RADIO COMMUNIC
HANDBOOK

FIFTH EDITION

RADIO SOCIETY OF GREAT BRITAIN

Published by the Radio Society of Great Britain
Cranborne Road, Potter's Bar, Herts EN6 3JE

ISBN 0 900612 58 4

First published 1938

Fifth edition (in hardback) 1976, 1977

This paperback edition 1982

Reprinted 1983

Reprinted 1984

Reprinted 1986

Reprinted 1987

Reprinted 1988

Printed in Great Britain
at the University Printing House, Oxford
by David Stanford
Printer to the University

ACKNOWLEDGEMENTS

The Radio Society of Great Britain expresses its gratitude to the following who have contributed chapters:

E. J. Allaway, MB, ChB, MRCS, LRCP, G3FKM
The late W. H. Allen, MBE, G2UJ
A. C. Carr, MA, MRCP, G3OSU
C. L. Desborough, G3NNG
R. G. Flavell, FRMetS, G3LTP
H. L. Gibson, CEng, MIEE, G2BUP
J. P. Hawker, G3VA
A. M. B. Holloway, G3VUQ
J. Hum, G5UM
I. Jackson, G3OHX
H. D. James, MIQA, MIEEE, G3HZP
G. R. Jessop, CEng, MIERE, G6JP
P. D. Martin, BSc, G3PDM/W1
M. H. McFadden, BSc, CEng, MIEE, GI3VCI
L. A. Moxon, BSc, CEng, MIEE, G6XN
A. D. Patterson, BA, CEng, MIERE, GI3KYP
H. W. Rees, G3HWR
C. Sharpe, G2HIF
R. F. Stevens, G2BVN
G. M. C. Stone, CEng, MIEE, MIERE, G3FZL
R. G. B. Vaughan, G3FRV/VK6RV
J. L. Wood, G3YQC

The Society is also grateful to the following who provided extra material for, or assistance with, certain chapters:

B. R. Arnold, G3FP
D. A. Evans, G3OUF
J. E. W. Groom, G8FQM
P. J. Hart, BSc, G3SJX
J. S. K. Hitchins, G4FGN
P. I. Klein, W3PK
J. P. Martinez, G3PLX
J. C. Pennell, G3EFP
R. O. Phillips, G8CXJ
R. F. G. Thurlow, G3WW
L. L. Williams G8AVX

Acknowledgement is also made to the authors of articles published in *Radio Communication* from which extracts have been made.

The 100W solid-state linear amplifier design in Chapter 6 was taken from the Mullard technical publication TP1337, *Transistors for single sideband amplifiers*, by kind permission of Mullard Ltd.

Production of the fifth edition was undertaken by J. K. Bayles, D. E. Cole (draughtsman), R. J. Eckersley, J. Hum, P. Linnett, and J. E. Tebbit, FISTC.

Updates

pp7.1, 7.2	The UK 70MHz and 1·3GHz allocations are now 70·025–70·500MHz and 1,240–1,325MHz respectively.
p10.2	The 1981 IARU Region 1 Conference recommended that rtty speeds of 50, 75 and 100 bauds should be encouraged.
p10.18	The UK rtty calling frequency is now 70·3MHz.
p14.1	Additional channels S10, S11 and S13–19 have been allocated for fm simplex use.
pp14.30, 14.31	UK repeaters operating to IARU Region 1 specifications require an initial toneburst but other stations may then carrier re-access indefinitely.
p19.6	The 1981 IARU Region 1 Conference recommended that moonbounce operation should take place in the segments 144·000–144.015, 432·000–432·015 and 1,296·000–1,296·015MHz. The 144MHz ms reference frequencies are now 144·100 (cw) and 144·400 (ssb); operation can take place up to 26kHz higher than the reference frequency. The 144MHz beacon band is now 144·845–144·990MHz.
p19.7	The Federal Republic of Germany has been allocated DAA–DRZ and the Republic of Korea DSA–DTZ and HLA–HLZ. The Democratic People's Republic of Korea has been allocated HMA–HMZ.
p20.11	There are now also amateur-satellite allocations in the 18 and 24MHz, and 1·3, 2·4, 5·6, 47, 76, 142 and 241GHz amateur bands.
p22.1	Associate members of the RSGB are now allocated BRS numbers.

January 1982

CONTENTS

FOREWORD

By M. MILI

Secretary-General, International Telecommunication Union

IN keeping with the rapid advances in technology during the past five years the number of stations in the amateur service has correspondingly grown, and it is now estimated that by 1980 this will reach one million. This is in contrast with the position in 1939 when the first edition of this handbook was printed. Successive editions have reflected the current interests of the amateur movement and the work that has been done, which has made an appreciable contribution to the progress of radio communication.

Regular communication is now effected on frequencies which only a short time ago would have been beyond the capabilities of radio amateurs. Their progress in microwave techniques has been rapid and the results achieved and the equipment used has owed little to their professional colleagues. The continuing interest in space techniques has led to the creation of the amateur satellite service in 1971 and the Oscar satellites are now an accepted and exciting part of the amateur radio scene. With the limited power and aerial systems at his disposal the operator must make the fullest possible use of the propagation characteristics of the various frequencies which are used. A great deal of useful scientific research has been carried out, the results of which are reflected in this handbook and which will aid radio amateurs in making the most out of those parts of the radio frequency spectrum allocated for their use.

In parallel with the scientific aspects of the amateur service are the opportunities offered for international co-operation. The ability to create friendships with persons in all parts of the world in the pursuit of a common interest is a factor in the maintenance of goodwill which is shared by few other activities. With the development of amateur radio there is an ever increasing range of subjects on which there is a high degree of worldwide liaison. The international body of the amateur service—the International Amateur Radio Union—exists for the purpose of furthering this co-operation. This body is recognized as a participating observer at the conferences of the International Telecommunication Union at which the tasks of frequency allocation and conditions of use are fulfilled.

The amateur radio service enjoys a special place in the body of services recognized by the ITU for it is one of the oldest radio services in existence. It is defined in the Radio Regulations as "a service of self-training, intercommunication and technical investigations carried on by amateurs, that is, by duly authorized persons interested in radio technique solely with a personal aim and without pecuniary interest".

The service is therefore recognized as having two missions—first, to instruct and to take part in the training of those who, in any capacity, bear responsibility for the operation of radio services. Second, to engage in disinterested research in order to deepen our knowledge of scientific matters. The role of radio amateurs in technical training seems to be little known for all its great importance. The ITU is engaged on a vast programme of technical co-operation to aid developing countries to expand their telecommunication services. In this programme training plays a vital role. There is no doubt that the development of an amateur radio service in the countries concerned makes a substantial contribution to the execution of this immense task, and a contribution moreover that costs the community so little.

RSGB—RADIO COMMUNICATION HANDBOOK

The field for further contribution to the science of radio communication is as great today as ever. There are many persons who have found that experience as a radio amateur has served them well in a professional career. Whether the reader's interest lies in the amateur or professional fields he will find this handbook to be indispensable in following his chosen area of study and experiment.

M. MILI

PREFACE

By PAT HAWKER, G3VA

Author of "Amateur Radio Techniques", "A Guide to Amateur Radio" etc

BY the time any book has reached its fifth edition—with total printings exceeding 250,000 copies—it may be safely assumed that it fulfils a real need in a manner that appeals to the amateur radio enthusiast, and indeed also to many students and those concerned professionally with radio communication.

The Society does not revise this handbook annually or even frequently, but when the technology has changed sufficiently to justify it. This means that each new edition (five in almost 40 years) is virtually a new book, each chapter being completely revised and up-dated, often indeed entirely rewritten.

The problems that this practice presents should not be underestimated. The future in radio communication increasingly depends upon semiconductor devices but at present (and rightly so) amateur stations still continue to make wide use of thermionic devices where these are reliable and cost-effective. So the amateur still needs to understand valves or tubes, while learning to cope with an increasing range of bipolar and field effect transistors, integrated circuits and digital techniques, as well as the more exotic semiconductor devices such as Gunn diodes.

To include the new while retaining the best and most valid of the old is a process that makes impossible demands on authors and editors alike. No matter how ruthlessly the blue pencil is wielded, the sheer amount of information that a keen amateur expects to find in a standard book of reference grows and grows.

This edition has consequently been split into two volumes for convenience in handling; the division does not represent any fundamental rearrangement in subject matter compared with earlier editions. Like those editions it is written almost entirely by active radio amateurs for both active and would-be active amateurs. It does not assume that every amateur always builds his own equipment but it does take it for granted that the radio amateur continues to be keenly interested in the technical aspects of the equipment he uses.

In recent years, amateur radio has developed in Europe along rather different lines from North America, with perhaps a little more interest in *why* as well as *how*. This ensures that this handbook is not just a British version of the popular American handbooks—though we owe a real debt to those in North America who originally conceived and pioneered technical handbooks of a style very different from the traditional engineering textbooks.

Practical down-to-earth information—certainly every amateur needs that if he is to continue to represent a balanced combination of designer, constructor, purchaser and operator of modern radio equipment. However, he also needs to absorb a feeling and an instinct for what is and is not really important. One of the attractions of radio communication is the way in which fundamental ideas and techniques remain valid in the midst of innovation.

It is perhaps just as well that one can still, even in a computer age, put a lot within the covers of a book.

PAT HAWKER

CHAPTER 1

PRINCIPLES

OPERATING an amateur radio station is a fascinating hobby, and the enjoyment can be greatly increased by an intimate knowledge of the equipment and the ability to modify it experimentally. For success in such home construction there are two essentials. An understanding of the theoretical principles involved allows the amateur to get the best out of his equipment by good design, tailored to his exact needs. A practical grasp of components and their limitations is also needed, allowing him to choose suitable components for any purpose and if necessary replace a specified type by another already to hand, perhaps with minor circuit adjustments.

In this introductory chapter a practical section describing common components and the choice between them is followed by a section outlining the basic theory underlying each piece of equipment in the amateur station. Subsequent chapters are devoted to a detailed discussion of each piece of equipment with practical examples. Readers new to electricity and radio should turn to p1.9.

COMPONENT SELECTION

Equipment breakdown can be frustrating, and it is worth using the best quality components in amateur construction. With care these can be found quite cheaply on the "surplus" market. This section gives the relevant advantages of different types of component. Note that it is always worth testing components before use.

Wire

Copper is the most often used conductor, being strong, cheap and of low resistance. Enamelled copper wire is used in coils and transformers because the insulating enamel takes up little extra room. Wire is sold in Standard Wire Gauge sizes (swg), and the size needed to carry a given current is listed in Chapter 23—*General Data*. Available wire can be roughly sized by comparing it with fuse wire or counting the number of turns per inch in a close-wound coil.

Enamelled wire should not be re-used if it has been kinked, as this cracks the enamel.

Radio frequency (rf) currents travel only along the surface of a wire. For this reason transmitter coils are often made from silver-plated wire. Where high currents are involved hollow tube can be used to reduce the cost of material, while some high-*Q* coils use Litz-wound wire in which many fine strands increase the surface area.

TABLE 1.1

Approx swg	Turns per inch	Fuse wire of same size
20	25	—
25	45	15A
30	70	10A
35	105	5A

Resistors

Resistors are very reliable components, lasting usually for many years, but, being present in such large numbers, they are the commonest cause of equipment failure. Ordinary resistors are moulded carbon, painted or encased in ceramic. Their resistance decreases on heating and tends to increase with ageing, so high-stability types are preferable.

Resistors less than ½in long are usually rated at ⅛ or ¼W power dissipation, while those ¾in long or more will dissipate 1–2W. There is however a maximum voltage rating of 150V for the small carbon resistors, 250V for medium-sized ones, regardless of power. When soldering the miniature types use a heat sink clip, for excessive heat will permanently alter their resistance.

Wire-wound resistors have higher power ratings (usually stamped on the resistor) but get very hot at full power, so they should be kept away from coils, etc. Unlike carbon types their resistance is greater when hot.

Low-cost resistors are usually made within 10 per cent tolerance, which means that a nominal 1,000Ω resistor may have an actual resistance of anything from 900 to 1,100Ω as measurement will quickly show. Resistors are therefore sold in preference values at intervals of twice the tolerance. For 10 per cent resistors this is the "E24" series of 20 per cent steps, ie 10, 12, 15, 18, 22, 27, 33Ω, etc, and subsequent decades likewise.

The nominal value of resistance is marked either with a series of coloured rings to be read from one end, or with the R-code. In the *colour code* the first two rings indicate digits and the third is a multiplier (see **Table 1.3**). A fourth ring of gold or silver indicates a tolerance of ±5 per cent or ±10 per cent respectively; where no fourth ring exists the tolerance is ± 20 per cent.

In the *R-code* the first letter marks the decimal point and indicates any multiplier (R = Ω, K = kΩ, M = MΩ). The second letter gives the tolerance as F (1 per cent), G (2 per

TABLE 1.2

Resistor type	Tolerance (per cent)	Power rating (W)	Advantages
Moulded carbon	±10 (or 20)	½–2	Cheap
Carbon film	±5	⅛–½	Better stability
Metal oxide (metal film)	±1–2	¼–½	Small, high stability, low temperature variation
Wire-wound	±5–10	1–100	High power

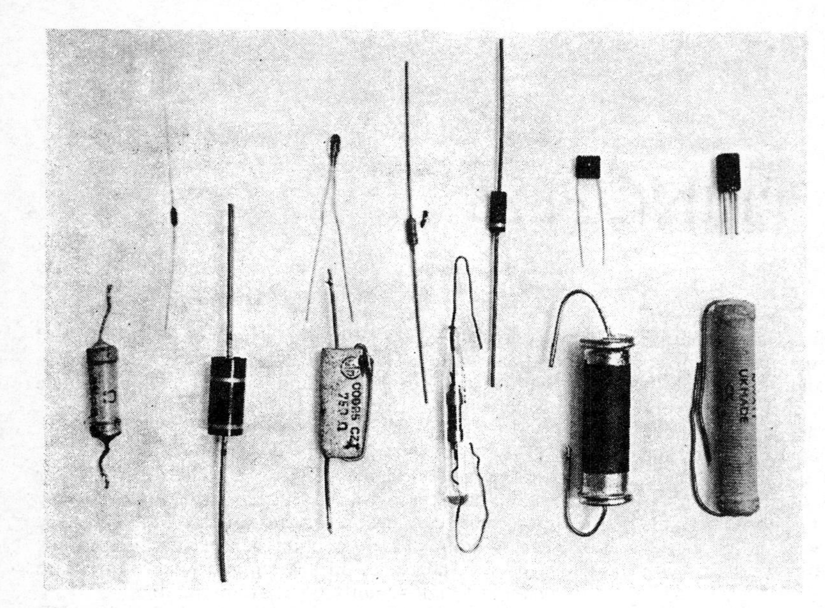

Fixed resistors. Top row, left to right: subminiature 10mW type, bead-type ntc Thermistor, two low-power carbon types, photoresistor, pcb-mounting type. Bottom row, left to right: two 1W carbon types, resistor with fusible link, glass-encapsulated high-resistance type, high-power rod-type ntc Thermistor, high-power wirewound resistor.

cent), J (5 per cent), K (10 per cent) or M (20 per cent). Thus the resistors described at the foot of Table 1.3 would be marked M39J, 1kM and 4R7K respectively.

There are various special types of resistor. The *fusible resistor* is used as a safety resistor: when a certain current is exceeded for a few seconds a soldered joint melts and springs apart to protect equipment from overload. The *voltage-dependent resistor* has a resistance which falls as the applied voltage reaches 50V or so, and is used to absorb voltage spikes. An *ntc Thermistor* has a cold resistance about 50 times larger than its resistance when hot. The bead type is mounted on a transistor heat sink to adjust the bias current with operating temperature, while the rod type is heated by its own power dissipation. The opposite function is performed by the *ptc Thermistor* (positive temperature coefficient, or Silistor), the resistance of which rises rapidly on heating and can protect transistors from thermal runaway.

Potentiometers ("pots", variable resistors) are usually of carbon film construction. The track often gets dirty and worn, causing crackles and unreliability, but can sometimes be renovated with switch cleaning fluid. The *log-law* (logarithmic) potentiometer has a graduated carbon film, so that on rotating the spindle the resistance increases very slowly at first. This type is used as a volume control, because the ear responds logarithmically to sound power; the terminals must be connected in the correct sense or an antilogarithmic effect is produced. Carbon potentiometers are limited to ½W power dissipation, while wire-wound types can be obtained for powers up to 50W. Tolerance is usually 10 per cent or 20 per cent. Other types of potentiometer include *preset* (set by screwdriver), miniature *skeleton presets*, ganged or concentric pairs, and types with a full 360° rotation or many turns. The photographs show a selection of the various types of resistor commonly in use, including most of those mentioned above.

TABLE 1.3

Colour	Digit	Multiplier
Black	0	× 1
Brown	1	× 10
Red	2	× 100
Orange	3	× 1,000
Yellow	4	× 10,000
Green	5	× 100,000
Blue	6	× 1,000,000
Violet	7	
Grey	8	
White	9	
Gold		× 0·1

Examples: orange/white/yellow/gold bands indicate 390kΩ ±5 per cent; brown/black/red bands indicate 1kΩ ±20 per cent; yellow/violet/gold/silver bands indicate 4·7Ω ±10 per cent.

Potentiometers (variable resistors). Left to right: the type commonly used as a volume control with an on-off switch built-in, skeleton preset for pcb mounting, screened preset for chassis mounting.

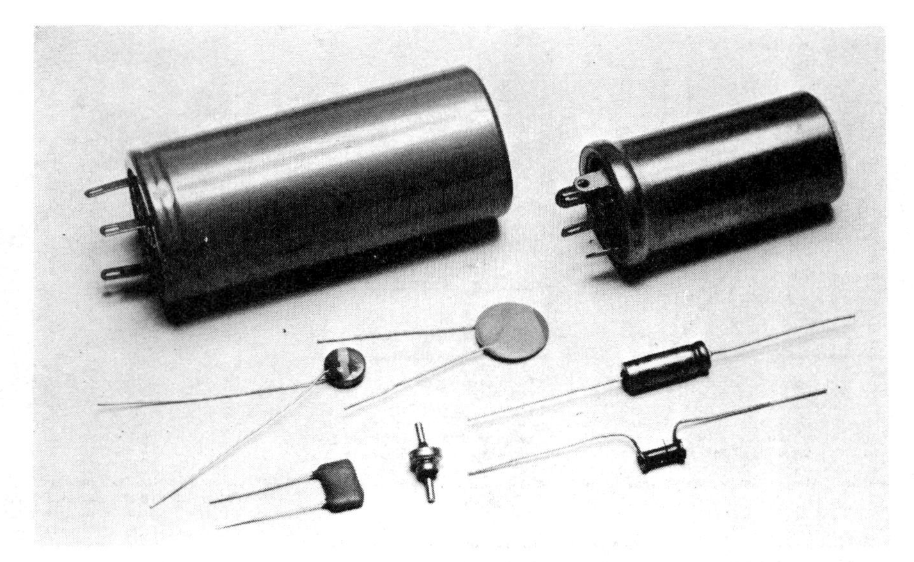

Fixed capacitors. Top row: two large multi-section electrolytic types. Middle row: two disc ceramic types and a miniature electrolytic type as used in transistor circuits. Bottom row, left to right: silver mica, feed-through and tubular ceramic types.

Capacitors

A typical capacitor contains two layers of metal foil separated by a layer of insulation (the dielectric), rolled up in a tubular case. The nominal capacitance is usually stamped on the case, and tolerance is often +50 per cent −20 per cent. The resistor colour code or R-code may be used to indicate value (eg yellow/violet/orange = 47,000pF, 6R8 = 6·8μF). Also marked is the *working voltage*, which is the maximum dc voltage the device will withstand in either direction: the ac rating is about half this. In the case of electrolytic capacitors dc may only be applied with correct polarity, and the ac (ripple) component must not exceed about 10 per cent of the working voltage.

Capacitors usually fail by short circuiting, and leakage current should be checked before use. Ordinary capacitors should leak much less than 1μA. Aluminium electrolytics have a leakage current given approximately by:

$$\text{Leakage current} = \frac{C(\mu\text{F}) \times V(\text{wkg})}{6}\ \mu\text{A}$$

while tantalum electrolytics leak about one third as much. Other important properties are the *temperature coefficient* of capacitance change, θ (which may be positive or negative), the *power factor*, which is a measure of the power loss introduced by the capacitor, and the physical size of the component. **Table 1.4** gives some common types and their uses.

Electrolytic capacitors have a foil electrode immersed in an electrolyte solution or paste, which forms the other electrode. Passing a small forward current builds up a thin layer of insulating deposit on the foil, so thin that huge capacitances (eg 100,000μF) can be accommodated in a small can. Unfortunately this layer of deposit allows a substantial leakage, and is easily destroyed by excessive voltage, heat, or a voltage of reversed polarity. During a long period of disuse the layer may fragment and require reforming by charging slowly through a 10kΩ resistor.

Electrolytic capacitors have a very high power factor, so an excessive ac ripple current will generate internal heat and damage the capacitor. For the same reason they do not pass rf well, and rf decoupling necessitates a ceramic capacitor in parallel. Where an electrolytic capacitor is used for af coupling a dc bias voltage must be provided to prevent damage by reverse voltages (**Fig 1.1**): an exception is the tantalum type, which will stand up to 0·5V of ac signal without bias.

Tuning capacitors are air-spaced variable capacitors, often with vanes shaped to give a "log-law" increase on rotation. The broadcast receiver type has two or more sections *ganged* on one spindle, and each section has a capacitance of about 300pF with the moving vanes earthed via the spindle. If a lower capacitance is required vanes can be carefully removed; hf types have a wider gap, giving both lower capacitance and higher breakdown voltage. A reduction gear on the spindle will provide mechanical *bandspread*.

Trimmers are small preset adjustable capacitors, either air-spaced (capacitance up to 30pF) or of the mica compression type (to 750pF). If one is connected in parallel with a tuning capacitor the minimum capacitance of the circuit can be preset, ie the upper limit of the frequency range covered can be adjusted. Another preset capacitor in series with the tuning capacitor (a *padder*) will set the maximum

TABLE 1.4

Function	Type	Advantages
Capacitors for af/i.f. coupling	Paper/polyester	High voltage, cheap
	Polycarbonate	High capacitance (to 10μF), compact
RF coupling	Ceramic	Small, cheap but lossy
	Polystyrene	Very low loss, low leakage but bulky
	Silver mica	Close tolerance
RF decoupling	Ceramic disc or feed-through	Very low inductance
Tuned circuits	Polystyrene	Tolerance 1–2 per cent very low loss, θ negative (—150 ppm)
	Silver mica	High stability, low loss, tolerance 1 per cent, low θ (+50 ppm)
	Ceramic	Available with range of θ (+75, 0, —150, —750 ppm) for temperature compensation, but lossy

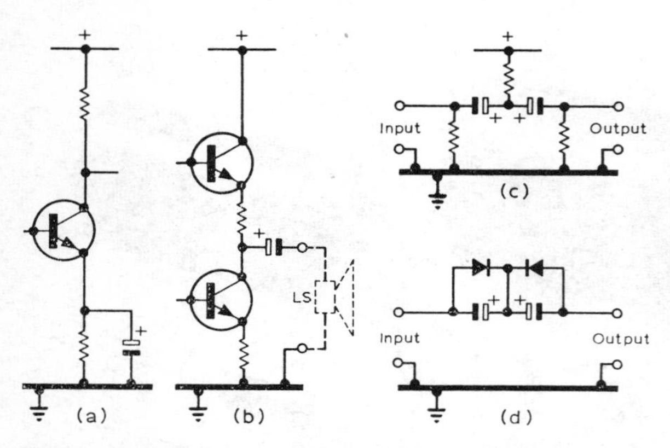

Fig 1.1. Electrolytic capacitors must be biased to avoid reverse voltage. When coupling or decoupling af, bias is often provided by the circuit (a,b). Otherwise it is provided externally (c) or by signal rectification (d) (negative plate shown solid). Note that in the last method the effective capacitance is halved because the capacitors are in series.

Variable capacitors. Top left: split-stator variable type for push-pull circuits. Top right: ganged two-section log-law tuning capacitor. Bottom left: tubular ceramic trimmer. Bottom right: miniature solid-dielectric tuning capacitor.

capacitance and thus the lower frequency limit. Where two ganged capacitors tune the aerial and the local oscillator, careful trimming and padding ensures that the circuits *track* across the whole range so that their frequencies always differ by the same amount (see **Fig 1.2**).

Inductors

Coil design depends chiefly on the frequency in use, and also on the permissible losses. Most short-wave coils are wound of fine wire (such as 32swg) on a paxolin, polystyrene or ceramic former to form a compact lightweight component. Thicker wire is needed for a high-current coil such as a transmitter tank coil, and also where low loss (high Q) is important, as in aerial loading circuits, etc. At very high frequencies losses are minimized by omitting the former and using thick wire to form a self-supporting coil.

For maximum Q coils should be air cored, single-layer and space-wound (ie with a small gap between each turn to reduce self-capacitance). Where space permits, further improvement is obtained by making the coil slightly shorter than its diameter, as this optimum shape uses the shortest length of wire to achieve the desired inductance.

At low frequencies coil inductance can be increased a hundredfold by a *ferromagnetic core*. Solid iron is unsuitable because it becomes heated by a flow of eddy currents, causing large power losses from the coil. These can be reduced by using a core of insulated iron laminations, giving an inductance of several henrys in ac mains chokes and transformers. A typical *ac supply transformer* has a primary winding with several taps near one end (for 200V, 220V, 240V supplies), and will draw about 100mA from the supply with no load on any winding. Incorrectly applying the ac mains to a winding designed for a lower voltage causes a loud buzz often accompanied by rapid heating, although the same effect can be caused by shorted turns inside the transformer. The principal secondary winding is often centre-tapped for use with a full-wave rectifier. If there is another single lead isolated from the windings, this is usually a Faraday screen which when earthed prevents mains interference passing from primary to secondary windings by capacitative coupling (Fig 1.66).

At audio frequencies smaller cores suffice. Typical *speaker transformers* have a turns ratio of about 40:1 (valve type) or 5:1 (transistor type), and the primary may be centre-tapped for push-pull circuits. The valve types are plentiful, and can be used for various impedance-matching applications (**Fig 1.3** shows one example).

Above audio frequencies the losses in a laminated core become too great, and it must be replaced by a *dust core* or *ferrite core* which are available in various shapes. Different grades of ferrite produce greater or smaller inductance, but for amateur purposes any grade can be used, the number of turns required being found by trial and error. Particularly

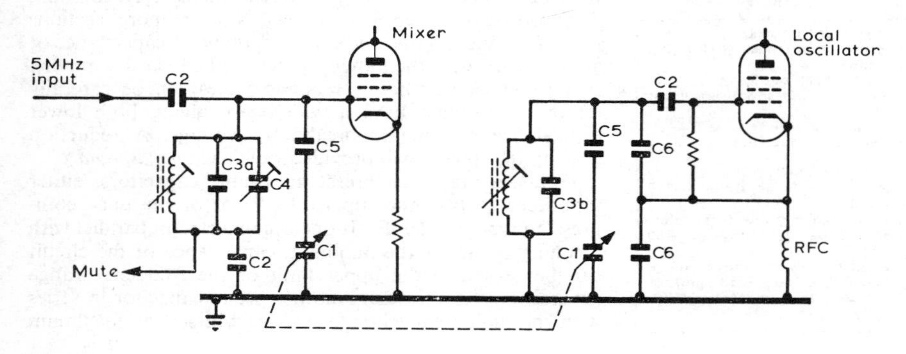

Fig 1.2. Part of the well known G2DAF receiver to show tracking arrangements. C1, ganged tuning capacitors. C2, dc blocking. C3a, b, capacitors setting the highest tuned frequency (C3b was a negative-temperature-coefficient type to offset thermal drift). C4, trimmer for fine adjustment of maximum frequency. C5, padder setting the lowest tuned frequency. C6, capacitative tap for the Colpitts oscillator.

Various inductors. Top left: rf coil on paxolin former and loudspeaker transformer. Bottom left: self-supporting silver-plated vhf coil and Ferroxcube-potted high-Q coil. Centre: exploded view of mains transformer, showing two-part laminated iron core, coil former and aluminium cover. Top right: i.f. transformer with and without its screening can. Bottom right: small toroid transformer for use at rf.

useful are *ferrite slugs*, which can be screwed towards the centre of a coil such as an i.f. transformer for fine tuning adjustment, and the *ferrite toroid* which, when used as a transformer core, gives very wideband inductive coupling with minimal capacitative link between the windings.

For an rf *choke* one quarter-wavelength of wire on a small air-cored former gives an impedance of about 100kΩ.

Charts and formulae for use in the design of coils are presented in Chapter 23, while the inductance of an available coil can be checked with a transmitter and an oscilloscope, as in **Fig 1.4**. Small capacitors can also be checked by this method.

The resonant frequency of a complete tuned circuit can be measured with a signal generator or a *grid dip oscillator* (gdo): as the gdo is tuned past the resonant frequency power is drawn from its probe into the tuned circuit and causes a dip in the meter current, allowing the frequency to be read off the calibrated scale to within about 5 per cent.

A *relay* is a magnetically operated switch, and the "Post Office" relay has served generations of amateurs, being cheap and plentiful. It has a large iron-cored coil designed to operate on 50V dc but can be run on lower voltages or from an ac supply: **Fig 1.5** shows one method. Various one- or two-way switches can be assembled on it according to need. Relays are often used for transmit/receive switching, and for aerial rf switching *coaxial relays* are available. Several different types including the Post Office model are shown in the photograph on the next page.

SEMICONDUCTOR DEVICES

Semiconductor devices are very small, long-lasting, mechanically rugged and cheap compared to valves. They are also very versatile—a typical silicon power diode will rectify a kilowatt or a few milliwatts of supply with equal ease. Early devices were made of germanium and were easily damaged by heat or overload, but silicon types are much less vulnerable and preferred for most purposes, except for work at very low voltages.

Diodes are designed for several different uses such as signal detectors, power rectifiers, regulators and amplifiers. The first group, *signal diodes*, are rated at about 25V and need to have a low capacitance. The following types are commonly available:

Germanium point contact (OA91)—low capacitance but a high forward voltage drop (perhaps 1V at 20mA)

Fig 1.3. A small loudspeaker used as a microphone needs its impedance matched to the amplifier input. To match a 3Ω speaker to a transistor input (about 3kΩ) requires a transformer with a ratio of about 30:1.

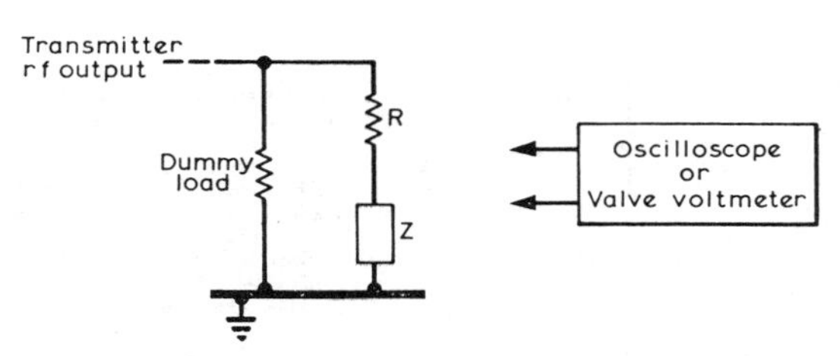

Fig 1.4. To measure the impedance of a coil (or capacitor) approximately, rf current from a transmitter is passed through it in series with a carbon resistor (R). R is chosen so that the measuring device indicates equal voltage drops across R and Z: then the impedance is R ohms, and R = $2\pi fL$ (or $1/2\pi fC$).

Various relays. The large one on the left is the Post Office type, still widely used, and the others are more modern types, some of which are suitable for mounting on a pcb.

Gold bonded (1N141) —low voltage drop, high current (say 100mA)
Silicon junction (OA202) —very low leakage, high breakdown voltage
Hot-carrier (HP2800) —excellent at vhf, low noise, low capacitance (low carrier storage), high efficiency

For signal detection all these are superior to valve diodes, which require higher signal voltage (eg 10V to pass 1mA). The germanium type will pass an adequate current at 100mV or even lower.

Power rectifiers are usually silicon junction diodes designed to stand a peak inverse voltage (piv) of up to 1kV, and a mean forward current of approximately 1A. The *avalanche diode* (avalanche-protected diode) is built to withstand small overloads of reverse voltage without permanent damage, so long as excessive current is prevented by an adequate series resistance. A simple method of testing diodes to assess the piv is shown in **Fig 1.6**. Diodes must *not* be used in series to withstand higher voltages unless special precautions are taken. For very high voltages miniature selenium rectifiers are convenient, typically rated at 6kV at 1mA. For high-current applications diodes can be used in parallel, although they should ideally be matched; germanium rectifiers dissipate less power and may be rated at higher currents (10 or 20A), albeit at a lower piv.

The *zener* diode (occasionally called an avalanche diode) breaks down at a low reverse voltage (usually between 5 and 15V) which remains remarkably constant whatever reverse current is applied. A more accurate version is the *reference diode*, supplied at an accurately stated breakdown voltage with low temperature variation, but at a higher price. There is also an ingenious method of making a transistor function as a rather leaky zener device but with an adjustable breakdown voltage **(Fig 1.7)**.

Any diode when reverse-biased has a small capacitance which decreases as the reverse voltage rises, and it can therefore be used as a remote tuning device (see Chapter 3—*Semiconductors*). The *Varicap* or *Epicap* diode is particularly designed for this purpose, with a stated capacitance swing from, say, 10pF down to 4pF as the voltage rises by 10V or so. Any rectifier diode will perform the same function, and a little experimenting may be useful, eg an OA202 varies from about 10pF to 3pF at a reverse voltage of 0-10V. Zener diodes give a much greater capacitance change, say 500pF to 150pF from 0-5V.

A voltage-variable capacitor can also be used as a *parametric* amplifier or frequency multiplier at extremely high frequencies (up to 10GHz or so), and diodes made for this purpose are called *varactors*. Input powers accepted by the types available vary between 4 and 40W and they have efficiencies of 25 to 50 per cent.

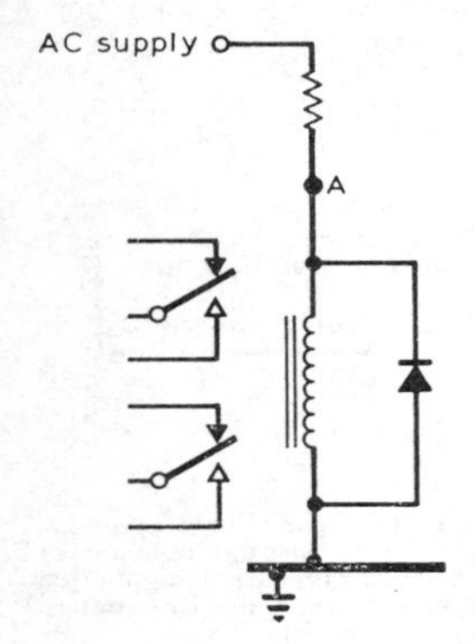

Fig 1.5. A relay can be powered from an ac source without "chatter" by fitting a shunt diode. Between positive half-cycles the relay coil continues to draw forward current from the diode, since an inductance opposes current change. A series diode can be added at A to reduce heat dissipation in the resistor.

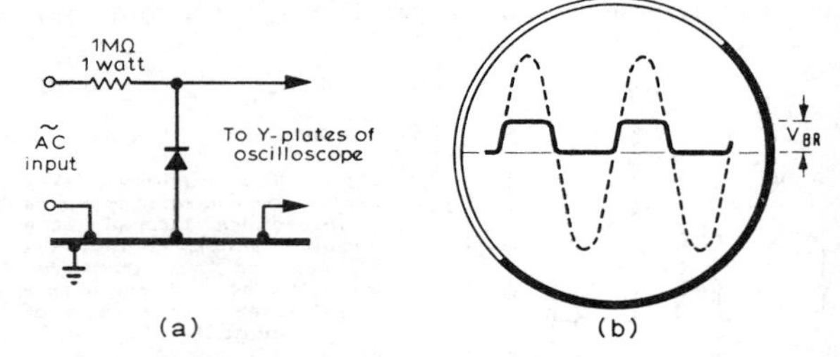

Fig 1.6. To test silicon diodes a large ac voltage is applied through a high series resistance which protects the diode. The oscilloscope trace indicates the diode breakdown voltage (piv) at which the diode starts to leak about 0·1mA. For zener diodes a low voltage, about 12V ac, is adequate.

Breakdown Devices

Certain devices have a high resistance until the applied voltage reaches a threshold level, when the resistance falls rapidly: the resistance remains low until forward conduction ceases, and then the device recovers its high resistance. The *four-layer diode* (Shockley diode) breaks down (triggers) at a set forward voltage around 20V, and the *diac* is a two-way version. The *silicon controlled rectifier* (scr, thyristor) is a four-layer diode with a control electrode, the *gate*, and breaks down when a small current is passed through the gate. High-voltage thyristors (piv up to 1kV) are used for supply regulation, while low-voltage versions function as switches. The *triac* is a two-way thyristor, permitting current flow in either direction when triggered, and the *quadrac* is a refinement incorporating a diac in the gate lead. The *unijunction* transistor has no gate electrode, but the breakdown voltage can be set by external biasing. See Chapter 3 for details.

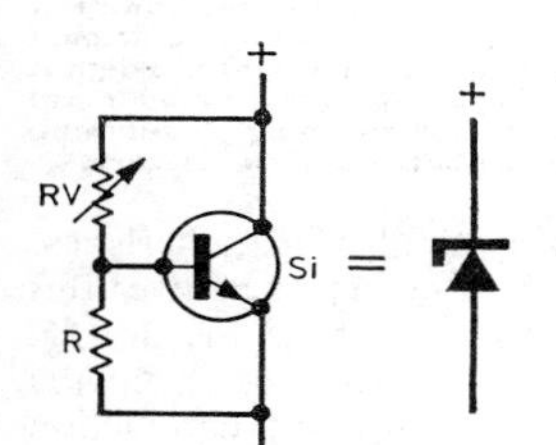

Fig 1.7. A constant-voltage device using a silicon transistor. When its emitter-base voltage reaches 0·6V, ie the applied voltage reaches (1 + RV/R) × 0·6V, the transistor conducts and maintains the voltage at this level.

Transistors

Ordinary transistors are *bipolar*, having two pn junctions separated by a very thin layer called the *base*. Choice of transistor for a given purpose is dictated by the following considerations:

Polarity—Silicon npn transistors are often cheaper than pnp types, and use a positive supply voltage.
Voltage—The collector must be rated at twice the supply voltage or more, when using an inductive load.
Frequency—In most common-emitter circuits the transistor needs a cut-off frequency (f_T, f_1 or f_α) of at least 100 times the operating frequency in use.
Power—Small transistors will dissipate 200-300mW, so the product of collector current and voltage in use must be less than this. Power transistors must be rated far above the actual power to be supplied, because their official rating (P_D, P_O, P_∞ etc) is usually referred to a constant case temperature. No practical heat sink achieves this. As a very rough guide, a sheet of aluminium 15cm square with the transistor mounted firmly in contact with the metal in the centre will dissipate a quarter of the rated power.

These are all simply minimum requirements, so a high-performance transistor (such as the 2N3375) could be used for *any* application within its generous limits of voltage, frequency and power.

Transistors have many other characteristics, but these show little variation between modern types and so rarely govern the choice for a particular application. Examples are the current gain β or h_{fe} (generally 50 to 100), leakage current (a few μA in common-emitter circuits), emitter reverse voltage (about 5V), etc.

Several methods of transistor construction are in use. The *alloy* transistor (alloy-junction, alloy-diffused) can be made to stand high voltage and power dissipation, but most have a poor frequency response. An example is the 2N3055, rated at 100V 115W with f_T = 1MHz. A variation with better frequency response is the germanium *drift* or *micro-alloy diffused* transistor (madt), in which the base is doped more heavily on one side.

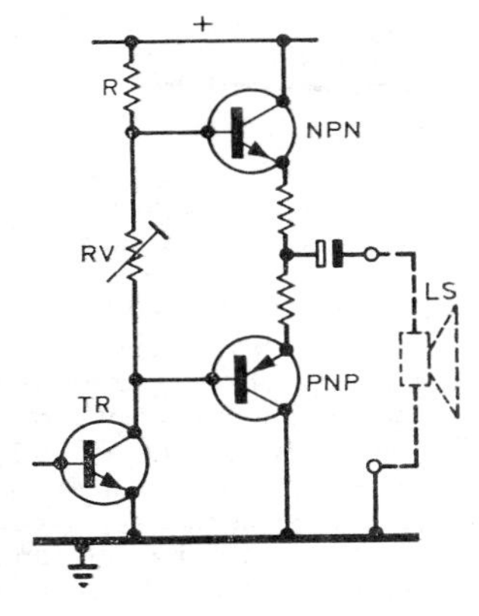

Fig 1.8. This transistor output stage uses complementary power transistors in push-pull. A driver transistor TR generates signal voltage across a load R, and the power transistors act as emitter followers giving a low-impedance output. Without applied signal a small dc current flows, controlled by the preset resistor RV (which may include temperature compensation).

Much higher frequency response can be obtained from silicon *planar* construction. The base and emitter are deposited as thin films on the surface of a silicon chip, which forms the collector. A refinement is the planar *epitaxial* transistor, where the chip is heavily doped at its base to lower the collector resistance. An example of an epitaxial transistor is the 2N709A, rated at 15V 0·3W with f_T = 800MHz; the very thin base layer gives good frequency response but a low voltage rating. A rather similar process in germanium transistor construction produced the *diffused-base* and *double-diffused mesa* transistors, also with high frequency response.

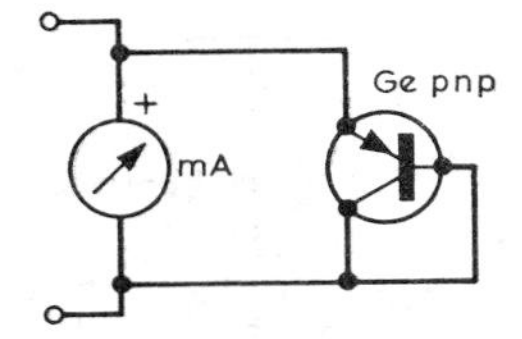

Fig 1.9. The simplest meter protection circuit uses a germanium transistor shunt which limits the movement voltage to about 200mV, corresponding to a current overload of about 10 times the scale maximum. Transistor leakage introduces a small inaccuracy, the meter reading low.

Matched pairs of transistors selected for their similar characteristics are available for differential amplifiers, etc. For best matching the transistors are formed on one chip side by side, and are known as *dual* transistors (eg the 2N2223). *Complementary* transistors have similar ratings but opposite polarity (eg TIS90 npn, TIS91 pnp), and are used in push-pull amplifiers to avoid the need for transformers (**Fig 1.8**).

The *symmetrical* transistor was a germanium type designed to function at voltages right down to about 10mV. Any germanium transistor will work at 100mV or so, and this property can be used for meter protection (**Fig 1.9**).

Any type of transistor should be tested before use as this takes only a few moments, and a simple tester is shown in **Fig 1.10**. Most small silicon transistors should leak less than

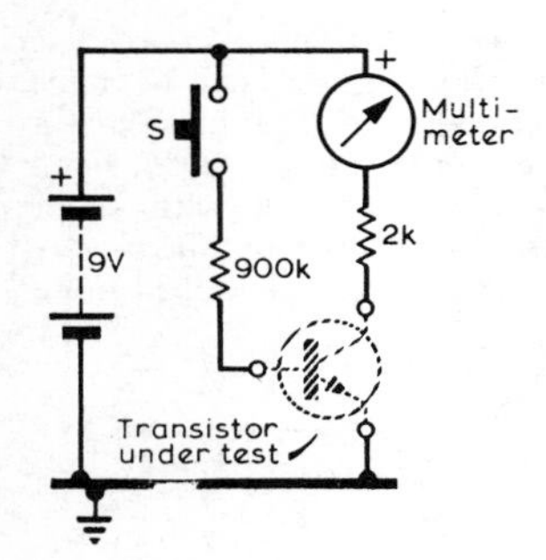

Fig 1.10. Testing transistors: first the leakage current I_{ceo} is measured with the multimeter, then pressing switch S gives a current increase in microamps of 10 times the gain (10β, or more exactly 10 × hFE, the dc gain).

10μA and have a gain of 50 or more. Another transistor tester is described in Chapter 18—*Measurements*.

Field-effect Transistors

The field-effect transistor (fet) has a narrow silicon *channel* surrounded by a *gate* electrode, which controls the number of mobile current carriers in the channel and therefore its resistance. In use it has high input and output impedances, rather like a pentode valve, and when coupled to a high-impedance input circuit it generates less noise than a comparable bipolar transistor when used as a simple amplifier. It will also function as a square-law detector (see page 1.41), and as a variable resistor since the channel small-signal impedance varies from about 1kΩ up to several megohms depending on the gate voltage.

In the *junction-gate* fet (jfet) the gate and channel meet at a pn junction, and a dc reverse bias voltage is applied to isolate the gate. An example is the MPF102 series of *n-channel* JFETS which in the absence of gate voltage pass several milliamps at 10–20V. A negative gate voltage depletes the channel of carriers and lowers the current, which stops entirely at the gate "pinch-off" voltage of about −5V.

Other FETS have the gate insulated from the channel by a fine layer of metal oxide, the *insulated-gate* FETS (IGFETS) or *metal-oxide-silicon* FETS (MOSFETS). One type functions in just the same way as a jfet with similar characteristics, except that the gate voltage need not be kept negative: this type comprises the *depletion* MOSFETS, such as the 3N155. There is a second type, the *enhancement* mosfet, eg the 3N157, in which no channel current passes until conduction is enhanced by a forward gate voltage above a certain threshold of a few volts. MOSFETS may be n-channel or p-channel.

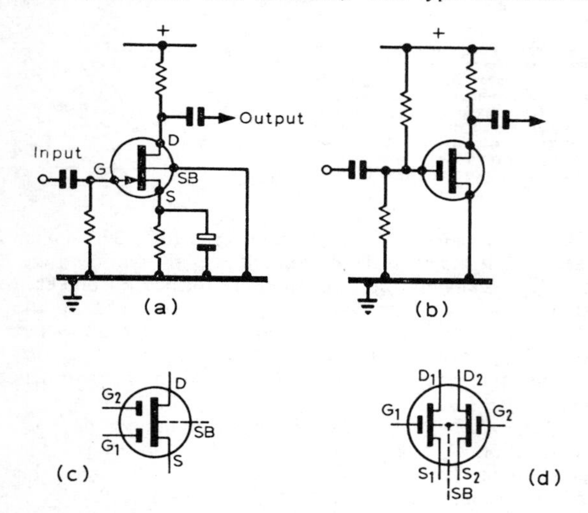

Fig 1.11. Types of fet: (a) n-channel jfet with simple biasing network (SB = substrate lead), (b) n-channel enhancement-type mosfet with its positive bias voltage supply, (c) dual-gate fet, (d) dual fet.

Integrated circuits. The black rectangular types are in the common 'dual-in-line" package; the 8-pin one on the left is an operational amplifier, and the 14-pin and 16-pin versions on the right are digital logic circuits. The other frequently-used package is the round metal can shown on the left, which affords better shielding. A subminiature "flat-pack" package is shown at the bottom of the picture.

Because FETS have a very high input impedance (about 10^{10}Ω for a mosfet) the gate can easily be damaged by hum voltage while handling and installing. To prevent this they are supplied with a substrate lead wrapped closely around the other leads, and this should not be removed until the fet has been soldered into the circuit. Some devices also have internal zener diodes to limit the gate voltage, and are known as *gate-protected* FETS. There may be a diode junction between substrate and channel, a source of confusion when testing FETS.

In a *dual-gate* mosfet the channel passes through two gates in succession: **Fig 1.11(c)**. The second gate may be earthed, giving excellent input/output isolation, or an agc voltage applied to control the gain of the device, or a modulating signal applied. They must not be confused with *dual* MOSFETS, which contain two matched single-gate MOSFETS formed on the same silicon chip.

Integrated Circuits

A monolithic *integrated circuit* (ic) contains several interconnected transistors and resistors formed on the surface of a single silicon chip. Small capacitors may be included, and external tags are provided for connecting large capacitors, inductors, variable resistors etc. *Linear* ICS respond to a continuously variable signal, while *digital* ICS respond only to "on" or "off" states and are mainly confined to computers.

The most common linear ic is the *operational amplifier*, a direct-coupled amplifier with high input impedance providing a voltage gain of 100 to 10,000 in most cases. Operational amplifiers function from dc up to about 100MHz, and some are designed as i.f. amplifiers with provision for

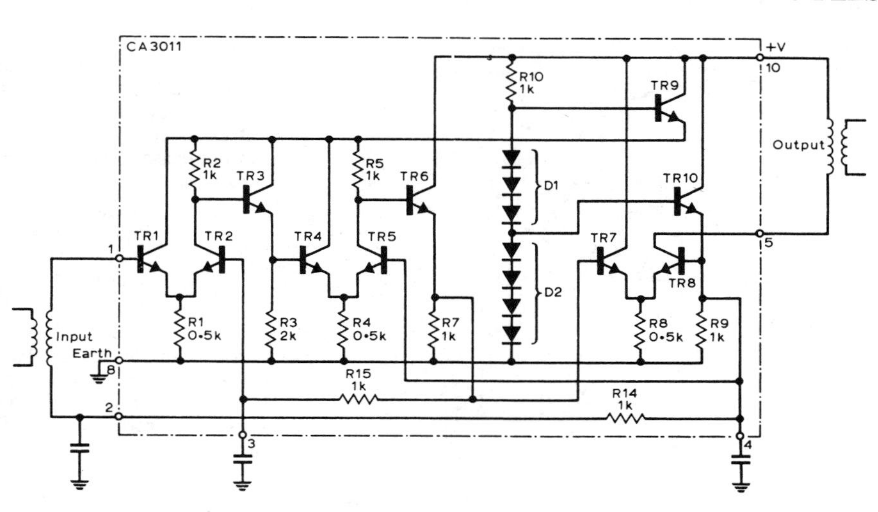

Fig 1.12. Circuit diagram of a typical integrated circuit, the RCA CA3011 wide-band amplifier, showing pin numbers and external connections. Transistors TR2, TR5 and TR8 are common-base signal amplifiers and the others are emitter followers. The diodes and R10 provide two stabilized voltages, transformed to low impedance by TR9 and TR10; further control of dc bias is achieved by using "long-tailed pairs" with direct-coupled emitters (TR1, 2, 4, 5, 7 and 8) and via two dc negative feedback loops (pins 3 and 4). Because the device uses no conventional common-emitter stages it gives good amplification at fairly high frequencies (up to 20MHz). Overall current gain is 5,000.

agc. Other linear ICs function as modulators, audio amplifiers, agc generators etc. **Fig 1.12** is the circuit of an RCA CA3011 ic showing the 10 transistors and associated components.

VALVES

Thermionic valves are categorized by the number of their electrodes, eg a *pentode* has five (anode, cathode, and three grids), and their function is fully described in Chapter 2—*Valves.* A few special terms are mentioned here.

A *beam* valve such as a beam tetrode has an extra pair of plates which focus the electron beam onto the anode, so as to minimize secondary emission from the anode. In the *beam-deflection* valve the electron stream is deflected from one anode to another by suitable voltages applied to the plates, and this is useful in ssb generation. A *cold-cathode* valve does not need a heater supply: the cathode coating is activated by an avalanche effect stimulated by the voltage on a nearby accessory anode.

A *Compactron* is a small unit with a 12-pin base containing two conventional valves. A *Nuvistor* is a very small rigid valve of metal and ceramic construction for vhf work. The *klystron* is a microwave device: in the reflex klystron an electron beam resonates in a metal cavity producing self-oscillation.

The *neon* or voltage regulator contains two electrodes and neon gas. No current flows until a certain *striking* voltage is reached, when the gas ionizes and the voltage across the electrodes drops to a fairly constant value regardless of the current passed, typically 75V or 150V. It is used as a shunt regulator, and also in the relaxation oscillator (p 1.35).

The *thyratron* was the thermionic forerunner of the thyristor. It is a triode containing gas, and acts as a switch. When a heavy negative bias is applied to the grid no current flows, but once the bias is relaxed anode current ionizes the gas, which maintains current flow regardless of grid voltage until the anode supply is interrupted.

METERS

The common type of dc meter has a moving-coil movement with a resistance of 10Ω or so, and requires about 1mA to produce full-scale deflection (fsd). It will function as a voltmeter (by adding a series resistor) or ammeter (with shunt resistor), but for general test measurements the basic movement should be sensitive, say 0·3mA fsd or better (ie 3,000Ω/V or more). Many low-cost instruments have a sensitivity of 30,000Ω/V.

Most ac voltmeters comprise a dc voltmeter with a built-in bridge rectifier and a suitably recalibrated dial. Unfortunately alternating current cannot be measured so easily, as the bridge rectifier has a considerable and non-ohmic resistance. It can be measured with a moving-iron meter, but only at low frequencies. To measure rf currents (usually the output to the transmitting aerial) a thermocouple meter can be used, in which the current heats a tiny filament. This heats a thermocouple, which in turn passes current through a dc meter. These meters have a small ohmic resistance, but the filament is easily destroyed by overload.

An effective method of measuring high voltages is to apply them directly to an oscilloscope Y-plate, having previously calibrated the screen with a known voltage such as the mains supply. The scale is linear, the impedance very high, and the frequency response excellent. Details of test equipment and methods will be found in Chapter 18.

THEORY OF ELECTRICITY

Electricity

The ancient Greeks noticed that rubbing a piece of amber (*elektron*) made it attract hairs and dust by some mysterious force, but it took 3,000 years to establish the explanation. Matter is composed of atoms, which contain many *electrons.* Some of the electrons on the surface of a piece of amber can be rubbed off, leaving it *charged,* so that it attracts the electrons on other bodies such as hairs, which then stick to it.

Many materials can be charged, such as glass, rubber and plastic. In each case the charge remains on the surface where it is placed and cannot cross the material, which is called an *insulator.* Metals, however, are quite different. They contain many free electrons which can move about within the metal, giving it special properties. If one end of a metal rod is charged, these electrons will *conduct* the charge right along the rod. Carbon, tap-water and wet objects are also *conductors* of electricity.

If electrons are continually removed from one end of a

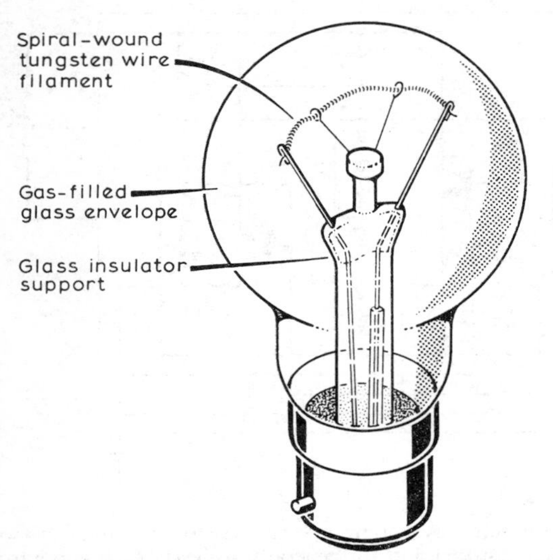

Fig 1.13. In a light bulb the conductor is a fine tungsten wire supported by a glass rod insulator. The wire is heated to white heat by the passage of electrons, and is suspended in an inert gas to prevent oxidation.

metal wire (by a battery) and placed at the other end, a *current* of electrons will flow along the wire. The stronger the battery, the larger the current which flows. This was discovered by Ohm, who established that in any given conductor the flow of current is proportional to the electrical force applied to its ends. Mathematically *Ohm's law* is written

$$\frac{\text{Potential difference (in volts)}}{\text{Rate of current flow (in amps)}} = \text{The } \textit{resistance} \text{ of the conductor}$$

$$\text{or} \quad \frac{V\text{ (voltage)}}{I\text{ (current)}} = R\text{ (resistance).}$$

The resistance of a conductor is measured in ohms (Ω), defined as the number of volts (V) required to push a current of one ampere (A) along the conductor. Consider a light bulb (**Fig 1.13**), in which the ac mains supply is connected to the ends of a fine wire filament. Suppose the mains (240V) passes a current of 0·5A through this filament. By simple proportion it would take 480V to push 1A through, so the filament has a resistance of 480Ω. Or, mathematically,

$$R = \frac{V}{I} = \frac{240\text{V}}{0{\cdot}5\text{A}} = 480\Omega$$

Using Ohm's law in this way can sometimes give inaccurate results. One example is that when electrons are forced along a wire they generate heat (this makes the light bulb glow), and electrons have difficulty in passing along a hot wire. The bulb resistance is therefore not a constant, and is in fact about ten times higher when the bulb is lit.

A *battery* is a column of cells joined end to end (in series) (**Fig 1.14**). Each cell contains chemicals, usually zinc and acid, which generate about 1·5V, so six cells in series make a 9V battery. If this battery were connected to each end of a short piece of wire (ie "shorted"), a large current would flow and quickly exhaust the battery. If the battery is connected to a high-resistance element such as a carbon rod (a *resistor*) a small current would flow. Carbon is a poor conductor, so the rod presents considerable resistance to the passage of current, perhaps many thousands of ohms, thereby preventing an excessively large current flow.

Because it is inconvenient to write out large numbers, unit prefixes are used to indicate size in electrical quantities, as follows:

mega-	(M)	= 1,000,000, ie 10^6
kilo-	(k)	= 1,000, ie 10^3
milli-	(m)	= 1/1,000, ie 10^{-3}
micro-	(μ)	= 1/1,000,000, ie 10^{-6}
nano-	(n)	= 10^{-9}
pico-	(p)	= 10^{-12}

For example, a 9V battery connected across a 1kΩ resistor will pass 9mA, and a small PP3 battery can keep this up for 36h or so. Batteries are, however, an expensive source of electricity: the same power from the ac mains supply would cost about 1,000 times less than the cost of a PP3.

"High-power" batteries are designed to stand a heavy current drain such as that from an electric motor, and mercury and manganese batteries will also pass a high current for a short time. Nickel-iron batteries are suitable for emergency equipment, as they can stand unused for many years without deteriorating. Mercury batteries give a particularly constant voltage, after a short running-in period, useful as a reference voltage. Chapter 16—*Power Supplies* gives a fuller description of common types of battery.

For electricity to flow there must be a complete *circuit* of conductors for the electrons to travel along. To stop the current the circuit need only be broken by a small gap, eg by a *switch*. When the switch is closed electrons from the battery negative terminal (−) pass round the circuit and are recaptured by the positive terminal (+), as in **Fig 1.15**.

Electrons actually travel along wire very slowly, at only a few feet per hour. However, closing a switch causes immediate current flow around the circuit: each conductor

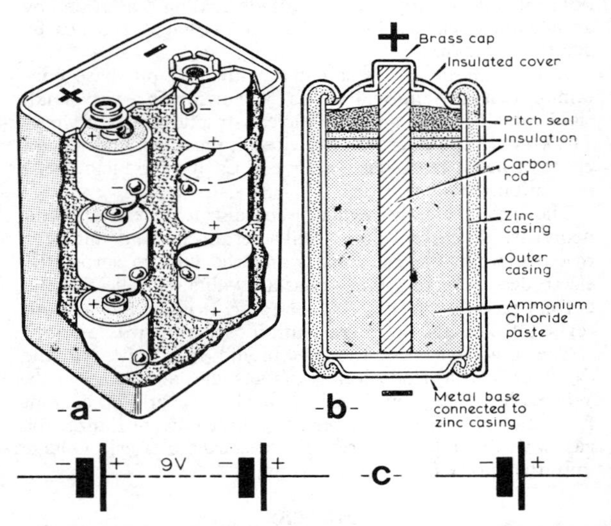

Fig 1.14. A 9V battery contains six cells in series, one arrangement being as in (a). (b) A single common dry cell: the zinc case slowly dissolves in the hydrolysed ammonium chloride, and hydrogen liberated at the carbon anode is absorbed by a layer of oxidants. (c) Circuit symbols.

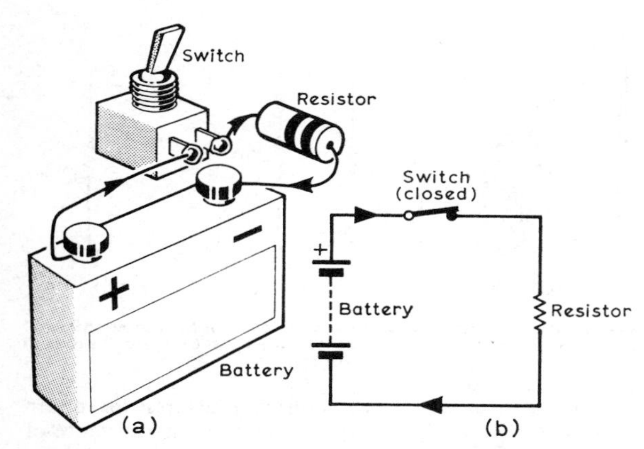

Fig 1.15. When the terminals of a battery are joined by a complete circuit of conductors current flows, conventionally from positive to negative terminals (a), although electrons actually flow in the opposite direction. Circuit symbols are shown in (b).

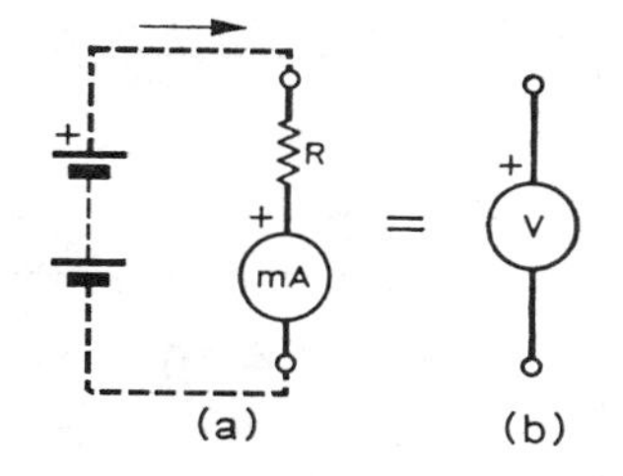

Fig 1.17. To measure the current flowing through a resistor a milliammeter is connected in series with it (a). If the resistance is known the applied battery voltage can be calculated: the resistor is mounted inside the milliammeter to form a "voltmeter". The circuit symbol is shown in (b).

already contains electrons, so pushing more in at one end immediately forces some out of the far end.

Before electrons were discovered, scientists believed that electricity emerged from the copper or carbon terminal of batteries, which they named the positive terminal, and was absorbed by the zinc or negative terminal. This theory proved wrong, as electrons actually travel from negative to positive, but it is still customary to indicate the direction of "conventional current" flow rather than electron flow.

Meters

A method of observing the passage of electric current was devised by Oersted. He showed that current passing down a wire would deflect a nearby compass needle. The larger the current, the more it will deflect the compass needle, and this is the principle of the *ammeter* (**Fig 1.16**). Rather than pivot the magnet, modern meters have a pivoted wire loop or coil suspended between the poles of a fixed magnet. This assembly is known as the *movement*. They are quite *sensitive*, a typical *milliammeter* requiring only 1mA to produce full scale deflection (fsd) of the pointer. To measure the current passing through a resistor, for example, the meter is connected in series with it: **Fig 1.17**.

No one has discovered an equally simple way to measure voltage. It is usually measured indirectly, by applying the voltage to a known resistance and measuring the current passed. A *voltmeter* is simply a milliammeter containing a suitable series resistor inside the case. As drawing current may alter the voltage to be measured, a voltmeter should have a high resistance so that it draws very little current. It must therefore be very sensitive so that the small currents can be accurately measured. If several different series resistors are provided, with a selector switch, the meter can be used to measure a range of voltages.

For example, a high-quality voltmeter may contain a 50μA fsd movement (ammeter) and choice of two series resistors, of 200kΩ and 2MΩ. The voltage producing fsd will be, from Ohm's law,

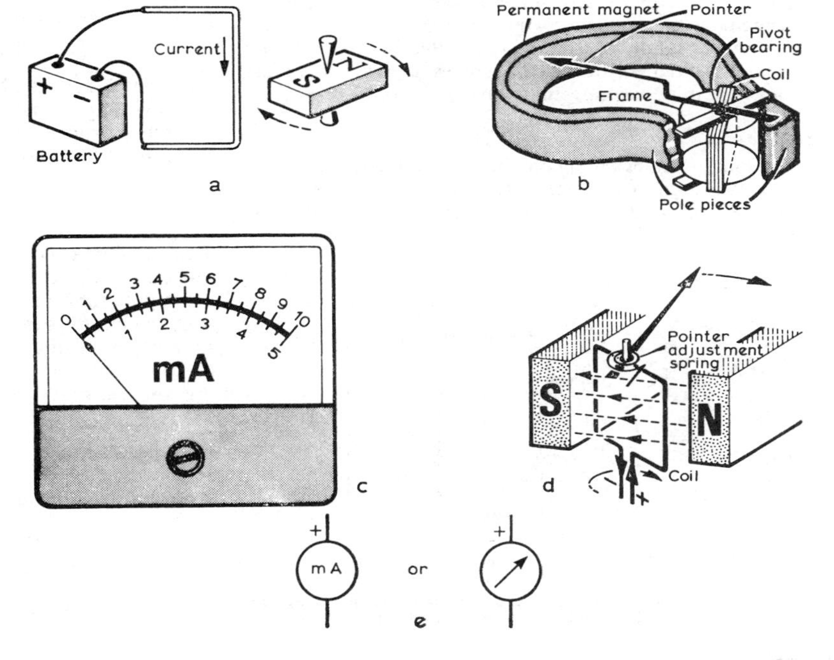

Fig 1.16. The principle of the ammeter. Oersted observed that a current will deflect a nearby magnet (a). In the moving-coil ammeter (b,c) the current passes between the poles of a large magnet and makes the coil twist, because each side of it is pulled in opposite directions (d). (e) Ammeter circuit symbol.

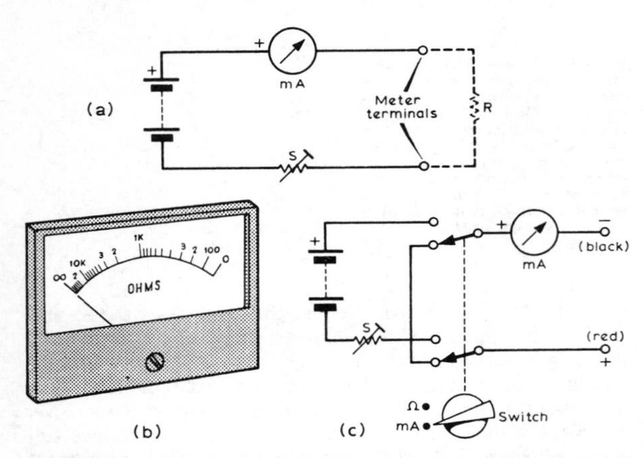

Fig 1.18. An ohmmeter (a) is a milliammeter containing a battery and preset series resistor S, which is adjusted to give full scale deflection with the terminals shorted (zero ohms). A resistance R is then measured by observing the fall in meter current on a reciprocal scale (b). In the multimeter the battery and series resistor are switched out when measuring current and voltage, but this reverses the terminal polarity (c).

$$\text{(i) } 50\mu A \times 200k\Omega = 10V$$
$$\text{(ii) } 50\mu A \times 2M\Omega = 100V$$

The meter therefore has two ranges across its scale, 0 to 10V and 0 to 100V.

The sensitivity of a voltmeter is defined as its resistance divided by the voltage for fsd. On either range the above meter has a sensitivity of 20,000Ω/V. Note that a voltmeter is not placed in series in a circuit as an ammeter would be, but directly across the voltage source to be measured.

A milliammeter can be used to measure resistance, by incorporating a battery and a series resistor (to protect the meter from overload). Such an *ohmmeter* is shown in **Fig 1.18**. In the *multimeter* version, selector switches allow measurement of voltage, current or resistance on several ranges. A peculiarity of these meters is that when measuring resistance the red (+) terminal becomes *negative*: this must be borne in mind when testing semiconductor devices.

Resistance

The total resistance of a combination of different resistors can be measured or calculated. If two resistors R1 and R2, having resistances R_1 and R_2, are joined in series (**Fig 1.19**)

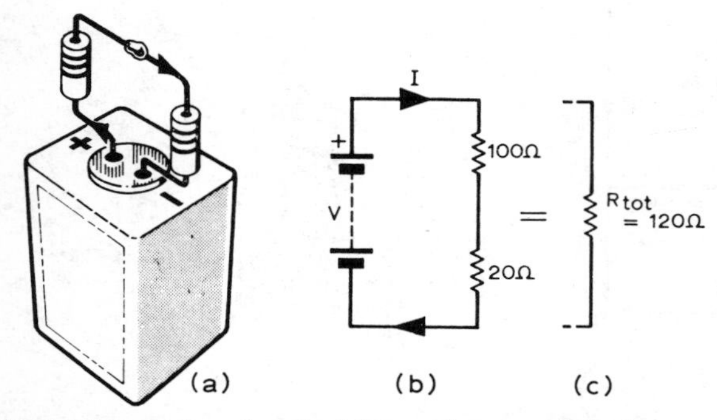

Fig 1.19. Two resistors in series (a) cause the same voltage drop as a larger resistor (c) equal to the sum of their resistances.

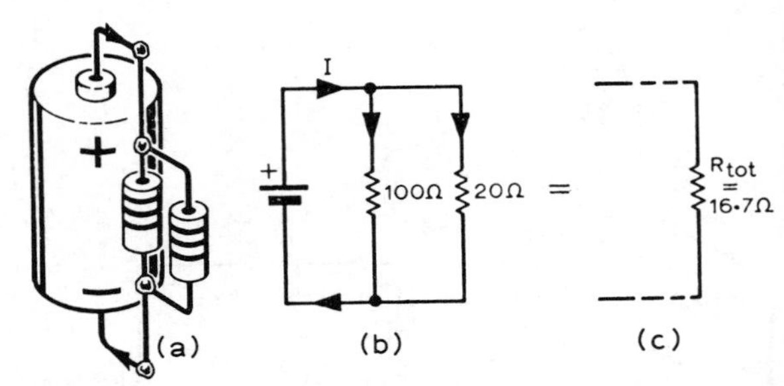

Fig 1.20. Resistors in parallel allow more current to pass, so the total resistance is smaller than that of either of the two individual resistors.

the voltage required to pass 1A through each resistor in turn is evidently $R_1 + R_2$, and is by definition the resistance of the combination.

$$R_{tot} = R_1 + R_2 \text{ (in series)}$$

If the resistors are joined not in series but side by side (in *parallel*, or shunting each other) calculation is more difficult. Imagine a 1V battery connected across them (**Fig 1.20**). From Ohm's law, any resistance R will pass a current $1/R$ amperes from this battery, so the resistance of the combination is given by:

$$\frac{1}{R_{tot}} = I_{tot} = I_1 + I_2 = \frac{1}{R_1} + \frac{1}{R_2}$$

or

$$R_{tot} = \frac{R_1 \times R_2}{R_1 + R_2}$$

Thus the total resistance in this case is a smaller resistance than either individual resistor.

Electric power passing into a resistor is converted into heat. The power is proportional to both voltage and current,

$$W \text{ (in watts)} = V \times I$$

One watt generates about 0·25 calories of heat per second, and this is as much as most carbon resistors will stand without burning. Applying Ohm's law gives alternative expressions for power:

$$W = \frac{V^2}{R} \text{ or } W = I^2 \times R$$

Thus doubling the voltage across a resistor generates four times as much heat.

Suppose for example that a valve heater designed to draw 0·3A from a 6V supply is to be powered from the ac mains supply. The mains voltage must be dropped from 240V to 6V. If a "dropper" resistor is to be used it must pass 0·3A at this voltage, so it must have a resistance of:

$$R = \frac{V}{I} = \frac{234}{0 \cdot 3} = 780\Omega$$

It must also withstand the heat generated,

$$W = V \times I = 234 \times 0 \cdot 3 = 70W$$

so it would have to be a large well-ventilated wire-wound resistor. It would be possible to use a 75W light bulb, since this has a resistance when hot of:

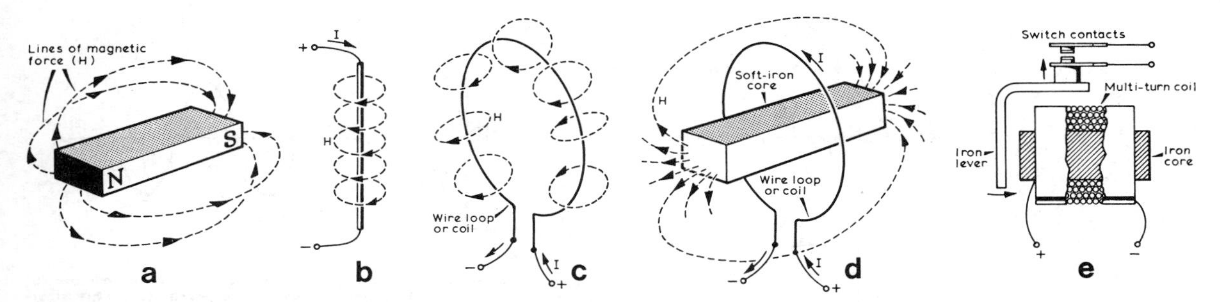

Fig 1.21. A bar magnet (a) is surrounded by lines of force, the force which would act on the north pole of another magnet brought near it. A wire carrying current (b) also generates a magnetic field, of strength proportional to the current passing. A coil or loop of wire (c) has a field like a bar magnet, and will magnetize a soft-iron or ferrite core (d): note the increased field strength due to the core. In the relay (e) this magnet pulls a lever to operate a switch.

$$R = \frac{V^2}{W} = \frac{240^2}{75} = 770\Omega$$

However, because its resistance was much lower at switch-on, the bulb would allow several amperes to flow for a moment before it lit up, and this surge of current might damage the valve heater.

Magnetism

A *bar magnet* is a hard metal bar with the interesting property that it will push the pole (end) of another magnet away from one end of itself towards the other end. This force around the magnet is called a *magnetic field* (**Fig 1.21**), and the stronger the field the harder another pole will be pushed along the "lines of force".

Since a wire carrying current deflects a nearby compass needle (which is actually a pivoted magnet), the current must generate a magnetic field. If the wire is bent into a circular coil the magnetic field resembles that of a bar magnet. This field will affect a soft iron *core* placed in the centre of the coil, making it a powerful magnet whenever current passes around the coil. The arrangement is called an electromagnet or solenoid, and it is used in the *relay* to operate a switch, and in the *earphone* to vibrate a metal diaphragm (**Fig 1.22**).

Induction

When a wire is moved across a magnetic field a voltage is *induced* in the wire, and current will flow if there is a complete circuit. This is the principle of the electricity *generator* or *dynamo*. A coil is rapidly rotated near a powerful magnet, and a large voltage appears at the ends of the coil. Another example of induction is the *ribbon microphone*, in which sound waves agitate a wire "ribbon" near a magnet generating tiny electric signals resembling the sound waves.

This effect also works in reverse. When a current is passed along a wire near a magnet a force is exerted and the wire tries to move across the magnetic field. In an ammeter this causes the coil to rotate. In the *electric motor* several coils are provided at different angles to form the armature: each is energized in the field of a permanent or electro-magnet (field coil) in turn and the armature thus spins round. A *loudspeaker* has a coil mounted on a paper cone near a magnet so that currents agitate the coil and cone, thus producing sound waves in the air: **Fig 1.23.**

When current passes through a meter the needle may swing irritatingly back and forth before giving a steady reading. One ingenious method of damping these oscillations uses the tiny voltages which they generate in the meter coil:

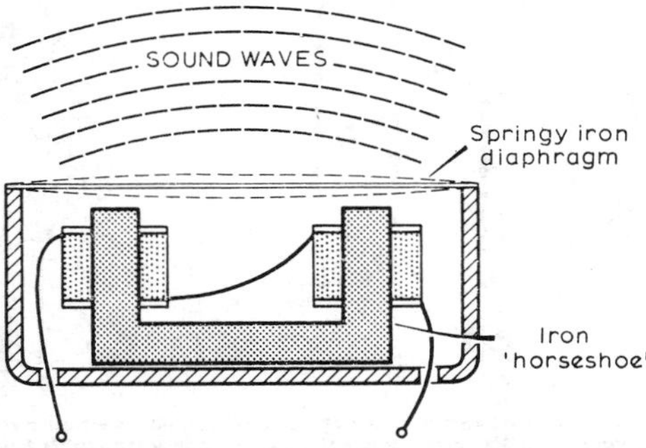

SOUND WAVES
Springy iron diaphragm
Iron 'horseshoe'

Fig 1.22. In the earphone or headphone, signal currents energize a U-shaped electromagnet which attracts an iron diaphragm, so that it vibrates and generates sound waves. In practice the electromagnet is also permanently magnetized to avoid distortion.

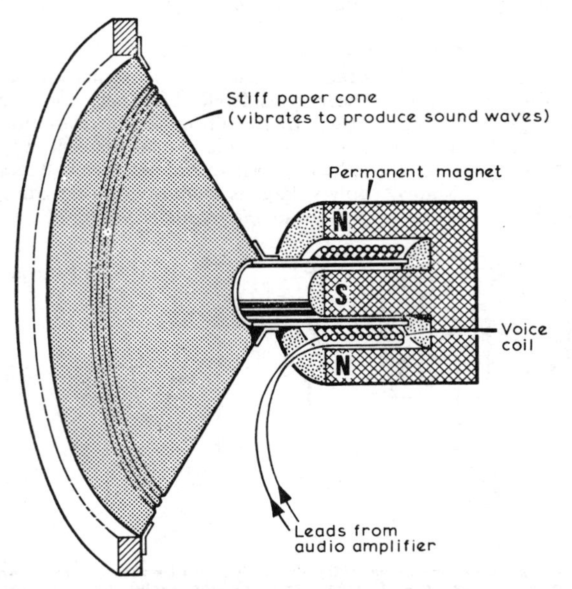

Fig 1.23. The loudspeaker has a coil suspended near a powerful magnet. Signal currents in the coil agitate an attached cone thereby generating sound waves.

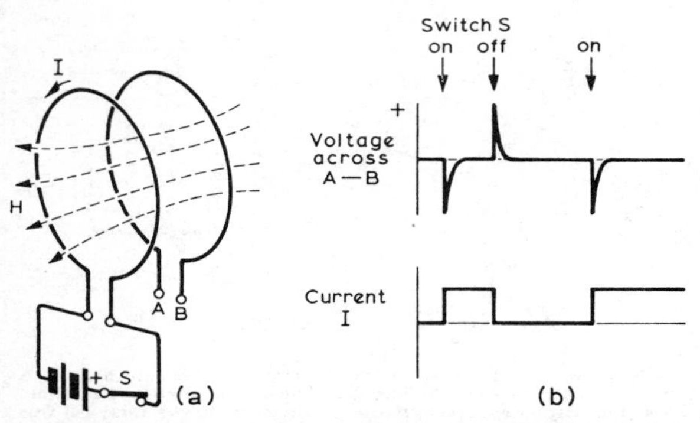

Fig 1.24. To demonstrate magnetic induction. Closing switch S starts a sudden current in the primary coil and a magnetic field springs up around it (a). The movement of this field momentarily induces a voltage in the secondary coil. Opening the switch causes another voltage spike as the field collapses (b). In a suitable coil the voltage spikes may be far larger than the battery voltage, eg a car ignition coil may produce several thousand volts.

a resistor is connected across the coil and absorbs power from the voltages, so that the oscillations rapidly die away. A resistor of six times the movement resistance gives adequate *damping* without excessive loss of sensitivity. Other meters use a metal coil former, which because it forms a closed circuit provides damping due to eddy currents circulating within it.

Voltage can also be induced in a stationary wire or coil by a moving or varying magnetic field, as in Faraday's original experiment (**Fig 1.24**). A sudden change of current in one coil moves its magnetic field, inducing a momentary current in the second coil. Similarly, the oscillating magnetic field in a radio wave induces tiny voltages in any conductor exposed to it, such as a receiving aerial or coil. A suitable core increases the inductance of the coil, giving a larger induced voltage, as in the *ferrite rod aerial.*

Alternating Current (AC)

Until now only steady voltages and currents (direct current or dc) have been considered. However, consider the generator again (**Fig 1.25**): during each revolution (cycle) the coil cuts across the magnetic field first one way and then the other, producing positive and negative voltages alternately. Mains supply voltage is therefore of alternating polarity, and causes a flow of *alternating current* (ac). In

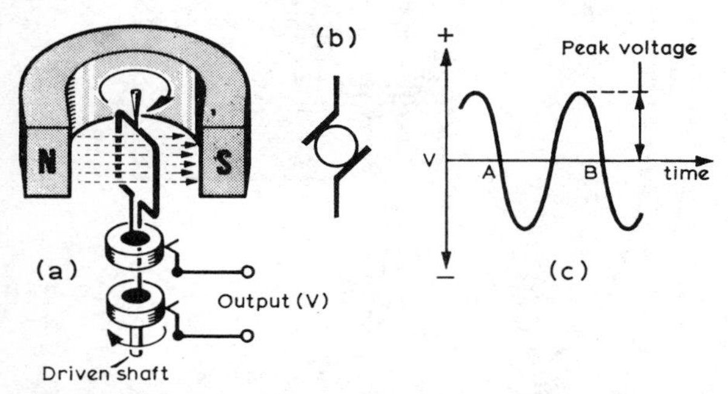

Fig 1.25. An ac generator (alternator) is a coil rotated between the poles of a magnet (a). During each revolution (cycle) one side of the coil passes first the north pole, then the south pole of the magnet, generating a voltage first positive and then negative alternately (c). (b) Circuit symbol of a voltage generator.

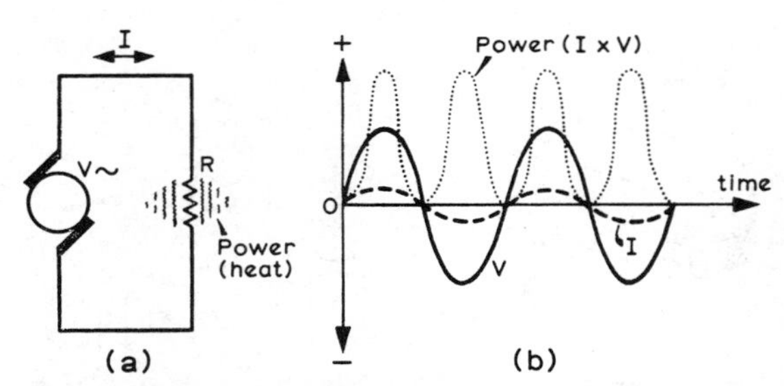

Fig 1.26. When an alternating voltage is applied to a resistor (a) current flows alternately backwards and forwards (b). This alternating current (ac) generates heat in the resistor proportional to the voltage-current product at each instant.

Great Britain ac supply generators are driven at a carefully synchronized speed which ensures the supply *frequency* is 50 cycles per second. In recent years the term "cycles per second" has been superseded by the unit "hertz" and this or its abbreviation "Hz" is used throughout this book.

Mains voltage reaches peak values of +340V and −340V alternately, but for the rest of each cycle it is smaller than this. Consequently it gives less power than a steady dc supply of 340V would give. It is in fact as *effective* as a dc supply of 240V, so it is described as "240V ac" (more accurately, 240V rms—see below).

Mains voltage applied to a resistor causes a flow of alternating current (**Fig 1.26**). The power generated is $V \times I$ or V^2/R, the dotted curve. The actual power dissipated in the resistor will be the mean of this square term, V^2/R, and from the graph this amounts to half its peak value, $\frac{1}{2}V_{pk}^2/R$. An ac supply is therefore as effective as a dc voltage of $\sqrt{\frac{1}{2}V_{pk}^2}$ which would dissipate the same power in the resistor. This is the *root mean square* (rms) voltage, and for a sinewave supply is:

$$V_{rms} = \sqrt{\tfrac{1}{2}V_{pk}^2} = \frac{V_{pk}}{\sqrt{2}} = 0{\cdot}7\ V_{pk}\ \text{approximately}$$

Capacitance

Consider two metal plates connected to a battery (**Fig 1.27**). The battery removes a few electrons from plate A, leaving it positively charged, and pushes them on to plate B.

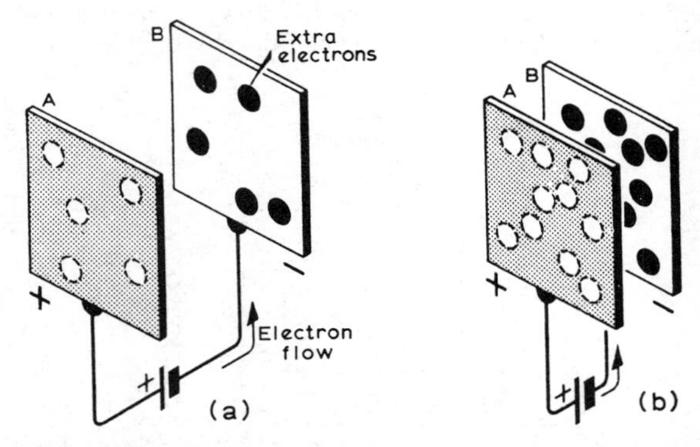

Fig 1.27. The quantity of electricity which flows when two plates are charged by a 1V battery is called their capacitance. If the plates are close together they have a much larger capacitance, because the positive charge on one plate attracts electrons onto the other plate. (+ indicates electrons removed from plate; − indicates electrons added).

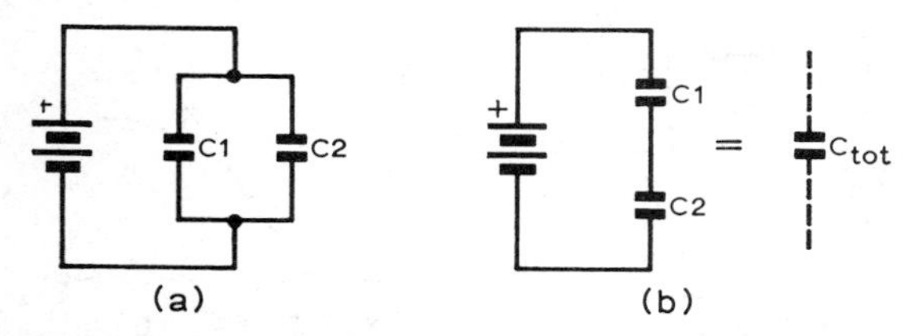

Fig 1.28. Two capacitors in parallel (a) accommodate more charge per volt applied, so the total capacitance is the sum of each. In series (b) the applied voltage is divided between the capacitors, so a smaller current flows during charging and the overall capacitance is smaller than that of either capacitor.

Plate B is now negatively charged and repels further electrons so strongly that current stops. The plates thus have a very small *capacitance* for electricity. Now imagine the plates brought very close together. Electrons approaching plate B are still repelled by its negative charge but they will also experience attraction by the positive plate in very close proximity, which partly overcomes the repulsion. The battery can now push many more electrons on to plate B, so the plates have a much larger capacitance.

Such a device is called a *capacitor*, and its capacitance (measured in *farads*) is the number of amperes for one second needed to charge it up to one volt. If it charges at I amperes for t seconds,

$$\text{Capacitance } C = \frac{Q \text{ (quantity)}}{V} = \frac{I \times t}{V} \text{ farads (F)}$$

The larger and closer the plates, the greater is their capacitance.

One practical difficulty is that the capacitance of air-spaced plates is minute. The common *tuning capacitor*, for example, despite many interleaved plates, has a capacitance of only 500pF per section (picofarad: 10^{-12} farads) or less. In fixed capacitors a capacitance up to a few microfarads can be achieved with fine metal foil separated by thin paper or plastic; for larger capacitances electrolytic construction is used (see p 1.3).

Capacitance can be varied by moving the plates together or apart. Rotating the spindle of a tuning capacitor enmeshes a greater area of the plates, thereby increasing the capacitance. A *compression trimmer* is adjusted by squeezing the plates together with a fixing screw. In the *condenser microphone* sound waves move one plate back and forth and the varying capacitance is used to generate signal currents.

Capacitors may be used in combination (**Fig 1.28**). Two capacitors joined in parallel simply act as another capacitor with a larger area of plates. The charging current to reach 1V will be the sum of the currents to each capacitor, so

$$\text{Total capacitance } C_{tot} = C_1 + C_2 \text{ (in parallel)}$$

When two capacitors are placed in series the charging voltage is divided between them, so they will accommodate less energy.

$$V_{tot} = V_1 + V_2$$

Therefore

$$\frac{It}{C_{tot}} = \frac{It}{C_1} + \frac{It}{C_2}$$

and

$$\frac{1}{C_{tot}} = \frac{1}{C_1} + \frac{1}{C_2} \text{ in series.}$$

When a capacitor is charged through a series resistor (**Fig 1.29**) it charges quite slowly, and as the voltage across the resistor falls during charging the current becomes progressively smaller. At first the current is simply V/R, and if this were maintained it would fully charge the capacitor in t seconds, where:

$$V = \frac{It}{C} = \frac{\frac{V}{R} \times t}{C}$$

so $t = RC$, the *time constant* of the circuit. Because the current falls, the capacitor is in fact only two thirds charged by this time.

If the capacitor after charging is quickly discharged by a neon valve or unijunction transistor it will commence recharging, be discharged again, and generate a saw-tooth waveform: Fig 1.29(d). This arrangement is the *relaxation oscillator*, and is used to provide the timebase for an oscilloscope.

Once the capacitor in Fig 1.29(a) is fully charged, current ceases—the capacitor blocks dc current. In contrast, an ac voltage will cause continual charging and discharging currents (**Fig 1.30**), thereby passing an alternating current. If a mixed ac and dc voltage is applied to a *blocking capacitor*, only the ac component passes through: in this way a signal voltage (ac) can be separated from a power supply voltage (**Fig 1.31**).

A larger capacitor will pass larger charging and discharging currents when the same ac voltage is applied, so it offers less *impedance* to ac. Impedance, denoted by the symbol Z, is similar to resistance, thus:

$$\frac{\text{Voltage (ac)}}{\text{Current (ac)}} = \text{Impedance}$$

However, this impedance depends on the frequency of

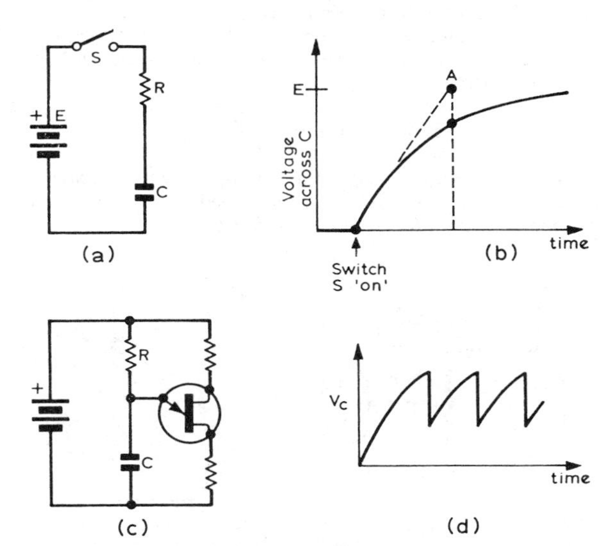

Fig 1.29. A capacitor charging through a resistor (a) draws a high current at first. As the capacitor voltage builds up the current tails off (b). If the capacitor is then suddenly discharged by a breakdown device such as a unijunction transistor (c) or neon tube it will recharge repeatedly, forming a saw-tooth wave (d).

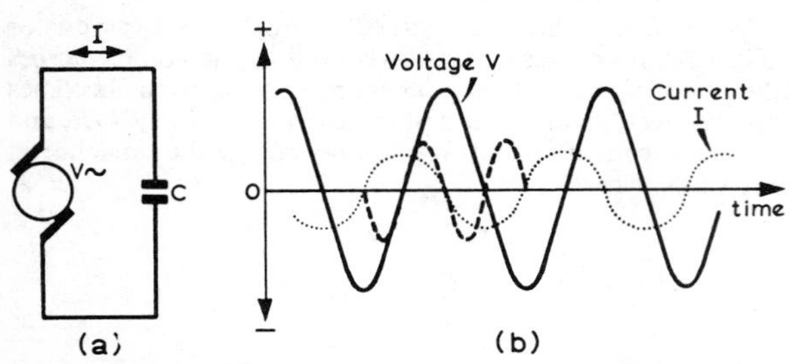

Fig 1.30. An alternating voltage applied to a capacitor (a) alternately charges and discharges it, current flowing alternately back and forth (b). Thus a capacitor does not block the passage of alternating current, but presents a definite impedance to it, measured in ohms. No heat is dissipated because the capacitor alternately stores power and returns it to the supply (dashed curve).

the alternating voltage, higher frequencies being impeded less in the case of a capacitor. At a frequency of f Hz,

$$\text{Impedance } Z = \frac{1}{2\pi fC} \text{ ohms}$$

A capacitor thus behaves rather like a resistor towards ac and obeys Ohm's law. One important difference is that virtually no power is dissipated as heat.

Close examination of Fig 1.30(b) shows that voltage and current peaks do not coincide, but are separated by a phase shift of a quarter-cycle (90°). This has several effects, one of which is to prevent power loss as heat, because the mean power dissipated over one cycle $V \times I$ is zero (dotted curve). A capacitor can be used as a cold mains "dropper", lowering the voltage without generating heat: Fig 1.31(b). For example, a 1·4μF capacitor will provide 100mA for an indicator bulb since

$$I = \frac{V}{Z} = \frac{240}{1/2\pi fC} = 240 \times 2\pi \times 50 \times \frac{1{\cdot}4}{10^6} = 0{\cdot}1\text{A approx}$$

In practice a little heat does appear, because the capacitor has a small "ohmic" resistance as well as capacitance, and this wastes power. The *power factor* is very small, perhaps 1 in 10^4, where

$$\text{Power factor} = \frac{\text{Resistance}}{\text{Impedance}}$$

Strictly speaking, the non-ohmic impedance of a capacitor or coil is termed *reactance*, while impedance refers to the combination of reactance and resistance.

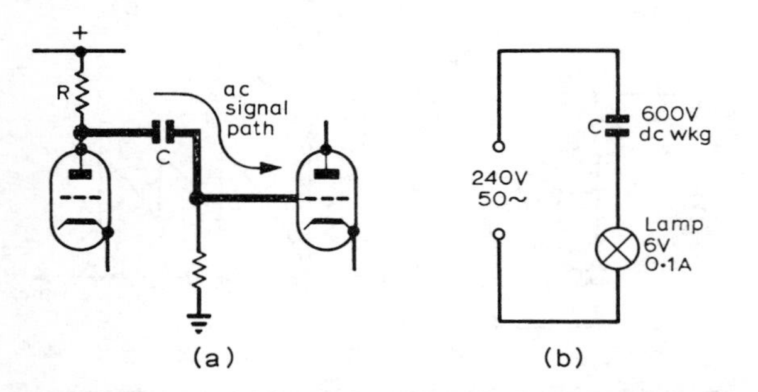

Fig 1.31. A blocking capacitor passes alternating currents (signal) while maintaining isolation from the dc supply. (a). Another use for a suitable capacitor is as an ac mains dropper (b), almost no heat being generated.

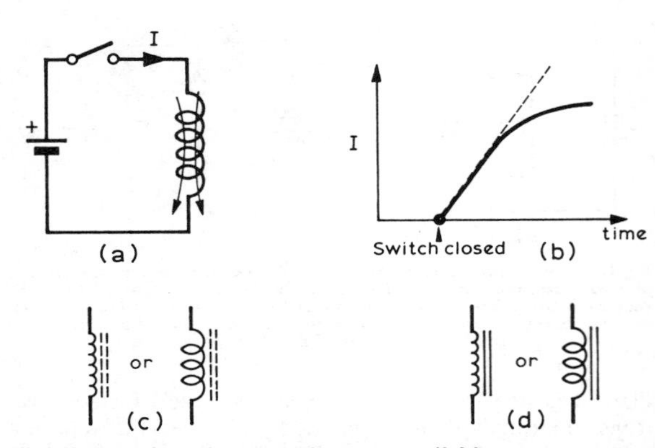

Fig 1.32. Applying a battery voltage to a coil (a) causes a current which builds up slowly, delayed by opposing induced voltages at each turn, until the current is finally limited by the resistance of the coil wire (b). The length of the delay (or self-inductance) can be increased by winding more turns on the coil, or by using a core of iron dust, ferrite (c) or iron strips (d).

Inductance

Consider a conductor wound into a helix (**Fig 1.32**). As current begins to flow round the turns a magnetic field arises and moves past all the turns inducing *opposing* voltages in them. This slows the current down very greatly at first. However, such *self-induction* occurs only when the magnetic fields are moving; once a steady current is established the fields stop moving and the coil simply behaves as a piece of wire (of low resistance).

When a battery is connected to a coil, current slowly increases as the self-induction effect is overcome, until the wire resistance limits further increase. The larger the coil the greater the induction and the more slowly the current will rise. The *inductance* in *henrys* (H) of the coil may be defined as the length of time taken to build up current of 1A from a 1V battery, neglecting wire resistance.

When a steady current through a coil is suddenly switched off the magnetic field collapses and a huge voltage "spike" appears across the coil (**Fig 1.33**): in a car ignition circuit this provides a spark at the sparking plugs. Switching off any coil (such as a relay or a motor) is likely to produce a spark at the switch, causing radio interference and also damage in some instances.

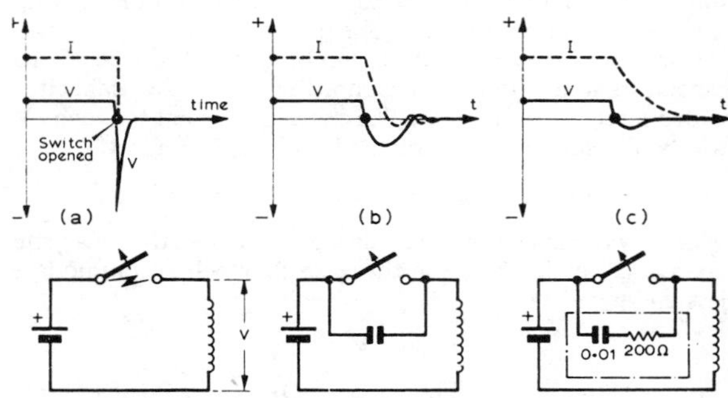

Fig 1.33. Sudden cessation of the current in an inductance causes a large transient "spike" (a) which may arc across switch contacts and cause radio interference. A capacitor (usually 0·01—1·0μF) connected across the switch allows the current to decay slowly at switch-off, preventing arcing (b). Commercial suppressors often contain a resistor and capacitor in one capsule (c).

Both these problems can be avoided by fitting a capacitor suppressor to absorb the voltage spike: Fig 1.33(b). Where a transistor is used to switch off the current to a coil it needs protection from the voltage spike. This is achieved by the use of a diode connected as shown in **Fig 1.34**.

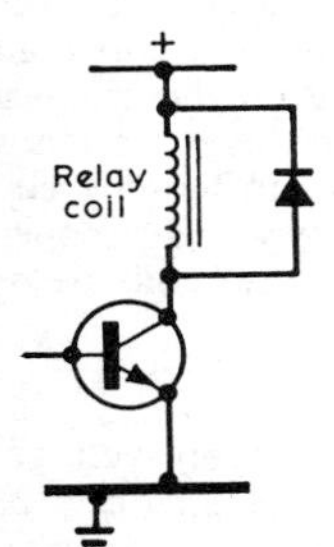

Fig 1.34. At switch-off the relay coil would develop a high voltage spike liable to damage the transistor. As the spike is of opposite polarity to the supply voltage it can be removed by a diode.

If an alternating voltage is applied to a coil only a little current flows, because by the time a small current begins to flow in one direction the supply changes polarity, and the current has to decrease and flow the other way (**Fig 1.35**). The more rapidly the supply alternates the less time the current has to build up, so the coil presents a greater impedance to high frequencies. It behaves like a resistance of:

$$Z = 2\pi f L \text{ ohms, at a frequency } f \text{ Hz}$$

Mathematically, the voltage across a coil is proportional to the rate of change of current,

$$V = L\frac{dI}{dt}$$

Alternating current usually has the waveform of a sine wave, with current

$$I = I_{peak} \sin(2\pi f t)$$

Therefore

$$Z_L = \frac{V}{I} = \frac{L.dI/dt}{I} = 2\pi f L \text{ ohms}$$

Once again current and voltage are 90° out of phase, but unlike a capacitor the coil current lags behind the applied voltage. As before, no power is dissipated in a coil save that due to the resistance of the wire.

An *af choke* is a large coil with an iron core to give it a very high inductance, up to 20H. It is used for smoothing (**Fig 1.36**), since the dc supply passes easily through while ac ripple encounters high impedance. Chokes are rarely used in

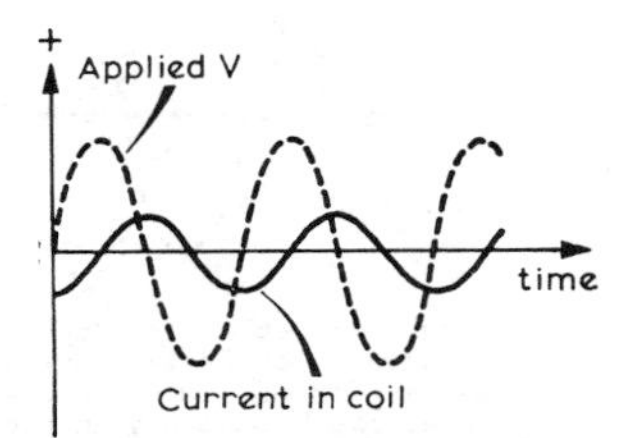

Fig 1.35. Alternating voltage applied to a coil causes a current which slowly increases during positive half-cycles and slowly decreases on negative half-cycles. The current peaks are thus delayed 90° by the coil's self-inductance, and no power is dissipated as the mean voltage-current product is zero.

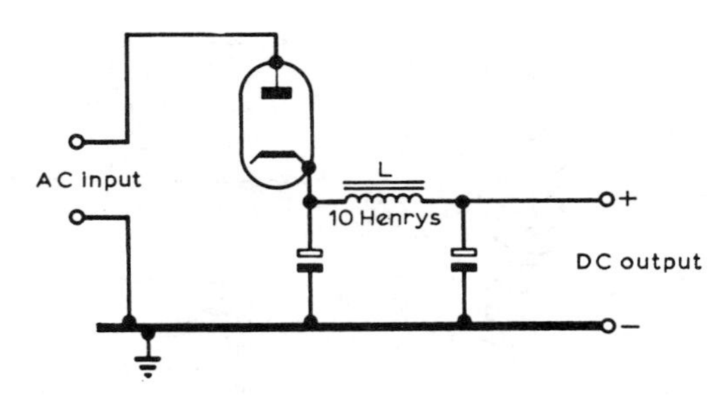

Fig 1.36. A smoothing choke presents a high impedance to mains ripple, about 3kΩ, but allows dc to pass with little voltage drop.

PSUS supplying transistorized equipment, because these usually require a high current and therefore a thick-wire choke, which would be inconveniently large and heavy.

A much smaller inductance will pass mains or audio frequencies, but will block radio frequencies (rf) of several megahertz. Such an *rf choke* can be made by winding a quarter-wavelength of wire on a small air-cored former, giving an impedance of about 100kΩ at that wavelength. At very high frequencies a few turns is enough, and even a ferrite bead threaded on to a supply lead will prevent leakage of rf currents along it. In the typical transmitter pa stage shown in **Fig 1.37** the rf choke L blocks rf current changes while allowing dc supply current and audio modulations to reach the valve anode. The rf signal is thus diverted via a "coupling" capacitor to the tank circuit and transmitting aerial.

Transformers

When two coils are wound on the same core any current in one coil (the primary) induces a voltage in the other coil (the secondary). This is called *mutual induction*, see **Fig 1.38**. The larger the secondary coil the larger is the voltage induced in it, this voltage being

$$V_s = nV_p$$

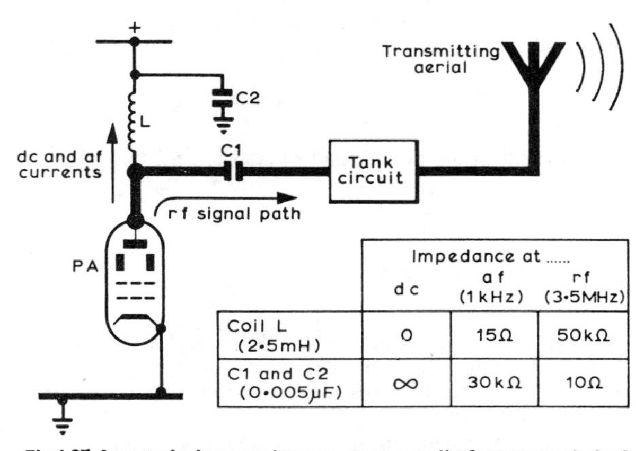

	Impedance at		
	dc	af (1kHz)	rf (3·5MHz)
Coil L (2·5mH)	0	15Ω	50kΩ
C1 and C2 (0·005μF)	∞	30kΩ	10Ω

Fig 1.37. In a typical transmitter pa stage a radio-frequency choke L blocks rf current changes while allowing dc supply current and audio modulations to pass to the valve anode. The rf signal current instead passes via a coupling capacitor C1 into the tank circuit; any rf which does leak past the choke is smoothed out by a capacitor C2. Both capacitors block both dc and audio frequencies.

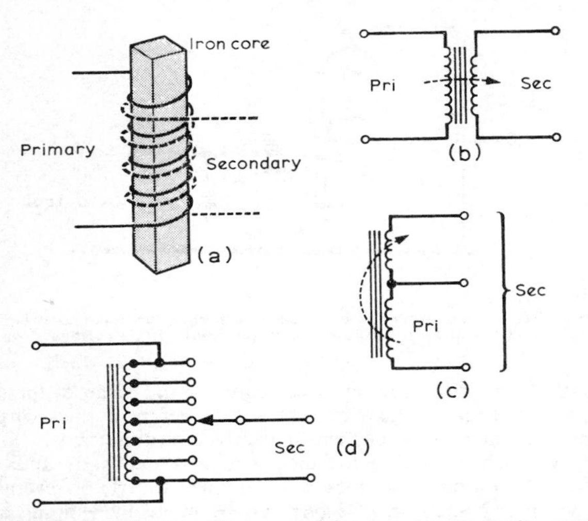

Fig 1.38. A mains transformer contains two or more coils on a common iron core (a): in the circuit symbol (b) a curved arrow may be used to indicate mutual induction between the coils. An auto-transformer contains one tapped coil (c), so that part of the coil is common to both primary and secondary circuits—as the currents in each are of opposite polarity the total current in the common section is actually less than elsewhere, and thick wire is unnecessary. The "Variac" transformer (d) supplies an adjustable output voltage.

where n is the turns ratio (secondary turns : primary turns) and V_p is the voltage applied to the primary.

Thus mains voltage can be *transformed* into any desired voltage; a 40:1 turns ratio yields 6V from a 240V mains supply, or a 1:3 ratio will step up the mains to 720V. Such manipulation is possible only with an alternating supply, and it was for this reason that mains supplies throughout the world are ac, although many (including those in Great Britain) were originally dc.

Although a step-up transformer increases the available voltage, there is no power gain. Current drawn from the secondary induces a "resistive" current in the primary coil in phase with the supply voltage, drawing as much power from it as is being supplied by the secondary.

Two currents flow in the primary winding, one inductive (with a 90° phase shift) and one resistive (0° shift), given by:

$$I_p\,(90°) = \frac{V}{2\pi fL} \qquad I_p\,(0°) = nI_s$$

If the secondary supplies a load of resistance R,

$$I_s = \frac{V_s}{R}$$

so

$$I_p\,(0°) = n\frac{V_s}{R} = n^2\frac{V_p}{R} = \frac{V_p}{R/n^2}$$

In other words, a resistance of R/n^2 is *reflected* into the primary winding. By suitable choice of turns ratio the secondary load can be transformed to any desired resistance for matching purposes.

The inductive current in the primary depends on the inductance of the transformer, and limits the maximum input voltage. If this voltage is exceeded the transformer will buzz and become hot, even with no current drawn from the secondary. A shorted turn inside the transformer would have a heavy current induced into it, thus producing the same effects.

If high frequencies are applied to an iron-cored coil or transformer, power is wasted in the core, which is heated by a flow of eddy currents. *Laminated cores* (insulated iron strips) minimize eddy currents, and they can be further reduced by using non-conducting magnetizable core material such as iron dust or *ferrite* compounds. At higher frequencies the core must be small to reduce power loss; it often takes the form of a small ferrite *slug* which can be screwed into the coil to adjust its inductance. At very high frequencies (vhf) even this is unsatisfactory and air-cored coils are used.

Rectification

Radio equipment needs a steady dc supply voltage. To produce this, ac mains voltage must be *rectified* in a *power supply unit* (psu) to form dc. The heart of the psu is a device which allows electrons to flow one way only, a *diode*. There are two types, semiconductor and thermionic.

Silicon is a *semiconductor*: it has no free electrons when pure, but can be *doped* with impurity to provide either free electrons (n-type alloy), or free positive current carriers, *holes* (p-type alloy). Either type alone acts as a simple resistance, but in the silicon diode the two types meet at a central junction (**Fig 1.39**). When the diode is *forward biased* electrons and holes are driven to this junction and will carry current across it. If a voltage of opposite polarity is applied, however, current stops because the junction is depleted of these carriers. Consequently when an ac voltage is applied

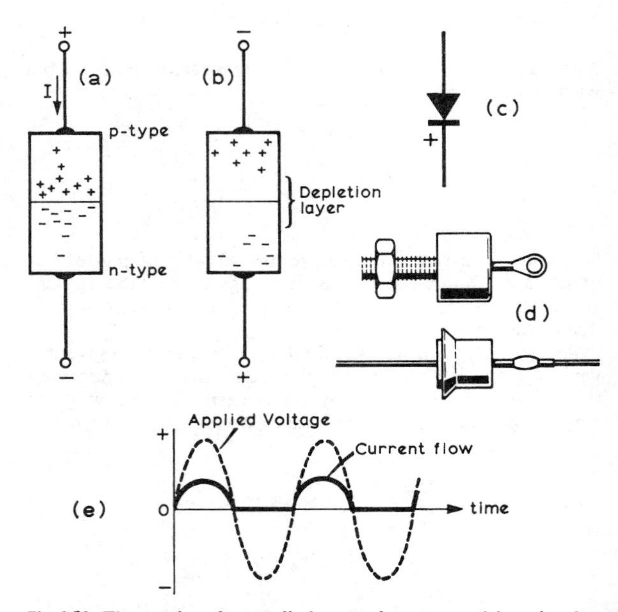

Fig 1.39. The semiconductor diode contains a central junction between two alloys, one with negative current carriers (electrons) and the other with positive carriers (holes). Forward bias forces carriers to the junction (a) and they combine, carrying current across it. Reverse bias draws them away from the junction leaving a depletion layer (which is an insulator), so current ceases (b). (c) Diode symbol: the n-type end is marked + because it will be connected to the positive end of the load resistance. (d) Top-hat and stud-mounting types. (e) Applying an alternating voltage results in forward current pulses on alternate half-cycles, or half-wave rectification.

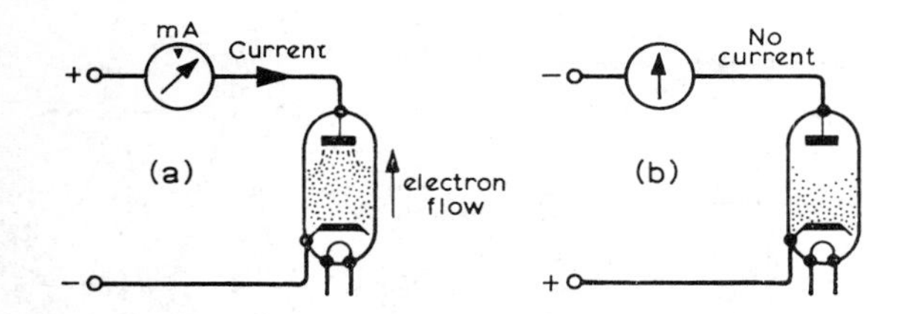

Fig 1.40. The valve diode has a heated cathode which emits electrons into the vacuum. Forward bias attracts these to the anode and current flows (a), but as no electrons can leave the cold anode reverse bias produces no current flow (b).

current flows only on forward half-cycles. This subject is dealt with more fully in Chapter 3—*Semiconductors.*

In the thermionic or *valve* diode two electrodes are mounted in a vacuum and one is heated so that it emits electrons into the vacuum (**Fig 1.40**). When the *cathode*, the hot electrode, is negatively charged electrons are repelled towards the *anode* and conduct current across the vacuum. Reversing the polarity repels electrons away from the anode, so no current can flow. Chapter 2—*Valves* covers the subject in more detail.

A diode produces unidirectional pulses unsuitable for powering equipment, as they would produce a loud mains hum and distortion. To smooth these pulses into a steady dc voltage a large *reservoir capacitor* is connected across the supply. It is charged up by each pulse and maintains a fairly steady supply voltage, with only a small variation or *ripple.* Power supply design is considered in Chapter 16—*Power Supplies.*

Basic psu design is instructive and quite simple. Suppose it is wished to supply 100mA to an apparatus at 9V dc. As shown in **Fig 1.41(b)**, the ac input voltage peaks must slightly exceed the output, so an input of 10V peak (7V rms) would be suitable. A reservoir capacitor of 10V working voltage and a very large capacitance (to give good smoothing), say 1,500μF, is required. Also required is a rectifier diode rated at 100mA (the mean current drawn: the manufacturer will have designed it for much larger current pulses to charge the capacitor). This diode must be rated at 20 peak inverse volts (piv), since the input voltage varies down to −10V while the output voltage remains at +9V. Of course, all these are minimum specifications—components with higher ratings will do equally well.

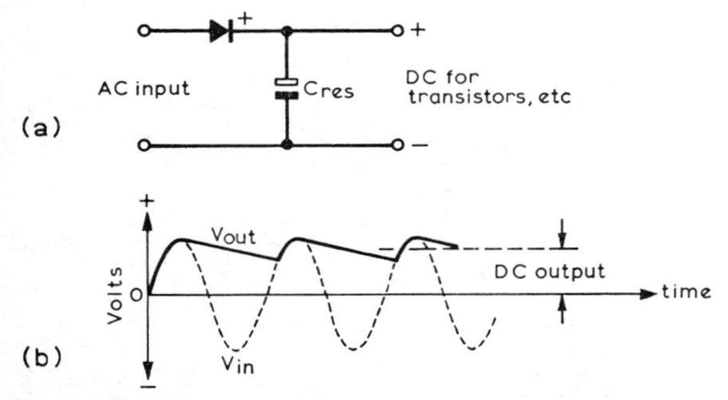

Fig 1.41. The half-wave rectifier contains a large reservoir capacitor which stores the current pulses (a). Its charge is topped up by supply peaks to maintain a steady dc voltage (b) with a small variation or ripple. Between peaks the capacitor slowly discharges to provide output current. The special capacitor symbol indicates an electrolytic type.

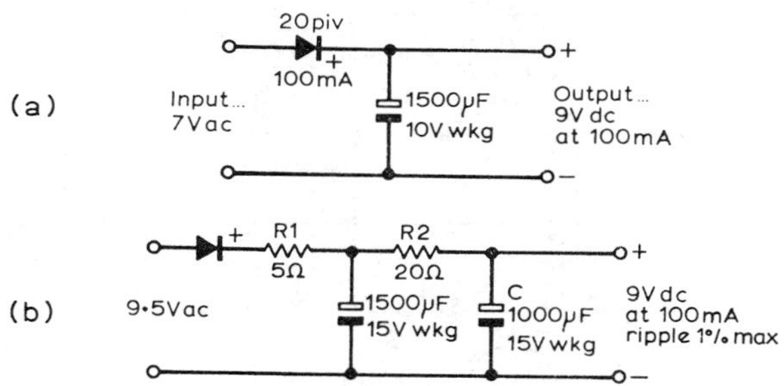

Fig 1.42. Without smoothing, the simple half-wave rectifier (a) gives an output of 9V dc with a ripple component of 0·5V ac. The ripple can be reduced by a smoothing network (R2,C) to only 0·06V, but a higher input voltage is needed to offset the loss in R2 (b). The series resistor R1 protects the diode and capacitor from very high charging current peaks, and further reduces ripple.

The ripple from this simple psu (**Fig 1.42(a)**) can be calculated easily. Between cycles the capacitor discharges 100mA for nearly one fiftieth of a second. Its voltage will fall by:

$$V = \frac{It}{C} = \frac{0{\cdot}1 \times 0{\cdot}02}{0{\cdot}0015} = 1{\cdot}4\text{V approx}$$

The output voltage therefore fluctuates by ±0·7V, so the ripple voltage is 0·7V peak (0·5V rms).

To reduce this ripple further, a more sophisticated smoothing circuit is needed: Fig 1.42(b). The ripple is divided across an extra resistor and capacitor. The capacitor has a low impedance to ac (about 3Ω), so the ripple will be reduced to:

$$\text{Output ripple } V' = 0{\cdot}5\text{V} \times \frac{3}{R_2 + 3} = 60\text{mV approx}$$

A single-diode rectifier like the above makes use of only half the supply cycle. *Full-wave* rectification can be achieved with a transformer or a diode bridge (**Fig 1.43**), giving better stability and less ripple. Another refinement is to stabilize the voltage, maintaining a constant output voltage despite supply or load changes: the simplest method is the shunt regulator (**Fig 1.44**), or a more complex type as shown in Fig 1.88.

Sometimes there is a need to generate a dc supply from a smaller ac voltage. Two half-wave rectifiers in series as shown in **Fig 1.45(a)** will double the voltage, or the voltage can be increased almost indefinitely by a *voltage multiplier*, as shown in **Fig 1.45(b)**. Although the circuit arrangement can be used to obtain millions of volts it is more commonly used in the eht circuits of oscilloscopes.

SIGNAL CIRCUITS

Sound travels through the air as small compression waves, and the frequency with which waves arrive at the ear is sensed as "pitch". The lowest audible tone has a frequency of about 30Hz and the highest about 20kHz, while most of the energy in speech lies in the range 300–3,000Hz. A microphone will convert these sound waves into small voltages carrying all the information of the original speech. Such a voltage is often called a *signal*, and some examples are shown in the photograph. They are alternating voltages, in this case at the same frequencies as the sound waves, ie *audio frequency* signals (af).

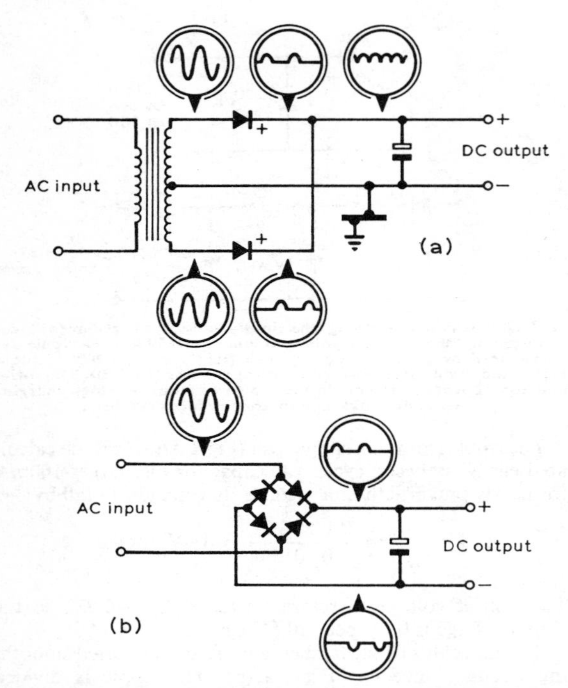

Fig 1.43. The usual full-wave rectifier (a) contains a centre-tapped transformer which supplies positive half-cycles from each end of the winding alternately. The bridge rectifier needs no transformer (b), and is used in ac meters: a disadvantage is that neither output terminal is at earth potential with respect to the input.

A popular microphone is the *crystal* type (**Fig 1.46**), in which sound waves cause vibration in a piezoelectric crystal, a crystal which generates voltage when it is compressed or deformed. Louder sounds generate larger signal voltages, up to about 100mV ac. A crystal gramophone pickup works

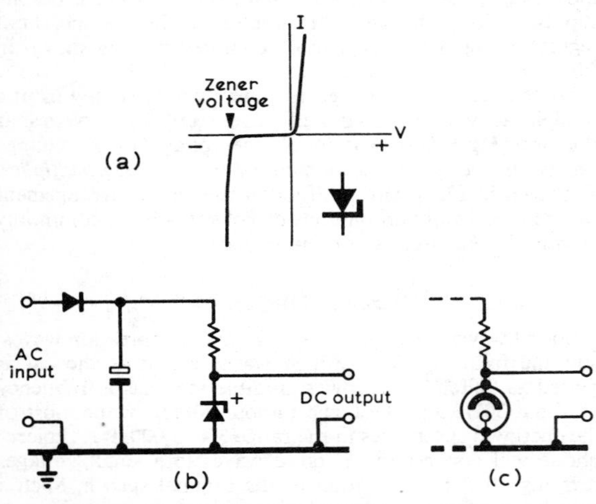

Fig 1.44. A zener diode conducts at a fixed reverse voltage, its zener voltage (a). A simple stabilized psu (b) can be constructed by shunting excess power away via a zener diode, limiting the output to the reverse breakdown voltage of the diode, usually 5 to 15V. Above 50V a neon regulator would replace the zener diode (c).

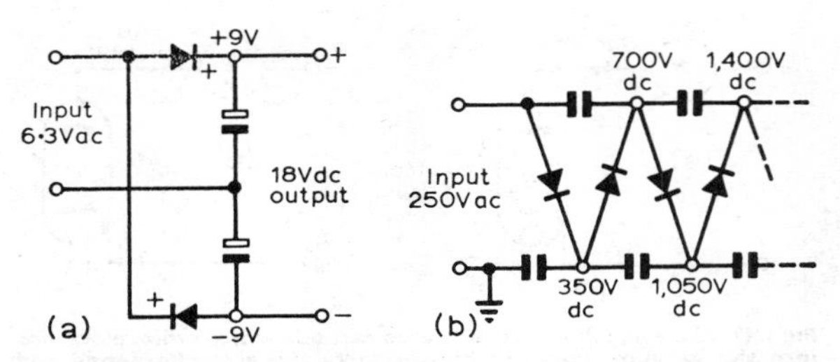

Fig 1.45. A voltage doubler consists of two half-wave rectifiers in series (a). With a voltage multiplier (b) any desired voltage can be achieved by a cascade effect. In each circuit capacitors with isolated cans are needed.

similarly, the crystal being agitated by the vibrations of the gramophone stylus.

Signals can be modified in several ways. Audio signals, for example, can be *amplified* to a high power, enough to operate a loudspeaker. This might require about 50mW for comfortable listening volume. (This is for speech—to produce loud bass notes for music reproduction much more power is needed.) Signals can also be *attenuated* (made smaller): two resistors in series will divide the signal voltage between them (**Fig 1.47**) so that the output is only a fraction of the input voltage. The signals can also be *filtered*, by passing them through resistors and a capacitor for example (**Fig 1.48**). The capacitor allows high frequencies (treble) to pass more easily than others, so the combination gives more attenuation at some frequencies than at others (a non-uniform *frequency response*).

Resonance

A coil and a capacitor connected in parallel have some interesting properties. If the capacitor is momentarily charged by an external voltage it discharges through the coil, but the coil then continues to pass current until its magnetic fields have collapsed; see **Fig 1.49**. By this time the capacitor is charged in the reverse direction, so it again discharges through the coil, and once more becomes charged by the coil current. The whole cycle is repeated many times, and the larger the coil and capacitor the longer each cycle takes. The frequency of this *resonance* is given by

$$f_{res} = \frac{1}{2\pi\sqrt{LC}} \text{ Hz}$$

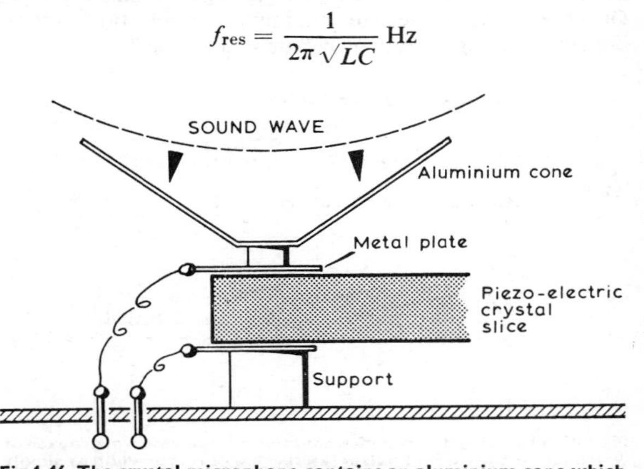

Fig 1.46. The crystal microphone contains an aluminium cone which is agitated by the sound waves. This makes the crystal vibrate, and small voltages from it are picked up by the metal contacts on each side.

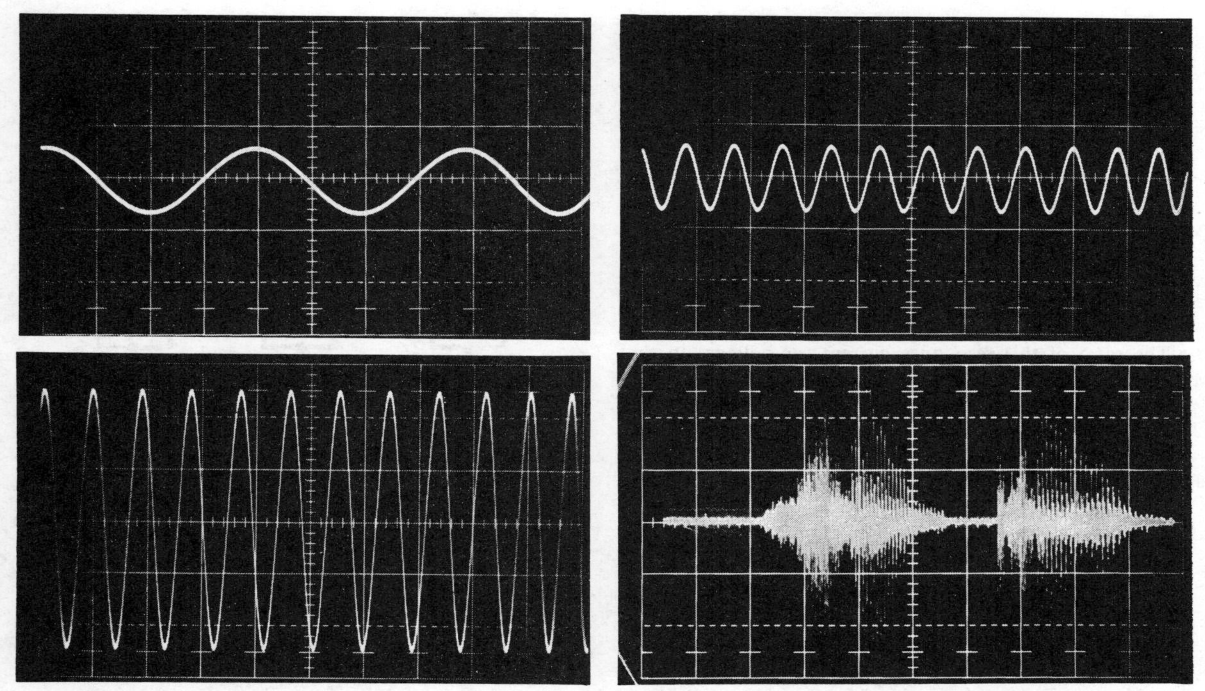

A microphone converts sound waves into signal voltages, small ac voltages which alternate at the same frequency as the sound waves. These photographs show oscilloscope traces of signals produced by whistling (i) middle C, (ii) a note two octaves higher, (iii) the same note but more loudly, and (iv) the signal from a human voice calling "CQ".

When an ac voltage is applied to the circuit the result depends on the applied frequency (**Fig 1.50**). If the applied voltage alternates at exactly the resonant frequency a large current builds up inside the circuit and little external current can pass through. The circuit is said to be *tuned* to this frequency, at which it has a very high impedance or *dynamic resistance* of perhaps 100kΩ. A small current is drawn from the applied source while a much magnified current circulates inside the circuit. Such a parallel-tuned circuit can be used to select signals at its resonant frequency, or to reject them (**Fig 1.51**). The use of a variable capacitor allows the circuit to be tuned over a range of frequencies.

Fig 1.47. To attenuate a signal it is applied to two resistors in series: the signal voltage is divided proportionately between them (a). The potentiometer allows the degree of attenuation to be varied, the output voltage being proportional to R (b). Because the human ear is most sensitive at low volume levels, a volume control potentiometer for an audio stage should have a non-linear (log-law) resistance scale (c) if uniform changes in loudness over its range are to be achieved.

A coil and capacitor in series also form a tuned circuit, but with almost opposite properties to that mentioned above. At resonance the circuit has almost zero impedance and draws a large current from the supply. For this reason it is called an acceptor circuit. It is often used to magnify the output voltage in a vhf amplifier (**Fig 1.52**), because at resonance the high current generates a very large voltage across both coil and capacitor, providing voltage gain.

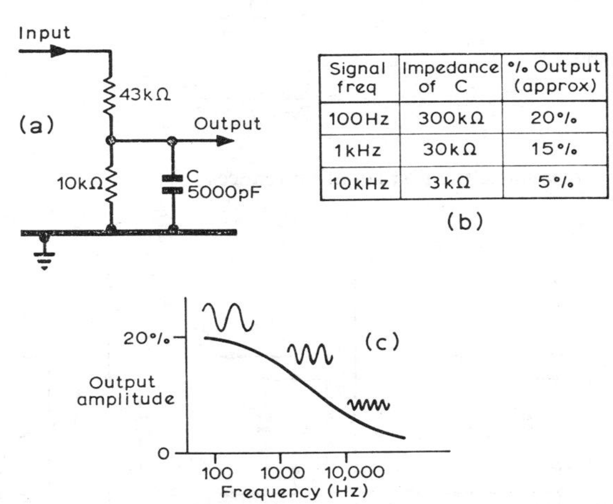

Signal freq	Impedance of C	% Output (approx)
100Hz	300kΩ	20%
1kHz	30kΩ	15%
10kHz	3kΩ	5%

Fig 1.48. A treble filter (a). The capacitor impedance changes with frequency and so varies the attenuation (b). Its low impedance at high frequencies gives a falling frequency response, ie the treble frequencies are filtered off.

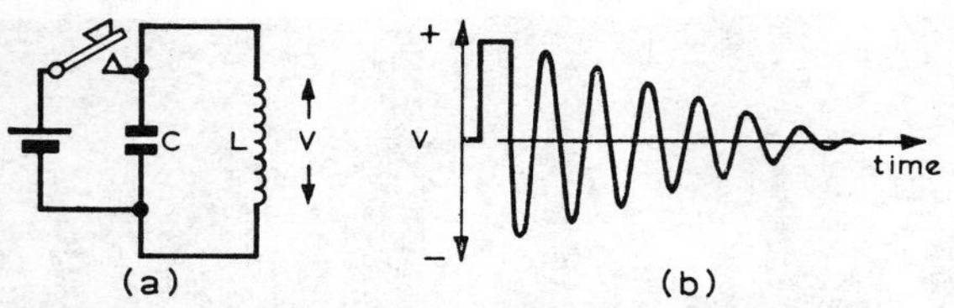

Fig 1.49. A coil and capacitor form a resonant circuit (a). A charge placed on the capacitor flows through the coil, which continues to pass current until almost the whole charge has accumulated on the other plate of the capacitor. This reversed charge then flows back through the coil, and current oscillates back and forth at the circuit resonant frequency, slowly losing amplitude. After Q cycles the current has fallen to about 5 per cent of its initial value, due to resistive losses.

The ratio of this resonance current (or voltage) to that supplied externally is a measure of the "goodness" of the circuit, and is known as the *magnification factor* (Q).

The magnification factor produced by a resonating circuit is theoretically infinite. In practice the ohmic resistance of the coil wire (and a small loss in the capacitor) limit Q to between 10 and 300 times magnification. Building high-Q tuned circuits is an art for which amateurs are justly famous. The coil must be of the shortest possible length of thick wire for minimum resistance, which necessitates a short squat coil (wider than its length). It should generally be a single-layer air-cored coil of as large an inductance as is feasible, and the capacitor should be a low-loss type.

The effect of ohmic resistance on Q can be calculated. Consider a series-tuned circuit of L, C and a small ohmic resistance r, which is passing a current I at frequency f. The total voltage across the circuit is the sum of that across each element,

$$V = IZ_L - IZ_c \quad (+Ir \text{ at } 0°)$$

The subtraction occurs because coil and capacitor voltages are of opposite phase. This voltage falls to a minimum when $Z_L = Z_c$, ie resonance occurs at a frequency such that:

$$\pi f_{res}L = \frac{1}{2\pi f_{res}C}$$

whence

$$f_{res} = \frac{1}{2\pi\sqrt{LC}}$$

At resonance the first equation reduces to:

$$V = I_{res}\,r$$

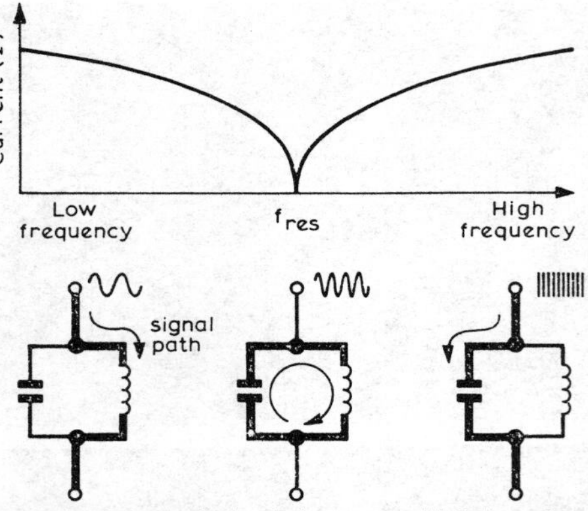

Fig 1.50. A parallel-tuned circuit has a very high impedance at its resonant frequency, fres, because a small applied current excites a huge circulating current inside the circuit. Current at a lower frequency can pass easily through the coil, while frequencies higher than fres pass through the capacitor.

so the dynamic resistance is simply r. The voltage across the coil at resonance will be:

$$V_L = I_{res}Z_L = V\frac{Z_L}{r}$$

The voltage magnification, Q, is therefore:

$$Q = \frac{Z_L}{r} = \frac{2\pi f_{res}L}{r}$$

Equally,

$$Q = \frac{Z_c}{r} = \frac{1}{2\pi f_{res}Cr}$$

As the ohmic resistance of a tuned circuit is almost entirely due to the coil, amateurs often refer loosely to the "Q" of a coil, but the intended resonant frequency must be specified (since Q varies with frequency).

For a parallel-tuned circuit similar calculations show that the current magnification is again

$$Q = \frac{Z_L}{r} = \frac{Z_c}{r}$$

but in this case the impedance at resonance (the dynamic resistance) is

$$R_D = \frac{L}{Cr}$$

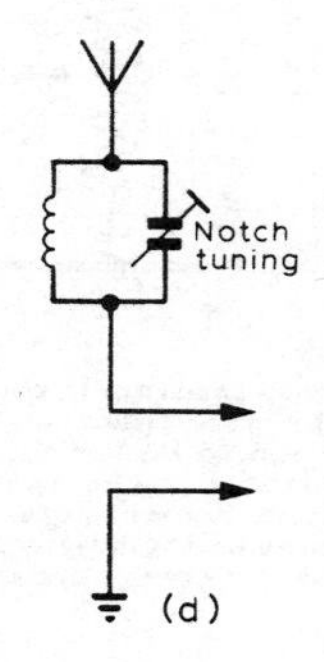

Fig 1.51. Signals can be selected with a parallel-tuned circuit by shorting unwanted frequencies to ground (a), transforming the desired signal to a high voltage (b), or by generating a high current at resonance (c). An unwanted local signal can be attenuated by a "wave-trap" placed between aerial and receiver (d).

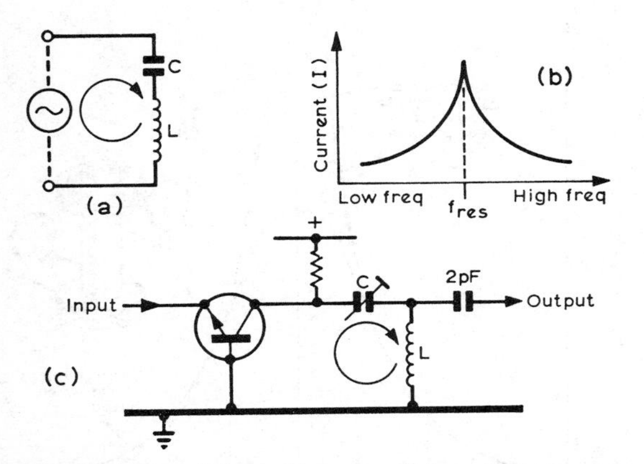

Fig 1.52. A series-tuned circuit (a) passes a high current at its resonant frequency (b), which generates a very large voltage across the coil. This is used to provide voltage gain in vhf amplifiers (c), the magnified voltage being coupled to the next stage by a small capacitor. (Biasing not shown).

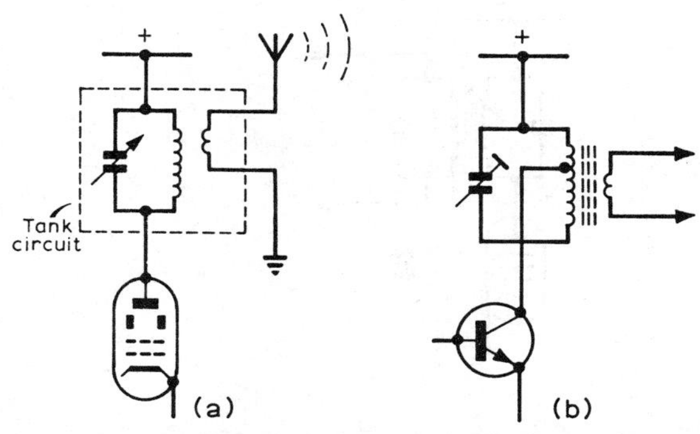

Fig 1.53. When a tuned circuit is heavily loaded (ie closely coupled to a resistive load) oscillations in it are damped by the load, and it becomes less selective. In the tank circuit of a transmitter the aerial must be quite loosely coupled to the pa to minimize this effect (a). To reduce loading in transistor i.f. transformers the collector impedance is placed across only part of the primary winding (b). Other examples of loose coupling to reduce loading are shown in Figs 1.51 (b) and (c).

Any load which draws power from a tuned circuit will damp down the oscillations and reduce Q. For example, a transmitter tank circuit (**Fig 1.53(a)**) is loaded by the aerial; close aerial coupling will give a very low Q, but loose coupling will transfer little power to the aerial. The best compromise (for maximum aerial power) lowers Q to about 12 and needs careful adjustment, preferably by a variable loading control in a pi-tank circuit. The low Q has certain advantages, making the pa stage more efficient and easier to tune.

Bandwidth

Because a tuned circuit contains some resistance it will not only resonate at its exact resonant frequency but also at any nearby frequency, the magnification falling as the frequency departs from f_{res} (**Fig 1.54**). The 70 per cent *bandwidth* of the circuit is the range of frequencies which will pass through it with at least 70 per cent of the full magnification (ie at half power or more compared to a signal at f_{res}). It can be shown that this range is

$$\text{Bandwidth}\, f = \frac{f_{res}}{Q}\ \text{Hz}$$

At high frequencies a tuned circuit makes a very poor filter. For example a circuit with a Q of 100 tuned to 7·05MHz will pass all frequencies between 7·01 and 7·09MHz—almost the whole of the 40m amateur band!

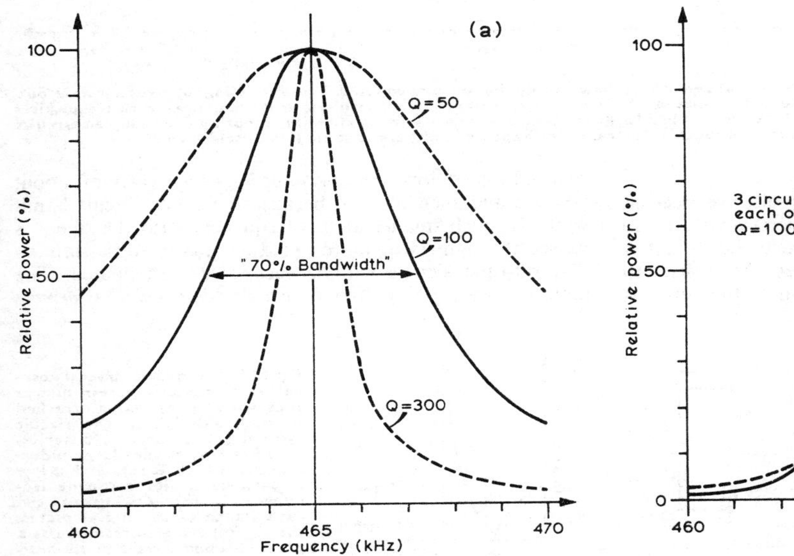

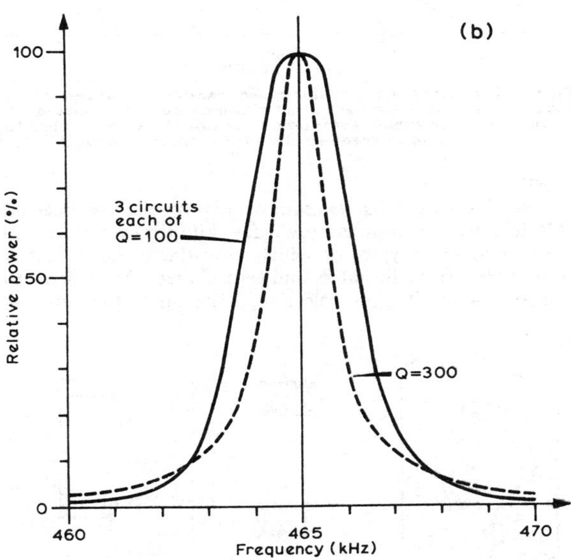

Fig 1.54. Tuned circuits with high magnification Q make better filters (a). The "70 per cent bandwidth", the bandwidth at half power (or 70 per cent voltage), is inversely proportional to Q. For receivers a better filter consists of several tuned circuits in succession (b), giving a broader "nose" and better attenuation of distant frequencies.

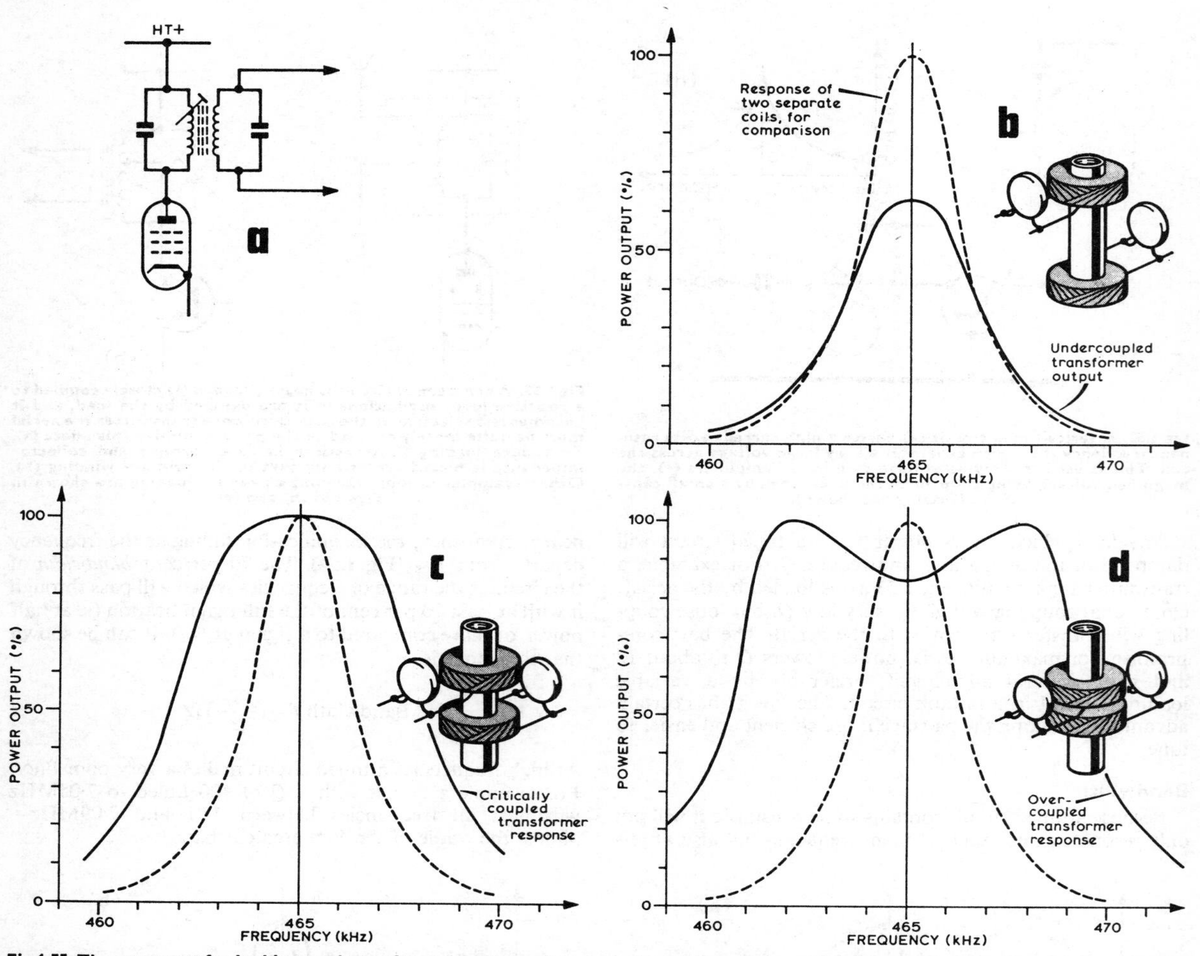

Fig 1.55. The response of a double-tuned transformer (a) depends on how closely the coils are coupled. Loose coupling (b) gives a low output, the bandwidth being similar to that of two uncoupled circuits (dotted curve). Critically coupled circuits give a broad peak with steep skirts (c) suitable for telephony reception, since a.m. signals have a bandwidth of 10kHz or more. For high fidelity reception over-coupled circuits give an even broader bandwidth (d), although a central notch appears and there is some loss of selectivity.

Filters

Today's crowded bands require a highly selective receiver, able to separate stations only a few kilohertz apart. There are several filter types, of which only the tuned circuit is cheap and easily adjustable, and several methods are used to compensate for its poor selectivity. The signal frequency is reduced to a low *intermediate frequency* (i.f.) of about 500kHz and then filtered, because the tuned circuit bandwidth is much smaller at this frequency. Also the signal is passed through six or more tuned circuits in succession, so that off-tune signals will be successively attenuated. The tuned circuits can be paired to form *double-tuned transformers*

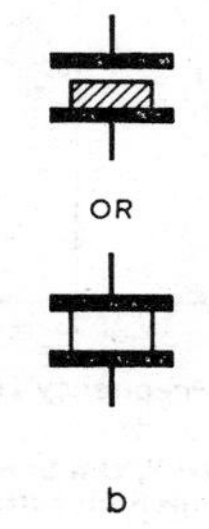

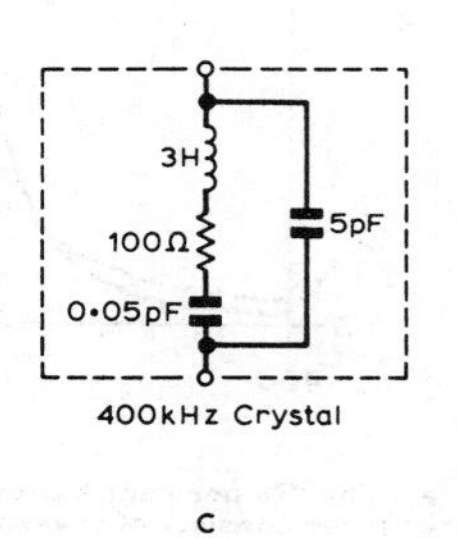

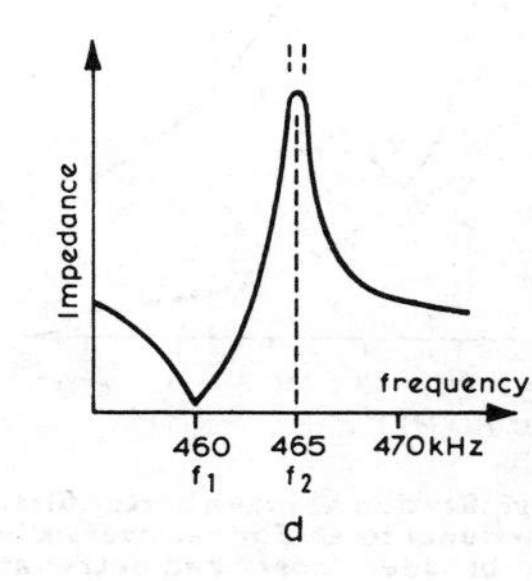

Fig 1.56. The quartz crystal resonator (a) contains a thin quartz slice with metal plates attached to each side, as in the circuit symbol (b). It acts as a series-tuned circuit of very large inductance and high Q (c), with a low impedance at its resonant frequency f_1 (d). Owing to the capacitance of the metal plates, the crystal can also resonate as a parallel-tuned circuit at its anti-resonant frequency f_2, with an impedance of several megohms.

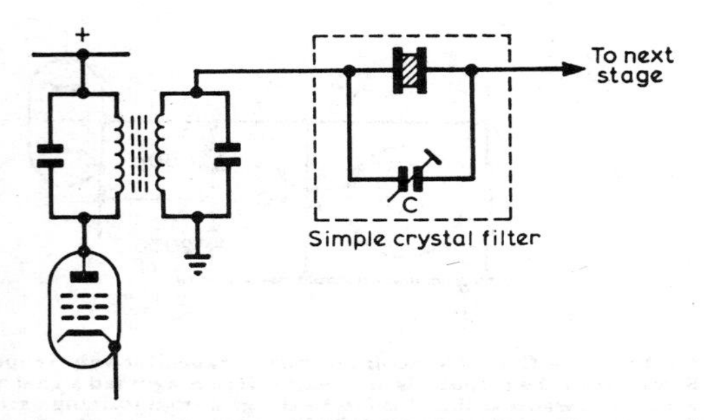

Fig 1.57. In this simple crystal i.f. filter the crystal resonates in its series mode. The resonant frequency can be adjusted slightly by a small trimmer C, about 30pF for a 465kHz filter.

(Fig 1.55): the tuned circuits interact to produce better rejection of adjacent frequencies, especially when they are *critically coupled.*

Unfortunately tuned circuits of high Q interact so strongly that critical coupling occurs with them wide apart, when little signal would reach the secondary coil. Practical i.f. transformers are therefore a compromise, using coils with an uncoupled Q of about 100. They are slightly under-coupled to make alignment easier.

Other types of filter act by converting the electric signal into mechanical vibrations. In the *crystal filter* **(Fig 1.56)** the signal is applied to a piezoelectric quartz crystal and makes it vibrate. At a certain frequency the crystal "rings" and its impedance drops sharply to almost zero, giving the effect of a "Q" of 100,000 or so. The thinner the quartz slice the higher its resonant frequency, but a 0·1 mm slice (which resonates at about 20MHz) is as thin as is practicable. For higher frequencies the quartz can be resonated at an odd harmonic (*overtone*) of its fundamental resonant frequency, and overtone crystals are cut to facilitate vibration in this mode.

Crystal filters have superb stability, the frequency remaining almost unchanged during years of correct use. It does

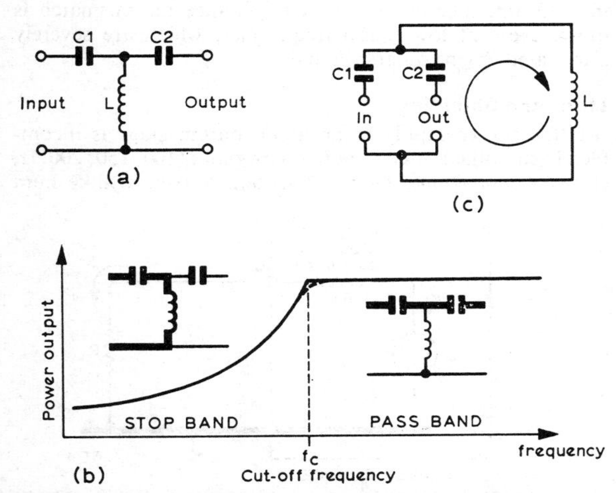

Fig 1.58. A high-pass T-filter (a) allows high frequencies to pass through the capacitors, low frequencies being shunted away by the coil (b). At the cut-off frequency fc resonance occurs (c), giving the curve a fairly sharp edge (depending on the circuit Q).

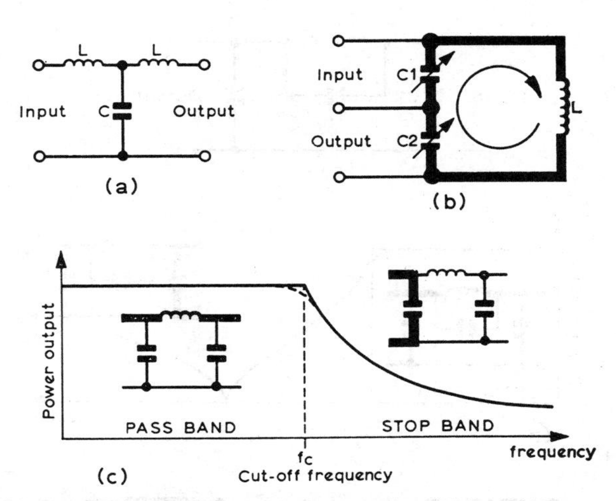

Fig 1.59. Low-pass T-filter (a) and pi-filter (c), the latter being preferred as it needs only one coil. Low frequencies pass through the coil, while high frequencies are removed by the capacitors. Resonance occurs at the cut-off frequency, and the circuit forms a capacitatively-tapped transformer (b): using variable capacitors any desired output impedance (eg a transmitting aerial) can be matched to the input impedance.

vary a little with temperature, perhaps by 50 parts per million over a day. Quartz filters are expensive but can be found on the surplus market. A very simple crystal filter circuit is shown in **Fig 1.57**, and many others in Chapters 4 and 6.

Other filters include the ceramic *transfilter*, a piezoelectric device with high and low impedance terminals, and also the *mechanical filter* which is excellent but expensive and delicate.

Wideband Filters

Filters can be deliberately constructed to pass a wide range of frequencies, for example any signal above a certain cut-off frequency f_c. An example of such a ***high-pass*** **filter is** shown in **Fig 1.58** and a *low-pass* filter in **Fig 1.59. These** filters are tuned to resonate at the cut-off frequency (***constant-k*** *filters*): signal rejection is good throughout most of the *stop band* (the band of unwanted frequencies) but low near the cut-off frequency.

Another type is the *m-derived filter*, which contains circuits tuned to more than one frequency. It can be designed to give a very sharp cut-off and excellent rejection near the band edge, but rejection is poor elsewhere in the stop band **(Fig 1.60)**. M-derived filters are simple to build, but design and alignment are more complex.

Capacitive Tapping

When a coil is *tapped* by connecting a terminal part-way along the coil it becomes an auto-transformer, and any voltage across one part appears in greater magnitude across the whole coil. If the coil and a capacitor form a tuned circuit at resonance the coil can still be tapped directly, or it can be as effectively tapped by dividing the *capacitor* into two parts, since the coil and capacitor bear the same voltage at resonance **(Figs 1.61 and 1.62)**. The circuit behaves like an auto-transformer, and signals applied across one capacitor appear enlarged across the whole circuit.

An advantage of capacitive tapping is that the effective position of the tap can be adjusted by varying the capacitors.

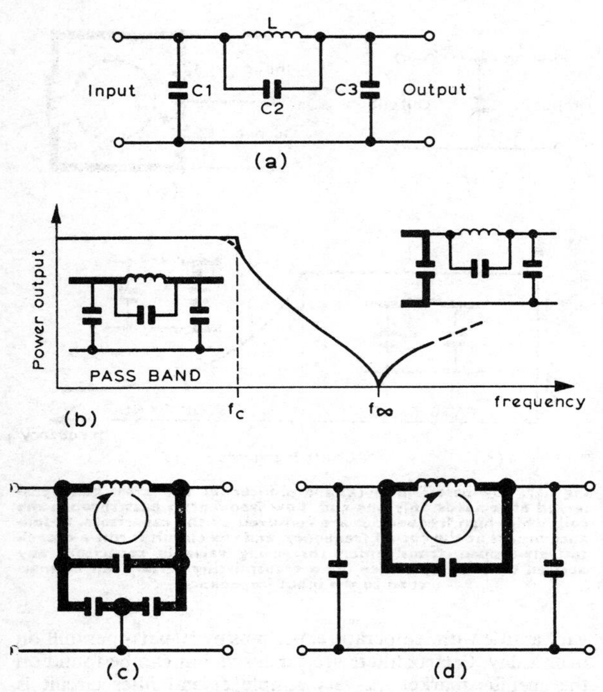

Fig 1.60. An m-derived low-pass filter (a) resonates at two adjacent frequencies. At the cut-off frequency fc it acts as a tuned transformer (c), while at its infinite rejection frequency f∞ it forms a notch filter (d). This gives the m-filter a steep slope near the edge of the pass-band (b). Unfortunately higher frequencies leak through C2, so attenuation is poor in the rest of the stop-band.

This is the principle of the popular pi-tank transmitter circuit, in which varying the tap position alters the loading on the output valve so that it can be matched to the aerial resistance: Fig 1.59(b).

Matching

No power source will supply an infinite current. When a heavy current is drawn from a source its output voltage falls, as though it contained an internal *source resistance* in series (**Fig 1.63**). Maximum power will be drawn from the source by a load resistance equal to this ohmic source resistance. This applies to any signal source, such as an aerial, tuned circuit or pa valve; the maximum signal power is extracted by matching the source resistance with an equal

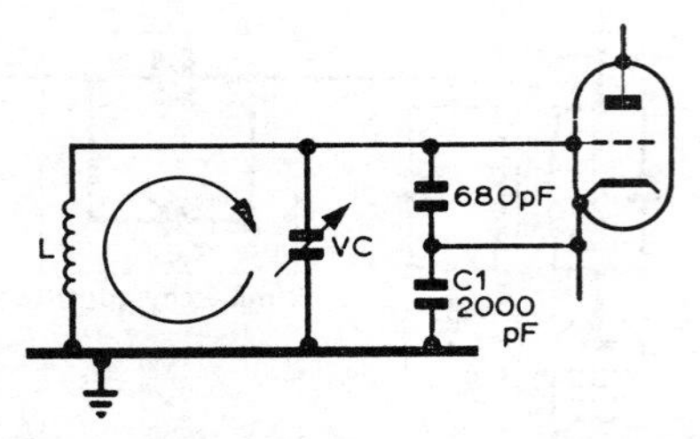

Fig 1.62. The Colpitts oscillator uses a capacitatively tapped coil. Signal from the cathode is applied to C1; a magnified signal appears across the whole coil and is fed to the grid, maintaining oscillation. A separate tuning capacitor VC allows the circuit to be tuned without altering the tap position.

load. The reason is simply that a larger load would draw less current, while a smaller load resistance will cause a large drop in output voltage, in either case lowering the power output.

A signal source will often have additional non-ohmic impedance due to internal inductance or capacitance, which will limit the current drawn. This internal impedance can be tuned out by resonating it with an external coil or capacitor, because this forms a series-tuned circuit which has zero impedance at resonance. Complete matching thus requires separate matching of ohmic and non-ohmic source impedance.

For example, a mobile aerial at hf acts as a signal source with a small internal resistance and capacitance (**Fig 1.64**). Very little current can be drawn through the capacitance (which has an impedance of several thousand ohms) unless it is resonated with a *loading coil* to give zero impedance. Aerial current is then only limited by the aerial's ohmic resistance, and for maximum power transfer the receiver input stage must present an equal resistance. The aerial is then fully matched.

Source impedance usually varies with the frequency in use, so matching arrangements have to be readjusted at each change of transmission frequency. If a mismatch occurs less power is transferred. For example, when a crystal microphone (which has a high source impedance) is plugged into an amplifier of low input impedance the mismatch is most severe at low (bass) frequencies, which are severely attenuated, giving a harsh sound.

Hum and Shielding

If the power supply to an audio output stage is incompletely smoothed, mains supply harmonics (100, 150, 200Hz, etc) cause *smoothing hum* at the loudspeaker. Unlike hum

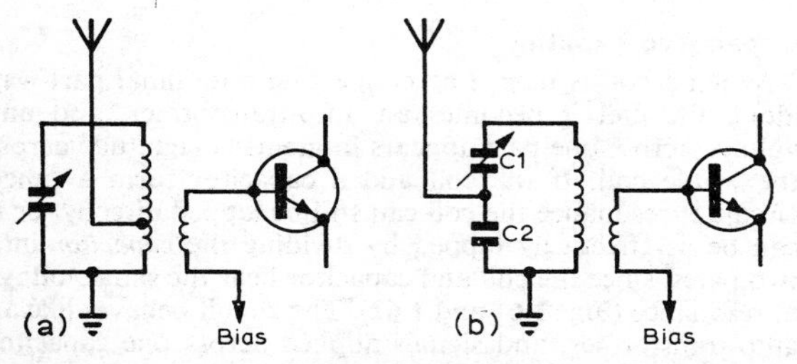

Fig 1.61. In this 144MHz converter front end (a), the need for a tapped coil can be avoided by using a capacitative tap (b). To load the circuit lightly the tap is placed near the earthy end of the coil by making C2 larger (of lower impedance) than C1.

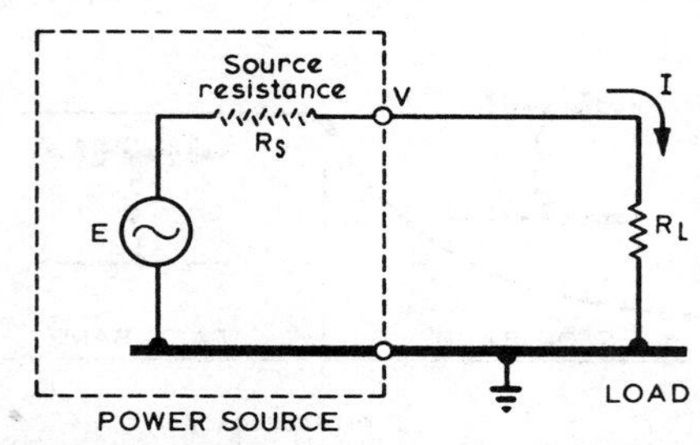

Fig 1.63. Matching. The power drawn from a source is limited by its internal resistance Rs. Maximum power will be delivered if the load RL is of exactly the same resistance, ie a load matched to the source resistance.

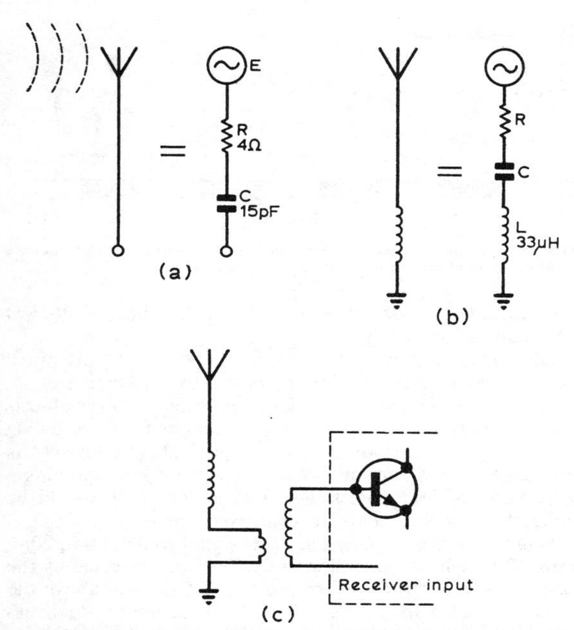

Fig 1.64. A short receiving aerial such as a 7MHz mobile whip acts as a signal source E in series with a small resistance and capacitance (a) The capacitance can be tuned with a loading coil to form a series-resonant circuit of very low impedance (b). For maximum signal transfer the receiver input impedance of 75 Ω or so is matched to the aerial resistance by a transformer (c) or a tap on the loading coil.

from other sources it is not affected by the volume control, but can be reduced by fitting an extra smoothing capacitor.

More common is *induced hum.* Any wire or conductor carrying a small signal to a sensitive amplifier is liable to pick up induced voltages from nearby mains cables and their alternating magnetic fields. This adds hum to the signal. Input wiring should be carefully shielded with earthed metal braid and kept away from ac supply lines, transformers and the heater supply line. The latter should be made from twisted flex: the two wires carry equal and opposite currents, so the resultant magnetic field is almost zero.

Wires carrying rf currents are surrounded by very powerful magnetic fields, which easily introduce signal voltage into neighbouring circuits. This may produce heterodyne whistles ("birdies") or parasitic oscillation, or the signal may leak out of the apparatus along supply wires and cause television interference. Stages passing large rf currents therefore need shielding, usually by enclosing the whole stage in a wire mesh cage. Wires entering or leaving the cage pass through a filter such as a feed-through capacitor or a ferrite bead to prevent rf leakage.

If one wire or strip of chassis provides a return path for both a signal and an alternating current (eg heater supply), the large ac current adds an *earth-return hum* to the signal. Separate earth return paths are necessary. Other causes of hum are valve heaters affecting the electron stream (this effect can be balanced out as in **Fig 1.65**), a loose transformer core buzzing, and earth-loop hum. *AC mains interference* is a series of clicks and crackles from the mains supply: it crosses the mains transformer by capacitive coupling, which can be prevented by a *Faraday screen* (**Fig 1.66**).

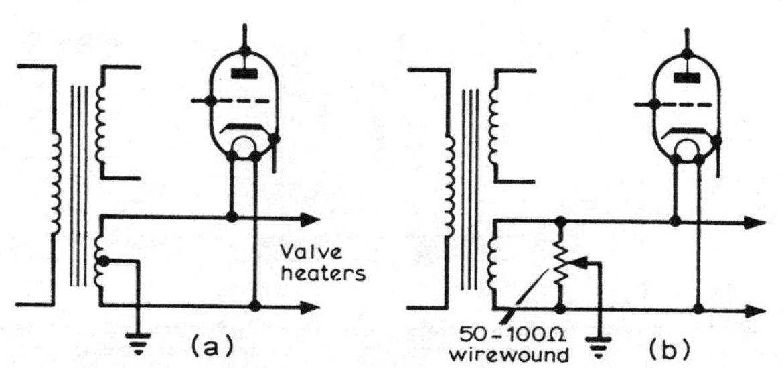

Fig 1.65. Hum from valve heaters can be reduced by a balanced voltage supply from a winding with an earthed centre-tap (a), so that effects from each end of the heater cancel out. Even better is a "hum-bucking potentiometer", which can be adjusted for minimum hum (b).

RADIATION

A conductor carrying alternating current radiates or transmits energy in the form of *electromagnetic waves* (radio waves). These waves travel at the speed of light (300,000,000m/s) through air and vacuum and nearly as fast through most solid materials. When they meet another conductor, such as an aerial, they are absorbed and induce a current in it similar to the original current in the transmitting aerial.

After a one-second transmission the distance between first and last waves is 300,000,000m. If n waves were produced in that second, the distance between waves (*the wavelength*) is evidently:

$$\text{Distance} = \frac{300{,}000{,}000}{n} \text{ metres, or } \frac{3 \times 10^{10}}{n} \text{ cm.}$$

Thus a transmission at a frequency n of 28MHz has a wavelength of 10·7m.

Short electromagnetic waves (of high frequency) tend to travel in straight lines and are quickly absorbed by solids: the shortest are light waves, X-rays and gamma-rays (**Fig 1.67**). Radiation from an aerial is efficient only at lower frequencies, over the range 10^5 to 10^{10}Hz, and these are

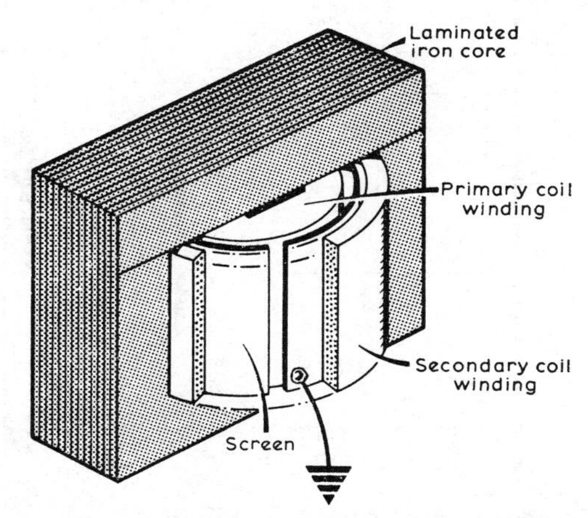

Fig 1.66. A Faraday screen is an earthed sheet of copper plate or wires between the primary and secondary windings of a transformer. As copper is non-magnetic mutual induction is unaffected, and no current flows because it does not form a complete turn. It prevents mains interference reaching the secondary winding by capacitive coupling.

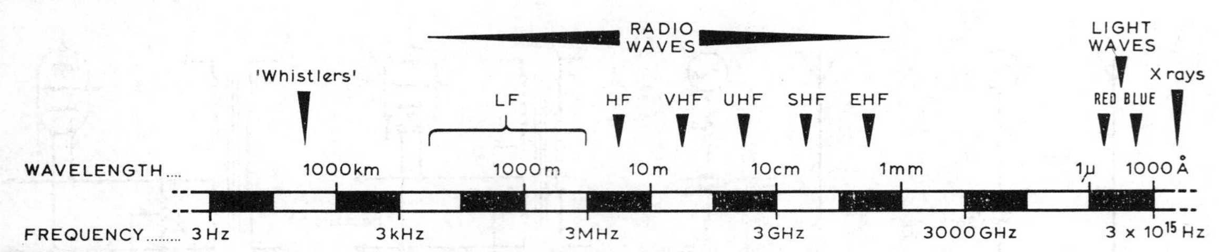

Fig 1.67. The electromagnetic spectrum, showing the central portion used for radio communication. Longer waves are difficult to radiate, and shorter waves do not reach distant stations. The central portion is shown in more detail in Fig 1.68.

termed *radio frequencies* (rf). **Fig 1.68** shows the rf spectrum together with the amateur and broadcast bands.

Use of the various bands depends how the waves travel (their *propagation*), which depends on the frequency. In the *super high frequency* (shf) bands (9cm, 5cm, 3cm, 12mm) signals travel in a straight line almost like light waves, and can be similarly focused on to the aerial by a curved dish reflector. The bands are therefore used for *line-of-sight* communication.

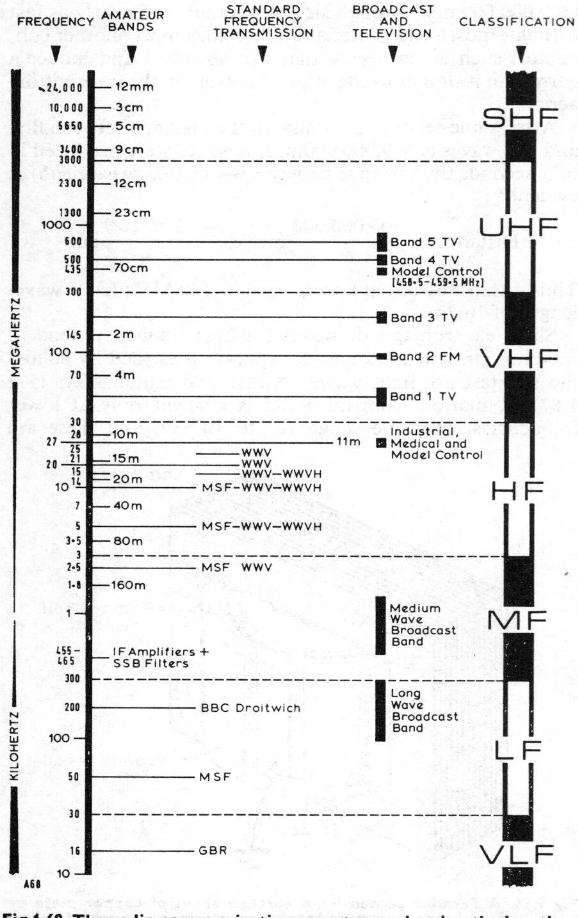

Fig 1.68. The radio communications spectrum, showing the location of the amateur bands. Amateur television is restricted to certain parts of some bands, and telephony is transmitted in the high-frequency portion of each band.

On *very high frequency* (vhf) bands (4m and 2m) direct *ground-wave* communication is possible over 100 miles or so. These frequencies are also very well reflected by large objects such as mountains, the moon and meteor trails, allowing occasional reception over much longer paths. Television is broadcast at uhf and vhf because a video signal contains a wide band of frequencies (3–6MHz wide) which would be difficult to accommodate at lower frequencies.

Short wave or *high frequency* (hf) signals (80m, 40m, 20m, 15m, 10m) behave quite differently. They are reflected by the *ionosphere*, a layer of ionized gas about 100 miles above the earth, and also by the earth's surface, so that round-the-world *sky-wave* communication is possible under ideal conditions **(Fig 1.69)**. Propagation depends on the time of day, however. Sunshine disturbs the ionosphere, so that signals are reflected better on the dark side of the globe, and bands are *open* mainly in the evening and at night. The ionosphere also shows seasonal variations, and a slow change over the years due to sunspot activity. The effect of all this is to confront the amateur tuning in at a particular time with some open bands (open for communication in certain directions) and some *dead* ones: a dead band is evidenced by lack of static as well as signals, since static too travels round the world by ionospheric reflections. Charts predicting the conditions and open times on each band appear monthly in *Radio Communication.*

Sky-wave propagation introduces some odd effects. One is *skip*: a transmission clearly audible thousands of miles away may not be received at 50 miles, this being too close to catch the reflected signals. Movements in the ionosphere may cause *fading*, a periodic fluctuation in signal strength at the receiver. At times ionospheric reflection may separate different frequency components of the signal, particularly an amplitude-modulated signal, producing *phase distortion* (harsh garbled reproduction). Together with the loud background noise and intense interference from other stations, these difficulties make successful communication on the hf bands a task requiring considerable skill and patience.

Low-frequency signals (lf and mf, 10km to 100m wavelength) tend to curve round the earth's surface for 100 miles or more. This ground-wave reception is undistorted and reliable, and is used by sound broadcasting stations and marine services. Further information can be found in Chapter 11—*Propagation.*

Aerials

When a radio wave meets a metal surface, free electrons in the metal rush back and forth under the influence of the wave's electric and magnetic fields, generating a small signal

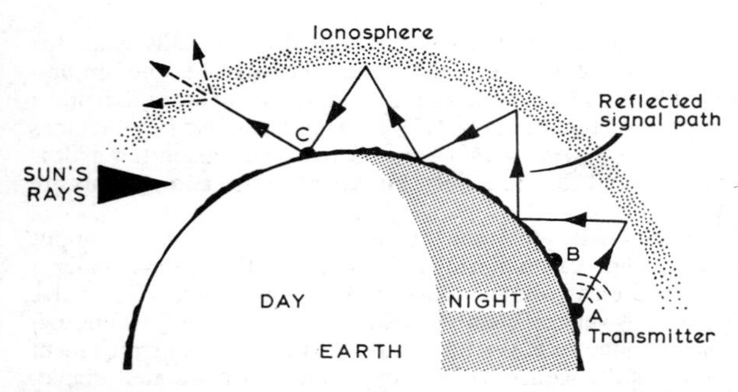

Fig 1.69. Short waves from transmitter A are reflected by the ionosphere and the earth's surface allowing communication over thousands of miles, although the signal may "skip" over a nearby station B so that it cannot be received there. Ionospheric reflection is poor over the daylight side of the earth and activity on the amateur bands thus depends on the time of day, most bands being busiest during the evening.

voltage. The object of aerial design is to fashion the conductor so that the maximum signal power can be drawn from it into the receiver. An excellent aerial for vhf work is the *half-wave dipole* of Hertz (**Fig 1.70**). The radio wave meets a straight rod one half-wavelength long and induces signal current in it, greatest at its centre. The rod is broken at this point and the two poles connected to the receiver input. This half-wave aerial has a low impedance (70Ω) which is entirely resistive and almost independent of the rod's thickness, making aerial matching very simple. For maximum signal strength the aerial must be mounted parallel to the approaching wave-front and parallel to its electric field (the direction of *polarization*).

Any aerial exhibits the same properties when radiating as when receiving. When a signal voltage is applied to the centre of a half-wave dipole current flows as though it were a 70Ω resistor in phase with the applied voltage. In aerials of

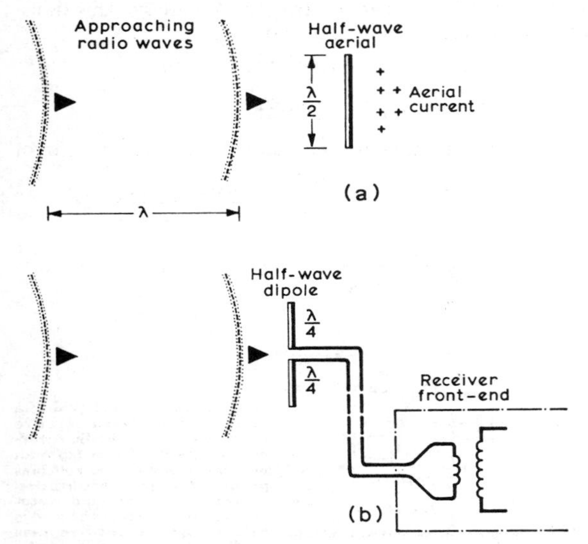

Fig 1.70. When radio waves meet an aerial half a wavelength long, signal currents are induced in it, these being a maximum at its centre (a). In the half-wave dipole a gap at the centre diverts this large current to the receiver input (b).

most other lengths current would be out of phase, ie most aerials have inductance or capacitance needing careful matching if they are to function efficiently.

Unfortunately half-wave dipoles become inconveniently large at hf and shorter aerials are much less efficient. The effective length can be doubled by Marconi's method. Wave reflection by the earth produces an image of the vertical aerial, and signal current passes from aerial to image (ie the ground): **Fig 1.71**. The difficulty here is to obtain a low-resistance ground connection, sometimes avoided by providing an artificial ground reflector of horizontal wires (the *ground-plane* aerial). Many ingenious aerial arrangements are discussed in Chapters 12 and 13.

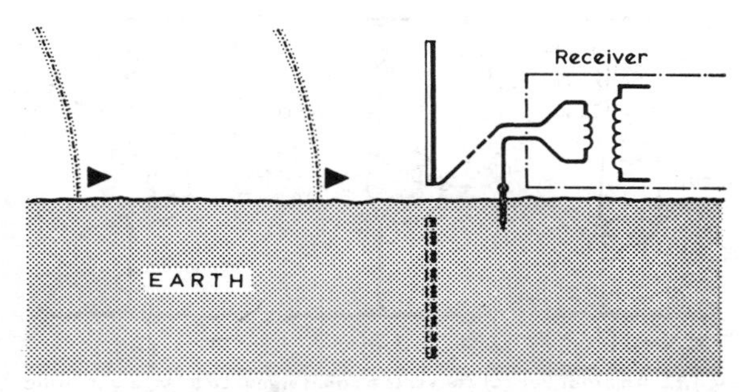

Fig 1.71. For longer wavelengths a Marconi vertical aerial is used. It acts as a dipole by virtue of the image produced by ground reflection.

AMPLIFICATION

An early radio was the *crystal set*, consisting of an aerial, a tuned circuit to select the station, and a crystal *detector* to convert the rf signal to audio frequencies (see p1.40). Only the small amount of power collected by the aerial was available at the headphones, and modern receivers only became possible with the invention of *active* devices or *amplifiers* (valves, transistors and certain diodes). These absorb power from a local source (battery or mains) and add part of it to the signal, which is amplified to a larger voltage or current but retains its original form.

Several simple amplifiers are shown in **Fig 1.72**. They all draw a few milliamps from a dc power supply: about 10V for transistors, 200V for valves (valves also require a low-voltage heater supply). These amplifiers produce an output signal at increased voltage and current, appearing as fluctuations in the current drawn from the supply. The current is usually supplied through a load resistance, which develops the output signal voltage across it: being ac this can be drawn off via a *coupling* capacitor. The amplifiers differ in their input requirements. The bipolar transistor requires appreciable signal current: the other devices require very little input current but a larger signal voltage. This circuit *configuration* (common-cathode or common-emitter) amplifies both input current and voltage—where only one is to be amplified, other configurations may be more suitable.

Valves have obvious disadvantages. They are large, breakable, expensive, get hot and use too much power for battery operation. They are of quite complicated construction, designed for a specific purpose and operating conditions, and have a limited life. Nevertheless they will in most cases handle larger voltages and power levels than presently

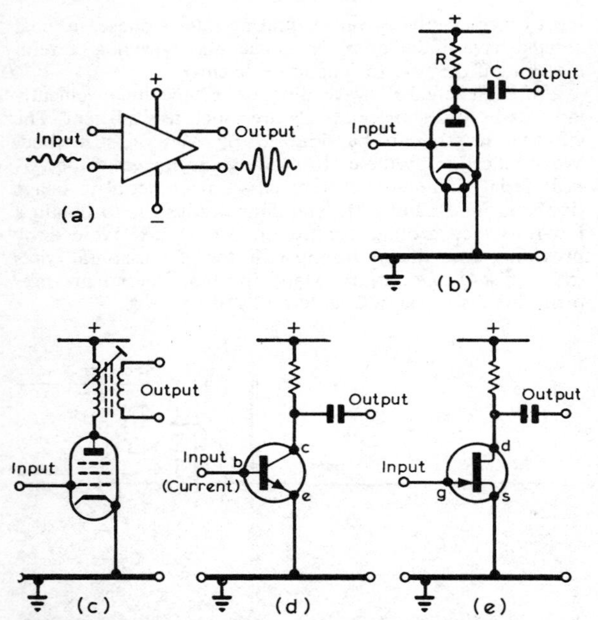

Fig 1.72. An amplifier (a) converts a small signal to a large one using power from an external source. Examples are the triode valve (b), pentode (c), bipolar transistor (d) and fet (e). At low frequencies RC coupling is used between amplifying stages (b), but radio frequencies may be transformer-coupled (c).

available transistors, and for this reason are almost always used for transmitter output stages in amateur equipment.

In contrast, a transistor is a very simple device. A good-quality transistor can be used for almost any purpose, and the manufacturer's data are only needed if it must perform near the limits of its capability. Below its maximum voltage, power and frequency it will function efficiently over a wide range of conditions.

Much more comprehensive details on valves and transistors will be found in Chapters 2 and 3. Biasing techniques are considered here as they illustrate several basic principles.

Valve Bias

In addition to the signal (ac) input, an amplifier also requires a dc input, the *bias*. Valves need a bias voltage applied to the grid (whereas transistors need a bias current), since a valve amplifier will only function normally while the grid is kept at a lower voltage than the cathode. The simplest method is to apply a negative voltage to the grid from a battery or rectifier: **Fig 1.73(a)**. This *fixed bias* is sometimes used in the *power amplifier* (pa) valve of a transmitter as it is reliable and easy to adjust, but several extra components are used.

Self-bias is generated by the valve itself from the input signal. The cathode and grid act as a diode, developing a rectified bias voltage across the *grid leak* resistor. The valve functions very efficiently because there is always just enough bias voltage to cover the input signal, and this arrangement will simultaneously *detect* an amplitude-modulated signal. Self-bias is not used for audio amplifiers as it produces distortion. If it is used for a pa stage great care must be taken that the input rf signal (the *drive*) does not fail, allowing the valve to run without bias and burn out.

Auto-bias (cathode bias) is produced by a cathode resistor R_k. The valve current passes through R_k and generates a positive voltage at the cathode; the grid is earthed through a large resistor, and is thus at a lower voltage: **Fig 1.73(c)**. As an example of the calculation, consider the E88CC triode, which is designed to pass 15mA of anode current and needs a grid bias of −1V. From Ohm's law,

$$R_k = \frac{V}{I} = \frac{1}{0{\cdot}015} = 68\Omega \text{ approx}$$

Because the valve current carries signal variations the bias voltage also fluctuates, and this reduces the gain. To avoid this a cathode *bypass* (decoupling) capacitor is fitted, which maintains a steady bias voltage.

Transistor Biasing

Bipolar transistors require a steady bias current across the base-emitter junction. The simplest method is to provide it directly from the supply line via a large resistor: **Fig 1.74(a)**. Calculation is simple: suppose a BFY90 transistor is to pass 10mA collector current (I_c) from a 9V battery. This transistor has a stated gain of about 90, so

$$I_c = 90 \times I_b \quad (I_b = \text{the base current})$$

$$I_b = \frac{10\text{mA}}{90} = 0{\cdot}11\text{mA}$$

From Ohm's law, this current is provided by a bias resistor of

$$R_b = \frac{V}{I} = \frac{9\text{V}}{0{\cdot}11\text{mA}} = 82\text{k}\Omega \text{ approx}$$

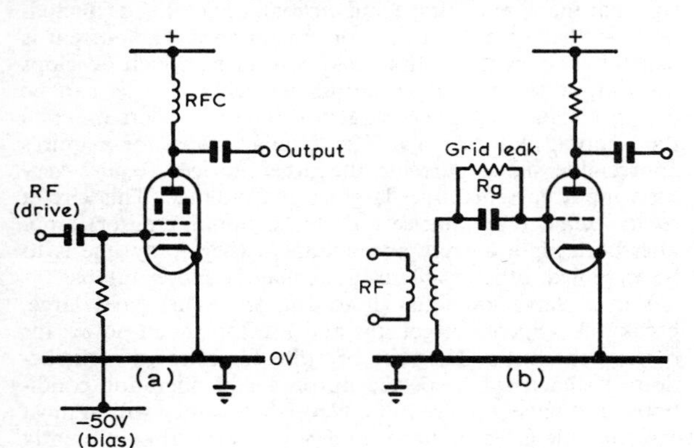

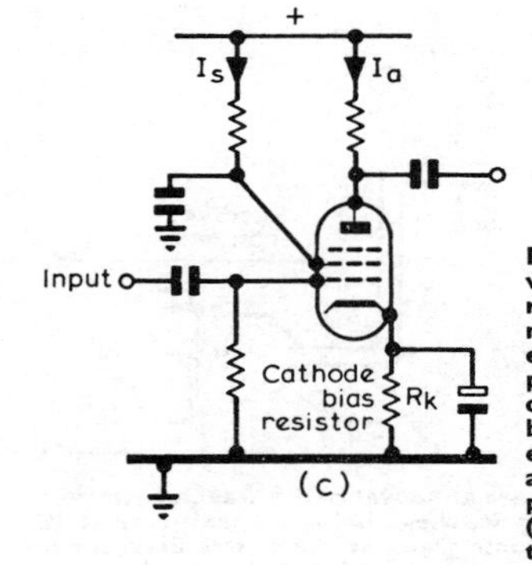

Fig 1.73. The control grid of a valve must be biased negative relative to the cathode. Alternatives are fixed bias from an external source (a), self-bias produced by grid rectification of the signal (b), and auto-bias(c). In the latter the grid is earthed and a positive bias applied to the cathode by passing the valve current (from anode and screen grid) through a cathode resistor Rk.

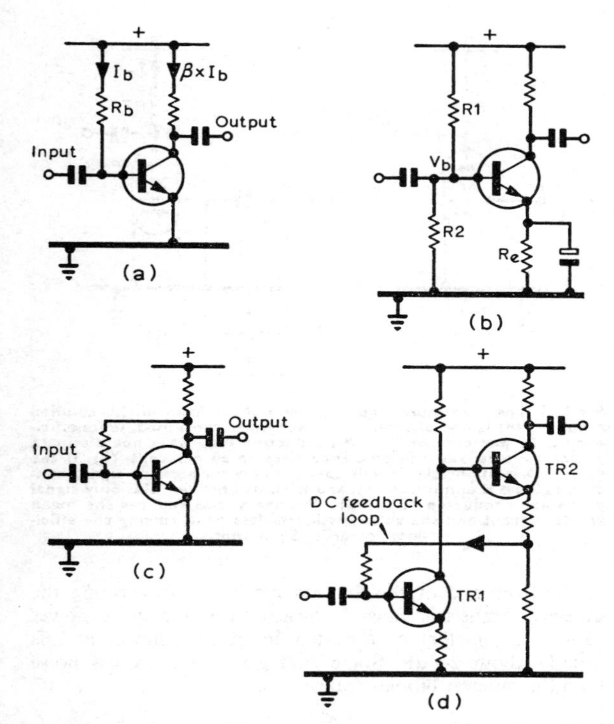

Fig 1.74. Bipolar transistors require a steady base-emitter bias current, which can be supplied directly through a large resistor Rb (a). In the voltage bias method a divider network R1,R2 maintains a steady base voltage which develops a predetermined current through the emitter resistor Re (b). Negative feedback can be used to control the bias current (c), and in direct-coupled amplifiers this is essential (d) in view of the high gain: any increase in TR1 bias current is neutralized by a drop in the current supplied to it from TR2. (Bypass capacitors omitted for clarity).

In practice the transistor gain may be anything from 40 to 200, so the calculation is only a rough guide. The resistor is fitted, the collector current (or voltage) is then measured, and the resistor changed as necessary until the desired current flows.

Apart from the slight effort of changing resistors direct biasing is a good simple method. Occasions when it is unsuitable are when using germanium transistors (which have high leakage currents) and when the base resistor value needed would be very large, as resistors of several megohms tend to age badly.

In the *voltage bias* method the transistor base is maintained at a suitable voltage by a divider network, and the desired current is developed across an emitter resistor R_e: see **Fig 1.74(b)**. This method suits large-scale manufacturers as no measurement is necessary, the desired current flowing regardless of transistor gain. The circuit uses extra resistors, the values of which must be calculated, and a large bypass capacitor to avoid loss of gain. The same gain is obtained as with direct biasing, but the maximum output voltage is reduced.

Using the same example of the BFY90, first a suitable base voltage is chosen. It should be much larger than the internal emitter-base voltage of the transistor (0·6V for silicon), say 3V. Next, the divider current is chosen to be much larger than the transistor base current; in this case the base current is about 0·1mA so a reasonable divider current would be 1mA. Therefore:

$$R_1 = \frac{9\text{V} - 3\text{V}}{1\text{mA}} = 6\text{k}\Omega$$

$$R_2 = \frac{3\text{V}}{1\text{mA}} = 3\text{k}\Omega$$

The emitter current is almost the same as the collector current, which is to be 10mA. Since the emitter voltage is:

$$\text{Emitter voltage} = \text{Base voltage} - 0{\cdot}6\text{V} = 2{\cdot}4\text{V}$$

the emitter resistor needed is:

$$\text{Emitter resistor } R_e = \frac{V}{I} = \frac{2{\cdot}4\text{V}}{10\text{mA}} = 240\Omega$$

Transistor gain does not enter the calculation, except to give a rough estimate of base current. The gain is still important, however, as it determines the maximum amplification available from the stage.

Another method of transistor biasing is to regulate the bias current using dc feedback, as shown in **Figs 1.74(c) and (d)**. This is chiefly used in direct-coupled amplifier chains, where the very high overall gain necessitates careful control of the bias current.

Gain

The effectiveness with which a voltage on a valve grid controls the anode current is called the *slope* (g_m) of the valve. If 1V on the grid alters the anode current by 10mA the slope is 10mA/V, a typical figure. The *voltage gain* is the change in output voltage per volt of input, and for an ideal valve this would be simply

$$\begin{aligned}\text{Voltage gain} &= \text{Anode voltage change per grid volt}\\ &= R_L \times \text{current change per grid volt}\\ &= R_L \times g_m \ (R_L = \text{anode load resistance})\end{aligned}$$

In practice a fall in anode voltage lowers the anode current, the valve behaving as a resistance (r_a) to ac. This reduces the gain with large anode loads, so that with an average triode there is little point in having a larger load than 5kΩ. The approximate maximum gain is then:

$$\text{Gain} = R_L g_m = 5\text{k}\Omega \times 10\text{mA/V} = 50$$

Pentode valves have a much higher anode resistance and will accept a load up to 1MΩ, giving a gain of over 1,000. The exact gain is:

$$\text{Gain} = \frac{g_m r_a R_L}{r_a + R_L}$$

A transistor absorbs current and power from the input signal, and the *current* gain depends on its construction:

$$\text{Current gain} = \frac{I_{out}}{I_{in}}$$

The current gain in common-emitter use (β) is typically 100 at low frequencies.

Transistor gain falls at vhf because the electrons (or holes) scarcely have time to cross the base layer during one half cycle; in fact the current gain falls to unity at the cut-off frequency (f_1), typically 200–500MHz. Amplification can still be achieved at vhf by using the common-base configuration (**Fig 1.75**), often with a series-tuned load to magnify the output voltage, as in Fig 1.52(c). Although there is a net current *loss* there will be a *power gain* of perhaps 10.

Power gain is usually measured on a logarithmic scale of *decibels* (dB):

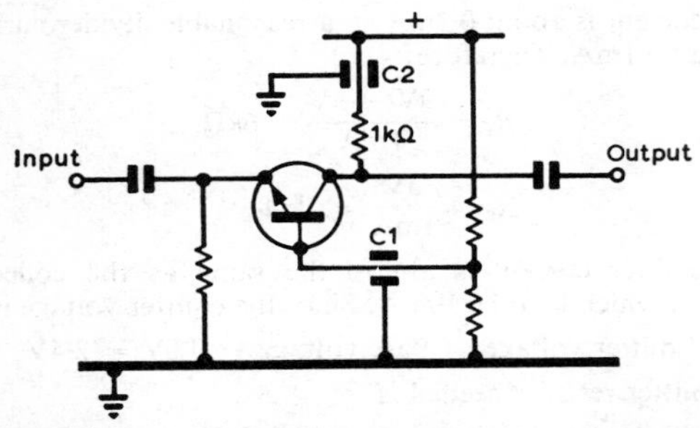

Fig 1.75. A common-base transistor amplifier for vhf (part of a 70cm pre-amplifier). The collector current is smaller than the input current due to losses in the base at vhf, but flows through a large resistance, giving an increased output voltage and a net gain in signal power. C1,C2 are decoupling feedthrough capacitors: C1 effectively earths the transistor base, which is thus common to both input and output circuits.

$$\text{Power gain} = \frac{P_{out}}{P_{in}}$$

$$\text{Log power gain} = \log \frac{P_{out}}{P_{in}} \text{ bels} = 10 \log \frac{P_{out}}{P_{in}} \text{ dB}$$

If input and output impedances are equal, doubling the voltage also doubles the current, giving a power gain of four times (6dB). For a transistor,

$$\text{Power gain} = \frac{I_0{}^2 R_L}{I_i{}^2 R_{in}} = (\text{current gain})^2 \frac{R_L}{R_{in}}$$

The common-base input impedance may be 25Ω or less, so a large load will compensate for the current "gain" of about 0·5.

A common-emitter transistor amplifier begins to lose gain at a much lower frequency than the common-base type. A transistor with rated gain of 200 and f_t = 300MHz will lose half its gain at only 1·5MHz (= f_t/β), despite the impressive frequency rating. (The cut-off frequency may be listed as f_1, f_t, f_α, or f_{ab}, depending on the method of measurement. The differences are unimportant.)

Noise

Electricity is not a fluid but discrete particles, and the current through an amplifier or resistor varies slightly from moment to moment as random numbers of electrons pass. This generates a noise voltage. In general low currents and low resistances produce least noise. A low source impedance also reduces amplifier noise, about 1kΩ being best for transistor circuits. In a radio receiver most of the noise is due to the first stage (*front-end*), because this noise is amplified by each subsequent stage and gives a hiss or roar, limiting the reception of weak signals. The input stage must therefore be a low-noise device.

In a valve each electrode contributes to the noise. Valve noise is usually quoted as the *equivalent noise resistance*, the input resistance which would generate an equal thermal noise. Typical values are 300Ω for a triode, 1,500Ω for pentodes, and 100,000Ω or more for multigrid mixer valves. This accounts for the popularity of triodes as mixers and rf amplifiers.

Transistor noise depends greatly on impurities and surface effects, so different specimens of the same transistor vary enormously in how much noise they produce. It is worthwhile trying several samples of a low-noise transistor

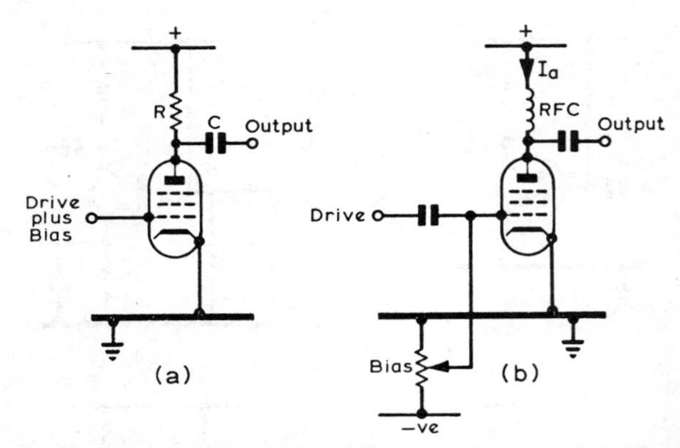

Fig 1.76. The maximum signal power output from an RC-coupled amplifier (a) is one quarter of the dc power consumed, ie its efficiency is 25 per cent or less. An inductive load does not dissipate power as heat, and the efficiency rises to 50 per cent, (b). If the valve is heavily biased it will conduct only on positive signal half-cycles (Class B amplification), and with still heavier bias only signal peaks are conducted (Class C). The heavy bias reduces the mean anode current and the valve dissipates less heat, raising the efficiency towards 75 per cent.

in the front-end of the finished receiver and selecting the quietest. Transistor noise is measured as the noise power gain (*noise factor*) at a stated input impedance, and is usually about 2–4dB. Some FETS generate even less noise than the quietest bipolar transistors.

Efficiency

Only part of the power consumed by a valve from the dc supply is recovered as signal output, the rest being wasted as heat. UK amateurs may only supply a maximum 150W of dc to the pa stage of the transmitter, so its efficiency is of vital importance. The efficiency of an amplifier depends on the input power,

$$\text{Efficiency} = \frac{\text{Output power (signal)}}{\text{Input power (dc)}}$$

which in turn depends on the average current drawn. An ordinary linear amplifier (ie *Class A*) passes a high mean current and has an efficiency of less than 50 per cent: half the power is wasted and makes the valve very hot. An RC coupled amplifier is even less efficient **(Fig 1.76).**

Suppose a transmitter pa stage is a single 807 valve operating in Class A, drawing 50mA from a 500V supply. The power supplied is:

$$P_{in} = 500\text{V} \times 0{\cdot}05\text{A} = 25\text{W}$$

The anode current can vary from 0–100mA, so the maximum output current is:

$$I_{out} = \pm 50\text{mA peak} = 35\text{mA rms of rf signal}$$

The maximum anode voltage swing (with an inductive load) is from 0–1kV, so:

$$V_{out} = \pm 500\text{V peak} = 350\text{V rms}$$

The maximum output power is therefore:

$$P_{out} = 350\text{V} \times 0{\cdot}035\text{A} = 12{\cdot}5\text{W}$$

$$\text{Efficiency} = \frac{12{\cdot}5}{25} = 50 \text{ per cent (at maximum output),}$$

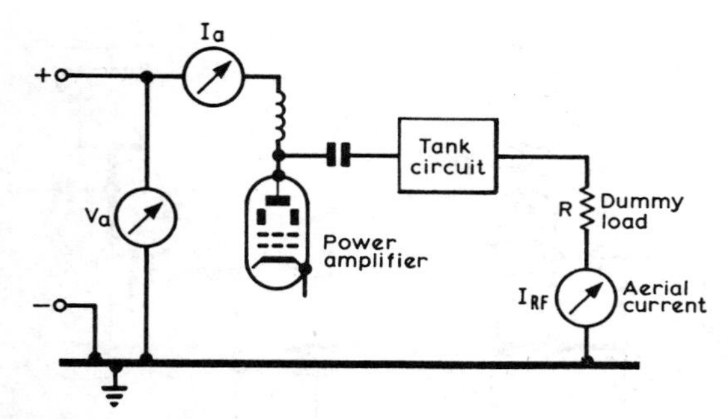

Fig 1.77. The efficiency of a pa is the ratio of its signal output power (I_{rf}^2 × load resistance) to the dc input power (Va × Ia). Direct measurement of the rf aerial current requires a thermocouple ammeter.

and to achieve this the valve would require a load resistance of:

$$R_L = \frac{350\text{V}}{0{\cdot}035\text{A}} = 10\text{k}\Omega$$

In practice the valve cannot be operated down to zero anode volts, so the maximum efficiency is less than 50 per cent. Also, when the signal falls below its maximum level the output power drops, so the overall efficiency on a speech transmission may be only 10 per cent or less.

For greater efficiency the valve is cut off between peaks by a large grid bias, so that it amplifies only signal peaks (*Class C*). These peaks excite resonance in the anode tank circuit, which restores the rest of the waveform. Not only is the mean anode current reduced (increasing efficiency towards 75 per cent) but it falls when the signal amplitude drops, so that during quiet passages input power falls and the efficiency remains high. PA efficiency can be measured as shown in **Fig 1.77.**

Class C amplification would distort an audio signal beyond recognition. Audio power amplifiers (eg an a.m. modulator) use two valves (**Fig 1.78**) or transistors in *push-pull*, amplifying half cycles alternately The valves are biased to cut-off during negative half-cycles (*Class B*) to increase efficiency, but the transition from one valve to the other between half cycles introduces some *cross-over* distortion. If necessary this can be reduced by allowing each valve to pass rather more than half the cycle (*Class AB*).

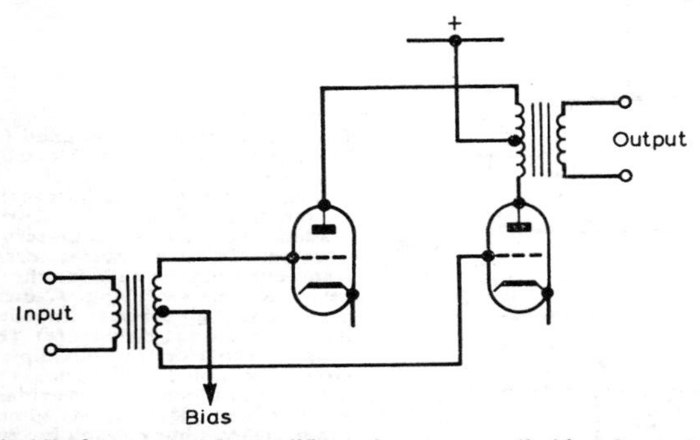

Fig 1.78. In a push-pull amplifier valves are usually biased near cut-off. Positive signal half-cycles are delivered to each valve alternately by the centre-tapped transformer, and output half-cycles are re-combined in the anode transformer.

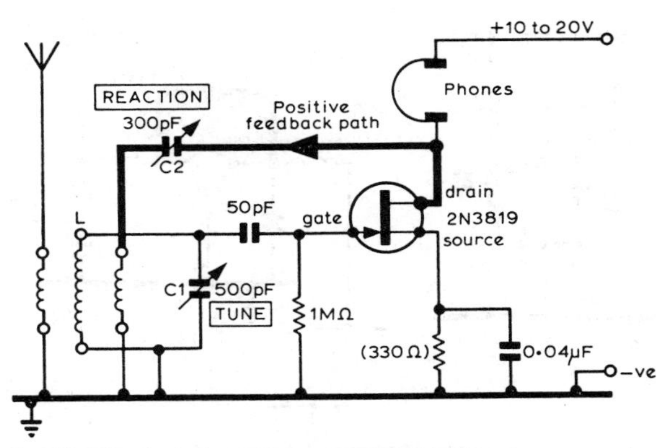

Fig 1.79. This single-fet radio receiver by G3UMP uses regeneration (reaction) to achieve sufficient amplification and selectivity. Aerial current excites oscillations in the tuned circuit L, C1 and these are amplified by the jfet. Part of the output is fed back to the input via the reaction control C2 (which acts as a variable impedance) and greatly increases the output at the resonant frequency. Positive feedback path shown in bold.

Positive Feedback

Suppose the output of an amplifier is partly coupled to its input. The input signal will be amplified, return along this *feedback* path, and be amplified over and over again. The total gain will be very large; indeed a single valve or transistor with positive feedback (*reaction, regeneration*) becomes powerful enough to function as a complete radio set (**Fig 1.79**). A filter is included in the feedback path, so that the signal passes through the filter many times, giving high selectivity. The same principle is used in the *Q-multiplier* to convert a tuned circuit into a very selective filter with a *Q* of 1,000 or more (**Fig 1.80**).

Adjusting the amount of positive feedback is always difficult. With a fraction too much feedback the stage gain becomes infinite and the amplifier bursts into spontaneous oscillation, maintained by the feedback signal.

Oscillators

An oscillator is an amplifier with generous positive feedback via a filter, so that it can oscillate only at the filter frequency. The filter may be a tuned circuit, forming a

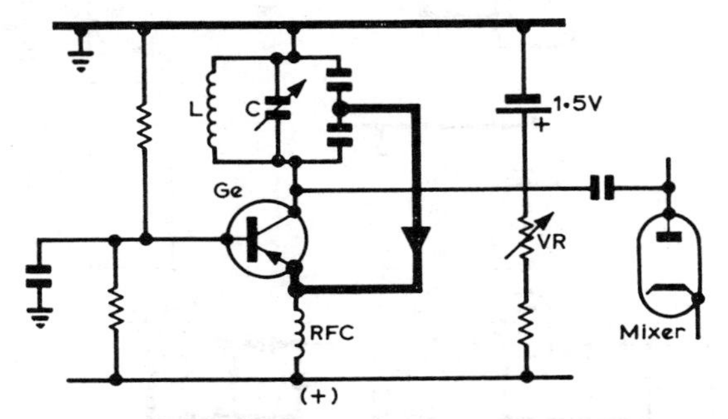

Fig 1.80. A simple Q-multiplier. The i.f. signal is applied to the tuned circuit L, C which shunts away unwanted frequencies. As resistive losses in L and C are overcome by positive feedback to the transistor the circuit has an almost infinite impedance at resonance and acts as a very narrow filter. Amplification, and therefore filter width, are controlled by the resistor VR. Positive feedback path drawn in bold.

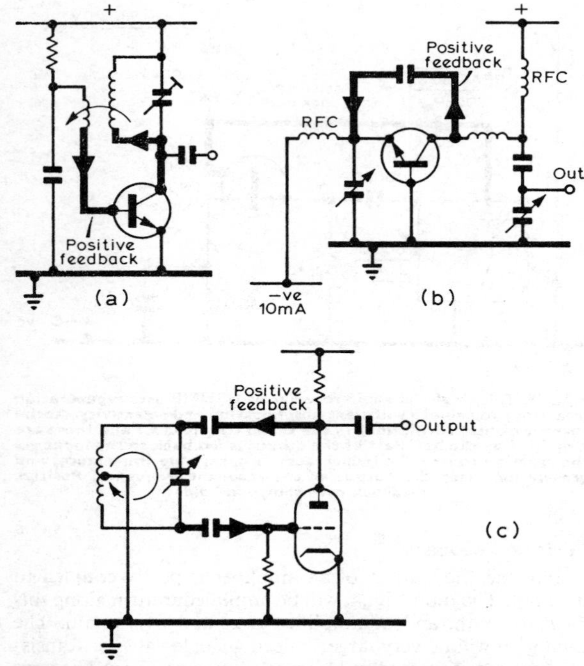

Fig 1.81. In an oscillator input signal is provided from the output by positive feedback. Simple oscillators are the transformer-coupled (a), the emitter-coupled for vhf work (b), and the Hartley oscillator (c) which uses the tapped tuned circuit as a transformer to provide feedback of the correct polarity.

variable frequency oscillator (vfo), or a quartz crystal (*crystal oscillator*), or both. The crystal oscillator has much better frequency stability, varying by only one part in 10^6 per day due to temperature and supply voltage variations, but it can

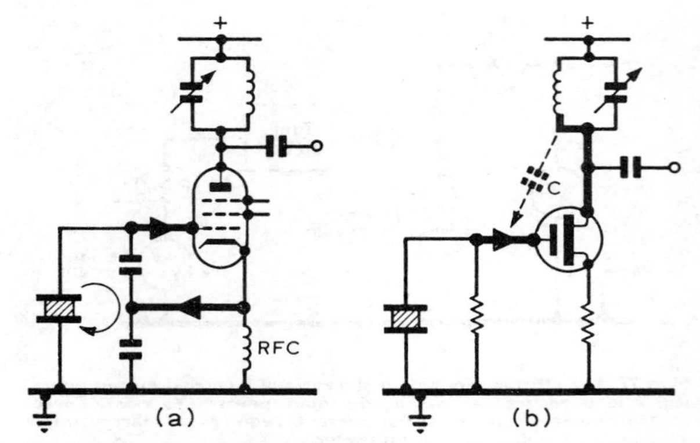

Fig 1.83. Parallel-mode crystal oscillators. The "Harmonic" oscillator (a) is a Colpitts circuit with a crystal replacing the parallel-tuned circuit. The anode circuit may be set to select the crystal fundamental frequency or one of its harmonics (overtones), with adequate output power. In the Miller circuit (b) signal is fed back via the internal capacitance of the transistor or valve, the crystal again acting as a parallel-tuned circuit.

only oscillate at or near the crystal frequency. Many efforts have therefore been made to improve the stability of the versatile vfo.

The simple transformer-coupled oscillator has many defects, one being that the output voltage varies with frequency. This is partly overcome by providing feedback from a tap on the tuned circuit, as in the *Hartley* oscillator and others (**Fig 1.81**). Another defect is that the following stage affects (*pulls*) the oscillator frequency: to minimize this *electron-coupled oscillators* are arranged so that the output is not taken from part of the feedback path (**Fig 1.82**).

Slow change of oscillator frequency (*drift*) is caused by coil and capacitor variation as the temperature alters, and also by changes in the internal capacitance and resistance of

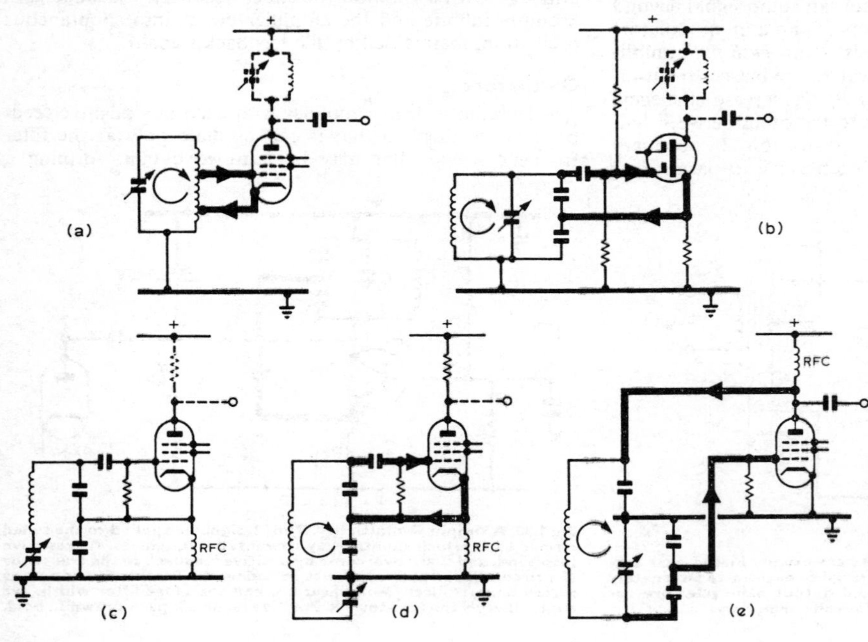

Fig 1.82. Oscillators designed for frequency stability. (a) "Electron-coupled" Hartley circuit to isolate the feedback path from the output circuitry: note the lightly-loaded tuned circuit. (b) Electron-coupled Colpitts mosfet oscillator, avoiding the need for three-terminal coil switching. A dual-gate fet will give better isolation from the output circuit. (c) The Clapp oscillator is a Colpitts circuit modified to reduce the tuned circuit loading, as is evident when it is redrawn (d). Its tuning range is small but enough for any amateur band. (e) The Vackar or Tesla oscillator is a similar modification with a wider tuning range.

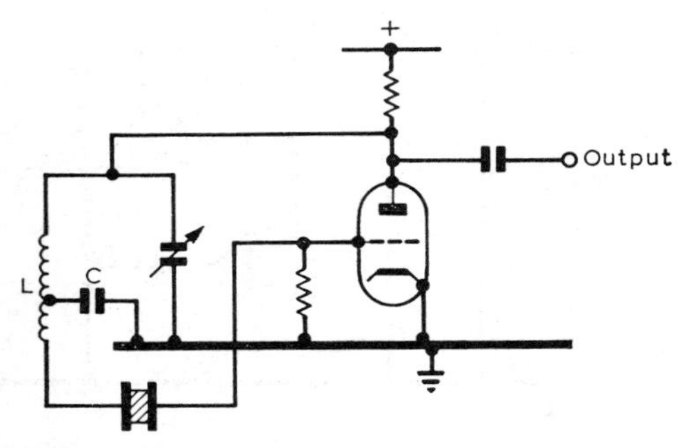

Fig 1.84. Most vhf oscillators use a crystal in its series-mode. The Squier circuit is a Hartley with an added crystal low-impedance filter (compare with Fig 1.81b), designed to operate at an overtone of the crystal frequency. Capacitor C is for dc blocking only.

the valve or transistor. The latter effect can be minimized by using a lightly tapped high-Q tuned circuit as in the *Clapp* and *Vackar* (Tesla) oscillators, which at the same time lowers the harmonic content of the output. A stability of 1 in 10^4 per day can be achieved.

Practical considerations when choosing an oscillator circuit are the coil requirements (so a capacitatively tapped circuit may be preferred) and the number of connections which have to be switched to change bands.

Crystal oscillators work at an almost fixed frequency, a trimmer providing fine adjustment. In oscillators designed for 20MHz or below, the crystal resonates at its fundamental frequency and forms a high-impedance tuned circuit, ie these are *parallel-mode* oscillators. The crystal replaces the coil and capacitor of a vfo, eg the Colpitts oscillator becomes a "harmonic" oscillator, **(Fig 1.83)**. A special type is the *Miller* oscillator, in which feedback occurs via the internal capacitance of the valve or transistor. It is not quite so stable but gives a high output, and the frequency can be varied over a somewhat larger range.

Above 20MHz crystal overtone oscillators are usually *series-mode*, and give a stable output at low amplitude. Practical arrangements are discussed in Chapters 6 and 7: a simple example is the *Squier* oscillator, which is the Hartley circuit with an added crystal filter **(Fig 1.84)**. VHF oscillator construction is difficult, and an oscillator at lower frequency followed by frequency multipliers usually gives better results.

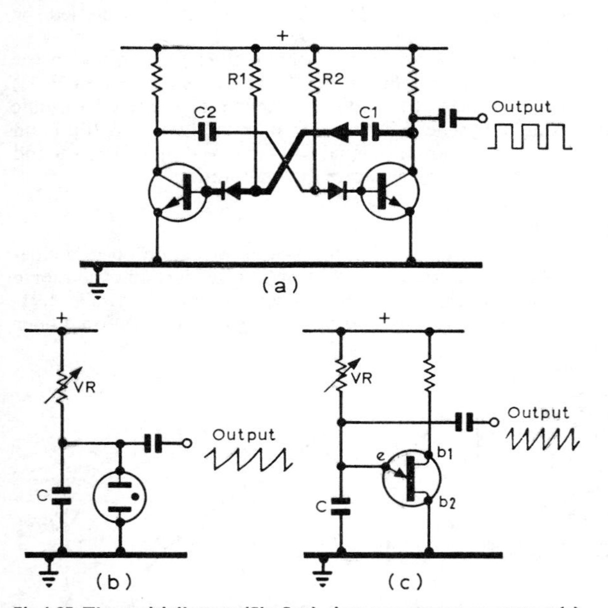

Fig 1.85. The multivibrator (flip-flop) gives a square-wave output (a). When one transistor conducts, the base of the other transistor is driven negative until the coupling capacitor has discharged: then it starts to conduct and the first transistor is cut off. The diodes protect the transistors from excessive negative base voltage, and the frequency is set by the time-constants R1C1 and R2C2. A relaxation oscillator gives a triangular waveform by repeatedly discharging a capacitor through a breakdown device such as a neon tube (b), thyristor, four-layer diode or a unijunction transistor (c). Frequency depends on the time-constant VR.C, which sets the charging time.

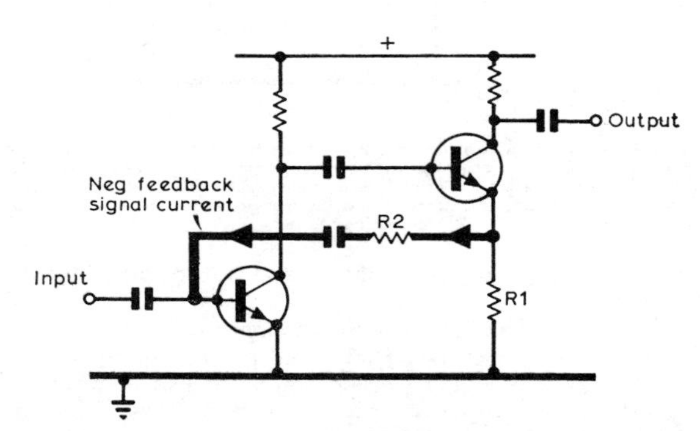

Fig 1.86. The gain of this amplifier is controlled by a negative feedback signal current (shown in bold), which cancels most of the input signal and leaves a gain R2/R1, regardless of individual transistor gains and operating frequency. (Capacitors are for coupling and dc blocking, and biasing is not shown).

Some oscillators are designed to produce a non-sinusoidal waveform. The *multivibrator* generates a square wave rich in odd harmonics of the fundamental frequency **(Fig 1.85)**. *Relaxation oscillators* produce a roughly triangular or *saw-tooth* waveform, such as that used for the timebase of an oscilloscope. A capacitor is slowly charged to a preset voltage, at which a neon tube or unijunction transistor discharges it and the cycle starts again. Microwave signals are sometimes generated by a similar arrangement using a tunnel diode or a Gunn diode.

Negative Feedback

If a fraction of the output from an amplifier is *subtracted* from its input, the gain is reduced. This negative feedback is used in mass-produced transistor circuits because it produces a predetermined gain, regardless of wide variations in individual transistors **(Fig 1.86)**. Advantages are constant gain over a wide frequency range, and reduced distortion. However, the low total gain necessitates extra stages of amplifica tion.

RF negative feedback was formerly used to *neutralize* the internal capacitance of valves or transistors **(Fig 1.87)**, to prevent instability (parasitic oscillation as a Miller oscillator). Modern devices rarely need neutralization except at uhf, since pentodes, triode-cascodes and dual-gate FETs all have low output-to-input capacitance.

AF negative feedback is used in high-fidelity amplifiers to reduce distortion. New frequencies produced by distortion

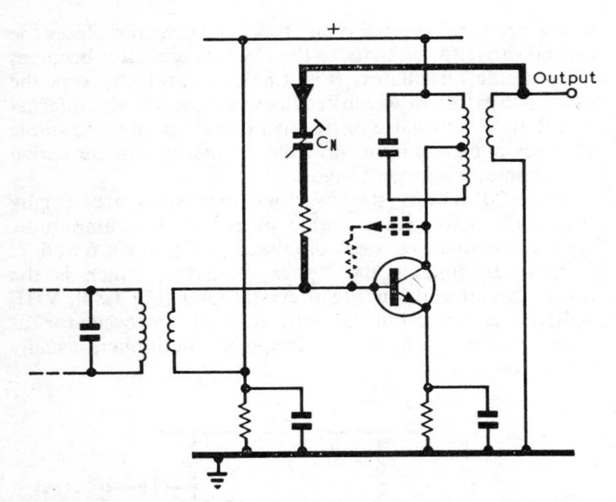

Fig 1.87. Neutralization. An rf amplifier is liable to oscillate due to positive feedback across its internal capacitance (shown dotted), like a Miller oscillator. This feedback may be neutralized by an equal negative feedback current of opposite polarity, adjusted by the capacitor C_N.

return to the input in antiphase, and so are largely self-cancelling.

DC negative feedback is used to *stabilize* dc currents or voltages in amplifiers (see p1.31) and power supply units. **Fig 1.88** shows a simple voltage-stabilized psu in which a fraction of the output voltage is compared with a zener diode voltage. The difference generates a current which controls the output voltage, in this case by a series transistor, thus compensating for any change in output voltage. Other methods of stabilization are described in Chapter 16—*Power Supplies.*

THE TRANSMITTER

The amateur transmitter may be considered in five sections. The *oscillator* generates a steady signal, *frequency multipliers* convert this to the transmission frequency, a *modulator* impresses the intelligence (speech) upon this carrier, the *power amplifier* magnifies the signal to a high power, and the *aerial matching* (tank) circuit feeds most of this power to the aerial. See **Fig 1.89.**

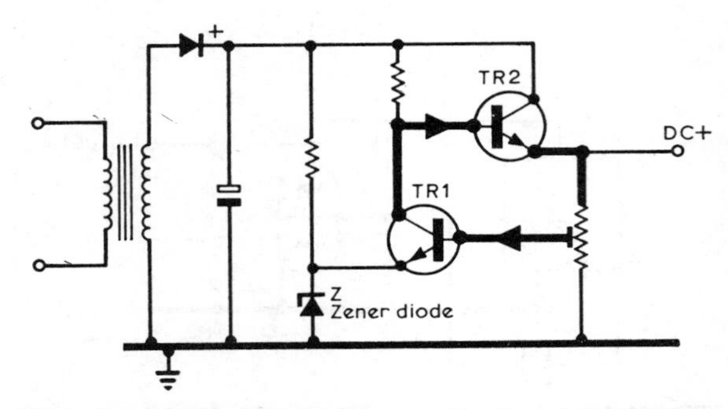

Fig 1.88. In a series-stabilized psu the output voltage is regulated by dc negative feedback. Any fall in output lowers the negative feedback current; the drop is amplified by TR1 and causes a rise in TR2 base voltage, which in turn counteracts the falling output voltage.

Frequency Multiplier

When a sine-wave signal passes through a non-linear stage such as a specially biased amplifier the waveform becomes distorted. This distorted signal is actually a mixture of new frequencies, *harmonics* of the original signal frequency (*f*), with frequencies 2*f*, 3*f*, 4*f*, 5*f*, etc. Harmonics can be selected by a tuned circuit in the amplifier output; if for example the second harmonic (2*f*) is selected, the stage becomes a *frequency doubler*. A typical transmitter might have a vfo tunable from 1·7 to 2·0MHz with four doublers which can be switched in, allowing transmission directly on the 1·8MHz band and by successive doubling on the 3·5MHz, 7MHz, 14MHz and 28MHz bands. If a tripler is added the 21MHz band can also be covered.

Another example of harmonic distortion is found in the *crystal calibrator* (**Fig 1.90**). Signal at exactly 100kHz is heavily distorted to produce components at every harmonic of that frequency. If the output is applied to a top-band receiver, for example, signal will be picked up at 1·8, 1·9 and 2·0MHz, allowing accurate dial calibration.

Modulation

The simplest way of impressing information on to a sine-wave *carrier* is to interrupt it by *keying*, forming the morse code characters in continuous-wave carrier (cw): **Fig 1.91.** Such *telegraphy* gives very reliable communication under

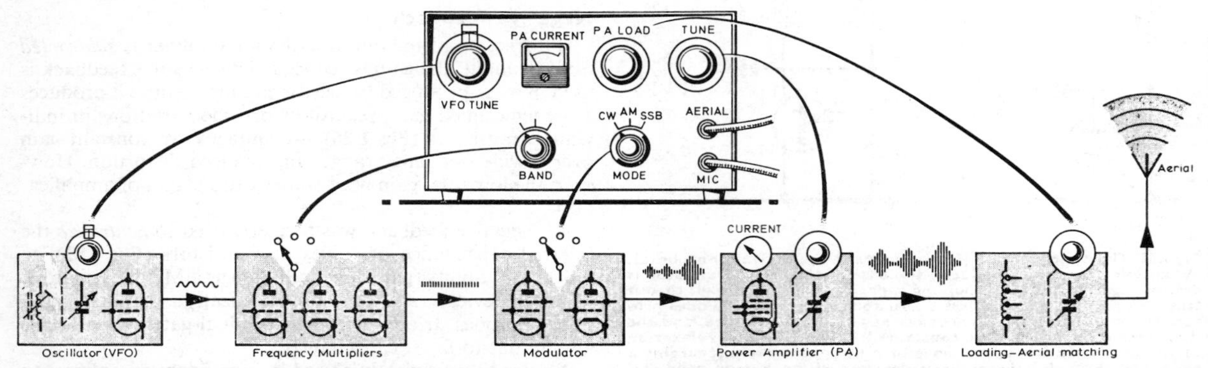

Fig 1.89. The stages in a transmitter. Signal is generated by the oscillator, multiplied to the desired frequency, modulated by the speech waveform, amplified to a high power and fed to the aerial via an impedance-matching unit, the "tank" circuit.

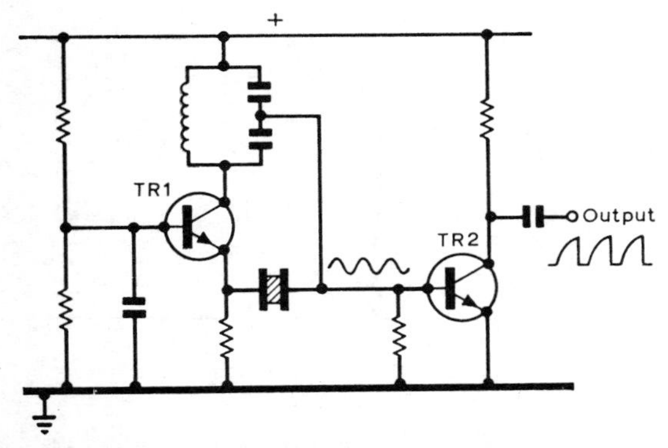

Fig 1.90. In this simple crystal calibrator the output of the emitter-coupled oscillator TR1 is heavily distorted by TR2, which has no bias supply. This distortion generates many harmonics, up to 10MHz from a 100kHz crystal.

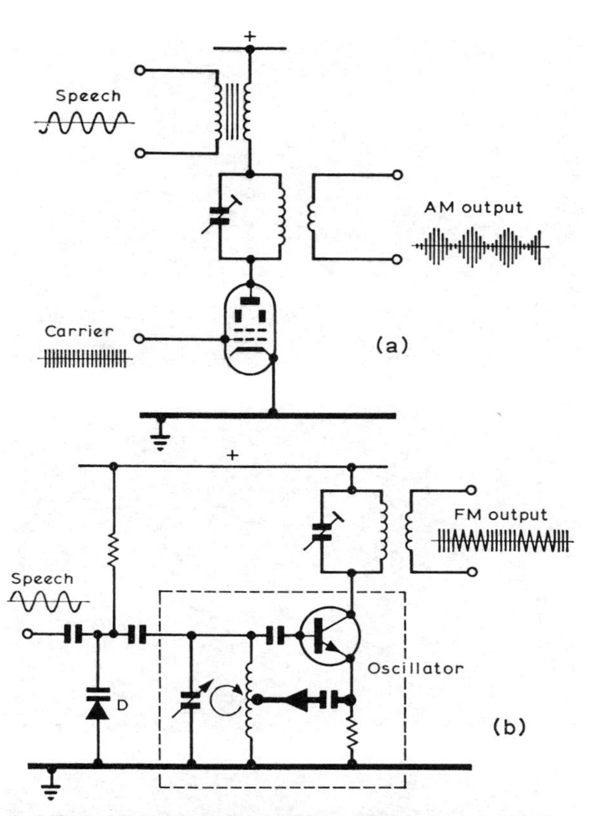

Fig 1.92. A simple amplitude modulator uses the speech signal to vary the ht voltage supply to the valve: this alters the valve gain and produces a carrier of fluctuating amplitude (a). Frequency modulation is conveniently achieved by applying the speech signal to a reverse-biased diode, causing its capacitance to vary slightly (b). The diode is placed across the tuned circuit of the transmitter vfo, in this case an electron-coupled Colpitts oscillator, and varies the output frequency. (Biasing not shown).

even the worst conditions, but is rather slow and impersonal. To transmit speech (*telephony*) the voice frequencies must modify or *modulate* the carrier waveform, and there are several available methods.

When voice and carrier signals are passed together through a non-linear stage, two new frequencies called *intermodulation* products are generated by the distortion (as well as harmonics of each signal). One is at the sum of the two input frequencies and the other at the difference frequency. The output contains unchanged carrier surrounded by two *sidebands* of intermodulation products, an upper sideband of sum frequencies and a lower sideband of difference frequencies. If the signal is examined on an oscilloscope the carrier is found to be *amplitude modulated* (a.m.) by the speech signal. Most broadcast radio stations use a.m., as it is very easy to recover the voice signal in a receiver *demodulator*, and slight detuning of the receiver produces little distortion.

Close examination of an a.m. signal shows that half the power is being used to transmit a pure carrier frequency, which carries no information. Attempting to modulate more strongly cuts off the carrier entirely on voice peaks, causing *overmodulation distortion* (splatter), which interferes with transmissions on adjacent channels. The carrier can be suppressed by other methods, however, leaving only a *double-sideband* signal (dsb). As both of these sidebands carry the same information, even more power can be saved

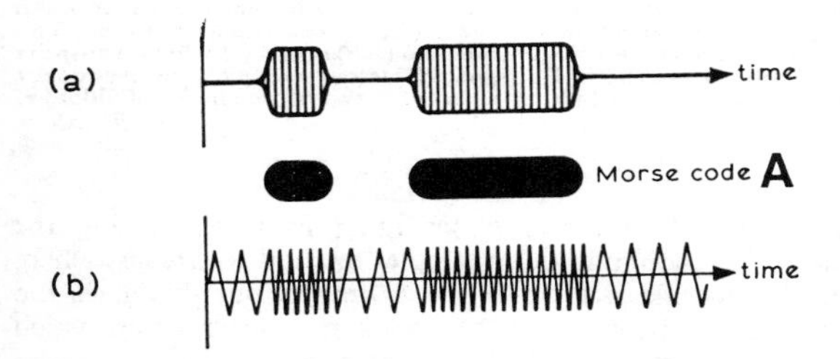

Fig 1.91. Continuous-wave (cw) telegraphy morse symbols are transmitted as bursts of rf carrier (a), the ends being rounded off to avoid key-clicks. Frequency-shift keying (fsk) is an alternative method using an unbroken carrier (b) which alternates between two fixed frequencies during keying.

by filtering off one of them and transmitting a *single sideband* signal (ssb), with the added advantage that the signal bandwidth is halved and takes up less room on the frequency band.

SSB modulation offers a four-fold gain in efficiency compared to a.m. transmissions, where half the power is divided between two sidebands. However, the equipment is complex and requires careful adjustment to suppress the carrier in the transmitter, and to regenerate it at exactly the same frequency in the receiver. There is also a loss of listening comfort as receiver tuning is very critical and some distortion usually remains.

Some relief from both distortion and background noise is offered by another method of modulation, *narrow-band frequency modulation* (nbfm). Speech signal is applied to a voltage-variable capacitor (a diode) in the oscillator producing small fluctuations in the frequency of the carrier, which is transmitted at constant amplitude **(Fig 1.92(b))**. Most background interference appears as amplitude variations of the carrier, so it can be filtered off by amplitude *limiters* in the receiver without affecting the fm signal. See Chapters 5 and 9 for details.

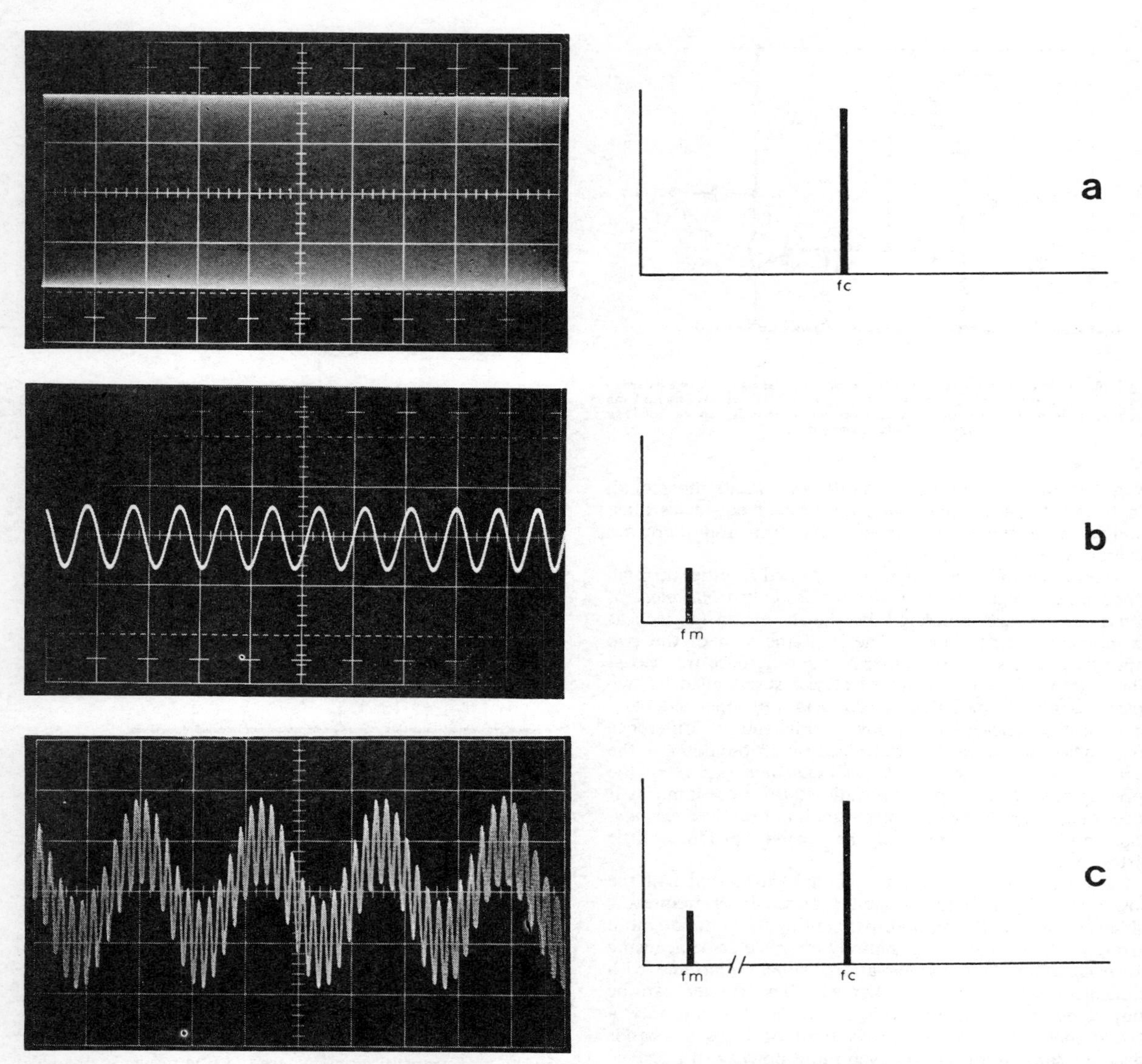

Modulated waveforms as displayed on an oscilloscope. The corresponding spectrum analysis is shown alongside, the horizontal axis being frequency (increasing going right) and the vertical axis being amplitude. Adding a pure rf carrier (a) to an audio (speech) signal (b) produces a signal (c) which cannot be transmitted, since it contains audio frequencies which will not radiate. Multiplying a carrier by an audio signal produces an amplitude-modulated (a.m.) signal (d) which contains carrier and two sidebands at adjacent frequencies. An audio signal which is too powerful produces overmodulation distortion (e), the extensive sidebands splattering across nearby channels. The carrier and one sideband of an a.m. signal can be suppressed by special techniques to leave a single-sideband (ssb) signal (f) which transmits the same information using a quarter of the power. Frequency modulation impresses the audio signal as fluctuations in carrier frequency rather than amplitude (g). Note that for clarity the oscilloscope time base has been speeded up in parts (c) and (g) to show individual rf oscillations.

Transmission Power

A speech signal shows great variations in amplitude, accented syllables causing large peaks. If the transmitter is adjusted to accept these peaks without overmodulating (a.m.) or exceeding the legal power limit (ssb) the *average* transmitted speech power will be small, perhaps 5 per cent of the legal maximum. It can be increased by *speech compression*: an automatic volume control device in the speech amplifier selectively amplifies quiet passages, raising the mean speech amplitude. A simpler method is *clipping*, where the loud peaks are cut off by a limiter, but this gives the voice a harsh sound and generates many harmonics, which must be filtered off before transmission.

The rf power of an ssb transmission drops to zero between words, and the dc input to the pa also drops to zero if it is operating in Class B or C. Therefore the legal power

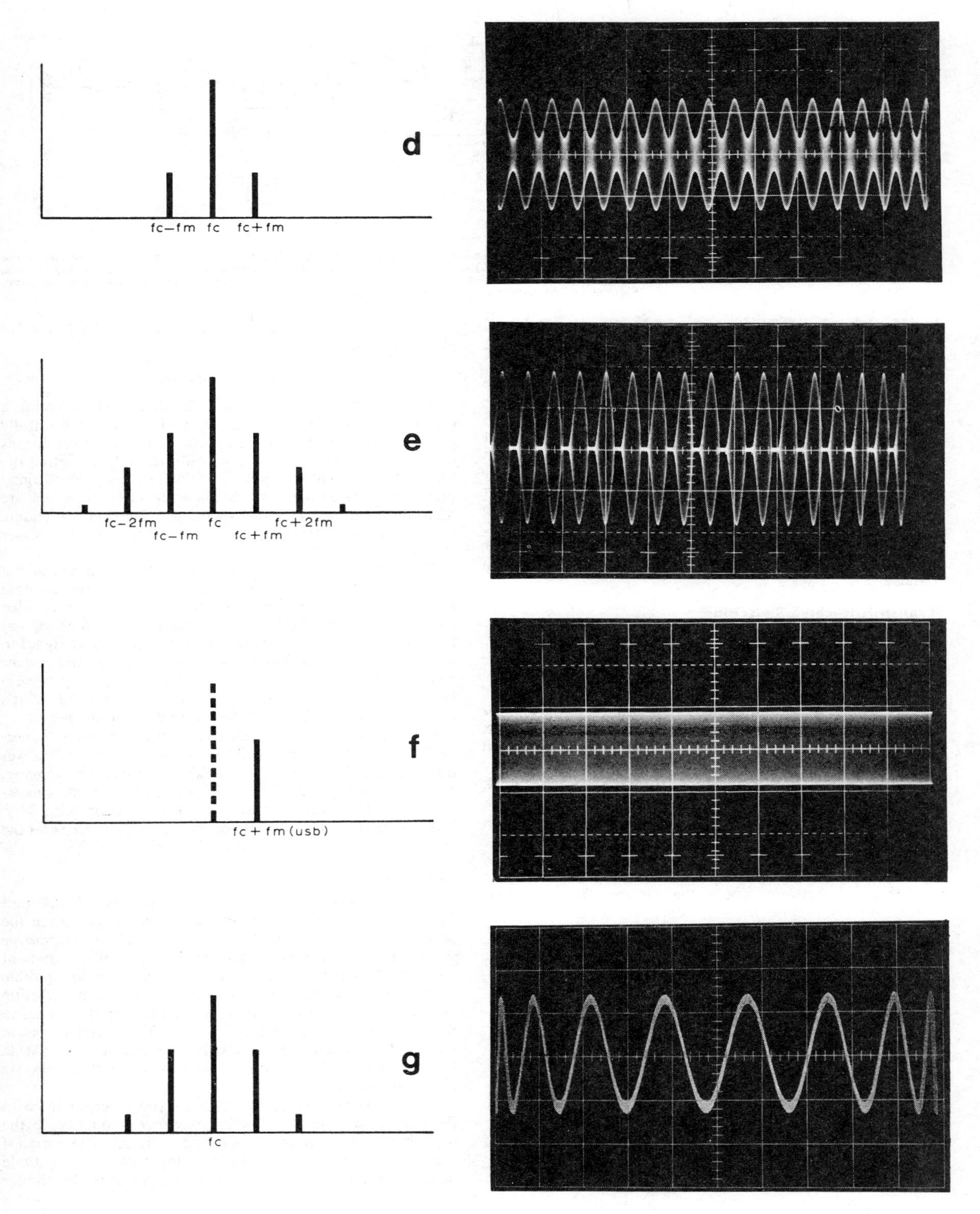
d
fc−fm
fc
fc+fm
e
fc−2fm
fc−fm
fc
fc+fm
fc+2fm
f
fc+fm(usb)
g
fc

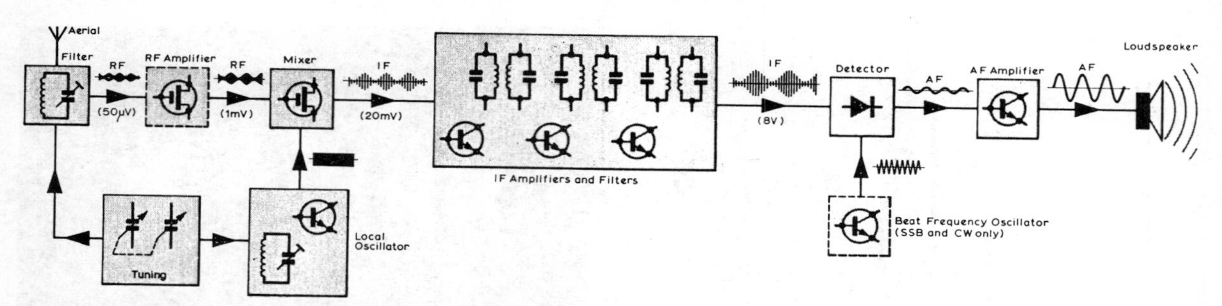

Fig 1.93. In a superheterodyne receiver the aerial signal is mixed with another rf signal from an oscillator to produce a beat note or heterodyne at a lower frequency, about 0·5MHz. This intermediate-frequency (i.f.) signal can be easily amplified and filtered to remove interfering signals, and the speech modulations are then recovered by a detector. SSB and cw signals must be combined with another i.f. signal before detection. Figures in brackets are typical signal voltages.

limit for ssb is not defined in terms of dc power input, but as a maximum rf power actually transmitted of 400W *peak envelope power* (p.e.p.).

This corresponds to the envelope power of a fully modulated a.m. transmission from a 150W input pa. About two-thirds of the 150W is radiated as rf under optimum conditions, giving 100W mean aerial power. At 100 per cent modulation the envelope voltage and current vary from zero to twice their mean values, so the peak envelope power is 400W. In the a.m. transmission this occurs as carrier (200W) and sidebands (100W each), but the ssb transmission reaches 400W in the sideband.

Legal power limits in Europe are generally similar to those in Great Britain, but in the USA the limit is 1,000W on most bands.

Transmit/Receive Switching

When changing from reception to transmission the transmitter must be energized, the receiver muted, and the aerial transferred from receiver to transmitter. This can be accomplished by a single switch operating a relay: press a single *ptt* (push-to-talk) button and the microphone is "live". For convenience the button may be mounted on the microphone, the so-called *mocs* or mox (microphone-operated control switching) operation. Still further refinement is provided by operating the change-over relay from the speech signal, ie *vocs* or vox (voice-operated control switching) with automatic return to the receive position a short while after speech ceases.

For telegraphy the corresponding refinement is *break-in* operation. Pressing the key mutes the receiver and starts the transmission; between words the receiver functions and gives early warning of any interference. See Chapter 8 for further details.

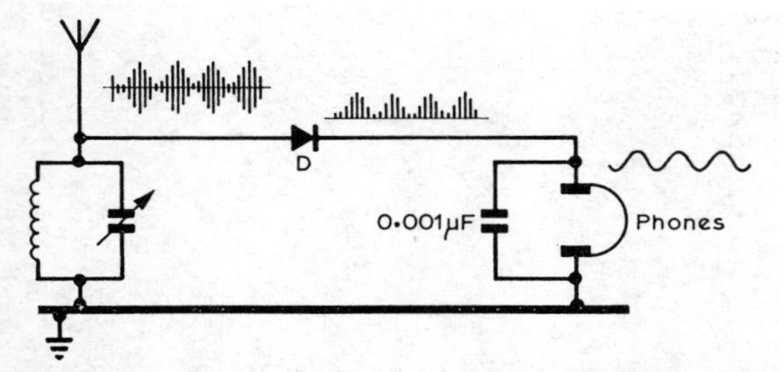

Fig 1.94. An early radio receiver the crystal set, contained only a tuned circuit and a crystal diode. A diode performs envelope detection by rectifying the a.m. signal: after smoothing only the speech modulations and a dc component remain.

THE RECEIVER

A good receiver performs an incredible task, selecting a tiny signal from thousands at the aerial, keeping faithfully to the same frequency for hours, smoothing out fluctuations of a million or more in signal amplitude, and converting the signal to intelligible sound, neither wasting any signal power nor introducing much noise. Modern receivers are dauntingly complex and expensive. However, adequate results and great pleasure can be obtained from much simpler "home-brewed" constructions.

The essentials of a receiver are filters, amplifiers and a detector (demodulator). Of these, filters pose the greatest problem. It has already been shown that a tuned circuit (or half a dozen tuned circuits) is not nearly selective enough at radio frequencies. One solution is to convert the rf signal to a much lower *intermediate frequency* (i.f.) in a *superheterodyne* receiver. This also avoids the problem of tuning several filters: any desired signal is converted to a fixed i.f. (often 465kHz) so that subsequent filters need no adjustment.

Fig 1.93 shows the progress of a signal through a "superhet". The incoming signal is converted to the much lower intermediate frequency by a *mixer* or *frequency changer*. At this frequency it can be amplified with ease, and a succession of double-tuned transformers provides good selectivity. The resultant i.f. signal is demodulated by a *detector* and the modulations passed to a loudspeaker or headphones.

Detector

As mentioned earlier, when two signals of different frequencies are distorted by any non-linear stage one of the new frequencies generated is a "beat note" or *heterodyne* signal at the difference frequency. For example, signals at 465 and 466kHz would combine to produce an audible 1kHz tone. An a.m. signal can be detected by *any* distorting device because each sideband heterodynes with the carrier to produce audio frequencies. Thus a 7MHz transmission amplitude-modulated by a 1kHz tone consists of: 7·001MHz (upper sideband), 7·000MHz (carrier), and 6·999MHz (lower sideband).

At the receiver detector, each sideband heterodynes with the carrier to generate a 1kHz difference frequency, ie the modulation is recovered as an audio signal. The simplest non-linear device is the diode, and **Fig 1.94** shows a diode detector in the simple *crystal set*. Audio passes to the phones

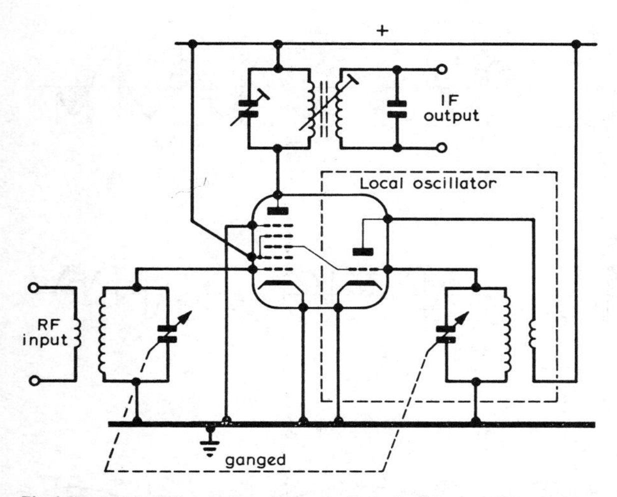

Fig 1.95. A triode-heptode mixer consists of a local oscillator (the triode section) with its output internally connected to the second control grid of the heptode. Incoming signal is applied to the first control grid, so the electron stream is modulated by both voltages and the resulting anode current is roughly proportional to their product. This contains a heterodyne frequency which is selected by the anode transformer. The other grids serve only to isolate input and output circuits from each other. To change receiving frequency both incoming and local oscillator signals are varied together, using ganged tuning capacitors on one spindle (indicated by the dotted line). (Biasing not shown).

while the various higher frequencies are shunted away by a capacitor.

To detect an ssb signal the carrier must first be replaced by a signal from an oscillator in the receiver. The mixture can then be detected by a single diode as above, or by a *balanced detector* or a *product detector*, described below.

Detection of morse (cw) signals requires that the bursts of carrier be made audible in some way. The simplest method is to add a signal at a frequency slightly above the i.f. from a *beat-frequency oscillator* (bfo) and pass the combination through any detector, generating a heterodyne note during each burst of carrier.

Mixer (Frequency Changer)

The frequency of the incoming signal is lowered by *mixing* in a non-linear stage with another rf signal from a *local oscillator* (lo). This produces an ultrasonic difference frequency (superheterodyne) signal with the same speech modulations as the original signal. Signals from several stations will reach the mixer, and by tuning the local oscillator one signal is selected to heterodyne at the desired i.f. The others then produce lower or higher heterodyne frequencies which will be rejected by the i.f. filters.

The *multigrid* valve is a popular mixer, and **Fig 1.95** shows a triode-heptode version. Electron flow through the heptode is controlled successively by two control grids, one carrying incoming signal and one coupled to the local oscillator (the triode section). The anode current is roughly proportional to the *product* of the two grid voltages, and this non-linear process generates the difference (heterodyne) frequency.

Mathematically, if the incoming signal has the form $V \cos At$ and the local oscillator signal is $V' \cos Bt$, the anode current is of the form:

$$I = I_0 \cos At . \cos Bt$$
$$= \tfrac{1}{2} I_0 [\cos (A + B)t + \cos (A - B)t]$$

which are the sum and difference frequencies.

The chief defect of the multigrid mixer is noise—it is so noisy that quiet signals would be missed unless they were amplified before reaching the mixer, and an rf amplifier stage is usually provided. However, this highlights another defect of the multigrid (and most other mixers), the tendency to *cross-modulate*. A strong interfering signal on a nearby frequency after amplification may reach the multigrid mixer at such high voltage that the valve becomes cut off on peaks, and the interfering signal modulates the desired carrier. This interference cannot be removed by any subsequent stage.

Suppose the input to a mixer is:

$$V = X \cos At \text{ (desired carrier with modulation X)}$$
$$+ Y \cos Bt \text{ (unwanted signal)}$$
$$+ \cos Ct \text{ (output from local oscillator)}$$

The non-linear mixer output can be written as follows:

$$\text{Output } I = a + bV + cV^2 + dV^3 \text{ etc}$$

The square term generates the desired difference frequency:

$$V^2 = 2X \cos At . \cos Ct + \text{other terms}$$
$$= X \cos (A - C)t \text{ etc}$$

Higher order terms produce cross-modulation, eg

$$V^4 = 12XY^2 \cos Bt . \cos At . \cos Ct + \text{other terms}$$
$$= 3XY^2 \cos (A - C)t \text{ etc}$$

This latter is a signal at the desired i.f. but bearing modulations from the interfering signal, ie the interfering modulations appear on the desired carrier.

Cross-modulation can thus be avoided by using a *square-law detector* such as a depletion fet with a low gate bias (**Fig 1.96**). This device gives an output proportional to the square of the input voltage (over a small range), which generates a heterodyne but no cross-modulation unless driven outside this range.

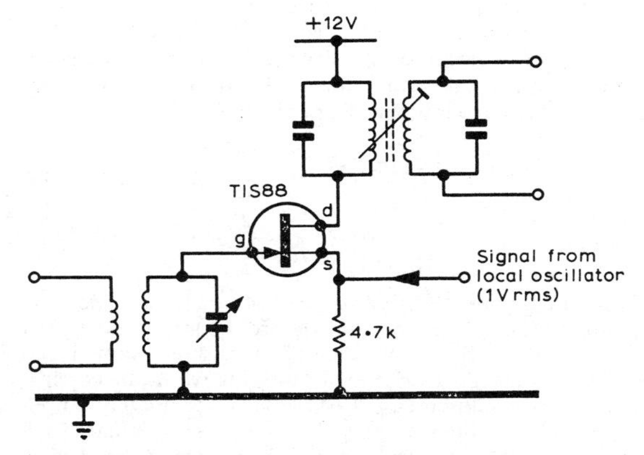

Fig 1.96. A fet square-law detector operates on a curved portion of the device's response characteristic, at which the drain current is proportional to V_{in}^2. Cross-modulation is absent unless the fet is driven off this portion by a very strong signal.

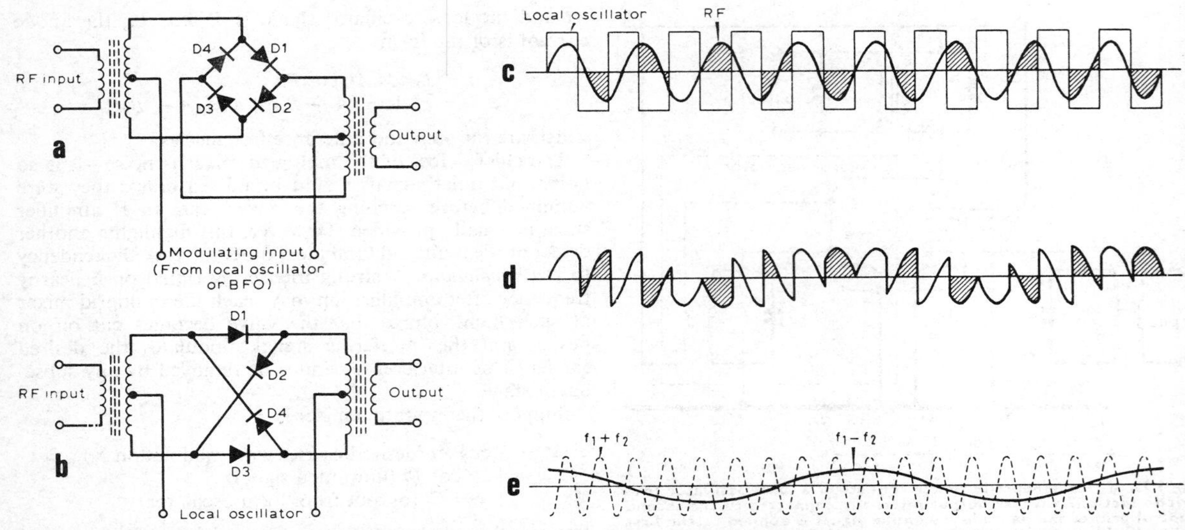

Fig 1.97. The ring-diode circuit (a) is a double-balanced modulator, both input signals cancelling at the output to leave only heterodyne products. In (b) the same circuit is redrawn to show the switching action of the diodes. When the local oscillator applies a positive half-cycle (ideally a square wave), diodes D1 and D3 conduct (c). On negative half-cycles these diodes are cut off, and signal passes through D2 and D4, arriving at the output transformer with reversed polarity. The resulting waveform (d) consists entirely of sum and difference frequencies (e) and harmonics.

Other types of mixer are quieter than the multigrid type but produce less output. In the *triode mixer* signal is fed to the grid and the local oscillator is coupled to the cathode, or vice versa. This type is more reliable than the multigrid one, which has a short working life.

Since the advent of hot-carrier diodes interest has centred on the *balanced mixer*, in which one or both input frequencies are removed from the output by cancellation, leaving only intermodulation products. The *ring diode* circuit **(Fig 1.97)** will function as mixer, modulator or product detector: the local oscillator signal is injected at high voltage and the diodes act as switches. Diode mixers are compact and reliable but give no amplification, and do add a little noise to the signal.

All of these mixers introduce *second-channel interference*: each station tends to appear twice on the band, and these *images* interact to produce howls ("birdies") as the receiver is tuned. This is because the mixer converts two frequencies to the same i.f., one from above the local oscillator frequency and one from below. For example:

Desired signal = 28·00MHz } heterodyne = 465kHz i.f.
Local oscillator = 28·465MHz }
Interfering station = 28·93MHz } heterodyne = 465kHz

Second-channel interference can only be avoided by filtering off the unwanted signal before the mixer stage, ie at radio frequency. Since the separation between wanted and interfering signals is twice the i.f., filtering would be easier if a high i.f. was chosen (say 1·6MHz). Unfortunately amplification and i.f. filtering then becomes less efficient, so a compromise must be reached. An alternative is the *double-conversion* receiver, which uses two intermediate frequencies (see Chapter 4—*HF Receivers*).

Automatic Gain Control

A weak signal reaching the receiver may fluctuate in strength from 1μV to 1V as atmospheric conditions vary during the transmission. To prevent great variations in listening volume the receiver is fitted with *automatic gain control* (agc, avc). Signal reaching the detector is rectified and applied as a negative bias voltage to the i.f. amplifiers. Strong signals provide a heavy bias, reduce the gain, and so emerge only a little louder than weak signals.

Refinements include avoiding any attenuation of weak signals (*delayed* agc), suppression of background noise in the absence of signal (*quiet* agc or *squelch*), preventing sudden noise spikes from causing attenuation of the signal (*slow-rise* agc), and preventing noise increase during short pauses in transmission such as between sentences (*hang* agc).

Static interference often occurs as very short "spikes" of high amplitude, and these are unpleasant for the listener. A *noise limiter* is a gate applied to the i.f. signal which limits its maximum amplitude, so that noise spikes are clipped off at a preset level. In the automatic noise limiter (anl) the width of the gate is controlled by the steady signal voltage, so that clipping of any sudden increase occurs (see Chapter 4—*HF Receivers*).

Transceive Operation

The vfo of a transmitter is quite similar to the local oscillator of a superhet, although the lo operates slightly off the station frequency. In the *transceiver* one oscillator functions as combined vfo and lo, so that once the receiver has been *netted* (tuned to the exact frequency of an incoming station) the transmitter is already set to reply on the same frequency. To allow for errors and drift, the receiver can usually be tuned over a small range by a receiver independent tuning (rit) control without disturbing the transmission frequency.

CHAPTER 2

ELECTRONIC TUBES AND VALVES

MODERN electronic tubes and valves have attained a high degree of reliability and are available in many forms for a wide variety of common and specialist applications. Although for many purposes semiconductors are superseding the electronic tube, the latter retains many important advantages in its ability to withstand severe overload without serious damage, its electrode isolation and the high power gain it provides at very high frequencies.

FUNDAMENTALS

Emission

In most types of evacuated electronic tube the emission of electrons is produced by heating the cathode, either directly by passing a current through it, or indirectly by using an insulated heater in close proximity. The quantity of electrons emitted is governed by the construction and surface coating of the cathode and the temperature to which it is heated. This is known as *thermionic emission.*

Emission may also be produced when electrons impinge on to a surface at a sufficient velocity. For example, electrons emitted from a hot cathode may be accelerated to an anode by the latter's positive potential. If the velocity is high enough electrons will be released from the anode. This is known as *secondary emission.*

The emission of electrons from metals or coated surfaces heated to a certain temperature is a characteristic property of that metal or coated surface. The value of the thermionic emission may be calculated from Richardson's formula:

$$I_s = A_1 T^2 e^{-b_1/T}$$

where I_s is emission current in amperes per square centimetre
A_1 is a constant of the emitting substance
T is absolute temperature in degrees Kelvin (°K)
b_1 is a constant depending on the material of the emitting surface,
or from a similar formula developed by Dushman.

Electron Flow

Electrons are negatively charged. When in an evacuated tube an electron leaves a parent molecule, as for example during emission from a cathode, the molecule becomes more positively charged. If an electrode such as an anode is placed near to the cathode and is charged positively with respect to it, the electrons released by emission from the cathode will be attracted to the anode. As the electrons traverse the space between one electrode and another they may collide with gas molecules (because no vacuum can be perfect) and such collisions will impede their transit. For this reason the residual gas left inside the evacuated envelope must be minimal. An electronic tube which has been adequately evacuated is termed *hard.*

If however a significant amount of gas is present the collisions between electrons and gas molecules will cause it to ionize. The resultant blue glow between the electrodes indicates that the tube is *soft*. This blue glow should not be confused with a blue haze which may occur on the inside of the envelope external to the electrode structure: this is caused by bombardment of the glass, and in fact indicates that the tube is very hard.

Space Charge

When electrons travel from cathode to anode in useful quantities they form a cloud in the space between the electrodes. The electric charge associated with this cloud is known as the space charge. It tends to repel the electrons leaving the cathode because it carries the same polarity. However, if the anode potential is sufficiently high the effect of the space charge will be overcome and electrons will flow from the cathode, the flow being completed by an external circuit back to the cathode. As the anode potential is raised, the electron flow or current will increase to a point where the space charge is completely neutralized and the total emission from the cathode reaches the anode. The flow can be further increased only by raising the cathode temperature.

Cathodes

Although several types of cathode are used in modern valves, the differences are only in the method of producing thermionic emission. The earliest type is the *bright emitter* in which a pure tungsten wire is heated to a temperature in the region of 2500–2600°K. At such a temperature emission of 4 to 40mA per watt of heating power may be obtained.

Bright emitters are still employed in high power transmitting valves for broadcasting but the only common amateur use is in diodes for applications such as noise generators. The life of a pure tungsten filament at full operating temperature is limited by evaporation of the tungsten, failure occurring when about 10 per cent has been evaporated.

Dull emitters are directly heated thoriated tungsten cathodes which produce greater emission than bright emitters and require less heating power. In a dull emitter, a small quantity of thorium oxide is introduced into the pure tungsten wire. A process known as carburization is used to create an outer skin of tungsten carbide on the wire which facilitates the reduction of the thorium oxide to metallic thoria, stabilizes the emission and increases the surface resistance of the cathode to gas poisoning. Typical emission efficiency is in the region of 30–100mA per watt of heating power at an operating temperature of 1900–2100°K. This type of cathode is relatively fragile and valves should not be subjected to shocks or sharp blows.

Provided the operating temperature is correctly maintained long life may be expected. In particular, the rated voltage or current should be closely controlled. Operation at constant filament power will give the longest life.

Oxide coated cathodes are the most common type of thermionic emitter found in both directly and indirectly heated valves. In this type, the emissive material is usually some form of nickel ribbon, tube or thimble coated with a mixture of barium and strontium carbonate, often with a small percentage of calcium. During manufacture, the coating is reduced to its metallic form and the products of decomposition removed during the exhaustion process. The active ingredient is the barium which provides much greater emission than thoriated tungsten at lower heating powers. Typically, 50–150mA per watt is obtained at temperatures of 950–1050°K.

Although the emission efficiency of oxide coated cathodes is high and large currents may be drawn, they are less able to resist the poisoning effects of gas or ion bombardment. This type must not be operated under temperature limited conditions.

In certain valves that are subject to back bombardment of the cathode, such as magnetrons, some form of protected cathode coating/material is necessary. Such cathodes are known as impregnated, the active coating material being mixed with nickel or tungsten powder; this mixture is then coated on the cathode surface.

An *indirectly-heated cathode* is a metal tube or sleeve, or in some cases is of thimble shape, having a coating of emissive material on the outer surface. The cathode is heated by radiation from a metal filament, called the *heater*, which is mounted inside the cathode. The heater is electrically insulated from the cathode. The emissive material is generally the same as that employed for filamentary oxide-coated cathodes and operates at about the same temperature. The cathode may be made of pure nickel or of special alloys, depending upon the purpose of the valve. The heater is normally made of tungsten or molybdenum-tungsten alloy.

The life of valves with oxide coated cathodes is generally good provided the ratings are not exceeded. Occasionally there is some apparent reduction in anode current due to the formation of a resistive layer, between the oxide coating and the base metal, which operates as a bias resistor.

In *cold cathode* valves, such as gas stabilizers, the cathode is an activated metal or coated surface.

Anodes

In most electronic tubes the anode takes the form of an open-ended cylinder or box surrounding the other electrodes, and is intended to collect as many as possible of the electrons emitted from the cathode; some electrons will of course be intercepted by the grids interposed between the cathode and the anode.

The material used for the anode of the small general purpose type of valve is normally bright nickel or some form of metal coated black to increase its thermal capacity. Power dissipated in the anode is radiated through the glass envelope, a process which is assisted when adequate circulation of air is provided around the glass surface. In some cases a significant improvement in heat radiation is obtained by attaching to the valve envelope a close-fitting finned metal radiator which is bolted to the equipment chassis so that this functions as a worthwhile heat sink.

Higher power valves with external anodes are cooled directly by forced air, by liquid, or by conduction to a heat sink, as follows:

Forced air: this method of cooling requires a blower, preferably of the turbine type, capable of providing a substantial quantity of air at a pressure high enough to force it through the cooler attached to the anode.

Liquid cooling calls for a suitable cooler jacket to be fitted to the anode; generally, this method is confined to large power valves. If water is used as the coolant care must be taken to ensure that no significant leakage occurs through the water by reason of the high voltage used on the anode.

Conduction cooling carries the heat from the anode to a suitable heat sink via a heat conducting insulator forming part of the tube envelope or by attachment of a heat conducting block connecting the anode to the heat sink. The latter is preferable, for the heat conductor block forms part of the external equipment, but a heat conductor that is part of the tube envelope is lost when failure occurs and the tube is discarded.

Conduction-cooled tetrodes: (left) requiring separate heat sink (centre picture), and (right) with heat conduction direct from envelope to chassis. *(Photo by courtesy of the M-O Valve Co. Ltd.)*

In certain uhf disc seal valves a different form of conduction cooling is used, the anode seal being directly attached to an external tuned-line circuit that functions as the heat-sink radiator. Needless to say, it must be suitably isolated electrically from the chassis.

Whatever the type of electronic tube, and whatever the method used to cool it, the limiting temperatures quoted by the makers, such as bulb or seal temperature, should be adhered to if a reasonable life is to be expected.

Grids

The electron flow from cathode to anode may be controlled by the introduction of one or more electrodes known as grids, the number of such electrodes depending on the purpose for which the tube or valve is required.

Electronic tubes are classified by a generic title based on the number of active electrodes they contain, as shown in **Table 2.1**.

Mechanically, the grid electrode takes many forms, dictated largely by power and the frequency of operation. In small general purpose valves the grids are usually in the form of a helix (molybdenum or other suitable alloy wire) with two side support rods (copper or nickel) and a cross-section varying from circular to flat rectangular, dependent on cathode shape.

TABLE 2.1

number of electrodes	generic title	number of grids
2	diode	none
3	triode	1
4	tetrode	2
5	pentode	3
6	hexode	4
7	heptode	5
8	octode	6

In some uhf valves the grid consists of a single winding of wire or mesh attached to a flat frame fixed directly to a disc seal.

In high performance tubes where close clearance between cathode and grid is required for high mutual conductance, the grid winding is made on to a frame consisting of support rods and metal straps, the helix being wound under considerable tension (about half its breaking strain). This method of construction is generally confined to Grid No 1 (the control grid), where the minor axis of cross-section is decided by the support rod diameter.

In beam tubes such as klystrons, travelling wave tubes and cathode ray tubes, the electron flow is concentrated through a single hole in the electrode plates, the potential applied to a plate having the effect of varying the tube characteristics. For higher power tubes the grid may be in the form of a "squirrel cage".

Primary emission from a grid due to heating must be minimized to obviate affecting the tube's operation. To inhibit emission at the normal operating temperature various types of coating or plated surface perform both this function and at the same time increase the working surface of the grid. It is particularly important to minimize grid emission in transmitting valves, where the grid may be operating in a positive mode. Similar operating conditions apply also to Grid No 2 (the screen grid), more especially in audio output and transmitting types.

Both No 1 and No 2 grids are sometimes fitted with radiation fins to cool them in the interests of holding primary grid emission at a satisfactory level.

Where a variable gain facility is required, as in i.f. amplifiers, the No 1 (control) grid is given a variable helix pitch to enable it to handle changes from high gain to relatively low gain. Pitch variation is conveniently introduced by providing suitable gaps in the winding around the centre of the cathode system.

TUBE TYPES

Diodes

The simplest form of electronic tube is the two electrode diode, consisting simply of an emitting surface (heater-cathode or filament) and an anode. The current which can be drawn from the anode is governed by the type of emitting cathode employed, its temperature and its spacing from the anode.

For a given operating cathode temperature a point of saturation is reached beyond which no further increase in current can be obtained unless the cathode temperature is increased.

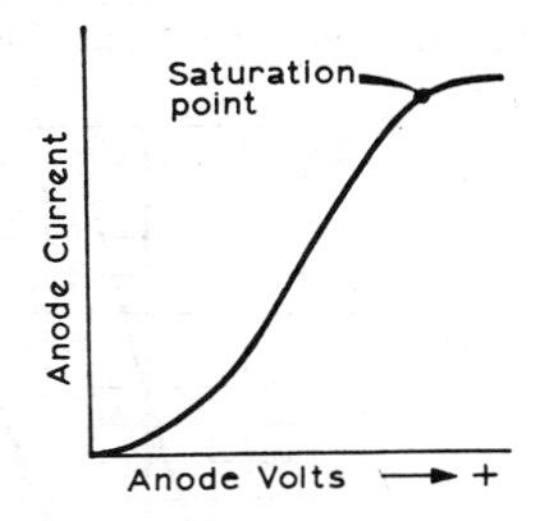

Fig 2.1. Typical diode characteristic, showing emission limitation for a given filament temperature.

Diodes have a wide variety of applications, from low level rf signal detection to power rectification up to very high voltages.

To provide a low impedance characteristic, where the voltage drop across the diode is virtually independent of the current drawn, gas filling is used, either mercury vapour or an inert gas such as xenon. The resultant low voltage drop helps to reduce the anode dissipation.

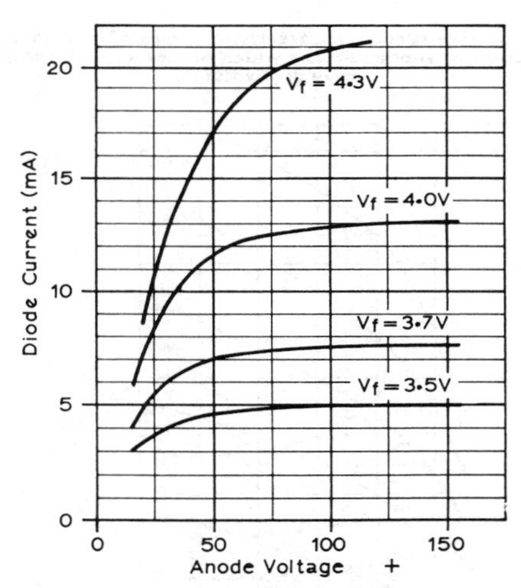

Fig 2.2. Diode saturation curves, showing the effect of different filament voltages and hence temperatures.

Triodes

By introducing a grid between the cathode and anode of an electronic tube the electron flow may be controlled. This flow may be varied in accordance with the voltage applied to the grid, its value being decided by the geometry of the grid and in particular the amplification factor and mutual conductance required from the valve.

Varying the potential applied to the grid modifies the space charge, but because the grid has an open mesh only partial interception of the electrons occurs, the majority of them remaining available for acceleration to the high potential anode. Electron collection by the grid is low if its potential is low, or negative, but increases significantly as the voltage is made more positive.

It is important to recognize that a small general purpose triode or pentode used as an audio amplifier with a negative potential on its grid may, if operated as a positively driven amplifier, be called upon to withstand more grid dissipation than the designer intended, and could be less reliable than a valve designed specifically for the purpose.

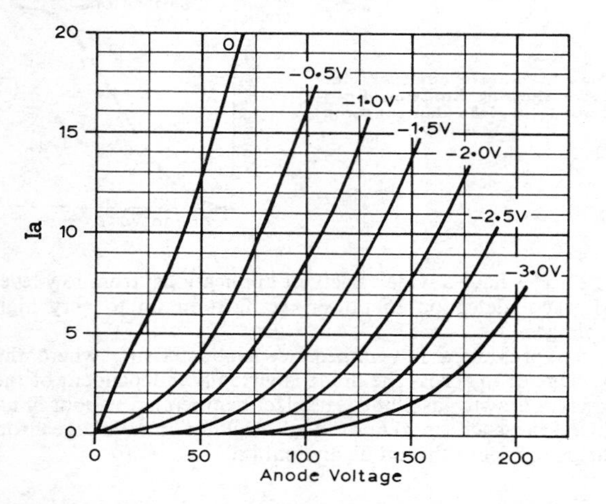

Fig 2.3. Negative region characteristic curves of a triode, showing the reduction in anode current which occurs with increase of negative grid voltage.

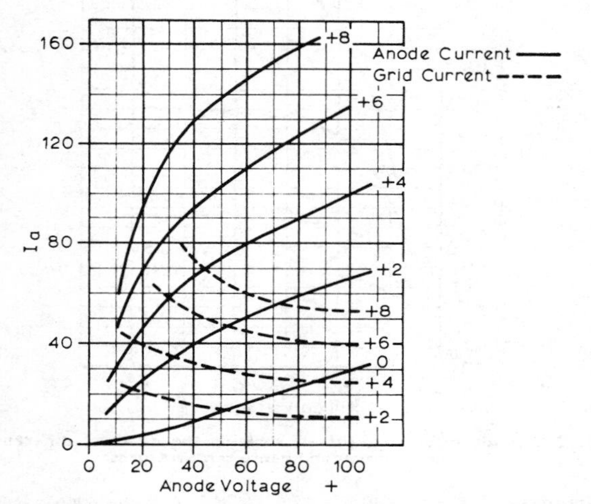

Fig 2.4. Positive region triode characteristic curves, showing how enhanced values of positive grid voltage increase anode current flow.

The grid voltage (grid bias) for a small general purpose valve may be obtained by one of several methods:

A separate battery;

A resistor connected between the cathode and the chassis (earth) so that when current flows the voltage drop across it renders the cathode more positive with respect to the chassis (earth), and the grid circuit return becomes negative with respect to cathode;

A resistor connected between the grid and the chassis (earth). When the grid is so driven that appreciable current flows (as in an rf driver, amplifier or multiplier), the grid resistor furnishes a potential difference between grid and chassis (earth), and with cathode connected to chassis a corresponding negative voltage occurs at the grid. A combination of grid resistor and cathode resistor is good practice and provides protection against failure of drive, which a grid resistor alone would not give.

Contact potential. A high value of resistance (1–10MΩ) is used as the grid resistor. A small current flow through it will provide sufficient negative voltage for small signal applications (eg, an input audio amplifier).

Characteristic curves which show graphically the relationship between anode current and anode voltage for various values of grid voltage are given, one illustrating the curve produced when negative grid voltages are applied, the other those for positive grid voltages, together with the corresponding grid currents.

Tetrodes

A tetrode ("four electrode") valve is basically a triode with an additional grid mounted outside the control grid. When this additional grid is maintained at a steady positive potential a considerable increase in amplification factor occurs compared with the triode state; at the same time the valve impedance is greatly increased.

The reason for this increased amplification lies in the fact that the anode current in the tetrode valve is far less dependent on the anode voltage than it is in the triode. In any amplifier circuit, of course, the voltage on the anode must be expected to vary since the varying anode current produces a varying voltage-drop across the load in the anode circuit. A triode amplifier suffers from the disadvantage that when, for instance, the anode current begins to rise due to a positive half-cycle of grid voltage swing, the anode voltage falls (by an amount equal to the voltage developed across the load) and the effect of the reduction in anode voltage is to diminish the amount by which the anode current would otherwise increase. Conversely, when the grid voltage swings negatively the anode current falls and the anode voltage rises. Because of this increased anode voltage the anode current is not so low as it would have been if it were independent of anode voltage. This means that the full amplification of the triode cannot be achieved. The introduction of the screen grid, however, almost entirely eliminates the effect of the anode voltage on the anode current, and the amplification obtainable is thus much greater.

A screen functions best when its voltage is below the mean value of the anode voltage. Most of the electrons from the cathode are thereby accelerated towards the anode, but some of them are unavoidably caught by the screen. The resulting screen current serves no useful purpose, and if it becomes excessive it may cause overheating of the screen.

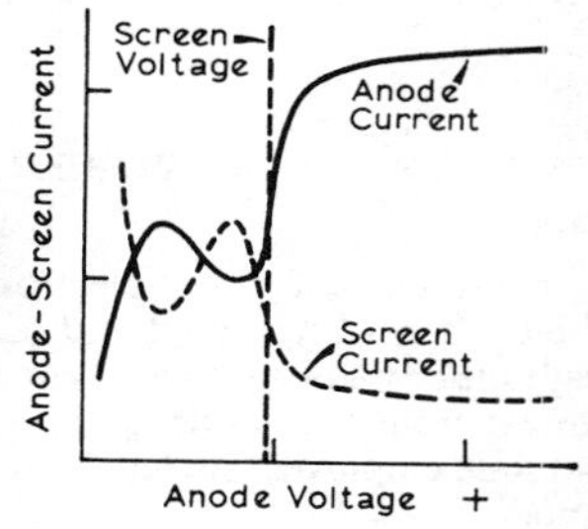

Fig 2.5. Characteristic curves of a pure tetrode (often termed "screened grid"), showing the considerable secondary emission occurring when no suppression is used.

If in low voltage applications the anode voltage swings down to the screen voltage or lower, the anode current falls rapidly while that of the screen rises, due to secondary emission from the anode to the screen. It should be noted that the total cathode current is equal to the sum of the screen and anode currents.

Another important effect of introducing the screen grid (No 2) is that it considerably reduces the capacitive coupling between the input (control) grid and the anode, making possible the use of stable, high gain rf amplification. To utilize this facility additional shields are added to the grid (electrically connected) so that the input connection cannot "see" the anode or its supports. With such a structure it is possible to reduce the unit's capacitance by a factor of almost 1000 compared with the triode. Adequate decoupling of the screen at the operating frequency by the use of a suitable external by-pass capacitor is essential.

In another type of tetrode known as the space charge grid tetrode the second grid is positioned between the usual control grid and the cathode. When a positive potential is applied to this space charge grid, it overcomes the limiting effect of the negative space charge, allowing satisfactory operation to be achieved at very low anode potentials, typically 12–24V.

Pentodes

To overcome the problem presented by secondary emission in the pure tetrode a third grid may be introduced between the screen and the anode and maintained at a low potential, or connected to the cathode. Anode secondary emission is overcome and much larger swings of the anode voltage may be realized. This third grid is known as the suppressor grid (G3).

Other methods which achieve the same effect are:

(a) Increasing the space between screen grid and anode, as in the Harries Critical Anode-distance valve;

(b) Fitting small fins to the inside surface of the anode;

(c) Fitting suppressor plates to the cathode to produce what is known as the "kinkless tetrode", which is the basis of the beam tetrode suppression system.

In some special types of pentode where it is necessary for application reasons to provide two control grids, the No 3 (suppressor) grid is used as the second and lower sensitivity control for gating, modulation or mixing purposes. Units of this type need to have a relatively high screen grid (No 2) rating to allow for the condition when the anode current is cut off by the suppressor grid (No 3).

Beam Tetrodes

A beam tetrode employs principles not found in other types of valve; the electron stream from the cathode is focused ("beamed") towards the anode. The control grid and the screen grid are made with the same winding pitch and they are assembled so that the turns in each grid are in optical alignment: see **Fig 2.6.** The effect of the grid and screen turns being in line is to reduce the screen current compared with a non-beam construction. For example, in a pentode of ordinary construction the screen current is about 20 per cent of the anode current, whereas in a beam valve the figure is 5–10 per cent.

The pair of plates for suppressing secondary emission referred to above are bent round so as to shield the anode from any electrons coming from the regions exposed to the influence of the grid support wires at points where the focusing of the electrons is imperfect. These plates are known as *beam-confining* or *beam-forming plates*.

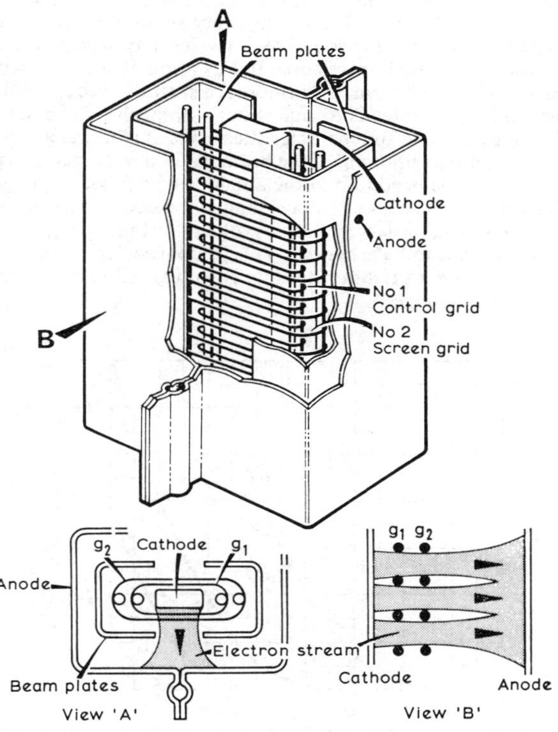

Fig 2.6. The general arrangement of a modern beam tetrode, showing the aligned grid winding and the position of the beam forming plates. View "A": looking vertically into a beam tetrode. View "B": showing how the aligned electrode structure focuses electrons from cathode to anode.

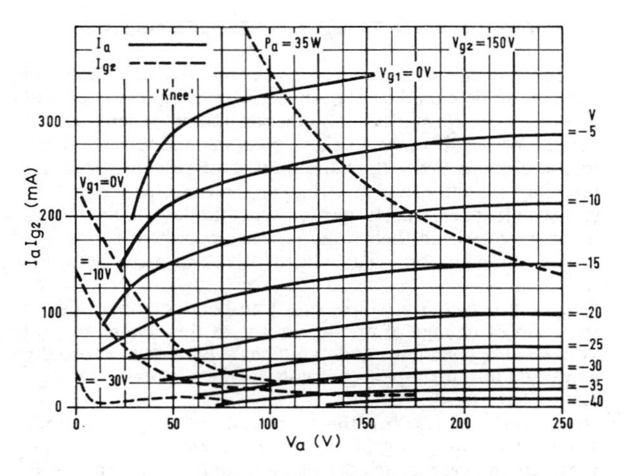

Fig 2.7. Characteristic curves of a beam tetrode. Anode secondary emission is practically eliminated by the shape and position of the suppressor plates.

Beam valves were originally developed for use as audio-frequency output valves, but the principle is now applied to many types of radio-frequency tetrodes both for receiving and transmitting. Their superiority over pentodes for af output is due to the fact that the distortion is caused mainly by the second harmonic and only very slightly by the third harmonic, which is the converse of the result obtained with a pentode. Two such valves used in push-pull give a relatively large output with small harmonic distortion because the second harmonic tends to cancel out with push-pull connection. Fig 2.7 shows the characteristic curves of a beam tetrode.

By careful positioning of the beam plates a relatively sharp "knee" can be produced in the anode current/anode voltage characteristic, at a lower voltage than in the case of a pentode, thus allowing a larger anode voltage swing and greater power output to be achieved. This is a particularly valuable feature where an rf beam tetrode is to be used at relatively low anode voltages.

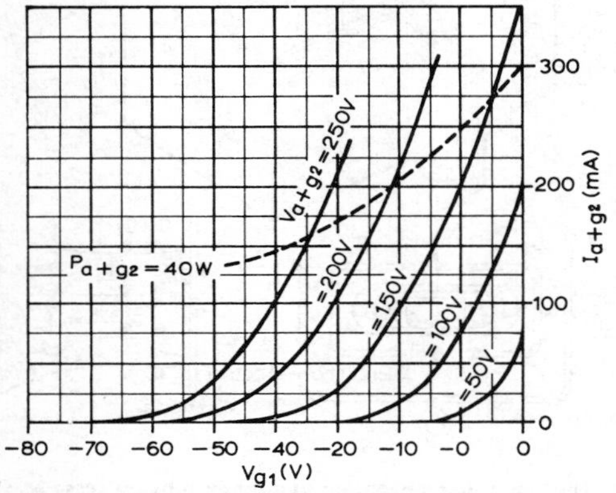

Fig 2.8. Characteristic curves of the beam tetrode connected for use as a triode.

VALVE AND TUBE CHARACTERISTICS

Technical data available from valve and tube manufacturers includes static characteristics and information about typical operating performances obtainable under recommended conditions. Adherence to these recommendations—indeed, to use units at lower than the quoted values—will increase life and reliability, which at higher than quoted values can easily be jeopardized. In particular, cathodes should always be operated within their rated power recommendations.

The following terms customarily occur in manufacturers' data:

Mutual Conductance (slope, g_m, transconductance): this is the ratio of change of anode current to the change of grid voltage at a constant anode voltage. This factor is usually expressed in milliamperes per volt or micromhos (1mA/V = 1000 micromhos).

Amplification Factor (μ): this is the ratio of change of anode voltage to change of grid voltage for a constant anode current. In the case of triodes classification is customarily in three groups, low μ, where the amplification factor is less than 10, medium μ (10 to 50) and high μ (greater than 50).

Impedance (r_a, ac resistance, slope resistance): when the anode voltage is changed while grid voltage remains constant, the anode current will change, an Ohm's Law effect. Consequently, impedance is measured in ohms.

The relationship between these three primary characteristics is given by:

$$\text{Impedance (ohms)} = \frac{\text{Amplification Factor}}{\text{Mutual Conductance}} \times 1000$$

$$\text{or, } r_a = \frac{\mu}{g_m}$$

It will be noted that the mutual conductance and the impedance are equal to the slopes of the I_a/V_g and I_a/V_a characteristics respectively.

Inner and Outer Amplification Factor (μ): in tetrodes and pentodes it is customary to quote the amplification factor of the device as if it were a triode, using the screen (G2) as the anode, or screen and anode connected together. This is known as inner μ or inner amplification factor. Similarly, where the screen (G2) is considered as a control grid in conjunction with the anode, the ratio of change of anode volts to the change of screen voltage for a constant anode current is known as the outer μ or outer amplification factor.

Electrode Dissipation

The conversion from anode input power to useful output power will depend upon the tube type and the operating conditions. The difference between these two values, known as the anode dissipation, is radiated as heat. If maximum dissipation is exceeded overheating will cause the release of occluded gas, which will poison the cathode and seriously reduce cathode emission.

The input power to be handled by any valve or tube will, in the limiting case, depend on the class of operation. Typical output efficiencies expressed as percentages of the input power are:

class A 33%
class AB1 60–65%
class AB2 60–65%
class B 65%
class C 75–80%

Considering a valve with a 10W anode dissipation, the above efficiencies would give outputs as follows (assuming there are no other limiting factors such as peak cathode current):

class A 5W
class AB1 and AB2 15 to 18·5W
class B 18·5W
class C 30–40W

Control and Screen Grid Dissipations

The control grid (No 1) and screen grid (No 2) due to their more fragile construction have a lower thermal capacity than the anode, and their maximum ratings must therefore not be exceeded.

The control grid is heated considerably by radiation from the cathode and therefore its dissipation rating is quite small; any significant overheating will cause grid primary emission (grid emission).

The screen grid, being shielded to a large extent from the cathode by the control grid, can accept a higher temperature rise before primary emission ensues. In an aligned grid tetrode any grid wire which becomes distorted by overheating may become red hot by drawing excessive current.

Noise in Valves

Microphonic Effects: These are simple mechanical effects producing modulation of the wanted signal by variation in the relative positions of the parts of a valve under mechanical vibration, shock, etc.

Resistive Noise: An ideal resistance R (Ω) at an absolute temperature T(°K) has current fluctuations generated within it by the random motion of its free electrons. The mean square values of voltage and current so produced are described below under "Equivalent Noise Resistance".

Flicker Noise: This is so called because it describes the comparatively large amplitudes of the noise output pulses produced by a valve at low frequencies. The amplitude is usually related to frequency according to a 1/f law and becomes noticeable compared with shot noise below a threshold frequency which may range from 100Hz to 50kHz, depending upon the valve type and the operating conditions.

Valves have been designed to produce a minimum of flicker noise when operated under specified conditions. For very exacting low frequency applications it is necessary to select valves for first stage operation and to pay special attention to other noise sources such as carbon resistors.

An approximate expression for the anode current fluctuation $\overline{\delta I_a^2}$ due to flicker noise, for a valve with an oxide coated cathode, is given by:

$$\overline{\delta I_a^2} = \frac{10^{-13}}{f} \times B \times g_m \text{ (amperes)}^2$$

where g_m is the valve mutual conductance in A/V.

Partition Noise: In a pentode or other multigrid valve, the division of the cathode current between the anode, screen and other grids is not constant at all instants of time. There are fluctuation components in both the anode and other electrode currents which add to the total fluctuation in the anode current itself. This added fluctuation component in the anode current is termed partition noise.

At all but low frequencies, where flicker noise predominates, the partition noise component in the output is greater than the shot noise; consequently triodes are generally preferred to pentodes in low noise amplifiers for most purposes.

In general, the greater the number of current taking electrodes in a valve, the greater the effect of partition noise. Thus, multigrid mixers are especially noisy and so should be preceded by as much signal frequency amplification as possible.

Induced Grid Noise: At very high frequencies, the finite transit times of electrons through the cathode-to-grid space enable the grid to extract energy from the electrons. This effect means that the high frequency components of the anode current fluctuations already present due to shot noise, themselves induce current fluctuations in the grid circuit, which then remodulate the electron stream, giving rise to another fluctuation component.

This transit time loading of the grid circuit has the effect of connecting a shunt conductance across the amplifier input circuit. This is discussed later.

Equivalent Noise Resistance: This is defined as the ideal resistance, held at a temperature of 290°K (17°C) which, when placed in series with the grid lead of an ideal noiseless valve, produces the anode current fluctuation due to shot or partition noise effects obtained from a given actual valve of the same characteristics.

The value of the equivalent noise resistance, R_{eq}, can be calculated as follows:

(a) For a triode (considering shot noise only)

$$R_{eq} = \frac{3{\cdot}5}{g_m}\,\Omega$$

(b) For a pentode or tetrode (considering shot and partition noise only):

$$R_{eq} \simeq \left(\frac{I_a}{I_a + I_{g2}}\right) \times \left(\frac{3{\cdot}5}{g_m} + \frac{20\,I_{g2}}{g_m^2}\right)$$

The equivalent noise resistance explains the noise behaviour of valves fairly well over the range 50kHz to about 20MHz. Below 50kHz, flicker noise effects become increasingly important and above 20MHz induced grid noise predominates.

Transit Time Conductance and Induced Grid Noise: Above 20MHz the induced grid noise effects predominate, imposing a shunt conductance across the input circuit of an amplifier. This shunt conductance, G_e, can be calculated as follows:

$$G_e = \frac{5g_m(af)^2}{V_1 \times 10^{15}} \times \left(1 + \frac{3{\cdot}3\,b/a}{1 + (V_a/V)^{\frac{1}{2}}}\right) \quad \text{(A/V)}$$

where: $V_1 = 5{\cdot}69 \times 10^3 \times a^{4/3} \times J_c^{2/3}$ (V)

g_m = mutual conductance of the valve under the given operating conditions (A/V).

a is the grid-cathode spacing (cm).

b is the grid-anode spacing (cm).

f is the frequency (Hz).

V_a is the anode voltage (V).

J_c is the cathode current density of the valve under the given operating conditions (A/cm²).

This relationship assumes the following:

(a) The valve geometry is planar.
(b) The initial velocities of the electrons leaving the cathode surface are zero.
(c) The emission is space charge limited.
(d) The grid plane can be considered as equipotential.
(e) The μ value is large.
(f) The signal voltages on grid and anode are small.
(g) The transit angles through the cathode-to-grid and grid-to-anode spaces are small.

Hence: $G_e \propto f^2$

The shunt conductance, G_e, applied across the amplifier input terminals, generates current fluctuations as if it were an ideal resistance held at a temperature of about 1400°K.

The transit time shunt conductance G_e should not be confused with the shunt conductance placed across the input terminals of an amplifier containing a valve, either in the grounded cathode or grounded grid connection.

In the case of the grounded cathode connection, the interaction between the valve and the cathode lead inductance produces a conductance G_c, placed across the amplifier input, of value given by:

$$G_c = g_m\,(4\pi^2 f^2)\,L_c\,c_{g-k} \text{ (A/V)}$$

where: g_m is the mutual conductance of the valve in A/V.

f is the operating frequency in hertz.

L_c is the cathode lead inductance in henries.

c_{g-k} is the grid-cathode capacitance in farads.

Here again, $G_c \propto f^2$.

G_c does not contribute any noise to the amplifier but it does serve to apply extra damping to the input of the amplifier, which is not usually desirable.

In the case of the grounded grid connection, a shunt conductance term G_{in} is applied across the input, taking into account the valve, the transit time conductance G_e and the load applied to the output of the valve circuit.

G_{in} is given by:

$$G_{in} = G_e + \frac{g_m}{1 + 1/(r_a G_L)}$$

where: g_m is the valve mutual conductance.
G_L is the external load shunt conductance.
r_a^{-1} is the valve anode conductance.

To minimize the noise factor of an amplifier, the effects of the equivalent noise resistance and transit time conductance, considered as noise sources, must be made as low as possible at the operating frequency, always bearing in mind the importance of sufficient gain to eliminate the effects of second stage noise contributions.

Hum: When a cathode is heated by ac the current generates a magnetic field which can modulate the electron stream, and a modulating voltage is injected into the control grid through the inter-electrode capacitance and leakages: additionally there can be emission from the heater in an indirectly-heated valve.

When operating directly heated valves such as transmitting valves with thoriated tungsten filaments, the filament supply should be connected to earth by a centre tap or a centre tapped resistor connected across the filament supply (a "hum bucking" resistor).

The hum is usually expressed as an equivalent voltage (in microvolts) applied to the control grid. Valve hum should not be confused with hum generated in other circuit components.

Electrode Primary Emission

In an indirectly heated valve it is possible for the heater to become contaminated with emissive material and so emit electrons. This is known as *heater emission.* The emitted electrons will be attracted to any electrode which is positive with respect to the heater, and such an electrode may be the cathode or the screen grid or anode; it can even be the control grid, which although generally negative to the cathode can become sufficiently positive to one end of the heater during at least part of the ac cycle of the heater supply. This explains why it is sometimes recommended that the centre-tap of the heater supply should be connected to a positive point in the circuit to reduce hum. The grid can thereby be maintained at a negative potential with respect to the heater.

Grid primary emission is a condition in which a grid commences to emit electrons itself and, figuratively speaking, competes with the cathode. The effect is produced by the heating of the grid which may be caused by an excessive flow of grid current, by the close proximity of the hot cathode or by radiated heat from the anode. The effects are accentuated if the grid becomes contaminated with active cathode material, which can happen if the valve is appreciably over-run even for a relatively short time.

The control grid, the screen and the suppressor grid are all subject to these effects. They are avoided by keeping the grid-cathode resistance low and by avoiding excessive heater, anode or bulb temperature (ie by not over-running the valve).

Two examples of the effects of control-grid primary emission can be given:

(a) In a small output valve the anode current rises steadily accompanied by distortion due to grid current flowing in the high-resistance grid leak in such a direction as to oppose the grid bias.

(b) In a power amplifier or frequency multiplier in a telegraphy transmitter the drive diminishes when the key is held down, accompanied by rising anode current.

Primary screen emission: in an oscillator or amplifier when the screen voltage is removed, for example by keying, the output is maintained often for quite long periods if the valve is already hot.

Anode primary emission is similar to grid emission but occurs when the anode attains a sufficiently high temperature to emit electrons. This effect occurs mostly in rectifiers and causes breakdown between anode and cathode.

Secondary Emission

When an electron which has been accelerated to a high velocity hits an electrode such as a grid or anode, electrons are dislodged and these electrons can be attracted to any other electrode having a higher potential. This effect is termed *secondary emission.* Under controlled conditions one electron can dislodge several secondary electrons, and a series of secondary-emitting "cathodes" will give a considerable gain in electrons. This principle is used in the electron-multiplier type of valve.

Conversion Conductance

The term *conversion conductance* is used in regard to frequency changers to represent the ratio of the output current of one frequency to the input voltage of another frequency. As applied to the mixer of a super-heterodyne receiver, for example, the conversion conductance is the current in the anode circuit at intermediate frequency (measured in microamperes) divided by the input voltage to the grid at signal frequency. The symbol commonly used is g_c and it is generally measured in microamperes per volt.

Conversion or Translation Gain

Conversion or translation gain is the ratio of intermediate frequency output voltage to radio-frequency input voltage. It can be obtained from the conversion conductance if the dynamic resistance and other parameters of the i.f. transformer used in the mixer anode circuit are known.

Cathode-interface Impedance

When a valve is operated for long periods, particularly with low cathode current or at complete cut-off, the mutual conductance steadily falls and so also does the available peak emission. This effect is due to the growth of a film between the metallic cathode and its emissive coating. This film possesses an impedance—the *cathode interface impedance*—which may be represented by a resistance with capacitive shunt connected in series with the cathode and acting as an automatic bias resistor. The rate of growth of interface resistance is considerably affected by the material of the cathode and is accelerated by high temperatures resulting from excessive heater voltage. Since the cathode-interface resistance is normally of the order of a few hundred ohms it has a most serious effect on valves having a high slope and a short grid base because the normal cathode resistor is likely to be comparable with this value. The effect of the parallel capacitance is to make the drop in performance less noticeable as the frequency is increased.

Contact Potential

A small potential difference exists between electrodes of dissimilar materials in a valve irrespective of any externally applied potentials. This is known as the *contact potential.* In a simple diode there is a potential difference between the anode and the cathode which causes a current to flow in any external circuit from anode to cathode. The magnitude of this contact potential depends on the cathode material, the type of emissive coating, the anode material and any contaminating film present upon its surface. Its value (anode to cathode) is between +1V and −0·5V but it is most frequently positive; it is affected by cathode temperature and varies throughout the life of the valve. In a triode or any other

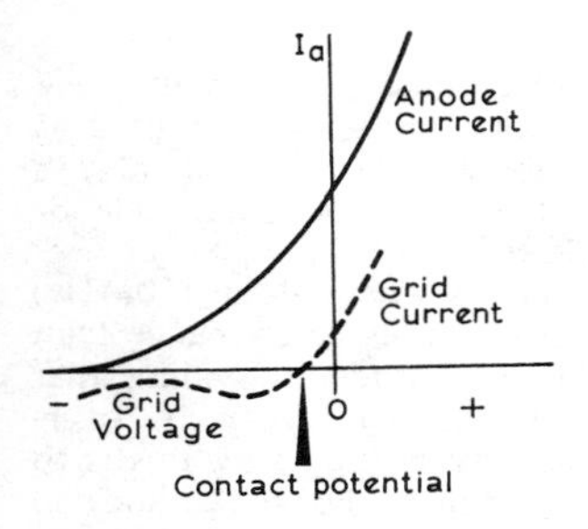

Fig 2.9. The diagram illustrates the *grid* contact potential point —always a negative voltage.

valve with a control grid a potential difference exists similarly between the control grid and cathode and between other electrodes. All electrodes except the control grid and possibly the suppressor grid can be ignored in practical applications, because the current due to it is small compared with other currents flowing in the circuit. The contact potential is effectively in series with the control grid, with the result that if no external grid bias is applied grid current will flow in the external circuit in the same manner as in a diode. If an increasing external grid bias is applied, a value of negative bias is reached when the grid current ceases and this is a measure of the contact potential for this particular valve. In order to operate the valve satisfactorily it is necessary in most cases to increase the bias still further to a point where the maximum positive signal input does not swing the grid more positive than the contact potential.

VALVE APPLICATIONS

Amplifiers

When an impedance is connected in series with the anode of a valve and the voltage on the grid is varied, the resulting change of anode current will cause a voltage change across the impedance. The curves at **Fig 2.10** illustrate the classifications of valve amplifier operating conditions, showing anode current/grid voltage characteristics, and the anode current variations caused by varying the grid voltage.

Class A: The mean anode current is set to the middle of the straight portion of the characteristic curve. If the input signal is allowed either to extend into the curved lower region or to approach zero grid voltage, distortion will occur because grid current is caused to flow by the grid contact potential (usually 0·7 to 1·0V). Under class A conditions anode current should show no movement with respect to the signal impressed on the grid. The amplifier is said to be linear.

Typical valves used in many linear amplifiers. *(Photo by courtesy of the M-O Valve Co. Ltd.)*

Class AB1: The amount of distortion produced by a non-linear amplifier may be expressed in terms of the harmonics generated by it. When a sine wave is applied to an amplifier the output will contain the fundamental component, but if the valve is allowed to operate on the curved lower portion of its characteristic, ie, running into grid current, harmonics will be produced as well. Harmonic components are expressed as a percentage of the fundamental. Cancellation of even harmonics may be secured by connecting valves in push-pull, a method which has the further virtue of providing more power than a single valve can give, and is to be widely found in audio amplifiers and modulators.

Class AB2: If the signal input is increased beyond that used in the class AB1 condition peaks reaching into the positive region will cause appreciable grid current to be drawn and the power output to be further increased. In both the class AB1 and AB2 conditions the anode current will vary from the zero signal mean level to a higher value determined by the peak input signal.

Class B: This mode, an extension of class AB2, uses a push-pull pair of valves with bias set near to the cut-off voltage. For zero signal input the anode current of the push-pull pair is low, but rises to high values when the signal is applied. Because grid current is considerable an appreciable input power from the drive source is required. Moreover, the large variations of anode current necessitate the use of a well regulated power (ht) supply.

Class C: This condition includes rf power amplifiers and frequency multipliers where high efficiency is required without linearity, as in cw, a.m. and nbfm transmitters. Bias voltage applied to the grid is at least twice, sometimes three times, the cut-off voltage, and is further increased for pulse operation. The input signal must be large compared with the other classes of operation outlined above, and no anode current flows until the drive exceeds the cut-off voltage. This could be for as little as 120° in the full 360° cycle, and is known as the conduction angle. Still smaller conduction angles increase the efficiency further, but more drive power

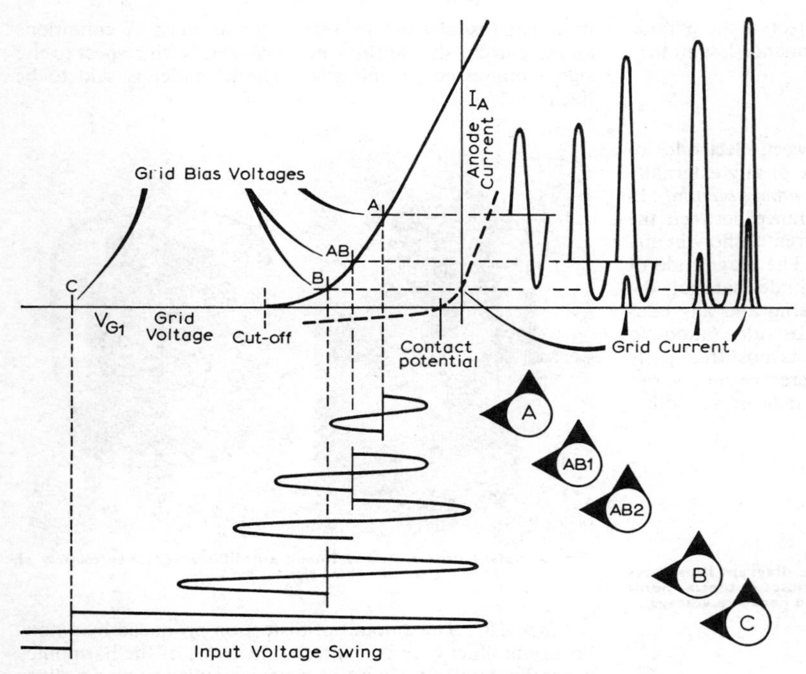

Fig 2.10. The valve as an amplifier: the five classes of operation.

is then needed. Pulse operation is simply "super class C"; very high bias is applied to the grid and a very small angle of conduction used.

Grid Driving Power

An important consideration in the design of class B or class C rf power amplifiers is the provision of adequate driving power. The driving power dissipated in the grid-cathode circuit and in the resistance of the bias circuit is normally quoted in valve manufacturers' data. These figures frequently do not include the power lost in the valveholder and in components and wiring or the valve losses due to electron transit-time phenomena, internal lead impedances and other factors. Where an overall figure is quoted, it is given as *driver power output*. If this overall figure is not quoted, it can be taken that at frequencies up to about 30 MHz the figure given should be multiplied by two, but at higher frequencies electron transit-time losses increase so rapidly that it is often necessary to use a driver stage capable of supplying 3–10 times the driving power shown in the published data. The driving power available for a class C amplifier or frequency multiplier should be sufficient to permit saturation of the driven valve; ie, a substantial increase or decrease in driving power should produce no appreciable change in the output of the driven stage. This is particularly important when the driven stage is anode-modulated.

Passive Grid

In linear amplifiers the driver stage must work into an adequate load, and the use of the passive grid arrangement is to be recommended. A relatively low resistance (typically 1kΩ) is applied between grid and cathode with a resonant grid circuit, where appropriate. This arrangement helps to secure stable operation, but should not be used as a cure for amplifier instability.

Grounded Cathode

Most valves are used with the cathode connected to chassis or earth, or where a cathode-bias resistor is employed it is shunted with a capacitor of low reactance at the lowest signal frequency used so that the cathode is effectively earthed. In modulated amplifiers two capacitors, one for rf and the other for af, must be used.

Grounded Grid

Although a triode must be neutralized to avoid instability when it is used as an rf amplifier this is not always essential if an rf type of tetrode or pentode is employed. However, at very high frequencies (above about 100MHz) a triode gives better performance than a tetrode or pentode, providing that the inherent instability can be overcome. One way of achieving this is to earth the grid instead of the cathode so that the grid acts as an rf screen between cathode and anode, the input being applied to the cathode. The capacitance tending to make the circuit unstable is then that between cathode and anode, which is much smaller than the grid-to-anode capacitance.

The input impedance of a grounded grid stage is normally low, of the order of 100Ω, and therefore appreciable grid input power is required. Since the input circuit is common to the anode-cathode circuit, much of this power is, however, transferred directly to the output circuit, ie, not all of the driving power is lost.

Grounded Anode

For some purposes it is desirable to apply the input to the grid and to connect the load in the cathode circuit, the anode being decoupled to chassis or earth through a low-reactance capacitor. Such circuits are employed in cathode followers and infinite-impedance detectors.

Neutralizing Amplifiers

Instability in rf amplifiers results from feedback from the anode to the grid through the grid-to-anode capacitance and is minimized by using a tetrode or pentode. At high frequencies, particularly if the grid and/or anode circuit has high dynamic impedance, this capacitance may still be too large for complete stability. A solution is to employ a circuit in which there is feedback in opposite phase from the anode circuit to the grid so that the effect of this capacitance is balanced out. The circuit is then said to be *neutralized.*

A typical arrangement is shown in **Fig 2.11.** Here the anode coil is centre-tapped in order to produce a voltage at the "free" end which is equal and opposite in phase to that at the anode end. If the free end is connected to the grid by a capacitor (C_n) having a value equal to that of the valve grid-to-anode capacitance (C_{g-a}) shown dotted, any current flowing through C_{g-a} will be exactly balanced by that through C_n. This is an idealized case because the anode tuned circuit is loaded with the valve anode impedance at one end but not at

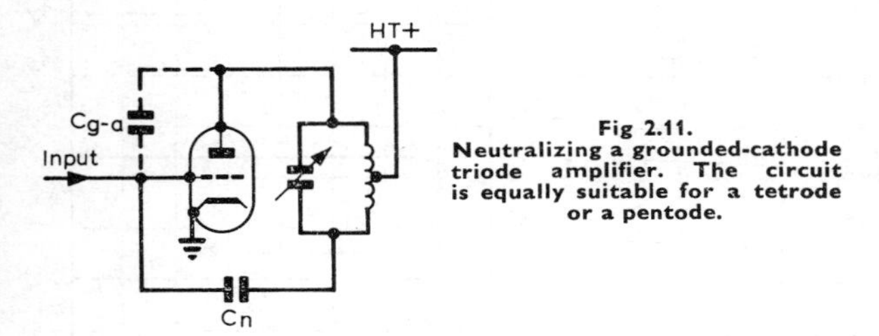

Fig 2.11. Neutralizing a grounded-cathode triode amplifier. The circuit is equally suitable for a tetrode or a pentode.

the other; also the power factor of C_n will not necessarily be equal to that of C_{g-a}. The importance of accurate neutralization in transmitter power-amplifier circuits cannot be overstressed, and will be achieved if the layout avoids multiple earth connections and inductive leads—copperstrip is generally preferable to wire for valve socket cathode connections.

Double Tetrodes for VHF/UHF Service

Early designs of double tetrode contained two complete electrode systems in one envelope. In current designs there is a common cathode and a common screen grid, with separate control grids and anodes positioned on each side of the cathode. In the design illustrated at **Fig 2.12** crossover neutralizing capacitors are built in. The form of construction shown helps to reduce the effects of circulating currents and lead inductance, especially in cathode and screen.

Fig 2.12. Cut-away view of a vhf double tetrode, the QQV06-40A
r, r′—electrode support rods; c, c′—neutralizing capacitors; a, a′—anodes; B—beam plate; M—mica electrode supporting plate; k—cathode; g_1—control grid; g_2—screen grid; S—internal screen. *(Photo by courtesy of Mullard Ltd.)*

Suitably designed class C amplifier and frequency multiplier valves will give satisfactory operation up to about 600MHz.

CALCULATION OF OPERATING CONDITIONS FOR RF AMPLIFIERS

In a tuned amplifier the anode and grid voltages are of sine-wave form and in phase opposition. The anode current does not flow continuously, but in a series of pulses whose duration varies from 40° to more than 180° of each complete cycle of 360°.

The grid current flows for a shorter duration, since this only occurs when the grid is positive relative to the cathode.

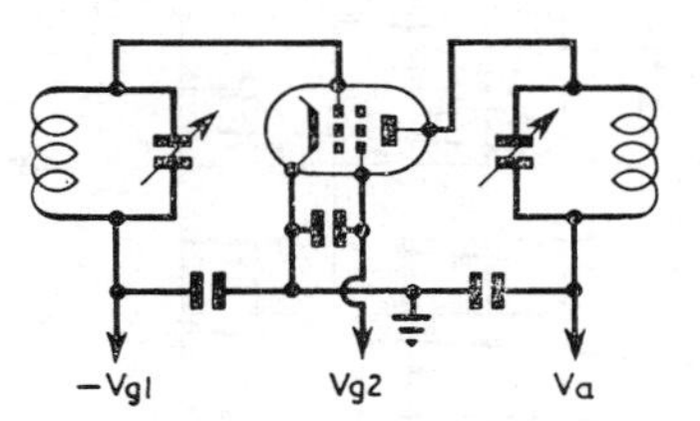

Fig 2.13. Basic circuit of a tuned amplifier.

Figs 2.13 and **2.14** show the basic circuit and phase relationships, respectively. It will be seen that the peak values of anode and grid currents occur when the anode voltage is low and the grid voltage is at its maximum positive value. The design methods given here are based on the location of this point on the valve characteristic curves and the translation of the peak values into rms and mean values, by applying factors derived from a Fourier analysis of sine and sine squared pulses of appropriate angles of flow. This method is very much quicker and only slightly less accurate than the alternative of plotting load lines on constant current characteristics.

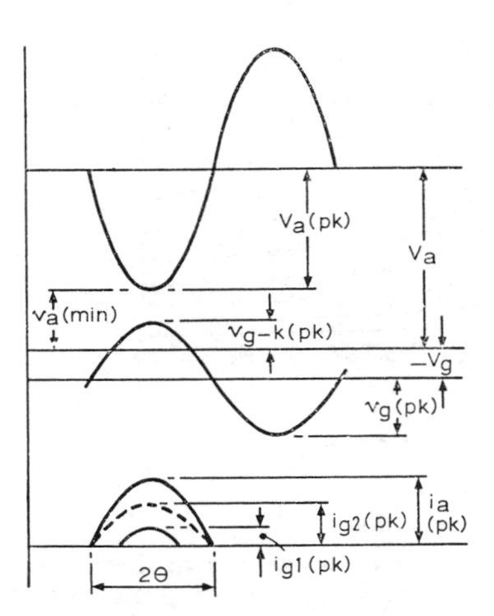

Fig 2.14. Anode/grid phase relationship in the tuned amplifier.

The method is best illustrated by a typical example; in this case a transmitting tetrode type TT21 (7623) has been used. The valve has a rated continuous anode dissipation of 37·5W. Its characteristics measured at I_a = 140 mA are: mutual conductance (g_m) = 11mA/V, and inner amplification factor ($\mu_{g_1-g_2}$) = 8. The relevant valve curves are shown in **Figs 2.15, 2.16, 2.17** and **2.18.**

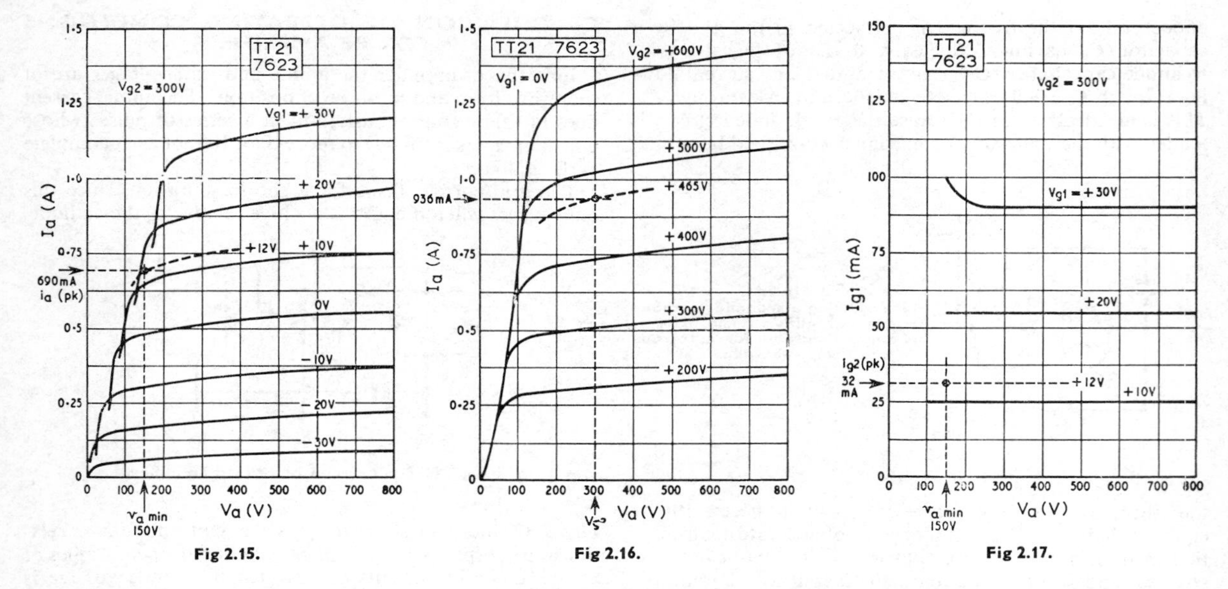

Fig 2.15.

Fig 2.16.

Fig 2.17.

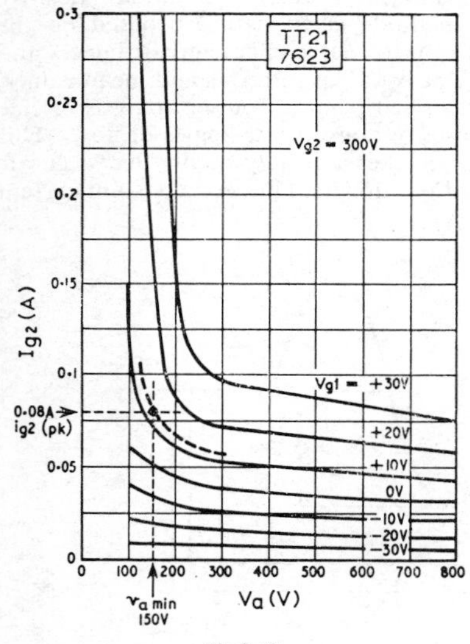

Fig 2.18.

Figs 2.15, 2.16, 2.17, 2.18. Characteristic curves of a TT21 (7623) transmitting tetrode.

Class C Telegraphy

A typical angle of anode current flow (2) for class C telegraphy is 120°. Smaller angles give increased efficiency, but at the expense of increased peak emission demand, greater driving power and possibly shorter valve life. Larger angles are sometimes used when power output is more important than efficiency.

The design factors required for calculations are F_1, F_2, F_3 and F_4. These can be obtained from the curves in **Fig 2.19** for an angle of θ of 60°. These are:

$$F_1 = 4{\cdot}6 \qquad F_3 = 2{\cdot}0$$
$$F_2 = 1{\cdot}8 \qquad F_4 = 5{\cdot}8$$

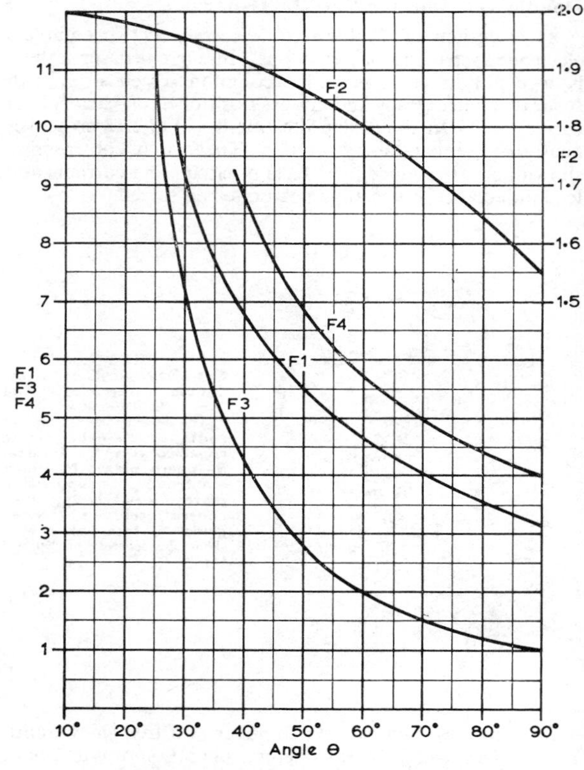

Fig 2.19. Design factors for Class C telegraphy (see text).

The design formulae are:

Peak Anode Current $i_{a(pk)} = F_1 \times I_a$ (1)

Peak Anode Voltage $v_{a(pk)} = V_a - v_{a(min)}$ (2)

Power Output $P_{out} = \frac{F_2}{2} \times I_a \times v_{a(pk)}$ (3)

Grid Voltage (Triodes)

$$-V_g = \frac{V_a \times F_3}{\mu} + (v_{g-k(pk)}) \times (F_3 - 1) \quad (4a)$$

Grid Voltage (Tetrodes)

$$-V_g = \frac{V_{g2} \times F_3}{\mu_{(g1-g2)}} + (v_{g1-k(pk)}) \times (F_3 - 1) \quad (4b)$$

Peak Grid Voltage $v_{g1(pk)} = V_{g1} + (v_{g1-k(pk)})$ (5)

Calculate ratio $\frac{V_g}{v_{g(pk)}}$ and from curve in Fig 2.20 read F_5 and F_6

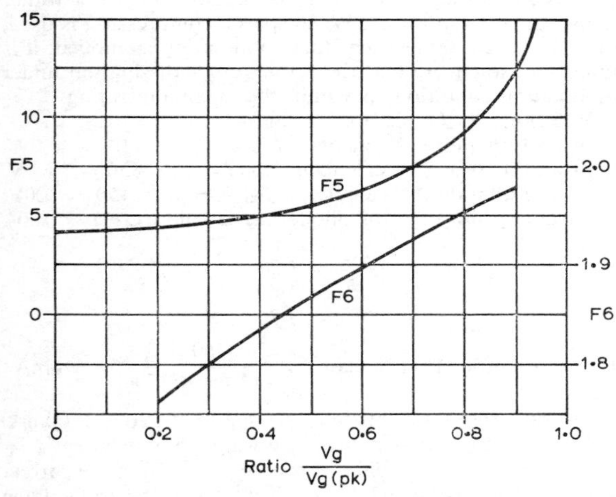

Fig 2.20. Designf actorsf or Class C telegraphy (see text).

Grid Current $I_g = \frac{i_{g(pk)}}{F_5}$ (6)

Grid Dissipation $p_{g1} = \frac{I_g \times F_6 \times (V_{g-k(pk)})}{2}$ (7)

Driving Power $P_{dr} = p_{g1} + (V_g \times I_g)$ (8)

Screen Current $i_{g2} = \frac{i_{g2\,(pk)}}{F_4}$ (9)

Screen Dissipation $p_{g2} = V_{g2} \times I_{g2}$ (10)

Output Impedance $Z_a = \frac{v_{a(pk)}}{F_2 I_a}$ (11)

In order to choose a value for anode input which will exploit the ratings of a chosen valve, an estimated efficiency may be assumed. Alternatively, the input may be fixed by other considerations, such as available power supplies or licence regulations.

A reasonable efficiency for a class C amplifier, at frequencies up to 30MHz, is 75 per cent. Hence, for the valve chosen, which has an anode dissipation rating of 37·5W:

$$\text{Anode input} = \frac{37{\cdot}5}{1 - 0{\cdot}75} = 150\text{W}$$

At an anode voltage of 1000 this corresponds to a dc anode current of 150mA.

From Equation (1) calculate $I_{a(pk)} = 4{\cdot}6 \times 150 = 690$mA. Next locate the current on the valve's anode current (I_a), anode voltage (V_a) characteristic (Fig 2.15) at a low value of anode voltage, just inside the knee of the curve; this corresponds to an anode voltage of 150V and a grid voltage of +12V.

From Equation (2), calculate $v_{a(pk)} = 1000 - 150 = 850$V.

From Equation (3), calculate

$$P_{out} = \frac{1{\cdot}8}{2} \times 0{\cdot}15 \times 850 = 115\text{W}$$

The anode dissipation is the difference between anode input and power output.

$$p_a \text{ (dissipation)} = 150 - 115 = 35\text{W}$$

This dissipation is sufficiently close to the maximum rating and can be accepted for the rest of the calculation. If the figure had been greater or considerably lower than the rated maximum, a new design should be made using a different power input, angle of flow or minimum anode voltage $V_{a(min)}$.

The chosen valve is a tetrode and from Equation 4(b) calculate grid voltage:

$$-V_g = \frac{300 \times 2}{8} + 12 \times 1 = -87\text{V}$$

From Equation (5) calculate $v_{g(pk)} = 87 + 12 = 99$V

Calculate: $\frac{V_g}{v_{g(pk)}} = \frac{87}{99} = 0{\cdot}88$ and from Fig 2.20 read values of F_5 and F_6.

These are 11·7 and 1·975, respectively.

From the grid current (I_g), anode voltage (V_a) curves of the TT21 (7623) a peak grid current of 32mA occurs at V_a = 150V and V_{g1} = +12V

From Equation (6) calculate $I_g = \frac{32}{11{\cdot}7} = 2{\cdot}75$mA

From Equation (7) calculate

$$p_{g1} = \frac{2{\cdot}75 \times 1{\cdot}975 \times 12}{2} = 32{\cdot}5\text{mW}$$

From Equation (8) calculate

$$P_{dr} = 32{\cdot}5 + (2{\cdot}75 \times 87) = 273\text{mW}$$

The driver stage should produce considerably more than this minimum power in order to allow for losses in the coupling system.

From the screen grid current (I_{g2}), anode voltage (V_a) curves of the TT21 (7623), a peak screen current of 80mA occurs at V_a = 150V and V_{g1} = +12V

From Equation (9) calculate $I_{g2} = \frac{80}{5{\cdot}8} = 13{\cdot}8$mA

From Equation (10) calculate $p_{g2} = 300 \times 13{\cdot}8 = 4{\cdot}15$W. This dissipation is within the maximum rating of 6W and is acceptable.

From Equation (11) calculate $Z_a = \dfrac{850}{150 \times 1{\cdot}8} = 3{\cdot}16\text{k}\Omega$

It is now possible to design a pi-coupler to match 3·16kΩ to the impedance of the load.

Anode-modulated Amplifiers

Anode-modulated amplifiers are designed in a similar manner to that given for class C telegraphy, but checks must be made to ensure that the required conditions at the modulation crest are met.

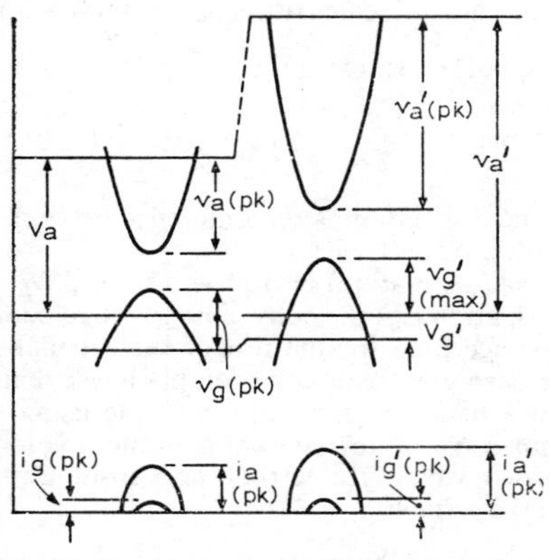

Fig 2.21. Phase relationship at the carrier and modulation crest for an anode-modulated Class C amplifier.

At the modulation crest, the anode and screen voltages will be increased but the bias will be unchanged; hence the angle of anode current flow will increase. Typical values are between 150° and 180°. In making a design, it is necessary to assume an angle and later check the accuracy of the assumption.

In the following equations, values at the crest of modulation are indicated by ('), thus θ' may be between 75° and 90°.

Since the amplifier is assumed to be linear, then:

$$P'_{out} = 4\,P_{out} \quad (12)$$

$$v'_{a(pk)} = 2\,v_{a(pk)} \quad (13)$$

Hence $$v'_{a(min)} = 2v_{a(min)} \quad (14)$$

By using Equation (3) rearranged, the anode current at modulation crest can be calculated from—

$$I'_a = \frac{P'_{out} \times 2}{F'_2 \times v_{a(pk)}} \quad (15)$$

and from Equation (1)

$$i'_{a(pk)} = F'_1 \times I'_a$$

Normally, the positive grid voltage may be assumed to have the same value as calculated at the carrier.

The peak working point corresponding to $i'_{a(pk)}$, $v'_{a(min)}$ and $v_{g1-k(pk)}$ must be located on the anode current (I_a), anode voltage (V_a) curves.

In the case of a tetrode, a value of the screen voltage must be found which satisfies these conditions. In triodes, it may be found that a different (usually greater) value of $v_{g-k(pk)}$ is required to satisfy $i'_{a(pk)}$ and $v'_{a(min)}$.

The grid current at the modulation crest is usually significantly less than at the carrier. By using some grid leak bias, the angle of flow can be increased to 180°, requiring less bias, and hence making available an increased positive grid excursion. An alternative is to supply sufficient modulation to the driver stage to provide the required positive excursion.

For convenience of illustration, it will be assumed that the foregoing class C telegraphy design is now to be modulated, but it should be noted that this will not necessarily give a practical result, since the anode dissipation rating may be exceeded during modulation.

It is usual practice to quote anode dissipation ratings at carrier (unmodulated) conditions of two-thirds of the maximum valve rating. This is based on the assumption that the average power dissipation will be increased by 1·5 times when modulation is applied. In the valve used for the example, the anode dissipation under modulation must be reduced to $\dfrac{37{\cdot}5}{1{\cdot}5} = 25$W.

In practice, however, with speech waveforms of relatively high peak to mean ratio, it is satisfactory to use a rather higher dissipation rating. When speech compression is used, or continuous 100 per cent tone modulation is applied, it is important to ensure that the actual anode dissipation under modulation conditions is within the maximum rating.

Returning to the previous design—

From Equation (12) calculate $P'_{out} = 4 \times 115 = 460$W.

From Equation (13) calculate $v'_{a(pk)} = 2 \times 850 = 1700$V.

From Equation (14) calculate $v'_{a(min)} = 2 \times 150 = 300$V.

Assuming an angle of anode current flow $(2\theta') = 150°$, then:

$$F'_1 = 3{\cdot}75$$
$$F'_2 = 1{\cdot}69$$
$$F'_3 = 1{\cdot}35$$

From Equation (15) calculate $I'_a = \dfrac{460 \times 2}{1{\cdot}69 \times 1700} = 320$mA

From Equation (1) calculate $i'_{a(pk)} = 3{\cdot}75 \times 320 = 1200$mA.

In order to obtain a peak working point where $i'_a = 1200$mA at $v'_{a(min)} = 300$V, it is necessary to find the correct value of screen voltage, it being assumed that the grid voltage for the carrier conditions is still available (+12V).

From the I_a/V_a curves for the valve at various screen voltages when $V_{g1} = 0$, it is now necessary to predict the screen voltage required to produce $i_{a'(pk)} = 1200$mA at $v'_{a(min)} = 300$V and $v'_{g1} = +12$V.

From the TT21 data, the mutual conductance (g_m) at $I_a = 140$mA is 11mA/V, therefore, at $i'_{a(pk)} = 1200$mA, the mutual conductance will increase by:

$$\left(\frac{1200}{140}\right)^{\frac{1}{3}} = 2{\cdot}046 \text{ which gives 22mA/V approx.}$$

From this it follows that the anode current at $V_{g1} = +12$V will be $12 \times 22 = 264$mA greater than the value at $V_{g1} = 0$V.

The point on the curve that now has to be found is for $V'_a = 300$V, I_a $1200 - 264 = 936$mA. This corresponds to a screen voltage of 465V.

The screen voltage should therefore be increased by slightly more than 1·5 times when the anode voltage is doubled by modulation. The modulation transformer should be designed to provide this screen modulation point either by a tap on the main winding or by additional winding.

The assumed angle of flow can be checked to see if it is realistic, by calculation of the bias from Equation (4b).

$$-V'_{g1} = \frac{465 \times 1{\cdot}35}{8} + 12 \times 0{\cdot}35 = -82{\cdot}5\text{V}$$

This is close enough to the original value of −87V for a practical design.

In practice, the regulation of the driver source, the change of grid current when the screen voltage is raised, and the method of obtaining the bias, will modify the available positive grid voltage at the crest, but the calculation gives sufficient guide as a practical starting point.

Class AB and Class B Linear Amplifiers

In class AB and class B linear service, the amplifier is required to handle modulated waveforms without distortion. The amplification of single sideband suppressed carrier signals is the most usual example.

In a class B amplifier, the angle of flow of anode current is close to 180°. An acceptable design can be made using the procedure given for class C telegraphy but with $\theta = 90°$.

In practice, however, such amplifiers are operated with some standing anode current ($I_{a(o)}$) in the absence of a signal, as a means of improving the linearity.

Class AB amplifiers invariably operate at significant standing anode current. Design curves based on angle of flow are therefore inconvenient; curves based on the ratio of mean anode current under driven conditions to standing anode current are more useful.

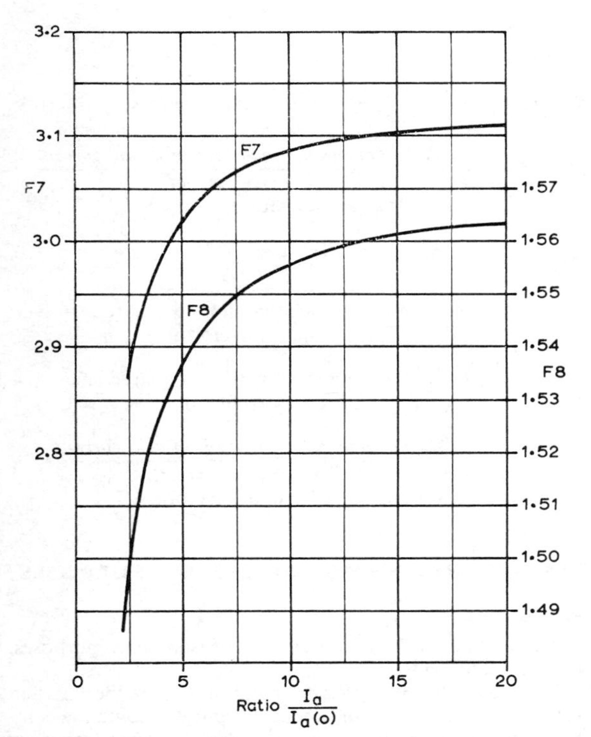

Fig 2.22. Design factors for Class AB and Class B linear amplifiers (see text).

The curves given in **Fig 2.22** are suitable. In these, F_7 corresponds to F_1 and F_8 to F_2; from which, under these new conditions:

Peak Anode Current $i_{a(pk)} = F_7 \times I_a$ (16)

Power Output, $P_{out} = \frac{F_8}{2} \times I_a \times v_{a(pk)}$ (17)

In a typical class AB amplifier driven to maximum peak envelope power the valve will have an anode efficiency of about 70 per cent. The anode dissipation is a maximum at some value of drive less than the maximum. The anode dissipation at maximum drive must therefore be less than the maximum rating, say, 80 per cent.

Taking the same example as used for the class C calculations, the TT21 (7623), an anode dissipation of 30W is a suitable starting point. In a final design, the values must be chosen so that, taking into account the peak to mean ratio of the modulation waveform, excessive anode dissipation does not occur.

Taking anode dissipation as 30W and anode efficiency of 70 per cent, then:

$$\text{Anode Input } P_{in} = \frac{30}{1 - 0{\cdot}7} = 100\text{W}.$$

Decide on the anode voltage; in this case, take $V_a = 1000$V; then the anode current $I_a = 100$mA.

Next, it is necessary to decide the zero signal (standing) anode current $I_{a(o)}$; this depends on a compromise between efficiency and intermodulation distortion. Generally a current corresponding to about 66 per cent of the rated anode dissipation is typical, from which

$$I_{a(o)} = \frac{2}{3} \times 37{\cdot}5 = 25\text{mA}.$$

$$\text{Then } \frac{I_a}{I_{a(o)}} = 4$$

from Fig 2.20 $F_5 = 2{\cdot}99$ and $F_6 = 1{\cdot}53$

and from Equation (16) $i_{a(pk)} = 2{\cdot}99 \times 100 = 299$mA.

Locate this current on the I_a/V_a characteristic curve to find the value of $v_{a(min)}$. To preserve linearity it is important that this point shall not be in the curved part of the knee characteristic.

From the curve a value of 100V is suitable.

Hence:

$$v_{a(pk)} = 1000 - 100 = 900\text{V}$$

and from Equation (17) $P_{out} = \frac{1{\cdot}53}{2} \times 0{\cdot}10 \times 900 = 69$W.

Anode dissipation $p_a = 100 - 69 = 31$W.

The calculation of driving power (if any) and anode load impedance follow the same procedure as for class C telegraphy. The bias will, however, be decided by the chosen value of $I_{a(o)}$. The approximate value can be taken from the characteristic curve, but in practice should be set to give the required value of $I_{a(o)}$.

The intermodulation of linear amplifiers is frequently assessed by using a test signal consisting of two or more signals (tones) of equal amplitude. The average power output will decrease as the number of tones is increased in the test signal as shown in **Fig 2.23**.

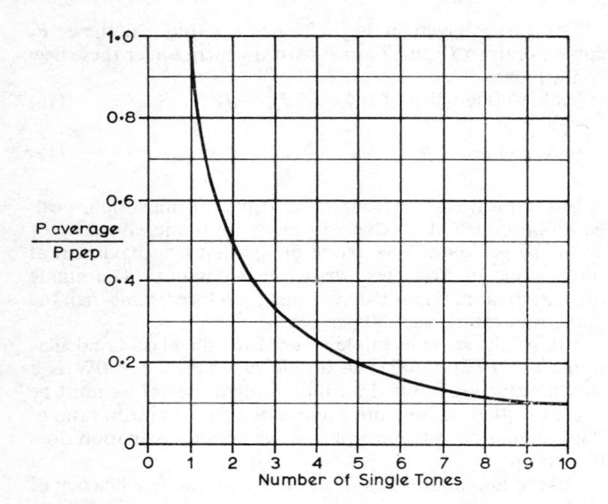

Fig 2.23. The assessment of intermodulation in linear amplifiers.

In the usual case of a two-tone test signal, and assuming ideal linear characteristics, the relation between single and two-tone conditions is:

$$I_{a\,(two\text{-}tone)} = \frac{2}{\pi} I_{a\,(single\ tone)}$$

Average input power:

$$P_{in\,(two\text{-}tone)} = V_a \times I_{a\,(two\text{-}tone)}$$

Average output power:

$$P_{out\,(two\text{-}tone)} = \tfrac{1}{2} P_{out\,(single\ tone)}$$

Grounded Grid Operation

All the preceding designs are based on the assumption that the signal is applied to the grid and the cathode earthed (grid drive or common cathode connection). Sometimes the signal is applied to the cathode and the grid earthed (grounded grid or cathode drive connection).

This arrangement has the advantage of improved stability usually without neutralizing. It has the disadvantage that much greater driving power is required than that needed for grid drive connection, but some of the driving power is recovered in the output circuit.

The driving power

$$P_{dr} = (V_g \times I_g) + p_{g1} + \left(\frac{v_{g1(pk)} \times F_2 \times I_a}{2}\right)$$

The drive power which appears in the output is

$$\left(\frac{v_{g1(pk)} \times F_2 \times I_a}{2}\right)$$

In the case of a tetrode, there is a small additional driving power which is not recovered in the output; this occurs due to the product of peak drive voltage and the fundamental component of the screen current. It is usually sufficiently small to be ignored.

Frequency Multipliers

Frequency multipliers are class C amplifiers in which the anode circuit is tuned to a harmonic of the drive frequency, and may be designed in the same way as a class C amplifier. In general, smaller angles of flow are used, as this tends to increase the harmonic output.

The factor F_2, which in the amplifier design gives the ratio of peak fundamental to dc anode current, is replaced by a factor giving the ratio for peak harmonic to dc anode current. These factors for harmonics up to the fifth are shown in Fig 2.24.

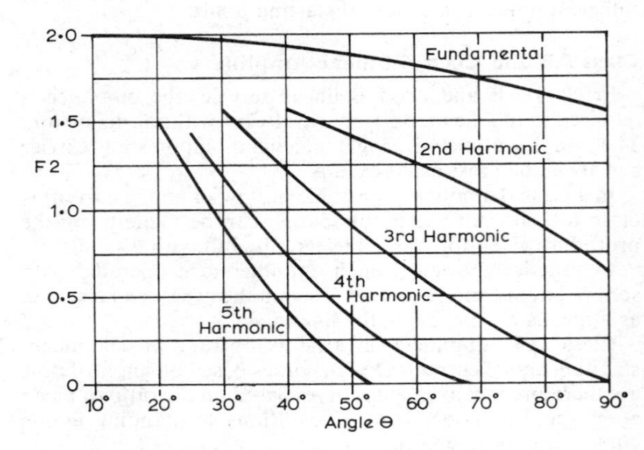

Fig 2.24. Design factors for frequency multipliers (see text).

Factors

F_1 and F_7 — $\frac{\textit{Peak anode current}}{\textit{DC anode current}}$ (assuming sine waveform)

F_2 and F_8 — $\frac{\textit{Peak fundamental component of anode current}}{\textit{DC anode current}}$ (assuming sine waveform)

F_3 — $\frac{1}{1 - \cos\theta}$

F_4 — $\frac{\textit{Peak screen current}}{\textit{DC screen current}}$ (assuming squared sine waveform)

F_5 — $\frac{\textit{Peak grid current}}{\textit{DC grid current}}$ (assuming squared sine waveform)

F_6 — $\frac{\textit{Peak fundamental component of grid current}}{\textit{DC grid current}}$ (assuming squared sine waveform)

VALVES AND TUBES FOR SPECIAL PURPOSES

Noise Diodes

Noise diode valves, useful for measurement purposes, operate as follows.

When a diode is operated in a condition where the emission is temperature-limited, the anode current will contain a component of noise which is readily calculable. The frequency spectrum of the noise would be infinite unless it was limited at the higher frequencies by the shunt capacitance of the valve and its holder and by the electron transit-time between the cathode and the anode.

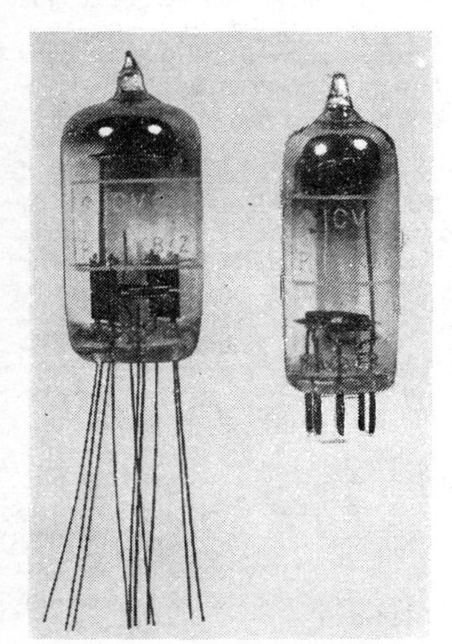

Typical noise diodes: (left) wire-in type and (right) plug-in type. *(Photo by courtesy of the M-O Valve Co. Ltd.)*

The noise output is controlled by adjustment of the anode current. This is achieved, while maintaining the temperature limited condition, by variation of filament temperature. A directly heated pure tungsten bright emitter filament is used and the filament current adjusted to give the required anode current. An indirectly heated cathode or a coated or thoriated filament cannot be used since the emissive properties would be destroyed by operation in a temperature-limited condition.

Coaxial noise diode (for insertion in the line). *(Photo by courtesy of the M-O Valve Co. Ltd.)*

Noise diodes are specially designed for the purpose of noise generation and if required for uhf circuits they are frequently of coaxial construction.

At frequencies above 1000MHz gas discharge noise diode tubes may be inserted across a waveguide feeder system to inject noise. These tubes are similar to fluorescent lighting tubes but without the surface phosphor because no light output is needed. Those in the 4-10W range are used for test purposes.

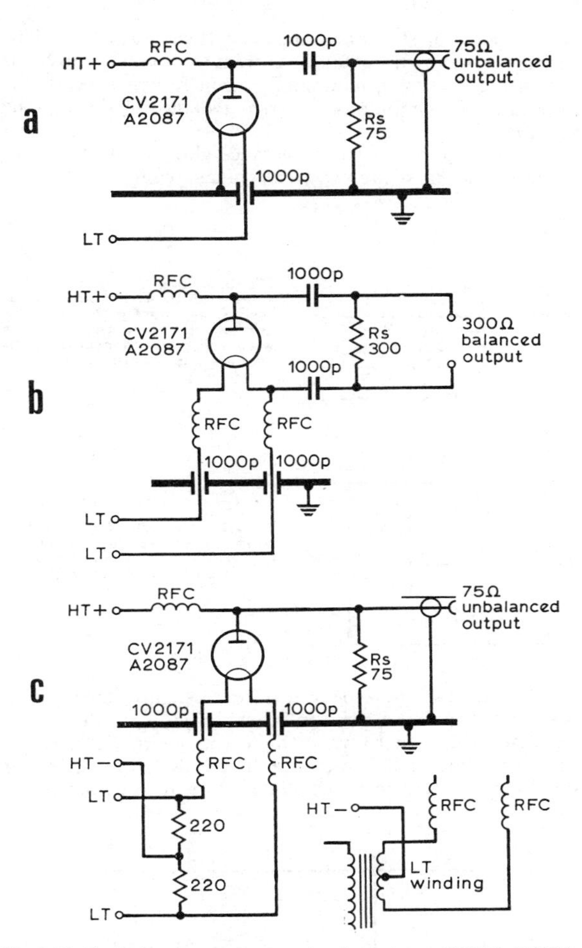

Fig 2.25. Typical noise generator circuits using a CV2171 (A2087): (a) for coaxial output, (b) for balanced output and (c) with earthed anode.

Voltage Stabilizers

Shunt type stabilizers: These may be of the cold glow discharge gasfilled type for use at up to 150V, or the corona discharge gasfilled tube for 20kV or higher, each having a cylindrical electrode system. Provided that the current passing through the device is kept within the published rating any voltage variation across the tube will be small. Particularly with corona discharge stabilizers the current must not be exceeded, otherwise the discharge may change from the "corona mode" to the "glow mode" and the voltage will fall considerably.

Corona discharge stabilizers are used for stabilizing the final anode voltage of a cathode ray tube where the currents involved are at most a few hundred microamperes.

With glow discharge stabilizers the current passing through the valve may approach 50-60mA, depending on type. The striking voltage is usually significantly higher than the operating voltage, and an additional trigger electrode is often provided. Glow discharge stabilizers may be used singly or in series where higher voltages are to be controlled, so long as some means of striking is included (**Figs 2.26, 2.27**). The resistor R is chosen so that the current through the valve(s) is within the maximum-minimum current rating, making due allowance for the total load current also passing through the resistor.

For some purposes close tolerance glow discharge valves are available; these are known as voltage reference tubes.

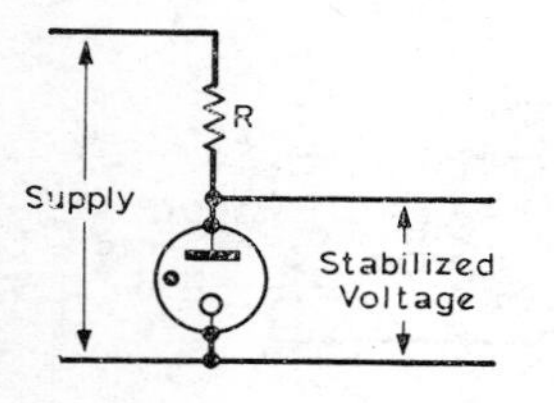

Fig 2.26. The connection of a shunt stabilizer. The series resistor R is chosen so that the current through the tube is within the limits necessary to provide the stabilized voltage.

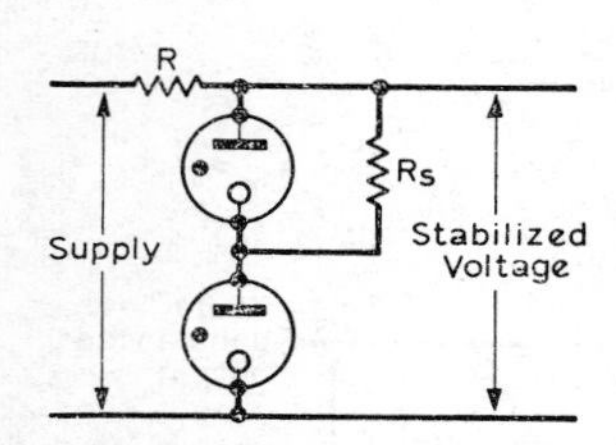

Fig 2.27. The use of two stabilizers in series to provide a higher voltage. Rs is the common striking resistor.

Where the current to be drawn from a high voltage supply is greater than that which can be stabilized by the devices so far described, a hard valve high voltage triode may be employed (**Fig 2.28**), where a glow discharge reference stabilizer is placed in the cathode return to provide steady bias.

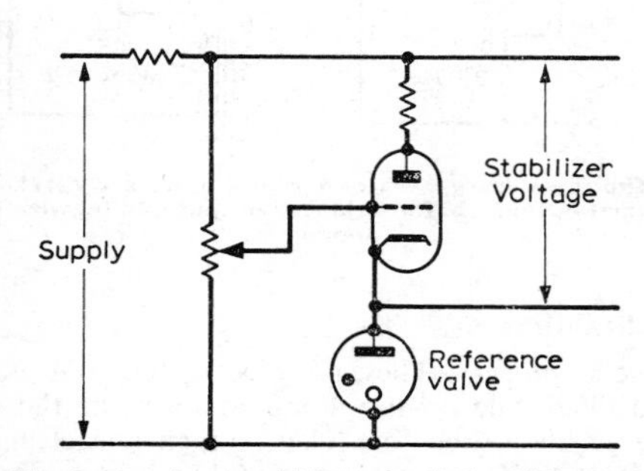

Fig 2.28. Hard valve shunt stabilizer circuit for higher voltages.

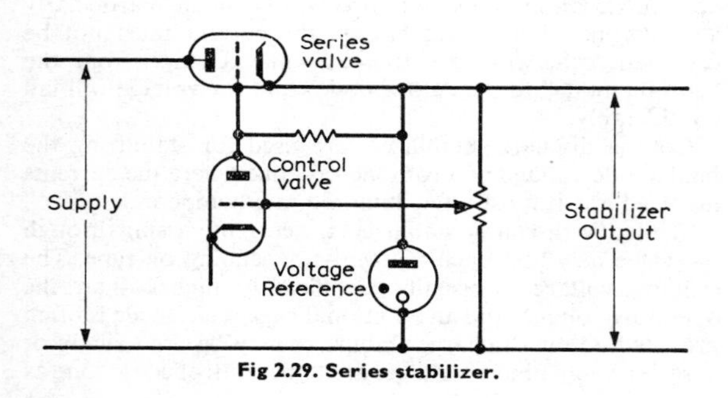

Fig 2.29. Series stabilizer.

Series stabilizers: For high current supplies where variations of voltage are to be absorbed a low impedance valve is connected in series with a control triode and a voltage reference tube (**Fig 2.29**). As voltage changes occur the control valve applies a varying bias to the series valve.

Special low impedance valves are available for this application, but many standard power output tetrodes triode-connected operate satisfactorily.

Deflection Beam Tubes

In certain types of valve the electrode construction is so arranged that the electron flow may be deflected from one anode to another by deflecting electrodes (**Fig 2.30**). The

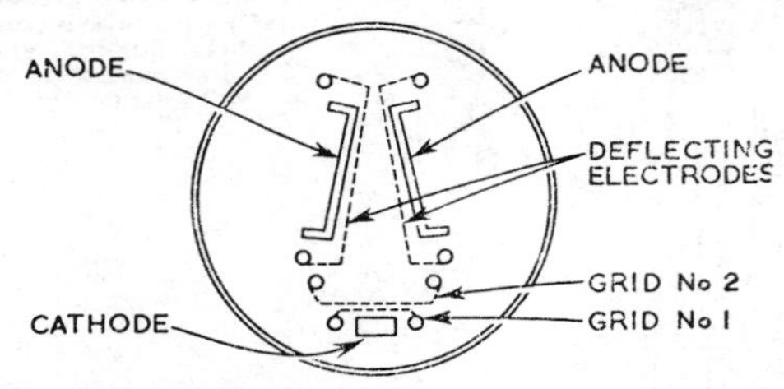

Fig 2.30. Arrangement of electrodes in the 7360 beam deflection valve.

type 7360 is shown; another type, the 6AR8, has an additional suppressor grid interposed between the accelerator (G2) and the deflecting electrodes. Each of these valves is especially suitable for use as a balanced modulator for generating single or double sideband signals, and achieves significantly better carrier suppression than that given by conventional double triodes.

Nuvistors

The Nuvistor is a receiving valve of sub-miniaturized construction in a metal-ceramic envelope, and is especially applicable at vhf by reason of its low internal capacitance and high mutual conductance. It is marketed in simple triode form (6CW4, 6DS4), as a grounded grid triode (**Fig 2.31**) and as a tetrode (7587) which is similar in appearance to the grounded grid version but is provided with a screen grid (G2).

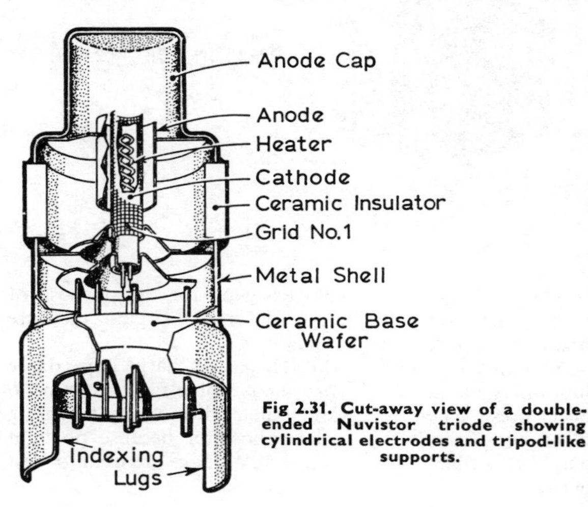

Fig 2.31. Cut-away view of a double-ended Nuvistor triode showing cylindrical electrodes and tripod-like supports.

Disc Seal Valves

In the disc seal triode (**Fig 2.32**), characteristically of high mutual conductance, the electrode spacing is minimal. The "top hat" cathode contains the insulated heater, one side of which is connected to the cathode and the other brought out coaxially through the cathode sleeve connection. The fine-wire grid stretched across a frame emerges through the envelope by an annular connection. Because the clearance between grid and cathode is very small, the cathode surface is shaved during construction to provide a plane surface.

The anode also emerges via an external disc for coaxial connection. On larger disc seal valves the anode may form part of the valve envelope.

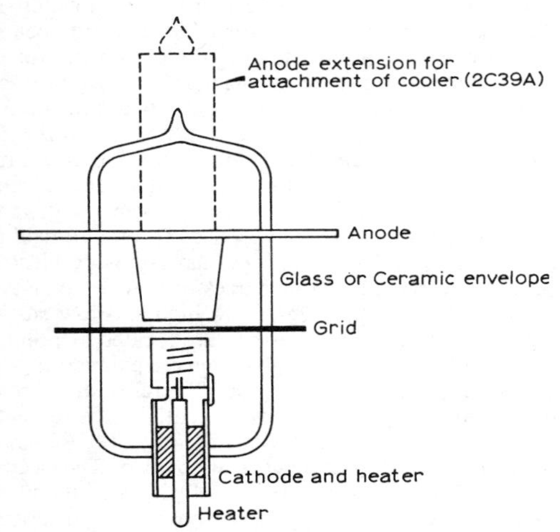

Fig 2.32. General form of a disc seal valve.

Disc seal valves are available for power dissipations of a few watts to 100W with forced air cooling and outputs in the frequency range 500-6000MHz. It should be noted that maximum power and frequency are not available simultaneously.

A 2C39A disc seal transmitting valve.

(Photo by courtesy of S.T. & C. Ltd).

Disc seal valves although intended for coaxial circuits may be effectively employed with slab type circuits. Important points to be observed are (*a*) only one electrode may be rigidly fixed, in order to obviate fracture of the seals (which is more likely to occur in the glass envelope types); and (*b*) except in the case of forced air anode cooling the anode is cooled by conduction into its associated circuit. With a shunt-fed circuit thin mica insulation will function both as a capacitance and a good transmitter of heat.

Certain sub-miniature metal-ceramic envelope types such as the 7077, although of the generic disc seal form, require special sockets if they are used in conventional circuits. Many of them give significant output at the lower shf bands.

Cathode Ray Tubes

A cathode ray tube contains an electron gun, a deflection system and a phosphor coated screen for the display.

The electron gun, which is a heated cathode, is followed by a grid consisting of a hole in a plate exerting control on the electron flow according to the potential applied to it, followed in turn by an accelerating anode or anodes.

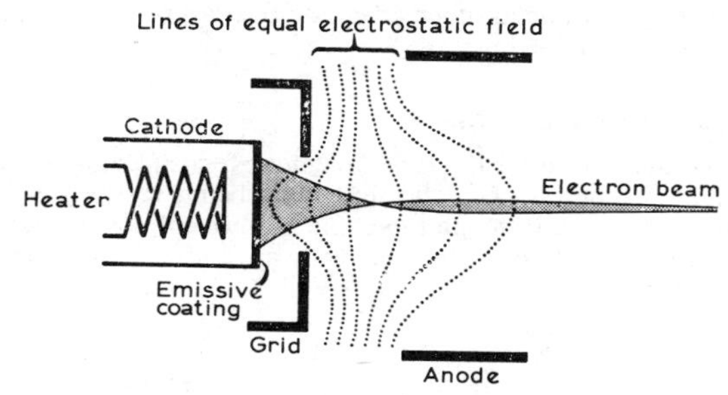

Fig 2.33. Diagram of the electron gun used in cathode ray tubes, travelling-wave tubes and klystrons.

The simplest form of triode gun is shown at **Fig 2.33**. The beam is focused by the field between the grid and the first anode. In tubes where fine line spots and good linearity are essential (eg, in measurement oscilloscopes) the gun is often extended by the addition of a number of anodes to form a lens system.

Beam focusing may be by either electrostatic or electromagnetic means.

Oscilloscope Tubes

Electrostatic focusing is used in oscilloscope tubes, the deflection system consisting of pairs of plates to deflect the beam from its natural centre position, depending on the relative potentials applied. Interaction between the two pairs of deflection plates is prevented by placing an isolation plate between them (**Fig 2.34**).

After deflection the beam is influenced by a further accelerating electrode known as the *post deflection accelerator* (PDA) which may take the form of a wide band of conducting material on the inside of the cone shaped part of the bulb, or

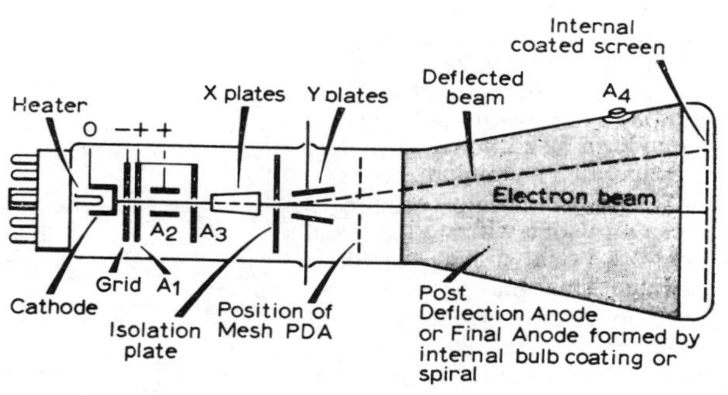

Fig 2.34. Diagrammatic arrangement of a cathode ray tube with electrostatic focusing and deflection.

of a close-pitch spiral of conducting material connected to the final anode. For some purposes when it is important to maintain display size constant irrespective of the final anode voltage, a mesh post deflection accelerator is fitted close to the deflecting plates.

If a double beam is required the electron flow is split into two and there are two sets of deflection plates. Alternatively, two complete systems are enclosed in one tube. Although a common X-deflection plate may be fitted the advantage of two complete systems lies in providing complete alignment of the timing (horizontal) deflection. By setting two systems at an angle to one another adequate overlap of each of the displays is provided.

Radar and Picture Tubes

In radar and television tubes magnetic focusing and deflection are common, the deflection angle being vastly greater than with an oscilloscope tube. Such tubes employ a simple triode or tetrode electron gun with an anode potential of 20kV or more.

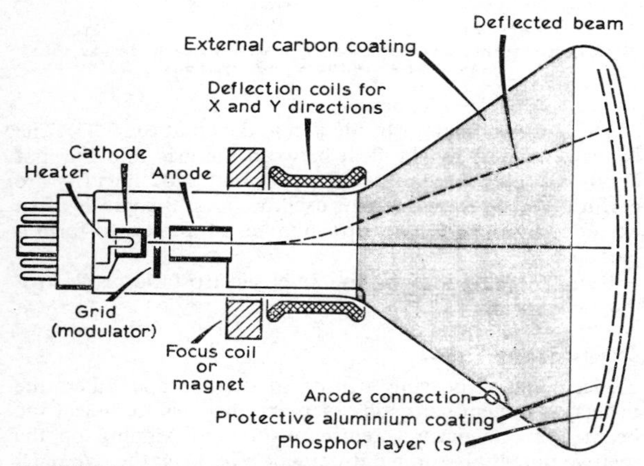

Fig 2.35. Diagrammatic arrangement of a cathode ray tube having magnetic focusing and deflection.

Screen Phosphors

Many types of phosphor are used for coating the screens of cathode ray tubes, their characteristics varying according to the application. In all of them the light output is determined by the final anode voltage used, but where this exceeds about 4kV the phosphor is protected against screen burn by a thin backing layer of evaporated aluminium.

Oscilloscope tubes require a phosphor with a wide optical band to give a bright display for direct viewing. This phosphor, yellow-green in colour, extends into the blue region to enable direct photographs to be taken from the display.

CR Tube Power Supplies

Unlike the valve, the cathode ray tube's current requirements are low, the beam current being only a few tens of microamperes.

For oscilloscope work the deflector plates need to be at earth potential, cathode and other electrodes consequently being at a high negative potential to earth. In magnetically focused tubes the cathode may be at earth potential and the final anode many kilovolts above.

Power supplies for either type of tube need to be of very high impedance for safety reasons, and the short circuit current should not exceed about 0·5mA.

Klystrons

A *klystron* is a valve containing an electron gun similar to that used in a cathode ray tube, from which a narrow beam of electrons is projected along the axis of the tube and focused through small apertures across which one or more oscillatory circuits are connected; these circuits are in the form of hollow toroidal chambers known as *rhumbatrons*. The beam of electrons is velocity-modulated by the application of an rf field, for example by passing the beam through small apertures having an rf voltage between them. If this velocity-modulated beam is now passed through a field-free space, known as a *drift space*, the faster-moving electrons will begin to overtake the slower ones so that, at some point along the beam, alternate regions of high and low electron density will exist. The rhumbatrons are located at points of maximum density. If two rhumbatrons are coupled together by means of a feedback loop, or if a reflector electrode maintained at a slightly negative potential with respect to the cathode is mounted at the end of the tube, the energy will be reflected back into a single rhumbatron. The electrons become bunched due to their being velocity-modulated, and since the energy is in phase by reason of the distance of travel to the reflector and back again, sustained oscillation will take place. Klystrons are employed as oscillators, the reflector type being preferred for low-power work, eg as a local oscillator in a superheterodyne receiver.

Travelling-wave Tubes

A *travelling-wave tube* consists of a wire helix supported in a long glass envelope which fits through two waveguide stubs or coaxial couplers used as input and output elements. At one end of the tube is a conventional electron-gun assembly which directs an electron beam to the collector at the other end (**Fig 2.36**). The rf signal input is coupled into the helix

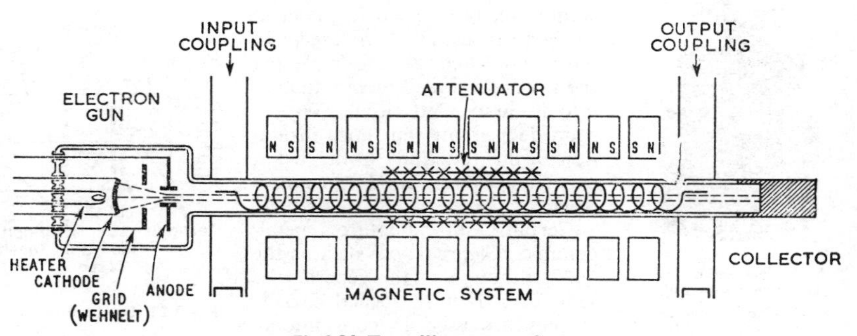

Fig 2.36. Travelling-wave tube.

and will run round the turns of the helix at roughly its normal velocity, but its axial velocity may be only about a tenth of that value, depending on the pitch of the helix. If the electron beam is sent along the centre of the helix and focused by a magnet it will be charge-density modulated by the voltage pulse (ie amplitude-modulated) owing to the interaction between the magnetic field of the current in the helix and the electrons. This modulation grows in amplitude along the length of the helix roughly according to the square of the axial distance from the beginning of the helix. By suitable design the output waveguide coupling will extract more energy than was put in by the input coupling and a considerable power gain can be achieved. For example, travelling-wave tube amplifiers with a bandwidth of 500MHz centred on frequencies in the range 1700–8500MHz are available. Outputs of 5–10W and power gains of 30–40dB are possible.

To prevent self-oscillation, it is necessary to introduce some attenuation between the input and output, and to match the input and output impedances to the tube. Serious mismatch at the output is likely to damage the tube.

The Magnetron

A *magnetron* is a diode with a cathode and anode in a cylindrical assembly. In a simple magnetron the anode is split into two parts; more commonly it comprises multiple cavities which resonate at the operating frequency. The whole assembly is mounted in a magnetic field parallel to the axis of the electrodes. The magnet may be an integral part of the valve (the *package magnetron*) or a close fitting separate assembly.

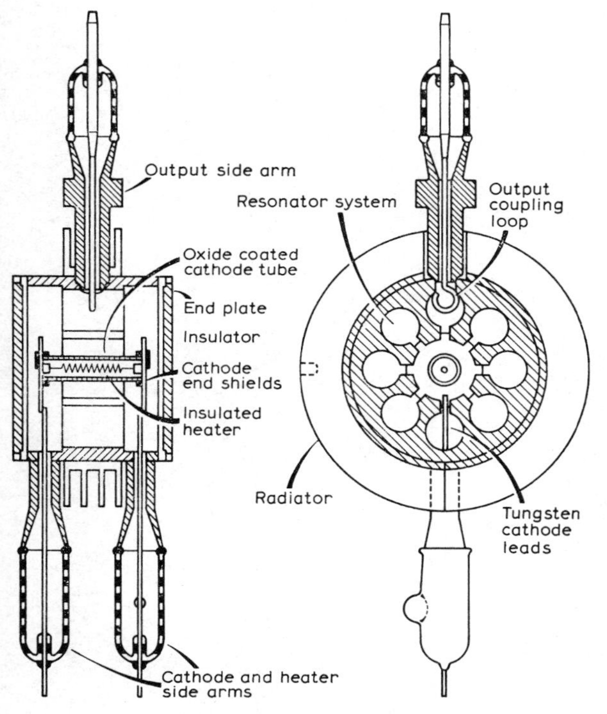

Fig 2.37. General arrangement of a magnetron.

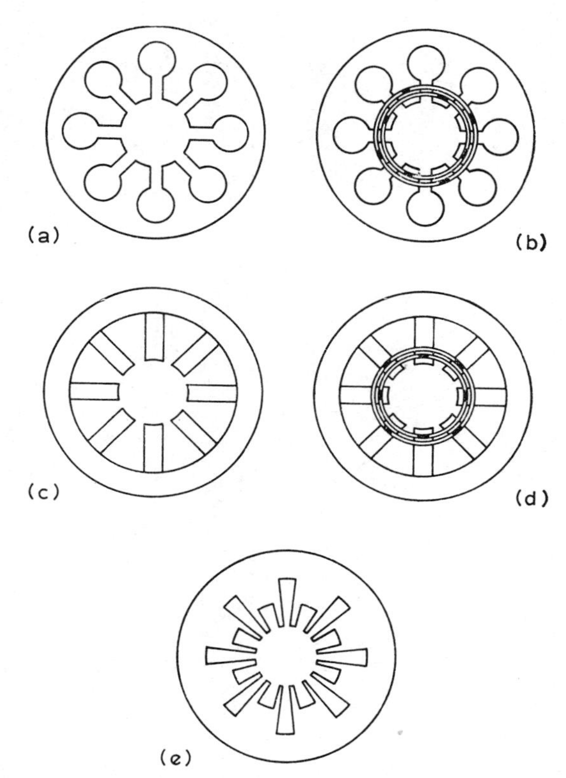

Fig 2.38. Various forms of anode cavity in the magnetron: (a) unstrapped cavity, (b) strapped cavity, (c) unstrapped vane-block, (d) strapped vane-block, and (e) rising-sun block.

The axial magnetic field causes the path of the electrons leaving the cathode to be curved. At a certain field intensity the electrons fail to reach the anode and return to the cathode in circular orbits. The time taken for an electron to complete its orbital journey decides the frequency of oscillation. The energy associated with the moving electrons is given up in the space around the cathode, transferred to the resonant cavities and coupled out into an associated waveguide by loop and probe or by a windowed slot cut in the back of one of the cavities.

The anode cavity can take a number of different forms and these are illustrated in Fig 2.38.

CHAPTER 3

SEMICONDUCTORS

SEMICONDUCTOR is the generic name given to all devices which are manufactured from materials having a resistivity some way between that of an insulator and a conductor. They perform functions in electronic circuits similar to those of thermionic valves, and are most commonly made from silicon or germanium. Although it was not until 1948 that the first transistor was made, such intensive development has taken place since that there are now very few electronic applications in which the solid-state device has not entirely superseded the ordinary valve.

The semiconductor combines the practical advantages of small physical size with ruggedness and reliability; and unlike the thermionic valve, it requires no heater supply. Its electrical characteristics are virtually independent of age; it will operate from a supply rail which is measured in volts rather than in tens or hundreds of volts, and as conduction occurs entirely within the semiconductor materials from which it is manufactured, it may be readily protected by encapsulation without the need of vacuum seals.

The uses of the solid-state device are unlimited, and extend from the generation of power in the microwave region to switching applications of all kinds. As a linear amplifier, which is its most common function, the semiconductor device is able to provide low noise gain in any frequency band ranging from a few gigahertz to dc; it can exhibit an input impedance which may be as low as a few ohms or higher than 10^9 megohms, and in different circuit configurations, it may be used either as a voltage or current generator. Complex monolithic arrays comprising many semiconductor devices and associated components on a single chip of material are commonplace in digital and analogue computer applications, and as the integrated circuit, requiring only a minimum of additional components in order to function as a complete stage, these arrays are playing an ever increasing role in modern communications technology.

Almost all semiconductor devices fall broadly into one of three categories: diodes, bipolar and field effect transistors, and although the functions and characteristics of each are quite different, the principles underlying their operation are basically those of the junction diode.

This device is, in fact, already familiar to most radio amateurs both as a modern development of the old crystal detector and as a replacement for the selenium power rectifier. For these reasons, it has been readily accepted without any real understanding of the fundamental physics governing its behaviour. The bipolar and field effect transistor, however, have neither a familiar forerunner nor any simple equivalence to the thermionic valve. They demand, therefore, a basic understanding of the principles upon which they operate before they can be employed successfully as circuit elements. Unfortunately few texts on this subject meet the needs of the average radio amateur, and it is in an attempt to satisfy these requirements that this chapter introduces the reader to some of the fundamentals of semiconductor physics without resorting to advanced electron flow theories.

THEORY OF SEMICONDUCTORS

A single isolated atom consists of a positively charged inner nucleus and one or more outer electrons whose total negative charge exactly balances that of the nucleus. Each element has a different number of electrons, all of which are bound to the nucleus and possess energy levels that fall into discrete bands. For example, silicon has fourteen electrons distributed through three specific bands containing two, eight and four electrons respectively (Fig 3.1).

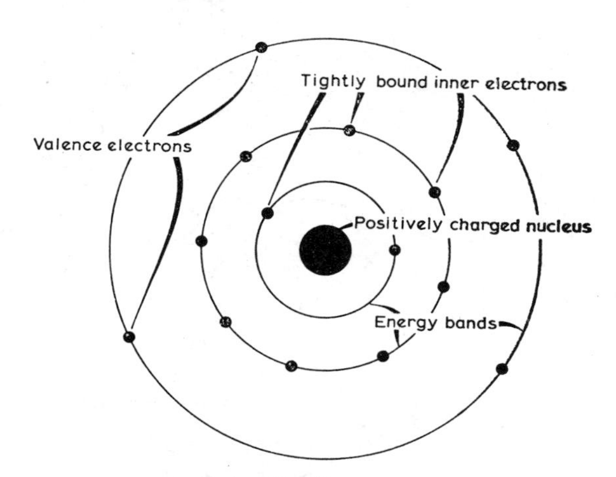

Fig 3.1. The silicon atom.

While the electrons in the lower energy bands are tightly bound to the nucleus, those occupying the higher energy levels are only loosely bonded. These electrons, which are called valence electrons, determine the characteristic properties of the element and enter into chemical reactions with other elements. By forming covalent bonds (pairs of shared valence electrons) with their neighbours, the individual atoms are able to link into regular lattice structures from which crystals of the element are formed.

The atoms of both silicon and germanium each have four valence electrons, and thus form crystals which have a tetrahedron lattice that is typical of a tetravalent substance. **Fig 3.2** is a two dimensional representation of such a structure with only the nucleus and valence electrons shown for reasons of simplicity.

The quantum theory states that the electrons in an atom can occupy only a finite number of discrete energy levels or bands, and that no intermediate levels are possible. The energy spectrum for an isolated atom can thus be represented as in Fig 3.3.

The energy band which contains the valence electrons is called the valence band. It is separated from the next highest allowable energy band (the conduction band) by a forbidden region of energies which the electron cannot occupy. This region, or energy gap, corresponds to the minimum energy that is required to raise a valence electron into the conduction band. It is also the amount of energy the electron must gain or lose in making the transition between the two bands.

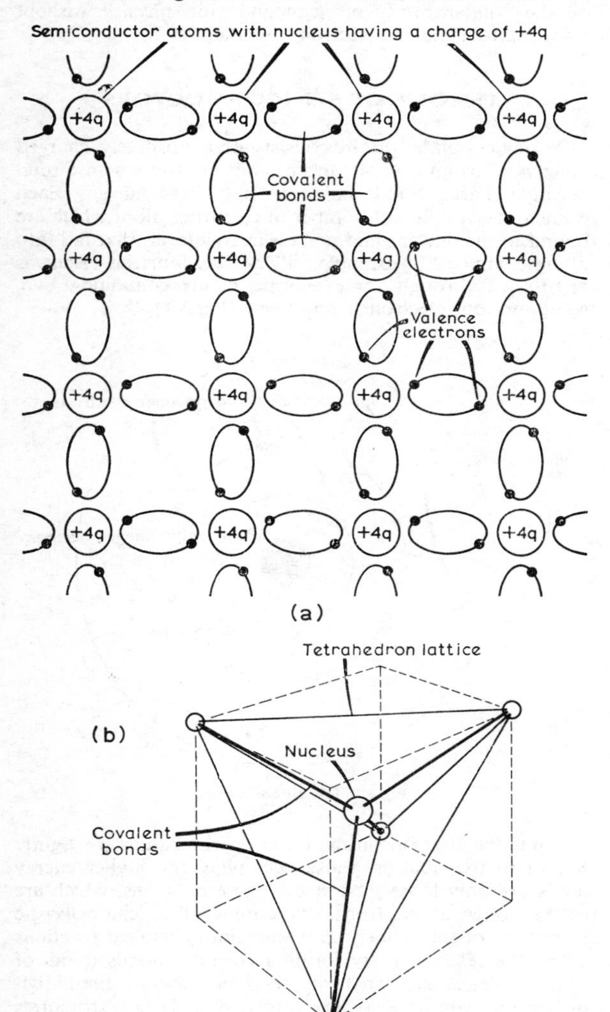

Fig 3.2. (a) Two dimensional diagram of a tetravalent semiconductor crystal. (b) Pictorial representation of unit of crystal structure in silicon.

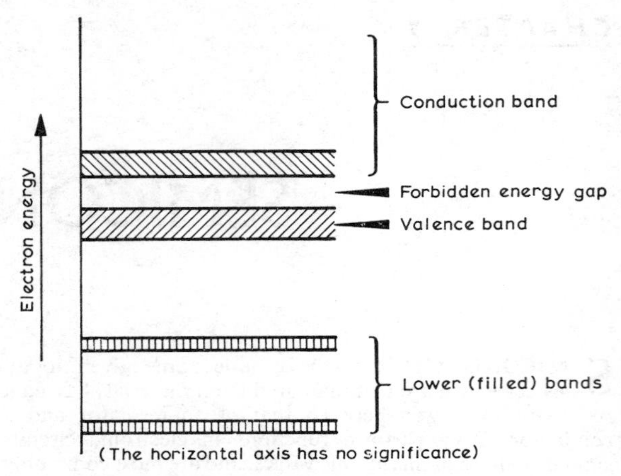

Fig 3.3. Energy levels in an isolated atom.

In a crystal, the loosely bonded valence electrons are not confined to any particular atom, but are free to wander randomly through the entire lattice keeping within the valence band of energies. Some valence electrons, however, acquire sufficient energy through thermal excitation to break their bonds, thus enabling them to surmount the forbidden energy gap and to move into the conduction band. Similarly, electrons within the conduction band can lose energy and fall back into the valence band. There exists, in fact, a dynamic equilibrium between the two states with a fixed proportion of free electrons available for conduction at any given time and for any particular temperature above absolute zero.

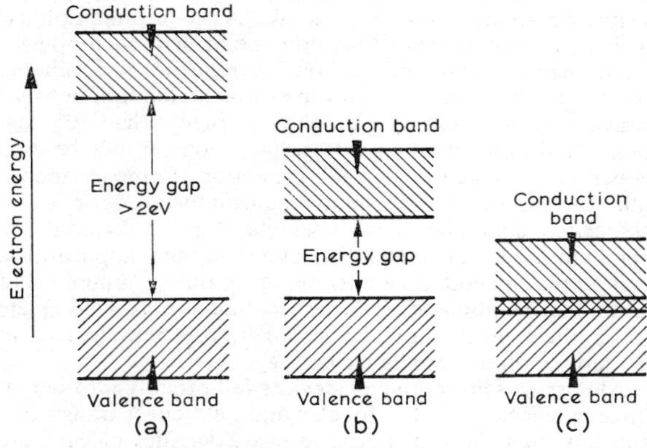

Fig 3.4. Energy band diagram for (a) an insulator, (b) a semiconductor, (c) a conductor.

The fraction of valence electrons able to make the transition is proportional to $e^{-Eg/kT}$, where k is Boltzman's constant and T the absolute temperature. The amount of energy, Eg, which these electrons must gain in order to cross the energy gap varies with different materials. It is large in a good insulator and negligible, or non-existent, in a conductor (**Fig 3.4**). Semiconductors occupy an intermediate position having a forbidden energy gap between the valence and conduction bands equal to 0·72eV for germanium and to 1·03eV for silicon.

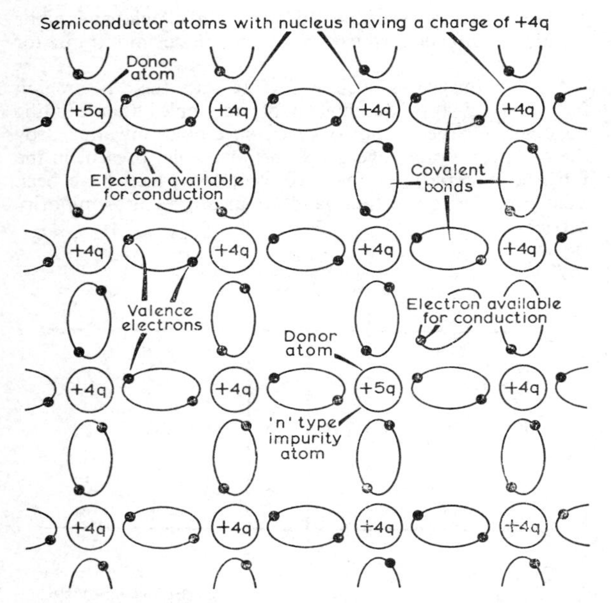

Fig 3.5. Semiconductor crystal structure with "n" type impurity atoms.

When a valence electron moves into the conduction band, it leaves a vacancy in the valence band. This situation is equivalent to regarding the space as being filled by a positively charged "hole", ie a fictitious particle of mass and shape similar to the electron, but having a single positive charge. Neighbouring electrons will, therefore, experience an attraction tending to capture an electron and fill the space. When this occurs, the hole effectively moves into the space vacated by the electron which it captured, and thus by successive motion of different electrons, it is able to move over long distances within the crystal lattice. Eventually, a drifting hole will meet a free electron and both hole and electron disappear. Normally this movement of holes and electrons is random, so that the net effect in any direction is zero and the crystal as a whole is electrically neutral.

If a voltage is applied between two points on a conductor or semiconductor, the random movement of holes and electrons becomes a general drift of electrons towards the positive pole and holes towards the negative pole. Free electrons in the conduction band are accelerated readily so that a significant flow takes place with only a very small applied field or voltage. At the same time, the removal of an electron from the valence band permits the acceleration of electrons within the band, and hence by the mechanism described above, the hole is similarly accelerated in the opposite direction.

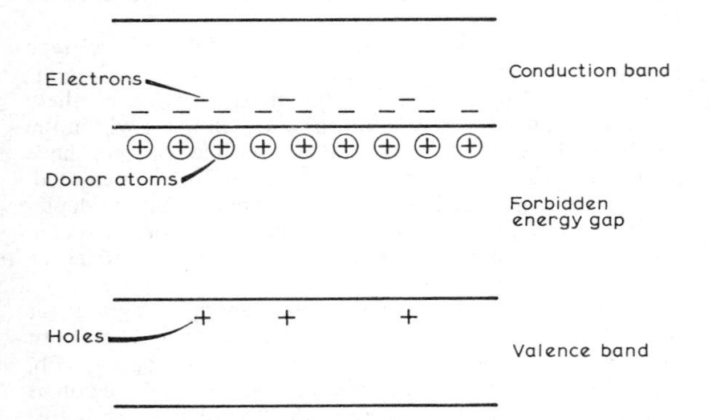

Fig 3.6. Energy band diagram for extrinsic "n" type semiconductor.

It is important to realize, however, that because holes are only vacancies left by parting electrons, they are not, in fact, influenced directly by the applied field. The migration of holes towards the negative pole is much less direct, and therefore slower, than that of the electron towards the positive pole. As will be seen later, this effect has a bearing upon the performance of the semiconductor diode and transistor.

In general, a flow of current in a semiconductor can be regarded as an acceleration of electrons in the conduction band and of positively charged holes in the valence band. For every electron that has been raised to the conduction band, there is a corresponding hole in the valence band, since in the pure, "intrinsic" semiconductor free holes and electrons cannot originate from any source other than by the breaking of a covalent bond.

The conductivity of the material from which transistors and other semiconductor devices are made, however, is much greater than that of the intrinsic silicon or germanium. The larger number of free current carriers (both holes and electrons) is produced by the controlled addition of a trace of an impurity element into the lattice structure. These impurities, which are taken from group three or five of the periodic table, are introduced into the pure crystal during the manufacture of the semiconductor material. The amount of doping necessary to achieve the required degree of conductivity is about one impurity atom per 10^8 atoms of silicon or germanium, and material which has been doped in this way is called an "extrinsic" semiconductor.

When the impurity element is pentavalent, it provides surplus electrons which cannot fit into the tetravalent lattice of the semiconductor crystal **(Fig 3.5).** The atoms of such an element are called "donor" atoms. Only four of the five valence electrons form covalent bonds with the intrinsic semiconductor, leaving the fifth electron only loosely bonded to its parent atom. This electron is readily excited into the conduction band, and in consequence at room temperature, nearly all the impurity atoms become ionized and positively charged. The total number of electrons in the conduction band at any given time is exactly equal to the sum of the donor atoms plus the holes in the valence band **(Fig 3.6).** Material which has been doped in this way is referred to as being an "n" type semiconductor. The elements that are used most frequently as pentavalent impurities are arsenic, phosphorus and antimony.

If a trivalent impurity is added to the intrinsic semiconductor, a similar effect occurs, but in this case, each atom introduces a hole **(Fig 3.7).** By having one less valence electron with which to complete the fourth covalent bond in the tetravalent lattice, the impurity atom acquires an extra electron from the intrinsic semiconductor. These atoms are therefore termed "acceptor" atoms. As a result of taking up the extra electron, the impurity atom becomes negatively charged and the semiconductor atoms are left with a concentration of holes in the valence band **(Fig 3.8).**

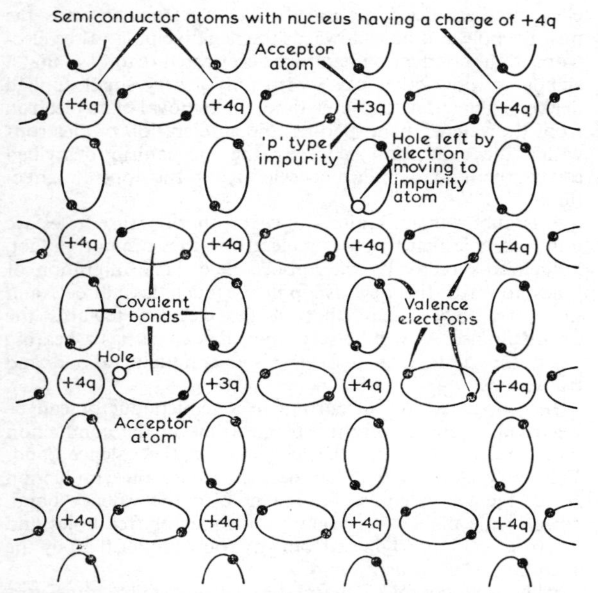

Fig 3.7. Semiconductor crystal structure with "p" type impurity atoms.

Only a relatively small number of electrons are available for excitation by thermal energy to the conduction band since all must originate in the intrinsic material. The increase in conductivity resulting from the doping is therefore primarily by the holes introduced with the trivalent impurity. The total number of holes in the valence band is exactly balanced by the sum of the negatively charged acceptor atoms and the electrons available in the conduction band. Material doped in this way is designated a "p" type semiconductor; the commonly used trivalent impurities being boron, aluminium, gallium and indium.

The P-N Junction

Holes and electrons normally co-exist within the same crystal of semiconductor, but in proportions which, at a given temperature, depend upon the impurities present. If a trivalent element is introduced into one part of the crystal, and a pentavalent element into the remainder, the junction,

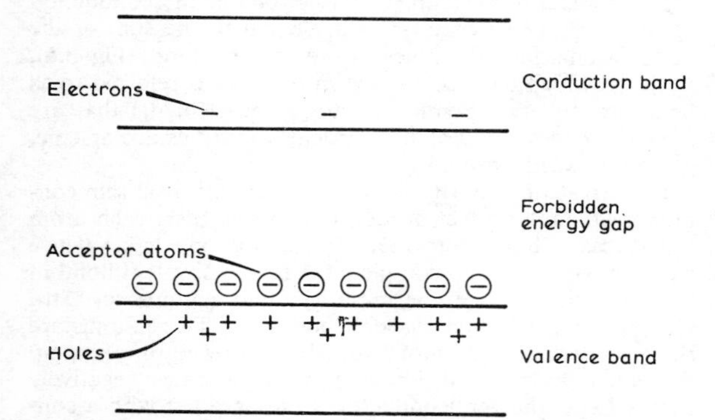

Fig 3.8. Energy band diagram for extrinsic "p" type semiconductor.

or interface, between the "p" and "n" regions exhibits certain phenomena which make the diode and transistor possible.

Fig 3.9 illustrates such a crystal with the "p" region indicated by a predominance of free holes that are the positively charged majority carriers (ie holes introduced by the acceptor atoms together with those holes present in the intrinsic material). A few free electrons which have been excited to the conduction band are also present as minority carriers.

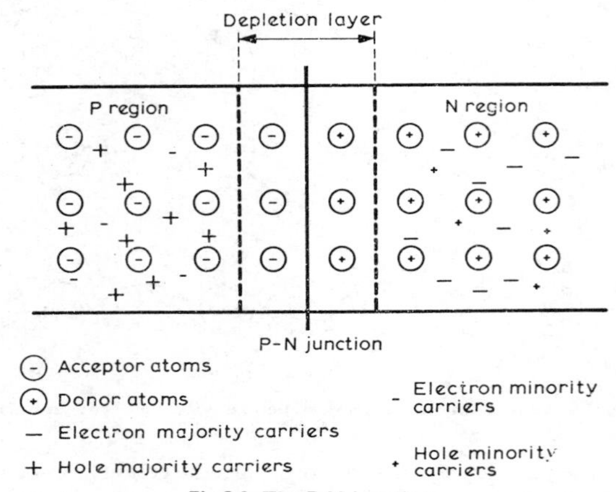

Fig 3.9. The P-N junction.

The "n" region is similarly indicated by a concentration of negatively charged majority carriers (ie electrons introduced by the donor atoms together with those electrons from the intrinsic material that have been raised to the conduction band). The minority carriers present are the few free holes created in the valence band as a result of electrons from the intrinsic semiconductor being excited to the conduction band.

It is apparent that because of the large number of free holes in the "p" region and of free electrons in the "n" region, a tendency will exist for some holes and some electrons to diffuse across the junction. In each case, electrons and holes adjacent to the junction will therefore neutralize each other to produce a local area devoid of both negatively and positively charged current carriers. This area is called the "depletion layer".

Also distributed within their respective "p" and "n" type regions are the immobile, ionized donor and acceptor atoms. Over the bulk of each region, the charge carried by these atoms is balanced by free electrons and holes, but within the depletion layer itself, where these mobile carriers have neutralized each other, this charge is left uncompensated. The net effect is to build up an electric field within the depletion layer, and thereby to create a potential barrier of such a polarity as to oppose the tendency for further free carriers to diffuse across the junction.

Fig 3.10 shows the corresponding energy diagram for the P–N junction. The energy gap, Eg, is the same on both sides of the junction, but due to the potential barrier, Eb, the potential energy of the electrons in the "p" region is increased with respect to those in the "n" region. Thus free majority carriers (electrons) from the "n" region must acquire additional energy equal to that of the potential

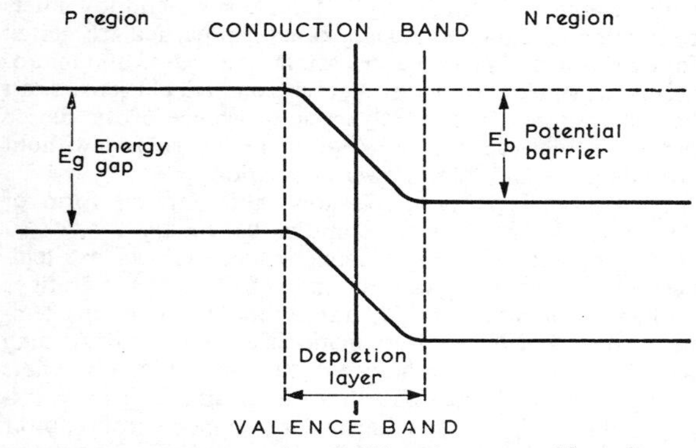

Fig 3.10. Energy diagram for a P-N junction at zero bias voltage.

barrier in order to cross the junction. The same conditions apply to the majority carriers (holes) in the "p" region, hence any transitions due to thermal excitation that might take place in one direction are exactly balanced by transitions in the reverse direction. As a result, the net current flow is zero as might be expected when the applied voltage across the junction is zero.

If an external potential is applied to the junction, and its polarity is such as to assist the migration of holes and electrons across the depletion layer (ie the "p" region is connected to a positive voltage and the "n" region to a negative voltage), a forward current will flow. The application of the voltage reduces the height of the potential barrier, thus facilitating the diffusion of both holes and electrons over the barrier **(Fig 3.11a).** The magnitude of this current will increase rapidly with the applied voltage, and in fact, will follow the exponential law shown in the upper right quadrant of the graph in **Fig 3.12.**

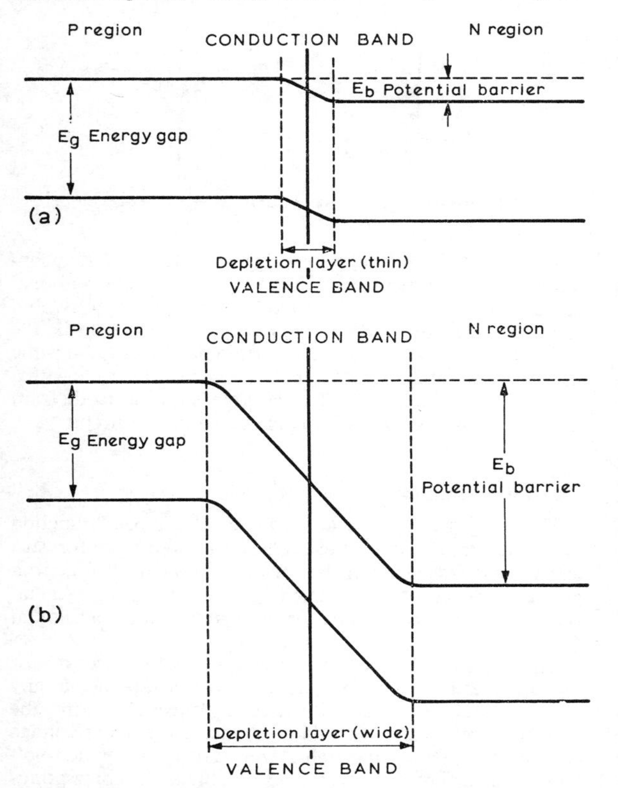

Fig 3.11. (a) Energy diagram for P-N junction at forward bias voltage. (b) Energy diagram for P-N junction at reverse bias voltage.

When a reverse field is created by connecting the "p" region to a negative voltage and the "n" region to a positive voltage, it augments the internal potential difference existing across the junction **(Fig 3.11b).** It thus opposes more strongly any tendency for the majority carriers from either region to cross the junction. The minority carriers, however, are assisted by this field, and because they are few in number, they reach a saturation flow with applied voltages of only a few millivolts. This reverse current is often referred to as the leakage current, and is illustrated in the lower left quadrant of Fig 3.12.

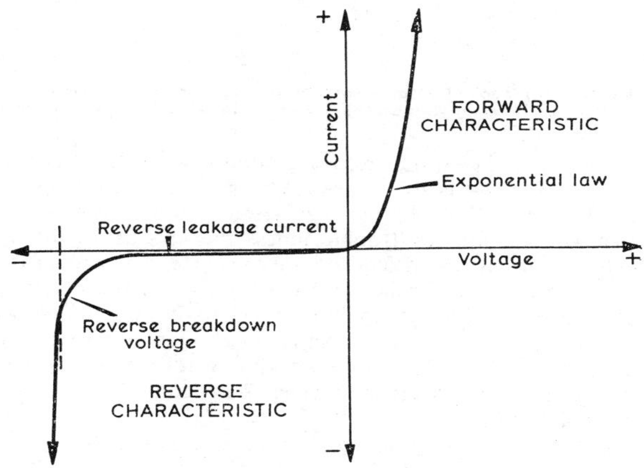

Fig 3.12. Characteristics of a P-N junction.

The minority carrier leakage current is due entirely to the presence of thermally generated free holes and electrons in the intrinsic semiconductor. At room temperatures, it seldom exceeds a few tens of picoamps in a silicon junction and a few microamps in germanium. In both materials, however, the leakage current approximately doubles for each 8°C rise in the temperature of the junction, and due account of this factor must always be taken into consideration when designing solid-state circuits in which significant rises in temperature may be expected. The de-rating curves quoted by the manufacturers of semiconductors should always be observed, but in general it is good practice to ensure that the equilibrium operating temperature of a silicon device does not rise above 120/140°C, and that of a germanium device is kept below 40/50°C.

The non-linear properties of a P–N junction are very apparent, and make it ideally suited as a rectifier. **Fig 3.13** is a typical characteristic of a semiconductor junction diode. The forward voltage drop at 20mA is about 0·8V for silicon and 0·3V for germanium; at 100mA these figures increase to 0·9V and 0·4V respectively.

The maximum reverse voltage that may be applied to a junction diode varies with its design and impurity content, but if exceeded, the resulting flow velocity of the minority carriers will become sufficient to ionize the neutral atoms in

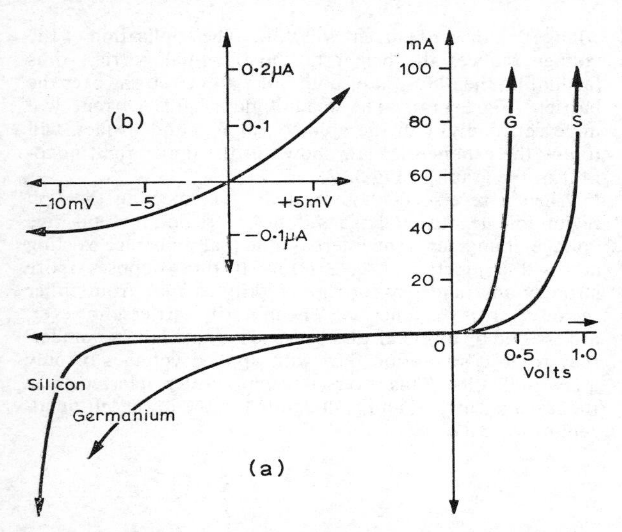

Fig 3.13. (a) Typical voltage/current characteristics of germanium and silicon junction diodes. (b) Detail of characteristics in the zero region.

the depletion layer, thus releasing additional charge carriers. This process leads to "avalanche" breakdown, and causes the leakage current to rise very steeply with consequences that are usually catastrophic unless the impedance of the external circuit is sufficient to limit the current flow to a safe value. Once a controlled breakdown occurs, the junction diode offers only a very small resistance to the current flow so defining a precise voltage virtually independent of the current magnitude. A P–N junction working in this mode is often used as a voltage reference device, and is commonly termed a "zener" diode.

A further property of the P–N junction is that the width of the depletion layer increases with the reverse voltage applied. The capacitance which must exist between the "p" and "n" regions, and is a function of the depletion layer width, therefore decreases as the reverse voltage is increased. This effect makes possible solid-state variable capacitors which find useful application in high frequency circuits for frequency control and parametric amplification.

CHARACTERISTICS OF THE SEMICONDUCTOR DIODE

The preceding section describes the elementary physics of the P–N junction, and while the electrical characteristics of the practical solid-state diode may vary according to the application for which it was designed, the basic properties of the P–N junction remain a common factor in each type.

There are perhaps more variants of the basic semiconductor diode than there are of any other single group of solid-state devices, and the following paragraphs describe the characteristics of some of the more important types. Typical applications will be found elsewhere in this handbook.

Semiconductor diodes offer combinations of characteristics not possible in the thermionic diode, and at the same time they allow the circuit design engineer complete freedom from the disadvantages which so often accompany the use of a heater or heater supply. The diodes can be manufactured with exceptionally small physical dimensions and low inter electrode capacitances, thus making them particularly suited for detection and mixing applications at all radio and microwave frequencies. In power rectification high efficiencies can be achieved, and the low dynamic resistance of the device permits a high peak current rating to be realized without exceeding a very modest power dissipation.

The photograph on p3.7 shows the physical form of several solid-state diodes in common use today.

The characteristics of all semiconductor diodes are temperature dependent, and the measurements and ratings quoted in manufacturers' literature apply only at the temperature at which they were made or are specified. At high temperatures it may be necessary to de-rate the permissible forward current, and sometimes the maximum inverse voltage. Such de-rating is particularly important in power rectification where the diode losses may be sufficient to raise the junction temperature appreciably.

It must not be assumed, however, that the temperature dependence of the semiconductor diode in any way reduces the reliability of the device when it is operating within its correct ratings. Indeed the failure rate of a solid-state device in correctly designed applications is considerably less than that of its thermionic counterpart, and emphasis is placed upon this point solely because it is so often neglected by inexperienced designers. Moreover the mechanism of semiconductor failure due to overload conditions is catastrophic, and usually results in the device being damaged beyond further use.

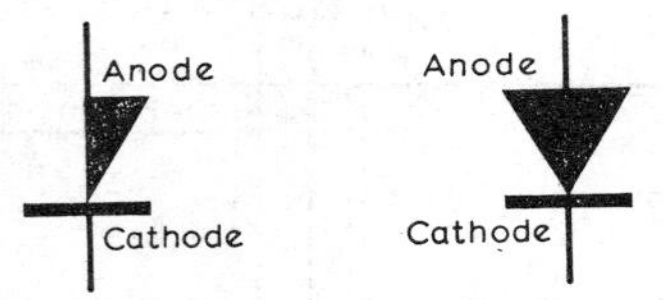

Fig 3.14. Alternative circuit symbols for the junction diode.

Fig 3.14 shows the general symbol for the semiconductor diode, or rectifier, with the direction of conventional current flow indicated by the arrow. The line of the symbol towards which the arrow points is often referred to as the "cathode" by analogy with the thermionic diode, and it is the usual practice of manufacturers to mark this electrode with a red spot or a white band. The cathode is the electrode from which the positive output is taken in a rectifying circuit.

The Junction Diode

The form of the characteristic of the typical junction diode can be seen from Fig 3.13, and although the forward and reverse resistance in the various types will diverge to a greater or lesser extent from this characteristic, the general shape will remain the same for both silicon and germanium devices.

The application of a forward voltage to a semiconductor diode causes a current to flow, which, in the absence of any external limiting resistance, rises exponentially with the voltage. Unlike the thermionic diode, the current/voltage characteristic passes through a true zero (zero applied voltage, zero current), and rises considerably faster corresponding to a much lower forward resistance. The forward current

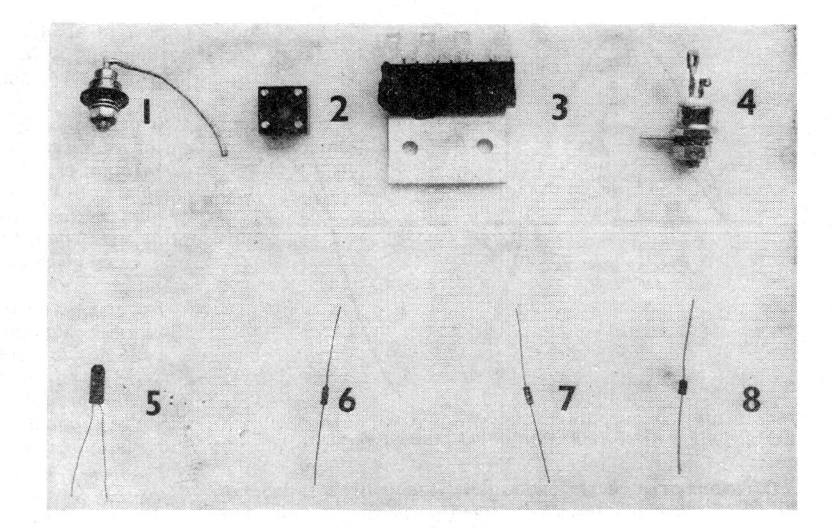

A selection of typical diodes. 1: 6A germanium power rectifier. 2 and 3: four-diode bridge rectifiers. 4: 500V scr. 5: general purpose germanium diode. 6: 5·1V 278mW zener diode. 7 and 8: high-voltage silicon power rectifiers.

for a voltage drop across the diode of between 0·5V and 1·0V ranges from a few milliamps for certain types to several amperes for others. At high forward currents, the characteristic approaches a straight line parallel to the current (y) axis. There is no saturation effect.

As a first approximation, the voltage drop across a typical device in more than a few milliamps of forward conduction is usually taken as being a constant, the exact value of which is a function of the diode type. This voltage in a germanium diode is approximately 0·3V and in silicon, about 0·75V. Below these voltages, the forward resistance of the diode rises rapidly to become less significantly different from the reverse characteristic as the zero region is approached.

An applied reverse voltage produces a small leakage current as shown in Fig 3.13(b). This leakage very rapidly reaches a saturation value of a few microamps at voltages of only a few tens of millivolts, then rises very slowly until the "turn over", or reverse breakdown, voltage is reached. Beyond this point, the characteristic exhibits a sudden increase in the reverse current for a relatively small increase in the reverse voltage. Under these conditions, the back resistance of the diode falls to a few ohms, and unless the current is limited by external means, the diode dissipation would exceed its rated value. The consequences of even a short duration overload in this region invariably results in permanent damage to the junction due to overheating. The maximum peak inverse voltage rating of a semiconductor diode is, therefore, of fundamental significance and should be strictly observed. Excess voltages lasting for even a few microseconds may be fatal.

In the zero region (-10mV to $+50$mV), the diode characteristic is approximately linear, and while this is normally not significant when the applied voltage is greater than a few volts, it considerably reduces the efficiency of the device as a rectifier of small signals. Within this range, both the forward and reverse resistances of the junction are high, and considerable non-linearity can be introduced between the applied signal voltage and the rectified output.

Optimum linearity and good detector efficiency are achieved when the applied voltage is large compared with the zero region of the diode characteristic, and when the ratio of the load resistance to the forward resistance of the diode is high. In practice, the reverse breakdown voltage imposes a limitation on the magnitude of the applied voltage, and the reverse leakage current places an upper limit on the permissible value of load resistance. When the diode is used as a conventional a.m. detector, load values of between 10kΩ and 100kΩ will usually prove satisfactory for all signal levels greater than one volt. Certain types of germanium diode will operate with good efficiency and linearity with somewhat lower applied voltages, but as the leakage current is generally greater than that of the silicon diode, they should be designed to work into loads which do not exceed about 20kΩ. The "hot carrier" diode (qv), which has a very low reverse leakage current and a forward voltage drop intermediate between silicon and germanium, is an ideal device for detector applications at all frequencies into the shf range.

The rectification efficiency of a diode in a specific circuit at a given peak input voltage is

$$\eta = \frac{\text{dc output voltage}}{\text{peak input voltage}} \times 100 \text{ per cent}$$

Efficiencies of over 75 per cent can be readily achieved.

One extremely important aspect of diode performance not apparent from the dc characteristic curves is that of hole, or charge, storage. The semiconductor diode usually consists of a strongly "p" type anode and a slightly "n" type cathode, thus when the device is conducting normally, holes are injected into the cathode. As described earlier, these holes do not possess the mobility of the electrons, therefore when the polarity of the applied voltage is reversed, those remaining within the "n" material migrate back to the anode. Thus the current through the diode does not fall immediately to the leakage value as soon as the polarity is changed, and a pulse of reverse current in excess of the normal leakage value flows for a short time until the depletion layer is swept clear of the current carriers. This phenomenon, commonly referred to as hole, or charge, storage is shown diagrammatically in **Fig 3.15.**

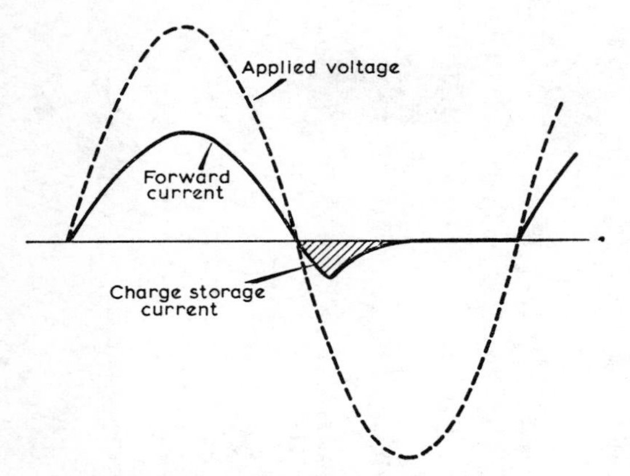

Fig 3.15. Charge storage in a junction diode.

The storage of charge within the diode causes it to behave as if a small capacitance is connected across the junction, and as such clearly has an adverse effect upon the operation of the diode at high frequencies.

The effects of hole storage can be expressed quantitatively as the amount of charge stored for a given forward current, or it may be quoted as the time required for the transient current to decay to a certain specified value; usually 10% of the peak figure. This time is called the "recovery time" of the device. Fast diodes suitable for operation at vhf have a storage charge of not more than a few picocoulombs and have a recovery time expressed in nanoseconds (10^{-9} s). The comparable figures for the power rectifier are typically one to three orders of magnitude greater.

As the charge stored in the depletion layer of the diode requires a finite time to dissipate, so a similar time is required for the build up of the charge following the application of a forward voltage. During this period the current through the diode is greater than in the steady state condition thus causing a turn-on transient to appear on the leading edge of the forward current waveform **(Fig 3.16).**

The Point Contact Diode

The point contact diode is the modern form of the original crystal ("cat's whisker") detector used in the earliest radio receivers, and although it was developed before the basic junction diode, it nevertheless depends for its operation upon the same physical principles as the latter.

While the point contact diode is less robust than the junction diode both electrically and physically, it has several important advantages over the junction diode. These are particularly evident at very high frequencies where a low inter-electrode capacitance and electron transit time are important.

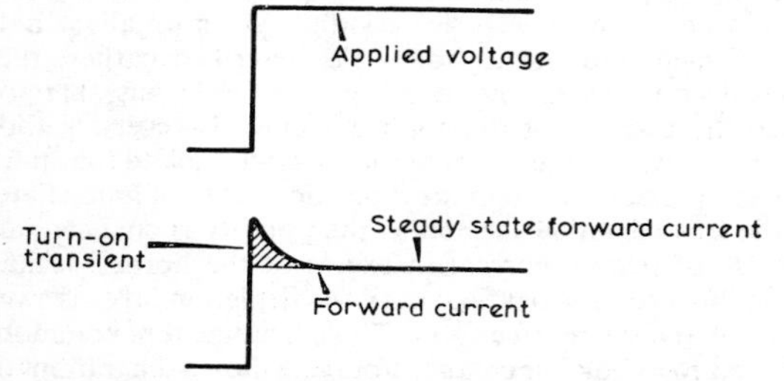

Fig 3.16. Forward current transient of a junction diode.

Structurally the point contact diode comprises a fine, springy tungsten wire which presses against an "n"-type germanium crystal **(Fig 3.17).** During manufacture, a very small area of "p" material is formed at the point of contact of the tungsten wire, thus creating a P–N junction of minute proportions on the face of the germanium crystal. Rectification takes place at the interface between the "p" and "n" materials.

The characteristics of the point contact diode resemble those of the junction diode over much of its working range, but the device exhibits several important differences as the forward and reverse voltages are increased. In the forward direction the voltage drop is similar to that of a junction diode of like material, but owing to the small interface area, the maximum current rating is much reduced.

The reverse leakage current initially follows the theoretical figure for an ordinary junction, but soon begins to rise more rapidly well before the turn-over or breakdown voltage is reached. The maximum peak inverse voltage (piv) rating is, therefore, generally much less than that of the junction diode. This figure may be as low as 8·0 to 10·0V for very fast diodes, and rarely exceeds 70V in any point contact type. The reverse capacitance and charge storage are considerably lower than in junction diodes hence they are widely used at all radio frequencies. The temperature sensitivity of point contact and junction diodes is similar.

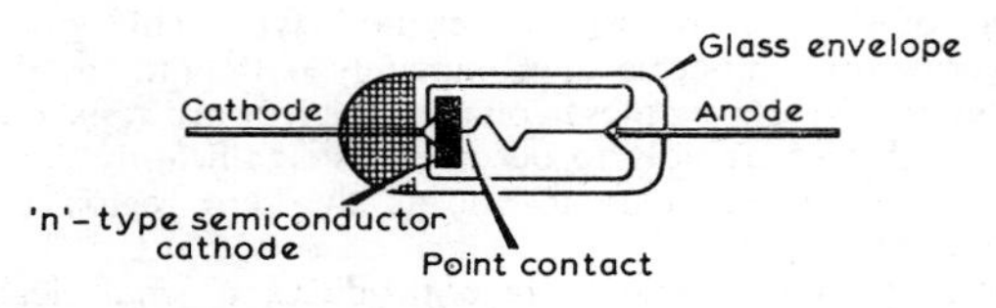

Fig 3.17. Construction of a point contact diode.

The Gold Bonded Diode

In its construction, the gold bonded diode is essentially a point contact diode in which the tungsten wire is replaced by one made of gold. The result is a small area junction diode in which both the forward and back resistance of the device are improved.

This allows higher forward current ratings to be achieved, and considerably improves the front to back impedance ratio. The low capacitance characteristics of the ordinary point contact device are retained, and while the maximum reverse voltage ratings are similar, the reverse leakage current is rather less.

The diode is particularly suited to the detection of small rf signals because conduction starts at a lower forward voltage than with most other types of diode, and the higher back resistance allows the load to be increased, thereby increasing its efficiency as a rectifier.

The Hot Carrier Diode

The hot carrier, or Schottky, diode is one of the more recent developments in semiconductor technology. The device consists of a metal to silicon junction in which the current flow is predominantly by means of electron majority carriers. This reduces the minority carrier lifetime to less

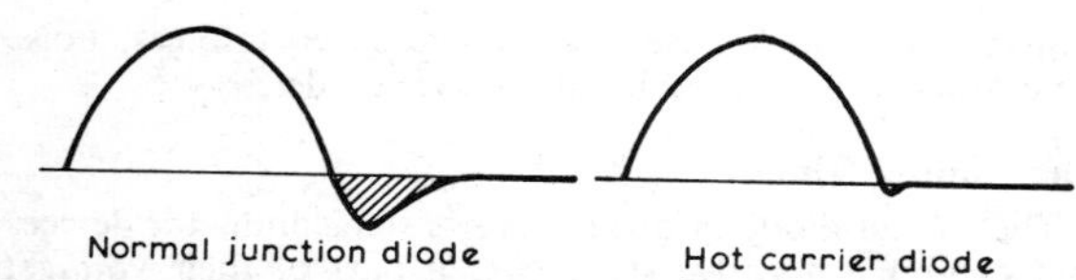

Fig 3.18. Comparison of charge storage effects in a P-N junction diode and a hot carrier diode.

than 100 picoseconds, and virtually eliminates the charge storage effects which such carriers produce in ordinary P–N junctions. **Fig 3.18** compares the charge stored in a typical junction diode with that in a hot carrier diode.

The performance of the hot carrier diode approaches the ideal, and its combination of superior speed and forward conduction is unparalleled in any other type of diode. The barrier potential at which forward conduction commences may be varied by the choice of metal in the metal-to-silicon junction. This is clearly illustrated in the characteristics of two hot carrier diodes given in **Fig 3.19.** The left hand curve in which conduction begins to rise rapidly for applied voltages greater than 0·25V is similar to that of a germanium device, while the right hand curve exhibits a barrier potential of more than 0·5V and is typical of a silicon junction.

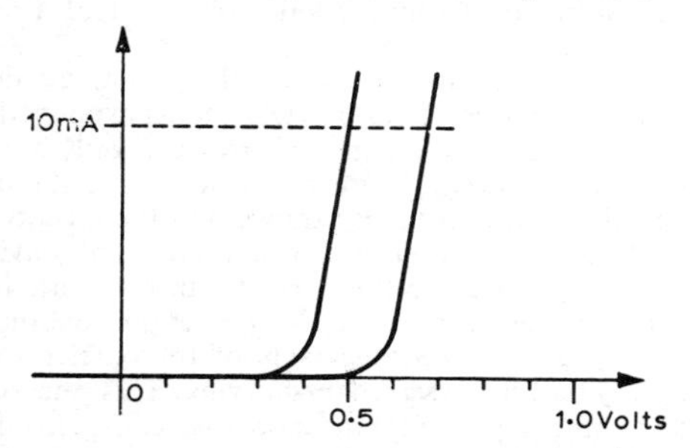

Fig 3.19. Comparison of barrier potentials in two hot carrier diodes.

The properties of the hot carrier diode enable an excellent noise figure to be realized in rf mixer applications, while in terms of conversion loss the device is not equalled by any of the more commonly used point contact diodes. The absence of minority carrier charge storage effects allows the hot carrier diode to fulfil the stringent specifications of microwave applications.

Hot carrier diodes may also be used to advantage in all circuits which require the ultimate in performance and reliability. In pulse operation, the device is ideal for clamping, sampling gates, pulse shaping and general purpose usage requiring fractional picosecond switching times. It makes an excellent small signal detector and balanced modulator through the spectrum from low frequencies well into the microwave region.

The Zener Diode

In the foregoing sections the performance of the semiconductor diode has been evaluated primarily upon its rectifying characteristics, and upon its behaviour at forward and reverse voltages within its maximum ratings. The characteristics of the P–N junction, however, exhibit an important feature outside this working range which can be exploited for voltage limiting, voltage stabilization or reference applications. Such devices have been given the generic name "zener", and may be regarded as being the solid-state equivalent of the gas-filled voltage stabilizer valve.

The circuit symbols for the zener diode are illustrated in **Fig 3.20.** For consistency, they are drawn with the arrow indicating the direction of conventional forward current through a normal diode. In use, the positive electrode, which is indicated similarly to the cathode of a rectifying diode, is returned directly or indirectly to the more positive pole of the applied voltage (ie the diode is reversed biased in normal use). A zener diode must never be connected across a supply voltage greater than its rated working voltage without there being sufficient resistance in series with the device to limit the current flow to a safe value.

Fig 3.20. Symbols for the zener diode.

Fig 3.21 shows the typical zener diode characteristic, and it will be seen that it resembles very closely that of the normal rectifying junction diode. Unlike this device, however, the zener diode is used in the reverse current region where the applied voltage is sufficient to cause avalanche breakdown. The knee in the silicon junction reverse characteristic is more sharply defined than in the germanium junction, and over a wide range of currents the curve runs almost parallel to the negative y axis. The dynamic impedance of the diode in this region is not more than a few tens of ohms, and it thus determines a fairly precise voltage across the junction which is nearly independent of the current through it.

By exercising careful control over the doping and manufacturing processes, the turnover voltage can be made to requirements within the range 3V to over 150V. The dynamic resistance of the zener varies over the range; rising towards the upper and lower limits, and passing through a minimum value of only a few ohms between 5V and 7V.

In practice, two factors determine the current range over which the zener diode can be used. The lower limit is set by

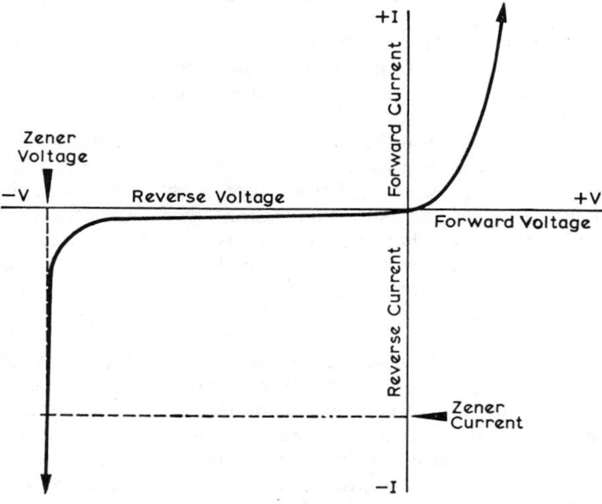

Fig 3.21. Zener diode characteristic.

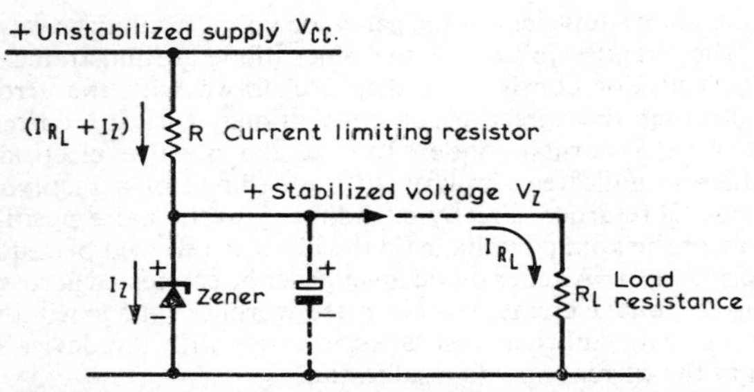

Fig 3.22. The basic circuit using a zener diode to provide a stabilized voltage from an unstabilized supply rail.

the increase in the slope resistance round the knee in the characteristic, and the upper limit by the maximum power dissipation that can be permitted within the device. As the incremental resistance in most zeners is small, the power dissipation will be approximately $V_z.I_z$, where V_z is the zener voltage and I_z is the current through it. Hence for a family of zener diodes of a given physical size, the maximum current can be taken as being inversely proportional to V_z.

Zener diodes possess a normal forward voltage/current characteristic.

The most frequent use of the zener diode is in generating a simple reference voltage or stabilized supply rail. The basic circuit for this application is given in **Fig 3.22.** Since the device has a finite dynamic resistance, R_z, good stabilization of the zener voltage against variations of the unstabilized supply rail can be achieved only when the ratio $\frac{R}{R_z}$ is large. The values of the parameters, V_{cc}, I_{R_L} and I_z should, therefore, determine a value of the current limiting resistance, R, greater than a few hundreds of ohms; ie at least ten to twenty times R_z. The value of R may be calculated by applying Ohm's Law as follows:

$$R = \frac{(V_{cc} - V_z)}{(I_{R_L} + I_z)}$$

where V_{cc} is minimum value of the unstabilized supply rail
V_z is stabilized rail voltage to load; zener voltage.
I_z is current in zener diode. (Typically not less than 1mA for small devices having a 250mW maximum dissipation).

When R has been determined, a maximum value of I_z should be calculated for the maximum value of V_{cc}, and the product $V_z.I_{z(max)}$ should not exceed the manufacturer's power rating for the device.

The basic circuit in Fig 3.22 has only limited application, and is not recommended when the load current is large and/or is subject to wide variations; nor is the simple circuit able to provide adequate stabilization against variations of supply rail, V_{cc}, when the over-voltage (ie the voltage by which V_{cc} is greater than V_z) is small compared with the zener voltage. Examples of circuits for use in these circumstances may be found in the chapter dealing with power supplies.

Less familiar applications of the zener include that of providing a dc voltage drop between two points in a circuit without the need to decouple in order to permit a low impedance path at signal frequency. A zener diode of the correct rating placed in the cathode circuit of a thermionic valve will, for example, provide a fixed bias voltage, and can replace the usual bias resistor and decoupling capacitor. Zeners may also be used as voltage surge limiters, noise generators, and as variable capacitance diodes.

The Tunnel Diode

The tunnel diode is one of several semiconductor devices that exhibit a negative slope over a part of their voltage/current characteristic. **Fig 3.23** plots this relationship in a typical tunnel diode, and the negative slope is clearly indicated as that region in which the current through the device decreases as the voltage across it increases.

The anomalous behaviour occurs only over that part of the characteristic near the origin, and beyond the valley, the curve rapidly assumes the exponential law of the conventional junction diode.

The commonly accepted symbol for the tunnel diode is shown in **Fig 3.24.**

Tunnel diodes are P–N junctions manufactured from very heavily doped semiconductor which is usually either germanium or gallium arsenide. The heavy doping reduces the width of the depletion layer to about 100 angstroms (10^{-6} cm) as compared with approximately 10,000 angstroms (10^{-4} cm) for a conventional junction. It is this difference in width that accounts for the major divergence in the current flow mechanism of a tunnel diode from that of a conventional semiconductor device.

According to the laws of classical physics, an electron cannot penetrate a potential barrier, and as explained in the theory of semiconductors, the electron can only surmount the barrier if its energy is greater than that of the barrier. However, the quantum theory shows that there is, in fact, a small but finite probability that an electron, having insufficient energy to climb the barrier, can penetrate it if the barrier is sufficiently narrow. Moreover in so doing, the electron emerges on the other side of the barrier with the same energy that it possessed on entering. This phenomenon is called "tunnelling", and is the basic mechanism from which the tunnel diode derives its name.

Fig 3.25 shows the energy band diagrams for the tunnel diode. Under zero bias conditions conduction band electrons in the "n" region and valence band electrons in the "p" region have similar energy levels. It is possible, therefore, for these electrons to cross the junction without

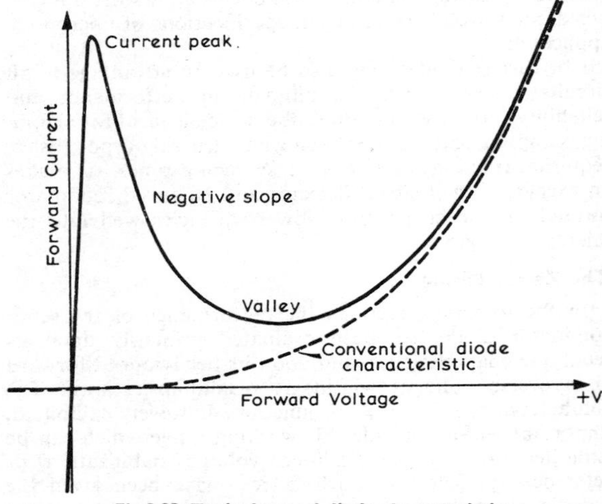

Fig 3.23. Typical tunnel diode characteristic.

changing their energies by tunnelling through the forbidden energy gap at this level. There is, in fact, a constant interchange which is exactly balanced in each direction so that the net current flow is zero.

When a reverse bias voltage is applied, all the energy levels on the "p" side are increased in relation to those on the "n" side. There are, therefore, many unoccupied energy levels in the "n" region conduction band readily available to the valence band electrons in the "p" region, so that a large current of electrons flows from the "p" side to the "n" side. The current flow in the reverse direction remains as in the zero bias condition since the "n" region conduction band electrons remain opposite the "p" region valence band electrons.

Fig 3.24. Symbol of tunnel diode.

The tunnelling phenomenon is an exponential function of the electric field across the barrier, and as a result the reverse current increases very rapidly with the reverse voltage, thus making the device highly conductive for all values of reverse bias.

Under forward bias voltage conditions, the energy levels of the "p" region are depressed in relation to those of the "n" region. If this applied voltage is small, the "n" type conduction band electrons are opposite the unoccupied energy levels in the "p" type valence band while the "p" type valence band electrons appear opposite the forbidden energy gap. Electrons are, therefore, able to cross from the "n" region conduction band into the "p" region valence band, but not in the reverse direction. The net effect is a current flow in the forward direction which reaches a maximum value at the peak in the tunnel diode characteristic.

When the forward voltage across the junction is increased further, the energy bands in the "p" region are depressed to an even greater extent so that eventually both the conduction and valence band electrons become opposite the forbidden energy gap. Progressively less current flows across the junction as the applied voltage passes through this region, and the negative slope in the tunnel diode current/voltage characteristic is established.

Ultimately, when the trough of the valley is reached, tunnelling ceases completely. At the same time, the height of the potential barrier is reduced to a level that permits the flow of electrons over it by the conventional mechanism. The tunnel diode thus behaves as an ordinary junction diode, and the current/voltage relationship gradually assumes the normal exponential curve for all further increases in the applied bias.

The exceptionally thin depletion layer between the "p" and "n" regions renders the tunnel diode very susceptible to damage through overload. For these reasons, the device should always be selected to have a working current in the range required by the application, and at no point in its operational cycle should the current be allowed to rise above three to four times the rated figure. Tunnel diodes are manufactured to operate at specific currents within a 1mA to 1A range, and exhibit their negative slope typically between 150 and 300mV.

Because of the very short distances the electrons are required to travel across the junction, the tunnel diode is especially suited to applications in the uhf/shf range. In the communications field, their most common use is that of an oscillator. As a low noise device, they also have application as amplifiers operating at frequencies above those at which other semiconductor devices fail.

The Back Diode, or Tunnel Rectifier

The back diode, or tunnel rectifier, is a normal tunnel diode in which the rectifying properties have been emphasized by a diminution of the negative resistance characteristics. It is highly conductive in the reverse direction, but for small forward voltages it exhibits a characteristic similar to that of the conventional junction rectifier; ie it presents a relatively high resistance to the flow of forward current while the applied voltage is less than the barrier potential.

The major differences in the current/voltage characteristics of the tunnel rectifier and conventional rectifier are shown in **Fig 3.26.** Whereas over the normal working range, the conventional diode conducts heavily in the forward direction and only very slightly in the reverse direction, the tunnel rectifier exhibits only a small reverse resistance and allows very little current to flow in the forward direction until the barrier potential is exceeded.

It is possible, therefore, for a tunnel rectifier to provide efficient rectification at much smaller signal voltages than the conventional junction rectifier, although the polarity requirements are opposite. Tunnel rectifiers have very high speed capabilities, and over their limited voltage range have superior characteristics for microwave applications.

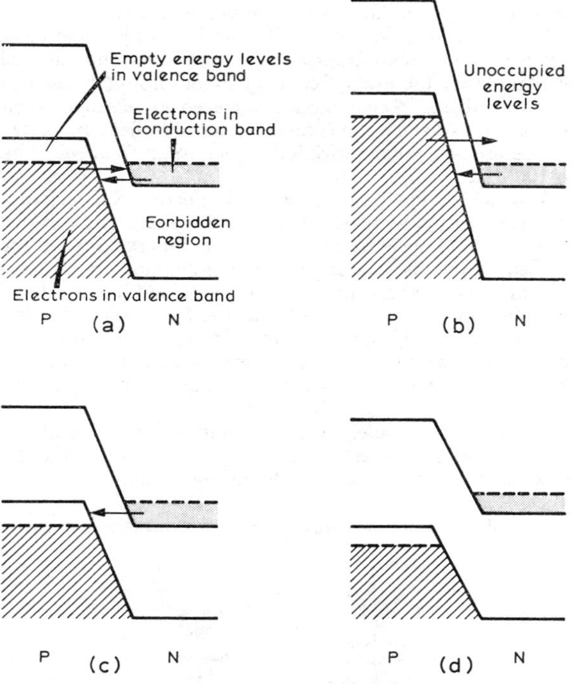

Fig 3.25. Energy band diagrams for a P-N tunnel diode junction at (a) zero bias voltage, (b) reverse bias voltage, (c) peak voltage, (d) valley voltage.

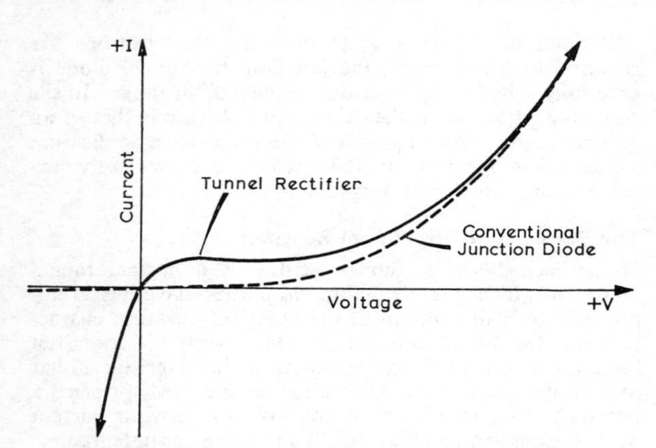

Fig 3.26. Characteristic of a back diode, or tunnel rectifier.

The Varactor Diode

In describing the basic physical properties of the simple P–N junction, reference has been made to the voltage dependent capacitance which is present across the depletion layer. Diodes that exploit this effect, and are primarily intended to be used in this mode rather than as rectifying junctions are termed "varactor", or variable reactance, diodes.

The symbol is shown in **Fig 3.27,** and a typical capacitance/voltage curve is given in **Fig 3.28.** The junction capacitance, which is clearly non-linear with the applied voltage, varies as V^{-n}, where V is the reverse bias voltage and n lies between 1/2 and 1/3 according to the diode type. All P–N junction devices, in fact, exhibit this phenomenon to a greater or lesser extent, and for many simple applications the ordinary rectifying junction diode will be found adequate.

The capacitance range possible varies from a few picofarads for small high frequency diodes to over one hundred picofarads for the lower frequency diodes normally used as power rectifiers. Zener diodes operated at less than their reverse breakdown voltage also exhibit this effect and often can be used as voltage variable capacitors at frequencies up to 100MHz.

At dc and at low frequencies, the limits of the capacitance swing are set by the reverse breakdown voltage and forward conduction. However, at uhf, where the polarity of the applied voltage is reversed before the charge carriers have had time to diffuse away from the junction and to recombine, the diode may be driven well into the forward conduction region without appreciable current flowing. This effectively reduces the width of the depletion layer and further increases the available capacitance swing in any particular diode.

Fig 3.29 shows the equivalent circuit of a varactor diode. The junction is represented by C_J and R_J in parallel, and the bulk resistance of the semiconductor by a small series resistance, R_S.

In a silicon junction, where the reverse leakage current is

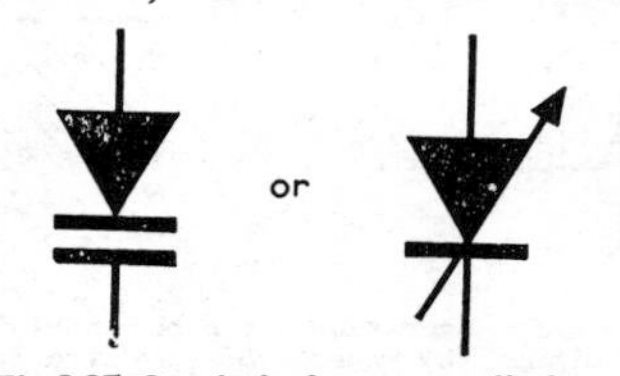

Fig 3.27. Symbol of varactor diode.

small enough to be neglected, R_J approaches 10^6 ohms, and the performance of the junction as a varactor at a given frequency is determined primarily by R_S and C_J. As the frequency is increased, the time constant R_SC_J becomes correspondingly more significant and thus accounts for the deterioration of the device's behaviour with this parameter.

Just as Q, or the figure of merit, for any conventional reactance is given by X/R_S; where X is the value of the reactance at the given frequency, and R_S is the resistance in series with it, so is the Q of a varactor diode similarly determined as the ratio of its capacitative reactance, X_J, and its bulk resistance, R_S.

Modern varactor diodes are able to function well into the microwave region, and to operate as efficient devices handling rf powers rated in the tens of watts.

Varactor diodes fall essentially into two categories; the abrupt junction and the graded junction device. The former, which are similar to the ordinary rectifying P–N junction, are used primarily in applications that require a simple capacitance variation as a function of the applied voltage. Such applications include reactance modulators, automatic frequency control, and parametric amplification. Some power devices for frequency multiplication also fall into this category. Varactor diodes in which the doping profile (map of the impurity concentration) is graded are known as "step recovery" diodes. They are most frequently employed as harmonic generators and comb generators in the lower powered applications, and as frequency multipliers in solid-state microwave transmitters.

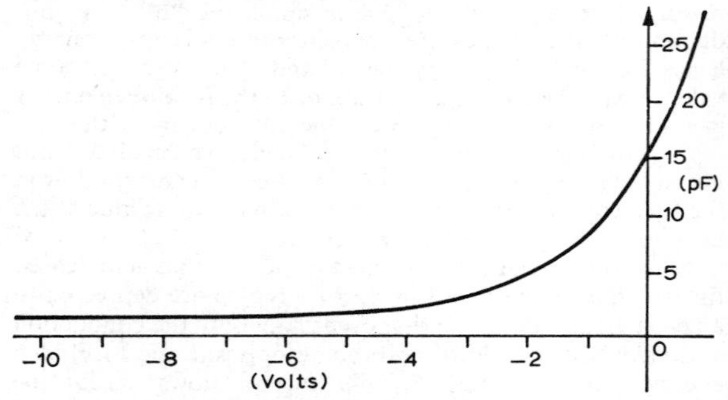

Fig 3.28. Capacitance/voltage characteristic of a typical varactor diode.

The step recovery varactor diode exploits the charge storage effects present in normal P–N junctions by carefully controlling the doping of the semiconductor material in the regions adjacent to the junction. When a P–N junction is forward biased, the majority carriers cross the depletion layer and enter the opposite side to become minority carriers. They then diffuse away from the junction and recombine. The time required for recombination is about 10^{-8}s, thus if the bias is suddenly reversed before all the charge carriers have recombined, those remaining will flow back across the junction and form a short current pulse in excess of the normal reverse bias leakage current. This is illustrated in Fig 3.16.

The delay in recombination represents stored charge, and has the effect of a capacitance which is additional to that appearing across the depletion layer under steady state conditions. In the abrupt junction the stored charge dissipates relatively slowly, and thus imposes considerable limitations on the behaviour of the junction as a rectifier at very

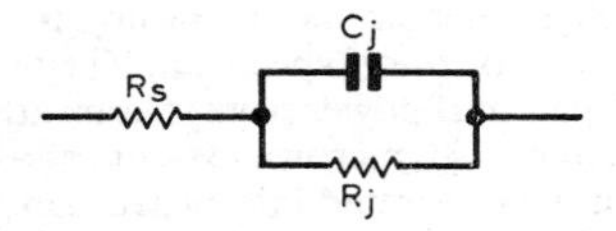

Fig 3.29. Equivalent circuit of a varactor diode.

high frequencies. However in the graded junction, the diffusion time of the carriers is modified by the doping profile so that the charge storage current cuts off very sharply as in **Fig 3.30.** This phenomenon, from which this type of varactor takes its name, results in the production of a great deal of additional harmonic energy that can extend far into the microwave spectrum.

The diagrams in **Figs 3.31(a)** and **3.31(b)** are examples of two uses of the varactor diode. The first shows an abrupt junction device in a typical automatic frequency control circuit where the local oscillator of a fm receiver is held "on tune" by the error voltage derived from the Foster-Seeley discriminator. Any frequency drift in the local oscillator results in a dc signal which automatically trims the tuning capacitance (of which the varactor is a part) in such a direction as to correct the drift.

The second example shows a step recovery varactor diode in a uhf transmitter application where the output frequency is twice that of the input. The network, L1, C1, C2, matches the varactor to a source of rf energy, and the second harmonic is selected and matched to the 50Ω load through a similar LC network tuned to twice the input frequency. The varactor is correctly biased in operation by the resistance, R, (typically 50kΩ to 1·0MΩ), and no external power supply is required. Efficiencies of greater than 50 per cent can be readily achieved. It is also worth noting that since the output power is a linear function of the input, it is possible to multiply the carrier frequency of an amplitude modulated signal without destroying the original modulation characteristics.

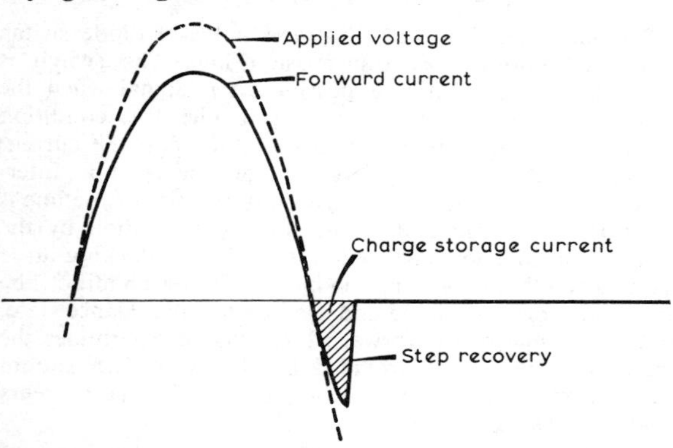

Fig 3.30. Forward and reverse current of a step recovery varactor diode.

The P-I-N Diode

The P–I–N diode is a two terminal semiconductor device in which thin, highly conductive "p"-type and "n"-type regions are separated by a region of intrinsic, or nearly intrinsic, material. This is shown diagrammatically in **Fig 3.32.** The dc characteristics of the P–I–N diode closely resemble those of a conventional P–N junction, but the different construction gives the device several properties that are important technically. For example, the P–I–N diode is able to withstand a high reverse bias thus making it suitable for use as a power rectifier or other high voltage application.

The intrinsic region also allows a wider separation of the "p" and "n" regions, so reducing the self-capacitance of the device under reverse bias conditions. It can store almost unlimited amounts of charge in the intrinsic region, and when forward biased to between 50 and 100mA the diode exhibits practically zero impedance (less than one ohm). The P–I–N structure therefore possesses considerable advantages for a step-recovery varactor diode.

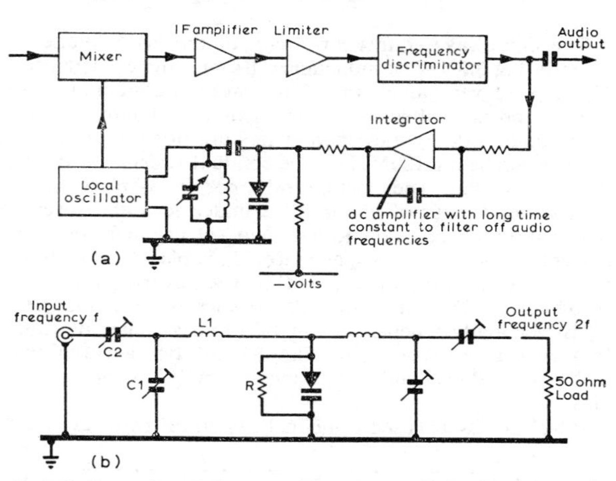

Fig 3.31. Examples of the uses of a varactor diode. (a) Automatic frequency control. (b) Frequency multiplier.

Normal P-N junctions often have a slow response time because the diffusion time of the holes in the field-free, bulk semiconductor material is comparatively long (tens, or even hundreds of microseconds). In the P-I-N diode very little field-free bulk material exists because the "p" and "n" regions are very thin. The field due to an applied bias permeates the intrinsic region, so that when holes (and electrons) are injected, or are generated by the incidence of external radiation (eg as in the photojunction), they are quickly swept clear and are collected by the electrodes. The response time of the P-I-N diode can, therefore, be reduced to a few tens of nanoseconds.

The dynamic impedance can be determined over several decades by controlling the dc bias current through the device. The precise dependence of the high frequency impedance upon such factors as current density, transit time and carrier lifetime is beyond the scope of this book. It is sufficient for most practical purposes, however, to consider the forward biased diode as a current controlled linear resistance at all frequencies above about 50MHz, and to represent it in the reverse direction by a small capacitance and a large resistance in parallel.

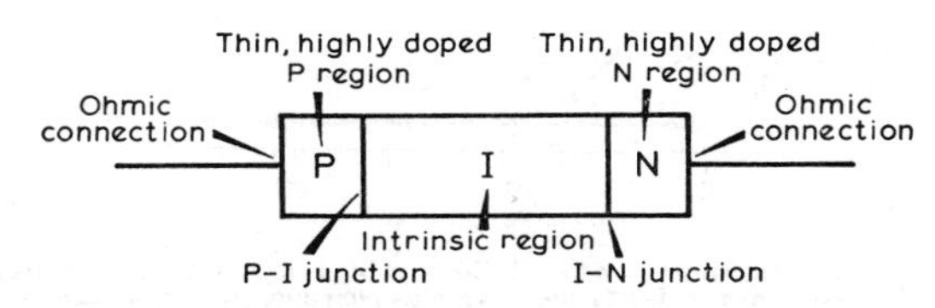

Fig 3.32. The physical form of the P-I-N diode.

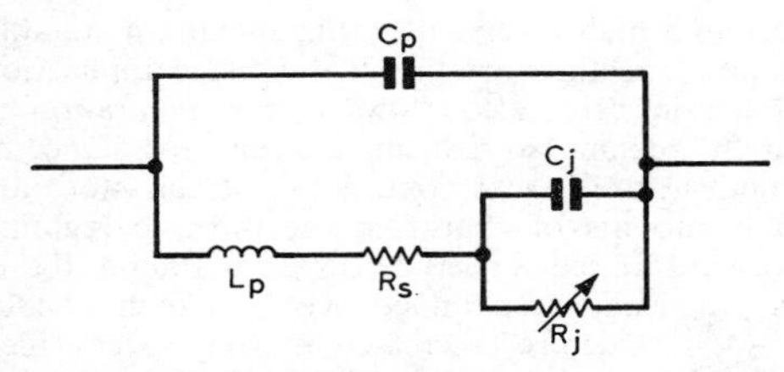

Fig 3.33. Microwave equivalent circuit of the P-I-N diode.

The microwave equivalent circuit of the P-I-N diode is given in Fig 3.33. The inductance, L_p, and the capacitance, C_p, depend very much upon the package containing the diode, and are affected by the external mounting and connexions; usually these parameters are not significant at frequencies below 100MHz. The residual resistance, R_s, is less than an ohm. The junction resistance, R_j, asymptotically approaches a limit of about 10kΩ under zero and reverse bias conditions, and falls to less than 1·0Ω at high forward currents. The junction capacitance, C_j, is bias dependent at low frequencies; typically at 1MHz it varies from 0·7pF at zero bias to 0·05pF at minus 50V. At microwave frequencies C_j is essentially independent of bias and assumes the low frequency, reverse bias value. Fig 3.34 plots typical variations of the rf resistance and capacitance of the P-I-N diode with bias.

P-I-N diodes are used principally as on/off switches or as current controlled attenuators operating in the uhf and microwave portions of the rf spectrum. They are capable of controlling all levels of rf power from microwatts to hundreds of watts, pulsed or cw. At microwave frequencies two or more P-I-N diodes can be integrated into broad-band 50Ω systems that are ideally suited for application as pulse or amplitude modulators, phase shifters in phased aerial arrays, and as T and R switches. Both the on and off switching times are typically only a few tens of nanoseconds; insertion losses are less than 3dB and isolation of 40 to 60dB is readily achieved.

As a modulator or switch, the P-I-N diode is generally mounted as a shunt element across a transmission line or waveguide. At signal frequencies above 100MHz, the diode current remains constant through the signal cycle, and the device presents an essentially constant impedance to the signal even if the diode is reverse biased during an appreciable portion of the cycle.

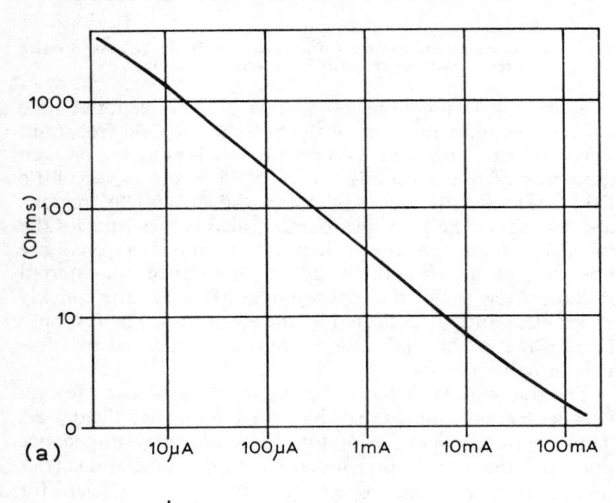

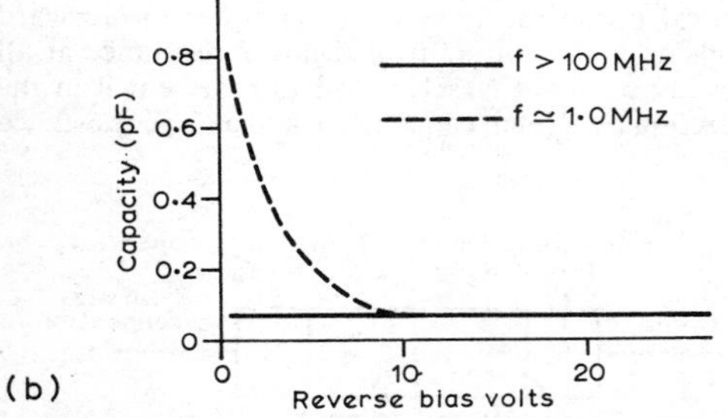

Fig 3.34. RF resistance of a P-I-N diode vs bias current (log-log scale). (a) RF resistance of P-I-N diode vs bias current. (b) Capacitance of the P-I-N diode vs reverse bias voltage.

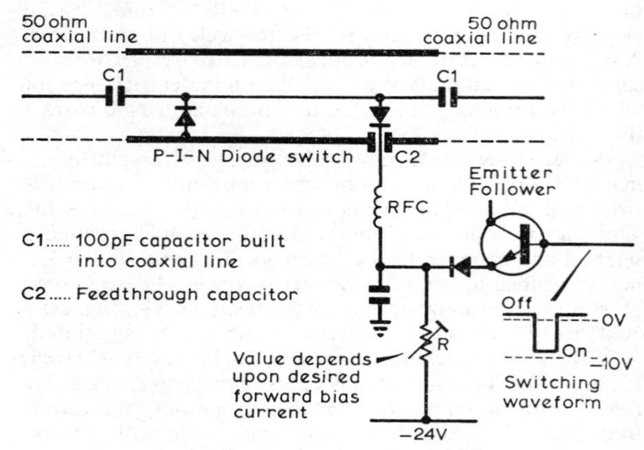

Fig 3.35. The P-I-N diode switch.

The circuit in Fig 3.35 illustrates a P-I-N diode switch connected into a 50Ω transmission line. The switch is considered to be "on", or passing an rf signal when the diodes are zero or reverse biased. Under this condition the diodes are in a high impedance state, and the current through them is typically less than one microamp. Interaction with the normal transmission properties of the line is therefore a minimum, and any noise contribution by the switch is negligible. The switch is "off", or blocking an rf signal when the diodes are conducting. In this condition the diodes are forward biased and are in a low impedance state. The magnitude of the forward bias current determines the degree of attenuation provided by the switch; maximum attenuation, or full isolation being obtained for bias currents of about 100mA.

The Impatt Diode

The impatt diode is one of a number of semiconductor devices that make use of two phenomena; avalanche breakdown and transit time. The name impatt describes the method of operation and is an abbreviation of IMPact Avalanche and Transit Time. As a generator of microwave energy, it is sometimes referred to as the Read oscillator.

The device comprises a P-N junction in which the depletion layer has a doping profile such that a narrow avalanche region and a wider drift region are created within its boundaries. In operation the diode is biased in the reverse direction and the resulting field appearing across the depletion layer has an intensity distribution that reaches a high maximum value immediately to the "n" side of the junction.

This region of high field intensity occurs only over a very thin section; falls away steeply on the "p" side of the junction, and decreases much less rapidly on the "n" side. This is illustrated by the field intensity diagram in **Fig 3.36**.

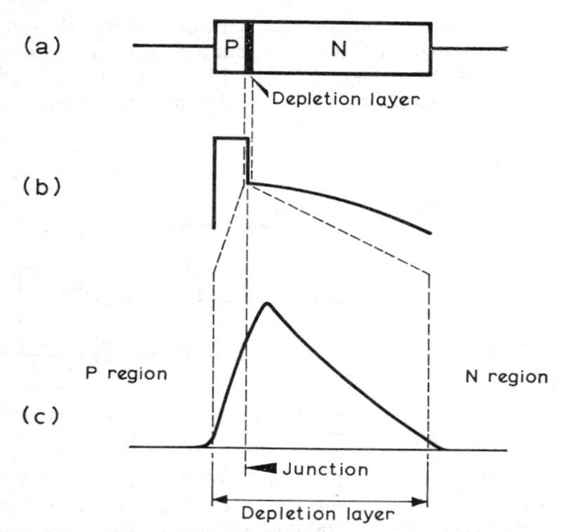

Fig 3.36. (a) The impatt diode. (b) Doping profile of the impatt diode. (c) Field distribution in the depletion layer of the impatt diode.

If the bias voltage applied to the diode is sufficiently large, avalanche multiplication takes place in the region of high field. This mechanism is similar to that which causes the reverse bias breakdown in a normal P-N junction. Over the remainder of the depletion layer, avalanching is inhibited by the lower impurity level in the semiconductor material. The current carriers drift under the influence of the reduced field intensity and reach insufficient velocity to cause multiplication.

When charge carriers are generated by the avalanche mechanism, they enter the drift region and propagate through it to produce a potential across the diode which can be coupled into an external circuit to produce oscillations. It can be shown mathematically that the impatt diode operating in this mode has a total impedance which is equivalent to a small positive resistance, a complex reactance and an active resistance in series. Since the reactance is predominantly capacitative and the active resistance is negative for small transit angles (the transit angle is defined as being $2fT_d\pi$, where f lies within a limited range in the microwave spectrum, and T_d is the carrier transit time through the drift region), it only requires the addition of an external parallel inductance, usually in the form of a cavity, to produce an oscillator.

The impatt diode is essentially a microwave device capable of generating cw powers of the order of a few hundreds of milliwatts; pulsed powers of up to a few tens of watts are possible. The obtainable power decreases with frequency, but quite useful amounts of energy can be generated at frequencies greater than 50GHz.

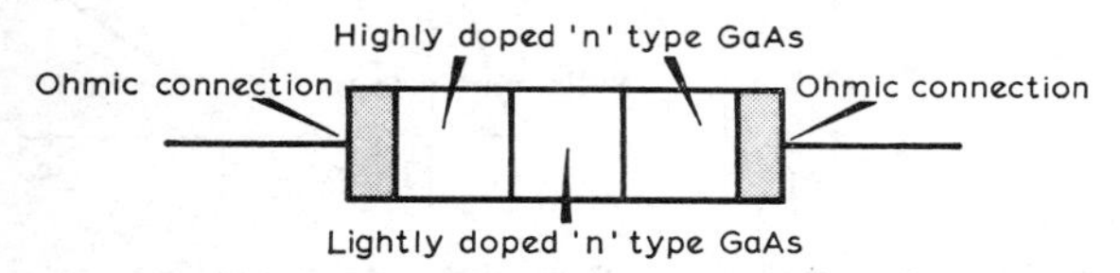

Fig 3.37. The Gunn diode; general form of construction.

The Gunn Diode

All the devices described so far depend for their operation upon phenomena associated with the basic P-N junction. The Gunn diode, however, does not possess any rectifying properties, and it is classified in this section because it is a two terminal device that exhibits negative conductive properties.

Fig 3.37 illustrates the general form of the construction which makes up a Gunn diode. The active part of the device is the centre region of "n"-type semiconductor which is sandwiched between two more highly doped "n"-type regions of similar material. Connections are made to the diode through ohmic contacts attached to the outer "n"-type regions. In operation an electric field is established across the central "n" region by applying a voltage to the ohmic contacts.

As a consequence of this field, the Gunn diode exhibits a negative conductive phenomenon that has a different mechanism from the more familiar negative resistance effects created by other active semiconductor devices. This phenomenon is known as the Gunn Effect. It occurs in the bulk of semiconductor materials which show energy minima in two conduction bands that are separated by an energy gap of a few tenths of a volt.

The distribution of the relevant energy bands within the atom are shown clearly in **Fig 3.38**. E_a and E_b are the minimum energy levels attained by electrons in the lower and upper conduction bands respectively, and E_g is the energy gap between them. One of the most suitable semiconductor materials that shows this effect is gallium arsenide; it is, therefore, the usual material from which Gunn diodes are manufactured.

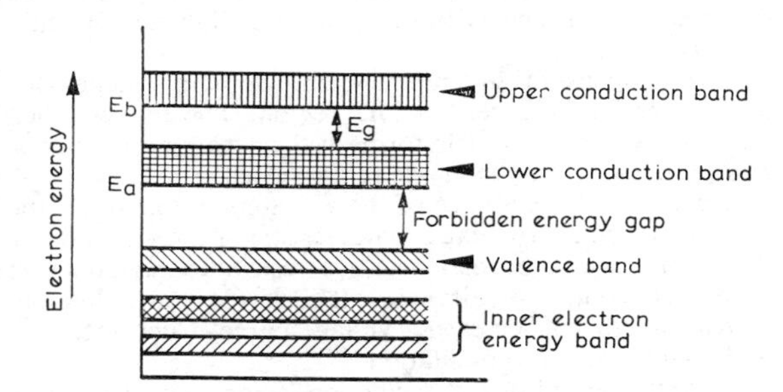

Fig 3.38. Energy band diagram for GaAs atom.

At low fields almost all the free electrons are in the lower conduction band, and the diode exhibits a normal increasing voltage/increasing current characteristic. This corresponds to region A in **Fig 3.39**. As the field is raised above a certain threshold level, more electrons are able to transfer to the higher conduction band, and in doing so are said to become "hot". The effective mass of these electrons is thus increased and as a result there is a significant decrease in their mobility. In gallium arsenide, for example, the ratio of the electron mobility in the lower band to that in the upper band is approximately 60:1.

As the upper conduction band becomes more populated, and the majority of the electrons available for conduction

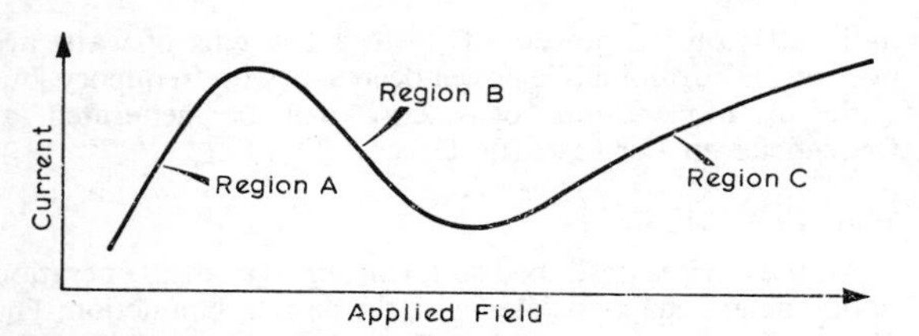

Fig 3.39. Current vs applied field characteristic of a Gunn diode.

becomes less mobile, the resistance of the diode increases. This corresponds to region B in Fig 3.39, where the current falls as the applied voltage raised. At still higher fields, almost all the free electrons are excited to the upper conduction band, and the positive slope of the current/voltage characteristic is re-established, though at a different value. This is region C of Fig 3.39.

The values of field at which the Gunn diode functions occur in region B. The uniform field distribution set up in region A becomes unstable in region B, and space-charge accumulation layers form towards the negative end of the central "n"-type region. The reduction of electron mobility at a point in the bulk semiconductor causes a local increase in the field. This, in turn, excites more electrons to the higher energy level where their mobility is reduced and the field in the vicinity of the disturbance is increased still further. Thus by a cumulative action, a concentration, or "domain", of hot electrons is built up which then propagates through the semiconductor towards the positive end.

The growth of the domain does not go unchecked, however, for since the total voltage across the device is constant, any increase in the field at the positive side of the disturbance must be accompanied by a decrease in field strength at the negative side. Ultimately, the field on the negative side becomes so small that no new excitation of high mobility electrons to low mobility electrons can take place. The disturbance then begins to decay until equilibrium is again restored.

This happens at the moment the domain reaches the positive end of the central "n" region. The process then starts again; a new domain forms at the negative end, grows, and propagates, and the cycle repeats itself.

Because of the existence of these "running domains", the terminal dc current/voltage characteristic of the Gunn diode does not show a normal negative resistance region as in other semiconductor devices. It behaves more like an increasing current/increasing voltage curve which saturates and then breaks into oscillation.

Since the period of this oscillation is approximately equal to the transit time, T, of a domain through the active region of the device, it follows that the frequency of the oscillations, f, is given by:

$$f = \frac{1}{T} \quad \text{or} \quad \frac{v}{d}$$

where v is the drift velocity of the domain, and d is tne path length through the active region.

The drift velocity is about 10^8mm/s, thus for a diode with an active region of 100 to 10 microns, the frequency lies in the 1·0 to 10·0GHz range. Gunn diodes are, therefore, attractive sources of microwave power. CW operation with powers up to at least 100mW and pulsed powers of 100W are attainable with efficiencies as high as several per cent.

A simple microwave transmitter incorporating a Gunn diode is illustrated in Fig 3.40. The device is matched into a length of waveguide that is terminated by a horn aerial. The oscillation is supported with a voltage that is sufficient to cause the domain formation described above.

The Unijunction or Double-base Diode

This is another semiconductor device which exhibits a negative slope in its characteristic, but unlike the tunnel diode, it operates at slower speeds and with applied voltages that are more compatible with other semiconductor devices.

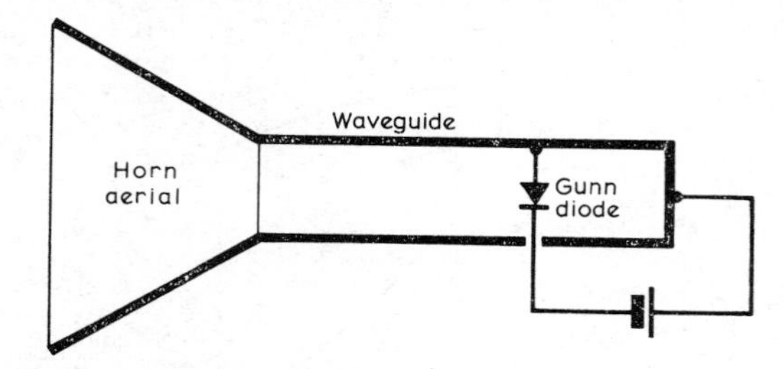

Fig 3.40. Schematic diagram of a simple Gunn diode microwave transmitter.

The construction and symbol of the unijunction are illustrated in Fig 3.41(a) and (b). It consists of a bar of lightly doped "n"-type silicon with ohmic (non-rectifying) electrodes welded to each end. A short distance along the length of the bar, a P-N junction is formed which is referred to as the "emitter". The ohmic electrodes are termed "bases", although their functions are not analogous to that of the base of a bipolar transistor.

In use, a voltage V_{bb} is applied across these electrodes such that base$_2$ is the more positive, Fig 3.41(c), and a small current is allowed to flow through the bar. A fraction of the applied voltage, $\left(\frac{R_1}{R_1 + R_2}\right)V_{bb}$, will appear at the emitter

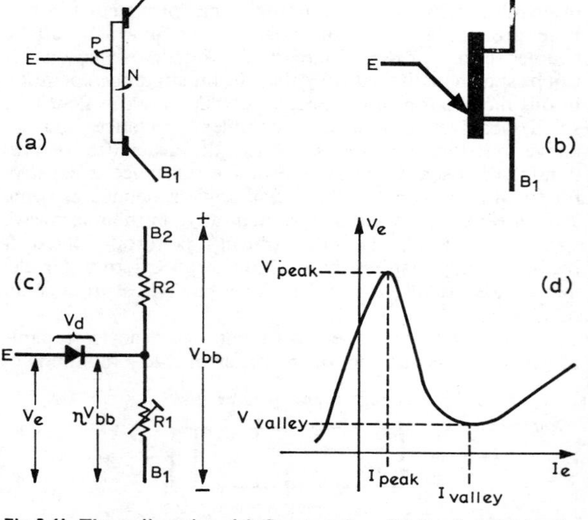

Fig 3.41. The unijunction. (a) Construction. (b) Symbol. (c) Equivalent circuit. (d) Characteristic.

junction, and if the emitter is returned to a voltage lower than this fraction of V_{bb}, it will be reversed biased and no current will flow.

The fraction, $\frac{R_1}{R_1 + R_2}$, is termed the intrinsic stand-off ratio, η, thus when the emitter is made more positive than ηV_{bb}, a current will flow to the lower electrode, base$_1$. The emitter-base$_1$ voltage, V_e, under these conditions, is therefore given by

$$V_e = \eta V_{bb} + I_e \frac{R_1 R_2}{R_1 + R_2} + V_d$$

where I_e is the emitter current; R_1 and R_2 the resistance of the "n" region from emitter to base$_1$ and base$_2$ respectively; and V_d is the normal forward junction diode voltage drop.

When the emitter junction is biased in the forward direction, holes are injected into the "n" region and subsequently swept down to the base$_1$ electrode. The density of holes in the "n" region is proportional to the emitter current, I_e, hence, as the space-charge neutrality is maintained at all times, the conductivity of this section of the bar is correspondingly increased (ie, R_1 decreases).

The fraction of V_{bb} across the lower part of the bar therefore decreases, biasing the junction further in the forward direction, and injecting more holes into the "n" region. This action is cumulative and results in the device exhibiting the increasing current/falling voltage, or negative slope, characteristic shown in **Fig 3.41(d)**.

It is interesting to compare the characteristics of the unijunction with those of the tunnel diode. Whereas the latter is essentially a current operated device: ie the voltage across the diode is a function of the current; the unijunction is voltage operated. Initially presenting a high impedance to any current flow, the resistivity of the emitter-base$_1$ section of the device falls to a low value, and the emitter current rises rapidly once the applied voltage exceeds the critical potential at which the emitter junction becomes forward biased.

The operating speed of the unijunction is considerably slower than that of the tunnel diode, and fall times of between one and 10μs are typical. The range of devices currently available have "trigger" voltages between 3 and 20V, and are able to pass emitter currents up to about 50mA.

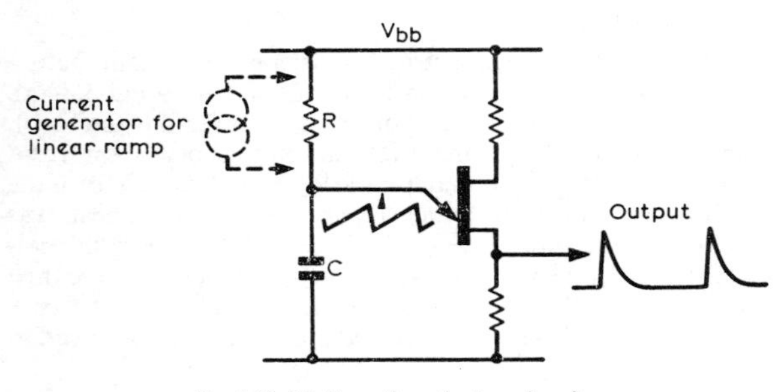

Fig 3.42. Unijunction timing circuit.

The basic application of the unijunction is that of a voltage sensing device, and uses in the purely communications field are few. In association with a simple RC network, it can form a useful timing circuit, or ramp generator, as shown in **Fig 3.42**.

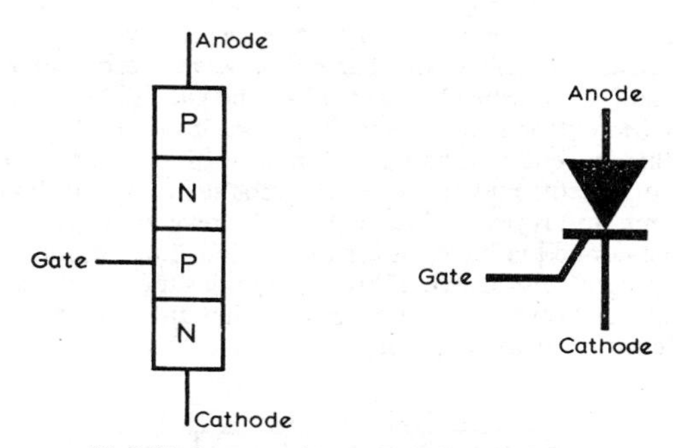

Fig 3.43. Construction and symbol of the scr.

The Silicon Controlled Rectifier

The scr, or thyristor, is essentially a four layer P-N-P-N device which can exhibit both a high and low resistance condition in its forward voltage/current characteristic. Shown diagrammatically and symbolically in **Fig 3.43**, it will be seen that there are three electrodes, which by analogy with the gas filled thyratron are termed anode, cathode and gate (triggering electrode).

Unlike the thyratron, however, which requires a voltage pulse to trigger it from the high resistance state to the low, the scr requires a current (or more correctly, a charge) to be injected into the control electrode to achieve the change of state.

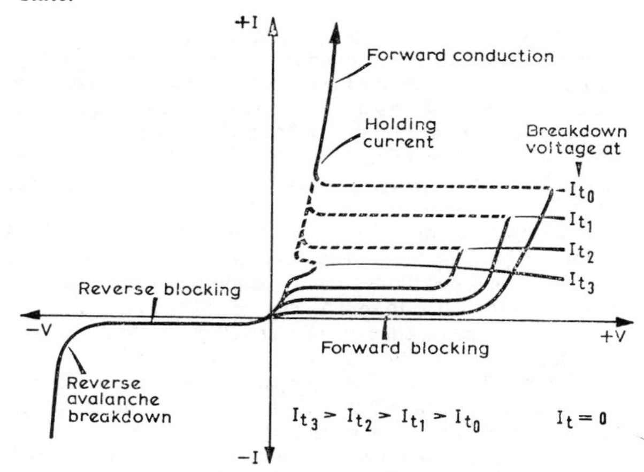

Fig 3.44. Typical characteristics of an scr.

The voltage/current characteristic of the device in the reverse direction is similar to that of a normal single P-N junction; but in the forward direction no current flows unless the anode voltage is raised above a critical value. This figure is itself a function of the current flowing into the gate electrode.

Above the critical voltage, the forward resistance of the scr falls rapidly to a low value, and the forward characteristic becomes similar to that of a normal junction diode. These characteristics are illustrated in **Fig 3.44**. Within the limits set by the voltages at which forward and reverse breakdown occurs, the preferred mode of operation is that of a normal silicon junction diode in which conduction in the forward direction can be initiated by injecting a current or charge into the gate or control electrode.

As with the thyratron, once the scr is triggered into the low resistance state, the current flow between anode and cathode is not subject to control by the gate. The device can only be returned to its high resistance condition by cutting off this current externally. In practice, this implies a reduction of the current to a value below the sustaining, or holding, current, and is preferably achieved by reversing the polarity of the applied voltage for a short period. In ac applications, this presents no problem, but where the supply is dc, special circuits have to be used to ensure that the current can be switched off prior to retriggering.

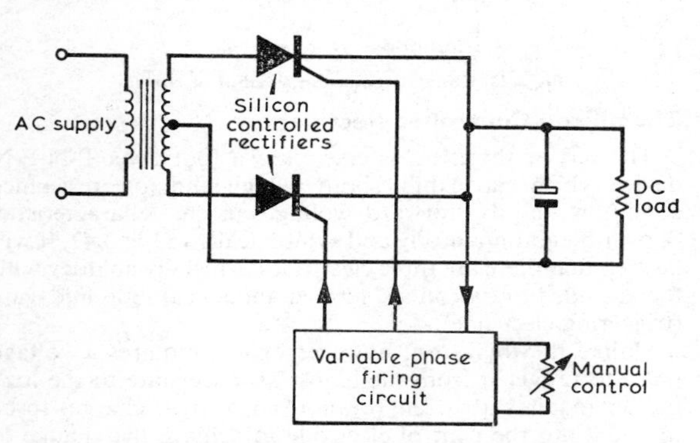

Fig 3.45. Phase controlled dc supply using SCRs.

Because the forward voltage drop across an scr is comparable with that of a simple junction diode, very high currents can be handled without excessive dissipation. Devices capable of passing currents of a few hundreds of amperes are not uncommon, and as it is possible to manufacture the scr with a thick base region, breakdown voltages, referred to zero triggering current, in excess of 1000V are possible.

The large area of the P-N junctions, and the thick base regions in the scr, made necessary by the high current and voltage ratings, inevitably reduce the speed of the device. It is therefore quite usual to find a delay of 250ns to 2·5μs occurs between the leading edge of the triggering waveform and the main anode/cathode breakdown.

Furthermore the scr has inherent limitations in the rate at which the forward current may be allowed to build up and at which the forward voltage may be re-applied after the main current has been switched off by the removal or reversal of the anode voltage. These limitations vary according to the particular scr, and in order to avoid the possibility of permanent damage, it is advisable to consult the manufacturer's data sheet before selecting an scr for a particular application.

The scr finds common usage in power applications such as might be required in motor speed control, or in high current switching by electronic means. SCRs may also be used in high power invertors. A development of the scr which exhibits a forward characteristic to both positive and negative applied voltages is known as the triac.

The circuit in **Fig 3.45** illustrates the use of silicon controlled rectifiers in a typical single phase, full-wave application suitable for such purposes as battery charging and dc motor control. By replacing the manual control by voltage feedback from across the load, the circuit can be extended to make a voltage regulated supply.

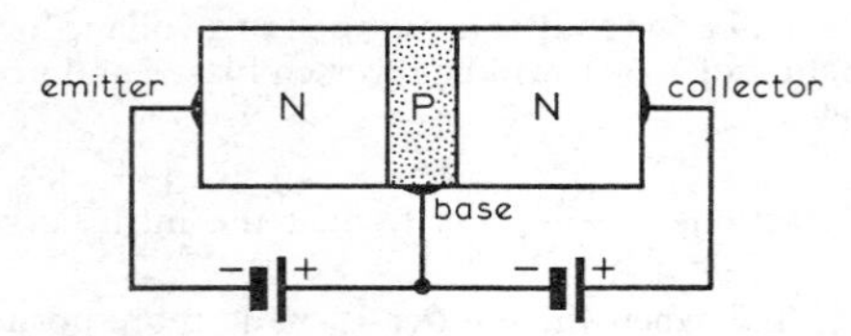

Fig 3.46. N-P-N sandwich formed by two P-N junctions mounted back-to-back.

THE BIPOLAR TRANSISTOR

The bipolar junction transistor is basically two P–N junctions, of the type described earlier, mounted back to back as shown in **Fig 3.46.** This structure may be in the form of either a N–P–N or P–N–P sandwich, and both types operate in essentially the same way, except that the holes and electrons as majority carriers are interchanged. The centre region is always very much thinner than the outer two and is termed the "base", while the outer two are termed "emitter" and "collector". It is interesting to note that the bipolar transistor was invented by Bardeen and Brattain in 1948.

N-P-N Transistor Operation

The collector-base junction is reverse biased and a thin depletion layer is produced at this junction. Any free electrons near this depletion region are drawn into the collector as a result of the applied voltage and hence negligible current can flow. The collector junction therefore acts as a high resistance.

The emitter-base junction is forward biased and thus electrons are injected into the base from the emitter as majority carriers. However, because the base region is very thin, most of these injected electrons will diffuse across the base and be drawn into the collector region as minority carriers to produce the collector current. This process is very nearly independent of the collector voltage provided it is greater than a few times KT/e (K = Boltzman's constant, T = temperature (°A) and e = the charge on an electron), this being approximately 26mV at room temperature. The emitter to-base circuit has therefore a low resistance—that of a conducting diode.

The standard symbols for N–P–N and P–N–P junction transistors are given in **Fig 3.47.**

Construction

The most common construction used in early manufacture resulted in the alloy junction transistor shown in **Fig 3.48(a).** In this process, a thin wafer of "n" type material (the base), had a bead of "p" type material (the emitter) placed on it and melted so that the two materials alloyed together to form the emitter-base junction. Similarly, another "p" type bead was melted on the other side of the base wafer to form the collector and complete the P–N–P structure. Contacts were then welded on to the collector and emitter beads, and the base connection made via a ring welded around the emitter bead to

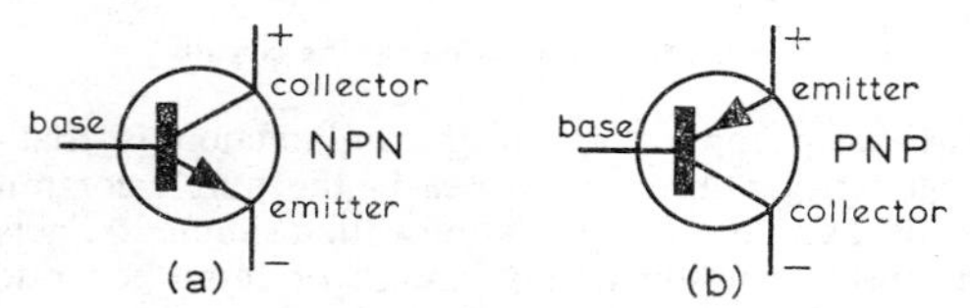

Fig 3.47. Transistor symbols.

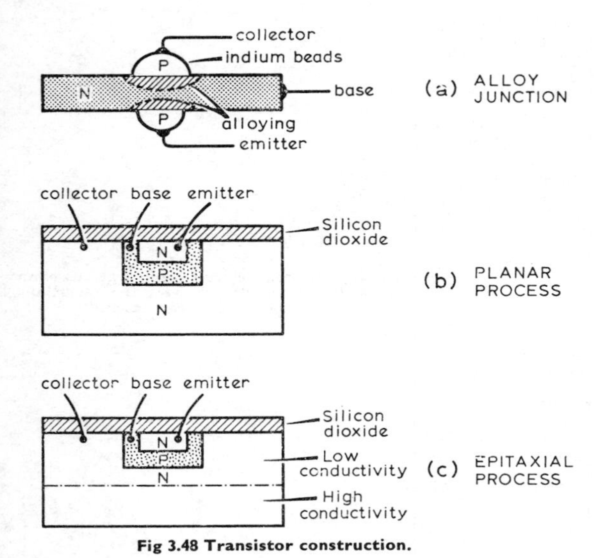

Fig 3.48 Transistor construction.

reduce the internal base resistance. Due to the difficulty in controlling precisely the alloying process, this construction was only suitable for lf transistors. Variations and improvements on this method of fabrication included the grown junction, diffused and mesa transistors, but all of these types are now virtually obsolete and the majority of present-day transistors are made by the planar and epitaxial processes.

The planar construction is shown diagrammatically in cross section in **Fig 3.48(b)** and is produced as follows. A section of lightly-doped "n" material is coated with a passive layer of silicon dioxide by heating it in oxygen. Part of this layer is then etched away and a "p" material is diffused in to form the base, and the surface repassivated. Part of the surface over the "p" region is again etched away and a second "n" impurity is added to form the emitter. Finally the surface is recoated with silicon dioxide.

The planar device does suffer one disadvantage in that the collector region must be lightly doped, and hence its conductivity is low. Thus a relatively high saturation resistance results—ie an appreciable voltage drop occurs across the collector material, which in turn means that the ht voltage cannot be used to the full because appreciable power will be dissipated in the device.

This limitation of the planar device is overcome by the addition of another step in the process known as epitaxy, which is shown in **Fig 3.84(c)**. Basically the process is as described for the planar transistor, except that the "n" material forming the collector consists of a very thin layer of lightly doped material which has been deposited on to a bulk "n" material of high conductivity. Thus the expitaxial planar transistor is capable of exhibiting a low collector saturation resistance and, at the same time, a high collector-base breakdown voltage. It is also readily made with values of f_t (gain/bandwidth product) well in excess of 1GHz. These advantages have made the planar and planar-expitaxial processes the almost-universal method of manufacture in use today for both discrete transistors and complex monolithic integrated circuits.

The construction of high-power planar rf transistors, however, requires a more stringent chip geometry. First, as the frequency of operation is increased, the current tends to flow from the edges of the emitter contact rather than uniformly over the surface and, second, lead inductances, stray capacitances and transit time become significant. To eliminate the effect of this current-crowding effect at the emitter edge, high periphery-to-area ratio is required of the emitter. This may be achieved in two ways—the interdigitated construction in which the emitter is constructed in a comb shape, or the multi-emitter version which consists of many small emitter junctions in parallel. Typical examples of the multi-emitter construction are the 2N3866 (16 emitters) and 2N3632 (312 emitters).

Transistor Parameters

In the previous paragraph, the injection of electrons into the base from the emitter was described. If now the unity current flowing in the emitter is considered, then the resultant current that flows in the collector (about 98 per cent of this input) is α, which means that $(1-\alpha)$ must have flowed through the base terminal as base current. This may be defined mathematically as $I_b = I_e - I_c$ and the base-to-collector current, β or h_{fe}, as $\alpha/(1-\alpha)$. Typical values of β range from 30 for power transistors carrying 15A to 500 for some small-signal devices operating at a few milliamps. Thus it can be seen that if the input current is fed to the base rather than the emitter, the transistor can provide a large current gain and also a high power gain.

A typical characteristic curve for a N–P–N transistor in the grounded-emitter configuration is shown in **Fig 3.49**, and it can be seen that the curves are very similar to those of the pentode valve. Two important points: the collector current I_c is plotted against input base *current* and not voltage, and the "knee voltage" is below 300mV which enables the full ht voltage to develop across the transistor load.

So far only the dc operation of the transistor has been dealt with, but there is obviously a limit to the frequency at which a given device may be used. The main limitation to this frequency arises from the mean time required for the electrons to cross the base width—the transit time t. For a current i flowing through the base there is a stored charge q, given by $q = i \times t$. This charge is of course neutralized by an equal charge of the excess holes in the base region. As the current i changes, there must be a change in the stored charge, and this implies that there is a storage capacitor associated with the movement of electrons from emitter to collector. To

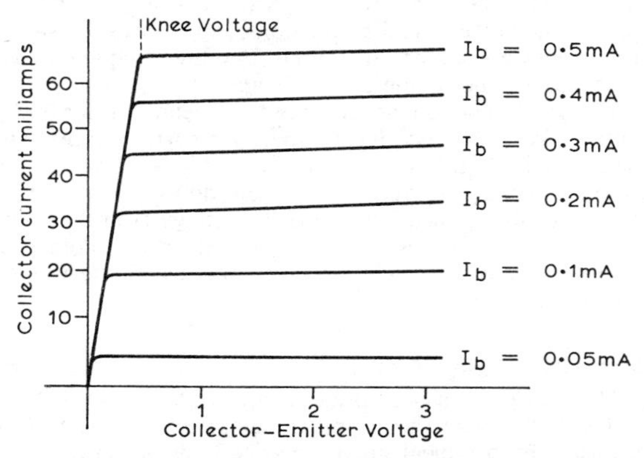

Fig 3.49. Typical collector characteristics for an N-P-N transistor in the common-emitter mode.

Configuration	Input resistance	Output resistance	Current gain
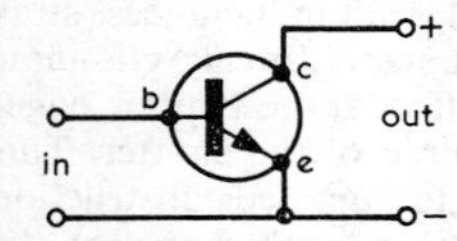 COMMON EMITTER	$(\beta \times r_e)+r_b$ approx 1·4kΩ	Slope of V_c/I_c curve approx 10kΩ to 100kΩ	Varies with types approx 30 to 200
	The common emitter amplifier has the highest power gain of the three configurations		
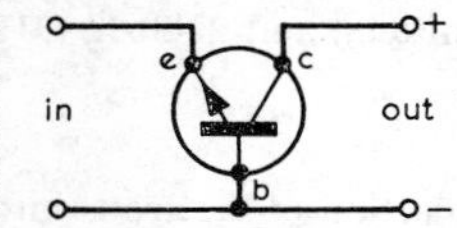 COMMON BASE	$r_e = \frac{26}{I_c} + \frac{r_b}{1+h_{fe}}$ approx 13Ω for I_c=2mA	Very high approx 1MΩ or greater	0·98
	The common base amplifier has lower power gain than the common emitter stage but the gain does not fall-off so rapidly at the higher frequencies		
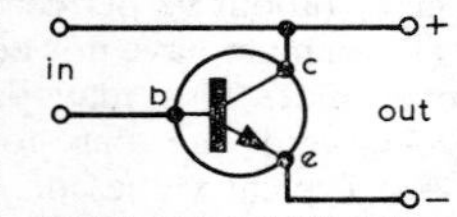 COMMON COLLECTOR (Emitter Follower)	$\beta \times (r_e+R_L)+r_b$ approx 100kΩ for R_L = 1kΩ	$r_e = \frac{26}{I_c}$ approx 13Ω for I_c = 2mA	Same as common emitter approx 30 to 200

Fig 3.50. Basic circuit configurations of an N-P-N transistor showing the variations between parameters.

charge and discharge this capacitance, a base current is required; the higher the frequency and the longer the transit time, the greater the base current required.

This means that, for a given collector current, the current gain of the transistor falls as the frequency of operation is increased. Two ways of defining this effect can be used: (i) the alpha cut-off is that frequency at which α has fallen to $1/\sqrt{2}$ (70 per cent) of its low-frequency value and is denoted by f_α (or in a few instances f_c): (ii) f_1 is that frequency at which the base-to-collector current gain has dropped to 1.

A third alternative often used is f_t, or gain/bandwidth product $\beta' f'$, where β' is the base-to-collector current gain (h_{fe}) at f', this being the frequency above that where the active base-emitter resistance ceases to be important.

Of these three definitions, f_α is always the highest frequency and hence there may be an advantage in using the transistor in the grounded-base configuration when operating near its frequency limit.

As stated earlier, although the base width is normally very small, an internal *base* spreading resistance $r_{bb'}$ is present. This resistance, together with the storage and depletion capacitances of the junctions, introduces unwanted time constants which also limit the high-frequency performance and noise figure of the transistor.

The P–N junction as described in detail earlier has a finite leakage current due to majority carriers only. This manifests itself in the junction transistor as a very small collector leakage current I_{ceo} which flows without input current being applied. Values of I_{ceo} range from less than 1nA in silicon devices to 1mA in certain germanium transistors. As a general rule, however, the leakage in modern silicon transistors may be ignored.

Circuit Configurations

The transistor may be used in three basic ways, as shown in **Fig 3.50**, where the various input and output resistances are listed for a typical small-signal N–P–N silicon transistor operating in the 2mA collector current region.

The internal emitter resistance r_e is that of the conventional P–N junction and is equal to (KT/e) × (1/I_c) which at room temperature approximates to 26/I_c Ω where I_c is in milliamps. The internal base resistance r_b is a constant value and is normally between 30 and 100Ω.

Practical Design of a Standard Circuit

A practical circuit in the common-emitter configuration, as shown in **Fig 3.51**, will now be considered. This is applicable to most of the low-level stages used in communication and low-frequency amplifiers. The arrows indicate the direction of *conventional* current flow and *not* electron flow.

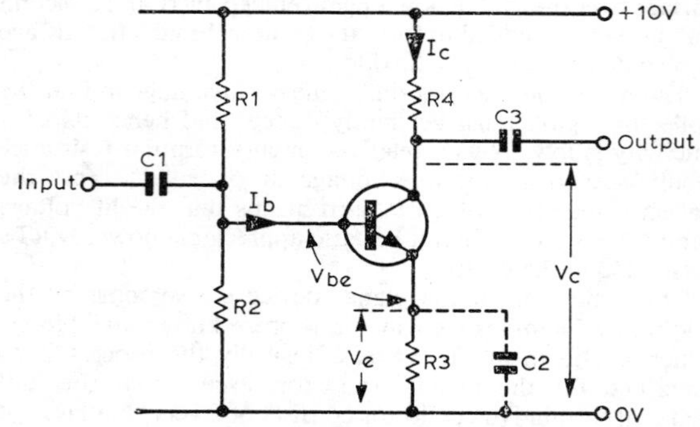

Fig 3.51. The common-emitter amplifier.

The base of the transistor is fed from a potential divider network R1 and R2 such that the current through R1 and R2 is much greater than the base current I_b, and hence the voltage at the base is held constant for small changes in base current. A typical value of bleed current is 8–20 times the base current.

The following data for an N–P–N silicon transistor will be assumed. At I_c = 1mA, h_{FE} = 100 minimum and ht available = 10V. Hence the base current will be $\frac{1}{100}$ mA and the current through R1 + R2 will be $\frac{1}{10}$ mA.

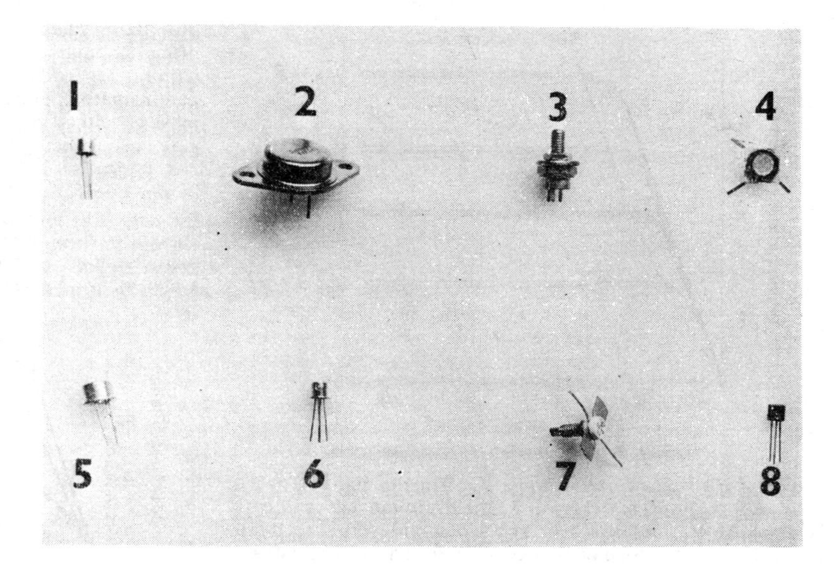

A selection of typical transistors. 1: small signal uhf bipolar. 2: audio power bipolar. 3: vhf power bipolar. 4: uhf power bipolar. 5: general purpose bipolar. 6: zener-protected dual-gate mosfet. 7: microwave power bipolar. 8: vhf jfet.

It is normal to take an arbitrary value of V_e of between $\frac{1}{10}$ and $\frac{1}{5}$ of the supply voltage for stability reasons. Should the collector current increase for any reason (remember collector leakage doubles with every 10°C rise in temperature), then V_e will increase while the base potential remains constant. Hence this will tend to reduce V_{be} and offset this collector current increase.

If V_e is taken as 1·5V the value of R3 is $1{\cdot}5/I_c = 1{\cdot}5\text{k}\Omega$ and the potential at the base will be 1·5V + V_{be}. For silicon transistors V_{be} is 0·7V, giving a base potential of 2·2V. V_{be} for germanium transistors is 0·25V.

Hence $R2 = 2{\cdot}2\text{V}/\frac{1}{10}\text{mA} = 22\text{k}\Omega$ and $R1 = (10 - 2{\cdot}2\text{V}) \div \frac{1}{10}\text{mA} = 78\text{k}\Omega$.

It must be remembered that these two resistances are both effectively shunting the input and will therefore affect the loading of the previous stage.

The choice of collector load will depend upon the application of the amplifier, but for the resistive load R4 shown, a value may be taken such that V_c is about $\frac{1}{2}(V_{ht} + V_e)$, 5·75V in this case, to give maximum output voltage swing without distortion. R4 is then given by 5·75V/1mA = 5·75kΩ and in practice the nearest standard value of 5·6kΩ would be used.

The voltage gain that can be expected from this amplifier will be $h_{fe} \times (R4/R_{in})$ where R_{in} from **Fig 3.50** is approximately 2·6kΩ (this assumes a value of C2 is used such that its reactance at the lowest input frequency is small compared with r_e). This gives the minimum voltage gain of the amplifier as 215.

Although the *voltage* gain of this resistive load amplifier has been calculated it must be remembered at all times that the transistor is a current-operated device.

The emitter follower is a special case of voltage amplifier with a gain of almost unity, but with an impedance transformation from a high value at the input to a low value at the output. It must be pointed out, however, that if a 1V peak output is expected at the emitter into a 50Ω load, then the standing dc emitter current must be greater than $\frac{1}{50} \times$ 1,000mA, ie 20mA.

THE FIELD-EFFECT TRANSISTOR

The Junction FET

The field-effect transistor was invented in 1928, well before the bipolar transistor, but was not actually produced until about 1960 due to manufacturing difficulties.

Basically the junction fet (jfet) consists of a bar of doped silicon which, when a voltage is applied across it, acts as a resistor, ie the current flowing through the bar is related to the applied voltage and the resistance of that bar. The terminal from which the electron current flows is termed the "source" and the terminal into which it flows is called the "drain". Consider the case of an "n" type bar with two "p" type regions diffused in as shown in **Fig 3.52** but leaving a channel between the two regions. These two regions are normally joined together electrically and are termed the "gate". The P–N junctions formed by the gate behave in exactly the way described on p 3.4, in that a depletion layer is formed around the junction when a reverse bias is applied.

If the bias is sufficiently high, the depletion layers will meet across the bar and increasing the drain-source voltage V_{DS} will have little effect on the drain current flowing. The drain current (I_D) flow in the channel also sets up a reverse bias across the gate surface which will increase with increasing drain-source voltage. This reinforces the effect of the bias, so that the depletion regions meet. The drain-source potential that causes the depletion layers to meet is

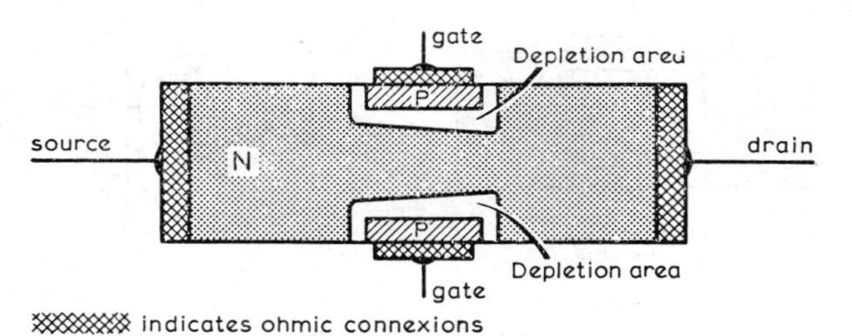

Fig 3.52. FET construction.

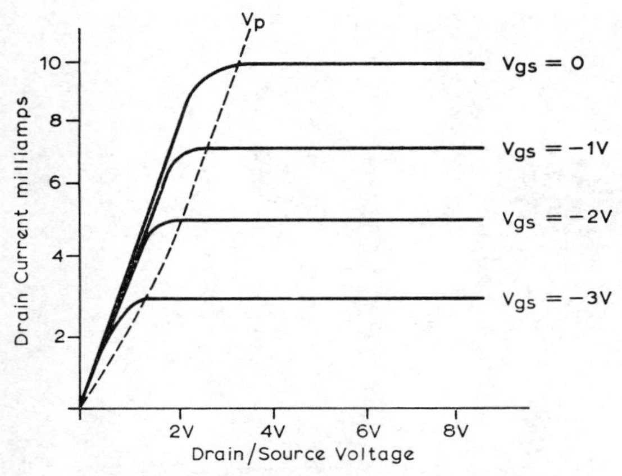

Fig 3.53. Typical junction fet characteristics.

termed the "pinch off" voltage V_P. Thus in **Fig 3.53** it can be seen that at low values of I_D the drain current is linearly related to V_{DS}. As, however, the depletion layers spread into the channel the slope of I_D decreases until V_P is reached, and the drain current saturates and stays relatively constant. Avalanche breakdown will occur beyond the normal operating range of V_{DS}: in this case I_D rises very steeply with V_{DS} and damage to the device will result. The standard symbols for both P- and N-channel devices are shown in **Fig 3.54.**

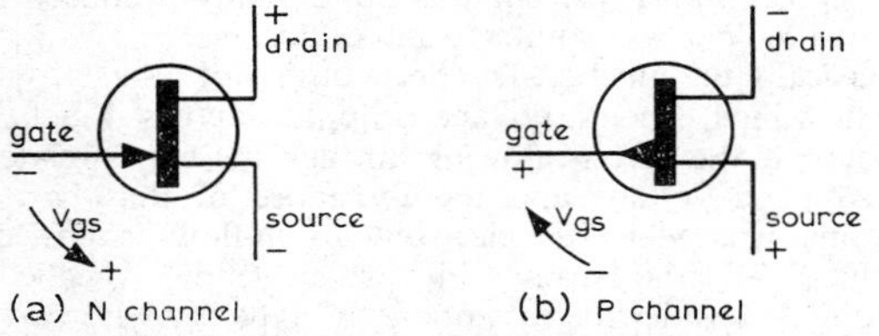

Fig 3.54. Junction fet symbols and polarities.

The Insulated-gate FET

The igfet differs from the junction fet in that there is no actual semiconductor junction between the gate and the drain-source channel, although gate control of the channel current is still achieved. The construction of the igfet is shown diagrammatically in **Fig 3.55(a)** and the standard symbol is shown in **Fig 3.55(b)**. The device consists of a conventional doped channel between source and drain which may be either "n" or "p" type material, though more commonly "n" type. The gate is normally a metallic electrode and is insulated from the channel by a thin layer of silicon

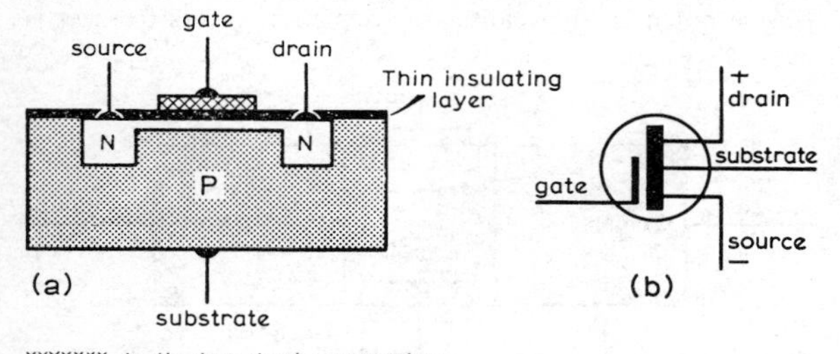

Fig 3.55. (a) N-channel igfet construction; (b) symbol and polarities for an N-channel depletion mode igfet.

dioxide or silicon nitride. (*Note:* Insulated-gate devices are often popularly called MOSFETS (metal-oxide-silicon fet) without regard to whether oxide or nitride is used as the gate insulator.) Thus, because no junction is produced, the gate may be either positive or negative to the source without gate current being produced.

A typical set of characteristic curves for an igfet is shown in **Fig 3.56** for an N-channel device, and it can be seen that the gate bias voltage may be positive or negative for drain current to flow. The curves for $V_{gs} = 0$ and negative values are analogous to the junction fet. This type of device, in which I_D flows when $V_{gs} = 0$, is known as a *depletion-mode* igfet.

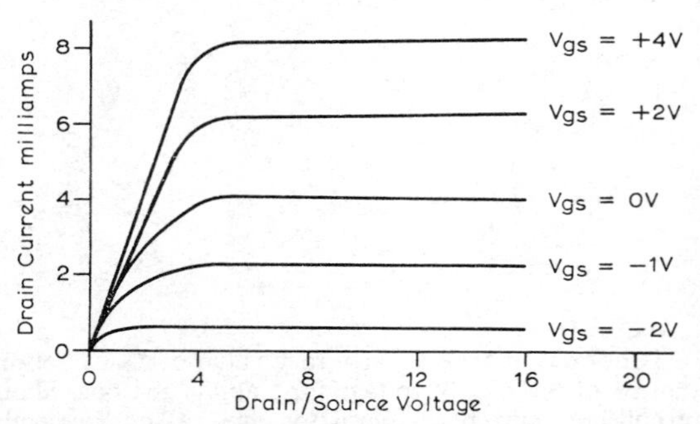

Fig 3.56. Typical N-channel depletion-mode igfet characteristics.

It is also possible to construct an insulated-gate device in which *no* drain current flows when $V_{gs} \leqslant 0$. Such a fet is known as an *enhancement-mode* fet and is shown in **Fig 3.57.** Two "p" type regions are diffused into an "n" type silicon chip or substrate to form source and drain regions such that a P-N junction exists between each region and the substrate. Obviously no drain-source current can normally flow, as one of these P-N junctions will always be reverse biased, irrespective of the drain-source polarity. If a voltage, negative to the source and substrate, is now applied to the gate, a field will be set up below the gate which will attract holes, thus linking the source and drain regions by a hole conduction path. Thus the more negative V_{gs} is made, the more the hole conduction between source and drain is enhanced. This condition is shown in **Fig 3.58** and typical curves are shown in **Fig 3.59.** Insulated-gate FETS usually have gate leakage currents of only a few picoamperes.

This P-channel enhancement-mode igfet is the basic element in most large-scale integrated circuits.

The igfet may also be manufactured with two independent insulated gate regions on the one substrate. Normally, only the depletion mode fet is utilized for this dual-gate construction. Either gate may be used to produce the typical

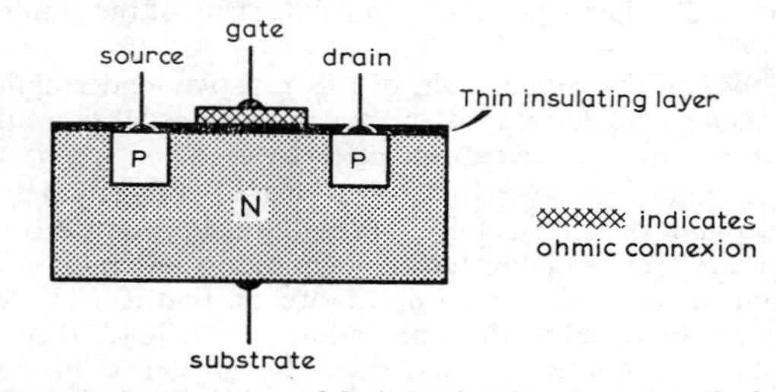

Fig 3.57. Construction of P-channel enhancement-mode igfet.

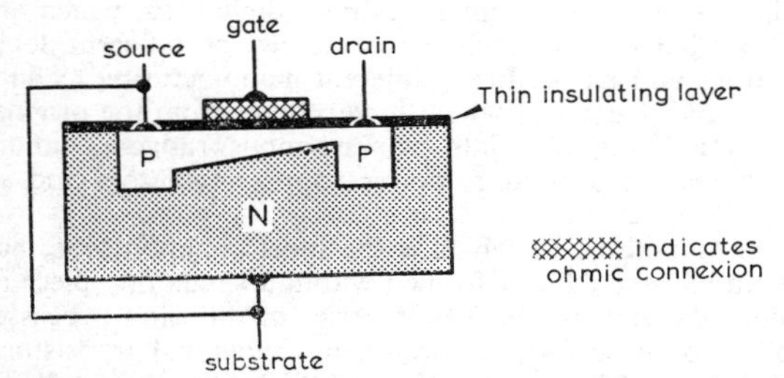

Fig 3.58. Enhancement-mode igfet with Vgs greater than Vds, showing the "p" region extended between drain and source to provide a conduction path.

curves shown in Fig 3.56, provided that the other gate is held at a slightly more positive bias voltage than the source.

FET Parameters

The majority of amateur applications of the jfet require the transistor to be operated in the saturated or "pinched off" region, where the drain current I_D is determined by V_{gs} and is virtually independent of V_{DS} (Fig 3.53). The main parameter is therefore the transconductance curve of I_D against V_{gs}, from which g_{fs} (the equivalent of g_m for a pentode valve) may be deduced. Also, this curve enables the static bias conditions of a circuit to be determined. **Fig 3.60** shows a typical transconductance curve for an N-channel fet, where the curve follows the formula

$$I_D = I_{DSS}(1 - \frac{V_{gs}}{V_p})^2$$

ie it follows a square law. (I_{DSS} is defined as that drain current flowing at zero gate-source voltage).

In small-signal, high-frequency applications the fet has two major advantages over the bipolar device: (i) the square-law characteristic which gives an improvement of at least 10 times in input-voltage handling ability, with the consequent reduction in the generation of spurious signals; (ii) the change of operating point with agc action has a much reduced detuning effect on the input and output circuits.

The most obvious of all fet applications is the high input impedance amplifier stage working in the range dc to 100MHz. A typical simple circuit is shown in **Fig 3.61** for an N-channel device. For the circuit to be repeatable, the value of I_{DSS} for the fet must be known. The value of I_D must be chosen to be below this value so that the device dissipation is not exceeded. For example, if I_{DSS}(min) = 4mA, ie worst case, then a value of I_D = 3mA may be used. (As g_{fs} is proportional to I_D a high value of I_D should be chosen.)

The value of R_g should not be so high that the product of

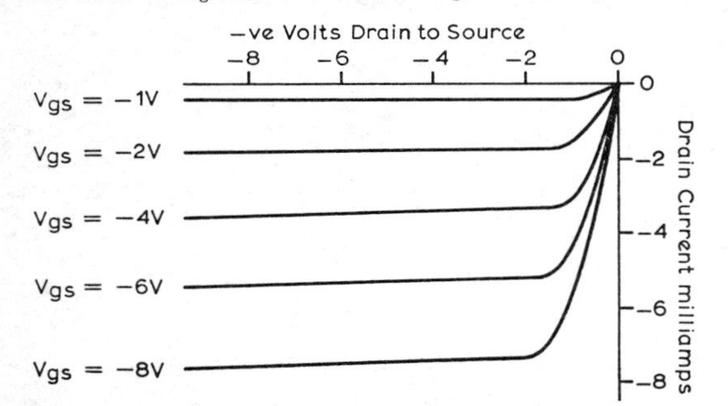

Fig 3.59. Typical P-channel enhancement-mode igfet characteristics.

R_g and gate leakage (normally less than 10^{-9}A) becomes significant.

The value of V_{gs} for I_D = 3mA is taken from the data sheet and R_s calculated from $R_s = V_{gs}/I_D$. R_s may, however, have to be found empirically if an exact value of I_D is required.

If it is assumed that a minimum of 4V is required across the fet at all times for linearity, then (V_{ht}–4)V ie 11V is available for the peak-to-peak drain voltage swing, and V_D should "sit" at $(4 + \frac{11}{2})$V or +9·5V. R_L is then given by $(15-9{\cdot}5)I_D$ or 1·8kΩ to the nearest preferred value.

The gain of such an amplifier is approximately equal to the product of R_L and g_{fs} provided that C_s fully decouples the source. To ensure this latter condition is met, the reactance of C_s must be at least 10 times less than $1/g_{fs}$ at the lowest operating frequency.

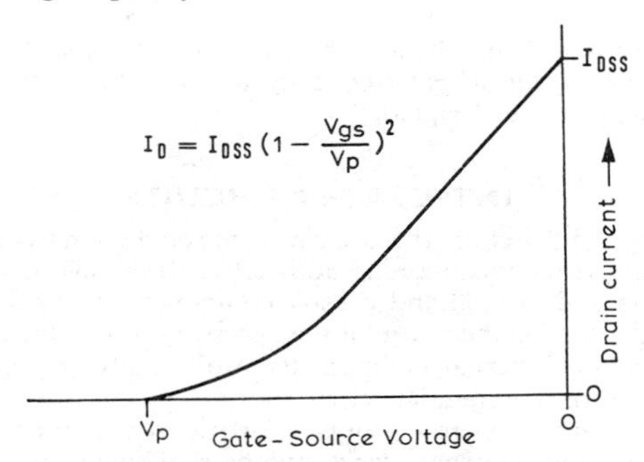

Fig 3.60. Transconductance curve for an N-channel igfet.

The dual-gate fet is widely used as a small-signal amplifier or mixer for frequencies up to 500MHz. A typical circuit is shown in **Fig 3.62**. The potential divider chain comprising R1R2 is used to forward bias gate 2 by between 1V and 3V —the actual value depending on the device used. In some cases it may be necessary to forward bias gate 1 also.

The circuit may be used as an unneutralized rf amplifier with the input applied to gate 1 under the following conditions: (i) that the impedance from the signal source is about 1kΩ at gate 1, and (ii) that gate 2 and source are decoupled at the signal frequency. This implies that R3 should have a value of at least 10kΩ to avoid reducing the wanted signal.

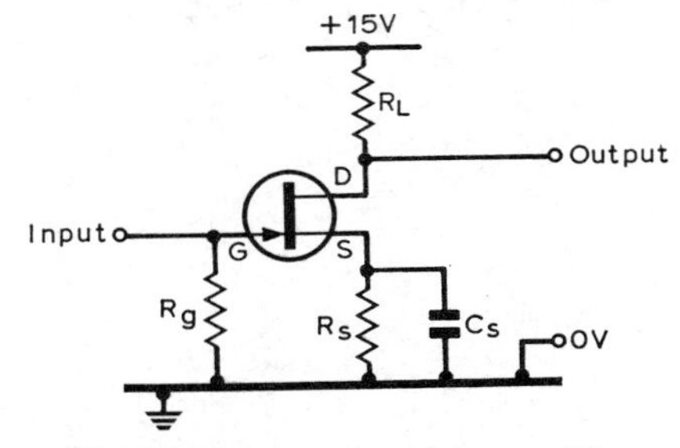

Fig 3.61. High input impedance amplifier.

To preserve a high Q value for the tank circuit, the drain may have to be tapped into L. (The typical output resistance of the 3N200 is only 6·5kΩ at 100MHz.)

The circuit shown in Fig 3.62 may also be used as a mixer. In this condition, gate 1 will always require the smaller signal of up to 300mV rms and gate 2 an oscillator signal of about 1V rms. Because of the device characteristics, the

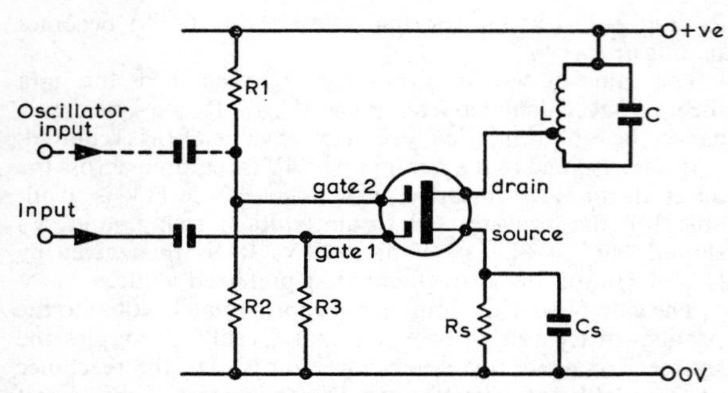

Fig 3.62. Dual-gate igfet amplifier or mixer.

input to gate 1 may be an ssb or modulated signal for conversion and the output frequency will also be of the same form as this gate 1 signal.

INTEGRATED CIRCUITS

Integrated circuits (ICS) are electronic circuits in a compact package, usually containing both active devices (transistors, MOSFETS and diodes) and passive components (resistors and capacitors), together with all the necessary interconnections. Only a few external components are usually required in order to form a complete circuit.

Originally ICS were developed for use in digital and analogue computers where large numbers of identical circuits had to be packed into a small space. Digital computers required logic circuits to have either a high or low voltage at their output in response to various patterns of high or low voltages at their inputs, while analogue computers required a basic building-block amplifier (the operational amplifier) which could simulate certain mathematical functions by the suitable choice of an external feedback network. Thus two main families of ICS were developed: digital ICS, which are now available with a choice of scores of different logic functions and half-a-dozen different manufacturing technologies, and linear ICS, which have grown from the original operational amplifier into an enormous range of audio amplifiers, radio circuits, power supply regulators and so forth.

The monolithic ic, which is the most common type, has the entire circuit block formed within a single tiny piece of silicon (the chip or slice) by a series of diffusion processes similar to those used in producing individual transistors. Other types of ic, the thin-film and thick-film varieties, have the passive components and interconnections formed as a thin film on an insulating substrate (eg glass) and then have microminiature discrete devices bonded in later. This type of construction allows more flexibility than the monolithic one, but is more costly to produce and is usually reserved for small quantity specialized applications.

Most of the early ICS used planar bipolar transistors as the active elements, but modern units often contain MOSFETS, which take up less room on the silicon chip, thus allowing more complex circuits. One particular variety, known as the complementary-symmetry mos (cmos) family, uses large numbers of both N-channel and P-channel MOSFETS in a special configuration which allows quite complex logic circuits to be fabricated without any other type of component being used.

The manufacture of monolithic ICS is usually carried out by diffusing large numbers of identical circuits onto a small (about 3cm diameter) slice of silicon, and then cutting this into individual circuits by a scribing process. Present-day techniques now permit the manufacture of ICS containing many thousands of active and passive devices, and this is termed large-scale integration (lsi).

Further discussion of this topic is beyond the scope of the present work, but some notes on the use of ICS are given on p 4.11.

CHAPTER 4

HF RECEIVERS

AMATEUR hf operation imposes stringent requirements on the receiver; the operator needs to find and hold extremely weak signals in the presence of many much stronger signals. Additionally the hf receiver may also, in conjunction with a converter, form the heart of a vhf receiver with its different modes and conditions. The performance requirements of an amateur receiver are thus every bit as exacting (in some respects even more exacting) than those demanded for professional communications services—although the amateur usually has to achieve this within a restricted budget.

In recent years professional general-purpose *communications receivers* (a type of receiver which stemmed originally from the amateur service) have become significantly more complex and more costly, and it is often no longer possible for the amateur to follow parallel lines. So he has to seek alternative means of achieving comparable results by accepting some compromises. The amateur receiver, whether factory-built or home-constructed, has usually to be designed within severe cost restraints, but by careful appreciation of just what performance characteristics are the most necessary for amateur operation—and often by limiting reception to amateur bands rather than general coverage—extremely satisfactory results can be achieved at costs moderate in comparison with those of professional communications receivers.

The radio amateur needs a communications receiver able to provide good intelligibility from signals which may easily differ in input voltage by 10,000 times and occasionally by up to a million times (say from 0·3μV from a weak distant station to 0·3V from a near neighbour) in restricted and often highly congested bands. To tune and listen to ssb and cw using narrowband selectivity the receiver must be stable—at least over periods of say 15min—to within just a few tens of hertz, even at 30MHz; it must be capable of being tuned with this degree of precision, though it is not essential to have this degree of readout or calibration accuracy. A single receiver will often be used to receive many different transmission modes—ssb, cw, a.m., nbfm, rtty etc—each calling for different selectivity and demodulation characteristics.

The main requirements for a good hf receiver are:

(1) High *sensitivity*, coupled with a wide *dynamic range* and good *linearity* to allow it to cope with both the very weak and very strong signals that will appear together at the input; it should be able to do this with the minimum impairment of the *signal-to-noise ratio* by receiver *noise*, *cross-modulation*, *blocking*, *intermodulation*, *hum* etc.

(2) Good *selectivity* to allow the selection of the required signal from among other (possibly much stronger) signals on adjacent or near adjacent frequencies. The selectivity characteristics should "match" the mode of transmission, so that interference susceptibility and noise bandwidth should be as close as possible to the intelligence bandwidth of the signal.

(3) Maximum freedom from *spurious responses*—that is to say signals which appear to the user to be transmitting on specific frequencies when in fact this is not the case. Such spurious responses include those arising from image responses, breakthrough of signals, harmonics of the receiver's internal oscillators, and those arising from *reciprocal mixing*.

(4) A high order of *stability*, in particular the absence of short-term frequency drift or jumping.

(5) Good read-out and calibration of the frequency to which the set is tuned, coupled with the ability to reset the receiver accurately and quickly to a given frequency or station.

(6) Means of receiving ssb and cw, normally requiring a stable beat frequency oscillator preferably in conjunction with product detection; it is also an advantage (though still uncommon) for the set to include a demodulator suitable for effective reception of narrow-band frequency modulation.

(7) It must incorporate sufficient amplification to allow the reception of signals of under 1μV input; this implies a minimum voltage gain of about one million times (120dB), preferably with effective automatic gain control (agc) to hold the audio output steady over a very wide range of input signals.

(8) Sturdy construction with good quality components and with consideration given to problems of access for servicing when the inevitable occasional fault occurs.

A number of other refinements are also desirable: for example it is normal practice to provide a headphone socket on all communications receivers; it is useful to have ready provision for receiver "muting" by an externally applied voltage to allow voice-operated, push-to-talk or cw break-in operation; an S-meter to provide immediate indication of relative signal strengths; a power take-off socket to facilitate the use of accessories; an i.f. signal take-off socket to allow use of external special demodulators for nbfm, fsk, dsbsc etc.

In recent years, significant progress has continued to be made in meeting these requirements—although we are still some considerable way short of being able to provide them over the entire signal range of 120dB at the ideal few hertz stability. The gradual introduction of more and more semiconductor devices into receivers has brought a number of very useful advantages, but has also paradoxically made it more difficult to achieve the highly desirable wide dynamic range. Professional users now often require frequency readout and long-term stability of an extremely high order (1Hz stability is needed for some applications) and this has led

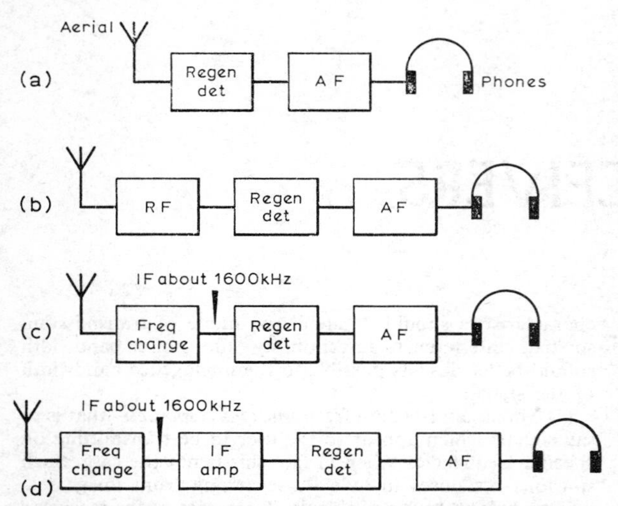

Fig 4.1. Simple receivers intended primarily for Morse reception. (a) Two-stage "straight" receiver with high-gain regenerative detector. (b) The addition of a tuned rf stage improves sensitivity and selectivity (this arrangement is often known as a trf receiver). (c) The simplest form of superhet. (d) The addition of an i.f. stage will greatly increase the gain and selectivity.

to the use of frequency synthesized local oscillators and digital read-out systems; although these are effective for the purposes which led to their adoption, they are not necessarily the correct approach for amateur receivers since unless very great care is taken, a complex frequency synthesizer not only adds significantly to the cost but may actually result in a degradation of other even more desirable characteristics.

So long as continuous tuning systems with calibrated dials are used the mechanical aspects of a receiver remain very important; it is perhaps no accident that one of the outstanding early receivers (HRO) was designed by someone whose early training was that of a mechanical engineer.

It should be recognized that receivers which fall far short of ideal performance by modern standards may nevertheless still provide entirely usable results, and can often be modified to take advantage of recent techniques. Despite all the progress made in recent decades receiver designs dating from the 'thirties and early 'forties are still capable of being put to good use, provided that the original electrical and mechanical design was sound. Similarly, the constructor may find that a simple, straightforward and low-cost receiver can give good results even when its specification is well below that now possible. It is ironical that almost all the design trends of the past 30 years have, until quite recently, impaired rather than improved the performance of receivers in the presence of strong signals!

BASIC TYPES OF RECEIVERS

Amateur hf receivers fall into one of two main categories: (1) "*Straight*" regenerative and *direct conversion* receivers in which the incoming signal is converted directly into audio by means of a demodulator working at the signal frequency; (2) single and multiple conversion *superhet* receivers in which the incoming signal is first converted to one or more intermediate frequencies before being demodulated. Each type of receiver has basic advantages and disadvantages.

Regenerative Detector ("Straight" or trf) Receivers

At one time receivers based on a regenerative (reaction) detector, plus one or more stages of af amplification (ie 0-V-1, 0-V-2 etc), and sometimes one or more stages of rf amplification at signal frequency (1-V-1 etc) were widely used by amateurs. High gain can be achieved in a correctly adjusted regenerative detector when set to a degree of positive feedback just beyond that at which oscillation begins; this makes a regenerative receiver capable of receiving weak cw and ssb signals. However this form of detector is non-linear and cannot cope well in situations where the weak signal is at all close to a strong signal; it is also inefficient as an a.m. detector since the gain is much reduced when the positive feedback (regeneration) is reduced below the oscillation threshold. Since the detector is non-linear, it is usually impossible to provide adequate selectivity by means of audio filters. Only by the most careful design is it possible to design a receiver of this type able to cope with modern band conditions, although for cw operation it can still provide a useful low-cost receiver for portable operation.

Simple Direct-conversion Receivers

A modified form of "straight" receiver which can provide good results, even under modern conditions, becomes possible by using a linear detector which is in effect simply a frequency converter, in conjunction with a stable local oscillator set to the signal frequency (or spaced only the audio beat away from it). Provided that this stage has good linearity in respect of the signal path, it becomes possible to provide almost any desired degree of selectivity by means of audio filters. This form of receiver (sometimes termed a "homodyne") has a long history but only in the past decade has it been widely used for amateur operation since it is more suited (in its simplest form) to cw and ssb reception than a.m. The direct-conversion receiver may be likened to a superhet with an i.f. of 0kHz or alternatively to a straight receiver with a linear rather than a regenerative detector. In a superhet receiver the incoming signal is mixed with a local oscillator signal and the intermediate frequency represents the difference between the two frequencies; thus as the two signals approach one another the i.f. becomes lower and lower; if this process is continued until the oscillator is at the same frequency as the incoming signal, then the output will be at audio frequency; in effect one is using a frequency changer or translator to demodulate the signal. Because high gain cannot be achieved in a linear detector, it is necessary to provide very high af amplification. Direct-conversion receivers can be designed to receive weak signals with good selectivity, but in this form do not provide true single-sideband reception (see later); another problem often found in practice is that very strong broadcast signals (eg on 7MHz) drive the detector into non-linearity and are then demodulated directly and not affected by any setting of the loca oscillator.

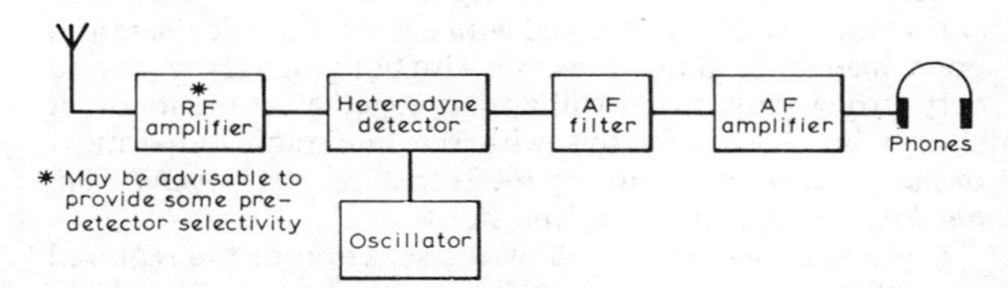

Fig 4.2. Outline of simple direct-conversion receiver in which high selectivity can be achieved by means of filters.

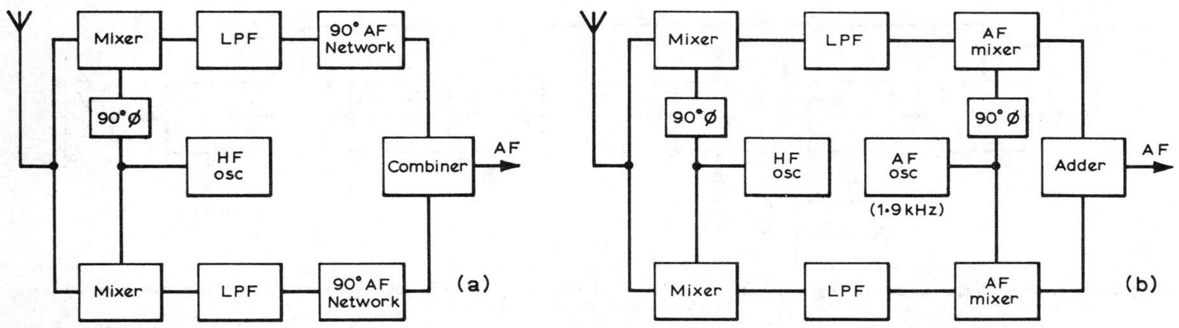

Fig 4.3. (a) Block outline of two-phase ("outphasing") form of ssb direct-conversion receiver. (b) Block outline of "third method" (Weaver or Barber) ssb direct-conversion receiver.

A crystal-controlled converter can be used in front of a direct-conversion receiver so forming a superhet with variable i.f. only.

Two phase and "Third-method" Direct-conversion Receivers

An inherent disadvantage of the simple dc receiver is that it responds equally to signals on both sides of its local oscillator frequency, and cannot reject what is termed the "audio image" no matter how good the audio filter characteristics; this is a serious disadvantage since it means that the selectivity of the receiver can only be made half as good as the theoretically ideal bandwidth. This problem can be overcome, though at the cost of additional complexity, by phasing techniques similar to those associated with ssb generation. Two main approaches are possible, see **Fig 4.3**. Fig 4.3(a) shows the use of broadband af 90 degree phase-shift networks in an "outphasing" system, and with care can result in the reduction of one sideband to the extent of 30–40dB. Another possibility, still to be exploited, is the polyphase ssb demodulator which does not require such critical component values as conventional ssb phase-shift networks. Fig 4.3(b) shows the "third method" (sometimes called the "Weaver" or "Barber" system) which requires the use of additional balanced mixers working at af but eliminates the need for accurate 90 degree af networks. The "third method" system, particularly in its ac-coupled form (*The Radio & Electronic Engineer* Vol 43, No 3, March 1973, pp209–215) provides the basis for high-performance receivers at relatively low cost, although suitable designs for amateur operation have yet to be developed.

Super-regenerative Detector Receivers

The high gain of a regenerative detector can be further increased by introducing a voltage at a supersonic frequency in such a way that rf oscillation ceases every half cycle of this second frequency, termed the quench frequency. This quenching voltage can be generated in a separate stage or, more commonly, within the regenerative detector stage. Although extremely high gain in a single stage is possible, this form of reception introduces high interstation noise and has poor selectivity, while additional complications are needed for cw reception and to avoid the emission of a rough signal capable of causing interference over a wide area. Super-regenerative receivers are suitable for both a.m. and fm reception but are today seldom used on hf; the principle however remains of interest.

Superhet Receivers

The vast majority of amateur receivers are based on the superhet principle. By changing the incoming signals to a fixed frequency (which may be lower or higher than the incoming signals) it becomes possible to build a high-gain amplifier of controlled selectivity to a degree which would not be possible over a wide spread of signal frequencies. The main practical disadvantage with this system is that the frequency conversion process involves unwanted products which give rise to spurious responses, and much of the design process has to be concentrated on minimizing the extent of these spurious responses in practical situations.

A *single-conversion* superhet is a receiver in which the incoming signal is converted to its intermediate frequency, amplified and then demodulated at this second frequency. Virtually all domestic broadcast receivers use this principle, with an i.f. of about 455–470kHz, and a similar arrangement but with refinements is to be found in many communications receivers. However, for reasons that will be made clear later in this chapter, some receivers convert the incoming signal successively to several different frequencies; these may all be fixed I.F.S: for example the first i.f. might be 9MHz and the second 455kHz and possibly a third at 85kHz. Or the first i.f. may consist of a whole spectrum of frequencies so that the first i.f. is variable when tuning a given band, with a subsequent second conversion to a fixed i.f. (this is a system widely used in current practice). There are in fact many receivers using double or even triple conversion, and a few with even more conversions, though unless care is taken each conversion makes the receiver susceptible to more spurious responses. The block diagram of a typical single conversion receiver is shown in **Fig 4.4**. **Fig 4.5** illustrates a double-conversion receiver with fixed I.F.S, while **Fig 4.6** is representative of a receiver using a variable first i.f. in conjunction with a crystal-controlled first local oscillator (hfo).

As the degree of selectivity provided in a receiver increases it reaches the stage where the receiver becomes a single-sideband receiver, although this does not mean that only ssb signals can be received. In fact the first application of this principle was the single-signal receiver for cw reception where the selectivity is sufficient to reduce the strength of the "audio image" (resulting from beating the i.f. signal with the bfo) to an insignificant value, thus virtually at one stroke halving the apparent number of cw stations operating on the band (previously each cw signal was heard on each side of the zero beat). Similarly double-sideband a.m. signals can be received on a set having a carefully controlled passband as

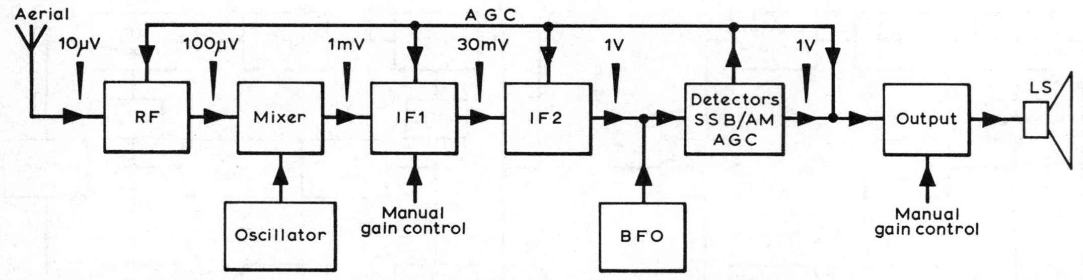

Fig 4.4. Block outline of representative single-conversion superhet receiver, typical of many used for amateur operation and showing typical levels of signal at various stages.

though they were ssb, with the possibility of receiving either sideband should there be interference on the other. This degree of selectivity can be achieved with good i.f. filters or alternatively the demodulator can itself be designed to reject one or other of the sidebands, by using phasing techniques similar to those sometimes used to generate ssb signals and for two-phase direct-conversion receivers. But most receivers rely on the use of crystal or mechanical filters to provide the necessary degree of sideband selectivity, and then use heterodyne oscillators placed either side of the nominal i.f. to select upper or lower sidebands.

It is important to note that whenever frequency conversion is accomplished by beating with the incoming signal an oscillator lower in frequency than the signal frequency then the sidebands retain their original position relative to the carrier frequency, but when conversion is by means of an oscillator placed higher in frequency than the carrier, the sidebands are inverted. That is to say an upper sideband becomes a lower sideband and vice versa (see **Fig 4.8**).

RECEIVER SPECIFICATION

The performance of a communications receiver is normally specified by manufacturers or given in equipment reviews or stated in the various constructional articles. It is important to understand what these specifications mean and how they relate to practical requirements in order to know what to look for in a good receiver. It will be necessary to study any specifications with some caution, since a manufacturer or designer will usually wish to present his receiver in the most favourable light, and either omit unfavourable characteristics or specify them in obscure terms. The specifications do not tell the whole story: the operational "feel" may be as important as the electrical performance—the "touch" of the tuning control, the absence of mechanical backlash or other irregularities, the convenient placing of controls, the positive or uncertain action of the band-change switch and so on will all be vitally important. Furthermore there is a big difference between receiver measurements made under laboratory conditions, with only a single locally generated signal applied to the input, and the actual conditions under which it will be used, with literally hundreds of amateur, commercial and broadcasting signals being delivered by the aerial in the presence of electrical interference and possibly including one or more "block-busting" signals from a nearby transmitter.

Sensitivity

Weak signals need to be amplified much more than strong ones in order to provide a satisfactory output to the loudspeaker or headphones. There are however limits to this process, and simply increasing amplification does not provide a solution. The limits are set by the noise generated internally within the set and external noise picked up by the aerial; noise within the set will be most significant when it arises from the early stages of a receiver and is amplified by the later stages; this noise is under the control of the designer. External noise, including electrical interference and other "site" noise, also sets severe limitations at hf beyond the control of the designer. The weakest signal which can be satisfactorily used for communications purposes is today seldom governed by the amount of overall amplification available in a receiver but by how weak a signal can be heard above the general noise level, which will in turn be set by the noise bandwidth of the receiver. Ultimate sensitivity is thus a function both of the internal noise and of the bandwidth necessary to receive the signal: usable sensitivity by site noise which varies with frequency.

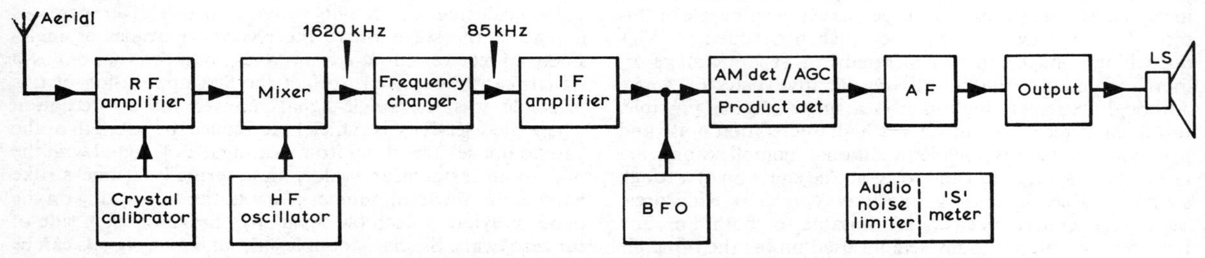

Fig 4.5. Block outline of double-conversion communications receiver with both I.F.s fixed.

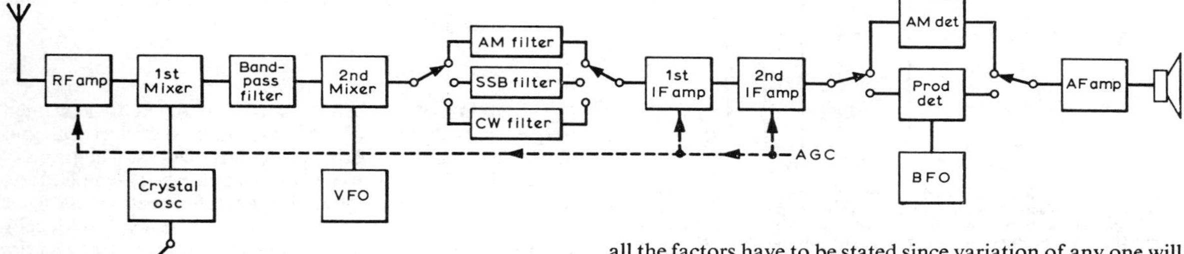

Fig 4.6. Block diagram of a double-conversion receiver with crystal-controlled first oscillator—typical of many current designs.

In practice, the sensitivity of receivers is usually specified in one (or both) of two ways: *noise factor* and *signal-to-noise ratio* (snr) or more precisely *signal-plus-noise-to-noise ratio*.

Noise factor defines the maximum sensitivity of a receiver without regard to its passband (bandwidth) or input impedance, and will generally be determined almost entirely by the first stage of the receiver. Because of the atmospheric and man-made noise always present on hf there is little need for a noise factor of less than about 10dB on bands up to and including 14MHz; a noise factor of about 6–8dB may be useful on 21 and 28MHz in quiet sites. However it can be an advantage to have a first stage with a noise factor significantly less than these figures since this will allow the use of an electrically short aerial, or alternatively will allow a more complex and "lossy" narrowband filter to be placed between the aerial and first stage; on the other hand good low-noise performance and a wide dynamic range seldom go together. It may be better to use a low-gain pre-amplifier on 21 and 28MHz with a basic receiver having an overall noise factor of about 10dB.

Signal-to-noise ratio indicates the minimum input voltage at the aerial input of the receiver needed to provide a stated signal-plus-noise-to-noise ratio when the receiver is used with its appropriate bandwidth. The input voltage will depend on the input impedance of the receiver, the noise bandwidth under which the receiver is operating, and the signal-to-noise ratio specified for the output signal. A given receiver thus gives different results on different transmission modes (which require different bandwidths); further, some manufacturers specify for 10dB snr output, others 6dB snr which will clearly result in a lower input signal for equivalent performance. For accurate assessment of a receiver's performance all the factors have to be stated since variation of any one will affect the others.

Typically a high-grade modern receiver may provide a 6dB signal-to-noise ratio with an 0·5μV ssb input signal, or less than 0·2μV cw signal with a narrowband filter; a comparable figure for a.m. signals modulated 30 per cent might be 2 or 3μV. In many locations the weakest signals receivable even on a high class receiver will be about 0·5μV cw and about 1–2μV ssb.

A simple but effective (and severe) test of sensitivity which can be applied to any well screened receiver is to remove the aerial and replace it by a resistor equal to the receiver's input impedance; if the receiver noise peaks up when the first rf circuit is tuned through resonance on the highest frequency bands (often possible by means of a panel-mounted control) then it is reasonably certain that the receiver's sensitivity cannot be further improved.

Selectivity

The ability of a receiver to separate stations on closely adjacent frequencies is determined by its *selectivity*. The limit to usable selectivity is governed by the *bandwidth* of the type of signal which is being received. For high-fidelity reception of a double-sideband a.m. signal the response of a receiver would need ideally to extend some 15kHz either side of the carrier frequency, equivalent to 30kHz bandwidth; any reduction of bandwidth would cause some loss of the information being transmitted. In practice for average mf broadcast reception the figure is reduced to about 9kHz or even less; for communications-quality speech in a double-sideband system we require a bandwidth of about 6kHz; for single-sideband speech about half this figure or 3kHz is adequate; for cw, at manual keying speeds, the minimum

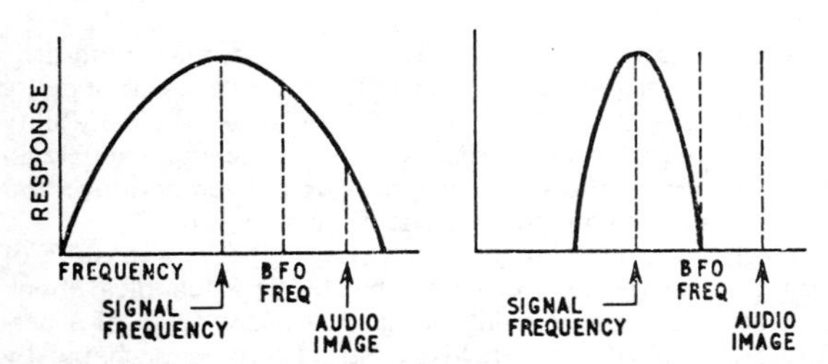

Fig 4.7. How a really selective receiver provides "single-signal" reception of cw. The broad selectivity of the response curve on the left is unable to provide substantial rejection of the audio "image" frequency, whereas with the more selective curve the audio image is inaudible and cw signals are received only on one side of zero-beat.

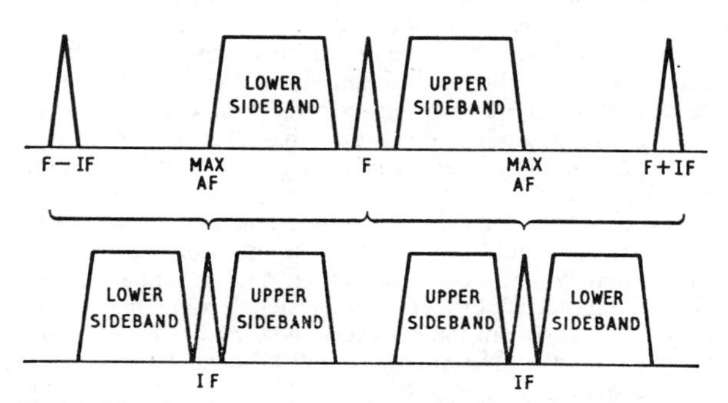

Fig 4.8. A local oscillator frequency lower than the signal frequency (ie f − i.f.) keeps the upper and lower sidebands of the intermediate frequency signals in their original positions. However when the local oscillator is placed higher in frequency than the signal frequency (f + i.f.) the positions of the sidebands are transposed. By incorporating two oscillators, one above and the other below the input signal, sideband selection is facilitated (this is generally carried out at the final i.f. by switching the bfo or carrier insertion oscillator).

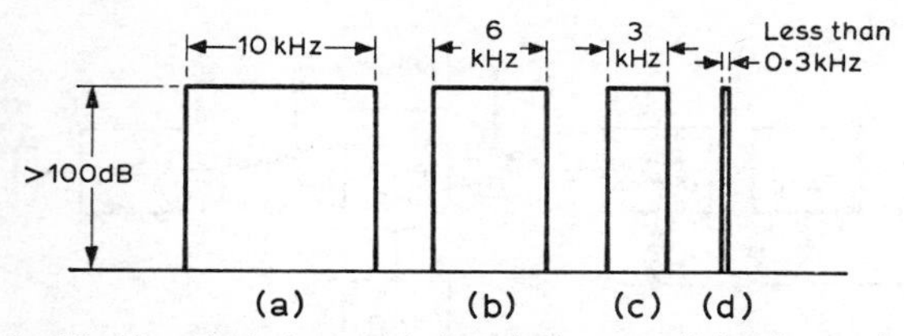

Fig 4.9. The ideal characteristics of the overall bandpass of a receiver are affected by the type of signals to be received. (a) This would be suitable for normal broadcast reception (dsb signals) permitting af response to 5kHz. (b) Suitable for a.m. phone (af to 3kHz). (c) For ssb the bandpass can be halved without affecting the af response (in the example shown this would be about 300 to 3,300 Hz). (d) Extremely narrow channels (under 100Hz) are occupied by manually keyed cw signals but some allowance must usually be made for receiver or transmitter drift and a 300Hz bandwidth is typical—by selection of the bfo frequency any desired af beat note can be produced.

theoretically possible bandwidth will reduce with speed from about 100Hz to about 10Hz for very slow Morse; in practice however the stability of the receiver or transmitter will seldom allow a bandwidth of much less than 100–300Hz to be used. Ideally, again, we would like to receive just the right bandwidth, with the response of the receiver then dropping right off as shown in **Fig 4.9**, to keep the *noise bandwidth* to a minimum. Although modern filters can approach this response quite closely, in practice the response will not drop away as sharply over as many dB as the ideal.

To compare the selectivity of different receivers, or the same receiver for different modes, a series of curves of the type shown in **Fig 4.10** may be used. There are two ways in which these curves should be considered: first the bandwidth at the *nose*, representing the bandwidth over which a signal will be received with little loss of strength; the other figure—in practice every bit as important—is the bandwidth over which a powerful signal is still audible, termed the *skirt* bandwidth.

The nose bandwidth is usually measured for a reduction of not more than 6dB, the skirt bandwidth for a reduction of one thousand times on its strength when correctly tuned in, that is 60dB down. These two figures can then be related by what is termed the *shape factor*, representing the bandwidth at the skirt divided by the bandwidth at the nose. The idealized curves of Fig 4.9, which have the same bandwidth regardless of signal strength, would represent a shape factor of 1; such a receiver cannot be designed at the present state of the art, although it can be approached by some ssb filters; the narrower cw filters, although much sharper at the nose, tend to broaden out to about the same bandwidth as the ssb filter and thus have a rather worse shape factor. Typically a high-grade modern receiver might have an ssb shape factor of 1·2 to 2 with a skirt bandwidth of less than 5kHz.

It should be noted that such characteristics are determined when applying only *one* signal to the input of the receiver; unfortunately in practice this does not mean that a receiver will be unaffected by very strong signals operating many kilohertz away from the required signal and outside the i.f. passband; this important point will be considered later in this chapter.

It therefore needs to be stressed that the effective selectivity cannot be considered solely in terms of static characteristics determined when just one test signal is applied to the input, but rather in the real life situation of hundreds of signals present at the input: in other words it is the *dynamic selectivity* which largely determines the operational value of an amateur hf receiver.

Stability

The ability of a receiver to remain tuned to a wanted frequency without *drift* depends upon the electrical and mechanical stability of the internal oscillators. The primary cause of instability in an oscillator is a change of temperature, usually as the result of internally generated heat. With valves, even in the best designs, there will usually be steady frequency variation of any oscillator using inductors and capacitors during a period of perhaps 15 minutes or more after first switching on; transistor and fet oscillators reach thermal stability in a few seconds, provided that they are not then affected by other local sources of heat.

After undergoing a number of heat cycles, some components do not return precisely to their original values, making it difficult to maintain accurate calibration over a long period. Many receivers include a crystal controlled oscillator of high stability providing marker signals (for example every 100kHz) so that receiver calibration can be checked and brought into adjustment.

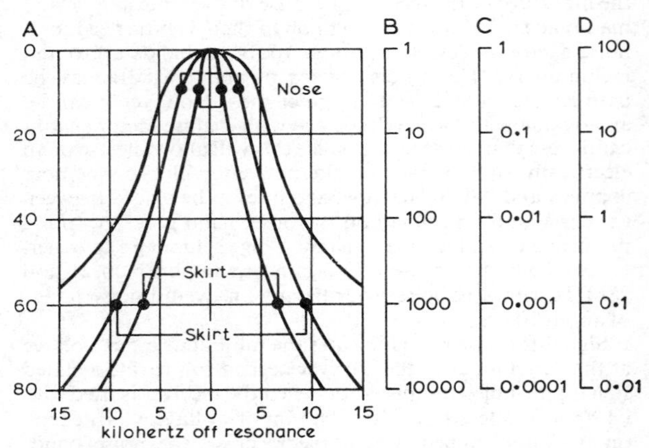

Fig 4.10. The ideal vertical sides of Fig 4.9 cannot be achieved in practice. The curves shown here are typical. These three curves represent the overall selectivity of receivers varying from the "just adequate" broadcast curve of a superhet receiver having about four tuned i.f. circuits on 470kHz to those of a moderately good communications receiver. A, B, C and D indicate four different scales often used to indicate similar results: A is a scale based on the attenuation in decibels from maximum response; B represents the relative signal outputs for a constant output; C is the output voltage compared with that at maximum response; D is the response expressed as a percentage.

Receiver drift can be specified in terms of maximum drift in hertz over a specified time, usually quoting a separate figure to cover the warming-up period. For example a high-stability receiver might specify drift as "not worse than 200Hz in any five hour period at constant ambient temperature and constant mains voltage after one hour warm up".

Mechanical instability, which may appear as a shift of frequency when the receiver is subjected to mechanical shock or vibration, cannot easily be defined in the form of a performance specification. Sturdy construction on a mechanically rigid chassis can help. The need for high stability for ssb reception has led to much greater use of crystal-controlled oscillators and various forms of frequency synthesis.

Spurious Responses

One of the major defects of superhet receivers is that the same station may be received at more than one position of the tuning dial, or alternatively that the various oscillators within the receiver provide signals which may be tuned in at various points on the dial, as though they were external signals. This means, in practice, that strong broadcasting and commercial hf stations may be heard as though they were operating within the amateur bands. This question of spurious responses will be discussed in detail later; the most important is usually "image" reception of signals spaced twice the intermediate frequency away from the required signal, and in most receivers image reception increases with increasing frequency. Whereas for example on 3·5MHz noticeable "image" signals would be unusual except in the very simplest superhet designs, on 28MHz it is quite common, even with a fairly good receiver, for the "image" signal to be reduced perhaps by only 30dB. Ideally, since we are concerned with a range of signals which may differ in strength by up to one million times, spurious responses need to be reduced by 120dB; such a figure is seldom if ever achieved at 30MHz.

Typically the image rejection of a receiver might be specified as "greater than 75dB relative to 1μV below 10MHz; greater than 60dB relative to 1μV above 10MHz".

Internally generated spuriae might be specified as "below receiver noise over range 1·5 to 30MHz except at *x*MHz and *y*MHz, where they will not exceed 1·5μV equivalent signal applied to aerial socket". It is extremely difficult in a multi-conversion superhet to avoid these "birdies" altogether, and the problem then resolves itself to ensuring that they do not fall in frequently used portions of the amateur bands. This is largely a question of choice of i.f.(s).

Cross-modulation, Blocking and Intermodulation

Even with a receiver that is highly selective to one signal down to the —60dB level, there remains the problem of coping with numbers of extremely strong signals. When an unwanted signal is transmitting on a frequency that is well outside the i.f. passband of the receiver, it may unfortunately still affect reception as a result of cross-modulation, blocking or intermodulation.

When any active device such as a transistor or valve is operated with an input signal that is large enough to drive the device into a non-linear part of its transfer characteristic (ie so that some parts of the input waveform are distorted and amplified to different degrees) the device acts as a "modulator", impressing on the wanted signal the modulation of the strong signal by the normal process of mixing. When a very strong signal reaches a receiver the broad selectivity of the signal frequency tuned circuits, and often any tuned circuits, including i.f. circuits, prior to the main selective filter, means it will be amplified, along with the wanted signal, until one or more stages are likely to be driven into a non-linear condition. It should be noted that the strong signal may be many kilohertz away from the wanted signal, but once cross-modulation has occurred there is no means of separating the wanted and unwanted modulation. A strong cw carrier can reduce the amplification of the wanted signal by a similar process which is in this case called *desensitization*, or in extreme cases *blocking*.

In these processes there need be no special frequency relationship between wanted and unwanted signals. However a further condition arises when there are specific frequency relationships between wanted and unwanted signals, or between two strong unwanted signals; a process called *intermodulation*, see **Fig 4.11.**

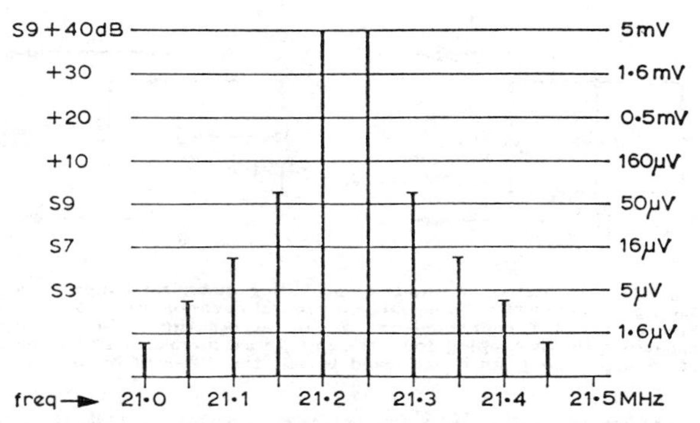

Fig 4.11. Intermodulation products. This diagram shows the effect of two very strong signals on 21,200kHz and 21,250kHz reaching the front-end of a typical modern transceiver and producing spurious signals at 50kHz intervals. Note the S9 signals produced as the third-order products f1 + (f1 — f2) and f2 — (f1 — f2). Three very strong signals will produce far more spurious signals, and so on.

Intermodulation is closely allied to the normal mixing process but with strong signals providing the equivalent of local oscillators; unwanted intermodulation products (IPS) can result from many different combinations of input signal.

It should be appreciated that cross-modulation, blocking and intermodulation products can all result from the presence of extremely strong unwanted signals applied to any stage having insufficient linearity over the full required dynamic range. The solution to this problem is either to reduce the strength of unwanted signals applied to the stage, or alternatively to improve the dynamic range of the stage.

Clearly the more amplifying stages there are in a receiver before the circuits or filter which determines its final selectivity, the greater are the chances that one or more may be overloaded unless particular care is paid to the *gain-distribution* in the receiver (see later). From this it follows that multiple conversion receivers are more prone to these problems than single-conversion or direct-conversion receivers.

It is often difficult for an amateur to assess accurately the performance of a receiver in this respect; undoubtedly many multi-conversion superhets and many receivers based on bipolar transistors fall far below the desired performance. Even high-grade valve receivers are likely to be affected by S9 + 60dB signals 50kHz or more away from the wanted signal.

The susceptibility of receivers to these forms of interference depends on a number of factors: notably the type of active device used in the front end of the receiver; the pattern of gain distribution through the receiver and how this is modified by the action of agc. On receivers suffering from this problem (often most apparent on 7MHz where weak amateur stations may be sought alongside extremely strong broadcast stations) considerable improvement often results from the inclusion of an attenuator in the aerial feeder, since this will reduce the unwanted signals to an extent where they may not cause spuriae before reducing the wanted signal to the level where it cannot be copied.

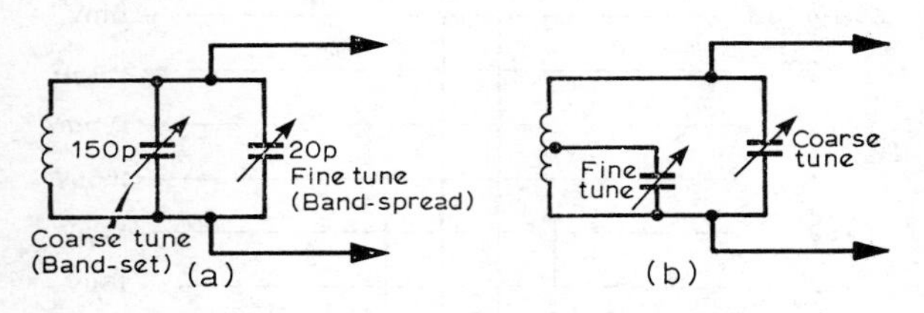

Fig 4.12. Alternative methods of providing electrical bandspread tuning to reduce the tuning rate on general coverage receivers. (a) Small capacitor connected in parallel across the main tuning capacitor. (b) By tapping down the coil almost any required degree of bandspreading can be achieved to suit the different bands.

If sufficient dynamic range could be achieved in *all* stages before the final selectivity filter, then selectivity could be determined at any point in the receiver (for example at first, second or third i.f. or even at af).

In practice some semiconductor receivers are unable to cope with undesired signals more than about 20–30dB stronger than a required signal of moderate strength, though their "static" selectivity may be extremely good. The overall effect of limited dynamic range depends upon the actual situation and band on which the receiver is used; it is of most importance on 7MHz or where there is another amateur within a few hundred yards. Even where there are no local amateurs, an analysis of commercial hf signals has shown that typically, out of a total of some 3,800 signals logged between 3 and 29MHz at strengths more than 10dB above atmospheric noise, 154 were between 60–70dBμV, 72 between 70–80dBμV, 36 between 80–90dBμV and 34 between 90–100dBμV. If weak signals are to be received satisfactorily the receiver needs to be able to cope with signals almost 100dB stronger (ie up to about S9 + 50dB). This underlines the importance of achieving front-end stages of extremely wide dynamic range or alternatively providing sufficient pre-mixer selectivity to cut down the strength of all unwanted signals reaching the mixer (preferably to well under 100mV).

Another form of spurious response is where a strong signal breaks through into the i.f. stages of the receiver. For example, with a simple single-conversion receiver having an i.f. of 470kHz, strong coastal stations and ships operating around 500kHz may be heard; these signals can usually be eliminated by providing additional pre-mixer tuned circuits. With receivers having a high (first) i.f. or with the first i.f. tunable over a range of frequencies, it becomes increasingly difficult to keep strong hf signals from breaking through into the i.f. stages. Whereas ideally one would like an i.f. rejection (compared with the wanted signal) of almost 120dB, most amateurs would be well satisfied with 80–100dB of protection; in practice many receivers with variable first i.f. do not provide more than about 40–60dB protection.

Automatic Gain Control (AGC)

The specification of a communications receiver will usually provide information on the operation of the agc circuits indicating the change of audio output for a specified change of rf input. For example a really good system might show only a 3dB rise in audio output for an rf input change from 1μV to 100mV. Equally important for the use of agc on ssb and cw (where there are no steady carriers on which to base the operation of the system) are the attack and release times, that is to say how long a signal has to be present before the the agc system responds to it, and for how long after it has gone will the system continue to limit gain; when this release time is significantly long, one has what is termed "hang" agc. Generally one requires a very fast attack time (preferably 20 milliseconds or less); and a long effective decay time (200ms to 1 second), although it is also useful to have available a shorter release time (say 25ms) for "fast agc", for use with a.m. signals. Modern "hang" agc systems are often based on two time constants to allow the system to be relatively unaffected by noise pulses while retaining fast attack and hang characteristics.

Tuning Rate

To tune accurately and rapidly to an ssb signal one needs to be able to adjust the receiver's frequency to within about 25Hz, and accurate tuning of this order is also highly desirable for narrowband cw reception. This accuracy is unlikely to be achieved with normal mechanical tuning systems unless each complete revolution of the tuning knob represents only a moderate shift in frequency. Typical figure for a good receiver would be about 5–10kHz per revolution, although some operators prefer even lower figures. Equally important is a tuning mechanism with a smooth action, free from backlash. A useful feature of receivers having a variable first i.f. is that an equal tuning rate can be achieved on all bands; this facility can also be provided on single conversion receivers by using pre-mixer frequency synthesis (see later). On many older receivers much higher tuning rates will be found, particularly on the higher frequency bands, and it is often beneficial to add a further step-down tuning mechanism in the form of a slow-motion dial operating on the normal tuning control of the receiver.

DESIGN TRENDS

After the "straight" receiver, because of its relatively poor performance and lack of selectivity on a.m. phone signals, had fallen into disfavour in the mid-'thirties, came the era of the superhet communications receiver. Most early models were single-conversion designs based on an i.f. of 455–470kHz, with two or three i.f. stages, a multi-electrode triode-hexode or pentagrid mixer, sometimes but not always with a separate oscillator valve. This approach made at least one rf amplifying stage essential in order to raise the level of the incoming signal before it was applied to the relatively noisy mixer; two stages were to be preferred, since this meant they could be operated in less critical conditions and provided the additional pre-mixer rf selectivity needed to reduce "image" response on 14MHz and above. Usually a band-switched LC hf oscillator was gang-tuned so as to track with two or three signal frequency tuned circuits, calling for fairly critical and expensive tuning and alignment systems. These receivers were often designed basically to provide full coverage on the hf band (and often also the mf band), sometimes with a second tuning control to provide electrical band-spread on amateur bands, or with provision (as on the HRO) optionally to limit coverage to amateur bands only. Selectivity depended on the use of good quality i.f. transformers (sometimes with a tertiary tuned circuit) in conjunction with a single-crystal i.f. filter which could easily be adjusted for varying degrees of selectivity and included a phasing control for nulling out interfering carriers.

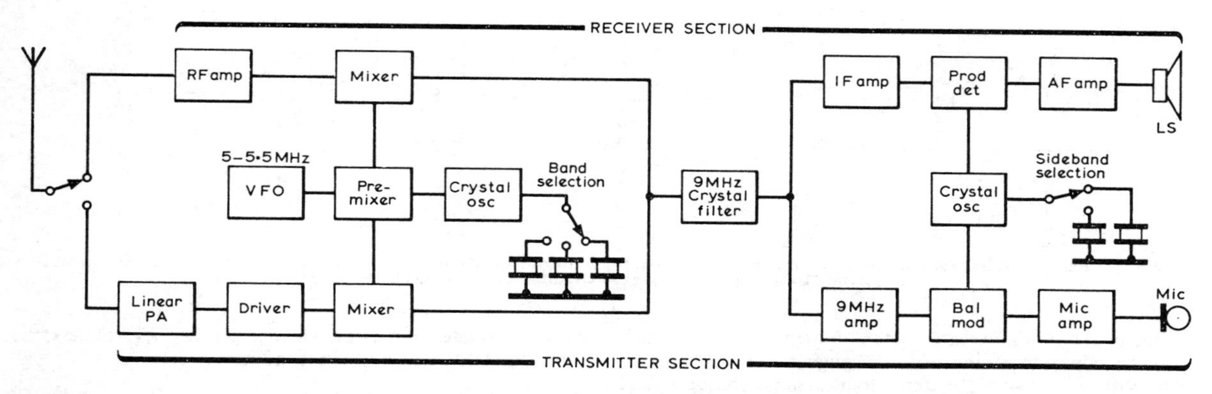

Fig 4.13. Block diagram of a typical modern ssb transceiver in which the receiver is a single conversion superhet with 9MHz i.f. in conjunction with the pre-mixer form of partial frequency synthesis.

Later, to overcome the problem of image response with only one rf stage, there was a trend towards double or triple conversion receivers with a first i.f. of 1·6MHz or above, a second i.f. about 470kHz and (sometimes) a third i.f. about 50kHz.

With a final i.f. of 50kHz it was possible to provide good single-signal selectivity without the use of a crystal filter.

The need for higher stability than is usually possible with a band-switched hf oscillator and the attraction of a similar degree of band-spreading on all bands has in the past two decades led to the widespread adoption of an alternative form of multi-conversion superhet; this, in effect, provides a series of integral crystal-controlled "converters" in front of a superhet receiver (single or double conversion) covering only a single frequency range (for example 5,000–5,500kHz) This arrangement provides a fixed tuning span (in this example 500kHz) for each crystal in the hf oscillator. Since a separate crystal is needed for each band segment, most receivers of this type are designed for amateur bands only (though often with provision for the reception of a standard frequency transmission, for example on 10MHz); more recently some designs have eliminated the need for separate crystals by means of frequency synthesis, and in such cases it is economically possible to provide general coverage. The selectivity in these receivers is usually determined by a bandpass crystal filter, mechanical filter or multi-pole ceramic filter, separate filters being used for ssb, cw and a.m. reception (although for economic reasons sets may be fitted with only one filter, usually intended for ssb reception). In this system the basic "superhet" section forms in effect a variable i.f. amplifier.

In practice the variable i.f. type of receiver provides significantly enhanced stability and lower tuning rates on the higher frequency bands, compared with receivers using fixed i.f., though it is considerably more difficult to prevent breakthrough of strong signals within the variable i.f. range, and to avoid altogether the appearance of "birdies" from internal oscillators. Very many current receivers are based on this system and with careful design a high standard of performance can be achieved; the use of multiple conversion (with the selective filter further from the aerial input stage) makes the system less suitable for semiconductors than for valves, particularly where broad-band circuits are employed in the front-end and in the variable i.f. stage.

There is now a trend back to the use of fixed i.f. receivers either with single-conversion or occasionally with double-conversion (provided that in this case an effective *roofing filter* is used at the first i.f.). A roofing filter is a selective filter intended to reduce the number of strong signals passing down an i.f. chain without necessarily being of such high grade or as narrow-band as the main selective filter. To overcome the problem of image reception a much higher first i.f. is used; for amateur band receivers this is often 9MHz since effective ssb and cw filters at this frequency are now available. This reduces (though does not eliminate) the need for pre-mixer selectivity; while the use of low-noise mixers makes it possible to reduce or eliminate rf amplification. To overcome frequency stability problems inherent in a single-conversion approach, it is possible to obtain better stability with fet oscillators than was usually possible with valves; another approach is to use mixer-vfo systems (essentially a simple form of frequency synthesis) and such systems can provide identical tuning rates on all bands, though care has to be taken to reduce to a minimum spurious injection frequencies resulting from the mixing process.

To achieve the maximum possible dynamic range, particular attention has to be given to the mixer stage, and it is an advantage to make this a balanced, or double-balanced (see later) arrangement using either beam deflection valves, double-triodes, Schottky (hot-carrier) diodes or FETS (particularly power FETS).

A further significant reduction of spurious responses may prove possible by abandoning the superhet in favour of high-performance direct-conversion receivers (such as the Weaver or third-method ssb direct-conversion arrangement); such designs however are still only at an early stage of development.

Many modern receivers are built in the form of compact transceivers functioning both as receiver and transmitter, and with some stages common to both functions (**Fig 4.13**). Modern transceivers often use semiconductor devices throughout the receiver stages, although valves continue to be used in the power stages of the transmitter. Dual-gate fet devices are generally found in the signal path of the receiver. Most transceivers have a common ssb filter for receive and transmit; this may be a mechanical or crystal filter at about 455kHz but current models more often use crystal filters at about 3,180, 5,200 or 9,000kHz, since the use of a higher frequency reduces the total number of frequency conversions necessary.

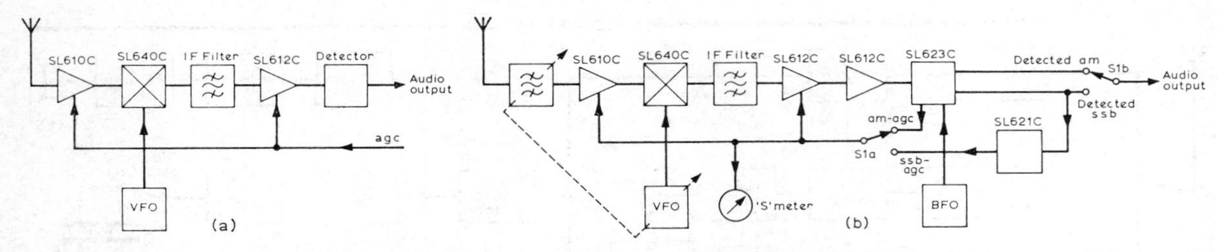

Fig 4.14. (a) Basic superhet receiver based on integrated circuits. (b) A more typical communications receiver for a.m. and ssb reception using the SL600 range of integrated circuits manufactured by Plessey Ltd.

One of the fundamental benefits of a transceiver is that it provides common tuning of the receiver and transmitter so that both are always "netted" to the same frequency. It remains however an operational advantage to be able to tune the receiver a few kilohertz around the transmit frequency and vice versa, and provision for this "incremental" tuning is often incorporated; alternatively some amateurs prefer to use a separate external vfo so that the two frequencies may be separated when required.

The most critical aspect of modern receivers is the signal handling capabilities of the early ("front-end") stages. Various circuit techniques are available to enhance such characteristics: for example the use of balanced ("push-pull") rather than "single-ended" signal frequency amplifiers; the use of balanced or double-balanced mixer stages; the provision of manual or agc actuated aerial-input attenuators; and careful attention to the question of gain distribution.

An important advantage of modern techniques such as linear integrated circuits and wide-band fixed-tuned filters rather than tunable resonant circuits is that they make it possible to build satisfactory receivers without the time-consuming and constructional complexity formerly associated with high-performance receivers. Nevertheless a multi-band receiver must still be regarded as a project requiring considerable skill and patience.

ACTIVE DEVICES FOR RECEIVERS

The amateur is today faced with a wide and sometimes puzzling choice of *active devices* around which to design a receiver: valves, bipolar transistors, field effect transistors including single- and dual-gate MOSFETS and junction FETS; special diodes such as Schottky (hot-carrier) diodes, and an increasing number of *integrated circuits*, many designed specifically for receiver applications. Each possesses advantages and disadvantages when applied to high-performance receivers, and most recent designs tend to draw freely from among these different devices.

Valves

In general the valve is bulky, requires additional wiring and power supplies for heaters, generates heat, and is subject to ageing in the form of a gradual change of characteristics throughout its useful life. On the other hand it is not easily damaged by high-voltage transients, is available in a wide variety of types for specific purposes to fairly close tolerances; is capable of handling small signals with good linearity (and special types can cope well with large signals). It is only recently that semiconductor receivers can approach the dynamic range of a good valve receiver; nevertheless the future now lies with the semiconductor.

Bipolar Transistors

These devices can provide very good noise performance with high gain, are simple to wire and need only low voltage supplies consuming very little power and so generating very little heat (except power types needed to form the audio output stage). On the other hand they are low-impedance, current-operated devices making the interstage matching more critical and tending to impose increased loading on the tuned circuits; they have feedback capacitances that may require neutralizing; they are sensitive to heat, changing characteristics with changing temperature; they can be damaged by large input voltages or transients; provide significant noise sidebands when used as an oscillator. Their main drawback in the signal path of a receiver is the difficulty of achieving wide dynamic range (except multi-emitter rf power types) and satisfactory agc characteristics. On the other hand bipolar transistors are suitable for most af applications. They are not recommended for rf/mixer/oscillator/i.f. applications, unless care is taken to overcome their limitations, for example by using relatively high-power types. The bipolars developed for catv (wire distribution of tv) such as the BFW17, BFW17A, 2N5109 etc, used with heat sinks, can form excellent rf stages or mixers.

Field Effect Transistors

These devices offer significant advantages over most bipolar devices for the low-level signal path. Their high input impedance makes accurate matching less important; their near square-law characteristics make them comparable with variable-mu valves in reducing susceptibility to cross-modulation; they can readily be controlled by agc systems. The dual-gate form of device is particularly useful for small-signal applications, and forms an important device for modern receivers. They tend however to be limited in signal handling capabilities. Special types of high-current field effect transistors have been developed capable of providing extremely wide dynamic range (up to 140dB) in the front-ends of receivers. A problem with fet devices is the wide spread of characteristics between different devices bearing the same type number, and this may make individual adjustment of the bias levels of fet stages desirable. The good signal handling capabilities of mosfet mixers can be lost by incorrect signal or local oscillator levels.

Integrated Circuits

Special-purpose integrated circuits use large numbers of bipolar transistors in configurations designed to overcome many of the problems of circuits based on discrete devices.

Because of the extremely high gain that can be achieved within a single ic they also offer the home-constructor simplification of design and construction. High-performance receivers can be designed around a few special-purpose linear ICs, or one or two consumer-type ICs may alternatively form the "heart" of a useful communications receiver. It should be recognized however that their rf signal handling capabilities are less than can be achieved with special purpose discrete devices and their temperature sensitivity and heat generation (due to the large number of active devices in close proximity) usually make them unsuitable for oscillator applications. They also have a very wide spread of characteristics which make it desirable to select devices from a batch for critical applications.

Integrated-circuit Precautions

As with all semiconductor devices it is necessary to take precautions with integrated circuits, although if handled correctly high reliability may be expected.

Recommended precautions include:

(1) Do not use excessive soldering heat and ensure that the tip is not at significant potential to earth (due to mains leakage).
(2) Check and recheck all connections several times before applying any voltages.
(3) Keep integrated circuits away from strong rf fields.
(4) Keep supply voltages within $\pm$ 10 per cent of those specified for the device on well-smoothed supplies.

Integrated circuit amplifiers can provide very high gains (eg up to 80dB or so) within a single device having input and output leads separated by only a small distance: this means that careful layout is needed to avoid instability, and some devices may require the use of a shield between input and output circuits. Some devices have earth leads arranged so that a shield can be connected across the underside of the device.

Earth returns are important in high gain devices: some (eg SL610) have input and output earth returns brought out separately in order to minimize unwanted coupling due to common earth return impedances, but this is not true of all devices.

Normally ic amplifiers are not intended to require neutralization to achieve stability; unwanted oscillation can usually be traced to unsatisfactory layout or circuit arrangements. VHF parasitics may generally be eliminated by fitting a 10Ω resistor in series with either the input or output lead, close to the ic.

With high-gain amplifiers, particular importance attaches to the decoupling of the voltage feeds. At the low voltages involved, values of series decoupling resistors must generally be kept low so that the inclusion of low-impedance bypass capacitors is usually essential. Since high-Q rf chokes may be a cause of rf oscillation it may be advisable to thread ferrite beads over one lead of any rf choke to reduce the Q.

As with bipolar transistors, ic devices (if based on bipolars) have relatively low input and output impedances so that correct matching is necessary between stages. The use of a fet source follower stage may be a useful alternative to stepdown transformers for matching.

Maximum and minimum operating temperatures should be observed. Many linear devices are available at significantly lower cost in limited temperature ranges which are usually more than adequate for operation under normal domestic conditions.

Because of the relatively high temperature sensitivity of bipolar-type integrated circuits, they are not suitable as free-running oscillators in high-performance receivers.

For the very highest grade receivers discrete components and devices are still required in the front-end since currently available integrated circuits do not have comparable dynamic range. The ic makes possible extremely compact receivers; in practice miniaturization is now limited—at least for general purpose receivers—by the need to provide easy-to-use controls for the non-miniaturized operator.

SPURIOUS RESPONSES

A most important test of any receiver is the extent to which it receives signals when it is tuned to frequencies on which they are not really present, so adding to interference problems and misleading the operator. Every known type of receiver suffers from various forms of "phantom" signals, but some are much worse than others. Unfortunately, the mixing process is inherently prone to the generation of unwanted "products" and receiver design is concerned with minimizing their effect rather than their complete elimination.

Spuriae may take the form of:

(a) External signals heard on frequencies other than their true frequency.
(b) Carriers heard within the tuning range of the receiver but stemming not from external signals but from the receiver's own oscillators (*birdies*).
(c) External signals which cannot be tuned out but are heard regardless of the setting of the tuning knob.

In any superhet receiver tunable signals may be created in the set whenever the interfering station or one of its harmonics (often produced *within* the receiver) differs from the intermediate frequency by a frequency equal to the local oscillator or one of its harmonics. This is reflected in the general expression: $mf_u \pm nf_o = f_i$ where m, n are any integers, including 0, f_u is the frequency of the unwanted signal, f_o is the frequency of the local oscillator, and f_i is the intermediate frequency.

An important case occurs when m and n are 1 giving $f_u - f_o = f_i$. This implies that f_u will either be on the frequency to which the set is correctly tuned (f_s) or differs from it by twice the i.f. ($2 \times f_i$), producing the so-called "image" frequency. This is either $f_u = f_s + 2f_i$ (for cases where the local oscillator is higher in frequency than the wanted signal, or $f_u = f_s - 2f_i$ (where the local oscillator is lower in frequency than the wanted signal).

For an example take a receiver tuned to about 14,200kHz with an i.f. of 470kHz and the oscillator high (ie about 14,670kHz). Such a receiver may, because of "image", receive a station operating on $14{,}200 + (2 \times 470) = 15{,}140$kHz. Since 15,140kHz is within the "19-metre" broadcast band, there is thus every likelihood that as the set is tuned around 14,200kHz strong broadcast signals will be received.

To reduce such undesirable effects pre-mixer selectivity must be provided in the form of more rf tuned circuits or rf bandpass filters, or by increasing the Q of such circuits, or alternatively by increasing the frequency difference between the wanted and unwanted "image" signals. This frequency separation can be increased by increasing the factor $2f_i$, in other words by raising the intermediate frequency, and so allowing the broadly tuned circuits at signal frequency to have more effect in reducing signals on the unwanted "image" frequency before they reach the mixer.

It should also be noted from the general formula $mf_u \pm nf_o$ that "image" is only one (though usually the most important) of *many* possible frequency combinations that can cause unwanted signals to appear at the intermediate frequency, even on a single conversion receiver. The problem is greatly increased when more than one frequency conversion is employed.

Even with good pre-mixer selectivity it is still possible for a number of strong signals to reach the mixer, drive this or an rf stage into non-linearity and then produce a series of intermodulation products as spurious signals within an amateur band (see Fig 4.11).

The harmonics of the hf oscillator(s) may beat against incoming signals and produce output in the i.f. passband; strong signals may generate harmonics in the receiver stages and these can be received as spuriae. Most such forms of spuriae can be reduced by increasing pre-mixer selectivity to decrease the number of strong signals reaching the mixer, by increasing the linearity of the early stages of the receiver, or by reducing the amplitude of signals within these stages by the use of an aerial attenuator.

Very strong signals on or near the i.f. may break directly into the i.f. amplifier and then appear as untunable interference. This form of interference (although it then becomes tunable) is particularly serious with the variable i.f. type of multiple conversion receiver since there are almost certain to be a number of very strong signals operating over the segment of the hf spectrum chosen to provide the variable i.f. Direct breakthrough may occur if the screening within the receiver is insufficient or if signals can leak in through the early stages due to lack of pre-mixer selectivity. For single conversion and double-conversion with fixed first i.f. it is common practice to include a resonant "trap" (tuned to the i.f.) to reject incoming signals on this frequency. The multiple conversion superhet contains more internal oscillators and harmonics of the second and third (and occasionally the bfo) can be troublesome. For amateur-bands-only receivers every effort should be made to choose intermediate and oscillator frequencies that avoid as far as possible the effects of oscillator harmonics ("birdies").

Because of the great difficulty in eliminating spurious responses in double and triple conversion receivers, the modern designer tends to think more in terms of single-conversion with high i.f. (eg 9MHz). Potentially the direct-conversion receiver is even more attractive, though it needs to be fairly complex to eliminate the "audio-image" response. It must also have sufficient linearity or pre-detector selectivity to reduce any envelope detection of very strong signals which may otherwise break through into the audio channel, regardless of the setting of the heterodyne oscillator. The direct-conversion receiver can also suffer from spuriae resulting from harmonics of the signal or oscillator and this needs to be reduced by rf selectivity.

Gain Distribution

In many receivers of conventional design, it has been the practice to distribute the gain throughout the receiver in such a manner as to optimize signal-to-noise ratio and to minimize spurious responses. So long as relatively noisy mixer stages were used it was essential to amplify the signal considerably before it reached the mixer. This means that any strong unwanted signals, even when many kilohertz

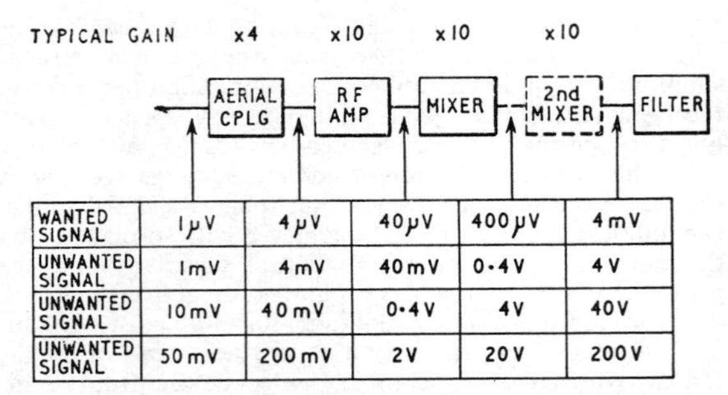

Fig 4.15. This diagram shows how unwanted signals are built up in high-gain front-ends to levels at which cross-modulation, blocking and intermodulation are virtually bound to occur.

from the wanted signal, pass through the early unselective amplifiers and are built up to levels where they cause cross-modulation within the mixer, see **Fig 4.15.** Today it is recognized that it is more satisfactory if pre-mixer gain can be kept low to prevent this happening. Older multi-grid valve mixers had an equivalent noise resistance as high as 200,000Ω (representing some 4–5µV of noise referred to the grid). Later types such as the ECH81 and 6BA7 reduced this to an enr of about 60,000Ω (about 2·25µV of noise) while the enr of pentode and triode mixers is lower still (although these may not be as satisfactory for mixers in other respects). **Table 4.1** shows the enr of valves which are widely used in rf amplifier and mixer stages.

For example the 6J6 has a mixer enr of under 2,000Ω, while the 7360 beam-deflection mixer has extremely wide dynamic range and an enr of only 1,500Ω. When the enr is less than about 3,000Ω no pre-mixer amplification is needed in an hf receiver.

The noise contribution of semiconductor mixers is also low—for example an fet mixer may have a noise factor as

TABLE 4.1

Equivalent noise resistance of typical valves

Valve	enr
RF Amplifiers	
6AC7	720
6AC7 (triode connected)	220
6AG5	1900
6AK5	1880
6AK5 (triode connected)	385
6BA6	3520
6BQ7A	390
6BZ6	1460
6CB6	1440
6F23	670
6F24	370
6J6	470
6K7	16,400
6SG7	3300
6SH7	2850
6SJ7	5840
6SK7	10,500
6U8 (pentode)	2280
6U8 (triode)	295
12AT7	380
12AU7	1140
12AX7	1560
ECC84	*420
ECC85	500
EF42	750
EF50	1400
EF54	700
EF85	1500
EF91	1200
EF183	490
EF184	300
KTW61	5000
Z77	1000
Mixers	
6AK5	7520
6BA6	14,080
6BA7	60,000
6BE6	190,000
6J6	1880
6K8	290,000
6L7	255,000
6SA7	240,000
6SB7Y	62,000
6U8 (pentode)	*9300
6U8 (triode)	2000
12AT7	2400
12AU7	7280
ECF86 (pentode)	*2700
ECH81	66,000
7360	1500

* *Calculated*

Notes: The above values are normally makers' optimum figures and the enr will rise sharply when the mutual conductance is lowered by increasing the bias or due to valve ageing.

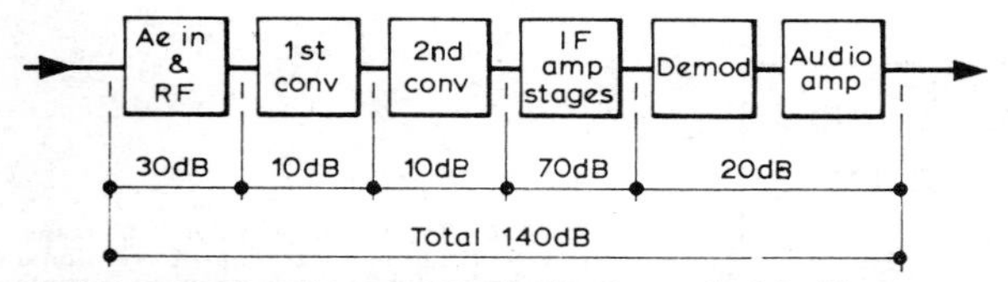

Fig 4.16. Representative gain distribution in a typical double-conversion receiver designed to optimize signal-to-noise rather than to achieve optimum dynamic range.

low as 3dB—so that generally the designer need no longer worry unduly about the requirement for pre-mixer amplification to overcome noise problems. Nevertheless a signal frequency stage may still be useful in helping to overcome image reception by providing a convenient and efficient method of coupling together signal frequency tuned circuits, and when correctly controlled by agc it becomes an automatic large-signal attenuator. Pre-mixer selectivity limits the number of strong signals reaching the mixer.

Fig 4.16 shows a typical gain distribution as found in many double conversion superhets. **Fig 4.17** shows a more modern design in which the signal applied to the first mixer is much lower than before.

Choice of I.F.

Choice of the intermediate frequency or frequencies is a most important consideration in the design of any superhet receiver. The lower the frequency, the easier it is to obtain high gain and good selectivity and also to avoid unwanted leakage of signals round the selective filter. On the other hand, the higher the i.f. the greater will be the frequency difference between the wanted signal and the "image" response, so making it simpler to obtain good protection against image reception of unwanted signals and also reducing the "pulling" of the local oscillator frequency. These considerations are basically opposed, and the i.f. of a single-conversion receiver is thus a matter of compromise; however in recent years it has become easier to obtain good selectivity with higher-frequency bandpass crystal filters and it is no longer any problem to obtain high gain at high frequencies. The very early superhet receivers used an i.f. of about 100kHz; then for many years 455–470kHz was the usual choice; many modern designs use between about 3 and 9MHz, and ssb i.f. filters are now available to 40MHz. Where the i.f. is *higher* than the signal frequency the action of the mixer is to raise the frequency of the incoming signal, and this process is now often termed *up-conversion* (a term formerly reserved for a special form of parametric mixer). Up-conversion, in conjunction with a low-pass filter at the input, is an effective means of reducing i.f. breakthrough as well as image response.

A superhet receiver, whether single- or multi-conversion, must have its first i.f. outside its tuning range. For general-coverage hf receivers tuning between say 1·5 and 30MHz, this limits the choice to below 1·5MHz or above 30MHz. To reduce image response without having to increase pre-mixer rf selectivity (which can involve costly gang-tuned circuits) professional designers are increasingly using a first i.f. well above 30MHz. This trend is being encouraged by the availability of vhf crystal filters suitable for use either directly as an ssb filter or more often with relaxed specification as a *roofing filter*.

The use of a very high first i.f. however tends to make the design of the local oscillator more critical (unless the Wadley triple-conversion drift-free technique is used). For amateur-bands-only receivers the range of choice for the first i.f. is much wider, and 3·395MHz and 9MHz are typical.

A number of receivers have adopted 9MHz i.f. with 5·0–5·5MHz local oscillator: this enables 4MHz–3·5MHz and 14·0–14·5MHz to be received without any band switching in the vfo; other bands are received using crystal-controlled converters with output at 3·5 or 14MHz.

Selective Filters

The selectivity characteristics of any receiver are determined by filters: these filters may be at signal frequency (as in a straight receiver); intermediate frequency; or audio frequency (as in a direct conversion receiver). Filters at signal frequency or i.f. are usually of bandpass characterstics; those at af may be either band-pass or low-pass. With a very high first i.f. (112–150MHz) low-pass filters may be used at rf.

A number of different types of filters are in common use: LC (inductor-capacitor) filters as in a conventional i.f. transformer or tuned circuit; crystal filters; mechanical filters; ceramic filters; RC (resistor-capacitor) active filters (usually only at af but feasible also at i.f.).

The significance of "nose" and "skirt" bandwidths, and "shape factor" was noted earlier. In many receivers, the overall shape factor is about 3 to 10. Very high-grade filters can have shape factors better than 1·1. A filter for ssb reception should have a nose bandwidth of about 2·2 to 2·5kHz; for cw about 300 to 500Hz.

LC filters By using high-Q inductors it is possible to construct effective ssb filters only up to about 150kHz, but LC filters (i.f. transformers) are generally used only in support of other forms of filter. **Fig 4.18** shows the conventional use of LC filters to form inter-stage coupling. The slope and shape of the characteristics are affected by the degree of coupling between the resonant circuits. The effective Q of

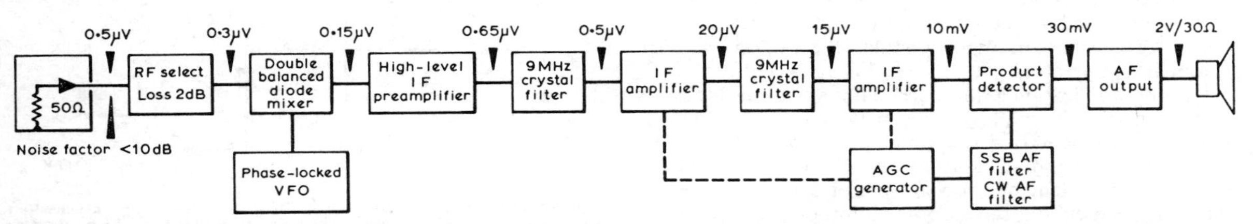

Fig 4.17. Gain distribution in a high-performance semiconductor single-conversion receiver built by G3URX using 7 integrated circuits, 27 transistors and 14 diodes. 1μV signals can be received 5kHz off tune from a 60mV signal, and the limiting factor for weak signals is the noise sidebands of the local oscillator, although the phase-locked vfo gives lower noise and spurious responses than the more usual pre-mixer vfo system.

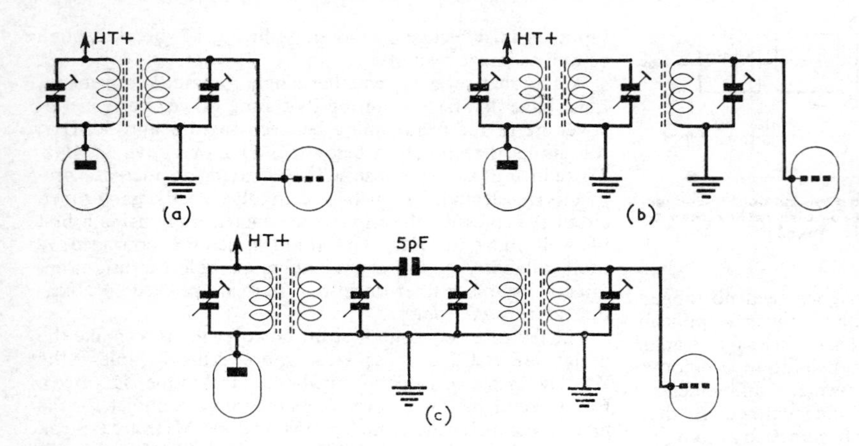

Fig 4.18. Typical inter-valve i.f. transformer couplings. (a) The conventional double-circuit i.f. transformer. (b) A triple tuned i.f. transformer provides a useful improvement in skirt selectivity. (c) If a triple tuned i.f. transformer is not available two conventional i.f. transformers can be used to provide four tuned circuits per stage.

filters of this type may be increased considerably by the technique of *Q-multiplication*: this depends on the high Q obtainable from a tuned circuit when it is operated near the point of oscillation. Some modern amateur receivers use Q-multiplication to sharpen up the selectivity for cw while using an ssb filter; this is a more economical approach than using separate ssb and cw crystal/mechanical filters.

Crystal filters A quartz crystal is the equivalent of a series/parallel tuned circuit of extremely high Q. During the period 1935 to about 1960 single-crystal filters were commonly fitted in communications receivers, with a sharply peaked resonance usually between 455 and 470kHz, but occasionally about 1,600kHz. Such filters can have a nose bandwidth of only a few hundred hertz, though the shape factor is not very good; variable selectivity can be achieved by varying the impedance into which the filter operates, a low impedance providing extremely narrow bandwidths. Although this form of filter is still very well suited to cw reception, it is not ideal for ssb, although satisfactory results can be achieved by the use of a treble-rise af network ("stenode" reception). Largely as a result of the increasing use of ssb, it is now more common to use multiple-crystal filters providing bandpass characteristics with centre frequencies up to 9MHz or so. Crystals may be grouped to form one filter or distributed in pairs over a number of stages.

Crystal Filters

The selectivity of a tuned circuit is governed by its frequency and by its Q (ratio of reactance to resistance). There are practical limits to the Q obtainable in coils and i.f. transformers. In 1929, Dr J. Robinson, a British scientist, introduced the quartz crystal resonator into radio receivers. The advantages of such a device for communication receivers were appreciated by James Lamb of the American Radio Relay League and he made popular the i.f. *crystal filter* for amateur operators.

For this application a quartz crystal may be considered as a resonant circuit with a Q of from 10,000 to 100,000 compared with about 300 for a very high grade coil and capacitor tuned circuit. From Chapter 1, it will be noted that the electrical equivalent of a crystal is not a simple series or parallel tuned circuit, but a combination of the two: it has (a) a fixed series resonant frequency (f_s) and (b) a parallel resonant frequency (f_p). The frequency (f_p) is determined partly by the capacitance of the crystal holder and by any added parallel capacitance and can be varied over a small range.

The crystal offers low impedance to signals at its series resonant frequency; a very high impedance to signals at its parallel resonant frequency, and a moderately high impedance to signals on other frequencies, tending to decrease as the frequency increases due to the parallel capacitance.

While there are a number of ways in which this high Q circuit can be incorporated into an i.f. stage, a common method—providing a variable degree of selectivity—is shown in **Fig 4.19**. When the series resonant frequency of the crystal coincides with the incoming i.f. signals, it forms a sharply tuned "acceptor" circuit, passing the signals with only slight insertion loss (loss of strength) to the grid of the succeeding stage. The exact setting of the associated parallel resonant circuit, at which the crystal will offer an extremely high resistance, is governed by the setting of the *phasing control* which balances out the effect of the holder capacitance. Such a filter can provide a nose selectivity of the order of 1kHz bandwidth or less, while the sharp rejection notch which can be shifted by the phasing control through the pass band can be of the order of 45dB. **Fig 4.20** shows the improvement which can be obtained by switching in a filter of this type in a good communications receiver (a simple method of switching the filter is to arrange the phasing trimmer to short-circuit at one end of its travel). Inspection of the response curve will show that the improvement in the skirt selectivity is not so spectacular as at the nose, and even with a

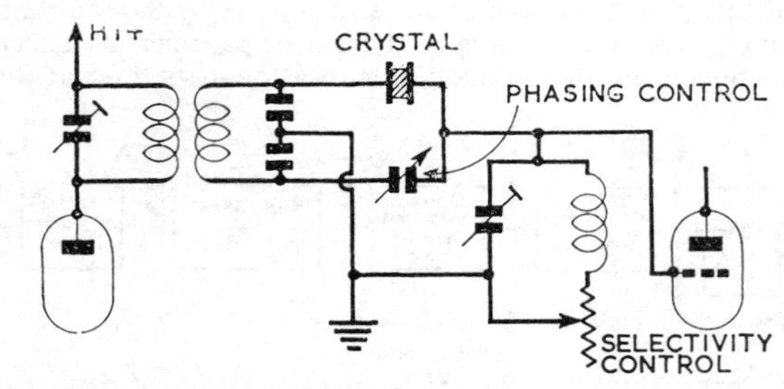

Fig 4.19. Variable selectivity crystal filter. Selectivity is greatest when the impedance of the tuned circuit is reduced by bringing the variable resistor fully into circuit. For optimum results there must be adequate screening to prevent stray coupling between the input and output circuits which would permit strong off-resonance signals to leak round the filter.

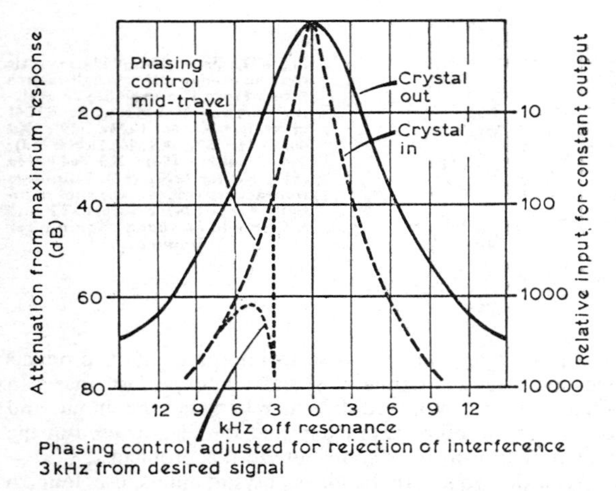

Fig 4.20. A graph showing the improvement in selectivity which can be obtained by the use of a crystal filter of the type shown in Fig 4.19.

well-designed filter may leave something to be desired in the presence of strong signals.

The degree of selectivity provided by a single crystal depends not only upon the Q of the crystal and the i.f. but also upon the impedance of the input and output circuits. The lower these impedances are, the greater will be the effect of the filter, though this will usually be accompanied by a rise in insertion loss. To broaden the selectivity curve and make the Q of the filter *appear* less, it is only necessary to raise the input or output impedances. In Fig 4.19 the input impedance may be lowered by detuning the secondary of the i.f. transformer. The output impedance will depend upon the setting of the variable resistor which forms the selectivity control. With minimum resistance in circuit, the tuned circuit will offer maximum impedance, which can be gradually lowered by bringing more resistance into circuit. Maximum impedance corresponds with minimum selectivity.

A disadvantage of the single crystal filter is that when used in the position of maximum selectivity, it may introduce considerable "ringing", rendering it difficult to copy a weak cw signal: this is due to the tendency of a high Q circuit to oscillate for a short period after being stimulated by a signal, producing a bell-like echo on the signal.

Since the minimum nose bandwidth of a single crystal filter may be as low as 100–200Hz it is not possible to receive a.m. signals through the filter unless its efficiency is degraded by a high impedance or by "stenode" tone correction. If this is done, such a filter will often prove most useful for telephony reception through bad interference, though at some cost to the quality of reproduction. Even the most simple form of crystal filter, consisting of a crystal in series with the i.f. signal path, without balancing or phasing, can be of use.

Bandpass Crystal Filters

As already noted, the sharply peaked response curve of a single-crystal filter is not ideal, and has by modern standards a relatively poor shape factor: improved results can be achieved with what is termed a *half-lattice* or bandpass filter. Basically, this comprises two crystals chosen so that their series resonant frequencies differ by an amount approximately equal to the bandwidth required; for example about 300Hz apart for cw, 3–4kHz apart for a.m. telephony or 2kHz apart for ssb, see **Fig 4.21(b)**. This form of filter, developed in the 'thirties, has in recent years come into widespread amateur use. Although it has a much improved slope over the single crystal filter, in its simplest form there will still be certain frequencies, just outside the main passband, at which the attenuation is reduced. Unless a balancing trimmer is connected across the higher frequency crystal the sides of the response curve tend to broaden out towards the bottom. As the capacitance across the crystal is increased, the sides of the curve steepen, but the side lobes tend to become more pronounced. Capacitance across the lower frequency crystal broadens the response and deepens the trough in the centre of the passband.

To eliminate the "humps" in the response curve additional crystals may be included in the filter; these may be up to about six in number, **Fig 4.22**. Examples of bandpass filter response characteristics are shown in Fig 4.21. Alternatively, additional filter sections may be incorporated in the i.f. section in cascade. An advantage of using several cascaded filters is that less critical balancing and adjustment are needed. Provided that there is no leakage of i.f. signals around the filter due to stray capacitances or other forms of unwanted coupling (an important consideration with all selective filters)

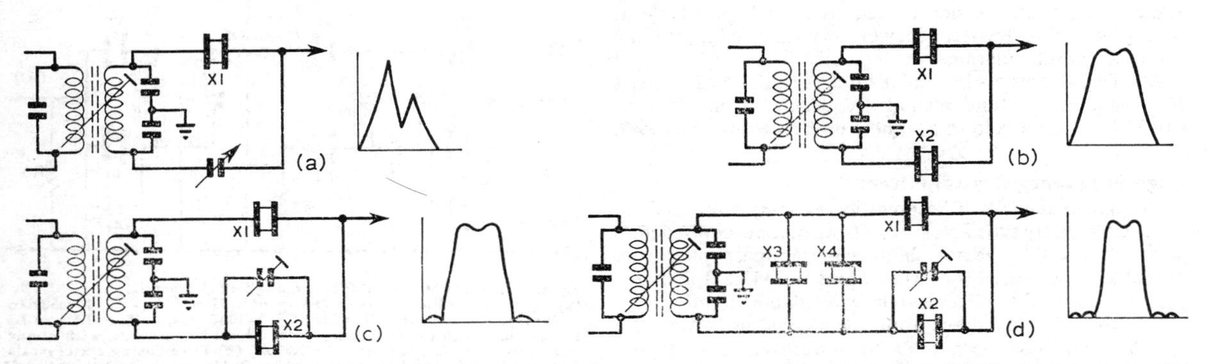

Fig 4.21. Half-lattice crystal filters showing the improvement in the shape of the curve which can be obtained when the crystals are correctly balanced or when extra crystals are used to reduce the "humps". Typical crystal frequencies would be X1 464·8kHz, X2 466·7kHz, X3 463kHz, X4 468·5kHz.

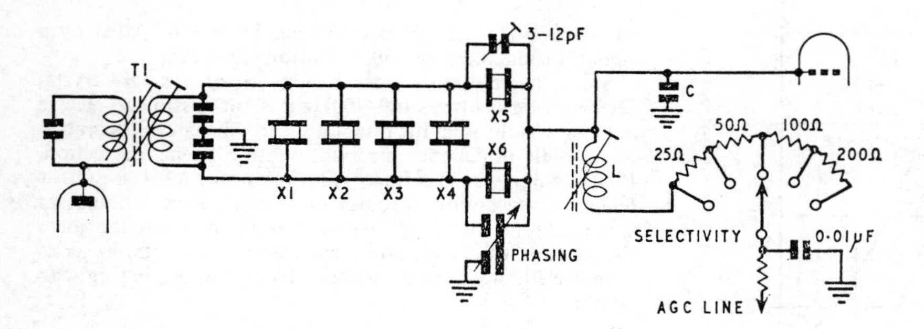

Fig 4.22. Up to about six crystals may be used in a single half-lattice filter. Here is a variable selectivity unit for 465kHz using FT241 crystals. X1 461·1kHz (49); X2 462·9kHz (50); X3 468·5kHz (53); X4 470·4kHz (54); X5 464·8kHz (51); X6 466·7kHz (52). Numbers in brackets refer to channel numbers for FT241 crystals. T1 and L/C should be tuned to mid-filter frequency.

extremely good shape factors of the order of 1·5 can be achieved with about three cascaded filters on about 460kHz. This system is used in the receiver described on p4.50. There will be an insertion loss of the order of 6dB per section.

To provide a sharp variable rejection notch, it is advantageous to incorporate some form of bridged-*T* filter or *Q* multiplier in order to be able to eliminate steady heterodyne interference.

To reduce susceptibility to cross-modulation or blocking, crystal and other selective filters should be placed as near as possible to the front end of the receiver.

Where the first fixed i.f. is too high for a crystal filter to be completely effective, it is nevertheless advantageous to include a preliminary filter at this point, with the more selective filter later in the receiver (but still at the earliest practicable stage).

Crystals for MF Bandpass Filters

For mf bandpass filters, most amateurs use surplus type FT241 crystals. These are in two groups labelled in frequency and with a channel number. The fundamental frequencies are from 370 to 540kHz. The first group are marked in frequencies from 20 to 27·9MHz with channel numbers up to 79. The actual crystal frequency may be found by dividing the frequencies marked on the holder by 54. As the channels are spaced 0·1MHz apart, the crystal separations are 1·85kHz. The second group of FT241 crystals are marked from 27 to 38·9MHz in 0·1MHz steps, with channel numbers from 270 to 389; the fundamental frequency can be found by dividing the marked frequency by 72. The spacing of this group is about 1·49kHz. FT241 crystals of similar channel numbers are likely to show some slight spread of frequencies. Should the precise frequencies required not be available, it is possible to raise the frequency of a crystal by edge grinding and to lower it by plating.

AN/TRC-1 crystals in similar holders to the FT241 series are available in fundamental frequencies from 729 to 1,040 kHz and marked in channel numbers from 70 to 99·9.

High Frequency Crystal Filters

At high frequencies it becomes increasingly more difficult to obtain entirely satisfactory results with home-construction although excellent factory-built ssb bandpass filters are available (at appreciable cost) at 9 and 10·7MHz. Filters for use as "roofing filters" are similarly available at vhf.

Effective hf crystal filters using FT243 crystals between 5·5 and 6·5MHz have been built by a number of amateurs; typical designs are shown in **Fig 4.23**.

It should be noted that crystal filters are not always linear and passive and can give rise to intermodulation products when subject to high-level signals; imd performance can sometimes be improved by interchanging the input and output connections. It is likely that the ferrites used in the matching transformers contribute to this problem.

Typically good ssb bandpass crystal filters use four to six crystals. A more recent form of hf and vhf filter is the compact *monolithic crystal filter* (mxf). Such filters consist of a quartz wafer on which pairs of metal electrodes are deposited on opposite sides of the plate. Complete filters may occupy only a TO-5 transistor capsule, and can be designed for resonant frequencies up to the uhf region.

Ladder Crystal Filters

For home-construction a particularly useful alternative to the half-lattice arrangement is the *ladder filter*, which can provide excellent ssb filters at frequencies between about

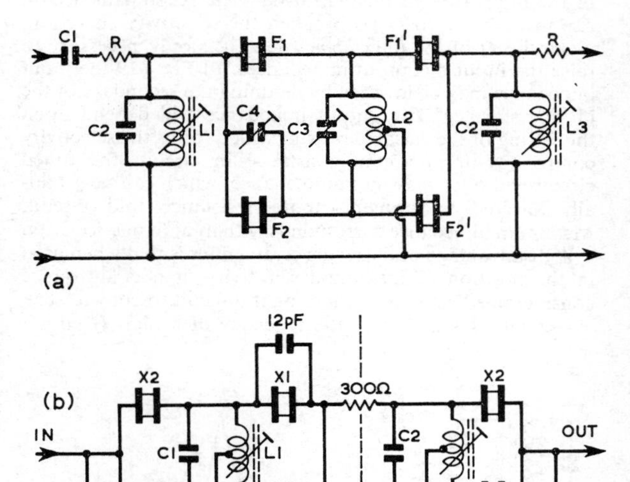

Fig 4.23. Typical hf crystal filters using FT243 crystals. (a) Filter using four crystals (F1 = F1′; F2 = F2′ = F1 + 1,500 to 2,500Hz); C1, 2,000pF; C2, 47pF; C3, 3–30pF; C4, 3–10pF. L1 and L3 to resonate at filter frequency with C2. L2 to resonate with about 15pF setting of C3. R should be 2,000Ω. (b) Filter using six crystals and capable of "nose" passbands of 2·4–3kHz and —60dB skirt bandwidth of 6–7kHz. The sets of X1 and X2 crystals should be separated by 1·7kHz in series resonant mode. R1, R2 and R3 are 560 to 820Ω.

4–11MHz. This form of crystal filter uses a number of crystals of the same (or nearly the same) frequency and so avoids the need for accurate crystal etching or selection. Further, provided it is correctly terminated, it does not require the use of transformers or inductors. Plated crystals such as the HC6U or 10XJ types are more likely to form good ssb filters, although virtually any type of crystal may be used for cw filters.

A number of practical design approaches have been described by J. Pochet, F6BQP, ("Technical Topics", *Radio Communication* September 1976 and *Wireless World* July 1977) and in a series of articles "Some experiments with hf ladder crystal filters" by J. A. Hardcastle, G3JIR (*Radio Communication* December 1976, January, February and September 1977).

Fig 4.25 outlines the F6BQP approach. By designing for lower termination impedances and/or lower frequency crystals excellent cw filters can be formed. A feature of the ladder design is the very high ultimate out-of-band rejection that can be achieved (75–95dB) in three- or four-section filters. For ssb filters at about 8MHz a suitable design impedance would be about 800Ω with a typical "nose" bandpass of 2–2·1kHz.

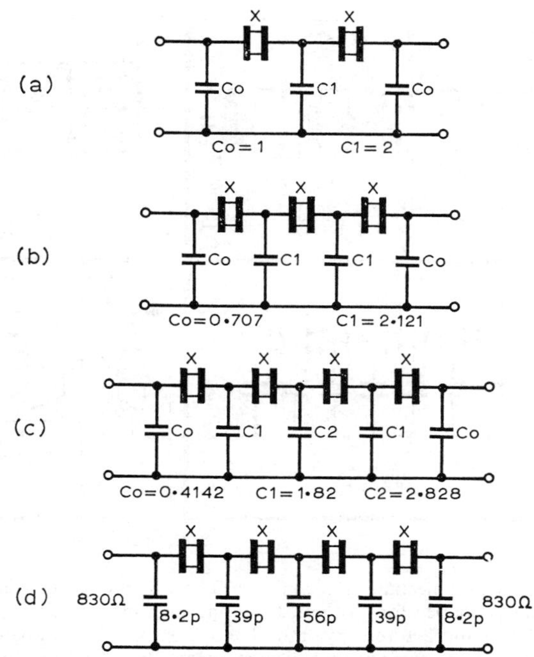

Fig 4.25. Crystal ladder filters, investigated by F6BQP, can provide effective ssb and cw bandpass filters. All crystals (X) are of the same resonant frequency and preferably between 8 and 10MHz for ssb units. To calculate values for the capacitors multiply the coefficients given above by $1/(2\pi fR)$ where f is frequency of crystal in hertz (MHz by 10^6), R is input and output termination impedance, and 2π is roughly 6·28. (a) Two-crystal unit with relatively poor shape factor. (b) Three-crystal filter can give good results. (c) Four-crystal unit capable of excellent results. (d) Practical realization of four-crystal unit using 8,314kHz crystals, 10 per cent preferred-value capacitors and termination impedance of 820Ω. Note that for crystals between 8 and 10MHz the termination impedance should be between about 800 and 1,000Ω for ssb. At lower crystal frequencies use higher design impedances to obtain sufficient bandwidth. For cw filters use lower impedance and/or lower frequency crystals.

Ceramic Filters

Piezoelectric effects are not confined to quartz crystals; in recent years increasing use has been made of certain ceramics, such as lead zirconate titanate (PZT). Small discs of PZT, which resonate in the radial dimension, can form economical selective filters in much the same way as quartz, though with considerably lower *Q*. Ceramic i.f. "transfilters" are a convenient means of providing the low impedances needed for bipolar transistor circuits. The simplest ceramic filters use just one resonator, but numbers of resonators can be coupled together to form filters of required bandwidth and shape factor. While quite good nose selectivity is achieved with simple ceramic filters, multiple resonators are required if good shape factors are to be achieved. Some filters are of "hybrid" form using combinations of inductors and ceramic resonators.

Examples of ceramic filters include the Mullard LP1175 in which a hybrid unit provides the degree of selectivity associated with much larger conventional i.f. transformers; a somewhat similar arrangement is used in the smaller Toko filters such as the CFT455C which has a bandwidth (to −6dB) of 6kHz. A more complex 15-element filter is the Murata CFS-455A with a bandwidth of 3kHz at −6dB, 7·5kHz at −70dB and insertion loss 9dB, with input and output impedances of 2kΩ and centre frequency of 455kHz. In general ceramic filters are available from 50kHz to about 10·7MHz centre frequencies.

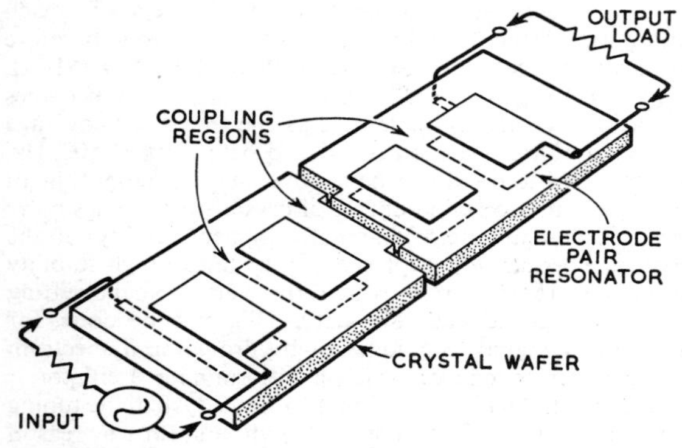

Fig 4.24. Monolithic form of hf crystal bandpass filter (mxf).

Ceramic filters tend to be more economical than crystal or mechanical filters but have lower temperature stability and may have greater passband attenuation.

Mechanical and Miscellaneous Filters

Very effective ssb and cw filters at intermediate frequencies from about 60 to 600kHz depend on the mechanical resonances of a series of small elements usually in the form of discs: **Fig 4.27**. The mechanical filter consists of three basic elements: two magneto-striction transducers which convert the i.f. signals into mechanical vibrations and vice versa; a series of metal discs mechanically resonated to the required frequency; disc coupling rods. Each disc represents a high *Q* series resonant circuit and the bandwidth of the filter is determined by the coupling rods. 6–60dB shape factors can be as low as about 1·2, with low passband attenuation.

Other forms of mechanical filters have been developed which include ceramic piezoelectric transducers with mechanical coupling; they thus represent a combination of

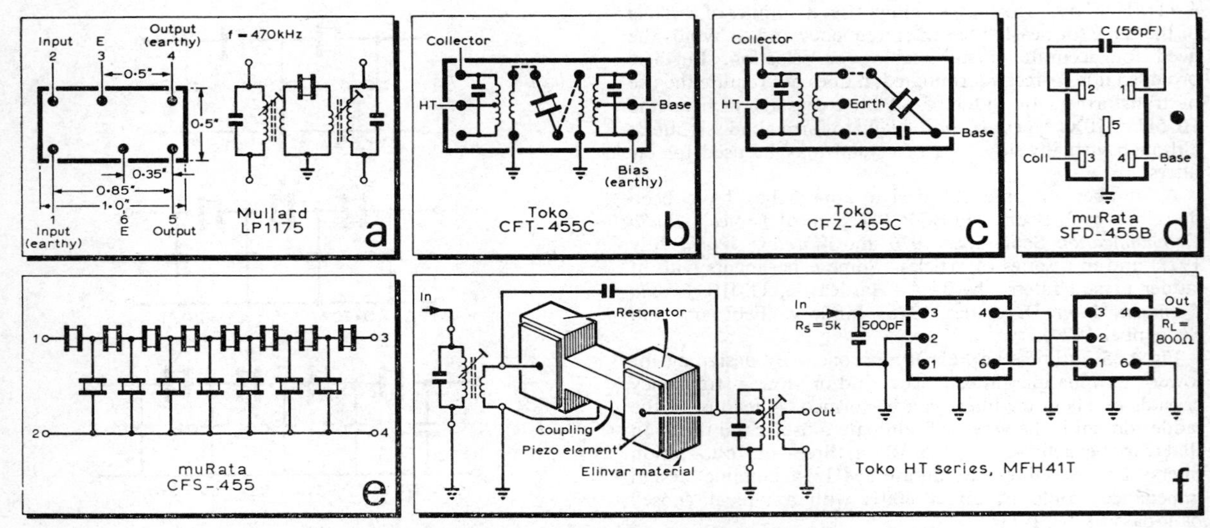

Fig 4.26. Representative types of ceramic filters.

ceramic and mechanical techniques. These filters may consist of an H-shaped form of construction; such filters include a range manufactured by the Toko company of Japan. Generally the performance of such filters is below that of the disc resonator type, but can still be useful.

Surface acoustic wave filters are being developed for possible bandpass filter applications where discrete-element filters have previously been used, including i.f. filters.

STABILITY OF RECEIVERS

The resolution of ssb speech and the reception of a cw signal requires that a receiver can be tuned to, and remain within, about 25–30Hz of the frequency of the incoming signal. At 29MHz this represents a tolerance of only about one part in a million. For amateur operation the main requirement is that this degree of stability should be maintained over periods of up to about 30 min. Long-term stability is less important for amateurs than short-term stability; it will also be most convenient if a receiver reaches this degree of stability within a fairly short time of switching on.

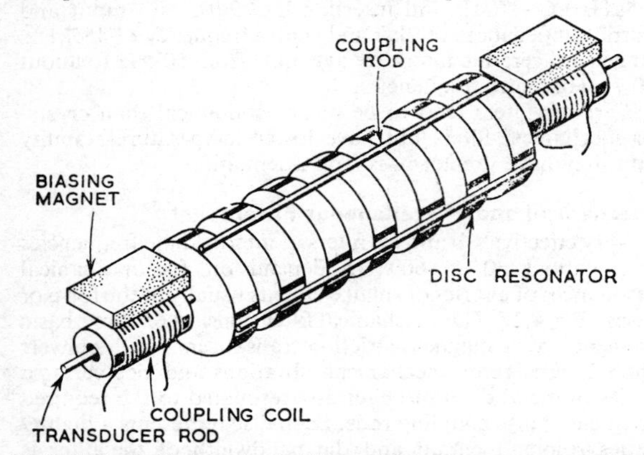

Fig 4.27. The Collins mechanical filter. I.F. signals are converted into mechanical vibrations by a magneto-strictive transducer and passed along a series of resonant discs, then finally reconverted into electrical (i.f.) signals by a second magneto-strictive transducer. The bandwidth of the filter is governed by the number of resonant discs and the design of the coupling rods. Very good shape factors can be achieved but the maximum frequency of such filters is usually about 500kHz.

It is extremely difficult to achieve or even approach this order of stability with a free-running, band-switched variably tuned oscillator working on the fundamental injection frequency, although with care a well-designed fet oscillator can come fairly close. This has led (in the same way as for transmitting vfo units) to various frequency-synthesis techniques in which the stability of a free-running oscillator is enhanced by the use of crystals. The following are among the techniques used:

(1) Multi-conversion receiver with crystal-controlled first oscillator and variable first i.f. This very popular technique has a tunable receiver section covering only one fixed frequency band; for example 5·0 to 5·5MHz. The oscillator can be carefully designed and temperature compensated over one band without the problems arising from the uncertain action of wavechange switches, and can be separated from the mixer stage by means of an isolating or buffer stage. For the front-end section a separate crystal is needed for each tuning range (the 28MHz band may require four or more crystals to provide full coverage of 28·0MHz to 29·7MHz).

(2) *Partial synthesis* The arrangement of (1) becomes increasingly costly to implement as the frequency coverage of the variable i.f. section is reduced below about 500kHz. or is required to provide general coverage throughout the hf band. Beyond a certain number of crystals it becomes more economical (and offers potentially higher stability) if the separate crystals are replaced by a single high-stability crystal (eg 1MHz) from which the various band-setting frequencies are derived (**Fig 4.28**). This may be done, for example, by digital techniques or by providing a spectrum of harmonics to one of which a free-running oscillator is phase-locked. It will be noted that with this system the tuning within any band still depends on the vfo and for this reason is termed *partial* synthesis.

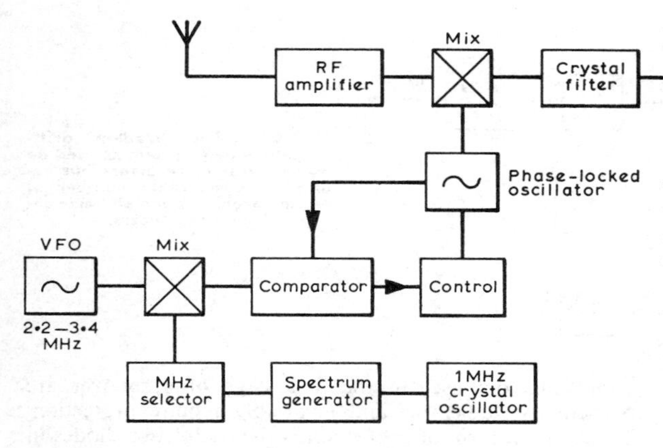

Fig 4.28. Partial frequency synthesis using fixed range vfo with MHz signals derived from a single 1MHz crystal.

(3) *Single-conversion receiver with heterodyne (pre-mixer) vfo* In this system of partial synthesis, the receiver may be a single-conversion superhet (or dual-conversion with fixed I.F.S) (**Fig 4.29**). The variable hf injection frequency is obtained using a heterodyne-type vfo, in which the output of a crystal-controlled oscillator is mixed with that of a single-range vfo, and the output is then filtered and used as the injection frequency. The overall stability will be much the same as for (1) but the system allows the selective filter to be placed immediately after the first mixer stage. However to reduce spurious responses the unwanted mixer products of the heterodyne-vfo must be reduced to a very low level and not reach the mixer. As with the tunable i.f. system, this arrangement results in equal tuning rates on all bands. The system can be extended by replacing the series of separate crystals with a single crystal plus phase-locking arrangement as in (2).

(4) *Fixed i.f. receiver with partial frequency synthesis* An ingenious frequency synthesizer (due to Plessey) incorporating an interpolating LC oscillator and suitable for use with single- or multiple-conversion receivers having fixed intermediate frequencies is outlined in **Fig 4.30**. The output of the vfo is passed through a variable-ratio divider and then added to the reference frequency by means of a mixer. The sum of the two frequencies applied to the mixer is selected by means of a bandpass filter and provides one input to a phase comparator; the other input to this phase comparator is obtained from the output of the voltage-controlled oscillator after it has also been divided in the same ratio as the interpolating frequency. The phase comparator can then be used to phase lock the vco to the frequency $mf_{ref} + f_{vfo}$ where m represents the variable-ratio division. If for example f_{ref} is 1MHz and f_{vfo} covers a tuning range of 1–2MHz and the variable ratio dividers are set to 16, then the vco output can be controlled over 17,000kHz to 18,000kHz; if the ratio divider is changed to 20 then the tuning range becomes 21,000kHz to 22,000kHz and so on. The use of two relatively simple variable ratio divider chains thus makes it possible to provide output over the full hf range, with the vfo at a low frequency (eg 1–2MHz).

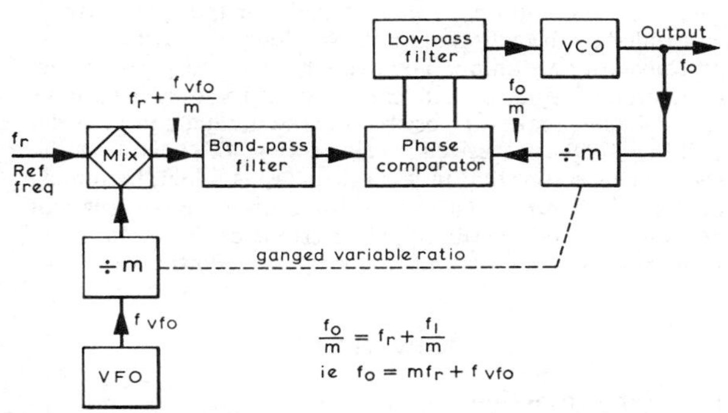

Fig 4.30. A digital form of partial synthesis developed by Plessey and suitable for use in single-conversion receivers.

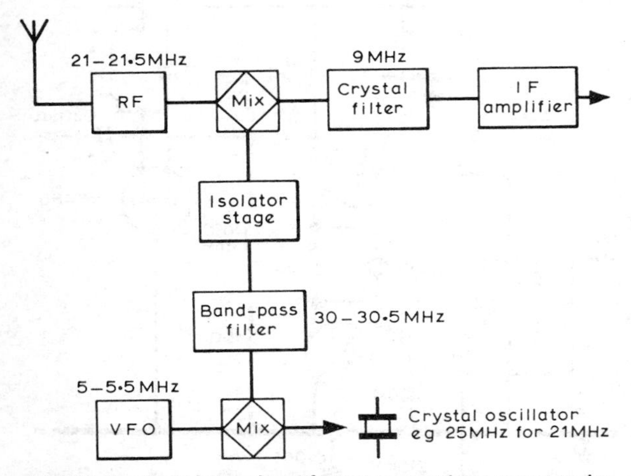

Fig 4.29. Pre-mixer heterodyne vfo system provides constant tuning rate with single conversion receiver. Requires a number of crystals and care must be taken to reduce spurious oscillator products reaching the main mixer.

(5) *VXO Local Oscillator* For reception over only small segments of a band or bands, a variable-crystal oscillator (vxo) can be used to provide high stability for mobile or portable receivers. As explained in Chapter 6, the frequency of a crystal can be "pulled" over a small percentage of its nominal frequency without significant loss of stability. The system is attractive for small transceivers.

(6) *Drift-Cancelling Wadley Loop* A stable form of front-end for use with variable i.f.-type receivers is the multiple-conversion Wadley loop which was pioneered in the Racal RA17 receiver (**Fig 4.31**). By means of an ingenious triple mixing arrangement a variable oscillator tuning 40·5 to 69·5MHz and a 1MHz crystal oscillator provides continuous tuning over the range 0·5 to 30MHz as a series of 1MHz segments. Any drift of the variable vhf oscillator is automatically corrected. Although the system has been used successfully in home-constructed receivers, it is essential to use a good vhf band-pass filter (eg 40MHz ± 0·65MHz) if spurious responses are to be minimized. Further since it involves multiple conversion in the signal-path it is difficult (especially in all-semiconductor form) to achieve wide dynamic range.

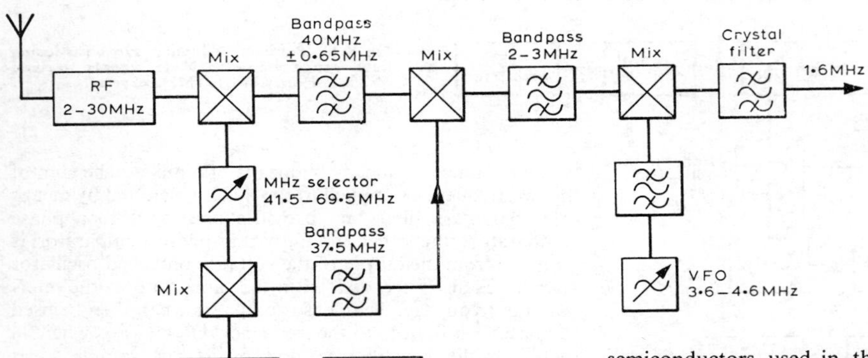

Fig 4.31. The Wadley drift-cancelling loop system as used on many Racal hf receivers but requiring considerable number of mixing processes and effective vhf bandpass filters.

(7) *Full Frequency Synthesis* In this system all the required injection frequencies are derived from one or more high-stability crystal oscillators in a series of discrete steps which may be spaced at 1kHz intervals or less (some professional receivers provide 10Hz steps). Sometimes the required frequency is set up on a series of dials, or the synthesizer is controlled by the action of knob switching to provide the feel of conventional tuning. The availability of low-cost digital integrated circuits, which can be used to form a variable ratio divider, makes it possible for the amateur to consider building fully synthesized receivers (**Fig 4.32**). However it should be stressed that unless great care is taken, there may be excessive "noise" (jitter) on the output signal, spurious outputs and the possibility of interference from the high-speed pulses formed within the frequency synthesizer.

CIRCUITRY

Receiver Protection

Receivers, particularly where they are to be used alongside a medium- or high-powered transmitter, need to be protected from high transient or other voltages induced by the local transmitter or by build-up of static voltages on the aerial. Valve receivers may suffer burn out of aerial input coils; semiconductors used in the first stage of a receiver are particularly vulnerable and invariably require protection. The simplest form of protection is the use of two diodes in back-to back configuration. Such a combination passes signals less than the potential hill of the diodes (about 0·3V for germanium diodes, about 0·6V for silicon diodes) but provides virtually a short circuit for higher voltage signals. This system is usually effective but has the disadvantage that it introduces non-linear devices into the signal path and may occasionally be the cause of cross- and inter-modulation.

The mosfet devices are particularly vulnerable to static puncture and some types (eg RCA 40673) include built in zener diodes to protect the "gates" of the main structure. Since these have limited rating it may still be advisable to support them with external diodes or small gas filled transient suppressors.

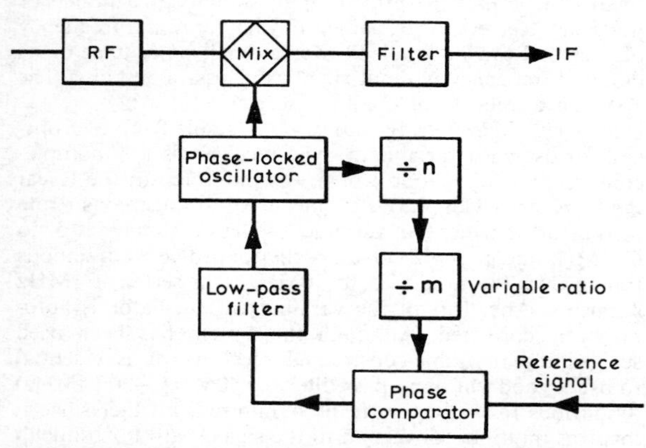

Fig 4.32. Frequency synthesized local oscillator using digital techniques—typical of the approach now used in many professional hf receivers.

Input Circuits and RF Amplifiers

It has already been noted that with low noise mixers it is now possible to dispense with high gain rf amplification. Amplifiers at the signal frequency may however still be advisable to provide: pre-mixer selectivity; an agc controlled stage which is in effect a controlled attenuator on strong signals; to counter the effects of conversion loss in diode mixer stages.

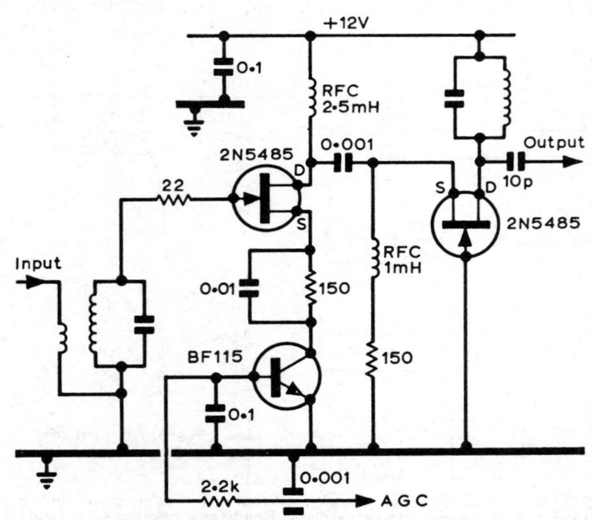

Fig 4.33. Cascode rf amplifier using two jfets with transistor as agc control element.

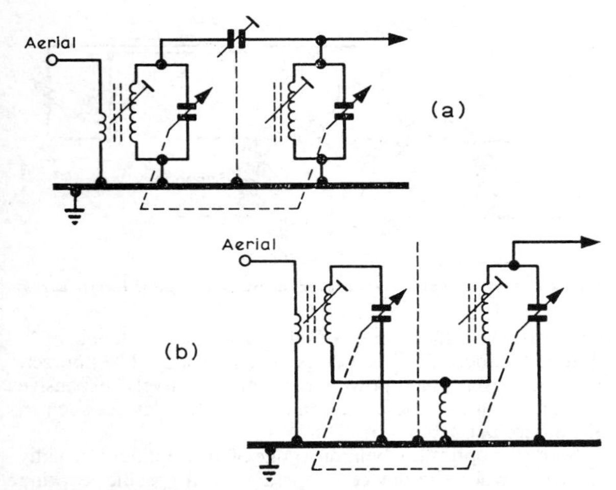

Fig 4.34. Typical rf input circuits used to enhance rf selectivity and capable of providing more than 40dB attenuation of unwanted signals 10 per cent off tune.

In practice semiconductor rf stages are often based on junction FETS as shown in **Fig 4.33** or dual-gate MOSFETS, or alternatively integrated circuits in which large numbers of bipolar transistors are used in configurations designed to increase their signal handling capabilities.

Tuned circuits between the aerial and the first stage (mixer or rf amplifier) have two main functions: to provide high attenuation at the image frequency; to reduce as far as possible the amplitude of all signals outside the i.f. passband.

Most amateur receivers still require good pre-mixer selectivity; this can be achieved by using a number of tuned circuits coupled through low-gain amplifiers, or alternatively by tunable or fixed band-pass filters that attenuate all signals outside the amateur bands (**Fig 4.34**). The most commonly used input arrangement consists of two tuned circuits with screening between them and either top-coupled through a small value fixed capacitor, or bottom-coupled through a small common inductance. Slightly more complex but capable of rather better results is the minimum-loss Cohn filter; this is capable of reducing signals 10 per cent off-tune by as much as 60dB provided that an insertion loss of about 4dB is acceptable. This compares with about 50dB (and rather less insertion loss) for an undercoupled pair of tuned circuits. The Cohn filter is perhaps more suited for use as a fixed band-pass filter which can be used in front of receivers

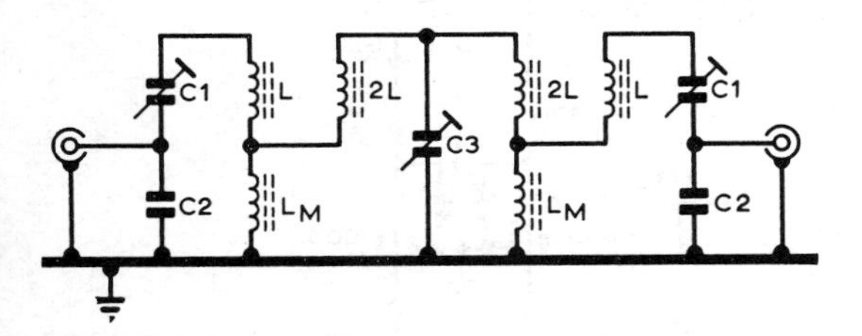

Fig 4.35. Cohn minimum loss bandpass filter suitable for providing additional rf selectivity to existing receivers. Values for 14MHz L 2·95μH, 2L 5·9μH, Lm 0·27μF, C1 22pF with 25pF trimmer, C2 340pF, C3 10–60pF (about 34pF nominal). Values for 3·5MHz L 8μH, 2L 16μH, Lm 2·4μH, C1 150pF + 33pF + 5–25pF trimmer, C2 1nF, C3 150pF + 10–60pF trimmer.

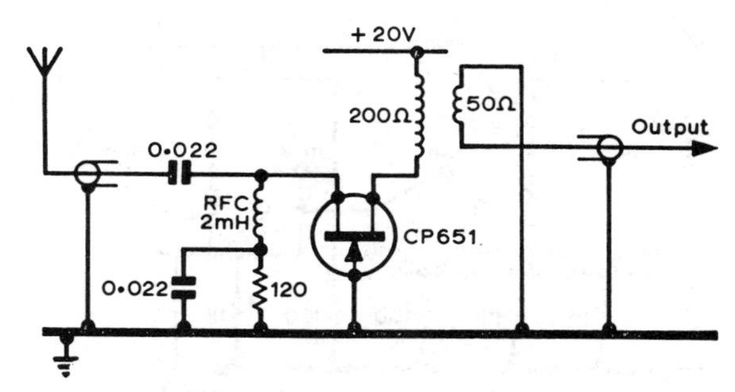

Fig 4.36. Broadband rf amplifier using power fet and capable of handling signals to almost 3-V p-p, 0·5 to 40MHz with 2·5dB noise figure and 140dB dynamic range. Drain current 40mA. Voltage gain 10dB.

having inadequate rf selectivity. **Fig 4.35** shows suitable values for 3·5 and 14MHz filters.

Broadband and untuned rf stages are convenient in construction but can be recommended only when the devices used in the front-end of the receiver have wide dynamic range. An example shown in **Fig 4.36** is a power fet designed specifically for this application and operated in the earthed-gate mode suitable for use on incoming low-impedance coaxial feeders.

Unless the front-end of the receiver is capable of coping with the full range of signals likely to be received, it may be useful to fit an attenuator working directly on the input signal. Such attenuators are particularly useful in front of integrated-circuit and mosfet amplifiers. **Fig 4.37** shows simple techniques for providing manual attenuation control; **Fig 4.38** is a switched attenuator providing constant impedance characteristics.

Various forms of attenuators controlled from the agc line are possible. **Fig 4.39** shows a system based on p-i-n diodes; **Fig 4.40** is based on toroid ferrite cores and can provide up to about 45dB attenuation when controlled by a potentiometer. **Fig 4.41** shows an adaptation with mosfet control element for use on agc lines although the range is limited to about 20dB.

The tuned circuits used in front ends may be based on toroid cores since these can be used without screening with little risk of oscillation due to mutual coupling.

It is important to check filters and tuned circuits for non-linearity in iron or ferrite materials. Intermodulation and cross-modulation can be caused by the cores where the flux level rises above the point at which saturation effects begin to occur.

A wideband amplifier placed in front of a mixer of wide

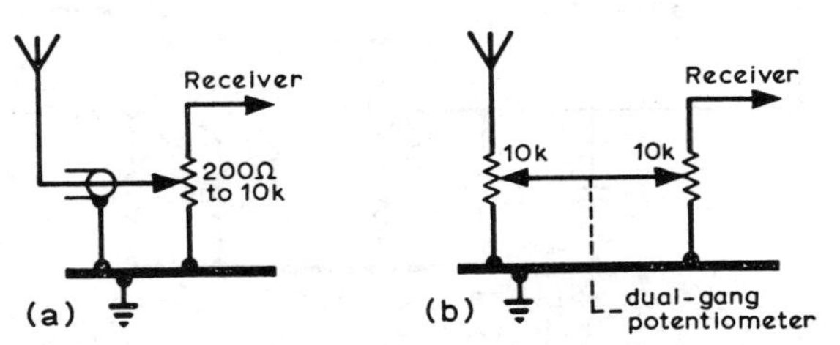

Fig 4.37. Simple attenuators for use in front of a receiver of restricted dynamic range. (a) No attempt is made to maintain constant impedance. (b) Represents less change in impedance.

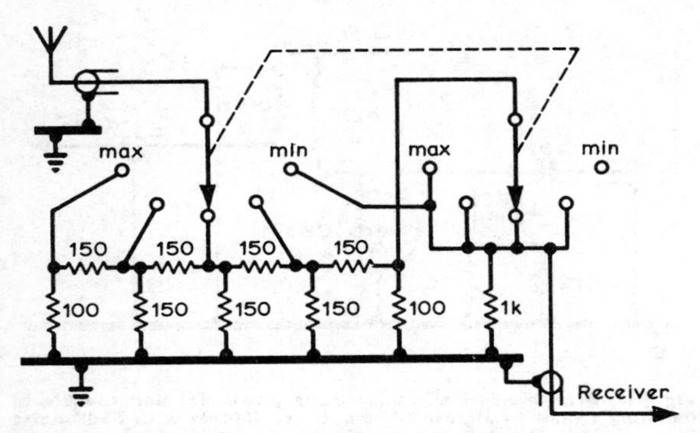

Fig 4.38. Switched aerial attenuator for incorporation in semiconductor receivers.

dynamic range must itself have good dynamic range. Dynamic range can be defined as the ratio of the minimum detectable signal (10dB above noise) to that signal which gives a barely noticeable departure from linearity (eg 1dB gain compression).

Amplifiers based on power FETS can approach 140dB dynamic range when operated at low gain (about 10dB) and with about 40mA drain current. This compares with about 90 to 100dB for good valves (eg E810F), 80 to 85dB for small signal fet devices; 70 to 90dB for small-signal bipolars. The dynamic range of an amplifier can be increased by the operation of two devices in a balanced (push-pull) mode.

Fig 4.36 shows an amplifier with a dynamic range

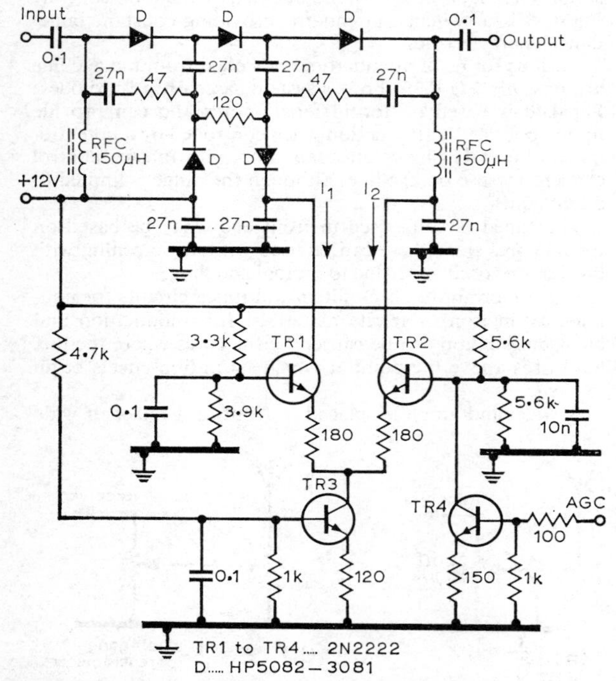

Fig 4.39. Five p-i-n diodes in a double-T arrangement form an agc-controlled attenuator. The sum of the transistor collector currents is maintained constant to keep input and output impedances constant.

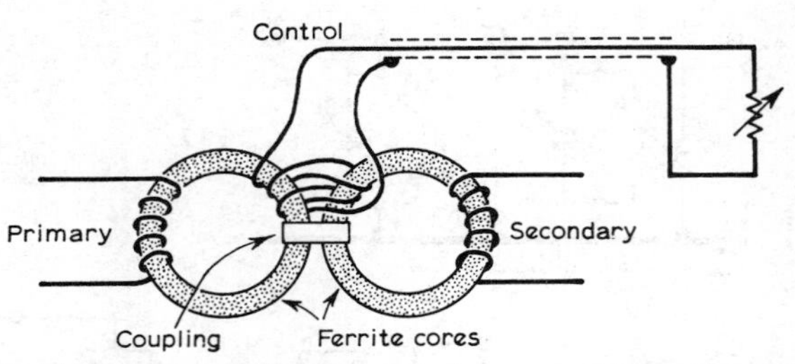

Fig 4.40. Basic form of rf level control using two toroidal ferrite cores.

approaching 140dB and suitable for use in front of a double-balanced Schottky diode mixer. It should be appreciated however that power FETS are relatively expensive devices, although these are some lower cost devices such as the Siliconix E310.

Since the optimum dynamic range of an amplifier is usually achieved when the device is operated at a specific working point (ie bias potential) it may be an advantage to design the stage for fixed (low) gain with front-end gain controlled by means of an aerial attenuator (manual or agc-controlled). Attenuation of signals ahead of a stage subject to cross-modulation is often beneficial since 1dB of attenuation reduces cross-modulation by 2dB.

For semiconductor stages using FETS and bipolar transistors the grounded-gate and grounded-base configuration is to be preferred. Special types of bipolar transistors (such as the BF314, BF324 developed for fm radio tuners) provide a dynamic range comparable with many junction FETS, a noise figure of about 4dB and a gain of about 15dB with a collector current of about 5mA. Generally the higher the input power of the device the greater is likely to be its signal handling capabilities: overlay and multi-emitter rf power transistors or those developed for catv applications can have very good intermodulation characteristics.

For general purpose small-signal unneutralized amplifiers

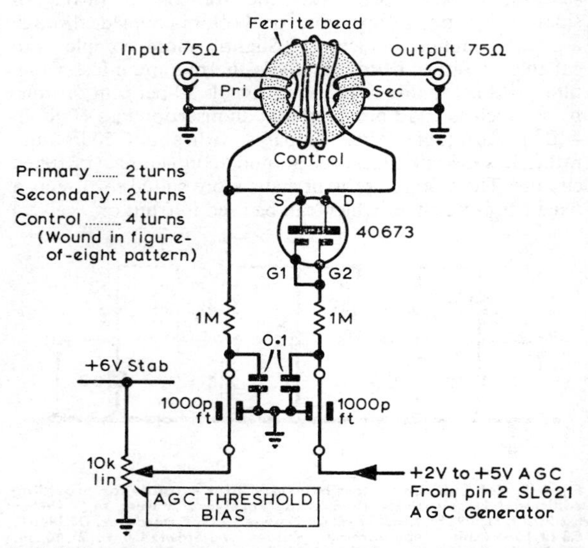

Fig 4.41. Automatic aerial attenuator based on the technique shown in Fig 4.40.

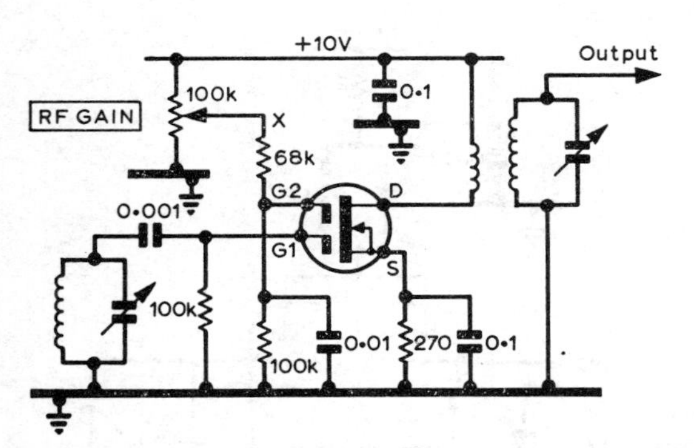

Fig 4.42. Typical dual-gate mosfet rf or i.f. amplifier. G2 is normally biased to about one-third of the positive voltage of drain. In place of manual gain control point X can be connected to a positive agc line.

the zener-protected dual-gate mosfet is probably the best and most versatile of the low-cost discrete semiconductor devices, with its inherent cascode configuration. If gate 2 is based initially at about 30 to 40 per cent of the drain voltage, gain can be reduced (manually or by agc action) by lowering this gate 2 voltage with the advantage that this then *increases* the signal handling capacity at this stage. This type of device can be used effectively for rf, mixer, i.f., product detector, af and oscillator applications in hf receivers.

Mixers

Much attention has been given in recent years to improving mixer performance in order to make superhet designs less subject to spurious responses and to improve their ability to handle weak signals in the presence of strong unwanted signals. In particular there has been increasing use of low-noise balanced and double-balanced mixers, sometimes constructed in wideband form.

Mixers operate either in the form of switching mixers (the normal arrangement with diode mixers) or in what are termed "continuous non-linear" (cnl) modes. Generally switching mixers can provide better performance than cnl modes but require more oscillator injection. The concept of "linearity" in mixers may seem a contradiction in terms since in order to introduce frequency conversion the device *must* behave in a highly non-linear fashion in so far as the oscillator/signal mixing process is concerned and the term "linearity" refers only to the signal path.

For valve receivers a breakthrough was the appearance of the *beam-deflection mixer* (7360, 6JH8) which can be used in unbalanced or balanced arrangements as a switching-type mixer, with a noise performance that makes it suitable for use as the first stage of an hf receiver. As for all switching mixers there is a requirement for appreciable oscillator power. The 7360 can provide exceptional performance in its ability to handle input signals of the order of 2·5V, with conversion gain of 20dB and noise figure of only 5dB.

Because of their near square-law characteristics field effect devices make successful mixers provided that care is taken on the oscillator drive level and the operating point (ie bias resistor); preferably both these should be adjusted to suit the individual device used (**Figs 4.43, 4.44**).

Low-noise mixers capable of wide dynamic range (**Fig 4.45**) and handling signals of the order of a volt or more include:

Power FETS or catv bipolars in balanced and double-balanced configurations.
Beam deflection valves (7360, 6JH8).
Parametric up-converters

Wide but rather lower dynamic ranges can be achieved with:

Balanced triode mixers.
Balanced and double-balanced mixers using Schottky (hot carrier) diodes.

Signal levels of the order of 10–50mV (taking into account the necessary pre-mixer gain) can be handled by a number of devices including multi-electrode valves, double-balanced

TABLE 4.2

Basic mixer arrangements

Characteristic	Single-ended	Single-balanced	Double-balanced
Bandwidth	several decades possible	decade	decade
Relative intermodulation density	1	0·5	0·25
Interport isolation	Little	10–20dB	> 30dB
Relative oscillator power	0dB	+ 3dB	+ 6dB

TABLE 4.3

Device comparisons

Device	Advantages	Disadvantages
Bipolar transistor	Low noise figure High gain Low dc power	High intermodulation Easy overload Subject to burnout
Diode	Low noise figure High power handling High burn-out level	High lo drive Interface to i.f. Conversion loss
JFET	Low noise figure Conversion gain Excellent in performance Square law characteristic Excellent overload High burn-out level	Optimum conversion gain not possible at optimum square law response level High lo power
Dual-gate mosfet	Low im distortion AGC Square law characteristic	High noise figure Poor burnout level Unstable

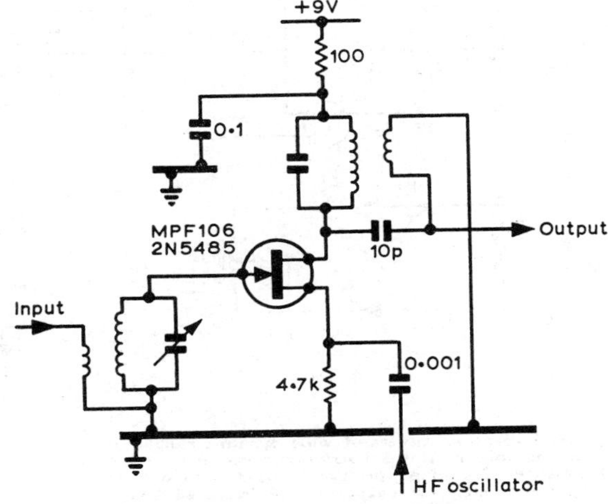

Fig 4.43. Mixer using junction fet with oscillator fed to source.

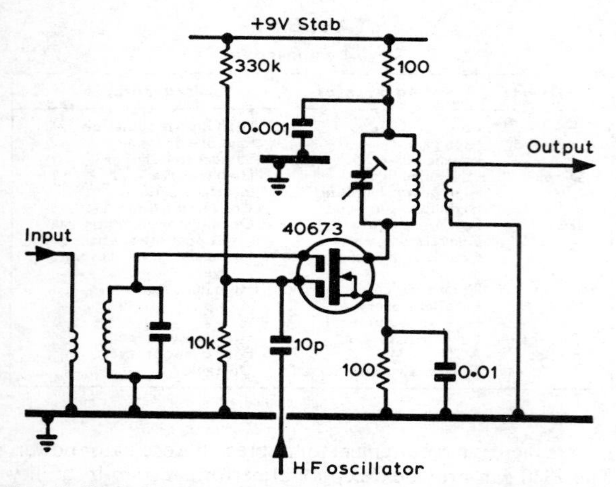

Fig 4.44. Typical dual-gate mosfet mixer—one of the best "simple" semiconductor mixers providing gain and requiring only low oscillator injection, but of rather limited dynamic range.

Fig 4.46. Double-triode mixer designed for good strong-signal performance rather than maximum sensitivity and suitable for use as a second mixer, using such valves as 12AU7, 12AT7.

mixers in integrated circuit form, field-effect devices such as JFETS, MOSFETS, dual-gate MOSFETS etc.

Optimum performance of a mixer requires correct levels of the injected local oscillator signal and operation of the device at the correct working point. This is particularly important for fet devices. Some switching-mode mixers require appreciable oscillator power.

Junction FETS used as mixers can be operated in three different ways: (1) rf signal applied to gate, oscillator signal to source; (2) rf signal to source, oscillator signal to gate; and (3) rf and oscillator signals applied to gate. Approach (1) provides high conversion gain but requires high oscillator power and may result in oscillator pulling; (2) gives good freedom from oscillator pulling and requires low oscillator power, but provides significantly lower gain; (3) gives fairly high gain with low oscillator power and may often be the optimum choice. For all fet mixers careful attention must be paid to operating point and local oscillator drive level. For most applications the dual-gate mosfet mixer (Fig 4.44) probably represents the best of the "simple" arrangements.

Among semiconductor devices, the trend has been towards the use of Schottky (hot-carrier) diodes in the diode ring double-balanced configuration using wide-band ferrite toroid transformers (**Fig 4.47**); such an arrangement can have a dynamic range of over 100dB with a noise figure of 6·5 to 7dB, but has a conversion loss of about 5 to 6dB; it can cope with signals of up to about 300 to 500mV. Double-balanced mixers of this type need care in construction if optimum performance and rejection of unwanted outputs are to be achieved. For optimum performance of a receiver on 28 and 21MHz, such a mixer should be preceded by a low gain rf amplifier, but quite acceptable results can be obtained using the mixer as the first stage.

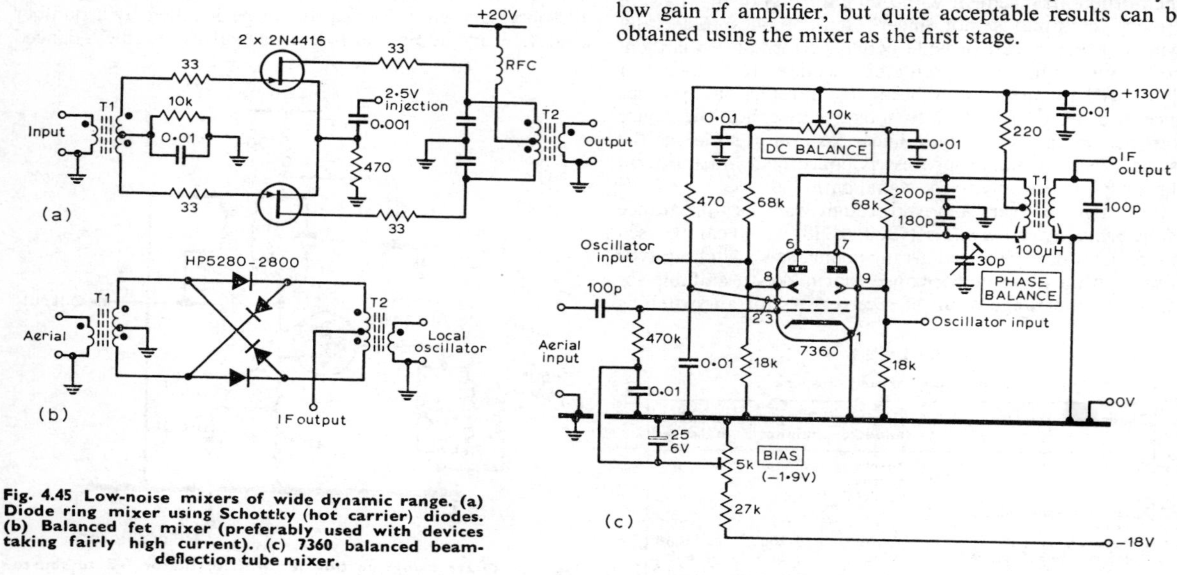

Fig. 4.45 Low-noise mixers of wide dynamic range. (a) Diode ring mixer using Schottky (hot carrier) diodes. (b) Balanced fet mixer (preferably used with devices taking fairly high current). (c) 7360 balanced beam-deflection tube mixer.

TABLE 4.4

Mixer Products

1 Unbalanced Mixer

Local Oscillator f_o

	f_o	$2f_o$	$3f_o$	$4f_o$	$5f_o$
f_s	$f_o \pm f_s$	$2f_o \pm f_s$	$3f_o \pm f_s$	$4f_o \pm f_s$	$5f_o \pm f_s$
$2f_s$	$2f_s \pm f_o$	$2f_o \pm 2f_s$	$3f_o \pm 2f_s$	$4f_o \pm 2f_s$	$5f_o \pm 2f_s$
$3f_s$	$3f_s \pm f_o$	$3f_s \pm 2f_o$	$3f_o \pm 3f_s$	$4f_o \pm 3f_s$	$5f_o \pm 3f_s$
$4f_s$	$4f_s \pm f_o$	$4f_s \pm 2f_o$	$4f_s \pm 3f_o$	$4f_o \pm 4f_s$	$5f_o \pm 4f_s$
$5f_s$	$5f_s \pm f_o$	$5f_s \pm 2f_o$	$5f_s \pm 3f_o$	$5f_s \pm 4f_o$	$5f_o \pm 5f_s$

2 Balanced Mixer—half the number of mixer products

	f_o	$2f_o$	$3f_o$	$4f_o$	$5f_o$
f_s	$f_o \pm f_s$	$2f_o \pm f_s$	$3f_o \pm f_s$	$4f_o \pm f_s$	$5f_o \pm f_s$
$2f_s$	—	—	—	—	—
$3f_s$	$3f_s \pm f_o$	$3f_s \pm 2f_o$	$3f_s \pm 3f_o$	$4f_o \pm 3f_s$	$5f_o \pm 3f_s$
$4f_s$	—	—	—	—	—
$5f_s$	$5f_s \pm f_o$	$5f_s \pm 2f_o$	$5f_s \pm 3f_o$	$5f_s \pm 4f_o$	$5f_o \pm 5f_s$

3 Double-Balanced Mixer—one quarter the number of mixer products

	f_o	$2f_o$	$3f_o$	$4f_o$	$5f_o$
f_s	$f_o \pm f_s$	—	$3f_o \pm f_s$	—	$5f_o \pm f_s$
$2f_s$	—	—	—	—	—
$3f_s$	$3f_s \pm f_o$	—	$3f_o \pm 3f_s$	—	$5f_o \pm 3f_s$
$4f_s$	—	—	—	—	—
$5f_s$	$5f_s \pm f_o$	—	$5f_s \pm 3f_o$	—	$5f_o \pm 5f_s$

Note a product such as $2f_o \pm f_s$ is known as a third-order product, $3f_s \pm 3f_o$ as a sixth-order product and so on.

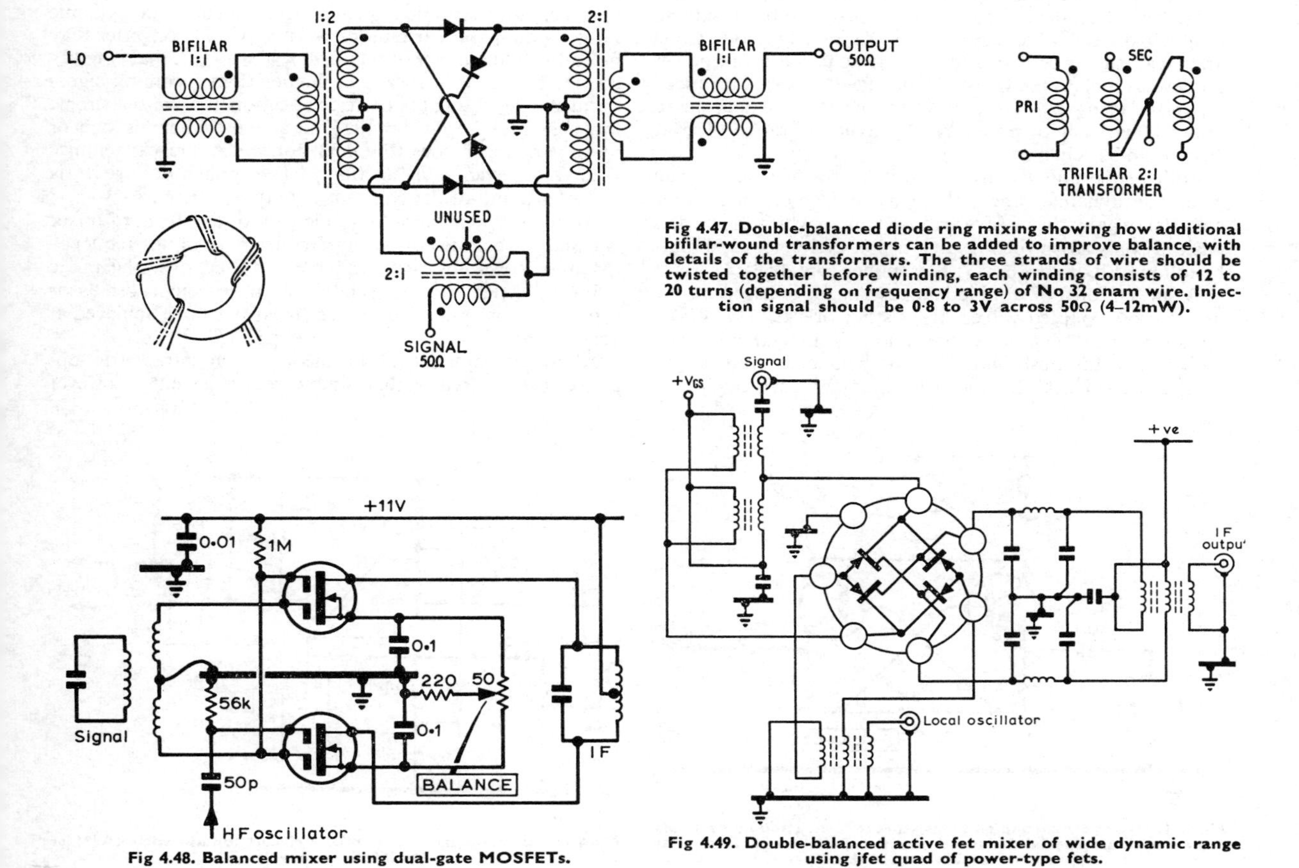

Fig 4.47. Double-balanced diode ring mixing showing how additional bifilar-wound transformers can be added to improve balance, with details of the transformers. The three strands of wire should be twisted together before winding, each winding consists of 12 to 20 turns (depending on frequency range) of No 32 enam wire. Injection signal should be 0·8 to 3V across 50Ω (4–12mW).

Fig 4.48. Balanced mixer using dual-gate MOSFETs.

Fig 4.49. Double-balanced active fet mixer of wide dynamic range using jfet quad of power-type fets.

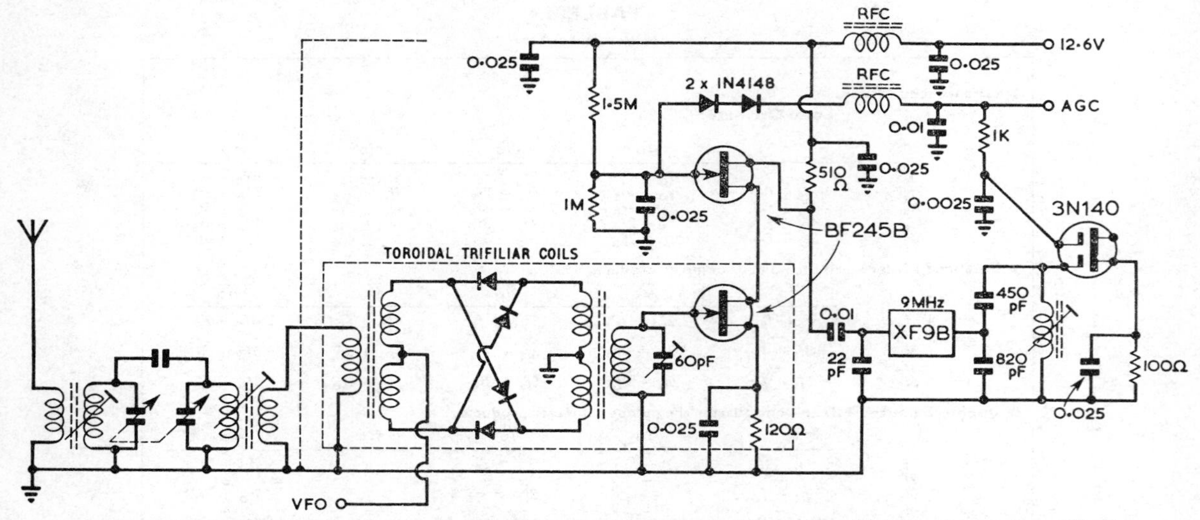

Fig 4.50. Balanced mixer and cascode first i.f. used in compact all-semiconductor receiver. Note absence of rf amplification.

More recently it has been shown that wider dynamic range, coupled with conversion gain rather than loss, is possible using field effect devices as switching mixers in balanced and double-balanced wide band mixers. Extremely good strong and weak signal performance can be achieved using power fet devices such as the U310, E310, CP640 series. A balanced fet mixer using 2N4416 FETS can cope with signals up to about 800mV; with power FETS a dynamic range of over 100dB can be achieved.

Both *active* and *passive* fet switching mode mixers can give wide dynamic range; the active mixer provides some conversion gain, the passive arrangement in which the fet acts basically only as a switch results in conversion loss and must be followed by a low-noise i.f. amplifier (**Fig 4.50**).

A special form of double-balanced mixer is based on the use of cross-coupled valves, transistors or FETS (**Fig 4.51**); an advantage of this configuration is that there is no requirement for push-pull drive or balanced input/output transformers. This technique, using bipolar transistors, forms the basis of the SL640 integrated circuit double-balanced mixers.

A technique capable of providing extremely wide dynamic range is the parametric up-converter. An up-converter may be considered a mixer of the form $f_{signal} + f_{osc}$, that is to say the i.f. output will always be higher than incoming signal frequency, enabling good image rejection even with a simple low-pass filter input circuit. The active element is one or more varactor diodes (**Fig 4.53**) for which the capacitance C characteristic, as a function of the applied voltage V, is very sharp and of the constant CV^n type, where $n > 1$.

Although the maximum usable gain of this form of mixer is equal to the ratio of output frequency to input frequency (Manley-Rowe law) gain can be stabilized over the range 1·5 to 30MHz at about 6dB. With sufficient oscillator ("pump") power excellent mixer linearity can be achieved at very low noise.

Although complicated in theory, such parametric up-converters are reasonably simple to implement. However

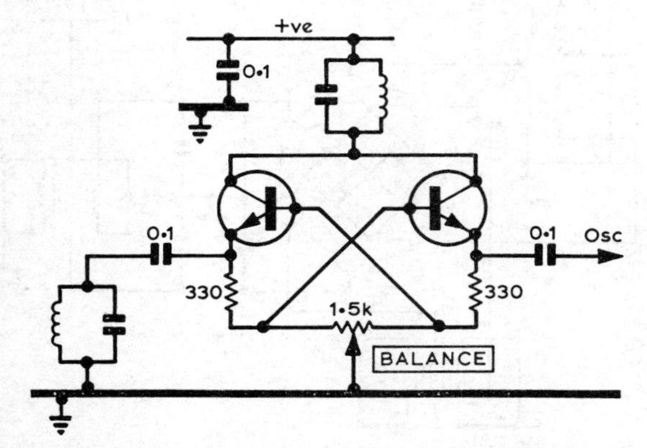

Fig 4.51. Use of cross-coupled transistors to form a double-balanced mixer without special balanced input transformers. A similar approach is used in the SL641 mixers.

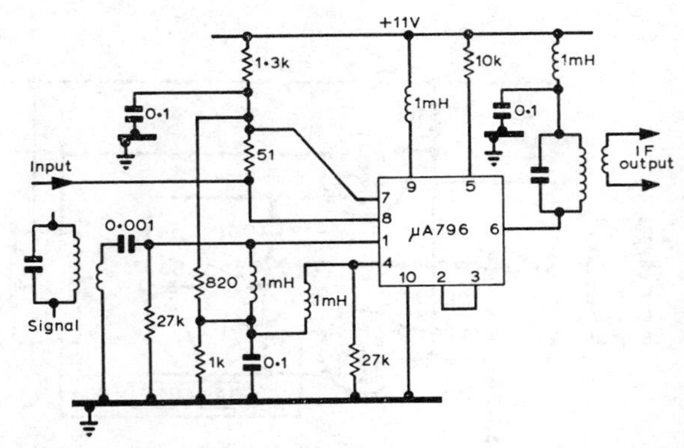

Fig 4.52. Double-balanced i.c. mixer circuit for use with μA796 or MC1596G etc devices.

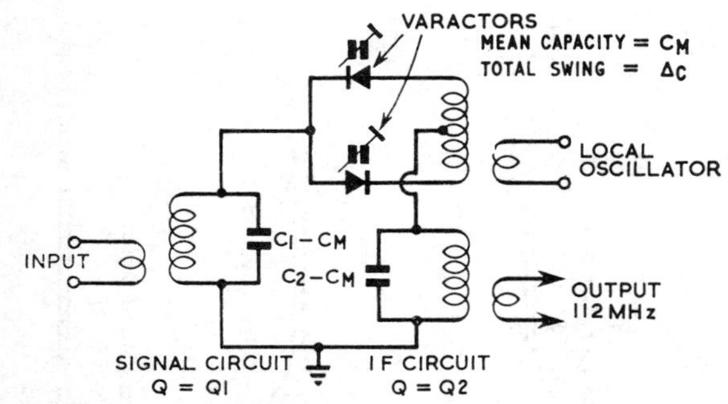

Fig 4.53. Basic circuit of parametric up converter used with i.f. of the order of 112MHz.

if the upper limit of the receiver is 30MHz it is advisable for the first i.f. of an up-conversion receiver to be at least three and preferably five times this frequency, implying an i.f. of 100 to 150MHz. While such receivers are produced for professional purposes, the difficulty of providing a good roofing or ssb filter at such frequencies has so far limited their amateur application: it is also difficult to keep the noise sidebands of a vhf pump source sufficiently low to achieve in practice the benefits that are theoretically possible.

The HF Oscillator(s)

The frequency to which a superhet or direct conversion receiver responds is governed not by the input signal frequency circuits but by the output of the local oscillator. Any frequency variations or drift of the oscillator are reflected in variation of the received signal; for ssb reception variations of more than about 50Hz will render the signal unintelligible unless the set is retuned. Some of the design techniques by which stability can be improved have already been outlined, but for most amateur receivers stability is still largely determined by one or more free-running oscillators.

The overall stability of a single conversion model thus depends on the high frequency oscillator (hfo) and to a lesser extent (since it is usually at a lower frequency) on the bfo; that of a double conversion model on the stability of all three oscillators, and so on. Generally it is more difficult to achieve good stability as the frequency increases, which is one reason why so many receivers use a crystal controlled hfo, while designing the variable oscillator at lower frequency.

The prime requirements of an oscillator used for heterodyne conversion are: freedom from frequency changes resulting from mechanical vibration or temperature changes; sufficient output for maximum conversion efficiency; low harmonic output (particularly important in second and third oscillators); no undue variation of output throughout the tuning range. The methods and circuits used to achieve these results are substantially the same as for transmitter VFOS.

Where a vfo is to be used in conjunction with the variable-i.f. or pre-mixer technique it need cover only a single range; for example 1MHz or commonly 500kHz. The absence of range switching makes it possible to design this for high-stability, along the lines associated with a transmitter vfo.

Fig 4.54 shows a fet vfo which, if all recommended precautions are taken, is capable of providing an extremely stable source. The following precautions should be noted:

(1) Design with genuine Vackar configuration, ie $C/(C4 + C6) \simeq C3/C2 \simeq 6$.
(2) Mount in strong box (eg diecast).
(3) Use high quality variable capacitor (eg Jackson U101)
(4) C2 should be air-spaced and adjusted for the minimum capacitance that allows the circuit to oscillate freely.
(5) Variable capacitors should preferably be effectively cleaned before construction (eg ultrasonic bath if at all possible).
(6) Temperature compensation should be provided. For example by means of an Oxley "Tempatrimmer" or the lower-cost "Thermotrimmer".
(7) C1, C3 and C6 should be silver-mica types firmly stuck (eg by Araldite) to convenient firmly mounted components or chassis.
(8) R1 should be 2W type for minimum heating, and of low inductance.
(9) The use of good buffer/isolating amplifier is essential; the two-stage arrangement shown should prove entirely satisfactory.
(10) Use in conjunction with a well-stabilized power supply (for example using a zener diode with constant-current fet). Disc ceramic bypass capacitors should be used liberally along the supply rail to prevent feed-back.
(11) L, C1, C2, C3, C4, C6, R1 and fet source connection

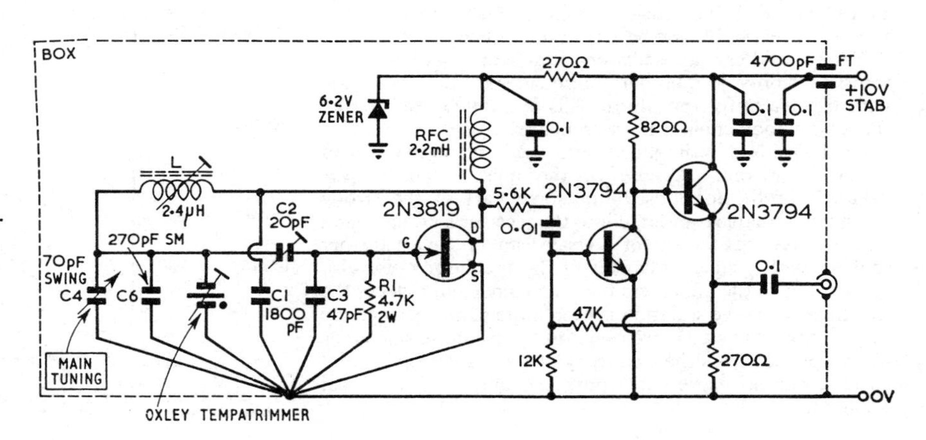

Fig 4.54. High-stability fet Vackar vfo covering 5·88 to 6·93MHz.

should have a common earthing point (for example one of the fixing screws of C4).

(12) Ceramic coil formers preferred but iron dust-cores can be used to facilitate calibration adjustment; ferrite cores should be avoided for this application.

(13) Keep wiring leads short and use stiff wire (16 or 18swg) for all interconnections affecting the oscillator tank coil.

Oscillator Noise and Reciprocal Mixing

A single conversion superhet requires a variable hf oscillator which should not only be stable but should provide an extremely "pure" signal with the minimum of noise sidebands. Any variation of the oscillator output in terms of frequency drift or sudden "jumps" will cause the receiver to detune from the incoming signal. The need for a spectrally pure output is less readily grasped, yet it is this feature which represents a practical limitation on the performance of modern receivers. This is due to noise sidebands or jitter in the oscillator output. Noise voltages, well known in amplifiers, occur also in oscillators, producing output voltages spread over a wide frequency band and rising rapidly immediately adjacent to the wanted oscillator output.

The noise jitter and sidebands immediately adjacent to the oscillator frequency are particularly important. When a large interfering signal reaches the mixer on an immediately adjacent channel to the wanted signal, this signal will mix with the tiny noise sidebands of the oscillator (the sidebands represent in effect a spread of oscillator frequencies) and so may produce output in the i.f. pass-band of the receiver: this effect is termed *reciprocal mixing*. Such a receiver will appear noisy, and the effect is usually confused with a high noise factor.

At hf the "noise" output of an oscillator falls away very sharply either side of the oscillator frequency, yet it is now recognized that this noise may be sufficient to limit the performance of a receiver. Particular care is necessary with some forms of frequency synthesizers, including those based on the phase-locking of a free-running oscillator, since these can often produce significantly more noise sidebands and jitter than that from a free-running oscillator alone.

Synthesizers involving a number of mixing processes may easily have a noise spectrum 40 to 50dB higher than a basic LC oscillator. A tightly controlled phase-locked oscillator with variable divider might be some 20 to 30dB higher than an LC oscillator, but possibly less than this where the vco is inherently very stable and needs only infrequent "correction".

For LC and crystal oscillators, it appears that field effect transistors provide minimum noise sidebands; valves next; with bipolar transistors third. This is another reason why the fet is a good choice for an oscillator.

Fig 4.55 shows in simplified form the basic mechanism of reciprocal mixing. Because of the noise sidebands (or "skirts") of the local oscillator some part of the strong unwanted signal is translated into the receiver's i.f. passband and thus reduces the snr of a weak wanted signal; further (not shown) a small fragment of the oscillator noise also spreads out to the i.f., enters the i.f. channel and reduces the snr. In practice the situation is even more complex since the very strong unwanted carrier will itself have noise sidebands which spread across the frequency of the wanted signal and degrade snr no matter how pure the output of the local oscillator.

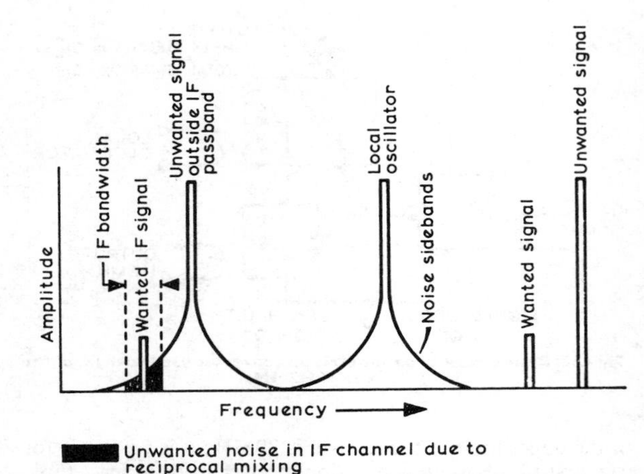

Fig 4.55. Showing mechanism whereby reciprocal mixing degrades the snr of weak signals in the presence of strong signals due to the noise sidebands of the local oscillator.

Fig 4.56 indicates the practical effect of reciprocal mixing on high performance receivers; in the case of receiver "A" (typical of many high-performance receivers) it is seen that the dynamic selectivity is degraded to the extent that the snr of a very weak signal will be reduced by strong unwanted signals (1 to 10mV or more) up to 20kHz or more off tune, even assuming that the front-end linearity is such that there is no cross-modulation, blocking or intermodulation. Reciprocal mixing thus tends to be the limiting factor affecting very weak station performance in real situations, although intermodulation or cross-modulation characteristics become the dominant factors with stronger signals.

It is thus important in the highest-performance receivers to pay attention to achieving low noise sidebands in oscillators, and this is one reason why the simpler forms of frequency synthesizers, which often have appreciable jitter and noise in the output, must be viewed with caution despite the high stability they achieve. The three major forms of basic oscillator noise are: (1) low frequency (lf) noise which predominates very close to carrier but is insignificant beyond about 250Hz; (2) thermal noise which predominates between about

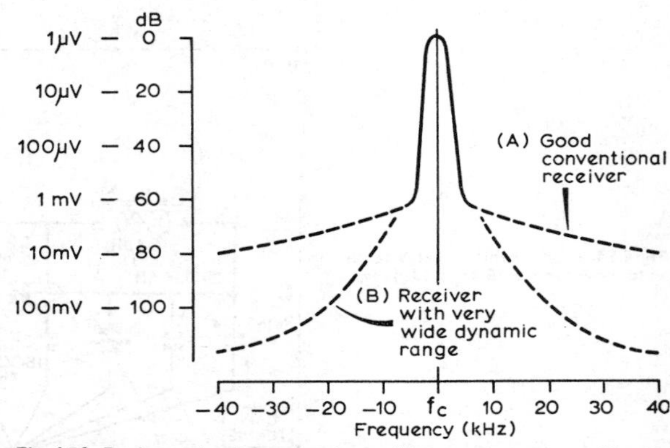

Fig 4.56. Reciprocal mixing due to oscillator noise can modify the overall selectivity curve of an otherwise very good receiver.

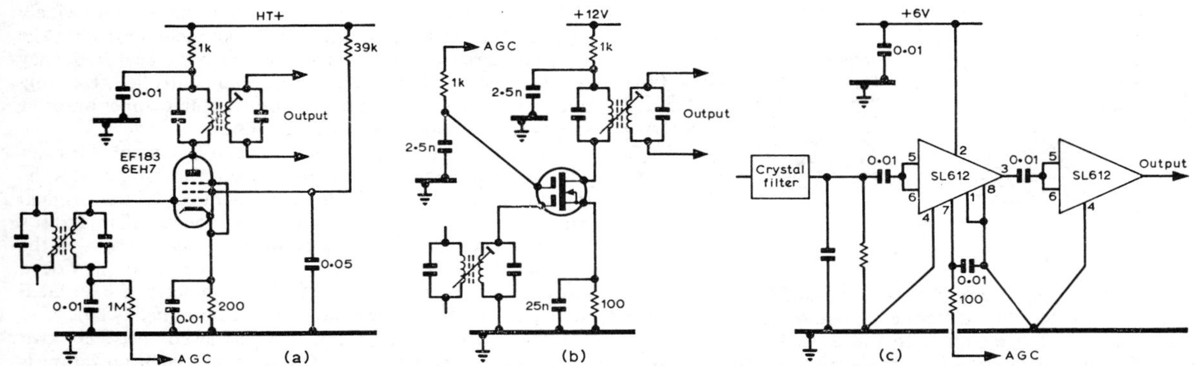

Fig 4.57. Typical automatic gain controlled i.f. amplifiers using (a) valves (b) dual-gate MOSFETs and (c) integrated circuits.

250Hz to about 20kHz from carrier; and (3) "shot" noise attributable to noise current and more or less evenly spread at all frequencies. Since the shot noise of an oscillator spreads across the i.f. channel care is needed with low noise mixers to limit the amount of this *injection-source noise* that enters the i.f. amplifier. Balanced and double-balanced mixers provide up to about 30dB rejection of oscillator noise. Another technique is to use a rejector "trap" (tuned circuit) resonant at the i.f. between oscillator and mixer.

Optimum oscillator performance calls for the use of a fet with a high forward transconductance and for a high unloaded tank-circuit Q.

For switching mode mixers and product detectors the optimum oscillator output waveform would be a square wave but this refinement is comparatively rare in practice.

I.F. Amplifiers

The i.f. remains the heart of a superhet receiver, for it is in this section that virtually all of the voltage gain of the signal and the selectivity response are achieved. Whereas with older superhets having significant front-end gain, the i.f. gain was of the order of 70–80dB, today it is often over 100dB.

Where the output from the mixer is low (possibly less than 1μV) it is essential that the first stage of the i.f. section should have low noise characteristics and yet not be easily overloaded. Although it is desirable that the crystal filter (or roofing filter) should be placed immediately after the mixer, the very low output of diode and passive fet mixers may require that a stage of amplification takes place before the signal suffers the insertion loss of the filter.

Similarly it is important that where the signal passes through the sideband filter at very low levels the subsequent i.f. amplifier must have good noise characteristics. Further, for optimum cw reception, it will often be necessary to ensure that the *noise bandwidth* of the i.f. amplifier *after* the filter is kept narrow. The noise bandwidth of the entire amplifier should be little more than that of the filter. This can be achieved by including a further narrow-band filter (for example a single-crystal filter with phasing control) later in the receiver, or alternatively by further frequency conversion to a low i.f.

To achieve a flat agc characteristic it may be desirable for all i.f. amplifiers to be controlled by the agc loop; and it is important that amplifier distortion should be low throughout the dynamic range of the control loop.

The dual-gate mosfet (**Fig 4.57(b)**) with reverse agc on gate 1 and partial forward agc on gate 2 has excellent cross-modulation properties but the control range is limited to about 35dB per stage. Integrated circuits with high performance gain controlled stages are available (**Fig 4.57(c)**).

For valve amplifiers frame grid valves such as the EF183 provide a control range of over 50dB and cope well with large signals (**Fig 4.57(a)**).

Where a high-grade ssb or cw filter is incorporated it is vital to ensure that signals cannot "leak" around the filter due to stray coupling; good screening and careful layout are needed.

In multiple conversion receivers, it is possible to provide continuously variable selectivity by arranging to vary slightly the frequency of a later conversion oscillator so that the bandpass of the two i.f. channels overlap to differing degrees. For optimum results this requires that the shape factor of both sections of the i.f. channel should be good, so that the edges are sharply defined.

With double-tuned i.f. transformers, gain will be maximum when the product kQ is unity (where k is the coupling between the windings). I.F. transformers designed for this condition are said to be critically coupled; when the coupling is increased beyond this point (over-coupled) maximum gain occurs at two points equally spaced about the resonant frequency with a slight reduction of gain at exact resonance: this condition may be used in broadcast receivers to increase bandwidth for good quality reception. If the coupling factor is lowered (under-coupled) the stage gain falls but the response curve is sharpened, and this may be useful in communications receivers.

DEMODULATION

For many years, the standard form of demodulation for communications receivers, as for broadcast receivers, was the *envelope detector* using valve or semiconductor diodes; occasionally for superhet applications the *regenerative detector* has been used, based on circuits used in straight receivers. Envelope detection is a non-linear process (part mixing, part rectification) and is inefficient at very low signal levels. On weak signals this form of detector distorts or may even lose the intelligence signals altogether. On the other hand *synchronous* or *product detection* preserves the signal-to-noise ratio, enabling post-detector signal processing and

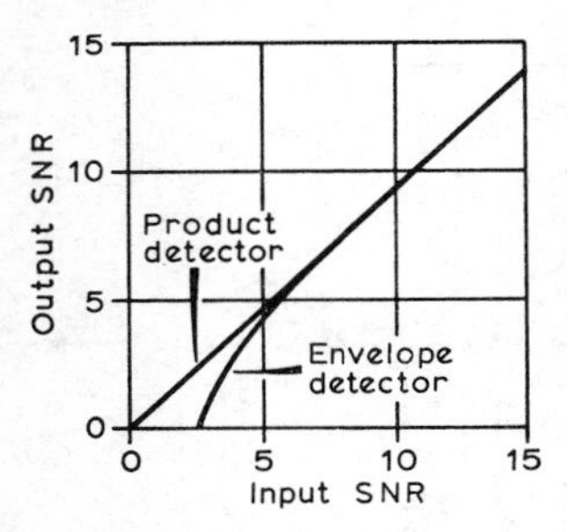

Fig 4.58. Synchronous (product) detection maintains the snr of signals down to the lowest levels whereas the efficiency of envelope detection falls off rapidly at low snr, although as efficient on strong signals.

audio-filters to be used effectively (**Fig 4.58**). Synchronous detection is essentially a frequency conversion process and the circuits used are similar to those used in mixer stages (**Figs 4.59** to **4.61**). The i.f. or rf signal is heterodyned by a carrier at the same frequency as the original carrier frequency and so reverts back to the original audio modulation frequencies (or is shifted from these frequencies by any difference between the inserted carrier and the original carrier as in cw where such a shift is used to provide an audio output between about 500 and 1,000Hz).

It should be noted that a carrier is needed for both envelope and product detection: the carrier may be radiated along with the sidebands, as in a.m., or locally generated and inserted in the receiver (either at rf, or i.f.—usually at i.f. in superhets, at rf in direct conversion receivers).

Synchronous or product detection has been widely adopted for ssb and cw reception in amateur receivers; the injected carrier frequency is derived from the beat frequency oscillator, which is either LC or crystal controlled. By using two crystals it is possible to provide selectable upper or lower sideband reception (**Figs 4.62** and **4.63**).

The use of synchronous detection can be extended further to cover a.m., dsb-sc, nbfm and rtty but for these modes the injected carrier really needs to be identical to the original carrier not only in frequency but also in phase: that is to say the local oscillator needs to be in *phase-coherence* with the original carrier (an alternative technique is to provide a strong local carrier that virtually eliminates the original a.m. carrier—this is termed *exalted carrier detection*).

Phase coherence cannot be achieved between two oscillators unless some effective form of synchronization is used. The simplest form of synchronization is to feed a little of the original carrier into a local oscillator so forcing a phase lock on a free-running oscillator; such a technique was used in the *synchrodyne* receiver. The more usual technique is to have a phase-lock loop: **Fig 4.64.** At one time such a system involved a large number of components and would have been regarded as too complex for most purposes; today however complete phase-lock loop detectors are available in the form of a single integrated circuit, both for a.m. and nbfm applications.

Apart from the phase-lock loop approach a number of

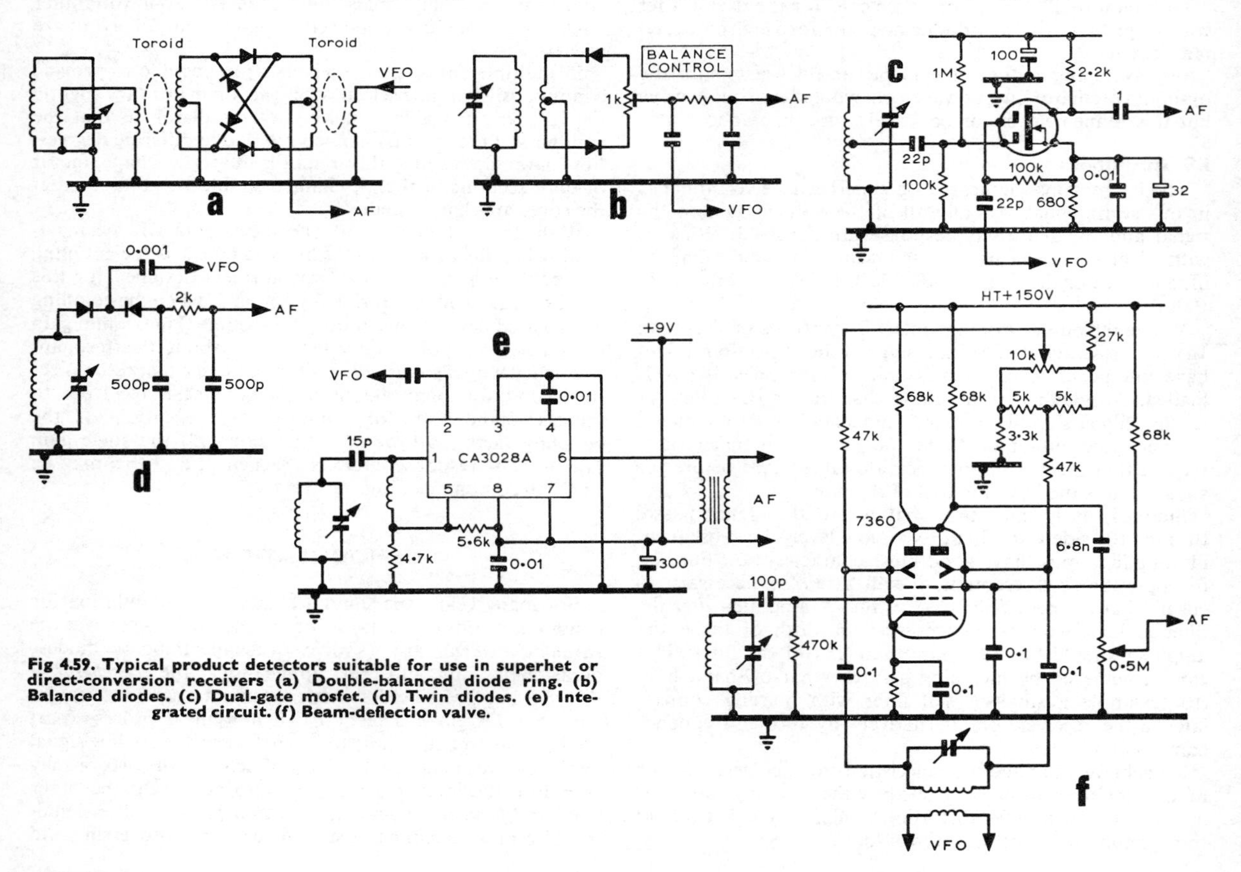

Fig 4.59. Typical product detectors suitable for use in superhet or direct-conversion receivers (a) Double-balanced diode ring. (b) Balanced diodes. (c) Dual-gate mosfet. (d) Twin diodes. (e) Integrated circuit. (f) Beam-deflection valve.

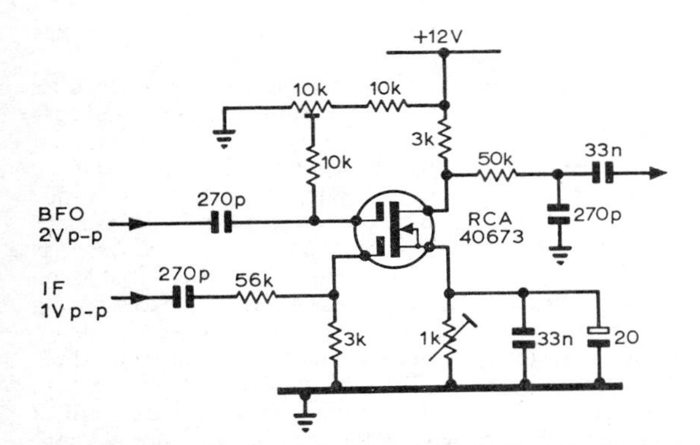

Fig 4.60. Dual-gate mosfet product detector including bias adjustment controls for optimum results.

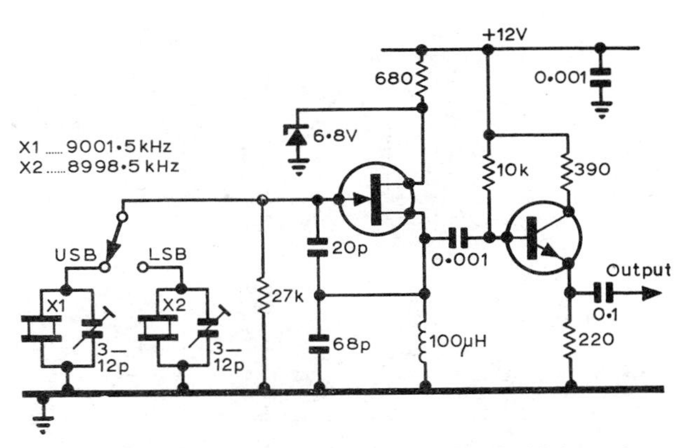

Fig 4.62. BFO suitable for use with 9MHz i.f. and with ic product detectors. Manual switching of crystals to select upper and lower sidebands.

alternative forms of synchronous multi-mode detectors have been developed. One interesting technique which synthesizes a local phase coherent carrier from the incoming signal is the *reciprocating detector*.

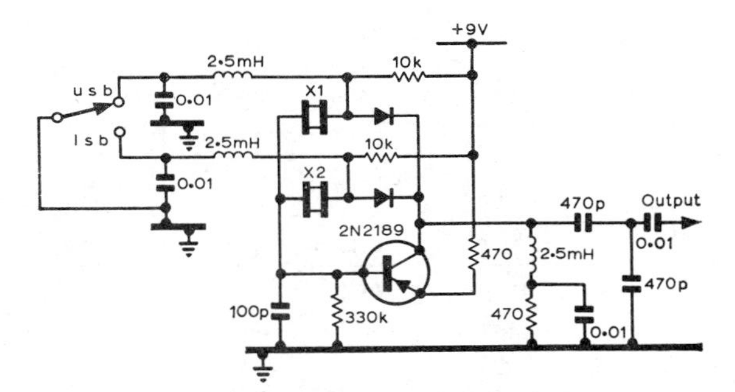

Fig 4.63. BFO with diode-switched sideband selection.

Automatic Gain Control

In hf operation incoming signals may be constantly varying by 20 to 30dB or even more. The basic function of an agc system is to hold the af output from the receiver as constant as possible despite these wide variations of signal level, by decreasing the overall gain of the receiver in such a way as to compensate for the signal fluctuations.

With signals having a continuous carrier (ie a.m. or fm) the control may be conveniently based on the strength of the carrier by rectifying a portion of the amplified i.f. signal so as to obtain a dc control voltage which is then applied as bias to one or more earlier stages. An increase in signal increases the bias causing the overall gain to be reduced; a decrease in signal causes the bias to be reduced and gain to increase. The application of agc to front-end stages may be *delayed* until the signal reaches a level sufficient to provide good signal-to-noise ratio.

The more stages controlled by the agc system and the greater the maximum gain within the controlled section, the greater can be the efficiency of the agc system; that is to say the more constant will be the audio output over a wide range of signal strengths. However for other reasons it may be desirable that a particular stage should operate at a fixed point on its transfer curve; this is often the case with *mixer* stages which are seldom made subject to agc action.

With semiconductor receivers, the limited signal-handling capacity makes it advantageous to apply at least some part of the gain control ahead of the receiver, that is to say by controlling the signals reaching the set. This can be done by incorporating an *agc-controlled rf attenuator* at the low-impedance aerial feeder connection.

For ssb and cw reception there is no continuously available

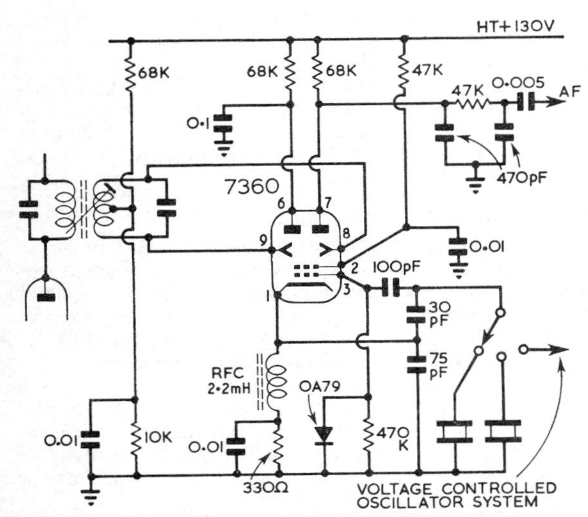

Fig 4.61. A self-excited 7360 product detector. The OA79 diode prevents the control grid from going positive. Note that this is not a fully balanced arrangement as output is taken from one anode only.

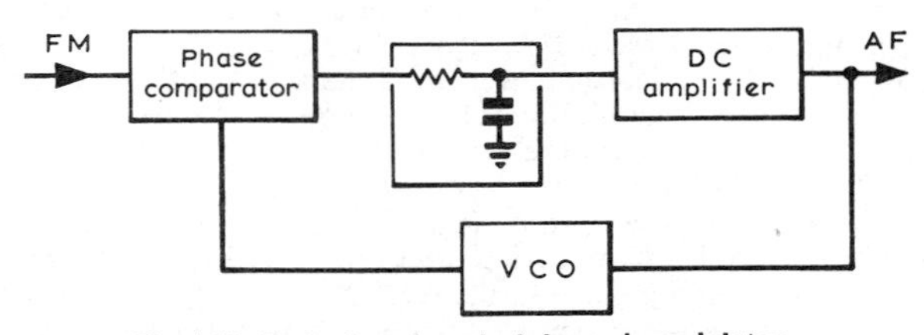

Fig 4.64. Basic fm phase-lock-loop demodulator.

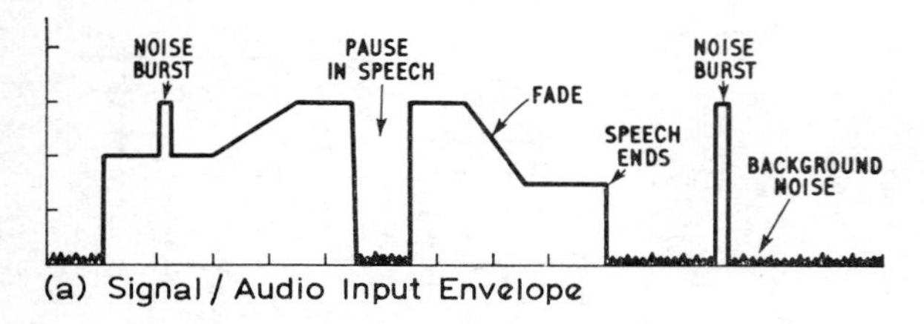

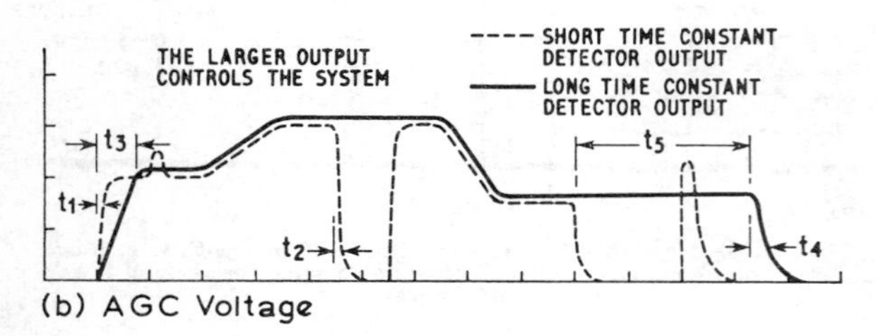

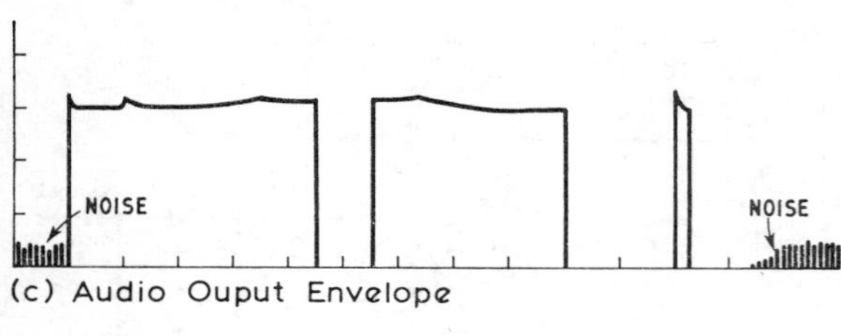

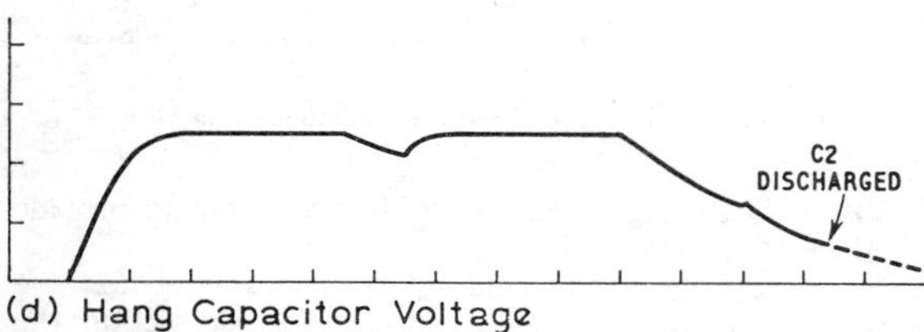

Fig 4.65. The behaviour of dual-time constants agc circuits under various operating conditions, where t1 is the fast detector risetime, t2 the fast detector decay time, t3 the slow detector risetime, t4 the slow detector fall time, and t5 the hang time. This approach is used in the SL621 ic or can also be implemented using discrete devices.

carrier which can be measured to provide the agc; many modern sets are controlled by signals in the af section of the receiver (but others use i.f. derived agc). Such signals can also be "averaged" over a short period to provide an indication of signal strength by an S-meter.

For ssb and cw operation the agc loop should have a very fast attack time (20ms or less) and a long effective decay time (0·2 to 1s), Short rise times are obtained with low impedance agc detectors; long decay times may be achieved by having a large capacitance across the agc line to provide a long time constant. However this approach means that noise pulses can mute the receiver for periods of seconds; to be preferred are "hang" systems which are fast acting in both rise and decay times but which maintain a steady agc voltage for a finite period after the actuating signal has gone, then rapidly increase the receiver gain. With "hang" systems background noise does not increase during short pauses in speech or cw transmission; however, these systems may be a disadvantage when tuning across the band.

Effective "agc generators" providing hang characteristics are available in integrated circuit form (eg Plessey SL621) (**Fig 4.66**).

With valves, gain control is normally achieved by shifting the bias conditions of the amplifying stages towards cut-off; that is to say the more negative the agc control line becomes the lower will be the receiver gain. This *reverse agc* system is satisfactory so long as the amplifying stages have variable-mu characteristics; it is far less satisfactory for devices having sharp cut-off characteristics. This is particularly true of bipolar transistors and to a lesser extent with some field effect transistors.

Fortunately there is an alternative technique called *forward agc* which depends upon the "high-frequency knee" region of transistor characteristics where when the emitter current is increased the amplifying properties begin to deteriorate. Transistors intended for forward agc operation have this region made larger than usual. Forward agc provides better signal handling characteristics than reverse agc.

A problem with either form of agc is the relatively large variation in input and output admittance that occurs when the operating point of a bipolar transistor is changed.

When a valve amplifier is agc controlled, the screen grid is usually fed through a high value resistor (about 22 to 47kΩ). This ensures that the screen voltage rises as the negative bias to the valve increases, so providing a longer grid base for the valve and reducing the likelihood of it being driven into non-linearity.

To prevent the bfo from reducing the gain in receivers using i.f.-derived agc, the agc detector may be fed from an agc amplifier stage before the injection of the bfo. In some designs the take-off point may be made early in the receiver, before the final selectivity characteristics are shaped, so that selectivity characteristics of the agc system are broader than those of the signal path. While this might seem undesirable in that the gain will be reduced by strong signals outside the pass band, in practice this is often beneficial since the strong adjacent-channel signals would otherwise cause splatter or cross-modulation. Some designs have more than one agc loop: one of broad selectivity and the other of similar selectivity to that of the signal path.

For simple receivers, particularly those intended primarily for cw reception, it may be advisable to dispense altogether with agc and rely on manual control and af output limiting; a poor agc system may prove worse than no agc.

Noise Limiters and Blankers

The hf spectrum, particularly above 15MHz or so, is susceptible to man-made electrical impulse interference stemming from electric motors and appliances, car ignition systems, thyristor light controls, high-voltage power lines and many other causes. A similar problem arises on 1·8MHz due to the operation of Loran navigation systems.

These interference signals are usually in the form of high amplitude, short-duration pulses covering a wide spectrum of frequencies. In many urban and residential areas this man-made interference sets a limit to the usable sensitivity of receivers and may spoil the reception of even strong amateur signals.

Because the interference pulses though of high amplitude are often of extremely short duration, a considerable

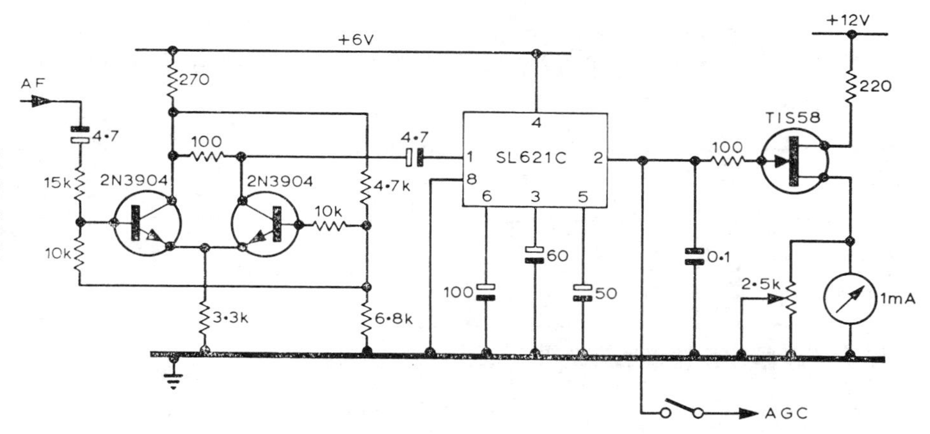

Fig 4.66. Use of SL621 agc generator with emitter coupled clipper and resistive network to keep audio level at about 10mV drive. These extra precautions are not needed when using this device in conjunction with other SL600 range ICs.

improvement can be obtained by "slicing" off all parts of the audio signal which are significantly greater than the desired signal. This can be done by simple af limiters such as back-to-back diodes. For a.m. reception more elegant noise limiters develop fast-acting biasing pulses to reduce momentarily the receiver gain during noise peaks. The ear is much less disturbed by "holes of silence" than by peaks of noise. Many limiters of this type have been fitted in the past to a.m.-type receivers.

Unfortunately, since the noise pulses contain high frequency transients, highly selective i.f. filters will distort and broaden out the pulses. To overcome this problem, *noise blankers* have been developed which derive the blanking bias potentials from noise pulses which have not passed through the receiver's selective filters. In some cases a parallel broadly-tuned receiver is used, but more often the noise signals are taken from a point early in the receiver. For example, the output from the mixer goes to two channels: the signal channel which includes a blanking control element which can rapidly reduce gain when activated; and a wideband noise channel to detect the noise pulse and initiate the gain reduction of the signal channel. To be most effective it is necessary for the gain reduction to take place virtually at the instant that the interference pulse begins. In practice because of the time constants involved it is difficult to do this unless the signal channel incorporates a time delay to ensure that the gain reduction can take place simultaneously with or even just before the noise pulse. One form of time delay which has been described in the literature utilizes a PAL-type glass ultrasonic television delay line to delay signals by 64μs. It is however difficult to eliminate completely transients imposed on the incoming signal.

One possible approach which has been investigated at the University College, Swansea, is to think in terms of receivers using synchronous demodulation at low level so that a substantial part of the selectivity, but not all of it, is obtained after demodulation. This allows noise blankers to operate at a fairly low level on af signals.

A control element which has been used successfully consists of a fet gate pulsed by signals derived from a wideband noise amplifier. The noise gate is interposed between the mixer and the first crystal filter, with the input signal to the noise amplifier taken off directly from the mixer. An example of noise blanking will be found in the "advanced hybrid receiver" described later.

AF Stages

The af output from an envelope or product detector of a superhet receiver is usually of the order of 0·5 to 1V, and many receivers incorporate relatively simple one- or two-stage audio amplifiers providing about 2W output. On the other hand the direct-conversion receiver may require a high-gain audio section capable of dealing with signals of less than 1μV.

Provided that all stages of the receiver up to and including the product detector are substantially linear many forms of post-demodulation signal processing are possible: for example bandpass or narrowband filtering to optimize signal-to-noise ratio of the desired signal, audio compression or expansion; the removal of audio peaks, af noise blanking, or (for cw) the removal by gating of background noise. Audio phasing techniques may be used to convert a dsb receiver into an ssb receiver (as in two-phase or third-method ssb demodulation) or to insert nulls into the audio passband for the removal of heterodynes. Then again, in modern designs the agc and S-meter circuits are usually operated from a low-level af stage rather than the i.f. derived techniques used in a.m.-type receivers.

It should be appreciated that linear low-distortion demodulation and af stages are necessary if full advantage is to be taken of such signal processing, since strong intermodulation products can easily be produced in these stages. Thus, despite the restricted af bandwidth of speech and cw communications, the intermodulation distortion characteristics of the entire audio section should preferably be designed to high-fidelity audio standards. It is sometimes suggested that the greatest sources of audio distortion in amateur stations are the use of old pre-war headphones and surplus tv loudspeakers.

Audio filters may be passive using inductors and capacitors, or active, usually with resistors and capacitors in conjunction with op-amps or FETS (**Figs 4.67** and **4.68**). Many different circuits have been published covering af filters of variable bandwidth, tunable centre frequencies and for the insertion of notches. The full theoretical advantage of a narrowband af filter for cw reception may not always be achieved in operational use: this is because the human ear can itself provide a "filter" bandwidth of about 50Hz with a remarkably large dynamic range and the ability to tune from 200 to 1,000Hz without introducing "ringing".

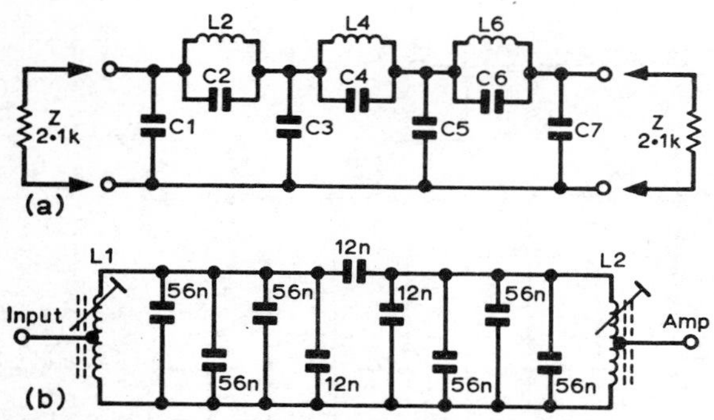

Fig 4.67. (a) Phone and (b) cw af filters suitable for use in direct-conversion or other receivers requiring very sharply defined af responses. The cw filter is tuned to about 875Hz. Values for (a) can be made from preferred values as follows: C1 37·26nF (33,000 + 2,200 + 1,800 + 220pF) C2 3·871nF (3,300 + 560pF); C3 51·87nF (47,000 + 4,700 + 150pF); C4 19·06nF (18,000 + 1,000pF); C5 46·41nF (39,000 + 6,800 + 560pF); C6 13·53nF (12,000 + 1,500pF); C7 29·85nF (27,000 + 2,700 + 150pF). All capacitors mica or polyester or styroflex types. L2 168·2mH (540 turns); L4 124·5mH (460 turns); L6 129·5mH (470 turns) using P30/19·3H1 pot cores and 0·25mm enam wire. Design values based on 2,000Ω impedance.

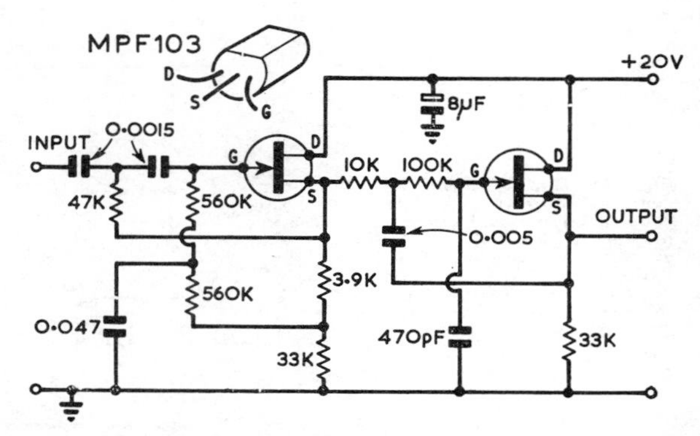

Fig 4.68. "Active" bandpass af filter for amateur telephony. The —6dB points about 380 and 3,200Hz, —18dB about 160 and 6,000Hz.

For this reason the experienced operator may himself be providing the advantage of an af filter.

Examples of both high- and low-gain af amplifiers will be found in the complete receivers described later in this chapter.

Power Supplies

HF receivers for home use are normally designed for operation from ac mains supplies, even when fully-transistorized, and use double-wound mains transformers to give full isolation between chassis and the mains supply. The voltage and current ratings of the supplies depend on the active devices and it is useful if there is sufficient in reserve to allow the operation of ancillary units from the receiver supply. Valves usually require a heater supply of 6·3V at several amps plus a well smoothed ht rail from 130 to 250V at 60–120mA. The advent of silicon power diodes made possible a significant reduction in the heat formerly associated with rectifier valves, and their use is general: precautions should be taken to prevent "mush" arising from the sudden switching action of the diodes, but this can normally be suppressed by connection of capacitors or capacitor-resistor combinations across the diodes. Some stages need a fairly closely regulated ht supply (oscillators, mixer screens) and this is usually achieved by the use of gas-filled voltage regulator tubes. For other stages the constant load represented by the receiver provides adequate regulation. The stability of valve oscillators depends as much on heater as ht regulation and some receivers incorporate various forms of heater current stabilization; one solution is to run the oscillator heater from a well-regulated dc supply.

For semiconductors a number of different supply rails may be needed for different devices, from 18 to 20V down to 5V for ttl logic devices, and sometimes both positive and negative rails are needed for ICs. Electronic regulation of the supply is to be preferred, in view of the fluctuating load of Class B af stages. For oscillators, simple zener diode regulation is not usually sufficient: in the absence of a closely regulated supply rail a solution here is to use a fet as a constant-current diode in conjunction with a zener diode: see **Fig 4.70.**

RF filtering on the input side of the power supply is helpful in reducing electrical interference and rf currents entering the receiver via the mains supply: components used in such applications should always be specifically rated for ac applications. Transients on the mains supply can damage semiconductor devices, and surge and transient protection is advised on receiver power supplies.

OSCILLATOR TRACKING AND ALIGNMENT

The superhet principle depends upon the provision of a locally generated signal which differs from the input signal by a predetermined fixed amount which is the i.f. Where both the oscillator and the mixed input circuits are tuned by a single control (ie ganged tuned circuits), these must *track* over the tuning range so as to maintain, as closely as possible, this specified frequency difference at all settings of the tuning control. For example if the waveband is 13,000 to 22,000kHz and the receiver's i.f. is 3,300kHz, then the rf circuits must

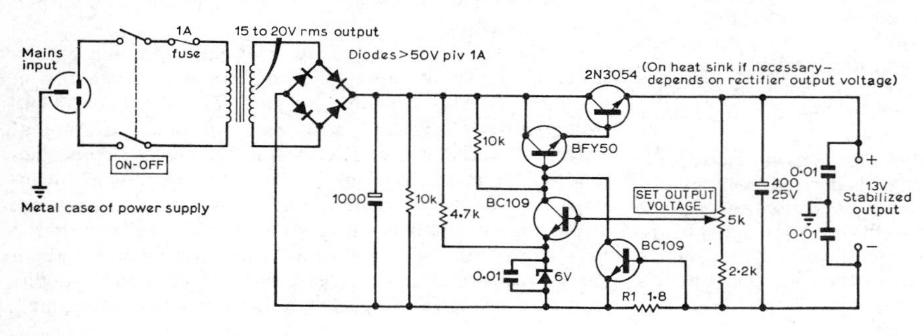

Fig 4.69(a). Power supply for semiconductor receivers with stabilized output.

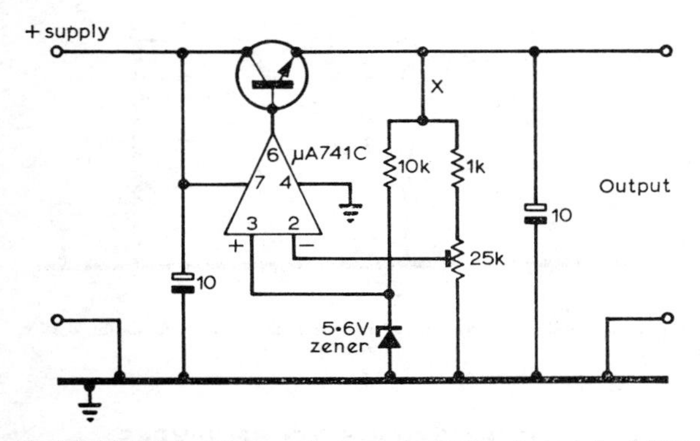

Fig 4.69(b). High-stability series voltage stabilized supply for 6·0V ic devices. Note the transistor is not needed where the output current is not greater than 10mA and in this case the op-amp output (pin 6) is connected directly to point X. The pin numbers refer to eight-lead TO5 or dual-in-line packages.

tune 13,000 to 22,000kHz in step with an oscillator tuning 16,300 to 25,300kHz. Because the resonant frequency of a tuned circuit depends on the square root of the product of inductance (L) and capacitance (C), it is not possible to obtain these two tuning ranges unless both L and C are reduced in the correct proportions for the oscillator. In practice this may be done by having fewer turns on the oscillator coil or by adjustment of the core; capacitance swing may be reduced by the use of fixed *padding* capacitors, effectively in series with the variable capacitor, while fine adjustment and the effective minimum capacitance is usually achieved by having small preset capacitors (*trimmers*) across the main tuning capacitor.

Even with correct scaling of L and C between oscillator and rf circuits, it is not possible to obtain exact tracking over the full sweep of a ganged tuning capacitor. The best compromise provides three accurately aligned frequencies, one towards each end of the tuning span, and one in a fairly central position. The practical effects of such errors are insignificant with small tuning ranges (for example in amateur-bands-only receivers) but become increasingly important when attempting to provide general coverage with only a few wavebands.

Many older single-conversion receivers used four-gang capacitors, tuning at the same time three rf resonant circuits and the oscillator—an approach which is expensive to implement; modern practice, especially with variable-i.f. type receivers, is to gang tune the oscillator with only the input circuit to the variable i.f. mixer (and even this may be eliminated by using band-pass filters in the input circuit so that tuning is then carried out without any ganging). The rf tuned circuit(s) are controlled independently by means of a pre-selector control; or may consist of sub-octave bandpass

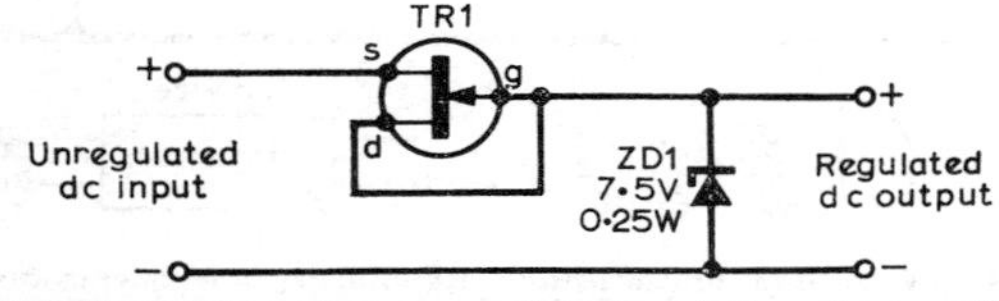

Fig 4.70. Use of fet to improve voltage regulation of an oscillator stage. For low currents TR1 can be MPF102 or almost any general purpose fet.

filters so that no rf tuning is needed. These systems enormously reduce the problem of tracking and alignment but unfortunately may also reduce the pre-mixer attenuation of strong signals.

The performance of many superhets is governed to a considerable extent by the accuracy with which all the various tuned circuits (i.f., rf, variable i.f. etc) are *aligned*, that is adjusted in accordance with the design requirements. In most equipment there is some tendency for circuits gradually to drift out of alignment (although modern sets are generally extremely stable in this respect) and so occasionally to require re-alignment.

The i.f. transformers need to be carefully set to the centre frequency of any crystal or mechanical filters, preferably using a signal generator with attenuator.

RF circuits are normally re-aligned with the aid of a signal generator or gdo. The appropriate trimmer capacitors are adjusted while injecting a frequency about 10 per cent of the tuning span down from the hf band edge and the appropriate cores in a similar manner about 10 per cent of the tuning range up from the low-frequency band edge.

A useful accessory when first adjusting equipment is the gdo as this can check the tuning ranges while the set is being constructed. A convenient aid when adjusting equipment using accessible air-cored coils is the trimming wand, consisting of an insulated non-metallic rod about 6in long with a small piece of dust-iron or ferrite core at one end and a piece of brass at the other. Inserting the brass end into a coil lowers its effective inductance, whereas ferrite or iron dust increases it. It is thus possible to determine simply whether a circuit is tuned low or high. Another tip when dealing with an air-cored coil is that its inductance can be increased by winding one or more layers of magnetic audio recording tape over the coil.

It should be emphasized that the complete re-alignment of a complex multi-band hf receiver requires skill and patience and at least some test equipment. The newcomer should not rush blindly into altering the setting of trimmers or cores; even the more experienced constructors should seek to obtain maker's alignment instructions before tackling such a task on a factory-built set.

DIGITAL TECHNIQUES

The availability of digital-type integrated circuits, including low-cost ttl logic, is encouraging the use by amateurs of a number of techniques that formerly would have required unduly complex circuitry.

Among these techniques are:

Digital Direct Read-out Systems

By using an add-on or built-in digital frequency counter off-set by the amount of the receiver's i.f. it is possible to display the frequency to which the set is tuned directly on numerical display (Nixie-type) tubes or light-emitting-diode matrices. This is achieved by continuously measuring the frequency of the local oscillator and may supplement or replace calibrated tuning dials.

Digital Frequency Synthesis

To improve frequency stability, professional receivers now often replace the tunable free-running local oscillator with a frequency synthesizer deriving its outputs from one

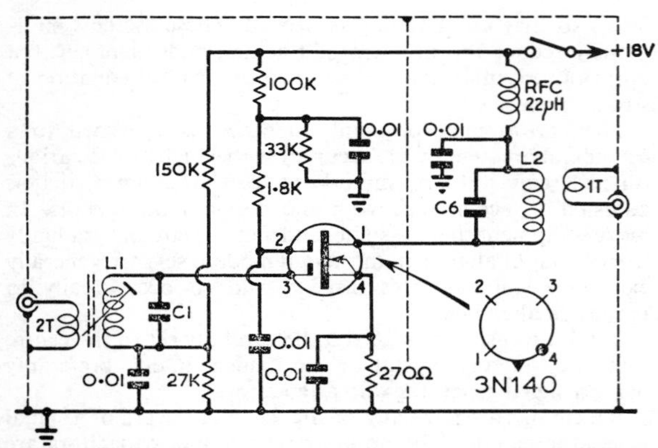

Fig 4.71. Dual-gate mosfet amplifier. Values for 28MHz C1 8pF, C6 10pF, L1, L2 1·6 to 3·1μH. Values for 21MHz C1 22pF, C6 22pF, L1, L2 no change. Similar circuits can be used with zener-protected MOSFETs.

or a few crystals. Such synthesizers may be built into the receiver in the form of variable-ratio-divider type synthesizers and the receiver set to any required frequency by means of decade switches, or provided with rotary switching having much the same "feel" as a vfo-tuned receiver. Tuning may be in steps of 1kHz, 100Hz or 10Hz or even less. To "clean up" the signals derived from digital systems they are used to phase-lock a free-running oscillator. It should be noted however that a phase-locked oscillator has an inherent jitter and that digital synthesizers involve high-frequency pulses which must be carefully screened from the signal path of the receiver.

Stabilized VFO

Digital techniques may be used to stabilize a free-running vfo by continuously "sampling" the frequency over pre-determined timing periods and then applying a dc correction to a varactor forming part of the vfo tuned circuit. The timing periods can be derived from a crystal oscillator and the technique has been shown to be capable of holding a reasonably good vfo to within a few hertz. Here again some care is needed to prevent the digital pulses, with harmonics extending into the vhf range, from affecting reception.

Digital Calibrators

The most widespread and simplest use of digital technique is to provide many more "marker" calibration signals than would normally be available from a crystal calibrator.

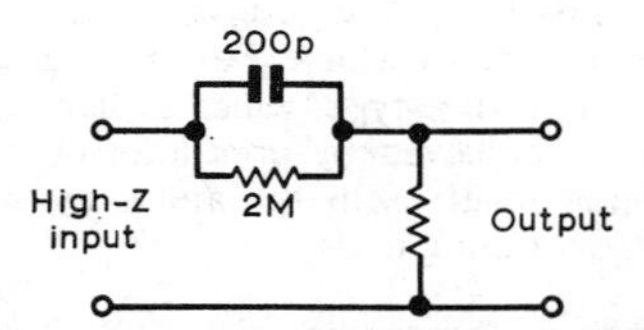

Fig 4.72. The single-crystal filter can be used effectively for phone reception by incorporating af tone correction to remove the "wooliness" of the heavily top-cut speech. A simple network such as the above provides top lift that restores intelligibility when used with the response curve of a typical single-crystal filter.

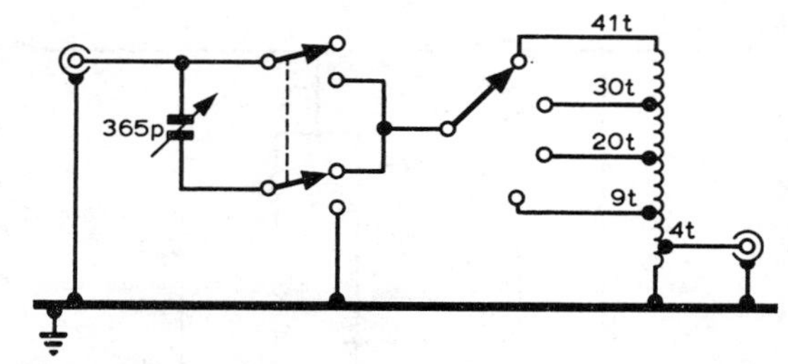

Fig 4.73. Typical aerial tuning and matching unit to cover 0·55 to 30MHz.

MODIFICATIONS TO RECEIVERS

While the number of amateurs who build their own receivers from scratch is today in a minority, many buy relatively low-cost models or older secondhand receivers and then set about improving the performance. Old, but basically well-designed and mechanically satisfactory valved receivers can form the basis of excellent receivers; often rather better than is possible by modifying some more recent low-cost receivers. The main drawback of the older receivers is their long warm-up period making it difficult to receive ssb signals satisfactorily until the receiver has been switched on for perhaps 15 or 20 minutes.

Some of the older models using relatively very noisy mixer stages may be improved on 14, 21 and 28MHz by the addition of an external *pre-amplifier* and such a unit may also be useful in reducing image and other spurious responses. However high-gain pre-amplifiers should not always be used indiscriminately since on a low noise receiver they will seriously degrade the signal-handling capabilities without providing a worthwhile improvement of signal-to-noise ratio. Receivers having low noise but poor signal handling capabilities can more often be improved by the fitting of a switched, adjustable or agc-controlled aerial attenuator: such an attenuator is likely to prove of most use on the 7MHz band where the presence of extremely strong broadcast

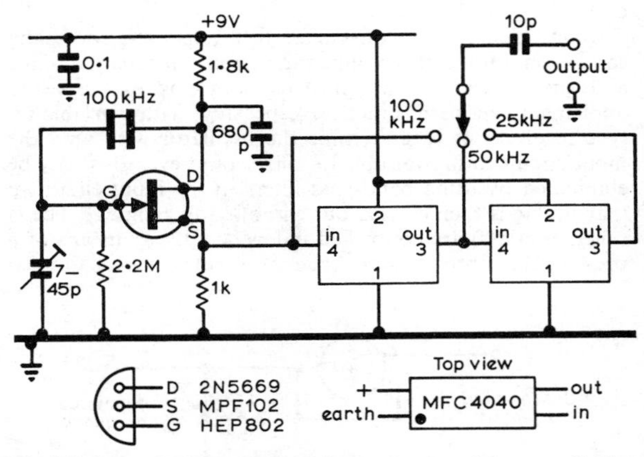

Fig 4.74. Crystal calibrator using electronic organ frequency divider integrated circuits and suitable for battery operation. Providing 100, 50 and 25kHz markers throughout the hf range. Suitable for external unit or for incorporation into the receiver.

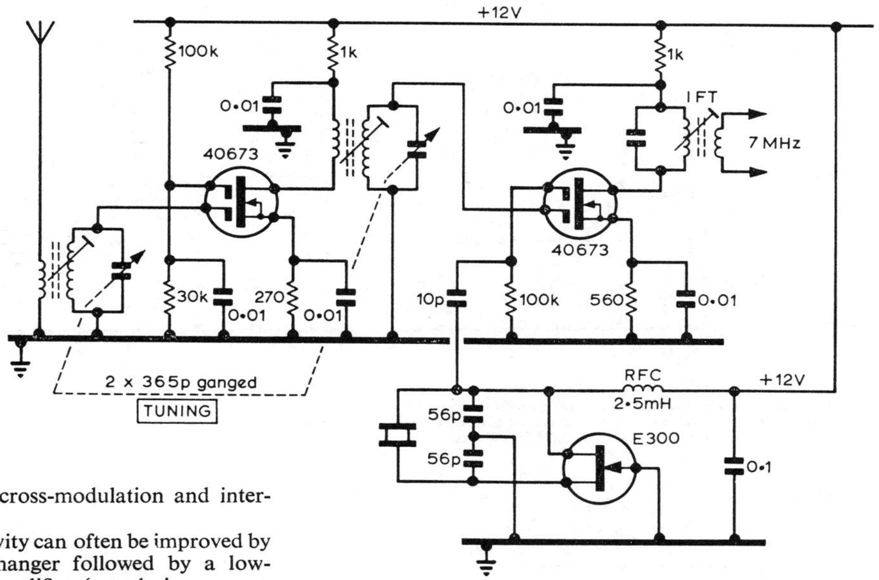

Fig 4.75. Representative hf converter using semiconductors for reception of 3·5, 14, 21 and 28MHz on receiver tuning 7MHz, or as required. Since the oscillator harmonics are sufficient to drive mixer only two crystals, 10·5 and 14MHz, will provide reception on the four bands.

signals often results in severe cross-modulation and intermodulation.

A receiver deficient in selectivity can often be improved by adding a second frequency changer followed by a low-frequency (50 to 100kHz) i.f. amplifier (a technique sometimes known as a Q5-er); or by adding a crystal or mechanical filter, or by fitting a *Q*-multiplier. CW reception can be improved by the use of narrow-band audio filters, although the degree of improvement may not always be as much as might be expected theoretically because of the ability of an experienced operator to provide a high degree of discrimination.

Older receivers having only envelope detection may be improved for ssb and cw operation by the fitting of a product detector; or for nbfm reception by adding an fm discriminator.

Many older receivers use single rather than bandpass crystal filters (and the excellence of the single crystal plus phasing control for cw reception should not be underestimated) and these often provide a degree of nose selectivity too sharp for satisfactory a.m. or ssb phone reception: speech may sound "woolly" and virtually unintelligible due to the loss of high and low frequency components. However since the response curve of such filters is by no means vertical, the addition of a high degree of tone correction (about 6dB/octave) can do much to restore intelligibility and the combination then provides an effective selectivity filter for ssb reception. The tone correction circuit shown in **Fig 4.72** is suitable for use in high-impedance circuits and can be adapted by using higher C lower R for low-impedance circuits.

The addition of an aerial matching unit between receiver and aerial can improve reception significantly in those cases where appreciable mis-match may exist (for example when using long-wire aerials with receivers intended for use with a 50 or 70Ω dipole feeder) (**Fig 4.73**).

A common fault with older receivers is deterioration of the Yaxley-type wave-change switch and/or the connection to the rotor spindle of the variable tuning capacitors; such faults may often cause bad frequency instability and poor re-set performance. Improvement is often possible by the careful use of modern switch cleaning lubricants and aerosols.

A simple accessory for older receivers (or those modern receivers not already incorporating one) is a crystal calibrator providing "marker" signals derived from a 100kHz or 1MHz crystal. While a simple 100kHz oscillator will usually provide harmonics throughout the hf range, the availability of integrated circuit dividers makes it practicable to provide markers which are not direct harmonics of the crystal. For example 10kHz or 25kHz or even 1kHz markers can be provided using ttl decade divider logic or divide-by-two devices, as in **Fig 4.74.**

A receiver deficient in hf oscillator stability on the higher frequency bands may still form the basis of a good tunable i.f. strip when used on a low frequency band in conjunction with a crystal controlled converter. Again, when the basic problem is oscillator drift due to heat this can sometimes be reduced by fitting silicon power diodes in place of a hot-running rectifier valve, or by adding temperature compensation to the hfo. A more drastic modification is to replace an existing valve hfo with an internal or external fet vfo: see **Fig 4.76**. Excessive tuning rate can sometimes be overcome by fitting an additional or improved slow-motion drive. A receiver with a good vfo can be modified for really high stability performance (better than about 20Hz) by means of external "huff and puff" digital stabilization using crystal-derived timing periods.

Receivers not initially designed for ssb operation can be improved by fitting a product detector with crystal-controlled bfo, and possibly adding a good mechnical or crystal band-pass filter.

In brief, the excellent mechanical and some of the electrical characteristics of the large and solidly built receivers such as the AR88, HRO and SUPER-PRO which featured single conversion with two tuned rf stages, are not always equalled in modern "cost-effective" designs. It may prove well worth spending time and trouble to up-grade these vintage models into receivers which can be excellent even by modern standards.

The following summary indicates some common faults with older models and ways in which these can be overcome.

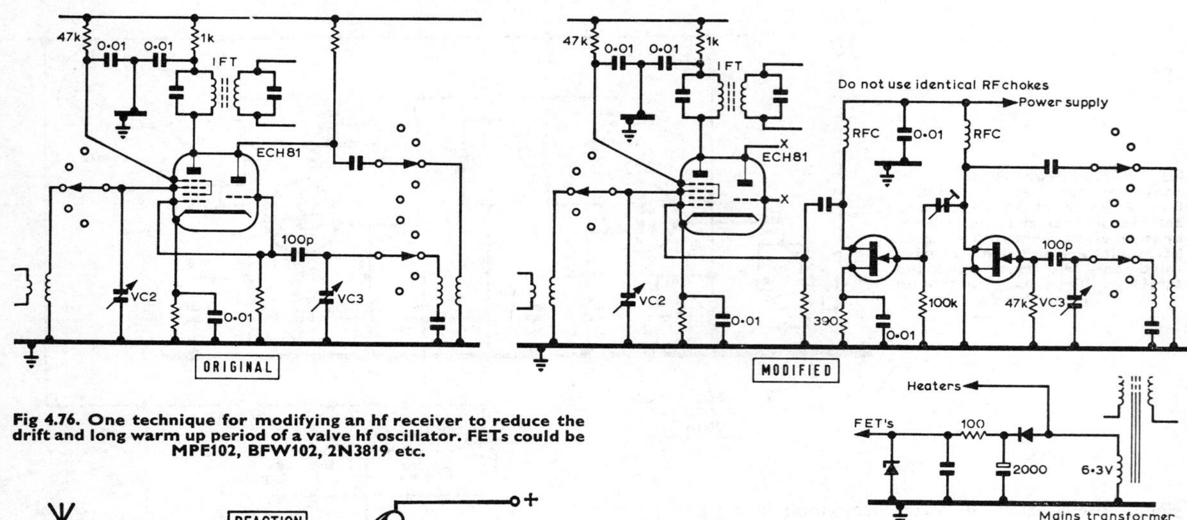

Fig 4.76. One technique for modifying an hf receiver to reduce the drift and long warm up period of a valve hf oscillator. FETs could be MPF102, BFW102, 2N3819 etc.

Fig 4.77. A single-fet receiver.

Poor sensitivity: Due to atmospheric noise this usually only degrades performance on 21 and 28MHz and then only on older valve models. Sensitivity can be improved by the addition of a pre-amplifier, but gain should not be more than is necessary to overcome receiver noise. Note that the sensitivity of a receiver may have been impaired by poor alignment, or by mismatched aerials, or due to the ageing of valves.

Image response: This can be reduced by additional pre-mixer selectivity, often most conveniently by means of a low-gain pre-amplifier with two or more tuned circuits. It is also possible to use a pre-tuned filter such as the Cohn minimum loss filter for particular bands.

Stability: This is a direct function of the oscillators within the receiver. Excessive drift and frequency "jumping" may be due to a faulty valve or band-change switch, or to incorrect adjustment of any temperature-compensation adjustments. Drift can sometimes be reduced by reducing the amount of heating of the oscillator coil by fitting heat screens, or by the addition of temperature compensation. But often with older receivers it will be found difficult to achieve sufficient stability on the higher frequency bands. In such cases considerably greater stability may be achieved by using the receiver as a variable i.f. system on one of the lower frequency bands, with the addition of one or more crystal-controlled converters for the higher frequency bands. It is worth noting that all oscillators (not only the first "hfo") may be the cause of instability (eg second or third frequency conversion oscillator or even the beat frequency oscillator).

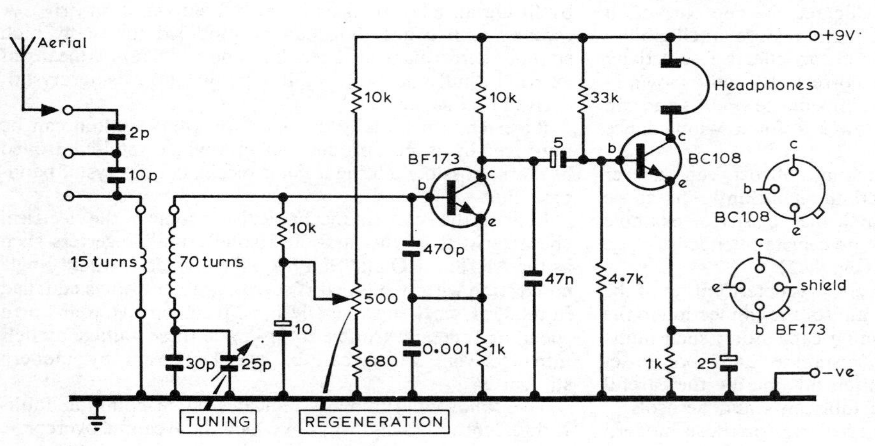

Fig 4.78. A simple "straight" receiver intended for 3·5MHz ssb/cw reception and using Clapp type oscillator to improve stability.

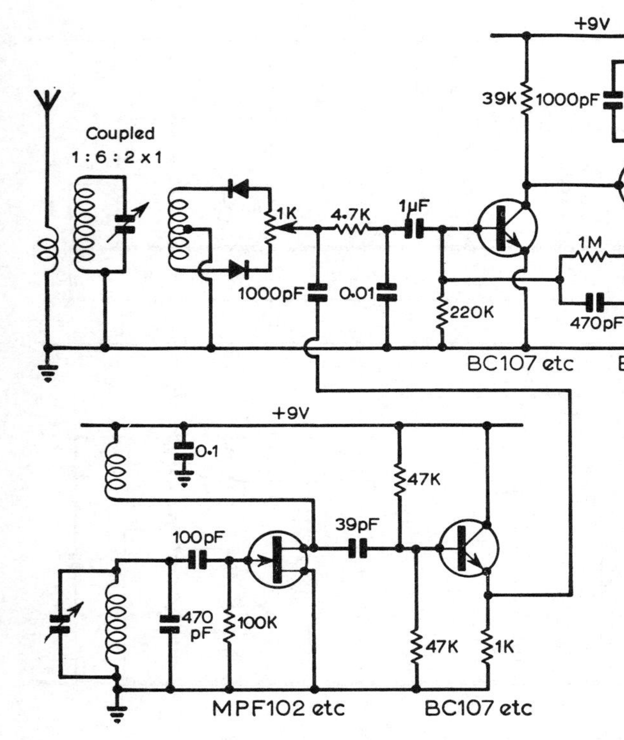

Fig 4.79. Simple direct-conversion receiver using balanced-diode demodulator (the 1kΩ pot adjusts balance).

Tuning rate: The tuning rate of some older but still good receivers tends to be too fast for easy tuning of ssb and cw signals. Often this problem can be overcome by the fitting of an additional slow-motion drive on the main tuning control. Alternatively the receiver may be used, as mentioned above, as the variable i.f. section with a converter since the tuning rate on lower frequency bands may be satisfactory. Performance on ssb may be improved also by fitting a product detector where only envelope detection is built-in.

Selectivity: It is possible to improve the selectivity of a receiver by fitting an external low i.f. section, or by fitting a (better) crystal filter, or a *Q*-multiplier. Many ssb receivers make little provision for narrow-band cw reception and in such cases it may be possible to include a single-crystal filter with phasing control in one of the later i.f. stages, or to add a *Q*-multiplier or audio filter.

Blocking and intermodulation: Performance of many semiconductor (and some valve) receivers can be improved by the addition of even a simple aerial attenuator for use on 7MHz in the presence of extremely strong broadcast signals.

CONSTRUCTION OF RECEIVERS

For a number of years most amateurs have used factory-built hf receivers. Yet the constructor who is prepared to spend time and effort can build extremely good receivers for quite modest sums. In doing so he will find that the design and construction of a receiver will teach him far more about radio than he is likely to learn in any other way. The constructor can take advantage of new techniques and devices in a far shorter lead time than it takes for their incorporation in production models. The factory design must usually cater for all possible applications, whereas the constructor can build a no-compromise receiver to suit his own interests. Either by construction or modification the amateur can provide himself with a station receiver, or a portable receiver or transceiver that can bear comparison with the best available models, and in doing so prove that the communications receiver is not a "black box" labelled "Not to be Opened".

Simple receivers and converters can be built in an evening or two, but a more advanced model may take several months of work and adjustment.

Simple Receivers

Figs 4.77 to **4.81** show some ideas for simple, low-cost receivers capable of bringing in many amateur signals when tuned and operated carefully. **Fig 4.77** is perhaps the simplest of all: a single-fet regenerative "straight" receiver able to receive cw/ssb/a.m. signals. It can be improved by adding a stage of af amplification. **Fig 4.78** represents a 3·5MHz regenerative receiver using two bipolar transistors with a Clapp-type oscillator to improve stability.

However the high-gain regenerative detector has fallen from favour on the grounds that it is easily overloaded making it difficult to obtain full advantage from any form of af selectivity. Nevertheless a well-constructed regenerative receiver can still be effective for cw operation, particularly for portable operation.

The regenerative receiver has been largely displaced by the direct-conversion approach which offers the simplicity of the traditional "straight" receiver without its inherent limitation of poor selectivity on strong signals. This does not mean that all direct-conversion receivers are better than those using regenerative detectors—only that this approach has few of the inherent limitations of either the "regenerative" receiver or indeed of the superhet, though for truly effective single-sideband reception (by which is meant the removal of the audio image and not reception of ssb stations) the design becomes of a complexity comparable with many superhet receivers.

The main drawback with direct-conversion receivers stems from limitations of simple product detectors, since these may be incapable of preventing breakthrough into the af stages of strong broadcast signals, particularly on the 7MHz band: such signals will be heard as "untunable" signals demodulated by "envelope" detection. This effect should not be observed with well balanced demodulators.

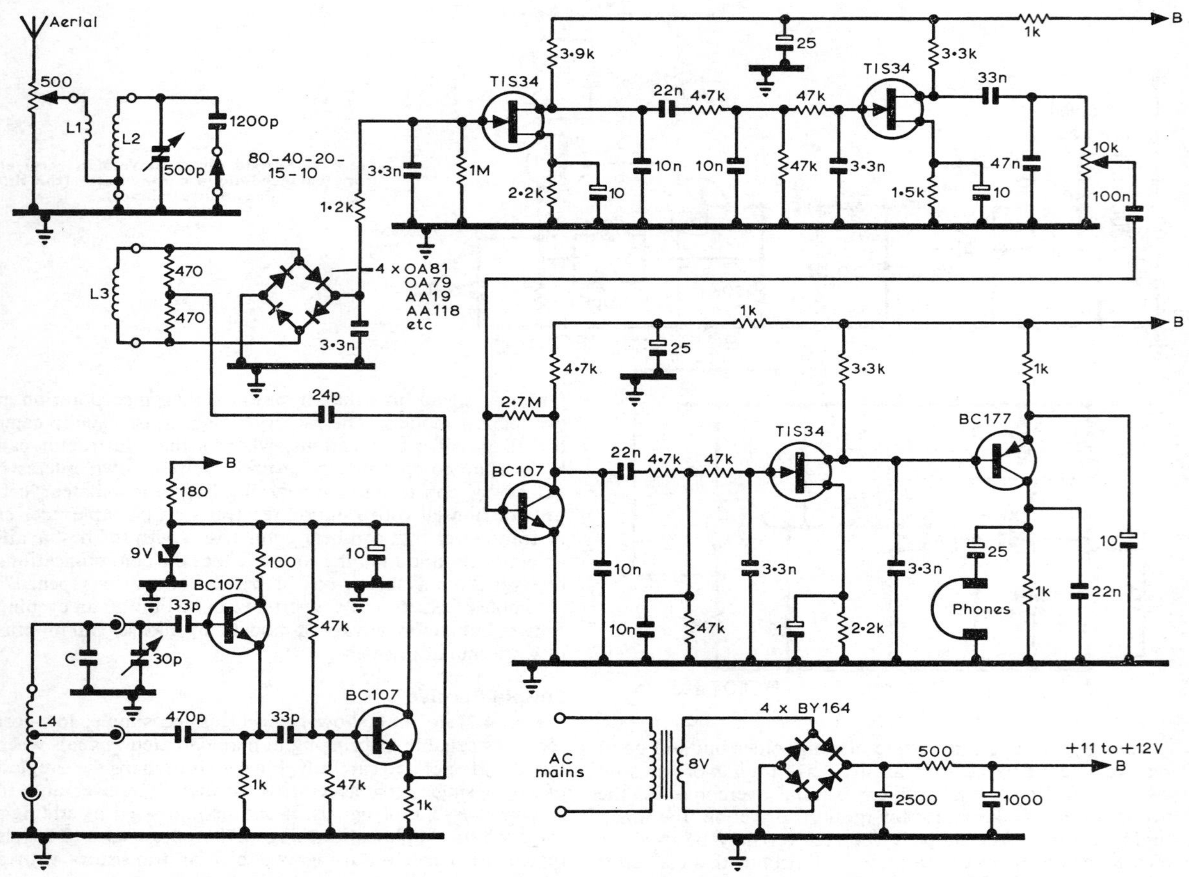

Fig 4.80. A multi-band direct conversion receiver using diode ring demodulator and plug-in coils for oscillator section.

Fig 4.79 shows an extremely simple design: with the aid of plug-in coils it can be used on several bands, although the oscillator may not be sufficiently stable for effective reception of ssb signals above about 7MHz.

Fig 4.80 is a multi-band dc receiver based on a diode ring demodulator designed originally by a Dutch listener. Several integrated circuits can be used in dc receivers including the CA3028A and SL640.

For the home constructor, a first hf superhet represents an important milestone. This is not so much because of any inherently greater complexity in its construction. Rather it is because in the construction of a superhet, the basic assembling and wiring represents only the first parts of the complete work; equal effort and skill may be required in the adjustment of the tuned circuits and the calibration of the receiver.

With a superhet one may complete the set correctly and yet be unable to receive any signals until a certain stage in the alignment is reached. This can present difficulties to the constructor who does not have access to an adjustable calibrated oscillator of the signal generator or grid dip type and who has to rely on "blind" adjustment of the circuits. It must be stressed that the alignment of a complex single or multi-conversion superhet without an adjustable oscillator can be a difficult operation calling for patience and not a little luck. Where a calibrated oscillator covering the i.f. and hf ranges is available the difficulties are greatly reduced. Again, the use of fixed-tuned ceramic i.f. filters means that i.f. alignment is virtually completed on assembly.

Coils and Wavechanging

The method adopted to alter the tuning coils when changing bands has considerable influence on the general construction of a receiver. The main methods in use are:

(a) Plug-in coils, either separate or combined into a single plug-in assembly for each band.

(b) All coils mounted on the chassis, with a Yaxley type wafer switch to select the coils required for each band.

(c) The coils mounted in a rotating turret assembly, so that although the coils are contained in the receiver only those required for one band are connected in circuit at one time.

(d) Individual front-ends for each band, usually constructed in the form of interchangeable sub-units.

Plug-in coils have the great merit of simplicity and high efficiency; they can be recommended for experimental work and for reducing the constructional complexity of receivers. The coils are usually wound on low-loss moulded formers,

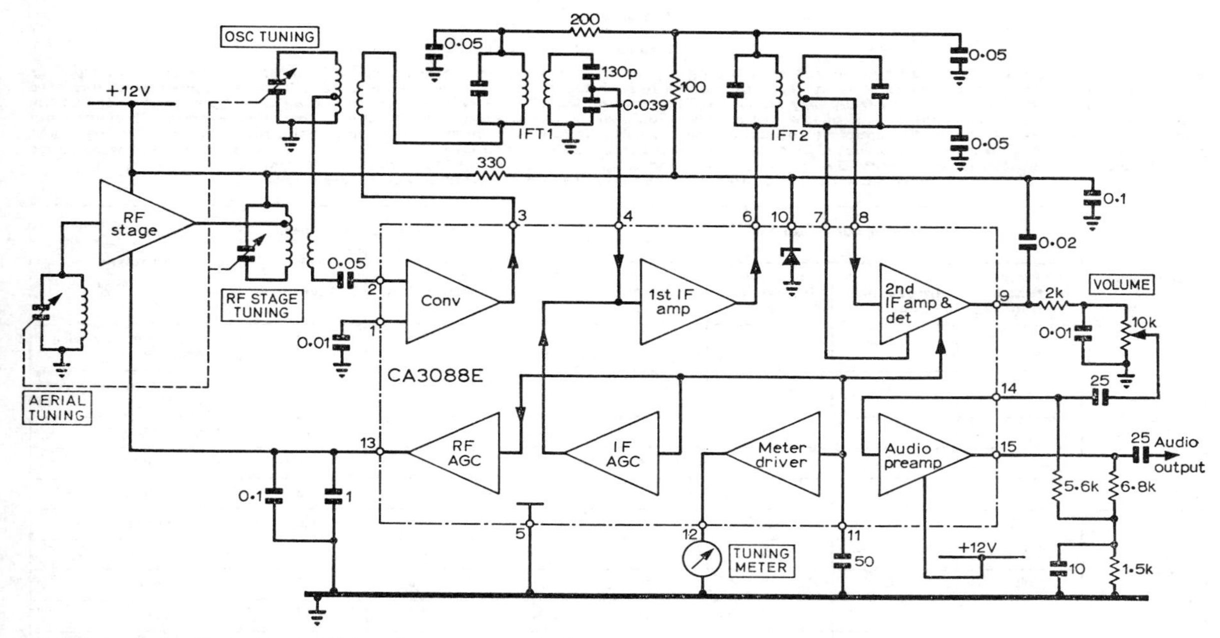

Fig 4.81. Typical outline of how a high-performance a.m. receiver can be based on the CA3088E integrated circuit—only the rf amplifier and audio output functions are external to the integrated circuit. For ssb/cw reception a bfo would be needed and could be coupled into pin 8 of the ic. Other useful consumer ICs include Mullard TBA570, SGS TBA651 etc.

the bases of which carry pins designed to make good electrical contact with sockets mounted on the receiver chassis. The coils thus occupy only a small space on the chassis and this permits short rf wiring. The main disadvantages are a degree of operating inconvenience, and the difficulty of varying the inductance for alignment purposes, since plug-in formers with variable cores are not always available.

In those cases where the coils are mounted on the chassis, it is usual to wind them on small formers with variable brass or ferrite cores. The wavechange switch should have sufficient contacts to allow coils not in use to be short-circuited, otherwise unwanted resonances and absorption effects may occur. The coils are often mounted below the chassis, either directly underneath or to one side of the ganged tuning capacitor. The switch should be mounted to keep wiring short; similarly, the highest frequency coils should be placed nearest to the switch. For low losses on the highest frequencies, the wavechange switch insulation should be ceramic (particularly important where a single wafer carries sets of contacts at both earth and ht positive); otherwise, since wafer losses form only a small proportion of the whole, Paxolin type insulation will normally prove reasonably satisfactory. The switch action should be positive and the contacts preferably kept coated with silicone lubricant to prevent oxidization. Electrolube No. 1 (Green) is a useful proprietary lubricant for switches.

Turret coil assemblies are electrically very efficient but mechanically difficult to arrange. Stray capacitance and inductive coupling can be kept low, and the main receiver wiring simplified. The various coils are usually mounted on a framework which is rotated by the action of the wavechange knob. Sets of contacts fitted along strips holding the coils press against spring contacts connected to the main circuit. Television turret tuners have been successfully adapted for this purpose.

The severe restrictions on layout, the stray capacitances and couplings represented by switch wiring and long term unreliability of multi-contact rotary switches can be reduced or eliminated by the use of diode electronic switching provided that care is taken to minimize problems arising from the introduction of non-linear devices into the signal path.

The interchangeable front-end or converter unit for each band is particularly well suited for use with a tunable first i.f. The unit generally comprises a broad-band rf stage, first mixer and crystal-controlled hf oscillator. This system avoids the difficulties of switching hf circuits and allows each unit to be designed for optimum performance on the particular band concerned. It is however, a relatively expensive form of construction, amounting to a single-range receiver with a series of fixed-tuned converters.

Table 4.5 provides a guide to typical coil windings. For use with ganged circuits, care should be taken to wind coils as similar as possible; coils with adjustable cores permit accurate matching of inductances. Single-layer coils for the hf bands can easily be wound by hand; ready-made coils are also available.

Toroid Cores

Increased use is being made of small toroid cores; since the magnetic field is virtually closed, inductors made in this way are largely self-screening and can be used in close proximity to the chassis or other components. To preserve the closed magnetic field the inductors should be wound

TABLE 4.5

Guide to Coil Windings

Figures are given as a guide only and are based on a tuning capacitor with a maximum capacitance of 160pF. The maximum frequency limit will depend largely on the value of stray capacitances, while such factors as closeness of turns, lengths of lead, position of dustcore (where used) will materially affect the frequency coverage. The reaction winding for trf receivers should be close to the lower end of the main winding. The aerial coupling coil, if of low impedance, should be wound over the earthy end of the main winding. If of medium impedance or for intervalve use, the coupling coil should be spaced a little way from the lower end of the main winding. Where an ht potential exists between windings, care should be taken to see that insulation is adequate. Small departures from the quoted wire gauge will not make any substantial difference. Generally, reaction and coupling windings can be of moderately fine wire.
The number of turns for intermediate ranges can be judged from the figures given.

Diameter of former	Number of turns				SWG (main winding)	Approximate frequency range (MHz)		Remarks
	Main tuned winding	Low impedance aerial	Medium impedance aerial	Reaction		Minimum	Maximum	
1½in ribbed air-core	3	1	2	2	20	13·5	31	Turns spaced two wire diameters
	5	1	2 or 3	2 or 3	20	11·5	23	Turns spaced one wire diameter
	9	2	4	3	22	6·5	14	Slight spacing
	17	3	5	4	24	3·4	6·8	Close wound
	42	6	10	10	30	1·6	3·3	Close wound
¾in ribbed air-core	8	2	3	3	24	16	30	Close wound
	18	3	5	5	24/26	7	16	Close wound
	40	6	10	10	30	3·5	8	Close wound
⅜in dust-iron core	8	2	3	3	26	13·5	31	Close wound
	14	3	5	4	28	7·0	15	Close wound
	26	5	8	6	30	3·2	7	Close wound
	40	6	10	9	32	1·6	3·6	Slightly pile wound

symmetrically and it is not good practice to trim the coils by spreading or compressing the turns.

Approximate windings can be determined from: turns $= 1{,}000\ \sqrt{(L/\mathrm{A1})}$ where L is the wanted inductance in millihenrys and A1 is the factor "millihenrys per thousand turns" which has to be obtained from manufacturer's data (occasionally stamped on the core) or from trial windings. Conventional gdo checking is not usually possible due to the closed magnetic loop. **Fig 4.82** shows a technique providing adjustment where it is not important to preserve a completely closed loop.

When trimming by subtracting turns a useful accessory is a small crochet hook. Resonant circuits using toroids plus disc-ceramic capacitors may be trimmed by reducing the value of the capacitor by grinding the capacitor on a sanding disc or grinding wheel (up to 50 per cent reduction in capacitance is usually possible provided that the grinding is longitudinal and does not smear metal dust between the plates). Exposed plate edges should be sealed by wax or polystyrene dope.

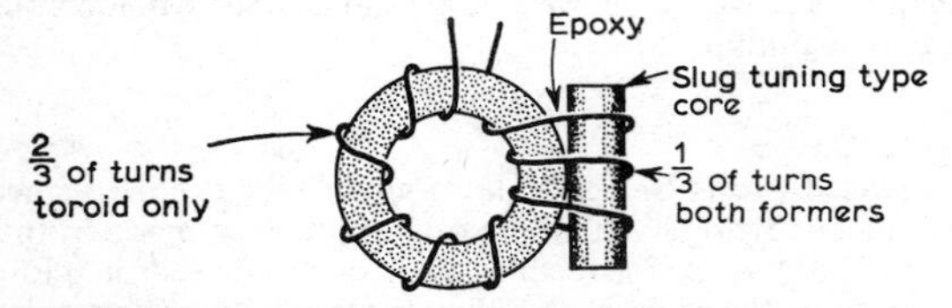

Fig 4.82. Tunable toroid technique. About 10 per cent variation in inductance can be achieved.

One cannot tap part of a turn on a toroid form; any time the wire passes through the central hole, it is effectively *one* turn. Try to wind turns just tight enough to keep them sliding around the core, but not too tight. Coils have a higher figure of merit if wound loosely but crossovers should be avoided. Avoid the use of the epoxy type adhesive to hold windings in position on small coils since this increases distributed capacitance and reduces the number of turns, lowering the figure of merit. Nylon screws are the best mounting medium and will hold turns in place. Low impedance link turns can be wound over other windings but space the link winding evenly around the core for optimum results.

Tuning Mechanisms

The tuning drive arrangements can make or mar the performance of a receiver, particularly on ssb and cw signals. Points to consider are smooth continuous action and absence of backlash in the reduction gearing. Some operators find a moderately heavy flywheel action of assistance.

Assuming a total bandwidth of the order of 500kHz and a 180 degree tuning span, a reduction ratio of 100 : 1 would be needed to give a tuning rate of about 10kHz per revolution. This would generally be considered a good tuning rate, but where the receiver is intended primarily for cw or ssb reception it may be advantageous to reduce the tuning rate still further, to around 5kHz or even 2·5kHz per revolution. With very low tuning rates it will be found advisable to fit a small handle to the tuning knob to facilitate turning from one end of a band to the other. With receivers having a fairly high tuning rate (on many well-known models this may

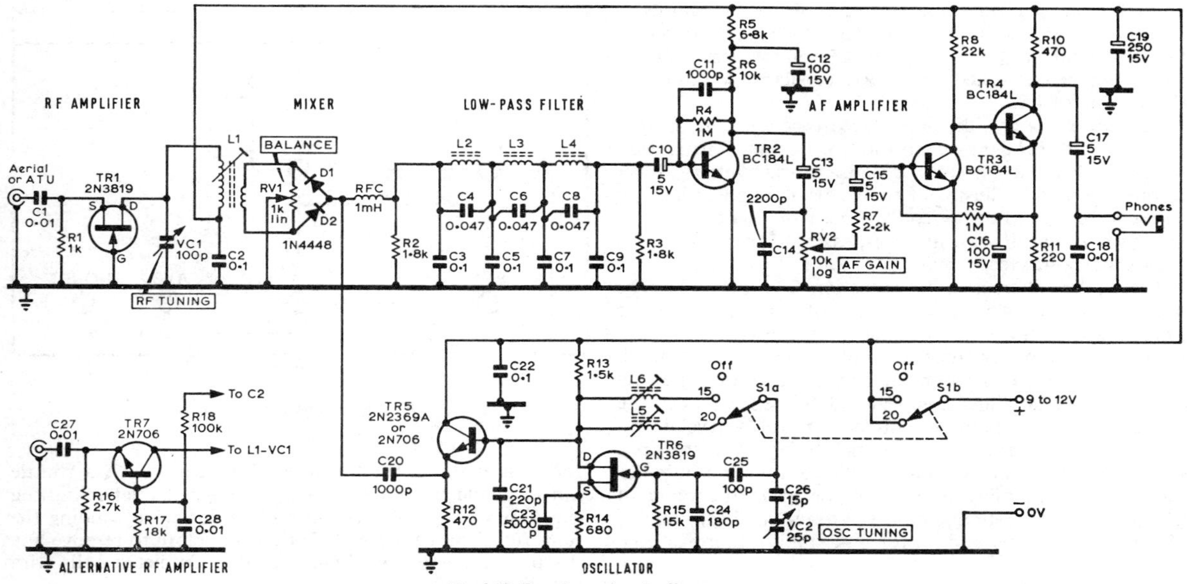

Fig 4.83. Receiver circuit diagram.

exceed 100kHz per revolution and considerably more on the 28MHz band) it is sometimes possible to fit a further reduction drive on to the existing tuning spindle.

The knob with which the tuning is normally done (the bandspread knob on a general coverage receiver) should have a fairly large diameter (at least 2 inches), to facilitate careful setting and should be mounted in the most convenient position for the operator. A very suitable drive mechanism for incorporation in a home-built receiver is the Eddystone type 898.

Permeability Tuning

While it is the usual practice to tune hf circuits with a variable capacitor, it is equally possible to do this by varying the inductance by means of sliding cores. This system, known as permeability tuning, offers considerable advantages in the maintenance of correct LC ratios and permits closer aerial input coupling. Since the screw principle can be adopted to alter the position of the cores it makes possible almost unlimited mechanical bandspread without the problems of gear reduction drives.

Diode Capacitors

A reverse-biased junction diode (and to a lesser extent a point-contact diode) has a capacitance which varies with a change in the reverse voltage. This characteristic can be used to provide what in effect is a variable capacitor tuned by adjusting the dc bias potential applied across it, if necessary from a physically remote position. Silicon junction diodes developed for use as capacitors are generally termed *varactors*, or by various trade names such as "Varicaps". One typical varactor diode (Hughes HC7005) undergoes a change of capacitance of from 24pF to 81pF as the reverse bias is reduced from 21 to 1V. Although the capacitance continues to increase below this figure, the *Q* factor tends to fall and temperature coefficient degrades. The *Q* of such devices is considerably below that of conventional capacitors but is sufficiently high to allow their direct use in various tuning applications in hf receivers; the relatively low *Q* allows ganged tuning without calling for extremely close tolerances in capacitance characteristics.

Because of the low *Q* and problems arising from introducing non-linear devices into the signal path, electronic tuning is not often used in the front-end of high-performance receivers, but can be useful for such purposes as bfo adjustment.

TWO-BAND DIRECT-CONVERSION RECEIVER

Introduction

The receiver is a direct conversion type intended for the reception of ssb and cw amateur stations. It is simple to construct, inexpensive to build, yet capable of performing as well as many of the more complex and more expensive superheterodyne receivers. Although intended as a project for the novice it should also prove an interesting project for the more experienced constructor, for use as a stand-by receiver for home station or portable use.

It is designed for the 14 and 21MHz bands but may be constructed as a single-band receiver by omitting the band-change switch and one oscillator coil. It may be further simplified by omitting the af filter and rf amplifier but at the cost of performance.

Circuit Description

A field effect transistor was chosen for the rf amplifier as it is less prone to cross modulation than the bipolar transistor, and a common gate configuration was used to provide a low input impedance for the aerial or atu as it does not require neutralizing. A bipolar transistor rf amplifier is also shown for those who wish to use it. No band switching is required for either amplifier, as the tuning covers both bands. The rf amplifier will increase the gain of the receiver and improve the signal-to-noise ratio but may be omitted if a lower performance is acceptable, in which case a two-turn link is wound on the cold end of L1 for the aerial, or the aerial may be connected via a 47pF capacitor connected to the hot end of L1.

The mixer is a two-diode balanced type with the balance potentiometer brought out to the front panel. This is done so that when strong a.m. broadcast transmitters break through, a small adjustment to the balance potentiometer will remove them. Both silicon and germanium diodes in the mixer work well. It was, however, easier to find two silicon diodes that were closely matched.

The audio filter is a three-section LC network. The inductors can be wound on ferrite pot cores or on ferrite rings. The cut-off frequency of the filter is 3kHz.

In the af amplifier three low-noise transistors are used. The gain control is between the first and second stages. By doing this the amplifier is not at full gain except when the gain control is at maximum. This keeps the noise level down.

The local oscillator employs a field effect transistor in a stable Vackar circuit. An emitter follower is used as a buffer stage to prevent pulling of the oscillator during the tuning of the front end. No retuning has been found necessary during 15 minutes or more, and no form of stabilization has been used.

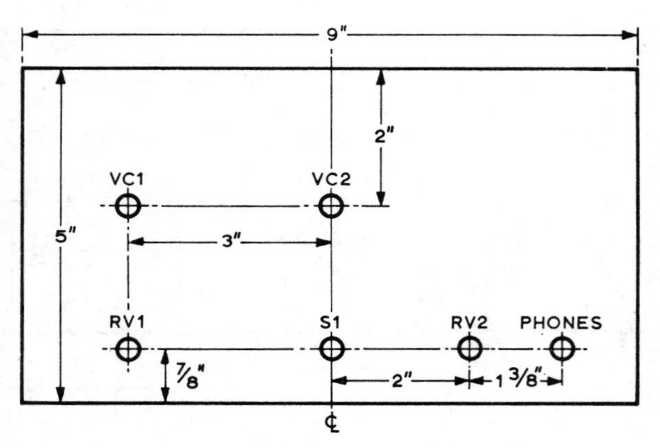

Fig 4.84. Front panel layout.

Construction

This receiver has been constructed, in single and dual band versions, in various cases from die-cast boxes to ex-surplus equipment cases. The choice is left to the constructor. Only one will be described here for the benefit of the novice, a simple front panel and chassis housed in a metal or wooden case. If the wooden case is used it should be lined with aluminium cooking foil. The front panel is a 9in by 5in sheet of aluminium or steel, and the chassis is 9in by 5in by 2in deep.

First mark out and drill the front panel, **Fig 4.84** and chassis, then mount the rf tuning capacitor, balance potentiometer, af gain control and phones socket. The oscillator tuning capacitor is mounted on the top side of the chassis and a 3-to-1 reduction drive (or better still a 10-to-1) is mounted in line with it on the front panel. The remainder of the oscillator components, including the band switch if required, are mounted underneath the chassis. See the photographs for top and bottom layouts.

Start with the rf stage, keeping all wiring as short as possible. The leads from the two-turn link on L1 are passed through the chassis and connected to the balance potentiometer. Mount a tag strip or terminal posts to take the diodes and rf choke.

The next stage is the oscillator. All the components of this stage are mounted on a small tag strip behind the coils. Ensure that all the components and wiring in the oscillator are ridged. The last to be wired in is the fet. A three-sided screen is then placed around the oscillator section and fixed to the front panel. A screened lead is brought out through the screen to the mixer tag strip.

The filter is the next item to be constructed and it is better to do this before putting it into the receiver, **Fig 4.85**. As pot cores were rather expensive it was decided to use ferrite rings, Mullard Type FX1593 wound with 300 turns of 38swg enamelled wire, for the inductors. A simple way of winding them is to make a shuttle, wind on 30 turns of wire, and pass it through the centre of the ring. Such a shuttle can be made from a 3½in length of ⅛in diameter knitting needle with a ¼in slot cut in each end. Before winding the rings rub them gently with fine emergy cloth to remove any sharp edges. After winding, coat the completed coils with varnish and leave to dry.

The coils, capacitors and resistors R2 and R3 are mounted on a paxolin board 1½in by 2½in. The coils are held in place with cotton passed through small diameter holes drilled in the board. A 6BA solder tag is placed under both mounting holes on the board and a 16swg bus wire soldered between them to which the capacitors are connected. The board is mounted on two pillars, earthing both ends of the bus wire. A lead is taken from the filter to the mixer tag strip. The output of the filter is connected via a screened lead to the input of the af amplifier. The capacitors may be soldered to pins inserted in the board or their leads may be passed through small holes in the board and connected with tinned copper wire. If constructors wish to use ferrite pot cores for the inductors, Mullard LA2400 or LA1114 may be used and should be wound with 350 turns of 34swg enamelled wire. The inductance should be about 60mH.

The af amplifier is constructed on a 2¾in by 2½in paxolin board. Again, small holes are drilled in the board to take the component leads which are connected on the underside with tinned copper wire. Transistors type BC109 may be

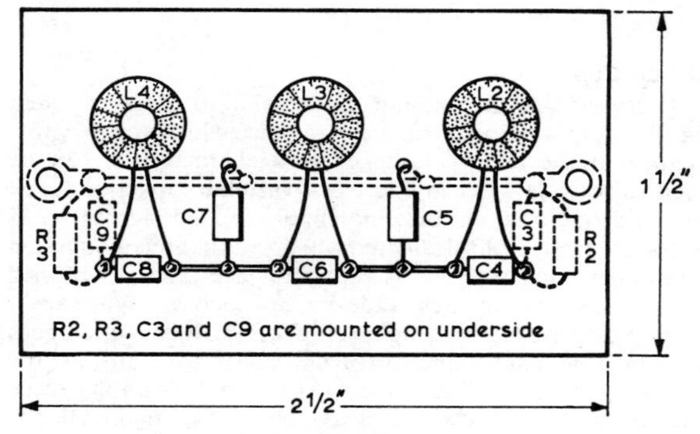

Fig 4.85. Filter layout.

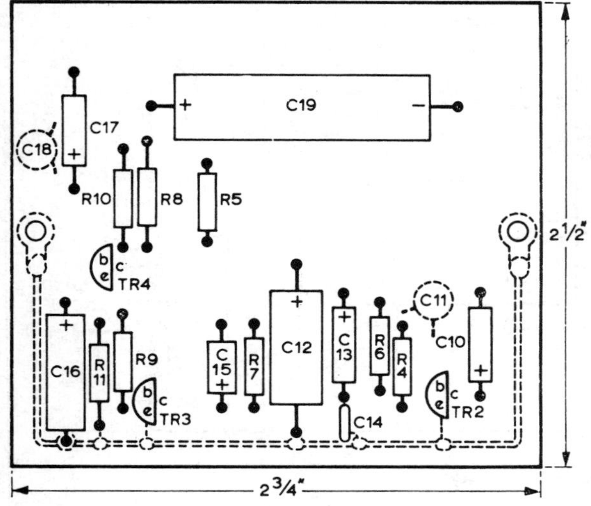

Fig 4.86. AF amplifier layout.

used but the plastic type are to be recommended as they are less likely to be affected by the rf from the oscillator. The BC184L was used in this design. The af amplifier board has a 16swg bus wire which is used for the common (negative) line. This is bent round at the ends and is soldered to the 6BA tags through which the fixing screws are passed and is earthed through the mounting pillars. The layout is shown in **Fig 4.86.**

The atu, although not essential, is a very worthwhile addition, particularly when the receiver is to be used with random lengths of aerial wire. The circuit is shown in **Fig 4.87** and was constructed in a die-cast box, RS Components Type 994.

The knobs and dials used were Bulgin but here the constructor has a free choice. The reduction drive was a Jackson epicyclic ball type with a reduction of 10 to 1.

One section of the bandchange switch was used for the ON/OFF switch but whether one uses this or a separate switch is again a matter of choice.

Setting Up

First check and then double check all wiring. Then connect two 6V batteries (Type PP1) in series to the receiver, negative to earth. Set the balance potentiometer RV1 to mid-way, and the oscillator tuning capacitor VC2 to full mesh. Switch to Band 1 and with the aid of a communications receiver (bfo on) adjust the tuning slug in L5 for zero beat at 14MHz. Switch to Band 2 and adjust the tuning slug in L6 for zero

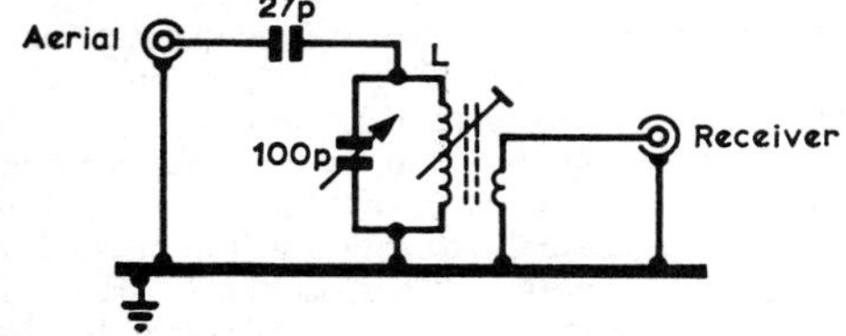

Fig 4.87. Aerial tuning unit. L — 12 turns of 28swg enamelled on ⅜in diameter former with 2-turn link and tuning slug.

Under-chassis view.

Top view of chassis.

TABLE 4.6

Components list

R1 1kΩ	C1 0·01μF	C24 180pF sm
R2 1·8kΩ	C2 0·1μF	C25 100pF sm
R3 1·8kΩ	C3 0·1μF	C26 15pF
R4 1MΩ	C4 0·047μF	C27 0·01μF
R5 6·8kΩ	C5 0·1μF	C28 0·01μF
R6 10kΩ	C6 0·047μF	VC1 100pF
R7 2·2kΩ	C7 0·1μF	VC2 25pF
R8 22kΩ	C8 0·047μF	TR1 2N3819 fe
R9 1MΩ	C9 0·1μF	TR2 BC184L
R10 470Ω	C10 5μF/15V	TR1 2N3819 fet
R11 220Ω	C11 1,000pF	TR2 BC184L
R12 470Ω	C12 100μF/15V	TR3 BC184L
R13 1·5kΩ	C13 5μF/15V	TR4 BC184L
R14 680Ω	C14 2,200pF	TR5 2N2369A/2N706
R15 15kΩ	C15 5μF/15V	TR6 2N3819 fet
R16 2·7kΩ	C16 100μF/15V	TR7 2N706
R17 18kΩ	C17 5μF/15V	D1 and D2 1N4448 or other
R18 100kΩ	C18 0·01μF	general purpose silicon diodes.
RV1 1kΩ lin	C19 250μF/15V	Germanium diodes such as
RV2 10kΩ log	C20 1,000pF	OA79, GEX-66 may be used.
All resistors 10%	C21 220pF polyester	RFC1 1mH
¼W	C22 0·1μF	SW1 2-pole 3-way wafer
	C23 5,000pF sm	

Coil details

L1 12 turns of 28swg enamelled close-wound with 2-turn link wound on cold end
L5 12 turns of 28swg enamelled close-wound
L6 7 turns of 28swg enamelled close-wound
L1, L5 and L6 wound on ⅜in diameter formers slug tuned
L2, L3 and L4 300 turns of 38swg enamelled wound on Mullard ferrite rings Type FX1593

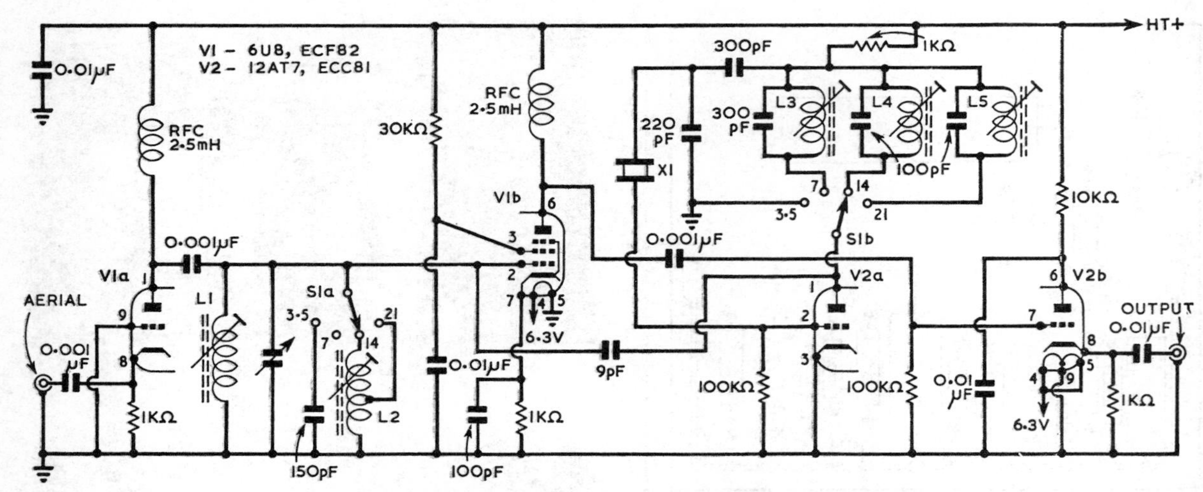

Fig 4.88. Circuit diagram of the high stability converter.

beat at 21MHz. Do not tune the coil slugs with a metal screwdriver, use a non-metal trimming tool, such as a large knitting needle filed to a blade.

The next step is to connect an aerial and tune the front end. Do this with the receiver switched to Band 1 and the oscillator at about 14MHz. Set VC1 to about 90 per cent of full mesh and adjust the tuning slug of L1 for the maximum output of the headphones. Switch to Band 2 and rotate VC1 again for maximum output. Make a note of where the rf tuning is for each band. If a communications receiver or wavemeter is not available for tuning it is no problem to tune the oscillator until a station is heard.

Conclusion

Although simple, this receiver should not be regarded as a gimmick or as an inexpensive substitute for the superheterodyne. It is not by any means claimed to be the ultimate in receivers but it does perform well and is an ideal project for the beginner interested in dx reception.

LOW-COST HIGH STABILITY CONVERTER

Many older communications receivers provide adequate performance on the lower frequency bands but lack stability and/or sensitivity and good bandspread tuning on the higher frequency bands. One satisfactory answer to this problem, often providing an additional useful lease of life to be given to receivers acquired second-hand, is to use the basic receiver as a tunable i.f. system on, say, 3·5MHz in conjunction with a sensitive crystal-controlled front-end converter for bands above 3·5MHz.

Many suitable designs can be evolved from the circuits presented earlier in this chapter; basically such a converter will usually comprise an rf amplifier (possibly omitted if a low-noise mixer is used) mixer, crystal-controlled oscillator and possibly a cathode follower to facilitate low-impedance coupling to the main receiver.

Most circuits require a separate crystal for each band, or alternatively reverse the receiver tuning direction on one or more bands (for example by using a 10·5MHz crystal to provide 3·5MHz output on 7MHz and on 14MHz, or a 17·5MHz crystal frequency for 14 and 21MHz). However an economical and simple converter capable of good performance using only a single 3500kHz crystal has been described by F. Johnson, ZL2AMJ—see **Fig 4.88**.

This two-valve unit functions as a crystal-controlled converter on the 7, 14 and 21MHz amateur bands and as a straight pre-amplifier on 3·5MHz. The converter provides a broad-band output on the 3·5MHz band, and all tuning is carried out on the main receiver. It should prove particularly effective in conjunction with an older receiver having good bandspread tuning and stability on the 3·5MHz band. The selectivity will depend entirely upon that of the main receiver when operating on 3·5MHz. One disadvantage is that the unit produces a strong spurious marker signal on 7,000kHz which can block the first few kilohertz of the 7MHz band. This can be minimized or eliminated if the crystal is slightly lower in frequency than 3,500kHz as described later.

The converter comprises the triode section of a 6U8 (ECF82) triode-pentode working as a grounded-grid amplifier. The pentode section of the functions as mixer with a low noise contribution. One half of a 12AT7 (ECC81) functions as crystal oscillator and the other section as a cathode follower output stage following the mixer. There are no tuned circuits at the broadband i.f. of about 3·5–4MHz and consequently no ganged tuning circuits are used.

For 3·5MHz reception the oscillator is switched off and the remaining stages act as a pre-amplifier. On other bands the crystal oscillator frequency is always 3·5MHz lower than the band being received, so that the signal frequency rises as the main receiver is tuned higher across the 3·5MHz band. For 7MHz the crystal oscillates on its fundamental frequency of 3,500kHz, for 14MHz on its third "overtone" of about 10·5MHz and for 21MHz on its fifth "overtone" of 17·5MHz. Note that for overtone operation the output is not precisely at the exact harmonic of the crystal fundamental, although close to it, so that a small correction will be needed to the original receiver calibration. Otherwise the 3·5MHz calibration applies in terms of tens and hundreds on all bands, and the tuning rate will be the same on all bands. Thus 7·1,

14·1 or 21·1MHz signals should be received on about 3·6MHz and 7·2, 14·2 or 21·2MHz on 3·7MHz, etc.

In order to maintain this approximate 3·5MHz calibration on all the other bands, the crystal needs to be fairly accurate on 3,500kHz. However if this is not regarded as vital, it is often possible to use a crystal some kHz lower in frequency. For example a 3490 kHz crystal would tune 7000–7300kHz as 3510–3810kHz, 14,000–14,350kHz as 3530–3880kHz and 21,000–21,450kHz as approximately 3550–4000kHz. This tolerance would allow the use of a cheap surplus crystal provided that it is sufficiently active to oscillate readily on its fifth overtone.

Note that only two adjustable coils are used to tune the four bands with L1 left in parallel with L2 on 14 and 21MHz.

Construction and layout are not particularly critical provided that rf leads are kept short, and output circuits placed well away from input circuits.

All coils are wound on ¼in diameter slug-tuned unshielded formers using 30swg enamelled wire: L1 42 turns; L2 26 turns, tapped at 17 turns from the "earthy" end; L3 35 turns; L4 13 turns; L5 8 turns, spaced over ⅜in. Formers could be 0·3in diameter with a slight reduction in numbers of turns—this can easily be determined if a calibrated grid dip oscillator is available. The formers can be mounted around the Yaxley-type bandswitch.

Alignment

After completing and checking all wiring, the oscillator circuits must be adjusted so that the crystal oscillates on the correct overtone frequencies. This can be done most readily with a grid-dip oscillator used as an absorption wavemeter, though it should be possible to achieve satisfactory results by checking the output on the main receiver if this will tune to 10·5 and 17·5MHz and care is taken not to be misled by harmonic output. With the absorption device check and adjust the cores of L3, L4 and L5 until the oscillator is operating on the required frequencies.

Then set the converter to 7MHz and connect it to the main receiver which is set to 3·5MHz. Without switching on adjust L1 for coverage of 7–7·5MHz in conjunction with the 35pF tuning capacitor by using the grid dip oscillator in the usual way (this can be done with incoming signals without a gdo but the process is more difficult). Switch to 3·5MHz and check that the rf tuning range now covers 3·5–4·0MHz (or just the European section of the band—3·5–3·8MHz—if this is all that is required). If the tuning range is incorrect, it may be necessary to replace or parallel the 150pF tuning capacitor with other values. Once the 3·5MHz tuning range has been established, switch to 14MHz and adjust L2 so that the tuning range covers 14–14·5MHz. It should then be possible to tune the 21MHz band without any further adjustment (the full tuning range will probably exceed the amateur band but this is of no consequence). In practice it will probably prove necessary to change the converter tuning only when tuning right across the band; when searching around only a small section of the band the control can be left unaltered.

Once the tuning ranges have been adjusted, the unit should be ready for use. Signals can be peaked with the converter tuning control when necessary, otherwise all tuning is done on the main receiver. If the gain of the hf converter should be excessive on some signals a gain control can be fitted on the cathode follower stage.

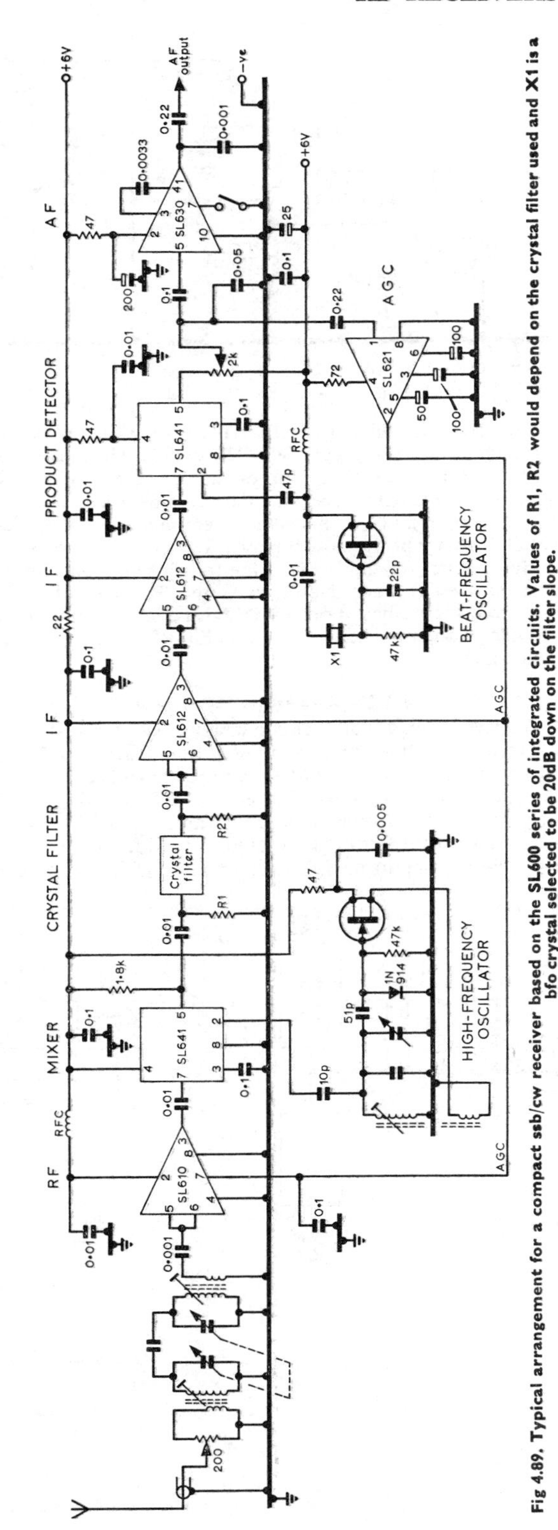

Fig 4.89. Typical arrangement for a compact ssb/cw receiver based on the SL600 series of integrated circuits. Values of R1, R2 would depend on the crystal filter used and X1 is a bfo crystal selected to be 20dB down on the filter slope.

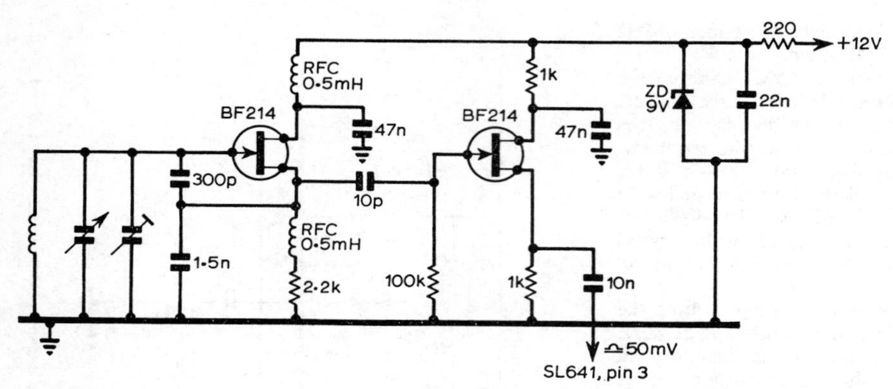

Fig 4.90. A 4·95 to 5·55MHz vfo suitable for use with ic mixers or for general applications.

One possible difficulty is that there may be i.f. breakthrough from strong local 3·5MHz signals when receiving on the other bands. Screening of the converter and the use of coaxial cable with screened connectors between converter and receiver may be sufficient to overcome any difficulties. Should this not be the case, it may in some cases be necessary to fit a further tuned circuit in the form of an aerial tuning unit between the aerial and converter. Provided that the main receiver is efficient on 3·5MHz this combination should provide excellent sensitivity and stability on all the bands concerned.

RECEIVER USING LINEAR INTEGRATED CIRCUITS

Fig 4.89 shows a typical circuit arrangement for a compact ssb receiver built around the Plessey SL600 series of linear integrated circuits. The SL610 is an agc controlled rf amplifier; the SL641 a double-balanced mixer followed immediately by the ssb crystal filter (for example a KVG 9MHz filter). Two SL612 devices are used in the i.f. amplifier, the first being agc-controlled. The second SL641 functions as a product (synchronous) detector with the af output fed both to the SL630 af preamplifier and the SL621 agc generator. This "hang" agc system is capable of holding the audio output constant to within 4dB for signal input variations of over 100dB. Dual detectors in the SL621 provide two different agc time constants; the long time constant keeps the agc output voltage nearly constant during brief speech pauses while noise pulses are sampled by the second detector and activate a trigger circuit to prevent the system from attempting to follow noise pulses. **Fig 4.90** shows an alternative vfo design.

Another SL600 receiver, forming part of a transceiver, is described in Chapter 6—*HF Transmitters*.

COMPACT HIGH-PERFORMANCE RECEIVER

To illustrate many of the points made in the earlier sections of this chapter brief circuit details are included of a compact all-semiconductor receiver of modern design, built by Jukka Vermaasvuori, OH2GF.

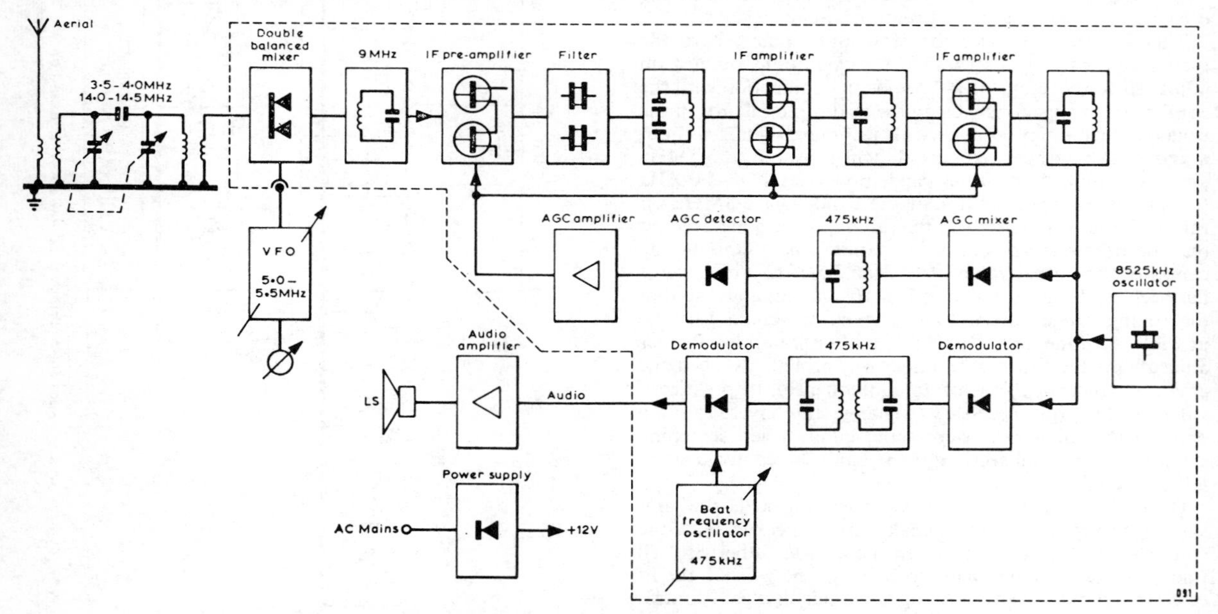

Fig 4.91. Block diagram of the compact receiver.

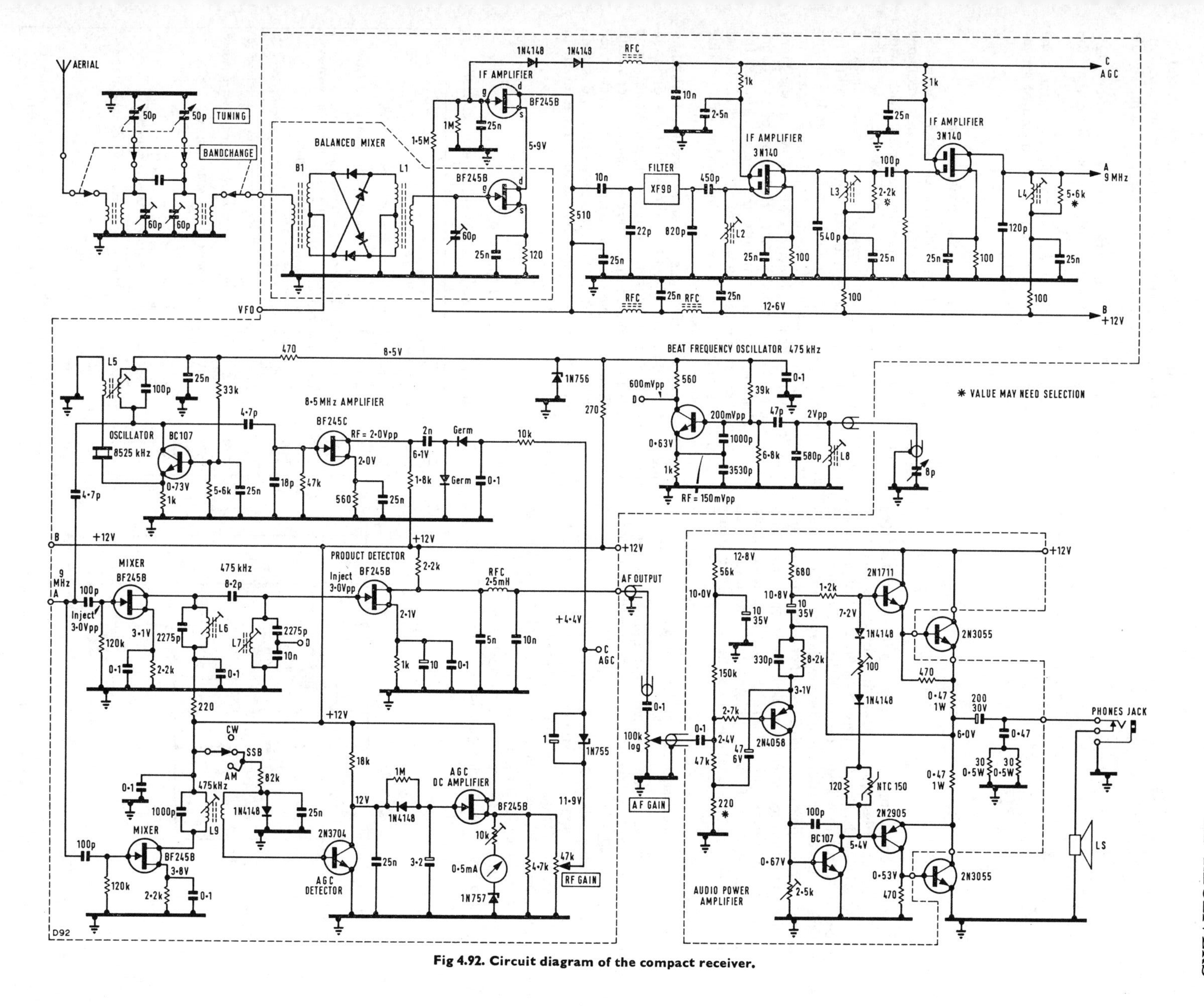

Fig 4.92. Circuit diagram of the compact receiver.

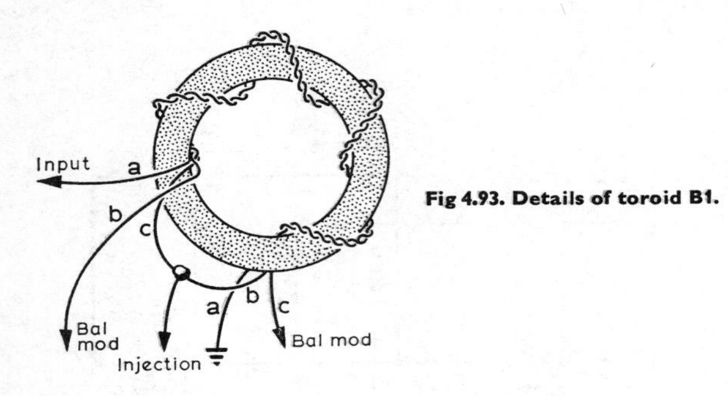

Fig 4.93. Details of toroid B1.

When incorporating a vfo tuning 5·0 to 5·5MHz (eg those shown in Figs 4.54 or 4.90) the receiver tunes the 3·5 and 14MHz bands. For operation on other bands it would be necessary either to redesign the oscillator and twin rf tuned circuits to cover the additional frequencies, or alternatively the receiver could be used with a crystal-controlled converter.

Basically the receiver is a double-conversion model with fixed I.F.S of 9MHz and 475kHz, using a double-balanced diode mixer and no rf stage. No integrated circuits are used and all devices in the signal path are either hot carrier diodes or single- or dual-gate FETS.

The following are among the general principles adopted in the design and construction of the receiver:

(a) To improve dynamic range no rf amplifier is used.
(b) Selectivity characteristics are determined near to the front-end.
(c) A double-balanced mixer has been used as first mixer to reduce spurious responses, increase linearity and to decrease the need for front-end selectivity.
(d) Demodulation is removed from 9MHz to 475kHz to overcome bfo leakage problems; this second frequency conversion however is only done after the agc-controlled stages when signal level is nearly constant.
(e) The agc loop is derived from i.f. rather than af, and the agc characteristics have been chosen for ssb reception. No envelope detection is included, and a.m. signals are received as ssb.
(f) In order to permit operation from a car battery, 12V is used as the working voltage (but note that the receiver is negative earthed).
(g) All stages from rf mixer to af are placed on the same printed circuit board to form the main unit of a two-band (3·5 and 14MHz) set.

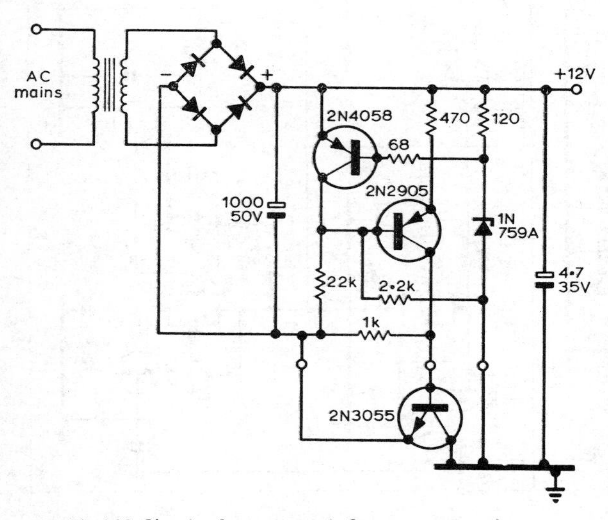

Fig 4.94. Circuit of power supply for compact receiver.

AN ADVANCED HYBRID RECEIVER

This is an advanced high-performance single-conversion receiver with an i.f. of 1·62MHz and partial frequency synthesis, providing high stability coupled with wide dynamic range. It features the use of 7360 beam deflection valves for the mixer and product detector functions; a phase-locked frequency synthesizer with high-stability fet interpolation vfo, noise blanking, hang agc and with the crystal filter distributed over several i.f. stages. The receiver was designed by Peter G. Martin, G3PDM/W1, as the Mark 2 version of one described in the 4th edition of this handbook. Full constructional details are not given since the prototype receiver uses a number of components no longer readily available, but a study of this design will provide many ideas for advanced rf receivers.

The prototype receiver was unit constructed so that building blocks could be removed for laboratory work or modification. The circuit is broken down into nine units and these will be discussed briefly in turn. The complete receiver circuit diagram (**Fig 4.95**) is given in eight parts; for clarity, switched tuned circuits are shown for one band only, and switch sections are denoted X. A list of special parts is given in **Table 4.7**.

TABLE 4.7

Details of special components in the G3PDM receiver

C501-3	See Table 4.8
D401, 501-4	Brush-Clevite BA111
L101	Amateur band coils (see Table 4.8)
L302, 4, 6	See text
L501	Amateur band local oscillator coils (see Table 4.8)
L502	1·2μH (Electroniques BP21)
L601	2·4μH. Silver-plated coil on Cambion 3354-6 coil former with Carbonyl TH slug
T101	Amateur band aerial coils (see Table 4.8)
T102, 204	1·62MHz i.f. transformer. Primary 100μH centre-tapped, secondary 100μH.
T201, 2, 3	1·62MHz ift, primary and secondary 100μH
T401	5000Ω:3Ω output transformer
T501	5·88–6·38MHz wideband coupler (Electroniques WBC-5 with 33pF capacitors reduced to 22pF)
T502	Primary tuned to 6·13MHz, turns ratio 2:1 (Modified Electroniques WBC-5)
T701, 2, 3	1·35–1·65MHz wideband couplers (Modified Electroniques WBC 1·8)
T901	100V rms at 130mA and 6·3V rms at 3A outputs
T902	16·3V 0·5A output
VC101, 201-4	Tubular ceramic trimmers
VC102	Rotary ceramic trimmers
VC501, 2	Rotary ceramic trimmers (See Table 4.8)
VC601	75pF swing, double-bearing variable capacitor
VC602	Air-spaced miniature trimmer
VC603	Temperature compensating capacitor (Oxley Tempatrimmer)
X201, 3, 5	1620·2kHz filter crystals
X202, 4, 6	1618·4kHz filter crystals
X401, 2	See text
X501	See Table 4.8
Note:	"Electroniques" components are no longer marketed but are possibly available on the second-hand market.

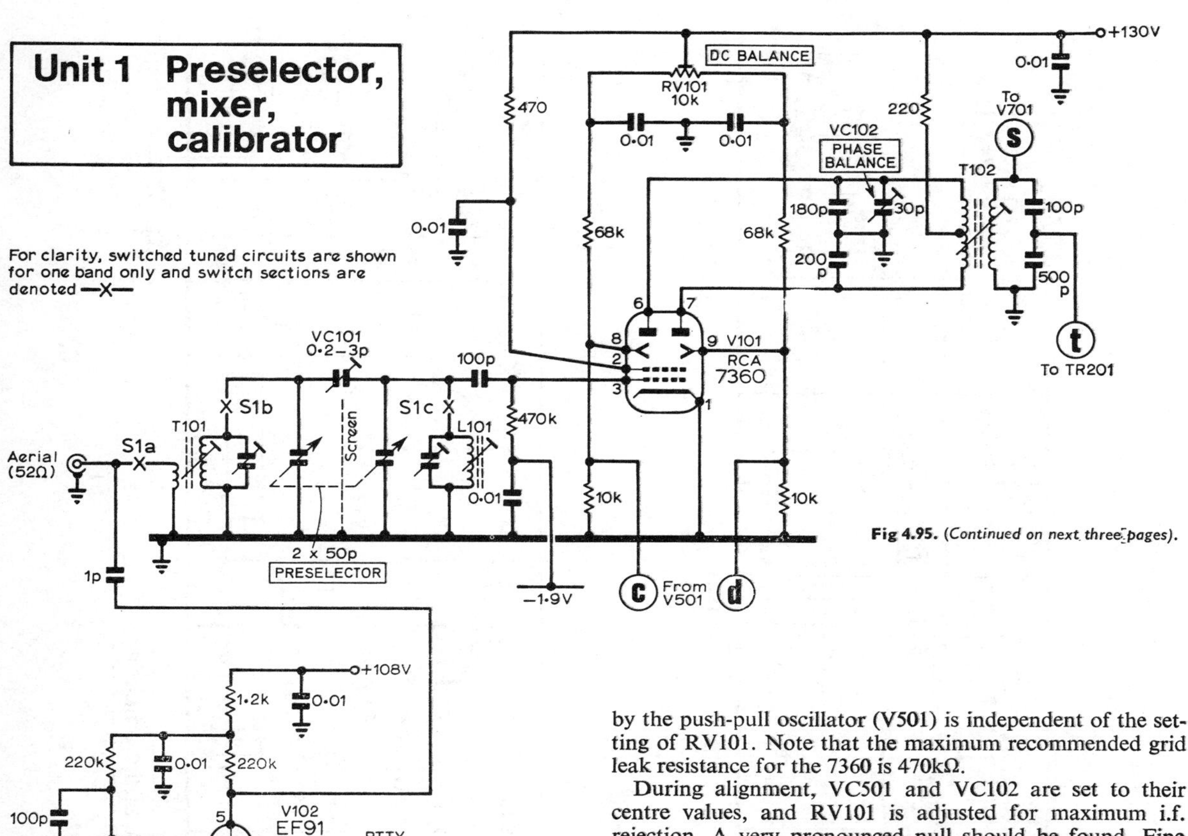

Fig 4.95. *(Continued on next three pages).*

by the push-pull oscillator (V501) is independent of the setting of RV101. Note that the maximum recommended grid leak resistance for the 7360 is 470kΩ.

During alignment, VC501 and VC102 are set to their centre values, and RV101 is adjusted for maximum i.f. rejection. A very pronounced null should be found. Fine adjustment can be made with VC102, if necessary. VC501 is adjusted to null the local oscillator radiation, using a general coverage receiver as a monitor. Readjustment of RV101 and VC102 will be necessary.

The 100kHz crystal calibrator is entirely conventional; a semiconductor device could be substituted.

Unit 1: Preselector, Mixer and Calibrator

Top-capacitance coupled tuned circuits are used in the preselector. The circuits are tuned by a two-gang 50pF variable capacitor, and coupled through a 0·2–3pF ceramic trimmer. The two sets of coils are screened from each other to prevent spurious coupling.

During alignment, the coupling coefficient is optimized by alternately reducing VC101 and re-peaking the tuned circuits, until the S-meter shows a slight loss of signal.

The inherently balanced nature of the 7360 mixer is preserved by the use of push-pull oscillator injection and a balanced output transformer (T102).

The circuit is balanced to dc by RV101, and to rf by VC102. VC501 in the synthesizer unit is used to equalize the amplitudes of the two anti-phase injection signals. The ends of RV101 are bypassed to rf so that the rf impedance seen

Unit 2: Product Detector and I.F. Amplifier

TR201 is the fet noise gate, which open-circuits the signal path when pulsed by the noise blanker (Unit 7). The receiver i.f. chain consists of three frame-grid pentode amplifiers, each preceded by a half-lattice crystal filter with resistive loading. Unbypassed cathode resistors are used on V201 and V202 to reduce the capacitive load on the first two filters.

The 7360 product detector (V204) is fed in push-pull from T204. The cathode of the 7360 is bypassed to rf and audio frequencies to reduce cross-modulation and noise modulation. The product detector output passes through a pi-section rf filter before reaching the main low-pass audio filter (Unit 3).

The filter crystal frequencies are 1,618·4 and 1,620·2kHz. Trimmers across the higher frequency crystals provide adjustment of the infinite attenuation frequencies of each section.

The overall gain of the i.f. strip is in excess of 120dB. To maintain stability and prevent deterioration of filter performance, thorough decoupling of supply lines is imperative.

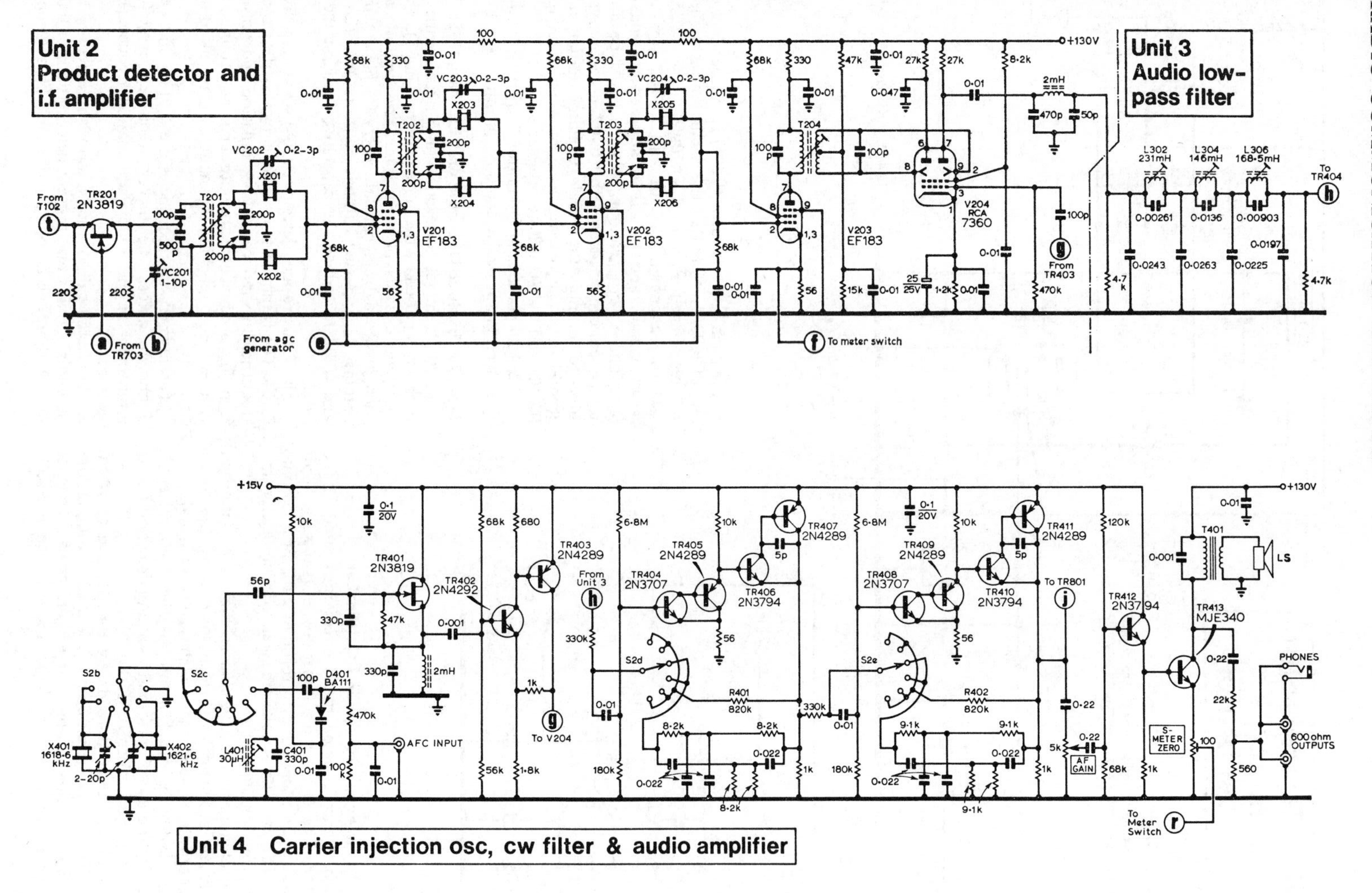

Unit 2
Product detector and
i.f. amplifier
Unit 3
Audio low-
pass filter
Unit 4 Carrier injection osc, cw filter & audio amplifier

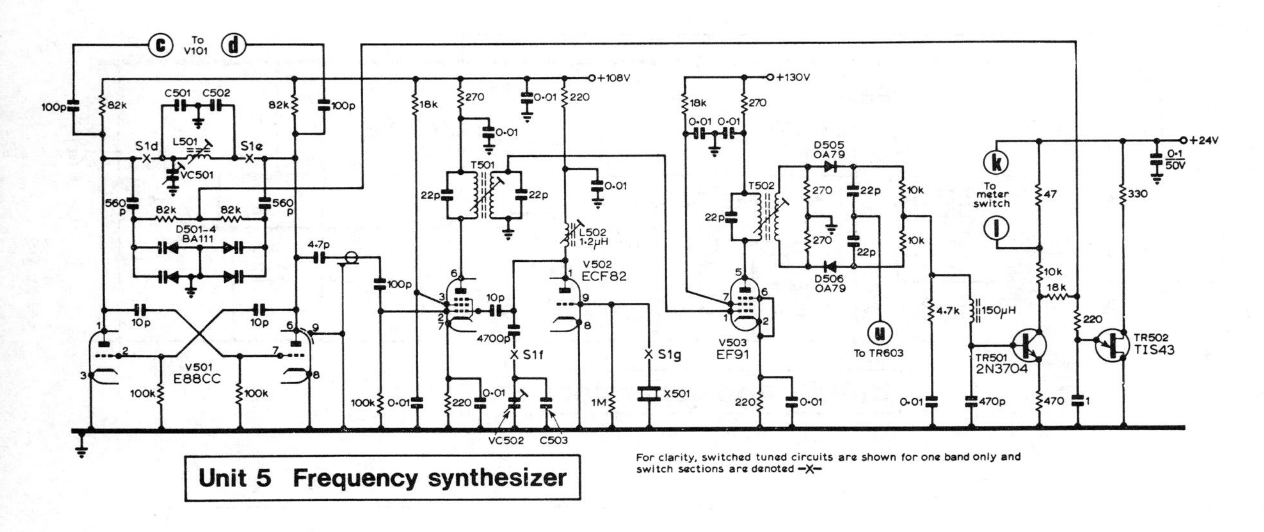

For clarity, switched tuned circuits are shown for one band only and switch sections are denoted –X–

Unit 5 Frequency synthesizer

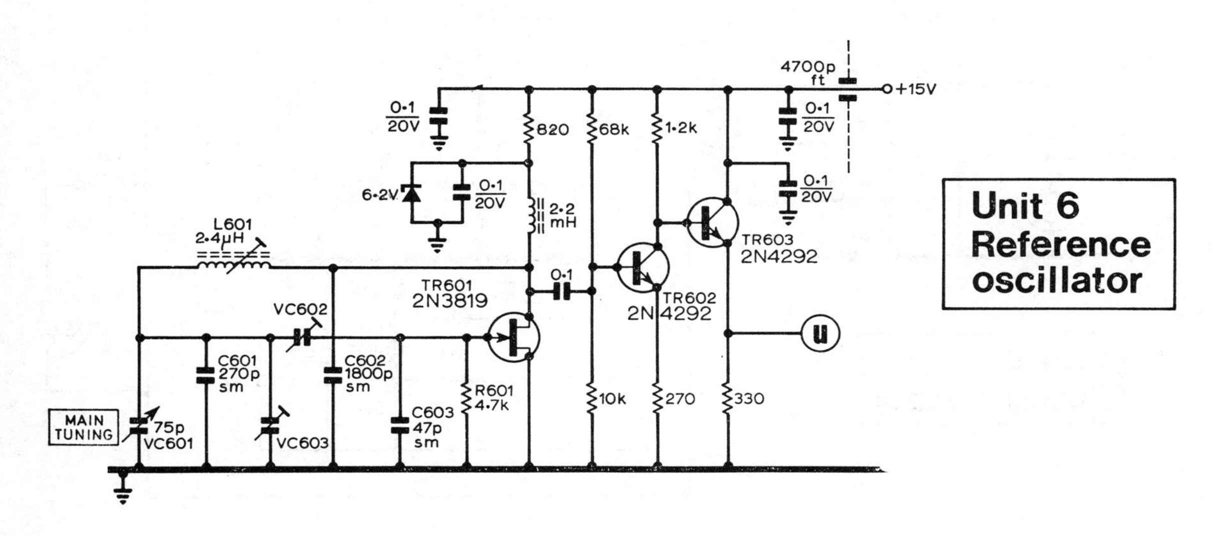

Unit 6 Reference oscillator

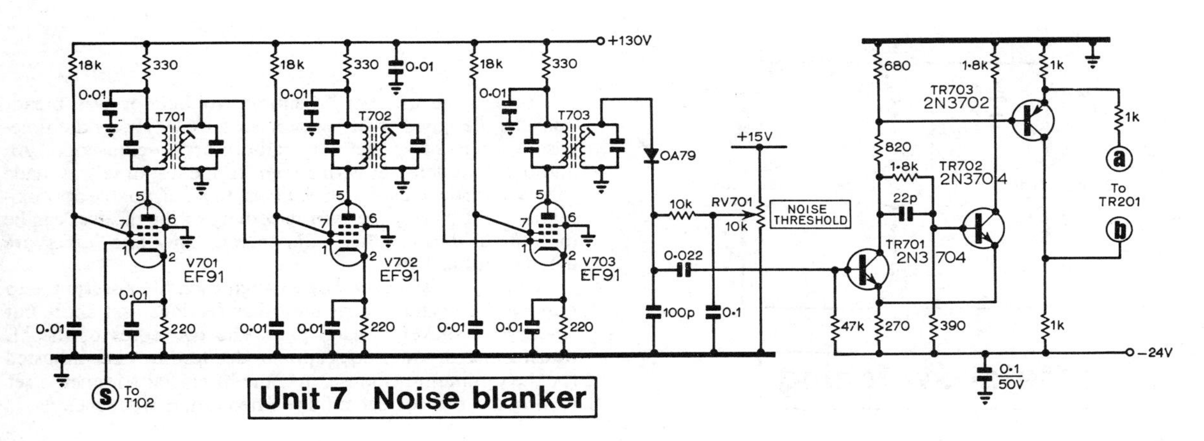

Unit 7 Noise blanker

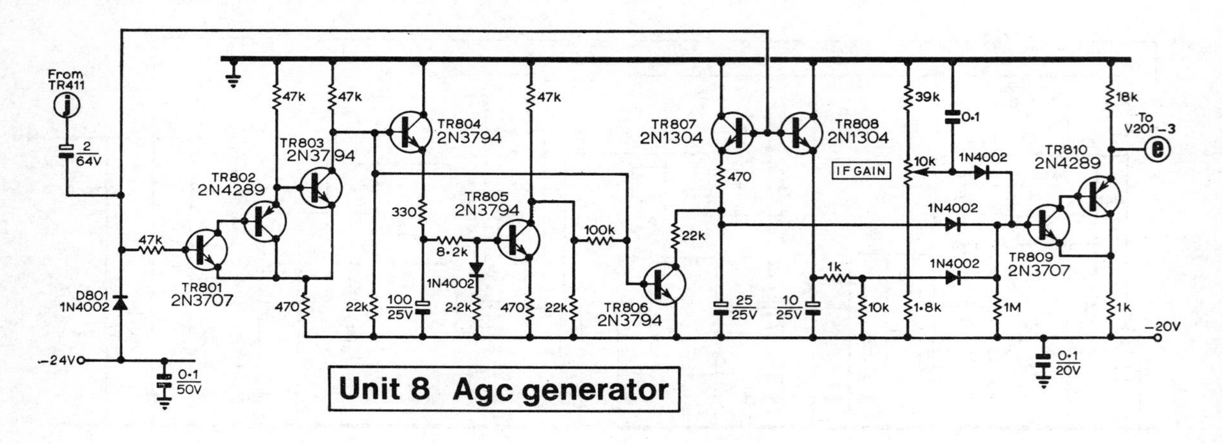

Unit 8 Agc generator

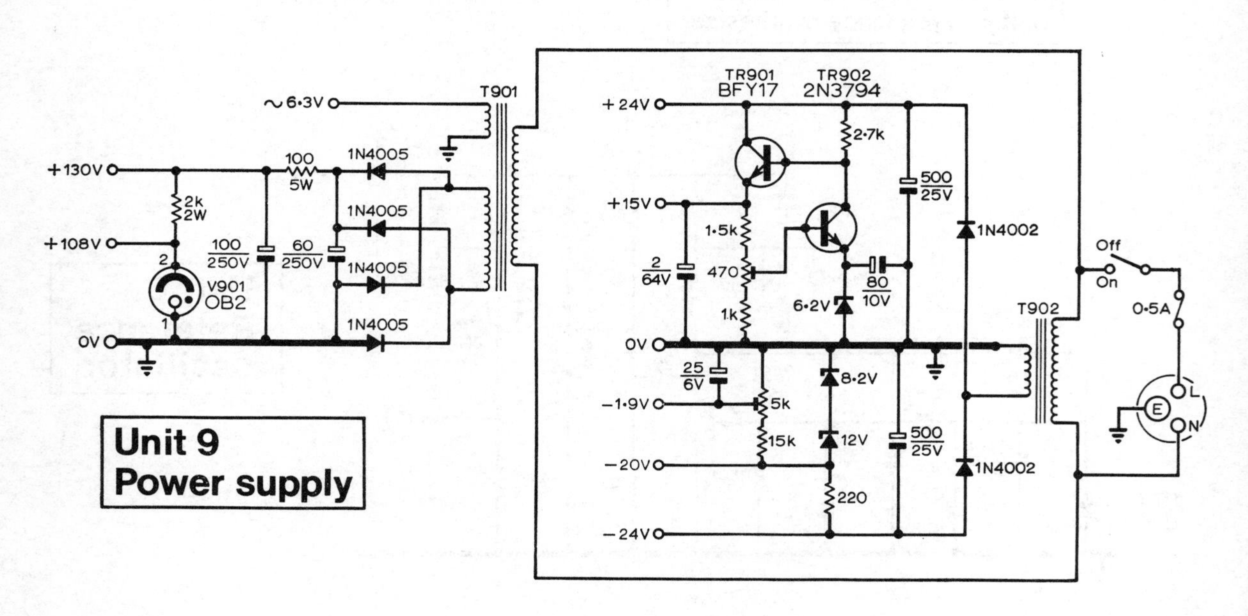

Unit 9 Power supply

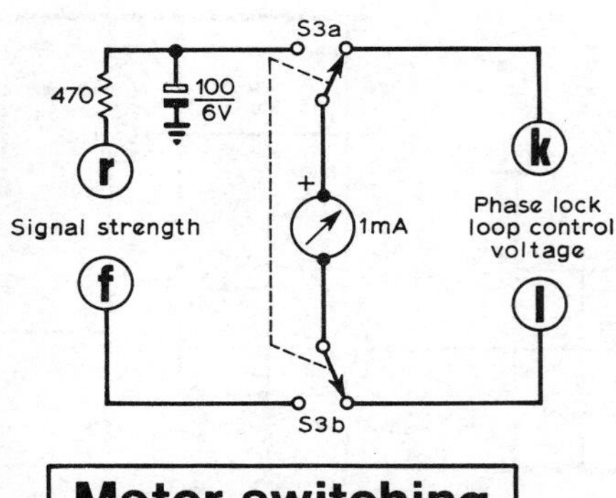

Meter switching

To align the i.f. strip, a simple wobbulator was breadboarded, designed to be swept across 1·62MHz by the timebase output of a standard laboratory oscilloscope. An audio output detector with a logarithmic response was made to give a vertical display in decibels, to a reasonable approximation. In this way the skirt selectivity of the receiver can be monitored during filter adjustment. The two units are detailed in **Fig 4.96**.

The only satisfactory way to align each filter section is to remove the crystals from the other sections and short out *one* crystal socket in each pair. The two cores of the i.f. transformer driving the section under test are then adjusted for the optimum response, and the hf crystal trimmer is set to provide the correct infinite attenuation frequencies.

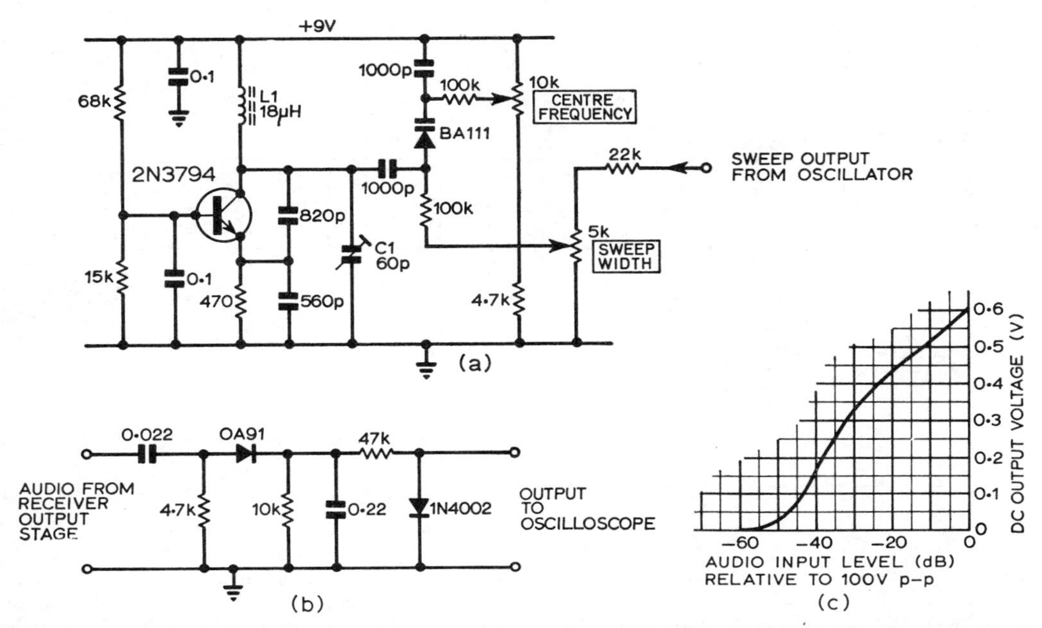

Fig 4.96. Circuit used to display filter response curves on an oscilloscope during receiver alignment. (a) Sweep frequency generator with adjustable centre frequency and sweep width. L1 is 13 turns of wire on a Mullard FX1595 ferrite ring. C1 is used to set the correct operating frequency. For most purposes direct coupling between the oscillator and the receiver under test is not needed. (b) Audio detector used to give a roughly logarithmic display of filter attenuation. The actual detector characteristic is shown in (c).

Unit 3: Audio Low Pass Filter

This elliptic function filter was described earlier. In the prototype, inductors are wound on Mullard LA2303 pot cores. L302, L304 and L306 have 650, 520 and 560 turns of 38swg wire respectively.

Unit 4: Audio Amplifier, CW filter and CIO

The carrier insertion oscillator is a fet Hartley circuit. Carrier crystal frequencies depend on the i.f. strip response achieved, but were 1,608·6 and 1,621·6kHz in the prototype.

In the rtty mode, the oscillator is tuned by L401, C401 and the capacitance of the varactor diode D401. Automatic frequency control for unattended rtty operation is available by using feedback from a teleprinter terminal unit to control the varactor bias. The carrier voltage is 5V peak-to-peak at the source of TR401, and 8V peak-to-peak at the collector of TR403, in any position of switch S2.

In the two-section cw filter, twin-T networks provide frequency selective feedback around amplifiers consisting of two complementary pairs (TR404–7 and TR408–411). In each section, the four 0·022μF capacitors and four resistors must be matched to better than 1 per cent (preferably 0·1 per cent). The RC products of the two sections are staggered by about 10 per cent, to provide a cw bandwidth of around 180Hz. In the ssb and rtty modes, the twin-T networks are replaced by R401 and R402. These are selected to give a constant audio output level when switching from one mode to another. The 5pF capacitors from base to collector on TR407 and TR411 are for high frequency stabilization.

The audio output stage consists of an emitter-follower (TR412) driving a high voltage power amplifier (TR413)

TABLE 4.8

Frequencies and component details for the phase-lock synthesizer and preselector. Inductors are described by the original Electroniques part numbers.

Band (m)	Signal frequency range (MHz)	Local oscillator range (MHz)	Crystal (X501) frequency (MHz)	T101	L101	L501	C501, 2 (pF)	C503 (pF)	VC501 (pF)	VC502 (pF)
160	1·8–2	3·12–3·62	9·5	LZ1·8	BP1·8	OS1·8/16	47	100	40	40
80	3·5–4	5·12–5·62	11·5	LZ3·5	BP3·5	OS3·5/16	0	68	40	40
40	7–7·5	8·62–9·12	15	LZ7	BP7	OS7/16	68	22	40	40
20	14–14·5	15·62–16·12	22	LZ14	BP14	OS14/16	68		20	20
15	21–21·5	22·62–23·12	29	LZ21	BP21	OS21/16	33	0	20	20
10A	28–28·5	29·62–30·12	36							20
10B	28·5–29	30·12–30·62	36·5	LZ28	BP28	OS28/16	0	0	20	20
10C	29–29·5	30·62–31·12	37							20
10D	29·5–30	31·12–31·62	37·5							20

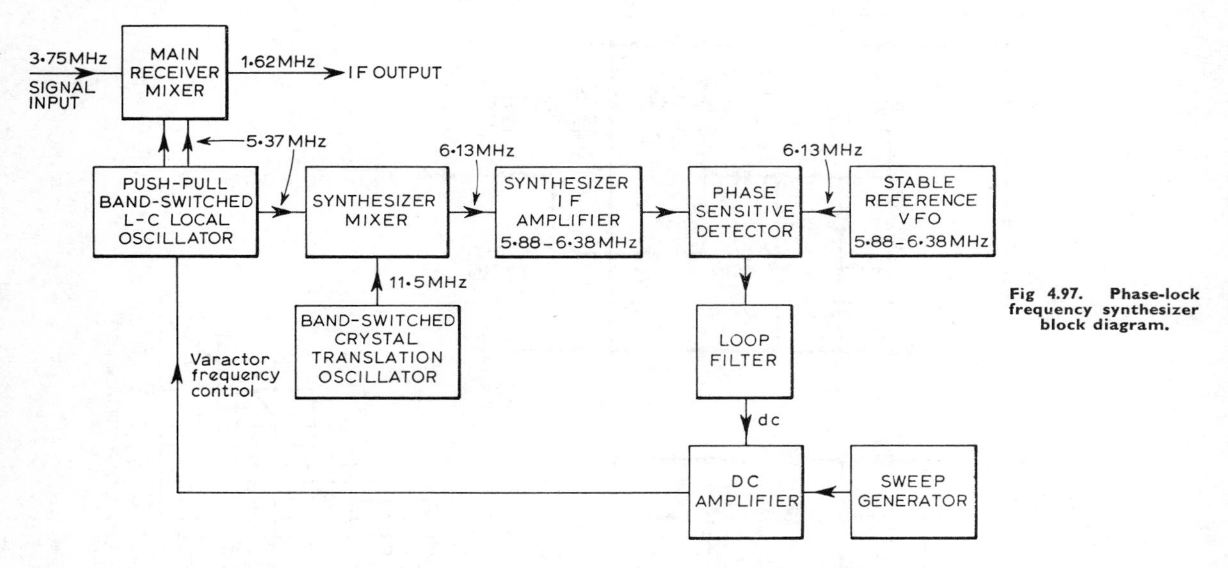

Fig 4.97. Phase-lock frequency synthesizer block diagram.

with a degenerative emitter resistor to reduce distortion. The Motorola MJE340 specified is a low-cost plastic-encapsulated device with a 300V collector-base rating. Three low-level outputs are provided: for headphones, an rtty terminal unit, and other accessories. T401 transforms the loudspeaker impedance to a 5kΩ load for TR413.

A high-voltage output stage was used to avoid the necessity for a low-voltage high-current supply in the receiver.

Unit 5: Frequency Synthesizer

The synthesizer unit is assembled in a $7\frac{1}{4}$ by $4\frac{1}{2}$ by 2in die-cast box, with screens between each section. V501 is the band-switched push-pull local oscillator driving the main receiver mixer, and the remaining circuitry is used to phase-lock this oscillator to the reference vfo (Unit 6); see **Fig 4.97**.

Although the use of a phase-lock loop in amateur equipment might be discouraging to some constructors, it is reassuring that when the prototype receiver was first switched on, the frequency synthesizer was the only unit to function correctly! Phase-lock systems are notorious for loop instability, but provided some simple ground rules are observed, few problems arise.

The pentode section of V502 mixes the outputs of the push-pull oscillator and the crystal-controlled translation oscillator to provide a signal in the reference vfo range (5·88–6·38MHz). The triode section of V502 is a Miller-type crystal oscillator operating on frequencies between 9·5 and 37·5MHz. Overtone operation is used above 20MHz. The entire frequency range of this oscillator can be tuned by a single inductor (L502) and switched capacitors (C503 and VC502). See **Table 4.8**. Wideband couplers are used between the mixer, synthesizer i.f. amplifier (V503), and phase-sensitive detector (D505–6).

The psd diodes must be matched for forward resistance and reverse-bias capacitance. If this is not carried out with sufficient accuracy, a trimming capacitor will be needed across one of the diodes. "Matched" pairs of germanium diodes supplied by one retail outlet have been found to be hopelessly unmatched.

The several filtering networks associated with the psd and dc amplifier (TR501) determine the phase-lock loop frequency response, and hence its stability. Circuit values should not be changed without a thorough understanding of the consequences. If phase-lock is lost, the collector voltage of TR501 increases towards +24V. This causes the unijunction sweep generator (TR502) to sweep the varactor bias voltage between zero and +20V. The push-pull oscillator frequency sweeps in sympathy until a beat note is detected at the psd output, and phase-lock is achieved.

To align the synthesizer, the reference oscillator is disconnected from the psd. The receiver is switched to the 14MHz band, and L501 is adjusted so that the local oscillator is sweeping over the range 15·62–16·12MHz. The crystal oscillator output on 22MHz is peaked by VC502. If a high-frequency oscilloscope is connected across one of the 270Ω resistors in the secondary of T502, the unijunction sweep generator will cause the synthesizer i.f. strip to wobbulate itself. T501 and T502 can be visually aligned to give a flat-topped response over the synthesizer i.f. range of

TABLE 4.9

Summary of receiver performance achieved

Frequency coverage:	All amateur frequencies between 1·8 and 30MHz, in 500kHz bands. A further three switch positions are provided or reception of MSF and other stations.
Tuning rate:	9kHz per knob revolution on all bands.
Sensitivity	SSB: Better than 0·5μV for 6dB signal-to-noise ratio. CW: Better than 0·2μV for 6dB signal-to-noise ratio.
Power output:	1·5W rms continuous rating.
Selectivity:	SSB: 2·2kHz at −6dB, 3·9kHz at −60dB, 6·8kHz at −120dB. 6:60dB shape factor: 1·77. CW: 130Hz at −6dB, 660Hz at −30dB, 6·5kHz at −120dB.
Image rejection:	Better than 50dB.
IF rejection:	Better than 85dB on 160m band, and better than 100dB on other bands.
Spurious responses:	All spurii below a 1μV equivalent level, except as noted in the text.
Transient responses:	IF recovery time of 470μs.
AGC characteristics	1·5dB rise in audio output for 132dB increase in signal strength above 1·5μV. Attack time 20ms, hang time 1s and decay time 200ms.
Frequency stability:	500Hz warm-up drift in first 60s, thereafter ±2Hz in any 30min period.
Resetability:	After 12-hour turn-off period, stabilizes within 10Hz of original frequency.

5·88–6·38MHz. If the oscilloscope is now connected to the psd output, any signal seen is due to imbalance in the psd diodes. This should be less than 50mV peak-to-peak. Reconnection of the reference oscillator should cause phase-lock, and the local oscillator should be tunable over its full range by varying the reference oscillator frequency between 5·88 and 6·38MHz.

The synthesizer should function correctly on each amateur band once the local oscillator sweep range is set to overlap the required operating range. VC501 should be set on each band to minimize oscillator radiation from the receiver. A meter switch is included on the rear panel of the receiver so that the varactor diode bias can be monitored. On each band L501 should be adjusted so that the bias voltage is between +5V and +17V over the receiver tuning range.

Unit 6: Reference Oscillator

The reference oscillator was designed according to the principles laid down by Vackar. The oscillator capacitor values are selected so that

$$\frac{C602}{VC601 + C601} = \frac{C603}{VC602} = 6 \text{ approximately.}$$

The tuning indicator L601 is a silver-plated coil on a highly stable ceramic former with an iron dust core mounted on a sprung screw thread. The core material is high stability Carbonyl TH powdered iron, with a temperature coefficient of permeability of +0·0015 per cent/°C. VC602 is an air-spaced trimmer capacitor used to set feedback just above the level needed to sustain oscillation. The oscillator dc supply is double-stabilized and decoupled at several points to reduce feedback and spurious a.m. or fm on the output. C601, 2, 3 are silvered mica capacitors secured to surrounding solid objects by epoxy resin. VC603 is a 6·5pF bi-metallic temperature-compensating capacitor with an adjustable coefficient of ±2,000ppm/°C.

TR602 and TR603 form a buffer/isolator to reduce oscillator loading effects such as frequency pulling and to increase the oscillator output to a level suitable for driving the psd (8V peak-to-peak).

If the oscillator is run away from sources of heat, a medium-term stability of ±2Hz should be achieved without adjustment of VC603. Warm-up period is about one minute, corresponding to the thermal time-constant of TR401. Warm-up frequency drift should be less than 500Hz.

Once the oscillator is placed inside the receiver cabinet, heat from the valve circuits necessitates adjustment of VC603 to minimize thermal drift. This is a time-consuming process, but with care it is possible to null out the effects of temperature changes inside the receiver cabinet. The problem is reduced considerably by running the receiver valve circuits from a +130V ht supply, and by using a cabinet designed to maximize ventilation of heat sources. In the prototype receiver, the hottest part of the cabinet stabilizes less than 15°C above ambient.

The reference oscillator tuning range is set to 5·88–6·38 MHz by judicious adjustment of L601 and C601.

Unit 7: Noise Blanker

The noise silencer circuit shown has not been tested in the current receiver, but is based on the design used successfully in the earlier model. An extra stage of amplification is shown so that the noise gate will be operated by lower level impulses, and the pulse shaping circuit has been modified. The envelope detector is ac coupled to the Schmitt trigger pulse squarer so that the receiver is not muted by strong signals inside the noise silencer passband. The three noise amplifiers (V701–3) are wideband-coupled by T701–3, and the detector threshold is set by the front-panel control RV701. The phase-splitter (TR703) provides inverted blanking pulses which are coupled to the noise gate output through VC201 to eliminate switching transients in the main receiver signal path.

The noise silencer is aligned by connecting a fast pulse generator to the aerial socket, feeding in 1μs pulses with a repetition frequency of about 100Hz, and adjusting the wideband couplers so that the output pulse resembles the input pulse. It is important that the output pulse should have the fastest possible rise-time, even at the expense of pulses being appreciably lengthened. VC201 is set so that the gate output is free of switching transients.

Unit 8: The AGC Generator

The agc circuit has already been described in detail. The agc threshold can be varied by means of the voltage at the anode of D801, which should be between −20 and −26V. No setting-up adjustments are necessary.

Unit 9: The Power Supply

The power unit must provide eight separate supplies for the receiver circuits. T901 provides 6·3V ac at 3A rms for valve heaters, and 100V rms at 130mA for bridge rectification to provide the main +130V ht line. An OB2WA stabilizer (V901) gives a +108V supply for valve oscillators and some other voltage-sensitive circuits. T902 gives 16·3V rms at 0·5A for two half-wave rectifiers providing +24V and −24V supplies. TR901–2 form a series regulator giving a stabilized +15V supply for the reference oscillator, and zener diodes give −20V for the agc generator. The adjustable −1·9V supply biases the main receiver mixer.

G2DAF MARK 2 RECEIVER

A double-conversion (variable first i.f.) receiver based on valves has been described by G. R. B. Thornley, G2DAF. This is a later version of an advanced design which was built by about 1,000 amateurs; by reducing pre-mixer gain and by using push-pull balanced stages in the front-end, it achieves a wide dynamic range while using conventional valves. A block diagram is shown in **Fig 4.98.**

The design targets for this receiver were:

(1) A high degree of bandspread, constant on all bands, and a slow tuning rate.
(2) High stability and dial setting accuracy, including freedom from slow drift and absence of frequency shifts produced by agc action.
(3) 10dB signal to noise ratio for less than 0·5μV aerial input on all bands.
(4) Selectivity 2·5kHz at 6dB down; not more than 5kHz at 60dB down.
(5) Second channel rejection not less than 80dB down.
(6) I.F. breakthrough rejection not less than 80dB.
(7) Self-generated spurious responses below threshold noise level on all bands.
(8) Two-speed agc system suitable for a.m., cw or ssb reception having audio rise less than 6dB for 60dB change in signal input.
(9) Effective a.m. noise limiter.

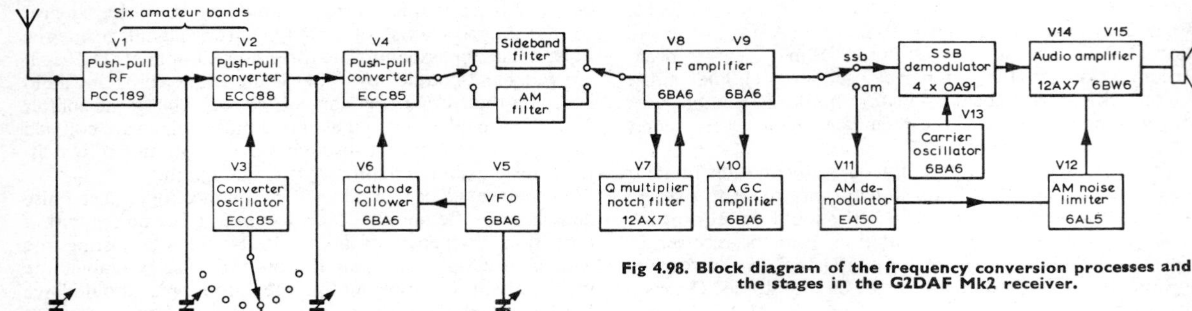

Fig 4.98. Block diagram of the frequency conversion processes and the stages in the G2DAF Mk2 receiver.

(10) Stability such that under conditions of constant ambient temperature and constant mains voltage drift will not exceed 100kHz/hour after 30 minutes warm-up period.

(11) Front-end dynamic range not less than 100dB.

(12) *Q*-multiplier heterodyne rejection filter; built in 100kHz calibrator; facilities for sideband switching and for a.m. reception.

Although high and low sideband filters were used in the original model, it is unlikely that many amateurs would wish to adopt this relatively high-cost approach, and an alternative i.f. circuit using the Kokusai mechanical filter has been developed.

Circuit Description

It will be noted that the circuit for chassis section B shows two Collins mechanical filters for ssb reception. The F455Z-4 is a high sideband filter and the F455Z-5 a low sideband filter, either side of the same 455kHz carrier frequency. This enables sideband switching to be undertaken without error caused by shift of carrier frequency. At the time the Mark 2 receiver was being constructed, the author was developing a phasing exciter and required a selectable sideband receiver that could be used as a precision measuring instrument to determine the degree of sideband suppression obtained by various experimental phasing units.

The two Collins F455Z-4 and F455Z-5 filters are costly items. For normal communication use, sideband switching filters are an extravagance and not necessary, so an alternative i.f. circuit using the relatively inexpensive Kokusai filter is also included. CW operators may wish to use the filter switching facility for a second narrow-band Kokusai unit. The receiver will be described as it was built, because the two Collins filters and associated i.f. transformers are clearly visible in the photographs.

Chassis Section A—Fig 4.99 (a)

Push-pull rf amplifier V1 (PCC189) is operated at unity gain (or less) and is therefore completely stable on all bands. Valve type PCC189 was chosen because it has variable-mu characteristics and can be controlled from the agc line; additionally the RF GAIN variable resistor in the cathode return circuit enables the valve to be used as a variable aerial attenuator—a useful feature under very heavy commercial interference conditions. Although the rated heater voltage is 7·6V, in this application the PCC189 is completely satisfactory when connected to the common 6·3V heater supply.

Primary windings on T6 to T10 provide the correct impedance load; the secondaries feeding the grids of the first converter valve V2 (ECC88). Signal input to the converter is in push-pull and oscillator injection in parallel.

When switched to the 3·5MHz or 7MHz band, the swing of the signal frequency tuning capacitor will resonate the front end circuits to the tunable i.f. of 5 to 5·5MHz. Under these conditions the converter V2 will behave as a tuned-anode tuned-grid oscillator. To maintain stability the valve is cross-neutralized by the two 2 to 8pF trimmers.

Conversion oscillator V3 (ECC85) is a Butler circuit operating as a fundamental or overtone oscillator.

All signal frequency and oscillator coils, and the required crystal, are selected by the 11-bank, single-pole, 8-way RANGE switch S1 to S11. T5 and T10 each cover the 21MHz band and the three sections of the 28MHz band, therefore the last four contacts of S1, 2, 3, 4, 5, 6 and 7 are linked together. In regard to T11, 12, 13, 14, 15, 16 and 17, each coil is pre-tuned with its own fixed mica capacitor and dust core. One coil is fitted for each band from 1·5MHz to 21MHz, one coil for 28MHz and one coil for 28·5MHz; the eighth position of the RANGE switch allows an additional oscillator coil and crystal to be used if 29 to 29·5MHz coverage is required.

The secondary centre taps of T1, 2, 3, 4 and 5 are connected together and taken to the 100kΩ grid resistor; secondary centre-taps of T6, 7, 8, 9 and 10 are connected together and taken to the rfc; cold ends of T11, 12, 13, 14, 15, 16 and 17 primary are taken to the 1·5kΩ resistor and 0·01μF by-pass capacitor; and cold ends of T11 to T17 secondaries are then connected together and taken to the 0·01μF by-pass capacitor and 4·7kΩ ht feed.

Signal frequency tuning is affected by the 4-gang 7 to 75pF variable capacitor. The centre-tap of the push-pull signal

TABLE 4.10

First conversion crystal frequencies

Band (MHz)	Crystal freq (MHz)	Mode	Output freq (MHz)
1·5	7·0	Fundamental	7·0
3·5	9·0	Fundamental	9·0
7·0	12·5	Fundamental	12·5
14·0	19·5	3rd overtone	19·5
21·0	26·5	3rd overtone	26·5
28·0	33·5	3rd overtone	33·5
28·5	34·0	3rd overtone	34·0

All crystals must be ordered for series resonance operation stating mode of operation.

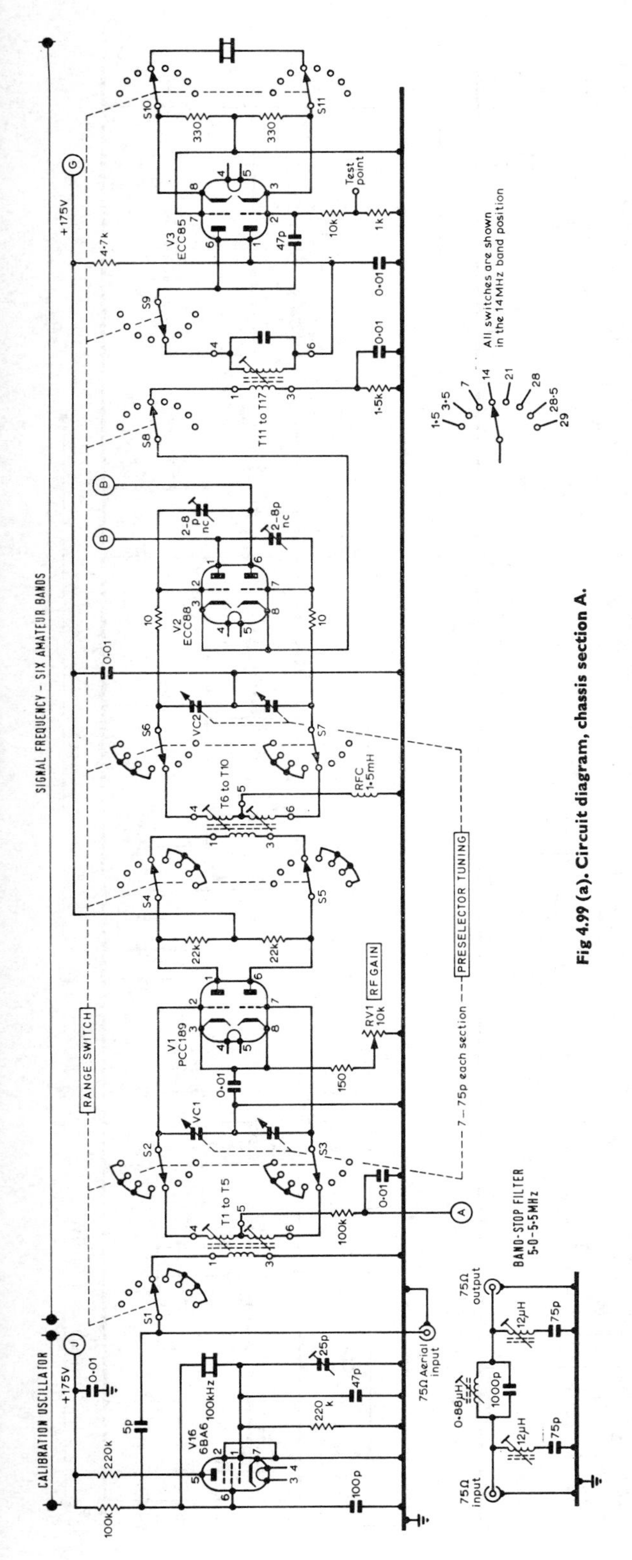

Fig 4.99 (a). Circuit diagram, chassis section A.

frequency coils must not be by-passed to rf but allowed to "float", circuit balance being obtained by the equal sections of the split stator tuning capacitors.

Chassis Section B—Fig 4.99 (b)

Output of the first converter is fed into T18 and T19, and in push-pull to the grids of the second converter V4 (ECC85). Oscillator injection to the two cathodes strapped in parallel is derived from cathode follower V6 (6BA6) driven by vfo V5 (6BA6).

These three stages constitute the tunable i.f. covering the range 5 to 5·5MHz and resonated by a 5-gang variable capacitor of 3 to 35pF each section—KILOHERTZ main tuning (VC3, 4 and 5).

The vfo is arranged as a Colpitts parallel-tuned oscillator to obtain a high frequency stability, together with a level amplitude of output voltage across the 500kHz range, and tunes on the high side (ie 5-5·7MHz plus 455kHz) 5·455 to 5·955MHz.

T18 and T19 are coupled by a low impedance link around the electrical and mechanical centre of each main winding. The centre tap of each coil is allowed to "float" and must not under any circumstances be bypassed to rf.

The tunable i.f. of 5 to 5·5MHz is converted by V4 to the second i.f. of 455kHz. Switch banks S12, 13 and 14 select the appropriate mechanical filter, or top-coupled i.f. transformers, for high sideband/low sideband/a.m. reception.

Q multiplier V7 (12AX7) is connected to points "C" "C" by 75Ω coaxial cable, thus forming the connecting link between the pole of S14 and the grid of V8.

I.F. amplifier V8 (6BA6) is capacitance coupled to the second i.f. amplifier V9 (6BA6), both valves being agc controlled. The S-meter is connected to G2 of V9 instead of the customary cathode position, in order to obtain better S-meter scale linearity. Both i.f. amplifiers have negative current feedback in order to reduce Miller effect due to agc action. Additional i.f. transformers T22, 23 and T25, 26 were incorporated in order to avoid the possibility of distorting the F455Z-4 and F455Z-5 response curves.

A proportion of the 455kHz i.f. is fed to the agc amplifier V10 (6BA6) whose output feeds the OA5 rectifier and the OA5 gate circuit. AGC release time is selected by the OFF-FAST-SLOW switch S19a and S19b. The envelope detector V11 (EA50) is fed from the i.f. transformer T27. S17 selects high or low sideband output and feeds the ring demodulator.

Chassis Section C—Fig 4.99 (c)

Audio output of V11 feeds through the 100kΩ variable resistor, NL SET, into the 47kΩ diode load. V12 (6AL5) is a conventional Dickerts twin-diode envelope-following a.m. noise limiter. This is very effective against Loran interference when operating in the top section of the 1·5MHz band. The noise limiter is switched in or out of circuit by S20, part of the OPERATION switch.

Carrier insertion oscillator V13 (6BA6) is operated in a modified Colpitts circuit to allow the valve cathode to be grounded to rf and eliminate the possibility of carrier leakage along the 6·3V heater rail getting back into the i.f. amplifier and agc amplifier stages. Oscillator operation is controlled by S18a—part of the main A.M.-L-H sideband switch.

Amplitude modulated output from V12 circuitry, or sideband output from the ring demodulator, is selected by S18b and feeds via the AF GAIN control into the two-stage

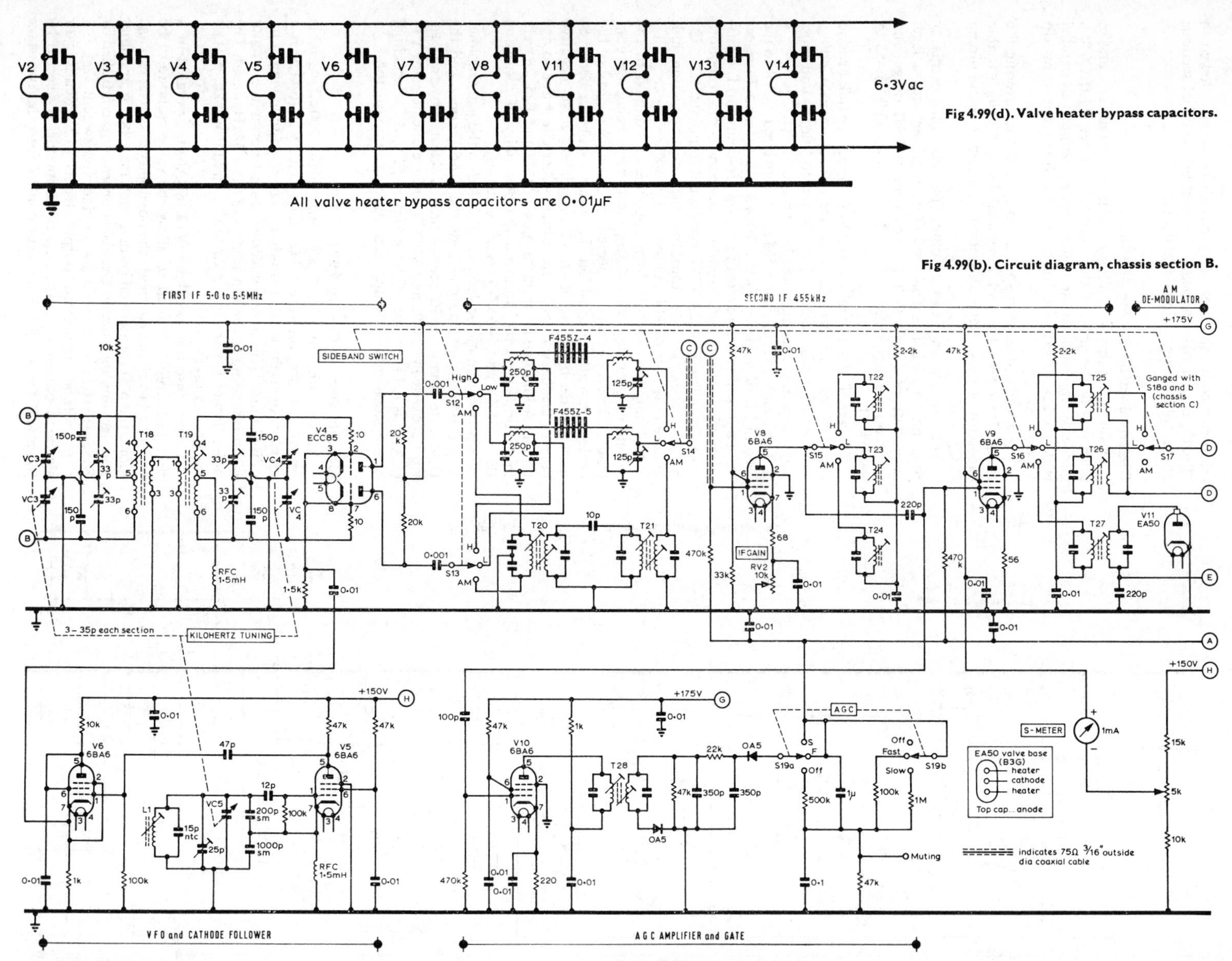

Fig 4.99(d). Valve heater bypass capacitors.

Fig 4.99(b). Circuit diagram, chassis section B.

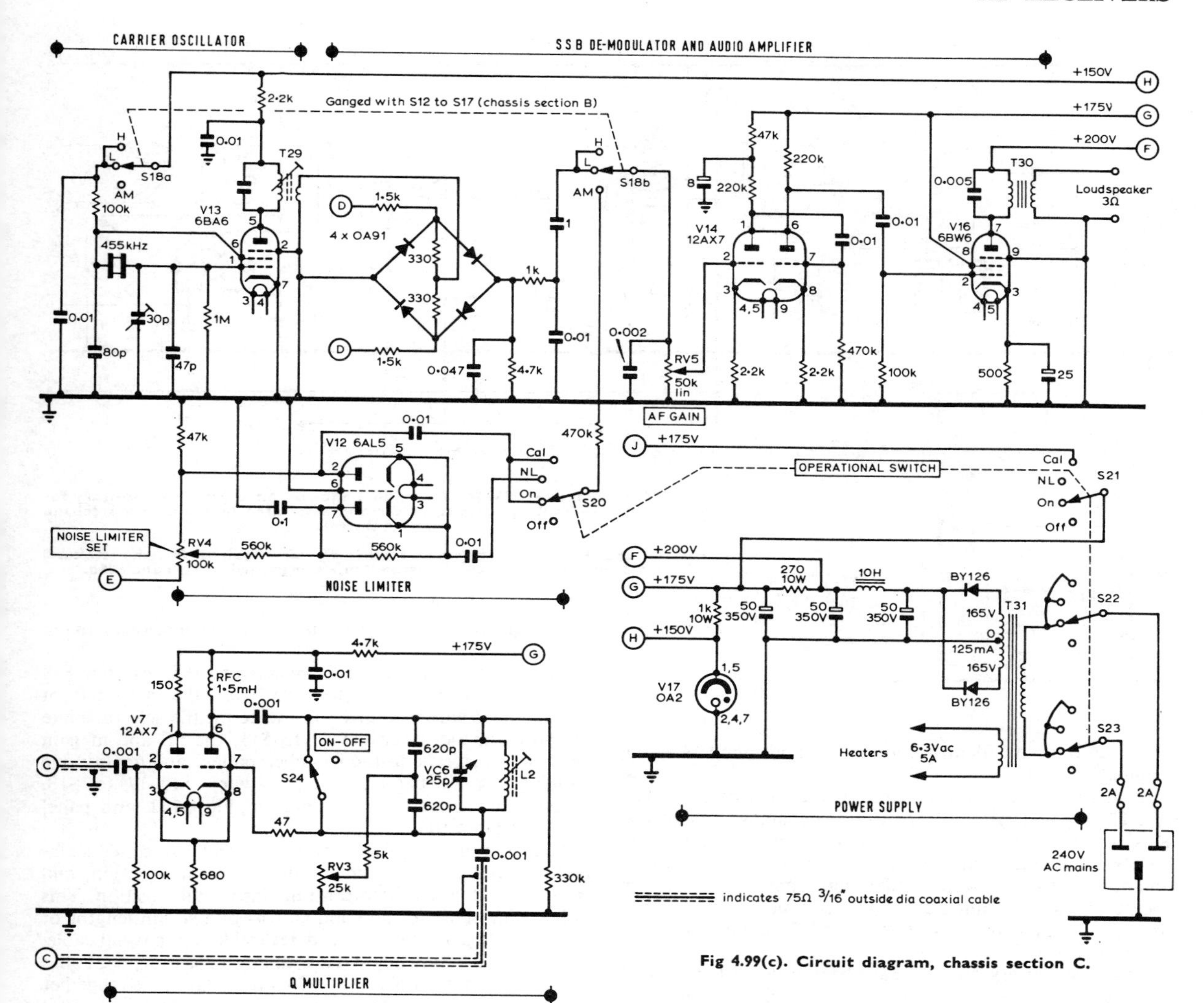

Fig 4.99(c). Circuit diagram, chassis section C.

audio amplifier V14 (12AX7) and finally to the output valve V15 (6BW6). Output transformer T30 matches the required 5kΩ anode load of V15 to a standard 3Ω loudspeaker.

The calibration oscillator V16 (6BA6) uses a 100kHz crystal in a conventional circuit arrangement, and is switched on or off as required by S21—part of the OPERATION switch.

Power supply provides ht outputs of 200 and 175V, and 150V regulated by V17 (OA2) stabilizer valve, using silicon diodes in a full-wave circuit, and provides 6·3V ac for all valve heaters. The receiver is switched ON or OFF by S22 and S23 sections of the OPERATION switch.

I.F. Rejection Filter

The receiver i.f. breakthrough rejection will be at its lowest level when using 3·5MHz or 7MHz, the amateur bands nearest to the tunable i.f. range. With the receiver tuned to 3·75MHz or 7MHz, the measured i.f. rejection is in fact 62dB. Under good propagation conditions the level of the commercial teleprinter transmissions in the 5 to 5·5MHz range can be very high and the author is of the opinion that 62dB rejection is not good enough. This in fact was the reason for stipulating in "required specification": "i.f. breakthrough rejection not less than 80dB". To obtain this figure with a margin in hand, a band-stop filter is included with the theoretical circuit shown in Fig 4.99 (a).

The filter is constructed as an outboard unit, intended to be connected between the 75Ω aerial feeder and the 75Ω receiver input terminal, so as to allow the filter to be taken out of circuit and the aerial to be connected directly when it is desired to receive MSF on 5MHz for vfo stability checks or to set the 100kHz calibration oscillator on frequency.

Alternative I.F. Amplifier

An alternative simplified i.f. amplifier using a Kokusai MF-455-10CK mechanical filter is shown in **Fig 4.100**. To avoid confusion the switch banks and i.f. transformers used

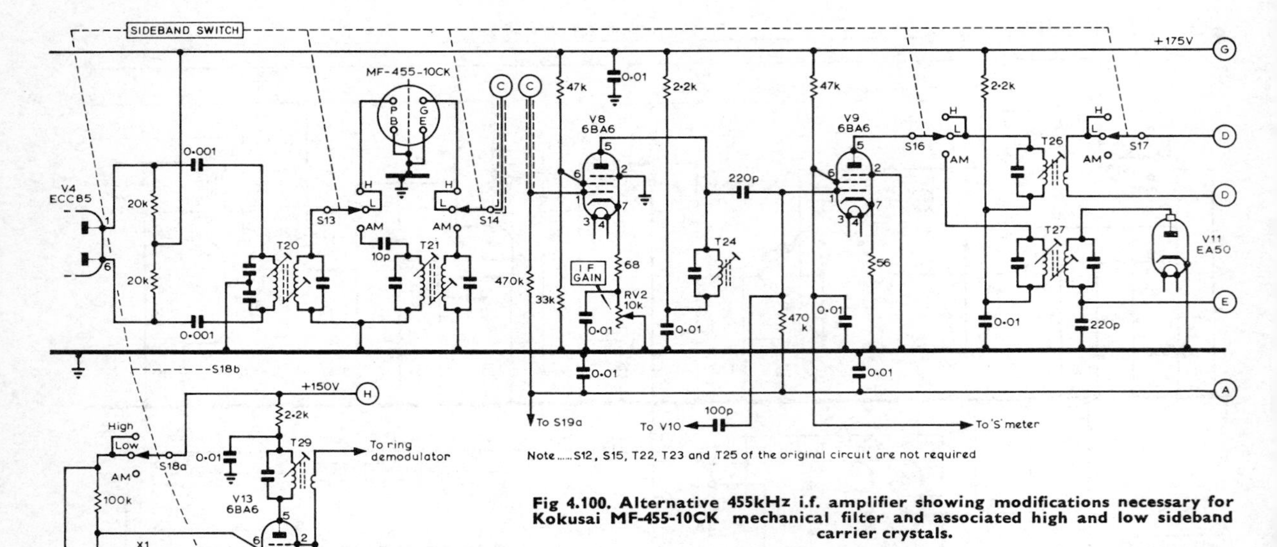

Fig 4.100. Alternative 455kHz i.f. amplifier showing modifications necessary for Kokusai MF-455-10CK mechanical filter and associated high and low sideband carrier crystals.

Left: **X1 and X2 carrier crystals: frequencies as required for the Kokusai MF-455-10CK filter. S25 additional switch wafer ganged to S18a and S18b.**

are given the same identification letters and numbers as the original circuit.

Switch banks S12 and S15 and i.f. transformers T22, T23 and T25 are no longer required. For sideband switching, the Kokusai filter requires two carrier crystals—nominally 300Hz either side of the filter 6dB points—and this requires an additional switch wafer in the input circuit of the carrier oscillator V13, to automatically select the correct carrier frequency when the A.M.-LOW-HIGH sideband switch is operated.

As the carrier crystals will have been ordered for parallel resonance with the usual 30pF shunt capacitance, the 30pF trimmer capacitor can also be omitted.

The Kokusai MF-455-10CK filter together with matching carrier crystals are obtainable from KW Electronics Limited, Dartford, Kent, UK.

Construction

The chassis was made up of four separate box sections of 16 swg aluminium to give a total size of 16in by 12in by 2½in deep, with a 17½in by 9in front panel. The box sections were machine pressings, true to size, having sharp radius corners. Metalwork for this design is available from North West Electrics, 769 Stockport Road, Levenshulme, Manchester, UK.

The Eddystone 898 dial drive assembly has the driving boss almost central behind the panel cut-out, and in order to bring the shaft of VC5 into line the complete vfo assembly is built into the 6in by 3in "box" and this unit is raised 2in above the normal chassis top face. The tunable i.f. capacitors VC3 and VC4 are driven from this shaft by means of two 1½in diameter drive pulleys and a spring tensioned nylon cord.

In order to clear the lead flywheel of the Eddystone 898 drive, the panel is set 1½in away from the chassis front apron—this space also gives clearance for the selector plate of the main RANGE switch S1 to S11, the i.f. and af gain variable resistors, and the sprocket pulleys and chain drive ganging the sideband and a.m. switch S12–17 to S18. Chassis layout is shown in detail in **Fig 4.102**, and panel layout in **Fig 4.103**.

The *Q* multiplier notch filter, V7 and associated components, is built on a small sub-chassis 3in by 2¼in, and is wired and tested before fitting into final position. This unit is connected to the main chassis by two 8in lengths of 3/16in outside diameter standard 75Ω television coaxial cable.

The signal frequency 4-gang tuning capacitor, VC1 and VC2, is in fact two Polar type C28-142 units coupled together after a rear shaft extension had been fitted to the unit nearest to the panel. Polar capacitors were used because they were available, but any standard 4-gang variable of either 75pF or 100pF maximum capacitance would be equally suitable. Both Polar capacitors are mounted on a piece of hardwood ⅜in thick to lift the drive shaft and obtain a more pleasing panel layout.

It will be observed from the chassis underside photograph that holes were drilled for six aerial input coils and six

Fig 4.101. Preselector tuning control, showing how the receiver panel should be engraved to show the control knob positions for each of the amateur bands. This illustration assumes that clockwise rotation reduces the tuning capacitance value.

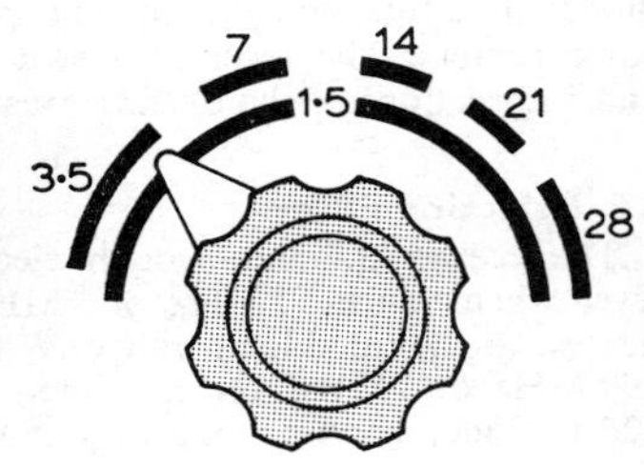

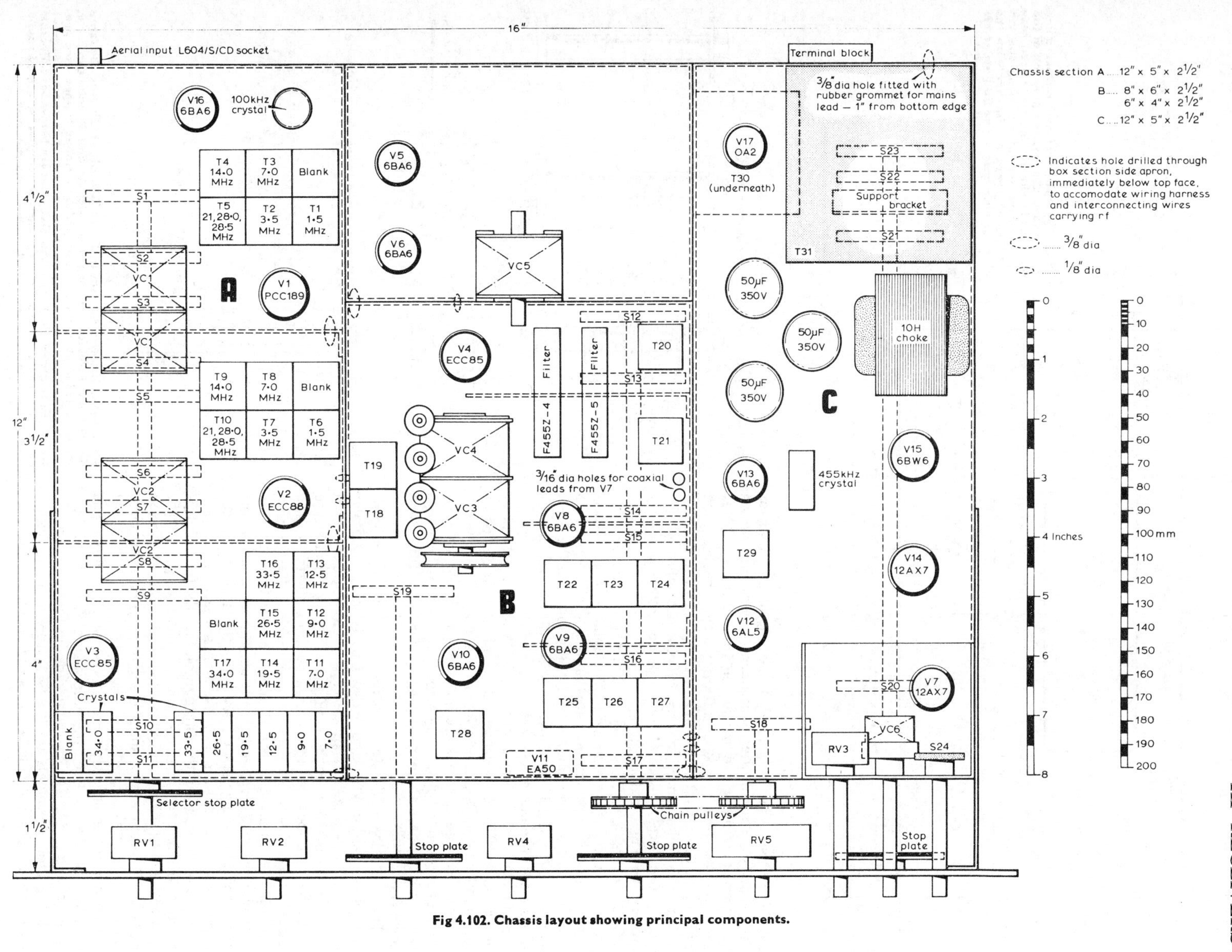

Fig 4.102. Chassis layout showing principal components.

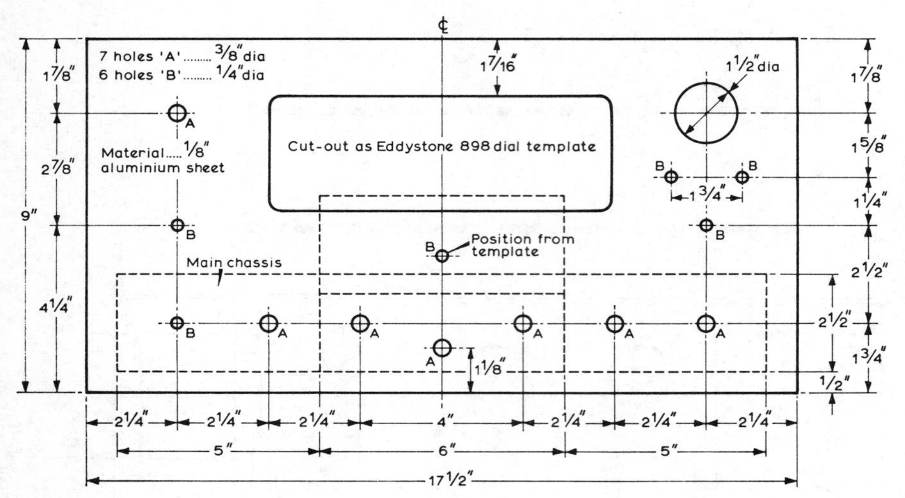

Fig 4.103. Panel layout. Dotted line shows the relative position of the main chassis and the raised rear centre section to bring the vfo tuning capacitor in line with the driving boss of the Eddystone 898 dial.

converter input coils. In fact only five of the six positions for each stage were used, after experimental work had shown that it was possible to use one coil only to cover the 21MHz band and the three 500kHz sections of the 28MHz band without adversely affecting the receiver performance. To mount each coil it is necessary to drill nine holes through the chassis top face, and this work can be greatly eased by initially making a drilling jig.

The conversion crystals used were HC-6/U type with ½in pin spacing and are mounted on 2-pin holders. When ordering crystals it is important to state that the 7, 9, 12·5 and 19·5 MHz crystals are for fundamental operation, and the 26·5, 33·5 and 34MHz for third overtone operation, and all for *series* resonance operation.

Standard Electroniques G2DAF Mark 1 Receiver oscillator coils were used for T11 to T17 and modified as required by the addition of a low impedance primary winding. As Electroniques coils are no longer available, winding details are given for coils using Neosid or Aladdin 2¼in high cans and standard 0·3in diameter formers with 0BA dust cores **(Fig 4.104)**.

The 5 to 5·5MHz i.f. section is tuned by a 4-gang 3–35pF variable capacitor VC3 and VC4. This is in fact a Polar type C28-142 2-gang of maximum capacitance 75pF modified to a 4-gang of 35pF each section, by sawing through the stator bars—in-between the ceramic support pillars—with a model makers' saw. Any other make of standard 4-gang variable can be used, provided only that it will fit into the available chassis space, and that it has the same law as the 35pF variable tuning the vfo (VC5), also that the constructor is prepared to accept a non-linear dial calibration.

It will be noted from the photograph that the four Philips beehive trimmers of 3–33pF are adjacent to the Polar C28-142 capacitor VC3 and VC4—they are in fact supported by soldering the trimmer support leg to the brass capacitor frame.

In order not to degrade the filter response, the input side must not be able to "see" the output side. A Kokusai filter must be mounted so that it straddles the 4in-long cross screen, with the input terminals on the S13 side, and the output terminals on the S14 side. A rectangular notch is cut into

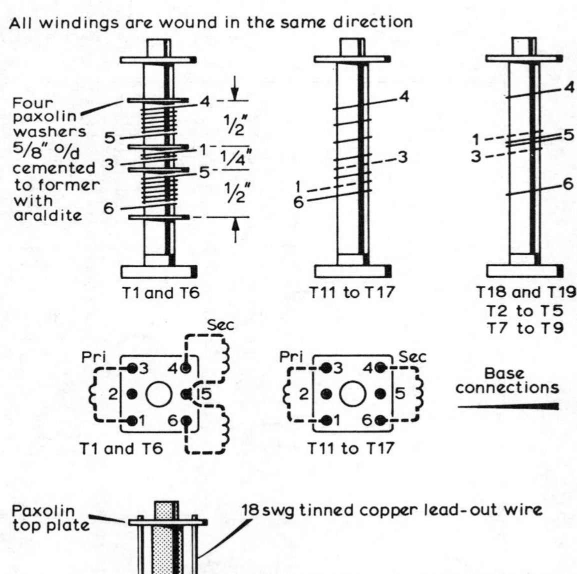

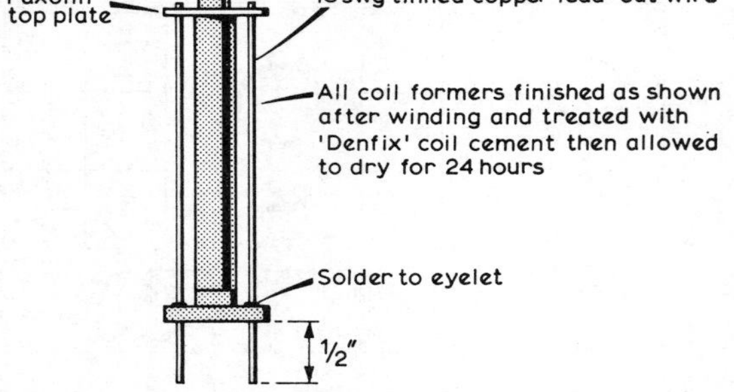

Fig 4.104. Coil assembly details, T11 to T17. All windings in the same direction. All coil formers are fitted with Aladdin Paxolin top plate. Made-up lead-out wires as shown of 18swg tinned copper wire pre-stretched and cut to 4in lengths. After soldering to eyelet, lead-out wires are trimmed flush with top plate and ½in below base. Enamel covering is removed from ends of each winding—end looped round lead-out wire and soldered, as required.

TABLE 4.11

Resonant circuit component details

Component	Freq or Band (MHz)	Primary Winding	Secondary Winding
T1	1·5	4t 38swg enam	100t-ct-100t 33swg enam scramble wound in 2 bobbins (see Fig 4.104)
T2	3·5	2·75t 24swg enam	62t-ct-62t 38swg enam
T3	7·0	1·75t 24swg enam	35t-ct-35t 32swg enam
T4	14·0	1·75t 24swg enam	20t-ct-20t 24swg enam
T5	21 to 28·5	1·25t 24swg enam	10t-ct-10t 24swg enam
T6	1·5	7t 38swg enam	As T1
T7	3·5	2·25t 24swg enam	As T2
T8	7·0	2·25t 24swg enam	As T3
T9	14·0	3·75t 24swg enam	As T4
T10	21 to 28·5	1·75t 24swg enam	As T5
T11	7·0	9t 36swg enam	36t 36swg enam with 65pF sm capacitor across winding.
T12	9·0	6t 28swg enam	24t 28swg enam with 65pF sm capacitor across winding
T13	12·5	4·5t 24swg enam	18t 24swg enam with 50pF sm capacitor across winding
T14	19·5	3t 24swg enam	12t 24swg enam with 35pF sm capacitor across winding
T15	26·5	2·5t 24swg enam	10t 24swg enam with 20pF sm capacitor across winding
T16	33·5	2·5t 24swg enam	8t 24swg enam with 10pF sm capacitor across winding
T17	34·0	2·5t 24swg enam	8t 24swg enam with 10pF sm capacitor across winding
T18	5 to 5·5	32t-ct-32t 32swg enam	5t 32swg enam
T19	5 to 5·5	5t 32swg enam	32t-ct-32t 32swg enam
Band-stop filter	5 to 5·5	12μH coil. 45t 36swg enam 0·83μH coil. 10t 24swg enam spaced to ⅜in long.	Electroniques or similar 0·3in dia former 1⅛in long with 0BA dust core.

Component	Freq or Band (kHz)	Description
T20	455	Denco IFT11-465 i.f. transformer with 65pF primary capacitors removed and replaced with two 130pF sm capacitors
T21, T27, T28	455	Denco IFT11-465 i.f. transformer
T22, T23, T24	455	Denco IFT11-465 i.f. transformer with top pie and resonating capacitor removed
T25, T26, T29	455	Denco IFT11-465 i.f. transformer with top pie and capacitor removed and replaced with 75t 36swg enam scramble wound against primary pie
T30	Wharfedale output transformer type GP8 used on 36 : 1 ratio tappings	
T31	Secondary : 165-0-165V 125mA Secondary: 6·3V 5A Primary 10-0-200-220-240V 50Hz ac.	

All coils T1 to T19 wound on Neosid or Aladdin 0·3in dia formers with 0BA dust cores (T1 to T10, T18 and T19 two dust cores in each former) and in 2¼ × 1 3/16 × 1 3/16in seamless aluminium cans. All windings close-wound single layer except T1 and T6. Primary of T2, 3, 4, 5, 7, 8, 9 and 10 wound over centre of secondary winding. Primary of T11 to T17 wound over cold end of secondary winding (see Fig 4.104). Secondary to T18 and primary of T19 wound over centre of main winding.

Component	Freq or Band (kHz)	Description
L1	5455 to 5955	14·5t 22swg enam close wound on ½in dia ceramic former, with dust core
L2	455	Standard pot core Q multiplier coil on 0·3in dia former, with dust core
S1, 2, 3, 4, 5, 6, 7, 8, 9, 10, 11	Yaxley Paxolin SP/8W 11-bank bandchange switch	
S12, 13, 14, 15, 16, 17	Yaxley Paxolin SP/3W 6-bank sideband switch	
S18a, S18b	Yaxley Paxolin 2P/3W 1-bank sideband switch	
S19a, S19b	Yaxley Paxolin 2P/3W 1-bank agc switch	
S20, 21, 22, 23	Yaxley Paxolin 1P/4W 4-bank operational switch	
S24	1P/2W 1-bank rotary on/off switch (Q multiplier)	
S25 (see Fig 4.100)	Yaxley Paxolin 1P/3W 1-bank (ganged to S18a and S18b)	

each of the two 3in long screens to clear the valveholders, and these are positioned so that a line drawn between pins 3–4 and pin 7 is parallel to the screen, and pin 1 is on the input side and pin 5 on the output side of the screen.

VFO tuning capacitor V5 is a Polar type C28-141 of 35pF maximum. Both the C28-142 and the C28-141 have a straight line capacitor law (semi-circular rotor plates) and this, together with the LC ratio used over the vfo tuning range of 5,455 to 5,955kHz, will give almost a linear dial calibration without the need for tracking capacitors.

The tuning capacitor should be of rugged construction with two rotor bearings, and the coil former ceramic with a dust core mounted on a screwed brass rod fitting into a pressure-loaded clutch or a bush with a locking nut to prevent end or side "float" within the winding.

Switch S18 is required to be screened from the 455kHz i.f. stages. It is therefore mounted on the chassis section C and ganged to the sideband selector switch S12-17 by sprocket pulleys and chain drive. The author was fortunate in having suitable sprockets drilled for a ¼in shaft available, but believes that standard Meccano items (normally drilled for ⅛in shafts) have sufficiently large centre bushes to accommodate opening out to ¼in.

In order to conserve space, the mains transformer T31 is assembled on a C core and was wound to the author's specification by R. F. Gilson Ltd, St George's Works, 11a St George's Road, Wimbledon, London SW19. Overall dimensions are 3 5/16in by 3¼in by 3⅜in high. The two silicon rectifiers and the mains fuses are mounted on the transformer top plate.

Any small box with a close-fitting lid—or 18 or 16swg aluminium, brass, copper or tinned iron—large enough to accommodate the three coils, capacitors and the two Belling-Lee L604/S/CD coaxial sockets, is suitable for the construction of the 5 to 5·5MHz band-stop filter. In order not to degrade the filter performance it is important to fit two close-fitting cross-screens so that the coils and associated capacitors of each section cannot "see" each other. The filter is connected to the receiver by a short 75Ω coaxial fly lead with Belling-Lee L734/P/AL plugs at each end.

Alignment

A serviceman's signal generator is really the minimum requirement, and if an oscilloscope is also available each chassis section can be individually aligned before final assembly. Assuming that this has not been done and that the

Front panel

Top view of the chassis. Note the positioning of the four beehive trimmers next to VC3 and VC4 (see text).

View of the underside of the chassis.

TABLE 4.12

Conversion oscillator output voltages

Output freq (MHz)	Output volts (rms)	Remarks
7·0	2·2	
9·0	1·8	
12·5	1·5	Measurement made at cathode of V2
19·5	1·65	
26·5	1·85	
33·5	1·65	
34·0	1·65	

signal generator was only available for a short amount of time—after final construction was completed—the most satisfactory alignment procedure is to start at the back and finish at the front. That is from V9 to the aerial, with an Avo or similar test meter on the 100V ac range connected across the primary of the audio output transformer T30 as an output indicator. The loudspeaker should be left connected so that the audio quality can also be monitored.

The author assumes that anyone undertaking the construction of the G2DAF Mark 2 Receiver will have sufficient experience to make detailed alignment instructions unnecessary.

A noise generator is the test instrument that will tell "the truth and nothing but the truth" about the receiver's performance. It is simple to build, costs little money, and can be constructed by most amateurs with parts to a large extent already available in the junk box. With the noise generator connected to the aerial input terminal, it is very rewarding to watch the improvement possible by a little further "tweaking" of the front-end circuits on each range. It should be possible to obtain a 7 to 8dB noise factor on each of the six amateur bands. The noise generator will also give a "figure of merit" that will enable the constructor to directly compare his results with those of the circuit designer, or to compare the performance of his home constructed "baby" with that of a comparable receiver made by a commercial manufacturer. The subject of noise generator measurement, detailed alignment procedure plus general fault-finding and construction hints, is covered in full in "Alignment of a G2DAF-type receiver, single sideband," *RSGB Bulletin*, February, March, April, June, August and November 1967.

TABLE 4.13

DC Voltage check

Valve	Stage	Type	Pin No 1	2	3	4	5	6	7	8	9	
V1	RF amp	PCC189	85	—	1·7	H	H	85	—	1·7	—	
V2	1st conv	ECC88	140	—	5	H	H	140	—	5	—	
V3	Conv osc	ECC85	73	—	1·8	H	H	90	—	2	—	All measurements made with rf and i.f. gain controls at maximum. Range switch at 14MHz. Sideband switch on H. Q multiplier off. Aerial input disconnected
V4	2nd conv	ECC85	125	—	2·2	H	H	125	—	2·2	—	
V5	VFO	6BA6	—1	—	H	H	30	60	2·2			
V6	Cath foll	6BA6	8	6	H	H	100	100	6			
V7	Q mult	12AX7										
V8	1st i.f. amp	6BA6	—0·25	—	H	H	170	80	5			
V9	2nd i.f. amp	6BA6	—0·1	—	H	H	180	62	5·5			
V10	AGC amp	6BA6	0·2	—	H	H	175	70	1·75			
V11	A.M. de-mod	EA50	anode —0·4									
V12	Noise limiter	6AL5	—	—	H	H	—	—	—			
V13	Car osc	6BA6	—2·5	—	H	H	145	60	—			
V14	Audio amp	12AX7	80	—	0·7	H	H	100	—	0·8	H	
V15	Audio output	6BW6	—	—	9	H	H	—	190	182	—	Anode current = 24mA
V16	100kHz cal osc	6BA6	—3	—	H	H	45	90	—			
V17	Regulator	OA2		—			150					

(— in front of figure indicates negative reading)

Measuring instrument Avo Model 8

TABLE 4.14

Final performance data

Band (MHz)	Input for 10dB S + N/N ratio (μV)	Input for 20dB S + N/N ratio (μV)	Noise Factor (dB)	Image rejection (dB)	I.F. break-through rejection (dB)	White noise output: Preselector off resonance (volts rms)	White noise output: Preselector on resonance (volts rms)
1·5–2·0	Less than 0·25μV on all bands, ssb operation	Less than 0·75μV on all bands, ssb operation	6·5	108	88	5V on all bands ssb reception	11V or more on all bands, ssb reception
3·5–4·0			7	112	94**		
7·0–7·5			7	98	94**		
14–14·5			7	92	95		
21–21·5			7	90	97		
28–28·5			7	88	98		
28·5–29			7	88	98		

Image rejection—second conversion process = > 120dB (Receiver tuned to 3·75MHz with preselector correctly resonated. Input freq 2·84MHz)

Cross-modulation While receiving a wanted signal of 1μV, an interfering signal modulated 30 per cent at 400Hz of strength 100mV and 15kHz away, has no effect on the wanted signal, ssb reception.

Dynamic range Greater than 120dB, ssb operation

Stability After initial warm-up period of 10 to 15min, drift less than 100Hz/h under normal ambient conditions

Notes (1) **Measurement made with band-stop filter in use.

(2) White noise measurements made with Avo meter on 100V ac range, connected across output transformer (T30) primary. No aerial input. All gain controls at maximum. Sideband switch in lsb position. Agc off.

Alignment of the band-stop filter is undertaken by adjusting the dust cores of the vertical legs for maximum attenuation at 5·5MHz, and the horizontal arms for maximum attenuation at 5MHz.

Muting

Modern practice is to control a transmitter and receiver by means of a relay with either voice (vox) or press-to-talk operation. Breaking ht or cathode connections usually causes clicks and thumps, and the most satisfactory control method is a source of negative voltage applied to the grids of the controlled valves. In many cases there is already a source of negative voltage provided for the pa bias supply, and the available potential is usually suitable for muting requirements.

The control voltage can be any value from about 30 to 100V and is connected to the receiver MUTING terminal. During transmission periods the applied voltage carries the agc line negative to cut-off and disables the controlled rf and i.f. stages. As the gate circuit reservoir capacitance is returned to the muting line, the agc attack and release time constants remain unaffected.

Conclusion

For those amateurs who have a thermionic diode probe valve voltmeter, the conversion oscillator (V3) output volts are given in **Table 4.12**. The output voltage from the vfo should be approximately 1·4V rms, measured at the cathode pin of V6. DC voltage checks at each valve pin are given in **Table 4.13**. Every constructional project undertaken by the author has been concluded by measuring and recording final performance figures. These figures are invaluable for future reference or for future comparison, and are given in **Table 4.14**.

It is hardly likely that the construction of an ambitious receiver such as that described in this article will be undertaken lightly, but the effort will be found interesting and instructive and a refreshing change from the more usual radio activity.

Finally, it is undeniable that a receiver is the most used and the most important piece of apparatus in an amateur station. Its construction can be guaranteed to result in a pride of possession and a pleasure of operation that will give greater reward than that experienced with any previous amateur radio activity.

CHAPTER 5

VHF AND UHF RECEIVERS

IN a vhf receiver it is possible to realize a performance superior in terms of sensitivity and signal-to-noise ratio to that normally obtained on the lower frequencies where, owing to various forms of both man-made and natural interference, a limit is imposed beyond which any attempt to recover signals is fruitless. In vhf reception there is no appreciable atmospheric noise with the exception of that originating in lightning discharges in the immediate vicinity of the receiver, or from charged rain which is an infrequent occurrence. Band occupation, while increasing steadily, is still (even during a popular contest on the 144MHz band), not as high as to create the sort of problem as that normally found on certain of the lower-frequency allocations.

The limiting factor in a good location is extra-terrestrial noise and, at frequencies up to at least 250MHz, receivers can be designed which will respond to signals only slightly above this level.

Interference from the ignition systems of cars is more noticeable on the higher frequencies, and in fact becomes objectionable from about 20MHz up to around 300MHz or so, depending upon the offending vehicle. However, such forms of impulsive interference may be greatly reduced by the use of some form of noise limiter.

Noise Factor

For these reasons the performance of vhf receivers is usually (and more usefully) specified in terms of *noise factor*, which may be defined as:

$$\frac{\text{Aerial noise} + \text{Receiver noise}}{\text{Aerial noise}}$$

and measured as the noise power present at the output of the receiver. With the theoretically perfect receiver, producing no noise in itself, this equation becomes 1/1, ie a noise factor of 1 or 0dB. The noise factor of a practical receiver that does itself generate unwanted noise is a measure of the amount by which it falls short of perfection.

It is customary to regard each stage in the receiver as a noiseless amplifier, and to consider the noise that it does in fact generate as originating in a fictitious resistor in its input circuit. The value of this resistor is such that the noise voltage which would be developed across it is multiplied by the gain of the valve or transistor, the result being the real noise present in the stage output circuit. The subject of receiver noise is discussed more fully in Chapter 15—*Noise* and the measurement of noise factor in Chapter 18—*Measurements* to which the reader is referred for a more extensive explanation.

Cross Modulation

This is an effect inherent in the receiver and manifests itself as the modulation from a strong local transmission superimposed on other signals. It has no connection with spurious radiation from the stronger station, and is caused by a lack of linearity in the receiver circuits—mainly in the first mixer—which causes modulation of one signal by the other. Improved selectivity after the stage in which such an effect has occurred will do nothing to separate one transmission from the other, and the only cure is prevention by improving the design of the circuit. Bipolar transistors are inherently more likely to produce cross modulation than are valves, but the *field effect transistor* (fet) is potentially better than a valve in this respect because of its superior square-law characteristic.

Those stations situated well away from others operating on the same band may never experience cross modulation but, when situations arise where the effect is significant, a receiver designed to prevent this form of interference will be found more satisfactory in use than one possessing a superior noise factor.

Intermodulation need not, of course, be caused only by amateur signals. Very strong signals may be received from vhf or uhf broadcast and television stations or from local mobile radio services. Although individual transmissions may not be strong enough to cause difficulty, the sum of several such transmissions may overload the receiver. For example, a receiver in South London now has to withstand the assault of ten vhf and uhf broadcast stations with a total erp of over 4MW, not counting several local stations with a power of a mere 10kW. Since the bandwidths of the broadcast transmissions are much wider than that of most amateur receivers, the intermodulation products are often not identifiable as such, the complex intermodulation from several stations appearing as a general increase in aerial noise without obvious source. In many cases of apparently "noisy" locations the noise is in fact generated in this way in the receiver. Wherever there is a significant increase in received noise when the aerial is connected the circumstances should be checked by repeating the test in the early morning when most of the broadcasting stations are off the air.

FRONT-END STAGES

The requirements of the early stage or stages in a vhf receiver are (a) low noise content, (b) power gain and (c) absence of cross modulation.

The first of these requirements implies that noise due to

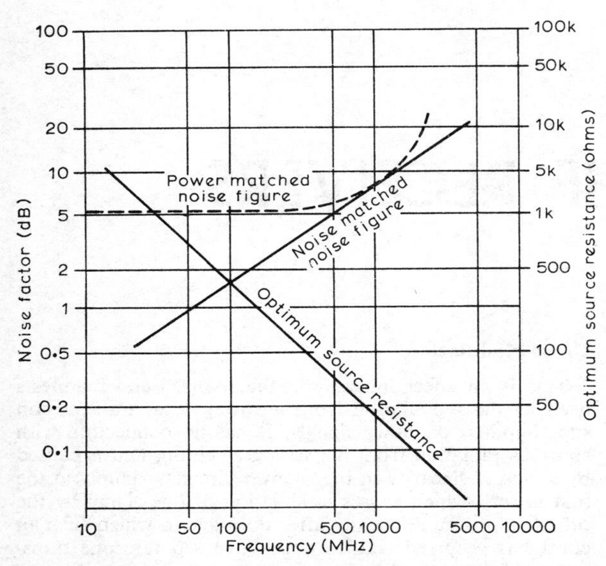

Fig 5.1. Noise factor and optimum source resistance plotted against frequency for a typical grounded grid triode (type 7077, g_m 9mA/V Ia 6·4mA).

the fictitious resistor in its input circuit, known as the *equivalent noise resistance*, must be as low as possible.

Since a pentode valve has an equivalent noise resistance which may be several times higher than that of a triode, owing to the added effect of *partition noise*, triodes are used exclusively for low-noise input stages in valve receivers. Bipolar or field effect transistors specially designed for such an application are readily obtainable.

The second requirement needs some explanation because sheer power gain is not sufficient in itself or even desirable. In a multi-stage receiver with, say, eight amplifying stages, the noise originating at the input of the first stage will be amplified by eight stages, that in the second by seven and so on. If the effective noise voltages are donated by V_1, V_2, $V_3 \ldots V_8$ and the stage gains by G_1, G_2, $G_3 \ldots G_8$, the total noise present at the output of the receiver will be $V_1 (G_1, G_2, G_3 \ldots G_8) + V_2 (G_2, G_3 \ldots G_8) + V_3 (G_3 \ldots G_8)$ and so on.

It follows, therefore, that if G_1 is high, say 20 or more, only the noise due to the first stage is of importance, and provided that the remaining stages are reasonably efficient they will contribute little to the overall noise. In cases where the gain of the first stage is low due to an unsuitable circuit, valve or transistor for the frequency concerned, or if a considerable bandwidth is required, or when both of these conditions exist, the noise contributed by the second or even the third stage may become important.

However, the function of the rf amplifying stage is to provide just sufficient gain so that the noise due to the mixer —and this is inherently greater when any device is performing this function, rather than acting as a straightforward amplifier—is small compared with the desired signal. If the rf stage has too much gain, there is the risk that a very strong local signal might induce cross modulation by overloading the mixer. There is obviously an argument here for some form of gain control on the rf amplifier. Automatic gain control on this stage is not altogether desirable as it might well spoil the weak-signal performance of the receiver, unless means for delaying the application of the control voltage is incorporated. The provision of a manual control of first-stage gain could, however, be a worthwhile addition if a number of strong local stations are in the area.

Input Impedance Matching

When the impedance of the feeder is matched to the input impedance of the first stage, an optimum power gain is achieved and the stage is said to be *power matched.* The input impedance of the stage should be as high as possible, the aerial impedance as seen at the end of the feeder being stepped up by transformer action and the signal voltage thus increased. This condition does not, however, result in the lowest noise factor. When the step-up ratio between feeder and input circuit, and the tuning of that circuit, are adjusted for optimum noise factor in contrast to maximum power gain, the stage is said to be *noise matched.* A practical application of this procedure appears later in this chapter. **Fig 5.1** shows the variation with frequency of the noise factor and optimum source impedance of a certain grounded-grid triode valve. It will be seen that when the valve is power matched the noise factor is constant at 5·5dB up to about 400MHz, whereas below that frequency a higher source impedance (over-coupling) produces considerable improvement in the noise factor. The noise factor does not diminish above 400MHz because the optimum source impedance is then comparable with, or may be less than, the cathode input impedance ($1/g_m$ for this valve is $1{,}000 \div 9 = 110\Omega$).

It is possible, with careful circuit design, to arrange the input circuit of either a valve or transistor rf stage so that resonance, optimum noise factor and matching all coincide. Useful information on this and many other aspects of vhf receiver design is contained in *Design of Low-Noise Transistor Input Circuits* by William A. Rheinfelder, and published by Iliffe. A technique recommended in that publication is the use of inductive coupling for the aerial input circuit rather than tapping the feeder on to the coil, and a form of bridge circuit is suggested which enables the above optimum matching condition to be achieved. The basic circuit is shown in **Fig 5.2,** from which it will be seen that two series-connected variable capacitors are employed to provide a tap on the inductances in both the input and the output circuits. In practice it is possible, once the correct ratio of capacitances

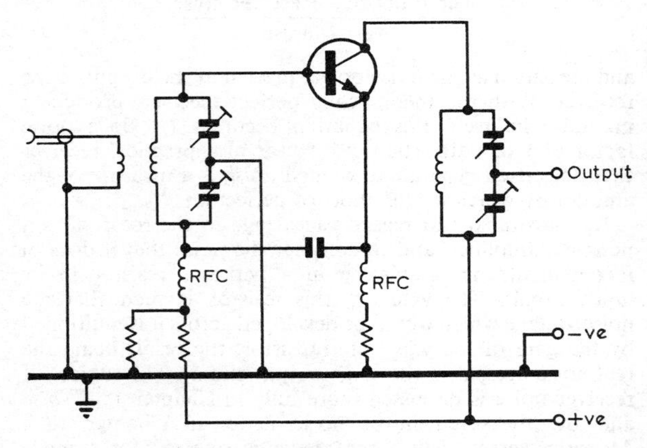

Fig 5.2. Basic circuit for obtaining optimum matching conditions in a transistor rf amplifier.

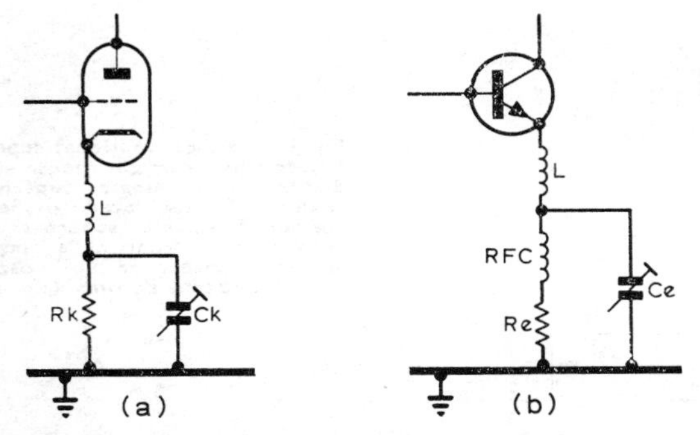

Fig 5.3. Method of reducing degeneration due to (a) cathode and (b) emitter lead inductance in an rf amplifying stage. L/C_k or L/C_e are resonant at the centre frequency of the band.

has been found, to replace one of the trimmers in each case by a fixed capacitor, and to use the remaining trimmer to tune the circuit to resonance.

One limiting factor to the gain which may be realized in an rf amplifier is degeneration due to cathode (or emitter) lead inductance. The author of the above book suggests how this may be overcome by the use of a series tuned circuit in the cathode (or emitter) lead as shown in **Fig 5.3**.

The noise factor of a valve is affected by the cathode temperature owing to the existence of the space charge in the grid/cathode region. The extent of the space charge is critically dependent upon the temperature of the cathode, and consequently there will be considerable increase in noise if the heater current is reduced by even a very small amount, eg by only five per cent. It is most important, therefore, to ensure that the correct heater voltage is applied to valves, particularly in the early stages of the receiver.

Adjustment of the value of the grid bias in relation to the contact potential of the grid is a means of improving the noise factor. If the grid bias of a valve is steadily reduced below the cut-off point the noise diminishes until a point is reached where grid current begins to flow; beyond this point the noise increases rapidly. **Fig 5.4** shows the relation between noise factor and grid bias at a constant anode current for two grounded-grid triodes. The general shape of the curves is the same but the minimum noise factors occur at different grid bias values. The optimum bias will vary with the type of valve and different specimens of the same type will also show some variation. It is often worth experimenting to obtain the optimum bias, but care should also be taken to keep the anode current constant at the recommended value by means of an adjustable series ht dropping resistor.

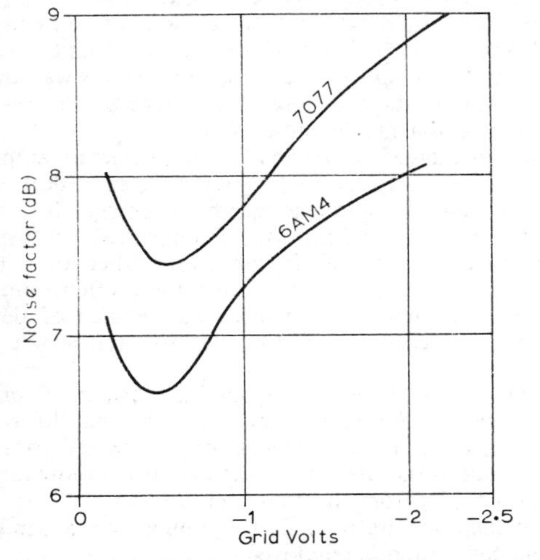

Fig 5.4. Noise factor plotted against grid bias for grounded grid triodes, the anode current being held constant (7077 at I_a 6·4mA at 450MHz; 6AM4 at I_a 10mA at 250MHz.

The noise factor of a transistor amplifier is also dependent upon the emitter current, and as the spread of characteristics for a given type may be quite wide it is well worthwhile trying the effect of variation in bias voltages if it is desired to achieve the best possible noise factor.

Circuit Noise

Noise due to devices other than valves or transistors is produced solely by the resistive component; inductive or capacitive reactances do not produce noise. Coils have negligible resistance in the vhf ranges but the leakage resistance of capacitors and insulators is important, and it is therefore imperative to select good-quality components, including such items as valveholders and switches. Attention should also be given to the use of low-noise resistors. Differences in noise as high as 25dB have been observed between a standard carbon resistor and a high-quality deposited carbon or metal film type when dc was present.

Circuit noise may be regarded as including noise caused by regeneration, and although it is common practice to enhance the gain of low-frequency circuits by applying regeneration, this should be avoided at all costs in vhf receivers because of the additional noise produced. Common causes of regeneration are:

(a) Insufficient decoupling of supply leads and particularly heater and cathode circuits. In the range 20–30MHz 0·01μF non-inductive capacitors are a minimum value for this purpose. On vhf, however, value of capacitance is less important than the physical size and properties of the capacitor itself and, when following published designs with results that do not come up to expectations, it is as well to check that the actual *type* of component has been used, otherwise the optimum results might be realized by a somewhat different value in another make or type of capacitor.

(b) Incomplete neutralizing of triode (or transistor) amplifiers. An adjustment that merely ensures that the circuit does not actually oscillate is not sufficient. Accurate neutralizing is essential, the appropriate methods being described later in this chapter.

(c) Insufficient or wrongly-placed screening between the input and output circuits of an amplifying stage. In valve circuits where a screen is required it should be mounted close-fitting across the valveholder, so that the input and output connections are on opposite sides of the screen with the latter firmly attached to the chassis on either side of the valveholder and to its central spigot.

(d) Circulating rf currents in the chassis due to multi-point earthing.

(e) An attempt to achieve an excessive voltage gain from an amplifying stage. By tapping down the anode, collector or drain of a valve, bipolar transistor or fet respectively on its tuned circuit, the voltage gain falls more rapidly

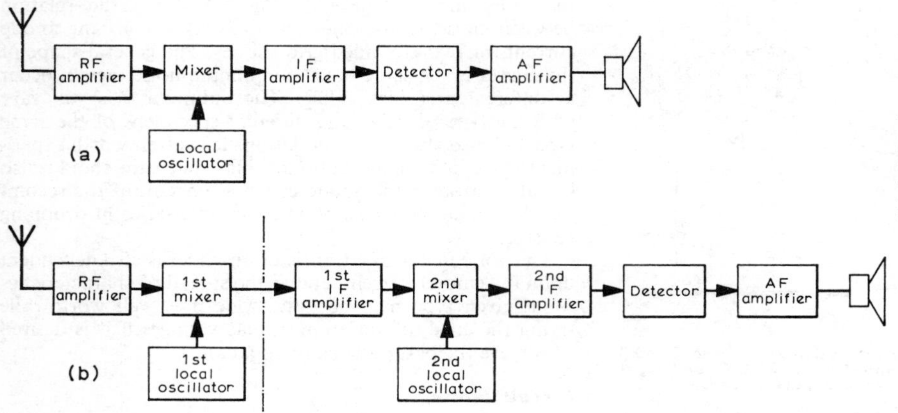

Fig 5.5. Two conventional superheterodyne configurations. (a) Self-contained single superhet with tunable oscillator. (b) Self-contained double superhet or converter in front of a single superhet. Either or both oscillators may be tunable.

than does the power gain and stability depends upon voltage gain. A further advantage is to be expected from such an intentional mismatch because selectivity will be improved, thereby reducing the bandwidth of the circuit and with it the noise.

(f) Poor earth contacts to the chassis. For low-noise circuits a brass or copper chassis should be employed; aluminium is not the best chassis material because it is extremely difficult to make a lasting low-resistance connection to it. Tinplate, on the other hand, is quite satisfactory as it is easy to make good soldered earth connections provided a reasonably high-wattage iron is used.

Effect of Bandwidth

Before leaving the subject of noise, there is one further point which must be considered. If the noise factor of a receiver is measured with a noise generator it will be found to be independent of receiver bandwidth. This is because the noise from the generator is of the same nature as circuit noise, so that if the bandwidth is doubled the overall noise is also doubled. For the reception of a signal of finite bandwidth, however, the optimum signal-to-noise ratio is obtained when the bandwidth of the receiver is only just sufficient to accommodate the signal, and any further increase in bandwidth results merely in additional noise. The signal-to-noise ratio at the receiver, therefore, depends on the power per unit bandwidth of the transmitted signal.

To illustrate this point, suppose that the receiver produces 1μV of noise for each 10kHz of bandwidth and that an amplitude-modulated transmitter radiates a signal 10kHz wide and produces 10μV at the receiver. The signal-to-noise ratio is therefore 10, provided that the receiver uses a bandwidth of 10kHz. If the bandwidth of the transmission is reduced to 5kHz and the radiated power remains the same, the input to the receiver will still be 10μV, but if the bandwidth of the receiver is also reduced to 5kHz, only 0·5μV of noise will be accepted and the signal-to-noise ratio will increase to 20.

The use of a bandwidth exceeding 5kHz for telephony and considerably less for cw reception is therefore undesirable when a good signal-to-noise ratio is the prime consideration. Also, an unstable transmission, requiring a large-bandwidth receiver, results in a degraded signal-to-noise ratio, so there is clearly a great advantage to be gained by the employment of stable transmitters and, of course, receiver local oscillators.

CHOICE OF RECEIVER CONFIGURATION

Receivers using other than superheterodyne techniques are rare on vhf or uhf. Simple designs based on the super-regenerative detector are still met occasionally but will not be considered further here.

Superheterodyne receivers for vhf or uhf may have one, two or even several frequency changers and any or all of the oscillators may be fixed or tunable. This leads to a variety of configurations to be considered and **Fig 5.5** illustrates some of these. Fig 5.5(a) is a conventional superheterodyne such as may be found on hf. The i.f. should be rather higher than is usual in hf receivers to reduce second-channel interference (see Chapter 4). With the advent of crystal filters at frequencies around 10MHz which provide desirable selectivity, this configuration can be expected to become very popular.

In Fig 5.5(b) is shown a double superhet. This can be a purpose-built complete receiver or alternatively the portion to the left of the dotted line can be added to an existing hf band receiver, in which case the additional unit is called a *converter*. Where an hf receiver is used in this way the converter provides the full range of facilities on the new band for the cost of a simple add-on unit.

A disadvantage in using an hf receiver as an i.f. amplifier is that when preceded by a converter with, say, 30dB of gain, it is very liable to overloading on strong signals. To some extent this problem can be circumvented by using an inexpensive receiver of limited gain rather than a high performance receiver, but because these often omit many of the refinements of selectivity and choice of detectors required for serious operation, this solution is not a very satisfactory one.

A much better idea is to build a *tunable i.f. amplifier* containing all the refinements of a normal hf receiver, except that the tuning is deliberately restricted to a single range of a few megahertz and the gain is kept low to permit additional gain from the converter.

Two methods are possible for tuning over a band when double conversion is employed:

(a) The oscillator in the converter may be fixed in frequency and may therefore be crystal controlled, tuning being effected by variation of the first i.f. (ie by tuning the main receiver) over a band equal in width to that of the vhf band to be covered.

(b) The oscillator in the converter may be variable in frequency and the main receiver tuning set at the frequency chosen for the first i.f.

Choice of the First I.F.

In any superheterodyne receiver it is possible for two incoming frequencies to mix with the local oscillator to give the i.f., these being the desired signal and the image frequency. A few figures should make the position clear. It will be assumed that the converter is to cover the 144 to 146MHz band, and that the first i.f. is to be 4 to 6MHz. The crystal oscillator frequency must differ from that of the signal by this range of frequencies as the band is tuned and could therefore be 144 − 4MHz = 140MHz or, alternatively, 144 + 4MHz. Assuming that the lower of the two crystal frequencies is used, a signal on 136MHz would also produce a difference of 4MHz, and unless the rf and mixer stages are selective enough to discriminate against such a signal, it will be heard along with the desired signal on 144MHz. From the foregoing it will be appreciated that the image frequency is always removed from the signal frequency by twice the i.f., and is on the same side as the local oscillator.

It should be noted that even if no actual signal is present at the image frequency there will be some contributed noise which will be added to that already present on the desired signal. It is usual to set the rf and mixer tuned circuits to the centre of the band in use so that on the 144MHz band they should be at least 2MHz wide in order to respond to signals anywhere in the band. This bandwidth only represents approximately 1·4 per cent of the mid-band frequency and it is not surprising that appreciable response should still be obtained over the image frequency range of 134 to 136MHz unless something is done to restrict it. Naturally, the higher the first i.f. the greater the separation between desired and image frequencies, but as low an i.f. as 4 to 6MHz is feasible provided some attempt is made to restrict the bandwidth of the converter by, for example, employing two inductively-coupled tuned circuits between the rf and mixer stages, thus providing a bandpass effect.

The choice of first i.f. is also conditioned by other factors. First it is desirable that no harmonic of the oscillator in the main receiver should fall in the vhf band in use, and second there should be no breakthrough from stations operating on the frequency or band of frequencies selected for the first i.f.

Many hf receiver oscillators produce quite strong harmonics in the vhf bands and, although these are high-order harmonics and are therefore tuned through quickly, they can be distracting when searching for signals in the band in question. The problem only exists when the converter oscillator is crystal controlled as freedom from harmonic interference is then required over a band equal in width to the vhf band to be covered. This also applies, of course, to i.f. breakthrough.

As it is practically impossible to find a band some hundreds of kilohertz wide which is unoccupied by at least some strong signals it is necessary to take steps to ensure that the main receiver does not respond to them when an aerial is not connected. Frequencies in the range 20 to 30MHz are often chosen since fewer strong signals are normally found there than on the lower frequencies, but this state of affairs may well be reversed during periods of high sunspot activity. With the greatly increased use of communication receivers which cover the amateur bands only, the sole range on which a continuous coverage of 2MHz is available is 28 to 30MHz, and for these reasons this is probably the most popular choice of first i.f.

Preventing Breakthrough

The main receiver should be switched on without an aerial connected and with rf and i.f. gains at maximum. Signals may be picked up on the aerial terminal itself, in which case it may be necessary to replace it with a coaxial socket. They may also be picked up on the mains wiring, in the power supply and/or its connecting cables, or on the loudspeaker leads if these last two items are separate from the receiver. In order to discover where the unwanted signals are entering the receiver, the following procedure may be carried out with the under-chassis wiring exposed:

(1) Short circuit the aerial input to the nearest earthy point *inside the receiver*. If signals drop in strength a coaxial socket, as previously mentioned, should help. Leave the short in position for the time being.

(2) With a few yards of *insulated* wire as aerial, attach the bared end to some form of insulated rod and touch the wire on either side of the mains input in turn. If signals increase a capacitor *of at least 1,000V working rating* between that point and chassis should be fitted. A capacitance of between 1,000pF and 0·01μF should be tried—for the higher frequencies the *lower* value may be found to be more effective.

(3) Continue touching the "aerial" on various points on the supply wiring and bypass any points to chassis where the signal increases. Particular attention should be paid to the agc line at different points, but when bypassing this it is important to avoid using such a value of capacitance as to upset the time constant of the agc system.

(4) Having removed the short circuit on the aerial input and, if necessary, fitted a coaxial socket, the receiver should have appreciably less unwanted signal pick-up. It should be remembered that the converter will contribute a certain amount of noise so that a small residual breakthrough may well be masked by this.

(5) With the converter connected through a short length of coaxial cable it may be found that breakthrough is again in evidence. Should this be due to pick-up on the mains wiring, action may be taken as already described, but it may be found that, for instance, signals are coming in on the supply leads (from a battery or mains unit) feeding power to a transistor converter. In one such instance a bypass capacitor made no improvement and recourse had to be made to a simple filter consisting of a parallel-tuned circuit resonant at the centre of the first i.f. band and placed in series with the unearthed supply lead. For 5MHz operation the most satisfactory inductance for the purpose was found to be a toroidally wound coil on a small ferrite ring of about 1in in diameter resonated by a capacitance of around 20pF. The toroid was found to have a greater bandwidth than a solenoid and had the added advantage of not requiring screening. It is worth noting that such a filter may be adjusted by placing it in series with the aerial on the main receiver and varying the number of turns on the coil until it effectively eliminates

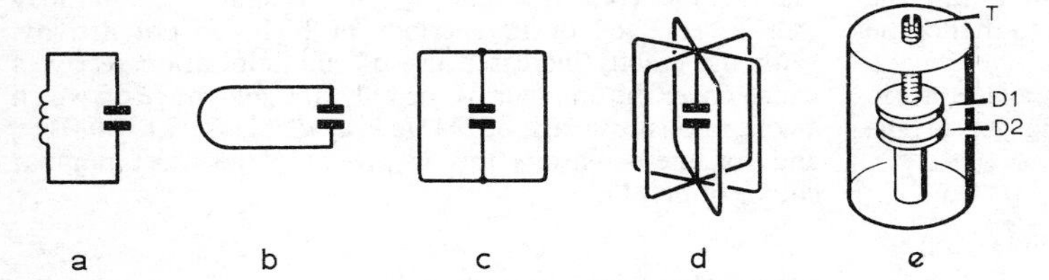

Fig 5.6. Progressive development of tuned circuits from a coil to a cavity as the frequency is increased.

signals over the required frequency range. The tuning of the filter will then be approximately correct when it is installed in the supply lead.

(6) Should first i.f. signals be picked up on the vhf aerial and pass through the converter, they may be eliminated by the use of a $\lambda/4$ shorted stub made from coaxial cable and connected across the converter input in parallel with the aerial feeder. The actual length of the stub will depend upon the velocity factor of the cable used, but should lie between 13·5 and 17·5in for the 144MHz band. The exact length can readily be ascertained by connecting as stated, say, 18in of *open-circuited* cable. This will act as a short circuit to in-band signals and small pieces of the cable are then cut off until a signal around the *low* end of the band is made as weak as possible. Obviously the final cut will make the cable too short—the strength of the signal will start to rise again—but if preliminary adjustment is made at the *low* frequency end of the band this will be no disadvantage, as the resonant frequency of the stub will then be nearer to that desired, ie mid-band. When the correct length has been found, the inner conductor of the stub is soldered to the braid at the free end and signals should then return to their normal strength—the insertion loss is very low—but first i.f. signals will "see" a virtual short circuit across the converter input.

TUNED CIRCUITS

Tuning is readily achieved at hf by "lumped circuits" in which the inductor and capacitor are substantially discrete components. At vhf the two components are never wholly separate, the capacitance between the turns of the inductor being often a significant part of the total circuit capacitance. The self-inductance of the plates of the capacitor is similarly important. Often the capacitance required is equal to or less than the necessary minimum capacitance associated with the wiring and active devices, in which case no physical component identifiable as "the capacitor" is present and the circuit is said to be tuned by the "stray" circuit capacitance.

As the required frequency of a tuned circuit increases, obviously the physical size of the inductor and capacitor becomes smaller until they can no longer be manipulated with conventional tools. For amateur purposes the limits of physical coils and capacitors occur in the lower uhf bands: lumped circuits are often used in the 432MHz band but are rare in the 1·3GHz band.

Distributed Circuits

Fig 5.6 illustrates how progressively lower inductances are used to tune a fixed capacitor to higher frequencies. In Fig 5.6(b) the "coil" is reduced to a single hairpin loop, this configuration being commonly used at 432MHz. Two loops can be connected to the same capacitor as in Fig 5.6(c). This halves the inductance and can be very convenient for filters. Fig 5.6(d) represents a multiplication of this structure and in Fig 5.6(e) there is in effect an infinite number of loops in parallel, ie a cylinder closed at both ends with a central rod in series with the capacitor. If the diameter of the structure is greater than its height it is termed a "rhumbatron" otherwise it is a "coaxial cavity".

The rhumbatron is very suitable for disc-seal valves such as microwave triodes or klystrons, while the coaxial cavity is more convenient for use with pin-based valves and semiconductors which are wired into the circuit. The capacitive element can be made adjustable as a tuning element by threading one of the rods T which passes through a threaded end plate; a disc D1 is fixed to the lower end of the rod so that it is parallel to a similar disc D2 attached to the central support.

The simple hairpin, shown at Fig 5.6(b), is a very convenient form of construction: it can be made of wide strip rather than wire and is especially suitable for push-pull circuits. It may be tuned by parallel capacitance at the open end, or by a series capacitance at the closed end.

In a modification of the hairpin loop, one side is a sheet of metal which can be part of the chassis, the strip being spaced from the metal, and such an arrangement is known as a "strip-line". When the strip is very close to the metal sheet but insulated from it by dielectric, the result is called "microstrip". This can be produced very easily from good-quality double-sided printed circuit board by etching or cutting with a sharp razor blade.

Because a cylindrical cavity usually requires the use of a lathe in its construction, a simpler form of a trough which can be made with hand tools is frequently used: see **Fig 5.7 (a)**. The open side may be closed with a well-fitting lid if required for screening—the omission of the lid only reduces the Q slightly. One end of the trough is closed by a plate to which the inner line is secured. The three edges of the plate must make a good low-resistance connection with the trough, preferably by being soldered. Tuning may be effected by a disc-type capacitor or a small trimmer connected between the inner and outer of the line.

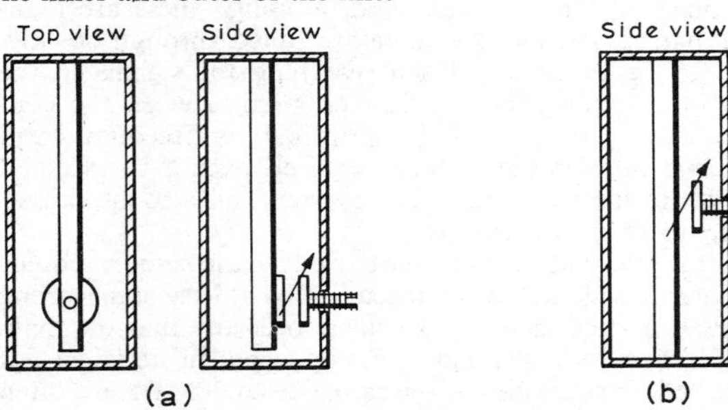

Fig 5.7. Trough line tuned circuits.

Another convenient form of trough line is shown in **Fig 5.7(b)**. The line is electrically $\lambda/2$ long with a central tuning capacitor and is closed at both ends. This structure is very rigid and examples of its use will be found later in this chapter. Note that in this case the tuning capacitor is effectively divided between two lines and has twice the capacitance to be calculated in the next section.

The dimensions of a cavity or trough line can be calculated from the basic equation:

$$l = \frac{\lambda}{2\pi} \tan^{-1} \frac{1}{2\pi f Z_0 C}$$

where l is the length in the same units as the wavelength, f is the frequency in Hz, C is the total capacitance in farads and Z_0 is the characteristic impedance in ohms. This formula is more conveniently rewritten as:

$$l = \frac{30{,}000}{2\pi f} \tan^{-1} \frac{10^6}{2\pi f Z_0 C}$$

where l is now in centimetres, f is in megahertz, C is in picofarads and Z_0 is in ohms.

It can be seen from this formula that Z_0C occurs only in one portion of the equation, so any value of the two parameters could be chosen at any specified frequency. In practice, considerations outside the control of the designer determine the choice of Z_0. The capacitance may be fixed by a valve or transistor mounted at the end of the line, or the diameter of the line determined by the geometry of that device. **Fig 5.8** is a set of graphs showing the lengths of the line plotted against frequency (MHz) for various values of Z_0C.

The characteristic impedance of a coaxial line is given by the formula $Z_0 = 138 \log_{10} b/a$, where b is the inside diameter of the outer tube and a is the outside diameter of the inner tube or rod. For a square trough line with round inner the formula is:

$$Z = 138 \log_{10} b/a + 6{\cdot}48 - 2{\cdot}34A - 0{\cdot}48B - 0{\cdot}12C$$

where b is now the width inside the trough. The correction factors are given by:

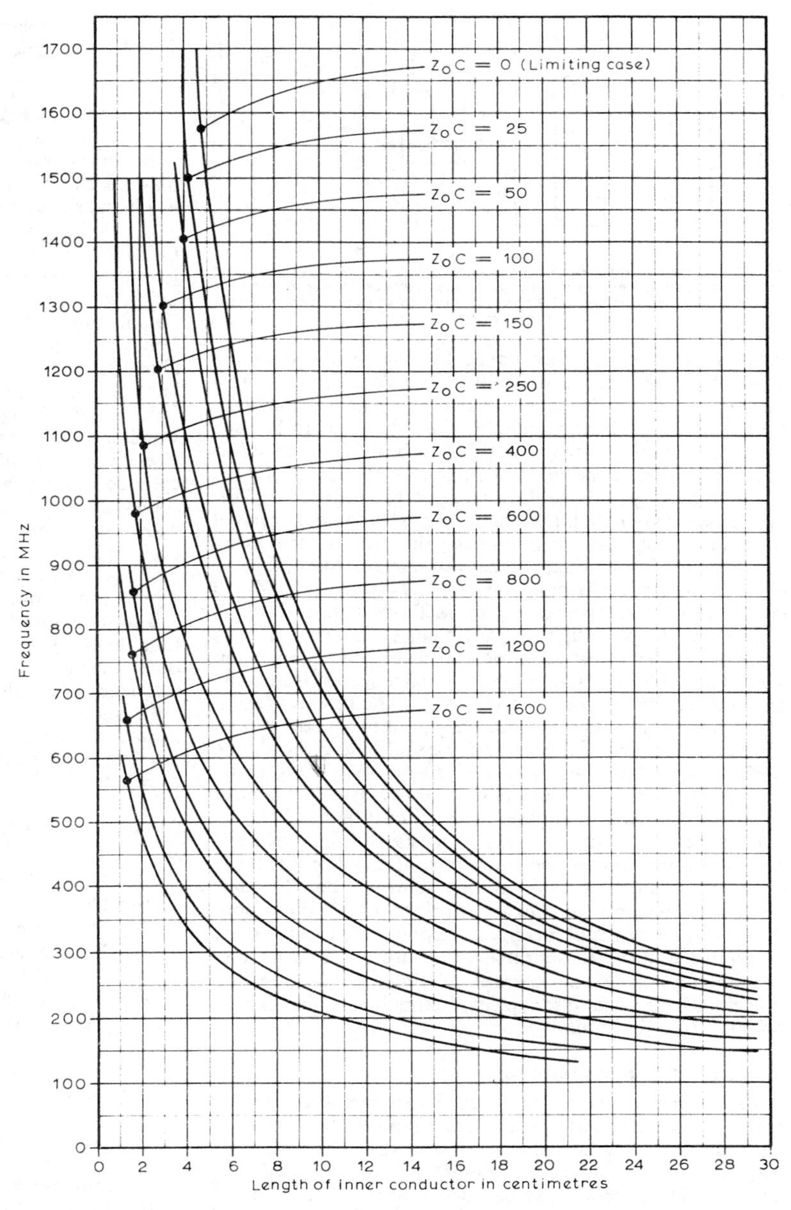

Fig 5.8. Chart plotting frequency against length of inner line for various values of characteristic impedance multiplied by the total capacitance.

$$A = \frac{1 - 0{\cdot}405(b/a)^{-4}}{1 - 0{\cdot}405(b/a)^{-4}} \qquad B = \frac{1 - 0{\cdot}163(b/a)^{-8}}{1 - 0{\cdot}163(b/a)^{-8}}$$

$$C = \frac{1 - 0{\cdot}067(b/a)^{-12}}{1 - 0{\cdot}067(b/a)^{-12}}$$

These factors need only be evaluated for very low impedance lines where the inner is comparable in diameter to the outer.

In practice, the following approximate formula will serve for lines of impedance greater than 50Ω:

$$Z = 138 \log_{10}(b/a) - 3{\cdot}42$$

Coaxial cavities are most readily constructed of lengths of standard-sized brass or copper tubing, and the choice of the ratio of outer to inner dimensions will of course be limited by the range of sizes available. Optimum Q for an unloaded cavity occurs when the characteristic impedance is 77Ω. In

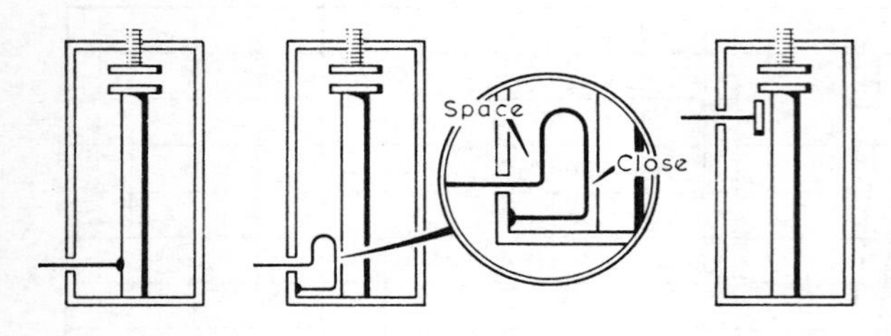

Fig 5.9. Three methods of coupling into a line or cavity. (a) Tapping the inner conductor. (b) Boot-shaped coupling loop at low impedance end of inner conductor. (c) Probe at high impedance end of inner conductor.

order to reduce the capacitance loading or the damping effect of a low-resistance device it is possible to tap down to a "lower" point on the inner conductor (ie nearer the zero-potential end). When this is done the value of C to be used in the above formulae can be taken to be: $C\, l_1/l_2$ where l_1 is the distance from the low-potential end to the tap and l_2 is the total length of the line. This result is satisfactory for taps down to about half the length. Taps should not in general be taken below this point due to the risk of provoking spurious modes and unwanted resonances. Should it be essential to tap down more than this it is usually better to couple into the cavity by one of the methods described below.

Due to the variation in capacitance between various samples of the same valve or transistor, some provision should be made for varying the electrical length of the cavity. In principle this can be done by varying the physical length of the line by means of a sliding short, but in practice a high standard of mechanical precision is essential and the sliding surfaces must be plated with a hard non-tarnishing metal such as rhodium. Such techniques are not usually available to the amateur and so another method must be sought.

A more convenient method is to use a small trimmer capacitor consisting of two discs, one mounted on the inner conductor and the other carried on a screwed stem, the latter being mounted in a threaded bush fitted in the outer conductor. Some means of locking or gripping the thread of the adjustable element is desirable to ensure a sound contact and smooth adjustment. A graph relating the diameters and capacitance ranges of disc-type capacitors is given in Chapter 7—*VHF/UHF Transmitters*.

Input and Output Coupling

Fig 5.9 shows the three methods of coupling power into or out of a resonant cavity or line: (a) a tapping may be made on to the inner line near the closed end, (b) a coupling loop may be inserted near the inner line towards the closed end, (c) a capacitance probe may be inserted near the open end of the inner line.

In the first method coupling is not very easy to adjust. It can be used in trough lines, however, by sliding a valve top-cap along the line, and also in strip-line where it is possible to unsolder the tap and move it. The second method is more versatile since the degree of coupling can be varied both by varying the size of the loop and by rotating it with respect to the cavity. The loop can also be a series or parallel tuned with a capacitor. When very tight coupling is required the loop should be "boot-shaped" as shown in Fig 5.9(b). The third method is generally only suitable for high-impedance circuits since the reactance of the effective series capacitor is quite high. The method can be used to vary the coupling to a relatively high-resistance valve or transistor in strip-line and microstrip circuits.

THE LOCAL OSCILLATOR

It has already been mentioned that the local oscillator may be either tunable or fixed in frequency. Each system has its merits and these will now be considered.

It is clear that if the final bandwidth of the complete receiver is to be the same as that used on the lower frequencies, the stability of the local oscillator must be as good as before, even though the frequency is much higher. For the 70 and 144MHz bands it is possible to build tunable oscillators of reasonable stability to operate at frequencies either above or below the signal frequency (by the value of the first i.f.), but apart from some inevitable drift there is also a rapid small change in frequency known as "scintillation" which mitigates against a note better than about T8. In valve circuits part of this failing is due to the ac heating of the cathode and therefore transistor oscillators are capable of somewhat better performance than valves in this respect. If a wider bandwidth than optimum can be tolerated, a tunable oscillator can be fairly satisfactory for a.m. or fm telephony, but a cw transmission always sounds worse than T9 and ssb signals can be difficult to resolve.

The same disadvantages apply, but with greater force, to tunable oscillators on the 432MHz band, even if the circuit is designed to operate at one-half or one-third of the injection frequency and followed by frequency multipliers.

Crystal-controlled local oscillators can, however, be used in any vhf or uhf converter, and the advantages and disadvantages of the two types of oscillator may be summarized as follows.

Tunable oscillators

(*a*) *Advantages*

(i) A directly calibrated dial is possible.
(ii) The cost of a crystal is eliminated.
(iii) The circuitry is simple.
(iv) There is less likelihood of harmonic interference from the main receiver.
(v) Only one clear channel for the first i.f. is required on the main receiver.

(*b*) *Disadvantages*

(i) Long-term oscillator drift makes dial calibration unreliable.
(ii) Warm-up drift can be troublesome.
(iii) It is difficult to obtain a better than T8 note, but not impossible.
(iv) Reception of ssb signals is unsatisfactory.
(v) They cannot readily be used remotely with mechanical tuning.

Crystal controlled oscillators

(*a*) *Advantages*

(i) Accurate logging of stations on the main receiver dial is possible.
(ii) A T9 note for cw is assured.
(iii) SSB reception is satisfactory.
(iv) Negligible short term or warm-up drift.
(v) Absence of controls permits remote operation of the converter.

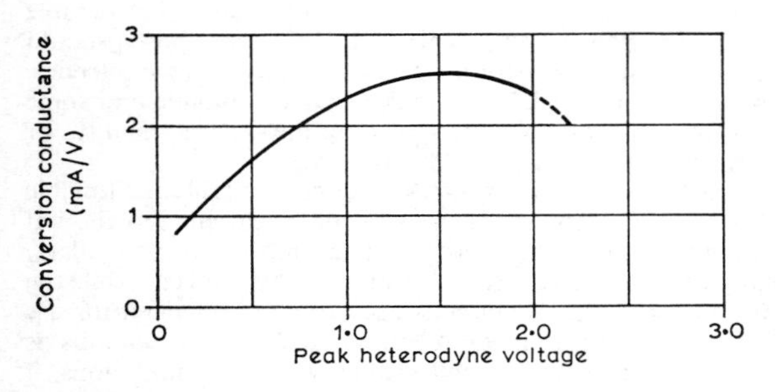

Fig 5.10. The relation between conversion conductance and oscillator (heterodyne) voltage in a typical triode mixer (6AM4, V_a 125V, Rk 220Ω).

(*b*) *Disadvantages*

(i) More expensive in valves or transistors (for 432MHz and above) and crystal.

(ii) Possibility of additional self-generated whistles, particularly if frequency multiplication is necessary.

(iii) More adjustments and therefore more difficult to align in the first place.

The question of oscillator injection level arises whatever the type of oscillator used. As already stated, if the rf stages have sufficient gain and low noise factor, the noise contributed by later stages tends to be less significant, but this assumes reasonable efficiency in the mixer stage. **Fig 5.10** shows a typical conversion conductance curve for a mixer valve stage and indicates how the i.f. output varies as the injection voltage is changed, the signal being held constant; it will be noticed that a considerable loss of gain results if inadequate injection is used. Similar considerations apply in the case of transistors but the optimum oscillator voltage is far lower and generally easier to realize in practice.

The injection voltage in a triode valve mixer is found by measuring the current flowing through the mixer grid resistor due to the oscillator injection, ie the difference in grid current with and without the oscillator operative. The maximum value of oscillator voltage normally to be expected is about 3V peak, and with a grid resistor of 50kΩ this corresponds to a grid current of only 60μA, so a sensitive meter is required for this measurement.

It must not be supposed that a voltage of this magnitude is easy to achieve; on the contrary, to produce such a voltage at the mixer grid can be quite difficult. **Fig 5.11** shows a typical triode mixer circuit in which the inductance L1 is tuned to resonance at, say, 70·3MHz and the signal voltage developed across it fed through a 30pF capacitor to the mixer grid. Suppose L2 is carrying oscillations at 50MHz, coupled through a 10pF capacitor to the grid of the mixer (V2). At 50MHz the inductance L1 is in effect a short circuit to earth, and at this frequency the 50MHz oscillations therefore "see" the mixer grid circuit as shown in **Fig 5.12**. The 10pF and 30pF capacitors (C2 and C1) in the mixer grid circuit become a potential divider such that only one-quarter of the original 50MHz voltage at the anode of V3 appears at the grid of V2.

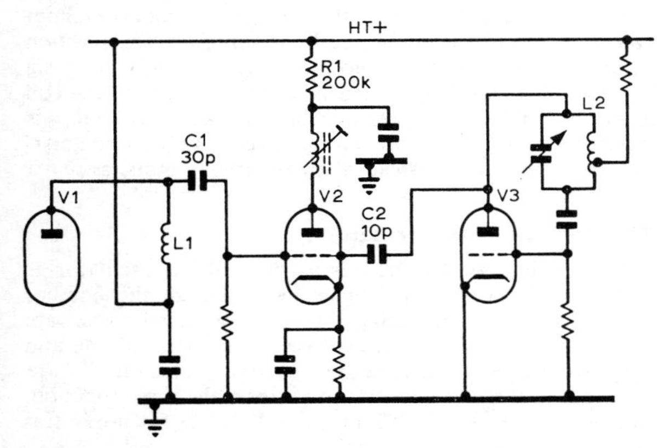

Fig 5.11. Typical triode mixer (V2) and associated oscillator (V3).

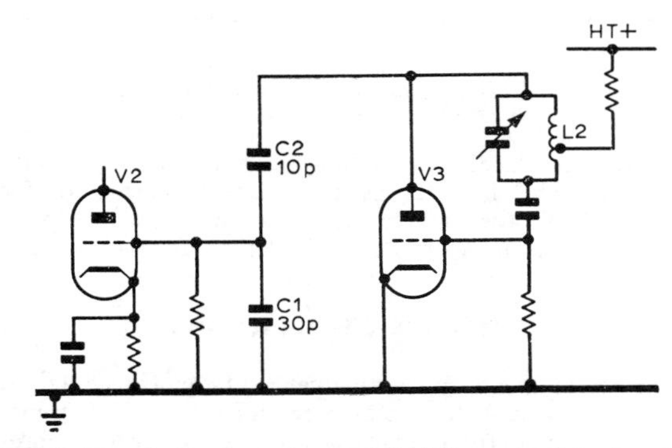

Fig 5.12. The triode mixer circuit of Fig 5.11 arranged to show the potential divider formed by the oscillator coupling and rf to mixer stage coupling capacitors at the injection frequency.

An obvious course would be to increase the value of C2, but signal voltages at the grid of the mixer "see" this capacitor between grid and earth, since L2 appears a short circuit at signal frequency. Thus only three-quarters of the signal voltage at the anode of V1 appears at the grid of V2 with the existing capacitor ratio; if C2 was increased in value, even less voltage would be available at the mixer grid.

Experimental adjustment of the values of C1 and C2 may be made (with appropriate re-adjustment of L1 and L2) to effect the best compromise. Capacitances of 30 and 10pF respectively are of the correct order of magnitude because an rf voltage of 10 to 20V is commonly obtained at the anode of V3, and one-quarter of this voltage at the mixer grid is usually adequate.

The required value of oscillator injection voltage depends upon the anode potential of the mixer valve; the lower the anode voltage the lower the required injection voltage. It is this fact that has made the use of relatively high values for R1 (Fig 5.11) commonplace.

If R1 is replaced temporarily by a variable resistor of about 250kΩ which can be adjusted while listening to a weak signal, it will be found that the optimum value of mixer anode potential is readily obtained. The value of resistance thus determined (which is not unduly critical) may attenuate strong signals slightly, but this is of little consequence in relation to the marked improvement effected in the response to weak signals.

Capacitive coupling between oscillator and mixer is not essential, and there are certain advantages to be had from using inductive coupling for the injection voltage. Part of a transistor converter circuit is shown in **Fig 5.13**, from which

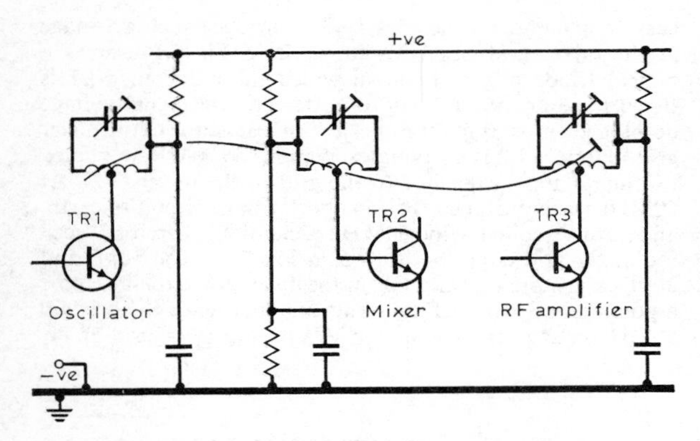

Fig 5.13. Inductive coupling between mixer and local oscillator in a transistorized converter.

it will be seen that the earthy ends of the local oscillator TR1 and mixer TR2 inductances are adjacent (spacing of the order of 0·25in for 70MHz) while the "hot" ends of the rf and mixer coils are spaced by a similar distance.

TUNABLE OSCILLATORS

The requirements for a tunable oscillator are similar to those for a variable-frequency oscillator designed for transmitter control, ie robust construction to guard against vibrations, the employment of only the best-quality components (including valveholders and insulation), short and rigid wiring, and self-supported coils of stout wire. The operating frequency of the oscillator may be either above or below the signal frequency, but to ensure the best stability it is usual for it to be on the *low* frequency side of the signal. Where the choice of such a frequency leads to interference with a local television channel, operation on the *high* side of the signal may be necessary, although a small change in intermediate frequency sometimes enables the oscillator frequency to be placed to avoid such trouble.

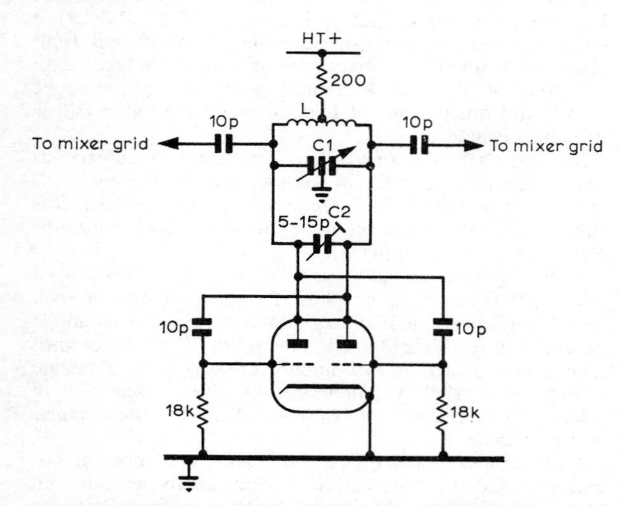

Fig 5.14. The kalitron oscillator. L is 2 turns of 16swg $\frac{5}{16}$in diameter for 120MHz. C1 may be 7·5pF per section.

The valve employed will, of course, have considerable influence on the results obtained, and the type chosen should be of robust construction with short leads to the electrodes. All-glass types on B7G or B9A bases are suitable and some advantage may be obtained by using the "special quality" or "trustworthy" ("ruggedized") variety.

There are three main types of circuit suitable for tunable vhf local oscillators: (a) the kalitron push-pull, (b) the vhf Colpitts, and (c) the Hartley. The kalitron is particularly applicable to the requriements of a balanced mixer, while the Colpitts and Hartley circuits are primarily intended for use with single-ended mixer circuits. Each of the circuits is capable of producing good results, within its limitations, if well constructed.

The Kalitron Oscillator

This circuit, shown in **Fig 5.14**, is well suited to twin triode valves.

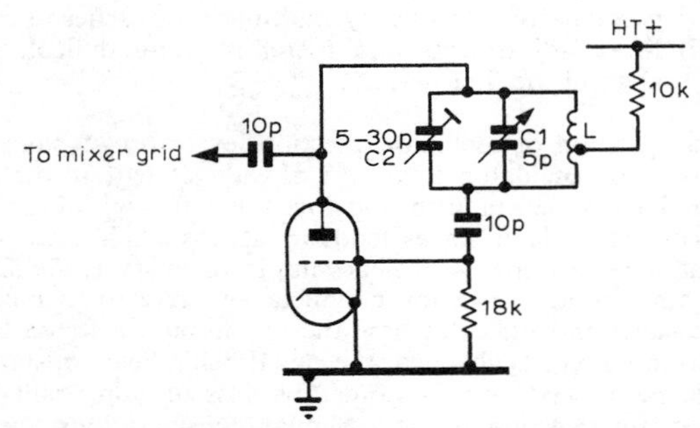

Fig 5.15. The vhf Colpitts oscillator circuit. For 120MHz L should be two turns of 16swg $\frac{3}{8}$in diameter.

For either 70 or 144MHz converters the oscillator tuning capacitor C1 may be a small split-stator type with one fixed and one moving plate per section. The wire used for the coil should not be thinner than 16swg, and the coils should be mounted directly on the soldering lugs of the capacitor. The wires to the anode tags of a high-quality (preferably ptfe) valveholder should be equally robust. Careful proportioning of the parallel capacitance C2 and the size of the coil will enable the required bandspread to be obtained.

With an ht supply of 100V the circuit produces about 18V p-p across the anode coil, which gives an injection voltage of the correct order to the mixer with 10pF coupling capacitors. It should be noted that the ht supply to the coil is through a low-value resistor; if a choke is used at this point the frequency of oscillation is very likely to be determined by the choke instead of by the tuning inductance.

The VHF Colpitts Oscillator

This circuit (see **Fig 5.15**) is a true Colpitts oscillator as employed on the lower frequencies, although the familiar pair of capacitors providing a tap on the tuned circuit are missing in the vhf version because the grid-to-cathode and anode-to-cathode capacitances of the valve itself are sufficient at these frequencies to perform the same function. The 6C4, 6CW4 or 12AT7 are suitable valves for use in this circuit.

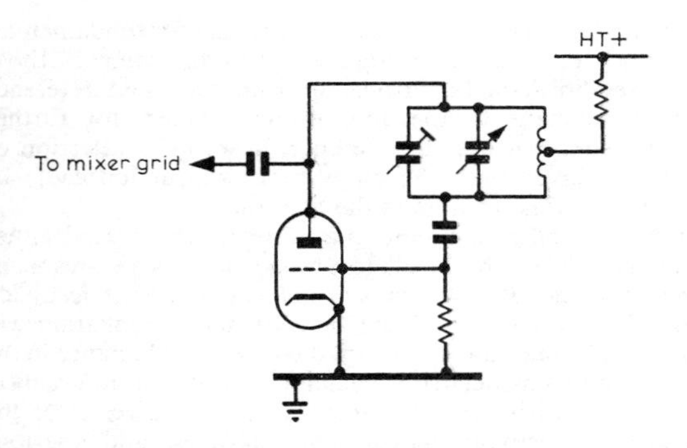

Fig 5.16. The Hartley oscillator. Similar values to those shown in Fig 5.15 apply.

The Hartley Oscillator

The Hartley oscillator circuit shown in **Fig 5.16** differs from the Colpitts (Fig 5.15) only in the position of the ht feed to the tuning coil, but it generally gives rather more output.

Transistor Tunable Oscillator

The constructional requirements for a transistor tunable oscillator differ in no way from those for valves, but due to the small size of the transistor a more compact and therefore more rigid layout is often possible. The frequency stability of a transistor oscillator is very dependent upon the applied voltage, and it is good practice to employ a zener diode (see Chapter 3) to ensure a constant-voltage supply.

Due to heating of the small device, high rf output and stability do not go together and it may be necessary to trade stability against output to achieve optimum results.

A Hartley-type circuit suitable for the local oscillator in either a 70 or 144MHz converter appears in **Fig 5.17.** With the 2N706 transistor shown the value of the emitter resistor should be regarded as nominal—a lower value will increase the output at the expense of some degree of stability—but in a typical case, with an emitter resistor of 500Ω, the output at 65MHz varied from 0·5V for a supply voltage of 4·5V up to 2·5V when 12V was applied. The output at 130MHz was approximately 0·7V at 12V input, and if more output should be required some reduction in the value of the emitter resistor would be called for.

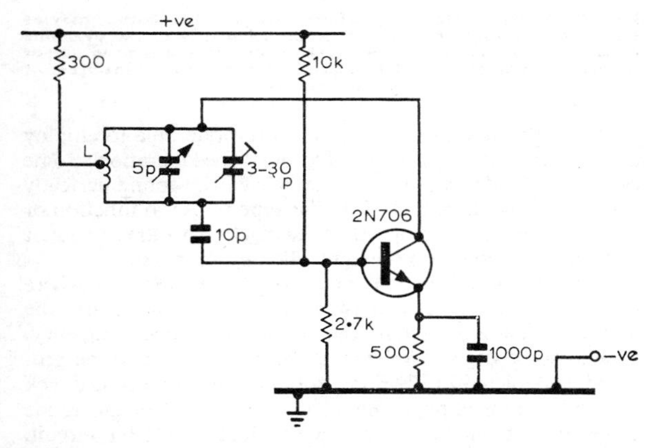

Fig 5.17. Hartley circuit for a tunable local oscillator. L should be 3 turns 0·5in diameter of 16swg for 65MHz or 2½ turns for 135MHz. Consumption approximately 6mA at 12V for the transistor shown.

CRYSTAL-CONTROLLED OSCILLATORS

In a crystal-controlled oscillator/multiplier chain for a converter, the aim should be to produce the least number of frequencies in order to avoid unwanted heterodynes or "birdies". This means the use of overtone crystals and ideally no frequency-multiplier stages. Such an arrangement is quite feasible for 70MHz where the crystal frequency would lie between 42 and 66MHz for I.F.S ranging from 28 to 4MHz respectively, but for the 144MHz band (crystal frequencies between 116 and 140MHz) it may be thought that the cost of such crystals would be unjustified. In that case, and certainly for the 432MHz allocation, some frequency multiplication would be necessary, but the aim should still be to start with as high a frequency crystal as possible.

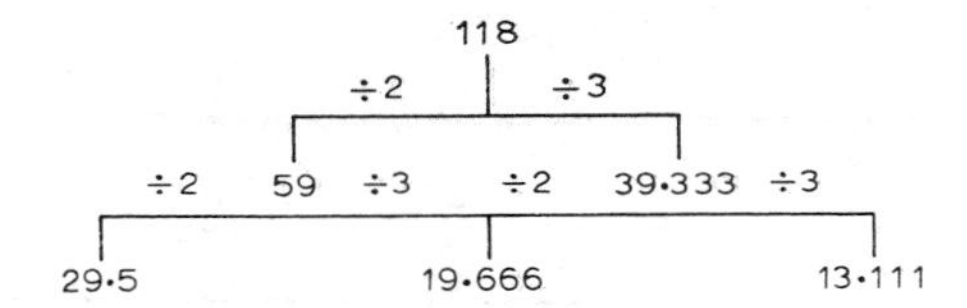

Fig 5.18. Type of chart used for determining the fundamental crystal frequency required in an oscillator multiplier chain.

Many designs utilizing crystals requiring high multiplication factors have been described in the literature and, provided certain precautions are taken to avoid frequencies which have harmonics falling inside the band concerned or frequencies which beat with harmonics of the oscillator in the main receiver, satisfactory results can be obtained.

Before deciding upon the frequency of a crystal for this service a chart such as that shown in **Fig 5.18** should be prepared. As an example, assume that it is found that if the local oscillator in the main receiver covers the range 26·5–28·5MHz, no harmonics from it will fall into the 144MHz band. If the main receiver has an i.f. of 500kHz and its oscillator operates above the signal frequency, a first i.f. of 26–28MHz will be satisfactory. Since 26MHz will be the tuning position for a signal on 144MHz, the injection frequency for the converter will be 118MHz (ie 144 — 26) and reference to the chart will show how this may be obtained.

It is normal practice to use an overtone oscillator in which the crystal vibrates mechanically at approximately three or five times its fundamental frequency, depending upon the configuration of the circuit. There is no output at the fundamental frequency of the crystal, and the harmonics present are related only to the overtone frequency. For example, a 10MHz crystal operating on its third overtone would generate no energy below 30MHz and behave like a crystal of that fundamental frequency. It should be noted that overtone operation *only* takes place on *odd* multiples of the fundamental frequency of the crystal.

Crystals specially manufactured for overtone operation are commercially available up to over 100MHz and suitable circuits are supplied by the makers. The cost of such crystals may seem high when compared with surplus fundamental-mode types but, unless the frequency of operation is very high or the specified tolerance very close (which is unnecessary for most amateur purposes), careful consideration should be given to the purchase of an overtone crystal. It should be borne in mind that such a crystal will give a greater certainty of trouble-free operation and, in the case of one of the high frequency crystals oscillating at 60MHz or more, the greater cost may well be offset in the saving in valves or transistors and other components, and the smaller space required for the converter.

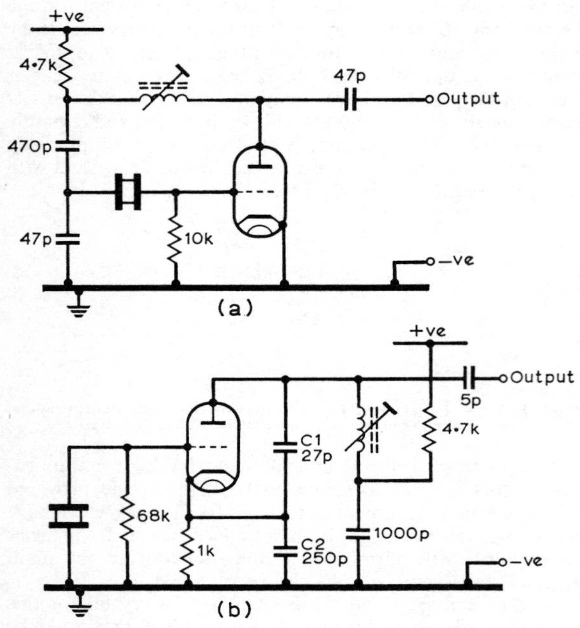

Fig 5.19. (a) The Robert Dollar overtone crystal oscillator circuit. (b) Anode-cathode feedback overtone crystal oscillator circuit.

In the example under consideration an ideal crystal frequency would be 59MHz, necessitating only a doubler stage to arrive at the required injection frequency of 118MHz.

For those who prefer to use the cheaper crystals other choices must be made with the aid of Fig 5.18. It would be undesirable to start at 13·111MHz since the second harmonic would fall at 26·222MHz. This is within the tuning range of the main receiver and would result in a spurious carrier or "birdie". A frequency of 29·5MHz appears to be a possibility and a 5·9MHz crystal, provided it would operate on its fifth overtone, would satisfy the requirements.

Alternatively a start could be made at 19·65MHz, obtained from a 6·55MHz crystal operating on its third overtone, the output frequency being doubled and trebled to produce the required 118MHz.

Double triode valves are used sometimes as oscillators in vhf converters, and among those suitable are the 12AT7, 6BQ7A and ECC88. The triode section of the 6U8/ECF82 or the ECF804 is also suitable for use as an overtone oscillator.

The 6BQ7A and ECC88 have an internal screen between the two anodes and are therefore particularly suitable for use when one section is to be employed as a mixer.

Oscillator circuits for converters, whether for fundamental-frequency or overtone crystals, are just the same as those employed in crystal-controlled transmitters, and reference should be made to the appropriate chapters for further information. For ease of reference, however, a selection of circuits is given here and many more will undoubtedly be found from time to time in the literature.

Crystal oscillator circuits may operate the crystal either in the parallel (high-impedance) mode or in a series-resonant circuit. In the latter the crystal forms part of the feedback loop so that if it is replaced by a capacitor oscillation will take place at a frequency determined by the inductance in the circuit and its associated capacitances. This can, in practice, happen even with the crystal in position as a result of the capacitance between the crystal electrodes, and spurious (uncontrolled) oscillations may be generated unless the circuit is properly adjusted.

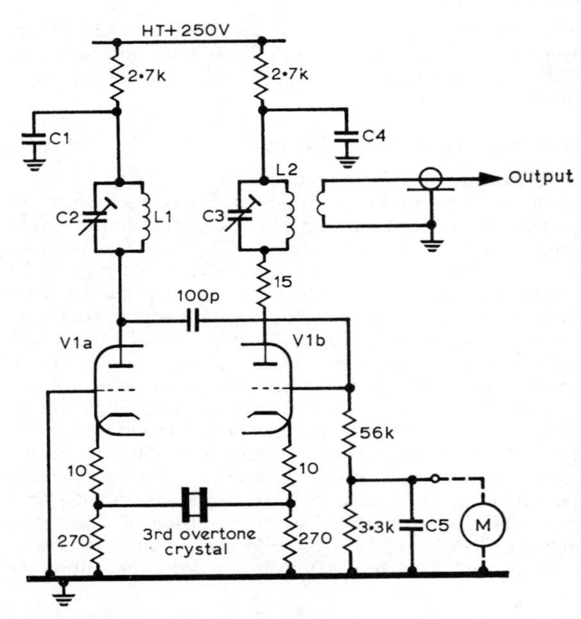

Fig 5.20. Butler overtone crystal oscillator circuit. Output may be taken at two, three or four times the frequency of the overtone crystal. L1, C2 and L2, C3 tune to the overtone and desired output frequencies respectively. C1, C4 and C5 are bypass capacitors of normal values.

Generally speaking it is much more preferable to employ crystals specifically intended for overtone operation: some essentially fundamental-mode crystals will be found perfectly satisfactory while others of similar type refuse to function or require an excessive amount of feedback, with the attendant possibility of oscillation uncontrolled by the crystal.

The simplest of the valve overtone circuits is that where the crystal is connected between grid and cathode and the anode circuit is tuned to the desired overtone frequency. Should some degree of feedback be required it may be provided as depicted in **Fig 5.19.** Both circuits rely on feedback due to incomplete bypassing of the "cold" end of the anode tuned circuit: in (a), which is the Robert Dollar circuit, feedback is applied via the crystal which operates in the series mode, while in (b) parallel mode operation is employed with feedback to the cathode across an only-partially bypassed cathode resistor. Particularly in the latter circuit, oscillation at the fundamental frequency of the crystal will occur when the anode circuit is not tuned to overtone resonance.

The values of feedback capacitors shown in Fig 5.19(b) are suitable for valves with a modest mutual conductance, but for triodes of higher g_m,C1 may have to be in the range 5 to 15pF and C2 80 to 100pF. Should oscillation at the fundamental frequency of the crystal still persist when the anode circuit is correctly tuned, a reduction in the value of the grid resistor to 20 or 30kΩ or even lower has been found effective in curing the trouble.

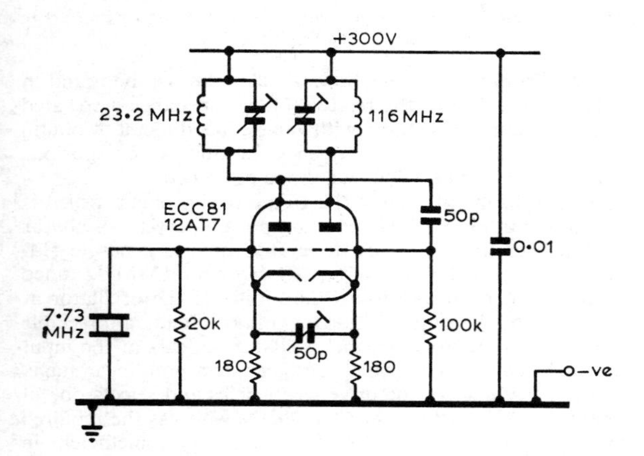

Fig 5.21. The PA0TP crystal oscillator/multiplier circuit.

The Butler circuit shown in **Fig 5.20** is particularly useful in receivers and may be employed either as an overtone or as a fundamental oscillator with provision for taking an output at two, three or four times the frequency of the crystal. Although two stages are required, the arrangement is very satisfactory when good stability is the main criterion. For use as a fundamental oscillator, L2 is omitted and the anode of V1b bypassed to earth by a suitable capacitor, the output then being taken from a link winding coupled to L1.

When setting up the oscillator a voltmeter may be connected (as shown with dotted lines in the diagram) and the tuned circuits adjusted for maximum deflection. For highest stability the coils should both have a low Q value, and if necessary damping resistors may be connected across them for this purpose.

Another twin-triode oscillator/multiplier circuit capable of a high order of frequency multiplication is due to PA0TP and appears in **Fig 5.21**. The crystal and tuned circuit frequencies shown would be suitable for a converter with a 28 to 30MHz i.f.

A very satisfactory transistor overtone oscillator, suitable only for use with overtone crystals, is that shown in **Fig 5.22**. Almost any transistor capable of oscillating at a frequency of 100MHz or higher could replace the 2N708 shown. For pnp types, such as the OC170 or 2N1742 etc, the polarity of the supply voltage would of course have to be reversed. The 3–30pF trimmer C1 is not essential but provides a ready means of controlling the amount of output. With other types of transistor it is worthwhile experimenting with other

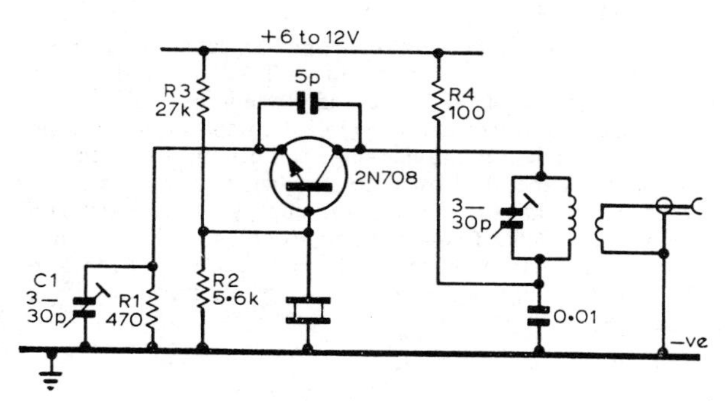

Fig 5.22. Transistor overtone oscillator circuit suitable for use with either third or fifth overtone crystals.

values of R1 and R3 for optimum results. There is little point in changing the value of R2.

FREQUENCY MULTIPLIERS

Unless the operating (overtone) frequency of a crystal is that required for injection into the mixer, some frequency multiplication will have to be arranged, and circuits to do this will be found among the designs for converters described later in this chapter.

While basically similar to the frequency-multiplying circuits in transmitters, the power output necessary for the satisfactory operation of a mixer is far smaller than would be required for transmitting applications, and in consequence greater multiplication factors per stage can be realized.

It is always good practice to start with a reasonably high frequency crystal requiring fewer stages of multiplication, rather than the reverse, because there is a distinct possibility that one of the unwanted multiples of the crystal frequency might reach the mixer with sufficient power to convert (perhaps inefficiently) a strong local signal outside the amateur band and cause its appearance within the range of the tunable i.f.

An example may make this clear. An overtone crystal is used to produce a frequency of 28MHz, and this is multiplied by five in a subsequent stage to provide injection for the mixer at 140MHz. A tunable i.f. of 4 to 6MHz then covers the 144 to 146MHz band. Although the output of the multiplier stage is tuned to five times its input frequency, there may be sufficient output at 84MHz (multiplication of three) to beat with a broadcast station in the 87·5 to 100MHz band to produce an i.f. within the range tuned. Radio 2 from Wrotham on 89·1MHz would appear at 89·1 – 84 = 5·1MHz. Thus a wideband fm station would appear at an apparent frequency of 145·1MHz, making reception of wanted signals near this frequency very difficult.

Other spurious responses may appear at other harmonics of the crystal frequency. If, for example, a crystal frequency of 35MHz had been chosen then the harmonics at 70 and 105MHz would give rise to a completely different set of possible spurious responses. The optimum choice of oscillator frequency will vary with geographical areas, and receivers which perform well in some areas may be found to be unsatisfactory in others. Designers of equipment for portable operation should give particular attention to this point.

High-Q Break

In addition to possible unwanted injection frequencies, all local oscillators generate noise. The magnitude of this noise depends on the operating conditions of the last frequency multiplier and the Q of the final multiplier tuned circuit. This final multiplier stage can be considered as an rf amplifier at the local oscillator frequency, and if its output tuned circuit has too low a Q, the stage will have appreciable gain at the signal frequency and so add to the noise input to the mixer. If, as is so often the case with uhf receivers, there is little or no rf gain in front of the mixer, oscillator noise can be quite serious.

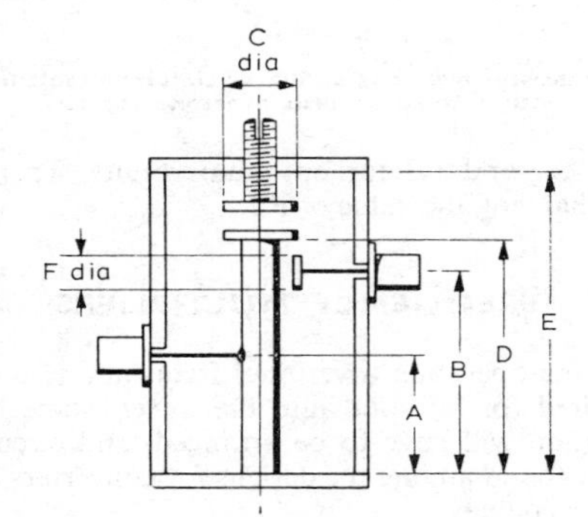

Fig 5.23. Essential dimensions for high Q breaks.

	23cm	30cm	70cm
Inside diameter of outer conductor*	43/64in	43/64in	43/64in
Outside diameter of inner conductor	$\frac{1}{4}$in	$\frac{1}{4}$in	$\frac{1}{4}$in
Dimension A (end to tap on inner)	$\frac{5}{16}$in	$\frac{5}{16}$in	$\frac{3}{8}$in
Dimension B (end to coupling probe)	$1\frac{1}{4}$in	$1\frac{3}{4}$in	$5\frac{3}{4}$in
Dimension C (diam of trimmer)	$\frac{3}{8}$in	$\frac{3}{8}$in	$\frac{3}{8}$in
Dimension D (length of inner conductor)	$1\frac{3}{8}$in	2in	6in
Dimension E (length of outer conductor)	$1\frac{3}{4}$in	$2\frac{3}{8}$in	$6\frac{1}{2}$in
Dimension F (diam of probe)	$\frac{1}{4}$in	$\frac{1}{2}$in	$\frac{1}{2}$in

*$\frac{3}{4}$in od tube with 20swg wall.

Oscillator noise and unwanted injection frequencies can both be reduced to insignificance by interposing a filter between the local oscillator and the mixer. This filter may consist of a single high-Q tuned circuit called a *high-Q break* or two loosely coupled tuned circuits.

Examples of high-Q breaks for 420, 1,000 and 1,296MHz appear in **Fig 5.23**. The capacitive probe connects to the source of rf (ie the output of the last frequency multiplier), and the coupling loop supplies filtered rf via a coaxial cable to the mixer. Such filters have a typical attenuation of 15dB at a frequency 3·5MHz away from the operating frequency.

Choice of Multiplier Valves or Transistors

Valves suitable for frequency multiplication are double triodes, such as the 12AT7, small rf pentodes of the 6AK5 or EF91 (6AM6) class, and triode pentodes designed for vhf frequency-changer service, such as the ECF80 (6BL8), ECF82 (6U8) and ECF804. High-slope valves designed for wideband amplifiers such as the E88CC or E180F are particularly good because of their low drive requirements.

Since the factor that limits the frequency of operation of a valve is usually transit time in the grid/cathode space, many valves can be used as frequency multipliers at higher frequencies than is possible if they are used as amplifiers.

Most vhf transistors make efficient frequency multipliers, and they may be operated in zero-bias Class B or with bias derived from the rf drive. Several examples will be found in this chapter. Due to the relatively low impedance associated with transistors compared with valves, it is difficult to obtain high-Q tuned circuits in frequency multipliers and spurious outputs of the type discussed above often occur.

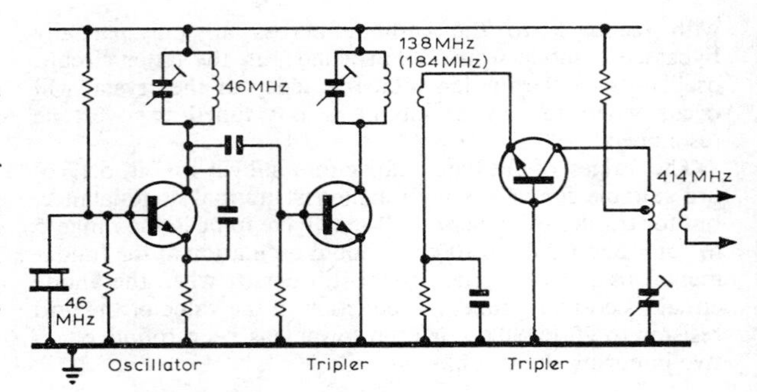

Fig 5.24. Typical local oscillator multiplier chain showing an error in alignment.

This difficulty also gives rise to another problem which is illustrated in **Fig 5.24**. Here the output of a crystal oscillator at 46MHz is multiplied by three and three to generate 414-MHz for a 432MHz converter. By error the 138MHz tuned circuit is tuned to the fourth harmonic of the oscillator at 184MHz, but because of the low Q of this circuit, appreciable fundamental power from the oscillator appears at the input to the last stage. The tripler, designed as a non-linear stage, mixes harmonics of the drive frequencies and selects one of these, eg $(184 \times 2) + 46 = 414$MHz which is the required output. The final multiplier, however, is very inefficient in this mode of operation, and requires an inordinate amount of rf drive. Much time can be wasted adjusting this apparently inefficient stage when the fault lies in an earlier one.

The cure for these and similar problems is to check all tuned circuits with a grid-dip oscillator to verify that they are all operating at the required frequencies.

Diode Multipliers

The last frequency multiplier of a uhf local oscillator often presents difficulties, and the straightforward solution is to use a valve or transistor similar to that used in the rf stage. If, however, such a device cannot be used in the rf stage for economic reasons this approach is not helpful in the case of the local oscillator. An inexpensive but relatively inefficient multiplier can be constructed using a point-contact or Schottky-barrier diode as a simple non-linear element. Approximately 1mW is required to produce the required current in a mixer diode; to generate this in a diode multiplier some 10 to 50mW of input will be required, depending on the order of multiplication required.

Quite thorough filtering is required after a diode multiplier since the harmonic output is only a small fraction of the fundamental, and a high-Q break should be included in the design as a matter of course. The current through the diode is typically 20 to 50mA, and a low-impedance diode is required with the minimum of circuit resistance (some workers have reported that the insertion of a meter to measure the diode current has significantly reduced the harmonic output). Satisfactory results have been obtained with 1N64, GEX66 and 1N914 diodes, with a preference for the latter.

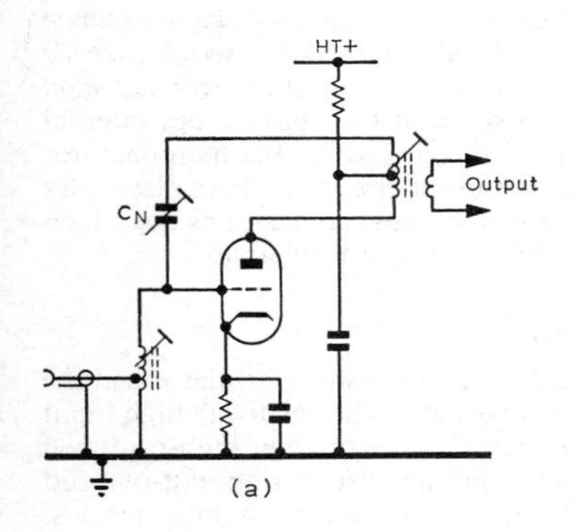

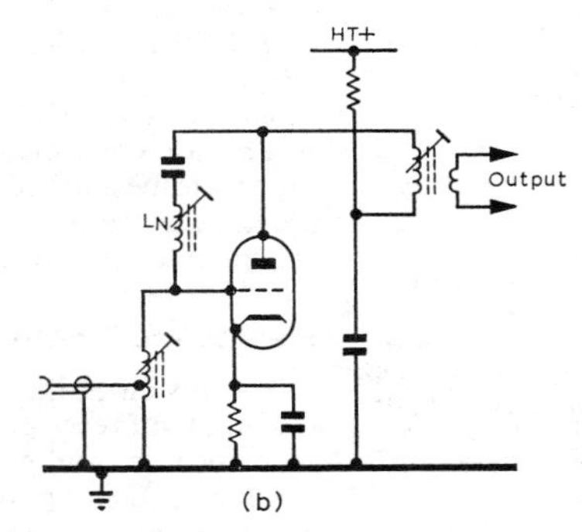

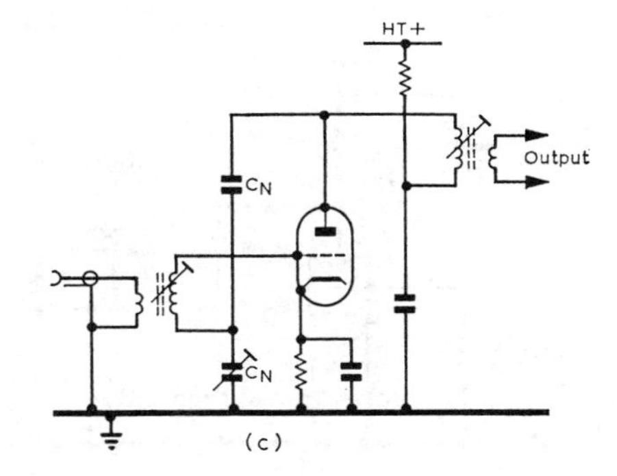

Fig 5.25. Neutralized triode rf stages.

RF AMPLIFIERS

Most devices (valves or transistors) used as vhf amplifiers can be considered to have three signal connections and can be operated with any one of these common to input and output circuits. Fig 3.50 shows transistor arrangements. Cathode and emitter followers are not normally used at vhf as they are very prone to self-oscillation but any of the other configurations will be encountered. Taken in isolation none of these arrangements has overriding qualities but when considered in context each has special merits which may be disadvantages in some other application.

Although a large number of circuits for rf stages have been described from time to time, many of them are only slight modifications of the better-known arrangements. So far as the newcomer to vhf work is concerned, the simpler the circuit the greater the chance of success. Whichever arrangement is chosen, it is strongly recommended that a copper or brass chassis be used to ensure the best possible connection for earth returns. This gives a consequent reduction in circulating currents in the chassis and therefore better stability and usable gain.

Common-Cathode Amplifiers

Fig 5.25 shows the arrangement of typical common-cathode amplifiers. Neutralization is necessary to reduce regeneration via the grid-anode capacitance and this can be produced in one of several ways. In Fig 5.25(a) the anode inductor is decoupled at a tap and Cn is taken back to the grid. In Fig 5.25(b) the feedback capacitance is tuned to a high-impedance parallel circuit by Ln. In Fig 5.25(c) the two neutralizing capacitors form a bridge circuit, the other two arms of which are the grid-anode and grid-cathode capacitances of the valve. This circuit is more convenient to adjust than that of Fig 5.25(a) since one terminal of Cn is at chassis potential, thus eliminating hand capacitance effects.

Examples of converters using rf stages with and without neutralization are given later in this chapter, including modern bipolar and field effect transistor designs.

The Common-Grid Amplifier

In the common- or grounded-grid configuration it is the grid, rather than the cathode, which is maintained at zero rf potential, thus providing a screen between the cathode and anode. The latter form the input and output electrodes respectively.

The input impedance of a common-grid triode is equal to $1/g_m$ where g_m is the mutual conductance of the valve in *amps* per volt, so for a g_m in the range 6–15mA/V this would imply input impedances between 167 and 67Ω. Due to this low input impedance the common-grid amplifier is by nature a wideband device and is well suited as an rf stage for the 432MHz band. In populated areas, however, this wide bandwidth can be a mixed blessing since stations such as uhf broadcast stations outside the amateur band may be amplified and cause non-linearity in later stages.

Cascode Amplifiers

In the circuit shown in **Fig 5.26** V1 is a common-cathode amplifier having an anode load which is the input resistance of V2, the latter being a common-grid amplifier. Since the gain of a common-cathode amplifier is $g_m R_L$ and in this case R_L is $1/g_m$, V1 has unit voltage gain between its grid and anode. V1 provides gain by virtue of the impedance step-up in the input transformer.

Fig 5.27 shows two practical designs. Fig 5.27(a) is used for valves requiring 200V or more on the anode, while Fig 5.27(b) is suitable for valves requiring about half that voltage. As can be seen, the arrangement of Fig 5.27(b) uses fewer rf components and the simpler design is more likely to give stable operation. Lp tunes the capacitance at the anode of the first stage and the (often considerable) cathode-to-heater capacitance of the second stage. Since it is heavily damped by the input resistance of the second stage, Lp appears to be not critical if adjusted for maximum gain. It has a marked effect on noise figure, however, and should be set up with the aid of a noise generator. Similarly, since the first stage has only unity gain the neutralizing has no apparent effect

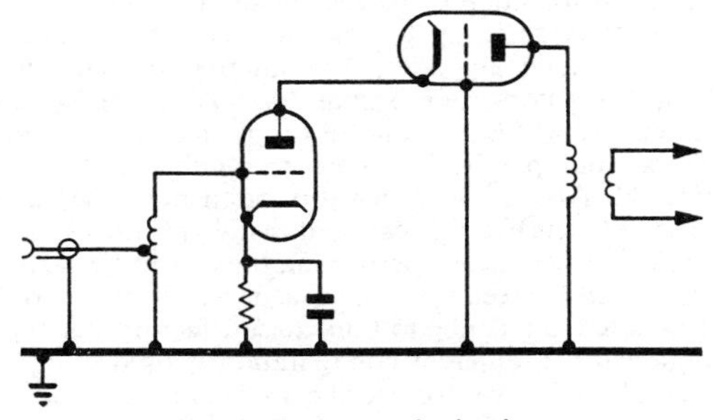

Fig 5.26. Basic cascode circuit.

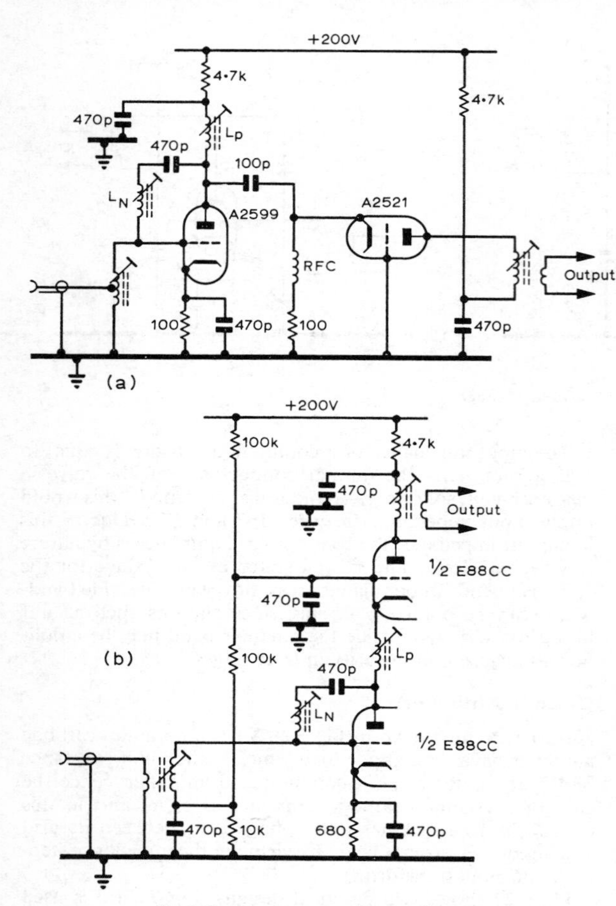

Fig 5.27. Practical cascode amplifiers.

on signal strength. This adjustment is also important for best noise performance.

Valves and Transistors for VHF Operation

Because the internal capacitance and inductance of devices used as vhf and uhf amplifiers may be a significant part of the total circuit values, the device manufacturer may make special provision for certain kinds of circuit configuration. Thus a valve designed for common-grid operation may have several grid connections to reduce the inductance to that electrode, while another ostensibly similar valve may have multiple cathode leads to improve its performance as a common-cathode amplifier. The internal parasitic impedances in transistors are too small to make this specialization worthwhile at vhf but certain uhf types may have wide tape leads or multiple leads where the inductance must be minimized. AF and hf transistors often have the collector connected to the external can, but many vhf transistors have separate collector leads, permitting the can to be earthed to provide better screening. Where a device is designed for a particular circuit configuration, unsatisfactory results may be obtained if a different configuration is used.

Examples of valves for specific applications are: *push-pull common cathode*, 6J6 (now obsolescent); *single common cathode*, 6CW4, 6AJ4, PCC89, A2599 (6CT4); *single common grid*, EC91, 6AM4, PCC89, A2521 (6CR4); *series cascode* 6BQ7, E88CC, PCC89, *single common grid with common cathode mixer*, ECC85. In some of the above types internal screens are connected to electrodes which the manufacturer expects to be earthed in the usual circuit. If these electrodes are not earthed but are used for signal connections the screen will radiate, and instability is almost inevitable.

Field Effect Transistors

The advent of the field effect transistor (fet) has eased the design of vhf receivers in two ways. The relatively high input resistance at the gate permits reasonably high-Q tuned circuits, providing protection against strong out-of-band signals such as from broadcast or vehicle mobile stations. Also the drain current is quite exactly proportional to the square of the gate voltage; this form of non-linearity gives rise to harmonics (and the fet is a very efficient frequency doubler) but a very low level of intermodulation (cross modulation).

The use of a square-law rf stage is not, however, as straightforward as it first appears. Second order products are still present, such as the sum of two strong signals. For example, an rf stage operating in a television Channel 2 or 3 area with insufficient aerial input filtering would generate signals on the 144MHz band by mixing signals in Broadcast Band II or the adjacent mobile band with the local television signal. It follows, of course, that if two fet amplifiers are operated in cascade a bandpass filter is required between them to ensure that the distortion products generated in the first stage are not passed to the next stage where they will be re-mixed with the wanted signal. A development of the fet is the metal-oxide-semiconductor fet (mosfet), in which the gate is insulated by a very thin layer of silica. The gate therefore draws no current and a high input resistance is possible, limited only by the losses in the gate capacitance. These devices may be damaged by static charges, and must be protected against aerial pickup during electrical storms, and also from rf from the transmitter feeding through an aerial changeover relay with an excessively high contact capacitance. Recent FETS have protective diodes incorporated in the device which limit the input voltage to a safe level, and these devices are thus much more rugged.

In the dual-gate mosfet the drain current is controlled by two gates and various useful circuit improvements result. If it is desired to apply gain control to a conventional transistor stage the control voltage is applied to the same electrode as the signal, but the result can be a reduction in power-handling capability showing up as a cross modulation. The dual-gate fet avoids this problem, and automatic or manual gain control can be applied to gate 2 without reducing the signal-handling capability at gate 1. **Fig 5.28** shows how such an rf stage for a converter is arranged. When a strong local station causes intermodulation at the mixer or an early stage of the main receiver, the rf gain can be reduced until interference-free reception is again possible.

Pre-amplifiers

As available devices improve and new circuit designs are published, it will become apparent that a receiver which may have been considered a first-rate design when built is no longer as good as may be desired. Specifically, a receiver using valves designed for vhf tv broadcast use (or early types of transistor) is not as sensitive as is now readily achievable, although the local oscillator may be quite satisfactory. The sensitivity of such a receiver can be improved without radical redesign by means of an additional separate rf amplifier, usually referred to as a *pre-amplifier*. Such an amplifier should have the lowest possible noise figure and just sufficient gain to ensure that the overall performance is satisfactory. **Fig 5.29** shows the improvement to be expected from a pre-amplifier, given its gain and the noise factor of the pre-amplifier alone and the main receiver.

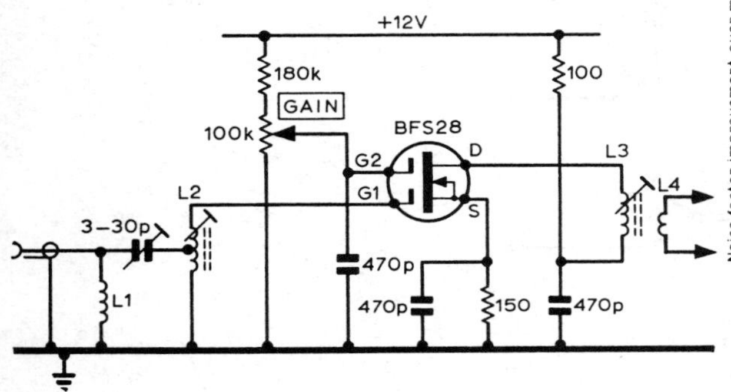

Fig 5.28. Dual-gate mosfet rf amplifier with gain control. L1 3 turns 20swg, 0·25in inside diameter 0·3in long; L2 5 turns 20swg tinned copper wire on 0·3in former 0·4in long tapped at 1½ turns and tuned with dust core; L3 6 turns 20swg on 0·3in former tuned with dust core and coupled L4; L4; 2 turns 20swg on same former closed spaced to capacitor end of L3.

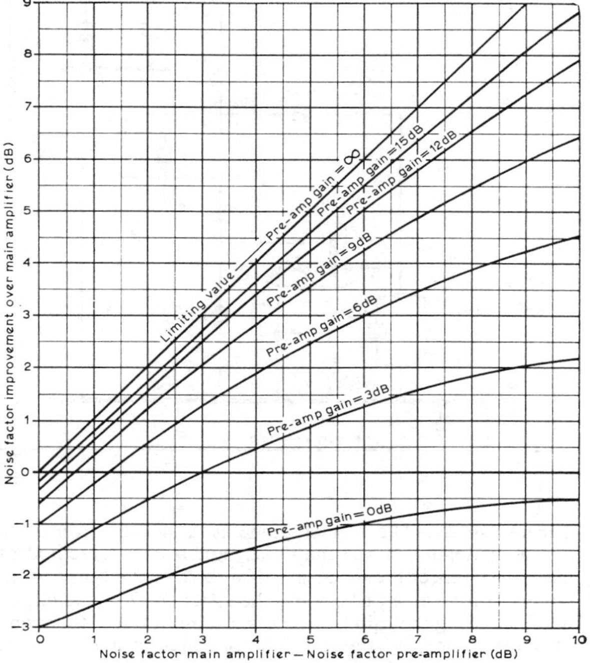

Fig 5.29. Receiver noise figures.

An example will suffice to show the application. An existing 145MHz receiver has a measured noise figure of 6dB and is connected to its aerial via a feeder with 3dB loss. It is desired to fit a pre-amplifier at the mast head; what is the performance required of the pre-amplifier?

Suppose a BF180 transistor is available; this has a specified maximum noise figure of 2·5dB at 200MHz and will be slightly better than this at 145MHz. The main receiver and feeder can be treated as having an overall noise figure of $3 + 6 = 9$dB, and from the graph if the pre-amplifier has a gain of 10dB the overall noise figure will be better than 4·1dB. Increasing the gain of the pre-amplifier to 15dB will only reduce the overall noise figure to 3·6dB, and may lead to difficulty due to the effect of varying temperatures on critical adjustments. The addition of so much gain in front of an existing receiver is also very likely to give rise to intermodulation from strong local signals. If it is desired to operate under such conditions it is essential that provision is made for disconnecting the pre-amplifier when the local station is on.

Fig 5.30 shows a masthead pre-amplifier suitable for 144MHz with the controls at the main station. Two feeders are required—the original feeder is used for the transmitter and a second feeder is used for the receiver (note that this feeder may have a slightly higher loss since this may be overcome by the pre-amplifier). RLA is fitted in the masthead unit and is energized *on receive*, and R1 and the zener diode reduce the relay supply to 12V for the amplifier. Note that even when the relay requires 12V this zener diode is still required to absorb the transient back-emf from the relay coil. RLB is the original station transmit-receive relay and S1 enables the main receiver to be quickly connected to this in an emergency.

Other RF Amplifiers

Many sophisticated amplifying devices have been developed for uhf at various times, but they are becoming of less interest to amateurs as transistors are developed with adequate performance. Two of the more important of these are included here for completeness, but further information should be sought in more specialized works.

Parametric Amplifiers. As the name suggests, these operate by virtue of a varying circuit parameter, usually capacitance, which is made to store and release signal energy without adding noise.

The most valuable parametric device is the variable-capacitance diode or varactor. This type of diode has a junction capacitance which can be varied in a controlled manner by varying the voltage across the diode. In use the diode is connected across a tuned circuit at the input to the receiver and its capacitance varied with a "pump" oscillator. This is an oscillator at a frequency much higher than the operating frequency and, to ensure that the varactor capacitance is varied in correct phase with respect to the signal being amplified, a third tuned circuit, "the idler", is also connected across the diode. This is usually adjusted to the difference between the pump and signal frequencies, and its presence makes the pump frequency non-critical. The pump frequency should be high for the lowest noise performance: 3GHz is suitable for a 432MHz amplifier but 10GHz is necessary for a 1·3GHz amplifier.

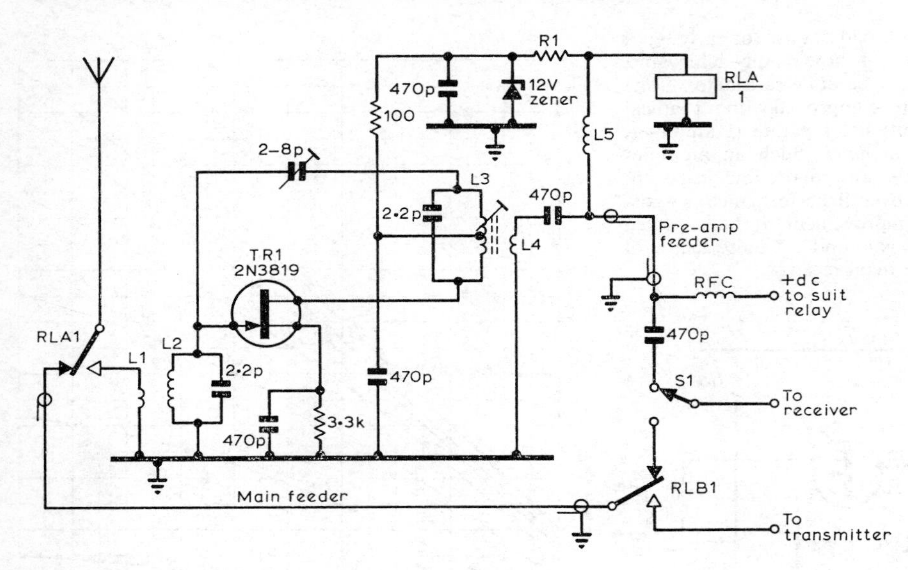

Fig 5.30. Masthead pre-amplifier for 144MHz. L1 2 turns 22swg wound within cold end of L2; L2 4 turns 22swg on ⅜in mandrel about 0·4in long (adjust to tune); L3 5 turns 22swg on 0·3in former slug tuned and centre tapped; L4 2 turns 22swg wound over centre of L3.

Tunnel Diodes. The tunnel diode is a negative-resistance device analogous to a Q multiplier, and a typical method of use is to connect the diode across a part of the tuned input circuit of a receiver, so raising the effective Q of the circuit and increasing the gain from the aerial input to the following rf stage. Since the tunnel diode operates at a current of approximately 1mA and its major noise contribution is due to shot noise, low-noise amplification results.

In practice it is desirable to incorporate some form of dc stabilization of the operating point. This is because negative resistance devices are very prone to instability unless such a precaution is taken. In any case it is essential to avoid any high-Q parasitic resonances due to chokes, inductive leads on capacitors etc, since oscillation at unexpected frequencies will certainly result. One consequence of the very low operating power of tunnel diode amplifiers is that they are very liable to overload, and except in the most remote areas it is advisable to make provision for disconnecting the amplifier when strong signals are liable to be received.

MIXERS

The common forms of mixers use diodes, triodes and most types of transistors.

Diode Mixers

A diode operates non-linearly either around the bottom bend, where the current through the diode is proportional to the square of the applied voltage (see **Fig 5.31(a)**), or by the switching action between forward conduction and reverse cut-off, as shown in **Fig 5.31(b)**. In the first case it is often necessary to forward bias the diode with dc to obtain an optimum working point. A bias of 100 to 200mV is typical but will vary with the type of diode. The maker's data sheets should be consulted for the optimum working conditions.

The second type of mixer is used where a high overload level is required. Signals approaching one tenth of the local oscillator power can be handled without distortion, and the local oscillator level is limited only by the power-handling

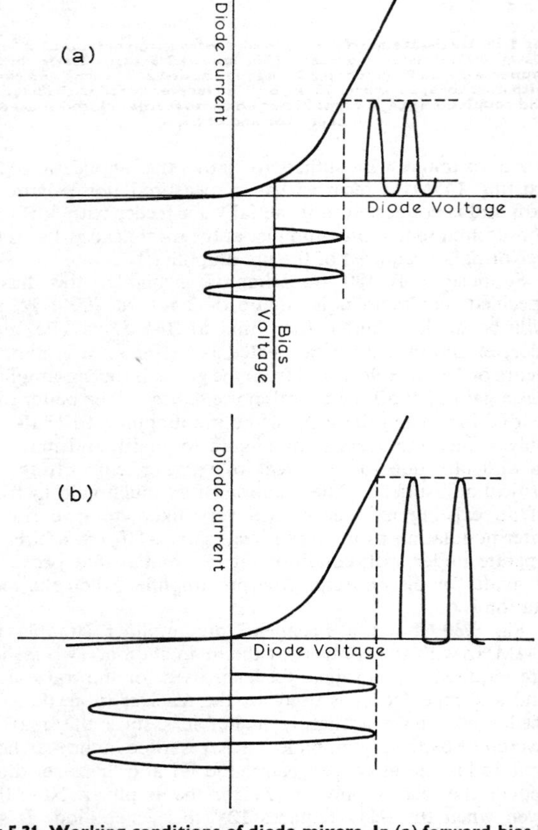

Fig 5.31. Working conditions of diode mixers. In (a) forward bias is required to provide an optimum working point. In (b) the local oscillator power is higher and no bias is required.

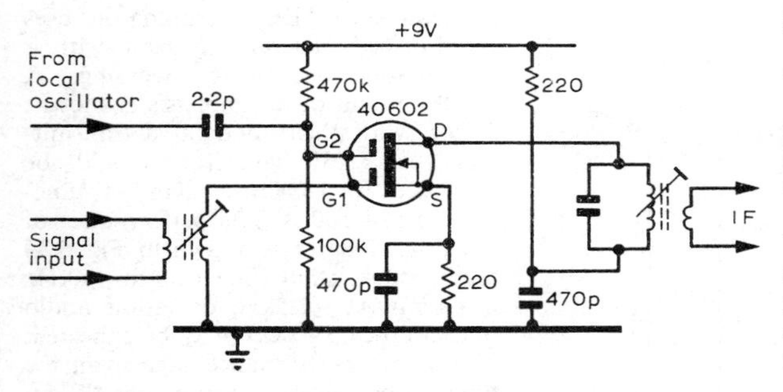

Fig 5.32. Dual gate mosfet mixer.

capacity of the diodes. The noise generated in the mixer rises with increasing diode current, however, and this sets a limit to the usable overload level if maximum sensitivity is required, although it is possible to adjust the local oscillator power from time to time to select a compromise between sensitivity and overload capacity.

The diode mixer is necessarily "lossy", and a loss of between 3 and 6dB may be obtained in practice. It is therefore essential that the stage following the mixer has the lowest possible noise figure.

Balanced Mixers

Noise components from the local oscillator can be reduced to a low level by means of balanced mixers; these are similar to the balanced modulators used in ssb transmitters, differing only in the use of low-noise receiver diodes. With modern diodes it is not usually necessary to provide adjustment of balance at the local signal for best noise performance, but some adjustment is desirable to reduce local-oscillator radiation from the aerial. In many cases it will be necessary to include a simple filter in the aerial lead to reduce this radiation; this can conveniently be the same filter as is used to reduce the second-channel response. Note that the balanced mixer does not provide any improvement over single-ended mixers in respect of noise originating in the diodes due to a high local oscillator level.

Diodes suitable for receiver mixers are all semiconductor types, and they include the well-known metal whisker-semiconductor junction and the more modern alloyed metal-semiconductor or Schottky-barrier junction. The latter are the only types used in mixers designed specifically to have a high overload capability. The older point-contact diodes are very noisy when the local oscillator power is such as to cause reverse breakdown, and this occurs at quite a low level in this type of diode.

Triode Mixers

A triode valve can be used as a mixer in two essentially different ways. The non-linear characteristic necessary for mixing results from biasing the valve to operate either on the grid-current bend (leaky grid) or the anode bend. In the anode-bend method the application of the local oscillator voltage will drive the anode current upwards; in the leaky grid method it will drive the anode current down. Provided that the recommended oscillator injection voltage applied is correct, the resultant anode current will be the same using either method and the conversion gain and noise factor will also be the same. For anode-bend operation, valves having indirectly heated cathodes can be provided with automatic bias. Those uhf valves with a common heater and cathode connection will, however, require a separate heater supply.

Triode mixers may be operated in the common-cathode or common-grid configuration but the leaky grid circuit is not often used with a common-grid stage due to the difficulty of providing a satisfactory bypass connection for the grid.

It is essential that the mixer provides no gain at the i.f. between the input electrode and the output since this will make the prevention of i.f. breakthrough very difficult. The input electrode should always have a low impedance to earth at the i.f. A vhf tuning coil is best for this but a fairly low inductance choke will sometimes serve.

Transistor Mixers

The above remarks on diode and triode mixers can be applied to transistor mixers. The emitter-base junction of the transistor can be used as a diode mixer with the amplified i.f. taken from the collector; or the non-linearities of the transfer characteristic can be used by biasing the device for a low value of oscillator current and applying local oscillator power to the base-emitter junction so as to increase the collector current. Often in practice both mechanisms contribute to mixing action.

Multiplicative Mixers

Fig 5.32 shows the use of a dual-gate fet as a mixer. The signal input is applied to gate 1 as for an rf amplifier, but instead of just a decoupled bias supply to gate 2, the decoupling is omitted and the oscillator voltage is applied to this gate. The output at the drain is then controlled by the two inputs and contains the usual intermodulation products between them. This form of mixer is important since the signal-handling capability of gate 1 is not reduced by the presence of local oscillator signal and quite a high overload level results. Next to the diode mixer the dual-gate mixer has the highest overload level of any semiconductor mixer. (None of these devices approaches the signal-handling capability of a well-designed valve mixer.)

The dual-gate fet mixer is very easy to set up since the various bias oscillator and signal level adjustments can be separate and the tuned circuits do not interact.

FREQUENCY MODULATION

The basic principles of frequency modulation are discussed in Chapter 9. It has a number of advantages over amplitude modulation for contacts over moderate distances and is now very popular for vhf mobile operation. The action of the "capture effect" greatly reduces the effect of weak interference since the annoying heterodyne typical of a.m. interference is often not heard. The principal disadvantage is the existence of a "threshold" limiting the sensitivity of a typical nbfm receiver to about 13dB above that of the corresponding a.m. receiver. For local contacts, however, this is an advantage since the range is well defined and the frequency can be re-used by another station outside this range without risk of mutual interference.

FM can be received after a fashion by detuning an a.m. receiver and using the skirt selectivity (or lack of it) to produce a slope detector. This gives quite poor results and should be considered only as a temporary measure. The

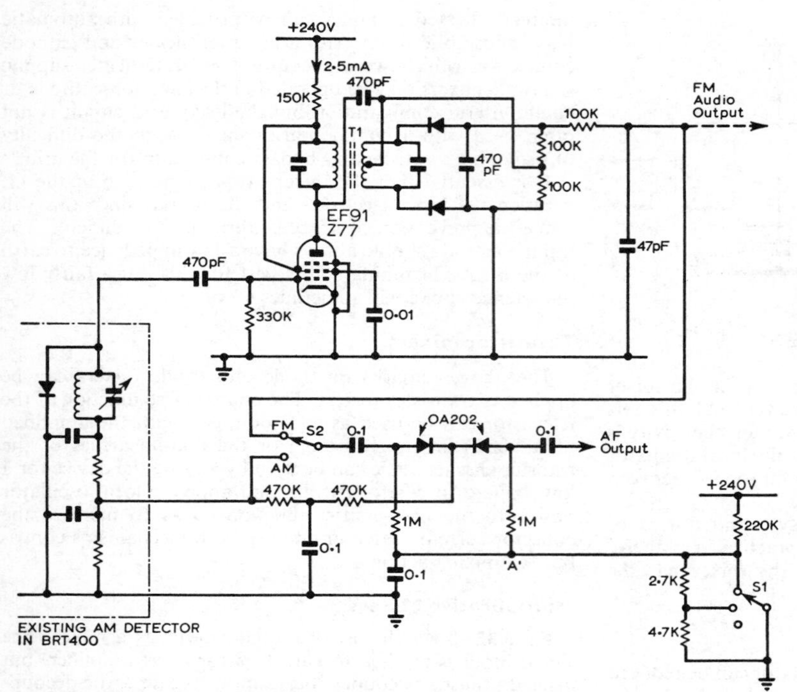

Fig 5.33. Circuit diagram. I.F. transformer is 455kHz, centre tapped.

following designs are intended as add-on units to existing a.m. receivers to provide something more approaching optimum performance.

Fig 5.33 shows a unit intended to be fitted to a valve receiver; it can be mounted on a bracket inside the receiver cabinet near the existing a.m. detector and takes power from the receiver supplies. The limiter is fed from the same point as the a.m. detector, the final i.f. transformer being retrimmed if necessary. The discriminator uses the Foster-Seeley arrangement with a centre-tapped secondary winding. A small amount of de-emphasis is shown; for use with phase-modulated transmitters the 47pF capacitor should be increased to 1,000pF, giving a time-constant of 100μs. The performance of the discriminator is shown in **Fig 5.34** and is substantially linear up to $\pm$4kHz deviation. At $\pm$2·5kHz deviation, audio output up to 5V peak may be expected. The limiter performance is given in the table. Since even "narrow band" fm requires a bandwidth of at least 11kHz for distortion-free reception, the bandwidth of the receiver will usually be set at its maximum when receiving fm; receivers with permanently connected crystal filters are not suitable for this use.

The noise limiter is not an essential part of the discriminator but is of interest because of the technique of using the dc output of the a.m. detector to set the limiter clipping level on fm. If point "A" were earthed, the diodes would be cut off until the negative bias reached about 0·5V and would severely clip speech. Clipping is avoided by applying several volts of additional bias derived from the ht rail; the preset potentiometer is adjusted so that there is no significant distortion on weak signals.

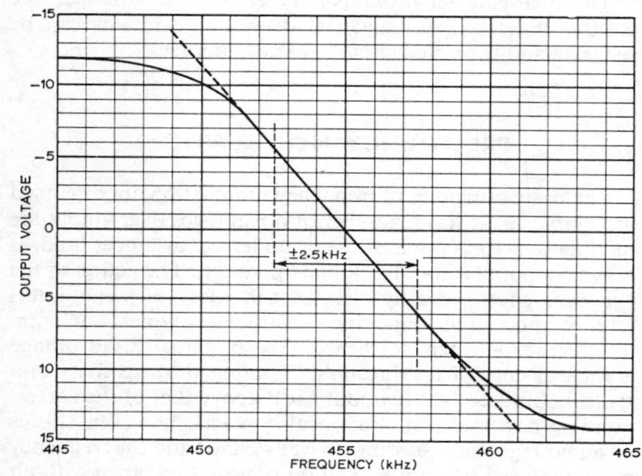

Fig 5.34. Curve taken with 4V peak at the limiter grid.

At reduced signal of	Output at 5kHz deviation	At reduced signal of	Output at 5kHz deviation
4V	10·1V	0·2V	5·7V
1V	9·8V	0·1V	3·0V
0·5V	8·9V	0·05V	1·5V

A Pulse Counting Discriminator Unit

The advantages of this type of discriminator over other designs are the absence of any coils and the possibility of using the device as a deviation measuring instrument in conjunction with an ac valve voltmeter. An example of this type of discriminator designed by G3JGO is shown in **Fig 5.35.**

The i.f. signal at 450 to 470kHz is applied to the longtailed-pair limiter TR1 and TR2. The variable resistor adjusts the limiting threshold and is set for best a.m. rejection. The limited output is used to switch TR3 on and off, giving a square wave of nearly 12V amplitude, which provides a constant quantity of charge via C into the emitter of TR4. Hence the emitter current of TR4 is directly proportional to frequency, and in this way variations in frequency produce a corresponding voltage variation across R. As shown, the circuit is set for about 500kHz $\equiv$ 10V, ie 20mV/kHz, so $\pm$2·5kHz deviation produces $\pm$50mV peak. This is amplified about 50 times by TR5 and TR6 to give an output suitable for feeding into the receiver af stages.

The circuit can be used at any i.f. up to 500kHz without adjustment, but for much lower I.F.S, eg 100kHz, C can with advantage be raised to increase the af output. Conversely, if C is reduced for higher I.F.S then the af output will fall, and a mixer converting to a lower i.f. may be preferable.

The voltage produced across R is given by:

$$V = vCRf$$

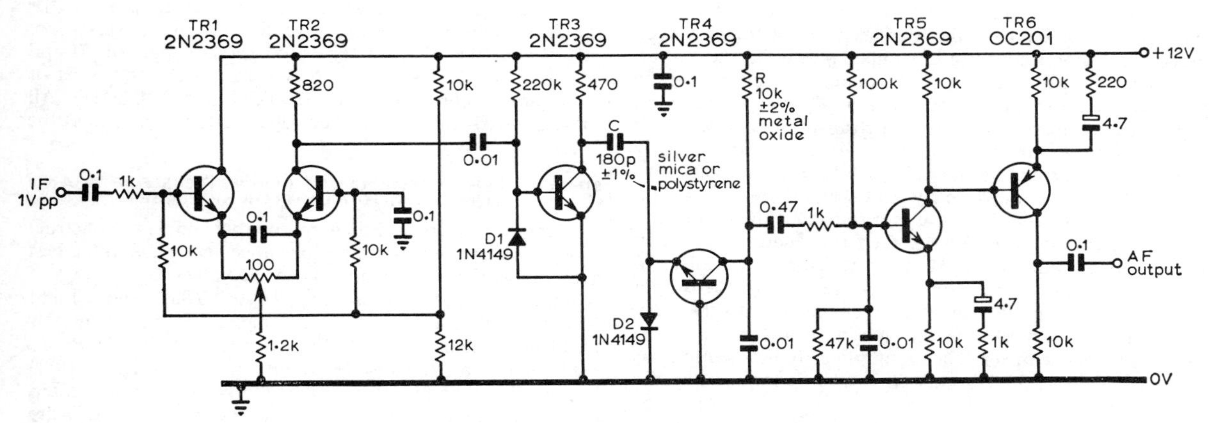

Fig 5.35. A pulse counting discriminator by G3JGO.

where v = output pulse amplitude of TR4 = 12V − 0·2V and f = the instantaneous intermediate frequency.

Hence a frequency deviation of $\pm f$ produces a peak-to-peak voltage of $2vCRf$.

As all these quantities are either easily measured or can be defined with accurate components, an ac peak millivoltmeter can be used to measure the deviation at TR4 output.

In practice, however, it may be more convenient to find the dc output across R when a crystal-controlled bfo is fed in and obtain its millivolts/kilohertz value this way. The ac gain of the amplifier TR5, TR6 should then be measured so that a less-sensitive ac voltmeter can be used at the output of TR1. The meter will then not be upset by the residual i.f. voltages.

Note that TR5 and TR6, as well as amplifying the signal, also serve as an af limiter since although the discriminator output is only 50mV peak of audio, noise voltages can produce outputs of 9·5V when a signal is near the i.f. limiter threshold. Although the circuit may work without TR5 and TR6, their omission will create the need for some other form of limiter.

Integrated Circuits

Both of the above fm detectors are unsuitable for use with selective receivers. It is necessary to increase the bandwidth and the most convenient method of doing this is to take off the input to the fm adaptor early in the receiver i.f. chain where the signal has passed through fewer tuned circuits. This results in a loss of gain which has to be made up in the fm adaptor; with the introduction of integrated circuits with multi-stage limiters this gain can be obtained in one package together with the components of the detector itself. **Fig 5.36** shows an ic limiter/discriminator using the Fairchild μA754C, originally designed for television i.f. systems, and includes an af amplifier which is not usually needed if there is already an af stage in the receiver.

The quadrature tuned circuit C1,L1 is tuned to the receiver i.f. and can be a spare bfo coil assembly or one half of a standard i.f. transformer. If the latter is used, the tuning capacitor should be removed from the unused winding to obviate the risk of absorption in this winding distorting the discriminator characteristic. The input impedance of the ic is 10kΩ which is a little low to connect across an existing

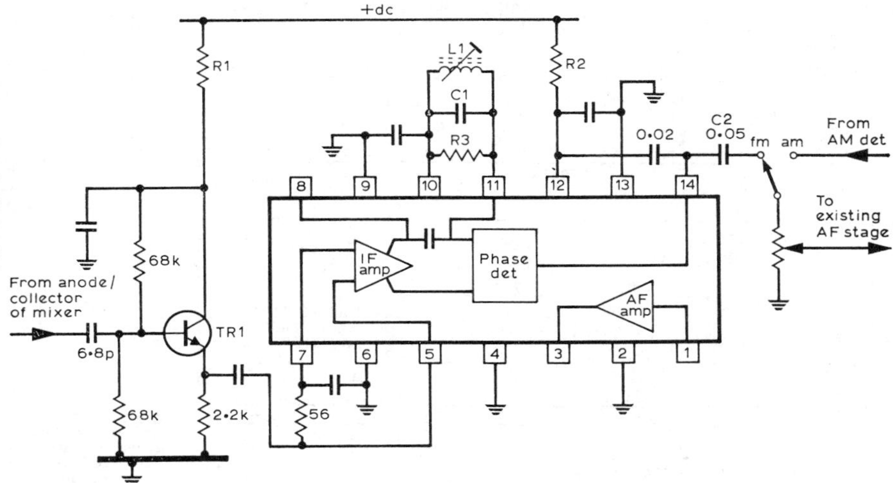

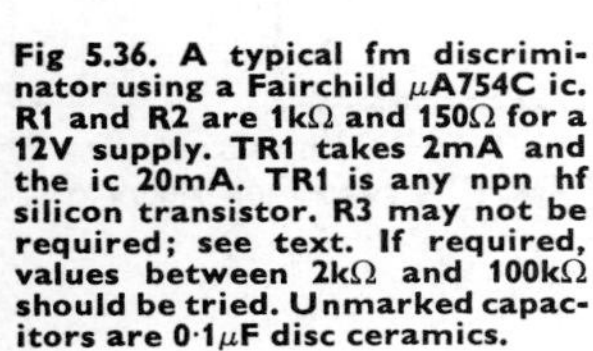
Fig 5.36. A typical fm discriminator using a Fairchild μA754C ic. R1 and R2 are 1kΩ and 150Ω for a 12V supply. TR1 takes 2mA and the ic 20mA. TR1 is any npn hf silicon transistor. R3 may not be required; see text. If required, values between 2kΩ and 100kΩ should be tried. Unmarked capacitors are 0·1μF disc ceramics.

high-impedance winding, and so an emitter follower is provided at the input. If the ic is used with a transistorized receiver and connected at a low-impedance point the emitter follower will not be required. The quadrature circuit can be tuned for minimum af output on a *weakly* modulated a.m. signal. (Since many amateur a.m. transmitters suffer from incidental pm at high modulation levels, the use of a heavily modulated transmission may give a false tuning position.)

If the tuning seems extremely critical and there is distortion on modulation peaks, and if it is certain that the bandwidth of the i.f. circuits is adequate, then the quadrature circuit may need an additional damping resistor in parallel. If, however, it is found that this merely reduces the af output without improving the tuning characteristics, then the main i.f. circuits are too selective and the input to the ic must be moved nearer the mixer. The i.f. amplifier/limiter section has a voltage gain of 10,000 and great care is required in the layout to prevent instability; several blank pins are provided in the package and these must be earthed to screen active pins. The 0·1μF ceramic disc capacitors must also be connected to earth at the pins shown in the diagram to prevent regeneration due to coupled earth loops.

TYPICAL VHF AND UHF DESIGNS

The best sensitivity at vhf and uhf is now to be obtained with designs based on semiconductors; however, there are still many individuals who are experienced in the techniques associated with valves and who find in practice that *they* get better results with these devices. Therefore, possibly for the last time, a range of valve designs of converter are included in this chapter. These comprise a pair of 70 and 144MHz converters using a Nuvistor rf stage and a set of three similar converters for 432, 1,296 and 2,304MHz. All these are well-tried designs developed towards the end of the "valve era".

SIMPLE HIGH PERFORMANCE CONVERTERS FOR 70 AND 144MHz USING NUVISTOR RF STAGES

These converters, although reasonably simple to construct even by a beginner, give a performance very close to the best available without using any expensive valves.

As many operators use the RCA AR88D and similar communication receivers as a tunable i.f., relatively low frequencies are used: 2·1–2·7MHz (for the 70MHz converter) or 2–4MHz (for the 144MHz converter) where the AR88D has excellent bandspread and stability. A different intermediate frequency can be employed if desired and details are given later on this point.

Precautions have been taken to eliminate breakthrough, which can be troublesome at such low I.F.S and these (which include the use of a screen over the oscillator crystal) have proved effective.

Basic Circuitry

Both converters use an earthed-cathode, capacitance-neutralized, 6CW4 Nuvistor rf amplifier inductively coupled into a triode mixer, the latter having a suitable i.f. tuning

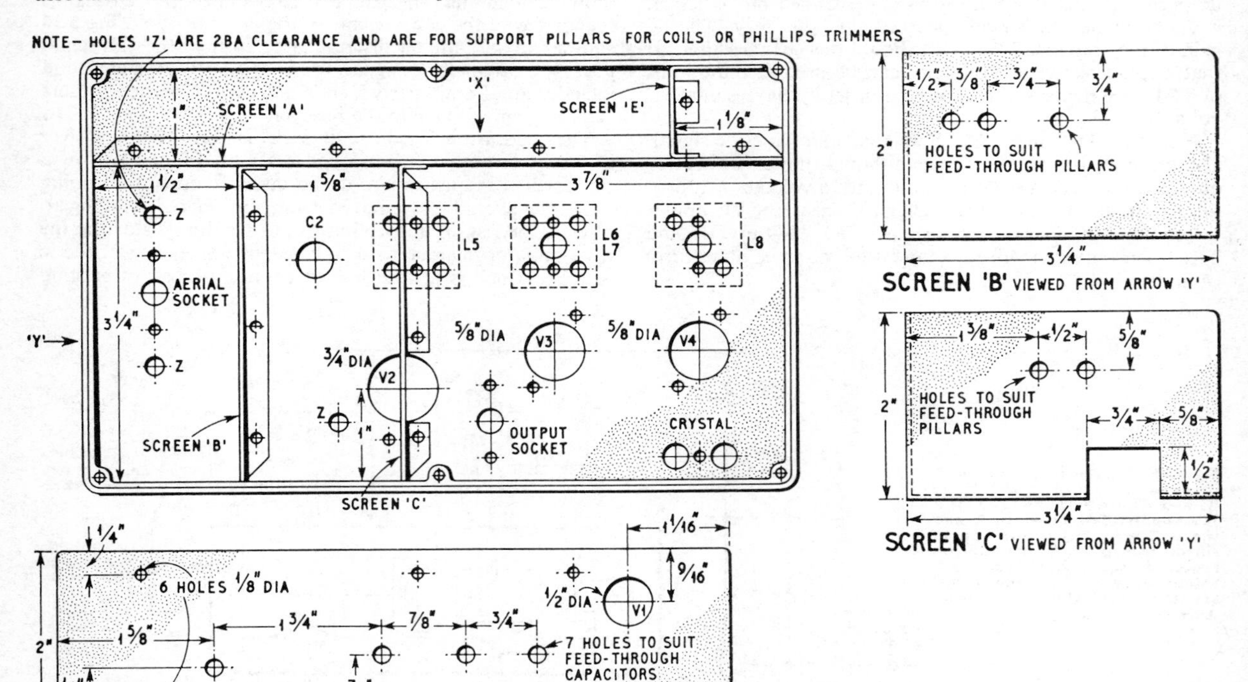

Fig 5.37. Mechanical details of the 70MHz Nuvistor converter. The valveholder for the Nuvistor valve and any solder-in type feed-through capacitors should be fitted to the screen before the assembly is mounted on the lid of the die-cast box.

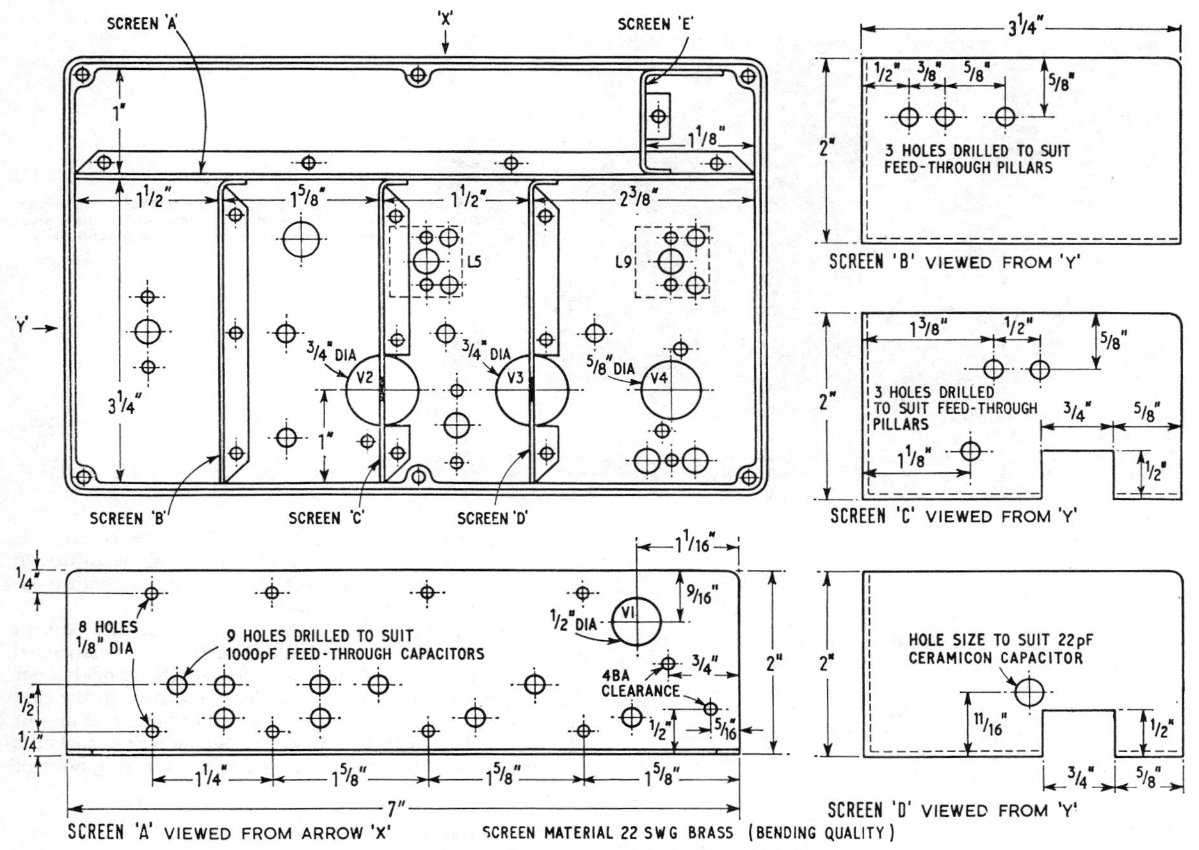

Fig 5.38. Mechanical details of the 144MHz converter.

coil in its anode circuit. To provide a low impedance for use with a coaxial cable, the output is fed to the main receiver through a triode cathode follower.

While the converters are basically similar, the crystal oscillator/multiplier stages differ. In both instances the oscillator valve is an EF91 (6AM6) in a simple Colpitts circuit. This is to be preferred to the Squier oscillator which may prove difficult to use, particularly with surplus crystals.

In the case of the 70MHz converter, an 8·500MHz FT243 crystal is used, the oscillator anode circuit beng tuned to the fourth harmonic of the crystal frequency. The output of this stage is capacitance coupled to the grid of the EF91 doubler stage, the anode of which is tuned to a frequency of 68·0MHz, hence producing an i.f. of 2·1–2·7MHz.

For the 144MHz converter, a 7·100MHz FT243 crystal is used but the oscillator anode circuit is tuned to the fifth harmonic of the crystal. If V4 fails to oscillate, the value of the 2·2kΩ cathode resistor should be reduced; poor-activity crystals may require a value as low as 100Ω. The subsequent ECC81 (12AT7) double triode is connected in cascade and doubling in each half produces a final frequency of 142MHz which results in an i.f. of 2–4MHz.

In both converters the output from the local oscillator is link coupled to the mixer stage, and more than sufficient output is available to ensure efficient mixing. No significant spurious responses from the crystal oscillator chain have been found within the tuning range of the converters.

Mechanical Construction

Each converter is built on the lid of an Eddystone 7½in by 4½in No 845 die-cast box. This ensures easy access for assembly and wiring. The construction of the interior screens, with dimensions, is shown in **Figs 5.37** and **5.38.** If solder-in type feedthrough capacitors are used, these, together with the Nuvistor holder, should be soldered into position before attempting to bolt the screen sections together and to mount the assembly on the lid.

The only component mounted through the body of the box is the Bulgin P360 three-way miniature power input socket, a flying lead from which carries the power to the heater and ht feedthrough capacitors mounted on screen "E". This screen also serves to support RFC1 and its associated 0·01μF ceramic capacitors. RFC2 is soldered directly to the heads of the appropriate feedthrough capacitors. Other aspects of construction may be seen from the photographs, but it should be noted that in the photograph of the top of the completed 144MHz converter the aforementioned power socket is shown at the wrong end of the box.

An interior view of the 70MHz converter. Note the use of ceramic feed-through insulators for supporting the air-spaced coils.

Components and Wiring

Almost all the components used in both converters are standard types which are readily available. All valveholders should be of the low-loss variety, preferably ptfe. A special holder, Cinch Type 133-65-10-001, is required for the 6CW4 Nuvistor and is generally available from dealers stocking this type of valve. As the 6CW4 has a very low anode-grid capacitance, it is essential that the neutralizing capacitor C_N should have a minimum capacitance not greater than 1pF. A Wingrove and Rogers or Plessey miniature 1–10pF air-spaced trimmer is suitable. The Philips concentric type should not be used due to its relatively high minimum capacitance.

All $0{\cdot}01\mu$F capacitors are high-*K* ceramic types. The 1,000pF feedthrough ceramic capacitors can either be the solder-in or nut-secured types. All other fixed capacitors are ceramic.

All resistors are $\frac{1}{4}$W except the $8{\cdot}2$kΩ ht feed for V1, which has a $\frac{1}{2}$W rating. A number of KLG feedthrough insulators are used both for support and feedthrough purposes, resulting in a neat and rigid assembly.

All wiring associated with the rf circuitry should be kept as short as possible. Decoupling and earth returns associated with each valve are made to solder tags which are held under the securing nuts for the valveholder concerned. In the case of the Nuvistor, tags 8 and 10 are wired to the frame lugs of the holder. The earth lead from the Bulgin power socket is connected to a solder tag under one of the securing bolts of screen "E".

The 70MHz Nuvistor Converter

The circuit is shown in **Fig 5.39** and mechanical details in Fig 5.37. Coil details are given in **Table 5.1.**

To align the converter, disconnect the ht supply from V1,

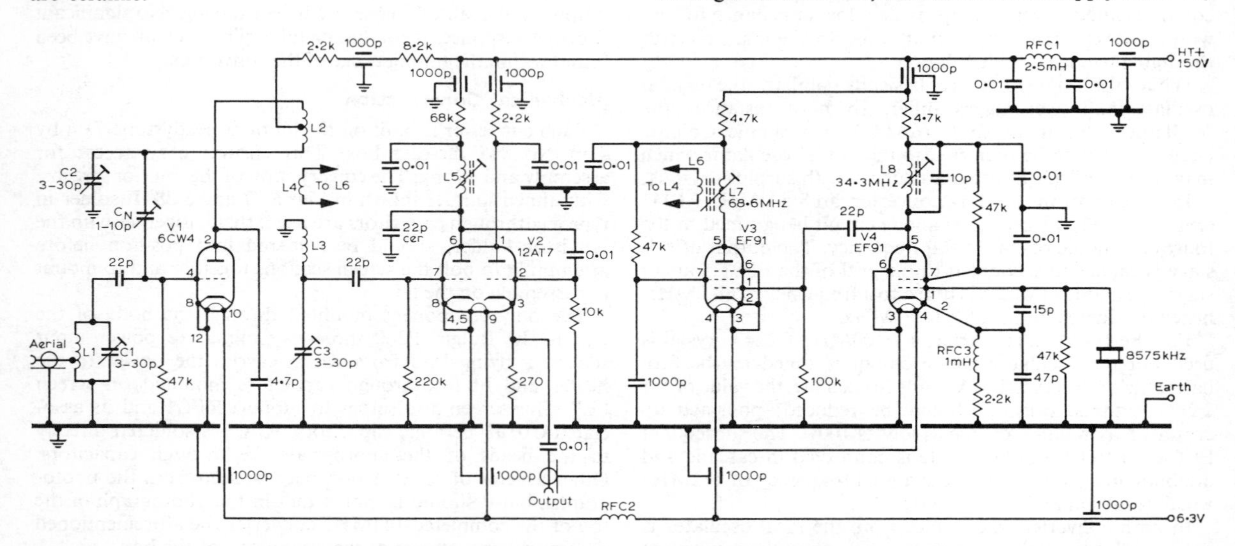

Fig 5.39. Circuit diagram of the 70MHz Nuvistor converter. A 22pF ceramic capacitor should be connected from pin 6 (anode) of V2 to earth.

TABLE 5.1

Inductor details for the 70MHz converter. Tunable i.f. 2·1 to 2·7MHz, crystal 8,500kHz

L1	8 turns 18swg enam wound on $\frac{7}{16}$in mandrel, length 1in, tapped 3 turns from earth end, air spaced
L2	13 turns 18swg enam wound on $\frac{7}{16}$in mandrel, length $1\frac{1}{4}$in, tapped $5\frac{1}{2}$ turns from C_N air spaced.
L3	$8\frac{1}{2}$ turns 18swg enam wound on $\frac{7}{16}$ in mandrel, length $\frac{7}{8}$in, air spaced
L4	2 turns 18swg enam wound on $\frac{7}{16}$in mandrel and placed between L2 and L3
L5	Maxi-Q i.f. transformer type IFT 11/1·6, with silver mica capacitors in can removed and primary and secondary windings in series (top of lower layer winding to bottom of upper layer winding)
L6	2 turns 22swg pvc covered tinned copper wire wound at power supply end of L7
L7	10 turns 26swg enam wound on $\frac{1}{4}$in by $1\frac{3}{8}$in former, slug tuned (Aladdin type with can).
L8	10 turns 26swg enam wound on $\frac{1}{4}$in by $1\frac{3}{8}$in former, slug tuned (Aladdin type with can)
RFC1	2·5mH rf choke
RFC2	51in 18swg enam wound on $\frac{1}{4}$in mandrel, close spaced
RFC3	1mH rf choke

remove the cans from L7 and L8 and first adjust L8 for maximum indication on the rf checking meter—**Fig 5.40.** Then adjust L7 in a similar manner. The tuning range of each coil is such that only the required harmonic should be selected, but it is wise to check the actual frequencies with an absorption wavemeter or gdo. When this has been done the cans should be replaced on L7 and L8, and the two cores re-adjusted for maximum rf output from the coupling L4. The output should be connected to the communications receiver, tuned to 2·4MHz. L5 will be approximately correct. Adjust C3 for maximum hiss; two positions will be found, the one with the smaller capacitance being the correct one.

A strong signal is then required (from a local transmitter or signal generator) and should be fed to the aerial socket. Adjust C1 and C2 followed by C3 and L5 for maximum output, followed by adjustment of C_N, the neutralizing capacitor, with an insulated screwdriver, for *minimum* output. (In practice this is usually found to be near the minimum capacitance of C_N). This procedure should be repeated several times as there is some interaction between the adjustments.

The initial adjustments are now complete and ht may be reconnected to V1. If there is a tendency for oscillation, C_N is incorrectly adjusted; the minimum position referred to is very sharp and to a certain extent adjustment is interdependent with C2 and C1. No difficulty should be experienced, however, in obtaining the correct setting. For final adjustments a noise generator is desirable (see Chapter 18—*Measurements*) but if this is not available, C1, C2, C3 and L5 (also L7 and L8) should be adjusted for maximum output on a local signal. Finally, the capacitance of C1 should be increased slightly so that the circuit is detuned towards 70MHz and the output just drops. This is near to the optimum position for the best signal-to-noise ratio. In practice a noise factor of better than 2·5dB should be obtained.

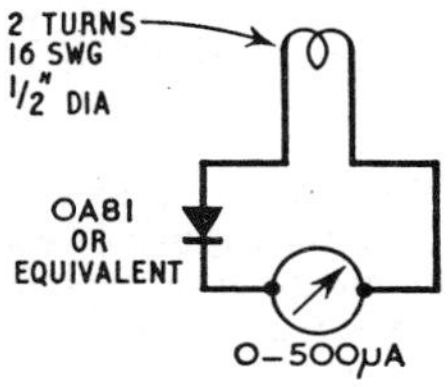

2 TURNS
16 SWG
1/2" DIA
OA81
OR
EQUIVALENT
0–500μA

Fig 5.40. RF checking meter.

The 144MHz Nuvistor Converter

The circuit of the 144MHz converter is shown in **Fig 5.41** and the mechanical arrangements in Fig 5.38. **Table 5.2** gives details of the coils and chokes.

The adjustment of this converter is very similar to the 70MHz unit except that the communications receiver is set to 3MHz, corresponding to a signal frequency of 145MHz. First, power is applied with the ht disconnected from V1. The can is removed from L9 and the slug is adjusted for maximum rf indication followed by C5 (rf indication being

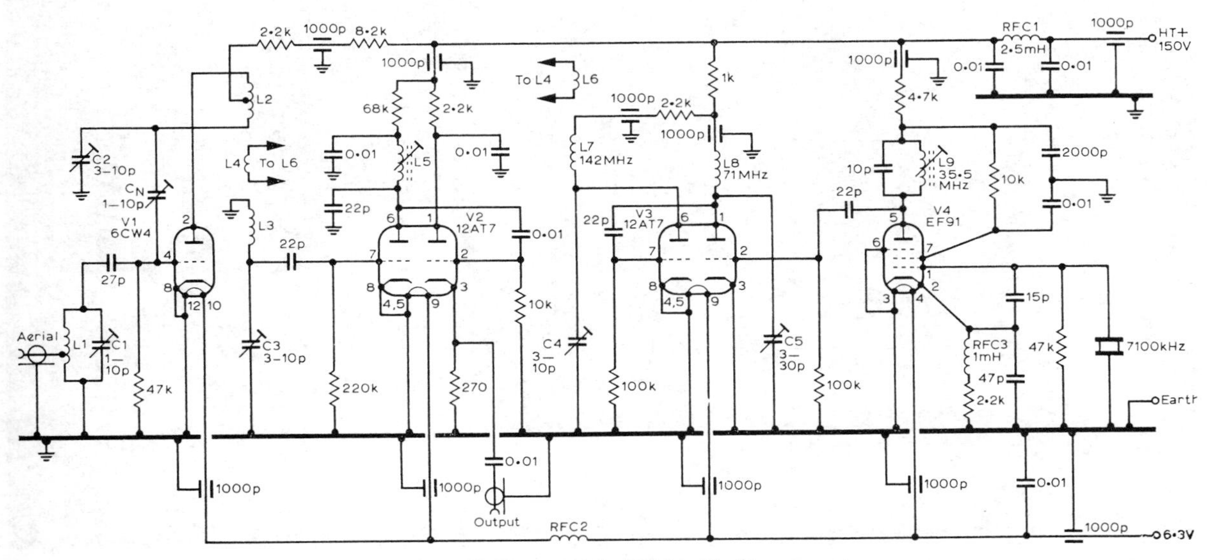

Fig 5.41. Circuit diagram of the 144MHz Nuvistor converter.

TABLE 5.2

Inductor details for the 144MHz converter. Tunable i.f. 2 to 4MHz, crystal 7,100kHz

L1	4 turns 18swg enam wound on $\frac{1}{4}$in mandrel, length $\frac{7}{8}$in, tapped $1\frac{1}{2}$ turns from earthy end, air spaced
L2	8 turns 16swg enam wound on $\frac{7}{16}$in mandrel, length 1in, tapped $3\frac{1}{2}$ turns for C_N, air spaced
L3	4 turns 16swg enam wound on $\frac{7}{16}$in mandrel, length $\frac{5}{8}$in, air spaced
L4	1 turn 16swg enam wound on $\frac{7}{16}$in mandrel, interwound at earthy end of L3, air spaced
L5	32swg enam wound on $\frac{1}{4}$in by $2\frac{1}{2}$in former (Aladdin type with can), 1 layer $1\frac{1}{2}$in long and 1 layer $\frac{3}{4}$in long. Tissue paper interleaving, secured with polystyrene cement. Tuned by two slugs. 3·3kΩ resistor in parallel
L6	as for L4 but mounted adjacent to the ht end of L7, air spaced
L7	$3\frac{1}{2}$ turns 16swg enam wound on $\frac{7}{16}$in mandrel, length $\frac{3}{4}$in, air spaced
L8	7 turns 16swg enam wound on $\frac{7}{16}$in mandrel, length 1in, air spaced
L9	10 turns 26swg enam wound on $\frac{1}{4}$in by $1\frac{3}{8}$in former slug tuned (Aladdin type with can)
RFC1	2·5mH rf choke
RFC2	26in 18swg enam close wound on $\frac{1}{4}$in mandrel, self-supporting, air spaced
RFC3	1mH rf choke

The 144MHz converter. The screening can for the crystal has been removed for this picture.

observed at L8 and L7 respectively), and then C4. Again, the tuning range of each tuned circuit is such that only the required harmonic should be selected, but it is wise to check the actual frequencies with an absorption meter or gdo, if available. L5, which is very flat in its tuning, should be centred on 3MHz. C1, C2 and C3 are then adjusted for maximum output on a strong local signal. C_N is adjusted for *minimum* output with an insulated screwdriver, the correct setting being very critical. When ht has been reconnected to V1 the converter is ready for use. Final adjustment should again, if possible, be carried out with a noise generator.

Measured under laboratory conditions the noise factor

An underside view of the 144MHz converter.

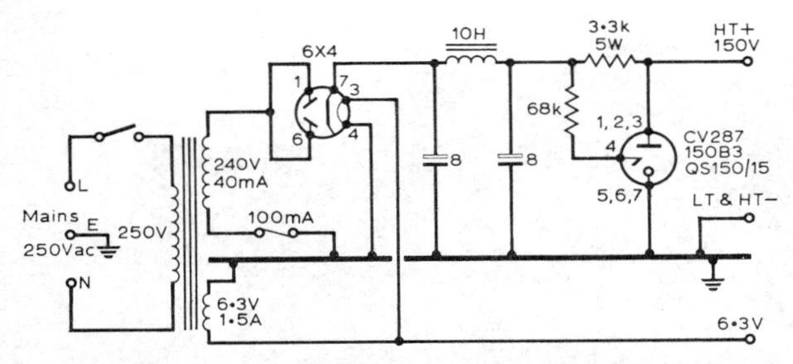

Fig 5.42. A power supply suitable for use with either the 70MHz or 144MHz converter. A semiconductor rectifier of suitable rating could be substituted for the 6X4 valve.

of the original model was 2·9dB, protection against i.f. breakthrough better than 80dB, the image response better then 20dB down at 144MHz and the overall gain 23dB.

Power Supply

A power supply which delivers 6·3V at 1·5A and 150V at 30mA is adequate—a suitable circuit is shown in **Fig 5.42.** The components are all of a modest size and the complete unit can be built into a 3½in by 4½in Eddystone No 650 die-cast box. A transmit/receive relay can be added (if required) to mute the converter while a transmitter is operating, by removing the ht supply.

Wiring Details

The orientation of the valveholders should be as follows: V2, V3, pins 1 and 9 are on the screen "A" side, V4, pins 1 and 7 are on the screen "A" side when viewed from the underside of the chassis. Earthing of the valveholder pins is important: all pins to be earthed are connected to the centre spigot of the valveholder. For V2 and V3, these are connected by as short a wire as possible, ie from the nearest earthed pin, to 6BA solder tags under the valveholder retaining nuts nearer screen "A".

In the case of V4 the earth tag is mounted under the nut nearer the crystal holder. Screens "C" and "D" are also soldered to the centre spigots of their related valveholders.

VALVE CONVERTERS FOR 432, 1,296 AND 2,300MHz

These converters were developed from the famous K6AXN design for 1,296MHz, and they use very similar local oscillator chains and crystal diode mixers. They can be built on a brass plate to replace the lid of a die-cast box or as a chassis in a conventional cabinet. In the previous design for 70 and 144MHz the Nuvistor stage was mounted on a bracket and the remainder of the circuit could tolerate the relatively long leads to the soldering tags on the chassis. In the three designs to be described the higher frequencies call for shorter leads and exactly reproducible earth points; a brass plate is therefore used and if a die-cast box is used for screening its lid is discarded.

Basic Circuitry

In each case a Squier overtone crystal oscillator using half of a 12AT7/ECC81 is followed by three frequency multiplier stages using the other half of the oscillator valve and two 6AK5/EF95 pentodes. In the 432MHz converter this produces the required injection frequency but in the two microwave converters further frequency multiplication is required. The 432MHz converter includes an rf stage, but since no satisfactory and inexpensive valve was available for the higher bands there are no rf stages in the other two converters; this makes for flexibility since separate pre-amplifiers can be added later which can be improved or replaced without disturbing the main converter.

Mechanical Construction

Figs 5.43, 5.44 and **5.45** show the chassis layouts. The brass plate for the chassis can be made either to replace the lid of an Eddystone 7½ by 4½in die-cast box or as a sub-assembly in a "Lectrokit" cabinet. Some constructors may prefer to fit the power connector on the side of the box as in the vhf converters.

The Local Oscillators

With the exception of V3b of the 432MHz converter, which is a tripler, all the valve frequency multipliers are doublers. This has been done to ensure that ample rf power is available at each stage and to reduce the risk of inadvertent selection of wrong harmonics. All the tuned circuits up to the grid of the last pentode should be preset with a gdo, and if a gdo for the required uhf is available, the anode of the last pentode can also be preset. Final adjustments should be made for maximum grid current in the following stage or for current through the appropriate diode in the case of the final pentode in the 432 and 1,296MHz converters.

The ht supply to the oscillator and first multiplier stage is stabilized at 150V. The low input to the oscillator reduces heating of the crystal and the stabilizer not only reduces the effect of power supply variations, but eliminates the last trace of hum, producing a very clean note. The stabilizer can be fitted on the converter chassis or in the power supply unit. In the latter case it is necessary to add a 100Ω resistor and 0·01μF capacitor to decouple this line and to prevent radiation in the television band. The value of the dc dropping resistor (R27 in the circuits) is calculated from the formula:

$$R = \left(\frac{V_{\text{ht}} - 150}{30}\right) \text{k}\Omega$$

R22 is the screen resistor of the second pentode multiplier, and it is adjusted for each individual multiplier chain to

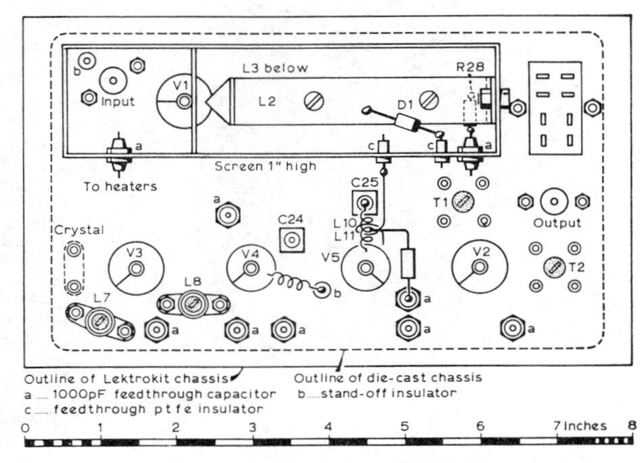

Fig 5.43. Chassis layout of the 432MHz converter.

Underchassis view of the 432MHz converter. In this example, crystal X1 was mounted in a B7G holder (shown at the extreme left of the picture), and not as shown in Fig 5.43.

Underchassis view of the 1,296MHz converter.

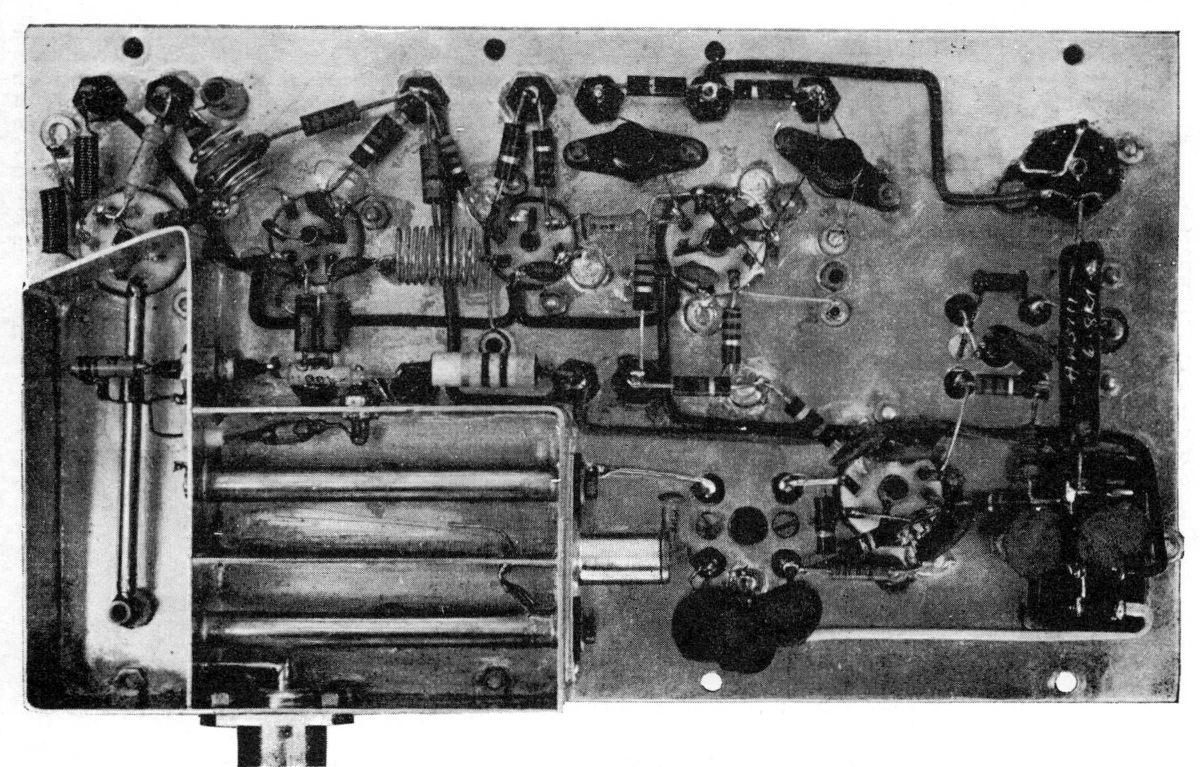

Underchassis view of the 2,300MHz converter.

provide the required drive to the next stage. **Table 5.3** gives typical starting values.

The I.F. Pre-amplifier

Since there is necessarily a signal loss in the diode mixers, the overall noise figure is critically dependent on the noise figure of the next stage. A cascode amplifier is therefore included, and details of coils for various I.F.S are given in **Table 5.4**.

In two of the photographs a ferrite pot assembly can be seen near the i.f. output. This is a hybrid used to feed output to two receivers without interaction between either. If this refinement is not required it can of course be omitted. The winding details and connections are given in **Fig 5.46.**

Choice of Intermediate Frequencies

The designs cater for two philosophies with regard to intermediate frequencies. In one case the local oscillator

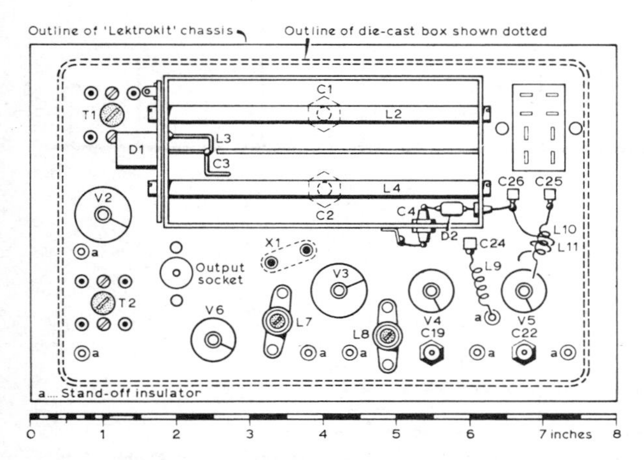

Fig 5.44. Chassis layout of the 1,296MHz converter.

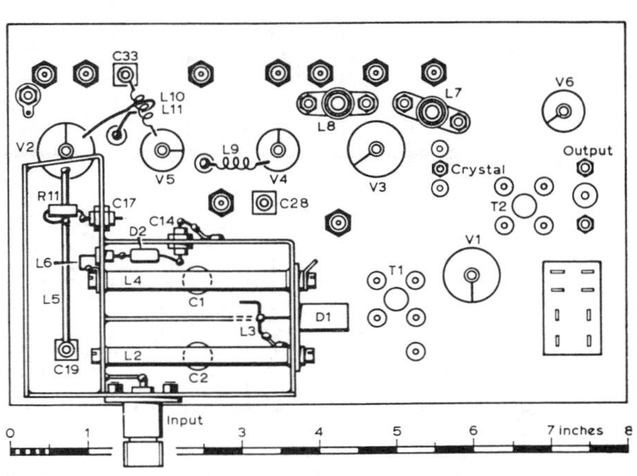

Fig 5.45. Chassis layout of the 2,300MHz converter.

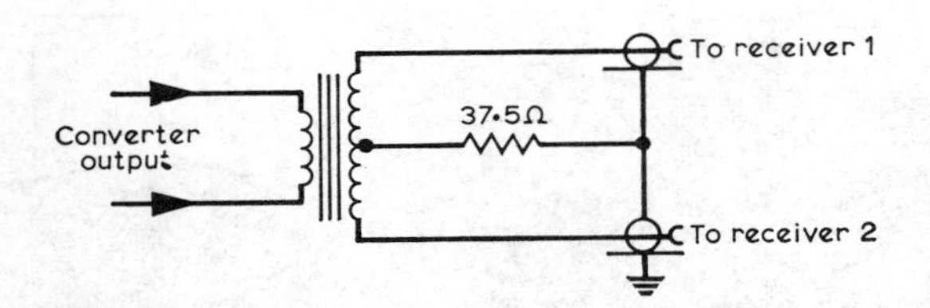

Fig 5.46. Hybrid transformer for supplying two receivers from one converter. The ferrite pot is a Mullard FX2238 or equivalent. Primary: 7 turns 26swg in a single layer. Secondary: 5 + 5 turns 26swg, also in a single layer over primary, with a layer of insulating tape between. The 37·5Ω resistor can be conveniently made with two 75Ω 5 per cent resistors in parallel.

frequency is a "round number" so that the received frequency is easily derived from the reading of the main receiver dial. This has the disadvantage that if one main receiver is to be used with a number of converters and switched rapidly between uhf bands it may be necessary to switch the range of the main receiver as well as to retune.

In the second case all the I.F.s are the same, reducing the amount of adjustment required at the main receiver when changing bands, but increasing the risk of breakthrough from strong signals on bands on other than the one of immediate interest. This last is often a serious problem when strong signals on 432MHz appear as weak signals in the higher bands. The selected common i.f. must be high to reduce local oscillator noise in the highest band required. I.F.s below 10MHz are not suitable for the microwave bands and if the 2,304MHz band is to be included then the common i.f. should preferably be above 20MHz.

The 432MHz Converter

Fig 5.47 is the circuit diagram of this converter and Fig 5.43 the chassis layout. Coil details are given in **Table 5.5.** The grounded-grid rf stage is screened by the trough assembly

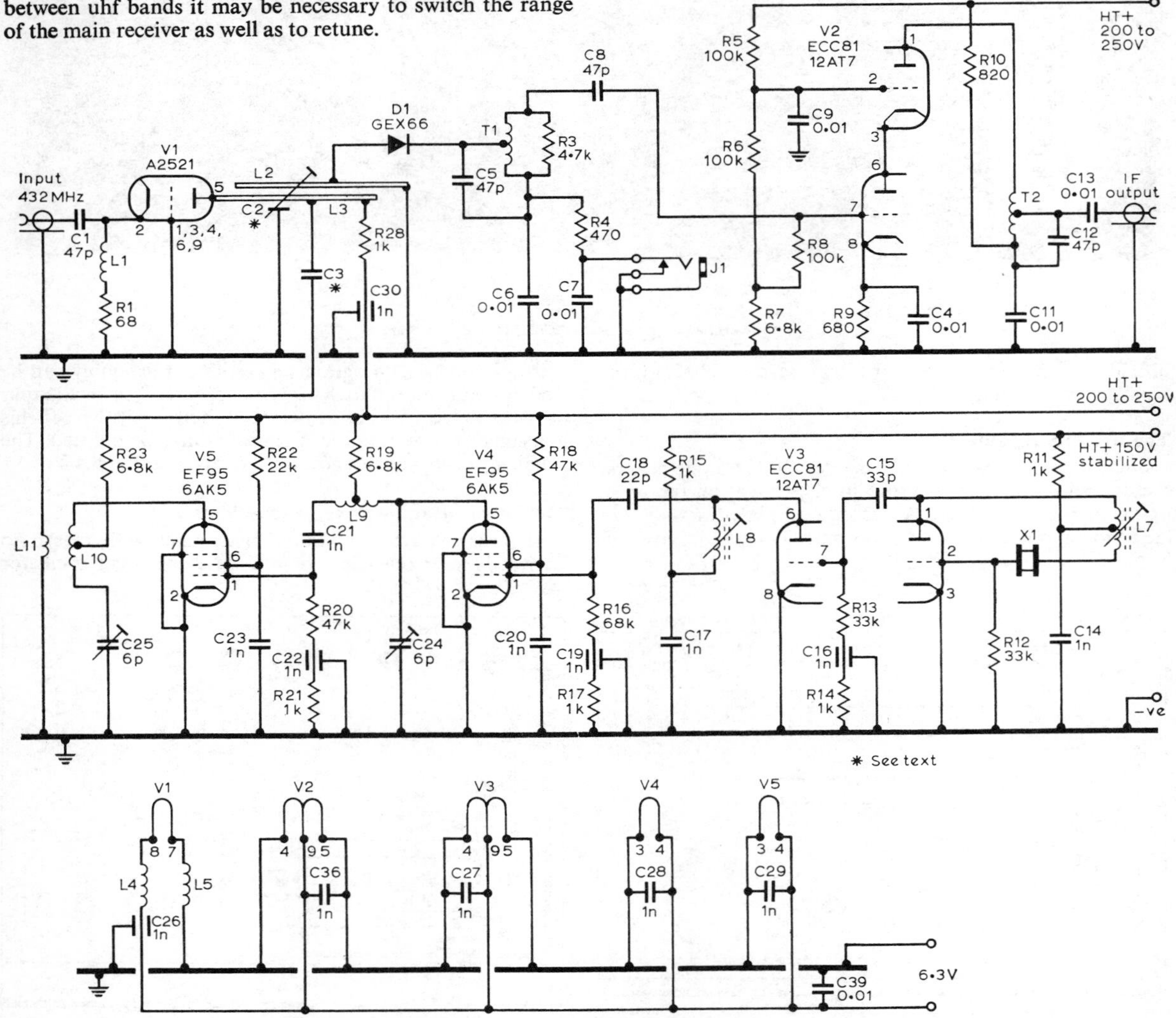

Fig 5.47. Circuit diagram of the 432MHz converter. See text for C2, C3 and R27, and Table 5.4 for details of T1, T2 and associated tuning components.

TABLE 5.3

Selection of R22 in the uhf converters

Band (MHz)	Typical R22 (kΩ)	Current D2	Current D1
432	47	200–300μA	—
1,296	22	5mA	500μA
2,300	22	40mA	500μA

and an additional screen is fitted across the valveholder. It is important that this is soldered to the centre spigot of the valveholder and to tags 3, 4 and 6 of the A2521. Tags 1 and 9 are bent inwards and also soldered to the centre spigot. If other valves are used in place of the A2521 then the corresponding grid tags are dealt with in the same way.

The anode line of the rf stage is split and **Fig 5.48** shows a detailed view of its construction. The two parts are clamped together with two nylon screws threaded into nuts soldered to the under side of the anode line. The two lines are insulated with a thin sheet of polythene; the thickness of this sheet is not critical and any piece of domestic wrapping material will do. The circuit is tuned by a disc capacitor, and this must be screwed through the nut fixed to the top of the chassis as shown before the line is fixed in place. The disc may be made by soldering a coin to a length of studding or by filing a screw head down to a thin disc.

The GEX66 diode used in the original design is now obsolete but any modern silicon or germanium mixer diode should serve since the noise performance of this converter is determined by the high-gain rf stage. If other diodes are used it may be necessary to experiment with the position of the tapping point of the diode. In this case it is important to dismantle the line assembly before moving the tap, or the insulator will melt with the heat: if care is taken it is not necessary to unsolder the anode connection to the rf amplifier.

Valves other than the A2521 can be used in the rf stage, and most modern uhf tuner types such as the EC8010 will give satisfactory results. It is important to orientate the valveholder so that the anode connection is on the centre line of the trough. If the holder has tags which come out at an angle to its radius it will be necessary to rotate it slightly for the most convenient connection. Fixing holes for the valveholder should therefore not be drilled until the trough walls have been soldered in place.

TABLE 5.4

I.F. transformers for the valve uhf converters. All coils wound on 0·3in formers

I.F. (MHz)	T1	T2	C12 (pF)	R2 (kΩ)	R3 (kΩ)
12–14	40 turns 30swg tapped 15 turns from cold end	40 turns 30swg tapped 10 turns from cold end	47	—	4·7
16–18	Primary 8 turns 30swg close wound Secondary 37 turns close wound and adjacent to primary	Primary 37 turns 34swg close wound Secondary 4 turns 34swg close wound spaced $\frac{1}{16}$in from primary	330	4·7	3·9
20–30	Primary 7½ turns 34swg close wound Secondary 15 turns 34swg pile wound $\frac{5}{32}$in long and adjacent to primary	Primary 18 turns 40swg close wound Secondary 4½ turns 40swg adjacent to primary	47	2·2	1·5

TABLE 5.5

Inductor details for the 432MHz converter

	Tuned to (MHz)	Winding
L1	—	20 turns 26swg close wound ¼in inside diam, selfsupporting
L2, L3	432	see Fig 5.48
L4, L5	—	15 turns as L1
L7	35	21 turns 26swg on 0·3in former tapped 3 turns from crystal
L8	70	8 turns as L7
L9	140	7 turns 22swg $\frac{5}{16}$in inside diam ½in overall, selfsupporting, centre tapped
L10	420	4 turns 22swg $\frac{5}{16}$in inside diam ½in long, selfsupporting, centre tapped
L11	—	1 turn 22swg sleeved, link in centre of L10

The 1,296MHz Converter

This converter is an "anglicized" version of the original design by K6AXN. Valves and other components more readily available in Europe are specified, but otherwise the original arrangement has been closely followed. **Fig 5.49** shows the circuit. The frequency multiplier up to the anode of the second pentode has already been described. D2 triples from 426·6 to 1,280MHz which is selected by the trough line tuned by C2. C3, the coupling capacitor into the mixer, is a short length of 22swg wire lying alongside the inner of the trough and adjusted to give the specified mixer current. The aerial connection is made directly on to the inner of the other trough line, and the position of the tap should be set for the best signal-to-noise ratio on a weak

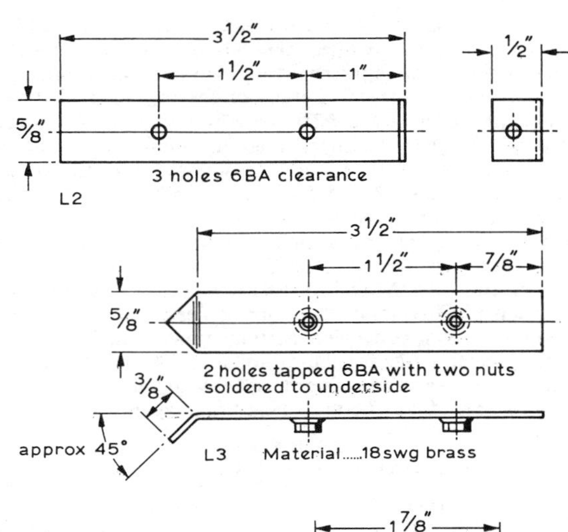

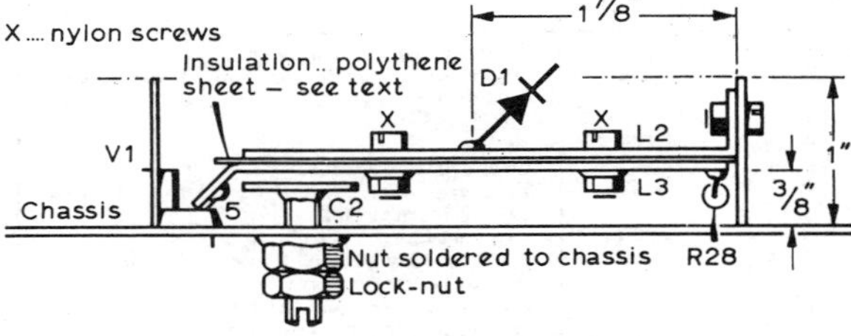

Fig 5.48. Details of L2 and L3 of the 432MHz converter. These strips must be flat and free of burrs where the polythene sheet is clamped.

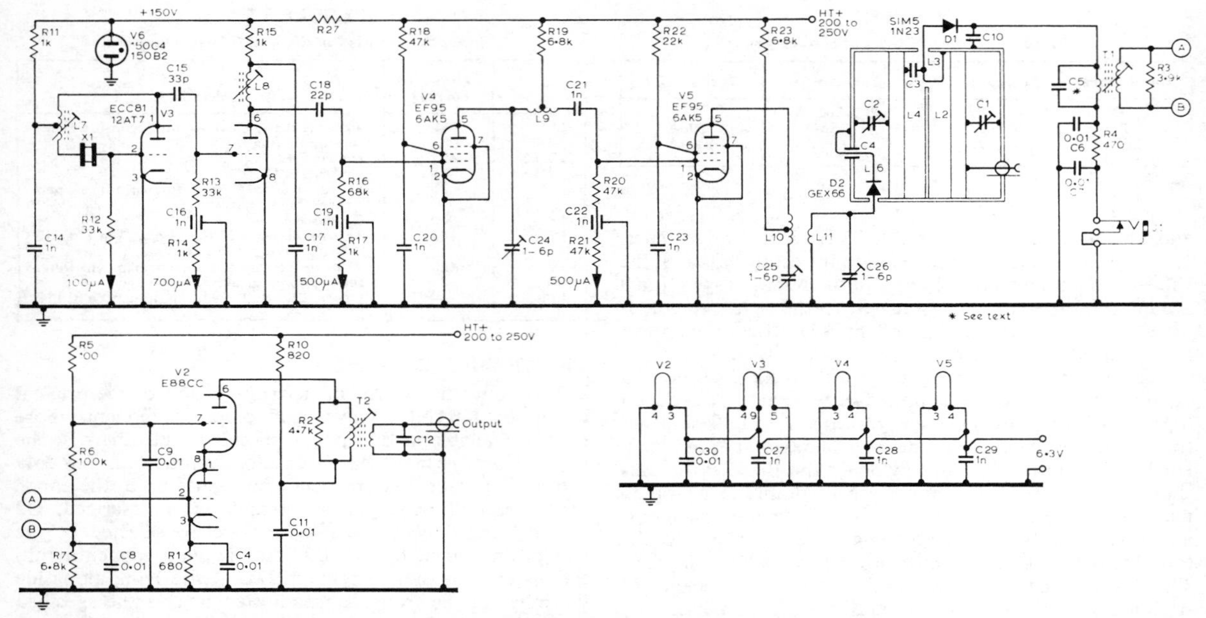

Fig 5.49. Circuit diagram of the 1,296MHz converter. See text for C5, C10, C31, C32, R27 and Table 5.4 for details of T1, T2 and associated tuning components.

signal. A starting position ⅜ to ½in from the end is suitable. In the prototype a ¼in valve top cap was used to provide a readily movable tap.

The general layout of the converter is shown in the photograph and Fig 5.44, while **Table 5.6** gives the coil details. The troughs, 1in wide internally and 1in deep, are fabricated from 20swg brass sheet. Details of the septum and diode mount are given in **Fig 5.50**. The trough components are temporarily jigged together using chromium-plated instrument screws (which do not solder) screwed into the tapped ends of the lines L2 and L4. The assembly may then be soldered directly to the 18swg brass chassis using a small gas flame. 0BA brass nuts soldered to the top of the chassis act as threaded bearings for C1 and C2, which are 0BA brass screws with the end filed flat to increase the available capacitance. 6BA solder tags are soldered directly to the chassis to provide convenient earthing points, and in particular short earth paths for the cathodes of V4 and V5 (pins 2 and 7). PTFE valveholders are used for all valves except the stabilizer. The diode mount is held firmly against the trough wall by 6BA screws fixed in the ends of the lines L2 and L4, the bolts being insulated by ptfe or fibre washers.

A polythene sheet of about 0·01in thickness is fitted between the mount and the trough. The thickness was chosen to produce a capacitor of value 40–51pF. This capacitor, C10, which is both the rf bypass at 1,296MHz and also the

TABLE 5.6
Inductor details for the 1,296MHz converter

	Tuned to (MHz)	Winding
L1	—	Optional loop coupling to L2
L2	1,296	4·25in ¼in diam brass rod, ends tapped 6BA mounted ⅜in clear of chassis
L3	—	see Fig 5.51
L4	1,280	As L2
L6	—	Lead of D2 connected to C4
L7	40	18 turns 24swg on 0·3in former tapped 2 turns from crystal
L8	80	9 turns as L7
L9	160	8 turns 22swg 3/16in inside diam ⅜in long, selfsupporting, centre tapped
L10	320	5 turns 22swg 3/16in inside diam ¼in long, self supporting, centre tapped
L11	320	1½ turns 22swg sleeved, wound into centre of L10

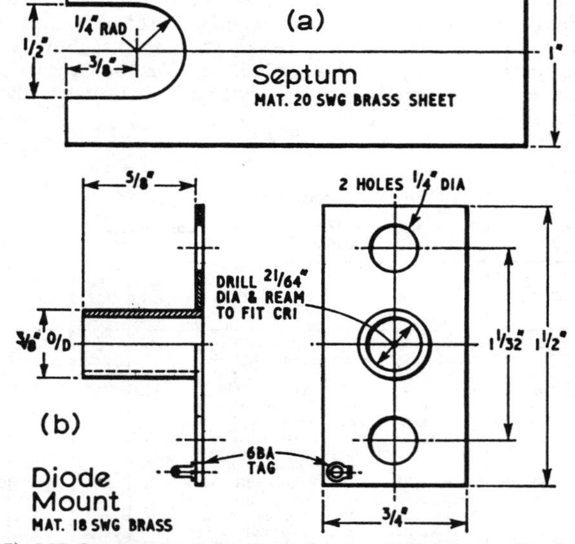

Fig 5.50. Constructional details of septum and diode mount for the 1,296 and 2,300MHz converters. The septum length of 2·4in shown is for the 2,300MHz converter; for 1,296MHz this length would be 4·25in.

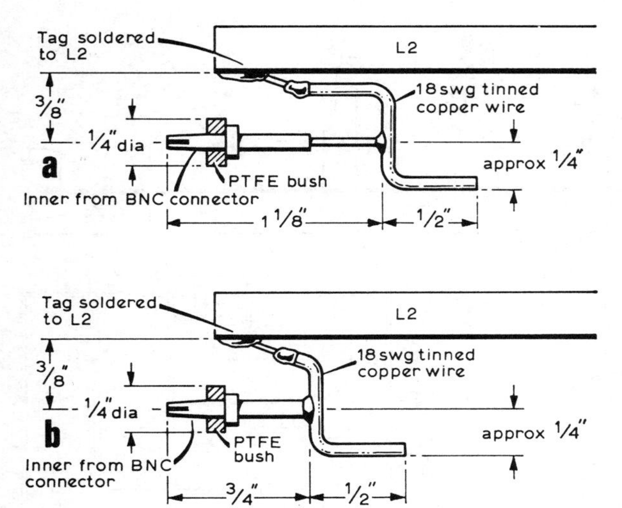

Fig 5.51. Connection to mixer diode D1 using inner of a BNC socket. (a) 1,296MHz; (b) 2,300MHz.

tuning capacitor for the primary of T1, is padded to the value specified in **Table 5.4** by C5. Nylon screws should not be used here since it is not possible to apply enough pressure to the plate, and the capacitance will drift in time. The connector to the mixer diode D1 is the inner of a BNC socket. The ptfe insulation from this socket is also used in locating the inner within D2 (see **Figs 5.51** and **5.52**).

2,300MHz Converter

In this converter signals at 2,300MHz are mixed with the output of a crystal oscillator chain at 2,280MHz, the i.f. produced being amplified by a wideband head amplifier. The upper half of **Fig 5.53** shows the i.f. head amplifier and the trough-line assembly which comprises the aerial and mixer circuits and the final stages of the multiplier chain. The lower half shows the crystal oscillator and the multiplier chain to 253MHz which is practically identical to the other two designs. Component details are given in **Table 5.7**.

Signals at 2,300MHz applied via the aerial connector and the input loop L1 are tuned by L2, C1 and coupled to the mixer diode D1 by L3. The 2,280MHz output of the multiplier diode D2 is also coupled to D1 via C3. Intermediate frequency signals produced by D1 are fed to the i.f. head amplifier V1. Wideband transformers T1 and T2 respectively match D1 to V1, and provide a 75Ω output to the main receiver.

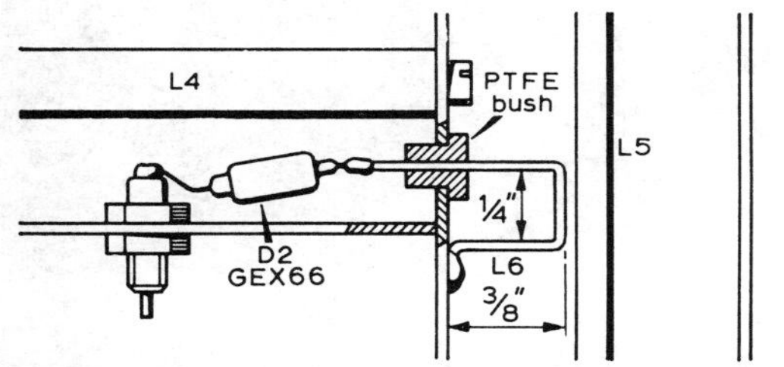

Fig 5.52. Method of connecting diode D2. In the 1,300MHz converter C4 is 1·26in from the end of the trough. In the 2,300MHz unit the distance is 1·0in.

TABLE 5.7

Component details for the 2,300MHz converter

C3, C4, C5, C6, C8, C9, C11, C34, C39	0·01μF disc ceramics
C5	value up to 10pF to pad C10 to 51pF
C10	capacitor formed between diode mount and trough wall. Dielectric polythene approximately 0·010in thick selected to give capacitance 40–51pF
C11, C14, C15, C17, C18, C20, C21, C23, C24, C26, C30, C32 C35	1000pF feedthrough
C1, C2	capacitor formed between 0BA brass bolts and L2, L4
C16, C27, C29, C31, C36, C37, C38	1000pF disc ceramics
C19, C28, C33	ceramic tubular trimmers 0·5–5pF Erie 3116A
D1	SIM5 or CV2155 (SIM2 and CV2154 similar but reversed polarity)
D2	GEX66
R10, R13	metering resistors 10–100Ω matched ±2 per cent

The oscillator and earlier stages of the multiplier chain are shown in the lower part of Fig 5.53. The output of the second 6AK5 is fed via L10 and L11 to the trough-line assembly. This assembly incorporates two further stages of multiplication: a grounded-grid triode V2 tripling to 760MHz, followed by a diode multiplier D2, the 2,280MHz output of which is tuned by L4, C2. The positive end of D2 is decoupled through C14 and earthed with a link which may be removed for metering during aligning. In connecting the valveholder for V2, pins 3, 4 and 6 and the centre screen are soldered directly to the trough wall, and pins 1 and 9 to the solder tag of the valveholder centre screen. The adjustable tap on L5 consists of a connector removed from a Paxolin octal valveholder. Other construction details are given in Figs 5.45, 5.51 and 5.52.

Alignment of the converter should present few problems provided that the critical dimensions of the trough circuitry, ie the lengths of L2, L4 and L5, are within ± 1/32in. With the valves plugged in, the anode circuits of V3a, V3b, V4 and V5 are tuned to the frequencies given in **Table 5.8** using a gdo. HT may then be applied and the oscillator and multiplier circuits peaked by metering test points 4, 5, 6 and 3 in that order. The currents measured should be similar to those given in **Table 5.9**.

TABLE 5.8

Inductor details for the 2,300MHz converter

	Tuned to (MHz)	Winding
L1	—	1/2in 18swg tinned copper wire spaced 1/8in from L2
L2	2,305	2·40in 1/4in diam brass rod, ends tapped 6BA. Mounted 3/8in clear of chassis
L3	—	see Fig 5.51
L4	2,280	as L2
L5	760	2·2in 1/8in diam brass rod, mounted 1/4in clear from chassis, tapped approximately 5/8in from anode end
L6	—	Formed from wire end of D2. See Fig 5.52
L7	31·7	25 turns 30swg enam copper wire close wound on 0·27in diam former
L9	126	10 turns 22swg tinned copper wire wound on 5/16in mandrel air-spaced to 9/16in, centre-tapped
L10	253	6 turns ditto 7/16in long, centre-tapped
L11	—	1 turn 22swg pvc covered wire wound over the centre of L10
L12, L13		10in 24swg enam copper wire on 1/8in diam mandrel, air-spaced

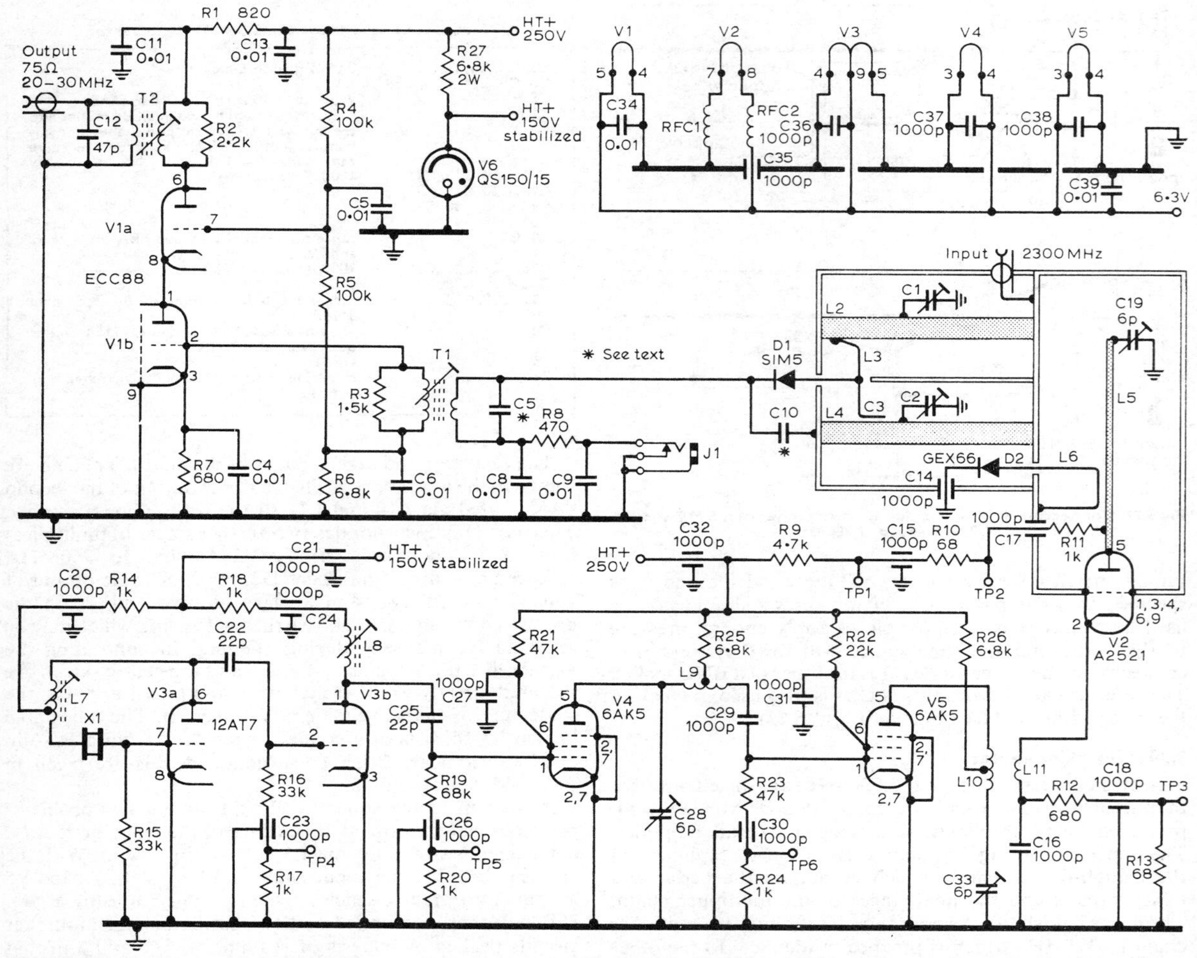

Fig 5.53. Circuit diagram of the 2,300MHz converter. See text for C5, C10, C2, C3, R27 and Table 5.4 for details of T1, T2 and associated tuning components.

The drive applied to the grid of V2 should be about 1mA, and is the difference between the cathode and anode currents measured from the voltage drop across the matched resistors R13 and R10 respectively. C19, together with the tap on L5 and the coupling link to D2 (L6), should then be adjusted to produce the maximum current in D2 (up to a maximum of 40mA). The tuning point of C2 is strongly dependent on the position of D2: that shown in **Fig 5.52** was found to be the optimum and resulted in the 0BA tuning bolt being unscrewed 2½ turns from being in contact with L4. Finally, the

TABLE 5.9

Meter between		Current (mA)
D1	chassis	0·6
D2	chassis	40
TP1	TP2	11·3
TP3	chassis	12·3
TP4	chassis	−0·9
TP5	chassis	−0·5
TP6	chassis	−1·3

Top view of the 2,300MHz converter.

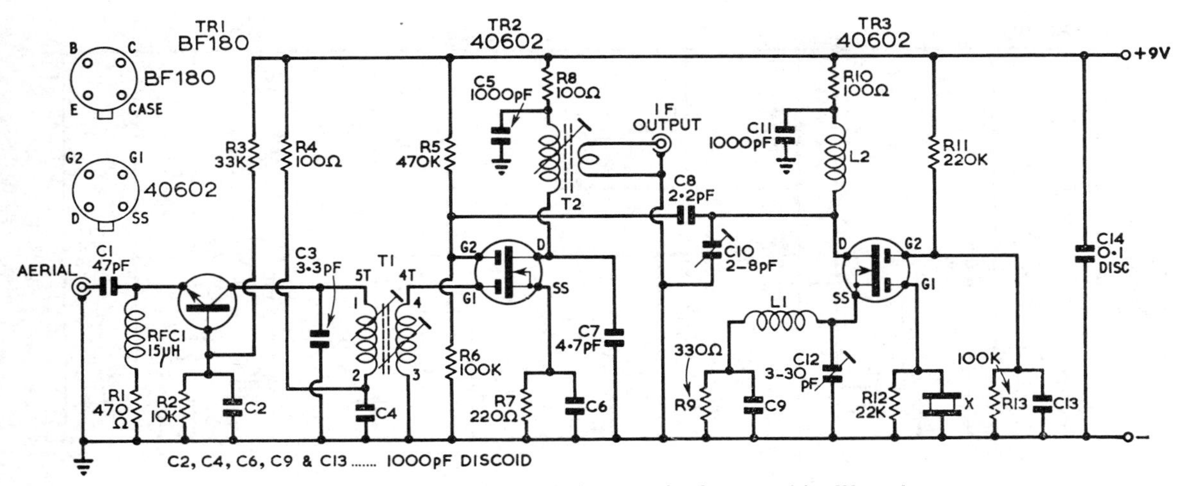

Fig 5.54. 144MHz mosfet converter for operation from a positive 9V supply.

small capacitive coupling (C3) between L3 and L4 is adjusted to give a mixer current of 0·5–1mA. Final peaking should be done after L2 has been tuned to 2,305MHz: an approximate position for C1 is ¾ turn from being in contact with L2.

144MHz MOSFET CONVERTER

This 144MHz converter was produced to fill the need for a very simple, cheap and sure-fire design. It was never intended to feature a limiting sensitivity although, using modern devices, it has been found possible to achieve a noise figure below 2dB. Using only three transistors, the insertion gain of the converter is about 30dB (spread from 28 to 35dB) and the rejection of small-signal spurii (including i.f. interference but not image) is never worse than 60dB.

The converter is basically intended for negative-chassis, positive-supply operation from a 9V supply. This has two main advantages over the other arrangement, namely that it facilitates running the unit from the ht supply of a valve receiver through a dropping resistor if so desired and also that the input circuit of the mixer may easily be very effectively decoupled at the i.f. by connecting the return lead of the input gate circuit directly to chassis. However, positive-chassis operation may be required when, for instance, working from old-style car electrical systems and, indeed, a small bonus is gained in that a few decoupling components may be saved this way. Therefore, an "inverted" circuit diagram and wiring plan are given (**Figs 5.56** and **5.57**). The performance of this version of the design is the same in all respects as that of the original, apart from a very slight increase in i.f. breakthrough, for the reasons just given.

Circuit Arrangements

Two available American surplus crystal frequencies are 69·992 and 70·992MHz (fifth overtones) giving output tuned I.F.s of about 4 to 6MHz and 2 to 4MHz respectively, each with a calibration error of 16kHz.

A single dual-gate mosfet operates as oscillator and doubler, delivering the required 1V rms to the second gate of the mosfet mixer (**Fig 5.54**). Both of these stages use the inexpensive RCA40602. The signal is fed into the first gate of the mixer, an arrangement that provides the maximum conversion slope, together with good signal oscillator isolation. The gate 1 and gate 2 dc voltages wrt source-substrate are −0·9 and +0·6V respectively, giving an average drain current of 4mA—close to the optimum for a suitable compromise between mixer linearity combined with allowable input-voltage handling capacity and low noise, characterized mainly by a high conversion slope of about 2·5mA/V.

The rf stage is standard in design, with untuned input in common-base and double-tuned output transformer coupling into the mixer. The operating conditions of the BF180 are 8·5V at 2·0mA.

The local oscillator utilizes an apparently unconventional circuit which is, however, in reality only an ordinary tuned Pierce arrangement with the tuned circuit connected to the drain and a feedback path from drain to gate 1 via the crystal. This circuit has been "turned upside-down" by grounding the drain at oscillator frequency instead of the source in order that a second-harmonic tuned circuit may be connected from drain to ground. This has only a small series impedance at oscillator frequency, thus allowing the drain

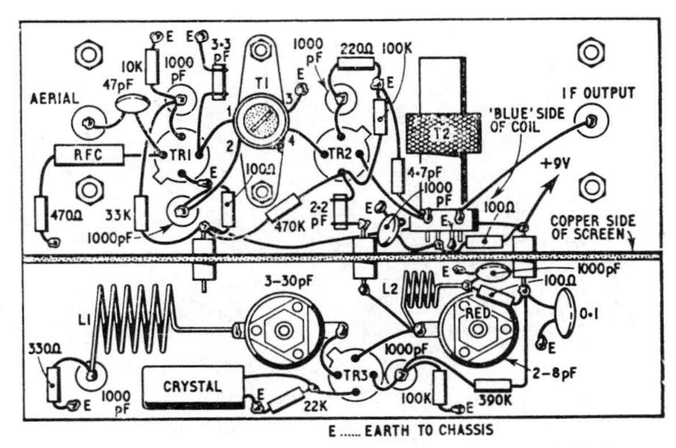

Fig 5.55. Wiring plan for the converter shown in Fig 5.54.

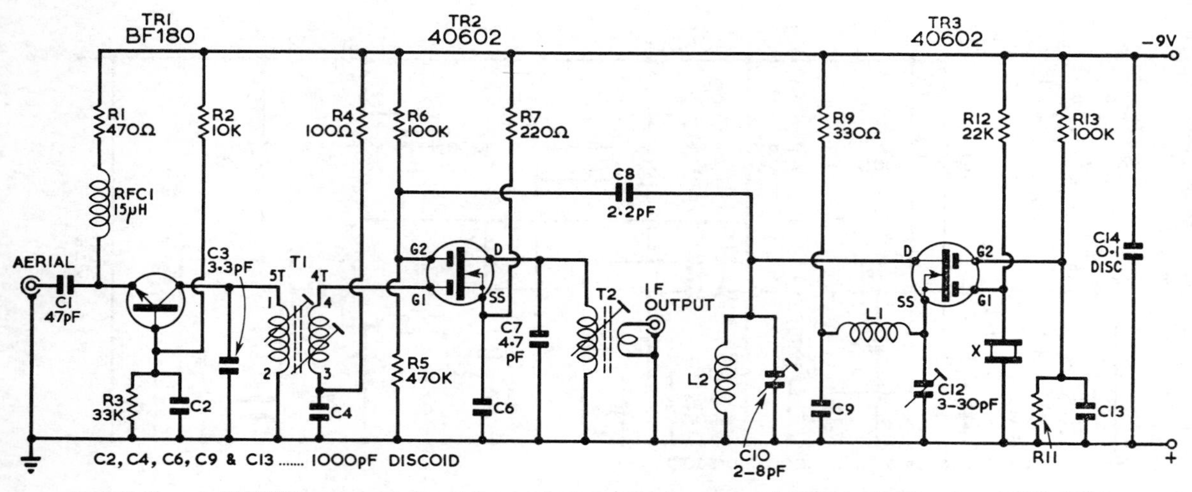

Fig 5.56. "Inverted" 144MHz converter circuit diagram for operation from a negative 9V supply. The value of R11 is 220kΩ.

to participate fully in the local-oscillator action. This latter condition is essential in this type of circuit when using a mosfet as gate 2 is not capable of operating effectively as a "sub-anode", as does the screen grid in a valve tritet circuit, because, of course, the insulated gate does not intercept any of the dc current in the channel.

The operating conditions of the mosfet have been arranged to favour operation as a doubler, by optimizing the square-law transfer characteristic from gate 1 to drain. This means that operation as a trebler or quadrupler from lower-frequency crystals will not be so favourable and, if this type of operation is desired, an extra multiplier stage will probably have to be added.

Substitution of Components

Almost any high-frequency, depletion type of n-channel dual-gate mosfet may be used for TR2 and TR3, including the earlier and more expensive 3N140 series of RCA and probably the Mullard BFS28.

It is rather difficult to substitute for the BF180 in the rf stage, TR1, because of its exceptional performance, as already mentioned. The particular requirement is for an npn silicon planar transistor with a cut-off in the region of 500 or 600MHz and very low collector-emitter feedback capacitance. Genuine Mullard BF180s have very low feedback capacitances and are normally stable in the common-base arrangement employed, even with the aerial loading removed. However, there do seem to be some BF180s of unknown manufacture which have much larger feedback and these may require shunt damping of the collector coil (T1 primary) to attain stability even with the aerial connected. Resistors from about 33kΩ down to about 10kΩ are required; use the highest value that will keep the stage stable. In really difficult cases, short-circuiting the emitter choke RFC1 usually helps, at the expense of a small increase in noise figure.

The miniature 15μH choke may be replaced by a home-made one, if desired. About 35 turns, 36swg enam close-wound on an ⅛in dia resistor, of value not less than 10kΩ, will serve.

The output i.f. transformer T2 can use almost any tuned primary type of transformer, with a turns ratio to the un-tuned winding of between 7 to 1 and 15 to 1. Primary inductances required are about 420μH for a 2 to 4MHz i.f. and 140μH for the 4 to 6MHz i.f. The stray capacitance of the tuned winding should of course be small.

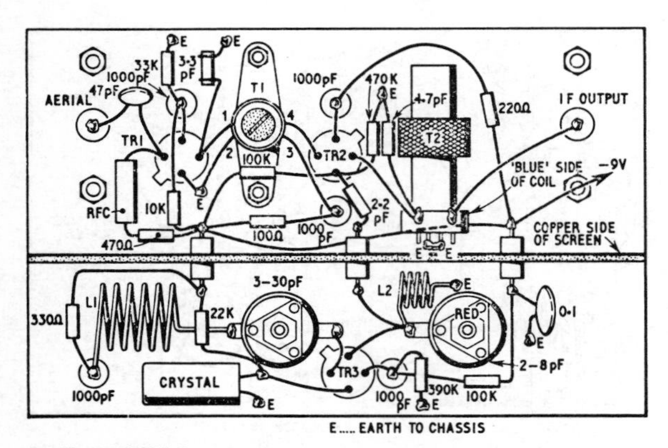

Fig 5.57. Wiring plan for the converter shown in Fig 5.56.

Construction

The converter is intended to be made on a simple flat-plate type of chassis, with a dividing screen mounted at right angles to it (**Fig 5.58**). The chassis and screen may be constructed from any suitable material, including aluminium or brass sheet or the cheap and effective copper laminate. Wherever possible components are soldered directly to the chassis. The prototype converters were made from the last-named material, which also facilitated the use of solder-in discoidal bypass capacitors in critical positions. These components have a much lower inductance than conventional feed-through capacitors and are greatly to be preferred at vhf.

Note that the position of one of the discoidal capacitors is changed if the positive-ground version of the converter is being built; hole position "1" in Fig 5.58 is correct for the

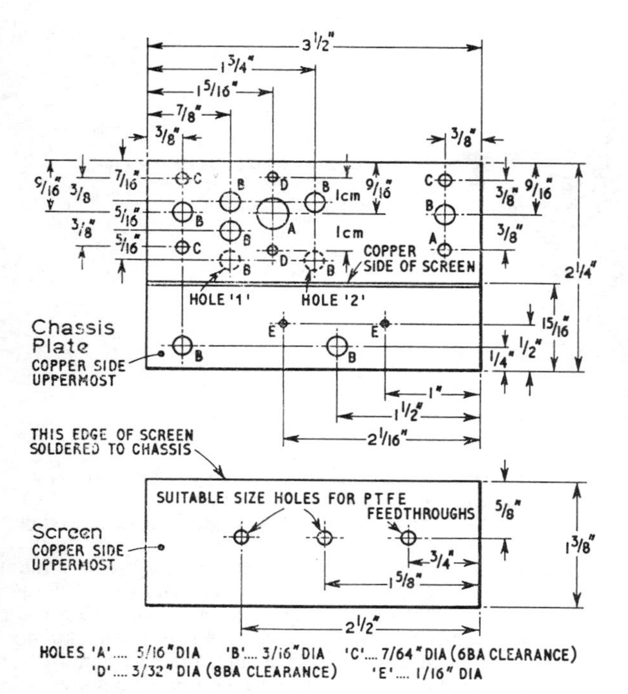

Fig 5.58. Mechanical details of the converter chassis.

normal negative-ground system but hole position "2" should be used for the positive-ground arrangement.

The rf stage and mixer are situated on one side of the screen (if laminate is in use, on the copper-clad side), with the oscillator on the other. Three ptfe feedthroughs are mounted in the screen, two for carrying supply voltages and the centre one for accommodating the local-oscillator coupling lead. Standard BS3061 coaxial sockets serve as input and output connectors. The BF180 has its case isolated from its electrodes and is conveniently mounted in a hole in the chassis, with the case lead independently grounded. It should *not* be soldered into the hole of course! The 40602 mosfet has the source and substrate tied to the case internally and so the two MOSFETS are supported in the wiring to avoid grounding the cases. The major support for each is the very

Component layout.

TABLE 5.10
Inductor details

L1	6½ turns 22swg enam, close-wound, air-spaced (wound on 5/16in drill)
L2	5 turns 22swg enam, close-wound, air-spaced (wound on ¼in drill)
T1	primary 5 turns 22swg enam, close-wound on ¼in drill, and then sprung on to Aladdin former; secondary 4 turns 22swg enam, close-wound on ¼in drill and then sprung on to Aladdin former (order of winding leads indicated on circuit diagram, "1" being nearest to the chassis)
T2	Osmor type QO9 coil for 2 to 4MHz i.f. Osmor type QO8 coil for 4 to 6MHz i.f.

Note: Osmor coils are no longer in production but suitable alternatives should readily be found.

short wire connecting to the bypass capacitor, from the mixer source lead and the oscillator gate 2 lead.

The output i.f. transformer is a standard, cheap commercial item, actually intended for use in valve receivers. It is tuned by a small capacitance to obtain the required low operating Q and therefore large bandwidth, combined with a large resonant impedance. It is supported by fixing one of its leads and also the spring clip at the top of the former to the chassis. The inter rf stage-mixer transformer is wound on a single Aladdin ¼in dia former. The windings are first formed on a ¼in rod and then sprung on to the former, one at each end. The coils are each tuned with a standard vhf type dust slug. The shiny black Neosid material (Grade 910) is best but Grade 901 (yellow) or even 900 (violet) will do. The numbers against the coil leads in the diagram refer to their order on the coil former, "1" being nearest to the chassis.

The oscillator-multiplier stage is confined entirely to one side of the screen. L1 is tuned to the overtone frequency by the 30pF Mullard concentric trimmer and L2 to the second harmonic by the 8pF trimmer. Both L1 and L2 are air-spaced, and supported by their respective trimmers at one end and by grounded components at the other. The centre (earth) connections of the trimmers are pushed into small holes in the chassis (if brass or laminate) prior to being

TABLE 5.11
Components list

C1	47pF ceramicon	
C2, 4, 6, 9, 13	1000pF discoidal feed-through capacitor	Erie type CDFT/100.
C5, 11	1000pF disc ceramic	
C3	3·3pF ceramicon	
C7	4·7pF ceramicon	
C8	2·2pF ceramicon	
C10	2 to 8pF Philips trimmer	
C12	3 to 30pF Philips trimmer	
C14	0·1μF Transcap (disc ceramic)	
R1	470Ω	
R2	10kΩ	
R3	33kΩ	
R4, 8, 10	100Ω	
R5	470kΩ	
R6, 13	100kΩ	Note: R8, R10, C5 and C11 are not required for the positive ground version
R7	220Ω	
R9	330Ω	
R11	220kΩ	
R12	22kΩ	
T1	¼in dia polystyrene Aladdin former with two dust cores (see text)	
T2	Osmor Type "Q" Coil (see Table 5.10)	
TR1	BF180	
TR2, 3	40602 dual-gate mosfet	
X1	approx 70 or 71MHz HC-18/U miniature overtone crystal according to i.f.	
RFC1	15μH miniature rf choke, Painton or STC, or similar component wound on high-value resistor	

The crystal oscillator is clearly shown on the lower left.

soldered, for extra mechanical support. The HC-18/U miniature wire-ended crystal is soldered directly to the chassis at one end. When soldering to the crystal wires, *it is essential to use a thermal shunt* as, otherwise, the crystal connections inside the can may drop off.

The 0·1μF Transcap C14 is a precautionary measure, to help prevent i.f. interference being picked up on the power leads; it was not found necessary in the prototype converter and so is not shown in the photographs.

Alignment of the Converter

Lining up the converter is extremely straightforward, only four tuned circuits being involved. C12 should be adjusted until the oscillator starts. A check with a wavemeter will confirm the frequency although there is little chance of incorrect operation with the circuit component values shown. Next, C10 is tuned until the second harmonic appears (at about one-half of maximum capacitance). Between 0·7 and 1·2V rms should be measurable at the mixer gate 2. A setting of C12 will be found (at about two-thirds of maximum capacitance) which will combine reliable oscillator starting with a satisfactory oscillator injection level. Then, peak the i.f. coil T2 for maximum noise at the i.f. band centre. Tuning the secondary of T1 will be found to result in two noise peaks corresponding to the wanted signal range and the image. Select the higher frequency of the two and then adjust the primary of T1 for an enhancement of the noise peak. Minor trimming adjustments may now be made on a signal generator or actual signals, and the alignment is complete.

Some constructors have experienced difficulty with the local oscillator arrangement employed, often due to crystals of low activity being employed. Correct operation of the local oscillator will almost certainly be obtained if the following steps are carried out:

(1) If the converter is of the printed circuit variety, remove the common connection of the crystal, R12 and the mosfet gate from the board and "stand it up" in the wiring. Many crystals will not tolerate the extra few picofarads to earth of the board connection when used in this circuit.

(2) Ensure that the oscillator coils are wound with the correct gauge of wire. If they are close-wound with thinner wire in error, the inductance will of course turn out to be much too high. Several converters have failed to operate simply because L1 had too high an inductance. It is probably safer to pull out L1 to a length of about $\frac{7}{16}$in or so to ensure resonance, instead of leaving it close-wound.

(3) If the oscillator still will not perform, reduce the value of R9 from 330Ω to 180Ω, or even to 150Ω. To a certain extent the operating conditions of TR3 are a compromise between good oscillator operation, requiring linear gain and a high drain current, and good doubler performance, which needs a square-law mode and therefore low drain current. The quoted value of source resistance leads to a set of operating conditions in which the linear gain of the 40602 can be a bit on the low side and 180Ω produces a rather better compromise.

(4) The final stage of the proceedings involves removing the oscillator gate 2 and feed resistors R11 and R13 from the discoidal bypass capacitor C13 and, instead, bypassing the gate 2 and its resistors directly to the source with an extra 1,000pF disc ceramic capacitor. This produces a large increase in feedback which will accommodate even quite poor crystals that will not oscillate in other circuits. This circuit arrangement was not used in the original design as the degree of feedback so obtained usually causes instability with a crystal of average activity.

Modifications to Converter for 70MHz Operation

The following changes are required to convert the basic 144MHz design to a 70MHz one:

(1) Replace the 70 or 71MHz crystal by a 33·0 or 34·0MHz crystal to obtain band coverage with 4·025 to 4·700MHz and 2·025 to 2·700MHz tuned I.F.S, respectively.

(2) Replace the 2 to 8pF trimmer C10 by a similar 3 to 30pF component.

(3) Replace C3 by a 22pF tubular ceramic or polystyrene capacitor.

(4) Add C15, a 15pF tubular ceramic or polystyrene capacitor, from the mixer gate 1 to ground.

(5) Add C16, a 68pF tubular ceramic or polystyrene capacitor, in shunt with the oscillator trimmer, C12.

No modifications will be required to the coils.

The performance of the converter on 70MHz will be found to be very similar to that of the 144MHz unit. The correct settings of the trimmers are C10, three-quarters way in and C12, half way in.

432MHz HYBRID RING CONVERTER

This converter design by L. Williams, G8AVX, is built round a hybrid ring mixer using hot-carrier diodes. The hybrid ring mixer is commonly used for frequency conversion at frequencies above 1GHz where its dimensions become practical.

This type of mixer has many properties which render it attractive to the home constructor. It is simple to make, and its wideband characteristics make local oscillator injection much easier than in the case of cavity mixers, and also render tuning devices and precision construction unnecessary. When used with hot-carrier or Schottky diodes, which may now be obtained at reasonable cost, excellent performance combining low mixer noise, cross-modulation and conversion loss may be anticipated.

The principal obstacle in realizing the potential of this circuit on 432MHz is simply physical size. A conventional hybrid ring with air dielectric for 432MHz would be more than 300mm in diameter. The solution is to fold the ring in such a way as to make its dimensions more acceptable without prejudicing its desirable characteristics.

Since not all readers will be familiar with the hybrid ring mixer principle, it will be briefly described. The hybrid ring is a transmission line in the form of a closed ring $3\lambda/2$ in circumference, with four ports, each separated by $\lambda/4$. Mixer diodes are connected to two ports located $\lambda/4$ on each side of the local oscillator port; the signal is injected into the remaining port which is $\lambda/2$ from the local oscillator port and $\lambda/4$ and $3\lambda/4$ respectively from the two mixers.

The mixers are thus fed in phase by the local oscillator and in antiphase by the signal, as in some other types of lower frequency balanced mixer. This arrangement has the further advantage of providing a considerable degree of isolation between signal and local oscillator circuits.

In home construction the transmission line is usually an air-spaced strip line, suitable low-loss uhf dielectric materials being difficult to obtain and expensive.

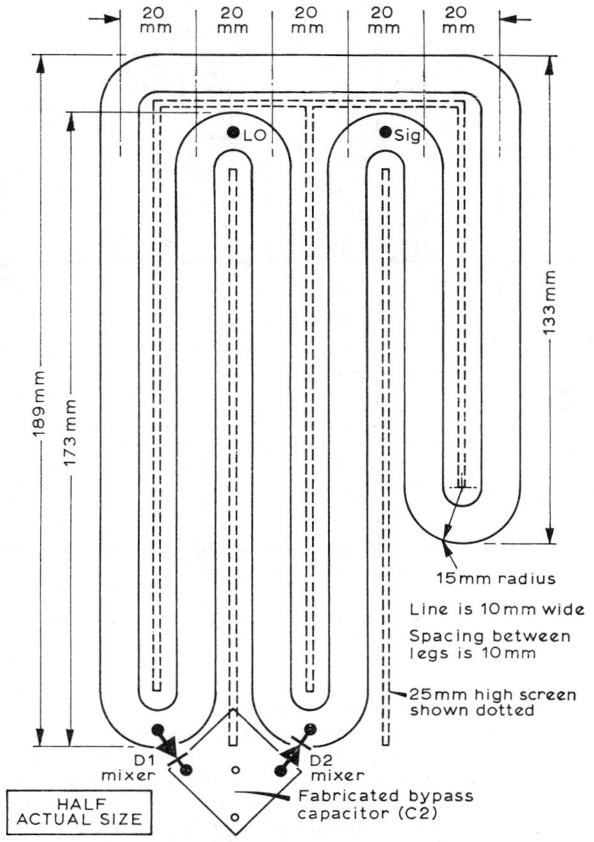

Fig 5.59. Folded hybrid ring mixer for 432MHz.

Folded Hybrid Ring Mixer

The ring (**Fig 5.59**) is folded into four parallel $\lambda/4$ sections, plus a $\lambda/2$ section which is folded and bent to close the ring. By these means the ring is reduced to a manageable rectangle and the mixer ports which are electrically $\lambda/2$ apart are brought into close physical proximity. Screens between adjacent legs of the folded ring are necessary.

The ring may be cut from any suitable light-gauge sheet metal: brass, copper or aluminium. Holes of 10mm diameter are drilled to form the inside radius of all bends and are then connected by shears or fretsaw. The ring is finished by filing to size.

The prototype ring was made from double-sided $\frac{1}{32}$in pcb with the copper on both sides connected at each port so that no potential difference existed between the upper and lower conductors. The board was therefore not used as a dielectric.

Holes are drilled at the port positions and the ring is mounted 10mm above an earth plane, which may be sheet metal or copper laminate.

In the prototype, coaxial sockets were used for the signal and local oscillator ports and small stand-off insulators provide support elsewhere.

Screens 25mm high are fixed to the earth plane between the adjacent legs of the folded ring. A specially-fabricated signal-frequency bypass capacitor for the mixers is made by fixing a 25mm square of single-sided $\frac{1}{32}$in pcb to the earth plane (*copper side up*) with a thin film of epoxy resin. This capacitor has a value of approximately 40pF.

The inter-line screen is undercut so that it may be positioned to allow minimum connection lengths to the mixer diodes. Motorola MBD102 diodes in miniature plastic "L packs" were used in the prototype; these are supplied with leads about 4mm long and were soldered between the ring and the bypass capacitor. Other types of uhf hot-carrier diode mixer in glass or strip-line packages would be equally suitable.

The hybrid ring mixer may be used as it stands by connecting an aerial and a local oscillator to the appropriate ports and taking the i.f. from the bypass capacitor. When the i.f. signal is fed direct from the mixer into a very good communications receiver quite acceptable results will be obtained, although the lack of gain will be apparent.

I.F. Amplifier

In the prototype an i.f. amplifier using a low-cost 2N5245 fet is employed after the mixer: see **Fig 5.60**. Because the low noise factor of the mixer makes very weak signals readable given sufficient gain, and because little selectivity is obtained

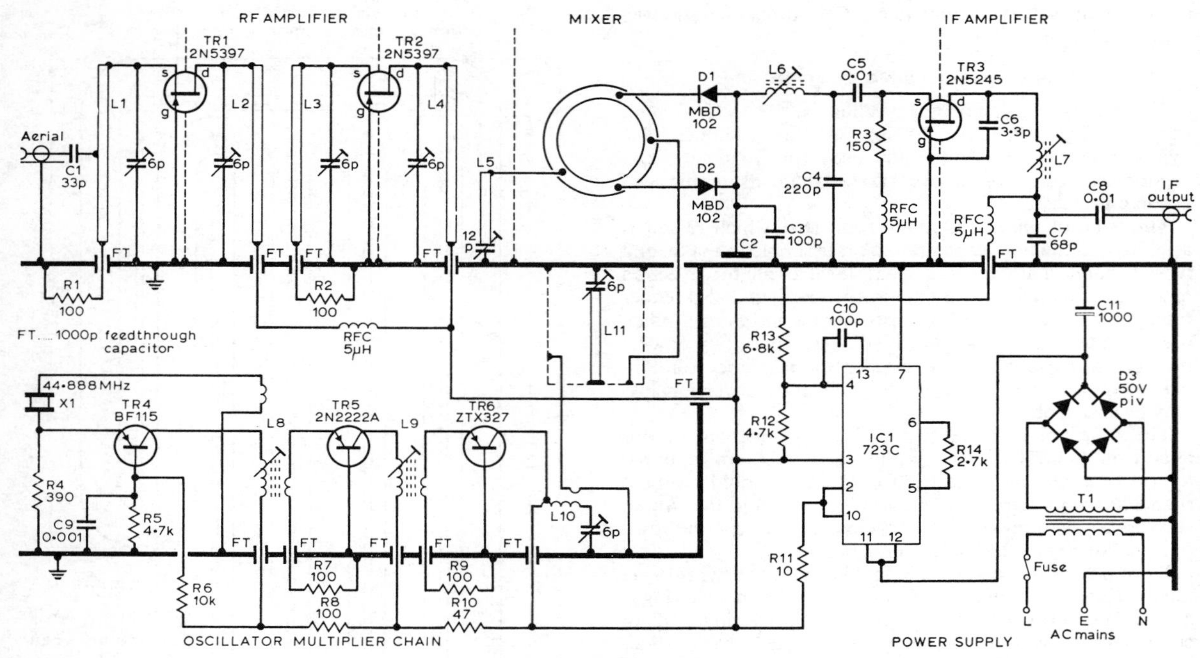

Fig 5.60. Circuit diagram.

before second conversion, the noise factor and cross-modulation performance of this first i.f. stage is important.

The i.f. impedance of the mixer is low and the grounded-gate configuration is the most suitable for direct connection to it. In practice a pi-network was used to provide a slight impedance change combined with an extra i.f. tuned circuit and low-pass filter function. The fabricated capacitor is padded to the required value by adding a silver mica type in parallel. The circuits shown are designed for an i.f. of 28–30MHz, but may be easily modified for other frequencies.

The first pi-circuit will have very flat tuning and is not at all critical. The drain circuit is matched to a 75Ω output by another pi-network, this time designed for a loaded Q of about 15. It is important that this i.f. output is terminated in the correct impedance or the required bandwidth will not be obtained. The design procedures for pi-networks given in Chapter 6 may be used to design networks for other intermediate frequencies or impedances if required.

The i.f. amplifier shown provides about 12dB voltage gain and makes the converter much less dependent upon the tunable i.f. performance. It is especially desirable if the converter is located some distance from the tunable i.f.

Local Oscillator

The local oscillator used in the prototype has a 44·888MHz overtone crystal oscillator, followed by two grounded-base tripler stages. This arrangement has the merit of simplicity. Other crystal frequencies and multiplication factors are of course possible. There are, however, two important differences between the hybrid ring and bipolar/fet mixers in respect of local oscillator injection.

Crystal oscillator multiplier chains have outputs at other multiples of the starting frequency as well as the desired injection frequency. With narrow-band mixers this is of less consequence, but the hybrid ring mixer has little discrimination against these unwanted injection frequencies. If inadequate pre-mixer selectivity is present, strong out-of-band signals, for example uhf tv, will mix with higher injection harmonics, resulting in products falling within the desired i.f. band. To prevent this a high-Q filter is interposed between the output of the multiplier chain and the mixer. In the prototype this is a $\lambda/4$ line with the input and output loose-coupled by short loops on opposite sides of the line. The tuning of this line is very sharp and if a piston trimmer having a finer adjustment thread than the Mullard type used in the prototype is available it should be used.

The other important requirement of the hybrid ring is mixer drive power. Reference to **Fig 5.61** shows that although

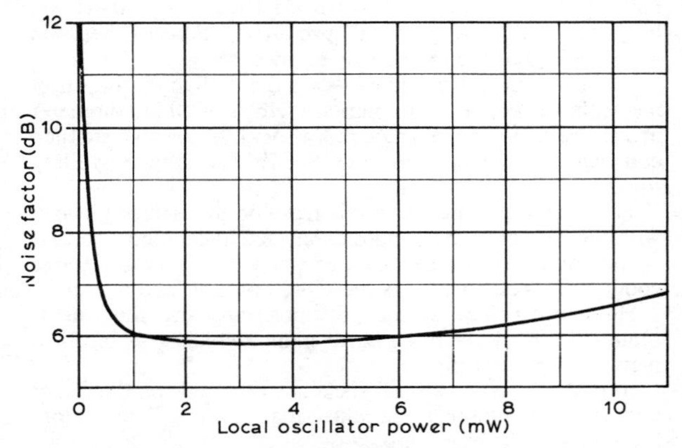

Fig. 5.61. Noise factor v local oscillator power for typical uhf hot-carrier diode mixer.

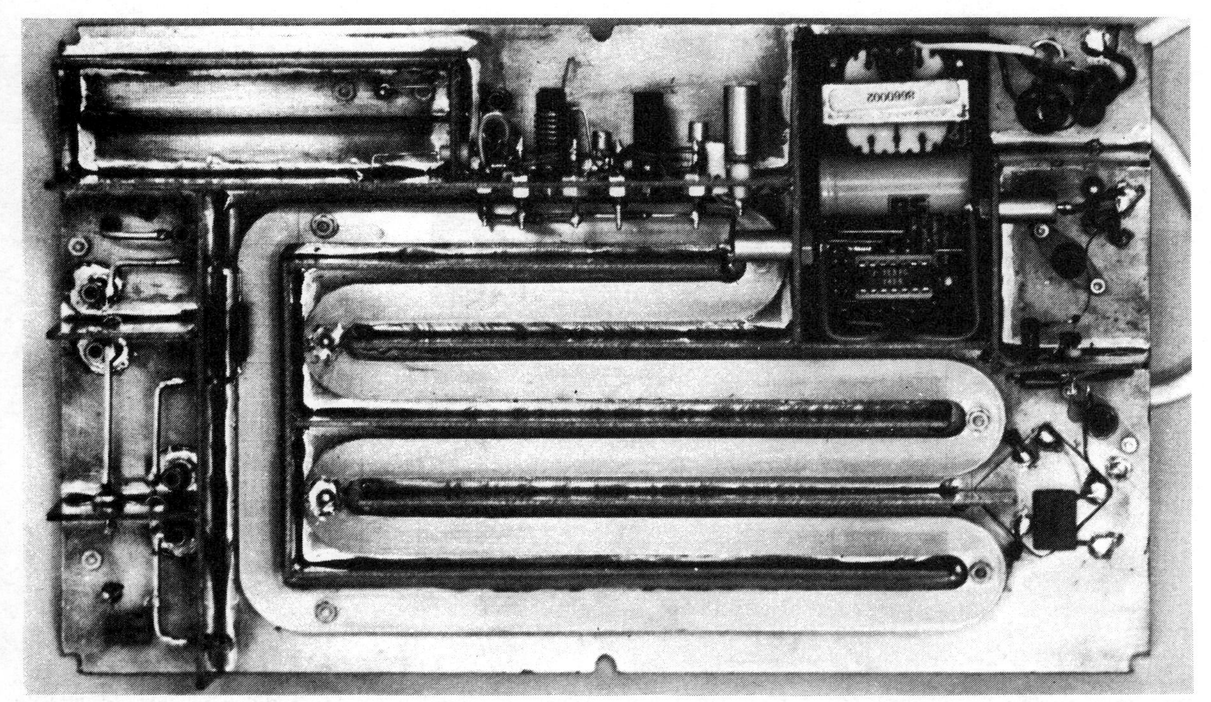

General view of the 432MHz hybrid ring converter. Left: the rf amplifier, with L1, C1 at the bottom. Top left: the oscillator chain and its high-Q break. Top right: the power supply and i.f. amplifier. Centre: the hybrid ring itself, with the fabricated bypass capacitor on the right.

drive level is not at all critical 1–10mW per diode will be required for the best performance. This is quite a high power when compared to the requirements of a fet or bipolar transistor mixer. As a rough guide, allowing for the final tripler efficiency, filter and coupling losses, the final tripler must be driven to approximately 100mW dc collector input. It is common practice to use tv front-end transistors such as the BF180 as the final stage of oscillator chains for uhf converters. These are not suitable at the level required for hot-carrier diodes. Their forward agc characteristic results in increased collector current causing reduced gain.

A low-cost plastic uhf power amplifier type, the Ferranti ZTX327, gave excellent results in the prototype when driven to about 8mA collector current.

The collector current may be monitored by a voltmeter across the emitter resistor. A temporary Lecher line strung across the shack makes an excellent absorption wavemeter for selecting the desired harmonic from each stage when tuning up the oscillator multiplier chain. For an injection frequency of 404MHz the nodal points will be 371mm apart.

RF Amplifier

Although the hot-carrier diodes give the mixer a very low noise factor (typically 6dB), it will benefit from a low-noise rf amplifier. A pre-mixer gain of not more than 20dB is desirable, and if the already good mixer performance is to be enhanced very good amplifier devices are required. FETs were preferred to bipolar devices due to their better cross-modulation performance. Two stages operating in grounded-gate configuration were used for simplicity in preference to a single grounded-source stage. This would have provided sufficient gain but would have required neutralization, which can be very critical to adjust for optimum noise and gain performance.

Unfortunately devices characterized for noise factor and gain a 1432MHz are not cheap. The prototype used the 2N5397, which has typically 10dB gain and 3dB noise factor in the circuit given. Other types have been substituted with success. The cheaper 2N4416 works well with a slight reduction in gain, and so does the low-cost 2N5245, with, however, a noticeable reduction in gain and increase in noise factor.

The tuned circuits in the rf amplifier are simple lengths of 14swg copper wire bent as required. L2 and L3 (also L4 and L5) are parallel spaced 12mm apart. The FETs are fitted in holes in the interstage screens so that good screening between the source and drain circuits is obtained. The gate and case terminals are soldered direct to the screen with lead lengths not exceeding 1mm.

No adjustment of the rf amplifier is required other than to peak all the tuned lines for maximum signal output.

Power Supply

The converter is designed to operate from a 12V negative-earth system and is therefore suitable for operations from a vehicle battery. The total current drain of approximately 50mA makes operation from dry batteries rather expensive. A mains psu for fixed-station operation is therefore included in the assembly. This psu uses the 723 ic regulator which gives excellent regulation at load currents up to 65mA with a minimum of external components and at a modest cost.

Close-up view of the fabricated bypass capacitor and associated components.

Construction

Although all the sections described were built into a single structure, each was built as a separate circuit with coaxial sockets for the signal terminals, inter-section connections being made with short coaxial links. If desired the sections may be permanently wired together and the expense of plugs and sockets avoided. The modular section arrangement has, however, much to recommend it. For example, the rf stages may be removed and the aerial feeder connected direct to the mixer when an aerial pre-amplifier is used.

A second oscillator multiplier chain allows an instant band change (there is no mixer tuning to adjust). Thus the converter can be rapidly switched to tv, satellites, or the French 70cm band. For the experimenter, comparisons may be readily made by plugging in experimental sections or taking signals out for other converters. If it is required to experiment with mixer injection, tv aerial attenuator pads may be inserted between the oscillator and mixer without disturbing the circuit.

Since the sections in the converter are independent their physical locations are not critical and no precise dimensions need be maintained. The only important features are adequate screening and filtering of the supply input to each section. The two cylindrical feed-through capacitors in the psu side screens are professional pi-section filters but ordinary 1,000pF feed-through capacitors will be adequate.

The prototype is housed in a large die-cast box, although an inverted aluminium chassis or any other shallow metal box will serve equally well.

All the components used are available from the usual mail-order houses except the MBD102, which was obtained from Jermyn Industries, and the 2N5397, which is available from Siliconix distributors.

All metalwork and assembly can be carried out with the normal hand tools found in any amateur workshop. Almost all the prototype was made from copper laminate off-cuts, cut with a small hacksaw and soldered together.

No special test instruments are required for setting up. Some form of absorption wavemeter for setting up the oscillator multiplier chain will be helpful, but if one is not available a signal should be found on the band and all trimmers adjusted for maximum S-meter reading. Provided the inter-stage screening is adequate that is all that is required to obtain a very high performance.

Conclusion

The prototype has been used for over a year under various conditions at the author's QTH, which is in line-of-sight with Sutton Coldfield, 10km away. It is the only converter tested on this site which is completely free of tv cross-modulation effects. The noise factor has not been measured but it is sufficiently low to make externally generated noise the

TABLE 5.12

Components list

R1, 2, 7, 8, 9	100Ω	
R3	150Ω	
R4	390Ω	
R5, 12	4·7kΩ	
R6	10kΩ	
R10	47Ω	
R11	10Ω	
R13	6·8kΩ	
R14	2·7kΩ	
All resistors are ¼W		
C9	0·001μF	ceramic disc
C10	100pF	polystyrene
C11	1,000μF	25V electrolytic
FT	1,000pF	feed-through (discoidal for rf amplifier)
Trimmers	6pF	Mullard ceramic tubular except that for L5, which is 12pF
C1	33pF	silver mica
C2	40pF	see text
C3	100pF	silver mica
C4	220pF	silver mica
C5, 8	0·01μF	ceramic disc
C6	3·3pF	ceramic tube
C7	68pF	silver mica
TR1, 2	2N5397	
TR3	2N5245	
TR4	BF115	
TR5	2N2222A	
TR6	ZTX327	
IC1	723C	ic regulator
D1, 2	MBD102	hot-carrier diodes
D3	50V piv rectifier bridge	
X1	44·688MHz 3rd overtone series-resonant	
T1	mains transformer, 15V 75mA secondary	
L1, 3	40 mm	14swg
L2, 4	30 mm	14swg
L5	25mm	14swg
L6	7t 20swg close-wound on 7mm former	
L7	34swg close-wound, winding length 8mm, on 7mm former	
L8	12t 34swg close-wound on 7mm former. 3½t 34swg coupling to TR5 close to cold end. 1½t thin bell wire wound over for crystal feedback	
L9	7t 22swg, spaced own diameter on 7mm former. 2t thin bell wire coupling	
L10	2t 20swg 12mm long, 5mm i.d. Coupling 1t thin bell wire at cold end. Collector tapping at ¾t	
L11	12mm wide strip 90mm long in 25mm square. Trough line with 15mm long coupling loops	
RFC	5μH (Radiospares 1A tv choke)	

limiting factor on most very weak signals when the converter is used with a tunable i.f. having an s/n ratio of 10dB at 2μV.

G8AMU MULTIMODE ALL-BAND RECEIVER

A general-purpose receiver providing for all modes of reception in any vhf or uhf band with provision for a wide range of refinements to be added at will is described below. Full constructional and alignment details are not provided but the design should provide useful ideas for the more experienced constructor.

The receiver was evolved with the following points in mind:

(1) Capable of receiving a.m., fm, ssb and cw.
(2) High sensitivity with low noise figure.
(3) Good cross-modulation and blocking performance.
(4) Good frequency stability and accurate dial setting.
(5) Free from spurious responses.
(6) Effective noise limiter.
(7) Good agc with fast or slow decay time.
(8) Self-contained, requiring only power and aerial inputs.
(9) 240V mains or 12V dc operation for portable use.
(10) Using modern, readily available devices and components.
(11) Capable of being updated without radical mechanical changes.

Principle of Operation

The complete receiver system is shown in the block diagram, **Fig 5.62.** The basic receiver is tunable over a range of 28–30MHz, this frequency being chosen for the first i.f., rather than a low frequency, because of the ease with which a flat 2MHz-wide band could be achieved without the use of large tuning capacitors, or the problem of tracking the rf and mixer circuits. Also the need for double conversion converters does not arise, due to the adequate separation of signal and oscillator frequency at least up to 1,296MHz. For frequencies higher than 1,296MHz converters should be fed into the 144MHz converter, making this frequency band the first i.f.

The selectivity is provided in the 10·7MHz crystal filter, which was the most costly part of the receiver. This filter is an fm type with a very flat top response; although suitable for a.m. and fm reception it is a little wide for serious ssb or cw work, but sufficient room has been left for a ssb filter to be added and switchable from the front panel. By using an i.f. at 10·7MHz, the image signals will be 21·4MHz away and out of the required tuning range.

The remaining i.f., detector and af circuits are fairly conventional with the outputs of the a.m. detector, fm discriminator and product detector switched into the audio stages. All supply voltages are stabilized and are derived from a 15kHz oscillator, the output of which is also stabilized against any voltage fluctuation of the input supply.

The converters for 70MHz, 144MHz and 432MHz are of conventional design, with their outputs at 28–30MHz switched into the receiver. High frequency crystals are used in the converters to reduce the number of multiplier stages and to ensure that the fundamental or harmonic frequencies do not appear in the passband of the tunable i.f.

The following controls are available on the front panel:

AF gain	Mode switch	Mains on/off
RF gain	AGC fast/slow	Internal ls on/off
Tuning	Noise limiter	S-meter
Band switch	Ext power on/off	

70MHz and 144MHz Converters

These converter circuits and printed circuit boards are identical apart from the crystal frequency, and the tuned circuits (see **Fig 5.63**).

The rf amplifier, TR37, is a dual-gate mosfet used in common source configuration. Bandpass capacitance coupled input and output tuned circuits are used to give selectivity and good rejection of out of band signals. The output of the rf amplifier is fed into gate 1 of the mixer, another dual-gate mosfet, and the output is taken from a low impedance link winding coupled to the tuned circuit L20 in its drain supply.

TR38 is an overtone oscillator operating at 42MHz for the 70MHz converter and at 58MHz for the 144MHz converter.

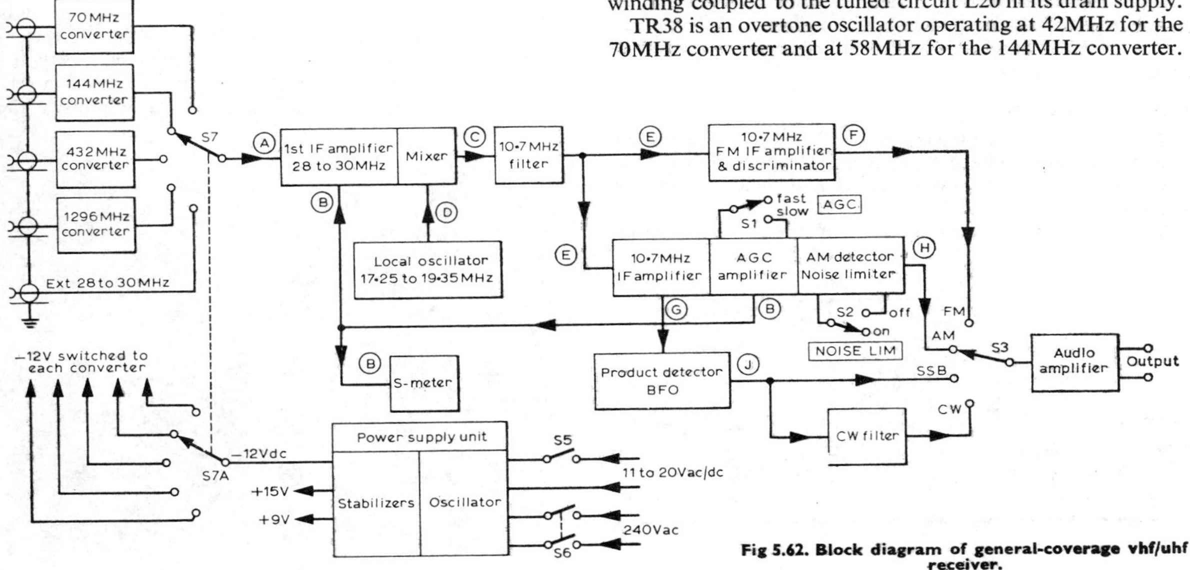

Fig 5.62. Block diagram of general-coverage vhf/uhf receiver.

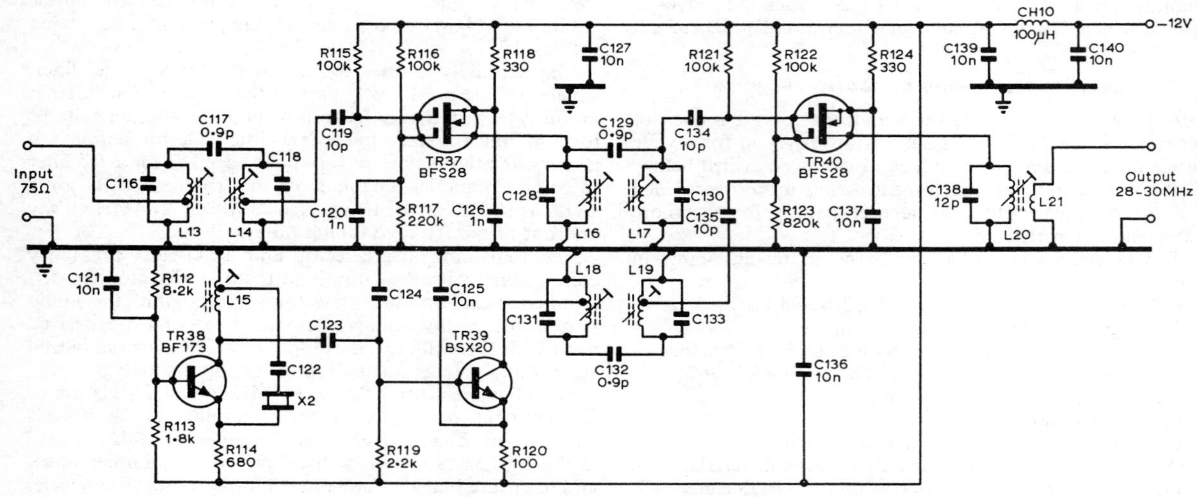

Fig 5.63. 70MHz and 144MHz converters.

The oscillator is capacitance coupled to TR39 which is an amplifier for the 70MHz converter and is used as a doubler to 116MHz for 144MHz. The output of the oscillator chain again uses capacitance-coupled bandpass tuned circuits to reduce to a minimum any other multiple of the oscillator frequency reaching gate 2 of the mixer.

Coupling between the tuned circuits is very small and in the region of 0·9pF; this can be made up from two 1·8pF capacitors in series, or a short length of twisted pvc-covered wire.

The converters use a negative power supply for the npn transistors so that the tuned circuits can be taken direct to the chassis, which eliminates any possible instability caused by poor rf decoupling. An additional BF180 rf amplifier, TR41, was added to the 144MHz converter because the noise figure of the BFS28 was not acceptable, Fig 5.64.

If the use of converters for the microwave bands is contemplated, then this amplifier should be arranged to operate on the 144MHz band only and a similar amplifier incorporated in each converter.

432MHz Converter

Two grounded base rf stages, TR43 and TR44, are used with emitters and collectors tapped into λ/4 tuned lines. Inductive coupling is used between the rf and mixer stages (see Fig 5.65).

The local oscillator chain consists of three stages, starting with an overtone oscillator, TR42, at 67·333MHz. The second stage, TR45, is tripling to 202MHz and the final stage doubling to 404MHz with TR46. All interstage tuned circuits are coupled by mutual inductance employing tapped inputs and outputs.

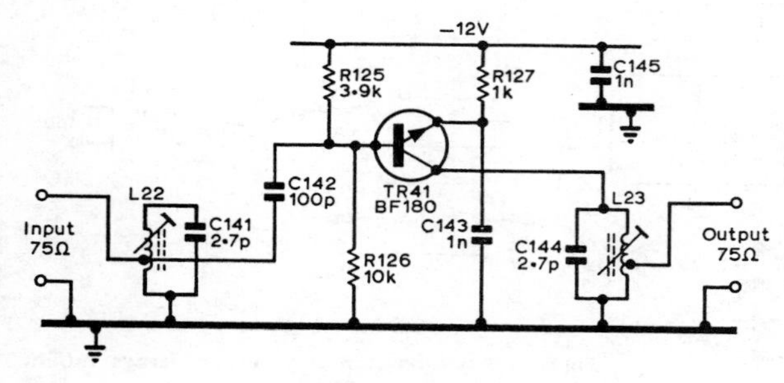

Fig 5.64. 144MHz rf amplifier.

Oscillator injection at 404MHz is also inductively coupled to the fet mixer TR47 by L34, and the output at 28MHz is taken from a link wound over L36 in the drain of the mixer TR47.

The converter is built on double-sided copper laminate board with interstage screening, and the rf and mixer transistors are mounted in slots cut in the screens dividing input from output.

28–30MHz Tunable I.F. Amplifier and Mixer

The purpose of the i.f. stage is to overcome mixer and subsequent i.f. noise (particularly when used with a converter having no rf amplifier), and to provide front end selectivity by enabling more tuned circuits to be added. The rf and mixer circuits are not tunable over the band, but a flat response with

70MHz and 144MHz converters.

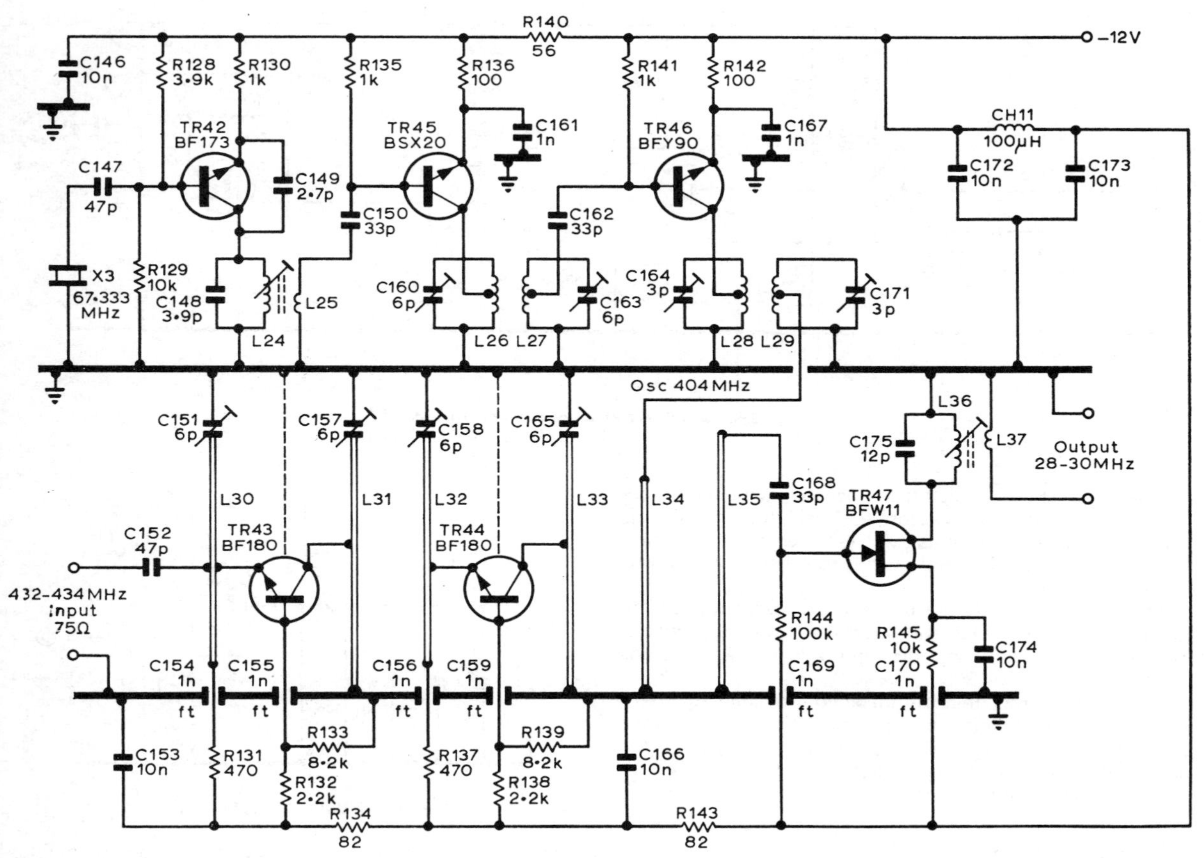

Fig 5.65. 432MHz converter.

sharp cut-off at band edges is achieved, by using capacitance coupled, bandpass stagger-tuned circuits (see **Fig 5.66**).

The rf amplifier TR2 has forward gain control characteristics which can reduce the stage gain by as much as 50dB under maximum signal conditions. To achieve this, the agc voltage must be positive going to increase the collector current of the rf stage. As the agc provided for the i.f. stages is negative going, TR1 is used as an inverter and the zener diode D1 delays the agc voltage operating the rf stage until the gain of the i.f. amplifier has been reduced. The gain of the rf stage is not reduced until the agc voltage has fallen from its no-signal level of +2V to approximately 1·2V, causing the collector voltage of TR1 to rise to 7·5V above the base voltage of TR2, when D1 will conduct and increase the current through the rf stage, thus reducing its gain.

The mixer TR3 is a dual-gate mosfet, which is an excellent device for low cross-modulation, blocking and spurious response performance, due to its square-law characteristics and its very high isolation between gates, which also reduces to a minimum the pulling of the local oscillator under large signals.

The local oscillator signal of 17·25–19·35MHz is fed into gate 2 of the mixer and the rf signal of 28–30MHz into gate 1, and the difference of 10·7MHz is extracted from the mixer's drain by L5.

Local Oscillator

The most stable oscillator tried with a high output in the region of 6V p-p to drive gate 2 of the mosfet mixer was the fet Colpitts (see **Fig 5.67**). The oscillator covers a frequency range of 17·25–19·35MHz, which gives 50kHz above and below the 2MHz tuning range required. Good temperature stability was achieved by using silver mica capacitors and

432MHz converter.

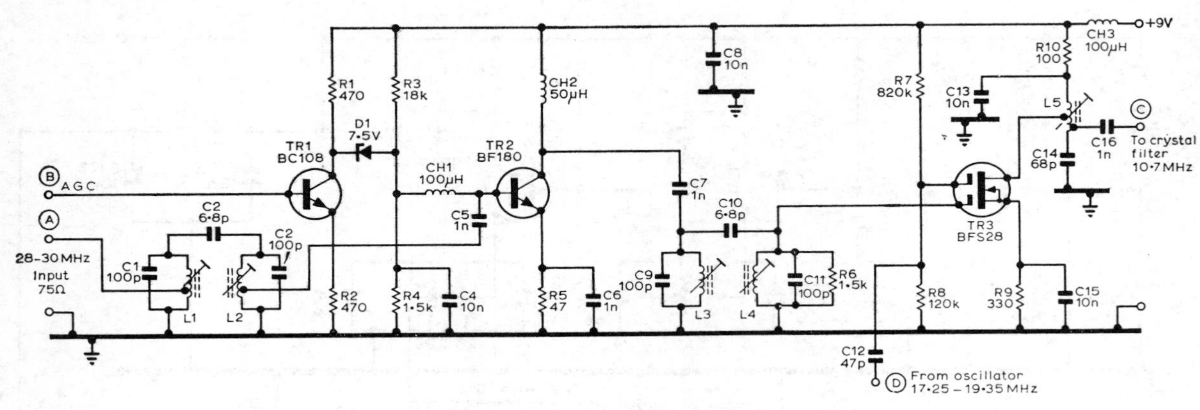

Fig 5.66. 28–30MHz rf amplifier and mixer.

a 39pF ceramic C27, which has a negative temperature coefficient of −150 ppm/°C across the oscillator coil L6. It is supplied with +15V from its own stabilized supply to eliminate any frequency shift and fm due to other circuits affecting the supply voltage.

The 4th, 8th and 24th harmonics of the oscillator can fall in the 70MHz, 144MHz and 432MHz bands respectively, but adequate screening and decoupling of the supply leads eliminates this.

The oscillator was built on tinned copper laminate board mounted on aluminium for mechanical strength, and then bolted into the base of a diecast box, which is fitted to the back of the dial. The tuning capacitor is of high quality with ball-bearing races fitted at each end.

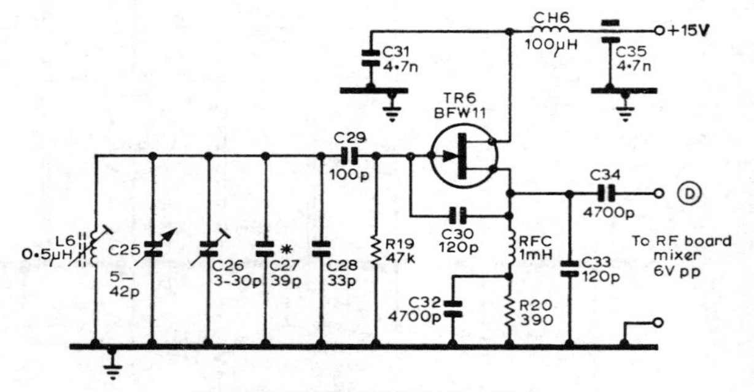

Fig 5.67. 17·25–19·35MHz oscillator

Crystal Filter Buffer Board

The i.f. filter (Fig 5.68) must provide sufficient selectivity to discriminate against stations operating on adjacent frequencies, and still have a sufficiently broad response so that the outer sidebands of a desired fm signal are not distorted. The filter used is a 10·7MHz fm type which has a very good flat top response with about 1dB of ripple and is 80dB down at ±12·5kHz. Although the bandwidth is rather large for ssb reception it is very convenient for shf operation since a moderate amount of frequency drift can be tolerated.

Transistor buffers are used on each side of the crystal filter to avoid the mixer and i.f. circuits loading the filter and spoiling its response, and so that the correct terminating impedance for the filter can be used. These values are given in the manufacturers' specifications, and for the Nikko Denshi filter type B10F12A which was used they are 1kΩ and 10pF. The values for C20 and C22 are 8·2pF and 6·8pF added to the output capacitance of TR4 and the input capacitance of TR5 making an effective value of 10pF.

TR5 is a fet source follower with its output taken to the

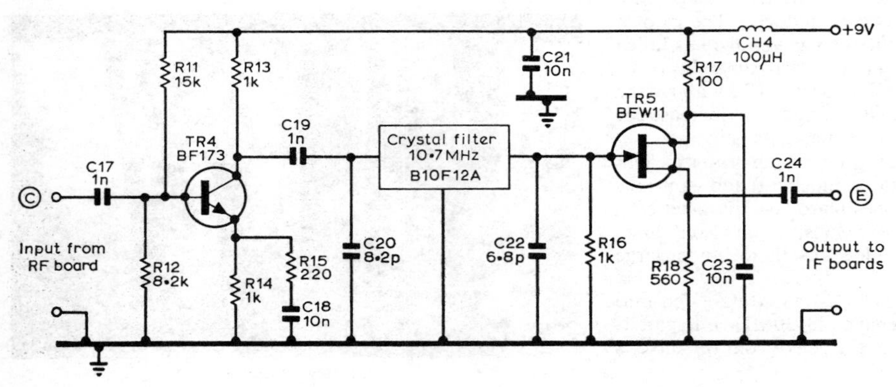

Fig 5.68. 10·7MHz crystal filter buffer board.

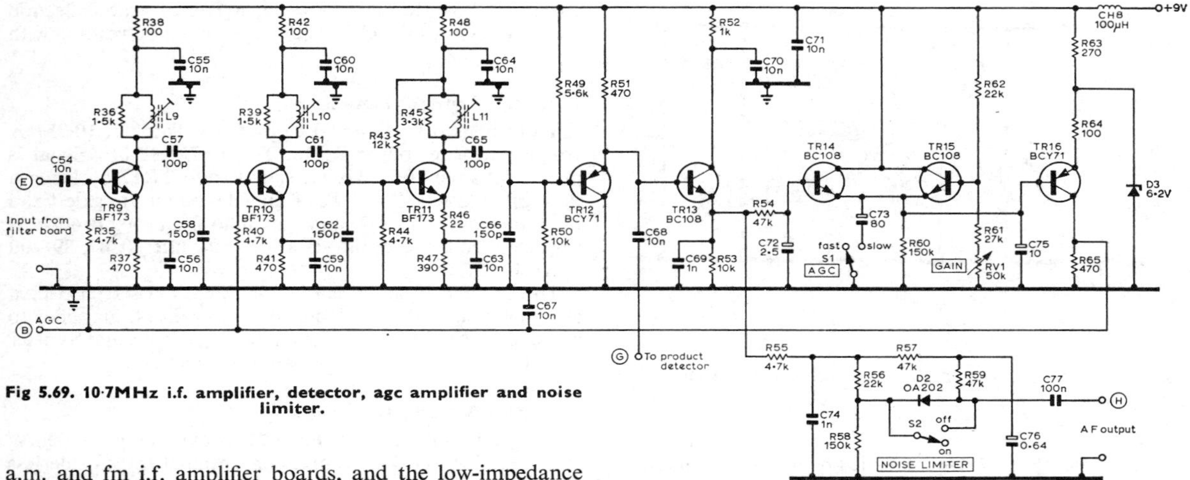

Fig 5.69. 10·7MHz i.f. amplifier, detector, agc amplifier and noise limiter.

a.m. and fm i.f. amplifier boards, and the low-impedance output enables other specialized amplifier/detector boards to be added in parallel if required.

10·7MHz A.M. I.F. Amplifier, Detector, AGC and Noise Limiter

The main i.f. amplifier (Fig **5.69**) uses three stages, tuned at 10·7MHz, and has a gain in the region of 100dB. The first two stages, TR9 and TR10, are agc controlled. TR12, an emitter follower stage, supplies the i.f. signal to the product detector and also to TR13, an emitter follower peak envelope detector for a.m. signals. The detector signal from the emitter of TR13 is taken via the usual low-pass filter to a conventional series-type noise limiter, D2, and then to the audio stages. The noise limiter can be switched off by bypassing the series diode D2 with the switch S2.

The agc voltage is obtained by comparing the dc component in the detected signal with a reference voltage. The comparator, TR16, is stabilized at 6·2V with a zener diode D3, and the agc voltage is taken from the collector of TR16 which is at approximately +2V under no-signal conditions. As the voltage at the base of TR14 rises with an increasing signal level, the emitter voltage will also rise, causing the collector current of TR16 to fall, thus reducing the agc voltage across R65. The agc normally has a fast attack and fast decay, but when C73 is connected across the emitter resistor R60 of TR14 and TR15 with S1, it will produce a slow decay of the agc voltage suitable for ssb and cw. RV1 is the rf and i.f. gain control which manually overrides the agc voltage by causing the base voltage of TR15 to rise, which in turn will increase the base voltage of TR16, thus reducing its collector current and the agc voltage to the level set by RV1.

FM I.F. Amplifier and Discriminator Board

TR7 and TR8 are used as a two-stage high-gain, low-noise, tuned amplifier at 10·7MHz (**Fig 5.70**). As most fm detectors are sensitive to changes in the amplitude of the received signal, they must be preceded by a limiter that removes all amplitude fluctuations caused by fading and noise spikes; this is provided by an integrated circuit IC2 which is a four-stage high-gain fm/i.f. limiting amplifier using long-tailed pair stages which clip symmetrically. Thus when a certain signal level is reached further increases in input signal produce no change in the output. The limiting amplifier also has very low fm distortion and high a.m. rejection in the region of 60dB.

The fm agc, which is derived from the a.m. detector board, the rf gain control acting to give a variable agc delay, is only applied to the rf stage to reduce overloading of the mixer with very large signals. The output of the integrated circuit

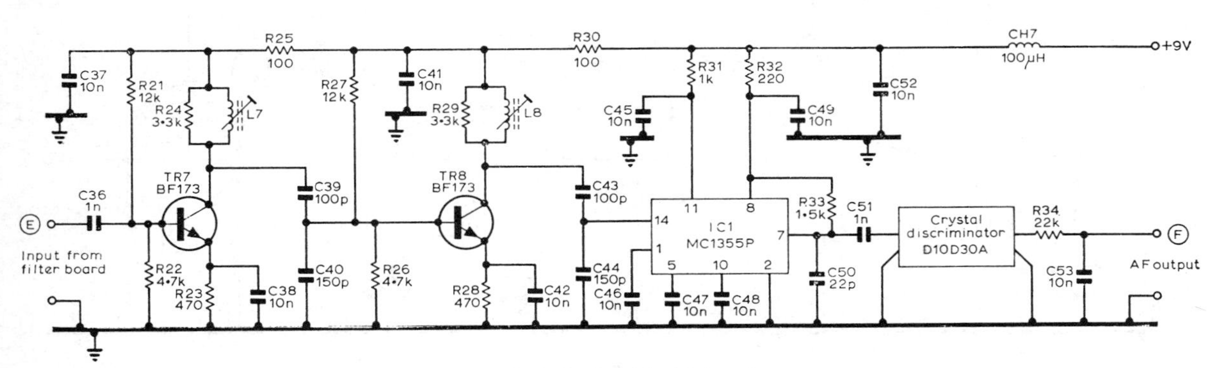

Fig 5.70. 10·7MHz nbfm i.f. board.

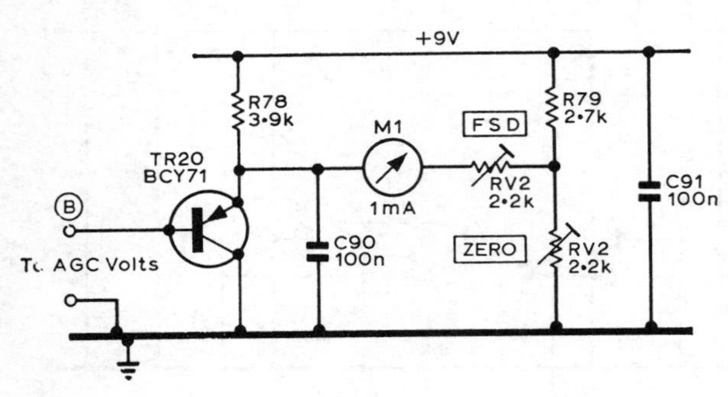

Fig 5.71. S-meter.

limiter is coupled into a crystal discriminator and then to the audio stage.

S-meter

The S-meter (**Fig 5.71**) is controlled by the agc voltage, and under no-signal conditions the voltage at the emitter of TR20 is equal to that of the junction of RV3 and R79. As the agc voltage falls with a received signal the emitter of TR20 will rise above that of the RV3, R79 junction causing current to flow through the meter M1. Full scale deflection is set by RV2, while RV3 is adjusted for a zero reading with no input signal.

BFO and Product Detector

TR17 is used as a crystal Colpitts oscillator at 10·7MHz coupled to a fet buffer TR18. The 10·7MHz bfo signal is mixed with the 10·7MHz i.f. signal from TR12 by using a dual-gate mosfet TR19. The i.f. signal is applied to gate 1 and the bfo signal to gate 2, and the audio frequency product is extracted from the drain via a low-pass filter (C87, 89 and R77): **Fig 5.72.**

The bfo and product detector is fitted in a screened box, and mounted well away from the high-gain i.f. amplifier to prevent the 10·7MHz bfo signal affecting the agc and S-meter circuits. The agc can thus be used on ssb and cw.

AF Amplifier and CW Filter

The audio output stage (**Fig 5.73**) provides about 500mW into a 3Ω loudspeaker from a conventional transformerless Class B amplifier, using a complementary pair of germanium transistors TR25 and TR26. The preset trimmer RV5 is adjusted to give a quiescent current of approximately 10mA with no input signal.

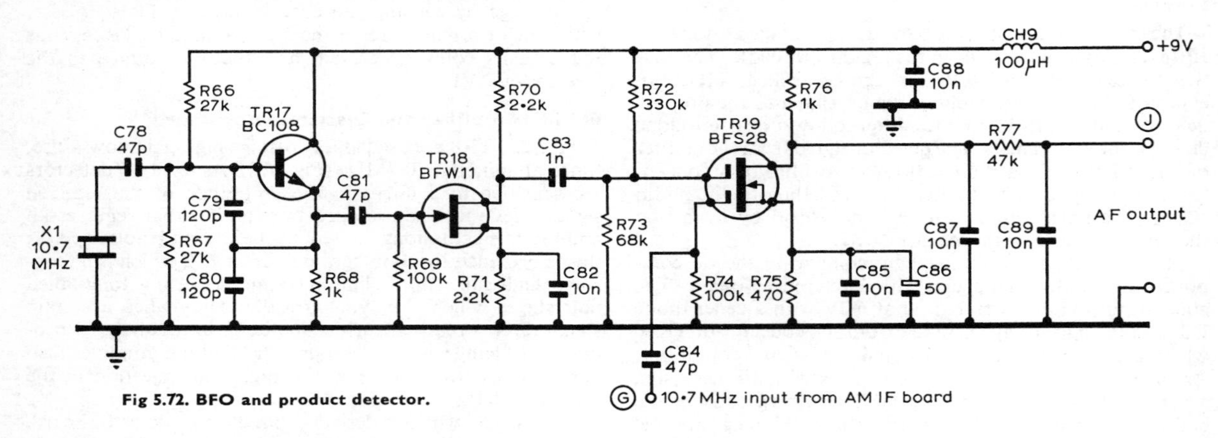

Fig 5.72. BFO and product detector.

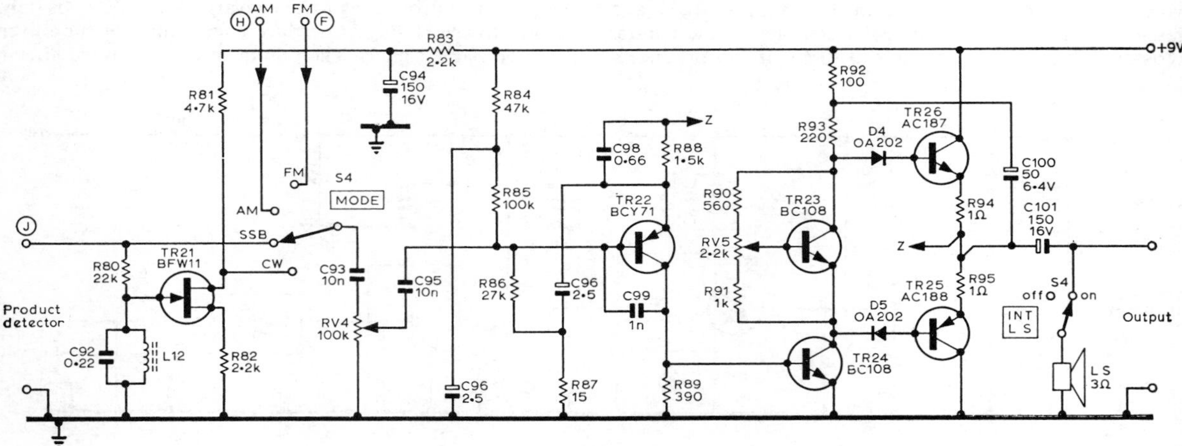

Fig 5.73. Audio amplifier and cw filter.

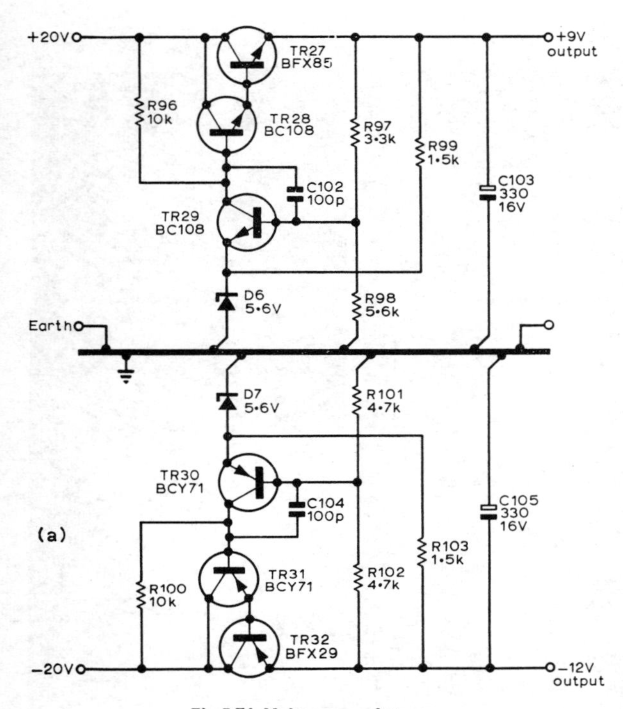

Fig 5.74. Voltage regulators.

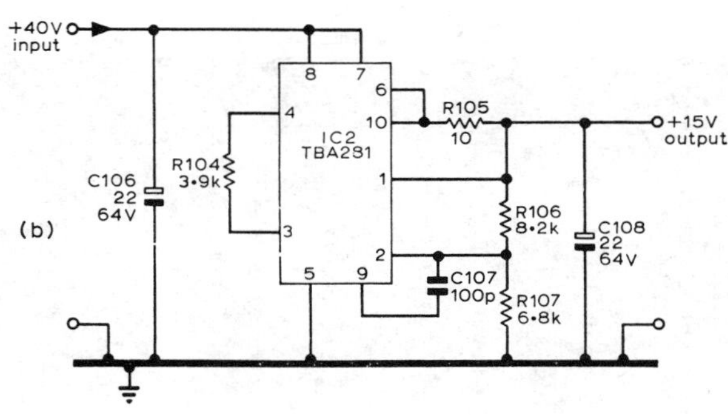

The cw filter has a bandwidth of about 100Hz at a frequency of 800Hz, but this frequency could be changed to suit the operator by adjusting the value of L12 or C93. The width of the filter can be decreased by reducing the value of the series resistor R80. TR21 is a single-stage fet amplifier to make up for the attenuation of the filter and to keep the audio output at the same level when the cw mode is used.

Power Supply

Three stabilized voltages are used for the receiver: +9V for all the main boards, +15V for the local oscillator and −12V for all converters. The +9 and −12V supplies **(Fig 5.74(a))** are derived from conventional stabilized power-supply circuits, while the +15V supply uses a monolithic voltage regulator IC2 **(Fig 5.74(b))**.

The input to the power supply **(Fig 5.75)** may be a little unusual; it can be supplied with its own 12V ac from the receiver transformer or from an external supply of 11 to 20V ac or dc positive or negative earth. All supply voltages are fed through a bridge rectifier D8 to a simple series regulator supplying 8V. The psu is floating from earth enabling any polarity supply to be connected to its input since the bridge rectifier remains in circuit.

The regulated 8V dc supplies a 15kHz oscillator TR35 and TR36, and the output of the secondary winding of oscillator transformer T2 is rectified with D12 and D15 giving +20V and D13 and D14 giving −20V which supplies the +9V and −12V regulators. D16, 17 and C112 form a voltage doubler giving +40V, which is used to supply the +15V regulator using IC2.

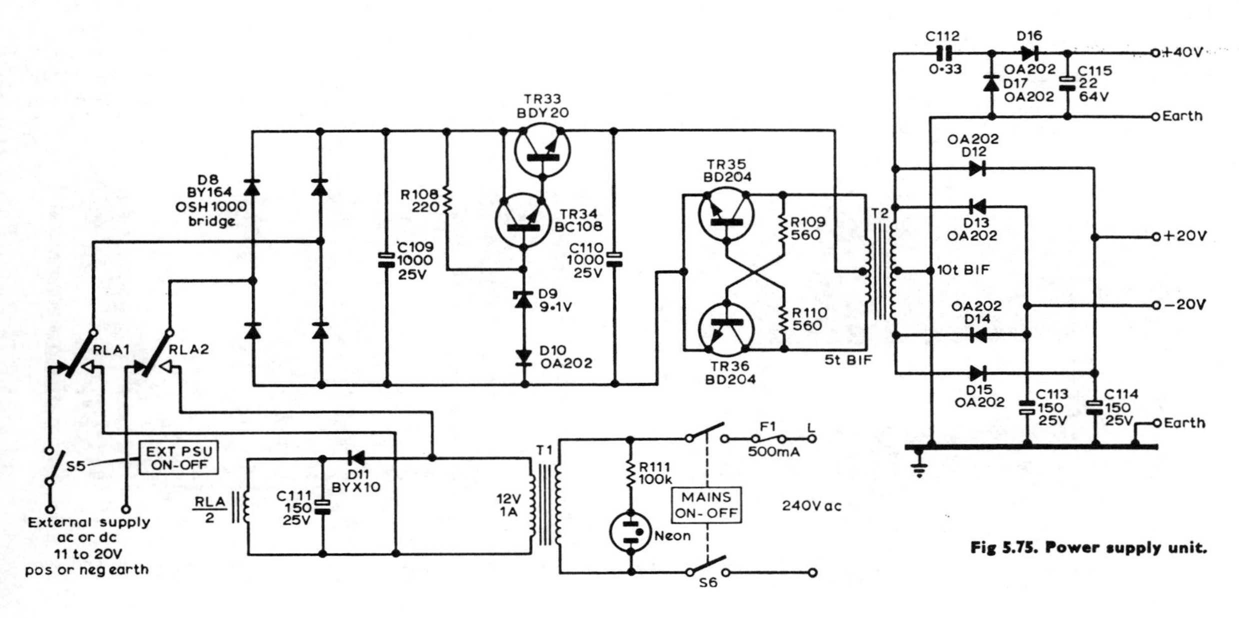

Fig 5.75. Power supply unit.

Underchassis view of general-coverage receiver, showing (left to right): af board and cw filter; 10·7MHz a.m. i.f. agc detector and noise limiter, 10·7MHz fm i.f. board; rf/mixer board; and power supply. The B10F12A 10·7MHz crystal filter is mounted on the board at the top.

Relay 1 is normally mains-energized, but if the supply fails the relay automatically connects an external supply of either polarity to the bridge rectifier D8.

General Construction

The complete receiver measures 15in by 9in by 6½in. The ront and back panels are constructed from 10swg aluminium and are connected by four horizontal bars of ⅜in square-section plated mild steel 8¾in long, the ends of which are drilled and tapped for 4BA screws, for fixing the panels. A platform of 14swg aluminium is fitted between the front and back panels by two 15in by ⅜in bars to make a very strong chassis. The top, bottom and side panels are fitted around the chassis with 4BA screws tapped into the plated steel bars.

Front view with top removed. The small unit at the back with two coil cans is the 144MHz rf amplifier.

Top view, showing local oscillator (with cover removed) and converters mounted on the sides and back.

TABLE 5.13

Component details

CAPACITORS

All ceramic except those stated below.

C26	3–30pF	Bee Hive trimmer
C27	39pF	
C28	33pF	1 per cent sm
C29	100pF	1 per cent sm
C30, 33	120pF	1 per cent sm
C35	4·7pF	Bush mounting lead-through
C92	0·22μF	Polyester foil
C98	0·66μF	Polyester foil
C112	0·33μF	Polyester foil
C117, 129, 132	0·9pF 2 × 1·8pF in series	
C151, 157, 158, 160, 163, 165	0·8 to 6·8pF	Tubular ceramic trimmer
C154, 155, 156, 159, 169, 170	1nF	Solder-in lead-through
C164, 171	0·8 to 3·8pF	Tubular ceramic trimmer

1·8pF to 330pF	Mullard 632 series ceramic
390pF to 4700pF	Mullard 630 series ceramic
1nF to 22nF	Mullard 629 series ceramic

RESISTORS

All resistors type Mullard CR16 0·2W carbon except those listed below

R94, 95	1Ω	CR25 0·33W
R108	220Ω	3W ww
RV1	50kΩ	Linear
RV2, 3, 5	2·2kΩ	0·05 trimmer pot
RV4	100kΩ	Log

ZENER DIODES

D1	BZY88 7·5V
D3	BZY88 6·2V
D6, 7	BZY88 5·6V
D9	BZY88 9·1V

CRYSTALS

X1	10·7MHz Type HC/18/C
X2	42MHz (70MHz converter) 58MHz (144MHz converter) Type HC/18/U
X3	67·333MHz Type HC/18/U
Xtal filter	10·7MHz Type Nikko Denshi B10F12A or Cathodeon BF4129
Xtal discriminator	10·7MHz Type Nikko Denshi D10D30A or Cathodeon BF4781

CHOKES

CH1, 3, 4, 6, 7, 8, 9, 10, 11	100μH
CH2	50μH
CH5	1mH

COILS

All coils wound on 4mm coil formers with dust cores and screening cans unless otherwise stated. Type Neosid 722/1

L1	8t 22swg tap 2t
L2	8t 22swg tap 3t
L3, 4	8t 22swg tap 3t
L5	30t 34swg tap 3t and 15t
L6	7t 20swg 0·4in former with dust core (0·5μH)
L7, 8, 9, 10, 11	30t 34swg
L12	500t 30swg LA1216 former (180mH)
L24	8t 34swg
L25	2t 34swg over L24
L26, 27	5t 18swg ¼in dia ⅜in long
L28, 29	3t 18swg ¼in dia ⅜in long
L30, 32	3in 14swg tap ¾in from cold end
L31, 33	3in 14swg tap 2in from cold end
L34	2·5in 20swg pvc covered
L35	3in 16swg
L36	28t 32swg
L37	5t 32swg over L36
T2	5t bifilar primary 10t bifilar secondary 26swg FX2241 former

70MHz and 144MHz CONVERTER COMPONENTS

Coils	70MHz	144MHz
L13	9t 24swg tap 2t	8t 22swg tap 1t
L14	9t 24swg tap 5t	8t 22swg tap 3t
L15	15t 26swg tap 3t	15t 26swg tap 3t
L16, 17	9t 24swg	5t 22swg
L18, 19	15t 26swg tap 5t	5t 22swg
L20	26t 32swg	26t 32swg
L21	3t 26swg	3t 26swg
L22, 23	Not used	8t 22swg tap 1t

Capacitors	70MHz	144 MHz
C116, 118	18pF	2·7pF
C122	22pF	6·8pF
C123	15pF	27pF
C124	47pF	100pF
C128, 130	18pF	6·8pF
C131, 133	15pF	12pF

All coils wound on Neosid 4mm 722/1 formers with dust cores and screening cans

MISCELLANEOUS

S1, 2, 4, 5, 6	Miniature toggle switch
S3	4-way single-pole switch
S7	6-way 2-pole switch
T1	12V 1A heater transformer
RLY1, 2	Miniature relays with two c/o contacts
Dial	Eddystone Type 898
M1	0–1mA S-meter

A 4in elliptical loudspeaker is fitted into the top of the case.

The receiver was constructed as a number of sub-units so that each section could be set up and tested in turn before installing it into the receiver. This method also simplifies the drawing of the printed circuit boards and enables any board to be replaced in the future with a new or improved design without upsetting the rest of the receiver's calibration.

Each of the main receiver printed circuit boards is fitted to the underside of the chassis on 6BA pillars with vertical aluminium screens between them. The product detector was fitted in a screened brass box and mounted as far away as possible from the i.f. board to eliminate any of the 10·7MHz bfo power from reaching the i.f. and operating the agc circuit. The top side of the chassis is used to accommodate the local oscillator, converters and part of the power supply, leaving ample room for any other converter or unit to be added if required. Special care should be taken that the local oscillator is mechanically as well as electrically stable.

CHAPTER 6

HF TRANSMITTERS

THE purpose of a transmitter is to generate radio frequency power which may be keyed or modulated and thus employed to convey intelligence to one or more receiving stations. This chapter deals with the design and adjustment of that part of the transmitter which produces the rf signal, while the methods by which this signal may be keyed or modulated are described separately in other chapters. Transmitters operating on frequencies below 30MHz only are discussed here. Methods of generating rf power at frequencies higher than 30MHz are described in Chapter 7—*VHF/UHF Transmitters*.

One of the most important requirements of any transmitter is that the desired frequency of transmission shall be stable within close limits to permit reception by a selective receiver, and to avoid interference with other amateurs using the same frequency band. Spurious frequency radiations, capable of causing interference with other services, including television and broadcasting, must also be avoided. These problems are considered in Chapter 17—*Interference*.

The simplest form of transmitter is a single stage self-excited oscillator coupled directly to an aerial system. Such an elementary arrangement has, however, three serious disadvantages:

(*a*) The limited power obtainable with adequate frequency stability;
(*b*) The possibility of spurious frequency radiation, and
(*c*) The difficulty of securing satisfactory modulation or keying characteristics.

In order to overcome these disadvantages, the oscillator must not be called upon to supply power to the following stage. Since frequency multipliers are essentially class C stages (ie, they require driving power), and since power amplifiers are usually operated in class C these stages must, in the interests of efficiency, be isolated from the oscillator by a buffer stage, which is capable of supplying this power without loading the oscillator.

Since the hf amateur bands are harmonically related, the oscillator is generally designed to operate on the lowest frequency band. In order to achieve output on the higher frequencies, it is necessary to employ one or more stages of *frequency multiplication*. An advantage gained from this arrangement is that by operating the oscillator at a relatively low frequency, it is easier to achieve the necessary frequency stability. A single multiplier stage may produce twice the fundamental frequency—*doubler*, three times—*tripler*, or more. Usually, however, the frequency multiplication in one stage is limited to three or four, owing to the rapid diminution in output power as the order of frequency multiplication is increased. It is undesirable to feed the aerial directly from a multiplier stage since harmonics, other than the wanted signal, are produced and may be radiated simultaneously. Where attenuation of unwanted signals is of the utmost importance, the power amplifier must not be driven directly from a frequency multiplier stage. The coupling between one stage and the next may be accomplished in a variety of ways, some of which permit physical separation between the various units. Others have the advantages of simplicity, harmonic rejection and so forth. Methods of coupling are described later.

The power supply for the transmitter may be obtained from conventional circuits such as those described in Chapter 16—*Power Supplies*. It is common practice to use separate power supplies for the low and high power sections of the transmitter, first because the voltages used in the two parts often differ greatly and, secondly, because interaction between the various stages can make the adjustment more difficult when a common supply is used. Unwanted modulation effects may also be introduced. The basic principles of transmitter circuit arrangement are illustrated by the block diagram in **Fig 6.1.**

THE CRYSTAL OSCILLATOR

The simplest method of achieving the high degree of frequency stability is by using a quartz crystal to control the frequency of the oscillator. Such an oscillator when correctly designed and adjusted remains the most frequency stable device available to the amateur.

The quartz crystal, cut in the form of a suitably dimensioned plate, behaves like a tuned circuit of exceptionally high Q-value and may therefore be connected in the oscillator circuit as would a normal frequency-determining tuned

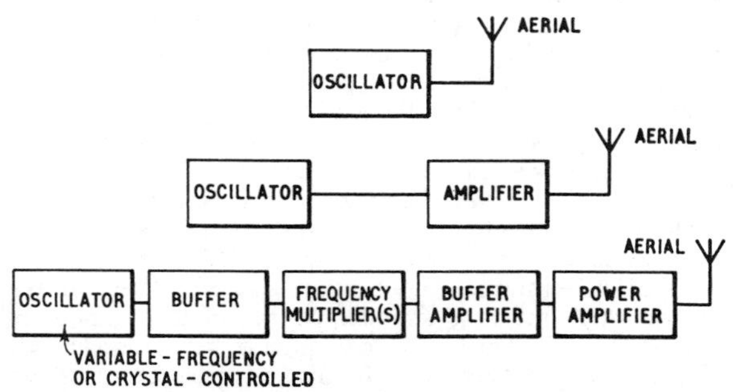

Fig 6.1. Block diagram of typical transmitters showing three possible combinations of oscillators, amplifiers and frequency multipliers. Many other arrangements are in common use, depending on the power output required and the number of frequency bands on which the transmitter is to operate.

circuit. The action of the quartz crystal depends on the piezo-electric effect explained in Chapter 1—*Principles.*

In the natural vibration of a quartz plate when used in such an oscillator unit, a small but appreciable amount of heat is generated by the internal frictional effects. The temperature rise may cause the otherwise very stable frequency to drift from its nominal value. This frequency drift can be quite serious, especially when the fundamental frequency is multiplied several times in the subsequent stages of the transmitter.

If the rf current through the crystal is allowed to become excessive, the vibrations of the quartz plate may become so large that fracture occurs.

In order to limit the rf current and the consequent temperature rise and to reduce the danger of crystal breakage, the power output of an oscillator must be kept at a relatively low value. Certain types of crystal have low temperature coefficients which reduce the amount of frequency drift likely to be encountered, but as these crystals are often very thin there is an increased possibility of damage. The power output of an oscillator using such a plate is therefore rather less than with other types of crystal.

There are several common types of crystal cuts, some of these, particularly the AT- and BT-cuts, being commonly used for hf transmitter control by reason of their low temperature coefficients. The X- and Y-cuts have relatively high coefficients. Some examples of the temperature coefficients for various crystal cuts are given in **Table 6.1.**

TABLE 6.1

Characteristics of various cuts of quartz crystals.

Type of Cut	Normal Frequency Range (MHz)	Temperature Coefficient (Hz/MHz/°C)
X	1 - 5	−20
Y	1 -10	+75
AT	0·5- 8	0
BC	1 -20	−20
BT	1 -20	0
GT	0·1- 0·5	0

Although the temperature coefficients of the AT- and BT-cuts are given in Table 6.1 as zero, this is true only at certain temperatures; at other temperatures the coefficient may have a small positive or negative value. Only the GT-cut has a true zero temperature coefficient over a wide range of temperature (0–100°C) but its use is unfortunately restricted to low-frequency applications. For exceptionally high-precision operation over long periods, the crystal may be installed in a thermostatically controlled oven, although in amateur transmitters sufficient frequency stability can almost always be obtained by isolating the crystal from any part of the transmitter which becomes heated in normal operation, such as valves, resistors and power transformers.

The equivalent circuit of a crystal contains a resistance *R* in series with one of its branches. This represents the mechanical resistance to vibration and is related to the energy which must be supplied to the plate to maintain it in a state of vibration; a low value of resistance implies a high degree of "activity". If this resistance becomes too great, it will be difficult or impossible to maintain oscillation. The crystal is then said to be "sluggish" or inactive. The fault is occasionally found to be due to the contamination of the crystal surfaces. These surfaces must therefore be kept scrupulously clean and dry and without any trace of grease. When necessary crystals can be cleaned with carbon tetrachloride or even with soap and water.

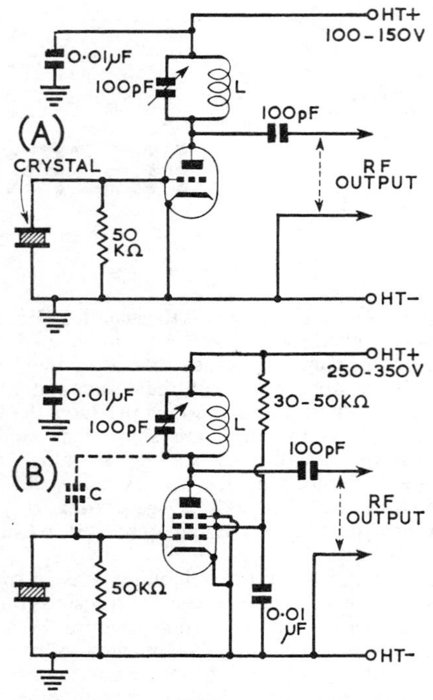

Fig 6.2. Two examples of simple Pierce-Miller oscillator circuits. For operation on 3·5MHz, L should have an inductance of approximately 23μH; for 7MHz, 7·5μH.

Crystal Oscillator Circuits

Almost any small receiving triode, tetrode or pentode valve may be used in a crystal oscillator. Solid state oscillators are described on page 6.6. The use of a small valve is recommended since the power output must be limited in order to reduce the possibility of overheating and crystal fracture.

The simplest possible circuits using a triode or pentode in Pierce-Miller arrangements are shown in **Fig 6.2.** In these circuits the reactance of the anode circuit causes feedback through the anode-to-grid capacitance with sufficient amplitude and *in the correct phase* to maintain oscillation of the crystal.

As the anode tuning capacitance is reduced from its maximum value a point will be reached at which there is a sharp drop in the anode current. This denotes that the conditions for oscillation are being fulfilled and that the crystal is vibrating. At the lowest part of the "dip" the crystal will be vibrating at maximum amplitude. Continuing towards minimum capacitance, the feed current begins to rise again more gradually and the power output drops. When the oscillator is loaded, for example by the input resistance of a following stage, the dip in anode current is less marked than with the unloaded valve. **Fig 6.3(a)** shows how the feed current can be expected to vary with the tuning adjustment.

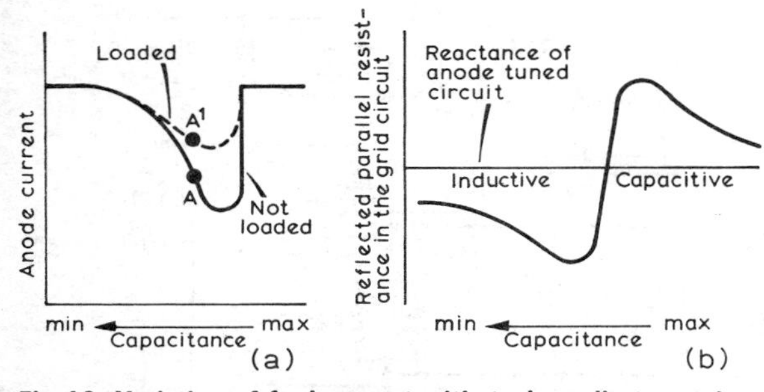

Fig 6.3. Variation of feed current with tuning adjustment in a crystal oscillator. When the circuit is not loaded the dip in anode current is very pronounced, but it becomes less marked when a load is applied. For stable operation the tuning should be adjusted to correspond approximately to the point A (or A[1] when loaded).

Oscillation is not, however, just a case of the magnitude of anode circuit impedance. Oscillation occurs in the vicinity of the resonant condition of the tuned circuit but only when it is inductively reactive at the operating frequency. The reflected impedance in the grid circuit, due to the Miller effect, then has a negative resistive component, ie power is fed back to sustain the losses of the grid circuit. When the anode circuit is capacitively reactive, the reflected impedance has a positive resistive component and oscillation is not possible. At the exact point of resonance the resistance is zero. This explains the sudden "flop" of the crystal oscillator as the anode circuit is tuned from an inductively reactive condition to a capacitively reactive one. The reflected resistive component is shown in **Fig 6.3(b).**

With a triode valve (Fig 6.2(a)) in which the interelectrode capacitance between anode and grid is likely to be relatively high, and so gives more than adequate feedback, the anode voltage should be limited to between 100–150V for BT and similar low temperature coefficient cuts, with the power output level between 0·5–1W. If X cut crystals are employed, it is permissible to increase the applied voltage to 250V, and the power output to 2–3W. However, when maximum stability is required, the power in the crystal oscillator should be kept to the lowest possible value consistent with adequate drive to the following stage; if the desired conditions cannot be achieved a buffer stage should be used after the crystal oscillator.

With the pentode or tetrode version of the simple crystal oscillator shown at **(b)** in Fig 6.2 the ht supply can be increased without risk of damage to the crystal and the corresponding power output can be increased to a possible maximum of 5W with a suitable valve. This is because a pentode or tetrode has greater gain, allowing a reduction of the rf crystal current for a given output power. A further advantage is that the amplitude of the output can be controlled by varying the screen voltage with a suitable potentiometer. If there is very effective screening between the grid and anode circuits, the small capacitor *C* of 1–2pF may be required to provide sufficient feedback to maintain oscillation.

The rf current through an X-cut crystal may be monitored by means of a small low-current lamp in series with it, such as a 60mA lamp. Alternatively the current may be measured with a suitable thermocouple meter of about 60mA full-scale deflection. AT-cut and BT-cut crystals are not suitable for such high currents and no convenient method is available to the amateur for current monitoring purposes. The most satisfactory technique, therefore, is to ensure that the condition whereby excessive current is obtained cannot arise by limiting the power input to the stage. A convenient method of quickly checking the power output of an oscillator, in the absence of better test equipment, is to couple a low-power lamp to the anode-circuit inductor (or "tank" circuit) with a single-turn loop of wire: see **Fig 6.4.** The loop will absorb sufficient power from the anode coil to cause the lamp to light.

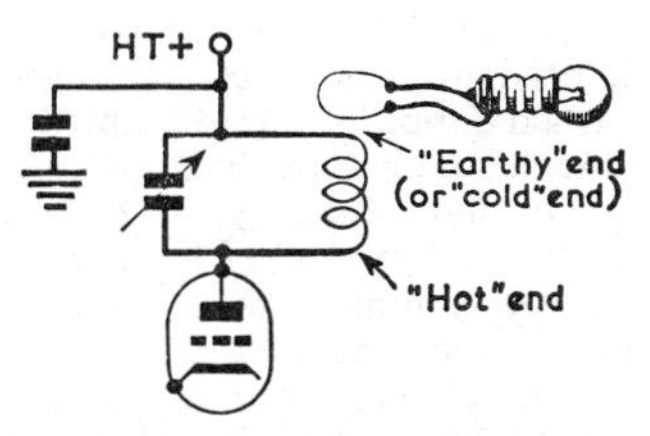

Fig 6.4. A dial-lamp with a single turn loop of wire soldered to it will glow brightly when coupled to the coil in a tuned circuit in which rf power is present.

The Pierce-Colpitts Oscillator

In the Pierce-Colpitts oscillator, shown in **Fig 6.5,** the crystal operates at a frequency just below the parallel resonance frequency and its equivalent inductance resonates with the anode-to-earth and grid-to-earth capacitances. The simple untuned circuit shown in Fig 6.5(a) will oscillate with crystals of any frequency and is therefore convenient where widely different frequencies are to be used for various channels. However, if the crystal happens to have a spurious

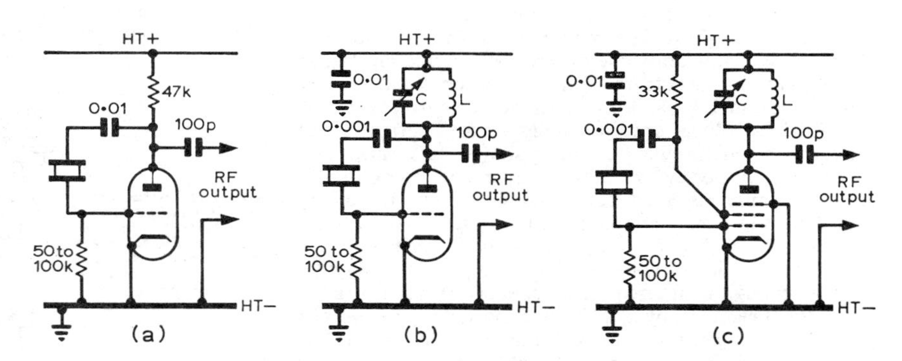

Fig 6.5. The Pierce-Colpitts crystal oscillator. (a) shows the simplest form of untuned oscillator In (b) the resonant circuit enables the fundamental frequency or harmonic output to be selected. The electron coupled circuit shown at (c) is especially useful for developing a large harmonic output.

TABLE 6.2

Harmonic output voltage compared to crystal frequency and the order of the harmonic when employing a 6AG7 valve in the circuit of Fig 6.5 (c).

Harmonic	Fundamental Crystal Frequency			
	3·7MHz	5MHz	11MHz	14MHz
Second	117	142	43	41
Third	58	71	26	25
Fourth	27	21	17	15
Fifth	21	16	10	9
Sixth	12	12	5	4·3
Seventh	7	7	3	—

These voltages were obtained using an anode voltage of 250 volts.

mode of high activity at a frequency differing from the desired frequency, this may be accidentally excited. The arrangement shown at (b) avoids this possibility by the use of a tuned circuit *LC* to select the desired frequency and at the same time gives appreciably greater output. An electron-coupled circuit based on the untuned Pierce-Colpitts oscillator is shown at (c). Here the screen grid acts as the "anode" of the oscillator and electron coupling takes place through the earthed suppressor grid from the screen to the actual anode. It is dangerous to tune the output tank circuit *LC* to the crystal frequency, since the crystal current may then be excessive. Any desired harmonic can, however, be selected by suitably tuning the anode circuit. Measurements taken on a typical pentode valve indicate that harmonics up to the sixth are of useful magnitude and **Table 6.2** shows the voltages that can be expected.

The Tritet Oscillator

The triode-tetrode (tritet) circuit was intended mainly for the production of harmonics of the crystal frequency, but is now really only of classical interest. A complete practical circuit is shown in **Fig 6.6**: it must not be used with modern miniature crystals owing to the danger of fracture.

An analysis shows it to be a form of tuned-anode/tuned-grid oscillator. In this case the crystal constitutes the resonant grid circuit and L1C1 is the resonant feedback circuit, the effective anode being the screen grid (as in the electron-coupled oscillator). Besides the feedback which occurs in the oscillator section, coupling takes place from the main anode circuit since current through this circuit passes also through the tank circuit L1C1 in the cathode lead. Feedback is therefore least when the stage is not loaded.

The circuit is adjusted by first tuning the cathode capacitor C1 from its maximum value downwards; oscillation begins when L1C1 is tuned to the fundamental crystal frequency and is indicated by a drop in anode current. The anode circuit L2C2 is then tuned for minimum feed current at the chosen harmonic frequency (which should be verified with an absorption wavemeter); at this setting the power output is at its maximum. By increasing C1 slightly above the setting which gives minimum anode feed current, the crystal current will be reduced and the frequency stability improved.

The value of L1 is important; it must be so chosen that C1 is large enough at the required resonance frequency to act as a low-reactance bypass for the harmonic frequencies which may be generated in the anode tank circuit L2C2. A recommended arrangement is a 200pF variable capacitor in parallel with at least a 100pF fixed capacitor. The addition of this fixed capacitance has the advantage that it will limit

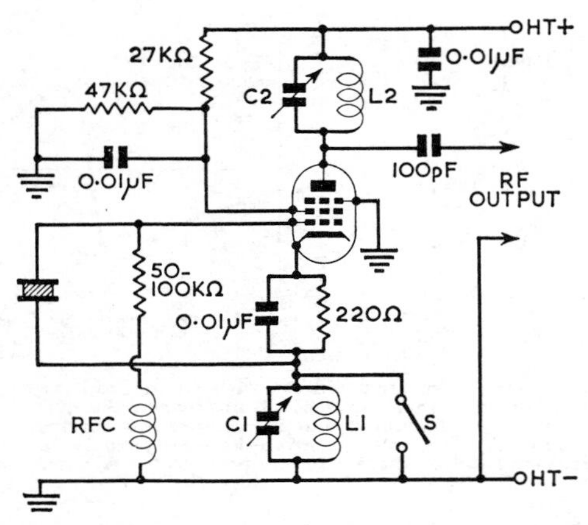

Fig 6.6. The tritet oscillator. The oscillator feedback circuit L1C1 is tuned to approximately the crystal frequency, while the anode tank circuit L2C2 selects the desired harmonic frequency. Care is needed to avoid tuning L2C2 to the crystal frequency. If output is required at the fundamental the switch S must be closed to short-circuit L1C1. Many different types of valve may be used, eg 6AG7 or 5763.

the tuning range and prevent the selection of a harmonic frequency with its associated heavy crystal current and the possibility of overloading caused by oscillation between the two tuned circuits. Output at the fundamental frequency is obtainable by short-circuiting L1C1. The circuit is then identical with that shown at (B) in Fig 6.2. An important point to note is that the anode circuit must never be tuned to the fundamental frequency of the crystal without L1C1 being short-circuited; otherwise there will be excessive crystal current owing to self-oscillation between the anode and cathode circuits.

Greater power output can often be obtained by the use of a beam tetrode rather than a pentode. However, if the latter is preferred the suppressor grid should have an independent connection which must be directly earthed: a pentode in which the suppressor grid is internally joined to the cathode will not be satisfactory. Valves with good screening between anode and grid circuits are preferable.

If it is inconvenient to allow the cathode of a tritet oscillator to be at a radio-frequency potential with respect to earth, the circuit may be rearranged as shown in **Fig 6.7.** The circuit then operates in exactly the same way as before, except that the electron coupling to the true anode is accomplished without permitting the output current to pass

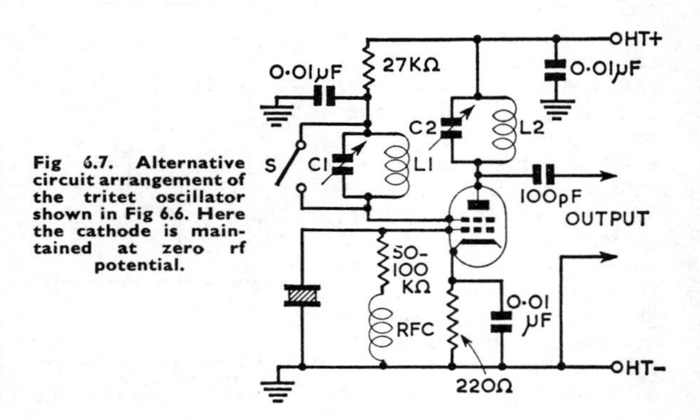

Fig 6.7. Alternative circuit arrangement of the tritet oscillator shown in Fig 6.6. Here the cathode is maintained at zero rf potential.

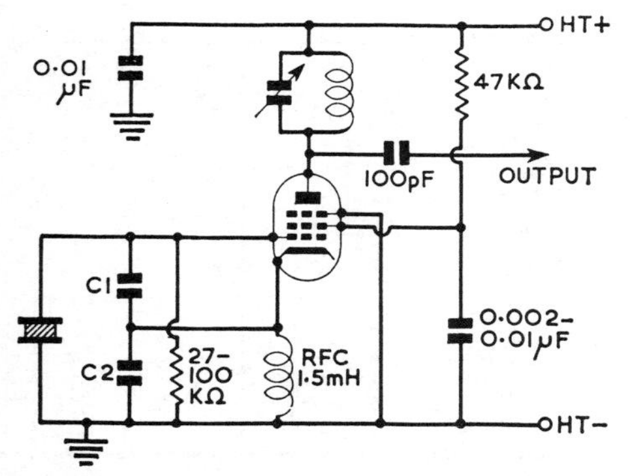

Fig 6.8. Harmonic crystal oscillator. Ample power output is available on the fundamental and harmonic frequencies. Recommended valves: EF91, 6AU6, EL91, 6AM5 or similar.

Crystal Frequency	C1	C2
2-4MHz	30pF	100pF
4-16MHz	22pF	47pF

through the oscillator tuned circuit L1C1. At the harmonic frequencies, since the output current is not passed through the oscillator tuned circuit L1C1, slightly greater output can be expected than with the first arrangement. The usual precautions to avoid danger to the crystal should be observed.

Colpitts Oscillator

The circuit of the "harmonic" crystal oscillator, shown in **Fig 6.8**, appears to resemble that of the tritet oscillator. In fact, however, it is a modification of the Colpitts oscillator described on page 6.8; the "anode" of the oscillator section proper (ie the screen) is at zero rf potential and feedback takes place through the capacitors C1 and C2. The anode circuit is electron-coupled to the oscillator section, and good power output at either the fundamental or a harmonic frequency can be obtained. No cathode tuning circuit is required, the purpose of the rf choke being to provide a dc return path for the cathode current while at the same time

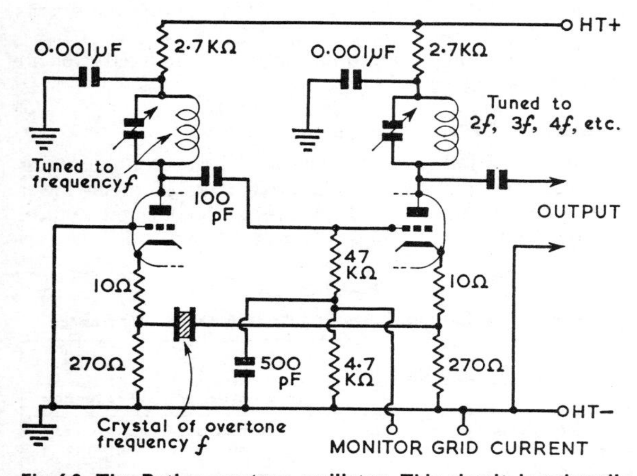

Fig 6.9. The Butler overtone oscillator. This circuit is primarily intended for use with crystals designed for overtone operation.

allowing the cathode to take up the correct rf operating potentials.

Overtone Oscillators

The oscillator circuits described above produce vibration of the crystal plate in its fundamental mode. Any required harmonic output is then obtained by the non-linear operation of the valve, the harmonics being selected in the tuned-anode tank circuit. Harmonic output can alternatively be obtained by making a crystal oscillate at an "overtone" of its fundamental. Theoretically the foregoing circuits could be adjusted to give overtone oscillation of the crystal itself, but it would in general be necessary to inhibit the tendency of the crystal to vibrate at the fundamental frequency; this may be difficult to accomplish with the circuits described.

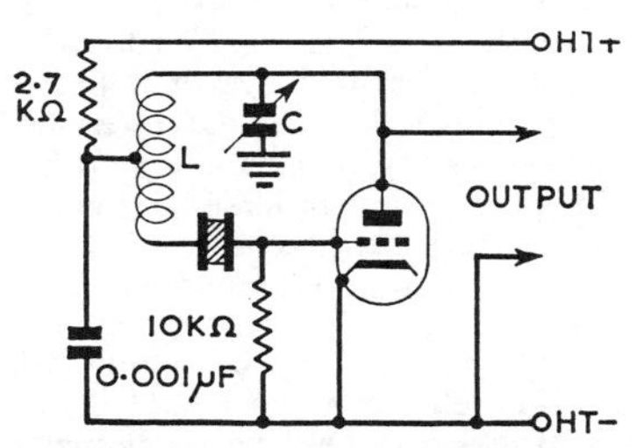

Fig 6.10. The Squier overtone oscillator. The circuit should resonate at the overtone frequency and most small triodes are suitable. The valve may be one half of a twin triode such as a 12AT7. Note that C resonates only with the upper portion of L.

Usually, circuits which assist the overtone mode of operation cause oscillation at the series-resonance frequency (of the overtone mode) and are typified by the Butler, Squier and Robert Dollar circuits: these are shown in **Figs 6.9, 6.10** and **6.11** respectively. Overtone oscillation is more often required for vhf applications where a large frequency-multiplication factor is wanted, and further information on these circuit types is therefore given in Chapter 7—*VHF/UHF Transmitters*. Nevertheless, the possibility of using overtone modes to assist in the avoidance of interference (for example, with television) at certain frequencies is suggested. The separation of the harmonics produced (compared with operation of the crystal at its fundamental frequency) is increased; the harmonics which are produced are multiples of the overtone frequency.

When such an oscillator is adopted, care has to be taken in the design of the transmitter as a whole to ensure that none of the unwanted harmonics are eventually radiated.

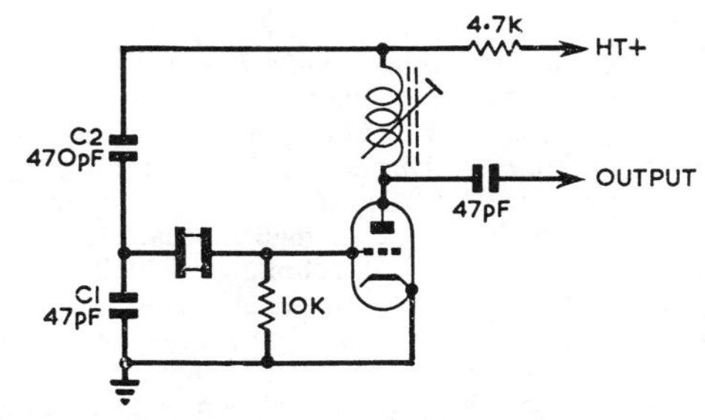

Fig 6.11. The Robert Dollar overtone oscillator. This circuit does not require the use of a tapped coil, and may be used with the same valves as the Squier oscillator.

Varying the Crystal Frequency

There are two methods of varying the frequency of a crystal oscillator. One is mechanical and the other is electrical.

The mechanical method requires an alteration of the dimensions of the crystal plate or its mass in some way. This generally involves the grinding of the surface to *increase* its resonant frequency or the loading of the vibrating part to cause a *reduction* of the frequency. Such changes to the plate must of course be regarded in general as permanent and irreversible. Crystal grinding, mentioned later in this chapter, requires great care and mistakes made through lack of skill can easily have disastrous results. The limit to the mechanical loading of a crystal is usually set by a gradual decrease of activity as material is added; for this reason a maximum reduction of about 25kHz is possible with a 3·5MHz plate. Suitable loading material for a crystal is cold soft solder. It is gently and gradually rubbed into the two surfaces as evenly as possible near the plate centre; the activity should be checked at regular intervals, and in the event of a decrease in activity being noted no further loading should be attempted. Another material found suitable for this purpose is ordinary pencil-lead.

In the electrical method, a reactance is connected in series with the crystal when the circuit is operated at the series-resonance crystal frequency, or in shunt with it for the parallel mode. The effects of such added reactances are summarized in **Table 6.3.** This electrical method is often useful for making slight frequency changes such as may be required to avoid temporary interference. The maximum adjustments possible are very small; for instance in the 7MHz band the change will not be more than a few kilohertz. The limits of variation are set either by instability or by a complete lack of oscillation.

Mixer Crystal VXO Circuits

An increase in the frequency deviation obtained by varying crystal frequencies can be achieved by employing two separate crystal oscillators, one of which is tuned by a reactance which makes its frequency go lf; the other is tuned by a reactance which moves the frequency in an hf direction. Such oscillators used in a circuit similar to that of **Fig 6.12** will permit greater deviation to be secured, compared with a basic frequency crystal.

If two oscillators are used, one on 8·5MHz, the other on 12MHz, and assuming that the 8·5MHz crystal is changed lf by 1kHz per MHz, and the 12MHz crystal hf by 1kHz per MHz, the maximum alteration in frequency will be 12 + 8·5 ~ 20kHz.

General

In many of the oscillator circuits described, the grid is connected to earth through a simple resistor. In some cases, an increase in output may be obtained by fitting an rf choke between the grid of the valve and the resistor but this must be done with care however because, in certain circumstances and particularly with a valve of high mutual conductance, self-oscillation may occur if the choke is of high *Q*.

If the crystal fails to oscillate in some circuits it is possible for the anode current to rise to such a high level that the valve is destroyed. This can be avoided by fitting a resistor in the cathode circuit to provide sufficient bias to limit the anode current to a safe value in case of crystal failure. It is essential to bypass such a resistor for rf.

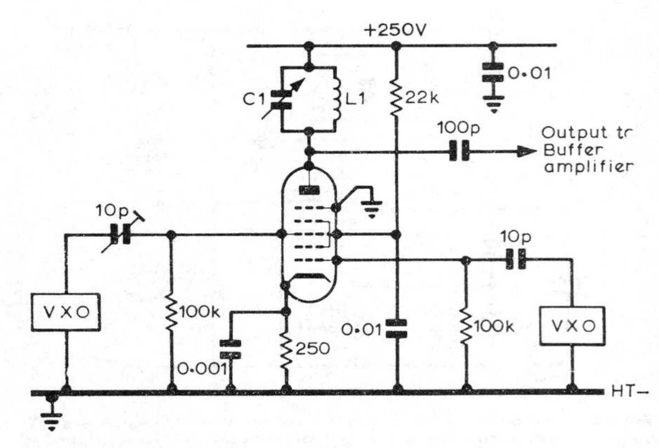

Fig 6.12. Low-level mixing circuit for a vxo showing typical values for a 6SA7. The values of L1 and C1 are similar to those for a normal tank circuit.

When using pentode or tetrode valves as oscillators, the screen voltage should be derived from a potential divider rather than the usual series dropping resistor. For best stability, however, both anode and screen voltages for oscillator valves should be obtained from a stabilized source. Electronically regulated supplies are preferable to those depending on simple voltage regulator tubes.

Keying of any oscillator is seldom satisfactory but when it is necessary the crystal oscillator must receive special attention. First, the crystal itself must be of high activity: an inactive crystal is likely to produce chirp due to changes in frequency as oscillation commences. Secondly, the oscillator must be lightly loaded and the crystal current kept as low as possible. DC keying was popular at one time, the usual method being to do so in the cathode lead. This is a dangerous practice from the point of view of the lethal voltage that may exist across the key as well as overstressing the heater/cathode insulation. A better method is to employ grid block keying in which a high negative voltage sufficient to bias the valve to cut-off is applied under key-up conditions. When the key is closed, the negative supply is short-circuited to earth, thus allowing the oscillator to function normally. A high resistance in series with the negative supply limits the current under key-down conditions and protects the operator.

Information on keying and control methods is given in Chapter 8—*Keying and Break-in*.

Solid State Crystal Oscillators

The basic triode crystal oscillator circuit configurations may be transistorized with the advantages of much lower power consumption and heat dissipation. A few typical examples are shown in **Figs 6.13** to **6.19.**

TABLE 6.3

Effect on Crystal Frequency of Adding External Reactances

Reactance Added	Series-resonance Frequency	Parallel-resonance Frequency
Capacitance in series	Increased	Unaltered
Capacitance in parallel	Unaltered	Lowered
Inductance in series	Lowered	Unaltered
Inductance in parallel	Unaltered	Increased

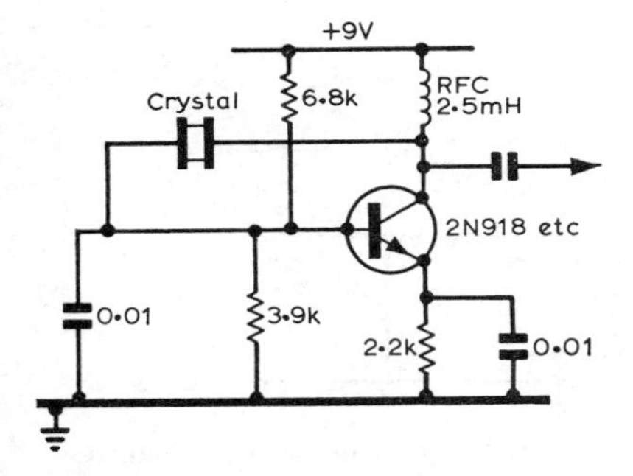

Fig 6.13. Oscillator suitable for a wide range of crystals.

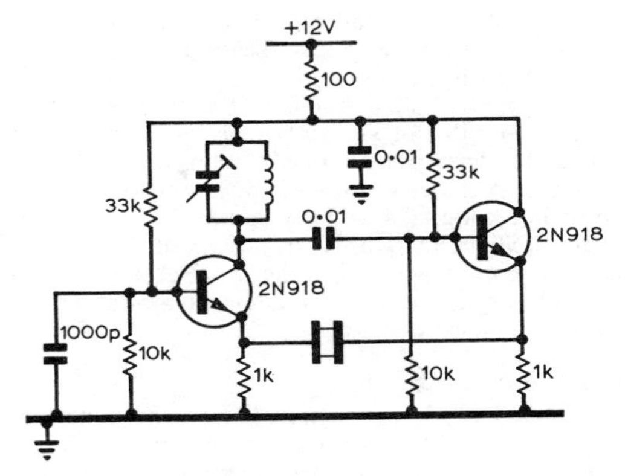

Fig 6.17. Butler oscillator.

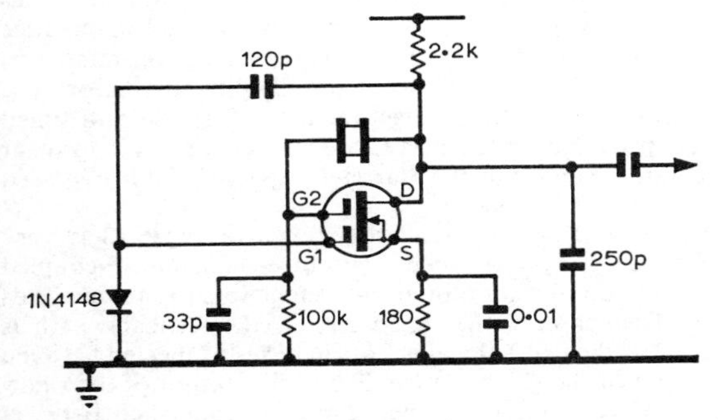

ig 6.14. High stability oscillator with agc loop. Values shown are for the 10MHz range.

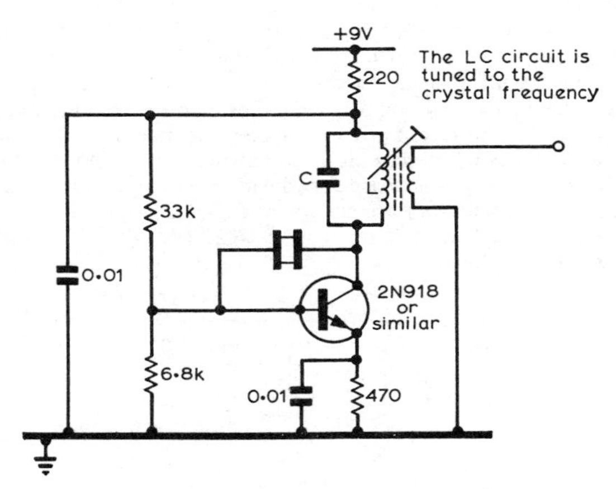

Fig 6.18.

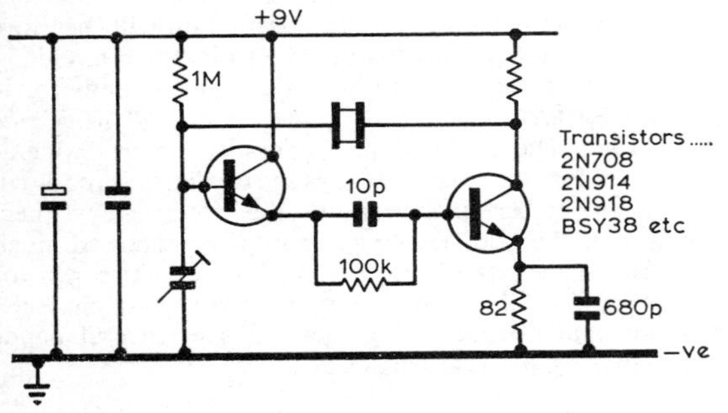

Fig 6.15. A 1–20MHz crystal oscillator.

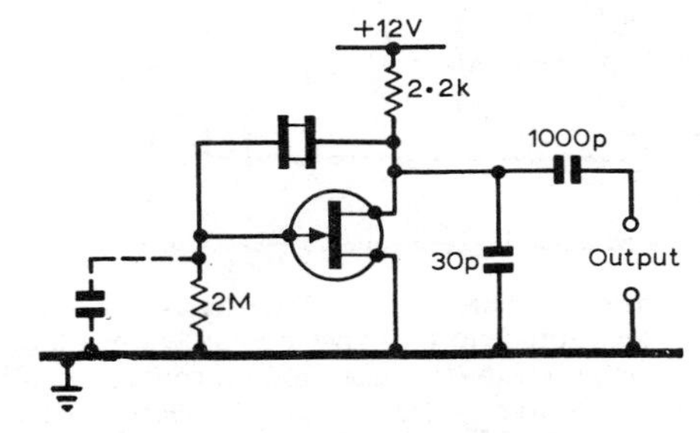

Fig 6.19. An fet oscillator for FT241 crystals (400–500kHz).

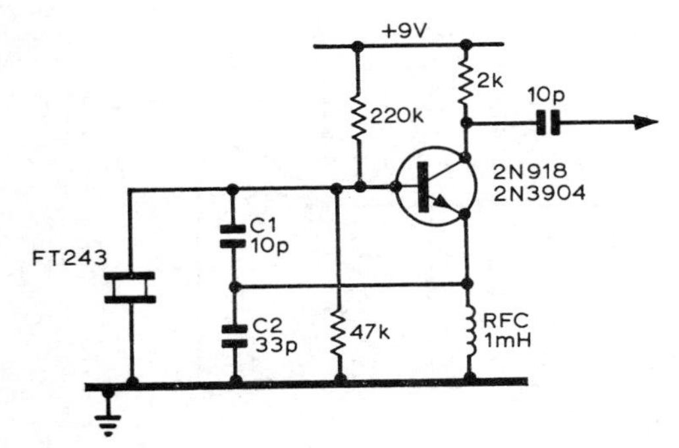

Fig 6.16. Crystal oscillator suitable for 10MHz operation.

VARIABLE FREQUENCY OSCILLATORS

Although the crystal oscillator is the most simple, stable and satisfactory device for providing accurate frequency control, it suffers from the obvious disadvantages of fixed-frequency operation, particularly on congested amateur bands where the operating conditions change from minute to minute. A *variable frequency oscillator* (*vfo*) is now essential for most if not all of the hf amateur bands. The frequency-determining circuits of such an oscillator consist of ordinary inductances and capacitances, either or both of which may be variable. Special care must be taken to eliminate all possible cause of unwanted frequency variation.

A vfo of good design should give a frequency stability comparable with that of a crystal oscillator. Some idea of the stability desirable may be obtained by considering the following example. A sudden frequency variation of 50Hz when operating in a congested cw band can cause the signal to be lost or its intelligibility to be severely reduced if the narrowest receiver passband is in use: on the 21MHz band this variation would correspond to only one part in 400,000 of the transmitted frequency.

Basic Types of VFO

Many different vfo arrangements have been developed, but when analysed they are found to be modifications of a few basic circuits. The modifications are introduced with practical aspects in view; eg to allow the earthing of one end of the tuned circuit or to improve the stability.

An oscillator is essentially a tuned amplifier in which some of the output voltage is fed back to the grid 180 degrees out of phase with respect to the anode voltage. The appropriate feedback may be obtained either by mutual inductance between the anode and grid circuits or by capacitive coupling between parts of the circuit at which the current phase and voltage relationships exist for positive feedback to occur. Four basic oscillators are shown in **Figs 6.20** (the Hartley), **6.21** (the tuned anode tuned grid), **6.22** (the Colpitts) and **6.23** (the Franklin).

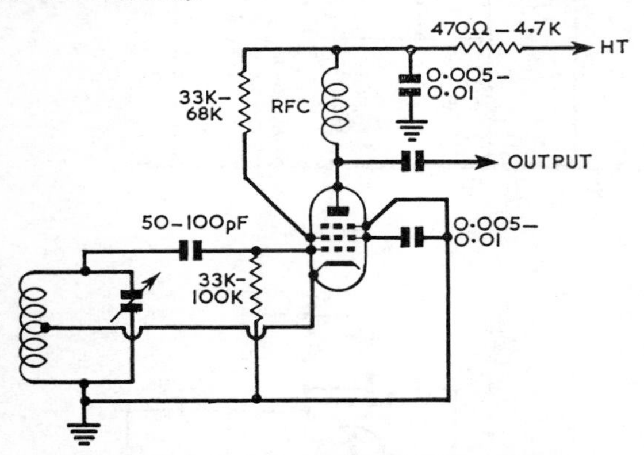

Fig 6.20. Basic electron-coupled Hartley oscillator.

In the Hartley oscillator, Fig 6.20, feedback is achieved by connecting the cathode of the valve to a tap on the tuned grid circuit. As the anode current also flows through the cathode of the valve, this arrangement effectively allows part of the rf output current to flow through the tuned circuit in phase with the grid circuit current. Alternatively the oscillator can be considered as having grid and anode connected to opposite ends of a tapped tuned circuit, at which points the voltages are in antiphase. The degree of feedback is determined by the position of the tap and will clearly be zero if it is at either end of the coil, or maximum if near the centre. In practice, it is normal for one quarter to one third of the total turns to be included between cathode and earth.

The tuned anode tuned grid oscillator, Fig 6.21, is similar to the Pierce-Miller crystal oscillator of Fig 6.2(a). When the anode and grid circuits of the tatg (sometimes known

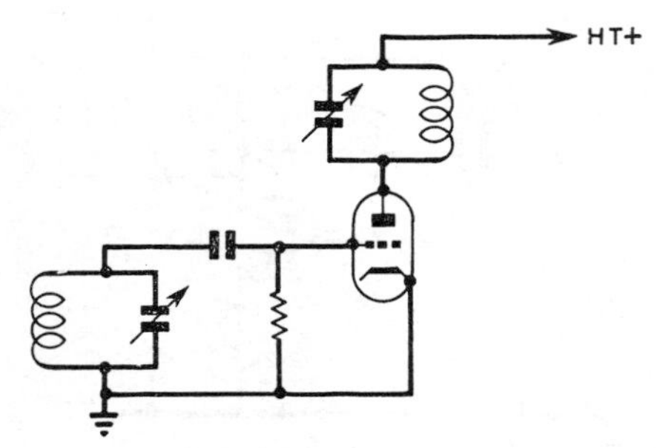

Fig 6.21. Tuned anode tuned grid oscillator.

as a tptg) are tuned to the same frequency, feedback occurs through the internal capacitance of the valve from anode to grid. This simple oscillator is no longer used in practice but its principle of operation should be borne in mind.

The cathode of the valve in the Colpitts oscillator, Fig 6.22, is connected to a capacitive tap across the grid tuned circuit provided by C1 and C2. The feedback arrangement is therefore similar to the Hartley. Capacitor C2 is bypassed for dc by the rf choke.

The basic Franklin oscillator shown in Fig 6.23 is very similar to a conventional resistance-capacitance coupled amplifier with the addition of two capacitors to provide feedback. Disregarding the tuned circuit, the feedback path is formed by the two 5pF capacitors in series connected between the input of the left-hand triode and the output of the right-hand valve. A parallel tuned circuit is connected between these two capacitors and earth. At frequencies other than the resonant frequency of this tuned circuit there is a short-circuit to earth across the feedback path but at resonance there is a high impedance to earth. Oscillation will therefore only occur at the resonant frequency of the tuned circuit.

The circuit of the so-called cathode coupled oscillator is shown in **Fig 6.24** which first appeared in *CQ Magazine* in April 1960. The circuit is very tolerant of high *C* operation, a useful feature where a wide range oscillator is required. However, under certain conditions the circuit may squeg, particularly if the grid coupling capacitor is increased much beyond the suggested 10pF. A small non-inductive resistor of 10 to 50Ω in the grid will prevent this and enable a nearly constant output to be obtained over several bands without altering the grid capacitor.

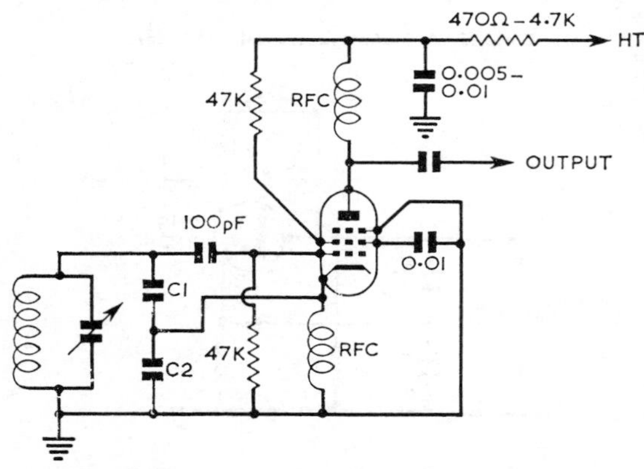

Fig 6.22. Basic electron-coupled Colpitts oscillator.

The ht bypass C3 should be connected directly between the "free" anode of the valve and the earthing point of the oscillator to ensure low harmonic output. For harmonic suppression, the series inductance of the capacitor is more important than its actual value and it is worth trying different physical types. For example, a cylindrical paper capacitor, particularly if in an earthed metal casing, may be found superior to a silvered mica type whereas with different makes the opposite might be true. Hi-*K* ceramic types seem to be unsuitable for this application. A combination of different types sometimes proves satisfactory. A 5kΩ resistor in the "free" anode circuit may also prove helpful in eliminating harmonics.

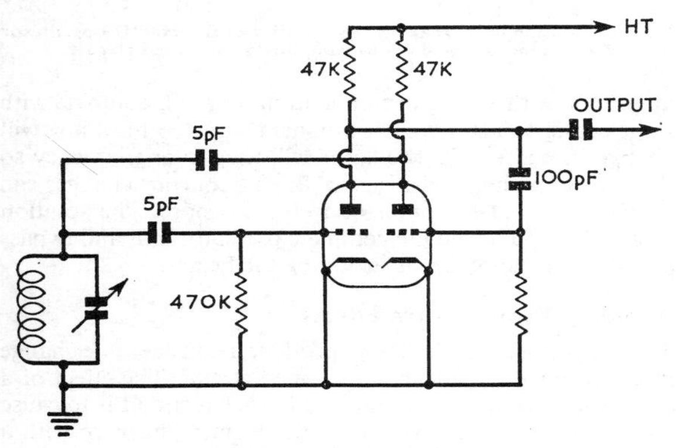

Fig 6.23. Basic Franklin oscillator.

In these basic types of oscillator or in any of their modifications the frequency stability is limited primarily by the stability of the frequency-determining tuned circuit and also by the effect of the valve and other components associated with it: other factors such as the operating voltages and currents and the loading of the oscillator are likewise important. All these design problems require close and careful study if satisfactory performance is to be achieved.

Tuned-circuit Stability

To ensure the highest stability over both long and short periods the primary requirement is that the frequency-determining tuned circuit shall be in every way of a high standard. Mechanically, it must be perfectly rigid since any vibration or displacement of the various parts of the tuned-circuit assembly will cause variations in the effective inductance and capacitance and therefore in the oscillator frequency. To overcome these effects, the whole oscillator (in particular the items associated with the tuned circuit) must be made as mechanically rigid as possible. It is preferable to construct the oscillator in a die-cast box, which gives sufficient strength for most amateur purposes and at the same time screens the vfo from external fields. The wiring should be of stout gauge (16swg or thicker).

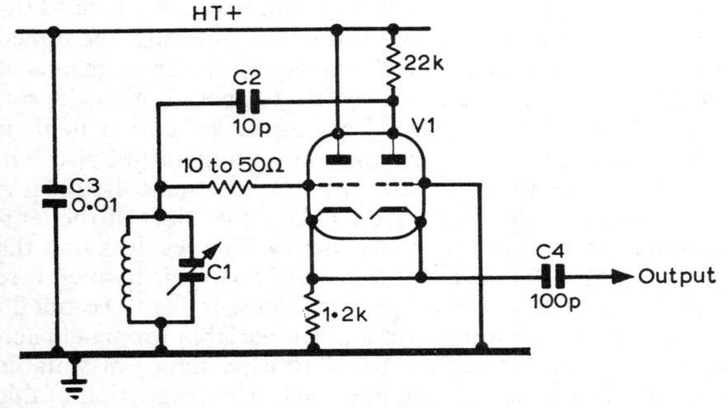

Fig 6.24. The so-called cathode coupled oscillator. V1 may be a 6J6 or 6SL7GT. It will be seen that the circuit is very similar to the standard Franklin oscillator.

The tuning capacitor must be of low-loss construction, preferably with ceramic insulation for the stator supports and with bearings for the rotor which prevent any tendency towards wobble or backlash. The electrical connection to the rotor should be made separately through a spring contact—or a pig tail—in order to avoid the introduction of spurious resistance paths through the bearings, which can cause frequency scintillation—small random changes in frequency. Since the capacitance of a variable capacitor depends not only on the area of the fixed and moving plates exposed to each other, but also on the spacing between them, it is advisable to employ a tuning capacitor with widely spaced plates. Like any other item, heat will cause expansion in the tuning capacitor, and the resulting capacitance change will be far less in a unit with widely spaced plates compared to one in which the plates are physically close. The capacitor must also be selected with regard to the nature of the dielectric supporting the fixed plates from the main frame, for consideration will show that this dielectric is directly across the tuned circuit when this is a parallel configuration, and across the capacitor itself in a series tuned arrangement. High grade ceramic or preferably ptfe must be used for these supports. Materials such as paxolin and Bakelite are quite unsuitable. To reduce rf losses to the minimum, it is desirable that the plates themselves should be good conductors viz copper or aluminium.

The coil should be wound in such a manner as to make any movement of its turns virtually impossible. Apart from the use of grooves in the former, sufficient rigidity can be achieved by winding the wire under tension.

There is an increasing tendency to employ physically small coils in oscillator circuits, and to fit these with ferrite cores in order to achieve the desired inductance. Such a procedure is satisfactory so long as the ferrite material is not subject to a high coefficient of expansion, and the core itself is firmly locked in place. If such coils are used and frequency drift is noted above that which may reasonably be expected, a change of ferrite core type may well result in a substantial improvement in frequency stability.

Changes in the values of the tuned circuit components are not the only source of frequency drift. Instability in the values of other components or changes in the interelectrode capacitances of the valve may produce capacitance changes across the tuned circuit or result in an alteration in the working point of the valve, both of which can cause the operating frequency to be unstable.

Circuit Q

An important factor in the maintenance of frequency stability is the loaded Q of the frequency controlling circuit, the stability being improved with increase of Q. This is because in a high Q circuit the phase change of the voltage existing during oscillation across the circuit varies with frequency more rapidly than in low Q circuits. Thus, in the high *Q* case, random phase changes in the oscillator feedback path are cancelled out by smaller changes of frequency than in the low *Q* case.

The loaded *Q* value of the tuned circuit may be improved by reducing the losses in its components, the chief loss usually occurring in the coil. Therefore coil formers, which may be ribbed, should be made of low loss material (eg ptfe, polystyrene or ceramic) and the wire must be copper; Litz wire is useful at frequencies below 1MHz. A specific shape can be found which gives highest *coil Q*.

The coil should be situated at least 1–1·5 coil diameters away from metal screening or other components to prevent loss of energy and subsequent lowering of *Q* due to induced eddy currents and dielectric losses. Other items associated directly or indirectly with the tuned circuit such as padding capacitors, insulators and so forth must be of high grade material, as any dielectric losses will be reflected in a reduced circuit *Q*. Capacitors should be of air or mica dielectric except for any necessary temperature-compensating ceramic capacitor.

In the practical oscillator the frequency controlling circuit must be coupled to the oscillator valve. The effect of this is to reduce the overall *Q* of the circuit by an amount depending on the degree of coupling. The coupling should therefore be small. For a given amount of loading, however, the *Q* of a tuned circuit is proportional to the *C*/*L* ratio, which should consequently be as great as possible. In fact the *C*/*L* ratio is usually chosen as a compromise between greatest ratio and the ability to tune over the required frequency range using components of convenient value.

Apart from the improvement in loaded *Q* value obtained through the use of high *C*/*L* ratio in the oscillator tuned circuit, the effect of changes in the valve interelectrode capacitances due to extraneous influences is considerably reduced, as the ratio of valve capacitance to total circuit capacitance is very small and as a result is less capable of varying the resonant frequency.

Reduction of oscillator *loading* as a means of improving the circuit *Q* is exemplified by the Franklin oscillator (see Fig 6.23). In this oscillator the two valves are extremely loosely coupled to the tuned circuit through two very small capacitors.

In another type of oscillator the connections from the valve electrodes are tapped down the coil (example: **Fig 6.25**) thus reducing the loading effect of the valve on the circuit and the influence of the interelectrode capacitance on the resonant frequency.

All other things being equal, the most stable oscillator will be that in which the loosest coupling exists between the tuned circuit and the valve, or a circuit in which the capacitance across the tuned circuit is extremely large in proportion to any likely external capacitance changes.

The valve, because of its non-linearity, will generate harmonics of the oscillator frequency. These can intermodulate with each other producing fundamental components in random phase relationship with the original fundamental component. The presence of these spurious components will produce instability of the frequency but their effect can be considerably reduced by employing high *Q* circuits which by virtue of their selectivity prevent them being fed back to the grid circuit.

It is also possible for the frequency to change if fundamental components are fed back to the oscillator from high power stages of the transmitter. Since these will change in amplitude with keying or modulation they will combine with the existing fundamental component and the resultant will change in phase. The oscillator will then alter frequency so that the loop phase shift is zero. Such frequency changes can sometimes be of the order of 100Hz or more. The solution is effectively to screen the complete oscillator unit and bypass and filter all supply leads including the heater.

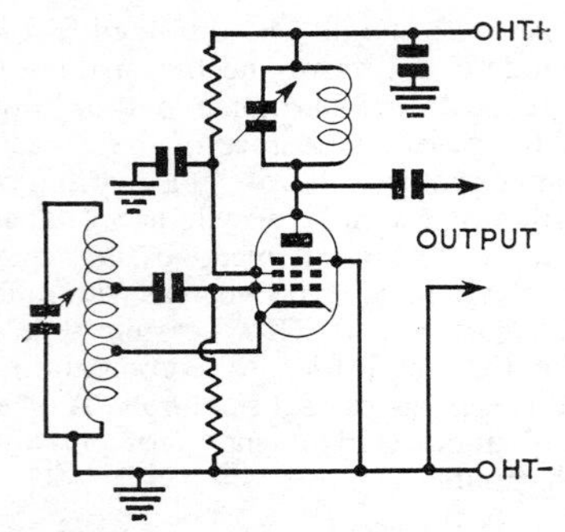

Fig 6.25. Method of increasing the stability of the Hartley oscillator by tapping the grid connection down the tuned circuit.

Avoiding Temperature Effects

Temperature variations can produce a considerable change in the resonant frequency of a tuned circuit. The effect of a temperature change upon the coil in such a circuit is to cause an expansion or contraction of the former and wire with a consequent change in inductance and resistance values. Both of these values normally increase with the temperature.

The various parts of the tuning capacitor assembly will likewise contract and expand and the amount of the resultant change in capacitance will depend on the type of construction. In general it is found that the resonant frequency of an ordinary tuned circuit comprising a parallel-connected coil and capacitor has an effective negative temperature coefficient of 50–100 parts/million/°C.

The most obvious method of avoiding the frequency drift caused by temperature variation is to remove the tuned circuit from all sources of heat. The tuned circuit should be located in a part of the equipment well away from valves (which can radiate a large proportion of their wasted power) and high-wattage resistors. It is preferable to place the circuit within its own compartment, and if this is brightly polished on the outside any unavoidable heat will be reflected away from the surface.

The second method, which is in common use owing to the difficulty of preventing all heat from affecting the tuned circuit, is to compensate the frequency variation by means of capacitors having a suitable negative temperature coefficient. When these are connected across the tuned circuit in their correct capacitance proportions, any temperature rise will tend to cause them to increase the resonant frequency, whereas in the remainder of the circuit the effect of the temperature rise is to lower the frequency. The result is that the frequency stays almost constant, ie the overall temperature coefficient of frequency is reduced almost to zero. To obtain a required tuning range with a given variable tuning capacitor it is necessary to connect a fixed capacitance of suitable size across the tuned circuit, and a selected proportion of this capacitance may conveniently be of the negative temperature coefficient type. The compensating capacitors themselves are

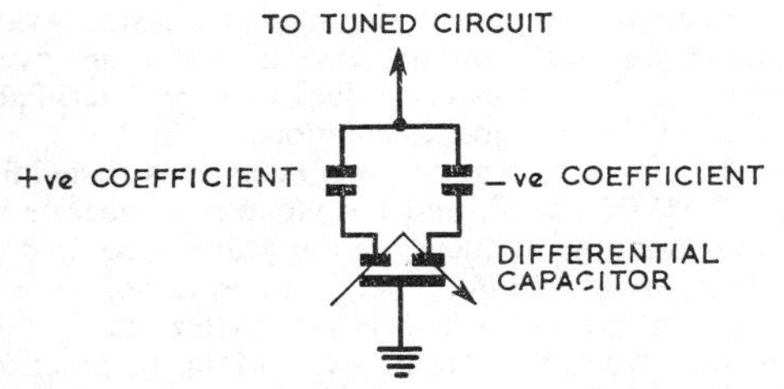

Fig 6.26. Temperature compensating circuit employing a differential capacitor.

usually of the tubular wire-ended ceramic type and are available in almost any nominal value of capacitance with several different values of temperature coefficient.

Selection of the correct value of compensating capacitor can be a laborious task. An ingenious solution to the problem is shown in the circuit of **Fig 6.26** in which a positive temperature coefficient capacitor and a negative coefficient capacitor of the same value are connected to the two arms of a differential capacitor. It is convenient to make both fixed capacitors the same value and by varying the differential capacitor the temperature cofficient may be adjusted precisely. The total capacitance "seen" by the tuned circuit will vary as the differential capacitor is adjusted (max. in mid-position) but this can be compensated by adjusting the band-set capacitor. The "Tempatrimmer" manufactured by Oxley Industries is a commercially available component performing the same function as the arrangement of Fig 6.26. Furthermore, if the oscillator is operated at a low power level there will be less difficulty in preventing a temperature rise in the tuning circuit due to heat from the valve. There will also be less heat developed in the tuning circuit itself (arising from rf losses), but in any case this is likely to be quite small compared with the amount of heat which reaches it from the valve.

Power-supply Variations

Variations in the supply voltages to the oscillator valve can alter its interelectrode capacitances and other parameters. Details of suitable methods of stabilizing the power supply voltages are given in Chapter 16—*Power Supplies*. Changes in heater voltage generally produce negligible variation of frequency, but their effects should not be ignored.

Loading Effects of Following Stages

The loading effect on the oscillator of the buffer or frequency-multiplying stages which follow it causes reactive and resistive components to be reflected into the frequency-determining circuit of the oscillator. The reactive components cause direct frequency variations while the effect of resistance across the resonant circuit is to decrease its Q, thereby reducing the inherent frequency stability. Moreover, rf voltages may be fed back from the succeeding stage. These voltages, the amplitudes of which depend upon the power output, and the tuning adjustments of other parts of the transmitter, will affect the calibration and the stability of the vfo.

For maximum stability the coupling to the following stage must therefore be as loose as possible consistent with the required drive being obtained.

The only way in which frequency variations due to changes in loading can be overcome is by the use of adequate buffer amplifier stages after the vfo. It is not possible to remove all loading from the vfo since it must supply some power to the following stage, but it is possible to arrange the loading on the vfo so that the frequency stays sensibly constant. The following stages then supply the required drive power.

It is strongly recommended that the vfo should be followed by an untuned class A buffer amplifier or a cathode follower, as shown in **Fig 6.27.** Both of the circuits shown in Fig 6.27 present a very high impedance to the oscillator stage, and feedback from subsequent stages is at a satisfactorily low level.

The circuit is that of a pentode amplifier operating in class A giving a high degree of isolation. For maximum efficiency, the output from the anode to the following stage should use the shortest possible lead length, and for preference be no longer than the wires on the 100pF coupling capacitor. To ensure that the stage operates under class A conditions when connected to the vfo the connection between the earthy end of the 470kΩ grid resistor and the chassis should be broken, and a 0–500μA meter inserted. With both the vfo and the buffer stage operating, the meter should be checked for any signs of grid current. If current is present, then the 100pF coupling capacitor in the grid circuit must be reduced in value to the point where grid current ceases.

Alternatively the cathode voltage of the class A amplifier

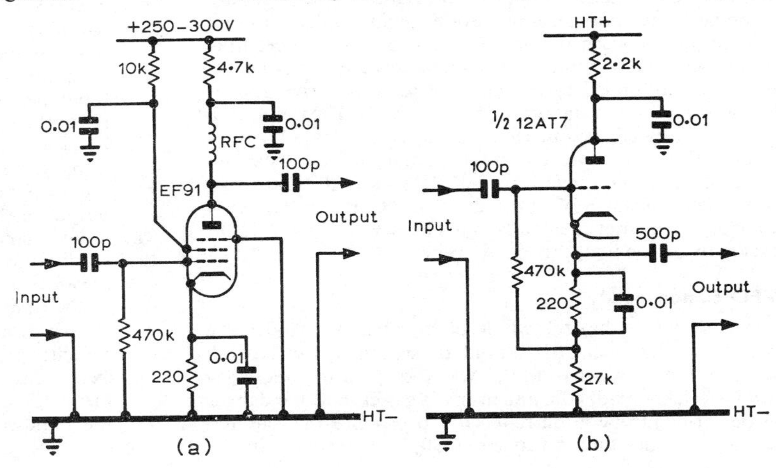

Fig 6.27. Methods of isolating the vfo from loading effects. The untuned class A amplifier shown at (a) provides a useful amount of gain besides affording adequate isolation. In the cathode follower (b), the impedance presented to the previous stage is very high while the output impedance is very low (eg about 100Ω): the voltage gain is, however, less than 1. In view of the low output impedance, the cathode follower is well suited for feeding into a coaxial line.

may be monitored and the coupling from the oscillator reduced to the point where disabling the oscillator causes no change in amplifier cathode voltage.

Although the gain of the cathode-follower circuit shown at Fig 6.27(b) is less than unity, the degree of isolation between input and output circuits is extremely high. The output impedance is quite low (approximately equal to $1/g_m$, where g_m is the mutual conductance at the operating point of the valve used); a typical value is 100Ω. It is therefore possible to feed the output directly and without serious mismatch into the low-impedance coaxial cable which can then supply the main transmitter at a point remote from the vfo unit.

VFO Design Recommendations

The rules for the avoidance of frequency drift and other variations in a vfo may be summarized as follows:

(*a*) Ensure complete rigidity and strength of the mechanical structure and wiring.
(*b*) Protect the tuned circuit from the effect of heat sources such as valves and resistors.
(*c*) Construct the tuned circuit and associated oscillator components of material having the lowest rf losses.
(*d*) Avoid draughts across the tuned circuits.
(*e*) Operate the oscillator at the lowest possible power level.
(*f*) Provide stabilized ht to the valve.
(*g*) Use a buffer amplifier and/or cathode follower to minimize the loading effect of the subsequent stages.
(*h*) Screen and decouple all the circuits to prevent the pick-up of rf energy from higher-power stages.
(*i*) Operate the oscillator on a sub-multiple of the output frequency or alternatively use an oscillator-mixer system.
(*j*) Use heat dissipating valve screens.
(*k*) Operate the oscillator or tuned circuit at a fixed temperature (above maximum ambient) by means of an oven and thermostat.

Selecting the Frequency of the VFO

It is usually advantageous to operate a vfo on a relatively low frequency since, first, the stability in terms of frequency is improved as a consequence of the diminished effect of the various factors causing frequency variations and, secondly, it can facilitate multi-band operation. Owing to the harmonic relationship between the various amateur bands the output of an oscillator working on a suitable low frequency can be frequency-multiplied to produce the required drive voltage in one or more of the higher-frequency bands. The multiplication is usually effected in separate stages, but often it can be arranged within the vfo unit itself. In many amateur stations a vfo covering the range 1·75–2·0MHz is found to be a very convenient means of providing multi-band frequency control. It is not generally satisfactory to operate the oscillator on the final carrier frequency.

VFO Bandspread

It is desirable that the whole of the amateur band to be covered by the oscillator should be spread across the full width of the tuning dial. In this way the operating frequency can be quickly read, adjustment made easier, and the danger of out-of-band operation reduced. To produce the required bandspread, the tuning capacitance is usually made in the form of fixed (or pre-set) capacitance plus a small variable capacitance just sufficient to vary the frequency over the desired range. The necessary values may be determined by experiment or from simple calculations.

If for example a transmitter were being constructed for the 28MHz–29·7MHz band, and the vfo was to operate in the 3·5MHz region, the frequency multiplication required in the transmitter would amount to eight times (3·5MHz × 8 = 28MHz). Thus to cover 28MHz–29·7MHz the vfo would need to tune from 3·5MHz to 3·7128MHz. In practice this would probably be adjusted to 3·49MHz to 3·72MHz to give about 10kHz overswing at each end. Similar calculations can be made for other bands.

One of the most satisfactory types of slow-motion drive for a bandspread tuning capacitor is the friction type with a ratio between 6 : 1 and 9 : 1 or the gear type. Frequency changes can then be made swiftly and the setting accuracy should be adequate, provided that a suitably marked dial is used. The dial may show ordinary scale divisions (0–100 or 0–180) but it is far more convenient to have a dial which has been directly calibrated. The frequency can then be read instantly and the risk of error that is always present in the use of a separate calibration chart is avoided. A particularly suitable slow motion drive and dial for the de-luxe transmitter is the Eddystone type 898.

Appraisal of a Newly Constructed VFO

As the vfo is the heart of the transmitter it is important that it should be thoroughly tested before the transmitter is put on the air. The following is intended as a guide to the procedure to be adopted when testing a vfo for the first time.

(*a*) The unit should be installed in position in the transmitter or chassis of which it is intended to form part.
(*b*) The buffer stage should be connected and suitably dummy loaded.
(*c*) It may be useful to monitor ht current and this may be conveniently done by measuring the voltage across a series resistor in the ht lead. In the absence of such a resistor it will be necessary to break into the ht lead and measure the current directly.
(*d*) Apply power and observe that the ht current is of a reasonable value.
(*e*) A check may be made that the unit is oscillating by touching the grid lead of the valve and observing that the anode current increases.
(*f*) It is now necessary to determine the approximate fundamental frequency of oscillation; for this purpose a calibrated absorption wavemeter should be used. Find the strongest signal using the absorption wavemeter and tune well to either side of this signal in order to ensure that it is the fundamental output of the oscillator.
(*g*) If necessary, adjust the oscillator frequency to the required value determined by the absorption wavemeter, using the band-set capacitor. The position of the tuning capacitor should be appropriate to the chosen frequency, ie it should be fully meshed if the chosen frequency is maximum lf and vice versa.
(*h*) Depending on the accuracy of the absorption wavemeter used, it may be possible to use it to determine the bandspread to a first approximation; otherwise a crystal calibrator must be employed.

(*i*) It may now be convenient to determine the stability of the oscillator; the oscillator should be tuned to zero beat with the signal from the crystal calibrator and the drift measured over a period.

(*j*) Should the circuit include adjustable temperature compensation based on the Oxley "Tempatrimmer" capacitor then adjustments may be made to reduce the drift to acceptable limits. A hairdryer is a useful tool at this juncture since accurate setting of the temperature compensation may be a long and tedious process without accelerated warm-up.

(*k*) The components normally employed in vfo circuits, for instance air spaced capacitors for tuning, band-set and band-spread, silvered-mica capacitors for the fixed values and valves all have positive temperature coefficients. A small amount of negative temperature coefficient compensation is therefore usually required. If capacitors employing other dielectrics are used note should be taken of their temperature coefficients so that appropriate compensation can be arranged.

(*l*) The purity of note should next be checked by listening to the vfo on a receiver. The presence of supply mains frequency modulation would normally indicate cathode heater leakage in the oscillator or buffer valve and ×2 mains frequency modulation would indicate inadequate smoothing of the ht supply. Modulation at other frequencies is probably attributable to the presence of parasitic oscillations.

(*m*) A useful check for parasitic oscillations, apart from a wide range sensitive absorption wavemeter, is to monitor the ht current to the vfo while tuning over its full tuning range. Sudden changes in ht current usually indicate the presence of parasitics.

(*n*) Should evidence of parasitic oscillations be found, reference should be made to page 6.36.

(*o*) Having achieved satisfactory operation of the vfo to this stage of testing, accurate calibration may be carried out using the crystal calibrator.

As a practical criterion of stability, a fundamental 5MHz vfo which drifts no more than 50Hz/h under average shack conditions may be considered to be good.

Frequency Measurement

If a digital frequency meter is not available then the following beat method may be employed.

While it is not essential to measure the frequency of the beat note between the vfo and the reference standard, it does permit some assessment of the effects of alterations.

If an oscilloscope and a variable frequency audio oscillator are available they can be used to determine the frequency of the beat notes (see Chapter 18—*Measurements*). Not all workshops have a variable frequency audio oscillator and an oscilloscope. In such cases a somewhat less accurate method can be adopted. This involves the use of a meter arranged in a circuit in such a manner that its deflection is proportional to the applied frequency, so allowing the beat frequency to be read directly.

Although not of extremely high accuracy, a simple direct reading frequency meter does have the advantage that the frequency change taking place over a given period can be instantly seen without the manipulation of controls. The circuit of such a frequency indicator is given in **Fig 6.28.** This instrument is provided with three ranges, the full scale deflections of which are 0–100Hz, 0–1kHz, and 0–10kHz.

The transistor functions as a limiter amplifier, the output of which is differentiated by one of the three capacitors associated with the range switch, resulting in a train of positive and negative going pulses. The negative pulses are clipped by the diode D1, the remaining positive pulses being fed to the metering circuit where they are integrated by the capacitor Cx. The voltage thus developed across Cx is linearly proportional to frequency and the meter measuring the voltage may therefore be calibrated directly in hertz.

It is most essential that the input to this circuit be sufficient to allow the transistor to be driven into limits, as otherwise the meter readings may be incorrect. Approximately 2V peak-to-peak is needed. A minor disadvantage of this device is that the indications are not accurate below about 5 per cent of the frequency range in use and thus while the lowest range claims 0–100Hz, frequencies lower than 5Hz should not be taken seriously. Consistent accuracy is maintained on all readings which are above 10 per cent of full scale deflection.

To calibrate the instrument the range potentiometers should first be set to maximum resistance. On Range 1 the output from the 6·3V winding of a mains transformer should be applied to the input and the range potentiometer adjusted so that the meter reads half scale. Range 2 can be calibrated against either the BBC tuning signal of 440Hz or the 1000Hz tone duration transmitted by MSF. An audio generator is needed to calibrate Range 3.

VFO CIRCUITS

Many different types of circuit are in common use, each one having its own particular merits. Many of the arrangements described here will be found satisfactory for use in amateur equipment.

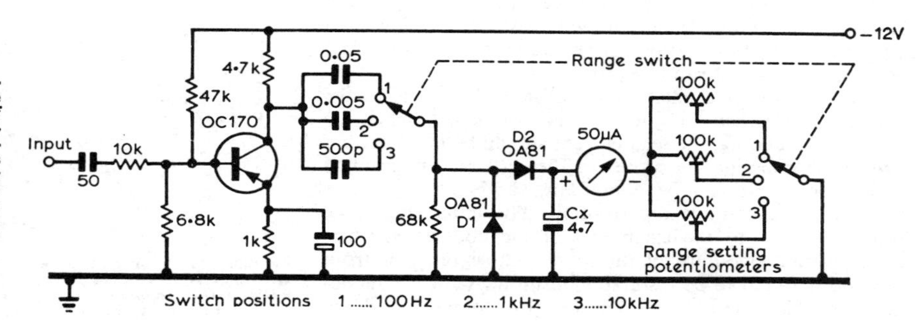

Fig 6.28. A direct reading frequency indicator suitable for the determination of the frequency of beat notes in the range 5Hz to 10kHz. The 50µA meter may be replaced by a 100µA meter; if this is done the 100kΩ range setting potentiometers must be reduced in value to 50kΩ each. A silicon transistor such as BC107 may be substituted if the supply polarity is reversed.

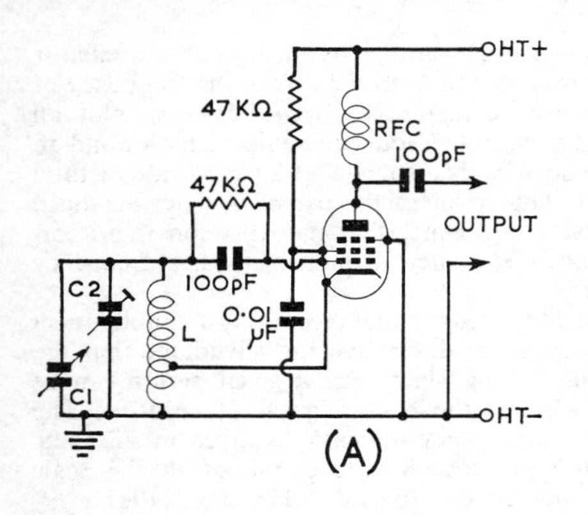

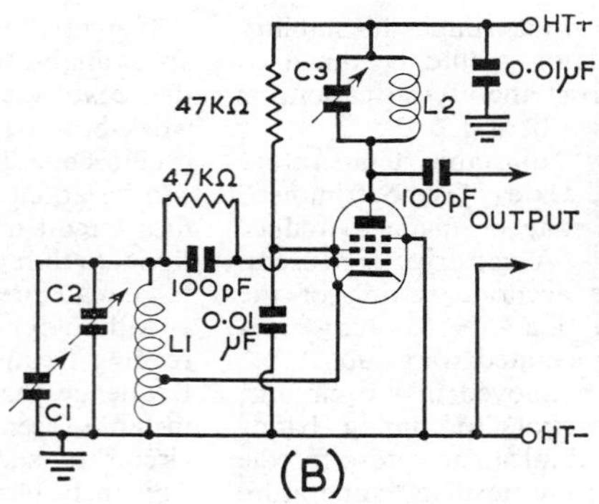

Fig 6.29. Two versions of the Hartley circuit. In (A) the output works into an rf choke, while in (B) a tuned circuit is employed resulting in a higher output level. For operation in the 3·5MHz range, L1 will be about 4μH, C2 (composed of a combination of fixed and variable capacitors in parallel to achieve the desired value), will total about 500pF, and C1, the tuning capacitor, 150pF. An EF91 (6AM6) or similar small pentode with separate suppressor grid is suitable. The ht voltage must be stabilized.

The Electron-coupled Oscillator

In this oscillator a pentode valve is used in such a way that the screen and control grids correspond to the anode and grid respectively of a triode. Electron flow through the valve is controlled by the action of this triode oscillator and the anode current, ie the current in the output circuit is therefore caused to vary in an oscillatory manner, although there is no direct coupling between the output circuit and the oscillator circuit. A marked reduction in the loading effect on the frequency determining circuit is thus obtained. A popular version of this circuit is shown in **Fig 6.29**. The anode circuit may be untuned as shown at (A) or alternatively the desired output frequency, which may be the fundamental or a harmonic, may be selected by means of the tuned circuit as shown at (B).

Generally the highest order of harmonic that can be used is the second or third (because of the very small amplitude of higher orders) and since frequency pulling (ie interaction of tuning adjustments of the two circuits) is considerably less with harmonic operation than with the output at the fundamental frequency, it is desirable to use harmonic output. A valuable property of the electron-coupled oscillator circuit is that the frequency variation for a given change in anode voltage is found to be of opposite sign to that caused by a similar change of screen voltage as shown in **Fig 6.30**.

If the relative values of these two voltages are carefully chosen the effect of one electrode can be made to balance that of the other over quite a large range when the two voltages are supplied from a common source. An important feature to note is that a pentode which has its suppressor grid internally connected to the cathode is unsuitable for the eco type of circuit since the cathode is at rf potential to earth. If the same rf potential is allowed to appear on the suppressor grid, direct capacitive coupling between the oscillator section and the output anode circuit will exist. A valve with a separate connection to the suppressor grid must therefore be chosen so that it can be earthed.

The presence of oscillation is indicated by a rise in the anode current when the grid terminal is touched. When the circuit is oscillating, the feed current should be not less than half the non-oscillating value in the case of an untuned anode circuit. It can be altered by varying the gridleak value and the cathode tap position. The latter should be placed as near to the earth end of the coil as possible; usually about one-third of the total number of turns from this end is found to be a suitable position. A further refinement to this circuit is to tap the grid connection down the tuning inductance as shown in Fig 6.25, thus reducing the loading thereon and improving the *Q*-value. The electron-coupled oscillator gives probably the highest output of all the vfo circuits consistent with high stability and good tone quality. It can therefore be recommended safely to the newcomer to vfo techniques.

The Gouriet Clapp Oscillator

The circuit of the Clapp oscillator, which was originally devised by G. C. Gouriet of the BBC, is shown in **Fig 6.31.** It closely resembles that of the Colpitts oscillator. The important difference is that a small capacitor C1 is inserted in series with the tuning inductance L1 (which is made correspondingly larger so as to resonate at the required frequency), while the capacitors C2 and C3 constitute the feedback system and are of the order of 1000pF each. Because they are connected across the valve electrodes, any capacitance variations within the valve itself become negligible in comparison. The whole arrangement may be compared with a crystal oscillator, since as previously explained the crystal has an equivalent circuit consisting of a very high inductance in series with a small capacitance in one arm in parallel with a second arm comprising a relatively large capacitance. The inductive reactance of the first arm in combination with the capacitive reactance of the second arm then causes the circuit to oscillate in the parallel-resonant mode. In a Clapp

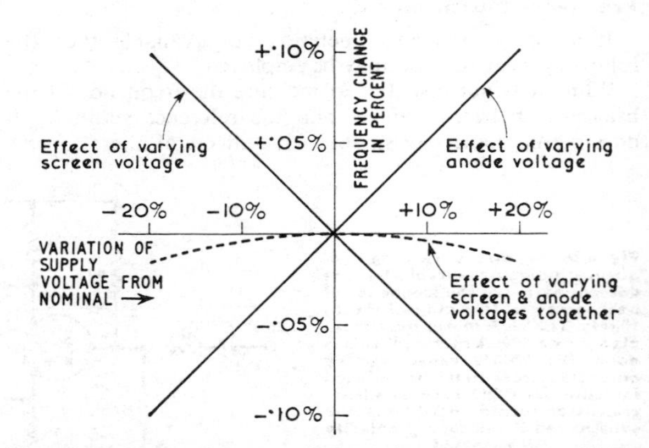

Fig 6.30. Variation of frequency of an electron-coupled oscillator with electrode voltages. When the screen and anode voltages are varied simultaneously the effect on the frequency is very much less than that caused by varying either of them alone.

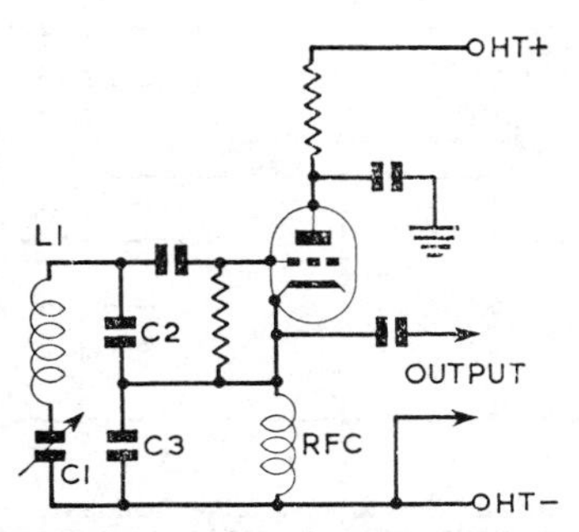

Fig 6.31. Fundamental circuit of the Gouriet-Clapp oscillator, from which its similarity to the Colpitts oscillator may be seen.

oscillator the first arm of the resonant circuit consists of L1 and C1 in series, while the second arm is the series combinations of C2 and C3.

The tuning control is effected by the small variable capacitor C1. Satisfactory performance can only be obtained over a frequency range of about 1·2 : 1 because the power output diminishes rapidly as the series capacitance is reduced. This range is, however, more than enough to cover any of the amateur bands.

A practical circuit for use on the amateur bands is shown in **Fig 6.32**. The triode oscillator portion comprises the cathode, grid and screen of a pentode and the output is derived from the tuned circuit connected to the anode proper which is electron-coupled to the oscillator section. For the highest frequency stability, the valve selected for use in this circuit should have the largest possible ratio between mutual conductance and inter-electrode capacitance.

The ratio of the normally equal capacitances C1 and C2 to C3 plus C4 should be as high as possible consistent with reliable oscillation. If oscillation is not maintained over the desired range, C1 and C2 must be reduced in value or the valve should be substituted by one of higher mutual conductance: it may also help to increase the Q of the coil if possible.

The oscillator can be adapted for frequency modulation by connecting a reactance valve to the cathode or across C3.

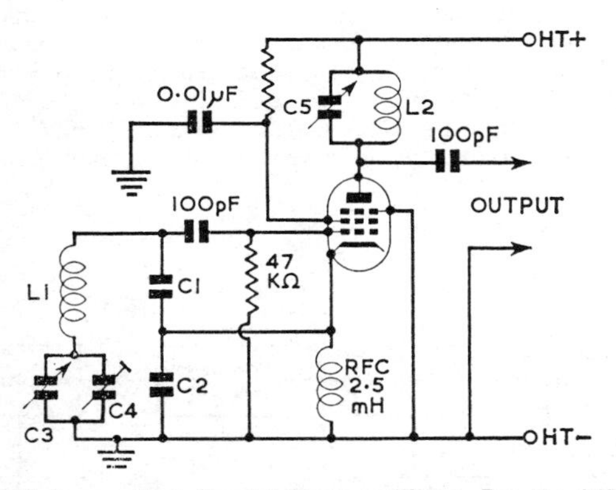

Fig 6.32. A practical Gouriet-Clapp oscillator. For the 1·8MHz band, typical values are: C1, C2, 2200pF; C3, tuning, 50 pF; C4, bandset, 80pF; C5, output tuning, 200pF; L1, 70µH; L2, 11µH. If the oscillator is required to operate in the 3·5MHz band, these values should be halved. Suitable valves are the EF91, 6AU6 etc. The ht voltage must be stabilized.

The Vackar Oscillator

Whereas the Clapp oscillator is limited in use to a relatively small frequency range, the Vackar (Tesla) oscillator provides a fairly constant power output over a much wider range (about 2·5 : 1). It has similar good frequency stability and low harmonic content.

The circuit adapted for amateur use is shown in **Fig 6.33**.

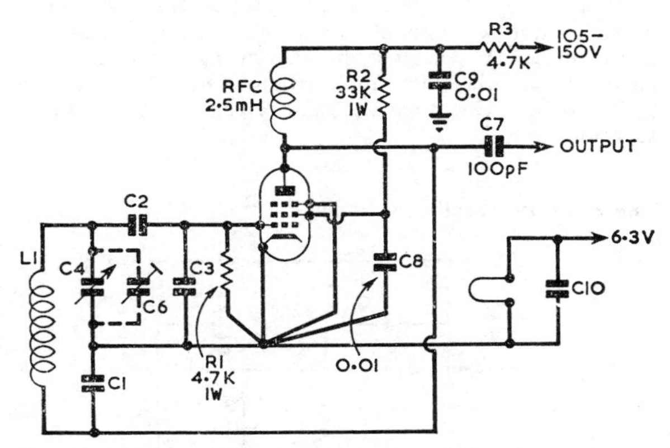

Fig 6.33. Practical Vackar oscillator for frequencies in the range 1·5MHz to 15MHz. See Tables 6.4 and 6.5 for component details not given on the diagram. The ht voltage must be stabilized.

TABLE 6.4

Tuning range	L1 Swg enam	L1 Turns close wound	C1	C2	C3	C4
Amateur bands						
1·8– 2MHz	34	70	556pF	556pF	4700pF	15–250pF
3·5– 3·8MHz	28	45	500pF	300pF	2700pF	10–100pF
7·0– 7·1MHz	26	30	200pF	200pF	1800pF	10–25pF
14 –14·35MHz	24	15	100pF	100pF	1000pF	10–35pF
Special Frequencies						
8MHz	26	25	200pF	200pF	1800pF	+
9MHz	26	20	200pF	200pF	1800pF	+
10MHz	24	25	140pF	140pF	1800pF	+
11MHz	24	20	140pF	140pF	1000pF	+

All coils are wound on $\frac{5}{16}$in diameter formers fitted with ferrite cores, the winding starting at the foot and progressing towards the top of the former. In the Special Frequencies section, no value is specified for the tuning capacitor C4 as this depends on the desired frequency coverage. A $\frac{1}{4}$in former may be substituted if the number of turns is increased by 10 per cent.

TABLE 6.5

Tuning range given by full excursion of core	L1 Swg enam	L1 Turns close wound	C1	C2	C3	C4
1·5– 2·5MHz	34	70	556pF	556pF	4700pF	+
2·3– 3·3MHz	34	45	556pF	556pF	4700pF	+
3·2– 4·5MHz	28	45	500pF	400pF	2700pF	+
4·3– 6·3MHz	28	35	300pF	300pF	2700pF	+
6·1– 8·8MHz	[illegible]	30	200pF	200pF	1800pF	+
7·8–11·0MHz	26	20	200pF	200pF	1800pF	+
10·5–15·0MHz	24	20	100pF	100pF	1000pF	+

All coils are wound on $\frac{5}{16}$in diameter formers fitted with dust iron cores. No value is quoted for C4 as this will depend on the frequency coverage required. Since adding capacitance at C4 will decrease the frequency, a coil is chosen which covers the highest frequency required within the range of its core, and the circuit tuned in a low frequency direction by a suitable capacitor at C4.

Operation at the fundamental frequency on all the hf bands is practicable. In the simplest arrangement tuning is effected by the variable capacitor C4, but a greatly improved performance is achieved if a split-stator capacitor is used, the other section being connected across C1. The capacitor C6 serves as a band-setting control. Typical coil and capacitor values are given in **Tables 6.4** and **6.5.**

As in the Clapp oscillator, maximum frequency stability is obtained by the selection of a valve with the highest possible ratio of mutual conductance to inter-electrode capacitances. A reactance modulator may be added for narrow-band frequency modulation. Cathode keying is found to be satisfactory although screen keying is usually superior.

The Franklin Oscillator

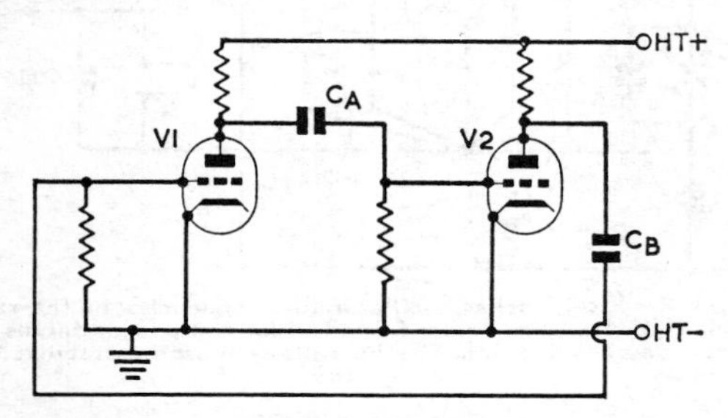

Fig 6.34. Basic circuit to explain the operation of the Franklin oscillator. See text.

The basic circuit principle of the Franklin oscillator is shown in **Fig 6.34.** Here two triodes V1 and V2 are connected as a conventional resistance-capacitance coupled amplifier. A rise of potential applied at the grid of V1 will appear on the grid of V2 as a fall in potential, and again at the anode of V2 as a rise in potential though now of course considerably greater than its original magnitude. If sufficient feedback is provided through C_B to the grid of V1 the whole amplifier will go into oscillation at a frequency determined by the natural time constant of the circuit, which will depend mainly on the values of the resistances and the capacitances C_A and C_B. This is called a multivibrator.

To convert the arrangement into a constant-frequency oscillator of controllable frequency, some means must be provided whereby the feedback or the loop gain is insufficient to maintain oscillation except at the desired frequency. This is done by inserting a parallel resonant circuit LC between the junction of C1 and C2 and earth in the feedback path from V2 to V1: see **Fig 6.35**. This circuit has a low impedance except at its resonant frequency, and at any other frequency the feedback path is virtually short-circuited.

If C1 and C2 are reduced to a minimum possible value for maintaining oscillation, the frequency of oscillation will depend almost entirely on the natural frequency of the LC circuit. In practice suitable values are found to be about 1pF, thus giving a high degree of isolation of the tuned circuit from the remainder of the circuit. Variations in valve capacitance can only slightly affect the tuning; further, the *Q* of the tuned circuit can be maintained at a high value since the load presented to it by the valve is small.

No tappings are required on the coil, and the tuned circuit

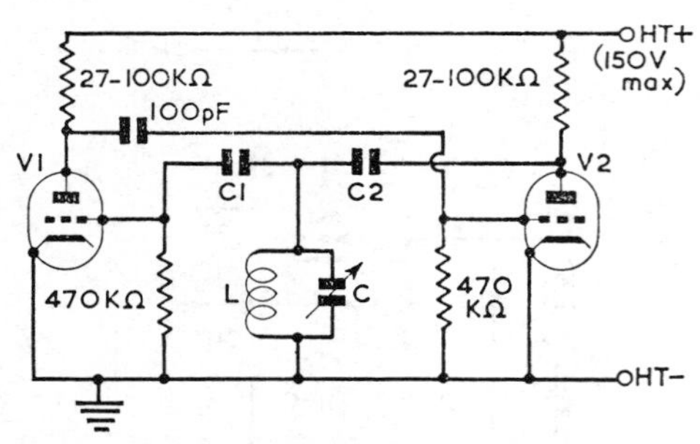

Fig 6.35. A practical Franklin variable frequency oscillator. The tuned circuit is coupled to the valves by the capacitors C1 and C2, both of which have very small values, typically 1–5pF depending on frequency. For optimum stability, C1 and C2 should be the lowest values needed to maintain adequate oscillation. Although separate triodes are shown, twin triodes such as the 12AT7 and the 12AU7 are satisfactory. The ht supply must be stabilized.

is earthed at one end. Both of these features are of great constructional advantage.

The circuit may be used also as a crystal-controlled oscillator by substituting a crystal for the LC circuit. Owing to the high amplification available, oscillation is easily obtained with relatively inactive crystals at the frequency of parallel resonance.

The power output obtainable is quite low and extra amplification may be needed to provide adequate drive. Usually this type of oscillator will be found to have poor keying characteristics and the keying should therefore take place in one of the later stages.

To achieve the maximum frequency stability the values of C1 and C2 should be reduced as far as possible consistent with stable oscillation. The two capacitors should be varied together so as to be always approximately equal.

Mixer Oscillators

In a mixer type of vfo the output voltage of a crystal oscillator is mixed with the output voltage of a variable-frequency oscillator of much lower frequency in a suitable valve mixer circuit. The sum or difference of these frequencies may be employed to control the transmitter frequency. The degree of frequency stability attainable by this method on any particular band is superior to that normally achieved by the use of the straightforward type of vfo, but the additional complexity of the mixer arrangement has prevented it from becoming widely adopted. The principle of its operation is shown in block diagram form in **Fig 6.36.**

The reason for the improved stability is explained by

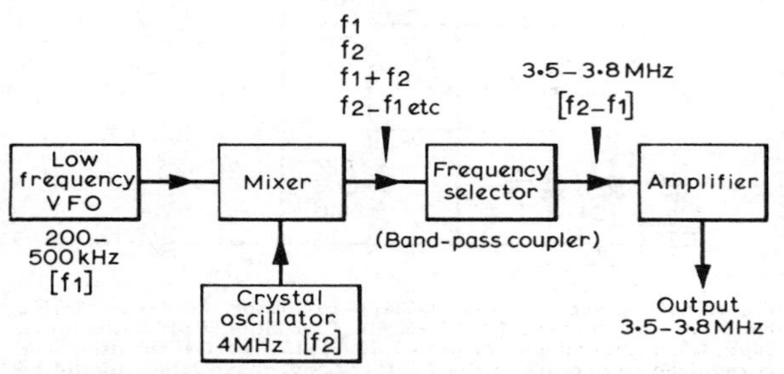

Fig 6.36. Block diagram of a mixer-type vfo. The frequencies shown would be suitable for producing an output in the 3·5MHz band.

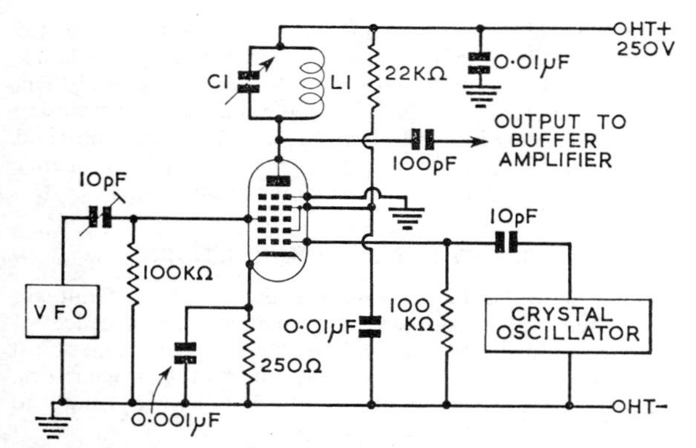

Fig 6.37. Low-level mixing circuit for a vfo showing typical values for a 6SA7. The values of L1 and C1 are selected as for a normal tank circuit.

considering the following example. Suppose that the crystal frequency (f_2) is 4·0MHz and that the vfo tunes from 200 to 500kHz (f_1). By extracting the difference frequency ($f_2 - f_1$) from the mixer by means of a frequency selector the output would cover the amateur band 3·5–3·8MHz. If the vfo is reasonably well designed its frequency should be constant within the limits of $\pm$ 6Hz at 300kHz, which is only one part in 25,000; at the output from the mixer, ie in the range 3·5–3·8MHz, this represents a stability of the order of one part in 300,000. For comparison, a typical vfo of straightforward design having an output in the same band would probably have a stability of about one part in 50,000.

In the example given, the stability multiplication factor is 12. This figure could be increased by using an even lower frequency for the vfo but this would necessitate a crystal frequency nearer to the required output frequency, and a point is eventually reached at which it is difficult to separate them in the succeeding amplifier circuits. The best compromise must therefore be found. The figures given in the example should prove satisfactory for the 3·5MHz band: a crystal frequency of 7·8MHz in conjunction with a vfo operating over a range of 500–800kHz should be satisfactory for the 7MHz band.

A low-level mixer circuit similar to that used in superheterodyne receivers is recommended: either a triode-hexode arrangement or a heptode such as the 6SA7 shown in **Fig 6.37** is suitable. The anode circuit L1C1 is tuned to the centre of the required frequency band and is coupled to a tuned buffer amplifier. Any of the orthodox coupling methods may be used instead of the capacitive coupling indicated.

In another mixing arrangement, the crystal-oscillator voltage is applied in the same phase to the screen grids of a push-pull mixer while the vfo voltage is applied in opposite phases to the control grids, as shown in **Fig 6.38**. Alternatively heptode or hexode mixer valves may be used with the vfo voltage applied in push-pull to the control grids and the crystal-oscillator voltage in parallel to the oscillator grids. The crystal-oscillator voltages are then balanced out in the output circuit, leaving only the vfo frequency and the sum and difference frequencies. These are relatively easy to separate in the tuned anode circuit and in the following buffer amplifiers.

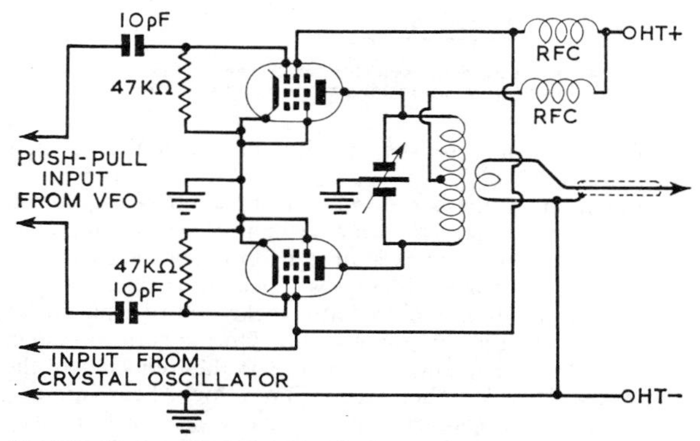

Fig 6.38. Push-pull mixing circuit for a vfo using screen injection of the crystal-oscillator voltage. The vfo voltage is applied in push-pull either through a suitably tapped transformer arrangement or from a push-pull oscillator direct. The crystal-oscillator voltage is balanced out in the centre-tapped output tank circuit. Suitable valves are 6AK5 and 6BH6.

Ideal keying characteristics may be more easily obtained with this type of oscillator unit. Neither of the two frequencies initially generated falls within the amateur bands and thus the two oscillators may be left running continuously without fear of causing interference; it is only when mixing takes place that other frequencies are produced. If the mixer itself is suitably keyed, the output in the required band will be interrupted at a low power level and chirp can be avoided, provided of course that the effect of the changing load on the variable-frequency oscillator is negligible.

Phase Locked VFO

The phase locked vfo is relatively more complicated than those already described, but it has advantages in transmitters where the carrier frequency is high and large excursions in frequency are required. In the phase locked system, the basic arrangement of which is shown in block form in **Fig 6.39**, a low frequency oscillator directly governs the frequency of a high frequency variable oscillator in such a manner that the stability of the high frequency variable

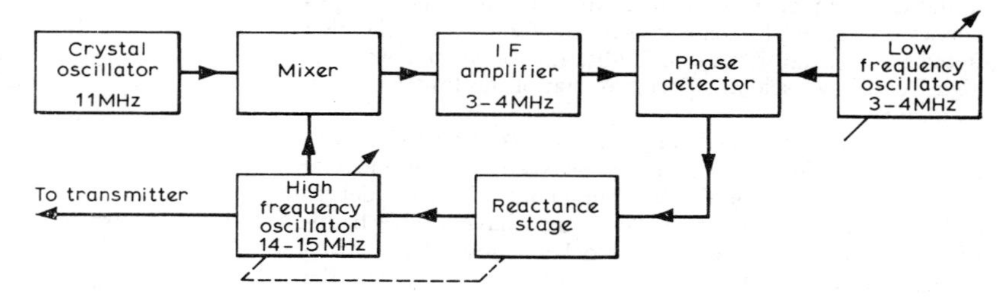

Fig 6.39. Block diagram of a phase locked vfo. As the frequency of the lf oscillator is varied it affects the reactance valve which in turn tunes the hf oscillator. When the i.f. is equal to the lf oscillator, the system is said to be locked; under these conditions any change in the hf oscillator produces an output from the phase detector due to the difference between i.f. and the lf oscillator frequencies which, via the reactance stage, automatically corrects the hf oscillator.

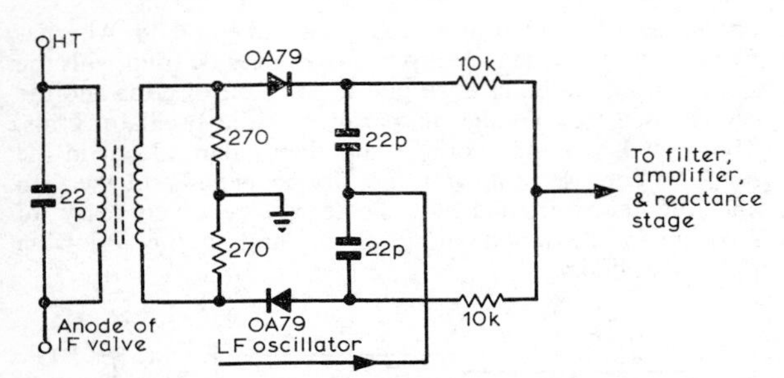

Fig 6.40. A phase detector suitable for use at 6·0MHz. It should be followed by a smoothing circuit and dc amplifier.

oscillator is equal to the stability of the low frequency oscillator.

In many ways the control circuit of the phase locked vfo resembles the automatic frequency control system employed in some vhf receivers to overcome drift in the local oscillator. There are of course differences since, in a receiver, the object is to maintain the oscillator locked on to a particular frequency, while in the phase locked vfo the control system has to respond to intentional changes in the frequency of the carrier oscillator and then assume a new locked condition.

In Fig 6.39, a relatively high frequency variable frequency (usually operating at half the carrier frequency) has its actual operating frequency controlled by a reactance valve, while its output, in addition to being fed to the transmitter, is also fed to a mixer. A frequency from a crystal controlled source is also fed to this mixer with the result that an intermediate frequency is produced, the frequency range of which will be either the sum or the difference between the upper and lower limits of the high frequency oscillator and the crystal oscillator. This i.f. is fed to a phase detector, as is the output from a low frequency oscillator covering the same range as the i.f.

The basis of the system is the phase detector **(Fig 6.40)**. Only when the two frequencies are precisely the same will their phase difference remain constant; under all other conditions there will be a changing phase difference. The voltages from the i.f. amplifier and the lf oscillator are fed to the phase detector in such a manner that the output from the phase detector fed to the high frequency oscillator via a reactance stage adjusts the frequency of that oscillator until the output from the i.f. amplifier is exactly in phase with that of the low frequency oscillator, ie until the i.f. is of precisely the same frequency as that of the low frequency oscillator. Any drift in the high frequency oscillator will immediately reflect as a change in the intermediate frequency, and hence as a change in phase in the phase detector; this will in turn apply a correction signal to the reactance modulator to restore the hf oscillator frequency to that which produces an i.f. equal to the low frequency oscillator. Not only does the low frequency oscillator physically control the frequency of the high frequency oscillator, but in addition, the stability of the high frequency oscillator becomes that of the low frequency oscillator.

At the moment of switching on, there is almost bound to be a very considerable difference between the intermediate frequency and that of the low frequency vfo. In order to lock the frequency, the phase loop must have a wide bandwidth. Once the frequency is locked, however, a wide bandwidth is not needed. One way of overcoming the difficulties caused by the wide bandwidth is to arrange the circuit in such a manner than upon switching on the bandwidth is wide enough to ensure capture and lock, and then, after a predetermined time, for the circuit automatically to reduce its bandwidth. Modern integrated circuits and solid state techniques have made highly complex frequency synthesizers available to the amateur at reasonable cost.

SOLID STATE OSCILLATORS

Solid state oscillators possess some obvious advantages over their valve counterparts: smaller size, absence of heaters, good mechanical stability and long-term electrical stability. The transistor possesses certain characteristics, however, which must be taken into account when using it to maintain an oscillator circuit:

(*a*) The low input and output impedances.
(*b*) The gain characteristics are frequency dependent.
(*c*) The gain characteristics are also temperature dependent.
(*d*) The crystal or tuned circuit may be excessively damped due to the low values of biasing resistors which are necessary.
(*e*) The considerable spread in the characteristics of any given type of transistor makes individual bias adjustment advisable.
(*f*) Because of the very low thermal capacity of the transistor, it is subject to relatively rapid changes in temperature and must therefore be provided with an adequate heat sink.
(*g*) Transistors are electrically fragile and care must be taken not to exceed the maximum permitted voltages, currents, and temperatures.

The criteria applied to the oscillatory circuit, described under valve circuits, apply equally to the solid state oscillator.

Most of the well-known valve oscillator circuits may be adapted to transistors, taking into account the difference in the parameters of the two devices. Examples of such circuits are shown in **Figs 6.41, 6.42** and **6.43.**

The circuit of a transistor bridge type vfo is shown in **Fig 6.44**. The attraction of this circuit is that external conditions do not affect the tuned circuit and, by carefully balancing the bridge, the effects of temperature and voltage changes on the transistors can be reduced to the point where they practically have no effect. Under such conditions, it is the reaction of the tuned circuit itself to temperature changes which will determine the ultimate stability.

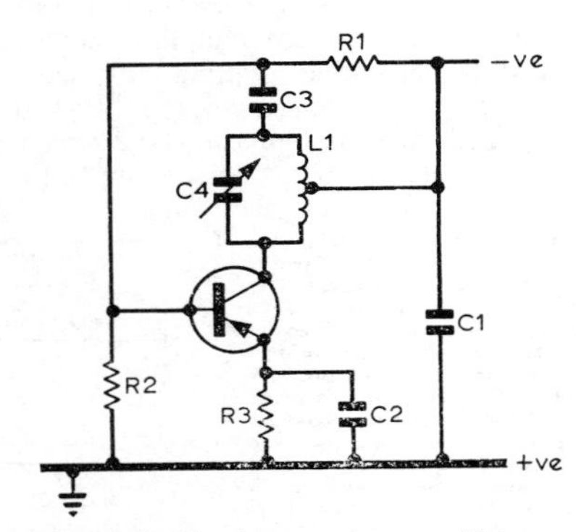

Fig 6.41. Hartley transistor oscillator.

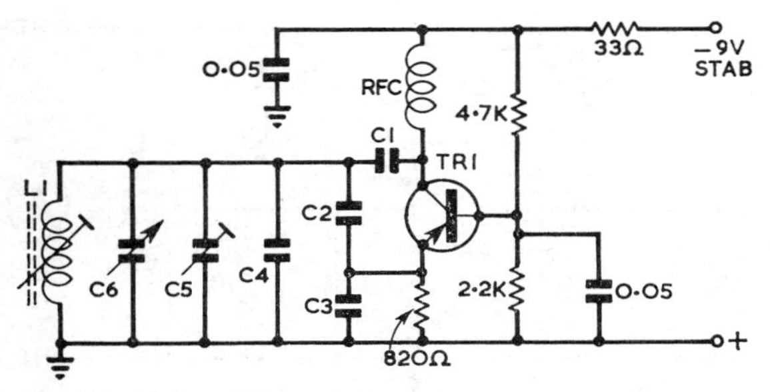

Fig 6.42. High stability transistor high C Colpitts oscillator.

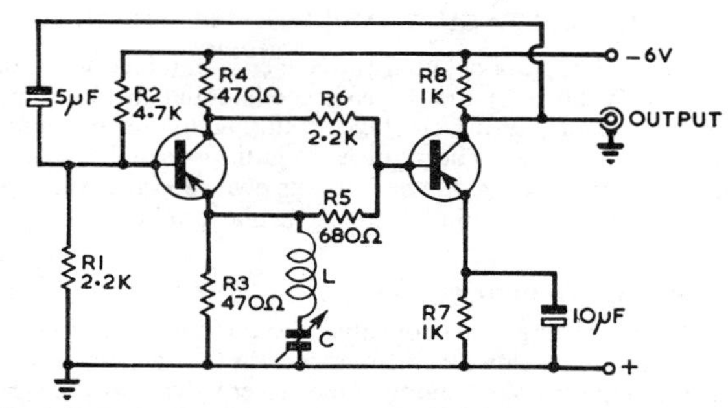

Fig 6.44. Typical example of a bridge type transistor hf oscillator suitable for use as a vfo.

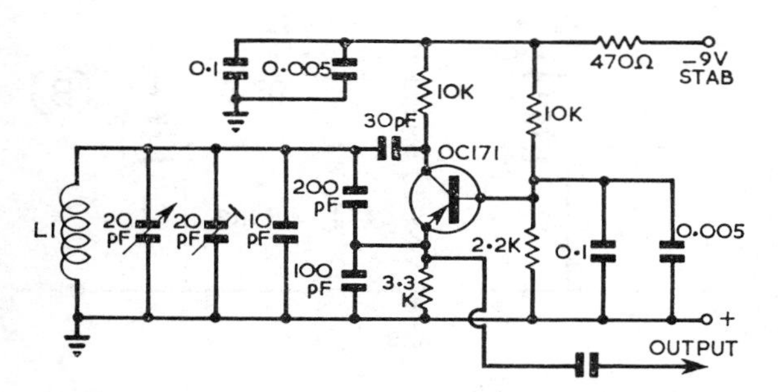

Fig 6.43. Colpitts vfo for 30MHz. L1, 4½ turns of 16swg enam. wire close wound on ¼in diam. former fitted with ferrite core.

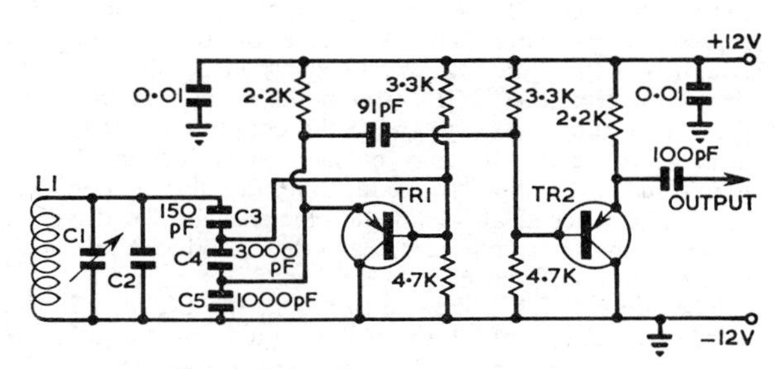

Fig 6.45. The "synthetic rock" transistor vfo designed by W3JHR. The values shown are suitable for the 5MHz range.

The resistors R3, in parallel with which is the series tuned circuit, R4, R5 and R6 form the bridge circuit which is connected between the input and output of a two stage amplifier. At the resonant frequency of the tuned circuit, R3 is virtually shorted out, and the bridge changes from a negative feedback condition to a positive feedback circuit, and thus oscillation occurs at the resonant frequency of the series tuned circuit. The feedback level may be adjusted by changes in the value of R5 or R6. Increasing R5, or decreasing R6, will result in a higher level of feedback.

The "synthetic rock" circuit due to W3JHR **(Fig 6.45)** illustrates the principle of employing a capacitive divider network to match the transistor to the oscillatory circuit in order to prevent the low input impedance of the transistor damping the circuit excessively.

Adequate buffering is essential in the transistor oscillator, as in its valve counterpart, but the transistor buffer stage is limited in its effectiveness by virtue of its essentially low input impedance and also by its "transparency"; that is, its input characteristics are effected by changes in output load.

With the advent of the field effect transistor, all the advantages of a solid state device, combined with characteristics directly comparable with the valve are available. The fet is therefore the ideal device for oscillator applications: it has high input impedance, it is much less transparent than the bipolar transistor and it may therefore replace valves in conventional oscillator circuits, taking account of the operating voltages. See **Fig 6.46**. For the same reasons FETS make excellent buffer amplifiers.

The fet, although exhibiting pentode drain/gate (anode/grid) characteristics, does not have the effect of the screen grid for input/output capacitive isolation and thus the drain/gate (anode/grid) capacitance is comparable with that of a triode valve. In most straight amplifier applications the fet must be neutralized.

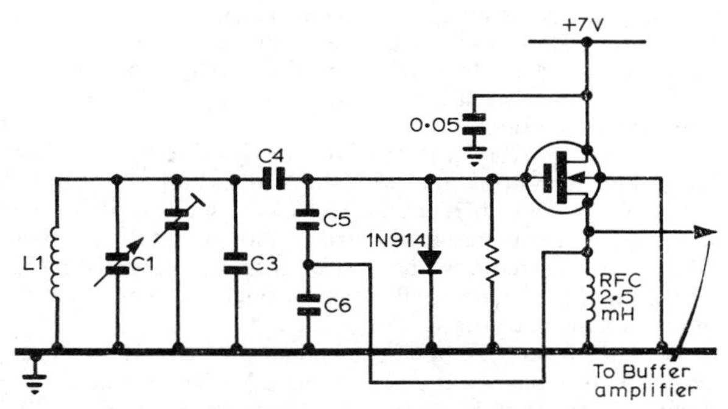

Fig 6.46. Circuit diagram of a mosfet vfo. Values for the 3·5–4·0MHz version are L1 17 turns 20BS, 16 tpi 1in dia; C1 100pF; C2 25pF; C3 100pF silver mica; C4 390pF SM; C5 680pF SM C6 680pF SM. The transistor is a 3N128 or equivalent, and the gate resistor is 22kΩ.

The fet possesses a convenient square law relationship between gate voltage and drain current, thus giving the device a linear relationship between gate voltage and g_m. This permits amplitude control of the oscillator by rectifying the output and feeding it back to control the gate bias voltage, thus allowing class A operation with increased stability due to reduction in the harmonic currents flowing in the device. It should be noted that an igfet cannot draw gate current corresponding to grid current in a valve. The gate voltage must be limited by a diode suitably connected between gate and source as shown in the above diagram.

INTERSTAGE COUPLING METHODS

Correct coupling between two stages is obtained when the input impedance of the driven stage is equal to the output impedance of the driver stage, resulting in the maximum transfer of power. The required condition can be secured in several ways, each of which has its own special advantages that make it more suitable for particular applications.

Capacitive Coupling

Various arrangements of capacitive coupling are shown in **Fig 6.47**. In its simplest form, shown in (A), a capacitor C3 is connected from the anode of the driver valve V1 to the grid of the next stage V2 in order to transfer the rf voltage from the driver tank circuit L1C1. The capacitor C4, shown dotted, represents the grid-to-earth capacitance of V2 while R2 represents the equivalent resistance of the driven grid circuit. Both C4 and R2 are effectively in parallel with the driver tank circuit L1C1 through the rf bypass capacitor C2 across the ht supply, and if the coupling capacitor C3 is made too great, V1 may be incorrectly loaded. As the load is increased, the *Q* of the tank circuit falls and a point is eventually reached at which the efficiency of the driver is lowered; moreover harmonics are then produced and may give rise to interference.

The effect of C4 is likewise to reduce the LC ratio in the anode tank circuit of V1 and if it is large it must be allowed for in the design. A method of improving the LC ratio of the tank circuit is to tap the grid of V2 into the coil as shown in (B); the *Q* of the tank circuit is also increased and often the impedance matching between stages can be achieved more accurately with consequent greater power transfer. Besides adding to the mechanical complexity, however, the tapped-coil arrangement shows a tendency to generate parasitic oscillations.

The resistor R1, which is the dc grid return for V2, is in effect also in parallel with the tank circuit of V1 and if it were not for the rf choke (RFC) in series with it this resistance would present an extra (though small) rf load. The rf choke reduces the power loss in the resistor by reducing the rf current flowing through it, while the dc path remains unchanged.

As shown in (C), a single-ended driver valve can be coupled to a push-pull stage V2 and V3. In this case, the tank coil of V1 is centre-tapped and is tuned by the split-stator capacitor C1. The rf voltages at the ends of the coil are equal in amplitude and 180 degrees out-of-phase and are therefore suitable for feeding directly to the grids of V2 and V3. The rotor of the split-stator capacitor must be earthed by the shortest route possible, preferably direct on to the chassis. If this is not done, the inductance of the connection may be sufficiently great to prevent harmonics from being adequately bypassed to earth and interference through the inadvertent radiation of these harmonics may result.

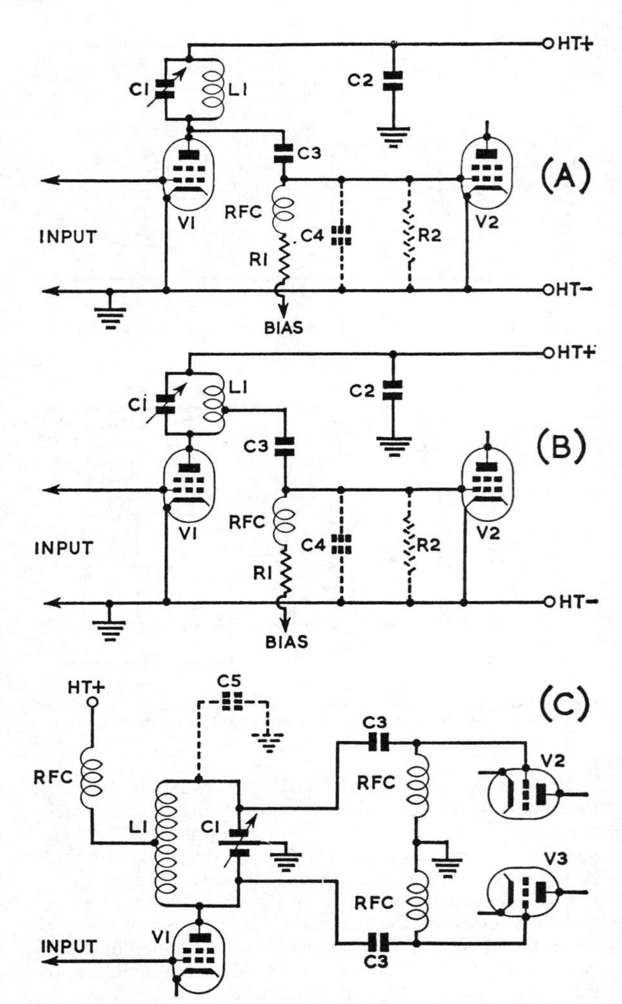

Fig 6.47. (A) Simplest capacitance coupling method which usually results in a high degree of harmonic transfer from V1 to V2. With this circuit accurate matching between V1 and V2 is difficult to obtain. (B) Provided the impedance of the anode of V1 is higher than that of the grid of V2, tapping the coupling capacitor down L1 may permit an acceptable match to be secured. (C) The circuit of (A) re-arranged to drive valves in push-pull. This circuit has the same general characteristics as that shown in (A). To obtain equal drive, the capacitor C5 may be needed to balance the anode to earth capacitance of V1. The grid circuits of V2 and V3 are symbolic only, no biasing arrangements being shown.

Inductive Coupling

Inductive coupling between stages has several distinct advantages as compared with capacitive coupling. The basic circuit arrangement is shown in **Fig 6.48.**

To understand these advantages, consider the two tuned circuits L1C1 (primary) and L2C2 (secondary), loosely coupled by a small degree of mutual induction between the coils, and suppose that each of the circuits resonates separately at a frequency *f*. If an rf voltage of variable frequency is applied to the primary of this arrangement and the resulting output voltage across the secondary is measured, the variation of output voltage with the frequency of the applied voltage will be in the form shown in **Fig 6.49**, curve A: i e, the circuit would possess considerable selectivity with a single peak at the frequency *f*. If the coils are now brought closer together, thus increasing the coupling between windings, the curve obtained on repeating the measurement will become flatter and the output voltage at the peak will be slightly higher, as depicted in curve B. Further tightening of the coupling causes the response to become still flatter until, after a certain critical value of the coupling has been reached (curve C), two peaks separated by equal amounts on opposite sides of the central frequency appear, with a pronounced dip at the centre (curve D). The value of coupling just below that which causes the appearance of the two peaks is known as the *critical coupling*. When the circuits are in this condition, it is found that the output voltage remains virtually constant over quite a wide frequency range, often as much as 5 per cent of the central frequency, and also that on either side of the flat top of the curve the voltage falls away somewhat more rapidly than in the single-peaked curve obtained with looser coupling (such as curve A or curve B). A pair of resonant circuits having critical coupling, and therefore exhibiting a flat-topped characteristic, are referred to as a *bandpass* or *wideband* coupling.

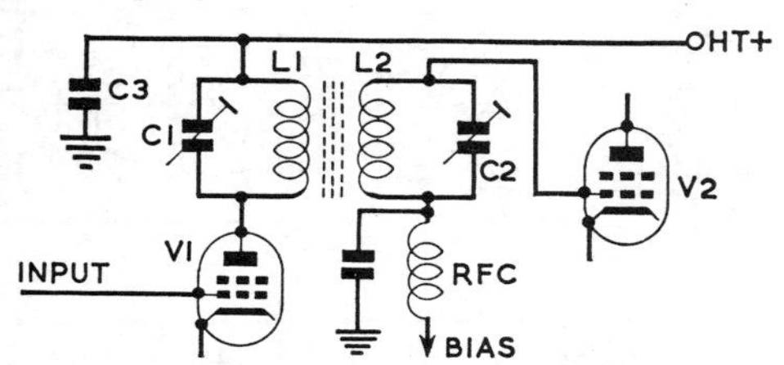

Fig 6.48. Inductive coupling. A band-pass effect can be obtained by suitably close coupling between L1 and L2. L1C1—anode tank circuit of driver stage; L2C2—grid tank circuit of driven stage; C3—rf by-pass capacitor.

The bandwidth (ie the frequency range over which the output voltage remains within specified limits, usually 3dB) of such a critically coupled arrangement is about 1·4 times that of a single tuned-circuit coupling. The overall bandwidth of a series of stages coupled by double-tuned circuits, as might be used as a drive unit, narrows much less rapidly as the number of stages is increased, compared with single tuned-circuit coupling. The same applies to double-tuned circuits irrespective of whether they are coupled by mutual inductance, common inductance or capacitance, or by link coupling as described in the following section.

The practical advantage gained therefore by the use of a properly adjusted bandpass system is that after the initial setting up of the circuits, the drive to the final stage can be made to remain virtually constant over the width of an amateur band without the necessity of any tuning adjustments.

The circuit shown in Fig 6.48 is a simple practical bandpass coupling arrangement. To avoid capacitive coupling effects which would modify the bandpass characteristic in an undesirable and unpredictable way, the grid end of the secondary winding should be situated at the opposite end from the anode end of the primary. It is quite a simple matter to measure the frequency-response characteristics (as typified by Fig 6.49) merely by using the vfo as the source of variable-frequency input voltage and observing the variations in the grid current of the driven stage. The voltage produced by the vfo (or by an untuned buffer stage following it) can be assumed to be reasonably constant over a limited frequency range. The correct degree of inductive coupling can then be found from a progressive series of measurements. If the bandwidth obtained with critical coupling is found to be too small, it is usually possible to broaden it by increasing the LC ratio of the tuned circuits.

Fig 6.49. Variation of output voltage with frequency for the inductively coupled circuits shown in Fig 6.48. The curve C indicates the conditions known as "critical coupling" the output voltage being constant over approximately a 5 per cent variation in frequency.

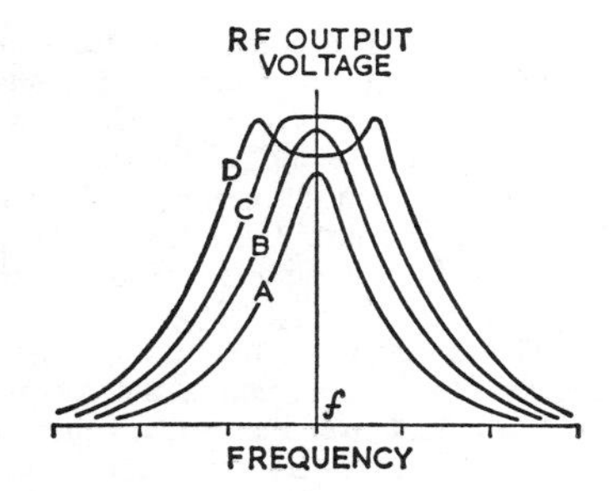

Sometimes it can happen that at first only one voltage peak is observed, although in fact the over-coupled case shown in curve D is being obtained. The other peak may be lying outside the range of measurement.

Bandpass Couplers

The use of bandpass couplers in a transmitter not only reduces the possibility of radiating spurious signals, but brings about a worthwhile reduction in the number of controls required for the rf driver section. While such couplers have these advantages, their use does invariably require somewhat higher power to be developed in the driver stages than that needed in the case of single tuned circuits. This however is a small price to pay for the increase in operating convenience.

The construction of a bandpass coupler is shown in **Fig 6.50** and winding details in **Table 6.6.** No additional capacitance is required across the windings except on 1·8 and 3·5MHz. The capacitance provided by the valves is sufficient.

The couplers are constructed on standard $\frac{5}{16}$in diameter i.f. former assemblies fitted with screening cans. To assist in the winding, a length of adhesive paper has fitted to it a narrower piece in such a manner that part of the length of the narrower strip is in contact with the adhesive of the

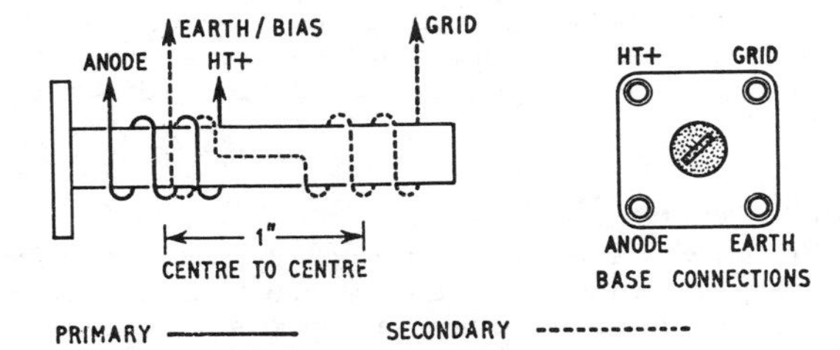

Fig 6.50. General construction of wideband couplers for the amateur bands 1·8–28MHz. Standard $\frac{5}{16}$in diameter formers are employed complete with screening cans. Both windings are in the same sense. Further details are given in Table 6.6.

TABLE 6.6

Amateur Band	Number of Turns Primary	Link	Secondary	Wire	Capacitance
1·8MHz	65 + 60 + 55	16	60 + 55 + 50	40 swg	10/15pF
3·5MHz	33 + 32 + 31	9	30 + 29 + 28	32 swg	5pF
7MHz	25 + 25	5	24 + 22	32 swg	Nil
14MHz	24	3	24	32 swg	Nil
21MHz	18	1	18	28 swg	Nil
28MHz	13	2	12	28 swg	Nil

The primaries and secondaries of the three lower frequency ranges are layer wound. All wires are enamelled. Each winding is fitted with a dust iron core.

wider strip. The combined pieces are then wrapped round the former with the adhesive of the narrower piece in contact with the former, and the adhesive of the wider section exposed. The winding is then made over the exposed adhesive. On the three lower frequency bands the windings are layer wound. Once one layer has been completed, a layer of tape should be wound over it before starting the next layer.

The actual winding should commence with the primary. When this is completed, the secondary should be started with the link over the primary first, and then progress on to the main secondary winding. Four 18swg tinned copper spills should be affixed, one to each terminal eyelet, and the windings terminated as indicated in the diagram.

As it may be necessary to adjust the position of the link winding on the 28–30MHz coupler in order in achieve a level response, it should not be too tightly wound over the primary in the first instance.

While these couplers are primarily intended for use with 6BW6 valves, there is no reason why they should not be employed with similar types so long as the mutual conductance and the grid anode capacitances of the substitute do not differ very greatly.

In the actual alignment, the primary (lower) core should be adjusted to give the required drive at the lower end of the band, and the secondary (upper) core for the same value of drive at the high frequency end of the band. The adjustments will tend to interact to some extent, and it will be necessary to go back and forth to arrive at a condition where the drive is reasonably level from one end of the band to the other. If it is suspected that a particular winding cannot be tuned sufficiently in an hf direction a brass core can be inserted in place of the iron core to confirm that there is excessive inductance. In some cases, it may be found that the use of a brass core will provide the required tuning range, and that modification to the winding is not necessary.

These couplers will provide more than adequate drive from a 6BW6 driver stage to a 6146 pa either single ended or push-pull. On the lower frequencies it may be necessary to fit a variable potentiometer in the screen supply of the driver valve in order to reduce the grid drive to the correct value.

Link Coupling

The practical disadvantage of simple bandpass inductive coupling between a pair of tuned circuits is that it may be difficult to arrange for the necessary mechanical adjustment. This can be overcome by using a variable link coupling between the two coils.

A small coil having a very few turns of wire is coupled to the driver tank circuit, thus transforming the output from a high impedance to a low impedance. A flexible transmission line transfers the energy at this low impedance to another similar coil coupled to the tuned input circuit of the driven valve where it is again transformed to the high impedance value required by its grid: see **Fig 6.51.** The degree of coupling is adjusted by altering the position of either of the link coils or by altering the number of turns in them.

The link coils should be situated at the "cold" or "earthy" ends (ie the ends having zero rf potential) of the tank coils since the capacitance coupling which exists between the link and the tuned circuits will otherwise result in spurious tuning effects and the transference of unwanted harmonics.

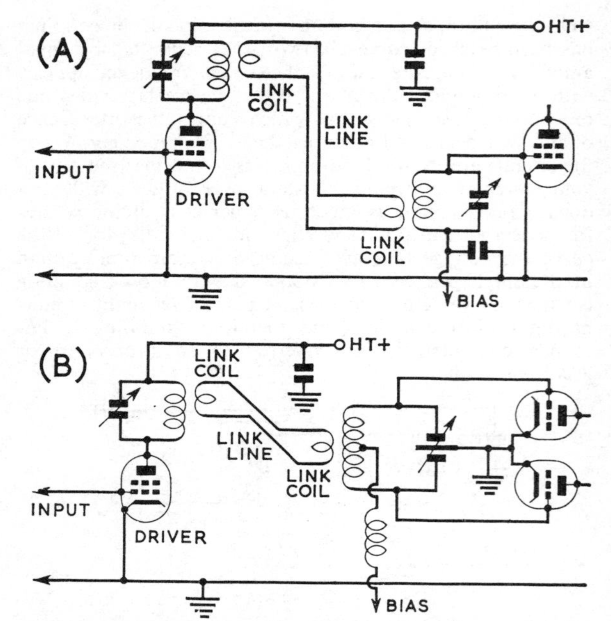

Fig 6.51. Link coupling. The arrangement at (A) is used for coupling two single-ended stages, while (B) shows how a single-ended driver can be coupled to a push-pull amplifier. The link coils must always be placed at the "cold" end of the tuned-circuit inductances to avoid unwanted capacitive coupling. Suitable numbers of turns for the link coils are as follows:

Frequency (MHz)	1·8	3·5	7	14	21	28
Number of turns	6-8	4-5	3	2	1	1

In certain circumstances it may be desirable to earth one side of the link line.

It is also beneficial to earth one side of the link circuit because this eliminates the possibility of stray capacitive coupling between the tuned circuits through the link.

The link line may in theory be of any length, but in practice unless it is limited to about 2ft the resulting reactance presented by the link may make adequate coupling difficult. However, if the line is initially adjusted to have a "flat" characteristic, this difficulty should not arise. The link reactance may also be tuned out by making the link circuit resonant; this is usually accomplished by the addition of capacitance in parallel or in series with one of the link coils. The method is not always satisfactory since the correct adjustment varies with the operating frequency and the losses in the line and the radiation from it are increased. A further advantage of link coupling is that it simplifies the problem of coupling a single-ended stage to a push-pull (or push-push) amplifier, as shown at (B) in Fig 6.51. Balanced low-impedance line, such as flat-twin cable of 75Ω impedance, is suitable for this purpose; coaxial cable (outer braid earthed) can also be used and may be desirable in order to reduce radiation from the line.

Parallel Anode Feed

If desired, the dc anode voltage of the driver valve can be removed from the associated coupling components in any of the coupling arrangements so far described. The

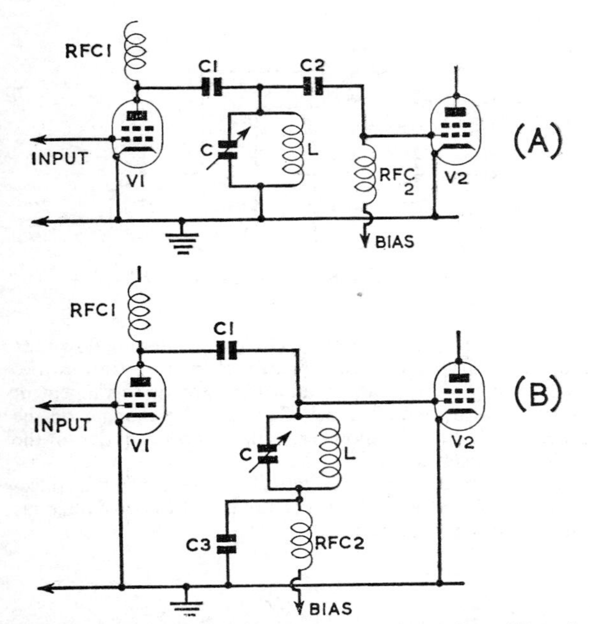

Fig 6.52. Parallel anode feed in capacitance-coupled amplifiers. In these arrangements the resonant circuit LC is the anode load. In (A) the bias for V2 is applied through the grid choke RFC2 whereas in (B) it is applied in series with the tuned circuit. C1, isolating coupling capacitor (0·001–0·01μF); C2, inter-stage coupling capacitor (100pF); C3, bypass capacitor (0·001–0·01μF); LC, tank circuit of driver stage V1.

basic principles are shown in **Figs 6.52** and **6.53.** The choke RFC1 presents a high impedance over a reasonably wide range of frequencies including the operating frequencies, and the necessary rf voltage is developed across the tuned circuit to which it is coupled through the isolating capacitor C1.

Apart from the increased factor of safety achieved by removing the anode voltage from the tuned circuits, the peak voltage across the tuning capacitor is reduced and one of smaller voltage rating may therefore be used—a matter of some importance in high-power driver stages or power amplifiers. The capacitor C1 should be rated to withstand the dc anode voltage plus the peak rf voltage and should preferably have a mica dielectric.

RF Chokes

As mentioned above, the rf choke must present a high impedance at the operating frequency. If the latter is variable over a very wide range as in a multi-band transmitter covering 1·8–29·7MHz appreciable losses may occur in the choke owing to the effects of self-capacitance and perhaps resonance. The difficulty may be partially overcome by the use of two or more different types of chokes connected in series. A good compromise for an all-band choke is one having an inductance in the region of 2·5mH, but its self-capacitance should be low.

Where rf chokes are present in both the grid and the anode circuits of the valve amplifier, it is advisable to make them of different inductance values since otherwise there is a possibility of low-frequency parasitic oscillations being

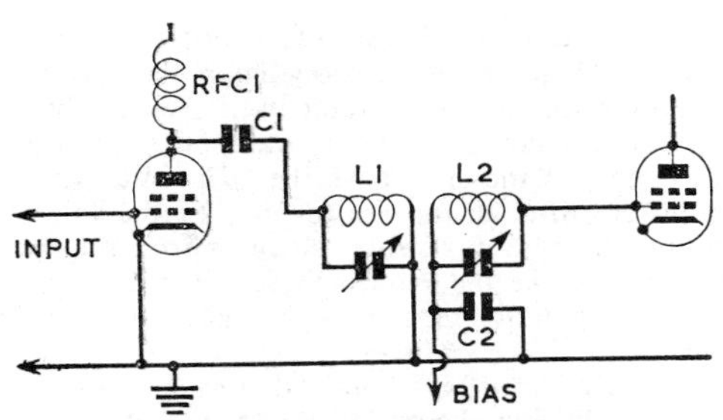

Fig 6.53. Parallel anode feed in an inductively-coupled amplifier. C1, isolating capacitor (0·001–0·01μF); C2, bypass capacitor (0·001–0·01μF); L1, L2, mutually coupled coils.

generated by feedback in the tuned-anode/tuned-grid circuit thus formed.

Interstage Pi-Network Coupler

The pi-network coupler is often used for delivering power to an aerial or an aerial feeder but it is equally suitable for coupling together two intermediate stages as shown in **Fig 6.54.** The impedances of the two stages can be matched by adjusting the tuning capacitors C1 and C2. The properties of pi-network couplers are further discussed on page 6.39 and the method of finding the coil and capacitor values is given. When used as a matching device between two class C stages the following values may be taken:

Load required by driven stage

$$R_1 = \frac{(0{\cdot}57 \times V_{ht})^2}{P}$$

where V_{ht} = dc anode voltage
P = output power in watts (input power to driven stage).

Input resistance of the driven stage

$$R_2 = \frac{6{\cdot}22 \times P}{I_g^2} \times 10^5$$

where P = total power to grid circuit of the driven stage
I_g = dc grid current in milliamps.

The total power P to the grid circuit depends upon the biasing arrangement used. For example, if the bias is derived by current through a grid resistor and this resistor is not isolated from the grid by an rf choke then it must be added in parallel with the resistance R_2 derived above. In addition.

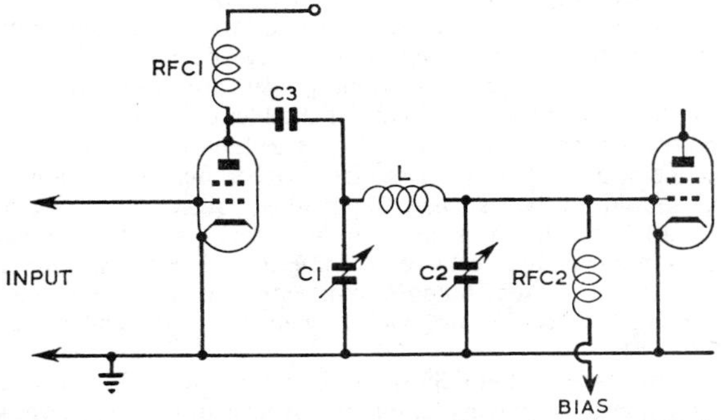

Fig 6.54. Inter-stage pi-network coupler. By varying C1 and C2 the impedances of the two stages can be matched; see text for design data. The blocking capacitor may be 0·001–0·01μF.

the value of P used in the formula is not only the driving power required for the valve alone since the driver has to supply power for the bias circuit itself. Where the bias is derived from the flow of grid current through a grid resistor the power loss in the circuit will be I_g^2R watts where I_g is the dc current through the resistor and R_g the value of the resistor in ohms. If the grid resistor is isolated from the grid by an rf choke the shunting effect can be ignored but the bias power $I_g^2R_g$ must still be included in P in the formula.

In the case of fixed bias obtained from a source isolated from the grid by a choke or tuned circuit, the power loss in the bias circuit is I_gV_g where I_g is the dc grid current and V_g the bias voltage. Again there is no shunting effect.

The formula for R_1 may be expressed in terms of anode supply voltage and anode current. If the efficiency is assumed to be 70 per cent when operating at the fundamental, 50 per cent when doubling and 30 per cent when tripling the expressions are:

Fundamental

$$\frac{(0{\cdot}57 \times V_{ht})^2}{0{\cdot}7\dfrac{(V_{ht} \times I_a)}{1000}} = 460\frac{V_{ht}}{I_a}\ \Omega$$

Second Harmonic

$$\frac{(0{\cdot}57 \times V_{ht})^2}{0{\cdot}5\dfrac{(V_{ht} \times I_a)}{1000}} = 650\frac{V_{ht}}{I_a}\ \Omega$$

Third Harmonic

$$\frac{(0{\cdot}57 \times V_{ht})^2}{0{\cdot}3\dfrac{(V_{ht} \times I_a)}{1000}} = 1080\frac{V_{ht}}{I_a}\ \Omega$$

The pi-network coupler also acts as a low-pass filter and therefore helps to prevent the transfer of undesirable harmonics from the driver to the following stages. In fact, the harmonic power output from a transmitter stage using such a coupler is less than that produced by a normal tank circuit of the same Q by a factor of $1/n^2$ where n is the order of the harmonic.

Transistor Interstage Coupling

In coupling transistor stages to succeeding stages or to output loads attention must be paid to proper matching. Because they are relatively low voltage devices the current swings involved in generating power are large and this implies low resistance loads. Tuned circuits are normally employed in the coupling networks to tune out stray capacitance and to reduce harmonics. To be effective the loaded Q should not be less than 10 and this requires an equivalent parallel circuit resistance of not less than 1000Ω with reasonable values of circuit components. For example at 3·5MHz a parallel tuned circuit with a parallel load resistance of 1000Ω has a Q of 10 when the tuning capacitance is 450pF.

Reference to **Fig 6.55** shows how the low input resistance of a power amplifier is transformed to a suitably high value to meet the driver tuned circuit Q requirements by link coupling. The low resistance required by the collector to

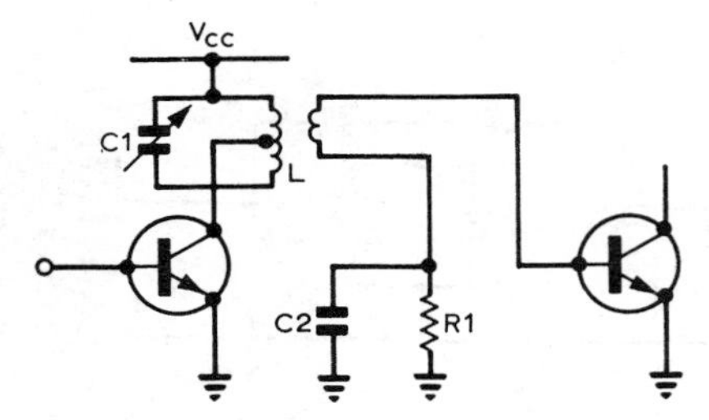

Fig 6.55. Transistor interstage coupling.

produce the specified power is achieved by tapping down the inductor. The output capacitance of a transistor varies considerably with voltage and the tapping arrangement described also minimizes the effect of these changes on the tuned circuit. **Fig 6.56** shows capacitive transformation of the pa input resistance.

The resistance (R_C) which must be presented to the collector can be found if the power (P) required to drive the pa is known using the formula:

$$R_C = \frac{V_{cc}^2}{2P}$$

The position of the inductor tap can be calculated easily from the formula $N^2 = \dfrac{R_T}{R_c}$ where R_T is the tank circuit parallel loading resistance. The tap is made at a point $1/N$th of the total turns from the cold end.

It is usually not easy to find the input resistance of the pa stage and some trial and error may be required with the link coupling or capacitance transformation arrangement.

At vhf (say above 50MHz) the coupling coefficient in coils is poor and the coil tap position cannot be calculated. Trial and error is again necessary.

There are a number of interstage networks which can be used at the higher frequencies to avoid the coupling coefficient difficulty. Some of these are shown in **Fig 6.57.** The values of capacitance and inductance involved are such that compact circuits result at vhf. The R_2 in the formula is the resistive part of the transistor input impedance and is sometimes described in manufacturers' data sheets as $r_{bb'}$. The capacitance C_o is the collector output capacitance and because it varies over the collector voltage swing an average value should be used in the formula which is twice the C_{ob} value found in the manufacturers' data.

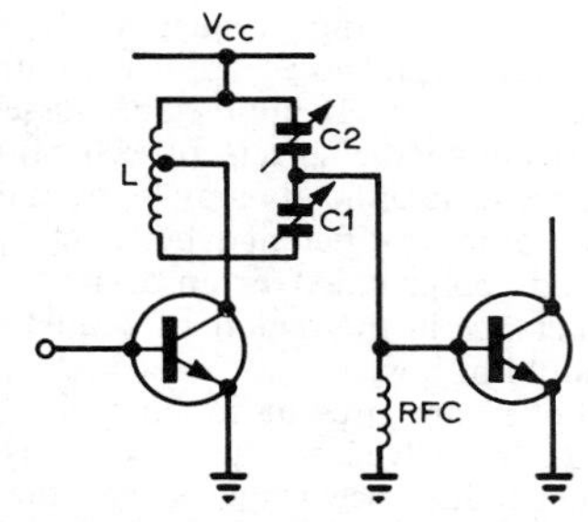

Fig 6.56. Transistor interstage coupling.

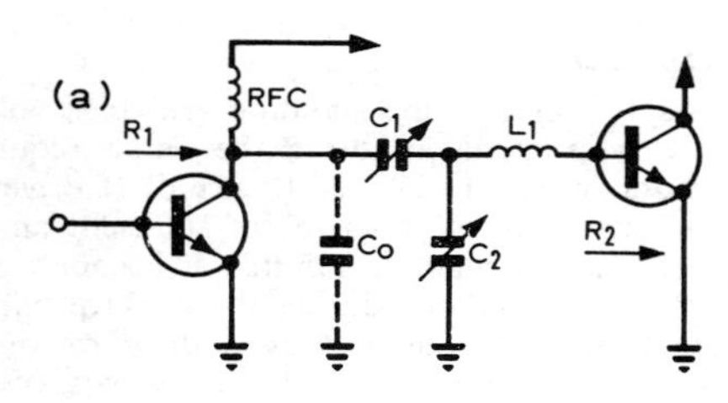

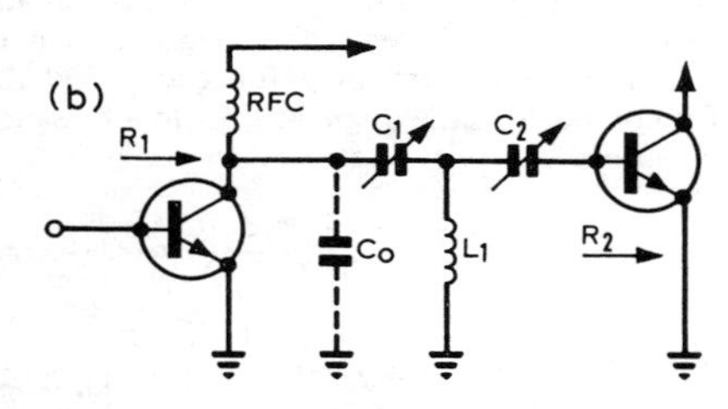

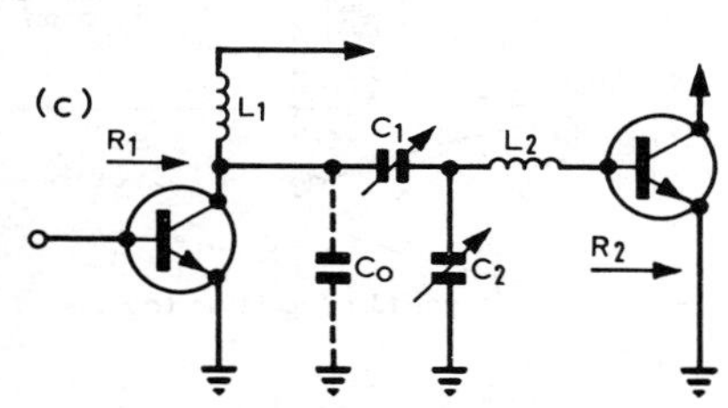

Fig 6.57. Interstage coupling networks.

Formulae for Fig 6.57(a)

For $R_1 > R_2$

(1) $X_{L_1} = Q_L R_2$

(2) $X_{C_1} = X_{C_0}\left[\sqrt{\dfrac{(Q_L^2+1)R_2}{R_1}} - 1\right]$

(3) $X_{C_2} = \dfrac{R_2(Q_L^2+1)}{Q_L} \cdot \dfrac{1}{\left[1 - \sqrt{\dfrac{R_1 R_2 (Q_L^2+1)}{X_{C_0}^2 Q_L^2}}\right]}$

Formulae for Fig 6.57(b)

For $R_1 > R_2$

(1) $X_{L_1} = \dfrac{R_2(Q_L^2+1)}{Q_L} \cdot \dfrac{1}{\left[1 + \sqrt{\dfrac{R_1 R_2}{X_{C_0}^2} \cdot \dfrac{Q_L^2+1}{Q_L^2}}\right]}$

(2) $X_{C_1} = X_{C_0}\left[\sqrt{\dfrac{R_2(Q_L^2+1)}{R_1}} - 1\right]$

(3) $X_{C_2} = Q_L R_2$

Formulae for Fig 6.57(c)

For $R_1 > R_2$

(1) $X_{L_1} = \dfrac{R_1}{Q_L}$

(2) $X_{L_2} = \dfrac{R_2}{Q_L} \cdot \dfrac{\left[\sqrt{\dfrac{R_1}{R_2}} - 1\right]}{\left[1 - \dfrac{R}{Q_L X_{C_0}}\right]}$

(3) $X_{C_1} = \dfrac{R_1}{Q_L} \cdot \dfrac{\left[1 - \sqrt{\dfrac{R_2}{R_1}}\right]}{1 - \dfrac{R_1}{Q_L X_{C_0}}}$

(4) $X_{C_2} = \dfrac{R_1}{Q_L} \cdot \dfrac{\sqrt{\dfrac{R_2}{R_1}}}{\left[1 - \dfrac{R_1}{Q_L X_{C_0}}\right]}$

FREQUENCY MULTIPLIERS

A power output at a multiple of the generated frequency is often required in amateur transmitters to obtain, for example, multi-band operation or merely to gain the advantage of the higher frequency stability which results from operating the oscillator on a relatively low frequency.

Frequency multiplication is made possible by the distortion of the anode-current rf waveform which occurs in a class C amplifier stage. As described on page 6.26, the current in such an amplifier flows only for a short time during the cycle in the form of pulses. It can be shown that such a series of pulses is composed of a number of sine waves of frequencies, f, $2f$, $3f$ and so on (where f is the fundamental or input frequency), the amplitude of these *harmonics* as they are called gradually diminishing with increasing order. Their relative amplitudes depend upon the shape of the pulse, which is determined by the operating conditions of the valve.

By inserting a parallel-tuned circuit in the anode circuit of the valve, resonant at the required harmonic frequency and therefore offering a high impedance at this frequency, the passage of the harmonic current produces a corresponding voltage-drop across the circuit. The tuned circuit in this way acts as a selector for the appropriate frequency.

In order to increase the energy of the harmonic current contained within the pulse, and therefore the efficiency of the valve as a frequency-multiplying device, it is desirable to reduce the angle of flow of current (defined on page 6.26) to values somewhat below the normal class C amplifier levels, eg 90 degrees or less. This is accomplished by increasing the grid bias to beyond normal class C operation and driving the valve with a high rf grid voltage.

The maximum power output obtainable at various harmonic frequencies in a frequency-multiplier stage, relative to the output at the fundamental frequency are—

Second harmonic	..	55 per cent
Third harmonic	..	35 per cent
Fourth harmonic	..	25 per cent

Push-push Doubler

To improve the efficiency of frequency multiplication at the *even* harmonic frequencies a push-push doubler is often used: see **Fig 6.58.** The anodes of the two valves are connected in parallel to the normal tank circuit, whereas the

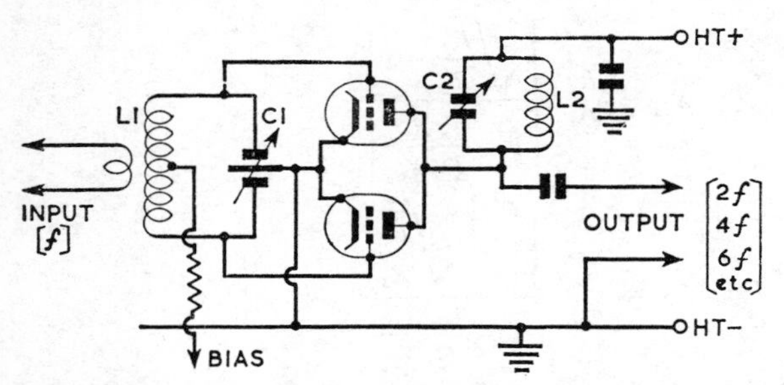

Fig 6.58. A push-push frequency multiplier. Output from this circuit is only obtained on even order harmonics, ie second, fourth, etc.

grids are driven in push-pull. The tuned circuit thus receives pulses of current alternately from the two anodes and the harmonic energy supplied to the tuned circuit is thereby doubled. When operated as a doubler, such a stage is capable of giving efficiencies approaching that of a straight class C amplifier; at the higher *even* harmonics the output is correspondingly increased.

Push-pull Frequency-Multipliers

To increase the efficiency of a multiplier stage at *odd* harmonics a true push-pull circuit such as is shown in **Fig 6.59** may be used, both the grids and the anodes being connected in push-pull. In this case, the even harmonics are self-cancelling in the anode circuit and the odd harmonics are reinforced.

POWER AMPLIFIERS

As already explained, the generation of rf power must be at a low level to ensure complete stability. The power level is raised to the desired value by means of rf amplifiers. The amplification must take place without the introduction of any spurious or harmonic frequencies which if radiated might cause undue interference to other services. According to the terms of the Amateur (Sound) Licences the maximum dc power input to the anode circuit of the final amplifier must be limited to certain specified values, depending on the frequency band in use and other factors, and therefore in order to radiate the greatest possible power for the limited input the anode-circuit efficiency of the amplifier (ie conversion of dc power to rf power) must be as high as possible. For this reason and also on the grounds of economy, the two classes of rf power amplifier in common use for cw and a.m. operation are the class B and class C types described below.

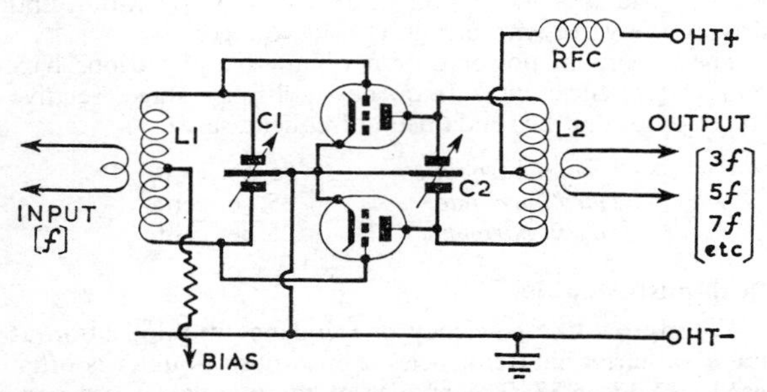

Fig 6.59. A push-pull frequency multiplier. Output from this circuit is only obtained on odd order harmonics, ie, third, fifth, etc.

Class C Operation

In a class C amplifier the standing grid bias voltage is increased to approximately twice the value required to cut off the anode current in the absence of rf drive to the grid. The rf drive voltage must be of sufficient amplitude to swing the grid potential so far that it becomes positive with respect to the cathode during the positive half-cycle. The resultant anode current is thus in the form of a succession of pulses, flowing only during a part of the rf drive cycle; the "angle of flow" (which measures its duration) is usually about 120 electrical degrees in a class C amplifier, ie about one-third of a full cycle (360 degrees), whereas for comparison the angle of flow in a class B amplifier is 180 degrees.

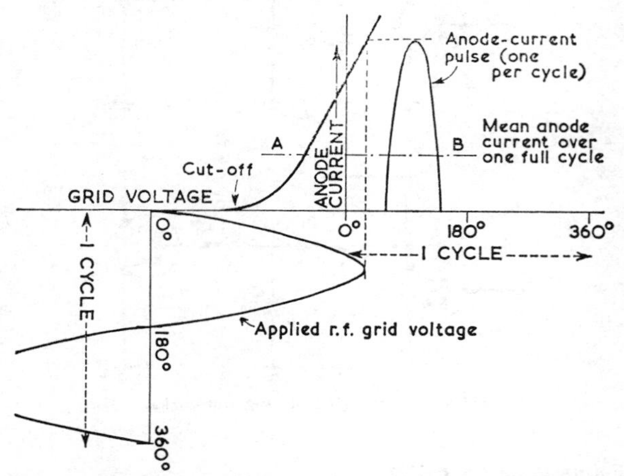

Fig 6.60. Class C operation. The valve is biased by a voltage which is not less than twice that required to give anode current cut-off under dc conditions. The signal voltage applied to the grid must be of sufficient amplitude to swing the grid positive with respect to the cathode, causing grid current to flow. The resulting anode current flows in the form of a single pulse for each grid cycle.

Fig 6.60 shows the conditions existing in a class C amplifier. The broken horizontal line AB at about one-third of the peak pulse height represents the mean anode current during one cycle, as would be indicated by a moving-coil milliammeter.

Although the anode current flows in the form of pulses, the rf voltage developed across a resonant circuit connected to the anode is approximately sinusoidal. This is due to the "flywheel" action of the resonant circuit, ie its tendency to allow the oscillations to continue after the removal of the current pulse.

As the angle of flow of anode current is reduced the efficiency increases. This is accomplished by increasing the negative grid bias and increasing the drive, which results in increased peak amplitude of the anode current pulses and increased peak anode current. Because the pulses are narrowed the power output will fall, but due to the increased efficiency it will fall proportionately less than the anode input power. The increase in efficiency achieved by narrowing the angle of flow to less than 120 degrees is small and is rarely worthwhile. Taking into consideration the increased drive and the higher peak current which the valve must provide, the efficiency to be expected at 120 degrees is about 75 per cent.

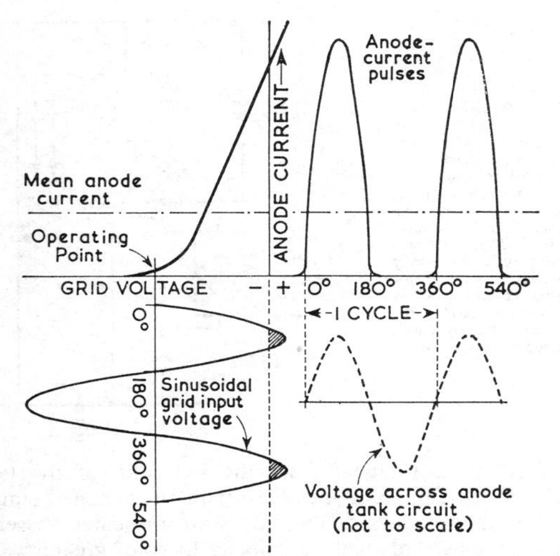

Fig 6.61. Class B operation. The anode-current pulses are approximately of half-cycle duration and appreciable grid current flows owing to the positive incursions of the grid drive voltage.

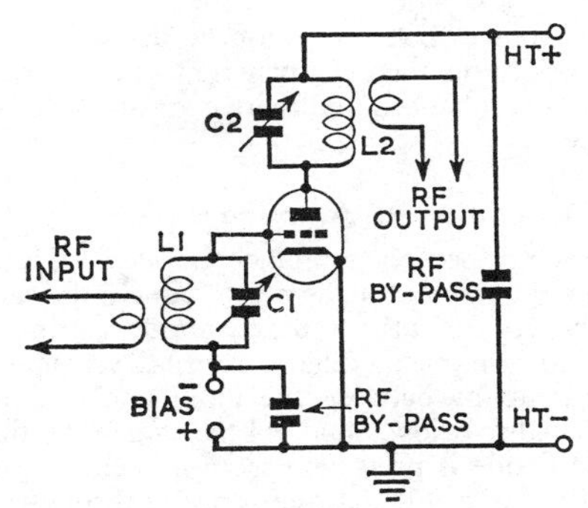

Fig 6.62. Basic triode amplifier circuit.

Class B Operation

In the class B amplifier, the anode-current pulses flow for approximately 180 degrees, ie half of the rf cycle as indicated in **Fig 6.61.** The negative grid bias is sufficient to reduce the anode current in the absence of drive only to the cut-off point, and the applied grid voltage swing drives the grid slightly positive once in each cycle, as represented by the shaded areas.

Owing to the larger angle of flow the anode efficiency is less than in class C operation, but even so it may reach 60–65 per cent for cw operation. The rf voltage in the output circuit is proportional to the rf grid excitation voltage; ie the power output is proportional to the square of the grid excitation voltage. This class of amplifier may therefore be used for the amplification of amplitude-modulated signals (including single sideband signals). Since the anode efficiency varies with the grid excitation level, the mean efficiency for the amplification of modulated signals is considerably less than the value of 60–65 per cent quoted above and is dependent on the nature of the amplified signal.

Linear amplifiers operating in class B are considered in detail later in this chapter.

Triode RF Amplifiers

The fundamental circuit of an rf amplifier using a triode is shown in **Fig 6.62.** The grid is connected to a tuned circuit which is coupled to the oscillator or driver valve by any of the means previously suggested, while the output is derived from a resonant anode circuit, which is tuned to the same frequency as the grid.

It will be noticed that this basic circuit corresponds to that of a tuned anode/tuned grid oscillator. Unless something is done to prevent it, the amplifier will oscillate at a frequency determined by the tuned circuits due to the feedback from anode to grid through the inter-electrode capacitance. Only in pentode or tetrode valves designed specifically for rf work will the anode/grid capacitance be small enough to prevent the occurrence of self-oscillation. An 807 for example has an anode/grid capacitance of about 0·25pF and will not generally oscillate on any amateur hf band unless the layout is poor. Poor layout can cause an increase in effective anode-grid capacitance due to wiring, use of poor quality valveholders, etc, or may permit sufficient inductive coupling to exist between anode and grid tank circuits to cause self-oscillation.

While it is considered advisable to neutralize pentodes and tetrodes, all triodes, without exception, must be neutralized when employed as straight amplifiers in the grounded cathode mode as Fig 6.62. The subject of neutralization is dealt with in more detail later in this chapter.

When adjusting a triode amplifier for the first time, the following is a guide to the procedure which should be adopted:

(*a*) Connect the stage to a suitable dummy load.

(*b*) Reduce the neutralizing feedback to minimum, ie set the capacitor NC to minimum or reduce inductive coupling to minimum as appropriate.

(*c*) Without applying ht to this power stage, apply drive to the grid circuit and tune for maximum grid current. Reduce the drive coupling as necessary to maintain the grid current at the required value.

(*d*) While observing the grid current tune the anode tuning capacitor over its full range: as it passes through resonance a sharp dip will be observed in the grid current. Increase the neutralization until this dip is just reduced to minimum. The stage is now neutralized.

(*e*) Reduced ht may now be applied and the tuned circuits and the drive level readjusted.

(*f*) If operation is satisfactory under reduced ht then full ht may be applied.

(*g*) Check the stability of the stage by removing the drive ("key-up" in a cw transmitter) and noting that the anode current falls to the desired quiescent value and remains so for all settings of the anode tuning capacitor.

It will be found that the rf power output increases as the drive is increased. The power output continues to rise to a point beyond which no further increase is achieved: the amplifier is then said to be saturated. The drive power should be set slightly higher than the level at which saturation commences; at this point, the anode current can be adjusted to the value corresponding to the required power input by increasing the load coupling. Slight readjustment of the tank tuning capacitor may then be required. Finally, the excitation should again be varied to ensure saturation. It is inadvisable to allow the grid excitation to rise much higher than the value at which saturation occurs since not only is the grid dissipation thereby increased but the production of harmonics is enhanced and there is no significant increase in efficiency.

Tetrode or Pentode RF Amplifiers

In a tetrode or pentode amplifier the screen grid is held at a constant potential and it therefore reduces the capacitance between the control grid and the anode to an extremely low value. If suitable auxiliary external screening is provided, the coupling between the input and output circuits can be reduced to a low value. At the same time the electric field of the anode is prevented from influencing the ability of the grid to control the current flowing through the valve, and consequently much less power is required to drive a tetrode or pentode than a triode; ie a greater power gain is available.

Because the characteristics of a tetrode or pentode valve tend to make it operate like a constant-current device, very high voltages may occur in the anode tank circuit if it is tuned to resonance in the absence of an output load or if it is inadequately loaded. No attempt should be made therefore to explore the anode-current dip with such valves except with greatly reduced power input.

It is more satisfactory to supply the screen through a series voltage-dropping resistor from the anode supply than directly from a fixed source, since in this way the screen is protected from the danger of excessive current if the control grid should be overdriven. The series dropping resistor also protects the screen from excessive dissipation in the event of the load being removed from the anode circuit. Moreover, by feeding the screen from the anode ht supply the possibility of damage owing to the application of screen voltage without the normal anode voltage is diminished.

If, however, it is considered that overdriving and underloading are not likely to occur, it will be found easier to obtain full output from the valve when the screen voltage is fixed at its correct value by the use of a potential divider across the ht supply or a separate constant-voltage source. The high power-sensitivity of pentodes and tetrodes and the fact that neutralization is not always required are the main reasons for their popularity as rf amplifiers. The small drive requirements enable a complete multiband transmitter to be built within a single screened container, and the possibility of tvi resulting from the leakage of rf power is thus greatly reduced.

Push-pull Operation

A pair of identical valves in a push-pull circuit will produce double the power output obtainable from a single valve. Although this is also true of a pair of valves connected in parallel there are several theoretical advantages to be gained from the inherent balance which exists in the push-pull arrangement. If the valves are similar in characteristics the second harmonic and other even harmonics are balanced out in the common tank circuit. Similarly, the current set up by one valve in the common bypass circuits are cancelled by those of opposite phase set up by the other. Since these currents can become quite large, particularly at the higher frequencies, increased stability is attainable by the use of push-pull operation. In practice, however, it is unlikely that a perfect balance will be found to exist.

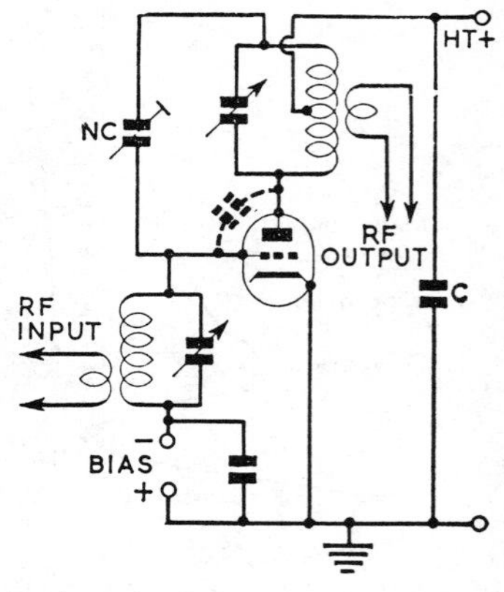

Fig 6.63. Anode neutralization. In this arrangement the centre tap of the tank coil is earthed through the rf bypass capacitor C. A typical value for the neutralizing capacitor NC is 3-30pF. When the circuit is correctly balanced, the peak voltage across NC is approximately the same as that present at the anode of the valve.

A further advantage is that the LC ratio of the tank circuits may be made higher than with single-ended stages, since the capacitances of the valves are presented in series across the coils. This feature is likely to be of greater value at frequencies of 21MHz and above where the desired LC ratio is difficult to achieve.

Parallel Operation

The output power obtainable from a push-pull or parallel arrangement is the same, provided that the correct impedances are presented to the anode and the grid circuits by use of correct LC ratios in the tank circuits. It may prove impossible to satisfy this condition when certain valves of high anode-to-earth capacitances are used in the parallel arrangement above about 21MHz, but nevertheless this circuit is often considered preferable to the push-pull arrangement, owing to the simpler construction involved and in view of the fact that the theoretical advantages of the push-pull circuit mentioned earlier are not easily obtained in practice. The parallel circuit also lends itself more easily to use with a pi-section aerial coupling network.

NEUTRALIZATION

The anode-to-grid capacitance of a valve provides a path for the feedback of energy from the anode to grid; if this feedback is of sufficient amplitude oscillation will occur. This can be overcome by deliberately feeding back a voltage of equal amplitude but opposite phase. This process is known as *neutralization*.

Although it is often considered that pentodes and tetrodes do not require neutralizing for operation in the hf range (up to 30MHz), it is usually preferable to arrange for neutralization, even when good layouts are employed. A valve may appear to be stable under constant carrier conditions but when it is amplitude modulated the rise in feedback through the anode-to-grid capacitance may be sufficient to cause either momentary or continuous self-oscillation. The former is the more difficult to diagnose because the distortion of the signal may be confused with a fault in the modulator. It is even more difficult to find when self-oscillation occurs only on the peaks of modulation.

Anode Neutralization

In this method, a centre-tapped anode tank circuit is used in place of the simple inductor-and-capacitor so that an out-of-phase voltage is made available for neutralizing the

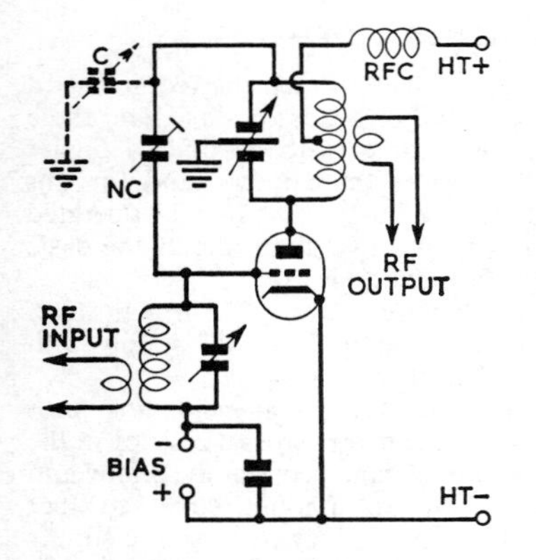

Fig 6.64. Anode neutralization using a split-stator capacitor to provide the zero-potential centre tap in the tank circuit. The circuit also shows a capacitor C for balancing the anode-to-earth capacitance of the valves.

feedback from the anode. **Fig 6.63** shows such an arrangement in which the anode tuning inductor is centre-tapped and is fed at this point from the ht line, thereby producing an rf voltage at the end of the coil opposite to the anode equal in value to the anode voltage but 180 degrees out-of-phase with it. To neutralize the internal feedback due to the anode-to-grid capacitance (shown dotted), the neutralizing capacitor NC should therefore be equal in value to this inter-electrode capacitance; in most valves this is quite small—less than 10pF—but its effect is important and an accurate balance is required.

With this circuit a certain amount of undesirable feedback is usually present when the operating frequency is higher than about 7MHz owing partly to the difficulty of tapping the coil at its exact radio-frequency centre and partly to the inevitable inductance of the leads connected to the neutralizing capacitor.

An improvement on this arrangement is shown in **Fig 6.64.** Here a split-stator capacitor is used to provide the zero-potential centre-tap in the tank circuit. The voltages to earth of the opposite ends of the coil are determined by the relative capacitance values of the two sections of the capacitor and an accurate balancing of the two voltages is thus more easily obtained. The balancing of the circuit may be further improved by the addition of the small variable capacitor C shown dotted in Fig 6.64. This serves to balance the anode-to-*earth* capacitance of the valve. It may be found especially

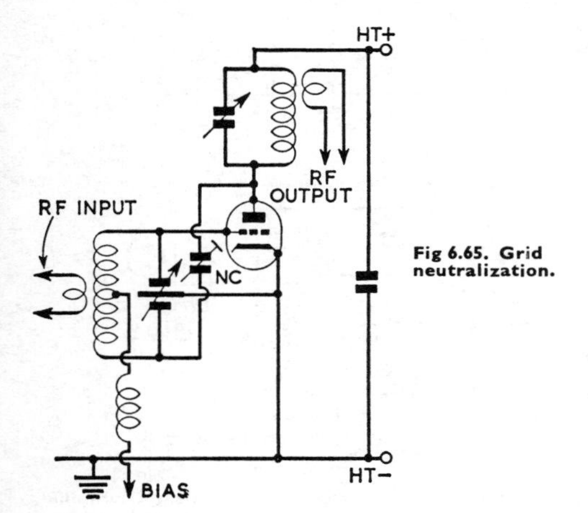

Fig 6.65. Grid neutralization.

beneficial whenever the neutralizing adjustment shows a tendency to vary from one band to another although with the split-stator circuit this does not normally occur.

Grid Neutralization

A less commonly used circuit is shown in **Fig 6.65.** This system is analogous to anode neutralization, the only important difference being that the voltage-phase reversal is achieved in the grid circuit. Its chief advantage is that it allows a normal anode tank circuit to be retained and the balanced circuit transferred to the grid where working voltages are lower and components smaller.

Link Neutralization

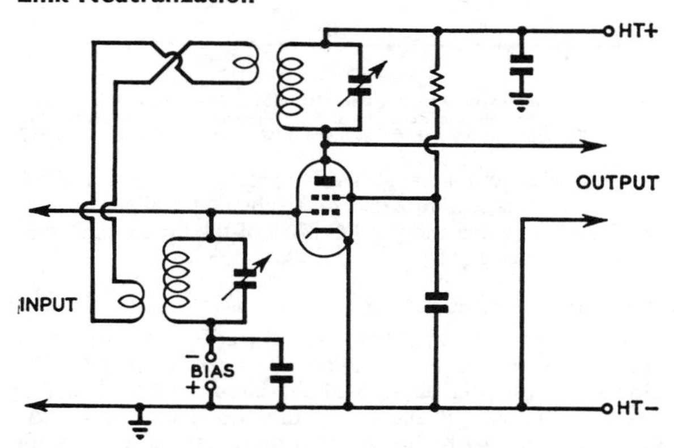

Fig 6.66. Link neutralization. This arrangement is particularly suitable for pentode or tetrode circuits where the required amount of neutralization is relatively small.

The required feedback voltage may be coupled from the anode tank circuit to the grid tank circuit by means of a link line connecting the two, as shown in **Fig 6.66.** The phase is determined by the polarization of the two coils, and if incorrect the connections to one or other of the coils should be reversed. The method is useful when a suitable neutralizing capacitor is not available or cannot easily be fitted into the layout. It is also suitable for neutralizing a tetrode or pentode amplifier which has been built without provision for neutralization and subsequently has been found slightly unstable.

As the method depends on magnetic coupling rather than capacitive feedback, it is only suitable for use on a single band unless separately positioned links are switched for each frequency range.

Series-capacitance Neutralization

Another neutralizing circuit using a capacitance potential divider is shown in **Fig 6.67.** A small capacitor NC is connected between the valve anode and a large capacitor in series with the grid tank circuit or the anode tank circuit of the preceding valve. The relative values of NC and C may be selected from the following expression:

$$\frac{NC}{C} = \frac{C_{ga}}{C_{in}}$$

where C_{ga} = anode-to-grid capacitance
C_{in} = input capacitance of the valve plus stray capacitance across the input circuit.

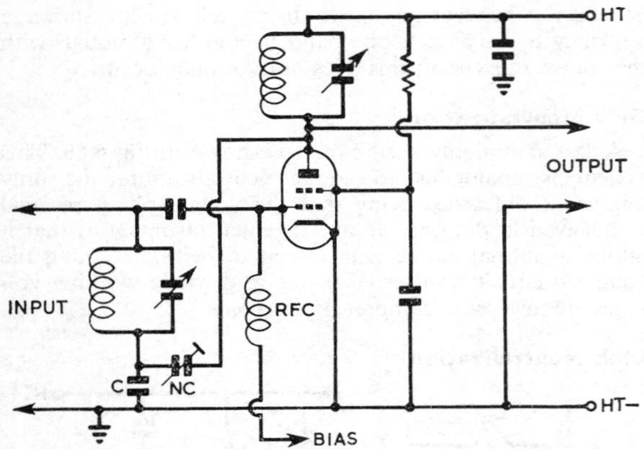

Fig 6.67. Series-capacitance neutralization. The balancing voltage is derived from the potential divider comprising the capacitors NC and C. Typical values for an 807 are NC = 5pF and C = 1000pF.

It is advisable not to use a value greater than 5pF for NC because this capacitance is effectively in parallel with the output of the valve and the LC ratio of the tank circuit may be reduced undesirably.

Push-pull Neutralization

In a push-pull amplifier, neutralization can be easily applied, as shown in **Fig 6.68,** since a centre-tapped coil is already an essential feature of the circuit. Perfect balance can be obtained; if the connecting leads are kept short and equal in length the neutralization setting remains the same over a wide range of frequency including perhaps several of the amateur bands. It is common practice to use split-stator capacitors in both the grid and the anode circuits and to earth the rotors, preferably by the shortest and thickest possible connections to ensure that harmonics are effectively bypassed.

Neutralizing Capacitors

Neutralizing capacitors are usually 20pF for triodes and 10pF for tetrodes and pentodes. They are subjected to high voltages and air spaced trimmers of high voltage rating are normally used. Tubular ptfe dielectric capacitors have a suitable voltage rating combined with small size.

Methods of adjusting neutralizing circuits are described on page 6.42.

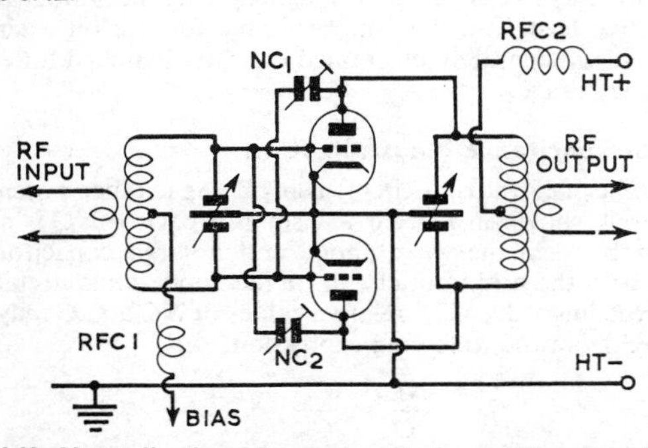

Fig 6.68. Neutralization of a push-pull rf amplifier. Due to the symmetry of the arrangement, excellent balance is usually easy to obtain: one setting of the neutralizing capacitor will hold over several amateur bands provided that the connecting leads are kept short.

GRID BIASING METHODS

There are three main methods by which the required grid bias voltage may be developed for a class C stage, and these are illustrated in **Fig 6.69**. Some methods offer a higher degree of protection to a valve than others, and for this reason, it is not unusual to find that a valve may be operated under conditions which employ more than one of the basic methods simultaneously. The methods are:

(*a*) *Fixed bias*, applied directly to the grid through an rf choke from a battery or suitable power supply with the positive terminal earthed.

(*b*) *Cathode bias* (sometimes known as *automatic bias*) obtained from the voltage drop across a resistor in the cathode lead. Since the voltage between the anode and cathode is reduced by an amount equal to that developed across the cathode bias resistor, the supply voltage has to be increased accordingly to arrive at the optimum operating conditions. This method is therefore generally limited to valves requiring a low anode to cathode voltage.

(*c*) *Grid resistor* or *drive bias* in which the cathode and grid of the valve function as a diode. The applied rf

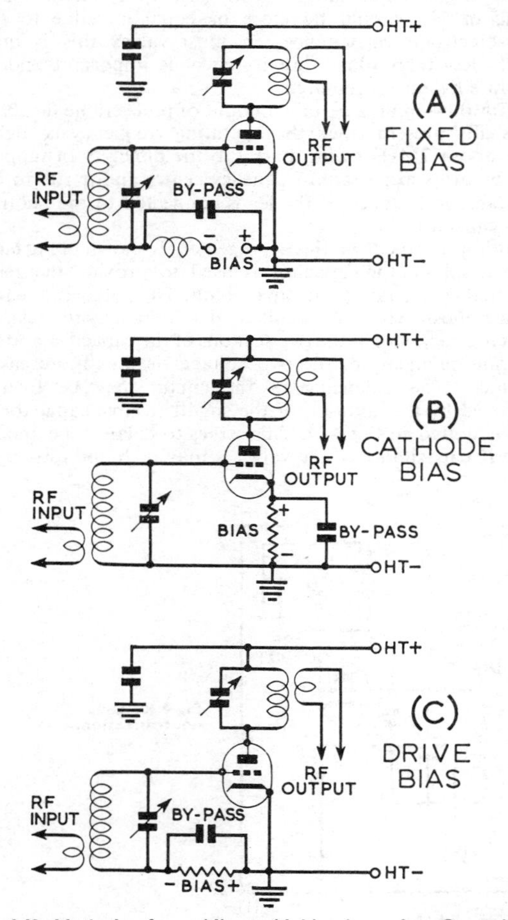

Fig 6.69. Methods of providing grid bias in a class C amplifier.

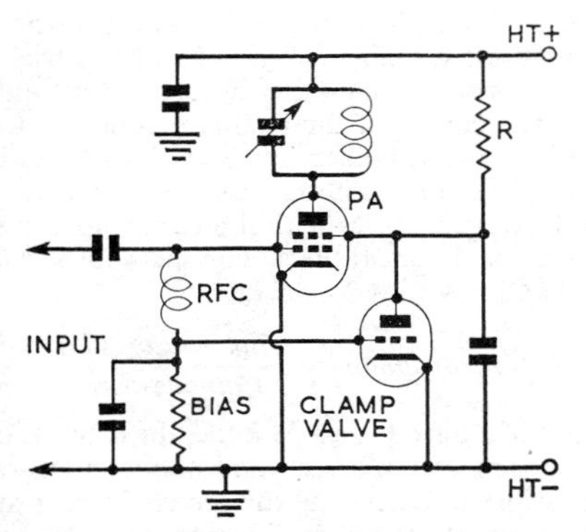

Fig 6.70 . Basic arrangement of a clamp valve to limit the anode current of the power amplifier to a safe level in the event of failure of the drive.

voltage developed across the tuned circuit causes a dc current to flow through a resistor between grid and ground, the voltage drop across which is equal to the desired bias.

Clamp Valve

If the grid bias is derived solely from the flow of rf current through the grid resistor, the valve will take excessive anode current when the drive is removed. To prevent this occurring in a tetrode or pentode amplifier, a clamp valve connected as shown in **Fig 6.70** may be used. During normal operation the pa grid resistor or drive bias also provides bias for the clamp valve and thus renders it ineffective. On removal of the excitation, the bias on the grid of the clamp valve becomes zero and its anode/cathode resistance falls to a relatively low value. The resulting increased voltage-drop across the screen resistor *R* reduces the screen voltage of the pa stage sufficiently to limit the anode current to a safe value.

The clamp valve chosen must, however, have a grid base sufficiently short to ensure cut-off under normal drive conditions, a requirement not difficult to meet with the 6BW6, EL84 or a similar type if the bias derived across the grid resistor is greater than about 25V.

This method will not reduce the screen grid voltage to zero since, for current to flow through the clamp valve, there must always be a positive potential on its anode, and hence on the screen of the power amplifier. Successful operation largely depends on the slope and current capabilities of the clamp valve. In practice it is better to use a tetrode or pentode with its screen grid fed from a fixed source rather than a triode.

It is worth mentioning that bipolar transistors have characteristics ideally suited to this requirement. Under saturated conditions the collector-emitter voltage is extremely low, usually not more than 1 or 2V even for relatively high power types. Transistors are now available with suitably high collector-emitter voltage ratings for many applications.

The current which has to be passed by the clamp valve anode should be based on the total current which could flow through the screen resistor if it were shorted to earth. In the case of a QV06-20 or 6146 operated at 600V, the screen resistor is 50kΩ, and therefore the maximum current is 12mA. To allow a generous margin, the clamp valve should be capable of passing three to four times this current and in this case a 6BW6 or EL84 valve would be suitable. Where a pentode or tetrode is used, it is essential to derive its screen grid voltage from a potential divider. Care should be taken to ensure that the screen dissipation of the valve is not exceeded when the anode voltage falls to a value which is well below that of the screen grid. If the supply to the screen grid of the clamp valve is from a potential divider connected across a high voltage source, the series resistor will usually ensure sufficient limiting action.

A clamp valve can be used as a convenient method of controlling the power developed in a power amplifier stage.

Since the anode current, and hence the output power, of a tetrode or pentode is directly related to the value of the screen grid voltage there will be a corresponding variation in the power output if it is varied. Where a clamp valve is employed to protect the power amplifier, such a facility can be arranged with very little change in the circuitry. A suitable circuit arrangement is shown in **Fig 6.71.** Instead of

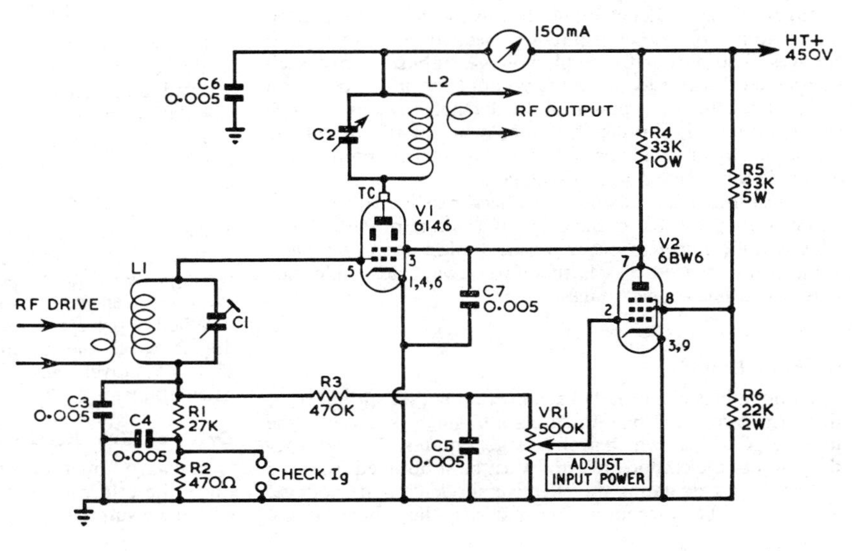

Fig 6.71. Clamp circuit modified to provide control of the dc input to the power amplifier. Possible overdrive to the grid of the amplifier is discussed in the text.

being connected direct to the grid of the pa stage, the grid of the clamp valve V2 is connected to the slider of VR1 which functions as the power input control. To avoid affecting R1, the total value of R3 and VR1 is substantially higher. VR1 permits the bias on V2 to be varied from zero to about −35V so that, when the bias on V2 is maximum, the valve is cut off and the screen of V1 is at its normal operating voltage. As the value of the bias on V2 is reduced by means of VR1, the increased current flow through R4 lowers the voltage on V1 and hence the input.

With this arrangement it is possible for the grid of V1 to be driven too heavily under reduced anode input conditions. In such a case, the power reduction must be limited or the grid drive reduced. The clamp valve can form the basis of a simple amplitude modulation system so that the minimum of additional circuitry is required to add a modulation capability to a clamp valve controlled cw transmitter. Details of such a system are contained on page 9.4.

Fixed Bias

The simplest method for the provision of fixed bias is a battery. For two reasons, however, its use may be found unsatisfactory:

(*a*) As the battery ages its resistance increases and an additional bias gradually develops owing to the grid current flowing through the battery.

(*b*) Grid current, which flows in a direction opposite to the normal discharge from a battery, tends to charge up the battery the voltage of which may be caused to rise well above the nominal value.

These drawbacks can be avoided by using a small mains-operated power unit incorporating the usual rectifier and ripple filter. The chief requirement in such a supply is that the output resistance must be low enough to prevent excessive auxiliary bias being developed across it by the flow of grid current. One method of ensuring this is to place a low-value bleeder resistor across the output terminals, in which case the power supply must be designed to supply a considerably larger current than is necessary for the actual biasing purposes. The difficulty is best met by the use of a voltage-stabilizing Zener diode or valve which will maintain the output voltage constant irrespective of grid-current changes. The output from this voltage stabilizer must then be applied directly to the valve without the inclusion of an external resistor or a potential-divider network. In practice the minimum bias voltage obtainable by the use of voltage regulator valves will be 60–80V but Zener diodes are available for a wide range of voltages.

A most important advantage of fixed bias is that the valve will be fully protected in the event of a failure of the rf drive. This is quite likely to occur during tuning operations, and some form of fixed bias is therefore always advisable particularly in high-power stages.

Cathode Bias

Cathode bias is normally used in conjunction with fixed or drive (grid resistor) bias for a class C stage because of the high value of bias which such a stage requires. The voltage-drop across the cathode resistor must be subtracted from the ht supply in calculating the effective anode-cathode voltage, and it may be uneconomical to derive the whole of the required bias from the voltage-drop across the cathode resistor. The most valuable feature of cathode bias is that it automatically protects the valve by preventing the flow of excessive anode current if the auxiliary sources of bias fail. The cathode resistor, which must be bypassed by rf, is generally chosen so that, without auxiliary grid bias, sufficient voltage is developed across it to limit the anode dissipation to within the maximum ratings. The method of calculation follows from Ohm's law ($R = E/I$):

$$\text{Cathode resistor (ohms)} = \frac{\text{Bias voltage } (V) \times 1000}{\text{Cathode current } (mA)}$$

In the case of a tetrode or a pentode, the cathode current is equal to the sum of the screen and anode currents. Where it is not possible to determine the screen current at a given anode current, due to the wide difference in screen and anode currents in such valves, it is permissible to base the calculation on the required anode current limit alone and to disregard the screen current. This will result in a somewhat higher value resistor, providing a higher level of actual bias and an anode current below the maximum level fixed.

For example, consider a tetrode in which the anode current is to be limited to 70mA, and which at this anode current has a screen current of 10mA. Assume that the characteristic curves show that a bias of 10V is required to the applied ht voltage to produce these currents, then:

$$\text{Cathode bias resistor (ohms)} = \frac{10 \times 1000}{70 + 10} = 125\Omega$$

Disregarding the screen current:

$$\text{Cathode bias resistor (ohms)} = \frac{10 \times 1000}{70} = 140\Omega \text{ (approximately)}$$

The *actual* bias developed across the resistor calculated by disregarding the screen current will be:

$$E \text{ (volts)} = \frac{I \text{ (total current in } mA) \times R \text{ (bias resistor)}}{1000} = \frac{80 \times 140}{1000} = 11{\cdot}2\text{V}$$

This will be seen to be within the 20 per cent tolerance likely to be encountered due to variations between the marked and actual value of the resistor fitted where the calculation included the screen current.

The power dissipated in the cathode resistor is given by I^2R, where I is the total current in amps; thus—

$$\text{Power} = (0{\cdot}08)^2 \times 125 = 0{\cdot}8\text{W}$$

Since this resistor is designed to protect the valve, its rated wattage should be appreciably higher than its working wattage: twice the working wattage is adequate and in the example given the resistor would be rated at 2W.

When a drive voltage is applied, the cathode current is augmented by the current flowing in the grid circuit but since this is relatively small it can safely be ignored in these calculations.

Drive or Grid Resistor Bias

In class C operation the rf grid drive is large enough to cause the grid potential to become positive at peak values, with the result that a rectified current flows in the grid circuit.

If a resistor is inserted into the grid return lead the voltage drop across this resistor due to the grid current causes the grid to be biased more negatively. This provides an automatic biasing action, the amount of bias varying with the drive power and thus to some extent adjusting the actual peak drive voltage to the operating requirements of the valve.

Since failure of the drive would result in the loss of the whole of the bias if this were derived solely from a grid resistor, it is recommended that grid resistor bias should only be used in conjunction with sufficient fixed or cathode bias to limit the anode dissipation to a safe value should the drive fail. To illustrate the method of choosing the relative proportion of fixed, cathode and grid resistor bias, the following example is given.

EXAMPLE: *An 807 is to be operated under maximum power input conditions (ICAS rating) with a combination of grid-resistor bias and fixed bias. It is required to calculate the value of the grid and the fixed bias voltage.*

The relevant operating conditions obtained by reference to the published resistor data for the valve are:

Anode voltage	..	750V
Anode current	..	100mA
Anode dissipation	..	30W
Grid bias ..	..	−45V
Grid current	..	4mA
Driving power	..	0·3W

First it is necessary to calculate the minimum bias required to limit the anode current to a safe value in the absence of rf drive. At 750V the maximum current permissible for a dissipation of 30W would be 40mA. By referring to the characteristic curve, the bias needed is found to be approximately −20V. With the rf drive applied an additional bias of 25V is therefore required to raise it to the specified −45V. However, if the fixed bias is to be obtained from a mains-driven power unit having an internal resistance of, say, 1000Ω, there will be an extra voltage developed due to the flow of grid current through this internal resistance: since the grid current is expected to be 4mA, this extra voltage drop will be 1000 × 0·004 = 4V. The fixed bias supply unit (nominally supplying 20V) will therefore develop an effective bias, under drive conditions, of 24V. This means that the series grid resistor which is to be relied upon to provide the additional bias to bring the total up to 45V need only contribute 21V. For a grid current of 4mA, a resistor of 21/0·004 = 5250Ω would thus be required.

If the fixed bias were to be obtained from a 20V battery (the internal resistance of which is negligibly small), the extra bias to be contributed by the series grid resistor would be 25V and its value would have to be increased to 25/0·004 = 6250Ω.

In either case the total resistance in the grid circuit is 6250Ω and the total resistive voltage drop amounts to 25V.

In practice the amplifier may be tested initially with battery bias in order to find the grid current. The value of the grid resistor can then be calculated from the following formula and the battery replaced by it:

$$\textit{Grid resistor in ohms} = \frac{\textit{Grid bias voltage} \times 1000}{\textit{Grid current (mA)}}$$

In the case of a 6146 valve with an anode voltage of 600 and a screen voltage of 150, the grid current and grid bias voltages are 2·8mA and 58V negative respectively. Thus where the bias is derived solely from the grid resistor, this resistor will need to have a value of:

$$\frac{58 \times 1000}{2 \cdot 8} = 20\text{k}\Omega \text{ (approximately)}$$

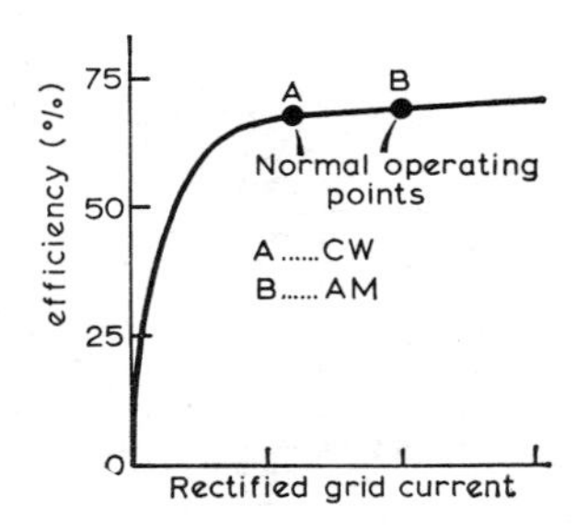

Fig 6.72. The anode efficiency of a class C power amplifier initially rises rapidly with a small increase of grid current up to about 65 per cent, after which the increase of efficiency with drive becomes relatively slight.

Valve manufacturers specify a driving power for valves operated under class B or class C conditions but this power takes no account of circuit losses, and is simply that needed by the valve itself. Working on the assumption that the transfer efficiency between the driver and the power amplifier will be no worse than 33⅓ per cent the driver should be capable of delivering three times the power needed by the driven valve. In the case of the 6146, the drive power required is 0·2W and the driver should be capable of delivering at least 0·6W.

Grid-drive Requirements

Grid Current. In normal class C operation the rectified grid current is sufficiently small to cause only a moderate rise in the temperature of the grid, but if the valve is over-driven the temperature rise may become great enough to produce grid emission and possibly cause permanent physical damage to the grid itself. Where the valve is a pentode or tetrode over-driving may also result in excessive screen dissipation. Another danger is the production of undesirable harmonics.

A curve drawn to show the relation between grid current and the anode efficiency of a class C pa stage will have the form shown in **Fig 6.72.** The efficiency initially rises quickly for a small increase of grid current from zero after which it remains substantially constant over quite a wide range.

No advantage accrues from increasing grid current beyond point A for cw work. Greater drive (eg point B) is required if the stage is to be amplitude modulated.

Drive Power. The drive power required is made up of:

(*a*) the power required to drive the grid current against the bias voltage,

(*b*) the power dissipated in the grid of the valve and,

(*c*) the loss in the coupling arrangement.

The power in the bias source is the product of the grid current times the total bias voltage from all sources and is the same no matter how the bias voltage is obtained. The power dissipated in the grid of the valve is approximately equal to the product of grid current times the peak positive grid excursion which is often about the same as the power in the bias source. The loss in the coupling components will depend on the quality of the components used and on the layout and this is usually at least as great as the total drive power specified for the valve. The drive power available should be at least five times the power lost in the bias sources. Valves in parallel or push-pull require twice as much drive as a single valve of the same type.

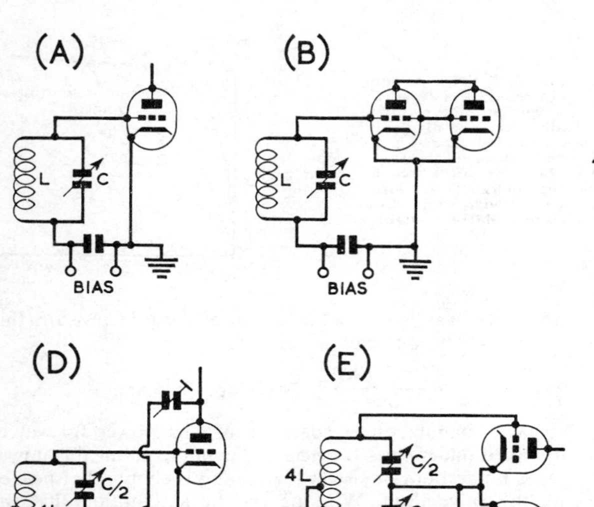

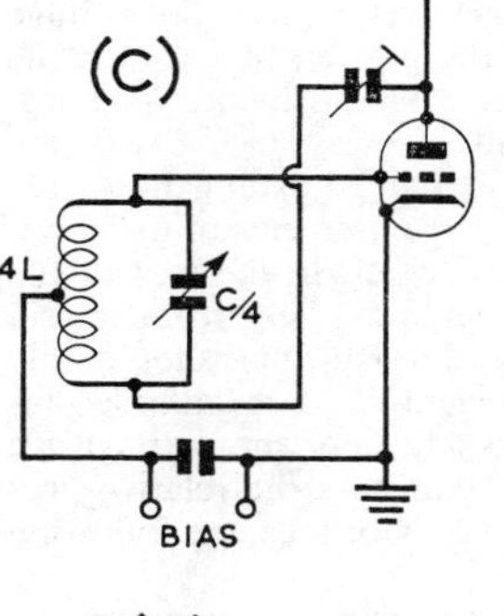

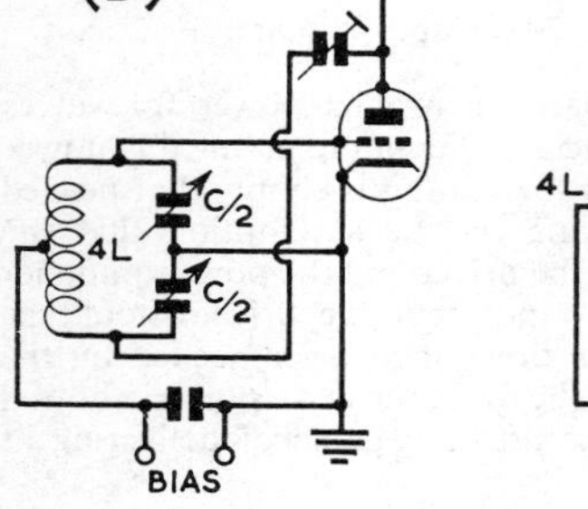

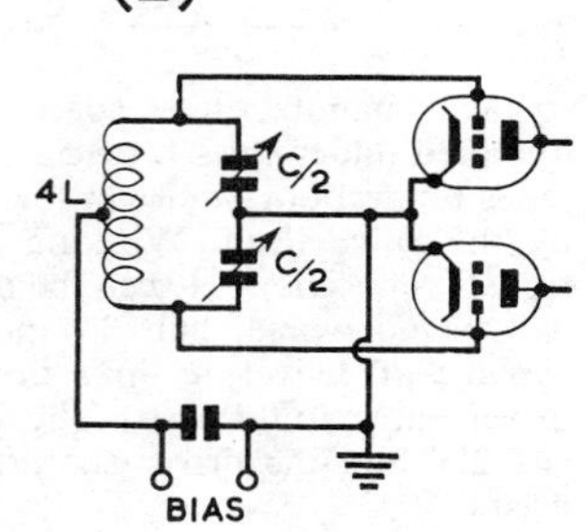

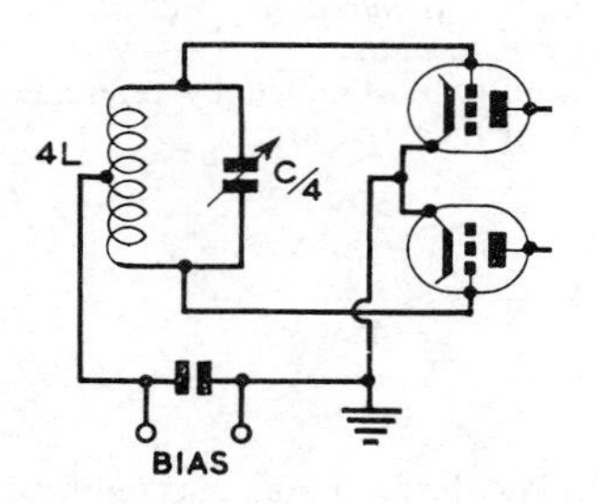

Fig 6.73. Grid tank circuit arrangements. The values of C and L can be obtained from Fig 6.74.

In an amplitude modulated output stage the drive power and bias must be large enough to accommodate modulation peaks during which the anode voltage may rise to twice its normal maximum value in the unmodulated condition and the drive power requirement correspondingly increased.

Grid Tank Circuits

A grid tank circuit should be designed to have a Q value of about 12, for reasons which are outlined later in the discussion on *Anode Tank Circuits*. The required Q value is obtained by adjusting the LC ratio of the tank circuit, although the relative values of L and C are also dependent on the type of circuit in use. **Fig 6.73** shows a selection of the more popular arrangements.

The optimum value of capacitance for a grid tank circuit in any of these arrangements can be found directly from the abac in **Fig 6.74.** To use this abac the data required are

(*a*) the operating frequency,
(*b*) the grid-drive power and
(*c*) the grid current.

In the case of push-pull circuits, the grid-drive power and the grid current will be double the value for a single-ended circuit.

EXAMPLE: *A single QV06-20/6146 valve is to be operated on the 3·5MHz band under class C conditions. The driving power required is 0·2W, and grid current 2·5mA. Find the grid circuit tank capacitance and the inductance of the associated coil.*

Using Fig 6.74 join the values 2·5mA on the grid current scale and 0·2W on the grid driving power scale by a straight-edge. This intersects the X_C/X_L scale at $1{\cdot}5 \times 10^3\Omega$. Join this value to the 3·5MHz marker on the frequency scale. The tank circuit capacitance may now be read off the scale marked C—in this case 30pF.

The capacitance found is that of the total across the coil, and therefore includes the stray capacitances of the circuit and the input capacitance of the valve. As these are likely to range between 10pF and 15pF, the net value to be provided is 30pF − 15pF = 15pF. In practice a 25pF variable capacitor would be fitted, to provide sufficient variation to compensate for all usual values of stray capacitance while still giving the total of 30pF required.

Once the capacitance has been determined, the inductance may be calculated from:

$$L = \frac{X_L\,(ohms)}{2\pi \times 3{\cdot}5}\ \mu\text{H}$$

Which in the example corresponds to:

$$L = \frac{1{\cdot}5 \times 10^3}{2\pi \times 3{\cdot}5} = 71\mu\text{H}$$

When applying these calculations to push-pull circuits it should be borne in mind that the capacitance value arrived at will be the total capacitance across the complete coil, and therefore the capacitance on each side of centre will be of twice the calculated value; the value of inductance calculated by use of the formula assumes that the coil is tuned by the total capacitance.

As the operating frequency is raised it will be found that the optimum value of C derived from the abac in Fig 6.74 becomes smaller, and it may easily happen that it is less than the input capacitance of the valve. In such circumstances, the circuit will have to be allowed to operate with more than the optimum capacitance (ie a lower LC ratio): the only serious disadvantage will be a lowering of the efficiency of power transfer from the driver to the grid circuit of the pa stage. To offset this there will be the advantage of a reduction in the harmonic content of the grid-voltage waveform and a reduced likelihood of interference due to harmonic radiation.

Anode Tank Circuits

The correct design of an anode tank circuit must fulfil the three following conditions:

(*a*) The anode circuit of the valve must be presented with the proper resistance in relation to its operating conditions to ensure efficient generation of power.

(*b*) This power must be transferred to the output load without appreciable loss.

(*c*) Owing to the nature of the anode-current pulses, the Q of the circuit when it is loaded by the output load must be sufficiently high to maintain a good "flywheel" action and produce a close approximation to a sinusoidal rf voltage waveform.

Maximum power is generated in the valve anode circuit when it is presented with a load appropriate to its operating conditions. The optimum anode load resistance (R) for class C operation is:

$$R = 460 \frac{V_{ht}}{I_a}$$

where V_{ht} is anode supply voltage and I_a is anode current in milliamps.

The efficiency of the tank circuit depends on the ratio of its dynamic resistance when unloaded to its value when loaded. Because the dynamic resistance is proportional to Q, the efficiency depends on the ratio of the unloaded Q to the loaded Q.

For a given Q value the dynamic resistance can be raised or lowered by changing the reactance of the coil or capacitor, ie by altering the LC ratio. In a good tank circuit the unloaded Q is likely to be of the order of 100-300, while the loaded Q will be much lower. Reducing the loaded Q will reduce the circulating currents in the tank circuit, and because of the diminishing heating effect it will be practicable to use smaller components. However, if the Q of the tank circuit under load is reduced too severely the "flywheel" effect becomes too small, and the harmonic content of the output increases as a consequence of the non-sinusoidal waveform. A compromise between the two requirements which has been generally accepted for use in amateur transmitters is represented by a loaded Q value of 12.

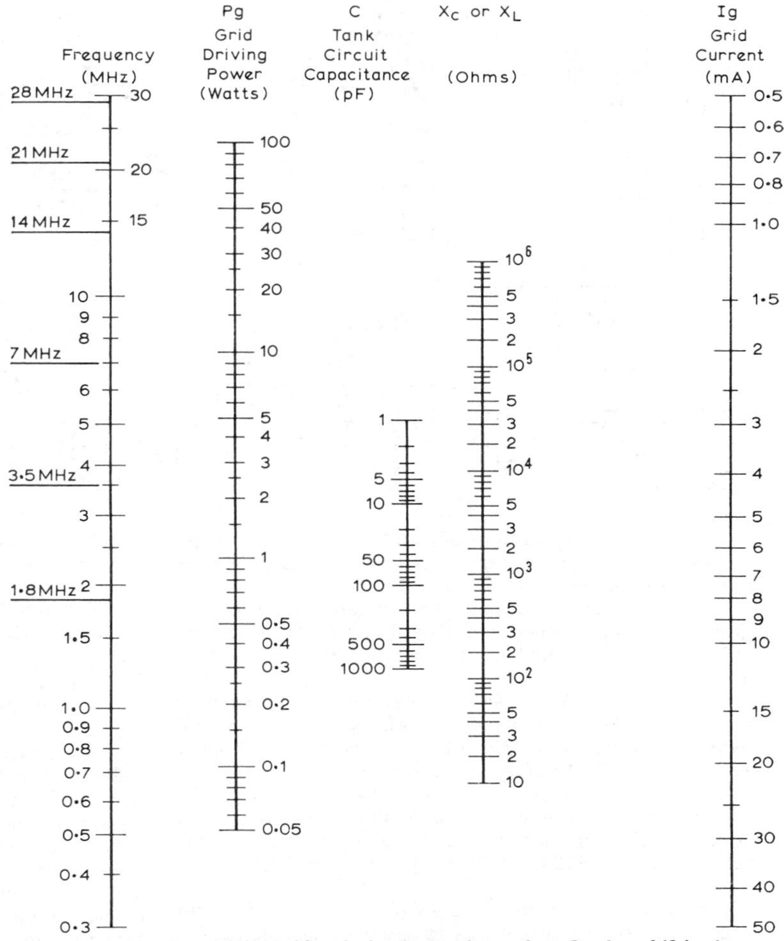

Fig 6.74. Abac for determining grid tank circuit capacitance for a Q value of 12 in the circuit arrangements shown in Fig 6.73. For push-pull and parallel connections, the appropriate values of grid current and power are those for the two valves taken together.

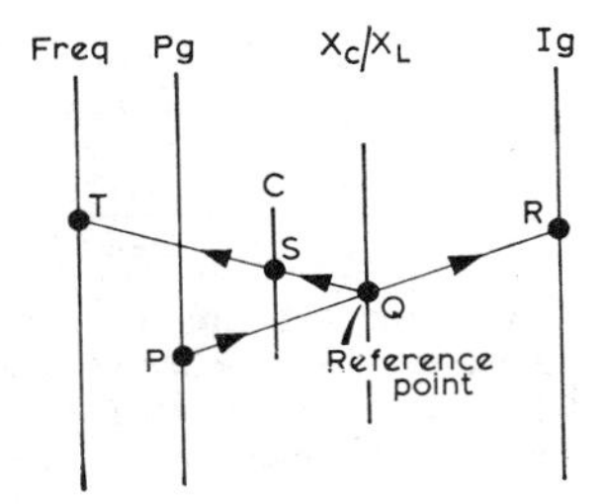

Key to Fig 6.74. Join the selected values of Pg and Ig by a line PQR. Note the point Q on the X_C/X_L scale. Join the point Q to the appropriate frequency T on the extreme lefthand scale. The required value of C is given at the point S. The corresponding value of L is given by the reactance value X_L at the point Q divided by 6·28 × frequency in MHz.

A slight increase of Q above 12 will in fact slightly lower the efficiency, but at higher Q values two other factors become significant. First, distortion can occur if the stage is modulated or if it is amplifying a modulated carrier (under class AB or B conditions) owing to the reduction in the impedance of the tank circuit at the sideband frequencies relative to the carrier frequency, but the Q would have to be greater than 200 to cause sideband clipping at 1·8MHz. The effect is to give an apparent lack of modulating power at the higher audio frequencies. Secondly, the frequency band over which substantially constant power output can be obtained without re-tuning the tank circuit will become narrower, and the serviceable band width then obtainable may be much less than is required for normal frequency changes within the permitted frequency band. The second effect will be much more significant than the first since the modulating frequencies are low in comparison with the carrier frequency.

The correct operating Q value can be obtained by adjusting the LC ratio or the loading but it must be remembered

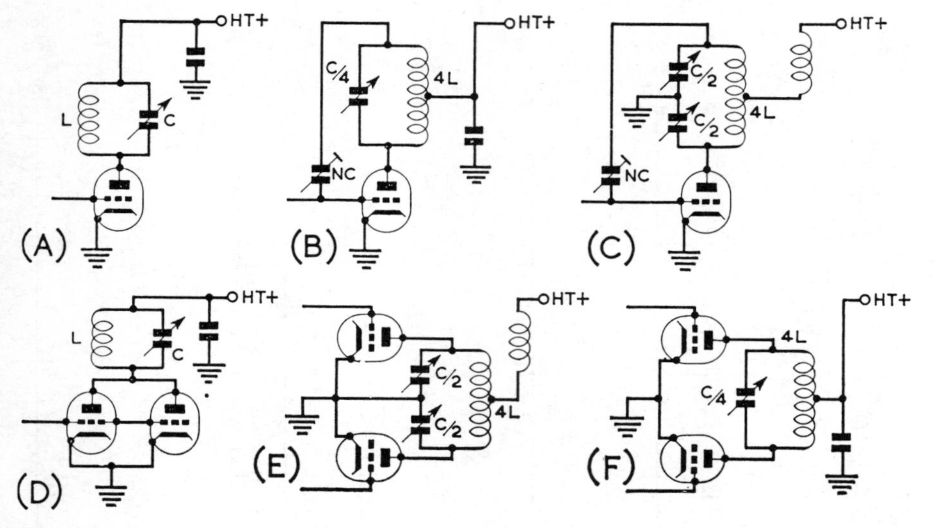

Fig 6.75. Anode tank-circuit arrangements. The values of C and L corresponding to a loaded Q of 12 can be obtained from Fig 6.76.

that the optimum proportions of L and C also depend on the particular form of anode tank circuit which is used. Several popular arrangements are shown in **Fig 6.75.**

An abac for the direct estimation of capacitance for any given operating frequency will be found in **Fig 6.76**: the only other data required are the operating anode voltage V_a and the anode current I_a.

EXAMPLE: *Find the anode tank capacitance and inductance required at 7MHz for a pair of 5763 valves connected in parallel in a class C output stage: the anode current for a single valve is 50mA and the ht supply is 300V.*

Using Fig 6.75, join the values 100mA and 300V on the anode-current and anode-voltage scales respectively by a straight-edge. This line intersects the X_L/X_C scale at the reference point of 120Ω. Join this point to the 7MHz-band marker on the frequency scale. The optimum tank-circuit capacitance is then read off as 180pF. The anode-to-earth capacitance of the two valves (plus stray capacitances) is about 15pF, leaving about 165pF to be provided by the actual tank capacitor. In practice, a 200pF unit would be used.

The inductance is given by

$$L = \frac{X_C}{2\pi f\,(\text{MHz})}\,\mu\text{H}$$

$$= \frac{120}{6{\cdot}28 \times 7} = 2{\cdot}75\mu\text{H}$$

As in the grid circuit, the desired LC ratio cannot always be achieved in practice with some valve types at the higher frequencies and a lower LC ratio must then be accepted with some loss of efficiency.

PARASITIC OSCILLATION

Under certain conditions, the amplification of an rf drive voltage may be accompanied by the production of oscillations within the amplifying stage itself. The frequency of these *parasitic oscillations* is usually quite unrelated to the normal carrier frequency, and therefore interference may be caused to other services outside the amateur bands. They may be found to occur at a relatively low frequency or at a very high frequency.

Low-frequency Parasitics

Low-frequency parasitics are likely to be generated when the rf chokes in the grid and anode circuits happen to resonate at approximately the same frequency. The use of two similar chokes in these positions should be avoided. It is better to use a grid resistor in place of one of the chokes. The frequency of oscillation is normally below about 1–2MHz and the reactance of the tank coils to such low frequencies is therefore very small except on the lower-frequency bands.

When drive (grid resistor) bias is employed, it is usually possible to dispense with the rf choke in the grid circuit and to connect the grid resistor directly to the tank circuit or grid, although this may mean increasing the loading on the driver to make up for the higher losses. When the grid circuit is tuned by a split-stator capacitor, direct connection of the grid resistor to the tap on the coil usually produces no measurable effect. However, if a clamp valve is used in conjunction with a push-pull input circuit, or if split-stator tuning of the grid circuit is used for neutralizing purposes, a grid choke must be employed. In such cases different types of choke may be used in the grid and anode circuits or the chokes may be damped by series or parallel resistors.

A general coverage receiver with a short aerial may be employed to detect low frequency parasitic oscillations. Alternatively, a sensitive wavemeter can be coupled to the suspected parts of the circuit. There will be no mistaking a parasitic when it is found: they are invariably highly unstable and any change in frequency is usually unrelated to vfo tuning or crystal frequency.

The most difficult type of parasitic oscillation to trace is one that occurs only as the applied voltage reaches its peak during amplitude modulation. The effect on the signal is to make it sound rough. As this is similar to the effect produced by a slightly regenerative stage it is not possible to tell whether it is due to regeneration or to parasitic oscillation. The only approach is to neutralize the stage and then to look for parasitic oscillation if distortion persists.

VHF Parasitics

The causes of vhf parasitic oscillation are somewhat more subtle than those of the lf parasitics and is not at first obvious how they can be produced. They are more common in tetrode and pentode amplifiers where the valve has a relatively high mutual conductance; often valves such as the 807, TT21 and 6146 are viewed with suspicion for this reason but the prevention of parasitic oscillation is quite simple provided that the cause is understood and the proper precautions are taken.

A tuned circuit is formed by both the grid and anode circuits as shown by the heavy lines in **Fig 6.77** and is likely to have a resonant frequency in the range 100-200 MHz. At these frequencies the bypass and other capacitors and their connecting leads act as inductive reactances. The capacitance which resonates with them is that of the grid and anode circuits and other stray capacitances. If the resonant frequencies of the grid and anode circuits are not greatly different from each other, oscillation may occur. Similar conditions apply if the lead to the screen grid is resonant. By reducing the stage gain at the frequencies concerned, or alternatively by separating the vhf resonance frequencies of the grid and anode circuits, oscillation can be prevented. By making connections to the tank circuits with short leads, preferably of dissimilar wire or copper strip, the inductance of the wiring may be reduced sufficiently to increase the resonant frequency to a value so high that the amplification afforded by the valve is insufficient to maintain oscillation.

A frequent cause of parasitic oscillation is the reactance at vhf of bypass capacitors intended for use at hf. For example, if the ht line in Fig 6.77 were decoupled at vhf at the point where the anode tank circuit joins it, in addition to being decoupled in the usual way for hf, the vhf resonant circuit would be limited to ABC instead of ABCDE.

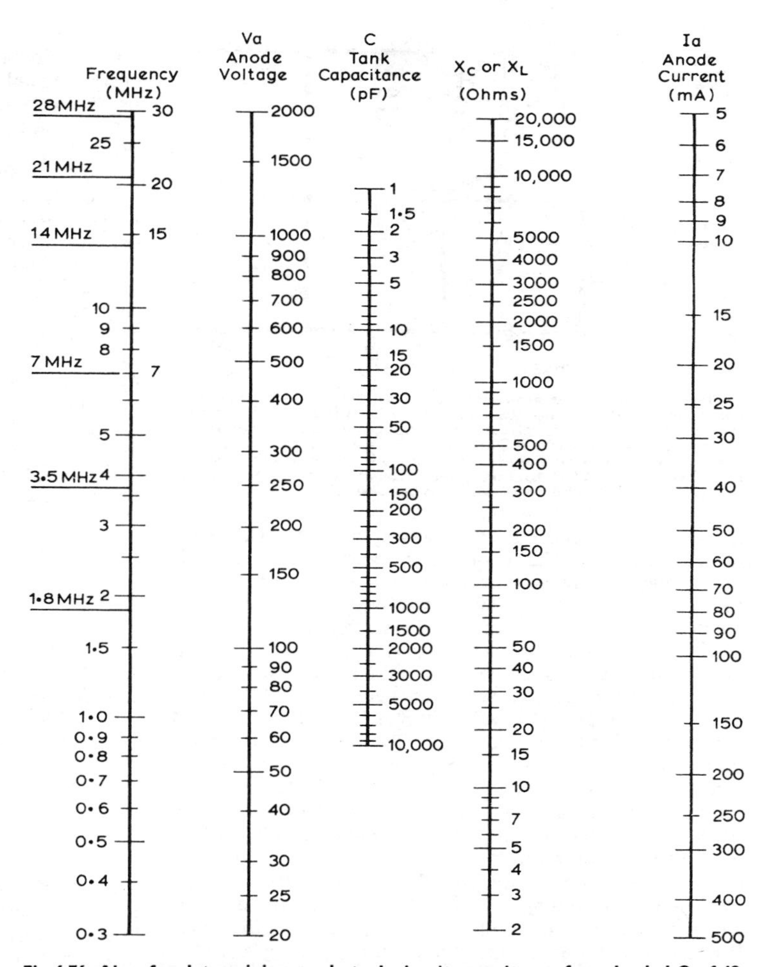

Fig 6.76. Abac for determining anode tank-circuit capacitance for a loaded Q of 12 in the circuit arrangements shown in Fig 6.75. For push-pull and parallel connections, the appropriate value of anode current is that for the two valves taken together.

Key to Fig 6.76. Join the selected values of V_a and I_a by a line PQR. Note the point Q on the X_C/X_L scale. Join the point Q to the appropriate frequency T on the extreme left-hand scale. The required value of C is given at the point S. The corresponding value of L is given by the reactance value X at the point Q divided by 6·28 × frequency (in MHz).

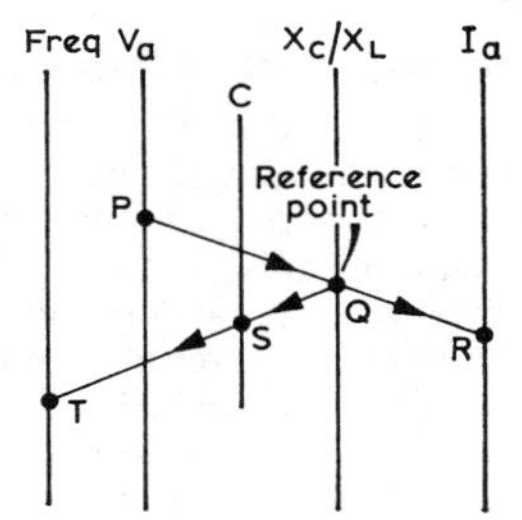

The length of the cathode lead of a valve has a direct bearing on its ability to oscillate: the lead should therefore be made as short as possible. If the lead must be long, or has included in it a bypassed resistor for biasing, a vhf bypass capacitor should be fitted with minimal leads from the cathode terminal on the valveholder to an earth point as near to it as possible. This capacitor should have a value of 470pF and should be a type designed for bypassing purposes. Unfortunately, the addition of a larger capacitor in parallel in order to obtain sufficient decoupling at the working frequency defeats the object of the 470pF capacitor unless the resulting parasitic circuit is damped by a low value (47Ω) resistor close to the valveholder.

The screen grid requires special attention and it should be the rule to connect a 470pF capacitor with minimal leads between the screen terminal of the valveholder and the cathode connection. This capacitor must be fitted between the screen grid and the cathode: to connect it from screen grid to earth would not prevent parasitic oscillation. A similar bypass capacitor should be fitted from the ht end of the anode tank circuit to earth with the shortest possible leads.

One of the most effective vhf parasitic stoppers is a vhf choke in series with the anode lead to the valve: a low value

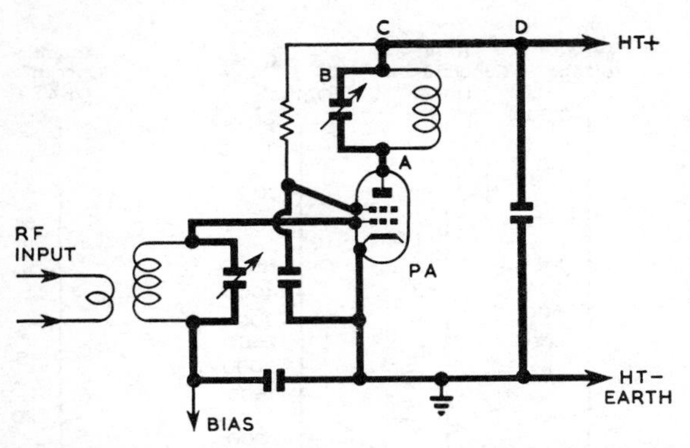

Fig 6.77. Typical power-amplifier circuit showing how vhf parasitic oscillations may be generated. The heavy lines indicate the respective vhf grid and anode circuits, the inductance being constituted by the connecting leads. A neon lamp can be used to indicate the presence of such oscillation; if it is moved, for example, from A, round the path ABCD to earth, the intensity of the glow will be found to diminish progressively.

(5–10Ω) wire wound resistor functions well as a vhf choke for this purpose. This choke must be positioned directly on the anode terminal with no leads. If this is not effective, the screen grid lead should be removed and fitted with a ferrite bead, the lead reconnected and the bead positioned as close to the screen grid connection as possible. If oscillation persists, the anode choke should be removed, and a carbon resistor with a value of between 10–47Ω fitted directly to the grid pin with minimal leads, the other end being connected to the grid circuit. The lowest value resistor needed to stop the oscillation should be used because it will introduce a power loss into the grid circuit. The positions for parasitic stoppers, the nature of the stopper, and the order in which they should be tried is shown in **Fig 6.78.**

Small vhf chokes can be constructed by taking approximately a quarter of a wavelength of wire at the parasitic frequency, and winding it on a low loss former so that it has a length which is at least three times as great as its diameter. If it is possible to make the choke self-supporting it will be measurably more efficient. For parasitic oscillations in the range 100–200MHz the wire length to be wound should be 18in, and for the 200–300MHz range 9in.

Identification of Parasitics

The presence of either type of parasitic oscillation will render the tuning characteristics of the amplifier erratic and will reduce the efficiency at the operating frequency. In some cases, the valve is overloaded owing to the parasitic oscillation and may be damaged as a result. Apart from the possibility of interference by unwanted radiation, these parasitics should therefore be eliminated to allow the full efficiency to be obtained. Before checking for parasitic oscillation it is desirable to carry out "cold" neutralization of the stage first by observing the grid current as the anode circuit is tuned through resonance with the ht switched off.

The tendency of the pa stage to oscillate can be investigated in the following way. The rf drive to the amplifier should be removed and the grid bias adjusted so that the anode current does not exceed the safe maximum value. If self-oscillation at a frequency corresponding to the actual tuning circuits is taking place, owing to lack of screening or neutralization, an adjustment of either the grid or anode tuning capacitors will alter the frequency (which will be heard on a receiver possibly as a rough unstable note): the anode current will also vary with the tuning adjustment. Such self-oscillation must first be eliminated by improving the screening arrangements and by ensuring that the amplifier is properly neutralized.

If the parasitic oscillation is of the low frequency type, the tuning capacitors will have very little effect on its frequency; a neon lamp held near the anode circuit will glow with a darkish red colour at any part of it, including the apparently "earthy" end. In the case of vhf parasitics, the neon lamp will glow with a reddish-purple colour when held near the anode, but the intensity of the glow will diminish as the neon is moved round the "parasitic" anode circuit (ABCDE in Fig 6.77). Sometimes parasitics are more readily generated when the tank capacitors are at their maximum values, and the capacitors should therefore be varied throughout their full ranges to ensure all possible conditions are examined. A sharp upward "flick" of the anode current while this is being done indicates that the stage requires neutralizing.

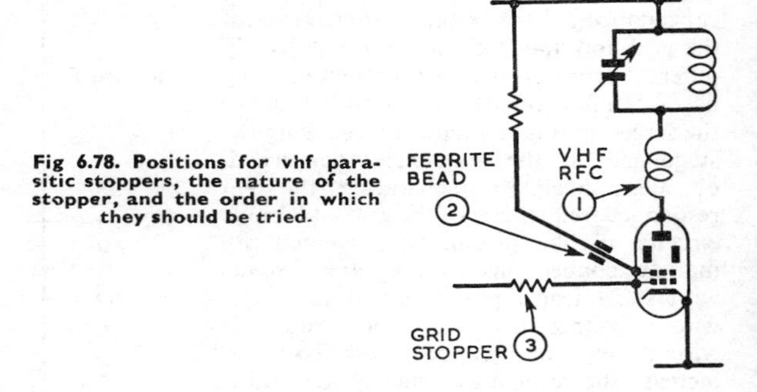

Fig 6.78. Positions for vhf parasitic stoppers, the nature of the stopper, and the order in which they should be tried.

In clearing parasitic troubles, it often helps to recognize that the spurious oscillation adopts a different mode of operation from that at the operating frequency. For example, in a push-pull stage a low frequency oscillation caused by the rf chokes could well occur due to the valves operating in parallel in the spurious mode. Similarly, when the valves are connected in parallel, a vhf parasitic could occur in a push-pull mode. Suppressing devices which affect the parasitic but do not affect normal operation can sometimes be devised.

OUTPUT COUPLING CIRCUITS

The output coupling circuit of the tank circuit of a transmitter is as important as the output circuit itself. There is no point in having potential energy available if it cannot be used.

Link Coupling

A quite commonly used method of transferring the power available in the anode circuit is by means of a link winding. This consists of a few turns of wire, usually of the same gauge as the winding on the coil, placed adjacent to the earthy end of the coil and coaxial with it. In practice, and to allow the position of the link winding to be moved relative to the tank coil, it may be wound fractionally larger so allowing it to slide up and down the coil former.

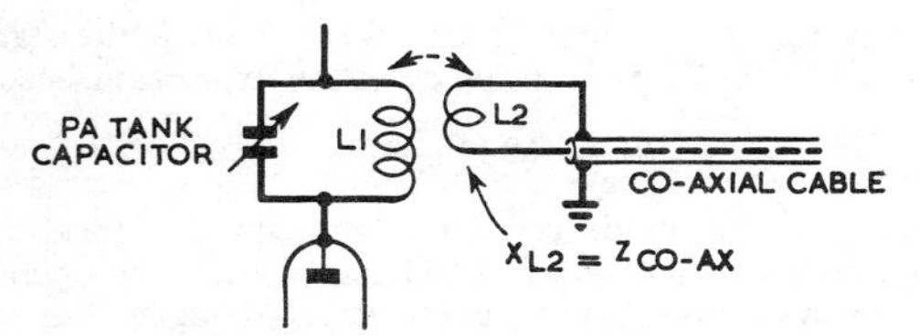

Fig 6.79. Untuned link output circuit.

Before winding the link, it is necessary to know the impedance into which the link will feed. The turns on the link and the turns on the tank coil constitute an impedance transformer. Just as it is necessary to match the anode tank circuit to the pa impedance for optimum power, so it is necessary to match the link to the impedance it has to feed if the most efficient transfer is to take place between the tank circuit and the link.

The load resistance which is placed across the link coil is transformed into the primary circuit to produce the load resistance for the valve or transistor.

The usual practice is to feed the power from the pa by means of a link connected to a coaxial cable which feeds an aerial array with an impedance which matches that of the coaxial line or an aerial matching device. The arrangement is illustrated in **Fig 6.79.**

The ideal conditions are not always easy to meet in practice. If the link is accurately matched to the line, it is frequently found that as it is coupled to the pa tank coil, the leakage inductance of the link detunes the pa circuit. In order to restore the pa current to its loaded value, the tuning capacitor needs to be adjusted to compensate for this leakage inductance. As the coupling between the link and the pa coil is increased (as they are moved physically closer) the detuning becomes more pronounced. However, the effect is usually more irritating than serious.

Tuned Link Coupling

As an alternative to the simple link, a tuned link system can be employed, and in many instances this can show distinct advantages. The arrangement is shown in **Fig 6.80.**

Provided that the far end of the coaxial cable is correctly terminated, the input impedance will be substantially resistive and equal in value to the impedance of the cable. By the use of a series tuned circuit this impedance can be matched. While the Q will be low, this is no particular disadvantage, for if the Q were high, the tuning of the link would require adjusting with excursions in frequency.

Table 6.7 gives the value of capacitance which, in association with a suitable coil, will produce a link circuit with a Q of two. The link should be wound so that when it is lightly

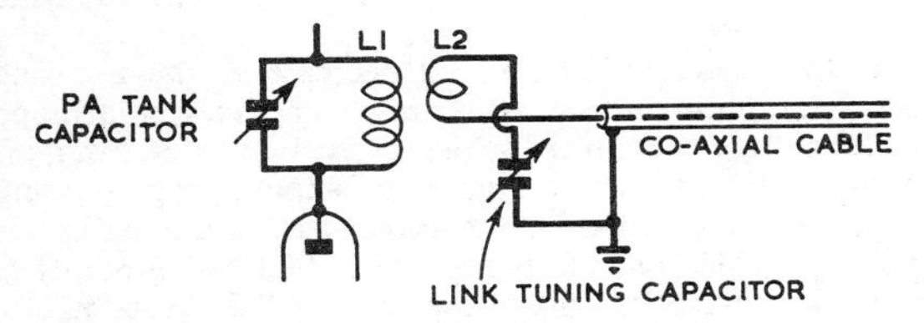

Fig 6.80. Tuned link output circuit. The link L2 together with the capacitor forms a series tuned circuit resonant at the operating frequency. The link capacitor should tune through the value given in Table 6.7.

TABLE 6.7

Values of the link tuning capacitor in Fig 6.80 with which the link L2 must be made to resonate at the operating frequency. The values given should not be exceeded at resonance, but a value of capacitor should be fitted which is somewhat higher than the maximum figure given.

Amateur Band	Impedance of Coaxial Cable 52Ω	75Ω
1·8MHz	900pF	600pF
3·5MHz	450pF	300pF
7MHz	230pF	150pF
14MHz	115pF	75pF
21MHz	80pF	50pF
28MHz	60pF	40pF

coupled to the pa tank circuit, the anode current of the pa rises as the link tuning capacitor is tuned through the value specified in the table. Once the coil has been adjusted to produce this condition, the link capacitor should be left at the value which produces the peak in the anode current of the pa. The link coil should be moved towards the tank coil until the desired degree of loading is obtained. It is probable that the coaxial line, even if correctly terminated, will exhibit some reactive element. As the frequency is changed, the link tuning capacitor may therefore require slight adjustment. The degree of trimming that this capacitor will require will depend on the extent of the reactive component.

Due to the low Q of the tuned link, fairly tight coupling between the link and the tuned circuit will be required. However, as it is resonant, and any reactance can be tuned out, the tank capacitor will show less deviation from the unloaded dip position as the coupling is increased compared to the untuned link method. This is to be expected in a double tuned circuit using mutual inductance. Any change experienced is usually due to increases in stray capacitive coupling as they approach each other.

Pi-Network Coupler

The pi-network is a circuit used to couple the pa anode to the output load in a manner which transfers the load resistance into the anode circuit with a value which is optimum for efficient power generation. It uses comparatively few components.

The commonly used pi-network arrangement is shown in **Fig 6.81.** The resistance R_1 is the valve anode and is the input resistance of the network when properly adjusted.

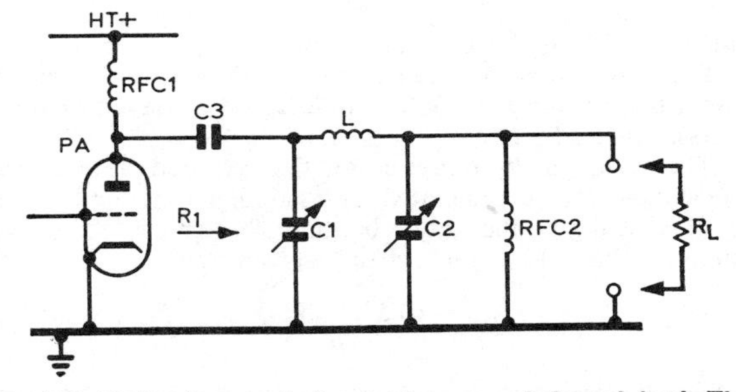

Fig 6.81. Basic pi-network for feeding an unbalanced load. The resistance presented to the valve anode is R_1 (see text). The choke RFC2 is added to prevent high dc voltage appearing at the output terminals in the event of an insulation breakdown in C3.

R_1	X_L	XC_1	XC_2
300	37	25	13
400	47	33	15
500	57	42	16
600	67	50	18
700	76	58	20
800	86	67	21
900	95	75	23
1000	104	83	24
1200	123	100	26
1400	141	117	29
1600	159	133	31
1800	177	150	33
2000	195	167	36
2200	212	183	38
2400	230	300	40
2600	247	217	42
2800	265	233	44
3000	282	250	46
3500	325	292	52
4000	367	333	57
4500	409	375	63
5000	451	417	69
5500	493	458	76
6000	534	500	83

(a) $R_L = 75\Omega$

R_1	X_L	XC_1	XC_2
300	35	25	10
400	45	33	12
500	54	42	14
600	63	50	15
700	73	58	16
800	82	67	18
900	91	75	19
1000	100	83	20
1200	118	100	22
1400	136	117	24
1600	153	133	27
1800	171	150	29
2000	188	167	31
2200	205	183	33
2400	222	200	35
2600	239	217	37
2800	256	233	40
3000	273	250	42
3500	315	292	48
4000	356	333	55
4500	397	375	64
5000	437	417	75
5500	477	458	89
6000	515	500	110

(b) $R_L = 50\Omega$

Fig 6.82. Reactance tables for pi-network couplers.

It is usual to design the network to have a loaded Q of 12 which is a compromise between the need to suppress harmonics for which a high loaded Q is required and the circuit efficiency which requires a high ratio of unloaded Q to loaded Q.

To meet these requirements the capacitors C1 and C2 and the inductance L should have the reactance values derived from the following formulae:

$$XC_1 = \frac{R_1}{Q} \qquad \ldots (1)$$

$$XC_2 = \sqrt{\frac{R_1 R_L}{(Q^2 + 1) - \frac{R_1}{R_L}}} \qquad \ldots (2)$$

$$X_L = \frac{QR_1 + (R_1 R_L / XC_2)}{Q^2 + 1} \qquad \ldots (3)$$

when Q is the loaded Q of the circuit and R_L is the output load resistance.

For any particular frequency the value of C and L may be obtained using the reactance formulae:

$$X_L = 2\pi f L \qquad \ldots (4)$$

$$X_C = \frac{10^6}{2\pi f C} \qquad \ldots (5)$$

where f is MHz, C is pF and L is μH.

Reactance tables derived from equations (1), (2) and (3) for load resistances of 50Ω and 75Ω and a loaded Q of 12 are shown in Fig 6.82.

The value of R_1 representing the required anode load resistance can be calculated by assuming that the peak rf voltage swing at the anode is about 80 per cent of the dc supply voltage V_{ht}. It will then be given by—

$$R_1 = \frac{(0{\cdot}57 \times V_{ht})^2}{P}\ \Omega \qquad \ldots (6)$$

where P is the power *output* in watts. It is much simpler, however, to measure the power input and if an assumption is made with regard to the anode efficiency the expression can be converted into a more useful form. Thus if the stage is assumed to be 70 per cent efficient the expression becomes

$$R_1 = 460 \times \frac{V_{ht}}{I_a}\ \Omega \qquad \ldots (7)$$

where I_a is the anode current in milliamps. Either of the expressions (6) or (7) may be calculated from the operating data given in the valve manufacturer's catalogue.

EXAMPLE: *A single ended QV06-20/6146 is to be operated on 3·5MHz at an ht of 500V and an anode current of 100mA. Find the constants of a pi-network with a Q of 12 for a load impedance of 75Ω.*

From equation (7)

$$R_1 = 460 \times \frac{500}{100} = 2300\Omega$$

Referring to Fig 6.82(a) and interpolating between R_1 = 2200 and 2400Ω the value of XC_1 is 191Ω, XC_2 is 39Ω and X_L is 221Ω. Using formulae (4) and (5) gives L = 10μH, C1 = 238pF and C2 = 1166pF.

The capacitors C1 and C2 are variable units with maximum values 25 to 50 per cent greater than those calculated. It is also a common practice to use a fixed capacitor in parallel with C2 to reduce the value of the required variable capacitor. In this example a suitable fixed value would be 500pF with 1,000pF variable. Additional charts for the design of pi-network tank circuits are given in the *Radio Data Reference Book* (RSGB).

The pi-network coupler has one important advantage over the parallel tank circuit. As it functions as a low pass filter, it shows a high degree of attenuation to harmonics of the frequency to which it is tuned, but to secure the maximum degree of suppression, the components have to be carefully placed. The capacitors C1 and C2 should be mounted directly on the chassis with C1 as close to the pa valve as possible. The outer braiding of the coaxial cable should be connected to the same point as the earthing of the rotor of C2, and the inner connector should be shielded right up to the point where it is connected to C2.

The earth connection through the chassis between C1, L and C2 should be of good conductivity. Since the chassis will carry the circulating current care must be taken that stray coupling to earlier stages does not exist. An excellent arrangement would be to connect the capacitors together with copper strip and connect them to earth at one point only.

In operation the pi-network offers the particular attraction that any reactance on the coaxial feed line can usually be tuned out.

The amount of capacitance used at positions C1 and C2 will be modified by the stray circuit capacitances, but because these are normally small when compared to the values of C1 and C2, and since it is usual to provide a somewhat higher variable capacitor than that of the calculated value, stray capacitances can usually be ignored within the frequency range 1·8–30MHz.

As the values of C1 and C2 decrease as the frequency increases, multiband operation can sometimes be arranged by designing the coupler for the lowest frequency band, and providing the coil with a number of suitable tapping points.

This will not give as high efficiency as that attainable with individual coils for each band, for as the tapping points are bridged the working part of the coil will "see" a number of shorted turns, and these will reduce the efficiency. Alternatively a variable inductance can be used and C1 switched for each band.

PA Choke Considerations

Owing to the tendency of an rf choke to resonate at various frequencies, depending on its design, some caution is necessary when using a parallel-fed tank circuit. A choke suitable for an all-band high-power parallel-fed pa stage is illustrated in **Fig 6.83**. This has been designed so that no resonances are present within the amateur bands; nevertheless the possibility of an unexpected resonance should be borne in mind since a metal chassis or a screening partition in the vicinity of the windings can alter the characteristics appreciably. Resonance will be noted by a loss in efficiency and by overheating of the choke windings. The choke former should be made from a material with good electrical properties and capable of withstanding without softening the temperature which it may acquire by being mounted close to the pa valve. A suitable material is Tufnol, or ptfe.

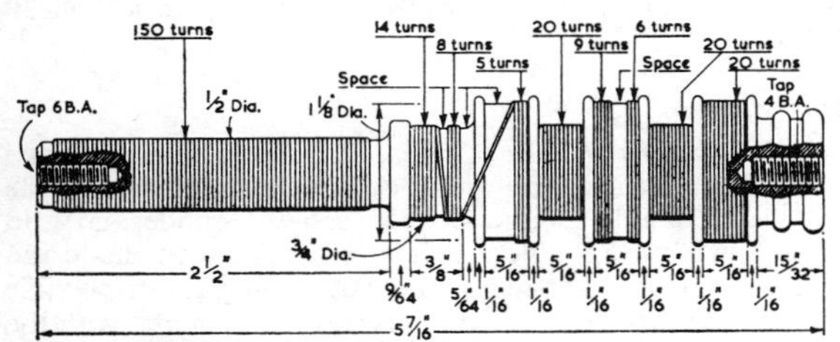

Fig 6.83. Construction of an all-band rf choke suitable for a parallel-fed pa stage. The former may be made of Tufnol or ptfe. The wire used is 30 swg dsc.

Pi-networks in Push-pull

It is practical to apply the advantages of a pi-network to valves operating in push-pull and this is shown in **Fig 6.84.**

In this the calculations are made for one half of the circuit according to the values given on the diagram, and similar components fitted to each half.

Pi-output Safety Precautions

If the blocking capacitor shown as C3 in Fig 6.81 develops a short circuit, there would be a possibility of the load assuming a dangerously high potential if it had no dc return. To prevent such a situation, the rf choke RFC2 should always be fitted, and the ht supply to the pa valve correctly fused. With certain aerial systems, this choke will provide a path for electrostatic voltages built up on the aerial. To avoid the transmitter enclosure assuming a dangerous potential under electrical storm conditions, it should always be connected to an efficient earth.

Pi-network Neutralization

Pa stages employing pi-networks may be neutralized by means of the arrangements illustrated in Figs 6.65 and 6.67. Cross-over neutralization may be used with push-pull pi-network stages. Contrary to popular opinion on the subject it is always best to neutralize pa stages using modern valves.

Adjusting the Pi-network Coupler

To tune a pi-network coupling the following procedure will be found satisfactory. First, with a dummy load connected, set the "loading" capacitor (C2 with Fig 6.81) to maximum capacitance and tune the anode capacitor (C1 in Fig 6.81) for minimum pa anode current, with a low power input to avoid possible damage to the valve. Next connect the aerial load or its equivalent and decrease the capacitance of C2 slightly from its maximum value; this capacitor C2 can conveniently be regarded as a "loading" control. Then C1 should be readjusted for minimum anode current. The procedure is then repeated until a value of C2 is found at which the output into the load is a maximum and at the same time a dip in anode current of 10–15 per cent can still be obtained by varying C1, the dip being coincident with maximum output.

An alternative, and possibly more satisfactory, method is to monitor the output voltage or current and adjust the pi-network for the required output with normal input.

If the pa loading is found to increase as the value of C2 is increased, the ratio between the impedances R1 and R2 is too great to allow correct matching to take place. In this event, it is preferable to modify the load impedance R2, usually by connecting a capacitance in series; its optimum value will best be determined by experiment. This capacitor should be suitably designed to carry the relatively heavy output-load current and one having a good mica dielectric should prove satisfactory.

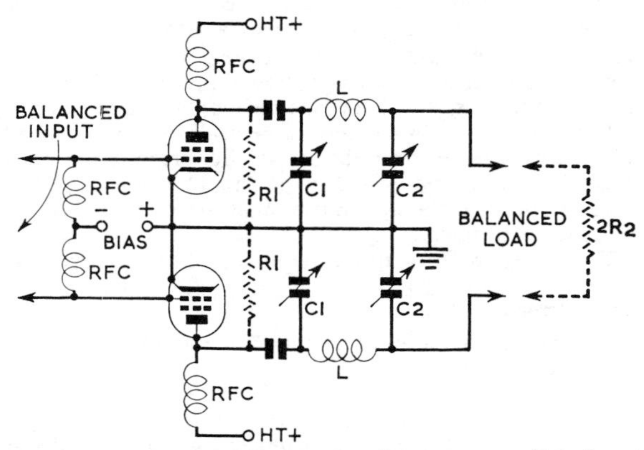

Fig 6.84. Twin pi-network for a push-pull output stage. If preferred the pairs of capacitors C1, C1, and C2, C2 can be replaced by single capacitors, the earthed centre connections then being ignored.

ADJUSTING POWER AMPLIFIERS

To avoid possible damage to the valve in a pa stage it is desirable to make the preliminary tuning adjustments with the anode and screen voltages removed. If a grid dip oscillator is available, it is a simple matter to tune the anode and grid circuits to the frequency of operation before any power is applied.

If a grid dip oscillator is not employed, normal bias should be applied to the valve. If link coupling is used, both links should be adjusted to give rather less than maximum coupling. A milliammeter in the grid circuit of the pa stage, **Fig 6.85,** will indicate when optimum coupling has been obtained. The procedure is to tune the drive anode tank circuit to give minimum anode current in the driver, and, with the links set as above, the pa grid circuit is tuned to give maximum drive as indicated by the grid milliammeter. The link coils may then be readjusted to give tighter coupling,

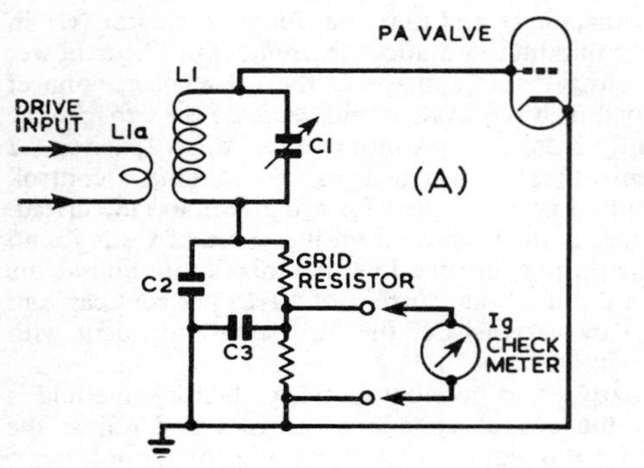

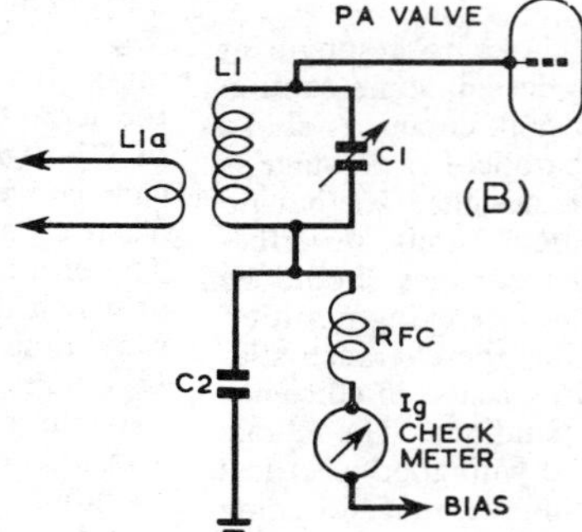

Fig 6.85. Grid current meter positions. The use of a grid current meter is essential during the tuning-up of a pa and should preferably be a permanent feature in any transmitter.

followed by further slight retuning of the driver anode and pa tank circuits, if necessary, to give maximum grid current. The latter should have a value 30–40 per cent greater than the specified value for the valve since it will fall by approximately this amount when the anode voltage is applied later. The frequency to which the grid circuit is tuned should be checked with an absorption wavemeter.

Neutralization

The stage must now be neutralized. Variation of the anode tuning capacitor through resonance should have no influence on the grid current, provided that the stage is correctly neutralized. Neutralization should be carried out at the highest frequency on which the amplifier is to operate.

There are two main methods of neutralizing an amplifier: with the ht removed from the anode and screen of the valve ("cold" neutralization) and with the ht applied ("hot" neutralization). Cold neutralization is to be preferred on the grounds of safety.

The procedure for neutralizing an amplifier is as follows:

Cold Method

(*a*) Disconnect the ht from the anode and screen of the valve.
(*b*) Disconnect any output load from the valve.
(*c*) Set the neutralizing capacitor to minimum capacitance.
(*d*) Set the anode tuning capacitor to the point at which it influences the grid current.
(*e*) Increase the neutralizing capacitor slightly.
(*f*) Vary the anode tuning capacitor to determine whether neutralization has been achieved.
(*g*) If not, repeat steps (*b*), (*c*) and (*d*) until the stage is correctly neutralized.

An alternative method of cold neutralization is to feed a signal to the anode of the valve and adjust the neutralizing capacitor for minimum indication on a valve voltmeter connected to the grid circuit.

Sometimes it is not possible to eliminate entirely interaction between the two circuits but this does not necessarily mean that the circuit will not function properly. However, the greatest care should always be taken to achieve the lowest possible degree of interaction.

In a push-pull amplifier the two neutralizing capacitors should be varied together (ie with similar values) so that an approximate balance is maintained during the process. The final correct balance may be found to occur with slightly different settings of the two capacitors but if the values are greatly different the balance may be upset when the operating frequency is changed to a different band. A well designed and constructed amplifier, when correctly neutralized, should remain so over the full range of frequencies 1·8–30 MHz.

Hot Method

This is carried out in a similar manner to that described for cold neutralization with the same object: to achieve the lowest possible interaction between the anode and grid circuits of the amplifier. As it is performed with anode and screen voltages applied to the valve, the greatest possible care must be taken to avoid coming into contact with the high voltage supplies. It is a good rule to keep one hand in your pocket while carrying out this type of adjustment. Before making any adjustment, the power should be switched off and the capacitors allowed to discharge: the ht line should then be shorted to earth. (See Chapter 19 for further advice on safety precautions.)

A final check of the neutralization should be made by applying a low anode voltage, removing the grid drive, and adjusting the bias so that it produces a standing anode current corresponding to about half the maximum permissible anode dissipation. If the neutralizing setting is incorrect, self-oscillation is almost certain to occur. This can be detected in various ways, eg by the glow in a neon lamp held near the "hot" parts of the anode circuit, by the lighting of a flash-lamp bulb connected to a loop of wire when coupled to the tank coil, or by an absorption wavemeter. Anode-current variations will also be observed when the grid or anode tank capacitor is tuned through resonance.

If it appears impossible to achieve proper neutralization the cause of the difficulty may lie in the presence of parasitic oscillation. Methods for suppressing such oscillations are described on page 6.36.

When the amplifier has been neutralized, a suitable load should be coupled to the anode tank coil; the grid drive may then be applied though still with reduced anode and screen voltages. This load may consist of an ordinary domestic electric lamp of suitable power rating or preferably a carbon resistor of suitable rating to dissipate the power output (the Electrosil type H39, for example).

When ht is applied to the pa the anode tuning capacitor should be quickly tuned for maximum dip in the anode current to avoid damaging the valve. It can be assumed that the circuit is correctly loaded if the dip in anode current when the tank circuit is tuned through resonance is 10–15 per cent. If a lamp is used as the load it should be connected to the coil by the shortest practicable leads, the wires being soldered directly to the lamp-cap contacts.

After the amplifier has been tuned at reduced power, full power may be applied. A rough estimate of the power output may be made by observing the brilliance of the lamp. Finally the anode efficiency and dissipation should be calculated to ensure that the valve is being operated within its ratings.

If it is found that varying the anode tank capacitor about the resonance position causes the grid current to change sharply at any point, it is a sign that feedback is taking place and closer attention should be paid to the screening or to the neutralizing adjustments. Besides being troublesome in other ways such feedback can cause intermittent spurious emissions in phone operation and produce effects similar to those of over-modulation.

Sometimes the tuning adjustment that produces the maximum output from the power amplifier does not coincide with the tuning adjustment for minimum anode current. This can be due to faulty neutralization but it can sometimes be traced to a variation in the screen voltage as the circuit is tuned through resonance, and in this event the remedy is to stabilize the screen voltage.

TRANSISTOR POWER AMPLIFIERS

Transistors which are suitable for low and medium power pa stages are now more readily available at prices attractive to the amateur.

They have the advantage of considerable longer life expectancy than their valve counterparts although care is necessary to ensure that the quoted maximum voltages are not exceeded even under transient conditions and they are not very tolerant of the overloads which may occur during adjustment.

Detailed and specific information on devices and their recommended circuitry is generally available from the manufacturers in the form of application reports and circuit manuals. This literature can be of great value to the serious experimenter.

Output Matching

The problem of matching the output load resistance which is normally of the order of 50Ω to the collector of the pa transistor is very similar to that of interstage coupling. In general however the required collector load is of a lower value than the output load which is the opposite state of affairs to the interstage case.

Fig 6.86 shows coupling circuits employing a tap on the collector coil. The design procedure is straightforward and uses well known circuit techniques. The peak value of the collector rf voltage is approximately equal to the supply voltage V_{cc} since the collector can be assumed to bottom at the negative peaks. The rms value of collector voltage is then $\frac{V_{cc}}{\sqrt{2}}$ and the power (P_o) developed in the collector load resistance (R_c) is $P_o = \frac{V_{cc}^2}{2R_c}$. Knowing the power which is to be produced the collector resistance may be found from $R_c = \frac{V_{cc}^2}{2P_o}$. As discussed under *Transistor Interstage Coupling* (page 6.24) a collector tank circuit parallel loading resistance (R_T) of 1000Ω is reasonable and this enables the position of the coil tap to be established using the formula $N^2 = \frac{R_T}{R_c}$ where N is the ratio of total turns to turns from the tap to the earthy end.

It is usual to design for a loaded Q of 10 as a compromise between circuit efficiency and harmonic suppression and by using the formula $Q_L = \omega C_T R_T$ the circuit total tuning capacitance C_T may be found. The inductance is derived from $L = \frac{1}{\omega^2 C_T}$

The next step in the design is to devise a circuit which will make the load resistance R_L appear like a resistance R in parallel with the collector tank circuit.

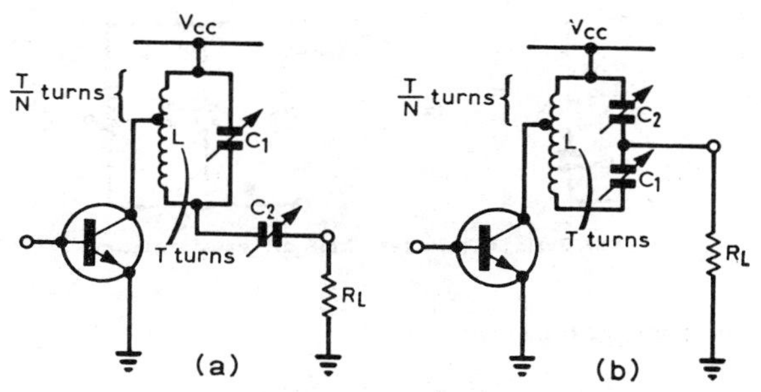

Fig 6.86. Transistor output coupling circuits.

Formulae for Fig 6.86(a)

(1) $R_c = \frac{V_{cc}^2}{2P}$

(2) $X_L = \frac{R_T}{Q_L} = \frac{N^2 R_c}{Q_L}$

(3) $X_{C_2} = R_L \sqrt{\frac{R_T}{R_T} - 1}$

(4) $X_{C_1} = \frac{R_T}{Q_L} \cdot \frac{1}{\left(1 - \frac{X_{C_2}}{Q_L R_L}\right)}$

Formulae for Fig 6.86(b)

(1) $R_c = \frac{V_{cc}^2}{2P}$

(2) $X_L = \frac{R_T}{Q_L} = \frac{N^2 R_c}{Q_L}$

(3) $X_{C_1} = \frac{R_T Q_L}{(Q_L^2 + 1)} \left[1 - \frac{R_L}{Q_L X_{C_2}}\right]$

(4) $X_{C_2} = \frac{R_L}{\sqrt{\frac{(Q_L^2 + 1) R_L}{R_T} - 1}}$

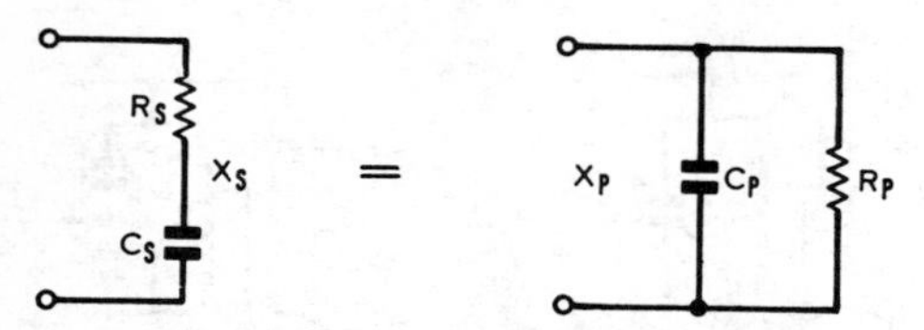

Fig 6.87. Series to parallel RC circuit transformation.

Formulae for Fig 6.87

$$X_s = \frac{1}{2\pi f C_s} \qquad X_p = \frac{1}{2\pi f C_p}$$

$$R_p = \frac{R_s^2 + X_s^2}{R_s} \qquad X_p = \frac{R_s^2 + X_s^2}{X_s}$$

The matching circuit of Fig 6.86(a) makes use of a series to parallel transformation by connecting a capacitor C2 in series with the load so that the equivalent parallel circuit has a resistance R_T. The capacitance of the equivalent parallel circuit requires to be subtracted from the total tuning capacitance calculated above in order to find the value of the tuning capacitance C1.

Fig 6.87 shows the method of calculating the component values of the equivalent parallel circuit. The formulae shown in Fig 6.86 give the same answers as the step by step method outlined above. The diagram in Fig 6.86(b) and the accompanying formulae show a capacitance transformation of the output load.

As with interstage coupling the tapped inductor ceases to be effective at vhf and other networks should be used.

Some suitable networks are shown in **Figs 6.88, 6.89** and **6.90**. When adjusting these circuits the variable capacitor C2 may be regarded as a loading control and C1 as the tuning capacitor. Final adjustments are carried out by varying C2 slightly in a direction which increases the power output and then re-tuning with C1.

Biasing

In class B operation no standing bias is provided. Unless a transistor has a forward bias current applied between its emitter and base, no current will flow through it until such a bias is provided. Thus a transistor can be used in class B simply by connecting it so that there is no forward bias applied.

Fig 6.91 is the circuit of a typical transistor power amplifier in which the drive is applied to the base, and the base matched to the preceding stage by a link winding. While one end of the link winding is bypassed for rf, the dc return is via a resistor R to earth; the emitter is also earthed. Under static conditions, ie without drive applied, the transistor rests in the class B condition, and the only collector current which flows is that due to leakage through the transistor. When drive is applied progressively, the transistor conducts on alternate half cycles. As the level of the drive is increased, diode rectifying action takes place between the base and emitter, and a biasing voltage is developed, the level of which will be in proportion to the drive power and

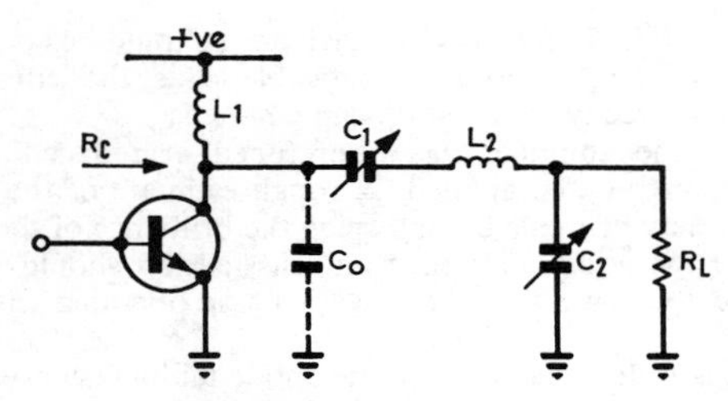

Fig 6.88. Output matching (1).

Formulae for Fig 6.88

(1) $X_{C_1} = Q_L R_c$

(2) $$X_{C_2} = \frac{R_L}{\sqrt{\dfrac{R_L(Q_L^2+1)}{R_c Q_L^2}}}$$

(3) $$X_{L_1} = \frac{Q_L R_c}{\left[\dfrac{Q_L R_c}{X_{C_o}} + 1\right]}$$

(4) $$X_{L_2} = Q_L R_c \left[1 + \frac{R_L}{Q_L X_{C_2}}\right]$$

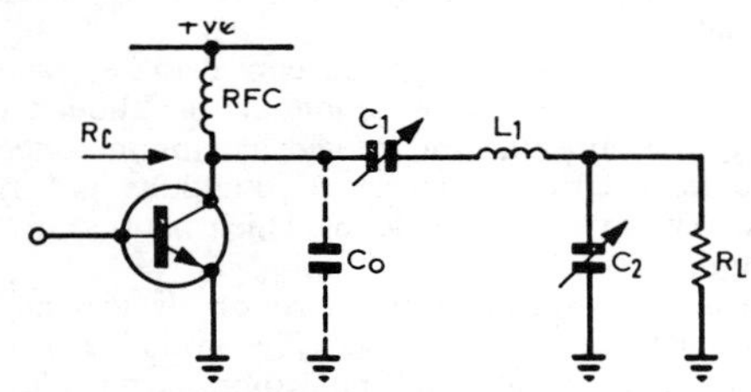

Fig 6.89. Output matching (2)

Formulae for Fig 6.89

(1) $$X_{C_1} = \frac{Q_L X_{C_o}^2}{R_c}\left[1 - \frac{R_c}{Q_L X_{C_o}}\right]$$

(2) $$X_{C_2} = \frac{R_L}{\sqrt{\dfrac{(Q_L^2+1)}{Q_L^2}\cdot\dfrac{R_c R_L}{X_{C_o}^2} - 1}}$$

(3) $$X_{L_1} = \frac{Q_L X_{C_o}^2}{R_c}\left[1 + \frac{R_L}{Q_L X_{C_2}}\right]$$

---

the value of R. Thus by varying the drive power, the transistor can be moved progressively from class B to class C operation.

As it is not usual to find that manufacturers specify precise conditions for transmitter service, except for semiconductor devices specifically produced for this purpose, a certain amount of work may be needed in designing a transmitter employing general purpose transistors. So far

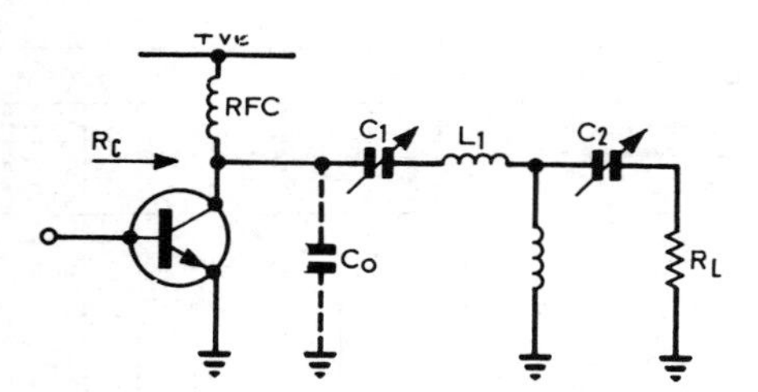

Fig 6.90. Output matching (3). The unmarked coil is L2.

Formulae for Fig 6.90

(1) $X_{L_1} = \frac{Q_L X^2_{C_o}}{\sqrt{R_c R_L}} \left[- \frac{\sqrt{R_c R_L}}{Q_L X_{C_o}} \right]$

(2) $X_{L_2} = X_{C_o} \sqrt{\frac{R_L}{R_c}}$

(3) $X_{C_1} = \frac{Q_L X^2_{C_o}}{R_c} \left[1 - \frac{R_c}{Q_L X_{C_o}} \right]$

(4) $X_{C_2} = \frac{R_L}{Q_L} \left[\frac{Q_L X_{C_o}}{\sqrt{R_c R_L}} - 1 \right]$

as the bias level is concerned, it is important to ensure that the base-emitter voltage specification is not exceeded, particularly the reverse voltage rating, and this should be continuously monitored as the drive is increased. Overdriving will quickly ruin a transistor.

Bypass Capacitors

Due to the lower impedances encountered in transistor circuits compared with those in valve circuits, any residual impedance in bypass capacitors must be correspondingly lower. It will be found therefore that the bypass capacitors employed in transistor transmitter circuits are larger relative to frequency than those in similar valve circuits. It is essential that the type of capacitor is selected with care, since increasing the value of the capacitance may be accompanied *by an increase* in reactance in some cases. This is of particular importance so far as the capacitor across R in Fig 6.91 is concerned, for if this shows reactance at the operating frequency it will either upset the matching of the link to the previous stage, or necessitate an increase in the drive to the pa transistor, or both.

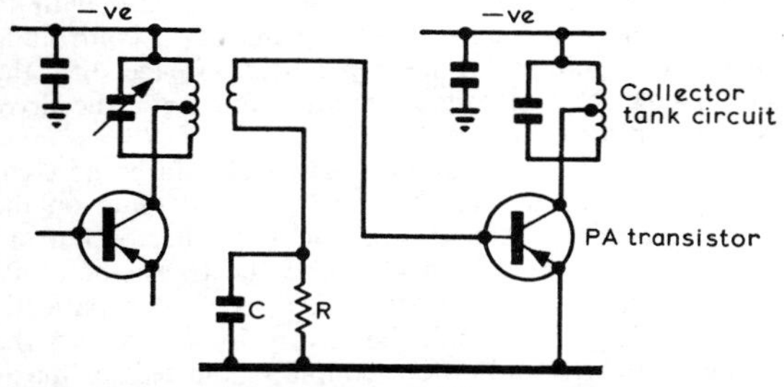

Fig 6.91. Representative common-emitter connected transistor pa stage.

Voltage Ratings

The collector voltage of a grounded emitter transistor class C pa stage reaches a value approximately twice the supply line voltage at the peak of the rf cycle. Under conditions of 100 per cent amplitude modulation it is obvious that the maximum collector voltage becomes approximately four times the supply rail voltage. To allow an adequate margin of safety the transistor chosen for operation under these conditions should have a quoted maximum permissible collector to emitter breakdown voltage of say five times the supply rail.

Transistor Neutralization

Like valves, transistors have internal capacitances, and for the same reasons, these may have to be neutralized depending on their magnitude and the frequency of operation. Transistors also have internal conductance which sometimes requires neutralization. If the neutralizing of a transistor is incorrect, there may be a loss of gain (degeneration) at one frequency setting of the output circuit while at another it may be highly regenerative.

The neutralizing system which appears to be the easiest to adjust is illustrated in **Fig 6.92.** It was originally designed for use with triode valve amplifiers and was later used in cascode rf amplifiers.

In the circuit of Fig 6.92, the inductance is arranged to resonate with the collector-to-base capacitance of the transistor at the operating frequency. At resonance, since $X_C = X_L$, the internal capacitance of the transistor will be neutralized.

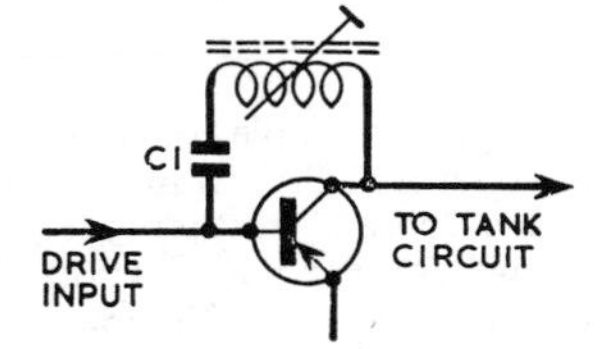

Fig 6.92. Neutralizing circuit for a transistor pa.

Transistor Heat Dissipation

The manufacturer's data on a particular transistor usually shows the maximum junction temperature which is permissible and the designer must take steps to ensure that this is not exceeded. The design of the cooling arrangement is based on the concept of thermal resistance. When heat flows in a heat conducting medium a difference in temperature is created between two points in the medium, and the thermal resistance between the two points is the temperature difference in degrees C divided by the heat flow in watts.

This is analogous to current flow producing a voltage drop in an electrically conducting medium.

Adequate cooling is normally achieved by mounting the transistor on an appropriate heat sink—three heat conducting paths in series are then connected between the transistor junction and the ambient air. The total thermal resistance is the sum of the thermal resistances from junction to case, case to sink and sink to air or expressed symbolically

$$\theta_{J\text{-}A} = \theta_{J\text{-}C} + \theta_{C\text{-}S} + \theta_{S\text{-}A}$$

The temperature difference between the junction and the

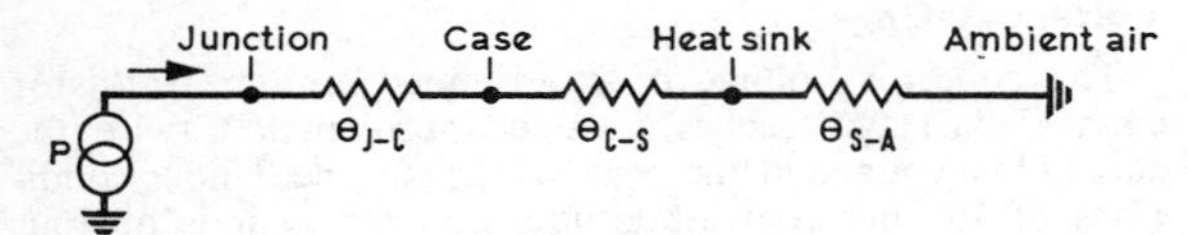

Fig 6.93. Thermal equivalent circuit.

ambient air can be found by multiplying the total thermal resistance by the power dissipated ie:

$$T_{J-A} = P\theta_{J-A}$$

Thermal resistances of heat sinks to free air are available from the manufacturers. The case to heat sink thermal resistance θ_{C-S} can be made low by the use of mica washers and silicone grease and often it can be neglected in calculations. The transistor data sheets usually quote the junction-to-case thermal resistance θ_{J-C}.

Fig 6.93 shows the thermal equivalent circuit of a transistor mounted on a heat sink and **Table 6.8** shows the thermal resistance of some transistor cases to free air. The table allows an assessment to be made of the dissipation possible without the use of a heat sink.

TABLE 6.8

Case	θ_{C-A} (°C/W)
TO–18	300
TO–46	300
TO–5	150
TO–39	150
TO–8	75
TO–66	60
TO–60	70
TO–3	30
TO–36	25

As an example of the power dissipation calculation consider a 2N3055 transistor mounted on a heat sink which has a thermal resistance of 2·5°C/W. According to the manufacturer's specification the junction-to-case thermal resistance is 1·5°C/W and the washer mounting arrangement may be assumed to have a thermal resistance of 0·5°C/W. The total series thermal resistance is:

$$\theta_{J-A} = 1{\cdot}5 + 0{\cdot}5 + 2{\cdot}5$$
$$= 4{\cdot}5°\text{C/W}.$$

The maximum junction temperature $T_{J(max)}$ of the 2N3055 given in the data sheet is 200°C so for an ambient temperature of say 50°C the maximum power which may be dissipated is:

$$P_{max} = \frac{T_{J(max)} - T_A}{\theta_{J-A}}$$

$$= \frac{200 - 50}{4{\cdot}5} = 33{\cdot}3\text{W}$$

It is a wise precaution to design for a heat sink which will enable the total dc power input to a transistor pa to be dissipated although during normal operation it is required to dissipate only the difference between dc power input and rf power output.

Heat sinks may be constructed from plates of copper or aluminium and often the equipment chassis itself may be used for this purpose. The thermal resistance of such arrangements depends primarily on surface area and to a lesser degree on plate thickness and mounting arrangement (ie horizontal or vertical). The nomograph of **Fig 6.94** enables the thermal resistance of plates of area 1 to 80 square inches to be determined by laying a straight edge horizontally across the chart at the required area and reading off the thermal resistance under the appropriate *Thickness and mounting position* column for the metal in question.

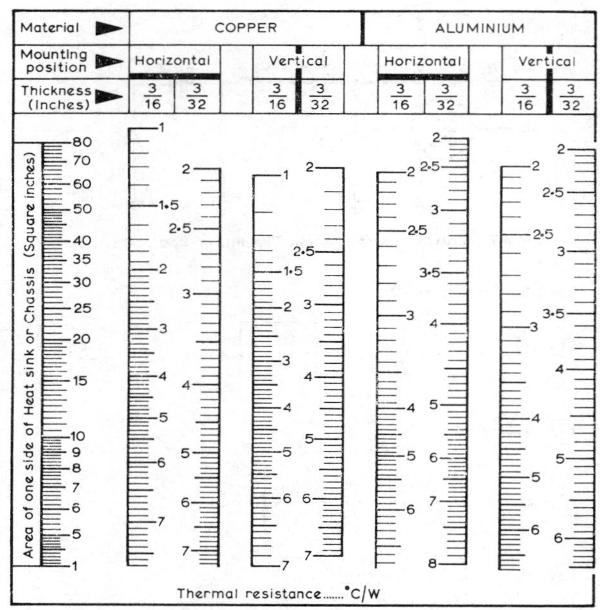

Fig 6.94. Thermal resistance as a fraction of heat-sink dimensions. (from *Electronic Design*, 16 August 1961)

Adjusting Transistor Power Amplifiers

It is essential to load the output circuit of a transistor power amplifier during the tuning procedure. Unless this is done, the peak collector voltage may exceed the collector-base rating and destroy the transistor. A suitable low power dummy load is shown in **Fig 6.95** although it is of course quite possible to use a high wattage carbon resistor of the type used as a dummy load for valve transmitters.

An indication of output power can conveniently be obtained by using a reflectometer in the feedline to the load as a forward power meter. Suitable indicators of this sort are described in Chapter 18—*Measurements*. Initial testing should be carried out with low supply voltage and with no drive. Some designs incorporate easy means of controlling the drive. For example in the transmitter on page 6.51 the potentiometer RV2 permits a smooth control of the drive to be made in the cw mode.

Under the zero drive condition there should be no signs of instability, such as may be shown by a reading on the reflectometer or a tendency for the pa collector current to change as a hand is brought near the transistor or its associated tuned circuits. In the absence of such parasitic oscillations the drive should be slowly increased and the interstage networks and output coupling adjusted for maximum output power. By this process maximum supply voltage can be applied eventually and the design power

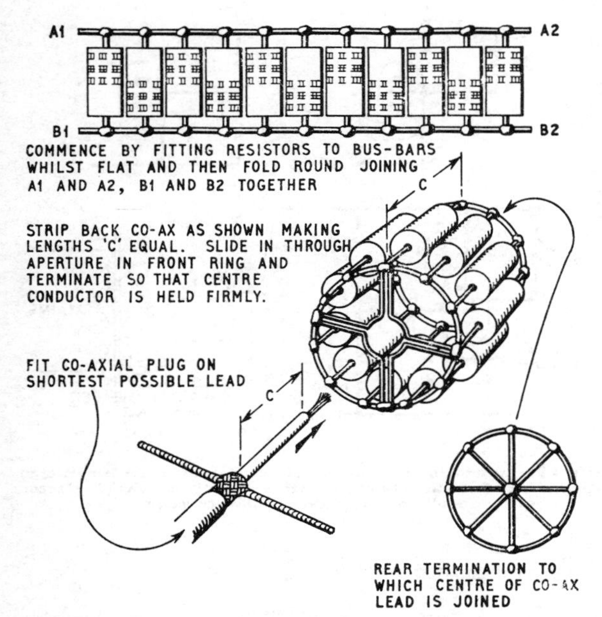

Fig 6.95. A 75Ω dummy load. For 10W dissipation, 11 × 820Ω, 1W, 10 per cent carbon resistors should be fitted to form the "cage"; for 50W, 25 1800Ω, 2W, 10 per cent carbon resistors.

output safety achieved. The tuning up process should result in a smooth increase in power at each stage and any abrupt changes may be indicative of instability.

POWER RATINGS OF RF AMPLIFIERS

The maximum permissible ratings at which a valve may be operated are specified in the valve manufacturers' published data. In the case of some USA valves, the manufacturers specify two different ratings: (i) the CCS or *Continuous Commercial Service* rating, in which long life and consistent performance are the chief requirements, and (ii) the ICAS or *Intermittent Commercial and Amateur Service* rating in which high output from the valve is a more important consideration than long life. The term "intermittent" applies to ON periods (key down) not exceeding five minutes followed by OFF periods (key up) of the same or greater duration. This must be remembered if it becomes necessary, as for instance, while testing to keep the key closed so that the valve is then in *continuous* operation.

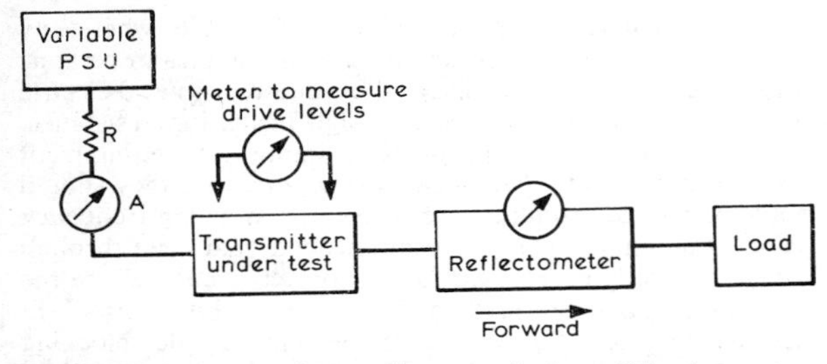

Fig 6.96. Typical set-up for tx alignment. Resistor R is chosen to limit the current to a safe value during alignment and may then be removed.

The following are the main features to be borne in mind when considering the power-handling capabilities of a valve.

Power Input

The power input to the valve is the dc power P_i fed to the anode circuit from the ht supply. It is given by—

$$P_i = \frac{I_a \times V_{ht}}{1000}\,\text{W}$$

where I_i = anode current (mA)
V_{ht} = anode voltage (V)

Power Output and Anode Efficiency

Not all of the ht power supplied to the valve is converted into rf energy; part is expended in heating the anode. The power-handling capacity of the valve is limited by the maximum permissible dissipation of heat by the anode through radiation and conduction. The anode efficiency is given by—

$$\text{Efficiency} = \frac{P_o}{P_i} \times 100 \text{ per cent}$$

where P_o = rf power output
P_i = dc power input to the anode circuit

The maximum rf power output obtainable from a valve may be predicted from a knowledge of the anode dissipation. For example, consider a valve with a maximum permissible anode dissipation of 12W, operating in class C, the efficiency being about 70 per cent. Of the dc power input, 70 per cent will appear as rf energy and 30 per cent as heat which must be dissipated by the valve anode. This amount of heat is limited by the valve rating to 12W. Thus—

$$\textit{Maximum dissipation} = 12\text{W} = \frac{30}{100} \times \text{dc input power}$$

Therefore—

$$\textit{Maximum power input} = 12 \times \frac{100}{30} = 40\text{W}$$

provided the maker's ratings are not exceeded.

For two such valves operating under the same conditions, either in parallel or in push-pull, the maximum power input would be 80W. If the anode efficiency is less than 70 per cent, the maximum power output would be correspondingly lower.

In a frequency multiplier, the anode efficiency will be substantially lower, and the maximum permitted dc input will be correspondingly reduced. In a frequency doubler, the efficiency is likely to be 50 per cent, leaving 50 per cent to be dissipated in the anode. In a tripler, as the output efficiency is likely to be 30 per cent, the anode power loss will be 70 per cent.

To arrive at the maximum permitted dc input for a frequency multiplier, the equation above should be modified so that the power loss figure is substituted as follows:

$$\text{Maximum dc input} = \text{Anode dissipation (W)} \times \frac{100}{A_{pl}}$$

where A_{pl} is the percentage anode dissipation, ie doubler 50 per cent, tripler 70 per cent.

Screen Dissipation

All the dc power supplied to the screen circuit is expended in heating the screen. Since the screen is constructed of relatively thin wire mesh, the valve may be permanently damaged if the permissible dissipation is exceeded even for a relatively short period.

When the screen grid is fed with ht from a separate supply provision must be made for it to be removed immediately if the anode supply fails. If screen grid voltage is applied in the absence of anode voltage, the screen will be burnt out.

Screen Voltage Supply

There are three common methods of supplying the screen grid voltage for a tetrode or pentode valve:

(*a*) by a voltage dropping resistor in series with the anode supply;

(*b*) from a potential divider connected across the anode supply.

(*c*) from a separate supply.

In the case of the series resistor, the voltage regulation of the screen grid is relatively poor due to variations in the anode supply voltage with variations in the valve loading, and consequential variations in the screen grid current. The advantage of a series resistor method is that since the value of the series resistor is normally high, this limits the maximum current of the screen grid, and so gives some degree of automatic protection. Some valves, however, will draw negative screen current; in such cases a series resistor alone is useless.

In the case of a supply derived from a potential divider across the ht supply, the current through the network is substantially higher than that taken by the screen grid, and this provides a measure of voltage regulation.

Methods (i) and (ii) are illustrated in **Fig 6.97(a)** and **(b).**

If the screen voltage is applied through a series dropping resistor, as shown in Fig 6.97(a) the calculation of the correct resistor value is very simple. Consider, for example, the requirements for a typical pa valve:

Screen voltage = 250V
Screen current = 6mA
Supply line voltage = 750V

The series resistor must drop 750 − 250V (ie 500V) for a current of 6mA. Therefore its resistance must be—

$$R1 = \frac{750 - 250}{0{\cdot}006} = 83{,}000\Omega$$

Under these conditions the power dissipated in the resistor will be (750−250) V × 6mA = 3W.

Fig 6.97(b) shows the voltage supply to the screen grid derived from a potential divider. The design procedure is best explained by taking the same example. The resistor R1 in this case carries the screen grid current and the current through R2. Assume that the total current through R1 is three times the screen grid current (ie 18mA). Then R1 can be calculated:

$$R1 = \frac{750 - 250}{18} = 27{\cdot}8\text{k}\Omega$$

The resistor R2 carries the total current minus the screen grid current (ie 12mA). So R2 can be calculated:

$$R2 = \frac{250}{12} = 20{\cdot}8\text{k}\Omega$$

The power dissipated in each resistor is easily calculated using the formula $P = \dfrac{V^2}{R}$.

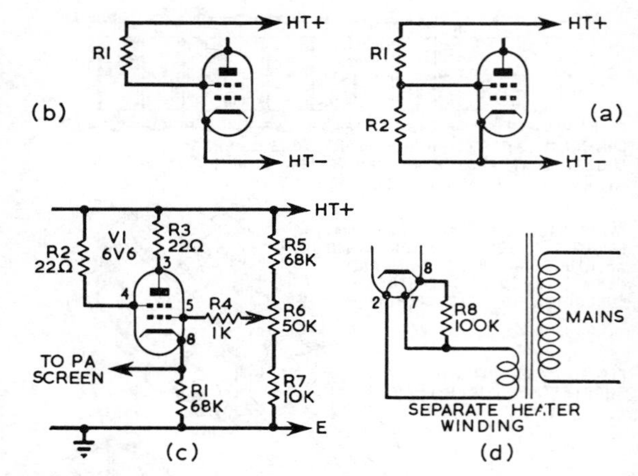

Fig 6.97. (a) and (b). Calculation of resistances in screen voltage circuits. (c) Cathode follower pa screen supply circuit. (d) Heater supply for the cathode follower valve.

$$\text{Power dissipated in R1} = \frac{(500^2)}{27800} = 9\text{W}$$

$$\text{Power dissipated in R2} = \frac{(250)^2}{20800} = 3\text{W}$$

Resistors of the nearest preferred value should be chosen with wattage ratings of double the above calculated power dissipations. The higher current consumption of the potential divider method is the price paid for improved regulation.

Fig 6.97(c) shows a cathode follower circuit which provides a variable screen grid voltage with very good regulation. The output resistance of a cathode follower is low, being approximately $1/g_m$ so current variations affect the voltage output only slightly. The output voltage can be easily adjusted by varying the control grid voltage of the cathode follower. Most of the common triodes and pentodes may be used, though it should be remembered that the valve selected must be capable of carrying more than the pa screen current. A separate heater supply should be provided for the cathode follower with the cathode connected to one side of the heater through a 100kΩ resistor. This eliminates the danger of breakdown of the heater to cathode insulation.

The cathode follower circuit may be modified to perform a number of other functions; see, for example the series gate modulator in Chapter 9—*Modulation.*

Capacitor Ratings

A fixed capacitor which is used in rf circuits where any loss of power must be avoided should have a mica or ceramic dielectric. These types have the required low dielectric loss, low inductance values and high insulation resistance. Low inductance is a particularly important requirement of decoupling and other bypass capacitors since they should have negligible reactance over the entire working frequency range (and above). It is recommended that feed-through or other disc types be used for low-power circuits where the ht supply does not exceed about 500V. For normal hf transmitter use, the values of decoupling, dc blocking and rf coupling capacitors lie within the range 1000pF to 0·01μF.

Fig 6.98. Grid-tank capacitor voltage ratings. The peak voltage, V, across each section is equal to the bias voltage plus 25 per cent of the applied dc anode voltage.

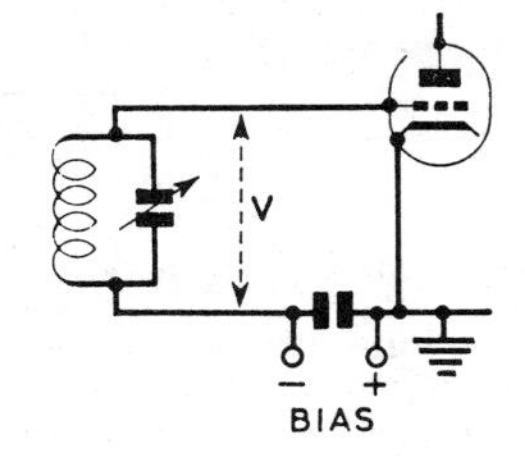

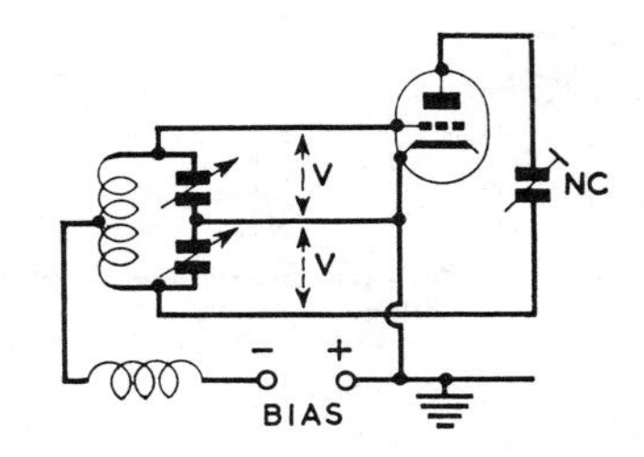

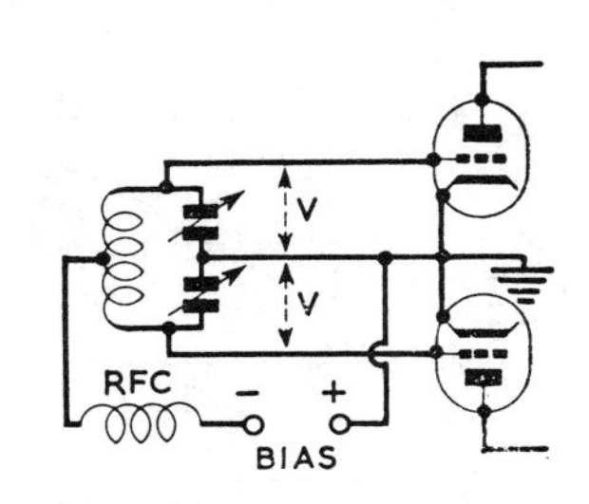

The inductance of capacitors and hence their resonant frequency can be modified by mounting them so that their leads enclose as small an area as possible. With tubular types, mounting them in earthed tubes will reduce the inductance. If a capacitor is series resonant with its leads at or near to the operating frequency, it is ideal for bypassing the screen grid and cathode, particularly at vhf.

The self-resonant frequencies of a number of British fixed capacitors commonly used in amateur transmitters are given in **Table 6.9.**

TABLE 6.9

Self-resonant frequencies of capacitors commonly used in amateur transmitters.

Capacitor Type	Maker	Lead Length	Frequency
0·1μF 350 volts paper foil (4702A)	Dubilier	½in	3·4MHz
0·1μF metalized paper (W99)	Hunts	¼in	5·95MHz
0·01μF ceramic disc (P∠)	Erie	¼in	17·8MHz
2200pF polystyrene	GEC	¼in	30MHz
0·001μF metalized paper (W99)	Hunts	¼in	43·6MHz
0·001μF ceramic disc (NY)	Erie	¼in	53MHz

To allow a margin of safety, the voltage rating of a capacitor should be at least 50 per cent greater than the peak voltage which is likely to occur across it, and its size should be adequate to carry the rf current without overheating.

Grid Coupling Capacitors. The peak applied voltage is approximately the sum of the grid bias voltage of the driven stage and the dc anode voltage of the driver stage; the voltage rating of a grid coupling capacitor should therefore be about 50 per cent higher than this peak value. Since some portion of the circulating current passes through the capacitor the current rating must be adequate.

Grid Tuning Capacitors. The peak voltage across the grid tank capacitor depends on the circuit arrangement and is indicated in **Fig 6.98.** In some cases the grid bias voltage must be added to the drive voltage.

Screen Bypass Capacitors. The voltage rating should be 1·5 times the anode voltage, or 2·5–3 times this value when anode-and-screen modulation is used. The current rating must also be considered because some or all of the circuit current can flow through the bypass capacitor and C_{a-SG} in series.

Neutralizing Capacitors. The neutralizing capacitor is subjected to peak voltages equal to the sum of the peak grid voltage and the peak anode voltage, amounting in the case of an unmodulated amplifier to about two to three times the dc anode voltage, and for a modulated amplifier about twice this value.

Anode Decoupling and Blocking Capacitors. In cw transmitters the peak voltage across the anode decoupling and blocking capacitors is equal to the dc anode voltage; in telephony operation it rises to at least twice this value on modulation.

Anode Tank Capacitors (including Pi-networks). The peak voltage across an anode tank capacitor depends on the arrangement of the circuit, eg single-ended, split-stator or push-pull. The values given in **Fig 6.99** assume that the amplifier is correctly loaded. In circuits where the dc supply voltage also exists across the capacitor, as at (C) and (F), twice the rating is required, and it is therefore an advantage in high-power stages to remove this voltage from

Fig 6.99. Anode-tank capacitor voltage ratings. The peak voltage may be taken as 80–90 per cent of the ht supply voltage.

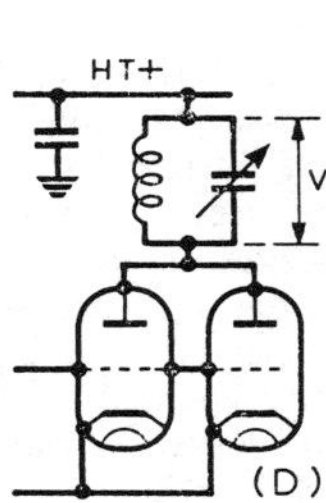

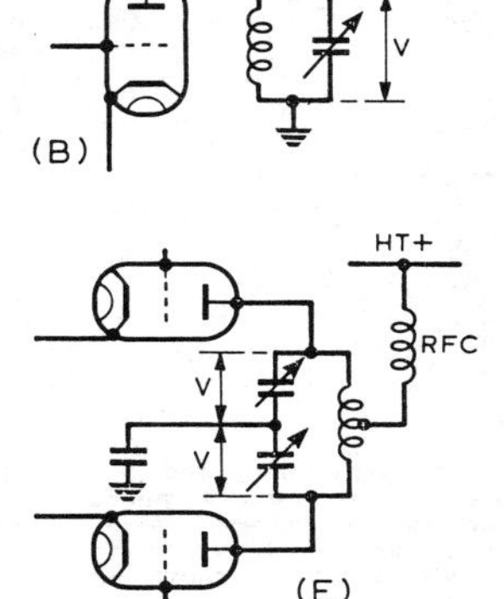

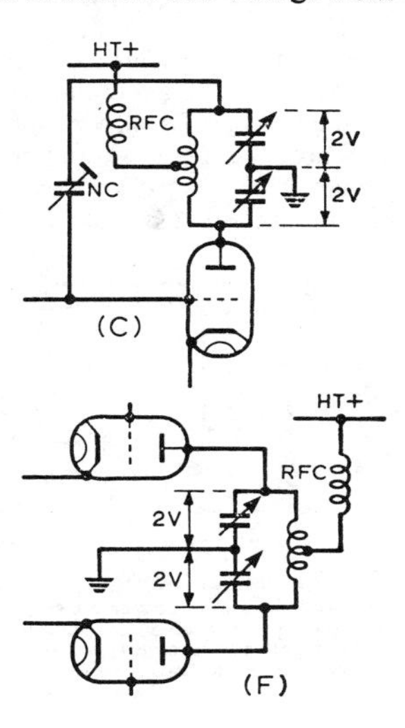

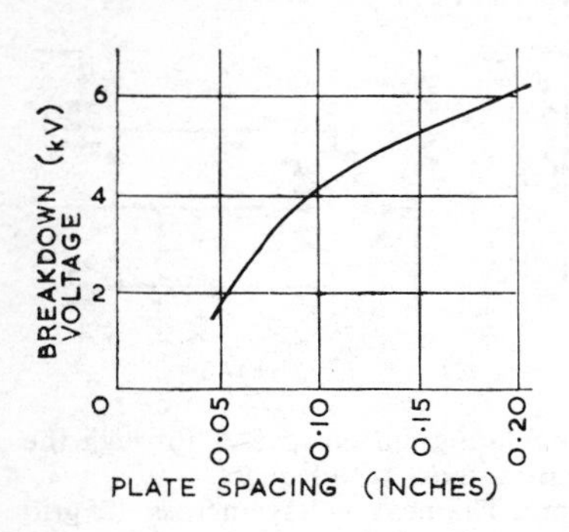

Fig 6.100. Breakdown voltage of air-spaced capacitors.

the capacitor by the use of parallel feed or other suitable arrangements. The application of 100 per cent modulation doubles the peak voltages occurring in any of the above circuits.

Although the necessary plate-spacing of an air-spaced capacitor for a given peak voltage rating depends to some extent upon constructional features such as the shape of the plates and the rounding of their edges, a useful guide to the voltage at which break-down occurs for different values of spacing will be found in Fig 6.100.

Pi-network Output (Loading) Capacitors. The peak voltage E occurring across a pi-network output capacitor is given by—

$$E = \sqrt{2P_oR_2}\ \text{V}$$

where P_o = power output of the amplifier (W)
R_2 = resistive component of the load or aerial impedance.

With a low-impedance load (eg a coaxial line not exceeding 100Ω) the peak voltage is not likely to be higher than about 300V, even in a 150W modulated transmitter, and a receiving-type variable capacitor should be quite satisfactory. In many cases it may be convenient to use a good-quality three or four gang receiving-type capacitor to obtain a maximum capacitance of 1500–2000pF, which may be necessary on 1·8MHz and 3·5MHz. The sections may be switched out progressively on changing to operation on the higher-frequency bands.

EXAMPLE. *A 150W phone transmitter is operating into a 75Ω line through a pi-network coupler. What is the peak voltage across the output capacitor?*

Since the peak power input occurring on modulation peaks is 4 × 150 = 600W, and if the anode efficiency of the pa stage is assumed to be 70 per cent, the peak power output will not exceed 420W. The peak voltage occurring across the pi-network output capacitor would therefore be—

$$E = \sqrt{2 \times 420 \times 75}$$
$$\doteqdot 250\text{V}$$

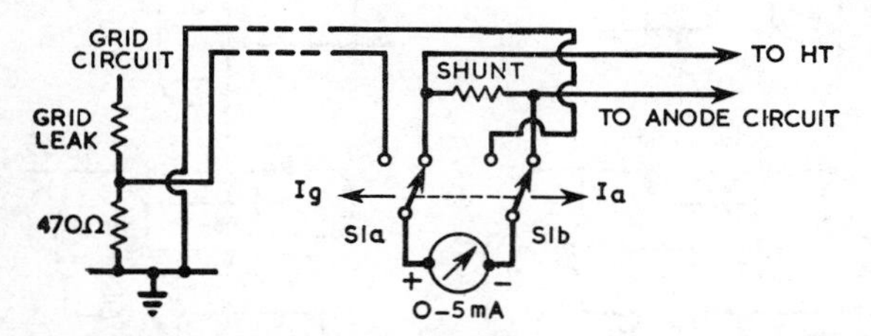

Fig 6.101. Meter switching circuit measuring the anode and grid currents in a low power pa stage. The switch S1 should be arranged to have a spare earthed position between the two ranges to avoid arcing and damage to the meter. In medium and high power transmitters it is advisable to meter the cathode current of a pa rather than the anode current.

Pi-tank Tuning Capacitor

The tuning capacitor of a pi-tank circuit in a cw or ssb transmitter must be rated at not less than 1·25 times the dc voltage of the anode, and for a.m., at not less than 2·5 times the dc anode voltage.

METERING

Although it is necessary to meter the current and/or the voltage at several different parts of a transmitter to ensure that all the various stages are operating correctly, an economy of instruments is made possible by the fact that one single reading (of the anode current in the pa stage, for example) is usually all that is required for noting quickly whether the transmitter is operating normally. The simplest arrangement—to measure grid and anode current—is illustrated in **Fig 6.101.**

A single instrument of about 1mA fsd in conjunction with a multi-way switch to allow connection to the appropriate points is therefore often found most suitable. Such an arrangement is shown in **Fig 6.102.**

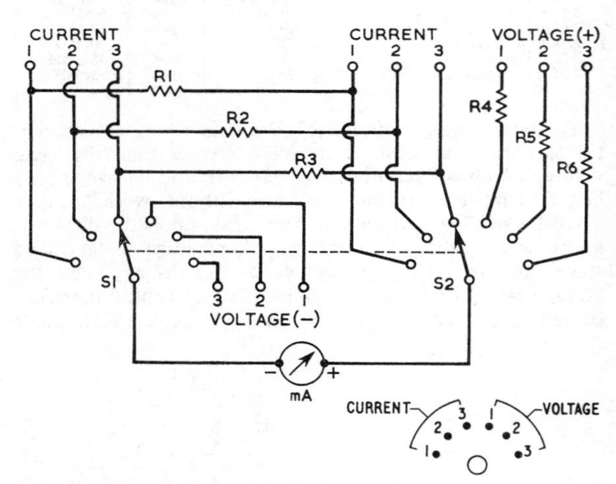

Fig 6.102. A simple switching system enabling one meter to be used for measuring three different currents and three different voltages. The switches S1 and S2 can be of the ganged-wafer type. The resistors R1, R2, R3 (current range shunts) and R4, R5, R6 (voltage-range series resistors) are selected according to the required ranges and the full-scale current of the meter.

The resistors R1, R2 and R3 in Fig 6.102 are the shunt resistors necessary to give a convenient range-multiplication factor for the current to be measured. They are left permanently in the circuit to be monitored but owing to their low resistance values, the operation of the circuit is not affected. On the voltage ranges, the series resistors R4, R5 and R6 will likewise be selected to give a convenient full-scale deflection for the meter. The actual values are calculated as suggested in Chapter 18—*Measurements*.

The selector switch should be of the wafer type, with

ceramic insulation for the higher voltage circuits; the contacts should be of the break-before-make type to avoid a short-circuit when changing from one circuit to the next. For voltages above about 300V it may be necessary to use double spacing between adjacent contacts to prevent flash-over, and the circuit should be arranged, where possible, so that adjacent contacts are at approximately similar potentials.

As an alternative to switching, closed-circuit jacks are sometimes used. These are jacks which keep the circuit closed except when the plug is inserted. The advantage of the method is that an independent external meter may be used but this advantage is outweighed by the simplicity and convenience of the switched meter arrangement. It must be remembered that spurious radiation is reduced when the meter and the associated wiring are completely contained within the transmitter enclosure.

TRANSMITTER DESIGN SUMMARY

The following summary is intended to help in the design and construction of amateur transmitters for use on the hf (1·8–28MHz) bands:

(1) If the transmitter is to be vfo controlled, the vfo should be constructed as a separate unit and subsequently fitted to the main transmitter assembly.

(2) A vfo must operate satisfactorily and be thoroughly checked before it is fitted to the transmitter. Final testing should be carried out in the complete transmitter.

(3) Only the highest quality components, rigidly mounted, and positioned for minimum temperature excursions should be used in the construction of a vfo.

(4) Where temperature compensation is needed, the adjustments should be made after the complete transmitter has reached its current operating temperature.

(5) A vfo must be followed by a suitable buffer amplifier.

(6) Frequency multiplier stages and driver amplifiers should be operated at a power level which is just sufficient to provide the required drive to the following stage.

(7) In the interests of reducing the radiation of unwanted harmonics of the vfo or crystal oscillator, capacitance coupling between the driver and the pa should be avoided.

(8) In high power designs, bandpass couplers between the stages significantly reduce the level of the harmonic output. While this is also true of low power transmitters, such couplers may not be needed to reduce harmonic radiation to an acceptable level in view of the lower power.

(9) A pi-network tank circuit tends to limit harmonic radiation.

(10) In certain areas where the television signals are weak an aerial coupler should be employed as an aid to harmonic reduction whatever type of pa tank circuit is employed.

(11) Provision for neutralizing should be included in all pa stages.

(12) Power amplifiers should be thoroughly checked for low frequency or high frequency parasitic oscillation.

(13) In power amplifiers employing grid-leak bias, some form of safety bias should always be included.

(14) It is preferable to construct the transmitter as a single unit in the interests of reducing stray radiation, but care must be taken to provide sufficient ventilation.

(15) If the transmitter is constructed as two units, one of which is the power supply, all leads leaving the transmitter enclosure must be fully decoupled at the point at which they leave the cabinet.

(16) The physical layout should be as near to a straight line as possible to ensure maximum separation between the master oscillator and the power amplifier.

(17) Both the vfo and the pa should be thoroughly screened. All power supply leads, including heater circuits, to the vfo should be decoupled at the point where they enter the vfo box. This will help to prevent feedback from the power amplifier.

(18) Tuned circuits should be positioned so that they do not exhibit mutual coupling one to the other, unless this is intended.

(19) Always ensure that the ht supplies are properly fused.

(20) In the interests of personal safety, never make adjustments to the transmitter, and especially the power amplifier, while ht is applied.

TRANSMITTER CONSTRUCTION

The circuits given in this section are representative of current amateur constructional practice and provide a broad indication of present trends in the design of high frequency amateur transmitters. While layout plans are not given for all the projects described, no particular difficulties should be encountered provided the layout follows a logical sequence.

Some of the circuit techniques used may also be of interest to the more advanced constructor who is designing his own equipment. Care should be taken when using sections of circuits in this way to ensure that voltage levels and impedances match the rest of the circuit.

A SOLID STATE MULTIMODE TRANSMITTER FOR 1·8MHz

This transmitter provides facilities for cw and either amplitude or double sideband with suppressed carrier modulation (dsb) and has given reliable service with good reports since it was built some two years ago. The quiescent current being less than 200mA at 12V makes it suitable for portable operation with, if necessary, no more than a pair of 6V "lantern" dry batteries for power supply.

Eyebrows may be raised at mention of dsb modulation, bearing in mind the theoretical requirement that the receiver must supply the missing carrier in correct *phase* when receiving this mode, but experience has shown that for fully 95 per cent of the contacts the distant operator has assumed, until told, that it was normal ssb, while the remainder—with less selective receivers?—have commented on the unusual form of modulation.

The reasons for the adoption of dsb rather than ssb were less complexity and cost, in as much as no expensive filter was required. Clearly, without synchronous demodulation dsb can only operate at 50 per cent of the efficiency of ssb, which would result in a 3dB loss of signal, ie about half an S-point.

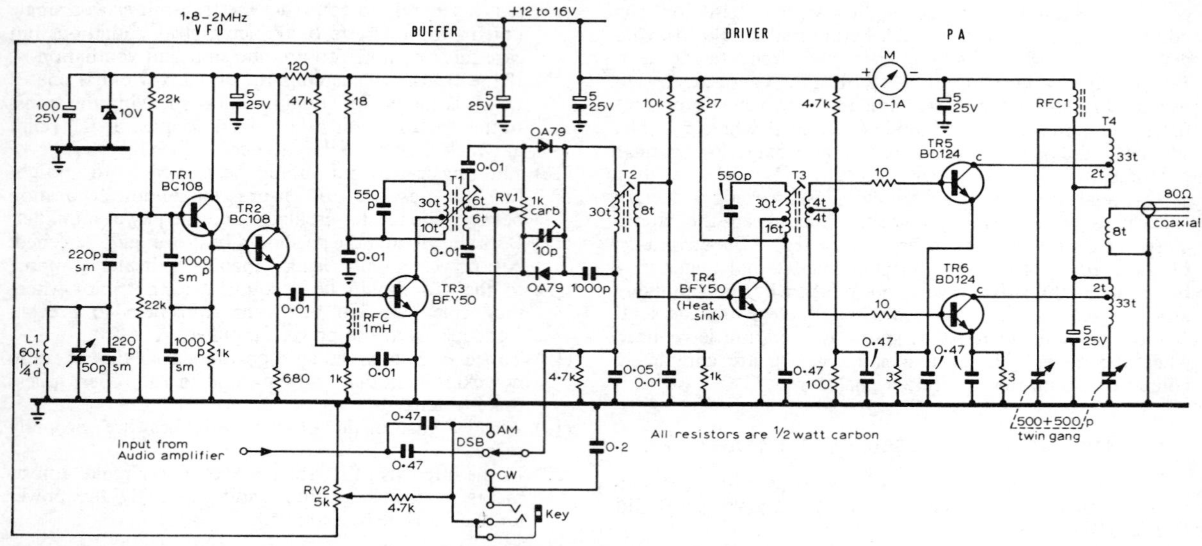

Fig 6.103. Circuit diagram of the transmitter. The 10pF trimmer shown connected in parallel with one of the diodes in the balanced modulator should be placed across the diode having the least capacitance.

The Circuit

From the circuit diagram **Fig 6.103** it will be seen that the Colpitts vfo, comprising TR1, with TR2 as a directly coupled emitter follower, drives the buffer amplifier TR3 in the collector circuit of which is the primary winding of T1. The secondary of this transformer is connected to a diode balanced modulator which eliminates the carrier. The following stage is the driver for the push-pull output pair of transistors TR5 and TR6. In the absence of modulation the pa standing current should lie between 50 and 100mA and is adjusted to a minimum by RV1. The necessary bias to provide a carrier for cw or a.m. operation is controlled by RV2 so that on cw the driver and pa stages pass a maximum current of 800mA. On a.m. RV2 is set so that the rf output, as indicated on a swr bridge, is half that obtained on cw. Only half the audio input from the modulator is required on a.m. as opposed to that necessary for dsb.

The two positive voltage terminals may be "commoned" if the power supply has sufficiently good regulation to withstand the fluctuating current drawn by the pa stage without seriously varying the voltage on the vfo and buffer amplifier stages. Otherwise the low current required by these may be obtained from a battery or other suitable source.

The circuit of the four-transistor modulator will be found in **Fig 6.104.** The resistance/capacitor network between the second and third stages is designed to tailor the audio frequency response of the unit so that output is 3dB down at 300 and 2,000Hz as compared with mid-frequency.

Construction

The transmitter itself was built on a tinplate chassis 8½in long by 3½in wide with a ¼in turn-up all round to add rigidity. The layout of the main components is indicated in **Fig 6.105.** L1, T1, T2 and T3 were mounted in cut-down i.f. cans salvaged from an old television receiver. Winding details, together with those for T4, are set out in **Table 6.10** and in **Fig 6.106** respectively.

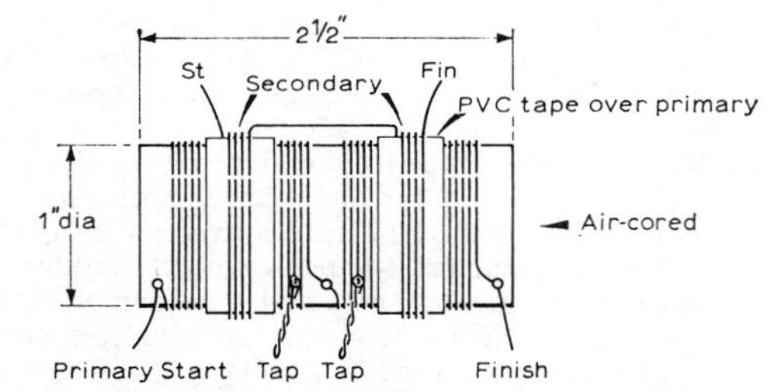

Primary..... 66 turns (33+33) of 22 swg wire tapped at 2 turns in positions shown
Secondary..... 8 turns (4+4) of 22 swg wire wound over primary

Fig 6.106. Constructional details of pa tank circuit, T4.

TABLE 6.10

L1	60 turns of No 34 swg enam. close wound.
T1 (primary)	40 turns of No 34 swg enam. close wound and tapped at the tenth turn from the "cold" end.
T1 (secondary)	6 + 6 turns of No 34 swg enam. bifilar wound over centre of primary winding.
T2 (primary)	30 turns of No 34 swg enam. close wound.
T2 (secondary)	8 turns of No 34 swg enam. close wound over "cold" end of primary winding.
T3 (primary)	40 turns of No 34 swg enam. close wound, tapped at tenth turn from the "cold" end.
T3 (secondary)	8 turns of No 34 swg enam. centre tapped, wound over centre of primary winding.
All above coils are wound on standard 8mm diameter formers with iron dust cores and all in ⅜in by ⅜in aluminium screening cans. A singler layer of pvc or Sellotape separates the primary from the secondary winding in the three transformers.	
T4 (see Fig 6.106)	On former 1in diameter by 2½in long. Primary, in two sections each of 33 turns of No 22 swg enam. with ¼in space between sections and each tapped 2 turns from the centre tap. Secondary, two sections each of 4 turns of No 22 swg enam. wound over centre of each section of primary winding to make an 8-turn coil. A single layer of pvc or Sellotape separates primary and secondary windings.
CH1	30 turns of No 22 swg enam. on length of ferrite aerial rod, mounted in aluminium can 1in diameter by 1½in high.

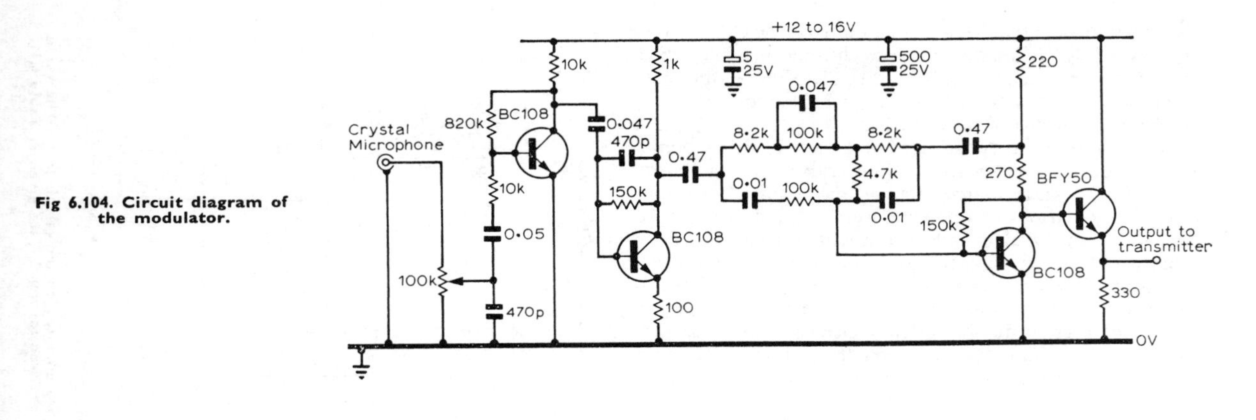

Fig 6.104. Circuit diagram of the modulator.

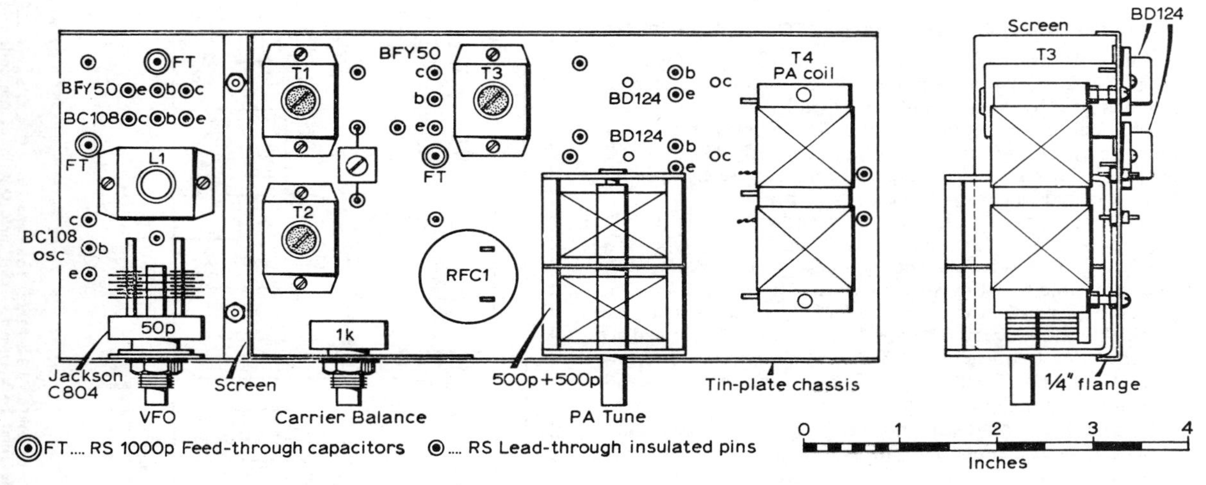

Fig 6.105. Plan view and right-hand elevation of the transmitter chassis showing the positions of the main components.

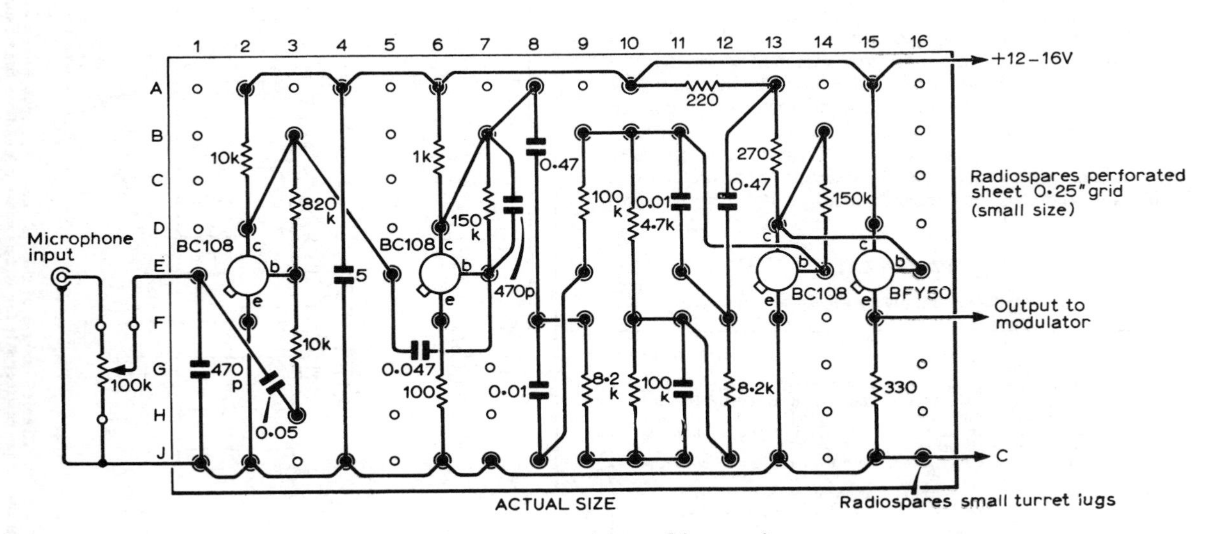

Fig 6.107. Layout of the modulator section.

All components requiring insulation from the chassis were mounted on R.S. Components lead-through insulated pins. The output transistors (TR5 and TR6) were mounted underneath the chassis at the points indicated and insulated therefrom by means of mica washers.

The modulator was constructed on a piece of R.S. Components perforated sheet with a 0·25in grid, with small turret tags from the same source fitted in the required positions to which the various components were secured. This board measured 4¼in by 2⅜in and was mounted separately from the transmitter unit. **Fig 6.107** indicates the layout employed.

Although it is sometimes found that transistorized transmitters are prone to cause tvi, there has been no difficulty in making the present design perfectly clean in this respect by attention to screening, the employment of a low-pass filter in the aerial feeder and the decoupling of all external leads.

VFO CONTROLLED VALVE TRANSMITTER FOR 1·8–2MHz

This 10W transmitter has a built-in vfo covering 1·8–2MHz and provision for high level anode and screen modulation. Cw operation is also possible.

Although an ht supply of 250V is suggested, the transmitter may be used on any voltage between 250 and 300 provided the screen dropping resistor is correctly chosen. The measured efficiency of the pa is 70 per cent.

The circuit of the transmitter is shown in **Fig 6.108.** V1 (EF91/6AM6) functions as a Tesla-Vackar type vfo (see Fig 6.33) and is followed by a class A buffer stage V2. The grid circuit of this stage is purely resistive, R4 being provided to permit the connection of an external meter to check the grid current. The anode circuit of V2 is tuned to the operating frequency by L2 C13. If this circuit is resonated at 1·9MHz it will be found that drive to the pa will remain substantially constant between 1·8 and 2MHz. C13 may therefore be a pre-set trimmer. R7 provides a test point for checking the anode current during the alignment procedure.

The pa V3 (6BW6) is conventional. Drive to the grid is via C14 and to improve the efficiency a 2·5MHz rf choke is connected in series with the grid resistor R9. The anode tank is a simple parallel tuned circuit C19 L3, the output of which is taken by a link, L3a, proportioned to match 75Ω coaxial cable. A 1·5MHz rf choke is used in the anode circuit to prevent rf feeding into the metering circuit or into the modulator.

M1, a 0–5mA meter, switched by S4, is used to measure grid and anode currents. To avoid damage to the meter when switching, there is an unused position between the two

TABLE 6.11

Inductor details for Fig 6.108

L1	70 turns 34 swg enam. close wound on ¼in former fitted ferrite core.
L2	90 turns 34 swg enam. close wound on ¼in former.
L3	26 swg enam. wire close wound for a distance of 1½in on a 1½in diam. former. Over one end bind two layers of pvc tape. This is the "cold" end of the pa coil, the anode being connected to the other end.
L3a	Starting at the end of L3, wind 8 turns 26 swg over the pvc band on L3.
T1	Centre tapped output transformer. Primary 8kΩ to 10kΩ centre tapped, rated at 50mA dc. Secondary not used.

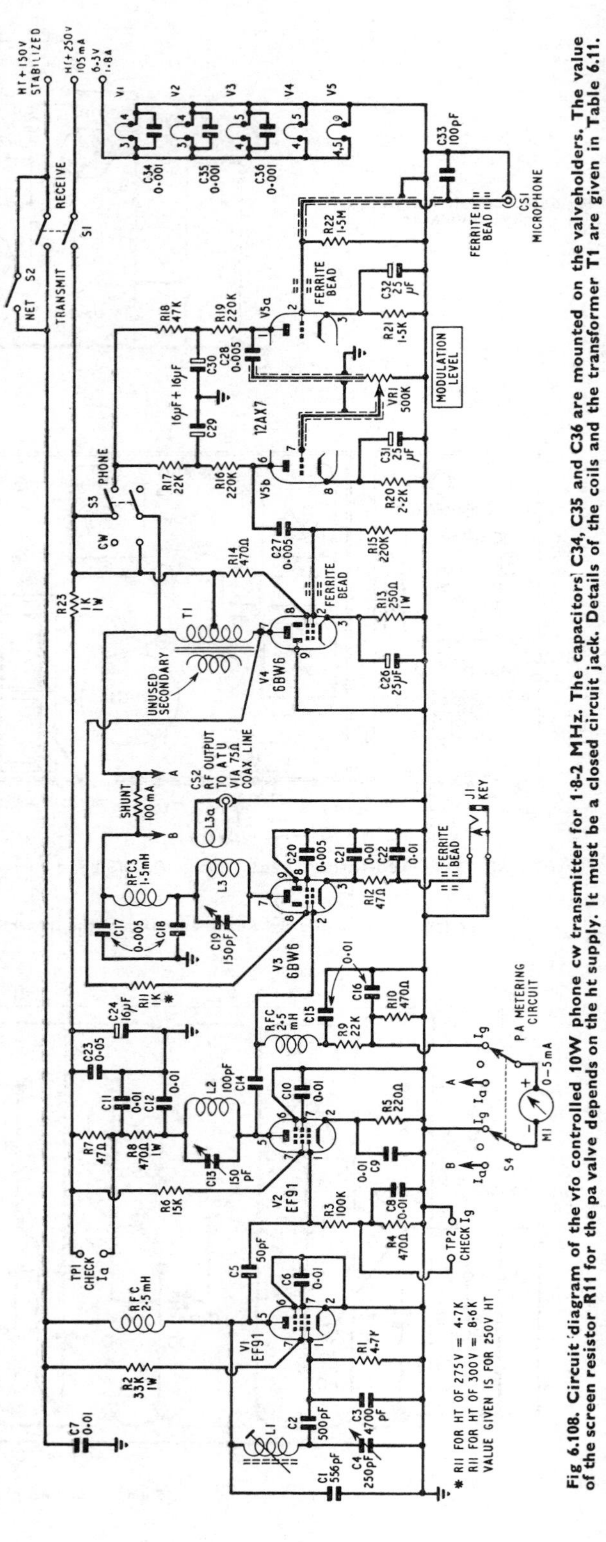

Fig 6.108. Circuit diagram of the vfo controlled 10W phone cw transmitter for 1·8-2 MHz. The capacitors C34, C35 and C36 are mounted on the valveholders. The value of the screen resistor R11 for the pa valve depends on the ht supply. It must be a closed circuit jack. Details of the coils and the transformer T1 are given in Table 6.11.

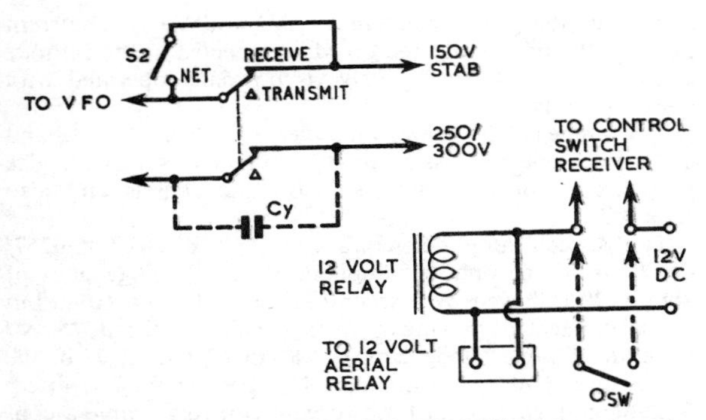

Fig 6.109. Substituting a relay for S1 in Fig 6.108. If there is arcing across the main contacts of the relay, a 0·001–0·01μF capacitor should be fitted at Cy. The smallest satisfactory value should be used. A manual override switch SW may be fitted to permit the transmitter to be switched on independently of the receiver.

current readings. When switched to read anode current, the meter is connected across a shunt permanently wired into the anode circuit which modifies the range to read 100mA full scale. The leads to S4 should be screened to reduce the possibility of stray coupling.

The modulator valve V4 is also a 6BW6 and is coupled to the pa by T1, a standard centre tapped output transformer which is used as a 1 : 1 modulation transformer. As the impedance of the pa is almost equal to the recommended load for the modulator valve, a good match is obtained. T1 may be relatively small provided that the winding can carry the currents drawn by the pa and modulator valves.

V4 is driven by a two-stage speech amplifier, with the gain control between V5a and V5b. Ferrite beads are fitted to grids of V4 and V5 to prevent any stray rf picked up on the modulator leads being fed to these stages.

The transmitter is controlled by three switches; the main send/receive switch S1 controls power to the vfo, rf and modulator sections, S2 allows the vfo to be switched on for netting and switching from a.m. to cw is provided by S3. In the cw position the modulator is disabled and a short circuit is placed across the modulation transformer to protect it from dc surges.

Many modern receivers incorporate a send-receive switch

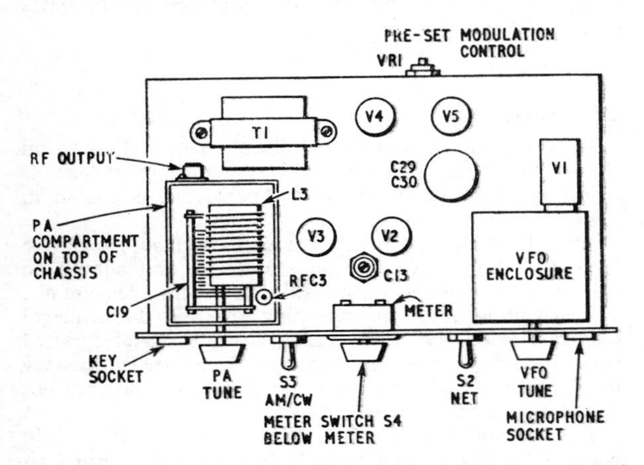

Fig 6.110. Layout of the top of the chassis of the vfo controlled 10W transmitter.

intended to control both receiver and transmitter. A suitable control circuit for use with such a switch is shown in **Fig 6.109.**

Construction

The transmitter should be laid out in a straight line with vertical screens between each section and the coils in each section arranged so that they are at right angles to those in the adjacent sections. A screen is placed around the under-chassis rf circuits so that, with the bottom plate in place, they are effectively within a screened box. The layout is shown in **Fig 6.110.**

The vfo should be constructed as a separate unit. Most of the pa components are on top of the chassis, C49, L3 and RFC3 being contained in a screened compartment to the rear of which is fitted the coaxial output socket CS2.

To avoid earth loops which might give rise to hum or instability, the screened cable which runs from the microphone socket CS1 to the grid of V5a should have its outer screening earthed only at the socket, the screening at the grid end being left disconnected. The grid resistor R22 should be mounted directly on the grid pin of V5a and earthed by the shortest route.

Alignment

After the vfo has been constructed, it should be tested as a separate unit. When it has been fitted to the chassis and the wiring checked, V2, V3 and V4 are fitted, S3 set to the cw position and S4 to read pa grid current. Ht to the pa is disconnected by removing the connection from T1. Power is then applied by closing S1. C13 is tuned to produce maximum grid current (approximately 3mA) and the frequency 1·8–2MHz) checked with an absorption wavemeter.

With the drive to V4 still applied, C19 should be tuned through its range to see whether the pa grid current varies. If it does, the stage must be neutralized or the decoupling and screening improved. The power supply may now be switched off and the ht reconnected to the pa. When this has been done, S4 should be set to read the anode current to V4. C19 should be at maximum capacitance. The power supply may now be switched on again and C19 quickly tuned for maximum dip in anode current. Next open S1. Lightly couple an absorption wavemeter to L3, close S1 and confirm that the frequency is in the 1·8–2MHz range. Open S1 again. A 75Ω dummy load should now be connected to the rf output socket CS2, S1 closed and C19 tuned for maximum dip.

To avoid exceeding a dc input of 10W, the anode current must not exceed the following values under loaded conditions:

at 250V ht the current should be between 35–40mA,
at 275V between 32–36mA and
at 300V between 29–33mA.

If the anode current exceeds these values, L3a should be moved away from L3 to reduce the coupling.

V5 may now be fitted and a high impedance microphone connected to CS1. VR1 should be at minimum gain. S1 may then be closed and the gain control advanced about half way. Speaking into the microphone should produce a marked increase in the brilliance of a flashlamp bulb loosely coupled to L3 by a three turn coil. The microphone should be kept away from the pa tank circuit to avoid rf pickup which might cause feedback.

The transmitter is now ready for connection to a suitable aerial through an aerial tuning unit.

A TT21 (7623) TRANSMITTER FOR 14, 21 AND 28MHz

The transmitter illustrated below employs a TT21 in the pa stage and is capable of 60W peak input on a.m. and 80W input when using narrow band frequency modulation or cw. A.m. is provided by series gate screen modulation of the pa.

In the interests of tvi prevention, the rf circuits of the transmitter are completely screened without reliance on a cabinet so that the unit can be used either as a table top transmitter or installed in a rack with other equipment without any significant increase in unwanted radiation. The more complicated units are completely removable from the main chassis so that construction and servicing are simplified.

The circuit of the complete transmitter, including the built-in power supply, is shown in **Fig 6.111.**

The vfo employs a screened rf pentode V1 (EF80/Z719/6BX6) operating as a high *C* Colpitts electron coupled oscillator covering 3·5–3·65MHz. Provision for frequency modulation of the oscillator is made by use of a silicon voltage variable diode CR1 across part of the tuned circuit, a potentiometer providing the required bias so that a suitable linear portion of the diode operating curve can be selected. In this case the bias is set to +1·0 to +1·25V which will give approximately equal capacitance change for both sidebands.

The second stage, V2 (Z77/EF91/6AM6), acts either as a buffer stage or as a crystal oscillator, selected by S1. The anode circuit of this stage is tuned to 3·5MHz and coupled to the next stage by a wideband coupler.

Valve V3 is a straightforward frequency doubler and feeds the next stage via a 7MHz wideband coupler. The final stage of the exciter V4 is operated as a doubler from 7 to 14MHz, as a trebler from 7 to 21MHz and as a quadrupler from 7 to 28MHz. The valve chosen (a type N78) gives somewhat better performance than the more usual 5763, due to its higher slope and lower drive requirement. Coupling between this stage and the pa is by means of a pi-network which provides harmonic attenuation while keeping the impedance of the grid circuit of the power amplifier relatively low, so helping to maintain the stability of the pa valve. The circuit is adjusted by tuning for maximum grid current to V5. When netting to a frequency, ht to the screens of this stage and of the power amplifier is switched off by S6. For cw operation, the cathodes of V2 and V3 are keyed.

The whole of the exciter is built as a unit so that it can be removed from the main chassis by removing four fixing screws and unsoldering the interconnections to the main chassis.

The final stage V5 uses a type TT21 (7623) tetrode, which is operated at 60W peak input on a.m. and at 80W input on cw and nbfm. The anode circuit is a pi-network with the inductance tapped at suitable points for three bands by S3. Across the output are connected a tv Band 1 series tuned trap circuit and an rf voltmeter which is a useful means of correctly tuning the output circuit, the power output being arbitrarily indicated on the meter on the righthand side of the front panel. By means of the switch S4, this meter also reads the grid current to V3, V4 and V5. The current input to the pa is indicated on M1 connected in the cathode circuit of the TT21. The valve is normally operated with 600V ht on its anode.

The whole of the power amplifier is built at the lefthand end of the main chassis and is completely screened. The grid circuit for the valve is below the chassis and also screened.

The speech amplifier comprises V6 (Z729/EF86/6267/6F22) as a microphone amplifier with a voltage gain of about 120 followed by a second stage V7 (also a EF86 but triode connected) having a voltage gain of about 25. V7 feeds the double triode clipper V8 (ECC82/12AU7) in the anode circuit of which a simple low pass af filter is fitted. Suitable adjustments of the two gain controls enable a considerable amount of clipping to be applied which is desirable when using nbfm.

The speech amplifier/clipper is built into an Eddystone $4\frac{1}{2}$in × $3\frac{3}{4}$in × 2in die-cast box (**Fig 6.112**) which is attached to the top of the main chassis at the front right hand corner. The output from this unit is switched by S5 either to the series gate screen modulator valve, V9 (12BH7), or to the reactance modulator CR1 (EW76 or SX761). When the switch is in either the cw or fm positions V9 operates as a clamp valve for the pa stage (see **Fig 6.113**).

An alternative series gate clamp circuit to provide for adjustable carrier level when using nbfm is shown in **Fig 6.114.**

The power supply is conventional and is provided by a 650V ht transformer and separate heater transformers. The main rectifiers, V13 and V14, are indirectly heated type CV4044/6443 half-wave valves, the pair providing the total current of about 225mA. Smoothing of the supply is by a single 5H choke and a 16μF paper capacitor (two 8μF units in parallel). The pa is fed directly but all other stages receive ht from the main potentiometer made up of a fixed resistance in series with three gas stabilizer tubes V10 (S130) and V11 and V12 (both type QS150/45). The striking electrodes are connected to suitable points to provide quick ignition.

To avoid breakdown between the heater and cathode of the screen modulator valve V9, a separate transformer is used. The heater must not be earthed. A small bias supply using a 0–125V transformer (T4) is provided for the series gate modulator.

Construction

The complete transmitter is built on a 17in × 13in × 3in chassis to which a standard 19in × $8\frac{3}{4}$in front panel is rigidly fixed by panel brackets. The layout is shown in **Fig 6.115.**

After the holes for fixing the chassis to the front panel have been drilled the hole for the vfo-exciter unit should be made. This is probably most readily done with a tension file. The mounting of the pa screening box and the principal components on the main chassis is quite straightforward and should not present any difficulty. Some care is necessary in placing the fixing holes for the pa so that they are not fouled by operating shafts.

The vfo and exciter unit is built on to a 10in × $5\frac{1}{4}$in plate of $\frac{1}{8}$in thick aluminium. A screening box is fitted on the underside of this plate while the vfo tuned circuit is

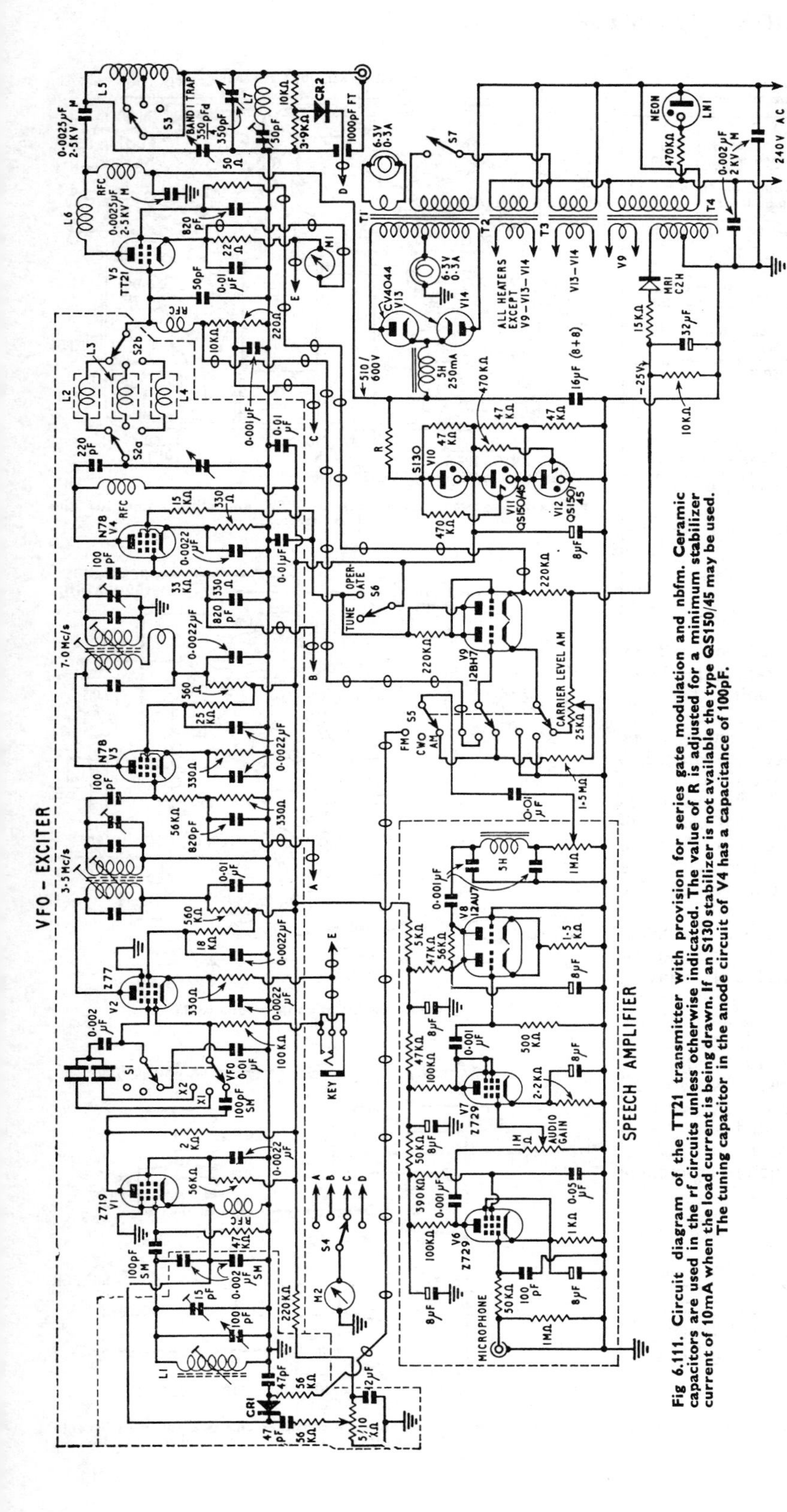

Fig 6.111. Circuit diagram of the TT21 transmitter with provision for series gate modulation and nbfm. Ceramic capacitors are used in the rf circuits unless otherwise indicated. The value of R is adjusted for a minimum stabilizer current of 10mA when the load current is being drawn. If an S130 stabilizer is not available the type QS150/45 may be used. The tuning capacitor in the anode circuit of V4 has a capacitance of 100pF.

built into an Eddystone die-cast box measuring 4½in × 3¾in × 2in and attached to the top near the front end by its long side with the removable plate at the rear.

The speech amplifier/clipper is built into a similar die-cast box and mounted upside down so that the valves and other components are on the bottom of the box which is attached to the main chassis by its normal four fixing screws. The lid is not used. Considerable care is necessary to position the microphone socket (a Belling and Lee type L722/S) so that it

(*a*) is as close as possible to the corner of the box but with sufficient clearance for attachment on the front panel.

(*b*) does not foul the handle on the front panel.

Both of these units are electrically connected to the rest of the transmitter by soldered connections on the distribution terminal strip in the main chassis and are readily movable for servicing and testing.

Details of the coils are given in **Table 6.12.**

The primary and secondary windings of the 3·5 and 7MHz wideband couplers are wound in the same sense, the other ends of the coils being connected to anode (top) and grid (bottom) with the earthy ends of the two coils at the centre. Coupling between the two windings is as follows:

3·5MHz Top coupled through a 10pF capacitor.

7MHz Extra turns of the primary are wound over the earthy end of the secondary winding. This method was found to give more output than top capacitor coupling.

So that both the anode and grid circuits of these units can be adjusted from the top of the chassis a small trimmer is connected across each of the secondaries in addition to the core. The anode coil has a dust iron slug.

The driver anode coils are wound on the same type of former as those used for the transformer couplings. The 14 and 21MHz coils are wound on the top half of the former. The inside diameter of the 28MHz coil is about $\frac{7}{16}$in. All the coils and couplers are individually screened in cans.

Adjustment and Operation

After completion of the constructional work, the two smaller units should be adjusted before being fixed on to the

TABLE 6.12

Inductor Details for the TT21 Transmitter

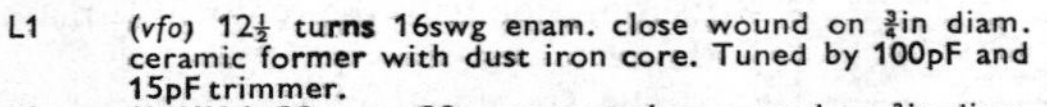

L1	(*vfo*) 12½ turns 16swg enam. close wound on ¾in diam. ceramic former with dust iron core. Tuned by 100pF and 15pF trimmer.
L2	(*14MHz*), 20 turns 20swg enam. close wound on ¾in diam. former with dust iron core (wound at top of ift former from type 373 unit).
L3	(*21MHz*), 15 turns 18swg enam. close wound on ¾in diam. former with dust iron core (as for L2).
L4	(*28MHz*), 14 turns 16swg close wound ¾in diam. (self supporting inside ift can from type 373 unit).
	(*pa*), 11 turns 14swg enam. spaced two wire formers 1¼in diam. self supporting, tapped from anode end at 8 turns for 21MHz and at 4 turns for 28MHz.
L6	(*parasitic suppressor*), 4 turns 16swg enam. spaced one wire diameter, 5/16in inside diam. self supporting.
L7	(*Band 1 trap*), 7 turns 18 swg spaced one wire diameter ½in inside diam. self supporting and tuned by 50pF.

Wideband Couplers

3·5MHz—primary and secondary both 45 turns 30swg enam. close wound on ½in diam. former (ift former from type 373 unit), both tuned by 100pF, dust iron core in primary and brass core in secondary, coupling 10pF.

7MHz—primary and secondary both 34 turns 26swg enam. close wound on ½in diam. former (ift former from type 373 unit), primary tuned by 22pF, secondary tuned by 30pF, dust iron core in primary, brass core in secondary, coupling 3 turns extra on primary wound over earthy end of secondary (see text).

main chassis. It is desirable that an alternative power supply giving about 300V at 60mA and heater voltage should be employed while adjusting the vfo exciter unit rather than the much more dangerous 650V ht source.

First, the vfo should be set up and its tuning range adjusted, preferably with the help of a receiver switched to the 10m band. With the components specified the vfo covers only 1·2MHz (28–29·2MHz) of this band as it was felt that the more open tuning for 15m and 20m was worth the sacrifice of the higher frequency band. If preferred a larger tuning capacitor could be used so that the whole band is covered but this might result in some loss of drive at the band edges.

The stability of the vfo is reasonably good. Provided, however, that the main tuning capacitor and the fixed capacitors are of good quality it should be satisfactory. Having adjusted the vfo range, S1 should be set to select one of the crystals and the operation of the crystal oscillator checked. Next, with V3 in position and using its grid current at point *A* on Fig 6.111 as an indication on M2, adjust the tuning of the primary and secondary of the 3·5MHz band-pass coupler until a satisfactory frequency response is obtained. The procedure should be repeated with V4 in position, using the grid current indicated at point *B*. Check that there is no instability in these stages before proceeding

TABLE 6.13

Mode	Audio Gain Control	Clipper Control	Reactance Diode Bias	PA Anode Current
AM (Series Gate)	Half to maximum	Maximum	—	40–50mA (no sig.) 110–120mA (max. sig.)
NBFM *	Half to maximum	Quarter to half	+ 1 to +1·25V	110–120mA
CW	—	—	—	150–160mA

*When the transmitter is wired so that S5 permits the carrier level to be adjusted on both series gate a.m. and on nbfm.

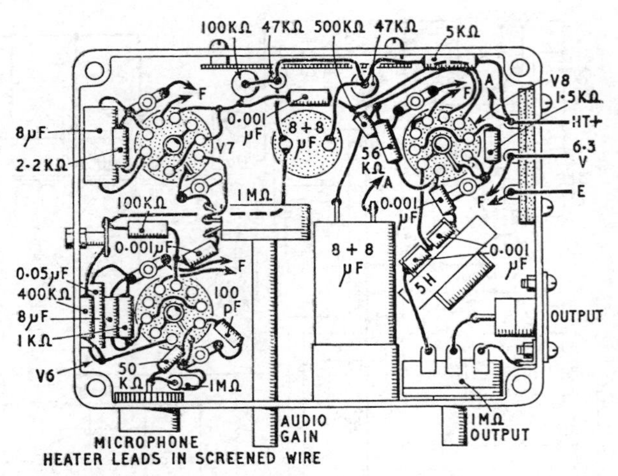

Fig 6.112. Layout of the speech amplifier and clipper unit for the TT21 transmitter.

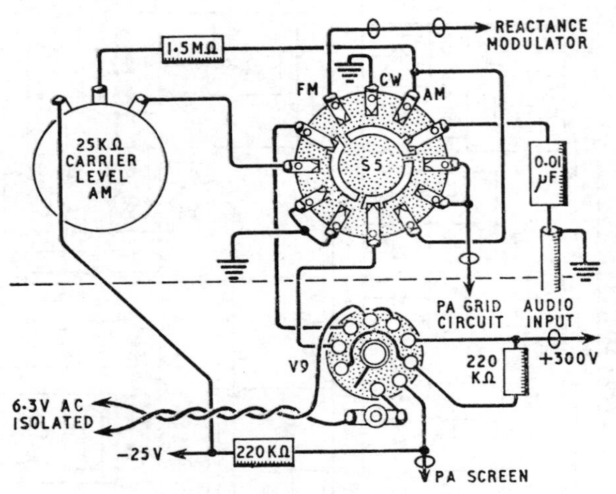

Fig 6.113. Details of connections to S5.

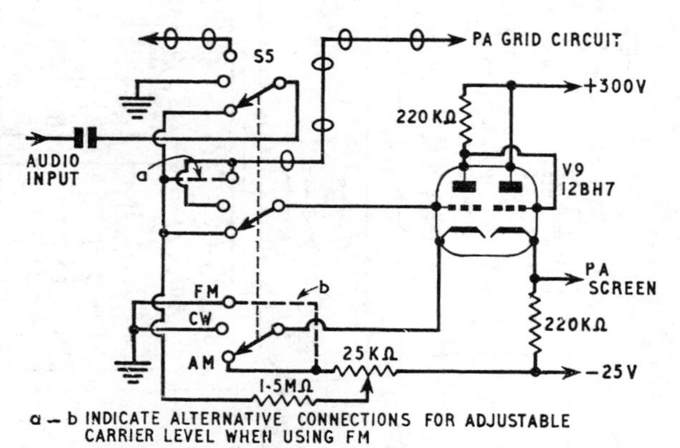

Fig 6.114. Alternative series gate clamp circuit providing adjustable carrier level when using narrow band frequency modulation.

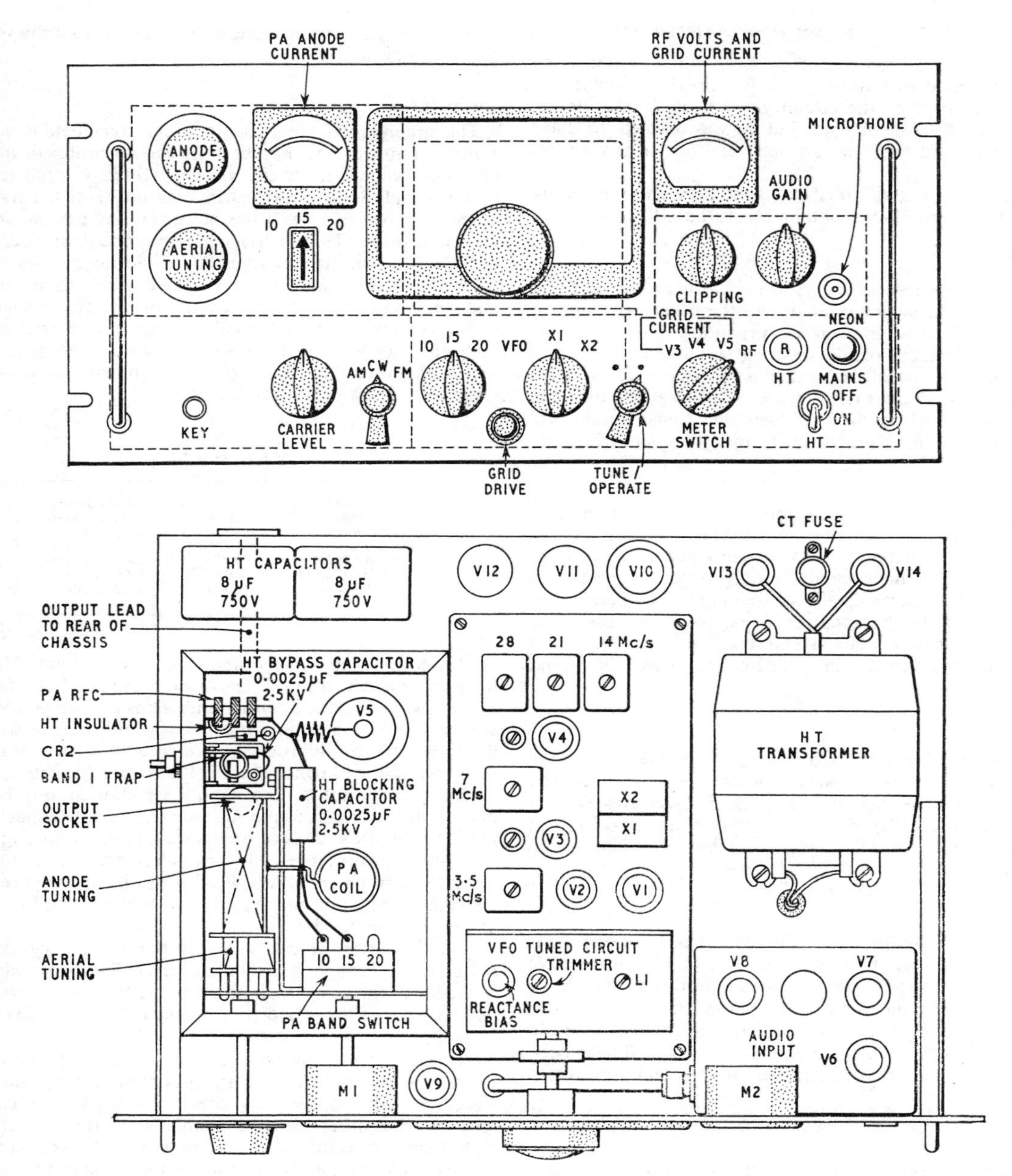

Fig 6.115. Layout of the front panel and chassis.

to adjust the anode circuits of V4. The total anode and screen input to the first three stages should be between 40 and 55mA at 300V.

If, as recommended, this preliminary tuning is carried out with an external power supply, some form of temporary rf voltmeter will be required to check the output from V4.

When adjustment of the exciter stages is complete, the unit may be connected to the main power supply and the complete transmitter tested.

The pa anode circuit is adjusted in the usual manner for a pi-network: tune for maximum dip as indicated on M1 and then increase the loading until V5 is drawing 120mA. Final adjustment should be made by adjusting the pa tuning controls for maximum output as indicated on M2 when switched to position *D*.

Satisfactory speech quality and modulation depth or deviation is obtained with the settings of the audio controls shown in **Table 6.13.**

THE WIRRAL SIX-TEN NFD TRANSMITTER

This collates a number of systems which produce an extremely stable transmitter with full break-in facilities.

No doubt part of the circuitry, particularly the vfo/co, could have been simplified but it was wished to have good bandspread and with an open dial on all bands from 160 to 10m.

A glance at **Fig 6.116** shows that only five valves are used, all of them readily obtainable in the UK.

VFO CO

A 12AT7 operates as a crystal oscillator in one half and as a variable frequency oscillator in the other—the sum or difference of their frequencies providing output in the 160 to 10m amateur bands. In order to allow each band to occupy all or most of the dial, the vfo has three switched tuned circuits. If constructors are not concerned over the problem of good bandspread then one tuned circuit will suffice as is common practice with many transmitters.

Mixer

Mixing takes place in the anode of the mixer valve, a 6BA7, and it is in this stage that keying takes place. The vfo/co runs continuously thus avoiding the risk of chirp. Drive on the amateur bands is only produced when the mixer is conducting which allows listening through under key up conditions. In the event that leakage does take place then the cathode of the crystal oscillator can be keyed simultaneously, though in practice this was not found necessary.

Buffer

The buffer valve is a 6CH6 which gives ample drive to the pa. Indeed, drive is embarrassing on the bands between 160m and 20m. Despite the fact that a screen was erected across the base of the 6CH6, such that its anode could not "see" its grid, this stage was unstable and had to be neutralized.

PA

The power amplifier is a 2E26 which has an anode dissipation greater than that required to run a genuine 10W dc input. With 250V on the anode, however, it can be kept within the limit using the constants shown. This valve was chosen since it has such excellent characteristics on the hf bands and has a top cap anode. A TT11 would have been quite suitable but its performance on 15m and 10m is not as good.

TR Switch

An EF80, triode-connected, acts as a transmit-receive switch and receiver muting. Under key up conditions the valve conducts and couples the pa pi-network to the input of the receiver. It is so effective that signals can only be heard in the receiver when the pa pi-network is resonant. Under key down conditions the valve is cut off by the rectification of a little rf taken from the pa. A potentiometer adjusts this bias such that it is possible to monitor the note of the transmitter in the receiver—this being absolutely essential during fast operating. The bias available lies between −10V and zero volts dependent on the potentiometer setting. This bias is fed to the avc system of the receiver and experience has shown that fast avc is necessary in order to achieve full break-in working.

Circuitry

The circuit is self-explanatory though some details require a little comment. As has been already mentioned, the vfo has three switched tuned circuits in its grid in order to achieve an open dial. Some constructors may not trust the stability of the switched vfo but it has not given any trouble and all reports have been T9. The designer used various surplus crystals of HC6/U mounting and adjusted the vfo frequencies to suit them. All crystals were not available on the surplus market and some had to be bought at retail prices. The frequencies of crystals and vfo tuned circuits are shown below, but it is realized that constructors may use crystals available to them with suitable modifications to the vfo in order to bring the resultant mixing in the amateur bands.

TABLE 6.14

Amateur band	Crystal Freq	VFO Swing
1·8–2MHz	6978kHz	4978–5178kHz
3·5–4MHz	9000kHz	5000–5510kHz
7·0–7·2MHz	1994kHz	4978–5178kHz
14·0–14·350MHz	9000kHz	5000–5510kHz
21·0–21·450MHz	15,940kHz	5000–5510kHz
28·0–28·5MHz	23,000kHz	5000–5500kHz

The coils in the mixer anode are wound on Aladdin formers, rescued from old television receivers. Each former carries two coils, one at either end. Thus only three formers are used, resulting in a saving in space and short leads. The lf bands (160, 80, 40m) coils were wound at the top of the formers and the hf bands (20, 15, 10m) wound at the bottom.

Ideally coils not in use should be shorted out but in practice this was not found necessary. The only capacitance used to bring these coils to resonance was that resulting from self capacitance of the coils, inter-electrode capacitance of the 6BA7 and circuit strays. By this means tuning is reasonably flat such that output is fairly constant over the full width of the amateur bands.

A similar arrangement is used in the driver stage though the coils are brought to resonance by a 5 to 50pF variable capacitor in the grid circuit of the pa. A 50kΩ potentiometer in the screen of the 6CH6 controls the drive available on the pa grid.

The pa is quite conventional except for the plug-in coils in the pi-network. Coil changing is timewasting and constructors may wonder why they were used in a contest type of transmitter. Several members of the Wirral club had often expressed some doubt about the efficiency of shorted coils in a pa pi-tank and felt that plug-in coils would be preferable for a low power transmitter. Further it is not always desirable to terminate at 80Ω impedance, particularly for end fed aerials on the lf bands. By using plug-in coils, inductances and capacitances of the correct value could be used in order to feed the aerials direct from the pi-tank, thus eliminating the use of a tuning unit. Television interference is not usually a problem on NFD.

The function of the transmit-receive switch and receiver muting arrangement has already been dealt with and all components are standard except for the transformer in the cathode circuit of the EF80. This is wound on a slab of

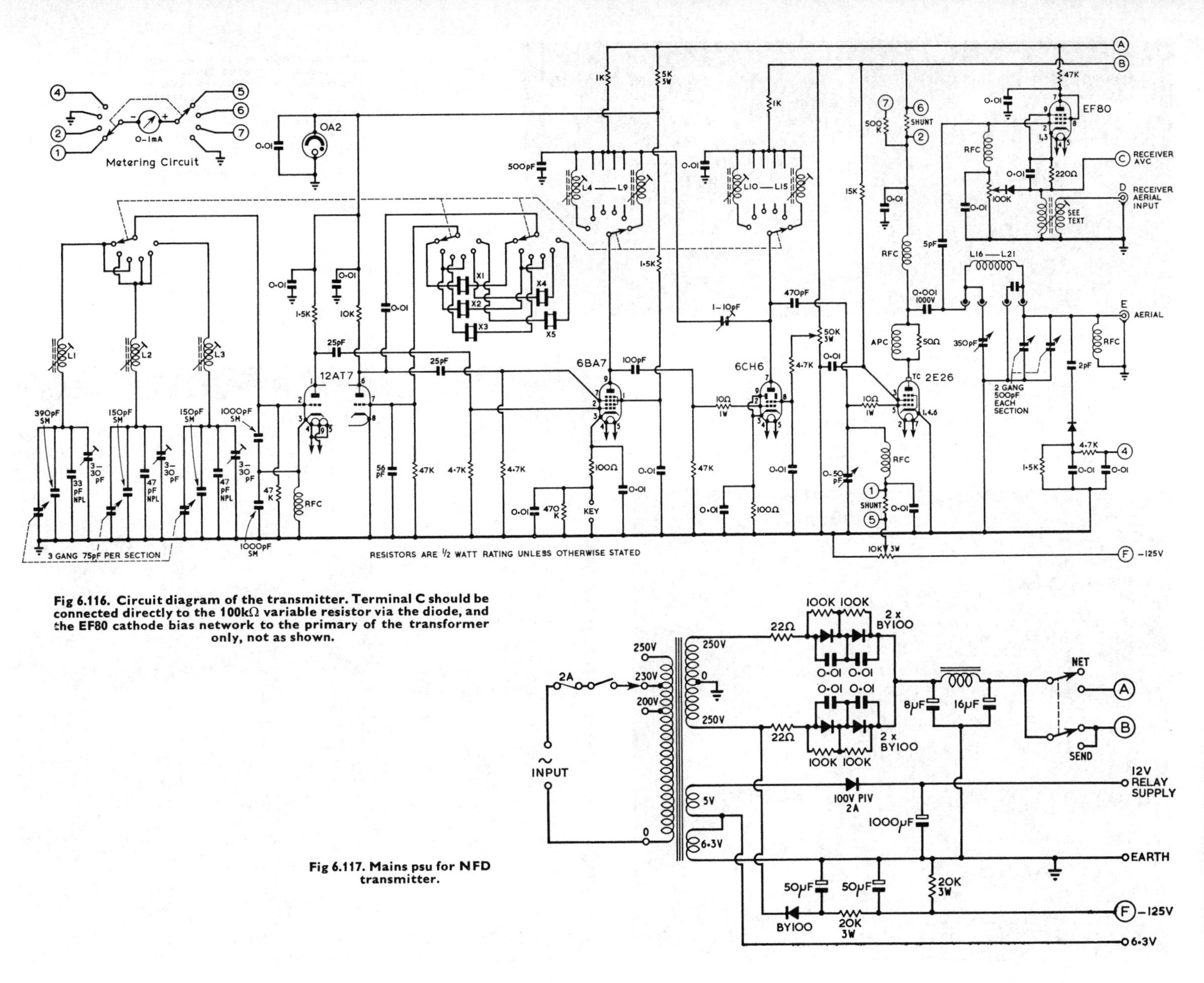

Fig 6.116. Circuit diagram of the transmitter. Terminal C should be connected directly to the 100kΩ variable resistor via the diode, and the EF80 cathode bias network to the primary of the transformer only, not as shown.

Fig 6.117. Mains psu for NFD transmitter.

Above: **Plan view showing the layout. On the right hand side can be seen the vfo three-gang tuning capacitor and behind this the six cans containing the coils for the mixer and driver stages. On the left hand side is the 2E26 pa and the EF80 TR switch.**
Photo by R. E. Foster

Above: **Underside view of the transmitter. The screening between sections can be seen with the vfo and co compartment on the lower right. Immediately above is the mixer stage and the top screen passes across the base of the 6CH6 driver stage effectively screening the grid circuit from the anode components. On the top left is the pa and tr switch. The rear apron contains (from left to right) aerial socket, avc mute socket, mute gain, services socket, pa bias potentiometer, relay socket and key jack.**
Photo by R. E. Foster

Right: **This shows the construction of the plug-in coils. They are made from Perspex tubing of 2in od and 3½in long, with a wall thickness of ⅛in. The base of the coil former is also made from Perspex, 3½in long, ¾in wide and ¼in thick. Four plugs on the coil enable extra capacitance to be placed across the loading capacitor and series capacitance inserted to the tuning capacitor in order to attain ease of resonance on the hf bands.**

TABLE 6.15

Coil Table

Coil Number	Function	Inductance	Coil Former	Dia.	No. of Turns	swg
L1	vfo 1·8/7·0 MHz	4·5μH	Cambion L55 series	0·375 in	21	26 close wound
L2	vfo 3·5/14/21MHz	6·4μH	Cambion L55 series	0·375in	30	26 close wound
L3	vfo 28MHz	6·6μH	Cambion L55 series	0·375in	30	25 close wound
L4	mixer 1·8MHz	105μH	Aladdin type former	0·3in	$\frac{15}{16}$in	40 enam. close wound
L5	mixer 3·5MHz	37μH	Aladdin type former	0·3in	$\frac{9}{16}$in	38 enam. close wound
L6	mixer 7·0MHz	14·4μH	Aladdin type former	0·3in	$\frac{9}{16}$in	34 enam. close wound
L7	mixer 14·0MHz	4·4μH	Aladdin type former	0·3in	28	26 close wound
L8	mixer 21·0MHz	1·9μH	Aladdin type former	0·3in	13	26 close wound
L9	mixer 28·0MHz	1·42μH	Aladdin type former	0·3in	6	26 close wound
L10	Driver 1·8MHz	130μH	Aladdin type former	0·3in	1in	40 enam. close wound
L11	Driver 3·5MHz	36μH	Aladdin type former	0·3in	$\frac{7}{16}$in	38 enam. close wound
L12	Driver 7·0MHz	9·3μH	Aladdin type former	0·3in	$\frac{7}{16}$in	34 enam. close wound
L13	Driver 14·0MHz	3·0μH	Aladdin type former	0·3in	24	26 close wound
L14	Driver 21·0MHz	1·65μH	Aladdin type former	0·3in	7	26 close wound
L16	PA 1·8MHz	48μH	Perspex tube × 3½in	2in	40	16 enam. close wound
L17	PA 3·5MHz	26μH	Perspex tube × 3½in	2in	30	14 enam. close wound
L18	PA 7·0MHz	10·4μH	Perspex tube × 3½in	2in	17	14 spaced 2in
L19	PA 14·0MHz	3·45μH	Perspex tube × 3½in	2in	9	14 spaced 2in
L20	PA 21·0MHz	1·37μH	Perspex tube × 3½in	2in	5	14 spaced 2in
L21	PA 28·0MHz	1·07μH	Perspex tube × 3½in	2in	4	14 spaced 2in

The rf transformer in the cathode circuit of TR switch EF80 consists of a slab of ferrite 2in long, ½in wide and ⅛in thick. Fill the slab with bifilar wound 26 swg enamelled wire. Form a suitable length of wire into a single pair. Wind on from one end so that the pair lie side by side, fill the slab completely and then varnish with polyurethane. When dry secure the whole to a suitable tagboard and earth one end of each wire. Other ends of wire are connected as shown, one to the cathode circuit and the other to the receiver aerial input terminal.

ferrite as used in small domestic transistor receivers. Details of the coils appear in **Table 6.15**.

Since the designer was not certain how the break-in facilities would work in practice a net/send/receive switch was fitted and provision for an aerial changeover relay. In fact these were found unnecessary and constructors are advised to arrange for netting simply by switching on all stages except the pa.

Only the pa is metered by means of a two pole four position switch as follows:

1	Grid current	0–5mA
2	Anode current	0–50mA
3	Anode voltage	0–500V
4	RF	

The rf reading is simply an indication of the feed to the aerial and consists of the rectification of a small amount picked up by means of a capacitor consisting of two insulated wires twisted together a few turns.

Complete constructional details are not given in this section since it is felt that few clubs will wish to take the trouble to build a pair of transmitters for NFD. In the event that some may do so the layout of components can be easily seen from the photographs. A cord drive was used for the vfo in order to keep costs down and the dial was home-made for the same reason. The size of chassis and front panel will depend largely upon what is to hand, but in the instance of the Wirral transmitter, the cabinet, panel and chassis were purpose made by the designer and his colleagues. No attempt was made to produce a transmitter of small physical dimensions simply for the sake of miniaturization. Rather was the size and layout suitable for available components.

Coils

Coil data for the transmitter is shown in the accompanying table. Polyurethane varnish is used to keep windings in place and here mention should be made of the method that was used to wind coils on Aladdin type formers.

The former is secured in the chuck of a wheel-brace by means of a length of 0BA studding and the wheel-brace clamped horizontally in a vice. One end of the coil is secured to one of the spill holes at the base of the former and the coil wound on neatly by rotating the former. The free end of the wire is then tied to the paxolin spill holder at the top of the former and the coil then coated with clear polyurethane varnish. Allow 24 hours for the varnish to dry and harden and then peel off excess turns and secure ends of the wire to the 18swg spills. The same method is used when winding two coils on one former. If different gauges of wire are used then join the wires temporarily at the centre of the former, varnish and make up as previously.

Power Supplies

Power supplies for the transmitter are quite modest and **Fig 6.117** shows the required information. Needless to say they are constructed on a separate chassis and are connected to the transmitter by a five core cable.

SINGLE SIDEBAND TRANSMISSION

Single sideband suppressed carrier telephony transmission (commonly called ssb) is a specialized form of amplitude modulated telephony and a brief examination of basic a.m. theory is therefore an essential preliminary to any description of the more advanced system.

A.M. Carrier and Sideband Relationships

The carrier of an a.m. transmission does not vary in amplitude; it is at all times of constant strength. The modulation introduced at the transmitter heterodynes the carrier and produces sum and difference frequencies. These are symmetrically disposed either side of the original carrier frequency and constitute two bands of side frequencies—those below the carrier form the lower sideband and those above the carrier form the upper sideband. The carrier by itself does not convey any intelligence; the intelligence is conveyed solely by the sideband frequencies.

Consider what happens when a carrier of 1000kHz is modulated 100 per cent by a pure audio tone of 2kHz. The energy propagated from the aerial would be three entirely separate and individual rf outputs on 998, 1000 and 1002 kHz (**Fig 6.118**). These three channels of rf energy would travel quite separately through the ionosphere and would eventually arrive at the receiving aerial and would then be accepted by the receiver, heterodyned by the local oscillator and converted to the final intermediate frequency of the receiver. If the i.f. passband was centred on 460kHz the i.f. amplifier would present three separate frequencies—462, 460 and 458kHz—to the detector. The combined effect of these three frequencies at the detector would in turn produce the modulation envelope. The way in which the modulation envelope is produced can be shown very simply by means of vector diagrams.

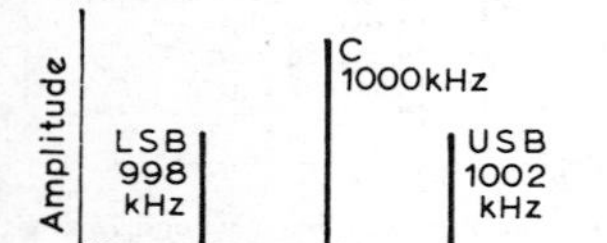

Fig 6.118. Amplitude frequency relationship of a.m. signal modulated with single tone of 2kHz.

The carrier vector is actually rotating at 460,000Hz with one sideband vector rotating at 458,000Hz and the other sideband vector rotating at 462,000Hz. Relative to the carrier, one sideband vector is rotating 2000Hz slower (lagging) and the other sideband vector 2000Hz faster (leading). If then the carrier is assumed to be stationary and is drawn as a vertical line whose length denotes the carrier voltage—let this be one unit in length—each sideband vector would be 0·5 unit in length (the power in one sideband is equal to 0·25 of the power in the carrier and as power is equal to E^2/R, the voltage is equal to 0·5 that of the carrier) and would be shown rotating round the carrier vector in opposite directions at the frequency of the modulation (**Fig 6.119(a)**). At some moment of time, the two sidebands will be in phase with each other and in phase with the carrier and the resultant vector length will be two units—the modulation crest. At 180° of rotation later they will again be in phase with each other but opposite in phase to the carrier and the resultant vector length will be zero—the modulation trough. (**Fig 6.119(c)**). At all other degrees of angular rotation the resultant voltage will be some in-between value. The modulation envelope recovered at the detector is shown in **Fig 6.120** and to the same scale, the audio output in **Fig 6.121.**

If, therefore, the resultant rf voltage output from a 100 per cent sine wave modulated transmitter is examined by means of an oscilloscope, the display will show the modulation envelope of the classical textbooks. The important point to understand is that the oscilloscope is not showing the true relationship of the carrier and the sidebands at all.

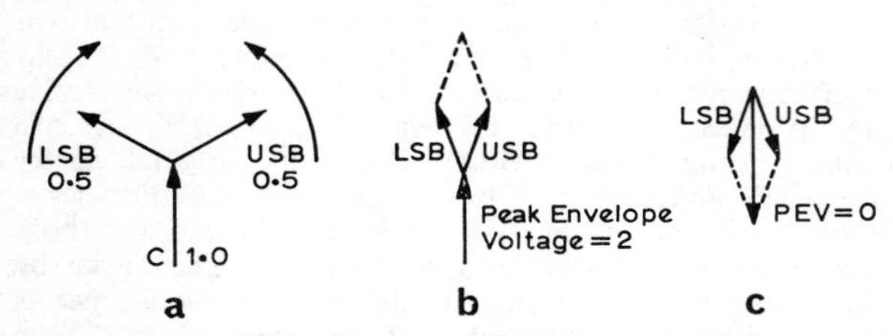

Fig 6.119. Vector presentation showing how the two sideband voltages combine with the carrier voltage to form the modulation envelope.

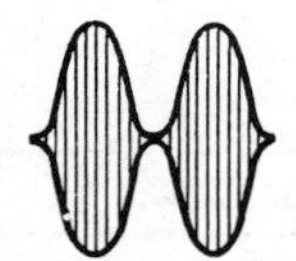

Fig 6.120. Modulation envelope recovered at the detector.

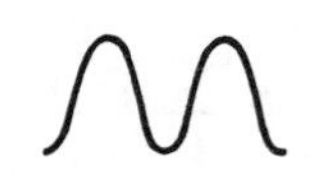

Fig 6.121. Resultant audio output voltage.

It cannot separate them and show each individual component. The display seen is the resultant effect in terms of voltage—in fact, the oscilloscope is showing a continuous panorama of voltage vector diagrams.

When the single tone modulating signal is replaced by the output of a speech amplifier, the picture grows more complex. It may, however, still be analysed into the original carrier, unchanged in amplitude and frequency, about which are displaced symmetrically the upper and lower sidebands. The sidebands will, of course, undergo continuous change in amplitude and frequency as the voice changes in inflexion and intensity, but they may never exceed half the amplitude of the carrier if over-modulation is to be avoided. Their maximum excursion in frequency is limited by the highest frequency component of the voice; for practical communications purposes this may be taken as 3kHz.

Fig 6.122 illustrates in diagrammatic form the relationship of the carrier and its sidebands under speech modulation; it is this representation, rather than that of Fig 6.119, which forms the foundation stone on which to build a solid understanding of ssb. There is, however, one noteworthy point which is more immediately evident from Fig 6.119 than from the alternative presentation. At the modulation crest the carrier and sideband voltages are in phase and add together; the voltages at the anode and screen of the modulated stage will also rise to twice the applied dc potentials, so the components in these circuits must be appropriately rated to avoid breakdown. As the transmitter peak envelope voltage (pev) at the crest of modulation is double that of the unmodulated carrier, and as output power is proportional to voltage squared, the peak envelope power (p.e.p.) will be increased four times. Assuming that the modulated stage is being operated under the usual amateur conditions of 150W dc input with an overall efficiency of 66 per cent, the unmodulated carrier output will be 100W and the peak envelope power output 400W.

Single Sideband Transmission

If the transmitter of Fig 6.118 were radiating a lower sideband ssb signal, the carrier on 1000kHz and the upper sideband on 1002kHz would be suppressed at the transmitter, and the only output to the aerial would be continuous rf energy on 998kHz. This would be converted by the receiver to an i.f. of 462kHz and fed to the detector. The carrier of 460kHz would be generated by a local oscillator (bfo) and also fed to the detector. A vector diagram would show the inserted carrier from the bfo as a vertical line with the sideband vector rotating around it at the modulating frequency; the modulation envelope would be recovered at the detector and the resultant audio output would be 2kHz. It will be seen that the original 2kHz tone input has been recovered, yet the ssb transmitter has radiated one continuous rf output on one frequency only, 998kHz.

If the tone modulation at the transmitter had been changed from 2 to 3kHz the sideband radiated would change from

998 to 997kHz. This would be converted by the receiver to an i.f. of 463kHz, combined in the detector with the local carrier on 460kHz and the resultant audio output would be 3kHz. Should the transmitter be modulated with both the 2 and 3kHz tone simultaneously two rf outputs would be radiated from the aerial on 997 and 998kHz. These would be converted by the receiver to two i.f. outputs of 463 and 462kHz. The resultant output recovered by the detector would be 3 and 2kHz—the original modulating frequencies.

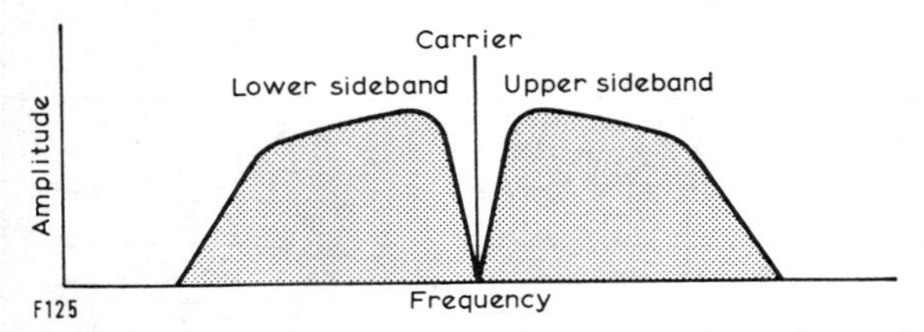

Fig 6.122. Amplitude/frequency relationships of carrier and sidebands with 100 per cent speech modulation.

When the tone modulation is replaced by the output of a speech amplifier the transmitted sideband becomes essentially a band of frequencies undergoing continuous changes in amplitude and frequency as the voice changes in inflection and intensity. Provided that the receiver i.f. passband is sufficiently wide, all frequencies within the received sideband will be amplified equally and fed into the detector. The locally inserted carrier merely serves as a datum line against which the sideband is demodulated and the resultant audio output is the original speech modulation.

COMMUNICATIONS EFFICIENCY

In an a.m. transmitter, the amplitude of each sideband is limited to half that of the carrier, so its maximum power will be one-quarter of the unmodulated output power. In the example already quoted of the 150W input amateur transmitter with an overall efficiency of 66 per cent, the unmodulated carrier output would be 100W, and the maximum power in one sideband would be 25W; both sidebands together would produce 50W. It is not, however, possible to specify the effectiveness of the transmission until something is known about the equipment on which it is being received. The sidebands necessarily occupy a band of frequencies 3kHz on either side of the carrier, so that if the receiver has a bandwidth of 6kHz they both add in phase and contribute to the total *talk power*. If a more selective receiver with a 3kHz bandwidth is used, only one sideband contributes to its output. Under the crowded conditions which obtain in the amateur bands, good selectivity is almost invariably a necessity, so that only 25W of the signal actually conveys intelligence to the listener. This is not a particularly profitable return for the generation of 100W of carrier power, not to mention the cost, weight and bulk of a high level modulator and associated power supply.

As the carrier remains constant in frequency and amplitude it conveys no intelligence from transmitter to receiver, and serves merely as a datum line against which the sidebands are demodulated. All the intelligence is conveyed in the sidebands, so that the carrier may be omitted and the signal demodulated perfectly clearly if the reference function is transferred to a local oscillator at the receiving end. The attractiveness of substituting for the 100W carrier the signal from a small oscillator valve in the receiver is obvious. The reduction of heterodyne interference within the amateur bands, which results from carrier suppression, may alone appear to be adequate justification for making the change, but there are other advantages.

The power amplifier which produces a 100W carrier must be capable of a peak output of 400W. With modifications the full capacity of the amplifier may be utilized to radiate intelligence conveying sideband energy. This does not necessarily mean that the peak envelope power output of the hypothetical amplifier will be 400W, because there are important differences between the operation conditions of anode-modulated class C amplifiers and the class AB or B linear amplifiers customarily employed in ssb output stages. The question of linear amplification will be dealt with in detail later in this chapter, but as a rough guide it may be said that a 100W a.m. stage should be able to produce about 200W p.e.p. output in ssb service. The whole of these 200W will contribute to the talk power at the receiving end, which compares very favourably with the 50W which produce speech output from a conventional a.m. transmission in a receiver of 6kHz bandwidth, and even more favourably if the receiver has a narrower bandwidth.

There is no difficulty at all in suppressing the carrier; a miniature double triode, or a pair of crystal diodes, together with a few resistors and capacitors are all that is necessary, leaving only the problem of how to get rid of one of the sidebands. It has already been said, however, that a receiver of 6kHz bandwidth is capable of combining the two sidebands so that both may contribute to the audio output. Furthermore, it matters little at the transmitting end whether the linear amplifier is producing single sideband or double sideband output, so long as the peak power remains substantially the same in both cases.

It is, however, desirable to eliminate one sideband in order to simplify the demodulation process in the receiver. Even if the local oscillator at the receiver can be made stable enough to simulate the precise frequency of the suppressed carrier, that alone is not sufficient. It must also be identical in *phase*, otherwise the two sidebands will not combine to produce anything approaching the original modulating waveform. The best way to utilize fully a double sideband suppressed carrier (dsb) signal is to tune it in on a wideband receiver to which is coupled an adaptor capable of synchronizing the local oscillator to the required frequency and phase by means of information derived from the sidebands themselves. The main disadvantage of this system of detection lies in its susceptibility to lose synchronization because of interference from random noise or from strong adjacent-channel signals.

If, however, one of the sidebands is eliminated as well as the carrier, reception becomes fairly easy. The local oscillator may wander quite a few hertz away from the correct frequency before intelligibility suffers appreciably and does not have to be locked in phase. The best place to eliminate the unwanted sideband is clearly at the transmitter, because the full capacity of the pa stage may then be used to amplify the wanted sideband. In addition the transmission will occupy a bandwidth of only 3kHz. Alternatively, the conversion from dsb to ssb may be effected at the receiving end. This is relatively simple because increasing interference has already led many amateurs to increase the selectivity of their receivers to a point at which it is possible to reject one-half of the dsb transmission and deal with the remainder

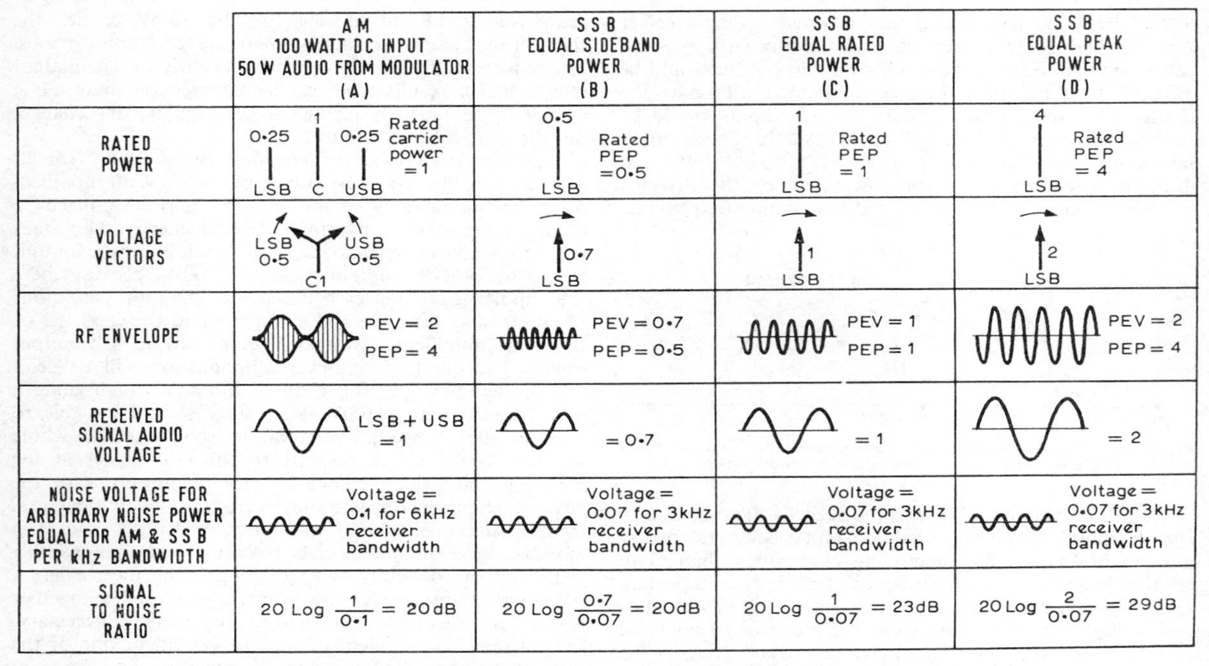

Fig 6.123. Power relationships for a.m. and ssb transmission. Single tone sine wave modulation.

as if it were a true ssb signal. When received in this way dsb is considerably more difficult to tune than ssb and is only half as efficient.

Half of the power radiated by the dsb transmitter is not used for reception and merely serves to increase the general level of interference. In this respect it is no worse than conventional a.m., although it has the advantage of having no resting carrier to cause a constant heterodyne whistle. When the fact that a dsb transmitter is simpler to construct than an a.m. rig of comparable power output is considered, the reasons for its former popularity become evident. It has now, however, been superseded almost completely by ssb.

Talk Power

Perhaps the clearest method—least open to misinterpretation—is to show the relative efficiency of the two systems in diagrammatical form where the powers and voltages concerned are to the same relative values. This method has been adopted in **Fig 6.123.** The basis of comparison given in column A is an a.m. transmission of 100W dc power *input* rating, modulated by 50W of audio. At the crest of the modulation cycle the peak envelope power (p.e.p.) is four times the carrier power—400W. The term peak envelope power is defined as the rf power developed at the crest of the modulation envelope.

The carrier of one power unit in value requires a half-power unit of audio for 100 per cent modulation (this is the maximum power than can be used; any greater audio input would produce overmodulation and distortion) and this produces two sidebands with 0·25 unit of power in each. As voltage is proportional to the square root of the power, the carrier voltage is 1 and the voltage of each sideband is 0·5. The rf envelope developed by the voltage vectors is shown, and for every 100 per cent modulation the peak envelope voltage (pev) is the sum of the carrier and the two sideband voltages, and this equals two units. This results in a p.e.p. of four units of power.

The rf signal is demodulated in the receiver and the diode detector develops an audio output voltage that is equivalent to the sum of the upper and lower sideband voltages. The noise power per kilohertz is an arbitrary value equal for a.m. and ssb. For a 20dB signal-to-noise ratio, the noise voltage would be 0·1 units for the 6kHz receiver bandwidth and the signal-to-noise ratio is then 20 log. the ratio audio voltage: noise voltage.

Column B shows the power and voltage relationships for an ssb transmission of equal *sideband* power to the a.m. transmission. The audio output from the ssb transmission (recovered by heterodyning the received signal with a locally inserted carrier) is 0·707 units in value. This represents a loss in detector output voltage of 3dB due to the ssb power being in one sideband. However, the reduction in receiver bandwidth gives a 3dB advantage so therefore the signal-to-noise ratio is the same for the two modes of transmission.

Column C shows the relationship for an ssb transmission of equal *rated* power (100W a.m. transmitter and a 100W p.e.p. input ssb transmission). It is seen that the audio voltage developed at the output of the diode detector is equal to the audio voltage of the a.m. transmission. The reduction of receiver bandwidth gives a 3dB advantage and the signal-to-noise ratio is 23dB.

Column D shows the relationship for an ssb transmission of equal *p.e.p. input*. It is seen that the ssb transmission gives a gain in detector output voltage of 6dB. This, in addition

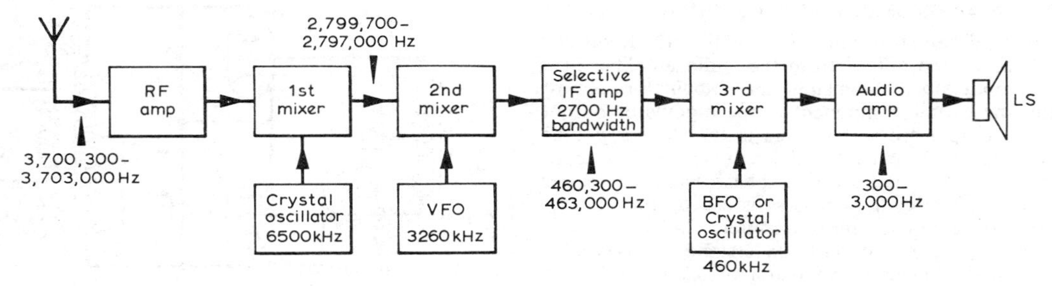

Fig 6.124. Frequency translation process in a typical double superhet receiver.

to the 3dB improvement in receiver noise output, gives a total signal-to-noise ratio of 29dB, a system gain for ssb of 9dB.

It will be noted that the carrier envelope voltage of the a.m. transmission serves no useful purpose other than to beat against the sidebands and demodulate the signal at the detector. Furthermore, that for both modes of transmission the audio voltage recovered at the detector is directly proportional to the total *sideband* voltage, and that the two transmitters, a.m. and ssb of equal *rated* power, will produce an equal receiver audio voltage.

SSB Advantages

There are no inherent difficulties in the construction of a ssb transmitter. In the filter type, the chassis, the layout and even the majority of the components are virtually identical to a selectable sideband receiver. The usual 75W modulator with its bulky driver and output transformers and separate 150W power supply is no longer required. The ssb transmitter can be built and used initially to gain operating experience with a low power output—say 50W peak. At a later date the power can be increased to the licensed maximum by building a suitable linear amplifier and ht supply and driving this with the existing transmitter.

It is quite true that ssb working demands a higher standard of frequency stability and more care in netting, but this should not be looked upon as a disadvantage peculiar to sideband working. First, as receiver selectivity is improved there will be a demand for better frequency stability from existing a.m. transmitters. Second, many of the methods developed by amateurs for ssb working are already being adopted to improve a.m. operation. These include exalted carrier detection, improved bandwidth control, better oscillator stability and press-to-talk or automatic voice control. Third, the power gain of 9dB with ssb operation (made up of 6dB gain at the transmitter and 3dB at the receiver) represents an equivalent power increase at the transmitter of eight times.

The advantages of ssb are, therefore:

(i) the rf spectrum required to transmit a given signal by means of ssb is exactly that of the original signal, thus maximum use can be made of the available rf spectrum.

(ii) since only essential signals are transmitted by ssb without a superfluous carrier or mirror image sideband, there is a considerable effective power gain.

(iii) most important of all, ssb systems are effected far less adversely by the transmission disturbances inherent in ionospheric transmissions than a.m., fm, or any of the double sideband systems.

The terms heterodyning and mixing have the same meaning in ssb transmission as in receiver practice and they are both frequency translation processes. All superhet receivers translate the frequency down from the required amateur band to the i.f. channel (or channels) and then to the final audio channel. This is shown clearly by the block diagram of a typical double superhet receiver given in **Fig 6.124.** If the third mixer is renamed product detector, the circuit arrangement is well-known and easily followed and understood. Suppose then that the translation process is reversed and the block diagram redrawn as in **Fig 6.125.** The mixing processes are exactly the same as in Fig 6.124 but the block diagram of Fig 6.125 is that of a ssb transmitter. In the receiver the translation process is *down* from the rf signal frequency to the audio frequency, and in the transmitter *up* from the audio frequency to the rf signal frequency. It is possible (in familiar terms) to follow mentally the progress of the signal intelligence from the microphone to the transmitting aerial, and the question, "How can you transmit voice frequencies without a carrier?" does not arise.

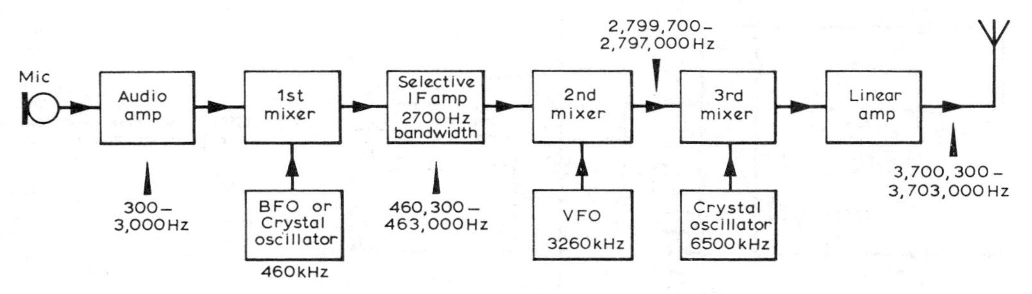

Fig 6.125. Frequency translation process in a typical ssb transmitter.

SUPPRESSING THE CARRIER

In a single sideband transmitter the initial carrier frequency is modulated by the output from the audio amplifier in a balanced modulator; balancing the modulator almost eliminates the carrier component in the modulator output circuit. The output from this stage is therefore a double sideband suppressed carrier signal. At the modulation crest, the two sideband voltages are in phase with the carrier voltage, ie, the modulation process is exactly the same as that of the conventional a.m. transmitter.

It is most important in order to understand clearly what is happening in the balanced modulator of a ssb transmitter to remember that when a carrier is "amplitude modulated" it does not in fact vary in amplitude at all. The output from the modulated stage, whether the balanced mixer in a ssb transmitter or the pa in an a.m. transmitter, contains three components. These are the upper sideband, the carrier, and the lower sideband. The carrier wave Fc is heterodyned by the audio wave Fm and this produces sum and difference frequencies $Fc + Fm$ and $Fc - Fm$. The modulation process is thus seen to be a frequency translation process and the modulator is therefore a converter or mixer. When an audio frequency modulates a radio frequency, the process is generally called *modulation*, but when a radio frequency modulates another radio frequency it is called *heterodyning*. The processes are, however, identical, and the general terms modulator, converter or mixer mean the same thing.

The sidebands, $Fc + Fm$ and $Fc - Fm$, contain the voice intelligence, so the carrier frequency can be attenuated or balanced out in the modulator output without affecting the sidebands in any way.

This can be done at high level by arranging a push-pull pa stage as a balanced modulator or at low level as in a typical ssb transmitter. In both cases the output is exactly the same—a double sideband suppressed carrier signal.

Balanced Modulators

Balanced modulators come in many and various forms employing a wide variety of devices and circuit techniques. In their simplest forms they can be adaptations of the bridge circuit and provide a high performance at low cost.

Valves have been used in a number of circuit configurations but they are mainly arrangements of the basic circuit shown in **Fig 6.126.** An rf voltage applied to the grids of V1 and V2 will appear in amplified form at the anodes. This will cause rf currents to flow in opposite directions through the two halves of L2. If the circuit is so adjusted that the signal induced in one-half of L2 is balanced exactly by that induced

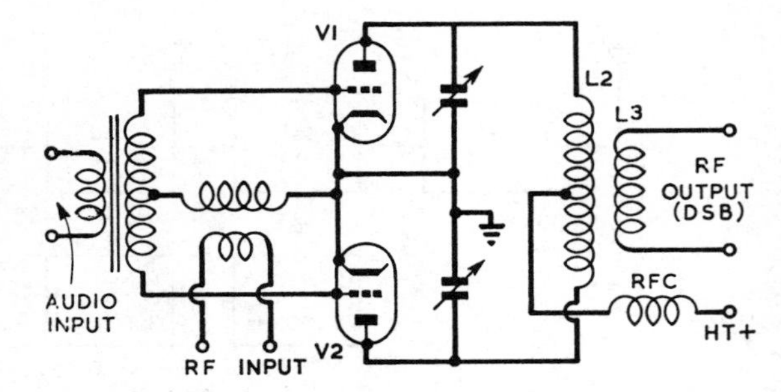

Fig 6.126. Basic circuit for twin-triode balanced modulator.

in the other half, no energy whatsoever will be passed on to L3.

Balance is governed primarily by the amplification factors of V1 and V2; ideally these valves should be identical, but this is difficult to achieve in practice and the best that can be done is to choose a pair whose characteristics match as closely as possible. The two units in a double-triode envelope often match fairly well, and if half-a-dozen valves are available at least one should be good enough for balanced modulator service. Almost any valve type can be made to give good results, but for ease of adjustment and long-term stability of balance, a low-mu triode such as the 12AU7, 12BH7 or GEC A2900 is recommended. The characteristics of high-mu types such as the 12AT7 are more prone to change as the valve ages. A valve which has seen enough service to permit its characteristics to stabilize is more likely to prove satisfactory than a brand new one; if a new valve must be used it should be aged artificially by operating it at normal heater voltage and a few milliamperes of anode current for 50-100 hours before it is tested for use in a balanced modulator. The importance of selecting the right valve cannot be overstressed. While it is possible to balance out large inequalities by external means, the circuit will drift out of balance quite rapidly, and will require continual readjustment.

Although a well-matched pair of valves will give about 25dB carrier suppression on their own, this is rarely enough, and additional balancing is required. If the remainder of the circuit permits, the simplest way is to vary the operating point of one or other of the valves by altering the standing bias. An example of this method is shown in **Fig 6.127.**

The basic arrangement may of course be varied widely to suit particular circumstances but there are certain pitfalls for the unwary. The anode circuit is balanced about earth

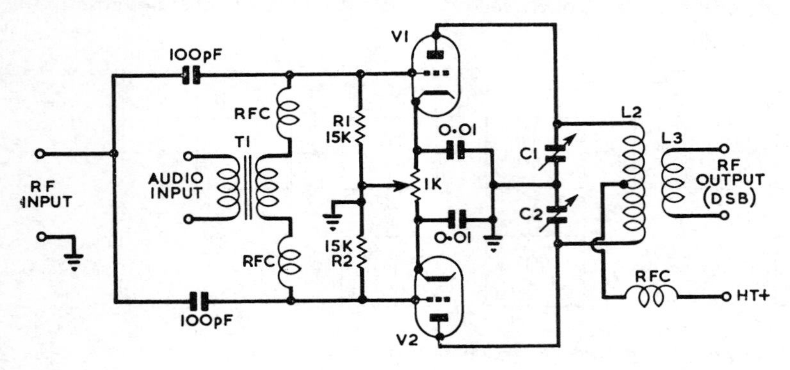

Fig 6.127. Balancing potentiometer adjustment of grid bias.

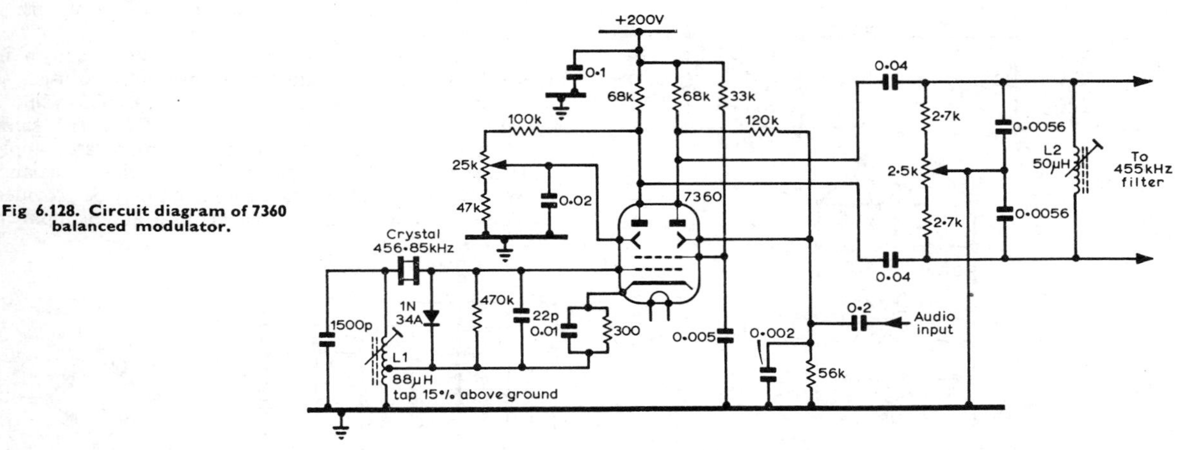

Fig 6.128. Circuit diagram of 7360 balanced modulator.

and this may be achieved by using a split stator tuning capacitor with earthed rotor or by decoupling to earth the coil centre tap, but not both. In the former case an rf choke as shown in Fig 6.127 will ensure that the coil tap floats in the required manner.

An extremely good balanced modulator circuit can be designed using the 7360 beam deflection valve. This is a special valve using a flat sheet electron beam and two shaped anodes so that the current to one or other anode may be controlled by deflecting the beam by means of crt type deflection plates. The beam current is controlled in the conventional fashion by a grid. A 7360 balanced modulator is shown in **Fig 6.128.**

This circuit is of particular interest because it uses the triode formed by the cathode, $grid_1$ and $grid_2$ as a crystal controlled carrier oscillator. The audio is applied to one of the deflection plates. A balancing potentiometer is connected to the other plate to provide a means of positioning the beam and hence achieving maximum carrier suppression. A further adjustment for carrier suppression is provided by a balancing potentiometer across the output tuned circuit. When properly adjusted this circuit is capable of a carrier suppression of 60dB. Like the crt the deflection tube is susceptable to external magnetic fields which offset the beam from the carefully balanced condition and can reduce the carrier suppression considerably. Nearby mains transformers can likewise introduce ac hum and proper positions of the valve and screening with mu-metal in some cases are necessary.

There are many possible variations of the diode modulator. They all stem from the four diode ring modulator which has given the telephone companies such excellent service in landline ssb work. The basic modulator shown in **Fig 6.129** is rarely used by amateurs but a number of derivatives have become popular.

The shunt connected circuit shown in **Fig 6.130** is capable of excellent results, and is suitable for providing input to filters operating at intermediate or higher frequencies. The key to successful operation lies in the selection of a pair of diodes whose characteristics match closely and in the correct transformation of impedances at both audio and radio frequencies.

Suitable diodes are germanium point contact types and matched pairs are available commercially eg Mullard type AA119. When installing them, an efficient heat shunt should be used to prevent the heat of the soldering iron causing the characteristics to change. In addition, the diodes should be located well clear of components which might generate enough heat to affect their performance. Many a mediocre balanced modulator owes its indifferent operation to the fact that it has been placed too close to a valve or a high wattage voltage-dropping resistor.

The circuit operates as a chopper with the diodes and

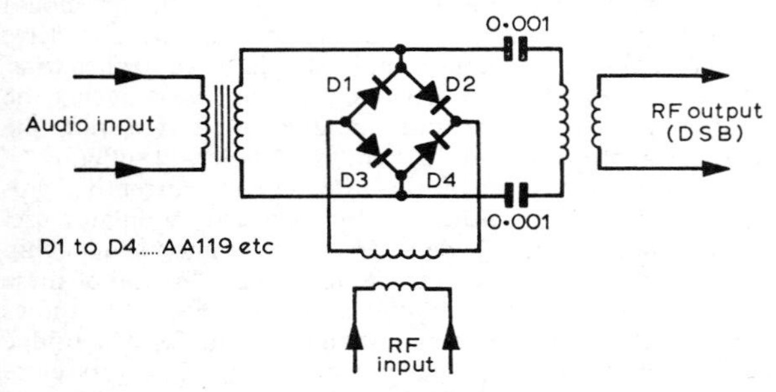

Fig 6.129. Diode bridge modulator.

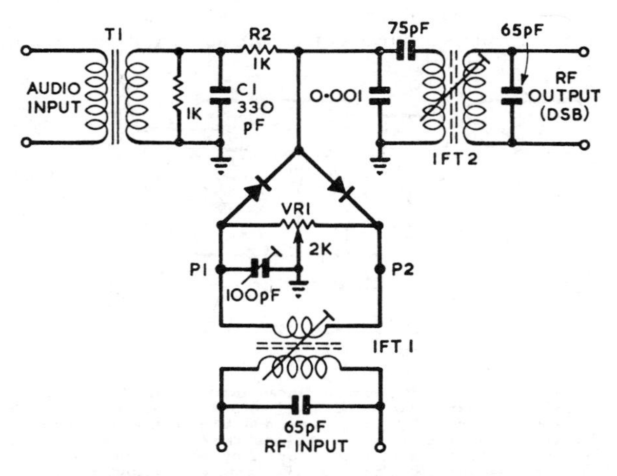

Fig 6.130. Shunt-type diode balanced modulator.

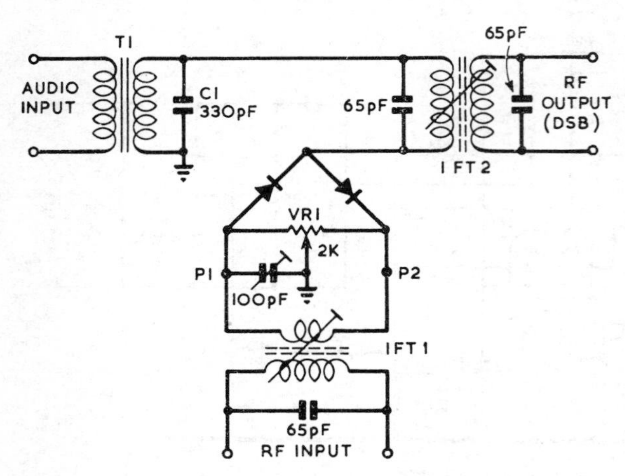

Fig 6.131. Series-type balanced modulator.

VR1 behaving as a fast switch which alternately connects and disconnects the right hand end of R2 to earth at the rf carrier frequency. The value of R2 requires to be a compromise between a low value which permits fast rise of the voltage at the junction of the two diodes at the instant of switch opening and a high value which decreases the loading on the previous circuit and increases the switching efficiency. The input to the i.f. transformer (IFT2) employs a capacitive impedance transformation arrangement so that the tuned circuit is not severely loaded. It also achieves a useful voltage step-up.

Carrier input must be at low impedance; a standard i.f. transformer with the original secondary removed and replaced with a scramble winding of 50 to 100 turns tightly coupled against the primary will be found satisfactory. If the potentiometer VR1 does not alone give sufficient carrier suppression, it may be supplemented by a small variable capacitor of about 100pF maximum value connected between earth and one or other of the points marked P1 and P2 in Fig 6.130. If the balancing capacitor is needed, its correct placement and adjustment may quickly be determined by trial and error.

For minimum distortion, the carrier voltage should be about ten times as great as the peak audio voltage. Several

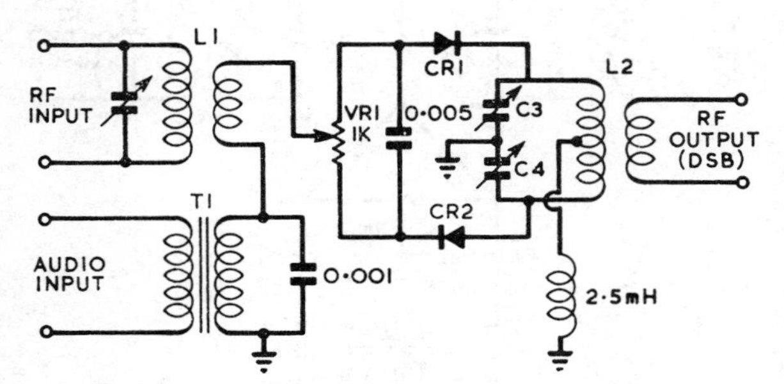

Fig 6.132. Modified ring type balanced modulator. T1 is an audio transformer with low impedance secondary winding; a ratio between 1 : 8 and 1 : 12 is recommended. L1 is a three or four turn link wound over the cold end of the input tank inductor and adjusted to give 3 or 4V rms across its output. Any matched pair of diodes may be used for CR1 and CR2. The input and output rf circuits should resonate at the operating frequency.

volts of rf and a fraction of a volt of audio are the usual operating levels.

The series connected balanced modulator is shown in **Fig 6.131.** Almost any audio transformer may be used, as long as it is bypassed at carrier frequency by C1, while a standard i.f. transformer will serve for IFT2. The same considerations as in the shunt-connected modulator apply to the carrier input transformer and to capacitative balancing. With either modulator the audio input may be provided by a cathode follower, emitter follower or ic audio amplifier.

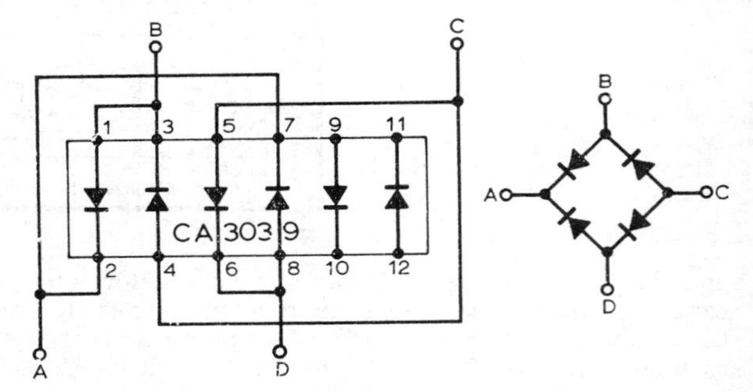

Fig 6.133. The RCA CA3039 connected as a modulator bridge.

Although the series and shunt balanced modulators are suitable for use at i.f. and rf to feed filter units, a pair of them may not be connected conveniently in parallel in the manner required for the generation of ssb output by the phasing method. The modified ring circuit designed by D. E. Norgaard, W2KUJ, and reproduced in **Fig 6.132** overcomes this difficulty. Almost all the commercial phasing-type transmitters made in the USA which employ diode balanced modulators make use of this arrangement. An ordinary loudspeaker transformer with a 15Ω secondary will be suitable for T1, while L1 is a three or four turn link wound over the cold end of the carrier oscillator coil, and adjusted to give about 4V of rf across VR1. The push-pull tank circuit resonates at the desired output frequency. The 2·5-mH choke must not be omitted.

In describing balanced modulators, the convention of rating a circuit in terms of "so many decibels of carrier suppression" has been followed. This may be misleading to the uninitiated, because in fact the circuits work the other way round. To be strictly correct, the rating of a balanced modulator should be expressed as the number of decibels by which the output at maximum audio drive exceeds the minimum resting output in the absence of modulation. The point of making this distinction is to show that for the highest degree of carrier suppression any balanced modulator should be driven to its maximum undistorted output. The drive to the final stages of a transmitter should be controlled *after* the ssb signal has been generated and not by reducing the audio input to the balanced modulator; if it is not, the carrier suppression of the system as a whole will suffer.

There are a number of integrated circuits currently available which are very attractive for balanced modulator use. The simplest are the diode arrays on a common silicon chip which are offered by several manufacturers. Typical of these is the CA3039 which contains six diodes. **Fig 6.133** shows four of the diodes externally connected to form a bridge which can be incorporated in a modulator. The very close matching of the diodes and the fact that they share a common

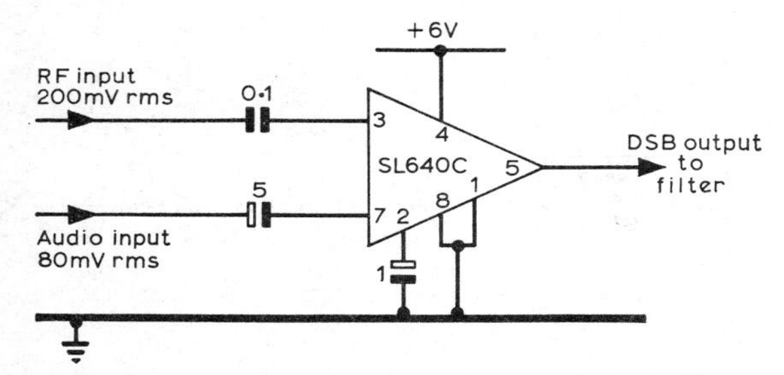

Fig 6.134. A balanced modulator using the Plessey SL640C.

temperature environment makes this device very suitable for inclusion in circuits where good balance is required.

More complex integrated circuits are also available which have been designed to act as modulators, mixers, product detectors etc. They are in fact analogue multipliers which means that they provide very high attenuation of both the carrier and modulating signal. Anyone interested in mathematics can confirm that if $E_1 \sin \omega_1 t$ and $E_2 \sin \omega_2 t$ are multiplied together that the only sinusoidal terms contained in the result are

$$\frac{E_1E_2}{2} \cos (\omega_1 + \omega_2)t \text{ and } \frac{E_1E_2}{2} \cos (\omega_1 - \omega_2)t$$

showing that only the sum and difference (ie upper and lower sideband) frequencies exist at the output of such a multiplier. In the balanced modulator application neither the carrier nor the modulating frequencies would exist at the output of a perfect multiplier. In practice of course this infinite attenuation is not achieved because of errors in the multiplication process and carrier breakthrough. Examples of this type of balanced modulator are the Plessey SL640C shown in the circuit in **Fig 6.134** and the Motorola MC1496G shown in **Fig 6.135**. The carrier suppression quoted for the MC1496G is −50dB in a typical case and −40dB for the SL640C. An advantage of these balanced modulators over other types is that they produce very low harmonic output levels and they provide useful gain. The voltage gain is of the order 2–3 in both cases.

Because of their general purpose nature the MC1496G and SL640C type of device are particularly useful in transceivers where they may for example perform the functions of balanced modulator on TRANSMIT and receiver mixer or product detector on RECEIVE.

SIDEBAND ATTENUATION

The double sideband signal generated by any of the balanced modulators described is of little value for communication purposes, so it has to be turned into ssb by attenuating one of the sidebands. No matter what system may be used, the unwanted sideband is not eliminated completely; it is merely attenuated to the extent at which its nuisance effect becomes negligible. A filter attenuation of 30–35dB has come to be regarded as the minimum acceptable standard. With care, suppression of 50dB or more is attainable but it is debatable whether there is any practical advantage in striving after greater perfection.

The unwanted sideband may be attenuated either by phasing or filtering. The two methods are totally different in conception, and will be discussed in detail.

Although the filter method provides the classic way to generate a single sideband signal and is now used almost exclusively, the phasing method will be described first because of its popularity in the early days of amateur ssb.

The Phasing System

The phasing system of single sideband generation can be simply explained with the aid of vector diagrams. **Fig 6.136(a)** shows two carriers A and B, of the same frequency and phase, one of which is modulated in a balanced modulator by an audio tone to produce contra-rotating sidebands A1 and A2, and the other modulated by a 90° phase shifted version of the same audio tone. This produces sidebands B1 and B2 which have a 90° phase relationship with their A counterparts. The vector presentation is similar to Fig 6.119 and the carrier vector is shown dotted since the carrier is

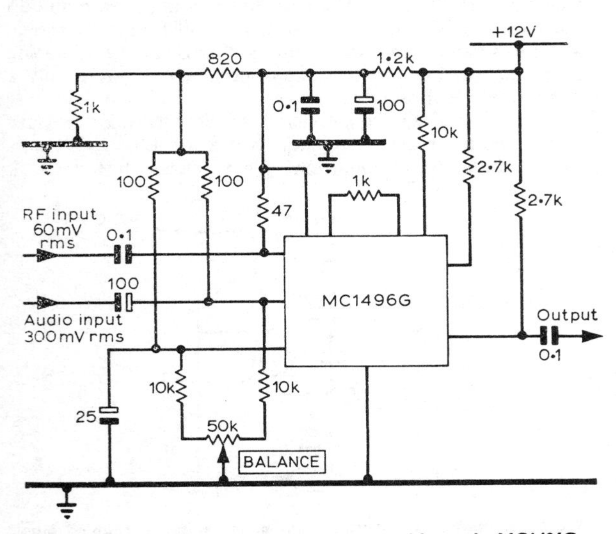

Fig 6.135. A balanced modulator using the Motorola MC1496G.

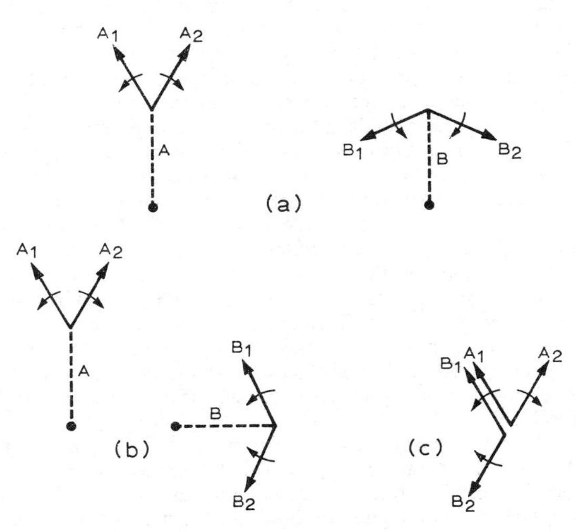

Fig 6.136. Phasing system vectors.

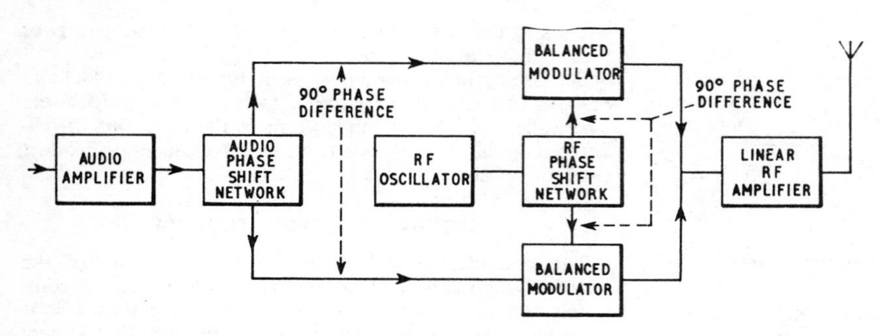

Fig 6.137. Phasing type transmitter.

absent from the output of the balanced modulators. **Fig 6.136(b)** shows the vector relationship if the carrier B is shifted in phase by 90° and **Fig 6.136(c)** shows the addition of these two signals. It is evident that sidebands A2 and B2 are in antiphase and therefore cancel whereas A1 and B1 are in phase and are additive. The result is that a single sideband is produced by this process.

A block diagram of a phasing type transmitter is shown in **Fig 6.137**, from which will be seen that the output of an rf oscillator is fed into a network in which it is split into two separate components, equal in amplitude but differing in phase by 90°. Similarly, the output of an audio amplifier is split into two components of equal amplitude and 90° phase difference. One rf and one af component are combined in each of two balanced modulators. The double sideband suppressed carrier energy from the two balanced modulators is fed into a common tank circuit. The relative phases of the sidebands produced by the two balanced modulators are such that one sideband is balanced out, while the other is reinforced. The resultant in the common tank circuit is an ssb signal. The main advantages of a phasing exciter are that sideband suppression may be accomplished at the operating frequency, and that selection of the upper or lower sideband may be made by reversing the phase of the audio input to one of the balanced modulators. These facilities are denied to the user of the filter system.

If it were possible to arrange for absolute precision of phase shift in the rf and af networks, and absolute equality in the amplitude of the outputs, the attenuation of the unwanted sideband would be infinite. In practice, perfection is impossible to achieve, and some degradation of performance is inevitable. Assuming that there is no error in the amplitude adjustment, a phase error of 1° in either the af or the rf network will reduce the suppression to 40dB, while an error of 2° will produce 35dB and 3.5° will result in 30dB suppression. If, on the other hand, phase adjustment is exact, a difference of amplitude between the two audio channels will similarly reduce the suppression. A difference between the two voltages of 1 per cent would give 45dB attenuation, an error of 2 per cent would result in 40dB, and 4 per cent 35dB approximately. These figures are not given to discourage the intending constructor, but to stress the need for *high precision workmanship and adjustment* if a satisfactory phasing-type ssb transmitter is to be produced.

There are many ways of obtaining rf phase shifts, but the two most satisfactory are those which use a pair of loosely coupled tuned circuits, and the so called low-*Q* network which uses a combination of resistance, inductance and capacitance. The former is shown diagrammatically in **Fig 6.138** and is based on the property of coupled circuits which provides that if both are tuned to resonance the voltage induced in the secondary will lag behind that appearing across the driven primary by 90°. In practice it may prove difficult to tune both circuits to the precise point of resonance because of mutual interaction, but if this defect is encountered it may be easily overcome. The recommended method is to start by detuning the secondary completely, or to shunt it by a resistive load of about 200Ω so that it cannot react upon the primary. The primary is tuned to resonance, using a valve voltmeter or other rf indicator, and is then detuned to the high frequency side until the voltage drops to 70 per cent of that indicated at resonance. The secondary is next brought to resonance (with its shunt removed) and finally detuned to the low frequency side until the indicated voltage drops to 70 per cent. The voltage across the two circuits will then differ in phase by precisely 90°. Because of losses, it is unlikely that the voltage appearing across the secondary will be exactly the same amplitude as that across the primary; it is therefore inadvisable to use capacitive coupling from the hot ends of the windings. Link coupling, as shown in Fig 6.138, is preferable because it allows compensation to be made for inequalities in amplitude.

The early amateur phasing transmitters were designed for fundamental-frequency operation, driven directly from an existing vfo tuning the 80m band, and used a low-*Q* phase shift network. The low-*Q* circuit has the ability to maintain the required 90° phase shift over a small frequency range, and this made the network suitable for use at the operating frequency in single band exciters designed to cover only a portion of the chosen band.

Today, with the great increase in ssb activity, it is necessary to be able to change frequency over a range of 200kHz or more. Under these conditions the rf phase shift network

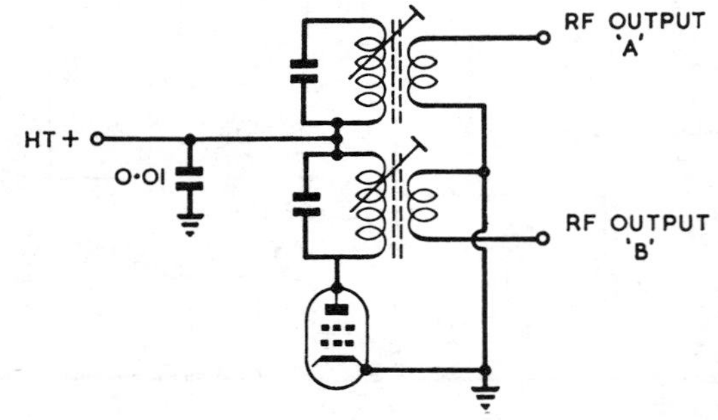

Fig 6.138. Method of obtaining 90° phase shift by coupled tuned circuits.

would be quite incapable of maintaining the required accuracy of phase shift and the available sideband suppression would deteriorate to a point at which the exciter was virtually radiating a double sideband signal. It is true that the plain RC network will provide a phase shift that is independent of frequency. Unfortunately experience over the years indicates that this simple method is not capable of providing the required sideband suppression. Additionally, a change of frequency into any type of rf phase shift network will cause a change in output amplitude ratio, and this in turn will unbalance the modulator and severely degrade the carrier suppression.

If amateurs are to enjoy the full benefits of ssb working and reduce adjacent channel splatter and heterodyne interference to a low level on the amateur bands, a sideband suppression of 30-35dB and a carrier suppression of 50dB should be considered the minimum acceptable standard.

Any operating method that is fundamentally incapable of maintaining this standard should not be used on the amateur bands. For this reason, the fundamental type of phasing unit is not recommended. For acceptable results, the rf phase shift must be operated at a fixed frequency outside the amateur bands. The ssb output from the balanced modulator is then heterodyned to the required bands by means of an external vfo. This method will be discussed in greater detail later in this chapter.

Audio Phase Shift Network

Although it is quite practicable to construct an audio phase shift network capable of covering from 50 to 5000Hz, the results scarcely justify the high cost involved. Experiments have shown conclusively that frequencies below 300 and above 3000Hz contribute little to the intelligence conveyed by the human voice, and that under conditions of high ambient noise there is a decided advantage in accentuating the higher frequency speech components and progressively attenuating those components which lie below 500 or 600Hz. The naturalness of the voice under this kind of treatment must necessarily suffer, but to a lesser degree than might be expected. Telephone lines have a frequency response no better than that mentioned and their quality is adequate for normal aural communication.

A straightforward circuit giving exactly 90° phase shift over the total audio spectrum is impossible to realize, but by restricting the range to the minimum required for satisfactory communication, a tolerably good network may be constructed from simple fixed capacitors and resistors. The voltage at one of the output terminals leads the applied af voltage by an amount which varies according to the frequency, while the output at the other terminal lags behind the applied voltage. The values of the components may be chosen so that the total phase shift between both output terminals approximates 90°.

Fig 6.139 shows a network particularly suitable for amateur use. This is similar in configuration to that used by W2KUJ in his famous "SSB Jr" which popularized single sideband in the USA, but the circuit values have been modified to enable more easily obtainable components to be used. The network operates over the range 300–3000Hz and if it is constructed of components of 1 per cent tolerance the maximum deviation from the ideal phase shift of 90° will be within 1·5°. Sideband suppression is therefore better than 35dB at the worst frequency within the range. This network should be fed from a source impedance of 500Ω,

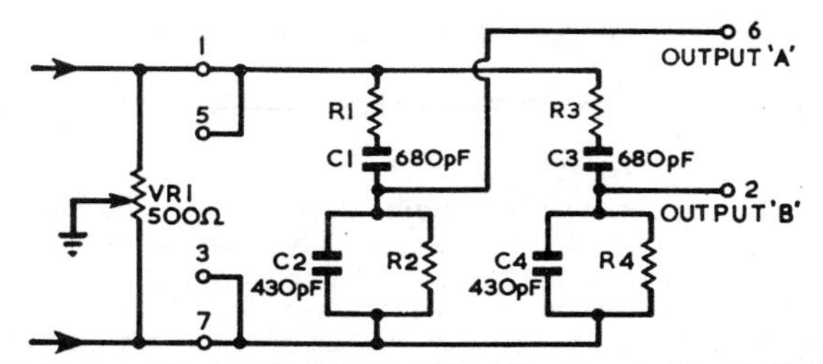

Fig 6.139. "SSB Jr." passive af phase shift network. C1, C3 680pF. C2, C4 430pF. R1 487·5kΩ (470kΩ + 15kΩ + 2·2kΩ in series); R2 770kΩ (500kΩ + 270kΩ in series); R3 125kΩ (100k + 15kΩ + 10kΩ in series); R4 198kΩ (150kΩ+47kΩ in series); VR1 500Ω. All capacitors are silver mica 1% tolerance. All resisistors are high-stability ± 1%, of half watt rating. The input and output connection numbers relate to the octal base pin numbers of the B & W 2Q4 P.S. network.

and the two af input voltages should have a precise amplitude ratio (measured to earth) of 2 : 7. This is obtained by the position of the slider of the balancing potentiometer VR1; it therefore follows that this slider will be some way off the central position. The lower of the two voltages should occur at the input terminal common to the two 430 pF capacitors, C2 and C4. The output terminals should be connected to a load of infinite impedance such as the grids of a pair of class A amplifier valves. (This audio phase shift network may be obtained commercially as the Barker & Williamson Type 2Q4 Phase Shift Network, with all components sealed in a metal valve envelope designed to plug into a standard octal valveholder.)

As the audio phase shift network is the heart of a phasing exciter, no effort should be spared in constructing it as closely as possible to specification. Close tolerance silvered mica capacitors and high stability resistors should be used exclusively.

If the speech amplifier used to drive the network described does not have its response restricted to the range over which the network is effective, objectionable distortion will result. An adequate degree of roll-off at low frequencies may be obtained by using low value coupling capacitors in the speech stages preceding the network. Insufficient high frequency attenuation will not necessarily make the signal as difficult to tune and read as poor low frequency cut-off, but it will cause splatter and interference to users of adjacent channels. This may be minimized by restricting the high frequency response in the audio amplifier in the manner shown in Chapter 9—*Modulation*.

A stage of class A amplification must be interposed between each of the output channels of the phase shift network and its associated balanced modulator for three reasons: to provide amplification, to offer the correct terminating impedance, and to isolate the network from the somewhat variable load presented by the balanced modulator.

The necessity of building these two amplifiers as really high fidelity devices is frequently overlooked, but it is vitally important. Unless the frequency response is excellent, the amplifiers may introduce phase-shifts on their own account and nullify the good work done by the phase shift network. If *RC* coupling is employed, large-value capacitors are essential, while audio transformers should be of proven high quality and wide frequency response. Shunt feed is recommended with transformer coupling to obviate the ill-effects of dc current flowing through the primary. Finally, the valves themselves should be operated on the strictly linear part of their characteristic curves and kept well within their ratings.

The Filter System

Since the objective is to transmit only a single sideband, it is necessary to select the desired sideband and suppress the undesired sideband. The relationship between the carrier and sidebands is shown in the diagram of Fig 6.122. Removing the unwanted sideband by the use of selective filters has the advantage of simplicity and good stability. The unwanted sideband suppression is determined by the attenuation of the sideband selecting filter, and the stability of this suppression is determined by the stability of the elements used in constructing the filter. This stability can be quite high because it is possible to use materials that have a very low temperature coefficient of expansion. Two commonly used materials are quartz crystal plates and small metal plates.

The filter system, because of its proven long term stability, has become the most popular method used by amateurs. At present three types of selective sideband filters are in common use:

(i) the low frequency crystal filter,
(ii) the high frequency crystal filter and,
(iii) the mechanical filter.

These three methods will be described in detail.

LF Crystal Filters

The crystal sideband filter is attractive to the amateur because it can be home constructed and the response characteristics and shape factor are under the constructor's control. Additionally, suitable crystals in the range 400–500 kHz have been available on the surplus market at low cost in the FT241 series of 54th and 72nd harmonic crystals. All crystals whose marked frequency commences with the figure 2 are 54th harmonic types, and all crystals whose marked frequency commences with the figure 3 are 72nd harmonic types. In each case the fundamental frequency of the quartz plate is the frequency marked on the box in MHz, divided by the harmonic series (ie 54 or 72).

Experience has shown that an audio frequency range of 300–3000Hz is satisfactory for voice communication and gives acceptable speech quality. The filter passband is therefore required to be 2700Hz and in the perfect filter the "slope" of the sides of the filter response curve would be vertical so that the bandwidth 60dB down would also be 2700Hz. In practice this ideal is not practicable and the sides of the filter response slope outward so that the bandwidth 60dB down is greater than the bandwidth at the 6dB points. The ratio, 60dB bandwidth to 6dB bandwidth, is the *shape factor* of the filter. A shape factor of two is a good figure to aim for, and if the filter can be made with a steeper response than this, so much the better—the sideband suppression will be further improved.

The circuit of a single half-lattice filter is shown in **Fig 6.140.** For improved sideband suppression half-lattice sections may be connected in cascade as shown in **Fig 6.141.**

The optimum crystal spacing for an audio bandwidth of 300–3000Hz is affected by the steepness of the passband *skirts*, (ie approximately 1·85kHz for a single half-lattice section, 2·2kHz for two sections, and 2·4kHz for three sections), and also by the inclusion or otherwise of a shunt

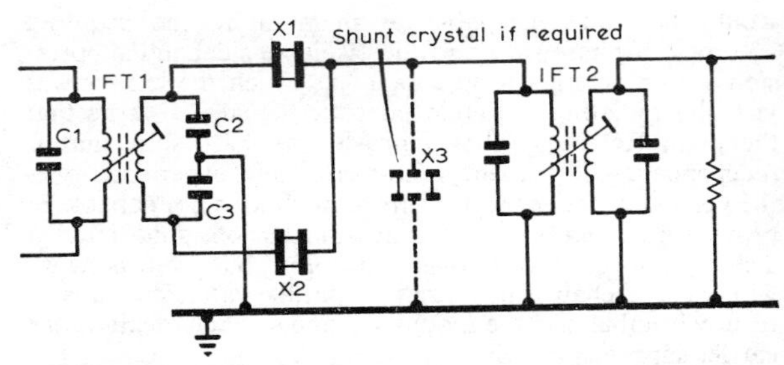

Fig 6.140. Single section half-lattice filter. To obtain the centre tap IFT1 is modified by replacing the original resonating capacitor by two capacitors of double the value (C2 = C3 = 2C1).

crystal which may be used to steepen the response on the carrier side. A simple half-lattice filter will give an unwanted sideband suppression of approximately 30dB, and if a better performance than this is required it will be necessary to use two or three filter sections. As a guide, the response of a single half-lattice filter is shown in **Fig 6.142(a)** and the response of the same filter with an additional shunt crystal is shown in **Fig 6.142(b).** A number of suitable crystal combinations for use between 400 and 500kHz are given in **Table 6.16.**

All FT241 crystals should be checked for frequency and activity by placing the crystal in series with the output from a BC221 frequency meter and the input to a diode probe valve voltmeter used on the 10V range. The BC221 should be *slowly* tuned across the required frequency range and at some setting the valve voltmeter pointer will swing over. This is the series resonant frequency of the crystal and if the activity is satisfactory the valve voltmeter will read about half scale. Any crystal giving less than 3 or 4V should be discarded. If the filter comprises two or three sections, those crystals on either side of the passband centre should be matched in frequency within a few hertz of each other, either by edge grinding or by plating.

The only satisfactory method of edge grinding is to hold the crystal plate (without dismounting from its support wires) in a jig so that it cannot possibly move under any circumstances, and then apply the grinding medium to it in the form of a rigid flat surface such as a 3in × 1in × ½in Carborundum slip stone of No 280 grit, obtainable from a local hardware dealer. The frequency may be raised a few hundred hertz by grinding the top edge.

Copper plating may be used to lower the frequency. A standard plating solution may be made by adding 15g of copper sulphate, 5cc of concentrated sulphuric acid and

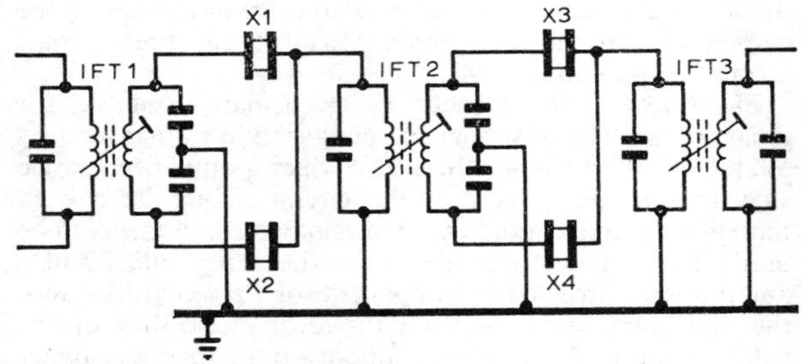

Fig 6.141. Double section half-lattice filter. X1 and X3 same frequency. X2 and X4 same frequency. Recommended crystal spacing is given in Table 6.16.

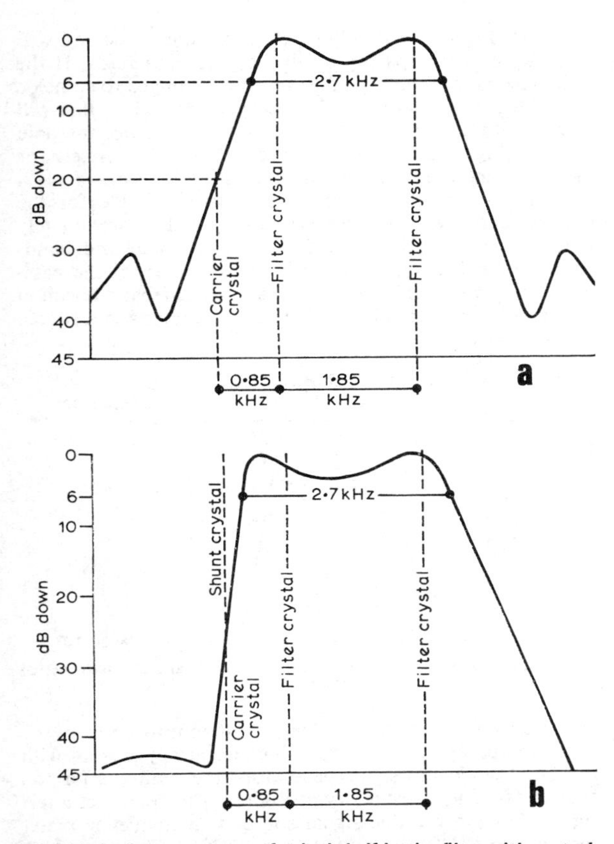

Fig 6.142. (a) Response curve of a single half-lattice filter with neutralizing. Note the side lobes just below the 30dB level. The steepness of the response can be improved by a small neutralizing capacitor of 2-3pF (two lengths of pvc insulated 22swg connecting wire twisted together for half an inch) across the pins of the highest frequency crystal in each half-lattice arm. Too much capacitance may bring up the side lobes to an unacceptable level. (b) Response curve of a single half lattice filter with shunt crystal. Note how the shunt crystal steepens the passband on the carrier side and improves the sideband suppression.

5cc of commercial alcohol to 100cc of distilled water. The solution should be poured into a glass vessel in which is placed a piece of clean copper wire bent so that it clamps the edge of the vessel and extends about 1in into the liquid. The wire should be connected in series with a 330Ω resistor to the positive terminal of a 1·5V dry cell. Both pins of the crystal are then connected together and wired to the negative side of the battery. When the crystal is dipped into the solution, a copper deposit will be formed which will lower its frequency. The exact variation may be determined by trial and measurement; if the process is accidentally carried too far, it may be reversed by changing the polarity of the dry cell and removing some of the copper. The process is not necessarily reversible with *all* crystals so caution is recommended.

When the frequency of the crystal has been increased by grinding, or lowered by plating, the unit should be thoroughly washed under running warm water, or alternatively washed in pure carbon tetrachloride, before being returned to service.

The most suitable type of i.f. transformers to use for

TABLE 6.16

Combinations of two-digit (54th harmonic) and three-digit (72nd harmonic) type FT241 surplus crystals suitable for sideband filters in the 400–500kHz intermediate frequency range.

Upper Sideband		
Shunt Crystal	Series Crystals	
	Lower	Upper
289	17	291 (18)
293	20	295 (21)
297	23	299 (24)
301	26	303 (27)
305	29	307 (30)
309	32	311 (33)
313	35	315 (36)
317	38	319 (39)
321	41	323 (42)
325	44	327 (45)
329	47	331 (48)
333	50	335 (51)
337	53	339 (54)
341	56	343 (57)
345	59	347 (60)
349	62	351 (63)
353	65	355 (66)
357	68	359 (69)

Lower Sideband		
Shunt Crystal	Series Crystals	
	Upper	Lower
291	18	289 (17)
295	21	293 (20)
299	24	297 (23)
303	27	301 (26)
307	30	305 (29)
311	33	309 (32)
315	36	313 (35)
319	39	317 (38)
323	42	321 (41)
327	45	325 (44)
331	48	329 (47)
335	51	333 (50)
339	54	337 (53)
343	57	341 (56)
347	60	345 (59)
351	63	349 (62)
355	66	353 (65)
359	69	357 (68)

The three-digit series of crystals shown in the third column of each of the above tables are recommended as giving optimum frequency response over a modulating range of 300–3,000Hz with an appropriate oscillator crystal. The two-digit series (shown in brackets) may be substituted in the event of the three-digit series being unobtainable, but the af response will be adversely affected.

coupling are the medium *Q*, high *L*/*C* ratio types; that is a *Q* of around 60 and a resonating capacitance value of not more than 100pF; the Maxi-Q miniature ift Type IFT.11/465 is very satisfactory. After alignment by adjusting all transformer dust cores for maximum response at the centre passband frequency, the filter response should be plotted and the bandwidth measured at the 6 and 60dB points to determine the shape factor that has been achieved. Normally the carrier crystal is placed at a frequency that is 20–25dB down on the filter response curve.

It should be appreciated, however, that its positioning is a compromise and it should finally be adjusted to obtain the best balance of audio quality determined by on-the-air reports.

Before embarking on the construction of a sideband filter it is most important to realize that surplus FT241 crystals are more than 30 years old, have deteriorated with time and may be several hundred hertz—in some cases more than 1kHz—off their original frequency. The *Q* of the crystal may be very low and in some cases the crystal may fail to oscillate at all.

Modern gold plated *AT* cut crystals are supplied in hermetically sealed B7G holders and have much better stability and output, and may be obtained as current equipment ground and calibrated to within 0·01 per cent of specified frequency; alternatively the Quartz Crystal Co Ltd will supply a complete set including the carrier crystal, for either a single section or a double section filter. Newcomers to ssb without previous experience of crystal manipulation and filter alignment are strongly advised that the initial additional cost of current production crystals is, in the long run, a worthwhile investment.

HF Crystal Filters

Many amateurs are natural experimenters and prefer, if at all possible, to make their own items of equipment. This applies particularly where there is a ready supply of low cost

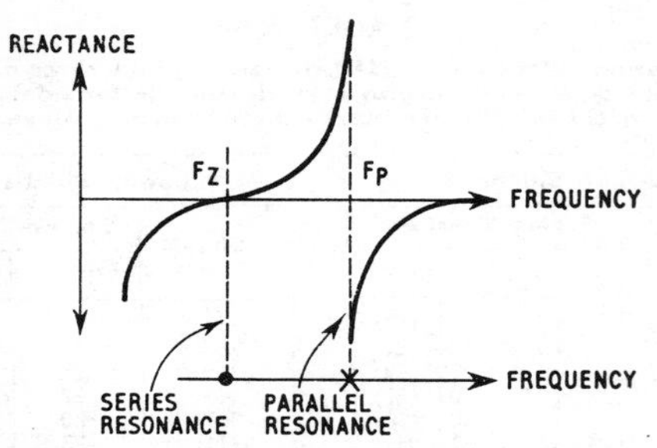

Fig 6.143. Reactance characteristics of a quartz crystal. The horizontal arrows denote an increasing frequency.

surplus "raw material". It is therefore not surprising that there has been increasing interest in the possibility of using the range of FT243 high frequency crystals in a bandpass filter with characteristics suitable for single sideband application. Experience has shown that it is in fact practicable to use these crystals and meet the requirements of the amateur ssb operator in regard to a filter giving a bandwidth of approximately 2·7kHz and a shape factor of two, with simple circuitry, the application of a little common sense and initiative, and the use of test equipment likely to be used by an amateur.

The main advantage of the crystal filter is the exceptionally high *Q* and stability of the quartz plates and these characteristics make it possible to obtain the required selectivity at high intermediate frequencies, avoiding the need for the usual double conversion that is generally necessary to obtain the required degree of image rejection.

It is well known that a quartz crystal has two resonant points very close together: these are the series resonance (the "zero" of impedance) and the parallel resonance (the "pole" of impedance) and this change of reactance or impedance is shown in **Fig 6.143.**

The circuit of a simple one-section half-lattice filter is given in **Fig 6.144** and it will be seen that the two crystals and the two halves of the inductance T1 form the legs of a bridge. Provided that the voltage across the coil from 1 to 2 is exactly the same as the voltage from 3 to 4, and provided also that the impedances of the two crystals A and B are equal there will be no voltage at the common connection (point 0).

For the requirement of a bandpass filter, crystals A and B are chosen to be different in frequency. Assuming the input frequency is at the zero (series resonance) frequency of crystal A the impedance balance of A and B is spoiled and there will be an output voltage between point 0 and the centre of the coil; this will also occur if the input frequency is at the pole (parallel resonance) frequency of crystal A. At the appropriate input frequencies the same thing will happen for crystal B, only the unbalance will be in the opposite direction. From this it follows that the filter passband will be as wide as the spacing of all the poles and zeros. If the series resonant frequency of crystal B is arranged to coincide with the parallel resonant frequency of crystal A this will theoretically give a perfectly flat passband of twice the pole zero spacing of each crystal; see **Fig 6.145.** Fortunately for the constructor the surplus range of FT243 crystals have a measured range of pole zero frequencies suitable for the requirement. Two of these crystals spaced approximately 2kHz apart will give a satisfactory single sideband bandwidth. A further improvement in the steepness of the passband skirts and therefore better unwanted sideband rejection is obtained by cascading two half-lattice sections in a simple back-to-back circuit.

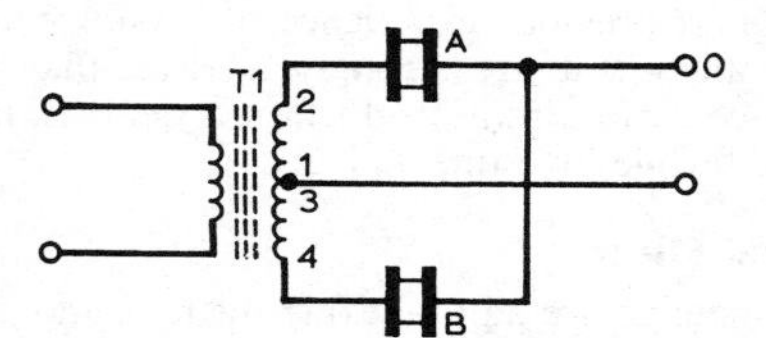

Fig 6.144. Basic single section half-lattice crystal filter. For optimum results the two halves of T1 must have very tight coupling.

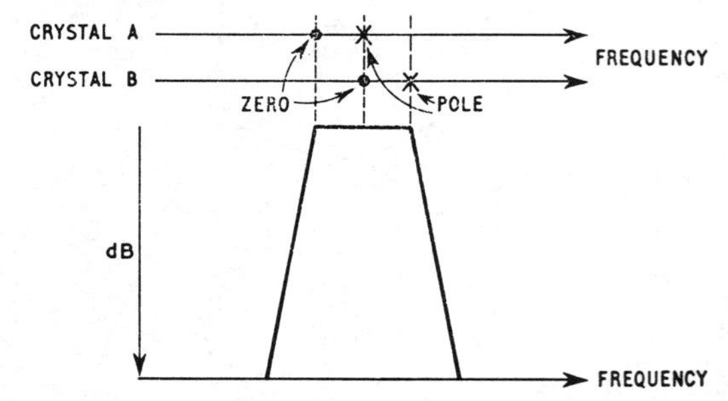

Fig 6.145. Diagram showing the theoretical passband of a half lattice crystal filter.

It is an impossible undertaking to attempt to measure the series and parallel resonances of high frequency crystals with the usual workshop signal generator—the tuning is far too coarse. The only satisfactory procedure is to construct a test rig so that the variable signal source is a frequency meter such as the BC221 used on its normal lf ranges. A suitable test rig is shown in **Fig 6.146.** It will be seen that the BC221 beats against a crystal controlled oscillator to produce the required frequency at the converter valve anode. The crystal to be tested is plugged into the holder whose input and output has a resistive padding network to avoid measurement error due to stray circuit capacitance (stray circuit capacitance has no effect on the series resonance but will affect the parallel resonance). At the parallel resonance the crystal offers a high impedance and the output voltage would be too small to be measured on a workshop valve voltmeter. Accordingly a class A amplifier is used, and the valve voltmeter probe is connected to the valve anode. This gives ample voltage to enable the pole and the zero frequencies to be accurately determined, and also enables the test rig to be used to plot the response of the complete filter, using the alternative connections shown.

Initially it is wise to purchase at least eight crystals—preferably 12—all of the same channel number. Identify each alphabetically and using the test rig, measure the pole and zero frequency of each crystal (the crystal zero frequency coincides with a sharp rise in the valve voltmeter reading and the pole frequency with a sharp dip). Record the frequencies on a list against the identification letters. It is also helpful to plot the pole and zero frequencies for each crystal and draw joining lines as shown in **Fig 6.147(a).** It is then possible to see easily the crystals which are most likely to make a satisfactory filter and to see the shift in frequency which is necessary. The crystals A and H in this case have equal

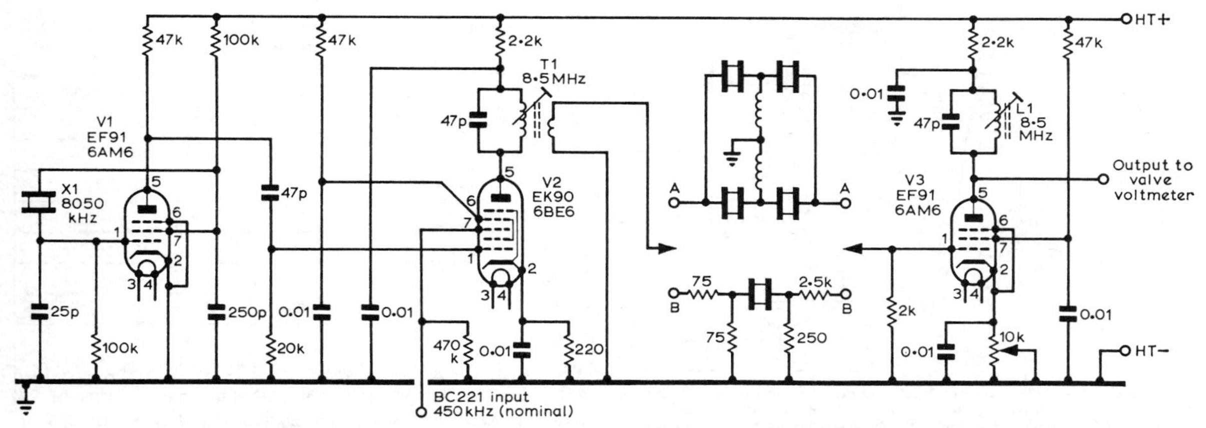

Fig 6.146. This is the test circuit used by G2DAF for plotting filter passbands and crystal frequency measurement. The BC221 frequency meter operates on its normal 430 to 470kHz range, and beats against the 8050kHz crystal controlled oscillator to give a precision output over the range 8480 to 8520kHz. For pole zero frequency measurement the arrowed connection points are taken to the B terminals, and for filter passband plotting they are taken to terminals A. The potentiometer in the cathode circuit of V3 enables amplifier gain to be set to give a convenient scale deflection on the valve voltmeter at the point of maximum filter response, ie 0dB.

pole-zero spacings and require no etching to align them. For the other two crystals of the filter E and C are chosen because they have closely similar pole zero spacings. These two are then etched until their zero frequencies correspond to the pole frequencies of A and H. The final pole and zero frequencies are shown on **Fig 6.147(b)**. When the BC221 is used on the range shown in this example the smallest readable division represents approximately 6Hz and the aim should be to match the crystal frequencies to within two divisions (12Hz). A suitable etchant is ammonium bifluoride with one part of a normal saturated solution diluted with two parts of water. The crystal is removed carefully from its holder and immersed in the solution for not more than

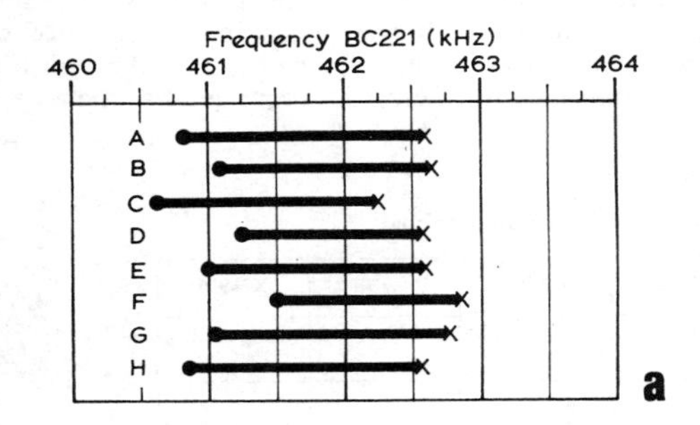

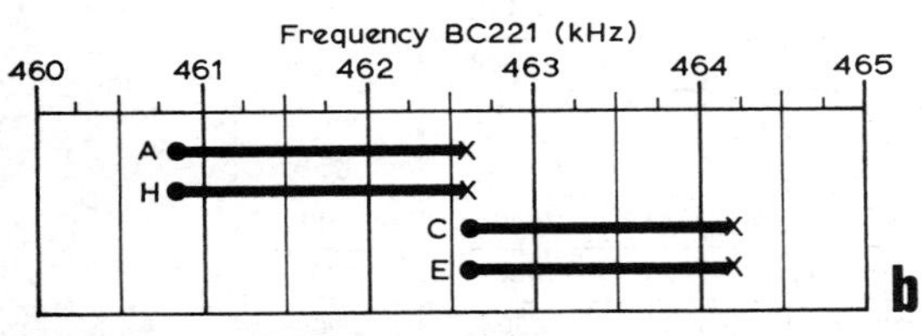

Fig 6.147.(a) The positions of the poles and zeros of eight crystals with nominal frequency 8512kHz. The test circuit beat oscillator frequency is 8050kHz. (b) Final positions of the poles and zeros of the four crystals selected for the filter.

one minute at the start. It is convenient to use a plastic tea strainer to hold the crystal and this is then dipped into the solution. At the end of the etching period the crystal must be carefully washed by holding the strainer under a running tap and then dried with paper tissue. Measure the new pole and zero frequencies and locate them on the graph paper. In this way progress can be closely monitored. (*Note: Ammonium bifluoride is highly corrosive and must be treated with extreme caution.*)

Finally, the filter coil is constructed by taking a length of 22swg pvc insulated connecting wire, doubled back on itself to form two parallel wires. This is then wound on a ferrite ring core to form nine double turns (total 18 turns) and the inner of one winding connected to the outer of the other to form a bifilar winding with the junction the centre tap. The main requirement of the inductance is very tight coupling between each half, together with a perfect electrical balance. In practice the grade of ferrite material does not appear to be critical. Filters have been successfully built using the ring from a Mullard LA4 pot core assembly (35mm outside diameter) and also the ring from a mains suppressor filter choke (1⅜in outside diameter) obtained from surplus sources.

To give some guidance as to the characteristics that may be expected, the plot of the filter passband and the pole and zero frequencies of a typical experimental filter is given in **Fig 6.148.**

The carrier crystal frequency is determined by plotting the filter passband and marking the 20dB down points. One of the remaining FT243 crystals is etched so that its *parallel* resonance frequency is at this frequency. Finally when the transmitter is completed and tested on the air the carrier crystal can be pulled by means of a 50pF trimmer across the oscillator grid circuit to obtain the best balance of voice quality.

FT243 crystals are mounted in ½in pin spacing holders, and are available over a wide range of frequencies in 25kHz steps from 5,700kHz to 8,650kHz, and in 33·333kHz steps from 5,706·666kHz to 8,340kHz. In general all crystals throughout the range are suitable for hf filter construction;

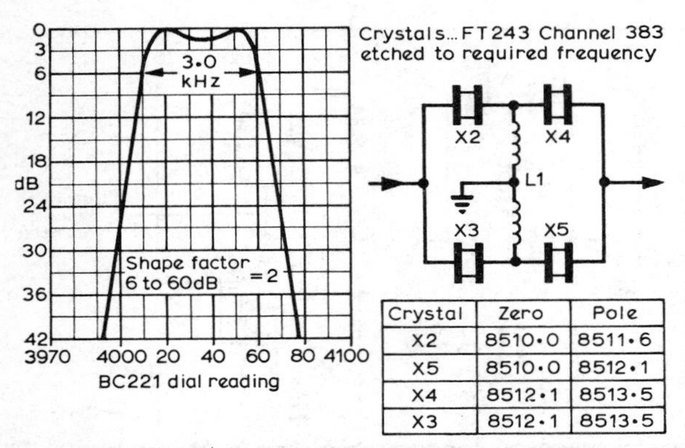

Crystal	Zero	Pole
X2	8510·0	8511·6
X5	8510·0	8512·1
X4	8512·1	8513·5
X3	8512·1	8513·5

Fig 6.148. The high frequency filter passband and pole zero frequencies of the filter in use at G2DAF. It can be seen that crystal X5 is the odd one out, with a greater pole-zero spacing than the others. This causes the slight asymmetry of the curve.

there is however one notable point. These crystals were made by a large number of different manufacturers for the USA services and the "cut" of the quartz plate is not always the same (ie certain crystals may have a pole zero spacing that is much greater than the average figure of about 2kHz; such crystals should be discarded as not suitable for bandpass filter use).

Commercially made hf filters are now available from many manufacturers. The best known are: McCoy Electronics Co, Holly Springs, Pa, USA, The Quartz Crystal Co Ltd, New Malden, Surrey, KVG West Germany (available from several UK sources) and Salford Electrical Instruments Ltd.

Table 6.17 shows the characteristics of the range of KVG filters and **Table 6.18** shows characteristics of two SEI Ltd filters which are suitable for ssb applications.

Mechanical Filters

Three types of mechanical filters are available in the UK at competitive prices that put them within the price range of many sideband workers. These are the Collins F455 FA-21 filter with a 6dB bandwidth of 2·1kHz and the Kokusai MF455-10K and MF455-15K with 6dB bandwidths of 2·0 and 3·0kHz respectively. The mechanical filter is a mechanically resonant device which receives electrical energy, converts it into mechanical vibration, then converts the mechanical energy back into electrical energy at the output. The mechanical filter consists basically of four elements:

(i) an input transducer which converts the electrical input into mechanical oscillations.
(ii) metal discs which are mechanically resonant,
(iii) coupling rods which couple the metal discs, and
(iv) an output transducer which converts the mechanical oscillations back into electrical oscillations.

Fig 6.149 shows the elements of the mechanical filter, and **Fig 6.150** shows the electrical analogy of the mechanical filter. In the electrical analogy the series resonant circuits L1 C1 represent the metal discs, the coupling capacitors C2 represent the coupling rods, and the input and output resistances R represent the matching mechanical loads.

The transducer, which converts electrical energy into mechanical energy and vice versa, may be either a magnetostrictive device or an electrostrictive device. The magnetostrictive transducer is based on the principle that certain materials elongate or shorten when in the presence of a magnetic field. If an electrical signal is sent through a coil which contains the magnetostrictive material as the core, the

TABLE 6.18

Two SEI filters for ssb applications

Filter type		QC1246AX	QC1246AA
Centre frequency		9·0MHz	5·2MHz
Application		SSB	SSB
Bandwidth (6dB down)		2·51kHz	2·5kHz
Insertion loss		< 3dB	< 3dB
Shape factor		(6 : 60dB) 1·7	(6 : 50dB) 1·7
Input-output Termination	Rt	500Ω	620Ω
	Ct	30pF	30pF

TABLE 6.17

KVG 9MHz crystal filters for ssb, a.m., fm and cw applications

Filter Type		XF-9A	XF-9B	XF-9C	XF-9D	XF-9E	XF-9M
Application		SSB Transmit	SSB	AM	AM	FM	CW
Number of Filter Crystals		5	8	8	8	8	4
Bandwidth (6dB down)		2·5kHz	2·4kHz	3·75kHz	5·0kHz	12·0kHz	0·5kHz
Passband Ripple		< 1dB	< 2dB	< 2dB	< 2dB	< 2dB	< 1dB
Insertion Loss		< 3dB	< 3·5dB	< 3·5dB	< 3·5dB	< 3dB	< 5dB
Input-Output	Zt	500Ω	500Ω	500Ω	500Ω	1200Ω	500Ω
Termination	Ct	30pF	30pF	30pF	30pF	30pF	30pF
Shape Factor		(6:50dB) 1·7	(6:60dB) 1·8	(6:60dB) 1·8	(6:60dB) 1·8	(6:60dB) 1·8	(6:40dB) 2·5
			(6:80dB) 2·2	(6:80dB) 2·2	(6:80dB) 2·2	(6:80dB) 2·2	(6:60dB) 4·4
Ultimate Attenuation		> 45dB	> 100dB	> 100dB	> 100dB	> 90dB	> 90dB

In order to simplify matching, the input and output of the filters comprise tuned differential transformers with galvanic connection to the casing.

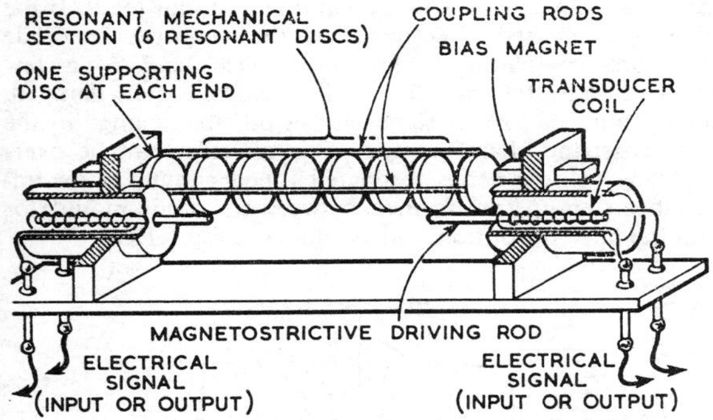

Fig 6.149. Constructional details of a mechanical filter.

electrical oscillation will be converted into mechanical oscillations. The mechanical oscillation can then be used to drive the mechanical elements of the filter. The electrostrictive transducer is based on the principle that certain materials, such as piezoelectric crystals, will compress when subjected to an electric current. The transducer not only converts electrical energy into mechanical energy and vice versa; it also provides the correct termination for the mechanical network.

In practice, filters between 50kHz and 600kHz can be manufactured. Since each disc represents a series resonant circuit, increasing the number of discs will increase the skirt selectivity of the filter. Manufacturing technique at present limits the number of discs to eight or nine in a mechanical filter. A six-disc filter has a shape factor of approximately 2·2, a seven-disc filter a shape factor of approximately 1·85, a nine-disc filter a shape factor of approximately 1·5.

The Collins Mechanical Filter

Collins mechanical filters have been in use for a number of years and are available in the range from 60kHz to 500kHz with 6dB bandwidths from 0·5kHz to 35·0kHz—in fact Bulletin 1007 obtainable from Collins Radio Co of England Ltd lists no fewer than 92 different types. The Type F455 FA-21 is designed specifically for the amateur market with a nominal 6dB bandwidth of 2·1kHz and a 60dB maximum bandwidth of 5·3kHz providing a shape factor of just over 2·5 : 1. It should be noted that although the shape factor is not so good as the more expensive 3·1kHz filter (shape factor 2 : 1) the slope is marginally better.

The F455 FA-21 filter is fitted in a rectangular case 2½in long, just over ½in wide, and ½in high, and is intended for horizontal mounting. The centre frequency is 455kHz nominal; 6dB bandwidth 2·1kHz nominal; 60dB bandwidth 5·3kHz maximum; passband ripple 3dB maximum; transfer impedance 5kΩ; resonating capacitance 130pF ± 5pF; transmission loss 9·5dB and spurious response attenuation (405–505kHz) 60dB minimum. The resonating discs are driven by magnetostrictive transducers employing polarized

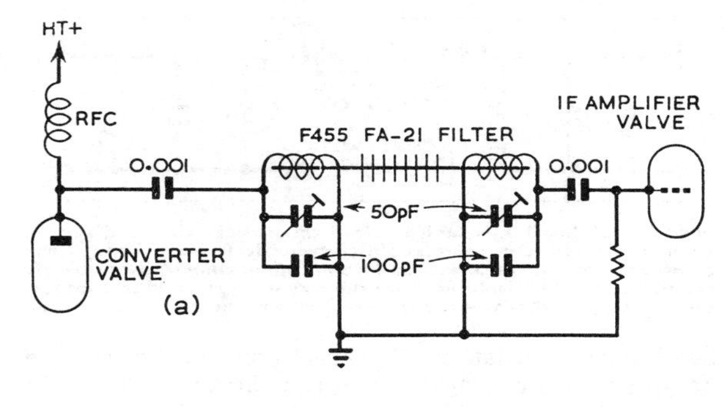

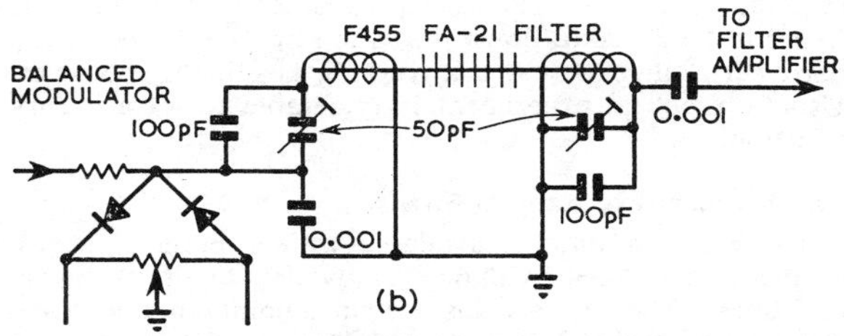

Fig 6.151. Basic circuits for the Collins mechanical filter type F455 FA-21 showing at (a) a parallel resonated transducer coil following a high impedance load, and at (b) a series resonated transducer coil following a low impedance load. The resonating capacitors are adjusted for maximum signal through the filter at the centre passband frequency.

biasing magnets; for this reason it is not permissible to have any dc at all flowing through the coil, and shunt feed to the valve anode is essential. The two transducer coils are identical, are isolated from the case, and are balanced to earth; either coil may therefore be arbitrarily designated as input or output and the filter can be coupled to the two anodes of a balanced modulator, or may be used to feed the two grids of a following balanced mixer.

Installation is simple, but it is important to provide screening between the input and output connections (a small cross screen underneath the chassis). The filter case must be earthed. Drive level should be kept below 2V rms. For best performance the input and output transducer coils should be carefully resonated at the filter centre frequency. The two basic circuits (*a*) filter following a high impedance load such as the mixer valve in a receiver, and (*b*) following a low impedance load such as the modulator in a transmitter are shown in **Fig 6.151.**

The positioning of the carrier crystal frequency is usually stated by the manufacturer. If it is not, the best procedure is to use a BC221 frequency meter as a variable carrier oscillator and get reports from other stations until the best balance of audio quality is obtained. The frequency is then obtained from the dial reading of the BC221 and a crystal ordered from the supplier, stipulating that the crystal is required for oscillation on the parallel resonance mode with

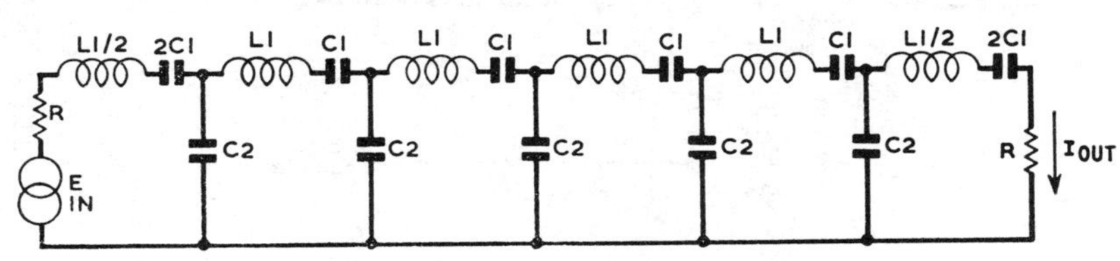

Fig 6.150. Electrical analogy of a mechanical filter.

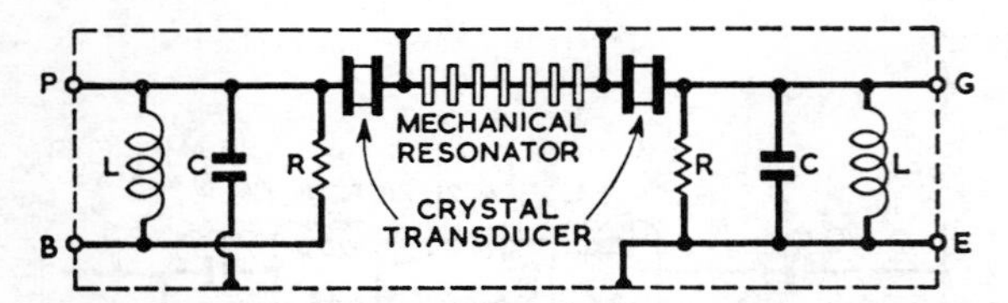

Fig 6.152. Kokusai mechanical filter theoretical circuit diagram. With this filter it is permissible to have dc flowing through the transducer coil L, (ie series fed ht to a mixer anode). Accordingly the "B" terminal is isolated from the screening can and it is important that the filter is wired into the circuit the right way round.

30pF shunt capacitance. Alternatively, an FT241 crystal can be edge ground or plated so that it oscillates in situ on the required frequency. The correct position will usually be found to be 20–30dB down the skirt response; however the narrower a filter the more critical the positioning becomes. Usually practical experiment is preferable to an arbitrary selection.

The Kokusai Mechanical Filter

The Kokusai filter is available with a nominal centre frequency of 455kHz, with 6dB bandwidths of either 2·0kHz or 3·0kHz. The case style is a spun aluminium cylindrical can approximately 2⅝in high and 1⅜in diameter with two threaded mounting studs and four insulated feed-through connecting tags. Mounting is intended to be vertical with the connecting tags going through the chassis.

This filter has the disc resonators excited by quartz crystal transducers; the circuit arrangement is shown in **Fig 6.152.** The general specification of the two filters is as follows: MF-455-10K, 6dB bandwidth 2·0kHz nominal; 60dB bandwidth less than 7·0kHz. MF-455-15K, 6dB bandwidth 3·0kHz nominal; 60dB bandwidth less than 9·0kHz (in practice the filters stocked in the UK are generally better than the manufacturer's quoted figures, the 60dB bandwidth rarely being greater than twice the 6dB bandwidth, ie a shape factor of 2 or better); maximum deviation from centre frequency ± 0·8kHz; passband ripple less than 4dB; temperature range 0°C to 70°C.

The manufacturer does not give any values for input or output terminal impedance but does state "Both input and output should be connected to high impedances. Since mechanical filters tend to be rather sensitive to capacitances across both input and output terminals care should be exercised not to lower the impedance by stray capacitance." It is clear from this that the Kokusai filter is designed to operate into relatively high impedance input and output loads. If the filter is connected directly across the output of a diode modulator presenting a low impedance load of a few hundred ohms, the transducer coil would be heavily damped, there would be loss of signal and a possible change in the filter passband characteristics. This difficulty can be overcome by using a series resonated tuned circuit, as an impedance step-up transformer, between the diode modulator and the filter transducer coil as shown in **Fig 6.153.**

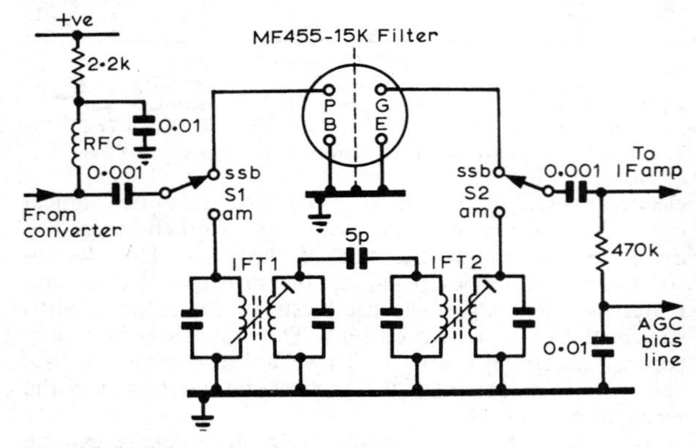

Fig 6.154. Suggested circuit for the use of the Kokusai filter in an existing communication receiver to improve the ssb selectivity. S1 and S2 can be switch banks fitted to the existing selectivity switch assembly, and should be of the type incorporating shorting plates. The shorting plate pole must be earthed. If the filter selectivity is not to be degraded, the two banks must be isolated from each other with an earthed cross screen. IFT1 and IFT2 are standard Maxi-Q IFT 11/465 i.f. transformers or similar.

Because of its small physical size and ease of mounting, this filter is particularly suitable for incorporation in the older communication receivers such as the AR88, HRO and CR100, etc, and it is relatively easy to arrange to switch the filter in or out of circuit so that the normal selectivity can be used for a.m. and the filter for ssb or cw. A circuit suitable for most receivers is shown in **Fig 6.154.**

The nominal centre passband frequency is 455·0kHz, but because of manufacturing tolerances the actual centre frequency may vary plus or minus 0·8kHz. For this reason, each filter is packed with a data sheet giving the filter serial number and the measured bandwidth at the 6dB and 60dB points for the filter concerned. Each bandwidth is given as plus or minus *X*kHz relative to the design centre frequency

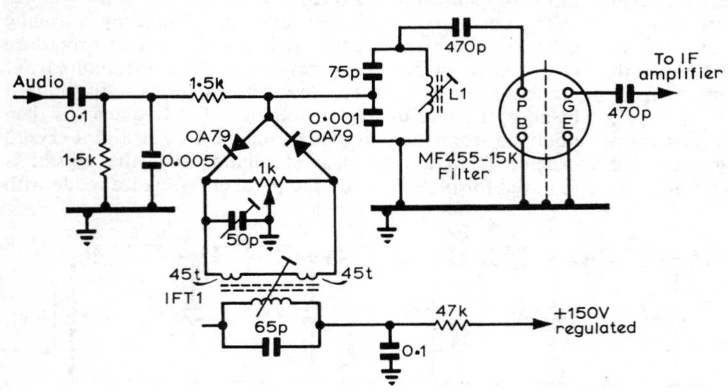

Fig 6.153. Circuit details showing the method of using the Kokusai mechanical filter following a low impedance diode modulator. L1 is the primary winding of a standard Maxi-Q IFT.11-465 i.f. transformer with the original 65pF resonating capacitor removed and replaced by the 75pF and 0·001μF capacitor shown. (The unwanted secondary pie is cut away with a sharp knife.) Feed a 1·5kHz tone into the microphone socket and resonate L1 for the maximum output at the anode of the filter amplifier valve.

KOKUSAI ELECTRIC Co., LTD.

MECHANICAL FILTER CHARACTERISTICS	Made by [illegible]	Checked by M.N.	Approvals R.F.
TYPE MF455–15K	SERIES No. CCK 0665	Date APRIL 13, 1973 Temp. 26°C	
CENTRE FREQUENCY	(+) − 0·4 kHz	GAIN PER STAGE	24·0 dB
BANDWIDTH AT 6dB ATTENUATION	+ 2·0 kHz − 1·2 kHz	INSULATION (More than DC 500V 200Mohm)	O.K.
30dB ATTENUATION (SSB)		457·52 kHz 453·29 kHz	
BANDWIDTH AT 60dB ATTENUATION	+ 3·3 kHz − 2·6 kHz		

Fig. 6.155. Example of the data sheet packed with each Kokusai mechanical filter.

of 455·0kHz—not relative to the actual filter centre frequency. To illustrate this more clearly, an example of the data sheet issued by the manufacturer is shown in **Fig 6.155.** It will be noted that the total passband width at the 6dB point is 3·2kHz but that this is quoted as two frequencies—each plus or minus relative to 455·0kHz and above this is a figure (in this case + 0·4kHz) indicating that the actual passband centre frequency is 455·0kHz + 0·4kHz = 455·4 kHz. At the bottom of the table the bandwidth is given at the 60dB points—again as two frequencies relative to 455·0kHz; these two figures added together give the total passband width at 60dB down. The ratio of the total bandwidth at 60dB to the total bandwidth at 6dB is the shape factor of the filter. In this case, it is 5·9/3·2 = 1·9 approximately.

From the 6dB and 60dB frequencies given, the filter passband could be plotted on squared paper: however, the graph would not be quite correct because in practice the two skirts are not straight lines—they are in fact slightly concave. To avoid this error the manufacturer also gives the two passband frequencies at the 30dB points. The procedure then is to get a sheet of one tenth of an inch squared paper and divide this vertically at 10dB per inch starting from the top at 0dB and horizontally at 1·0kHz per inch arranging that the 455·0kHz point is in the centre of the graph paper—this is then indicated clearly by a dotted centre-line. From the data sheet the 6dB and 60dB figures are plotted, the + (plus) figures to the right of the centre-line and the − (minus) figures to the left. By reference to the horizontal frequency scale, the two 30dB figures are marked on the graph at the 30dB down position. The 60dB, 30dB and 6dB points on each side of the centre-line are then joined together with a shallow curve and this is the correct passband for the filter. A complete plot of the filter characteristics already given is shown in **Fig 6.156.**

In the interests of acceptable voice quality the filter should pass frequencies down to 300Hz, ie 300Hz should be at the 6dB point. From this it follows that the carrier frequency must be 300Hz away from this position. A vertical line is therefore drawn on the graph 0·3kHz outside the 6dB point and where this line cuts the passband curve is the correct position for the carrier crystal—on the graph shown, this is a

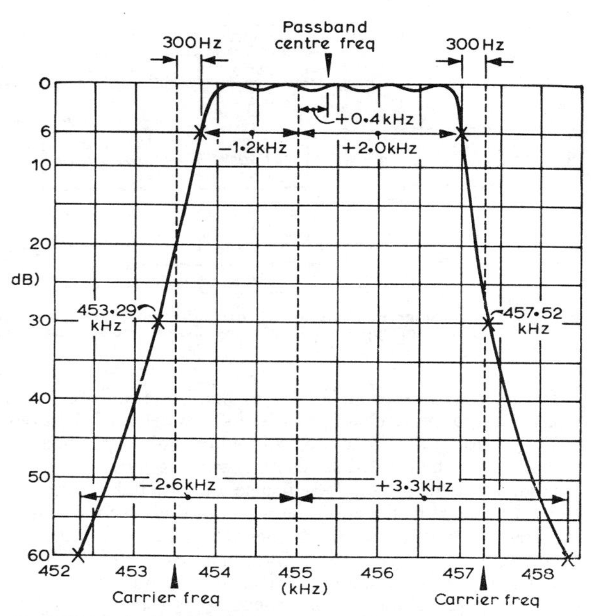

Fig 6.156. Method of plotting the passband curve for a Kokusai mechanical filter from characteristics given on the data sheet supplied and from this determining the required carrier crystal frequencies.

frequency of 457·3kHz. If two carrier crystals are required (ie sideband switching in a receiver) the second frequency is plotted in exactly the same way but at the other side of the passband. If the carrier crystal on the lower side of the passband is in use, the filter will pass the higher sideband. If the carrier crystal on the higher side of the passband is in use, the filter will pass the lower sideband.

Ordering Crystals

When ordering crystals it should be remembered that every crystal can be made to oscillate on either its series or parallel resonant frequency and these are not the same (at frequencies around 455kHz the difference may be 200Hz or more). It is not sufficient to quote only the required frequency; the manufacturer must know whether operation is required on the parallel or the series resonance, and this must be stated. As the parallel resonance is affected by the shunt circuit capacitance, this value should also be stated. (Normally in the United Kingdom 30pF is taken as a preferred standard value and the manufacturer will grind the crystal to oscillate at parallel resonance at the stated frequency with this capacitance, unless otherwise instructed.)

For amateur use it is not necessary to go into involved calculations in an attempt to determine circuit capacitance—the value of 30pF can be quoted and is quite near enough in practice. The manufacturer also has to know (i) the type of holder, ie 10XJ (½in pin spacing) or B7G; (ii) frequency tolerance (this is normally plus or minus 100ppm or 0·01 per cent; (iii) operating temperature range (it would be sufficient here to quote "normal room temperature" or "amateur equipment").

Carrier crystals are normally used in either Pierce, Colpitts or Miller oscillator circuits and in all these the crystal is excited at its fundamental parallel resonant frequency.

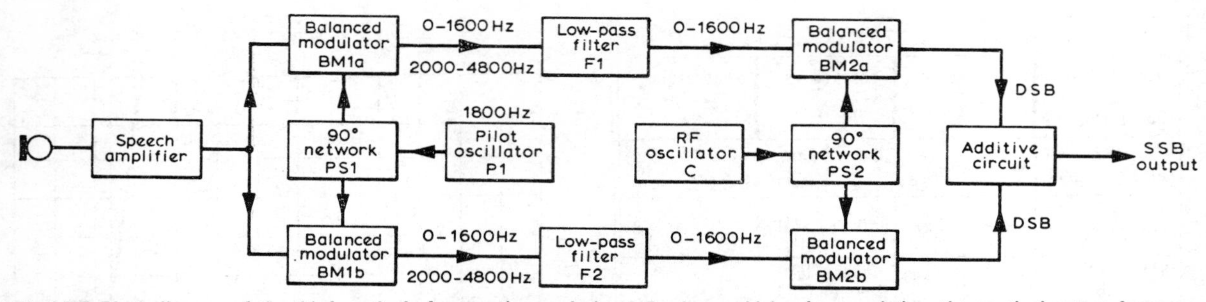

Fig 6.157. Block diagram of the third method of generating a ssb signal. For lower sideband transmission, the nominal output frequency will be fp + fc and for upper sideband transmission, the nominal output will be fc – fp. The nominal output frequency itself is not generated.

In the case of either mf or hf sideband filters, the crystals operate as coupling elements in the series arms of the half-lattice sections; therefore the required crystal is specified for operation on the quoted series resonant frequency.

High frequency crystals will also be ordered for use in some other part of the equipment and in this case the manufacturer will have to know whether the mode of operation is fundamental or overtone. Overtone oscillators always operate at series resonance and as this is not affected by shunt circuit capacitance it is unnecessary to state this. Often final conversion crystals are operated on their overtone to avoid generating spurious signals—this applies particularly to a double conversion amateur band receiver. The mode of operation can be readily identified by reference to the type of oscillator circuit—this will be a Butler, Squier or Robert Dollar in which the crystal is used as a *series* coupling of low impedance to the required oscillatory rf current.

Remember that in an harmonic oscillator—as distinct from an overtone oscillator—the crystal oscillates at its *fundamental* parallel resonant frequency. The harmonic generation is obtained by the non-linearity of the associated valve, the valve output circuit being tuned to resonance at the required harmonic frequency. This type of oscillator normally uses the Colpitts, Miller or electron-coupled circuit.

The subject of crystals and crystal oscillator modes of operation is dealt with in detail earlier in this chapter.

The Third Method

In addition to the methods already described, details have been published of a third method of generating a ssb signal. Because it is rather more complicated, insofar as it combines the principles of both phasing and filter-type equipment, the third method has not attracted much interest among amateurs. A brief description is however included because the system is said to preserve a higher degree of fidelity than a sharp-cutting filter, and to give better sideband suppression than a straightforward phasing exciter.

The third method combines the principles of both systems in such a way that a wideband audio phase-shift network is unnecessary, and although filters are used, their cut-off characteristics are not abrupt enough to detract from speech quality. The method is shown diagrammatically in **Fig 6.157.** The output of a pilot audio frequency oscillator P1 is applied to a differential phase-shift network, preferably of low Q. The two components emerging from the network are adjusted to be approximately equal in amplitude, and to differ in phase by exactly 90°. Oscillator frequency is not particularly critical; the system can be made to work at any point from 1·6 to 4·5kHz, but as will be seen later there is an incidental dividend in remaining within the range 1,600–1,800 Hz. The quadrature signals from the first phase-shift network PS1 are applied to two identical balanced modulators BM1a and BM1b. Audio excitation from a conventional speech amplifier is fed in parallel to the other inputs of BM1a and BM1b. Because of the low frequencies involved, it is possible to achieve an exceptionally high order of pilot carrier attenuation in the balanced modulators, and the output of each will consist of two sidebands at audio frequency, $Fp + Fm$ and $Fp - Fm$. Amplitudes and waveforms will be identical, but the sidebands from one balanced modulator will be 90° out of phase with those from the second. The higher-frequency sidebands are eliminated from both channels by a pair of identical low-pass audio filters. If a pilot frequency of 1,800Hz is assumed and a minimum modulating frequency of 200Hz, each filter will be required to pass all frequencies up to 1,600Hz, and to attenuate rapidly above 1,600Hz, reaching maximum rejection at 2,000Hz. It is comparatively easy to attain this performance with standard tolerance components, and high stability is not essential in either the pilot oscillator or the filter itself. The phase shift network PS1 is the only part of the circuit calling for precision adjustment, but a simple differential arrangement will take care of more drift than is likely to be met in any practical pilot carrier oscillator.

The quadrature audio signals issuing from F1 and F2 are applied to a further pair of balanced modulators, where they modulate quadrature rf carriers from PS1 exactly as in a conventional phasing transmitter. The rf excitation may be at the desired output frequency, or if multiband operation is contemplated, at a neutral frequency convenient for heterodyning into the required bands.

When the double sideband signals from the two balanced modulators are combined in an additive tank circuit, the wanted components reinforce one another, and the unwanted sideband balances itself out to an extent which depends upon the accuracy of the two 90° networks PS1 and PS2. It is claimed that attenuation exceeding 40dB is achievable.

The main advantage of using as low an audio frequency as possible for the pilot carrier lies in the resulting simplification of the low-pass filter design. In the numerical example quoted, the response of the filter is required to change from zero to full attenuation in the 400Hz between 1,600 and 2,000Hz; that is, in 25 per cent of the highest frequency to be passed. Had 3,600Hz been chosen as the pilot frequency, the slope of the filter response curve would have to be twice as steep in proportion if the same performance were to be

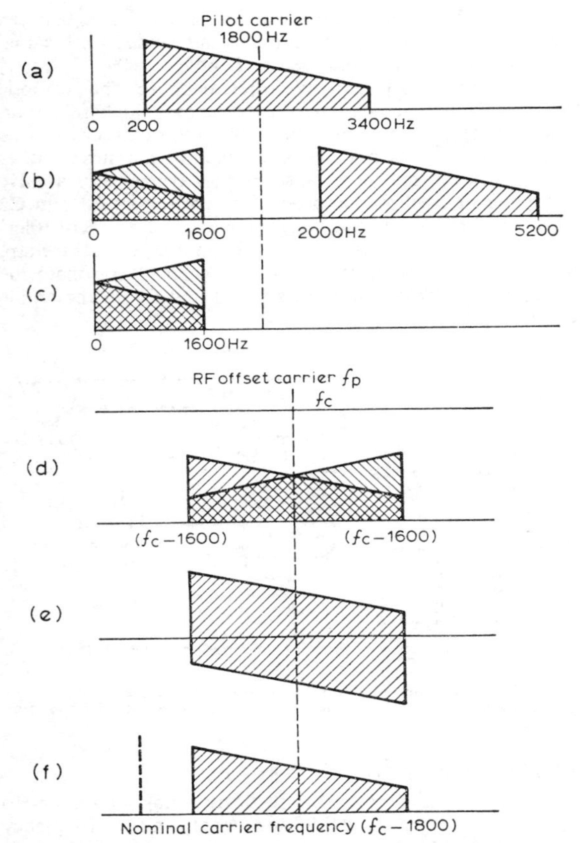

Fig 6.158. Frequency relationship in a third method transmitter; (a) af input spectrum to BM1a and BM1b; (b) af output from BM1a and BM1b for a pilot carrier frequency of 1800Hz. Note how the lower sideband is folded; (c) af output from lowpass filters F1 and F2; (d) rf output from upper modulator chain to Fig 6.156; (e) rf output from lower modulator chain in Fig 6.156; (f) Single sideband from additive tank circuit in which (d) and (e) are combined. By reversing the phase shift of the output from either BM2a or BM2b, upper or lower sideband transmission may be selected at will.

maintained. This would naturally call for more sophisticated filter design. An incidental advantage of a low pilot frequency stems from the fact that the modulating spectrum becomes "folded" upon itself for all audio frequencies exceeding that of the pilot oscillator. In the earlier numerical example, a tone of 1,000Hz would beat with the pilot carrier to give a difference frequency of 800Hz. If, however, the audio tone were 2,600Hz, the difference frequency would also be 800Hz, but the phase would be altered by 180°. Taking the system as a whole, it follows that imperfect sideband suppression manifests itself as inverted speech superimposed upon the wanted sideband and not as adjacent channel interference. The advocates of the third method claim that the listener's ear can discriminate more readily against this type of imperfection than against the type of interference caused by poor suppression in a conventional ssb transmitter. The validity of this argument is open to question, because many amateurs now use highly selective receivers capable of rejecting image-type interference. Selectivity at the receiving end affords no improvement to a poor third method transmission. There is however some rough justice in the fact that a badly adjusted third method transmitter causes more irritation to its owner than to the occupants of adjacent channels, while the reverse is usually true of a poor transmitter of more conventional design.

A fuller explanation of audio spectrum folding would require mathematical treatment beyond the scope of this Handbook. The description in the preceding paragraph in conjunction with the diagrams of **Fig 6.158** should, however, show in broad outline how the system operates. Readers who desire a deeper understanding of the third method are referred to the excellent article by D. K. Weaver in the *Proceedings of the I.R.E.* for December 1956.

A third method transmission is received in exactly the same way as ssb generated by either of the more common systems. As will be seen from Fig 6.158, the carrier insertion oscillator will have to be offset from the nominal frequency of transmission by an amount equal to the frequency of the pilot audio oscillator. The receiving operator will not, however, be conscious of this fact. Imperfect carrier suppression will show up, not as a steady heterodyne against which the carrier insertion oscillator may be zeroed, but as an audio tone at the frequency of the pilot carrier in the transmitter.

FREQUENCY CONVERSION

Multiband operation of an ssb transmitter normally involves the generation of the signal at a fixed frequency, and the provision of one or more heterodyne converters to enable output to be obtained in the required amateur bands.

The pentagrid valve, of which the 6BE6 is representative, is probably the best known type of frequency converter. The circuit of **Fig 6.159** is typical, and may be used either to raise the output from a filter in the i.f. range to the 80m band, or to change the transmitter frequency from one band to another. For linearity, the oscillator excitation to grid 1 should be close to 2·5V rms and the ssb signal at grid 3 should not be allowed to exceed 0·1V rms at peak input. Should the ssb output of the preceding stage exceed this value, a voltage divider should be interposed to reduce the excitation to the recommended level. The excitation could of course be reduced by turning down the audio gain but this would impair the ratio of sideband signal to resting carrier and nullify the good work done by the balanced modulator. It is

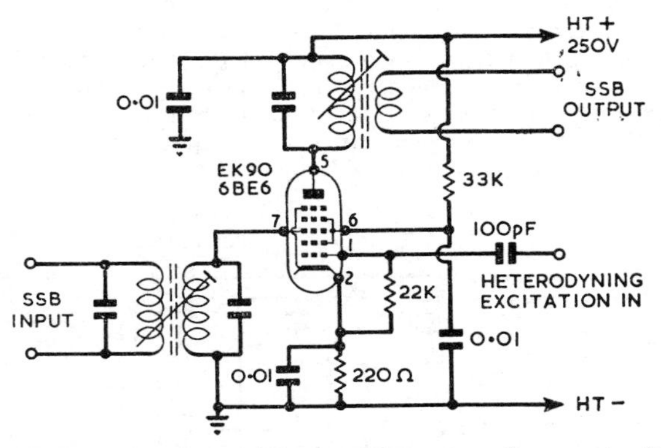

Fig 6.159. Simplest form of frequency converter using a pentagrid valve. The tuned circuits should resonate at the appropriate input and output frequencies.

a sound principle of ssb transmission that the audio gain control should be used solely to secure optimum operating conditions for the balanced modulator, and that the excitation of the converter or amplifier stages should be adjusted *after* the single sideband has been generated.

Apart from the low level of output, the only serious shortcoming of a simple pentagrid converter lies in its inability to discriminate against the unwanted energy from the heterodyning oscillator if the ssb signal is to be raised in frequency by a ratio of more than five or six to one. The following numerical example will make the reason clear. Assuming that the output of a 460kHz crystal filter is to be converted to 3,760kHz, the heterodyning oscillator will operate at 3,300kHz. A single tuned anode circuit of good quality (high *Q*) adjusted precisely to 3,760kHz, will present roughly 25 times the impedance to the wanted ssb signal as it will to the steady off-tune oscillator signal. The oscillator input voltage will, however, have been adjusted to at least 25 times the amplitude of the peak ssb input, so despite the selective effect of the output tank circuit, the steady oscillator voltage across it will be just about the same as the peak ssb voltage. Subsequent amplifier stages will, of course, afford further discrimination against the oscillator signal, but if the tuned circuits are wide-band or swamped in the interests of linearity, enough energy may leak through to produce an appreciable spurious signal outside the amateur bands. For conversion ratios exceeding three to one, it is desirable to employ a critically coupled pair of really high *Q* tuned windings in the anode circuit.

With appropriate changes to circuit values, any receiving type frequency converter may be substituted for the pentagrid in Fig 6.159. If the oscillator is crystal controlled, a valve envelope may be saved by using the triode section of a triode-hexode as oscillator, but it may prove difficult to adjust the excitation to the precise value for maximum linearity. A self-excited oscillator should preferably be isolated from the frequency converter by a buffer stage, otherwise pulling and frequency modulation may be experienced.

Although multi-element valves are convenient and efficient, comparable results may be obtained with double triodes such as the 12AU7, 12AT7 and ECC85. A representative circuit is shown in **Fig 6.160.** It may be operated at the same voltage levels as those recommended for the 6BE6 and has the same shortcomings.

When the conversion ratio exceeds three to one, many designers prefer to attenuate the oscillator voltage in the mixer output circuit by a system of phasing or balancing. The principle is identical to that of carrier elimination by balanced modulator. The classical balanced frequency converter circuit, widely used in ssb equipment, is shown in **Fig 6.161**. This is capable of attenuating the oscillator output by 20–25dB. Excitation may be any value between 2·5 and 10V rms, depending on the amplitude of ssb signal the converter valve is required to handle.

Balancing is affected by adjustment to the cathode potentiometer with oscillator drive but no sideband input, until the heterodyning voltage in the output circuit is at its lowest value. This may be conveniently measured either with a pick-up loop to an absorption wavemeter or a valve voltmeter or oscilloscope probe. The capacitors C1 and C2 in the grid circuit should be silver mica of 1 per cent tolerance, each of equal value to provide an accurate centre-tap, and with an effective value that will correctly resonate the tuned circuit. This also applies to C3 and C4 in the anode circuit.

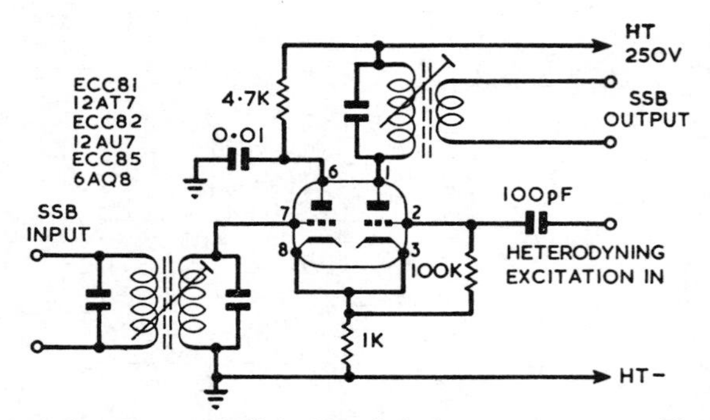

Fig 6.160. "Single-ended" double triode frequency converter. The tuned circuits should resonate at the appropriate frequencies.

Choice of Mixing Frequencies

It is the usual practice in ssb transmitting equipment to generate the initial single sideband signal at a fixed frequency and translate this signal to the required operating frequency by one or more frequency changing processes. The sideband signal *F2* is used to modulate a high frequency carrier *F1*, whose frequency is chosen so that either the upper or the lower sideband of the mixing process is on the desired operating frequency. As a result of this modulation process, the sideband signal will be translated to a new frequency that is either the sum of the carrier and sideband frequencies $F1 + F2$ or the difference between the carrier and the sideband frequency $F1 - F2$. Of further importance is the fact that if the lower sideband of the mixing process is selected, an inversion of the sideband occurs—an upper sideband signal will be converted to a lower sideband signal.

The two essential components of the translation system are the modulator (commonly called mixer) and the carrier (commonly called oscillator, injection, or heterodyne signal).

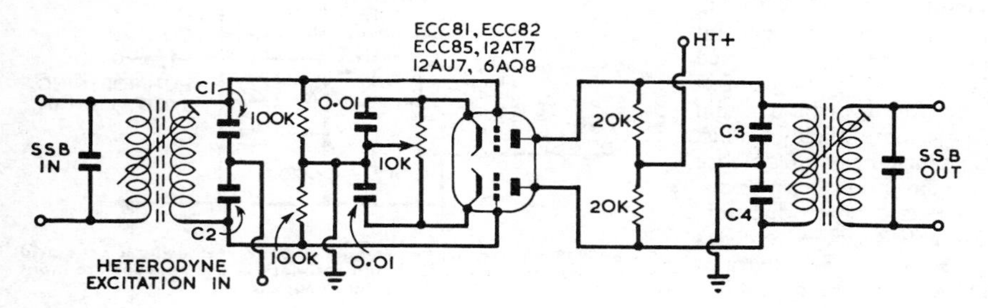

Fig 6.161 Double-triode balanced frequency converter.

In order to provide a mixer output waveform that is in linear relationship to the ssb input waveform, it is necessary to make the amplitude of the oscillator injection several times greater than the amplitude of the sideband input. A practical figure generally used is an oscillator injection voltage ten times (+ 20dB) greater than the maximum signal input.

If two frequencies $F1$ and $F2$ are fed into a mixer it is important to remember that the mixer output will contain not only the wanted $F1 + F2$ or $F1 - F2$, but additional frequencies composed of the following:

Funda-mental	*Second Order*	*Third Order*	*Fourth Order*	*Fifth Order*
$F1$, $F2$	$F1 \pm F2$	$2F1 \pm F2$	$3F1 \pm F2$	$4F1 \pm F2$
	$2F1$	$2F2 \pm F1$	$3F2 \pm F1$	$4F2 \pm F1$
	$2F2$	$3F1$	$2F1 \pm 2F2$	$3F1 \pm 2F2$
		$3F2$	$4F1$	$3F2 \pm 2F1$
			$4F2$	$5F1$
				$5F2$

Assuming the inputs were $F1 = 3{,}000$kHz and $F2 = 500$ kHz the mixer output would have these frequencies:

Funda-mental	*Second Order*	*Third Order*	*Fourth Order*	*Fifth Order*
	3,500	6,500	9,500	12,500
3,000	2,500	5,500	8,500	11,500
500	6,000	4,000	4,500	5,000
	1,000	2,000	1,500	1,000
		9,000	7,000	10,000
		1,500	5,000	8,000
			12,000	7,500
			2,000	4,500
				15,000
				2,500

(Wanted output $Fo = 3{,}500$kHz)

From these figures it is clearly seen that a train of frequencies has been generated, with a separation equal to $F2$ (the lowest input freq.).

The response or selectivity curve of an average tuned circuit of the kind likely to be used is shown in **Fig 6.162.** The two frequencies most difficult to eliminate will be the two either side of the wanted output Fo of 3,500kHz. These are the strong $F1$ heterodyning input of 3,000kHz and the third order product ($2F2 + F1$) of 4,000kHz. The curve shown is actually that of the 80m coil in position in an ssb transmitter. The Q of the circuit can be determined from the formula

$$\frac{Fo}{Fh - Fl} = \frac{\textit{centre frequency}}{\textit{bandwidth at 3dB points}}$$

In this particular case this gives the value of $Q = 35$. The response has been plotted with the coil and the associated capacitor in place on the chassis, and is affected by the shunt loading of the valve, the shunt ht feed and the following grid input resistance. It is considered to be more sensible to evaluate the function of the tuned circuit under actual operating conditions, rather than to consider the unloaded Q measured on a Q meter. The response shown is believed to be a fair average of tests taken on representative coils and at different frequencies.

The two frequencies it is desired to eliminate are 500kHz removed. Plotting the 500kHz points on the response curve gives an attenuation of 20dB and it is noted that the ratio $Fo : F2$ is 7 : 1. A tuned circuit with equal Q value, resonant at 7MHz would have the same response curve and is shown in Fig 6.162 by doubling the frequency scale. However, 500kHz off tune is now a ratio of 14 : 1 and the attenuation has dropped to 14dB. If the circuit were resonant at 1·75 MHz, 500kHz off tune would be a ratio of 3·5 : 1 and the attenuation would have increased to 25dB.

It is therefore clear that the lower the ratio $Fo : F2$ the better the attenuation from the tuned circuit. If two tuned circuits are used, coupled to give the highest possible selectivity, the total attenuation is twice that of a single tuned circuit and in this case would be $20 \times 2 = 40$dB.

Looked at the other way round it can be seen that if an attenuation of 40dB is required from two coupled tuned circuits in the mixer anode, the ratio of the lowest input frequency ($F2$) to the required output frequency (Fo) should not exceed a ratio of 7 : 1.

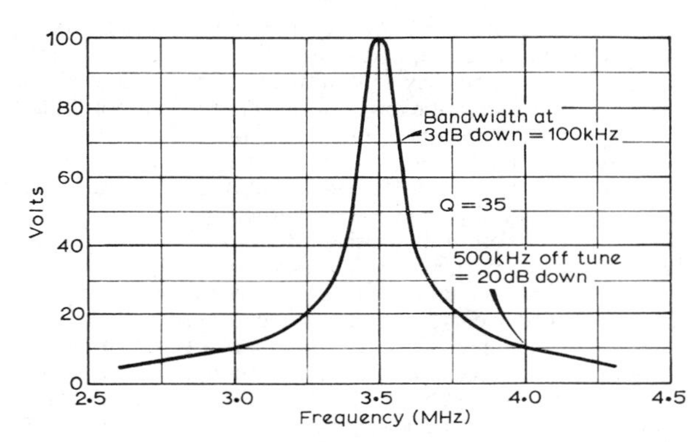

Fig 6.162. Resonance curve for an 80m coil. Frequency at Fo = 3·5MHz.

It is good design practice to aim for adequate suppression of all spurious conversion products and 60dB should be looked upon as a desirable target. With the usual initial ssb generation on 455kHz, a ratio of 7 : 1 sets the upper limit for Fo of $455 \times 7 = 3{,}185$kHz. It is possible to go to a ratio of 8 : 1 (ie direct from 455kHz into the 80m band; or direct from 80m to 10m) but it is safer and results in a cleaner output to keep within the ratio of 7 : 1. In practice coupling is often tightened beyond the "critical coupling" point to prevent too much loss of gain, and this would reduce the available attenuation to a slightly lower value than 40dB. This loss of attenuation could be offset by making the limit for the multiplying ratio 6 : 1. To sum up:

(i) In the interests of constructional simplicity it is desirable to keep the number of tuned circuits at the mixer output to two.

(ii) Assuming two coupled tuned circuits and normal Q values, the multiplying ratio of the frequency translation process should wherever possible not exceed 6 : 1.

(iii) A lower value than 6 : 1 will give greater attenuation of the unwanted products and therefore a cleaner output.

(iv) The lower multiplying ratio can be obtained either by increasing $F2$ (the ssb input), or reducing Fo (the wanted ssb output); or a combination of both.

Spurious Product Amplitude

In order to show how undesired frequencies are generated in a mixer stage, it is necessary to consider the case where oscillator and signal voltages are applied to the same grid of the mixer valve. The wanted sum or difference frequencies can only be generated if the anode current, grid voltage characteristics have some non-linearity or curvature. The components of the anode current will be the dc; signal; oscillator; signal second, third, fourth, fifth . . . harmonics; oscillator second, third, fourth, fifth . . . harmonics; and the sum and differences of the signal and oscillator.

To obtain the desired sum or difference product it would be necessary to use a valve in which the characteristic curve had only second order curvature. Unfortunately, all practical valves have characteristic curves having higher order curvature and this contributes additional unwanted frequency components into the output current. Sometimes the frequency of these unwanted components is sufficiently removed from the desired output frequency and they are easily filtered out, but often these frequencies are very near to the desired signal frequency and they will fall within the passband of the selective filter used in the mixer output circuit. The amplitude or strength of these undesired mixer products varies with valve type, and even with valve to valve of the same type. It is also critically dependent on the amplitude of the signal input, the bias point, and the amplitude of oscillator injection. It is therefore not surprising that valve manufacturers have not been particularly successful in designing valves having the desired second order curvature to the exclusion of any higher order curvature. In practice, the circuit designer must select his mixer valves by means of a series of experiments in which the amplitudes of these undesired mixer products are measured. A representative calculated determination of mixer product amplitude is shown in **Fig 6.163**.

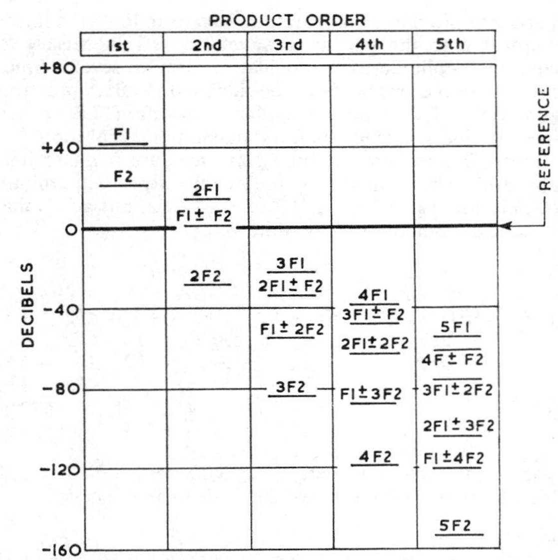

Fig 6.163. Calculated frequency products contained in the anode current of a 12AU7 triode mixer.

Consideration of this chart shows that there are several undesired products that are greater in amplitude than the desired signal and a considerable number that are weaker than the desired signal. Of further interest is the fact that as the product order increases, its amplitude decreases.

Avoiding Spurious Mixer Products

It is possible by an intelligent choice of the signal and oscillator frequencies to minimize the presence of undesired mixer products within the passband of the output circuits of the frequency translation system. The problem of frequency selection is relatively simple where the operating frequencies are fixed, but becomes increasingly complex where the operating frequency must be varied. In an attempt to simplify the problem, circuit designers make use of charts in which the frequency of the spurious mixer products is plotted with respect to the signal and oscillator frequencies. This type of chart is shown in **Fig 6.164**.

As an example of the spurious product problem, consider the case of an ssb transmitter for operation on the 80, 40 and 20m bands. The ssb signal has been generated at 455kHz using a mechanical or crystal filter. Coverage of the lowest frequency band 3·5 to 4kHz can be obtained by mixing the ssb signal with the output from a vfo tunable from 3·955 to 4·455MHz. As the required output is the low sideband of the translation process (ie the heterodyning frequency is *higher* than the wanted 80m output) there will be a sideband inversion. In order to give a low sideband signal in the 80m band, this means that the mechanical filter must be arranged to provide a high sideband output. It has become an accepted convention adhered to by all ssb stations, to transmit *low* sideband below 10MHz and *high* sideband above 10MHz as a result of a CCIR (International Radio Consultative Committee) recommendation.

The strong oscillator signal is only 455kHz removed from the wanted output frequency and must be filtered out by the following tuned circuits or else balanced out through the use of a balanced mixer. As the mixer cannot be expected to retain an accurate balance over a 500kHz tuning range, it is necessary to resort to a combination of both methods to obtain suppression of this spurious frequency of at least 60dB. At the higher operating frequencies required for the other amateur bands, it becomes increasingly difficult to suppress this product because the selectivity required in the tuned circuits is so high as to become impracticable. This difficulty can be overcome by using a second stage of frequency conversion.

The output from the first mixer at 3·5 to 4MHz is mixed with the output from a crystal oscillator. This could be on 3·5MHz, giving a tunable output of 7 to 7·5MHz for the 40m band; however, the second harmonic of the crystal will give a crossover with the wanted 7MHz output signal and in practice this would be avoided by using a lower frequency crystal—say 3·3MHz. Even with this choice of frequency, the second harmonic of the crystal at 6·6MHz is only 400kHz removed from the low-frequency end of the wanted output. The 7·4MHz second harmonic of the input signal (3·7 to 3·8MHz to give a band coverage of 7 to 7·1MHz) is only 400kHz on the other side of the wanted output frequency. A very high order of selectivity would be required to reduce these spurious signals to a satisfactory level. In practice the designer would overcome this problem by placing the heterodyning frequency on the high side of the wanted output frequency by using a crystal of 11MHz. The sideband input of 3·5 to 4MHz modulating the 11MHz carrier would produce a difference frequency output over the range

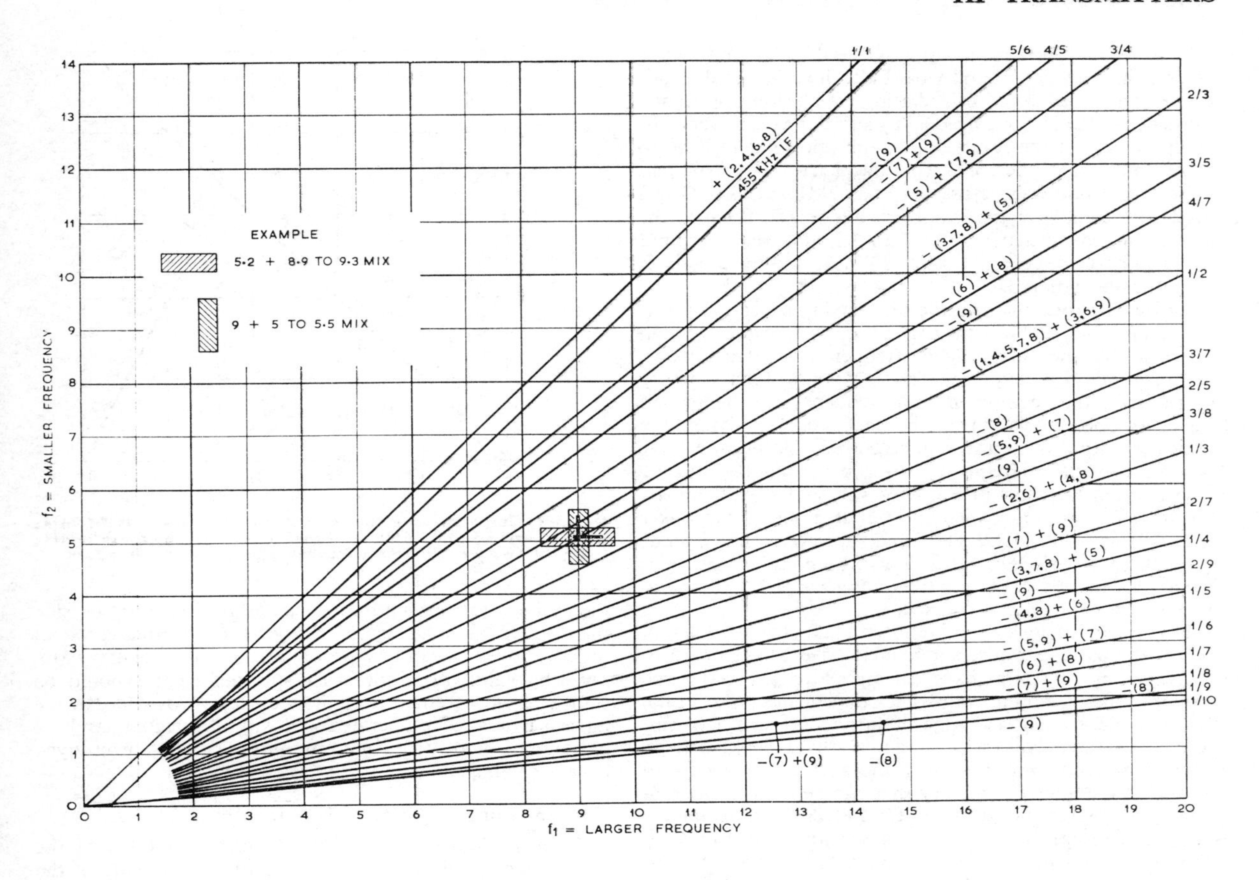

	$F2 \sim F1$								
Order	**1**	**2**	**3**	**4**	**5**	**6**	**7**	**8**	**9**
1 / 1		•20 •02		•13 •31		•24 •42		•35 •53	
1 / 2	10		•12 •30	31	32	•33 •51	52	53	•54 •72
1 / 3		20		•22 •40		42 51		•53 •71	
1 / 4			30		•32 •50		52	71	
1 / 5				40		•42 •60		62	
1 / 6					50		•52 •70		72
1 / 7						60		•62 •80	
1 / 8							70		•72 •90
1 / 9								80	
1 / 10									90
2 / 3			21		•23 •41		43	53	

	$F2 \sim F1$								
Order	**1**	**2**	**3**	**4**	**5**	**6**	**7**	**8**	**9**
2 / 5				41			•43 •61		63
2 / 7							61		•63 •81
2 / 9									81
3 / 4					32		•34 •52		54
3 / 5						42		•44 •62	
3 / 7								62	
3 / 8									72
4 / 5							43		•45 •63
4 / 7									63
5 / 6									54

• INDICATES SUM MIXING
OTHER DIFFERENCE MIXING

Fig 6.164. Spurious response chart.

(Reproduced, by permission, from the 2nd Edition of "Fundamentals of Single Sideband" (Published by Collins Radio Company, USA) where the subject is treated in greater detail.)

7·5 to 7MHz. Unfortunately this would also produce a reverse dial calibration and would also invert the sideband; ie the transmitter would be radiating a high sideband signal. The sideband could be corrected by switching carrier crystals at the filter, but the reverse dial calibration would have to be accepted. It will be noted from the chart of Fig 6.164 that there is a second order crossover at 3·67MHz; this is shown in greater detail by the graph of **Fig 6.165**. Because the 40m band only extends from 7 to 7·1MHz the ssb input (*F2*) will never be required below 3·9MHz and the crossover may be safely ignored.

As the 80m output from the first mixing process is a low sideband signal, and it is the convention to transmit the high sideband on the three higher frequency amateur bands, the heterodyning frequency for 14MHz will have to be higher than the required output frequency, ie the second translation process will invert the sideband, thus giving high sideband output from a low sideband modulating signal. A suitable conversion frequency for 20m would be 18MHz. Inspection of the chart in Fig 6.164 will show that there is a fourth order spurious when the signal input to the mixer (*F2*) is near to 3·5MHz. The band, however, only extends from 14 to 14·350MHz and this will be given by an input signal of 4 to 3·65MHz, and the tuning range 3·65 down to 3·5MHz will not need to be used.

There is an obvious temptation to include the 160m band. Where the basic exciter has a 500kHz coverage from 3·5 to 4MHz many amateurs have attempted to get 160m ssb output by heterodyning with some convenient crystal in the easily obtainable FT243 range of surplus crystals. Remembering the 160m band has a bandwidth of 200kHz from 1·8 to 2MHz, consideration of the factors involved will show that the lowest frequency that can be used for heterodyning is 5·5MHz and the highest is 5·8MHz. Unfortunately the mixing process will produce the second harmonic of the 80m ssb output and this will also be heterodyned by the crystal oscillator to produce a spurious distorted signal. (This is the third order difference product *2F2 — F1*.) When the vfo is altered this spurious signal changes frequency at twice the rate of the wanted signal and in the opposite direction. **Table 6.19** has been compiled to show this clearly.

Many existing exciters are peaked to give maximum output at the hf end of the 80m band and the logical choice of heterodyning crystal would be 5·65 or 5·7MHz. Under these conditions the spurious output is either on or very close to the wanted output. If any adjustment is made to the main tuning, either initially to find a clear channel or to avoid interference, the spurious output will move twice as fast and in the opposite direction. While it is appreciated that this output will not be as strong as the wanted one, it will most certainly be radiated by the aerial with sufficient strength to cause interference to stations in a wide area. It can at certain settings of the main tuning actually be on the wanted output frequency and put severe distortion on the transmitter signal.

It is clear from the examples shown that the design of a multi-band exciter is by no means a simple problem if spurious outputs are to be avoided—the correct choice of operating frequencies used in each of the frequency translation processes is of major importance. With a one or two band exciter the problem is relatively simple. As the requirement for multi-band operation increases—particularly if this is a six band coverage from 160m to 10m—the difficulties become more complex and can only be satisfactorily solved by an additional stage of frequency conversion together with the initial tunable ssb output on a neutral frequency outside the wanted amateur bands. In general spurious product orders above five can be ignored because the level of a sixth, seventh, eighth or ninth order spurious signal should be sufficiently low to have no nuisance value, provided that the mixer valve is being operated well within its signal handling capabilities and with the optimum level of heterodyne injection voltage.

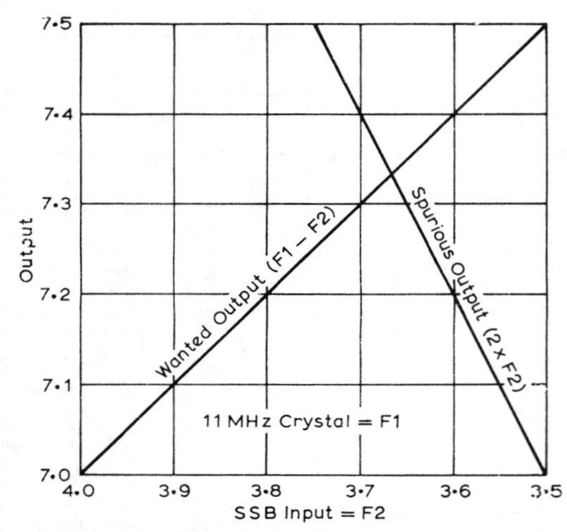

Fig 6.165. Graph showing the second order crossover at 7·34MHz when a basic 3·5 to 4·0MHz ssb signal is heterodyned by an 11MHz crystal to give a difference frequency output in the 40m band.

Heterodyning Oscillator Requirements

From any ssb transmitter the frequency stability of the output signal is dependent on the frequency stability of the heterodyning inputs to the frequency changer valves. The total frequency error is the arithmetic of the errors in all these oscillators. For the purpose of amateur requirements the crystal oscillator can be considered stable; the stability of the transmitted signal will then be directly dependent on the stability of the variable LC oscillator, the vfo.

It is common practice on the ssb bands for a number of stations to operate in a "net" (all stations accurately tuned to the same carrier frequency) and carry on a quick "backwards and forwards" normal type of conversation. Under

TABLE 6.19

Basic 80m ssb output	Second harmonic of basic 80m ssb output	Heterodyning crystal frequency	160m output	
			Wanted	Spurious
3·5MHz	7·0MHz	5·5MHz	2·0MHz	1·5MHz
3·6MHz	7·2MHz	5·5MHz	1·9MHz	1·7MHz
3·7MHz	7·4MHz	5·5MHz	1·8MHz	1·9MHz
3·6MHz	7·2MHz	5·6MHz	2·0MHz	1·6MHz
3·7MHz	7·4MHz	5·6MHz	1·9MHz	1·8MHz
3·8MHz	7·6MHz	5·6MHz	1·8MHz	2·0MHz
3·7MHz	7·4MHz	5·7MHz	2·0MHz	1·7MHz
3·8MHz	7·6MHz	5·7MHz	1·9MHz	1·9MHz
3·9MHz	7·8MHz	5·7MHz	1·8MHz	2·1MHz
3·8MHz	7·6MHz	5·8MHz	2·0MHz	1·8MHz
3·9MHz	7·8MHz	5·8MHz	1·9MHz	2·0MHz
4·0MHz	8·0MHz	5·8MHz	1·8MHz	2·2MHz

these conditions it is most important that all stations remain accurately on frequency with negligible drift. With a transmitter that is moving off channel because of frequency drift in the vfo speech can no longer be demodulated correctly at the receiver. Quite obviously vfo stability is highly valued by ssb workers. Ever since amateur sideband commenced, thousands of experimenters and enthusiasts have been searching for that very elusive object, the driftless vfo.

In regard to vfo stability it should be clearly understood that, notwithstanding the many claims that have been made by the authors of various "ultimate" circuits published in the past, there is no such thing as a driftless LC oscillator. A quartz crystal has a high degree of frequency stability because quartz is a material with a low temperature co-efficient. Replacing the crystal by building an equivalent series tuned circuit using L and C—as for instance in the so-called Clapp vfo—does not give the same standard of stability as a crystal, since it is not possible to manufacture standard coils and variable capacitors with the temperature co-efficient of natural quartz.

Certain circuits such as the Tesla/Vackar and the Clapp use large values of fixed swamping capacitance across the oscillator valve and claim that this reduces the effect of changing valve input capacitance during initial warm-up. An LC oscillator that drifted 2 or 3kHz during the first 10 or 15 minutes and was thereafter rock stable, would, however, be very acceptable to almost everyone—no amateur would mind switching on and waiting 15 minutes before he intended to transmit. In practice the major annoyance is the long term drift taking place continually over the transmitting period of a couple of hours or so. This slow drift is caused by the changing temperature of the two components that make up the frequency determining resonant circuit, the L and the C. Aside from the temperature rise of the air in the cabinet and the air in the transmitting room, a considerable amount of heat from the valves—though the valveholders, the screening cans and skirts—warms up the chassis and this in turns warms up the vfo tank coil and tuning capacitor.

Stability can be materially improved—not by a change to some "ultimate" circuit—but by re-building the vfo so that the coil and tuning capacitor and associated components are built as a self-contained unit mounted on a sub-platform made of some poor heat conducting material such a poly-styrene or Perspex. The "tuning unit" is then enclosed in a heat insulating "box" made from cellular polythene sheet ½in. or so in thickness. The vfo valve or valves are left in their original positions on the main chassis. As the object is to isolate the tank circuit elements from conducted heat, the variable capacitor shaft is coupled to the dial drive mechanism by a poor heat conducting flexible coupler such as the Eddystone Type 529 or Type 50.

That there is no simple solution to the problem of vfo stability is borne out by the fact that a number of the leading world communication equipment manufacturers have spent a considerable amount of time and effort in developing phase-locked or crystal synthesizer variable master oscillators and that these are now being supplied not only to the commercial market but to the amateur market as well.

The requirement for a home constructed amateur vfo are:

(i) constant amplitude of output voltage over a 500kHz tuning range;
(ii) low harmonic output;
(iii) "simple" circuitry using standard components and valves;
(iv) after an initial warm-up period of 10 to 15 minutes frequency drift not to exceed 10 parts per million (10Hz per MHz) per hour.

With reasonable care in construction and some experimental work to determine the required value of negative temperature co-efficient capacitor compensation, these parameters can be met with the standard tunable oscillators described earlier in this chapter.

THEORY OF LINEAR AMPLIFICATION

Most modern low-power ssb transmitters are capable of driving a high power linear amplifier. Much effort will have gone into the construction and adjustment of an exciter capable of an acceptable degree of carrier and sideband attenuation, so care must be taken to ensure that subsequent amplifier stages do not degrade the overall performance of the transmitter by introducing avoidable distortion. This makes it essential to operate all ssb amplifiers under strictly linear conditions.

RF Linear Power Amplifiers

The function of the power amplifier in a ssb transmitter is to raise the power level of the input signal without change, so that the envelope of the output signal is a faithful replica of the envelope of the input signal.

Radio frequency amplifiers are classified A, B, and C according to the angle of anode current flow—the number of degrees the anode current flows during the 360° rf cycle. The class A amplifier has a continuous anode current flow and operates over a small portion of the anode current range of the valve. This amplifier is used for amplification of small signals where low distortion is required. Its efficiency in converting dc anode power input into rf power output is low, of the order of 30 per cent, but this is not of great importance where small signals are concerned (see **Fig 6.166.**)

Class B amplifiers are biased to near anode cut-off so that anode current flows for approximately 180° of the rf cycle. Amplifiers operating with more than 180° of anode current flow, but less than 360°, are called class AB amplifiers. Both class AB and class B amplifiers are used in high power

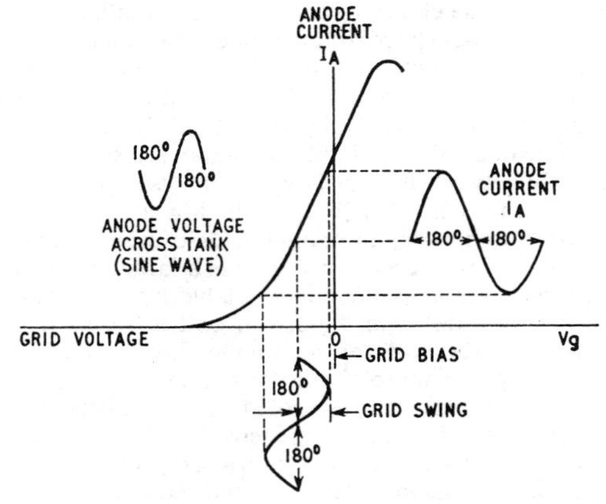

Fig 6.166. Class A operation of a valve (dynamic characteristics).

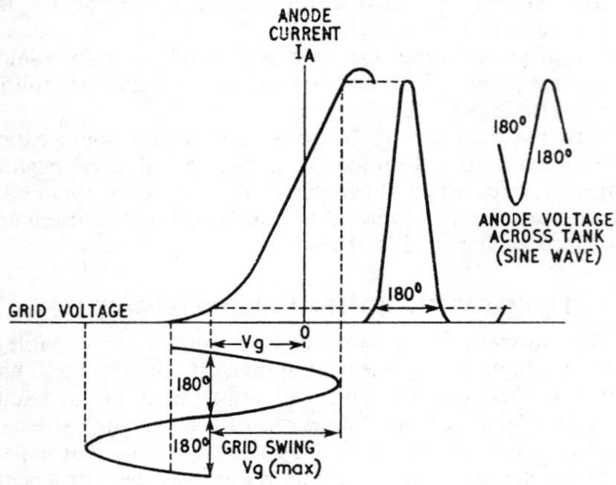

Fig 6.167. Class B operation (dynamic characteristics).

linear amplifier stages to obtain higher efficiency and maximum output power with low distortion. The amplifier efficiency depends on the operating condition selected and the type of valve used and is usually or the order of 50 to 66 per cent (see **Fig 6.167**).

Class C amplifiers are biased well beyond cut-off so that anode current flows for less than 180° of the rf cycle. The principal advantage of this mode of operation is high anode efficiency (of the order of 65–80 per cent); however, the class C amplifier is not suitable for ssb use because the amplifier is not linear and it will not respond to low level input signals (see **Fig 6.168**). The amplifier class can be followed by a number to indicate whether or not the valve is operated in the positive grid region over part of the rf cycle. Class AB1 indicates that the grid never goes positive and that no grid current is drawn; class AB2 indicates that the grid goes positive and that grid current is drawn. Class A amplifiers are nearly always operated without grid current and class C amplifiers are nearly always operated with grid current. It is therefore normal practice to refer to these as class A or class C without the necessity for further designation.

Choice of Valve

There is a wide choice of triode valves suitable for power amplifiers and these have the advantage of simplicity and low cost. Generally, they require a large amount of driving power and because of the considerable grid-to-anode capacitance the valve must be carefully neutralized. The amplification factor of triode valves suitable for amateur ssb application is generally between five for low mu triodes and 75 or more for high mu triodes. Usually only the low and medium mu valves are suitable for the requirements of power amplifiers and it is therefore necessary to provide a large grid swing to obtain the power amplification available from the valve.

With the tetrode valve the screen grid is an electrostatic shield between the grid and the anode and this reduces the grid anode capacitance to about one hundredth of that of the triode. It is still necessary to provide neutralization because of the higher amplification of the tetrode valve but the small value of feedback eases the neutralization problem. Because of the higher gain available, the valve requires a relatively low drive to obtain a high power output.

Pentode construction is also available in power amplifier valves and these give improved efficiency because the rf anode swing can be increased. The pentode, however, is more complex and expensive than the tetrode and in some cases requires additional supply voltage for the supressor grid. These disadvantages have limited the development of pentode power valves and at present the pentode has little advantage over the well-designed tetrode.

For linear power amplifier operation the following features are desirable:

(i) high gain;
(ii) good efficiency;
(iii) low grid to anode capacitance;
(iv) linear characteristics at all frequencies within the desired operating range.

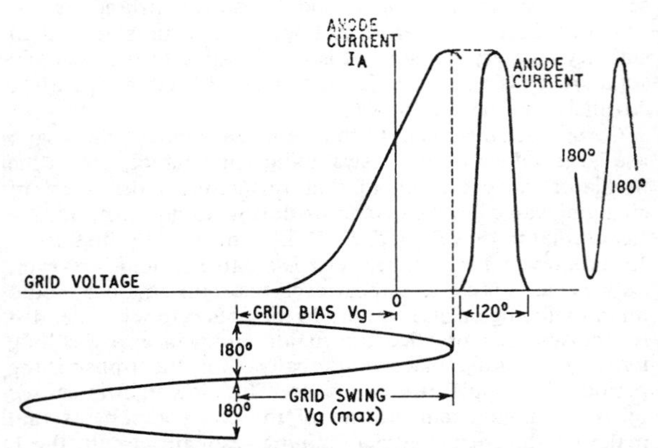

Fig 6.168. Class C operation (dynamic characteristics).

When a valve is operating in class A the degree of linearity is quite high but the efficiency is low—of the order of 25–30 per cent. By operating class AB1 the principal advantages of class A operation are retained while the efficiency is raised to between 50 and 65 per cent.

If operation is further advanced into the class AB2 region, the efficiency is improved only slightly, but any gain in this direction is more than offset by the more stringent requirements imposed upon the driver stage brought about by variation in driver loading as the amplifier is driven into the positive region. This can, and often does, result in a compression of the modulation peaks and non-linearity unless the driver is capable of providing the increased loading. Usually, in order to maintain driver regulation in class AB2 linear amplifier operation, a relatively large percentage of the driver output is dissipated in a swamping resistor so that when the grid is driven positive the relative increase in driver loading is small. After careful consideration of these factors one finds that class AB1 operation has much to offer.

Operating Methods

There are four basic methods of operating a linear amplifier as follows:

(i) Grid driven—anode neutralized;
(ii) Grid driven—grid neutralized;
(iii) Cathode driven—grounded grid;
(iv) Grid driven—passive grid.

These basic circuits are shown in **Figs 6.169, 6.170, 6.171**

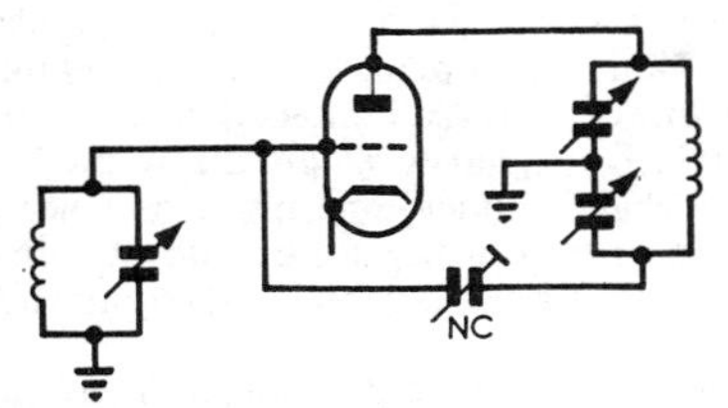

Fig 6.169. Grid driven anode neutralized amplifier.

and **6.172** and in each circuit configuration the valve can be either a triode or a tetrode.

At this stage it might be advisable to consider what exactly is meant by the term *linear amplifier*. To the high fidelity music lover a linear amplifier is intended to give superb quality with the lowest level of distortion. To the single sideband operator a linear amplifier can be added to an existing exciter and will give a more powerful signal without impairing the speech quality. In fact the hi-fi enthusiast and the sideband operator when they talk about a linear amplifier are talking about the same thing. In theory, a good hi-fi linear amplifier can be turned into a good low distortion sideband amplifier by replacing the audio input and output transformer by rf tank circuits. In addition, because the flywheel effect of the tuned circuits puts back the missing half cycle it is not even necessary to use two valves in push-pull. Finally, the operating parameters of anode, screen and grid supply voltages, anode load resistance, grid driving voltage and power output, supplied by the valve manufacturer for audio operation, apply equally for single sideband rf service. An example of this is given in **Table 6.20** for the valve type 811A.

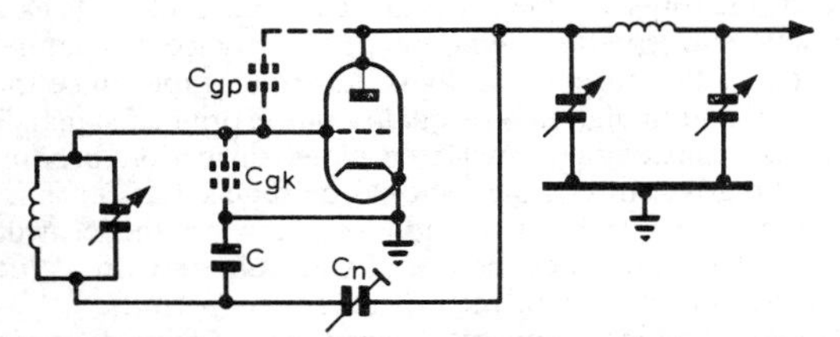

Fig 6.170. Grid driven grid neutralized amplifier. For neutralization, $C_{gp}/C_{gk} = C_n/C$. C may be any convenient value between 250 and 500pF.

Is Linearity Necessary?

For sideband service the envelope of the signal in the anode circuit must be an exact replica of the envelope of the exciting signal. This implies that the power gain of the stage must be constant regardless of the signal level. This desirable basis of operation can only be obtained if the amplifier is operated in a linear manner. Any non-linearity creates distortion products that appear both in the signal passband and in those channels adjacent to it.

A practical place to examine a sideband signal for linearity and quality is in the adjacent suppressed sideband channel, not in the frequency band of the signal itself.

The excellence of a sideband signal is judged by the amount of (or lack of) sideband splatter in nearby channels. Theoretically, a sideband signal should be just as wide as the voice passband of the equipment—3kHz—and no wider. If the output signal of a linear amplifier stage is a replica of the existing signal there will be no distortion products—however,

TABLE 6.20

811A Operating Data

	Class B Audio Service (Two valves)	Class B RF Service (SSB) (One valve) Grid Driven	Grounded Grid
Anode Voltage	1250	1250	1250
Grid Bias	0	0	0
Peak Grid Voltage	175	88	88
Zero Sig. Anode Current (mA)	54	27	27
Max. Sig. Anode Current (mA)	350	175	175
Load Resistance (ohms)	9200	4600	4600
Max. Sig. Grid Current (mA) *	26	13	13
Power Output (W)	310†	155†	141‡

*Varies from valve to valve. †Computed power output. ‡Measured including circuit loss.

The operating parameters of a class B amplifier stage remain the same regardless of whether the valve functions in audio or rf service. Grounded grid operation is similar, except that the exciter must supply additional feed through power. Since class B audio service requires two valves, all currents and anode load resistance must be halved for single valve rf service. Class B audio data is readily available for most valves and can be used for rf service as shown above.

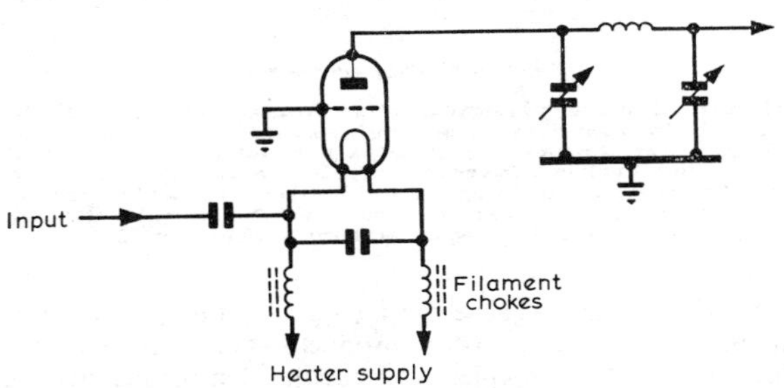

Fig 6.171. Grounded grid (cathode driven) amplifier.

valves are not perfect and the transfer characteristics of even the best linear amplifiers exhibit non-linearity at the extremes of the anode current swing. So long as the signal input is a single tone (such as inserted carrier, or a single tone into the microphone socket) departure from linearity has no effect but, if the signal input contains two or more tones, the non-linearity of the power amplifier will cause "mixing" of the signal source and will produce new additional sum and difference frequencies that were not present in the original input signal. These new frequencies, generated in the power amplifier, are known as intermodulation distortion products.

The standard method of testing a linear amplifier to determine the level of distortion products is the two-tone test, in which two radio frequencies of equal amplitude are applied to the amplifier and the output signal is examined for

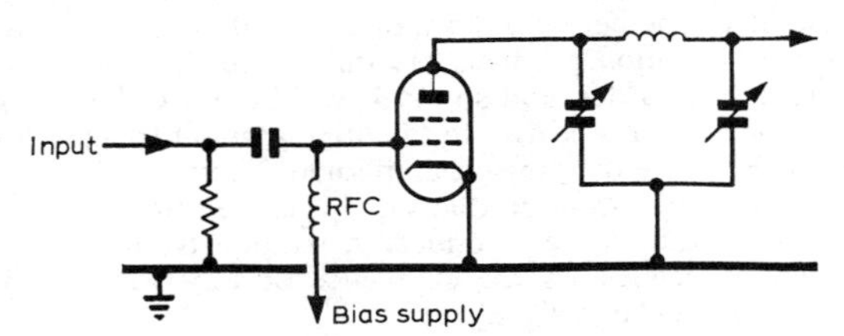

Fig 6.172. Grid driven passive grid amplifier.

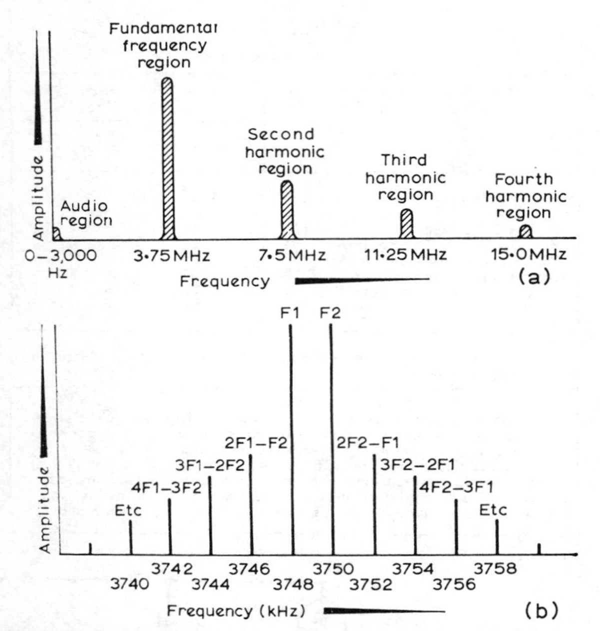

Fig 6.173. Odd order intermodulation products causing ssb distortion. The frequencies shown assume a transmitter with a carrier on 3750kHz (radiating lower sideband) and modulated by a 2kHz tone input. The amplifier is driven at carrier frequency (F2) by unbalancing the modulator or using carrier insertion. The audio input of 2kHz produces the second frequency of 3748kHz (F1). Controls are adjusted for equal amplitude of F1 and F2.

spurious products. Those output signals falling in the harmonic region—"even order" products—are attenuated to a low level by the amplifier tank circuits. Unfortunately, the "odd order" products fall close to the fundamental output frequencies and cannot be removed by tuned circuits. These are the distortion products that put back the signal on the unwanted sideband and in the case of a poorly designed or incorrectly operated linear amplifier, cause objectionable splatter.

Fig 6.173(a) shows the spectrum distribution of the products generated in a typical pa stage, while **Fig 6.173(b)** shows on an expanded frequency scale those intermodulation products within the amplifier passband that cause ssb distortion. In the example shown the two frequencies making up a typical two tone test are 3,748 and 3,750kHz. If the linear amplifier is perfect these will be the only frequencies appearing in the output. In practice, the amplifier is not perfect and there will be additional combinations of sum and difference frequencies generated by the non-linear transfer characteristics of the valve. These odd order products fall within the passband of the selective output circuits and will be radiated together with the wanted signal. The inside pair of intermodulation products are third order, the next fifth order, seventh order, and so on. It will be noted that those distortion products nearest to the original input frequencies, F1 and F2, have the greatest amplitude. These are the third order intermodulation products and it is the relative amplitude of the third order products in relation to the wanted signal that determines the excellence or otherwise of the transmitted sideband signal.

Fortunately as far as amateur operation is concerned, it is not necessary to have or be able to use elaborate test equipment. Many ssb operators have selectable sideband receivers with correctly calibrated S meters and they can check the relative signal level on the wanted and on the unwanted sideband, under tone input or voice conditions, and give a report of so many decibels down with a degree of accuracy that is reasonably high and quite adequate for amateur use.

The reduction of adjacent channel interference in the crowded amateur bands is ultimately of benefit to all. There is obviously little sense in aiming for a rejection of at least 35dB in the filter and then allowing a badly operated linear amplifier to put the unwanted sideband back again in the form of objectionable intermodulation distortion products. The aim therefore is a distortion product level at least as good as the filter (35dB down).

Reducing Distortion

Consideration has been given to the effect on the output signal of non-linearity in the power amplifier and how such non-linearity causes "mixing" and the generation of odd order intermodulation products. It will be noted that the frequency spacing of the distortion products is always equal to the frequency difference between the two original tones. A voice signal is made up of a multiplicity of tones—there will therefore under voice operating conditions be a multiplicity of intermodulation distortion products. These will be present in the transmitted signal and will be heard on the unwanted sideband as blurred and distorted speech that is completely unintelligible—in short as splatter.

When a linear amplifier is improperly adjusted or overdriven the spurious frequencies rise in amplitude and also extend far outside the original channel, and will cause unintelligible splatter interference in adjacent channels. Splatter of this type is usually of far more importance than the effect on intelligibility or quality of the original signal. To minimize unnecessary interference the distortion products falling in adjacent channels should be reduced as far is it is possible to get them. Common courtesy on the crowded amateur bands dictates the use of transmitters with as little distortion as the state of the art reasonably permits.

The first and most important means of reducing distortion in a single sideband linear power amplifier is to choose a valve with a good anode characteristic and choose the operating conditions for low odd order curvature. **Fig 6.174** shows the anode characteristic and the operating point that will allow class AB operation with no odd order distortion products. From point *A* to *B* the curvature is second order or a simple ($I_a = kV_g^2$) curve. From point *B*, the curve continues at the same slope in a straight line to point *C*. The zero signal operating point *Q* is located midway horizontally between *A* and *B*. It is also located directly above the point of projected cut-off, point *P*, where an extension of *CB* crosses the zero anode current line.

Small signals whose peak-to-peak amplitude is less than the horizontal distance between *A* and *B* operate on a pure second-order curve, resulting in no single sideband distortion. When the input signal becomes greater than *AB* it enters a linear region on both peaks at the same time and since the slope of *BC* is correct there is no change in gain of the fundamental components and no single sideband distortion will result at large signals either. The anode current at point *Q* determines the static anode current I_a of the valve

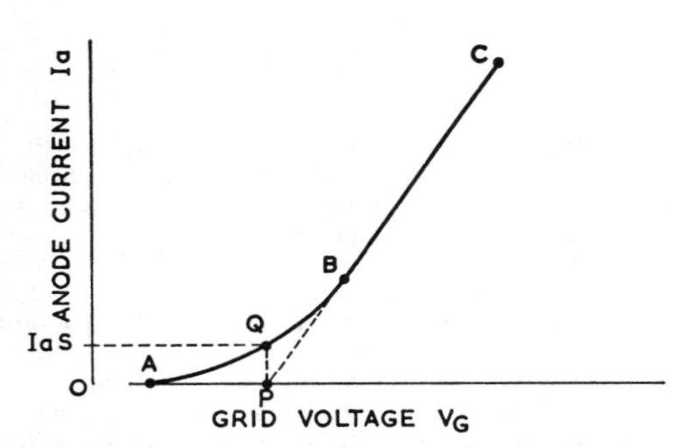

Fig 6.174. Ideal valve characteristics for class AB operation.

and, when multiplied by the dc anode voltage, determines the static anode dissipation.

Most valves have a characteristic similar to Fig 6.174, although *AB* is not a pure square law and the region from *B* to *C* is rather limited and seldom straight. In practice, however, an anode current/grid voltage curve can be plotted from the desired load line on a set of constant current curves, or obtained from the valve manufacturer. By projecting the most linear portion of this curve to intersect with the zero anode current line, the point of projected cut-off and therefore the grid bias and static anode current can be determined. This static anode current is the correct value for minimum distortion.

The screen voltage of the tetrode valve has a very pronounced effect on the optimum static anode current because the anode current of a valve varies approximately as the three-halves power of the screen voltage. For example, raising the screen voltage from 300 to 500V will double the anode current. The shape of the dynamic characteristic will stay nearly the same; however, the optimum static anode current for minimum distortion is now also doubled. In practice a limit is reached when the higher static anode current and therefore the higher static anode dissipation exceeds the rated anode dissipation for the particular valve in use. Should this condition arise it is necessary to make a choice between operating the valve at lower than optimum static anode current or alternatively reducing the screen voltage.

Although single sideband has now been in use for many years there is not yet any linear power amplifier arrangement that has shown itself to be superior either in performance or characteristics to an extent where it has become accepted as a standard particularly suitable for the requirements of ssb.

The type of linear amplifier that would be most suitable is dependent on circumstances that vary from one station to another. In fact, the choice must be made by the operator because it is inherently a personal one. It is possible, however, to give an indirect answer to the question by outlining a simple "design consideration" procedure as follows:

(i) Determine the ht supply voltage that is to be used. (This may be from an existing power pack, or may depend on the use of components that are already available, or it may be determined by considerations of safety, ie, many amateurs fight shy of using voltages much above 750.)

(ii) Determine the driving power that is available from the exciter. (This can be obtained from the manufacturers' data for the driver valve in use, or can be measured into a dummy load, or approximated by lighting a suitable lamp load.)

(iii) By reference to **Table 6.21** choose a valve suitable for the ht supply available.

(iv) By reference to **Table 6.22**, decide on the basic method of operation suitable for the driving power that is available.

TABLE 6.21

500V	500-750V	750–1250V	1500–2000V	2000–3000V
2E26 829B	6146 807 1625	TT21 807 1625 805 811A 4-65A 4X150A EL38 QV08-100	813 4-125A 4X150A 304TL 4X250B	304TL 4-250A 4-400A PL-6569

The values given in Tables 6.21 and 6.22 are approximations intended to serve as a guide. In Table 6.22 the "Driving Power" includes:

(*a*) The grid circuit and coupling losses,
(*b*) The input damping losses of the valve,
(*c*) The loss in the grid circuit swamping resistor,
(*d*) The grid current which may flow,
(*e*) In the grounded grid application, the percentage of driving power that appears in the output circuit as "feed through power." The driving power for the valve shown in the class AB1 column assumes a grid swamping resistor of 2,000Ω.

As the input impedance of a grounded grid amplifier is a function of the peak cathode current, the driving power required will be greater for four valves than for two. The number of valves in use is indicated by (× 4) or by (× 2). It is

TABLE 6.22
Basic Method of Operation

Driving Power (P.E.P.) output	Tuned Grid AB1	Tuned Grid AB2	Passive Grid	Cathode Driven (Grounded Grid)
5W	2E26 6146 4-65A 4X150A 807, 1625 QV08-100 TT21 EL38	829B		
10W	4-125A 813	807 1625 4-65A 813 4-125A		
25W	304TL	805 811A	813 4-65A 4-125A 4X150A	
50W		304L	805 811A 4-250A 4-400A	813 4-125A 4X150A 805 (×2) 811A (×2)
100W			304TL PL6569	304TL 811A (×4) 805 (×4) PL6569

assumed in all cases that where two or more valves are used in the amplifier, they are operated in parallel.

Having decided on the type of valve that is most suitable for the ht supply and driving power available, the valve manufacturers' data can be consulted for the p.e.p. output rating. If this is less than the total output required it will be necessary to run two, three or four valves in parallel. To give two examples:

(i) The choice of valve type is 6146 with a 750V ht supply. The required maximum signal power output (p.e.p.) is 200W. This will require four valves in parallel.

(ii) The choice of valve type is 813 with a 2,000V supply. The required output is 400W p.e.p. This can be obtained with two valves in class AB2.

Basic Circuit Considerations

As a guide to the choice of basic methods of operation given in Table 6.22, the main advantages and disadvantages can be summarized as follows:

Tuned Grid, Class AB1

Advantages

(*a*) Low driving power.

(*b*) As there is no grid current the load on the driver valve is constant.

(*c*) There is no problem of grid bias supply regulation.

(*d*) Good linearity and low distortion.

Disadvantages

(*a*) Requires tuned grid input circuit and associated switching or plug-in coils for multi-band operation.

(*b*) Amplifier must be neutralized.

(*c*) Lower efficiency than class AB2 operation.

Tuned Grid, Class AB2

Advantages

(*a*) Less driving power than passive grid or cathode driven operation.

(*b*) Higher efficiency than class AB1.

(*c*) Greater power output.

Disadvantages

(*a*) Requires tuned grid input circuit.

(*b*) Amplifier must be neutralized.

(*c*) Because of wide changes in input impedance due to grid current flow there is a varying load on the driver valve.

(*d*) Bias supply must be very "stiff" (have good regulation.)

(*e*) Varying load on driver valve may cause envelope distortion with possibility of increased harmonic output and difficulty with tvi.

Passive Grid

Advantages

(*a*) No tuned grid circuit.

(*b*) Due to relatively low value of passive grid resistor, high level of grid damping makes neutralizing unnecessary.

(*c*) Constant load on drive valve.

(*d*) Compact layout and simplicity of tuning.

(*e*) Clean signal with low distortion level.

(*f*) Simple circuitry and construction lending itself readily to compact layout without feedback troubles.

Disadvantages

(*a*) Requires higher driving power than tuned grid operation.

Cathode Driven

Advantages

(*a*) No tuned grid circuit.

(*b*) No neutralizing. (This may be necessary on 10m.)

(*c*) Good linearity due to inherent negative feedback.

(*d*) A small proportion of the driving power appears in the anode circuit as "feed-through power."

Disadvantages

(*a*) High driving power—greater than the other methods.

(*b*) Isolation of the heater circuit with ferrite chokes or special low capacitance wound heater transformer.

(*c*) Wide variation in input impedance throughout the driving cycle causing peak limiting and distortion of the envelope *at the driver valve*.

(*d*) The necessity for a high-*C* tuned cathode circuit to stabilize the load impedance as seen by the driver valve and overcome the disadvantage of (*c*).

(*e*) In practice, with the type of valves commonly used by United Kingdom amateurs, the power output appears to be the same or slightly less than that for passive grid operation. The active anode current flows through the cathode circuit and produces across the cathode impedance a voltage which decreases the exciting voltage. This corresponds to negative feedback, and it is possible that the loss due to this feedback can be roughly equal to the feed-through power—the net advantage is then zero.

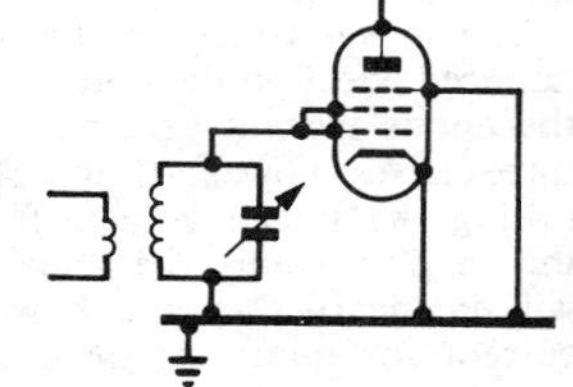

Fig 6.175. High-mu triode connection of a tetrode or pentode amplifier valve.

Beam power valves such as the 807, 1625 and 813 may be connected as high-mu triodes in the manner shown in **Fig 6.175**. The triode-connected tetrode usually requires a much higher excitation voltage than a conventional triode, but is otherwise capable of similar performance. It is important, however, to remember that this high excitation voltage appears between the grid and cathode, and there is no grid-bias voltage to be overcome before grid current can flow. As a result of the large driving voltage and high current flow, triode-connected tetrodes often operate with control-grid dissipation powers in excess of the manufacturers' ratings. This may happen inadvertently if the grid circuit has only a single current meter because up to 75 per cent of the combined grid current may be to the control grid in such a circuit. Should the valve be cathode driven in a grounded grid circuit, with insufficient anode circuit loading, the power formerly fed through the amplifier into the output circuit becomes available to heat the control grid to even higher temperatures. Such action can destroy the valve in a short time.

While it is possible to protect the valve with a grid-current overload relay of low coil resistance and sufficient sensitivity, it is hardly likely that a suitable component would be available to the amateur constructor. Conventional tetrode connection (either grid driven or cathode driven) will avoid the risk of damage to the valve and is the preferred method.

Push-pull or Parallel Operation

In theory two valves operated in linear amplifier service, either in push-pull or in parallel, will give the same maximum signal power output. However, there may be certain *practical* considerations that make one method more desirable than the other.

A single sideband amplifier must operate at all times in the most linear manner if objectionable intermodulation distortion is to be avoided. This means that the valve (or valves) must operate into a precise value of anode load: this may vary in value from a few hundred ohms up to several thousand ohms depending on the valve type, operating potentials and required maximum signal power output. The anode load (R_L) is then required to be stepped down in impedance to the design value of the amplifier output, usually 50Ω. From this it follows that the tank circuit is arranged as a step-down transformer.

As a precaution against the possibility of tvi it has become accepted practice to feed the transmitter output via a screened cable into a low-pass filter (capable of 40dB or more attenuation to any rf in the television channels) and then either directly or via an aerial tuning unit into the aerial system. As it is reasonably cheap and easy to obtain, coaxial cable is generally used, and modern transmitters are therefore designed to have a tank circuit with an unbalanced output.

Of further advantage is some simple means—preferably from a panel control—of varying the loading so that the amplifier operation can be adjusted while the output waveform is monitored on an oscilloscope.

The old convention of plug-in tank coils is now considered to be completely out-of-date, and modern transmitters are designed for rapid changes from band to band by means of switching. It obviously would be quite a complex electrical and mechanical operation to arrange switching for five push-pull anode coils and five output link windings.

For these reasons the pi-network circuit has become deservedly popular; the anode tuning capacitor need not be a split stator type, inductance change is easily effected by suitable taps on a single coil, and the required switch is a simple single pole type. It is easy to see why in modern transmitters the pi-network is almost universally used. As both the input and the output of the tank circuit are single ended, two or more pa valves are always operated in parallel.

Finding the Value of the Anode Load

A large number of valve types are suitable for linear amplifier use but in many cases manufacturers' figures for single sideband service are not available. It is therefore necessary for the constructor to work out the valve operating conditions from first principles. This is particularly important in regard to the value of R_L and the values of L and C in the tank circuit. The amplifier can only give its rated power output without distortion if it is working into the correct anode load corresponding to the dynamic operating conditions and the load line that has been selected.

One "rule of thumb" formula that can be used to find the value of anode load (R_L) where the makers' figure for single tone anode current is known, is $V_a/2 \times I_a$, where V_a is the dc supply voltage and I_a is the maximum signal anode current in amps. This formula cannot be more than an approximation because it does not differentiate between the different classes of working. The correct load for class

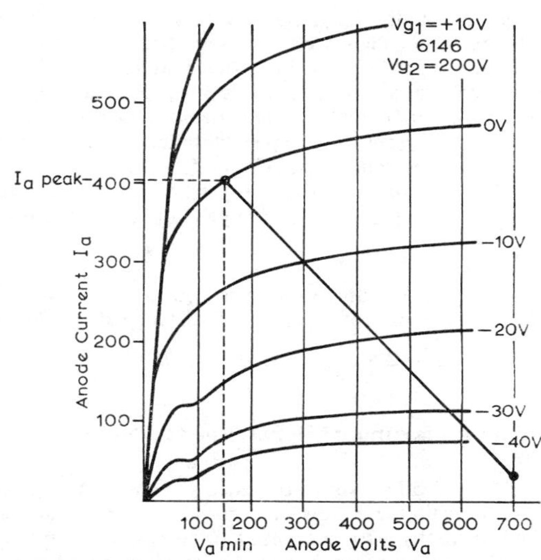

Fig 6.176. Characteristics of a 6146 operating with a zero signal current of approximately 30mA and anode supply voltage of 700V.

AB1 is not the same as the load for class AB2 or class B working.

The recommended formula to provide a much more accurate answer for the value of R_L is based on the operating conditions and takes into account the angle of anode current flow. The anode circuit is tuned to the fundamental frequency of anode current and when loaded by the output it presents a resistance R_L at this frequency.

The value of the resistance may be found as usual by dividing the voltage across it by the current through it, the voltage being the peak voltage and the current the peak value of the fundamental component of anode current. In class B operation the anode current flows for approximately half of each cycle (or 180°) and the fundamental component therefore has a peak value equal to half that of the anode current. In class AB operation where the angle of flow is greater than 180° but usually not greater than 200°, the fundamental component is not significantly different to that in the class B case. It is more convenient to calculate R_L in terms of the dc or average value of anode current because this is usually displayed on the panel meter and therefore readily accessible.

If the peak current is taken as k times the dc value (ie $I_{a\,peak} = \text{k}\,I_a$), for the class B case k is 3·14 (ie π) and for 200° operation it is 2·88. The formula for R_L can be written therefore as:

$$R_L = 2 \times \frac{(V_a - V_{a\,min})}{I_a \times \text{k}}$$

where V_a is the dc supply voltage, $V_{a\,min}$ the minimum value of anode voltage at the crest of the cycle and I_a the manufacturer's figure for maximum signal anode current.

Zero bias triodes such as the 805, 811A and TZ40 are designed to have a low zero signal anode current and be operated in class B. Conversely, many high slope tetrode or pentode valves are capable of high power output without grid current and are normally driven in class AB1. Examples of the latter are 6146, 4-125A, 4X150 and 813.

The V_a/I_a characteristics of a 6146 are shown in **Fig 6.176** and a load line drawn to correspond to typical class AB1 operation. Normally $V_{a\ min}$ is given a value slightly less than the recommended screen voltage for the valve. The actual anode load has a value twice that calculated from the line because as already mentioned the fundamental component of anode current has a peak value which is 0·5 $I_{a\ peak}$. The dc current to be expected under maximum signal conditions may be found by dividing $I_{a\ peak}$ by 2·88. In the case of the 6146 operating as represented in Fig 6.176 the peak current $I_{a\ peak}$ is approximately 400mA. The dc value is therefore $\frac{400}{2{\cdot}88} = 139\text{mA}$.

The anode peak voltage swing is 700 − 150V = 550V and the actual value of anode load is twice the load line resistance or $R_L = 2 \times \frac{550}{400} \times 1{,}000 = 2{,}750\Omega$.

Where the Maximum Signal Anode Current is not Known

A number of valve types—designed for some other application—are available as low cost initial equipment and are therefore attractive to the amateur for linear amplifier operation. In many cases, however, operating data for rf or audio use is unobtainable. In these cases it is necessary to design from first principles. The procedure can perhaps most clearly be shown by taking a specific example. An excellent valve available at low cost is the EL38, a high slope pentode developed as a television receiver line time base output valve capable of operation at high anode voltage and high peak anode current—characteristics particularly suitable for ssb amplifier use.

It will be assumed that the valve is to be operated in class AB1 with an expected efficiency of 60 per cent. Inspection of the manufacturers' data for limiting values gives V_a 800 volts, $V_{a\ peak}$ 8kV and V_{g2} 400V. (Suitable values of anode and screen supplies for amateur service would be 1,000 and 300V.) Since the 40 per cent power loss must equal the maximum rated anode dissipation of 25W, the total power input (100 per cent) must be 25 × 100/40 = 62·5W.

Dividing the maximum power input by the anode voltage gives a maximum signal average anode current I_a of 0·0625A and the peak value of anode current is then $I_{a\ peak}$ = 0·0625 × 2·88 = 0·18A since k is 2·88 for class AB operation.

The peak value of the fundamental component is half this value or 0·09A.

Assuming that the instantaneous anode voltage is allowed to swing down just below the value of the screen voltage $(V_a - V_{a\ min}) = 800\text{V}$ and $R_L = 800/0{\cdot}09 = 8{,}888\Omega$.

The p.e.p. output is calculated by multiplying together the rms value of anode voltage swing (ie $\frac{V_a - V_{a\ min}}{\sqrt{2}}$) and the rms value of the fundamental component of anode current (ie $\frac{I_{a\ peak}}{2} \times \frac{1}{\sqrt{2}}$)

The p.e.p. output becomes $\frac{800}{\sqrt{2}} \times \frac{0{\cdot}18}{2\sqrt{2}} = 36\text{W}$.

To check the figures the output power is subtracted from the input power, 62·5 − 36 = 26·5. This is slightly more than the rated anode dissipation but is quite satisfactory for amateur sideband use. (If a greater power output than this is required it would be satisfactory to increase the ht supply to say 1,250V.)

Four EL38 valves in parallel with a 1,000V ht supply would make an excellent linear amplifier running at 250W input and a p.e.p. output of 150W. The correct value of R_L would be one quarter the value of one valve—8888/4 = 2,222Ω.

The power output obtainable from a linear amplifier at a given anode voltage is determined by the peak anode current $I_{a\ peak}$; this in turn is determined by the minimum anode voltage $V_{a\ min}$. A large $I_{a\ peak}$ is required in order to obtain large output, and a small $V_{a\ min}$ in order to have a good efficiency. With $I_{a\ peak}$ and $V_{a\ min}$ determined by considerations of power output and anode efficiency, respectively, it is required that R_L have the value given by the formula

$$R_L = \frac{2 \times (V_a - V_{a\ min}).}{I_{a\ peak}}$$

If R_L is smaller than it should be, then $V_{a\ min}$ is thereby increased and both efficiency and output power suffer. If R_L is made too large, $V_{a\ min}$ becomes very small and with triodes this causes $I_{a\ peak}$, and therefore the output power, to decrease; with pentode and tetrode valves a virtual cathode forms if $V_{a\ min}$ is too small causing secondary emission from the anode to the screen, excessive screen current and a flattening of the anode current peaks. In the examples given $V_{a\ min}$ is determined by the operating voltage on the screen. In the case of triode valves $V_{a\ min}$ is limited to a value that is more positive than the peak rf grid voltage.

A summary of the operating conditions for four valves parallel would be as follows:

Operating Conditions for Four EL38

Dc anode voltage	1,000V
Dc screen voltage	300V
Zero signal dc anode current	80mA
Max. signal dc anode current	250mA
Effective load resistance	2,220Ω
Dc grid voltage	value required to give 80mA standing anode current
Maximum signal power input	250W
Maximum signal power output	150W (approximately)

The amplifier would be loaded and the drive adjusted with single tone input (audio tone or inserted carrier) to the maximum signal anode current of 250mA. On speech the pa anode current meter would not be allowed to swing beyond half this value to prevent overdriving and distortion. In general the linearity of a tetrode or pentode amplifier is improved by running the zero signal anode current as high as possible without exceeding the rated anode dissipation—80W is a good compromise value.

The Pi Network Tank Circuit

The Q of the anode circuit, of which the tank is a part, must be sufficient to keep the rf anode voltage close to a sine wave shape. Because of the greater angle of anode current flow the requirements for linear amplifier operation are less stringent than for class C operation. However, if the anode circuit Q is insufficient the rf waveform may be distorted resulting in low anode efficiency and also poor attenuation of the harmonics of the output signal. Too high a value of Q results in large circulating rf currents and power loss. A compromise value giving a good balance

TABLE 6.23

Former Diameter	Winding Length	Number of Turns	Wire Gauge and Spacing
(A) 1½in	2in	$8{\cdot}6\sqrt{\frac{L}{d}}$	18swg spaced 16 turns per inch.
(B) 1¾in	3¼in	$10\sqrt{\frac{L}{d}}$	16swg spaced 12 turns per inch.
(C) 2¼in	3¾in	$9{\cdot}5\sqrt{\frac{L}{d}}$	14swg spaced 8 turns per inch. (Eddystone former).

L = Inductance in μH
d = Former diameter in inches.

Former "A" is suitable for two 6146 valves or similar.
Former "B" is suitable for two TT21 valves or similar.
Former "C" is suitable for two QY3-125 valves or similar.

With the transmitter output connected to a 75Ω non-inductive dummy load, check that each band resonates with the correct values of C1 and C2; it may be necessary to adjust the tapping points to achieve this.

between the conflicting requirements and fully sufficient for sideband working is a Q of 12.

This is the recommended type of output circuit and once the correct value of R_L has been calculated as shown, the required circuit reactances can be read directly from the tables on p6.40.

Pi Coil Winding Data

While it is possible to wind a pi-tank coil and get it to resonate correctly on each band—with the calculated values of C1 and C2—by a process of trial and error, this is a time consuming method. A much more satisfactory procedure is a method of relating the inductance values in μH to actual turns of wire on a coil former of suitable diameter. Amateurs are indebted to R. G. Wheatland, G3SZW for providing the practical formulae* given in **Table 6.23**.

THE ADVANTAGES OF PASSIVE GRID OPERATION

The term "passive grid" is used to define a method of operating a linear power amplifier in which the grid input circuit is "passive" in regard to frequency, ie the normal coil and tuning capacitor are omitted and replaced with a non-inductive resistor. This method has a number of advantages making it particularly suitable for amateur operation.

Variation of Input Damping

It is of interest to consider the basic problems that are common to the normal grid driven method. The first is the variation of input damping. Throughout the drive cycle, in a radio frequency amplifier biased to class AB1, the input damping changes and these changes are reflected back to the driver valve, as shown in **Fig 6.177**.

A method of overcoming this distortion is to make the driver impedance as low as possible. This can be accomplished by connecting a swamping resistor across the grid circuit. The value of this resistor depends on the valve being used, the operating conditions, and possibility of grid current, but is usually of the order of 2,000Ω. The characteristics of certain valve types are such that it is advantageous

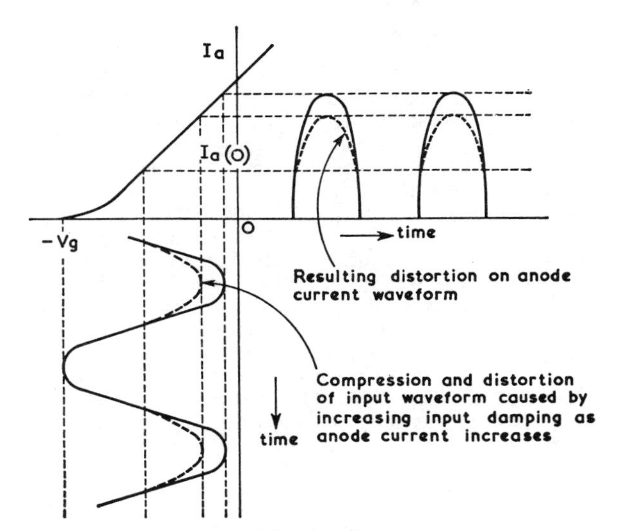

Fig 6.177. Linear amplification characteristics showing how variation of input damping causes distortion of the input waveform.

to drive the valve slightly into grid current. In this instance a swamping resistor will assist in masking the effects of input impedance change when grid current starts to flow.

Amplifier Stability

The second problem is that of amplifier stability. It is vitally important that there is no positive feedback in a linear power amplifier and most tetrode and pentode valves used for amateur sideband use require neutralizing. Because of the high power gain it is more difficult to neutralize correctly a class AB amplifier than a class C one. The variation in grid-cathode capacitance can also cause an amplifier that is perfectly stable and correctly neutralized under static conditions to become imperfectly neutralized and become unstable under voice peak conditions.

Heavy grid damping will prevent any feedback due to the grid-anode capacitance of the valve being of sufficient amplitude to cause instability, without any necessity to incorporate neutralizing circuits. In addition the low grid resistor will provide a constant load on the driver valve and will effectively damp out the effect of variation of input damping during the driving cycle caused by input capacitance change, and the further effect caused by the onset of grid current flow.

The Passive Grid Linear Amplifier

While it is possible to stabilize a triode amplifier by making the passive grid resistor of low enough value it is likely that in practice the required value would be so low, an excessive amount of driving power would be necessary in order to develop the necessary grid driving voltage across it. The passive grid method is therefore particularly recommended for tetrode and pentode amplifier valves.

The lowest value of the grid resistor that will normally be used is 75Ω—this would be a good match to an exciter with a pi-output network usually working directly into a 75Ω aerial load. The highest possible value of grid resistor will be that at which the amplifier becomes unstable due to positive

* "LC Calculations" by R. G. Wheatland, G3SZW, *RSGB Bulletin* August 1965.

feedback; this will vary from valve to valve but is estimated to be around 1,000Ω.

As a low value will give a greater measure of stability and a more constant load on the driver valve, its value in ohms will be determined by the required maximum signal grid driving voltage and the available power output from the driver valve. This is determined from the formula $R = V^2/P$, where V is the maker's value for the peak rf grid driving voltage and P is the p.e.p. output rating of the driver valve in use. It should be noted that as p.e.p. is an rms value, the value of V used in the formula should also be an rms value. However, the use of the peak voltage figure normally given in valve data ensures that in practice the required drive voltage is developed using only half of the available driver output. This gives a very desirable two to one margin of safety and ensures that the driver stage can never be overrun at any time, and that the level of intermodulation distortion products from this stage is always less than the distortion product level from the linear power amplifier.

PROBLEMS OF RATINGS

It is customary to rate a.m. transmitters on the power developed in the carrier and most published data on valves for a.m. telephony gives operating conditions which take into account the peaks which occur when 100 per cent modulation is applied. With single sideband operation it is not possible to have a similar form of rating since at zero modulation there is no output from the valve (this assumes complete suppression of the carrier).

Looking at the A.M. Modulation Envelope

It will be assumed that a transmitter set up to the maximum licensed input of 150W is anode modulated with 75W of audio (the usual amateur conditions). Further, that the overall efficiency of transfer of rf output into the load is 66·6 per cent. The load is a non-inductive resistor of 100Ω. If the vertical deflection plates of a cathode ray oscilloscope are connected across the dummy load the vertical deflection will be the measure of the rf voltage appearing across the load. In the unmodulated condition the rf output power is 100W, the voltage across the load is 100V and the current flowing through the load is 1A. The modulator is now driven to the 75W output condition by applying a 1kHz sine wave to the microphone input socket. This will fully modulate the rf carrier and the cathode ray tube trace will double in amplitude. If the oscilloscope horizontal time base is switched on and the speed adjusted to some multiple of the 1kHz modulating frequency, the rf modulation envelope will be displayed.

The carrier and the sidebands are quite separate from each other and can be received independently. As all the component frequencies of the transmitted wave are rf (two sidebands and a carrier) on different frequencies, they get "in and out of step with each other" and the resultant is the modulation envelope.

The diagram in **Fig 6.178** is a graph of voltage plotted against time—it would be equally valid if it were a graph of current through a 100Ω load plotted against time. It is convenient to think in terms of voltage because the oscilloscope is the only instrument that will show the transmitter output as a visual presentation, and an oscilloscope is a voltage operated device. Considering the single audio cycle of the modulation envelope shown in Fig 6.178, at the left

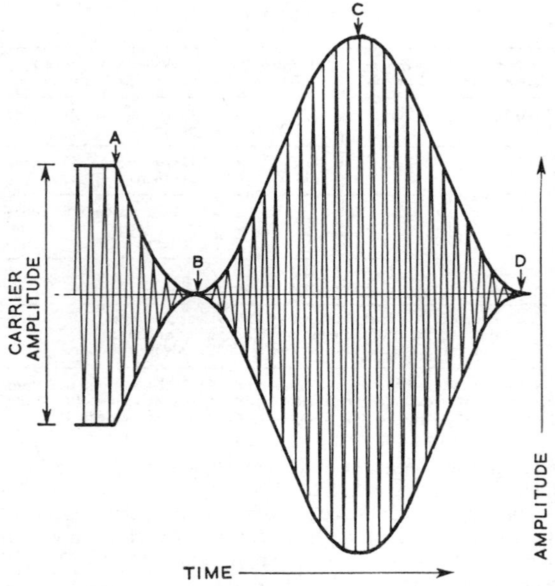

Fig 6.178. Graph showing how the cycles of rf produced by the resultant voltages of the carrier and the two sidebands are continually changing in amplitude from zero at the modulation trough at point B to twice the carrier value at the modulation crest at C.

hand side of the diagram at point A the carrier voltage is 100V and the current through the load is 1A. One-four-thousandths of a second later at point B, the rf cycles are zero—the voltage across the load and the current through it are also zero. During the next two-thousandths of a second the rf cycles increase in amplitude until they reach the crest of the envelope at point C. During this period of time the individual rf cycles occurring have twice the original carrier amplitude and the voltage across the load has doubled, therefore as $P = E^2/R$ the power output of each cycle of rf energy at the modulation crest is four times the unmodulated carrier power—this is the peak envelope of the transmission, the p.e.p. output, and this is 400W. It is also important to appreciate that the p.e.p. output is a real power rms value. The voltage across the load is 200V rms, and the current through it is 2A rms. *The product of these two is 400W of effective power.* One-two-thousandths of a second later the cycles of rf have reached point D in Fig 6.178, and the transmitter output is again zero.

The method of finding the effective (rms) value is to take the peak rf values at many points over a complete cycle of audio alternation, square these values and then find their average, finally taking the square root of the value thus obtained. The part of the diagram between points B and D in Fig 6.178 is one complete cycle of the modulating frequency and is already divided into 25 cycles of rf. It is then convenient to measure the individual lengths of these, square all the values obtained, add them together and divide by 25 to find the average and then calculate the square root. This value will be found to be exactly 1·225 times the carrier amplitude, ie the effective (rms) value of the individual cycles of rf energy occurring within each cycle at the modulating frequency is 1·225 times the unmodulated carrier value, therefore an rf ammeter indicating 1A of current

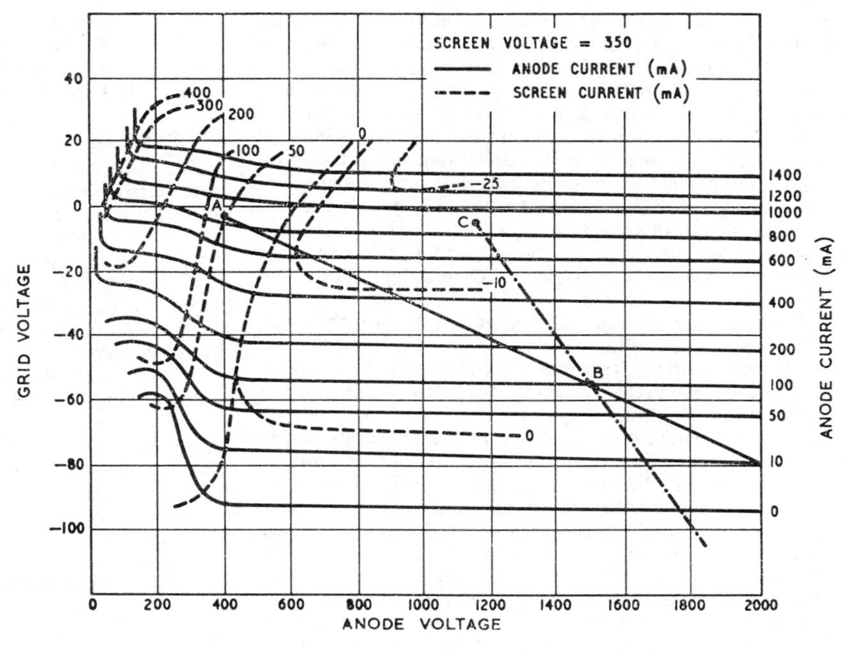

Fig 6.179. Typical constant current characteristic of the Eimac 4CX250B valve.

through the 100Ω load will be expected to indicate under 100 per cent modulation conditions, 1·225A, and this is exactly what it does do.

The output power is given by the formula $P = I^2R$, and substituting the values this becomes 1·225 squared × 100 = 1·5 × 100 = 150W.

The most satisfactory method of giving the operating conditions for valves designed especially for linear amplifier sideband service is that of p.e.p. input and output ratings.

The mean rf output power of 200W is equivalent to 400W p.e.p. output when using a two tone test signal and can be measured with an rf ammeter in series with a resistive dummy load of known value. The power dissipated in the load is I^2R; ie the current squared multiplied by the value (in ohms) of the load, therefore the current in a 75Ω load when running at maximum allowable output will be 1·63A. For a load of 50Ω the current will be 2A; other current readings can readily be converted to mean output power by the formula given above (see **Table 6.24**).

RF AMPLIFIER ADJUSTMENT AND LOADING

The tetrode amplifier has the advantage of high power efficiency, together with good linearity and low harmonic output. It has become deservedly popular over the years as a ssb power amplifier. For these reasons the following discussion in regard to rf amplifier adjustment and loading will be based on class AB1 tetrode operation.

Fig 6.179 shows a set of constant current characteristics for a typical tetrode valve. These curves show the dynamic characteristics of the valve—that is, the instantaneous values of anode and screen current for any given grid and anode voltage conditions. Constant current characteristics of this sort are often published for valves which are intended specifically for rf power amplifier use. This form of presentation demonstrates some important features of tuned power amplifiers and is useful when designing such amplifiers, including class B and class AB linears.

Typical operating conditions for the 4CX250B valve are: V_a 1500V; V_{g2} 350V; V_{g1} − 55V; I_a (*zero signal*) 100mA. These values show that with an ht supply to the anode of 1,500V and a screen supply of 350V, the valve will require −55V grid bias and will draw a standing anode current of 100mA. From these figures it is possible to make the zero-signal operating point of the valve, corresponding to V_a = 1,500 and V_{g1} = −55V; this is shown in Fig 6.179 by the point *B*. This is the point at which the valve rests with zero-signal rf grid drive.

As it is desired to operate in class AB1 and not draw grid current, the control grid must not be driven positive and the peak rf drive must be restricted to a value that is slightly less than the bias voltage—it will be assumed that the peak rf grid drive is held at 100V peak to peak (eg + 50V on the positive half of the driving cycle, and −50V on the negative half of the driving cycle). When the peak grid drive is applied, the first positive half cycle will carry the valve operating point along the line from *B* to *A* and back to *B* again. During this half-cycle the grid-voltage swing from −55V up to −5V and back again to −55V, has caused the valve anode current to swing from 100mA up to 1,000mA and

TABLE 6.24

Output Power of a SSB Transmitter using a Two Tone Test Input

50Ω dummy load (R)			75Ω dummy load (R)		
Current (I) (A)	Mean Power output (W)	P.E.P. output (W)	Current (I) (A)	Mean Power output (W)	P.E.P. output (W)
0·5	12·5	25	0·5	19	38
1·0	50·0	100	1·0	75	150
1·5	112·5	225	1·5	168·75	337·5
2·0	200	400	1·63	200	400

back to 100mA again. At the same time the anode voltage swings from 1,500V down to 400V. During the negative half cycle the control grid voltage will swing from −55V down to −105V and the valve operating point will move down the line from *B* to a point opposite −105 on the grid voltage scale and back to *B* again. The negative going grid voltage therefore swings the anode current down to cut-off for a small portion of the cycle, and the anode voltage continues rising up to 2,600V and back down again due to the fly-wheel action of the anode tank circuit. This half of the operating line need not be plotted and is not important because the valve does not "work" during the negative half cycle.

The operating line represents in graphical form the dynamic characteristics of the pa stage—that is the instantaneous anode current and voltage and the grid voltage at any point on the sine wave cycle. The line is straight because it relates a pure sine wave grid voltage to the anode voltage which is also a synchronous sine wave due to the flywheel effect of the tuned circuit. This operating line must not be confused with the more familiar load line and its slope does not represent resistance but rather the voltage gain of the pa stage, ie anode voltage swing : grid voltage swing. However a change in load resistance, with other conditions unchanged, results in a change of slope of the operating line. For example if the load resistance is decreased the operating line will rotate clockwise about the point B with the point A moving horizontally to the right until the line takes up a steeper position. This occurs because of the reduced anode voltage swing resulting from the lowering of the load resistance.

The load resistance is easily calculated from values obtainable from the diagram. The peak anode voltage is 1,500 − 400 = 1,100V. The peak value of anode current is approximately 1,000mA and since this is roughly of half sine wave shape the peak value of the fundamental component is 500mA.

The load resistance is therefore $\frac{1,100}{500} \times 1,000 = 2,200\Omega$.

Driving and Tuning

From inspection of Fig 6.179 it is now possible to make the following deductions:

(i) *Reducing drive will reduce the length of the operating line.* (If the rf grid drive is reduced to half, the grid voltage swings to only half the original peak-to-peak amplitude; the operating point *B* remains the same but the operating line is reduced to half its original length.)

(ii) *Detuning the tank circuit will tilt the operating line in a clockwise direction.* The operating line will therefore have *minimum* slope at resonance. Note, however, that the moving point will still intercept essentially the same anode current values. It is clear from this that the anode current in a tetrode is not a good indicator of resonance (very little dip).

It will be seen from Fig 6.179 that the constant screen current lines are concentrated in the upper left hand side of the graph and are tilted upwards at a steep angle. Also of note is the fact that the screen current consists of zero or even negative values in the off-resonance position (*C* −*B*), but at resonance is always positive. From this it is now possible to make the third deduction:

(iii) *A peak in screen current indicates tank circuit resonance.* As the operating line is confined vertically by the constant peak-to-peak amplitude of the grid driving voltage (two imaginary horizontal lines, one at −5V and one at −105V) during the rotation of the operating line while tuning, its length increases as resonance is approached and reaches a maximum at resonance. As point *A* penetrates the heavy screen current region the dc screen current rises and the screen current meter indicates a sharp peak at resonance.

Loading

Once the tank circuit is tuned to resonance it presents a pure resistive load to the valve. However the value of this load is affected by the coupling to the external load (the aerial); increased coupling lowers the value of R_L and the operating line assumes a steeper angle. At this steeper angle the operating line will not intercept the heavy screen current region to the same extent as it did before and the screen current will reduce. From this it is now possible to make the fourth deduction:

(iv) *Screen current will fall as loading is increased.* During the rf driving cycle the valve operating point is continually moving through many different instantaneous values of screen and anode current. The average of all these values is what the dc meter in the circuit reads. For a linear amplifier valve operated in class AB1 the dc meter reading is approximately one-third of the peak value of current at the top of the operating line.

Tuning-up Procedure

It is now clear that the screen current meter is by far the best indicator both of resonance and loading with a tetrode amplifier and should always be used in preference to the anode current meter. Of great importance is the realization that a linear amplifier loaded for maximum rf output indicated on an aerial ammeter or forward power meter is *not* sufficiently loaded to prevent flat topping on speech peaks. Loading must always be set to obtain a predetermined value of screen current under single-tone driving conditions to obtain as nearly as possible a given set of data sheet conditions as given by the valve manufacturer. If data sheet conditions are not available it is essential to examine the modulated envelope on an oscilloscope.

Due to the inertia of the movement the anode current meter cannot follow at syllabic rate. It is therefore normal on voice peaks for the anode current meter to read no more than half the true maximum value. This means that an amplifier should never be talked up to more than half of the maximum signal single-tone anode current reading.

Tuning-up should be undertaken with the transmitter output connected to a non-inductive dummy load and after matching the manufacturer's data sheet conditions the output should be transferred to the aerial feeder. A change in meter readings indicates a small amount of mismatch (standing wave on the line) and it should be possible to correct this with a further small adjustment to anode tuning and loading capacitors. The amplifier is now ready for speech operation, and after removing the carrier or the inserted tone and re-connecting the microphone, talk into the microphone in a normal operating voice level and adjust the exciter rf drive control for the highest level that is possible without drawing grid current on voice peaks or flat topping (check this with the oscilloscope).

The Two-tone Envelope

A single tone input to an ssb transmitter drives the linear amplifier at one frequency. The amplifier output is a pure cw signal exactly the same as the output of a telegraph

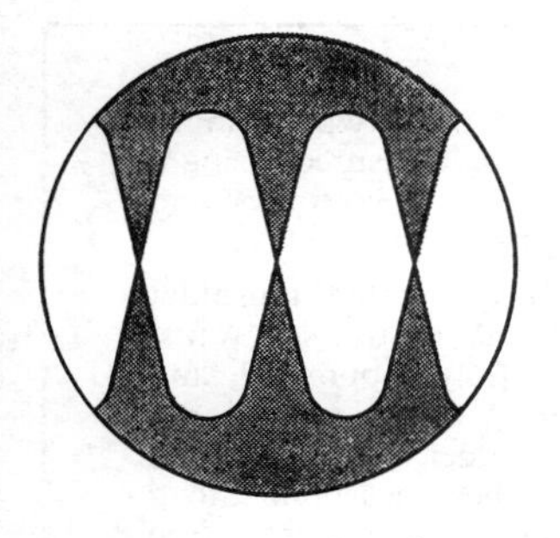

Fig 6.180. Oscilloscope pattern of a two-tone test signal used for aligning linear amplifiers.

transmitter under key-down conditions. As such it is possible to ascertain by meter readings the performance of the amplifier at maximum signal (p.e.p.) conditions. Information on the linearity of the amplifier is, however, lacking.

In order to study linearity thoroughly by observation of the amplifier output, some means must be provided which will vary the output level from zero to maximum signal with a regular pattern (**Fig 6.180**) that is easily interpreted. A simple means of obtaining an output signal is to use two audio tones of equal amplitude to modulate the sideband transmitter. The resultant (or beat between the two rf signals) produces a regular pattern which, when observed on an oscilloscope, has the appearance of a carrier 100 per cent amplitude modulated by a series of half sine waves, as shown in **Fig 6.181.**

Because the pattern is produced by adding two pure sine waves, it is known as the *two-tone test* signal. As it is merely a double sideband suppressed carrier signal in another guise, it may be generated in a phasing exciter by disabling one of the two balanced modulators and applying a single audio tone to the input of the speech amplifier. In the filter exciter, the same result may be produced by applying single tone modulation and reinserting carrier of exactly the same amplitude as the ssb signal passed through the filter. It is not necessary to go to the trouble of building an audio oscillator to give output at two separate frequencies.

To the sideband operator the two-tone envelope is of special importance because it is from the envelope that the power output from an ssb system is usually determined. An ssb transmitter is rated in p.e.p. output with the power

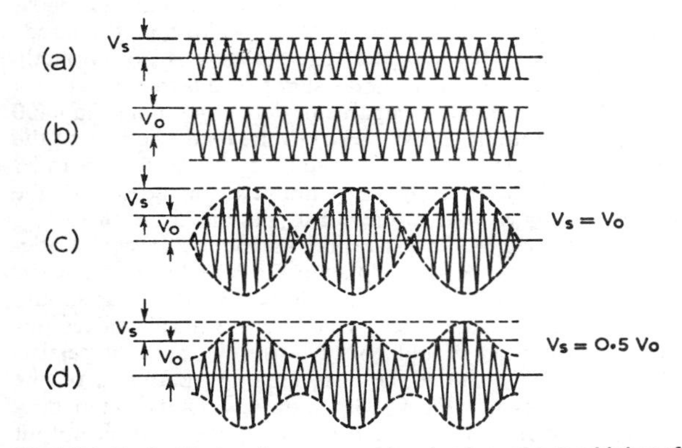

Fig 6.181. Typical heterodyne waves, showing how the combining of two waves of slightly different frequencies results in a wave which pulsates in amplitude at the difference frequency of the component waves, and how the wave shape of the envelope of the resultant wave depends upon the relative amplitudes of the two components. (c) Shows the familiar half-sine-wave two-tone test pattern.

measured with a two equal-tone test signal. With such a signal the actual watts dissipated in the load are one-half the p.e.p.

The generation of this two-tone envelope can be shown clearly with vectors representing the two audio frequencies as shown in **Fig 6.182.** When the two vectors are opposite in phase the envelope voltage is zero. When the two vectors are exactly in phase, the envelope value is maximum. This generates the half sine-wave shape of the two-tone ssb envelope which has a repetition frequency equal to the difference between the two audio tones.

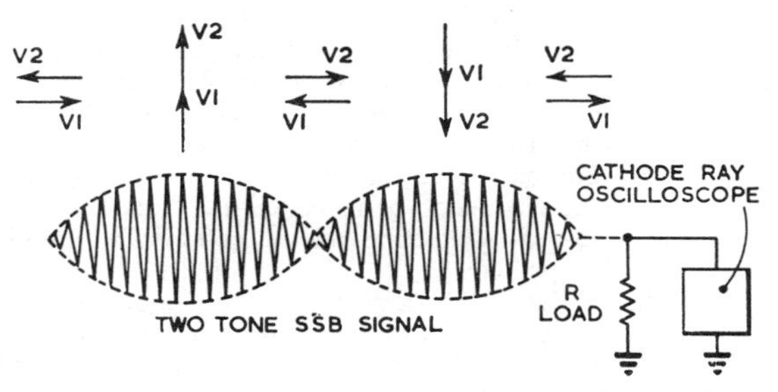

Fig 6.182. Power measurement from two-tone ssb test signal.

When the half sine-wave signal is fed into a load, an rms calibrated cathode ray oscilloscope across the load indicates the rms value of the peak envelope voltage. This cro deflection is equal to the in-phase sum of $V1 + V2$, where $V1$ and $V2$ are the rms voltages of the two tones. Since $V1 = V2$, the p.e.p. $= (2V1)^2/R$ or $(2V2)^2/R$. The mean power dissipated in the load must equal the sum of the power represented by each tone, $V1^2/R + V2^2/R = 2V1^2/R$ or $2V2^2/R$. Therefore, with a two equal-tone ssb test signal, the mean power dissipated in the load is equal to 0·5 of the p.e.p., and the power in each tone is equal to 0·25 of the p.e.p. The peak envelope power can be determined from the relationship, p.e.p. $= V^2(cro)/R$. The mean power can be determined from the relationship, $P_{mean} = \frac{1}{2}V^2(cro)/R$. Similar measurements can be made using an rf ammeter in series with the load instead of the cro across the load. The analysis can be carried further to show that with a three equal-tone ssb test signal, the power in each tone is one-ninth of the p.e.p., and the mean power dissipated in the load is one-third the p.e.p.; with a four equal-tone test signal the power in each tone is one-sixteenth of the p.e.p., and the mean power is one-quarter the p.e.p. and so on.

Practical Adjustment Procedure

When a linear amplifier has been built and given a preliminary check for accuracy of wiring, the following procedure is recommended:

(i) Apply reduced potentials to anode (and screen if applicable) and adjust the bias so that the valve is running at the maximum permissible anode current—this will be just within the maximum rated anode dissipation. Check that there is no trace of self-oscillation at the required operating frequencies, or of parasitic oscillation at vhf. Correct neutralization is infinitely more important in a class AB amplifier than in a class C stage used for telegraphy or A3 telephony. This does not apply to a passive grid amplifier where the heavy grid damping makes neutralization unnecessary.

(ii) Increase applied potentials to the values recommended

POWER MEASUREMENTS

Extract from the UK licence conditions governing ssb operation

Suppressed or reduced carrier single sideband systems

The radio frequency output peak envelope power under linear operation from an A3A or A3J transmitter must not exceed that from an A3 transmitter working at an overall efficiency of 66 per cent when supplied with the appropriate maximum permitted dc input. The output power shall be measured, using an oscilloscope, by the following process:

(i) Adjust the A3 transmitter output stage for class C working and apply a pure sinusoidal tone to the transmitter. With the dc input power limited to the maximum value appropriate to the frequency band concerned note the peak-to-peak deflection on the cathode-ray oscilloscope.

(ii) Adjust the transmitter for single sideband linear operation and replace the tone by speech; the maximum deflection on the cathode-ray oscilloscope, showing the rf output caused by the peaks of speech, should not be greater than twice the previously measured deflection obtained with tone input.

As an alternative the following method may be used:

Suppressed or reduced carrier single sideband operation

The radio frequency output peak envelope power must not exceed that from an A3 transmitter working at an overall efficiency of 66 per cent when supplied with the appropriate maximum permitted dc input power. The output power shall be measured, using a resistive dummy load, rf ammeter or voltmeter and oscilloscope, by the following method:

(i) Apply two non-harmonically related sinusoidal tones* of equal amplitude to the ssb transmitter, with the carrier fully suppressed, and adjust the input power to give a mean radio frequency output power under linear operation of 200W (see Note 1) when measured into a resistive load by means of an rf meter (see Note 2). Under this condition note the peak-to-peak deflection on the cathode-ray oscilloscope (see Note 3).

(ii) Replace the tone by speech; the maximum vertical deflection on the cathode-ray oscilloscope shall not be greater than the previously recorded deflection obtained with the two-tone input.

Note (1) 200W mean radio frequency output power in the case of those bands limited to a maximum dc input power of 150W; 66⅔ and 13⅓W for those bands limited to a maximum dc input power of 50W and 10W respectively.

Note (2) In the case of vhf and uhf measurements the rf meter may be replaced by a crystal rectifier and calibrated meter; for shf measurements a bolometer may be used.

Note (3) In the case of vhf, uhf and shf measurements, this use of an oscilloscope may not be practical. In this case the test may be limited to a measurement of the mean radio frequency output power as outlined in part (i) of the procedure.

* A two-tone oscillator is described in Chapter 18—*Measurements.*

by the valve manufacturer and adjust the grid bias voltage so that the valve is taking the recommended zero signal anode current. If this is not known, adjust the bias so that the standing anode current is just within the rated maximum anode dissipation. Make sure that the neutralization is effective and the amplifier perfectly stable, no matter how the grid and anode tuning controls may be set.

(iii) Couple an oscilloscope to the output tank inductor by means of a link and coaxial cable. Connect a non-inductive resistive load of 50 or 70Ω capable of dissipating the full mean power output to the output tank and fully mesh the loading capacitor: this corresponds to minimum loading.

(iv) Switch the exciter to the lowest band and with a 1kHz or 1·5kHz audio input from an audio signal generator into the microphone socket, adjust the audio gain and carrier injection controls so that the exciter drives the linear amplifier with two frequencies of equal amplitude (two-tone output). If there is no provision for carrier insertion, it will be necessary to use a two-tone oscillator—see Chapter 18.

(v) Adjust the exciter rf drive control, the amplifier anode tuning and loading together until an undistorted pattern of maximum amplitude is obtained at the maximum dc input permitted by the valve manufacturers' ratings for two-tone input conditions. If two-tone test data is not available, a value of 0·7 times the permitted dc maximum signal current for single-tone input may be employed. The correct tuning point for anode resonance is indicated by a sharp increase in screen current and a dip in anode current—this position of the anode tuning control should coincide with maximum rf output. If it does not, the maximum rf output occurs at some other position of the anode tuning control; this indicates that the valve is not correctly neutralized and the nc requires slight readjustment.

(iv) If the amplifier is capable of delivering more than 400 watts p.e.p. output (200W mean) to the load, the face of the oscilloscope should be calibrated at the 200W mean rf output level by inserting an rf ammeter in series with the resistive dummy load as previously described on p6.99.

(vii) Examine the "cross-overs" on the oscilloscope critically. They should be as sharp as in **Fig 6.183(a)**. If, however, they appear compressed as in **Fig 6.183(c)** the bias should be adjusted until they are correct. With suitable valves this type of distortion should not give trouble, but if it persists the bias supply should be suspected. A flattening of the envelope peaks as in **Fig 6.183(b)** indicates insufficient loading (loading is increased by *reducing* the capacitance of the output pi-network capacitor) or overdriving the amplifier beyond its maximum signal capabilities. This condition produces severe splatter and distortion on the transmitted signal and must be avoided.

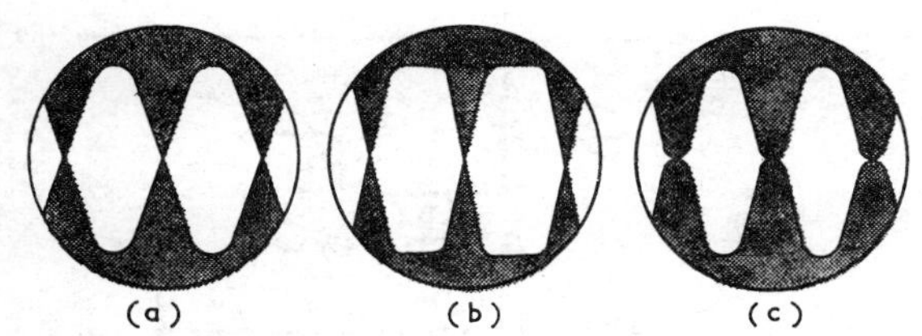

Fig 6.183. Oscilloscope patterns of the output of a linear amplifier with two-tone test input; (a) amplifier correctly adjusted; (b) peaks flattened because of insufficient anode loading or overdrive; (c) distortion at cross-over points because of incorrect bias voltage.

Whenever possible the manufacturers' figures for anode and screen current for maximum signal two-tone input conditions should be used and the drive and loading adjusted to obtain these readings on the amplifier panel meters. (This procedure is described in detail on p6.99.)

(viii) The exciter carrier injection control should be returned to its zero setting and the audio input from the audio signal generator removed. Insert the microphone and drive the linear amplifier under normal speech conditions to the maximum deflection on the oscilloscope previously observed under two-tone input conditions; while doing this the readings to which the anode and screen current meters kick should be carefully noted. Thereafter, the gain of the exciter should be controlled so that the recorded values are not exceeded, otherwise intermodulation distortion and splatter will result. As the tuning of a correctly loaded amplifier is rather flat, the anode current dip is not so pronounced as with a class C stage. The screen current on the other hand, will rise to an unmistakable peak as the anode circuit is tuned to resonance. It should be remembered that the anode and screen meters cannot follow the syllabic rate because of the inertia of the meter movement—an amplifier loaded to 200mA under single-tone conditions will show 140mA under two-tone input conditions and approximately 100mA under speech conditions, *for the same peak envelope power output*. Never talk the amplifier up to a peak current swing (as seen on the anode current meter) that is more than half of the loaded current (single-tone) value.

Summary of Basic Formulae

Symbols

The following symbols will be used in all formulae given below.

VOLTAGE

V_a .. .. .. DC anode voltage

$V_{a\,min}$ .. .. Instantaneous anode voltage at the peak of the rf driving cycle.

V_{g2} .. .. .. DC screen voltage.

V_{g1} .. .. .. DC grid voltage.

CURRENT

I_a .. .. .. DC anode current (as read on anode current meter).

$I_{a\,peak}$ .. .. Instantaneous anode current at the peak of the rf grid driving cycle.

$I_{a\,zero\,sig}$ .. .. Static anode current (resting anode current under zero drive condition).

I_{g2} .. .. .. DC screen current.

I_{g1} .. .. .. DC grid current.

RESISTANCE

R_L .. .. .. External anode load resistance.

R_{out} .. .. External aerial load resistance (normally 50 or 75Ω).

POWER

P.E.P. input .. Peak envelope power input. This is the rms power input at the crest of the modulation cycle.

P.E.P. output .. Peak envelope power output. This is the rms rf power output at the crest of the modulation cycle.

$P_{out\,mean}$ .. .. Average, or mean rf power output under modulating conditions.

P_a .. .. .. Anode dissipation.

P_{g2} .. .. .. Screen dissipation.

P_{g1} .. .. .. Grid dissipation.

MISCELLANEOUS

K .. .. .. Constant whose value depends on the angle of anode current flow. (For 180° $K = 3{\cdot}14$; for 200° $K = 2{\cdot}88$). A useful figure for quick calculation is a value of $K = 3{\cdot}0$.

Eff .. .. .. Anode efficiency.

Basic Formulae

P.E.P. input $= V_a \times I_a$

$I_{a\,peak} = I_a + K$

P.E.P. output $= I_{a\,peak} \dfrac{(V_a - V_{a\,min})}{4}$

$R_L = 2 \dfrac{(V_a - V_{a\,min})}{I_{a\,peak}}$

Single-tone Input Test Signal

DC anode current $I_a = \dfrac{I_{a\,peak}}{K}$

Anode Input Watts and p.e.p. $= \dfrac{I_{a\,peak} \times V_a}{K}$

Average Output Watts and p.e.p. $= I_{a\,peak} \dfrac{(V_a - V_{a\,min})}{4}$

Anode Efficiency (Eff) % $= \dfrac{p.e.p.\ output}{p.e.p.\ input} \times 100$

$= K \dfrac{(V_a - V_{a\,min})}{4 \times V_a}$

Two-tone Input Test Signal

DC anode current $I_a = \dfrac{2 \times I_{a\,peak}}{K^2}$

Anode Input Watts $= \dfrac{2 \times I_{a\,peak} \times V_a}{K^2} = I_a \times V_a$

Average Output Watts ($P_{out\,mean}$) $= I_{a\,peak} \dfrac{(V_a - V_{a\,min})}{8}$

$$\text{Peak Output Watts (p.e.p. output)} = \frac{I_{a\ peak}\ (V_a - V_{a\ min})}{4}$$

$$\text{Anode Efficiency (Eff) \%} = \left(\frac{K}{4}\right)^2 \times \frac{(V_a - V_{a\ min})}{V_a}$$

$$= \frac{P_{out\ mean}}{I_a \times V_a} \times 100$$

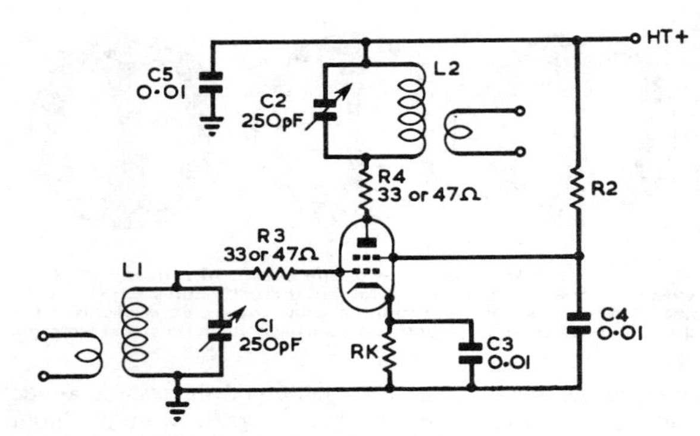

Fig 6.184. Typical class A rf linear amplifier. For operating conditions, see Table 6.25.

TYPES OF LINEAR AMPLIFIERS

There are many different linear amplifier circuits, all capable of giving satisfactory service. The final choice is purely a personal one and will depend on the individual requirements in regard to such things as maximum power output, availability of valves and components and preference in regard to the use of high voltage (2,000 to 3,000) or low voltage (700 to 1,250) from the main power supply.

The Class A Linear Amplifier

The class A linear, of which a representative diagram is given in **Fig 6.184**, will be seen to bear a strong family resemblance to the rf and i.f. amplifiers in receivers. Care must, however, be taken not to presume too far upon this relationship, otherwise trouble in the form of unexpected distortion may result. A true class A stage must be so biased and driven that the valve operates at all times on the linear part of its characteristic curve. The rf/i.f. amplifier is essentially a small-signal amplifier; that is, the signal applied to its grid is only a fraction of a volt, so the valve will operate linearly no matter how it is biased. Furthermore, the stage is not required to produce any power and works into the constant load provided by the grid of the following stage. In ssb service few class A stages work under such favourable conditions. They are usually called upon to produce a reasonably large voltage swing, and perhaps several watts of output power as well. If they are to drive a class AB2 stage, the loading will require some kind of stabilization to minimize fluctuations. There is, therefore, limited scope for small receiving valves in ssb linears. It pays to follow the lead of the designer of high fidelity audio equipment who customarily builds an amplifier of far higher power handling capability than is really needed. By underrunning valves, the problem of distortion is avoided.

The efficiency of a practical class A stage varies from zero under no-drive conditions to about 30 per cent at full input which means that the maximum output is limited to something less than one-third of the rated anode dissipation of the valve used. Provided that the internal insulation is built to stand it, operation at high voltage and low current will give greater efficiency and better linearity than low voltage and high current. It is impossible to list recommended operating conditions for all suitable valves under all circumstances, but **Table 6.25** gives a representative set of typical working conditions for the valves already mentioned. Suitable conditions for valves not included in the table may be deduced from the audio frequency power amplifier ratings given in manufacturers' characteristic sheets.

The construction of a class A stage follows common practice; input and output circuits should be well shielded from one another to prevent self-oscillation, and high frequency parasitics should be eliminated by the generous use of stopper resistors. If oscillation at operating frequency should prove troublesome, the capacitance bridge neutralizing arrangement will provide a sure remedy.

Class AB Amplifiers—Triodes

Although it would be technically possible to achieve sufficient output for amateur purposes by using class A amplifiers exclusively, the result would certainly be uneconomical. Greater efficiency is essential in any practical transmitter, and this may be obtained by operating the high level stages in class AB or class B. For the purposes of this chapter, the precise shade of difference between a class AB amplifier and a true class B stage is unimportant and will be ignored.

When examining the ratings of power valves, it is important to note that the efficiency and typical operating conditions occasionally quoted for rf amplifier service in valve data sheets refer to the amplification of carrier type a.m. waveforms and are not directly applicable to ssb use. The linear amplifier is a variable efficiency device: its efficiency in the absence of drive is zero, rising in a regular manner to between 50 and 75 per cent at maximum drive. The greatest

TABLE 6.25

Typical operating conditions of a representative selection of rf tetrodes and pentodes in class A linear amplifier service.

	6AU6	EF89	6CH6	6CL6	EL84	5763	6146	807
V_a	250V	250V	250V	300V	250V	300V	500V	500V
I_a	10mA	9mA	40mA	30mA	48mA	40mA	40mA	50mA
V_{g_2}	150V	100V	250V	150V	250V	225V	150V	200V
I_{g_2}	4·5mA	3mA	6mA	7mA	5·5mA	2·4mA	2mA	1·6mA
V_{g_1}	−1V	−2V	−4·5V	−3V	−7·5V	−7·5V	−22V	−15V
Rk	68Ω	180Ω	100Ω	82Ω	135Ω	175Ω	470Ω	280Ω

The fixed bias voltage and the cathode resistor in the table above are alternatives.

efficiency obtainable depends on a number of interdependent variables, as well as upon the valve itself, but 66 per cent may be taken as rough generalization. The figures given in manufacturers' literature relate to dc measurements under conditions of no modulation; on modulation peaks, input and efficiency double, and in troughs both drop to zero. In ssb service there is no carrier to provide a reference point against which to strike an average, so peak input and peak power handling capacity are the standards of measurement normally used. Conveniently enough, these standards are used by manufacturers themselves when rating valves for audio frequency linear amplifier service. When available, these audio ratings provide a more convenient guide as to what may be expected in an rf linear than any other published data.

In theory, any rf valve may be used as a class AB or class B linear, but there are a number of external factors which narrow the field considerably. Except in the special case of the zero-bias triode, fixed negative bias is essential, and if the stage is driven into grid current the voltage must have excellent regulation so that it does not vary over any part of the input cycle. Dry cells are inexpensive and may be used if the amount of grid current flowing through them is strictly limited. At high values of reverse current any battery will quickly develop enough internal resistance to cause the instantaneous bias voltage to fluctuate with input. The life of the cells under such conditions would be uneconomically short. Battery bias is therefore restricted to tetrode and pentode stages.

Most important of all is the effect of the amplifier on the preceding drive stage. With any valve which draws grid current over only a part of the input cycle, the onset of grid current will cause the load presented to the driving stage to fall from something approaching infinity to a fairly low value. Wide fluctuations in loading would have a disastrous effect on the linearity of the driver so some way has to be found to minimize the variation. The most practical method is to load the driver with a resistor chosen to dissipate at least ten times the peak load presented by the driven stage. In this way the load on the driver increases only 10 per cent at the onset of grid current, and distortion is kept within bounds. Resistive loading gives rise to no difficulty with tetrodes or pentodes, which require a watt or less of driving power, but rules out low-mu triodes, which would need far too large a driver stage to make them either practical or economical.

One possible way to use low-mu triodes is to bias them to cut-off and to adjust the drive so that they will not operate beyond the negative grid-voltage region; that is, in class AB1. The peak efficiency in this mode of operation is only 40–50 per cent, because the peak current which the valve can draw at zero grid voltage is relatively low. A high voltage bias supply has also to be provided. Zero bias triodes such as the 805, 811 and TZ40 are a much more favourable proposition. As grid current flows throughout the whole of the driving cycle, the load presented to the driver *during the time at which it is delivering power* is virtually constant.

Fig 6.185 shows a zero bias triode driven by a pure sine wave. Only that portion of the driving waveform which is depicted by the solid line can, however, result in rf output; when the driving signal enters the region represented by the dotted line the valve is cut off, and the positive excursion of the output waveform is provided by the flywheel effect inherent in the anode tank circuit. As the valve itself is cut off,

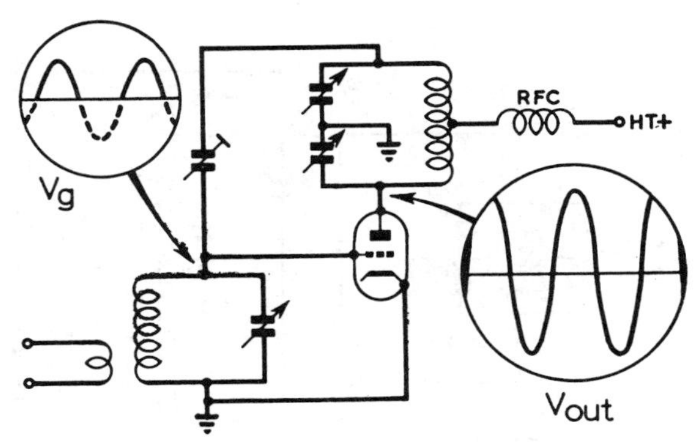

Fig 6.185. Waveforms at grid and anode of a class B zero bias linear amplifier.

it cannot produce intermodulation distortion, and it isolates its output circuit from any distortion which may occur in the driver stage because of the change in loading. Some distortion close to the cross-over point is, of course, inevitable, but practice has proved that this has negligible effect on the signal. It is the distortion at maximum point which causes trouble.

Triode valves have the disadvantage in rf amplifier service of requiring neutralization which is more difficult to accomplish in a class AB amplifier than it is in a class C amplifier because of the high power gain. In order to ensure that neutralization is effective and the amplifier stable on all bands, both the theoretical circuit and the physical layout must be symmetrical. For this reason the valves should preferably be operated in push pull as shown in **Fig 6.186** with a balanced tank circuit instead of the more usual parallel operation with a pi output network.

The 805 was originally designed as a high quality class B audio amplifier, and it is difficult to find a better valve for linear service at peak outputs up to 150W. A pair in push-pull (or parallel) will deliver 300W and require only 6W of driving power, which can be supplied fairly comfortably by a single 807 in class A. The circuit diagram is self-explanatory, but the factors governing the design of the various tank circuits will be examined in detail so that those wishing to adapt the design for use with different valve types may be able to do so without difficulty. The makers' recommended operating conditions for the two in push-pull are given in **Table 6.26**, from which it will be seen that the optimum load for highest undistorted output is 6,700Ω. An output circuit Q of about 15 is desirable which necessitates that the reactance of both the tank capacitor and the inductor at resonance shall be $\frac{6700}{15}$ or 430Ω approximately.

The required tank capacitance may easily be calculated from the formula $X_C = \frac{1}{2\pi fC}$. This is, however, the total value of C_3 and C_4 in series, so the value of each half of the split stator capacitor requires to be double that obtained by calculations. Suitable values for all amateur bands from 3·5 to 28MHz are given in **Table 6.27**, together with coil winding instructions based on the Eddystone 2½in diameter ceramic formers.

If a different type of valve is used, it is advisable to obtain the manufacturers' recommendation about anode loading,

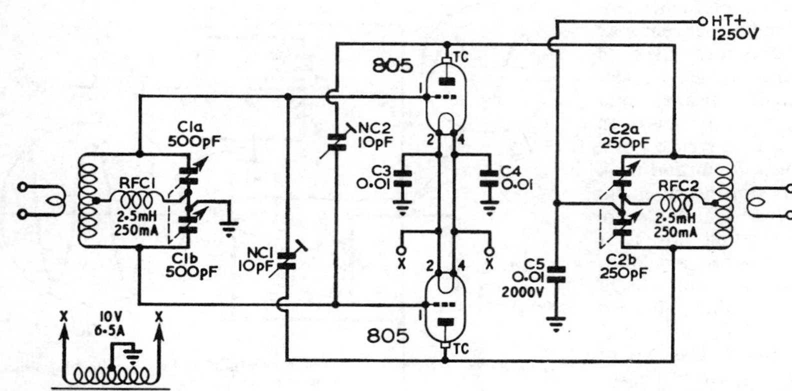

Fig 6.186. Zero bias triode linear amplifier employing two 805 triodes in push-pull. For circuit values not shown above, see Table 6.27.

and to calculate capacitor and inductor values in the manner demonstrated. Loading information is readily available for most of the valves likely to be encountered, but if it should prove difficult to locate for any particular valve, it can be calculated by the methods already discussed under the heading "Theory of Linear Amplification" on p6.89.

The calculation of circuit values for the grid tank may be performed in exactly the same manner as for the anode circuit, and the results will be shown in Table 6.27. Some manufacturers do not quote the grid load impedance in direct form, but for zero-bias valves this may be calculated from the stated grid driving-power and peak grid-to-grid input voltage by means of the simple formula

$$R_{grid} \simeq \frac{(Peak\ Input\ Voltage)^2}{2 \times Driving\ Power}$$

The relatively large values of capacitance for the grid circuit are neither the result of miscalculation nor misprint; they arise from the characteristics of the valve under consideration. A two-gang receiving type variable capacitor will probably be found more convenient than a standard transmitting component, for under no circumstances should the tabulated values of capacitance be reduced, because the proper operation of linear amplifiers in general is vitally linked with the choice of correct component values in the tuned circuits.

Class AB Amplifiers Employing Tetrodes

The tetrode or pentode linear amplifier has become deservedly popular because it is capable of a much higher power gain than comparable triode valves. In the passive grid circuit arrangement it does not require neutralization and construction is thereby simplified.

TABLE 6.26

Recommended operating conditions for two type 805 triodes in class B push-pull zero bias.

DC anode volts	1250V
Grid excitation—peak-to-peak	235V
Anode current—zero signal	148mA
Anode current—maximum signal	400mA
Grid driving power—maximum signal	6W
Anode input—maximum signal	500W
Power output—maximum signal	300W
Anode-to-anode load	6700Ω

PRACTICAL SSB DESIGNS

The following sections give the circuits and complete constructional details of a valve transmitter, a valve linear amplifier, an ic transceiver and two solid state linear amplifiers.

G2DAF TRANSMITTER MARK 3

Design Considerations

The basic requirements for the Mark 3 transmitter may be summarized as follows.

1. Reduction of frequency translation processes to two, in order to simplify the construction and present a design that would appeal to the relative newcomer to ssb.
2. A reduction in the total number of valves used, and simplification of the circuitry wherever this was possible without compromising the overall performance in any way.
3. Straightforward setting up and alignment procedure.
4. First-class carrier, sideband and intermodulation product suppression, together with natural speech quality.
5. No complicated constructional methods.

TABLE 6.27

Anode and grid circuit values for push-pull zero-bias class B amplifier using type 805 triodes.

Band	Anode		Grid	
	C	L	C	L
3.5MHz	200 + 200pF	20 turns CT	300 + 300pF	32 turns CT
7MHz	100 + 100pF	14 turns CT	150 + 150pF	20 turns CT
14MHz	55 + 55pF	8 turns CT	75 + 75pF	12 turns CT
21MHz	35 + 35pF	8 turns CT double spaced	50 + 50pF	12 turns CT double spaced
28MHz	25 + 25pF	6 turns CT double spaced	40 + 40pF	10 turns CT double spaced

Anode inductors are wound on Eddystone or similar 2¼in diam ceramic formers. The 21 and 28MHz coils should be double spaced, and because stray inductance has an appreciable effect at these frequencies, pruning may be necessary in individual cases.
Grid inductors are wound on ⅞in or 1in diam. formers, threaded 20 or 21 turns/in. Capacitor settings are "in use" values and include valve capacitances and circuit strays.

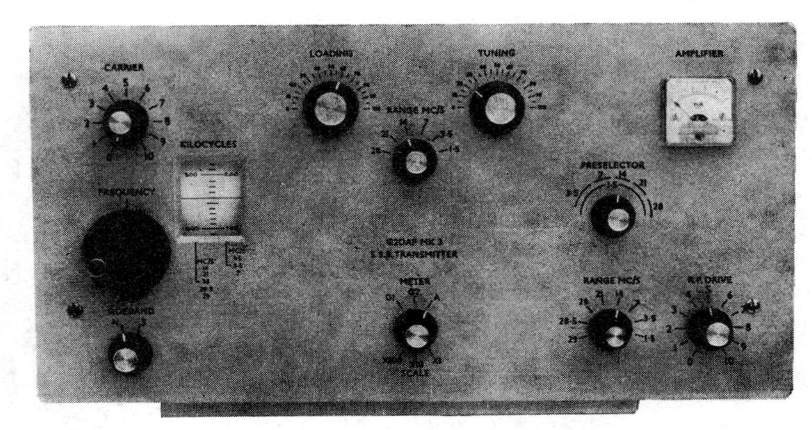

Front view of the G2DAF transmitter Mark 3.

6. The use of push-pull frequency conversion throughout, giving a high discrimination against breakthrough of the heterodyning frequency, and a clean output with a low order of distortion products.
7. Wherever possible, standard production easily obtainable valves and components, avoiding high-cost items.

Constructional features considered to be desirable were also itemized.

1. Unit construction.
2. Home-made coils using standard, readily available coil formers, dust cores and screening cans.
3. Simple press-to-talk control.
4. Clean layout with good accessibility and professional appearance.
5. Separate power supply.

A block diagram of the Mark 3 transmitter is shown in **Fig 6.187** and the circuit diagrams in **Figs 6.188–6.192**. Various aspects of the design in detail follow.

Audio Amplifier

Any residual hum modulation from the audio amplifier would unbalance the diode modulator and impair the carrier suppression. A common source of hum leakage is via the cathode heater insulation of the first amplifying stage. Accordingly an EF86, a valve specifically designed for the input stage of a low-noise audio amplifier and intended to be used with a crystal microphone, was chosen for V1.

The second audio stage, V2, uses a 12AT7 with one half as a voltage amplifier and the second half as a cathode follower, to give a low-impedance output to drive the OA7 diode balanced modulator.

Both valves have cathode resistors without bypass capacitors giving negative current feedback to each amplifying stage, and it will be noted that the coupling capacitors have a low value of 0·005μF, giving a low frequency roll-off to further improve the unwanted sideband suppression.

Carrier Oscillator

Because the final conversion process uses a heterodyning frequency above the 160, 80 and 40m bands, and below the 20, 15 and 10m bands, the transmitter automatically gives the correct sideband output for each of the six bands in use.* It is, however, an operating convenience to be able to switch sidebands to obtain suppression reports and on occasions to temporarily dodge interference.

The carrier oscillator uses the first half of a 12AT7 valve in

* It has become an accepted convention adhered to by all ssb stations, to transmit *low* sideband below 10MHz and *high* sideband above 10MHz as the result of a CCIR (International Radio Consultative Committee) recommendation.

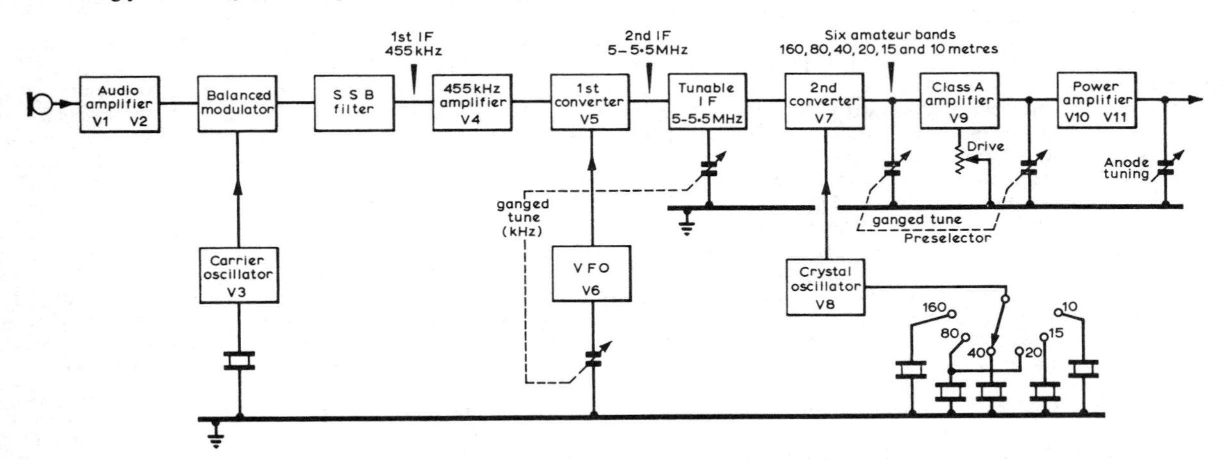

Fig 6.187. Block diagram of the frequency conversion processes and the stages in the G2DAF Mark 3 transmitter.

a modified Colpitts circuit, chosen to allow the cathode to be earthed to rf. This avoids any possibility of carrier leakage along the common heater wiring impairing the carrier suppression in the balanced modulator. This oscillator does not require a tuned anode circuit, and the customary 455kHz i.f. transformer that requires modification in order to provide a low-impedance balanced output is not required. Switch S1 is operated by a panel-mounted control knob and selects the required carrier crystal X1 or X2. The second half of the 12AT7 is used as a phase splitter to provide the push-pull rf drive to the diode balanced modulator. This method has proved in practice to have very good stability—an important requirement if the diode modulator is to hold its balance setting.

A proportion of the carrier oscillator output is taken via a coaxial cable coupling link to a panel-mounted carrier injection control (CARRIER). This control enables a predetermined amount of carrier to be fed round the mechanical sideband filter, for tuning up and netting purposes.

Balanced Modulator

There are many types of balanced modulators using either low-impedance semiconductor diodes, or high-impedance thermionic valves. The two-diode modulator using a pair of OA7 diodes has been chosen for its simplicity, long-term stability and proven performance. The diodes receive audio in parallel and rf in push-pull, the rf being balanced by the potentiometer RV2 and the 50pF preset phasing capacitor. The modulator can be balanced to obtain a carrier suppression of better than 40dB, and this together with the 20 to 25dB suppression in the following sideband filter gives an overall suppression of not less than 60dB. This figure represents a very acceptable amateur transmitter performance.

Sideband Filter

The major design requirement for the Mark 3 transmitter was simplicity of construction, with the less experienced amateur constructor very much in mind. Accordingly, crystal etching and grinding and the relatively highly-skilled half-lattice filter alignment and response curve plotting have been avoided by the choice of a mechanical filter on a nominal frequency of 455kHz, for unwanted sideband suppression. At the current market price the Kokusai MF455–15K filter represents excellent value for money, gives very acceptable performance and has proven long-term stability.

In regard to filter bandwidth (6dB points) the author is firmly of the opinion that 2·2kHz is too narrow, giving an un-natural "boxey" speech quality, and that a much more natural and pleasing transmission is obtained by a filter that will pass all voice frequencies up to 3kHz.

The Kokusai filter also has the advantage, compared to a two half-lattice crystal filter, of compact construction and uncomplicated mounting, enabling a simpler and smaller chassis layout to be adopted. The nominal centre passband frequency is 455kHz, but because of manufacturing tolerances the actual centre-frequency may vary $\pm 0{\cdot}8$kHz. For this reason, each filter is packed with a data sheet giving the filter serial number and the measured bandwidth at the 6dB and 60dB points for the filter concerned. Each bandwidth is given as $\pm X$ kilohertz relative to the design centre frequency of 455kHz—not relative to the actual filter centre

Top side of the transmitter.

frequency. As an example, the data sheet may quote, "bandwidth at 6dB attenuation, + 2kHz and − 1·2kHz".

For good speech quality it is desired to pass all audio frequencies down to 0·3kHz. It therefore follows that the two carrier crystal frequencies should be 0·3kHz outside the 6dB points. They would then be at + 2·3kHz and − 1·5kHz relative to 455kHz, giving actual frequencies of 457·3kHz and 453·5kHz for the example quoted.

When the correct carrier frequencies have been determined for the Kokusai filter that has been purchased, crystals should be ordered quoting the actual frequencies required, and stating that operation is required on the parallel resonant frequency with 30pF shunt capacitance. More detailed information on this topic can be found on p6.81.

Filter Amplifier

Owing to the low output from the diode modulator, and the insertion loss of the filter, the filter amplifier is required to give a high order of amplification. It must, however, do this with a high degree of stability—not only under static conditions but under speech conditions as well. Any tendency towards self-oscillation will completely ruin the sideband signal and must be avoided at all cost.

The amplifier valve for the filter is an EF85 with a medium g_m, and as a further aid to stability the grid resistor is reduced in value to 22kΩ. Because of the greater dynamic resistance of the anode tuned circuit at 455kHz there is adequate stage gain, and high slope valves of the EF80 or EF183 class should not be used.

Coupling to the following converter valve is by a standard Denco type IFT11–465 i.f transformer. The secondary is required to provide a push-pull output to the double-triode valve V5, and in order to avoid the difficulty of rewinding the secondary pi, the centre tap is obtained by a capacitance divider made up with two silver mica capacitors of double the original value—the grid returns of V5 being by suitable value resistors.

First Conversion

A sideband signal—initially generated at the low frequency of 455kHz—has to be raised in frequency to the required amateur band by a process of heterodyning. Unfortunately the converter output contains not only the wanted sum or difference frequency, but also the strong heterodyning input, together with the image frequency and many other unwanted spurious frequency combinations. See p6.84 for detailed analysis. The problem of obtaining a clean output from the converter can be greatly eased by bringing the ssb signal on to the required amateur band in a series of steps. This is the reason why the original G2DAF transmitter used three conversion processes with an intermediate i.f. of 2MHz.

In the Mark 3 transmitter the intermediate i.f. has been omitted in the interest of simplicity and to make the final alignment easier. This means that the initial sideband generated on 455kHz is translated to the first i.f. of 5–5·5MHz in one step—a frequency ratio of approximately 11 : 1. The strong vfo heterodyning frequency is only 455kHz removed from the wanted converter output—less than 10 per cent away. It is therefore much more important to ensure that there is no vfo breakthrough, and to this end *the design of the first converter stage and the discrimination in the tunable converter output circuits is the keystone of the Mark 3 transmitter design.*

As the initial sideband generation is on 455kHz and this frequency is less than the required tuning range of 500kHz, a wideband coupler cannot be used. The i.f. circuits will require to be continuously tuned. Two tuned circuits—coupled for optimum selectivity—will be required.

In order to obtain an output waveform in linear relationship to the sideband input waveform it is necessary to have an injection level to the converter valve of approximately 10 times (20dB) greater. Assuming the use of two tuned circuits at the converter output frequency of 5–5·5MHz and normal loaded-Q values, the attenuation to the heterodyning frequency 455kHz removed would not be sufficient to prevent vfo breakthrough. It is then vitally important to obtain additional rejection to the vfo output by a process of balancing in the double-triode converter valve.

Underside view showing chassis box sections and component wiring.

It will be noted that the converter is operated in push-pull, balance in the anode circuit being obtained by a split-stator tuning capacitor, with the centre tap of the anode coil not bypassed for rf but being allowed to "float". The wanted signal input is applied to the two grids in push-pull, but the vfo to the two cathodes strapped in parallel. The injection rf currents will flow in opposite directions in the push-pull anode tank circuit and cancel out, so that there will be almost no coupling of the vfo frequency into the secondary winding.

Cathode injection was adopted after a series of experiments had shown that this method gave a high conversion efficiency, good long-term stability, an inherent balance of better than 30dB, and did not require an embarrassingly high vfo output level. The converter stage is further simplified because the customary cathode-balancing potentiometer was found not to be necessary.

The two tuned circuits cover a range 5–5·5MHz, and they are gang-tuned to the vfo operating on the high side—5·455 to 5·955MHz. As the tuning range of 500kHz is less than 10 per cent of the tunable i.f. centre-frequency (5·25MHz) there is no tracking problem. In practice the vfo is adjusted to tune correctly at either end of the range by setting the core of L2 at the low end and the 50pF trimmer at the high end. Finally the vfo is set by the main tuning control to mid-frequency (5·705MHz) and the two dust cores of T2 and T3 adjusted to resonate at 5·25MHz. It will be found that the three tuned circuits hold tracking quite satisfactorily throughout the 500kHz tuning range.

Because the converter is a push-pull triode and it is also required to feed a second push-pull triode in the final converter stage, the whole of the tunable i.f. is balanced to earth, and this, together with the single ended vfo coil, necessitates a five-gang variable capacitor of 5–40pF each section. The five-gang capacitor used by the author was purchased many years ago from surplus sources and measures 6in long (excluding the spindle) by 2¼in high by 1¾in wide. The assembly screws have BA threads, indicating that the unit was made by a British manufacturer. Unfortunately the maker's name is not known, and five-gang capacitors are not shown in any of the current catalogues in the author's possession.

This should not, however, deter a prospective constructor, because there are at least three alternatives open:

1. Jackson Bros Ltd are able to supply the Type ME or Type E gang capacitor with a spindle extending through the back plate. This would enable standard two- and three-gang units to be coupled together, in line, to make up a five-gang tuning capacitor.

2. Many of the standard three-gang variable capacitors lying around in amateur junk boxes are of approximately 100pF maximum and have the stator plates supported by two ceramic pillars on each side. It is a relatively simple job to saw through the centre of each stator bar—in between the ceramic pillars—with a model-maker's saw. This procedure converts a three-gang into a six-gang of slightly less than half the original maximum capacitance. (The sawing process

"loses" the odd number stator plate.) If the original capacitance is more than 100pF—provided the value is known—it will be necessary to count the number of air gaps in each section and by simple division determine the capacitance per air gap. From this one can determine the number of air gaps for 40pF maximum per "half section" and saw out additional stator plates as necessary.

3. Wingrove & Rogers Ltd currently list the C28–141 (one-section) and the C28–421 (two section) obtainable with 18 air gaps per section, giving a capacitance range of 7–80pF. The one-section capacitor has $\frac{1}{4}$in spindles at both ends and may be ganged to the two-section capacitor in line. These tuning capacitors are of compact construction and would occupy a chassis space approximately 5in long, including the shaft coupler. The stator sections are supported by two ceramic pillars on each side and can be split with a model-maker's saw as already described. This procedure will result in a six-gang having 4–40pF (approximately) capacitance swing. Five sections will be wired up in the transmitter and one section left unused. (The Polar C28 type units were used in the surplus RF27 units, familiar to a large number of RSGB members.)

The capacitor manufacturer would no doubt be horrified to contemplate his precision tuning unit being operated on with a metal saw. Nevertheless, in practice variable capacitors modified in this manner continue to work very well with apparently no adverse distortion of the frame. The author has used successfully a Polar type C28–142 (75pF two-gang), modified with a saw to a 30pF four-gang, in an experimental push-pull Mark 3 receiver for more than five years.

The two tuned circuits are inductively coupled for optimum selectivity by low-impedance link windings positioned centrally round each coil in order to offer additional attenuation to higher frequency products.

It will be appreciated that the major function of the tunable i.f. is to pass the required ssb signal in the range 5–5·5MHz, and greatly attenuate all other signals, including the strong vfo injection. The rejection of these two tuned circuits must not be degraded by vfo leakage across the wiring. For this reason the coupling link between T2 and T3 is by a short length of $\frac{3}{16}$in outside diameter 75Ω coaxial cable, with the outer screening effectively bonded to chassis earth. The unscreened vfo coil L2 must be positioned as near to the gang tuning capacitor as practicable, and the rfc and L2 positioned as far away from T3 and V7 as is practicable.

Variable Frequency Oscillator

The long-term drift taking place continuously over the operating period of a couple of hours or so is a major annoyance. This slow drift is caused by the changing temperature of the two components that make up the frequency-determining resonant circuit, ie the inductor and capacitor. Apart from the temperature rise of the air in the cabinet and the air in the transmitting room, a considerable amount of heat from the valves—through the valveholders, the screening cans and skirts—warms up the chassis and this in turn warms up the vfo tank coil and tuning capacitor. Stability can be materially improved by omitting the screening cans of those valves in close proximity to the vfo resonant circuit, and in this design the valveholders of V5, V6 and V7 are standard B9G types *without* skirts. Additionally the holders are positioned in the chassis as far removed from the vfo coil, L2, as practicable. The variable capacitor is largely protected from conducted chassis heat by its positioning and by its physical size.

The vfo is an EF80 pentode, V6, arranged as a Colpitts parallel-tuned oscillator with capacitive swamping across the grid input, tuning being effected by one section of the five-gang main tuning capacitor. A small amount of temperature compensation is provided by the NTC capacitor (nominal value 15pF) mounted close to, and across, the coil L2.

White noise from the oscillator is reduced by incorporating an inductive anode load, L3. This has the further advantage of holding the valve anode voltage constant, unaffected by cathode current, and also increases the available drive to the converter V5.

For good vfo stability it is most important that the coil L2 can be adjusted initially to the precise inductance value required and that it will hold its setting over a long period of time. The dust core should be held in position by a screwed brass rod, running through the mounting bush of the coil former, and capable of being locked in position by either a spring-loaded clutch, or alternatively a locking nut, so that there is neither end- or side-float of the core within the winding. Ideally the vfo coil-former should be ceramic, with the coil winding put on under tension. Should difficulty be experienced in obtaining a suitable $\frac{3}{8}$in diameter ceramic former, the next best substitute is the baked-paper type of former as used in the coil-pack of the CR100 receiver. As this former is $\frac{5}{8}$in outside diameter and has a larger dust core, the winding turns will have to be reduced by about 25 per cent using a larger gauge of wire—22 or 24swg enamelled would be suitable.

Second Conversion

The second and final conversion process translates the 5–5·5MHz ssb signal into the required amateur band. It will be noted that this stage, V7, is also operated as a balanced converter with push-pull input and output circuits, and parallel cathode injection from the crystal-controlled heterodyne oscillator V8.

Switch banks S4 and S5 select the required coil for each band. As optimum performance is required, no attempt has been made to make one coil tune more than one band. There are, in fact, six separate coils—each with its own secondary winding—in order to control the optimum level of rf output into the following class A stage.

When the 80m or the 40m band coils are selected, the swing of the preselector tuning capacitor (PRESELECTOR) of 75pF each section is sufficient to tune across the incoming 5–5·5MHz i.f. signal. When this happens the converter valve momentarily becomes a tuned-grid/tuned-anode triode amplifier with sufficient feedback across the grid anode capacitance to behave as a class C oscillator. In order to prevent this possibility of unwanted regeneration, the valve is cross-neutralized by the two 2–8pF trimmer capacitors.

The anode circuit is balanced by the split-stator tuning capacitor—two Polar Type C28–142 of 75pF each section, with the common centre-tap of each coil not bypassed to rf and allowed to float, ht being fed via the 10kΩ resistor.

Five crystals are required in the heterodyne oscillator to cover the six amateur bands from 160m to 10m inclusive, and provision is made to select two additional crystals to give 10m band coverage in two more steps up to 29·5MHz, by the use of eight-way switch wafers. The lowest injection frequency is 7MHz for the 160m band and the highest 24·0MHz for the top of the 10m band—a frequency ratio of approximately

3·5 : 1. Additionally it will be appreciated that the crystal output will be greater on the lower frequency bands and less on the higher frequency bands. For these two reasons it is permissible to use a single anode coil, with the band-change switch S2 selecting the right value of capacitance to tune the output circuit to the required heterodyning frequency. With this arrangement the *L* : *C* ratio is lower on the low ranges and therefore the dynamic resistance of the tuned circuit and the oscillator voltage developed will also be lower. This is compensated by the greater output with the lower-frequency crystals used on the fundamental, and the smaller output with the 15m and 10m band crystals used on the second harmonic, and should result in a reasonably constant amplitude of output voltage throughout the six bands required.

The serious shortcoming of the single-ended converter lies in its inability to discriminate against the unwanted energy from the heterodyning oscillator if the ssb signal is to be raised in frequency by a ratio of more than approximately 4 : 1. For instance, when operating on the 10m band the heterodyning input will be 23MHz. A single-tuned circuit of high *Q*, tuned to 28MHz, will present approximately 10 times the impedance to the wanted signal as it will to the steady off-tune oscillator signal. However, the oscillator input voltage will be five to 10 times the amplitude of the peak ssb input (this is necessary to prevent distortion of the modulating waveform) so despite the selective effect of the output tank circuit, the steady oscillator voltage across it will be just about the same as the peak ssb voltage. Under these conditions it would be quite easy to tune up the following pa grid and anode circuits to the 23MHz output frequency instead of the required 28MHz ssb signal. Even if care were taken to ensure that all signal frequency circuits were correctly resonant in the 10m band, enough energy at 23MHz could leak through to produce a spurious signal outside the amateur band.

This, in fact, is the basic reason for accepting the additional complication of a balanced-converter anode circuit, and in order to obtain optimum performance at any frequency within the amateur band in use, continuous tuning right across the 500kHz range is needed. In practice, in order to avoid an additional panel control, the converter anode-tuning capacitor is ganged to the following class A amplifier/pa grid input tuning capacitor. This is the preselector tuning control and after setting the required transmitting frequency on the calibrated tuning scale, is used in much the same way as the preselector tune control on an amateur band double-superhet receiver.

Output Stages

A perfect single sideband transmitter would do exactly what the title describes and radiate a single sideband containing only the required voice modulating frequencies. The attenuation of the carrier and the unwanted sideband would be infinitely great. In fact, the transmitter output signal would be truly a single sideband signal. In practice this ideal state of affairs is not obtainable, and an amateur ssb transmitter is considered to be in the top class if the carrier suppression approaches 60dB and all signals in the vestigial sideband are 40dB down or better.

Most prospective constructors of ssb equipment will have sufficient knowledge to be aware that non-linearity in the final class A amplifier and class AB output stages produces intermodulation distortion products that appear close in on either side of the nominal carrier frequency and are not attenuated by the selectivity of the tuned circuits. Under correct operating conditions the distortion products on the wanted sideband are masked by the output signal. However, on the other side of the suppressed carrier they appear to the receiving station as a distorted (completely unintelligible) sideband that is much greater in amplitude than the true (clean and readable) remnant sideband that has not been completely suppressed in the mechanical bandpass filter.

Obviously there is no point whatsoever in using a filter that will attenuate the unwanted sideband to a level 45dB down, and then allowing this sideband to be put back again in the following stages in the form of distortion, and at a much higher level.

The class A amplifier stage V9 is required to provide an output of 50V peak to drive fully the pa valves. It is most important that while it is doing this it is operating in the most linear manner, over the straight part of its characteristic curve. However, the achievement of good linearity is made difficult in practice because it is also necessary to be able to control the total voltage amplification from this valve. In general, control of gain by alteration of bias or screen potential shifts the operating point and this is not compatible with a high order of linearity. A simple solution would be to put the RF DRIVE control in some other part of the transmitter circuit. There are, however, good reasons for controlling the amplification of the ssb signal in the band-switched circuits following the final conversion process. These will now be considered in detail.

All amateurs with past constructional or transmitting experience know that it is much more difficult to get voltage amplification and sufficient drive to the pa on 10m than it is on 160 or 80m. This occurs because the circuit losses become greater as the frequency goes higher. These losses are due to a number of factors.

1. the reduction in dynamic resistance of the tuned circuits;
2. greater losses due to absorption by stray coupling;
3. lower conversion efficiency of the converter valve;
4. increased input damping losses in the pa valves.

These losses add up to quite an appreciable amount, and taking the drive into the pa on 80m as a reference level, switching to the 10m band will show a drop of approximately 20dB. It follows that the total gain requirement in the transmitter—from the microphone input to the pa grids—can vary over a ratio of 10 : 1. Therefore, when changing bands it is necessary to have some manual control to set the total amplification to a value that will give the correct drive into the pa grids. This is the RF DRIVE control, and as the variation in circuit loss is taking place in the band-switched stages, a logical place for a drive control is in the circuitry associated with the class A amplifier stage following the final conversion process.

It is important to remember that in an ssb transmitter all stages in front of the final converter are running with a constant-peak sideband input and with a constant heterodyning output that is not affected by the position of the band-change switch. A good design will provide the optimum operating conditions to obtain the best signal to distortion ratio in the balanced modulator and the following converter stages. The audio gain control should never—repeat *never*—be used as a drive control. Its function is to enable the operator to set the audio level to suit the microphone in use and the characteristics of his voice to obtain optimum working conditions in the balanced modulator. Too much audio gain will cause over-modulation, and too little will impair the ratio of peak signal to resting carrier in the modulator, and the available carrier suppression will suffer. Once the audio gain control has been set correctly, it should not be touched again.

In the Mark 3 transmitter, linearity in the class A stage is improved by negative current feedback across the unbypassed cathode resistor, gain being controlled by varying the screen potential of the EF80 valve by means of a 25kΩ RF DRIVE control.

The 180W p.e.p. input pa stage comprises two 6146 valves operated without grid current in class AB1, using a conventional pi tank output circuit, switched for six bands. The intermodulation distortion product level in the pa stage is kept to the lowest practicable level by:

1. a "stiff" bias supply;
2. generous stabilization of the screen voltage;
3. correct choice of $L : C$ ratio in the pi tank circuit giving the optimum value of anode load;
4. a "stiff" ht supply with good *dynamic* regulation.

Under these conditions, with the pa correctly loaded, (ie a maximum signal—under single-tone conditions—screen current of 15mA for both valves) on-the-air reports indicate that the overall distortion product level from the transmitter is 40 to 45dB below the wanted sideband level. Note that this figure can only be obtained with correct driving of the pa—that is, strictly in class AB1. Overdriving is fatal to a clean signal, and 6146 valves have been specifically designed to be driven without grid current.

Muting

In addition to muting the filter amplifier valve, it is necessary to mute at least one stage in the output side of the transmitter to prevent feedback into the receiver. Accordingly, the muting bias is taken to the bottom end of the bias setting potentiometer in the pa bias supply. The two 6146 valves are therefore held at cut-off during transmitter standby periods, and allowed to take normal standing current when the press-to-talk button is depressed for transmit.

Circuit Description

Chassis Section A

The audio stage V1 (EF86) and the first half of V2 (12AT7), have cathode resistors without bypass capacitors giving negative current feedback to each amplifying stage. The second half of V2 is connected as a cathode follower to present the correct impedance to the OA7 balanced diode modulator. Although the EF86 introduces an additional valve type, it has been specifically designed as a low-noise af voltage amplifier, with characteristics exhibiting low microphony and low cathode-heater leakage, and is well worthwhile as the first stage in a high-gain audio amplifier intended to be used with a crystal microphone.

Output from the carrier oscillator V3a is connected to the 12AT7, V3b, operating as a phase splitter to provide a push-pull drive to the OA7 diode balanced modulator, and the carrier balance potentiometer RV2.

The switch, S1, SIDEBAND, selects the required carrier crystal (X1 and X2) to obtain final transmitter output on either the "Normal" or the "Suppressed" sideband as required. A proportion of the carrier oscillator output is fed via the 47pF capacitor to the panel-mounted carrier insertion

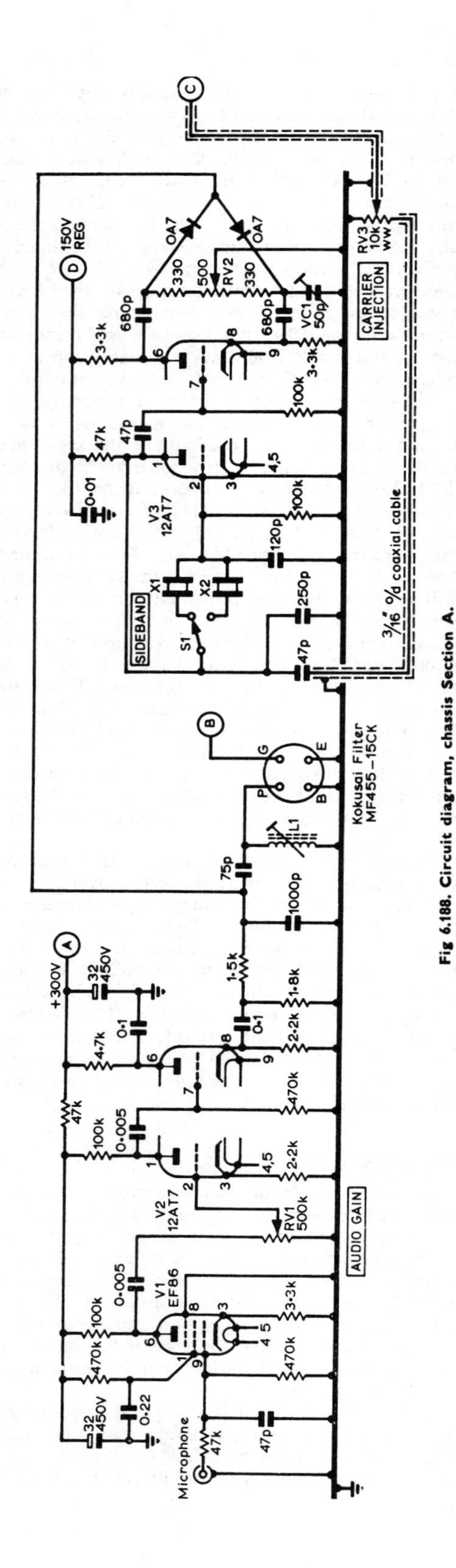

Fig 6.188. Circuit diagram, chassis Section A.

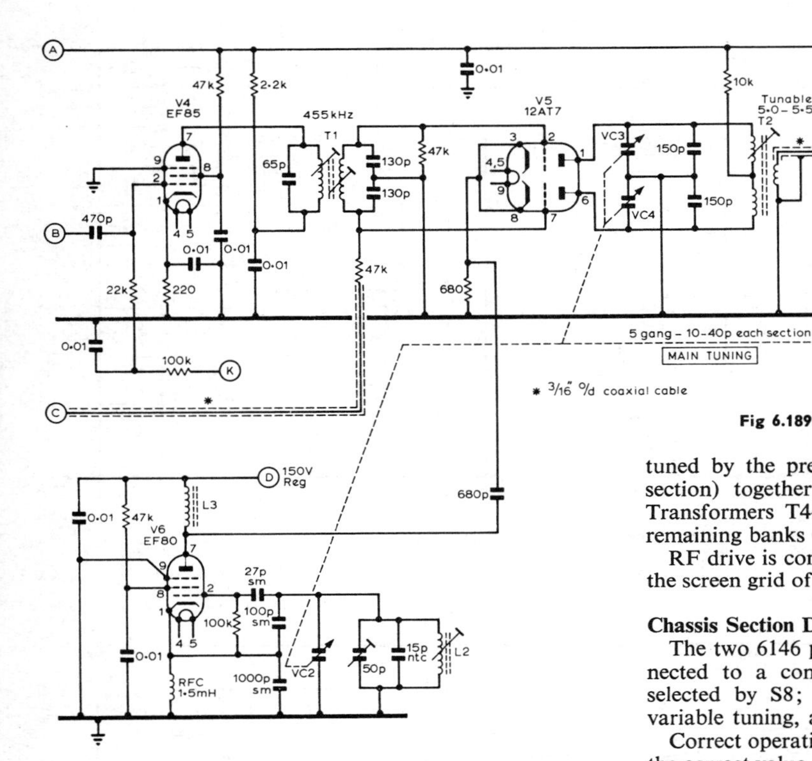

Fig 6.189. Circuit diagram, chassis Section B.

control RV3, and this enables an rf signal at carrier frequency to be fed round the sideband filter and be used for carrier insertion or netting purposes.

The double sideband suppressed carrier output from the balanced modulator is fed into the capacitor impedance-matching network of L1 to drive the Kokusai mechanical band-pass filter.

Chassis Section B

The low-level ssb output from the mechanical filter is amplified by V4 (EF85) operating in class A, and fed into a 455kHz i.f. transformer T1 feeding a push-pull output into the grids of the balanced converter V5 (12AT7). The heterodyning input from the vfo V6 (EF80) is fed to the two cathodes strapped in parallel.

Push-pull anode circuits of V5 are link coupled to push-pull grid input circuits of the final balanced converter V7 (12AT7).

Transformers T2 and T3 and the vfo coil L2 are the frequency-determining circuits tuned by a five-gang variable capacitor of 10–40pF each section, and make up the tunable i.f. covering the range 5–5·5MHz—this is the transmitter main tuning control FREQUENCY (KILOCYCLES).

Chassis Section C

The required injection frequency to convert the tunable i.f. to each of the six amateur bands is provided by a crystal-controlled switched oscillator V8 (EF80). Output circuit T10 is resonated to the correct frequency by a combination of fixed silver-mica capacitors and pre-set trimmers selected by S2 wafer of the five-bank range selector switch; the required crystal being selected by wafer S3.

The anode circuits of the final conversion valve V7 are tuned by the preselector capacitor (three-gang 75pF each section) together with the grid input of the pa valves. Transformers T4–T9 and coils L4–L9 are selected by the remaining banks (S4, S5, S6 and S7) of the range switch.

RF drive is controlled by varying the potential applied to the screen grid of the EF80 class A amplifier V9.

Chassis Section D

The two 6146 pa valves are strapped in parallel and connected to a conventional pi tank circuit. Coil taps are selected by S8; resonance being obtained by the 400pF variable tuning, and the three 350pF loading capacitors.

Correct operation of the pa can only be obtained by using the correct value of anode load (R_L). The L and C values for each of the amateur bands have been calculated for an R_L of 2,000Ω and an external load of 75Ω, and these are given in **Table 6.30**.

The VR150 stabilizer valve V12 provides 150V regulated for the vfo and the carrier oscillator, together with 200V regulated for the screens of the 6146 valves—obtained by tapping the ht feed made up with the 2·5kΩ and 1·5kΩ resistors in series.

All control functions for transmit-receive are controlled by the two-pole changeover relay having a high resistance coil energized by current obtained from the main 300V ht rail, and operated by a simple press-to-talk foot switch. The second pair of contacts marked AERIAL RELAY should be connected in series with the coil of an external low resistance relay of the GPO 600 type and an external 6–12V dc supply. The relay is shown in the non-energized receive position, the negative muting voltage to the receiver being short-circuited and the full 100V bias being fed to the grid return of V4 and, via the bias-setting potentiometer RV5, to the grids of the two 6146s; these three valves are therefore held at cut-off

TABLE 6.28

Final conversion crystal frequencies

Band	Crystal Freq (MHz)	Mode	Output Freq (MHz)
160	7·0	Fundamental	7·0
80	9·0	Fundamental	9·0
40	6·25	2nd harmonic	12·5
20	9·0	Fundamental	9·0
15	8·0	2nd harmonic	16·0
10	11·5	2nd harmonic	23·0
10	11·75	2nd harmonic	23·5
10	12·0	2nd harmonic	24·0

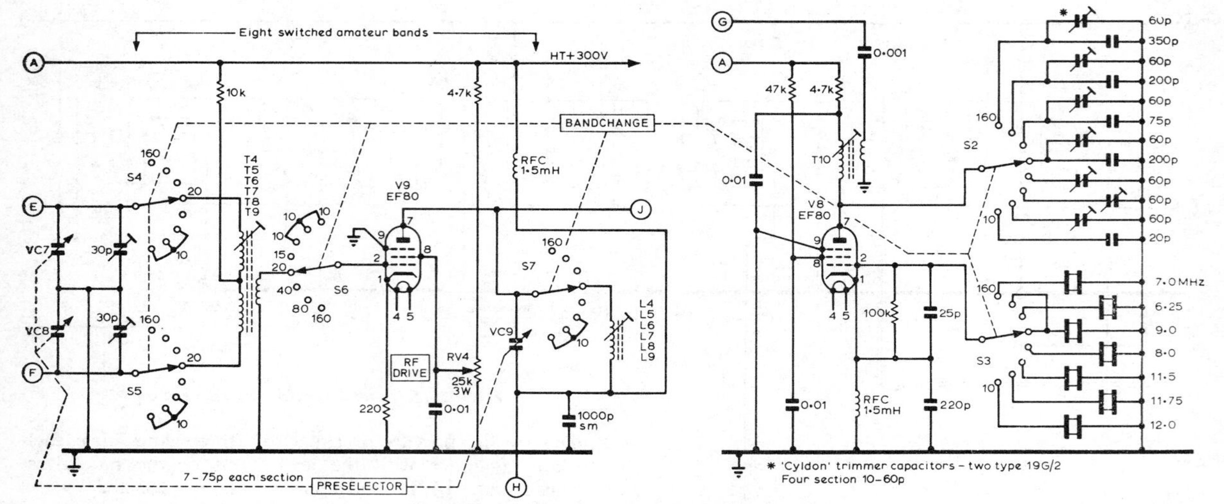

Fig 6.190. Circuit diagram, chassis Section C.

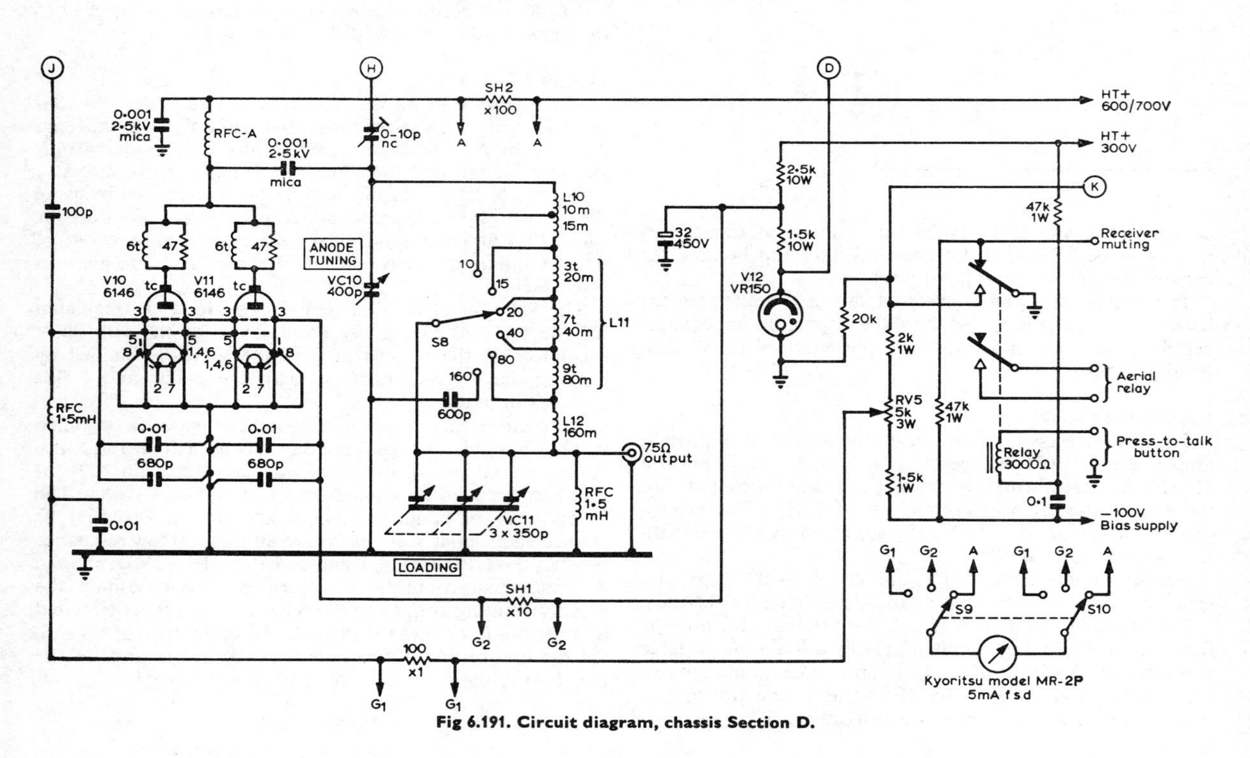

Fig 6.191. Circuit diagram, chassis Section D.

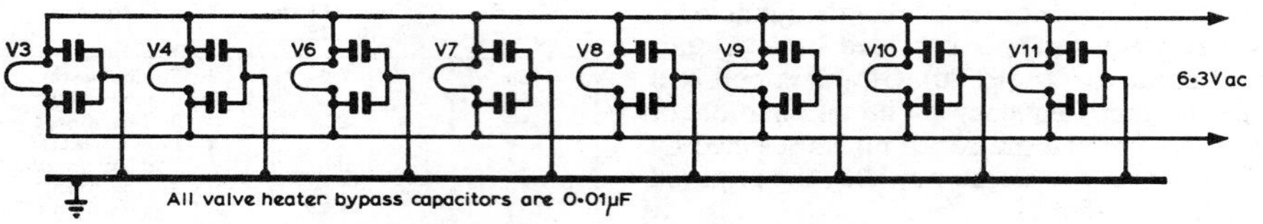

V3 V4 V6 V7 V8 V9 V10 V11

6·3Vac

All valve heater bypass capacitors are 0·01μF

Fig 6.192. Valve heater bypass capacitors.

TABLE 6.29

Resonant circuit component details

Component	Freq or band	Description
L1	455kHz	Half standard Denco type IF11–465 i.f. transformer.
L2	5·455–5·955MHz	28t 28swg enam, on ⅜in od Cambion former with 20063 slug (3·3μH).
L3	5·7MHz	20t 28swg enam, on 0·3in od former, core in centre of coil.
L4	160m	75t 38 dsc close-wound, single layer wire, 135pF, padding capacitor across coil, fit two dust cores in coil former.
L5	80m	78t 38 dsc, close-wound, single layer.
L6	40m	36t 36swg enam, close-wound, single layer.
L7	20m	20t 24swg enam, close-wound, single layer.
L8	15m	11t 22swg enam, close-wound, single layer.
L9	10m	9t 18swg enam, close-wound, single layer.
L10	10m & 15m	5½t 16swg tinned copper, spaced to 1¼in long, tap 3t from hot end, 1in od.
L11	20m, 40m & 80m	Total 19t 18swg tinned on 1¼in od ceramic former, grooved 18t per inch, wound 9t—3 groove gap—7t—3 groove gap—3t.
L12	160m	30t 22swg enam, close-wound on 1in od ceramic former.
T1	455kHz	Denco type IFT 11-465 with secondary capacitors modified.
T2	5–5·5MHz	32t—ct—32t 28swg enam, 5t link over centre.
T3	5–5·5MHz	32t—ct—32t 28swg enam, 5t link over centre.
T4	160m	Primary 75t—ct—75t 38 dsc close-wound, single layer, secondary 35t 38dsc over centre of primary wire, 135pF padding capacitor across primary.
T5	80m	Primary 78t—ct—78t 38dsc, close-wound single layer, secondary 40t 38dsc over centre of primary.
T6	40m	Primary 36t—ct—36t, 36swg enam, close-wound single layer, secondary 24t 38dsc over centre of primary.
T7	20m	Primary 18t—ct—18t 24swg enam, close-wound single later, secondary 20t 28swg enam, over centre of primary.
T8	15m	Primary 11t—ct—11t 22swg enam, close-wound, single layer, secondary 11t 28swg enam, over centre of primary.
T9	10m	Primary 9t—ct—9t 18swg enam, close-wound single layer, secondary 9t 28swg enam, over centre of primary.
T10	7–24MHz	Primary 13t 24swg enam, close-wound single layer, secondary 4t 24swg enam at cold end of primary.
All coils wound on Neosid or Aladdin 0·3in dia formers with 0BA dust cores, unless otherwise specified, and in 2¼in by $\frac{13}{16}$in by $\frac{13}{16}$in seamless aluminium cans.		
S1		"Normal-suppressed" sideband switch sp/2-way.
S2, 3, 4, 5, 6, 7		Yaxley sp/8-way six-bank band-change switch.
S8		Pi tank band change switch, h/duty ceramic, sp/6-way.
SH1		42t 38swg enam wound on 1W 10kΩ resistor.
SH2		6t 36swg enam wound on 1W 10kΩ resistor.
S9, S10		Yaxley sp/3-way two-bank ceramic, break-before-make.

All resistors ½W unless otherwise specified.

and there is no output from the transmitter. When the press-to-talk button is depressed the relay closes, allowing the 100V muting bias to cut off the receiver, and at the same moment of time to short-circuit the bias rail to earth, allowing V4 to conduct. The bias on the pa valves is now a proportion of the negative 100V supply, determined by the setting of the 5kΩ potentiometer RV5 which is now the centre part of a potential divider between the 100V bias rail and chassis earth. RV5 is adjusted until the two 6146 valves are taking a total of 50mA standing anode current.

Panel Meter

A panel meter having a basic 5mA movement is provided to monitor the operating parameters of the pa. Grid, screen or anode current can be selected as required by the two-pole three-way meter switch S9, S10.

Shunts SH1 and SH2 are adjusted to obtain a full-scale current of 50mA in the G2 position and 500mA in the A position. The constructional details of the shunts given in the circuit diagram will only be correct for the type and make of meter used in the original transmitter. Any other meter having a different internal resistance will require adjustments to SH1 and SH2 to obtain the correct × 10 and × 100 multiplying ratio: see Chapter 18—*Measurements*.

Construction

The chassis is made up of four separate box sections of 16swg aluminium to give a total size of 16in by 12in by 2½in deep, with a 17½in by 9in front panel. This enables the transmitter to be constructed and wired in units; in fact each unit can be individually tested and aligned before final assembly should this be considered necessary. These box sections are machine pressings, are true to size, have sharp radius corners, and are available in the required sizes shown on **Fig 6.193** from North West Electrics, 769 Stockport Road, Levenshulme, Manchester.

A pair of 6146 valves will give an appreciable output, and as a precaution against any possibility of stray rf fields the two valves and the components in the anode circuit are screened with a 16swg aluminium box measuring 6in by 5¾in by 5¼in high. This is, in fact, a U-shaped assembly with ⅜in lips top and bottom, permanently fixed to the main chassis with PK self-tapping screws. The pi tank switch and coil, the anode tuning and loading capacitors and the rf choke are all supported by the screening box. It is most important that there is a free flow of air throughout the pa compartment; accordingly a ring of $\frac{3}{16}$in diameter holes is drilled round each octal valve holder, and the box top and back are made from expanded aluminium mesh, also held in position by PK screws, and removable for access to the 6146 valves.

The chassis layout showing the principal components is given in Fig 6.193, and a panel layout with suggested positioning of the dial aperture and the control knobs in **Fig 6.194**. It is necessary to remember that the three ¼in diameter shafts from the pa box to the front panel control knobs have to pass over the top of the five-gang main tuning capacitor. The height of this capacitor will therefore determine the final vertical positioning for the LOADING, TUNING and RANGE-MC/S control knobs.

The Yaxley two-pole changeover switch bank S1 must be mounted reasonably close to the two carrier crystals X1 and X2, and the oscillator valve V3, and is supported by a small L-shaped bracket bolted to the chassis side apron. A clear space should be kept down the centre of this section to clear the switch control rod and support bearing before connection to the Eddystone flexible coupler and the shaft to the front panel control knob. In order to avoid the possibility of induced rf currents bypassing the balanced modulator and the filter it is important that the whole of the shaft length is insulated, and ¼in diameter polystyrene rod is recommended.

The details of the assembly and the positioning of the main

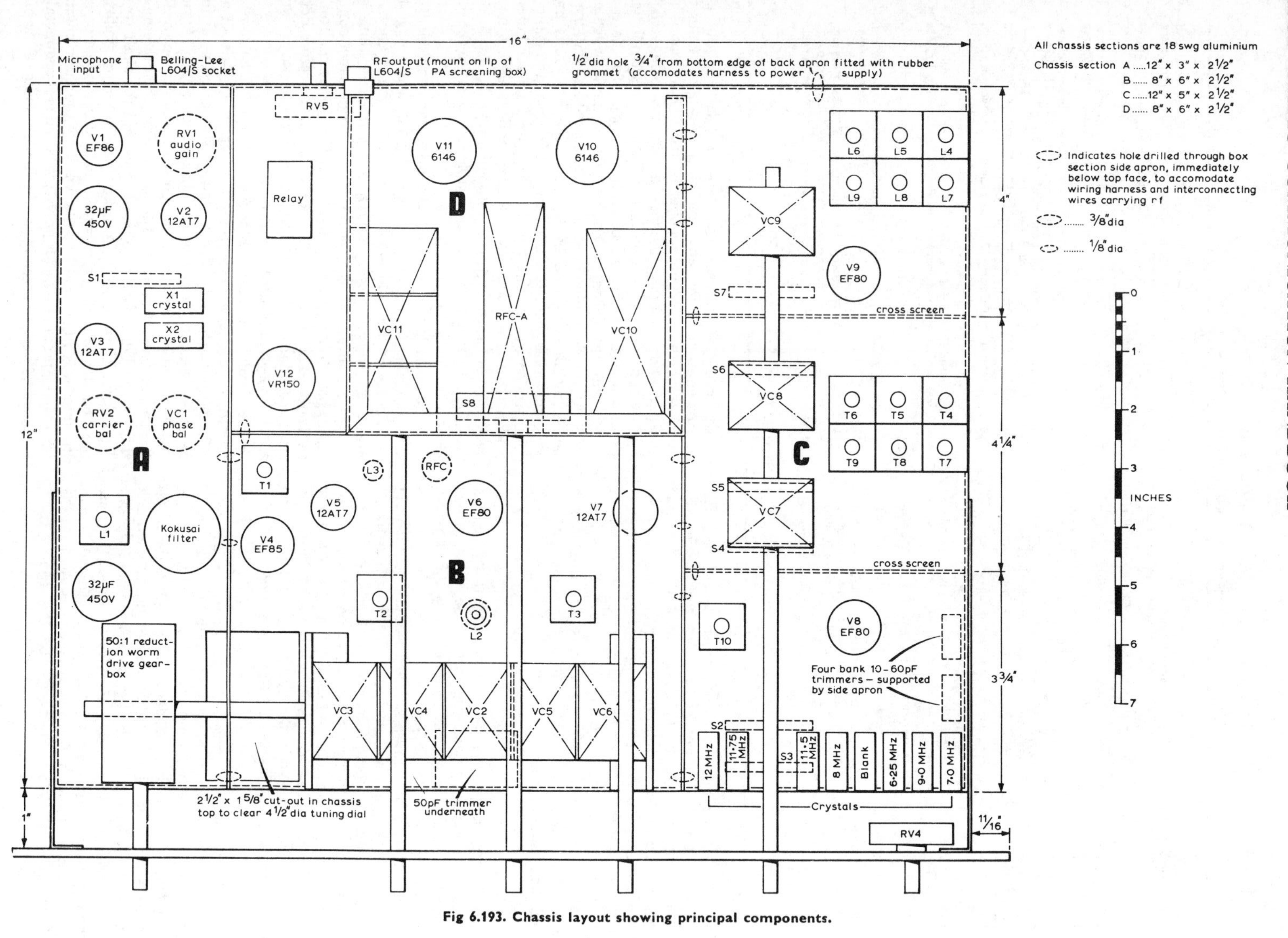

Fig 6.193. Chassis layout showing principal components.

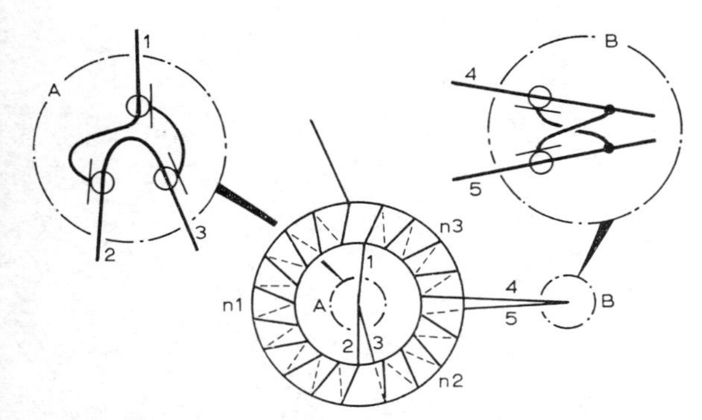

Fig 6.207. Winding of transformer T1 with interconnection detail shown enlarged.

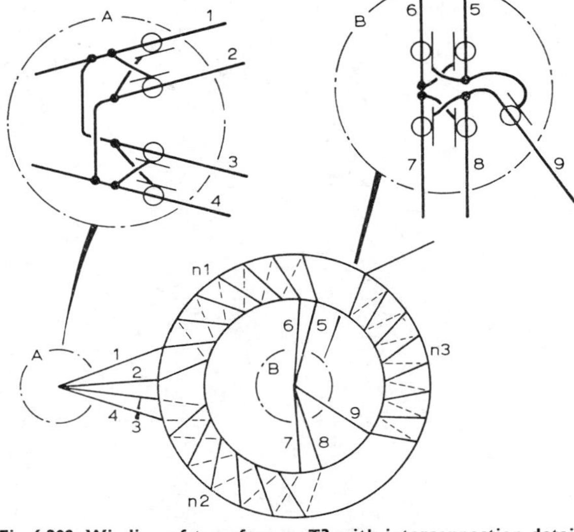

Fig 6.209. Winding of transformer T3 with interconnection detail shown enlarged.

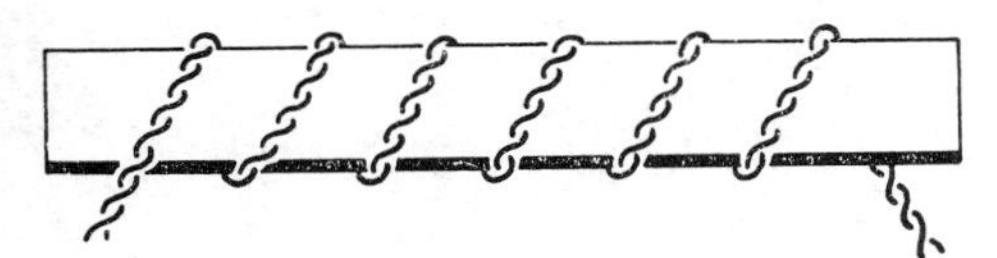

Fig 6.208. Winding of transformer T2.

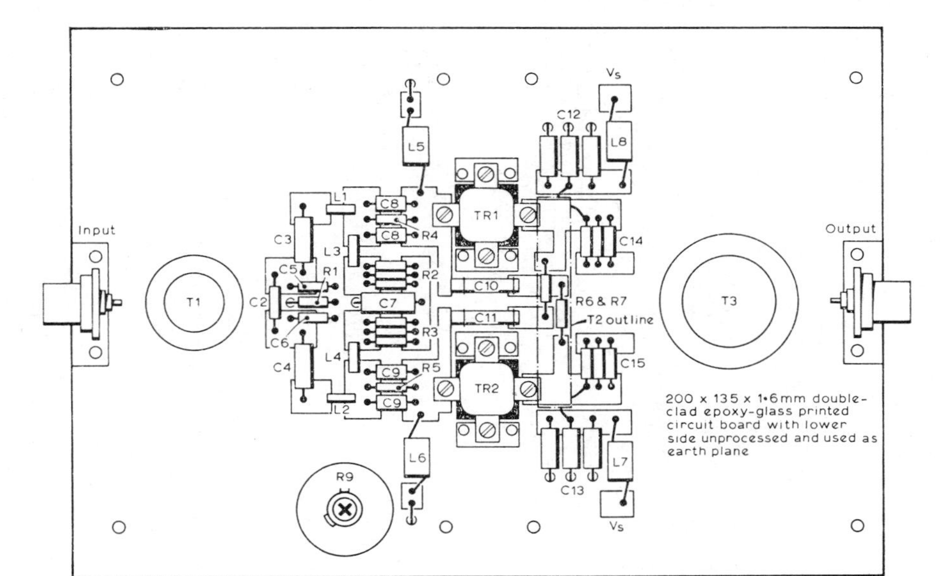

Fig 6.210. Main circuit board layout.

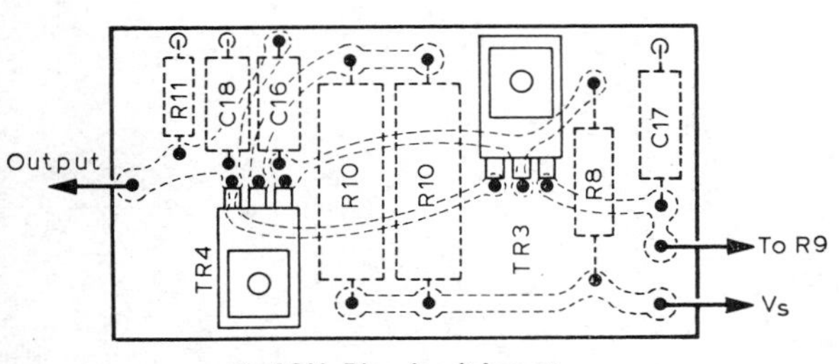

Fig 6.211. Bias circuit layout.

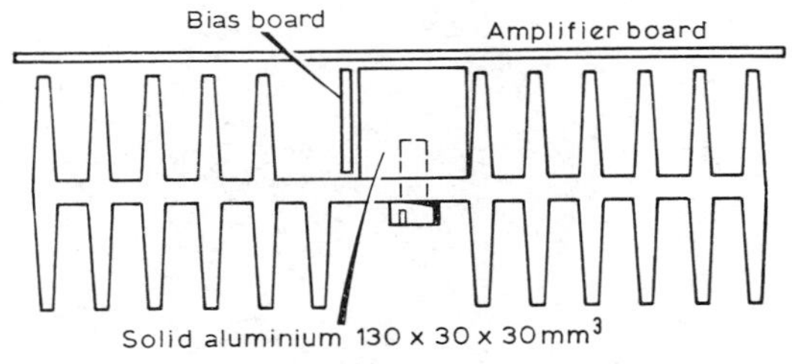

Fig 6.212. Heat sink assembly.

single-pole eight-way five-bank range switch assembly is shown clearly in the photograph of the underside of the chassis. These are standard Yaxley paxolin wafers. It will be noted that the use of individual coil cans on top of the chassis, and switch banks running underneath, gives a clean accessible layout together with very short connecting wires. Switch sections S2 and S3, together with the indexing mechanism, are supported by the chassis apron, sections S4 and S5 by the front cross screen, and S6 and S7 by the rear cross screen—these can conveniently be assembled before the screens are bolted in position in the main chassis, Section C.

Also shown in the under-chassis photograph is a 50pF air-spaced variable capacitor (Polar Type C28–141) fitted with a fine-toothed sprocket and spring-loaded "latch" as an indexing mechanism. This assembly is in a central position, bolted to the front chassis apron, and is the 50pF pre-set trimmer shown in the circuit diagram across the vfo coil L2.

A balanced bridge circuit is used for neutralization of the 6146 valves, and this requires that the cold ends of the six coils and the frame of the 75pF grid tuning capacitor are taken to a common bus-bar which is insulated from the chassis and bypassed with a 1000pF mica capacitor. The two remaining 75pF tuning capacitors have their frames at chassis earth. However, in order to keep the drive shafts in line, all three capacitors are mounted on an insulated plate made from dry hardwood, $\frac{3}{8}$in thick—the mounting screws for the grid tuning section being well countersunk into the wood to guard against an inadvertent short circuit to the chassis—and the shaft coupled by a length of $\frac{1}{4}$in internal diameter paxolin tubing. These three capacitors are Polar Type C28–141 (available in the surplus RF27 units). The frames of the two sections tuning the converter anode (V7) are connected to the chassis by short lengths of 18swg tinned copper wire.

The pi tank inductance is made up of three separate units. The 160m coil L12 is on a separate ceramic former at right angles to the main winding L11 and adjacent to its "cold" end, with the self-supporting 10m coil L10 positioned at the "hot" end to form the connecting link between L11 and the stator plates of the anode tuning capacitor.

The 400pF anode tuning capacitor was obtained from surplus sources. Overall dimensions are $3\frac{1}{4}$in long, $1\frac{7}{8}$in wide and $1\frac{5}{8}$in high—a convenient size to fit into the available space.

TABLE 6.30

Pi network data—two 6146 valves

Band	R_L = 2,000Ω XC1 = 200Ω	XL = 250Ω	XC2 = 46Ω
80m	220pF	11·0μH	900pF
40m	110pF	5·5μH	450pF
20m	56pF	2·7μH	225pF
15m	38pF	1·8μH	160pF
10m	28pF	1·4μH	115pF

The rotor vanes are semi-circular giving a straight-line capacitance law, and the air gap is 0·02in.

A standard pi-wound rf choke is unsuitable for use with a shunt-fed tank circuit and this component must be specially wound to have a low self-capacitance and no self-resonant points within the required amateur bands. There is insufficient space in the pa screening box for the full size pi tank choke described in this chapter. However, the smaller version actually employed appears to give a satisfactory performance. This consists of 250 turns of 36swg enamelled wire wound in unequal sections of 150, 60, 30 and 10 turns on a ceramic former, $\frac{3}{4}$in diameter and 4in long, with a $\frac{1}{8}$in spacing between each of the sections.

It is convenient to use a standard broadcast type three-gang variable capacitor of 350pF each section for the aerial loading capacitor. As this would not have a large enough value for use on the lowest amateur band, the 1·5MHz position of the pi tank band-change switch is used to bring into circuit a further fixed loading capacitor of 600pF. A standard single-pole, six-way, Yaxley *ceramic* switch S8 is supported by the front apron of the pa screening box and the switch in turn supports the coil L11 by means of the two ends and the three tapping connections. S8 and L11 are conveniently made up as a unit before fitting into position.

On the 160m band, power is reduced to the equivalent p.e.p. of a 100 per cent modulated 10W dc input A3 transmitter by switching the anode feed to the pa valves from the normal 700–750V line to the 275–300V rail feeding the remainder of the transmitter.

As the transmitter is controlled by a press-to-talk button (this may be a circular bell-push screwed to a block of wood and placed on the floor for foot operation) there is no relay

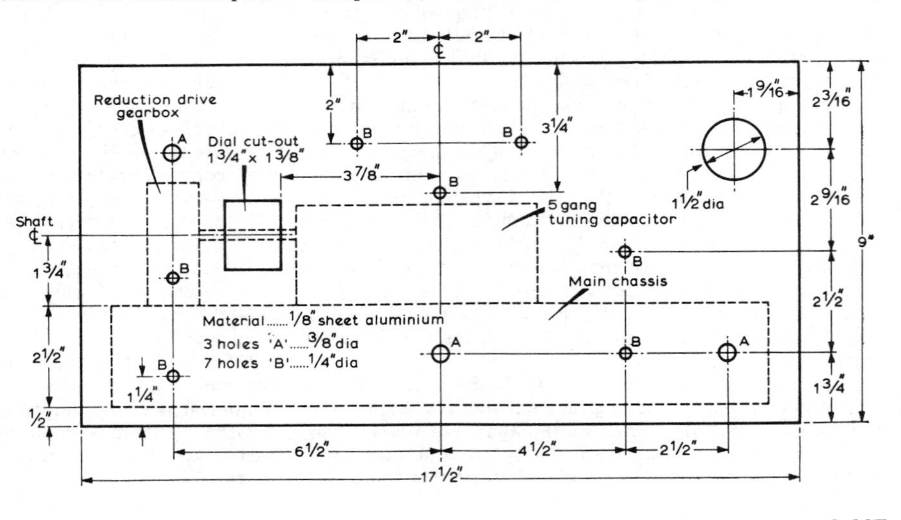

Fig 6.194. Panel layout. Dotted line shows the relative position of the main chassis, the five-gang tuning capacitor and the reduction drive gearbox. The shaft centre-line of the five-gang capacitor used by the author is $1\frac{3}{4}$in above the chassis top face; a different type of five-gang may require re-positioning of the dial cut-out and the drive gearbox tuning knob. Front panel material is $\frac{1}{8}$in thick aluminium.

Power supply unit.

clicking while the operator is actually talking. The type 3000 relay does not require a flexible rubber shock mount and may be screwed directly to the chassis.

The meter switch S9 and S10 is a single-pole, three-way, two-bank Yaxley type. This may be paxolin, but ceramic is preferred to eliminate any possibility of tracking. It is most important that this switch is the break-before-make type, otherwise ht would be momentarily connected to the 6146 grids as the switch poles moved over. If this type of Yaxley wafer is not available, standard make-before-break single-pole six-way banks can be used with each adjacent contact left blank (ie the three positions of the control knob would be at 60° instead of the customary 30°).

Any good quality moving coil meter of 5mA full scale deflection is suitable. However, the constructional details of the shunts SH1 and SH2—giving a multiplying ratio of × 10 and × 100—are only correct for the type of meter used by the author; this is a Kyoritsu Model MR–2P. The 1W carbon resistors used as formers for the shunt windings may be any available value of 1kΩ or greater.

The calibrated tuning dial can be seen clearly in the chassis photographs. This is constructed in a simple manner by cutting a piece of 20swg aluminium 9in by 1½in, and cementing it with Araldite to the rim of a standard 4½in diameter cord drive drum (Jackson Part No 4029 or similar). The calibrated scale is hand printed with Indian ink on a piece of glazed drawing paper 8½in by 1½in held in position by self-adhesive tape along either end. This enables the calibration points to be initially marked up in pencil and the paper scale removed for the final hand printing, without having to upset the alignment by removing the drum from the five-gang capacitor.

A worm-drive gearbox with a ratio of 50 : 1 is very convenient, because with a tuning range of 500kHz, one turn of the FREQUENCY knob represents a tuning rate of 10kHz. This ratio allows precise frequency setting without being too tedious when it is desired to traverse completely across the band. The gearbox used by the author was obtained from a scrapped Collins CT12 transmitter. Equally suitable is the 50 : 1 worm drive used in the BC221 frequency meter. In the event that these items are no longer available from surplus sources, a possible alternative is the 25 : 1 worm drive from a TU5B tuning unit, still available in many amateur junk boxes.

Alignment

An ambitious project such as the construction of an all-band ssb transmitter is only likely to be undertaken by an amateur with past constructional knowledge. For this reason it is felt that step by step detailed alignment instruction is not necessary.

In the interest of valve life the 6146s should be run with the power supply ht switch in the 275–300V position—at least until the anode and grid tuning has been checked and the valves correctly neutralized on all bands.

A serviceman's signal generator is really the minimum requirement, and, assuming that this is available, each chassis section can be individually aligned before final assembly. Assuming that this has not been done and that the signal generator is only available for a short space of time —after final construction is completed—the most satisfactory alignment procedure is to start at the back and finish at the front. That is from the pa tank coil back to the carrier oscillator, using the serviceman's generator as a signal source.

All oscillators should be checked for satisfactory operation. It is most important that the dust core of L2 is adjusted at the low end, and the trimmer VC1 at the high end to obtain the *correct* tuning of the vfo over the frequency range of 5,455–5,955kHz.

All coils are in screening cans and a grid dip oscillator cannot be brought up to the coil to check alignment, neither is it possible to use the usual pick-up loop feeding into an absorption wavemeter. This may cause difficulty when it is required to set each of the pre-set trimmers across the final conversion oscillator coil T10, because it is important on some ranges to know that the circuit is resonant on the correct harmonic. There are two ways to overcome this limitation:

1. A pick-up loop tucked into the wiring close to the valve anode will deflect a sensitive absorption wavemeter using a diode and 100μA meter; alternatively the "signal" can be fed into the station receiver via a length of coaxial cable.
2. A pick-up loop can be constructed by winding six turns of 24swg enamelled wire round the end of a 6in length of ⅛in diameter Systoflex sleeving, the loop connections then feeding back down the centre of the sleeving and terminating with a length of twin flex or coaxial cable. This "probe" can then be pushed down towards the dust core inside the former of the circuit to be checked.

The tunable i.f. must tune accurately over the range 5·0–5·5MHz. As it is relatively easy to inadvertently tune the frequency determining circuits to the strong vfo signal of 5·455–5·955MHz, it is advisable to disable the vfo by removing the oscillator valve V6 while the correct tunable i.f. alignment is being undertaken. The signal generator, set to 5·25MHz, should be connected to one of the grids of V5, the KILOCYCLES dial set to the centre band position, and the dust cores of T2 and T3 adjusted for maximum output. It will be

found that the ganging holds quite accurately right across the 500kHz tuning range and that (assuming the five-gang variable capacitor has semi-circular rotor plates giving a straight-line capacitance law) the dial calibration is linear right across the scale.

It will be noted that there are eight positions of the main range switch, one each for the five bands from 160 to 15m and three for the 10m band to give a 28–29·5MHz coverage. A separate signal frequency coil is used for each of the six amateur bands, T4–T9 in the anode circuit of V7, and L4–L9 in the anode circuit of V9. As the 10m coils (T9 and L9) cover the three sections of the 10m band, the respective switch contacts of S4, S5, S6 and S7 are strapped together.

The two Philips pre-set capacitors of 2–8pF are soldered directly to the valveholder pins of V7 and can be seen in the chassis underside photograph. They are initially fully unmeshed and their capacitance increased equally until neutralization is complete and V7 fully stable at all settings of the preselector tuning control (range switch set to the 3·5 MHz or the 7MHz band).

Alignment of the conversion oscillator V8 is initially undertaken with the range switch in the 29MHz position; the dust core of T10 is adjusted for maximum oscillator output at 24MHz. The range switch is then set to the 28·5 MHz position and the appropriate 60pF trimmer adjusted for maximum oscillator output at 23·5MHz. With the range switch in the 28MHz position the next 60pF trimmer is adjusted for maximum output at 23MHz . . . and so on throughout the eight amateur bands to 1·5MHz. On certain ranges the oscillator output is the second harmonic of the crystal frequency and it is important to ensure that the correct harmonic has been selected and that the oscillator is not inadvertently tuned to the fundamental or the third harmonic. At each adjustment the output signal frequency should be monitored using an absorption wavemeter or the station general coverage receiver.

The transformer T1 is a standard item purchased on the surplus market, believed to have been made originally by Parmeko and often referred to as a "table topper" type. An alternative for T2 is a centre tapped secondary winding 125–0–125V at 50mA with one DRM–1B rectifier (or similar) in the usual full wave circuit. A suitable type may be obtained from *Radio Communication* advertisers, or from RS Components Ltd through local radio retailers.

Power Supply

An rf power amplifier can only operate in a linear manner and give an undistorted output if it is provided with grid, screen and anode voltages that remain constant in potential and do not vary under differing load conditions. Additionally, the heavy anode current demand is at syllabic rate and the main ht power supply must have good regulation under dynamic conditions.

It will be appreciated that the carrier of an ssb transmitter is attenuated in the modulator by a process of balancing, and that normally the modulator will give an attenuation of 40dB or more. It will also be appreciated that any hum ripple, either on the ht supply to the carrier oscillator or to the audio valves, will unbalance the modulator and produce a hum-modulated carrier that cannot be removed by the balancing control RV2. A rough "carrier" will also be produced if there is any 50Hz mains ripple induced into the cathode of V2b from the 6·3V heater line. To overcome these difficulties it is good practice to provide (1) two-stage smoothing and a large value of reservoir capacitance in the 275–300V power supply, (2) a two-wire heater circuit that is balanced to earth.

Under-chassis view of the power supply unit.

In regard to the high voltage supply to the pa valves, the conditions are different because the anode current of a tetrode is almost independent of small changes in anode potential and hum ripple will not "anode modulate" the signal. However, dynamic regulation over the range 50–250mA is important because a fall in anode voltage at the moment of peak current demand would prevent the pa handling the peak signal and could cause flat-topping and distortion. In the 650–700V supply the customary smoothing chokes are therefore omitted and a large value of reservoir capacitance is placed directly across the rectifier output.

The bias supply is made "stiff" by the use of a generously rated mains transformer and rectifier stack, together with ample capacitance and a relatively heavy bleed current of approximately 20mA through the resistive load network.

A circuit diagram of the power supply used at G2DAF is given in **Fig 6.195.** All components can be mounted on an aluminium chassis 14in by 10in by 3in deep (see **Fig 6.196**). Points to note are:

(*a*) The heater voltage of 6·3V ac is balanced and "floating" above earth. This prevents hum being induced into the bias and ht lines in the cable harness between the power pack and the transmitter, and also ensures that the heavy heater current is not superimposing 50Hz ripple on the ht negative return connection.

(*b*) The bias supply is "stiff" and also provides a wide range of adjustment.

(*c*) The 32μF capacitor between the 200V rail and earth ensures, together with the bleed through the series

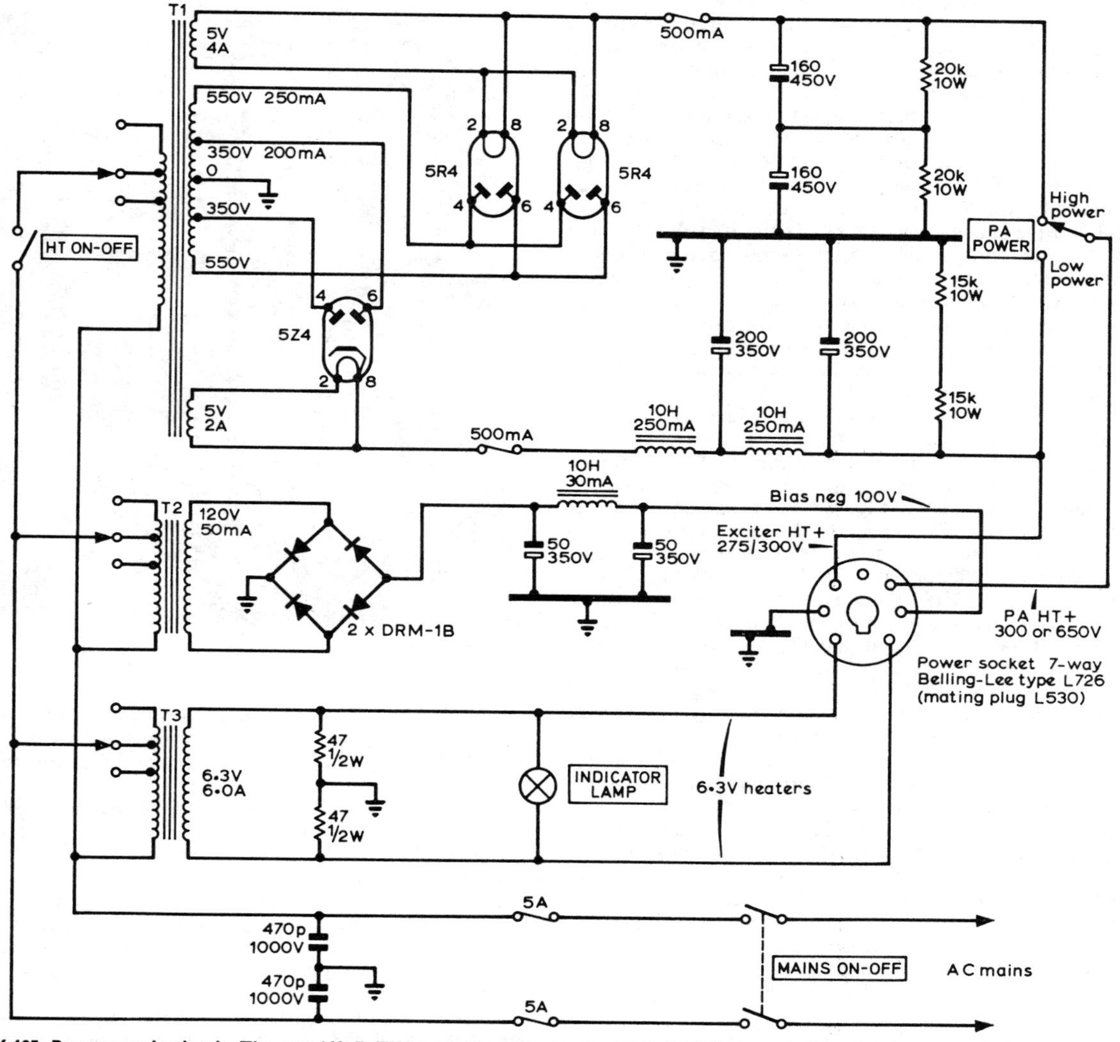

Fig 6.195. Power supply circuit. The two 160μF 450V capacitors are standard 100 + 60μF electrolytics, with the two sections strapped in parallel. T1 may be two separate transformers, operated from the common ht on/off switch.

resistors and the stabilizer valve, a "stiff" screen supply to the pa valves, with a low impedance to the demand that will vary at audio rate (see Fig 6.191).

(*d*) The high value of effective capacitance of 80μF across the 5R4 rectifier valves gives an output voltage of 650–700V and ensures excellent dynamic regulation.

S1 is provided so that the anode ht can be reduced to permit operation on the 160m band without exceeding the licensed power limit. This switch must never be used for tuning and loading purposes on the other bands. A 6146 valve fully driven with a low voltage on the anode and normal voltage on the screen will take excessive screen current, the rated screen dissipation will be exceeded and the valve may be destroyed.

Conclusion

For cw operation the transmitter is best controlled by the press-to-talk foot switch. A key jack socket of the closed circuit type can conveniently be connected in the cathode return of the final balanced converter valve V7—between the 680Ω resistor and chassis earth; this connecting point must be bypassed to rf by a 0·01μF capacitor. The keying characteristic is clean and free from chirp. The audio gain control should be turned to zero and the carrier control (RV3) advanced until the required amount of drive is available to the pa valves.

For the benefit of those amateurs who have—or can borrow—a thermionic diode probe valve voltmeter, **Table 6.31** gives details of the rf voltage readings of the conversion oscillator, and the sideband voltage readings at the grid

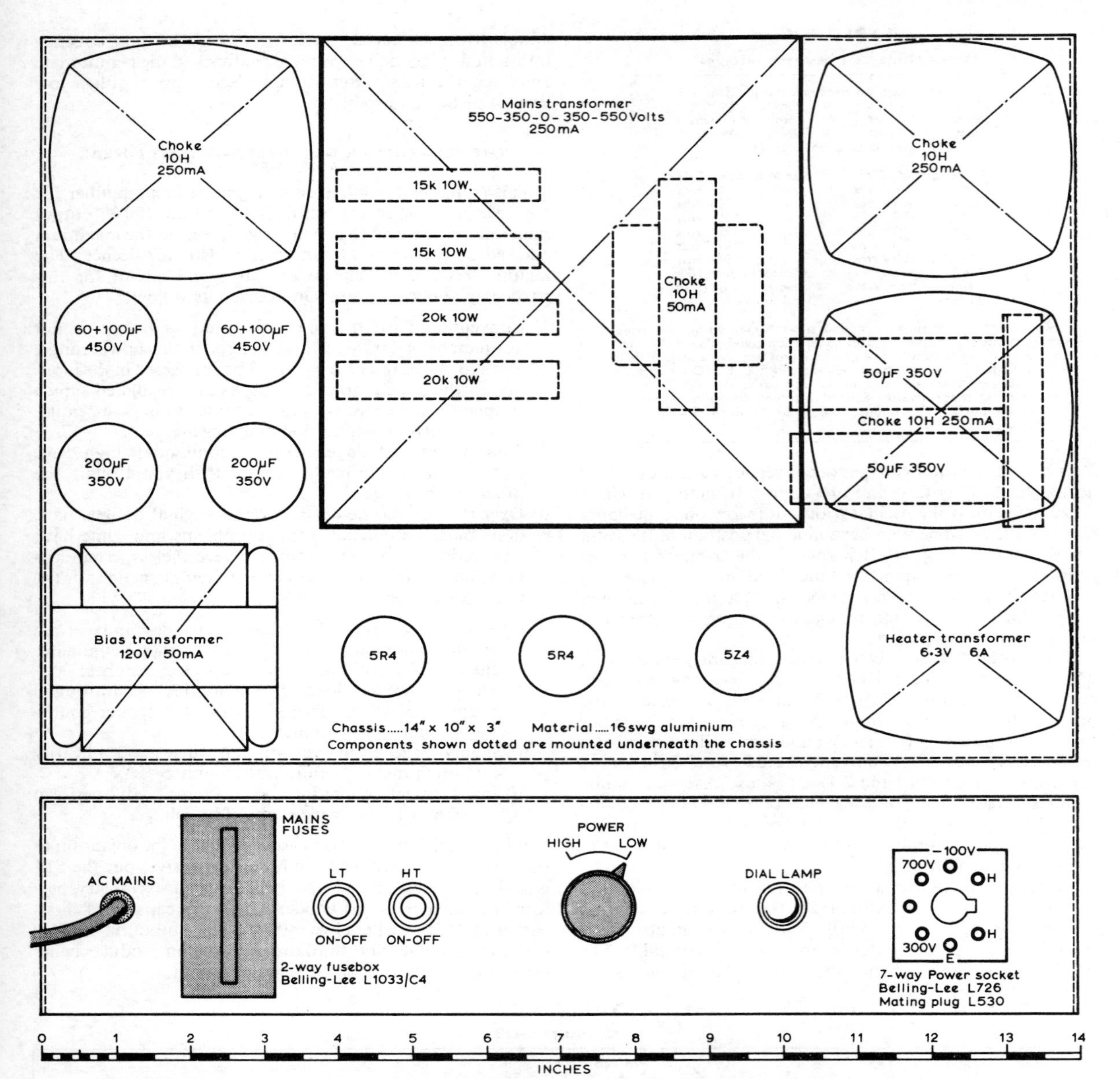

Fig 6.196. Power supply chassis and front apron layout showing the principal components. Material—16swg aluminium.

of the class A amplifier V9 and the available drive to the 6146 grids. The values quoted are rms values and were measured with a Salford Instruments valve voltmeter Type BW 211B. When measuring the voltages at the grid of the EF80 amplifier V9, it should be remembered that the input capacitance of the diode probe will de-tune the anode circuit of the converter V7 and give a false low reading. When the probe is in position each circuit must be brought back to resonance by adjustment to the preselector tuning control until maximum reading is obtained. Similarly, when measuring the voltage at the grids of the 6146 valves the diode probe should be connected to the V9 anode side of the 100pF coupling capacitor and the capacitors should be temporarily disconnected from the grids of V10 and V11. This is essential, otherwise any peak rf drive that exceeds the negative bias potential (about 50V) will cause the pa valves to draw grid current, the input impedance will fall to a few hundred ohms and severely damp the resonant circuit, thus giving a false low reading. Before taking the valve voltmeter reading the circuit must be correctly resonated, with the diode probe in position, by slightly unscrewing the dust core of the appropriate coil (L4 to L9) at the same time rocking the preselector tuning control to ensure that the anode circuit of V7 is also correctly resonated.

TABLE 6.31

Valve voltmeter rms check voltages

RF — **Conversion oscillator injection**

Band	Voltage	Band	Voltage
160m	3·4V rms	20m	4·1V rms
80m	4·1V rms	15m	4·4V rms
40m	4·8V rms	10m	5·0V rms

Measured at output of T10

SSB

Band	Volts at grid of V9	Volts at anode of V9
160m	3·4V rms	> 150V rms
80m	6·3V rms	> 150V rms
40m	9·8V rms	> 150V rms
20m	6·5V rms	120V rms
15m	7·4V rms	107V rms
10m	6·2V rms	92V rms
	Measured with V9 removed from valve holder	Measured with 100pF grid capacitor to V10 and V11 removed

Note: For all readings, sideband signal obtained by adjusting carrier injection potentiometer to position giving maximum valve voltmeter reading. (Audio gain control set to zero). KILOCYCLES tuning at mid band position, PRESELECTOR correctly resonated for maximum output on each band.

Measuring instrument: Salford Instruments valve voltmeter Type BW211B. (Thermionic diode probe into 50MΩ input impedance.)

The various output voltages of conversion oscillator V8 are not particularly critical and are quite satisfactory if within ± 20 per cent. If the oscillator output is low on some bands re-check the alignment and examine the position of the rotor vanes of the appropriate trimmer. If the trimmer is either fully meshed or fully unmeshed the circuit may not have fully reached the peak of the resonance point; adjust the value of the parallel fixed silver mica capacitor as necessary. If all else fails suspect a faulty crystal.

This transmitter has greater overall sideband gain than the Mark 2 version; due partly to the greater conversion efficiency of the cathode injection balanced converters, and partly to the use of a continuously tuned circuit in the anode of V7. On the lower frequency bands the rf drive control is normally set half on, and only increased beyond this position for the 15m and 10m bands. There is also reserve of audio gain (sufficient for a low level microphone). The correct setting of the pre-set audio gain control (RV1) when using a normal crystal microphone is between half and two thirds of the maximum position.

In regard to performance, the unwanted sideband suppression at 1kHz (measured at the transmitter output terminal) is better than 55dB. The low-impedance diode modulator makes the carrier balance particularly stable. The carrier suppression is better than 60dB. Due to the use of double triode balanced converters in each of the frequency translation stages and negative rf feedback in the penultimate amplifier, the transmitter gives a clean output with a low level of distortion products.

THE "CLASSIC"—A GROUNDED-GRID 813 LINEAR AMPLIFIER

This equipment is intended as a companion amplifier for any ssb transceiver in the nominal 50 to 100W output range and is designed to increase signal level to the maximum allowed in the terms of the amateur (sound) licence. The factors deciding the choice of valves and circuit for this particular application are summarized as follows:

1. Conventional grid-driven class AB tetrode linears do not adequately load transceivers (exciters) in this power range, and a matched resistive pad must be connected in the input line to the pa. Separate bias packs and regulated screen supplies are necessary and neutralization is essential, which involves complication and expense.
2. Passive grid AB1 stages readily accommodate high drive and do not require neutralizing, but they still retain the other disadvantages.
3. Over the last decade or so, several original designs have done much to overcome these problems and some have deservedly become very popular. Nevertheless, grounded-grid linear amplifiers have retained their attraction for the following reasons:

   (*a*) zero-bias operation of high-mu triodes and certain triode-connected tetrodes in class B mode eliminates the need for grid bias packs and screen supplies;
   (*b*) adequate exciter loading due to the "feedthrough" property of grounded-grid stages—the exciter contributes to amplifier output;
   (*c*) highly degenerative circuit ensuring good linearity and often making neutralization unnecessary;
   (*d*) few critical operating parameters; will perform satisfactorily over wide range of ht voltage.

It is unfortunate that many valves suitable for operation at the power level envisaged are rather expensive, but the 813 is still reasonably priced and is capable of exemplary performance as a zero-bias triode. A pair are capable of dissipating 250W and will deliver 400W p.e.p. output on 28MHz with excellent linearity; third order distortion products being better than −30dB below maximum output.

TABLE 6.32

DC voltage check

Valve	Stage	Type	Pin No 1	2	3	4	5	6	7	8	9
V1	1st audio	EF86	135	—	3·3	H	H	170	—	—	—
V2	2nd audio	12AT7	150	—	2·6	H		300	—	4·5	H
V3	Carrier osc	12AT7	55	Neg 1·5	—	H		150	—	3·6	H
V4	IF amp	EF85	3·7	Nil or —105	3·7	H	H	—	270	195	—
V5	1st converter	12AT7	255	—	4·2	H		255	—	4·2	H
V6	VFO	EF80	—	Neg 0·12	—	H	H	—	150	104	—
V7	2nd converter	12AT7	255	—	4·6	H		255	—	4·6	H
V8	Conversion Oscillator	EF80	2	—	2	H	H	—	297	150	—
V9	RF amp	EF80	3*	—	3*	H	H	—	305	0 to 250	—
V10	PA	6146	—	H	205	—	Neg 45 or	—	H	—	
V11	PA	6146	—	H	205		Neg 130	—	H	—	
V12	Regulator	VR150	Anode — 150								

Main ht + 310V
Pa ht + 670V at standing pa current of 50mA.
Bias —33 to —120 (limits of bias pot)
*DRIVE control at max position
1·5V DRIVE control at mid-way position.
H Denotes 6·3V ac heater connection.

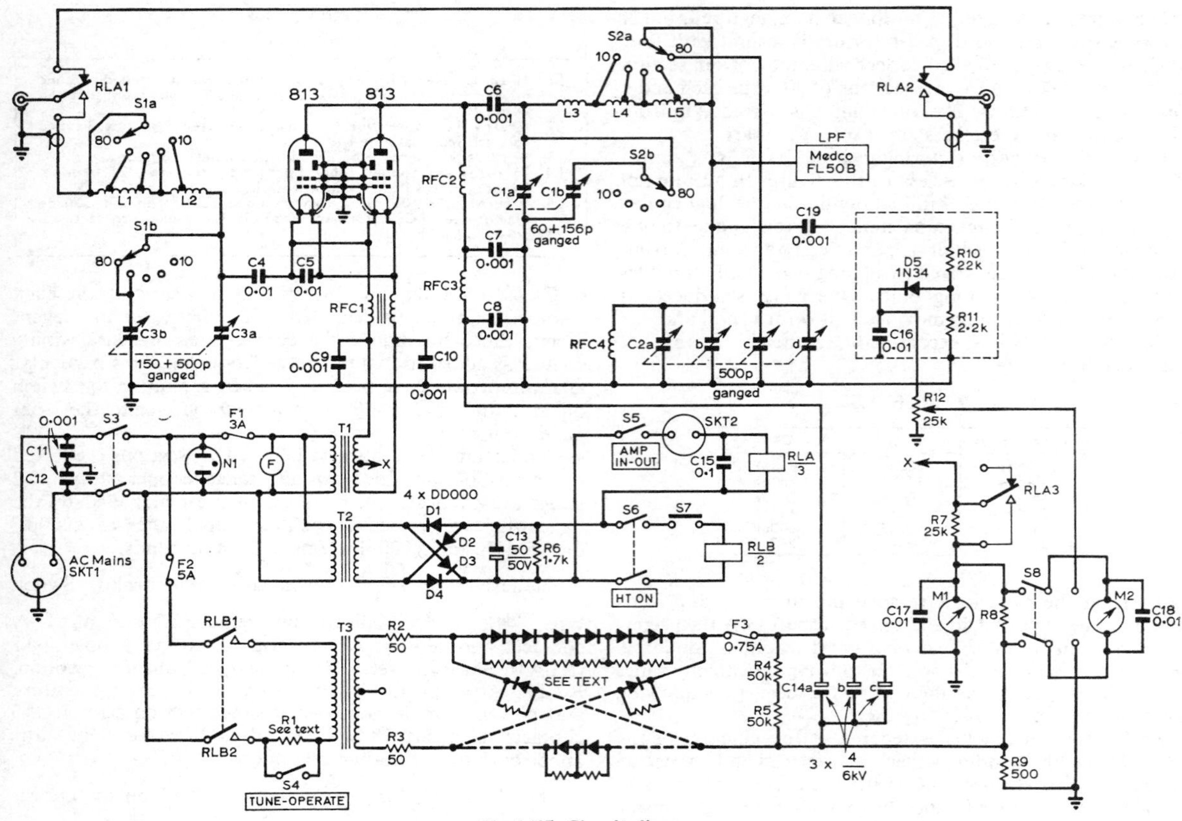

Fig 6.197. Circuit diagram.

RF Circuit Description

Details of the circuit are given in **Fig 6.197**. The drive impedance of the grounded-grid 813s is approximately 140Ω. This value is stepped down to 50Ω through a band-switched L network with manual tuning. This network could be omitted with exciters capable of loading into the higher impedance, but it is not recommended since the *Q* imparted by the network is valuable in suppressing waveform distortion arising from non-linear loading over the input drive cycle (see page 6.105).

The *Q* is very low and tuning so flat that a slow motion drive is not essential. However, a graduated dial allows resonance to be set quickly, and it can be left for operation over the whole telephony band.

An ordinary two-gang 500pF "broadcast" tuning capacitor is used to tune the network, the spacing being quite adequate for the low rf voltage encountered at this point. One gang, reduced to about 150pF by removing all but three of the moving vanes, is used on the 14–28MHz range; on the 3·5 and 7·0MHz bands the other gang is connected in parallel by the input bandswitch.

The filament choke, RFC1, consists of a 5in long ⅜in diameter ferrite rod close bifilar wound with 12swg enamelled copper wire over its full length. Regrettably, the specification of the ferrite material is unknown and it might be advisable for the intending constructor to equip himself with one or two rods of various grade material and select that which provides the most satisfactory all-band performance. The choke is bound with adhesive tape and dipped in copal varnish to secure the core.

The filament transformer is situated in the pa enclosure close to the valves with the terminal panel facing downwards through a cut-out in the chassis. It is unavoidably subjected to a hot blast from the fan and is protected by a ⅜in-thick slab of asbestos secured to the exposed side.

The valves are orientated so that the filament pins face one another. For maximum stability, the grids, screens and beam plates are earthed as directly as possible by soldering a "horseshoe" of copper braid around the relevant socket contacts and making direct connections to solder tags secured to each of the socket mounting bolts. Care should be taken to ensure that the ceramic material is not subjected to excessive compression, and the "double-nut" technique is advisable when securing the tags.

The metal shells surrounding the bottoms of the 813s are earthed in metal receptacles which are standard socket attachments. Should these items prove difficult to obtain, the spring clips found in certain types of fuse carrier may be put to good effect by bolting a number of them on to the chassis so that they butt firmly against the shells with the valves *in situ*.

The anode harness consists of short lengths of braid soldered to a ⅜in wide copper strap secured to the top of the anode choke, RFC2.

The anode choke used is no longer manufactured, but it is very easily reproduced. A ⅞in former is wound with four sections of 34swg enamelled copper wire, the bottom section is 145 turns, followed by two sections of 30 turns each and a top section of 15 turns. The top winding is spaced 3/16in from its neighbour and the others are spaced ⅛in apart.

The combined output capacitance of the 813s is almost 30pF, and strays are in excess of this figure. In an attempt to keep tank circuit Q within bounds on the higher frequencies, the minimum capacitance of the anode tuning capacitor is made as small as possible. To this end a twin-gang assembly was contrived by mating two 156pF variables taken from BC375 tuning units. One gang is reduced to approximately 60pF by removing eight rotor vanes, and it is used to cover all bands, except 3·5MHz when both gangs are used together.

TABLE 6.33

Frequency	C1	C2
3·5MHz	180pF	1,800pF
3·8MHz	170pF	1,700pF
7MHz	90pF	900pF
14MHz	45pF	450pF
21MHz	30pF	300pF
28MHz	23pF	230pF

To gang the capacitors remove the rotor bearing studs and bonding wipers, drill and tap each shaft 4BA then screw the rotors on to either end of a short length of studding. With a little patience it is possible to bring the moving vanes into radial alignment as the end plates meet, thus allowing the capacitors to be bolted together.

An aluminium bracket is secured at the junction of the two sections, the bonding wipers are replaced and connected to the bracket which serves as the main earth return, and additional support is provided by two pillars mounted inside the front panel.

The optimum anode load for the valves is 5,000Ω (R_L), and the desired output impedance 50Ω (R_{out}). A target loaded circuit Q of 20 was chosen, taking care to note that this value would be exceeded on the higher frequency ranges.

The ratio $\frac{R_L}{R_{out}} = 100$, and $\sqrt{100} = 10$, which is the reactance ratio of C1 to C2.

$$\text{Now, } X_{C1} = \frac{R_L}{Q} = \frac{5{,}000}{20} = 250\Omega$$

$$\text{and, } X_{C2} = \frac{X_{C1}}{10} = \frac{250}{10} = 25\Omega$$

The pi-tank capacitor values which equate these reactances are shown in **Table 6.33.** Bearing in mind the large extraneous capacitance shunting C1, it is evident from these figures that operating Q runs appreciably higher than datum on the 21 and 28MHz ranges. To minimize losses, large diameter coil material is employed and tank inductance is split into sections which are orientated to reduce undesirable absorption effects. Lengths of ⅜in-wide copper strapping are used as coil taps on 21 and 28MHz, and 12swg copper wire for the lower frequencies. More detail of the assembly may be had by relating **Table 6.34** to the photograph of the pa.

The bandswitch is modified by filing extra indexing detents and adding fixed contacts to bring the number of positions up to five. The rear wafer is used to shunt the two sections of C1 on 3·5MHz, and, since the shaft is earthed, the back rotor contact ring is removed from this wafer to prevent arcing. Since this halves the contact area an extra wiping contact is added to mate with the front ring. This particular alteration could have been avoided by mounting the switch on insulators and using an insulated coupling to the front panel control.

TABLE 6.34

Coil details

L1	10 turns 18swg tinned copper on 1⅜in diameter ribbed ceramic former.* 10 tpi with taps at 2 turns and 6 turns from junction with L2.
L2	6 turns ⅛in copper tube, 1½in inside diameter, 1¾in long. Tapped 3⅔ turns from cathode end.
L3	4¼ turns ¼in copper tube, 1⅝in inside diameter, 3in long.
L4	7¾ turns ⅛in copper tube, 1½in inside diameter, 2¼ in long. Tapped 3 turns from junction with L3.
L5	9 turns plus 10½ turns 16swg tinned copper on 2in outside diameter ex-BC375 former. 10tpi ¼in space between sections.

* "non-standard" item.

Connection between the switch and C1 sections is effected by means of ½in-wide low-inductance copper straps. C2 gangs are connected in pairs which are marshalled by the bandswitch to provide a maximum capacitance of 2,000pF on 3·5MHz and 1,000pF on the 7–28MHz bands.

The aerial relay, $\frac{\text{RLA,}}{3}$ was taken from a surplus "Switch type 78", and modified by removing the auxiliary contacts and replacing them with a normally open pair (RLA3). Since the relay is not energized during reception, R7 biases the pa to virtual cut-off to reduce temperature and prevent "diode noise" in the receiver; on SEND RLA3 contacts short out R7 and the valves draw 70mA of static anode current in zero-bias condition.

The 24V supply for $\frac{\text{RLA}}{3}$ is fed via SKT2 on the chassis back-apron to allow the transceiver send-receive relay to control the whole station; the transceiver may be operated "barefoot" by opening S5 (AMP IN-OUT), which inhibits the aerial change-over function and holds the pa cut-off. In both S5 positions the exciter "sees" the same impedance, therefore no retuning is needed and S5 behaves as an instant QRO/QRP selector.

It is appreciated that many intending constructors may find the surplus relay difficult to obtain, and in this case

Power amplifier. HT enters via the insulator to the left of C1; C7 and C8 are mounted on the chassis and RFC3 is suspended between them. C1(a) is the foremost gang. C6 and coil L3 are supported by an insulator mounted on C1 frame. L4 is supported off the rear end plate of C2 where it joins L5. The screening can to the right of L5 houses the rf output monitor; RFC4 may be observed nearby.

it is suggested that the range of aerial relays manufactured by the Dow Company may offer a suitable alternative.

Power Supply and Metering

The anode supply voltage may range from 1·5kV to 3kV with corresponding values of efficiency and output. The "Classic" ht voltage is 2·5kV on load and single tone efficiency ranges from 60 per cent on 3·5MHz to 50 per cent on 28MHz. On TUNE the anode voltage drops to 1·75kV or less, depending on loading.

The power transformer T3 is taken from a T1131 transmitter modulator supply which delivered 1,000V dc in a conventional full-wave centre tap circuit. A full-wave bridge comprising 24 BY100 silicon diodes is connected across the secondary in series with R2 and R3 for surge protection; each diode is shunted by a 150kΩ ½W resistor to even up inverse voltage.

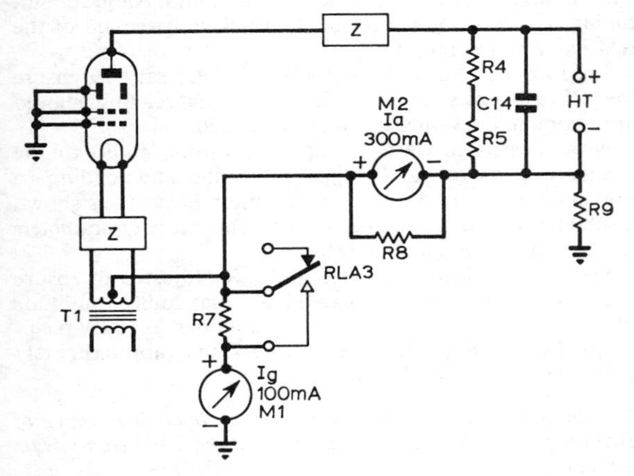

Fig 6.198. Simplified metering diagram.

The metal-cased smoothing capacitors are mounted above chassis on an insulating board so that any leakage to case does not appear across R9, M1 and possibly M2. Since the cases could assume high positive voltage under certain fault conditions, the tops are painted red to warn that a potential hazard exists. An HT ON panel neon has not been fitted, but one should be provided and connected at points Y-Y in Fig 6.197.

A 100mA meter, M1, reads combined grid and screen currents, while M2 is a 1mA instrument scaled 0–300mA which measures anode current when shunted across R8 by means of S8. In the alternate S8 position M2 monitors relative rf output, and to avoid personal risk and the possibility of meter insulation breakdown R8 is connected in the ht supply negative return rather than in the positive line. See **Fig 6.198.**

Constructional Features

The chassis layout is shown in the photograph. The pa and power supply are combined on a 17in by 16in by 2½in deep aluminium chassis in an assembly with detachable panels of top, bottom and rear, the bottom panel being equipped with castors. The top cover is finely perforated over the rf section and an air filter is attached to the rear panel in line with the valves. A hole 5in in diameter is cut in the partition which divides the interior to allow a fan to play on the valves from

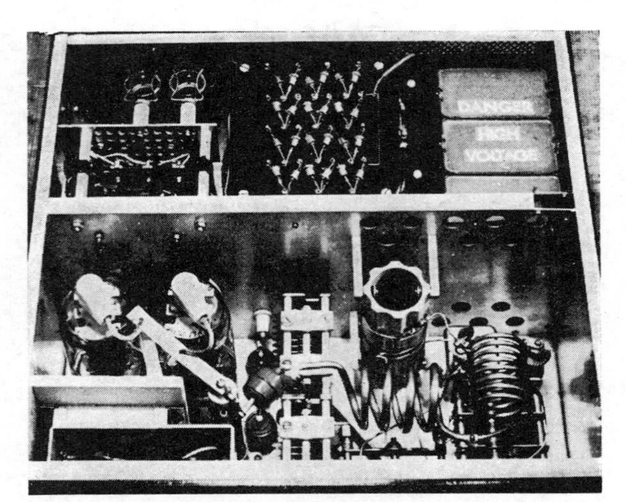

Chassis top view. HT bridge rectifier and fuse are mounted on the power transformer. Bleeder resistors R4 and R5 are to the left, R1 is alongside but not visible. The anode choke is mounted on 1½in insulators to level with the valve caps. Note the asbestos shield protecting the filament transformer.

immediately inside the power supply section. The ht bleeder resistors R4 and R5 are positioned on a perforated metal panel placed behind the fan in the path of cool air drawn through the filter. A safety interlock switch, S7, is let into the upper lip of the partition at the opposite end.

The sub-chassis view photograph shows T2 mounted on a shallow channel which also lends support to the bottom

Sub-chassis view. The aerial relay and MEDCO low pass filter are positioned along the left side with T2 near the centre. Input circuit L-net controls are grouped on a metal bracket near the front. L2 is mounted behind S1 on small ceramic insulators with L1 to the left RFC1 is located between the filament transformer terminals and the valve bases. C4, C5 and R7 are also visible in the same area.

TABLE 6.35

Component details

C1	60 + 156pF (ex-BC375, see text)
C2	4-gang 500pF, receiving type
C3	150 + 500pF, receiving type (see text)
C4, 5	0·01μF 600V dc wkg mica (ex-BC375)
C6, 7, 8	Erie Ceramicon 0·001μF 15kV dc wkg type CHU 410
C9, 10, 11, 12, 19	Disc ceramic 0·001μF 500V dc wkg
C13	Electrolytic 500μF 50V dc wkg
C14 (a, b, c)	Pyranol 4μF 6kV dc wkg (3 off)
C15	Paper 0·1μF 350V dc wkg
C16, 17, 18	Disc ceramic 0·01μF 350V dc wkg
D1, 2, 3, 4	Lucas DD000 or equivalent
D5	Diode, type 1N34 or similar
F1	3A Belling Lee 1¼in
F2	5A Belling Lee 1¼in
F3	0·75A Belling Lee 1¼in
LPF	Medco FL50B
R1	Bowl fire element 600W
R2, 3	Wirewound 50Ω 10W
R4, 5	50kΩ 85W wirewound
R6	Carbon 1·7kΩ 3W
R7	Wirewound 25kΩ 10W
R8	300mA shunt (see text)
R9	Wirewound 500Ω 10W
R10	Carbon 22kΩ 2W
R11	Carbon 2·2kΩ 1W
R12	Carbon variable 25kΩ
RFC1	Filament choke (see text)
RFC2	Labgear E5032
RFC3	2·5mH, 300mA
RFC4	1mH, 600mA (National R154U or similar)
S1	Ceramic 2-gang, 1-pole 5-way
S2	F and E. 2-gang 1-pole 5-way (modified)
S3	Dpst toggle 10A 250V ac
S4	Spst toggle 3A 250V ac
S5	Rotary wafer spst
S6	Dpdt toggle. 3A 250V ac
S7	Microswitch 1A dc
SKT1	Mains connector 3-pole 5A
SKT2	2-pole connector, female, 1A
T1	Filament transformer 10V 10A, tapped primary 220–250V ac
T2	24V 1·5A, tapped primary 220–250V ac
T3	1200–0–1200V ac. 300mA tapped primary representative values, see text)
RLA/3	Aerial relay, 24V dc coil (see text)
RLB/2	2-pole contactor, 10A at 230V ac, 24V dc coil
N1	Neon panel indicator 230V ac wkg

cover. Rectifier and smoothing components associated with T2 are supported on a terminal strip to the rear.

The valve socket filament contacts are interconnected by a pair of 12swg copper buss wires and the filament choke is suspended between these wires and the filament transformer terminals.

The 17in by 10½in 12swg aluminium front panel is dressed with a 16swg sub-panel to mask the unsightly heads of screws which secure the tank circuit components to the inner panel. A small box and a vertical channel are secured to the inner panel to shield the panel meters and wiring from the strong internal rf field.

Adjustment and Tuning

WARNING. *Before making* any *internal adjustment, switch off and disconnect the supply mains, then, using a well-insulated screwdriver, earth the ht line to confirm that the filter capacitors are discharged.*

A two-tone audio source, monitoring oscilloscope, and a 50Ω load with p.e.p. reading output meter are necessary to assess satisfactorily the performance of the amplifier and also to ensure that the terms of the licence are not subsequently exceeded. Modus operandi and the characteristic

TABLE 6.36

HT = 2·5kV   Output = 400W p.e.p (Two-tone drive)

Freq (MHz)	Grid current (mA)	Anode current (mA)	Input tuning (0–10 div)	Anode tuning (0–100 div)	Output loading (0–100 div)
3·5	40	170	8	50	85
7	40	200	2	50	80
14	42	220	7	18	68
21	50	295	3	12	50
28	58	300	2	8	40

display patterns to be expected have been thoroughly dealt with elsewhere in this chapter and will not be repeated here (see page 6.99).

Preliminary checks are necessary to confirm that the input circuit and pi-net resonate correctly on each band; they should be conducted at reduced power (S4 on TUNE), and with drive from the exciter held to a minimum. Anode circuit tuning and loading should be adjusted with the aid of the relative rf output monitor.

On 10m L3 should be deformed if necessary to ensure that C1 just begins to mesh at 29·7MHz. Dial readings should approximate the values given in **Table 6.36.**

Now, switch to OPERATE, apply two-tone audio to the exciter mic socket, and adjust pa tuning and loading to obtain 400W p.e.p. output with optimum linearity as shown by the familiar oscilloscope pattern. The related parameters to be expected are given in Table 6.36.

Speech may now be applied and level adjusted to ensure that speech peaks do not exceed the maximum amplitude of the two-tone oscilloscope pattern obtained at 400W p.e.p. output; grid and anode meters will kick up to approximately half the values shown in the table.

For the benefit of operators favouring an input impedance of 80Ω the procedure adopted in arriving at the L-net values is as follows:

Beginning with 28MHz, insert carrier and adjust the exciter into an 80Ω load with the aid of an swr bridge. Shift the load to the pa, apply ht and transfer the drive to the pa, retaining the bridge in circuit; make no further adjustment to the exciter except to regulate drive as necessary.

Now tune the L-net (C3) for maximum grid current and load the pa normally. Then, keeping C3 in tune, adjust the tap on L1 until maximum grid current coincides with minimum swr on the input line. (Remember to switch off pa ht and check for safety before each adjustment). Repeat the process on the remaining bands, tapping L2 as necessary.

This method should result in an swr of $<1\cdot5:1$ on 28MHz and $<1\cdot25:1$ on all other bands; without the L-net the figure may rise to as high as 3 : 1 on some bands.

AN SL600 SERIES SSB TRANSCEIVER

This section describes the i.f. and af signal circuitry of a single-sideband transceiver designed by the applications department of Plessey Semiconductors using their SL600 series integrated circuits. The unit consists of a single printed-circuit board which requires only the addition of a local oscillator, preselector, linear amplifier, volume control, microphone and loudspeaker to make a complete transceiver. The transceiver may be used at any frequency from a few kilohertz to 500MHz. The complete circuit is shown in **Fig 6.199** and is described below.

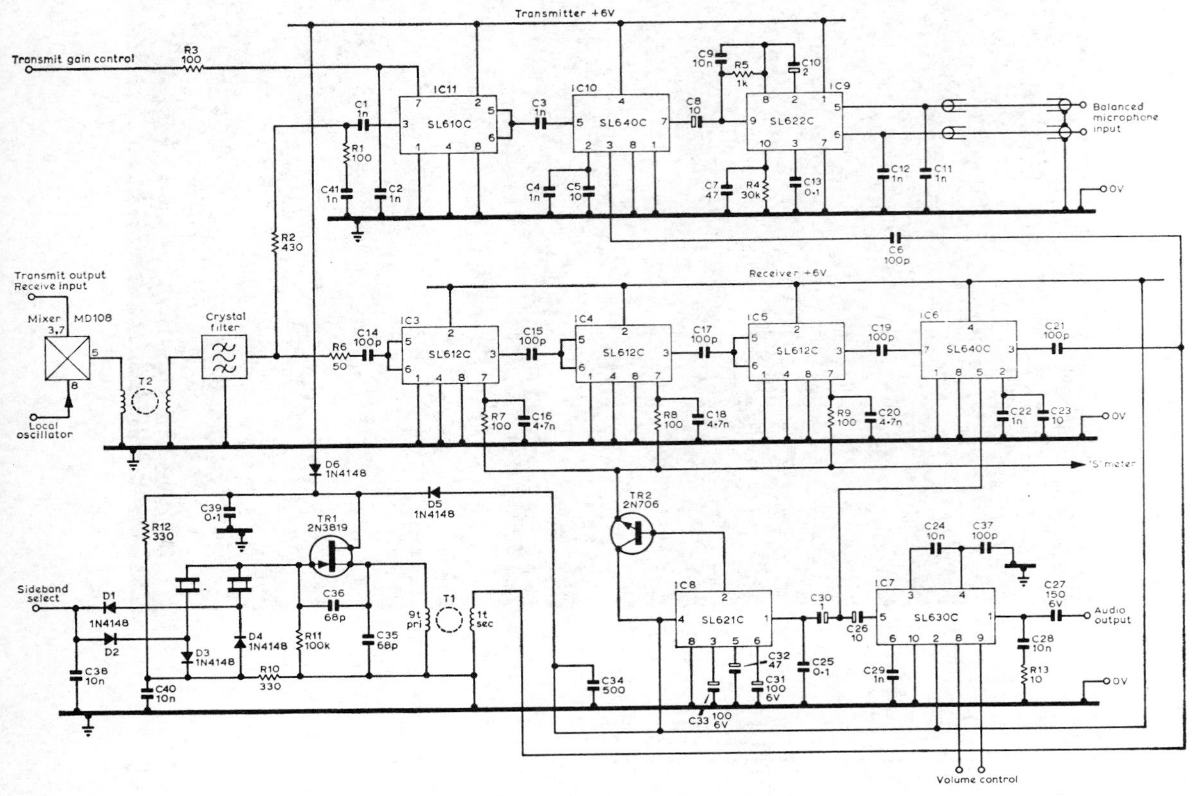

Fig 6.199. Circuit diagram.

Receiver

The receiver consists of a single-conversion superhet with a 9MHz i.f. In order to optimize its intermodulation performance there is no rf amplifier, and the incoming signal is fed directly to a hot-carrier diode ring mixer and then to the crystal filter. The i.f. sensitivity is such that at frequencies of 30MHz or less no rf amplification is required if a reasonable aerial is used (as it would be with a transceiver) but if the receiver is used at frequencies of over 30MHz, or with a less than ideal aerial, some rf gain may be necessary to obtain the necessary noise figure. The rf amplifier used should have the lowest gain consistent with the frequency and aerial to be employed and must have good large signal-handling capability if the receiver performance is not to be degraded.

The mixer is an Anzac MD108 hot-carrier diode ring. This was chosen for its conveniently small size, high performance and low cost, but doubtless similar devices from other manufacturers could be used. All the ports of this ring are 50Ω, and two have a frequency range of 5MHz to 500MHz while the third has a frequency range of dc to 500MHz. The input from the aerial is applied to this dc to 500MHz port via a preselector, and the local oscillator at a level of + 7dBm (500mV rms) is applied at pin 8. The mixer output from the last port passes a ferrite toroidal transformer to match it to the 500Ω input impedance of the filter. If other filters are used the impedance-matching transformer may need to be altered.

Once the signal has passed the crystal filter, a 2·4kHz bandwidth 9MHz filter with 90dB stopband suppression (the SEI QC1246AX), there is little further risk of cross-modulation or intermodulation. The i.f. strip consists of three cascaded SL612C i.f. amplifier circuits followed by an SL640C product detector. Without agc applied, each SL612C has 34dB gain and 15MHz bandwidth. A broadband i.f. strip of three SL612Cs has over 100dB gain and 15MHz bandwidth and can very easily become unstable. The circuit board layout used for this transceiver is critical if the i.f. strip is to be stable. It is relatively easy to make a three-stage broadband strip on double-sided printed circuit board if the component side is left as a plane of grounded copper, but on single-sided board the layout used in this description should be rigidly adhered to.

The beat frequency oscillator for the product detector is a fet crystal oscillator. It delivers about 100mV rms to the SL640C product detector and also supplies the carrier for the transmitter modulator. One of two crystals for upper or lower sideband is selected by diode switches.

The detected audio from the product detector drives an SL630C output stage, which is capable of providing about 65mW to headphones or a small loudspeaker and also drives an SL621C agc system. The SL630C has voltage-controlled gain so the volume control consists of a potentiometer providing the control voltage to the SL630C. If 65mW is insufficient output (it is worth listening to it before deciding as it is

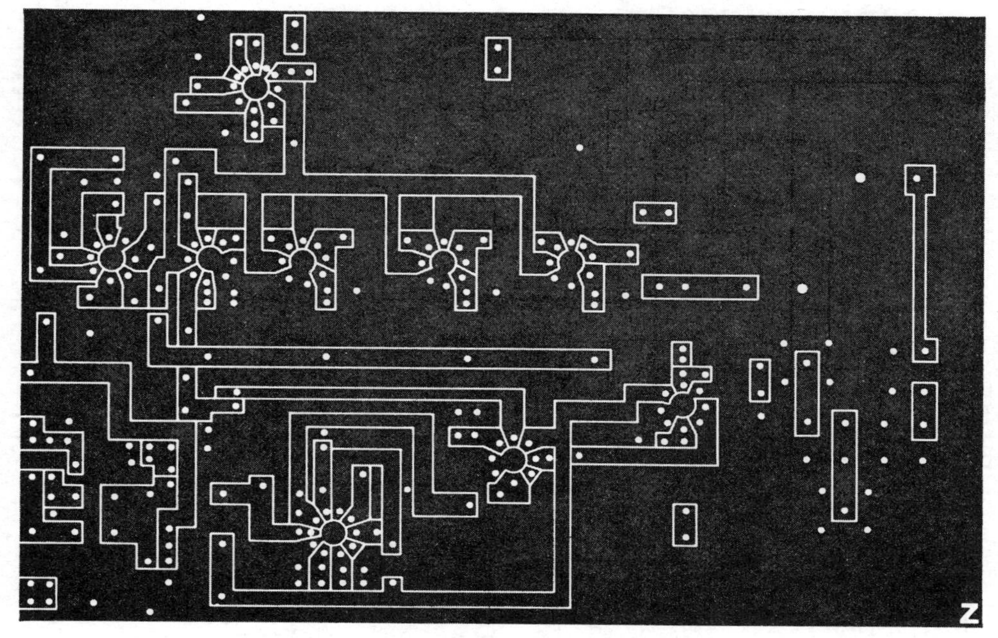

Fig 6.200. PCB pattern, actual size.

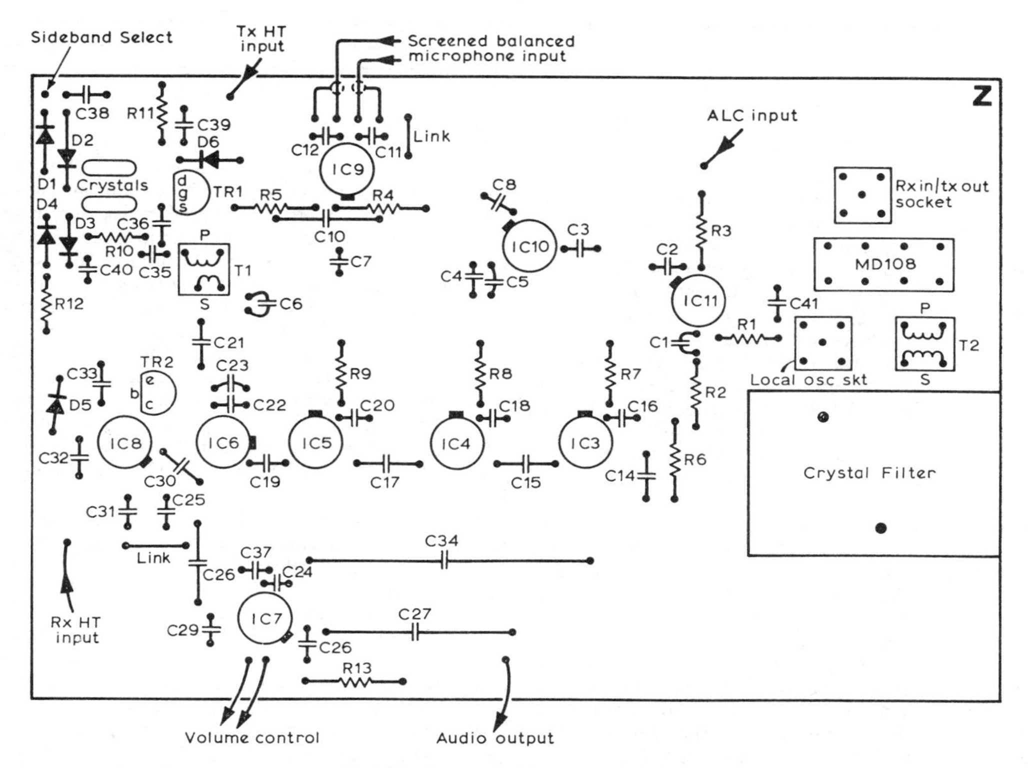

Fig 6.201. Component layout, actual size.

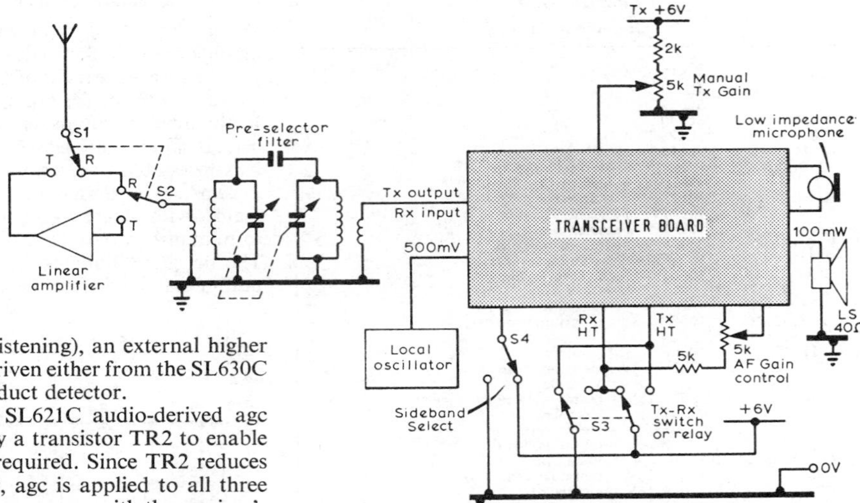

Fig 6.202. Block diagram of transceiver.

usually adequate for domestic listening), an external higher power audio amplifier may be driven either from the SL630C output or directly from the product detector.

The agc is provided by an SL621C audio-derived agc system. Its output is buffered by a transistor TR2 to enable an S-meter to be connected if required. Since TR2 reduces the available agc voltage swing, agc is applied to all three i.f. stages to ensure that the agc can cope with the receiver's 114dB dynamic range. If R7 is replaced by a germanium diode there will be a delay to the first stage agc which may improve the receiver noise figure very slightly on small signals—this is barely worthwhile. The capacitors C16, C18 and C20 are kept down to 4,700pF in order to retain the ignition suppression characteristics of the system.

Transmitter

The transmitter is also single conversion. It generates single-sideband at 9MHz by the filter method, using the same crystal filter as the receiver. The 9MHz ssb is then converted to the final frequency by the MD108 ring mixer with the unwanted product being removed by the preselector. This system entails no signal switching between the aerial side of the preselector and the transmitter/receiver side of the crystal filter on the change-over from receive to transmit. All the transmit/receive switching on the board is achieved by turning on the appropriate power line (transmit or receive) and grounding the unused line. The grounding of the unused line is most important and instability can result if it is not done.

The audio input from the microphone is amplified by an SL622C agc amplifier which will give a constant 100mV rms output for a 60dB range of input. If a single ended input is used rather than a balanced input this dynamic range is reduced to about 46dB. In most systems, 60dB input dynamic range is too large, 40dB being sufficient, so R5 has been included in the circuit. If 60dB is required R5 should be omitted and C9 reduced to 4,700pF.

The audio output from the SL622C goes to the SL640C double-balanced modulator. The carrier input to this modulator is fed by the bfo (which works on both transmit and receive since its power may be derived from either line via diodes D5 and D6). The output of the SL640C consists of double-sideband with low carrier leak (usually −40dB on signal) which is amplified by an SL610C which may have its gain controlled either by an alc signal derived from the transmitter linear amplifier or manually by a dc gain control. This amplified dsb is applied to the filter to yield ssb. Resistors R1 and R2 ensure a correct match to the filter both on transmit and receive.

The ssb output from the filter passes to the diode ring via the impedance-matching transformer and is mixed with the local oscillator to give the final transmitter frequency (and an image which is removed by the preselector). This is amplified by the linear amplifier and transmitted. The output from the preselector is about 70mV rms.

Construction

The system is built on a single-sided printed circuit board (**Figs 6.200** and **6.201**) with two wire links—one in the receive supply, the other in the transmit supply. If only a receiver is required the components R1 to R5 inclusive, C1 to C13 inclusive, C40, and the semiconductors IC9, IC10, IC11, D5 and D6 must be omitted, a wire link connected where D5 was, and a 500Ω resistor connected from the filter end of R6 to earth.

The layout of the board is critical and changes of printed circuit design will almost certainly lead to instability unless double-sided board is used. The design shown may be built on double-sided board quite safely.

The components used in the original are given in **Table 6.37.** Bead tantalum capacitors are used where possible for their small size but since they are hard to obtain in high capacitances at high voltages aluminium electrolytics have been specified in two places. The WeeCon capacitors specified may be replaced with other miniature high-K ceramic capacitors but the values of any components should not be altered without very good reason. The resistors are all $\frac{1}{8}$W 10 per cent types.

Transformer T2 is made on an ITT CR 071-8A ferrite core. In fact any small ferrite or iron dust toroid with cross-section greater than 3mm² and diameter between 7 and 12mm, capable of working at 9MHz, may be used. Square-loop materials, however, are not suitable. Four 5cm lengths of 26swg wire are twisted together, and two turns are wound on the core with the twisted wire. The ends are then opened and three windings are connected in series for the filter winding and the fourth is used as the winding connected to the diode ring. Transformer T1 is wound on a core of the same type and has a nine-turn primary and a single-turn secondary.

TABLE 6.37

Component List

Component	Value
IC3, 4, 5,	SL612C
IC7	SL630C
IC9	SL622C
IC6, 10	SL640C
IC8	SL621C
IC11	SL610C
R1, 3, 7–9	100Ω ⅛W hi-stab carbon film
R2	430Ω ⅛W hi-stab carbon film
R4	30kΩ ⅛W hi-stab carbon film
R5	1kΩ ⅛W hi-stab carbon film
R6	50Ω ⅛W hi-stab carbon film
R10, 12	330Ω ⅛W hi-stab carbon film
R11	100kΩ ⅛W hi-stab carbon film
R13	10Ω ⅛W hi-stab carbon film
C1, 2, 3, 4, 11, 12, 22, 29, 41	1nF Wee Con
C5, 8, 23, 26	10μF 6·3V tantalum bead
C6, 14, 15, 17, 19, 21, 37	100pF ceramic
C7, 32	47μF 6·3V tantalum bead
C16, 18, 20	4·7nF Wee Con
C10	2μF 16V tantalum bead
C13, 25, 39	0·1μF Wee Con
C27	150μF 6V electrolytic
C31, 33	100μF 6·3V tantalum bead
C9, 24, 28, 38, 40	10nF Wee Con
C30	1μF 35V tantalum bead
C34	500μF 16V electrolytic
C35, 36	68pF ceramic (see text)
D1–6	1N4148 or similar low capacitance switching diode
TR1	2N3819 or similar N channel jfet
TR2	2N706 or similar low-cost silicon npn transistor
Filter	SEI QC1246AX with sideband crystals (or KVG XF-9B)
Mixer	Anzac MD108 hot carrier diode ring mixer

The values of C35 and 36 may require slight adjustment under test conditions to ensure satisfactory oscillator operation. If this is done, both values should still be kept equal.

A small number of the transceivers built from this design suffer from apparent agc instability. The symptoms are generally motor-boating at certain signal levels.

The problem is not, in fact, due to the agc but to instability caused by i.f. feedback through the unused transmitter section of the circuit. It may easily be cured by connecting a single 0·1μF capacitor with low rf resistance between the transmitter section power supply rail and ground—as near as possible to the SL610C amplifier.

Installing this capacitor does not remove the necessity of grounding the transmitter power rail during reception, and vice versa.

Conclusion

A block diagram of the system's use in a single band transceiver is shown in **Fig 6.202**. Obviously it may be used in many different transceivers, the one in Fig 6.202 being the simplest.

This transceiver is probably the simplest which may be made using the SL600 Series but its performance is not compromised. It has a sensitivity of better than 0·5μV for 10dB s/n, it can handle signals of over 200mV rms at the diode ring with minimal intermodulation, and the board uses less than 500mW on transmit or receive. It has been designed so that anyone with basic technical competence but without previous experience in ssb transceiver design can build a successful ssb transceiver.

A complete kit of parts for this design is available from Amateur Radio Bulk Buying Group, Communications House, 20 Wallington Square, Wallington, Surrey.

A 30MHz SOLID STATE LINEAR AMPLIFIER

The theoretical principles of solid state linear power amplifiers are similar to those applying to their valve counterparts. In recent years hf power transistors with characteristics sufficiently linear for ssb applications have been developed. In the interests of efficiency these are normally operated in class B, and this mode of operation demands a small forward bias to establish a zero-signal collector current of typically 2–20mA. The design of the biasing circuit is important, mainly with regard to the low impedance which is necessary to minimize variations in the bias over the operating cycle. The design techniques relating to interstage coupling and output matching are as described earlier in this chapter.

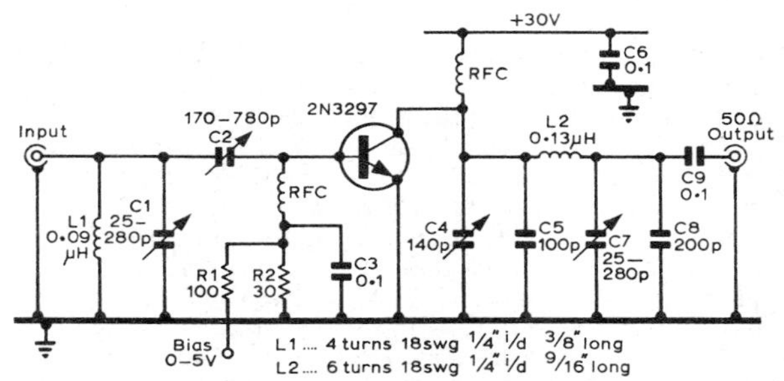

Fig 6.203. A 30MHz linear amplifier delivering 15 to 20W p.e.p. (*Motorola*).

The 30MHz amplifier shown in **Fig 6.203** is capable of delivering 15 to 20W p.e.p. with all odd order distortion products at least 30dB below the desired signal in a two-tone ssb test. The transistor is a 2N3297 npn silicon unit in a TO-3 (diamond) package.

The input circuit is a tapped network for which the primary function is impedance matching for maximum power transfer.

The transistor is bolted to the chassis for heat sinking, and two mica spacers isolate the collector electrically while providing the required thermal conductivity. Two spacers are used instead of one to reduce the capacitance between chassis and the collector, the latter being electrically connected to the transistor package. The total output capacitance of the 2N3297 with two spacers is about 100pF and the output circuit is designed to operate with this output capacitance in parallel with C4 and C5.

The forward bias required for linearity is supplied by the bias source in the base circuit and the impedance of the bias divider is selected for the best combination of linearity and collector efficiency. Collector current is 5mA with no signal at the input.

The performance data for this circuit is given in **Table 6.38**. More power output can be obtained at the expense of linearity. As shown, power output can be increased to 30W p.e.p. but in doing so the signal-to-distortion ratio is reduced to 22dB. Power output can also be increased by connecting transistors in parallel.

TABLE 6.38

Performance data of 30MHz linear amplifier

Power output (p.e.p)	3rd order distortion (dB)	5th order distortion (dB)	Power gain (dB)	Collector efficiency (per cent)
15	—32	—37	12	52
30	—22	—37	9·6	51

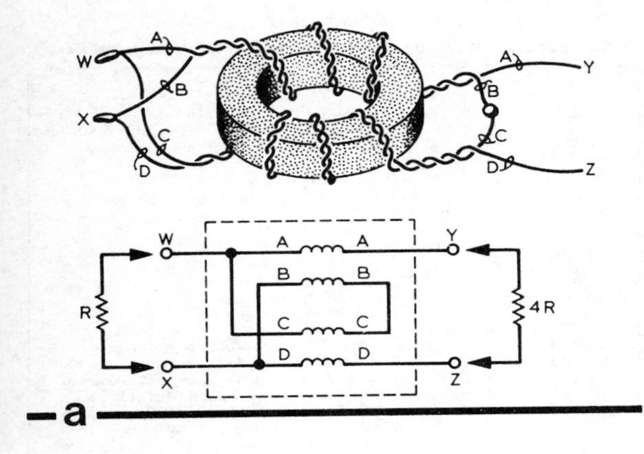

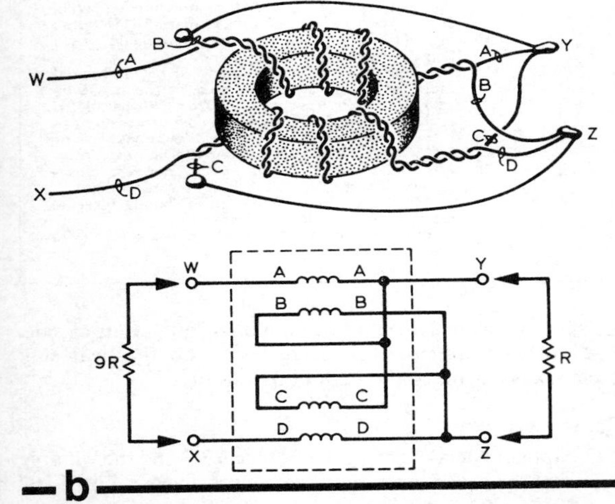

Fig 6.204. Two basic transmission line transformer configurations. (a) 1 : 4 transformer (b) 9 : 1 transformer.

The linearity of the amplifier of this type is optimum if it is slightly more heavily loaded than for the conditions of maximum power output. This entails tuning and loading the output pi-network in the usual manner to obtain maximum power output as indicated on a reflectometer. The loading capacitor is then decreased slightly and the tuning capacitor re-adjusted.

A 100W BROADBAND LINEAR AMPLIFIER

The solid state linear amplifier requires careful matching for satisfactory operation and this results in a more complicated tuning procedure than is necessary in valve amplifiers. Band changing is a particularly daunting task. This has prompted the development of broadband linear amplifiers which may be driven from ssb exciters at any frequency from 2 to 30MHz and which produce reasonably constant power output over this range with no adjustment.

The basis of this technique is the use of transmission line transformers as a means of obtaining impedance transformation over a wide range of frequencies. The transmission line transformer may be more familiar in its role as a matching device between a transmission line and an aerial and particularly in its simplest form, the balun, which is used to couple a balanced aerial to an unbalanced line. The use of ferrite ring cores greatly reduces the length of lines required in these transformers while still providing satisfactory transformer action. Operation over a wide frequency range is assisted by the fall in core efficiency which occurs as frequency increases, and this tends to keep the reactance of the line constant. Only certain transformation ratios, such as 1 : 1, 4 : 1 and 9 : 1 are available with this type of transformer. The transmission line which is wound on the ring core is usually in the form of miniature coaxial cable or a twisted pair of enamelled copper wires. Two basic configurations are shown in **Fig 6.204**, that shown in Fig 6.204(a) being utilized in the following design.

The circuit of the broadband amplifier, which is of Mullard design, is shown in **Fig 6.205**. It uses a matched pair of BLX14 transistors in class AB push-pull and can deliver up to 100W p.e.p. over the frequency range 1·6 to 30MHz with a maximum intermodulation distortion level of -30dB. The overall gain over the band is 16·75 $\pm$0·75dB and the

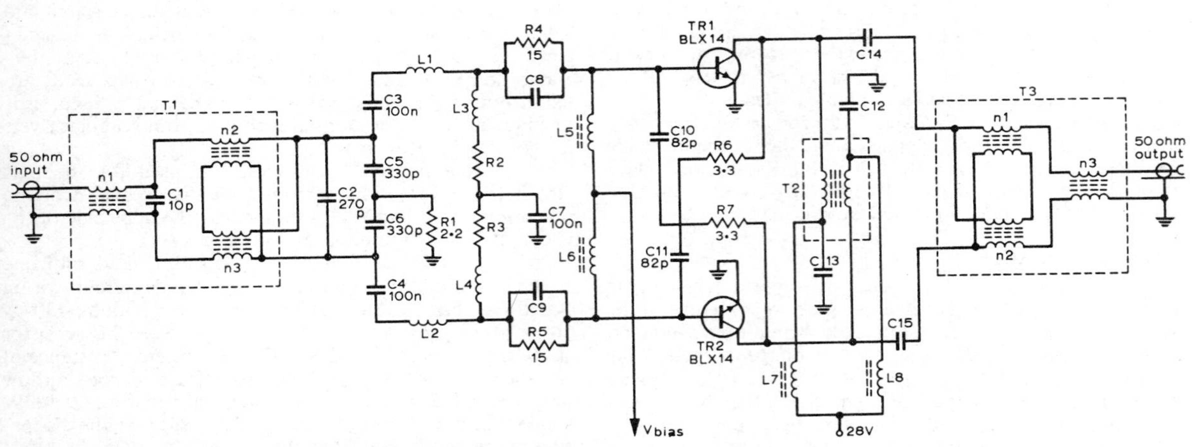

Fig 6.205. Circuit diagram of the 100W broadband linear amplifier (*Mullard*).

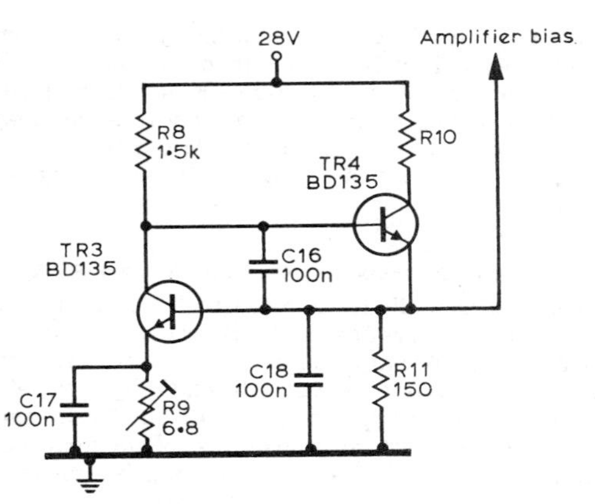

Fig 6.206. Bias compensation circuit.

minimum two-tone efficiency at 30MHz is 42 per cent. The vswr is less than 1·4 : 1.

Circuit Description

The input and output transformers T1 and T3 are both transmission line types with impedance transformations of 1 : 4 and 4 : 1 respectively. Each comprises a Ferroxcube toroid wound with 50Ω coaxial cable.

The BLX14 transistors are adjusted in class AB, each with a quiescent collector current of 70mA. A suitable temperature-compensated bias supply is described below. Variations in the gain and input impedance of the transistors over the band are compensated for by an LCR network between the input transformer and the push-pull pair. Blocking capacitors C3 and C4 also compensate for the shunting effect of the input transformer inductance on the LCR network.

The collector choke T2, wound on a Ferroxcube rod, provides tight coupling between the collectors of the transistors for even-order harmonics. This coupling, very nearly 100 per cent, together with the cross-neutralization provided by C10 and C11, results in reduced second-harmonic distortion.

Because of the practical difficulty of centre-tapping T2, two separate decoupling capacitors C12 and C13, each consisting of three parallel 100nF capacitors, are employed. Ferrite chokes L7 and L8 eliminate the possibility of in-band parallel resonance by reducing the resonant frequency of the collector choke circuit to a value below 1MHz.

Blocking capacitors C14 and C15 provide low-frequency compensation for the inductance of the collector choke and transformer T3. Resistors R6 and R7 in series with the neutralizing capacitors C10 and C11 prevent oscillation in the 100MHz region if the 50Ω output load is removed.

Bias Compensation

Bias compensation is needed to prevent changes in quiescent collector current due to variations in temperature. A suitable circuit is given in **Fig 6.206**. The output voltage is equal to the sum of the base-emitter voltage of the sensing transistor TR3 and the voltage across R9. The potentiometer R9 is used to adjust the quiescent current in the BLX14 transistors to the required value. The three capacitors C16, C17 and C18 prevent oscillation and provide a low-reactance path for rf signals entering the circuit from the output side. The sensing transistor TR3 is mounted on the heat sink as close as possible to the BLX14 transistor.

TABLE 6.39

Component details for the 100W broadband linear amplifier

Component	Details
R1	2·2Ω 0·3W
R2, 3	3 × 27Ω in parallel 0·3W
R4, 5	15Ω 0·3W
R6, 7	3·3 Ω 0·3W
R8	1·5kΩ 0·5W
R9	6·8Ω, 1W ww, ±10 per cent pot.
R10	2 × 120Ω, 5·5W ww, ± 10 per cent pot in parallel
R11	150Ω 0·3W
C1	10pF ceramic
C2	270pF polystyrene
C3, 4, 7, 16, 17, 18	100nF metallized polyester
C5, 6	330pF polystyrene
C8, 9	2 × 1·2nF in parallel, polystyrene
C10, 11	82pF ceramic
C12, 13	3 × 100nF in parallel, metallized polyester
C14, 15	3 × 22nF in parallel, metallized polyester
T1	PTFE jacketed 2·5mm outer diam 50Ω coaxial cable wound on a 4C6 Ferroxcube toroid, 23 by 14 by 7mm, Philips type 4322 020 91070, as shown in Fig 6.207. n_1 = 9 turns and n_2 = n_3 = 4·5 turns. Suitable alternative cores are Mullard toroids FX1582 (25·4mm diam) or FX1588, FX1589 (38mm diam).
T2	6 turns, twisted-pair 0·7mm enam copper wound on a 4B1 Ferroxcube tube, 7·7 by 4·6 by 50mm, Philips type 4322 020 34320, as shown in Fig 6.208. Suitable alternative cores are Mullard FX1159 tube or FX1099 rod. If these are unavailable use a length of 8mm diam ferrite rod.
T3	PTFE jacketed 2·5mm outer diam 50Ω coaxial cable wound on a 4C6 Ferroxcube toroid, 36 by 23 by 15mm. Philips type 4322 020 91090, paralleled as shown in Fig 6.209. n_1 = n_2 = 3·5 turns and n_3 = 7 turns. Two stacked toroids of type FX1588 or FX1589 can be substituted for the Philips toroid.
L1, 2	27nH; 2·5 turns, 0·7mm enam copper, close-wound, internal diameter 3·2mm, leads 3·2mm.
L3, 4	40nH; 2·5 turns, 0·7mm, closewound, internal diameter 4·2mm, leads 4·2mm.
L5, 6, 7, 8	3 turns, 0·6mm enam copper wound on FX1898 Ferroxcube core.

Construction

Component details are given in **Table 6.39**, and the winding details of the broadband transformers are given in **Figs 6.207 to 6.209**.

The rf circuit components are mounted on a glass fibre printed circuit board, dimensions 200 by 135 by 1·6mm (see **Fig 6.210**). The board is double-clad, that is, the lower side is left unprocessed and functions as the earth plane. Isolated parts on the upper side are used for mounting and interconnection. For components connected to earth, leads are fed through holes in the board. If leads cannot be used, or if low-inductance connections are required, 2mm tubular rivets are inserted in the board.

A separate printed circuit board is used for the bias circuit (**Fig 6.211**). This circuit, with the exception of the potentiometer R9 which is mounted on the upperside of the rf circuit board, is screwed on to the heat sink.

A length of finned aluminium heat sink is used with one of the fins removed to make room for the bias circuit wiring board (see **Fig 6.212**). A solid block of aluminium, 130 by 30 by 30mm, provides thermal coupling between the transistor studs and the heat sink. The estimated thermal resistance of the modified heat sink with a low-velocity forced airflow amounts to 0·5° C/W. Total maximum dissipation (at 100W single-tone and 28MHz) is about 80W, neglecting the dissipation of the bias circuit transistors.

CHAPTER 7

VHF AND UHF TRANSMITTERS

TRANSMITTERS for use in the amateur vhf and uhf allocations, the frequency spectrum above 30MHz, employ the same general principles as those designed for the lower frequency areas, that is to say, they utilize a stable signal generating source, a chain of frequency multipliers and a final power amplifying stage. There the resemblance ends: the design of transmitters for vhf/uhf (known for convenience as the metre-wave bands) calls for great attention to be given to such factors as stray capacitance, excessive lead inductance and the self-resonance of certain components. Moreover, transmitters for the metre-waves, unlike those intended for the hf spectrum, customarily cover one amateur allocation only, in the interests of maximum efficiency.

In the UK the vhf/uhf bands are those commonly known as 4m, 2m, 70cm and 23cm, of which the last three are in partial harmonic relationship **(Fig 7.1(a))**, a circumstance that permits the design of a three-band exciter capable of furnishing drive to multipliers for the next bands up. **Fig 7.1(b)** shows the proximity of these bands to broadcasting services.

In **Fig 7.2** the block diagram illustrates the stages required to realize multiband output. The earlier stages of the exciter unit would need to be either tunable or broadbanded enough to cover the whole range of output frequencies required.

CHOICE OF FREQUENCY

The actual number of stages required in the exciter section of a vhf/uhf transmitter is governed by the starting frequency of the oscillator and the final operating frequency. In between a number of doubling or trebling stages are provided. At lower frequencies quadrupling may be used. **Table 7.1** shows typical oscillator starting frequencies for four metre-wave bands, the multiplication factors required, and the final output frequencies. For convenience a 2MHz span is shown for the 70cm and 23cm bands, though in fact these bands extend considerably beyond these limits, and in their upper reaches accommodate amateur television transmissions. Voice communication in the 70cm band in the UK is confined to the area 432·000 to 433·500MHz.

In addition to the multipliers in a vhf/uhf transmitter chain additional amplifier stages are occasionally called for in such services as a buffer (isolation) amplifier between a variable frequency oscillator (vfo) and a subsequent stage, to prevent frequency pulling, or as an amplifier after the last multiplier to provide sufficient output to drive a large final amplifier.

The choice of the frequencies to be used in the oscillator and following frequency multiplier stages will be governed by the crystals or vfo which the constructor has available to him. Each of these two sources of frequency control may require a different starting frequency; a high starting frequency may be used with crystals (provided they are operated within their ratings), but adequate stability in a vfo becomes more difficult to achieve as the starting frequency is raised.

Choice of multiplier frequency is governed by a further factor: the need to ensure that no undue radiation occurs on unwanted frequencies. For this reason it is desirable to operate exciter stages at low power levels even if this necessitates

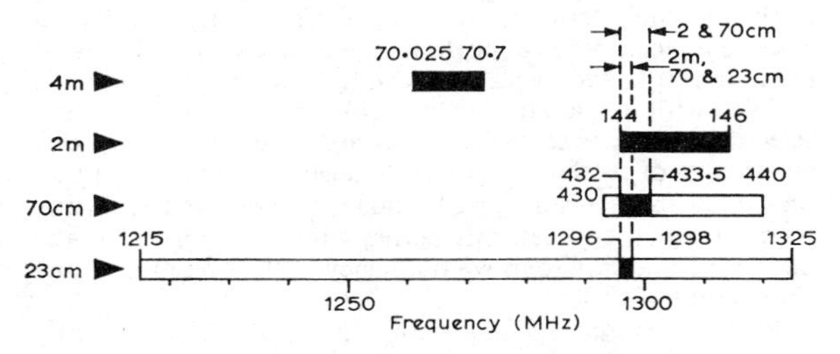

Fig 7.1 (a). The harmonic relationship of four of the UK metre-wave bands, the frequency extent of which is:
4m 70·025 to 70·7MHz
2m 144·000 to 146·000MHz
70cm 432·000 to 440·000MHz
23cm 1,215·000 to 1,325·000MHz
The communications sections are shown shaded. Other amateur services occupy the remainder, eg amateur television above 433·5 MHz. Fig 7.1(b) (below) shows the UK metre-wave bands in relation to television sectors

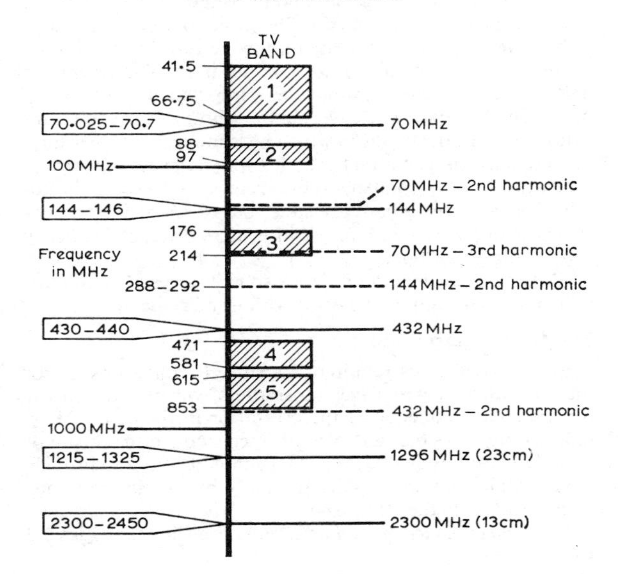

the addition of a driver-amplifier stage later in the design. Such an amplifier both raises the power level to the final stage and attenuates in its tuned circuits unwanted harmonics generated earlier.

Study of the figures given in Table 7.1 will assist in the selection of a starting frequency (crystal or vfo). For example, an oscillator at 72MHz will require no more than a single doubler after it to reach the 144MHz (2m) band. By adding a tripler stage 432MHz may be reached. A further tripler will deliver output in the 1,296MHz band. At such a high starting frequency separation of harmonics at the higher frequencies is an easy matter.

OSCILLATORS

In the design of a vhf/uhf transmitter for the amateur bands three basic types of oscillator are available, the traditional crystal controlled oscillator, the variable crystal oscillator (vxo) and the variable frequency oscillator. The advantages and disadvantages of each are as follows:

Crystal controlled oscillator: reliable frequency location, negligible short-term or warm-up drift and the ability to achieve a clean note (important when A1 cw is to be used) are important advantages provided by this form of oscillator, coupled with the fact that no control is needed, apart from a selection switch if several crystals are to be used. Two disadvantages of crystal control are that care to avoid output on unwanted frequencies must be taken (which applies to any form of signal source), but perhaps more important is the operator's inability to move frequency in the presence of on-channel interference.

Variable crystal oscillator (vxo): the small variable tuning range provided by a vxo permits the user to avoid on-channel interference when this occurs. Like the crystal oscillator, the vxo displays acceptable short-term and warm-up drift, and reliable frequency setting. A small disadvantage is that it is slightly more complex and requires care in the mechanical construction and layout. In some designs of vxo it is not possible to depart far from the nominal frequency of the crystal in use without endangering the quality of the emitted note.

Variable frequency oscillator: The choice of frequency range to be covered by a vfo for use on the metre-wave bands may be made by the operator to suit his particular operating habits and preferences. Almost certainly he will wish to employ a design capable of covering the whole of the communication areas of the bands in which he is interested rather than the restricted spots within the bands which the use of a crystal oscillator or vxo compels. This frequency freedom is obtained at a price: great care must be exercised in the construction of a vfo to achieve stability, a low order of frequency drift and a clean note.

Each of these forms of signal source—crystal oscillator, vxo and vfo—will now be dealt with in more detail.

The Crystal Oscillator

Crystal oscillators required for vhf/uhf converters or for the control of metre-wave transmitters will operate on the fundamental frequency of the crystal or in the case of harmonic generators on multiples of it, eg a converter crystal of a fundamental frequency of 35MHz could be used to provide the fourth harmonic at 140MHz simply by tuning the anode circuit to this higher frequency.

Parallel resonance crystal oscillators are shown at **Figs 7.3 to 7.6.**

TABLE 7.1
Multiplication factors required for four vhf/uhf bands

Multiplication Factor	4m Band 70.025–70.7	2m Band 144—146
x2	35.013–35.350	72–73
x3	23.342–23.566	48–48.6
x4 (x2x2)	17.507–17.675	36–36.5
x5	14.005–14.140	28.8–29.2
x6 (x2x3)	11.671–11.783	24–24.3
x8 (x2x2x2)	8.753– 8.837	18–18.25
x9 (x3x3)	7.781– 7.855	14.4–14.6
x10 (x5x2)	7.025– 7.070	16–16.2
x12 (x3x2x2)	5.836– 5.891	12–12.16
x16 (x2x2x2x2)		9– 9.125
x18 (x3x3x2)		8– 8.11
x20 (x5x2x2)		7.2– 7.3
x24 (x3x2x2x2)		6– 6.083

Multiplication Factor	70cm Band 432–434*	23cm Band 1296–1298**
x2		
x3	144–144.667	432–432.667
x4 (x2x2)	108. –108.5	
x5	86.4 – 86.8	
x6 (x2x3)	72 – 72.3	
x8 (x2x2x2)	54 – 54.25	
x9 (x3x3)	48 – 48.2	144 –144.2
x10 (x5x2)	43.2 – 43.4	129.6 –129.8
x12 (x3x2x2)	36 – 36.165	108 –108.16
x16 (x2x2x2x2)	27 – 27.125	81 – 81.125
x18 (x3x3x2)	24 – 24.1	72 – 72.1
x20 (x5x2x2)	21.6 – 21.7	64.8 – 64.9
x24 (x3x2x2x2)	18 – 18.0825	54 – 54.08
x32 (x2x2x2x2x2)	13.5 – 13.653	40.5 – 40.563
x36 (x3x3x2x2)	12 – 12.05	36 – 36.05
x40 (x5x2x2x2)	10.8 – 10.85	32.4 – 32.45
x48 (x3x4x4)	9 – 9.04125	27 – 27.04
x64 (x4x2x2x2x2)	6.75– 6.7813	20.25– 20.2813

* For convenience the band 432–434MHz is shown although in the UK there is little amateur voice communication above 433·5MHz.
** Communication Section.

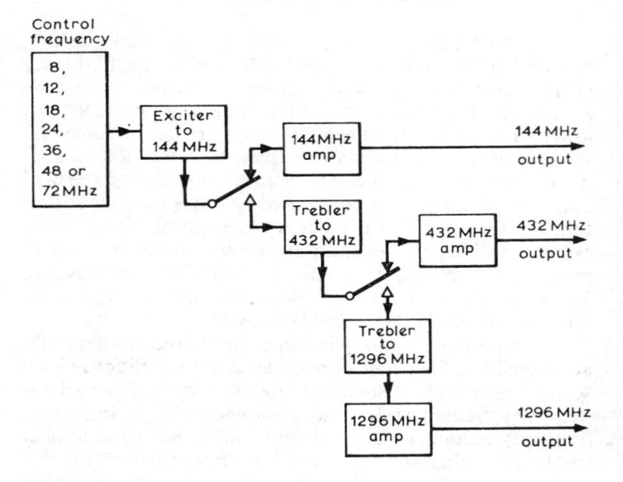

Fig 7.2. Block diagram to show how operation in the three harmonically related areas of the 2m, 70cm and 23cm bands may be achieved. Inter-unit switching may be by coaxial relays in the linking coaxial leads

Series resonance crystal oscillators include such circuits as the Squier **(Fig 7.7)**, suitable for fundamental frequency or overtone crystals but calling for care in setting up, and the Butler **(Fig 7.8)**, again suitable for fundamental or overtone use. In the Butler the output inductor in the anode of V1B is tuned to the second, third or fourth harmonic. Matched resistors should be used in each cathode circuit of the values shown. At **Fig 7.9** is shown a typical semiconductor harmonic generator in which the fifth harmonic is extracted from one set of inductors and the fifteenth from another pair, crystal connected in the emitter circuit of the transistor.

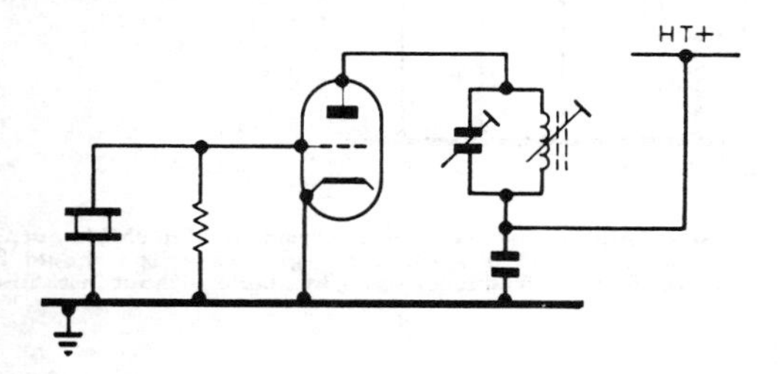

Fig 7.3. A standard crystal oscillator in which the anode circuit is tuned to the fundamental of the crystal. A triode valve is shown for the sake of simplicity

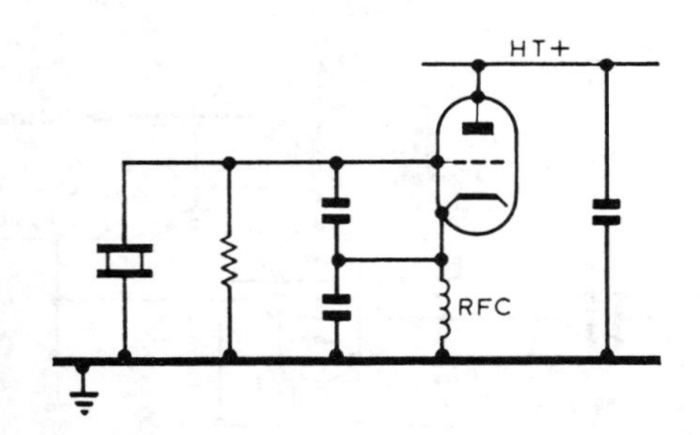

Fig 7.4. Basic Colpitts crystal oscillator. As a harmonic generator this circuit would require capacitance and inductance in the anode circuit tuned to f2, f3 or higher. Crystal starting is sometimes facilitated if a bypassed resistor of approx 380 Ω is included at the earth end of the cathode choke

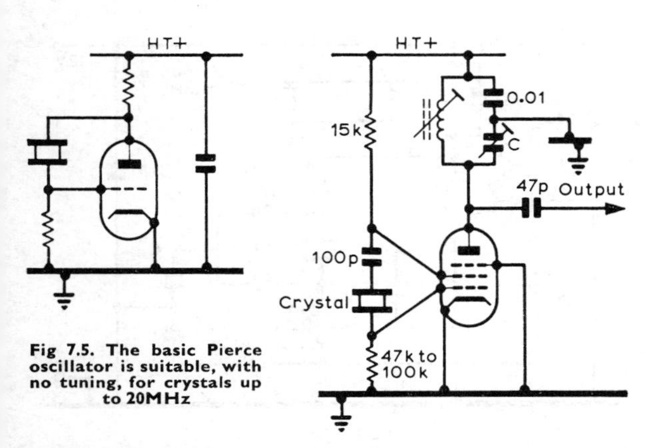

Fig 7.5. The basic Pierce oscillator is suitable, with no tuning, for crystals up to 20MHz

Fig 7.6. Modified Pierce oscillator in which the anode circuit is tuned to the required order of harmonic. A high slope valve such as the E180F should be used in this circuit

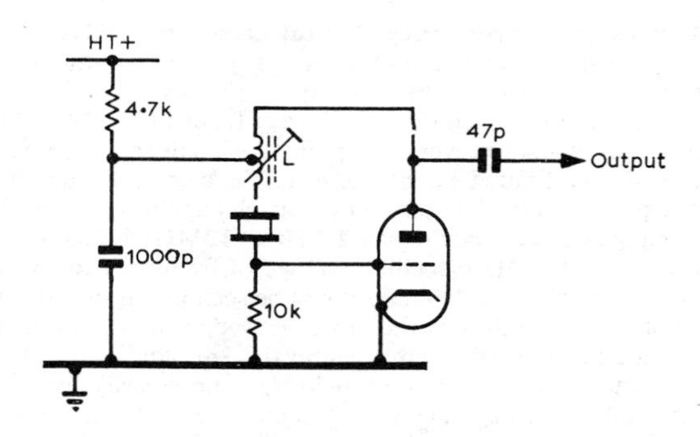

Fig 7.7. The Squier oscillator, in which the position of the tap on the inductor determines the amount of feedback: as the tap is moved from the grid end feedback is increased until eventually instability ensues. The circuit is suitable for use with fundamental and overtone crystals

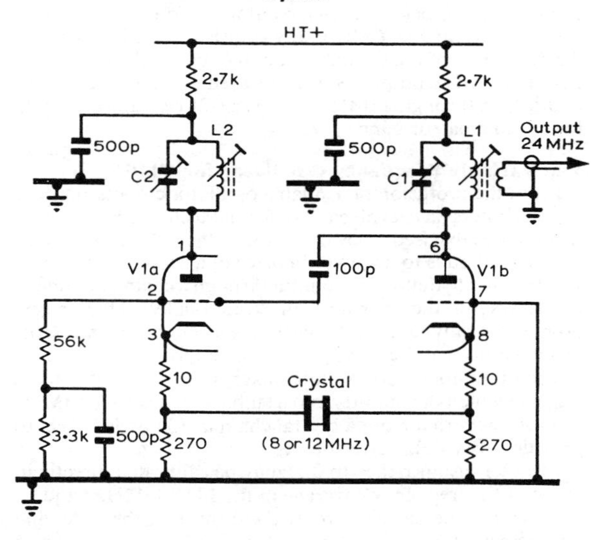

Fig 7.8. The Butler circuit: L2 is tuned to the crystal frequency and L1 to the required harmonic. Output may be taken from the cold end of L1 by means of a two-turn loop

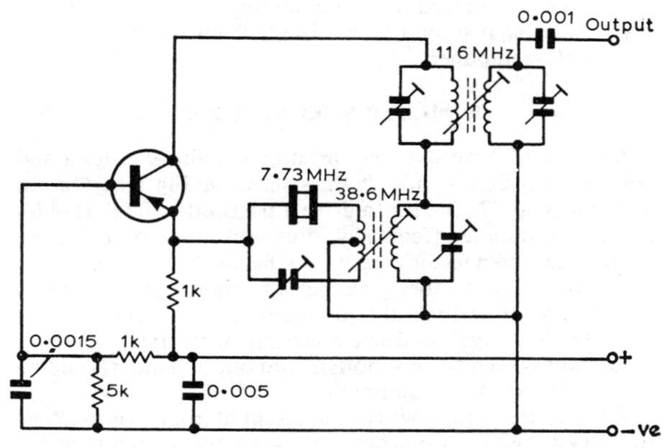

Fig 7.9. A transistor harmonic generator for series-resonant crystals. Any transistor rated to work in the 150MHz region will be suitable, PNP as shown or NPN if preferred

The Variable Frequency Crystal Oscillator (VXO)

A transmitter intended to operate anywhere within the 2m band would require an inordinate number of individual crystals to provide complete coverage from 144 to 146MHz. It is possible, however, to "pull" the frequency of many crystals to a limited extent; for example, a crystal which in a "frequency pulling" circuit showed itself capable of providing a change of frequency of say 20kHz at 12MHz fundamental would at 144MHz provide a change of frequency of twelve times this value, ie 240kHz. It will be seen that quite a small number of crystals may therefore be used in the vxo mode to provide full frequency coverage in the 2m band.

Crystals of similar type will not necessarily provide identical frequency excursions; some if pulled too far will produce a poor note and endanger the stability of the transmitter in which they are used.

Typical vxo circuits are shown at **Figs 7.10** and **7.11** for fundamental and harmonic output respectively, using any standard rf pentode of reasonable slope. A transistor equivalent is shown at Fig 7.11(b); any transistor rated to work at the frequency of the crystal in use may be employed. In this case the 9MHz output would be used to feed a chain of multipliers through 18MHz, 36MHz, 72MHz and 144MHz to the 2m amateur band.

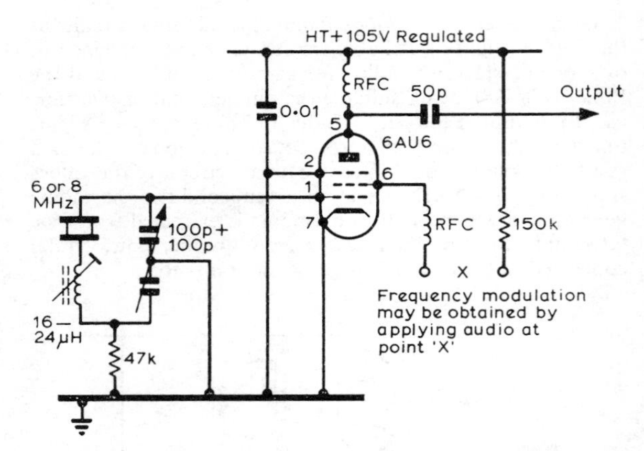

Fig 7.10. A simple vxo using an 8MHz crystal to give 8MHz output in the anode circuit. The split stator grid capacitor should be calibrated to show the frequency swing available without instability

The Variable Frequency Oscillator for VHF

From the profusion of available oscillator circuits amateur experimenters have evolved two distinct techniques for adapting the variable frequency oscillator to the vhf/uhf spectrum.

One method is to use a vfo in place of a crystal at, say, 8 or 12 MHz and to multiply its output through a chain of doublers or triplers into the 2m band. An exceptionally stable design, both electrically and mechanically, is essential, for any drift or instability will be multiplied accordingly.

The other approach uses the mixer/master oscillator technique in which the output from a stable vfo at say 6 or 8MHz is mixed with that from a crystal chain at 138 or 136MHz to provide a 144MHz drive source.

In either instance the vfo dial may be calibrated directly in terms of the frequency coverage in the 144-146MHz band.

A vfo for the 2m band must conform with the following requirements:

(a) the stability should be comparable with crystal control;
(b) spurious emissions, both in band and out of band, must be reduced to the minimum;
(c) the design should be realizable from ordinarily available components.

A MIXER-VFO FOR 2M

A design of mixer-type vfo meeting the above criteria and developed by G2UJ and G5OX is shown at **Fig 7.12.** Output is provided at 72-73MHz to drive a transmitter final doubler at half the radiated frequency, thus preventing output from the pa stage from feeding back into the low level mixer.

The following factors governed the choice of frequencies for the crystal and variable oscillators:

(a) the stability should be comparable with crystal control;
(b) spurious emissions, both in and out of band, should be reduced to the minimum.

With regard to (b) it was found essential to employ an overtone crystal with output only on the frequency required for injection into the mixer. In the final design this frequency was set at 55·525MHz and this in turn governed the choice of

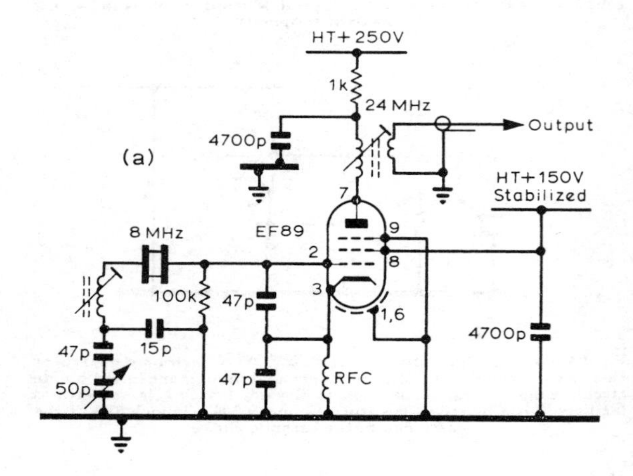

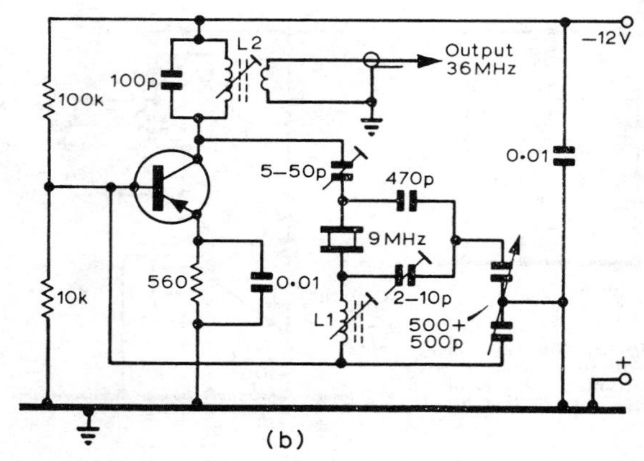

Fig 7.11. A vxo circuit for use with 6, 8, 9 or 12MHz crystals to give multiplied output: (a) valve version; (b) transistor version

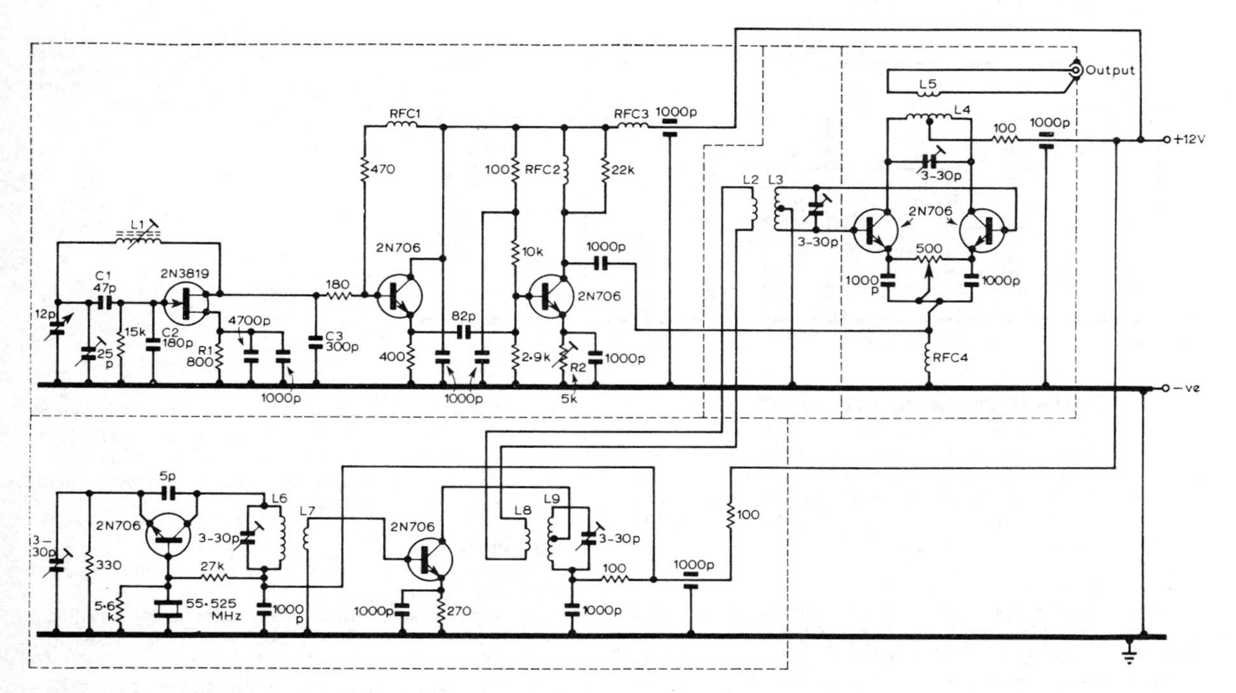

Fig 7.12. Diagram of a mixer vfo due to G2UJ/G5OX. The products of a high frequency crystal and a low frequency variable oscillator are added in the mixer to give output at 72-73MHz to feed the final doubler of a 2m transmitter

frequency coverage for the variable oscillator, a fet Vackar, namely 16·475 to 17·475MHz to provide the required 72MHz output. This output is applied through two buffer stages to the emitters of the push-pull mixer; the 500Ω variable resistor connecting the two emitters allows exact balance to be achieved.

The output of the 55·525MHz crystal oscillator and amplifier unit is coupled to the push-pull base inductor of the mixer pair, is added to the 17MHz input at the emitters and appears as 72MHz at the push-pull collectors. Precautions are taken to minimize unwanted harmonic output, for example, by careful selection of the source resistor R1 (the best value to reduce fourth and lower harmonic output may be found by experiment).

Direct coupling between the drain of the fet oscillator and the base of the first buffer will give far greater output from the latter stage without any deterioration in performance, in fact as much as 7V rf could be realized from the output of the second buffer amplifier. The optimum drive to the mixer will be found to lie around only 1·5V rf. Adjustment may be made by varying the value of R2, which should be replaced by a fixed resistor when the correct value has been ascertained. The output of the variable oscillator remains constant over 90 per cent of the band.

The variable oscillator and two stage amplifier may be built as a separate unit, the crystal and single stage amplifier as another unit, and the mixer as a third, all screened from one another and the individual supplies decoupled.

Setting up: The output from the crystal chain should be adjusted, by varying the link coupling, to be approximately 3V at the base of each of the mixer transistors. The circuit may be detuned by the valve voltmeter probe used to measure the rf voltage, and temporary retuning may be required. About 1V of rf should then be injected from the variable oscillator, its tuning range having been previously checked on an hf bands receiver. The collector circuit of the mixer is tuned to resonance in the band 72-73MHz by means of an rf indicator connected across the output link. Coupling should be quite light at this stage. The point of resonance may be checked by listening on a 2m receiver, to avoid the possibility that the circuit might be tuned to the crystal frequency, though with the considerable separation between the crystal and output frequencies this is unlikely.

With the trimmer across the mixer collector circuit at maximum capacitance it should be possible to detect a signal from the crystal frequency: this should be reduced to as low a value as possible by adjustment of the 500 Ω potentiometer in the emitter circuits. Next listen to the vfo output on the 2m receiver, if necessary disabling the converter rf stage to prevent overloading. This test will enable the presence of any gross spurious output to be discovered, and if any appears it will almost certainly be caused by excessive drive to the mixer from the variable oscillator.

A valve voltmeter measuring both dc and rf is invaluable in ascertaining transistor operating conditions and the rf output from the various circuits. Secondly, the use of a valve type grid dip oscillator *in its ordinary mode* can be dangerous to transistors as excessive rf voltage may easily be induced into the tuned circuit under test. A grid dip oscillator may be employed quite safely if, instead of using the indication of resonance of its built-in meter, a low reading milliammeter (say, 1mA fsd) is connected across the battery leads of the circuit being tested—battery disconnected. This auxiliary meter will give an indication of resonance with the coil of the gdo much farther away from the tuned circuit under test than would normally be the case and possible damage to the transistors thereby avoided.

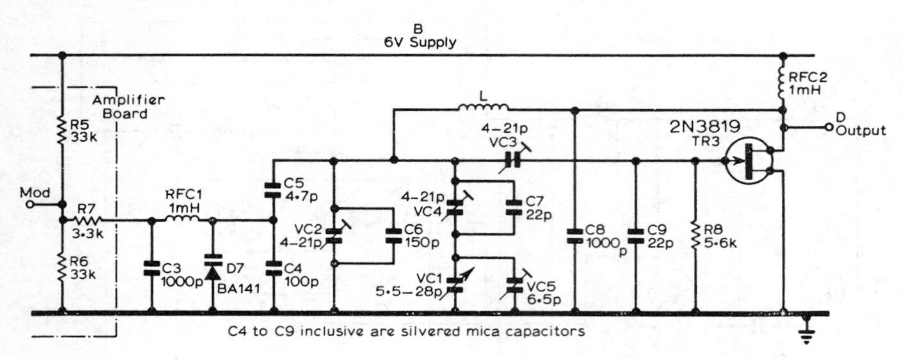

Fig 7.13. The 12MHz direct drive vfo due to G8CGA. Frequency modulation may be provided at the point "Mod" from an external audio unit

A DIRECT DRIVE VFO UNIT

The other approach to vfo design is to employ a unit **(Fig 7.13)** that may be used as a crystal substitute at 12MHz and will drive the 144MHz or 432MHz transmitter through a chain of multipliers. The following design (due to G8CGA) includes a small power supply unit and stabilizer (suitable equally for the mixer vfo already described).

The Oscillator

The oscillator operates at 12MHz and the circuit is shown in Fig 7.13. The trimmers VC2, VC3 and VC4 are all air dielectric variable types giving control of tuning range, feedback and frequency respectively, although all are interdependent. VC5 is a Tempatrimmer of 6·5pF capacitance, and provides an adjustable temperature coefficient from +2,000 to −2,000ppm/°C. Frequency modulation may be produced by a varactor but its effect is reduced by a small series capacitor and a large shunt capacitor, the overall effect being to produce 3kHz deviation at 2m for an audio signal of 1V peak.

The varactor is biased to one half of the stabilized oscillator supply voltage by a potential divider of 33kΩ + 33Ωk.

The deviation for a 1V audio signal can conveniently be checked by shunting the lower 33kΩ resistor with another of equal value. This changes the bias from 3V to 2V approximately; if the required frequency shift does not occur, either the series or shunt capacitors may be modified. Care is needed to prevent rf being fed into any associated audio amplifier.

The correct setting of the feedback capacitor VC3 is important in achieving good frequency stability. Only sufficient feedback to provide reliable operation over the band should be used. This can be achieved when the trimmer is about half meshed (approximately 12pF).

With the feedback set and VC2 and VC4 adjusted so that the required tuning range is just covered by the full range of VC1, the frequency drift should be examined with the Tempatrimmer in its mid position. A likely result is an increase of frequency of some 9kHz at 2m during the first two minutes, due mainly to heating in the transistor, followed by a similar drift over the following 20 minutes as the whole unit reaches a stable operating temperature. This second drift can be almost entirely eliminated by careful adjustment of the Tempatrimmer, and frequency stability from hour to hour of ± 100Hz is achievable.

During construction care must be taken that no connecting leads are in a state of mechanical stress: such leads tend to move with time and during heating cycles which cause changes of calibration. The coil is wound on a polystyrene rod and heavily doped with polystyrene cement. The fixed capacitors are anchored to solid objects with Araldite.

The output lead should also be clamped to prevent movement in the hole in the cast box.

The Amplifier

The oscillator output is only a few tens of millivolts and this must be raised to about 1V.

To do this a four-stage amplifier is used. The first amplifier, TR4, is a fet as a source follower feeding a bipolar transistor, TR5, with rf grounded base. The collector output from TR5 drives a conventional feedback amplifier, the resistor R14 controlling the current drive to the base of TR6; this may be any value from zero upwards and in the original design was 2·2kΩ.

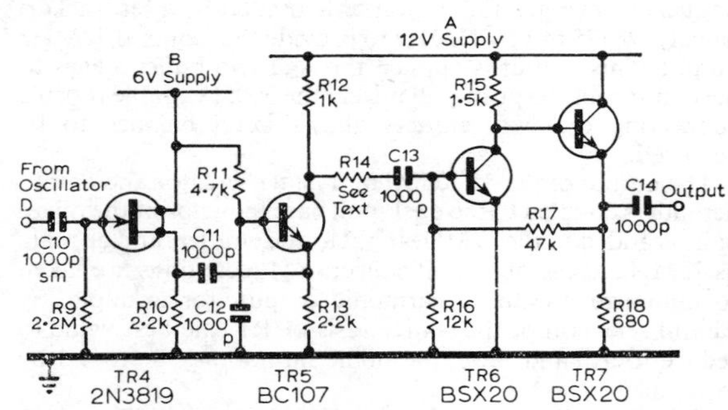

Fig 7.14. The four-stage amplifier unit following the G8CGA fet vfo incorporates one fet and three bipolar transistors

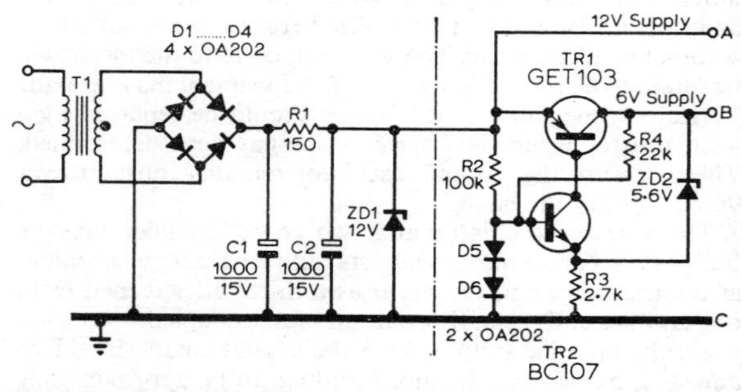

Fig 7.15. The power supply unit and stabilizer for use either with the direct-drive vfo or the mixer-vfo described earlier. For battery operation omit all to the left of the dotted line

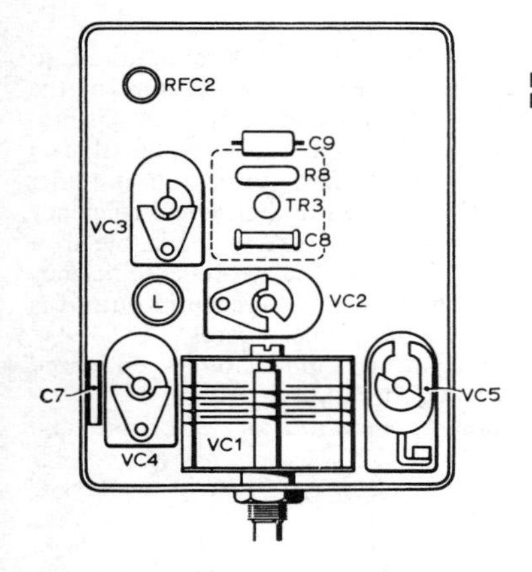

Fig 7.16. A suggested layout for the 12MHz vfo

TABLE 7.2
Direct drive vfo: components list

C1,	1,000μF 15V	R11	4·7kΩ
C3	1,000pF	R12	1kΩ
C4	100pF silvered mica	R13	2·2kΩ
C5	4·7pF silvered mica	R14	see text
C6	150pF silvered mica	R15	1·5kΩ
C7	22pF silvered mica	R16	12kΩ
C8	1,000pF silvered mica	R17	47kΩ
C9	22pF silvered mica	R18	680Ω
C10	1,000pF silvered mica	T1	Radiospares miniature mains—2, 6V secondaries in series
C11–14	1,000pF		
D1–6	OA202		
D7	BA141	TR1	GET103
L	17 turns close wound 18swg enamelled on ⅜in diam polystyrene rod (1·1μH)	TR2	BC107
		TR3	2N3819
		TR4	2N3819
		TR5	BC107
R1	150Ω	TR6, 7	BSX20
R2	100kΩ	VC1	Polar type C28. 6 gaps of 0·015in (5·5-28pF)
R3	2·7kΩ		
R4	22kΩ		
R5, 6*	33kΩ	VC2, 3, 4	Polar type C31, 9 gaps of 0·015in (4-21 pF)
R7*	3·3kΩ		
R8	5·6kΩ		
R9	2·2MΩ	VC5	Oxley Tempatrimmer 6·5pF
R10	2·2kΩ		
		RFC1, 2	1μH
		ZD1	12V
		ZD2	5·6V

* These resistors are mounted on the amplifier board

The Power Supply

The vfo is mains driven using OA202 diodes in a bridge rectifier followed by a capacitor-resistor-capacitor filter.

The source resistance of the supply plus the smoothing resistor adequately limits the current through the 12V zener diode to about 10mA, the current drain of the amplifier and oscillator being 24mA. The highly stable supply for the oscillator and first buffer amplifier uses a GET103 as a controlled series transistor, the output voltage being that of the zener diode plus the forward drop across one OA202 diode. Should it be desired to operate the oscillator from a battery supply, this stabilizer circuit will give substantially constant output until the battery has fallen to 6V. If a 9V battery is used to supply the entire vfo, the total battery drain is 22mA.

Construction

The oscillator must be rigidly built into a cast box 4½ by 3½ by 2in deep and the amplifier and power supply in a similar box. The two boxes may be fixed together, allowing a small space between them to reduce heat transfer from the power supply to the oscillator components. Fig 7.16 shows the approximate component layout of the oscillator unit; this may be varied to suit the components available. The tuning capacitor is bolted directly to the base of the box and all the air-spaced trimmers mounted off the base by 2BA bolts. TR3, R8, C8 and C9 are mounted on a piece of 0·15in matrix Veroboard about 1in square also mounted on a 2BA bolt. The 6V supply, audio and rf output leads pass through clearance holes in the walls of the two boxes, Araldited to prevent movement.

The amplifier and the stabilizer are mounted on a strip of Veroboard 3¾ by 1½in; an important connection is the rf ground to the base of TR5, which should be short and direct. The mains transformer, rectifiers, smoothing components and zener diode occupy the rest of the box and could of course be replaced by a battery for mobile operation.

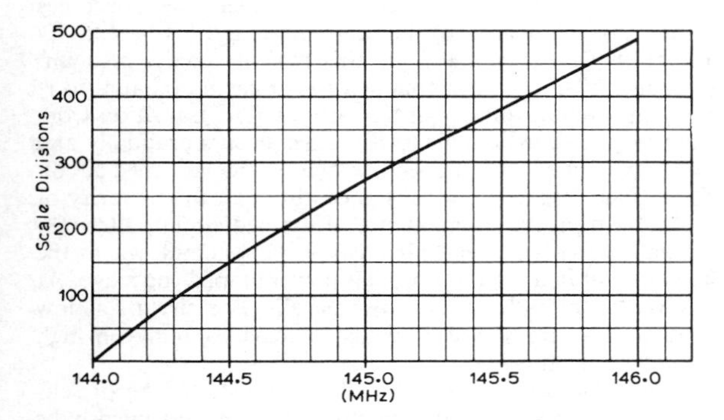

Fig 7.17. Calibration curve for the 12MHz vfo in terms of 2m frequencies

Calibration

Calibration should not be attempted until the vfo when vigorously tapped shows little transient change of frequency and no permanent change at all. It should also be heat cycled a number of times, preferably to a higher temperature than it will ever reach in practice (say 50° C), as this helps to relieve mechanical strains in the wiring and components. Screwing the lid on to the oscillator box has a considerable effect on frequency and the final adjustment of the trimmers for frequency and band spread will have to be done by cut and try. In the final stages the lid should be firmly screwed on at each corner.

FREQUENCY MULTIPLIERS

Because frequency generation sources usually operate at hf for reasons of stability and good output, they must be followed by a chain of multipliers in order to realize output in the vhf/uhf spectrum. These multipliers may use transistors or valves, according to the constructor's requirements. In the case of valves tetrodes or pentodes provide greatest sensitivity and efficiency; high sensitivity is desirable because a large amount of grid drive may not be available from the preceding stage, particularly as the frequency of operation is increased. As the frequency increases the efficiency of multiplying stages falls off and for this reason it is advisable that any quadrupling or tripling is confined to the early stages of the transmitter and that doubling only takes place in the later stages where the efficiency is lowest.

To minimize the generation of unwanted frequencies, which would increase the possibility of interference to other services, multiplying stages should be designed to give maximum output at the desired harmonic and minimum output at all other frequencies. This is achieved by the use of high *Q* tuned circuits, inductive or link coupling between stages, and operation at the lowest power level consistent with obtaining adequate drive to the succeeding stage.

At frequencies up to 145MHz single ended multiplier stages are usually employed. Suitable valves for low power operation are the EF91 (CV138, 6AM6), QV03–12 (CV2129, 5763) and E180F/6688. RF double tetrode valves are commonly used as push-pull frequency triplers and can also be used as push-push doublers with the two anodes connected in parallel.

Triode frequency multipliers are sometimes employed in low power fixed and mobile transmitters, and their use in crystal-oscillator/multiplier circuits has already been mentioned in the section dealing with overtone oscillators. At higher frequencies, for example as triplers from 144MHz to 430MHz, triodes such as the A2521 (6CR4), 5842 (417A), DET22 (CV273, TD03–10), EC8010, DET23 (CV354, TD03–5), and DET24 (CV397, TD04–20) will be found suitable. Varactor diodes and some uhf power transistors may also be used as frequency multipliers.

POWER AMPLIFIERS

Power amplifying stages may employ triode or tetrode valves, used either singly or in push-pull. Tetrodes have the advantage that the power gain is greater and a lower grid drive is required, an important factor at the higher frequencies where in many instances only a low output is available from the preceding driver stage.

For low and medium power 2m and 70cm transmitters undoubtedly the most popular valve is the rf power double tetrode, some versions of which are suitable for operation up to 600MHz. In the low power class with up to 10W anode dissipation, these valves are of miniature single-ended construction; examples are the Mullard QQV02–6 (6939) and QQV03–10 (6360). For higher power ratings up to a maximum of 40W anode dissipation a double-ended construction is exemplified by the Mullard QQV03–20A (6252/AX9910) and QQV06–40A. These valves incorporate internal neutralizing capacitances and no external neutralization is necessary provided that input and output circuits are shielded from each other.

Alternatively, grounded grid triodes for use as frequency multipliers, and amplifiers, offer somewhat simpler circuits. With such valves as the A2521 (6CR4), EC8010 (8556), DET22 (TD03–10) and DET24 (TD04–20) good power gain above 430MHz may be attained.

Single tetrodes are commonly employed in high power 2m transmitters, or a pair may be used in push-pull; suitable valves are the Mullard QY3–65 and QY3–125. For 70cm a typical high power amplifier consists of a 4X150A forced air cooled tetrode in a coaxial circuit; such an amplifier can be operated at the maximum input of 150W at 432MHz. Amplifiers for 23cm make use of planar electrode triodes in a coaxial circuit.

A list of valves suitable for operation as power amplifiers and in the later frequency multiplying stages of vhf and uhf transmitters appears in **Table 7.3.** Detailed information on these types and in many instances application notes may be obtained from the valve manufacturers.

Linear Operation

In some applications power amplifiers are required to operate in the Class C mode, driven into grid current by the preceding stage. In other applications linear operation is preferred, where the valve works over the straight (linear) portion of its characteristic curve. With an output amplifier operated under Class AB1 conditions a theoretical efficiency of 66% is possible and 50–55% is readily realizable in a practical design, whereas with Class C the theoretical efficiency is 75% and 50–70% achievable. The difference in output is unlikely to be significant except in special cases.

The advantages which a linear amplifier offers compared with a Class C operation may be summarized as:

(a) suitable for all modes of operation, ssb, a.m., nbfm and cw;
(b) harmonic generation much lower than in a Class C amplifier;
(c) drive requirements much lower.

From these facts it will be appreciated that there is a significant saving in cost and weight if the high level modulator needed with a Class C amplifier may be dispensed with. Further, reduced harmonic content will be passed by the amplifier, for it is by definition a linear device: harmonic output could well be 25dB or more below the fundamental. The stage will of course amplify any harmonics present in the input waveform presented to it. In the case of a Class C amplifier there will be generation of harmonics even with a pure input signal. Any reduction of harmonic output is an advantage in abating interference to television.

A further advantage of the linear amplifier is that its lower drive requirements may be provided readily by a transistor exciter, because as far as the amplifier is concerned, only voltage and not power is needed to operate it, although some power will be absorbed in the input circuit and valve capacitance: also, to provide a reasonably constant input load some power will be dissipated in an appropriate loading resistor.

TUNED CIRCUITS

At frequencies up to approximately 150MHz lumped circuits are efficient and are generally used throughout any vhf transmitter multiplying and amplifying stages. Above this frequency the efficiency of an inductor-capacitor tuned circuit falls off and use is made of parallel resonant lines up to frequencies of approximately 500MHz. At higher frequencies the spacing between parallel lines becomes appreciable compared to the wavelength and loss occurs by radiation from the lines; for this reason coaxial resonant lines in which the outer conductor is at earth potential to rf are employed at ultra high frequencies. The transition from one type of tuned circuit to the other occurs gradually and some overlap exists in the choice of a particular tuned circuit for a given frequency; for instance although the majority of 144MHz transmitters make use of lumped circuits throughout, a quarter wavelength line is sometimes employed as the tank circuit in a medium or high power amplifying stage. At 430MHz the tendency is to use parallel-line circuits at low and medium powers and coaxial or cavity circuits in high power transmitters.

The inductors used in lumped circuits should be of self-supporting air spaced construction, and can conveniently be wound with soft drawn enamelled copper wire, having a

TABLE 7.3

Make	Type No.	Class	Base	Cathode			Limiting Values			Max. freq. full ratings (MHz)	Output at full ratings (W)	Max. freq. reduced ratings (MHz)	Output at reduced ratings (W)
				Type	V	A	P_A	V_A	V_{G2}				
Mullard U.S.A. CV	ECC91 6J6 858	Double-triode	B7G	IH	6·3	0·45	2 × 1·5	300	—	80	3·5	—	1·0 at 250 MHz
Mullard	EC56	Disc-seal triode	Octal	IH	6·3	0·65	10	300	—	—	—	4000	0·5
Mullard	EC57	Disc-seal triode	Octal	IH	6·3	0·65	10	300	—	—	—	4000	1·8
M-O Valve Mullard CV	DET22 TD03-10 273	Disc-seal triode	—	IH	6·3	0·4	10	350	—	1000	2·8	3000	0·5
S.T.C. CV	33B/152M 1540	Double-triode	B9G	IH	6·3	0·92	15	375	—	300	28	—	—
M-O Valve Mullard CV	DET24 TD04-20 397	Disc-seal triode	—	IH	6·3	1·0	20	400	—	600	23	2000	3·5
S.T.C. CV	3B/240M 2214	Triode	B8G	IH	6·3	1·1	24	375	—	200	24	—	—
M-O Valve CV	ACT22 257	Disc-seal triode	—	IH	6·3	4·0	75	600	—	1000	90	—	—
Mullard CV	QQV02-6 2466	Double-tetrode	B9A	IH	6·3 12·6	0·8 0·4 }	2 × 3	275	200	200	6	500	5
Mullard CV	QV04-7 309	Tetrode	B9G	IH	6·3	0·6	7·5	400	250	20	7·9	150	6·3
Mullard U.S.A. CV	QQV03-10 6360 2798	Double-tetrode	B9A	IH	6·3 12·6	0·8 0·4 }	2 × 5	300	200	100	16	225	12·5
M-O Valve CV	TT15 415, 4046	Double-tetrode	B9G	IH	6·3	1·6	2 × 7·5	400	—	200	12	250	—
Mullard U.S.A. CV	QQV04-15 832A 788	Double-tetrode	B7A	IH	6·3 12·6	1·6 0·9 }	2 × 7·5	750	250	100	26	250	18
M-O Valve Mullard U.S.A. CV	TT20 QQV03-20A 6252 2799	Double-tetrode	B7A	IH	6·3 12·6	1·3 0·65 }	2 × 10	600	250	200	48	600	20
Mullard U.S.A. CV	QV06-20 6146 3523	Tetrode	Octal	IH	6·3	1·25	20	600	250	60	52	175	25
Mullard U.S.A. CV	QQV06-40A 5894 2797	Double-tetrode	B7A	IH	6·3 12·6	1·8 0·9 }	2 × 20	750	250	200	90	475	60
Mullard U.S.A. CV	QQV07-50 829B 2666	Double-tetrode	B7A	IH	6·3 12·6	2·5 1·25 }	2 × 25	750	225	100	87	250	60
Mullard U.S.A. CV	QY3-65 4-65A 1905	Tetrode	B7A	DH	6·0	3·5	65	3000	400	50	280	220	110
Mullard U.S.A. CV	QY3-125 4-125A 2130	Tetrode	B5F	DH	5·0	6·5	125	3000	400	120	375	200	225
Mullard S.T.C. U.S.A. CV	QV1-150A 4X150A 4X150A 2519	Tetrode—forced-air cooled	B8F special	IH	6·0	2·6	150	1250	300	165	195	500	140
M-O Valve ST&C U.S.A. CV	4CX250B 4CX250B 4CX250B 2487	Tetrode—forced-air cooled	B8F special	IH	6·0	2·6	250	1500	300	175	235	500	225 (at 2kV)
ST&C U.S.A. CV	2C39A 2C39A 2516	Triode Disc seal	— —	IH	6·3	1	100	1000	—	2400	20	—	—
M-O Valve CV	DET29 2397	Triode Disc seal	—	IH	6·3	0·5	10	450	—	3800	1·5	7000	—

gauge between 12 and 22 swg, the wire gauge increasing as the power rating of the stage and the frequency of operation increases. For power ratings in excess of 100W, it is advisable to use $\frac{3}{16}$ or $\frac{1}{4}$in od copper tube, preferably silver plated to minimize loss due to skin effect which can be appreciable at the frequencies under consideration As a vhf transmitter is essentially a one band device, an inductor should be soldered directly across the associated tuning capacitor; this eliminates any plug and socket connection which could cause losses if used in an rf circuit.

In tuned circuits associated with driver stages it is common practice to dispense with the tuning capacitor and to resonate the inductor with the stray and circuit capacitances. Tuning is effected by opening out or compressing the turns of the inductor.

Linear tuned circuits are most conveniently employed in push-pull stages and may be any multiple of a quarter wavelength (λ/4) long. This refers to the electrical length of the line, the actual length in practice being somewhat shorter due to the valve acting as an extension of the line, so that the added inductance of the leads and the inter-electrode capacitance result in a physical shortening of the line. A quarter wavelength linear anode circuit for a push-pull stage is shown in **Fig 7.19 (a).** Tuning may be effected either by varying the physical length of the line by adjustment of a shorting bar or by a variable capacitor C1 connected across the open circuited end of the line adjacent to the valve. To maintain circuit balance a split-stator capacitor should be used for tuning with the rotor left floating and not connected to earth. For the same reason the ht feed connection at the voltage antinode of the line should be left unbypassed. An electrical centre-tap on the circuit is provided by the valve inter-electrode capacitances.

A push-pull anode circuit employing a half wavelength line is shown in **Fig 7.19 (b).** The valve and tuning capacitance are connected at opposite ends and ht is applied at the electrical centre of the line, ie the voltage node. Half wavelength lines are sometimes employed in grid circuits at the higher frequencies where the valve leads and input capacitances are such that the natural frequency of the valve grid assembly is lower than the frequency of operation. All parallel-line circuits should be shielded to minimize loss due to rf radiation from the lines which becomes greater as the frequency of operation increases.

Linear tuned anode circuits can be used in single ended stages as shown in **Fig 7.19 (c).** The capacitor C2 is necessary to balance the output capacitance of the valve connected across the other half of the line.

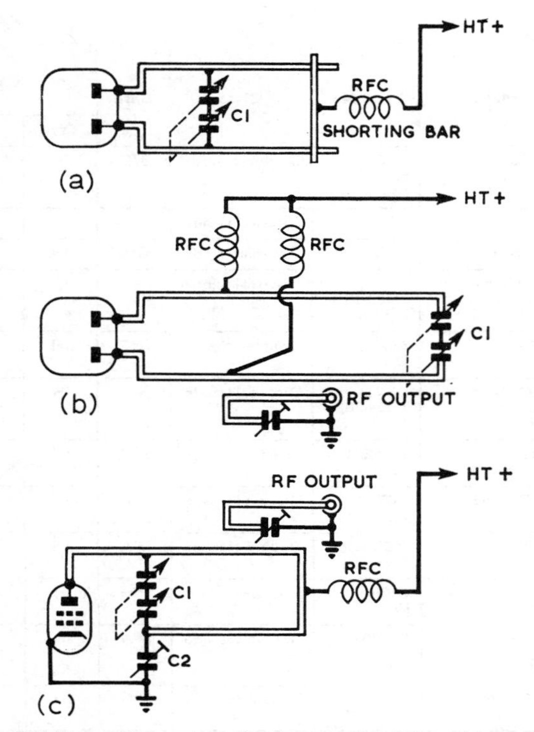

Fig 7.19. Transmitter output circuits using tuned lines: (a) quarter wavelength; (b) half wavelength with push-pull valve; (c) for use with a single ended stage, where C2 should be adjusted to balance the valve output capacitance

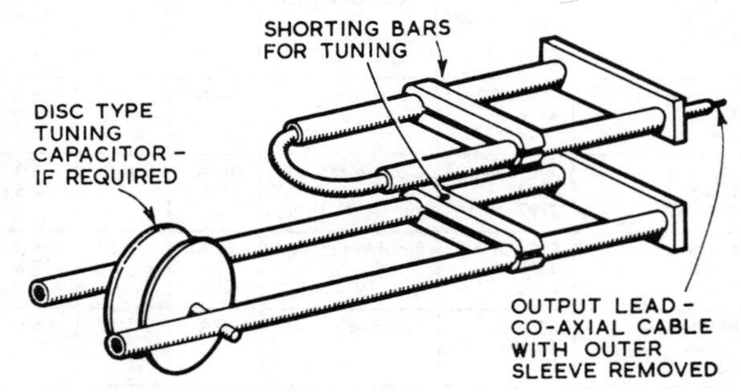

Fig 7.20. Balance-to-unbalanced aerial coupling

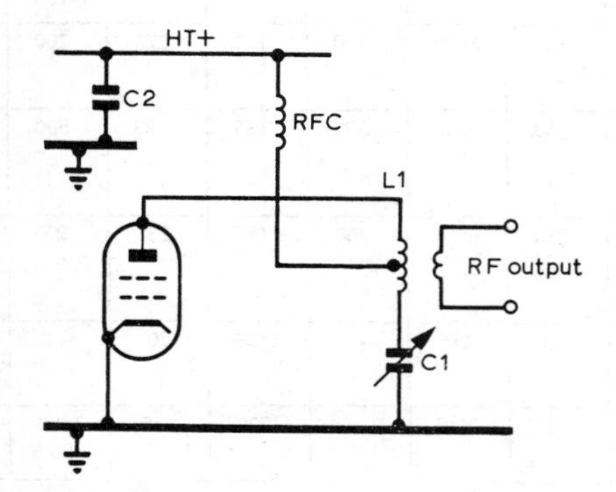

Fig 7.18. Series tuned anode circuit for use with single ended valves

Provision for adequate ventilation of the valve must be made to avoid excessive bulb temperatures being caused by enclosing this type of circuit.

Aerial Coupling to Parallel-line Tank Circuits

The usual method of coupling the output of the vhf transmitter to a low impedance coaxial feeder is by a one or two turn coupling winding located adjacent to the earthy end of the pa tank inductor. One side of the coupling circuit is invariably earthed, either directly or through a small preset capacitor which is adjusted to cancel out the inductive reactance of the coupling coil. With a parallel-line tank circuit a hairpin loop is employed, as shown in Fig 7.19 (b) and (c). This arrangement is satisfactory with the unbalanced circuit of Fig 7.19 (c) but when used with the push-pull anode circuit of Fig 7.19 (b) it results in some unbalance being reflected back into the tank circuit.

One method of overcoming this is to use a linear balance-to-unbalance and impedance transformer—balun **(Fig 7.20)**. The amount of coupling is determined by the distance between the balun and the anode lines, and also by the length of the balun lines which is dependent on the position of the shorting bar.

Coaxial or Cavity Circuits

Coaxial or cavity circuits are usually employed at 432MHz and above. At 1296MHz planar-electrode valves are incorporated in grounded grid circuits with quarter wavelength cathode and anode cavities, though three-quarter wavelength cavities are sometimes found. Use of a grounded grid

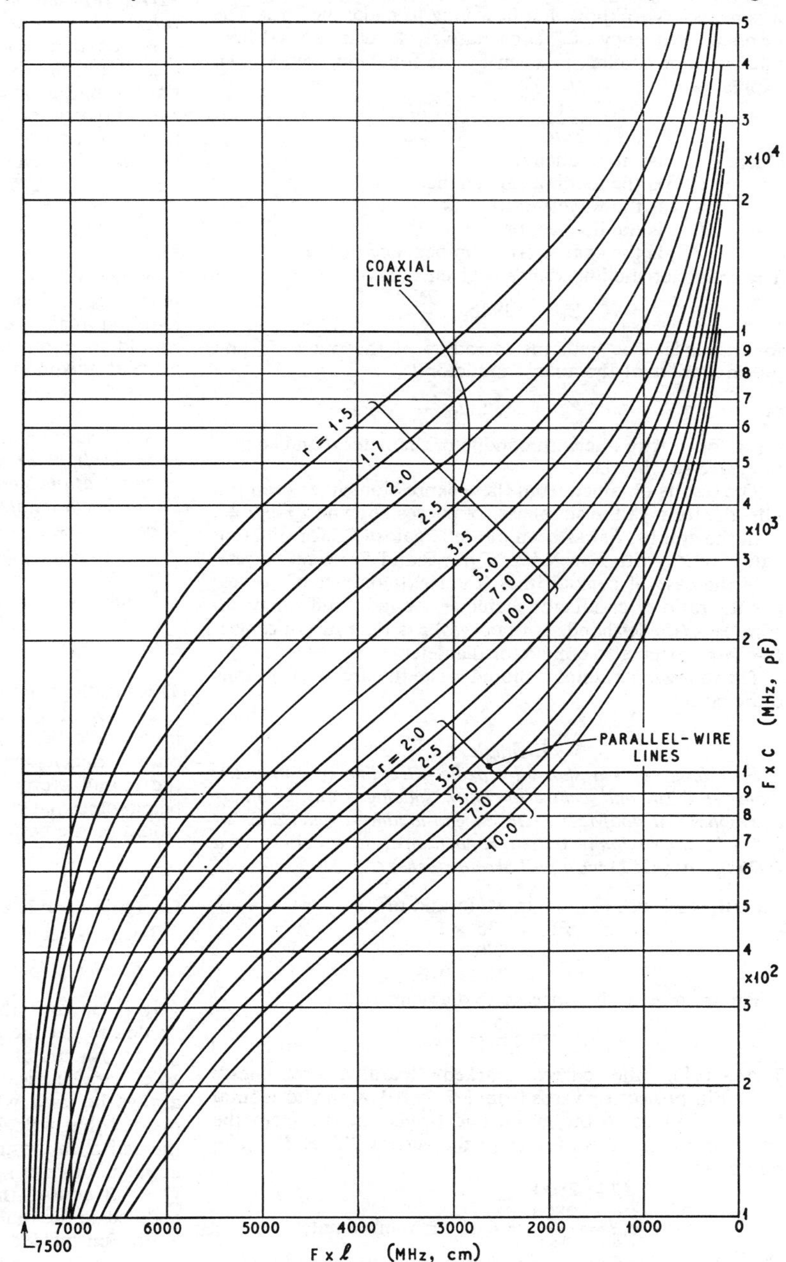

Fig 7.21. Resonance curve for capacitively loaded transmission line resonators

amplifier obviates the necessity for neutralization but as the amplifier and its driver stage are effectively connected in series, both stages must be modulated for telephony transmission. The audio power is not wasted, however, as the modulated rf output of the driver stages is added to the output of the amplifier.

Design of Transmission Line Resonators

When designing a resonator to be used as a tank circuit it is necessary to know first how long to make the lines. The resonant frequency of a capacitatively loaded shorted line, open-wire or coaxial, is given by the following well-known expression:

$$\frac{1}{2\pi fC} = Z_0 \tan \frac{2\pi l}{\lambda}$$

where F is the frequency
C is the loading capacitance
λ is the wavelength
l is the line length
Z_0 is the characteristic impedance of the line.

The characteristic impedance is given by:

$$Z_0 = 138 \log_{10} \frac{R_1}{R_2}$$

for a coaxial line with inside radius of the outer R_1 and outside radius of the inner conductor R_2

or
$$Z_0 = 276 \log_{10} \frac{2D}{d}$$

for an open wire line with conductor diameter d and centre-to-centre spacing D.

The results obtained from these expressions have been put into the form of the simple set of curves shown in **Fig 7.21.**

In the graphs, Fl has been plotted against FC for different values of Z_0, with F in MHz, C in pF and l in centimetres.

In the case of coaxial lines (the righthand set of curves) r is the ratio of conductor diameters or radii and for open-wire lines (the lefthand set of curves) r is the ratio of centre-to-centre spacing to conductor diameter.

The following examples should make the use of the graphs quite clear:

Example 1

How long must a shorted parallel-wire line of conductor diameter 0·3in and centre-to-centre spacing 1·5in be made to resonate at 435MHz, with an end-loading capacitance of 2pF (the approximate output capacitance, in practice, of a QQV03–20 (6252) push-pull arrangement)?

First, work out $F \times C$, in MHz and pF.

$$FC = 435 \times 2 = 870 = 8{\cdot}7 \times 10^2.$$

The ratio, r, of line spacing to diameter is:

$$r = \frac{1{\cdot}5}{0{\cdot}3} = 5{\cdot}0$$

Then, using the curves marked "parallel-wire lines," $r = 5{\cdot}0$in project upwards from $8{\cdot}7 \times 10^2$ on the horizontal "$F \times C$" scale to the graph and project across from the point on the graph so found to the vertical "$F \times l$" scale, obtaining:

$$Fl = 2800$$

$$\text{therefore, } l = \frac{2800}{435} = 6{\cdot}45 \text{ cm approximately.}$$

The anode pins would obviously absorb quite a good deal of this line length and if the lines were made 6 cm long with an adjustable shorting-bar they would be long enough.

Example 2

A transmission line consisting of a pair of 10 swg copper wires spaced 1in apart and 10cm long is to be used as part of the anode tank circuit of a QQV06–40 (5894) pa at 145 MHz. How much extra capacitance must be added at the valve end of the line to accomplish this?

For a pair of wires approximately ⅛in diameter spaced 1in r is about 8. Also $F \times l$ is equal to 145×10, ie 1450. Estimating the position of the "$r = 8$" curve for a parallel-wire line between "$r = 10$" and "$r = 7$," $F \times C$ is found to be about $1{\cdot}55 \times 10^3$, ie 1550. Hence the total capacitance C required is given by:

$$145 \times C = 1550$$
$$C = 1550 \div 145 = 10{\cdot}7\text{pF}.$$

Now the output capacitance of a QQV06–40A (5894) push-pull stage is around 4 pF in practice, so about 7 pF is required in addition. A 25 pF + 25 pF split stator capacitor should therefore be quite satisfactory giving 12 to 15 pF extra at maximum capacitance.

Example 3

A coaxial line with outer and inner radii of 5·0 and 2·0cm, respectively, is to be used as the resonant tank circuit (short-circuited at one end of course) for a 4X150A power amplifier on the 70cm amateur band. What length of line is required?

In this case:

$$F \times C = 435 \times 4{\cdot}6 = 2001.$$

Using the "$r = 2{\cdot}5$" curve for coaxial lines,

$$F \times l = 4620$$

Hence
$$l = 4620 \div 435 = 10{\cdot}6\text{cm approximately.}$$

This length would include the length of the anode and cooler of the 4X150A of course but, as in Example 1, a line 10cm long would be certain to be long enough, especially as the output capacitance used in the calculations is that quoted by the manufacturers for the valve, the effective capacitance being somewhat greater in practical circuits. A shorting bridge would be the best method of tuning the line to resonance.

Designing for Maximum Unloaded Q

The tank circuit efficiency is given by:

$$\textit{Efficiency} \text{ (per cent)} = \frac{\textit{unloaded } Q - \textit{loaded } Q}{\textit{unloaded } Q} \times 100$$

It is obvious that the highest possible unloaded Q is needed to get the greatest tank circuit efficiency. The Q is greater for radial and coaxial resonators than for comparable parallel-wire circuits and the former types should always be used where possible. It should perhaps be explained that unloaded Q is the Q of the tank circuit with the valve in position and all voltages and drive power applied, but with no load coupled up to it. The loaded Q is, of course, that measured when the load is correctly coupled to the tank circuit.

For unshielded parallel-wire lines, the unloaded Q is usually quite low because of power loss by radiation from the line, and the best value is obtained by using a small conductor spacing (low Z_o).

To obtain the best Q, the material of the line should be copper or brass, fairly smoothly finished, although a highly polished surface is not necessary. To improve its conductivity, the surface of a coaxial or radial line can be silver plated. In an industrial or city atmosphere, however, the silver plating is rapidly attacked by atmospheric gases and the surface conductivity suffers far more than does that of a copper or brass line. The best solution is to apply a "flash" of rhodium to the silver plating but this is rarely possible for the amateur. The Q of a coaxial line depends also upon the ratio of conductor diameters and, for fairly heavily capacitatively loaded resonators, which is true in most practical cases, this ratio should be between about 3 and 4·5 to 1.

Care should be taken that the moving contacts on the bridge (if one is used) are irreproachable: they should preferably make contact with the line a little way from the shorting disc, where the line impedance is somewhat higher than at the current antinode. It is better to use a large number of springy contact fingers rather than to use a few relatively rigid ones.

Attention to these points will often make all the difference between a reliable, satisfactory resonant circuit of high Q and one which possesses none of these qualities.

Disc-Type Capacitors

Parallel lines or concentric (coaxial) tuned circuits are conveniently tuned by means of a variable air capacitor comprising two parallel discs, but the calculation of the capacitance range of different size discs with differing spacing can be tiresome.

The chart in **Fig 7.22** gives the capacitance between two parallel discs of various convenient diameters with spacings between ¼in and 1/128in calculated according to the formula

$$CpF = \frac{0{\cdot}244 \times \textit{area (inches)}}{\textit{spacing (inches)}}$$

The diameter of the disc employed may be fixed by space considerations but it determines the minimum and maximum capacitance and the range available. Very close spacings should be avoided unless extremely accurate parallelism can be maintained, and must be avoided where high voltages exist, as in transmitters.

For example, a cathode concentric tuned circuit tuned by a disc capacitor is required for a 2C39A valve to be used as a tripler from 432 to 1296MHz. This valve has an input capacitance of 6·5 ± 1 pF. If the cavity has an outer diameter of 2⅜in and an inner conductor of ⅜in the ratio is 7:1 and at 430MHz the curves in Fig 7.21 show that if C is taken as 7·5 pF plus a minimum disc capacitance of, say 0·5 pF, ie 8 pF, the inner conductor length is 4·2cm or 1·65in. If a ½in disc capacitor is chosen then the range of tuning is from 0·2 pF at ¼in spacing to 1·5 pF at 1/32in spacing. As the drive voltage is of the order of 100V closer spacing must not be used. It is clear that a ½in disc will not do as the spread between top and bottom limit values is 2 pF so that ¾in discs will have to be used, giving a range of 0·4 to 3·0 pF which is sufficient for fixed frequency working; if it is necessary to cover a frequency band then a greater range is required. If at times the drive frequency were 420MHz then this would require a maximum capacitance across the tuned circuit of the original 8 pF multiplied by the square of the frequency ratio, ie $\frac{(430)^2}{(420)^2}$ or 1·05. This means a maximum disc capacitance of 2·4 pF and the ¾in diameter will still just do but ⅝in would not.

If it is found that the range of the capacitance required is larger than that given by the maximum size disc that can be accommodated, then the solution is to use two sets of discs located, for example, each side of a cavity and use one as tuning and one as a kind of band-set. Valve capacitance limits must be taken into account otherwise the replacement of a valve can result in the circuit no longer tuning. Where the tolerances in input and output capacitance are not known they can be taken as ± 30 per cent which would adequately cover most cases.

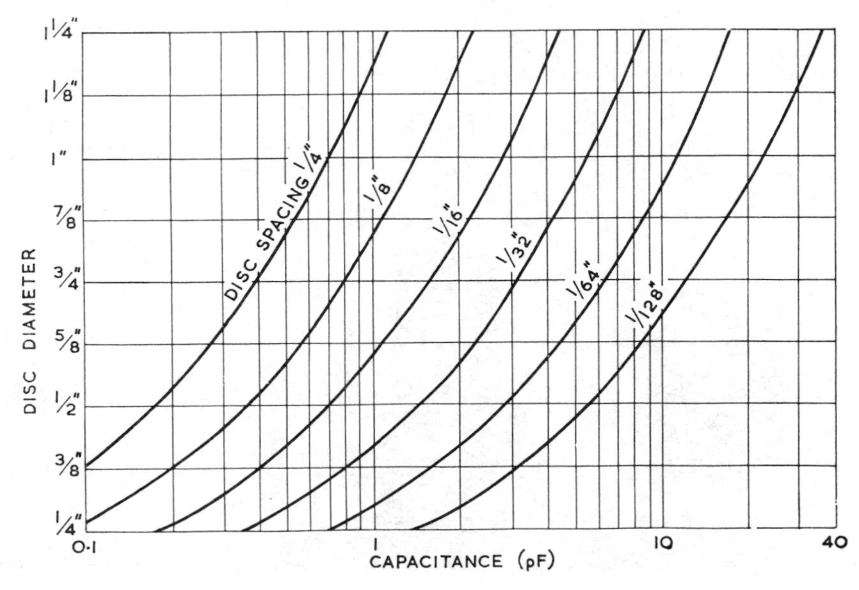

Fig 7.22. Capacitance between two parallel discs of various diameters

Neutralization

With triode valves and tetrodes operating below their self-resonant frequency the normal methods of neutralization apply, ie the application of an antiphase feedback voltage from anode to grid of the valve. The residual anode-to-grid capacitance of modern vhf tetrodes is quite low, in some instances being less than 0·1 pF. Provided that adequate screening exists between grid and anode circuits, operation of certain types of valve in the lower part of the vhf spectrum (up to 70MHz) is permissible without neutralization being required. At higher frequencies although the reactance of this capacitance is still not sufficiently low to cause oscillation, it will result in the stage being regenerative and neutralization will be necessary.

A convenient method of providing the small capacitance required for neutralizing push-pull double tetrode stages consists in using each valve anode as one plate of the capacitor, and a wire connected to the grid of the opposite valve as the other plate. The neutralizing capacitance is varied by bending each wire nearer or further away from the respective anode. In single ended circuits the capacitance is formed by a wire connected at one end to the grid and positioned so that the other end is adjacent to the appropriate side of the anode tank circuit **(Fig 7.23 (a))**. Heavy gauge enamelled copper wire (12 to 16 swg) should be used, and a metal tab may be attached at the end of the wire to increase the capacitance if that provided by the wire alone is insufficient.

When the valve is operated above the self-neutralizing frequency the neutralizing capacitor must be connected directly between anode and grid of the valve **(Fig 7.23 (b))**. Alternatively a variable capacitor may be connected between screen grid and earth and tuned with the screen lead inductance to form a series resonant circuit at the frequency of operation, thus providing a low screen-to-earth impedance and effectively placing the screen grid at earth potential with respect to rf **(Fig 7.23 (c))**. It must be remembered that this method of neutralization is frequency sensitive, and any considerable change in the operating frequency will necessitate re-adjustment of the variable capacitor.

In some double tetrode valves such as the QQV03–20A (6252) and QQV06–40A (5894) internal neutralizing is incorporated in the valve structure so that external neutralizing is unnecessary provided the circuit design and layout is good.

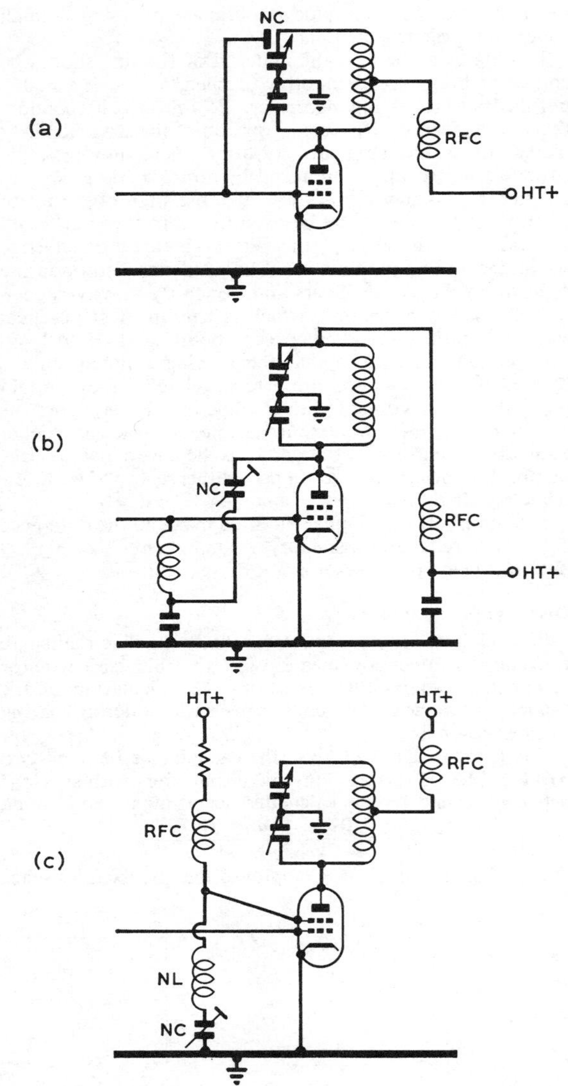

Fig 7.23. Methods of neutralizing single ended valve stages (a) "hot" grid to "cold" end of tank circuit; (b) "hot" anode to "cold" end of grid input; (c) screen grid neutralization

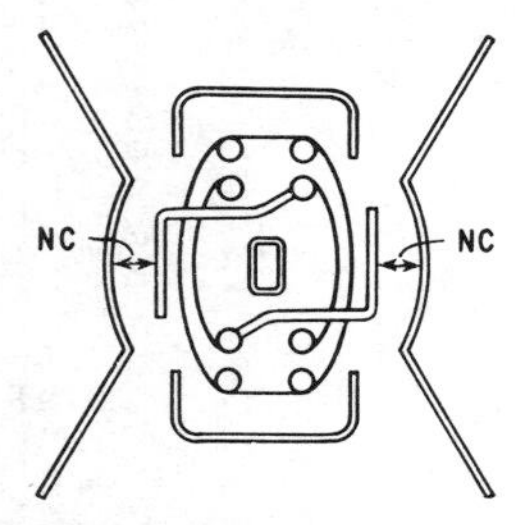

Fig 7.24. How internal neutralization is applied to the QQVO3/20A and QQV06/40A double tetrode transmitting valves

MODULATION

For telephony service a vhf transmitter may be modulated by one of several methods. For satisfactory modulation it is essential that there is adequate grid drive to the modulated amplifier and that the stage is correctly neutralized.

Experience has shown that it is desirable to maintain a high average level of modulation in a vhf transmitter and the use of speech clipping or compression is therefore recommended.

Instability in the modulator may be experienced due to rf feeding back into early stages of the speech amplifier. To eliminate this the microphone input circuit should be completely shielded and the effect of inserting grid stopper resistors, or ferrite beads threaded on to the grid lead close to the valve socket connection, particularly in low level speech amplifying stages, and the inclusion of an rf filter in the grid circuit of the first valve should be investigated.

Amplitude Modulation

Amplitude modulation can be applied to the pa valve anode by any of the usual methods and must also be applied to the screen by:

1. Series resistor from the modulated anode voltage;
2. Separate winding on the modulation transformer, with series resistor to the anode ht voltage as shown in **Fig 7.25(a).** The screen voltage dropping resistor must be by-passed to audio frequencies on the supply side. R is an rf stopper resistor.
3. A tapped modulation transformer: the screen dropper which must be connected in series with the tap must be by-passed to audio frequencies **(Fig 7.25(b)).**
4. A resistive potential divider across a standard modulation transformer: the values of the upper and lower resistor must be chosen to provide the proper screen voltage. **(Fig 7.25(c)).**

For optimum operation the screen voltage is critical, and can never remain reasonably constant when supplied from a series resistor, since any change in anode current results in a change in screen current. The circuit of **Fig 7.26** not only enables a definite value of screen voltage to be set, but also allows the correct amount of audio to be applied.

In operation RV1 controls the value of the screen voltage and the values of R1 and R2 must be chosen so that RV1

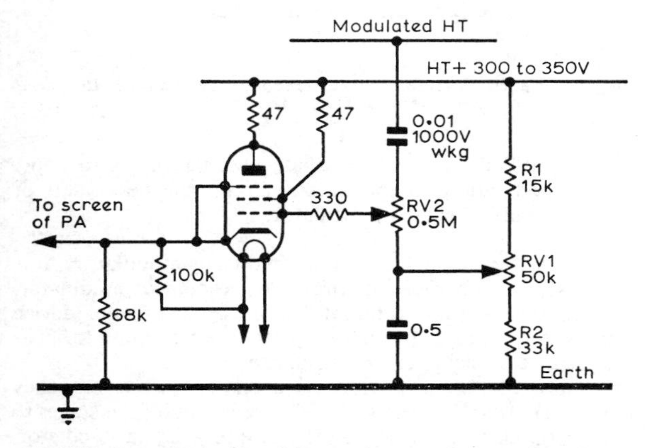

Fig 7.26. A method of applying optimum voltage and optimum modulation to the screen of a pa valve. Any small pentode such as the 6V6, 6CH6 or 5763 may be used

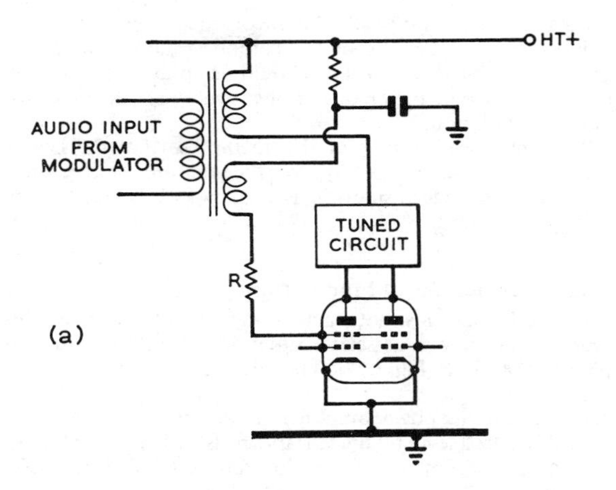

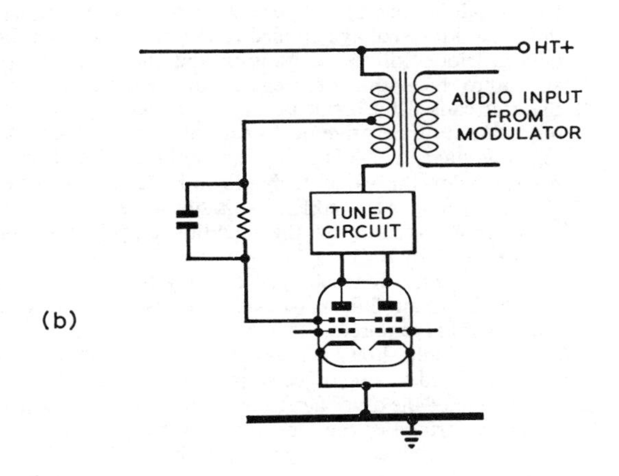

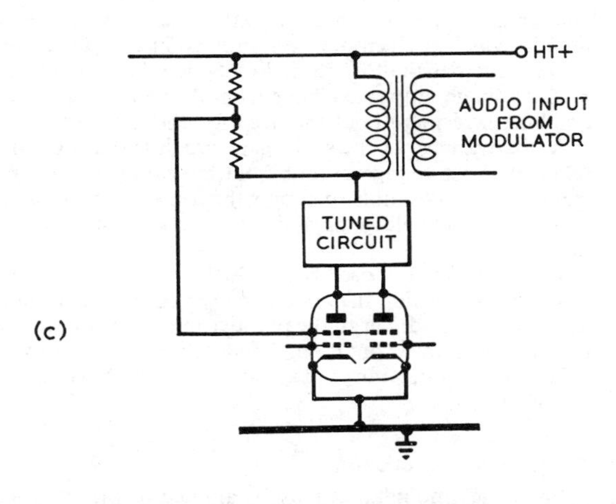

Fig 7.25. Methods of anode and anode-screen modulation of the final stage of a vhf transmitter: (a) screen supply by a separate winding on the modulation transformer (the resistor R is an rf stopper of typical value 47 Ω); (b) tapped modulation transformer; (c) by potential divider across the modulation transformer secondary winding

gives an adequate range. RV2 controls the amount of modulation applied to the grid of this cathode follower valve and hence to the pa screen grid. It should be adjusted so that the pa anode current meter does not move either up or down under modulation.

The only disadvantage of this method is that the heater supply to the valve must be from a separate source, since the heater–cathode insulation is usually insufficient for the voltages involved.

Screen (Series Gate) Modulation

Probably the most satisfactory method of screen modulation—that known as series gate modulation—is both simple and effective **(Fig 7.27)**. The advantages of this method are:

(a) The standing power amplifier anode current can be set to any desired level by the variable resistor RV1 in the cathode of the first section of the double triode series gate valve;

(b) Over-modulation on positive peaks cannot occur because the audio input voltage applied to the first section of the double triode will cause limiting and thus the voltage applied to the amplifier screen cannot rise too high;

(c) Splatter caused by break up of the carrier due to over-modulation on negative peaks cannot occur because the screen voltage, which is set by the position of RV1, cannot fall below this value. A simple and effective means of speech clipping is available by suitable adjustment of the audio input voltage by the standing amplifier screen voltage as set by RV1;

(d) It can be applied to existing cw or nbfm transmitters and to ssb linear amplifiers, with carrier inserted, when an a.m. signal is required.

With the circuit shown, the cathodes of the series gate valve are returned to a negative voltage. This enables the full screen voltage of the amplifier to be taken from the cathode of the second triode of the series gate valve, because of the voltage drop introduced by the valve itself.

An alternative arrangement would be to use a higher ht voltage and a lower impedance valve such as a triode pentode in which the pentode section is connected as a triode. A valve such as the ECL82 or ECL86 would enable a much higher voltage output available from the cathode to feed the amplifier screen, without the need for the negative voltage.

A disadvantage of the series gate circuit is that due to the relatively high voltage output from the cathode of the control valve, it is desirable to supply the heater from an isolated supply to avoid voltage breakdown between heater and cathode.

Whatever form is used it should be remembered that at peaks of modulation the anode and screen voltage is almost doubled and the driver to the pa valve must provide sufficient power under this condition. Modulation of the driver is sometimes useful, taking the audio from the screen grid of the amplifier.

Frequency Modulation

Frequency modulation may be applied to the transmitter carrier oscillator by either direct or indirect means:

(a) The direct method is the application of modulation to the frequency determining device, crystal or tuned circuit;

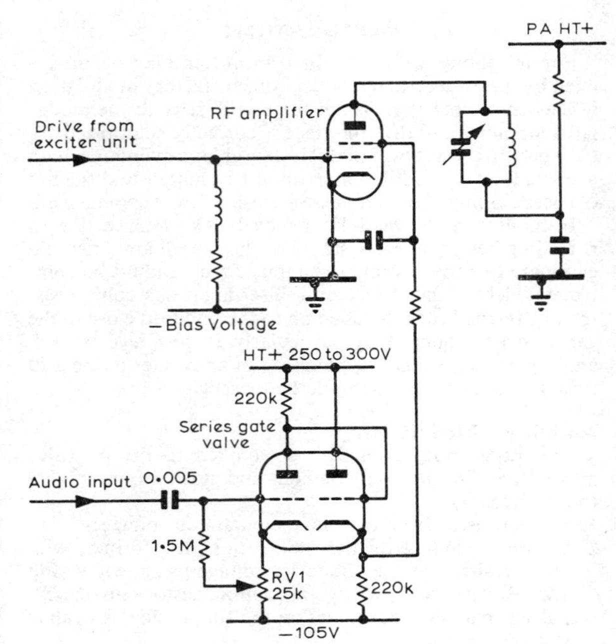

Fig 7.27. Series-gate modulation of a vhf transmitter pa stage from a doubler triode such as a 12BH7

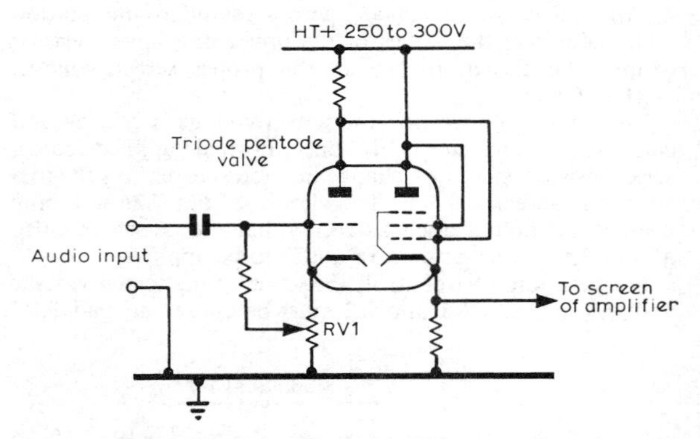

Fig 7.28. Another method of screen modulation of a pa by the series gate method, using a triode pentode

(b) The indirect method (phase modulation) is to apply modulation to the output circuit of the crystal or variable frequency oscillator.

Direct FM: To achieve direct frequency modulation, it is necessary to provide some form of reactance modulator, which may be either a thermionic or semiconductor device connected in parallel with the crystal or tuned circuit to act as a variable capacitance or inductance.

The oscillator is operated at a relatively low frequency such as 6–18MHz, followed by frequency multiplier stages to reach the final frequency. In the case of a crystal oscillator, the result obtained will be largely phase modulation because it is difficult to achieve sufficient frequency shift of the crystal

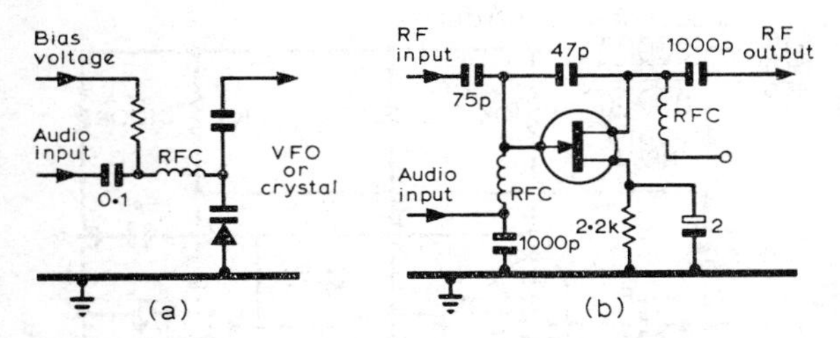

Fig 7.29. Basic methods of applying frequency modulation (a) direct fm, by the use of a varicap diode, and (b) indirect fm (phase modulation), in which a phase modulator is connected between the oscillator of a vhf transmitter and the following frequency multiplier

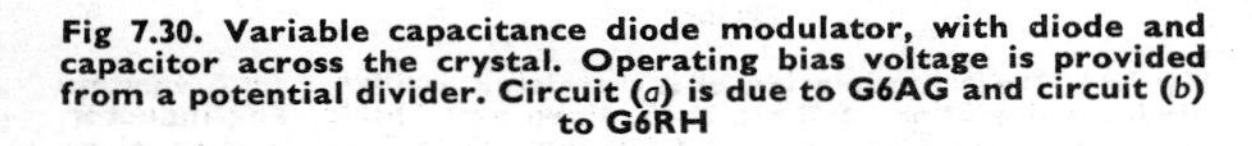

Fig 7.30. Variable capacitance diode modulator, with diode and capacitor across the crystal. Operating bias voltage is provided from a potential divider. Circuit (*a*) is due to G6AG and circuit (*b*) to G6RH

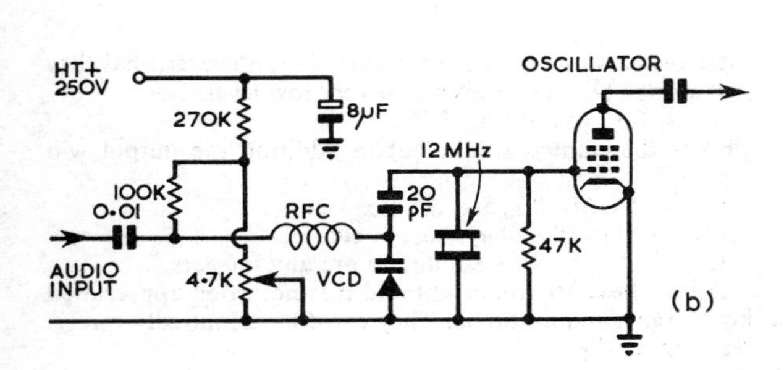

itself, although some additional shift can be attained when suitable inductance is placed in series with the crystal.

For most amateur purposes, a variable capacitance diode suitably biased will be suitable for this purpose. The operating point (bias) should be chosen such that the variations in capacitance, in both the upward and downward directions for the range of audio frequency voltage, are equal and linear.

Most variable capacitance diodes require a relatively small audio frequency voltage for a considerable change of capacitance. It is therefore usual to connect a small capacitor in series with the diode, feeding the audio frequency signal at the junction of these components **(Fig 7.29 (a))**. It is advisable to use an rf choke to feed in the audio frequency signal since supply voltage variations will cause frequency change in the oscillator, especially if it is a vfo. It is necessary to use some form of regulated supply both for the oscillator and the diode supply.

Indirect FM: Phase Modulation: In this method, the audio frequency modulation is applied to the output circuit of the oscillator. The amount of frequency deviation that can be achieved is dependent on the degree of circuit detuning that can be obtained. The circuit should have a reasonably high loaded Q; about 20 is usually satisfactory **(Fig 7.29 (b))**.

The phase shift that takes place when a circuit is detuned from resonance depends on the degree of detuning and on the Q of the circuit. The higher the Q, the smaller the amount of detuning needed to obtain a given number of degrees of phase shift.

Because in this type of modulation the actual frequency deviation increases with the audio frequency modulation, it is necessary to cut off frequencies above 3000Hz before modulation, otherwise unnecessary sidebands will be generated.

The following seven diagrams illustrate various methods of generating nbfm. See also Chapter 9—*Modulation*.

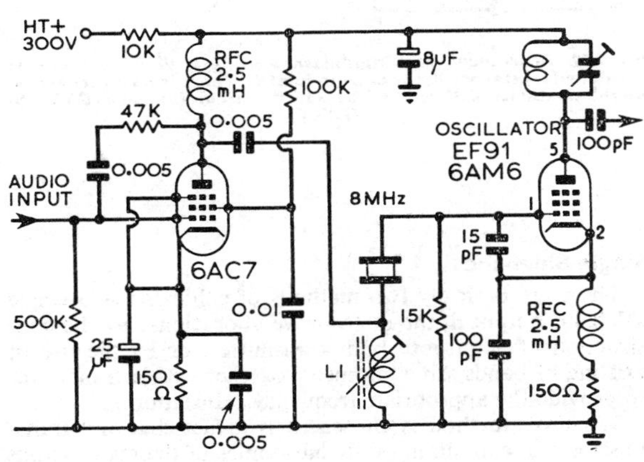

Fig 7.31. Reactance valve modulator by G6TA in which the reactance valve is placed across an inductor in series with the crystal. The inductance is varied by the reactance valve, thus causing frequency variation

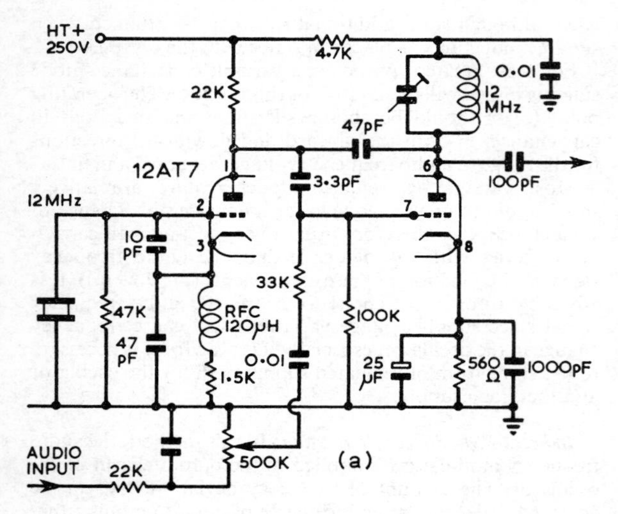

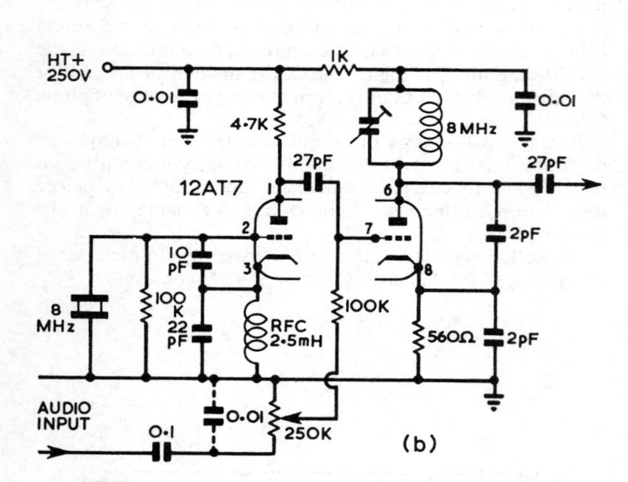

Fig 7.32. Reactance valve modulators using a double triode as combined crystal oscillator and modulator. There is no direct action on the crystal itself. Circuit (a) by G3EDD, circuit (b) due to G3AWS/G3SLF

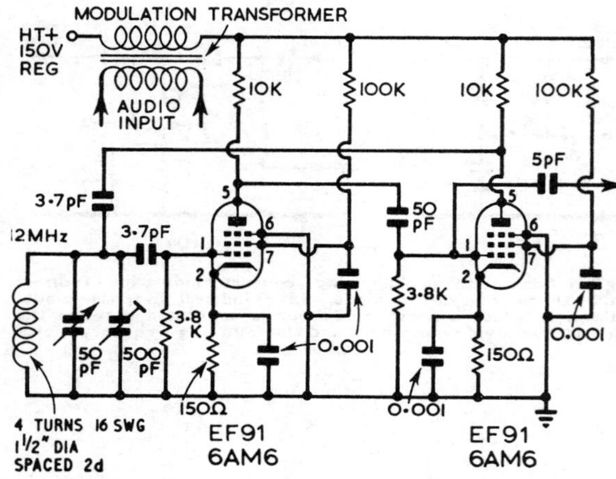

Fig 7.33. Direct frequency modulation of a Franklin vfo, due to G2HCG. The audio modulation is applied to the oscillator anodes via a small transformer

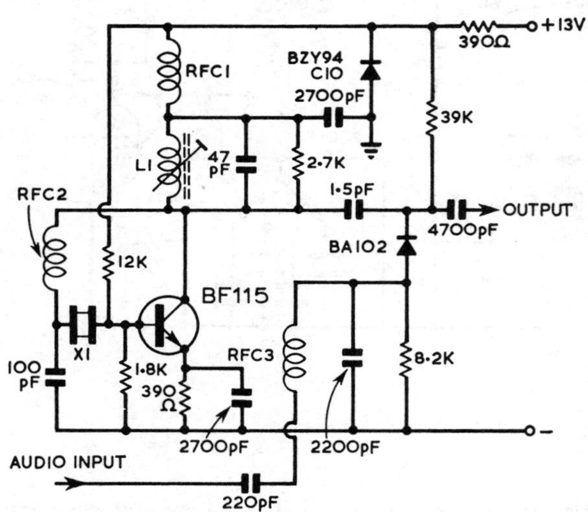

Fig 7.34. Phase modulation of a transistor oscillator by a varicap diode. A deviation of 5kHz is obtained at 145MHz. The inductor requires 31t of 26swg wire close-wound on a 7mm former, slug-tuned, when a 12MHz crystal is used

Single Sideband

There are basically two methods of achieving A3J single sideband output in the metre-wave allocations: by the construction of a purpose built transmitter, or by the use of existing hf bands ssb equipment together with a transverter to provide the appropriate frequency translation.

Whatever method is chosen, it is a fact that any transmission will contain not only harmonics of the main output frequency, but harmonics of any oscillator involved in the final frequency generation, and mixing products of any frequency present. It is essential to take steps to attenuate these outputs to a minimum. Remember, however, that they will always be present even if at very low level.

$nf_x + f_{nf} = f$ out **(Fig 7.35).**

This is the wanted result, but in addition the output will contain:

(1) $f_x, 2f_x, 3f_x, 4f_x, 5f_x, \ldots$ etc.

(2) $f_{nf}, 2f_{nf}, 3f_{nf}, 4f_{nf}, 5f_{nf}, \ldots$ etc.

(3) $nf_x \pm mf_{nf}$ where n and m are any integers.

All of these are important and it is not often appreciated how large n and m can be. They can (but admittedly rarely) reach three digits.

To take an actual case, as in **Fig 7.36:**
Not only does (2 × 58) + 28 = 144MHz, but
(3 × 58) − 28 = 146MHz
In this case, since post mixer selectivity is invariably low Q, the unwanted 146MHz transmission could be expected to be at a high level.

However, both Figs 7.35 and 7.36 are simplified. The actual generation of the hf ssb injection involves yet more frequencies, so that the full arrangements can be shown in **Fig 7.37.**

The total arrangement thus involves three crystal oscillators and one vfo. The calculation of potential spurious emissions from this is a task for a computer. It should not be forgotten that the crystal controlled injection frequency can be higher than the output frequency. Some problems concerning spurious frequencies can be solved in this manner, but the approach does result in inverted tuning direction and sideband transposition.

Measurements on transmitters where no deliberate attempt has been made to reduce spurious emissions show that 40dB spurious attenuation is typical. Thus if a local receiver receives an S9 + 40dB wanted signal from a vhf ssb transmitter, a spurious signal of S9 could also be received. Assuming 3dB per S point and an acceptable interference level of S2, the extra attenuation required is (7 × 3) dB ie 21dB. We are looking for approximately 60dB spurious attenuation in this case. However, local signals can well be in excess of S9 + 40dB and a case can be made for a target of 100dB spurious emission attenuation.

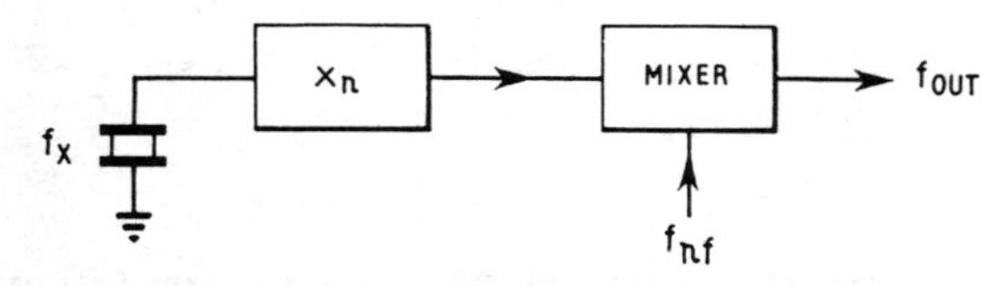

Fig 7.35. The basic mixing process in ssb generation

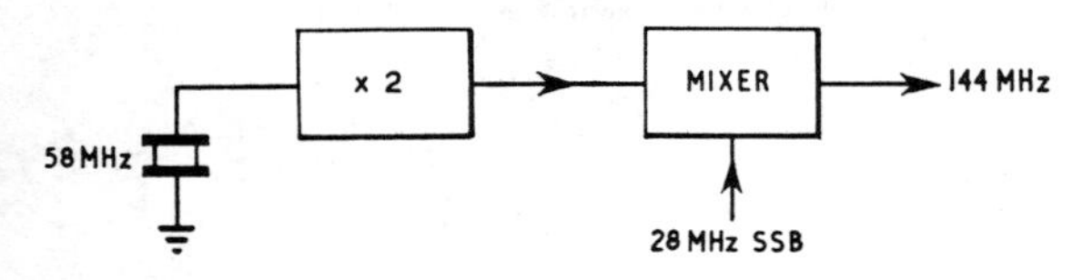

Fig 7.36. Crystal chain frequency and ssb input frequency into a mixer/transverter give 144MHz output. Reservations exist about the use of 28MHz ssb input injection

Only the professional engineer with access to specialized test gear and who has had experience of attempting to design to this target can truly appreciate the size of the task. It is possible however to indicate lines of approach which will at least show a major improvement on the 40dB figure.

Both hf and crystal frequencies should be chosen so that no spurious emissions produced by low order harmonics fall in the band or close to it. The crystal frequency should be as high as possible, but bear in mind the need to avoid use of a 58MHz crystal with 28MHz hf for instance.

Final Mixer: Although the double tetrode mixer has almost universal acceptance, a number of other approaches are open if mixing is kept at a low level. Both conventional and Schottky diode bridges, and balanced transistor circuits are possibilities. Whatever arrangement is decided, the aim should be to arrange for maximum rejection of the vhf injection frequency. Provision for balance adjustment should always be provided.

With a high level ssb input in which high power is dissipated in a resistive load, "hop over" effects can occur, in which appreciable ssb energy (with doubtful spectral purity) gets into both pre and post mixer circuits thus bypassing any selectivity in the ssb input. If this energy mixes with a second frequency to produce a third frequency within the bandwidth of either the oscillator or 144MHz circuits a spurious emission close to the wanted output frequency will result.

Crystal Oscillator and Multiplier: In order to provide the mixer with injection of excellent spectral purity, coupled tuned circuits should be used throughout, with low impedance link coupling.

HF SSB Input: The input level should be kept to a minimum. It is bad practice to use many watts of hf only to dissipate them in a resistor located on the converter chassis. On the other hand bad carrier rejection will result in the hf ssb transmitter if the level is turned down by reducing the audio drive. The best approach is to feed the input to the hf pa direct to the converter and it is often possible to obtain sufficient level by switching off the pa screen voltage and using the feed through voltage. Band pass filtering should be used in the converter hf feed in order to filter unwanted emissions.

Post Mixer Filtering: Band pass filtering should be provided immediately after the mixer. A series of low working *Q* tuned circuits in successive amplifiers is not considered sufficient.

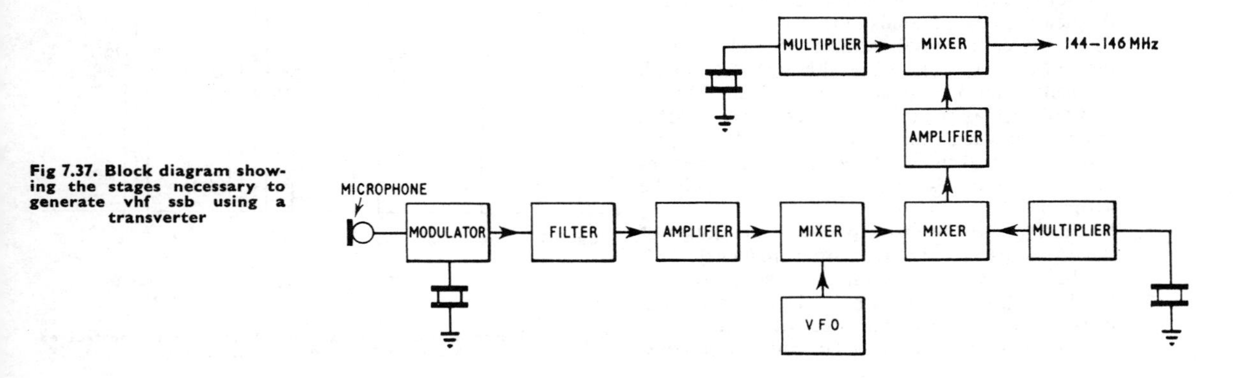

Fig 7.37. Block diagram showing the stages necessary to generate vhf ssb using a transverter

Aerial: Use an aerial with the narrowest acceptable gain-bandwidth characteristic. A broad band antenna may have appreciable gain at the local oscillator injection frequency or the image frequency. It is important not to confuse the vswr bandwidth with the gain bandwidth; they can be very different and the latter is rarely quoted.

Practical Design: A near perfect design can be unsuccessful if the engineering is poor, conversely an indifferent design can sometimes produce near acceptable results if the engineering is excellent.

The following points should be borne in mind:

(1) Every tuned circuit should be properly screened and preferably every stage should also be screened.
(2) Decoupling must be effective not only at the desired frequency but also on high order harmonics.
(3) Lt and ht feeds to individual stages must be thoroughly decoupled.
(4) In-line layout should be employed with the oscillator at one end and the output at the other.

The Final Solution? From the above discussion it becomes clear that there are serious problems using an hf ssb transmitter as the ssb generator. The best answer appears to be a purpose built vhf ssb transmitter along the lines of **Fig 7.38.**

The main problem is generating the ssb signal at a sufficiently high frequency. Filters at 10·7MHz are adequate, but 20MHz would be better. For variable frequency operation the injection frequency should be varied. Again, this is a problem if full 2MHz coverage of the 2m band is required, but if ± 50kHz on the sideband channel is acceptable, a vxo offers a simple solution.

Out of Band Radiations: Great care must be taken to ensure that no out of band spurious radiations occur. All rf circuits should be adequately screened and a bandpass filter with sufficient out of band attenuation should be included in the aerial feeder.

In Band Radiations: To avoid interference between 144 and 146MHz:

(1) Every effort should be made to keep the spurious radiations down to 90dB below the wanted signal.
(2) The following precautions should be taken when a transverter is used in association with an hf bands transmitter/transceiver as a ssb source.
 (a) Do not use the 28–30MHz tuning range because the fifth harmonic is in the band and the level of spurious signals is likely to be higher at the highest frequency of the hf bands unit.
 (*b*) The transverter crystal oscillator should be on as high a frequency as possible (although certain high frequencies must be avoided, eg 58MHz). The use of fundamental crystal oscillators below 30MHz must be avoided unless very special design precautions are taken. Any frequency that gives in-band signals of less than tenth order should be avoided.
 (*c*) Precautions must be taken to minimize radiation of the crystal oscillator chain output frequency. This can be done by using a balanced mixer—which can attenuate this component by 20dB or more—and by subsequent tuned circuit selectivity. In the latter case a minimum of *four* tuned circuits are required

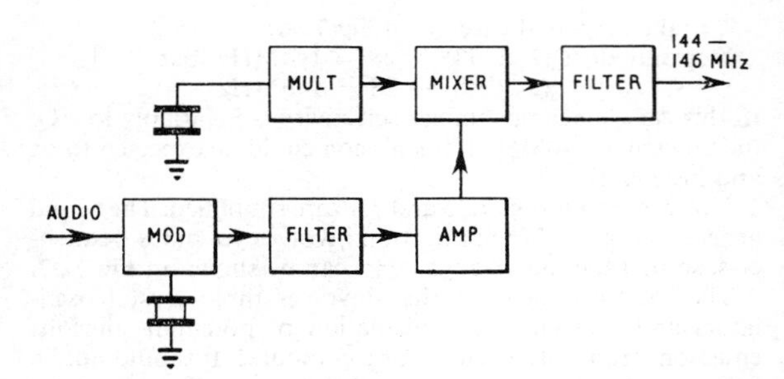

Fig 7.38. Suggested arrangement for a purpose-built vhf ssb transmitter

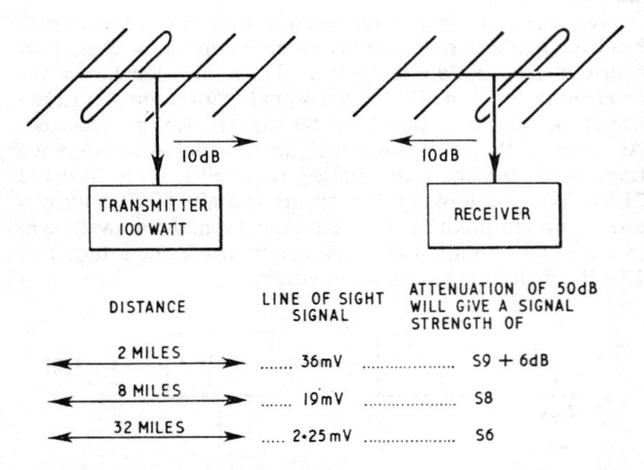

Fig 7.39. Diagram showing the signal strengths to be expected from a typical situation at various distances. A spurious transmission attenuated by 50dB from the desired signal will produce the strengths given in the third column

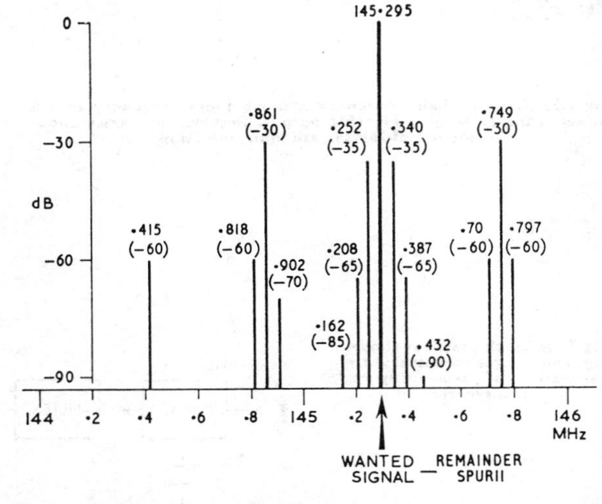

Fig 7.40. Spectrum analysis of a signal from a home constructed transceiver transverter combination

between the mixer and the aerial feeder. If two of these tuned circuits are coupled this should be done inductively.

(*d*) A frequency of 58MHz must be avoided in the transverter—either as the oscillator frequency or a multiplier stage output—when a 28 to 30MHz ssb feed is used.

Reason: $58 \times 2 + 29 = 145$MHz
$58 \times 3 - 29 = 145$MHz

Another undesirable combination is a 43·333-MHz crystal used with a 14MHz ssb feed. The third crystal harmonic is 130MHz and this with the addition of the feed frequency will give 144MHz output. However, the fourth harmonic is 173·333MHz and if twice the feed frequency is deducted from this a frequency of 145·333MHz will be obtained. This is another unwanted in-band spurious frequency.

(*e*) There must be an adequate degree of frequency selectivity between the hf feed and the transverter mixer as most hf ssb transmitters or transceivers will have unwanted frequency components in their output. Even those far removed from the nominal feed frequency can cause serious problems if fed without further attenuation to the transverter mixer. A satisfactory method is to insert a suitable filter and, if necessary, a combined power attenuator between the hf feed and the transverter. The dummy load, if used, should always be well screened from the transverter.

(3) Preference should be given to the use of equipment specifically designed for operation on 2m using a high frequency ssb generator. Recommended are 9, 10·7 or 25MHz as suitable for this purpose.

(4) Care must be taken in all amplifier stages following the final mixer to ensure that the operating conditions are such that harmonic distortion does not produce spurious frequency signals in the region of the unwanted sideband of level worse than −40dB relative to the wanted signal.

PRACTICAL TRANSMITTER DESIGNS FOR 144MHz

In the following pages a number of 144MHz transmitter designs to be described put into practice some of the technical concepts outlined in the foregoing section of this chapter. All of them may be readily reproduced by the constructor of average ability.

The first two designs adapt all-solid-state technology to the realization of one or two watts of rf generated from units intended for portable application or as exciters for following amplifier stages. The readily available transistors specified have a reasonably high voltage rating to reduce the liability to breakdown if they should happen to be operated into considerably mismatched loads.

DESIGN "A": A SIX-TRANSISTOR EXCITER UNIT

The first all-transistor transmitter to be described delivers 1W of rf from a pair of paralleled 2N2218 (or similar) devices. See **Fig 7.41.**

The crystal oscillator uses a 24MHz overtone crystal and is built on the underside of the chassis. The emitter biasing components, R1 and C1, are soldered direct to the chassis at the one end with the other ends soldered direct to the emitter of TR1 with no additional support. The normal base biasing resistors are R2 and R3. Feedback through the crystal is achieved by a centre tap on L1. Output from the oscillator stage is taken via C6 to the emitter of TR2. This transistor is connected in common base and its base lead should be cut to approximately ½in and soldered direct to the chassis. The bias resistor R4 is beneath the chassis and soldered direct to it. Reference to **Fig 7.42** should make the mounting of the transistors quite clear. Transistor TR2 doubles to 48MHz and output is taken via C10 to TR3 tripling to 144MHz. Tuning for TR3 is by two concentric trimmers C14 and C15 connected from TR3 collector to chassis. C15 has its centre connections soldered direct to the chassis and C14 is supported by soldering one of its outer connections to the adjacent feed-through insulator.

Refer again to Fig 7.41. Capacitor C16 which is connected in parallel with C15 is soldered below the chassis. Output from this stage is taken from the junction of C14 and C15 and by adjusting the two capacitors which in effect are tapped up the coil thus matching the impedance to the following stage. Transistor amplifiers of this type perform best when heavily

Fig 7.41. The six-transistor transmitter/exciter unit: circuit diagram overleaf, general view below

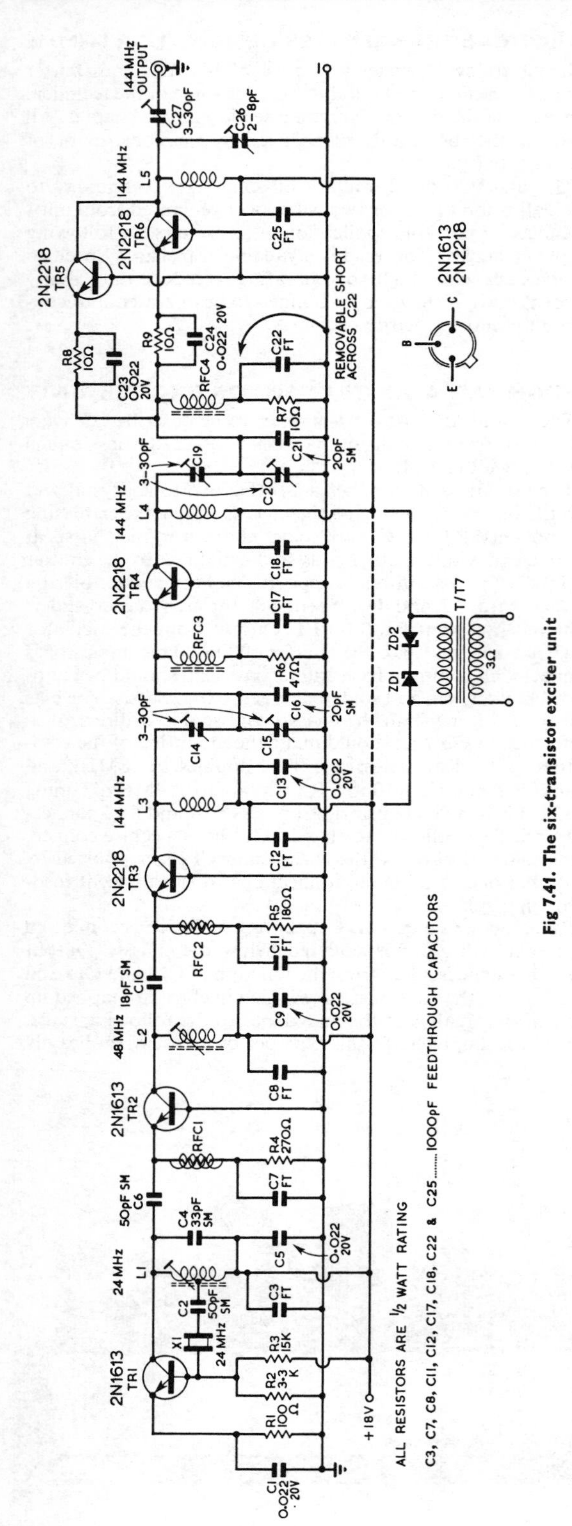

Fig 7.41. The six-transistor exciter unit

loaded and instability may result if the lower capacitor is screwed in too far.

TR4 is the driver stage and feeds TR5 and TR6, the power amplifiers, connected in parallel through separate emitters, thus preventing "current hogging" by one transistor. Should one of the power amplifier transistors become much hotter than the other increase the value of R8 and R9 slightly. This will reduce the output somewhat but slightly increase the efficiency. Another way to overcome this trouble is to try various pairs of transistors until they appear to run approximately at the same temperature. Testing with the finger is quite adequate. All the transistors in this transmitter run quite hot to the touch. To assist cooling TR5 and TR6 are fitted with small clip-on heat sinks. Silicon transistors can run quite safely to 200°C.

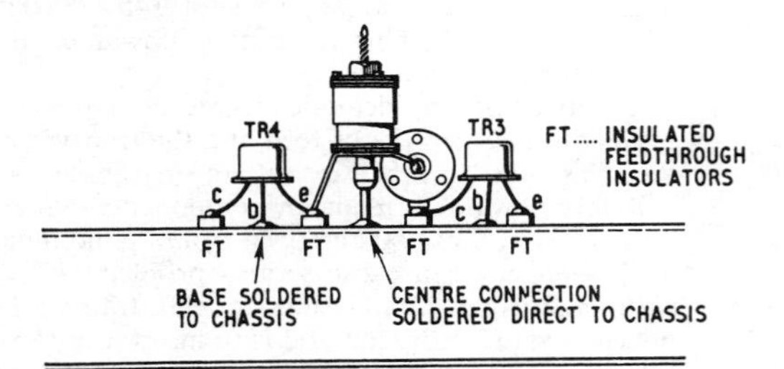

Fig 7.42. How transistors TR3 (tripler) and TR4 (first buffer amplifier) are mounted on the chassis in relation to their associated trimmers

The output stage has been designed to work into a 75Ω load and lamps which do not approximate to this resistance when hot may give a false indication of the output. A 6V, 60 mA type is probably best for initial tuning but it should be possible to light a 6V 0·1A bulb to the point of burn out when the circuit is peaked for maximum output.

Alignment

Alignment of the completed transmitter will be assisted by connecting a 6V, 60mA pilot lamp as a load across the output and by an absorption wavemeter tuning 24, 48 and 144MHz.

Unscrew all trimmers to the minimum capacitance position. Unscrew both slugs in L1 and L2 as far out as possible. Connect a 0 to 10V dc meter between C7 and the chassis. Apply positive 18V to the supply rail. Screw in the slug in L1 and adjust for maximum meter reading. This should be approximately 2V. Remove the meter and reconnect it

TABLE 7.4
Components list for the six-transistor exciter unit for the 144MHz band

RFC1, **RFC2**: 25μH.—90 turns of 36 swg enamel covered wire pile wound on a 1MΩ 1W resistor.
RFC3, **RFC4**: 3 turns of 23 swg on Radiospares ferrite bead, toroidal wound.
12 Lektrokit feed through bushes part No LK2121 }
12 Lektrokit soldering pins part No. LK3011 } Or Radiospares lead through insulators (fit 5/32 in hole).
L1 16 turns centre tapped 22 swg enamel covered wire on ¼in od former.
L2 8 turns 22 swg enamel covered wire on ¼in od former.
L3 5 turns 16 swg tinned copper wire ¼in id, ⅜in long.
L4 5 turns 16 swg tinned copper wire ¼in id, ⅜in long.
L5 4 turns 16 swg tinned copper wire ¼in id, ⅜in long.
1000pF feed through capacitors from Radiospares.

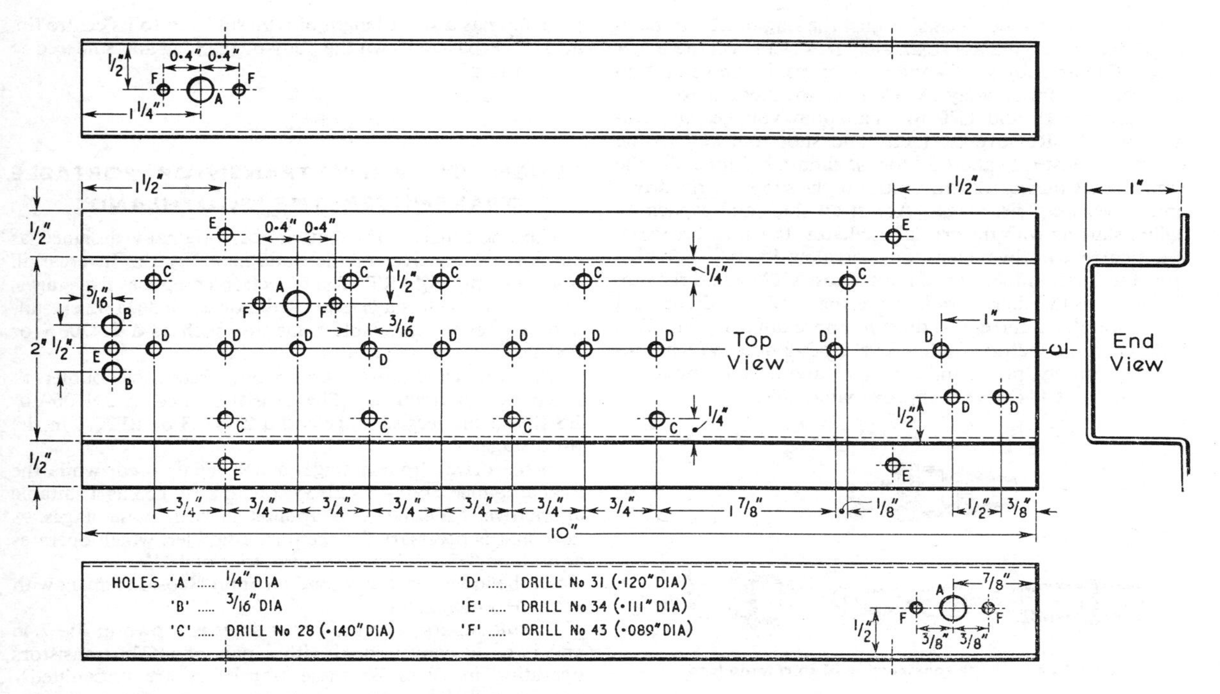

Fig 7.43. Drilling template for the six-transistor unit

Fig 7.44. Component layout diagram for the six-transistor unit

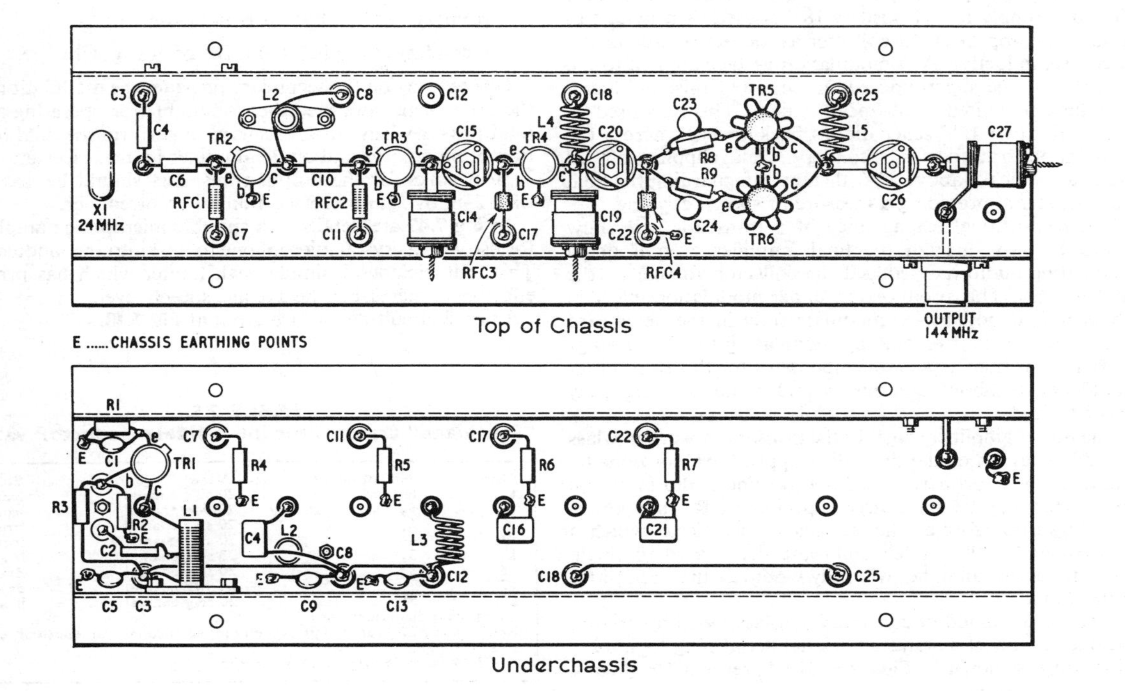

between C11 and the chassis. Adjust the slug in L2 for maximum meter reading, approximately 1·5V. Connect the meter across C17 and adjust C14 and C15 for maximum voltage on the meter, approximately 1V. Connect the meter across C22 and adjust C19 and C20 for maximum voltage, approximately 0·6V. Remove the meter and short out C22 to the chassis. Adjust C26 and C27 for maximum brightness in the lamp load. Connect a 200mA meter in the supply to the driver and power amplifier stages. Adjust all slugs and capacitors again, starting with the crystal oscillator, this time for maximum current in the meter, approximately 150mA. For high level modulation the short circuit across C22 should remain. Removal of the short should cause the combined driver and power amplifier current to drop to approximately half. This is the correct condition for low level modulation. With a positive 18V supply, power input to TR5 and TR6 is about 2W and output at 144MHz is approximately 1W.

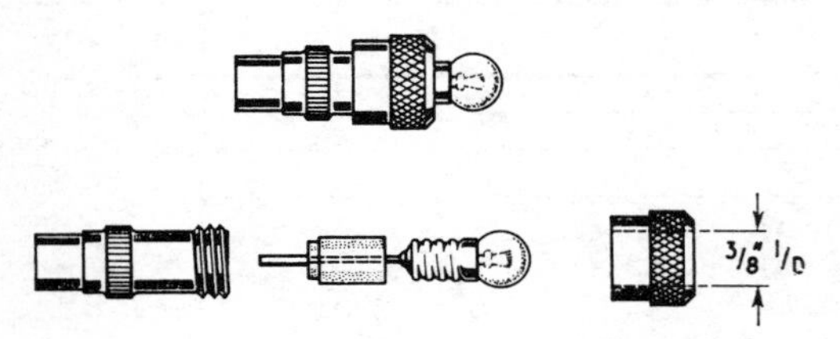

Fig 7.45. Shows the construction of an rf lamp load

Amplitude modulation may be applied to this little transmitter at the 144MHz driver stage as well as to the pa, TR5/6. It is most important that the maximum collector to base voltage rating (Vcbo) is at no time exceeded, in this case 60V. If a supply rail of positive 18V is used then twice this voltage can appear at the collector as the tuned circuits are, of course, inductive. Any modulation voltage applied to the collector will be superimposed on the top of this, and therefore must be limited to 24V peak to peak. This is assured by connecting two 12V zener diodes back to back across the modulation transformer secondary, thus clipping off all modulation peaks above 24V, thereby safeguarding the final transistors and providing a measure of speech clipping.

The feed-through capacitance of a transistor will allow power to pass through the final amplifier even if down modulating audio has reduced the collector voltage on the final to zero. This produces an under-modulation effect in which it is impossible to modulate fully in the downward direction. This is overcome by modulating the driver stage as well as the final. A suitable modulator for this transmitter would deliver about 2W output and could be completely transistorized.

The power amplifier stages in the transmitter work in Class B and low level modulation may be applied to them alone by removing the short across C22 and feeding audio in at this point. This may be via a large capacitor or R7 may be replaced by a transformer, the secondary resistance of which is approximately 10Ω. A few milliwatts from a small single ended transistor amplifier will fully modulate the transmitter at this point.

A suitable method of constructing a lamp load by drilling out one section of a standard coaxial aerial plug to hold a pilot lamp is shown in **Fig 7.45.** The lamp is a 6V 100mA type and has a short length of wire soldered to its centre tip, and this is passed down the body of the plug and soldered to the centre pin.

DESIGN "B": A FIVE-TRANSISTOR PORTABLE TRANSMITTER: THE "SOUTHLAND"

The "Southland" transmitter was originally designed as part of a transceiver of the same name by the Invercargill branch of the NZART. The rf section comprises five stages, the first of which is a crystal oscillator using 36MHz crystals and any readily available transistor such as a 2N706A or BSX20.

The second and third stages are both frequency doublers to reach the final frequency. The transistors used are 2N706A or BSX20 in the second stage and a 2N4123 or BFX44 in the third stage.

In the driver (fourth stage) a 2N3866 is used, while the final amplifier employs a 2N3553 with a BLY53 as a suitable alternative. Modulation is applied to both these stages. A heat sink is necessary for the final amplifier, which operates at an input of 2·5W to give an output of 1·5W.

All the inductors are wound on 5mm Neosid formers with suitable tuning slugs.

The modulator with the driver stage is shown in **Fig 7.46** and is quite conventional with a pair of OC28 transistors operating in Class B; these transistors are undoubtedly unnecessarily large for the output power required and if preferred may be replaced with an alternative smaller type.

The modulation transformer is a rewound standard 3W audio output type using the original former. Windings are as follows,

primary: 120 + 120 turns of 28swg wire.

secondary: 0/70/120/170 turns of 30swg wire.

One tapping on the secondary provides the modulation for the driver stage and the other is suitable for operating a 3Ω loudspeaker when the unit is used to perform the dual function of modulator and audio amplifier in a transceiver.

Heat sinks for the output transistors should be not less than 2½in by 1½in by 16swg aluminium or copper.

In **Fig 7.47** are details of a suitable microphone amplifier for use with a crystal microphone to feed into the modulator. This unit includes a simple peak limiter which has proved effective in increasing the average speech level.

Printed circuit details are given at **Fig 7.48.**

TABLE 7.5
"Southland" transmitter for 144MHz: inductor values

Coil	No. of turns	Wire	Slug
L1	8	20 swg enamel	Iron
L2	5½	20 swg enamel	Iron
L3	6½	20 swg enamel	Iron
L4	4½	20 swg enamel	Iron
L5	5½	20 swg enamel	Iron
L6	3¾	20 swg enamel	Brass
L7	2	20 swg enamel	Brass
L8	5½	20 swg enamel	Iron

all on 5mm diam. formers
RFC1 and RFC2 3 turns of 36 swg en looped through ½in length of ferrite beading.
RFC3 34 turns 30 swg en on ½W resistor.

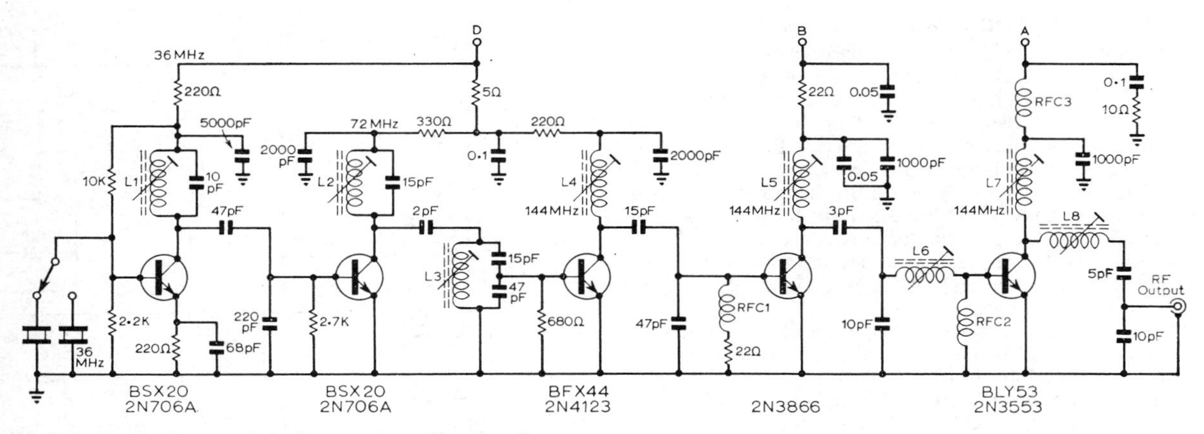

Fig 7.46. Circuit diagram of the five-transistor "Southland" transmitter

Fig 7.47. The modulator section of the "Southland" transmitter, showing microphone pre-amplifier (below) and driver-and-output stages (right)

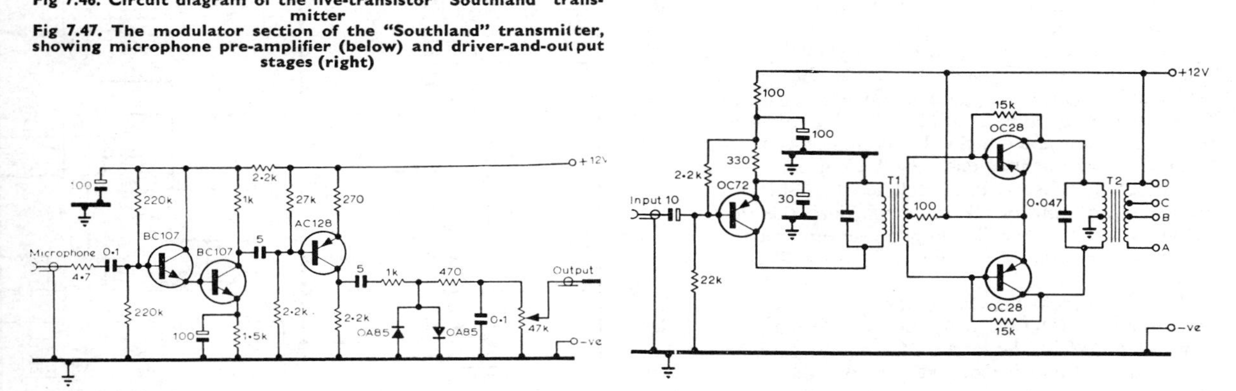

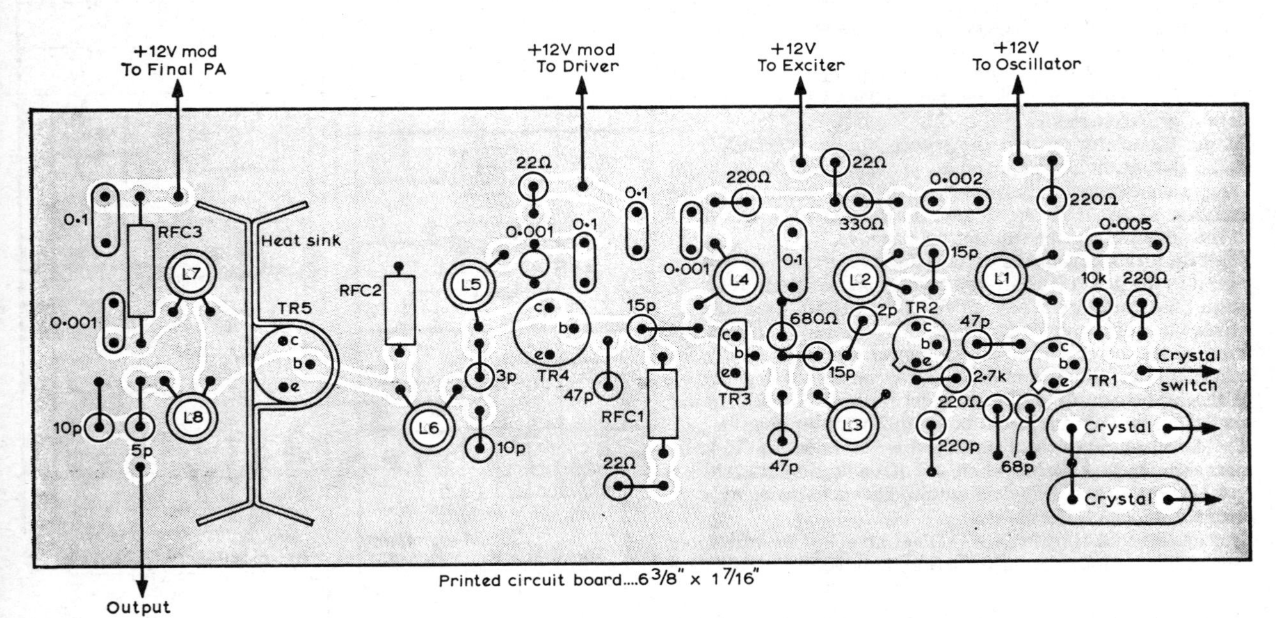

Fig 7.48. Printed circuit and etching details for the "Southland" transmitter

A 30W DUAL MODE TRANSMITTER FOR 144MHz

The use of relatively powerful narrow band frequency modulation has attractions for the amateur operator where other forms of modulation cause interference problems.

As described earlier, several methods of producing frequency modulation exist, but there seems little doubt that indirect phase modulation more readily complies with the true definition of narrow band fm. In the phase modulator the amount of phase shift (deviation) which can be obtained is dependent on the Q of the tuned circuit of the oscillator; the higher the Q the greater the deviation obtainable. An adequate tuned circuit can be made using standard coil formers with an appropriate dust core.

Many amateurs will not wish to confine their operations to one mode. Therefore facilities are provided in this 30W transmitter **(Fig 7.49)** to change from nbfm to series gate amplitude modulation at the touch of a switch. If the cw facility is preferred the ht feed to the modulator unit is switched out, and keying is effected via a relay inserted in the screen feed of V4.

In the unit described here a 12MHz crystal input is used. This is in the first half of a double triode. The second half of the double triode, the phase modulator, is tuned to the crystal frequency. Output to the following frequency multiplier is quite low and if higher levels are needed two separate pentodes should be used (eg, the 6AK5 or EF91).

The second stage (EF91) is tuned to 48MHz, using a double tuned transformer to couple into the third stage (E180F) with its anode circuit arranged in a balanced configuration to drive the following push-pull driver stage (QQV02/6).

The output from the driver stage is link-coupled to the power amplifier (QQV03/20A) but with 2W of drive available from the QQV02/6 the larger QQV06/40A or 7/50 may be used in the pa stage if required, provided the tuned circuits are adjusted to suit the different input capacitances presented by the larger valve. If a QQV03/20A is used the dc input will be about 30W.

Adjustable fixed bias is provided for the power amplifier so that, if desired, it may be operated in Class AB1 to reduce the emission of harmonics.

In the modulator section the speech amplifier (12AX7) with its output circuit arranged to provide suitable audio frequency characteristics may be switched either to the phase modulator V1 or to the series gate modulator V7/8, which feed the screen of the transmitter pa stage.

With the arrangements shown (Fig 7.49) the screen voltage supplied to the pa is relatively low: it may be raised by either returning the cathode resistor of V8 to the −100V available from the bias supply, or by connecting V8 anode to the higher voltage of the anode supply to the power amplifier. Under these conditions an input of 40/45W can be obtained. Note that the heater supply of V8 must be isolated because the potential on the cathode could be up to 250V above earth.

The construction of this transmitter is shown in **Fig 7.50.** Beneath the chassis the whole of the rf section is enclosed with a bottom cover. The power amplifier is enclosed by a U-shaped screen above the chassis.

The filament transformer (MT1) and relay for extension switching are mounted at the left and above the chassis; the

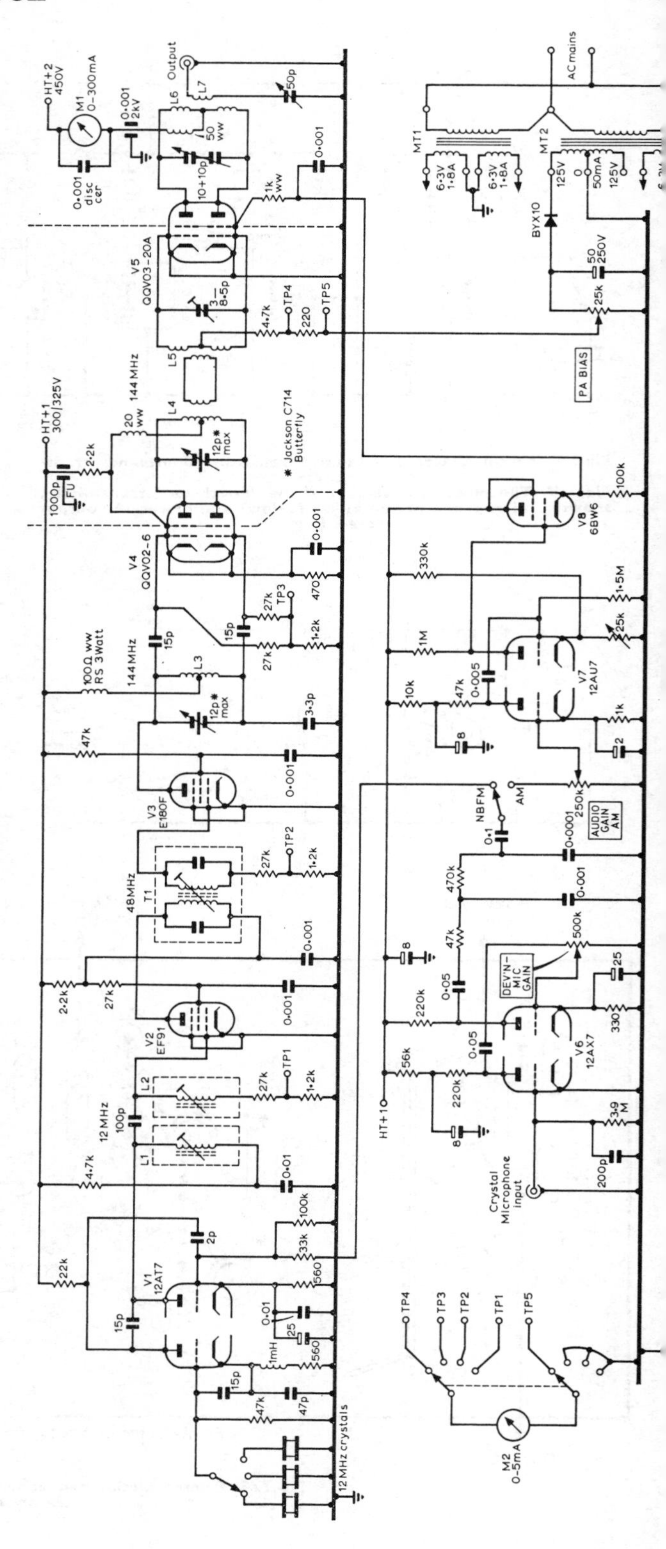

Fig 7.49. A standard five-stage 144MHz unit for use either as a driver for a high power final or as a small 30W transmitter for nbfm or series gate amplitude modulation

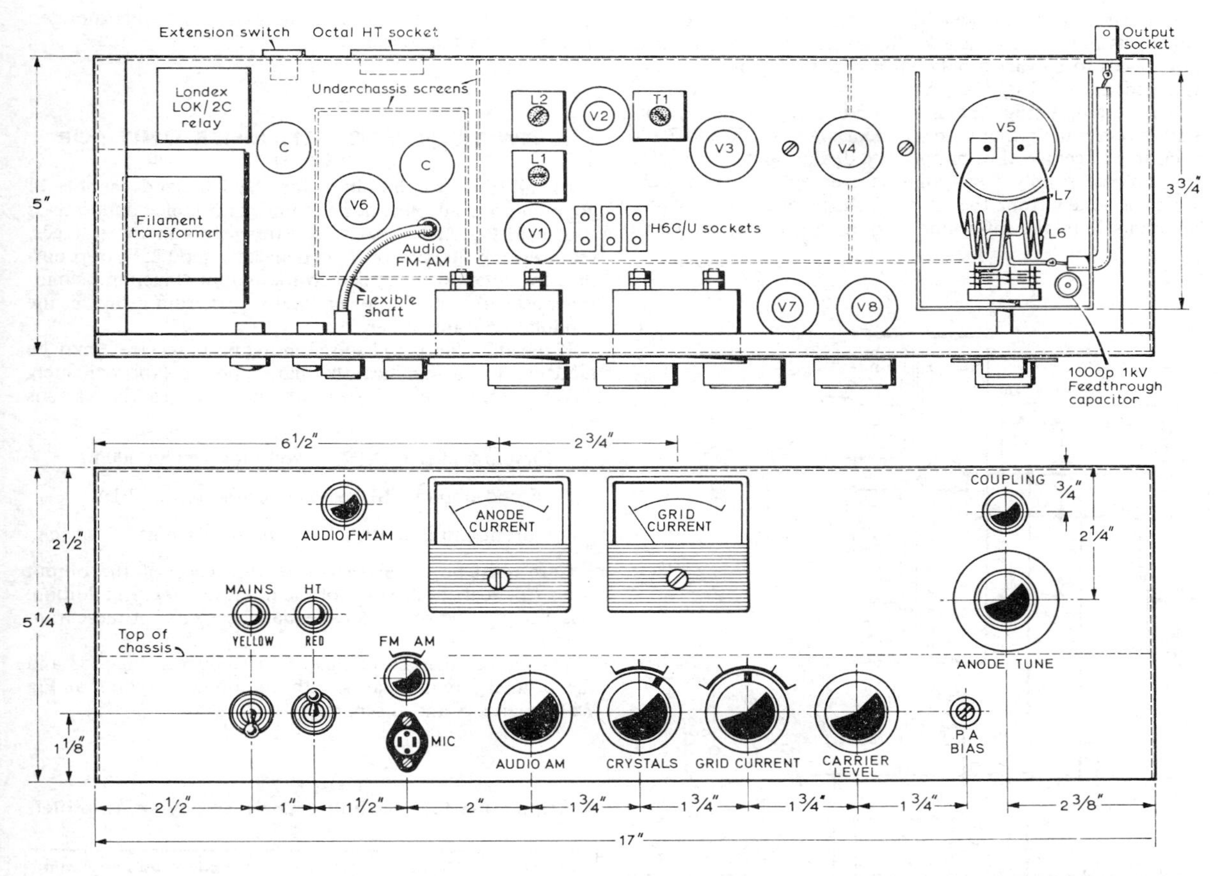

Fig 7.50. Front panel and top chassis layout for the 30W dual mode transmitter for 144MHz

bias and isolated heater supply transformer MT2 is mounted below the chassis.

The speech amplifier V6 and its components are enclosed in a separate screen **(Fig 7.51).** The gain control for this amplifier is connected by a flexible shaft to its front panel control. This may be left as a pre-set control if only nbfm is to be used.

The series-gate control valves V7/8 are mounted close to the front panel.

TABLE 7.6
Components list for the five-stage 144MHz exciter unit

L1, L2 25t 30swg on 7·5mm former close wound, with dust core, in standard ⅜in square can.
T1 7½t 24 swg on 7·5mm former close wound, primary and secondary closely coupled and an 18pF capacitor across each winding, all contained within a standard ⅜in square can.
L3 6t 18 swg 0·25 in inside diam., adjusted to length to resonate at 144MHz when tuned by 12pF trimmer.
L4 8t 18swg 0·25 in inside diameter, adjusted to length to resonate at 144MHz when tuned by the 12pF parallel capacitance, coupled by a single turn loop into L5 via either a short length of coaxial cable or twisted pair.
L5 4t centre tapped of 16swg 0·375 in inside diam. coupled from L4 with a single turn loop introduced into a gap 0·375 in in the centre of the coil.
L6 4t centre tapped of 14 swg 0·5in inside diam, with ½in gap in the centre to accommodate the aerial pick-up coil L7.
L7 one turn of 16 swg.
MT1 Radiospares "Hygrade" 6·3V at 1·8A and 6·3V at 1·8A.
MT2 Radiospares "Midget" 125-0-125V + 6·3V at 1·2A.

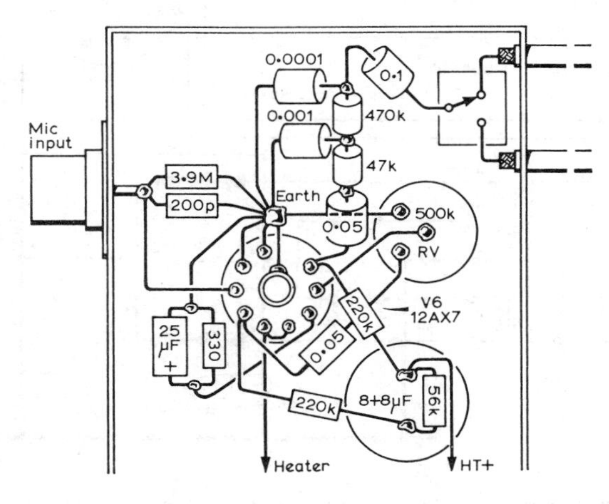

Fig 7.51. Layout of the speech amplifier stage V6

In the front panel layout (Fig 7.50) the power amplifier bias control is not fitted with a knob because once set the bias level needs to be varied only when operating conditions are changed, eg, to Class C or AB1.

The ht supply **(Fig 7.52)** is housed in a separate unit, consisting of a transformer with secondary taps at 350 and 250V each side of centre, thus providing a dual output source. The low voltage supply feeds all stages except the pa anode, which is taken from the higher voltage output. To avoid accidents both the input and output to this psu are taken from the transmitter unit, and the switching is incorporated on the front panel layout.

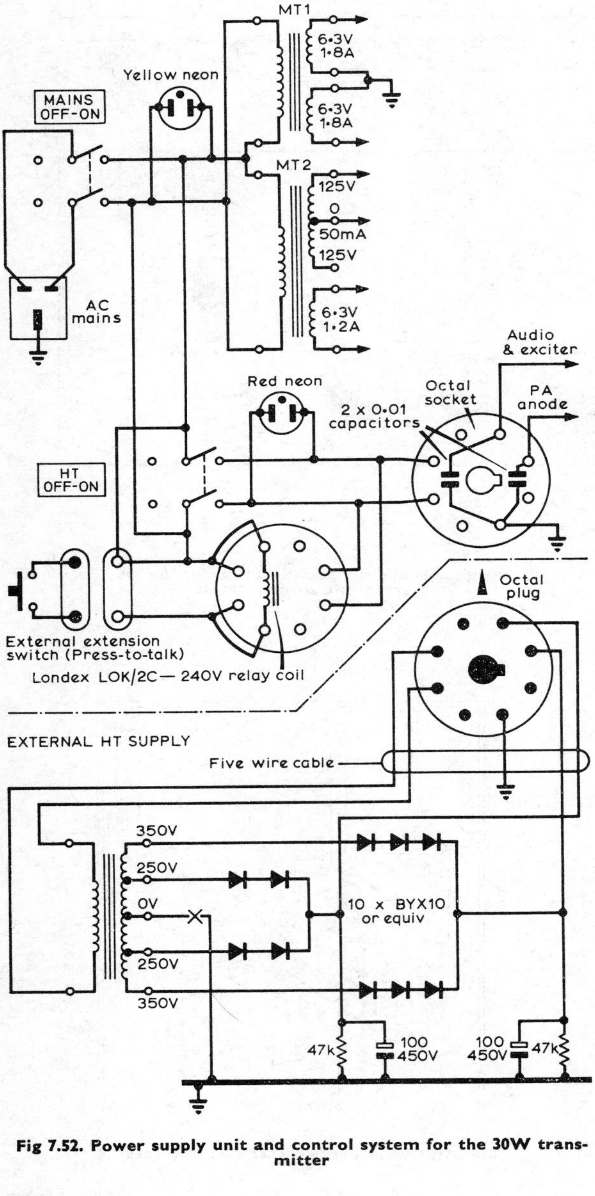

Fig 7.52. Power supply unit and control system for the 30W transmitter

A 25W TRANSISTOR AMPLIFIER UNIT FOR 144MHz

An all-transistor amplifier for the 2m band capable of giving a good account of itself under portable conditions is shown in **Fig 7.53.** It uses four transistors in three stages, requires only 300mW of rf yet delivers a good 25W of rf output. Modulation is applied to the output stage, and simultaneously to both the lower level stages from a tap on the modulation transformer.

If the amplifier is to be used in the nbfm service it may be adjusted for a significantly higher power (approximately 40W) output level. The maximum inputs to the various stages are:

First amplifier BLY97 collector current 400mA

Second amplifier BLY93A collector current 2·1A

Output amplifier BLY93A collector current 2·7A each.

Care must be taken to ensure that each of the output transistors carries its share of the input loading. In addition, it is essential to provide an adequate heat sink for these transistors.

The circuit diagram of this unit is shown at Fig 7.53 and the printed circuit boards with component layouts at **Fig 7.54,** boards A and B separately.

TABLE 7.7
Components for the four-transistor power amplifier unit

Inductor	Turns	Diam (inside)	Length	Wire size in mm
L1	2	5mm	9mm	1·25
L2	3	6·3mm	9mm	1·25
L3	2	8·3mm	5mm	1·25
L4	2	6·3mm	8mm	1·25
L5	3	6·3mm	9mm	1·25
L6	2	6·0mm	8mm	1·25
L7	2	6·0mm	8mm	1·25
L8	3	6·3mm	9mm	1·25
L9	2	8·5mm	6·5mm	1·60
RFC1	2½	0·4mm on FX1115 bead		
RFC2	1½	0·4mm on FX1115 bead		
RFC3/4	1½	0·5mm on FX1898 bead		
RFC5a, b, c	2½	0·5mm on FX1898 bead		

Capacitors: C4, 11, 19, to be 68nF Mullard type C280/P68K; C7, 14, 20 to be 100nF Mullard type C280/P100K; C1, 2, 8, 9, 15, 16, 24 and 25 to be 4-40pF Mullard type 809.08002. Other components may be standard types.

The printed circuit uses double sided fibre glass board. The circuit is etched on the top side and all the components are soldered to the top surface. Only the transistors and feed-through capacitors C5, C12 and C21 go through the underside. The method of securing the transistors with the heat sink attached to their fixing studs is illustrated in **Fig 7.55.** To avoid short circuits the through connection of the feed-through capacitors is cut and filed so that there is clearance between them and the attached heat sink. It may be an advantage to cover the underside of the capacitors with suitable insulation.

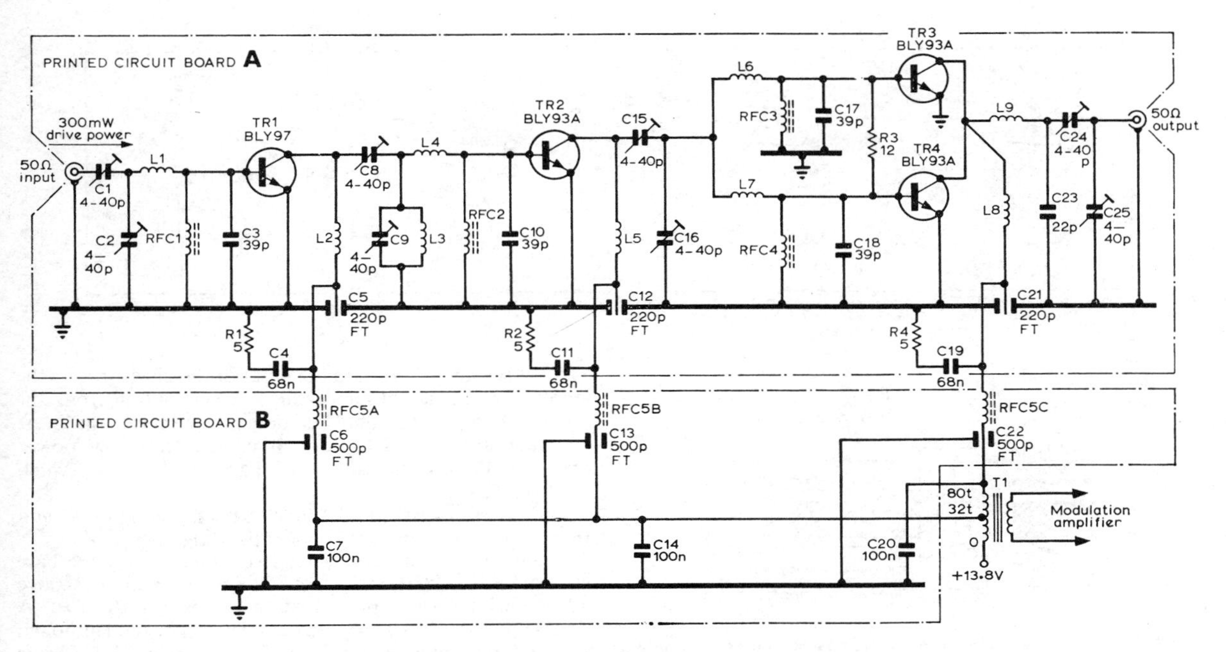

Fig 7.53. A four-transistor power amplifier unit capable of delivering 25W of rf under a.m. conditions, or 40W under nbfm conditions

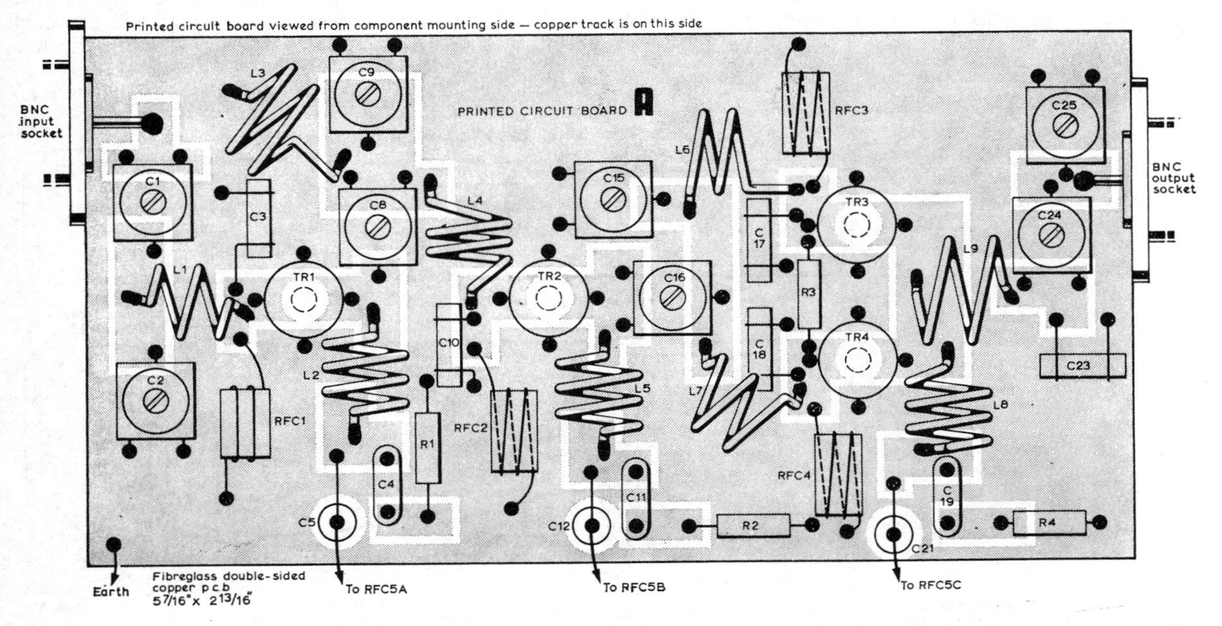

Fig 7.54. Printed circuit board layouts for the 25W transistor amplifier for the 2m band (a) for the upper section of the diagram at Fig 7.53 and (b) for the lower section overleaf

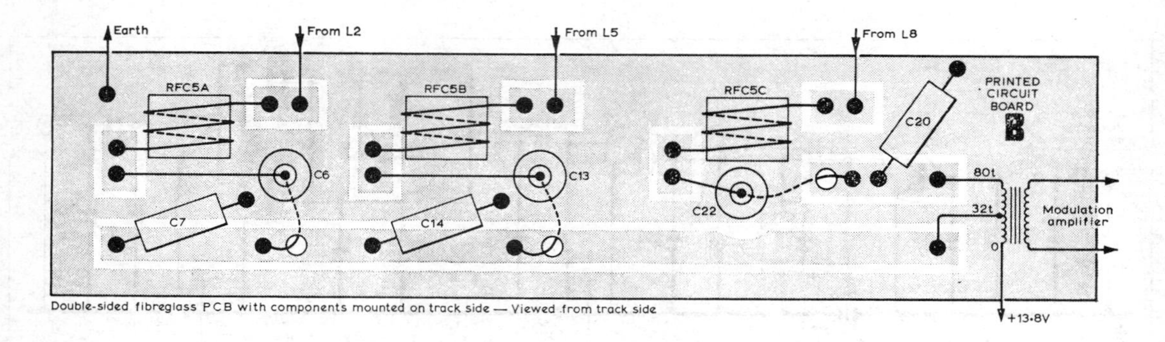

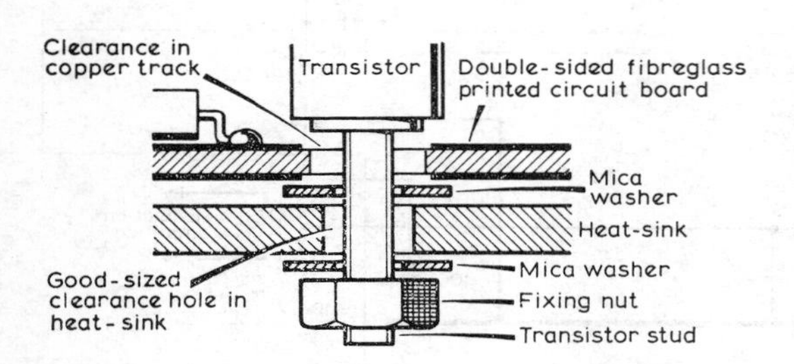

Fig 7.55. Method of securing transistors and heat sink when building the 25W transistor power amplifier for 144MHz shown on p 7.29

A 144MHz LINEAR AMPLIFIER FOR PORTABLE OR MOBILE USE

For constructors who prefer a valve rather than a transistor amplifier to add to an existing exciter unit the following design will be found to incorporate features that permit its operation under portable or mobile conditions from a 12V supply. This linear amplifier delivers 50W output for 2W of drive from a transceiver that may be used without modification. **(Fig 7.56).**

An rf sensing network in the input of the linear inserts the amplifier in the aerial line when it detects output from the exciter. Otherwise the input and output terminals of the linear amplifier are a straight-through connection. Low power operation requires nothing more than just simply to switch the linear off.

The active device is a QQV06/40A push-pull pa valve, and there is a built-in inverter supply operating from 12V.

Incoming rf from the exciter is rectified to provide base current for the 2N3053 which operates relays A and B; to give adequate input-to-output isolation two relays are used. A double-pole double-throw coaxial relay can be used although for economy small plug-in relays may be preferred. Contacts A1, B1 introduce the amplifier into the line. Contacts A2, B2 cause the dc/dc converter to start generating the plate and screen voltage. A simple transistor/zener regulator in the cathode circuit provides −27V of fixed bias for linear operation, and rf output is monitored by a meter connected to an anode power detector. When drive is cut off and relays A and B release, contact A2 shorts the BD130Y bases to stop the dc/dc converter. The normally closed contact of B2 could supply to a separate receiver pre-amplifier. The insertion loss is less than 0·5dB.

Although the primary application of this amplifier is as an

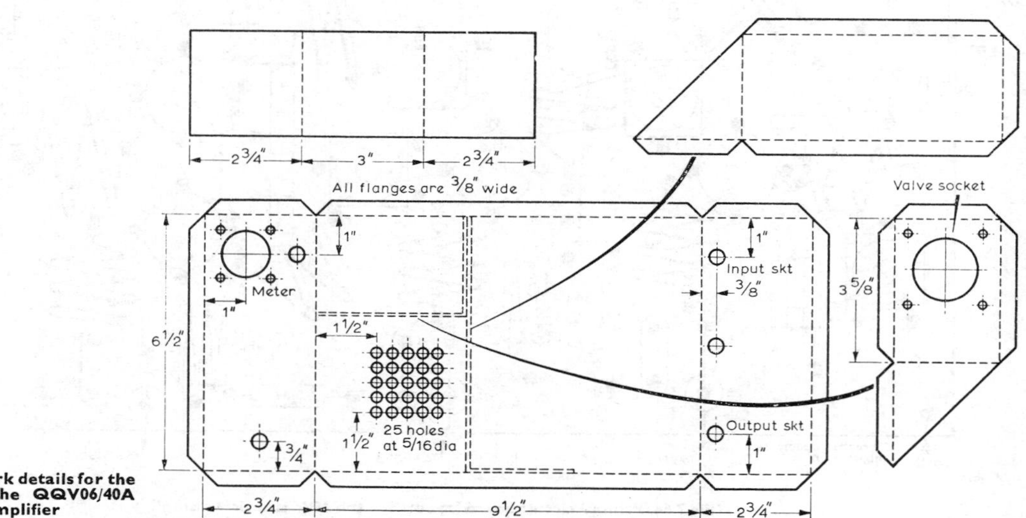

Fig 7.57. Metalwork details for the case to house the QQV06/40A power amplifier

Fig 7.56. A high power 2m pa unit for portable or mobile use. C1 2—16pF; C2 10 + 10pF; L1 4t ⅜in diameter; L2 4t ½in diameter. "M" is 0—1mA meter; D to be 1N4007 or equivalent. In the diagram "X" denotes feed-through insulator or high voltage feed-through capacitor. Winding data for the inverter toroid as follows:

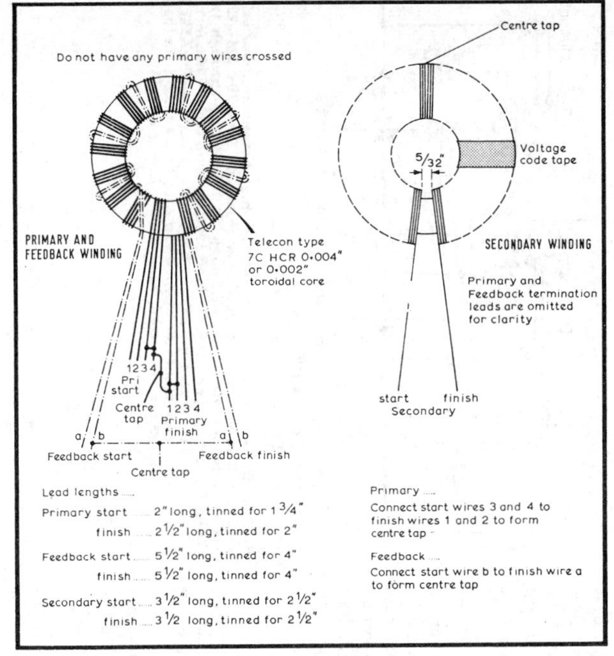

1. Use a Telecon type 7C HCR 0·004in or 0·002in toroidal core;
2. Seal the inside face of the toroid by applying a 4½in length of ½in wide No. 1409 Sellotape polythene electrical tape;
3. Wind the primary, which consists of 4 lengths of 18 swg self-flexing wire wound quadrafilar in one layer of 19 turns. Ensure that no wires cross. (If four 5ft lengths are used this will allow about 7in of leads at start and finish);
4. Wind on the feedback, which consists of two lengths of 24swg self-flexing wire bifilar wound 9 turns over the primary as shown.
5. Over the primary and feedback wind a layer of ½in wide 0·003in Melanex (approx 5ft 6in long). Each turn is to overlap the previous one by one-quarter or one third its width;
6. Wind on the secondary to the number of turns and wire gauge given for the particular transformer. The secondary should be progressively wound starting at the beginning of the primary. There should be no bunching of turns at any point on the circumference. There should be a gap of at least 5/32in between the start and finish of the secondary winding. The centre tap is to be half way around the coil.
7. Cut and tin all wires (on all transformers) to the lengths shown. When tinning the wires wipe the solder clean, as these wires have to fit through turret lugs.
8. Apply a band of coloured pvc or polythene electrical tape over the secondary to indicate the secondary voltage: 360V, orange; 460V, yellow; 500V, green.

NOTE: All windings are to be in the same direction as shown.

Secondary Turns: for 500V, 840 turns, centre-tapped, 30swg self-flexing wire.

a.m. or ssb linear, it will of course amplify nbfm. In the case of nbfm it can be operated in Class C if a 12kΩ 1W resistor is inserted between the centre of the grid coil and earth, and the input circuit of the amplifier adjusted to draw grid current and thus increase the bias. In this mode 50W of output can be obtained for 2W of drive.

The linear amplifier is housed in a case 6⅝in by 9⅝in by 2¾in. The 20 gauge aluminium metal work details are given in **Fig 7.57.** The power supply transistor mounting bracket is of 16 gauge metal. The internal bulkheads may be given an additional supporting flap if desired. The toroid transformer is mounted between two thick pieces of cardboard to reduce acoustic noise. Potting in Araldite (or equivalent) is not necessary.

Setting-up Procedure

(1) Switch on: check that the filament of the valve is alight;
(2) Using a gdo, tune the input coil to 145MHz with input link out;
(3) Short circuit the collector of the 2N3053 to ground and verify that the inverter starts: if not try reversing the base connections. If it still fails to start try reducing the 47Ω biasing resistor to 39Ω, 5W;
(4) Check that there is 500V on the anode, 250V on the screen, and 27V on the cathode of the valve;
(5) Check that there is no deflection on the meter;
(6) With about 500mW input, and the link withdrawn from the grid coil, tune the pa output for maximum deflection (should be about 1/10 on meter);
(7) Leaving the input link where it is, adjust the input link capacitor for a new maximum (say, one fifth);
(8) With swr bridge in input line push the link in for minimum swr (NB: not minimum reverse power);
(9) When adjusted for good linear operation with an a.m. exciter unmodulated output is 25W, which is approximately four fifths of the full scale deflection on the meter.

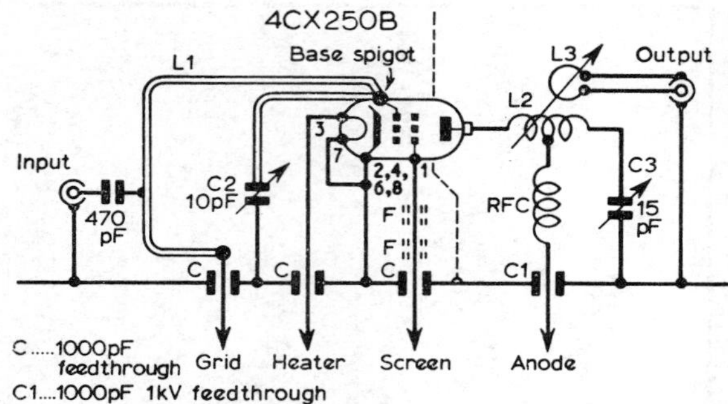

Fig 7.58. Circuit of the compact 144MHz high power amplifier using a 4CX250B valve cooled by a blower (Plannair Type 2PL 321—284C Mk 3). C denotes 1000pF feed-through, C1 1000pF feed-through 1kV, C2 10pF Jackson type C804, C3 15pF Jackson type C804, F is a ferrite bead, RFC 50 Ω wirewound 5W, L1 copper strip loop (see Fig 7.59), L2 3½t coil ⅜in inside diameter of ⅛in diam copper, L3 1t coil ⅜in inside diameter, insulated, valve socket Eimac or AEI

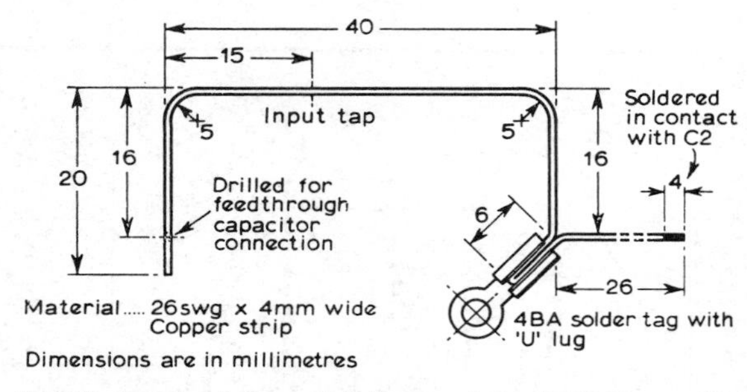

Fig 7.59. Details of the grid inductor of the 4CX250B amplifier

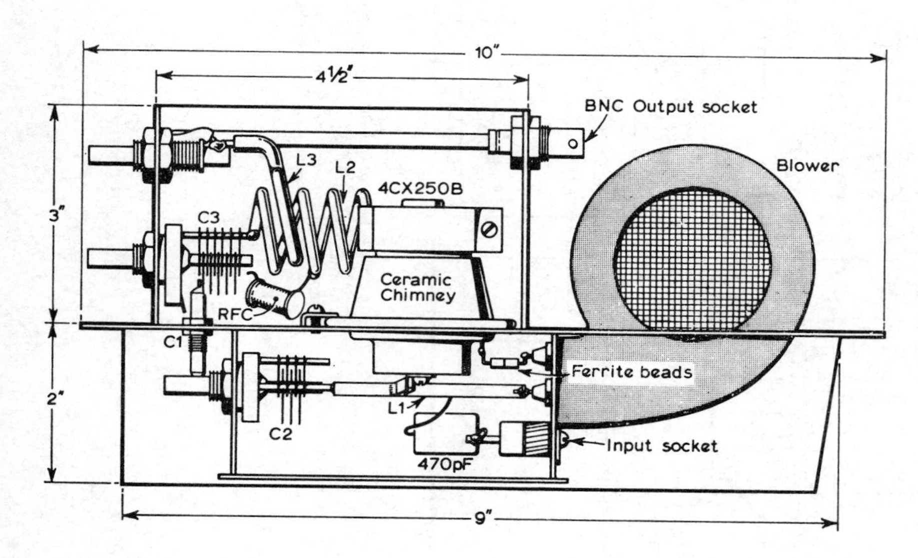

Fig 7.60. Side view of the compact 144MHz amplifier

A COMPACT GENERAL PURPOSE 150W POWER AMPLIFIER FOR 144MHz

The compact "add on" 144MHz amplifier at **Fig 7.58** is capable of operating under varying conditions, such as Class AB1 for linear amplification of an ssb signal or Class C for telegraphy. Using a single 4CX250B, it requires a power supply unit delivering 800V at 300mA, a bias line of approximately −130V variable, and an lt supply of 6·3V at 2A. A separate screen supply of 250V should be rated to deliver about 10mA, and there should of course be a supply for whatever blower is employed (quite a modest sized blower will suffice when the valve is run at the 150W power input level of the UK licence).

The circuit shows the simplicity of the arrangement, which is due to G6JP. The grid inductor L1 consists of a copper loop tuned by the capacitor C2 with the input tapped down from the grid. The anode circuit is a series tuned half wave circuit with the ht feed joint at the centre of the inductor L2. The output coupling consists of a well insulated single turn, which is adjustable.

The grid input circuit below the chassis must be fully enclosed to ensure the cooling air passes through the valve cooler. The valve and its anode circuit is enclosed by the box above the chassis with its top left open. If the amplifier is to be built into a cabinet, free airflow from the top of the valve must be provided and an adequate air inlet is necessary.

The blower used causes very little vibration, and no shock absorbent mounting is needed. It is mounted so that the air outlet is directed into the grid circuit enclosure under the chassis, by a suitable cutout in the chassis. The blower is mounted by a simple U-shaped strap around its motor.

The input socket used is a standard tv type but that used for the output is a bulkhead mounting BNC.

Performance

Anode voltage	750	800	V
Anode current	200	200	mA
Screen voltage	250	250	V
Screen current	5	8	mA
Grid voltage	−100	−100	V
Grid current	6·6	8	mA
Drive power	2·6	3·0	W
Output power (load)	90	100	W

A SINGLE SIDEBAND TRANSVERTER FOR 144MHz

Broadly, there are two methods of generating single sideband (A3J) for the metre-wave bands: either by means of equipment built specially for the purpose, or by a transverter in which ssb from an existing hf transmitter is mixed with the output of a crystal oscillator chain to provide signals on 144·3MHz.

In any mixing process there exists the danger of generating families of unwanted signal products as well as the wanted one, as discussed earlier in this chapter. Consequently, in the design now to be described, where 28MHz is mixed with the output from a 116MHz oscillator chain, enough selectivity is built into the system to ensure that only a low level of unwanted products appears in the output.

The block diagram at **Fig 7.61** shows how the 28–30MHz ssb output from an existing hf bands unit is mixed with the 116MHz oscillator output to give full coverage of the 144–146MHz amateur band. Although it might appear that the 116MHz frequency could be arrived at by the use of a lower frequency crystal multiplied up, this is an unwise procedure: it cannot be too strongly recommended that the fixed oscillator should originate at the final frequency using a fifth overtone crystal.

It is also important that the variable frequency ssb drive to the transverter should be taken from the driver stage of the ssb unit in preference to the final amplifier. Most hf band transmitters provide facilities for low level output of this kind.

The local oscillator (116MHz fifth overtone crystal and a BSX20 or equivalent transistor) is shown at **Fig 7.62.** It is mounted on copper laminate and enclosed in a screening box. The output is taken from a coaxial socket to a separate two-stage Class A valve amplifier **(Fig 7.63)** to raise the output

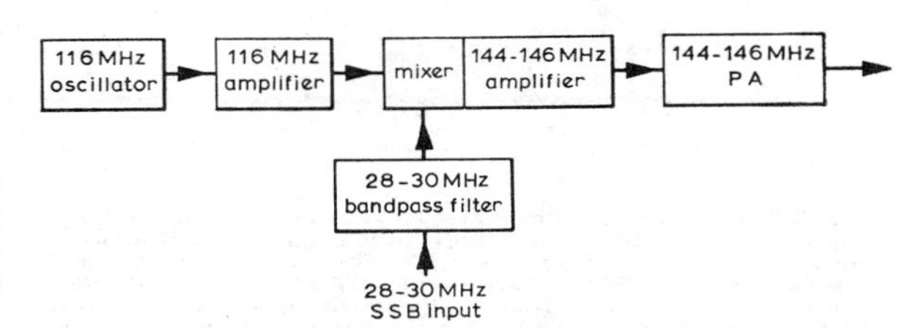

Fig 7.61. Block diagram of 144MHz transverter in which ssb input at 28MHz to 30MHz is mixed with the output from a 116MHz oscillator chain to deliver output in the 2m band. A bandpass filter is interposed between the ssb generator and the mixer

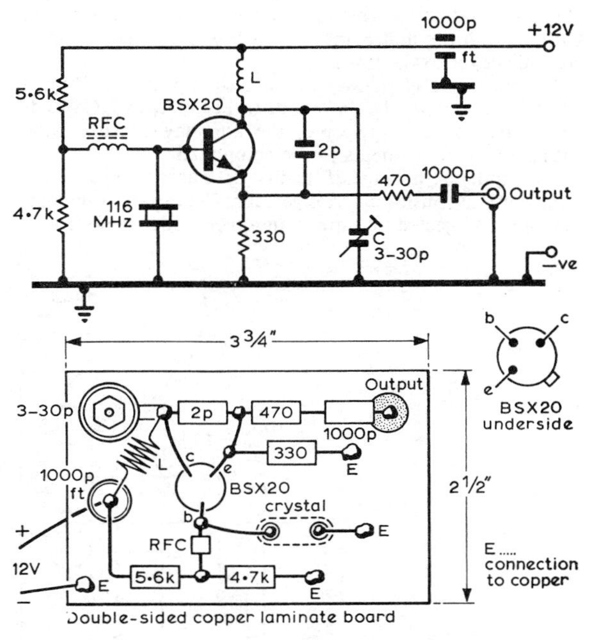

Fig 7.62. The circuit diagram and the electrical layout of the 116MHz oscillator. The inductor L requires 4t of 22swg on ¼in former, the winding to be ⅜in long. The choke rfc has 2t 30swg wound on a ferrite bead. Construction is on copper laminate board

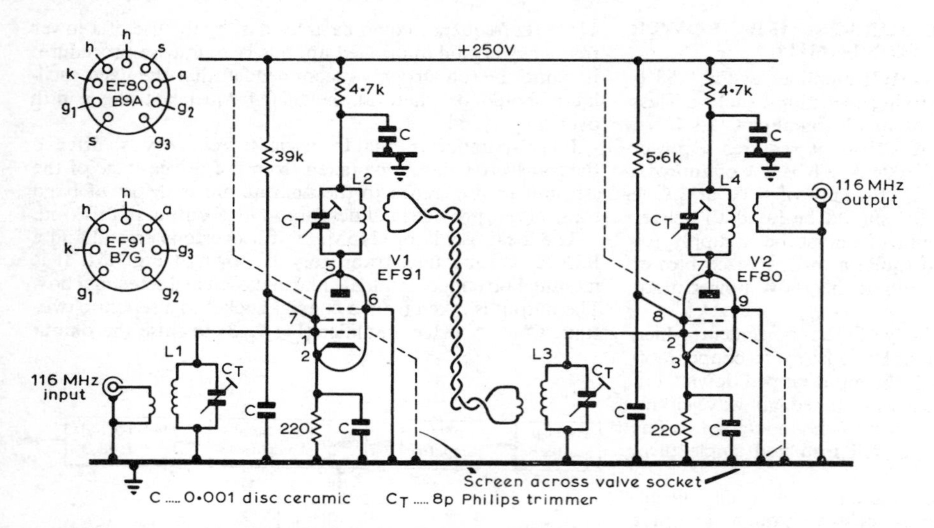

Fig 7.63. The 116MHz amplifier accepts the output from the transistor crystal oscillator and delivers it to the input of the 3/10 mixer (below). V1 is an EF91 or equivalent, V2 EF80 or equivalent screened pentode/tetrode, C 1000pF disc-ceramic, CT 8pF Philips trimmer. Inductor values are: L1 $4\frac{1}{2}$t $\frac{3}{4}$in long, 20swg self supporting, $\frac{1}{2}$in diam; L2 $4\frac{1}{2}$t $\frac{1}{2}$in long, 20swg self supporting, $\frac{1}{2}$in diam; L3 4t $\frac{3}{4}$in long, 20swg self supporting, $\frac{1}{2}$in diam; L4 6t $\frac{5}{8}$in long, 20swg self supporting, $\frac{1}{2}$in diam. Link coils 2t each at cold end of L2 and L3

level to suit the mixer requirements. Both stages have tuned grid and anode circuits with link coupling between the stages. To ensure stability a screen is fitted across each valve socket. Output from the unit is taken by a link coupling through a coaxial socket to the third unit.

The third unit (**Fig 7.64**) contains the mixer and Class A amplifier tuned to the 2m band. In both stages QQV03/10 valves are used: if preferred, the 2/6 may be substituted with suitable adjustments to the component values.

The input circuit stage of the first 3/10 accepts 116MHz in a conventional push-pull arrangement. The variable frequency ssb signal is injected into the common cathode connection.

From the output of V1 mixer a standard push-pull circuit at 144–146MHz is inductively coupled to the untuned input circuit of the 3/10 amplifier stage.

Between the ssb exciter output and the 3/10 input a band-pass filter covering the frequency range 28–30MHz is interposed to minimize unwanted products. With careful adjustment the filter presents a very low insertion loss. The link couplings to it should be adjusted to ensure a level adequate for optimum mixing. See **Fig 7.65.**

The final unit is a Class AB linear amplifier using a QQV06/40A or 7/50 as shown at **Fig 7.66.** Alternatively, the 4CX250B shown elsewhere in this chapter could be used as the final.

Fig 7.64. The mixer is a 3/10 accepting 28MHz input at the cathode and 116MHz at the push pull grids to produce 144MHz single sideband, which is then amplified by the following 3/10. The inductors are: L1 one turn 18swg insulated $\frac{5}{8}$in diameter; L2 4t as above to make $\frac{5}{8}$in length; L3 as L2; L4 2t 18swg $\frac{5}{8}$in diameter, $\frac{5}{8}$in long; L5 4t 18swg $\frac{5}{8}$in diameter, $\frac{5}{8}$in long; L6 one turn 18swg insulated, $\frac{5}{8}$in diameter; C 1,000pF disc ceramic, and CT 20 + 20pF

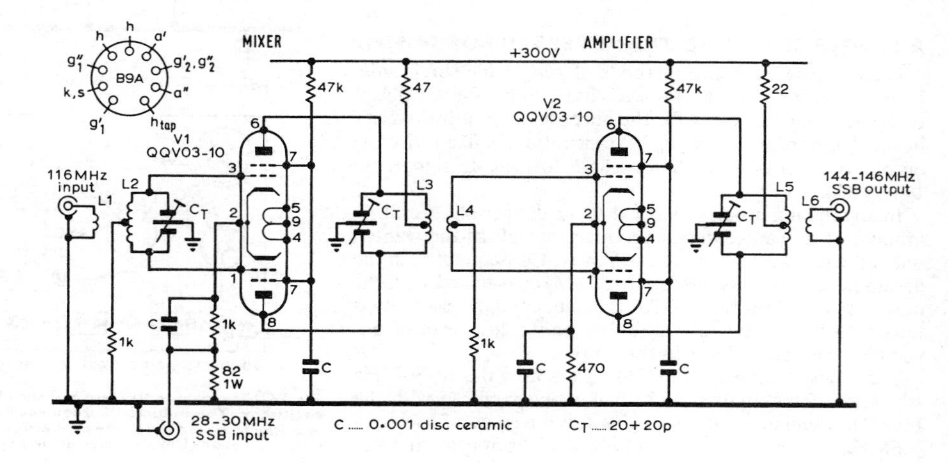

Setting up procedure: The oscillator tuned circuit LC (Fig 7.62) is adjusted to 116MHz using a gdo. When the supply voltage is applied the oscillator current is likely to be about 15mA; when the tuned circuit is retrimmed, at resonance with the crystal a sharp dip to 11–12mA will occur. Check that the frequency is 116MHz.

Next, tune all the circuits of the amplifier (Unit 2, Fig 7.63) to 116MHz. Connect the oscillator output to the amplifier, apply ht and retrim all the circuits for maximum output as shown by an rf indicator plugged in to the output socket of the amplifier unit.

Having set up the oscillator and amplifier stages to 116MHz adjustment of the mixer is the next operation. Correct adjustment of this stage is essential for satisfactory production of 2m ssb via the transverter. First adjust V1 mixer input circuit (Fig 7.64) for maximum delivery at 116MHz. Switch off the oscillator and its amplifiers, apply ht to the mixer and check its anode current (about 15mA). Now switch on oscillator and amplifiers; the anode current should rise to 40–50mA. Reduce couplings so that this falls to 30mA.

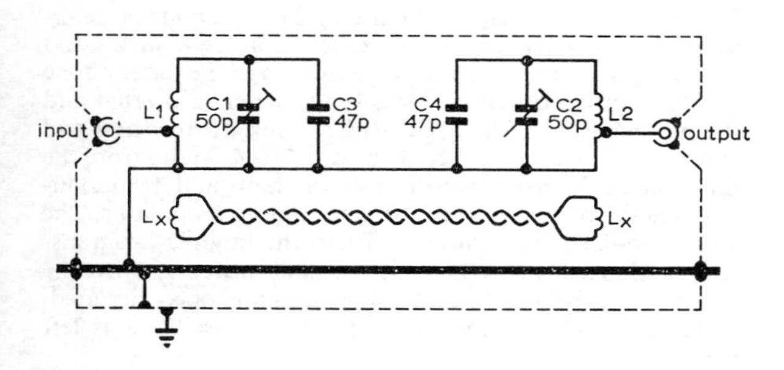

Fig 7.65. The 28-30MHz bandpass filter, bottom coupled by a single turn link. Both L1 and L2 have 7 turns of 20swg wound on a ⅜in former to make ⅞in length. Each is tapped one turn up from the earthy end. Each variable capacitor is 50pF shunted by 47pF silver mica

Now resonate the mixer anode circuit and the following amplifier circuits to 144MHz and insert the rf indicator into the output socket of the 3/10 amplifier (V2, Fig 7.64).

Now connect the ssb output to the socket in the cathode circuit of the mixer, insert a little carrier so that the output is about 0·5W and observe the rise in the anode current of the mixer. This should be about 1mA and not more than 2mA. Apply voltage to the amplifier and adjust both mixer and amplifier anode tuning to give maximum reading in the output indicator. The output should be in the region of 1–2W. At this point check the mixer/amplifier stages for self oscillation. Do this by switching off oscillator and ssb drive and tune both anode circuits while watching for changes in anode current. Adjust couplings for maximum output with both oscillator and ssb drive connected and ensure that there is no grid current present in the amplifier stage. With either the oscillator or the ssb input switched off the output indicator should show no output.

The linear amplifier (Fig 7.66) follows normal practice. With 750V on the anodes the standing anode current should be set at 35 to 40mA by means of the bias control potentiometer. Full carrier insertion from the ssb exciter to give an increase of 1 to 2mA in the mixer anode current will drive the anode current of the linear amplifier up to about 200mA at resonance. Do not run at this level for anything but very short periods or the valve will be damaged. With normal speech the anode current of the linear amplifier will peak up a little over 100mA and the grid current may peak up to 0·5mA; it should not be allowed to exceed 1mA.

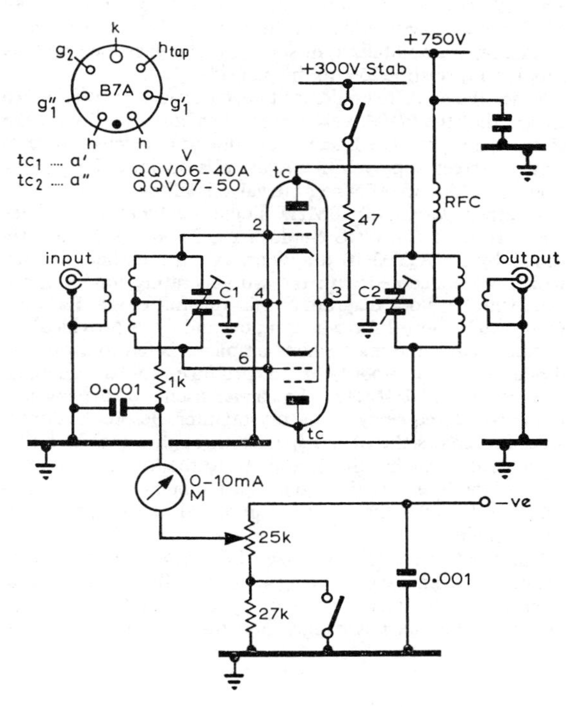

Fig 7.66. The power amplifier is a QQV06/40A or 7/50. The grid meter M should provide a reading of 10mA. The choke RFC may be a 50Ω wirewound resistor capable of carrying the full anode current of the valve. For linear operation the screen supply will need to be stabilized, and biasing facilities provided (bottom left hand corner) to set the negative grid voltage at the appropriate value. The grid input and anode output capacitors are 12pF + 12pF balanced variables, and the associated inductors will require approximately four turns of 16swg copper wire ⅝in diameter wide spaced to give coverage over 144-146MHz

These tests should be made using a dummy load; the couplings should be reduced to a minimum required to ensure that all tuned circuits are "on the nose".

As a linear amplifier in the quiescent state, ie with no drive, may act as a very effective noise generator some means must be used to prevent this during listening periods: either a relay operated by the send/receive switch on the ssb unit can be used to open the screen supply to the amplifier or an additional resistor in the bias supply can be introduced and switched to increase the negative bias and thus cut off the anode current of the linear amplifier.

SINGLE SIDEBAND ON 70MHz

The principles described above for generating ssb (A3J)on the 2m band may be applied to the other metre-wave bands in use in the UK. For 70MHz (the British amateur's lowest metre-wave allocation) a transversion method suggested by G3KQR makes use of a high level mixer (5763) which accepts the output from a crystal chain at 49MHz in the grid circuit and the output from a 21MHz single sideband source in the cathode circuit across a 68Ω cathode resistor, to produce 70MHz ssb in the 5763 anode. This is fed to the balanced input of a Class A buffer amplifier such as a QQV03/10 or 2/6. This delivers 5W of rf into a final linear amplifier (QQV06/40A) operated with 600V on the anode, 250V to the screen from a separate stabilized ht supply, and a variable grid bias supply to furnish −25V to the pa grids.

When the pa valve holder is fitted, care must be exercised to ensure that the 6/40A valve internal shield is level with the chassis platform. The linear is disabled on "receive" by cutting the screen supply with a relay; this obviates the generation of "white noise" when the valve is idling.

Another design of 70MHz single sideband transmitter which is due to G3WOS employs a QQV06/40A linear preceded by a QQV03/10 amplifier. All the earlier circuits, which are concerned with the ssb generating function, are solid state. A block diagram of this transmitter is at **Fig 7.67.** One switch control for ssb, a.m., cw and fm is provided.

Audio from the microphone amplifier is fed to a double-diode balanced modulator to produce double sideband centred around 1·4MHz. The carrier oscillator driving it is matched, in frequency, to the crystal filter. Before the double sideband enters the filter it passes through a mosfet buffer stage which matches the modulator to the low input impedance of the filter. As the signal is attenuated when it passes through the filter, the output is amplified in a single transistor amplifier.

The 1·4MHz ssb is now mixed in a balanced mixer with the output from a vfo covering 4·9 to 5·9MHz, producing an intermediate frequency range of 6·3 to 7·3MHz. The tuned circuit in the mixer is ganged and tracked with the vfo to reduce any spurious mixer products. This ssb component at 6·3 to 7·3MHz passes to a second mixer where it encounters the output of a crystal chain at 63·7MHz. The resultant 4m single sideband signal passes through an amplifier with a high-Q tuned circuit in the collector, and thence to a QQV03/10 buffer amplifier in Class A which feeds the QQV06/40A final amplifier.

TABLE 7.8
The multi-mode transmitter for 70MHz: coil winding data

L1 40t 30g on ¼ in slug-tuned former
L2 primary 40t 30g on ¼in slug-tuned former secondary 10t 30g
L3 primary 10t + 10t bifilar wound on ¼in slug-tuned former total 20t 30g
secondary 2t + 2t 30g
L4 primary 8t + 8t 18g ⅜in dia.
secondary 4t
L5 24t 24g on ⅜in slug-tuned former
L6 10t 18g ⅜in dia.
L7 6t 18g ⅜in dia.
L8, 5t 18g ⅜in dia.
L9 L11 primary 2t 20g ⅜in dia.
secondary 8t + 8t 18g ⅜in dia
L10 primary 8t + 8t 18g ⅜in dia.
secondary 4t
L12 primary 8t + 8t 18g 1in dia.
secondary 1t adjustable coupling

The mode in which the final amplifier operates is determined by the transmission type selected; in the a.m., cw and fm modes an audio oscillator is connected to the microphone amplifier to insert carrier. On ssb no carrier is inserted and the pa operates in Class AB1 with stabilized screen and grid supplies. The pa anode is connected to 700V. Audio from the microphone is disconnected from the a.m. and fm modulators and switched through to the balanced modulator. The a.m. modulator ht is turned off and the modulation transformer shorted out. When the transmit switch is open the driver is deprived of ht and the pa screen is open-circuited, power is removed from all stages but the oscillator is left running.

In the cw mode all audio circuits are shorted out and the carrier insertion oscillator is connected to the balanced modulator. The pa remains in the Class AB1 condition. The

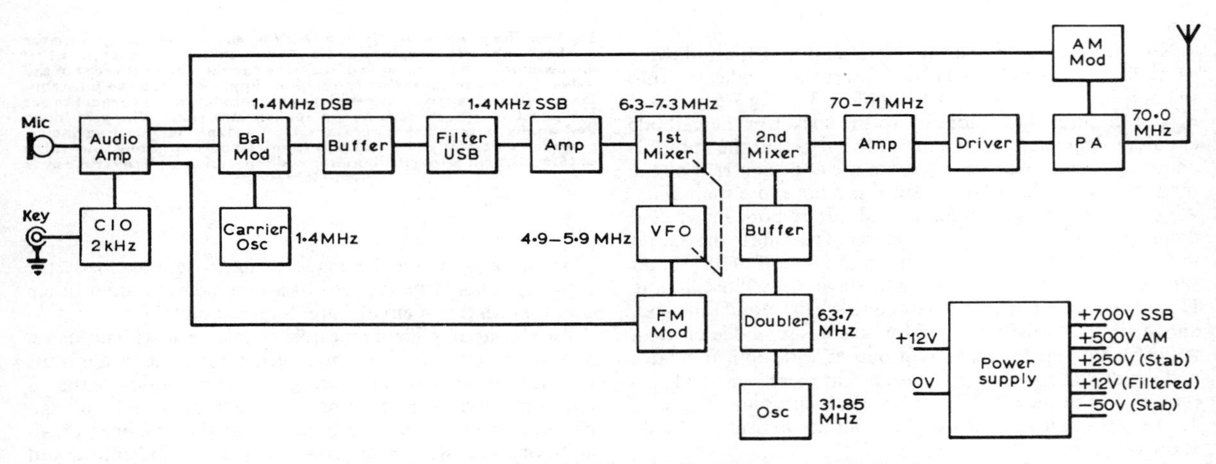

Fig 7.67. Block diagram of a multi-mode transmitter for the 4m band. For portable use a 12V inverter psu may be employed, as shown

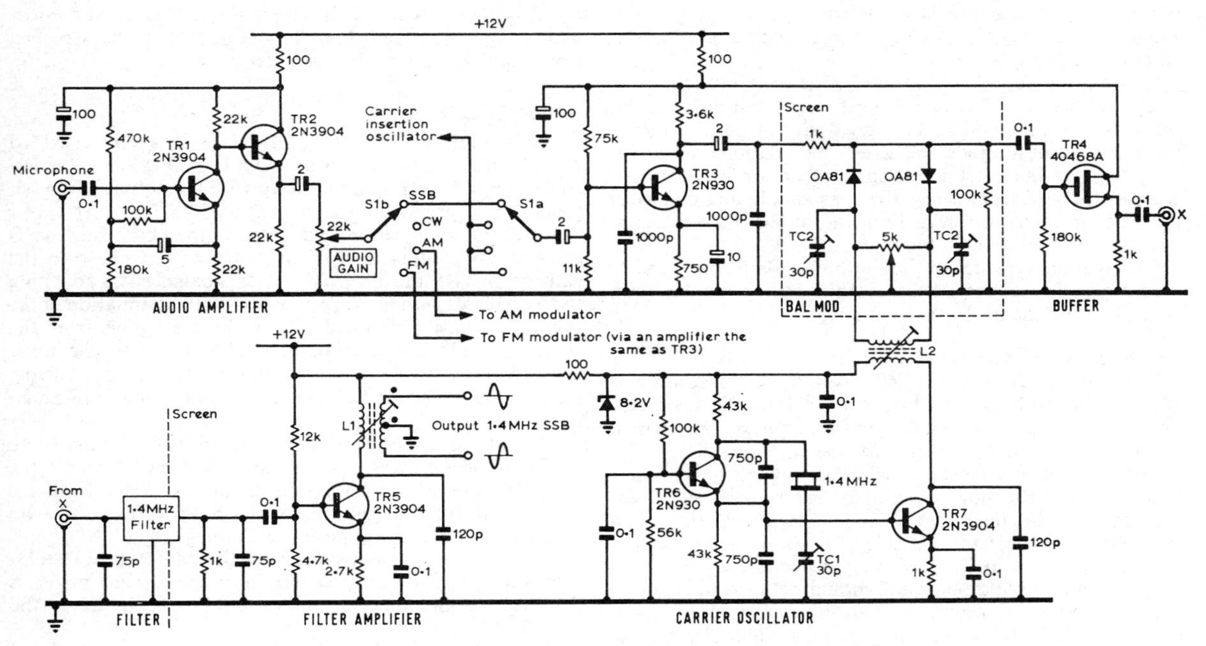

Fig 7.68. The ssb generator unit of the 4m multi-mode transmitter

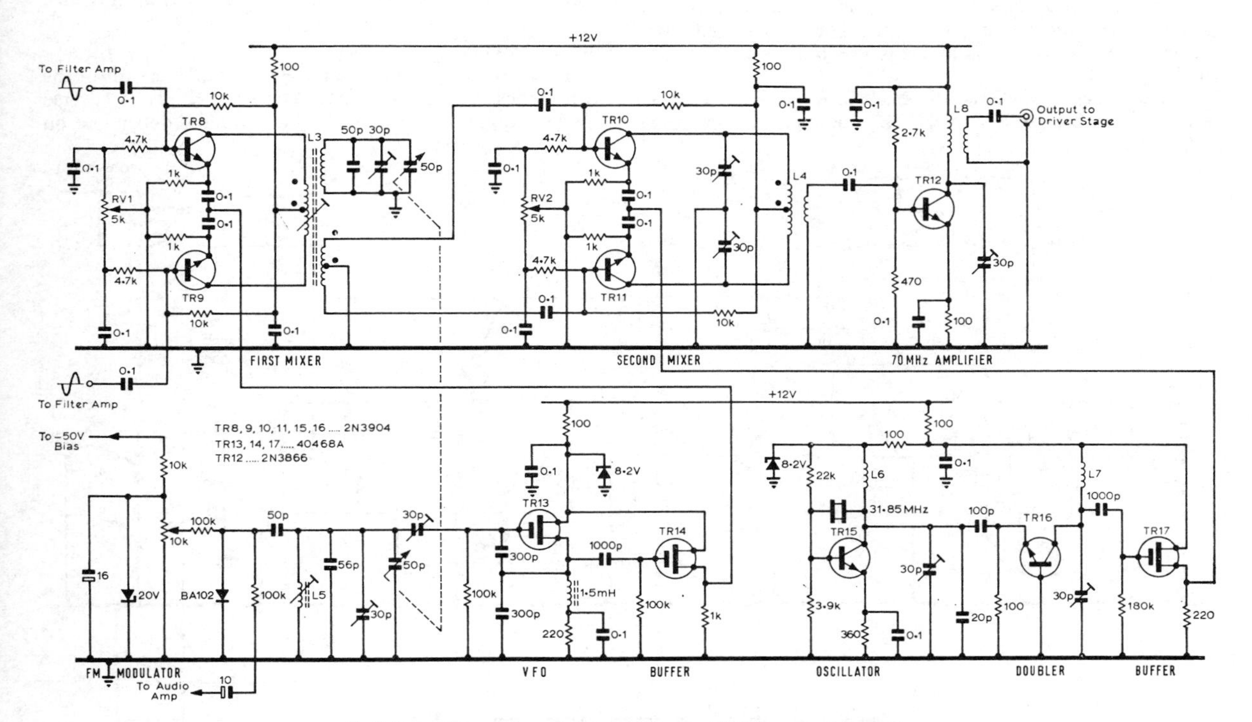

Fig 7.69. In the multi-mode transmitter for 70MHz output from the filter-amplifier in Fig 7.68 is applied to the balanced mixer shown here and fed to a second balanced mixer to which the output of the crystal oscillator chain TR15, 16 and 17 is applied

output of the cio is normally shorted out by a small relay, but when the morse key is pressed this short circuit is lifted and a carrier thus generated.

On a.m. the pa mode is Class C. One relay drops the pa anode voltage to 500V and changes the screen grid supply from +250V stabilized to a supply derived from the anode voltage to furnish anode-and-screen modulation. A second relay removes the short across the modulation transformer, changes the grid from a negative bias supply and connects grid to earth via a resistor. A third relay applies ht to the a.m. modulator, to which audio is routed.

On fm audio is switched to the fm modulator on the vfo. Audio gain for all modes is governed by a common gain control before the microphone amplifier.

The SSB Generator

The ssb generator, **Fig 7.68,** uses a Cathodeon filter BP4128/USB with a carrier frequency of 1·4MHz. The transistors TR1, TR2 form a high impedance input stage to match microphone impedance to microphone amplifier TR3.

The amplifier is connected via a 2μF capacitor (to maintain good low frequency response) to the twin diode balanced modulator, where the audio is chopped by the carrier oscillator TR6. The mosfet buffer stage TR4 prevents the modulator being loaded by the low input impedance of the filter. For maximum performance and minimum in-band ripple it needs to be fed from, and into, an impedance of 1kΩ and capacitance of 75pF. This is obtained on the input by the use of a 1kΩ resistor in the source lead of TR4.

The filter amplifier (a) restores the level of the signal after 6dB attenuation in the crystal filter, and (b) acts as a tuned phase splitter to drive the first mixer. The carrier oscillator frequency is adjusted by TC1 for the best overall audio quality.

The unit delivers 500mV of ssb, before overload, with a 3dB bandwidth of 300Hz to 3·1kHz, the maximum in-band ripple being less than 0·2dB. To ensure good carrier and sideband suppression the balanced modulator is totally screened from the filter and carrier oscillator.

Mixer Circuits

On the mixer board the 1·4MHz ssb is converted to 70MHz vfo controlled ssb in two mixers, **Fig 7.69.** The vfo covers 4·9 to 5·9MHz and is mixed with the output of the ssb generator to produce an i.f. of 6·3 to 7·3MHz. Each mixer balances out its injection frequency. In the first mixer this is achieved by feeding the output of the vfo in phase to the emitters of TR8 and TR9. The mixer is balanced by adjusting the base biases with RV1. As much as 30dB attenuation of the injection is easily obtained. The antiphase signals from the ssb exciter are fed into the bases of TR8 and TR9. The mixer is tuned by a variable capacitor ganged with the vfo, allowing a high-Q coil to be used to increase the rejection of unwanted mixer products.

The antiphase output of the first mixer is fed straight into the bases of the transistors in the second mixer, where the variable i.f. of 6·3 to 7·3MHz mixes with 63·7MHz from the crystal chain TR15, 16 and 17, to give full coverage of the 4m amateur band. The second mixer is tuned by a 30 + 30pF split stator capacitor across the collectors of TR10 and TR11. The 4m single sideband signal from the second mixer is delivered to the tuned amplifier TR12 which drives the QQV03/10.

The mixer and oscillator unit is adequately decoupled and screened. The supply lines to the two oscillators are zener stabilized to ensure stability when the battery voltage drops.

The FM Facility

An epicap diode connected across the tuned circuit of the vfo provides the nbfm facility. When the diode is reverse biased it exhibits a capacitance whose value is dependent on

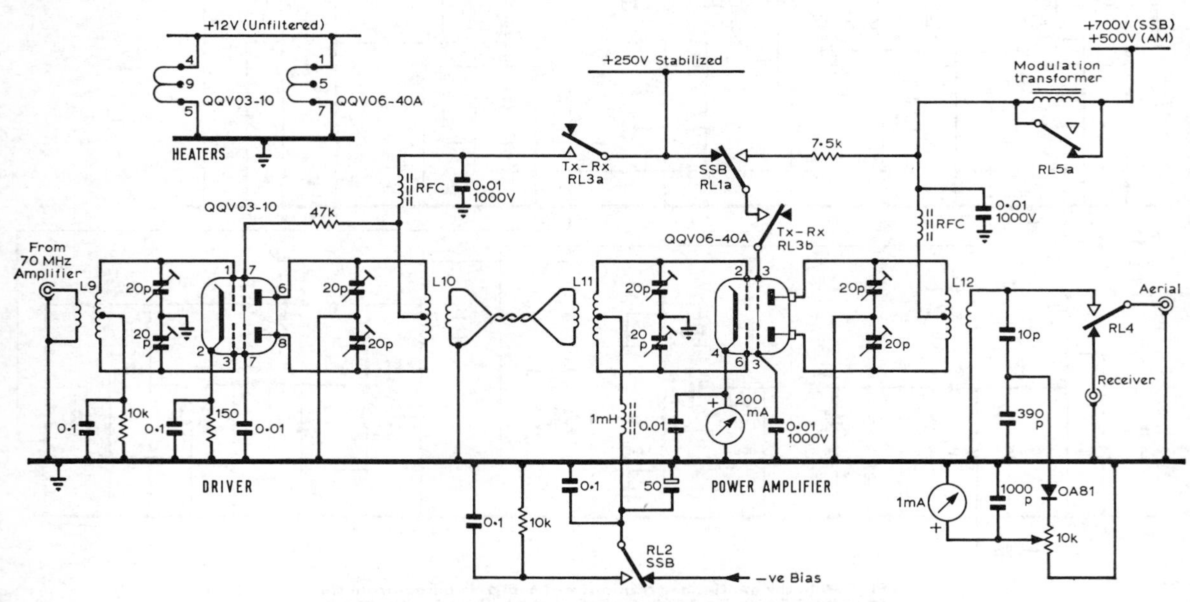

Fig 7.70. The driver and output stages of the multi-mode 70MHz transmitter

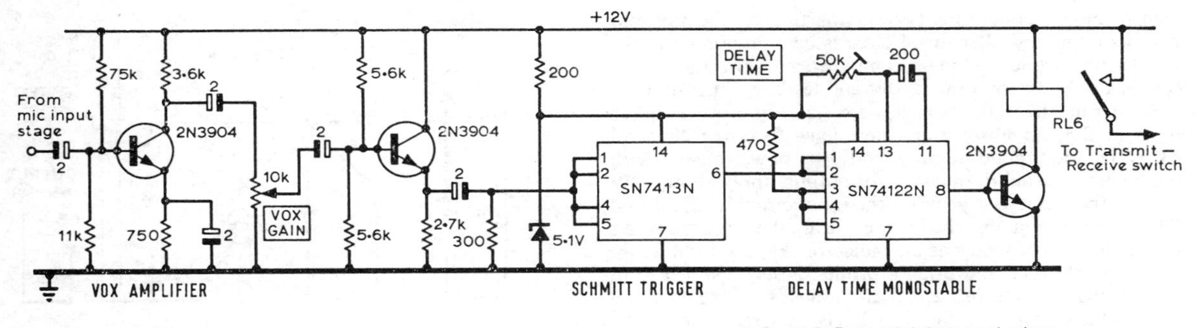

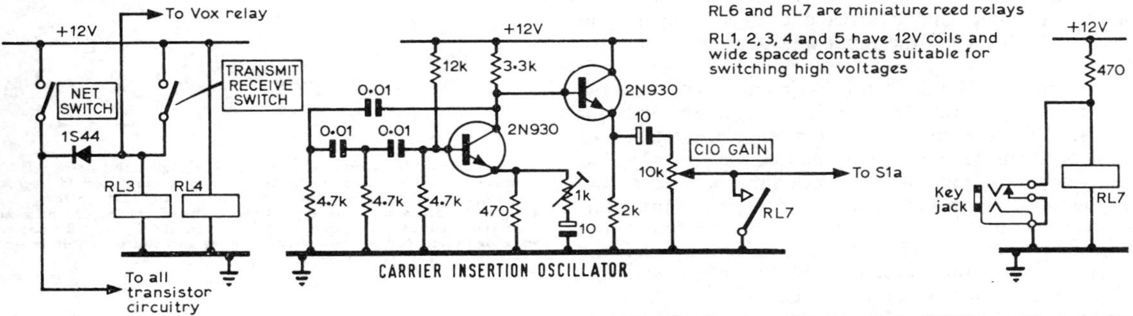

Fig 7.71. Miscellaneous circuitry of the 70MHz multi-mode transmitter

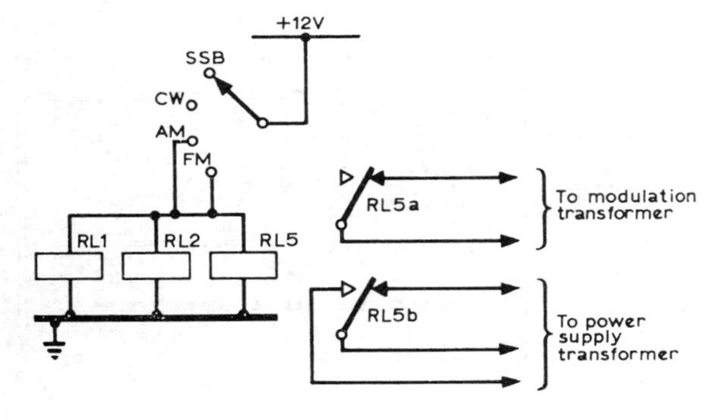

the level of bias. Choice of a suitable bias voltage is by means of the 10kΩ potentiometer connected to the diode (see **Fig 7.69**) and correct deviation achieved by on-air tests.

Driver and Power Amplifier Circuits

The QQV03/10 driver operates in Class A for all transmitter modes. The operating conditions of the QQV06/40A pa are controlled by relays RL1 and RL2; RL1a switches the pa screen between a stabilized 250V for ssb and a resistor to the ht supply for a.m. to provide anode-and-screen modulation. RL2 applies a negative bias to the control grid for ssb and shorts it to earth via a resistor for a.m. The relay connections RL3a and RL3b are controlled by the transmit/receive switch so that during "receive" the driver is switched off and the screen supply is removed from the pa. This ensures a low standby current. The power output is monitored by a 1mA meter driven by a capacitative voltage divider connected across the aerial output socket.

Before the aerial there are five circuits tuned to 70MHz to ensure reasonable attenuation of any unwanted products. As all the valve circuits are tuned to the same frequency good screening is essential between the grid and anode coils if unwanted feedback is to be avoided.

TRANSMITTERS FOR 432MHz

Much of the communication which takes place on the 2m band may usefully be transferred to 70cm, an allocation especially attractive for local and medium distance contacts at high signal levels. In the UK most of this communication takes place in the area 432–433·5MHz.

One of the simplest transmitters for 432MHz consists of a modulated tripler such as a QQV03/20A adequately driven by an existing 144MHz transmitter and modulated on screen and anode. The solid state varactor tripler is also much favoured, and two designs will now be described:

VARACTOR TRIPLER FOR 432MHz: DESIGN "A"

In the varactor tripler unit shown at **Fig 7.72** input at 144MHz is applied to the input socket at the left and 432MHz emerges from the output socket. The varactor itself is not directly modulated: speech intelligence is provided from the 2m drive source, either a.m. or nbfm.

The unit is built in a copper box 6in long by 1½in square, brazed or soldered up from 16 to 18 swg sheet. A partition

2½in from one end forms two compartments, the varactor with its input and idler tuned circuits being mounted in the smaller one, while the longer compartment forms the output cavity. **Fig 7.73** shows the layout and leading dimensions.

Setting up the multiplier requires, in addition to a source of rf at 2m, an absorption wavemeter to cover 288 and 432MHz, and a good dummy load or rf power meter. With the load connected to the varactor output, apply about 10W of rf to the input and peak up the input circuit (C2, L1 and C1) for maximum absorbed power, using the 2m transmitter pa anode current meter as an indicator. Next set the wavemeter to 288MHz, and bring a probe, loosely coupled to it, near to the junction of L2, C3 and adjust C3 for maximum idler voltage at this point (ie L2, C3 in series resonance). These first two steps can be carried out more rapidly if a gdo is used initially to set the input and idler to the correct frequencies.

Next, peak up C4 and C5 for maximum rf out, using the wavemeter to confirm that this is on 432MHz and not 576MHz. With the cavity dimensions shown, the plates of C5 will be about 1/16in apart. Now increase the power level in steps, readjusting the trimmers each time because of the changing varactor capacitance which forms part of the input, idler and first output tuned circuits. It will be found that correct adjustment of the idler circuit produces a peak in the output power, but this is not a reliable way of adjusting the idler since a number of spurious peaks will occur if C3 is varied over its range, owing to interaction between the various tuned circuits. Once the idler is set to series resonance by the method suggested above, it is best left alone, and

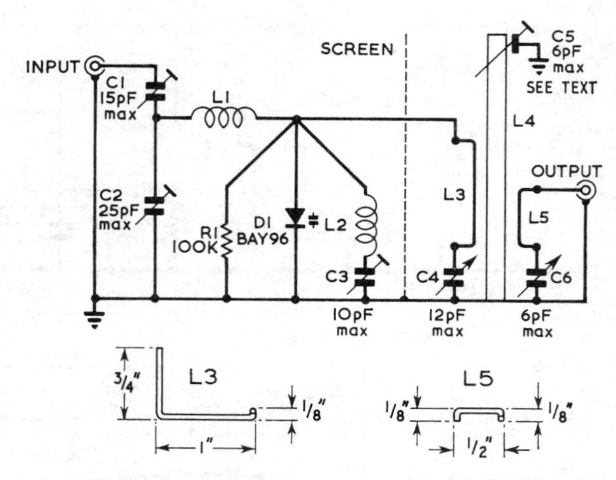

Fig 7.74. Circuit diagram of the 144-to-432MHz tripler using a BAY96 varactor to deliver approximately 9W of modulated rf at 70cm. Component values: L1 6t 18swg at 3/16in diam and ½in long; L2 3t 14swg 3/16in diam and ¾in long; L3 18swg shaped as shown and spaced 3/32in from L4. L4 ¼in od copper tube 3/16in id 4⅛in long; L5 coupling inductor fabricated as shown from 18swg and spaced 3/16in from L4

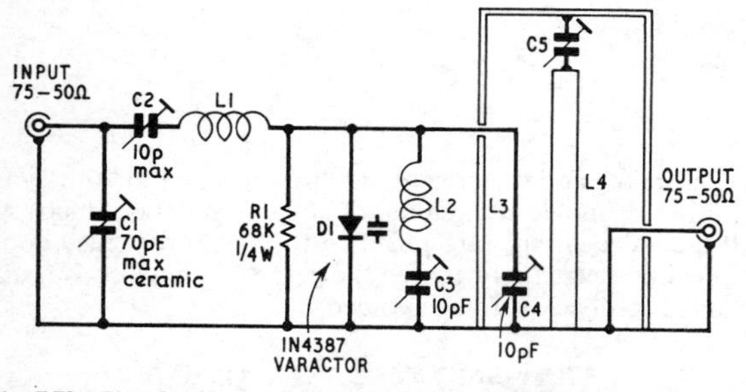

Fig 7.72. Circuit diagram of the 144-to-432MHz varactor tripler using a 1N4387 device. Component values: C1 70pF max ceramic; C2 10pF max; C3, C4 10pF beehive; C5 two 1in diam discs approx 1/16in apart; D1 the 1N4387 varactor diode; L1 3½t 16swg ½in diam to make approx 0·2 μH; L2 2t 16swg ⅜in diam to make approx 0·06μH; L3 2in of 16swg; L4 brass tube ⅜in od and 3½in long; R1 68kΩ ¼W

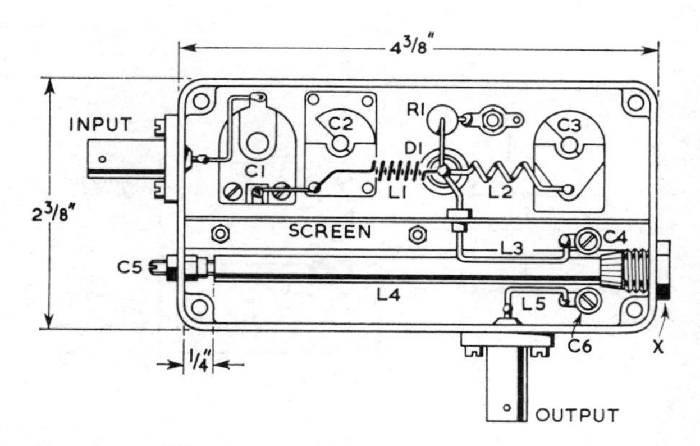

Fig 7.75. Mechanical layout of the BAY96 varactor tripler. The diode D1 is bolted directly to the base of the diecast box. At "X" is a Radiospares spindle lock, the inner part of which provides fixing for the tuned line at the earth end

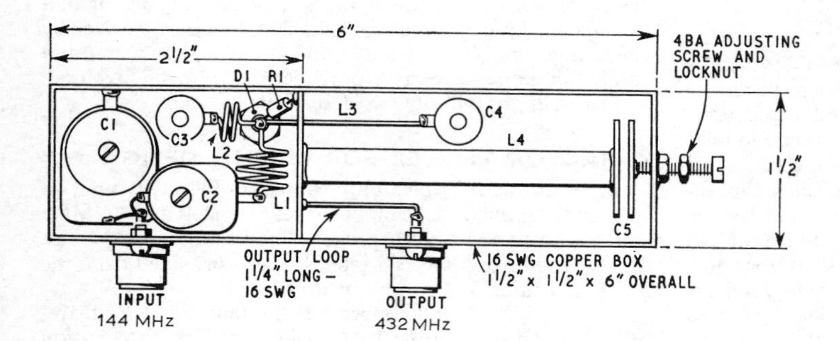

Fig 7.73. Mechanical layout of the varactor tripler, Design "A"

subsequent alignment confined to the input and output circuits. Using a reflectometer adjust C1 for optimum match to the feeder cable, again at the proper power. In practice, adjustment of C1 so that the 2m transmitter pa does not detune as the coupling loop is brought in is satisfactory.

After a short period of operation at full power, check the varactor stud temperature. At the maximum allowable rf input level of 40W, the 1N4387 will produce about 25W of rf output and will, therefore, be dissipating 15W as heat; at this dissipation the maximum permissible stud temperature is 100°C. The box shown in Fig 7.73, if made of 16 swg copper, will have a thermal resistance of about 5°C/W in free air, and at 15W will therefore hold the varactor stud at 100°C in an ambient of 25°C. This, of course, is on the limit of the device rating, and is uncomfortably hot in any case, so it is recommended that the tripler unit be mounted to a thick copper or brass bracket bolted directly on the varactor stud.

The power levels quoted above are cw, and in fact the varactor may limit, due to reverse voltage breakdown, at a power level slightly above the manufacturer's rated maximum. This means that although the 1N4387 or BAY96 will produce 25W of cw power, the peak power level attainable will not be much more than this. With the 1N4387 it is necessary to reduce the input power from 40 to 20W to produce reasonably symmetrical 100 per cent a.m. This diode, which is a graded-junction step recovery device, will multiply an a.m. signal very well with good linearity, but the abrupt junction types such as the BAY66 detune more rapidly with changes of drive level, and would probably be slightly less satisfactory for a.m. use.

Phase or frequency modulation is cheaper to achieve and more suitable (because of voltage breakdown and hence peak power limitations) for varactor and transistor transmitters. If it is necessary to use a.m. with a varactor tripler, it is advisable to tune it up with a 100 per cent sinewave modulated signal at the normal power level while watching the demodulated envelope on an oscilloscope. Look for peak clipping, and for discontinuities in the scope pattern due to dynamic detuning as the applied power varies, over the modulation cycle. These can be eliminated by adjusting the input power level and finding a compromise setting for the input tuned circuit. These complications are unnecessary, of course, with nbfm.

VARACTOR TRIPLER FOR 432MHz: DESIGN "B"

An alternative design of 144 to 432MHz varactor tripler is shown at **Fig 7.74.** The dimensions of the inductor lines are such that the unit may be accommodated within a standard 4⅜in by 2⅜in diecast box. A high-Q filter consists of a tuned ¼-wave line L4 to which the output of the varactor is coupled and from which low impedance output via L5 is passed to the aerial.

When 15W of modulated rf are applied to the input socket, 9W of output at 432MHz may be expected when the unit is properly set up. Good quality a.m. speech may be passed through this multiplier provided that the peak percentage modulation at the drive source is held to about 80 per cent. Detailed adjustments for optimum output are as Design "A".

A PURPOSE-BUILT EXCITER FOR 432MHz →

The following design for a 432MHz exciter unit employs five valve stages terminating in a QQV02/6 straight amplifier delivering 4W of output, and thus suitable for driving a larger

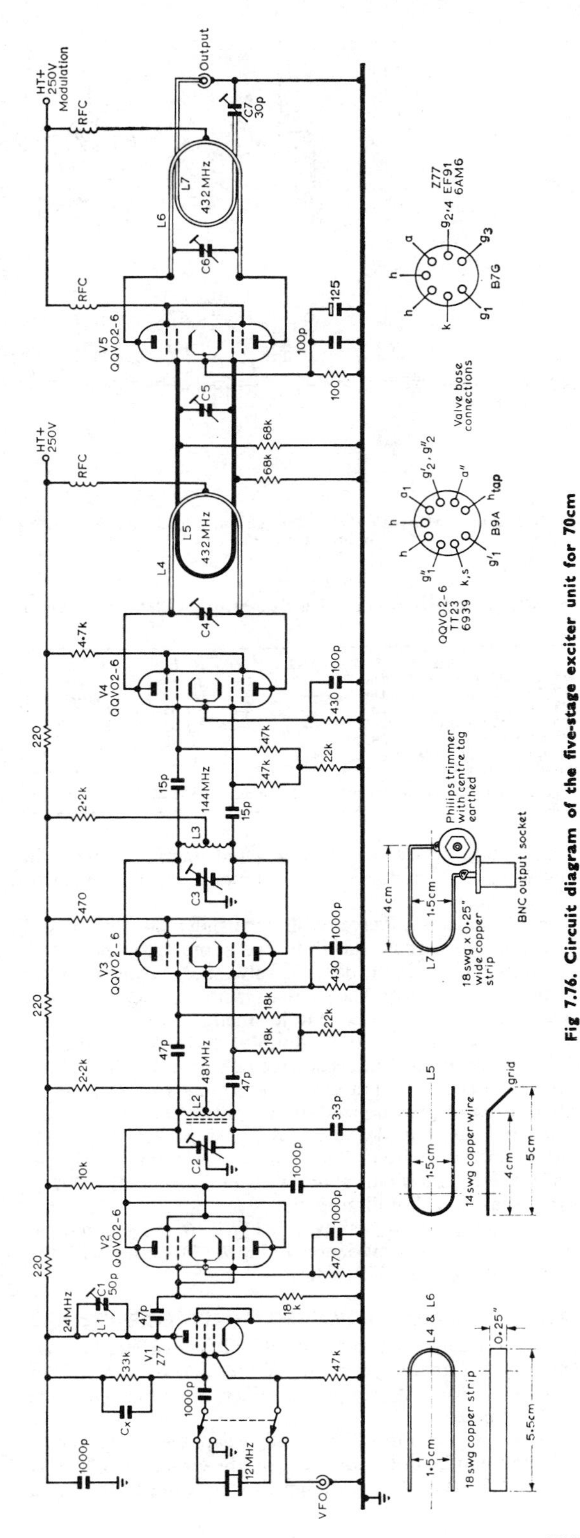

Fig 7.76. Circuit diagram of the five-stage exciter unit for 70cm

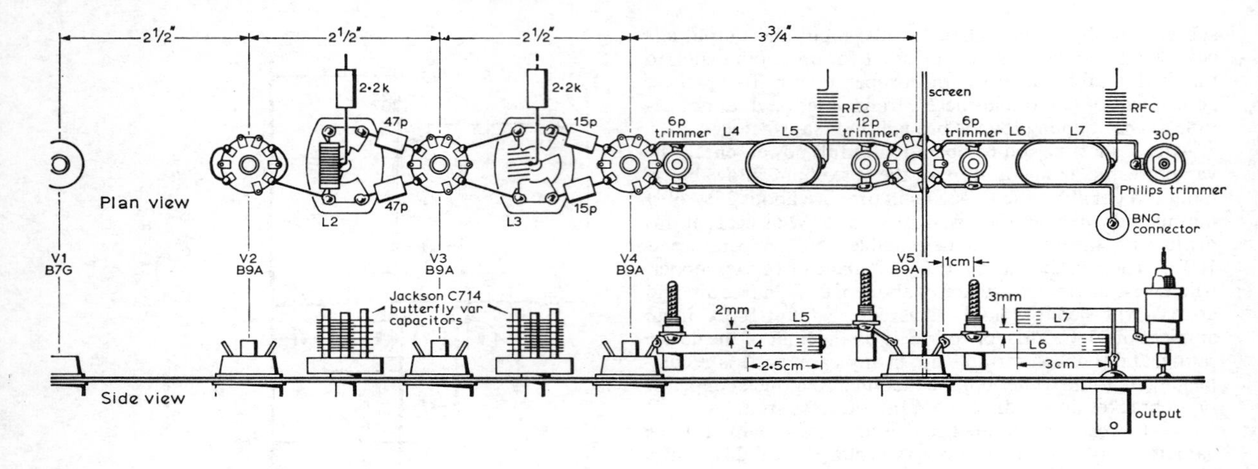

Fig 7.77. Mechanical layout of the 70cm all-valve exciter unit showing inductor details

TABLE 7.9
Operating conditions in the five-stage 432MHz exciter unit

	Va	VG2	IG2	VC	IC	IG1
V1 Z77/EF91	230	150	—	—	—	—
V2 QQVO2/6	220	160	—	–15	32	negligible
V3 QQVO2/6	235	235	—	–15	35	2 × 1·3
V4 QQVO2/6	250	240	—	–15	50	2 × 1·2
V5 QQVO2/6	240*	240	1·3	–6·6	53	2 × 0·6

Voltage supply 250V
Power output 4W

* Note that CCS rating for amplitude modulation is 200V.

amplifier stage or used as a low powered 70cm transmitter in its own right, if anode-and-screen modulation is applied to this final stage.

The first stage is a Pierce oscillator using a 24MHz crystal and the anode circuit tuned to the same frequency. For oscillation to occur in the pentode valve specified the load (ie, G2) must be capacitive. The small capacitor Cx across the G2 dropper resistor gives precise adjustment.

The second stage, which doubles to 48MHz, uses a QQV02/6 with both sections connected in parallel. The effectively single ended output circuit is arranged in a balanced configuration to provide push-pull input to the third valve.

The third stage triples to 144MHz using a QQV02/6 in a standard push-pull circuit. The fourth stage is similar, tripling to 432MHz. The final stage uses a further QQV02/6, operating as a push-pull amplifier.

The circuit diagram is shown in **Fig 7.76** and the general layout of the major components in **Fig 7.77** together with the side view of the 432MHz section.

A TRIPLER AND AMPLIFIER UNIT FOR 432MHz

A classic design of 70cm output unit which has held its popularity for many years utilizes a QQV03/20A tripler driving a QQV03/20A straight amplifier, the latter operating at 24W dc input.

A scaled-up version of this design capable of delivering considerably more output power uses the larger QQV06/40A valve in each stage. In the design shown at **Fig 7.78,** due to G8AVX, fine tuning of the tripler stage anode line inductors is accomplished by a copper or brass paddle which is rotated from the front panel so that it just clears the valve. The pa stage is fine-tuned by a dielectric paddle. In each stage initial resonance—to the top end of the 432–433·5MHz range—is achieved by bending the "flags" attached to each output line to or from one another. The tripler anode lines are ¼-wave; those for the amplifier grid input are ½-wave, tuned across the ends by a 0·5 to 3pF trimmer.

The choke feeding the tripler anode lines is a 50Ω wire-wound resistor. The chokes feeding the amplifier grid lines have 6in of wire close-wound on a 50Ω resistor; they should be positioned at a "cold" point on each line. This is readily found by running a screwdriver along the line (minimum deflection on the grid current meter corresponds to "cold" point).

The screens of both stages are each fed through variable resistors, to set the input power; the resistor feeding the tripler effectively operates as a drive control to the amplifier. The amplifier grid current should be set to 3mA with the bias resistor network provided in the circuit. The anode current to the tripler to give this input to the amplifier will be between 85 and 130mA.

The input to the amplifier at 400V is adjusted to 150mA at resonance by means of the variable screen resistor and loading.

As is shown in the illustrations, the tripler-amplifier circuit is completely screened. The power supply components are fitted between the front panel and the tripler-amplifier compartment.

The tuning control shafts for the tripler and amplifier are brought out to the front panel and the output coupling control is accessible from the back.

The U-shaped paddle used for tuning the tripler anode circuit consists of a piece of copper or brass ⅝in wide bent to form a U-shape slightly larger than the valve bulb diameter. This is fixed at its centre to an insulator block, suitably drilled for attaching to the tuning control (see **Fig 7.79 (a)**). It should be noted that this control has a limited range and the actual position of the anode lines on the valve anode pins is used to set the centre of the range required.

Fig 7.79 (b) gives details of the tripler anode lines and tuner.

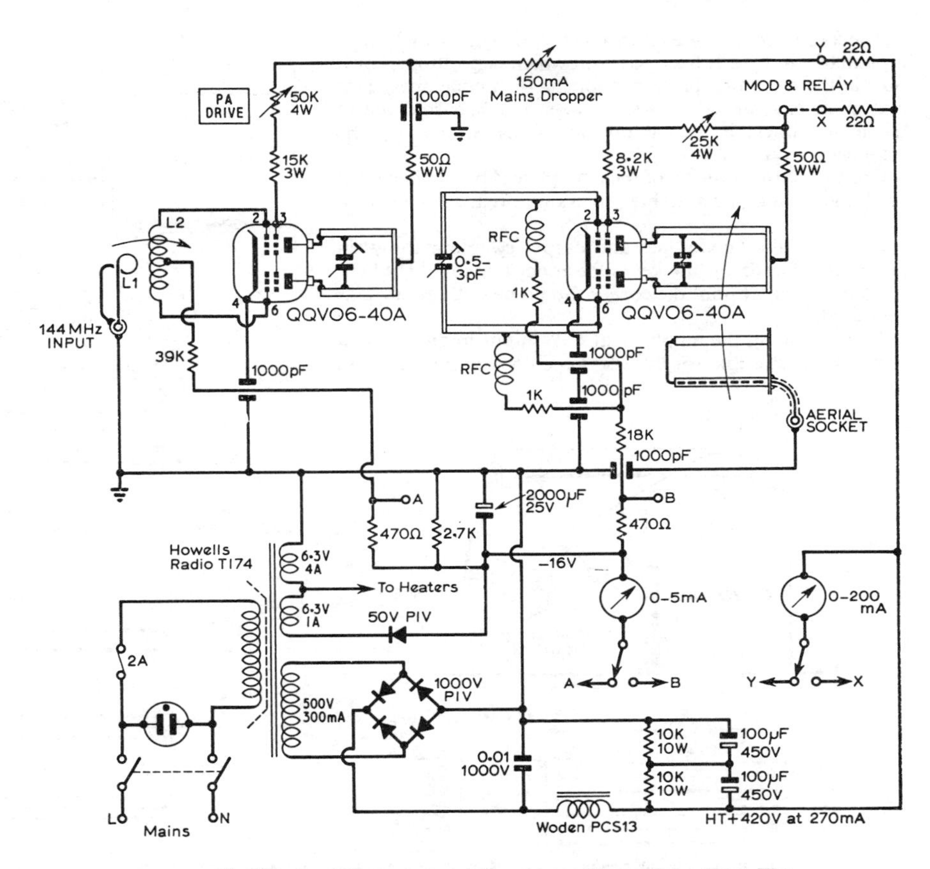

Fig 7.78. A medium power tripler-and-amplifier unit for 70cm. The inductor L2 is self-resonant at 144MHz, ie approx five turns ½in diam air spaced. The single-turn coupling loop L1 which furnishes drive from an existing 2m transmitter is positioned at the centre of L2 to deliver optimum input so that when the tripler output is tuned to 432MHz a grid current of 3mA is realized in the pa grid input circuit

Top view of the tripler-and-amplifier transmitter for 432MHz

Rear panel removed to show detail of line construction

The amplifier tuning as mentioned earlier is by dielectric; a plate of ⅛in PTFE is fixed to a rod which is positioned by a hole in the mounting block supporting the anode lines. This PTFE plate is arranged to pass between the flags soldered to the anode lines; its actual position is determined by the frequency required.

Control of the position of the PTFE plate is by standard dial cord and spring and suitable pulleys (see rear view illustration).

Details of the anode lines and output coupling are given in **Fig 7.81.** The fixing bracket of the anode line should be bent to allow for expansion; the corner should be well rounded, as indicated.

When running the amplifier at 60W input in an enclosed screen it is desirable to provide some cooling, and a small

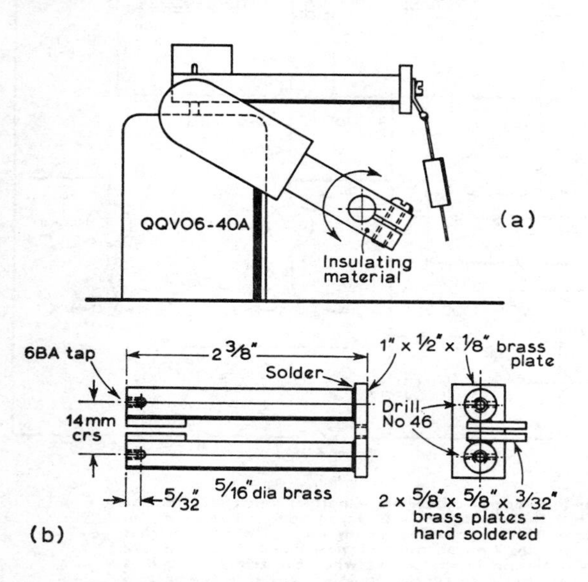

Fig 7.79. The tripler anode circuit and metal paddle fine tuner details

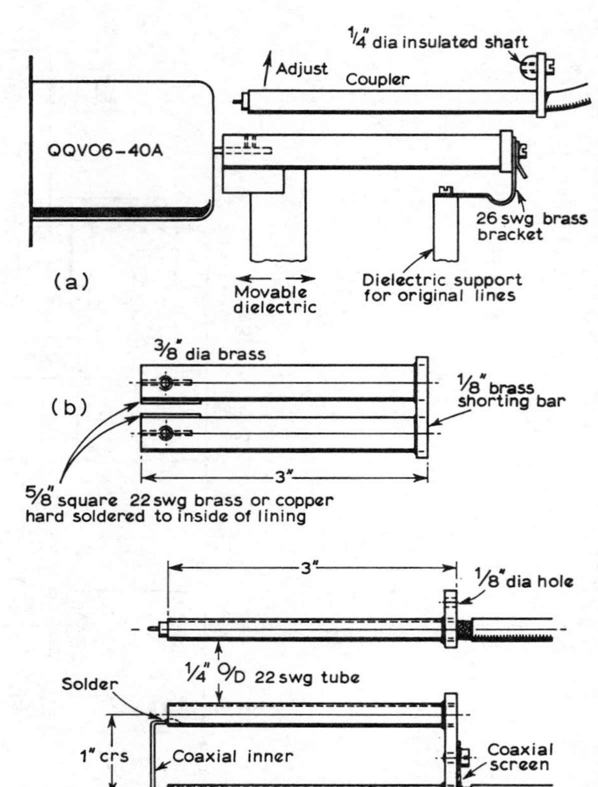

Fig 7.81. Amplifier anode lines construction. Fine tuning is by movable dielectric paddle. The lower diagram shows the construction of the balance-to-unbalance aerial coupling inductor

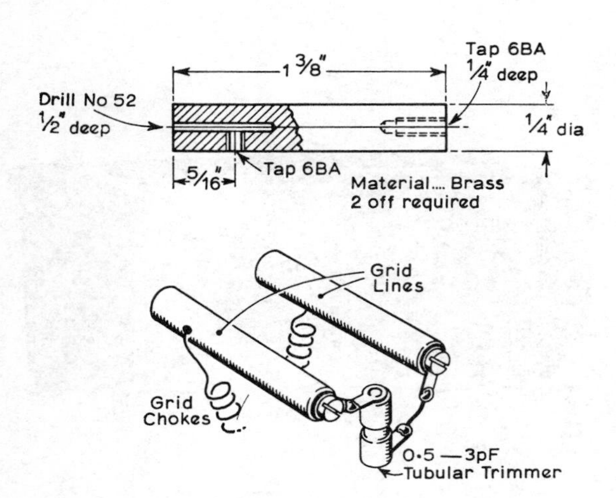

Fig 7.80. Amplifier grid lines construction

blower is attached to the rear panel of the enclosure. A number of ventilating holes in the base plate of the amplifier screen are located directly under the valve, with the air-flow from below.

Chassis details are shown in **Fig 7.82.** Only the major items are given; such detail as fixing holes for the screen or valve sockets are left to the constructor.

A HIGH POWER AMPLIFIER FOR 432MHz

Where it is desired to use the full permissible 150W dc input on 70cm a larger pa stage than those hitherto described will be necessary. A single 4CX250B fulfils this requirement excellently. The operating conditions which apply to this valve were given earlier under "A compact general purpose 150W power amplifier for 144MHz". The circuit diagram at Fig 7.58 for 2m applies broadly to the 70cm design except that substantial changes in the inductor geometry are dictated by the higher operating frequency.

In this 432MHz amplifier the anode and grid circuits are coaxial lines tuned by disc capacitors with inductive input and output coupling. The unit is contained in a box 8¼in long by

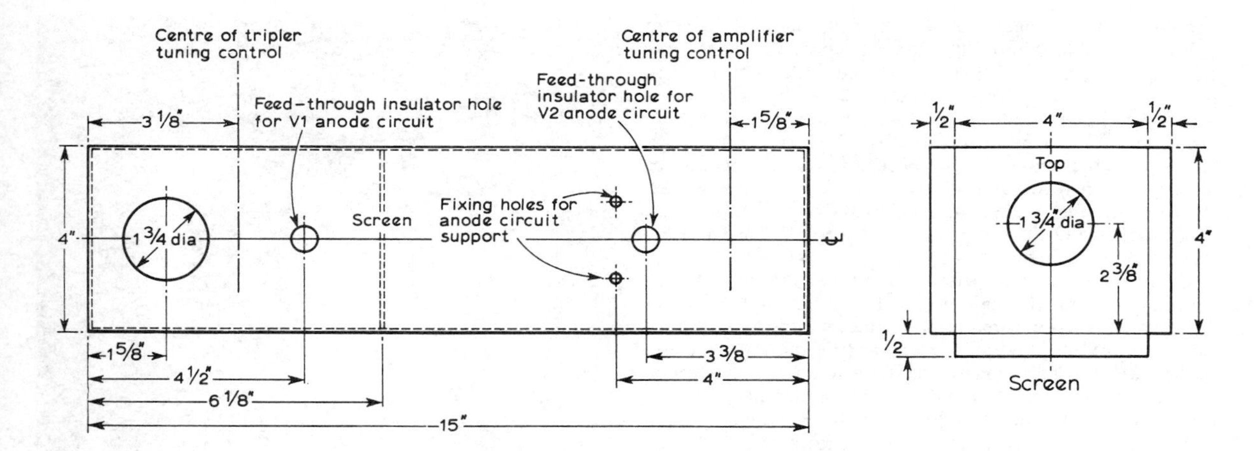

Fig 7.82. Metalwork for the tripler-and-amplifier unit

3½in square (inside) made of 18swg copper or brass. The lid is bent down on the long sides so that it makes good contact with the box, and the two end plates are similarly constructed. The end plate to which the anode inductor is attached is made easily removable for valve changing.

A screen across the centre of the box carries the valve socket, and isolates the anode and grid circuits from one another. The anode inductor (**Fig 7.84**) is made of 1¾in diameter (outside) tube with 20swg wall thickness. At one end a 2¾in diameter flange is fitted and at the other, eight equidistant slots are cut for a distance of 1in from the end. After these slots have been cut, the open end should close down to a diameter of 1·6in to make good contact to the valve cooler. The inner edge of the fingers should be chamfered to assist fitting to the valve. The fixed plate of the anode tuning capacitor is attached centrally to one of the end fingers by soldering; the position of this is arranged to be opposite to its moving plate.

The end plate to which the mounting flange is attached has a hole to match the inside diameter of the anode line for the air outlet. Insulation between the line and the end plate

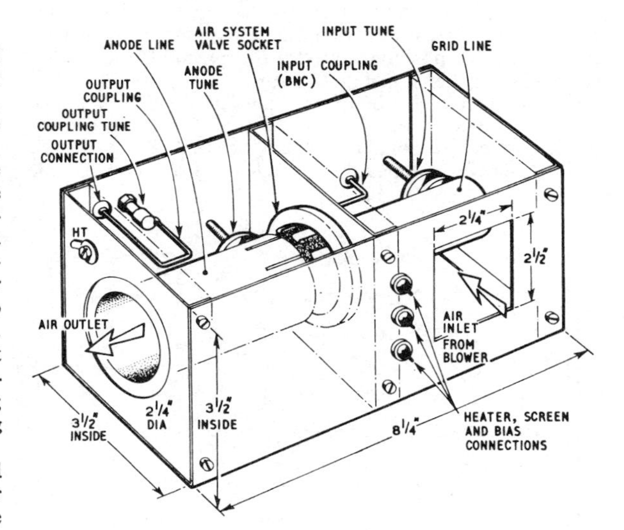

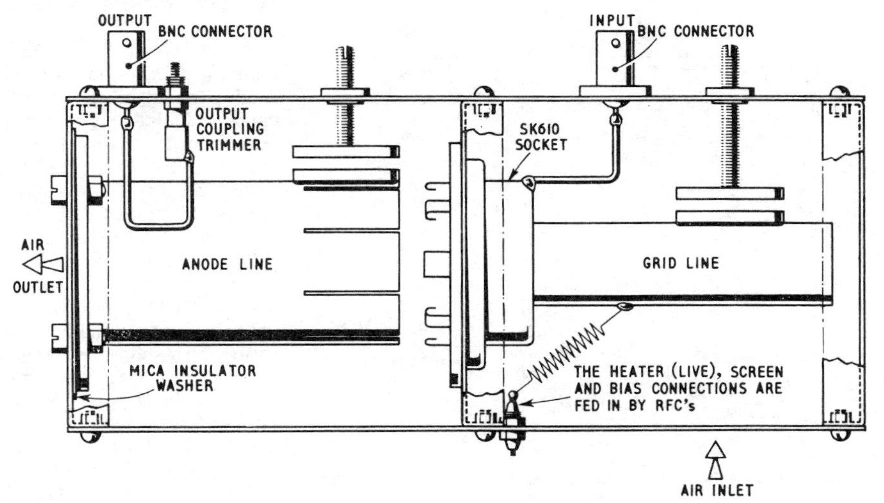

Fig 7.83. The 432MHz high power amplifier. Above right: general arrangement; right: top view

should be a mica ring, which with its high dielectric constant will provide a substantial by-pass capacitor. For safety it is desirable to cover the hole in the end plate with an open mesh to prevent accidental contact with the live anode line.

The grid line is made of $\frac{7}{8}$in diameter tube of 20swg wall. One end is closed by a disc soldered into the tube and a clearance in the centre of this is needed so that the line can be directly attached to the grid connection of the valve socket. As with the anode line, the fixed plate of the grid tuning capacitor is fixed directly on to the tube at $2\frac{1}{16}$in from the valve end of the line (see Fig 7.84).

Connections to the heater (live side), screen and bias supplies are made through insulated terminals and rf chokes. The bias connection is made on to the grid line at a point of minimum rf voltage, approximately 0·8in from the valve end of the grid line. The ht connection to the anode is taken by one of the insulated screws used to fix the anode flange to the end plate. An rf choke should be fitted externally.

The valve socket is fitted to the central screen, with the screen by-pass capacitor flange of the socket on the anode side of the screen and fixed by the three clamps provided with the socket. The position and size of the input and output couplings are shown in Fig 7.83 (b).

Arrangements must be made to initiate the blower motor when the heater comes on. It is essential that both compartments are reasonably airtight so that air is forced through the valve anode cooler and out through the anode line.

Fig 7.85. The 4CX250B valve

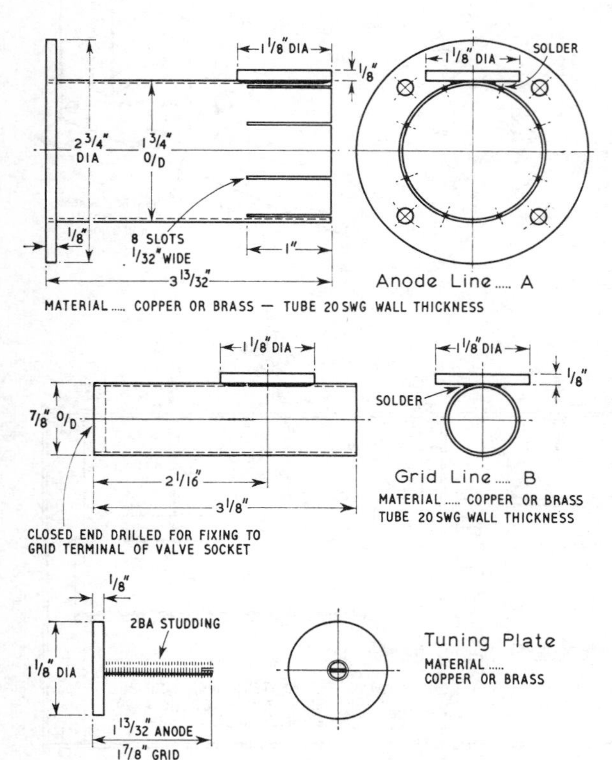

Fig 7.84. Mechanical details of the 432MHz anode lines (A), gridline (B) and tuning plate

SINGLE SIDEBAND ON 432MHz

Single sideband in the 432MHz band may be generated in a number of ways, adapting the principles already described earlier for 144MHz.

A method suggested by G2AIH has the merit of providing ssb output on two of the UK amateur bands, 70MHz as well as 432MHz. Output at 70MHz is mixed with the output from a crystal chain at 362MHz to deliver 432MHz input to the 70cm final class B amplifier. The p.e.p. output at 432MHz is approximately 35W; ssb output of approximately 1W is also available at 70MHz for driving an external 4m linear amplifier.

As will be seen from **Fig 7.86** amplified audio frequencies from the microphone are applied to a double-diode balanced modulator together with the rf output from a 498·6kHz carrier oscillator. The resultant output from the balanced modulator thus consists of the upper and lower sidebands of the (suppressed) carrier frequency. These two sidebands are then applied to a two-section half-lattice crystal filter, which rejects the upper sideband and passes the lower sideband only via a 498·6kHz amplifier to a second balanced modulator in the first frequency translator. Here the lower sideband signal is heterodyned with an rf voltage at approximately 2·65MHz to give an output frequency of 2·15MHz, and at the same time the lsb signal is converted to usb to conform with standard ssb practice for vhf and uhf transmission.

In the next frequency translator stage the 2·15MHz component is heterodyned with 9·47MHz in a balanced modulator to give a resultant output frequency of 11·6MHz usb. This 11·6MHz signal is then amplified and applied to a third frequency translator which converts the output frequency to 70MHz. The oscillator-multiplier chain in this translator

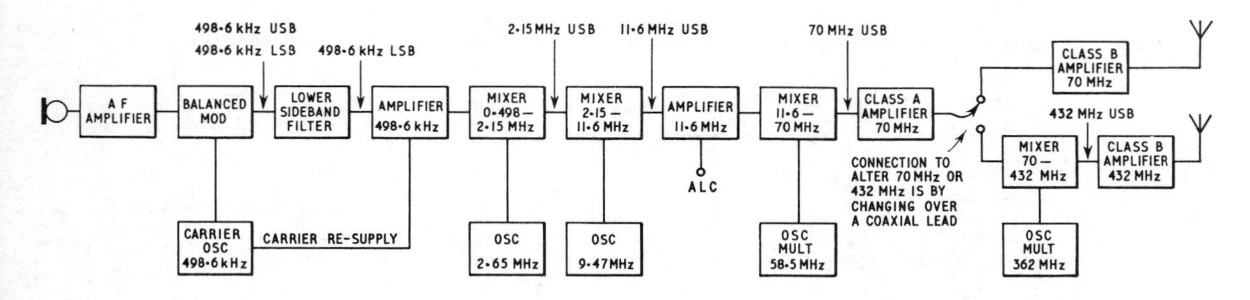

Fig 7.86. Block diagram of the G2AIH single sideband transmitter for 432MHz using ssb injection from a 70MHz source mixed with 362MHz from an oscillator-multiplier chain. The 70MHz ssb may be fed alternatively to a 70MHz pa unit when output is required in the 4m band

generates a frequency of approximately 58·5MHz, the actual frequency being determined by selection of any one of three crystals, which results in three spot frequencies being available for transmission in the 70MHz and 432MHz bands. In the prototype these crystals were for 6,525kHz, 6,540kHz and 6,550kHz, to deliver the required 58·5MHz at the end of the crystal chain. If continuously variable coverage is required instead of three spot frequencies, the crystal oscillator is substituted by a vxo or vfo at about 6,540kHz.

The 70MHz output from the balanced modulator in the third translator is amplified by a Class A amplifier, and can be applied either to an external 70MHz power amplifier for 4m operation or to the final frequency translator when 432MHz operation is required.

For 432MHz operation the 70MHz ssb signal is heterodyned with a frequency of 362MHz in the final frequency translator. The output stage is a 432MHz Class B amplifier (QQV03/20A).

The starting frequency in the 362MHz crystal chain is 40MHz; the output passes to a QQV02/6 mixer to which 70MHz ssb is also applied, delivering 432MHz ssb to the input of the QQV03/20A final.

Provision is made in the transmitter for injecting a voltage from the carrier oscillator (carrier re-supply) into the first mixer. This by-passes the first balanced modulator and crystal filter and enables a single sideband plus carrier transmission (A3H) to be made if desired.

To assist in obtaining good frequency stability, which is of paramount importance in ssb operation, oscillator output is derived from buffer stages in most instances; ht supplies to oscillators, balanced modulators and signal frequency amplifiers are regulated.

To ensure isolation between stages and to permit ease of any subsequent modifications (eg, introduction of a vfo as already mentioned) the prototype transmitter was constructed in the form of nine sub-assemblies:

(1) Audio frequency amplifier
(2) Carrier oscillator and balanced modulator
(3) Lower sideband filter
(4) Frequency translator 0·498 to 2·15MHz
(5) Frequency translator 2·15 to 11·6MHz
(6) Frequency translator 11·6 to 70MHz
(7) Frequency translator 70MHz to 432MHz
(8) 432MHz linear amplifier
(9) Control and power distribution panel.

Sub-assemblies (2), (3) and (4) above were mounted together to form a sideband generator assembly.

TRANSMITTERS FOR 1,296MHz

The two triplers now to be described accept 432MHz input and deliver 1,296MHz output. They may be used as small transmitters for 23cm or as driver units for a straight-through amplifier on this band.

DESIGN "A": A DIODE TRIPLER TO 23cm

A diode tripler configuration developed by G8AZM is shown as **Fig 7.87.**

The unit consists of the tripler unit itself followed by a simple filter to attenuate the other harmonics generated—a filter is necessary as the output at 1,728MHz, for example, is only 16dB below that at 1,296MHz. The tripler stage uses five 1N914 or 1S44 switching diodes which have a step recovery time of less than 90 picoseconds, although their capacitance/voltage ratio is small. Each diode is rated at nominally 200mW dissipation, but this can be greatly increased by using large thermal sinks as in the present design. By this means, up to 6W at 342MHz can be dissipated by the five diodes to produce up to 2W at 1,296MHz, a useful power level for the 23cm band. The maximum power input even for short times is about 9W.

The circuit is shown schematically in Fig 7.87 and the practical configuration in **Fig 7.88.** The five diodes are connected between two 1in × ⅝in plates to form a stack, one

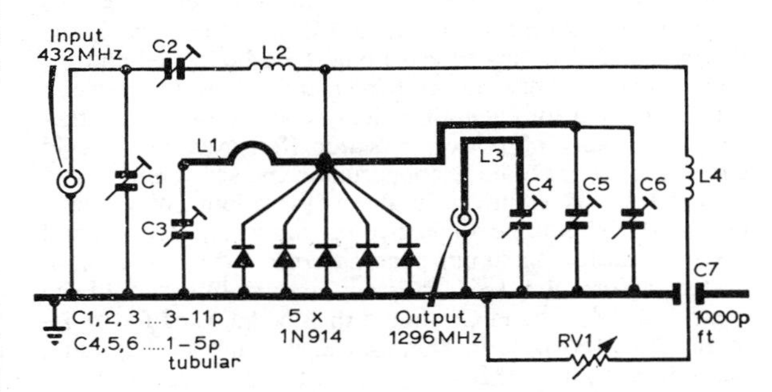

Fig 7.87. The G8AZM 70-to-23cm tripler using five diodes in parallel

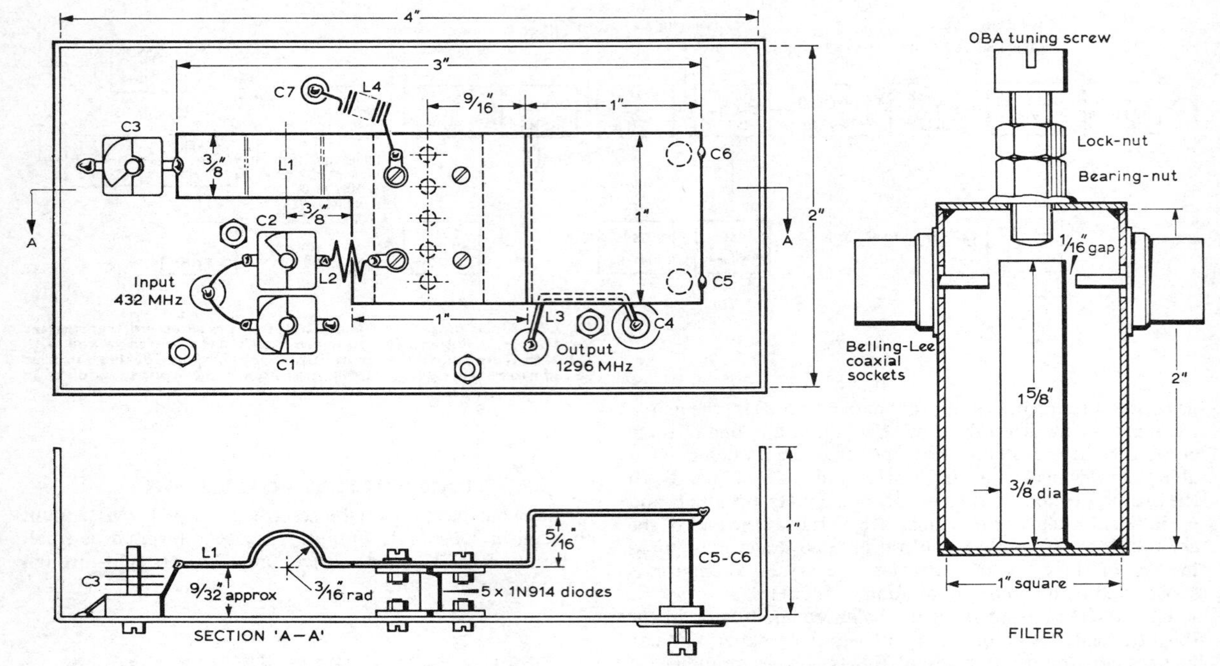

Fig 7.88. Practical configuration of the 23cm tripler and the associated filter

side of which is bolted to the chassis and the other to the line L1. The plates, the line and the chassis are made from copper (22swg) which provides high thermal conductivity. Great care must be taken in soldering in the diodes as their lead lengths should be as short as possible. To reduce the risk of damaging the diodes, the leads and the holes in the plates should be pre-tinned and a *hot* soldering iron used for the minimum time necessary to make each joint. The top and bottom surfaces of the stack should then be made flat by filing, and then with emery paper on a flat hard surface such as glass. The corresponding parts of the chassis and L1 should also be carefully flattened to ensure good thermal contact. The output loop is of 18swg wire running for ½in as near to L1 as possible, one end being connected to the output socket and the other to C4.

The filter consists of a box 1in square and 2in long fabricated from 16 or 18swg brass or copper sheet, containing a single resonant element.

The filter should first be tuned by connecting it to a receiver input and peaking preferably a weak distant 23cm signal: its insertion loss should be less than 1dB and therefore barely detectable. The tuning screw should then be locked and the filter connected between the tripler and a load (perhaps an aerial fed by a long length of lossy coaxial cable) together with some sort of power indicator (for example, a field strength meter or a diode connected in the aerial line).

With the bias resistor set at about 5kΩ, about 5W of power at 432MHz should be applied to the input, and C1, C2 and C3 adjusted for maximum reading from a voltmeter connected across the bias resistor: about 20V should be measured. C4, C5 and C6 should then be adjusted for maximum 1,296MHz output. The bias resistor and all the tuning capacitors should then be re-adjusted to maximize the output.

DESIGN "B": A VALVE TRIPLER TO 23cm

Fig 7.89 shows the circuit diagram of a simple tripler employing the grounded grid 2C39A valve. A drive of 10W is needed from a 432MHz transmitter if the full power output capability of this valve is to be realized.

The anode tuned circuit is a plate of copper or brass drilled as in **Fig 7.91** and fitted with contact finger material for connection to the anode of the valve. To reduce losses by radiation the anode circuit is enclosed in a standard diecast box, 4⅝in by 3⅜in by 2¼in. The lid is replaced by a copper or brass plate 1/16in thick with a U-shaped screen of the same material soldered to it.

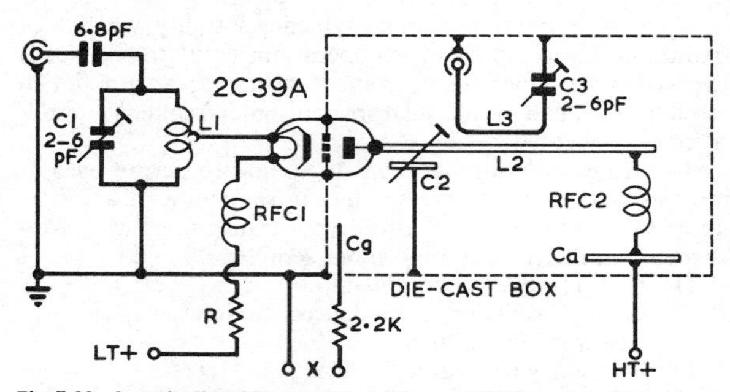

Fig 7.89. A strip-line 23cm tripler using a 2C39A valve. The input inductor L1 for 432MHz has 2t of ⅛in diameter strip ¾in long, cathode tap at ¾t from the end earth. L2 and L3, the anode strip line and output coupling loop, are shown in detail at Fig 7.91. Cg grid to chassis bypass capacitor using as insulation mica 0·004/6 in thickness Ca ht to chassis bypass capacitor, construction as Cg. RFC1 8t ¾in diam 18swg; RFC2 7t ⅛in diam 22swg; R to be adjusted to give 5·5V heater potential; C1 and C3 2-6pF concentric trimmer (Mullard C004/EA); C2 adjustable plate trimmer (see Fig 7.91); "X" is position for grid current meter

The plate forming the anode strip line is mounted $\frac{13}{16}$in above the chassis plate by two insulated pillars, and the ht supply to this is connected to a tag under one of the fixings, through the rf choke (RFC2) to the ht bypass capacitor centre fixing screw insulated under the plate, via the plate to the supply socket.

The grid, which is at rf earth potential, is connected to a 2in square plate insulated from chassis and fixed by three nylon and one metal screws. Note that the insulation is on the underside.

The screw is connected to the 2·2k resistor, which is in turn connected to the chassis plate. To read grid current this connection is lifted off earth and a meter connected between it and the chassis plate. The insulation of both the ht and grid bypass capacitors is 0·004/6in thick mica or other similar material.

The input circuit L1 C1 is fitted into the top left-hand corner close to the valve and the input socket. Contact to the valve anode and grid is made by phosphor bronze fingering $\frac{7}{32}$in long soldered to the anode and grid plates. The fingering should face the same direction.

The completed unit is operated with valve anode downwards, ie standing on the box in the normal manner.

Adjustment: For full power operation, a drive level of 10W is needed. The input circuit is tuned for maximum grid current which when correctly adjusted should be between 30 and 40mA. When the input has been properly adjusted apply ht of 250–450V to the anode and adjust the anode tuning capacitor and output coupling circuit for maximum output. The indicator described for the previous unit should be used to assist in tuning up the anode circuit.

With inputs up to 50W no forced cooling is needed, but if the valve is to be operated at its full input rating provision for forced air cooling must be made. Since the diecast box forms a cavity at the operating frequency, any holes which are cut in the sides of the box to allow cooling to be blown across the anode radiator should be covered with mesh to prevent disturbance of the rf field.

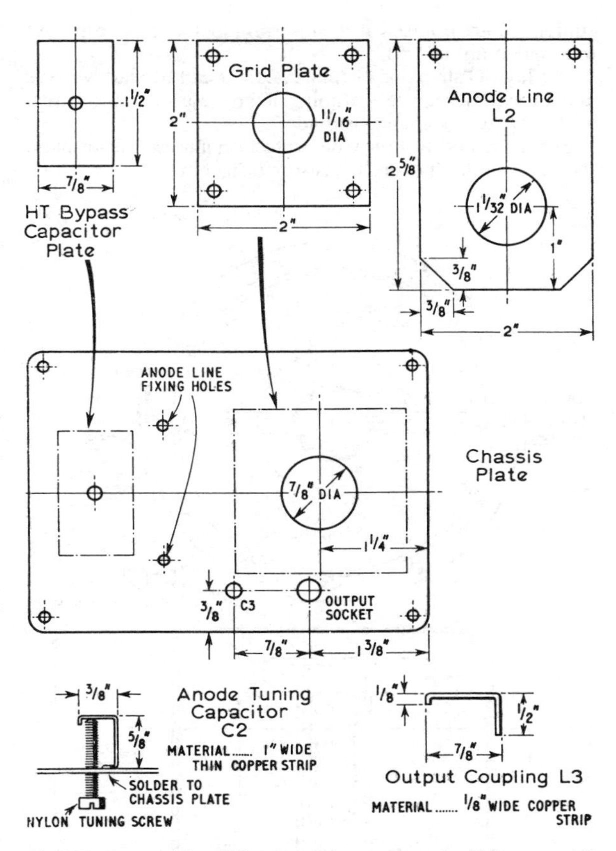

Fig 7.91. Constructional details of the anode and grid lines used in the 2C39A 70-to-23cm tripler

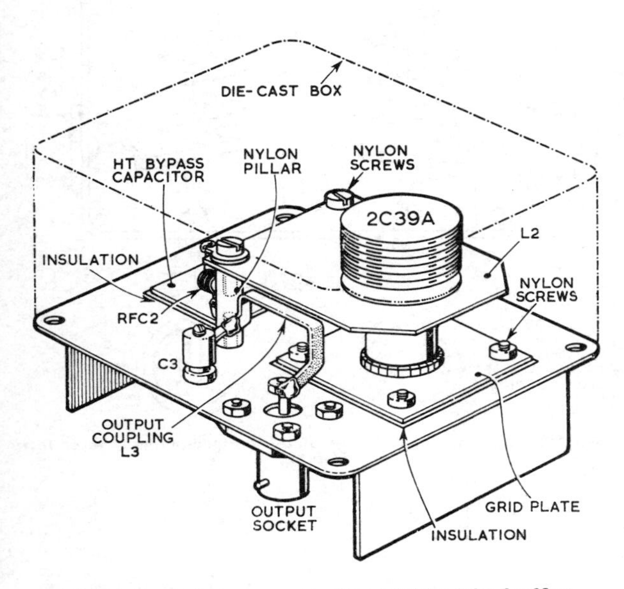

Fig 7.90. General arrangement of the 2C39A tripler for 23cm

For local (short range) working sufficient output can be obtained when the ht lead is connected to chassis (no anode voltage).

FILTERS

The amateur vhf and uhf allocations are situated among television, fm broadcasting, air traffic control and other services (see Fig 7.1 (b)). Special care needs to be exercised by the amateur to obviate interference to or from these services. Common practice with hf band equipment is to use a low pass filter at the transmitter and a highpass filter at the tv or other receiver input, but at vhf the method most likely to be effective is to use some form of bandpass filter for each of the bands used. It may be either a multi-element unit with a relatively wide band of, say, 5 to 7 per cent of the frequency in use, or a narrow band high-Q strip line version tuned to the actual transmitter frequency.

The bandpass filter is probably the more elegant method but its design and setting up without fairly elaborate test equipment are not easy if the insertion loss is to be restricted to a maximum of 1dB.

The high-Q strip line or tube line filter can be readily made and tuned. It requires retuning if any considerable transmitter frequency change is made.

Either method will provide protection against out-of-band radiation, notably from transistor transmitters.

A bandpass filter installed in the aerial feeder will not only restrict out-of-band radiation but will also reduce receiver desensitization caused by strong out-of-band signals.

Bandpass filters which are to pass significant rf power must be constructed with adequately rated components, eg, air spaced adjustable capacitors and large diameter wire for inductors.

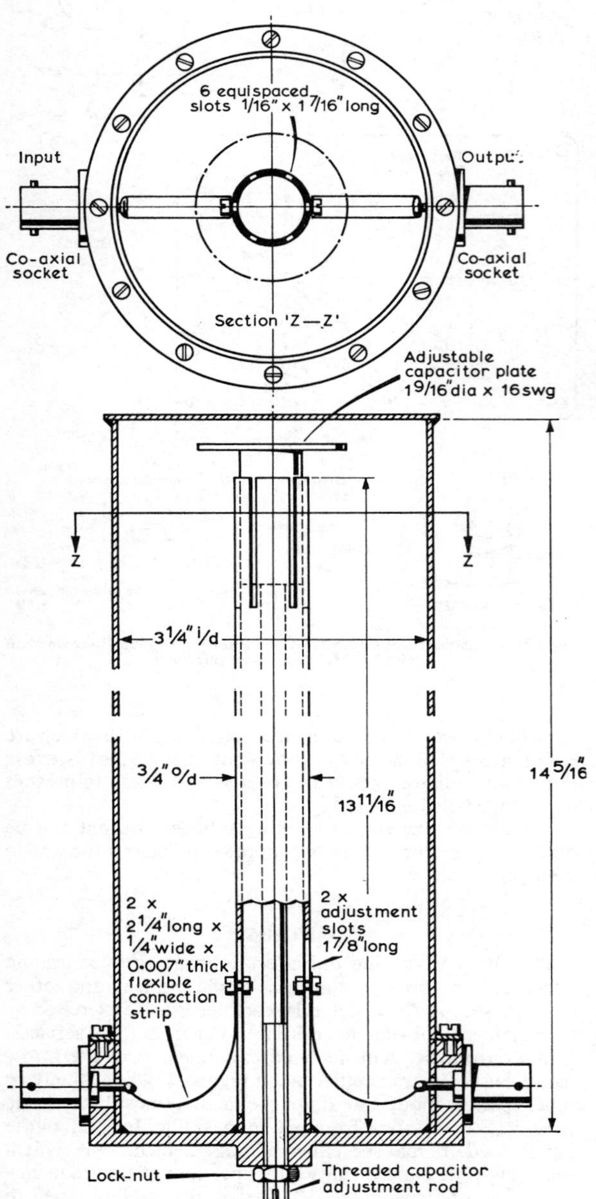

Fig 7.92. Constructional details of the high-Q filter for 144MHz

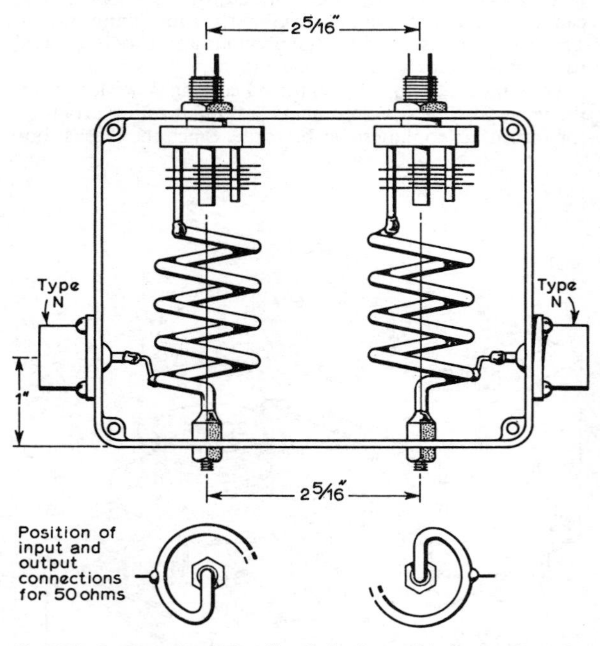

Fig 7.93. Constructional details of the lumped-inductor filter for 144MHz

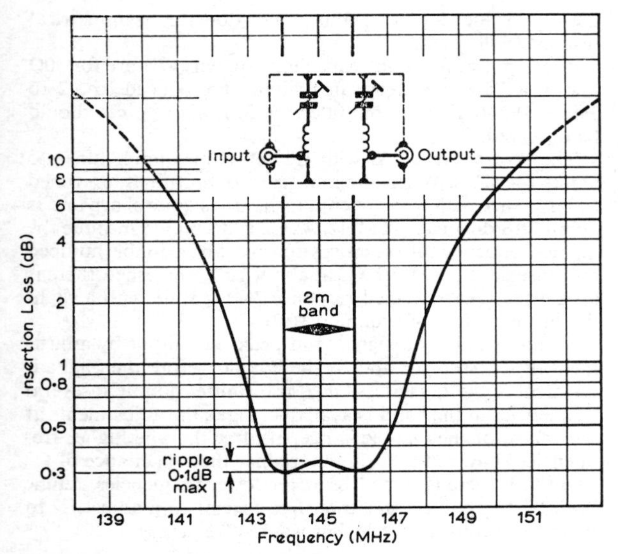

Fig 7.94. This graph shows the performance to be expected from the lumped inductor filter for 144MHz

A HIGH-Q FILTER FOR 144MHz

The high-Q filter shown at **Fig 7.92** consists of a copper or brass tube 3¼in inside diameter by 14$\frac{5}{16}$in tall, in which a single hollow inductor rod terminates at the top in a capacitor plate. The rod for adjusting this capacitor plate passes along the centre of the inductor and has a slotted base to accommodate a screwdriver head.

Versions for 70MHz or 432MHz may be constructed by scaling the length of the inductor accordingly. The coupling loops at the base must be symmetrical so that the input and output impedances remain the same.

The filter should be peaked on a steady signal for "reception" or for maximum transmitter output on "send" by adjusting the screwed end of the coaxial rod. Insertion loss should not be greater than 1dB.

A COMPACT FILTER FOR 144MHz USING LUMPED INDUCTORS

Rather simpler than the foregoing high-Q filter is the lumped inductor band pass filter shown at **Fig 7.93.** It employs two slightly over-coupled tuned circuits providing an adequate bandwidth (4MHz) to cover the whole of the 2m band, but with a negligible insertion loss of 0·4dB. Dimensions both for 144MHz and 432MHz are given.

Coupling to each inductor is by a sliver of solid wire. The tapping points shown are for 50Ω, but will need to be moved slightly higher up the coil for higher input impedances, and should be finally set by using a signal source in the 2m band. At the same time the top capacitors will require adjustment for optimum signal strength.

The filter is contained in an aluminium die-cast box 4¾in by 3¾in. The inductors are made from ⅛in diameter copper wound on a ⅞in mandrel; they consist of four full turns plus that required at the lower end to reach the central fixing. Care is needed to ensure that the coil spacing is correct; with the dimension given there is sufficient overcoupling to give a small ripple at the nose of the response curve. The method of fixing adopted was to solder the ⅛in diameter copper into a socket as used for the Erie type K1700 stand-off capacitor (see Fig 7.93), with the capacitor removed. Any alternative method of making a good electrical contact to the box would be suitable, bearing in mind the relatively high circulating current involved.

The tuning capacitors are Jackson type C804 trimmers 3·5 to 15pF with two plates removed from both the rotor and stator, leaving three fixed and three moving.

When the capacitor and coil are mounted and soldered together, the circuits should be rotated so that the tap points are opposite the input and output connectors.

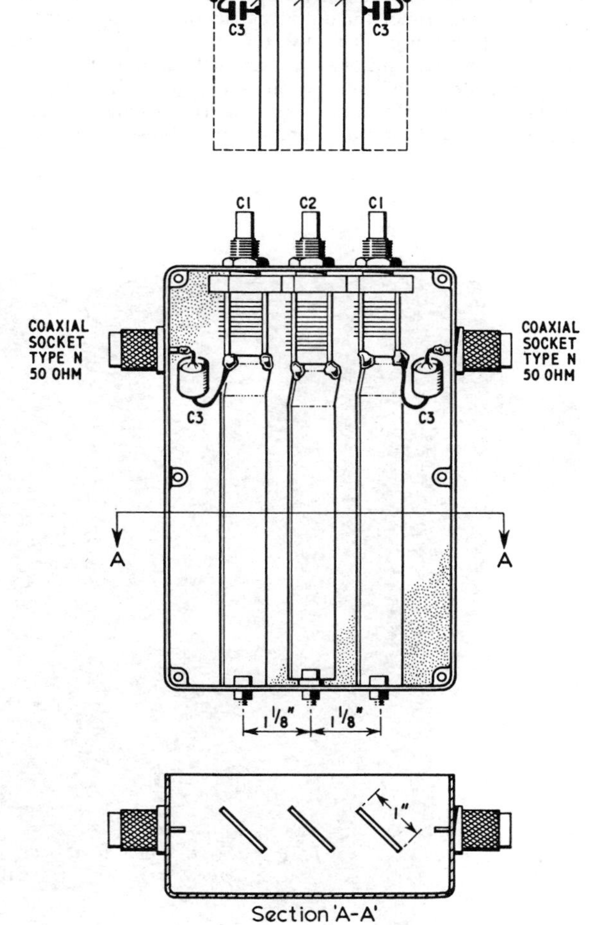

Fig 7.95. Mechanical layout and electrical circuit of the 144MHz strip-line band-pass filter for use with a high power 2m transmitter. The length of the central line is slightly shortened to allow for the rib in the cast box and the longer capacitor. Otherwise the line lengths should be 6¼in for 144MHz and 2¼in for 432MHz, both in 1in flat copper strip. Capacitor values at 144MHz are: C1 50pF, C2 60pF C3 4·4pF. At 432MHz they become: C1 50pF, C2 15pF and C3 5pF. Insertion loss is only 0·6dB at 144MHz and 1·2dB at 432MHz. For the higher frequency band a smaller 4¾in by 3¾in box is used

A HIGH POWER BAND PASS FILTER FOR 144MHz

A bandpass filter for transmitters delivering 100–120W output in the 2m band is shown at **Fig 7.95.**

Three shortened strip lines are tuned by capacitors at the top end. Input and output coupling is through small fixed capacitors to the top end of the outer lines. The third (central) line is free and couples the input and output circuits.

The strip lines are made of 1in wide copper at an angle of 45° (incidentally this angle will be provided naturally if the standard type of trimmer is used with the centres given). Fixing of the line to the end wall of the box is provided by 0BA brass cheesehead bolts with saw cut opened to allow the line to be brazed centrally to the end.

The top end of the line is bent to make contact with both the capacitor stator pillars, which are soldered directly to the line after assembly—a fairly large soldering iron will be needed for this. Input and output capacitors are taken directly from the connectors to the top of the respective line. Setting up is straightforward provided an output power measuring device of some form is available. Initially C2 should be set near maximum capacitance: this and the C1's are then adjusted for maximum output, taking care to keep the C1's value similar. If this is not done an asymmetrical response will result.

The input and output capacitors shown (C3) are for 50Ω line. For 75Ω line their value should be reduced to 3·2 to 3·4pF. When aligning the filter the following points should be observed:

The central strip line with C2 set at maximum tunes to about 144MHz. When the input and output lines are tuned to resonance with it the overall response is quite sharp and is about 3dB down at ±1MHz. As C2 is decreased in value the higher frequency cut-off moves upwards and a double humped response develops. With C2 at about 10% unmeshed there is an almost square shaped response 2MHz wide with a slight dip in the middle of around 0·1dB.

To adjust, apply a signal, and peak the output by adjustment of C1. Replace and fix the box lid before the final adjustment of the capacitors C1. If transmitter output at several frequencies in the band is available, make the adjustments at the centre of the wanted range. If after this the higher frequencies show reduced output, reduce the capacitance of C2 slightly. If there is some loss at the lower frequencies, retune both C1 to optimize the lower frequencies, and adjust C2 to maintain output at the higher wanted frequency.

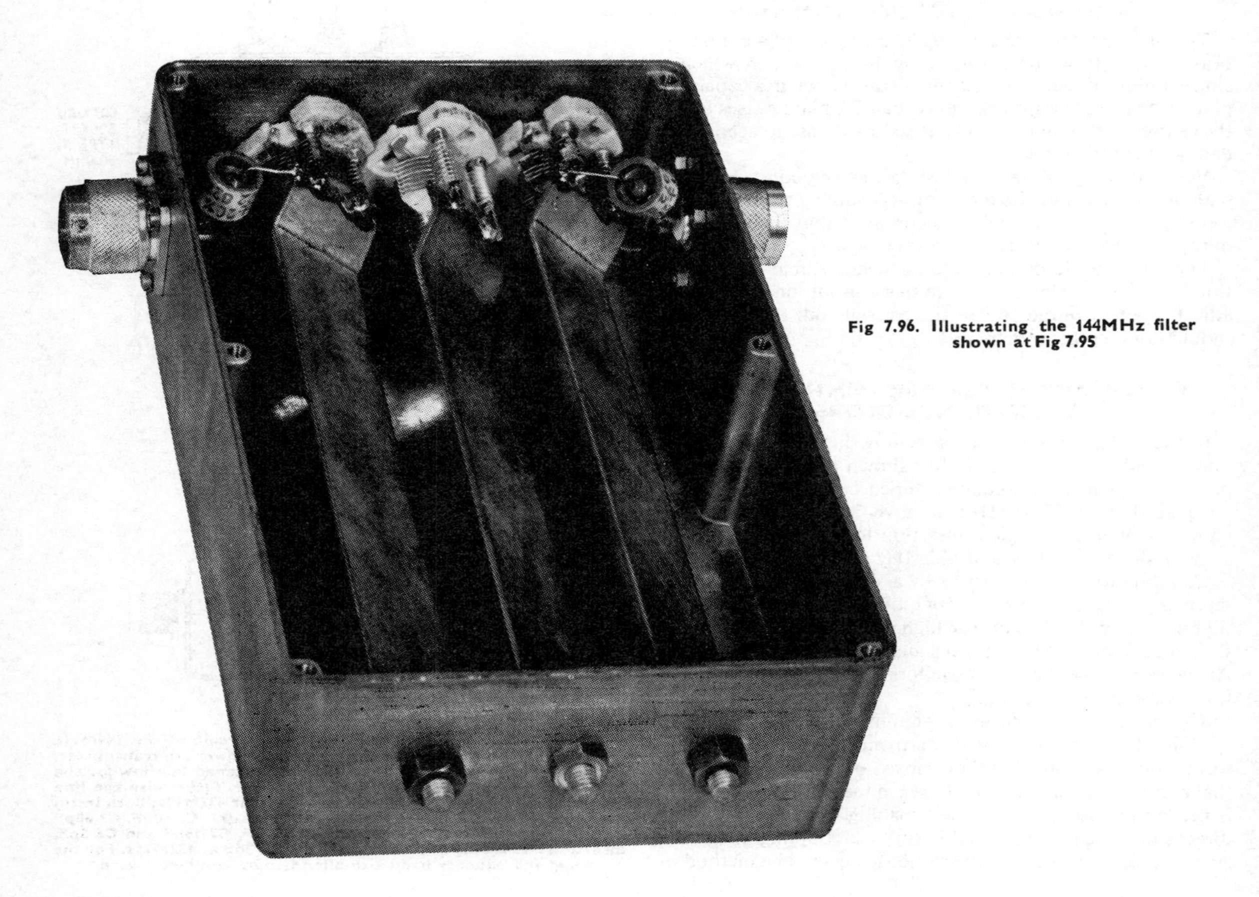

Fig 7.96. Illustrating the 144MHz filter shown at Fig 7.95

CHAPTER 8

KEYING AND BREAK-IN

TO impress intelligence on a carrier wave, it must undergo a process of modulation, and the simplest way of accomplishing this is on/off keying of a transmitter. This necessitates encoding the information in the form of specific trains of pulses of different lengths, to represent each character, the usual system being the morse code. The only method of keying the transmitter to be discussed in this chapter consists of changing the amplitude of the carrier; an alternative method, known as frequency shift keying, is to vary the frequency, but this offers very little advantage for normal morse communication purposes, and is usually applied solely to operating teleprinters. This subject is considered in Chapter 10.

The bandwidth required by a properly keyed signal is small, and should be directly related to the speed of sending involved. The optimum condition is not always easily attained, however, and the result of a maladjusted keying system is usually the radiation of spurious emissions over large parts of the frequency spectrum. It is also possible to obtain the opposite effect, however, through an attempt to avoid harmful keying transients; the result of this condition is to produce a very "soft" characteristic which will prevent the reading of morse signals at high speeds.

Morse code characters are made up of dots and dashes, each character consisting of a unique combination of these elements.

The standard morse code (International Telegraph Code No. 1) dictates that a space between the elements of a character shall be one dot length, and that the length of a dash and the interval between letters shall be three dot lengths. Fig 8.1, showing the length of a dot for a range of transmission speeds, should be used for reference when a keying waveform is being tailored, because it is important to provide the correct amount of softening of the key envelope in relation to the upper speed which is to be used: the rise or decay time should not exceed about one-third of a unit length at the highest speed, so that at 24 words/min, such times should not exceed about 15ms.

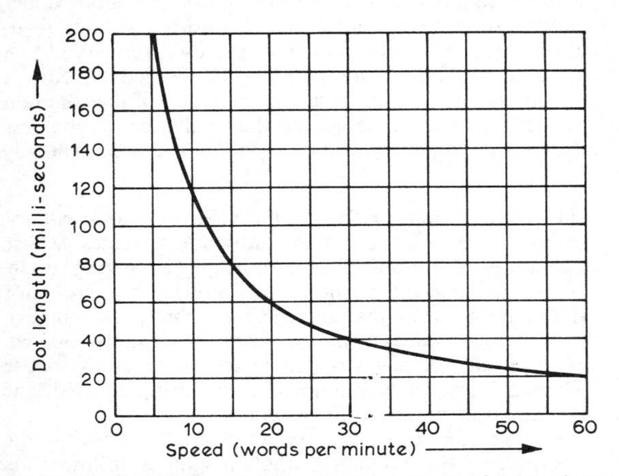

Fig 8.1. Graph showing dot lengths for a range of transmission speeds. Once the design figure of the maximum speed of sending for a transmitter has been decided, this graph can be used to select the appropriate time delays in keying filter circuits.

If a signal has its dots or dashes too long with respect to the spaces between them, it is said to be weighted. This term is used particularly in connection with keying relays, as a slow component would clip the dots of a transmission unless some weight is added to the switching pulses to compensate for it.

KEYING CHARACTERISTICS

There are two main components which affect keying characteristics: envelope shape and frequency stability. Any keying trouble such as key clicks, ripple, chirp and spacer-waves can be attributed to poor conditions in one of these areas.

Envelope Shape

The envelope of a keyed signal is the outline of the pattern that the signal would display on an oscilloscope. It can be observed by feeding the keyed rf signal on to the Y plates or vertical amplifier of a slow-scan oscilloscope, and setting the timebase in synchronism with the keying speed. The transmitter is best keyed by an automatic device producing a regular sequence of dots or dashes so as to obtain a steady display.

In general, if no precautions are taken, the pattern will be square and sharp, or "hard" (Fig 8.2(a)), and will radiate interference over a wide range of frequencies. The rise and decay times of the carrier must therefore be lengthened until the interference is no longer objectionable, but without impairing the intelligibility at high speed. This requires rise and decay times of 5–20ms, and the methods of achieving this will be described in the sections on different keying systems.

Key filter circuits rely for their operation on the time constant of a CR (or LR) circuit (see p 1.15). Filter circuits are used to limit the rate of change of the bias voltage to the keyed amplifier stage, so that the operating conditions of the stage do not switch abruptly from conduction to cut-off.

The characteristics of the pa power supply may contribute to the envelope shape, as the voltage from a power unit with

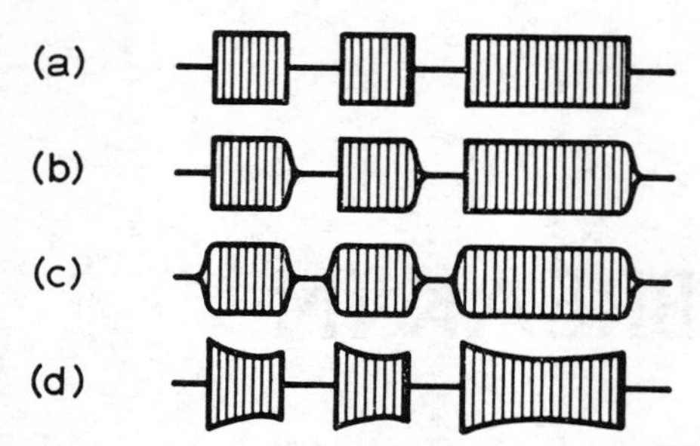

Fig 8.2. Keying envelope characteristics. (a) Click at make and break; (b) click at make, with click at break suppressed; (c) ideal envelope with no key clicks; (d) effect on keying envelope of poor power supply regulation.

poor regulation will drop quickly each time the key is closed, and rise when it is released. This can lead to the shape in **Fig 8.2(d)**. This effect is not necessarily undesirable, but it can be lessened by use of a choke-input filter in the pa supply, or by increasing the size of filter capacitors in the supply.

Envelope ripple can be caused by rf feedback, but poor power unit filters are generally responsible. If ripple is present on the pa supply, the carrier will be amplitude modulated at the ripple frequency, whereas on the oscillator supply it would probably cause both frequency and amplitude modulation. Clipping a large electrolytic capacitor on to various points in the transmitter will usually indicate the necessary treatment.

RF Clicks

Although clicks caused by a hard keying envelope are radiated with the signal, local rf interference may be caused by sparks at the key contacts, particularly if an appreciable current is keyed in an inductive circuit. This interference is usually removed by connecting a capacitor (typically 0·005μF) directly across the key or keying relay contacts (**Fig 8.3**), and in severe cases by also inserting an rf choke (1 or 2mH) in the live keying lead. The effectiveness of such treatment is judged by listening to the station receiver tuned to a frequency well removed from the transmitter frequency.

Some semi-automatic "bug" keys and keying relays tend to give more trouble with clicks than hand keys because of contact bounce or mis-shapen contacts. With bug keys, contact bounce can often be prevented by adjusting the position of the moving dot contact so that only about one-quarter of the contact surfaces are in use, and by mounting a small block of foam rubber inside the dot contact U-spring (**Fig 8.4**).

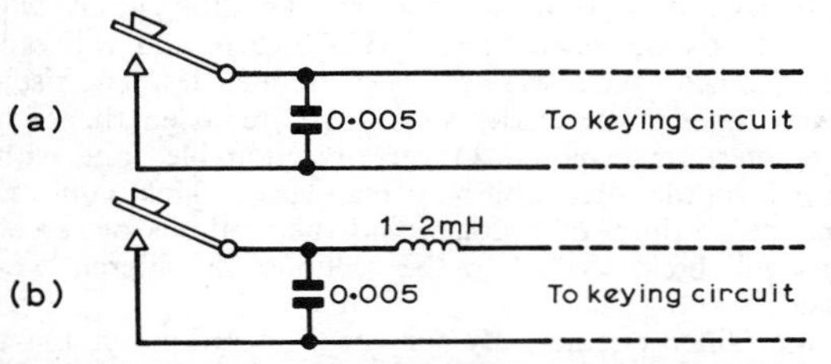

Fig 8.3. Circuits for suppression of interference caused by arcing at the key contacts. The rf choke will be necessary only in severe cases. The 0·005μF capacitor must be mounted directly across the key contacts. If an rf choke is used, it must be able to withstand the current flowing in the keyed circuits; in the case of cathode keying of a high power stage, this may be considerable, but then a high current choke is desirable so that a minimum of resistance is added to the cathode circuit.

Worn, pitted or dirty contacts are obviously undesirable, and great care should be exercised in treating them. Only a burnishing file should be used; anything rougher would leave a burred surface which would only cause further trouble.

In some transmitter designs, the power amplifier is clamp-valve controlled and an earlier stage keyed. If the pa is operated in Class C, it is capable of sharpening the softened signal from the buffer stage, thereby reintroducing clicks on the radiated signal. This is one reason for using sequential keying of more than one stage. The pa may also be triggered into low-frequency parasitic oscillations, again causing clicks. This is usually due to poor choice of rf chokes in the pa grid and anode circuits.

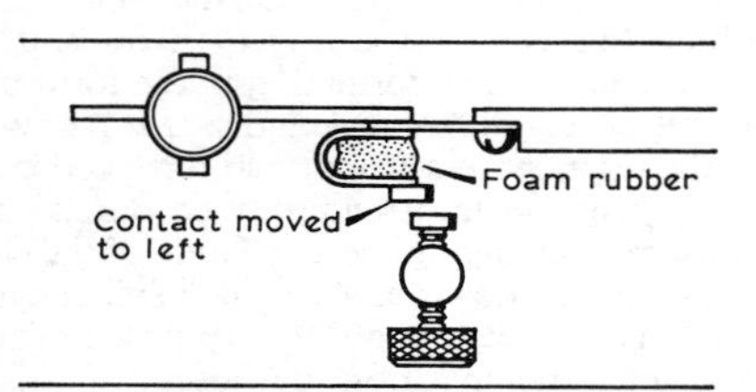

Fig 8.4. Method of reducing contact bounce from "bug" key dot contacts. Note the displacement of the two contacts.

Chirp

Chirp is a form of frequency instability occurring each time the transmitter is keyed, and is recognized by a change in beat frequency at the beginning and end of each character when the signal is monitored on a receiver. A signal with chirp is undesirable, as it is less easy to copy, and is not suitable for reception by narrow-bandwidth receivers. It is usually more prevalent in transmitters controlled by a vfo, and there are three principal causes, which are:

(*a*) *DC Instability*. DC instability occurs when a common power supply is used for the oscillator and the pa (or any circuit through which the current changes in sympathy with keying). No oscillator exhibits absolute stability under varying supply conditions, and therefore a voltage regulator should be incorporated in the oscillator supply. A separate oscillator supply might be needed to cure a difficult case, but improving the regulation of the common power unit and redesigning the oscillator to be less dependent on supply voltage variations should generally suffice.

(*b*) *Pulling*. Pulling refers to the effect on the oscillator frequency of one or more subsequent stages whose operating conditions change during the keying cycle. It can be expected if the stage following the oscillator draws input current (as would a Class C buffer or doubler), or if the early stages of the transmitter are tightly coupled. If the oscillator is on the same frequency as the pa, ie where no frequency multiplying or mixing is used, the likelihood of pulling is increased.

Pulling can invariably be treated by improving the isolation of the oscillator, and an emitter follower* is

* A cathode follower would be used in valve circuits.

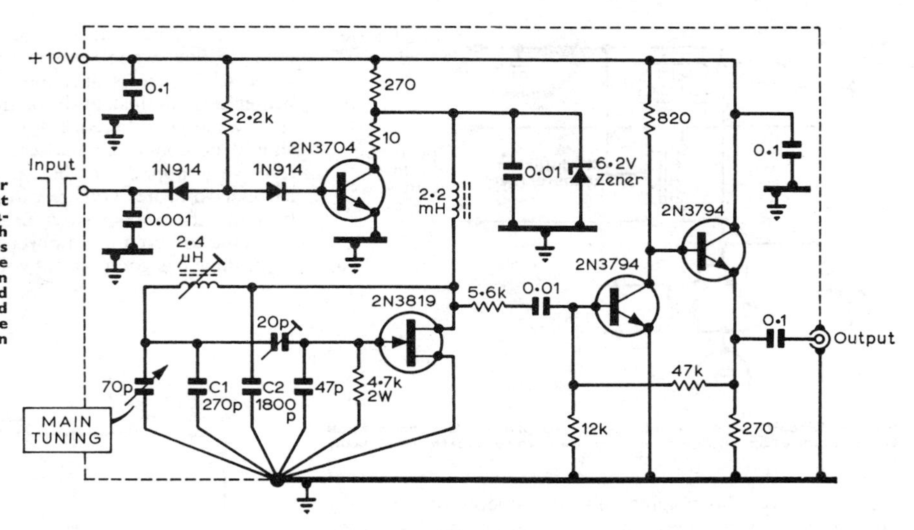

Fig 8.5. VFO and emitter follower designed to be stable against supply voltage and load fluctuations. Operating frequency with the component values shown is 5·88–6·38MHz, but this can be altered by changing L, C1 and C2 in inverse proportion to the desired frequency shift. The circuit around the oscillator permits it to be keyed for use in full break-in systems (see p 8.15).

highly recommended for this. The oscillator should be loosely coupled to it, and by careful design it should be possible to produce a unit whose output can be short-circuited without shifting the frequency by more than a few hertz. **Fig 8.5** shows a circuit combining a stable oscillator with an emitter follower.

It may be simpler to replace the frequency-determining components in the oscillator with values to halve the original operating frequency and to use a subsequent stage as a doubler. This may, however, reduce the drive to the pa to such an extent that a further buffer stage would be necessary.

(*c*) *RF Feedback*. RF feedback is a stray signal leaking back from a high-level stage to a previous stage, particularly a variable frequency oscillator. It may have an appreciable effect on the frequency of oscillation, depending on its strength and phase. The presence of feedback may be verified by noting the pulling which occurs on tuning the output circuits of the keyed stage through resonance. The feedback path may be either internal, by virtue of valve capacitances, or external, because of poor constructional layout.

Internal feedback may be treated by the methods outlined for pulling; isolation of the oscillator is of great importance. External feedback is only discovered after a transmitter has been built, and the commonest cause is the pa valve and circuitry being close to the oscillator section. Here also, pulling is most likely to occur when the pa runs on the same frequency as the oscillator, and the cure is either to resort to doubling, as before, or to screen both pa and oscillator. Sometimes it is sufficient to mount a metal plate between the two circuits.

It is recommended that any long leads carrying rf power near the oscillator be screened. In addition, the ht line must be bypassed to rf by means of series resistance and shunt capacitance.

Spacer Waves or Backwaves

The spacer wave is the small signal often radiated during key-up conditions. It is common for the spacer wave to be audible in the station receiver, but this does not mean that it is radiated far. If an appreciable spacer wave is radiated, it would make a signal difficult to copy.

Causes are (i) the keyed transmitter stage not being completely cut off, and (ii) leakage of rf through the pa valve capacitance.

The first is often due to insufficient bias being applied to the keyed stage. If drive is applied to the pa continuously, the bias required for cut-off during key-up conditions will be the static value obtained from data sheets plus the peak value of the drive voltage, which is considerable. In a cathode-keying system, a leaky capacitor in the keying filter may cause a spacer wave.

In valve circuits, the second effect increases with the grid-to-anode capacitance of the pa: it is considerably worse with triodes than pentodes. It may be cured either by neutralizing the stage in question or by keying more than one stage.

KEYING METHODS

There are many possible methods of keying, and the choice is largely one of practical convenience, personal preference and, particularly in break-in systems, suitability to the station as a whole. The important requirements of the system adopted are avoidance of the troubles outlined in the previous section, and the operation of all valves in the correct regions of their characteristics.

Almost any stage in the transmitter may be keyed, but there are good reasons for keying the final power amplifier rather than an earlier stage. For instance, if the oscillator is keyed, the requirements of a short time-constant to reduce chirp and a long time-constant to eliminate clicks conflict. If any stage before the pa is keyed, with softening, the pa may harden the keying, as described earlier in connection with key clicks. In some cases it is useful to key more than one stage sequentially, and this is covered in the section on break-in.

When keying the pa of a valve transmitter designed for anode modulated telephony it is necessary to short-circuit or remove the modulation transformer. If this is not done,

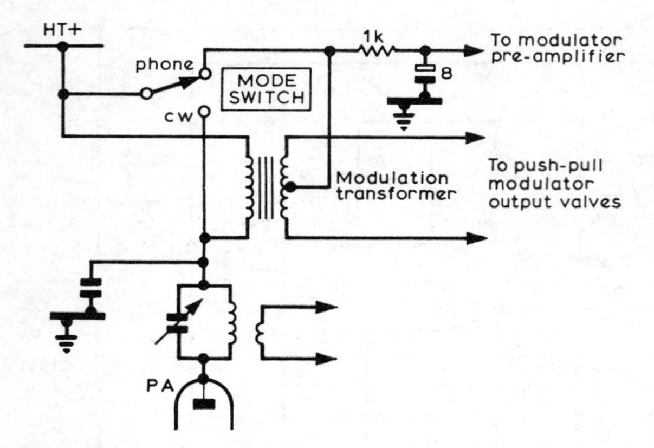

Fig 8.6. Circuit using a single switch to protect the modulation transformer of an a.m./cw transmitter and to control the modulator ht supply.

there is a risk of damaging this component with the very high voltages which may appear across its windings when the anode current of the keyed stage is switched. There is also a possibility of generating interference owing to the transformer "ringing".

If, in an anode modulation system, the modulator output valves are fed by the same ht supply as the pa, switching from one mode to the other may be achieved with a single pole changeover switch (**Fig 8.6**).

HT Supply Keying

A neglected but very reliable method, corresponding to anode modulation in a telephony transmitter (or anode-and-screen modulation if a tetrode or pentode is used), is to key the ht supply to the pa stage. In the circuit shown in **Fig 8.7** a filter (L, C, R) is incorporated to produce the desired shape of keying envelope. For reasons of safety, this method usually requires the use of a keying relay.

Cathode Keying

One of the most popular systems of keying is to insert the key in the cathode lead of one or more stages, often the pa alone. It has, however, three disadvantages.

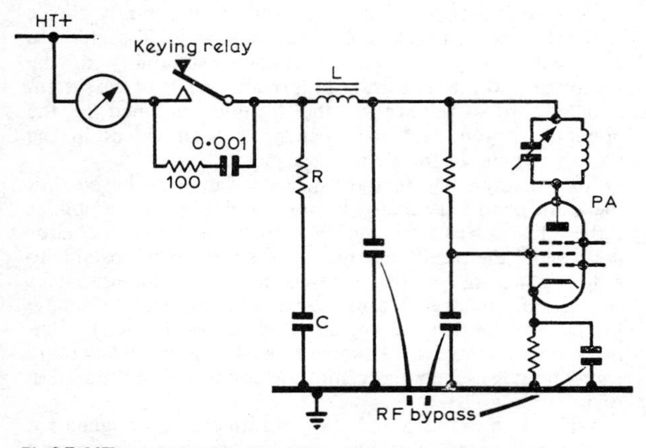

Fig 8.7. HT supply keying. An rf filter is connected across the keying relay contacts to suppress local clicks.

(i) Unless precautions are taken the cathode potential tends to float when the key is open and may rise to such a value that there is a serious danger of breakdown of the heater/cathode insulation of the valve. To avoid this excessive potential difference a resistor not exceeding 0·25MΩ should be connected between the cathode and the heater. If drive is applied to the valve when the key is open, the cathode-to-heater voltage will rise to the sum of the cut-off bias and the peak positive drive voltage, and it is therefore desirable to avoid any unnecessary rise by ensuring that the drive stage has good output voltage regulation.

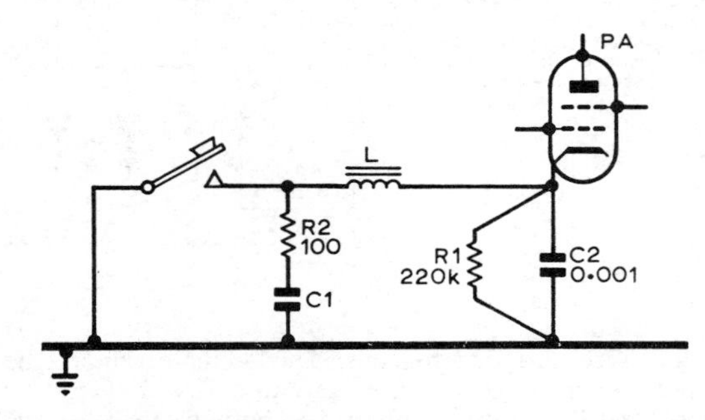

Fig 8.8. Cathode keying. The rise time is determined by L, a low-resistance choke of 1–5H. The decay time is determined by C1 which can have a value of 0·5–2·0μF. R1 is a safety resistor, and should not be more than 0·25MΩ.

(ii) Cathode keying is contrary to the practice, particularly in high level amplifiers, of avoiding any rf impedance in the cathode lead, as this is a common cause of parasitic oscillations. It is widely recommended, especially for vhf transmitters, that a short and substantial copper strip be soldered between the cathode connection and chassis. If cathode keying must be used in any stage handling appreciable rf currents, the keying lead must be by-passed to rf at the cathode pin. About four high-grade capacitors, typically 0·01μF each, with low internal inductance, should be used in parallel.

(iii) As the cathode circuit of a pa valve carries a high current, any key-click filter must include a series inductor, which will add its internal resistance to the cathode circuit thus introducing an unwanted additional bias. The operation conditions should be adjusted to allow for this.

The optimum component values of the click filter circuit are best found by experiment. When the key is depressed, capacitor C1 (**Fig 8.8**) is short-circuited through the key (the resistance R2 being included to limit the discharge current to a reasonable value) and the capacitor therefore plays no part in determining the rise-time of the carrier. Thus the first step is to choose a suitable inductance value to give an acceptable "make". The value of C1 may then be altered to give the desired decay time, or "break" characteristic.

Grid-block Keying

This system overcomes all three of the problems associated with cathode keying and is a very satisfactory method. It does, however, require a negative supply of up to −250V.

If a sufficiently large negative bias is applied to the control grid of a pa or other stage, no cathode current will flow even in the presence of drive voltage. For most tetrode or pentode

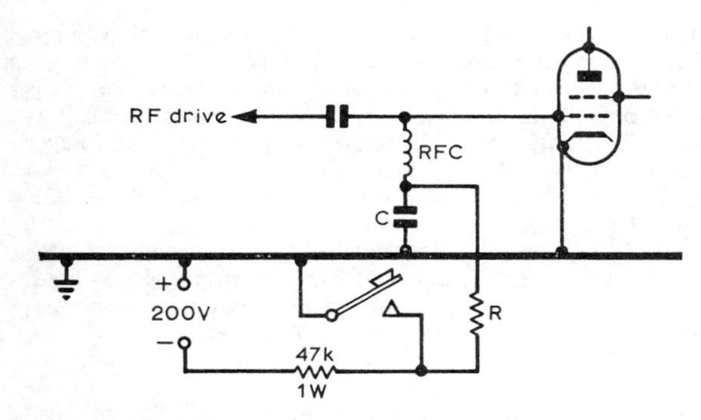

Fig 8.9. Grid block keying. The resistor R has the normal bias value, and the keying envelope is determined mainly by the time constant of R and C. The product of R (in kΩ) and C (in μF) should be about 5. The bias supply should be capable of supplying 4mA in the key-down condition and should have an open-circuit output of about 200V.

power amplifiers a bias of about 200V will be found adequate: for oscillators and other low-level stages, 50V or less is sufficient. As there is usually little or no grid current flowing in the stages to be keyed, this method lends itself to key-click filters using R and C components only. A suitable circuit is shown in **Fig 8.9**. The capacitor C is not only part of the softening circuit, but also serves to decouple any rf remaining at the earthy end of the rf choke. It should therefore be mounted near the keyed valve.

In the circuit shown the voltage across the key contacts will rise to 200V with the key up, but the impedance is high enough to prevent dangerous currents from flowing through the operator, and a hand key may be safely used. The voltage must not be allowed to exceed the specified maximum dc negative control-grid potential, or there is a danger of flashover, particularly in high-slope valves.

Grid-block keying is easily applied to several stages simultaneously, each of them having individual hardness characteristics. By adding a circuit to invert the switching of the blocking bias, the receiver muting may be controlled by a similar system operating on the agc line. This is discussed further in the section devoted to break-in.

Screen Grid Keying

Screen grid keying permits a pa of reasonable power to be keyed with contacts and filter components of low power rating. In its simplest form the method consists of breaking the positive supply line to the screen, but most pentodes and tetrodes do not completely cease to conduct when the screen is isolated and it is usually necessary to apply a negative bias of the order of 20–50V in the key-up condition to eliminate the spacer wave. The bias may be derived from a separate bias unit.

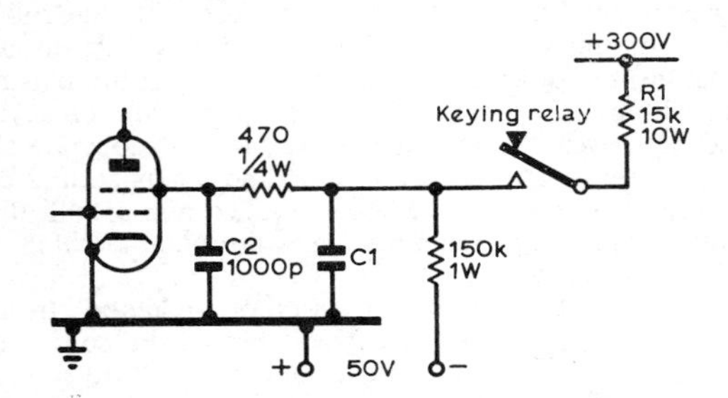

Fig 8.10. Screen-grid keying. The values shown are typical for a 100W amplifier, but may vary widely. R1 will be somewhat lower than would be required if it were a simple series dropper resistor. The capacitor C1 is chosen to give suitable hardness to the keying: a typical value is 0·25μF. C2 is the rf bypass capacitor.

A typical screen grid keying circuit suitable for 100W amplifiers is shown in **Fig 8.10**. This arrangement causes the "make" to be harder than the "break", although a measure of such disparity is often favoured.

In a differential keying system, involving the transmitter oscillator, screen grid keying of the oscillator is recommended, as a negative supply is unnecessary. This is particularly suitable for control by the back contacts of the key, as shown in **Fig 8.11**.

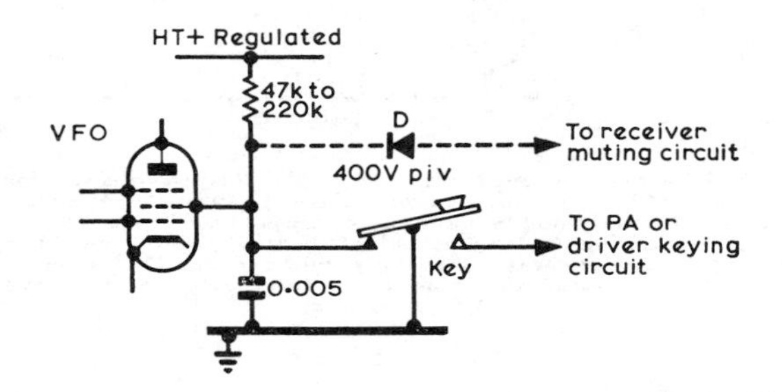

Fig 8.11. Simple circuit for obtaining differential keying of both oscillator and driver or pa using back and front contacts of the key. For full break-in operation, the receiver also may be controlled by the back contacts in the manner shown later in Fig 8.33, but diode D must be included to prevent the receiver muting circuit from reducing the vfo screen voltage. This method cannot be used with "bug" keys or many types of el-bug unless a keying relay is used to carry out the switching functions.

Suppressor Grid Keying

The output from a pentode amplifier may be reduced to zero by the application of a sufficiently high negative voltage to the suppressor grid, but this is seldom used as a method of keying. When the suppressor is biased to cut-off the anode current, heavy screen-grid current will flow and the screen dissipation may be exceeded. Moreover, the suppressor potential is a very insensitive form of control in most pentodes, and the keying voltage is therefore large.

The Application of Keying Methods

In a transmitter not including break-in circuits, it is usual to apply keying in one of two ways. Either the master oscillator and other low-level stages function continuously and the pa is controlled, or the oscillator and power amplifier are left on while an intermediate stage is keyed. The latter system has the advantage that low-level stages are easier to key properly but most Class C pa stages would need to be protected in some way under key-up conditions.

There are several ways of doing this, which are:

(a) A clamp valve may be added to the screen grid circuit of the pa.
(b) The pa can be redesigned to operate in a linear mode, such as Class B or AB, where its operating conditions are such that it passes safe currents, within the valve dissipation limits, when drive is absent. Although the pa efficiency would be lowered, harmonic output would be much reduced.

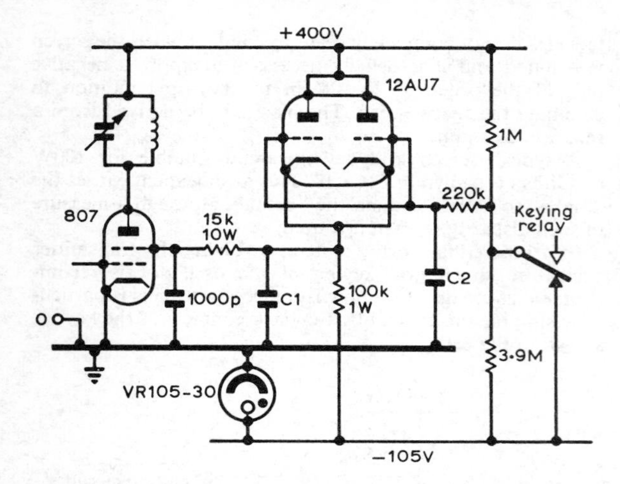

Fig 8.12. Valve keying of a screen grid supply. This circuit is suitable for inputs of 25–35W using a pa valve of the 807 class. The capacitors C1 and C2 determine the hardness at "break" and "make" respectively; suitable values for normal keying are C1 = 0·1μF and C2 = 0·005μF. A separate heater supply is required for the 12AU7 keyer valve, and the heaters should be strapped to the cathode through a 220kΩ resistor.

Keyer Valves

In a Class C amplifier every electrode, except the suppressor grid in a pentode, normally passes a current. This means that the keying circuit must be designed to deal with power and not merely voltage, and efforts must be made to prevent local interference being caused through arcing at the contacts or radiation from the key leads. A further problem is that in most circuits it is difficult to obtain the desired degrees of softness at both make and break with the same component values, and a compromise is necessary.

It is occasionally worthwhile to overcome these drawbacks by introducing an auxiliary valve used as a variable resistance in series with the supply of some electrode of a convenient valve in the transmitter, the key being used to control the grid bias on the auxiliary valve.

Fig 8.12 shows a circuit where a 12AU7 (with the two sections connected in parallel) is used to key the screen supply to an 807. This method uses a contact on the keying relay which is closed when the relay is unenergized. The steady current flowing through the contacts is only 0·5mA

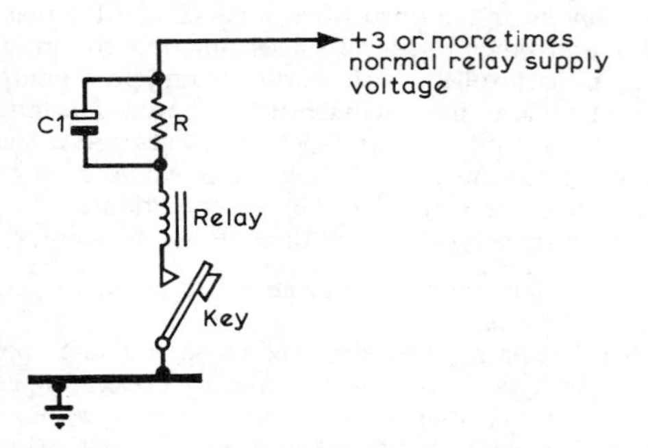

Fig 8.13. Circuit for speeding up the "make" action of a relay. The resistor R should be such that the steady-state relay current is about its rated figure, and C1 should be a few microfarads.

in the key-up condition, although the open-circuit voltage, ie, with the key depressed, is about 400V.

If the transmitter is controlled by an electronic key, it is often possible to dispense with relays in this arrangement as suitable voltages may be derived directly from the keyer circuit. With a 12AU7 keyer valve as in Fig 8.12, the two grids should be at +300V with the key down, and −100V with the key up.

There is little advantage in keying the master oscillator of a transmitter, unless break-in operation is required. Indeed, satisfactory oscillator keying is not easily achieved, and the system should be avoided if possible.

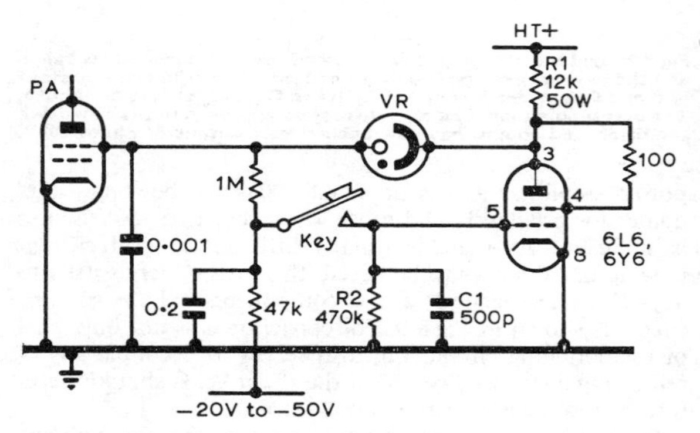

Fig 8.14. Circuit for using the clamp valve as a keyer valve. The negative supply may be derived from the pa grid circuit provided that drive is applied to the pa continuously. This circuit also is designed for valves similar to the 807. For other types, it will be necessary to modify the value of the screen dropper R1. The keying characteristics are determined by C1 and R2.

Keying Relays

In many systems, the voltages in the part of the circuit where the key is connected are low enough to preclude any possibility of shock to the operator, but if the voltages are higher than about 150V, or preferably 75V, a keying relay should be used. Such a relay should be of robust construction, rapid in operation, and its contacts and insulation must be adequate for the current and voltages in the keyed circuit. It should also be reasonably silent; if necessary a noisy relay can be quietened by mounting it on a sheet of rubber or foamed plastic.

Keying relays are particularly useful where several circuits are keyed simultaneously, but it is unfortunate that robust relays with high current and voltage ratings are usually neither fast nor quiet. Many relays take 10ms or more to act, and such devices would impose severe restrictions on keying speed. However, dry reed relays are fast, with operate and release times of 1 or 2ms, are quiet and compact, and can key currents of up to 1A. Mercury-wetted relays, with their freedom from contact bounce, may be found suitable for some applications.

It is possible to speed up the action of a relay on "make" by driving a large pulse of current through its coil at the beginning of each operation. The circuit of **Fig 8.13** shows a method of achieving this.

In a transmitter using clamp-valve efficiency modulation for telephony, it is possible to use the clamp valve itself as a keyer valve, as shown in **Fig 8.14**.

Many keyer valve systems have been devised and applied to power amplifier stages. In any circuit of this type, the keying envelope shape is determined from grid-block keying principles.

Changeover Switches and Relays

Switching the station between transmitting and receiving states usually involves several functions, and some of these may need to take place sequentially. Post Office key switches can be used, but relays are more satisfactory.

A multiple-section switch mounted on the front panel of the transmitter or receiver may necessitate long runs of cable carrying rf; it is advantageous to have a separate aerial relay, specially designed for the purpose and mounted at a convenient point on the chassis. When other units are involved (linear amplifiers and other accessories) it is more satisfactory to run a single line between them which will control their individual relay switching circuits. This may mean that equipment built at a later date can be accommodated more easily. It is also simpler to adjust the timing of relay circuits (by the use of capacitors, for instance) than multisection switches, and this is important for sequential switching.

Aerial switching can be avoided completely by using separate arrays for transmitter and receiver, but this is not recommended. Few situations allow the erection of two high-gain aerials, and so the communication efficiency of a station using this system is likely to be affected. If separate arrays are used, they should be as far removed from one another as possible, or large voltages may appear on the receiver aerial and resonance effects may drastically reduce the transmitting aerial efficiency. Separate aerials are necessary for full duplex operation, but where different bands are used, these may already be available.

A common problem is absorption of incoming signals by the transmitter pa and noise generated by the pa reaching the receiver. Both of these effects may be observed by swinging the output tuned circuit of the transmitter through resonance, and listening for fluctuations in signal strength and noise. The noise problem can be aggravated if the pa is not neutralized properly: it is also wise to ensure that it is completely cut off during receiving periods.

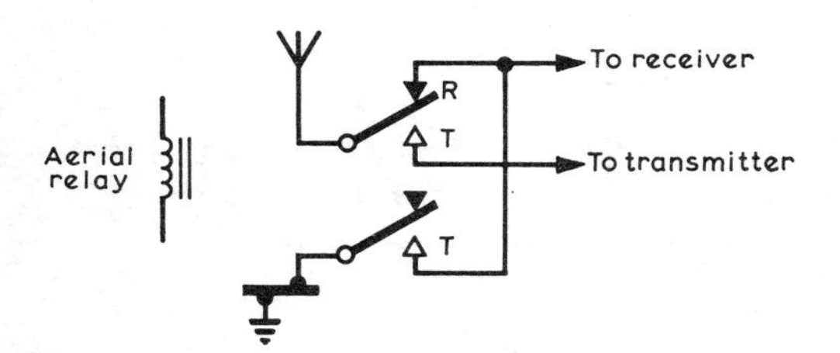

Fig 8.15. Circuits showing the use of the standard arrangement of contacts of aerial relays.

Aerial relays are found in many forms. Heavy duty devices with one changeover contact and one make contact are useful (**Fig 8.15**), but where better isolation and screening are required, coaxial relays are preferable.

Electronic aerial relays (T/R switches) are satisfactory, and these are described in the section on break-in, together with other methods of high-speed aerial switching.

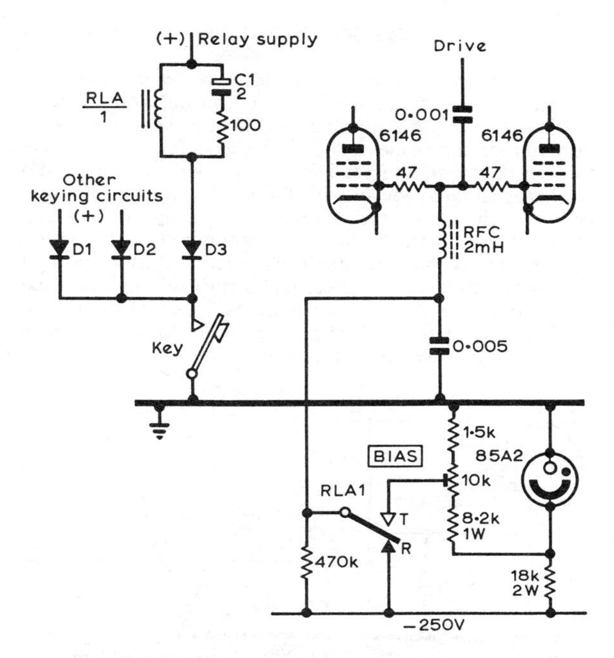

Fig 8.16. Dry reed switch circuit for linear amplifier control. Diodes D1, D2 and D3 (rated at 400 piv or more) ensure that different keying circuits do not interact: if other circuits have negative polarity, as in grid-block keying, the diodes, C1 and the relay supply must be reversed in polarity. R1 prevents the linear amplifier grid circuit from floating during armature travel. Although the circuit is designed for controlling valves of the 6146 class, the bias circuits could be redesigned to suit any valve type.

Control of Linear Amplifiers

Where the transmitting equipment includes a high power linear amplifier, it is desirable to bias such stages to cut-off during listening periods, either because of noise generated in the valves affecting receiver performance, or because the continuous anode dissipation rating is being exceeded. The problem of noise is more likely to occur in a system where an electronic T/R switch is used, and the final amplifier is connected directly to the aerial at all times.

A circuit which will switch the grid of a linear amplifier from operating bias to cut-off, regardless of the class of operation, is shown in **Fig 8.16**. It utilizes a dry reed relay, which switches the grid of the linear amplifier between TRANSMIT and RECEIVE states at a very high speed. The normal, or unenergized, state is that corresponding to the RECEIVE condition, where 200–300V negative is applied to the valve control grid.

The timing of linear amplifier switching is important. If the linear amplifier is switched on late, that is, after its exciter output has started building up, the envelope of the rf signal will rise sharply, causing spurious signals to be radiated. Similarly, there will be a sharp decay if the linear amplifier is turned off before the exciter output has fallen almost to zero. Hence, it is necessary for the circuit controlling the linear amplifier grid to switch on instantaneously, and off after a delay of 20ms or more. The circuit of Fig 8.16 switches to TRANSMIT in about 500μs, and back to RECEIVE in about 20ms, this delay being determined by the value of C1.

KEYING MONITORS

In the interests of good sending, it is desirable that an operator listen to his own morse transmission. This is particularly important when automatic keys are used. Several monitoring methods have received favour, especially the use of the station receiver, and the keying of an audio oscillator simultaneously with the transmitter by means of either dc circuits or the rf voltage appearing at the transmitter output. If the signal is not monitored in the station receiver, signal characteristics such as chirp, drift, ripple and key-clicks cannot be detected, and a note of constant pitch causes operator fatigue over long periods. RF-controlled oscillators tend to produce a tone whose frequency varies with transmitter power or frequency.

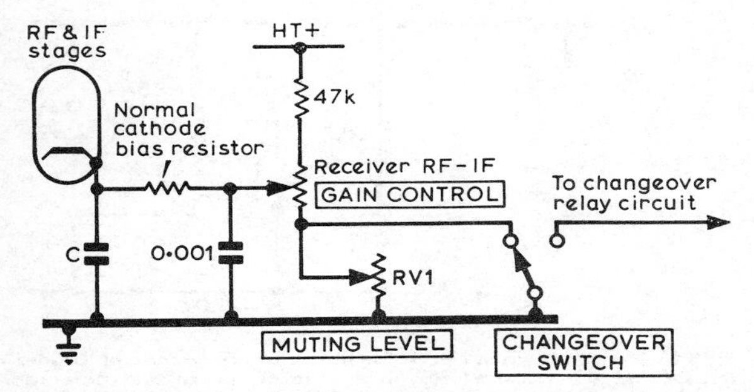

Fig 8.17. Method of muting the station receiver during transmission periods. RV1 should be 25–50kΩ. Needing only occasional adjustment it is usually mounted at the back of (or inside) the receiver cabinet.

When cw signals are generated in an ssb transmitter by the use of a tone oscillator, this can be used directly as a monitor by coupling the oscillator to the receiver audio stages. Tone generators are readily added to electronic bug keys using integrated logic circuits (see p8.9).

If a tone generator is used for monitoring, the station receiver must be completely silenced during transmission periods. This can be done by biasing the receiver rf or i.f. stages into a low-gain state. The circuit of **Fig 8.17** is typical.

The method of using the station receiver for monitoring is more popular, but there are three important conditions which must be satisfied.

(a) The signal appearing at the receiver must be small to prevent overload or even damage. This means, of course, that the receiver must be disconnected from the transmitting aerial, and it may be necessary to short its aerial terminal to earth. These operations may be carried out by the changeover relay or switch, or by another relay specially designed for aerial switching.

(b) The receiver gain must be further suitably reduced except when an exceptionally good agc system is available in the cw mode. The method shown in Fig 8.17 may be used.

(c) The sequence of switching the transmitter and receiver must be correct. If the transmitter oscillator and driver stages come on just before the receiver is muted, a burst of rf will reach the receiver. It is also undesirable for the receiver to be resensitized before the transmitter is switched off. This can be overcome by slightly setting the appropriate contacts on the changeover relay, or by using electronic delay circuits.

MORSE KEYS

The various devices for sending cw can be classified as straight keys, sideswipers, semi-automatic ("bug") keys, and electronic keys ("el-bugs").

Sideswipers are devices built with a horizontal arm having a contact on either side of it, and pivoted in a vertical plane. Operation involves the use of each contact alternately as the keying contact, so that dots and dashes are sent on one side and then on the other. Sideswipers are not popular, and tend to produce incorrectly spaced characters.

Semi-automatic or "bug" keys are usually purchased complete, as their construction includes precision machined parts. They are similar in function to sideswiper keys, except that whereas one of the contacts remains fixed and is used to provide dashes, the other initiates oscillations in a spring-loaded arm so that a sequence of dots is produced. With bug keys it is possible to send a wide range of speeds, depending on the position of a movable weight on the dot arm.

Electronic Keys

Electronic keys vary in complexity from those solely providing dot sequences like a bug key (**Fig 8.18**), to complex machines with typewriter keyboards, using logic circuits to create the correct morse characters. Most electronic keys produce a series of dots or dashes when a moving arm or paddle is held against one or other of two fixed contacts. Once a dot or dash has been started, it will be completed irrespective of what happens to the paddle, and a space of the correct length will be left before the next dot or dash can commence. The timing of the hand movements is thus very much less critical than when using a "straight" key.

Fig 8.19 shows the timing circuits of an electronic key based on five ttl integrated circuits, and designed by G3RUZ. The output of the key is very accurately timed, as the beginning and end of each dot or dash is locked to a stable pulse generator.

The keying speed is determined by a pair of monostable multivibrators (U2, U3) which are cross-coupled to form a low frequency pulse generator with a controllable period and well-defined start-up characteristics. Because U2 produces an asymmetrical output, the pulse generator frequency is divided by two in the counter stage U4A. This produces the symmetrical dot output of the key

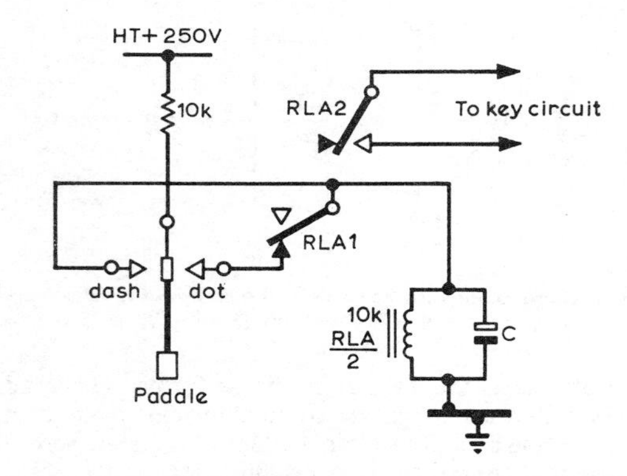

Fig 8.18. Circuit of a device with the same properties as a mechanical "bug" key, producing a sequence of dots electronically, at a speed determined by C (which can be made switchable). Contacts RLA1 are normally closed, contacts RLA2 are normally open.

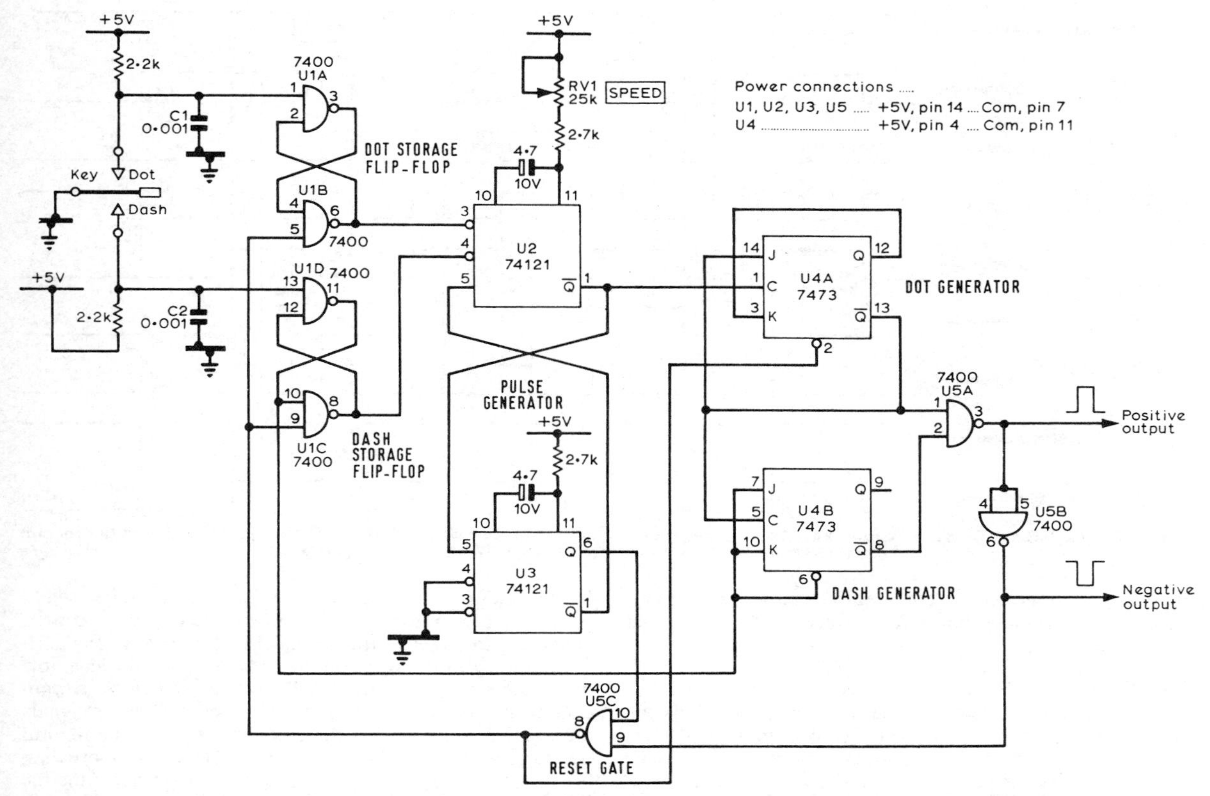

Fig 8.19. The timing circuits of an electronic key using ttl integrated circuits. The characters are accurately timed and self-completing. Capacitors C1 and C2 reduce the sensitivity of the key to external sources of interference.

and, in conjunction with a further counter stage (U4B), creates the dash output of the key. The keyer speed can be varied from about 10 to 60 words/min, by means of the potentiometer RV1.

If the dot contact of the key is closed momentarily, the dot storage flip-flop (U1A, U1B) is turned on (**Fig 8.20**). In rapid succession, U2 is triggered, U4A is clocked, and U5A produces a high output corresponding to a dot pulse. When U2 times out, U3 is triggered. When U3 times out, U2 is triggered again, U4A is clocked again, and the dot output from U5A goes to zero. Once again, U2 times out, U3 is triggered, and the dot storage flip-flop is reset by a low output from gate U5C. When U3 times out for the second time, U2 is not retriggered, as neither pin 3 nor pin 4 is low, unless the one of the key contacts is being held closed. Note that if the operator closes the dash contact immediately after the initial closure of the dot contact, a dash is not produced because the dash generator (U4B) is clocked at the very beginning of the dot generation cycle, when its J and K inputs are low.

A dash is produced by using the dash storage flip-flop (U1C, U1D) to enable the dash generator (U4B). This flip-flop is clocked on at the beginning of the cycle, and is clocked off by the leading edge of the second pulse from the dot generator (U4B) (Fig 8.20). The dash cycle is reset when both inputs to the reset gate are high, after the completion of a dash and the correct gap after it.

Fig 8.19 shows the timing circuits of the key, but does not include accessories such as a +5V power supply, a monitor, or a transmitter keying circuit. **Fig 8.21** shows examples of these.

The 7400 series of ttl logic will normally operate satisfactorily from a 4·5V battery, although it should preferably be operated by a regulated supply between 4·75 and 5·25V. This requirement is met by the integrated regulator type LM309H (National Semiconductor) shown in **Fig 8.21(a)**. The monitor in **Fig 8.21(b)** uses a keyed Schmitt trigger oscillator driving a simple loudspeaker amplifier. The circuit is based on the SN7413N dual 4-input NAND Schmitt trigger circuit. The transmitter keying circuit shown in **Fig 8.21(c)** is for general purposes where a dry reed relay can handle the keying circuit current and voltage. Where a high current relay is needed, a Darlington circuit should be used rather than reducing R1. The circuit of **Fig 8.21(d)** uses a high-voltage (300V) transistor to key the cathode of a low power pa valve directly. The base circuit of the MJE340 includes an RC network which controls the rate of rise and decay of cathode current in the keyed stage. This avoids the need for a large inductor in the cathode circuit, which would develop excessive voltages across the switching transistor. The circuit is designed to key stages running from a +250V supply, with a cathode current of about 50mA. The MJE340 should be mounted on a small heatsink for protection against temporary overloads. A similar circuit to that in

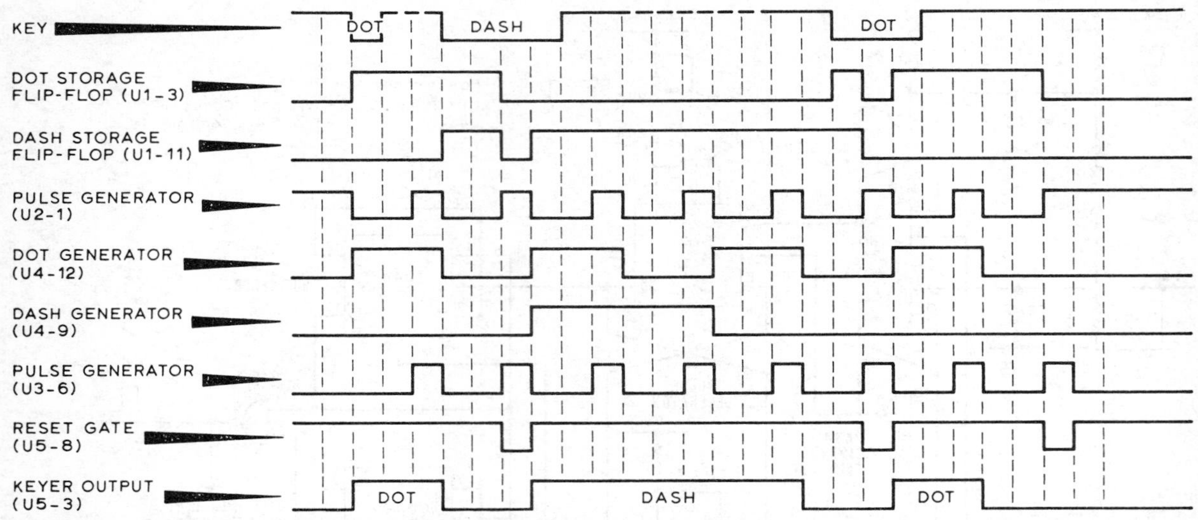

Fig 8.20. Timing diagram for the electronic key, showing the generation of the morse character "R". Even though the timing of the contact closures is poor, the keyer output produces correctly timed output pulses.

Fig 8.21(d), connected to the negative output of the key, could be used to mute the station receiver in a full break-in system (see later).

Tone Keying

Many applications exist for devices which permit keyed audio tones to control a transmitter operating in the cw mode. For instance, tape recordings can be used for slow morse transmissions, automatic CQ machines, and transmission at very high speeds.

Fig 8.22 shows a circuit which converts a keyed audio tone from one of two sources into dc pulses which can be used to control a transmitter. The output of the converter can be connected to either of the keying circuits shown in Fig 8.21.

Circuit operation relies on the properties of a retriggerable monostable multivibrator (U2), which generates a continuous output pulse provided that repetitive trigger signals are applied to its input with a period less than the natural period of the monostable circuit. If a 1kHz tone of more than 0·75V peak is applied to the A input of the converter (and the SOURCE switch enables the A channel), transistor TR1 will switch on and off at the tone frequency. The Schmitt trigger

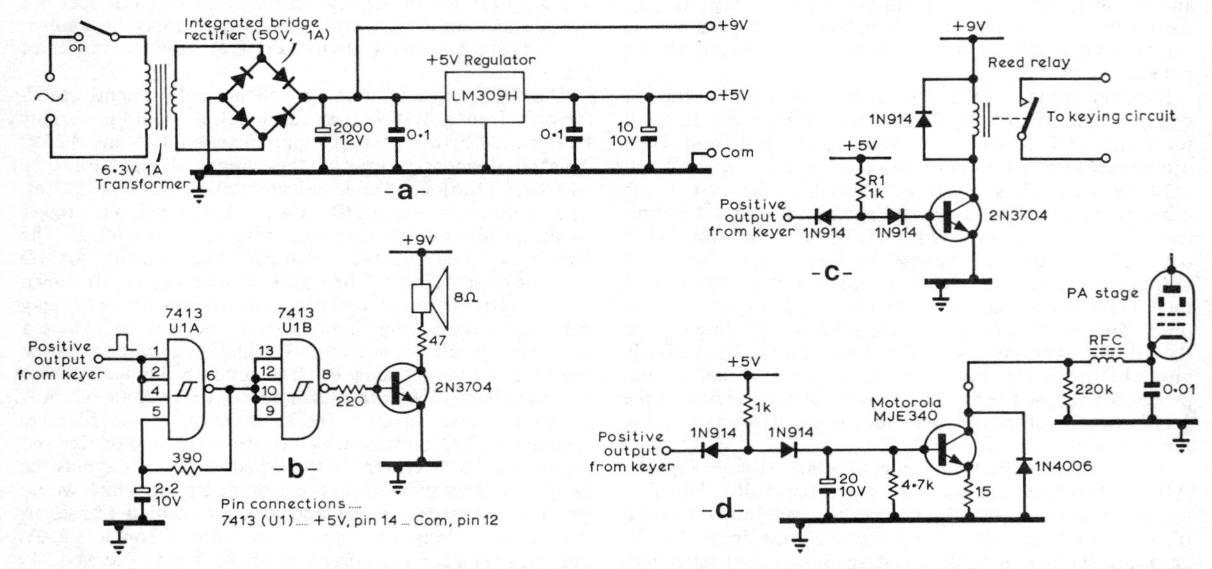

Fig 8.21. Ancillary circuits for the electronic key. (a) Simple +5V regulated power supply, also providing an unregulated +9V supply for relay and monitor circuits; (b) monitor circuit providing an audible 700Hz tone; (c) general-purpose transmitter keying circuit using a dry reed relay; (d) direct cathode keying of a small pa stage using a high-voltage transistor between the electronic key and the transmitter.

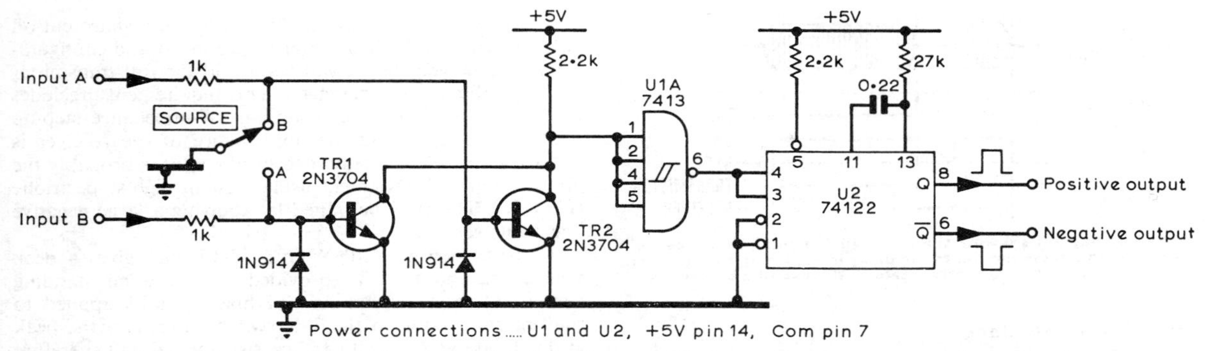

Fig 8.22. Tone keyer circuit which operates from two different signal sources. The keyer converts tones into dc keying signals, and operates at speeds up to 80 words/min.

circuit (U1) acts as a pulse squarer, and triggers the monostable circuit (U2). If the monostable period is greater than 1ms, U2 will produce a steady output whenever a tone is present. In effect, the converter provides a dc signal corresponding to the envelope of the input tone.

With the circuit values shown, the natural period of the monostable circuit is 2ms: this permits the use of tone frequencies as low as 500Hz. The converter will operate satisfactorily at keying speeds up to about 80 words/min.

BREAK-IN

Before describing full break-in systems, mention should be made of automatic changeover schemes which have some of the benefits of true break-in. **Fig 8.23** shows a circuit which uses the initial closure of the key contacts to actuate the main station changeover relay, in addition to keying the transmitter in the normal way. The station returns automatically to the receive state a fixed time after the last closure of the key contacts. The time delay is normally set to about 1s, and is defined by the capacitor C, the relay resistance R, and the relay drop-out voltage. In this way, the station is held in the transmit state during transmissions, without reverting to the receive state in the natural gaps between morse characters.

Full Break-in

The essential feature of a break-in system is that the operator is able to receive incoming signals in the spaces between his own transmitted morse characters so that duplex operation becomes possible. Much time is thereby saved, particularly in multiple contacts or when interference is present, and the advantages in contests or when hunting dx are obvious. Transmissions are interspersed with the sign "BK" and the callsigns are given at frequent intervals to conform with the conditions of the licence.

When using break-in, the normal changeover and keying functions are controlled by the key, and they must take place in the right sequence. The station should return to the receiving condition at the sensitivity level required by the operator between each dot and dash of the transmitted message.

It is not easy to install a good break-in system, one of the problems being that of keying the transmitter oscillator stage. This can be avoided in two ways: either the oscillator is screened so well that it is inaudible in the station receiver, or a mixer-type vfo with a keyed mixer is used. With either of these methods, the master oscillator can be made to run continuously. However, as it is difficult to screen a vfo to the extent required, and as mixer-VFOs are not common except in single-sideband transmitters, it is common to find break-in circuits involving oscillator keying.

In the section on changeover methods it was pointed out that the transmitter power amplifier output should be switched on after the receiver is desensitized, and off before it is turned back on. The receiver must be disconnected from the transmitting aerial (or the aerial must be switched from receiver to transmitter) just before radiation commences, and it must remain so until after the rf output has fallen to zero. The transmitter oscillator must be timed relative to the keyed stage in the same way. In fact aerial, oscillator and receiver may be switched together, and it is then only necessary to provide a delay before they return to the receive state, as the softening of the keyed amplifier stage automatically ensures that they are in the transmit condition before the rf output rises (**Fig 8.24**).

The delay time at the end of each morse character will naturally depend on the degree of softening used, but from 10 to 20ms is adequate for most situations. Three methods of obtaining this delay are in common use: utilization of the back contacts of a hand key, suitable timing of relay circuits, and application of pulse circuits such as the Schmitt trigger or monostable multivibrator. The first system precludes the use of bug keys and many el-bugs, the second demands one or more keying relays (with their own inherent disadvantages) and the third tends to be more complicated.

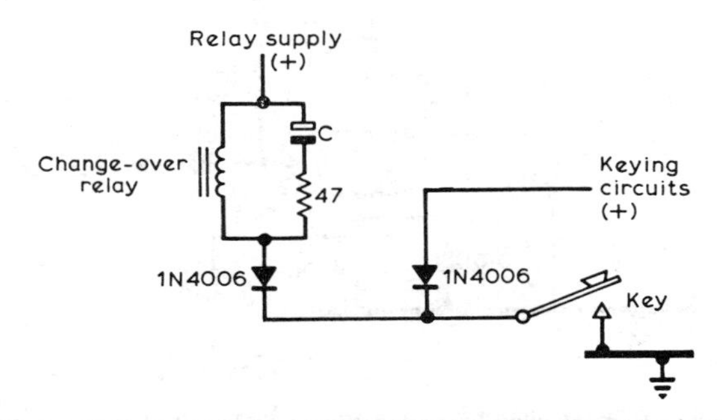

Fig 8.23. Automatic changeover circuit which holds the station in the transmit state for a fixed time after transmission ceases. The capacitor C is given approximately by 500/R μF, where R is the relay coil resistance in kilohms.

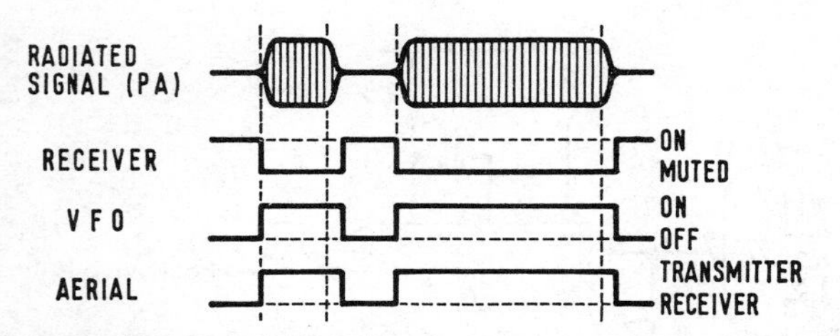

Fig 8.24. Diagram showing the required switching sequence for a typical break-in system. Note the delay in returning the vfo, aerial and receiver to the receive state at the end of a character.

Fast Aerial Switching

A simple expedient often used in a break-in system is to erect a separate aerial for the receiver, but the disadvantages of this method have been mentioned elsewhere. A single aerial may be switched by an accurately timed keying relay, but fast high-current aerial relays are scarce.

It is preferable to leave the transmitter connected permanently to the aerial, and to use a low-current reed switch or an electronic T/R switch to disconnect the receiver during transmission periods.

Electronic T/R switches may be divided into two classes: (a) those which require the application of some control voltage or current synchronized with the key, and (b) those which act as limiters and are entirely automatic. With either type, the transmitter is connected permanently to the aerial, the receiver being connected to it through the T/R switch. In the key-down condition the T/R switch must not absorb any appreciable fraction of the transmitted output power, nor must it permit any excessive amount of power to reach the receiver. The switch is usually arranged so that it presents a high impedance in the key-down condition, and is connected to the aerial at a point of low impedance, and thus of low voltage.

The T/R switches in class (a) have the disadvantage that they require a high negative blocking bias voltage although they reduce the problems of harmonic generation and tvi inherent in class (b). Both types suffer from the absorption of incoming signals by the transmitter pa stage and interference due to noise generated by the pa.

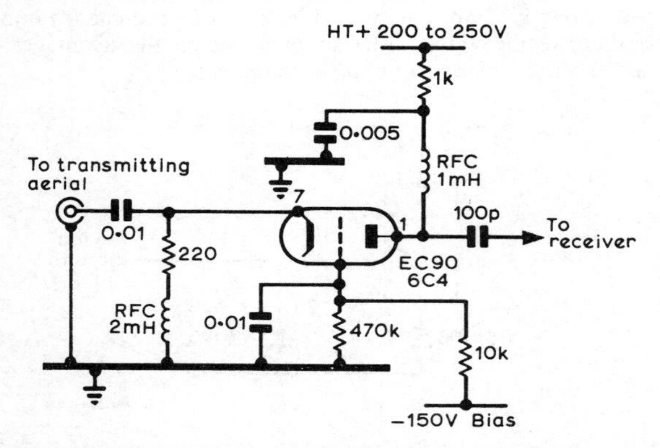

Fig 8.25. DC controlled T/R switch using an EC90 (6C4). The swr on the transmitting aerial feeder must be as low as possible, and the output power limited to about 75W. If it would be an advantage to use a circuit with appreciable gain, the rf choke in the anode circuit could be replaced by tuned circuits for the bands in use.

A switch in either class usually consists of a sharp cut-off valve, often in cathode follower or grounded-grid configuration, so arranged that it will withstand the full transmitter output without being damaged. This requirement precludes the provision of any useful amount of impedance step-up in its grid circuit, and the noise figure of the receiver is generally degraded. The best form of circuit is probably the grounded-grid arrangement using a small high-slope triode (**Fig 8.25**). This may be muted by applying a large negative bias to its grid.

A carrier power of 100W in a 75Ω circuit gives a peak voltage of about 125V (provided there are no standing waves); at least this amount of bias must be applied to prevent conduction in the T/R switch valve, and the peak grid/cathode voltage with the carrier applied will therefore be 250–300V. This is beyond the rating of most high-slope valves, but the majority of samples will withstand it without premature failure.

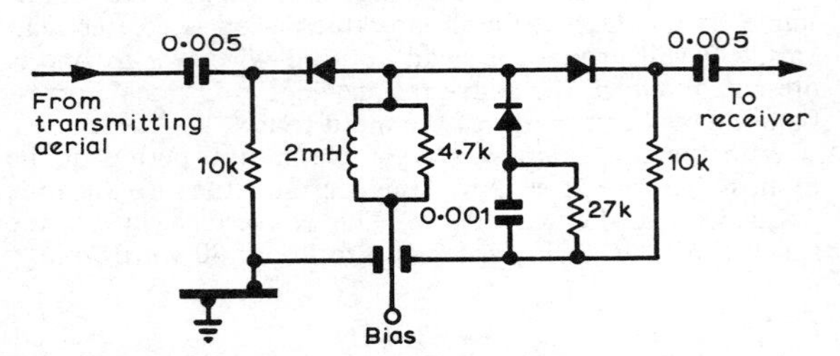

Fig 8.26. Diode T/R switch due to VE2AUB. The receiving bias is +30V, and the transmitting bias —350V, although the switch offers some protection without this. It may be convenient to reverse diode and bias polarities to suit switching systems already in other equipments.

The heater/cathode insulation is subject to voltages of the same order, and it may be necessary to insert rf chokes in the heater supply leads, each heater pin being bypassed to the cathode with a 1,000pF capacitor.

Other forms of amplifier may be preferred but it should be remembered that pentodes generate more valve noise than triodes, while triodes in circuits other than the grounded-grid circuit suffer from excessive breakthrough owing to the anode-to-grid or grid-to-cathode capacitance.

The circuit in Fig 8.25 is suitable for power up to about 75W. For powers up to 200W in a low impedance line with a reasonable swr, the semiconductor circuit of **Fig 8.26** is recommended. The silicon diodes may be any type with a very low capacitance at reverse voltages of about 30V, and with a peak inverse rating of 800V or more. The insertion loss is less than 1dB in the receive condition and up to 80dB in the transmitting state. The receiving bias should be about +30V, and the transmitting bias about −350V.

Fig 8.27 shows a circuit which is switched off by the voltage generated when rf signal causes current to flow through the grid bias resistor; it possesses useful gain and has a low output impedance. The design of the tuned circuit L1, C1 will depend on the bands to be covered.

Tunable T/R switches with gain may be beneficial to receivers of poor sensitivity, but they may lead to trouble with cross-modulation or overload as the pre-selectivity gain is increased. All T/R switches involving tuned circuits need adjustment when large frequency changes are made. The circuits working on the self-limiting principle should be well screened to reduce radiation of harmonics and should

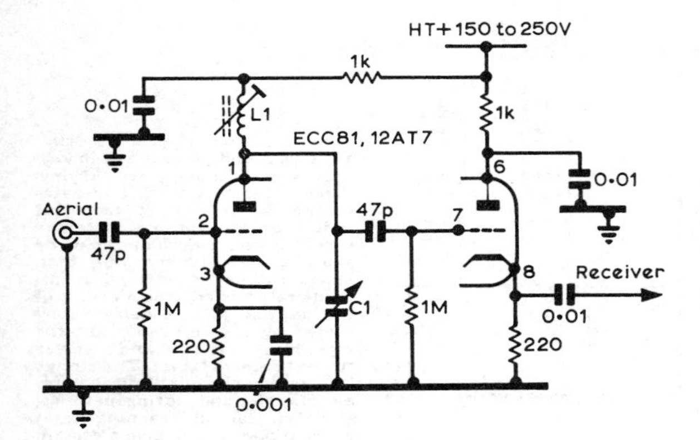

Fig 8.27. A tunable self-limiting T/R switch, with fair gain and low output impedance. The design of the tuned circuit L1, C1 will depend on the bands to be covered.

be connected to the aerial between the transmitter and low-pass filter or aerial tuning unit. Circuits using diode limiters may cause strong harmonics of a local broadcasting station to be heard in the receiver: rejection circuits tuned to its frequency would overcome this defect.

It is recommended that T/R switches be completely screened and mounted on the back of the receiver or transmitter cabinet. The coaxial lead to the receiver should be as short as possible to avoid resonance effects and to minimize pick-up of the transmitted signal.

Dry Reed T/R Switches

Electronic T/R switches can be made to act exceedingly fast, operating times of 50μs being readily attained. A system which is rather slower (yet 10 to 20 times faster than normal keying relays), but which avoids most of the problems peculiar to electronic devices, is one using dry-reed switches.

Fig 8.28 shows a circuit which ensures that the aerial is switched at the right times relative to the transmitter. Unlike previous circuits, it switches the aerial between the receiver and transmitter, and in times of less than 1ms. Dry-reed relays will switch only quite low currents, but if the switching operations are completed before the rf current appears, they can pass much higher currents through the closed contacts. The circuit shown will handle 700W into a well-matched 50Ω line.

Transmitter Keying Requirements

It has already been mentioned that it is necessary to key the master oscillator unless it is exceedingly well screened or unless it works on mixing principles, but it is almost impossible to achieve perfect keying characteristics from the vfo alone. It is usual to find the vfo and the pa or driver keyed differentially. This method requires the vfo to be keyed as "hard" as possible, with softening applied to the later stage. Furthermore, as shown by Fig 8.24, the vfo must be held on for a time while the keyed amplifier is being exponentially turned off: it must then be switched off sharply, to prevent radiation reaching the receiver.

This can be achieved by the use of a keying relay with at least two pairs of contacts, one pair switching the vfo and another the keyed amplifier stage. The contact springs should be set so that the vfo contacts close as soon as the armature starts to move, the amplifier contact closing later in its travel. On release the sequence is reversed so that the vfo stays on for a short time after the amplifier contacts have opened. A difficulty is that the time interval between the operations of the two pairs of contacts is limited to a few milliseconds, and therefore the keying cannot be made very soft.

The circuit in **Fig 8.29** is more flexible in this respect, and uses a high speed relay. It includes several silicon diodes (400 piv) and is rather limited in the types of keying which can be applied to the various units. For instance, the pa must be grid-block keyed, and the receiver muting circuits must involve switching positive voltages.

Muting the Receiver

When the key is depressed, the receiver must lose sensitivity quickly enough to prevent a click being produced from the leading edge of the transmitted signal: when the key is raised the gain should remain low until radiation has ceased and then rise rapidly.

If the receiver is used for monitoring, the amount of desensitization required will depend on the transmitter power, the type of aerial switching used, and the effectiveness of any screening. As it is generally insufficient to cut off the receiver rf stage alone, the i.f. stages are usually controlled with it. In valve receivers, it is often convenient to do this by breaking the cathode return of the rf and i.f. stages and replacing the normal rf/i.f. gain control with a MONITOR LEVEL control of higher resistance value. The normal control can be restored in parallel with this through the back contacts of the key or a pair of similar contacts on the

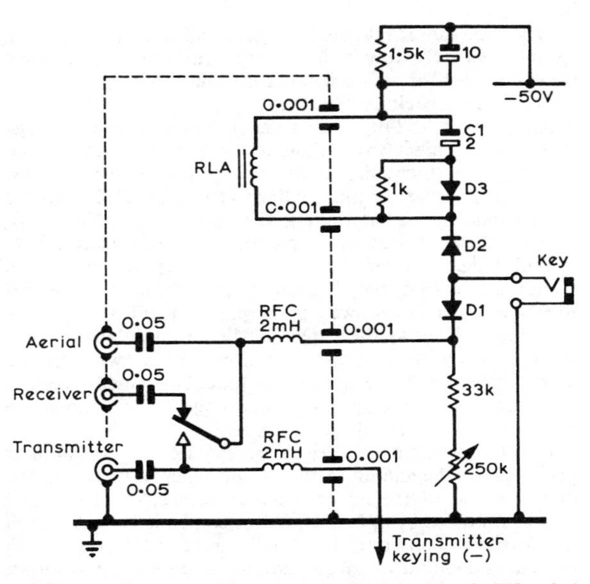

Fig 8.28. A dry reed switch used in a modified version of a T/R switch devised by VE3AU. Here sequential keying of the aerial and other circuits is obtained by (a) the use of the relay contacts to switch both dc and rf and (b) the release delay for the relay due to C1. Diodes D1 and D2 isolate the relay and transmitter switching circuits, and D3 together with the electrolytic capacitors provides the fast attack and slow release characteristics of the relay. As mentioned in connection with the diode T/R switch of Fig 8.26, it is simple to adapt the circuit for keying systems of the opposite polarity. The circuit shown is for negative keying lines: positive voltages may be controlled by reversing all three diodes, both electrolytic capacitors and the 50V supply.

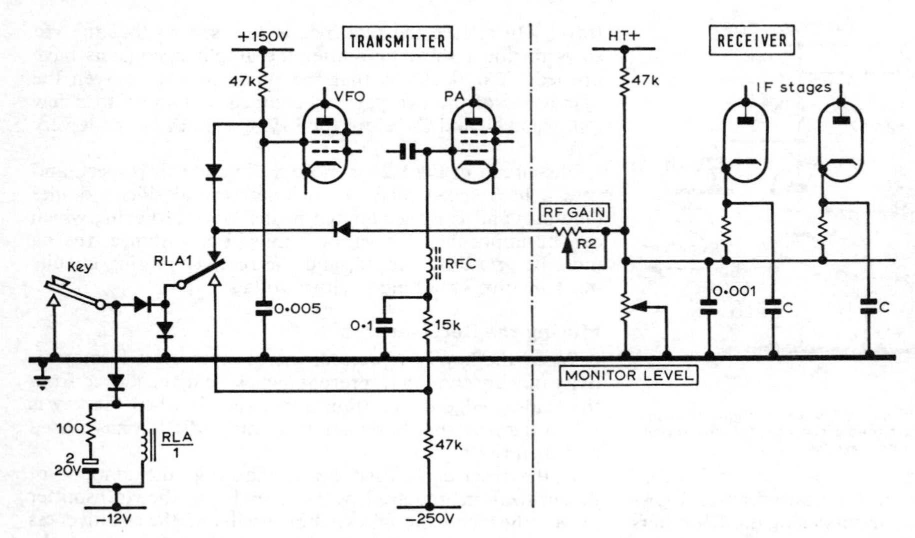

Fig 8.29. A complete break-in keying system with sequential keying. When the key is closed, it operates relay RLA. The relay contacts first switch on the vfo and mute the receiver, then switch on the pa by shorting out the blocking bias. When the key is released, it immediately initiates the blocking of the pa, and after a delay determined by the 2μF capacitor, the relay returns the vfo and receiver to the "receive" state. The circuit is suitable only for a limited number of keying and muting methods, although careful rearrangement of the diodes would give a certain amount of flexibility.

keying relay, adjusted to open at the same time or a little earlier than the instant when the vfo contacts close. Fig 8.29 shows a circuit of this kind.

In order to obtain a sufficiently rapid fall in sensitivity where the key is depressed, it may be necessary to reduce the values of the rf and i.f. cathode decoupling capacitors C in the receiver to the minimum required for proper working.

When the key is released, the gain will rise rapidly owing to the relatively low value of the normal gain control potentiometer R2 which forms the main discharge path for the cathode capacitors.

Another method of receiver desensitization, using the agc line, is convenient in conjunction with grid-block keying systems, as both circuits involve switching negative voltages. However, the receiver muting line must be isolated from the large capacitance determining the gain control time constants; otherwise the receiver recovery time will be excessive.

With either method of receiver muting, trouble may arise from the beat frequency oscillator in the receiver. If the bfo injection point is followed by a gain-controlled stage, the high bfo voltage at the detector will be modulated by the muting voltage, and for fast rates of muting this will produce an audible click or thump at the output. The remedy is either to inject the bfo voltage at a later stage or to remove the gain control from the stage in question.

Another problem may arise from the use of a crystal filter placed at the beginning of the i.f. amplifier chain, eg in the anode circuit of the mixer. In such a receiver, the crystal filter is followed by at least one gain-controlled stage, and in the key-down condition the signal voltage at the crystal may be many times that which corresponds to a reasonable output when the receiver is being operated at maximum sensitivity. Due to the very high Q value of the filter elements, this voltage cannot die away quickly, even after the transmitter output has fallen to zero, and if the gain of the following stages of the receiver is allowed to increase rapidly to normal after the key is released, this signal will be heard as a loud click in the receiver immediately after the instant of break. This can be avoided by controlling only the rf gain, if by doing so it is possible to reduce the overall gain sufficiently. Otherwise the receiver gain must be held down until the crystal filter has recovered.

Complete Break-in Systems

A common requirement is the need to add break-in facilities to commercial equipment without any drastic modifications, and this often means controlling voltages of both polarities. In this instance a keying relay must be resorted to, or its equivalent using the properties of diodes, or electronic switching circuits.

As has been shown in several instances already, it is possible to switch any number of voltages of the same polarity with a single key or relay contact by isolating each control line with a small diode, suitably connected (**Fig 8.30**). It is helpful therefore to use keying methods which have identically polarized control lines from stages which need to be switched simultaneously.

T/R switches of the self-limiting type should be used where possible as this reduces the number of circuits needing precise control. Otherwise the choice should be between diode and triode types requiring keyed bias voltages, and dry reed relay circuits.

If a switching voltage is available which is the wrong phase relative to the transmitted characters, it may be inverted by an auxiliary circuit being switched from saturation to cut-off. This principle may be developed to the extent of using flip-flops and other switching circuits which produce two complementary keying voltages. In this case, one is used to control the transmitter vfo and pa, and the other the receiver and T/R switch.

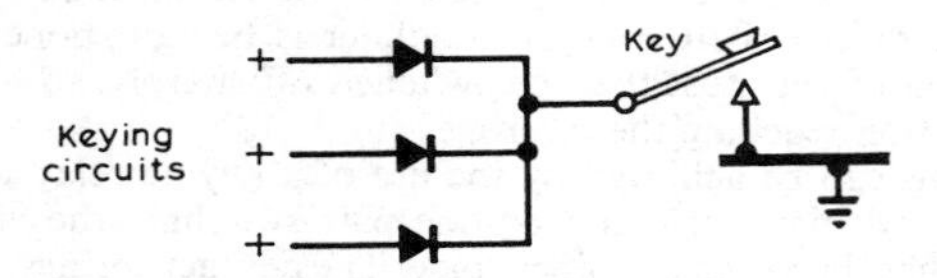

Fig 8.30. Method of switching several voltages of the same polarity with a single switch contact.

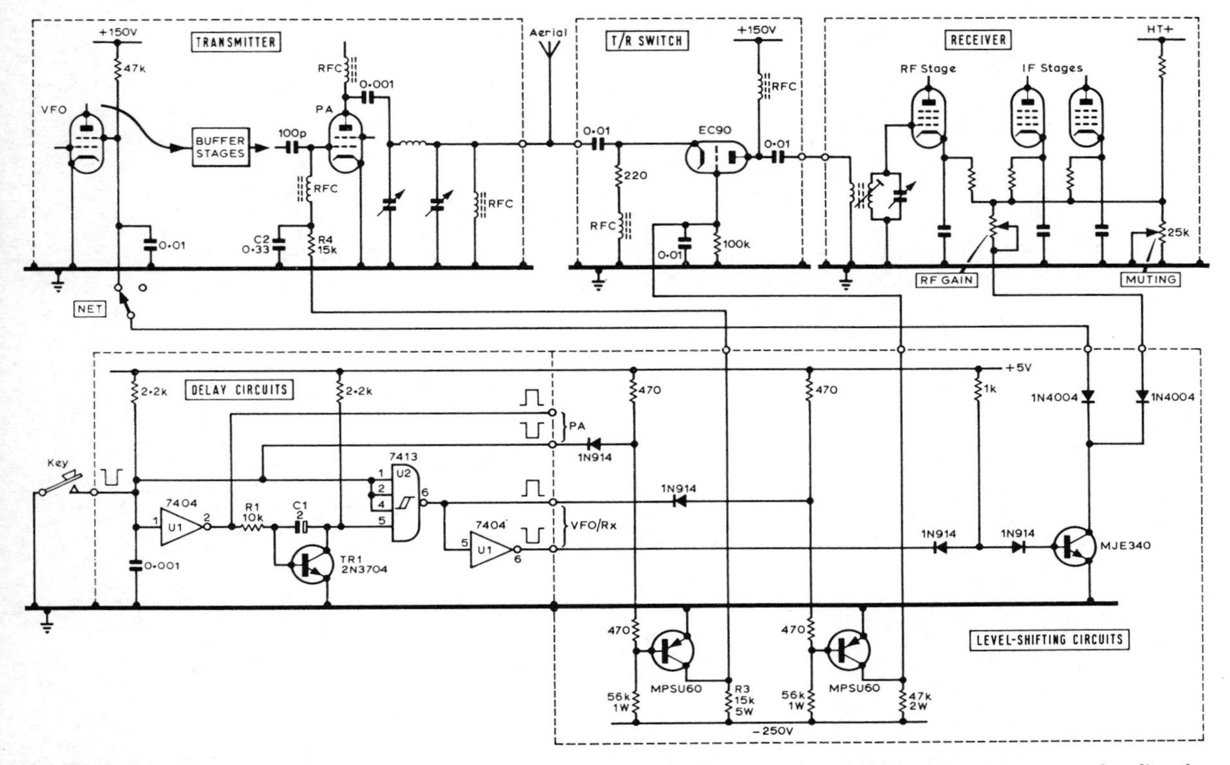

Fig 8.31. Example of a complete break-in system using electronic switching throughout. The circuit can be driven from a morse key directly, or from the negative output of either the electronic key in Fig 8.19 or the tone keyer in Fig 8.22. The delay circuit around TR1 and the Schmitt trigger circuit (U2) ensure that the various transmitter and receiver circuits are switched in the correct sequence.

Full Break-in Without Relays

Fig 8.31 shows the circuit of a full break-in system using electronic switching throughout. It can be driven directly from a morse key or from the negative output of the electronic key shown in Fig 8.19. The circuit timing corresponds to the recommended practice shown in Fig 8.24. Examples of both positive and negative switching circuits are shown for controlling the various receiver and transmitter circuits: these circuits can be interchanged to suit particular requirements.

In the ttl logic circuits, the network R1, C1 provides a signal which corresponds to the key contact closures, except that it is delayed. Schmitt trigger U2 provides a high output when the key line is low or the delay network output is low. Thus the output of U2 persists after the key circuit is broken, and is suitable for controlling the transmitter vfo, the T/R switch, and the receiver gain stages. The key circuit can be used directly to control the transmitter pa stage. Note that R3, R4 and C2 provide softening of the keying waveform in the usual way.

The transistors used in the level-shifting circuits are rated at 300V. If the supplies associated with these circuits are greater than about 250V, transistors with a higher breakdown rating should be used.

The advantages of this type of break-in system are that (a) it operates quickly, without "weighting" the morse characters; (b) there are no keying relays to produce noise, rf interference, or contact bounce problems; and (c) the relative timing of the switching signals is easily controlled by varying C1.

Note that the circuit can be converted from a full break-in mode of operation to an automatic changeover system (see p8.11) simply by increasing C1 from 2μF to about 50μF.

Level-shifting circuits similar to those shown in Fig 8.31 can be used to control transistorized receivers and transmitters. The lower voltages associated with transistor circuits often permit simpler switching circuits to be used. For example, the stable vfo shown in Fig 8.5 includes a control transistor which can be driven directly from a standard logic-level switching signal, such as that obtained from pin 6 of U1 in Fig 8.31.

CHAPTER 9

MODULATION SYSTEMS

A RADIO-FREQUENCY carrier wave may be used for the transmission of speech by varying its amplitude, frequency or phase in sympathy with the speech waveforms. Only one of these parameters should be deliberately varied in any one modulation system.

Apart from the use of single sideband working which is dealt with elsewhere, *amplitude* modulation is most commonly used in amateur radio. *Frequency* modulation is, however, increasingly used on the higher frequencies as it is less liable to produce interference to domestic entertainment equipment resulting from rectification in the audio circuits of transistor amplifiers. *Phase* modulation is used as a method of indirectly producing frequency modulation with which it is indistinguishable at the receiver, provided that the af response has been suitably modified in the transmitter. The frequency *deviation* of fm signals is normally confined to $\pm 2{\cdot}5$kHz which limits the bandwidth occupied to that of an equivalent a.m. signal and is normally receivable by slope detection on receivers lacking frequency discriminators. The full potentialities of fm are only obtained with receivers having amplitude limiters and narrow-band discriminators; integrated circuits for this purpose are readily available.

AMPLITUDE MODULATION

All modulation systems produce additional frequencies above and below that of the carrier wave, and these are known as *sidebands*. In an amplitude modulation system, the sidebands extend on either side of the carrier by the highest modulation frequency used so that a carrier of 1,000 kHz modulated by a signal containing frequencies up to 15kHz will extend from 985 to 1,015kHz. It is for this reason that the maximum modulation frequency should be limited to that needed for communication; 3kHz is usually regarded as a satisfactory maximum.

Modulation Depth

The amplitude-modulated wave is shown graphically in **Fig 9.1.** Here (a) represents the unmodulated carrier wave of constant amplitude and frequency which when modulated by the audio-frequency wave (b) acquires a varying amplitude as shown at (c). This is the modulated carrier wave, and the two curved lines touching the crests of the modulated carrier wave constitute the *modulation envelope*. The modulation amplitude is represented by either x or y (which in most cases can be assumed to be equal) and the ratio of this to the amplitude of the unmodulated carrier wave z is known as the *modulation depth* or *modulation factor*. This ratio may also be expressed as a percentage. When the amplitude of the modulating signal is increased as at (d), the condition (e) is reached where the negative peak of the modulating signal has reduced the amplitude of the carrier to zero, while the positive peak increases the carrier amplitude to twice the unmodulated value. This represents 100 per cent modulation, or a modulation factor of 1. Further increase of the modulating signal amplitude as indicated by (f) produces the condition (g) where the carrier wave is reduced to zero for an appreciable period by the negative peaks of the modulating signal. This condition is known as *over-modulation*. The breaking up of the carrier in this way causes distortion and the introduction of harmonics of the modulating frequencies which will be radiated as *spurious sidebands;* this causes the transmission to occupy a much greater bandwidth than necessary, and considerable interference is likely to be experienced in nearby receivers. The radiation of such spurious sidebands by over-modulation (sometimes known as *splatter* or *spitch*) must be avoided.

At 100 per cent modulation, the instantaneous amplitude of the carrier voltage or current is double and it follows that the instantaneous power output is four times that of the carrier. For linear modulation, it is necessary that the

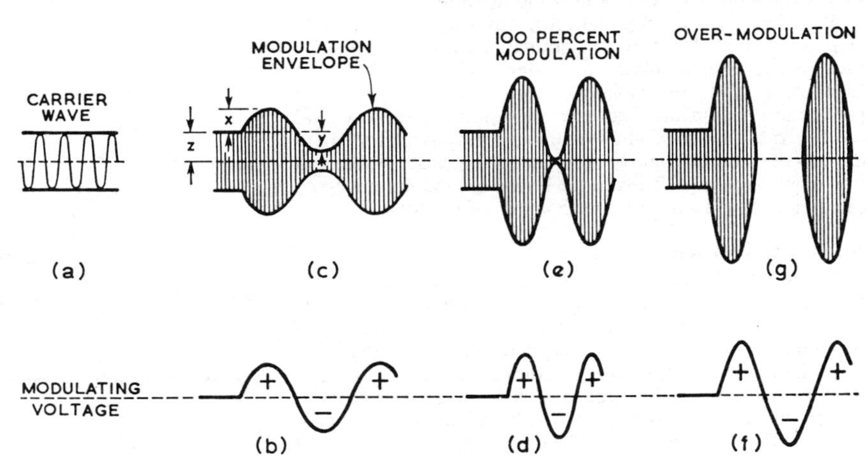

Fig 9.1. Graphical representation of amplitude modulation: (a) un-modulated carrier wave; (b) modulating signal; (c) modulated carrier wave; (d) (e) 100 per cent modulation; (f) (g) over-modulation.

modulated stage produces this power; correct adjustment of rf grid drive and load impedance are particularly important. For sinusoidal modulation, the average increase in output power for 100 per cent modulation is 1·5 times and it follows that the increase of 50 per cent in the output is supplied by the modulator. Thus to fully modulate a carrier of 150W input power will require a modulator capable of producing 75W of audio frequency power. The aerial current will, of course, only increase by $\sqrt{1 \cdot 5}$ or 1·226 times the carrier level for sine wave modulation but for speech waveforms which have not been processed, the rise in aerial current will be considerably less due to the "*peaky*" waveform of speech.

Table 9.1 shows the relative amount of audio power required to produce various depths of modulation and the corresponding increase of aerial current for sine wave modulation.

TABLE 9.1
Effect of Modulation on Aerial Current

Depth of Modulation (per cent)	Ratio: af power / dc power	Increase in aerial current (per cent)
100	0·5	22·6
90	0·405	18·5
80	0·32	15·1
70	0·245	11·5
60	0·18	8·6
50	0·125	6·0

AMPLITUDE MODULATION SYSTEMS

There are two basic systems of applying amplitude modulation to the carrier.

1. The rf power amplifier is operated in an efficient (normally class C) condition and modulation is achieved by supplying additional power from the modulator. *Anode modulation* is the principal example.

2. The rf power amplifier is operated in an efficient condition only at the crest of modulation; in the carrier condition, the efficiency is deliberately reduced to about one half of the possible value (typically to about 33 per cent) and modulation is achieved by increasing the efficiency of the rf amplifier towards the normal value. In this case, the increased output power resulting from modulation comes from the increased efficiency of the rf amplifier and not significantly from the modulator. These are known as *efficiency modulation* systems and include screen grid, suppressor grid and control grid modulation.

Whatever form of modulation is applied, it should only be to an rf stage driven from a source of constant frequency such as a crystal oscillator or a stable vfo followed by at least one unmodulated isolating stage. This is to prevent the production of spurious frequency modulation such as would result from the direct modulation of an oscillator. It is important to check that the modulated stages are stable and free of *parasitic oscillation* at all levels of modulation. Such oscillation may occur only on the peak of modulation.

Anode Modulation

Anode modulation of a class C rf stage is accomplished by superimposing the modulating voltage on the dc anode voltage. For full modulation, the anode voltage should be swinging from zero to twice its normal value and if the valve is a triode and the operating conditions are correct, the rf output will vary as the square of the anode voltage. The design information given in the valve chapter will enable correct operating conditions to be determined. If the pa valve is a pentode or tetrode, varying the anode voltage will have little effect on the anode current and current modulation will only be obtained if modulation is also applied to the screen grid voltage. The modulating voltage required at the screen-grid is sometimes given in the valve maker's data but may have to be found experimentally. A peak modulating voltage swing of two thirds of the dc screen voltage is typical. A peak modulating voltage equal to the screen voltage is sometimes used but this will result in over-modulation since the anode current rises more than linearly with screen voltage; however, if the grid drive is inadequate, the extra screen voltage swing will sometimes compensate and give acceptable results.

Although the *effective* power input is modulated, the *average* power input remains constant because each positive excursion of the anode voltage is immediately followed by a corresponding negative one. This means that the anode current, measured by a moving-coil dc meter, does not vary during modulation because such a meter measures average current. However, the aerial current is usually measured on a thermal meter which indicates the rms value and, as was shown earlier, the effective or rms value of the aerial current does depend upon the amount of modulation. Although the thermal ammeter in the aerial circuit should show a rise of current on modulation, the moving-coil "*average-current*" meter in the anode circuit should remain steady, and any movement of this meter during modulation indicates incorrect operation.

Transformer Modulation

The most straightforward means of anode modulation is to use a modulation transformer to couple and match the modulator into the modulated stage. The load impedance offered by the modulated stage is given by:

$$Zm = \frac{Va}{Ia} \times 1000 \text{ ohms}$$

where Va = anode voltage of rf stage in volts
Ia = anode carrier of rf stage in mA

and the transformer should have a ratio which will transform this into the load impedance required by the modulator. The circuit arrangement is shown in **Fig 9.2.** If the rf stage

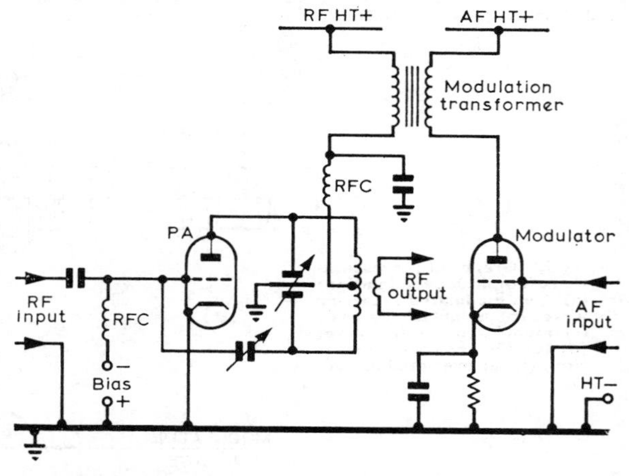

Fig 9.2. Anode modulation using transformer coupling. The system is sometimes referred to as "transformer modulation".

requires modulation of the screen grid a separate winding should be provided on the transformer secondary through which the screen voltage is fed. The screen-grid should be supplied from an ht rail of appropriate voltage and should not be derived from the anode supply through a series resistor.

If a transformer with two secondary windings is not available, the modulated screen voltage may be obtained from the modulated anode supply through a series resistor as shown in **Fig 9.3.** The screen bypass capacitance *C* must have a reasonably high reactance at audio frequencies so as not to bypass the modulating voltage. A potentiometer cannot be used for the screen supply and the method may not be used with valves which have negative screen current under some load conditions eg 4CX250B. In the case of Fig 9.3 the anode and screen currents should be added in determining the local impedance offered by the modulated stage.

Many tetrodes can be satisfactorily anode and screen modulated by the use of a small inductance in the screen lead to an unmodulated supply. The screen voltage is self-modulated by virtue of the small variations of screen current when the anode voltage is varied by modulation. A choke of 10–15H at 60mA would be typical. The arrangement is

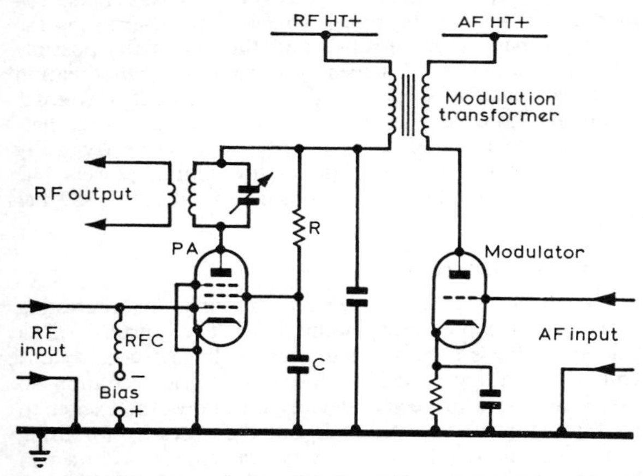

Fig 9.3. Anode-and-screen modulation. When the modulated stage consists of a pentode or a tetrode, the screen voltage as well as the anode voltage should be modulated by a proportionate amount in order to ensure linear modulation.

shown in **Fig 9.4.** The modulation depth obtained by this method is somewhat unpredictable.

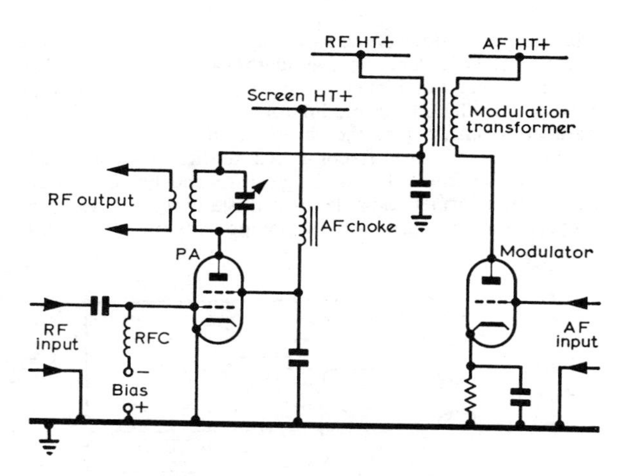

Fig 9.4. Anode modulation of a tetrode by the use of an inductance in the screen circuit.

Choke Modulation

Choke modulation employs a choke as the coupling impedance between the modulator and the rf stage, as shown in **Fig 9.5.**

The modulation choke must have an impedance at audio frequencies which is high compared with the impedance of the modulated rf stage when carrying the combined anode currents of that stage and the modulator. Obviously in this system both valves operate at the same ht voltage. The undistorted output voltage from the modulator which is developed across the choke must be less than the ht voltage and therefore it cannot modulate the rf stage to 100 per cent. For this reason the pa stage is run at a lower ht voltage than that of the modulator, the excess voltage being dropped across the resistor *R* which is bypassed at audio frequency by the capacitor *C*. The ht voltage on the pa stage should be reduced by this means to about 60–70 per cent of the modulator ht voltage. Since a push-pull arrangement in the modulator is obviously precluded, the modulator must therefore operate under class A conditions. This may necessitate the use of very large modulator valves, or several in parallel, and for this reason choke modulation is only economic for low-power operation, eg up to about 15W input to the rf amplifier.

Series Modulation

In this case the modulator and the pa valves are connected in series across one ht supply, as shown in **Fig 9.6.** Thus both will pass the same current, the method of adjustment being to vary the operating conditions (ie the amount of grid drive, the aerial loading and the grid bias of the pa stage) so that it is operating at a lower anode-to-cathode voltage than the modulator. This system has the advantage that one power unit feeds both the pa and the modulator, and that no modulation transformer or choke is required. Valves such as the 807 or TT21 can be used in both positions in a series-modulation system from a 1,000V supply. The

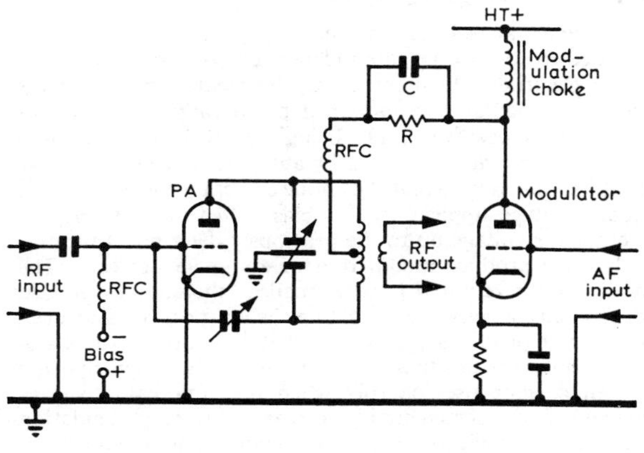

Fig 9.5. Choke modulation of a class C rf amplifier.

operating conditions of the pa stage should be such that there are 500–600V across the modulator and from 400–500V across the pa. For higher powers two such valves may be used in parallel as the modulator and two more either in parallel or push-pull as the pa. It is immaterial whether the pa or the modulator is connected to the positive supply provided that the valve at the positive end has a separate heater transformer capable of withstanding the total ht voltage and has grid circuit components rated for this voltage.

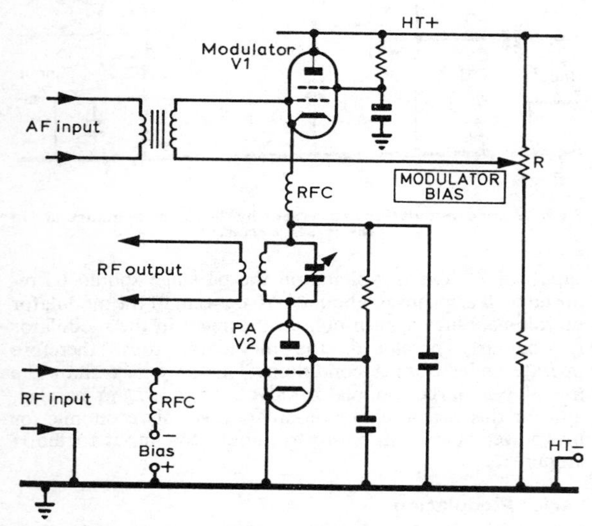

Fig 9.6. Series modulation. The relative positions of the af modulator V1 and the rf output valve V2 between the ht line and earth can be interchanged if desired. The appropriate negative bias on the control grid of V1 with respect to its cathode is obtained by adjusting the potentiometer R.

Efficiency Modulation Systems

Efficiency modulation may be obtained by modulating the voltage applied to any of the grids of a valve. Although little power is required from the modulator, the efficiency of the rf stage is low and full modulation is only attainable by critical adjustment.

Suppressor-grid Modulation

Pentode valves have a normal suppressor grid which may be used to control the gain independently of the voltages on other electrodes. In many cases, the greatest output at the peak of modulation is obtained by driving the suppressor grid slightly positive, +40V being typical. From this point, the *carrier condition* is found by applying a negative voltage to the suppressor so that the anode current is halved. A peak audio voltage applied to this bias sufficient to swing the suppressor voltage up to the previously determined positive value will produce modulation of about 95 per cent. The audio power required is very small as the suppressor grid draws only a very few milliamps when positive and even this small power can be avoided if the suppressor is not driven into the positive region. It should be noted that the screen current rises sharply when the suppressor is very negative and a screen supply of reasonably good regulation should be provided. A typical arrangement is shown in **Fig 9.7.** Beam tetrode valves in which beam deflecting plates

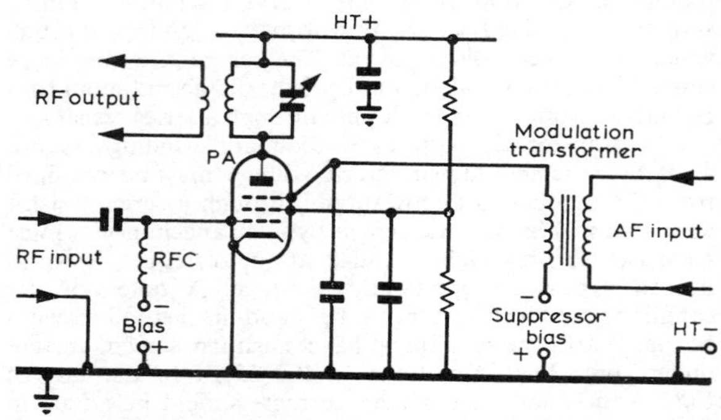

Fig 9.7. Suppressor-grid modulation. The bias on the suppressor grid needs to be carefully adjusted.

are used instead of normal grids cannot be successfully modulated by this means.

Screen-grid Modulation

The general circuit arrangement for *screen-grid modulation* is shown in **Fig 9.8.** The main drawback of screen modulation is that the screen current rises sharply as the screen is made more positive and this limits the possible upward excursion of screen voltage during modulation since the positive peaks of the modulated rf waveform would be flattened. The screen voltage at the carrier has, therefore, to be reduced to a rather low value (typically 100–125V) which results in rather low output power. The maximum modulation depth is generally about 75–80 per cent.

Clamp Modulation

The so-called *clamp-modulation* system is an interesting variation of screen-grid modulation. The purpose of a "*clamp*" valve is to reduce the screen voltage of a tetrode rf amplifier to a low value in the absence of grid excitation, so that it acts as a protective device and allows the exciter to be keyed for cw operation without the necessity for using fixed bias on the pa stage. A typical circuit is shown in **Fig 9.9.** The clamp valve V2 which may be of the KT66 or 6L6 class, is cut off by the operating bias of the rf stage V1 which is developed across R1. When the excitation is removed, there is no bias on V2 and hence it passes a large

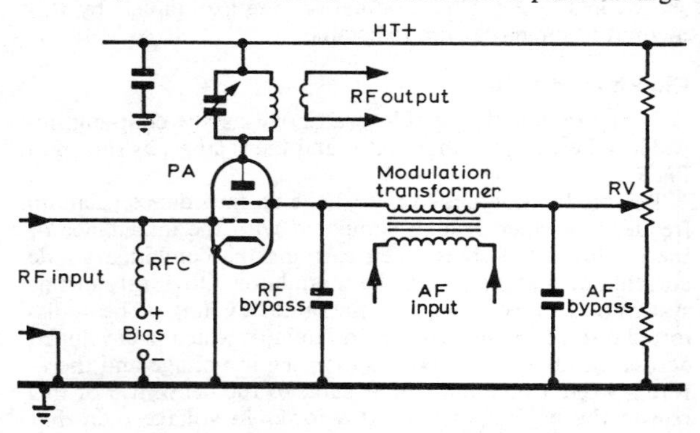

Fig 9.8. Screen modulation. The screen supply voltage must be obtained from a potentiometer RV to enable the necessary critical adjustment to be made for maximum depth of modulation.

anode current which develops a high voltage-drop across R2, the screen resistor of V1. This reduction in the screen voltage of V1 reduces the anode current to a very low and safe value.

For the purposes of modulation the grid of the clamp valve may be switched from the rf bias resistor R1 to the output of a speech amplifier V3 so that the screen of V1 receives the audio output voltage through V2.

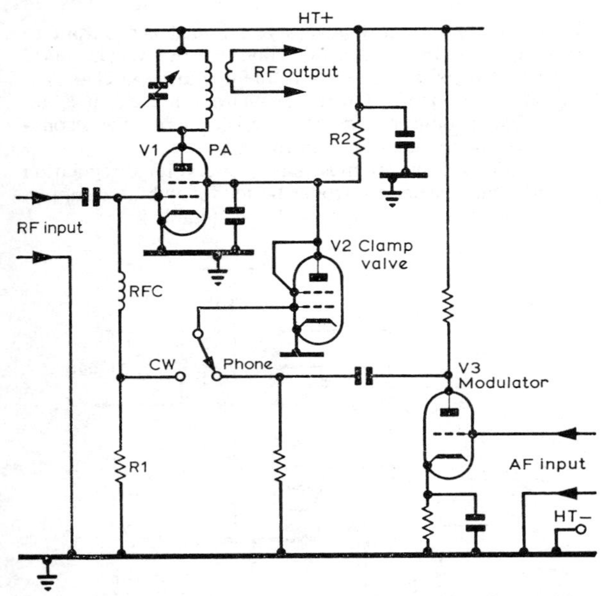

Fig 9.9. Clamp modulation. This is an elaborated form of screen modulation which enables the transmitter to be easily switched to cw operation.

When V2 is switched for telephony operation its bias should be arranged to be such that its anode voltage, ie the screen voltage of V1, is about one half of the value which it has in the cw position. Tuning-up can take place in the cw position, but when V2 is switched to the telephony position the carrier output of V1 will of course fall. When modulating, the audio gain should be such that the anode current of V1 on modulation peaks does not exceed about three-quarters of the normal cw value.

Series Gate Modulation

Series gate modulation is another form of screen modulation and is probably the most used of any system of efficiency modulation. The name *series gate* was given by the originator of the method but is not very helpful in understanding the operation.

The basic series gate modulator circuit is shown in **Fig 9.10.** This employs a twin triode valve, but there is, of course, no reason why two separate triodes should not be used. The circuit is essentially a voltage amplifier V1a direct-coupled to a cathode-follower V1b. Audio input is applied via C1.

When the slider of R2 is at the cathode end of its travel, V1a is operating without bias; there is, therefore, a large voltage drop across R3 and the anode voltage of V1a is reduced to a low value, of the order of 15V relative to the chassis. This voltage is direct-coupled to the cathode-follower V1b, and so about 95 per cent of this voltage or 14V appears across the cathode load R4. Thus the screen voltage of the rf amplifier is also 14V and the rf output is low. If the slider of R2 is moved away from the cathode end, the bias on V1a increases and the voltage drop across R3 is reduced. The anode voltage of V1a, and hence the grid voltage of V1b, then rises to the order of 250V. The screen voltage of the pa stage will therefore be increased to somewhat less than this as a result of the cathode-follower action of V1b. Thus the output of the pa stage will be increased towards its maximum.

The setting of R2 therefore makes it possible to vary the output of the rf stage from a low level to almost full cw ratings. If S1 is opened V1a is inoperative and the anode of V1a and the grid of V1b rise towards the anode supply voltage and the screen voltage of the pa stage is limited by the flow of grid current of V1b through R3. The screen potential is then the highest obtainable with this particular circuit arrangement. The maximum screen voltage may be set by the introduction of Rx. R3 and Rx then form a potentiometer across the ht supply which holds the grid voltage of V1b constant.

Modulation may be achieved by applying an af voltage to C1. Assuming that the bias on the grid of V1a is set (by R2) to −1V, the dc voltage applied to the screen of the pa is low and so the rf output is low. If an af voltage of 1V peak is now applied to the grid of V1a, it will be amplified by V1a and appear at the pa screen and so modulate the low rf output to a depth of approximately 95 per cent, the mean dc potential of the pa screen remaining constant. If the af voltage applied to V1a is increased, grid current will flow in V1a and a negative charge will build up on C1 proportional to the peak value of the af voltage. This additional dc bias applied to the grid of V1a will cause the potential on its anode and the grid of V1b (and of course the mean screen voltage of the pa stage) to rise, resulting in increased rf output from the pa stage. Since the period of time for which this potential is raised depends upon the time it takes for C1 to discharge through R1, the time constant of R1 and C1 is important and must be considerably longer than the time corresponding to the lowest audio frequency used.

The increased af voltage at the grid of V1a will appear as an amplified voltage at the pa screen relative to its previous value, but as the mean screen potential has also been raised the carrier is again modulated to a level of about 95 per cent.

The advantages of this method are:

(a) The standing pa anode current and hence the residual power output can be set to any desired level by the potentiometer R2.

(b) Over-modulation on positive peaks cannot occur because the audio input voltage applied to the first section of the double triode will cause limiting and thus the voltage applied to the pa screen cannot rise too high.

(c) Over-modulation on negative peaks sufficient to cause break-up of the carrier cannot occur because the screen voltage cannot fall below the value set by R2.

(d) It provides a simple method of adding an a.m. facility to an existing cw or nbfm transmitter and provides a power control for the existing facilities.

Fig 9.10 shows the cathode of the series valve returned to a negative voltage. This is done to provide an adequate screen voltage to the modulated stage since the available voltage is reduced by the voltage drop across the series valve. The need for a negative voltage may be avoided by using a fairly high

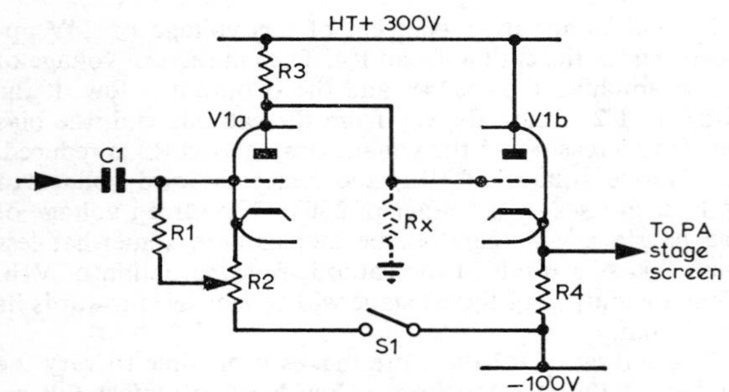

Fig 9.10. Basic "series gate" modulation circuit.

ht voltage to the series valve, by using a series valve of low impedance and by choosing a pa valve which requires a relatively low screen voltage. A suitable low impedance series valve may be obtained by connecting in parallel two triodes such as the two triodes of a 12AU7 or by triode connection of a pentode. A suitable example of the latter is shown in **Fig 9.11** where an ECL82 with the pentode section triode connected provides two triodes of different impedance. The 500pF capacitor C1 is the normal screen bypass capacitor and should be of low enough value not to attenuate the audio signal while the 100Ω resistor R1 is a parasitic stopper and should be mounted close to the pa valve holder. The modulator circuit is sometimes liable to oscillate and the use of ferrite beads on all grid leads is a worthwhile precaution. The audio frequency response of the series gate circuit is very wide and a filter to limit the audio bandwidth should be used.

The time constant of C2R2 is important as this determines the period that the increased screen voltage is retained once the af signal has disappeared. The values given in Fig 9.11 are generally regarded as satisfactory. There are considerable variations of opinion for the optimum level of standing output power. Some operators use a very low value but this presents difficulty in reception and is not recommended except when the maximum possible "speech power" is essential. A standing power output of about a quarter of the output under cw conditions is generally suitable.

Grid Modulation

The circuit arrangement for *grid modulation* is shown in **Fig 9.12** in which it will be seen that the audio frequency

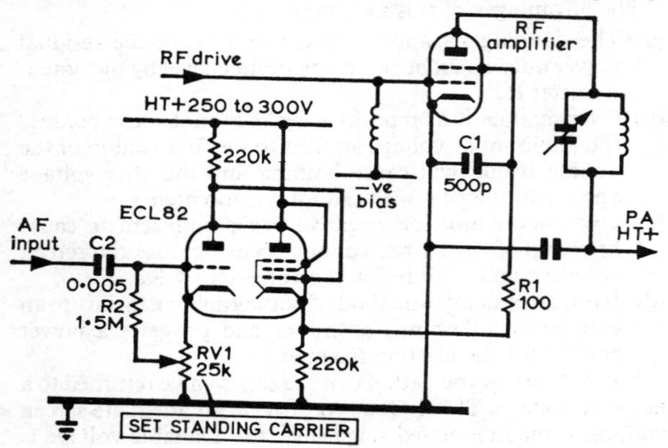

Fig 9.11. Practical "series gate" modulator.

voltage is applied in series with the rf voltage to the grid of the rf amplifier. It is more difficult to adjust the system for full modulation with low distortion than with most other systems.

The requirements are as follows:

(a) the rf drive must have good regulation in spite of the varying grid current load of the rf amplifier. This is generally arranged by providing much more drive than is really necessary and dissipating the excess in a resistive load.

(b) the modulator must have a low impedance output to avoid waveform distortion due to the varying grid level of the rf amplifier. A triode operating in class A is preferable but if a tetrode or pentode is used, it is advisable to connect a swamping resistor across the secondary of the modulation transformer.

(c) the grid bias supply must have good voltage regulation and should be derived from a battery or stabilized supply. A zener diode regulator would be suitable.

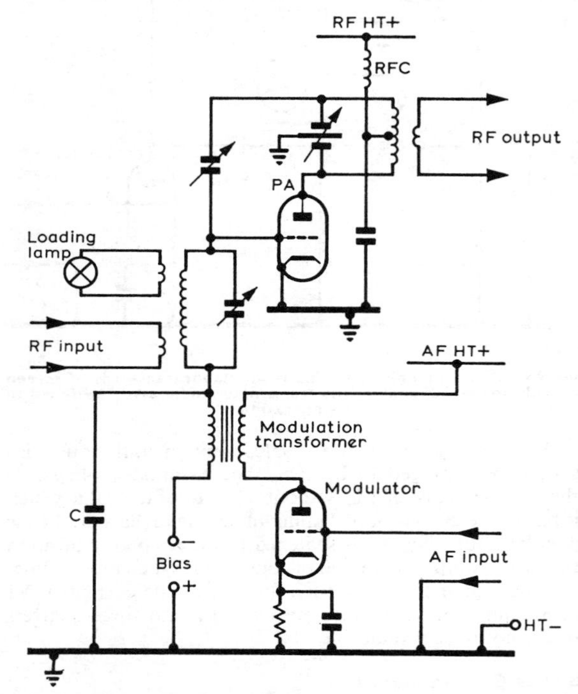

Fig 9.12. Grid modulation. The lamp is used to absorb an appreciable proportion of the grid-driving power and thus improve the grid drive regulation. The capacitor C serves as an rf bypass: it must be small enough to present a high reactance at the highest modulation frequency.

The power input to the pa valve must be limited so that the rated maximum anode dissipation of the valve is not exceeded bearing in mind that the average efficiency will not exceed 35 per cent. The adjustment generally consists of setting the grid drive and aerial loading until the appropriate amount of aerial current (or power output) increase is obtained without causing variation of the pa anode current when the grid voltage is varied. The audio frequency gain control is then adjusted to swing the bias by the amount previously determined. The modulation depth will usually be limited to about 90 per cent to prevent distortion.

Grid-leak Modulation

A form of grid modulation which is occasionally used for low power operation is called *grid-leak modulation*. If, in an rf amplifier which obtains its operating bias solely by the grid leak method, the value of the grid leak is varied, the grid bias and hence the output will also be varied. In practice this may be accomplished by using a valve as the grid leak **(Fig 9.13)**. The usual adjustments should be made to grid drive and aerial loading for approximately one half of the normal cw output. Under these conditions the aerial current should rise during modulation by the normal amount.

The power efficiency of this system is very low, being about 25–30 per cent, but on the other hand it is probably the most economical system since it can be made to function successfully with only a carbon microphone and its transformer and one small triode.

Cathode Modulation

Modulation may alternatively be applied to the cathode circuit of the pa as illustrated in **Fig 9.14**. This is known as *cathode modulation*. In effect both anode and grid voltages are varied simultaneously with respect to the cathode. Cathode modulation is therefore a compromise between anode modulation at high output efficiency and grid modulation at low output efficiency. By suitable adjustment, any desired proportion of each form may be achieved.

The ratio of the respective amounts of anode modulation and grid modulation is controlled by the value of the grid bias on the pa, which in turn will determine the amount of audio-frequency power required. The latter determines the amount of anode modulation, while the former determines the amount of control that a given value of modulating voltage will have. The grid bias, of course, will depend upon the value of the rf and af voltages. A greater range of adjustment in the grid circuit may be obtained if the grid return can be taken to a variable tap on the secondary winding of the modulation transformer; for this reason a multi-ratio type of transformer is preferable.

The setting-up of such a circuit can be fairly complicated because in addition to the normal adjustments, the proportion of grid and of anode modulation should be adjusted to obtain full modulation. However, once set up the system is capable of very good results.

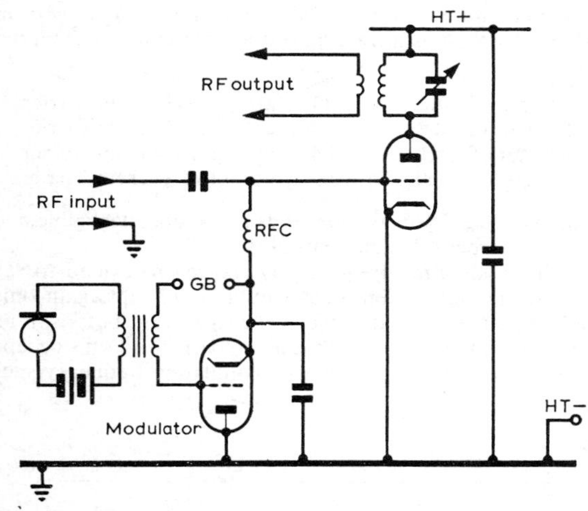

Fig 9.13. Grid-leak modulation. This system requires the minimum amount of audio equipment but the rf efficiency is low.

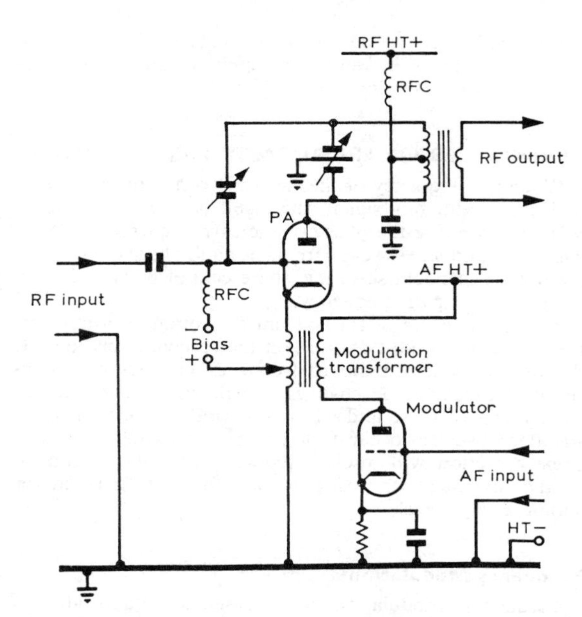

Fig 9.14. Cathode modulation. This method is a combination of (a) anode modulation with its high output efficiency and high-modulation-power requirement and (b) grid modulation with its low output efficiency and low modulation-power requirement.

Low-level Modulation

The modulation systems so far considered are known as *high-level* systems because the modulation is applied to the rf stage working at the highest power level. There is no reason why the modulation should not be applied to a low-power stage, the modulated output of which is then amplified. The *low-level* modulation system is used in ssb transmitters and is occasionally found in other amateur equipment. The modulated rf voltage can be amplified to increase the output power, but in order to avoid distortion of the modulated wave the rf amplifiers following the modulated stage must be of the linear type.

Modulation in Transistorized Equipment

Speech amplifiers and modulators frequently use transistors rather than valves. This does not alter any of the preceding discussion except perhaps that matching may be somewhat less critical in the case of transistor power amplifiers due to their low output impedance.

When the rf amplifiers which are to be modulated make use of transistors, a few special considerations apply. In order to retain the high efficiency of transistors, efficiency modulation systems are rarely used and the equivalent of anode modulation is usual. In the case of bipolar transistors there is significant output power fed from the penultimate rf amplifier to the aerial and it may be necessary to apply some modulation to this rf stage as well as full modulation of the ht supply to the pa stage if 100 per cent modulation is to be approached. A centre tap on the secondary of the modulation transformer will frequently supply a suitable modulation level to the penultimate amplifier.

PRINCIPLES OF FREQUENCY MODULATION

When the frequency of the carrier is varied in accordance with the modulating signal, the result is *frequency modulation* while varying the phase of the carrier current is called *phase modulation*. However, frequency and phase modulation are not independent since the phase cannot be varied without also varying the frequency.

The effectiveness of fm and pm for communication purposes depends almost entirely on the receiving methods. If the receiver will respond to frequency changes but is insensitive to amplitude changes, it will *discriminate* against most forms of noise and will ignore quite large changes in signal level such as occur during mobile operation. Although slope detection will resolve frequency modulation almost all the advantages are lost and some form of discriminator should always be used.

Frequency Modulation

Frequency modulation is represented graphically in **Fig 9.15**. When a modulating signal is applied, the carrier frequency is increased during one half-cycle of the modulating signal and decreased during the half cycle of opposite polarity. This is indicated in the diagram by the fact that the rf cycles occupy less time (high frequency) when the modulating signal is positive and more time (lower frequency) when the modulating signal is negative. The change in the carrier frequency (frequency deviation) is proportional to the instantaneous amplitude of the modulating signal. The total power does not change during modulation.

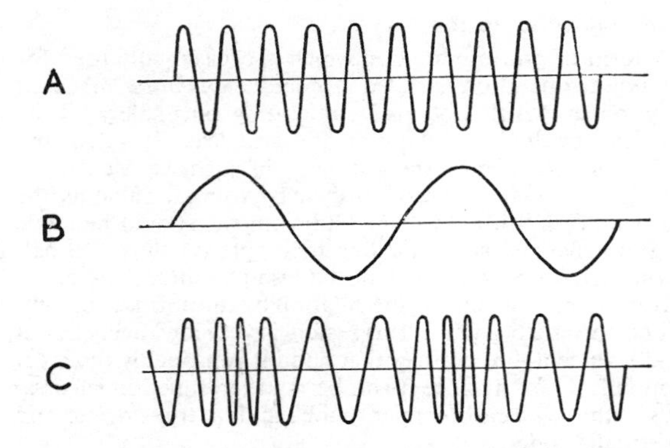

Fig 9.15. Graphical representation of frequency modulation. (A) Unmodulated carrier wave. (B) Modulating signal. (C) Frequency-modulated carrier wave (ie its frequency is varied in sympathy with B).

Sidebands in FM Systems

When a carrier is amplitude modulated by a signal of, say, 1kHz, two sidebands are produced spaced 1kHz either side of the carrier and, for 100 per cent modulation, the sidebands contain half the power of the carrier. When a carrier is frequency modulated by a 1kHz signal, an infinite number of sidebands are produced at 1kHz intervals on both sides of the carrier **(Fig 9.16)**. The power contained in the sidebands is obtained by a corresponding reduction in the carrier power since the total power is unchanged.

It is seen in Fig 9.16 that the sidebands remote from the carrier are small in amplitude although those next to the carrier are not necessarily the largest. It can be shown that the number of significant sidebands depends on the *modulation index* which is defined as:

$$\text{Modulation index} = \frac{\text{Deviation from carrier}}{\text{Audio frequency producing the deviation}}$$

Table 9.2 shows the value of carrier and sidebands as a fraction of the unmodulated carrier from which it will be seen that the number of significant sidebands increases with modulation index. In amateur practice in the United Kingdom the deviation should be limited to 2·5kHz and the maximum modulating frequency to 4kHz. This corresponds to a modulation index of 0·625 and approximating from Table 9.2 we get as a fraction of the unmodulated carrier:

Carrier	92%	
1st pair of sidebands	29%	at ± 4kHz from carrier
2nd pair of sidebands	4%	at ± 8kHz from carrier
3rd pair of sidebands	0·4%	at ±12kHz from carrier
4th pair of sidebands	0·03%	at ±16kHz from carrier

Thus the second pair of sidebands are almost negligible and the higher order sidebands entirely so.

If the modulation frequency is allowed to extend to say 7·5kHz, the modulation index would be 0·3 and again only the first pair of sidebands would be significant but, as in an a.m. system, these would occur at ±7·5kHz and thus occupy excessive bandwidth. For a low modulating frequency such

TABLE 9.2

Modulation Index	Carrier value	1st set of sidebands	2nd set	3rd set	4th set	5th set	6th set	7th set	8th set
0·0	1·0000								
0·2	0·9900	0·0995	0·0050						
0·4	0·9604	0·1960	0·0197	0·0013					
0·6	0·9120	0·2867	0·0437	0·0044					
0·8	0·8463	0·3688	0·0758	0·0102	0·0010				
1·0	0·7652	0·4401	0·1149	0·0196	0·0025				
2·0	0·2239	0·5767	0·3528	0·1289	0·0341				
3·0	−0·2601	0·3391	0·4861	0·3091	0·1320	0·0430	0·0114		
4·0	−0·3971	−0·0661	0·3641	0·4302	0·2811	0·1321	0·0491	0·0152	
5·0	−0·1776	−0·3276	0·0466	0·3648	0·3912	0·2611	0·1310	0·0534	C·0184

Where blank spaces are indicated, the values of the sidebands are insignificant. The negative sign indicates that the componen tis 180° out of phase with respect to the others. Values given refer to current or voltage, not power.

as 500Hz, the modulation index would be 5 and Table 9.2 shows that sidebands up to the 8th are significant but as they are only spaced at 500Hz intervals the bandwidth occupied is still only ±4kHz.

Although the sidebands are small beyond the highest modulating frequency, they are not entirely negligible from the viewpoint of possible interference with other services and an fm transmission should never be centred nearer than 100kHz from a band edge.

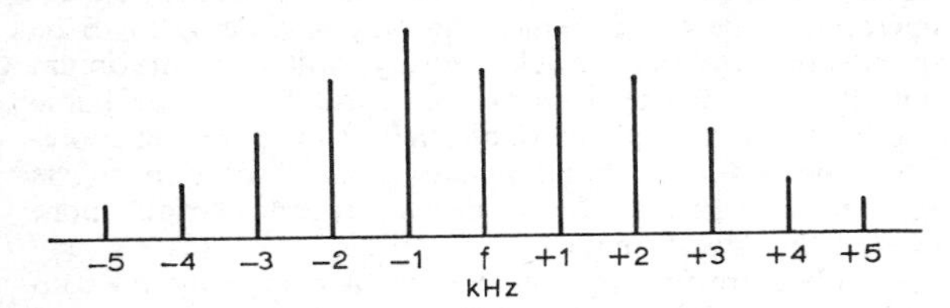

Fig 9.16. FM signal with 1kHz modulation.

Phase Modulation

Phase modulation is sometimes referred to as indirect fm since it is normal practice to modify the phase modulated signal so that the receiver cannot distinguish it from a direct frequency modulated one.

Consider **Fig 9.17** in which a vector OA represents the unmodulated carrier in an XY co-ordinate system which is rotating counter-clockwise at an angular velocity $\omega_c = 2\pi f_c$ where f_c is the carrier frequency. If sinusoidal phase modulation is now applied, the vector oscillates between the extreme positions OB and OC. The vector moves from OA towards OB at a rate determined by the modulating frequency while the angle between OA and OB depends on the amplitude of the modulating signal. On reaching OB, the vector must instantaneously stand still before reversing and returning towards OC where it again instantaneously stands still. While standing at OB or OC, the instantaneous frequency will be that of the unmodulated carrier since the vector is rotating at ω_c with the whole XY co-ordinate system. At the instant of passing through position OA from position OC the vector will be travelling at its maximum velocity and faster than the general rotation ω_c; hence at this instant the frequency will be at its maximum increased value. Similarly, when passing through OA from OB, the instantaneous frequency will be at its minimum value. The frequency deviation at the position OA depends on the rate of change of phase and hence depends on the modulating frequency itself as well as the amplitude of the modulating frequency. This shows the essential difference between phase and frequency modulation and in order that phase modulation should appear as frequency modulation at the receiver the amplitude of the modulating signal must be halved for each octave increase in frequency, ie a fall of 6dB/octave.

Comparison of FM and PM

Frequency modulation can only be applied to an oscillator but phase modulation may be applied to both oscillators and amplifiers. Frequency modulation applied to either a crystal oscillator or a vfo can be entirely satisfactory but it must follow that the oscillator is capable of changing frequency due to external influences and care is needed to prevent unintentional influences from causing frequency changes. Such problems can be avoided if phase modulation is used.

Available Deviation

In a phase modulation system with the required shaped audio response, the maximum linear frequency deviation which can be produced is limited to only a few hundred hertz. In a frequency modulation system rather greater deviation can be produced but often at the expense of frequency stability of the modulated oscillator. It is therefore necessary to produce the frequency modulated signal at a relatively low frequency and to increase the deviation by frequency multiplication. A multiplication of at least eight times is normally required.

Interference in FM Systems

Reception of an fm signal using a discriminator provides a measure of protection against an interfering a.m. signal. The amplitude variations of the interfering signal will be largely eliminated by the limiter but phase modulation will occur due to the interfering signal. The amount of phase modulation increases with the frequency difference between the signals but, of course, the amplitude of an interfering signal at significant frequency separation from the wanted signal is reduced by the receiver bandpass characteristics. The phase modulation produced by an interfering signal is usually no more than a few hundred Hz and the degree of interference decreases as the deviation of the wanted signal increases.

If two fm signals are received, the stronger signal assumes control and if the ratio of the signals exceeds about 2:1, the weaker signal almost completely disappears.

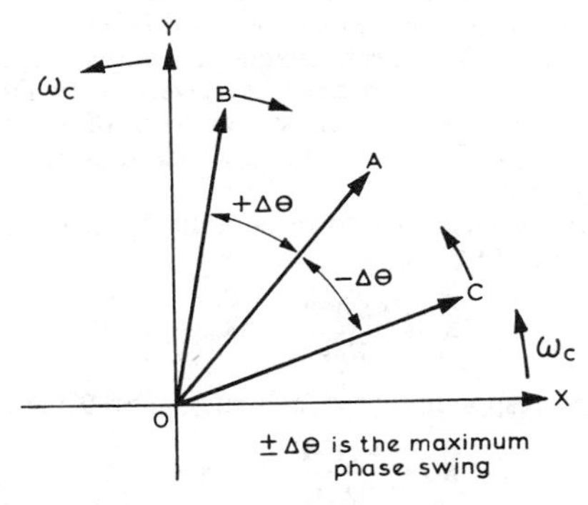

Fig 9.17. Vector representation of phase modulation.

MICROPHONES

The microphone is the starting point of any electrical system of speech communication, as it is the device which converts the energy of the sound wave of the human voice into electrical energy. The output voltage or current can then be amplified and used to modulate a radio transmitter.

Basic Acoustical Principles

A sound wave is a wave of alternating pressure and rarefaction spreading out from the source, in which the instantaneous air pressure at any fixed point varies above and below the mean atmospheric pressure by an amount which determines the intensity of the sound. The waveform of this pressure variation is a curve whose shape determines the characteristics by which one sound differs from another. For example, a pure sine wave of pressure results in a single tone described as "pure," the frequency determining the pitch. Different musical instruments all playing a note of the same pitch produce waves of varying harmonic content, the proportions of which determine the tonal quality or *timbre* of the sound. Harmonics (or overtones) similarly determine the characteristic differences between the male and female singing voices. The important parts of the wave-shape of speech sound are the harmonics of the fundamental pitch and the transient wave-shapes.

The pressure of a sound wave represents the variation about the mean pressure, and is usually measured in dynes per square centimetre. When referring to sinusoidal waves either the peak value or the rms value may be used as in other alternating quantities, the rms figure being more generally quoted.

In an electrical circuit an emf or electrical pressure produces an electrical current, the magnitude of which depends upon the electrical resistance. In an analogous manner an acoustical pressure-wave is accompanied by a movement of the air particles, which may conveniently be measured by the velocity of these particles. This velocity depends upon the *acoustic impedance* of the air space upon which the pressure is acting. This impedance depends in turn on the distributed mass and elasticity of the air and can be regarded as a "characteristic impedance", similar to the characteristic impedance of a cable or transmission line which is determined by the distributed inductance and capacitance.

The wavelength of an acoustic wave is derived in the same way as that for a radio wave, except that the wave velocity in air is approximately 344m/s compared with 300,000,000m/s for electromagnetic waves. This wave velocity is a constant figure for any given gas or liquid at a specified temperature and must not be confused with the particle velocity referred to above, which depends upon the acoustic pressure.

Thus, the wavelength corresponding to a frequency of 100Hz is given by:

$$\lambda = \frac{34{,}400}{100} = 344\text{cm}$$

and that corresponding to a frequency of 10,000Hz by:

$$\lambda = \frac{34{,}400}{10{,}000} = 3{\cdot}44\text{cm}$$

Microphones can be designed so that their electrical output voltage is proportional either to the acoustic pressure or to the particle velocity, thus giving rise to the terms *pressure microphone* and *velocity microphone.*

The pressure microphone, provided it is small enough to avoid complications due to reflection of the wave, will give a constant output no matter in what direction it is facing relative to the direction of travel of the wave. In practice, however, when the frequency is high enough for a half-wave to approximate to the size of the microphone, or higher, the microphone will be more responsive to sounds from a direction facing the microphone than to those from the sides or the rear. Thus, at 10,000Hz where the half-wave length is 1·72cm, the "front-to-back" ratio would be quite high for a microphone 3in in diameter, whereas at 100Hz, corresponding to a half-wavelength of 172cm, the same microphone would give the same output regardless of the direction of the sound.

The term *velocity* has no meaning unless it is associated with a direction as well as a magnitude, since sound waves consist of longitudinal movements of the air particles in the direction in which the wave is travelling. The velocity at any point is at a maximum in a direction facing the source, and zero in a direction at right angles to this. Thus a velocity microphone, ie one in which the output is designed to be proportional to the particle velocity, will have maximum output when facing towards or away from the source and zero output when at right angles to the source; moreover, the output is independent of any reflection effects referred to earlier, ie it is zero in the perpendicular direction, even at the lowest frequencies.

An ideal pressure microphone has a polar diagram consisting of a circle, while a velocity microphone has a diagram like a figure-of-eight, similar to that of a dipole aerial. These diagrams apply in any plane, which means that for a pressure microphone the solid polar diagram is a sphere, while that for a velocity microphone is two spheres touching at the point represented by the microphone.

Basic Electro-mechanical Principles

All practical microphones consist of a mechanical system in which some part moves in sympathy with either the acoustic pressure or the particle velocity, or a combination of both. The pressure microphone comprises a closed chamber containing air at normal atmospheric pressure, part of this chamber being made in the form of a diaphragm which, if it is small and light, will move in sympathy with the variations in air pressure produced by the sound wave outside the chamber above and below the mean pressure of the air inside the chamber.

In a velocity microphone, however, a very light membrane with both sides open to the air is used. When the plane of the membrane is at right angles to the direction of the air movements, it will move with the air, and its velocity will be almost the same as that of the air particles if it is light enough.

To convert a movement of the diaphragm of the pressure microphone or of the membrane of the velocity microphone into an electrical output either the resistance of a circuit may be varied, as for instance by the use of carbon granules, or electromotive forces may be produced by electro-magnetic induction or the piezoelectric effect.

Microphone Characteristics

Brief details are given below of the construction and characteristics of the more common types of microphone. Reference should be made to the manufacturers' literature for more detailed information.

The Carbon Microphone. In this microphone a small capsule filled with carbon granules is attached to a diaphragm, and the effect of the varying pressure of the sound waves is to cause a similar variation in the resistance of the capsule. When connected to a source of constant voltage the microphone thus produces a varying current. The normal resistance of the capsule is generally between 200 and 1,000Ω and the polarizing current should be in the

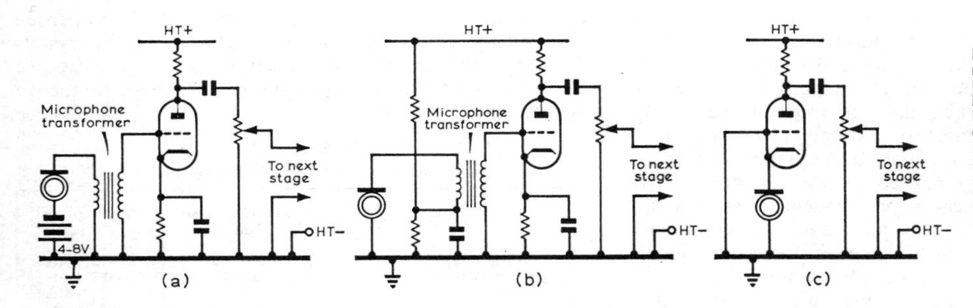

Fig 9.18. Typical carbon microphone input circuits. In (a) the polarizing current is obtained from a small battery: in (b) a potential divider across the ht supply is used as a substitute for the battery: in (c) the microphone is used to modulate the cathode current directly by acting as a variable cathode bias resistance.

range 5–40mA according to the particular design of capsule.

The source of current may be a small dry battery or a potential divider connected across the ht supply to the speech amplifier, as shown at (a) and (b) in **Fig 9.18.** To match the low impedance of the microphone circuit to the high impedance of the input grid circuit of the amplifier, a transformer with a large turns-ratio is required; any ratio between 1:30 and 1:100 should prove satisfactory.

Although these microphones are very sensitive, the frequency response is poor compared with that of most other types. An output of between 5 and 15V can be expected across the secondary of the transformer when speaking close to the microphone, depending on the transformer ratio and the polarizing current. To avoid distortion, the latter should be kept as low as possible consistent with the output required. It will be appreciated that a microphone which utilizes the change in pressure between a multitude of contact points cannot be as free from distortion as some of the more elaborate types described below.

The carbon microphone is particularly convenient for portable use as its high output is capable of fully driving valves such as the KT66 and 6L6 directly from the microphone transformer. An alternative method of using a carbon microphone is shown at (c) in Fig 9.18. Here the microphone acts as the cathode bias resistance of the first valve, the grid being earthed. This form of connection avoids the use of a transformer and a separate polarizing supply, and is capable of somewhat better quality, although the output may be lower.

The Transverse-current Microphone. This is another form of resistance microphone employing carbon granules. It has a mica diaphragm, usually rectangular in shape, behind which is a shallow chamber containing very fine carbon granules. The polarizing current passes between carbon electrodes which are placed at opposite edges of the chamber so that the granules form a thin layer of carbon through which the current passes from one edge to the other. As with the simple carbon microphone a polarizing current and step-up transformer are required. Because the diaphragm is heavily damped all over its surface by the carbon granules, the frequency response is much better than that of the capsule type, although the sensitivity is lower. An output in the order of 0·1–0·5V across the transformer secondary can generally be obtained.

The Ribbon Microphone. This type of microphone has an output proportional to the velocity of the air particles. The main difference between its operation and that of a pressure microphone is that it has a figure-of-eight polar diagram; it is thus necessary to align its axis with that of the direction of the sound source.

The most common form consists of a very thin strip of aluminium foil (about 1in long, 0·1in wide and 0·0003in thick) supported between the poles of a permanent magnet. As the ribbon vibrates it cuts the magnetic field between the poles, and so an emf is induced in the strip. A step-up transformer is normally placed within the microphone case to transform the very low ribbon impedance to 50–250Ω to make the microphone impedance suitable for working into a cable, and a further step-up transformer is used at the amplifier input. No polarizing current is required, and the frequency response can be very good, although the sensitivity tends to be rather low.

The Condenser Microphone. The condenser microphone consists of a thin conducting diaphragm which forms one plate of a capacitor whose capacitance varies when actuated by sound waves. A polarizing voltage of 100–200V is required in series with a very high resistance across which the output voltage is developed. The usual circuit arrangement is shown in **Fig 9.19** where *R* is the resistor across which the output voltage appears. Because the capacitance of any connecting cable will reduce the output, it is necessary for the first stage of the amplifier to be close to the microphone. Consequently the first stage of the amplifier is often built into the microphone housing. Owing to this complication and to the fact that the normal output is generally very low, condenser microphones are not commonly used in amateur stations.

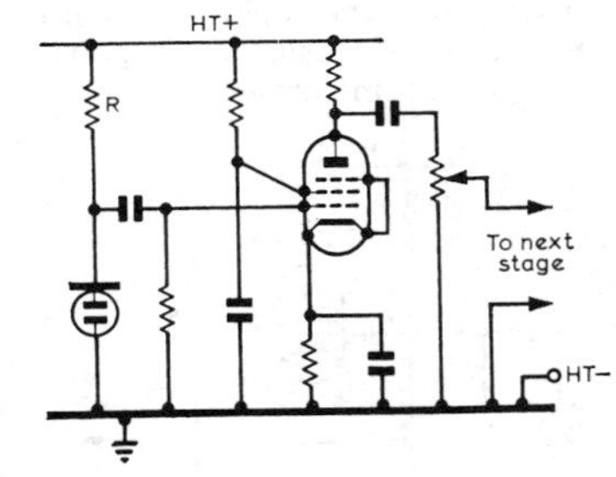

Fig 9.19. Condenser microphone input circuit. The resistance R must be very high and is usually of the order of several megohms.

The Moving-coil Microphone. This microphone is an electrodynamic type. Its construction is very similar to that of a moving-coil loudspeaker, except that the whole construction is very much smaller. The impedance of the coil is generally 20–50Ω. When the diaphragm is caused to vibrate by the sound pressure an emf is developed in the coil and no polarizing supply is necessary. A step-up transformer, however, is required to feed the grid of the first amplifier valve. A very good frequency response can be obtained, and this type of microphone is commonly used in broadcasting studios. The output from the secondary of a suitable transformer is usually not more than about 0·5V. A miniature moving-coil loudspeaker can sometimes be made to serve as a fairly satisfactory microphone, but its frequency response will not be as good as that of a specially designed instrument. The microphone transformer may be replaced by the use of an af transistor as a combined pre-amplifier and step-up matching device as shown in **Fig 9.20.** A low noise transistor should be used.

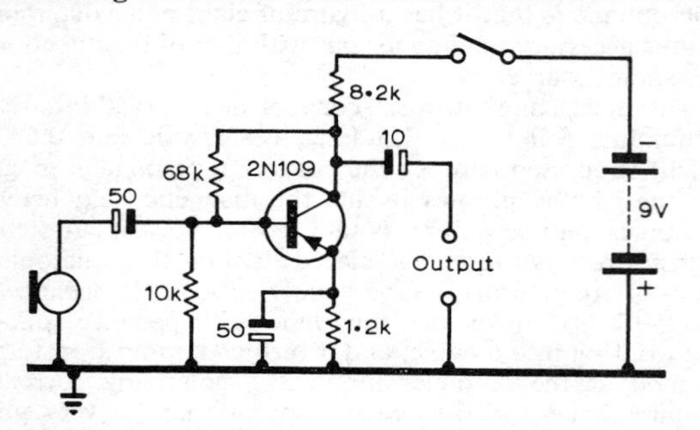

Fig 9.20. Use of an af transistor as a combined pre-amplifier and step-up matching device (*Electronics Illustrated*).

The Crystal Microphone. This type of microphone utilizes the piezoelectric effect whereby any mechanical stress applied to a suitably cut piece of certain materials such as quartz or Rochelle salt generates an emf between opposite faces. Such microphones are very popular amongst amateurs since their frequency response is usually very good, and neither a polarizing voltage nor a transformer is required.

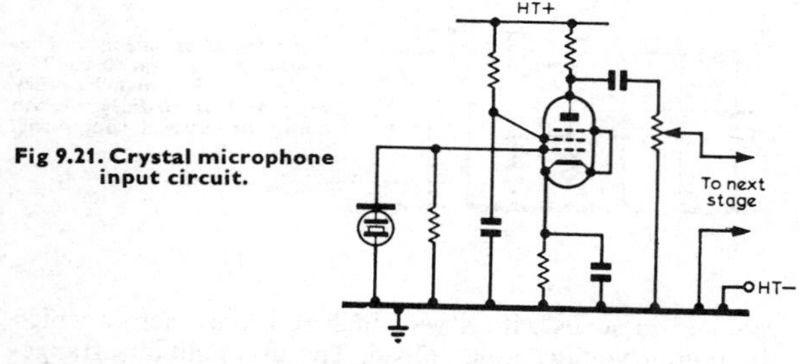

Fig 9.21. Crystal microphone input circuit.

They can be connected directly to the grid of the first amplifier stage through a moderate length of screened cable (up to about 8ft). A high resistance must, however, be connected across the output at the grid of the valve, as shown in **Fig 9.21.** This should be 2–5MΩ; any value lower than 2MΩ will impair the output at low frequencies.

There are two types of crystal microphone in general use. The most common consists of a diaphragm which acts directly on the crystal (normally Rochelle salt). In the second type, known as the "cell" type, the diaphragm is dispensed with, and the sound waves act directly on a pair of crystals which form the surface of a flat cell. Both types are capable of excellent quality, the latter type have a better frequency response but are less sensitive.

TABLE 9.3

Voltage Output of Various Microphones

Type	Transformer Ratio	Voltage at Grid
Carbon (capsule)	30–100	5 –15 volts
Transverse-current	20– 40	0·1–0·3 volts
Condenser	—	0·05 volts
Ribbon	10	0·03 volts
Moving-coil	30– 50	0·1–0·5 volts
Crystal (diaphragm)	—	0·05 volts
Crystal (cell)	—	0·01 volts

Relative Sensitivities

In commercial practice the output of a microphone is generally specified in decibels relative to a reference level of one volt per dyne per square centimetre of acoustic pressure. For amateur purposes, however, it is more convenient to be able to estimate the actual voltage output which is likely to be obtained from the various types of microphone described above when the microphone is held in the hand of the speaker and is actuated by normal speech. An approximate indication of typical output voltages is given in **Table 9.3.**

SPEECH AMPLIFIERS

The term *speech amplifier* is generally applied to the stages of af amplification between the low output voltage of the microphone and the power amplifier which is the modulator proper. The speech amplifier normally only contains voltage amplifiers but the final stage may be a power amplifier if the modulator is driven into grid current and hence requires significant power to drive it. Valves, transistors or integrated circuits are all suitable for use in speech amplifiers and the choice largely depends on the power supplies available from other parts of the equipment. If the microphone output is low, there is some advantage in using a transistor for the first stage as it is easier to keep the hum level imperceptible.

Voltage Amplifiers Using Valves

Audio frequency voltage amplifiers may be divided into two groups, viz resistance capacitance (RC) coupled amplifiers and transformer coupled amplifiers.

RC Coupled Amplifiers

The basic circuit of an RC amplifier is shown in **Fig 9.22.** The voltage amplification of such a stage is given by:

$$\frac{V\ out}{V\ in} = \frac{\mu\ RL}{(RL + ra)}$$

where μ = the amplification factor of V1
ra = anode resistance of V1
RL = external anode local resistance

It is assumed that the value of Rg, the following grid resistor is much greater than:

$$RL\ ra/(RL + ra)$$

which is generally the case.

It will be seen that in order to achieve high gain, RL must

be large compared with ra. If, however, RL is made too large, the dc voltage-drop across it will be excessive and a high dc voltage will be necessary in order to maintain adequate voltage at the anode of V1. In practice, RL is generally made 3–5 times as large as ra. Grid bias for V1 is provided by the voltage-drop across the cathode resistor R1, which is bypassed by C1, to prevent negative current feedback with resulting loss of gain.

When V1 is a pentode RL cannot satisfactorily be made greater than ra because the latter is normally of the order of several megohms and values of RL between 100,000 and 330,000Ω are normally used with pentode amplifiers. The value of RL may be increased to as much as 1MΩ if the screen voltage is made very low. The gain is increased but the valve will only handle very small input signals. The frequency response (ie the variation of gain over the frequency range) is determined principally by the values of three capacitances:

(a) The cathode bypass capacitor C1.
(b) The coupling capacitor Cc.
(c) The total input capacitance Cs of the following stage V2.

The first two affect the response at low frequencies and have negligible effect at high frequencies, while the third affects the high-frequency response.

The capacitance of the cathode bypass capacitor C1 should be such that its reactance at the lowest frequency at which amplification is desired should be small compared with the value of the cathode bias resistor. This capacitor also serves a useful purpose in that it bypasses ac hum voltages between the cathode and the heater; hence it is advantageous to use a somewhat larger capacitor than that dictated by low-frequency response considerations. For most practical purposes a 25μF capacitor is adequate.

At low frequencies, the reactance of the coupling capacitor Cc increases, and since Cc and Rg form a potential divider the voltage appearing across Rg, which is the input to the following stage, tends to fall. The relative gain at a low audio frequency, with respect to that at frequencies in the middle of the audio range, may be found from the expression:

$$\frac{Rg}{\sqrt{Rg^2 + \frac{1}{\omega^2 Cc^2}}} \times 100\%$$

where Rg and Cc are the values of the grid leak and the coupling capacitor respectively of the following stage and $\omega = 2\pi f$. In practice this means that the value of the coupling capacitor must be about 0·5μF for good low-frequency response. However, for speech, where good response below 300Hz is not essential, a value of 0·01μF is adequate for use with a grid leak Rg of 0·1MΩ or more.

At high frequencies the reactance of Cs falls, and as this capacitance is, in effect, in parallel with the output of the previous stage, appreciable high-frequency loss can occur if its value is too great. The capacitance Cs is composed of the input capacitance of the valve V2 and the various stray capacitances of the circuit components.

The input capacitance of an amplifier valve with a resistance load is:

$$C\ input = Cgc + (M + 1)\ Cga$$

where Cgc = grid/cathode capacitance of the valve
Cga = grid/anode capacitance of the valve
M = stage gain

This reflection of the grid-to-anode capacitance into the grid circuit is known as the *Miller Effect*, and in some circumstances it may make the input capacitance quite high. The relative amplification at high audio frequencies compared with that in the middle of the audio range may be calculated from the expression:

$$\frac{1}{\sqrt{1 + (\omega RCs)^2}} \times 100\%$$

where R = equivalent resistance of RL, ra and Rg in parallel
Cs = total shunt capacitance

This high-frequency loss can be very important in the design of high-fidelity amplifiers, but for speech communication where an upper frequency limit of about 3,000Hz is acceptable it is not usually of great significance.

Resistance-capacitance coupling is a cheap and convenient form of inter-stage coupling and is in almost universal use. It has the added advantage that it is not prone to pick up hum from stray magnetic fields, which can sometimes be troublesome with transformer-coupled circuits. It may be used with either triodes or pentodes, and in fact it is the only suitable form of coupling with high-μ valves because of the difficulty of making a transformer suitable for operating with the high load impedances associated with such valves.

Operating conditions for typical RC-coupled amplifiers are given in **Table 9.4** (triodes) and **Table 9.5** (af pentode).

Cathode Followers

If a valve is operated with the output load connected to the cathode instead of to the anode, the load is common to both grid and anode circuits and the arrangement is called a *cathode follower*: see **Fig 9.23**. Here the input voltage V_{sig} is applied between grid and earth while the output load R_k, across which the output voltage V_{out} is obtained, is

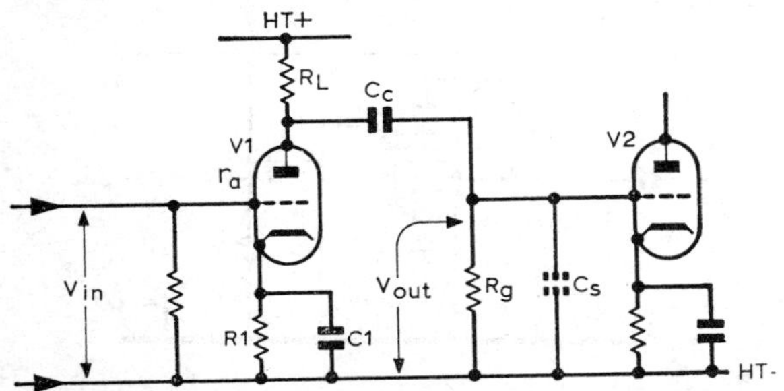

Fig 9.22. Basic circuit of a resistance-capacitance voltage amplifier.

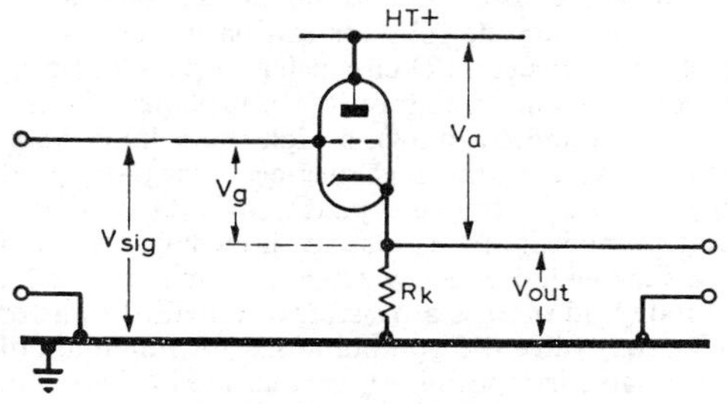

Fig. 9.23. Cathode follower.

TABLE 9.4

Valve	HT (V)	Rg (MΩ)	Rl (kΩ)	Cc (μF)	V out (V)	Gain
½·12AU7	180	0·1	2·0	·032	24	12
	180	0·22	2·8	·016	33	12
	180	0·47	3·6	·007	40	12
	300	0·1	1·9	·032	44	12
	300	0·22	3·0	·016	68	12
	300	0·47	4·0	·007	80	12
½·12AT7	180	0·22	2·6	·014	18	29
	180	0·47	2·6	·009	19	31
	180	1·00	2·7	·006	20	28
	300	0·22	1·2	·015	22	34
	300	0·47	1·2	·009	23	36
	300	1·00	1·25	·006	24	38
½·12AX7	180	0·22	3·0	·012	24	53
	180	0·47	3·5	·006	34	59
	180	1·00	3·9	·003	39	63
	300	0·22	2·2	·013	54	59
	300	0·47	2·8	·006	60	65
	300	1·00	3·1	·003	79	68

Symbols as in Fig 9.22 (R_L assumed to be 100kΩ)

TABLE 9.5

Valve	HT (V)	RL (kΩ)	R2 (MΩ)	Rl (kΩ)	Rg (kΩ)	V out (V)	Gain
EF86/Z729	200	100	0·39	1·0	330	40	106
	300	100	0·39	1·0	330	64	116
	200	220	1·0	2·2	690	36	176
	300	220	1·0	2·2	680	54	188

R2 is value of screen-dropper resistor, assumed to be bypassed to earth by 0·1μF capacitor. Otherwise symbols are as in Fig 9.22.

connected between cathode and earth. If at any instant V_{sig} is made more positive, the grid potential will become more positive and the anode current will increase: hence the cathode current will increase and the increased voltage-drop across R_k will cause V_{out} to increase in a positive direction also. Therefore the output voltage is in phase with the applied grid voltage: ie it "follows" the input voltage.

The amplification of a valve is concerned with the applied voltage between grid and cathode, and in the cathode follower this voltage V_g is not V_{sig} but $V_{sig} - V_{out}$. The higher the amplification of the valve the nearer V_{out} approaches V_{sig}. The voltage gain of the stage is:

$$\frac{V_{out}}{V_{sig}} = \frac{\mu R_k}{r_a + R_k(1 + \mu)}$$

where μ is the amplification factor of the valve and r_a is the anode impedance. It is evident that the stage gain must always be less than unity, although when μ is large and R_k is large compared with the anode impedance r_a the gain is nearly unity. Because there is no actual gain in voltage amplitude, the cathode follower can hardly be regarded as an amplifier. However, the circuit has certain advantageous properties, including a low output impedance (from about 50 to several hundred ohms), a high input impedance, low distortion (owing to the high degree of negative feedback through R_k) and negligible phase shift. One of its most important applications is as an impedance transformer where a very wide frequency range is required. The load in the cathode lead may be a resistor, or a loaded transformer.

For a given valve the optimum load for minimum distortion is the same irrespective of whether the load is connected in the anode or the cathode lead, but in the latter case, ie in the cathode follower, the distortion and output impedance are lower than in the former. It is also convenient at times to divide the load so that part of the load is in the anode and part in the cathode circuit. This arrangement gives a greater gain than the true cathode follower and results in less distortion and lower output impedance than the conventional amplifier arrangement.

If a tetrode or a pentode were used as a cathode follower it would behave as a triode because its anode, like the screen, would be at a constant potential. The relevant values of μ and r_a (anode impedance) are therefore those quoted for the valves connected as triodes.

Anode Followers

When a voltage amplifier with a uniform gain over a wide frequency range is required and when the gain is to be independent of changes in supply voltages, valve replacements, etc, it is convenient to use the *anode follower* principle. The anode follower circuit is essentially a single-stage amplifier in which a high degree of negative feedback from the anode to the grid is provided. A typical arrangement is shown in **Fig 9.24.** Here a calculated fraction of the output voltage is injected into the grid circuit by a potential divider consisting of the two resistors R1 and R2. The effect of this feedback is to reduce the stage gain to a value at which it is almost independent of any variations in the amplification actually produced by the valve. Thus, if the stage gain *without* feedback is A, the effective gain *with* feedback due to R1 and R2 is:

$$n = \frac{R_2}{R_1 + \left(\frac{R_1 + R_2}{A}\right)}$$

where the resistances are expressed in ohms. From this formula it is apparent that especially when A is large the gain depends mainly on the ratio R2/R1.

In a practical example, if the frequency characteristic of an amplifier *without* feedback were such that the gain fell by, say, 20dB at either end of the frequency range, this diminution in gain could be reduced to a much smaller amount, eg 3·5dB by the use of the anode follower principle. At the same time, however, the maximum gain would be lowered from perhaps 200 to something less than 10.

While it is the ratio of R2 to R1 that mainly determines the amount of feedback, they should both be reasonably high compared with R_a, the anode load impedance. Ordinarily R2 may be made equal to R_a.

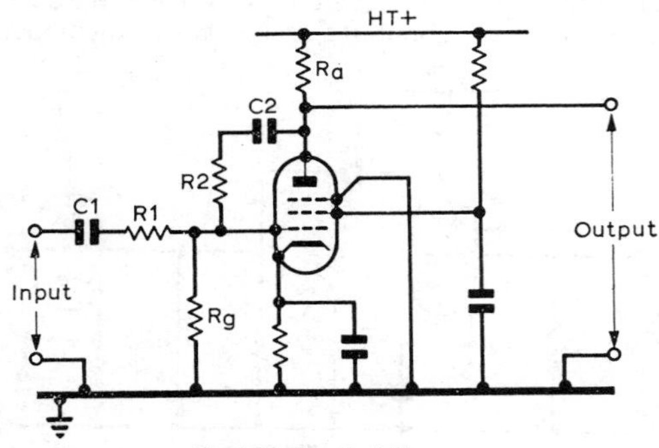

Fig 9.24. Anode follower.

The two capacitors C1 and C2 serve as dc blocking capacitors, but since they are associated with the feedback-determining resistors R1 and R2 it is important that the time constants R1C1 and R2C2 should be made equal so as to maintain a flat frequency response.

Provided that the grid leak R_g is not low compared with R2 it will not affect the performance. In practice it can be made approximately equal to R2.

The input impedance of an anode follower is determined by the value of the series grid impedance represented by R1. Because in general the input impedance is low, say 50,000 Ω, it is essential that the impedance of the source from which an anode follower is driven is relatively low. This would of course be so if a previous amplifying stage was used because this stage would have to supply an adequate output voltage of low distortion into a load of value R1: a valve with an impedance of the order of 10 per cent of R1 would be required.

The output impedance is $1{,}000(n + 1)/g_m$, where n is the stage gain *with* feedback and g_m is the mutual conductance in milliamperes per volt.

The output load need not be resistive as shown. If desired it could be in the form of a suitable choke or transformer.

Phase Splitting Circuits

There are a number of circuits capable of producing balanced voltages of opposite polarity for driving grid circuits in push-pull without the use of transformers, provided that the driven amplifier does not draw grid current. They all utilize forms of RC coupling and are essentially voltage amplifiers. Such circuits are generally referred to as phase splitting circuits.

Paraphase Amplifier. In this arrangement two similar triodes are used, the grid of the second valve being driven from the anode of the first valve to produce the oppositely phased output. The basic circuit is shown in **Fig 9.25**. Here V1 is a normal resistance-capacitance coupled voltage amplifier, its output appearing across R1 and R2 in series; this supplies one phase of the output. The part of this which appears across R2 drives the second valve V2 which also acts as an rc coupled voltage amplifier and supplies the opposite phase. The ratio of R1 to R2 is adjusted so that (R1 + R2)/R2 is equal to the voltage gain of V2, and R1 + R2 is made equal to R3. Then the output voltages of both valves will be equal and opposite. The operating conditions for each valve are calculated in the same way as for voltage amplifiers but the values of R1 and R2 are critical and are chosen to suit the voltage gain. A common cathode resistor can be used and its value should be half that for each valve alone. The bypass capacitor may be omitted with some loss of balance between the two phases at the higher frequencies.

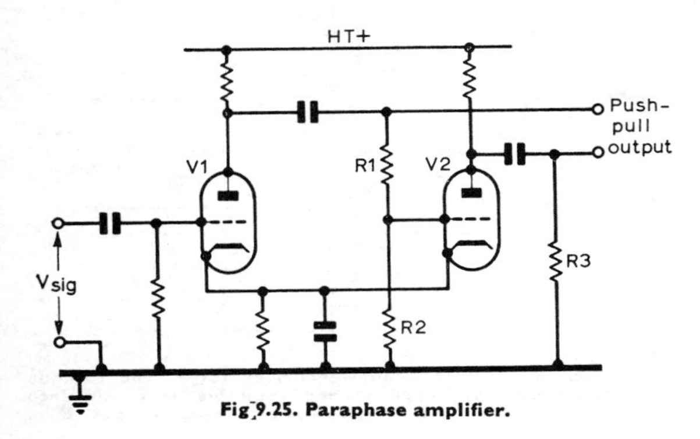

Fig 9.25. Paraphase amplifier.

Split-load Phase Inverter. This type uses a single valve having half the load in the anode circuit as shown in **Fig 9.26**. The two load resistors are R1 and R2. With respect to the load in the anode circuit the valve acts as a normal voltage amplifier, but with respect to the load in the cathode lead it behaves as a type of cathode follower. Since the current in both loads is the same, the voltages are equal if R1 = R2 and R3 = R4, but because of the large negative feedback through R2 the gain is reduced nearly to unity. Since the ratio of the feedback voltage to the output voltage is R2/(R1 + R2), and since R1 = R2 the feedback ratio is 0·5. In other words, an input signal V_{sig} of 1V will produce an output of 1V across both anode and cathode loads. Because of the feedback developed across R2, the output impedance in regard to R1 will be increased while the output impedance in regard to R2 will be decreased, and the source impedance from which the subsequent push-pull valves are to operate will therefore be unbalanced. Hence this type of phase splitter should not be used if the push-pull stage is driven into grid current at any time. The operating conditions should be determined as if the valve were working as an RC coupled amplifier with R2 omitted and R1 twice the value actually used. The effective output voltage will be half the value so obtained.

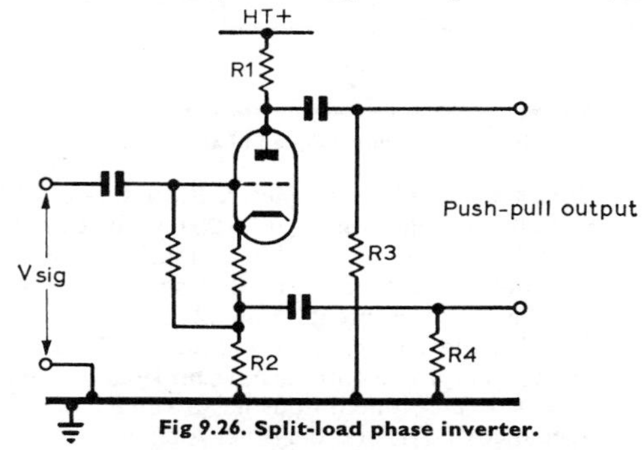

Fig 9.26. Split-load phase inverter.

The anode voltage is measured with respect to the cathode, not the earth connection.

Cathode-coupled Phase Splitter. In this arrangement two similar triodes are used, the input being applied to the grid of one of the triodes while the second valve receives its drive from a coupling resistor R_b in the common cathode load: see **Fig 9.27**. When the grid potential of the first valve V1 swings positively the current through V1 increases and the voltage at the point G becomes more positive with respect to earth; ie the earth line becomes more negative with respect to the point G and the grid of V2 which is coupled to earth through its capacitor C1 consequently becomes more negative with respect to its cathode. The two grids thus operate in opposite phases. The outputs are taken from the two anodes and are balanced in amplitude and impedance, and the voltage gain is about 25 per cent of that obtainable for each triode when used as a normal resistance-capacitance coupled amplifier. The operating conditions can be determined for each triode by assuming an anode load resistance of $R_a + 2R_b$ and a cathode-bias resistance of $2R_k$. The value of R_b should be approximately $\frac{1}{2}R_a$.

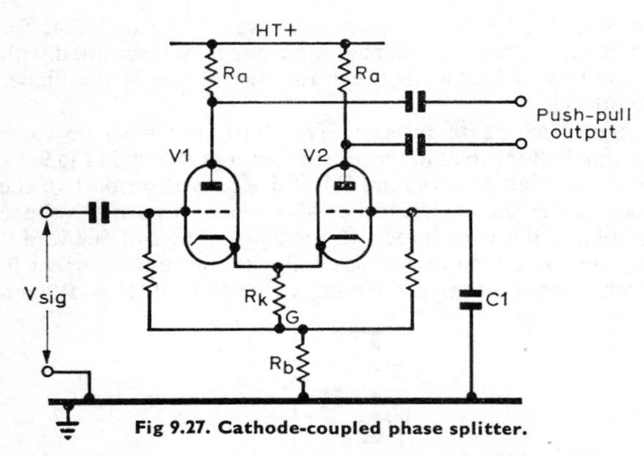

Fig 9.27. Cathode-coupled phase splitter.

Anode-follower Paraphase Amplifier. If an anode follower is designed to have a high degree of feedback, the gain will be unity and the output will be equal in voltage but opposite in phase to the input. A suitable circuit is shown in **Fig 9.28.** As in the anode follower circuit (Fig 9.24), if R1 = R2 and C1 = C2 and if the voltage gain without feedback is very high, the gain is unity. The outputs have impedances differing to some extent because one output has a value equal to the source impedance of V_{sig} and the other is equal to $2/g_m$. The circuit constants are derived as for an anode follower.

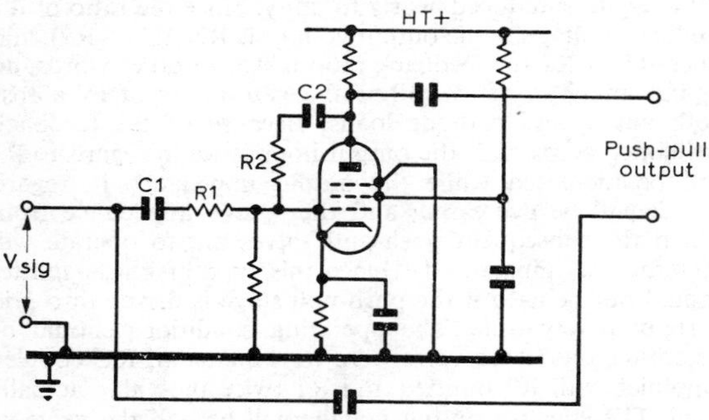

Fig 9.28. Anode-follower paraphase amplifier.

Balancing Adjustments

The amount of unbalance which may be permitted between the two halves of the output of a push-pull speech amplifier is dependent on the amount of distortion which can be tolerated. In high-fidelity work, accurate balancing is essential, but in amateur transmitting practice this is not so important, and the use of high-stability resistors with a tolerance of $\pm$ 2 per cent or even $\pm$ 5 per cent in the appropriate positions generally ensures an adequate degree of balance.

Nevertheless, it must not be deduced from this that it is unnecessary to check the balance of the input voltages to a push-pull modulator. This may be done by direct measurement of the two halves of the output by a high-resistance ac voltmeter (rectifier type), a valve voltmeter or oscilloscope.

Push-pull Amplifiers

A push-pull stage requires the grids of its two valves to be driven by identical voltages of opposite polarity. If the driven stage does not draw grid current, one of the RC coupling systems would normally be used. If grid current is drawn, the driver stage must be able to supply power and should have a low output impedance so that no wave form distortion is caused by the variation of grid current over the audio cycle. Transformer coupling meets this requirement.

Centre-tapped Transformers. The transformer-coupled amplifier using a transformer with a centre-tapped secondary, as shown in **Fig 9.29**, is the easiest way of obtaining a push-pull output. It can be used for supplying a push-pull grid circuit from a single-ended stage, a step-down ratio often being employed to obtain the required low impedance in the drive circuit when the push-pull stage is of the class B type. **Fig 9.30** shows another arrangement in which the input transformer is a step-up transformer driving the push-pull

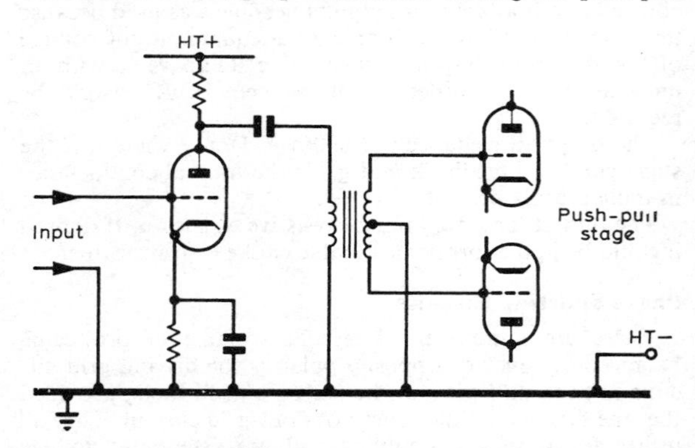

Fig 9.29. By using a transformer which has a centre-tapped secondary winding, a balanced output suitable for driving a push-pull stage is obtained.

stage in class A, while the following push-pull stage (not shown) is driven from the cathodes of the first stage through a transformer having centre-tapped primary and secondary windings. This utilizes the property of the low-impedance output of the cathode follower circuit, and hence it is not necessary for this transformer to have a step-down ratio for driving a class B push-pull output stage. It should be noted that the gain of cathode follower driver stage is only about 0·95.

A transformer with a centre-tapped primary is used almost universally to couple a push-pull power amplifier to its load.

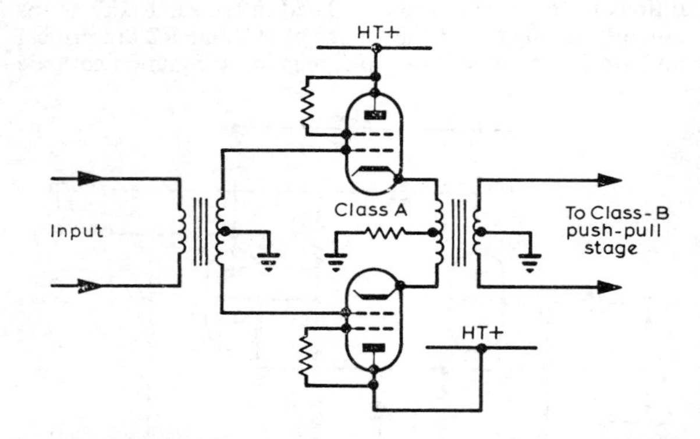

Fig 9.30. Push-pull cathode-follower driver stage. The cathode output circuit has a low impedance and is suitable for working into a step-up transformer for driving a class B push-pull output stage.

Cathode-coupled driver stages can be connected to the grid circuit of the output stage directly, without the use of push-pull transformers, in the manner shown in **Fig 9.31**. If the push-pull input to the driver stage is obtained from one of the circuits which follow, no transformers other than the modulation transformer are required.

Negative Feedback

Negative feedback is commonly used in audio frequency amplifiers, but it is not often applied to the modulators of amateur transmitters. Negative feedback consists of feeding back to an earlier stage a proportion of the output of the amplifier in such a phase relationship that the feedback voltage has an opposite polarity to the input voltage which produces it. Since the effect of the feedback voltage is thus to reduce the effective grid voltage of the earlier state, the overall result is to reduce the gain, but in return the distortion, the hum and the noise are reduced; in addition there are a number of other advantages of which the following are the most important:

(a) Less change in amplification due to changes in supply voltages and valve characteristics.
(b) Improvement of frequency response.
(c) Reduced harmonic and inter-modulation distortion.
(d) The ability to increase or decrease the output and input impedances of the amplifier, if desired.

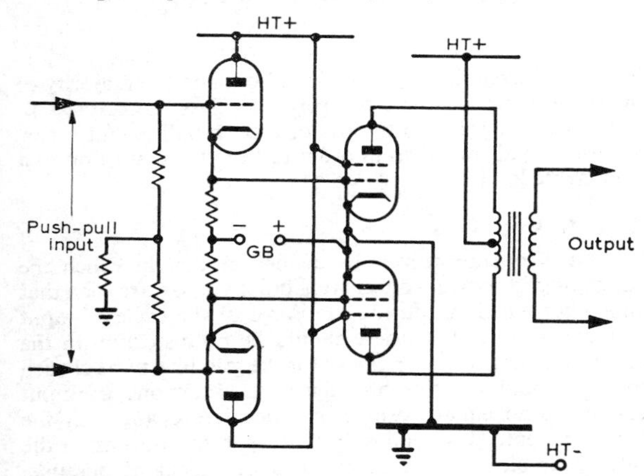

Fig 9.31. Push-pull cathode-follower driver stage using direct coupling from the cathode load resistances to a class B output stage.

It is this last property which can be of great value in modulator design, as it presents a convenient means of providing a low-impedance driver stage for a push-pull modulator working under class B conditions.

The feedback voltage may be obtained in various ways, the two most commonly used being:

(a) The provision of a separate secondary winding on the output transformer.
(b) A fairly high-resistance potentiometer across the output of one stage, the tapping point of which is returned to the input either of the same stage or a suitable earlier one.

A typical circuit showing the use of negative feedback in the push-pull driver stage is illustrated in **Fig 9.32**. The magnitude of the feedback voltage is governed by the

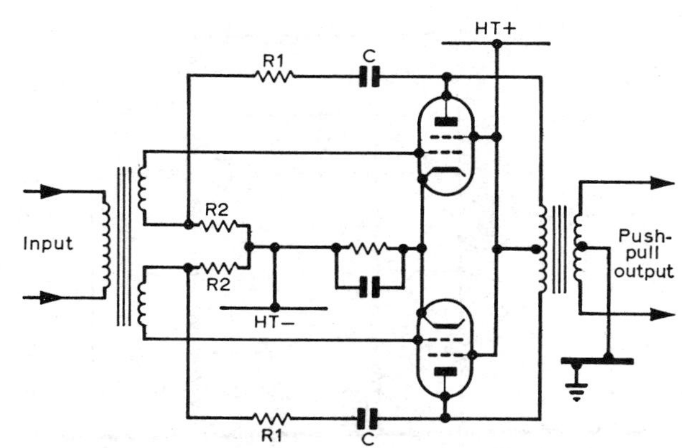

Fig 9.32. Application of negative feedback to push-pull driver stage.

proportions of the potential dividers R1R2, suitable values for which are 150,000 and 33,000Ω respectively. The two blocking capacitors C may have a value of 0·1μF and should be rated at about twice the ht voltage used. It will be seen that when a voltage at one of the grids is increased the corresponding anode voltage will be decreased, and hence a feedback voltage opposite to that of the input voltage will be added to the grid, thus providing negative feedback.

Overall Negative Feedback

The use of negative feedback need not be confined to the af equipment only, but may be applied over the whole chain from the radio frequency output of the transmitter right back to the first stage of the speech amplifier. The basic arrangement of such an application is shown in **Fig 9.33**. Here L is a small coil loosely coupled to the output of the transmitter. The voltage developed across this coil is then rectified, and the resulting voltage may be fed back to the speech amplifier.

Such a scheme has the advantage that it takes account of distortion appearing anywhere in the transmitter, a particularly valuable feature since most of the distortion occurs in the modulation transformer and the modulation system. It can also reduce hum introduced from the ht supply and from the ac used for the valve heaters. It should be pointed out, however, that negative feedback cannot reduce the distortion resulting from over-modulation once the rf output has been reduced to zero, any more than it can reduce the distortion in an ordinary af amplifier due to overloading which is severe enough to cut off the anode current.

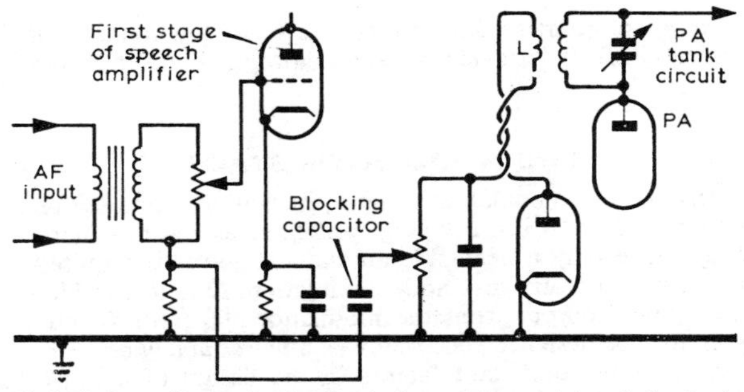

Fig 9.33. Overall negative feedback. Part of the modulated rf output voltage is rectified and fed back to the first stage of the speech amplifier.

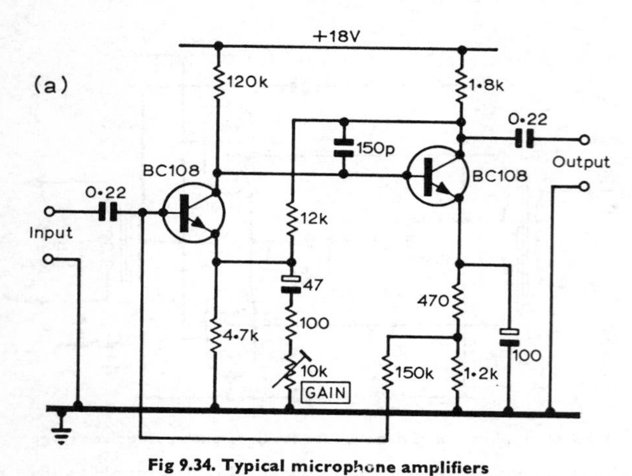

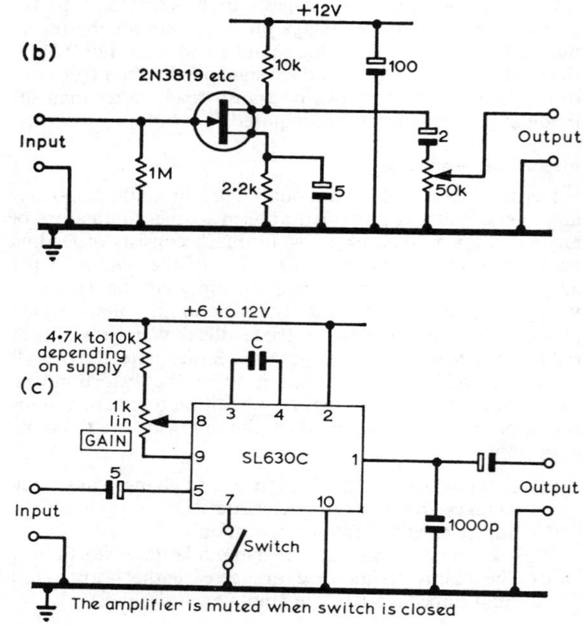

Fig 9.34. Typical microphone amplifiers

Voltage Amplifiers using Semiconductors

Transistors may readily be used in the voltage amplifying stages and the designs published for domestic high quality amplifiers may be used. A typical microphone amplifier is shown in **Fig 9.34(a)**. This will give an output of up to 2 volts with a voltage gain controllable between 13 and 40dB by means of adjustable feedback. The input impedance is about 120kΩ and the output impedance about 100Ω. Field effect transistors may be used with advantage when high input and output impedances are required. A typical circuit is shown in **Fig 9.34(b)**.

A number of integrated circuits are available intended for voltage amplifiers in domestic equipment. The SL630C is specifically intended for communications equipment and a typical circuit is shown in **Fig 9.34(c)**. The voltage gain is 46dB and the input impedance 1kΩ. If a high impedance microphone is used, a pre-amplifier of high input impedance such as shown in Fig 9.34(b) should be used. The maximum signal which the SL630C will handle without clipping is approximately 50mV rms. A capacitor connected across pins 3 and 4 defines the high frequency response which is 3dB down at a frequency f where:

$$f = \frac{16000}{C + 20} \text{ kHz (C in picofarads).}$$

A companion integrated circuit SL621C used in conjunction with the SL630C provides an automatic gain control system.

THE MODULATOR STAGE

In efficiency modulation systems, such as grid modulation, it is sufficient to use a voltage amplifier as the modulator stage since only a negligible amount of power is required. Other systems, such as anode modulation, require considerable power output from the modulator stage which must therefore be designed specifically as a *power amplifier*.

One of the dominant factors in the design of a power amplifier is the anode-circuit efficiency. It is uneconomical, in general, to operate such an amplifier in class A owing to the inherently low efficiency of this method, and the majority of modulators are arranged to operate in class AB or class B. This naturally leads to the use of a push-pull circuit incorporating a pair of valves in order to keep the distortion to a reasonable level.

Class A Amplifiers

Class A operation may be defined as that in which the values of grid bias and alternating input voltage are such that anode current flows during the *whole* of the cycle of input voltage. In order that there should be no distortion in the output waveform it is necessary for the grid bias to be chosen so that equal positive and negative excursions of input voltage cause equal positive and negative changes in the anode current. This implies that the operative portion of the valve characteristic should be as nearly linear as possible. The anode current in the presence of a signal remains at the same steady value as in the absence of a signal, since the equal positive and negative changes in anode current do not affect the average value. Thus there is a continuous dissipation of power in the valve, and hence the efficiency of a class A amplifier will always be low.

Class B Amplifiers

A class B amplifier is one in which the grid is biased approximately to the cut-off voltage, ie in the absence of any input signal the anode current is very low. Anode current therefore only flows for approximately half of the full cycle. In order to obtain high power-efficiency, class B amplifiers are commonly operated with a large input voltage, the peak value of which may be greater than that of the grid bias, causing grid current to flow over the part of the cycle where the grid potential becomes positive. Because this grid current flows for only part of the cycle, the driver stage must have good regulation to avoid distortion of the input wave by the loading effect of the grid current.

Because the anode current flows for only half of each cycle, class B operation at audio frequencies demands the use of two valves in push-pull in order that both halves of the input waveform may be reproduced without distortion. Such a push-pull class B amplifier can have a high anode efficiency since, in the absence of an input signal, the anode current is very low. However, owing to the variation in anode current with the input voltage, an anode power supply with good regulation is essential. Provided that the requirement of low grid-circuit impedance is met, class B operation enables full advantage to be taken of the portion of the anode-current/grid-voltage curve which lies in the positive grid region. This part of the curve is usually fairly linear. The lower portion of the characteristic is badly curved, but the curved portions of the two valve characteristics compensate each other when they are combined in the output circuit. Thus waveform distortion can be kept reasonably low, while still retaining high anode efficiency, so that class B amplification is well suited for use in high-power audio amplifiers.

Special valves with a high amplification factor, which are suitable for operating at zero grid bias, have been developed for this class of operation, a typical example being the DA42 which when used in pairs will give an output of up to 200W.

Class AB Amplifiers

In class AB operation, which is an intermediate stage between class A and class B, the grid bias is insufficient to reduce the anode current to zero, and hence anode current flows for more than half of the input cycle but for less than the whole cycle. In class AB1, the input signal is not permitted to be large enough to drive the grids positive, whereas in class AB2 grid current is allowed to flow, and consequently the driver stage must be capable of delivering appreciable power. Since the fixed grid bias is higher than in class A, larger signals can be handled without grid current, and the standing anode current is less. The efficiency and maximum power output are therefore greater than for class A. Operation in push-pull is, of course, still necessary to avoid distortion arising from the use of the curved part of the valve characteristic.

To summarize the foregoing, as the mode of operation is changed from class A through class AB to class B, the reduction in standing anode current and anode dissipation under no-signal conditions results in a considerable increase in the power-handling capacity and the anode efficiency. Against this, however, must be set the fact that both the ht and grid-drive requirements become more critical. A greater input voltage is required and, in the case of class AB2 and class B operation, the driver stage must be designed to supply the grid power and to have good regulation. Since the anode current varies over a wide range the anode power supply must also have good regulation.

As a general rule class B operation is liable to produce more distortion than class A, but it must be remembered that the overall distortion depends on many factors, and provided that the regulation of the power supplies and the driver stage is good, and particularly if negative feedback is employed, the overall distortion can be reduced to a very low level.

Ultra-linear Operation of Tetrodes and Pentodes

In ultra-linear or "distributed load" operation, the screen supply is obtained from taps on the primary winding of the output transformer as shown in **Fig 9.35**. As the position of the tap is varied so that the ratio a:b changes from zero to 100 per cent, the operating conditions of the output valves change from those of a tetrode to those of a triode.

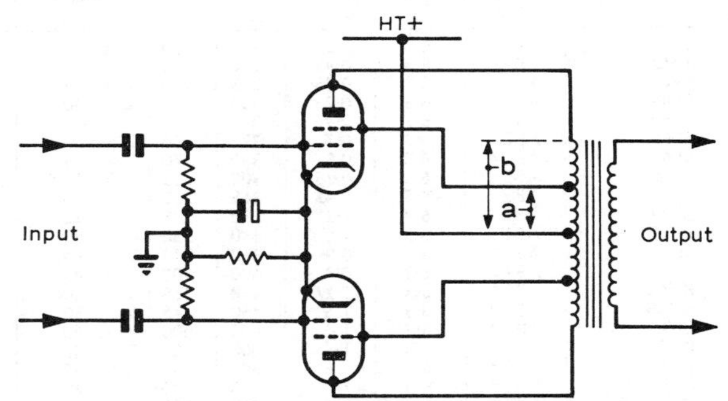

Fig 9.35. Basic arrangement of an ultra-linear output stage.

Ultra-linear operation thus provides a compromise between the high efficiency of tetrodes and the low distortion and uncritical load impedance of triodes. The ratio of a:b is not very critical and can be between 20 per cent and 50 per cent with an optimum value of about 40 per cent.

Ultra-linear operation is now almost universal in high fidelity audio amplifiers but its use is not common in amateur radio modulators. It has undoubted advantages for this service, ie the matching of the modulator to the rf stage is not so critical and as the screen supply is obtained automatically, a high-wattage screen potentiometer or stabilizer valves are not required. It should be noted that as the anode and screen voltages are effectively equal, the output valves must be so rated.

The Woden multi-ratio modulation transformer may be used for ultra-linear operation as the tapping points 2 and 5 on windings 1–4 and 3–6 respectively provide a ratio of about 40 per cent if the anodes are connected to 3 and 4 and a ratio of about 60 per cent if the anodes are connected to 1 and 6.

MODULATOR DESIGN AND CONSTRUCTION

To design any particular modulating equipment it is necessary to know (a) the voltage output of the microphone and (b) the maximum audio power required to fully modulate the transmitter. The power output stage should be chosen first and the preceding stages designed to give the necessary voltage gain.

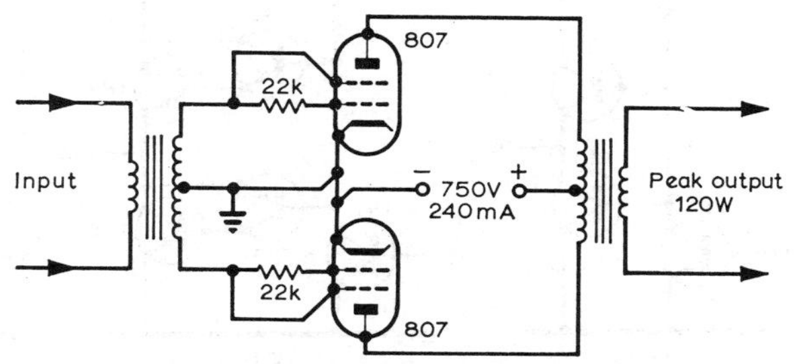

Fig 9.36. A pair of beam tetrodes, each with its two grids connected together, may be used as zero-bias triodes in a class B amplifier.

TABLE 9.6
Characteristics of Typical Modulator Valves

Type	Class	Heater (V)	Heater (A)	V_a	V_{g2}	I_a (No signal) (mA)	I_{g2} (No signal) (mA)	I_a (Max. signal) (mA)	I_{g2} (Max. signal) (mA)	Peak Input Voltage (grid-to-grid) (V)	Load Resistance (anode-to anode) (ohms)	Power Output (watts)
6AM5 ...	A	6·3	0·2	250	250	2 × 11	2 × 1·6	2 × 13	2 × 4·1	32	24,000	4
6AQ5 ...	AB1	6·3	0·45	250	250	2 × 35	2 × 2·5	2 × 39·5	2 × 6·5	30	10,000	10
6V6 ...	AB1	6·3	0·45	285	285	2 × 35	2 × 2	2 × 46	2 × 7	35	8000	14
EL84 ...	AB1	6·3	0·76	300	300	2 × 40	2 × 3	2 × 72	2 × 8	70	8000	17
6L6 ...	A	6·3	0·9	270	270	—	—	2 × 72·5	2 × 8·5	35	5000	18·5
6F6 ...	AB2	6·3	0·7	375	250	2 × 54	2 × 8	2 × 74	2 × 18	55	10,000	19
6L6 ...	AB1 *	6·3	0·9	360	270	2 × 44	2 × 2·5	2 × 50	2 × 8·5	45	9000	24·5
5763 ...	AB2	6·3	0·75	300	225	2 × 13·5	2 × 1·1	2 × 70	2 × 9	71	4500	25
6BW6 ...	AB2	6·3	0·45	315	285	2 × 35	2 × 2	2 × 77·5	2 × 8	80	5000	30
KT66 ...	AB1	6·3	1·27	390	275	2 × 52	2 × 2·5	2 × 62·5	2 × 9	70	8000	30
KT88 ...	AB1	6·3	1·6	330	330	2 × 85	2 × 7·5	2 × 95	2 × 20	80	3400	40
6L6 ...	AB1 *	6·3	0·9	360	270	2 × 44	2 × 2·5	2 × 102·5	2 × 8	72	6000	47
807 ...	AB1	6·3	0·9	600	300	2 × 40	2 × 3	2 × 75	2 × 9	60	10,000	47·5
KT66 ...	AB1 *	6·3	1·27	480	375	2 × 40	2 × 1·5	2 × 87·5	2 × 9·5	80	5000	50
KT88 ...	AB1 *	6·3	1·6	450	345	2 × 50	2 × 4	2 × 120	2 × 17·5	75	4000	65
EL37 ...	AB1 *	6·3	1·4	400	400	2 × 50	2 × 6	2 × 138	2 × 36	60	3250	69
KT77 ...	AB1 *	6·3	1·4	600	300	2 × 40	2 × 4	2 × 100	2 × 10	94	11,000	72
KT88 ...	AB1 *	6·3	1·6	600	330	2 × 50	2 × 3	2 × 125	2 × 16	70	5000	100
EL34 ...	AB2 *	6·3	1·5	800	400	2 × 25	2 × 3	2 × 91	2 × 19	67	11,000	100
807 ...	AB2 *	6·3	0 9	750	300	2 × 26	2 × 2·5	2 × 120	2 × 10	92	7000	120
807† ...	B	6·3	0·9	750	—	2 × 6	—	2 × 120	—	100	6650	120
6146 ...	AB1	6·3	1·25	750	195	2 × 12	2 × 1	2 × 110	2 × 13	100	8000	120
KT88† ...	B	6·3	1·6	750	—	2 × 10	—	2 × 140	—	360	6000	150
DA42† ...	B	7·5	1·2	1000	—	2 × 22	—	2 × 140	—	220	7000	175
TT21 ...	AB1 *	6·3	1·6	1250	300	2 × 28	2 × 1	2 × 130	2 × 13	71	15,000	200

* Fixed bias. The figures shown here relate to pairs of valves operating in push-pull. † Triode-connected

Typical Modulator Valves

Basic operating conditions of valves commonly used as modulators are shown in **Table 9.6.** It should be realized that valves operating under class AB2 or B conditions have a considerable change of grid and screen currents over the operating cycle and low impedance supplies are essential. These requirements can be avoided by using zero-bias triodes or tetrodes connected to give this characteristic. A typical arrangement of the latter is shown in **Fig 9.36**. This circuit is capable of an output of 120W and requires an anode supply of 750V at 240mA. At the same anode voltage the KT88 is capable of 150W in this circuit. Although neither grid nor screen dc supplies are needed, a low-impedance drive source is necessary capable of delivering about 5W.

Power Transistors for Modulators

A number of power transistors are available suitable for modulators up to 100W output while integrated circuit power amplifiers are available for lower powers.

For powers lower than about 4W, the simple complementary push-pull output stage of **Fig 9.37(a)** is the most popular using germanium transistors such as AC128/176. For higher powers, the quasi-complementary circuit of **Fig 9.37(b)** is most often used. The full complementary circuits of **Fig 9.37(c)** and **(d)** are becoming more popular as pnp and npn devices of equal reliability and cost become available.

Three basic input circuits are found and are shown in **Fig 9.38**. The parallel input arrangement of **Fig 9.38(a)**

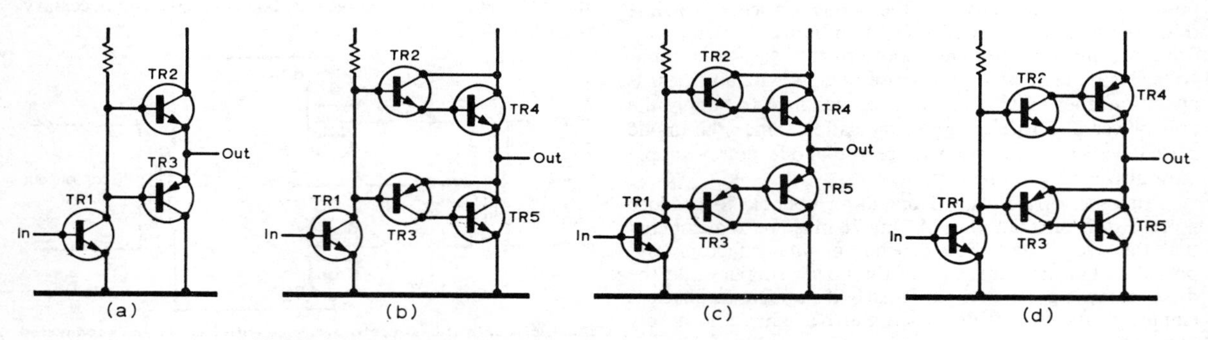

Fig 9.37. Various forms of complementary and quasi-complementary output stages.

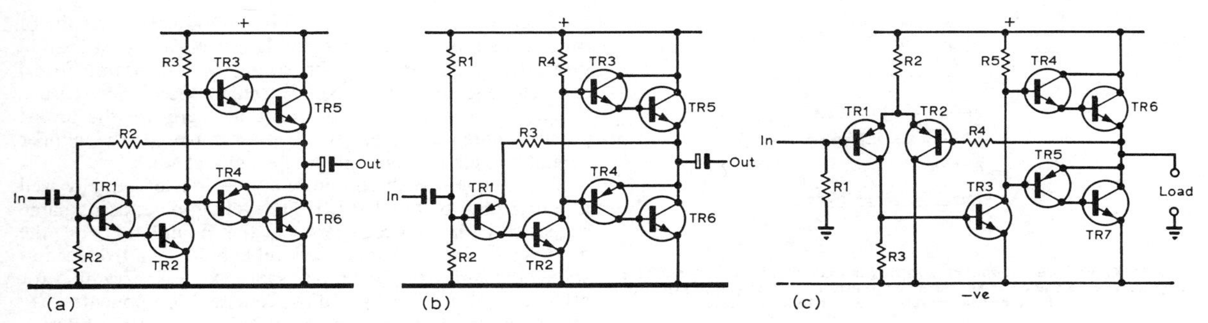

Fig 9.38. Input methods showing dc feedback paths; (a) parallel, (b) series, (c) differential.

suffers from rather low gain and very low input resistance but is simple and stable. The series arrangement of **Fig 9.38(b)** gives higher gain and input resistance but requires input devices of both polarities. **Fig 9.38(c)** shows a long tailed pair input and although it needs an extra active device, and a split ht supply, the mid point or centre line will follow earth potential if R1 and R4 are made equal. No output capacitor is then required.

The Modulation Transformer

The modulation transformer is a most important component, for its object is to superimpose the output of the modulator upon the dc supply to the rf stage in order to produce the required modulation: it must also convert the modulating impedance of the rf amplifier to the load impedance required by the modulator valves. To avoid distortion the iron core should be of generous proportions, and modulation transformers should always be used well within their ratings.

The choice of a suitable modulation transformer should receive careful attention. Two conditions must be fulfilled: (a) the ratio must be correct in order to transform the impedance of the modulated rf amplifier to the optimum load required for the modulator, and (b) the transformer must be capable of handling the maximum power output over the audio-frequency range without undue distortion. Multi-ratio modulation transformers, of which the Woden types UM1 and UM3 are excellent examples, are frequently used, since they enable adjustments of the matching conditions to be made. The use of miscellaneous transformers, such as mains transformers, which are not designed for audio work is not recommended as they rarely have the correct impedance ratio, and magnetic saturation of their cores is likely to occur even with relatively low values of direct current.

Modulator Power Supplies

The power supplies required for class A or AB1 modulators are relatively simple in view of the small variation of current drawn. A capacitor input smoothing circuit is generally adequate.

For class AB2 or B operation all the supply voltages must have good dynamic regulation in order that the voltages shall not vary by more than about 5 per cent in spite of the widely varying load. A choke input smoothing circuit is required and the use of low impedance semi-conductor rectifiers is recommended. If a separate screen supply is used, switching precautions should be taken to ensure that the screen voltage cannot be applied in the absence of the anode voltage. Alternatively a well regulated screen supply may be derived from the anode supply by utilizing the voltage drop across a gas-discharge stabilizer such as the VR150 or across high voltage zener diodes, such as a pair of BZX70C75s in series.

Speech Clipping and Volume Compression

Although a pure sine wave and a speech waveform may have the same peak value, due to the peaky nature of the speech waveform, the average value of the latter is lower than that of a sine wave. This is shown in **Fig 9.39**, where (A) represents a pure sine wave having just sufficient amplitude to modulate a given transmitter to a depth of 100 per cent. (B) is a typical speech waveform also with just sufficient amplitude to give 100 per cent modulation on peaks. In the latter case, if the audio level is conscientiously adjusted to give 100 per cent modulation on peaks which may not occur very often, the average modulation depth will be much less than 100 per cent, probably of the order of 30 per cent. If the

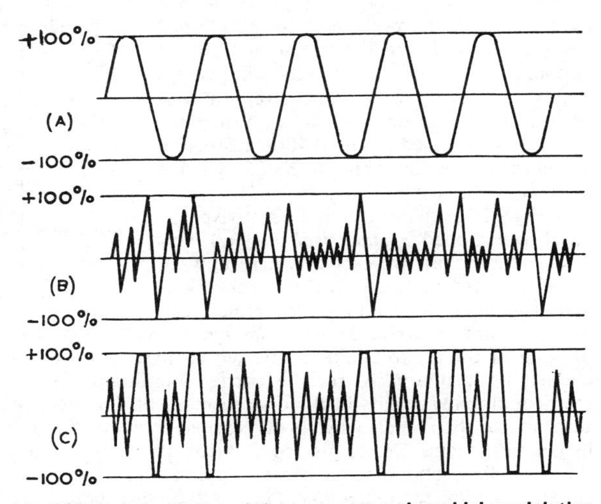

Fig 9.39. Speech clipping. (A) represents a sinusoidal modulation signal the peak amplitude of which is sufficient to modulate a given carrier wave to a depth of 100 per cent. (B) shows a typical speech waveform of equivalent peak amplitude to (A). Full 100 per cent modulation is reached relatively infrequently and the average modulation level is low. In (C) the same speech waveform has had its peaks removed by clipping, and the mean amplitude of the modulating voltage has been increased so that 100 per cent modulation is again reached. As compared with (B), there is greater average speech power in the clipped waveform (C).

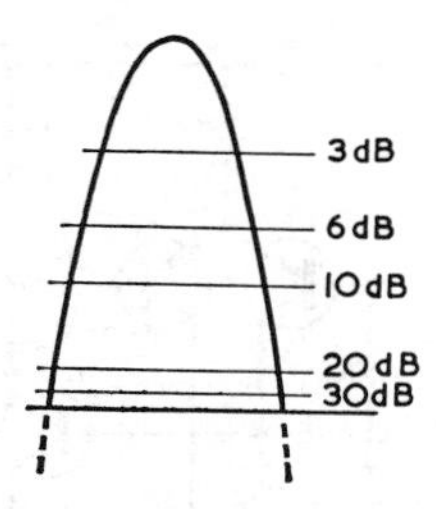

Fig 9.40. The degree of speech clipping is expressed in decibels. The diagram shows the corresponding amounts by which the maximum amplitude of the waveform must be reduced.

audio level is increased in order to increase the average modulation depth, then overmodulation will obviously occur on the peaks.

Principle of Speech Clipping

The process of speech clipping consists simply of increasing the audio input level and at the same time clipping or limiting the output level at the value corresponding to 100 per cent modulation as shown in Fig 9.39. Thus overmodulation is prevented and the average level is increased.

The degree to which this process can be carried is limited by the amount of distortion of the speech waveform that can be tolerated, and if taken to excess the quality of the speech will deteriorate and intelligibility will be diminished. Furthermore the squaring of the individual peaks as shown at (C) in Fig 9.39 would introduce very high audio-frequency components which will defeat the object of the operation by again increasing the bandwidth occupied. Any speech-clipping system must therefore be followed by a low-pass filter to restrict the audio bandwidth to the required 3kHz. The degree of speech clipping is defined as the ratio of the peak level to the clipped level, as shown in **Fig 9.40**. Small amounts of clipping, of the order of 4–6dB, will give an appreciable increase in audio strength with only a slight loss of naturalness. When the clipping is increased to about 10dB distortion becomes noticeable, while at 20–25dB some intelligibility is lost.

In Fig 9.39 both positive and negative peaks are shown clipped (*symmetrical clipping*). This is essential in an fm system where upward and downward modulation are not meaningful expressions, but in an a.m. system either peak may be clipped (*asymmetrical clipping*) provided that the phasing of the audio chain is adjusted so that it is the *un*-clipped peak that *increases* the carrier amplitude. Asymmetrical clipping results in the displacement of the effective zero line of the speech waveform. Normal intervalve couplings adjust themselves to give equal areas above and below the zero line, and asymmetrical clipping will upset this equality, giving a momentary shift in the zero axis. It is advisable to make all succeeding coupling capacitances small enough so that the time constant of the coupling circuits is short enough to allow rapid restoration of the normal zero position. It is quite possible for this momentary displacement of the zero axis, owing to the presence of ac intervalve couplings, to cause the intended level to be exceeded and therefore to permit over-modulation in spite of the speech-clipping action. A similar effect can occur even in symmetrical clipping circuits owing to the asymmetrical characteristic sometimes experienced in speech waveforms. It may be obviated by ensuring that there are the fewest possible couplings after the clipper stage, ie the clipper should be as late as possible in the audio chain.

In principle speech clipping may be effected equally well at either high or low level. High-level clipping, as its name suggests, is introduced between the modulator and the modulated rf stage, while low-level clipping is introduced at some point preceding the modulator. The necessary filter for attenuating the higher audio-frequency components may also be inserted at either high or low level, but the filtering action must, of course, always take place *after* the clipping.

Volume Compression

A volume compresser is an automatic gain control system applied to an audio amplifier. If the agc time constant is slow, say of the order of one second, the system will follow the average level of speech and will result in substantially constant modulation depth without the operator having to maintain a constant voice level and distance from the microphone. In the case of frequency modulation where excessive deviation can easily occur without any indication on the metering of the transmitter, some form of slow acting audio agc is highly desirable. If clipping of the audio signal is used, the agc has the advantage of producing a constant degree of clipping, thus avoiding distortion due to excessive clipping. If the time constant of the agc system is made very short (say a few milliseconds) it is possible to get an effect similar to clipping but this is difficult to achieve owing to the problems of over-shoot in the feedback loop. It is probable that such a system would not need a filter but since some means of limiting the audio response of the modulator to 3–4kHz is necessary to avoid occupying unnecessary bandwidth, a low pass filter is almost essential even in the absence of clipping. It is probable that a slow acting audio agc system followed by a moderate degree of clipping (say 10dB) and followed by a low-pass filter represents a good compromise.

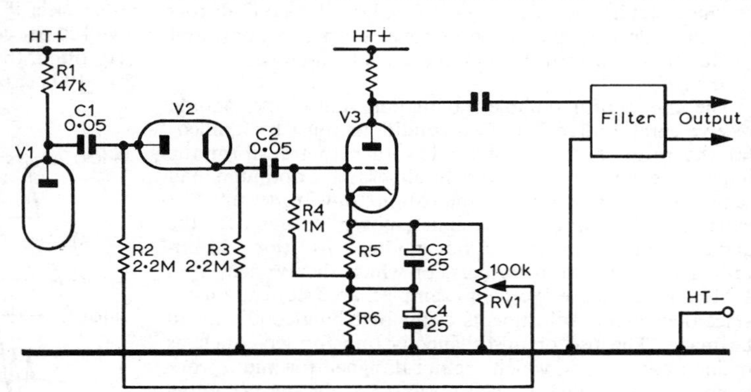

Fig 9.41. Low-level asymmetrical series clipper.

Frequency Response

A fairly sharp cut-off above 3–4kHz is necessary and has little effect on the naturalness of speech. It is sometimes

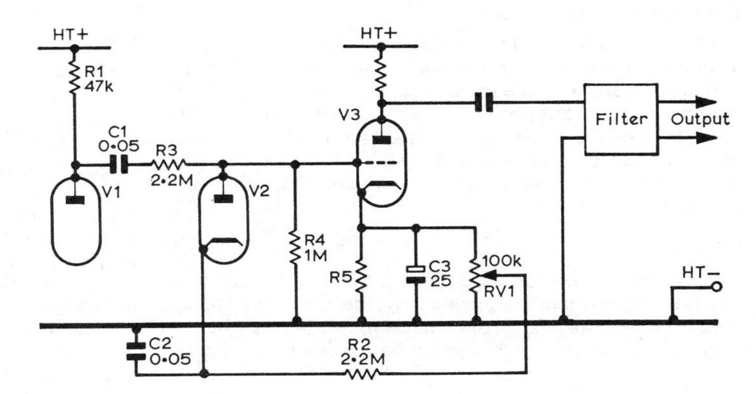

Fig 9.42. Low-level asymmetrical shunt clipper.

argued that a corresponding cut of frequencies below 300Hz helps to give a more balanced response; probably the voice characteristics of the particular speaker influence the advantage to be gained. It should not be assumed without trial that intelligibility is improved by a sharp cut below 300Hz but it is worthwhile ensuring that maximum response occurs in the region 500 to 1500Hz, the optimum range depending on the voice characteristics. This ensures that any clipping is effective over the frequencies carrying most intelligence.

When using fm it should be noted that many commercially made receivers incorporate substantial de-emphasis following their discriminators. Any transmission which does not employ corresponding pre-emphasis will lack brilliance and intelligibility may suffer. A rising response from about 1kHz to the cut-off at 3–4kHz is an advantage in this situation.

Low-level Clippers

The limiting of the peaks that occur in speech waveforms resembles the limiting of impulsive interference by a conventional noise limiter. The circuits employed are very similar and consist of a series or shunt arrangement of diodes suitably biased to give the desired clipping level. A typical low-level asymmetrical series clipper is shown in **Fig 9.41**. If the anode of V2 is held at, say, +5V by means of potentiometer RV1 signals at all amplitudes to 5V peak will be passed on unchanged, but any *negative* peak which exceeds −5V will cause the diode to become non-conducting and the peaks will be clipped.

The corresponding shunt diode clipper is shown in **Fig 9.42**. In this case the diode V2 is normally non-conducting, but it conducts when the *positive* peak voltage at its anode exceeds the positive bias applied to the cathode from RV1 and thus all positive peaks which exceed this value are bypassed by the diode.

A symmetrical shunt clipping circuit is shown in **Fig 9.43**.

Although thermionic diodes such as 6H6 are shown in these circuits, semi-conductor diodes can equally well be used. In general, a thermionic diode will give flat-topped peaks whereas semi-conductor diodes give a somewhat more rounded top. This is rarely a disadvantage since the flat-topped characteristic contains a wider range of undesired harmonics and makes the following filter design more difficult. It is worth remembering that germanium diodes conduct at a lower forward voltage than do silicon diodes and hence give somewhat greater clipping.

High-level Clippers

The high-level series-diode negative-peak clipper is the commonest arrangement. The circuit is shown in **Fig 9.44**. The diode V conducts only when its anode is positive with respect to its cathode. Therefore the modulated anode voltage cannot swing the anode of the rf amplifier negative. For use in a 150W transmitter, the diode V may be a rectifier of the 500V 250mA class with the two anodes connected together. It should be noted that the filament transformer T must have a primary-to-secondary insulation suitable for at least twice the anode voltage of the rf stage. Such a clipper used in conjunction with a good filter makes a very effective "splatter" suppressor. It has the advantage that relatively few extra components are required, although the filter components are likely to be rather bulky if entirely satisfactory performance is to be achieved.

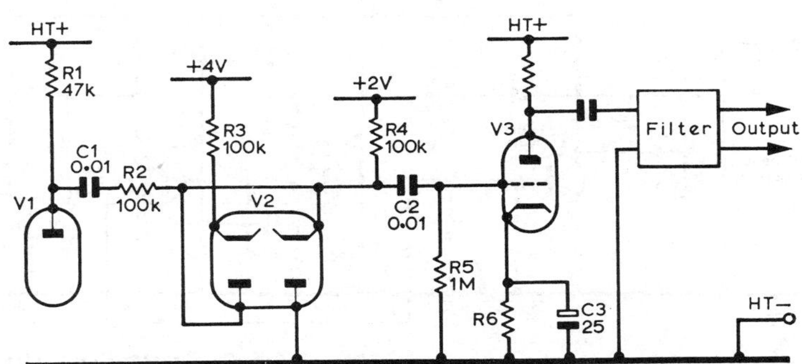

Fig 9.43. Low-level symmetrical shunt-clipper. The small positive bias voltages applied through R3 and R4 to the cathodes of the double-diode V2 can be conveniently obtained from taps on the cathode bias resistor R6 of the following amplifier stage V3.

Compression and Clipping

Fig 9.45 shows a circuit giving audio agc followed by symmetrical low level clipping and a low pass filter. The agc action is obtained by the use of a fet as a voltage controlled variable resistor. Substitution of germanium diodes for D2 and D3 would give a significant increase in clipping. This circuit, preceded and followed by any necessary voltage amplification would be very suitable as a speech amplifier for an fm transmitter.

Filters

Every speech-clipping stage must be followed by a filter to remove the harmonics generated by the action of speech clipping on the waveform. Such filters must obviously be put into the circuit after the clipping has taken place. Trans-

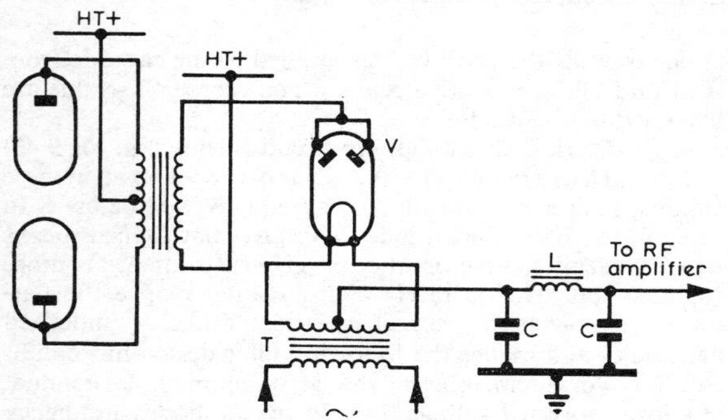

Fig 9.44. High-level series-diode negative-peak clipper. The transformer T must be insulated for at least twice the anode voltage of the rf stage. V may be a 500V, 250mA rectifier. Suitable values of L and C can be calculated from the basic filter design equations (see text).

mitters which only include volume compression do not introduce waveform distortion of the same type as clippers, but such transmitters, and even those without either clipping or compression can, with advantage, incorporate a low-pass filter to remove any modulation frequencies above about 3kHz, thus restricting the total transmitted bandwidth to about 6kHz.

Filter Design

The complete design of filters is outside the scope of this handbook, but the following information should enable the amateur to design and make satisfactory filters without very much trouble.

Before the design can be commenced, three factors must be known:

(a) The "cut-off" frequency, ie the frequency separating the regions of high attenuation and low attenuation.
(b) The amount of attenuation required in the attenuating band.
(c) The impedances between which the filter is to operate.

A recommendation of the IARU Congress (Paris), May 1950, is that the response of the modulator in an amateur telephony transmitter at 4kHz should be 26dB below the response at 1kHz. A suggested method of achieving this is by the use of two filters, one in the speech amplifier with an attenuation of 20dB at 4kHz and a second between the modulation transformer and the rf amplifier having an attenuation of 6dB, the second filter having a higher cut-off frequency than the first.

The operating impedance of the filter is not critical, and normally considerable variation is possible. In the case of a high-level filter, it should be designed to work from the modulating impedance of the rf amplifier.

The simplest form of low-pass filter is the pi-network illustrated in **Fig 9.46**. This is known as a "prototype" or "constant-k" filter. If the impedance of the two circuits which are coupled by the filter are assumed equal, the design equations are as follows:

$$R = \sqrt{\frac{1000L}{C}} \quad L = \frac{R}{\pi f} \quad C = \frac{1000}{\pi R f}$$

where R = terminal impedance (Ω)
L = inductance in filter (mH)
C = total capacitance in filter (μF)
f = cut-off frequency (kHz)

Thus L and C may be calculated for any chosen values of R and f. In a low-pass filter for the purpose described f could have a value of 3·3kHz.

Such a filter by itself has an attenuation curve which does not rise very steeply beyond the cut-off frequency, and would only give an attenuation of about 6dB at 4kHz and about 30dB at 10kHz. It would, however, be suitable for a high-level filter if a filter of higher attenuation is included in the speech amplifier.

The simplest method of increasing the attenuation of a filter is to connect several identical sections in cascade. Three sections of the type just referred to give an attenuation of about 20dB at 4kHz and about 60dB at 10kHz. Thus a three-section constant-k filter in the low-level amplifier plus

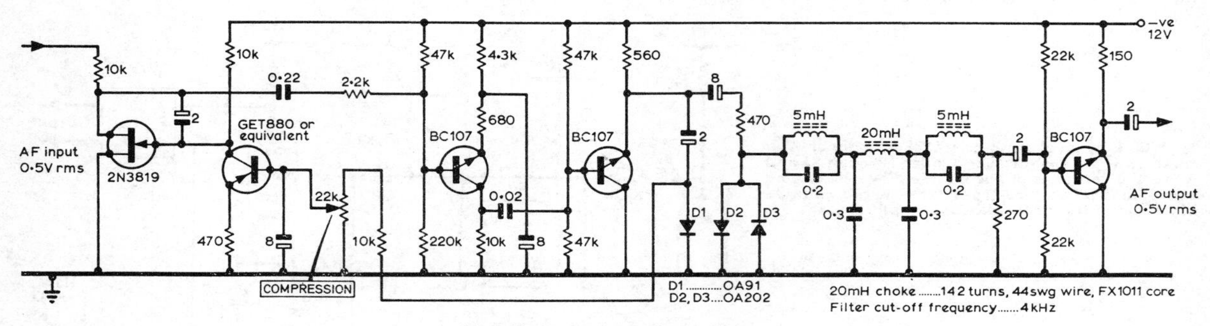

Fig 9.45. Compressor-clipper-filter.

a single section of similar type for the high-level filter would just about meet the specified requirement. If a steeper attenuation/frequency curve is desired, use may be made of what are called "m-derived" half-sections. The arrangement of a filter having a single constant-k section and an m-derived half section at each end is shown in **Fig 9.47**. The design equations for the m-derived half sections at the ends are:

$$L_1 = mL \quad C_1 = \frac{1 - m^2}{4m}C \quad C_2 = mC$$

The general attenuation characteristic of such section has a very sharp peak of attenuation followed by a reduced attenuation remote from the cut-off frequency, as compared

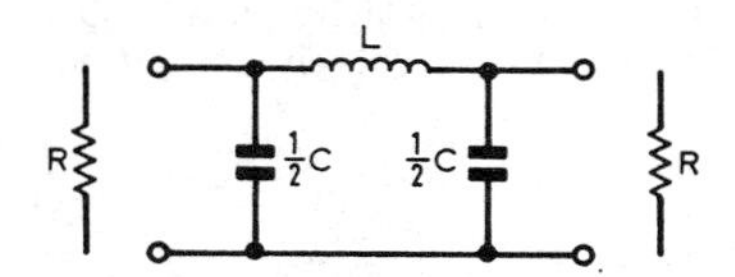

Fig 9.46. Low-pass prototype or constant-k filter. It is also referred to as a pi-network filter. Its function is to attenuate all voltages having frequencies above a certain value determined by the filter components. See text for design equations.

with the constant-k sections which have a slow but steadily increasing attenuation. The quantity m in the above equations is determined by the separation between the frequency of peak attenuation and the cut-off frequency. A convenient value of m for design purposes can be taken as 0·6, and these equations then become:

$$L_1 = 0{\cdot}6L \quad C_1 = 0{\cdot}27C \quad C_2 = 0{\cdot}6C$$

A filter consisting of a single constant-k section and two m-derived half-sections would give an attenuation of about 30dB at 4kHz.

It will be seen from the above equations that the value of inductance required is proportional to the terminal impedance of the filter. High values of inductance introduce difficulties owing to the presence of self-capacitance, and for this reason it is often preferable to avoid filters of higher impedance than, say, 10,000–15,000Ω. Such filters, however,

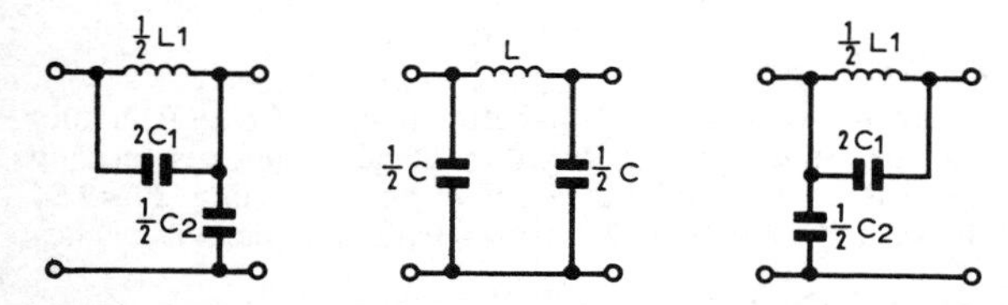

Fig 9.47. An improved low-pass filter having a more sharply-defined cut-off frequency than the simple filter shown in Fig 9.46. The diagram illustrates the three separate sections comprising a single-section "prototype" filter preceded and followed by an "m-derived" section. See text for design equations.

may not provide adequate load impedance for voltage amplifiers, and a means of using a low-pass filter with a terminal impedance as low as 500–1,000Ω is shown in **Fig 9.48**. The filter is connected between a cathode follower and a grounded-grid stage.

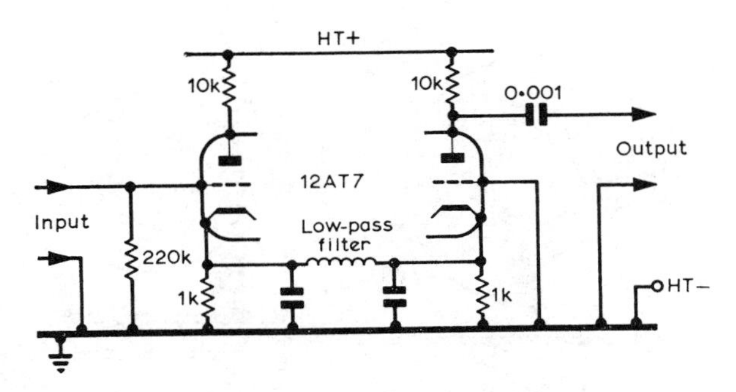

Fig 9.48. A suggested circuit arrangement for a low-impedance type of low-pass filter. The use of a cathode follower and a grounded-grid stage provides relatively low terminal impedances. The values of the filter components can be calculated from the basic filter design equations (see text).

Filter Construction

As the values of L and C required in a filter depend upon the choice of impedance and the cut-off frequency, these values may be selected in order to utilize existing components.

The layout of filters is not usually very critical. However, care should be taken to avoid undesired couplings between filter coils by suitable spacing. The mounting of such coils with their axes at right-angles will reduce the coupling between them, provided that the axis of one coil intersects the axis of another at its centre.

Capacitors. The capacitance values required rarely coincide with those easily obtainable, and series or parallel arrangements are usually necessary. Good quality paper capacitors are satisfactory, although mica ones are better, but if these are not readily available the extra expense involved is rarely justified, provided that the working voltage of the capacitors is adequate. In a high-level filter the voltage rating of the capacitors should be at least twice the anode voltage of the rf amplifier.

TABLE 9.7

Winding Data for Air-cored Inductances

Approximate Inductance (mH)	Number of Turns	Outside Diameter	
		Bobbin A	Bobbin B
80 100 120 150	2030 2240 2400 2620	2½in	1¼in
180 200 250 300	2800 2930 3200 3450	3in	1½in
350 400 450 500 600 700	3660 3850 4030 4200 4570 4840	3½in	1¾in
800 900	5160 5480	4in	2in
1000	5800	4in	2⅛in

For bobbin A the wire size is 32swg enamelled.
For bobbin B the wire size is 40swg enamelled.

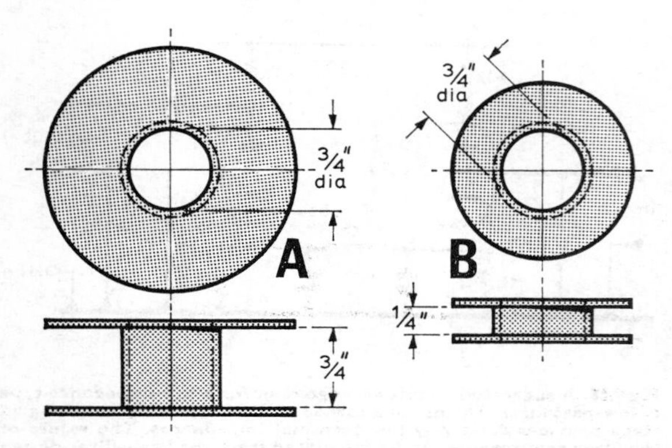

Fig 9.49. Dimensions of bobbins for inductances wound according to Table 9.7. The inductances are all air-cored.

Inductances. In the case of a high-level filter, the inductances must be capable of carrying the ht current of the modulated stage, and they therefore tend to be of larger dimensions than those intended for low-level filters where the current may be only a few milliamperes.

Either air-cored or iron-cored inductances may be used. Air-cored ones are cheaper and easier to design for high-level filters but iron-cored inductances using pot-cores are smaller, have less external field and are better screened from unwanted pick up. They are therefore a useful, if more expensive alternative for low-level filters.

Inductances with an accuracy of 5–10 per cent may be made by reference to **Table 9.7** which shows the number of turns of wire required to give the stated values of inductance when wound on one or other of the two bobbins illustrated in **Fig 9.49**. For the larger bobbin (A) the wire size is 32swg (enamelled), and inductances wound with this wire will carry 200mA without overheating, thus being suitable for high-level filters. The largest coil listed (1 henry) will take about 1lb of wire and will have a resistance of about 260Ω.

For the smaller bobbin (B) the wire size required is 40swg (enamelled), and the largest size shown will need about 2¼oz of wire and will have a resistance of about 850Ω. These inductances are suitable for low-level filters. In either case inductances of intermediate values may be found by simple interpolation.

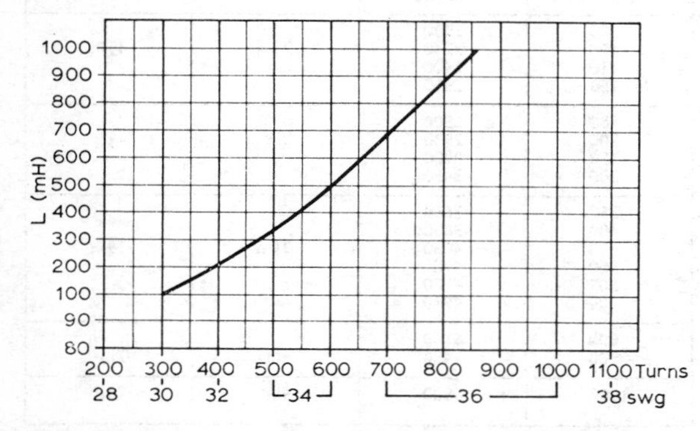

Fig 9.50. Approximate number of turns and gauge (swg) of enamelled wire on former type DT 2179 for inductances between 100 and 1,000mH with cores LA1218.

Many suitable pot-cores are available but a pair of 25mm cores type LA1218 are typical of those suitable. These cores require 26·9 turns for 1mH and **Fig 9.50** shows the number of turns and the gauge of wire to produce inductances up to 1H on these cores. The former should be as full as possible and hence the wire size should be no smaller than will allow the required number of turns to be wound. Inductances wound in this way would have a Q of about 400 and an inductance of 1H would have a dc resistance of about 100Ω.

Typical AF Filters

The filter shown in **Fig 9.51** is a single-section constant-k filter suitable for use at high level, feeding an rf amplifier having an input of 150W. Assuming an anode supply of 1,000V and a current of 150mA, the terminal impedance required is 1,000V/0·15A, which equals 6,700Ω. The values of the components shown will give a cut-off frequency of 3·3kHz.

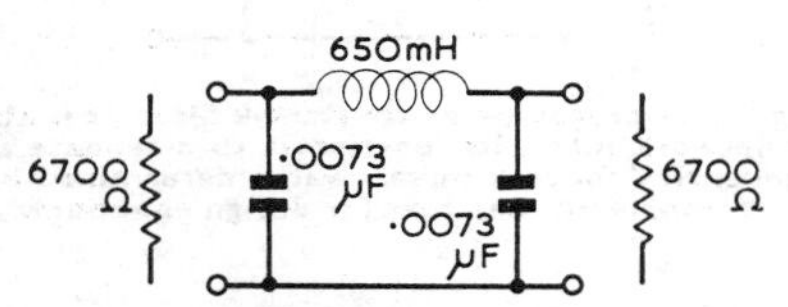

Fig 9.51. Single-section "prototype" filter suitable for feeding a 150W modulated rf stage.

Fig 9.52 shows a filter having a single constant-k section with m-derived end sections which would be suitable for use at low levels. It is designed for an impedance of 15kΩ and has a cut-off frequency of 3kHz. By careful design and construction, an attenuation of 50dB at 4kHz is attainable.

Resistance-capacitance Filters

A filter with a slower fall-off in its characteristic may be obtained from resistance and capacitance elements avoiding the use of wound inductances. A single section RC filter may be considered as a potentiometer composed of the resistance and the reactance of the capacitor. The ratio of the potentiometer therefore, is dependent on the frequency of the input voltage. A two-section filter is shown in **Fig 9.53** together with its loss/frequency characteristic.

Active Filters

Greater attenuation with better bandpass characteristics may be attained using RC filters if active devices (usually transistors) are used for coupling and matching. **Fig 9.54** shows an active low-pass filter with a typical frequency

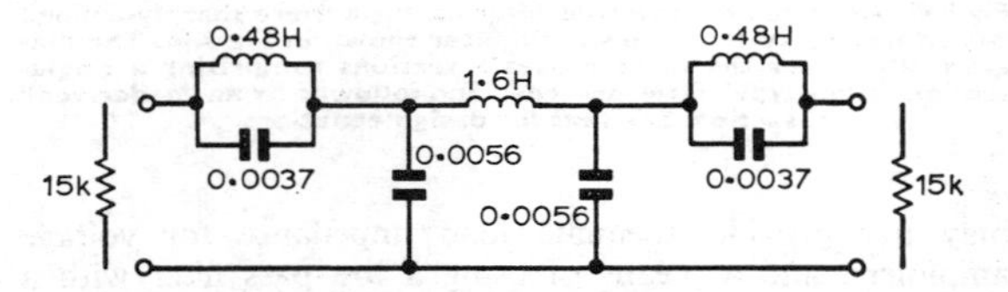

Fig 9.52. Low-pass filter suitable for use at low levels in a voltage amplifier having an output impedance of 15,000Ω. The cut-off frequency is 3kHz. The filter comprises a "constant-k" section with an "m-derived" section at each end.

response in **Fig 9.55.** The filter has an in-band attenuation of about 1·5dB and will handle an input signal of 2·5V peak. The load impedance following the filter should be at least 10kΩ.

TYPICAL FREQUENCY MODULATORS

A frequency modulated transmission may be produced either by frequency or phase modulation and the modulated stage may be a crystal oscillator, a vfo, or with phase modulation only, an amplifier. In all cases a voltage controlled reactance is required and this may be either a varicap silicon diode or a variable reactance valve modulator.

The Variable Capacitance Diode Modulator

The variable capacitance diode is a small area p-n junction which exhibits a change in junction capacitance with change of reverse voltage. This change of capacitance is due to a change in thickness of the "depletion layer" between the p and n regions. All p-n diodes have this characteristic and

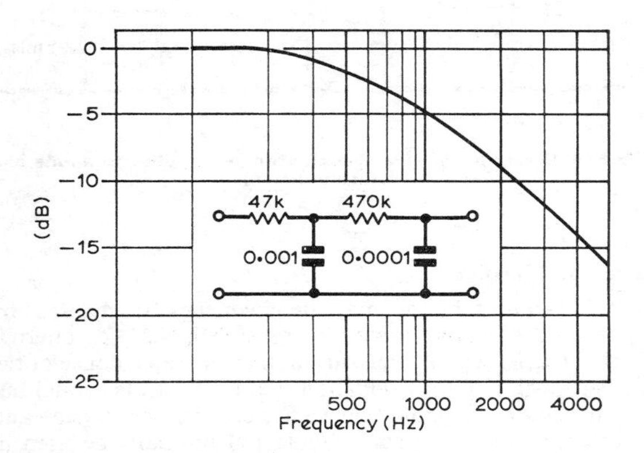

Fig 9.53. Typical two-stage RC filter and approximate loss/frequency characteristic.

many small silicon rectifiers may be used as modulators but in these the capacitance change with reverse voltage is not controlled in manufacture and the back resistance may be low enough to damp the circuit to which the diode is applied. Diodes of the OA200 class are normally successful but varactor or varicap diodes with controlled characteristics are to be preferred. The usual practice is to apply a few volts

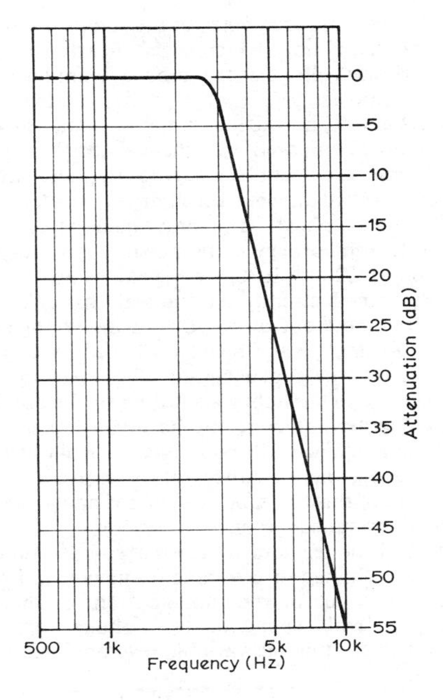

Fig 9.55. Frequency response of active filter shown in Fig 9.54.

of reverse dc bias so as to select a reasonably linear part of the diode characteristic and to superimpose the modulating voltage on this bias. It is important to ensure that the peak modulating voltage does not drive the diode into forward conduction as this would damp the circuit and result in amplitude modulation or even the cessation of oscillation on speech peaks.

The Variable-reactance Valve Modulator

The basic circuit of the variable-reactance or reactor type of modulator is shown in **Fig 9.56.** Essentially it consists of a valve connected across the tuned circuit of an oscillator in such a way that it behaves like a variable capacitance or inductance. In the circuit shown, a potentiometer consisting of a capacitor C2 and a resistor R in series is connected from anode to earth and also, of course, across the tuned circuit. The centre point is taken to the grid of the valve. Provided that the resistance of R is large compared with the reactance

Fig 9.54. Circuit of transistor active filter.

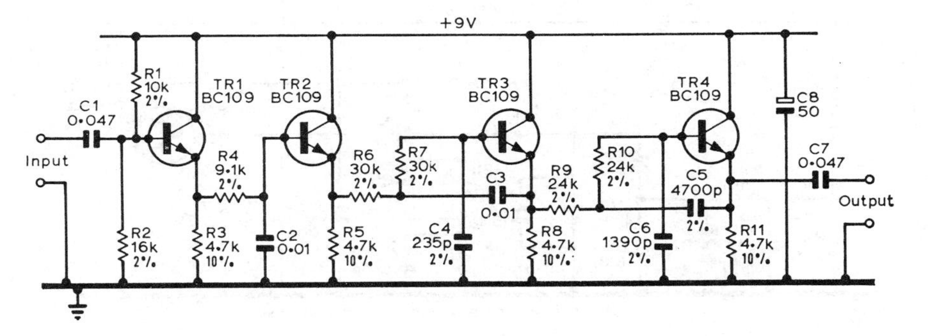

of C2 at the resonant frequency of the tuned circuit, the grid will be fed with a voltage which is very nearly 90 degrees out of phase with the voltage at the anode. A similar phase-shift but of the opposite sign may also be obtained by interchanging C2 and R, in which case the reactance of C2 would need to be large compared with the resistor R.

Since the anode current of a valve is in phase with its grid voltage, the current flowing through the reactance valve is 90 degrees out of phase with the voltage across the tuned circuit due to the 90 degrees phase-shift introduced by the combination of C2 and R. Since the valve current also flows through the tuned circuit, the current through the tuned circuit is 90 degrees out of phase with the voltage across it, which is equivalent to the result produced by connecting a reactance across the tuned circuit. This reactance may be either capacitive or inductive according to the configuration of the RC2 potentiometer. In the arrangement shown in Fig 9.56, the reactance will be inductive, while if R and C2 are interchanged, the reactance will be capacitive. In either case, the effect of this reactance will be to change the resonant frequency of the tuned circuit.

The value of the reactance thrown across the tuned circuit and hence the change in resonant frequency will obviously depend on the value of the anode current of the valve; in other words it will depend on the voltage applied to the grid of the valve. Thus the variable reactance valve presents

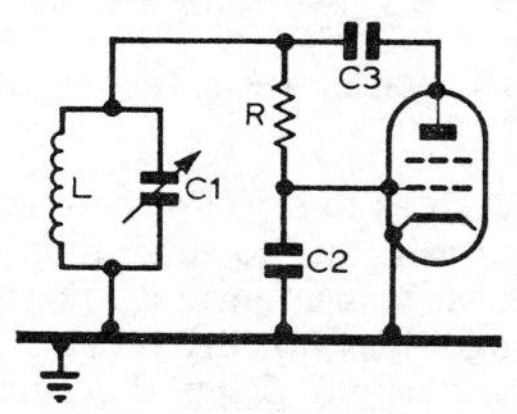

Fig 9.56. Basic circuit of a variable-reactance frequency modulator L-C1 is the tuned circuit of the master oscillator.

a simple means of varying the frequency of an oscillator in sympathy with an af voltage applied to the grid of the reactance valve. The amplitude of the grid voltage will govern the magnitude of the change in frequency; ie the deviation produced will be determined by the amplitude of the modulating signal and the rate at which the oscillator frequency is varied will be equal to the frequency of the voltage applied.

A reactance-valve modulator may be used with any type of oscillator. It is normally connected directly across the tuned circuit of the oscillator as shown in Fig 9.56, although in the case of a series-tuned circuit, such as the Clapp oscillator, it may be connected across the tuning capacitor.

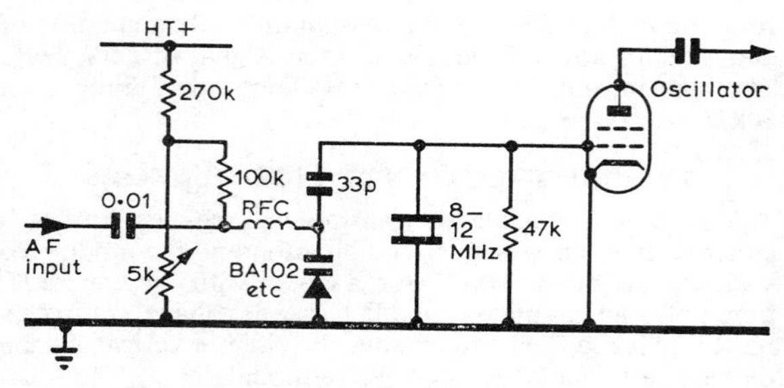

Fig 9.57. Direct fm of a crystal oscillator (dc supplies to anode and screen omitted).

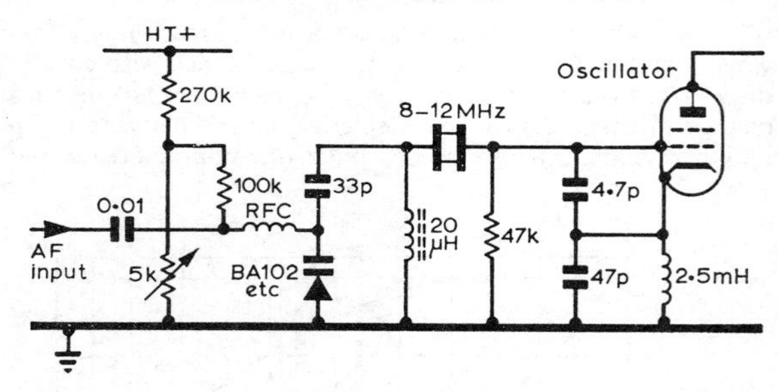

Fig 9.58. Direct fm of crystal oscillator (dc supplies to anode and screen omitted).

Typical Circuits

A crystal oscillator may be frequency modulated by applying a varactor across the crystal **(Fig 9.57)** or alternatively in series with it. The variation of the capacitance of the varactor will pull the oscillation frequency of the crystal but the deviation available will vary between one crystal and another. Overtone crystals should not normally be used in this way but this seldom arises because frequency multiplication is usually necessary to increase the deviation to a reasonable amount and this leads to the use of crystals in the 6–12MHz region or lower. These are normally fundamental mode crystals. Unless the shunt capacitance is exactly that for which the crystal was manufactured, it will not operate at precisely its marked frequency.

An alternative arrangement is shown in **Fig 9.58** in which a small inductance is placed in series with the crystal and this inductance is partly tuned out by the varactor across it. This arrangement gives rather better control of the deviation but

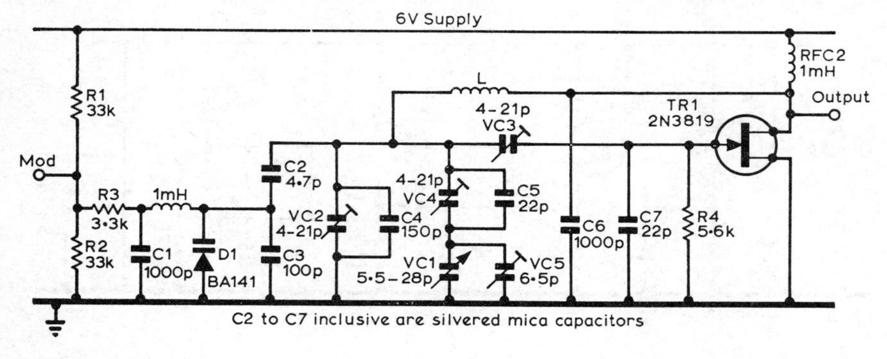

Fig 9.59. Oscillator circuit.

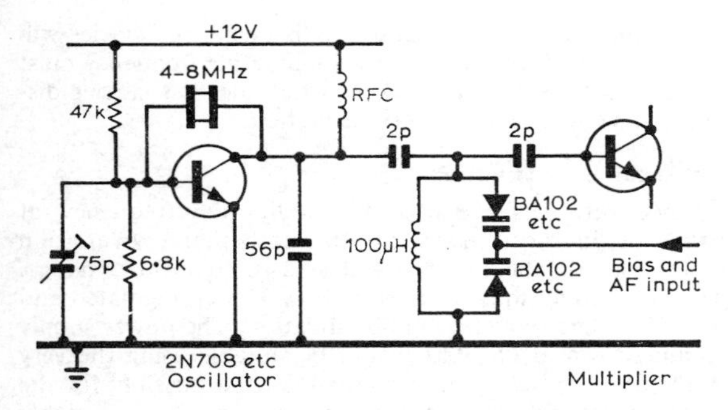

Fig 9.60. Phase modulation using varactors.

the final frequency is likely to be significantly removed from the marked value.

Direct frequency modulation of a vfo is shown in **Fig 9.59.** This is a Vackar oscillator operating around 12MHz. The varactor is across one of the frequency determining capacitors but its effect is reduced by a large shunt capacitance C3 and a small series capacitance C2. The overall effect is that a modulating signal of one volt produces a frequency shift of about 200Hz which corresponds to 2·4kHz deviation at 144MHz. Had the varactor been connected directly across VC2, adequate deviation would be obtained with a very small audio voltage but the vfo would be very sensitive to small voltages accidentally reaching the varactor and the frequency stability would suffer.

Fig 9.60 shows a phase modulation arrangement using a pair of varactor diodes to vary the tuning of a circuit loosely coupled between the oscillator and the following multiplier. **Fig 9.61** shows a valve crystal oscillator phase modulated by a reactance valve, the two valves being the two halves of a double triode; it is preceded by a typical speech amplifier and clipper. It should be remembered that the audio frequency response should fall at 6dB/octave when phase modulation is used in order that the transmission may give a normal audio response when resolved in a discriminator or by slope detection. This is in addition to a sharply falling response above 3kHz in order to limit the bandwidth occupied by the sidebands. A de-emphasis of 6dB/octave is simply obtained by applying the modulation signal across a resistor and capacitor in series and taking the output across the capacitor alone. This will give a 6dB/octave de-emphasis above the frequency at which the capacitive and resistive impedances are equal. The 22kΩ and 0·01µF components immediately following the low pass filter perform this function in **Fig 9.61.**

Setting up Procedure

When clipping is used, the audio gain control before the clipper should be used to set the clipping level and the gain control following the clipping to set the deviation. When audio agc is used, the control loop should be arranged to present to the clipper a signal which results in the desired

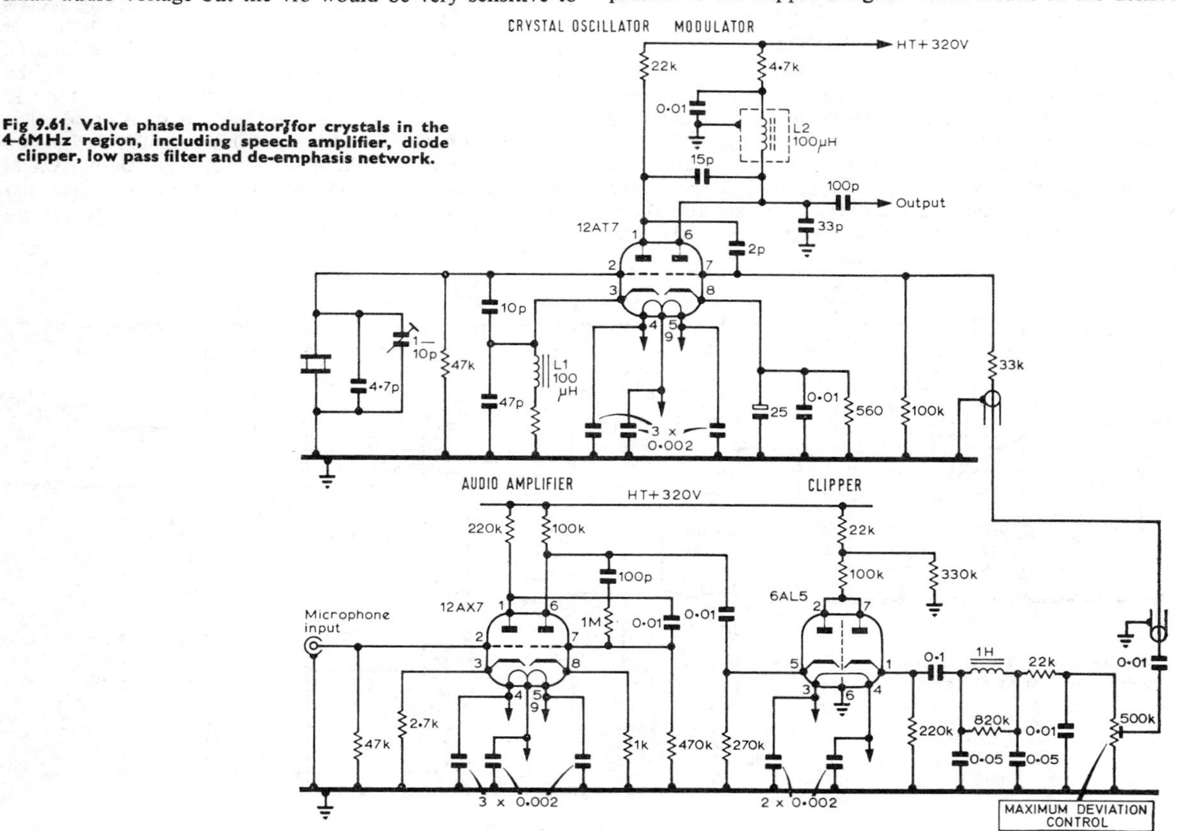

Fig 9.61. Valve phase modulator for crystals in the 4-6MHz region, including speech amplifier, diode clipper, low pass filter and de-emphasis network.

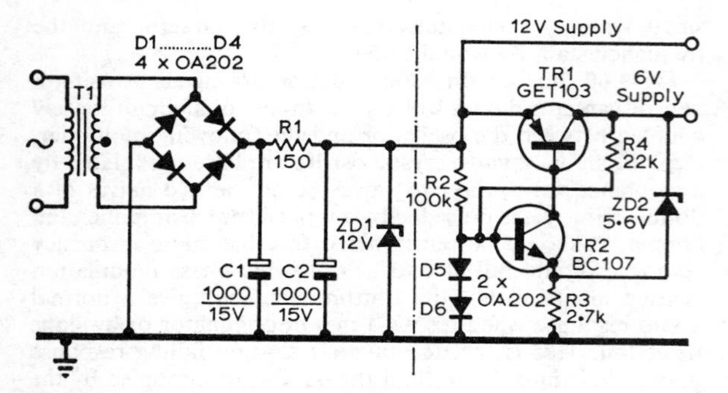

Fig 9.62. Power supply unit (for battery operation omit all to left of dotted line).

degree of clipping; 10–12dB is recommended. If direct fm is used with a varactor, the deviation may be set up by varying the dc bias on the varactor sufficiently to produce a frequency shift equal to the desired peak deviation and then to use an oscilloscope to ensure that the peak value of the audio waveform arriving at the varactor equals the change of bias previously determined. In other cases, the disappearing carrier technique may be used. This is based on the fact that the carrier has zero amplitude when the modulation index has values of 2·4, 5·5, 8·7 etc since all the energy then occurs in the sidebands. As

$$\text{modulation index M} = \frac{\text{deviation}}{\text{modulating frequency,}}$$

it follows that if a modulation frequency of 1kHz is applied, the carrier will disappear when the deviation is 2·4kHz, 5·5kHz etc. This disappearance may be detected on a receiver with a narrow filter or a Q multiplier. If the receiver is tuned accurately to the carrier frequency and the deviation slowly increased, the carrier will be seen to disappear on the S meter when the deviation is 2·4kHz and again at 5·5kHz. Should the receiver bandwidth be as great as ± 1kHz, the effect will not be seen because the first side frequencies will also be received. In this case, the modulating frequency must be increased to perhaps 2·5kHz when the first carrier disappearance will occur at 6kHz deviation.

Varactor Bias Supplies

Since varactors are used to modify the frequency of oscillators in direct fm systems, it follows that any variation of the bias on the varactor will lead to frequency change. The bias supply should therefore be very well regulated and well smoothed to avoid hum modulation. The power supply circuit shown in **Fig 9.62** would be suitable using the very stable 6V supply for the varactor bias and possibly for the oscillator supply when transistors are used. The less stable 12V supply would be suitable for subsequent amplifiers or multipliers.

TYPICAL AMPLITUDE MODULATORS

A 15W Modulator

This modulator will give an output of 15–17W using a power supply rated for 300V at 125mA. It would be adequate for a transmitter with an anode input to the final rf amplifier of 25W; such a transmitter might well operate from the same ht supply as the modulator. The circuit is shown in **Fig 9.63.**

The first stage uses a high gain pentode type EF86 RC coupled to a paraphase phase splitter (12AT7). The output stage uses a push-pull pair of EL84 valves operating in class AB1 with cathode bias. The grid, screen and anode stopper resistances should be mounted close to the valve holders. For cw operation, the modulator may be switched off by switch S1 which removes the ht from the modulator and shorts the secondary of the modulation transformer. The regulation of the ht supply need not be particularly good and a capacitor input filter is adequate. The output valves should operate at an anode current of about 46mA each and the modulation transformer should be designed to present an anode to anode load of 8,000Ω.

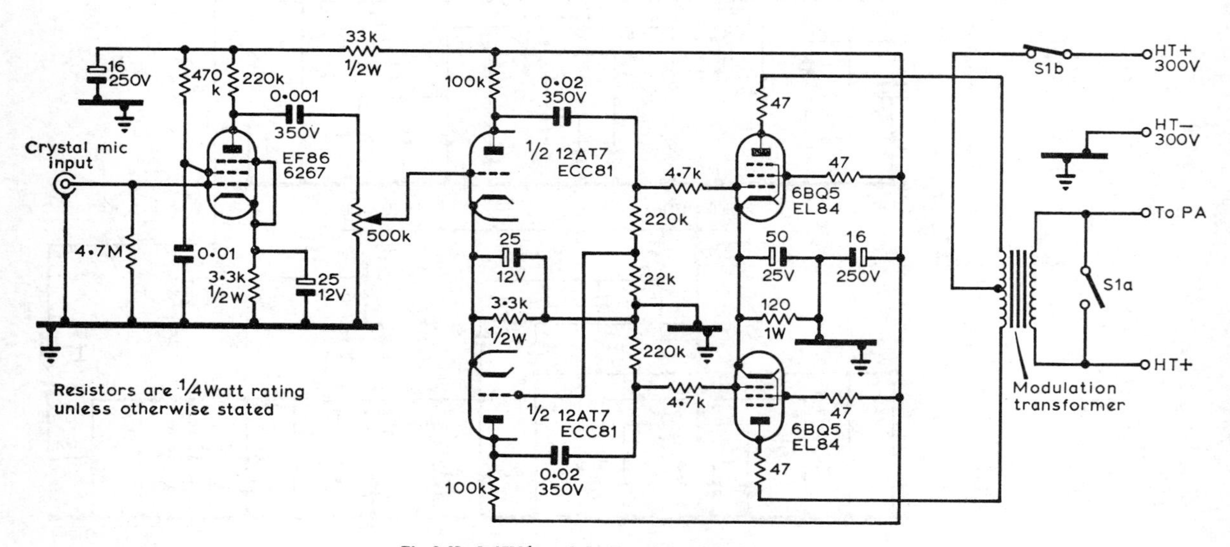

Fig 9.63. A 15W modulator using miniature valves.

A 100W Modulator

This modulator will give an output of 100W using four tetrodes type KT66 in parallel push-pull class AB1 in the output stage. The complete modulator and power supply shown in **Fig 9.64** can be assembled on a chassis 17 by 10 by 2in.

The output valves operate in the ultra-linear connection, the screen supply being obtained from taps on the primary of the Woden type UM3 modulation transformer. Grid and screen stoppers should be mounted close to each valve holder in the output stage. The total cathode current of the parallel pairs of output valves is measured by a 0–150mA meter which can be switched by S to either pair of valves.

The speech amplifier is coupled to the output stage by a push-pull transformer and consists of a triode-connected EF86, which is resistance-capacitance coupled to a twin-triode type 6SN7, the two sections of which are also RC coupled. The gain control (a 1MΩ potentiometer) is immediately after the first stage. The speech amplifier is generously decoupled at rf and af to prevent instability and is built into a small diecast box 5 by 3 by 2¼in.

The ht supply for the output stage (500V) is obtained from a conventional bi-phase half-wave rectifier arrangement which is composed of series-connected silicon diodes type SX635. A capacitor-input smoothing circuit is used and the output capacitor consists of two 160μF 450V working electrolytic capacitors in series. A 100kΩ resistor is connected across each to equalize the voltage across them. The ht supply for the speech amplifier is obtained from a resistive potentiometer connected across the 500V supply. The grid bias for the output valves is obtained from a small 200-0-200V transformer, one 6·3V secondary winding of which is connected to a 4V supply obtained from the main heater transformer. The bias is adjusted by means of the two 10,000Ω potentiometers.

The on-off switch S1 controls the mains input to both the ht and heater transformers while the supply to the ht transformer is also controlled by switch S2. For remote control operation, the secondary of the modulation transformer is shorted by S4.

The setting-up procedure is simple; before the ht is applied, the bias potentiometers are adjusted to give maximum bias, ht is then switched on and each bias is slowly reduced to give a standing cathode current of 70mA under no-signal conditions in each parallel pair of output valves. The values of bias voltage obtained should agree to within about 5 per cent.

If the bias voltages differ by more than this amount, the valves are not sufficiently well matched. A third valve paired with either of the other two will normally provide a pair sufficiently well-matched.

High-power Modulator

This modulator is capable of an output of 140W, obtained from a pair of tetrodes type TT21 operating in class AB1 at an anode voltage of 1,000V with a screen voltage of 300V and a grid bias of approximately −40V.

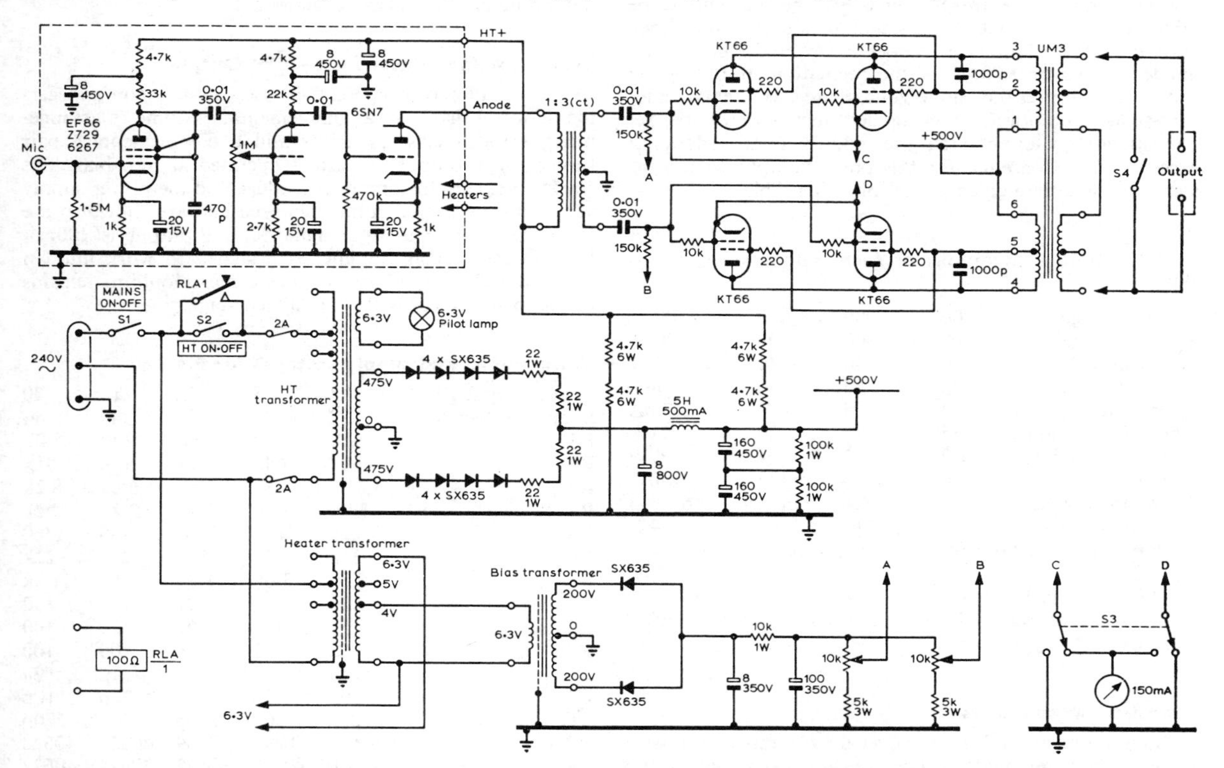

Fig 9.64. A 100W modulator. Anode-anode load impedance 4,000Ω. KT88 valves could replace the KT66 with negligible change of performance. Mixed types should not be used.

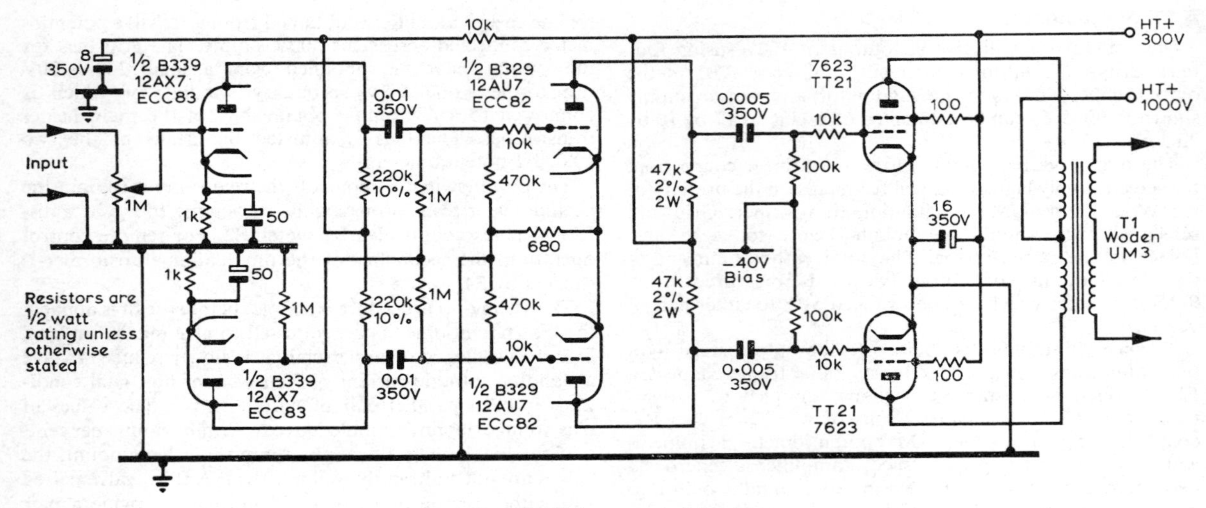

Fig 9.65. High power modulator.

As will be seen from **Fig 9.65**, the speech amplifier consists of two twin triodes; the first (12AX7) is a self-balancing floating para-phase phase inverter which feeds a push-pull voltage amplifier (12AU7).

In the interests of simplicity, no balancing arrangements for the output stage are provided, but the bias should be adjusted so that under zero-signal conditions the anode current of each output valve does not exceed 35mA at an anode voltage of 1,000V. If the regulation of the power supply is such that the anode voltage exceeds 1,000V under zero-signal conditions, the anode current must be set correspondingly lower so that the anode dissipation does not exceed 35W. Consequently the power output would be somewhat less than given in the following table.

The operating conditions of each output valve are as follows:

Anode voltage	1,000	1,000	1,000	1,000V
Screen voltage	300	300	300	300V
Grid voltage*	−40	−40	−40	−40V
Anode current	35	56	80	92mA
Screen current	4	5	9	17mA
Anode dissipation	35	35	30	22W
Screen dissipation	1·2	1·5	2·7	4·9W
Output†	0	40	100	140W
Anode load (a–a)		16,800	16,800	16,8004
Distortion		0·6	1·4	4%

* Set to give anode current of 35mA/valve at anode volta of 1,000V.

† Measured in resistive load of 7,000Ω.

Input to first stage (gain control at max, output of 100W) = 25mV.

Transistor Modulators

Transistors are available to cover the full range of powers required from modulators. The choice between valves and transistors is largely decided by available power supplies and consideration of space. Although transistor modulators may be considerably smaller than those using valves, the need for substantial heat sinks often makes high power modulators larger than at first expected. The two designs that follow may be arranged to cover output powers from 3W to 70W depending on the transistors chosen.

Transistor Modulator for 3–20W Output

The circuit shown in **Fig 9.66** uses both silicon and germanium transistors in a complementary symmetry arrangement, using a class A driver and a complementary pair (npn/pnp) of output transistors operated at zero bias. The transistors, certain component values and the power supply vary with the power output required and are shown in the following table. The input signal for full output is 100mV in each case. Each amplifier is designed to work into an 8Ω load and the modulation transformer should match this to the load offered by the modulated stage.

Components and supply voltages for Fig 9.66

Power output (W)	3	5	12	16	20
Supply voltage (V)	20	24	36	40	44
R1	91k	51k	16k	10k	8·2k
R3	68k	68k	91k	91k	91k
R5	2·7k	3·3k	7·5k	6·8k	8·2k
R7	3·9k	3·9k	2·7k	2·7k	2·2k
R8	620	620	390	430	360
R9	33k	27k	18k	27k	22k
R10	5·6k	3·6k	1·8k	1·8k	1·3k
R13	120	110	91	120	100
R14	150	110	91	120	100
R16	22	27	56	100	100
C1	0·1µ	0·25µ	1µ	1µ	2µ
C3	10p	5p	10p	22p	10p
C6	100p	150p	220p	470p	270p
TR4	40611	40616	40389	40625	40628
TR5	40610	40615	40622	40624	40627
TR6	40609	40614	40050	40623	40626

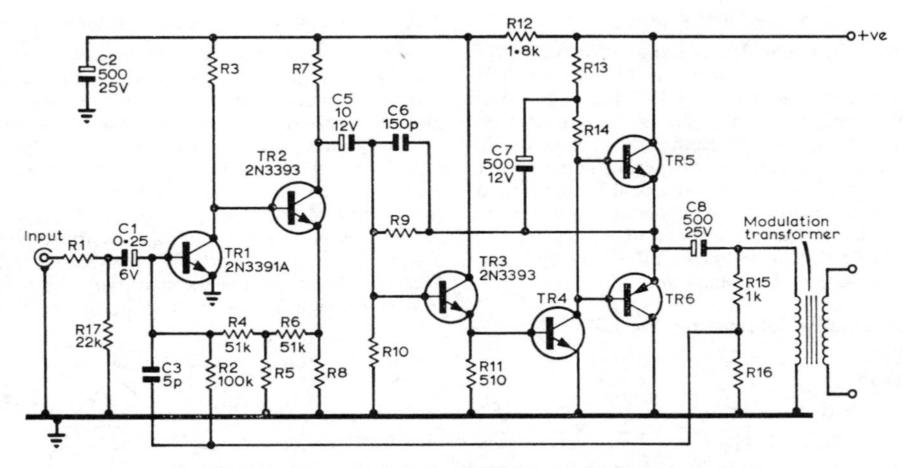

Fig 9.66. Transistor modulator for 3–20W. See table for component values, transistor types and supply voltage. (Note: R17 is omitted except for amplifiers of 3 and 5W output.)

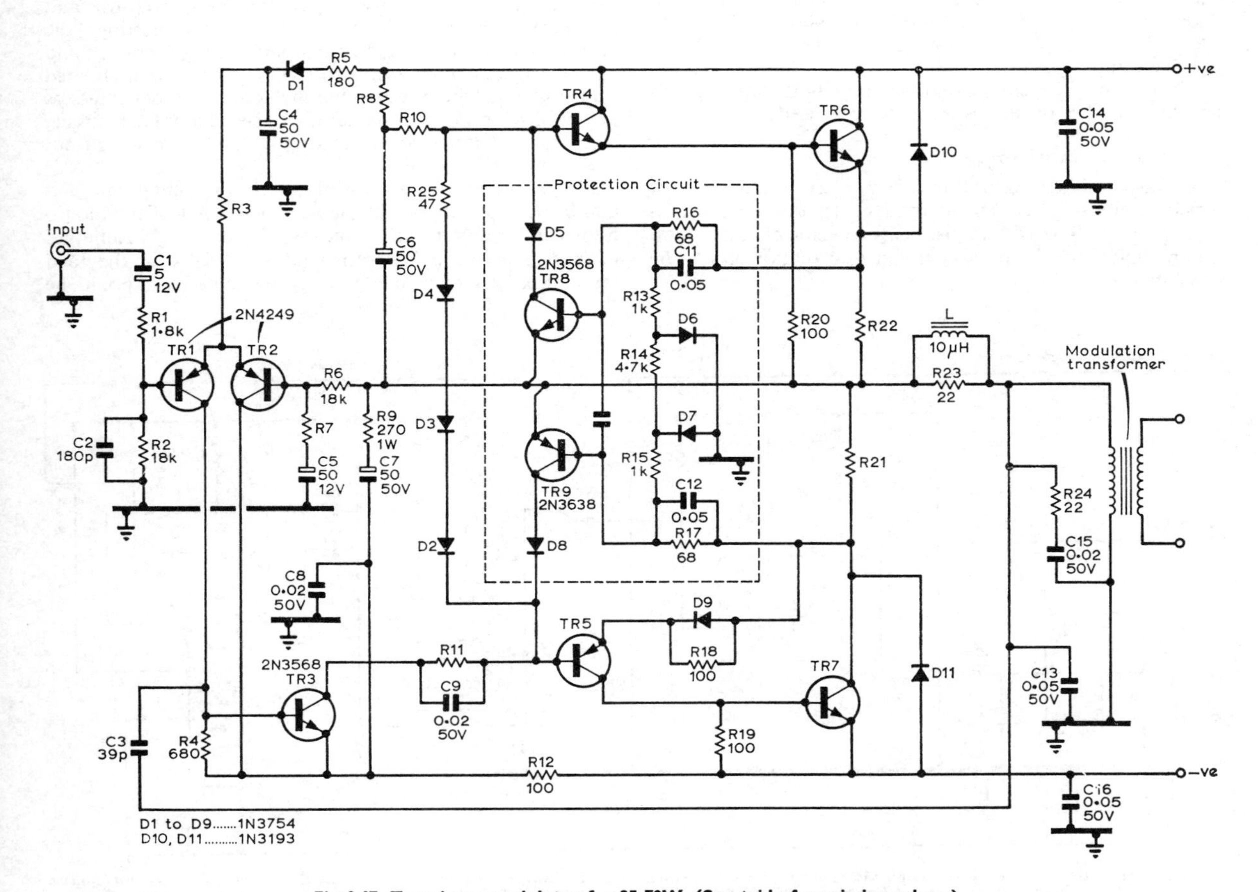

Fig 9.67. Transistor modulator for 25–70W. (See table for missing values.)

Transistor Modulator for 25–70W Output

The circuit shown in **Fig 9.67** uses the quasi-complementary arrangement with silicon transistors throughout. The class B npn/pnp driver stage is followed by the npn output transistors. A short circuit at the output is likely to damage the output transistors and the protection circuit shown within the dotted box is recommended. The performance of the amplifier under normal load conditions would be unchanged if the protection circuit were omitted. These circuits require an input of about 700mV for full output and are designed to work into a load of 8Ω. The following table lists the components not shown on the circuit diagram.

Component and Supply Voltages for Fig 9.67

Power output (W)	25	40	70
Supply voltage (V)*	26	32	42
R3	12k	15k	18k
R7	680	560	470
R8	1,800	2,200	2,700
R10	2,200	2,700	3,300
R11	270	390	470
R21†	0·43	0·39	0·33
R22†	0·43	0·39	0·33
TR4	2N3568	40635	40594
TR5	2N3638	40634	40595
TR6	40632	40633	40636
TR7	40632	40633	40636

† 5W rating

* Note from the circuit diagram that both positive and negative supplies of the values shown are needed.

Alternative Transistors

A considerable range of transistors are available, many of which could be used as alternatives to those shown in Figs 9.66 and 9.67 with adjustments to some circuit values. The manufacturers' data and circuit information should be consulted.

Adjustment of Amplitude-modulated Transmitters

Before tests on the complete transmitter are attempted, the performance of the modulator itself should be checked. For this purpose, it should be operated into a resistive load able to dissipate the power output of the modulator continuously and, of course, also of the correct load impedance. An oscilloscope connected across the load or across part of it may be used both to examine the output waveform and to measure the output voltage. If the oscilloscope does not have a calibrated shift voltage, an ac voltmeter will be needed suitable for use at speech frequencies. Many multi-range meters maintain their accuracy to a sufficiently high frequency. The modulator should be driven

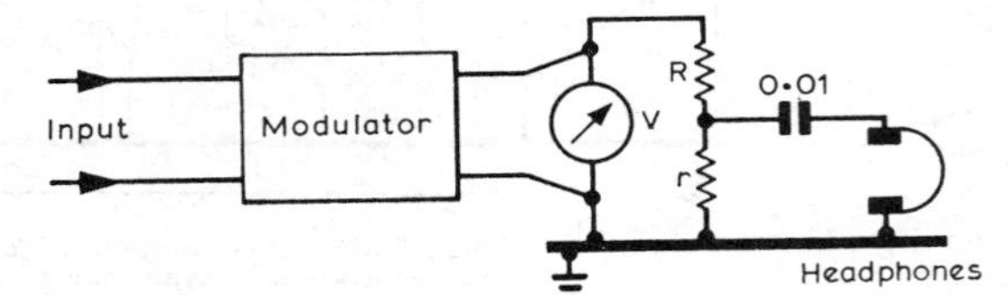

Fig 9.69. A simple method of checking the audio performance of the modulator by a direct listening test. Special care must be taken to ensure that no dangerously high voltage ever reaches the headphones.

from a sine-wave source of reasonable waveform and preferably variable frequency. During this testing, the performance of any filters or clipping arrangements should be investigated. It is useful to switch on the unmodulated transmitter, working into a dummy aerial, in order to assess any rf pick up in the modulator or speech amplifier, taking care that test leads do not cause pick-up which would not occur in actual operating.

Simpler methods are possible when an oscilloscope and audio oscillator are not available. The output of the modulator can be monitored by means of a pair of headphones connected across a low resistance in series with the load resistance, as shown in **Fig 9.69**. The value of r should be

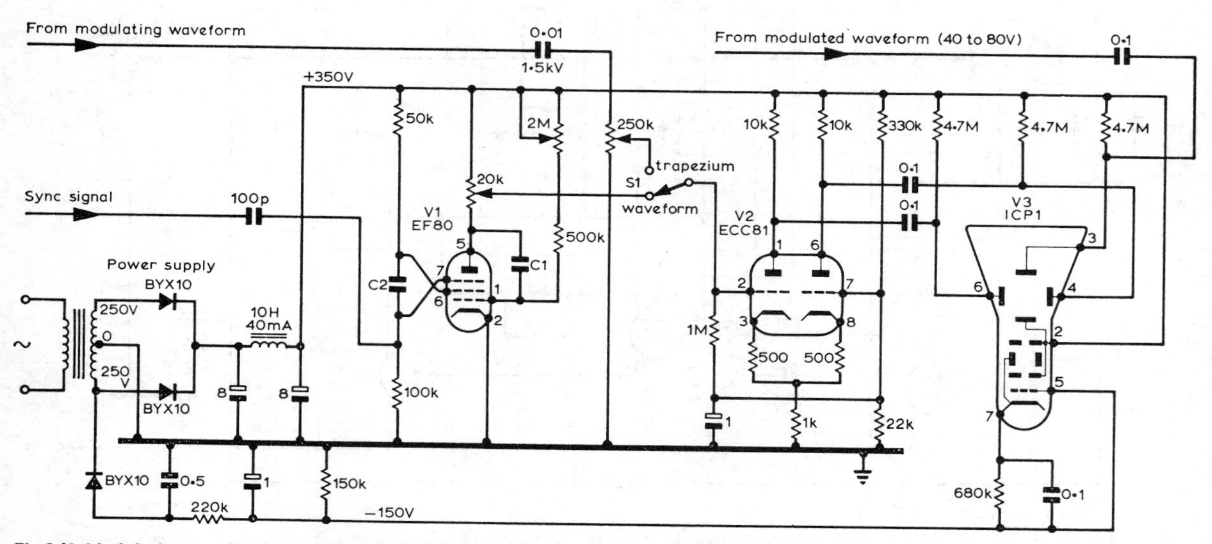

Fig 9.68. Modulation monitor employing a ICPI tube. The unit is suitable for building into a transmitter. The timebase speed depends on the value of C1 which may be 300pF (1·2–6kHz), 1,000pF (400Hz–2kHz) or 3,000pF (120–600Hz). C2 should be five times the value of C1.

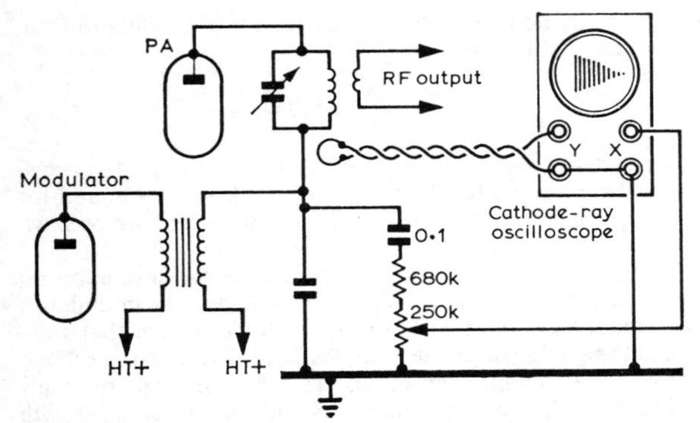

Fig 9.70. Trapezium test for checking the modulated output from the rf power amplifier. Here rf voltage is applied to the Y-plates and af voltage is applied to the X-plates. No timebase is required in the oscilloscope.

adjusted to give a comfortable signal strength in the headphones; generally 2–3 Ω is sufficient. As a safety precaution it is very important that one side of the headphones should be earthed, and that the resistor r should be of a very reliable type and very securely connected across the headphones. Such a simple testing method will enable any serious distortion to be heard when the modulator is driven from a test source or from an audible signal reproduced by the microphone.

Testing the Complete System

The rf sections of the transmitter should be operated into a dummy aerial and investigated for correct operating conditions. In particular any trace of instability should be corrected by the adjustment of neutralizing if fitted or even the introduction of neutralizing or increased decoupling or screening. Instability is bound to be made worse when modulation is applied and hence must be eliminated from the cw condition. When sine-wave modulation is applied, the power output should increase and should be one and a half times the carrier power at 100 per cent modulation. If this cannot be achieved, the operating conditions should be checked. Incorrect load coupling or inadequate drive are the most likely causes of inadequate increase of output at 100 per cent modulation. Remember that the modulated amplifier must produce four times the carrier output at the instant of crest modulation and the working conditions must make this possible; the design information in the valve chapter should be consulted. With efficiency modulation systems it is rarely possible to achieve 100 per cent modulation without distortion.

Measurement of Modulation Depth

The cathode ray oscilloscope is the best instrument for measuring the depth of modulation. There are two distinct methods, one of which requires a timebase.

A very simple oscilloscope using a cathode ray tube with a small diameter screen is shown in **Fig 9.68**. It is suitable for mounting alongside the meters in a transmitter and includes a simple timebase so that both presentations may be used. V1 is a transitron-Miller timebase while V2 provides symmetrical deflection and also amplifies the modulating waveform.

The method which does not require a timebase is illustrated in **Fig 9.70**. A small fraction of the modulated output voltage is fed to the Y-plates of the oscilloscope by a pick-up loop loosely coupled to the output tank circuit while the af modulating voltage is fed to the X-plates. With a steady modulating signal a trapezoidal or wedge-shaped pattern will be produced on the screen. **Fig 9.71** shows the variations in this pattern for various degrees of modulation. The modulation depth is calculated as follows:

$$m = \frac{P - Q}{P + Q} \times 100 \text{ per cent}$$

where P and Q are the lengths of the vertical sides of the pattern. P and Q can be measured with reasonable accuracy by means of a pair of dividers. For 100 per cent modulation a triangular pattern is produced.

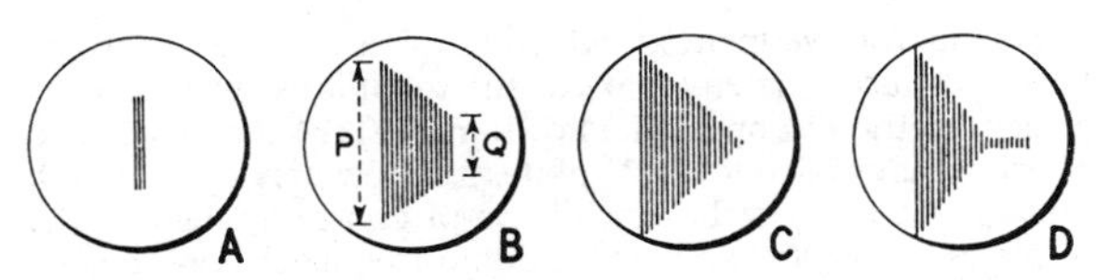

Fig 9.71. Typical oscilloscope patterns obtained by the method shown in Fig 9.70. The vertical line in A shows the unmodulated-carrier amplitude. In B the carrier is modulated 50 per cent while in C it is modulated 100 per cent. Where the sloping edges of the pattern are flattened, as in D, the carrier is over-modulated.

The oscilloscope pattern can also be used to examine the overall performance in other respects besides the actual modulation depth. If the operating conditions are correct so that there is no distortion in the modulation process, the sloping sides of the trapezoidal pattern will be perfectly straight: any curvature of these sides indicates non-linear modulation. Typical patterns for various fault conditions in both anode- and grid-modulation systems are shown in **Figs 9.72** and **9.73** respectively. The audio voltage applied to the X-plates may be derived directly from the audio input to the amplifying system of the transmitter instead of from the output of the modulator, in which case any distortion in the audio part of the system will become apparent.

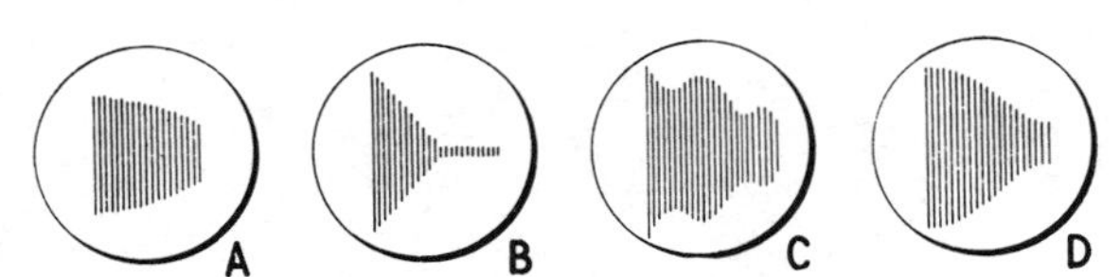

Fig 9.72. Fault conditions in anode modulation: (A) insufficient drive; (B) over-modulation; (C) instability in power amplifier; (D) incorrect matching of modulator to pa.

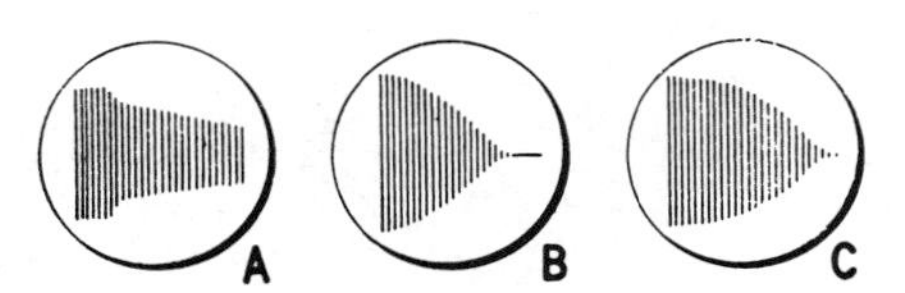

Fig 9.73. Fault conditions in grid modulation: (A) excessive drive; (B) over-modulation; (C) power amplifier insufficiently loaded.

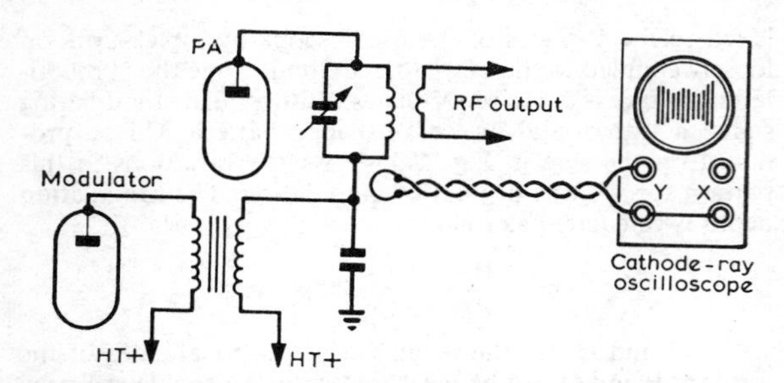

Fig 9.74. Alternative method for checking the modulated output from the rf power amplifier. The test signal is applied to the Y-plates, while the X-plates are supplied with a linear timebase voltage of any suitable frequency.

The alternative method, which requires the use of a time-base, produces the envelope of the modulated wave on the screen. In the arrangement shown in **Fig 9.74** the modulated rf voltage is applied to the Y-plates as in the previous method, and the normal timebase which is of course applied to the X-plates is adjusted so that the audio-frequency components of the envelope are suitably displayed. A rectangular patch of constant height will be observed when there is no modulation, and this will show the appropriate change of shape during modulation. The type of patterns displayed on the cathode-ray tube are shown in **Fig 9.75**. By measuring the height R corresponding to a modulation peak and the height S of the unmodulated carrier, the depth of modulation can be calculated directly:

$$m = \frac{R - S}{S} \times 100 \text{ per cent}$$

The patterns resulting from incorrect operation of the transmitter are more difficult to interpret in this method, and for this reason the trapezium test is usually found more satisfactory.

Whichever method is used the results are easier to interpret when a constant sinusoidal input is applied to the modulator. It should be noted that both the methods require that sufficient rf signal reaches the Y-plates of the cathode ray tube to give a reasonable deflection. This becomes increasingly difficult as the frequency increases and for vhf use a crt with side arm connections is required in order to keep the series inductance and shunt capacitance to reasonable values.

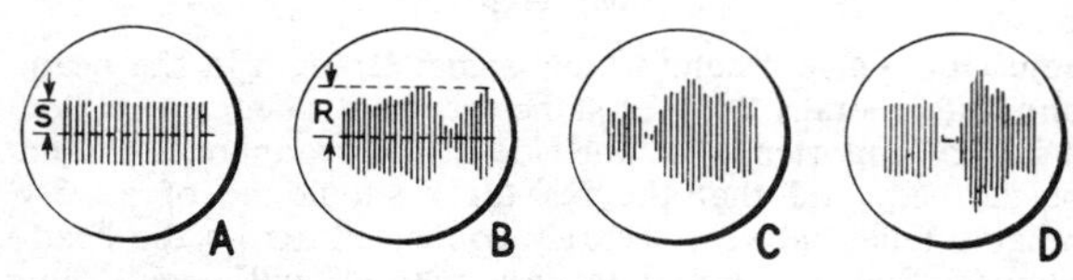

Fig 9.75. Typical oscilloscope pattern obtained by the method shown in Fig 9.74: (A) unmodulated carrier; (B) 50 per cent modulation; (C) 100 per cent modulation; (D) over-modulation. Since the time-base frequency is independently fixed the patterns obtained with a normal speech input are constantly varying in position on the horizontal axis.

CHAPTER 10

RTTY

RADIOTELETYPE (rtty) is a facet of amateur radio which has gained increasing support and interest since World War II. RTTY is a little different from other facets of amateur radio in that it is necessary to have sophisticated mechanical devices in the amateur station to operate in the rtty mode, and for this reason development has tended to follow the availability of surplus equipment. Unhappily very few amateurs have the facilities for the home construction of teleprinters and ancillary mechanical equipment.

From the outset of the science of telegraphy man has been concerned with providing a visible record of information transmitted between two points. It may come as a surprise to many people to learn that printing telegraphs were in existence as far back as 1832. With the speed-up in the processes of commerce and industry the increasing requirements for transfer of complex information between world centres brought about concentrated development work in the new field of line telegraphy. Such was the progress in machine telegraphy that by the early 1930s a comprehensive network of cable teleprinter circuits existed, and attention turned to using the then new development of high frequency radio as a carrier for telegraph signals. At the start of World War II some progress had been made, but it is a sad fact that only the necessities of wartime communication gave the impetus to the development of some of the sophisticated techniques in use today.

Amateur rtty really got under way in the early 1950s in Great Britain and the USA, and today activity originates from well over 100 countries.

As a result of the dual development in the sphere of printing telegraph machines in the USA and Great Britain, the two terms "teleprinter" and "teletype" have come into common usage to describe telegraph machines. It should be made clear at this stage that *Teleprinter* is a registered trade mark of ITT-Creed, and *Teletype* of the Teletype Corporation. Understandably, USA amateurs tend to refer to teletype and their British counterparts to teleprinter.

FIRST PRINCIPLES

The teleprinter can in many ways be likened to the typewriter, and the keyboards for the two devices are very similar. However, the requirements for transmission of information between two teleprinters are somewhat complicated by the need to transmit machine function signals such as carriage return and line feed.

Figs 10.1 and **10.2** show typical teleprinter keyboards. A number of detailed variations exist between keyboards fitted to machines of different manufacture. In addition special purpose machines having non-standard keyboards are often found on the surplus market. Where uncertainty exists as to the suitability of equipment for amateur purposes, the advice of an experienced rtty operator should be sought.

Apart from the various keys used for machine functions, the major difference between the typewriter and teleprinter is the absence of a lower case type face on the teleprinter. In place of this the teleprinter has letters and figures cases which are selected by the appropriate machine function keys.

Telegraph Codes

It should be first of all clearly established that the various telegraph codes bear no relationship to the Morse code. Accepting the premise that only a single path for the transmission of information will exist between two machines, clearly any code which can transmit the necessary number of combinations must be time-sequential. A machine receiving signals from another must therefore be aware of the time scale on which the signals were originated so that it may correctly decode them. In practice a state of mechanical synchronization between the two machines is necessary. While nowadays sophisticated techniques exist in commercial practice for synchronization, the majority of telegraph circuits and all amateur practice utilize the *start-stop* mode. In order to minimize the mechanical problems of synchronization, a *start* signal is transmitted by the originating machine, followed by the information coding. The process is

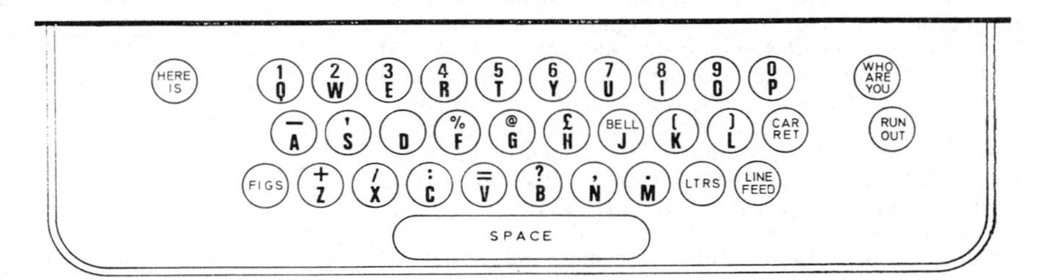

Fig 10.1. A typical three-row teleprinter keyboard.

Fig 10.2. A four-row teleprinter keyboard.

then completed by the transmission of a *stop* signal. In this way very accurate mechanical synchronization need only be held during the transmission of one character. Hence the most important difference between Morse and the telegraph codes—all character sequences in the telegraph codes are of identical length in time. Most readers will appreciate that the Morse character E takes less time to send than, for example, the character O.

The original telegraph codes due to Baudot and Murray have been slightly altered over the years, but the standard code now used for most commercial and all amateur purposes is the International Telegraph Alphabet No 2. **Fig 10.3** shows the time-sequential pulse signalling sequences used. Each letter, figure, or character starts with a space (off) pulse and terminates with a mark (on) pulse. The sequence of mark/space signals for the five signalling elements gives 32 binary combinations, each identifying a letter, figure or other character/function in the International Telegraph Alphabet No 2. While the *start* pulse or element is always of the same duration as each of the five signalling elements, some variation exists in the length of the *stop* pulse between different types of machine. The duration of the stop pulse can vary between one and two signalling elements according to the machine or convention in use.

Telegraph Speed

Prior to the start of international telegraph working separate development of telegraph machines had been carried out on both sides of the Atlantic. The result was that the majority of commercial operations in the USA standardized at one speed, and the UK and Europe standardized at another (slightly faster) speed. The differential in speeds and the consequent obvious compatibility problems have persisted to the present day. Only the modern requirement for greater information transfer rates and the obvious advantages of speed compatibility now that international telex working is an accomplished fact have led to international agreement on line speeds. The consequences of the original speed decisions linger on in amateur circles where the major source of equipment is surplus from the commercials. The proliferation of surplus USA equipment following World War II led to most international amateur working taking place at the USA speed. Domestic working in the UK (particularly on vhf) has been carried on at the CCIT/British standard speed in some cases, and many stations have found it necessary to provide for dual speed operation. It is difficult to visualize an early solution to this problem, particularly since much USA equipment cannot readily be modified for the UK speed. Luckily most British and European machines use mechanical speed governing devices which can readily be modified to the slower USA speed.

The standard unit of telegraph line speed is the baud. The line speed in bauds is defined as the number of pulses per second, and it should be emphasized that the variation in length of the *stop* pulse referred to earlier does not affect the baud rate.

For this reason it can be seen that two transmissions having identical baud rates can transmit a different number of characters in the same time purely because one of the transmissions uses a longer *stop* pulse than the other! For the same reason the equivalent speed in words *per minute* will also differ. **Table 10.1** relates line speed and the other characteristics of types of transmissions found in common usage.

Until early 1972 USA amateurs were only permitted to use 60 speed by their licensing authority. Although this restriction has now been removed, equipment availability in the rest of the world makes widespread operation on other speeds unlikely in the foreseeable future. In practice it is anticipated that most international working will continue at 45·45 bauds. In April 1975, at the Region 1 IARU conference in Warsaw, Region 1 voted in favour of 45·45 bauds for hf, vhf and uhf working. This means that amateurs worldwide have now standardized at a speed of 45·45 bauds.

Machine Speeds

From the foregoing the reader will appreciate that the time sequential nature of telegraph transmissions makes it essential that the sending and receiving machinery must be compatible in speed during the transmission of a character. Speed synchronization to tolerable limits may be achieved by adjusting motor speed (where possible) and viewing the governor or drive shaft stroboscopically by eye through a

TABLE 10.1

Characteristics of various common telegraphy systems

Speed	Baud rate	Units per second	Speed (wpm) (approx)	Start pulse length (ms)	Signal pulse length (ms)	Stop pulse length (ms)
60 speed (USA)	45·45	7·42	61·33	22	22	31
European/CCIT standard	50·00	7·50	66·67	20	20	20
75 speed (USA)	56·88	7·42	76·67	17·57	17·57	25
100 speed (USA)	74·20	7·42	100·00	13·47	13·47	19·18
75 baud (European/CCIT)	75·00	7·50	100·00	13·33	13·33	20

LETTERS		FIGURES
A		—
B		?
C		:
D		WHO ARE YOU
E		3
F		OPTIONAL
G		OPTIONAL
H		OPTIONAL
I		8
J		BELL
K		(
L		)
M		.
N		,
O		9
P		0
Q		1
R		4
S		'
T		5
U		7
V		=
W		2
X		/
Y		6
Z		+
CARRIAGE RETURN		CARRIAGE RETURN
LINE FEED		LINE FEED
LETTERS		LETTERS
FIGURES		FIGURES
SPACE		SPACE
LTRS	TAPE	FIGS

MARK ELEMENT ●

Fig 10.3. International Telegraph Alphabet No 2.

specially manufactured tuning fork. Alternatively a tachometer may be used to check shaft or motor speed. Motor speeds and also drive shaft speeds vary widely and the reader should consult handbook information for his machine. Where synchronous motors are fitted, machine speed can usually only be changed by altering gear ratios. Governor controlled machines lend themselves to speed changes (within a reasonable range) by adjustment of the governor and this can be carried out by electrical or mechanical means.

Telegraph Distortion

Telegraph signals sent by any transmission medium are subject to distortion which can take the form of pulse shortening or lengthening, or displacement of pulses in time

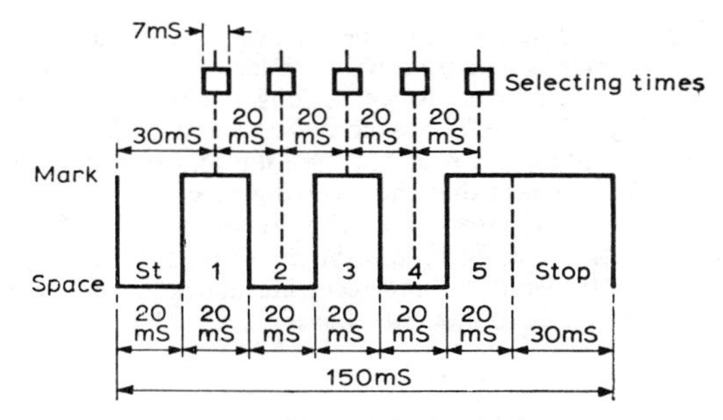

Fig 10.4. Letter Y showing ideal selection timing.

during transmission. In order to minimize these effects the receiving machine will normally examine the incoming signal for a very short period in the middle of the pulse or element time duration. **Fig 10.4** shows the selection timing diagram for a 7½ unit 50 baud machine.

A number of forms of distortion exist, any or all of which may be sufficient to cause the receiving machine to make the wrong decision during any or all of the five signalling elements, and result in an incorrect selection. Additionally, the transmitted signal may itself be distorted by incorrect machine adjustments. The reader will appreciate that the primary concern of any rtty operator must therefore be to minimize all controllable distortions so the maximum amount of unavoidable distortion due to the transmission medium can be tolerated.

Bias Distortion results in the lengthening of either mark or space pulses by a fixed amount and the shortening of the complementary space or mark pulses. **Fig 10.5(a)** illustrates lengthening of the mark pulses or marking bias. Spacing or marking bias distortion generally arises because of mechanical bias on transmitting contacts or receiving relay. Alternatively electrical bias can occur due to the lack of symmetry in electronic equipment. All forms of bias distortion can be eliminated by careful adjustment.

Characteristic Distortion is a more random effect occurring in the manner shown typically in **Fig 10.5(b)**. This type of distortion generally results from the failure of electronic circuitry to register a transition between the mark and space signalling conditions at the correct time following a long mark or long space condition. This effect is generally due to incorrect time constants in detector circuits, but can also occur as a result of transient disturbances in the transmission media, caused by the presence of the signal.

Fortuitous Distortion results from entirely random causes such as accidental irregularities in mechanical apparatus or external interference affecting the transmission media.

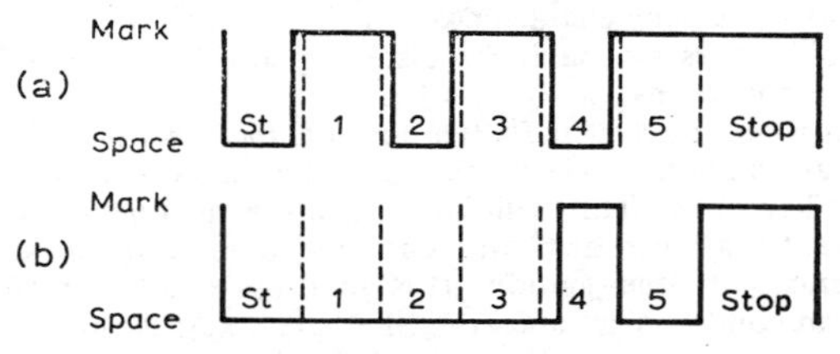

Fig 10.5.(a) Bias distortion, letter "Y" with marking bias (b) characteristic distortion, carriage return signal with characteristic distortion of element 4.

All the foregoing types of distortion can add or subtract in a manner which can cause a wrong selection to be made by the receiving machine. More serious problems can arise where telegraph signals are conveyed on a radio circuit, where deep fading can result in the total loss of say, a start element. This means that the receiving machine will regard the next space element as the start pulse and possibly take up part of the next character, resulting in a loss of synchronism between the transmitting and receiving machines which can take four or five characters to recover.

THE TELEPRINTER

It has been shown that teleprinters operate by the transmission and reception of a time-sequential series of pulses in a binary code using two signalling conditions which have been called mark and space. Before considering the problems and equipment associated with radio transmission and reception, it may be helpful to the newcomer to understand the way in which teleprinters are used on a landline circuit.

Indeed, an essential preliminary to successful rtty operation is operation on a dc loop or local basis. Prior to consideration of local loop operation it is as well to describe the facilities provided by the teleprinter keyboard. Not all the facilities listed exist on all machines, and in some cases they are not really applicable to amateur operation.

Keyboard Controls

Line Feed—each stroke of this key causes the paper in the local and distant machines to be fed up to the next printing line.

Carriage Return—this key returns the printing unit at the local and distant machines to the beginning of the line.

Space Bar—each stroke of the space bar moves the printing unit at the local and distant machines forward one space without printing.

Figure Shift—operate this key whenever you wish to change to the figures case, ie to print figures, symbols or punctuation marks.

Letter Shift—operate this key whenever you wish to change back to the letters case.

Run Out—this key provides for the continuous transmission of any selected printing character or any of the following functions:—carriage return; line feed; figure shift; letter shift, space and bell.

To run out on a particular character or function, operate the keys concerned in the following sequence:

(1) hold down the run out key.
(2) hold down the key whose character it is required to transmit continuously. Run out then commences.
(3) release the run out key. (Run out ceases).
(4) release the character key.

Avoid errors by making a distinct pause between each of these operations.

Alarm—to attract the attention of the operator at the distant station, operate the figures key once and then the bell key once. This will cause an alarm to operate at the distant station which will continue until another key is operated at either station. It is good practice to cut off the alarm condition by operating the letters key as this returns both machines to the letters condition for subsequent traffic. In amateur practice this key is often used at the end of a transmission.

Who-are-you? facility—operating the figures key and then the who-are-you? key will cause the distant machine to transmit its call sign back to the enquiring station. This assures the local operator that he is through to the correct station and that it is working properly even if it is unattended. This facility cannot, of course, be used via a simplex radio connection, and is often disconnected on amateur machines.

Here-is facility—operate this key once to transmit the home station call sign. It is important that this key should not be depressed again until the transmission of the call sign has been completed. In amateur practice it is often not possible to arrange for the device to be set to the operator's call sign and the facility is removed to prevent spurious character generation on receipt of a false who-are-you interrogation.

How to Use the Shift Keys

3-row keyboards: in the typical 3-row keyboard shown in Fig 10.1 most of the keys bear two characters, one a letter and the other a figure, symbol or punctuation mark. Two keys, one marked *Ltrs* and the other marked *Figs*, are provided to transmit special signals which enable the receiver to discriminate between the two characters associated with a particular printing key.

Operating the *Figs* (short for "change to figures") key causes the home and distant machines to condition themselves so that only figures, symbols and punctuation marks can be printed. Both machines will then remain in this state until the *Ltrs* key is operated.

Operating the *Ltrs* (short for "change to letters") key causes the home and distant machines to condition themselves so that only letters can be printed and both machines will remain in this state until the *Figs* key is operated again.

The carriage return, line feed and space keys may be used either in the letters or the figures condition of the keyboard.

4-row keyboards: The 4-row keyboard layout shown in Fig 10.2 differs from the 3-row in that it has a separate key for each character. Its mechanism is arranged so that when the letters key has been operated all its figure, symbol and punctuation-mark keys are locked; similarly when the figures key is operated all the letter keys are locked.

Operate the *figures* key whenever it is wished to print figures, symbols or punctuation marks and also when it is wished to use the bell and who-are-you? facilities. Operate the *letters* key whenever it is wished to print letters.

As with the 3-row keyboard, the carriage return, line feed and space keys may be used in either the *letters* or the *figures* condition of the keyboard.

Line Signalling Circuits

Figs 10.6 and **10.7** show the basic line signalling circuits for single current and double current signalling. In either circuit the earth return may be replaced by a physical connection.

Single current signalling has the advantage that only one signalling supply is required, but against this the space or open circuit condition leaves the receiving machine open to induced currents or spurious interference picked up on the telegraph line.

Double current signalling (sometimes called polar signalling) requires the use of two signalling supplies, but this added complexity is offset by the advantages of an unambiguous signalling condition for both mark and space. This renders the telegraph circuit relatively immune to induced interference. Double current signalling also permits the use

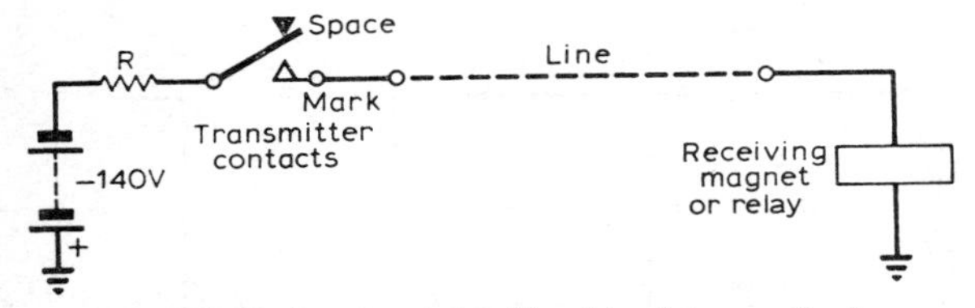

Fig 10.6. Single current signalling (simplex operation).

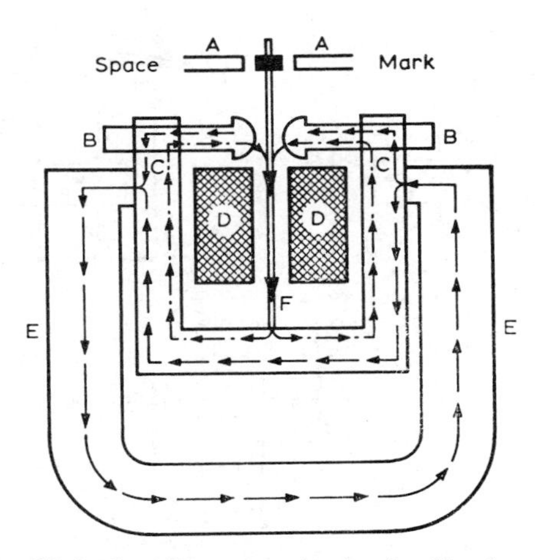

Fig 10.8. Simplified polar relay construction drawing. Flux due to the permanent magnet "E" is shown by the dashed line while flux paths due to current in coil windings "D" are indicated by the dot dash line. "A" are the contact assembly whilst armature "F" has its travel limited by adjustable pole piece screws "B".

of the balanced or polar receiving relay, which by the use of mechanical bias can be made much more sensitive than a single current telegraph relay.

Both of the circuits shown are for simplex or one way operation; somewhat more complex circuits utilizing the transmit/receive switch fitted to most teleprinters permit simplex operation in both directions over one signalling circuit. Even more complex arrangements provide half and full duplex operation. These arrangements do not normally concern the amateur radio operator.

The circuit shown in Fig 10.6 makes use of the flow of current in the line to attract the receiver magnet into the mark position. When no current flows in the line the receiver magnet is pulled into the space condition by a mechanical spring. This is known as a mechanical bias or neutral signalling circuit. It will be obvious that considerable mechanical energy may be necessary in teleprinters to operate the selection mechanisms, and therefore a high line current is necessary to secure reliable operation. An improvement on this system is to interpose a relay between line and teleprinter enabling the use of much reduced line current.

Fig 10.7 shows the differential or polar signalling arrangement, where a current flow in one direction indicates the mark condition, and the space condition is signalled by current flow in the reverse direction. This signalling arrangement is based on the use of the polar relay.

The Polar Relay and Selector Magnets

Fig 10.8 shows the construction of a typical polar relay, and similar principles are employed in teleprinter receiving magnets suitable for double current operation.

The basic difference between a polar relay and a conventional relay is that the polar type is centre stable without coil energization. The centre stable condition is obtained by setting up a balanced magnetic circuit using an armature

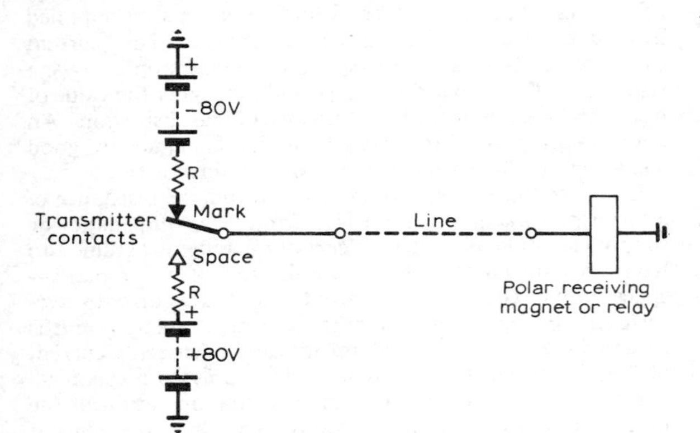

Fig 10.7. Double current signalling (simplex operation)

positioned in the centre of a permanent magnet yoke assembly. Electromagnet coils are also provided. The magnetic circuit is adjusted to give a condition where the armature floats between the pole faces of the permanent magnet. In this way it is only necessary to overcome the mass of the armature to deflect it to one pole face or the other. When released the armature will in theory return to the centre stable condition. As the armature has mass and cannot be perfectly balanced, it normally remains at one pole face or another until sufficient force is exerted to move it to the other pole face. The mechanical design is such that the mechanical effort required to obtain transition from one pole face to the other is the same for either direction, and the relay is said to be bias free. Balancing springs are often provided on teleprinter magnets in view of the relatively high mechanical energy required to operate the selection mechanism, but polar relays are normally magnetically balanced. This is done to make the electromagnetic force required to alter the state of the armature as small as possible and thus increase the sensitivity of the relay. With a polar relay, the contacts are normally mounted directly on the armature.

Two electromagnet coils are mounted symmetrically on either side of the armature. The application of current to either coil will generate a flux which will be additive with the flux due to the permanent magnet and deflect the armature to a pole face. A reversal of current flow in the coil will reverse the flux and cause the armature to deflect in the other direc tion. If the two coils are connected in series aiding and current flows, the action will be to aid the flux attracting the armature to one pole face, and tend to cancel the magnetic flux due to the other pole face of the permanent magnet. The result will be to aid the attraction to one pole face and thus further increase the sensitivity of the relay.

Polarized relays such as the Carpenter type are often encountered in telegraph practice. Like the double-current magnet they are completely balanced mechanically and side-stable. When the current is removed entirely from the coils, the contact remains switched according to the direction of the current prior to its removal. There is only one change-over contact, the fixed contacts being designated mark and

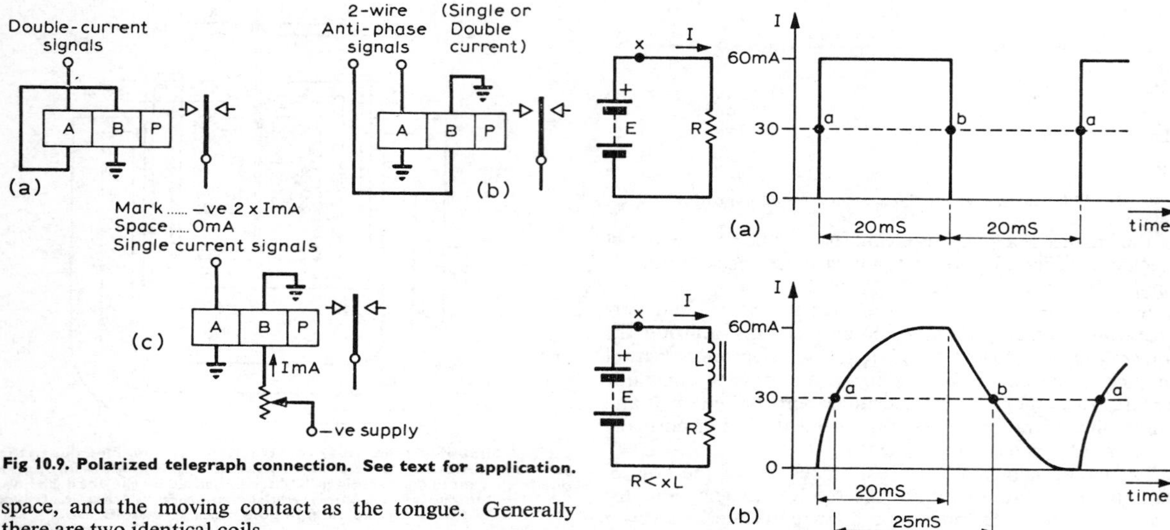

Fig 10.9. Polarized telegraph connection. See text for application.

space, and the moving contact as the tongue. Generally there are two identical coils.

Fig 10.9 shows a number of arrangements for telegraph relays. The circuit in (a) has coils A and B in series assisting, and requires the current in the coils to reverse direction for the contact to change over. It therefore accepts a normal double-current signal, and may be used to convert double to single-current signalling, or to convert a double-current signal working about one reference voltage to a double-current signal working about another reference voltage.

The circuit in (b) has each coil fed separately from a two-wire source. A positive current in coil A will operate the relay one way, and a similar current in coil B will change it over. Only one coil is energized at a time. It is used with circuits such as a bistable where one coil may be placed in series with each output. The relay then follows the switching of the bistable and allows the electronic circuitry to be interfaced with the telegraph circuitry.

Circuit (c) works entirely from single-current signals, and is the least sensitive of the three arrangements. A steady current is passed continuously through coil B, so that when no current flows in coil A, the relay will always switch one way. The signal current in coil A is set to be twice the current in coil B, so that the change in magnetic force acting upon the armature for operation from mark to space and from space to mark is always the same. Since the force arising from the continuous energization of coil B is always in opposition to the force arising from signal current in coil A sensitivity is reduced. However the arrangement is better than using a relay with a spring return for the armature, because operation in each direction is effected by identical change of magnetic force, and the relay retains its mechanical balance. This circuit can be used to convert single to double-current signalling.

All the circuit arrangements shown may be used for energizing teleprinter magnets.

Driving the Teleprinter Magnet

At first sight the energization of a teleprinter magnet seems to be a straightforward procedure, and the uninitiated tend to consider the magnet as being similar to an ordinary relay. Such a relay would have a coil voltage rating, and it would

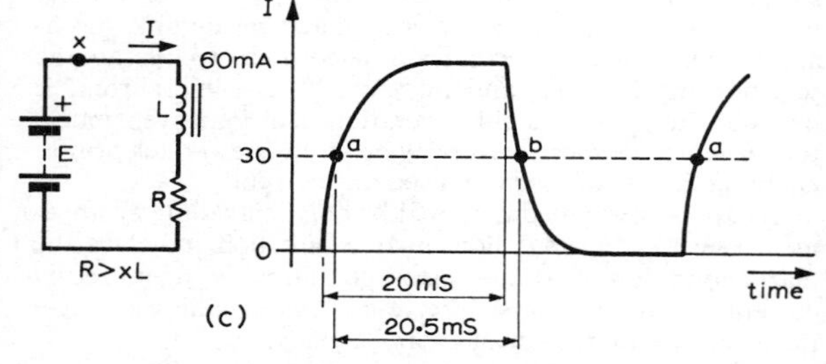

Fig 10.10. Marking bias distortion may be introduced by failure to drive a teleprinter magnet from a high resistance source. (a) shows the current waveform in a purely resistive circuit; a and b are the operate and release points. (b) shows the exponential build-up and decay of current in a series L-R circuit with R small in comparison with XL. Exponential build-up and decay of current may be minimized by keeping R as high as possible as shown in (c), thus avoiding pulse lengthening.

logical to expect the selector magnets in a teleprinter to have a similar rating. This is not the case, and hence we arrive at one of the pitfalls awaiting the newcomer to rtty. Selector magnets have a current rating and must be supplied from a current or high impedance source. This current source consists of a relatively high voltage applied via a resistance. The higher the voltage and the higher the value of resistance, the better for the avoidance of bias distortion. An understanding of this principle is fundamental to good results and the following explanation should assist.

Consider **Fig 10.10(a)** where R represents the resistance of a teleprinter magnet being driven from a low impedance dc supply via contacts x. The teleprinter magnet is assumed to have no inductance, and the application of a 1:1 mark—space ratio at contacts x will result in a square current waveform in the circuit. For the purpose of this explanation it is assumed that the magnet will operate and release at a current of 30mA. It can be seen that, subject only to mechanical delay which should not be significant, the operation of the magnet armature should faithfully reproduce the current waveform in the circuit and the bias introduced is zero.

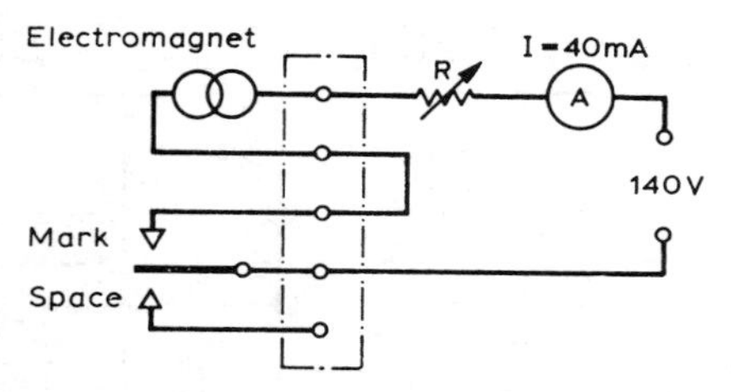

Fig 10.11. Simplified local loop circuit—single current keying.

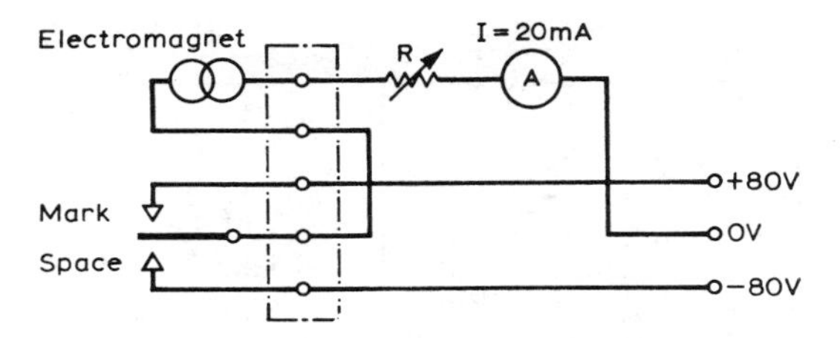

Fig 10.12. Simplified local loop circuit—double current keying.

A practical magnet will have a comparatively high inductance and thus the circuit must be modified to that shown in **Fig 10.10(b)**.

Due to the presence of series inductance the current waveform in the circuit is modified. Current in the circuit increases in an exponential manner at a rate determined by the L/R time constant. When contacts x break, the current decreases, again in an exponential manner.

It can be seen that the operate and release points of the armature have been displaced in time from the ideal and that the period of operation has been markedly affected. The result of this state of affairs is considerable telegraph distortion.

In order to minimize the effects of the coil inductance it is necessary to reduce the L/R time constant to a minimum in the magnet circuit. Since time is proportional to L/R, R should be at a maximum to minimize the time constant. However, a little thought will reveal that if R is increased to minimize the time constant, the circuit voltage must be increased to obtain the necessary operate current. In practice the magnet winding will have some resistance, but external series resistance is also used. A good empirical rule is to ensure that the series resistor is at least ten times the magnet winding resistance.

Where two or three machines are required to operate simultaneously, it is possible to put the magnet coils in series with an appropriate resistance.

The series resistance is often made variable to permit the setting up of the loop current to the correct value specified by the machine manufacturer.

Figs 10.11 and **10.12** show local loop circuits for single and double current keying using a single machine. The choice of single or double current loop operation within the amateur station is often dictated by the machines available, which may only be suitable for one or the other mode. Double current operation is to be preferred, although it requires more complex power supply arrangements. A word or two about the loop power supply may help at this stage: the reader might infer from the inclusion of a series resistor to provide a current source that regulation is unimportant. This is far from the case, particularly when one considers the reasons for poor transient regulation in power supplies. In many cases poor regulation occurs due to the inability of power supply reservoir capacitors to meet the demands made upon them. For this reason the supply reservoir time constants may well modify the current wave form in the loop circuit. This is clearly another way of introducing telegraph distortion. The telegraph supply must be capable of delivering the required current with no worse than a 10 per cent voltage drop.

Local Record

The circuits in Figs 10.11 and 10.12 will produce an electrical local record, that is, the transmitted signal is reproduced on the teleprinter receiver normally associated with the transmitter. This facility is necessary because most teleprinters consist of separate transmitting and receiving sections, and an electrical connection is required. A few types of teleprinter, such as the Creed 75, utilize mechanical local record and no electrical connection is necessary. However, when using mechanical local record the valuable self-checking obtained by the use of the electrical connection is lost. The electrical connection can include the station transceiver.

The problems of obtaining local record are complicated by the introduction of a second teleprinter or an auto transmitter into the local loop. Fortunately most teleprinters and auto transmitters incorporate transmit-receive contacts which can be used to solve this problem. A typical arrangement using two teleprinters and one auto transmitter is shown in **Fig 10.13**. Transmit-receive contacts in teleprinters are

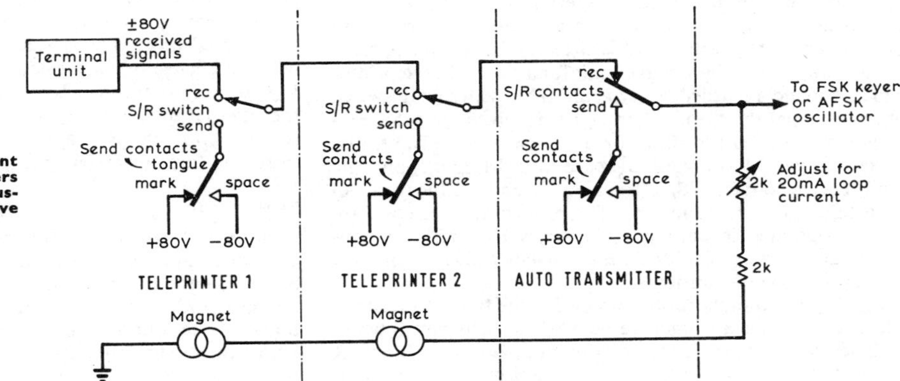

Fig 10.13. Typical double current loop circuit for two teleprinters and one auto-transmitter, illustrating the use of send-receive contacts.

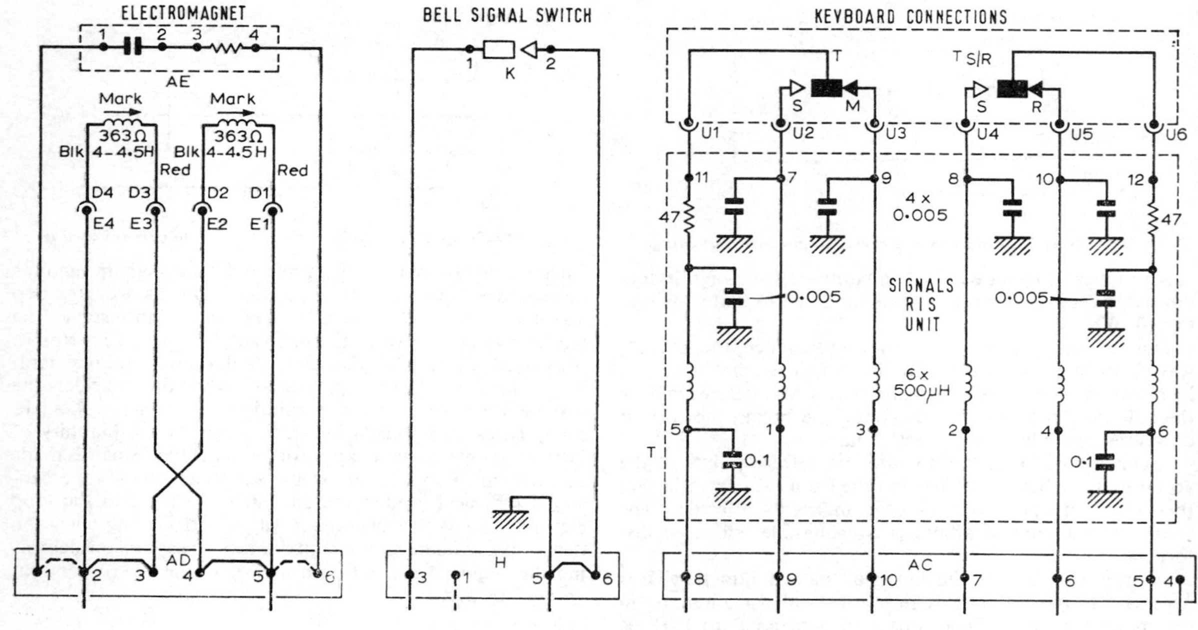

Fig 10.14. Signal circuits—Creed model 54 teleprinter.

arranged to change over from the normal receive condition to transmit immediately prior to the transmission of a character, and to restore immediately after that transmission. There are many variations of the basic circuit shown, but these variations must always be arranged to provide +80V, or the mark condition, when all the machines are quiescent. It will be noted that the first machine in the loop can always override the others, but this is not often a problem in amateur practice. Normally machines are connected into the series loop by plugging into closed circuit jack sockets, and this arrangement allows maximum flexibility.

Radio Interference Suppression

Up to this point the circuits presented to the reader have been simplified by the omission of many components which are provided for the suppression of radio interference. The switching, at comparatively high speeds, of inductive circuits can generate serious radio interference unless measures are taken to suppress the problem at source. Normally all machines are fitted with suitable circuitry for this purpose.

Fig 10.14 shows the signal circuit for a Creed 54 teleprinter which can be regarded as typical. The bell signal switch operates when the "bell" character is received, and the contacts may be used for alarm or other purposes. The transmitter contacts T (tongue), S (space) and M (mark) and the send/receive contacts TS/R (tongue), S (send) and R (receive) each have filter components. The electro-magnet coils are usually operated in parallel. The series RC circuit is connected across the coils, for double current working only, for the suppression of transients.

Fig 10.15 shows the motor circuit for the type 54 machine. Block Q contains a capacitive network to provide a low impedance path to earth for interference currents. Block R shows the centrifugal governor mechanism responsible for speed control of the series wound motor shown at W. 1mH chokes and an RC network are mounted in the governor assembly, which operates by placing a short circuit across series governor resistor L. B is the motor start switch assembly which allows the motor to be started by an incoming transmission. Following a period in excess of 90 seconds without incoming signals, the motor will be shut down automatically until further signals are received.

OTHER RTTY EQUIPMENT

There are a number of other pieces of equipment often found in the amateur rtty station and details are given below.

Auto Transmitter

A motor driven automatic transmitter which accepts punched or perforated paper tape and converts this to telegraph signals. Only the five signalling elements are punched on the tape, the start and stop elements are inserted by the automatic transmitter. The auto, as it is colloquially known, will send telegraph information at the full machine speed. Fig 10.3 shows the standard tape punching arrangement for the International Telegraph Alphabet No 2, the black representing holes in the paper. The small holes between the third and fourth elements are sprocket feed holes.

Perforator

Hand or keyboard perforators are available to punch or perforate paper tape for automatic transmission. Reference is often made to the chad or punching. Where the tape is punched and the hole is clean the tape is said to be chadless. Some perforating machinery only cuts the paper for some ⅞in of the circumference of the hole, allowing the chad or perforation to remain attached to the tape. Perforators are

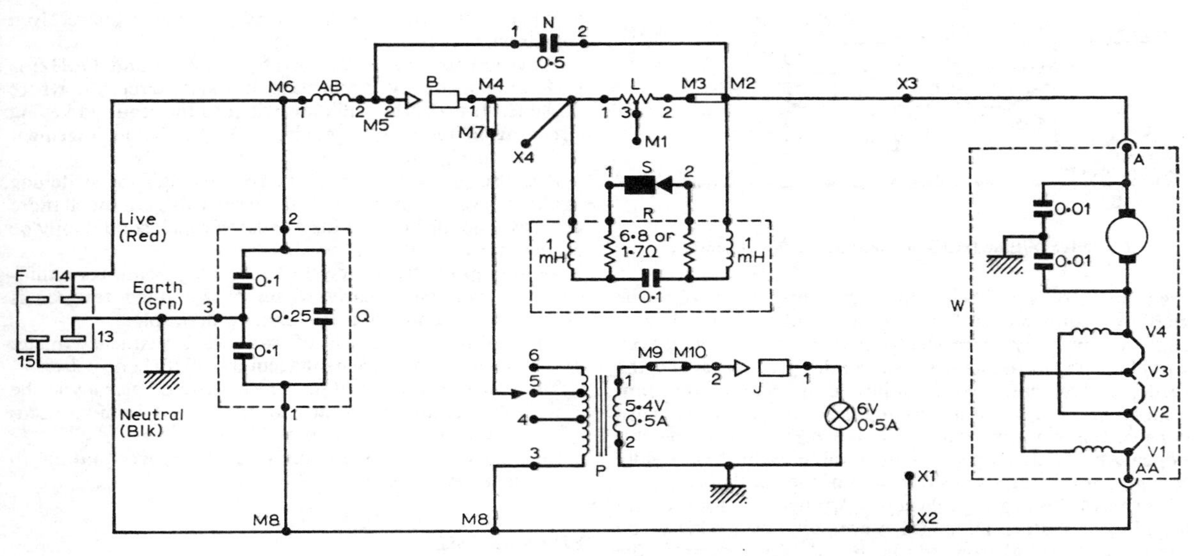

Fig 10.15. Motor circuit—Creed model 54 teleprinter.

available which print the character on the tape as well as punch the holes. In many cases these machines produce chadded tape to enable the printed character to remain legible.

Reperforators perform a similar function to perforators with the exception that they accept electrical signals in the same way as a teleprinter. Reperforators are commonly attached to teleprinters so that they may be used to prepare paper tapes for subsequent automatic transmission, either from keyboard initiated information or from incoming signals.

TYPES OF TRANSMISSION

There can be a problem in transmitting telegraph information over a radio path with the minimum of distortion. This problem is complicated by fading and interference, but with careful design good results can be obtained with simple equipment.

The simplest method of transmitting information is to use on-off keying and the cw operator will readily appreciate this mode. However, the key-up or space condition is completely open to interference and it is preferable to use polar signalling as for line operation. This is accomplished by transmitting mark and space as two separate frequencies.

The two frequencies can take the form of audio tones modulating the carrier of a conventional a.m. or fm transmission, and this mode is known as afsk (audio frequency shift keying). Alternatively transmissions can be made by changing the frequency of a carrier (frequency shift keying).

On the hf bands frequency shift keying is accomplished either by changing the frequency of an existing transmitter oscillator in response to telegraph modulation or by connecting a two-tone audio signal to the audio input of an ssb transmitter. Using a product detector or simply the conventional bfo as employed for cw reception it can be seen that audio frequencies can be reconstituted at the receiver for connection to a telegraph terminal unit.

A number of techniques may be used for the reception of rtty signals, but possibly the most popular in amateur circles is the audio frequency terminal unit. Either afsk, or fsk signals converted to afsk in the receiver detector circuits, can be utilized. No special modifications to the receiver are necessary and the terminal unit is a convenient add-on feature to a station used for other modes of emission.

A few years ago it was also common practice in the commercial fields to use intermediate frequency telegraph adapter units for reception of fsk signals. The received signal was extracted from the receiver i.f. stages and applied to the adapter unit. Many excellent adapter units of this type have reached the surplus market. The disadvantage is the requirement for an i.f. output from the station receiver, often at low impedance. Very few amateur station equipments have this facility available, and many operators are reluctant to modify their receivers. For this reason the majority of rtty operators now use the audio frequency type of terminal unit.

AFSK Transmission

The afsk mode is commonly employed on the vhf bands, generally over comparatively restricted distances. The standard tone frequencies used are 1445Hz (mark) and 1275Hz (space) giving a frequency shift of 170Hz. The change to these tone frequencics and the narrow shift afsk standard were agreed by IARU Region 1 in Warsaw, April 1975.

FSK Transmission

On the amateur bands up to 30MHz fsk transmission is normally employed and fsk is also used for longer distance working on vhf. Two main methods of generating fsk are used and the simplest method will be considered first. The general requirement is that it shall be possible to frequency shift the transmitted signal by either 850Hz or 170Hz at the radiated frequency, dependent upon whether wide or narrow shift is in use. Most stations have facilities for both, although narrow shift is rapidly becoming standard on all bands.

Any modification made to transmit fsk must presuppose a

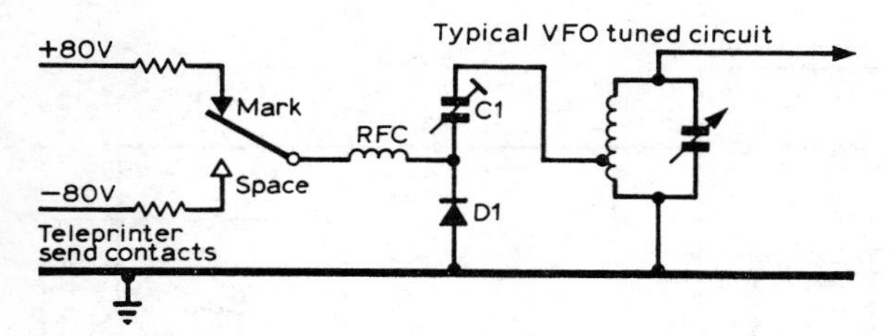

Fig 10.16. Outline frequency shift keying arrangement.

high order of stability in the oscillator concerned. The stability requirements for rtty, particularly with narrow shift, are possibly more stringent than for ssb operation. Modifications to oscillators also require great care, and the reader is recommended to utilize only high grade components and to take all the usual precautions associated with construction work on oscillators. It is essential that all components associated with the tuned circuits must be rigidly mounted inside the screened section of the oscillator.

Fig 10.16 shows a suitable keying arrangement for use with a typical oscillator. Diode D1 is normally reverse biased by the mark potential from the keyboard. On receipt of the space potential D1 switches to conduction and places capacitor C1 in parallel with existing circuit capacitances, thus lowering the frequency by the required amount of frequency shift. With the circuit in the mark condition it will be necessary to readjust the original oscillator trimming capacitor to restore the original calibration. This operation should be carried out only after reference to any handbook or alignment information. A reliable frequency standard should be employed.

Where frequency multiplication of the keyed oscillator takes place to obtain the radiated frequency, remember to allow for this when setting the frequency shift at the oscillator frequency. However it is preferable to set the shift at the radiated frequency.

When an oscillator is heterodyned or mixed to obtain the output frequency it may be necessary to reverse the diode to obtain the correct keyed output frequencies. The reason for this is that a frequency inversion may take place in the mixing process. If this is so a decrease in frequency at the oscillator will give an increase in radiated frequency.

The normal amateur convention on all bands is that the highest radiated frequency represents the mark and the lowest radiated frequency the space, and the diode orientation should be selected to give this condition. In the nature of things not all amateur operators adhere to this convention and most stations are equipped with reversal switches on the

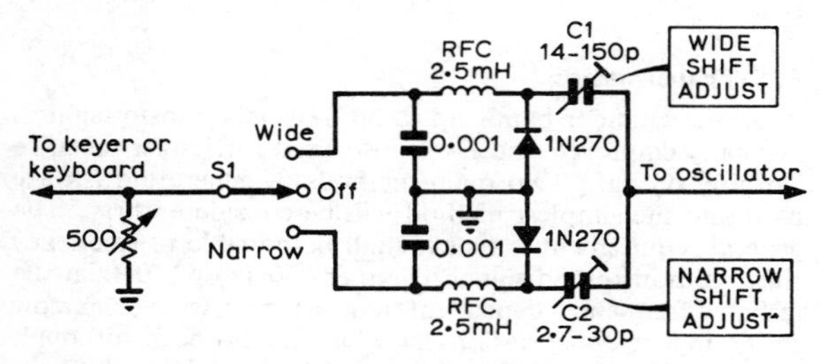

Fig 10.17. Switched frequency shift keyer for wide/narrow shift operation. S1 may be mounted remotely.

terminal unit to reverse the keying polarity of the received signals when necessary.

Provision for dual shift operation (850Hz and 170Hz) is preferably made by duplicating the shift circuitry, where possible. It is then possible to switch to the required keying circuit at dc, remotely if required. A practical arrangement is shown in **Fig 10.17.**

It is also possible to select the required shift by switching in alternative capacitors at C1 but clearly this is a much more extensive modification to the oscillator which should only be carried out when there is no alternative.

Setting up the frequency shift accurately is always a difficult problem, particularly where only limited testgear is available. A number of methods are practicable:

(i) Utilize some form of calibrated frequency meter having a measurement accuracy of 10Hz or better.
(ii) Listen to the signal on a receiver and heterodyne the receiver audio output with the output from an accurately calibrated audio oscillator.
(iii) Enlist the co-operation of another rtty station to measure the shift.

When on SSB

With the growing predominance of ssb equipment many operators find it very convenient to generate a quasi-fsk signal by applying a keyed tone audio frequency input to the audio stages of an ssb transmitter. Provided care is taken this is a quite acceptable procedure, and the same two-tone audio oscillator can be used for vhf operation. A number of precautions are necessary, however, and these are outlined below.

(i) In order to avoid the generation of excessive spurii, the carrier suppression and intermodulation distortion performance of the ssb equipment must be exceptionally good. Additionally, great care must be taken not to overdrive the equipment.
(ii) Many ssb equipments cannot be safely operated at full input on rtty; this is particularly true of equipment using colour television sweep tubes as output amplifiers. No attempt should be made to exceed the a.m. carrier input limitation or if no such rating is available the equipment manufacturer should be asked for a continuous carrier input rating. The input limitation should, of course, apply whatever method of fsk generation is used.
(iii) The accepted standard for narrow shift tone frequencies are 1275Hz (space) and 1445Hz (mark). If for some reason wide shift is used the mark tone should be 2125Hz. All these three frequencies will go through ssb filters without the problems previously associated with higher audio tones. The above audio standard will give correct sense fsk when these tones are fed into a usb transmitter. Care must be taken to avoid audio harmonics.

AFSK Oscillators

The principal requirements of an afsk oscillator can be summarized as:

(i) good waveform and low harmonic distortion to minimize spurious radiated components, and
(ii) minimum transient distortion and overshoot during keying transitions from the mark to the space tone.

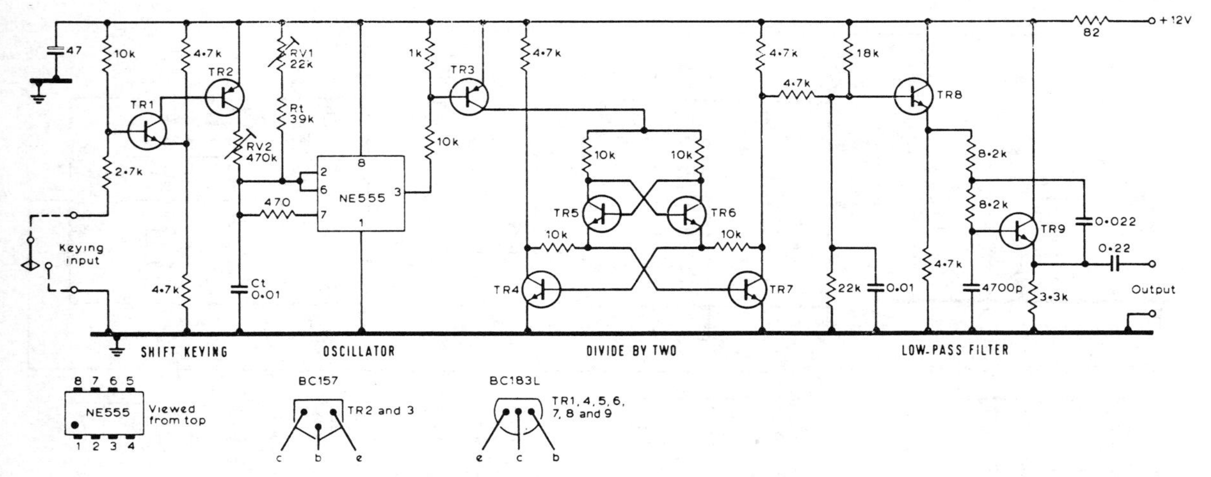

Fig 10.18. Two-tone afsk oscillator for 1275/1445Hz.

The tone keyer shown in **Fig 10.18** generates a sine wave tone on 1445Hz in the idle or mark condition, and when the input line is taken low, the frequency changes to 1275Hz for the space condition. It can thus be used for afsk operation into an a.m. or fm transmitter, or for fsk operation by feeding the unit into a usb transmitter.

The NE555 ic is connected as a multivibrator, the period of oscillation being determined by Ct and the network RV1, RV2 and Rt. In the space condition, with TR1 and TR2 turned off, the period is determined by Rt in series with RV1, and in the mark condition, TR2 switches in RV2 to raise the frequency. The output from pin 3 is a train of narrow negative-going pulses, rich in harmonics and unsuitable for transmission, and so the NE555 is actually made to oscillate on twice the required frequency. The pulses are taken via TR3 to a divide-by-two circuit consisting of TR4-7. The output of the divide-by-two circuit is a perfect square wave containing far less harmonics. This square wave is filtered by TR8 and TR9 which are connected as an active filter with a cut-off frequency of about 1900Hz, and this removes the harmonics, leaving a sine wave of about 500mV rms.

The unit was designed for use from a 12V supply, but will function well from 9 to 15V, and the frequency is substantially independent of the supply voltage over this range. A ceramic capacitor must not be used for Ct because of its temperature coefficient. Where wide temperature fluctuations may be found, Rt should be a metal film resistor and Ct a polystyrene capacitor so that their temperature coefficients will cancel. To set the frequencies, first adjust the space tone to 1275Hz with RV1, with the keying input grounded, and then set the mark tone to 1445Hz with RV2, with the keying input open-circuited.

TERMINAL UNITS

The reception of rtty signals requires a means of translating a frequency shifted signal into telegraph information for connection to the teleprinter system, the necessary device being known as a terminal unit. General amateur practice today is to use an audio frequency unit as an extra to the normal station receiving equipment. The station receiver is then set up to produce an audio frequency shift (afsk) output at the standard rtty audio frequencies, which are 1275/2125Hz for 850Hz wide shift and 1275/1445Hz for 170Hz narrow shift.

The requirements for the design of a terminal unit are:

(a) Selection of the appropriate audio tone frequencies together with a bandwidth appropriate to the reception of telegraph information, and the rejection to the greatest possible degree of other audio frequencies present in the receiver output.

(b) The provision of envelope detection facilities which will recover the telegraph information from the modulated audio frequencies.

(c) The provision of circuitry which will compensate for the wide variations in signal level which occur as a result of fading under hf band conditions. This fading may take the form of "flat fades" or level variations affecting both mark and space signals or selective fading affecting the level of one or other signal. Both types of signal fading can occur simultaneously on the amateur bands. Under commercial conditions a terminal unit might have to tolerate up to 50dB of selective fading between mark and space signals, and amateur conditions may require even better performance. A detailed discussion of design theory is beyond the scope of this handbook and the reader is referred to the references at the end of the chapter for further reading matter on this subject.

Terminal units in use on the amateur bands fall into two groups:

(i) Limiterless or a.m. units, or

(ii) Limiter-discriminator or fm units.

The limiterless two tone or a.m. unit utilizes a linear amplifier and envelope detector for the mark and space audio frequencies. The outputs from the two envelope detectors are combined in an "assessor" circuit which is arranged to compensate for level differences between the two tones. Separate filters are provided for both mark and space frequencies with bandwidth kept to the minimum necessary to convey telegraph intelligence. Separate linear amplifiers can be used with agc control to minimize the effect of signal level variations.

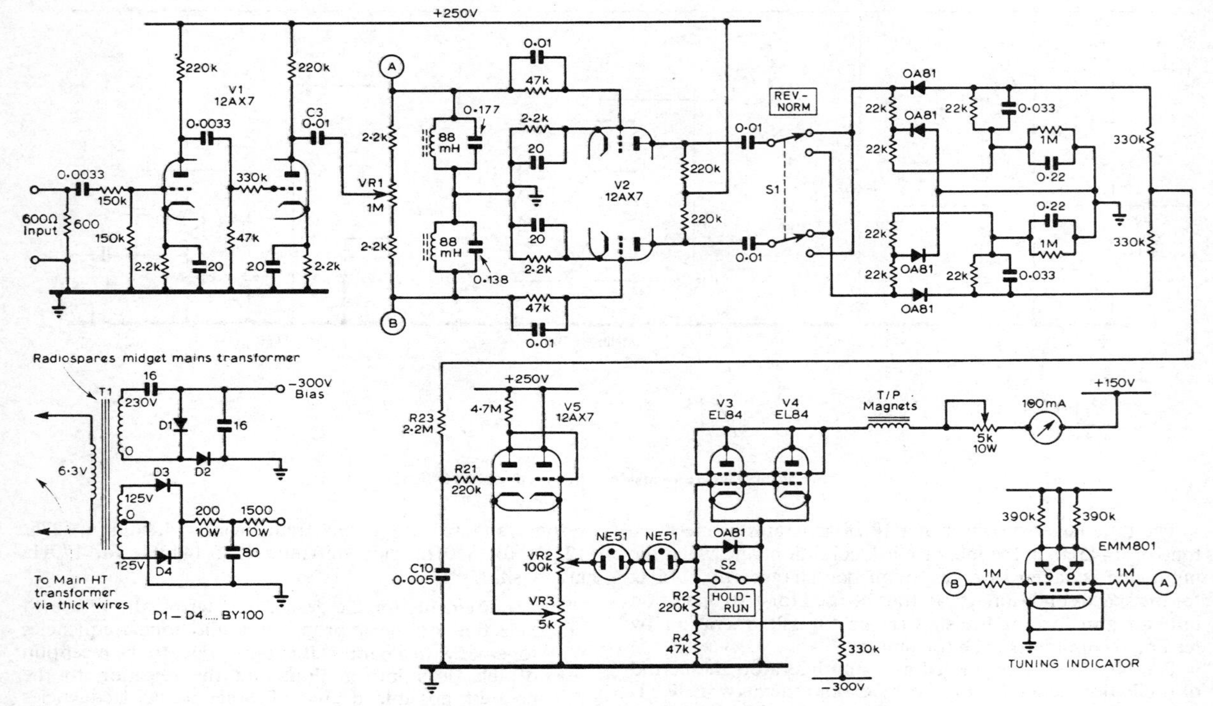

Fig 10.19. The DL6EQ terminal unit.

Limiter-discriminator or fm terminal units are the commonest type used in amateur service and consist of a limiting amplifier followed by a discriminator circuit. The limiting amplifier is often preceded by a filter having sufficient bandwidth to accept the mark and space frequencies together with the bandwidth required for telegraphic information.

Many learned authorities have argued the merits and demerits of the two systems, including the CCIR, which eventually issued a report stating that for very long distance working the two tone or limiterless method was slightly superior. In many respects the arguments between the two systems are analogous to the fm/a.m. dispute. With the use of narrower shifts in amateur practice the marginal theoretical advantage of the two tone system is largely cancelled out by the simpler techniques used in the fm system.

Both systems require the use of a low pass filter following the detector to minimize the effect of in-band noise and similar effects. Terminal unit designs are completed by a limiter or slicer and the necessary dc amplification to make the output suitable for driving the teleprinter magnet circuit.

DL6EQ TWO TONE LIMITERLESS DESIGN

Fig 10.19 shows the circuit of the DL6EQ terminal unit in use at many stations. It is characterized by simplicity and the use of non-critical components. V1 is a class A preamplifier stage having 600Ω input impedance and suitable for direct connection to the phone patch sockets provided on many modern receivers and transceivers. Alternatively the input may be connected in parallel with the loudspeaker circuit using a suitable matching transformer. As this is a limiterless type of terminal unit care should be taken not to exceed 0·2V rms input. V2a/b is a tuned audio amplifier with the grid circuits resonant to the mark/space frequencies of 1275Hz and 1445Hz respectively. Potentiometer VR1 is set to balance the af voltages at the anodes of V2 with 100mV rms inputs at the mark and space frequencies. The tuning capacitors across the 88mH inductors may require adjustment to resonate exactly at the correct tone frequencies, and this may conveniently be accomplished by padding with capacitors of suitable value.

For wide shift (850Hz) operation the 0·138μF capacitor will need to be decreased to approximately 0·064μF. S1 is provided as a reverse-normal switch and reverses the sense of the tone inputs to the detector circuits. Provision is made in the detector circuits for differences in the signal amplitudes between mark and space to be corrected and this terminal unit will operate correctly (although with increased telegraph distortion) in the complete absence of one tone input. The detector output is coupled to a 12AX7 (V5) connected as a limiter stage. Limiter potentiometer VR3 is set to obtain a symmetrically limited waveform when the terminal unit is driven from a local two tone oscillator keyed with a 1:1 mark-space ratio signal. The tone oscillator output level should be reduced until the 12AX7 just limits for this adjustment.

Output from the limiter cathode is taken to the magnet driver stage V3/V4 via two neon trigger tubes. Any small neon having a striking voltage around 35V will serve in this position. Adjust VR2 with S2 open so that the neons just strike reliably and then increase the setting slightly to give a safety margin. S2 is provided to hold the magnet driver stage in the mark or magnet operated condition and is used to inhibit printer operation while tuning. The driver anode

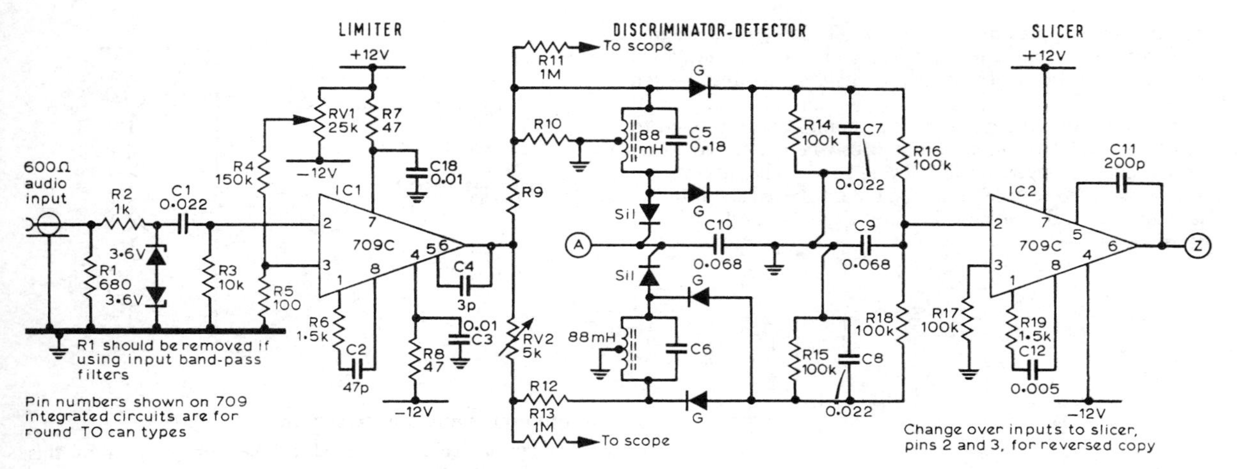

Fig 10.20. The Mainline ST5 terminal unit.

current is adjusted to a suitable value for the machines in use. Most single current machines require a magnet current of 40mA.

MAINLINE ST5 TERMINAL UNIT

Based on a design by Irvin M. Hoff, W6FFC, but modified for use on the new IARU Region 1 tone standards, this terminal unit uses two 709 ic operational amplifiers as limiter and slicer/trigger stages, together with a full-wave discriminator and a double current keyer using MJE340 high-voltage transistors. The use of high-gain operational amplifiers means a frequency shift as low as 5Hz can be copied.

Fig 10.20 shows the limiter, discriminator and slicer stages, and **Table 10.2** gives the component values for narrow and wide shift operation. The use of a 709 as limiter gives nearly 90dB of limiting with a threshold at signal input levels around 200μV. A simple RC high-pass filter is used in the input to offer some discrimination against hum. The limiter is followed by a discriminator/detector for 1275/1445Hz or 1275/2125Hz operation, using full-wave detection for greater efficiency. The discriminator can be used on both narrow and wide shift by using a combination of four 0·033μF capacitors for C6 (narrow shift) and then switching out two for wide shift. The unbalance of the discriminator

TABLE 10.2

Component values for narrow and wide shift in Fig 10.20

Component	Narrow shift (1275/1445Hz)	Wide shift (1275/2125Hz)
R9	2·7kΩ	1·5kΩ
R10	27kΩ	8·2kΩ
R12	2·7kΩ	2·2kΩ
C6	0·132μF	0·066μF

resulting from not using the wide-shift resistor values is not serious, but it will cause the tuning meter (see later) to read higher on 2125Hz on mark on wide shift. This is however a small price to pay for the convenience of switched operation.

Another simple RC filter, this time low-pass, removes tone frequency ripple from the discriminator output to the slicer stage. If reception of reversed tones is required the connection to pins 2 and 3 of IC2 may be reversed by means of a switch (not shown).

The double current keyer **(Fig 10.21)** uses two 2N706 transistors in a bistable flip-flop circuit which drives two MJE340 high-voltage keying transistors. The latter are used because of their comparative immunity to the high-voltage spikes generated by the teleprinter inductance, although

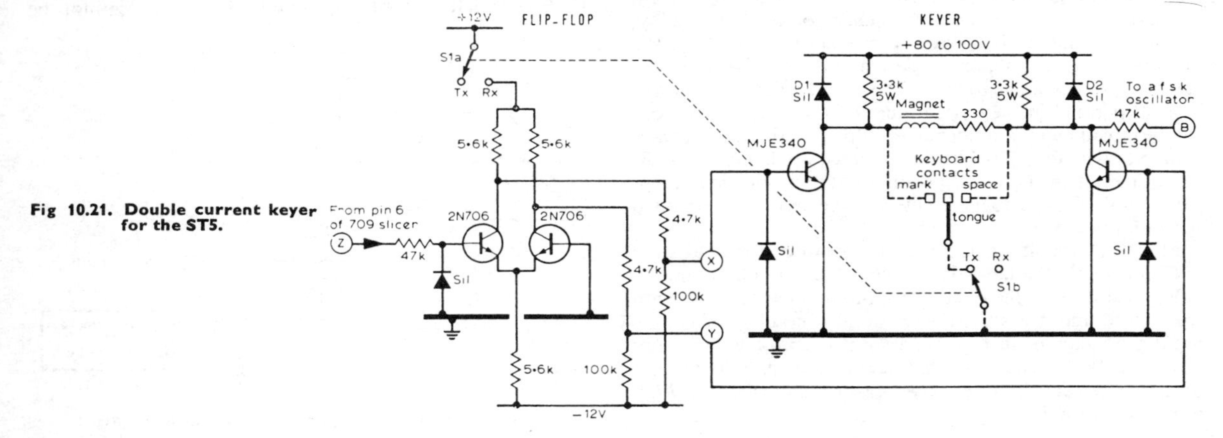

Fig 10.21. Double current keyer for the ST5.

diode spike suppressors are also fitted across the magnet resistors. No heat sinks are necessary for the MJE340 devices. The TRANSMIT/RECEIVE switch is a dpst switch S1a, b.

The input of the keyer circuit at point Z is connected to pin 6 of the 709 slicer ic. Point B can be connected to an afsk oscillator keyer, positive being mark. The 80V supply should be rated to at least 25mA, but no special stabilization is required and voltages in the range 80–120V will be satisfactory.

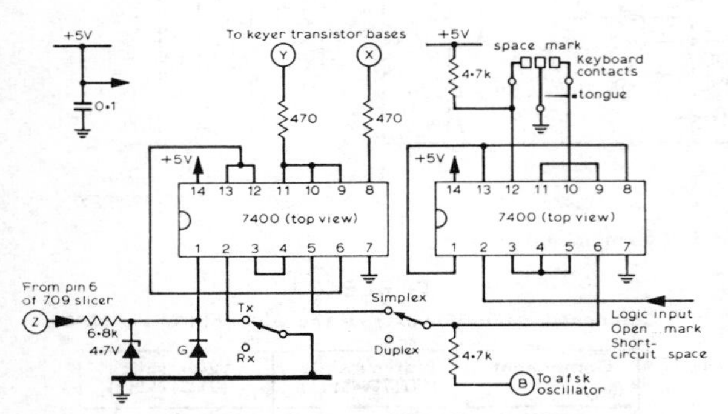

Fig 10.22. Logic keyer driver for the ST5.

Fig 10.22 shows an inexpensive ttl keyer driver which can be substituted for the 2N706 bistable if desired. A 5V stabilized supply is required, but the circuit has a number of advantages over the simpler driver. Instead of being connected across the magnet, as in Fig 10.21, the input from the printer keyboard is fed into a bistable circuit which removes any contact bounce, and compensates for slight maladjustment of the contacts themselves. A logic input is provided, and grounding this changes the system from mark to space, thus allowing single-current keying inputs. This logic input can also be used to connect other ttl-compatible devices, such as a memory, into the system.

The SIMPLEX/DUPLEX switch set to the DUPLEX position allows the received signal to be printed and the keyboard on logic input to be passed to the afsk oscillator for transmission without local copy. This position also allows the reception of one's own transmission for the receiving system and terminal unit, thus acting as a check on one's own copy.

The input has a 4·7V zener diode across it to limit the positive-going voltage, and a germanium diode to limit the negative-going voltage to 0·3V. The 4·7kΩ resistor on the key contact permits the keyboard plug to be withdrawn without the bistable having a tendency to "flop" to the space condition. No switching of the keyboard is required, and it may be left in circuit in the mark position without affecting reception. Similarly the logic input may be left open-circuited when not in use.

Make sure that if the ttl driver is substituted for an existing 2N706 bistable the latter is disconnected at points X and Y.

Fig 10.23 shows the circuit of a tuning meter for use with the ST5 terminal. Such a meter can also be of considerable use during alignment. In this design it is connected to point A on Fig 10.20. The transistor dc amplifier permits the use of a comparatively insensitive meter, and offers a high impedance at point A.

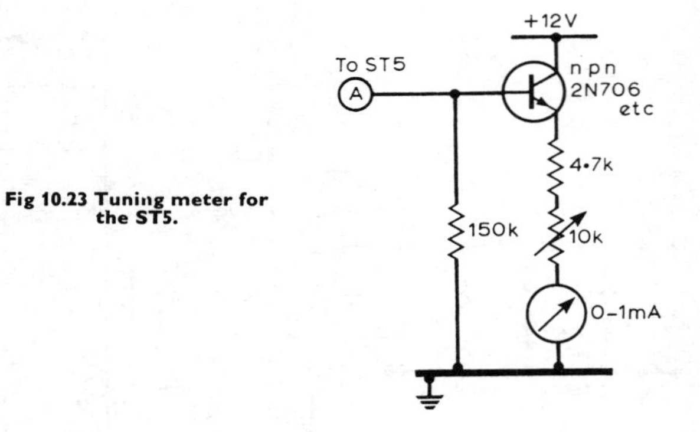

Fig 10.23 Tuning meter for the ST5.

Components and Construction

No special care is needed in the construction of this terminal unit other than the usual precautions necessary when dealing with high-gain amplifiers. All resistors are ¼W except where indicated. All capacitors should be polyester or polycarbonate in order to preserve circuit Q. Paper types should not be used.

Provision should be made in the component layout to allow small-value padding capacitors to be added to C5 and C6 during alignment (see next section).

All diodes marked "G" and "Sil" are small germanium and silicon types respectively. RV1 and RV2 are small preset potentiometers. C2, C4 and C11 in Fig 10.20 can be small disc, mica or similar types. The pin numbers of IC1 and IC2 are those of the round-can TO-style packages. The 2N706 transistors in Fig 10.21 may be replaced by any small silicon npn type with similar characteristics.

Alignment

With the input signal disconnected, connect a test meter between pin 6 on IC1 and earth, and adjust RV1 for zero voltage. With a high impedance meter (or the tuning meter described above) connected to point A, tune a tone input connected on to the terminal unit from mark to space frequency, and adjust RV2 to obtain equal meter indication on both frequencies. For dual shift users of mainly narrow shift, it is recommended that the balance be obtained on the narrow shift tone frequencies and the wide shift frequency of 2125Hz be left out of adjustment if balance cannot be

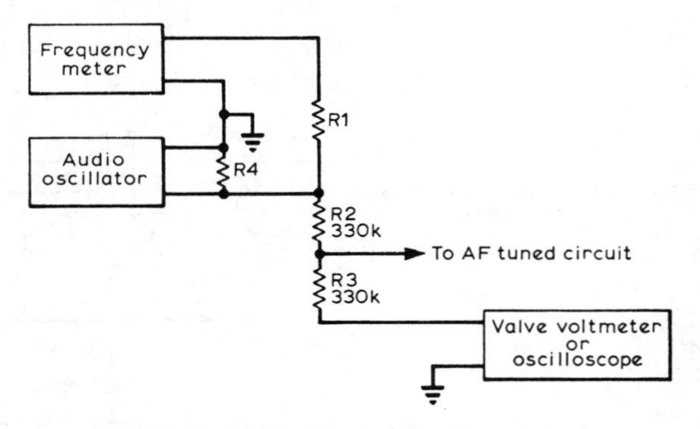

Fig 10.24. Method of alignment for af tuned circuits.

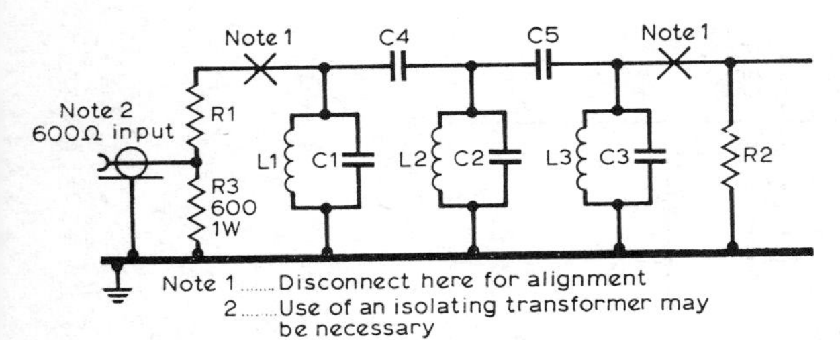

Fig 10.25. Bandpass filter for 170Hz or 850Hz shift.

achieved on all three tones. A compromise may be needed if both types of shifts are regularly used, but the resultant error is usually small.

Because the toroid inductors actually used may not be exactly 88mH, best results will be obtained if C5 and C6 are padded to the correct valve using the test circuit of **Fig 10.24.** The input impedance of the valve voltmeter or scope should be not less than 1MΩ, and the frequency meter or af oscillator calibration should be accurate to within 5Hz.

BANDPASS INPUT FILTERS

The performance of most terminal units can be considerably improved by the inclusion of a bandpass input filter between the receiver and the input to the terminal unit. A design suitable for use with the ST5 terminal on narrow or wide shift is shown in **Fig 10.25,** and the component values are given in **Table 10.3.**

The 850Hz (1275/2125Hz) filter can of course be used for 170Hz shift on 1275/1445Hz as well and, though the results will not be as good as those obtained with the narrower bandwidth filter, they will still be a great improvement. The inductors are the commonly available 88mH toroids, and the two windings are in series for 88mH and in parallel for 22mH. R1/R3 is a matching attenuator and R2 terminates the filter in the correct impedance.

Good-quality polycarbonate or polyester capacitors should be used—ceramic or paper types are not recommended. As with the terminal, provision should be made in the component layout for the addition of small-value padding capacitors during alignment.

The input of the filter should be connected to a 600Ω socket on the receiver. If there is no such output, 35 turns of approximately 32swg enamelled wire should be wound (in a similar manner to the existing windings) onto L1 in Fig 10.25, and connected to the 3Ω loudspeaker output of the receiver.

TABLE 10.3

Component values for narrow and wide shifts in Fig 10.25

Component	Narrow shift (1275/1445Hz)	Wide shift (1275/2125Hz)
L1, 2, 3	22mH	88mH
C1	0·47 + 0·15μF	0·15μF
C2	0·47 + 0·022μF	0·01μF
C3	0·63μF	0·15μF
C4	0·1μF	0·047μF
C5	0·1μF	0·047μF
R1	1·6kΩ	3·3kΩ
R2	2·2kΩ	3·3kΩ

The 680Ω resistor R1 in Fig 10.20 should be removed if the filter is used with the ST5 terminal. If both wide and narrow shift filters are incorporated then a simple switch may be used to change from one to the other by switching both inputs and outputs.

Aligning the Filters

The following procedure can be used to adjust the filters.

1. Disconnect the input and output, and also resistors R1, 2 and 3.
2. Short L2 out temporarily.
3. Adjust the first section to resonate at mid-band (narrow shift) or just above 1275Hz (wide shift).
4. Repeat step 3 for the last section.
5. Adjust the centre section to resonate at mid-band (narrow shift) or just below 2125Hz (wide shift).
7. Remove all short circuits and restore the input and output resistors and wiring.

Alignment may be conveniently carried out by padding the inductor/capacitor combination using small-value capacitors, and the circuit shown in Fig 10.24 can be used for the necessary measurements.

If other terminal units than the ST5 are used, the values of R1 to R3 may need adjustment.

TUNING INDICATORS

Nearly the most important piece of ancillary equipment in the rtty station is the tuning indicator. We have already seen that the station receiver is adjusted so as to translate the incoming rtty signal to audio frequency tones for application to the terminal unit. It is necessary to set the received tones to within 10Hz for narrow shift operation, and presupposing that the resetting accuracy of the receiver is sufficiently good, some convenient means of aligning the receiver output correctly is needed.

A number of methods exist, one of which is the tuning meter already described for use with the ST5. The meter circuit normally gives equal deflection on mark and space, and the receiver is adjusted to give no change in the meter deflection for a keyed signal. The disadvantage of this arrangement is that no direct indication of incorrect shift is available, and most serious rtty operators agree that some form of oscilloscope display is more satisfactory. The simplest arrangement is to apply the mark and space frequencies to the X and Y deflection plates respectively. The oscilloscope amplifier gains are then adjusted to give equal vertical and horizontal deflections. The inputs to the oscilloscope are normally derived from the terminal unit discriminator tuned circuits, or from the mark/space amplifiers in a two-tone or a.m. unit. The receiver tuning is adjusted to give equal vertical and horizontal deflections on a keyed signal. Once again the disadvantage with this form of display is that there is no direct indication of frequency error in the incoming signals. Signals having abnormal shifts cannot be detected except by observing the amplitude of the X and Y displays as compared with a standard signal.

An alternative form of oscilloscope display giving an indication of tone frequencies is the phase shift indicator. A typical circuit is shown in **Fig 10.26.**

The phase-shift monitor scope uses signals taken directly

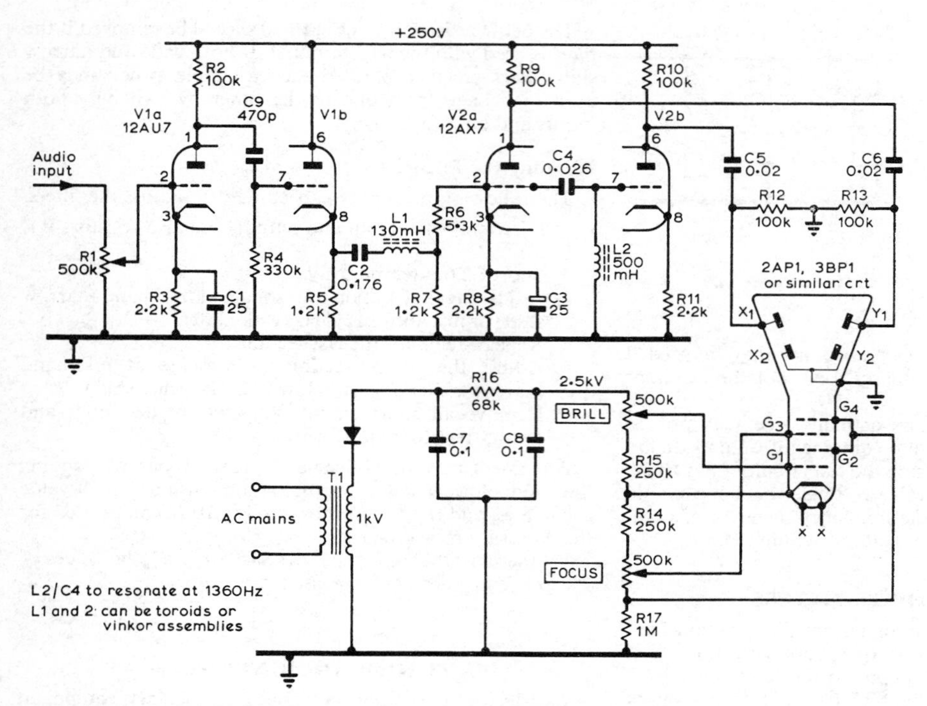

Fig 10.26. Phase shift monitor oscilloscope.

from the receiver output. Where a bandpass filter precedes the terminal unit, the monitor scope feed can with advantage be taken between the filter and the terminal unit.

The basis of the monitor is a series LC circuit resonant at the mean frequency of the rtty tone spectrum—normally 1360Hz (narrow shift).

When two sine waves are applied to the X and Y deflection plates of an oscilloscope the resultant display is dependent upon the phase and amplitude relationship between the sine waves.

As will be seen from Fig 10.26 two signals are derived from the series circuit. The phase/amplitude of the emf across the series circuit is compared with that across the inductor. The resultant display is a straight line rotating about the centre of the tube. The angle of rotation about the datum (equivalent to the mean frequency) is proportional to the frequency difference between the datum and input frequencies. For a keyed rtty signal a cross display is obtained. However, it will be realized that incorrect shifts and similar faults can be observed with this display, and in addition the ability to continuously check the outgoing transmission is very valuable.

For wide shift working it is possible to switch in an alternative circuit series-resonant at 1700Hz, giving a more precise display. With the mixture of shifts currently in use the wide shift network calibrated for use with both 850 and 170Hz shifts is normally adequate. Calibration, accomplished with accurate tone frequencies may be by lines drawn on the face of the crt, or by specially prepared graticule. **Fig 10.27** shows a number of displays commonly obtained.

In use, it is only necessary to align the crt display with the marks on the tube face by means of the receiver tuning and to turn on the printer. Not only is the shift of the incoming signal (or of one's own transmitter) shown instantly, but the correct tuning point also. The angle between the arms of the "X" gives the shift, while the whole display rotates on the tube face according to the correctness of tuning or otherwise. At the same time one can instantly detect from the position of the mark signal whether the station is being received with the correct orientation.

SETTING UP AN RTTY STATION

There are, of course, many different methods of getting going on rtty and undoubtedly the assistance of someone already knowledgeable in the field is of great value. In the nature of things such a person is not always available and it is the purpose of this chapter section to guide the newcomer and to describe a typical simple rtty station.

The first requirement is naturally a teleprinter and the availability of suitable machines varies from country to country. In general it is wise to choose a type of machine which is used in commercial service by some authority in your country. In this way you are more likely to find technicians experienced in teleprinter maintenance within the amateur or allied fraternities. Try also to obtain handbooks and maintenance instructions; these can be invaluable.

A reperforator attachment or separate tape perforator and auto-transmitter are not essential, but desirable if they can be obtained. Be sure that any machines you obtain can be operated at amateur speeds and have standard keyboards. Many special purpose machines, not suitable for amateur working, appear on the surplus market and are often purchased by enthusiastic amateurs unaware of the pitfalls.

When you obtain a suitable machine you will often find it to be in an unserviceable condition and here is where the assistance of a technician is valuable. If no such person is

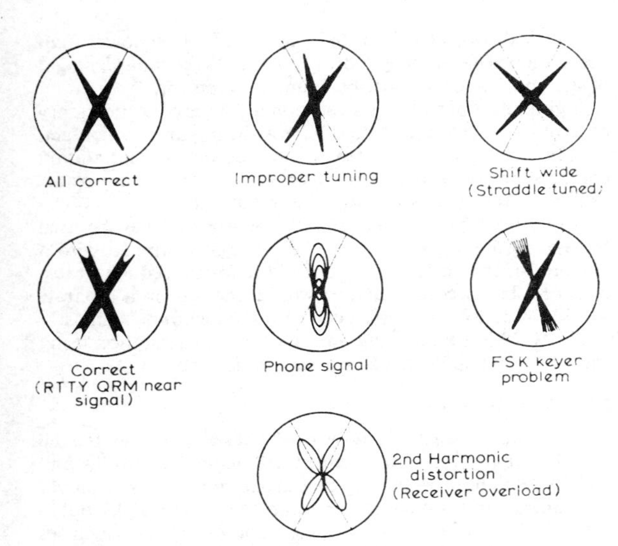

Fig 10.27. Examples of oscilloscope displays.

available then you are recommended to carry out the maintenance procedures detailed in the handbook. Initial efforts should be directed towards obtaining good local loop copy—do not attempt to adjust a machine on the air. The following general comments may be helpful although much depends on the type of machine in use.

Check:

(i) Mains tappings on motor.

(ii) All electrical connections with special attention to plugs and sockets.

(iii) Slip rings on governor for cleanliness—fine grade steel wool may be used to clean if necessary.

(iv) Contacts in governor for cleanliness and correct action—remove pits in contacts and dress up with diamond file. In bad cases it may be necessary to change the contacts.

(v) Governor and motor brushes for excessive wear; replace if necessary.

(vi) Clean motor commutator and check for excessive wear—in bad cases the commutator may require skimming. Remove any accumulation of carbon dust.

(vii) Remove any accumulated dust, paper lint, or excess grease. Excess grease can often be the cause of trouble when bearings become clogged. Methylated spirits will often soften solidified grease sufficiently to allow removal. When the machine has not been used for a lengthy period it may be necessary to repack ball-bearings with fresh grease. It is always worth while carrying out the maker's lubrication instructions.

Following the initial checks and lubrication, turn the motor by hand and ensure that no mechanical obstruction occurs. Manual operation of the magnet to the space condition should enable the receiver clutch and typehead mechanism to be checked. Only when you are satisfied that the machine can be freely turned over by hand should any attempt be made to apply power. Prior to the connection of the machine into a local loop circuit, check whether your machine requires single or double current drive and arrange your circuitry accordingly. Some machines can be set up for single or double current working, and in this case a choice should be made having regard to standardization with other machines in use and other factors.

Motor speed should be carefully set using the procedure given in the machine handbook, preferably using a stroboscopic fork. It is well worth while to carry out some preliminary electrical tests with the object of eliminating telegraph distortion. Unfortunately few amateurs have access to the necessary equipment, but it is possible to make a few simple checks which will eliminate bias distortion. Telegraph relays are a common source of bias distortion, and their adjustment is critical. The mark and space contacts mounted on the tips of screws with fine threads are made adjustable. Ideally they should be adjusted dynamically using a test set as shown in **Fig 10.28**. They can be set out of circuit approximately as follows:

(i) Back off both mark and space contacts until they are clear of the tongue contact.

(ii) Slowly advance one contact until the tongue switches away from it.

(iii) Advance the other contact until the tongue switches away from it.

(iv) Repeat the process, advancing each contact in turn to make the tongue just switch, until the tongue travel is correct for the type of relay.

(v) Fine up the adjustment with the contact screws, backing off one as the other is advanced by the same amount until the tongue remains whichever side it is put. Use a small screwdriver tip to toggle the tongue back and forth, judging that the force required to toggle it is the same for each direction. Telegraph mechanics become highly skilled at this process, but it takes some practice.

The contact travel of a telegraph relay when correctly set should be of the order of 0·002in, and this can be measured with a set of engineer's feeler gauges. Often the contact screws are calibrated around their knobs or heads, relative to the threads. When in good adjustment the relay should change over with 0·5mA flowing through either coil. The contacts will have some form of spring mounting to absorb the impact when the contact changes over, so that the contact will not bounce away.

Transmitter contacts should be mechanically adjusted in accordance with handbook instructions. A further electrical check may be carried out by connecting the teleprinter in a local loop circuit together with a series milliameter. The continuous transmission of the combination RYRYRY

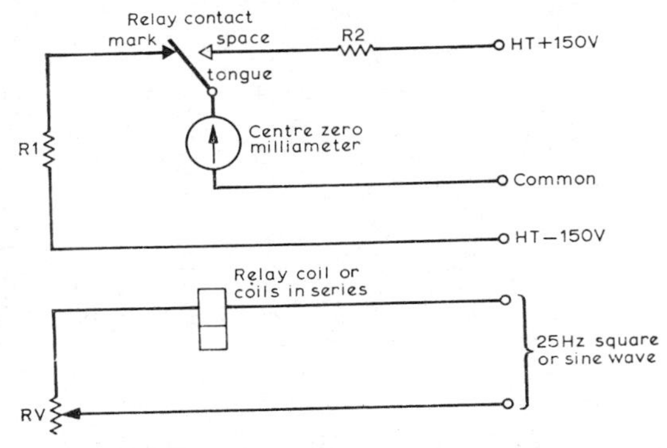

Fig 10.28. Bias test set for polarized telegraph relays. R1 and R2 are selected to give full scale deflection in mark or space direction when the contact is operated by hand. RV is used to set the coil current.

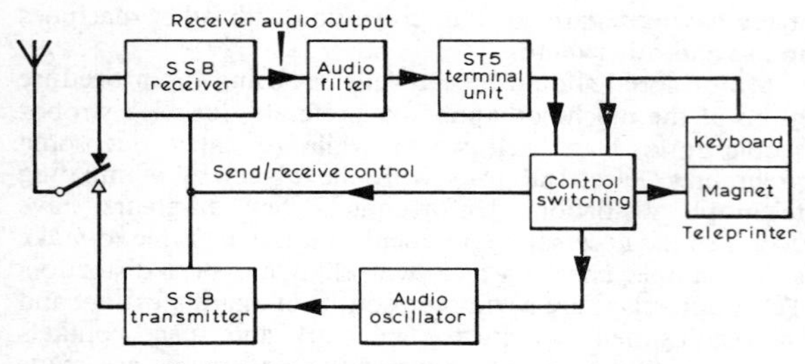

Fig 10.29. Block diagram of a rtty station.

using the keyboard should produce a 1 : 1 mark/space ratio. For a double current loop the meter should indicate zero, and in single current loops half the loop current. The same check can be applied to transmit contacts on an auto-transmitter, in this case a test tape carrying the test combination RYRYRY should be used.

Bias adjustments on a terminal unit can be set up by using a locally keyed signal derived from correctly adjusted transmit contacts.

Following the initial procedures listed above, efforts should be directed towards obtaining satisfactory operation on a local basis—and here the help of a knowledgeable amateur can be very useful. This is particularly the case where the machine requires extensive adjustments. However, the newcomer should not be discouraged because many rtty operators have succeeded by careful study of the handbook and lots of patience.

A word or two on the operating environment might be useful. It should be remembered that a teleprinter is like any other light machinery, and should be kept in warm dry conditions. The machine should always be covered when not in use and a plastic sheet is preferable for this purpose. Dust and/or a humid atmosphere can very quickly ruin any machinery.

Many operators prefer to mount their teleprinter on a trolley rather than occupy valuable bench space. This course of action allows the printer to be mounted a little lower than the standard table height of 30 inches. A keyboard height of around 24 inches will be found convenient.

A good deal of thought on station layout is often necessary to avoid an advanced state of chaos during an rtty contact. It should be possible to reach the receiver and all station controls from the operating position. Where an auto transmitter is used this should also be to hand.

If two teleprinters are available one machine can be used to take incoming copy, while the other, if equipped with a perforator, may be used to punch tape for subsequent transmission. The second machine would operate on a separate local loop to avoid interference with the incoming signals.

Fig 10.29 is a block diagram of a typical rtty station using the ST5 terminal described earlier in this chapter.

RTTY—where to find it

Most amateur rtty activity on hf takes place in the 80 and 20m bands, typical frequencies being 3·590MHz and 14·090MHz, both $\pm$10kHz. In Europe, most activity on vhf is found around 144·6MHz (fsk) and 145·3MHz (afsk), but in the UK 70·56MHz is also utilized. The uhf rtty frequencies are 432·6MHz (fsk) and 433·3MHz (afsk).

When receiving rtty signals for the first time, find a strong amateur rtty signal and make sure that the terminal unit is set to the correct shift. Set up the incoming signal levels and frequencies by adjustment of the receiver with reference to the display on the monitor scope, and there should then be no trouble in copying the signal.

References

Further and more detailed information on rtty can be found in the following publications:

The Teleprinter Handbook, D. J. Goacher, G3LLZ, and J. G. Denny, G3NNT, Radio Society of Great Britain.

Telegraphy, J. W. Freebody, Sir Isaac Pitman & Son, London.

In addition most English and USA amateur radio journals publish regular information on developments in the rtty field.

CHAPTER 11

PROPAGATION

ANY reader who consults this chapter in the belief that it contains the key to all his propagation problems is due for a disappointment, because the hard truth is that few such problems can be solved precisely, even in retrospect, for the simple reason that there are too many unknown variables involved. Whereas we know, for example, the relationship between the voltage across a carbon resistor and the current flowing through it and can draw a graph through just two points which will save us the trouble of having to work out any of the others in between, a line, such as that which appears to relate the field strength of a 300MHz signal with distance from the transmitter to a similar degree of precision has, in all probability, come about as a best-fit curve through a broad spread of points measured over quite a considerable period.

Thus many of the graphs which follow must be regarded as displaying statistical trends rather than hard-and-fast equalities, and they can no more predict an isolated S-meter reading than can a knowledge of the mean monthly temperature for August tell one whether or not to wear a jacket to the office sports.

With that reservation it is the purpose of this chapter to describe some of the various processes which play a part in determining the progress of a radio wave from the transmitting aerial to the receiver. The object has been to treat the material as a single topic, with no clear-cut divisions for uhf, vhf, hf, etc, because there are no definite limits to most of the issues involved. In the space available there is room only to indicate the broad outlines of a very large subject.

The RSGB, in common with a number of other national societies, conducts a continuing programme of research into certain aspects of a more specialized nature than can be dealt with here and, in that connection, it should be noted that the field of radio propagation is one in which radio amateurs can still usefully contribute something to complement the work of professionals.

A study of radio propagation should properly begin with a section about the sun, for it is there that most of the direct influences on our signal paths have their origin.

But to start there in this account would mean trying to explain the relationships between causes and effects without first defining the nature of the effects.

Therefore it seems preferable to examine first the make-up of our earth's atmosphere where the action takes place, as much to introduce and define the regions of interest as to show the scale on which they are arranged, and then to review certain fundamental considerations and the various modes by which radio waves may be propagated before attempting to describe the sun and its radiations, and the part they play in determining the properties of the medium through which radio waves pass.

The Earth's Atmosphere

There is a certain amount of confusion surrounding any attempt to identify various portions of the earth's atmosphere, and this has come about as a result of there being no obvious natural boundaries, as there are between land and sea. Workers in different disciplines have different ideas about which functions ought to be separated and it is particularly awkward for the radio engineer that he has to deal with the troposphere and stratosphere, which are terms from one set of divisions, and the ionosphere, which is from another.

The diagram **Fig 11.1** shows the nomenclature favoured by meteorologists and now widely accepted among physicists generally. It is based on temperature variations, as might be

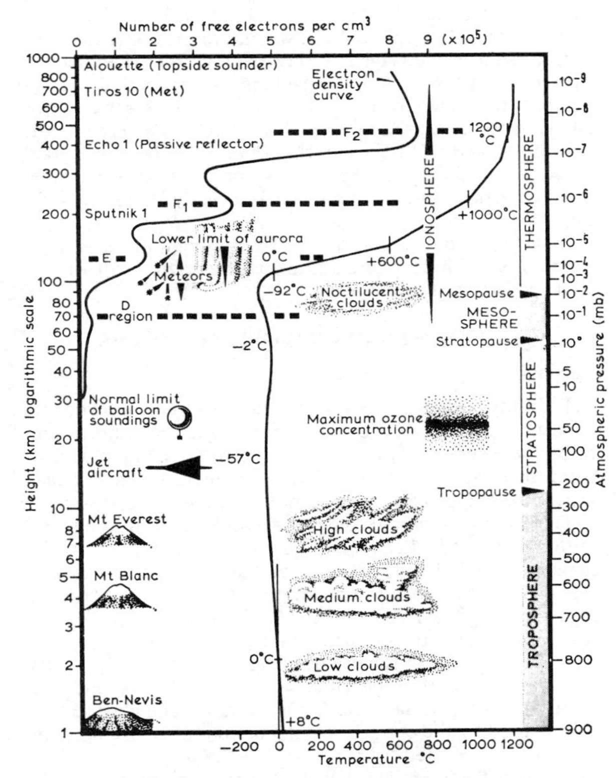

Fig 11.1. Some features of the earth's atmosphere. The height scale is logarithmic beginning at 1km above sea level. The equivalent pressure scale on the right is not regularly spaced because the relationship between pressure and height depends on temperature, which does not change uniformly with height.

expected from its origin. (Note, incidentally, that the height scale of kilometres on the diagram is a logarithmic one).

In the *troposphere*, the part of the atmosphere nearest the ground, temperature tends to fall off with height. At the *tropopause*, around 10km, (although all these heights vary from day to day and from place to place), it becomes fairly uniform at first and then begins to increase again in the region known as the *stratosphere*. This trend reverses at the *stratopause*, at an altitude of about 50km, to reach another minimum at the *mesopause* (ca 80km) after traversing the *mesosphere*. Above the mesopause temperatures begin to rise again in the *thermosphere*, soon surpassing anything encountered at lower altitudes, and levelling off at about +1200° C around 700km, where we must leave it in this survey.

The *ionosphere*, which has been defined as "The region above the earth's surface in which ionisation takes place, with diurnal and annual variations which are regularly associated with ultra-violet radiation from the sun, and sporadic variations arising from hydrogen bursts from sunspots" (Chambers' Technical Dictionary), overlaps the thermosphere, mesosphere and part of the stratosphere, but for practical purposes may be considered as lying between 60 and 700km. The name ionosphere is perhaps misleading because it is the number of free electrons, rather than the ions they have left behind, which principally determines the electrical properties of the region. The electron density curve in the diagram shows a number of "ledges", identified by the letters D, E, F1 and F2, which are the concentrations of free electrons described as "layers" later on, when dealing with propagation in the ionosphere. They tend to act as mirrors to transmissions of certain wavelengths, while allowing others to pass through. The lowest ledge is generally referred to as the D-region, rather than the D-layer, because, as we shall see, its principal role is one of absorption rather than reflection, and its presence is usually easier to infer than it is to observe.

It may be found helpful to refer to this diagram again when features of the ionosphere and troposphere are dealt with in later sections of this chapter.

FUNDAMENTAL CONSIDERATIONS

Radiation

The transmitted signal may be regarded as a succession of concentric spheres of ever-increasing radius, each one a unit of one wavelength apart, formed by forces moving outwards from the aerial. These hypothetical spherical surfaces, called *wave-fronts*, approximate to plane surfaces at great distances.

There are two inseparable fields associated with the transmitted signal, an *electric field* due to voltage changes and a magnetic field due to current changes, and these always remain at right-angles to one another and to the direction of propagation as the wave proceeds. They always oscillate in phase and the ratio of their amplitudes remains constant. The lines of force in the electric field run in the plane of the transmitting aerial in the same way as would longitude lines on a globe having the aerial along its axis. The electric field is measured by the change of potential per unit distance, and this value is referred to as the *field strength*.

The two fields are constantly changing in magnitude and reverse in direction with every half-cycle of the transmitted carrier. As shown in **Fig 11.2**, successive wave-fronts passing a suitably-placed second aerial induce in it a received signal

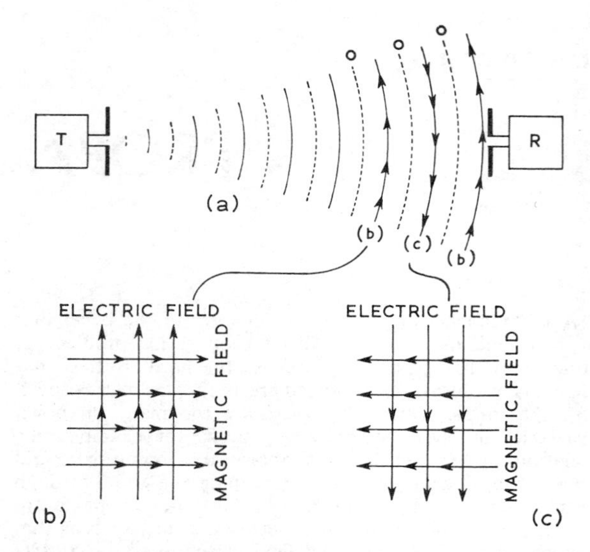

Fig 11.2. The fields radiated from a transmitting aerial. (a) The expanding spherical wavefront consists of alternate reversals of electric field, with which are associated simultaneous reversals of the magnetic field at right angles to it, as shown in (b) and (c). The dotted arcs represent nulls. The lower diagrams should be interpreted as though they have been rotated through 90° of arc, so that the magnetic field lines are perpendicular to the page.

which follows all the changes carried by the field and therefore reproduces the character of the transmitted signal.

By convention the direction of the electric lines of force defines the direction of *polarization* of the radio waves. Thus horizontal dipoles propagate horizontally polarized waves and vertical dipoles propagate vertically polarized waves. In free space, remote from ground effects and the influence of the earth's atmosphere, these senses remain constant and a suitably aligned receiving aerial would respond to the whole of the incident field. When the advancing wave-front encounters the surface of the earth or becomes deflected by certain layers in the atmosphere a degree of cross-polarization may be introduced which results in signals arriving at the receiving aerial with both horizontal and vertical components present. Circularly polarized signals, which contain equal components of both horizontal and vertical polarization, are receivable on dipoles having any alignment in the plane of the wavefront, but the magnitude of the received signal will be only half that which would result from the use of an aerial correctly designed for such a form of polarization (it must not be overlooked that there are two forms, differing only in their direction of rotation). Matters such as these are dealt with in detail in Chapter 13—*VHF/UHF aerials.*

Field Strength

As the energy in the expanding wave-front has to cover an ever-increasing area the further it travels, the amplitude of the signal induced in the receiving aerial diminishes as a function of distance. Under "free-space" conditions an inverse square-law relationship applies, but in most cases the nature of the intervening medium has a profound, and often very variable, effect on the magnitude of the received signal.

The intensity of a radio wave at any point in space may be

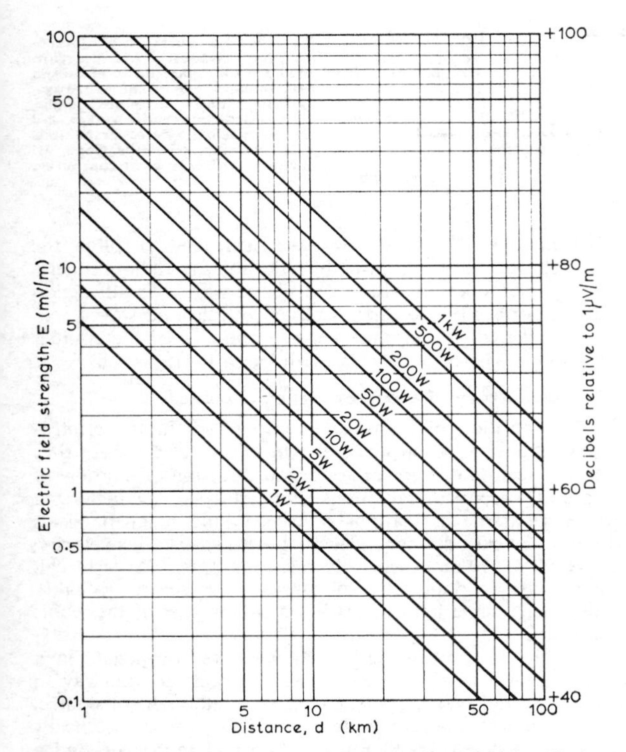

Fig 11.3. Field strength from an omnidirectional aerial radiating from 1kW to 1W into free space. This is an application of the "inverse distance" law.

expressed in terms of the strength of its electric field at right-angles to the line of the transmission path. The units used indicate the difference of electric force between two points one metre apart, and **Fig 11.3** (which should not be used for general calculations as it relates only to free-space conditions) has been included here to give an indication of the magnitudes of fields likely to be met with in the amateur service in cases where a more-or-less direct path exists between transmitting and receiving aerials and all other considerations may be ignored. In practice signal levels very much less than the free-space values may be expected because of various losses en route, so that field strengths down to about 1μV/m need to be considered.

Signal Input

In many calculations dealing with propagation the parameter of interest is not the field strength at a particular place but the voltage which is induced by it across the input of a receiver. If a half-wave dipole is introduced into a field and aligned for maximum signal pick-up, the open-circuit emf induced at its centre is given by the expression:

$$e = \frac{E\lambda}{\pi} \qquad \ldots (1)$$

where e = emf at the centre of the dipole, in volts
E = the incident field strength in volts/metre
λ = the wavelength of the transmitted signal in metres.

When connected to a matched feeder correctly terminated at the receiver the input voltage available will be half this, or e/2. By substituting frequency for wavelength the equation may be reduced to the more practical form:

$$V_r = \frac{47{\cdot}8}{f} \,.\, E \qquad \ldots (2)$$

where V_r = microvolts of signal across the receiver input
f = frequency of the transmitted signal in MHz
E = the incident field strength in μV/m.

It should be noted that for a given frequency the first term becomes a constant factor.

By a further rearrangement

$$E = \frac{f}{47{\cdot}8} \,.\, V_r \qquad \ldots (3)$$

which enables the field to be estimated from the magnitude of the received signal in cases where the receiver is fed from a perfectly matched half-wave dipole.

It is an advantage to work in decibels in calculations of this nature. If a standard level of 1μV/m is adopted for field strength and of 1μV for signal level it is a simple matter to take into account the various gains and losses in a practical receiving system. **Fig 11.4** shows the terminated voltage at the receiver (V_o) in terms of decibels relative to 1μV for a normalized incident field strength of 1μV/m at the transmitter frequency, derived from expression (2). Then,

$$V_r(dB) = V_o(dB) + V_i(dB) + G_r(dB) - L_{fr}(dB) \quad (4)$$

where $V_r(dB)$ = input to receiver in dB relative to 1μV
$V_o(dB)$ = input in dB relative to 1μV, for 1μV/m incident field strength (from graph)
$V_i(dB)$ = the actual incident field strength in dB relative to 1μV/m
$G_r(dB)$ = the gain in dB of the receiving aerial relative to a half-wave dipole
$L_{fr}(dB)$ = loss in the feeder line between aerial and receiver.

Example. The incident field strength of a 70MHz transmission is 100μV/m (or 40dB above 1μV/m). A 3-element Yagi having a gain of 5dB over a half-wave dipole is connected to the receiver through 100ft of coaxial cable which introduces a loss of 2dB. From this, $V_o = -3{\cdot}5$, $V_i = +40$, $G_r = +5$ and $L_{fr} = +2$ (the fact that L_{fr} is a loss is allowed for in Eqn (4)), and $V_r = -3{\cdot}5 + 40 + 5 - 2$dB
$= 39{\cdot}5$dB relative to 1μV,
or a receiver input voltage of 94μV.

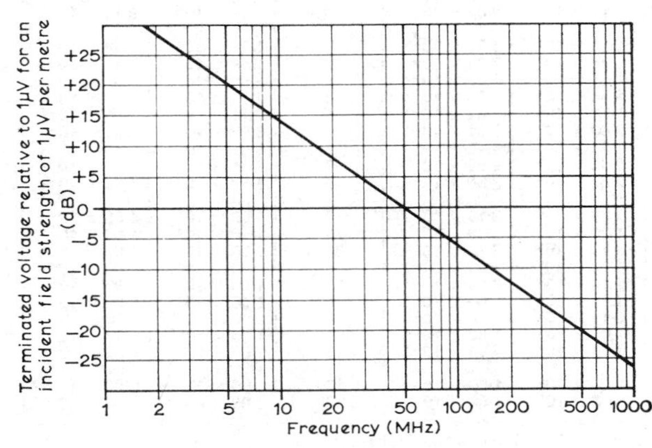

Fig 11.4. The relationship between the incident field strength on a half-wave dipole and the voltage at the receiver end of a correctly matched perfect feeder connected to it.

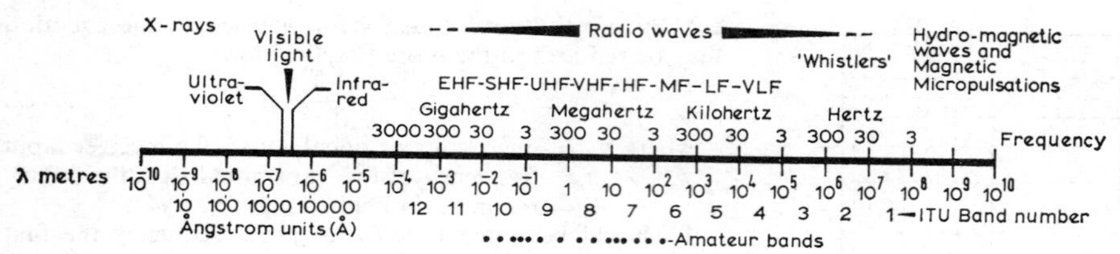

Fig 11.5. The spectrum of electromagnetic waves. This diagram shows on a logarithmic scale the relationship between X-rays, "visible" and "invisible" light, heat (infra-red), radio waves and the very slow waves associated with geomagnetic pulsations, all of them similar in basic character.

Table 11.1, a skeleton decibel table to slide rule accuracy, will be found useful in converting the rather unfamiliar looking values of decibels met with in radio propagation work into their corresponding voltage or power ratios.

MODES OF PROPAGATION

Introduction

There are four principal modes by which radio waves are propagated. They are:

Free-space waves, which are unaffected by any consideration other than distance,

Ionospheric waves, which are influenced by the action of free electrons in the upper levels of the earth's atmosphere,

Tropospheric waves, which are subject to deflection in the lower levels by variations in the refractive index structure of the air through which they pass, and

Ground waves, which are modified by the nature of the terrain over which they travel.

Free-space waves propagate from point-to-point by the most direct path. Waves in the other three categories are influenced by factors which make them tend to follow the curvature of the earth, either by reflection as with ionospheric waves, refraction as with tropospheric waves, or diffraction at the surface of the earth itself as with ground waves.

Wavelength is the chief consideration which determines the mode of propagation of earth-based transmissions.

TABLE 11.1

Skeleton decibel table to slide-rule accuracy

Combine by multiplication — Voltage ratios — up	down	Combine by addition — dB	Combine by multiplication — Power ratios — up	down
1.01×10^{0}	9.89×10^{-1}	0.1	1.02×10^{0}	9.77×10^{-1}
1.02×10^{0}	9.77×10^{-1}	0.2	1.05×10^{0}	9.55×10^{-1}
1.03×10^{0}	9.66×10^{-1}	0.3	1.07×10^{0}	9.33×10^{-1}
1.05×10^{0}	9.55×10^{-1}	0.4	1.10×10^{0}	9.12×10^{-1}
1.06×10^{0}	9.44×10^{-1}	0.5	1.12×10^{0}	8.91×10^{-1}
1.07×10^{0}	9.33×10^{-1}	0.6	1.15×10^{0}	8.71×10^{-1}
1.08×10^{0}	9.23×10^{-1}	0.7	1.17×10^{0}	8.51×10^{-1}
1.10×10^{0}	9.12×10^{-1}	0.8	1.20×10^{0}	8.32×10^{-1}
1.11×10^{0}	9.02×10^{-1}	0.9	1.23×10^{0}	8.13×10^{-1}
1.12×10^{0}	8.91×10^{-1}	1	1.26×10^{0}	7.94×10^{-1}
1.26×10^{0}	7.94×10^{-1}	2	1.58×10^{0}	6.31×10^{-1}
1.41×10^{0}	7.08×10^{-1}	3	2.00×10^{0}	5.01×10^{-1}
1.58×10^{0}	6.31×10^{-1}	4	2.51×10^{0}	3.98×10^{-1}
1.78×10^{0}	5.62×10^{-1}	5	3.16×10^{0}	3.16×10^{-1}
2.00×10^{0}	5.01×10^{-1}	6	3.98×10^{0}	2.51×10^{-1}
2.24×10^{0}	4.47×10^{-1}	7	5.01×10^{0}	1.99×10^{-1}
2.51×10^{0}	3.98×10^{-1}	8	6.31×10^{0}	1.59×10^{-1}
2.82×10^{0}	3.55×10^{-1}	9	7.94×10^{0}	1.26×10^{-1}
3.16×10^{0}	3.16×10^{-1}	10	1.00×10^{1}	1.00×10^{-1}
1.00×10^{1}	1.00×10^{-1}	20	1.00×10^{2}	1.00×10^{-2}
3.16×10^{1}	3.16×10^{-2}	30	1.00×10^{3}	1.00×10^{-3}
1.00×10^{2}	1.00×10^{-2}	40	1.00×10^{4}	1.00×10^{-4}
3.16×10^{2}	3.16×10^{-3}	50	1.00×10^{5}	1.00×10^{-5}
1.00×10^{3}	1.00×10^{-3}	60	1.00×10^{6}	1.00×10^{-6}
3.16×10^{3}	3.16×10^{-4}	70	1.00×10^{7}	1.00×10^{-7}
1.00×10^{4}	1.00×10^{-4}	80	1.00×10^{8}	1.00×10^{-8}
3.16×10^{4}	3.16×10^{-5}	90	1.00×10^{9}	1.00×10^{-9}
1.00×10^{5}	1.00×10^{-5}	100	1.00×10^{10}	1.00×10^{-10}
1.00×10^{10}	1.00×10^{-10}	200	1.00×10^{20}	1.00×10^{-20}

Example: 39·5dB above 1μV (voltage ratio).
39·5dB = 30 + 9 + 0·5dB
Combining equivalents by multiplication
$= (3.16 \times 10^{1}) \times (2.82 \times 10^{0}) \times (1.06 \times 10^{0})$
= 94 times 1μV, or 94μV.

The Spectrum of Electromagnetic Waves

The position of man-made radio waves in the electromagnetic wave spectrum is shown in **Fig 11.5**, where they can be seen to occupy an appreciable portion of a family of naturally-occurring radiations, all of which are characterized by inseparable oscillations of electric and magnetic fields and travel with the same velocity in free space. This velocity, 2.99790×10^{8}m/s (generally taken as 3×10^{8}m/s in calculations) is popularly known as "the speed of light" although visible light forms but a minor part of the whole range.

At the long wavelength end the waves propagate in a manner which is similar in many respects to the way in which sound waves propagate in air, although, of course, the actual mechanism is different. Thus, reports of heavy gunfire in the 1914–18 war at abnormal ranges beyond a zone of inaudibility revealed the presence of a sonic sky-wave which had been reflected by the thermal structure of the atmosphere around 30km in height, and this has a parallel in the reflection of long wavelength radio sky-waves by the atomic structure of the atmosphere around 100km in height, which also leads to a zone of inaudibility at medium ranges.

Radio waves at the other end of the spectrum show characteristics which are shared by the propagation of light waves, from which they differ only in wavelength. For example, millimetre waves, which represent the present frontier of practical technology, suffer attenuation due to scattering and absorption by clouds, fog and water droplets in the atmosphere—the same factors which determine "visibility" in the meteorological sense.

The radio wave portion of the spectrum has been divided by the International Telecommunication Union into a series of bands based on successive orders of magnitude in wavelength.

In **Table 11.2** an attempt has been made to outline the principal propagation characteristics of each band, but it must be appreciated that there are no clear-cut boundaries to the various effects described.

Wave Propagation in Free-Space

The concept of "free-space" propagation, of a transmitter radiating without restraint into an infinite empty surrounding space, has been introduced briefly in the section on field strength where it was used to illustrate, in a general way, the relationship between the strength of the field due to a transmitter and the distance over which the waves have travelled.

TABLE 11.2

A survey of the radio frequency spectrum

ITU band No	Metric name of band and limits by wavelength	Alternative name of band and limits by frequency	UK amateur bands by frequency (and usual description based on wavelength)	Principal propagation modes	Principal limitations
4	Myriametric 100,000–10,000m	Very low frequency (vlf) 3–30kHz	—	Extensive surface wave. Ground – ionosphere space acts as a waveguide	Very high power and very large aerials required. Few channels available
5	Kilometric 10,000–1,000m	Low frequency (lf) 30–300kHz	—	Surface wave and reflections from lower ionosphere	High power and large aerials required. Limited number of channels available. Subject to fading where surface wave and sky wave mix
6	Hectometric 1,000–100m	Medium frequency (mf) 300–3,000kHz	1,800–2,000kHz (160m band) (also known as "Top band")	Surface wave only during daylight. At night reflection from decaying E-layer	Strong D – region absorption during day. Long ranges possible at night but signals subject to fading and considerable mutual interference
7	Decametric 100m–10m	High frequency (hf) 3–30MHz	3·5–3·8MHz (80m), 7·00–7·10MHz (40m), 14·00–14·35MHz (20m), 21·00–21·45MHz (15m), 28·00–29·70MHz (10m)	Short-distance working via E-layer. Nearly all long-distance working via F2-layer	Daytime attenuation by D-region, E and F1-layer absorption. Signal strength subject to diurnal, seasonal, solar-cycle and irregular changes
8	Metric 10m–1m	Very high frequency (vhf) 30–300MHz	70·025–70·7MHz (4m) 144–146MHz (2m)	F2 occasionally at lf end of band around sunspot maximum. Irregularly by sporadic-E and auroral-E. Otherwise maximum range determined by temperature and humidity structure of lower troposphere	Ranges generally only just beyond the horizon
9	Decimetric 1m–10cm	Ultra high frequency (uhf) 300–3,000MHz	430–440MHz (70cm), 1,215–1,325MHz (23cm) 2,300–2,450MHz (13cm)	Line-of-sight, modified by tropospheric effects	Atmospheric absorption effects noticeable at top of band
10	Centimetric 10cm–1cm	Super high frequency (shf) 3–30GHz	3·400–3·475GHz (9cm) 5·650–5·850GHz (6cm) 10·000–10·500GHz (3cm) 24·000–24·250GHz (12mm)	Line-of-sight	Attenuation due to oxygen, water vapour and precipitation becomes increasingly important
11	Millimetric 1cm–1mm	Extra high frequency (ehf) 30–300GHz	—	Line-of-sight	Atmospheric propagation losses create pass and stop bands. Background noise sets a threshold
12	Decimilimetric (sub-millimetric) 1mm–0·1mm	— 300–3,000GHz		Line-of-sight	Present limit of technology

It is only recently, with the advent of the Space Age, that we have acquired a practical opportunity to operate long-distance circuits under true free-space conditions as, for example, between space-craft and orbiting satellites, and it may be some considerable time before radio amateurs have direct access to paths of that nature. Earth-moon-earth contacts are becoming increasingly popular, however, and reception of satellite signals commonplace, and for these the "free-space" calculations apply with only relatively minor adjustments because such a large part of the transmission paths involved lie beyond the reach of terrestrial influences.

In many cases, and especially where wavelengths of less than about 10 metres are concerned, the "free-space" calculations are even applied to paths which are subject to relatively unpredictable perturbations in order to estimate a convenient (and often unobtainable) ideal—a bogey for the path—against which the other losses may be compared.

The basic transmission loss in free space is given by the expressions:

$$L_{bf} = 32{\cdot}45 + 20 \log f(\text{MHz}) + 20 \log r(\text{km}) \quad (5)$$

$$L_{bf} = 36{\cdot}6 + 20 \log f(\text{MHz}) + 20 \log r(\text{miles}) \quad (6)$$

where r is the straight line distance involved.

If transmitter and receiver levels are expressed in either dBW (relative to 1 watt) or dBm (relative to 1 milliwatt)—it does not matter which providing that the same units are used for both—with other relevant parameters similarly given in terms of decibels it is a relatively simple matter to determine the received power at any distance by adding to the transmitter level all the appropriate gains and subtracting all the losses.

Thus:

$$P_r(\text{dBm}) = P_t(\text{dBm}) + G_t(\text{dB}) - L_{ft}(\text{dB}) - L_{bf}(\text{dB}) + G_r(\text{dB}) - L_{fr}(\text{dB})$$

where P_r = Received power level (dBm or dBW).
P_t = Transmitted power level (dBm or dBW)
G_t = Gain of the transmitting aerial in the direction of the path, relative to an isotropic radiator
L_{fr} = Transmitting feeder loss
L_{bf} = Free-space transmission loss
G_r = Gain of the receiving aerial in the direction of the path, relative to an isotropic radiator
L_{fr} = Receiving feeder loss

The free-space transmission loss may be estimated approximately from **Fig 11.6** which perhaps conveys a better idea of its relationship to frequency (or wavelength) and distance than the nomogram generally provided. Unless a large number of calculations have to be made it is no great hardship to use the formula for individual cases, should greater accuracy be desired. It should be noted that aerial gains quoted with respect to a half-wave dipole need to be increased by 2dB to express them relative to an isotropic radiator.

Example. A 70MHz transmitter radiates 100W erp in the direction of a receiver 550km away. The receiving aerial has

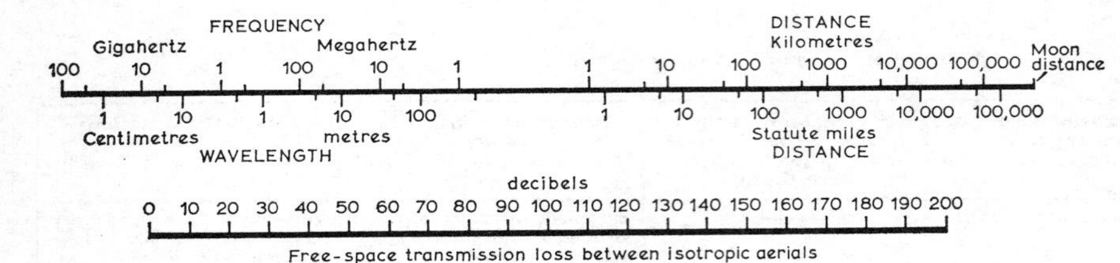

Fig 11.6. The effects of frequency (or wavelength) and distance on the free-space transmission loss between isotropic aerials. The length on the diagram between points corresponding to frequency and distance on the upper scale corresponds to the free-space transmission loss measured from the zero decibel mark on the lower scale.

a gain of 5dB over a half-wave dipole (2dB more over an isotropic radiator), and there is a 2dB loss in the feeder. In this case the effective radiated power is known, which takes the place of the terms P_t, G_t and L_{ft}.

$P_{erp} = 100W = 10^5mW = 50dBm$
$L_{bf} =$ 124 from Fig 11.6 or by use of the expressions (5) or (6)
$G_r =$ 7dB over an isotropic radiator
$L_{fr} = 2dB$

Then $P_t = P_{erp} - N + G_r - L_{fr}$
$= 50 - 124 + 7 - 2$
$= -69$dBm, or 69dB below 1mW
$= 12{\cdot}6 \times 10^{-6}$mW

If this power is dissipated in an input impedance of 70 ohms the voltage appearing across the receiver V_r is $\sqrt{P_r Z_{in}}$, in this case $\sqrt{12{\cdot}6 \times 10^{-9} \times 70}$ which is 94μV. This example deals with the same situation which was considered earlier in connection with field strength and offers an alternative method of calculating signal input.

Wave Propagation in the Ionosphere

It has been shown that a transmitted signal may be considered as consisting of a succession of spherical wave fronts, each one one wavelength apart, which approximate to plane surfaces at great distances. At certain heights in the upper atmosphere concentrations of negatively-charged free electrons occur, and these are set into oscillatory motion by the oncoming waves, which causes them to emit secondary wavelets having a phase which is 90 degrees in advance of the main wave. It is only in the forward direction that the original waves and their dependent wavelets combine coherently and their resultant consists of a wave in which the maxima and minima occur earlier than in the projection of the originating wave—to all intents and purposes the equivalent of the wave having travelled faster in order to arrive earlier. The amount of phase advancement is a function of the concentration of electrons and the change of speed is greatest at long wavelengths, decreasing therefore as the signal frequency increases.

The advancing wave-front, travelling, let us say, obliquely upwards from the ground, meets the layer containing the accumulation of free electrons in such a way that its upper portion passes through a greater concentration of charge than does a portion lower down. The top of the wave-front is therefore accelerated to a greater extent by the process just outlined than are the parts immediately below, which results in a gradual swing-round until the wave-front is being returned towards the ground as though it had experienced a reflection.

The nearer the wave-front is to being vertically above the transmitter the more quickly must the top accelerate relative to the bottom, and the more concentrated must be the charge of electrons. It may be that the density of electrons is sufficiently high to turn even wave-fronts propagating vertically upwards (a condition known as *vertical incidence*), although it must be appreciated that deeper penetration into the layer is required as the propagation angle becomes steeper.

Consider the circumstances outlined in **Fig 11.7**, where T indicates the site of a transmitting station and R1, R2 and R3 three receiving sites. For a given electron concentration there is a *critical frequency*, *fo*, which is the highest to return from radiation directly vertically upward. Frequencies higher than this will penetrate the layer completely and be lost in space, but their reflection may still be possible at *oblique incidence* where waves have to travel a greater distance within the electron concentration. This is not the case at point A in the diagram, so that reception by sky-wave is impossible at R under these circumstances, but at a certain angle of incidence to the layer (as at point B) the ray bending becomes just sufficient to return signals to the ground, making R2 the nearest location relative to the transmitter at which the sky-wave could be received. The range over which no signals are possible via the ionosphere is known as the *skip distance*, and the roughly circular area described by it is called the *skip zone*. Lower angle radiation results in longer ranges, as to point R3 from a reflection at C, until further extensions are prevented by the curvature of the earth.

For oblique incidence on a particular path (eg the ray from T to R3 via point C in the ionosphere) there is a *maximum usable frequency* (*muf*), generally much higher than the critical frequency was at vertical incidence, which is approximately equal to $f_o/\cos\phi$, where ϕ is the angle of incidence of the ray to the point of reflection, as is shown in the figure. The limiting angle which defines the point at which reflections first become possible is called the *critical wave angle*, and it is this function which determines the extent of the skip zone. The relationship between the two is shown in **Fig 11.8** where the two curves relate to ionospheric layers at approximately 120km and 400km above the ground.

This mechanism is effective for signals in excess of about 100kHz, for which the concentration of electrons appears as a succession of layers of increasing electron density having

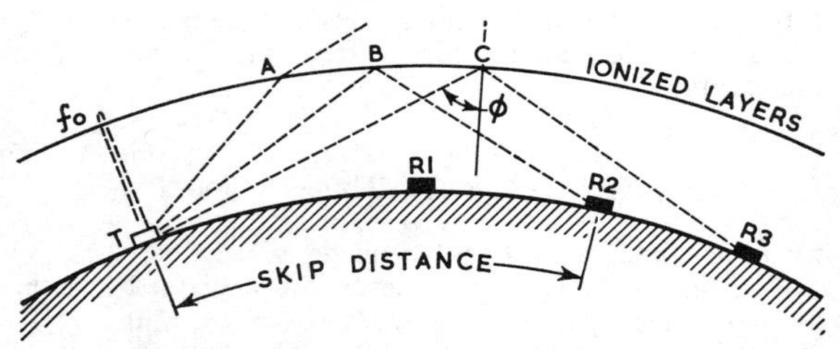

Fig 11.7. Wave propagation via the ionosphere. T is the site of a transmitter, R1, R2 and R3 are the sites of three receivers. The significance of the ray at vertical incidence (dotted) and the three oblique rays is explained in the text.

Fig 11.8. Relationship between angles of take-off and resulting path-length. The two ionospheric layers are presumed to be at heights of 120km and 400km respectively.

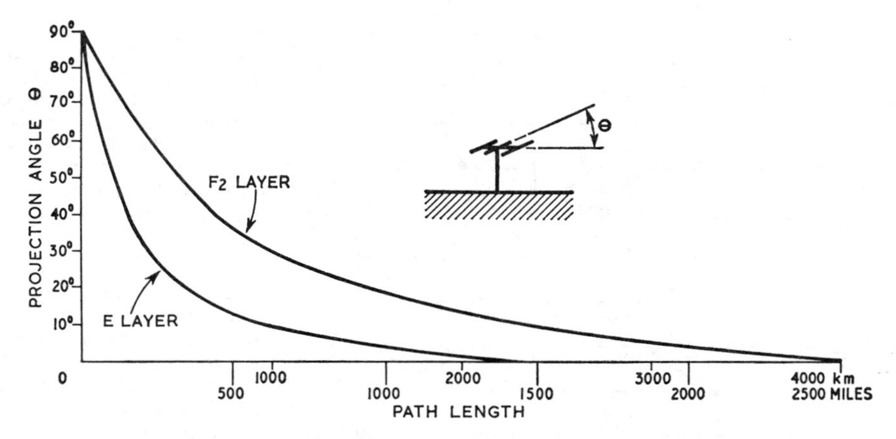

the effect of progressively bending the rays as the region is penetrated.

Below 100kHz the change in concentration occurs within a distance which is small compared to the wavelength and which therefore appears as an almost perfect reflector. Waves are propagated in that way over great distances by virtue of being confined between two concentric spheres, one being the lower edge of the layer and the other the surface of the earth.

During the hours of daylight the quantity of free electrons in the lower ionosphere becomes so great that the oscillations set up by incident waves are heavily damped on account of energy lost by frequent collisions with the surrounding neutral air particles. Medium wave broadcast band signals are so much affected by this as to have their sky-waves completely absorbed during the day, leading to the familiar rapid weakening of distant stations around dawn and their subsequent disappearance at the very time when the reflecting layers might otherwise be expected to be most effective. The shorter wavelengths are less severely affected, but suffer attenuation nevertheless.

A certain amount of cross-polarization occurs when ionospheric reflections take place so that the received signals generally contain a mixture of both horizontal and vertical components irrespective of which predominated at the transmitting aerial.

Ionospheric scatter propagation does not make use of the regular layers of increased electron density. Instead forward scattering takes place from small irregularities in the ionosphere comparable in size with the wavelength in use (generally around 8m or about 35MHz). With high powers and very low angles of radiation, paths of some 2,000km are possible and this mode has the advantage of being workable in auroral regions where conventional hf methods are often unreliable, but only a very small proportion of the transmitted power is able to find its way in the desired direction.

Wave Propagation in the Troposphere

The *troposphere* is that lower portion of the atmosphere in which the general tendency is for air temperature to decrease with height. It is separated from the *stratosphere*, the region immediately above, where the air temperature tends to remain invariant with height, by a boundary called the *tropopause*, at around 10km. The troposphere contains all the well-known cloud forms and is responsible for nearly everything loosely grouped under the general heading of "weather".

Its effect upon radio waves is to bend them, generally in the same direction as that taken by the earth's curvature, not by encounters with free electrons or layers of ionization, but as a result of successive changes in the refractive index of the air through which the waves pass. In optical terms this is the mechanism responsible for the appearance of mirages, where objects beyond the horizon are brought into view by ray-bending, in that case resulting from temperature changes along the line-of-sight path. In the case of radio signals the distribution of water-vapour also plays a part, often a major one where anomalous propagation events are concerned.

The refractive index, n, of a sample of air can be found from the expression $n = 1 + 10^{-6}N$, where N is the refractivity expressed by:

$$N = \frac{77{\cdot}6P}{T} + \frac{4810\,e}{T^2}$$

P being the atmospheric pressure in millibars
e the water vapour-pressure, also in millibars, and
T the temperature in degrees absolute.

Substituting successive values of P, e and T from a standard atmosphere table shows that there is a tendency for refractive index to decrease with height, from a value just above unity at the earth's surface to unity itself in free space. This gradient is sufficient to create a condition in which rays are normally bent down towards the earth, leading to a similar state of affairs as that which would result from the radio horizon being extended to beyond the optical line-of-sight limit by an average of about 15 per cent.

The distance, d, to the horizon from an aerial of height h is approximately $\sqrt{2ah}$, where a is the radius of the earth and h is small compared with it. The effect of refraction can be allowed for by increasing the true value of the earth's radius until the ray paths, curved by the refractive index gradient, become straight again. This modified radius a′ can be found from the expression:

$$\frac{1}{a'} = \frac{1}{a} + \frac{dn}{dh}$$

where dn/dh is the rate of decrease of refractive index with height. The ratio a′/a is known as the *effective earth radius factor*, *k*, so that the distance to the radio horizon becomes $d = \sqrt{2kah}$.

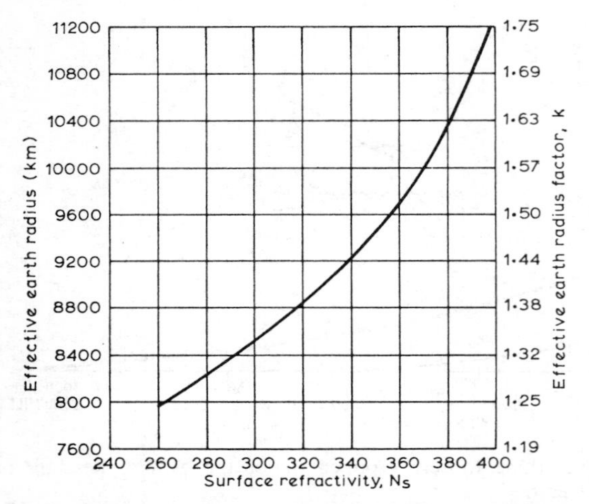

Fig 11.9. Effective earth radius corresponding to various values of surface refractivity. A curved ray-path drawn over an earth section between terminals may be rendered straight by exaggerating the earth's curvature. An average value often used is equal to 1·33 times the actual radius; this is sometimes referred to as "the four-thirds-earth approximation". *(Based on CCIR Report 244.)*

An average value for k, based on a standard atmosphere, is 1·33 or 4/3 (whence a common description of this convention, "the four-thirds earth"). This leads to the very simple relationship, often used when refraction effects in the troposphere must be taken into account, $d = \sqrt{2h}$, provided that in this case d is expressed in miles and h in feet, the various factors conveniently cancelling out. Notes relating to the construction of path profiles using this convention will be found in a later section of this chapter. When the four-thirds earth concept is known to be inappropriate an estimate of a suitable value for k can be obtained from the surface refractivity Ns (obtained from **Fig 11.9**), using for N the value obtained from ground-level readings of pressure, temperature and vapour pressure.

Waves of widely separated length are liable to be disturbed by the troposphere in some way or other, but it is generally only those shorter than about 10m (over 30MHz) which need be considered. There are two reasons for this; one is that usually ionospheric effects are so pronounced at the longer wavelengths that attention is diverted from the comparatively minor enhancements due to the refractive index structure of the troposphere, and the other that anomalies, when they occur, are often of insufficient depth to accommodate waves as long as, or longer than, 10m. For example, it might be that the decrease of refractive index with height becomes so sharp in the lower 100m of the troposphere, that waves are trapped in an *atmospheric duct*, within which they remain confined for abnormally long distances. The maximum wavelength which can be trapped completely in a duct of 100m thickness is about 1m (corresponding to a frequency of 300MHz), for example, so that the most favourable conditions are generally found in the vhf, uhf bands, or above. The relationship between maximum wavelength λ and duct thickness t is shown in the expression:

$$t = 500\ \lambda^{2/3}$$

where both t and λ are expressed in centimetres.

At centimetre wavelengths signals propagated through the troposphere suffer rapid fluctuations in amplitude and phase due to irregular small-scale variations in refractive index which give rise to continuous changes known as *scintillations*, and they are also attenuated by water in the form of precipitation (rain, snow, hail, etc) or as fog or cloud. As **Table 11.3** shows, this effect increases both with radio frequency and with either the rate of rainfall or the concentration of water droplets. Precipitation causes losses by absorption and by random scattering from the liquid (or solid in the case of ice) surfaces and this scattering becomes so pronounced as to act as a "target" for weather radars which use these *precipitation echoes* to detect rain areas.

At even shorter wavelengths resonances occur within the molecules of some of the gases which make up the atmosphere. Only one of these approximates to an amateur band, the attenuation at 22·23GHz due to water vapour. The principal oxygen resonances at around 60 and 120GHz are well beyond the reach of amateur activities, but are likely to prove to be limiting factors where professional work at millimetre wavelengths involves paths passing through the atmosphere, as opposed to outer space working.

Wave Propagation Near the Ground

Diffraction is an alteration in direction of the propagation of a wave due to change in velocity over its wave-front. Radio waves meeting an obstacle tend to diffract around it, and the surface of the earth is no exception to this. Bending comes about as a result of energy being extracted due to currents induced in the ground. These constitute an attenuation by absorption, having the effect of slowing down the lower parts of the wave-front, causing it to tilt forward in a way which follows the earth's curvature. The amount of diffraction is dependent on the ratio of the wavelength to the radius of the earth and so is greatest when the waves are longest. It also depends on the electrical characteristics of the surface, namely its relative permeability (generally regarded as unity

TABLE 11.3

The attenuation in decibels per kilometre to be expected from various rates of rainfall and for various degrees of cloud intensity.

Frequency band	Precipitation (mm/h)					Fog or cloud water content (g/m³) (at 0°C)		
(GHz)	100	50	25	10	1	2·35	0·42	0·043
3·400–3·475	0·1	0·4	0·02	—	—	—	—	—
5·650–5·850	0·6	0·25	0·1	0·02	—	0·09	—	—
10–10·5	3·0	1·5	0·6	0·2	0·01	0·23	0·04	—
21–22	13·0	6·0	2·5	1·0	0·1	0·94	0·17	0·02
24	17·0	8·0	3·8	1·5	0·1	1·41	0·25	0·03
48–49	30·0	17·0	9·0	4·0	0·6	4·70	0·84	0·09

100mm/h = tropical downpour; 50mm/h = very heavy rain; 25mm/h = heavy rain; 10mm/h = moderate rain, 1mm/h = light rain. 2·35g/m³ = visibility of 30m; 0·42g/m³ = 100m; 0·043g/m³ = 500m.

TABLE 11.4

Typical values of dielectric constant, conductivity and depth of wave penetration for various types of surface at various frequencies.

Type of surface	Dielectric constant ϵ	Conductivity σ (mho/m)	Depth of penetration δ (m)			
			1MHz	10MHz	100MHz	1,000MHz
Sea water 0°C	80	4–5	0·25	0·08	0·02	0·01
Fresh water 10°C	84	1×10^{-3} to 1×10^{-2}	11	9	4	0·2
Very moist soil	30	5×10^{-3} to 2×10^{-2}	5·5	3	2	0·3
Average ground	15	5×10^{-4} to 5×10^{-3}	21	16	16	16
Very dry ground	3	5×10^{-5} to 1×10^{-4}	95	90	90	90

for this purpose), its dielectric constant, ε (epsilon), and conductivity, σ (sigma). This diffracted wave is known as the *surface wave.*

Moisture content is probably the major factor in determining the electrical constants of the ground, which can vary considerably with the type of surface as can be seen from **Table 11.4**. The depth to which the wave penetrates is a function also of frequency, and the depth given by the δ (delta) value is that at which the wave has been attenuated to 1/e (or about 37 per cent) of its surface magnitude.

At vhf and at higher frequencies the depth of penetration is relatively small and normal diffraction effects are slight. At all frequencies open to amateur use, however, the ground itself appears as a reflector, and the better the conductivity of the surface the more effective the reflection **(Table 11.5).** It is this effect which makes generalizations of wave propagation near the ground difficult to make for frequencies greater than about 10MHz where the received field is, more often than not, due to the resultant of waves which have travelled by different paths.

Further aspects of propagation near the ground will be dealt with in the section on multiple-path propagation where it will be necessary to consider the consequences of operating with aerials at heights of one to several wavelengths above the ground.

MULTIPLE-PATH EFFECTS

Introduction

The preceding descriptions of the various modes of propagation do not necessarily paint a very realistic picture of the way in which the signals received at a distant location depend on the radio frequency in use and the distance from the transmitter, excluding ionospheric components. The reason for this is that the wave incident on the receiving aerial is rarely only the one which has arrived by the most direct path but is more often the resultant of two or more waves which have travelled by different routes and have covered different distances in doing so. If these waves should eventually arrive in phase they would act to reinforce one another, but should they reach the receiving aerial in antiphase they would interfere with one another and, if they happened to be equal in amplitude, would cancel one another completely.

These alternative paths may arise as a result of reflections in the horizontal plane (as in **Fig 11.10(a)**, where a tall gasholder intercepts the oncoming waves and deflects them towards the receiving site) or in the vertical plane (as in **Fig 11.10(b)**, where reflection occurs from a point on the ground in line-of-sight from both ends of the link). If the reflecting surface is stationary, as it ought to be in the two cases so far considered, the phase difference (whatever it may be) would be constant and a steady signal would result.

It may happen that the surface of reflection is in motion, as it would be if it was part of an aeroplane flying along the transmission path. In that case the distance travelled by the reflected wave would be changing continually and the relative phases would progressively advance or retard through successive cycles (effectively an increase or decrease in frequency—the *Doppler effect*), leading to alternate enhancements and degradations as the two waves aid or oppose one another. This performance is one which is particularly noticeable on television receivers sited near an airport, and even non-technical viewers can instantly diagnose as "aircraft flutter" the fluctuations in picture brilliance which result. Any "ghost" image on the television picture is evidence of a second transmission path, and the amount of its horizontal displacement from the main picture is a measure of the additional transmission distance involved. So, with the reflection from the moving aircraft the displacement of the ghost picture will change as the path length changes, and its brilliance will reach a maximum every time the difference between the direct path and the reflected path is exactly a whole number of wavelengths. An analytical treatment of the appearance of aircraft reflections on pen recordings of distant signals has appeared in the pages of *Radio Communication*; see reference [1].

Because the waves along the reflected path repeat themselves after intervals of exactly one wavelength, it is only

TABLE 11.5

The number of decibels to be subtracted from the calculated free-space field in order to take into account various combinations of ground conductivity and distance. The values are shown in each case for 1·8, 3·5 and 7·0 MHz. Vertical polarization is presumed.

Distance (km)	Free-space field in dB rel to 1μV/m for a 1kW transmitter	Sea, $\sigma = 4$			Land $\sigma = 3 \times 10^{-2}$			Land $\sigma = 3 \times 10^{-3}$			Land $\sigma = 10^{-3}$		
		1·8	3·6	7·0	1·8	3·5	7·0	1·8	3·5	7·0	1·8	3·5	7·0
3	100	1	1	1	1	3	12	8	18	30	19	28	36
10	90	1	1	1	6	8	22	18	29	42	30	38	47
30	80	1	2	2	8	18	34	28	41	51	40	48	56
100	70	2	3	7	18	33	51	44	56	68	58	63	73
300	60	10	12	23	43	61	88	68	85	—	78	—	—
1,000	50	48	55	—	—	—	—	—	—	—	—	—	—

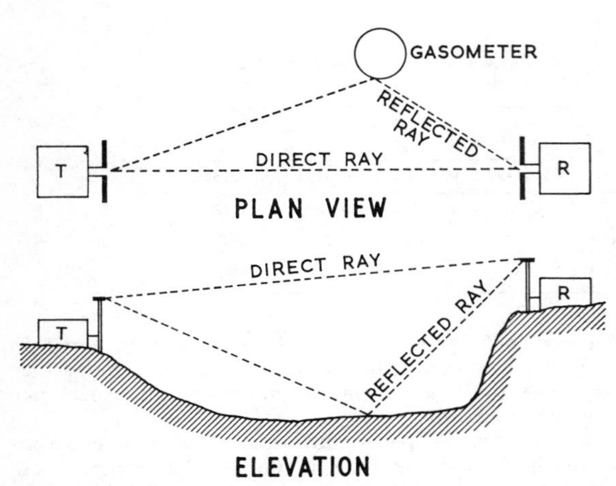

Fig 11.10. Multipath effects brought about by reflections in (a) the horizontal plane, and (b) the vertical plane.

the portion of a wavelength "left over" which determines the phase relationship in comparison with the direct-path wave. This suggests that relatively small changes in the position of a receiving aerial could have profound effects on the magnitude of the received signal when multiple paths are present, and this is indeed found to be the case, particularly where the point of reflection is near at hand.

Ground Wave Propagation

It should now be evident why it was not possible to generalize on the relationship between distance and received signal strength when dealing with propagation near the ground. The surface wave, influenced by the diffraction effects considered earlier, is only one of the possible paths. If the spacing of the transmitting and receiving aerials is such that they are not hidden from one another by the curvature of the earth there will also be a *space wave*, made up of a direct wave and a ground-reflected wave, as suggested by **Fig 11.11**. The combination of this space wave and the diffracted surface wave form what is called the *ground wave*, and it may sometimes be difficult (and often perhaps unnecessary) to try to separate this into its three components.

Beyond the radio horizon the direct and reflected rays are blocked by the bulge of the earth, and the range attained is then determined by the surface wave alone. This diffracted wave is strong at low frequencies (including the amateur 160m band) but becomes less so as the carrier frequency increases and may be considered negligible at vhf and beyond. When occasional signals are received well beyond the horizon the dominant mechanism may be forward scatter.

The strength of the reflected component of the space wave depends largely on the conductivity and smoothness of the ground at the point of reflection, being greatest where over-sea paths are involved, and least over dry ground and rock. An extensive treatment of the various factors concerned will be found in the Society's journal [2]. If perfect reflection is presumed, the received field strength due to the interaction of one reflected ray with the direct ray can be estimated from the expression:

$$E = \frac{2E_o}{d} \cdot \sin\left(2\pi \frac{h_1 h_2}{\lambda d}\right)$$

where E is the resultant received signal strength,

E_o is the direct-ray field strength,

h_t and h_r the effective aerial heights above a plane tangential to the earth at the point of reflection,

d is the distance traversed by the direct ray, and

λ the wavelength, all units being consistent (eg metres).

It can be seen that the magnitude of the received signal depends on the relative heights of the two aerials, the distance between them, and, of course, the frequency.

This relationship suggests that doubling the aerial height has the same effect on the received signal strength as halving the length of the path. In view of the respective distances involved it will be appreciated that an increase in the height of one of the aerials has a greater effect than a comparable horizontal movement towards the transmitter.

The Effect of Varying Height

A few moments of experiment with two pieces of cotton representing the two alternative ray paths will provide a convincing demonstration of the effect of altering the height of one or both of the aerials. An increase in height of (say) the receiving aerial has little effect on the length of the direct path, but the ground-reflected ray has to travel further to make up the additional distance and it therefore arrives at a later point on its cycle. The consequence is even more pronounced when it is realized that at very low angles of incidence, when the two aerials are at ground level, the indirect ray may well have experienced a phase change of 180 degrees upon reflection so that the two components arrive roughly equal in magnitude but opposite in phase, so tending to cancel. As the height of one or both aerials is increased the space wave increases in magnitude and the field becomes the vector sum of the diffracted surface wave and the space wave. At even greater heights the effect of the surface wave can be neglected, while the intensity of the space wave continues to increase.

In practice these considerations apply only to aerials carrying vhf, uhf and above. This is because it is not practicable to raise aerials to the necessary heights at the longer wavelengths, and in any case the reception of ionospherically propagated waves imposes different requirements as regards angle of arrival.

As with other functions which depend on earth constants for their effectiveness there is a marked difference between over-land and over-sea conditions. There is more to be gained by raising an aerial over land than over sea, for high frequencies than for low, and for horizontally-polarized waves rather than vertically polarized ones.

The ratio of the received field at any given height above ground to the field at ground level due to the surface wave alone (presuming that the two components of the space wave have cancelled one another) is known as the *Height-gain Factor*. This can be expressed either as a multiplier, or

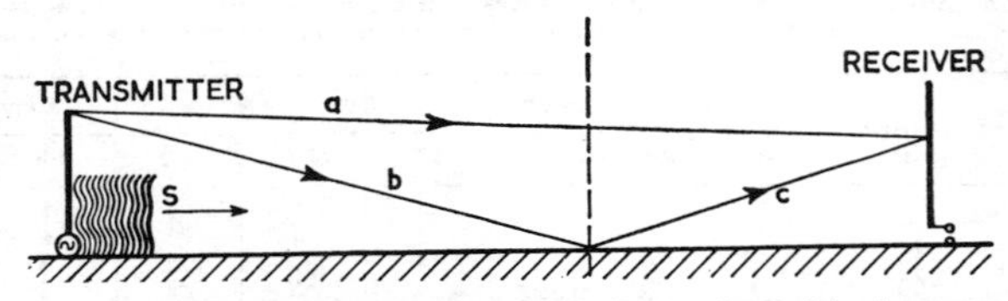

Fig 11.11. The diffracted surface wave S together with the space wave, composed in turn of a direct wave (a) and a reflected wave (b) and (c).

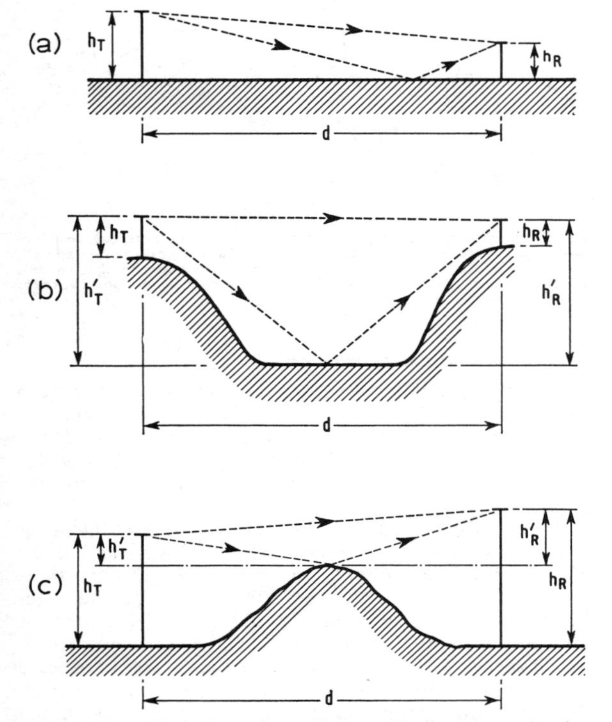

Fig 11.12. The effect of terrain on direct and indirect rays, showing the effective heights of the aerials when undertaking height-gain calculations.

as the corresponding equivalent in decibels, using the voltage scale of relationships.

Over flat ground there is little to be gained from raising aerials for frequencies below about 3MHz, unless it is to clear local obstacles, but it should not be overlooked that it may be desirable to raise aerials no matter what their frequency of operation for reasons unconnected with height-gain benefits—to increase the distance to the radio horizon, for example.

The result of changing the receiving aerial height is by no means as predictable as some authorities would have us believe, and the subject is still a matter thought worthy of further investigation at some research establishments. The following figures summarize the gains to be expected after raising a receiving aerial from a height of 3m to a height of 10m above the ground, according to a current CCIR report [3], primarily concerned with television broadcasting frequencies but relevant nevertheless:

50–100MHz	Median values of height-gain 9–10dB
180–230MHz	Median values of height-gain 7dB in flat terrain and 4–6dB in urban or hilly areas
450–1,000MHz	Median values of height-gain very dependent on terrain irregularity. In suburban areas the median is 6–7dB, and in areas with many tall buildings 4–5dB

A simple rule-of-thumb often adopted by radio amateurs is to reckon on a height-gain of 6dB for each time that the aerial height is doubled (eg if 12dB at 3m height, then expect 18dB at 6m, 24dB at 12m, etc), but the presence of more component waves than the two considered can lead to wide departures from this relationship, particularly in urban areas.

If the terrain is not flat the result of altering the aerial height depends largely on the position in the vertical plane of the reflection points relative to the two terminals. Thus in **Fig 11.12** the situation shown in (a) corresponds to the one already considered. Should the two aerials be sited on hills, or separated by a valley, as at (b), there will be large differences in path length between the direct and ground-reflected rays which alterations in aerial height will do little to alter so that elevating it is unlikely to have very much effect on the received signal strength. On the other hand, the presence of high ground between transmitter and receiver, as at (c), may make communication between them difficult at low aerial heights, and in that case there would be a great deal to be gained from raising them. In cases (b) and (c) the two aerials should be considered as having effective heights of h'_T and h'_R respectively when dealing with height-gain calculations.

If the intervening high ground has a relatively sharp and well-defined upper boundary, such as would be the case with a mountain ridge, the receiving aerial height at which signals cease might be much lower than would be expected from line-of-sight considerations, even when refractivity changes are taken into account. This is because of an effect known as *knife-edge diffraction*, which often enables 2m operators situated in the Scottish Highlands (to cite an instance) to receive signals from other stations which are apparently obscured from them by surrounding mountains.

The Effect of Varying Distance

The effect on the field strength of varying the distance between transmitter and receiver is shown in **Fig 11.13**, where, again, the result is due to the interference between

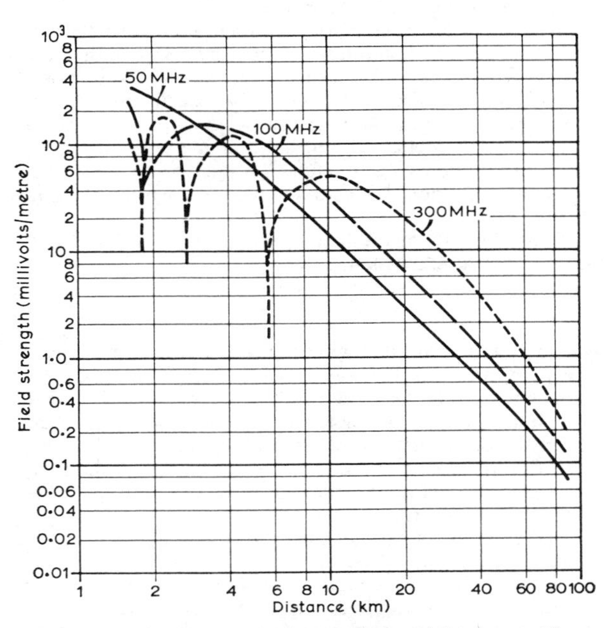

Fig 11.13. Relationship between field-strength and distance at vhf and uhf.

the direct wave and the ground-reflected wave, passing through successive maxima and minima as the path difference becomes an exact odd or even number of half-wavelengths. (It must be remembered that a low-angle ground-reflection itself introduces a phase change of very nearly 180 degrees). The spacing of the maxima (which are greater in magnitude than the free-space value) is closer the higher the frequency of operation and the shorter the path for a given frequency. The most distant maximum will occur when the path difference is down to one half wavelength; beyond that the difference tends towards zero and the two waves progressively oppose one another, the field rapidly falling below the free-space estimate. If the aerial heights are raised the patterns move outwards.

As with all these matters involving ground reflection there is a difference between the behaviour of horizontally and vertically polarized waves, and the foregoing description favours the former. The reflection coefficient and phase shift at the reflection point vary appreciably with the ground constants when vertical polarization is employed. In practice, whichever is used, the measured field strength may vary considerably from the calculated value because of the presence of other components due to local reflections. A fairly reliable first estimate for vhf and uhf paths up to about 50km unobstructed length is just to allow for a possible increase or decrease of 10dB on the free-space figure.

Fresnel Zones

In all the explanations so far it has been presumed that reflection at a surface occurs at the point which enables the reflected ray to travel the shortest distance.

Because the surfaces considered in radio propagation work are neither plane nor perfect reflectors, the received waveform is the resultant of signals which have been reflected from an area, the size of which is determined by the frequency and by the separation of the terminal aerials, so that the individual reflected path lengths differ by no more than half a wavelength from one another. The locus of all the points surrounding the direct path which give exactly half a wavelength path difference is described by an ellipsoid of revolution having its foci at the transmitting and receiving aerials respectively. A cross-section of this volume on an intersecting plane of reflection encloses an area known as the *first Fresnel zone*.

The radius, R, of the first Fresnel zone at any point, P, is given by the expression

$$R^2 = \frac{\lambda d_1 d_2}{d_1 + d_2}$$

where d_1 and d_2 are the two distances from P to the ends of the path, and λ is the wavelength, all quantities being in similar units. The maximum radius occurs when d_1 and d_2 are equal. Thus, on an 80km path at a wavelength of 2m the radius of the first Fresnel zone at the mid-point is 200m, and this represents the clearance of the line-of-sight ray (corrected for refraction) necessary if the path is to be considered unobstructed.

Higher orders of Fresnel zone surround the regions where similar relationships occur after separations of one-wavelength, two wavelengths, etc, but the conditions for reflection in the required direction rapidly become less favourable and Fresnel zones other than the first are rarely considered.

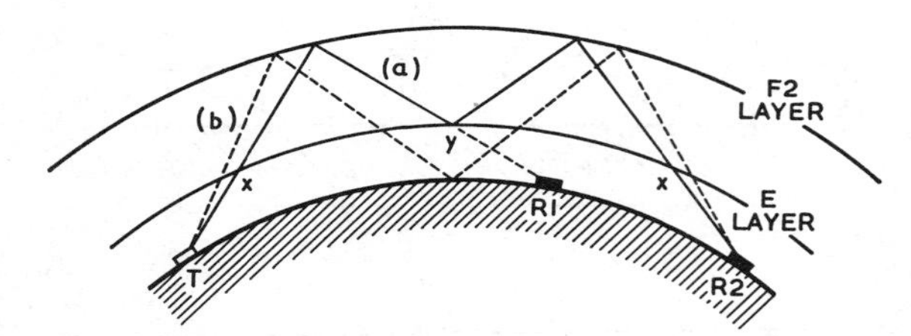

Fig 11.14. Fading due to multipath reception. In case (a) the frequency in use is higher than the E-layer muf at the points marked "x", but lower at "y", where the ray will suffer reflection. In case (b) the frequency is higher than the muf of the E-layer at all four contacts with it.

Ionospheric Multi-path Effects

The multiple-path effects so far considered have been mainly associated with vhf, uhf and shf where very low angles of radiation and reception are generally involved, and the wavelengths concerned are sufficiently small to enable optimum heights and favourable positions to be found for the aerials.

When ionospherically propagated signals are of interest it is generally sufficient to regard the ground wave as a single entity, without attempting to separate it into its three components. When multi-path effects occur (as they frequently do) they may be between the ground wave and an ionospheric wave, between two ionospheric waves which may have been propagated by different layers, or, in the case of long range transmissions, by signals which have followed different paths entirely in different directions around the earth's curvature, perhaps in several "hops" between the ionosphere and the ground. Whatever the cause the result is inevitably a fading signal.

The ionosphere is not a perfect reflector; it has no definite boundaries and it is subject to frequent changes in form and intensity. These deficiencies appear on the received signal as continual small alterations in phase or frequency as the effective path length alters and, when only a single ionospheric wave is present, can pass almost unnoticed by the average hf listener, who is remarkably tolerant of imperfections on distant transmissions. However, when a second signal from the same source is present, which may be either the relatively steady ground wave or another ionospheric component, these phase or frequency changes become further emphasized by appearing as changes in amplitude as the waves alternately reinforce and interfere, and by distortion of modulated signals if the various sideband frequencies do not resolve back into their original form.

Pen recordings of signal strength generally show very clearly the period of fading which results when two modes of propagation begin to interact, continuing until the second predominates. This could occur as a result of circumstances similar to those shown in **Fig 11.14**, where the receiver at R2 may receive signals either by double-hop off the ground (as in path (b)) or double-hop off an intermediate layer, path (a), depending on the relationship between the signal frequency and the maximum usable frequency at point y. The transition between one propagation mode and another is generally accomplished within a relatively short time.

Occasionally very long-distance transmissions may be heard with a marked echo on their modulation. As the two signal components responsible have obviously travelled routes of markedly differing length they probably require very different azimuths at both ends of the path. This effect is most noticeable on omnidirectional broadcast transmissions and is minimized by the use of narrow-beam aerials for transmission and reception.

Fading

Fading is generally, though not exclusively, a consequence of the presence of multiple transmission paths. For that reason it is appropriate to include a summary of its causes here, although more properly some of the comments belong elsewhere in this chapter.

Fading is a repetitive rise and fall in signal level, often described as being *deep* or *shallow* when referring to the range of amplitudes concerned, and *slow* or *rapid* when discussing the period. It is sometimes *random*, usually *periodic*, but occasionally *double periodic*, as when a signal with a rapid fade displays slow changes in mean level. Generally the fading rate increases with frequency because a particular motion in the ionosphere causes a greater phase shift at the shorter wavelengths.

At vhf and uhf the fading rate is often closely associated with the pattern of atmospheric pressure at the surface, tending to become slower during periods of high pressure. This can be particularly noticeable on a pen recording of signal strength taken while a ridge of high pressure moves along the transmission path, the slowest rate occurring as the ridge crosses the mid-point.

Interference fading, as its name implies, is caused by interference between two component waves when one or both path lengths are changing, perhaps due to fluctuations in the ionosphere or troposphere, or due to reflections from a moving surface. The period is relatively short, usually up to a few seconds. Fast interference fading is often called *flutter*. Auroral flutter comes in this category, being caused by motion of the reflecting surfaces.

Polarization fading is brought about by continuous changes in polarization due to the effect of the earth's magnetic field on the ionosphere. Signals are at a maximum when they arrive with the same polarization as the receiving aerial. Period again up to a few seconds.

Absorption fading, generally of fairly long period, is caused by inhomogeneities in the troposphere or ionosphere. Period up to an hour, or longer.

Skip fading occurs when a receiver is on the edge of a skip zone and changes in muf cause the skip distance to shorten and lengthen. Highly irregular as regards period of fade.

Selective fading is the name given to a form of fading characterized by severe modulation distortion in which the path length in the ionosphere varies with frequency to such an extent that the various sideband frequencies are differently affected. It is most severe when ground and sky waves are of comparable intensity.

Scintillations are rapid fluctuations in amplitude, phase and angle-of-arrival of tropospheric signals, produced by irregular small-scale variations in refractive index. The term is also used to describe irregular fluctuations on hf signals transmitted through the ionosphere from satellites and other sources outside the earth.

Diversity Reception

The effects of fading can be countered to a certain extent by the use of more than one receiving system coupled to a common network which selects at all times the strongest of the outputs available.

There are three principal versions in common use:

Space diversity, obtained by using aerials which are so positioned as to receive different combinations of components in situations where multi-path conditions exist,

Frequency diversity, realized by combining signals which have been transmitted on different frequencies, and

Polarization diversity, where two receivers are fed from aerials having different planes of polarization.

It is unlikely that any of these systems have any application in normal amateur activities, but they may be of interest in connection with research projects relating to the amateur bands.

SOLAR AND MAGNETIC INFLUENCES

The Sun

Our sun is at the centre of a complex system consisting of nine major planets (including ours), five of which have two or more attendant moons, of several thousand minor planets (or asteroids), and of an unknown number of lesser bodies variously classed as comets or meteoroid swarms.

It is a huge sphere of incandescence of a size which is equivalent to about double our moon's orbit around the earth, but, despite appearances to the contrary, it has no true "edge" because nearly the whole of the sun is gaseous and the part we see with apparently sharp boundaries is merely a layer of the solar atmosphere called the *photosphere* which has the appearance of a bright surface, preventing us from seeing anything which lies beneath.

Beyond the photosphere is a relatively cooler, transparent layer called the *chromosphere*, so named because it has a bright rose tint when visible as a bright narrow ring during total eclipses of the sun. From the chromosphere great fiery jets of gas, known as *prominences*, extend. Some are slow-changing and remain suspended for weeks, while others, called *eruptive prominences*, are like narrow jets of fire moving at high speeds and for great distances.

Outside the chromosphere is the *corona*, extending a distance of several solar diameters before it becomes lost in the general near-vacuum of interplanetary space. At the moment of totality in a solar eclipse it has the appearance of a bright halo surrounding the sun, and at certain times photographs of it clearly show it being influenced by the lines of force of the solar magnetic field.

The visible sun is not entirely featureless—often relatively dark *sunspots* appear and are seen to move from east to west, changing in size, number and dimensions as they go. They are of interest for two reasons: one is that they provide reference marks by which the angular rotation of the sun can be gauged, the other that by their variations in number they reveal that the solar activity waxes and wanes in fairly regular cycles. The sun's rotation period has been found to vary with latitude, with its maximum angular speed at the equator. The mean rotation period with respect to the stars is 25·38 mean solar days, but a more useful figure is the *mean synodic rotation period* of 27·2753 days, which is the time required for the sun to rotate until the same part faces the earth taking into account the earth's steady motion around it. We shall have occasion to note these periods again when dealing with the recurrence of abnormal magnetic activity on the earth.

The sun's rotations have been numbered since 1853

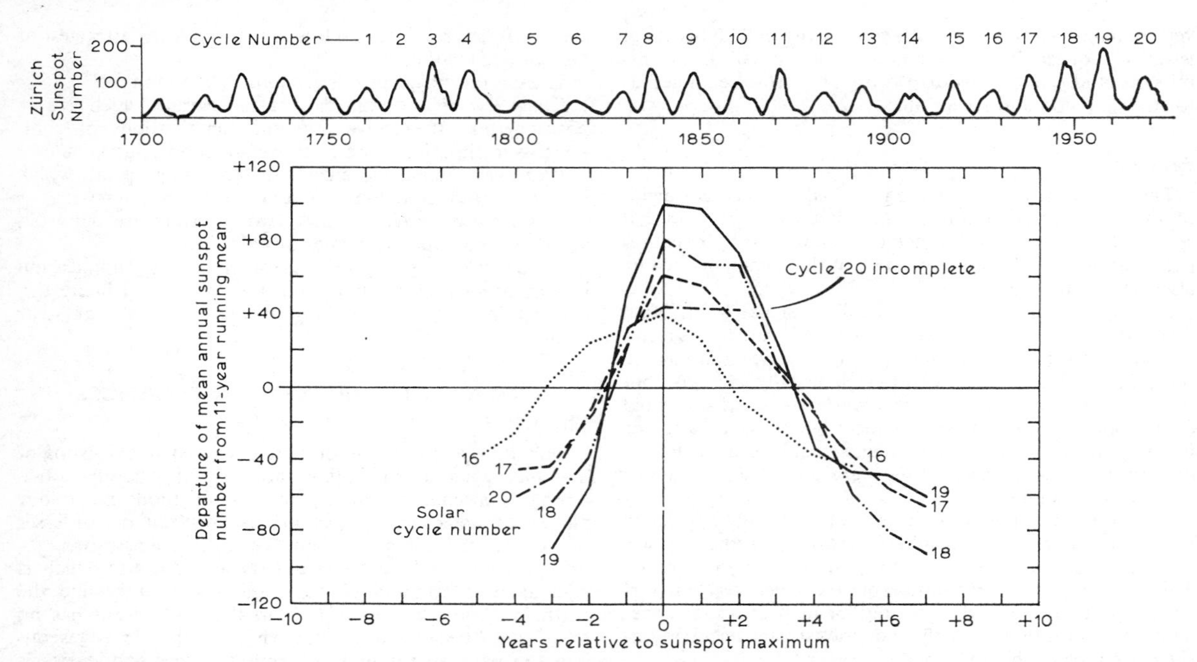

Fig 11.15. Monthly mean values of Zurich sunspot numbers. The recurrence pattern is a very complex one and every cycle remains a mystery until it becomes certain that the next has begun.

(Rotation 1651 commenced in 1977 on 28 January), and observations of solar features are referred to an imaginary network of latitudes and longitudes which rotates from east to west, as seen from the earth. A long-persistent feature which first appears at the *east limb* is visible for about 13½ days before it disappears from sight at the *west limb*. For many purposes it is more convenient to refer the positions of noteworthy features to a related, but stationary set of co-ordinates, the *heliocentric latitude and longitude*, in which locations are described with respect to the centre of the visible disc. An important statistic relating to sunspots is the time of *Central Meridian Passage*.

Sunspots are the visible manifestations of very powerful magnetic fields and adjacent spots often have opposite polarities. These intense magnetic fields also produce *solar flares*, which are emissions of hydrogen gas and are also responsible for the ejection of streams of charged particles and X-rays.

Sunspot numbers have been recorded for over 200 years and it has been found that their totals vary over a fairly regular cycle occupying around 11 years. The numbers at *sunspot maximum* vary considerably from one cycle to the next as can be seen from **Fig 11.15**, but there are often periods at *sunspot minimum* when there are no spots at all.

The sun also emits electromagnetic waves which cover a large portion of the radio section of the spectrum. There is a close correspondence between measurements made at a frequency of 2,800MHz and the physical counts of sunspots. The sun at radio frequencies is a larger body than the one outlined by the visible disc. Radio telescopes detect features which are usually situated in the vicinity of the corona, which accounts for the fact that they sometimes reveal disturbances beyond the limbs which travel across the disc at a faster rate than the spots beneath.

The Solar Wind

The solar corona was described in the last section as extending outwards until it became lost in interplanetary space. In fact it does not become lost at all, but turns into a tenuous flow of hydrogen which expands outwards through the solar system, taking with it gases evaporating from the planets, fine meteoric dust and cosmic rays. It becomes the solar wind.

Near the sun the corona behaves as a static atmosphere, but once away it gradually accelerates with increasing distance to speeds of hundreds of kilometres per second, the gas particles taking about nine days to travel the 150,000,000km to the earth, carrying with them a magnetic field (because the gas is ionized) which assumes a spiral form due to the sun's rotation. It is the solar wind, rather than light pressure alone, which is responsible for comets' tails flowing away from the sun, causing them to take on the appearance of celestial wind-socks.

The existence of the solar wind was first detected and measured by space vehicles such as the Luniks, Mariner II and Explorer X, which showed that its speed and turbulence are related to solar activity. Regular measurements of solar wind velocity are now routine.

There is thus a direct connection between the atmosphere of the sun and the atmosphere of the earth. In the circumstances it is hardly surprising that solar events, remote though they may at first seem, soon make their effects felt here on earth.

The Earth's Magnetosphere

It is well known that the earth possesses a magnetic field, for most of us have used a compass, at some time or another, to help us to get our "bearings". We know from such experiences that the field appears to be concentrated at a point somewhere near the north pole (and are prepared to believe that there is another point of opposing polarity somewhere near the south pole). Popular science articles have familiarized us with a picture of field lines surrounding the earth like a section of a ring doughnut made up of onion-like layers.

Because the particles carried by the solar wind are charged their movement produces a magnetic field which interacts with the geomagnetic field. A blunt shock-wave is set up, called the *magnetosheath*, and the wind flows round it rejoining behind where the field on the far side is stretched in the form of a long tail, the overall effect being reminiscent of the shape of a pear with its stalk pointing away from the sun. The region within the magnetosheath, into which the wind does not pass, is called the *magnetosphere*.

On the earthward side the magnetosphere merges into the ionosphere. Inside the magnetosphere there are regions where charged particles can become trapped by geomagnetic lines of force in a way which causes them to oscillate back and forth over great distances. Particles from the solar wind can enter these regions (often called *Van Allen belts* after their discoverer) in some way as yet not perfectly understood.

The concentration of electrons in the magnetosphere can be gauged from the ground by observations on "*whistlers*", naturally-occurring audio-frequency oscillations of descending pitch which are caused by waves radiated from the electric discharge in a lightning flash. These travel north and south through the ionosphere and magnetosphere from one hemisphere to another along the magnetic lines of force. The various component frequencies propagate at different speeds so that the original flash (which appears on an ordinary radio receiver as an "*atmospheric*") arrives at the observer considerably spread out in time. The interval between the reception of the highest and lowest frequencies is a function of the concentration of electrons encountered along the way.

These trapped particles move backwards and forwards along the geomagnetic field lines within the Van Allen belts and some collide with atoms in the ionosphere near the poles where the belts approach the earth most closely. Here they yield up energy either as ionization or illumination and are said to have been *dumped*. These dumping regions surround the two poles forming what is called the *auroral zones*. The radius of the circular motion in the spirals (they are of a similar form to that of a helical spring) is a function of the strength of the magnetic field, being small when the field is strong. Electrons and protons perform their circular motions in opposite senses, and the two kinds of spiralling columns drift sideways in opposite directions, the electrons eastward, the protons westward, around the world. Because of the different signs on the two charges these two drifts combine to give the equivalent of a current flowing in a ring around the earth from east to west.

This ring current creates a magnetic field at the ground which combines with the more-or-less steady field produced from within the earth. We shall see later the sort of effects that solar disturbances have on the magnetosphere, the ionosphere, and the total geomagnetic field.

The Quiet Ionosphere

With the stage set to follow the antics of the ions and electrons deposited by dumping we must pause again, this time to examine the normal day-to-day working of the ionosphere, which is dependent for its chemistry on another form of incoming solar radiation.

The gas molecules in the earth's upper atmosphere are normally electrically neutral, that is to say the overall negative charges carried by their orbiting electrons exactly balance the overall positive charges of their nuclei. Under the influence of ultra-violet radiation from the sun, however, some of the outer electrons can become detached from their parent atoms, leaving behind overall positive charges due to the resulting unbalance of the molecular structure. These ionized molecules are called *ions*, from which of course stems the word ionosphere.

This process, called *disassociation*, tends to produce layers of free electrons, brought about in the following manner. At the top of the atmosphere where the solar radiation is strong there are very few gas molecules and hence very few free electrons. At lower levels, as the numbers of molecules increase, more and more free electrons can be produced, but the action progressively weakens the strength of the radiation until it is unable to take full advantage of the increased availability of molecules and the electron density begins to decline. Because of this there is a tendency for a maximum (or peak) to occur in the production of electrons at the level where the increase in air density is matched by the decrease in the strength of radiation. A peak formed in this way is known as a *Chapman layer*, after the scientist who first outlined the process.

The height of the peak is determined not by the strength of the radiation but by the density/height distribution of the atmosphere and by its capability to absorb the solar radiation (which is a function of the uv wavelength) so that the layer is lower when the radiation is less readily absorbed. The strength of the radiation affects the rate of production of electrons at the peak, which is also dependent on the direction of arrival. The electron density is greatest when the radiation arrives vertically and it falls off as a function of zenith distance, being proportional to cos χ, where χ is the angle between the vertical and the direction of the incoming radiation. As cos χ decreases (ie when the sun's altitude declines) a process of recombination sets in, whereby the free electrons attach themselves to nearby ions and the gas molecules revert to their normal neutral state.

Experimental results have led to the belief that the E layer (at about 120 km) and the F1 layer (at about 200km) are formed according to Chapman's theory as a result of two different kinds of radiation with perhaps two different atmospheric constituents involved.

The uppermost layer, F2 (around 400km), which normally appears only during the day, does not follow the same pattern, and is thought to be formed in a different way, perhaps by the diffusion of ions and electrons, but there are still a number of anomalies in its behaviour which are the subject of current investigation. These include the *diurnal anomaly*, when the peak occurs at an unexpected time during the day; the *night anomaly*, when the intensity of the layer increases during the hours of darkness when no radiation falls upon it; the *polar anomaly*, when peaks occur during the winter months at high latitudes, when no illumination reaches the layer at all; the *seasonal anomaly*, when magnetically quiet days in summer (with a high sun) sometimes show lower penetration frequencies than quiet days in winter (with a low sun); and a *geomagnetic anomaly* where, at the equinoxes, when the sun is over the equator, the F2 layer is

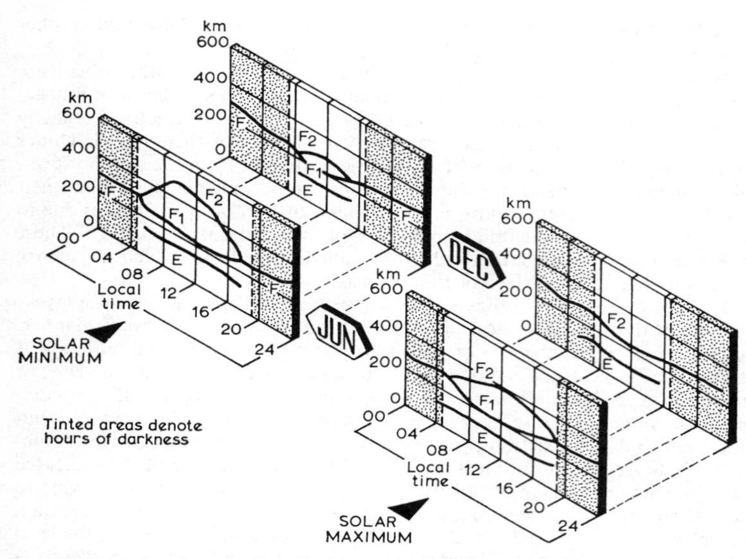

Fig 11.16. Typical diurnal variations of layer heights for summer and winter at minimum and maximum states of the solar cycle.

is continuously varied in frequency from the medium-wave broadcast band through the hf bands to an upper limit of about 20MHz, but beyond if conditions warrant. Reflections from the various layers are recorded photographically in the form of a graph called an *ionogram*, which displays *virtual height* (corresponding to radar range) as a function of signal frequency. *Critical frequencies*, where the signals pass straight through the layers, are read off directly. There are many such equipments in the world; the one serving the United Kingdom is located at the Appleton Laboratory (formerly the Radio and Space Research Station) at Slough, which is also the site of one of the World Data entries for the ionosphere.

The two sets of diagrams **(Fig 11.16** and **Fig 11.17)** summarize the forms taken by the diurnal variations in height and critical frequency for two seasons of the year at both extremes of the sunspot cycle. The actual figures vary very considerably from one day to the next, but an estimate of the expected monthly median values of maximum usable frequency and optimum working frequency between two locations at any particular year, month and time of day can be obtained from predictions published several months in advance by a number of research establishments.

The critical frequencies of the E and F1 layers are a function of R, the smoothed Zürich relative sunspot number (which is predicted six months in advance for this purpose), and the cosine of the zenith distance χ, the angle the sun makes with the local vertical, and thus, to a first approximation

$$f_oE = 0{\cdot}9\ [(180 + 1{\cdot}44R)\cos\chi]^{0{\cdot}25}$$

(usually to within 0·2MHz of the observed values),

and $$f_oF1 = (4{\cdot}3 + 0{\cdot}01R)\cos^{0{\cdot}2}\chi$$

which is less accurate because of uncertainty in the value of the exponent which varies with location and season.

The F2 layer is the most important for hf communication at a distance, but is also the most variable. It is subject to geomagnetic control which impresses a marked longitudinal effect on the overall world pattern, causing it to lag behind the sub-solar point so as to give maximum values in critical frequency during the local afternoon.

The F2 critical frequency f_oF2 varies with the solar cycle, as shown in **Table 11.6,** which shows monthly median values for Slough, applicable to sunspot numbers of 0 and 150. In recent years it has been found possible to predict the behaviour of the F2 layer by extrapolation several months

most intense at places to the north and south, separated by a minimum along the magnetic dip equator. It is thought that topside sounding from satellites probing the ionosphere from above the active layers may help to resolve some of these anomalies in F2 behaviour.

The Regular Ionospheric Layers

The various regular ionospheric layers were first defined by letters by Sir Edward Appleton who gave to the one previously known as the *Kennelly-Heaviside layer* the label E because he had so marked it in an earlier paper denoting the electric field reflected from it, and to the one he had discovered himself the letter F, rather than call it the *Appleton layer*, as some had done. To the band of absorption below thus naturally fell the choice of the letter D, although this was generally referred to as a region rather than a layer because its limits are less easy to define.

From comments already made it will be appreciated that the regular ionospheric layers which these letters define exhibit changes which are basically a function of day and night, season, and solar cycle.

Most of our knowledge of the ionosphere comes from regular soundings made at vertical incidence using a specialized form of radar called an *ionosonde*, which transmits short pulses upwards using a carrier which

TABLE 11.6

F2-layer critical frequencies at Slough. The two rows show mean median values of f_oF2 in megahertz for three-months weighted mean sunspot numbers of 0 and 150.

	Jan	Feb	Mar	Apr	May	Jun
$R_3 = 0$	5·3	5·1	4·75	4·80	5·10	5·03
$R_3 = 150$	12·17	12·42	11·67	9·88	8·23	7·70

	Jul	Aug	Sep	Oct	Nov	Dec
$R_3 = 0$	4·72	4·75	4·90	5·69	5·58	5·32
$R_3 = 150$	7·73	7·68	8·81	11·26	12·93	12·53

TABLE 11.7

MUF (maximum usable frequency) factors for various distances, assuming representative heights for the principal layers.

Layer	Distance 1km	2km	3km	4km
Sporadic E	4·0	5·2	—	—
E	3·2	4·8	—	—
F_1	2·0	3·2	3·9	—
F_2 winter	1·8	3·2	3·7	4·0
F_2 summer	1·5	2·4	3·0	3·3

ahead, using an index known as IF2, which is based on observations made at about ten observatories.

The *maximum usable frequency* which can be used on a particular circuit can be calculated from the critical frequency of the appropriate layer by applying the relationship

$$\text{muf} = f_o \sec i$$

where i is the angle which the incident ray makes with the vertical through the point of reflection at the layer. The factor sec i is called the *muf factor*; it is a function of the path length if the height of the layer is known. **Table 11.7** shows typical figures obtained by assuming representative heights. The *optimum working frequency* is the highest of those available which does not exceed the muf.

There is a lower limit to the band of frequencies which can be selected for a particular application. This is set by the *lowest usable frequency*, luf, below which the circuit becomes either unworkable or uneconomical because of the effects of absorption and the level of radio noise. Its calculation is quite a complicated process beyond the scope of this survey.

It is often useful to be able to estimate the radiation angle involved in one- and two-hop paths via the E and F2 layers. **Fig 11.18,** again prepared for average heights, accomplishes this. It is a useful rule-of-thumb to remember that the maximum one-hop E range is about 2,000km, and the maximum one-hop F2 range about 4,000km, both, of course, involving very low angles of take-off.

Irregular Ionization

Besides the regular E, F1 and F2 layers there are often more localized occurrences of ionization which make their contribution to radio propagation. They generally occur around the heights associated with the E layer and often the effects extend well into vhf, although the regular E layer can never be effective at frequencies of 30MHz or more.

Sporadic E, (Es), generally appears as an intensification in ionization in the form of a horizontal sheet about 1km thick and some 100km across, at a height of around 100–120km. It appears in a random manner but occurs more often and with greater intensity during the summer and more often

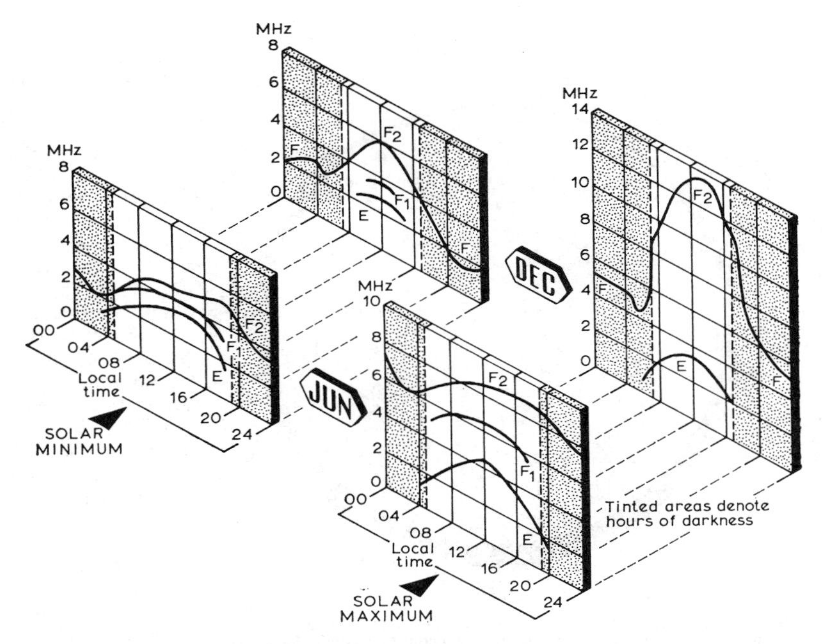

Fig 11.17. Typical diurnal variations of layer critical frequencies for summer and winter at the extremes of the solar cycle.

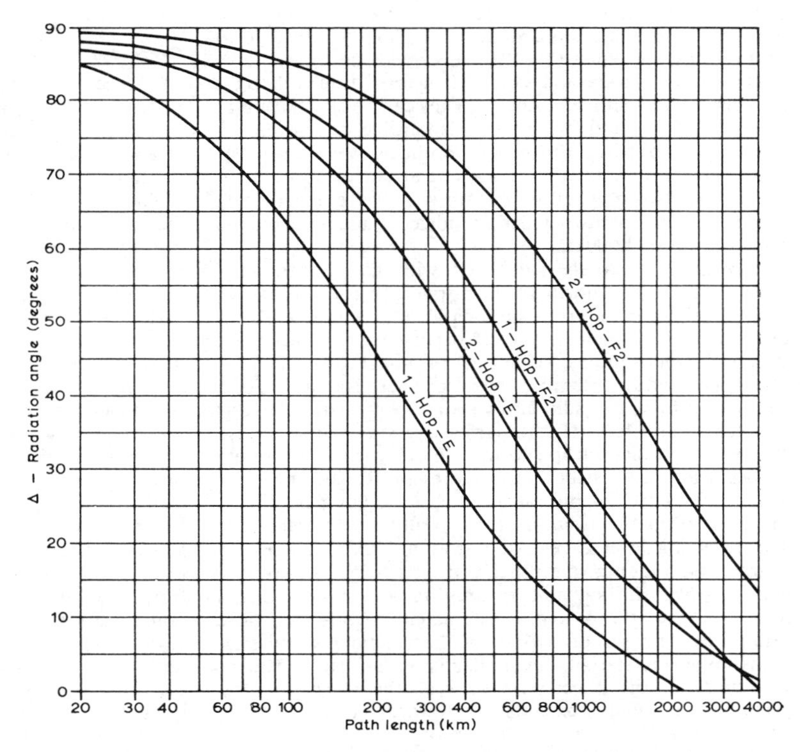

Fig 11.18. Radiation angle involved in one-hop and two-hop paths via E and F2 layers. (*From NBS publication "Ionospheric Radio Propagation", p191.*)

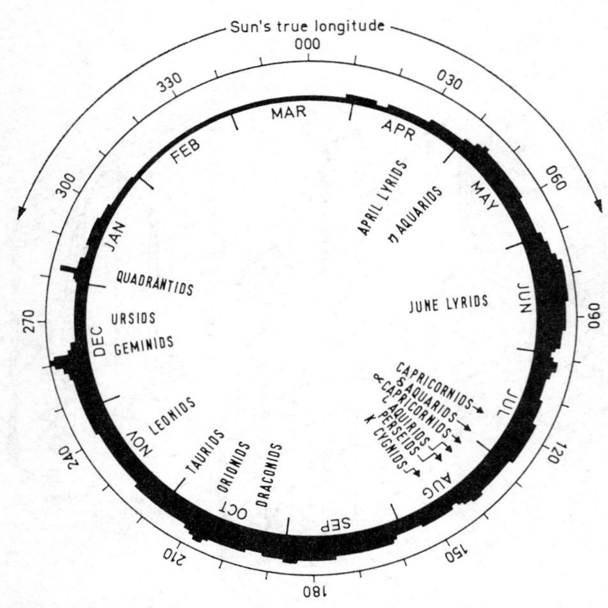

Fig 11.19. Seasonal variation of meteor activity, based on a daily relative index. Prepared from tables of 24h counts made by Dr Peter Millman, National Research Council, Ottawa. The maximum rate corresponds to an average of about 300 echoes per hour, corresponding to an equivalent visual magnitude of 6 or greater.

during the daytime. It is rare between 0000 and 0600 local time, and normally peaks around 1000. Clouds of Es tend to drift towards the equator in the temperate zone and to move westward in the equatorial zone. Their formation is thought to be associated with wind shears. They can be effective up to about 90MHz, and are often responsible for severe tv interference from the European Continent. VHF propagation from sporadic E rarely occurs at distances of less than 500km, but two-hop reflections are possible, either from a single cloud or from two which are adjacent, making possible ranges of up to 4,000km.

Auroral E is closely correlated with geomagnetic disturbances. In the northern hemisphere most reports of auroral echoes come from stations located within 1,100km of the zone of maximum auroral occurrence. Reports from stations further south coincide with auroral storms, when the displays extend further from the pole. There is a fairly good correlation between radio and visual aurora, particularly at between 50 and 100MHz. Radar observations show diffuse echoes when quiet auroral forms are observed and discrete echoes when rayed structures are present. Radar ranges are usually 400–1,100km, but ranges of up to 4,700km have been reported, probably resulting from multiple-hop reflection. The reflecting level is generally around 110km. There is a characteristic fading rate ranging from 50–250 per second, with Doppler shifting and an unmistakable tone. "Openings" on the 10, 4 and 2m amateur bands generally exhibit two phases of activity with bistatic transmission paths favouring stations on similar latitudes, each beaming well north of the direct path between them. The activity generally gradually extends southward, often ceasing abruptly while the longest paths are open.

Meteoric ionization is caused by the heating to incandescence by friction of small solid particles entering the earth's atmosphere. This results in the production of a long pencil of ionization extending over a length of 15km or more, chiefly in the height range 80 to 120km. It expands by diffusion and rapidly distorts due to vertical wind shears. Most trails detected by radio are effective for less than one second, but several last for longer periods, occasionally up to a minute and very occasionally for longer. There is a diurnal variation in activity, most trails occurring between midnight and dawn when the earth sweeps up the particles whose motion opposes it. There is a minimum around 1800 local time, when only meteors overtaking the earth are observed. The smaller *sporadic meteors*, most of them about the size of sand grains, are present throughout the year, but the larger *shower* meteors have definite orbits and predictable dates. **Fig 11.19** shows the daily and seasonal variation in meteor activity, based on a 24-hour continuous watch. Intermittent communication is possible using meteoric ionization between stations whose aerials have been prealigned to the optimum headings of 5 to 10 degrees to one side of the great-circle path between them. Small bursts of signal, referred to as *pings*, can be received by meteor scatter from distant broadcast (or other) stations situated 1,000–1,200km away (eg, for the south of England, Gdansk on 70·31MHz.)

Using the technique known as *ionospheric scatter* signals can be propagated over distances ranging between 1,000 and 2,000km at 30–60MHz by scattering from a volume in the lower D region of the ionosphere common to both transmitting and receiving aerials. The useful bandwidth is restricted to less than 10kHz and high power is necessary, but some useful immunity from fade-outs is claimed.

Geomagnetism

The earth's magnetic field is the resultant of two components, a *main field* originating within the earth, roughly equivalent to the field of a centred magnetic dipole inclined at about 11° to the earth's axis, and an *external field* produced by changes in the electric currents in the ionosphere.

The main field is strongest near the poles and exhibits slow secular changes of up to about 0·1 per cent a year. It is believed to be due to self-exciting dynamo action in the molten metallic core of the earth.

The field originating outside the earth is weaker and very variable, but it may amount to more than 5 per cent of the main field in the auroral zones, where it is strongest. It fluctuates regularly in intensity according to annual, lunar and diurnal cycles, and irregularly with a complex pattern of components down to *micropulsations* of very short duration.

Certain observatories are equipped with sensitive *magnetometers* which record changes in the field on at least three different axes, the total field being a vector quantity having both magnitude and direction. In the aspect of analysis which is of interest to us the daily records, called *magnetograms*, are read-off as eight *K-indices*, which are measures of the highest positive and negative departures from the "normal" daily curve during successive three-hourly periods, using a quasi-logarithmic scale ranging from 0 (quiet) to 9 (very disturbed). The various observatories do not all use the same scale factors in determining K-indices; the values are chosen so as to make the frequency distributions similar at all stations. Most large magnetic disturbances are global in nature and appear almost simultaneously all over the world. Another, but similar, indicator is the *planetary index*, Kp, which is graded on a finer scale of 0 to 9+

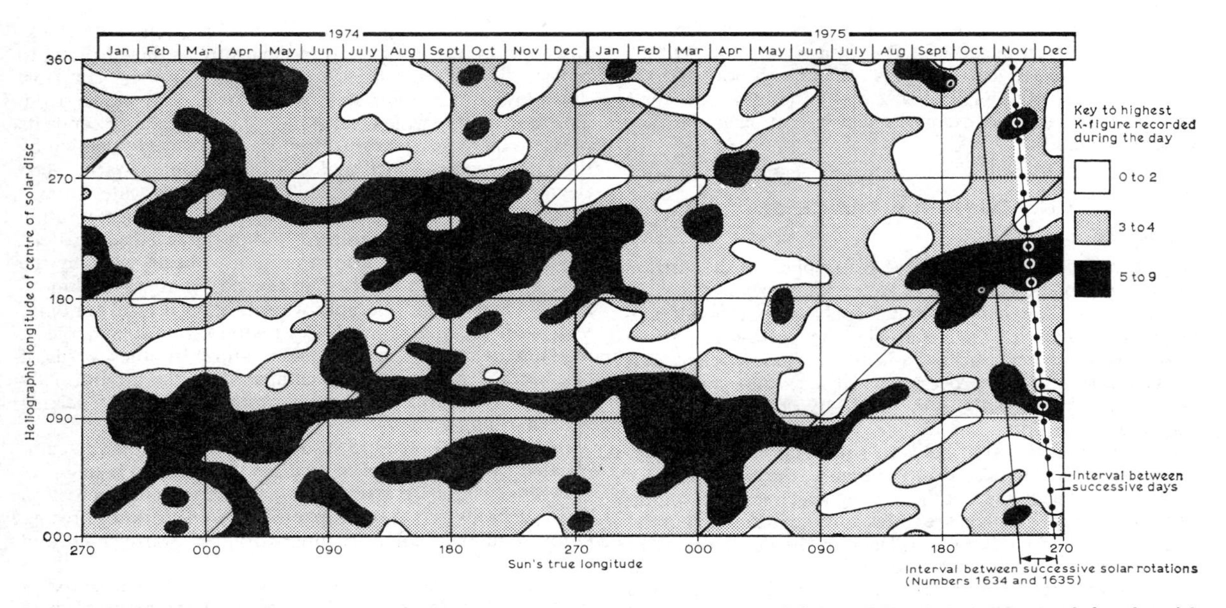

Fig 11.20. Geomagnetic activity diagram. The black areas indicate sequences where a K-figure of 5 or more was recorded at Lerwick Observatory. A horizontal line on this diagram denotes a recurrence period of 27 days, linked to the sun's rotation period as seen from the earth; the diagonal lines show the slope associated with a 25-day rotation period, such as the sun has in relation to a fixed point in space.

(in thirds), and is formed by combining the K-figures from 12 selected observatories. K (or Kp) indices of 5 or more may be regarded as indicative of magnetic storm conditions.

There is a tendency for 27-day recurrences in abnormal geomagnetic activity, linked to the solar synodic rotation period, but these are by no means as consistent as some authorities suggest. The chart shown in **Fig 11.20,** of a form which has been developed as part of the RSGB's Scientific Studies programme, clearly shows the patterns of recurring magnetic activity over a two-year period, 1974/75. The original diagram on which it is based was prepared by plotting the highest K-figure for each day on the spot determined by the longitude of the sun's central meridian facing the earth (thus a measure of the solar rotation), and a parameter called the sun's true longitude, which indicates the position of the earth around its orbit. Successive rotations build up a raster of daily figures in the way shown by the dots along the sloping right-hand edge, and the resulting chart should really be considered as being cylindrical, with the upper and lower edges brought together. The black areas surround the days when a K-figure of five or more was recorded—magnetic storm days—and the unshaded areas enclose relatively quiet days when the K-index was two or less. 27-day recurrences are clearly marked on this section of the record but occasionally there are periods when there is a marked tendency for storms to recur after an interval which appears to be linked to the sun's rotation period relative to the stars—indicated by the slope of the diagonal across the diagram. A good example of this occurred in 1971/72.

Periods of high geomagnetic activity are usually centred around the time of the equinoxes, with relatively quiet periods near the solstices. However, a *sudden commencement* (sc) is likely to occur at any time, linked to an event on the sun. There is an 11-year cycle in activity which tends to lag the solar cycle by a year or so.

Sudden Ionospheric Disturbances

We are now in a position to review the sequence of events which begins with the occurrence of a suitably-positioned major flare on the sun. When this takes place there is an emission of electromagnetic radiation covering a very wide range—X-rays, ultra-violet, visible light and radio waves between 3cm and 10m in length—all of which reaches the earth in about 8 minutes. The X-rays and uv light cause immediate increases in the D-layer ionization, leading to *short-wave* (*or Dellinger*) *fade-outs* which may persist for anything up to two hours. They frequently affect the E layer also and occasionally the F layer. At this time some magnetometers in high latitudes may record a short fluctuation known as a *crotchet*. Other effects observed are a sudden enhancement of atmospherics (s.e.a.), a sudden absorption of cosmic noise (scna), and sudden phase anomalies (s.p.a.) on very low-frequency transmissions.

This is followed after a few hours by the arrival of *cosmic ray particles*, and perhaps the onset of *polar-cap absorption* (pca).

The main stream of particles arrives after an interval of 20–40 hours and consists of protons and electrons borne by the solar wind. When they reach the day side of the earth's magnetosphere they compress it, causing some of the particles oscillating within the Van Allen belts to spill out into the ionosphere along the night side of the auroral zone, where they manifest themselves as visible displays of aurora, also causing a strong polar electro-jet to flow in the lower ionosphere. Changes in the make-up of the trapping regions leads to variations in the circulating ring-current which leads to violent alterations in the strength of the geomagnetic field, bringing about the sudden commencement, which is the first indication of a magnetic storm.

Associated with the magnetic storms are ionospheric storms, and both may persist for several days. The most prominent features are the reduction in F2 critical frequencies (f_oF2) and an increase in D-region absorption. During the storm period signal strengths remain very low and are subject to flutter fading. The effects of an ionospheric storm are most pronounced on paths which approach the geomagnetic poles.

TROPOSPHERIC PROCESSES

Introduction

Because it is all around us the troposphere is the portion of the atmosphere which we ought to know best.

We are dealing here with the sun's output of electromagnetic radiation which falls in the infra-red portion of the spectrum, between 10^{-6} and 10^{-5} metres wavelength, is converted to heat (by processes which need not concern us here) and distributed about the world by radiation, conduction and convection.

At this point our link between solar actions and atmospheric reactions breaks down, because the very variable nature of the medium, and the ease by which it can be modified both by topographical features and the differing thermal conductivities of land and sea, leads to the development of air masses having such widely contrasting properties that it becomes impossible to find a direct correlation between day-to-day climatic features and solar emissions. We must accept the fact that in meteorology "Chance" plays a powerful role and look to functions of the resulting weather pattern for any relationships with signal level, without inquiring too deeply into the way in which they may be connected with events on the sun.

Pressure Systems and Fronts

The television weatherman provides such a regular insight into the appearance and progressions of surface pressure patterns that it would be wasteful of space to repeat it all here. Suffice it to record that there are two closed systems of isobars involved, known as *anticyclones* and *depressions* (or, less-commonly nowadays, *cyclones*) within or around which appear *ridges* of high pressure, *troughs* of low pressure (whose very names betray their kinship), and *cols*, which are slack regions of even pressure, bounded by two opposing anticyclones and two opposing depressions.

The most important consideration about these pressure systems, in so far as it affects radio propagation at vhf and above, is the direction of the vertical motion associated with them.

Depressions are closed systems with low pressure at the centre. They vary considerably in size, and so also in mobility, and frequently follow one another in quick succession across the North Atlantic. They are accompanied by circulating winds which tend to blow towards the centre of the system in an anti-clockwise direction. The air so brought in has to find an outlet, so it rises, whereby its pressure falls, the air cools and in doing so causes the relative humidity to increase. When saturation is reached, cloud forms and further rising may cause water droplets to condense out and fall as rain. Point one: Depressions are associated with rising air.

Anticyclones are generally large closed systems which have high pressure in the centre. Once established they tend to persist for a relatively long time, moving but slowly and effectively blocking the path of approaching depressions which are forced to go round them. Winds circulate clockwise, spreading outwards from the centre as they do, and to replace air lost from the system in this way there is a slow downflow called *subsidence* which brings air down from aloft over a very wide area. As the subsiding air descends its pressure increases, and this produces dynamical warming by the same process which makes a bicycle pump warm when the air inside it is compressed. The amount of water vapour which can be contained in a sample of air without saturating it is a function of temperature, and in this particular case if the air was originally near saturation to begin with, by the time the subsiding air has descended from, say, 5km to 2km, it arrives considerably warmer than its surroundings and by then contains much less than a saturating charge of moisture at the new, higher, temperature. In other words, it has become warm and dry compared to the air normally found at that level. Point two: Anticyclones are associated with descending air.

In addition to pressure systems the weather map is complicated by the inclusion of *fronts*, which are the boundaries between two air masses having different characteristics. They generally arrive accompanied by some form of precipitation, and they come in three varieties: warm, cold and occluded.

Warm fronts (indicated on a chart by a line edged with rounded "bumps" on the forward side) are regions where warm air is meeting cold air and being forced to rise above it, precipitating on the way.

Cold fronts (indicated by triangular "spikes" on the forward side of a line) are regions where cold air is undercutting warm. The front itself is often accompanied by towering clouds and heavy rain (sometimes thundery), followed by the sort of weather described as "showers and bright intervals".

Occluded fronts (shown by alternate "bumps" and "spikes") are really the boundary between three air masses being, in effect, a cold front which has overtaken a warm front and one or the other has been lifted up above the ground.

It is perhaps unnecessary to add that there is rather more to meteorology than it has been possible to include in this brief survey.

Vertical Motion

It is a simple matter of observation that there is some correlation between vhf signal levels and surface pressure readings, but it is generally found to be only a coarse indicator, sometimes showing little more than the fact that high signal levels accompany high pressure and low signal levels accompany low pressure. The reason that it correlates at all is due to the fact that high pressure generally indicates the presence of an anticyclone which, in turn, heralds the likelihood of descending air.

The reason that subsidence is so important stems from the fact that it causes dry air to be brought down to lower levels where it is likely to meet up with cool moist air which has been stirred up from the surface by turbulence. The result then is the appearance of a narrow boundary region in which refractive index falls off very rapidly with increasing height—the conditions needed to bring about the sharp bending of high-angle radiation which causes it to return to the ground many miles beyond the normal radio horizon. Whether you regard this in the light of being a benefit or a misfortune depends on whether you are more interested in long-range communication or in wanting to watch an interference-free television screen.

The essential part of the process is that the descending air must meet turbulent moist air before it can become effective as a boundary. If the degree of turbulence declines the boundary descends along with the subsiding air above it and, when it reaches the ground all the abnormal conditions rapidly become sub-normal, and a sudden drop-out occurs. Occasionally this means that operators on a hill suffer the disappointment of hearing others below them still working dx they can no longer hear themselves. Note, however, that anticyclones are not uniformly distributed with descending air, nor is the necessary moist air always available lower down, but a situation such as a damp foggy night in the middle of an anticyclonic period is almost certain to be accompanied by a strong boundary layer. Ascending air on its own never leads to spectacular conditions. Depressions therefore result in situations in which the amount of ray bending is controlled by a fairly regular fall-off of refractive index. The passage of warm fronts is usually accompanied by declining signal strength, but occasionally cold fronts and some occlusions are preceded by a short period of enhancement.

To sum up, there is very little of value about propagation conditions which can be deduced from surface observations of atmospheric pressure. The only reliable indicator is a knowledge of the vertical refractive index structure in the neighbourhood of the transmission path.

Radio Meteorological Analysis

It remains now to consider how the vertical distribution of refractive index can be displayed in a way which gives emphasis to those features which are important in tropospheric propagation studies.

Obviously the first choice would be the construction of atmospheric cross-sections along paths of interest, at times when anomalous conditions were present, using values calculated by the normal refractive index formula. The results are often disappointing, however, because the general decrease of refractive index with height is so great compared to the magnitude of the anomalies looked for that, although they are undoubtedly there, they do not strike the eye without a search.

A closely-related function of refractive index overcomes this difficulty, with the added attraction that it can be computed graphically, and easily, directly from published data obtained from upper-air meteorological soundings. It is called *potential refractive index*, K, and may be defined as being the refractive index which a sample of air at any level would have if brought adiabatically (that is, without gain or loss of heat or moisture) to a standard pressure of 1,000mb; see references [4], [5] and [6].

This adiabatic process is the one which governs (among other things) the increase of temperature in air which is descending in an anticyclone, so that, besides the benefits of the normalizing process (which acts in a way similar to that whereby it is easy to compare different sized samples of statistics when they have all been converted to percentages) there is the added attraction that the subsiding air tends to retain its original value of potential refractive index all the time it is progressing on its downward journey. This means that low values of K are carried down with the subsiding

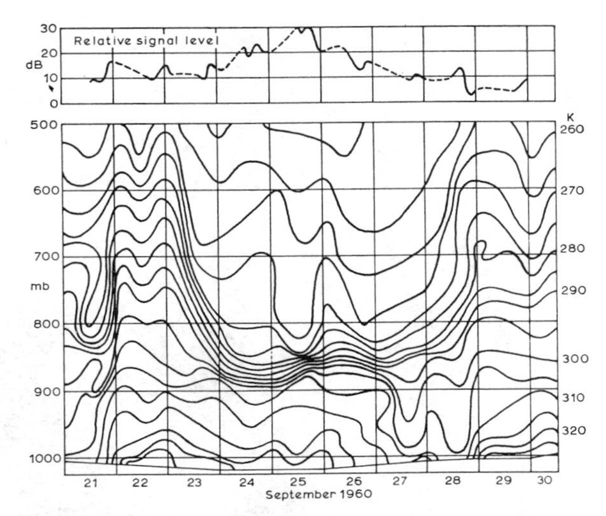

Fig 11.21. The relationship between variations in potential refractive index in the atmosphere and signal strengths over a long-distance vhf tropospheric path. (*With acknowledgements to J Atmos Terr Phys, Pergamon Press.*)

air, in sharp contrast to the values normally found there. A cross-section of the atmosphere during an anticyclonic period, drawn up using potential refractive index, gives an easily-recognizable impression of this.

Fig 11.21 is not a cross-section, but a time-section, showing the way in which the vertical potential refractive index distribution over Crawley, Sussex, varied during a 10-day period in September, 1960. There is no mistaking the downcoming air from the anticyclone and the establishment of the boundary layer around 850mb (about 1·5km). Note how the signal strength of the Lille television transmission on 174MHz varied on a pen recording made near Reading, Berkshire, during the period, with peak amplitudes occurring around the time when the layering was low and well-defined, and observe also the marked decline which coincided with the end of the anticyclonic period. Time-sections such as these also show very clearly the ascending air in depressions (although the K value begins to alter when saturation is reached) and the passage of any fronts which happen to be in the vicinity of the radiosonde station at ascent time.

GEODESY

Map Projections

Maps are very much a part of the life of a radio amateur; yet how few of us ever pause to wonder if we are using the right map for our particular purpose, or take the trouble to find out the reason why there are so many different forms of projection.

The cartographer is faced with a basic problem, namely, that a piece of paper is flat and the earth is not. For that reason his map, whatever the form it may take, can never succeed in being faithful in all respects—only a globe achieves that. The amount by which it departs from the truth depends not only on how big a portion of the globe has been displayed at one viewing, but on what quality the map-maker has wanted to keep correct at the expense of all others.

Projections can be divided into three groups: those which

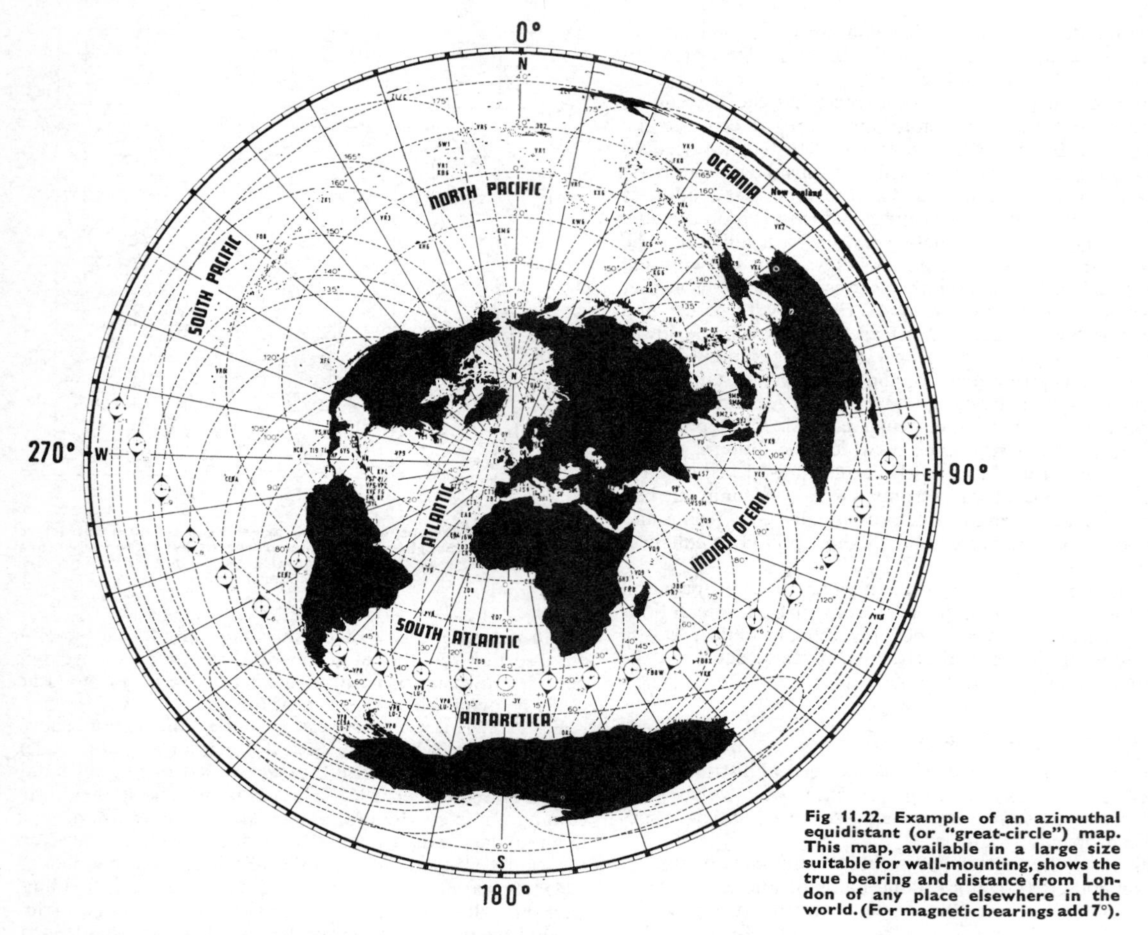

Fig 11.22. Example of an azimuthal equidistant (or "great-circle") map. This map, available in a large size suitable for wall-mounting, shows the true bearing and distance from London of any place elsewhere in the world. (For magnetic bearings add 7°).

show areas correctly, described, logically enough, as *equal-area projections;* those which show the shapes of small areas correctly, known as *orthomorphic* or *conformal projections;* and those which represent neither shape nor area correctly, but which have some other property which meets a particular need.

The conformal group, useful for atlas maps generally, weather charts, satellite tracks, etc, includes the following:

Stereographic, where latitudes and longitudes are all either straight lines or arcs of circles, formed by projection on to a plane surface tangent either to one of the poles (*Polar*), the equator (*Equatorial*), or somewhere intermediate (*Oblique*). Small circles on the globe remain circles on the map, but the scale increases with increasing radius from the centre of projection.

Lambert's conformal conic, where all meridians are straight lines and all parallels are circles. It is formed by projection on to a cone whose axis passes through the earth's poles.

Mercator's, where meridians and parallels are straight lines intersecting at right-angles. The meridians are equidistant, but the parallels are spaced at intervals which rapidly increase with latitude. It is formed by projection on to a cylinder which touches the globe at the equator. Any straight line is a line of constant bearing (a *rhumb-line*, which is not the same thing as a *great circle*, the path a radio wave takes between two given points on the earth's surface). There is a scale distortion which gets progressively more severe away from the equator, to such an extent that it becomes impossible to show the poles, but most people accept these distortions as being normal, because this is the best-known of all the projections.

Transverse Mercator is a modification of the "classical" system, and is formed by projection on to a cylinder which touches the globe along selected opposing meridians. It therefore corresponds to an ordinary Mercator turned through 90 degrees, and is of value for displaying an area which is extensive in latitude but limited in longitude. A variant is the *Universal Transverse Mercator*, which forms the basis of a number of reference grids, including the one used on British Ordnance Survey maps.

The equal-area group is used when it is necessary to display the relative distribution of something, generally on a world-wide scale. It includes the following:

Azimuthal equal-area projection, having radial symmetry about the centre, which may be at either pole (*Polar*), at the equator (*Equatorial*), or intermediate (*Oblique*). With this system the entire globe can be shown in a circular map, but there is severe distortion towards its periphery.

Mollweide's homolographic projection, where the central meridian is straight and the others elliptical.

Sinusoidal projection, where the central meridian is straight and the others parts of sine curves.

Homolosine projection, which is a combination of the previous two, with an irregular outline because of interruptions which are generally arranged to occur over ocean areas.

The final group includes the two following, which are of particular interest in propagation studies:

Azimuthal equidistant, centred on a particular place, from whence all straight lines are great circles at their true azimuths.

The scale is constant and linear along any radius. Well-known as a "Great-circle map", **Fig 11.22.**

Gnomonic, constructed by projection from a point at the earth's centre on to a tangent plane touching the globe. Any straight line on the map is a great-circle. Because the size of the map expands very rapidly with increasing distance from the centre they do not normally cover a large area. Often produced as a skeleton map on which a great-circle can be drawn and used to provide a series of latitudes and longitudes by which the path can be replotted on a more detailed map based on a different projection.

Great-circle Calculations

The shortest distance between two points on the surface of the earth lies along the great-circle which passes through them. On a globe this great-circle path can be represented by a tightly-stretched thread joining the two locations.

It is sometimes useful to be able to calculate the great-circle bearing, and the distance of one point from another, and the expressions which follow enable this to be done.

First label the two points A and B.

Then let L_a = latitude of point A

L_b = latitude of point B

L_o = the difference in longitude between A and B

C = the direction of B from A, in degrees east or west from north in the northern hemisphere, or from south in the southern hemisphere.

D = the angle of arc between A and B.

It follows that

$$\cos D = \sin L_a . \sin L_b + \cos L_a . \cos L_b . \cos L_o$$

D can be converted to distance, knowing that

1 degree of arc = 111·2km or 69·06 miles

1 minute of arc = 1·853km or 1·151 miles

Once D is known (in angle of arc), then,

$$\cos C = \frac{\sin L_b - \sin L_a . \cos D}{\cos L_a . \sin D}$$

Note:

1. For stations in the northern hemisphere call latitudes positive.
2. For stations in the southern hemisphere call latitudes negative.
3. Cos L_a and cos L_b are always positive.
4. Cos L_o is positive between 0 and 90°, negative between 90° and 180°.
5. Sin L_a and sin L_b are negative in the southern hemisphere.
6. The bearing for the reverse path can be found by transposing the letters on the two locations.

TABLE 11.8

A method of converting QTH-locator groups to latitude and longitude.

LONGITUDE

First letter	Mid-square longitude
A	01°00'E
B	03°00'E
C	05°00'E
D	07°00'E
E	09°00'E
F	11°00'E
G	13°00'E
H	15°00'E
I	17°00'E
J	19°00'E
K	21°00'E
L	23°00'E
M	25°00'E
N	27°00'E
O	29°00'E
P	31°00'E
Q	33°00'E
R	35°00'E
S	37°00'E
T	39°00'E
U	11°00'W
V	09°00'W
W	07°00'W
X	05°00'W
Y	03°00'W
Z	01°00'W

2nd figure	Increment of longitude Final letter F, G, H	A, E, J	B C, D
1	−58'E	−54'E	−50'E
2	−46'E	−42'E	−38'E
3	−34'E	−30'E	−26'E
4	−22'E	−18'E	−14'E
5	−10'E	−06'E	−02'E
6	+02'E	+06'E	+10'E
7	+14'E	+18'E	+22'E
8	+26'E	+30'E	+34'E
9	+38'E	+42'E	+46'E
0	+50'E	+54'E	+58'E
1	+58'W	+54'W	+50'W
2	+46'W	+42'W	+38'W
3	+34'W	+30'W	+26'W
4	+22'W	+18'W	+14'W
5	+10'W	+06'W	+02'W
6	−02'W	−06'W	−10'W
7	−14'W	−18'W	−22'W
8	−26'W	−30'W	−34'W
9	−38'W	−42'W	−46'W
0	−50'W	−54'W	−58'W

LATITUDE

Second letter	Mid-square latitude
A	40°30'N
B	41°30'N
C	42°30'N
D	43°30'N
E	44°30'N
F	45°30'N
G	46°30'N
H	47°30'N
I	48°30'N
J	49°30'N
K	50°30'N
L	51°30'N
M	52°30'N
N	53°30'N
O	54°30'N
P	55°30'N
Q	56°30'N
R	57°30'N
S	58°30'N
T	59°30'N
U	60°30'N
V	61°30'N
W	62°30'N
X	63°30'N
Y	64°30'N
Z	65°30'N

Figures	Increment of latitude Final letter A, B, H	C, G, J	D, E, F
01–10	+28¾'N	+26¼'N	+23¾'N
11–20	+21¼'N	+18¾'N	+16¼'N
21–30	+13¾'N	+11¼'N	+08¾'N
31–40	+06¼'N	+03¾'N	+01¼'N
41–50	−01¼'N	−03¾'N	−06¼'N
51–60	−08¾'N	−11¼'N	−13¾'N
61–70	−16¼'N	−18¾'N	−21¼'N
71–80	−23¾'N	−26¼'N	−28¾'N

Examples:

(1) YM70C
Long 03°00'W — 58'W = 02°02'W
Lat 52°30'N — 18¾'N = 52°11¼'N

(2) MB34H
Long 25°00'E — 22'E = 24°38'E
Lat 41°30'N + 05¼'N = 41°35¼'N

It is advisable to make estimates of the bearings on a globe, wherever possible, to ensure that they have been placed in the correct quadrant.

The QTH Locator

One of the basic requirements for an active interest in the practical aspects of propagation is a knowledge of the approximate whereabouts of the other station.

A form of position reporting which is popular on the Continent and used also in this country is the QTH locator. Although based on latitude and longitude the subdivisions have been made in such an illogical way that it is impossible

TABLE 11.9

D (miles)	Horizontal scale (inches)	D²	D²/2 = h	Vertical scale (inches)
5	0·5	25	12·5	0·125
10	1·0	100	50·0	0·500
15	1·5	225	112·5	1·125
20	2·0	400	200·0	2·000

to convert directly from one scale to the other. However, the method shown in **Table 11.8** provides a relatively simple way of converting a QTH locator group into its equivalent latitude and longitude, and the two examples should be sufficient to demonstrate the method if the column headings are found to be inadequate.

Plotting VHF/UHF Path Profiles

As mentioned in the section on wave propagation in the troposphere it is usual to draw path profiles for vhf/uhf working by presuming that standard refraction can be represented by regarding the earth's radius as being four-thirds of its normal value. The use of this convention then allows the convenience of being able to represent ray paths by straight lines.

Because the various factors conveniently cancel it is common practice to work in terms of miles for the distance and feet for the height, taking advantage of the simple relationship which results.

Suppose that a profile for a path 40 miles in length is required. It is usual to work from the centre outwards, so that, in this case, distances will extend from 0 to 20 miles in each direction. Decide on suitable horizontal and vertical scales—let us suppose here that 1 inch on paper will be equivalent to 10 miles and 100 feet respectively. Then, using the formula $h = D^2/2$, the figures given in **Table 11.9** result.

These scaled values should then be plotted as shown in **Fig 11.23** to provide the "sea-level" datum along the path, above which must be extended the actual land profile as determined from contours on a map, transferring the altitudes vertically upwards at the appropriate distances. The heights of the transmitting and receiving aerials must then be plotted, using the same scale. If it is possible to draw a straight line between both aerials, T-R in the diagram, without striking the intervening ground, the path should be an unobstructed line-of-sight one under normal conditions of tropospheric refraction. However, in cases of near-grazing it may be necessary to consider the extent of the first Fresnel zone before a clear decision can be made.

THE BEACON NETWORK

The establishment of an extensive network of beacon stations for amateur propagation research is one of the common aims of a number of European societies, and one in which the RSGB has been able to take a leading role. Beacons are now in operation in the UK on most bands from 28MHz to 10GHz, and details are given frequently in *Radio Communication.*

Many operators will tèstify to the value of these stations as reliable indicators of band conditions, particularly when the level of activity is low, and those who co-ordinate the amateur radio programme of research will welcome regular observations of any of those listed, or of similar beacon transmissions originating outside Great Britain, particularly during periods of auroral-E, sporadic-E, or other irregular modes of propagation.

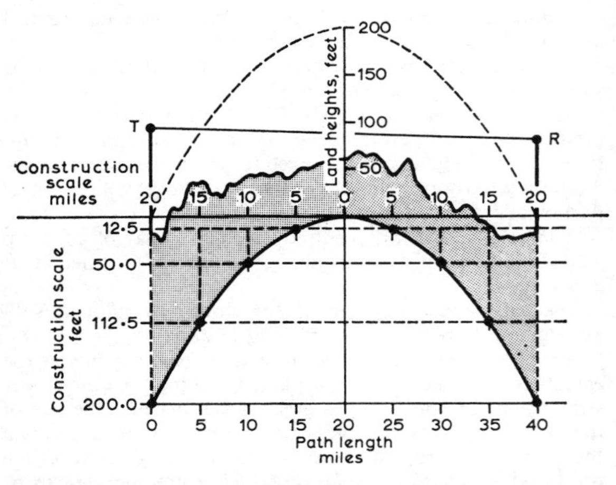

Fig 11.23. Construction of a "four-thirds-earth" profile. Land heights are measured upward from the lower curve. On this diagram rays subjected to "normal" variations of refractive index with height may be represented by straight lines.

REFERENCES AND BIBLIOGRAPHY

[1] "Flare spot", P. W. Sollom, *Radio Communication* December 1970, p820; January 1971, p20; February 1971, p92.

[2] "The ground beneath us", R. C. Hills, *RSGB Bulletin* June 1966, p375.

[3] Report 239–2 in "Propagation in non-ionized media", Study Group 5, CCIR XIIth Plenary Assembly, New Delhi, 1970, Volume II, Part I, ITU, Geneva, 1970.

[4] *VHF-UHF Manual,* 3rd edn, D. S. Evans and G. R. Jessop, RSGB, London, 1976, Chapter 2.

[5] "The application of potential refractive index in tropospheric wave propagation", R. G. Flavell and J. A. Lane, *J Atmos Terr Phys,* Vol 24, 1962, pp47–56.

[6] "Further radio-meteorological analysis using potential refractive index", R. G. Flavell, *J Atmos Terr Phys,* Vol 26, 1964, pp44–49.

General Works

"Transmission and propagation", *The Services' Textbook of Radio,* Vol 5, HMSO, London, 1958, Chapter 14.

Reference Data for Radio Engineers, ITT, Chapter 26.

Sun, Earth and Radio, J. A. Ratcliffe, World University Library, London, 1970.

The Ionosphere

Ionospheric Radio Propagation, National Bureau of Standards Monograph 80, NBS, Washington, 1965.

"Ionospheric propagation", Study Group 6, CCIR XIIIth Plenary Assembly, Geneva, 1974, Vol 6, ITU, Geneva, 1975.

The Troposphere

Radio Meteorology, National Bureau of Standards Monograph 92, NBS, Washington, 1966.

"Propagation in non-ionized media", Study Group 5, CCIR XIIIth Plenary Assembly, Geneva, 1974, Vol 5, ITU, Geneva, 1975.

CHAPTER 12

HF AERIALS

IN setting up a link for radio communication between two stations, certain specific items of equipment must be provided at each end of the circuit. At the sending end there must be a transmitter which imposes the signal intelligence upon a carrier wave at radio frequency and amplifies it to the required power level. At the other end a receiver is required which will again amplify the weak incoming signal, and then decode from it the original intelligence.

The signal passes from one station to the other as a wave propagating in the atmosphere, but in order to achieve this it is necessary to have at the sending end something which will take the power from the transmitter and launch it as a wave, and at the other end extract energy from the wave to feed the receiver. This is an *aerial* and, because the fundamental action of an aerial is reversible, similar aerials can be used at both ends. The aerial then is a means of converting power flowing in wires to energy flowing in a wave in space, or is simply considered as a coupling transformer between the wires and free space.

FUNDAMENTAL PROPERTIES

Many of the fundamental properties of aerials are common to their use in any part of the radio frequency spectrum and in free space, ie when the influence of the ground and surrounding objects can be neglected, a piece of wire which is one half-wavelength long will have the same directional radiation characteristics and appear as exactly the same load to a transmitter whether it be a half-wavelength at 1MHz or at 1,000MHz. All that matters is that the wire should maintain a constant relationship between its physical length and the wavelength used. Strictly speaking, the wire diameter should also be scaled with the wavelength since the ratio of length to diameter determines the sharpness of resonances, but this does not affect the basic radiating or receiving properties. The classification into aerials for hf and aerials for vhf in this handbook is one based on the differing practical requirements for aerials for use in the two frequency ranges, taking account of physical size limitations and also the widely differing mechanisms of propagation which dictate different forms of aerial. Much of what is to be said in this chapter about the fundamental characteristics of aerials is equally applicable to both hf and vhf, and it is only later in the chapter that consideration is given to aerials specifically designed for the hf bands. At that point the reader with vhf interests should pick up the story of specialized vhf aerials in Chapter 13—*VHF/UHF Aerials.*

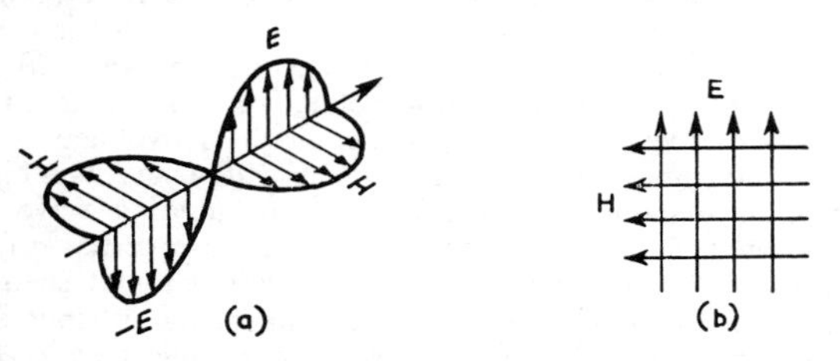

Fig 12.1. Instantaneous representation of a travelling radio wave: (a) along the path of travel and (b) for the wave approaching the observer.

Wave Motion

An understanding of aerial behaviour is very closely linked with an understanding of basic propagation, and some aspects cannot be dissociated from it.

In Chapter 11 it is explained that radio waves are propagated as an expanding electromagnetic wavefront whose intensity decays as it moves further out from the source. The wavefront is formed of electric and magnetic fields which exist at right angles to one another in the plane of the wavefront, and each field may be represented by parallel lines of force having lengths proportional to the strength of the field. The intensity of the wavefront is usually varying sinusoidally with time, as the waves expand like ripples on a pond, and the peak value of the wave decays as the wave moves further and further away from its source. The arrangement of one cycle of the wave along its direction of travel at a particular moment in time is represented in **Fig 12.1**, as it appears in general, and also as it appears to the observer whom it is approaching.

Some imagination is necessary to appreciate fully the structure of the wave as it propagates through space as a whole. The behaviour of the separate but related electric and magnetic fields is more easily visualized if one imagines them as oscillations travelling along a slack string which is being excited at one end. An observer at a point along the string will see approaching him a succession of crests and troughs which pass him along the string. A complete electromagnetic wave is then two such strings identically located but oscillating at right angles and with the crests on each string passing the observer at the same instant.

The distance between successive peaks in the intensity of the electromagnetic wave as it moves along its direction of propagation is called the *wavelength* and is customarily measured in metres. The wave moves through space with the velocity of light, or more exactly light moves through space with the velocity of electromagnetic waves since light is really em radiation in a narrow band of extremely high frequencies. This velocity is approximately 3×10^8 metres per second (186,000 miles per second), and is known as the *velocity of propagation*. As stated above, from the point of view of a fixed observer, the wave will pass by as a succession of peaks and troughs. The portion of a wave between successive peaks (or troughs) is called one *cycle*, and the number of such *cycles* which pass the observer in one second is known as the *frequency* of the wave, usually measured in hertz (Hz), where one hertz denotes one cycle per second.

Since the wave travels with a known velocity, it is easy to establish a relationship between the frequency and the wavelength, and if f is the frequency in hertz (Hz) and λ (lambda) the wavelength in metres, then

$$f \times \lambda = 3 \times 10^8$$

For example, a wavelength of 160 metres corresponds to a frequency of $1{\cdot}87 \times 10^6$Hz or 1·87MHz.

Referring again to the wave illustrated in Fig 12.1 an observer facing the wave would "see" the field as shown in Fig 12.1(b) which would be alternating at the frequency f. It is this alternating field that excites a receiving aerial into which it delivers some of its power. The arrows indicate the conventional directions of the electric and magnetic fields, E and H, relative to the direction of motion. If the polarity of one component along either E or H in the diagram is reversed, then the wave is receding instead of approaching the observer. The components E and H cannot be separated; together they are the wave and together they represent indestructible energy. They wax and wane together and when the energy disappears momentarily from one point, it must re-appear further along the track; in this way the wave is propagated.

Travelling and Standing Waves

The wave illustrated in Fig 12.1 is unrestricted in its motion and is called a travelling wave. As long as they are confined to the vicinity of the earth all waves must, however, eventually encounter obstacles. In order to understand what happens in these circumstances, it is convenient to imagine that the obstacle is a very large sheet of metal, since metal is a good conductor of electricity and, as will be seen, an effective barrier to the wave (**Fig 12.2**). In this case, the electric field is "short-circuited" by the metal and must therefore always be zero at the surface. It manifests itself in the form of a current in the sheet (shown dotted) like the current that is induced in a receiving aerial, and this current re-radiates the wave in the direction from whence it came.

In this way a wave is reflected from the metal but due to the relative directions of field and propagation, the components of forward and reflected waves appear as in Fig 12.2(b). It will be seen that at the surface of the metal the two electric fields are of opposite polarity and thus cancel each other out, but the magnetic fields are additive.

If the two waves are combined the resultant wave of Fig 12.2(a) is obtained in which the electric and magnetic maxima no longer coincide, but are separated by quarter wave intervals. Such a combined wave does not move, because the energy can alternate in form between electric field in one sector and magnetic field in an adjacent one. This type of wave is called a *standing wave* and exists near reflecting objects comparable in size with the length of the wave. If a small receiving aerial and detector were moved outwards from the front of the reflector the output of the receiver would vary with distance and thus indicate the stationary wave pattern.

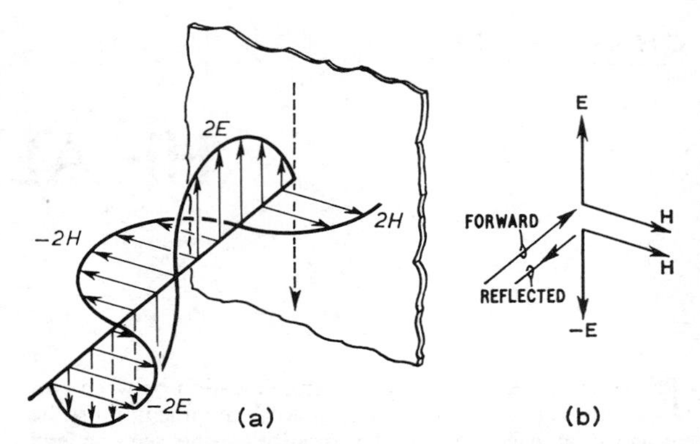

Fig 12.2. Standing waves due to short-circuit reflection at a metal surface, with analysis into forward and reflected wave components.

In practice radio waves more usually encounter poor conductors, such as buildings or hillsides, and in this case some of the power from the wave is absorbed by or reflected from the obstacle. Later in this chapter, aerials will be described which incorporate conductive elements deliberately used as reflectors. It will also be shown that oscillatory currents in aerials or feeders can be reflected to produce standing waves.

A travelling wave moving away into space represents a flow of energy from the source. The transmitter can be regarded as being at the centre of a large sphere, with the radiated power passing out through its surface. The larger the sphere the more thinly will it be spread, so that the further the receiver recedes from the transmitter, the smaller will be the signal extracted by means of a receiving aerial of given size from a given area of the wave front.

Resonant Aerials

If an oscillatory current is passed along a wire, the electric and magnetic fields associated with it can be considered as a wave attached to the wire and travelling along it as far as it continues, and if the wire finally terminates in, say, an insulator, the wave cannot proceed but is reflected. This reflection is an open-circuit reflection and the wire produces standing-wave fields complementary to those generated by a short-circuit reflection as in Fig 12.2, with corresponding voltages and currents. **Fig 12.3** shows two typical cases where the wire is of such a length that a number of complete cycles of the standing wave can exist along it. Since the end of the wire is an open circuit, the current at that point must be zero and the voltage a maximum. Therefore at a point one quarter wavelength from the end, the current must be a maximum and the voltage will be zero. The positions of maxima are usually known as current (or voltage) *anti-nodes* or *loops* and the intermediate positions as *nodes* or *zeros*. At positions of current loops, the current-to-voltage ratio is high and the wire will behave as a low impedance circuit. At voltage loops the condition is reversed and the wire will behave as a high impedance circuit. A wire carrying a standing wave as illustrated in Fig 12.3 exhibits similar properties to a resonant circuit and is also an efficient radiator of energy, most of the damping of the circuit being attributable to this radiation. This is a *resonant* or *standing wave aerial* and the majority of the aerials met with in practice are of this general type. The wire need not be of resonant

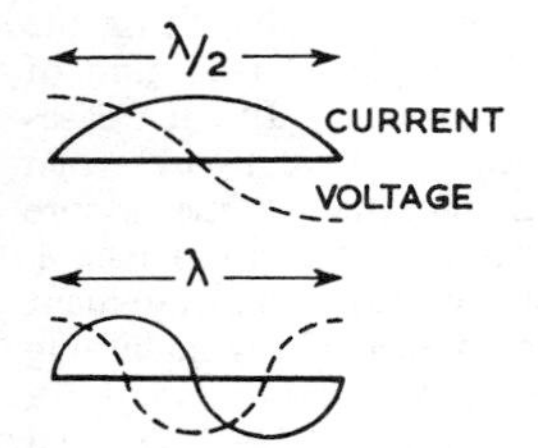

Fig 12.3. Standing waves on resonant aerials, showing voltage and current variation along the wire. The upper aerial is λ/2 long and is working at the fundamental frequency; the lower is a full-wave or second harmonic aerial.

length, but it is then equivalent to a detuned circuit and unless brought into resonance by additional capacitance or inductance it cannot be energized efficiently.

The length for true resonance is not quite an exact multiple of the half-wavelength because the effect of radiation causes a slight retardation of the wave on the wire and also because the supporting insulators may introduce a little extra capacitance at the ends. An approximate formula suitable for wire aerials is:

$$\textit{Length (feet)} = \frac{492(n - 0{\cdot}05)}{f}$$

where n is the number of complete half-waves in the aerial and f is the frequency in megahertz.

Radiation

The actual physics of the radiation of energy from the wire is an involved matter, which is best left to theoretical text-books. It suffices to say here that the current flowing up and down the wire gives rise to a magnetic field around the wire, while the charges in motion (which constitute the current) carry with them an electric field. Due to the reversing nature of the current, the two fields are mutually supporting and expand outwards from the wire, carrying with them energy from the exciting current. There exists in the immediate vicinity of the wire an oscillating field known as the *induction* field (similar to that surrounding an induction coil or a magnet), but this decays in strength rapidly as the distance from the wire increases. At a distance of $\lambda/2\pi$, or approximately one-sixth of a wavelength, it is equal in strength to the *radiation* field, but beyond one or two wavelengths has fallen to a negligible level. Radiation takes place from any elevated wire carrying a radio frequency current unless prevented by screening or cancelled by an opposing field of equal magnitude. At any given frequency and for a given direction relative to the wire the field strength produced at a distant point is proportional to the current multiplied by the length of wire through which it flows.

Directivity

The radiation field which surrounds the wire is not uniformly strong in all directions. It is strongest in directions at right angles to the current flow in the wire and falls in intensity to zero along the axis of the wire; in other words the wire exhibits *directivity* in its radiation pattern, the energy being concentrated in some directions at the expense of others. Later in this chapter it will be explained how directivity may be increased by using numbers of wires in arrays. These aerial arrays are called *beams* because they concentrate radiation in the desired direction like a beam of light from a torch. Because a number of wires or elements are needed to create the additional directivity, beam aerials usually require more space than simple ones, and this limits the extent to which they can usefully be employed at the longer wavelengths. When space is limited to an average garden, useful beam action tends to be obtainable only with aerials operating on 14MHz or higher bands, those for 1·8MHz, 3·5MHz and 7MHz being generally limited to simple arrangements involving single wires. In the vhf bands, the wavelength is sufficiently short that the various elements of the aerial may be made rigid and self-supporting and the aerial system may therefore be quite elaborate.

Dipoles

One of the most commonly used words in aerial work is *dipole*. Basically a dipole is simply some device (in the present context an aerial) which has two "poles" or terminals into which radiation-producing currents flow. The two elements may be of any length, and a certain amount of confusion sometimes arises from the failure to state the length involved. In practice it is usually safe to assume that when the word dipole is used by itself, it is intended to describe a half-wavelength aerial, ie a radiator of electrical length λ/2, fed by a balanced connection at the centre. Any reference to gain over a dipole is assumed to refer to this λ/2 dipole. When reference to another form of dipole is intended, it is usual to state the overall length, eg a *full-wave dipole*, *short dipole*. etc. A short dipole is less than λ/2 in length but needs to be tuned to resonance by the addition of inductance, usually in the centre, or some form of capacitive end-loading as discussed later. Shortening has little effect on radiation pattern but, if carried too far, leads to poor efficiency and excessively narrow bandwidth.

Loops containing between two-thirds and one-and-a-half wavelengths of wire have radiation patterns very similar to those of half-wave dipoles.

A further reference sometimes encountered is to the *monopole* or *unipole*. This is an unbalanced radiator, fed against an earth plane, and a common example is the ground plane vertical described later in this chapter.

Gain

If one aerial system can be made to concentrate more radiation in a certain direction than another aerial, for the same total power supplied, then it is said to exhibit *gain* over the second aerial in that direction. In other words, more power would have to be supplied to the reference aerial to give the same radiated signal in the direction under consideration, and hence the better aerial has effectively gained in power over the other. Gain can be expressed either as a ratio of the powers required to be supplied to each aerial to give equal signals at a distant point, or as the ratio of the signals received at that point from the two aerials when they are driven with the same power input. Gain is usually expressed in decibels: a table of conversion from voltage and power ratios will be found in the preceding chapter (Table 11.1). Gain is of course closely related to directivity, but an aerial can be highly directive and yet have a power loss as a result of energy dissipated in aerial wires and surrounding objects. As we shall see later, this is the main reason why it is impossible to obtain high gains from very small aerials.

It is important to note that in specifying gain for an aerial, some reference to direction must be included, for no aerial can exhibit gain simultaneously in all directions

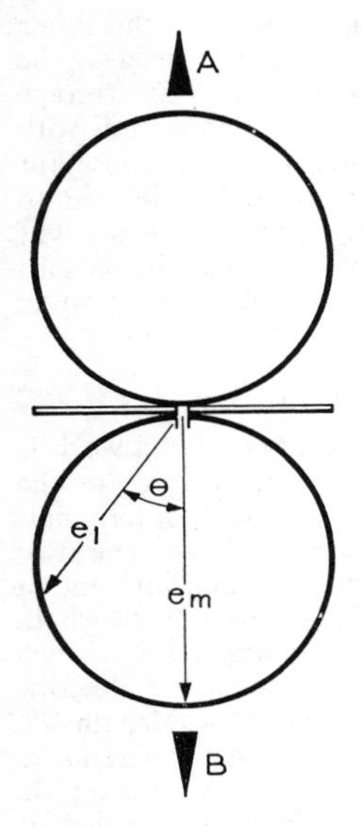

Fig 12.4. Directive polar diagram of a λ/2 (or shorter) dipole shown in section: e_l and e_m represent the relative field strength in the directions indicated.

relative to another aerial. The distribution of radiated energy from an aerial may be likened to the shape of a balloon filled with incompressible gas, with the aerial at the centre. The amount of gas represents the power fed to the aerial, and the volume of the balloon can only be increased by putting in more gas. The balloon may be distorted to many shapes, and elongated greatly in some directions so that the amount of gas squeezed in those directions is increased, but this can only be achieved by reducing the amount of gas in some other part of the balloon: the total volume must remain unaltered. Likewise the aerial can only direct extra energy in some required direction, by radiating less in others.

The gain of an aerial is expressed in terms of its performance relative to some agreed standard. This enables any two aerials to be directly compared, since if two aerials have for example a gain relative to the standard of 6dB and 4dB respectively, the first has a gain of 6 − 4 = 2dB relative to the second. It is unfortunate that two standards exist side by side and will be encountered in other references to aerials. One standard often used is the theoretical *isotropic* radiator, which radiates equal power in all directions, ie its solid polar diagram is a sphere. This is a strictly non-practical device which cannot be constructed or used, but has the advantage that the comparison is not complicated by the directional properties of the reference aerial. The other standard is the "half-wave dipole" which has in its own right a directional pattern as illustrated in **Fig 12.4.** This is a practical aerial which can be built and is therefore a more realistic basis for comparison, but it should be noted that gain expressed *relative to a λ/2 dipole* means by inference relative to the maximum radiation from the dipole, ie in directions A or B of Fig 12.4. The polar diagram consists of two circles and the relative gain in any direction relative to the wire, such as that of e_l, may be found by describing a circle, filling in e_l and e_m as shown, and using a rule to measure their relative lengths. This is known mathematically as a cos θ pattern because the ratio of e_l and e_m is given by the cosine of the angle θ between e_l and e_m.

It will be found in practice that gain is frequently expressed in dB without reference to the standard employed. This can lead to a disparity of 2·15dB in claimed results, being the relative gain of the two standards employed (the gain of a λ/2 dipole relative to the isotropic source). In such cases it is safer to assume the more conservative figure when comparing different aerial performance unless one is sure that the same standard aerial has been assumed in each case.

Because direction is inevitably associated with a statement of gain, it is usually assumed in the absence of any qualifying statement that the gain quoted for any aerial is its gain in the direction of its own maximum radiation. Where the aerial system can be rotated, as is often the case on 14MHz and higher frequencies, this is not so important, but when the aerial is fixed in position the superiority it exhibits in one direction over another aerial will not hold in other directions, because of the different shapes of the two directivity patterns. Aerial A may have a quoted gain of 6dB over aerial B, but only in the directions which favour the shape of its radiation pattern relative to that of aerial B. There is an important distinction between transmitting gain and effective receiving gain. In the first case it is required to maximize the power transmitted and in the second case we have to maximize the signal-to-noise ratio, and the two gains will be the same only if there are no power losses and noise is isotropic, ie arriving equally from all directions. In the hf bands, the useful receiver sensitivity is limited by external noise which is usually well above the receiver noise level, and, so long as this remains true, signal-to-noise ratio is unaffected by losses in the aerial system. Typically, with a low-noise receiver, aerial losses could reduce the power transmitted by up to 10dB or more before starting to affect the performance adversely when the same aerial is used for reception.

Radiation Resistance and Aerial Impedance

When power is delivered from the transmitter into the aerial, some small part will be lost as heat, since the material of which the aerial is made will have a finite resistance, albeit small, and a current flowing in it will dissipate some power. The bulk of the power will usually be radiated and, since power can only be consumed by a resistance, it is convenient to consider the radiated power as dissipated in a fictitious resistance which is called the *radiation resistance* of the aerial. Using ordinary circuit relations, if a current I is flowing into the radiation resistance R, then a power of I^2R watts is being radiated. As depicted in Fig 12.3 the rms current distribution along a resonant aerial or indeed any standing wave aerial is not uniform but is approximately sinusoidal. It is therefore necessary to specify the point of reference for the current when formulating the value of the radiation resistance, and it is usual to assume the value of current at the anti-node or maximum point. This is known as the current loop, and hence the value of R given by this current is known as the *loop radiation resistance*: in practice the word loop is omitted but inferred.

A half-wave dipole has a radiation resistance of about 73Ω. If it is made of highly conductive material such as copper or aluminium, the loss resistance may be less than 1Ω. The conductor loss is thus relatively small and the aerial provides an efficient coupling between the transmitter and free space. The effective loss resistance per half-wavelength of copper conductor varies with frequency and with size of conductor as shown in **Fig 12.5**; for comparison with the radiation resistance of a dipole these figures must be divided by two, since the current distribution along the dipole is

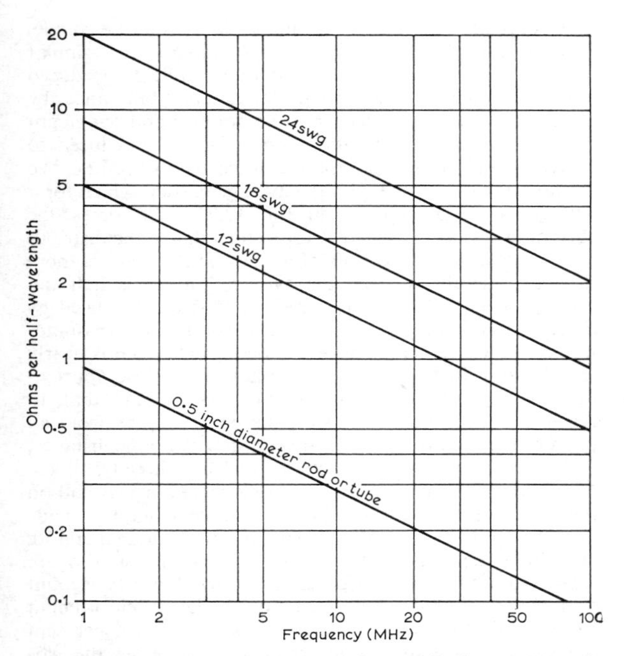

Fig 12.5. RF resistance of copper conductors. Resistance per unit length is proportional to the circumference of the conductor and the square root of the frequency.

more or less sinusoidal and the mean-square current is therefore only half the current in the centre. Approximately, for non-magnetic materials, the resistance varies as the square root of the resistivity and inversely as the surface area. Values obtained with the aid of Fig 12.5 should be increased by about 25 per cent for aluminium and 65 per cent for aluminium alloy. For magnetic materials the losses are much higher.

When the aerial is not a resonant length, it behaves like a resistance in series with a positive (inductive) or negative (capacitive) reactance and requires the addition of an equal but opposing reactance to bring it to resonance, so that it may be effectively supplied with power by the transmitter. The combination of resistance and reactance, which would be measured at the aerial terminals with an impedance meter, is referred to in general terms as the aerial *input impedance*. This impedance is only a pure resistance when the aerial is at one of its resonant lengths.

Fig 12.6 shows, by means of equivalent circuits, how the impedance of a dipole varies according to the length

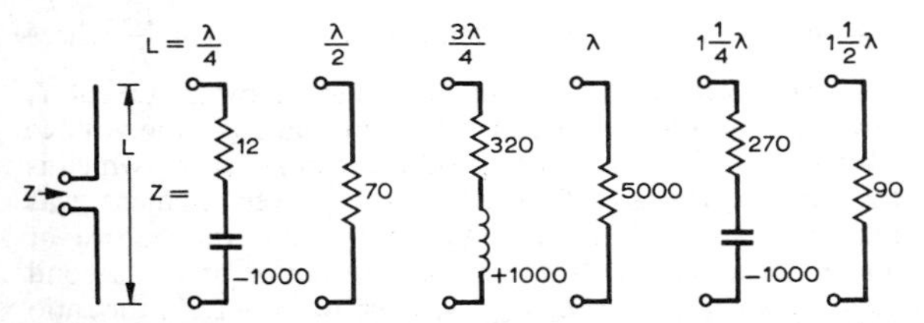

Fig 12.6. Typical input impedance (Z_i) value for dipoles of various lengths. The values for L = λ/2 and 3λ/2 are always approximately as shown but values for other lengths vary considerably according to the length/diameter ratio of the aerial. The values given are typical for wire hf aerials. Note how the reactance changes sign for each multiple of λ/4.

in wavelengths. It will be seen that the components of impedance vary over a wide range.

The input impedance of the aerial is related specifically to the input terminals, whereas the radiation resistance is usually related to the current at its loop position. It is possible to feed power into an aerial at any point along its length so that the input impedance and the loop radiation resistance even of a resonant aerial may be very different in value, although in this case both are pure resistances. Only when the feed point of the aerial coincides with the position of the current loop on a single wire, will the two be approximately equal (**Fig 12.7(a)**). If the feed point occurs at a position of current minimum and voltage maximum, the input impedance will be very high, but the loop radiation resistance remains unaltered (**Fig 12.7(b)**). For a given power fed into the aerial, the actual feed point current measured on an rf ammeter will be very low, but because the input impedance is high, the power delivered to the aerial is the same. Such an aerial is described as *voltage fed*, because the feed point concides with a point of maximum voltage in the distribution along the aerial. Conversely an aerial fed at a low impedance point, usually a current maximum, is described as current fed.

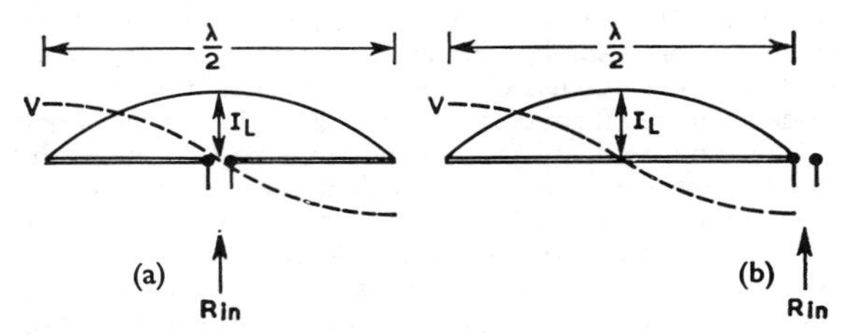

Fig 12.7. The input resistance of a λ/2 dipole is low at the centre and high at the end, although the loop radiation resistance is the same in each case. In (b) R_{in} is twice the R_e of Table 12.10.

The input impedance of a current-fed half-wave dipole consisting of a single straight conductor is approximately equal to 73Ω, though somewhat dependent on height (Fig 12.86), and will be much the same value irrespective of the size of wire or rod used to fabricate the dipole. The input impedance of a voltage fed half-wave dipole is very high, and its precise value depends not only upon the loop radiation resistance of the dipole, which is independent of the method of feed, but also the physical size of the wire used. The dipole wire between the current loop and the current zero at the ends may be considered to act as a quarter-wavelength transformer between the loop radiation resistance at the centre, and the input impedance at the end (**Fig 12.8**). As a transformer, the wire must exhibit a certain characteristic impedance Z_0, and the transformer ratio will depend upon the value of Z_0, which in turn depends upon the ratio of the conductor diameter to its length. If the end impedance is R_e, and the radiation resistance R_r, then approximately $R_e = \frac{Z_0^2}{R_r}$: the value of R_r is fixed by the current distribution, and hence the value of R_e will change as Z_0, or the wire length/diameter ratio changes. Typical values of R_e for different ratios are given in Table 12.10; the behaviour of transformers is dealt with in more detail in the section on transmission lines.

In certain cases, even though a half-wave dipole is centre fed at the current loop position, the input impedance may

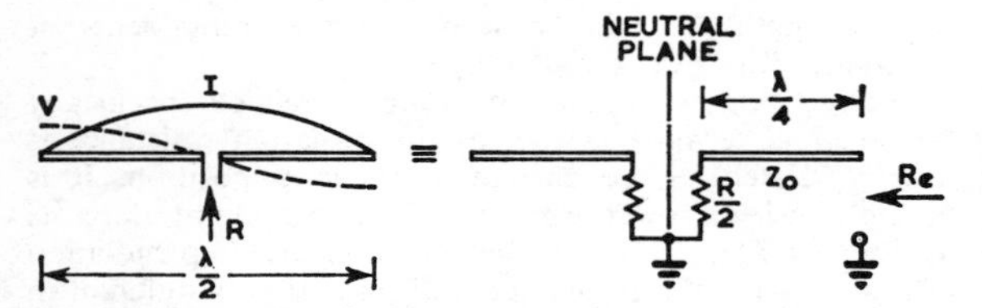

Fig 12.8. The end impedance of a λ/2 dipole can be derived by considering one half of the centre impedance transformed along λ/4 of transmission line formed by the dipole wire.

differ considerably from the value of 73Ω. If the half-wave dipole is folded (see p12.33), a transformation again occurs, to produce a different input impedance. If the dipole is the driven element of a multi-element array such as a Yagi, then the input impedance will also be modified (usually reduced) because of the presence of the other elements. If a wire is made up to be 33ft long and centre-fed, it would be resonant (λ/2) at 14MHz, three-quarter wavelength (3λ/4) at 21MHz and full wave (λ) at 28MHz. At 14MHz it is easy to supply power into the 70Ω input resistance, but at full wave resonance, ie 28MHz, the high value of 5,000Ω requires the use of some form of transformer. Again at 21MHz where the impedance is complex and evidently high, special arrangements are necessary. These problems are discussed later in connection with tuned feeders and aerial couplers and in relation to specific aerials.

This same aerial on 7MHz would only be λ/4 long, with a low resistance of 12Ω and a high capacitive reactance. It would require a *loading coil* of about 23μH inductance (1,000Ω reactance) for resonance at 7MHz. This loading coil will have appreciable loss resistance, typically about 6Ω so that only two-thirds of the power from the transmitter reaches the aerial. The remainder is expended as heat in the coil so that the efficiency is 67 per cent and signal reports will be degraded by about one-third of an S-unit. With 18Ω total input resistance and a reactance of 1,000Ω, the Q of the circuit (ratio of reactance to resistance) would be approximately 50, which is just sufficient to permit reasonable coverage of the band without retuning. On 3·5MHz the same aerial would have a resistance of only 3Ω and a reactance of 2,000Ω. The loss resistance of the loading coil would be 12Ω, the efficiency only 20 per cent, and it would be impossible to shift frequency over more than about 25kHz without having to re-tune. The losses could be at least halved by using big coils, probably needed in any case for dissipating the heat, but the bandwidth would then be reduced to a mere 15kHz.

Greater efficiency may be achieved by using a length of open-wire line (or tuning stub) in place of the loading coil. For 7MHz the required length of line is just over λ/8 and the total loss resistance is likely to be about 1Ω. Due to transformer action by the line, the radiation resistance is reduced to 6Ω, but the efficiency is about 85 per cent, which is reasonable. The Q is about 86 which is low enough for coverage of the phone or cw bands (UK) but not both. The total length of wire in aerial and stub is rather greater than that normally required for resonance; in effect, wire has been subtracted at the high Z_0 of the aerial wire and added at the lower Z_0 of the stub, the formula on p12.27 being applicable in each case.

Starting from the known value for a λ/2 dipole, the radiation resistance of other aerials may be deduced from knowledge of the gain and current distribution. For example, knowing that the directivity of a λ/4 dipole is very similar to that of a λ/2 dipole, it follows that the fields produced must be similar for the same power radiated but, since the length is halved the current must be doubled and one might therefore expect the radiation resistance to be reduced to 17·5Ω or rather more than the 12Ω indicated in Fig 12.6. We are however concerned with the average current which for a half-sine-wave as depicted in Fig 12.7 is $2/\pi$ times the current in the centre, whereas for a short dipole the distribution represents the two ends of a sine wave which are more or less triangular so that the average current is half the centre current. The radiation resistance is thus reduced by $(\pi/4)^2$ to 10·8Ω. The remaining slight error lies in the assumption that the two aerials have identical gain, which is nearly true because in both cases the aerials are a close approximation to point sources of radiation. This means that, to reach any given point in space, the contributions from all parts of the radiator have to travel nearly the same distance, arrive in almost the same phase, and are directly added. Nevertheless, if we look at a λ/2 dipole from a nearly end-on position, we find contributions from near its two ends arriving almost half a wavelength apart and therefore cancelling each other. These components of the field are small and can for most purposes be ignored, but they account for a reduction of about 4 per cent in the relative field strength from the shorter aerial. This is equivalent to a 4 per cent reduction of current and allowance for this brings the estimated radiation resistance up to 12Ω as tabulated. The variation of radiation resistance with dipole length, for different types of loading, is plotted in Fig 12.129, p12.88.

The possibility of approximating each element by a point source is particularly helpful towards understanding the operation of beam aerials as described later in this chapter.

Mutual Impedance

If two aerials are in proximity there is mutual coupling between them just as there would be between a pair of tuned circuits, except that instead of mutual reactance there is a *mutual impedance* which usually includes both resistance and reactance. It plays an important role in the operation of small beam aerials such as those normally used for the amateur hf bands, causing a big reduction of input impedance so that relatively large currents flow in the elements. The following example illustrates the calculation of input impedance for closely-spaced two-element arrays and provide some insight into a problem which becomes extremely complex for practical systems involving additional elements.

Mutual impedance Z_m is defined as the ratio $-\frac{E_2}{I_1}$ where E_2 is the voltage induced in one element by a current I_1 flowing in another. It should be noted that E_2 is the voltage which would be required to produce the current I_2 (which is the current actually flowing) if the driven element was removed, and the minus sign arises from the nature of magnetic induction. If Z_2 is the impedance of the second element we have $I_2 = E_2/Z_2$, but since $E_2 = -I_1 Z_m$ the ratio of the currents, $(-I_2/I_1)$ is given by Z_m/Z_2. The first element has, in addition to the driving voltage E_1, a voltage $-I_2 Z_m$ induced from the second element and putting these facts together we find that the driving point impedance (E_1/I_1) is

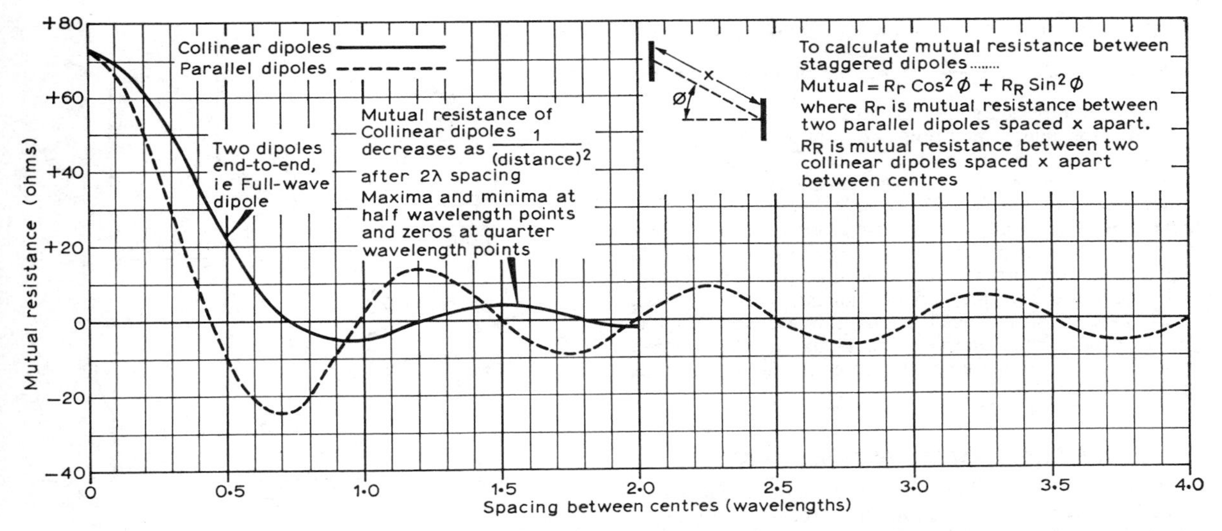

Fig 12.9. Variation of mutual resistance between parallel and collinear λ/2 dipoles as the spacing is altered.

given by $Z_1 - \dfrac{Z_m{}^2}{Z_2}$, Z_1 being the self-impedance of the first element and usually equal to the radiation resistance R, any reactive component being tuned out.

In the case of Z_2, the situation is slightly more complicated because the element is usually detuned slightly, which, as explained later, increases the gain and favours one direction of propagation rather than the other so that the aerial system is now said to possess a *front-to-back ratio*. For best performance a phase-shift of about 30° from the antiphase condition is required and for the simple case where Z_m is a pure resistance R_m (which happens for a convenient spacing of about λ/8 between dipoles) the phase-shift in radians is roughly equal to X/R where X is the reactance inserted in series with the radiation resistance R. (Instead of using a coil, the detuning effect can be achieved by altering the length of the element slightly, which is usually more convenient). In this situation, which though relatively simple is of considerable practical importance, the driving point impedance is given by

$$\frac{E_1}{I_1} = R\left(1 - \frac{R_m{}^2}{R^2 + X^2}\right)$$

This is the resistive part of what is really a complex impedance, but the reactive portion is neutralized by normal tuning procedures and need not concern us further. For other spacings Z_m is reactive and the algebra becomes quite involved, although this need not worry the experimenter unduly since beams are usually tuned experimentally, and it is mainly the theoretical difficulty which increases.

If there are more than two elements the impedance reflected from, say, the nth element into the first is given by $\dfrac{I_n}{I_1}\cdot Z_{mn}$ where I_n and Z_{mn} denote the appropriate values of current and mutual impedance respectively. In this case, to calculate the driving point impedance the above expression would have to be elaborated to take into account the relative phases of the currents, and the mutual impedances of all the elements and their images in the ground, Fig 12.21 on p12.14. This would have to be repeated for each element, leading to a complex set of simultaneous equations soluble only by computer. We can however effect a remarkable simplification by assuming that all the currents are of equal amplitude and either equal or opposite in phase in which case all the coefficients $\dfrac{I_n}{I_1}$ are equal to 1 or minus 1 and we have

$$\frac{E_1}{I_1} = R \pm R_{m2} \pm R_{m3} \pm R_{m4} \pm \text{etc},$$

the reactance terms (which add in the same way) being eliminated by the normal tuning process. The mutual resistances of parallel or collinear half-wave dipoles may be obtained from **Fig 12.9** and added or subtracted as required; thus for two collinear dipoles with their ends almost touching (centres λ/2 apart) the input resistance of each is 73 + 22 = 95Ω which reduces the current in each, and therefore the field at a distant point by 1·2dB compared with what it would be with no interaction. On the other hand for similarly phased parallel dipoles spaced by 0·7λ we have 73 − 24 = 49Ω, and the field is increased by 1·8dB. This procedure is particularly useful in dealing with arrays of widely-spaced elements; thus in the absence of mutual resistance the gain obtainable from stacking any number n of identical radiators (regardless of whether these are dipoles or elaborate beam aerials) is precisely equal to n, as explained on p12.16. For two elements, the basic gain of 3dB is modified in the case of the above examples to 3 − 1·2 = 1·8dB and 3 + 1·8 = 4·8dB respectively. A more dramatic illustration of the relation between gain and mutual impedance is provided by the 8JK beam, Fig 12.14 on p12.11, in which close-spaced parallel elements are fed in anti-phase. For a spacing of 0·15λ we have R_m = 61Ω so that allowing for the phase reversal the radiation resistance of each becomes 73 − 61 = 12Ω and the total power radiated for a current of I in each element is $24I^2$ compared with $73I_0{}^2$ for a single dipole. Equating these powers, we find that $I = 1{\cdot}71I_0$. Along a line at right angles to the elements the fields radiated do not quite cancel out because one has a start of 0·15λ over the other so that there is a 54° phase difference. The resultant field is therefore $2 \times 1{\cdot}71 \sin \left(\frac{54}{2}\right)^{\circ}$ times that of a dipole, ie 1·55 times, which implies a gain of 3·84dB.

In practice, simple addition and subtraction of mutual resistances can usefully be applied only to a few simple combinations of two or four elements, ie to symmetrical arrangements in which each element "sees" an identical pattern of other elements and their ground images. Otherwise different elements will experience different reflected impedances, and currents will be unequal unless the phases and amplitudes are independently adjusted in every case. In the case of close-spaced beams at reasonable heights, ground images, being relatively remote, have negligible effect on impedances compared with those due to interactions between the elements.

PRINCIPLES OF DIRECTIVE ARRAYS

All practical aerials are directive to some extent since, as we have seen, there is no radiation from a wire directly along its axis. When there is more than one half-wavelength in an aerial or if an aerial system is built up from a number of separate radiators it can be represented as a number of *point sources*, each representing one standing wave current loop. The total radiation can then be regarded as the resultant of a number of components, one from each point source. In any given direction these components may have to travel different distances, so that they do not arrive at a distant point in the same relative phases as they had in the wire. They can, therefore, augment or oppose each other. It is, however, possible to combine a number of elementary aerials so that their radiation accumulates in some favoured direction, at the expense of radiation in other directions. This gives a much stronger signal than a single aerial would give in the required direction for the same power input. For these reasons aerials are said to be *directive* and to have *gain*. The concept of gain was discussed on p12.3, and as we have seen, directivity and gain are closely interrelated. There are many ways of constructing directive arrays and the choice between them depends largely on the space and height available as well as basic requirements such as whether it is desired to achieve equally good performance in all directions or concentrate on one or more particular directions. Rotary beams meet the requirements for coverage of all directions with a gain of about 5–6dB [1] and comparable results may be obtained by switching between two or three fixed reversible beams, but higher gains require larger beams so that rotary construction becomes impossible, and the azimuthal patterns of fixed arrays become narrow so that a large number of separate arrays are necessary if it is desired to achieve uniform coverage of all compass directions.

For long distance work a low angle of radiation in the vertical plane is needed and depends primarily on aerial height and the nature of the surrounding contours [2] as discussed later in this chapter. In general the angle should be as low as possible and usually this means that the aerial must be as high as possible. Assuming a height of about $\lambda/2$ or greater to the centre of the elements, horizontal and vertical polarization give very similar low angle performance. Typically, for a height of $\lambda/2$ and radiation angle of 5° the ground reflected wave interferes destructively to produce a loss of 6dB but this changes to an enhancement

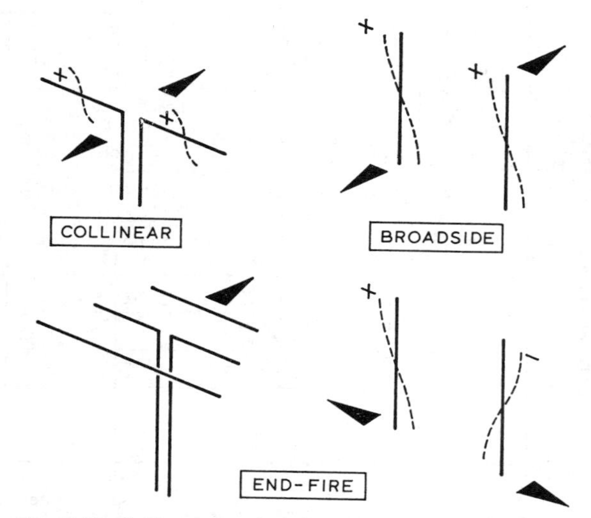

Fig 12.10. Collinear, vertical broadside and end-fire arrays, illustrating the terms. The arrows indicate directions of maximum radiation and the voltage standing waves on three of the aerials show their relative polarity or phase. The lower left aerial is a three-element parasitic array and is unidirectional.

of 3dB for a height of $3\lambda/2$ and 6dB for a height of 3λ. It is important to note that these figures are independent of the free-space gain of an aerial or the way it is obtained. However, due to interaction between the aerial and its image (see for example Fig 12.9 and the section on mutual impedance, p12.6) the current in the reference dipole will vary somewhat with height, and could account for a variation in the apparent gain of up to 1·2dB. It is also to be noted that with vertical polarization cancellation is incomplete leaving a small but useful amount of low-angle radiation [2] which is independent of height.

Long wire aerials have very distinctive radiation patterns, as will be seen later, but these patterns have several main beams or lobes and hence the gain is not so great as when most of the transmitter power is concentrated into one main beam. In order to achieve the latter condition, two or more elements, such as half wave dipoles, are combined in special ways and the combination is called an *array*. There are three general classes of array—the *broadside array* in which the main beam is at right angles to a plane containing the elements, the *end-fire array* in which the main beam is in the same direction as the row of elements and the *collinear array* which employs co-phased elements arranged end-to-end, radiation being a maximum at right-angles to the line of the elements; the three classes are illustrated in **Fig 12.10**. It should be noted that the term end-fire refers to the layout of the array and not to the ends of the wires, although the term is sometimes applied to a long wire, because its direction of maximum radiation tends towards the direction of the wire.

Radiation Patterns and Polar Diagrams

From the point of view of effective gain it is immaterial whether this comes from horizontal directivity, vertical directivity or both, but the practical usefulness of a fixed array using horizontal collinear elements is restricted by the relatively narrow beam width in the horizontal plane. To illustrate the radiation pattern of an aerial, *polar diagrams* are used in the form of curves, the radius of which in any direction represents the relative strength of signals in that direction.

The radiation from an aerial occurs in three dimensions and therefore the radiation pattern is best represented by the surface of a solid object. A polar diagram is any section of the solid shape and a large number of sections may be necessary to reduce the aerial radiation pattern to two dimensions. In practice it is necessary to be content with two polar diagrams taken in the principal planes, usually the horizontal and vertical, and giving the two cross-sections of the main beam.

Where the polar diagram has a definite directional form, the angle between the directions where the power radiated is half the value at the point of maximum gain (−3dB) is called the *beam width*. These points are marked on Fig 12.99 on p12.65.

To avoid confusion when discussing radiation, directions in the horizontal plane are referred to as *azimuth*; angles above the horizontal, in the vertical plane, are called *wave angles* or directions in *elevation*. Confusion often arises additionally when the expression horizontal (or vertical) polar diagram is used, unless it is made clear by a statement of the polarization of the aerial with respect to the Earth's surface. When reference is made to the polar diagram of an aerial in free space, the terms horizontal and vertical have no meaning, and the more precise descriptions of *E-plane* and *H-plane* polar diagrams are to be preferred. These are unambiguous, since the direction of the electric and magnetic fields around the aerial is a function only of the direction of current flow. The electric field (or *E*-plane) is parallel to the direction of current and therefore usually parallel to the radiating wire. The magnetic field (or *H*-plane) is at right angles to the current and therefore normal to the radiating wire. The polar diagram of the half-wave dipole illustrated in Fig 12.5 is then an *E*-plane diagram: the *H*-plane diagram of the dipole will be a circle. Such an aerial is then said to possess *E*-plane directivity, and is omni-directional in the *H*-plane.

No matter what name is used to describe the polar diagram, it should be remembered that these radiation patterns are for long distances and cannot be measured accurately at distances less than several wavelengths from the aerial. The greater the gain, the greater the distance required.

Construction of Polar Diagrams

Fig 12.11 is a plan of two vertical aerials A and B, spaced by one half-wavelength, and carrying equal in-phase currents. A receiver at a large distance in the direction *a* at right-angles to the line of the aerials (ie *broadside*) receives equally from both A and B. Since the two paths are equal the received components are in phase and directly additive, giving a relative signal strength of two units. In the direction *b* the signal from A travels one half-wavelength further than that from B. The two received components are therefore one half cycle different, or in antiphase, and so cancel to give zero signal. In an intermediate direction *c* the two paths are effectively parallel for the very distant receiver, but one component travels further by the distance *x*. In this case the components have a relative phase $\psi = 360x/\lambda$ degrees. The addition for this direction must be made *vectorially* using the "triangle of forces" also illustrated in Fig 12.11, in which the

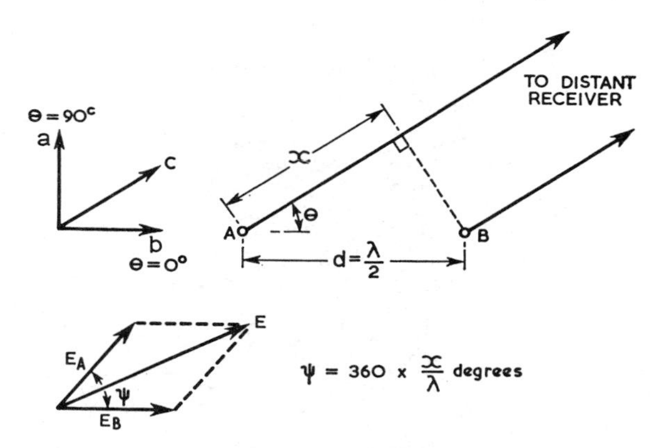

Fig 12.11. The radiation pattern of two sources can be calculated by vector addition of the respective field contributions at various angles of azimuth. For simple arrays this can be converted into a trigonometrical formula.

two equal lines representing signal strength from A and B are set at the phase angle ψ which corresponds to the path-difference x; the total or resultant signal is equal to the relative length of the line E which completes the triangle. This process could be carried out for all directions, but is tedious and it is usual therefore to convert the operations to a trigonometrical formula from which the result can be calculated more quickly.

In the general case of two aerials carrying equal but de-phased currents and spaced a distance d apart, the appropriate formula for the H-plane polar diagram is:

$$E = 2E_0 \cos\left(\frac{\pi d}{\lambda}\cos\theta + \frac{\phi}{2}\right) \quad \ldots \text{(i)}$$

where θ is the angle measured as shown in Fig 12.11, ϕ is the electrical phase difference between the currents in the two aerials, and E_0 is the value of the field from one aerial alone in the direction θ. In the particular case illustrated in **Fig 12.12**, $\theta = 0$, E_0 is constant for all directions of θ, and $d = \lambda/2$, so that the shape of the polar diagram can be plotted from the expression:

$$\cos\left(\frac{\pi}{2}\cos\theta\right)$$

The absolute magnitude of the field E will depend upon the actual value of the current flowing in each aerial and is of no direct importance when calculating the shape of the pattern. It comes into account only when the relative patterns of two aerial systems are being considered and the question of gain arises.

The polar diagram found in this way is a figure-of-eight, the curve marked $\lambda/2$ in Fig 12.12(a). If on the other hand the two aerials had been in antiphase, the resultant signal would have been zero in the direction *a* and maximum in direction *b*, and the polar diagram as in Fig 12.12(b).

In that case, $\phi = 180°$ (or π) in expression (i) above, and the polar diagram is given by

$$\cos\left(\frac{\pi}{2}\cos\theta + \frac{\pi}{2}\right)$$

$$= \sin\left(\frac{\pi}{2}\cos\theta\right)$$

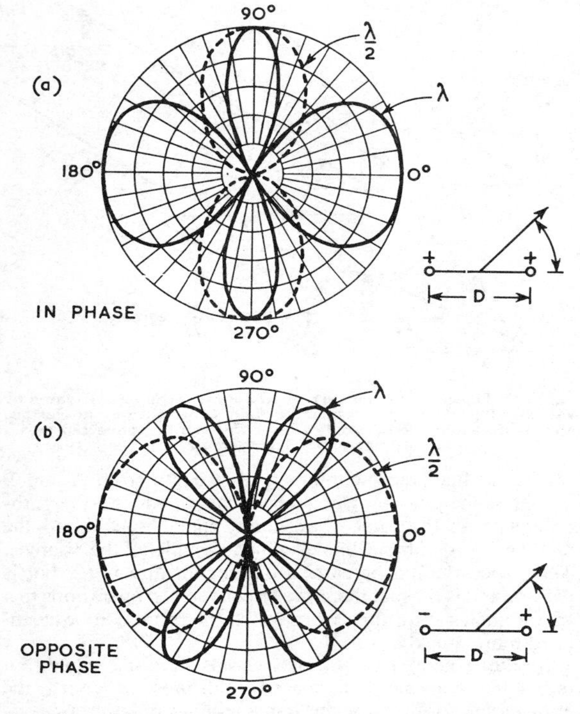

Fig 12.12. Horizontal polar diagrams of two vertical aerials spaced λ/2 and λ. The upper diagram is for aerials in phase, the lower diagram for antiphase connection. If separate feeders are used the pattern can be changed by reversing polarity of one feeder. Directions 90°, 270° are broadside: 0°, 180° end-fire. Note that the end-fire diagrams are broader than the broadside ones.

This expression for the antiphase aerials is very similar in form to that for the in-phase aerials, and the change from cosine to sine reveals that the general pattern is turned through 90° of azimuth. This is confirmed by the dotted curves of Fig 12.12.

A similar procedure is used if the aerials do not carry equal currents, or if they have an arbitrary phase relation. In this case the sides of the triangle would be drawn in lengths proportional to the two aerial currents and the relative phase of these currents would be added to the angle ψ which arises from the path differences.

In Fig 12.11 the path difference x and corresponding phase-shift ψ are of course proportional to the spacing between the aerials, which in the example above was one half-wavelength. If the spacing is one wavelength then the phase difference changes twice as rapidly with change of direction, and so there are twice as many maxima and minima to the pattern, giving the curves marked λ in Fig 12.12. These patterns are said to have four *lobes*.

With increased spacing between the two aerials, more lobes appear, two for each half-wave of spacing, and the pattern becomes like a flower with many petals. This type of pattern is not very useful, but if the intervening space is filled with aerials spaced λ/2, one pair of lobes grows at the expense of all the others, giving a sharp main beam with a number of relatively small *minor lobes*. This is the basis of stacked beam arrays. It should be noted, however, that the beam is only developed in the plane in which the array is extended; a broadside array of vertical aerials or collinear array of horizontal aerials all in phase produces a beam which is narrow in the horizontal plane, but its vertical pattern is the same as that of a single aerial. It is therefore necessary to extend a stacked array in the vertical direction if a sharp vertical pattern is required. In an end-fire array where the aerials are spaced and phased to give the main lobe along the line of aerials, the vertical and horizontal patterns are developed simultaneously by the single row of radiators because the array is extended simultaneously in both of these planes. The azimuth pattern of an end-fire array is always broader than that of a broadside array of the same length; this effect can be noted in the patterns of Fig 12.12. Note that for the purpose of this discussion the interaction of horizontal and vertical planes is assumed to be along the direction of propagation.

Although the trigonometrical formula for the polar diagram of a large array may be very complex, and therefore laborious to plot in full, it is always readily possible to find the directions of minimum radiation or *nulls* by solving the equation for θ when E is made equal to zero.

Unidirectional Patterns

The patterns so far considered are symmetrical: a somewhat different combination of aerials gives a unidirectional pattern. The two dipoles in **Fig 12.13** are connected a quarter-wave apart along a common feeder, with a wave entering from the left. In the direction away to the right it does not matter whether radiation leaves via the first aerial or the second; it takes the same time to reach its destination. Thus to the right the components of radiation are additive because they are in phase. This is shown by the upper set of broken lines. To the left, however, the wave from the end dipole has to travel further than that from the first dipole by a distance equal to twice the spacing between them. This extra journey is one half-wavelength and hence the two components differ in phase by 180° and cancel. Because of the λ/4 spacing the currents in the two aerials have a phase difference of 90° and are said to be in *quadrature*.

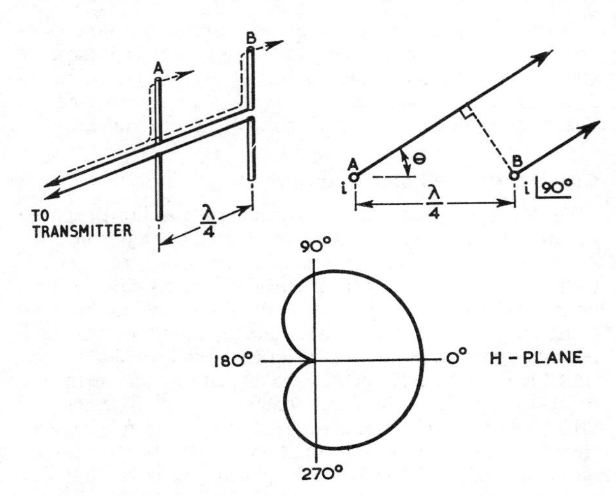

Fig 12.13. Illustrating the fundamental principle of the driven reflector. The two wave components have equal path lengths in one direction, and therefore add, but in the other direction the paths differ by λ/2 and the components cancel. The polar diagram is the geometrical figure called a cardioid (heart shape).

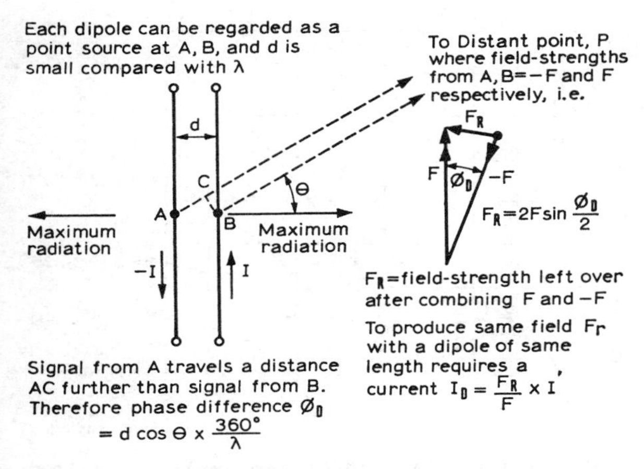

Fig 12.14. Mechanism of 8JK aerial.

The actual shape of the pattern can again be determined by substituting the appropriate values for d and ϕ in expression (i) given earlier, to give

$$\cos\left(\frac{\pi}{4}\cos\theta - \frac{\pi}{4}\right)$$

This fundamental principle is often used in broadside arrays in order to make the radiated beam unidirectional by placing a second set of aerials a quarter-wave behind the main set. For small beams such as those suitable for rotary use, much smaller spacings are used, the pattern being cardioidal so long as the phase shift differs from 180° by an amount exactly corresponding to the spacing between the elements, eg 135° for λ/8 spacing. However, as demonstrated on p12.12 improved gain and *directional characteristics* result from a phase shift somewhat less than that required for a cardioid pattern.

Close-spaced Arrays

Beam aerials may be classified into two types: (*a*) those in which the beam is formed by in-phase addition of fields which the various elements produce in the wanted direction and (*b*) those in which energy is concentrated in a wanted direction by less-effective cancellation of opposing fields. The simplest example of the second type is provided by the 8JK which is an end-fire array using a closely-spaced pair of dipoles driven in antiphase and operating as illustrated by **Fig 12.14**. Due to restrictions of space and the mechanical requirements of rotary beam construction the majority of amateur hf beams use closely-spaced elements and it will be shown that those with two elements can be evolved from the 8JK by the introduction of a small phase-shift.

The nearly-antiphase excitation of elements is an inherent feature of all close-spaced beams. If such elements are fed in the same phase the radiation resistances are additive so that the current in each is reduced, the total current tends to stay constant, and the radiation pattern and gain is the same as that of a single dipole. This follows also from the fact that for any given point in space, signals from each element arrive in nearly the same phase, a difference of say, 45° being insignificant when the fields are adding. On the other hand if the elements are fed in antiphase a narrow beam can be formed because as shown in Fig 12.14 the extent to which cancellation fails to take place in any given direction is proportional to the apparent spacing of the elements as viewed from that direction. This varies as cos θ, so that a second cos θ pattern is superimposed on the basic dipole pattern giving $\cos^2\theta$ in the end-fire direction and cos θ in the broadside direction, a concentration of energy which produces a gain of 4dB and is independent of the dimensions of the beam. Arrays characterized by this property are known as "supergain" arrays since for a given gain they are much smaller than "normal" arrays using the additive principle. Unfortunately since the increased field strength is attributable to the difference between opposing fields, each of these must be relatively large. This implies large currents in the element so that for a given radiated power the radiation resistance must be relatively low; bandwidth is therefore decreased and much larger diameter conductors are needed to prevent energy from being dissipated as heat instead of radiation.

This subtractive method of beam forming can only be applied with close spacing, since otherwise there must be some direction in which the field is additive and thus relatively large.

If the phase of the current in one element is advanced by an amount equal to the spacing, ie 45° for a spacing of λ/8, the total phase-shift is doubled for the direction of advancement and zero for the opposite direction so that a cardioid pattern results exactly as described in the previous section for λ/4 separation and 90° phase-shift, Fig 12.13, although different arrangements are needed to produce the phase-shift (eg Figs 12.108, 12.110). One highly desirable consequence of this is an increase of four times in radiation resistance, with the gain remaining the same as for an 8JK system. Better performance is obtained with somewhat smaller values of phase-shift and **Fig 12.15** is a set of universal curves showing how gain, front-to-back ratio (nominal) and radiation resistance vary with phasing. It should be noted that as phase-shift is reduced the single null of Fig 12.13 splits into two in different directions as shown by **Fig 12.16** so that, despite the reduction of nominal front-to-back ratio, the average discrimination against interference tends to be improved; in fact if interference were equally likely to arrive from all directions in space, adjustment for maximum gain would by definition ensure minimum interference. The variation of backward field strength with bearing and phase-shift is plotted in **Fig 12.17.**

In the discussion so far both elements have been driven so that the currents are equal. More often only one element is driven, the other being excited *parasitically* from the driven element via the mutual impedance, phase being adjusted by tuning the parasitic element which may be operated either as a *director* or *reflector*. Parasitic operation has little effect on gain, but, depending on the type of element, there is a tendency for nulls to be less pronounced because the currents are unequal so that there are no directions for which the fields are equal and opposite. The effect of unequal currents on gain and front-to-back ratio is shown in **Fig 12.18**.

The choice of director or reflector operation is immaterial if the mutual impedance is purely resistive, which occurs in the case of dipoles for a spacing of about λ/8; at larger spacings, or in the case of loop elements, the mutual impedance includes capacitive reactance which increases the current in a reflector and reduces it in a director so that directors are then relatively inefficient. This is particularly

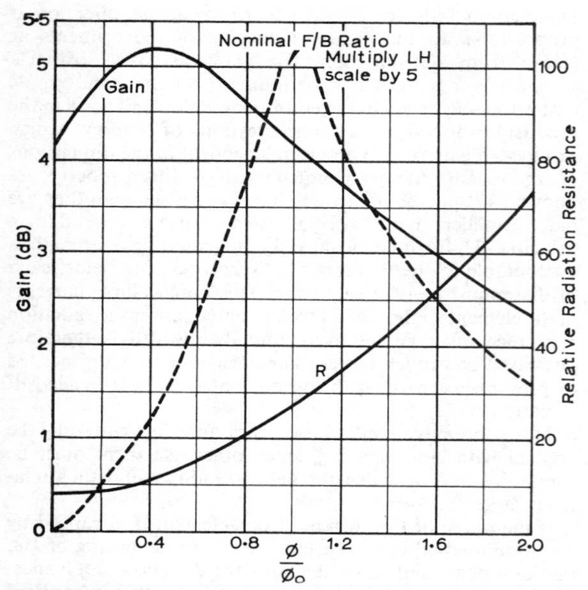

Fig 12.15. Variation of gain, radiation resistance and F/B ratio with ϕ/ϕ_0 where ϕ_0 is the phase shift corresponding to the distances between the elements and ϕ is the electrical phase shift relative to the antiphase condition. Resistance scale is correct in ohms for each of a pair of dipoles spaced $\lambda/8$ (equal currents are assumed).

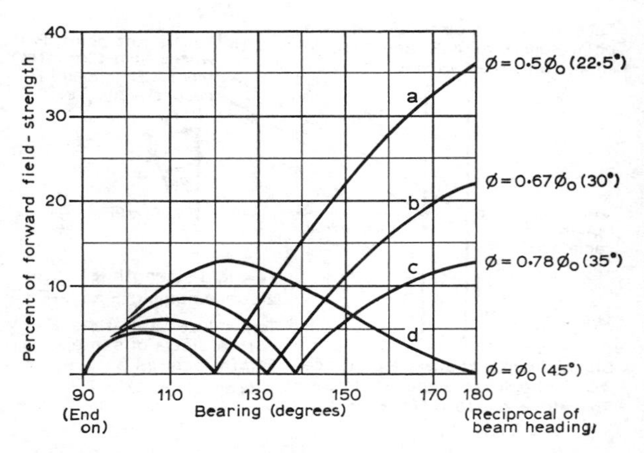

Fig 12.17. Variation of backward field strength with bearing and phase shift. Angles in brackets apply for S = $\lambda/8$. Curve (a) corresponds to maximum gain and curve (d) to maximum nominal back-to-front ratio. Curve (b) provides near-maximum gain and better rejection of interference than (d) for all angles less than 147°.

marked with loop aerials (eg the quad or delta loop, Figs 12.115, 12.120) with reflector operation providing almost equal currents and deep nulls, whereas directors provide much reduced gain and back-to-front ratio.

Parasitic beams frequently employ a reflector and one or more directors but as shown on p12.75 the increase of gain obtainable in practice, with close spacing, is limited to about 1dB although the theory of supergain aerials predicts a possible gain proportional to the square of the number of elements. This is unfortunately subject to very precise adjustment of current amplitude and phases, and as the size is reduced or the number of elements increased, bandwidth decreases and losses increase at an astronomical rate.

Point Sources

We have already made use of the fact that aerial elements can be represented by point sources, and although it is always admissible to use more than one point per element the need only arises if this radiates in more than one plane or if it contains separate concentrations of current with a distance of more than about $\lambda/4$ between them. In the latter

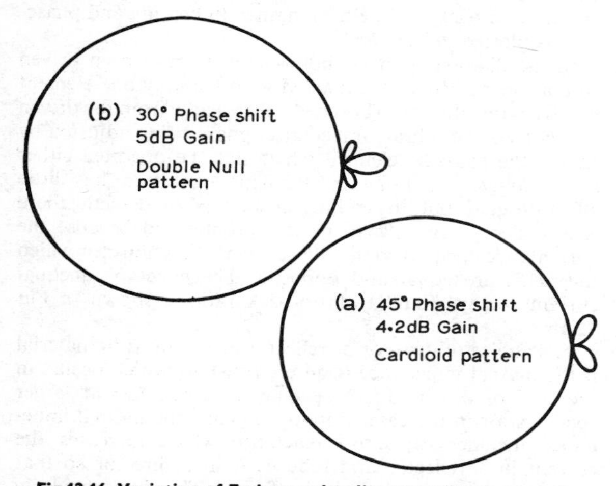

Fig 12.16. Variation of E-plane polar diagram with phase angle.

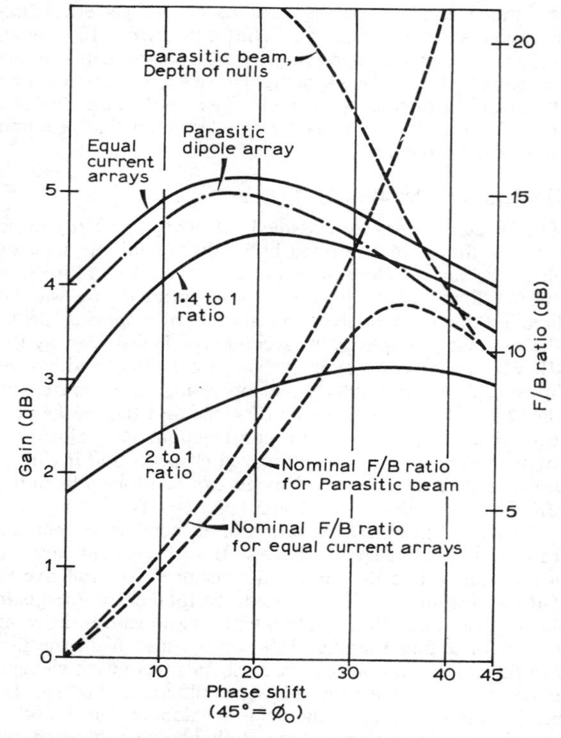

Fig 12.18. Gain and F/B ratio for two-element arrays with unequal currents, compared with equal current arrays. Spacing is $\lambda/8$.

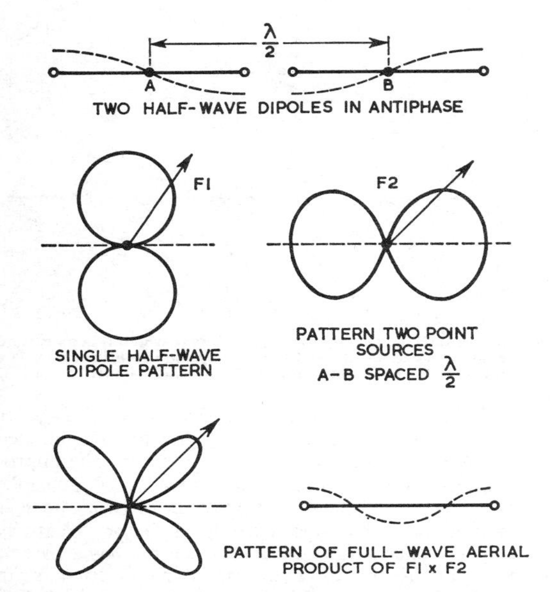

Fig 12.19. The array factor. Each dipole has a pattern like F1. Two sources in opposite phase at A and B would have a pattern like F2. For any direction (eg the arrows) the chord F1 is multiplied by that of F2 for the same angle. When this is done for all angles the pattern shown at the bottom is obtained for the complete aerial.

case radiation from different parts of the element could arrive in phase for one direction in space and yet be appreciably different in phase for another direction, in which case the radiation pattern will differ from that of a point source, and if the reduction of radiation in one direction accounts for a significant amount of energy there may be enough gain in another to invalidate the point-source assumption. An important borderline case is provided by the quad loop (Fig 12.115) the top and bottom of which could be regarded as individual point sources separated by λ/4; with a single loop this slightly reduces the radiation in the vertical plane and provides a small stacking gain, about 1dB. With loops assembled into a beam, the stack approximates to a pair of Yagis which as discussed later require a separation of the order of λ/2 instead of λ/4 before they can be regarded as stacked effectively, so that even in this case no appreciable error is to be expected from the point source approximation.

Array Factor

The radiation pattern of a large aerial array can often be calculated by breaking down the array into units of which the individual pattern is known, and then combining the patterns of the units together. For the purposes of combining the unit patterns, each is assumed to be a "point source" of radiation, located at its physical centre, such as the aerials A and B in **Fig 12.19,** and an expression F_1 derived for the pattern of the individual sources. The final pattern F of the whole aerial is then obtained by multiplying F_1 by the pattern F_2 of two antiphase point sources spaced λ/2 so that

$$F = F_1 \times F_2$$

The expression F_1 is known as the *unit* pattern and F_2 as the *array factor*.

For very large arrays of dipoles this technique can be repeated by successive breaking down of the array into smaller and smaller units. The application of the array factor to the determination of the *E*-plane polar diagram for a pair of collinear antiphased half-wave dipoles is shown in Fig 12.19, where F is the final pattern, F_1 that of the dipole and F_2 the array factor or pattern of the set of point sources. F_1 is unity for the azimuth pattern of vertical aerials. If each aerial were fitted with a reflector, then another multiplying factor would be included to represent for example the cardioid (heart shape) of Fig 12.13.

A particular application of array factor is the conversion of *H*-plane to *E*-plane polar diagrams for arrays made up of parallel half-wave dipoles. Once the *H*-plane pattern has been established by measurement or calculation, the *E*-plane pattern is obtained merely by multiplying the *H*-plane pattern by the dipole factor F_d (the *E*-plane pattern of a single half-wave dipole) so that:

$$E\text{-plane} = H\text{-plane} \times F_d$$

Since the half-wave dipole possesses *E*-plane directivity normal to its axis the effect of converting the *H*-plane diagram is to sharpen it in the directions at right angles to the axis of the dipoles, at the expense of radiation in the directions along the axis of the dipoles. This is illustrated in **Fig 12.20** which shows the *E*- and *H*-plane diagrams for an array of two parallel dipoles with reflectors.

Vertical plane diagrams are found in the same way, by treating the aerial and its image in the ground as an array of two sources spaced by twice the height of the real aerial.

Vertical Radiation Patterns over Earth

The relationship between the sign of the aerial and its image for the separate cases of horizontal and vertical polarization is shown in **Fig 12.21**. In the case of the horizontally polarized aerial, the image is of opposite sign or phase, and the resultant field along the surface of the reflecting plane will always be zero. In the case of vertically

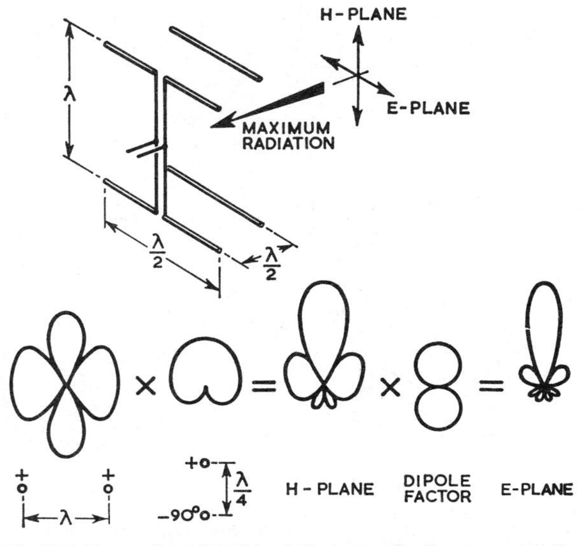

Fig 12.20. Conversion of H-plane to E-plane polar diagrams using the principle of array factor. Note reflector spacing is λ/4, not λ/2.

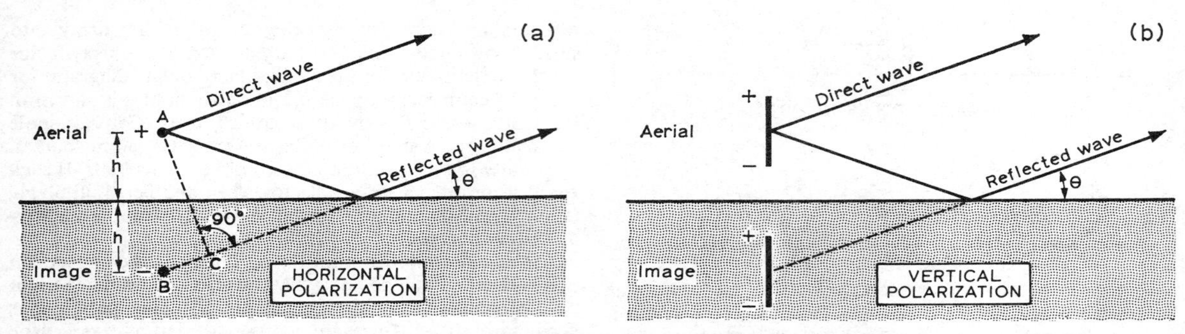

Fig 12.21. Images of aerials above a perfectly-conducting ground plane. The vertically polarized aerial produces an image which is in phase and supports radiation along the surface. The horizontally polarized image is in antiphase and cancels the radiation along the earth's surface. The path length from the image is greater by the distance 2h sinθ, resulting in a phase difference of $4\pi h \sin\theta/\lambda$ radians.

polarized aerials over perfect ground the image would be in phase and the field a maximum at low angles.

For horizontal polarization, there are maxima in the vertical plane whenever $2h \sin \theta$ is equal to an odd number of half-wavelengths thus producing a phase reversal, and at these angles the direct wave and ground-reflected waves reinforce each other, giving 6dB gain compared with the free-space pattern. On the other hand there is almost-zero radiation when $2h \sin \theta$ is equal to zero or an even number of half-wavelengths. Typical vertical plane polar diagrams are given in **Fig 12.22 (c)** and **(d)**, and it will be evident from the above discussion that these represent the array factor for any two sources fed in antiphase and spaced by $2h$. Vertical polar diagrams for any horizontal array may therefore be obtained by multiplying its free space polar diagram by the appropriate pattern from Fig 12.22, with the nose of each lobe corresponding to a voltage multiplication of two, and it is important to note that this multiplying factor is entirely independent of the type of aerial. If the array has a narrow free-space radiation pattern in the direction corresponding to the vertical plane this would suggest that it derives its free space gain from restriction of radiation in this plane rather than by a narrow azimuthal beam width, and the narrow vertical pattern does not imply any additional gain due to

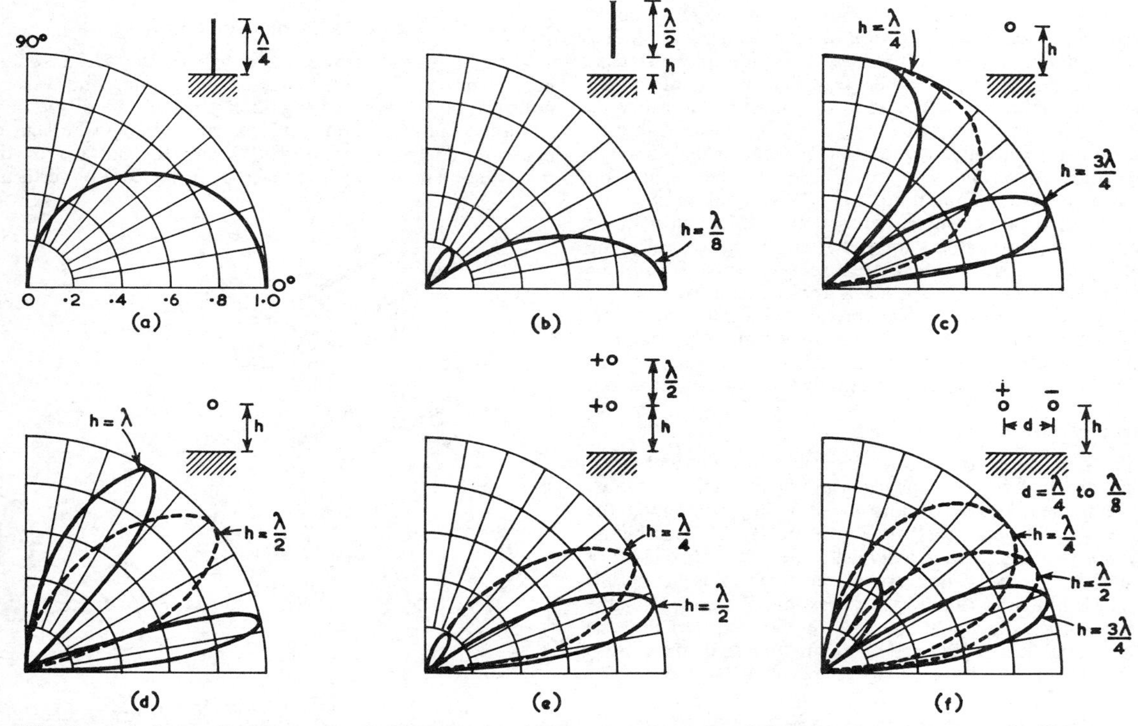

Fig 12.22. Vertical plane radiation patterns. (a, b) Vertical aerials over a perfect "earth". (c, d) Horizontal dipoles and collinear arrays. (e) Broadside horizontal arrays. (f) End-fire horizontal systems, eg W8JK. Only half of each pattern is shown: they are symmetrical about the vertical axis, unless reflectors are also used. Diagrams (a) and (b) hold for all azimuth directions: the remainder are for broadside direction only. The beamwidth in azimuth depends on the length of the arrays, as shown in Fig 12.99.

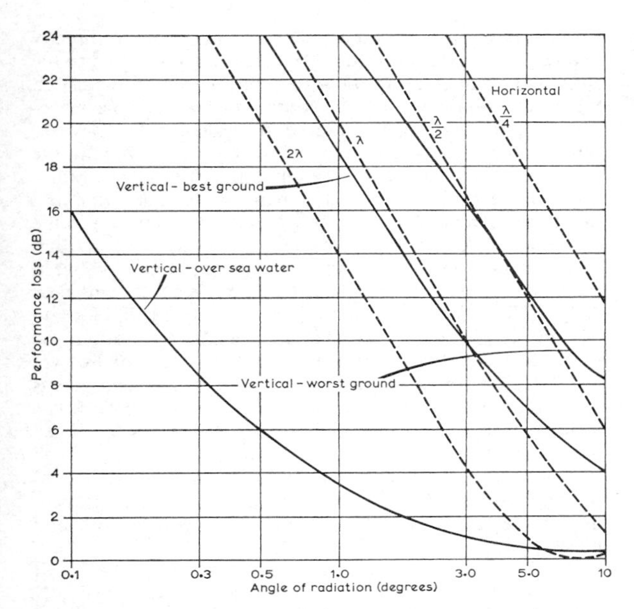

Fig 12.23. Comparison of short horizontal and vertical radiators at hf, assuming flat open country. "Zero loss" occurs with in-phase addition of the direct wave and a reflected wave of equal amplitude. Aerial heights are indicated in wavelengths for horizontal polarization (dotted curves). Vertical polarization curves are calculated for low height and a frequency of 7MHz; performance deteriorates slightly as frequency increases.

a lowering of the angle of radiation in the presence of the ground. The effect of the ground is almost fully accounted for by the appropriate array factor which is identical for all horizontal aerials at the same height, but there is another small factor resulting from the variation of radiation resistance with height, Fig 12.86. This can increase or decrease the current, causing a loss of 1·2dB for a dipole at a height of 0·4λ or a gain of 0·9dB if the height is 0·55λ, etc. It is to be expected that this effect will be neutralized in the case of close-spaced beams with antiphase elements such as the 8JK and its derivatives since each element "sees" two images of opposite phase and almost equidistant.

Effect of Earth Conductivity

So far the ground has been considered as a perfect conductor which is a useful approximation in the case of horizontal polarization but far from true for vertical polarization except for the special case of sea water when it is valid for elevation angles greater than about 3°. Due to a phenomenon which is known as the Brewster effect there is an almost complete sign reversal of the image at the angles of elevation most important for dx, ie a few degrees, so that vertical and horizontal aerials centred at the same height tend to give very similar performance. At the "Brewster angle" which is of the order of 10–20°, depending largely on ground conductivity and dielectric constant, the reflection coefficient falls to a low value and at higher angles the performance starts to decrease on account of the dipole factor. Even this however is an oversimplification since the reflection coefficient at low angles is appreciably less than unity, and leaves some useful radiation however low the aerial. This can be extremely useful at low frequencies where heights of even λ/2 are difficult to achieve, and **Fig 12.23** compares the low-angle performance of low-height vertical aerials above various types of ground with that of horizontal aerials at heights between λ/4 and 2λ. At the lower frequencies in the hf band vertical polarization tends to give stronger ground wave signals but, whichever polarization is used, ground wave signals will normally be stronger when received on an aerial of similar polarization.

Fig 12.22 includes some typical vertical patterns for horizontal broadside and end-fire arrays in addition to dipoles and confirms that for a given mean height the lower lobes are almost identical in all three cases, the high angle lobes being of course greatly reduced due to the unit patterns in the case of vertical aerials. The vertical patterns, although for a perfect earth, can be usefully applied at high angles (ie well above the Brewster angle). The modification of the vertical radiation pattern caused by this effect, which applies to vertical polarization only, is illustrated for a typical case by the dotted curve of **Fig 12.24**. A more searching treatment of this aspect is to be found in Reference [20].

Total Radiation Patterns over Earth

All the preceding references to polar diagrams in the *E*- or *H*-planes of an aerial are derived for radiation of the appropriate polarization. For aerials in free space, the diagrams are entirely valid for there is no external influence to produce any effective change of polarization. In free space there is literally no radiation off the ends of a dipole, and the *E*-plane diagram correctly falls to zero in those directions. Similarly, for a horizontal array along a direction at right angles to the elements the aerial and its image can be regarded as identical point sources and the polar diagrams computed from the appropriate array factor as described, though it should be appreciated that the diagram so obtained is relative to the centre of the system, ie a point midway between the aerial and its image, zero elevation being relative to this point and not to the actual aerial, ie zero field coincides with ground level.

If a dipole is disposed horizontally above an earth plane acting as a reflector, there will then be some radiation off the end even at zero elevation as may be appreciated by holding a pencil parallel to the surface of a mirror and observing the image from the end-on direction at various heights above the mirror. Closing one eye, the pencil shrinks to a dot but the image, though perhaps greatly reduced in length, remains visible. Neglecting ground losses the amount of radiation is determined by the usual cos θ formula and is therefore proportional to the apparent length of the image as viewed in this way. Since this image appears to be in the

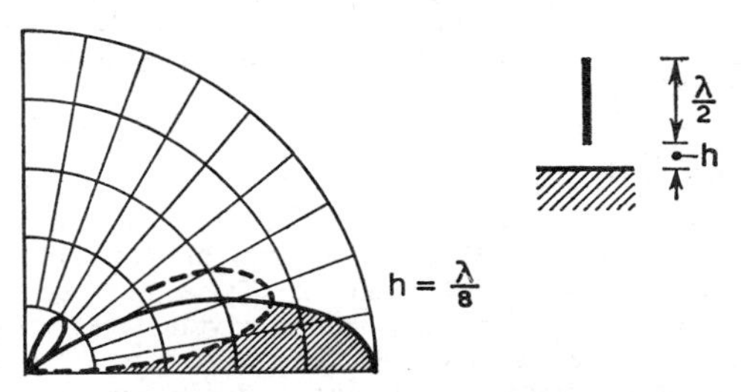

Fig 12.24. Effect of ground conductivity and Brewster reflection on vertical radiation pattern of a vertically polarized aerial over typical ground.

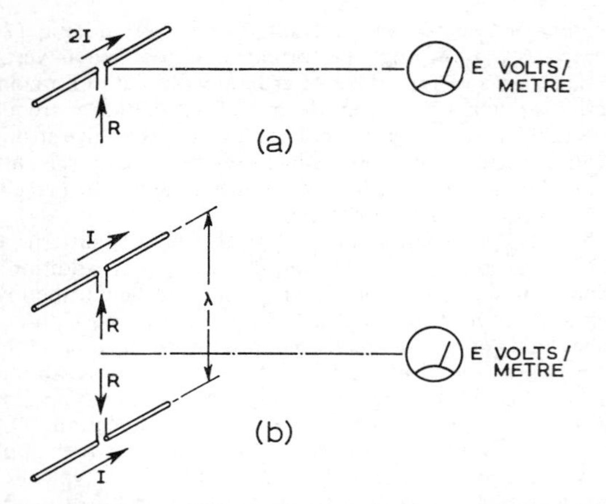

Fig 12.25. Stacking two dipoles gives an effective gain of 3dB (ignoring mutual effects) because the input power is proportional to the square of the current and the distant field to the total current.

vertical plane if the mirror is horizontal, the radiation is vertically polarized.

Ground wave propagation normal to the aerial wire is of course horizontally polarized but for intermediate directions the plane of polarization will be inclined, with completely vertical polarization off the ends of the aerial.

There is considerable high-angle sky-wave radiation "off the end" of a horizontal dipole. For example applying the cos θ formula it is evident that there will be only 6dB discrimination against signals arriving at an elevation angle of 30° from an endwise direction as compared with signals in the main lobe, neglecting the effect of the ground which due to the higher angle is more likely to favour short-skip interference than a wanted dx signal. In between these two directions the plane of polarization changes gradually from horizontal to one inclined at 30° to the vertical, remaining of course always normal to the wave direction and in the plane of the wire. Due to the varying effect of the ground as the ratio of vertical to horizontal polarization changes, it is difficult to compute accurately the vertical plane radiation pattern for directions well away from the main lobe. These directions in general are of minor importance but it should be noted that for short skip (ie high angle) signals there will be considerable reduction of directivity. On the other hand it may often be possible to increase the strength of a wanted signal, or reduce interference, by switching to another aerial at a different height. It is useful to note that the dipole tends only to exhibit reasonable horizontal directivity at the lower wave angles, and will tend to become omnidirectional for the higher angles associated with short skip propagation.

Stacking Gain

The general relationship between gain and directivity was explained on p12.3, the effect of mutual coupling on aerial currents on p12.6, and the procedure for building up the directional pattern from individual elements has now also been covered. It remains to assemble these factors together to establish the gain of arrays of various types. The achievement of gain with collinear, broadside and widely-spaced end-fire arrays depends upon the fact that the distant field, which excites the receiving aerial, is directly related to the radiation current, whereas the power fed to the aerial is proportional to the square of the radiation current. **Fig 12.25** shows two half-wave dipoles stacked one wavelength apart, each with a radiation resistance R. They are centre fed so that the radiation resistance is nominally the same as the input impedance, and each carries a loop radiation current I. For one dipole, the distant field is E due to a radiation current $2I$, and the power required to be fed to the dipole to achieve this is $(2I)^2 \times R = 4I^2R$. If the second dipole is now introduced and the current is split between them so that each has a current I and gives a distant field $E/2$, the fields will be in phase broadside to the dipoles and will add to give the same field strength E as before, but the power delivered to each dipole is now I^2R, so the total power is $2I^2R$. For the same distant field, the power required is only half that of a single dipole, so a power gain of 3dB has resulted from stacking the two dipoles. What has happened is that the fields radiated by the individual dipoles have added in phase in the desired directions but have either cancelled or added less effectively in others which results in a sharpening of the radiation pattern in the desired direction, and hence gain.

By simple extension this argument leads to the result that n elements provide a gain of n times, but this is slightly in error since it fails to take account of mutual impedance which as demonstrated earlier may increase or decrease the element currents somewhat even with wide spacing. Typically a reduction of 1·2dB is found for a collinear pair and an increase of 1·8dB for an optimally spaced broadside pair so that the effective gains in these two cases will be 1·8dB and 4·8dB respectively. The latter figure is the maximum gain obtainable from two elements fed in phase, and the required spacing (0·7λ) is known as the *optimum stacking distance*. As discussed on p12.11 however, somewhat greater gain is obtained from close-spaced elements with nearly-antiphase excitation. In the example of the two dipoles spaced one wavelength apart, the mutual resistance is very nearly zero, and the simple figure of 3dB is correct.

The principle of stacking gain may be extended to more than two radiators, and to radiators each of which is highly directive in its own right, but wherever many elements are involved, consideration must be given to the mutual coupling between each element and all other elements, in turn. The derivation of the optimum stacking distance for such systems becomes very complicated and varies with the type of array, but in general the higher the directivity and intrinsic gain of each unit in the array, or the larger the number of elements, the greater is the optimum stacking distance between them. Two broadside dipoles require a spacing of 0·7λ, four are better spaced at 0·8λ intervals, eight at 0·9λ intervals and so on. On the other hand to get maximum gain from a fixed length of array, instead of a fixed number of elements, the optimum stacking distance tends to remain constant regardless of the length. Typically, a broadside or end-fire array 4λ in length made up of λ/2 dipoles would provide about 12·5dB gain in each case, the number of elements required being 7 and 13 respectively [4]. It will be noted from the broadside examples that there is little or no change in optimum stacking distance as the array is lengthened.

When a fixed number of beam arrays are stacked, the optimum distance increases roughly as the square root of the power gain and for small Yagi arrays with 6dB gain (p12.75) it is equal to about 1·2λ though not unduly critical.

This aspect is covered more fully in Chapter 13—*VHF/UHF Aerials*.

TRANSMISSION LINES

Three separate parts are involved in an aerial system: the radiator, the feed line between transmitter and radiator, and the coupling arrangements to the transmitter. Wherever possible, the aerial itself should be placed in the most advantageous position, and a feed line used to connect it to the transmitter or receiver with a minimum of loss due to resistance or radiation. In some circumstances (eg in temporary installations or to simplify multi-band operation) the feed line is omitted and the end of the aerial is brought into the station and connected directly to the apparatus.

By the use of transmission lines or *feeders*, the power of the transmitter can be carried appreciable distances without much loss due to conductor resistance, insulator losses or radiation. It is thus possible to place the aerial in an advantageous position without having to suffer the effects of radiation from the connecting wires. For example, a 14MHz dipole 33ft in length can be raised 60ft high and fed with power without incurring appreciable loss. If, on the other hand, the aerial wire itself were brought down from this height to a transmitter at ground level, most of the radiation would be propagated from the down-lead in a high angular direction. An arrangement of this nature would be relatively poor for long distance communication.

Types of Line

There are three main types of transmission line:

(*a*) The single wire feed arranged so that there is a true travelling wave on it.

(*b*) The concentric line in which the outer conductor (or sheath) encloses the wave (*coaxial feeder*).

(*c*) The parallel wire line with two conductors carrying equal but oppositely-directed currents and voltages, ie balanced with respect to earth (*twin line*).

Single wire feeders are inefficient and now seldom used since it is impossible to prevent them acting to some extent as radiators, and the return path, which is via the earth, introduces further losses. In the two other types, the field is confined to the immediate vicinity of the conductors and there is negligible radiation if proper precautions are taken. Commercially available types of feeder are dealt with in Chapter 13 (*VHF/UHF Aerials*) with some emphasis on their vhf applications. In this chapter the discussion is confined to the mode of working and application to frequencies below 30MHz but the fundamental principles are the same.

The wave which travels on a transmission line is fundamentally the same as the free-space wave of Fig 12.1 but in this case it is confined to the conductors and the field is curved about the conductors instead of being linear as in that diagram. Concentric and two-wire lines and their associated fields are illustrated in **Fig 12.26**. In the concentric line the current passes along the centre conductor and returns along the inside of the sheath. Due to skin effect at high frequencies the currents do not penetrate more than a few thousandths of an inch into the metal; hence with any practical thickness of the sheath there is no current on the outside. The fields are thus held inside the cable and cannot radiate.

In the twin line the two wires carry "forward and return" currents producing equal and opposite fields which effectively neutralize each other away from the immediate vicinity of the wires. When the spacing between wires is a very small fraction of the wavelength, the radiation is negligible provided the line is accurately balanced. In the hf range, spacings of several inches may be employed, but in the vhf range small spacing is important.

Characteristic Impedance

In the earlier section on the behaviour of currents in radiating wires, it was stated that a travelling wave is one which moves along a wire in a certain direction without suffering any reflection at a discontinuity. The same applies to transmission lines, although in this case the presence of reflections and therefore standing waves does not cause radiation if the line remains balanced or shielded. If the line were infinitely long and free from losses a signal applied to the input end would travel on for ever, energy being drawn away from the source of signal just as if a resistance had been connected instead of the infinite line. In both cases there is no storage of energy such as there would be if the load included inductance or capacitance and the line, so far as concerns the generator of the signal, is strictly equivalent to a pure resistance. This resistance is known as the *characteristic impedance* of the line and usually denoted by the symbol Z_0. Suppose now that at some distance from the source we cut the line; what has been removed is still an infinitely long line and equivalent to a resistance Z_0 so if we replace it by an actual resistance of this value the generator will not be aware of any change. There is still no reflection, all the power applied to the input end of the line is absorbed in the terminating resistance, and the line is said to be matched.

Again because no reflections occur at the end of a correctly matched line, the ratio of the travelling waves of voltage and current, $V/I, = Z_0$. This enables the load presented to the feeder by the aerial to be in turn presented to the transmitter, without any change in the process. This is irrespective of the length of line employed, since the value of the characteristic impedance Z_0 is independent of the length of the line. In order to achieve maximum efficiency from a transmission line, it should be operated as close to a matched condition as possible, ie the load presented by the aerial

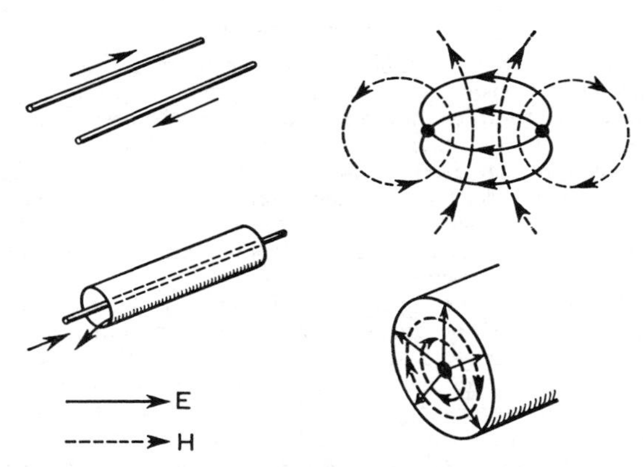

Fig 12.26. Concentric and two-wire transmission lines, with cross-sections of their wave fields. The field directions correspond to waves entering the page.

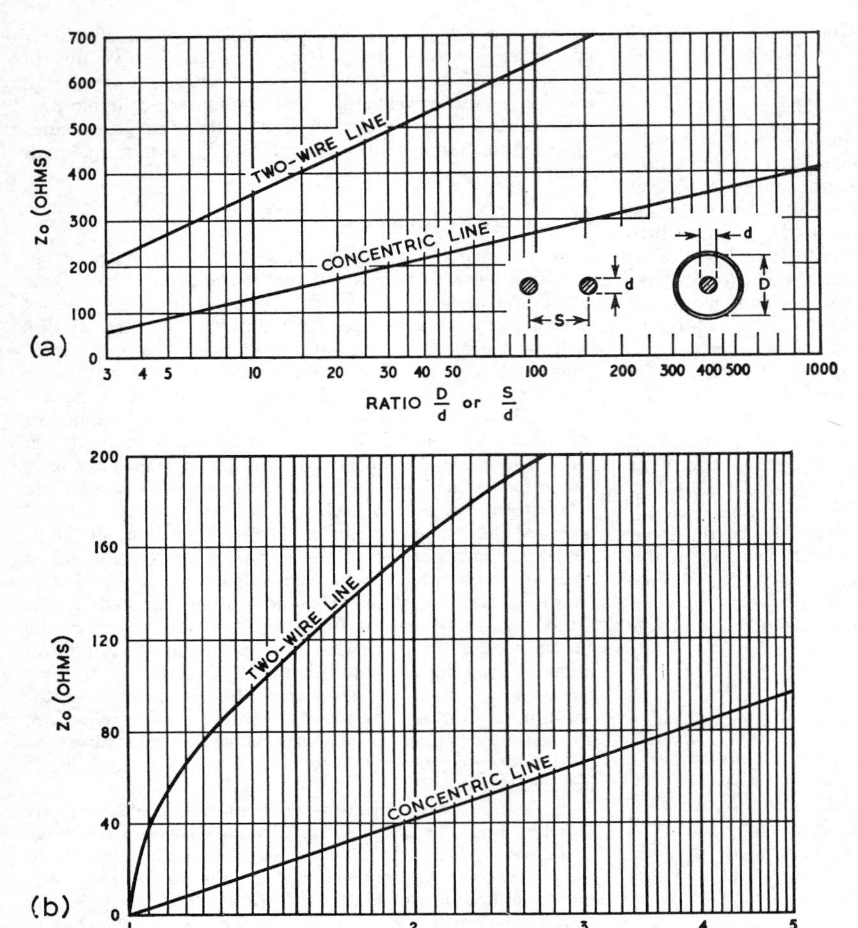

Fig 12.27. Chart giving characteristic impedances of concentric and two-wire lines in terms of their dimensional ratios, assuming air insulation. When the space around the wires is filled with insulation, the impedance given by the chart must be divided by the square root of its dielectric constant (permittivity). This ratio is called the velocity factor, because the wave velocity is reduced in the same proportion.

characteristic impedance values between about 50 and 120Ω. Twin lines have higher impedances; in practice, between 80 and 600Ω. **Fig 12.27** and **Fig 12.28** show the characteristic impedance of coaxial and two-wire lines in terms of the dimensional ratios, assuming air between the conductors. The formula for concentric lines of inner and outer diameters d and D respectively is:

$$Z_0 \text{ (ohms)} = 138 \log (D/d)$$

Thus if the diameter ratio D/d is 2·3:1 the logarithm of 2·3 is 0·362 and this multiplied by the constant 138 gives Z_0 = 50Ω. If the space between conductors is filled with insulating material with a *dielectric constant* ε (permittivity) greater than unity, the above value of Z_0 must be divided by the square root of the dielectric constant.

The usual material for insulation is polythene, which has a permittivity of 2·25. The square root of 2·25 is 1·5 so that a "solid" polythene cable has a characteristic impedance two-thirds of the value given by the formula. Many cables have a mixed air/polythene dielectric, and for these it is necessary to estimate the effect of the dielectric; this can be carried out by measuring the velocity factor as described later.

The characteristic impedance of two-wire lines of wire diameter d and centre spacing S is given by the approximate formula

$$Z_0 \text{ (ohms)} = 276 \log_{10} (2S/d).$$

should be arranged, either directly or by means of some impedance transformer, to present a good match to the line. The degree to which the load impedance can be permitted to depart from the characteristic impedance without introducing appreciable extra losses is however quite large as discussed in the section on attenuation. On the other hand bandwidth considerations or the need to avoid load variations which could damage transmitters may impose stringent matching requirements.

The characteristic impedance is determined by the dimensional ratios of the cross-section of the line, and not by its absolute size. A concentric line or cable with diameter ratio 2·3 : 1 and air dielectric, is always a 50Ω line, whatever its actual diameter may be. If it is connected to an aerial of 50Ω radiation resistance, then all the power available at the far end of the line will pass into the aerial and the impedance at the sending end of the line will also appear to be 50Ω.

Coaxial cables can conveniently be constructed with

For greater accuracy at small spacings use should be made of Fig 12.28. Appreciable errors may be introduced by bending and, during use, such lines should not be coiled with the turns touching.

As with concentric lines, an allowance must be made for the effect of the insulation. A thin coat of enamel or even a pvc covering will not produce any material change in lines with an impedance of 300Ω or less, but in the moulded feeder commonly known as 80Ω flat twin the electric field between the wires is substantially enclosed in the polythene insulator, which is thus effective in reducing the impedance to two-thirds of the value given by the formula.

When an open-wire line is constructed with wooden or polythene spacers it is again difficult to estimate the effect of the insulation, but since the proportion of insulator to air is relatively small, the effect is also small. A 600Ω line with spreaders spaced every few feet along it would normally be designed as if it were for about 625Ω, eg for 16swg wires

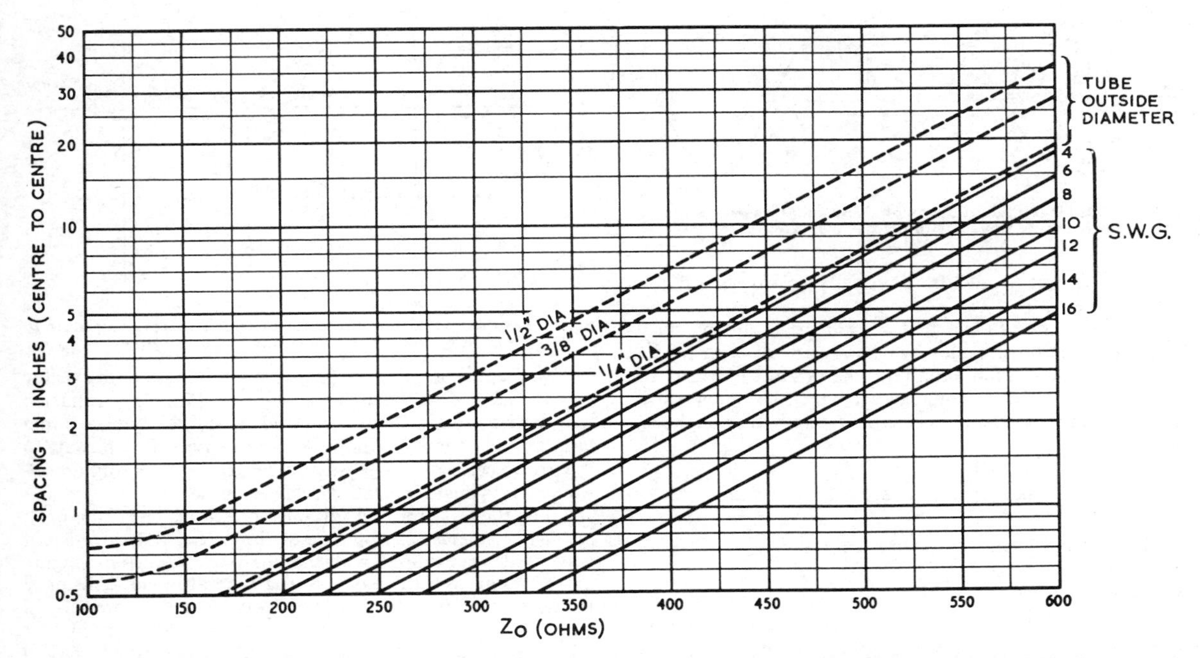

Fig 12.28. Characteristic impedance of balanced lines for different wire and tube sizes and spacings, for the range 200–600Ω. The curves for tubes are extended down to 100Ω to cover the design of Q-bar transformers. Air spacing is assumed.

(0·064in diameter) the spacing would be made 6in instead of 5in as given by the chart for 600Ω.

A transmission line can be considered as a long ladder network of series inductances and shunt capacitances, corresponding to the inductance of the wires and the capacitance between them. It differs from conventional L/C circuits in that these properties are uniformly distributed along the line, though applications are given later where short sections of line are used instead of coil or capacitor elements. If the inductance and capacitance per unit length, say, per metre, are known, the characteristic impedance is given by:

$$Z = \sqrt{L/C} \text{ (ohms)}$$

and the velocity of waves (v) on the line by:

$$v = 1 \div \sqrt{LC} \text{ metres per second}$$

The inductance and capacitance can both be determined from geometrical calculations based on the shape of the cross-section.

For the two-wire parallel line, they are obtained from the expressions:

$$L = 0{\cdot}921 \log_{10} \frac{2S}{d} \text{ microhenries/metre}$$

$$C = \frac{12{\cdot}05\varepsilon}{\log_{10} \frac{2S}{d}} \text{ picofarads/metre}$$

where ε is the dielectric constant of the material in the space between the conductors (= 1·0 for air).

Similarly for concentric lines:

$$L = 0{\cdot}46 \log_{10} \frac{D}{d} \text{ microhenries/metre}$$

$$C = \frac{24{\cdot}1\varepsilon}{\log_{10} \frac{D}{d}} \text{ picofarads/metre}$$

Velocity Factor

When the medium between the conductors of a transmission line is air, the travelling waves will propagate along it at the same speed as waves in free space. If a dielectric material is introduced between the conductors, for insulation or support purposes, the waves will be slowed down and will no longer travel at the free space velocity. The velocity of the waves along any line is equal to $1 \div \sqrt{LC}$; from the expressions for L and C given above, the value of C depends upon the dielectric constant of the insulating material. The introduction of such material increases the capacitance without increasing the inductance, and consequently the characteristic impedance and the velocity are both reduced by the same factor $\sqrt{\varepsilon}$. The ratio of the velocity of waves on the line to the velocity in free space is known as the *velocity factor*. It is as low as 0·5 for mineral or pvc insulated lines and is roughly 0·66 for solid polythene cables (ε = 2·25). Semi-airspaced lines have a factor which varies between 0·8 and 0·95, while open wire lines with spacers at intervals may reach 0·98.

It is important to make proper allowances for this factor in some feeder applications, particularly where the feeders are used as tuning elements or interconnecting lines in aerial arrays, or as tvi chokes. For example, if $v = \frac{2}{3}$, then a

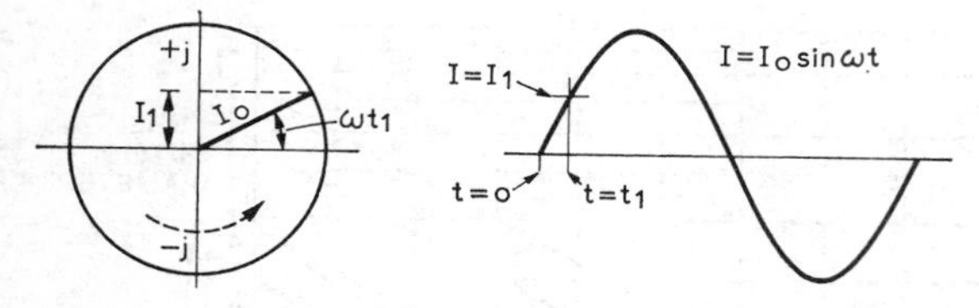

Fig 12.29. Graphical interpretation of a sinusoidal wave represented by a vector. A cosine wave is similar but displaced by quarter of a cycle.

quarter-wave line would be physically $\frac{1}{6}$ wavelength long ($\frac{2}{3} \times \frac{1}{4} = \frac{1}{6}$).

In practice, the velocity factor v can be found by short-circuiting a length of cable with about 1in of wire formed into a loop and then coupling the loop to a grid dip oscillator. The *lowest* frequency at which the cable shows resonance corresponds to an *electrical* length of one quarter-wave: then

$$v = \frac{f(\text{MHz}) \times \text{Length (feet)}}{246}$$

and should have a value between 0·5 and unity.

Standing Waves

It has already been stated that when a transmission line is terminated by a resistance equal in value to its characteristic impedance, there is no reflection and the line carries a pure travelling wave. When the line is not correctly terminated, the voltage-to-current ratio is not the same for the load as for the line and the power fed along the line cannot all be absorbed—some of it is reflected in the form of a second travelling wave, which must return along the line. These two waves, forward and reflected, interact all along the line to set up a *standing wave*.

The flow of power along the line can be interpreted as the progress along the line of a voltage wave and a current wave which are in phase, the product of which is the value of the power flowing.

If the voltage V at a point on the line is given by the expression:

$$V = V_0 \sin \omega t$$

and the current at the same point by:

$$I = I_0 \sin \omega t$$

then the amount of power flowing is the product of the rms voltage and current:

$$P = 0{\cdot}707\, V_0 \times 0{\cdot}707\, I_0 = 0{\cdot}5\, V_0 I_0$$

and is independent of the time or position on the line, varying only as the peak amplitude of the voltage or current wave is altered. Since the voltage and current waves can be expressed in identical terms, it is convenient to consider the current wave, bearing in mind that the discussion is equally applicable to the voltage wave.

The I and V waves are both of sinusoidal form, varying in amplitude at an angular rate $\omega = 2\pi f$ where f is the frequency at which the transmitter is generating the rf power. Such a wave may be represented graphically by a vector of constant magnitude (or length) rotating at an angular speed ω. The actual instantaneous value of the wave at

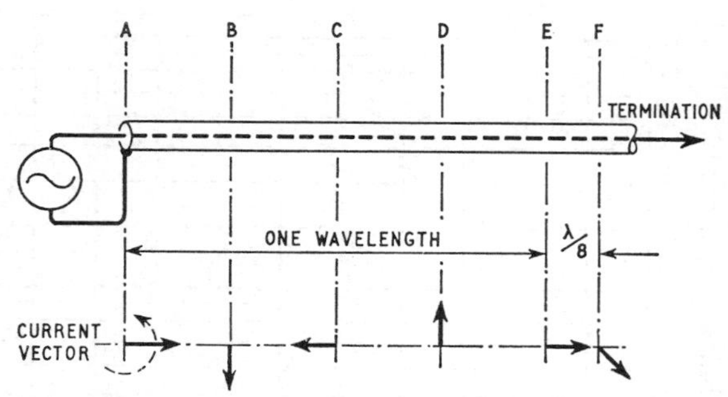

Fig 12.30. Initial variation of incident current vector, according to position along the line at any instant of time.

any moment is obtained by projecting the length of the vector on to a line passing through the origin (**Fig 12.29**). The nature and use of vectors is fully covered in any textbook dealing with ac waves but for the present purpose it is sufficient to regard a vector as a line with its length representing amplitude and its angle indicating relative phase.

Thus the distribution of current along the line *at a particular moment of time* due to the passage of the current wave is also sinusoidal, with the phase of the current lagging more and more with respect to the generator as the distance from it increases. It can therefore be represented by a whole series of vectors, each appropriate to a particular physical point on the line. This illustrates the physical meaning of the statement that a piece of line is "one wavelength" long, since this is the distance between adjacent points along the line at which the current is equal in amplitude and phase, ie the vectors are identical (**Fig 12.30**).

The vectors are drawn for a loss-less line, in which the current wave suffers no attenuation during its passage along the line. It is convenient to explain the principles of generation of standing waves by reference to the behaviour of such a line and subsequently to examine what happens when the line exhibits some amount of attenuation, as is the case in practice to a greater or lesser degree.

The current wave moving along the line will eventually reach the far end, and will then be influenced by the termination it meets. In order to establish the magnitude of this effect we must now look at the various loads that can exist. These may be divided broadly into three groups depending upon whether all, some, or none of the incident power is reflected:

(*a*) A resistive termination equal in value to the characteristic impedance of the transmission line. By definition there will be no reflection from such a termination, and all the current will flow into this load. Hence all the power delivered by the generator will be dissipated in this load (**Fig 12.31(a)**).

(*b*) A resistive termination of value other than in (*a*) above. This may or may not have a reactance associated with it. At such a termination reflection will occur, the actual amount of the reflected current being dependent upon the relative values of the resistive part of the load and the characteristic impedance of the line. In general, the greater the difference between these two, the larger the proportion of current reflected. Also, if the load resistance is greater than the characteristic impedance, a phase reversal occurs, ie the reflected current wave is 180° out of phase with the incident wave. In the most general case where a reactance is also involved, a further phase shift will occur, the amount depending upon the ratio of load resistance to reactance. Thus some of the generator power is dissipated in the resistive part of the load and some is reflected back along the line (**Fig 12.31(b)**).

(*c*) An entirely reactive termination. This includes the two extreme cases of a short circuit and an open circuit. Since there is no resistive component in the load, no power can be dissipated and therefore all must be reflected. This means that the whole of the incident current wave is reflected back down the line. There will be a phase change relative to the incident wave, the actual value of which will depend upon the reactance of the load.

In the limiting case of a short circuit, the current in the short circuit will have a maximum value and because there is no discontinuity the phase change is zero; in the case of an open circuit, no current will flow in the load, and this implies cancellation of the forward by the reflected current wave so that there is a phase change of 180° and the reflection coefficient has a negative sign. These are both special cases of the more general condition covered in (*b*).

In every case, apart from that of a perfect termination, some proportion of the incident current is reflected at the far end of the transmission line, and commences to flow back along the line at the same rate as the incident current flowing towards the end. This reflected current wave will, depending upon the circumstances of the termination, commence with an amplitude and phase both differing from the incident current wave. However, since the reflected wave is travelling back along the line, its value at any one moment in time may also be expressed in terms of a current vector, which is rotating in a clockwise direction, opposite to the incident current vector (since the waves are travelling in opposite directions) but nevertheless rotating at the same angular rate, since the frequency of the wave remains unaltered. This is illustrated in **Fig 12.31(b)–(d)** where the incident and reflected current wave vectors are shown at a precise moment of time for both the end of the line and a point one quarter-wavelength away from the end.

The reflected current wave is thus travelling back along the line, towards the generator supplying the power. At any point on the line the net current is represented in amplitude and phase relative to other points by the vector sum of the two waves as illustrated in **Fig 12.32**. At any point on the line the rf current is of course varying sinusoidally with time at the signal frequency and it is the maximum amplitude of this which is represented by the solid vectors in the lower line of the figure. The instants of time at which these amplitudes are reached for different points on the line are separated by fractions of a cycle which are represented by the relative angles of the vectors, and the variation of phase along the line is non-linear unless the reflected wave is zero.

The standing wave pattern may be observed by voltage or current meters attached to the line, or moved along the line, since the indications correspond to averages over many cycles. If on the other hand we were to measure the instantaneous values at one particular moment of time, the readings would be influenced as much by the phase variations as by the amplitude variations and would tell us little.

In Fig 12.32 the *i* and *r* vectors from the previous figure

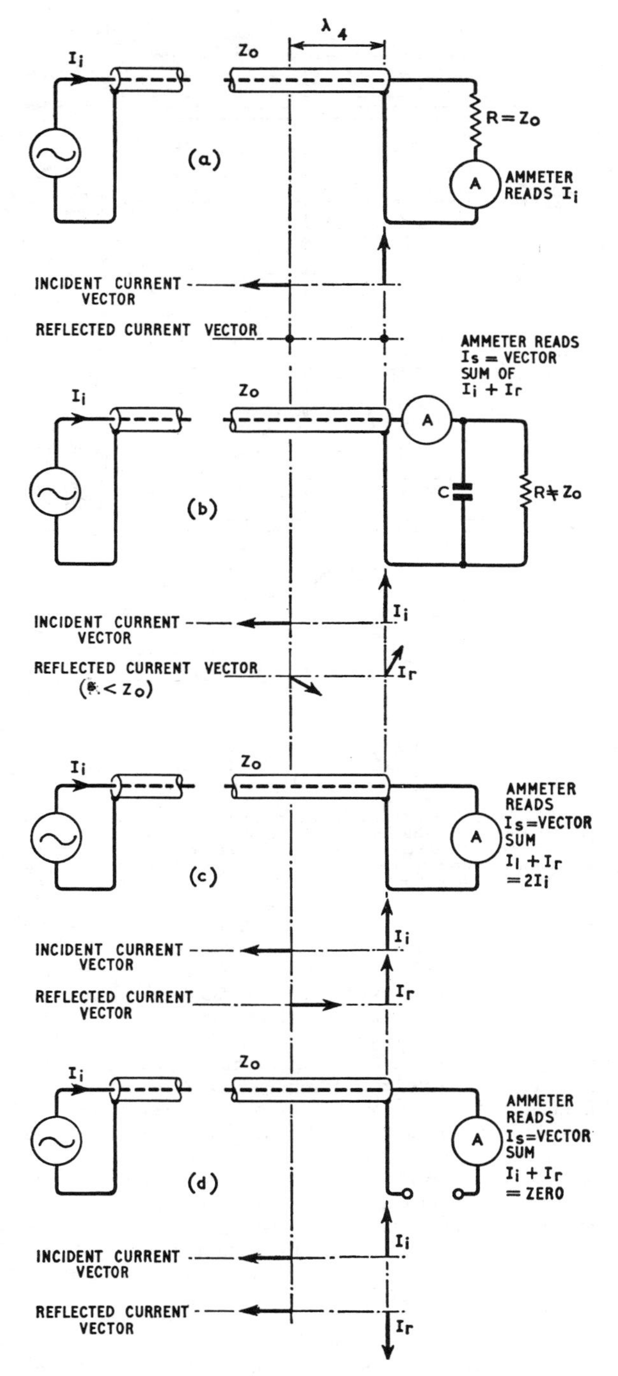

Fig 12.31. (a) Arrangement of incident and reflected current vectors in the vicinity of a correctly terminated load. (b) Arrangement of incident and reflected current vectors in the vicinity of a mismatch load of complex impedance. (c) Arrangement of incident and reflected current vectors in the vicinity of a short-circuit. (d) Arrangement of incident and reflected current vectors in the vicinity of an open circuit.

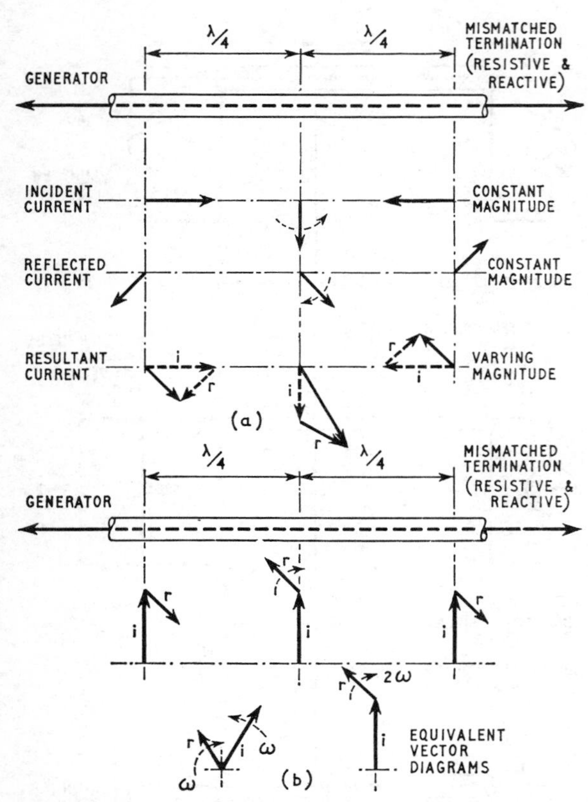

Fig 12.32. (a) Combination of incident and reflected currents along a typical section of the line. (b) Effect on (a) of maintaining the incident current vector stationary. The reflected current vector effectively rotates at twice the speed, to produce maxima and minima of current which have a crest-to-crest distance of λ/2.

have been re-drawn to show the phase variation of the reflected wave relative to the incident wave along the line, from which it will be seen that the same relative phases exist at any two points along the line separated by half a cycle, which means that maxima and minima occur twice per wavelength along the line. It follows that as the incident wave moves along the line at the velocity of propagation the reflected wave vector rotates relative to it at twice the signal frequency. Fortunately this harmonic frequency does not exist as such, the situation being analogous to that of two trains approaching each other at 50mph and passing therefore at a *relative* velocity of 100mph; in the same way the incident wave travelling outwards along the line "sees" the reflected wave rushing back past it at what appears to be twice its own velocity. The interaction of forward and reflected waves produces the familiar standing-wave pattern as typified by **Fig 12.33**.

The standing wave pattern is not a sine-wave, although it is approximately sinusoidal for low values of swr; it departs increasingly from this form as the swr increases, to an extent where in the limit of an open or short circuit termination it becomes a plot of the *amplitude* of a sine-wave, ie it resembles a succession of half sine-waves (**Fig 12.34(b)**). Because of the fact that in space the reflected current wave vector is effectively rotating at twice the generator frequency,

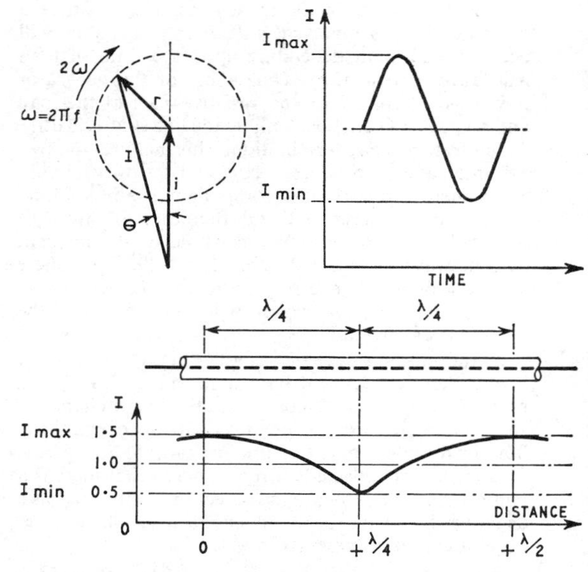

Fig 12.33. Development of the standing wave pattern due to the resultant of the incident and reflected current vectors. The variation of the resultant line current with distance is derived directly from the variations of incident and reflected current with time; the time variation is effectively removed by maintaining a constant incident current vector.

the positions of successive maxima and minima occur at quarter-wavelength intervals along the line. It is important to note that the voltage pattern though similar in other respects to the current distribution is displaced from it by λ/4 so that voltage maxima coincide with current minima and vice versa. This follows from the facts that voltage must be zero at a short circuit, where current is a maximum, and current must be zero at an open circuit. The voltage standing waves are shown in Fig 12.34 by dotted lines.

The distance from crest to crest of the standing wave is one half-wavelength at carrier frequency, this being also the distance between points of maximum amplitude of a travelling wave, and in both cases the voltages or currents at successive maxima are opposite in phase. In the case of a travelling wave, however, a meter inserted in the line "sees" a succession of peaks and troughs and records an average over many cycles, which is the same for any point in the line if losses are neglected. In general, for a standing wave, the conditions of voltage, current and therefore impedance repeat themselves every half-wavelength along the line, and use is made of this property for impedance matching.

The *standing wave ratio* (swr) k is the ratio of the maximum and minimum values of the standing wave (I or V) existing along the line. Also, by definition, the *reflection coefficient* r is the ratio of the reflected current vector to the incident current vector. Thus the maximum value of the standing wave will be $(1 + r)$ and the minimum value of the standing wave will be $(1 - r)$, the swr and reflection coefficient being related by the expression:

$$k = \frac{1 + r}{1 - r}$$

The value of r used in the above expression lies between zero (matched line) and unity (complete mistermination) but

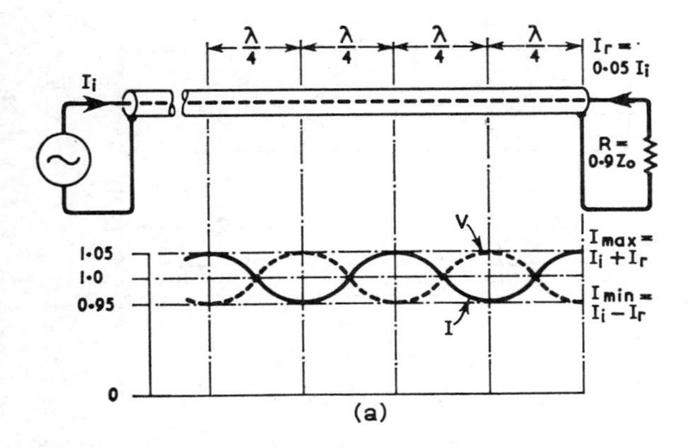

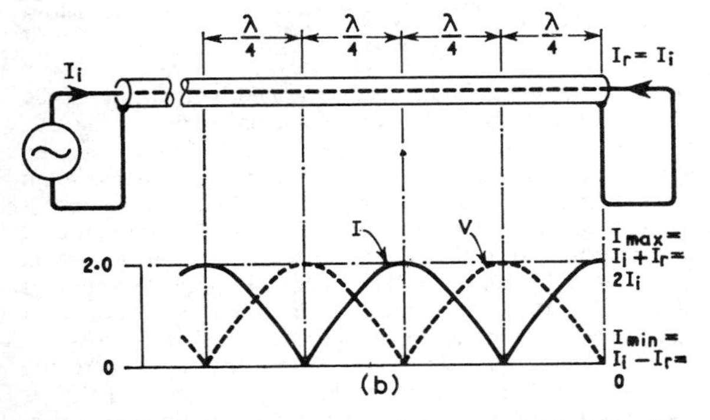

Fig 12.34. (a) Standing wave pattern for a near-matched line (R = 0·9 Z_0). (b) Standing wave pattern for a completely mismatched line (R = short circuit). The curve of Fig 12.33 is typical of the transition at medium swr from the near-sinusoid of Fig 12.34(a) to the rectified sinusoid of Fig 12.34(b).

it is sometimes expressed as a percentage. Typical values are given in **Table 12.1.**

As defined above, the standing wave ratio will have a range of values from unity (matched line) to infinity (complete mistermination). Occasional references will be found to values of swr in the range zero to unity. These refer to a scale of values obtained by inverting the expression given, and are exactly reciprocal to the more generally used figures, ie an swr of 0·5 : 1 is exactly the same as an swr of 2·0 : 1. This system is more frequently encountered in microwave work. Sometimes the designation vswr is used in place of swr to emphasize the normal convention of expressing it in terms of voltage or current, not power ratio.

Fig 12.34(a) shows the current standing wave pattern on a line which is terminated in a purely resistive load smaller than Z_0. The standing wave is a maximum at the load and therefore a minimum at a point λ/4 back from the load.

TABLE 12.1
VSWR in terms of reflection coefficient

Z/Z_o or Z_o/Z = vswr	Reflection coefficient percentage
1·0	0
1·5	20
2	33
3	50
5	67
9	80
∞	100

If a line is terminated in a mismatch R_L, which is entirely resistive, then the value of the swr is conventionally given by the simple relationship:

$$\text{swr} = \frac{R_L}{Z_0} \text{ or } \frac{Z_0}{R_L} \text{ whichever is larger}$$

This follows from a consideration of the power distribution at the resistive load. Considering the balance of power flow at the point of connection of the load to the line, the incident power, ie that associated with the forward travelling wave, is given by:

$$P_{in} = I_i^2 \times Z_0$$

and the reflected power, ie that associated with the backward travelling wave, is given by:

$$P_{ref} = I_r^2 \times Z_0$$

both from the earlier definition of $Z_0 = \frac{V}{I}$ along the line.

Assuming a positive value of reflection coefficient, ie R_L less than Z_0 so that I_i and I_r are in phase, the power in the load resistance R_L will be:

$$P_L = (I_i + I_r)^2 \times R_L$$

and must equal the difference between the incident and reflected power, so that

$$Z_0 (I_i^2 - I_r^2) = R_L (I_i + I_r)^2$$

and

$$\frac{R_L}{Z_0} = \frac{I_i + I_r}{I_i - I_r} = \frac{1 + r}{1 - r} = k \text{ (swr)}$$

If r is negative (ie R_L greater than Z_0) the expression inverts so that $k = \frac{Z_0}{R_L}$.

This relationship between R_L, Z_0 and the swr is a basic one which is exploited to good effect when using transmission lines as impedance transformers as discussed later.

The reflected current wave travelling back along the line from the mistermination at the load end will ultimately return to the generator which supplies the power. This generator will itself possess an internal impedance which may or may not match the Z_0 of the line. Subject to a suitable mismatch the reflected wave will be re-reflected at the generator and will travel forward again towards the load, as a new incident wave which will in turn be reflected at the load end, to exactly the same extent as was the original incident wave. This process of reflection at the load, and re-reflection at the generator will continue until all the power is absorbed in the load provided there are no line losses and the mismatch at the load is countered by the appropriate "mismatch" between line and generator. Part of the power in the load will have been delayed by several passages along the transmission line but the time interval, though sometimes long enough to cause ghost images in television reception, is of no significance for the narrow-band modulation systems used in hf communication.

The net effect of the passage up and down the line of travelling waves of current is to modify in turn the value of the incident and reflected currents. However, each contribution to the incident current due to re-reflection at the

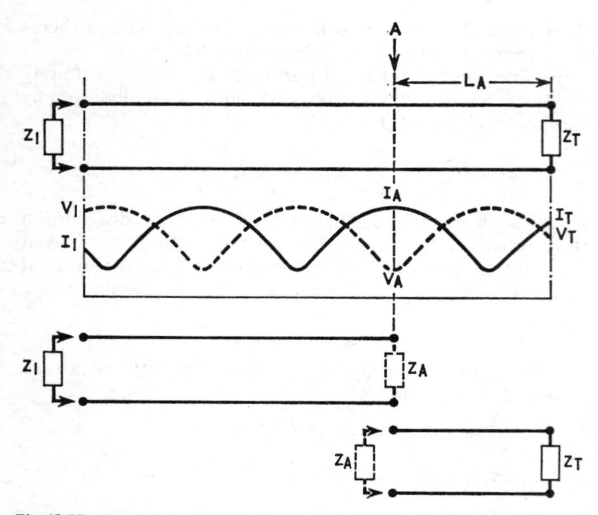

Fig 12.35. The impedance across the line at point A looking towards the load is such that if used to terminate the shortened line, the input impedance Z_I would be unaltered. The value of Z_A depends upon the ratio V_A/I_A and the phase difference between them: in this example it is less than Z_T or Z_I, and because it occurs at a point of current maximum is the lowest impedance at any position along the line.

generator is accompanied by a corresponding contribution to the reflected current, due to further reflection at the load. The overall effect of this is that the ratio of incident and reflected currents remains unaltered, and is dependent *only* upon the relative values of the characteristic impedance of the line and the impedance of the load. *The standing wave ratio along the line is dependent only upon the nature of the load at the far end, and no amount of alteration at the generator end can alter the magnitude of this standing wave.*

This is one of the most important aspects of the behaviour of transmission lines.

Since no power is lost in the lossless line, the current waves are not attenuated during their passage along the line, and hence the swr remains at a constant value for the whole length of the line.

Input Impedance

When a transmission line is operated in a mismatched condition, it may be necessary to know what value of impedance is seen across the input to the line. This is the load which is presented to the transmitter, and is termed the *input impedance* of the line.

The voltage and current along a misterminated line vary in the manner indicated by Fig 12.34 and depend as already described on the nature of the load impedance. At the far end of the line the ratio of voltage to current is of course equal to the load impedance and similarly the ratio at any other point along the line describes an impedance, which can be said to exist across the line at that point. If the line were cut at a point A and this value of impedance connected, the standing wave pattern between this point and the transmitter would be unaffected, and the input impedance would remain the same as before. Equally, the impedance across the line at the point of cutting is the input impedance seen looking back into the section of line which has been cut off. This is illustrated by **Fig 12.35**. The actual value of the impedance

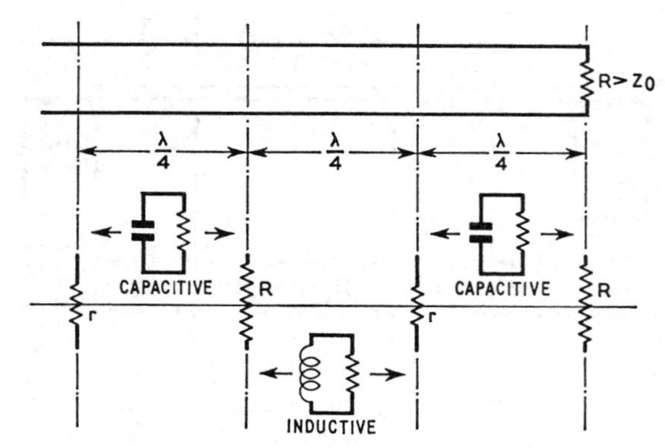

Fig 12.36. Variation of input impedance along a misterminated line with $R > Z_0$. The impedance is alternately inductive and capacitive for succeeding λ/4 sections. At the precise λ/4 points it is purely resistive.

at any point along the line is determined by the ratio of the actual line voltage and current at this point, taking due account of phase differences. Fig 12.34 shows that the apparent load impedance as observed at a voltage maximum is purely resistive and greater than Z_0. Equally, at current maxima the impedance will also be resistive, although of a lower value. At these points current and voltage are in phase, and in between them the current leads and lags alternately. This causes the impedance to become in turn capacitive and inductive along successive quarter-wavelengths of line, changing sign at the positions where the reactive term becomes zero in **Fig 12.36**. Note that if R is less than Z_0, the capacitive and inductive sections are interchanged.

The input impedance is dependent not only upon the load impedance, but also the length of line involved. It is a common fallacy to believe that by altering the length of a line, the match can be improved. The adjustment of line length may result in the presentation to the transmitter of a more acceptable load impedance so that it delivers more power but this indicates merely the provision of an insufficient range of adjustment at the transmitter. The standing-wave pattern remains completely unaltered.

In Fig 12.34 the case of a purely resistive termination is illustrated. In practice the far end termination may be inductive or capacitive. The effect of this is to cause the positions of voltage maxima to move along the line towards the generator by an amount determined by the nature of the reactive termination. If this is capacitive, as in **Fig 12.37,** there will then exist along the line a short region AB of capacitive impedance before the first voltage minimum. The reactive termination could be replaced by the input impedance of a further short length of line BC terminated in a suitable pure resistance: the length AB is then such as to make up the quarter-wavelength AC. The lower the capacitive reactance of the termination, the closer is the point of voltage minimum.

An important aspect of the input impedance of a mismatched line is its repetitive nature along the line. The conditions of voltage and current are repeated at intervals of one half-wavelength along the line, and the impedance across the line repeats similarly. *The input impedance is always equal to the load impedance for a length of line any exact number of half-wavelengths long* (*neglecting line losses*)

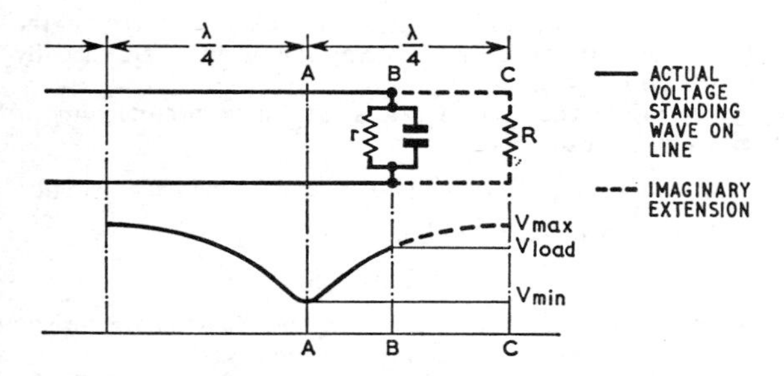

Fig 12.37. Derivation of the standing wave pattern on a line terminated in a capacitive load impedance. The identical pattern is provided by a hypothetical resistive load at a particular distance beyond the actual line load.

irrespective of its characteristic impedance. Such a length of line can be used to connect any two impedances together without introducing unwanted variations, and without regard to the Z_0 of the line being used. This property is the basis of the operation of "tuned" lines as aerial feeders, and its application is discussed later. Also, any existing feeder arrangement can be extended in length by any number of half-wavelengths without altering the swr or the load as seen at the input end. Because the repetitive nature of the line impedance is a function of frequency, being tied to intervals of one half-wavelength, the precise duplication of the load impedance at the input can only occur at a number of discrete frequencies for any specific length of line, for which the line is 1, 2, 3 . . . n half-wavelengths long. The extent to which the input impedance varies when the frequency is varied slightly determines the bandwidth over which the aerial system may be used without readjustment of the transmitter. For a given percentage change in frequency, there will be a greater variation of input impedance the longer the physical length of line, the change in electrical length (expressed in fractions of a wavelength) being greater, leading to a larger change in the voltage and current at the input to the line. This of course involves properties not only of the line itself but also of the load connected to it. Particular attention should be paid to the bandwidth aspects of lines when they are operating with high swr such as matching or tvi stubs, or as "tuned" feeders. In the case of tvi stubs which operate with a nearly infinite swr the effective Q is quite high and their useful bandwidths as filters is strictly limited, often to the extent of $\pm$ 2 per cent of nominal design frequency.

Line Transformers

It is possible to use short lengths of line with values of Z_0 different from that of the main feeder as impedance transformers for matching a wide range of load impedances to different values of line impedance. It has already been shown that the standing wave ratio $k = \frac{Z_0}{R_L}$ or $\frac{R_L}{Z_0}$ whichever is greater so that translating this into the symbols of Fig 12.36 we have:

$$k = \frac{R}{Z_0} = \frac{Z_0}{r} \text{ so that } r = \frac{Z_0^2}{R} \text{ and } Z_0 = \sqrt{Rr}$$

If the quarter-wavelength of line adjacent to the load in Fig 12.36 is replaced by a different line having a characteristic impedance Z_T the main transmission line will "see" a load impedance Z_T^2/R and by suitable choice of Z_T this can be made equal to Z_0 so that it looks like a matched load. The impedance transformation ratio $\frac{Z_0}{R}$ is therefore given by $\frac{Z_T^2}{R^2}$ and the required value of Z_T by $\sqrt{Z_0 R}$. The matching line is known in this case as a *quarter-wave transformer* and its operation is shown pictorially by **Fig 12.38**. It should be noted that R_1 and R_2 are interchangeable so that a $\lambda/4$ section of 150Ω line will match a resistance of 450Ω to a 50Ω line or alternatively for example a 50Ω impedance to a 450Ω open-wire line. This property is known as impedance inversion, an impedance of, say, one tenth of the matching line impedance Z_T being transformed to ten times Z_T and vice versa. This relationship holds good for any ratio of impedances.

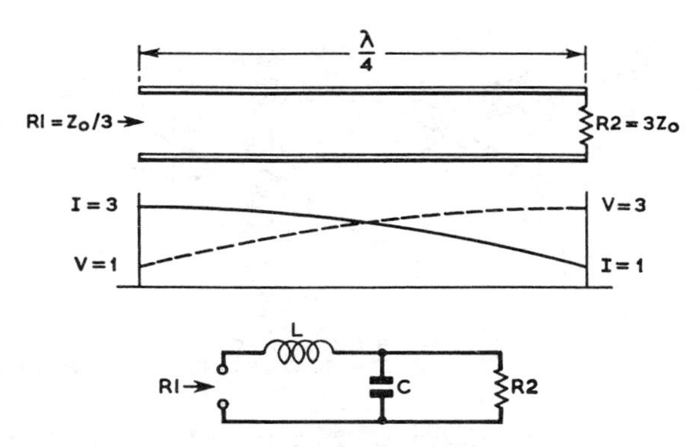

Fig 12.38. Illustrating the principle of the λ/4 transformer in the case of a 9:1 impedance ratio. The V/I ratio is inverted by the standing wave on the line with vswr = 3, giving R1 R2 = Z0². This corresponds to the series parallel action of a resonant circuit in which L/C = Z0². R2 is the parallel resistance and R1 the equivalent series resistance.

A transmission line can always be approximated by one or more inductances and capacitances, and the simple LC circuit shown in Fig 12.38 provides a close equivalent to the quarter-wave transformer; L and C are in resonance and ωL replaces the Z_T of the transmission line formula.

Apart from questions of possible convenience it is not always possible to find or construct matching lines with suitable characteristics, and lumped constants provide an important alternative.

As an example of the use of line transformers, **Fig 12.39** illustrates the matching of a broadside array of two dipoles into a 70Ω line. The ratio of the higher impedance to the Z_T of the transformer is closely analogous to the Q of a tuned circuit and the ratio in this case, 1·4, implies a large bandwidth. For large transformation ratios the bandwidth may be improved by the use of two quarter-wave sections in series, transforming to an intermediate value at their junction, which should ideally be the geometric mean of the end impedances. In such an arrangement the effective ratio of each transformer is reduced, which itself improves their bandwidth, and in addition the two tend to be mutually compensating with change of frequency to produce a further improvement, **Fig 12.40**. Use of this technique is somewhat restricted by the limited choice of cable impedances available.

The inverting properties of a quarter-wavelength of line can also be used to relate the voltage at the input to the

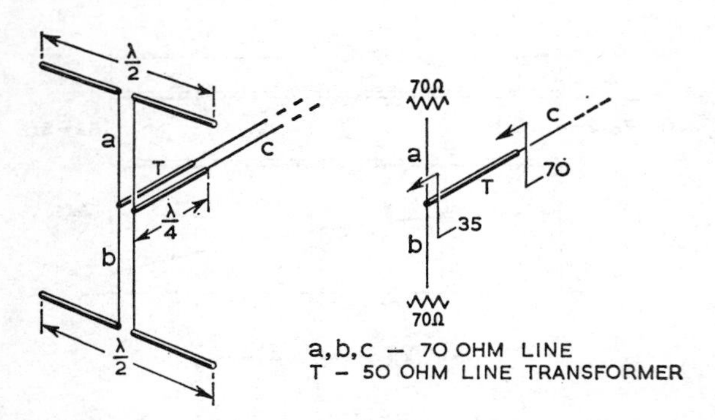

Fig 12.39. Use of a λ/4 50Ω line transformer to match two co-phased dipoles into a single 70Ω main feeder. The impedance at the junction of a and b is 70/2 = 35Ω, which is transformed up to 70Ω again.

current in the load (and vice versa). In Fig 12.38, the swr on the line can be defined in a number of ways.

$$\text{swr} = \frac{V_2}{V_1} = \frac{Z_0}{R_1} = \frac{Z_0 I_1}{V_1}$$

so that $V_2 = I_1 \times Z_0$
and similarly $V_1 = I_2 \times Z_0$

The voltage at the input end is determined *only* by the characteristic line impedance and the current at the other end, and does not depend upon the value of the load in which the current flows. Any change of load impedance alters the value of impedance across which this constant voltage is developed, and hence the power which must be supplied by the transmitter in order to maintain the current.

Stubs

A line is completely mismatched when the far end is terminated by either a short or open circuit. There is then no load resistance to dissipate any power, and a 100 per cent reflection of voltage and current takes place. The standing wave pattern is then that of Fig 12.34(b). Except at resonance, sections of line with open or short circuit terminations approximate to pure capacitance or inductance, although there is usually some resistance arising from losses in the conductors or dielectric material and these can be considerable as explained later unless the lines are air-spaced or very short. Such pieces of line are called *stubs* and can be used to compensate for, or match out, unwanted reactive impedance terms at or close to a mismatched termination, and allow the major part of the feed line to operate in a matched

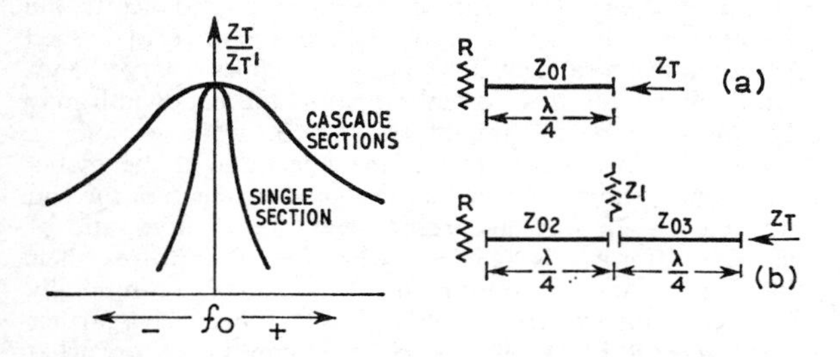

Fig 12.40. Bandwidth of single and double section λ/4 transformers. Z_T is the actual input impedance at each frequency, while Z_T' is the theoretical design frequency value. In the two-section case, Z_i is the geometric mean of R and Z_T.

condition. To minimize losses, and also to preserve the bandwidth of the aerial system, the lengths are usually restricted to λ/4 or less.

The input reactance of a loss free short circuited line is given by the expression:

$$X_{in} = Z_0 \tan\theta \text{ (inductive)} \simeq Z_0\,\theta \text{ when } \theta \text{ is small}$$

and for an open circuited line by:

$$X_{in} = -Z_0 \cot\theta \text{ (capacitive)}$$

where θ is the electrical line length and Z_0 its characteristic impedance as given for example by Fig 12.27.

Figs 12.41 and **12.42** illustrate the manner in which the input reactance varies with line length. For sections shorter than λ/4, the short circuited line is inductive, and the open circuited line capacitive, as may be appreciated from visualizing very short lengths of line which bear a close physical resemblance to inductors or capacitors. A short circuited line can also be made to behave like an open circuited line of the same electrical length (and vice versa) by adding a further λ/4 of line: this is another application of the inverting properties of a λ/4 section of line described previously. The behaviour of the two types of stub is completely complementary, and if two are joined together at their input ends they will make up between them precisely one quarter-wavelength when their respective reactances are equal and opposite in sign. This is closely analogous to a parallel tuned circuit, and a short-circuited λ/4 line is said to be a "resonant" section, possessing all the properties of the parallel tuned circuit. Near resonance the reactance-frequency slope of the resonant line is similar in shape to that of the lumped constant circuit, and a change of Z_0 is equivalent to a change in the L/C ratio. Such sections of line may be used as shunt chokes across other lines without affecting the latter, and may also at vhf be used as tuned circuits for receivers and transmitters, the physical lengths being then conveniently short.

In complementary fashion, a λ/4 open circuit line section behaves as a series resonant circuit, with a very low input resistance at the centre frequency. In the ideal case this is a

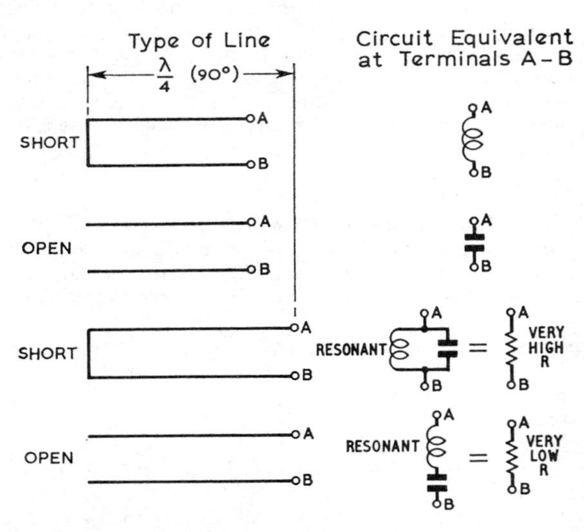

Fig 12.41. Circuit equivalents of open and short-circuited lines.

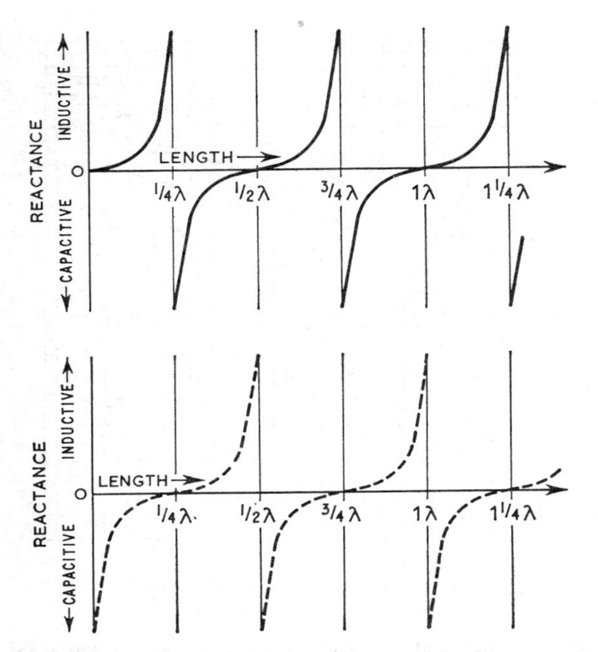

Fig 12.42. Variation of input reactance with length for short circuited (solid) and open circuited (broken) lines. The two sets of curves are identical and displaced by λ/4.

perfect short circuit: in practice the line losses degrade this to a finite but low value of resistance, which for all practical purposes can be considered a short circuit. The reactance-frequency slope of both open and short circuit resonant sections is shown in **Fig 12.43** which illustrates the effect of variations in the Z_0. For the open-circuited lines of Fig 12.41 operating near resonance but slightly detuned the reactance is given by the simplified formula:

$$X = Z_0\,\theta = 6{\cdot}28 Z_0 \cdot \frac{\Delta l}{\lambda}\ \Omega$$

where $\frac{\Delta l}{\lambda}$ is the fraction of a wavelength by which the line falls short of or exceeds the resonant length. This formula is valid for any length of line *or aerial system* observed at or near a current loop and given the radiation resistance (if any) and loss resistance it enables the bandwidth of the system to be roughly estimated. Thus for a quarter-wave line (or half-wave dipole which can be thought of as the λ/4 line "opened out") 5 per cent change of frequency produces a length error of λ/80. Z_0 for the λ/2 dipole may be as much as 1,000Ω so that X = 78Ω which is comparable with the radiation resistance and therefore indicates a half-bandwidth of the order of 5 per cent. For an open wire line Z_0 is typically 600Ω and from Fig 12.5 the loss resistance at 14MHz using 18swg is 2·4/2 = 1·2Ω. The reactance equals this resistance for an error in line length or a change in frequency of $5 \times \frac{1{\cdot}2}{62{\cdot}8}$ per cent, ie the total bandwidth is $\frac{12}{62{\cdot}8}$ per cent and the Q is therefore $\frac{62{\cdot}8}{12} \times 100 = 525$. If an additional half wavelength is added to the stub or a half-wave dipole is fed with λ/2 of open wire line, the reactance changes three times as fast, the loss resistance is multiplied by three, but radiation resistance stays the same so that for the dipole case the Q increases from 20 to 60 approximately whereas the Q of the stub remains constant. Strictly speaking, allowance should be made for a difference in Z_0 between the dipole and the feeder but to gain insight into the behaviour of such combinations of aerial and resonant feed line it is sufficient to assume an "average" Z_0 of the order of 800Ω. For lengths shorter than resonance, or for longer wavelengths, X is of course negative (capacitive) and for longer lengths or shorter wavelengths it is positive (inductive).

The above formula for X may be applied equally well to short inductive lines. For example it was found above that if a resonant λ/4 stub is shortened by 5 per cent the negative reactance is $\frac{6{\cdot}28}{80} = 0{\cdot}078\ Z_0$ so that the short portion removed must have a positive reactance of 0·078 Z_0. A convenient figure to remember is 55Ω per foot for 600Ω line at 14MHz and this can be adjusted proportionately for other impedances and frequencies multiplying also by the velocity factor (eg two-thirds) where this is applicable. This formula gives an error of minus 5 per cent for a change in length of λ/16 and minus 20 per cent for λ/8 which, incidentally indicates the danger of pursuing the analogy between lines and circuits too far. In the case of inductance for example, the reactance is strictly proportional to frequency and this process continues indefinitely, whereas the line reactance varies with frequency in a non-linear manner changing from positive to negative and back again repeatedly as discussed earlier.

Some practical applications of matching stubs are considered on p12.37 in the section dealing with impedance

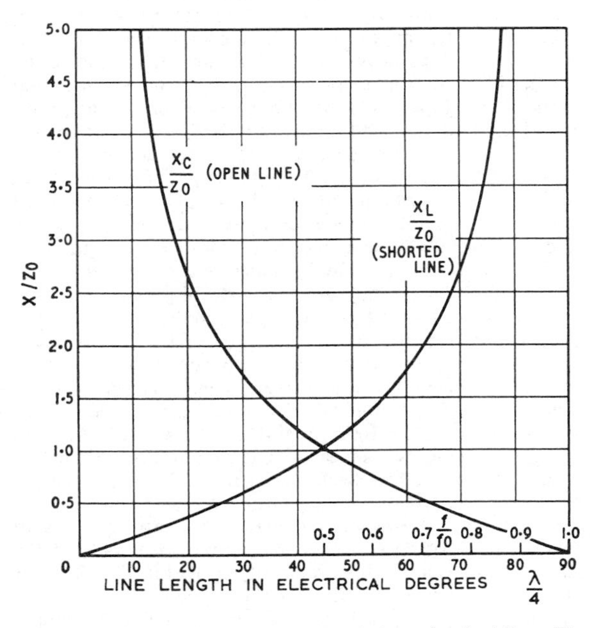

Fig 12.43. Reactance curves for open and short circuited lines. The quantity X/Z_0 multiplied by the characteristic impedance of the line equals the reactance at the input terminals. Longer lines alternate as capacitance or inductance according to Fig 12.42. The curves also indicate the manner in which input reactance varies with frequency for nominal λ/4 sections.

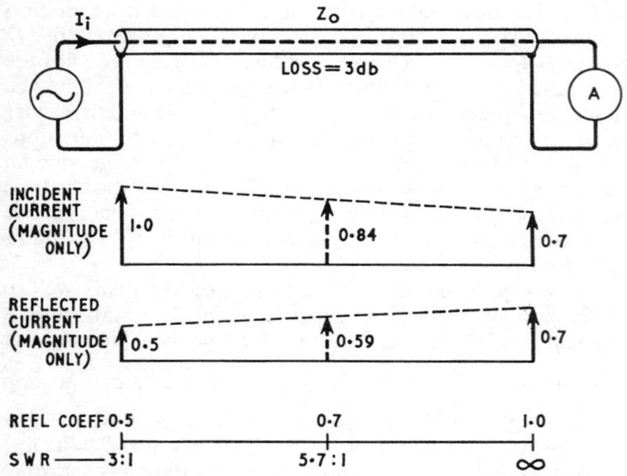

Fig 12.44. Effect of line attenuation on incident and reflected currents. The line is short circuited and the ammeter measures the current in the short circuit.

matching. Their use as single frequency filters is covered in detail in Chapter 17—*Interference*.

Attenuation and Loss

In practice all transmission lines have some loss associated with them. This loss may be due to radiation, resistive losses in the conductors, and leakage losses in the insulators; but however it arises, it is a function of the actual construction of the line and the materials employed. With the best of the moulded twin or coaxial lines supplied for amateur radio and television use this attenuation is less than 1dB per 100ft (when correctly matched) at frequencies below 30MHz. 1dB loss in 100ft means that at the far end of the feeder the amplitude of voltage or current is just under 90 per cent of that at the input end, and about 80 per cent of the power is delivered to the load. A second 100ft extension would deliver 80 per cent of the power left at the end of the first 100ft or 64 per cent of the original output. For each extra 100ft the factor is multiplied again by 0·8 but in decibel ratios 1·0dB loss is added. Open wire lines can be remarkably efficient: a 600Ω two-wire line made from 16swg copper wire spaced about 6in, carefully insulated and supported, has a loss of only a few decibels *per mile* in the hf range.

The *matched loss* of the line is quoted in the manufacturers' published information usually as n dB per 100ft. It increases more or less in proportion to the square root of the frequency at which it is being used, ie a line having a quoted matched loss of 0·5dB/100ft at 10MHz will have a matched loss of approximately 1·5dB/100ft at 90MHz. One effect of this loss is to improve the swr as measured at the input end of the line since at this point the reflected wave has suffered attenuation due to the outward and return journeys along the line whereas the outgoing wave is unattenuated. This can on occasion conceal the existence of a considerable mismatch at the load end of the line. This effect is illustrated in **Fig 12.44** for the extreme case of a short circuited line. Losses increase with increasing swr since they are proportional to the square of current or voltage. This means for example that the increased loss due to heating of the conductors at a current maximum is less than counterbalanced by reduced heating at the adjacent minimum, and similar considerations apply

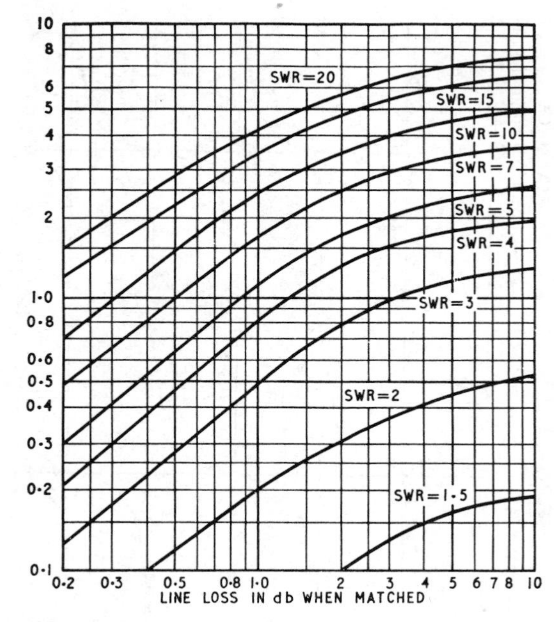

Fig 12.45. Curves of additional power loss for any given swr at the load and matched line loss. These losses are due to extra heating of the line with high values of swr and will be incurred in addition to the matched line loss regardless of whether or not the generator matches the line.

to the dielectric losses. Extra losses arising in this way depend only on the matched line loss and the swr at the load, not on the length of line, and may be obtained from **Fig 12.45**.

When a signal is fed into a lossy line terminated by a short circuit it arrives at the far end with reduced amplitude but in almost the same phase as if there were no losses. The reflected wave is initially of the same amplitude as the incident wave but in turn experiences attenuation on its way back to the generator as illustrated in Fig 12.44. In this case an infinite swr at the termination appears as a ratio of only 3 : 1 at the transmitter. If sufficient cable is available it is possible to use it in this way as a dummy load of reasonable accuracy and relatively low cost for use in the adjustment of transmitters; thus referring to Fig 12.45 a line with a matched loss of 3dB will have a total loss of at least 10dB when completely mismatched as in this example, the input swr will be only 1·22 : 1 and the load impedance presented to the equipment will be equal to $Z_0 \pm 20$ per cent. The cable, because of its considerable bulk, is able to dissipate a considerable amount of heat. This technique is particularly useful at vhf and uhf where the smaller diameter cables tend to be quite lossy. It should be noted that Fig 12.45 has in one respect been over-simplified, since the losses are not distributed uniformly along the line but are about three times as great per unit of length near the termination, where the swr is large.

In typical practical installations the improvement in swr at the transmitter due to line losses can be very large, leading to an over-optimistic assessment of performance as illustrated by the following example. In this case the matched-line loss (theoretical and measured) was 1·5dB and the swr at the transmitter was 2. The extra line loss due to this was estimated from Fig 12.45 to be only 0·25dB and from dx signal reports the aerial appeared to be working well until another aerial became available for comparison, when it became obvious that something was amiss. Measurements at the load end of the feeder revealed an swr of 5 thus increasing the total loss to an estimated 3·2dB and accounting, along the lines of Fig 12.44 for most of the "improvement" in swr at the transmitting end. (Further losses were attributable to feeder radiation as discussed later.)

As already indicated any losses in a transmission line are increased by the presence of standing waves, whereas in the case of the loss-less line all the power delivered to the line is eventually dissipated in the load, although, as we have seen, a certain proportion is delayed in time due to multiple reflections. When the mistermination of the line is not gross, the amount of power returned from the first reflection at the load is not large.

Assuming that the generator completely misterminates the line at the sending end then the power from this first load reflection will reappear at the load after a journey back down the line and up again during which time it is attenuated by twice the matched line loss (go and return). At this stage the major proportion of the attenuated reflected power is delivered to the load, and a small proportion is re-reflected; thus if 20 per cent is reflected on the first occasion only 4 per cent (less some attenuation) is reflected a second time and the nett power delivered to the load after only one reflection represents most of that delivered by the generator.

In practice, any reactive term present in the load offered to the transmitter by the line is removed by re-tuning of the output stage, and any variation in the resistive term is compensated by readjustment of whatever coupling device is used between the transmitter and the line. These adjustments ensure that the effective loading of the transmitter and therefore the power delivered to the aerial system remains constant despite variations of load impedance.

In other words the transmitter presents virtually a complete mistermination to the reflected power in the line, and the additional power lost due to the presence of reflections from the load is relatively small, arising only from the attenuation of the reflected wave. This point is frequently misunderstood and it is a common belief that once power has been reflected from a misterminating load this reflected power is lost. This case *only* occurs when the variation of line input impedance caused by reflections at the load is not compensated by tuning and matching adjustments at the transmitter. This is a situation of limited practical importance but can arise for example due to an insufficient range of adjustment or if attempts are made to obtain instantaneous comparisons (ie without retuning) between aerials which are not equally well matched. If the transmitter has been adjusted with a matched load and its output is then switched to another load with a known swr, the loss may be obtained from **Fig 12.46**; for example, with an swr of 3 there will be a loss of only 1·5dB or less than half an S-unit, though in the case of a linear amplifier the incorrect loading could lead to distortion and "splatter".

Matching adjustments are not usually provided in the case of receivers and it is sometimes erroneously thought that reception can be degraded by aerial mismatch. Maximum signal-to-noise ratio does not in general require a good match but a certain optimum degree of mismatch, though this consideration is usually of no importance for hf reception in view of the high external noise levels from galactic and other sources. Because these external noise levels tend to be a long

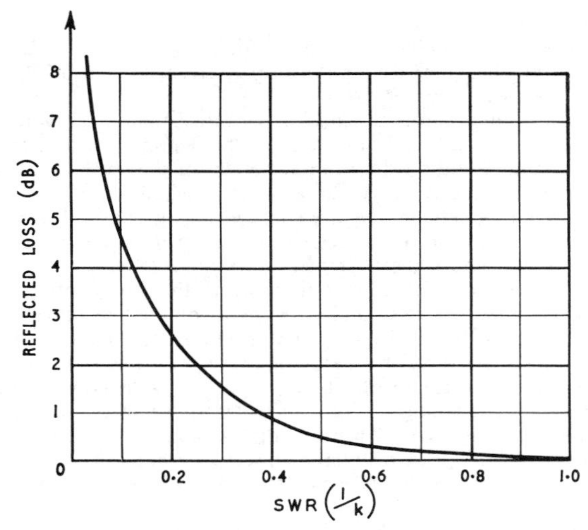

Fig 12.46. Curve of mismatch loss, ie the amount of power reflected from a mismatch for a given swr. This is also the curve of absolute power loss (when the generator matches the line, and therefore absorbs the reflected power) for lines of zero matched loss.

way above the internal receiver noise it is usually permissible to insert considerable attenuation in front of the receiver and this may indeed be beneficial in view of the prevalence of intermodulation and similar effects which contribute to interference and effective noise levels. This point is considered in more detail in Chapter 15.

As already noted the additional line loss at high values of swr for the normal transmitting situation is dependent only on the total matched line loss and the swr, and not influenced by the length of the line. From Fig 12.45 it will be noticed also that when the matched line loss is very high, the additional loss due to the presence of a standing wave tends to become independent of the matched line loss.

Cable loss may be determined very simply by measurement of swr at the input end with the far end short circuited. If for example the loss is 3dB the reflected wave arrives back at the input with an attenuation of 6dB so that the measured swr is (1 + 0·5)/(1 − 0·5), ie 3·0. For monitoring the condition of a feeder cable it should be possible to obtain the necessary mismatch by operating at a frequency for which the aerial is not designed, taking care of course not to overload the cable.

Power Handling

There is always a definite limit to the amount of rf power which can safely be transmitted along any form of transmission line. The limitation is set either by excessive heating of the conductors, which can lead to deformation or destruction of the insulation between them, or by voltage breakdown between the conductors, either in air or in the insulating material.

Open wire balanced lines will generally carry more power than any other form, there being adequate ventilation around the conductors and usually a relatively large spacing between them. Concentric lines usually have much lower impedances with small spacings between the conductors, and because the inner conductor must be maintained accurately in alignment there is often a considerable amount of dielectric material involved. The inner conductor is therefore well insulated thermally from the surrounding air and will rise to a much higher temperature for a given dissipation than its open wire counterpart.

In amateur practice, most lines other than the poorest quality coaxial cables are quite suitable for use up to the maximum authorized power, although for the uhf bands the smaller diameter cables should be avoided when carrier powers in excess of 50W are employed. This is usually done anyway, because the cables of lower power rating are also those with the highest attenuation per unit length.

The amateur should pay attention to the power rating of cables employed when a high swr exists on the line, either by accident or design. In such circumstances, the value of the current at maxima of the standing wave may rise to such an extent that local overheating will occur, with consequent damage. The voltage maxima may also approach the limiting value for flash-over. These conditions are unlikely to arise in a well designed installation except perhaps under fault conditions, a possibility which should not be overlooked.

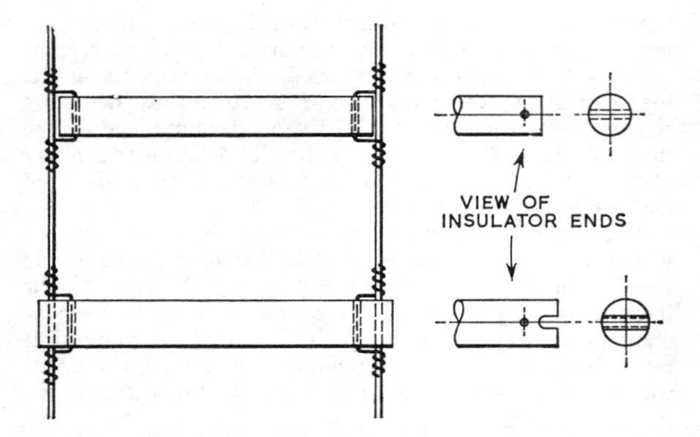

Fig 12.47. Spacing insulators for open wire line. The lower pattern with an end locating groove is to be preferred. In each case, the insulator is held in place by a 20swg binding wire passing through a hole in the insulator body.

Practical Feeders

The majority of commercially available coaxial cables have impedances of the order of 50Ω or 70–75Ω and twin feeder is available for impedances of 75, 150 and 300Ω. Two or four wire open line having impedances in the range 250–750Ω, though not as a rule commercially available, is readily constructed, providing the advantages of much lower cost and greatly reduced losses, typically 0·2dB per 100yd at 14MHz compared with 1·8 to 4dB for coaxial cable. Against this, coaxial cable and low-impedance twin-lead is often more convenient to use, with the latter providing a good match to the centre of a λ/2 dipole without the need for transformers or baluns. In general however practical beam aerials have relatively low impedances so that some form of transformer is needed between the aerial and line. In addition, if unbalanced feeders are used to feed balanced aerials or vice versa a suitable converter (balun) should be used as discussed later, and these devices can also be used as impedance transformers so that the choice of feeder is mainly dictated by personal preferences, practical convenience in individual cases, or in the case of commercial aerials the instructions of the manufacturer. If very long feeders are required, the use of open-wire line in conjunction with some form of impedance transformation at the aerial is normally essential.

Coaxial cables are usually made with a solid or stranded inner conductor and a braided wire sheath. Two-wire lines of 75–300Ω impedance are generally made with moulded polythene insulation and are reasonably flexible. The general properties of typical lines are listed in Chapter 13—*VHF Aerials*.

Almost any copper wire may be used for the construction of open wire lines and although 12–18swg is advisable for very long lines it can be inferred from Fig 12.5 on p12.5 that even with 24swg the losses only rise to 0·6dB per 100yd at 14MHz and the wire size can be decreased to 30swg before losses become comparable with those of low impedance lines. These figures do not take account of loss by radiation, commonly thought to be large in the case of open wire lines but typically (as discussed later) less than one-tenth of 1 per cent for lines with 6in spacing at 14MHz. Spacers are needed at intervals and subject to using the minimum number (depending on wire gauge, spacing, length of span, exposure to wind etc) almost any insulating material can be used provided it does not absorb moisture. Polythene rod is particularly recommended whereas it is inadvisable to use lossy materials such as bakelite or perspex for long lengths of line. Narrow spacings are inconvenient in so far as they require more spacers, and there is more risk of trouble from snow and ice. Rod spacers may be fixed as shown in **Fig 12.47.** Thin polythene line has excellent insulating and mechanical properties and may be used for supporting open-wire line, guiding it round obstacles etc, and also for supporting aerials without additional insulators; inexpensive lines, suitable for lengths of 100ft or more, may be constructed by using tensioned polythene cord as mechanical support for 22swg or finer wire gauges.

Although the above procedures are recommended, considerable liberties may be taken in the design of open wire lines without appreciably closing the performance gap which is in their favour relative to low impedance lines. The lead-in for example may use plastic insulated wire taken through holes drilled in a window frame without incurring measurable losses, but it is possible to go too far in this direction and in one instance a lead-in using twin plastic lighting flex pushed through a single hole in a wooden frame gave rise to a spectacular firework display; the situation was somewhat extreme as there was an accidental feeder short and 400W p.e.p. applied to the line, but this incident suggests a serious fire risk which could easily be overlooked. It is therefore strongly recommended that good insulation, preferably some sort of ceramic bushing, be used at this point unless the wire is brought through a *glass* window pane as shown in **Fig 12.48**. Replacement of glass by perspex for easier drilling is inadvisable as perspex is frequently subject to rf breakdown. Soft-drawn enamelled instrument wire can be used, although when the line is to be strained between insulators, it is better to use hard-drawn or cadmium-copper wire. Pre-stretching to about 10 per cent extra length by means of a steady pull will however appreciably harden soft-drawn copper wire and render it less liable to stretching in service.

Moulded 300Ω twin lines sometimes suffer noticeably from the effect of moisture in damp weather. This is most noticeable in the case of tuned lines, because water has a very high dielectric constant and affects the velocity factor of the line as well as its loss factor. Soot deposited on the

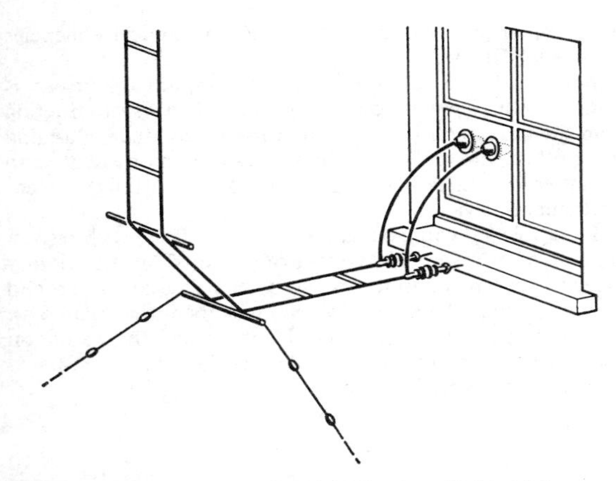

Fig 12.48. Typical layout of balanced line installation. A lower end fixing and a slow bend are included in this example. With polythene or artificial fibre cords (eg nylon or Terylene) there is no need for additional insulators.

surface will tend to retain moisture and accentuate these troubles. The remedy is to clean the line periodically and give it a dressing of silicone wax polish to repel any moisture. Tubular 300Ω line should be sealed so that water cannot enter the interior. Although the pvc outer jacket of a concentric cable is good enough to allow the cable to be buried, the open end of the cable is very vulnerable for it can "breathe in" moisture which does irreparable damage to the line. There is a variety of satisfactory sealing materials available, such as Bostik cement, Bostik sealing strip (putty) and Sylglas tape which is loaded with a silicone putty.

Particular attention should be paid to the adequate sealing of plugs and sockets used out of doors on coaxial cables. Even those intended for professional use (eg BICC Teleconnectors, US type C and BNC) are not entirely waterproof, and the usual Belling-Lee coaxial plug is virtually unsealed in its own right. Such fittings should be securely taped for at least 1in along the cable with a water repellent tape. The most effective and reliable sealing is obtained by a layer of self-amalgamating PIB tape, covered by a wrapping of a putty loaded fabric tape. (The former is manufactured by Rotunda Ltd, Denton, Manchester, and the latter, known as Densotape, by Winn and Coales Ltd, Denso House, Chapel Road, London S.E.27.)

In selecting an appropriate feeder for use with an aerial system, due allowance must be made for the physical layout involved. Any installation employing a balanced line should be arranged so that as far as possible the line is kept clear of masts, buildings and other obstructions. Failure to do this may sometimes result in severely unbalanced currents in the two conductors, which then become a source of undesirable radiation and adversely affect aerial performance. Unbalance can also be caused by abrupt changes of direction of the feeder, and to a lesser degree by slackness in the line which can allow it to swing about in the wind and to twist.

As far as possible, all balanced lines should be lightly tensioned in straight clear runs, with bends broken up into a slow change of direction. A practical layout is illustrated in Fig 12.48 in which it will be noted that the lower end of the feeder is tensioned to a secure anchorage, and flexible tails are used to connect to the feed-through insulators to relieve them of any strain.

Coaxial cables lend themselves to less obtrusive installation. If they are correctly terminated in an unbalanced load, no currents will flow on the outside of the outer conductor and they may therefore be laid close to any other surface or object (brickwork, tubular masts, etc) without affecting their behaviour in any way. It is only necessary to ensure that any bend is made on a sufficiently large radius to avoid distortion of the cable itself. Typically, bends in ¼in diameter television type cables may have as little as ½in radius, whereas those in larger cables such as UR67 should be of not less than 2in radius. The recommended minimum bending radius for cables can often be found in manufacturers' literature, but a useful rule of thumb is to use a radius equal to at least twice and preferably four times the cable diameter. It is easy to damage coaxial cables by excessively tight fastenings, such as staples driven hard into the cable support, and care should be exercised when making such fixings to avoid bruising the cable.

When plugs and sockets or other forms of termination are used on out-of-door coaxial cables, the cable should always be arranged in such a way that any water running down the outside is shed from the outer before it can run into the termination. The right and wrong ways for fixing such terminations are illustrated in **Fig 12.49.**

Radiation from Feeders

Radiation from feeders can arise in several different ways. **Fig 12.50** illustrates a typical aerial fed at the centre with a feeder which may be of any type, and it is assumed for the moment that the system is energized by inductive coupling into an aerial tuning unit with no earthing at the transmitter. If the feeder has an effective electrical length which (allowing for top loading by the dipole) is an even number of quarter-wavelengths, it will resonate as an aerial and can be used as such; in fact a 14MHz dipole with about 44ft of vertical feeder, voltage fed at the lower end, performs well as a low-angle radiator for 7MHz dx. On the other hand a length of

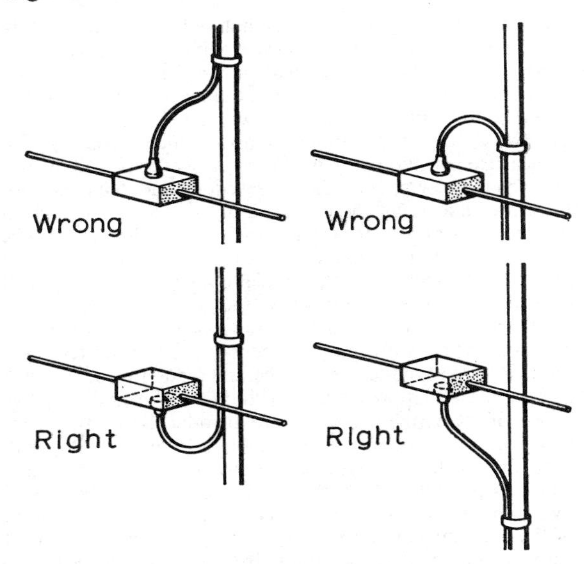

Fig 12.49. Method for terminating outdoor coaxial cables. The reverse loop in the cable prevents water from running back into the end termination of the cable.

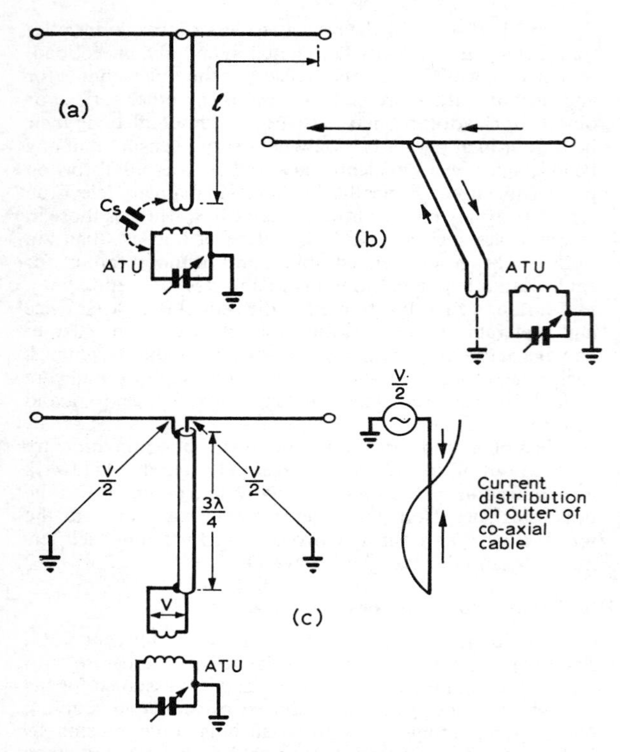

Fig 12.50. (a) If l is an odd number of wavelengths the feeder is energized as a vertical radiator via the stray capacitance C_S. (b) If not brought away at a right-angle the feeder is energized by unequal inductive or capacitive couplings as indicated by arrows. (c) Excitation of standing waves in coaxial feeder if current distribution is symmetrical. If feeder is at earth potential excitation can take place as at (a) or (b).

17 or 51ft would enable it to radiate efficiently on 14MHz though probably in an unwanted direction, or angle of elevation. With the arrangement shown the unwanted mode could be excited by capacitive coupling at the transmitter unless the coupling coil is balanced to earth and shielded. The two halves of the dipole can also induce currents in the vertical feeder (the outer in the case of a coaxial line) but given perfect symmetry these are equal and opposite and therefore cancel. On the other hand any asymmetry due to failure to bring the feeder away from the dipole at right angles or not connecting it to the exact centre of the dipole, or proximity of one half of the dipole to a tree or building, will lead to a net current in the down lead and hence to radiation in an unwanted and probably lossy mode with considerable potential for tvi. Earthing the feeder at the transmitter alters the length required for resonance and leads to a rather less predictable situation as the earth resistance and effective length of the earth connection are introduced into the problem. The safest course is probably to avoid earthing and keep the electrical length of the system as far from resonance as possible. If necessary the resonance can be shifted by a short counterpoise or earth connected to the centre point of the coupling coil as shown dotted or a short-circuited $\lambda/4$ stub. Avoiding resonance minimizes but does not necessarily eliminate the unwanted radiation at the fundamental frequency, and the problem though basically similar may prove much more intractable at the harmonic frequencies responsible for tvi.

In the case of balanced lines the unwanted current is induced in the same direction in both conductors, subtracting from the transmission line current in one conductor, adding in the other, and displacing the swr pattern in one relative to the other; this makes it impossible to obtain reliable measurements of swr.

If a dipole fed with coaxial line is symmetrical with respect to earth, half the terminal voltage of the aerial must obviously appear between the outer of the coaxial line at its top end and earth, Fig 12.50. The feeder must then appear like an "inverted ground plane" aerial for which the radiation resistance at resonance is about 18Ω and the power radiated in the unwanted mode will be

$$\frac{V^2}{4 \times 18} \text{ watts}$$

ie as much power is radiated from the feeder as from the aerial. More likely, symmetry will not be achieved and feeder radiation will take place as previously described. As before, the situation can be rectified, more or less, for the fundamental frequency by avoiding resonance but this does not necessarily eliminate harmonic radiation.

A long line will have many resonances which tend to overlap making it essential to use some form of balun as described later, but such devices have limited bandwidth and cannot be relied on to prevent the flow of harmonic currents on the outside of the feeder, so there is clearly no alternative to the suppression of harmonics by filtering at the source. It is, however, only the inner of the coaxial line which is normally filtered and the suppression of harmonic currents in the outer conductor may be incomplete. For these reasons it is sometimes found difficult when using coaxial feeder to eliminate tvi.

Balanced lines brought away from the aerial at right angles and cut to a non-resonant length with adequate harmonic filtering avoid these problems. Unscreened lines are in themselves potential radiators, but subject to accurate balancing this should be negligible for frequencies in the hf band; thus the power radiated for an rms line current of I amps and a spacing D is given by:

$$P_r = 160 \left(\frac{\pi D}{\lambda}\right)^2 I^2$$

D and λ being measured in the same units so that D/λ is the line spacing measured in wavelengths. This is the same power as would be radiated from the same current in a short dipole of length equal to the line spacing. For 600Ω line with 6in spacing and a current of 1A, the line radiation

at 14MHz is therefore $160 \left(\frac{\pi}{134}\right)^2 = 0{\cdot}09\text{W}$ for 600W

supplied to the aerial. The proportion of any harmonic power radiated by the line increases of course as the square of the harmonic number, but seems likely to remain insignificant. The figures given by the above formula must be multiplied by four to account for radiation from the terminal connectors, but even so for a 75Ω line with 0·08in spacing only about one millionth of the incident power is radiated as compared with the possibility of up to one-half as previously discussed for an unbalanced line.

This is still not the full story. A major advantage of coaxial line is that it can be laid anywhere, even buried underground; in contrast to this, low impedance twin line generates a field which can be detected externally and, though it spreads only a short distance, makes it advisable to keep the line clear of trees, buildings, etc, at least to the extent of mounting it on stand-off insulators. It is a disadvantage of open-wire lines that the balance is very easily disturbed and the presence of objects near the line can cause asymmetry and hence radiation; the use of open-circuited matching stubs is particularly dangerous and if their use is unavoidable they should be brought away from the line at right angles with the separate wires carefully equalized in length, and balance is particularly sensitive to proximity of the open ends of such stubs to surrounding objects. Even if the layout appears to be perfect, balance should be carefully checked, eg with simple indicators such as that illustrated in Fig 12.78.

IMPEDANCE MATCHING

In nearly every arrangement of aerial and feeder system there is a requirement to provide some means of impedance transformation, or *matching*, at a particular point between the aerial and the transmitter. It may take the form of the aerial coupler mentioned above, which transforms the varying input impedance for a tuned line system down to a suitable load value for the transmitter and helps to provide discrimination against harmonics or other spurious frequencies. It may be a network at or near the aerial terminals which transforms the aerial input impedance to the correct value to match a length of flat line back to the transmitter. In each case the function of the transformer is to alter the impedance on the aerial side of it to a suitable value for presentation to the transmitter or the line connecting it to the transmitter. The matching network whatever its construction has no effect on the impedance of the aerial and therefore cannot alter the swr on the line which connects it back to the aerial terminals. It is important to understand that the process of matching results in the elimination of standing waves only on the line between the point of application of the matching and the transmitter delivering the power (**Fig 12.51**).

It will be clear from the discussion of tuned line feeder systems that they are acceptable, carrying as they do a high standing wave, only when as in the case of open-wire line the inherent line loss and hence the additional loss due to the standing wave is very low. It is also evident as explained on p12.25 that the bandwidth of the aerial system reduces rapidly when the length of the tuned line is increased, the rate of change of reactance with change of frequency being proportional to the length of line whereas the radiation resistance is fixed, so that even in the narrow span of an amateur band re-tuning of the coupling circuits may be necessary as the frequency is changed. Any necessary matching should therefore be carried out as near as possible to the aerial. In a similar way, a reduction of bandwidth occurs if a large ratio of impedance is to be transformed, because high ratio transformers have smaller bandwidths. It is therefore best to avoid any large ratios between the impedances of aerials and feeders, and this also helps to reduce losses. This cannot always be done, especially if a long line is necessary, because it would then be preferable to employ a high impedance line in order to minimize losses. The importance of avoiding feeder radiation (sometimes referred to as the

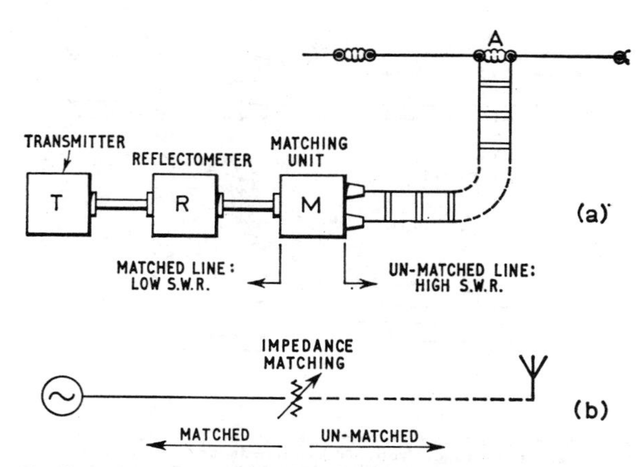

Fig 12.51. Application of impedance matching. Introduction of the matching unit can improve matters only on the section M-T. The feeder on the aerial side of the matching unit, M-A, remains "un-matched" with a high swr. It should therefore be as short as possible and of low attenuation (see text). The matching unit, often called an aerial coupler in this application, can be adjusted using a reflectometer to indicate minimum reflected power in the "matched" section of the line to the transmitter.

Marconi Effect) has been stressed in the previous section and in many cases the balun required to prevent this in the case of coaxial feeders can also act as a matching unit.

The technique employed for matching will depend upon the nature of the unmatched impedance and the physical details of the aerial. In some cases it is possible to modify the aerial itself to achieve the required match to the feed line, without significantly altering its characteristics as a radiating element. Examples of this are the use of folded wires, and the adjustment of reflector spacing and length in Yagi type beam aerials. Resonant aerials fed with high impedance balanced lines may usually be matched by tapping the line across a suitable portion of the aerial wire, eg by means of the T or delta matches described later, but in some cases it is necessary, or easier, to use a network consisting of lumped constant components or transmission line stubs interposed between the aerial terminals and the main feeder.

Folded Dipoles

The half-wave dipole is a balanced aerial and requires a balanced feeder. Normally this is a 70Ω line, but it is sometimes necessary to step up the aerial input impedance to a higher value; for example, where the feeder is very long and 300 or 600Ω line is employed to reduce feeder loss. Another case is that of a beam aerial in which the dipole input impedance is too low to match 70Ω line. A separate transformer to step the aerial impedance up to these values would reduce the bandwidth, but it has been found that by folding the aerial the step-up can be accomplished without any reduction of bandwidth. **Fig 12.52(a)** shows a full-wave wire folded into a half-wave dipole; this bears some resemblance to a loop aerial and has some useful properties in common with the quad and delta loops into which it can be formed by pulling the wire out to form respectively a square or triangular shape.

The relative direction of currents, at current maximum, is shown by arrows in the diagram. In a straight full-wave wire they would be of opposite polarity, but folding the full-wave

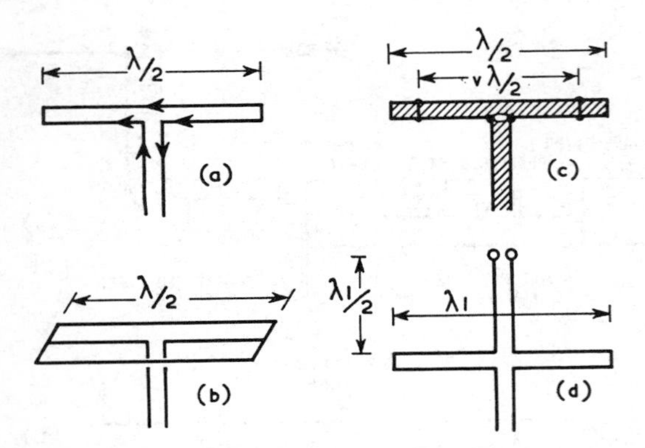

Fig 12.52. Folded dipoles, including (a) simple fold; (b) three-fold; (c) folded dipole made from 300Ω ribbon feeder; (d) arrangement for working on half- and full-wave resonance.

has "turned over" one current direction; thus the arrows in the diagram all point the same way. The folded dipole is therefore equivalent to two single-wire dipoles in parallel.

Since half the current flows in each wire, the radiation resistance referred to the centre of either is four times that of a simple dipole, ie 300Ω approximately, which is particularly convenient for matching to standard ribbon feeder, though the higher impedance and wider bandwidth of an open-wire line greatly eases the problems of resonant feed-lines, and this can be exploited to good effect for the development of multi-band beams as discussed later. Wider bandwidth is obtained with folded dipoles because the reactance change with varying frequency is approximately that of two quarter-wave stubs connected in series and is therefore rather less than twice that of a simple dipole, whereas the radiation resistance, as explained above, is increased by four times. Typically, a simple dipole well matched to low impedance line at the centre of the 14MHz band could show an swr as high as 1·4 at the band edges whereas the corresponding figure for a folded dipole fed with 300Ω line would be better than 1·2.

If the aerial is made three-fold (**Fig 12.52(b)**), it is equivalent to three radiators in parallel, the feed-point resistance is multiplied nine times and it would be used with a 600Ω line.

In some multi-element arrays the radiation resistance is very low, and folding the driven element is a useful aid to correct matching since with N wires the resistance is always raised N^2 times.

The folded dipole is usually regarded as a one band aerial, and will not normally operate on even harmonics of the frequency for which it is cut because the "folded" currents cancel each other. However, the alignment of currents is correct for the odd harmonics, so that a folded dipole for 7MHz can also be used on 21MHz, although the radiation pattern will have multiple lobes. On the other hand a 14MHz folded dipole with a half-wave resonant feeder will be nearly resonant on 21MHz and will provide some gain at the higher frequency, the radiation pattern being intermediate between those of half-wave and full-wave dipoles. With suitably tuned feeders, a 21MHz folded dipole may be used on 28MHz and 14MHz.

A folded dipole one wavelength long can be made to radiate but only if the centre of the second conductor is broken; in such a case the input impedance is *divided* by four. The centre impedance of a full-wave dipole is 4,000 to 6,000Ω; that of the folded version 1,000 to 1,500Ω. **Fig 12.52(d)** shows how such an aerial can be made to work at both fundamental and second harmonic frequencies by means of a length of open circuit twin line (called a *stub*) used as a "frequency switch". At the frequency at which the aerial is λ/2 long, the stub is one electrical quarter-wave long and, since its other end is open, the input end behaves as a short circuit and effectively closes the gap in the aerial. At the full-wave frequency the stub is one half-wave long and behaves as an open circuit.

The folded dipole can be conveniently made from 300Ω flat twin feeder as in **Fig 12.52(c)** because the spacing is not critical and the electrical length of the system, as a radiator, is not affected by the insulating material since the voltages have the same magnitude and phase in both wires and no current flows between them. The two halves of the dipole can, however, also be regarded as a pair of short-circuited stubs connected in series and, in this role, the normal velocity factor applies so that each stub is about 0·05λ longer than resonance. The capacitance being roughly 6pF per foot, this is equivalent at 14MHz to a reactance of about 1,200Ω across the feeder which increases the swr to 1·3. The velocity factor of ribbon feeder being about 0·8, the swr may be improved by placing the short circuits between the wires at this fraction of a quarter-wavelength from the

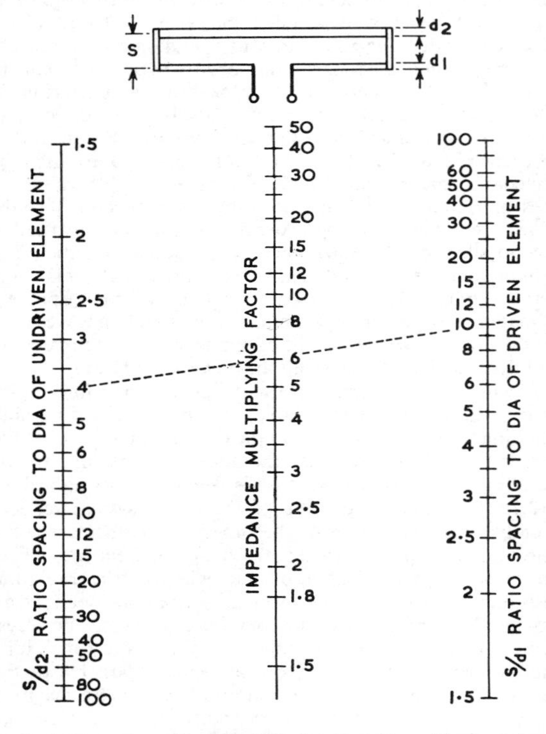

Fig 12.53. Nomogram for folded dipole calculations. The impedance multiplying factor depends on the two ratios of conductor diameter to spacing between centres, and is always 4:1 when the diameters are equal. A ruler laid across the scales will give pairs of spacing/diameter ratio for any required multiplier. In the example shown the driven element diameter is one-tenth of the spacing and the other element diameter one quarter of the spacing, resulting in a step up of 6:1. There is an unlimited number of solutions for any given ratio. The chart may also be used to find the step-up ratio of an aerial of given dimensions.

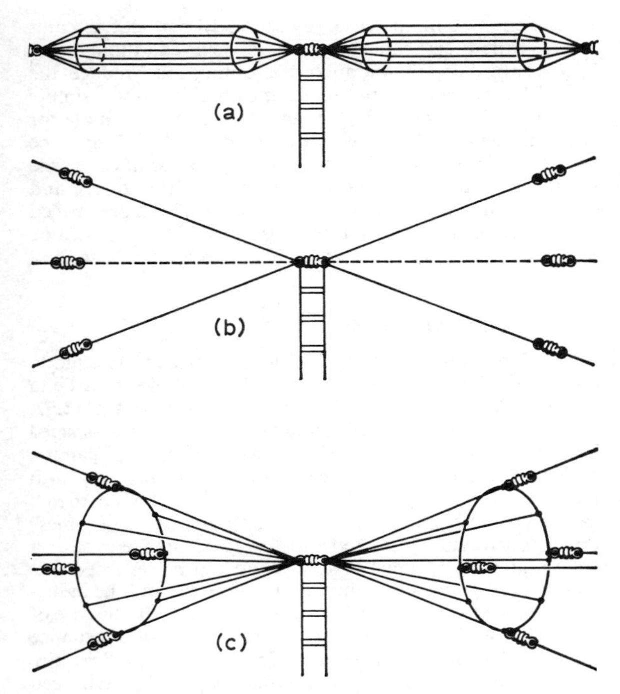

Fig 12.54. Broadband dipoles. (a) Cage. (b) Fan (two or three wire). (c) Conical. The semi-conical form is often used for broadband vertical monopoles over an earth plane.

centre rather than at the end—eg 27ft apart in a 33ft long 14MHz aerial.

Other ratios of transformation than four or nine can be obtained by using different conductor diameters for the elements of the radiator. When this is done, the spacing between the conductors is important and can be varied to alter the transformation ratio. The relative size and spacings can be determined with the aid of the nomogram in **Fig 12.53.**

These variations of the basic folded dipole do not lend themselves readily to multiband operation.

Aerial Transformers

In aerial systems comprising full-wave (or end fed half-wave) radiators, it is possible to obtain some control over the high input impedance by utilizing the radiator itself as a quarter-wave transformer, making use of its similarity to a $\lambda/4$ open-ended stub which has been unfolded. The principle was explained on p12.27. The loop radiation resistance of the radiator is principally a function of its length, and is independent of the conductor size. By varying the latter to alter the effective L/d ratio of the radiator, the Z_0 of the "transformer" can be altered and hence the input impedance at the voltage feed point. This is not practicable as an empirical adjustment after erection, but does permit some elementary matching to be built-in to the aerial at the design stage to restrict the extent of the initial mismatch to the open wire main feeder.

The effective Z_0 of the radiator as a transmission line being a function of the length/diameter ratio, and the length determined by the operating frequency, the diameter must be adjusted to control Z_0. This may be achieved by using a thicker wire, or more effectively by using several wires connected in parallel and spaced out to form a cage. Another alternative is to use several wires which are spread out from their common end at the feed point, to form a fan or semi-cone. These methods are illustrated in **Fig 12.54.** The curve in **Fig 12.55** shows the variation of input impedance of a full wave biconical dipole in terms of the cone angle: it is correct for solid cones and indicative of the general tendency for elemental cones comprising several wires. The values of input impedance for a fan dipole will be 50–100 per cent higher for the same angle.

The use of large diameter conductors or fans of wires improves the bandwidth of $\lambda/2$ and λ dipoles equally, since in accordance with line transformer theory, the Q is approximately equal to R/Z_0 or Z_0/R, whichever is the larger. The need for a reduction of Z_0 arises with $\lambda/2$ dipoles if it is required to combine large bandwidth with matching to a low impedance line (ie one matching the radiation resistance), otherwise the folded dipole makes better use of the extra conductors up to a maximum of three or four, beyond which further matching problems may arise. With full-wave dipoles however there are two separate problems, the bandwidth of the dipole itself and the constraints imposed by the high feed point impedance relative to the feeder Z_0 which may further restrict the bandwidth unless some form of cage or fan is adopted; for the amateur hf bands, however, this problem is unlikely to be acute unless the dipole forms part of a close-spaced beam system.

When fans of wires are used, the length required for resonance is reduced, typically by a factor of about 0·8. A simple explanation of this is implied by the $6{\cdot}28 Z_0\, \Delta l/\lambda$ formula for the reactance, p12.27. Thus if a low impedance (Z_{0L}) resonant line is shortened by an amount Δl, this subtracts a reactance proportional to $Z_{0L}.\ \Delta l$. If this length is now restored, using a line of higher impedance Z_{0H}, the added reactance is proportional to $Z_{0H}.\ \Delta l$ which is more than the amount removed, so that some shortening must be applied to one or other sections of the line, and the overall length for resonance is less than it was originally. In the case, for example, of a $\lambda/2$ fan dipole, Z_0 is higher where the wires are closer together, ie in the centre, which is the reference point for application of the formula. This demonstrates an important property of all aerial or aerial plus feeder systems involving a change of Z_0, enabling us to predict, for example, the well-known fact that for resonance a quad loop requires slightly more wire than a folded dipole, and other facts which appear to be less well-known, eg when a non-resonant

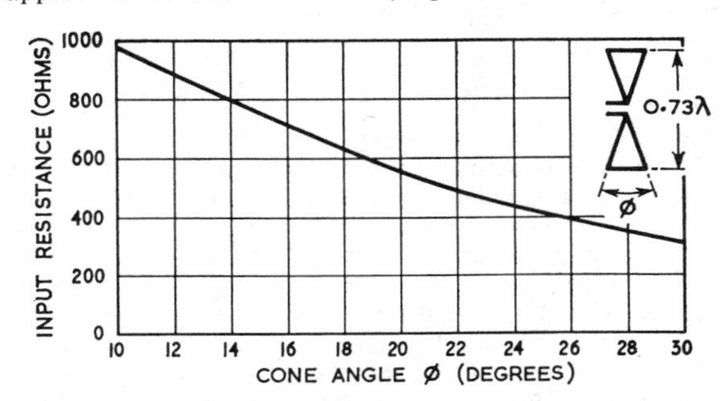

Fig 12.55. Input resistance of a full-wave bi-conical dipole as a function of the cone angle. The overall length should be 0·73 wave-length at the mid-frequency independent of the cone angle.

dipole is loaded by means of a stub the total length of wire required will be longer for a short dipole but shorter for lengths between λ/2 and λ, and harmonic resonances will not be quite coincident. Similarly a 14MHz quad or delta loop with a λ/2 resonant stub *nearly* resonates at 21MHz, the length of the whole system then being 3λ/2, but to obtain exact resonance the stub must be shortened by about 2ft.

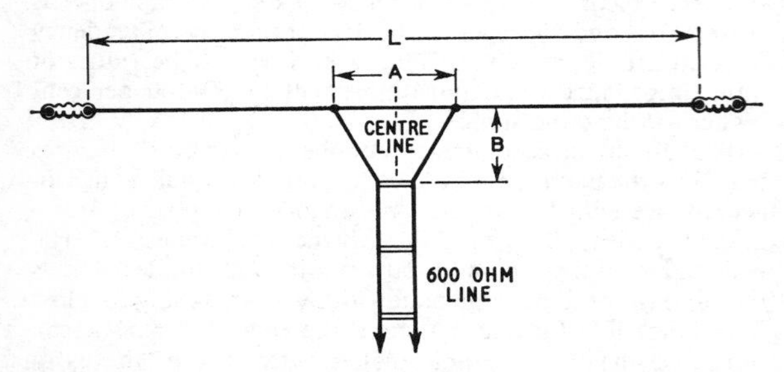

Fig 12.56. Use of the delta match with a λ/2 aerial fed with 600Ω open line.

Tapped Aerial Matching: The Delta Match

Pursuing the analogy between a resonant aerial and a tuned circuit, any point on it will have an impedance *relative to ground* varying from a low value (one-quarter of the radiation resistance) at points of maximum current, up to a very high value at points of maximum voltage. A single-wire feed-line (p12.17) if connected to the centre would cause a current to flow outwards in both directions which would of course be quite wrong, but when displaced to one side the line "sees" an inductive reactance one way and a capacitive reactance the other so that equal but opposite currents flow in the two directions and the required current distribution is set up. This is exactly analogous to connecting a wire to a tap on a tuned circuit and by choosing the right tapping point the impedance of the line can be matched. This is the principle of the single wire feed, which as discussed on p12.17 is now largely obsolescent due to wastage of power by radiation from the feeder and losses in the ground return path. These disadvantages are completely overcome by using *two* wires, tapped each side of centre as shown in **Fig 12.56**. This is called the *delta match*, and is a convenient way of connecting an open wire balanced line to a halfwave dipole or beam element. The precise dimensions *A* and *B* of the delta section may be obtained experimentally, but for the case of a 600Ω feed line are given approximately by:

$$A = \frac{118}{f(\text{MHz})} \qquad B = \frac{148}{f(\text{MHz})}$$

where *A*, *B* are measured in feet.

It might be thought that the delta section would radiate but using its resemblance to an 8JK aerial it is possible to arrive at an approximate figure of 0·002dB for the power loss by radiation. The previous formula is not applicable as the ratio of feeder length to spacing is less than specified and in any case most of the radiation from the delta is in the wanted mode. There is no need to make direct connection to the dipole, and the end of the feeder may be fanned out into a delta loop which provides inductive coupling to the aerial. Ordinary pvc insulated mains cable secured against the dipole with plastic tape has been found satisfactory, the dimension *A* being increased by about 10 per cent as compared with direct connection. It is believed, though not fully established, that inductive coupling may reduce the risk of line imbalance, but either method provides a simple and satisfactory method of matching between a dipole, or beam element, and a 600Ω line. These methods cannot be used in their simple form with low impedance lines, as the reactance of the loop is then excessive relative to Z_0 and must be removed by tuning. In the case of a close-spaced two- or three-element beam the dimensions *A* and *B* will be reduced by some 30 to 50 per cent compared with those for a dipole.

The Tee and Gamma Match

The dimensions *A* and *B* for the delta match are slightly interdependent but *B* is not critical and the whole of the delta may be folded upwards to form a tee-match as in **Fig 12.57**. The tee-match is particularly suitable for feeding close-spaced beam aerials using low-impedance twin-lead and the gamma match, also illustrated, is evolved from it for use with unbalanced lines. In each case there is some mismatch caused by the inductance of the coupling loop unless this is tuned out, eg as shown in Fig 12.57. In the case of a delta match, as in Fig 12.56, the loop can be regarded primarily as an extension of the transmission line and the inductive effect is small, but tuning is usually essential in the case of tee and gamma matches, particularly with low-impedance lines and low aerial impedances. The capacitors will require to be adjusted empirically for minimum swr on the main feed line: they will also need to be sealed adequately against weather and dirt accumulation, which would otherwise make them liable to voltage breakdown. The physical size and rating should be adequate to withstand the full line voltage with an adequate factor of safety.

As in the delta match, the exact dimensions of the matching section are a matter of experimental adjustment for minimum swr on the main feed line. For a half-wave dipole fed with 600Ω line, the approximate dimensions are given by:

$$A = \frac{180{\cdot}5}{f(\text{MHz})} \qquad B = \frac{114}{f(\text{MHz})}$$

where *A* is in feet and *B* in inches.

These formulae apply only when the extra conductor is of the same diameter as the aerial conductor. If a different size

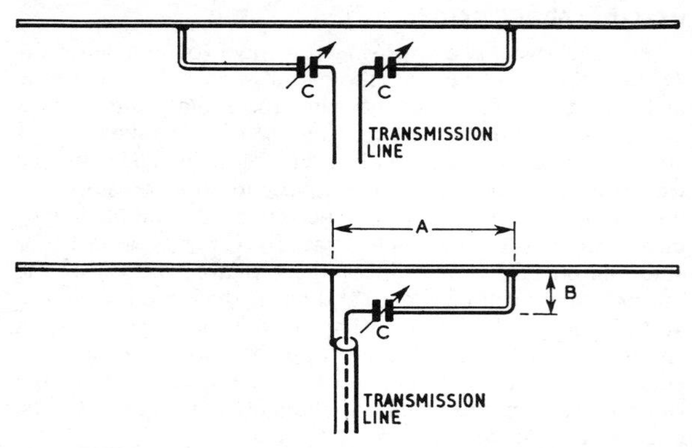

Fig 12.57. Tuning out residual reactance in the tee- and gamma-match systems with series capacitors. Capacitors with a maximum capacitance of 150pF are suitable at 14MHz. For higher frequencies proportionately smaller capacitances may be used.

of conductor is used for the matching section, an impedance transformation is obtained analogous to that achieved by varying the diameter ratio in a folded dipole.

The Clemens Match

With the gamma match, power is fed into one half only of the radiator, the other being excited by induction. This is inherently an unbalanced arrangement and does not fully meet the requirements for elimination of feeder radiation. A further disadvantage is the need to adjust the length of the matching section. An improved version has been developed which enables the tapping points to be fixed and also incorporates balance-to-unbalance transformation. It is known as the *Clemens match* and as shown in **Fig 12.58** can be constructed entirely of coaxial cable, the cable shield being attached to the metal aerial element using self-tapping screws.

This arrangement will cater for a very wide range of impedance transformations. The coaxial feeder cable is taken to the centre "neutral" point of the aerial, carried along one side to a distance of 0·05λ where the outer conductor is bonded to the aerial, the inner conductor being looped back to an equal distance the other side of centre and there joined to the driven element via a capacitor which is formed from the cable. The spacing between D and the driven element is variable from 0·01 to 0·02λ which adjusts the ratio, while the capacitor tunes out the transformer reactance and also helps with the tuning of the radiator. The dimensions given are those recommended for matching to a particular design of three-element Yagi aerial and may need modifying for other types of aerial by trial and error along the lines indicated above. The cable should be sealed against damp, Bostik sealing strip or Sylglas tape both being suitable for the purpose.

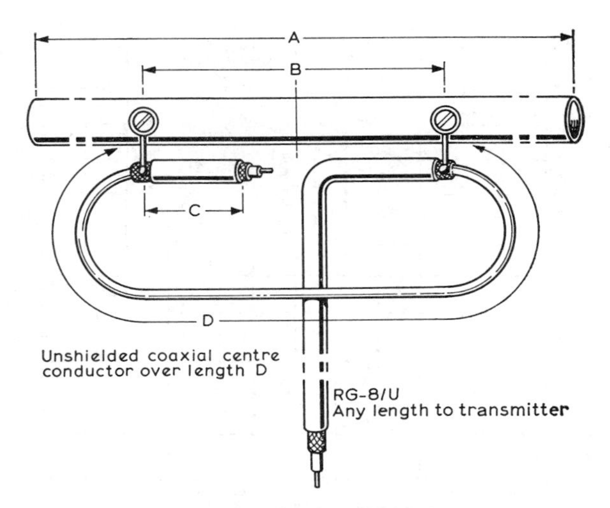

Fig 12.58. The Clemens transformer, a flexible arrangement which will match a wide range of dipole impedances into a concentric 50 or 70Ω line, and at the same time convert from balanced to unbalanced circuits. The suggested dimensions for a three-element aerial using this transformer are given below. L is the length of a λ/2 in free space (= 5,900/f inches at frequency f MHz). The third column gives the actual dimensions in inches for a 14·2MHz aerial.

Reflector length	L	412
Reflector-to-aerial spacing	0·3L	124
Aerial length (A)	0·95L	392
Aerial-to-director spacing	0·2L	82
Director length	0·90L	371
Spacing between aerial driving points (B)	0·2L	82
Length of unshielded coaxial centre conductor (D)	0·2L + 2in approx	84
Length of capacitive open stub (C)	0·039L	16

Line Transformers

Often the most convenient way of performing impedance matching between the aerial and the main feeder is by employing *line transformers*. These are made up as described on p12.27 from short sections of transmission line and are hence physically convenient to assemble and connect at the aerial terminals. They are also lighter than the equivalent coil and capacitor networks and more dependable when exposed to the weather.

The behaviour of short sections of transmission line with various terminations was explained on p12.26 where it was shown that quarter-wave sections can be considered equivalent to parallel and series tuned circuits. Shorter line sections behave like inductance or capacitance depending upon their length and whether they are open or short circuit at the far end.

Stub Matching

If the main feeder is connected directly to the aerial terminals the aerial will usually constitute an unmatched load, and a system of standing waves will be set up along the feed line as explained on p12.23. Moving away from the aerial, the impedance along the line will alter in sympathy with the pattern of voltage and current waves, as explained and illustrated in Fig 12.35. There will occur a position along the line *not more than* λ/2 *from the aerial* at which the resistive component of the impedance will equal the line characteristic impedance, but will have associated with it some degree of reactance. By connecting in parallel with the main line at that point a transmission line stub, which is either open or short circuited at the far end to give an equal but opposite reactance at its input end, the unwanted reactance on the main line can be cancelled or "tuned out", and the only term left is the resistive one which then presents a correct match to the remainder of the line. This is the principle of *stub matching*.

It is possible to calculate the length and position of the stub from a knowledge of the swr on the un-matched line and the position of a voltage maximum or minimum. The charts shown in **Figs 12.59** and **12.60** give the length of the stub and its distance from the point of maximum or minimum voltage *in the direction of the transmitter*. They are drawn specifically in terms of a stub of the same impedance Z_{01} as the main feed line. However a line of different impedance Z_{02} may be used to provide the stub, by determining the ratio $\frac{X}{Z_{01}}$ from Fig 12.43, and thence X for the stub length given by the stub-matching charts. The new ratio $\frac{X}{Z_{02}}$ may then be used to find a new stub length again from Fig 12.43. It may sometimes be advantageous to use a stub of lower Z_0 than that of a main feed line when the latter is relatively high, since the bandwidth of the match will then be improved. However particular care must be taken in using low-impedance lines as stubs in view of the possibility of high

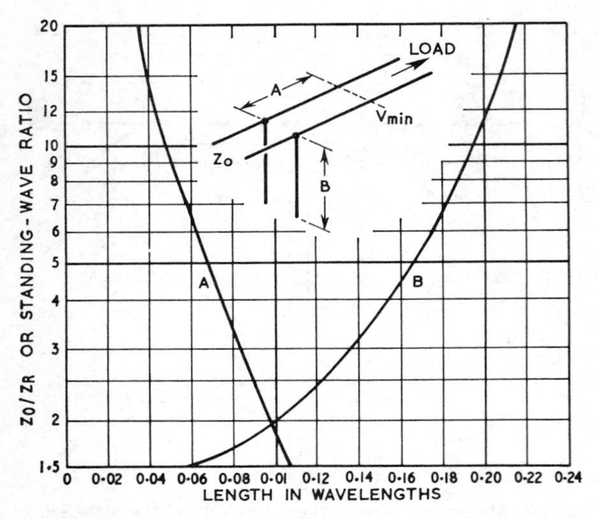

Fig 12.59. Position and length of an open stub as a function of Z_0/Z_R or swr when reference is at a voltage minimum.

losses; for example a 75Ω line with a loss rating of 1dB per 100ft will have a 2½ per cent loss per 10ft which is equivalent to a resistance of 2Ω in series with a λ/4 stub, or a Q of 40. If used as a line transformer with a voltage step up ratio greater than two, there will be a power loss exceeding 5 per cent. Attention is also drawn to the warning on p12.33 regarding the added danger of line unbalance when using open stubs.

The curves of Fig 12.59 apply when the point of reference is a voltage minimum, and those of Fig 12.60 when the reference is a voltage maximum. Since these voltage conditions along the line alternate at intervals of λ/4, either short circuit or open circuit stubs may be employed at the appropriate points on the line to achieve exactly the same match on the remainder of the line. Also because of the inverting properties of a quarter-wave section of line, the

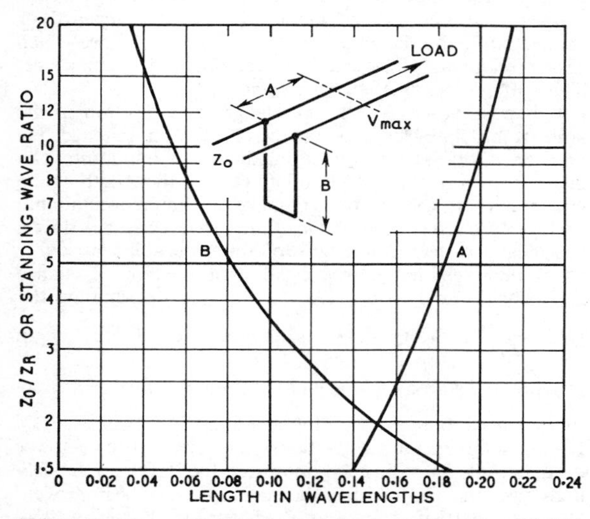

Fig 12.60. Position and length of a closed stub as a function of Z_R/Z_0 or swr when reference is at a voltage maximum.

open circuit stub may be replaced by a short circuit stub which is λ/4 longer, and vice versa. This system of stub matching is extremely flexible, and within certain limits permits the matching to be carried out at a point which is physically convenient, and with a stub which can be either open or short circuit to suit the particular mechanical layout. Generally speaking the use of short circuit stubs is preferable; apart from the problem of balance mentioned above, the electric field at the remote end of an open circuit stub is very high, and the "end effects" of the supporting insulator are difficult to assess accurately. Their action is to modify the electrical length of the stub and hence to "de-tune" it. In a short circuit stub, the remote end is accurately defined by the shortening bar and the length is not subject to the same end effects.

The voltage at the end is practically zero, and they can be tensioned or otherwise supported without the need for a strain insulator. There is also no objection to a dc connection from the centre of the shorting bar to earth, to provide lightning protection for the aerial and feeder system.

Another useful feature of stub matching is that it is not necessary for the aerial to be resonant, ie, cut to an exact number of quarter-wavelengths at the operating frequency. If the aerial is of an arbitrary length, the input impedance will be complex, and the series reactance necessary to tune the aerial is effectively provided by the short piece of line between the aerial and the first voltage maximum or minimum.

Quarter-wave Resonant Sections

When the standing wave ratio is high, the length of the stub together with its distance from the reference point add up to one quarter-wavelength, and because of this, such stubs are sometimes referred to as quarter-wave resonant line transformers. Typical stub matching arrangements are shown in **Fig 12.61(a)** and re-arranged in **Fig 12.61(b)** to illustrate the quarter-wave feature of this system of matching. The action is electrically identical, but it may sometimes be physically more convenient to mount a fixed λ/4 stub across the aerial input, and to adjust the tapping point of the main feeder along this stub for minimum swr on the main line.

Quarter-wave resonant line transformers can be used when the impedance ratio is fairly high (eg greater than four-to-one) and two varieties using short- and open-circuited stubs are shown. Both make use of the fact that there is a standing wave on the stubs, the main feeder being moved along the stub to a point where the voltage/current ratio becomes equal to the main line impedance. They are in effect tuned resonant circuits, with the feeder tapped into them. The open stub is employed when the feed point impedance of the aerial is too low for the line, ie, when the overall aerial length is an odd number of half-waves, and the closed stub is used for a high impedance feed, ie, when the aerial is a number of full waves long.

In practice the stub can be made to have the same impedance as the main line, though this is not essential. The line should be supported so that it remains at a fixed angle (preferably a right angle) to the stub. The tapping position is varied until the standing wave or the reflection coefficient on the main line is minimized. If a good match is not produced the length of the stub can be altered; should this not be successful, it is probable that the wrong type of stub is being used. It may be noted that the aerial itself need not be a resonant length, though the aerial-plus-stub must be; also, the stubs may be increased in length by one or more quarter-waves in order to bring the tap position nearer to ground level, changing from "open" to "short-circuit" or vice-versa with each quarter-wavelength addition. One method of adjustment requires a simple current indicator such as the one illustrated in Fig 12.78. If the current in a closed stub is less than the current on the aerial side of the junction, the length of aerial plus stub needs to be shortened, and vice-versa, the swr in the main feeder being unimportant for this part of the procedure. When the length of aerial plus stub is correct, the tapping point for the main feeder can be moved up and down the stub to obtain the minimum swr.

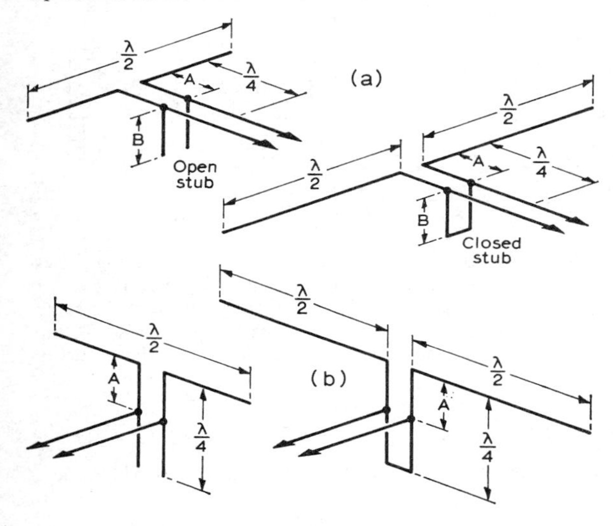

Fig 12.61. Use of matching stubs with various aerials.

The technique of stub matching is applicable both to balanced and unbalanced lines, and may be used with advantage on coaxial lines at vhf where the physical length of the stub is not great. Again because of the "end effect" of the open circuit stub, a short circuit version is to be preferred, which then completes the outer screen of the coaxial system and eliminates the possibility of unwanted signals being picked up on the inner conductor of an open circuit stub.

Series Quarter-wave Transformers

The impedance inverting properties of a λ/4 section of line were explained earlier in the section dealing with transmission lines and it was shown that a resistive load may be transformed to provide a match to some other impedance by one or more λ/4 sections of line having appropriate values of Z_0. If the load is partly reactive, it may be supplied with power from any convenient feed line and moving back along this line from the aerial, a position of voltage minimum may be located. Viewed from this point the load will appear to be a low value of resistance with no reactive component, and the next λ/4 section of line (moving towards the transmitter) may be designed to provide the required transformation. This may be merely a matter of "pinching in" the conductors of an open wire to reduce the Z_0, but it may also require a larger diameter of conductor. Such matching sections are

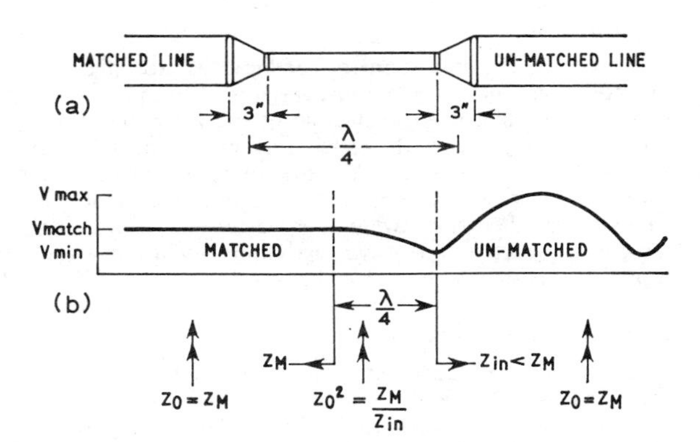

Fig 12.62. Series λ/4 transformer (Q-bar) using "pinch" matching. The voltage standing wave is a minimum at the point of commencing the λ/4 matching section, which is pinched together until the standing wave disappears on the following length of line. The "pinch" can be determined experimentally using string to hold the wires together. The string can then be replaced by permanent insulators.

often referred to as *Q-bar transformers*, and their action is illustrated by **Fig 12.62**. By measuring the swr in a line of known impedance it should be possible to deduce the value of matching line Z_0 required and design it according to the data given on pp12.18-12.19. As will be apparent from Fig 12.28 it is difficult to effect much change of Z_0, merely by reducing the spacing of a 600Ω line so that if a large ratio is required and no lines of suitable impedance are readily available, more drastic steps will be needed. These include the arrangements sketched in **Fig 12.63**, coaxial lines being connected in series with the outer conductors bonded, thus forming balanced lines of 100 or 150Ω, or connected in parallel to form unbalanced lines of 25 or 37Ω impedance. The four-wire modification of a 600Ω line divides the impedance by half approximately.

The bandwidth of an amateur-band hf aerial is usually determined by the aerial itself in connection with any resonant matching stubs, the ratio and therefore the effective loaded Q of λ/4 matching transformers being relatively low so that they do not introduce any further appreciable restriction of bandwidth.

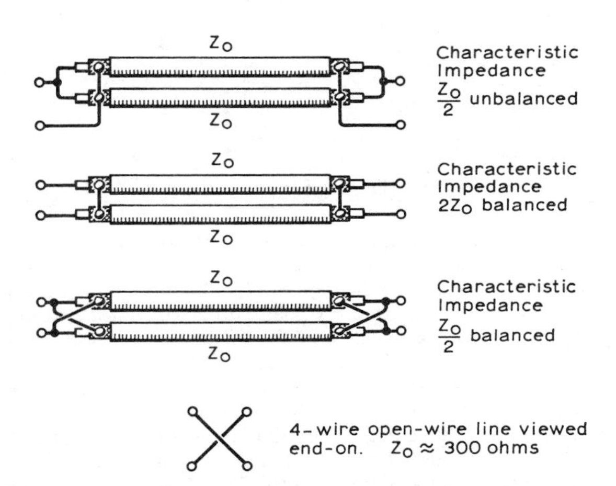

Fig 12.63. Methods of obtaining non-standard line impedances, eg for matching sections.

Tapered Lines

When the required transformer ratio is not high, for example when an 800Ω rhombic aerial is to be matched to a 600 or 300Ω line, it is practicable to use a simple tapered line. At the aerial end, the spacing is correct for 800Ω and gradually decreases to that for the lower impedance. This arrangement gradually converts the impedance from one value to the other. It is a wide-band device and, provided the taper is not less than one wavelength long, it will give a good, though not perfect, match.

Network Transformers

At ground level, particularly with coaxial feed lines, it is often more convenient to employ a coil and capacitor matching network. This could be similar to any of the couplers used near the transmitter (though probably simpler, eg as shown in Fig 12.38).

The ability to transform one value of resistance to another without loss is of course a common requirement in radio frequency circuits generally, not only in aerial systems. This general problem may be studied in terms of a "generator" supplying power to a load consisting of resistance and reactance, as in **Fig 12.64**. The generator may be a valve supplying power to an anode load, or it can be *any device from which power comes*, regardless of what may intervene in the form of tuned circuits or lines with or without losses. The important properties of generators are the available power and the source impedance. The available power, ie the maximum power obtainable from a generator, is expressed in terms of the open circuit voltage E_0 and the effective internal series resistance R_G and is equal to $E_0^2/4R_G$, any internal reactance being ignored for this purpose. The requirements for maximum power transfer are that the generator should see a "conjugate match", or in other words if the load consists of a resistance R_S in series with a reactance X_S and the generator has a series reactance X_G, we require $R_G = R$ and $X_G = -X_S$.

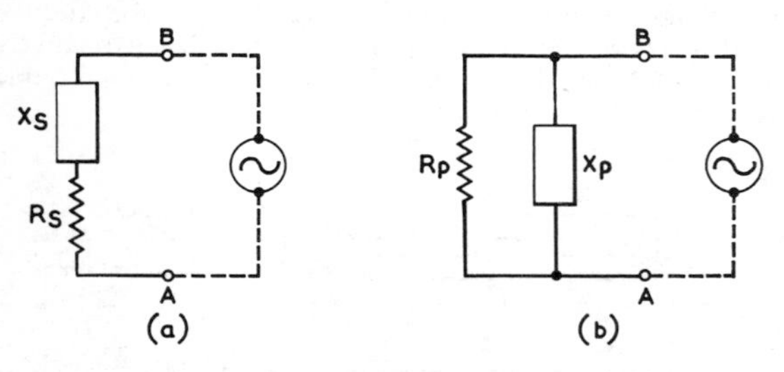

Fig 12.64. Equivalent impedance circuits.

Suppose now that the transmitter is intended to work into a 75Ω load subject to proper adjustment of tuning and matching controls. When the adjustment is completed successfully with a "dummy load" this will automatically ensure a conjugate match and the reader who is interested only in getting as much power as possible into the aerial need not be concerned further with what is happening inside the transmitter, provided he can make the aerial look like 75Ω and connect it in place of the dummy load. The problem next to be considered is how to make *any* impedance look like 75Ω using convenient "lumped" constants, ie capacitors and inductors. There is one further constraint: this must be achieved without losses, and if the components are loss-free (no internal resistance) all the available power must then go into the load for there is nothing else to absorb it. Usually there is a choice of many alternative ways of effecting the required transformations, and because in general the required components are reasonably efficient devices, low-loss transformation can usually be achieved, though it is advisable to avoid large currents in large inductors, or excessive voltages which produce leakage losses in capacitors. Losses are proportional to the square of these currents and voltages, and the loss resistance of a coil tends to be proportional to the inductance, ie Q values do not vary much assuming reasonable design practice.

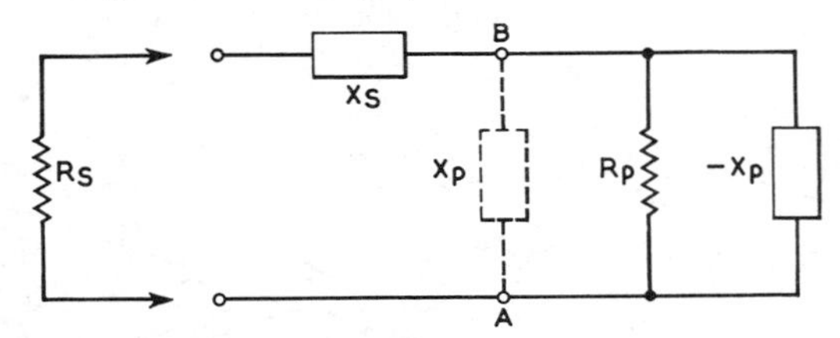

Fig 12.65. Transformation: stage 1.

Consider now two elementary networks involving series and shunt elements respectively as shown in Fig 12.64; for these two networks to be identical, that is present the same load to a generator connected across the terminals AB, the admittance seen at these terminals must be the same in each case, ie

$$\frac{1}{Rp} + \frac{1}{jXp} = \frac{1}{Rs + jXs} \qquad \ldots \text{(i)}$$

From this it is simple to derive the following two identities:

$$Rp = Rs\left(1 + \frac{Xs^2}{Rs^2}\right) \qquad \ldots \text{(ii)}$$

$$Xp = Xs\left(1 + \frac{Rs^2}{Xs^2}\right) \qquad \ldots \text{(iii)}$$

The "operator" j, (which is mathematically the square root of minus one) merely denotes that the quantity after it is a reactance and that, if added to a resistance, the addition must be done vectorially. This can be done by drawing a right angled triangle with the two shorter sides proportional to the two quantities, the magnitude of the sum then being represented by the length of the longer side.

Using these basic formulae it is possible to convert a network of series elements to its equivalent network of parallel elements (and conversely). Consider next the circuit shown in **Fig 12.65.** From an inspection of Fig 12.64 and equation (ii) above, it can be seen that it is possible to select a value of Xs to put in series with the resistance Rs, such that the shunt equivalent resistance is equal to Rp, but has in parallel with it a residual reactance Xp. If now a reactance of value $-Xp$ is placed across the terminals AB, the net effect will be to produce at those terminals a pure resistance equal to Rp, which is the requirement of the transformer. Then the basic form of the transformer is an L network comprising only reactive elements as shown in **Fig 12.66.** This also meets the terms of the original specification. If the ratio $R_2/R_1 = p$, it can be shown that:

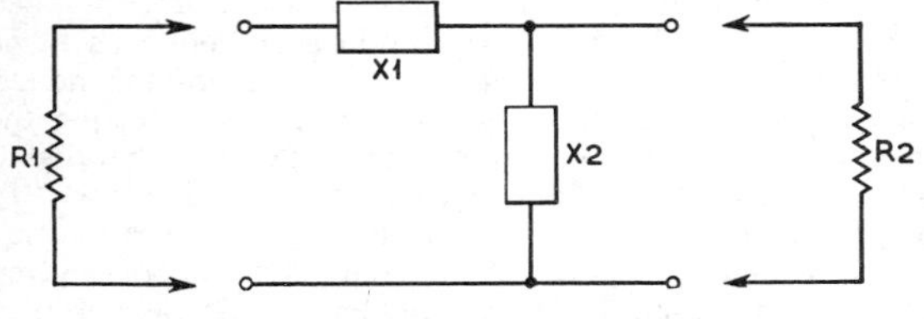

Fig 12.66. Transformation: stage 2.

$$X_1 = \pm R_1\sqrt{(p-1)} \qquad \text{. . . (iv)}$$

$$X_2 = \mp pR_1/\sqrt{(p-1)} \qquad \text{. . . (v)}$$

X_1 and X_2 must have opposite signs in all cases, or in more general terms, the network must consist of one inductive and one capacitive element. It is more usual in practice to make the series element inductive and the shunt element capacitive for reasons which vary with individual circumstances (eg a series dc path to the anode of a valve), but there is no basic objection to the reversal of this practice in cases where it may be more suitable.

Using the fundamental equations (iv) and (v) it is now possible to design any required network by a process of arithmetic as the following example will show:

Example

A single wire short aerial for 3·6MHz has a base impedance of 15 – j200Ω. It is required to match this to a feeder with a characteristic impedance of 75Ω (**Fig 12.67(a)**). This aerial is equivalent at its terminals to a resistance of 15Ω in series with a capacitance of 220pF. The steps of the calculation are as follows:

(*a*) Connect an equal and opposite reactance +j200 in series to tune out the aerial capacitance and leave only the resistive term of 15Ω: the transformation required is then from 15Ω resistive to 75Ω resistive (**Fig 12.67 (b)**).

(*b*) From equation (iv), X_1 (series term) $= \pm 15\sqrt{5-1}$
$= \pm\ 30\Omega$

(*c*) From equation (iv), X_2 (shunt term) $= \mp \dfrac{5 \times 15}{\sqrt{5-1}}$
$= \mp 37{\cdot}5\Omega$

Thus the transformer section required is as shown in **Fig 12.67(c)** and the complete arrangement in **Fig 12.67(d)**. To convert the reactance into physical components, since $X_L = 2\pi fL$ and $Xc = 1/2\pi fC$:

+ j230 is the reactance of 10μH at 3·6MHz.

– j37·5 is the reactance of 1,250pF at 3·6MHz.

If the coil has a Q of 200, ie a series loss resistance of 1Ω, this will be equivalent to an aerial loss resistance of 1Ω. Strictly speaking the matching network should be modified slightly, as if the aerial had in fact a resistance of 16Ω, and one-sixteenth of the transmitter power will be unavoidably lost as heat in the coil. This however is a loss in signal of 0·3dB only, and unlikely to have much impact on reports of signal strength.

The complete network is shown in **Fig 12.67(e)**). By making each element variable over a convenient range, it is possible to correct for the coil resistance and any inaccuracies in the original figure taken for the aerial base impedance, and also for variations in the working frequency.

A suitable protective cover for the network is a large air-tight can, used upside down, with components mounted on the lid. A metallized lead-through seal from an old capacitor or transformer can be soldered into the lid for the aerial lead. A sealed rf connector is preferable for the cable connection. The can should have a coat of paint, and Bostik putty can be used to seal all the joints; a small bag of silica-gel may be placed in the can to guard against condensation. The components of the network should be capable of handling the transmitter power; if fixed capacitors are used they should be of the foil-and-mica type.

Baluns

The majority of hf aerials are inherently balanced devices and equal voltages exist or should exist to earth from each input terminal. Exceptions include long-wire end-fed aerials often used on the lower frequency bands and verticals which are driven against either a ground plane or true earth.

It is not possible to connect an unbalanced feeder such as coaxial line to a balanced aerial and maintain zero potential on the outside of the line. In such cases currents are forced as

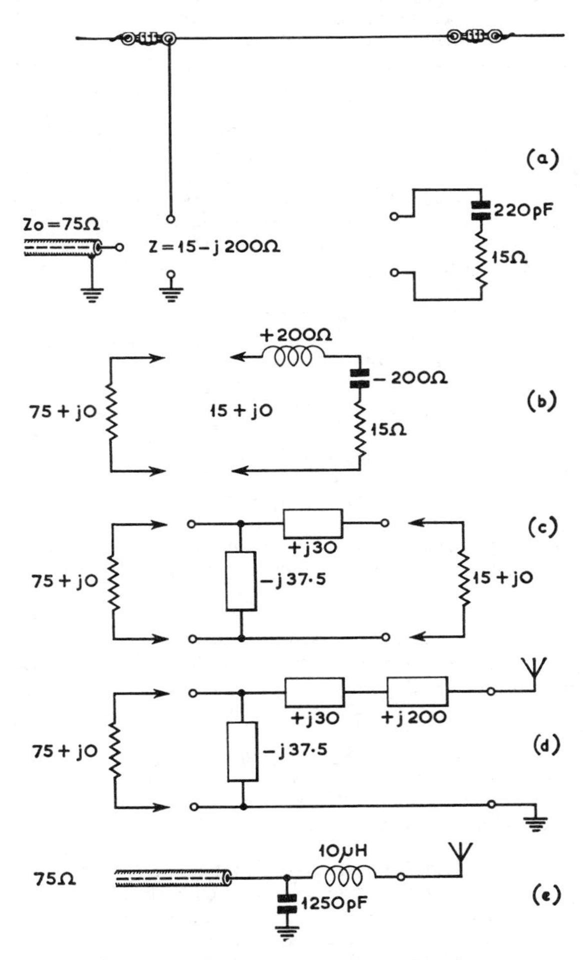

Fig 12.67. Matching of two arbitrary impedances.

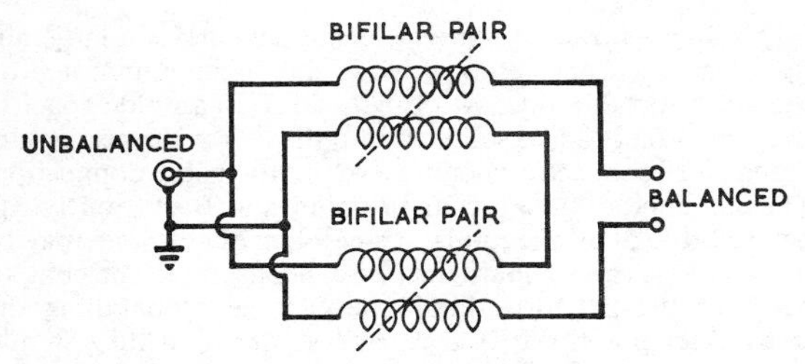

Fig 12.68. Pairing of bifilar choke coils.

explained on p12.32, to flow down the *outside* of the outer conductor by the voltages appearing at the aerial terminal to which it is connected. These currents give rise to unwanted radiation from the line itself since the field due to them cannot be cancelled out by the field due to the current flowing in the inner conductor, which is contained entirely within the outer conductor of the line and cannot penetrate beyond it.

The existence of the unwanted radiation field around the feeder will modify the pattern of the aerial, and possibly also its input impedance due to the coupling now existing between feeder and aerial, and will represent a loss of energy by radiation in undesired directions. As with an inadequately balanced twin wire line, it is also a potential source of tvi, bci and rf feedback problems within the station itself.

If it is required to drive a balanced aerial directly from an unbalanced line, then a choke or transformer can be employed to prevent the unwanted currents flowing back down the outside of the line. Such a device is known as a balance-to-unbalance transformer (*balun*) and often replaces the gamma match described earlier. The action of transmission line baluns depends principally upon the use of a quarter-wave section of line as a parallel resonant circuit to present a high impedance path to the unwanted currents. A number of balun designs based on this principle are described in Chapter 13, and although they will all work in theory at hf as well, their physical length precludes their convenient use below the vhf bands. A suitable balun for the longer wavelengths consists of a pair of bifilar wound choke coils connected as shown in **Fig 12.68.** These coils may be sealed up in a suitable waterproof housing mounted at the aerial feed point and taking the place of the usual centre insulator. The coaxial feeder may be soldered in place or taken into the housing through a weatherproof plug and socket, with a clamp or other device for taking the mechanical strain of the feeder.

The coils act as a choke for asymmetrical currents while having negligible effect on symmetrical currents, and there is a four-to-one impedance transformation, with the higher value on the balanced side.

Transformer-type baluns using ferrite or iron powder cores can be designed to operated over a wide frequency range such as 3 to 30MHz. The principle is the same as that of low-frequency transformers except that, due to the relatively low permeability of cores having sufficiently small rf losses, special precautions are needed to maintain a high coefficient of coupling. It is not possible to rely on the core alone to provide the necessary coupling, and use must be made of the fact that if two wires of equal length are laid side by side in close contact, and if the insulation between them has negligible thickness, then all the lines of magnetic force surrounding one wire must also go round the other. This by definition is unity coupling and a voltage applied across the ends of one wire will produce an identical voltage across the other. For efficient transformer action the inductance must be large enough to ensure that connecting the transformer has no effect on the input other than that due to whatever load may be connected to the secondary. In practice there will always be some leakage inductance which can be regarded as in series with the load, and for a given coupling coefficient the leakage inductance increases in proportion to the self-inductance so that, if this is adequate for 3·5MHz operation, it may sometimes be difficult to get a low swr on 28MHz. As with mains transformers there are losses in the windings and core, and reducing the inductance to improve swr at 28MHz can often lead to unacceptable core losses. In addition self-capacitance may have unpredictable effects on performance at the higher frequencies.

Despite these complexities it is not difficult to design small broadband baluns with losses of only a small fraction of a decibel, and acceptable swr. With efficient design baluns wound on toroid cores of only 2in diameter are capable of handling powers up to 1kW p.e.p., and rod-type baluns can be equally compact and efficient. The impedance ratios obtainable are somewhat restricted by the necessity for complete overlap between the windings but **Fig 12.69** shows how a trifilar winding may be used to obtain balun action with a 1:1 impedance ratio. Comparing this with a double-wound transformer (bifilar) it is found that the use of three wires instead of two tends to increase the leakage, but this is more than compensated as a result of the auto transformer connection whereby half the turns of each winding are shared with the other. For winding onto a toroid core the three wires may be laid side by side and bound tightly together with plastic tape so that they form a flat strip, or they may be bunched together and lightly twisted to keep them in position without air gaps. In the case of ferrite rods, the three wires may be wound side by side as a single layer winding without air gaps. Typically five to eight turns have been found suitable for ferrite cores, including 1in diameter "ferrite aerial" rod and also 2in diameter toroid rings. Although the wires must be wound as a tight bunch without spacing there is no objection to spaces between turns. Powder cores require rather more turns, a minimum of 14 being recommended in the case of the 1kW kits obtainable from TMP Electronic Supplies, Unit 27, Pinfold Workshops,

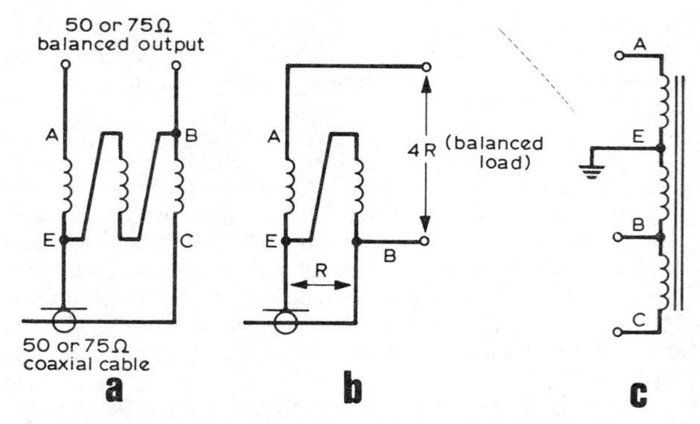

Fig 12.69. (a) Trifilar balun, 1:1 impedance ratio. (b) Bifilar balun, 4:1 ratio. (c) Trifilar balun redrawn as an auto-transformer to illustrate principle of operation. In bifilar case BC is omitted. The three (or two) windings are wound as one with the least possible spacing between wires but individual turns may be spaced out on the core.

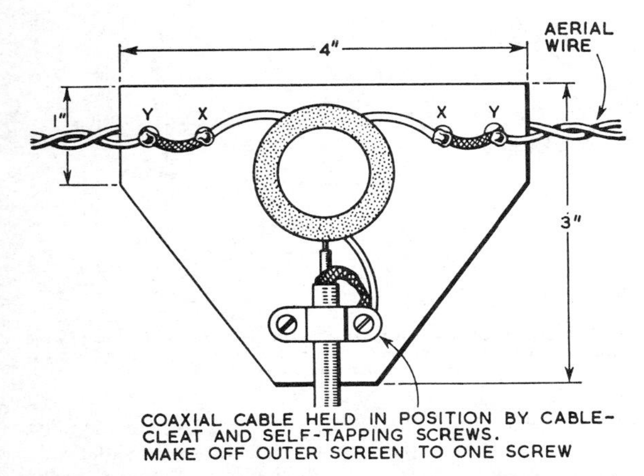

Fig 12.70. A triangular block of perspex makes a strong mounting for a balun positioned at the centre of a λ/2 dipole. The balun is anchored with its connecting wires and a coating of Araldite, the coaxial cable is held with a cable cleat, and the aerial wires are passed through holes Y. Braid is used to connect the output of the balun to the aerial wires to avoid straining the connections owing to flexing in the wind.

Pinfold Lane, Buckley, Clwyd CH7 3PL. In the case of 1:1 trifilar baluns there is a tendency for swr to increase appreciably with frequency, due to the leakage inductance, but considerable improvement can usually be obtained by use of a suitable parallel capacitance, typically about 50pF though the actual value must be found by trial and error. For powers up to the legal limit a voltage rating of 500 should be adequate for this capacitor except under extreme fault conditions which might in any case result in damage to the core. Alternatively, if the balun is used in, for example, the centre of a dipole, leakage inductance together with any lead inductances may be compensated by adjustment of dipole length for minimum swr.

Much smaller cores may be used for low power and receiving applications, and a 1:1 balun using 17 turns, on a 0·69in core type T–68–6 obtainable from the above source, had measured losses averaging less than 0·1dB, an swr better than 1·35 without capacitive compensation, and handled a power of 20W cw with very little temperature rise. Ferrite cores suitable for powers up to the legal limit include a pair of Ferroxcube FX 1588 toroid rings which should be stacked and bound with acetate, polyester or silk tape before winding with six trifilar turns. Ferrite aerial rod of ⅜in to 1in diameter is usually suitable, and rod-type baluns may also be purchased ready made from various sources. In the majority of cases the measured losses have been found to vary with frequency, from less than 0·1dB to a maximum of 0·2 to 0·4dB.

Bifilar winding auto-transformers connected as in Fig 12.69(b) provide 4:1 impedance transformation eg from 75Ω coaxial feeder to 300Ω twin lead. Because of lower leakage inductance it is much easier in this case to maintain low values of swr over the entire frequency range; conversely if multifilar windings are used to obtain impedance ratios other than 1:1 or 4:1 it is rather difficult to achieve good performance over a large range of frequencies. Despite recommendations in some quarters the writer has had no success with tapped windings, and this result is to be expected since the use of taps conflicts with the overlap requirement for minimizing leakage inductance, although in some cases this may be removed by tuning.

The baluns described above are all suitable for use with 50 or 75Ω coaxial feeder over at least the frequency range 3·5 to 30MHz. Subject to reasonable care in winding to ensure symmetry, balance should be within a few per cent.

If the balun forms part of an outside aerial system it must be protected from the weather, and two methods are illustrated. Perhaps the simplest method is to attach the toroid to a triangular sheet of perspex, and completely envelop the assembly in Araldite. The coaxial line can be readily secured with a sheet metal clamp bolted to the perspex; the toroid is held in position by the combined action of the mounting wires and Araldite. The dimensions and drilling points are shown in **Fig 12.70**; the inner holes X retain the balun lead-out wires, and the holes Y secure the aerial wires. The jumper between holes X and Y should preferably be a length of copper braid to prevent damage caused by flexing of the assembly in high winds. With auto-transformer windings all connections to the balun will usually be close together and holes marked X may need repositioning.

Another simple method of mounting is to enclose the balun in a short length of Marley plastic drainpipe, the ends being covered with discs of paxolin. The discs can be cut with a flycutter, the pilot holes being used to allow a brass screw to clamp the ends together, or alternatively used as fixing points to tie the unit to a mast. The coaxial cable is held tight by a rubber grommet in the centre of the tube wall, while the balanced winding can either be connected to the dipole by braid fly leads, or alternatively a small strain insulator cut from perspex sheet can be mounted within the unit; this possesses the added advantage that all joints are totally enclosed. The unit should be sealed with Evo-stik or Twin-pack Araldite. This method is readily adaptable for use with rod type baluns.

An inexpensive mount using a short length of plastic drainpipe.

METHODS OF FEEDING AN AERIAL

End Feeding

Any random length of aerial wire, whether resonant or not, may be energized by bringing one end of it into the station and connecting it to an earth or counterpoise through a tuned circuit, link coupled to the transmitter.

This method has the advantage of simplicity, and has often been used with great success, particularly in the early days of amateur radio, but cannot be generally recommended as it involves bringing the aerial right into the station; this means that radiation takes place in the vicinity of house wiring, brickwork etc, leading to unpredictable losses and added risk of tvi. The local radiated field in and around the shack may also give rise to feedback problems in audio and rf circuits, and will contribute little or nothing to the useful radiation, being generally absorbed in the fabric of the building in which the station is housed. If it is necessary to use this method, it is advisable to use a counterpoise rather than rely on an earth connection; earth wires, though in this case just as much part of the radiating system as the aerial wire, are usually attached to the building thereby introducing losses additional to those of the earth itself. Typically, a counterpoise consists of a quarter wavelength of wire at any convenient height, though preferably out of doors, but if the aerial is fed at or near a voltage point almost any random short length of wire is adequate. On no account should rf be allowed to find its own way to earth via housing wiring or other random paths.

Modern transmitter design tends to use a pi-circuit for the pa tank, designed to work into an output impedance within a limited range of low impedances, typically 40–150Ω. Since the end impedance of a current fed aerial, ie one which is an odd number of quarter-wavelengths long, is also low, and of the same order as the figures quoted, it would be quite possible to couple the aerial straight into the output socket of the transmitter and achieve a reasonable match by means of adjustments to the variables in the pa tank circuit. For the reasons which have been outlined above the use of this technique is deprecated, tempting as it may be to the amateur beginning on 1·8MHz with a simple transmitter and a short wire aerial. It is perhaps permissible in field day conditions, when the transmitter is in the open and can be connected in the direct path between the aerial and earth. In *all* other circumstances involving use of a random length of wire as an aerial a separate tuned circuit should be employed as an aerial coupling network. This network may be located in any suitable position and link coupled to the transmitter tank circuit by a low impedance coaxial line, to achieve adequate separation between the radiating part of the aerial and the building housing the station.

If the total aerial length is such as to cause the aerial section nearest to the transmitter to carry a high rf voltage, there may be considerable loss of power should it have to pass through or near to surrounding structures. On the other hand, an aerial tuned to an odd number of quarter-wavelengths and consequently of low input impedance can also be inefficient as a result of the comparatively high resistance of the earth connection. Unless a counterpoise is employed it is generally better therefore to use an intermediate length, for example, an odd number of eighth wavelengths so that the impedance will have some intermediate value, thereby reducing the current at the earth connection.

A suitable coupling circuit is illustrated in **Fig 12.71.** The inductor and capacitor may be similar to those in the transmitter output tank circuit. The link coil may vary from about four turns at 3·5MHz to one turn at 28MHz and may be tightly coupled into the earthed end of the transmitter tank coil, the aerial tap then being adjusted for correct loading of the transmitter. For good efficiency the resonance

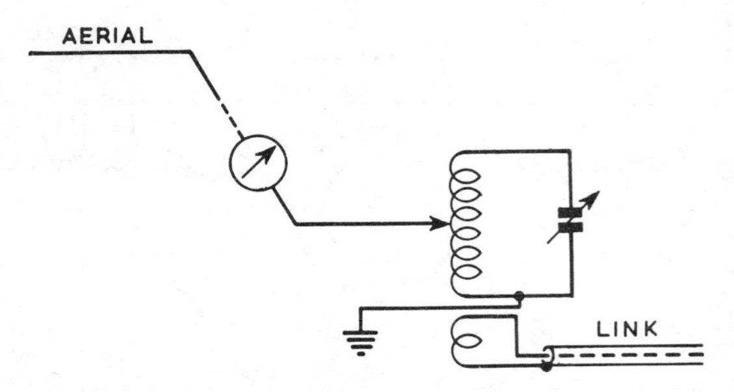

Fig 12.71. Aerial coupler for end-fed aerials. For some lengths of wire a series tuning capacitor and perhaps a separate coupling loop instead of a tap may be required.

of the aerial coupler should be fairly broad, correct loading combined with high Q being an indication of insufficiently tight coupling, and excessive dissipation of power in the tuned circuit. Alternatively a similar link coil may be used at the transmitter or, better still, the link may be fed from the output of a pi-network.

With a non-resonant length of aerial it will be found that, as the tap position is moved up the coil (from the "earthy" end) in order to draw more power, the coupler circuit will need retuning. This problem is considered later under "Aerial Couplers".

Earth Connections

When the earth connection forms part of the aerial system, as in the previous section, it should preferably be taken to a copper spike or tube several feet long, or to the nearest large earthed conductor such as a main water pipe. The joint should be heavily coated with bituminous paint to avoid corrosion. In this case *the earth connection of the electricity supply is dangerous* since rf voltages could lead to an insulation breakdown; also it could introduce noise into the receiver or be responsible for spreading interference to nearby television receivers. If the station is at the top of a building, the aerial coupler may fail to function properly on some frequencies because the earth lead is long enough to exhibit resonance. In such cases a "shortening" capacitor of 50–100pF may be inserted into the rf earth lead to de-tune it. This capacitor should be a variable air-spaced type.

Further information on earth connections will be found later in the section dealing with low frequency aerials.

Matched versus Resonant Feeders

Some feeders are described as matched lines and others as resonant lines. It is sometimes thought that these are two different kinds of line, resonant lines being usable with any load impedance whereas matched lines have to be operated at an swr better than some low value such as 1·4 to 1. A resonant line is in fact merely a badly mismatched line, a point being reached with increasing swr at which it becomes convenient to regard the aerial plus feeder as a resonant system rather than as a matched (or mismatched) resistive load.

If an aerial is to be used on one frequency only, there is rarely a good reason for not attempting to match it, eg by using a $\lambda/4$ resonant line or equivalent lumped-constant network to transform a high impedance down to a lower value. Referring to Fig 12.45, matching is not critical from the point of view of power transfer and even an swr of 3·0 only increases the feeder loss by a fraction of one dB unless the feeder is extremely lossy, but this degree of mismatch may have considerable nuisance value with a long feeder; for example a 2 per cent frequency change with a feeder three wavelengths long is equivalent to a change in line length of $\lambda/16$ and this would convert an impedance of $3Z_0$ into one of $(1{\cdot}4 + 1{\cdot}3j)Z_0$. Adjustments to obtain a conjugate match at the transmitter in the second case would have little effect on swr in the first case so that Fig 12.46 would be applicable and there would be a reflected power loss of 1·4dB, probably combined with flat-topping of the transmitter unless the tuning and matching are both readjusted.

When an aerial is required to work on several wavebands the matching problem usually requires a different solution for each band and the difficulties tend to escalate rapidly with the number of wavebands to be covered, except that an aerial fed with matched line will work on odd harmonics, eg a 7MHz aerial will work also on 21MHz. Since the feed-point is rarely accessible, it is often the lesser evil to forego any attempt to match the line at the aerial and use resonant lines, accepting a large swr on all bands and relying on adjustment of the atu to provide a match at the transmitter. If for example a 14MHz dipole is directly fed at the centre with 600Ω line the swr will be about 8·0. Because the line impedance is roughly the geometric mean of the current loop and voltage loop impedances, the same aerial will work as a full wave dipole on 28MHz with a comparable swr, and will perform rather similarly on 21MHz also. It will even work on 7MHz, though in this case the radiation resistance is very low (8Ω) and the swr will be very high, about 80 to 1. The loss in 100ft of matched 600Ω line is typically 0·07dB at 14MHz and extrapolating from Fig 12.45 a resonant feeder loss of the order of only about 0·3dB would be expected at the higher frequencies. The loss on 7MHz can be worked out with the aid of Fig 12.5 and for the same long lines comes to about 1·3dB assuming use of 12swg copper wire, the band-width being about 35kHz. On the other hand, with low impedance feeder the swr would be about 67 at 28MHz and extrapolating from Fig 12.45 the actual line loss will be at least 6dB, assuming a matched line loss of 1dB; moreover unless the power is very low, breakdown of the cable due to high voltages and overheating is to be expected. These examples underline the necessity of using open-wire line for resonant feeders if low-loss operation is required on even harmonics of the fundamental frequency or in any other circumstances involving large standing-wave ratios or high powers.

Tuned Lines

The input impedance at the lower end of a resonant line will depend upon the actual load impedance presented at the far end, and upon the length of feeder involved. In general, it will be quite arbitrary and the feeder is employed solely as a means of connecting the aerial to the transmitter without attempting to transform it to any particular value. It is therefore necessary to terminate the tuned line in a matching network which will transform the impedance at the lower end of the line into the optimum load resistance for the transmitter.

An aerial coupler for use with tuned open-wire lines is basically similar to the end-feed coupler shown in Fig 12.71 but differs in detail, and the tuned circuit will require to be of a series or parallel form depending on whether the impedance at the lower end of the line is lower or higher than the Z_0 of the open wire line, as illustrated in **Fig 12.72.**

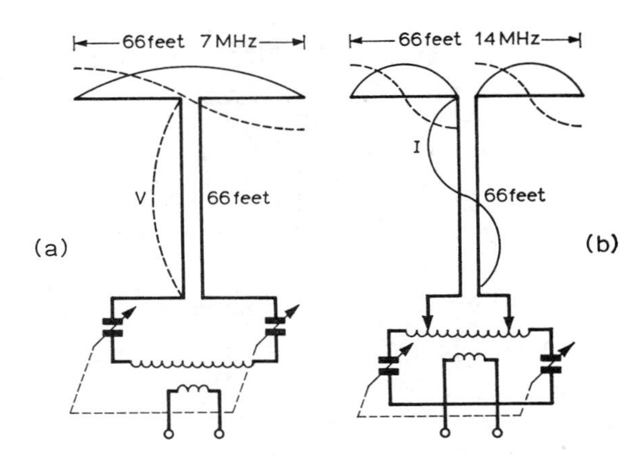

Fig 12.72. Centre-fed aerials using tuned feeders. The aerial to the left is λ/2 long and has a low input impedance; the feeder is also λ/2 long and thus repeats the low impedance, so that series tuning could be used. On the right is a similar aerial operating at twice the frequency of the other (full wave). The input impedance is high, and the feeder one wavelength long; parallel tuning would be best. The effect of feeder length is discussed in the text, and illustrated in Fig 12.73. The tuning capacitor sections should each be 50pF maximum for 14MHz and higher frequencies; proportionately larger values are needed for lower frequencies. The voltage rating of the capacitors should be the same as for those in the transmitter output tank circuit.

Without a knowledge of the exact aerial impedance, it is not possible to predict the precise value of the input impedance to the feeder or to determine the values of capacitance and inductance in the coupler. It is however possible to determine whether a series or parallel coupler is required, with the aid of **Fig 12.73** which is based on the total length of the system from the far end of the aerial to the near end of the feeder.

The impedance presented to the coupler will contain both resistive and reactive terms. The reactive term is neutralized by adjustment of the tuning control on the coupler (usually a variable capacitor) but provision must also be made for obtaining the correct transformation of the resistive term; this may be done by providing a range of tappings on the coil of the coupler unit, in order that the downcoming feeder may be tapped in at a position along the coil best suited to the value of the input resistance. A wide range of adjustment is desirable to accommodate very low impedances on taps near the coil centre, and very high impedances towards the outer ends of the coil. When the impedance is very low, corresponding to a current maximum on the line, it is possible to connect the feeder to a separate coupling coil of one or two turns and retain the parallel circuit, the coupling being adjusted to obtain the equivalent of adjustable taps, and many other combinations of coupling and tuning methods may be devised to suit particular circumstances, some examples being found later in this chapter. In general if the feeder input is of low impedance but largely reactive, series tuning is advisable, though it may be combined with parallel tuning which can be retained as the main variable.

When the feeder input resistance is of the same order as the feeder Z_0 a parallel coupler is necessary and the feeder should be tapped down the coil to obtain the best match to the transmitter. This situation implies that the feeder is

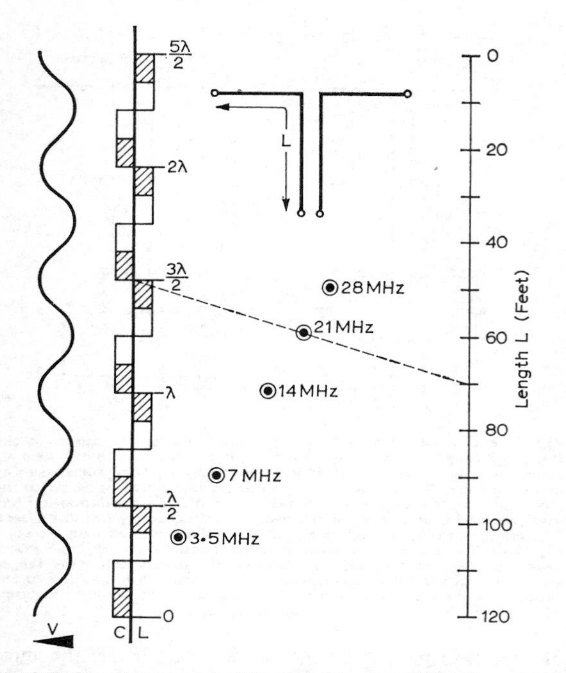

Fig 12.73. Standing wave chart for tuned feeders. A line through the length L of feeder plus half-aerial, and the frequency point, will show on the wavelength scale the nature of the input impedance. Rectangles to the left of the line are regions of capacitive impedance, those to the right inductive. The shaded areas correspond to high impedance input (high voltage) and call for parallel-tuned couplers (Fig 12.72(b)) and the blank areas to low impedances (high current) which require close taps on a parallel-tuned coupler, or a series circuit (Fig 12.72(a)). The chart may also be used for feeders alone in which case, as coded, it applies for the use of high resistance loads. When the load is lower than Z_0 the code must be reversed; the shaded areas indicate low input impedance, inductive to the left. The velocity factor of the feeder is not allowed for in the chart, and the physical length of the feeder must be divided by this factor when computing length L.

more or less matched to the aerial and alternative couplers for this case are discussed in the next section.

In general, very high impedances at the bottom of the feeder should be avoided by judicious selection of the length of feeder employed, using the chart of Fig 12.73. High impedances are often difficult to match into the parallel tuned circuit because of the tendency to "run out of coil" when selecting the correct tappings. The high impedance also has associated with it a high voltage which will appear across the components of the tuned circuit and requires disproportionately large components to avoid the danger of flash-over. This is particularly likely to occur between the vanes of the air spaced tuning capacitor during peaks of modulation, if the vane spacing is inadequate for the high voltages developed. If the tuned line has to be brought into the shack, regions of current maxima are preferred because rf voltages "getting loose" can cause feedback troubles, as well as losses in walls, window frames etc. Unfortunately achievement of this condition at the lowest frequency ensures high voltages at all the even harmonics and the best compromise will probably be found by aiming for a current maximum at the highest frequency, where the above troubles, if allowed to occur, are likely to be more serious.

It is desirable to reduce the length of resonant line to a minimum because of the bandwidth problem discussed earlier, and it is good practice if the tuner can be located say at the base of the mast or in a nearby garden shed, continuing from there to the shack with matched line. If this avoids the need for re-matching when altering frequency within a band, and allows reasonable access for band changing, it should be the method adopted. Otherwise the tuner should if possible be located at the point of entry into the building. It is inadvisable to extend open wire lines carrying high rf voltages for any appreciable distance inside a building, and particularly into the station itself, because of their tendency to radiate energy and thus aggravate the troubles mentioned above if any current unbalance exists; if such a course proves unavoidable, particular care must be taken to maintain balance and this will be assisted by reducing the spacing of the line.

The change of Z_0 will usually be unimportant with a resonant open-wire line but can be avoided by reducing the wire diameter proportionately, a course strongly recommended in the case of a matched line which also needs to be carefully balanced, though the situation is much worse with a large swr; this is mainly on account of the regions of high voltage, the field from which may extend unevenly into adjacent brickwork etc having poor insulating properties. If, however, a high-current section of line passes near another conductor and the layout is asymmetrical, this also may cause unbalance and additional losses, and both of these effects may aggravate tvi and bci problems.

Centre-fed Horizontal Wire

The 7MHz half-wave dipole shown in Fig 12.72 can be used on all bands from 10 to 80m subject to special care to minimize losses on 80m as was earlier shown to be necessary for a 14MHz λ/2 dipole used on 7MHz. The directional properties of this aerial on the various bands will be considered later, but for the moment the problem is to decide the best way to design a tuner unit which provides coverage of all bands and the ability to accommodate any desired lengths of feeder. From the previous example it follows that the swr will be about 8·0 or 9·0 on all bands except 3·5MHz where it will rise to about 60. For operation on this band it will be necessary to retune the coupler for any frequency change exceeding about 15–20kHz.

Fig 12.72 shows the aerial and its feeder (also 66ft long) and indicates the form of the standing waves on both aerial and feeder when used on 7 and 14MHz. Two different coupling circuits are shown for the transmitter.

Since the feeder is 66ft long, it will be a definite number of half wavelengths long on all the bands from 7 to 28MHz and the input impedance will therefore be the same as the aerial impedance for each band. On 7MHz for example, the impedance is low and the *series* circuit of Fig 12.72(a) would be used, the coil and tuning capacitors having about the same electrical and physical size as the transmitter tank circuit. The link coupling coil for the transmitter feeder connection may consist of three turns of insulated wire between the centre turns of the tuning coil. Alternatively the parallel tuned coupling circuit of Fig 12.72(b)) could be used, and the feeder tapping points set close to the centre of the coil. Similar arrangements could also be used for 21MHz with a size of coil suitable for that band. On 14 and 28MHz the impedance is high, and in these cases the parallel tuning network should be used because at high impedance the taps will be at the outer ends of the coil.

If the feeder is made only 33ft long it will be one quarter-wavelength on 7MHz, one half-wavelength on 14MHz, three-quarters of a wavelength on 21MHz and a full wavelength (four quarter-wavelengths) on 28MHz. The impedance at the input to the feeder will be the same as with the 66ft feeder on 14 and 28MHz but on 7 and 21MHz where the aerial impedance is low, the odd quarter-wavelength feeder will transform it to a high value on these bands also; thus, the parallel tuned coupler would be used on all bands.

With the aerials described the swr is high and consequently the coupling circuit will carry large voltages and currents. It is essential therefore to use good quality components. A suitable arrangement for multi-band working consists of a plug-in coil base with six contacts, two for the coil, two for the feeder and two for the transmitter link coil with a separate coil unit for each band. The tuning capacitor must have two sections in order to preserve feeder balance, as discussed later under aerial couplers. For 100W of rf power the feeder current may not exceed about 0·25A in each line when the impedance is high, but it may well exceed 1A for the low impedance cases, eg 66ft feeder on 7 and 21MHz. This is a disadvantage in that the same rf ammeters cannot be used in all cases, but when the current is low and the voltage high, a small neon lamp may be used instead as a voltage indicator.

Judicious choice of feeder lengths will avoid a high voltage feed point on most bands. A suitable length may be found with the aid of Fig 12.73. On this standing wave chart the total lengths may (if desired) include one half of the aerial so that the voltage wave illustrated starts from one aerial insulator. It will be seen that total lengths of 45 or 90ft bring the feeder input into the high current sections of the chart in nearly every case when using tuned feeders. It is not essential to make the aerial itself a resonant length; with the top increased to 84ft length, this arrangement is known as the "extended double Zepp" and provides a gain of 3dB on 14MHz as compared with 2dB for the length shown. The increased length has the advantage of doubling the radiation resistance on 3·5MHz, making it much more attractive for this band, but in both cases it may often be found preferable to connect the feeders in parallel and use the system as a top-loaded vertical operating against a counterpoise or another aerial, as discussed later.

Flat Lines

An important feature of flat lines lies in the fact that the input impedance of the line is equal to the load impedance and by definition to the characteristic impedance Z_0, *irrespective of the line length.* It follows from this that the route and length of feeder employed may be decided primarily by the physical layout of the aerial system and its position relative to the transmitter, though losses increase of course in proportion to the length. Because the matched loss of coaxial cables is considerably greater than for open-wire lines, such cables are usually employed as nominally flat lines, with a minimum standing wave and hence minimum additional loss. It is also of course advantageous wherever possible to operate open-wire lines in a matched condition as already discussed. The advantages of this are wider bandwidth and the associated convenience of operation, better balance and therefore less risk of feeder radiation or other possible complications, and the ability if required to use very long lines with negligible loss so that the aerial can be placed in the best location available. Though not common practice, presumably due to other difficulties, there is no technical reason to prevent someone living in a valley from installing an aerial on an adjacent hilltop, anything up to a mile or so distant. The loss could be held down to 3dB at 14MHz and the advantage could be many S-units.

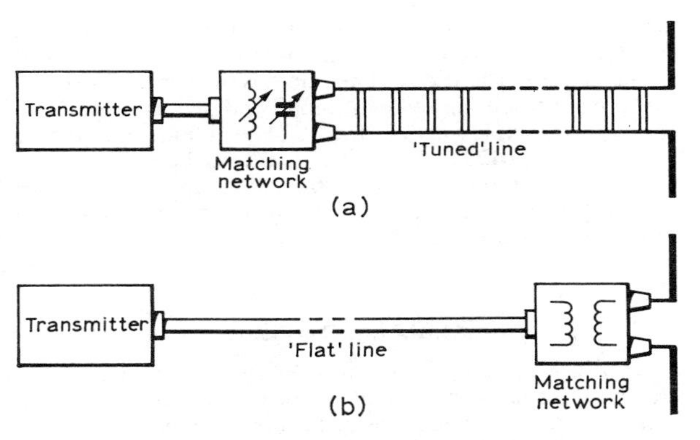

Fig 12.74. Transmission lines. (a) Tuned line. (b) "Flat" line.

It was shown in the section on transmission lines that any given line can be matched only by adjustment of the load impedance so that it equals the Z_0 of the line. Thus, unlike the tuned line method of aerial feeding, the matching arrangements must be incorporated *at the aerial end* of the flat line to obtain correct operation.

If the line is to be operated in a matched condition for its entire length, some means of matching must be incorporated into the actual aerial design, and, if the line employed is of the coaxial or unbalanced type, usually some form of unbalance-to-balance transformer (or *balun*) will also be required at the aerial terminals. Techniques of matching at the aerial are discussed in the earlier section on impedance matching, which also considers the design of baluns.

Because the use of flat lines dictates the need for impedance matching facilities at the aerial end of the line, it is possible to standardize on a characteristic impedance for flat lines for all applications, and arrange to match any particular aerial impedance which is encountered to that standard line impedance. The commonest coaxial line impedances are 75Ω and 50Ω. The former is used in Europe for domestic tv purposes, but the majority of amateur commercial equipment is designed for the lower impedance in conformity with American and Japanese domestic practice.

It is advisable to develop a station where possible around one or other of these standard values, rather than to attempt the simultaneous use of both. Transmitters and all auxiliary equipment such as tvi filters, swr indicators etc, can then be constructed to the selected standard impedance thus avoiding the need for duplication of components and simplifying the operation of the station. **Fig 12.74** illustrates some of the basic differences between tuned and flat line operation, the matching network in the case of a flat line being located at the aerial. In the case of the tuned line, the matching network is normally accessible so that it can incorporate tuning and matching adjustments, and it is use of a tuned line which usually makes this both possible and necessary. With a flat line the matching network, probably a balun with suitable impedance transformation, is not accessible and must therefore be a broad band device. Modern practice tends rightly or wrongly towards the use of matched unbalanced coaxial line wherever possible, though the relative merits of balance and unbalanced line are subject to some controversy and for further discussion the reader is referred to p12.32 and Reference [6]. The use of tuned lines is in general restricted, as previously discussed, to multiband aerial designs, a popular example being the G5RV multiband dipole described on p12.85. It is not, however, the only method of tackling the problem of multibanding, and for examples compatible with the use of matched lines the reader is referred to pp12.77 and 12.84–12.89.

AERIAL COUPLERS

In Fig 12.74 the flat line has been shown going directly to the transmitter but coupling such a line directly to the output circuit provides insufficient discrimination against harmonics or other possible spurious signals, and is an obsolete practice unsuited to modern conditions. At the very least a harmonic filter is required, and this may either be built into the transmitter or fitted externally but in close proximity. The normal output from a transmitter is through a coupling network and harmonic filter into a coaxial rf connector. In some cases, where the aerial is already fed by coaxial lines, it can be connected directly to the output of the filter, but more generally it is necessary to use an intermediate coupler or transformer. Even if the line is of the same impedance as the transmitter output (say 70Ω), it may be a balanced twin feeder and cannot therefore be connected to the unbalanced concentric transmitter output. It is therefore common practice to use an *aerial coupler* or *aerial tuning unit* (often abbreviated to atu), such as those illustrated in Figs 12.72 and **12.75,** even when the transmitter is working into a flat line as in Fig 12.74(b). An aerial coupler not only provides additional rejection of harmonics but also some protection against radiation of local oscillator, intermediate frequency and other spurious signals which because of their relatively low frequency, are not suppressed by the harmonic filter.

The aerial coupler is sometimes built as part of the transmitter, in which case when working with balanced lines it may be necessary to use an additional coupler at the point of entry into the building as discussed previously. In addition to the functions of tuning and matching as already discussed the atu may have to perform such tasks as conversion from a balanced feed line to an unbalanced transmitter outlet or vice versa, and electrical isolation of the aerial and feeder system from the transmitter and mains supply; connection as in Fig 12.74(b), though sanctioned by common usage, usually fails to meet the requirements discussed earlier for establishing a non-resonant length of feeder plus aerial and also implies reliance on the earth of the mains supply for lightning protection. As already stated, and discussed in more detail later in this chapter, this is **potentially dangerous.** If the house wiring is protected by some of the older varieties of earth-leakage trip, any ground connection to the aerial or feeder, or even leakage to ground, will short-circuit the trip and leave the entire mains installation unprotected unless the aerial is isolated by a suitable matching network or transformer.

The general design principles of couplers have already been discussed but as stated earlier, many practical variations are possible, and the following notes are intended to assist in choosing the optimum configuration for any particular situation, and suitable design parameters.

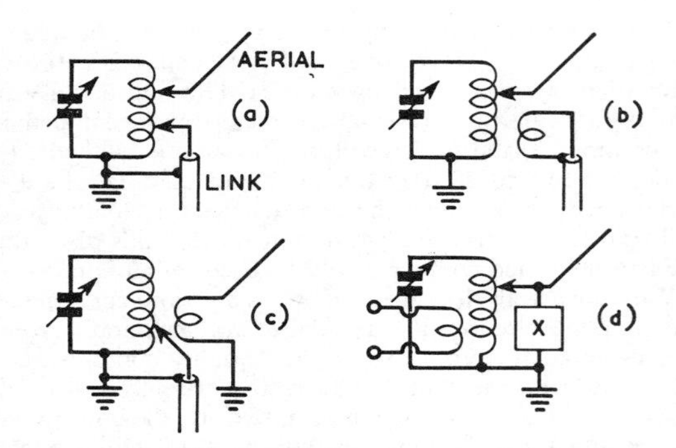

Fig 12.75. Various forms of aerial coupler for end-fed aerials. The use of reactance X for compensation of very reactive aerials or feeder loads is discussed in the text.

Unbalanced Aerials

Fig 12.75 shows some variations of an aerial coupler intended for unbalanced aerials or feeders. In cases (*b*) to (*d*) the inductance acts as a two-winding transformer, the capacitor serving to resonate the coil to ensure a high degree of coupling, to help the network to discriminate against unwanted frequencies and to assist in harmonic suppression. The auto-transformer arrangement of Fig 12.75(a), though often used, is deprecated on safety grounds as discussed above, although with a high impedance aerial this objection may be overcome by isolating it for dc with a small series capacitor not exceeding a few hundred pF. The link may have either a direct tap connection or a separate coil of a few turns interwound with the main inductance at the earthed end.

The tuning components may conveniently have a reactance of about 500Ω at the operating frequency. This may most conveniently be translated into practical figures by reckoning the capacitance as 1pF per metre of wavelength, and the inductance as 0·25μH per metre. For example, at 7MHz, approximately 40pF and 12μH could be used. There is nothing critical about these values, though for a very high impedance aerial load (as may occur when using tuned feeders) it is advisable to increase the inductance and use less capacitance. Coils of about 3in diameter wound with 14–26swg wire on a ceramic former will handle powers up to the legal limit without overheating. The tuning capacitor should be of good quality, have a plate spacing of $\frac{1}{16}$in or more and have adequate capacitance to tune to resonance at about half its maximum value. The link coil may be wound outside the tuning coil with about $\frac{1}{4}$in clearance and should be either self supporting or rely on spacers of good quality insulating material such as ceramic or polystyrene rod. Polystyrene rod may be notched to fit over the turns of the tuning coil, its ends being secured with tape or Araldite to the ends of the ceramic former, and the coupling winding may be embedded in the polystyrene by an application of heat from a soldering iron at the points of contact.

Couplers for 14–28MHz may use a self-supporting tuning coil wound with $\frac{1}{8}$in to $\frac{3}{16}$in diameter copper tubing. If a variable coupling is required it is necessary to use either a self-supporting tuning coil and a well-insulated coupling winding which can be slid in between the turns of the tuner, or alternatively split the tuning coil into two closely coupled halves wound on separate formers. The link may vary from four or five turns at 3·5MHz to one turn at 28MHz.

In adjusting the coupler the aim should be to achieve the sharpest possible tuning consistent with negligible loss of efficiency. This not only minimizes the radiation of harmonics or other spurious signals but gives valuable additional protection to the receiver against various forms of spurious signal and particularly intermodulation interference, which is usually caused by powerful signals in adjacent broadcast bands. With good construction the unloaded coil Q should be at least 300 which means that a loaded Q of 30 would result in a power loss of 0·4dB and an improvement in efficiency of only 0·2dB would result from loading twice as heavily, ie reducing the Q to 15. A loaded Q of 30 allows coverage of, say, all or most of the phone portion of the 14MHz band without retuning, whereas to achieve equal convenience of operation on 3·5MHz the Q must be reduced to 7 or 8 with correspondingly less rejection of harmonics and intermodulation interference. It is suggested that adjustment may be made by starting with maximum coupling and if necessary reducing the aerial tap until it is just possible to obtain adequate loading. The tap is then reduced, re-adjusting the atu and pa as necessary to keep the loading constant until the current or power going into the aerial just starts to decrease; if, however, there is no tvi problem and a larger bandwidth is required as is likely on 3·5MHz, the tap position may be increased.

Balanced Aerials

When the aerial or the feeder is balanced it is necessary to revert to a balanced circuit such as Fig 12.72(b) which has already been discussed in some detail. Similar components and methods of construction are suitable, the main differences being the provision of symmetrically arranged pairs of taps and the advisability of a two-gang capacitor with the two portions in series to ensure that the stray capacitances to ground are symmetrical.

Marconi Effect

To reduce the risk of feeder radiation due to the feeder wires being excited in parallel as a vertical radiator, as discussed on p12.32 and often known as *Marconi Effect*, the capacitance between the windings must be kept as small as possible and it is advisable to screen the coupling loop. This may be done by forming it from coaxial cable as illustrated in **Fig 12.76.** Trouble may still be experienced if the length of feeder plus aerial is near resonance and the easiest way to rectify this situation is likely to be the connection of an earth or counterpoise to the frame (ie centre of the split tuning capacitor. It is sometimes helpful to "damp" the unwanted resonance, eg by a 200Ω 2W resistor in series with the earth connection.

Matched Lines

When supplying a matched line the design requirements for the coupler are modified to the extent that no need arises for feeding into a high value of impedance and much smaller values of inductance may be used. Advantage should be taken of this, as it reduces the magnitude of rf voltages and it may often be possible to use a two-gang receiving-type capacitor, even at the legal power limit. Apart from this the risks of (*a*) feedback troubles and (*b*) excitation of the Marconi effect are reduced. By way of example, a balanced resonant feeder connected to a dipole can present an impedance as high as 6,000Ω; if this is connected across a

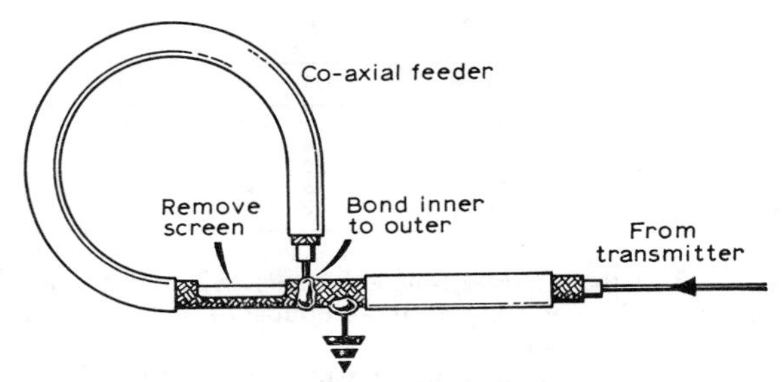

Fig 12.76. Screened coupling loop. Outer is sometimes partly removed as shown, to ensure that the current return path is inside rather than outside the sheath.

reactance of 500Ω having a Q of 300 there will be a 4 per cent wastage of power. In contrast, for a 600Ω line with an swr of 2 : 1 the reactance can be reduced in the ratio 6,000/1,200, ie to 100Ω for the same loss in the worst case and the voltages generated will be less than half. In practice working with experimental aerials a value of 150Ω has been found a good compromise, providing somewhat greater flexibility, but this is not critical and a reactance of 150Ω at 21MHz provides a coil that can be tuned over the range 14–28MHz, a second coil being required to cover 3·5–7MHz. For feeding low impedance balanced line a second coupling coil (same number of turns) may be used; this can consist of two coils in parallel to assist in achieving symmetry. The arrangement of Fig 12.75(c) may also be used by omitting the earth connection from the coupling coil, which should be located close to the earthy end of the tuning coil. If Marconi effect occurs it may be reduced by connecting an earth, not that of the transmitter casing, to the centre point of the coupling coil.

Fig 12.75(c) is also suitable for coaxial feeders, the coupling coil being connected to the line and the earth if required being connected to the outer of the line.

Table 12.2 overleaf gives suggested L and C values for feeding a balanced matched open-wire line. These should be found suitable for an swr up to at least 3. The turns are based on the assumption of 3in diameter and a length/diameter ratio of unity. For other diameters the turns must be adjusted inversely as the square root of the ratio of diameters, smaller diameters giving slightly larger losses which for a given coil are roughly proportional to the *loaded Q*. With so many variables depending on individual circumstances it is difficult to give any fixed rules, but couplers handling the full legal power limit should preferably not be less than 2in diameter and need not be more than 3in. For the purpose of the table the reactance has been allowed to vary slightly around 150Ω in order to arrive at a whole number of turns, but considerable variation in dimensions is admissible to suit available components, formers etc. The more robust varieties of two-gang capacitors salvaged from old broadcast receivers have been used successfully at powers up to the legal limit for balanced tuners, but must be in good condition, ie vanes must be evenly spaced, and those with ceramic insulation are preferable. Arcing may still be experienced but, except under fault conditions (eg excessive swr), can usually be cured by reducing the inductance, operating at a lower value of loaded Q, or both. In Table 12.2 the value of L for 3·5MHz has been increased slightly to allow the use of such capacitors.

Note the values of C are the actual values needed for the

TABLE 12.2

Band (MHz)	L (μH)	C (pF)	No. of turns
3·5	10·20	200	14
7·0	4·05	128	9
14·0	1·80	70	6
21·0	1·25	46	5
28·0	0·80	40	4

low end of the band with a perfect match; the maximum values should preferably be about double to provide a wide range of adjustment.

These values of capacitance are of course the effective totals, ie with a balanced circuit 200pF is obtained by connecting two 400pF capacitors in series. With four-gang capacitors a total of about 500pF can be obtained. For tuners feeding resonant lines at voltage maxima the number of turns should be increased by a factor of 1·7 to 2·0 and the capacitances reduced to a third or quarter respectively with provision for a much larger variation, say up to twice the design values, and receiver capacitors are not suitable. If the need arises the coupling may be increased somewhat in link circuits by tuning out the inductance of the coupling coil, reckoning this as about 0·25μH for a one-turn coil of 4in diameter, and directly proportional to diameter. This additional resonance is extremely flat and no adjustment is needed except from one band to another. When more turns are needed (up to three or four at 3·5MHz or two at 7MHz), they may be wound tightly together, the inductance increasing almost as the square of the turns. The ideal value of inductance for the coupling coil depends on the loaded *Q* of the atu as well as on the line impedance and, for values of 10 or more, single turn coupling coils have been found satisfactory for low impedance lines at 14MHz and higher frequencies. The coils are concentric and single layer with a spacing of ¼in or less, and may be constructed as previously described with the inner coil on a ceramic former and the outer coil wound over it using polystyrene rods located in end-plates of any convenient insulating material, with a clearance of at least ½in between the end turns and the end plates if these consist of a lossy material such as perspex or bakelite.

Tuning Compensation

When the aerial or feeder is coupled to the aerial tuning unit it is sometimes impossible to adjust the circuit for reasonably high loading. This occurs with tuned feeders having a high standing wave ratio in which case a low reactance load may be presented by the line. In such cases it is better to compensate the feeder reactance separately before transforming the resistance component, using a coil or capacitor connected in parallel with the feeder (see Fig 12.75(d)). If more capacitance than is available appears to be required for resonance, then capacitance loading is necessary; if the converse, inductive loading is the solution. A loading component of the same value as the coil or capacitor of the tuning circuit may be tried first.

Adjusting Matching Units and Couplers

The recommended way of adjusting the units just described requires an swr bridge or reflectometer. The layout of a typical setup is shown in Fig 12.51. The matching unit may be that of Fig 12.72(b), and a suitable reflectometer is described in Chapter 18.

The procedure is quite simple and should present no difficulties if followed carefully and in the correct sequence. The reflectometer should be capable of handling the power output of the transmitter and should be of the type which reads both forward and reflected power. For maximum convenience the instrument should employ two meters, enabling both powers to be observed simultaneously.

Connect up as shown in Fig 12.51 and adjust the transmitter power or the reflectometer sensitivity so that in the forward reading position the meter indicates well up the scale with the taps set at a trial position equidistant from the centre of the coil.

Now read the reflected power. Vary the tuning of the matching unit for the minimum possible reading on the meter.

If a low reading is not possible it will be necessary to adjust the tap positions, making certain that they are symmetrical about the centre of the coil.

If the coupling between the coils is variable then after tuning for lowest reflected power the coupling should be varied to obtain an even lower reading, repeating the tuning adjustment. Once again if a low reading is not possible the position of the taps should be changed and the whole procedure repeated.

Multiband Couplers

There are many variations of the aerial couplers described. For example, a pi-network similar to those described in Chapter 6 (*HF Transmitters*) for transmitter output circuits is often attractive because it can be arranged to cover as many as four amateur bands with the same set of components. The couplers of Figs 12.72 and 12.75 can usually be arranged to cover two adjacent amateur bands but where complete coverage is needed it may be necessary to make up a coil together with its link and taps as a plug-in unit for each band, though there may still be some difficulty if the capacitor has too large a minimum capacitance for 28MHz or insufficient maximum capacitance for 3·5MHz.

A popular multiband aerial tuning unit for 3·5–28MHz is shown in **Fig 12.77** and is known as a *Z-match coupler*. It is a compound network using two pairs of windings and is capable of matching the wide range of impedances which may be presented by a tuned aerial line. Coils L1 and L2 may each be 5 turns tightly coupled. L3 and L4 are 8 and 6 turns respectively. L1 and L3 may be about 2½in diameter and L2 and L4 about 3in diameter, with 0·25in spacing between turns.

The series capacitor C1 is 500pF maximum and it should be noted that it is "live" on both sides; the frame should be connected to the transmitter link cable and an insulated extension shaft provided. The other capacitor, C2, is the split stator type, 250pF per section. C2 tunes the coupler and C1 adjusts the load to the transmitter. A standing wave indicator (reflectometer) is again an aid to tuning. C1 and C2 are adjusted for minimum reflection and the transmitter can then, if necessary, be trimmed for maximum output.

Power Indicators

It is necessary to be cautious in interpreting the power radiated in terms of meter readings, because the current and voltage vary from point to point in an aerial system. Some suggestions as to what to expect may therefore be helpful.

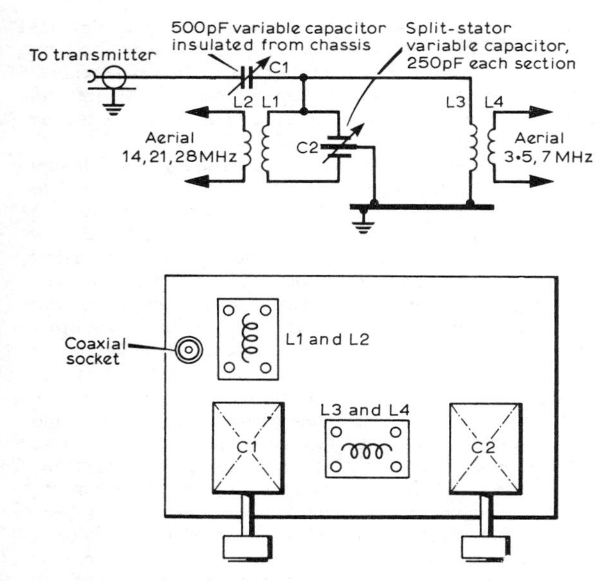

Fig 12.77. Circuit diagram of the Z-match coupler and suggested layout. C1 is the series capacitor and C2 the split-stator capacitor in the multiband tuning circuit.

A 100W dc input transmitter will deliver up to 1A into an 80Ω matched line, but only about 0·35A into a matched 600Ω feeder. In both cases this can be taken to represent true power. In a tuned line, however, the current may vary over a wide range and does not reveal the real radiated power, though it still serves as a guide for tuning and matching adjustments since the current at any point beyond that at which adjustments are made will be proportional to the square root of the true power, provided the adjustment is not of a nature to upset the balance of the system. RF thermo-ammeters are available with maximum readings of 0·5 to 5A, but usually have rather cramped and limited scale at low readings so that they cannot be used for widely differing currents. Care is necessary in cases where the current is greatly different in the same aerial for different bands, since these meters are easily damaged by overload. Such overloads are particularly prone to occur in the event of a fault developing in the aerial or feeder system.

Low-wattage lamps provide slightly more sensitive current indication and are cheaper to replace when destroyed by overload, but both devices are inconvenient to use as they usually require the line to be cut for their insertion, though if the current is high enough (eg at the centre of a dipole) they may be tapped across a few inches of conductor which then acts as an inductive shunt. Even this may be inconvenient and if the voltage is high enough, a neon lamp is a more handy though less precise indicator. A sensitive current meter which is portable and requires no direct connection to the line is illustrated in **Fig 12.78**; the line current induces a voltage in the loop which is detected by a silicon or germanium diode, and with a 4in square loop the sensitive microammeter provides a useful indication from a power level of only a few milliwatts in a flat open-wire line, enabling measurements to be made with very little radiation from the aerial. It is however advisable to use a little more power and space the loop at least ¼in away from the feeder in order to ensure a true current reading since the meter, acting as an "aerial", can respond slightly to line voltage and there is also a risk of unbalancing the line appreciably by its use. The loop can be used to indicate the current in each line separately, or it can be held so that it couples to both lines in which condition full scale deflection is obtainable even on an insensitive meter (3mA fsd) for about 500mW radiated power, ie a line current of only 30mA. The loop dimensions are not critical but should be comparable with the line spacing. These figures are of course only a rough guide and the instrument is not suitable for use as an accurate power meter. It can however be calibrated accurately in terms of *relative* power or voltage by coupling it loosely to the pa and plotting meter reading versus rf oscilloscope deflection, or against aerial current measured with an accurate meter. At very low currents readings are square law but for induced voltages of about 0·5V upwards a reasonably linear law is obtainable. The device should be mounted at the end of an insulated handle, the longer the better, to minimize body capacitance effects.

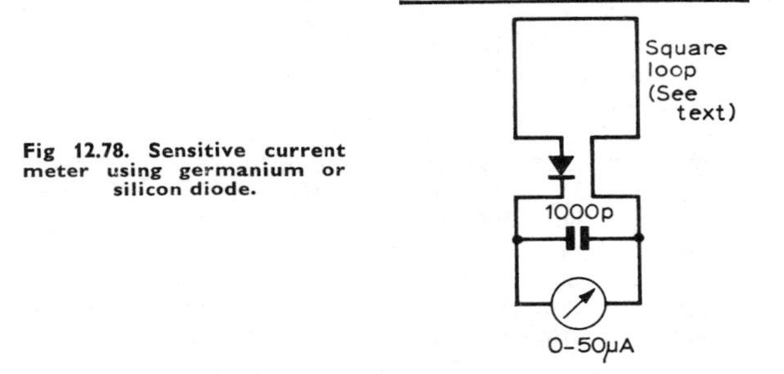

Fig 12.78. Sensitive current meter using germanium or silicon diode.

The current in a low-impedance twin feeder may also be detected by an external pick up device, eg a few inches of feeder taped to the main feeder, with one end shorted and the meter and rectifier of Fig 12.78 applied to the other end. As it stands the instrument can also be used as a sensitive detector of unwanted current flowing on the outside of a coaxial line, but measurements of currents in the line are not possible without damage to the line. For measurements on coaxial lines it is however almost as simple to construct a reflectometer on the lines of Fig 18.39. The reflectometer will be found much more useful as it provides an actual swr measurement for comparison with the swr on the transmitter side of the atu. It is an easy and inexpensive matter to construct two similar instruments, one for use on each side of the atu, using a single meter with a two-pole four-position switch, but the meter should be as sensitive as possible to allow adequate rf filtering of the leads by the use of series resistances, at least 10kΩ in each lead being recommended.

DESIGNS FOR PRACTICAL AERIALS

The first sections of this chapter have been devoted to the general aspects of aerials and transmission lines without any specific attempt to relate these to the particular requirements of any one amateur frequency band. The rest of the chapter deals mainly with practical aerial designs and has been divided into those primarily suitable for the low frequency bands (1·8MHz and 3·5MHz), and those which are more likely to have practical application for the higher frequency bands from 7MHz to 28MHz. A separate section is devoted to vertical aerials, which have a number of distinctive features relative to horizontal aerials, particularly the fact that in their simplest form, being extended only in the vertical plane, they have negligible requirements for horizontal area. There is of course considerable overlap between these sections and, to avoid duplication, descriptions are made fully comprehensive with cross-referencing as necessary.

LOW FREQUENCY AERIALS

Choice of Polarization

One feature of operation on the lower frequency bands is the widespread use of vertical aerials. This is due partly to ease of erection and partly to the important role of ground-wave propagation at these frequencies, particularly on 1·8–2·0MHz. Due to the low height, in relation to the wavelength, ground-wave propagation is very inefficient with horizontal polarization although better ground-wave signals will usually be obtained when both aerials have the same polarization regardless of whether this is horizontal or vertical. Vertical polarization is advantageous also for dx, since even at 3·5MHz it is usually not possible to erect a horizontal aerial at sufficient height to achieve good low-angle radiation.

Against this, the 3·5MHz band is particularly suitable for short and medium range communication via the *E*-layer in the daytime and the *F*-layer at night, including distances just beyond the ground-wave range which can usually be reached in daylight hours by nearly vertical incidence reflection from the *E*-layer, whereas for much of the time long-distance low-angle paths are useless due to excessive absorption in the ionosphere. Vertical aerials are relatively poor as high-angle radiators so that horizontal aerials, even at low heights, are usually better for short-range sky-wave working and reasonably satisfactory over medium distances. The directive properties of the horizontal aerial have comparatively little influence at the high angles appropriate to these distances.

Marconi Aerials

Most of the discussion so far of simple types of aerial has assumed horizontal polarization and although, in principle, these aerials can be erected vertically this is somewhat difficult in the case of say a half-wave dipole at 3·5MHz, which would require a mast about 140ft high! The usual procedure if vertical polarization is required is to use as long a vertical wire as possible and tune it as a monopole resonator working against some form of earth connection. The aerial, in conjunction with its image in the ground, then bears some resemblance to a shortened dipole. If the wire is a quarter-wavelength long with its base at ground level it can be thought of as a bisected half-wave aerial and its radiation resistance is about 35Ω, ie half that of a half-wave dipole. Such a height is not usually possible, even for 3·5MHz where it would be 66ft and hence it is necessary to use an electrically short aerial and to load it in such a way as to bring it to resonance, accepting the inevitably low value of radiation resistance; for example an aerial one-tenth of a wavelength high (50ft for 1·8MHz) has a radiation resistance of 4Ω, which would not matter if the earth were a perfect conductor, but this is far from being the case, and the resistance due to even a very elaborate earth system, together with the loss due to the current returning through the surface, may contribute 20Ω or more. The efficiency is thus low. The relative efficiency of various combinations of earth resistance and aerial height are considered in more detail in the following section on earth systems.

At higher frequencies where vertical aerials are used mainly for their convenience ground plane systems can be used to reduce earth loss, but at the lower frequencies these tend to be rather large and other methods of improving efficiency may have to be sought. Let us consider first a short vertical wire and assume access to the base only; this wire can only be brought to resonance by means of a series inductance which has some loss resistance, this being added to the other system losses. Apart from this, the current distribution in the wire, being the tip of a sine wave, is triangular and the mean current is only half the base current. If on the other hand the aerial can be tuned by capacitance between the top end and ground, and if this provides the main current path, the current distribution on the vertical wire will be nearly uniform; the mean current will then be almost equal to the base current, the required value of this being therefore halved for a given field-strength. This means that the radiation resistance is multiplied by four, and in addition the loading coil is eliminated. Capacitive end loading may be achieved in practice by various combinations of horizontal wires, and even when it is not possible to tune the vertical radiator fully by this means the amount of inductive base loading required may be greatly reduced. On the other hand it is sometimes possible in this way to increase the total effective length of the aerial to more than $\lambda/4$ thereby raising the impedance at the feed point, and hence even further reducing the current and the power loss in the earth connection.

The simplest form of top end loading is to add a horizontal top to make a T or inverted L aerial (**Fig 12.79(a)**). This loading should ideally be arranged to bring the current maximum into about the centre of the vertical portion—though it may sometimes be advantageous (eg if there are local obstructions) to bring it as high as possible when the loading is of T form. With L type construction it is undesirable to have too much current at the top where it is liable to result in wastage of power by useless radiation from the horizontal section, whereas with the T configuration the radiation from the two halves of the top is in antiphase and tends to cancel. For an L aerial the electrical length is given approximately by the usual formula, p12.3, the distance from the end of the wire to the centre of the current loop being about 130ft at 1·8MHz. For a T aerial the equivalent electrical length of the top portion can be reckoned as about two-thirds of the actual length for lengths up to $\lambda/4$, decreasing to half for a length of $\lambda/2$ when the centre of the T becomes the centre of a half-wave dipole. If the length obtainable by these means is insufficient, the next alternative is to increase the loading effect of the top by making it into a cage or "flat top" of two or three wires joined in parallel; this increases the capacitance and effectively lengthens the aerial. If this procedure is not practicable, the aerial may be loaded by including a coil (see **Fig 12.79(b)**) near the free end of the aerial. In this case the principle of operation is as follows; if the length of wire beyond the coil has a capacitive reactance to ground of $-X$, an inductive reactance of $X/2$ gives a net reactance of $-X/2$ which is equivalent to doubling the capacitance or, approximately, doubling the length of wire if this is shorter than about $\lambda/16$. The coil also has a

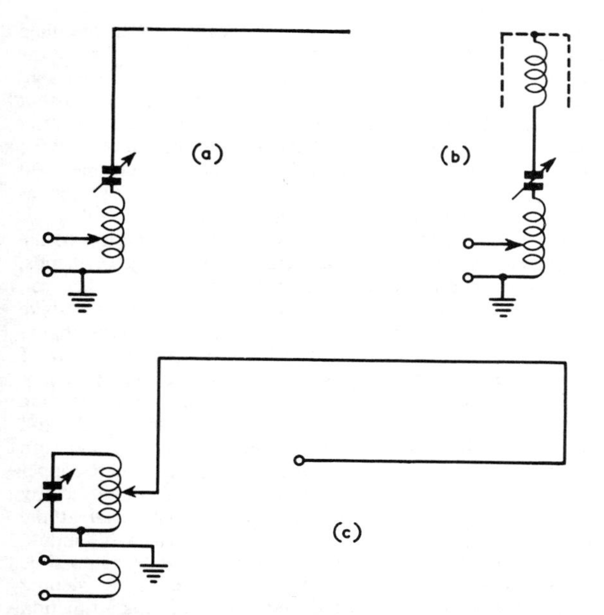

Fig 12.79. Aerials for low frequency bands. (a) λ/4 (inverted L) aerial with series tuning. (b) End loading to raise the efficiency of a short aerial. (c) Special extended aerial with high efficiency, detailed in text. The input current in aerials (a) and (b) may be 0·5A for 10W input; that in (c) is low, say 0·1A for 10W.

resistance $X/2Q$, part of which appears as a loss resistance in series with the radiation resistance, and thus limits the amount of this type of loading which can be usefully employed. As an example, a 34ft length of vertical wire top-loaded to resonate on 3·6MHz will have a radiation resistance of about 18Ω. The vertical wire has an inductive reactance of about 300Ω which can be tuned to resonance by 1,500Ω of capacitance in series with an additional 1,200Ω of inductance which will have about 4Ω loss resistance. This would require a coil of 53μH in conjunction with a capacitance of 32pF, which could be provided by 16ft of horizontal wire in place of the 70ft or more required without the coil. Being in series with the radiation resistance plus the earth resistance, the loss in the coil has relatively little effect on performance though the process cannot be carried very much further without loss of efficiency since halving the capacitance and doubling the inductance will roughly double the coil loss. On the other hand, if the capacitive top is long compared with the vertical wire it is immaterial electrically whether the coil is placed at the top or, more conveniently, at the base. In the above instance, whichever method of capacitive loading is used improves the efficiency by three to four times if the earth resistance is large. This example has been slightly oversimplified by ignoring some aspects of the current distribution but serves to illustrate the basic principle as well as the main practical aspects, and it will be obvious to the reader that there are many possible variations of this technique which find particular application in the field of mobile and indoor aerials.

Other methods of end loading may be devised; for example, folding the aerial back on itself with spacing of, say 6in.

Aerials for 1·8MHz with relatively high input currents are preferably series-tuned at the base, using about 250pF capacitance with say 30μH inductance. A suitable coil for 1·8MHz could be made with twenty turns of 16swg wire on a 3in diameter former, spaced to occupy a length of 3–4in. The transmitter link could be tapped across a few turns at the earthed end of the coil or the wire itself coupled to the transmitter output tank coil. About half the above values should be used for 3·5MHz.

To reduce the deleterious effect of an imperfect earth connection the aerial may have a total length of about $\frac{3}{8}$ or $\frac{5}{8}$ of a wavelength, the current in the vertical part being high in both cases though the earth current is considerably reduced. A very effective aerial is shown in **Fig 12.79(c)** where the main half-wave is folded into a U with one leg near the ground and adjusted in length until maximum current occurs at a point half-way up the right-hand vertical section. The down-lead makes the total length up to about $\frac{5}{8}$ of a wavelength, so that the impedance at the feed point is a few hundred ohms and capacitive, while the earth current is relatively low. The coupling circuit for this aerial should be parallel tuned but can use the same components as the previous examples.

The 1·8MHz aerials of Figs 12.79(a) and 12.79(c) can also be used effectively on 3·5MHz and higher frequencies, though their performance will not be accurately predictable.

Earth Systems

The above examples have demonstrated the importance of a low earth resistance when the ground acts as the return path for the flow of rf currents. As will be evident from this discussion, earth resistance tends to present a major problem on the low frequencies, particularly on "top band", as it is often difficult to dispose of a sufficient length of wire to achieve a radiation resistance larger than the earth resistance which is in series with it and represents wasted power. There are two lines of attack on this problem; increasing radiation resistance where possible by one or more of the methods outlined above, and decreasing the resistance of the earth connection. To illustrate the problems of earth resistance, **Fig 12.80** shows the current paths when a short vertical radiator is driven by a "generator" (ie transmitter), connected between the bottom of the radiator and ground.

The current distribution along the radiator possesses a maximum value at or near the ground, decreasing approximately sinusoidally (or in the case of a very short radiator, linearly) to the end of the radiator at which point it must be zero. Because of the capacitance of the radiator to ground, charges are induced in the earth surrounding it, giving rise to a circulating current which flows back to the generator.

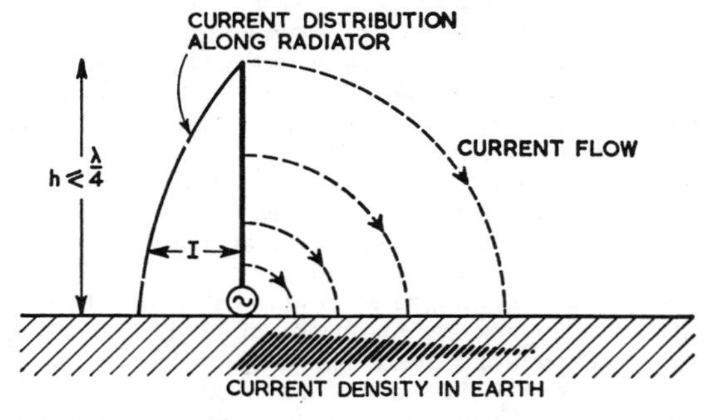

Fig 12.80. Current distribution on a short vertical radiator over a plane earth.

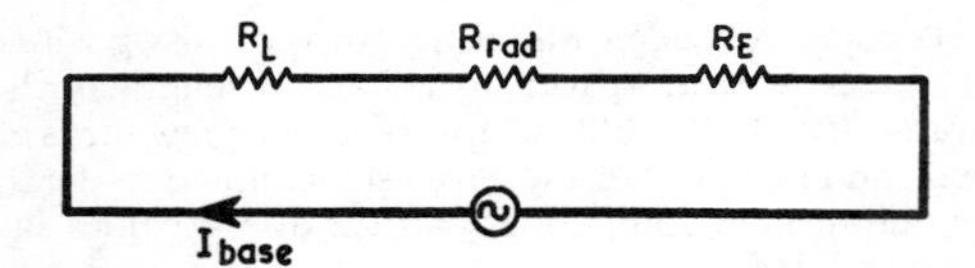

Fig 12.81. Equivalent circuit for a short vertical aerial over a plane earth. R_L, ohmic losses; R_{rad}, radiation resistance; R_E, effective earth resistance.

This current flows in the ground at a penetration which varies with frequency and near-surface conductivity, decreasing as both increase. The net effect is to include in the series circuit representing the aerial system a resistive loss term through which the aerial current flows, and which is dependent upon the near-surface conductivity in the neighbourhood of the aerial.

In order to determine the aerial efficiency we need to compare the power dissipated in the various resistive elements of the circuit; as illustrated by **Fig 12.81** these comprise the ohmic loss R_L of the aerial tuning components and the aerial conductor itself, the radiation resistance R_{rad}, and the earth loss resistance R_E.

Radiation resistance is a fictitious resistance which is included to account for the power dissipated by radiation, ie its value is determined by the useful radiated power and the current flowing in the aerial at the feed point. It does not exist as a physical resistance in the same way that R_L and R_E do. Of the three resistive elements dissipating power, only that in R_{rad} can be considered useful; that in R_L and R_E is wasted as heat, although it still has to be supplied by the transmitter. Since the same current flows through all resistances, the efficiency of the aerial system is given by:

$$\text{Efficiency } \eta = \frac{\text{power radiated}}{\text{power supplied}} = \frac{R_{rad}}{R_L + R_E + R_{rad}}$$

From Fig 12.5 the aerial wire resistance will contribute about 1Ω per quarter-wavelength and a tuning coil of 30μH with a Q of 200 will add a further 2Ω, making a total of 3Ω. The form of loading illustrated in Fig 12.79(b) if used will increase this figure, the extent to which it can be exploited being determined as already discussed by the point at which R_L becomes comparable with $R_E + R_{rad}$. The value of R_{rad} is fixed for a particular aerial by its physical size and shape, being typically 35Ω for a vertical quarter-wave wire, and as low as 9Ω for a vertical sixth-wave wire. Hence the shorter the aerial (electrically) the better is the earth system required to maintain a given efficiency. This can also be expressed by saying that for a given input power, the shorter aerial with lower radiation resistance gives rise to a larger feed current, and hence the earth loss must be reduced in proportion. The value of R_E varies widely, from as much as 300Ω for a simple earth in sandy soil, to as low as 2–3Ω for a comprehensive earth system in good soil.

Fig 12.82 shows how the aerial efficiency varies with effective electrical height for differing values of earth loss resistance. It is interesting to note that with a very good earth system ($R_E < 2\Omega$), an eighth-wave vertical radiator is almost as efficient as a quarter-wave radiator over the same earth, the figures being 95 per cent and 66 per cent respectively, or a difference of 1·6dB. Allowing for the difference in field strength due to the shape of the radiation patterns, the eighth-wave aerial would be 1·7dB down on the quarter-wave version at a distant point. If the earth loss were that of a simple "spike in the ground", typically 100Ω, the field strength from the shorter aerial would be 8·5dB down on the quarter-wave aerial, which itself would be only 27 per cent efficient. **Fig 12.83** shows the field strength relative to that produced by a quarter-wave vertical aerial with a perfect earth, for varying values of effective earth resistance, and for two different aerial heights. It is assumed that the field strength is measured at ground level, and the rate of attenuation with distance the same in both cases.

From what has been said it is clear that attention to the earth system of a base-fed radiator will pay large dividends, but the question now arises how the effective earth loss can in practice be reduced. The figure of 100Ω for a large spike in sandy soil is not unrealistic, and to reduce this figure to the 2–3Ω of a near perfect earth is beyond the scope of most amateurs, requiring as it would a massive system of radial earth wires, in excess of 120 in number and extending far out up to a wavelength from the base of the aerial—and all this in best quality agricultural ground as well! In practice, the best rule of thumb is to get as much copper wire into the ground as possible in the immediate vicinity of the aerial base, concentrating it near the aerial at the expense of the edges of the garden. The radial wires should be at least 16swg, and buried as near the surface as possible consistent with their remaining undisturbed by gardening activities since the depth of current penetration is a function of the near-surface resistivity. The inner ends of the wires should be joined to a heavy copper circular bus-bar (say 6swg wire in a 12in diameter loop). Brazing is to be preferred for resistance to corrosion, although soft soldering is adequate provided the joints are painted with a bitumen paint to seal out moisture. It is permissible to join the outer ends of the wires, but too much interconnection will result in large circulating currents being induced in the buried loops with a consequent increase in losses—the opposite in fact of the desired result (**Fig 12.84**).

The remarks in the previous paragraph apply mainly to amateur aerials for the lf bands having lengths of a quarter-wave or less and the current maximum at the base of the aerial. For those in the position to erect verticals (natural or loaded) with an effective length approaching a half-wave there is no need for an earth connection, and Fig 12.102

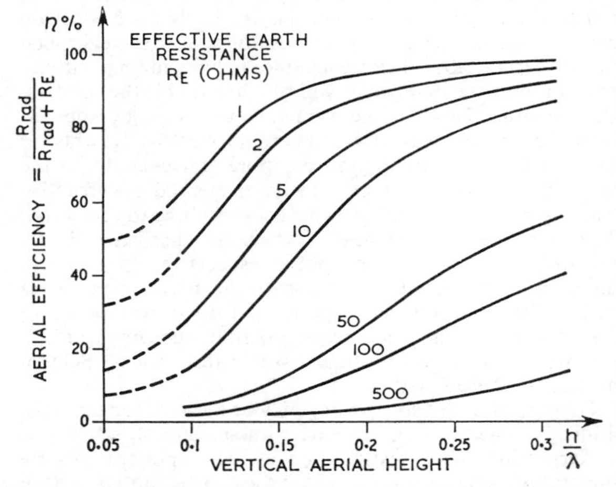

Fig 12.82. Variation in efficiency of a vertical aerial over a finite earth, for different heights of aerial and various earth resistances.

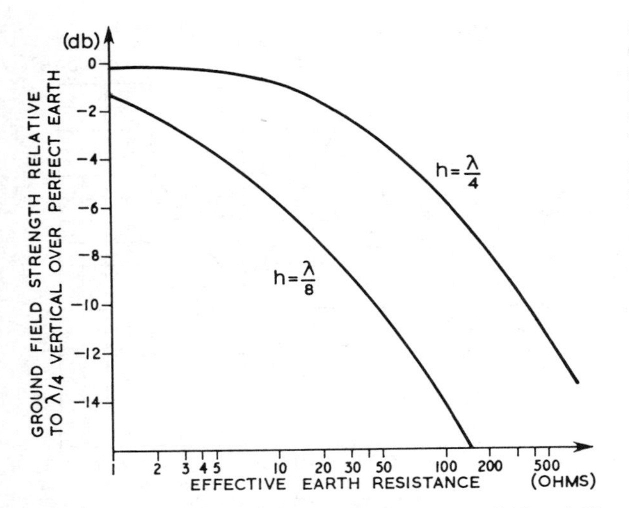

Fig 12.83. Comparative ground-wave field strength of λ/8 and λ/4 vertical aerials over an imperfect ground (neglecting ground-wave attenuation).

on p12.67 illustrates a 7MHz beam aerial based on the use of short end-loaded vertical dipoles. Used singly at the lowest possible height these elements have a radiation resistance of about 30–40Ω but this figure can be expected to vary considerably with height and ground conductivity. With direct scaling for 3·5MHz a height of over 50ft would be required but if a relatively longer span is accepted this could be considerably reduced for single elements, particularly if broad band operation is not essential.

It will be obvious from the above examples that every possible effort should be made to avoid reliance on earth return, and this is the main advantage of the various forms of ground-plane aerial (p12.81). Even if the radials are reduced to two (a reasonable compromise) a run of 133ft is required to accommodate a full size ground plane on 3·5MHz, but the vertical can be end-loaded by the use of a T configuration, the radiation resistance being halved when the height is reduced from λ/4 to λ/8, and there seems no reason why the radials should not also be of T form. There is also no reason why a vertical or sloping or inverted-L version of the W3EDP aerial described later should not be used. In general however these systems, like the previous example, tend to have too narrow a bandwidth for matched

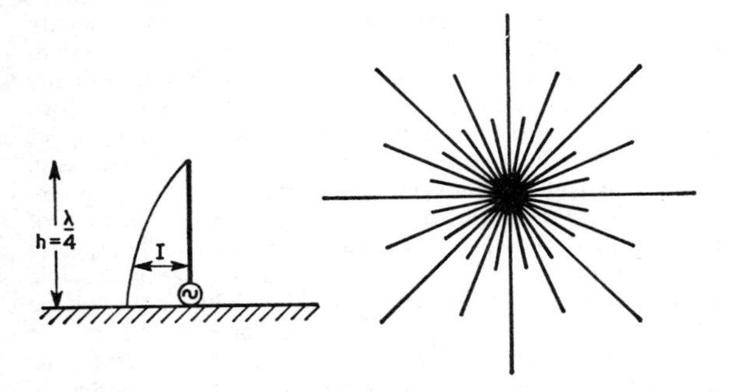

Fig 12.84. Typical radial earth system for a short vertical aerial. The earth currents are concentrated near the base.

coverage of the whole band, so that access for tuning may be essential. It has been pointed out that a particular virtue of the 3·5MHz band lies in its suitability for short and medium distance (high-angle) sky-wave communication for which horizontal polarization tends to be more suitable, but in both cases there is some value in being able to select either polarization at will, and it is worth noting that the horizontal dipoles described below can frequently also be used by connecting the feeder wires in parallel to act as a vertical radiator, with the dipole providing top loading or acting as an "inverted ground plane". This mode will usually be found better for dx except in the case of hilltop or hillside sites. Though not generally ideal for ground-wave propagation at the lower frequencies, horizontal polarization may give better ground-wave signals in those cases where the aerial at the other end is also horizontal; also, being a balanced load such aerials do not require an earth return, and even for ground wave communication they offer a better compromise in those cases where the earth system is very poor.

Finally, a word of warning about ground connections. It is sometimes suggested both in print and over the air that a rising water pipe is a good earth. This is not always true in practice since the amount of pipe in contact with the earth *in the vicinity of the aerial* may not be large, and the contact resistance will be indefinite because of corrosion deposits on the outer of the pipe. There will probably be indifferent contact between lengths of pipe at the screwed unions because of the sealing compounds used, and in any case modern practice is often to use plastic tubing which will negate the whole exercise. There is an obvious temptation in some cases to use the earth connection of the electricity supply for lf aerial systems but as previously noted *this may inject rf voltages into the supply and create a dangerous situation.* It may also introduce noise into the receiver or be responsible for spreading interference to nearby television receivers.

It is a mistake to bring a long "earth" lead into the radio room, which is often some distance from the point of connection to true earth, being in the limit on the upper floors of a building. The result of this is that the long earth lead will necessarily radiate since it carries the aerial feed current, and in consequence of the lead impedances the equipment in the shack will be up-in-the-air to rf with many consequent problems of filtering and feedback. A better arrangement is to install the atu in a box at ground level immediately adjacent to the earth mat, and connect it back to the shack with a low impedance coaxial line matched into the atu. This will not only isolate the shack from the radiating part of the aerial system, but will permit the vital vertical section of the aerial to be installed clear of obstructions which would otherwise degrade its performance.

Horizontal Dipoles

In an earlier section it was explained that the frequent need for high-angle propagation experienced on the lf bands is best met by the use of a horizontally polarized aerial. The appropriate lengths for half-wave dipoles for the 1·8MHz and 3·5MHz bands are given in **Fig 12.85.**

The current distribution along the dipole is roughly sinusoidal and is concentrated in the middle with little or no radiation from the ends of the wire. The aerial should therefore be supported in a such a way as to keep the centre

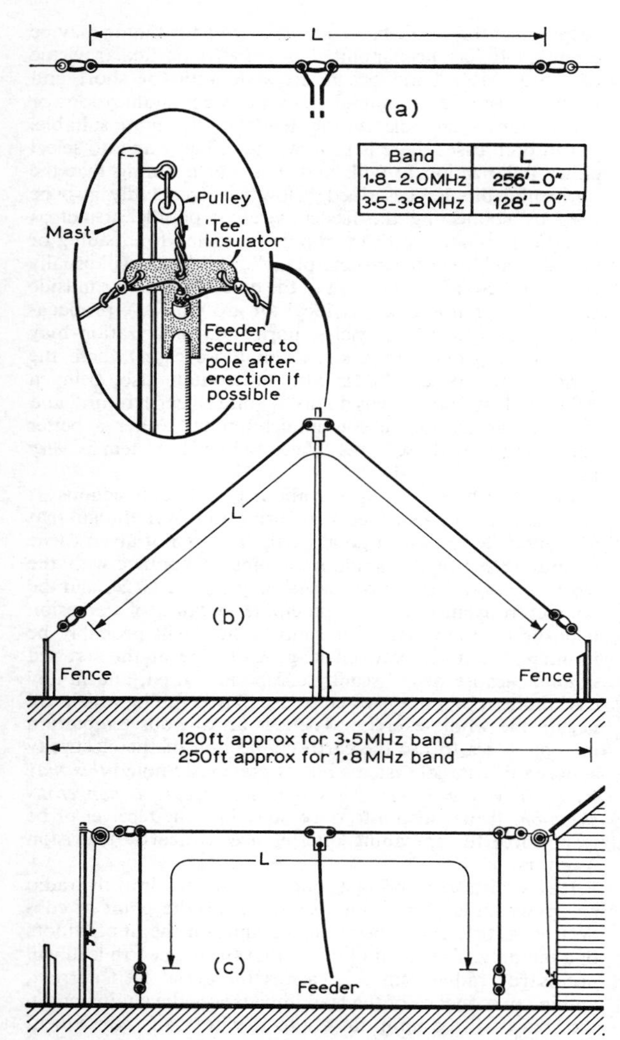

Band	L
1·8-2·0MHz	256'-0"
3·5-3·8MHz	128'-0"

Fig 12.85. Dipole horizontal aerials for 1·8 and 3·5MHz. The current is low at the ends and contributes little to the radiated signal. If alternative use as a vertical radiator is required, the feeder must be brought down well away from the pole which should be non-metallic.

region as high as possible, the ends being allowed to droop, or to hang down, depending upon the available space. Two possible arrangements for restricted gardens are shown in Fig 12.85. Shortened dipoles centrally loaded with a coil or tuning stub may also be used, with appropriate matching, their radiation resistance being given in Fig. 12.129.

The radiation resistance of a half-wave dipole tends to be a rather uncertain quantity when the aerial is less than a quarter-wave above ground, but is otherwise a function of height as shown in **Fig 12.86.** At low heights the resistance may rise as a result of excessive ground losses as suggested by the dotted line; this line is somewhat conjectural, being dependent on earth constants, and is intended only to serve as a warning against reliance on the solid curves for low heights. These curves which are classical, and appear in

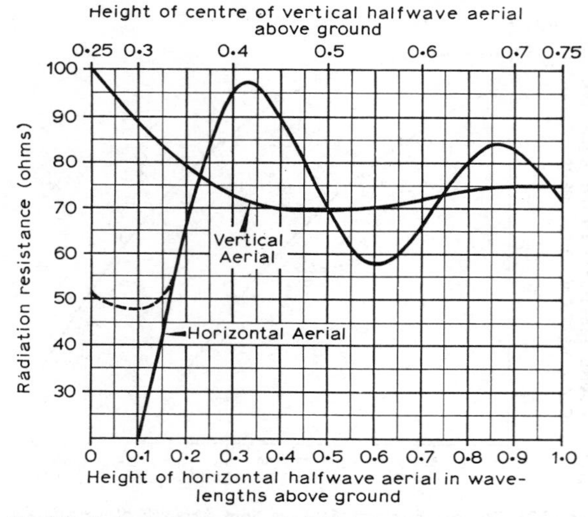

Fig 12.86. Radiation resistance of λ/2 horizontal and vertical dipoles as a function of height above a perfect earth. Dotted curve suggests likely behaviour of horizontal aerials over actual earth.

nearly all handbooks, have given rise to misunderstanding as they relate to an unreal situation namely a perfect earth, and this proviso is not always stated. In the case of the perfect earth and low heights, halving the height divides the radiation resistance by four so that the current is doubled and the low angle field remains constant, but as most amateurs are only too well aware, horizontal aerials, even if they are at a height of only 0·3λ say, are much more likely to bring in the dx than when they are lying on the ground! This is consistent with the fact that although the ground reflection coefficient for horizontal polarization approaches unity for *low angle* propagation, it drops to 0·6–0·8 for vertical incidence, suggesting appreciable absorption. Although the radiation resistance is shown as levelling off at 50Ω as height is reduced, both higher and lower values have been measured.

The W3EDP Aerial

Another approach to this problem, **Fig 12.87** avoids the necessity for a direct earth connection and is known by the callsign of its originator, W3EDP. In this design a short counterpoise wire is used in place of an earth, the aerial itself being suitably shortened to compensate for this addition.

A specific design which has been worked out experimentally for multiband use employs an aerial cut to a length of approximately 84ft, providing a simple and effective end-fed installation in cases where this length can be conveniently accommodated as a clear run from the radio room window. It is particularly useful when the radio room is on an upper floor and would involve a long and awkward earth lead. The counterpoise will vary from zero to 17ft in length, according to the frequency in use, and can be dropped out of the station window, or even accommodated indoors, provided that the far end is well insulated and as far as possible clear of earthed obstructions, since this will be a high voltage point during use. Ideally the counterpoise should run at right angles to the aerial, which may be bent if necessary, or can be sloping. An aerial of this type may be expected to radiate harmonics if these are present in the transmitter at

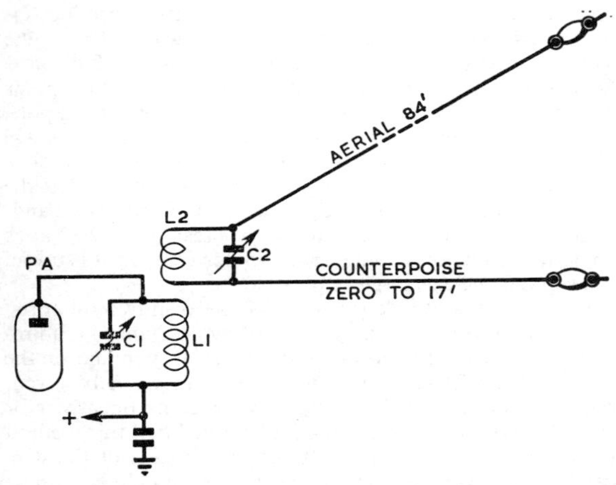

Fig 12.87. The W3EDP end-fed aerial using a counterpoise wire instead of an earth connection.

excessive strength, and in the form illustrated it is not compatible with modern practice. There is however no basic reason why it should not be link-coupled as shown in Fig 12.75(b) for an earthed system, and suitable harmonic filters can then be used in the coupling line. This aerial is of course subject to the warnings given earlier regarding the dangers of bringing long wire aerials into the shack but rf feedback and harmonic problems are generally much less serious on the lower frequencies. There is scope for much experiment with alternative lengths of aerial and counterpoise and in general it is much easier to achieve efficient radiation with systems of this type than those relying on earth connections.

HIGH FREQUENCY AERIALS

There is a wide choice of aerials for the higher frequency bands ranging from simple horizontal or vertical wire types to various kinds of small fixed or rotary beams and, for the fortunate few with plenty of space, large arrays which enable high gains to be obtained in one or more fixed directions. The choice depends on individual circumstances and preferences including the desire for single or multi-band operation, and the space available.

Simple Wire Aerials

Generally speaking almost any length of wire can be tuned to work as an aerial on any band although as we have seen it is difficult to make short lengths radiate efficiently on the lower frequencies. Longer wires on the other hand tend to have rather complex radiation patterns on the higher frequencies but, in the best directions, their performance for lengths up to 2λ is comparable with that of a dipole at the same mean height and average position though it may be degraded by, for example, the proximity of high-voltage points to various structures. The simplest forms of aerial include horizontal dipoles and loops, and these are capable of highly efficient multiband radiation if coverage is restricted to the three highest-frequency dx bands and they are cut to resonate in the region of 14–21MHz. It is not unduly difficult to work out matched feed systems for two bands, but efficient coverage on all three bands can be difficult to achieve if the radiator requires loading to make it resonate on 14MHz and the feeder length is considerable; otherwise resonant feeders can be used and the price paid for multiband operation is in terms of bandwidth, leading to the inconvenience of frequent retuning, rather than poor efficiency. When a simple wire aerial is long enough to cover lower .requency bands as well there is then a break-up of the radiation pattern at higher frequencies (p12.59) and the direction of maximum radiation on one band coincides with a null, or poor efficiency, on another, and if it is not desired to use resonant open-wire feeders with the possible disadvantages of high rf voltages around the shack, some more or less lossy compromise may have to be accepted.

Ideally all radiators, however simple, should be suspended as high as possible, and in the clear, using heavy gauge wire and good insulation but usually there are practical constraints such as the need to use existing supports, eg trees or chimneys. The aerial may have to be more or less invisible, or the span available may be much less than a half-wavelength at the frequency in use. Fortunately many "liberties" can be taken without compromising performance, but a lot of conflicting ideas are in circulation as to what is or is not permissible. The following guide-lines are aimed at providing optimum dx performance and, though inspired largely by theory in the first instance, have been for the most part confirmed by practical trials. Use has also been made of experimental results which do not lend themselves readily to theoretical treatment.

Height should be as great as possible unless the ground is sloping and the aerial must obviously be orientated for the wanted direction. The length of the radiator must not be less than about two-thirds the normal resonant dimensions if loading is by inductance at a current loop, or one-third if end-loading by capacitance is used. When electrically short radiators are used, greatly increased care must be taken to keep voltage maxima (eg the ends of radiators) away from lossy insulating material eg bricks, trees etc, thicker conductor must be used, and extreme precautions to avoid high-resistance joints at or near a current loop are advisable. As a rough guide it can be taken that the above figures for permissible shortening are valid for wire sizes of 12swg or thicker and spacings of not less than 3–4ft between the ends of the radiator and objects such as tree branches. There must be no other conductor within several feet of, and parallel to, the radiator. Regardless of aerial dimensions no other resonant conductor should be within a distance of about 0·7λ or if it is in front, similarly oriented and in the direct line of propagation, at least 2λ and possibly 3λ; there appears however to be no available data on the effect of metal supporting poles, which should be regarded with obvious suspicion unless the aerial is symmetrical in the sense that fields due to any one part of it in the vicinity of the mast are cancelled out by an equal and opposite field. The higher the radiation resistance the smaller the permissible wire gauge and the less important the proximity of trees etc. A quad loop (or even beam) can be "buried" in tree branches, and it has been found that twigs even if wet can make contact with a dipole at least half-way out from the centre without producing noticeable deterioration as judged by signal reports involving instantaneous comparisons between two aerials. Vertical radiators are however liable to be adversely affected by the close proximity of tree trunks, supports etc. Dipoles and full-size quad or delta loops may be constructed from very thin wire (as evidenced by Fig 12.5) provided the mechanical requirements are met. Insulation between points of high voltage,

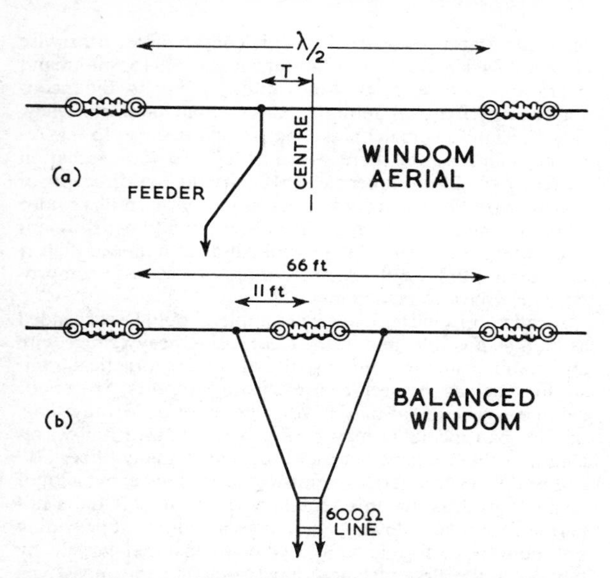

Fig 12.88. (a) The Windom single wire fed aerial. (b) Balanced version which minimizes feeder radiation and also works on multiple frequencies.

eg the centre of a dipole of length $\frac{3}{4}$ to $\frac{5}{4}$ wavelengths, or voltage antinodes on a resonant feeder must be good. Polythene cord or polystyrene rod are good insulators and fully adequate for all situations, but bakelite or perspex are not.

Considerable departure from "straight line" shape is usually admissible *provided* feeder balance is maintained and the inverted V dipole (Fig 12.85(b)) is a particularly important example, being probably the easiest of all aerials to support in an average situation, since it needs only one main support (though if only a short span is available additional poles of lower height will be required) and the two ends can be allowed to "droop" so that very little tension need be applied to the wire. It provides an excellent aerial for portable work since the centre can be supported from a single tree branch or by a "clothes prop" constructed from bamboo garden canes, the ends being strung out to points as far away as possible using light polythene cord. 24swg or thicker wire is satisfactory for this application.

It is commonly thought that the inverted V dipole is a compromise since it must produce vertically polarized radiation in an endwise direction, but although quite a strong ground-wave field can be measured in this direction it represents a negligible amount of wasted energy and the radiation pattern is hardly affected for apex angles down to 90°. This may be demonstrated by resolving each arm of the V into a pair of point sources of radiation as explained on p12.12, one horizontal, the other vertical, noting the resemblance of the two vertical sources to an "8JK" system, Fig 12.14. Intuitively one might expect the "centres of gravity" of the current distributions to be about one third of the way out from centre, but for greater precision it is not difficult to do a rough graphical integration to take account of the phase differences between the endwise fields produced by different portions of the current distribution. This gives the result that for currents I in the horizontal sources the 8JK system comprises currents $0{\cdot}97I$ with a physical separation of 48°, and the field strength produced is 0·35 times that of the normal mode. Allowing 4dB for the gain in the 8JK mode, the power wasted by it is 0·2dB only. Except in this inefficient endwise direction, the polar diagram is almost identical with that of a horizontal dipole, and the assembly of two such elements into a beam will almost completely cancel the "8JK" radiation. At an apex angle of 90° the radiation resistance is halved so that the minimum satisfactory wire diameter is doubled. The bandwidth however is almost halved, whereas for 120° apex angle the reduction is only about 25 per cent, so it is useful to keep the angle as large as possible.

If a horizontal dipole is suspended between two supports, the required wire gauge is determined by mechanical requirements and therefore dependent on the total span and on the weight of the feeder so that no general rules can be given, though 16swg is usually satisfactory assuming open-wire or other lightweight feeder. It should be noted that this method of construction requires two supports, the ends of the wire carry high rf voltages and should not be too close to these points, and there will be some sag in the centre where height is most needed. For comparison the inverted-V dipole loses effective height to the extent of about one third of the total "droop", ie 4ft and 2ft 9in respectively for apex angles of 90° and 120° at 14MHz. Approximate resonant lengths for straight horizontal dipoles centred on the higher frequency bands are as follows:

Band (MHz)	Length
7	66ft 0in
14	33ft 0in
21	22ft 0in
28	16ft 0in

Resonant lengths are influenced by various proximity effects and also by bends in the wire. Fig 12.85 indicates considerable shortening in the case of an inverted-V but the opposite effect has been observed, and in view of conflict between different sources of information the length should be determined experimentally, eg starting with an extra 5 per cent the ends of the dipole may be folded back on themselves symmetrically until minimum swr is obtained in the feeder.

Alternative methods of feeding are described on p12.36. To avoid Marconi effect (p12.49) it is usually reckoned that the feeder should be allowed to hang vertically from the feed point connection for at least a length equal to a quarter-wave in free space but as will be appreciated from earlier discussion, p12.32, the incidence of this effect depends not only on the disposition of the feeder but also on its length and the connections at the transmitting end.

The "Windom" and VS1AA Aerials

The principles of single wire fed (Windom) aerials, **Fig 12.88**, and the reasons for their obsolescence have been explained on p12.17. These objections can be overcome by balancing one Windom against another as in Fig 12.88(b). It is evident that the Windom principle is applicable to any multiple half-wave radiator and the feeder tap positions shown in **Table 12.3** for the fundamental frequency are quite close to being suitable for even harmonics. They apply equally to the balanced version.

The VS1AA version of the Windom achieves multiband operation by exploiting the principle that if the tap is placed one-third of the way along a standing wave current loop on the lowest frequency it will always be in the same position

TABLE 12.3
Single wire feeders

Band and aerial	Radiator length L (ft)	Tap distance T (ft)
3·5–14MHz half-wave	470/f MHz	66/f MHz
21–28MHz half-wave	460/f MHz	66/f MHz
14–28MHz full-wave	960/f MHz	170/f MHz

The values given are recommended for aerials constructed from 14 or 16swg wire.

on a current loop on all even harmonic frequencies. This system is most suitable for very long wires, the tap being about 22ft 6in from the end for all frequencies down to 7MHz, and even works quite well with long aerials on 21MHz (where the "one-third" rule breaks down), because the long wire places a substantial load on the end of the feeder. The essential difference between the VS1AA and the Windom lies in the use of very long wires whereby adjustment is less critical due to heavier damping by the radiation resistance, and this tends to overcome the difficulty that harmonic resonances are not exact multiples of the fundamental frequency.

The Double Windom

The VS1AA aerial also has the disadvantage that there is considerable radiation from the feeder. This can be overcome by making it into the balanced system shown in Fig 12.88(b) so that the feeder radiation is cancelled. This aerial can alternatively be regarded as a development of the centre-fed aerial with tuned feeders, the spacing of the taps helping to reduce the vswr on the feeder. The length of the arms of the V section of the line should be at least equal to the separation between the taps, preferably more, and the overall length from one end of the aerial into the station should be chosen as for the centre-fed aerial in order to simplify the tuning arrangements. Some difficulty may be experienced on 21MHz as the feeder system is effectively a resonant line and this, depending on the length, may influence the design requirements for the atu as discussed on p.12.45.

Asymmetrical Twin Feed

It will be realized that a long-wire aerial, ie one which is two or more half-wavelengths long, can be fed with low-impedance twin line at any current-maximum position, such as the point a quarter-wave from one end (**Fig 12.89**). In this case 70–80Ω twin line can be used and matches the aerial well enough, though the impedance at such a point in the aerial is somewhat greater than that of a single half-wave. This can be done with a 7MHz aerial, and at 21MHz the feeder is again at a current maximum position. Other bands have an even harmonic relationship, so that the feeder is badly matched.

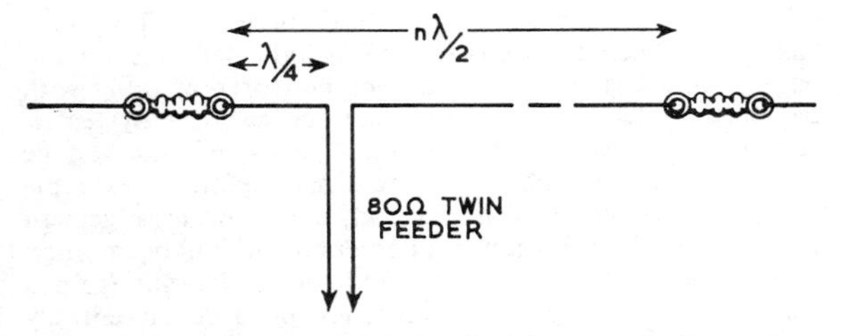

Fig 12.89. Asymmetrical twin line feed for a harmonic aerial.

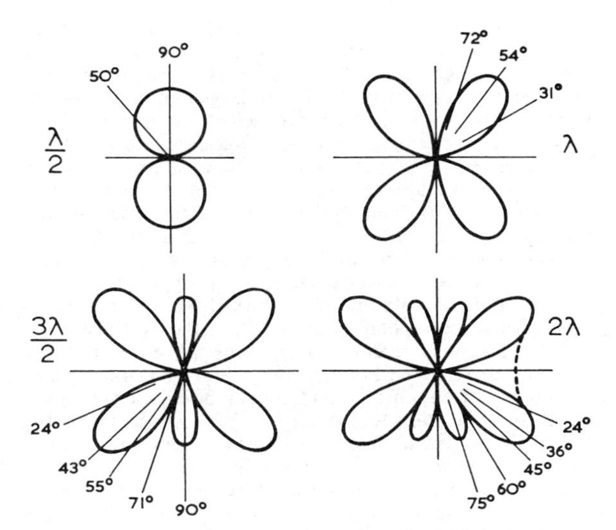

Fig 12.90. Theoretical polar diagrams for wire aerials up to 2λ in length. The angles of the main lobes and crevasses are shown, also the angles at which the loss is 3dB in the main lobe. The lobes should be visualized as cones about the wire. When the aerials are horizontal, radiation off the end can take place at useful wave angles, especially from very long wires, eg the broken line on the aerial. Details of long wire aerials are given in Table 12.4.

Long Wire Horizontal Aerials

The basic patterns of aerials of various harmonic lengths are shown in **Fig 12.90** and further information is given in **Table 12.4**. It will be seen that there are two lobes in the diagram for each half-wavelength of wire, and that those nearest the end-on direction are always strongest. It should be remembered that these diagrams are sections, and the reader should try to visualize the lobes as sections of cones about the aerial. The reason for this concentration of energy in the nearly-end on direction is immediately obvious if the aerial is visualized as a string of point sources alternating in sign. In the direction of the wire there is of course no radiation whereas at a small angle to the wire the radiation from all the sources adds up in phase. Viewed from a larger angle to the wire, for every current loop of one sign there is

TABLE 12.4
Properties of long wire radiators

Length (λ)	Angle of main lobe to wire	Gain of main lobe over half-wave dipole (dB)	Radiation resistance (Ω)
1	54° (90°)	0·4	90
1½	42°	1·0	100
2	36° (58°)	1·5	110
2½	33°	1·8	115
3	30° (46°)	2·3	120
4	26° (39°)	3·3	130
5	22° (35°)	4·2	140
6	20° (31°)	5·0	147
8	18° (26°)	6·4	153
10	16° (23°)	7·4	160

The number of complete conical lobes (see Fig 12.91) is equal to the number of half-waves in the aerial. The main lobe is the one nearest to the direction of the wire, and the figures in this table give its direction and gain. When a multiple full-wave aerial is centre-fed the pattern is like that of one half, but with more gain in the main lobe. The angles in brackets correspond to this case. When the aerial is terminated, or self-terminating, the radiation resistance is 30 to 50 per cent greater, and the main lobe slightly nearer to the wire.

another of opposite sign and almost at the same distance, so that radiation tends to be cancelled. At a small angle to the wire each source is of course a very weak radiator and although this is compensated by the large number of sources it follows that for a given gain the length of wire required is very much greater than if all the sources are assembled in a fully cooperative manner as they are with other types of array. The same applies to a wire terminated in a suitable resistance at the far end so that there are no reflections and no standing-wave pattern, except that in this case there is a travelling-wave and no radiation takes place in the backwards direction because radiation from any point is cancelled by radiation of opposite sign which started out half a cycle earlier but has half a wavelength further to travel.

Polar diagrams represent effectively the azimuth directivity of these aerials, but their general assessment must include the effect of height, and this can be done by considering them in relation to the vertical plane diagrams of Fig 12.22. There is, however, one very important feature: namely the existence of a major lobe of radiation at moderate angles of elevation in the endwise direction; thus the lobe at 36° to the wire in the 2λ pattern applies equally to the horizontal and vertical patterns and at this angle, depending on height, the ground reflection may reinforce the sky wave or possibly produce cancellation causing the lobe to split into two.

For a dipole, assuming comparable ground reflection in both cases, the end signal will be 6dB (one S point) less than the broadside at 30° elevation, but this effect is more marked with longer radiators, where the end pattern almost fills in for elevations of 15–30°, as shown for the 2λ case in Fig 12.90. In the case of a three wavelength long aerial (which would be about 100ft for 28MHz) the end radiation at a 15° wave-angle may be much greater than the broadside signal from a dipole at the same height. It should be noted however that in going from broadside to high-angle end radiation the sign of the ground reflection coefficient is reversed, the latter radiation being mainly of vertical polarization and the angle of maximum radiation being well above the Brewster angle except for very long wires. This means, for example, that for a height of λ/2 the broadside radiation at 30° angle of elevation will gain 6dB from the ground reflection whereas the end radiation at the same angle will be mainly cancelled.

The long wire aerial is the simplest form of directive aerial and is quite popular with those having sufficient space because it is at the same time a useful all-round radiator. Although the notch in the pattern in the broadside directions is quite noticeable, in practice for the even-multiple aerials the effect of the other crevasses is not so marked, chiefly because their direction varies with wave angle, so there is usually a "way through" for signals. It has the advantage of providing some discrimination against very high angle (short skip) signals but it should be noted that gain is appreciable only with very long wires and in the direction of their main lobes which are very sharp.

Table 12.4 gives important properties of wires up to 10 wavelengths long. It will be seen that a 10λ aerial has a main lobe at only 16° angle to the wire, with a gain of over 7dB. Remembering that these lobes exist all round the aerial, eg in the vertical as well as the horizontal plane, it will be realized that in the direction of the wire there is a vertical polar diagram with a lobe maximum at 16° elevation. This angle of radiation may be quite close to the Brewster angle in which case the reflection coefficient is small and, taking

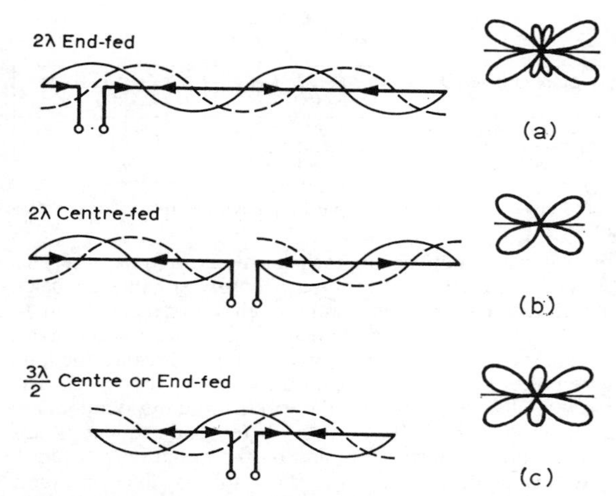

Fig 12.91. Illustrating the difference between feeding at a current maximum or a voltage maximum, and how centre-feed affects even- and odd-multiple λ/2 aerials differently. When the feed enters a current loop the pattern is the same as with end-feed, but when a voltage loop is entered, the pattern more nearly resembles that of one half only of the aerial.

into account the gain and the vertical polarization, a long wire aerial is likely to give more radiation "off its end" than a dipole at the same height. (It is this effect which sometimes causes confusion over the use of the term *end-fire* which refers to a particular type of array and not to the direction of a main lobe with respect to a wire.) It should however be noted that unless a wire is long enough to have appreciable gain it cannot differ much in main lobe performance from a dipole having the same height and polarization; this is because as we have seen on p12.13 the effect of the ground is accounted for by an array factor which, though dependent on ground constants, polarization, aerial height and angle of elevation, is independent of the number and disposition of the point sources of radiation comprising the aerial.

The radiation resistance (R_D) figures in Table 12.4 are the resistance at any one current maximum, say λ/4 from the end, and are representative free space values, fluctuating somewhat with height. However it is highly improbable that a long wire will behave in the same way as a dipole (Fig 12.86) since each current loop "sees" many image loops, some of one polarity and some of opposite polarity, instead of seeing only one loop of opposite sign. From Table 12.4 it will be seen that the resistance increases steadily as the size of the radiating system increases; this is a general rule for all large directive aerial systems.

Effect of Feed Position

The general subject of how to feed energy to a long wire has already been discussed and it was indicated that it could be fed with a single wire near one end or preferably with balanced feeders in the centre. In each case the system is capable of multi-band operation. A balanced line can be connected at any voltage or current maximum, for example a quarter-wavelength from one end, but in this arrangement the aerial works only on its fundamental and odd harmonics.

It should be noted that shifting the feed point from a current to a voltage maximum may produce an entirely different radiation pattern. The long wires described above

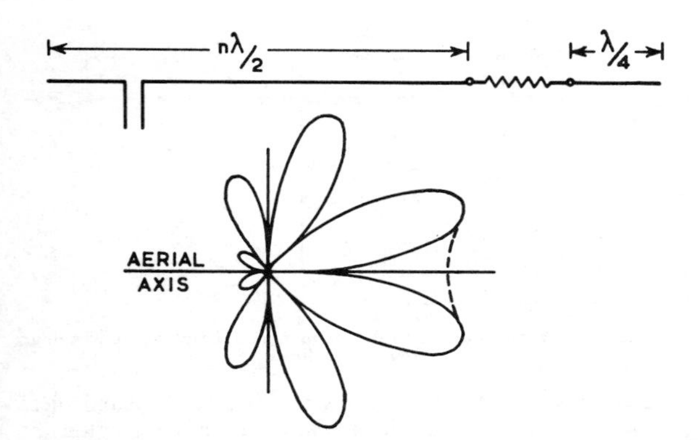

Fig 12.92. A 2λ terminated aerial and its radiation pattern, showing unidirectional effect. The end-to-end ratio depends on several factors but may be 3 to 5dB for a full-wave unterminated aerial and more than 10dB for a terminated aerial. The 500Ω terminating resistor is earthed to a λ/4 artificial earth. A multiple "earth" fan may be used to cover two or three frequency bands. These aerials do not need critical length adjustment.

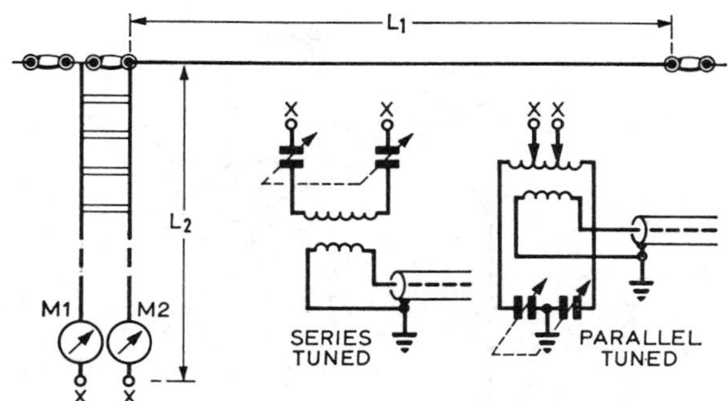

Aerial length L_1	Feeder length L_2	Band MHz	Tuned
135′ – 0″	45′ – 0″	3·5, 7, 14, 28	Series
135′ – 0″	45′ – 0″	21	Parallel
67′ – 0″	45′ – 0″	7, 14, 28	Series
67′ – 0″	45′ – 0″	21	Parallel

Fig 12.93. The end-fed Zepp aerial employs tuned feeders and an aerial coupler to achieve correct operation on each band. Currents in the meters M1 and M2 should be equal within ±10 per cent for balanced operation on the line.

were continuous with alternate positive and negative current loops along them. This is the situation with end-feed, single-wire feed, and feeding at any current loop, but if the aerial is fed with balanced line (**Fig 12.91**) at a voltage maximum an extra phase reversal is introduced at this point, as can be seen by sketching the standing waves. Thus a centre-fed three wavelength aerial becomes an array of two 3λ/2 aerials and has a pattern which is basically that of the 3λ/2 aerial multiplied by an array factor corresponding to the 3λ/2 spacing between the centre of the two halves. Since this particular array factor is very nearly the same as the pattern of each half, the final diagram is very like that of a single 3λ/2 aerial, but with emphasis on the broadside lobe. By comparison the 3λ end-fed aerial has a null in the broadside direction.

It should be noted that any asymmetrical feed arrangement will tend to be imperfectly balanced. The presence of Marconi effect should therefore be suspected and precautions taken as discussed on p12.32.

In practice a long wire fed at one end tends to radiate best from the opposite end, the pattern of a 2λ aerial tending to become like that of **Fig 12.92** (these patterns are ideal patterns). This tendency is greater the longer the wire, and occurs because the lobes to the right (in Fig 12.92) are due to the radiation from the forward wave along the wire, while those toward the left can only be due to the wave reflected from the far end. Because of loss by radiation, the reflected wave is weaker than the forward wave; thus a long wire tends to behave like a lossy transmission line. The one-way effect can be enhanced by joining the far end of the aerial to a quarter-wave "artificial earth" wire through a 500Ω (matching) resistor. This resistor must be able to absorb about 25 per cent of the transmitter power if the aerial is only 2λ long, but only about 10 per cent if it is very much longer. The power lost in this resistor is not wasted because it would all have gone in the opposite direction. Such an aerial is of course only correctly terminated on one band, though a fan of earth wires can be used, one for each band. It can be fed (for one band use) by means of an 80Ω feeder one quarter-wavelength from the free end as in Fig 12.92. For multi-band working a second fan at the free end may be used. It should be noted that although the unidirectional pattern does not increase the power radiated in the forward direction, it reduces noise and interference picked up from the reverse direction and thus improves reception. It also of course reduces the likelihood of causing interference to other stations during transmission.

The "Zepp" Aerial

A simple wire aerial can be made to work on a number of harmonically related bands by feeding it at one end. This will always present a high impedance and can be connected to open wire tuned lines. The operation of such lines is explained in the section dealing with transmission lines. Such an aerial is known as a Zeppelin or "Zepp" and is illustrated in its traditional form by **Fig 12.93**, which gives suitable dimensions for both the aerial and feeder. Despite its time honoured status this arrangement as illustrated is very uncertain in its behaviour [8]. Put bluntly it usually *does not work*, and the reason is the same one that we have met before in a different guise. In going from a balanced to an unbalanced system *or vice versa* a balun is essential. Without one, there is no guarantee that the aerial will work better than a random length of wire nor, to be fair, any certainty that it will not work, but the Zepp feed has been found particularly uncooperative in this regard. A suitable balun consists of a λ/4 short circuited stub, **Fig 12.94**. The end-fed aerial may now be seen clearly for what it is, namely a high impedance to ground connected across one half only of a resonant system which experiences relatively little damping so that the swr is extremely high. The fact that it does not appear so with the conventional Zepp feed is a measure of the inefficiency of the system and, as reference [8] states, simple connection of the aerial to one side of the line will not work—it is necessary to add a "transformer winding" in the form of a stub to "tell" the line it is balanced. The need to employ this stub which may not be readily accessible makes it more difficult to exploit efficiently the multiband properties inherent in this system but Fig 12.93 includes some suggested feeder lengths and tuning arrangements. These will not be affected by the addition of the stub.

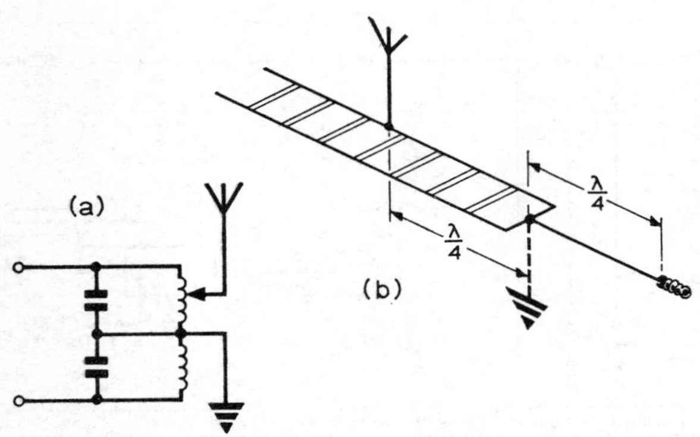

Fig 12.94. Unbalanced aerial—balanced line.

A length of 45ft will provide an intermediate impedance and work with close or medium spaced taps or series tuning on all bands except 21MHz. At the latter frequency a high impedance will be presented by the feeder. Open wire 600Ω line is essential in view of the high vswr. The aerial should be made just over one half-wavelength long at the lowest frequency, say 67ft for frequencies from 7MHz upwards (or 135ft if 3·5MHz is to be included) and adjusted in length until the feeder currents are equal at the *lowest* frequency. If the half-wavelength top cannot be erected for the lowe·' frequency band the free wire of the feeder (the one not connected directly to the aerial) may be disconnected at the transmitter end and the remaining wire plus the aerial used as an end-fed arrangement.

For the parallel coupler, the capacitor should have a maximum value of at least 100pF for operation down to 3·5MHz, and 25pF if the lowest band is 14MHz. The value of the corresponding capacitors in the series coupler is approximately one-half of these figures. The inductance value required can then be determined from a suitable abac relating frequency, *L* and *C*. The link coupling coil should have approximately one-tenth the turns of the main coil.

Alternatively the feeder may be adjusted to resonance and matched into a non-resonant line in the usual way. The length or termination of the balun stub, Fig 12.94, must be varied according to the band in use so that it presents a high impedance at the aerial.

The Extended Double Zepp

This aerial system is simple to erect and adjust and gives a gain of approximately 3dB over a halfwave dipole, at the price of a rather narrow azimuthal beam width.

The horizontal polar diagram, like that of a full-wave dipole, has main lobes which are at 90° to the run of the wire. In addition there are four minor lobes at 30° to the wire. At twice the frequency or more the pattern becomes similar to that of one and two wavelength aerials.

The system is a type of collinear aerial (see p12.64). The layout is shown diagrammatically in **Fig 12.95** and comprises the two lengths of wire each 0·64λ long, fed in the centre with an open wire line. Extension improves the swr which is reduced to about 6.

The table of Fig 12.95 gives the design length for the centre of different bands.

A feeder length of 45ft is again a good compromise and the coupling arrangements described for the ordinary Zepp

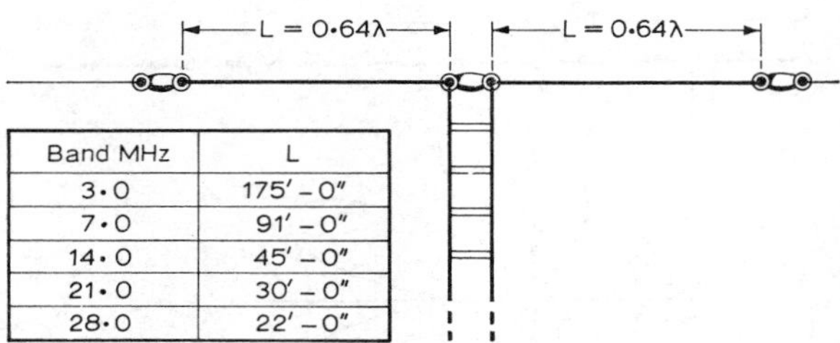

Band MHz	L
3·0	175′ – 0″
7·0	91′ – 0″
14·0	45′ – 0″
21·0	30′ – 0″
28·0	22′ – 0″

Fig 12.95. The extended double Zepp aerial employs centre feed with tuned lines.

are equally suitable, but as this is a balanced system there is no need for the additional stub. For other lengths the form of coupling (series or parallel) can be found from Fig 12.73. In this case the length L should include the half-top of the aerial.

300 Ohm Ribbon Aerials

A simple multiband aerial is shown in **Fig 12.96** and can be fed with 75Ω flat twin or 75Ω coaxial cable. L1 acts as a halfwave dipole on 7MHz and as three halfwaves on 21MHz; L2 functions as a halfwave dipole on 14MHz.

The lengths of L1 and L2 can be calculated from the formula 468/*f*(MHz), the answer being in feet. Dimensions for L1 and L2 are as follows when cut for mid-band operation:

L1 Length	Band	L2 Length	Band
246 ft	1·8MHz	128ft	3·5MHz
128ft	3·5MHz	66ft 5in	7MHz
66ft 5in	7 & 21MHz	33ft	14MHz
33ft	14MHz	22ft	21MHz
22ft	21MHz	16ft 3in	28MHz

Any combination of L1 and L2 can be chosen to cover the particular bands required.

F7FE All Band Dipoles

A way of connecting and laying out four separate dipoles to cover 3·5, 7, 14 and 28MHz has been worked out by F7FE. It uses two supporting masts and is fed by a single 75Ω flat twin feeder or coaxial cable. The principle of operation is that each dipole presents a high value of reactive impedance to the feed line at frequencies other than those at which it is resonant so that, on each band, most of the current flows only in the correct dipole. If coaxial cable is used, a balun is required.

Only four dipoles are used because the one cut for 7MHz works well on 21MHz.

The layout is shown in **Fig 12.97** and is more or less self-explanatory; each dipole is cut to mid-band using the formula 468/*f*(MHz).

The centre insulating plate is made of polystyrene sheet and has six holes drilled in it. The top pair take the 28 and

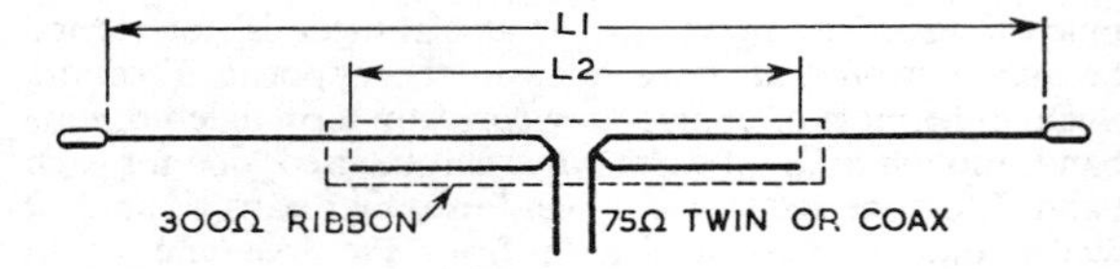

Fig 12.96. A two band 300Ω ribbon aerial. Dimensions are given in the text.

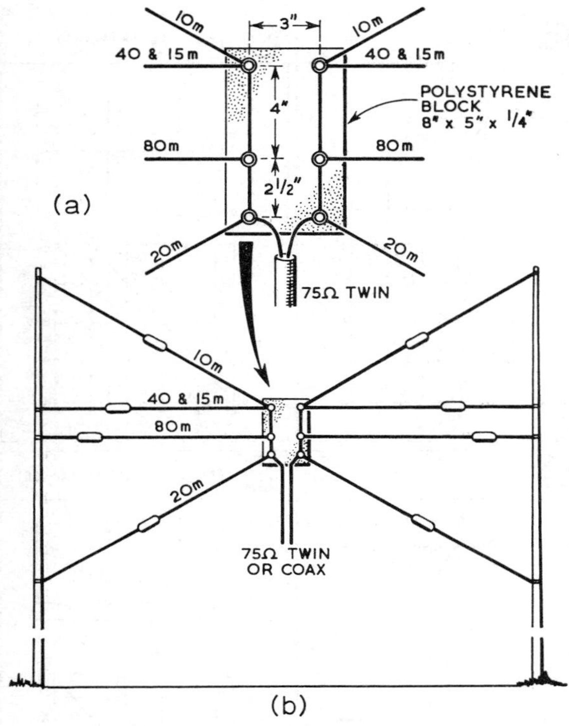

Fig 12.97. The F7FE all-band dipole aerial.

7MHz dipoles, the centre two the 3·5MHz dipole and the bottom two take the 14MHz dipole.

A short length of wire connects all the left-hand quarter-waves in a similar manner. The 75Ω feed is connected to the bottom end of these pieces of wire. Many variants of this system have been described and there is no reason why it should not be applied to an inverted-V dipole system.

BEAM AERIALS

Directive or beam aerials may be divided into two main classes: physically small beams using closely-spaced elements and wide-spaced beams or long-wire systems. The small beams can usually, if desired, be built as rotary systems and are particularly suitable for general purpose amateur use, whereas the larger beams, by virtue of narrow beam widths and requirements for large areas of land, find their main application in commercial point-to-point circuits. The smaller types of rhombic, Vee, and multiple-dipole arrays nevertheless play a substantial role in amateur communication and provide outstanding performance when there is sufficient space available combined with a primary interest in a small number of fixed directions.

There are other ways of classifying beams, eg as collinear, broadside, end-fire (Fig 12.10) and long-wire, small rotary beams being a special case of the end-fire category although as explained on p12.67 there is a radical change in the mechanism of beam-formation from additive to subtractive as the spacing of an end-fire array is reduced. Each of these types can further be subdivided into *parasitic* and *driven* arrays.

In broadside and collinear arrays the elements are connected together by phasing lines, so that they are all in one phase, but in the end-fire aerials the elements may be all connected to give a progressive phase change along the array (driven arrays) or there may be one driven dipole together with a number of nearly-resonant free *parasitic* elements which modify the local field of the radiator so that a unidirectional end-fire pattern is produced (*parasitic arrays*). Parasitic reflectors can also be added to broadside and collinear arrays to produce a unidirectional beam instead of the fore-and-aft pattern of a single row of elements. Driven arrays in general use resonant elements interconnected by tuned lines and are usually single-band, though by using traps or tuned feeders multiband arrangements are possible. Arrangement of elements (broadside or collinear) at the same height and along a line at right angles to the direction of propagation produces a narrow azimuthal pattern, whereas if sufficient height is available vertical stacking can be used to reduce high-angle radiation without affecting directivity in the horizontal plane, though due to the reduction of mean height for a given mast height it may also reduce very low-angle radiation. On the other hand end-fire systems provide gain by virtue of reduced beam width in both horizontal and vertical planes. Simple arrays suitable for fixed-direction amateur use include the lazy-H described below and the two-element collinear array (which is in fact the same thing as a full-wave dipole), preferably with the addition of reflectors, directors or both. Two or more close-spaced beams, if widely spaced from each other, may also be connected together to form high gain arrays.

Long wire aerials as described on p12.59 can also be assembled into arrays such as rhombics and V-beams, which, despite the disadvantage of requiring a relatively large area of land for a given gain, have many advantages, being easier to erect and adjust besides operating into a matched line over a wide frequency range without readjustment. Coverage can include as many as four amateur bands, though the optimum performance is limited to a 2 : 1 frequency range. Their patterns are not so well defined as those of dipole arrays, a difference very noticeable in reception, and, like the end-fire arrays, their vertical and azimuthal patterns are determined simultaneously by the height and length. Long wire arrays do not normally use reflectors but can be made unidirectional by terminating them with matching resistances.

Collinear and long-wire arrays have the disadvantage of producing a rather narrow azimuthal pattern which restricts the geographical coverage, a disadvantage in amateur work where even a small country like New Zealand may cover nearly 90° of bearing, and over near-antipodal distances there is no guarantee that signals will follow a great-circle path. This disadvantage is however somewhat mitigated in the case of rhombic aerials by rather high side-lobe levels which often produce good signals in directions other than those of the main beam. This in turn is offset by the disadvantages that interference may be received from or caused in the directions of the sidelobes.

Broadside Arrays—General

Fig 12.98 gives a selection of simple broadside arrays together with gain and average radiation resistance figures. A variety of feed connections are shown, which are interchangeable as discussed below, and it is apparent how to

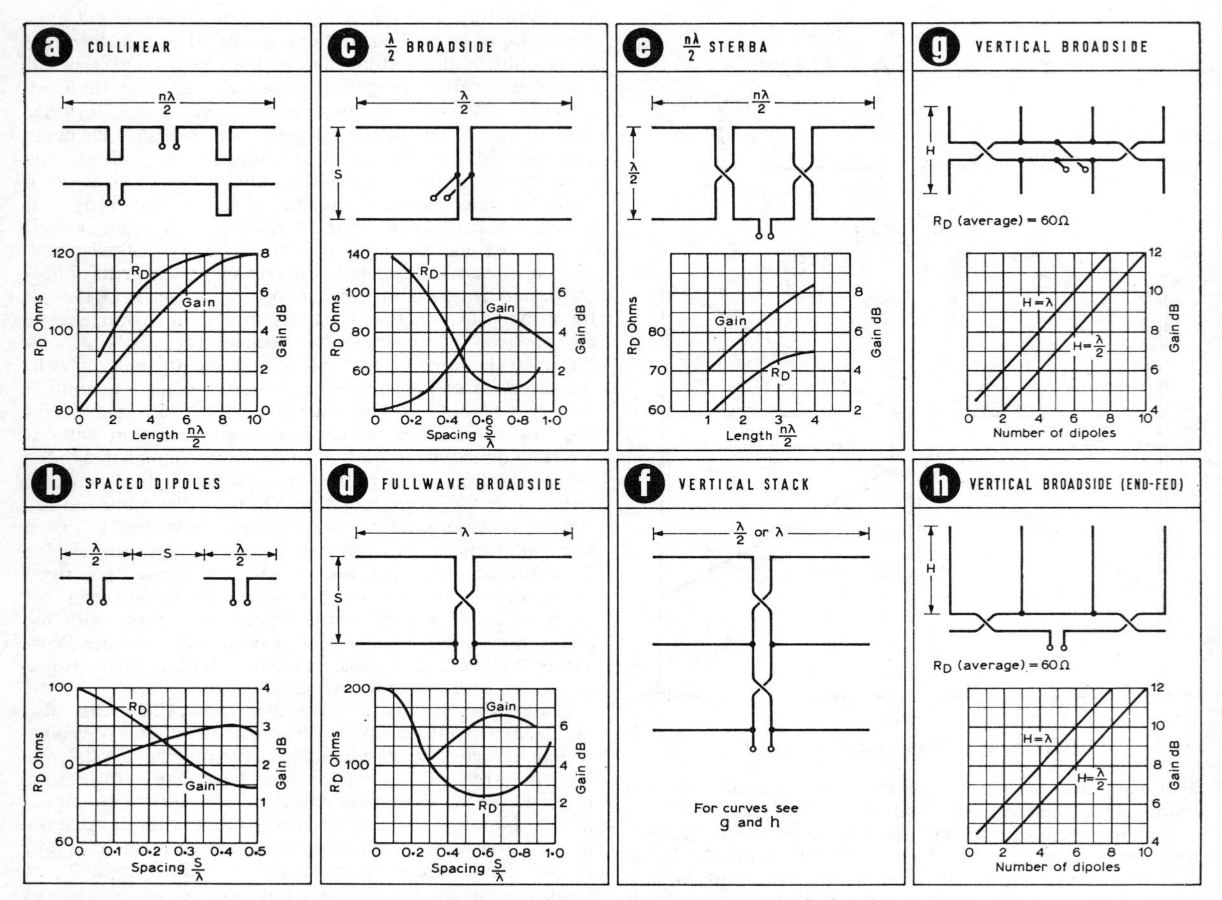

Fig 12.98. Four general types of broadside array. (a) Collinear arrays. (b) End-spaced dipoles. (c, d, e) Two-tier Sterba or Barrage arrays. (f) Pine tree or Koomans, stacked horizontal $\lambda/2$ or λ dipoles. (g, h) Vertically polarized broadside arrays. Gain figures are with reference to a free-space dipole, in terms of spacing or total length in half-wavelengths. Resistance figures are average over the array, and are added in series or parallel according to the feed arrangements, as described in the text. Various feed positions are shown, and details are also given in the text. The aerial in (c) can be arranged to give a broadside beam over a 2:1 frequency range, eg 14, 21 and 28MHz. In (g, h) two half-waves in phase ($H=\lambda$) are reckoned as "one dipole".

extend the arrays beyond the number of elements illustrated. For all these arrays the azimuth patterns can be estimated as a function of the length of the array from **Fig 12.99** and, in most cases, the broadside vertical patterns as a function of height from Fig 12.22.

In broadside arrays, the elements are all in phase and the interconnecting *phasing lines* or stubs must be adjusted to secure this condition. The spacing between elements and between the centre of the elements need not be one half-wave but can be varied up to about $\frac{3}{4}\lambda$, beyond which minor lobes in the pattern become too large to ignore. The choice of spacing is, however, in practice dependent on the type of phasing line used. The position of the feed-point depends on the input impedance required, or on requirements for multiband operation. A centre feed position should be used if possible, especially in long arrays, because power is being radiated as the currents travel along the array, and the more distant elements may not receive their proper share. Uneven power distribution can cause the beam to broaden and "squint".

Collinear Arrays

The collinear array is the simplest method of obtaining a sharp azimuthal pattern and is simply a row of half-wave radiators strung end-to-end. To bring all elements into phase it is necessary to provide a phase reversing stub between the high voltage ends of each pair, except where this position is occupied by the feeder.

The simplest form of collinear array is the centre-fed full-wave or *double Zepp* with a high impedance tuned line feed at the centre, and as we have seen (p12.62) this can be used on other bands; when the total length is only one half-wave the impedance is between 60 and 100Ω, but in the full-wave condition it is 5,000Ω or more. When the frequency is raised to the value giving three half-waves the impedance is about 100Ω, while at two full-waves it is about 3,000Ω. The high impedances can be lowered to between 1,000 and 2,000Ω by using a flat top of twin wires 2–3ft apart, joined in parallel, in order to improve the swr or permit the use of a 300Ω line; this does not alter the low impedance value.

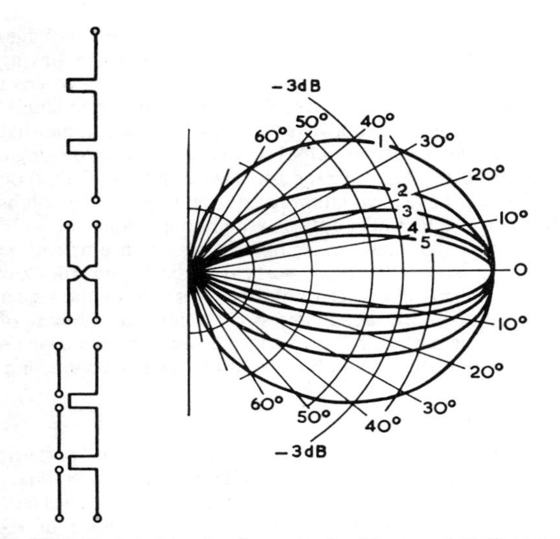

Fig 12.99. Horizontal polar diagrams of collinear and similar arrays, for one to five λ/2 overall length, showing half-power points (−3dB). Without a reflector the patterns are bi-directional. With a reflector the patterns are only very slightly sharper, because the forward pattern of a dipole and reflector in this plane is very little different from that of a dipole. Minor lobes are not shown but should be —14dB or lower in a well adjusted array. The pattern of a 1·25λ dipole is slightly sharper than that of a full-wave dipole, but has minor lobes at —10dB level at about 60°. Vertical patterns of vertically stacked dipoles are also represented by the upper half of the above curves.

Radiation is broadside for lengths up to about 1·3λ. At higher frequencies it becomes multi-lobed as shown in Fig 12.91(b).

Tuning and Matching Collinear Arrays

In order to match a full-wave dipole, a quarter-wave stub may be added at the centre and low impedance line connected into the end of the stub, or 300 or 600Ω line tapped on to the stub (Fig 12.61(b)). The array needs tuning when a stub is used, and for this purpose the stub is made a little too long and a moving short-circuit provided. It may be possible to couple a grid dip oscillator into the bottom of the stub to find resonance. Approximate dimensions (in feet) are given by 470/*f*(MHz) for the radiators and 240/*f* for the stub.

If a low impedance line is used the stub is left open, the line then being connected in place of the short circuit and moved along to find the position for minimum swr. Fig 12.98(b) is a variation in which the two halves are separated and is useful because the two equal-length low impedance feed lines can be connected in parallel, in or out of phase by means of a plug connector in the station, giving broadside or full-wave patterns at will. When the gap is λ/2 the pattern may be found by multiplying the λ-patterns of Fig 12.12 by sin ϕ where ϕ is the angle to the wire (see *Array Factors*).

Longer arrays may use half-wavelength elements with tuning stubs and a feed point either at the centre of one element (current maximum) or at a phase reversal point (voltage maximum). The impedance at the centre of any element rises rapidly with the number of elements but falls rapidly at the phase reversal position, the two meeting at about 1,200Ω which is the characteristic impedance of a typical aerial. These longer aerials should be tuned up section by section as described above, the total length of wire in feet between centres of shorting bars being 950/f (MHz).

Horizontal Broadside Arrays

These aerials, consisting of two horizontal arrays one above the other, give more gain than single arrays for the same width provided the height of the lowest dipole is not less than about half the height of the upper one. At low wave angles their azimuthal patterns are the same as for collinear arrays, but with λ/2 spacing the vertical pattern has only one main lobe whatever the height. In planning broadside arrays it is important to realize that at angles well below the main lobe of the upper dipole the lower element produces less field-strength than the upper one by an amount which, assuming flat ground, is *proportional to the ratio of the heights*. If therefore it receives an equal share of the power, the field-strength will be less than if the upper unit is used alone unless the above "half-height" rule is enforced. The benefit from the lower element is therefore unlikely to be appreciable, except at fairly high angles, unless the height of the lower element is at least λ/2, and decreases very rapidly with spacing between the elements, being only 1dB for λ/4 spacing even at the most favourable height.

Three Band Array

The simple array using two half-wave dipoles cut for the lowest frequency (Fig 12.98(c)) can be used as a broadside array over a 2 : 1 frequency band—eg 14, 21 and 28MHz. For this purpose the vertical spacing at the lowest frequency should be at least three-eighths wavelength. The phasing line should be 600Ω (not crossed) with the feed point at its centre. The swr on 600Ω main feeder will be between 6 and 10 and cannot easily be improved. There is nothing critical about any dimensions on this aerial. Methods of matching it to a line on any one frequency are considered later in this section.

Lazy-H

When the above array is one wavelength long it is called a *lazy-H*, the effective gain then being about 2dB higher. With the phasing line connected as in Fig 12.98(d) it must be electrically a half-wavelength long and crossed to restore phase. The impedance across the bottom end of the phasing line is then of the order of 3,000Ω and a λ/4 stub transformer using 500Ω line provides a good match to a low impedance twin feeder for single band operation. On the other hand, if it is connected as in **Fig 12.100** with a centre tap feed,

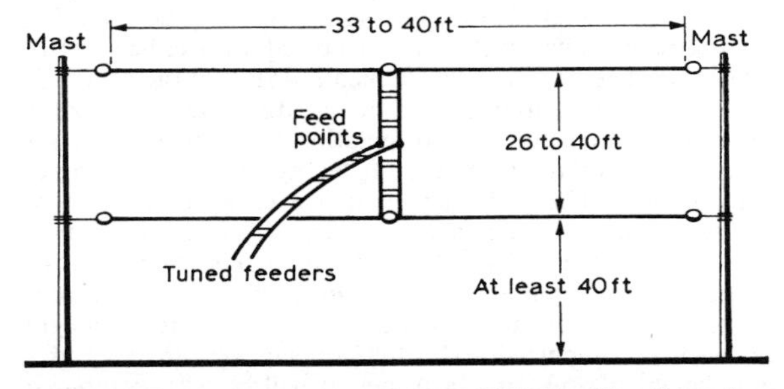

Fig 12.100. A lazy H array for 14, 21 and 28MHz. The dimensions given are a minimum for useful low-angle gain at the lowest frequency over flat ground.

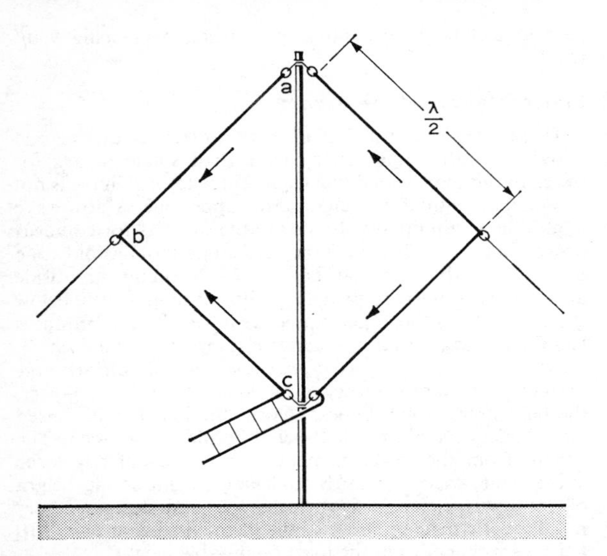

Fig 12.101. Bi-square aerial. Length of wire each side of the feedpoint is given by 960/f feet where f is the frequency in megahertz.

there is no need to cross the line as the two branches are always in phase. In this case, if $S = \lambda/2$, the 600Ω line connecting the dipoles becomes a pair of $\lambda/4$ transformers which reduce the feed-point resistance to around 70Ω so that for single-band operation a low impedance feeder can be used without additional transformation.

When a $\lambda/2$ broadside array for 14MHz, Fig 12.98(c), is used on 28MHz it becomes a lazy-H on the higher frequency and the gain is appropriately increased, ie from a matter of 2–4dB at 14MHz for the spacings suggested in Fig 12.100 to 6dB at 28MHz with an intermediate value at 21MHz. If the recommended spacings are not realizable, eg if the spacing is reduced to 15ft and the height of the lower element to 30ft, the array approximates to a single dipole at 37ft on 14MHz but the arrangement may still be considered worthwhile since most of the gain on 28MHz is retained.

The Bi-Square Aerial

A development of the lazy-H known as the *bi-square*, **Fig 12.101**, is particularly attractive as it can be supported from a single pole, although the gain is somewhat less than that of the original. Two can be mounted at right angles and switched to provide omnidirectional coverage, the aerial wires in this case acting also in part as guy wires. The radiation resistance is 300Ω so that it can be fed with either 300 or 600Ω line. The gain usually claimed is 4dB, but the radiation resistance in fact implies a gain of 3dB which is also the figure to be expected by extrapolation from the gain curves of Fig 12.98, taking into account the rather close spacing of the elements. The gain may be increased by a reflector or director, though problems then arise if it is desired to suspend two beams from the same pole. The gain, like the figures quoted for the lazy-H and other broadside arrays, is relative to a dipole at the same *mean* height, not the height of the top element; thus for a frequency of 14MHz and a pole height of 75ft, the mean height will be 52ft, compared with about 69ft (effective) for an inverted-V dipole mounted on the same pole, and for radiation angles below about 10° the bi-square will have a net gain reduced in the (voltage) ratio 69/52, ie to 0·5dB, but this is clearly worth having since it costs nothing, the bi-square being as easy to erect and easier to feed; these remarks relate however to single-band operation, the above example providing somewhat unpredictable results on 21MHz and very little low-angle radiation on 28MHz. Like the 14MHz inverted V, if used on 7MHz it will have a radiation resistance of only a few ohms but unlike the V it acts only as a high-angle radiator.

This array like many other high frequency beams can be excited as a very efficient vertical radiator on the low frequency bands if the feed point is accessible or the feeder is brought down to, say, an accessible point near the base of the mast, where it can be short-circuited and energized via an appropriate atu with suitable provision for band-switching.

Sterba Curtain

Longer broadside arrays have, of course, sharper patterns and greater gain, generally up to 4dB more than a collinear array of the same length. A six-element array with series feed (Fig 12.98(e)) has effectively the input impedance of six dipoles in series, say 500Ω, while an eight-element array fed at the base of the centre phasing line would resemble four full-wave centre fed aerials in parallel, about 800Ω. In either case, the vswr on a 600Ω line would be low enough without extra matching. The element length (feet) should be about 470/*f*(MHz) and the phasing line on these larger aerials may be 600Ω open wire or 300Ω ribbon of resonant length.

Phasing Lines

When the phasing lines are part of a series connection (eg Figs 12.98(d) and (e)) they must be electrically one or more complete half-wavelengths long. When the length is $\lambda/2$ they must be transposed to offset the phase reversal due to the wave travelling along the length of line, but if they are one wavelength long the phase is restored and the lines are not crossed. However, if the feeder is tapped into the middle of the phasing line as in Fig 12.98(c) the current divides in phase regardless of its length and no cross-over is needed in that phasing line.

The velocity factor of open wire lines is nearly unity and the length factor for a half-wave is 0·48–0·49λ; thus it is usually necessary to space the upper and lower rows by $\lambda/2$ in order to achieve a practical construction. When 300Ω ribbon is used advantage can be taken of its velocity factor of 0·8 and the aerials spaced vertically by a three-eighths wave with an electrical half-wave of ribbon (365/*f*) or a three-quarter wave with a full-wave of ribbon. On the other hand 300Ω ribbon is not so good for small arrays because in such applications it carries a high swr. The greater spacing gives about twice as much effective stacking gain *for the same mean height*.

Broadside Verticals and Stacks

Vertical patterns of these arrays are given in Fig 12.22 and horizontal patterns for two elements in Fig 12.12. The patterns are broader than those of the equivalent collinear arrays. The impedance of the individual dipoles has an average value of about 60Ω.

The stack of horizontal dipoles is really the same aerial rotated. Its azimuth pattern is that of a single horizontal element; the vertical pattern is not illustrated, but improves

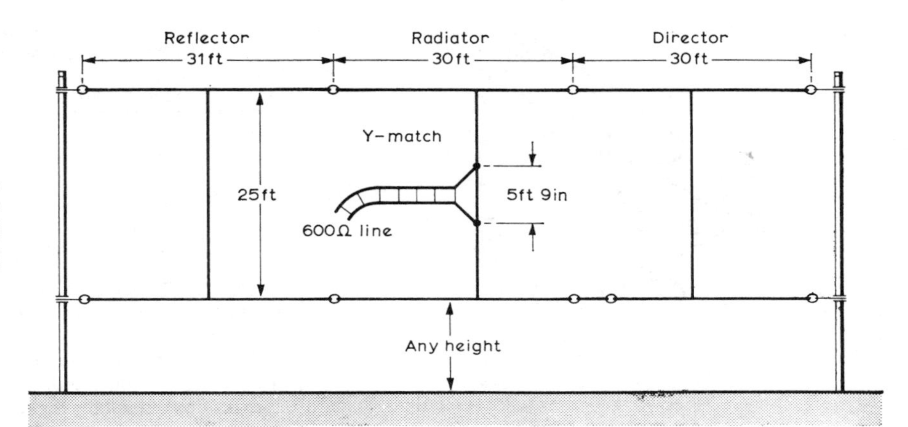

Fig 12.102. Low-angle aerial for 7MHz. The height is not critical. The reflector and director are tunable by adjusting the length of verticals or of the lower horizontal wires. All dimensions approximate.

relative to that given in Fig 12.22(e) as the number of elements increases. In these arrays advantage can be taken of the velocity factor of 80Ω line (0·67) or 300Ω lines (0·8) to make the phasing lines one wavelength long (uncrossed) and increase the spacing of the elements to $\frac{2}{3}\lambda$ or $\frac{4}{5}\lambda$, but the 80Ω line should not be used with full-wave dipoles because of the high vswr which would result.

End-fire Arrays

The vertical broadside array, Fig 12.98(g), may be converted into an end-fire array by reversal of alternate feeder connections so that aerials $\lambda/2$ apart are excited in opposite phases. Due to the space separation between the elements, their fields then add up in phase along the line of the array instead of broadside to it. This has the valuable feature of allowing the beam to be rotated 90° by a simple switching operation although, if there are many elements, the beams will be too sharp to provide all-round coverage without elaborate phase-shift networks. The end-fire connection offers the important advantage of broader azimuthal coverage enabling the beam to be used for communication with much larger geographical areas, and it also allows the beam to be mounted horizontally though it cannot then be switched back to operate in the broadside mode. The gain of a *horizontal* end-fire array is virtually identical with that of a *vertical* broadside array but the vertical end-fire array with $\lambda/2$ spacing has considerably less gain, by about one-third for up to four elements; the same gain as before may be obtained from a given length, but only by reducing the spacing to $3\lambda/8$ or less, ie increasing the number of elements.

When the spacing is less than $\lambda/2$ the phase shift between elements must also be reduced and this has the advantage of producing a unidirectional pattern but, being no longer a simple matter of phase reversal, gives rise to practical difficulties. The simplest way out of this problem is to use parasitic excitation of the radiators, this arrangement being known as a Yagi after one of its inventors. These can be made small enough to enable them to be supported from a single mast and rotated to produce some 5–6dB of gain in any desired direction.

Fig 12.102 is an example of a wide-spaced Yagi array for 7MHz which exploits the fact that vertical elements produce, albeit very inefficiently, some intrinsic low-angle radiation even when mounted at very low heights. Use of horizontal supporting wires to provide capacitive end-loading maintains the radiation resistance at a manageable value despite the use of very short radiators and leads to a simple construction which does not require tall supports. Any number of elements can be assembled on these lines depending on the space available, and for 5° radiation angle eight such elements mounted over average ground could be expected to give about the same performance as a 2-element beam at a height of 60–80ft. The space requirement for this number of elements would be about 0·25 acre, and the pole height about 30–35ft. Vertical arrays are not recommended for higher frequencies unless height is very restricted, less than 30ft, say, or unless the location is such that ground-reflection takes place from the surface of the sea.

LONG WIRE BEAM AERIALS

The V Beam Aerial

A long wire aerial two wavelengths long has a lobe of maximum radiation at an angle of 36° to the wire. If two such aerials are erected horizontally in the form of a V with an included angle of 72°, and if the phasing between them is correct, the two pairs of lobes will add fore and aft along a line in the plane of the aerial and bisecting the V. Remaining lobes do not act in this way and so this provides what is essentially a bi-directional beam, although minor lobes will occur away from the main beam.

Fig 12.103 illustrates the principle. If the waves on the wires were visible, they would be seen to flow in the directions of the arrows, to appear in phase from the front of the array; hence an anti-phase excitation is necessary, eg a balanced feed-line at the apex.

The directivity and gain of V-beams depend on the length of the legs and the angle at the apex of the V. The correct choice of this angle also depends on the length of the legs which are likely to be the limiting factor in most amateur installations, and this is the first point to be considered in designing a V-beam.

The correct angle and the gain to be expected in the most favourable direction is given in **Table 12.5.**

The horizontal polar diagram varies with the leg length and is virtually bi-directional along a line bisecting the apex. The shorter lengths show quite a broad lobe which becomes rapidly narrower as the number of wavelengths in each leg is increased.

The layout of a typical V-beam is shown in Fig 12.103(b).

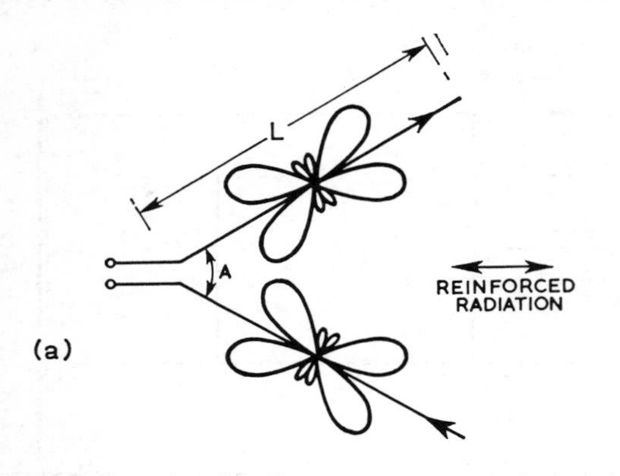

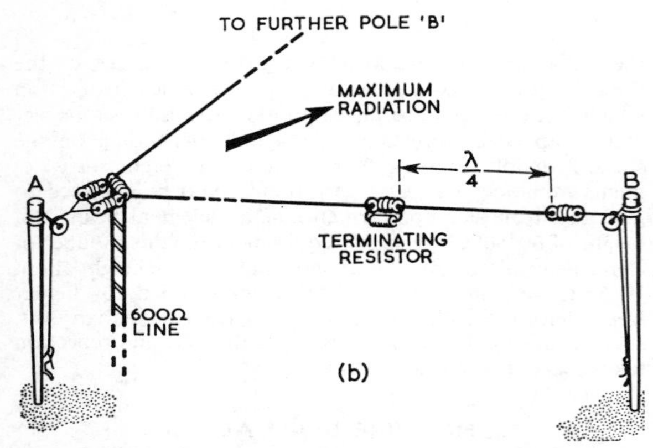

Fig 12.103. The V-beam aerial derived from two long wires at an acute angle A. Addition of resistors as shown in (b) results in a undirectional pattern.

Probably the best way of feeding a V-beam is by the use of tuned feeders as the impedance is high. If a V-beam is designed for a particular frequency it can then be used successfully at higher frequencies.

This aerial is too large for use by the majority of amateurs but where space is available it is attractive in view of its simplicity. Its other main limitation is the narrowness of the main lobes which restricts all round coverage unless a sufficient number can be erected to cover all the main land masses.

The V can be made unidirectional if it is terminated, for example, with the artificial earth described in connection with Fig 12.92. A suitable value of resistor would be 500Ω for each leg. The input impedance, in the resonant condition, may rise to 2,000Ω in a short V but will be between 800 and 1,000Ω in a longer or terminated aerial and thus 600Ω feed lines can be used. Alternatively a non-terminated V can be driven like the balanced Windom shown in Fig 12.88(b).

The Rhombic

Early difficulties of terminating a wire high in the air led to the development of the *rhombic* aerial in which a second V is added, so that the ends can be brought together. The same lobe addition principle is used but there is an additional complication, because the lobes from the front and rear halves must also add in phase at the required elevation angle. This introduces an extra degree of control in the design so that considerable variation of pattern can be obtained by choosing various apex angles and heights above ground.

TABLE 12.5
V beam aerials

Leg length (λ)	Gain (dB)	Apex angle
1	3	108°
2	4·5	70°
3	5·5	57°
4	6·5	47°
5	7·5	43°
6	8·5	37°
7	9·3	34°
8	10·0	32°

The rhombic aerial gives an increased gain but takes up a lot of room and requires an extra support. There are two forms of the rhombic; the resonant rhombic which exhibits a bi-directional pattern and the terminated rhombic which is non-resonant and unidirectional. The terminating resistance absorbs noise and interference coming from the back direction as well as transmitter power which would otherwise be radiated backwards; this means that it improves signal-to-noise ratio by up to 3dB without affecting signals transmitted in the wanted direction.

The layouts of the two forms are shown in **Fig 12.104.**

It will be evident that the resonant rhombic can be considered as two acute angle V-beams placed end on to each other. An advantage of the rhombic over a V is that it gives about 1-2dB greater gain for the same total wire length, and its directional pattern is less dependent on frequency. It also requires less space and is easier to terminate.

The use of tuned feeders enables the rhombic, like the V-beam, to be used on several amateur bands.

The non-resonant rhombic differs from the resonant type in being terminated at the far end by a non-inductive resistor comparable in value with the characteristic impedance, the optimum value being influenced by energy loss through

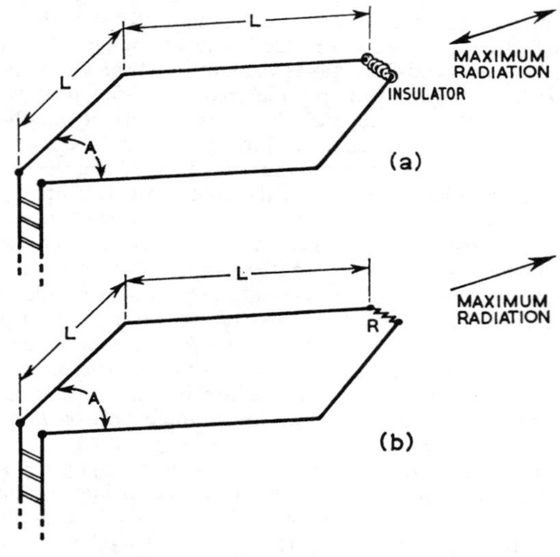

Fig 12.104. Terminated and unterminated rhombic aerials.

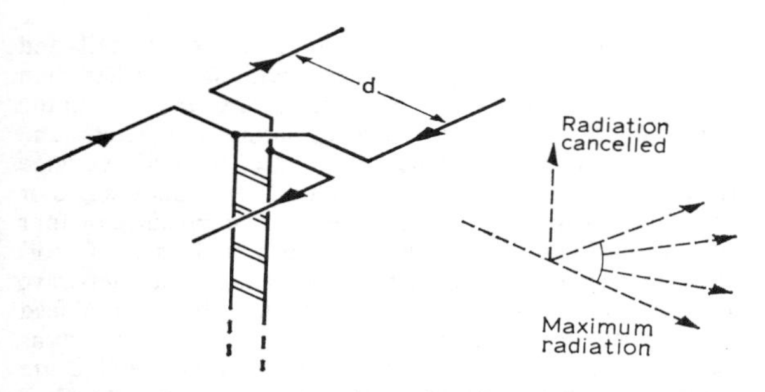

Fig 12.105. The basic W8JK flat-top array. The antiphase currents in the elements cancel any vertical radiation and reinforce in the plane of the array.

radiation as the wave travels outwards. An average termination will have a value of approximately 800Ω. It is essential that the terminating resistor be as near a pure resistance as possible, ie without inductance or capacitance and this rules out the use of wire wound resistors. The power rating of the terminating resistor should not be less than one third of the mean power input to the aerial. For medium powers suitable loads can be assembled from series or parallel combinations of say, 5W carbon resistors. Higher power loads may be constructed from a number of Morganite Type 702 resistors.

The terminating resistor may be mounted at the extreme ends of the rhombic at the top of the supporting mast or an open line of 800Ω can be brought down from the top and the resistor connected across this at near ground level. The impedance at the feed point of a terminated rhombic is 700–800Ω and a suitable feeder to match this can be made up of 16swg wire spaced 12in apart. Heavier gauge wire will need somewhat wider spacing, which can be determined by extrapolation from the curves of Fig 12.28.

The design of the rhombic aerials can be based on **Table 12.5** considering them to be two V-beams joined at the free ends. These figures will give a wave angle of the main lobe of approximately 15° in all cases, when the aerial height is one wavelength.

The design of V and rhombic aerials is quite flexible and both types will work over a 2 : 1 frequency range or even more, provided the legs are at least 2λ at the lowest frequency. For such wideband use the angle is chosen to suit the length *L* at the mid-range frequency. Generally the beamwidth and wave angle increase at the lower frequency and decrease at the upper frequency, even though the apex angle is not quite optimum over the whole range. In general leg-lengths exceeding 6λ should not be used because the beam is then too narrow to cater for random variations in the direction of propagation which does not always stick precisely to the great circle or a particular angle of elevation. The vertical radiation pattern of rhombic aerials is modified by the presence of the ground in exactly the same way as that of other horizontal arrays and for maximum very low angle radiation the height should be as great as possible.

A comparison with the figures given for tuned arrays in Fig 12.98 will show that, for the same total length of radiator, V and rhombic aerials give somewhat lower gain, but against this must be set the simplicity of construction, wideband properties and ease of feeding. They are applicable to amateur communication primarily in those situations where there is special interest in a particular geographical area extending over not more than 10–20° on a great circle map, and where there is room to accommodate not less than 2λ per leg at the lowest frequency which provides a gain of 8–10dB on three bands, compared with 5–6dB for, say, a tri-band rotary beam of considerably greater complexity. The gain increases in proportion to leg length and the coverage angle decreases roughly as the inverse square root of the leg length.

SMALL ARRAYS—DRIVEN

"Flat Top" or W8JK End-fire Array

These aerials use two anti-phase radiators side by side (**Fig 12.105**) and the theory has been explained on p12.11. Briefly in any given direction θ the fields produced do not quite cancel because the distance from the two elements is slightly different, by an amount proportional to $\frac{d}{\lambda}$ and also to cos θ. The radiation pattern is therefore similar to that of a dipole multiplied by cos θ in both planes. From this there arises a theoretical gain of 4dB but unfortunately the extent to which cancellation takes place is quite impressive and a gain of 4dB representing as it does the difference between two large quantities implies very large circulating currents. This translates into a radiation resistance of only 8·5Ω per element or just over 4Ω for the two in parallel assuming λ/8 spacing. Values for other spacings and for full-wave elements will be found in **Table 12.6**. The high currents in turn imply high voltages and consequent losses in insulators and in the surroundings generally so that the theoretical gain is never reached. It has been claimed that gains up to 3·5dB can be obtained with suitable precautions which should include a somewhat larger spacing (say 0·15–0·20λ), use of heavy gauge conductors, avoiding insulators as far as possible except near voltage nodes (though use of four end-insulators is unavoidable unless self-supporting elements are used), mounting in the open away from houses and trees, and using an open-wire matching stub as close as possible to the aerial so that high currents are confined to the elements themselves and the minimum amount of additional conductor necessary for matching. The use of resonant feeders for multiband operation, though attractive in principle, is therefore not possible without large losses unless the feeder length can be kept very short or the beam dimensions increased. The position may be eased considerably by use of folded dipole elements, accepting some restriction of multiband capabilities, though two-band 14/21 or 21/28MHz operation is

TABLE 12.6
Impedance values for W8JK aerials

Spacing S(λ)	L = λ/2		L = λ	
	Rd	Re	Rd	Re
0·1	6	40,000	10	50,000
0·15	12	20,000	20	25,000
0·2	20	12,000	30	16,000
0·25	33	7,500	50	10,000
0·3	46	5,500	65	8,000
0·4	64	4,000	100	5,000
0·5	85	3,000	125	4,000

Approximate theoretical impedances in ohms at two points of a W8JK array. Rd is the impedance at the centre of any λ/2 element and Re the impedance to earth of any free end. Figures are based on an aerial characteristic impedance of 700Ω. For typical wire elements Re will be roughly doubled.

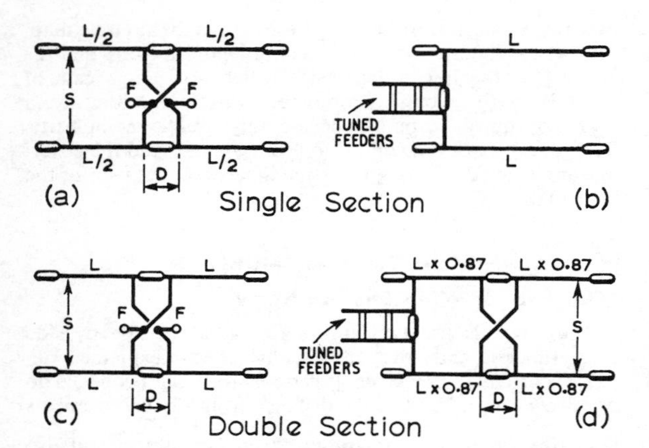

Fig 12.106. Single and double section W8JK arrays using centre and end feeding. Dimensions are not critical.

feasible. For single band operation the element lengths may be anywhere between about 0·5λ and 1·25λ, and are therefore not critical; shorter lengths involve even lower radiation resistances, whereas the longer length corresponds to the extended double-zepp giving higher gain and increased radiation resistance. Even longer lengths, or harmonic operation of the longer length, are feasible subject to the acceptability of long-wire type radiation patterns. Radiation resistance increases nearly as the square of the spacing which may be increased to 0·25λ, accepting about 0·5dB drop in theoretical gain. At 0·5λ spacing the 8JK becomes a wide-spaced end-fire array with 2·5dB gain and a feed point impedance in the region of 2,400Ω for full-wave elements.

A flat top beam comprising two half-wave dipoles spaced 0·15 to 0·2λ at 14MHz on the lines of **Fig 12.106** can be used as a full-wave array on 28MHz, where spacing would become 0·3 to 0·4λ, and although the vswr will be rather large on the lower frequency bands the aerial will also perform as a beam on 21MHz.

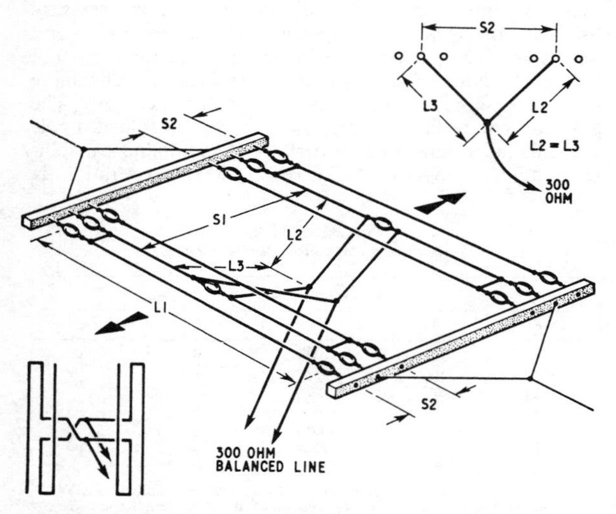

Fig 12.107. A practical single section W8JK design. The phasing lines can be made from 300Ω flat twin polythene cable if allowance is made for the velocity factor. Dimensions are given in Table 12.7.

Table 12.6 gives centre- and end-impedances for half- and full-wave arrays; it will be seen that with spacing less than 0·2λ power loss can easily occur, due to large currents in the wire and high voltages across the insulators, and matching to a line may be difficult. One way to deal with the extreme impedances is to reduce them by folding the radiators two or three times. This multiplies the low centre impedance by four or nine, and reduces the high impedance by the same factor. The two branch feed lines can then be used as quarter-wave transformers using suitable cable to achieve the required feed point impedance. Alternative methods of matching such as the use of transformers at the aerial end of the line are likely to prove difficult due to the high vswr.

A simple practical design for single band single section arrays is shown in **Fig 12.107** and **Table 12.7**. This utilizes 300Ω twin wire line for the phasing sections. By taking advantage of the impedance step-up provided by three-wire folded radiators, the array is arranged to present an acceptable match to a main feeder also made of 300Ω flat twin.

Phased Arrays

As explained on p12.11 an approximation to the performances of all close-spaced two-element beams may be derived by taking the 8JK as a starting point and assuming a phase-shift between the elements comparable with that corresponding to their spatial separation. Such phased arrays may be realized in practice by driving both elements, or one element only with parasitic excitation of the other. First the driven versions will be considered, and **Fig 12.108** shows how a generator may be connected to two equal resistances so that a phase-shift is introduced without disturbing the equality of the currents. The impedance of the two arms is of equal magnitude, therefore the currents are equal, but the phases are shifted in opposite directions, each by an angle given by $\tan \phi = X/R$. For small phase shifts ϕ (in radians) is equal to X/R. In the case of a close spaced beam the situation is in general greatly complicated by the mutual coupling but for the relatively simple case of $\lambda/8$ spacing the mutual coupling is a pure resistance R_m and the phase shift for each element is given by $\frac{X}{R + R_m}$ if the dipoles are connected in opposition, ie "8JK fashion", and $\frac{X}{R - R_m}$ if they are connected in parallel without phase-reversal, ie if to get the required nearly-antiphase condition we start from an in-phase connection. R is the intrinsic radiation resistance of each dipole considered in isolation. These formulae demonstrate the need to use the antiphase condition, adjustments being more critical in the ratio $(R + R_m)/(R - R_m)$ or about 15 times if the in-phase connection is used. Fig 12.15 on p12.12 shows the relation between phase-shift, gain, radiation resistance and nominal back-to-front ratio and we may decide from this to aim for

TABLE 12.7

Frequency	S1	S2	L1	L2, L3*
7·1MHz	20ft 9in	14in	65ft 10in	27ft 7in
14·2MHz	10ft 4in	12in	32ft 11in	13ft 9in
21·2MHz	6ft 4in	12in	22ft	9ft 3in
28·2MHz	5ft 2in	9in	16ft 7in	6ft 11in

*For air-spaced line. If flat polythene 300Ω twin is used, these lengths must be multiplied by 0·82 (velocity factor).

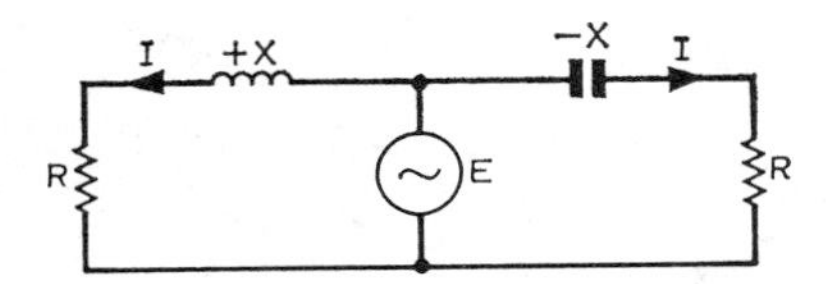

Fig 12.108. The ± X method of phasing. Total phase shift ϕ is given by tan (ϕ/2) = X/R or approximately by ϕ = 2X/R radians if this is less than unity. For parallel λ/2 dipoles spaced λ/8 and fed out of phase, R = 137·5Ω, the mutual resistance being added to the radiation resistance. (In-phase connection is unusable because the resistances are then subtracted and X becomes too critical). Note that the currents are equal only if the reactances are equal.

example at a phase angle of 151°, ie starting from the 8JK, 14½ degrees or 0·2 radian each side. Therefore, X = 0·2 (73 + 64) = 27·4Ω which from the formula on p12.26 is equal to the reactance of 6in of open wire line at 14MHz. In other words to obtain the required phase-shift we must first make sure the system consisting of the two resonators and the feeder between them is exactly resonant, and then shift the feed point *6in only* off-centre which subtracts the required reactance from one side and adds it to the other.

There are many types of driven array which use a phasing line, or unequal feeder lengths, to obtain the required phase-shift, and are based on the assumption that phase-shift is equal to the length of line expressed in electrical degrees. The method described above is in fact identical with the phasing-line method with the addition of a rigorous procedure for determining the correct length *and position* of phasing line, which turns out for the particular example chosen to be 1ft. For 45° phase-shift this becomes 18in instead of the usual 8ft, and must also be disposed symmetrically with respect to resonance. Phase-shift is in fact equal to line length *only* in the special case of perfect matching *at the aerial* and matching of course depends in turn on the phase-shift, so that any adjustment of one requires readjustment of the other. To account for the frequent success of phasing-line methods, Fig 12.15 shows the critical dependence of radiation resistance on phasing which could therefore be used primarily as a matching adjustment; element currents will then be equal, there will be a phase-shift, and therefore some gain and front-to-back ratio. By using folded-dipole elements the situation is greatly improved and with the aid of Fig 12.15 it can be shown that use of an 8ft phasing line, compared with the new correct value of 6ft, produces a loss of only 0·7dB in forward gain and the front-to-back ratio remains quite high (17dB) despite an swr of 4. The situation may be further improved by using low-impedance line as in some well-known arrangements, though ideally a Tee or gamma match (p12.36) should be used particularly if it is desired to work at the point of maximum gain. The plus and minus X method in conjunction with resonant feed-lines permits simple and precise adjustment at any one frequency from ground level, or even in the shack, and also allows multiband operation but, like other resonant multiband feeder systems, has the disadvantage of being frequency-sensitive; however acceptable band-coverage can be obtained on 14/21MHz or 21/28 MHz using folded-dipole elements designed for 14 and 21MHz respectively provided the feeder length does not greatly exceed $\lambda/2$ at the lowest frequency. Three-band coverage is feasible using single-wire 14MHz dipoles or 21MHz dipoles (single-wire or folded) but these designs are compromises involving bandwidth restriction or excessive spacing with some reduction in gain, or both.

There appears to be no reason to maximize the "nominal" back-to-front ratio, since this does not give the best average discrimination against unwanted signals, taking into account all directions, and it is better to operate closer to the condition for maximum gain; on the other hand larger phase-shifts produce greater bandwidth, sometimes assist matching, and are helpful in the design of miniature beams.

Simple Phased Arrays

The ± X method of phasing has been applied in a number of different ways to quad and dipole elements, including the Swiss quad (see below) which uses loops of slightly different sizes. The simplest arrangement is almost identical in appearance with the 8JK, Fig 12.105, the resonance requirement being met by shortening the elements to 27ft 6in with 8ft 4in spacing. The shortening of the elements unfortunately almost halves the radiation resistance, making it advisable to operate with phase-shifts well in excess of those required for maximum gain, optimum results being obtained with the main feeder displaced from the centre of the system by about 4½in in the required direction-of-fire, corresponding to about 135° phasing. A capacitance of 168pF bridged across the 600Ω line at 2ft from the feedpoint acts as a suitable "matching stub" but an swr of at least three must be expected at band edges, narrow bandwidth being of course a disadvantage of all centre-loaded short-dipole systems. If the aerial is not correctly resonated, displacing the feed-point causes more current to flow to the shortened (capacitive) side if the overall length of the elements and connecting feeder is too great, and the longer (inductive) side if the length is too short. This can be checked with any suitable current meter (eg that of Fig 12.78), thus providing a simple tuning indication.

Other versions use elements of standard length each fed with a half-wavelength of resonant feeder, the feeders being connected in parallel antiphase. For 14MHz the elements may be folded dipoles spaced about 9ft and fed by connecting a 50Ω line (balun essential if using coaxial cable) to points about 2ft either side of centre, permitting beam reversal by means of a switch or relay. For operation on 21MHz the feeders require shortening by 4ft in the case of 6in spacing between the wires of the dipoles, or 3ft 6in with the dipoles "opened out" to an average spacing of 3ft. The same feeder can be used, connected 14in off-centre. Shortening may be carried out electrically by series tuning with two centrally-located capacitors of 20–25pF, and in this case it has been found possible to obtain points for feeder connection which are a reasonable compromise for both bands although the beam now fires in the opposite direction on 21MHz because the physically longer side has become electrically shorter. Many variations on this theme are possible.

Phasing-line Systems

Examples of phasing-line systems include the "ZL Special" (**Fig 12.109**) and arrangements such as the "G8PO" in which separate feeders of equal length are brought down to a phasing-line and beam-reversing switch as in **Fig 12.110**. If the spacing is $\lambda/8$ and the matching arrangements at the aerial end of the feeder are such that a match is obtained for a satisfactory phase angle as indicated by Fig 12.15, then a phasing line of the appropriate electrical length will serve to establish this operating point. If not, there will, as discussed above, be at least some degree of current inequality unless

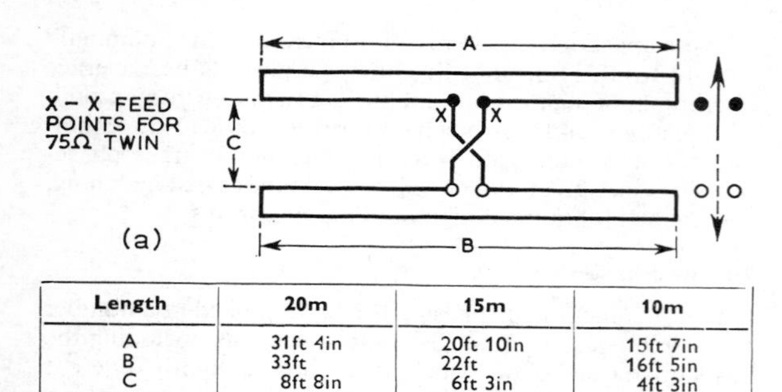

Length	20m	15m	10m
A	31ft 4in	20ft 10in	15ft 7in
B	33ft	22ft	16ft 5in
C	8ft 8in	6ft 3in	4ft 3in

Notes: VF for open wire 0·95.
VF for 300Ω feeder 0·85.
C dimensions given for 300Ω feeder.

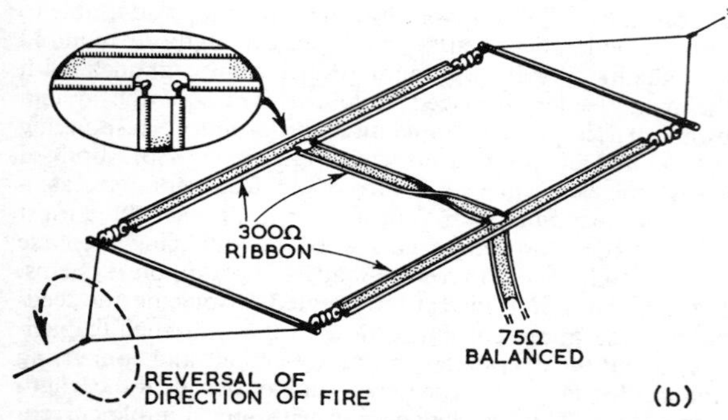

Fig 12.109. The ZL Special. Reversal of direction can be obtained either by feeding at the opposite dipole or by physical rotation of the aerial about a horizontal axis.

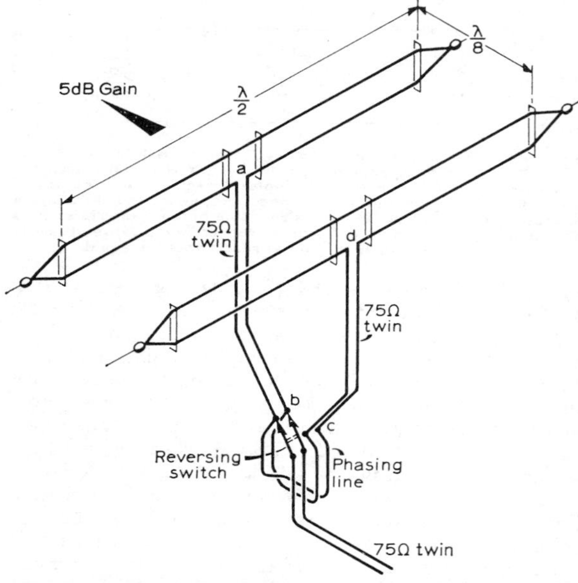

Fig 12.110. Driven beam with phasing lines. If feeders are perfectly matched ab, dc can be any length and phase-shift is equal to the electrical length of the phasing line. Mismatched feeders, eg open-wire line, may be used but feeder length is then critical and the length of the phasing line may be greatly reduced. Elements are connected in antiphase, in-phase operation (which reverses the beam direction) being permissible with matched feeders only.

the line lengths are disposed symmetrically either side of resonance, ie unless the feeder plus half the phasing line adds up to a resonant length, but so long as this symmetry is maintained it is possible to obtain any desired phase-shift combined with current equality by adjusting the line length. There may now however be a large swr in each feeder, and, although the extra loss due to this may be negligible, possibilities of overheating and voltage breakdown could arise at high power levels.

These procedures are strictly valid only for spacings of $\lambda/8$ or just over since the presence of a reactive term in the mutual impedance leads to interdependence of phase-shift and the effective resonant length of the elements. This does not mean that other spacings are unusable, but they may involve slightly more trial, error and patience for the achievement of optimum results.

Constructional details for one version of the "ZL Special" are illustrated in Fig 12.109. The dimensions apply to 300Ω line but open wire line may also be used. Various sources quote widely differing figures for the dimensions and performance, but as the phasing line is inherently mismatched and the reactances asymmetrical the behaviour is not readily amenable to calculation.

There seems no doubt that good results can be obtained from this system, but comparison of published designs suggests that a good deal of trial and error, such as varying the lengths of the elements separately, may be involved. The usual aim is to obtain a phase-shift of 135° with $\lambda/8$ spacing but if achieved, this is likely to give inferior performance to that of a parasitic beam, as will be appreciated from the above discussion and Figs 12.15, 12.17, and 12.18. On the other hand, it appears that errors in the adjustment of driven arrays may be to some extent self-compensating.

PARASITIC ARRAYS

The broadside arrays of an earlier section (pp12.64 to 12.66) radiate equally fore and aft. In order to make them unidirectional, reflectors are often added at a distance of $\lambda/4$ behind each dipole of the array, and this has the further advantage of increasing the gain by 3dB. The reflector may be driven but, as discussed above in the context of phased arrays, there are problems in adjusting arrays with driven reflectors so that the currents are equal in amplitude but different in phase, because one effect of the reactive part of the mutual impedance (p12.6) is to produce unequal reactances and radiation resistances in aerial and reflector. This difficulty can be overcome for example by using the $\pm X$ method, p12.70, but it is often simpler and equally effective (apart from some sacrifice of back-to-front ratio), to use a parasitic *reflector*, which picks up some of the power of the driven dipole and re-radiates it in the required phase.

From the discussion of phased, driven arrays, pp12.70 and 12.71, it might be thought that improved results could be obtained by closer spacing but this is not normal practice with large arrays, and taking due account of stacking requirements (p12.16) and the more critical design considerations which arise with close-spacing (as also with large numbers of elements) this course is not recommended.

Additional gain and improved front-to-back ratio may be obtained by the addition of parasitic directors, but these are not normally used alone since owing to the inductive component of the mutual impedance as discussed on p12.6 they are relatively ineffective at wide spacings, and close spacing is inadvisable as discussed above.

Adjustments are made sometimes for best nominal front-to-back ratio and sometimes for maximum forward gain. Referring again to Fig 12.15 the first condition tends to give slightly less gain but higher radiation resistance which may be desirable to preserve bandwidth. It is however important to realize that maximum gain *by definition* implies minimum interference if, as is true for the average case, interference is no more likely to be coming from one direction than another. This follows from the fact that maximum gain is obtainable only by achieving the maximum concentration in one direction at the expense of the average of all other directions. To select one particular direction which happens to be at $-180°$ and adjust for minimum interference from that one direction, neglecting the possibility that this could increase the signals picked up from or radiated in other unwanted directions is, to say the least, devoid of logic; on the other hand it is easier to adjust for a minimum than a maximum and it is evident from Figs 12.17 and 12.18 that if this course is adopted, the signal chosen should be on a bearing in the region of $-120°$ to $-150°$.

A field strength indicator may help in the adjustment of these arrays, but this does not always give the best result, because the local field of any aerial is different from the distant field, especially near the ground. If a field strength meter is used it should be connected to a short dipole, set up parallel to the array, as far away as possible and preferably at the same height. The greater the gain of the array, the greater the distance at which the field strength meter must be placed. With small beams (5–6dB gain) field-strength indicators at a distance of λ and the same height as the beam, or at more or less any height if the distance exceeds about 2λ, have been found to give an indication of optimum performance in fair agreement with reports from distant stations. Particular sources of error observed in such measurements have included the use of too short a dipole for the field strength meter, so that the harmonic sensitivity is many times greater than the fundamental sensitivity, and pick-up on insufficiently-filtered leads used for wiring the meter back to the point where the adjustments are being made. Polarization of the field strength measuring aerial must be correct, meters with short vertical rod aerials being useless for measurements on horizontal aerials. An important point is that the impedance of the driven aerial changes when the parasitic element is tuned, so that the power drawn from the transmitter also varies with this tuning. Thus, unless a power-flow monitor, not an ammeter but a reflectometer, is used the optimum aerial gain adjustment may not be obtained.

TABLE 12.8

Properties of two-element parasitic arrays

Spacing	Reflector				Director			
s/λ	X (Ω)	Gain (dB)	F/B (dB)	R_d (Ω)	X (Ω)	Gain (dB)	F/B (dB)	R_d (Ω)
0·05λ	—	—	—	—	−10	4·4	20	13
					+10	5·2	6·6	4
0·1λ	+30	5	6	16	−10	5·2	7·8	14
	+50	4	10	26	−40	4·2	17	43
0·15λ	+20	5	6	35	—	—	—	—
	+50	4	10	50	—	—	—	—

Theoretical properties of two-element parasitic arrays, which are largely realized in practice. The tuning of the parasitic element is given in terms of reactance X (see Fig 12.111). Of the two sets of figures given for each spacing, one is for maximum gain and the other maximum front/back ratio.

Two-Element Arrays

Two-element aerials with close-spaced elements can be constructed in fixed or rotating form for 14MHz and higher frequencies, and as suspended wire aerials for 7MHz. Rotating arrays are usually made of rigid metal tubing. When the array is fixed, for example supported between spreaders, good insulation should be used at the ends of the wires as the voltages there are relatively high, and serious loss can therefore easily occur. Polythene cord is inexpensive and has been found excellent for this purpose. In the case of fixed arrays, it is convenient to be able to reverse direction, and in principle this may be arranged by changing the parasitic element from reflector to director, using a U-section or stub in the centre which can be shorted out to change from the length of reflector to that of director. This stub is sometimes extended by one half or one full wavelength using 600 or 300Ω twin line, so that the position of the shorting link can be changed at ground level without lowering the aerial, but this arrangement is very restrictive of bandwidth, the change in reactance from tuning over the 14MHz band being more than enough to change a director into a reflector. As an alternative, any feeder may be loosely coupled into the parasitic element so that the tuning can be pulled one way or the other by reactances connected to the lower end or by altering the feeder length. The coupling should be only just sufficient for this purpose and the adjustment should then be non-critical, with negligible losses in the auxiliary feeder. This method has been tried successfully with 600Ω line terminated at the aerial by a small delta loop, with its base insulated and taped on to the aerial. The optimum size of loop depends on the radiation resistance but 3ft by 3ft is suggested as a starting point. It should be noted that changing from director to reflector operation alters the feedpoint impedance and it will not be possible to achieve very low values of swr in both cases without some form of compensation.

From **Table 12.8** it will be seen that $0{\cdot}1\lambda$ is a possible value of spacing for a combined reflector/director system; this is 7ft for 14MHz. At this spacing the feed point impedance at the centre of the driven element is between 15 and 30Ω and the aerial is fairly sharply tuned, but can still effectively cover the 14MHz band. To match this impedance to available transmission lines, a transformer is needed at the aerial. Suitable devices include the T-match for low impedance twin line, the gamma match or Clemens match (p12.37) for coaxial cable, and the delta match for high-impedance twin line. The transformers given later for three-element arrays may be used, the adjustments being practically the same.

Tuning the Parasitic Elements

Formulae for element lengths are frequently quoted, but are not useful unless the diameter of the elements is also specified. What controls the performance is the reactance of the parasitic element; that is, the amount by which the element is detuned from resonance, and this, in terms of length, depends on the length/diameter ratio. **Fig 12.111** shows reactance as a function of percentage change of length for various relative conductor diameters. The crossover point of the curves occurs when the radiator is a true free-space half wavelength long, the length in feet being given by

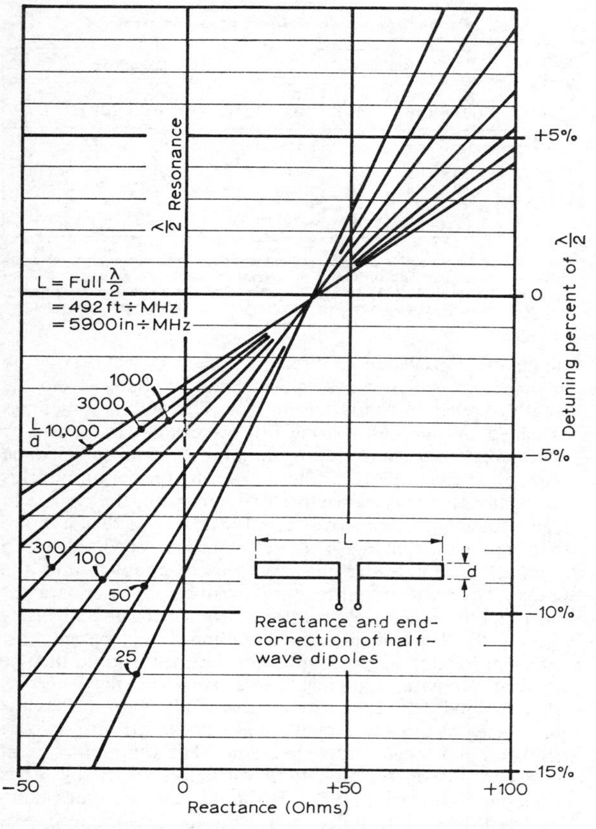

Fig 12.111. Tuning and reactance chart for λ/2 dipoles, as a function of the length/diameter ratio. A radiator exactly λ/2 long is "over tuned" by 42Ω and "end-correction", given as a percentage of the length, is necessary to bring it to resonance (zero reactance). The chart is useful for the construction of parasitic arrays and vhf dipoles. Each one per cent of length corresponds to five units in the factor 492/f or 60 in 5,900/f (f is in megahertz).

492/*f*(MHz). At this length the conductors all have a reactance of about + 40Ω, and to obtain resonance must be *end corrected* by the percentage which brings the reactance to zero. For example, the correction for a relatively "fat" aerial, with $L/d = 100$, is 5·5 per cent so that for 28·0MHz the full half-wave would be 17ft 6in and the end correction 1ft, making the resonant length 16ft 6in.

The best reflector has a positive reactance of about 40Ω, which brings its length to almost a true half-wave, independently of diameter. For 0·1λ spacing, best director action occurs at −10 to −40Ω reactance with a length some 5 to 10 per cent shorter than the full half-wavelength. For example, at 21·0MHz, $\lambda/2 = 492/f = 23{\cdot}4$ft. A 1in diameter half-wave radiator would have a L/d ratio of about 300, and an end-correction of −4·4 per cent, giving a resonant length of 22ft 4in, while a director of −30Ω reactance would be 7·4 per cent short of true half-wave, or 21ft 7in.

In practice it is always advisable to find the best adjustment experimentally, using the above theory as a guide, because the environment of the aerial has some effect. The parasitic elements can be tuned first, independently of the driven element, which may then be adjusted for optimum tuning and impedance match, as described later.

The following figures will be found satisfactory for *two-element* beams in most locations where the spacing using a director is 0·1 wavelength or where the reflector spacing is 0·15 wavelength.

Reflector length in feet	$\frac{500}{f(\text{MHz})}$
Driven element in feet	$\frac{475}{f(\text{MHz})}$
Director length in feet	$\frac{455}{f(\text{MHz})}$

From this it will be seen that the driven element is a little longer than the normal half-wavelength. The spacing of 0·1 wavelength, though usable, is somewhat on the narrow side for preserving a good swr throughout the 14MHz band, when operating in the maximum gain condition as recommended above, and 0·15 is too wide for efficient operation of a director, so that 0·125λ is a good choice for a reflector/director system. The curves of Fig 12.15 are applicable in so far as they indicate the relation between gain and front-to-back ratio, and the accompanying trend (though not the actual value) of radiation resistance; the null in the back direction for 135° phasing is however restricted to about 11·5dB due to current inequality and for the same reason performance deteriorates, relatively, for phase-differences less than 135°.

Three- and Four-element Parasitic Arrays

Slightly improved performance can be obtained by a combination of reflector and one or more directors, the usual arrangement having all the elements in one plane. Theoretically as mentioned on p12.12 the gain is proportional

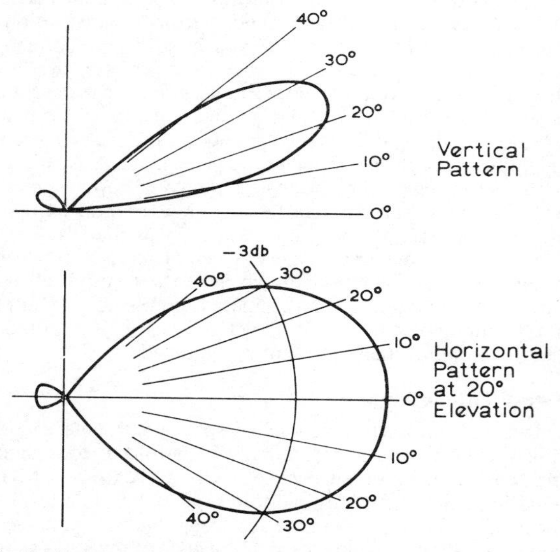

Fig 12.112. Vertical and horizontal radiation patterns of a three-element Yagi array spaced 0·15 + 0·15λ, and erected at a height of λ/2. These patterns, which were obtained from a scale model, do not vary much with other spacings of the elements. Note that the front/back ratio along the ground may be very different from that at 20° elevation. The vertical pattern, of which only the lower lobe is shown in the figure, depends on height, which should be as great as possible.

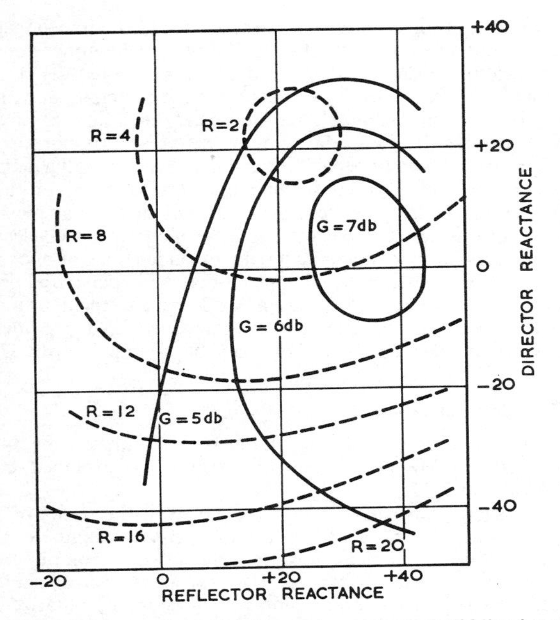

Fig 12.113. Contour chart showing contours of gain (solid lines) and input resistance (broken lines) as a function of the tuning of the parasitic elements. This chart is for a spacing of 0·15λ between elements, but is also typical of arrays using 0·2 + 0·1λ spacings.

to the square of the number of elements, but this requires very precise adjustment of all currents and phases together with acceptance of very narrow bandwidths and low radiation resistances which lead to severe losses. A four-element driven array on these lines has been described [3], but this had a gain of only 8·7dB out of a theoretical 10·1dB. The interdependence of amplitudes and phases prevent the achievement of comparable results with Yagi arrays. There is no advantage in using more than one reflector in an in-line array, though many directors may be used with advantage if they are widely spaced. For hf work one, or at the most, two, directors are the limit because of the size of the array, but in the vhf bands longer arrays with more than three elements may be built, and are described in Chapter 13.

The characteristics of two-element arrays can be calculated quite well, but three or more elements lead to such complexity that their analysis requires a computer. A vast amount of experimentation by amateurs and others has, however, resulted in comprehensive knowledge, and an infinite variety of prescriptions for element lengths and spacings.

The vertical and horizontal radiation patterns of a typical three-element Yagi array are given in **Fig 12.112**.

The maximum gain theoretically obtainable for parasitic arrays with equally-spaced elements is 7·5dB for three elements. These patterns have their half-power points at a beamwidth of about 55° in the *E*-plane (the plane of the elements) and 65–75° in the *H*-plane (at right angles to the elements). These high gains are, unfortunately, associated with a feedpoint resistance of only a few ohms, so that in practice the loss due to conductor resistance and environment may be appreciable in proportion. In addition, the tuning is very sharp, so the array will only work usefully over a small part of, say, the 14MHz band, while the line matching for a low vswr is very difficult.

The power loss due to conductor resistance, or high feeder vswr, is such that these high gains are never realized in practice. Better general performance is obtainable by designing the aerial for less gain, as may be appreciated with the aid of **Fig 12.113**. This is a "contour" map of gain and impedance as a function of parasitic tuning, for a Yagi with element spacings of 0·15. The lower right-hand corner of the chart is clearly the region to aim for, where the gain is still over 6dB while the feed resistance approaches 20Ω. This region of the chart corresponds to rather long reflectors (+ 40 to + 50Ω reactance) and rather shorter than optimum directors (— 30 to — 40Ω), and also is found to be the region of best front-to-back ratio.

Although the chart is for aerials with both spacings equal to 0·15λ, the resistance and bandwidth are both somewhat improved, with the gain remaining over 6dB, if the reflector spacing is increased to 0·2λ and the director spacing reduced to 0·1λ. Such an array would operate satisfactorily over at least half the 28–29MHz band or the whole of any other band for which it could be constructed.

The overall array length of 0·3λ is practicable on all bands from 14MHz upwards, though the 20ft boom required for a 14MHz array is rather heavy for a rotary array, and there is a temptation to shorten it. Spacings should not be reduced below 0·1λ for both reflector and director as the tuning will again become too critical and the transmitter load unstable as the elements or the feeder move in the wind. It should not be overlooked that in general for a given cost, and degree of practical difficulty, the lighter the beam the higher it can be erected; in terms of practical results this may be worth a lot more than additional elements.

The optimum tuning does not vary appreciably for the different spacings, and it is therefore possible to construct a practical design chart (**Fig 12.114**). An array made with its element lengths falling in the shaded regions of the diagram

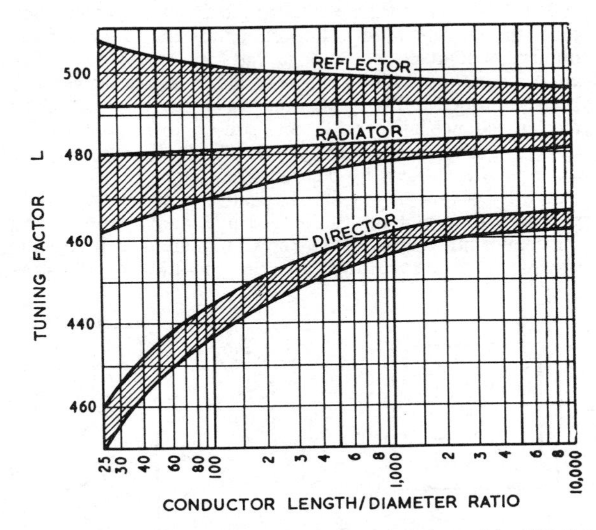

Fig 12.114. Design chart for Yagi arrays, giving element lengths as a function of conductor length-to-diameter ratio. The tuning factor L is divided by the frequency in megahertz to give the lengths in feet. These curves are for arrays of overall length 0·3λ, with reflector reactance +40 to +60Ω and director −30 to −40Ω, and give arrays of input impedance between 15 and 20Ω. Element lengths which fall within the shaded areas will give an array which can be used without further adjustment, though the front/back ratio may be improved by adjusting the reflector.

will give good performance without further adjustment. The length of the director may be decided in advance. The reflector may then be adjusted experimentally to improve the front-to-back ratio. It will be seen that the radiator is somewhat longer than a normal dipole; this is because the parasitic elements have a detuning effect on it. The addition of a second director also spaced 0·1 to 0·15λ will not materially affect the above recommendations.

The adjustment of the array can be carried out as indicated for two-element arrays, using in this case the nominal front-to-back ratio as the criterion. For this purpose the elements can be made with sliding tube extensions or, alternatively, a short variable stub can be fitted at the centre. The tuning can be carried out without reference to impedance matching, which is always the last operation. Adjustments should be made with the array as high and as clear as possible, since tuning alters near the ground, and performance should be checked with the array in its final position. It should be emphasized that attempts to adjust for maximum gain lead inevitably into the top right-hand corner of Fig 12.113 where bandwidth is low and matching difficult.

Impedance Matching of Yagis

The 15–20Ω aerial impedance resulting from the foregoing recommendations can conveniently be matched to 80Ω line by a single fold of the driven element. Since the impedance rises on either side of the optimum frequency, it is better not to match at this frequency, but to bring the impedance up to, say, 60Ω (vswr = 1·3 with maximum voltage at the aerial). The aerial will then hold within a vswr of two over a somewhat wider band.

It is possible to multiple-fold the driven element using, for example, a rigid central tube surrounded by a cage of wires joined to radial struts a few inches long at the free ends of the tubes. In this way the feed-point impedance can be raised to 300 or 600Ω. When there are several thin wires surrounding a tube, the step-up ratio is approximately the square of the number of conductors added, so that four wires would bring an 18Ω aerial up to 300Ω. Most of the examples which have been published are for maximum-gain Yagis, with a basic feed impedance of 6 to 8Ω, and use up to nine wires. To some extent this multi-folding improves the bandwidth, but this is limited largely by the parasitic elements.

It will be appreciated that basically two adjustments are necessary for matching to the feed line, namely tuning and impedance ratio. Most of the known matching arrangements provide only one of these, and the operation of adjusting the length of the radiator at the same time as the impedance ratio can be tedious and difficult. If however the dimensions are carefully established in accordance with Fig 12.114 it should only be necessary to carry out the matching adjustment to obtain reasonable performance, the T and gamma transformers of Fig 12.57 being suitable for this purpose. Alternatively high impedance feeder may be used with a delta match, but this can cause difficulty with disposal of the feeder which must be kept well clear of any metal mast. Provided insulated wire is used and the main run is well clear of the mast, accidental contact with a wooden mast at one or two points is not serious with a well-matched open wire line. Details of an alternative design for a three-element beam, centred around the Clemens match which also provides balance-unbalance conversion, will be found on p12.37.

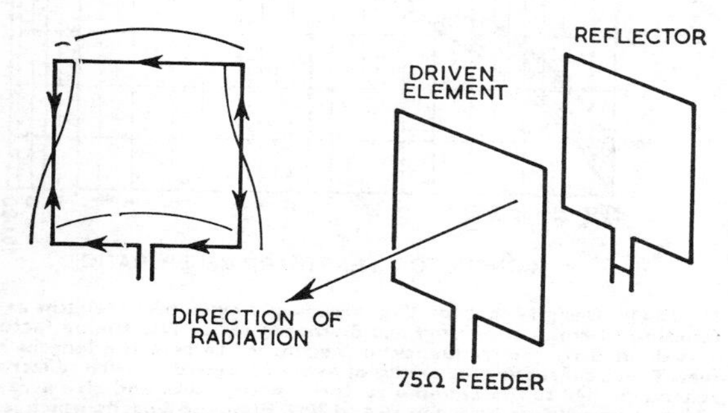

Fig 12.115. The basic cubical quad showing the current distribution and direction around the driven element. Radiation from the vertical sides tends to cancel as the currents are in opposition.

The Cubical Quad

This aerial is very popular with dx operators on the 14, 21 and 28MHz bands and was originally developed by W9LZX in 1942.

It usually consists of a driven element in the shape of a square, each side of which is a quarter-wavelength long. Behind this driven element is placed a closed square of wire to act as a parasitic reflector. The layout is shown in **Fig 12.115** and in this configuration it is horizontally polarized. Feeding in the middle of one of the vertical sides would change the polarization to vertical. The loops can alternatively be mounted with their diagonals vertical, the feed-point being at a current loop in both cases and the performance identical.

The currents in the top and bottom wires of any one loop are in phase and add to give broadside horizontally polarized radiation. Those in the vertical sides are in anti-phase and cancel in the broadside direction, giving rise only to a small amount of vertically polarized radiation off the side.

The radiation resistance and other properties of the cubical quad vary with the spacing between the driven element and the parasitic reflector in the same way as other small beams, and Fig 12.15 is applicable subject to use of the appropriate scale factors, ie 1·55 for the radiation resistance of each element if both are driven or 2·3 for the more usual arrangements using a driven element and reflector. These figures are derived from those for folded dipoles, dividing by two because the end portions of each half-wavelength of wire (which account for 30 per cent of the total field-strength from a dipole) have been virtually prevented from radiating, and reduced by a further 20 per cent to account for the slightly increased field-strength due to "stacking gain" which occurs because of the slight resemblance to a stacked pair of dipoles, the "point source" assumption being invalidated to this extent. The usual scale factors S for the required phase shift and S^2 for the radiation resistance, where S is the spacing in eighths of a wavelength, also apply. Element currents being nearly equal for spacings of 0·1 to 0·15λ, there is no need for further correction on this score as there is with dipole arrays.

The cubical quad presents a good match to 75Ω cable with an element-spacing of 0·15λ and the reflector tuning adjusted for ϕ/ϕ_0 = 0·8 but by shortening the reflector it should be possible to obtain a match to 50Ω cable together with an increase of 0·6dB in gain. A spacing of 0·125 is also satisfactory and provides a match to 50Ω cable assuming reversion to the lower gain condition, or alternatively an

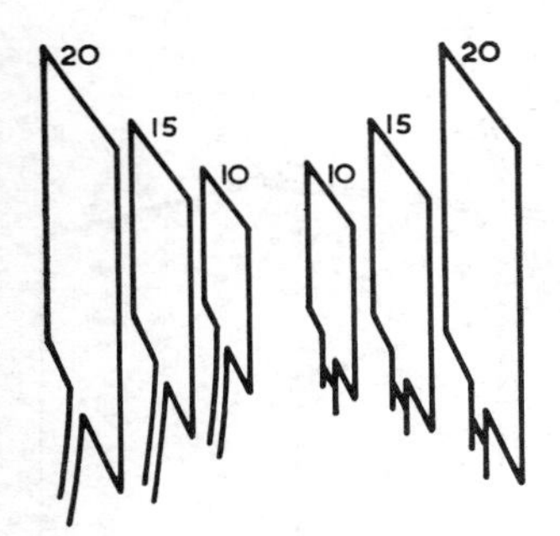

Fig 12.116. A three-band nest of cubical quads maintaining typical spacing for each band.

swr of about 1·5. The reader will appreciate however that many variations are possible.

There should be no difficulty in achieving a gain figure of 6dB if, for closer spacings, some restriction of bandwidth and either increased swr or more complex matching devices are acceptable. There is *no* justification for claims that the quad provides gain figures of 10dB or more, or that it achieves enhanced dx signal gain beyond the 6dB figure by concentration of energy at lower angles of elevation. The latter belief apparently arises from the fact that a quad element radiates less energy vertically upwards than a dipole, but this is in fact the origin of the 1dB superiority of quads over dipole arrays and effective gains appreciably in excess of 6dB can be achieved only in defiance of the laws of nature; practical aerials consist of point sources located above a reflecting plane, the distribution of energy in space and therefore the gain being determined entirely by the basic geometry of the situation. The relative phases and amplitudes of currents are part of this geometry and need to be "right", also the sources themselves have to comply with a few obvious rules in the interests of efficiency, but for a given geometry, polarization, and number of sources no type of aerial can be electrically superior to another. To appreciate why a quad loop approximates to a single point source it is convenient to imagine the top and bottom horizontal wires replaced by separate half-wave dipoles, ie two sources spaced by $\lambda/4$. From Fig 12.9 the mutual resistance between them is 40Ω so that each dipole has a radiation resistance of 73 + 40 = 113Ω. For equal power radiated therefore the current in each is 0·565 times that in a single dipole, the total field is increased by 13 per cent or 1·1dB only, and the loop is much closer in performance to one than two separate sources of radiation. Even this slight advantage is partly eroded when two elements are arranged as a beam, since an increase of gain increases the optimum stacking distance (p12.16). Despite these reservations the quad has several important advantages in addition to the slightly higher gain and possibility of feeding it from low-impedance *balanced* line without any kind of transformer. It is particularly easy to adjust for good results, the small span of the elements simplifies construction of rotary beams, and the near-equality of currents in the driven element and parasitic reflector produces sharp nulls in the radiation pattern which can often be used by rotation of the beam to suppress interference. This last item is a property which it shares with driven arrays. Another property of quad loops, deserving to be more widely known, is the possibility of using them with resonant feeders as multiband radiators, this being particularly valuable for the construction of fixed reversible beams covering two or three bands.

There are certain points of importance to note in the electrical construction of a cubical quad. Because of the variation of Z_0 round the loop, the wire-length formulae used for dipoles are not applicable and the overall length is slightly longer than it would be for an aerial in free space. For the driven element, each side of the square may be $248/f$(MHz) and fed with 75Ω flat twin, or coaxial with a balun. If an unbalanced coaxial cable feed is used there will, as with other aerials, be a distortion of the lobes caused by the imbalance. However, if suitable tuning stubs are used neither the dimensions nor the shape of the loop are in any way critical.

The size of a parasitic reflector element is usually made the same as that of the driven element but it is made electrically longer than the driven element by means of a short circuited stub which can be adjusted for maximum gain or front-to-back ratio. It will be found quite easy to adjust the stub for a front-to-back ratio better than 20dB.

Some amateurs have avoided using stubs to tune the reflector by making the overall length of the reflector longer than the driven element. This method works quite well in practice if a lengthening factor of about 2 or 3 per cent is used but if no adjustment is possible it is preferable to err on the high side.

The following table gives approximate dimensions suitable for the dx bands. If a stub is used it can be taken as a rough guide that 2ft of wire in the loop is equal to 3ft in the stub.

Band	Side of quad	Spacing	Side of quad reflector without stub
14MHz	17ft 6in	8ft 5in	18ft 0in
21MHz	11ft 8in	5ft 7in	12ft 0in
28MHz	8ft 8in	4ft 2in	8ft 11in

The foregoing description is for a single band quad but a two or three band quad can be made up by mounting the elements concentrically on a bamboo spider. The dimensions will be the same as in the above table if three quads are mounted as shown in **Fig 12.116**. If, however, the driven elements are mounted in the same plane and the reflectors similarly, there will be a change in the gain and the radiation resistance on the higher frequencies due to the increased spacing. If the spacing is optimum for 14MHz then the gain will be lowered and the radiation resistance will be increased for the 21 and 28MHz quads. At 28MHz the gain will drop to 5·0dB and the radiation resistance will increase to 140Ω.

A suggested constructional layout for a single band quad is shown in **Fig 12.117**. If a three band quad is to be constructed, the length of the horizontal boom should be reduced, and the bamboo arms angled outwards, so that the correct element-to-element spacing is maintained for each band.

The Swiss Quad

Quads may of course also use pairs of driven elements and the Swiss quad (**Fig 12.118**) described below provides improved mechanical construction without sacrificing performance [12]. The elements consist of two parallel squares having quarter-wavelength sides and spaced 0·075 to 0·1 wavelength, and the squares are fixed to the mast by bending the centres of the horizontal portions at an angle of 45° towards the fixing points, thus avoiding the need for any other supports. No insulators are required in the construction, a good point since the bamboo supports commonly used for quad loops can give trouble in wet weather.

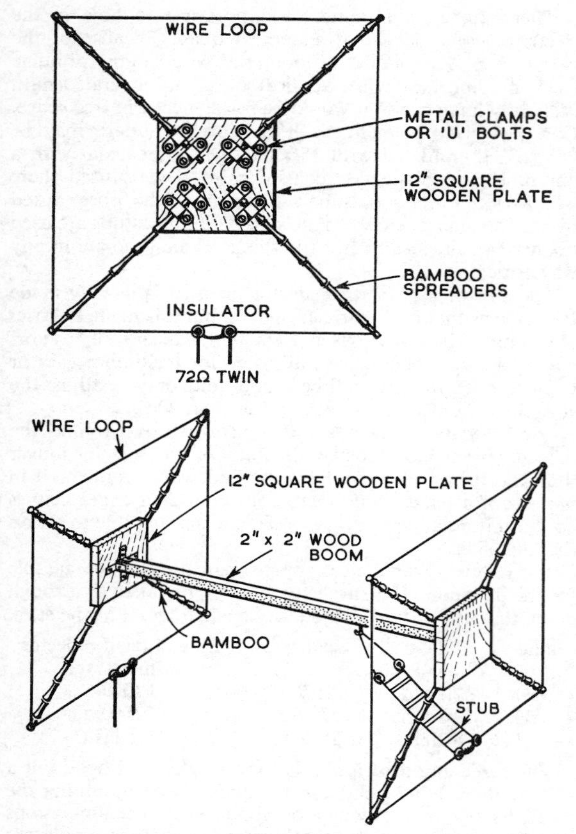

Fig 12.117. A suggested method of construction for a lightweight single band cubical quad.

The elements are fed in phase using a form of T-match as in **Fig 12.119**. This is readily modified to form a gamma match, but if coaxial cable has to be used it would be better to retain the T-match and use a balun to minimize feeder radiation. Phasing is obtained by the $\pm X$ method (p12.70), making one element shorter and the other longer than the resonant length. Despite the pinching-in, it has been possible to make the elements more or less full size overall; the pinch merely fulfils the general requirement for the lengthening of loop elements which with quads usually takes the alternative form of loading stubs. Fig 12.15 can be used as a rough design guide also for the Swiss quad but the narrow spacing reduces the radiation resistance so that a phase shift of 30° between the elements is likely to be a reasonable compromise, giving near-maximum gain. Adjusting the figures for a spacing of 0·1λ and the above scale factor of 1·6 gives a required phase-shift of 24°, and a radiation resistance of 21Ω. From the explanation of the 8JK aerial (p12.69) it follows that $(R - R_m)$ must be proportional to R and to S^2, from which $R_m = 104\Omega$ and the required reactance given by $(R + R_m)\phi/2$ is equal to 46Ω. Assuming $Z_0 = 800\Omega$ for the loop, this requires $\pm$ 2 per cent detuning. The original article specifies $\pm$ 2·5 per cent and quotes figures upwards of 30Ω for the radiation resistance, both figures being consistent with working slightly nearer to the condition of maximum nominal back-to-front ratio which

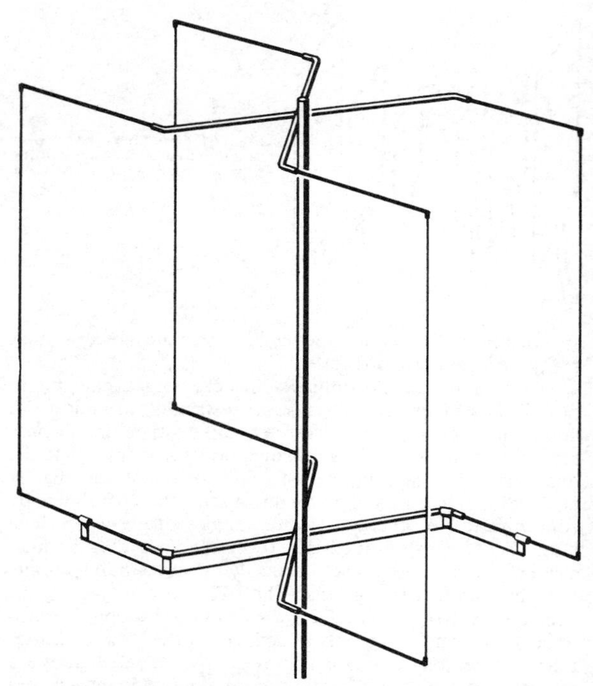

Fig 12.118. General arrangement of the Swiss quad.

could be helpful in view of the rather low radiation resistance. There is clearly considerable latitude in the design of this type of aerial but **Table 12.9** represents a proven design and is the obvious starting point for further experiments, being taken from reference [12] which may be consulted for general guidance on constructional methods. It should be noted that the "X" section in the middle contributes little to the wanted radiation because the wires are inclined at 45° to the direction of propagation and their average spacing is only half that of the main wires. The radiation from them is a clover-leaf pattern with some radiation at right angles, and indeed all four petals would become equal with "8JK" phasing. At 0·1λ spacing the "X" section will have negligible effect on the performance or the calculations, but if the spacing is increased the importance of the "X" increases rapidly, not only because of its own greater dimensions, but also because the width of the loops will

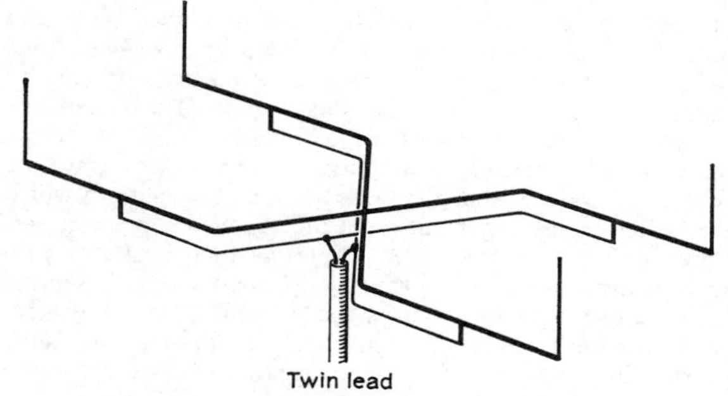

Fig 12.119. Balanced feed and matching system.

TABLE 12.9
A Swiss quad

Band and Frequency	Wave-length (m)	Aerial height (in)	Reflector width (in)	Director width (in)	Spacing (in) 0·1λ	Spacing (in) 0·075λ
10m 28,500kHz	10·52	116	121·5	110	41·3	31
15m 21,200kHz	14·14	156	164	148	55·5	42
20m 14,150kHz	21·20	234	246	222	83·5	62·5
40m 7,050kHz	42·60	470	493	443	168	126

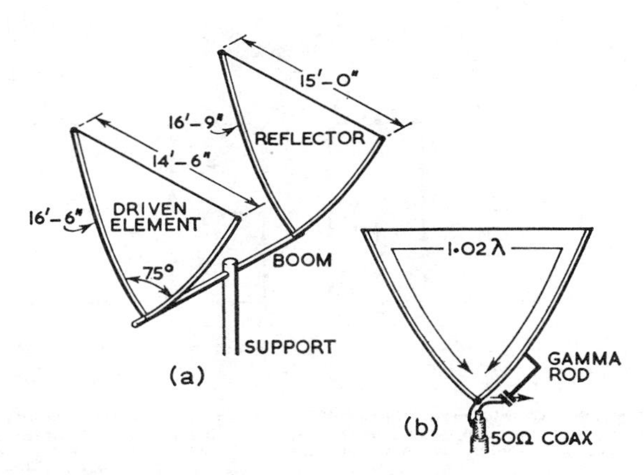

Fig 12.120. HRH delta-loop beam. Main dimensions shown for 21MHz, using 9ft spacing and 3ft gamma rod. For 28MHz the driven element has 12ft sides, 11ft top and a 75° angle; the boom spacing is then 6ft 6in and the reflector has 12ft sides and an 11ft 10in top.

shrink. This tends to reduce the radiation resistance as the inverse square of the width, thus offsetting and ultimately reversing the usual trend whereby an increase of spacing increases the radiation resistance. In the case of driven arrays using the reactance method of phase-shifting it is important to distinguish between the radiation resistances and the feedpoint impedance, since the reactances form a tuned circuit which acts as a transformer. But for this and the T-match, "looking" into the aerial from the feeder one would "see" the two radiation resistances in parallel, ie 10Ω, but this will be transformed by the reactances to 40 or 50Ω and the T-match provides further adjustment.

The Delta Loop

It has been mentioned that the shape of a quad loop is not critical, and it can in fact be distorted into a wide variety of shapes (even squashed flat so that it turns into a folded dipole), with little or no change in performance. One popular variation of the quad aerial is the delta loop, and experiments have been described [11] using a pair of triangular loops with 50ft of wire per loop as a triband beam for 14, 21 and 28MHz. Despite the incorporation of a rather complex and frequency-sensitive matching network, there was no noticeable change in tuning or matching on any of the bands when the shape of the loops was altered from inverted triangular (delta) to square. Designs have been worked out for 21 and 28MHz [13, 14] based on 2in diameter boom with 1in diameter side-members of aluminium tubing, the top ends being joined with wire as shown in **Fig 12.120**. The total length in feet is given by $1005/f$(MHz) for the driven element and $1030/f$ for the reflector. Directors if used should be 3 per cent shorter than the driven element but add very little to the gain. A spacing of 0·2λ is satisfactory. The wire at the top is in tension so that the sides of the triangle are bowed, the angle at the apexes, ie at the boom ends, being 75°. The gamma match could obviously be replaced by a T match with twin feeder, or a balun employed, as a precaution against feeder radiation.

Switchable Arrays

As an alternative to rotary beams, two or preferably three fixed reversible beams may be used to provide all-round coverage. Advantages include instant switching of beam direction and the ability to use a number of multiband systems which are difficult to adapt mechanically to rotary beam constructions; these are discussed in a later section. If only two horizontal beams are used, mutually at right angles, performance will be within about 1dB of maximum gain over four 60° segments but in between these it will fall off roughly to equality with a dipole. Two beams may be mounted on the same pole with a small vertical separation for mechanical clearance, provided the one not in use is detuned. This is achieved most easily with the aid of resonant feeders, the ends of which are accessible from ground level, and such an arrangement also facilitates multiband operation.

Beam reversal may be achieved in various ways. Switching a director to act as a reflector as described on p12.73 is one possibility but may have considerable effect on the swr. This may be overcome in the case of a three-element beam by using a switch or relay to shorten one or the other of the parasitic elements, an arrangement particularly recommended with loop or folded dipole elements having λ/2 resonant feeders which may enable both tuning and switching to be carried out at or near ground level. Any type of two-element beam can be made reversible by attaching feeders to both elements so that either can be used as a driven element, the other being tuned via the feeder as a parasitic element; alternatively both elements may be driven as described on pp12.11 and 12.72.

Multiple Arrays

It has been shown that unless considerable space or height is available, the effective gain of high-frequency beams cannot exceed (though it should equal) a total of 5 or 6dB. On the other hand, if enough height is available it may pay to stack two beams vertically, and assuming a large garden, farm or access to neighbouring fields a number of other possibilities arise; it is quite likely that due for example to uneven ground contours or local obstructions different aerial positions will favour different directions, and by using a number of switched beams it may be possible to obtain considerable advantage in certain directions. If for example the ground slopes steeply in one direction, the ground reflection reinforces the direct wave in that direction even for angles which are low relative to the horizon, so that optimum performance may be achieved with a height as low as 3λ/8 and closely approached with a height of only λ/4. To take full advantage of such slopes it is best to locate the aerial in a position some way below the top, as the ground area from

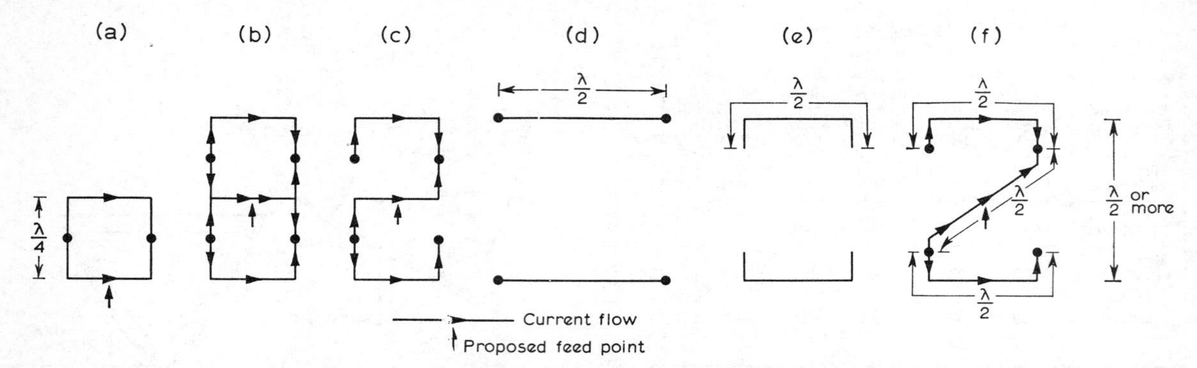

Fig 12.121. Development of the 9M2CP Z-beam: (a) conventional quad; (b) double quad—difficult to tune and little apparent advantage over (a); (c) open-ended double quad developed to eliminate possibility of double currents in centre element—it gave improved results over (b); (d) λ/2 dipoles spaced about λ/2 apart vertically; (e) as (d) but with ends bent inwards to permit joining diagonals with λ/2 wire; (f) the 9M2CP Z-beam as now evolved.

which reflection takes place extends for some distance behind the aerial when the height is low and the angle of radiation relative to the ground is large. For other directions, unless slopes are available for these also it will usually be best to mount the aerial as high as possible and preferably at the top of the slope.

If two rotary beams can be spaced apart by at least λ/2, a gain of about 3dB may be obtained by driving them both, with a phase difference such that the fields add up in phase in the required direction. This does not involve critical adjustments, as the gain drops by only 1dB for a power inequality of 10 times or a phase error of nearly 60°, assuming that the two aerials are equally good. If one aerial is xdB worse than the other it should be given xdB less power, the benefits obtained from addition of the inferior aerial, for performance differences of 3dB and 6dB, then being 1·8dB and 1·1dB respectively. Phasing may be achieved by switching additional short lengths of feeder in series with the feed to one or other of the aerials, which will produce a 2 : 1 mismatch if both aerials are perfectly matched initially; this can be corrected, eg by a λ/4 transformer (p12.25), if it is outside the range of adjustment provided on the atu. In principle the $\pm X$ method of phasing can also be used but, unless corrected, the matching will then vary with the amount of phase-shift. With balanced feeder a single λ/6 section of phasing line can be switched in as required to give either aerial a 60° lag compared with the other, and this with reversals gives six phases spaced 60°. The maximum phasing error is therefore 30° or, translating this into performance, 0·3dB. A 6-pole 6-way switch is required. With coaxial cable the arrangement shown in Fig 12.124 on p12.84 is suitable.

Z-Beam

An example of vertical stacking is provided by the 9M2CP Z-beam [15]. This is derived from the quad as illustrated in **Fig 12.121** and can incorporate a double delta 21MHz beam using the same supports, as illustrated by **Fig 12.122.**

On 21MHz there is a standard delta loop, with a similar delta pair mounted below the boom, as a mirror image of the upper pair. A common gamma tuning capacitor is used and is better and easier to adjust than separate capacitors. The gamma arms are of equal length and matching is similar to that found with a single delta loop beam.

For 14MHz, 5ft (or longer) wooden extensions are placed in the ends of the delta loop Vs to provide at least half-wavelength vertical spacing between upper and lower horizontals. The whole is laced round with nylon line, to form a framework for the 14MHz beam and also act as support for the lower Vs which otherwise tend to collapse inwards under their own weight. The radiator is made up of 1½λ of wire laced as shown in Fig 12.122, and fed with a gamma match at the centre. The length of wire is slightly shorter than 1½λ to account for two end effects only. A sliding loose noose of

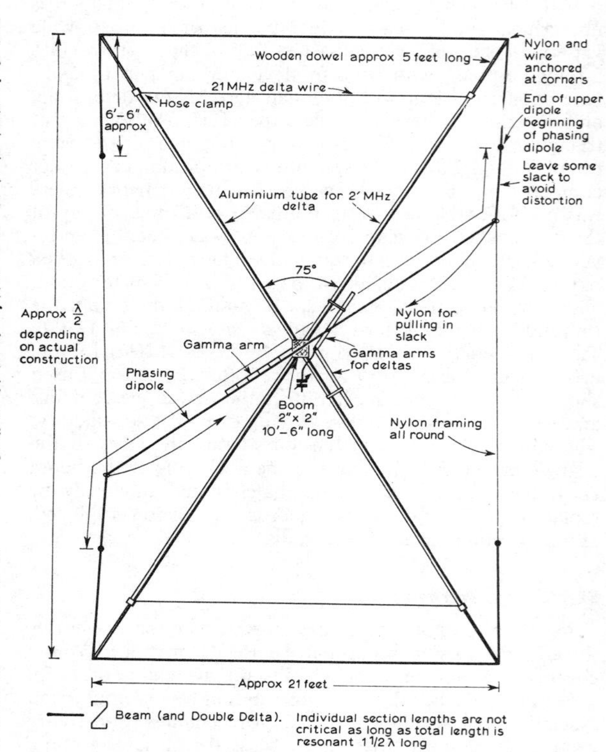

Fig 12.122. Constructional details of the 9M2CP 21MHz double delta and 14MHz Z-beam.

nylon is run round the vertical nylon and includes the radiator so that it can be adjusted to take up slack when in position. The reflector is cut to just under the full theoretical $1\frac{1}{2}\lambda$ and a tuning stub is put in the centre. This can readily be adjusted for maximum gain or best back-to-front ratio. The tuning is quite marked for best back-to-front ratio and this also provides what appears to be the best gain.

VERTICAL HF AERIALS

Vertical aerials have already been discussed in the context of the low frequency bands and in principle any of the low frequency arrangements may also be used on the higher frequency bands. In general, at 14MHz and higher frequencies reasonably good horizontal aerials can be erected so that the possibility of getting a small (though by no means useless) amount of low-angle radiation by using a poor but simple and cheap vertical aerial, loses much of its interest. To summarize a rather complex situation, vertical aerials tend to give a relatively poor account of themselves at these frequencies unless erected over sea water, or close to the sea and high enough to ensure that the reflected wave is accounted for mainly by sea reflection, in which cases they may afford a simple means of achieving efficient low-angle radiation [2]. If height or space are very restricted a vertical aerial may be superior in dx performance to a horizontal aerial and, assuming identical *mean* heights of $\lambda/2$ or greater with freedom from obstructions, both polarizations will give rather similar low-angle performance, although the vertical aerial is then usually more difficult to erect and feed. In the case of the higher frequency bands it is rarely necessary to use an earth connection and usually difficult in any case to achieve an electrically-short earth lead. Alternatives include conventional methods of centre feeding, possibly with some form of capacitive end loading to reduce the vertical dimensions (eg Fig 12.102), a counterpoise, eg on the lines of Fig 12.87, or a "ground plane", **Fig 12.123(a)** or (**b**). If however for some reason use of an earth is unavoidable, the aerial should if possible be folded as in **Fig 12.123(c)**. A high-voltage capacitor of the order of 2,000pF should be connected in the earth lead as a precaution against fault conditions involving the supply mains and its earth systems.

The Ground Plane

Problems of earth resistance could in principle be eliminated by erecting the vertical over a perfectly-conducting surface, such as a large sheet of copper. Moreover, if such a sheet were of infinite extent, the image-reversal property of vertical aerials could be fully exploited and efficient radiation achieved at angles down to zero elevation. Practical aerials have in fact been constructed using buried earth systems containing many miles of wire and achieving a dx performance difficult to achieve in any other way, but such systems are clearly beyond amateur resources and the ground-plane aerial, Fig 12.123, does not incorporate a ground plane in this sense. On the other hand a system of three or four radials as illustrated does provide a very low impedance path to earth, and is more efficient in this respect than a simple counterpoise as there is negligible loss of energy by radiation from it in unwanted directions; this follows from the fact that the fields produced at any point in space by the individual radials tend to add up to zero. The radials act

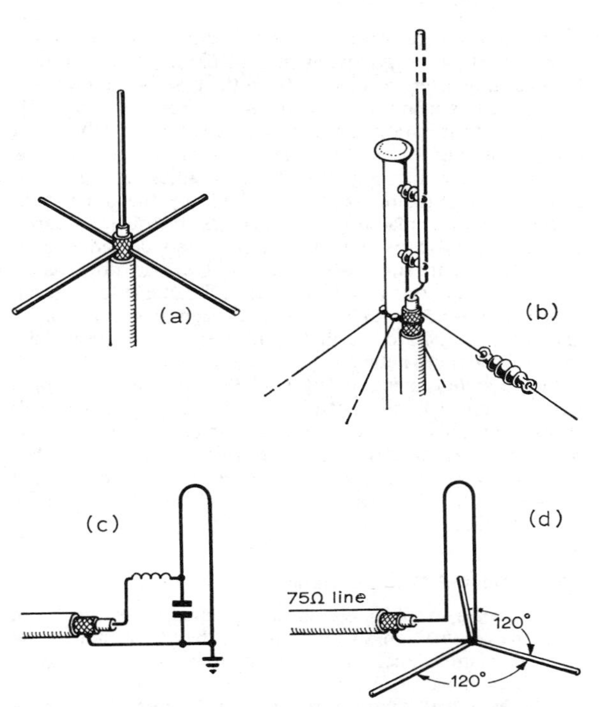

Fig 12.123. $\lambda/4$ vertical aerials. (a) Ground plane. (b) Ground plane with sloping wires which may be used as guy wires. (c) Folded wire with transformer (see text on impedance matching). (d) Folded ground plane. The cable size is exaggerated for clarity.

therefore as a zero-resistance earth connection, not as a substitute for a reflecting ground plane, and the radiator itself approximates to a point source of vertically-polarized radiation. As compared with a half-wave dipole the advantages are mechanical arising from a shorter vertical portion and an easier method of attaching the feeder. Being in effect one half of a dipole, twice the current is required to produce a given field and the radiation resistance remote from earth should therefore be $73/4 = 18{\cdot}25\Omega$, but measured values are usually a little higher. At very low height the aerial with its image forms a single dipole but there is nevertheless a gain of 3dB relative to a dipole in free space since the radiated energy is confined to a single hemisphere; this brings the radiation resistance up to about 36Ω, ie it is the same as that of an efficiently grounded $\lambda/4$ vertical, p12.54. For comparison, the lowest possible mean height of a full-length vertical dipole is $\lambda/4$ and its gain from ground reflection over perfectly conducting ground would be 6dB, as would also be that of a ground-plane aerial at the same effective height which, for this purpose, should be measured to the "centre of gravity" of the current distribution.

Matching the impedance of a ground plane to the widely used 50 or 75Ω cables presents a problem which is considered more fully in the section on impedance matching. Whereas the feed-point impedance of an earthed quarter-wave aerial may lie between 35 and 50Ω, that of a true ground plane may, as we have seen, be 20Ω or less, and feeder cables of this impedance are not available. One solution to this problem can be found by tilting the radials downwards as illustrated in Fig 12.123(b) since this raises the feed-point impedance towards that of the dipole, which the aerials would become if the radials were considered to be "tilted" directly downwards at 90°. Four guy-wires may be used, fitted with insulators at a quarter-wavelength from the base of the radiator, and spaced as nearly as possible at 90° from each other. If the angle of tilt is 45° the system will be a good match to 50Ω cable. Standing waves along the feeder can be measured as described later, the angle of tilt being adjusted slightly to arrive at a minimum swr. It is possible by the use of still steeper angles to match into 75Ω cables, but such an aerial becomes very nearly a vertical dipole and demands considerable height, since the radials will extend downwards for nearly a quarter-wavelength. If 75Ω cable is to be used, it is better practice to employ a horizontal ground plane and fold the radiator as in Fig 12.123(d), or to use some form of matching network.

In the arrangement of Fig 12.123(b) the radials become tightly coupled to the outside of the feeder and there will almost certainly be some radiation from it. This may be reduced by taking care to avoid resonant lengths of feeder, and if trouble is experienced, it may be necessary to coil up a few feet of the cable to act as a form of choke effective on the outer surface only.

The Extended Ground Plane

The aerials so far described are assumed to be very close to a quarter-wavelength long, and as has been stated, the feed-point impedance is invariably low. As the aerial is lengthened however, the feed-point impedance increases, and can be expressed in the form $R + jX$ where R is a pure resistance and X an inductive reactance which would represent a mismatch. Increasing the aerial length from a quarter to approximately one-third wavelength raises the resistive component to 75Ω, while the reactance X rises at the same time to between 300 and 400Ω. It is thus possible to obtain a good match directly into any feeder of impedance between 50 and 75Ω by lengthening the aerial by a suitable amount, while at the same time tuning out the additional inductive reactance thus introduced by means of an adjustable series capacitor. The adjustment will, of course, only be correct for one frequency band, where it provides a simple and effective alternative to the more sophisticated methods of impedance matching set out in the following section.

Arriving at the best length for the extended ground-plane aerial cannot be done reliably by calculation, since it is related to the form of earthing, the number and length of radials, etc, in actual use, and is dependent to a lesser extent upon the height of the system above ground, and the effect of surrounding objects. 0·375 wavelength is frequently regarded as a good compromise or starting point. However, if an swr meter is available the optimum length can be simply arrived at by trial and error. The standing-wave ratio is measured with the capacitor adjusted to give minimum reading. A small adjustment to aerial length is then made, and the ratio checked to see if it is better or worse; preferably the aerial should initially be too long, and the reduction should lead to a reduced swr. It is then only necessary to reduce further the length until by adjusting the capacitor the swr approaches unity. If sloping or movable radials are in use, the final adjustment can be made even more easily by repositioning these, or even shortening one of them slightly. Increasing the radial angle from the horizontal, or even reducing the length of one radial wire, is equivalent to slightly lengthening the aerial itself.

Multi-band Ground Plane

The principle of the all-band dipole, Fig 12.97, may be extended to the ground plane by using a set of verticals, one for each band joined in parallel at the base, and a separate set of radials for each band. Alternatively it may be found simpler to use traps (p12.84) in the vertical portion, in conjunction either with separate sets of radials or trapped radials.

The Loaded Vertical

Vertical or ground-plane aerials present little problem at the higher frequency bands where a self-supporting radiator up to 0·375 wavelength long or a wire supported by a mast can be erected. Not infrequently a metal mast or lattice tower insulated from earth at its base can be used effectively, as is common practice at broadcast frequencies. The loading of vertical aerials to permit the use of relatively short radiators has been discussed on p12.52 and aerials designed for the higher frequencies can often be pressed into service on the lower frequencies by the use of one or other of the loading devices described. Loading is common practice in mobile installations where the length of whip aerial is strictly limited by practical considerations.

Base loading is clearly the most convenient since the loading coils are accessible for changing bands, and can also be used as a coupling transformer on the lines shown in Fig 12.79. Because of relatively large coil losses, base loading is generally confined to adapting an aerial for use on the next lowest frequency band, because the inductance will not then be large enough to upset the current distribution seriously, or to introduce excessive losses. Thus a vertical aerial designed for $\frac{1}{4}$ or $\frac{3}{8}$ wavelength on 28 or 21MHz can be simply adapted for use on 14MHz; and more commonly a 14MHz vertical may be loaded for use on the 7MHz band. A 32ft vertical resonant at 7MHz is not unduly difficult to erect, and may be base-loaded effectively for 3·5MHz. The exact size of a base-loading inductance coil cannot be precisely stated, since in effect it is tuned by the self-capacitance of the aerial above it to earth, and this capacitance is dependent upon local conditions. As a guide however, it can be stated that a coil containing an equal length of wire to the aerial itself will prove rather too large for resonance in the next lowest band, but the aerial can be tapped down such a coil until resonance is achieved, and the unused portion then discarded. A roller-coaster coil is popular for base loading purposes. Whatever type is used, the coil should be of low-loss construction, and protected against the weather by a suitable waterproof housing.

The marked increase in mobile activity in recent years has led to the increase of numerous loaded whip aerials having demountable coils specially suitable for this work. Such coils are ideal as a basis for more permanent loaded vertical aerials for the hf bands. Mobile aerials are in many instances centre-loaded, which implies that the coil will be located part-way up the mobile system, and may be designed for mounting upon a section of 1in diameter aluminium tubing. The section of whip above this coil determines its resonance, and often takes the form of a 4ft to 5ft section of $\frac{5}{8}$in diameter tubing. Coils are available in interchangeable form for the various amateur bands.

Since in a mobile system of this kind maximum current is carried in the section below the loading-coil, there will be a very large reduction in loss resistance if this section is greatly lengthened. Making this change however has only a slight effect upon the resonant frequency, which tends to rise as the capacitance of the top section to earth is reduced, but is simultaneously lowered by the added inductance of the lengthened lower section. Excellent loaded verticals can therefore be constructed by mounting a standard coil and top whip section for the band required upon a mast of 1in diameter tubing, which may itself be of resonant length for a higher amateur band. The mast must be insulated at its base, either by the aid of a mounting insulator, or a short section of wood or bakelite rod or tube, and will be fed as already described. There is no objection to a small trimming-inductance at the base, since the frequency in the loaded-mode is likely to be a little high; should it be otherwise, the top section must be shortened with caution, as this length is very critical. An aerial of this kind will require to be guyed at a point just below the loading-coil, and at other points according to the actual construction and overall height chosen.

The aerial described is effectively top-loaded for the lowest frequency band, and will give useful low-angle radiation. Bearing in mind that a vertical radiator need not necessarily be quarter-wave, it is possible to excite such a system on several bands by altering the feeder termination. Where space is very limited, the feeder line reasonably short, and the base of the aerial accessible for adjustment, a vertical system on these lines can be a useful compromise providing a dx capability on several bands.

As a practical example, the vertical aerial may consist of from 32ft to 40ft of 1in diameter tubing, terminating at the base in a coil similar to that described in the context of Fig 12.79(b) with provision for tapping the feeder up or down the inductance, but without the series capacitor shown in that illustration. At the top this 1in tubing carries a loading-coil of the type used in 1·9MHz mobile installations, and carrying above it the usual whip section which will resonate towards the hf end of the 1·8 to 2·0MHz band. To operate on this band, the feeder is taken very nearly to the top of the base loading-coil, the remaining few turns of which resonate the system to the required part of the band while the larger portion of the coil forms a high-inductance shunt across the feeder termination.

For operation as a vertical dipole on the 3·5MHz band, the feed-point is moved down the coil until the electrical length of the whole aerial becomes three-quarter wavelength at the working frequency, when the system will be voltage-fed at the bottom of the vertical section. The inductance of the base loading-coil must be sufficient to ensure this condition. Where this is not so it will be of assistance to add a 100pF variable tank capacitor across the whole of this coil, and to tune it to resonance as indicated by maximum brightness from a neon lamp placed adjacent to the top of the loading coil. Maximum current will then occur at the top of the vertical section, immediately below the top-loading coil, and in a favourable position for effective radiation.

Operation on the 7MHz band makes use of the vertical section only as an extended ground plane, the top loading coil serving as a choke. The feedpoint will be very high on the base-loading coil, much as for 1·9MHz operation. The system can be made to resonate in the 14MHz band as a vertical half-wave by bringing down the feed-point as before until the electrical length to this point is three-quarter wavelength, the portion of the base loading coil above the feed-point resonating at 14MHz with its own self-capacitance.

Somewhat more advanced designs based on this conception have been described in which the optimum feed-point is selected by switching. It is recommended that the feeder line be taken underground in order to minimize stray radiation which may disrupt the low-angle pattern. In setting up such a system it is very advantageous to make use of an swr meter so that the best matching can be quickly found, but providing the feeder length is not excessive and that the transmitter loading is satisfactory, results are likely to prove reasonably acceptable.

Vertical Beam Aerials

Most of the horizontal beam aerials described earlier can in principle be used in a vertical position. Other things being equal, the pattern of a vertical beam is narrower in elevation but broader in azimuth than a horizontal beam so that all-round coverage may be obtained with a smaller number of fixed beams, and it is even possible to use for example a single driven element surrounded by parasitic elements suitably switched to give omnidirectional coverage.

Despite these attractive features, vertical beams have not achieved much popularity. It is usually easier mechanically to erect a horizontal beam than a vertical beam at the same mean height which would of course be a greater maximum height, and serious inefficiency may result from the proximity of vertical elements to vertical metal masts, or even wooden masts unless close proximity and high rf voltages can be avoided. There is also the obvious difficulty of bringing the feeder away at right angles, though some compromise may be acceptable if all possible precautions are taken to prevent feeder radiation. The most obvious line of attack on the feeder problem is of course to use lower-end Zepp feed but this usually leads to very poor results unless a balancing stub is used, Fig 12.94. Lack of general awareness on this last point may account for some of the unpopularity of vertical beams. Even with the balancing stub however, one problem remains, ie narrow bandwidth. From Fig 12.6 the impedance at the centre of a full wave dipole is typically 5,000Ω and but for the mutual impedance between the two halves, Fig 12.9, this would rise to 8,500Ω. The Zepp feed "sees" in effect one half of this across one half of the feeder, equivalent to 17,000Ω across the whole feeder, and this is multiplied by about 2–4 times in the case of a close-spaced beam aerial with likely values of spacing and phasing, so that the effective Q will be typically of the order of 80 and the bandwidth much less than the width of, say, the 14MHz band.

If the maximum height is severely restricted centre feed is less of a problem, the height of the feedpoint being probably comparable with the height of an upstairs window, allowing the feeder to be brought away at right angles all the way to the shack or at least for a considerable distance. Assuming sufficient ground area is available a vertical beam can in these circumstances be attractive and the 7MHz beam of Fig 12.102 scaled down to 14MHz has given useful results despite the very low height. Performance was however markedly inferior to that of horizontal beams at 50ft.

A very simple vertical array using two ground-plane radiators and providing all-round coverage with a gain of 3dB is illustrated in **Fig 12.124** overleaf. With $\lambda/2$ spacing and in-phase feed, the beam is bidirectional at right angles to the line joining the aerials whereas with antiphase feed the pattern is rotated through 90° and the array becomes end-fire with a broader radiation pattern. This arrangement is suitable for use also on the low frequency bands. There are of course many possible ways of combining the radiation from as many vertical radiators as space permits and in general a gain of n times can be realized from n radiators spaced upwards of half a wavelength. A particularly interesting arrangement suggested by GW3NJY [16] uses two pairs of grounded quarter-wave verticals located around a common centre, ie at the corners of a half-wavelength square. By feeding one diagonally-opposite pair in-phase and the other out-of-phase so that the two patterns coincide, then combining the two pairs with a phase shift of 90°, a unidirectional pattern is obtained which may be switched to provide all-round coverage. This idea should be equally applicable to dipole or ground-plane elements thereby avoiding earth losses.

It is difficult to give any definite rules governing the choice of horizontal or vertical polarization, because of wide variations in ground constants as well as individual circumstances, but some guidance [2] is provided by Fig 12.23 from which it may be inferred that a low vertical aerial over poor ground should be about equal to a horizontal aerial at a height of $\lambda/2$ for radiation angles of 2–8°, typical of those required for long distance communication. Over good ground the vertical may be appreciably better. It should be noted that whereas a horizontal aerial benefits from foreground reflections from a steep ground slope the vertical does not, because the image is tilted back into the ground. Fresh water or marshy ground is likely to be "good", but it is not an alternative to sea-water which owes its excellence to a conductivity which is greater, by orders of magnitude, than that of any likely alternative.

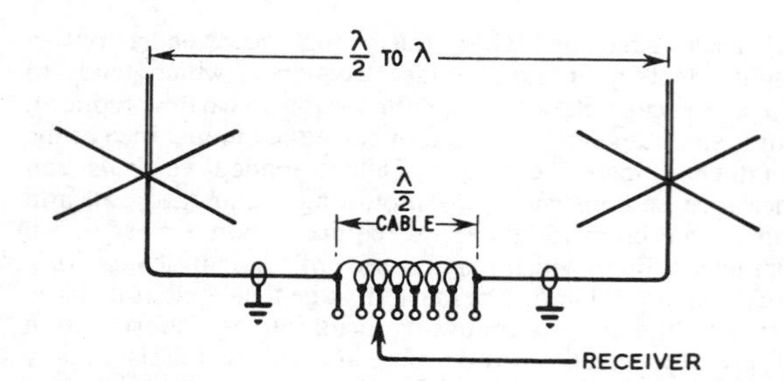

Fig 12.124. Special arrangement of two ground-plane aerials providing electrical steering of the patterns given in Fig 12.12. This gives a gain of 3dB and is often a great help for interference reduction in a receiver as the nulls in the radiation pattern are steerable and the vertical polarization tends to reject high-angle noise and interference.

MULTIBAND AERIALS

In the average location, an amateur will not have space for many aerials, and thus the problem of making one array work usefully on more than one amateur band is an important one. Throughout this chapter, care has been taken to indicate the multiband possibilities and limitations of the various types of aerial. This section presents a number of techniques specifically directed towards the solution of multibanding problems and includes examples of their use. The multiband properties of the simpler aerials, such as dipoles, or long wires fed in various ways, have already been discussed but there are always disadvantages—the radiation pattern changes, and usually there is a high swr on the feeder which can cause inconvenience and sometimes losses.

Multiband Dipoles and Ground Planes

The first approach to a multiband dipole is shown in **Fig 12.125(a)**. Dipoles are cut for the various bands and supported about one foot apart in the same plane, and then joined to a single 80Ω twin line. The theory is that the aerial which is in tune at half-wave resonance takes all the power; just as the coupling between resonant circuits is relatively small when they are tuned to widely different frequencies so the coupling between the dipoles can be ignored as a first approximation and the power radiated from the "wrong" dipoles will be a small fraction of the total, except in the case of a 7MHz dipole which will tend to produce a long-wire type of radiation pattern when the system is driven on 21MHz. One version of this, in which the 7MHz acts as the 21MHz radiator has already been described on p12.62. An exact theory taking account of all interactions is very complicated and some adjustment of lengths may be necessary to achieve an acceptable swr.

Another approach is to connect reactances into the aerial with one of two objects—(*a*) to cut off the aerial progressively for each frequency band, so that it is a dipole for each band, or (*b*) to use the reactance as a phase changer, so that at the higher frequencies the extensions behave like a collinear array. Method (*a*) will be described but method (*b*) is difficult to apply, and has only had limited use so far. **Fig 12.125(b)** shows the first method, using resonant *LC* units (*traps*). Starting from the feeder, the radiator is cut to length for the highest frequency, say f_3. Parallel circuits resonant at f_3 are then inserted, one for a ground-plane aerial, two for a dipole, and the aerial is then extended till it resonates at f_2, the next lower frequency. The procedure is repeated for a third frequency if required. Finally the lengths of the sections are re-trimmed for each band in the same order. For a dipole working on 7 and 14MHz, L and C values of 2μH and 50pF respectively are claimed to be

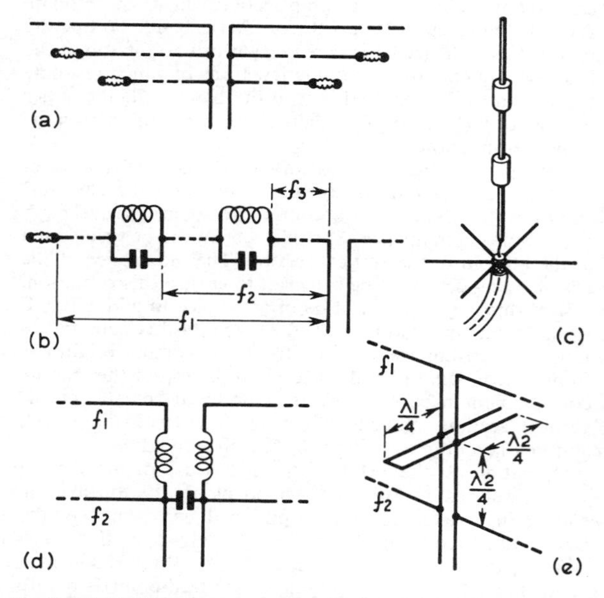

Fig 12.125. Multiband arrangements. (a) Parallel dipoles. (b) Multiband dipole. (c) Ground plane equivalent of (b). (d, e) Two "frequency switch" filters for feeding two aerials over one transmission line.

suitable for the 14MHz traps. Taking just-acceptable losses as the criterion, these values should be more or less proportional to wavelength and to the characteristic impedance Z_a of the aerial, assuming the basic element to be a half-wave dipole. **Table 12.10** gives examples of Z_a and the effective impedance R_e viewed between the *ends* of the dipole, this being equal of course to $\frac{Z_a^2}{R_r}$. Assuming a coil Q of 200 the two traps in series will have an impedance at 14MHz of $2 \times 200 \times 177 = 70{,}800\Omega$ and $2R_e$ for $1\frac{1}{4}$in tubing is 5,850Ω so that only 7·6 per cent of power (ie 0·35dB) is lost in the traps. At band edges the traps present a reactive shunt of 14,000Ω and the swr can be expected to rise to about 1·5 which is quite acceptable. In the case of a typical beam aerial however $2R_e$ is increased to about 24,000Ω and with the same traps the swr would rise to about six, causing considerable inconvenience even if the feeder cable and atu are able to cope with the situation. Sometimes damping is added to the traps to improve the swr but, for the above example, this would cause a loss of about 4dB for an swr of two at band edges. It is obvious therefore that coil Q must be kept high and the L/C ratio increased until the swr is acceptable. At lower frequencies than their resonance, the traps provide an inductive reactance which operates like the coil in Fig 12.79(b) to reduce the length of conductor required for resonance and, for the above example, the overall length required to obtain resonance on 7MHz will be about 52ft. This example may be adapted for other frequencies along the lines indicated, bearing in mind that the increased L/C ratios needed for beams will require additional shortening of the elements. In view of the complex electrical and mechanical requirements, the design of trapped beam aerials is not a task to be undertaken lightly and, if traps are purchased, it is important that their design should be related to the particular application and their suitability verified along the lines indicated above.

Replacing the tubing assumed for the above dipole example by thick wire will roughly double the losses, but these still remain quite small. A suitable coil would comprise eight turns of 16swg wire 2in diameter and 2in long supported by polystyrene strips cemented in position. The coil should be mounted over a long insulator and the capacitor may be mounted along the insulator. The whole assembly should be tuned to resonance by adjusting the coil with a grid dip oscillator, and then sealed into a polythene bag. The aerial described above will also work quite well on 21MHz. The required voltage rating of the capacitor can be obtained from the anticipated power level and the anti-resonant impedance which in turn depends of course on the aerial wire gauge. For this example a rating of 1,000V should cover all eventualities up to the "legal limit" of power.

The theory of operation described is approximate, and does not allow for the coupling which exists between aerial sections. It is therefore advisable to adjust the aerial during construction.

When the process is applied to a ground-plane aerial, **Fig 12.125(c)**, the physical construction is simpler, because the aerial can be made of self-supporting tube. The tube can be broken and supported by rod insulators over which the coil is wound; the trap circuits can then be well protected by tape, or even encased in Fibreglass. A "whip" can be used for the top section. Separate ground-plane radials may be provided for each band.

TABLE 12.10
Characteristic and anti-resonant dipole impedances

Ratio L/D	Char. imp. Z_a	End R_e ($\lambda/2$)	Centre R_c (λ)	Typical aerial
16,600	1,130	8,750	13,500	66ft of 18swg
2,500	920	5,750	8,900	16ft of 14swg
320	655	2,925	4,530	33ft of $1\frac{1}{4}$in dia tubing

This table gives the $\lambda/4$ characteristic impedance, Z_a, the end-impedance R_e of $\lambda/2$ dipoles and the centre impedance R_c of λ aerials, in terms of conductor length/diameter (L/D) ratio. R_e is based on a radiation resistance of 73Ω, and the full wave R_c on 95Ω per half-wave. Values may vary by 20 per cent in practice due to environment.

When it is required to use two aerials on different bands with a common feeder, this can be done by inserting a low pass filter immediately after the high frequency aerial (**Fig 12.125(d)**). The filter has a cut-off midway between the two frequencies: at this cut-off frequency the reactance of the capacitor is equal to the impedance (Z_0) of the line, while each inductor has half this reactance. An alternative form of filter (**Fig 12.125(e)**) uses stubs connected a distance $\frac{1}{4}\lambda$ from the first (f_2) aerial. One stub is $\frac{1}{2}\lambda_2$ (ie $2 \times \lambda_2/4$) and is open-circuit; this therefore presents a short-circuit on to the line at f_2 and thus the second stub has no effect on the first. The short circuit becomes a high impedance at the second aerial (f_1) by transformation. The second (shorted) stub tunes out the first at f_1, when the total length of the two stubs is $\lambda/4$. The length of line from the second stub to the first aerial is not important.

Fig 12.126 illustrates the use of tuned circuits as frequency switches. In Fig 12.126(a) a half-wave 14MHz dipole is fed with open-wire line, and a matching stub for this band is adjusted in the usual way, after which 21MHz traps are inserted as shown. The stub will then require to be shortened slightly. The process is repeated for 21MHz, the stub for this band requiring to be lengthened slightly when the 14MHz traps are in place. The process may be extended to achieve a matched line on any three bands but thereafter the method gets too cumbersome to be recommended. Because the traps are bridged across 600Ω line instead of the high impedance of the previous example, losses in them tend to be negligible and their design is greatly simplified. At 21MHz the length of the dipole is of course $3\lambda/4$ and it has a slight gain, about 1dB.

The G5RV 102ft Dipole

The 102ft dipole has been found an excellent compromise suitable for all hf bands and can be fed in either of the ways illustrated in **Fig 12.127**. In the upper diagram, tuned feeders (300 to 600Ω) are used all the way: in the lower version the high impedance feeder is 34ft long and is connected into a 72Ω twin or coaxial cable. At this junction, the aerial impedance is low on most bands, as can be checked with the aid of Fig 12.73 for a length of 34 + 51 = 85ft.

It has previously been suggested that there is an optimum height for this aerial ie a half or full wavelength above ground but, although in some cases these heights may provide better matching, the G5RV is no exception to the general rule that aerials should normally be erected as high as possible.

On 1·8MHz the two feeder wires at the transmitter end are connected together or the inner and outer of the coaxial cable joined and the top plus "feeder" used as a Marconi

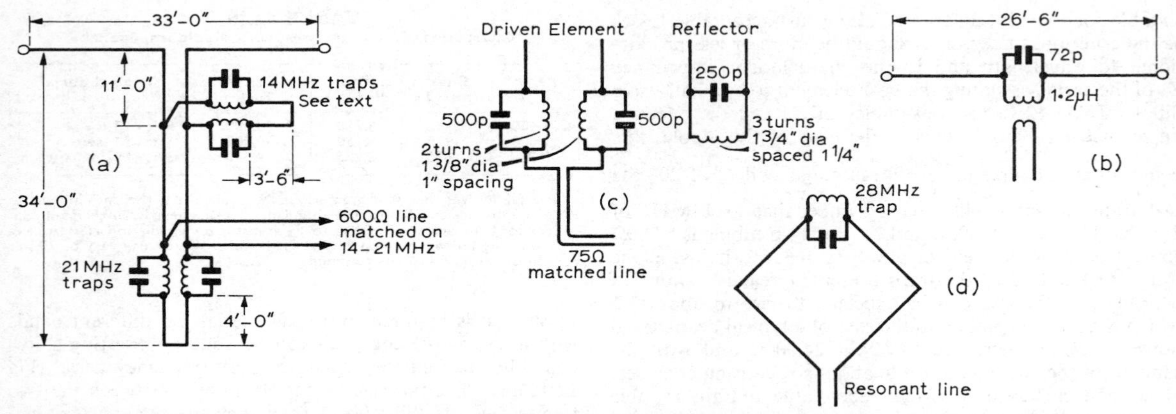

Fig 12.126. Examples of tuned circuits as frequency switches. (a) Illustration of principle enabling a dipole to be matched into an open-wire line on several wavelengths. (b) Shortened dipole which is resonant and matched on 14 and 21MHz. Component values depend on Z_a, being roughly correct for tubing of 0·7in diameter. (c) Driven element and reflector terminations for loops as Fig 12.128 (a), 16ft 8in square, L=70ft. This avoids the need for adjustment of stubs when changing bands. (d) Use of 28MHz trap in top corners allows a 14MHz quad to operate as a bi-square at 28MHz.

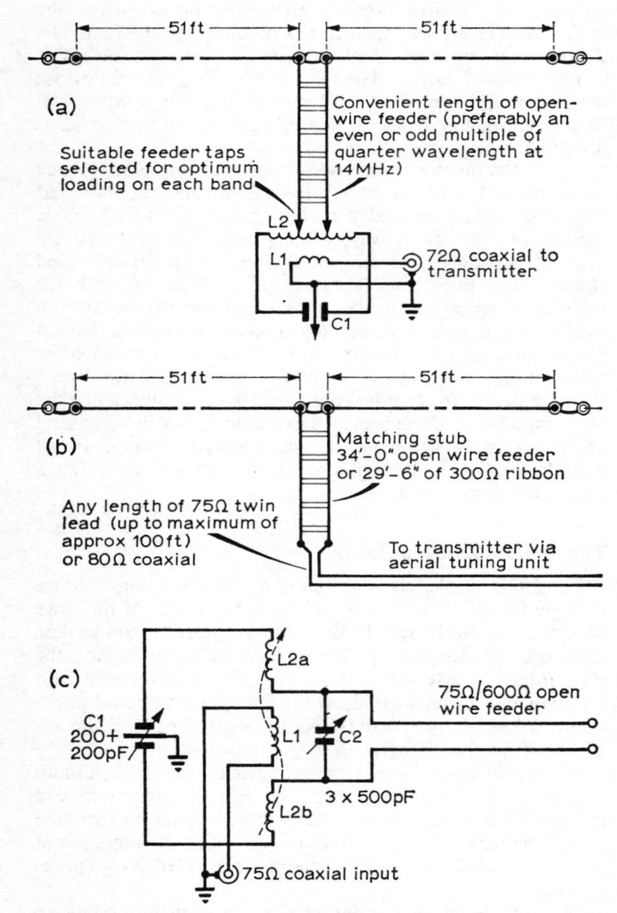

Fig 12.127. Two versions of a simple but effective multiband aerial for 1·8–30MHz. L1 is the coupling coil and C1 L2 form a resonant circuit at the operating frequency.

aerial with a series-tuned coupling circuit and a good earth connection.

On the 3·5MHz band, the electrical centre of the aerial commences about 15ft down the open line (in other words, the middle 30ft of the dipole is folded up). The aerial functions as two half-waves in phase on 7MHz with a portion "folded" at the centre. On these bands the termination is highly reactive and the atu must of course be able to take care of this if the aerial is to load satisfactorily and radiate effectively.

At 14MHz the aerial functions as a three half-wavelength aerial. Since the impedance at the centre is about 100Ω, a satisfactory match to the 72Ω feeder is obtained via the 34ft of half-wave stub. By making the height a half-wave or a full-wave above ground at 14MHz and then raising and lowering the aerial slightly while observing the standing wave ratio on the 72Ω twin-lead or coaxial feeder by means of an swr bridge, an excellent impedance match may be obtained on this band. If however low-angle radiation is required, height is all important and as most cables will withstand an swr of 2 or greater any temptation to improve the swr by lowering the aerial should be resisted.

On 21MHz, the aerial works as a slightly extended two-wavelength system or two full-waves in phase and is capable of very good results especially if open-wire feeders are used to reduce loss. On 28MHz it consists of two one-and-a-half wavelength in-line aerials fed in phase. Here again, results are better with a tuned feeder to minimize losses although satisfactory results have been claimed for the 34ft stub and 72Ω feeder.

When using tuned feeders, it is recommended that the feeder taps should be adjusted experimentally to obtain optimum loading on each band using separate plug-in or switched coils. Connection from the atu to the transmitter should be made with 72Ω coaxial cable in which a tvi suppression (low pass) filter may be inserted.

When using tuned feeders there is of course no particular merit in the length of 102ft. It should also be noted that the radiation pattern is of the general long wire type and the position of lobes and nulls will vary with length and frequency.

Multiband Parasitic Arrays: Frequency Switches

The principle of a stub or network used as a frequency switch, as in Fig 12.126(a), can be applied to parasitic arrays to enable them to work on more than one band. The range of frequency is limited to about 2 : 1 because, in terms of wavelength, the spacing will change, and at only one frequency can the optimum spacing be employed. For example a spacing of 0·25λ at 28MHz, which is rather higher than optimum, becomes 0·125λ at 14MHz, a spacing which gives good performance but rather more critical tuning.

Fig 12.126(b) shows an example of a tuned circuit acting as a frequency switch for a two-band dipole or beam-element. The tuned circuit resonates at 17·2MHz which is the geometric mean of 14 and 21MHz. It presents the correct value of loading inductance at 14MHz and acts as series capacitance of the appropriate value for tuning out the excess inductance of the dipole on 21MHz. The required value of inductance is proportional to the characteristic impedance of the aerial (Table 12.10), the indicated values being correct for tubing of 0·7in diameter approximately. Inevitably there is a large circulating current in the coil and therefore losses, the coil current being three times the dipole current on 14MHz and twice on 21MHz; in the first case, for a coil *Q* of 200, 4·5Ω of loss resistance appears in series with a radiation resistance of 50Ω and at 21MHz there are 3Ω of loss in series with 110Ω. These loss figures are acceptable for dipoles but not really good enough for a close-spaced 14MHz beam; however by tuning the switch a little higher and making the element a little longer, the loss can be reduced say to 3Ω, and the *Q* may be increased to about 600 by using some 2ft of shorted line as the inductance. This reduces the loss resistance to 1Ω in series with perhaps 15Ω of radiation resistance for a 14MHz beam, and on 21MHz there is less of a problem in view of the greater relative spacing and higher initial value of radiation resistance. The frequency switch may also be used as a matching transformer as indicated, and it should be possible by adjustment of mutual coupling to obtain better than 2 : 1 swr on both bands.

Fig 12.128 shows two simple ways of obtaining a multiband resonance. If the length is equal to two half-wavelengths on 14MHz it will be three on 21MHz, four on 28MHz and one on 7MHz, ie nominally resonant in every case, though slight adjustments are needed for the reasons given on p12.36. Of the two alternatives, the closed loop (a) is more generally useful, because the lower end is a point of current maximum on each band, and therefore in many cases accessible for tuning and matching into any non-resonant line. This is true in principle of the open loop; the end of the resonant feeder can be coupled loosely through a very small capacitance (eg by proximity) so that it presents a match into open wire line, but practical difficulties of adjustment are considerable. The open loop can be replaced by a dipole.

In the case of a 14MHz quad loop used on 21MHz or a 21MHz loop used on 28MHz, the current node slides up the side to a point about 11ft from the top centre, and vertically polarized radiation off the ends is less fully cancelled, but it has been estimated that loss from this cause is more than offset by higher intrinsic gain arising from the "oversize" elements, and as observed on skywave signals the radiation patterns obtained from a driven-element plus reflector are roughly comparable on the two frequencies. For two-band operation the loops should preferably be mounted with the diagonals vertical [11]. The loop length needs reducing

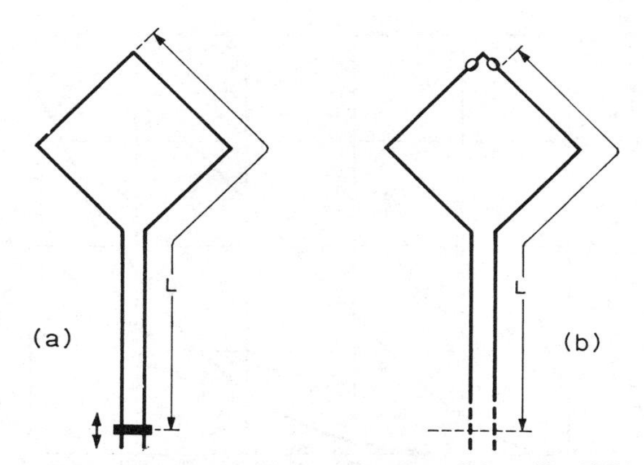

Fig 12.128. Examples of multiband resonators. If L = λ/2 at 7MHz, resonance occurs also at 14, 21 and 28MHz approximately, but small adjustments of L are needed. This applies to any size of loop.

slightly on 21MHz and this may be done for a reflector by connecting a quarter-wave open-ended stub about 2ft from the end; this has negligible effect on 14MHz. Alternatively, frequency switches may be used on the lines of Fig 12.126(c). In this case the traps have to effect only a small adjustment of line length and the values are not critical. The 14MHz quad also works, but not as a beam, on 7MHz and 28MHz but for lower frequencies it is probably better to excite one or more of the loops as a vertical, eg Zepp-fed (Fig 12.94) at the lower end as a λ/2 vertical on 7MHz or fed with coaxial cable against a ground-plane as a λ/4 vertical aerial on 3·5MHz. An example of a 14MHz quad used very successfully on 3·5 and 7MHz for long-haul dx has been described by G8PO [17] and, in general, any vertical bunch of wires provided it is not "nailed to the mast" is likely to be a useful candidate for consideration as a low-frequency aerial. On its second harmonic frequency a conventional quad or delta loop can be resolved into two end-fire two-element arrays, one of which fires vertically upwards so that half the energy is wasted, thereby reducing the gain approximately to that of a dipole. A single 14MHz quad loop can therefore be expected to work more or less equally well on 28MHz except that the direction of propagation and the polarization are both switched through 90° so that radiation is vertically polarized and "off-the-end". Very little gain is obtainable on 28MHz by exciting both elements of a close-spaced 14MHz quad, though a bidirectional pattern with 4dB gain can be obtained from feeding wide-spaced (λ/4 on 14MHz) elements in phase. A more efficient triband beam can be achieved by open-circuiting the top of the loop so that it turns into a bi-square (Fig 12.101) on the higher frequency, and this may be done by means of a frequency switch (Fig 12.126(d)) if narrow band operation is acceptable. If the loop is required to work on 14, 21 and 28MHz the *L*/*C* ratio must be kept low in order not to upset the 21MHz radiation pattern, values of 0·5μH and 62pF being suitable if the loop size is reduced to 16ft square. Identical treatment can be applied to a 14MHz folded dipole to obtain triband operation. Yet another alternative arises from the fact that a 21MHz loop retains a reasonable amount of radiation resistance, about 36Ω for a 12ft 6in square loop, when driven on 14MHz; this is coupled with the advantage of increased effective height since most of the radiation comes from the top half of the loop.

With all these arrangements the simplest method of feeding

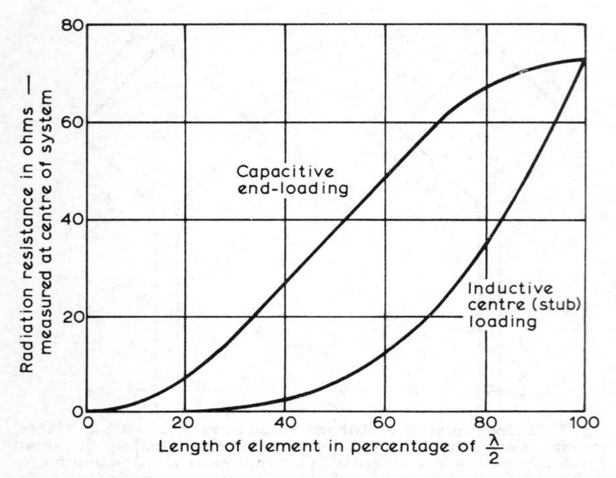

Fig 12.129. Radiation resistance of short dipoles with alternative methods of loading. For beams designed on the guide lines of Fig 12.15, the radiation resistance varies with length of element in the same ratio as that of single dipoles.

is to use resonant feeders on the lines of Fig 12.128, this being much easier with loops in view of the higher radiation resistances and the near-coincidence of resonances if the length is suitably chosen. It is possible even without the use of tuned feeders to devise elements with simultaneous resonances on two or three bands by using a variety of stubs and other loading devices [11, 18]. Triple resonance however, combined with efficient matching to an untuned line on all bands, is difficult to achieve without compromises of one sort or another, and the only systems currently enjoying much popularity employ trapped elements.

Many triband beams have been described using shortened elements for the lowest frequency, and **Fig 12.129** shows the radiation resistance of short elements for different types of loading. Typical elements are about 23ft in length with their natural resonances in or near the 21MHz band. These usually employ frequency switch techniques, on the lines of Fig 12.126(b), for resonating parasitic elements on two or more bands and either frequency switches or tuned feeders for the driven element. Making use of Figs 12.15 and 12.129 with appropriate scale factors and aiming for a gain of at least 4dB on 14MHz a radiation resistance (referred to a current loop, not the actual feed point) of 8Ω can be expected. From Fig 12.5, assuming 12swg and a feeder length of 50ft, the loss resistance referred to the same point is 2·5Ω and the power lost in the driven element is 1·2dB. Referring back to the discussion of Fig 12.126(b), a further loss of at least 0·5dB can be expected in the frequency-switching circuit of the reflector, leaving a nett gain of only 2·3dB. This calculation is of course extremely rough but serves to illustrate the basic principles and the extreme difficulty of obtaining effective gain on the lowest frequency, particularly as no account has been taken of the possibility of increased dielectric losses. In the case of a 21MHz quad loop, as described above, the situation is a little better, the radiation resistance being at least double.

The near-coincidence of resonances allows much more efficient frequency switching of the reflector, and a nett gain of 3·0 to 4·0dB with possibly an extra dB from the increased effective height for a given mean height can be expected. Alternatively resonant feeders to both elements can be used and both can be driven if desired. In all these cases tuning will be critical on 14MHz where full band coverage cannot be expected without retuning or accepting some loss, but the double feeder arrangement has the advantage of allowing all tuning to be carried out at or near ground level.

The most efficient method of feeding such beams is to tune both elements (dipole or loop) to resonance using the shortest possible stubs and match into these in the usual way with open-wire lines at 14MHz. In this case losses are negligible and bandwidth greatly improved but the adjustment must be correct prior to erection. At 21MHz the stub can be open-circuited by a frequency switch and at 28MHz it is harmless, allowing the feeders to be used as tuned lines in both cases.

Because of the relatively low radiation resistances of small beam aerials particular care must be taken to minimize losses in any tuned feeders by using the heaviest practicable wire gauge, good insulation, and avoidance of proximity of lossy objects other than in the vicinity of voltage nodes. The feeders must be as short as possible, except that it may be convenient to end them at a current loop for matching into a non-resonant line. In general, tuned feeders are more suitable for fixed than for rotary beams, although a two-feeder reversible beam may be arranged as sketched in **Fig 12.130** and has been found to allow sufficient rotation without harmful effects. A single-feeder version with, for example, separate parasitic elements for each band, presents no major problem but the feeder should be prevented from actual contact with the mast or metal guys. Joints, particularly those between copper and aluminium, should be suspect and one advantage of loop or folded elements lies in the possibility of making a dc resistance measurement which usually reveals the presence of any corrosion. If a resonant feeder is matched into a non-resonant open-wire line in the usual way, the sum of radiation resistance and loss resistance may be deduced approximately from the required length of matching stub; the reactance X_s of a short-circuited open-wire stub made of 600Ω line is quite close to $4fl$ Ω where f is the frequency in MHz and l is the length in feet, provided l is less than about 5ft at 14MHz and pro rata. The resistance is then given by $X_s^2/600$ and if this greatly exceeds the likely value of radiation resistance which may usually be inferred from Fig 12.15 or discussion in this chapter under the appropriate headings, the explanation must be excessive loss resistance or, possibly, incorrect

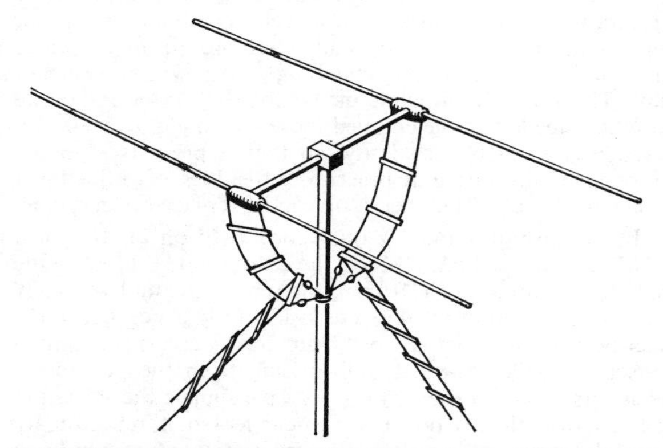

Fig 12.130. Arrangement to permit rotation of reversible beam with open-wire feeders. The top guy wires should be attached at or below the anchorage point of feeders. About 120° to 140° of rotation is sufficient in most cases.

phasing of beam-elements. Too low a resistance also indicates incorrect phasing but, if this is the cause, it will usually be obvious from the directional properties of the aerial. Although this procedure provides only very rough guidance it is capable of revealing major faults in design or maintenance, or alternatively inspiring confidence in the system.

Stacking

For coverage of more than two wavebands it is usually preferable to use two or more beams even if these have to be mounted on the same supporting structure. One example of this, the multiband quad, p12.77, has already been given and it is common practice also to use stacked dipole arrays. There is in general very little effective mutual coupling between parasitic elements having widely different resonant frequencies and separated by a foot or so but this is not necessarily true of driven elements which, if left with a random length of feeder attached to them, will usually resonate at some frequency other than the nominal resonance and possible coinciding with one of the other bands. This obviously spells trouble, and since the point has apparently not been widely appreciated, may well account for occasional failures to get satisfactory performance with stacked beams. It is a simple matter to test for interaction by short-circuiting any unused feeder and, if there is any change in performance, it can be assumed that the worst condition involves some interaction and the best almost certainly no interaction, though multiple resonances can occur, and a check with an additional termination such as an eighth wavelength of feeder should also be made. A 14MHz parasitic element will of course resonate on 28MHz but this is a full wave resonance and, assuming symmetry, any coupling to the 28MHz elements will be neutralized. On the other hand, depending on feeder length and termination, the 14MHz driven element can look like a pair of 28MHz half-wave dipoles in phase and the above termination test should be applied.

LIGHTNING PROTECTION

In order to achieve the most efficient radiation from an aerial, it should in general be erected in the clear and as high as possible. Its potential as a lightning hazard is thereby increased and serious consideration should be given to this problem which unfortunately is one of considerable complexity, so that no one set of simple rules can be devised to suit all situations. Some understanding of the general principles involved and, incidentally, the weakness of many common practices, is therefore desirable.

The object of lightning protection systems is achieved by providing a *safe* conducting path between ground and the atmosphere above the structure to be protected [24]. Radio aerial installations usually embody the main ingredients of a protective system but may tend to increase the hazard instead of reducing it, because of failure to meet certain minimum specifications for such systems. The majority of lightning strikes involve currents in the region of 2,000A to 100,000A, with an absolute maximum in the region of 220,000A. These are of short duration, eg a rise time of a few microseconds and decay time of a millisecond or less, though a complete lightning discharge may comprise a sequence of such strokes following the same path and lasting up to one second or more. Despite the short duration these currents cause intense heating if they pass through a bad joint in a metal conductor or poor insulators such as trees or brickwork where they may cause sudden generation of steam; in each of these cases the effect may be explosive, fires may be started, and there is danger to any individual near the path of the discharge. The current can pass safely to ground only if the following conditions are satisfied:

(1) The current path is of adequate conductivity and cross-section.
(2) The earth resistance is low enough.
(3) Other conductors in the vicinity of the lightning conductor are adequately isolated or bonded to it.
(4) There are no people or animals in the vicinity of the earth termination, where large potential gradients exist at the ground surface during a lightning strike.

The most frequent cause of damage to equipment is not a direct strike but voltage or current induced in aerials and mains wiring by lightning strikes in the vicinity, at distances up to several hundred yards. Static charges due to an accentuation of normal atmospheric stresses also come into this category. Protection against these effects may be obtained even with a high-resistance earth-connection but the presence of an hf aerial, despite adequate protection against static electricity, could in some cases increase the likelihood of a direct strike without providing the means for dissipating it.

Lightning danger varies enormously between different areas of the world and this must obviously enter into the assessment of what precautions should be taken. The BSI Code of Practice [25] puts forward a "points system" for deciding whether a building needs protection and the majority of private dwelling houses in the British Isles would appear to be exempt though many of them might be put at risk by erection of an hf beam or ground-plane aerial on the roof. Many aerials mounted on towers or tall masts would also appear to be in need of protection; on the other hand some aerials can be designed so that they themselves act as efficient protective systems. Short of full lightning protection it is recommended [26] that television aerials, etc, should be protected against atmospheric electricity by earthing with a conductor of not less than 1·5mm² cross-section, the outer conductor of a coaxial cable being regarded as suitable. An hf transmitting aerial should obviously receive at least this degree of protection. Aerials and masts may be earthed for this purpose by connection to an existing system of earthed metal work, eg suitable water pipes, the point of connection being as high as possible in the system. In the case of vhf or uhf aerials parasitic elements need not be earthed, but this may be advisable with the much larger elements used for hf beams.

Whether or not full protection is provided, the earth resistance should be as low as possible. For full protection the BSI Code of Practice [25] recommends a value not exceeding 10Ω and conductor cross-sections of the order of 60mm² though opinions differ on the latter figure; the official standards of some countries run as low as 28mm², and values as low as 5mm² have been described as adequate despite a small number of instances of damage to 60mm² conductors [24]. An earth resistance as low as 10Ω is often difficult to achieve but in such cases an annual dose of a solution of rock salt can be very effective. Low earth resistances are sometimes obtained by laying conductors in trenches near the surface, but it seems reasonable to suppose that this might tend, for a given resistance, to create dangerous potential gradients over a much larger area in the event of a strike.

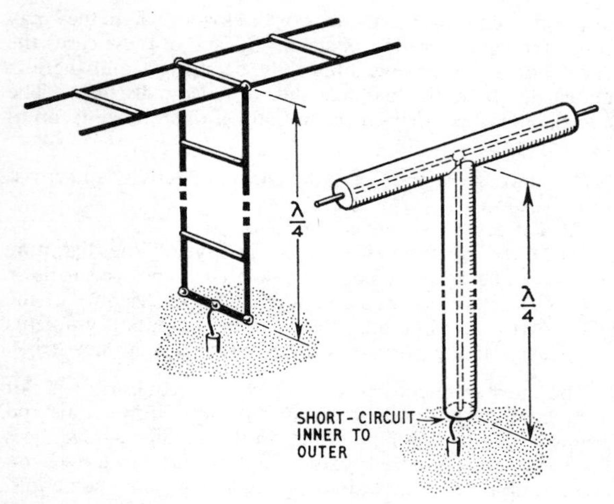

Fig 12.131. λ/4 lightning protection. The balanced line is approximately λ/4 in free space: the coaxial line is λ/4V where V is the velocity factor of the cable employed to make the stub.

The bonding of other conductors to the lightning protective system can raise complex problems and if it appears necessary, the references should be consulted. Isolation is the alternative, and for full protection requires separations of 1ft per ohm of earth resistance, plus 1ft for every 15ft of structure height to allow for the inductive voltage drop in the down conductor; however, for a slight increase in risk these distances may be halved. A rough idea of the zone of protection of a lightning conductor system is obtained by imagining a cone with a 90° apex angle extending down from the tip of the conductor and this may be extended by the efficient bonding to the conductor of, say, a "plumbers delight" hf beam erected over the roof. Most hf transmitting aerials however lie wholly or partly outside the protected zone and, because of inadequate conductor sizes, are unsuitable for its extension. In such cases the feeder (eg outer of coaxial cable), metal masts and parasitic elements should be connected to the highest convenient point on the lightning conductor. If there is no lightning protection system they should be *earthed outside the building*.

In the case of an aerial mounted on a tower or tall metal mast, all components including parasitic elements should be adequately earthed and unless a separate lightning conductor is fitted the tower must be free from high resistance joints (eg fully welded). Even if resistance is initially low, possibilities of corrosion following exposure to the weather must not be overlooked. Otherwise, if the lightning current is allowed to pass to earth via the tower, explosive disintegration and collapse could occur. The earth resistance must be as low as possible and a single 3ft earth spike as illustrated in earlier editions of this handbook, though adequate for the discharge of static electricity, is insufficient for lightning protection. Balanced feeders cannot be directly earthed except via resonant stubs as illustrated in **Fig 12.131.** Various spark-gap type lightning-arresters are suitable as an alternative [27].

The usual change-over switch for earthing an aerial and worse still other "executive" arrangements likely to require the handling of a feeder during a thunderstorm, such as removing it from the rig and plugging it into an earth connection, are not merely useless for lightning protection but could prove highly dangerous. Even in the event that an operator remembers to "switch to earth" before the outset of a storm, "any idea that they provide safety is illusory" [24]. This is because the switch is not capable of carrying lightning currents, nor is the gap sufficient for providing the required isolation which needs to be several feet as already indicated. It should be remembered that in the event of a lightning strike, even if the earth resistance is as low as 10Ω, voltages as high as two million could be lying around in the shack!

The BSI Code of Practice recommends where possible a common earth electrode for lightning protection and *all other services*, the lightning conductor being of course routed outside the building, and the installation must comply with regulations for the other services. It should be noted that an earth resistance of 10Ω is not low enough to provide in itself the fault protection required by wiring regulations and, if this is the best figure available, an earth-leakage trip will be fitted. Some older types are voltage-operated with a coil connected between true earth and the "earth connection" of the wiring system. If an earthed coaxial feeder is plugged into a transmitter or receiver to which the usual three-pin plug is fitted, the trip coil will be wholly or partly short-circuited and a dangerous situation involving the whole of the house wiring could be created.

CONSTRUCTING AERIALS

All the aerials described in this chapter can be constructed at home without much difficulty provided that suitable materials are selected. They fall approximately into two groups, those which are formed from wires under tension between fixed anchoring points and those which employ rigid or semi-rigid conductors or light frames which act as supports for flexible radiators. All rotatable beam aerials come in the second group.

Wire Aerials

The cardinal points of design of a fixed wire type aerial are:

(*a*) choice of conductor type and gauge,
(*b*) insulators,
(*c*) end anchorages for tensioning, and
(*d*) method of support.

The size of conductor to be employed depends upon the working tensions involved. The longer the span and the heavier the insulators, the greater will be the tension for a given amount of sag. Single copper conductors are cheapest, the choice of wire gauge being determined by the need to keep the loss resistance small compared with the radiation resistance and also by mechanical considerations. In general for "short dipoles", or for dipole beam elements, 14swg or thicker should be used in order to reduce losses and maintain bandwidth. Stranded wire, eg 7/029, is however slightly easier to handle. Thinner wires may be used for loops and folded dipoles even when these are part of a beam and, given a rough idea of the likely radiation resistance, Fig 12.5 provides the necessary guidance for meeting the electrical requirements. Wire should always be tensioned to avoid breakage due to flexing, but if anything more than light tension is required it is essential to employ hard-drawn copper to prevent stretch. For a 14MHz dipole the wire size can be as low as 30swg for a loss of only 0·5dB, and though liable to mechanical breakage such an aerial has the

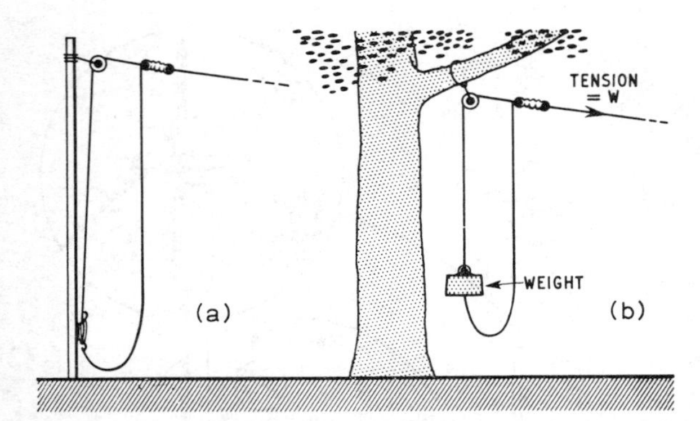

Fig 12.132. Halyard connections to poles and trees. The weight W is equal to the tension T required in the aerial wire. In both cases an endless loop is employed.

advantage of being much less conspicuous in areas where bye-laws or aesthetic requirements call for an invisible aerial. Wire aerials may be suspended between any available supports, using polythene line without additional insulation, lightly tensioned as in **Fig 12.132.** If only one support is available, or if a few feet of extra height can be obtained in this way, the dipole may be erected as an inverted-V.

A solid insulator is likely to be required in the centre of the dipole to provide mechanical anchorage for the feed line, and in the case of half-wave dipoles almost any insulating material is suitable. Losses at this point can however have a serious effect on the performance of full-wave dipoles or "short dipoles". Insulators can be obtained as proprietary items in glazed china or glass, or can be manufactured at home from polythene rod. It is possible to cast individual tension and centre feed-point insulators in *Araldite* resin, which has an acceptable power factor and low dielectric loss up to 30MHz. In all cases where a rigid insulator is not required for mechanical reasons, polythene cord will be found satisfactory.

The correct method of attachment of halyards and conductor wires to a thimble or insulator is shown in **Fig 12.133(a).** To form the PO splice in the stranded conductor, one strand of the free end of wire is untwisted back to the insulator, and then wrapped tightly around both wires to bind them together. A second strand is then untwisted back to the end of the first wrapping, and wrapped to continue the binding action. The process is repeated until all the strands of the free end have been used. This neat splice may also be used on galvanized stay wires for masts, and is very strong. It is aesthetically better and inherently safer as a permanent splice than the familiar Bulldog wire rope grips sometimes employed.

One method of securing the feed-point insulator for coaxial and for open-wire line is shown in **Figs 12.133 (b)** and **(c).** It is important to ensure that the weight of low impedance line is supported firmly by the insulator and not by the connections made to the aerial conductor. The latter should be wrapped to form a sound mechanical joint before soldering.

Aerial halyards should be of polythene, nylon, or other plastic material since conventional rope or cord tends to rot and in the meantime is much affected by weather conditions. They should always be reeved as an endless loop to avoid a climb to the top of the pole if an aerial or halyard

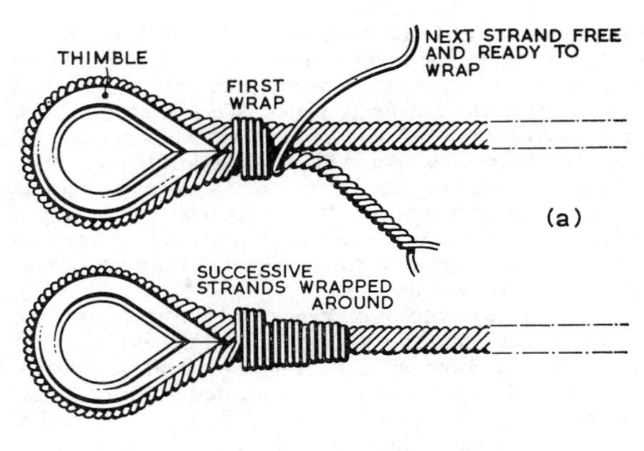

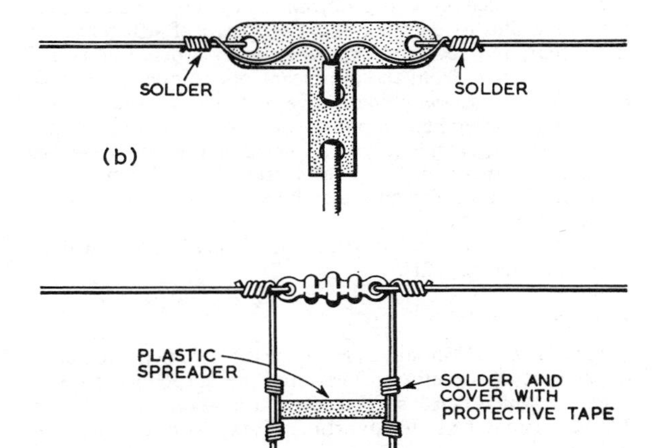

Fig 12.133. (a) The PO splice for securing stranded wire around a thimble or insulator. (b, c) Methods of supporting feeders from dipole centre insulators. The exposed end of the coaxial cable in (b) must be taped to exclude moisture. The "tee" insulator is described in the text.

connection fails: Fig 12.132(a). Some plastic cords are reinforced by a fine strand of wire and these must be avoided in view of possible effects on the aerial radiation pattern, apart from the fatal consequences likely to ensue if they come into accidental contact with overhead electric cables.

Wooden or metal poles, towers, or the walls or eaves of buildings are all suitable tension anchorages. In the latter case a long free end to the halyard is preferable to keep the aerial away from lossy brickwork and metal guttering. The branches or trunks of trees have disadvantages as halyard anchors as they move excessively in the wind and cause large variations of sag in aerial wires, and possibly failure under extreme tension. If they must be used, then some form of counterbalance on the lower end of the halyard is essential: Fig 12.132(b). The weight should be free to move up and down, and is equal to the tension in the aerial wire.

Beam Aerials

Various forms of rotary beam aerials can be constructed either from wires tensioned on a framework or from self-supporting tubes, remembering that wires should be of the heaviest convenient gauge, though 18 or 20swg is electrically satisfactory for quad or delta loops of normal size and at least λ/8 spacing. The framework can be made up from suitable timber which can be screwed together, or from bamboo poles clamped to end fixing plates. When timber is used, particularly for rotary Yagi type aerials, a hardwood is to be preferred. Softwoods should always be primed and painted for normal outdoor protection, or given a thorough creosote or polyurethane treatment. It is advisable, particularly if short elements are used, to keep any wood or bamboo supports well away from wires to ensure satisfactory performance in wet weather.

The most elegant form of Yagi aerial construction is the "plumber's delight", so called because all the elements and the supporting boom are made from tube sections. Since all elements in a Yagi are at zero rf potential at their centres, they may be joined to and supported by a metallic boom. This form of construction is usual in commercially available aerials, but often presents problems for the home constructor who may like to consider the alternative of using two or more fixed but reversible wire beams to provide all-round coverage. Apart from the important operational advantage of instantaneous switching between directions such arrays can often use lighter supports and are relatively cheap. Sometimes they can even be concealed among tree branches without incurring appreciable losses. Wire beams using inverted-V dipole elements may be supported from a single mast and spreader.

Many ideas for the construction of quad aerials will be found in reference [21].

Vertical Aerials

Vertical aerials should as far as possible be kept well clear of supporting structures. They may be suspended from a catenary which can also provide end-loading, eg as in Fig 12.102, or may be self-supporting. If stay wires are required polythene cord is suitable in view of its excellent insulating properties. Sometimes a metal mast is used in conjunction with a base insulator as a single-band radiator. The base insulator can be made from a soft drink or wine bottle, clamped around the lower part of the outside, and with the bottom end of the vertical tubing either inside or over the neck of the bottle. A hock bottle is probably the most suitable shape for this purpose. In all such cases it is essential to prevent the bottle filling up with rainwater.

Masts and Rigging

Although traditionally aerials have been supported by stayed or self-supporting wooden poles, usually painted or creosoted for weather protection, there is an increasing tendency to use steel or aluminium tubing with suitable coupling and stay attachment fittings. A most convenient mast kit can be built using nominal 2in diameter aluminium *scaffold* poles as the basis, and selecting fittings from a range manufactured by Gascoignes Ltd, Reading, Berks, under the trade name "Kee Klamps". A typical arrangement for a rotating lightweight tubular mast suitable for carrying light hf or vhf beams is shown in **Fig 12.134**. A simpler form using the same principle can be employed for fixed poles.

Alternative sources of tubes and clamp fittings include firms specializing in tv aerial installations, particularly in fringe areas of reception.

Metal guy wires if used should be broken up by insulators into non-resonant sections. It should be noted that metal masts, unlike guy wires, cannot be split up in this way and may have unpredictable resonances which can be excited by any asymmetry in the aerial system, so that the measures recommended (p12.32) for the prevention of feeder radiation may assume added importance; they should not be used to support vertical aerials with the possible exception of ground planes mounted at the top, although in this case the radials, if allowed to droop, could couple tightly into the mast. Asymmetry also exists in principle if the mast is used to support one end of a horizontal aerial which should preferably be several feet from the mast, though usually no serious difficulty is experienced.

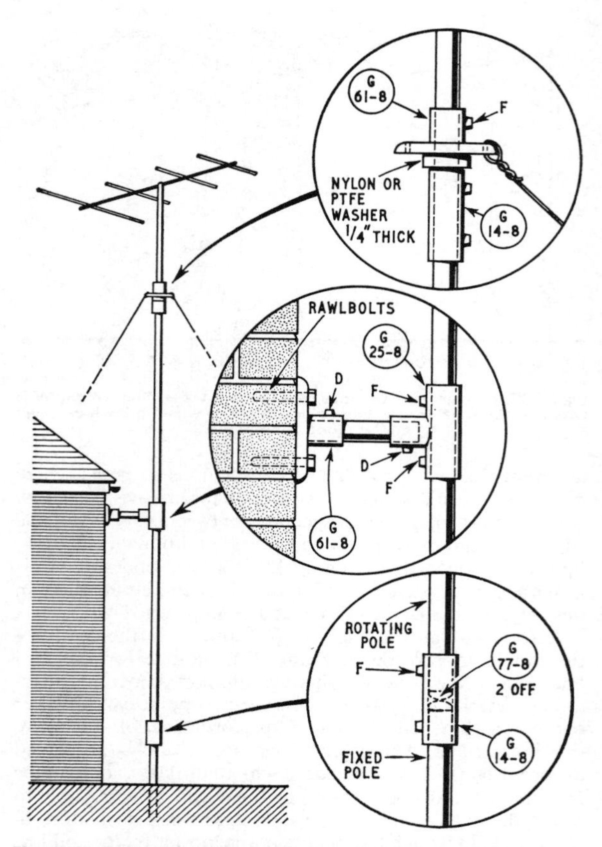

Fig 12.134. A light rotating mast using Kee-Klamp scaffold fittings. The clamping screws at F should be left slack and grease forced into the fitting at those positions. For extra security the stub pole can be drilled $\frac{1}{4}$in clearance at points D and $\frac{1}{4}$in BSW bolt substituted for the Allen screw, to penetrate the wall of the tube and form a lock against possible pull out. A simpler form of construction can be used for fixed poles.

Many tilt-over and telescopic towers are available as proprietary articles and the telescopic variety can be motor-operated. On the other hand wooden masts can be constructed from timber at low cost as in **Fig 12.135** and lend themselves readily to tilt-over methods of erection for easy replacement or maintenance of aerials.

Excellent articles on aerial masts and rigging have appeared in *Radio Communication* [22, 23] and will help the reader wishing to adopt a "professional" approach to aerial erection, or to benefit from the experience of others in evolving cheap and simple designs of aerial support while avoiding some of the more serious pitfalls. Every installation presents a different set of problems, and solutions have few features in common. Careful consideration must be given to any dangers to persons or property which could result from the collapse of a mast and beam during severe weather conditions and it is essential to err on the side of safety, even though this usually leads towards rather massive construction. On the other hand there are cases where collapse would be of minor consequence and repairs easily effected, for example where a light support is used for an inverted-V dipole or ground plane, and materials such as the pvc tubing available from builders merchants have been used successfully in these cases.

Guys are usually required, but the stranded galvanized clothes line often used has only limited life, particularly when the galvanizing has been damaged with pliers. Polypropylene rope (monofilament) is an excellent general-purpose material for guys and halyards, Terylene being slightly better but considerably more expensive. Nylon and polythene have been used successfully but stretch under tension and this can be disastrous when they are used as guys. Guy anchorages for light masts may consist of 2ft lengths of 2in angle-iron driven into the ground approximately at right angles to the line of pull of the guys. Particular attention should be paid to safety during the erection of masts, avoiding any danger of contact with power wires. Pairs of guy wires, carefully measured and attached to points at right angles to the line of erection will prevent the mast falling sideways. The "gin-pole" method permits the single-handed erection of masts up to 40ft or more in length. The gin-pole, which may conveniently be a ladder, is held vertically at the foot of the mast which is on the ground, and the free guy or guys from the top of the mast are passed over the top of the ladder and extended to a lower position where they can be easily secured. For single-handed erection a tackle may be rigged between the top of the mast and a suitable anchorage. A temporary crossbar should be clamped to the base of the ladder so that the mast sits on it during erection, being afterwards lifted off into a suitable socket [22]. In working out the required strength of guy-wires, it is convenient to estimate the effective surface area presented to the wind by the aerial, and add on the area of the top one-third of the mast. The wind drag is then given by $v^2/400$ where v is the wind velocity in mph. Depending on the location and other factors, v is usually taken as 80–100mph, and a factor of safety of six should be allowed for the breaking strain of the guys. Typical wind drag (100mph) is 5·8lb per foot length for 2in tubing, 0·17lb per foot for 14swg wire, 135lb for a three-element 14MHz Yagi and 260lb for a cubical quad, Fig 12.117 [23].

Planning Consent

In the UK all aerial supports erected at a height exceeding 10ft above ground level are subject to the provisions of the Town and Country Planning Acts. Before embarking upon any major aerial erection, it is advisable to consult with the Surveyor to the local Council and if necessary make formal application for planning consent. Failure to comply with the Town and Country Planning Acts could possibly lead to an injunction to dismantle the entire erection at a later date.

In other countries various zoning laws also apply in built-up areas, and the appropriate procedure should be ascertained. When purchasing or renting property, it is important to ascertain whether there are any covenants in the title deeds prohibiting the erection of outdoor aerials. This is particularly important in municipally owned estates and in areas where piped television is in use.

AERIAL MEASUREMENTS

The most important property of hf aerials is usually the effective dx performance which is difficult to measure directly as it involves not only the aerial but its entire surroundings and particularly the ground contours which may have markedly different effects even for aerials at say, opposite ends of a typical suburban garden. Even if the area is flat, power lines and trees can have marked effects in the case of horizontally and vertically polarized aerials respectively. Data on these effects are few, but in one case vertical aerials suspended from, on either side of and barely clear of a row of trees were found to be ineffective when "looking through" the trees. Local measurements of impedances and polar diagrams can be used to establish that an aerial is working approximately in the correct manner but when evaluating a new aerial it is advisable to obtain a considerable number of comparative signal reports before and after the change.

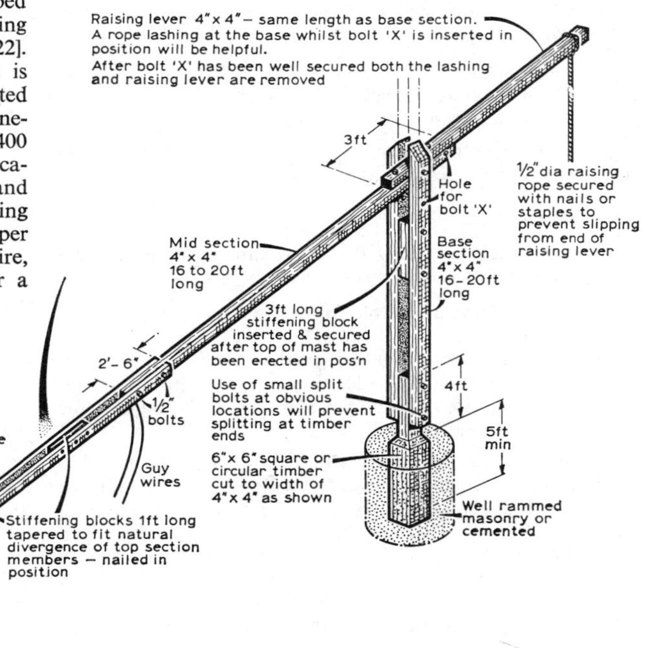

Fig 12.135. Constructional and raising details of the 55ft wooden mast. (*ARRL Antenna Book*).

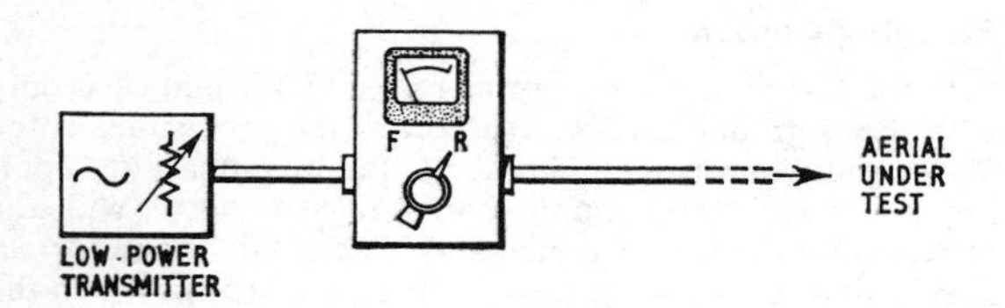

Fig 12.136. Adjustment of aerial impedance using a reflectometer. The output of the transmitter is set to maintain a constant indicated forward power. The aerial is then adjusted for minimum reflected power to match its impedance to the cable employed.

The yardstick for such comparisons may be provided by another amateur whose operating conditions are maintained constant, or another aerial (eg a dipole) at least $\frac{3}{4}\lambda$ away from the aerial under test and end-on to it, although these requirements may be relaxed if the aerial not in use *including any parasitic elements* can be completely detuned, eg by the open-circuiting of resonant feeders. It will be difficult to get consistent results unless both aerials have similar vertical plane radiation patterns which for horizontal aerials and low angles are a function only of height and ground contours. More consistent results have been obtained over long transequatorial paths such as G/VK or G/ZS, probably due to the prevalence of single modes of propagation (eg chordal hop) whereas transatlantic paths can sustain many different multihop modes, some of which may involve quite high angles of radiation. The "yardstick" station need not be particularly close, geographical separation up to 100 miles or more being usually less significant than a difference of say 2 : 1 in aerial height, though correlation may break down completely at times particularly if band conditions are changing. This "rule" also breaks down for near antipodal paths since places which are close together geographically may be on completely different great circle bearings, as will be obvious from looking at New Zealand on a great circle map centred on the UK.

Local measurement of aerial performance usually involves three particular properties of the aerial:

(*a*) Impedance,
(*b*) Polar diagram (in both planes), and
(*c*) Gain relative to a reference standard.

These will now be discussed in turn; see Chapter 18 for details of suitable test equipment.

Impedance

It is desirable to know the approximate value of input impedance of an aerial for two reasons:

(*a*) to determine a suitable form of matching transformer at the aerial,
(*b*) to verify as discussed below that the radiation resistance is of the right order since if too low this indicates faulty design or tuning, and if too high suggests the presence of excessive losses.

In the case of an aerial fed with open-wire line the position and length of matching stubs after following normal adjustment procedures should give the information required (Figs 12.59 and 12.60). With low impedance feeder and correct tuning of the aerial, swr should be a minimum around the middle of the band and, if the aerial is fed without a transformer, or with a transformer of known ratio, radiation resistance may be deduced from the minimum swr. An swr of 2·0 in 50Ω cable could result from an aerial impedance of 25 or 100Ω, but such ambiguities may be resolved by connecting say a 100Ω resistance across the aerial terminals. This would have little effect in the first case, but would improve the swr to unity in the second case.

Matching may be checked with an impedance bridge, noise bridge or reflectometer which should if possible in the first instance be connected close to the aerial, otherwise what is interpreted as a good match may in fact represent only a high value of feeder loss. Even if the feeder loss is low, as measured into a resistive load, additional loss and measurement errors may be caused by line imbalance or currents flowing on the outer surface of a coaxial feeder. To illustrate this by a practical example, a two-element beam fed with 120ft of coaxial feeder, without a balun, had an swr of 1·4 measured at the transmitter, and the feeder loss into a matched load was 1·5dB. Despite this the performance was very badly down and an improvement of about 6dB, as estimated from dx signal reports, was obtained by changing over to open-wire feeder with a matching section. Investigation revealed an swr in the first case of 4·0 *at the aerial*. This gave [19] a predicted swr at the transmitter, allowing for line-losses, of 2·0. Further reduction of swr and at least half the total loss was attributed to loss by radiation from the feeder.

After the performance of an aerial has been established, the reflectometer may be installed in the shack for permanent monitoring, but it should not be regarded as giving an infallible indication of correct operation and losses due for example to moisture getting into the feeder will improve rather than degrade the swr. **Fig 12.136** illustrates a typical test set-up using a reflectometer to indicate correct adjustment of the aerial for minimum reflected power in the line. When carrying out such adjustments it is important to maintain a constant forward power from the test oscillator, and this too can be verified by means of the reflectometer.

Adjustment of aerial impedance should ideally be carried out with the aerial in its operating position, since it is influenced by height above ground. Often this is not feasible and it may be necessary for reasons of accessibility to make adjustments at a height of 10ft or so with a final performance check at full height.

It is permissible, and usual, to leave the reflectometer connected in the transmission line during normal operation (with a reduction of sensitivity appropriate to the transmitter power employed). It is then possible to maintain a continuous check on the swr and detect aerial faults as they arise.

When aerials using tuned feeders are employed, the adjustment of impedance is transferred to the aerial coupler (see p12.48). Again a reflectometer may be used in a short piece of line of correct impedance for the transmitter output load, and the aerial coupler adjusted to give maximum swr on this piece of line (see Fig 12.74).

Polar Diagram

The horizontal polar diagram of an aerial can be measured to a reasonable degree of accuracy without having to raise the aerial to an excessive height above ground. A figure of 0·2–0·3 wavelength may be considered adequate, with a minimum of 8ft, to ensure adequate general clearance.

The horizontal polar diagram of small beam aerials depends on the number of elements and the relative phase and amplitudes of currents. In the case of two-loop or dipole

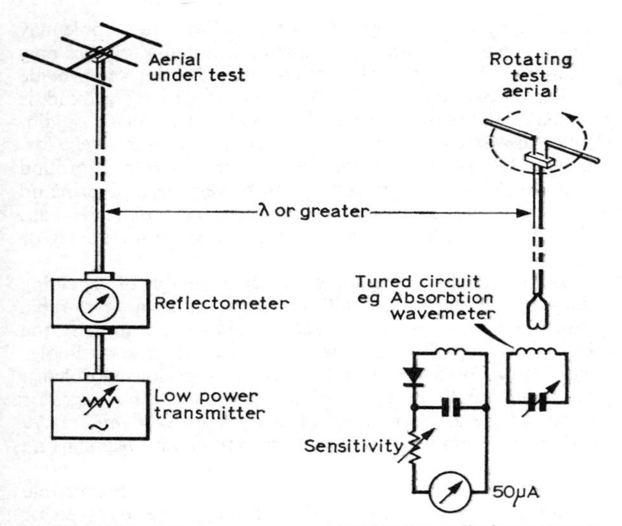

Fig 12.137. Test site for measurement of aerial radiation patterns. The site must be as free as possible from obstructions. The reflectometer is used in the forward position to monitor the transmitter output which must remain constant. The S-meter should first be calibrated using a signal generator into the receiver input.

elements a good idea what to expect can be gleaned from Figs 12.15, 12.17 and 12.18. With more elements the pattern is less predictable but a slightly narrower main lobe and reduction of other lobes can be expected. In both cases measurement of polar diagrams is useful for providing a rough check that the beam is operating correctly but it is important to realize that these diagrams are only rather loosely related to sky-wave performance. In particular, discrimination against short and medium range sky-wave signals "off the end" is quite poor because at an angle of incidence of 30°, for example, the vertical polar diagram is only 6dB down. Also, for directions "off the back" the effective element spacing for high-angle signals is different from what it is in the horizontal plane so that nulls will occur, if at all, in directions slightly different from those observed in the horizontal plane.

The test aerial must have the same polarization as the one being tested or, in some cases, it may be useful to make a measurement of both vertical and horizontal fields. This may reveal, for example, the presence of feeder radiation, but great care is needed in interpreting results if both polarizations are present; an inverted-V dipole for example radiates a very small amount of power in a vertically-polarized "8JK" mode, enough to produce a relatively strong ground-signal "off the end" although this has little effect on sky-wave performance.

Fig 12.137 illustrates a suitable test set-up for measuring horizontal polar diagrams. The distance between aerials should be as great as possible, otherwise the signal from one element travels significantly further than that from another and is thus subject to greater attenuation. A null which would be observable at greater distances is therefore partly filled in, and at the low heights usually favoured for such measurements the inverse distance-squared law tends to apply, so that if one element produced a field of 1V at a distance of λ an identical element spaced $\lambda/8$ behind it produces a field of 64/81V, and a null will be filled in to the extent of being only 14dB down. Observations under these conditions, though not useless, are not accurate either, and need careful interpretation. The greater the gain of the aerial being tested, the greater is the separation needed. At short distances accuracy is also affected by the presence of capacitive and inductive fields in addition to the radiation field. In the context of Fig 12.137 height is not critical but should be considerably greater than the spacing between the elements; even so there could be a significant change of tuning and radiation pattern when the aerial is raised to its full height. Depending on sensitivity the field-strength meter may be linear or square-law, and this can be checked against transmitter output as observed on an oscilloscope or other linear voltage indicator. The possibility of harmonic resonances which greatly increase the sensitivity of the test meter to harmonics should not be overlooked, and harmonic radiation from the transmitter must be prevented. This is particularly important if a short aerial is used for the test set.

Gain

Due to the complex role played by ground reflections it is not possible using ground-wave signals to make accurate measurements of effective dx gain of hf aerials. Although for many purposes flat ground can be regarded as a nearly perfect mirror with reflection taking place at a point as in Fig 12.21 the "point" is in fact a Fresnel zone of considerable extent depending on aerial height and the angle of reflection. For the reflection to be fully effective both aerials must "see" the whole of the relevant zone, which extends in the case of short ranges and low angles (as for ground-wave propagation) from the vicinity of the transmitting aerial to a long way beyond the measuring aerial. The allowance to be made for ground reflection must therefore depend on the polar diagram of the aerial under test and will be different for a beam and for the reference aerial. Errors will also occur because the mutual impedance between the aerial and its image will exert a different influence in the two cases and there will probably also be absorption or field distortion by nearby objects which will also be different for different aerials. These problems are all much less acute in the case of measurements on vhf and uhf aerials.

Even if ground-reflection can be regarded as taking place at a point there is *no limit to the possible error* of measurement, as the following example illustrates. A lazy-H with a vertical separation between elements of $0{\cdot}7\lambda$ is mounted at a fairly large mean height $n\lambda$ and a test aerial is placed at a distance $2n\lambda$ so that ground reflection takes place at an angle of 45°. Radiation from the two elements cancels at the point of reflection so that the test aerial experiences the true free-space gain of the lazy-H. If the lazy-H is replaced by a reference dipole, the ground reflection may (depending on the height) add or subtract from the direct signal giving a resultant field anywhere between zero and 6dB greater than the free space value, so that the lazy-H with its true free-space or dx signal gain of 4dB may have in this case an *apparent* gain anywhere between "plus infinity" and −2dB with respect to a dipole! Effects of this nature seem to be the most likely explanation of the large but impossible gain figures so often claimed and conversely perhaps the premature demise of some promising new ideas! At the very least this example emphasizes the need for extreme caution when interpreting the meaning of gain figures both measured and claimed. With this proviso gain measurements may be attempted with the arrangement of Fig 12.137 by replacing the aerial under test with a reference dipole and comparing the signal levels for the same value of radiated power. In the case of beams with tuned feeders it is a simple matter to open circuit all but one element which can then be used as the reference dipole.

Tested in this way close-spaced beams can be expected to show a gain of at least 3dB and sometimes as much as 6dB, but usually rather less than the theoretical value; if, however the radiation resistance and polar diagram of a two-element beam are correct, the gain must also be correct and any measurements which indicate otherwise should be ignored.

Comparisons between aerials based on the average of a sufficient number of signal reports must also be interpreted carefully, remembering that the comparison is not merely between a quad and a Yagi (or whatever the aerials may be) but between two different combinations of *aerial plus environment*. If, as usual, these "environments" include the other aerial, it becomes important to ensure that one aerial is not in the "line of fire" of the other, as the following example illustrates. A 14MHz fixed reversible beam for the north-south directions, which appeared to be working well, was swung round through 90° for comparison with another beam fixed on the long path to VK and found to be down by two S-units, the second beam being 80ft in front and having a good front-to-back ratio so that interaction was at first considered unlikely; it was suspected initially as a result of gain measurement on the lines indicated above, then confirmed by signal reports with the second aerial lowered to the ground, and finally substantiated on theoretical grounds. Any nearly-resonant element in the line of fire tends to generate an antiphase field, and the relative magnitude of this as a function of separation distance can be inferred from the mutual impedance or from the maxima of the mutual resistance curves, Fig 12.9, with due allowance for ground images. In this "parasitic" role a beam does not operate as such and has no front-to-back ratio, which exists only in respect of signals applied to (or taken from) its proper terminals. For an accuracy of ±1dB in comparisons based on average dx signal reports a separation of about 2·0 or 3·0λ appears to be required, but for aerials end-on to each other useful though not highly accurate comparisons can usually be made with quite small separations, eg 0·6λ between centres. The difficulties may however be aggravated, or eased, by environmental differences.

CHOOSING AERIALS

This chapter has attempted to explain aerials and the principles underlying their design: a large number of aerials have been illustrated together with many accessory devices. The beginner may well be confused by the range of information, and therefore need advice on the best way to start.

In particular he may well be discouraged by the complex and expensive installations often considered to be necessary for good results, and if he attempts without previous experience to build, say, one of the more exotic varieties of multiband rotary beam this is likely to breed further discouragement. It is likely that most pleasure will be obtained and more will be learnt by starting with a simple aerial which is bound to work, such as a one-band inverted-V dipole which can be "hitched up" in the centre to the highest point available and the ends allowed to droop (with an apex angle which should not be less than 90°) towards lower anchoring points. Alternatively if two high points are available the dipole may be suspended between them, or two inverted-V dipoles may be mounted at right angles with separate feed lines to provide switched coverage in all directions. If sloping ground is available this should be exploited, good results being obtainable "down the slope" with aerials at comparatively low heights [2]. If space is too limited for a dipole, a ground plane may be tried but results can be very disappointing on the higher frequencies and it is important that the base should be as high as possible, eg mounted on a chimney or rooftop.

Simple aerials such as these are often capable of providing daily ssb contacts with all continents and, in favourable cases, performance may be within one or two S-units of the best obtainable. For multiband operation the use of dipoles with resonant feeders should present no serious problems, but note should be taken of the particular advantages of quad or delta loops as described on pp12.76 and 12.79 and many beginners might find these to be an ideal starting point.

The next step could be the construction of fixed reversible beams using close-spaced pairs of any preferred type of simple radiator. If because of environmental constraints of one sort or another it has not been possible to erect an efficient single radiator, the possible gain from using a close-spaced pair may be as little as zero dB [20] but even if there is no increase of signal the important advantages of less interference received and caused can be fully realized provided *both elements are driven*. This requirement arises because if a parasitic element is lossy the current induced in it will be small compared with that in the driven element so that it is unable to exert much influence on the radiation pattern [20].

The next step may be the erection of a good triband rotary beam, and of the types in common use the easiest to construct and adjust correctly is probably the "triband quad" using three beams on a common support, eg Fig 12.116. There appears however to be no reason why equally good results should not be obtainable with other arrangements such as a pair of driven multiband dipoles (eg the F7FE type, Fig 12.97) or even reversible beams with resonant feeder systems using some arrangement (such as Fig 12.131) to permit at least 120° radiation. For single band operation Figs 12.109, 12.117 and Table 12.8 indicate suitable arrangements and if it is possible to erect several beams the choice includes fixed beams with directional switching and separate monoband rotary beams.

There are of course excellent multiband beams available from commercial sources, as well as some that are less good, and the beginner who thinks he may eventually want to purchase one of these would be well advised to take note of the ones which seem most consistently to produce the biggest signals from long distances.

Low frequency aerials have not been included in the above discussion owing to the complex variety of individual problems and solutions but the reader will have appreciated that many of the hf aerial systems can also be energized in various ways as lf radiators, as discussed on pp12.82 and 12.87.

Those fortunate enough to be able to choose a site, should pick one with plenty of clear space, preferably on a hilltop. With such a site, success comes more easily. Aerials designed for low angle propagation will increase the hours per day during which stations in, say, USA or Australia can be worked. It should be borne in mind that if the ground is sloping, particularly if it is a steep slope, very good results will be obtained very easily down the slope but it may be difficult or impossible to work against the slope. With more than one aerial, or even for multiband work, preset aerial couplers will be found a great advantage.

RECEIVING AERIALS

At the beginning of this chapter it was stated that the receiving problem is not quite the same as the transmitting one. The reciprocity theorem as applied to aerials is true in a limited sense but assumes that interference is isotropic, ie equally likely from all directions, whereas atmospheric noise arrives mainly at low angles, galactic noise from high angles, and man-made noise occurs in local concentration. Prevailing galactic noise levels ensure that useful sensitivity in the hf band is normally limited by external noise so that losses in aerials or feeder systems, up to a point, attenuate noise and signal equally and are much more likely to be acceptable for reception than transmission. The rule concerning the use of the transmitting aerial for reception is a good one; beam arrays, especially the unidirectional ones, can be a great help for reducing interference from directions not in the beam of the aerial. Long wire types are not so good in this respect, and often bring in more cosmic or atmospheric noise than other types: vertical aerials normally reduce the galactic background noise level, and help to discriminate against short-range interference—say from 500 to 1,000 miles, because of their selective vertical patterns. A properly constructed aerial with a good transmission line always helps to minimize man-made interference radiated by house wiring etc.

INTERFERENCE

Communication receivers often show signs of overloading when supplied with the full output of a very large array, such as a V or a rhombic. The effect is a large increase in background noise of a type which is clearly the interaction between high power stations in or near the working frequency; often a pair of stations will beat together to give an i.f. signal which does not respond to the rf tuning of the receiver. This is most likely to occur on the 7, 14 and 21MHz bands, and is recognized because it disappears more rapidly than the tunable signals if attenuation is introduced into the aerial lead and usually also if the rf gain is reduced. It may originate from powerful broadcasting stations in bands adjacent to the amateur bands which cause overloading of rf stages or, due to the presence of rf stages, overloading of the mixer.

It is difficult in a commercial receiver to scrap the rf stage but an attenuator can be readily inserted in the aerial lead and adjusted so that useful sensitivity is only just limited by external noise. This attenuator may take the form of a tunable filter, with considerable advantage if the source of the interference is not too close to the wanted frequency to be rejected by rf selectivity. Advice sometimes given to use a "smaller aerial" should be resisted because this sacrifices the normal interference-reducing properties of the big aerial. Also, due to differences in pattern between the transmitting and receiving aerials it could result in an "occupied" channel being inadvertently selected for transmission, or in failure to hear replies. For these reasons the importance of using the same aerial for transmission and reception cannot be overemphasized.

In order to prevent interference with broadcasting, special steps are taken in the transmitter to prevent over-modulation, key clicks and harmonic output. Loose connections in and around the aerial system can undo this work and rectify the transmitter output, causing serious interference of a harmonic type. Corroded or oxidized metal contacts make rectifiers, and can act as if a diode were connected in the aerial. Offending items include poor or broken solder joints, lead-through terminals, dirty rf connectors in the cable system, corroded earth connections, and loose metal joints such as those in gutters or rusty wire fences. Joints in the aerial should be carefully cleaned, made tight before soldering, and arranged so that the strain on the wire does not pull on a sharp bend. A solder joint may fracture in time through vibration, and should not therefore take strain: it should also be protected with tape, so that it is less likely to corrode. Regular inspection of the aerial system and its surroundings is necessary.

REFERENCES AND BIBLIOGRAPHY

[1] "Supergain Aerials", L. A. Moxon, G6XN, *Radio Communication*, September 1972.

[2] "Low-Angle Radiation", *Wireless World*, April 1970.

[3] "A New Approach to the Design of Super-directive Aerial Arrays", A. Bloch *et al*, *Proc IEE*, Part III, September 1953.

[4] *Radio Engineers Handbook*, F. E. Terman, McGraw Hill, 1943, pp799–802.

[5] *Ibid*, p193.

[6] "A 14 Mc Co-Ax Fed Dipole and TVI", F. G. Rayer, G3OGR, *RSGB Bulletin*, March 1965. See also "Why Coax?", E. M. Wagner, G3BID, *73*, November 1971.

[7] "The Determination of the Direction of Arrival of Short Radio Waves", H. T. Friis, C. B. Feldman, W. M. Sharpless, *Proc IRE*, Vol 22, 1934, pp47–78.

[8] "Aerial Reflections", F. Charman, BEM, G6CJ, *RSGB Bulletin*, December 1955.

[9] "Two-Element Driven Arrays", L. A. Moxon, G6XN, *QST*, July 1952.

[10] "Theoretical Treatment of Short Yagi Aerials", W. Walkinshaw, *JIEE*, Part IIIA, Volume 93, 1946, p 564.

[11] "Multiband Quads", L. A. Moxon, G6XN, *CQ*, November 1962.

[12] "The Swiss Quad Beam Aerial", R. A. Baumgartner, HB9CV, *RSGB Bulletin*, June 1964.

[13] "The HRH Delta-loop Beam", H. R. Habig, K8ANV and Lew McCoy, W1ICP, *QST*, January 1969. See also Technical Topics, *Radio Communication*, May 1969 and May 1973.

[14] "Evaluating Aerial Performance", L. A. Moxon, *Wireless World*, February and March 1959.

[15] "The 9M2CP Z-beam", Technical Topics, *Radio Communication*, August 1971.

[16] "Unidirectional Antenna for the LF Bands", Malcolm M. Bibby, GW3NJY, *Ham Radio*, January 1970.

[17] "The G8PO 'Guy Wire' array", J. E. Ironmonger, G8PO, *RSGB Bulletin*, July 1962.

[18] "More about the Minibeam", G. A. Bird, G4ZU, *RSGB Bulletin*, October 1957.

[19] "Measuring Cable Loss by SWR", O. J. Russell, *RSGB Bulletin*, November 1961.

[20] "Gains and Losses in Aerials", L. A. Moxon, G6XN, *Radio Communication*, December 1973 and January 1974.
[21] *All about Cubical Quad Antennas*, William I. Orr, W6SAI, Radio Publication Inc, Wilton, Conn.
[22] "Ropes and Rigging for Amateurs", J. Michael Gale, G3JMG, *Radio Communication*, March 1970.
[23] "Aerial Masts and Rotation Systems", R. Thornton, GM3PKV, and W. H. Allen, G2UJ, *Radio Communication*, August and September 1972.
[24] "The Protection of Structures against Lightning", J. F. Shipley, *JIEE*, Part I, December 1943.
[25] "The Protection of Structures against Lightning", British Standard Code of Practice, BSI, CP326; 1965.
[26] "The Reception of Sound and Television Broadcasting", BS Code of Practice, BSI, CP327.201; 1960.
[27] "Lightning and your Aerial", G. R. Jessop, G6JP, *Radio Communication*, January 1972.

CHAPTER 13

VHF AND UHF AERIALS

THE range of frequencies considered in this chapter extends from 30MHz to 3,000MHz. In accordance with the accepted terminology a distinction ought strictly to be made between the range 30–300MHz which is described as *very high frequency* (*vhf*) and the range 300–3,000MHz which is described as *ultra high frequency* (*uhf*). Over these two ranges, however, the problems of aerial design are often quite similar in character and to avoid unnecessary repetition the term *vhf* is used here to cover both of them. Some attention is given to aerials for *super high frequency* (*shf*), the range 3–30GHz, but this subject is largely beyond the scope of this chapter. Above 30MHz the wavelength of the radiation becomes short enough to allow efficient aerials of comparatively small dimensions to be constructed, since the effectiveness of an aerial system generally improves when its size becomes comparable with the wavelength that is being used. A special feature of vhf aerial design is the possibility of focusing the radiated energy into a beam, thereby obtaining a large effective gain in comparison with a half-wave dipole, which is the standard by which most vhf aerials are judged.

Power Gain and Beamwidth

A simple way to appreciate the meaning of aerial gain is to imagine the radiator to be totally enclosed in a hollow sphere, as shown in **Fig 13.1**. If the radiation is distributed uniformly over the interior surface of this sphere the radiator is said to be *isotropic*. An aerial which causes the radiation to be concentrated into any particular area of the inside surface of the sphere, and which thereby produces a greater intensity than that produced by an isotropic radiator fed with equal power, is said to exhibit gain relative to an isotropic radiator. This gain is inversely proportional to the fraction of the total interior surface area which received the concentrated radiation.

The term *gain* of any particular aerial system always applies in the direction of maximum radiation.

The gain of an aerial is usually expressed as a power ratio, either as a multiple of so many "times" or in decibel units. For example, a power gain of 20 times could be represented as 13dB (ie $10 \log_{10} 20$).

The truly isotropic radiator is a purely theoretical concept, and in practice the gain of beam aerials is usually compared with the radiation from a single half-wave dipole fed with an equal amount of power. The radiation pattern of even a single half-wave dipole is markedly non-uniform, and in consequence the power gain of such an aerial compared with the hypothetical isotropic radiator is about 64 per cent (ie 5/3 times or 2·15dB), but since the half-wave dipole is the simplest practical form of radiator it is generally acceptable as a basis of comparison.

The area of "illumination" is not sharply defined as shown by the shaded region in Fig 13.1 but falls away gradually from the centre of the area. The boundaries of the illuminated area are determined by joining together all points where the radiation intensity has fallen by half (ie 3dB): these are known as the *half-power points*. The gain of the aerial can then be determined by dividing the total surface area of the sphere by the illuminated area: eg if the total surface area were 100 sq cm, and the illuminated area bounded by the half-power points were 20 sq cm, the gain of the aerial would be five times or 7dB. The radiation in any particular plane can be plotted graphically, usually in polar co-ordinates; such a plot is called a *polar diagram*. A typical polar diagram is shown in **Fig 13.2**. The region of maximum radiation is called the *major lobe*. Unwanted radiation is occurring in other directions, and these regions, when small compared with the major lobe, are called *minor lobes*. All practical aerials exhibit such lobes and the aerial designer frequently has to compromise to obtain the optimum performance for any particular application. For example, an aerial may be designed for maximum *front-to-back ratio*; ie for minimum radiation in the direction opposite to the major lobe. To achieve such a condition it may be

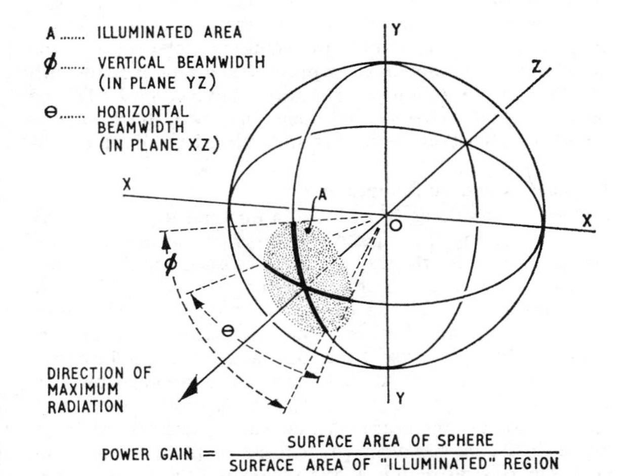

Fig 13.1. Radiation from an aerial. An isotropic radiator at the point O will give uniform "illumination" over the inner surface of the sphere. A directional radiator will concentrate the energy into a beam which will illuminate only a portion of the sphere as shown shaded.

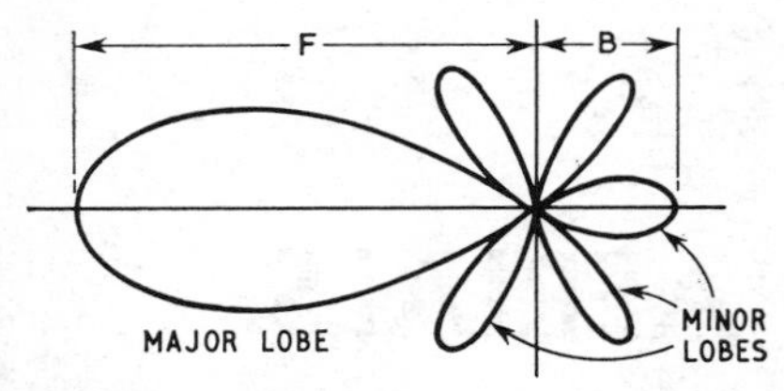

Fig 13.2. Polar diagram of a vhf beam aerial. The front-to-back ratio is represented by F/B.

necessary to sacrifice some gain in the major lobe (or forward radiation) with a possible increase in other minor lobes (or side lobes) and thus the designer will need to consider all the implications before finalizing any particular design.

In practice the radiation from an aerial is measured in a horizontal and a vertical plane. The *beamwidth* is the angle between the two half-power points in the plane under consideration. The vertical polar diagram is greatly influenced by the height of the aerial above ground; the higher the aerial the lower will be the angle of maximum radiation, and at the same time the effects of neighbouring objects such as houses and trees will be minimized. The important requirement is to place the aerial well clear of such objects and this frequently means as high as can be safely achieved. Any aerial which has the property of concentrating radiation into any particular direction is said to possess *directivity*.

Bandwidth

The performance of an aerial array generally depends upon the resonant properties of tuned radiators such as dipoles or other elements, and therefore any statement regarding its power gain or beamwidth will be valid only over a restricted frequency band. Beyond the limits of this band the properties of an aerial system may be entirely different. Hence it is useful to define *bandwidth* as that range of frequencies over which the power gain of the aerial array does not fall by more than a certain percentage as compared with the frequency at which maximum gain is obtained (eg a bandwidth of 15MHz for a 50 per cent reduction in power gain). Alternatively the bandwidth may be defined as the frequency band over which the standing wave ratio of the aerial feeder does not exceed a prescribed limit (eg bandwidth of 10MHz for a standing wave ratio not exceeding 2 : 1). The latter convention is the one generally used. It should be noted that these examples are not related to one another.

Capture Area or Aperture

Besides examining the action of a transmitting aerial array in concentrating the radiated power into a beam it is also helpful to examine the way in which the same aerial structure will effect the reception of an incoming signal. In this study it is convenient to introduce the concept of *capture area* or *aperture* of the aerial. This concept is frequently misunderstood, probably because it may appear to relate to the cross-sectional area of the radiated beam (as represented by *A* in Fig 13.1): it is in fact related to the *inverse* of the cross-sectional area of the beam inasmuch as an aerial which has a high gain usually has a sharply focused beam (ie one of small cross-sectional area) but at the same time the capture area of the aerial is large. The larger the capture area, the more effective is the aerial.

The actual size of the aerial system does not always give a reliable indication of the capture area. A high-gain array may have a capture area considerably greater than its frontal area determined by its physical dimensions.

The fundamental relationship between the capture area and the power gain of an aerial system is:

$$A = \frac{G_I \lambda^2}{4\pi}$$

where A is the capture area and λ is the wavelength (measured in the same units as A) and G_I is the power gain (arithmetically, not dB) relative to an isotropic radiator.

A half-wave dipole has a gain of $\frac{5}{3}$ relative to an isotropic radiator, and therefore this formula can be modified to give the capture area in terms of the gain of a half-wave dipole G_D instead of G_I simply by introducing the factor $\frac{5}{3}$, thus—

$$A = \frac{5}{3} \times \frac{G_D \lambda^2}{4\pi} = \frac{5G_D \lambda^2}{12}$$

This formula shows that if the wavelength is kept constant the capture area of an aerial is proportional to its gain, and therefore if an increase in gain results in a narrower beamwidth it must follow that a narrower beamwidth corresponds to a greater capture area (the term beamwidth being used here to signify both horizontal and vertical dimensions, ie in effect the cross-sectional area).

The formula also shows that for any given power gain the capture area is proportional to the square of the wavelength. For example, an aerial having a power gain of, say, 10 times relative to a dipole at 600MHz (0·5m) would have a capture area one-sixteenth of that of an aerial having a similar power gain at 150MHz (2m); to achieve equal capture area the gain of the 600MHz aerial would have to be 16 times greater than that of the 150MHz signal, ie 160 times relative to a dipole. This is unfortunate because it is the capture area of the aerial that determines its effectiveness in absorbing the incoming radiation: it means that as the wavelength is reduced it becomes increasingly important to design the aerial to have a higher gain in order to produce the same voltage at the receiver terminals and thus the same basic signal-to-noise ratio.

A very useful rule of thumb method of calculating the gain of a multiple array aerial system is that each time the aperture is doubled the gain is increased by approximately 3dB, although in practice the increase is usually a little less. For example, if two 5-element Yagi arrays are stacked, provided they are spaced so that their apertures just touch, then the overall gain will be increased from about 8 to 11dB. It is useful to remember this point when considering *stacking* later in the chapter.

These observations apply only to signals being received or transmitted in the direction of maximum gain. For directions other than the optimum the relationships become more complex.

Multi-radiator Arrays

High-gain aerial arrays can be built up from a number of individual radiators such as half-wave dipoles. To achieve the maximum gain, the spacing of these radiators should be such that their respective capture areas just touch. Where the individual radiators are themselves high-gain systems, such as Yagi arrays, the centre-to-centre spacing of each radiating system needs to be larger, since the individual capture areas are greater.

Reciprocity Theorem

The theorem of reciprocity states that any particular aerial gives the same performance either as a transmitting or as a receiving system. Practical aerial designs are therefore worked out in terms of transmission because the characteristics are more easily determined in this way, and the resulting aerials are assumed to have similar reception properties.

Angle of Radiation

The characteristics of propagation in the vhf ranges are determined principally by the influence of the troposphere, ie the part of the atmosphere extending from ground level up to a few thousand feet. There is rarely ionospheric propagation on frequencies higher than about 100MHz and thus any energy which is radiated at more than a few degrees above the horizontal is wasted. Similarly to transmit to a particular point it is unnecessary to radiate a broad beam in the horizontal plane. Generally, therefore, the aerial designer tries firstly to reduce the vertical beamwidth to avoid wastage of power into space and secondly to reduce the horizontal beamwidth according to the required ground coverage. A narrow horizontal beamwidth can however be a disadvantage because stations situated off the beam may be missed, when searching. Under conditions of high activity, on the other hand, a narrow bandwidth can be helpful in rejecting interference from unwanted stations in other directions. A compromise between gain and beamwidth must be made.

Height of the Aerial above Ground

Several considerations apply when deciding the height of the aerial above the ground for optimum performance:

(*a*) the aerial should be well above local screening from buildings and other obstacles;

(*b*) even if there is no screening, the earth will act as a plane reflector and will tend to direct the radiated energy into space rather than along the horizontal as is required at vhf. To minimize this effect, the aerial should be as high as possible (and the height should be judged in terms of wavelengths rather than feet);

(*c*) the higher the aerial the greater the feeder length and the more difficult and costly is its erection, especially in confined spaces. Thus, again, a compromise has to be made.

It is well to remember that even in an ideal location from the vhf point of view, such as a hilltop with no local screening, it is still necessary to erect the aerial as high as can be achieved within the resources available. This consideration is also very important when aerials are stacked as the effective height of the aerial is the height of the physical centre or electrical centre of gravity above ground. In general, the improvement to be obtained from increasing aerial height outweighs the increase in feeder loss. At an average amateur location the height/gain relationship is often better than 6dB for doubling of height. Only in the case of well-sited stations on high ground is such a rate of height gain not achieved.

Polarization

Radio waves are constituted from electric and magnetic fields mutually at right angles and also at right angles to the direction of propagation. The ratio of the electric component E to the magnetic component H in free space ($E/H = Z$) is known as the *impedance of free space* and has a value of about 377Ω. When the electric component is horizontal, the wave is said to be *horizontally polarized*. Such a wave is radiated from a horizontal dipole. If the electric component is vertical, as in a vertical dipole, the wave is said to be *vertically polarized*.

Sometimes the polarization is not exclusively horizontal or vertical and the radiation is then said to be *elliptically polarized* or, in the special case where the horizontal and vertical components are equal, *circularly polarized*. The effect of the addition of two components of the same kind (ie electric *or* magnetic) at right angles is to create a rotating field, the direction of rotation of which depends on the relative phase of the two components. Thus the polarization of the wave will appear to have either clockwise or counter-clockwise rotation, a feature which is important in the use of helical aerials. A dipole will receive an equal pick-up from a circularly polarized wave irrespective of whether it is mounted horizontally, vertically or in an intermediate position, but the strength will be 3dB less than if an aerial designed for circular polarization is used. This means that the full *gain* of a helix will only be realized when received by a helix. Horizontally or vertically polarized waves (or any other wave from linear sources) are known as *plane polarized* waves.

It has been found by some experimenters that in the vhf range horizontally polarized waves suffer less attenuation over long distances than vertically polarized waves, and this system is therefore often preferred. It has in fact been universally adopted for vhf dx communication in the British Isles and many other countries. Vertically polarized waves may be more suitable for other purposes such as short-distance or mobile communication and a simple ground-plane or similar aerial can then be used, as described in Chapter 14 (*Mobile and Portable Equipment*).

Tests have been conducted at frequencies of the order of 70MHz between fixed stations employing horizontally polarized beam aerials and mobile stations employing vertically polarized ground plane aerials which show that the loss in signal strength due to cross polarization is much less than might be expected due to the multiple reflections that occur from the objects that surround mobile radiators. The vertically polarized wave is transformed into one of random polarization and hence there is no difference in performance if either a horizontally or vertically polarized aerial is used at the fixed station. Further work needs to be done in this field in the uhf band to see whether the same results occur, although tests by the BBC on television Bands IV and V (470–582MHz and 606–960MHz) show less multi-path propagation from horizontally polarized aerials and hence less fading.

AERIAL FEEDERS

Before discussing vhf aerial design it will be helpful to review the methods of conveying power from the transmitter to the aerial. The feeder length should always be considered in terms of wavelengths rather than the actual length of the conductors. If the feeder length is short compared with the wavelength, the loss caused by its ohmic resistance and by the dielectric conductance is unimportant as also is the effect of incorrect impedance matching. For vhf operation, however, the aerial feeder is usually many wavelengths long,

TABLE 13.1

Characteristics of typical radio frequency feeder cables

Type of cable	Nominal impedance Z_0 (Ω)	Dimensions (in) Centre conductor	Dimensions (in) Over outer sheath	Dimensions (in) Over twin cores	Velocity factor	Approximate attenuation (dB per 100ft) 70MHz	145MHz	433MHz	1,296MHz	Remarks
Standard tv feeder ...	75	7/·0076	0·202	—	0·67	3·5	5·1	9·3	17	—
Low-loss tv feeder (semi-air-spaced) Super Aeraxial	75	0·048	0·290	—	0·86 approx.	2·0	3·0	5·5	10	Semi-air-spaced or cellular.
Flat twin	150	7/·012	—	0·18 × 0·09	0·71	2·1	3·1	5·7*	11*	*Theoretical figures, likely to be considerably worsened by radiation.
Flat twin	300	7/·012	—	0·405 × 0·09	0·85	1·2	1·8	3·4*	6·6*	
Tubular twin	300	7/·012	—	0·446	0·85	1·2	1·8	3·4*	6·6*	

UR No	Nominal impedance Z_0 (Ω)	Overall diameter (in)	Inner conductor (in)	Capacitance (pF/ft)	Maximum operating voltage (rms)	Approximate attenuation (dB per 100ft) 10MHz	100MHz	300MHz	1,000MHz	Approx RG equivalent
43	52	0·195	0·032	29	2,750	1·3	4·3	8·7	18·1	58/U
57	75	0·405	0·044	20·6	5,000	0·6	1·9	3·5	7·1	11A/U
63*	75	0·853	0·175	14	4,400	0·15	0·5	0·9	1·7	
67	50	0·405	7/0·029	30	4,800	0·6	2·0	3·7	7·5	213/U
74	51	0·870	0·188	30·7	15,000	0·3	1·0	1·9	4·2	218/U
76	51	0·195	19/0·0066	29	1,800	1·6	5·3	9·6	22·0	58C/U
77	75	0·870	0·104	20·5	12,500	0·3	1·0	1·9	4·2	164/U
79*	50	0·855	0·265	21	6,000	0·16	0·5	0·9	1·8	
83*	50	0·555	0·168	21	2,600	0·25	0·8	1·5	2·8	
85*	75	0·555	0·109	14	2,600	0·2	0·7	1·3	2·5	
90	75	0·242	0·022	0	2,500	1·1	3·5	6·3	12·3	59B/U

All the above cables have solid dielectric with a velocity factor of 0·66 with the exception of those marked with an asterisk which are helical membrane and have a velocity factor of 0·96. This table is compiled from information kindly supplied by Aerialite Ltd and BICC Ltd, and includes data extracted from Defence Specification DEF-14-A (HMSO).

and therefore both the loss introduced and the matching of the load to the feeder are of the utmost importance.

Two types of feeder, or transmission line, are in common use, the *unbalanced* or *coaxial feeder* and the *balanced pair:* the latter may be either of open construction or enclosed in polythene ribbon or tubular moulding. Each type has its own particular advantages and disadvantages.

In a coaxial cable, the radio frequency fields are contained entirely within the outer conductor and hence there should be no rf currents on the outside. This enables the cable to be carried in close proximity to other cables and metal objects without interaction or serious change of its cable properties which might cause reflections and thereby introduce appreciable loss. There is no loss by external radiation.

The open-wire or balanced feeder has a radiation loss which is dependent upon the ratio of the spacing of the wires to the wavelength and becomes more serious as the frequency is raised. However, if this spacing is less than 0·01 λ the radiation loss is negligible. The properties can also be severely changed by the close proximity of metal objects and the accumulation of ice or water on the separating insulation, and therefore much greater care must be taken in the routing of the feeder. Difficulties are often experienced when attempting to use this type of feeder with rotatable aerial arrays. It is, for instance, quite unsatisfactory to bind a ribbon feeder directly against a metal mast. When the feeder is kept well clear of metal objects, however, the loss tends to be less than that of coaxial cable unless the frequency is so high that the radiation loss is serious (ie 450MHz and above).

Generally speaking, the use of open-wire feeders is restricted to bands of 144MHz and below; coaxial cable is used for all frequencies up to about 3,000MHz. Above about 1,000MHz the loss in conventional types of flexible coaxial cable becomes prohibitive and rigid semi-air-spaced types of cable are then used. Above 3,000MHz waveguides are usually necessary for feeder runs of more than a few inches.

A list of typical aerial feeder cables obtainable in Great Britain is shown in **Table 13.1.** Further information can be obtained from the various manufacturers.

In the semi-air-spaced type of coaxial cable, the centre conductor is supported either on beads or on a helical thread: in some forms a continuous filling of cellular polythene is used as the insulator. This type of cable has a lower capacitance per unit length than the solid type and hence the *velocity factor* (ie the ratio of the wave velocity in the cable compared with that in free space) is higher, being about 0·88–0·98 for helical and bead types and about 0·8 for cellular polythene types compared with 0·66 for solid cable. The use of semi-air-spaced conductors allows cables to be designed with less attenuation for a given size, or with the same attenuation for a smaller size, but unfortunately the bead and helical types suffer from the disadvantage that moisture can easily enter the cable; special precautions must therefore be taken to ensure a good watertight seal at the aerial end if such cable is used. Suitable material for this purpose is Telecompound (a softened polyethylene compound manufactured by the Telecon division of BICC) or Bostik sealing strip.

Cellular polythene cables do, however, suffer from another disadvantage; the effective dielectric constant of the cellular polythene varies according to the number and volume of the air cells per unit length which unfortunately cannot be controlled with precision during manufacture. Thus the characteristic impedance may tend to vary along a length of

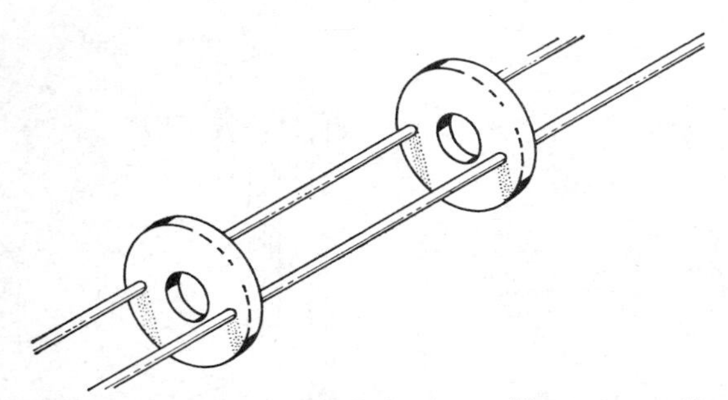

Fig 13.3. Open-wire line with disc insulators. The centre portions of the insulators are removed to minimize dielectric losses.

feeder and although this may not matter at the lower end of the vhf spectrum such cable is not recommended for frequencies above about 500MHz.

It is also important to exclude water even from the outer braiding of any coaxial cable. If water has once entered it is very difficult to dry it out, and the loss in the feeder becomes progressively higher as the copper braiding corrodes.

Open-wire Feeders

To obtain very low losses open-wire feeder line can conveniently be made from hard-drawn 16swg copper wire with separating insulators placed at intervals. The insulators should be made from polythene and be shaped in the form of a disc with the centre removed, as shown in **Fig 13.3**. This ensures that in places where the maximum electric stress occurs, ie between the conductors, the dielectric is air and not the solid insulating material, thus minimizing losses. To prevent excessive radiation loss the characteristic impedance should not exceed about 300Ω. Care must be taken to avoid sharp bends and also the close proximity of surrounding objects as stated earlier. The characteristic impedance of an open-twin line is given by—

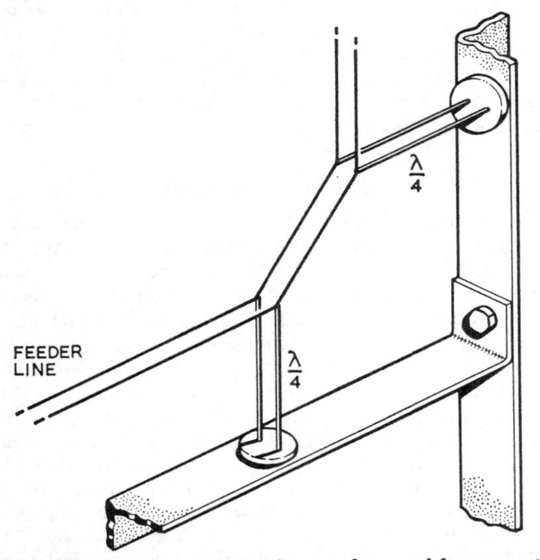

Fig 13.4. Quarter-wave closed stubs are often used for supporting open-wire feeders. When used in this way they are known as "metal insulators". The inductance of the short-circuiting conductors is minimized by making them in the form of large metal plates.

$$Z_0 = 276 \log_{10} (2D/d)$$

where D is the centre-to-centre spacing and d is the diameter of the wire (measured in the same units). A chart of characteristic impedance in relation to the conductor size and spacing is given in Chapter 12—*HF Aerials*.

Metal Insulators

A quarter-wave short-circuited transmission line presents a very high impedance at the open end and hence may be connected across an open-wire line without affecting the power flow in any way; such a device is called a *quarter-wave stub*. The stub so formed may conveniently be used as a support or termination for an open-wire line as shown in **Fig 13.4**. Since the stub must be resonant in order to behave as an insulator it can function only over a narrow range of frequency. Note that the characteristic impedance of the stub does not have to be the same as that of the line.

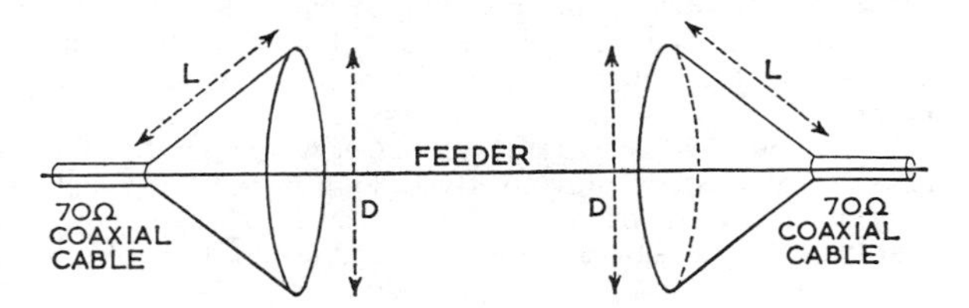

Fig 13.5. Surface-wave transmission line. The outer sheathing of each length of coaxial cable is terminated by an open cone. The feeder itself should be of 10-16swg copper wire, preferably enamelled; its length may be several hundred feet. The cone diameter D must be 0·6L, but L itself may be of any length greater than 3λ.

Surface-wave Transmission Lines

A type of feeder which becomes useful above about 400MHz is the surface-wave transmission line. The wave is directed on to a single conductor by means of a horn: see **Fig 13.5**. The dimensions of the horn are not critical but the angle should be correct, and for the best performance the side should be several wavelengths long. The single conductor wire should be covered with a thin dielectric, preferably polythene, although enamel or even an oxidized covering is usually sufficient. This covering layer minimizes radiation loss by reducing the effective diameter of the field surrounding the wire. Typical losses measured at a frequency of 3,300MHz (9cm) using horns 21in long and 13in in diameter and a 14swg wire feeder (bare but oxidized) are about 1·35dB per hundred feet plus about 0·4dB per horn, making a total of just over 2dB per hundred feet.

At 430MHz the attenuation of the field 8in away from the feeder is about 20dB. The feeder can therefore be run reasonably near to walls and other objects for short distances. If the feeder is a fairly long one, however, it would be desirable to maintain a spacing of several feet over most of its length. The radius of any bends should be kept greater than about one wavelength in order to avoid losses due to the tendency of the energy to be radiated into space whenever there is a sudden change in the direction of the feeder.

Balance-to-unbalance (Balun) Transformers

In most cases the aerial requires a balanced feed with respect to ground, and it is therefore necessary to use a device which converts the unbalanced output of a coaxial cable to the balanced output required by the aerial. This device also

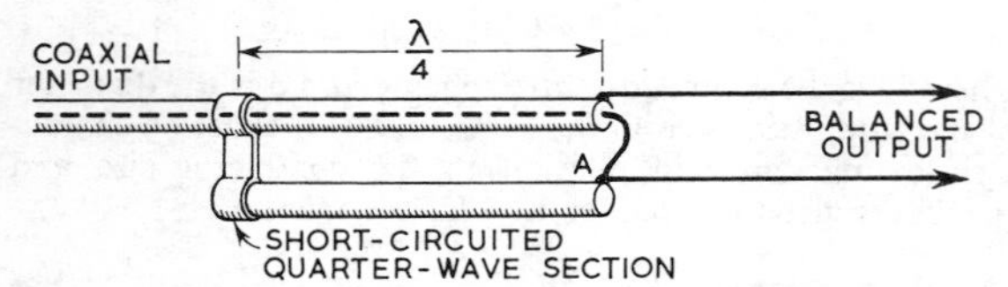

Fig 13.6. Quarter-wave open balun or Pawsey stub.

prevents the wave which has been contained within the cable from tending to "spill over" the extreme end and travel back over the surface of the cable. Whenever this occurs there are two important undesired effects; first the re-radiated wave modifies the polar diagram of the attached aerial, and secondly the outer surface of the cable is bound to have a radio frequency on it.

To prevent this, a balance-to-unbalance transformer (abbreviated to *balun*) is connected between the feeder cable and the aerial. The most simple balun consists of a short-circuited quarter-wave section of transmission line attached to the outer-braiding of the cable, as shown in **Fig 13.6**.

This is often known as a *Pawsey stub*. At the point *A*, the quarter-wave section presents a very high impedance which prevents the wave from travelling over the surface. Note that the $\lambda/4$ dimension given is in air and not in the cable.

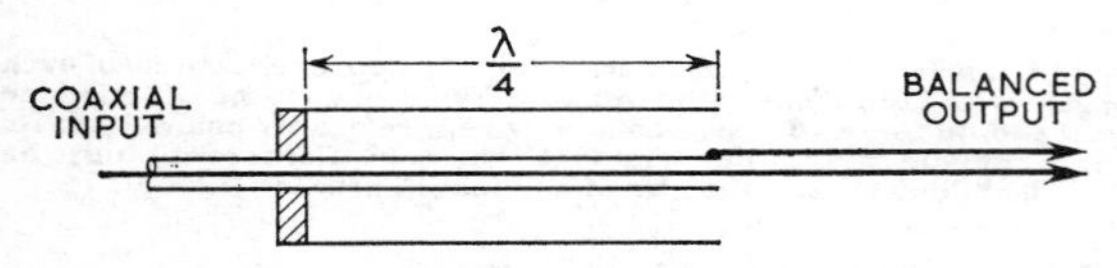

Fig 13.7(a). Coaxial-sleeve balun.

Several modifications to the simple balun are possible: for example, the single quarter-wave element may be replaced by a quarter-wave coaxial sleeve, thus reducing radiation loss: see **Fig 13.7(a).** To prevent the ingress of water and to improve the mechanical arrangement, the centre conductor may itself be connected to a short-circuited quarter-wave line acting as a "metallic insulator" as shown in **Fig 13.7(b).** The distance *d* should be kept small, and yet the capacitance between the sections should also be kept small otherwise the quarter-wave section will not be resonant at the desired frequency. A satisfactory compromise is to taper the end of the quarter-wave line, although this is by no means essential. In practice, at a frequency of 435MHz about ⅛in is a suitable spacing. The whole balun is totally enclosed, the output being taken through two insulators mounted in the wall.

A useful variation is that shown in **Fig 13.8**, which gives a 4 : 1 step-up of impedance. The half-wave loop is usually made from flexible coaxial cable, and allowance must

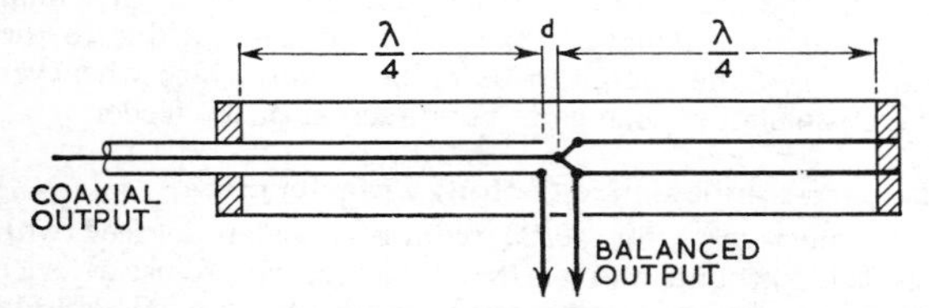

Fig 13.7(b). Totally enclosed-coaxial balun. The right-hand section acts as a "metal insulator".

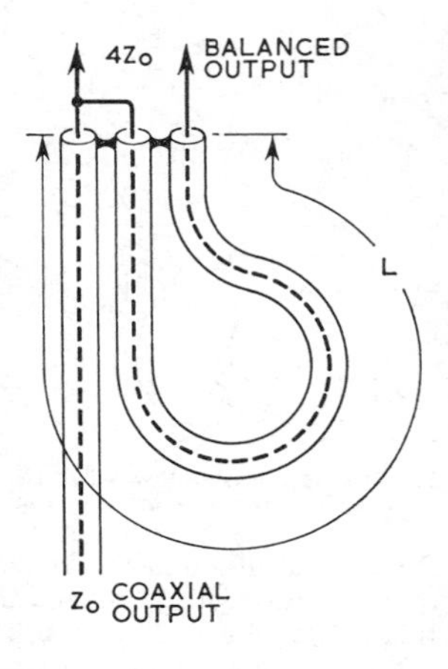

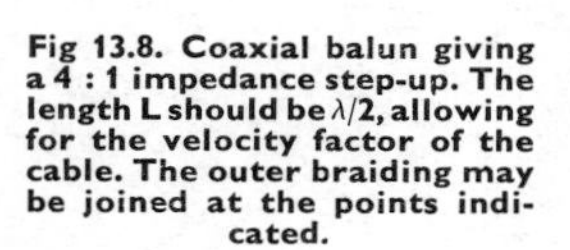

Fig 13.8. Coaxial balun giving a 4 : 1 impedance step-up. The length L should be $\lambda/2$, allowing for the velocity factor of the cable. The outer braiding may be joined at the points indicated.

therefore be made for the velocity factor of the cable when calculating a half-wavelength.

At frequencies above about 2,000MHz it may be inconvenient to mount the coaxial-sleeve balun close to a dipole radiator. In this case the sleeve can be mounted a short distance back from the end of the line. The characteristic impedance of the balun element is not critical.

A Pawsey stub may be constructed by attaching a piece of coaxial cable one physical quarter-wave long (the centre conductor being unused) to the braiding of the feeder cable. The two sections should be spaced sufficiently to ensure an air dielectric between them. If the two pieces lie closely alongside one another the resonant length must be reduced and an inferior dielectric introduced. It is important to note that since it is the electrical characteristics of the outer surface that are being used, there is no need to allow for the velocity factor of the cable. Coaxial-sleeve baluns should have an outer-to-inner diameter ratio of between 2 : 1 and 4 : 1.

The performance of these baluns will be frequency dependant as they rely for their operation upon the properties of resonant transmission lines. In all cases, however, these lines are effectively terminated in a low impedance and the bandwidth is comparatively large. All are quite suitable for use in the vhf/uhf amateur bands.

IMPEDANCE MATCHING

For an aerial feeder to deliver power to the aerial with minimum loss, it is necessary for the load to behave as a pure resistance equal in value to the characteristic impedance of the line. Under these conditions no energy is reflected from the point where the feeder is joined to the aerial, and in consequence no standing waves appear on the line.

When the correct terminating resistance is connected to any feeder, the voltage and current distribution along the line will be uniform. This may be checked by using a device to explore either the magnetic field (*H*) or the electric field (*E*) along the line. One such device, suitable for use with a coaxial feeder, is a section of coaxial line with a longitudinal slot cut in the wall parallel to the line. A movable probe connected to a crystal voltmeter is inserted through the slotted wall. This samples the electric field at any point, and the standing wave ratio may be determined by moving the probe along the line and noting the maximum and minimum

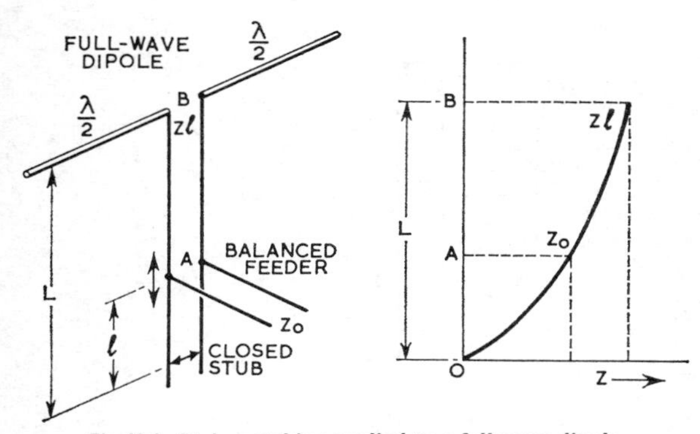

Fig 13.9. Stub matching applied to a full-wave dipole.

readings. The distance between adjacent maxima or between adjacent minima is one half-wavelength.

The fields surrounding an open line may be explored by means of an rf voltmeter, but it is much more difficult to obtain precise readings than with a coaxial line because of hand-proximity effects and similar disturbances.

Another device which measures forward and reflected waves is the *reflectometer* (see Chapter 18).

The term *matching* is used to describe the procedure of suitably modifying the effective load impedance to make it behave as a resistance and to ensure that this resistance has a value equal to the characteristic impedance of the feeder used. To make a *complex* load (ie a load possessing both resistance and reactance) behave as a resistance, it is necessary to introduce across the load a reactance of equal value and opposite sign to that of the load, ie the reactance is "tuned out". A convenient device which can theoretically give reactance values from minus infinity to plus infinity (ie large capacitance to large inductance) is a section of transmission line, either of length variable between zero and one half-wavelength having an open-circuited end, or alternatively of length a little greater than one half-wavelength having a movable short-circuit capable of being adjusted over a full half-wavelength.

Although there is no need to make the characteristic impedance of a stub equal to that of the transmission line, it may be desirable to do so for practical reasons.

In addition to tuning out the reactance, a match still has to be made to the transmission line. The impedance at any point along the length of a quarter-wave resonant stub varies from zero at the short-circuit to a very high impedance at the open end. If a load is connected to the open end and the power is fed into the stub at some point along its length the stub may be used as an auto-transformer to give various values of impedance, according to the position of the feed point. This is shown in **Fig 13.9**. The distance L is adjusted to tune the aerial to resonance and will be one quarter-wave long if the aerial is already resonant. The distance l is adjusted to obtain a match to the line. However, it is usually more convenient to have a stub with an adjustable short-circuit which can slide along the transmission line: see **Fig 13.10**.

In practice matching can be achieved entirely by the "cut and try" method of adjusting the stub length and position until no standing waves can be detected. The feeder line is then said to be *flat*. However, the frequency range over which any single-stub matching device is effective is quite small, and where wideband matching is required some other matching system must be used.

Stub-matching

When it is possible to measure the swr with good accuracy and to locate the exact position on the feeder at which the minimum current occurs, the length of the stub required to match the line and the correct point at which to connect it may be predicted and much of the time required by cut-and-try methods saved. The predictions may be read off from **Fig 13.11**. The application of the method is best suited to balanced feeders; adjustments are somewhat easier with open-wire lines.

The measurement of swr may be made when transmitting on low power:

(*a*) by inserting in series with the feeder an swr indicating or measuring device, near the point at which the stub is to be connected. For wide bandwidth this stub should be very near the aerial.

(*b*) by running along the feeder a calibrated probe which has arrangements to keep its coupling to the feeder constant while its position is moved along the line. It is helpful to have the feeder well tensioned. Either voltage- or current-probes can be used, having a thermocouple or crystal diode connected to a sensitive milliammeter. The probe must be *loosely* coupled to the line.

The position of a minimum is sharply defined, a maximum is flat. Even so, a minimum position is best found by "bracketing" it. The positions at which equal readings of current or voltage are obtained on either side of the minimum are found: the minimum position is taken to be midway between them. A voltage minimum can be assumed to be a quarter-wave distant from a current minimum. With the bracketing method even a torch bulb coupled to the line can be used to locate the I_{min} position with fair accuracy. Its coupling to the line must be the same in each position. The two positions at which the filament just begins to glow are noted.

It is very important that harmonic currents are not flowing in the line during measurements, and as filters do not work well unmatched, the probe should be sharply tuned to the

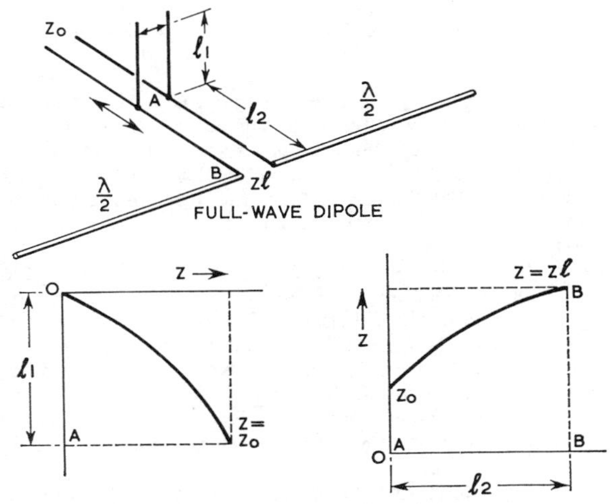

Fig 13.10. Stub matching with a movable short-circuited stub.

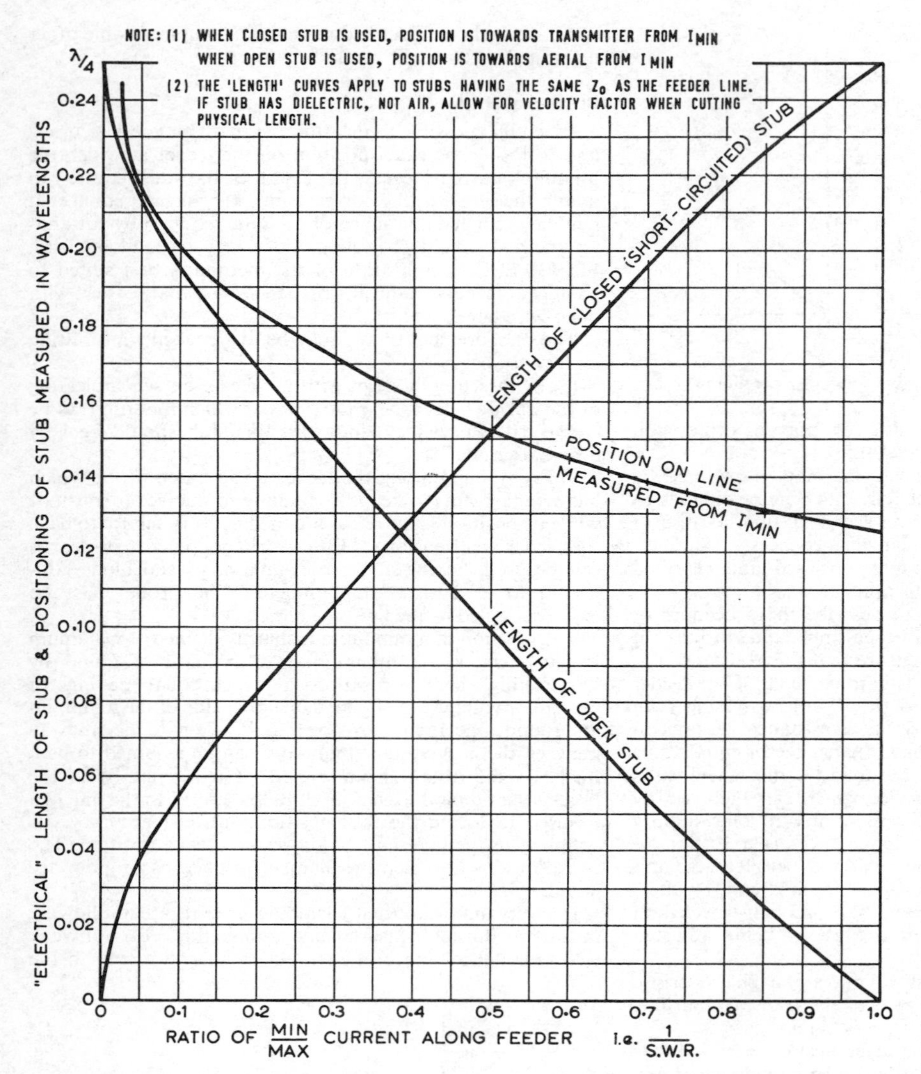

Fig 13.11. Chart giving details of stub lengths.

operating frequency. A field strength meter with suitable constant-coupling arrangements can be used as a probe. It may be calibrated by connecting a short circuit on the line to set up an infinite swr. The theoretical sinusoidal current pattern along the line is used to set the probe in positions of known current (ie current of known ratio relative to the maximum current).

eg (*a*) Set the probe at the I_{max} position and reduce the probe coupling, and adjust transmitter power or transmitter loading to get a full scale meter deflection.

(*b*) Locate the I_{min} position accurately by bracketing.

(*c*) Set the probe $\lambda/12$ from I_{min}. Note the meter reading. The actual current here is 0·5 I_{max} ($\lambda/12 = 30°$ phase. Sin 30° = 0·5).

(*d*) Set the probe $\lambda/8$ from I_{min}. Here the actual current is 0·707 I_{max}.

(*e*) Set the probe at other positions and calibrate similarly.

Any error due to a slightly incorrect position of I_{min} may be eliminated by repeating the readings on the opposite side of I_{min} and taking an average meter reading. Any appreciable asymmetry may indicate excessive probe coupling, variation in probe coupling, or harmonic currents.

If the predicted stub does not produce a perfect match immediately, note whether the point of attachment of the stub is now a position of I_{min} or I_{max} of the new standing wave on the "matched" line. If neither, then the stub is not tuning the aerial system to resonance. The first adjustment is therefore made to the length of the stub so that the point of attachment is resistive. Correct adjustment is when the position of I_{min} or I_{max} is coincident with the point of attachment. If this is now a point of I_{max}, the resistance is too low for a perfect match. If a closed stub is being used, lengthen it and simultaneously move its point of attachment nearer the aerial by the same amount. (These adjustments are equal only when the stub Z_o is the same as that of the feeder.)

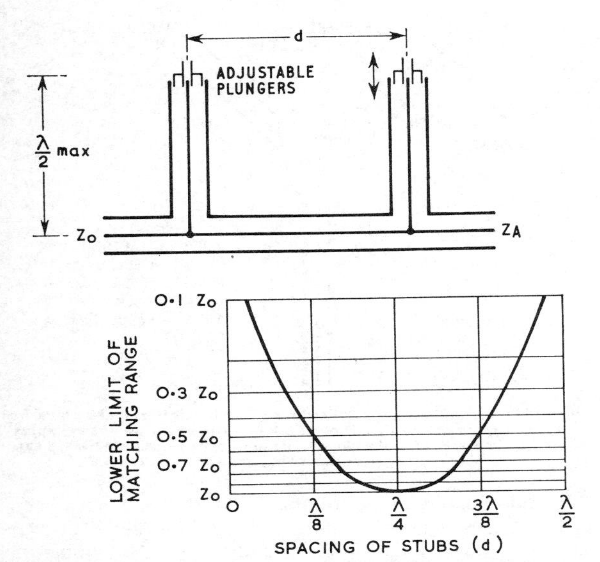

Fig 13.12. Two-stub coaxial tuner. The graph shows the lower limit of the matching range: the upper limit is determined by the Q of the stubs (ie it is dependent on the losses in the stubs). Z_o is the characteristic impedance of the feeder.

If the correct position is passed, the "matched" line will now have an I_{min} at the point of attachment, and the adjustment is to shorten the closed stub and simultaneously move it further from the aerial by the same amount. The total distance from the short circuit of the stub to the aerial remains the same to keep the system resonant. For an open-circuit stub the adjustments are in the opposite sense.

Stub Tuners

On a coaxial line it is impracticable to construct a stub having an adjustable position. However, two fixed stubs spaced by a certain fraction of a wavelength can be used for matching purposes: see **Fig 13.12**. The spacing usually employed is λ/8 or odd multiples thereof. With this spacing independent adjustment of the short-circuit plungers gives a matching range from 0·5 times the characteristic impedance of the transmission line (Z_o) upwards. As the spacing is increased towards λ/2 or decreased towards zero, the matching range increases, but the adjustments then become extremely critical and the bandwidth very narrow. The theoretical limit of matching range cannot be achieved owing to the resistance of the conductors and the dielectric loss; ie the Q is limited. To obtain the highest Q the ratio of outer-to-inner conductor diameters should be in the range 2 : 1 to 4 : 1 (as for coaxial baluns). An important mechanical detail is the provision of reliable short-circuiting plungers which will have negligible inductance and also ensure low-resistance contact. They can be constructed of short lengths of thin-walled brass tubing, their diameters being chosen so that when they are slotted and sprung they make a smooth sliding contact with both inner and outer conductors.

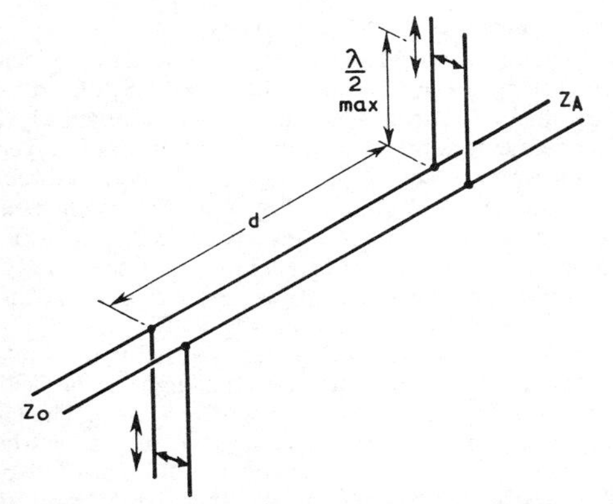

Fig 13.13. Two-stub open-wire tuner. With an open-wire line stubs should be mounted on opposite sides of the line as shown so as to avoid mutual coupling. The matching range can be seen from the graph in Fig 13.12.

The two-stub tuner may be applied to open transmission lines if it is inconvenient to have a movable stub. In this case the stubs must be mounted laterally opposite to each other to prevent mutual coupling: see **Fig 13.13**.

This type of tuner may, of course, be used for other purposes than to feed an aerial. For example, it will serve to match an aerial feeder into a receiver, or a dummy load to a transmitter. A greater matching range can be obtained by using a three-stub tuner, the stubs being spaced at intervals of one quarter-wavelength apart, as shown in **Fig 13.14**. The first and third stubs are usually ganged together to avoid the long and tedious matching operation which becomes necessary when adjustments are made to three infinitely variable stubs.

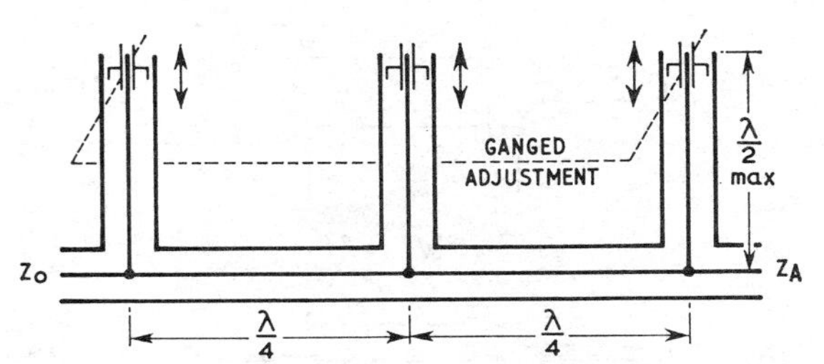

Fig 13.14. Three-stub tuner. This provides a greater matching range than a two-stub tuner. Z_o is the characteristic impedance of the feeder.

Quarter-wave Lines

An impedance transformation can be effected by using a certain length of transmission line of a different characteristic impedance from the feeder. This may be used to match a load to a transmission line. A special condition occurs when the length of the section of line is an odd number of quarter-wavelengths and the following formula then applies:

$$Z_t = \sqrt{Z_0 \cdot Z_1}$$

where Z_t is the characteristic impedance of the section of quarter-wave line and Z_0 and Z_1 are the feeder and load impedance respectively. For example, if Z_0 is 80Ω and Z_1 is 600Ω:

$$Z_t = \sqrt{80 \times 600} = 251\Omega$$

This matching section is useful for transforming impedance and is called a *quarter-wave transformer*: see **Fig 13.15**. A wider bandwidth match can be obtained by reducing the transformation ratio and using a number of quarter-wave sections of progressively changing impedance in cascade.

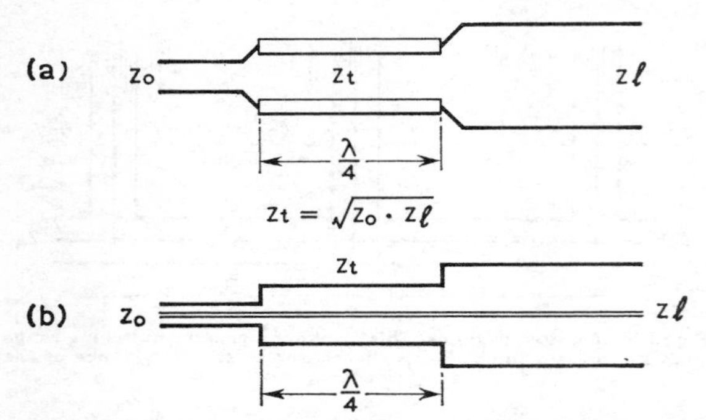

Fig 13.15. Quarter-wave transformers. (a) shows a construction suitable for open-wire lines, and (b) is the corresponding method for coaxial cables. Where a solid-dielectric section is used, due allowance must be made for the velocity factor.

It is often convenient to incorporate a quarter-wave transformer into a coaxial sleeve balun of the type shown in Fig 13.7(a). The length of the quarter-wave section will only be a physical quarter-wavelength if the inner coaxial transformer is air spaced, but as a quarter-wave transformer usually has an impedance different from that of standard coaxial cable it has to be fabricated anyway and thus this is no disadvantage. If a transformer is built into the balun shown in Fig 13.7(b), the characteristic impedance of the right hand quarter-wave section need not be changed since it is acting as a "metal insulator".

A section of tapered line can also be used to effect an impedance transformation, and an application of the principle is described later in this chapter. Again a quarter-wavelength section is only a special case, and to achieve a match in a particular installation the line lengths and the angle of taper should be varied until a perfect match is achieved. This form of matching device is often called a *delta match.*

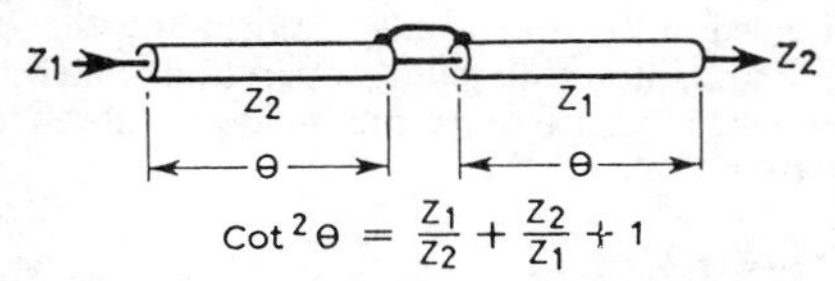

Fig 13.16. Transmission line transformer technique used to provide simple means of matching 50–75Ω coaxial cable.

Transmission Line Transformers

Other than the quarter-wave transformer described above, there is another arrangement that is very convenient as it allows any two cables of different impedance to be matched together simply by using matching lengths made up of the two cables concerned: see **Fig 13.16.** A simplified relationship for this type of transformer is

$$\cot^2 \theta = \frac{Z1}{Z2} + \frac{Z2}{Z1} + 1$$

For a 50/75Ω transformer this works out to an electrical length of 29·3° (0·079 λ) for each section of cable; the physical length must, as explained previously, take into account the velocity factor of the cable (typically 0·66 for solid polythene dielectric coaxial cable—physical length therefore is 0·053 λ).

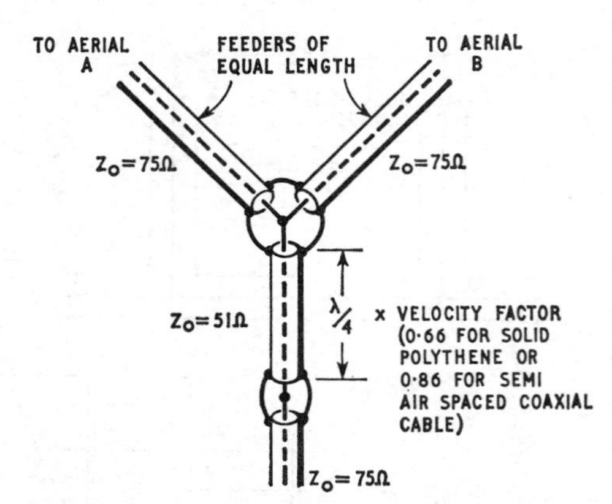

Fig 13.17. Arrangements for feeding two identical aerial arrays in phase using coaxial cables. Links between the braiding of the cables should be as short and of as low an impedance as possible. Braid can be used for this purpose.

Multiple Feed Arrangements

It is often required to feed two or more aerial arrays, such as stacked Yagis, in phase and there are a number of convenient ways of doing this using standard impedance coaxial cables. Details of one possible arrangement are shown in **Fig 13.17.** Others can be developed using similar principles.

AERIAL ARRAYS

The basic elements of vhf aerial arrays are usually half-wave or full-wave dipoles, depending upon the circumstances. The characteristics of such dipoles are described in Chapter 12—*HF Aerials.*

When elements are arranged together to form a beam aerial, the radiation pattern most interesting to amateurs is one having a large forward gain. However, by the study of beam-aerial polar diagrams it becomes apparent that in the course of building up a major lobe, other minor or subsidiary lobes may be created. No purpose would be served here by discussing these minor lobes further but it is well to remember that they exist and do not necessarily imply an inferior aerial array, although they do represent a loss of power from the main lobe.

Parasitic Arrays: The Yagi Array

By placing a *reflector*, usually a resonant element one half-wavelength long behind a half-wave dipole, the radiation can be concentrated within a narrower angle. By adding further elements somewhat shorter than one half-wavelength, called *directors*, at certain spacings in the forward direction, a further gain can be achieved. Any aerial array which employs elements not directly connected to the feed line, ie parasitic elements, is known as a *parasitic array.* If the arrangement consists of a dipole with a reflector and two or more directors it is known as a *Yagi array*: see **Fig 13.18.**

When compared with other aerial systems of similar size the Yagi array is found to have the highest forward gain, and it can be constructed in a very robust form. The effect of adding the reflector and director(s) is to cause the feed impedance of the dipole to fall considerably, often to a value of about 10Ω, and the matching is then critical and difficult to obtain. This, however, may be overcome by the use of a folded-dipole radiator. If the folded dipole has two elements

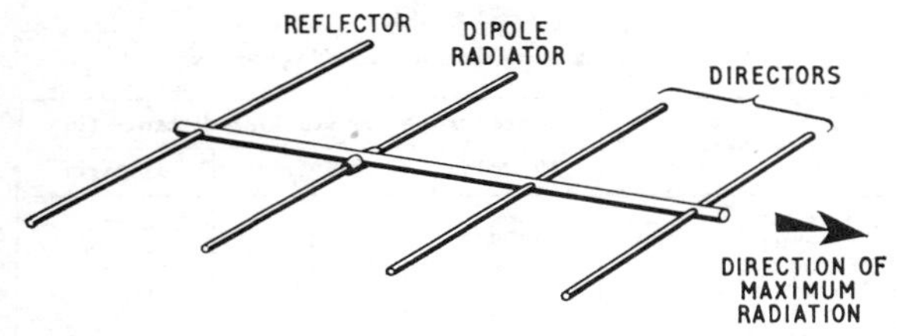

Fig 13.18. Yagi array (three elements). See Table 13.3 for typical dimensions.

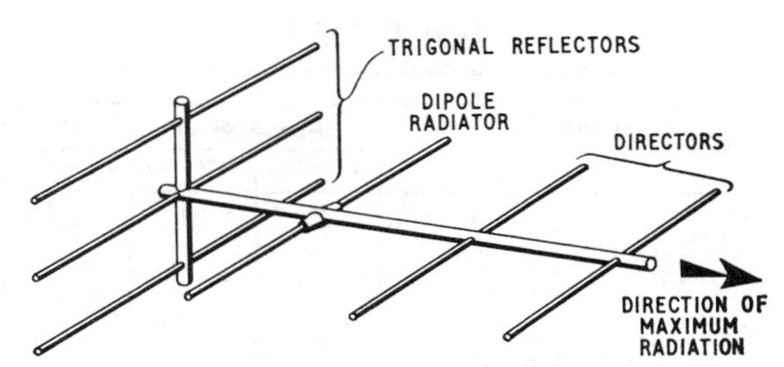

Fig 13.19. Yagi array with trigonal reflectors.

of equal diameter, a 4 : 1 impedance step-up is obtained. By varying the ratio of the diameters, different impedance step-up ratios become available.

The length of the dipole or folded dipole required for resonance depends not only on the frequency but also to a lesser extent on the ratio of the diameter of the element to the wavelength and the distance between the arms of the fold, the length required for resonance diminishing as the wavelength/diameter ratio is decreased: see **Table 13.2.** The overall length of conductor in the folded dipole remains approximately constant: as spacing is increased, the length of the radiating part is shortened accordingly.

The forward gain is not appreciably affected by a variation of reflector spacing over a range of $\lambda/8$ to $\lambda/4$: under these conditions the forward gain is approximately at its maximum value. A considerable change in feed impedance takes place when this spacing is varied and this may be used as a convenient form of adjustment. The reflector is usually $0{\cdot}5\lambda$ long although this should be reduced to about $0{\cdot}475\lambda$ for the closer spacing.

The length of the directors is usually made about $0{\cdot}43\lambda$ and the spacing approximately $0{\cdot}25\lambda$, but experiments have shown that where several directors are used, the bandwidth can be broadened by making them progressively shorter in the direction of radiation. The greater the number of directors, the higher the gain and the narrower the beamwidth. There is no advantage to be derived from using more reflectors spaced behind the first, but the front-to-back ratio may be improved somewhat by the use of additional reflectors as shown in **Fig 13.19**. These additional reflector elements should subtend a fairly wide angle at the farthest director to be effective. In practice, a trigonal reflector element/element spacing of about a quarter-wavelength is sufficient.

The gain obtainable from a Yagi array compared with a half-wave dipole is shown approximately by the curve in **Fig 13.20.**

The bandwidth for a standing-wave ratio less than 2 : 1 is about 2 per cent for close-spaced beams and about 3 per cent for wider spacing. Element lengths, particularly those of the directors, are very critical (ie within fractions of an inch), and ideally telescopic rods should be used to enable fine adjustments to be made. Each change of element length necessitates a readjustment of the matching either by moving the reflector or in the matching device itself.

Typical element lengths for spot frequencies in the 4m, 2m and 70cm bands are given in **Table 13.3.** The lengths are based on the assumption that the element diameter lies within the stated limits for the respective bands. Any departures from these diameter ranges will necessitate a change in the lengths of the elements; for a larger diameter the length will need to be decreased, and vice versa. Provision has been made for simple dipole, folded dipole or skeleton slot radiating elements. The latter radiator, which is described later in the chapter, is the feed system in a stacked pair of Yagi arrays, the dimensions of which are the same as for a single Yagi array given in Table 13.3.

Stacked and Bayed Yagi Arrays

To obtain a greater gain several Yagi arrays can be disposed horizontally or vertically or in both directions, the radiating elements being fed in phase. Normally, *stacking* refers to vertical addition and *baying* to horizontal. It is important to remember that with stacking the effective height of the aerial above the ground is half way between the lowest

TABLE 13.2

Resonant lengths of half-wave dipoles

(Wavelength / Diameter)	(Value of dipole length / Wavelength for resonance)	Feed impedance (Ω)
50	0·458	60·5
100	0·465	61·0
200	0·471	61·6
400	0·475	63·6
1,000	0·479	65·3
4,000	0·484	67·2
10,000	0·486	68·1
100,000	0·489	69·2

The dimensions used in calculating the ratios must be in similar units (eg both in metres or both in centimetres).
From *Aerials for Metre and Decimetre Wavelengths* by R. A. Smith.

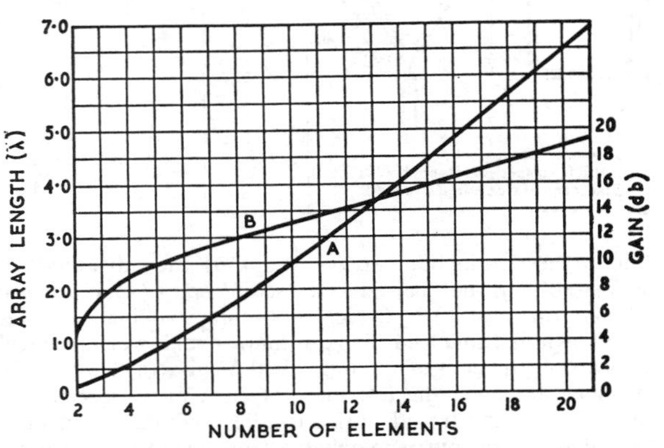

Fig 13.20. Design information for Yagi aerials. Curve A shows the optimum boom length in wavelengths for any number of elements. Curve B shows the maximum gain that can be expected when the design information of Curve A is used. *With acknowledgements to the ARRL.*

TABLE 13.3

Typical dimensions of Yagi arrays—parasitic elements

Element	Length of element (in)		
	70·3MHz	145MHz	433MHz
Reflector	85½	40	13¼
Director D1	74	35½	11¼
Director D2	73	35¼	11⅛
Director D3	72	35	11
Succeeding directors	1in less	½in less	⅛in less
Final director	2in less	1in less	¾in less
One wavelength (for reference)	168	81½	27¼
Diameter range for length given	½–¾	¼–⅜	⅛–¼
	Spacing between elements (in)		
Reflector to radiator	22½	17½	5½
Radiator to director 1	29	17½	5½
Director 1 to director 2	29	17½	7
Director 2 to director 3, etc.	29	17½	7

The above figures are based on a number of proved Yagi array designs. If the slot is used two sets of reflectors/directors are required, one mounted above the other thus forming a stacked array—see Fig 13.26. Match to the feeder can be effected by moving the radiator relative to the first director and the reflector relative to the radiator.

Typical dimensions of Yagi arrays—driven elements

	Length (in)		
	70·3MHz	145MHz	433MHz
Dipole (for use with gamma match)	79	38	12¾
Diameter range for length given	½–¾	¼–⅜	⅛–¼
Folded dipole (70Ω feed)	**Length (in)**		
	70·3MHz	**145MHz**	**433MHz**
l length centre/centre	77½	38½	12½
d spacing centre/centre	2½	⅞	½
Diameter	½	¼	⅛
	Length (in)		
	70·3MHz	**145MHz**	**433MHz**
a centre/centre	32	14	5¾
b centre/centre	96	43¾	15⅛
Delta feed sections (length 70Ω feed)	22½	12	4⅜
Diameter of slot and delta feed	½	⅜	⅜

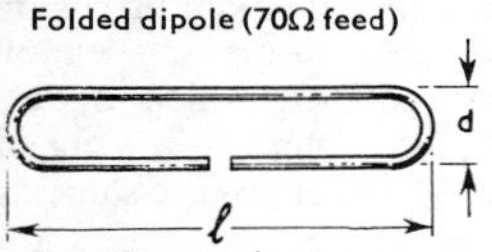

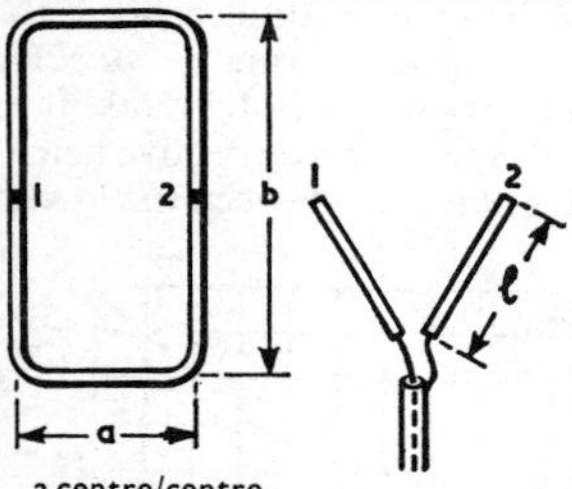

and highest individual arrays and thus the advantage of increased gain may be reduced—see *Height of Aerial above Ground* on page 13.3. Typical optimum stacking distances are shown in **Table 13.4**. The optimum stacking distance will depend on the gain of the individual Yagi arrays and is the distance at which the apertures of each array just touch. Where the vertical and horizontal beamwidths of an individual array are approximately the same, ie with five or more elements, the optimum stacking distance varies from about 0·75λ for an array of 4-element Yagis to about 2·0λ or more for 8-element Yagis. For practical reasons the spacing is

TABLE 13.4

Optimum stacking distance for Yagi arrays

Yagi array	Centre-to-centre stacking distance (in)		
	70·3MHz	145MHz	433MHz
3 element	84	—	—
4 element	108	82	—
8 element	—	120	—
10 element	—	132	44
14 element	—	—	66
24 element	—	—	—
4-over-4 slot fed	—	120	40
6-over-6 slot fed	—	147	49
8-over-8 slot fed	—	160	53

With acknowledgements to Jaybeam Ltd.

usually less than optimum and there is a consequent reduction in the total gain. It was stated earlier that the increase in gain obtainable by stacking two identical arrays in such a way that their capture areas do not overlap is simply two times (ie 3dB), but in practice this can sometimes be exceeded if there is a suitably favourable degree of coupling between the two arrays: for example, a pair of three-element Yagi arrays could be made to yield an increase of 4·2dB. However, the increase usually proves to be less than the theoretical 3dB and a figure of 2·2–2·5dB is all that can ordinarily be expected.

Theoretically the feed impedance of a stacked array is the feed impedance of an individual Yagi array divided by the total number of Yagis employed; in practice the feed impedance of stacked Yagis is slightly less than this because of interaction between each Yagi array, although this reduction is not so marked at the greater spacings.

Disadvantages of Conventional Yagi Arrays

Perhaps the most important disadvantage is that the variation of the element lengths and spacings causes interrelated changes in the feed impedance of a Yagi array. To obtain the maximum possible forward gain experimentally is extremely difficult because for each change of element length it is necessary to readjust the matching either by moving the reflector or by resetting a matching device. A method has been devised, however, for overcoming these practical disadvantages by the use of a radiating element in the form of a *skeleton slot*, this being far less susceptible to the changes in impedance caused by changes in the parasitic-element lengths. This development is due to B. Sykes, G2HCG.

A true slot would be a slot cut in an infinite sheet of metal, and such a slot when approximately one half-wavelength long would behave in a similar way to a dipole radiator. In contrast with a dipole, however, the polarization produced by a vertical slot is horizontal (ie the electric field is horizontal).

The skeleton slot was developed in the course of experiments to determine to what extent the infinite sheet of metal could be reduced before the slot aerial lost its radiating property. The limit of the reduction for satisfactory performance was found to occur when there remained approximately one half-wavelength of metal beyond the slot edges. Further experiments showed that a thin rod bent to form a "skeleton slot" of dimension approximately 5λ/8 × 5λ/24 exhibited similar properties to those of a true slot. The manner in which a skeleton slot functions can be understood by referring to the diagrams in **Fig 13.21**. Consider two half-wave dipoles spaced vertically by 5λ/8. Since the greater part of the radiation from each dipole takes place at the

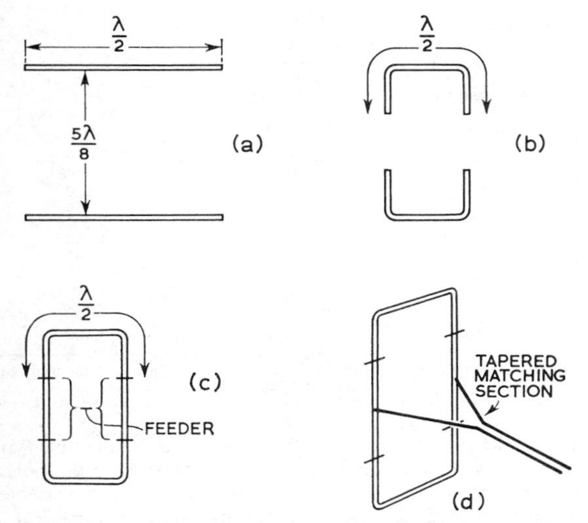

Fig 13.21. Development of a skeleton-slot radiator from two half-wave dipoles spaced ⅝λ apart.

current anti-node, ie the centre, the ends of the dipoles may be bent without serious effect. These ends may now be joined together with a high-impedance feeder, so that end-feeding can be applied to the bent dipoles. To radiate in phase, the power should be fed midway between the two dipoles. The high impedance at this point may be transformed down to one suitable for the type of feeder in use by means of a tapered matching-section transmission line (ie a delta match). Practical dimensions of a skeleton-slot radiator are given in **Fig 13.22.**

It is important to note that two sets of parasitic elements are required with a skeleton-slot radiator and not one set as required with a true slot. One further property of the skeleton slot is that its bandwidth is somewhat greater than a pair of stacked dipoles. A disadvantage of the basic slot-fed beam is that the spacing between upper and lower sets of parasitic elements is not optimum and thus the gain increase is less than 3dB compared with the equivalent single Yagi. This is increasingly so with slot-fed beams arrays having a greater number of parasitic elements.

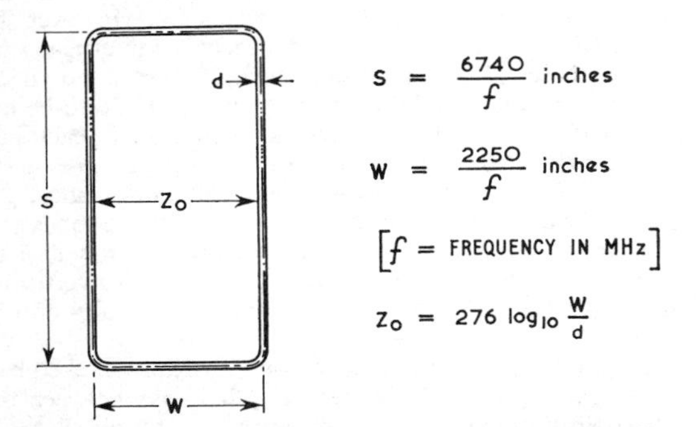

Fig 13.22. Dimensional relationships of a skeleton-slot radiator. Both S and W may be varied experimentally from the values indicated by these formulae. For small variations the radiation characteristics of the slot will not change greatly, but the feed impedance will undergo appreciable change and therefore the length of the delta matching section should always be adjusted to give a perfect match to the transmission line.

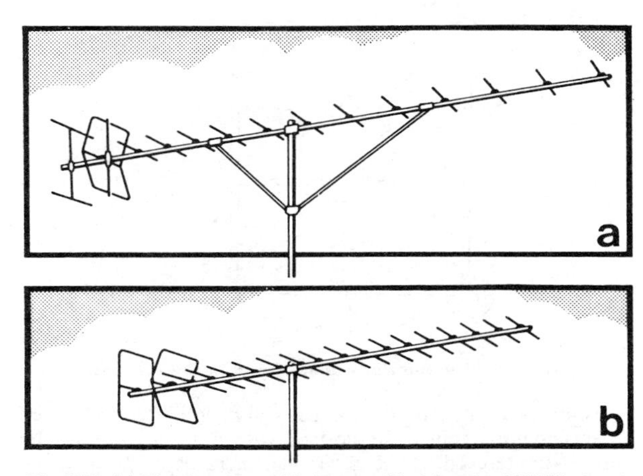

Fig 13.23. (a) The Parabeam 14-element Yagi for the 144MHz band. Gain is 15dB and the horizontal beamwidth between half-power points is 24°. (b) The Parabeam 18-element Yagi for the 432MHz band. Gain is 17dB and the horizontal beamwidth between half-power points is 28°. Note that the reflector is a pair of parasitic λ/2 elements in the 432MHz version.

Stacked Skeleton Slot Yagi Arrays

Skeleton slot Yagi arrays may be stacked to increase the gain but the considerations of optimum stacking distance previously discussed apply; in this case the centre-to-centre spacing of a pair of skeleton slot Yagi arrays should vary between 1λ and 3λ or more according to the number of elements in each Yagi array.

Each skeleton slot Yagi array may be fed by 72Ω coaxial cable, using equal lengths of feeder to some common feed point for the stacked array, and it would of course be desirable to use a balun at the point where the cable is attached to each array. A coaxial quarter-wave transformer can be used to transform the impedance to that of the main feeder. For example, if a pair of skeleton slot Yagi arrays, each of 72Ω feed impedance, is stacked, the combined impedance will be one-half of 72Ω, ie 36Ω; this may be transformed to 72Ω by the use of a quarter-wave section of 52Ω coaxial cable, allowance being made for its velocity factor. Larger assemblies of skeleton slot Yagi arrays can be fed in a similar manner by joining pairs and introducing quarter-wave transformers until only one feed is needed for the whole array.

The Long Yagi

The gain of the Yagi depends upon the number of directors and thus its length as already shown in Fig 13.20. The term long Yagi has been applied to arrays with, say, eight or more directors. Such arrays have become popular in recent years, largely due to the work of W2NLY and W6QKI. The construction of a typical long Yagi is described on page 13.23.

The Parabeam and Multibeam

Further developments of the skeleton slot have been conducted by Jaybeam Ltd to produce high gain Yagi type aerials having a wide bandwidth suitable for uhf television. These have also found ready application for communications purposes especially in the 144, 432 and 1,296MHz bands. The Parabeam, **Fig 13.23,** uses a skeleton slot radiator and reflector with conventional Yagi type parasitic directors, and

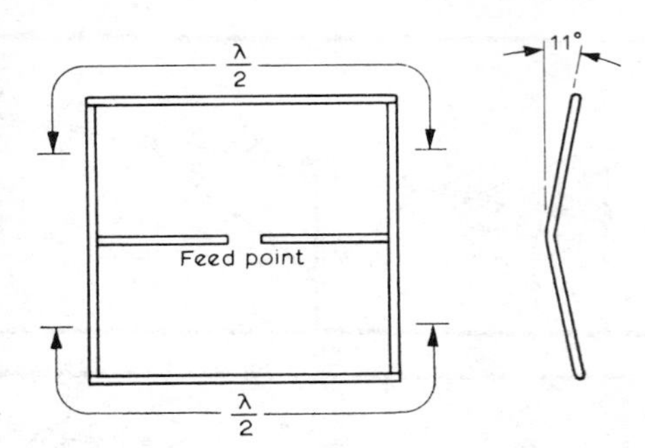

Fig 13.24. Skeleton slot radiator used in the Parabeam.

this increases the gain compared with a similar conventional Yagi having a folded dipole feed with multiple reflector by approximately 2dB. The skeleton slot radiator used in the Parabeam is arranged as shown in **Fig 13.24** and feed to the 300Ω point can be effected by using a 4 : 1 coaxial balun as shown in Fig 13.8. To obtain optimum launching into the director chain it has been found necessary to advance the horizontal radiating sections of the skeleton slot with respect to the 300Ω feed bars. This accounts for the rake seen in Fig 13.23 which is at an angle of 11° to the vertical.

The Multibeam is a further development of the Parabeam aimed at increasing the gain still further by replacing the conventional Yagi type directors with multiple director units, each unit consisting of four directors. Careful experiments showed that the gain increase from a 12-element Parabeam to a 42-element Multibeam (each multiple director counting as four elements) gave an increase of 5dB, a total of some 7dB over a conventional Yagi of the same length, a very useful gain obtained at the expense of some increase in mechanical complication. See **Fig 13.25** for details.

The bandwidth of both the Parabeam and Multibeam has proved to be very good, amounting to some 20 per cent of frequency for a voltage standing wave ratio of better than 2 : 1 and a gain variation of within 2dB. The gain of a 46-element Multibeam at 432MHz is approximately 19dB.

Stacked Dipole Arrays

Both horizontal and vertical beam widths can be reduced and gain increased by building up arrays of driven dipoles. This arrangement is usually referred to merely as a *stack* or

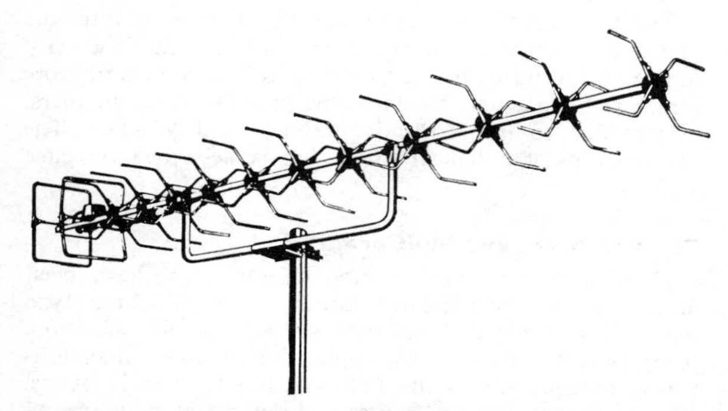

Fig 13.25. The Multibeam 46-element Yagi. Gain is 20dB and the horizontal beamwidth between half-power points is 24°.

TABLE 13.5

Resonant length of full-wave dipoles

(Wavelength / Diameter)	Value of (dipole length / Wavelength) at resonance	Feed impedance (Ω)
50	0·85	500
100	0·87	900
150	0·88	1,100
200	0·896	1,300
300	0·906	1,500
400	0·916	1,700
700	0·926	2,000
1,000	0·937	2,400
2,000	0·945	3,000
4,000	0·951	3,600
10,000	0·958	4,600

The dimensions used in calculating the ratios must be in similar units (eg both in metres or both in centimetres).
From *Aerials for Metre and Decimetre Wavelengths* by R. A. Smith.

sometimes as a *bill-board* or *broadside array*. Since this type of array is constituted from a number of radiating dipoles, the feed impedance would be extremely low if the dipoles were centre-fed. However, the impedance to earth of a dipole at its end is high, the precise value depending upon the ratio of its length to diameter, and it will therefore be more convenient to use a balanced high-impedance feeder to end-feed a pair of collinear half-wave dipoles, a system called a *full-wave dipole*. The length for resonance and the feed impedance in terms of wavelength/diameter ratio is shown in **Table 13.5**.

The full-wave dipoles are usually mounted with a centre-to-centre spacing, horizontally and vertically, of one half-wavelength and are fed in phase. Typical arrangements for stacks of full-wave dipoles are shown in **Fig 13.26**. Note that the feed wires between dipoles are one half-wavelength long and are crossed so that all dipoles in each bay are fed in phase. The impedance of these phasing sections is unimportant provided that the separators, if used, are made of low-loss dielectric material and that there is sufficiently wide separation at the cross-over points to prevent unintentional contact.

To obtain the radiation pattern expected, all dipoles should be fed with equal amounts of power (as indeed would be desirable in any multi-radiator array), but this may not be achieved in practice because those which are at the edges of the array have least mutual coupling to other dipoles and therefore have different radiation resistances. However, by locating the main feed point as nearly symmetrically as possible these effects are minimized. Hence it would be preferable for the aerial shown at (A) in Fig 13.26 to be fed in the centre of each bay of dipoles; the feeder to each bay must be connected as shown to ensure that the two resultant bays are fed in phase. If they were fed 180° out-of-phase the resultant beam pattern would have two major side lobes and there would be very little power radiated in the desired direction. The diagram (B) in Fig 13.26 shows two vertically stacked bays of full-wave dipoles fed symmetrically and in phase.

Diagram (C) in Fig 13.26 shows one vertically stacked bay of full-wave dipoles fed symmetrically half way between the centre pair of dipoles.

The spacing at the centre of each full-wave dipole should be sufficient to prevent a reduction of the resonant frequency by the capacitance between the ends. In practice this spacing is usually about 1in for the 144 and 432MHz bands.

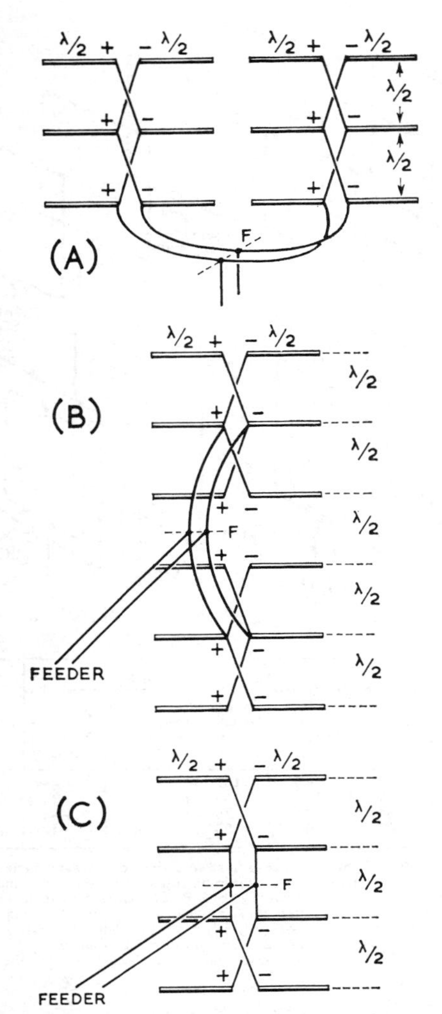

Fig 13.26. Typical stacks of full-wave dipoles. Note that the feed-point F is equi-distant from each bay of dipoles. Example of lengths are given in Table 13.5.

Matching can be carried out by the use of movable short-circuited open-wire stubs on the feed lines.

As with the Yagi array, the gain can be increased by placing half-wave reflectors behind the radiating elements at a spacing of 0·1–0·25λ, a figure of 0·125λ being frequently chosen. A perfect plane reflector will yield a gain of 6dB compared with the 3dB obtainable from a rod reflector. For the 432MHz band and for higher frequencies, a plane reflector made up of 1in mesh wire netting stretched on a frame can be used in place of the resonant reflector at a similar spacing. The mesh of the wire should be so orientated that the interlocking twists are parallel to the dipole. The wire netting should extend at least one half-wavelength beyond the extremities of the dipoles in order to ensure a high front-to-back ratio.

The half-wave sections of the full-wave dipole should be supported at the current anti-nodes, ie at their centres, either on small insulators or in suitably drilled wooden vertical members. Supports should not be mounted parallel to the elements because of possible influence on the properties of the aerial.

The bandwidth of this type of aerial is exceptionally large and its adjustments generally are far less critical than those of Yagi arrays.

For a stack having a wire-net reflector extending λ/2 beyond the extremities of the dipoles, the horizontal beamwidth θ, vertical beamwidth ϕ and power gain G (compared with an isotropic radiator) can be calculated approximately from the following formulae:

$$\theta = \frac{51\lambda}{a} \qquad \phi = \frac{51\lambda}{b} \qquad G = \frac{4\pi ab}{\lambda^2}$$

where a and b are the horizontal and vertical dimensions of the reflector respectively, both being expressed in the same units as the wavelength.

These formulae are true only for an array which is large compared with the wavelength, but are suitable as a criterion for judging aerials of any type provided the equivalent aperture or capture area is known.

Skeleton Slots in Stack

Skeleton slots can be used to replace vertically disposed pairs of half-wave dipoles. As the optimum vertical dimension for a horizontally polarized skeleton slot is approximately 5λ/8, it is no longer possible to use the vertical spacings shown for full-wave stacks. The slots are mounted vertically at a centre-to-centre spacing of one wavelength and fed through a tapered matching section, as for the skeleton-slot Yagi array, and are then connected to the phasing lines. Since the spacing between feed points is one wavelength there is no phase difference and it is unnecessary to transpose the phasing wires. The tapered matching sections should be adjusted to present an impedance of N times the desired feeder-cable impedance where N is the number of skeleton slots employed. The impedance resulting from the connection of all the feed points together will then equal the cable impedance.

A broadside array of skeleton slots may be built up by adding further slots horizontally at a centre-to-centre spacing of one half-wavelength.

Disadvantages of Multi-element Arrays

As the frequency becomes higher and the wavelength becomes shorter, it is possible to construct arrays of much higher gain although, as already described, the advantage is offset by the reduction in capture area when used for receiving. However, if the practice already described, that of using many driven or parasitic elements, either in line or in stack, is adopted, the complications of feeding become increasingly greater. Also, as the frequency increases the radiation loss from open-wire lines and from phasing and matching sections likewise increases, and it is then difficult to ensure an equal power-feed to a number of radiators. Preferably, therefore, the aerial should have a minimum number of radiating or other critical elements, such as resonant reflectors or directors. There are many aerials in this category but only those having immediate amateur application are described here.

The Cubical Quad

The cubical quad aerial has proved highly successful for vhf use especially indoors where its properties are almost unaffected by the close proximity of building structures, in

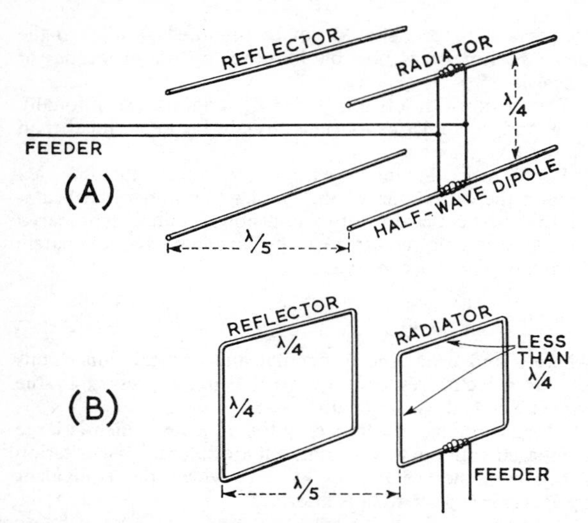

Fig 13.27. Development of a cubical quad aerial from a pair of dipoles and reflectors. In (A) the reflectors are λ/2 long and the dipole radiator lengths for resonance are approximately 0·38λ. In (B) the elements have been bent to form two squares, the combined reflector having sides 0·25λ long and the combined radiator having sides 0·23λ long. The two squares can be made equal in size by the method shown in Fig 13.28.

contrast to the Yagi array, the high *Q* elements of which are detuned with the result that its performance is markedly changed.

The basic cubical quad (**Fig 13.27**) can be constructed using copper wire of $\frac{1}{16}$in or $\frac{1}{8}$in diameter. The reflector square can either be made larger than the radiator by about 5 per cent in total length or alternatively can have its length extended electrically by inserting a small inductive stub at a current anti-node, ie short circuited and considerably less than a quarter-wave in length. A variation of the quad, developed by G2PU, known as the bi-square, has two symmetrically disposed inductive stubs located at current anti-nodes in the reflector which ensures current symmetry in the reflector and hence no beam tilt (see **Fig 13.28**). VHF bi-square aerials are usually constructed from $\frac{3}{8}$in diameter tube, although wire can be used.

The gain of a quarter-wave square cubical quad is 5·5 to

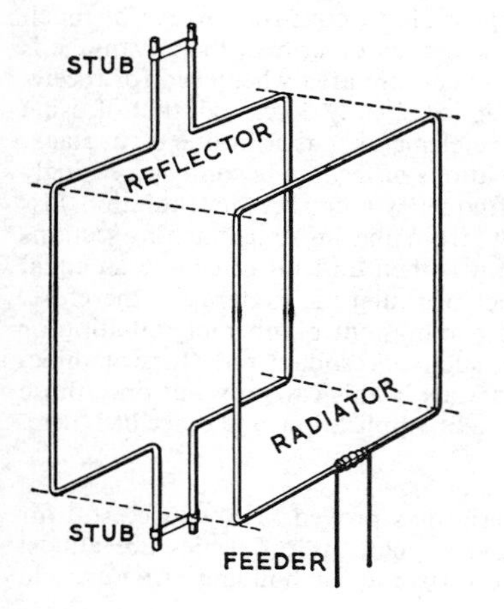

Fig 13.28. The bi-square aerial may be fitted with small stubs in the reflector element so that all the sides of both reflector and radiator can be made equal in length (approximately 0·23λ).

TABLE 13.6

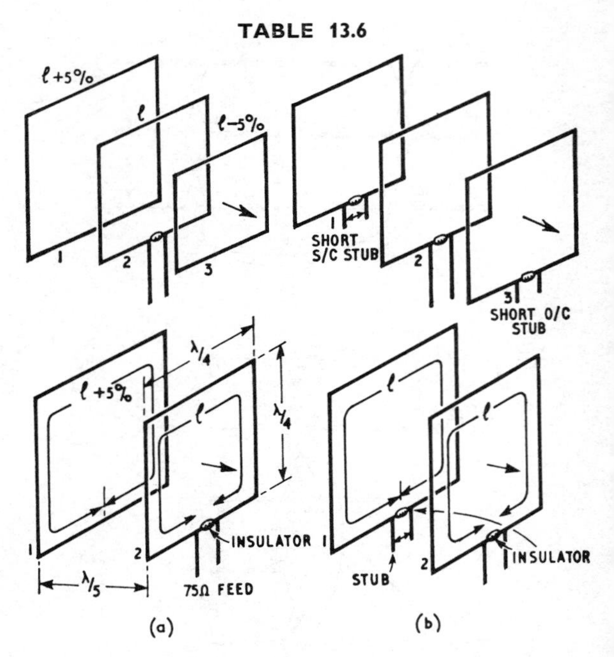

Band (MHz)	Reflector 1 total length* (in)	Radiator 2 total length* (in)	Director (if used) (in)	Approximate length of stubs if used (in) Reflector S/C	Director S/C	Spacing (in)
70 (a)	173	165	157	—	—	34
70 (b)	165	165	165	8	8	34
144 (a)	84	80	76	—	—	16
144 (b)	80	80	80	4	4	16

The total length of the radiator is just less than one wavelength in total (about 0·98 × λ). The stub in (*b*) should be adjusted in length for maximum forward gain. The gain is about 5½ to 6dB. Elements are made from $\frac{1}{8}$in diameter copper wire although the diameter is not critical. Insulators can be made from polystyrene or polythene about 1½in wide. Stub lengths should be adjusted for optimum performance.
*Each side is one quarter this length

6dB but this can be increased by using a director square in front of the radiator. This square can either be some 5 per cent shorter in total length than the radiator or have a short capacitive stub inserted, ie open circuited and considerably less than a quarter-wave. This increases the gain by about 1·5dB. Multiple directors are often added to increase the gain still further.

Two band arrays can conveniently be placed one inside the other as no significant interaction occurs. In the case of a quad without a director. the wires can be attached to bamboo spreaders originating from one central multiple X having eight arms. In the case of the quad with director, three separate bamboo X spreaders can be located on a single boom. In all, the array is non-critical, has a wide bandwidth and is ideal for home construction with limited facilities. It will pay to experiment both with stub lengths and reflector/radiator/director spacing to achieve the maximum gain with minimum feeder swr in any particular environment.

Typical dimensions for both quarter- and half-wave types of quads are shown in **Tables 13.6** and **13.7**.

TABLE 13.7

Half-wave cubical quad

Band (MHz)	Reflector total length (in)	Radiator total length (in)	Director (if used) (in)	Approximate length of stubs if used (in)		Spacing (in)
				Reflector S/C	Director O/C	
144	174	164	154	—	—	16
144	164	164	164	10	10	16

Gain without director 10dB; with director 12·5dB.

In Table 13.6 dimensions are given for typical quarter-wave types with and without stubs. The half-wave quad has an additional insulator half way along the top conductor of all elements of the squares shown in Table 13.6, ie impedance is high (between 2,000 and 4,000Ω) and this may be transformed down to 75Ω by means of a quarter-wave transformer.

Corner Reflector

The use of a metallic plane reflector spaced behind a radiating dipole has already been discussed. If this reflector is bent to form a V, as shown in **Fig 13.29**, a considerably higher gain is achieved. The critical factors in the design of such an aerial array are the corner angle and the dipole/vertex spacing *S*. The curves in **Fig 13.30** show that as α is reduced, the gain theoretically obtainable becomes progressively greater. However, at the same time the feed impedance of the dipole radiator (Fig 13.29) falls to a very low value, as can be seen from **Fig 13.31**: this makes matching difficult and hence a compromise has to be reached. In practice the angle α is usually made 90° or 60°; adjustments in a 60° corner are a little more critical although the maximum obtainable gain is higher. The final matching of the radiator to the line may be carried out by adjusting the distance *S*, which as seen from Fig 13.30 does not greatly affect the gain over a useful range of variation but causes a considerable change in feed impedance (see Fig 13.31). A two-stub tuner may also prove helpful in making final adjustments.

The length of the sides *L* of the reflector should exceed two wavelengths to secure the characteristics indicated by Figs 13.30 and 13.31, and the reflector width *W* should be greater than one wavelength for a half-wave dipole radiator. The reflecting sheet may be constructed of wire-netting as

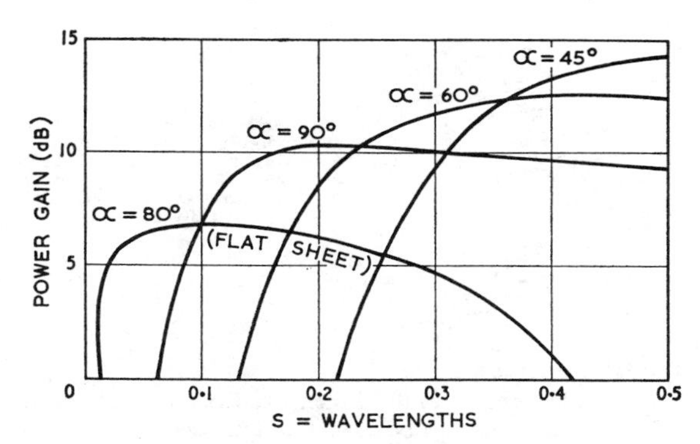

Fig 13.30. Theoretical power gain obtained by using a corner reflector with a half-wave dipole radiator: see Fig 13.29.

described previously or alternatively may be fabricated from metal spines arranged in a V-formation, all of them being parallel to the radiator: see **Fig 13.32**. The spacing between adjacent rods should not exceed $0{\cdot}1\,\lambda$ and the length of the rods is $0{\cdot}6\lambda$ for a half-wave driven element.

A useful approximation for the power gain *G* referred to a half-wave dipole is $G = 300/\alpha$, where α is the angle between the sides measured in degrees.

The maximum dipole/vertex spacing *S* included in the curves shown in Figs 13.30 and 13.31 is one half-wavelength. Spacings greater than this would require rather cumbersome constructions at lower frequencies, but at the higher frequencies larger spacings become practicable, and higher gains can then be obtained. This indicates that the corner reflector can become a specially attractive proposition for the 1,296MHz band, but the width across the opening should be in excess of 4λ to achieve the results shown. Dimensions for 60° (13dB gain) corner reflectors for 145, 433 and 1,296MHz are given in **Table 13.8**.

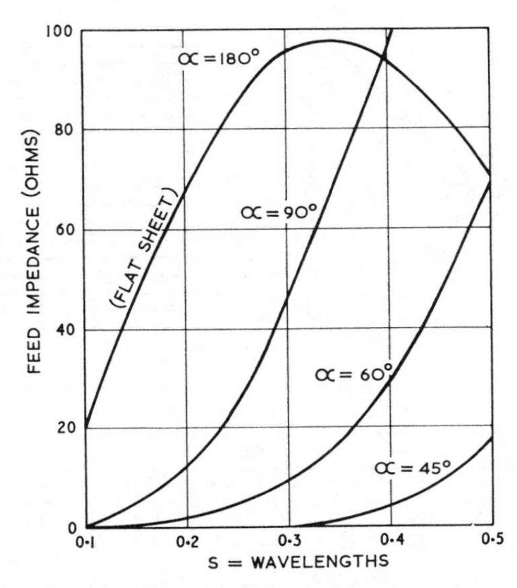

Fig 13.31. Feed impedance of a half-wave dipole provided with a corner reflector: see Fig 13.29.

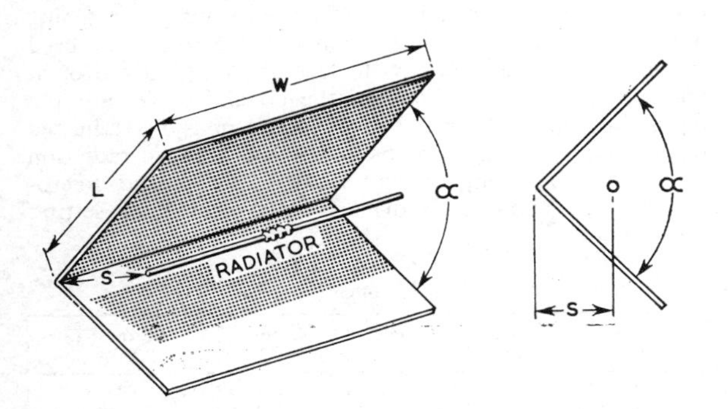

Fig 13.29. Corner reflector. The half-wave dipole radiator is spaced parallel with the vertex of the reflector at a distance S. Its characteristics are shown in Figs 13.30 and 13.31. For dimensions, see Table 13.8.

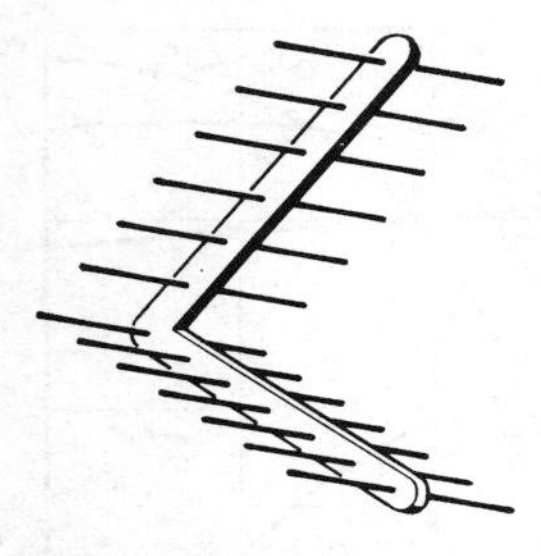

Fig 13.32. The corner reflector shown in Fig 13.29 can be modified by using a set of metal spines arranged in V-formation to replace the sheet-metal or wire-netting reflector.

Trough Reflector

To reduce the overall dimensions of a large corner reflector the vertex can be cut off and replaced with a plane reflector; such an arrangement is known as a *trough reflector:* see **Fig 13.33**. Similar performance to that of the large corner reflector can thereby be achieved provided that the dimensions of the trough do not exceed the limits indicated in **Table 13.9**.

The resulting aerial has a performance very little different to the corner-reflector type and presents fewer mechanical problems since the plane centre portion is relatively easy to mount on the mast and the sides are considerably shorter.

The gain of both corner reflectors and trough reflectors may be increased still further by stacking two or more and arranging them to radiate in phase, or alternatively by adding further collinear dipoles within a wider reflector similarly fed in phase. Not more than two or three radiating units should be used since the great virtue of the simple feeder arrangement would then be lost.

Trough Reflectors for 432 and 1,296MHz

Dimensions are given in **Table 13.10** for 432 and 1,296MHz trough reflectors. The gain to be expected is 15dB and 17dB respectively. A very convenient arrangement, especially for portable work, is to use a metal hinge at each angle of the reflector. This permits the reflector to be folded flat for transit. It also permits experiments to be carried out with different apex angles.

A housing will be required at the dipole centre to prevent the ingress of moisture and also, in the case of the 432MHz aerial, to support the dipole elements. The dipole may be moved in and out of the reflector to achieve either minimum standing wave ratio or, if this cannot be measured, for maximum gain. If a two-stub tuner or other matching device is used, the dipole may be placed to give optimum gain and the matching device adjusted to give optimum match. In the case of the 1,296MHz aerial, the dipole length can be adjusted by means of the brass screws at the ends of the elements. Locking nuts are essential.

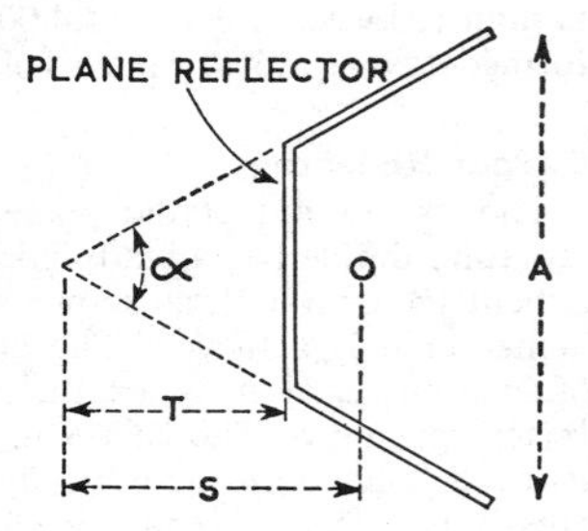

Fig 13.33. Trough reflector. This is a useful modification of the corner reflector shown in Fig 13.29. The vertex has been cut off and replaced by a simple plane section.

The reflector should be made of sheet aluminium for 1,296MHz but can be constructed of wire mesh (with twists parallel to the dipole) for 432MHz. To increase the gain by 3dB a pair can be stacked so that the reflectors are just not touching (to avoid a slot radiator being formed by the edges). The radiating dipoles must then be fed in phase and suitable feeding and matching must be arranged. A two-stub tuner can usefully be used for matching either for a single or double reflector system.

The Reflex Aerial

The reflex aerial, shown in **Fig 13.34**, comprises a radiating dipole located in front of a large reflecting sheet with a grating type of structure mounted in front of and parallel to the dipole. The total gain attainable with this aerial system is about 12dB.

The dipole has a feed impedance of about 120Ω and can be fed from a coaxial cable with a suitable matching device and balun. An arrangement that has been used successfully at 1,296MHz is to feed the dipole from open lines some 3½in long between the coaxial cable and the dipole (see **Fig 13.35**). An open circuited stub is then adjusted in position and length to give maximum forward radiation as detected by a dipole with diode detector and microammeter at about 20ft distant. A typical final setting

TABLE 13.8

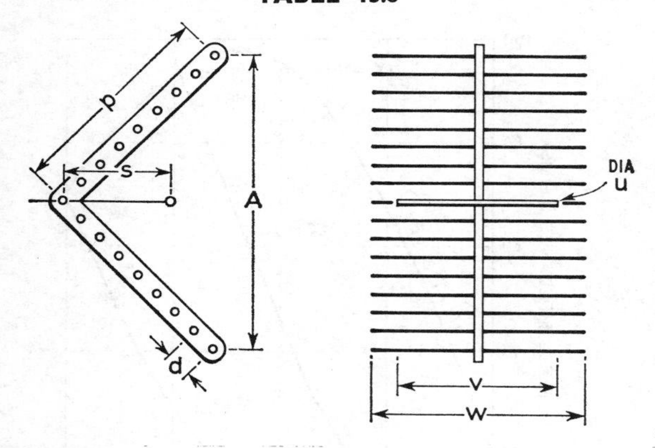

Band (MHz)	Dimensions (in)							
	p	s	d	v	w	A	u	λ
144	100	40	6	38	50	100	3/8	82
432	35	13¼	1½	12½	20	35	¼	27¼
1,296	12	4½	½	4	8	12	1/8	9 1/8

Dimensions for a 60° corner reflector aerial system giving a gain of about 13dB. The feed impedance of the dipole radiator is 75Ω. The apex may be hinged for portable work.

TABLE 13.9

Corner/trough reflector

Angle α	Value of S for maximum gain (λ)	Gain (dB)	Value of T (λ)
90°	1·5	13	1—1·25
60°	1·25	15	1·0
45°	2·0	17	1·9

This table shows the gain obtainable for greater values of S than those covered by Fig 13.30, assuming that the reflector is of adequate size.

TABLE 13.10

Trough reflectors for 432 and 1,296MHz

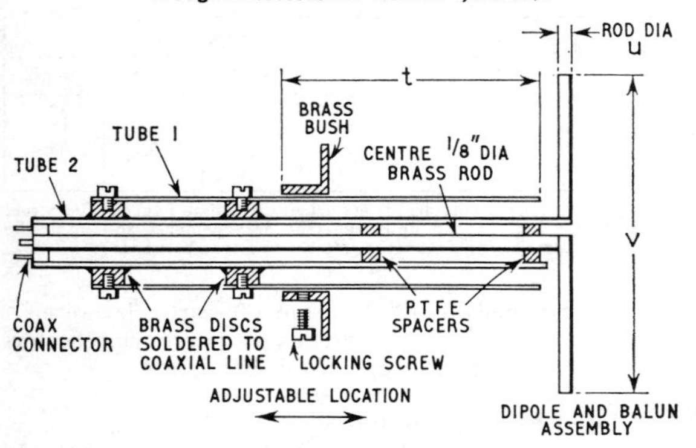

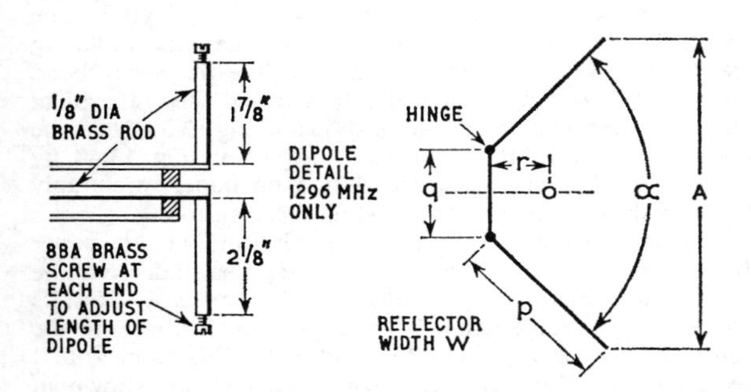

Band (MHz)	Dimensions (in) A	p	q	r	w	t	u	v	Tube 1 Brass or copper	Tube 2 Brass or copper	Gain (dB)
432	120	100	50	7½	60	6½	¼	12¾	1½ id 1/16 wall	⅝od 1/16 wall	17
1,296	45	36	9½	2½	20	2⅛	⅛	4¼	¾ id	⅜ od 20 swg	15

resulted in the stub about 1½in long at about 1in from the coaxial feeder. No balun was used. The performance of the aerial adjusted in this way proved to be very satisfactory.

Typical dimensions for the 432MHz and 1,296MHz bands are shown in **Table 13.11**. There is no advantage in increasing the reflector size beyond about two wavelengths square since above this size the oblique radiation from the dipole is so slight that little further gain can be achieved. However, the gain may be increased by adding further reflex aerials either horizontally or vertically as desired and feeding them so as to radiate in phase.

The Parabolic Reflector

The principle of the parabolic reflector as used for light waves is well known. Such a reflector may be used for radio waves, although the properties are somewhat different because the ratio of the diameter of the reflector to the wavelength is not extremely large, as it is in the case of light waves. If a radiator is placed at the focus of a paraboloid and if most of the energy is directed back into the "dish", a narrow beam will be produced. Assuming that the energy is uniformly distributed over the dish, the angular width of

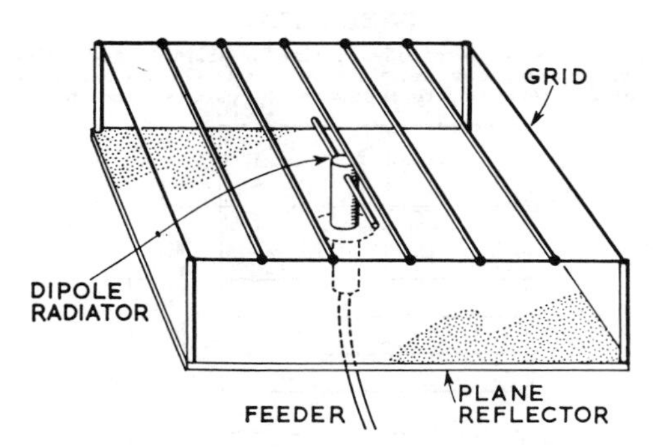

Fig 13.34. Reflex aerial. Note the grid-like structure in front of the dipole radiator. The plane reflector behind the radiator may be made of sheet metal or wire netting (1in mesh for 433MHz; ½in mesh for 1,296MHz).

the beam will depend on the diameter of the reflector A approximately according to the formula $\theta = 58/A$, where A is expressed in wavelengths. To avoid the obvious difficulties in making a parabolic reflector from sheet metal, a skeleton form of construction of wire netting stretched over wooden ribs may be found successful for frequencies up to 1,500MHz but as the wavelength becomes shorter the overall surface contour of the reflector must approach that of the true paraboloid more closely.

A suitable parabolic reflector for the 1,296MHz and 2,400MHz bands may be 1–3ft in diameter. The radiator element is usually a half-wave dipole having a resonant reflector in the form of a disc one half-wavelength in diameter mounted one quarter-wavelength in *front* of the dipole to reflect the radiation back into the parabolic dish. The centre of the dipole should be accurately positioned at the focus of the parabola. A balun transformer of the coaxial-sleeve type may be mounted on the feed stem.

A typical aerial assembly with a suitable feeder arrangement is shown in **Fig 13.36**. For this system a two-stub tuner will be found particularly convenient. The curve of the dish must not deviate from a true parabola by more than $\lambda/12$ at any point and thus the higher the frequency the more difficult it is to make a satisfactory reflector. Where wire mesh and ribs are used, the approximation to the true parabola must still be within $\lambda/12$ and the diameter of the holes in the mesh should also not exceed this amount. The radiator assembly can be based on the design given in Table 13.10 for the trough reflector. An alternative feed arrangement that has been used very successfully at 1,296MHz is shown in **Fig 13.37**.

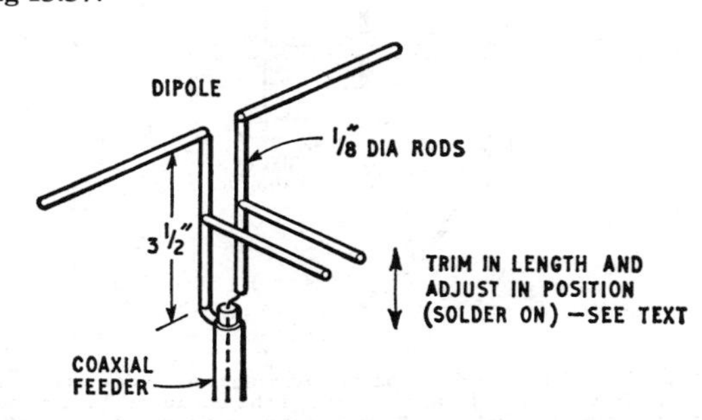

Fig 13.35. A possible feed arrangement for a 1,296MHz reflex aerial

TABLE 13.11

Typical dimensions of a reflex aerial for 432MHz and 1,296MHz. With these dimensions the gain should be 11-12dB. If the size of the reflector D is increased to 2λ square, the gain rises to about 16dB.

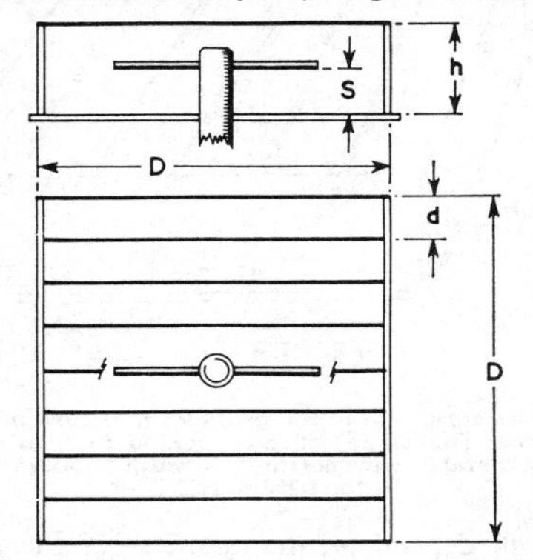

Band (MHz)	Dimensions (in)				rods diam (in)	Dipole	
	D	S	h	d		length (in)	diam (in)
432	30	7½	12	7½	½	12½	¼
1,296	18	2½	4	2½	5/32	4¼	⅛

Above 5,000MHz it is normal to use a waveguide feed either having a resonant dipole or alternatively a horn radiator.

A parabolic shape can be computed by plotting the curve $y = 4Sx$ (see **Fig 13.38**). If a suitable template is constructed from this formula, a complete dish can be fabricated using the template for reference purposes.

The gain of a parabolic reflector aerial system depends upon the diameter of the dish and the illumination of the dish from the feedpoint. Ideally, there should be a minimum radiation of energy from the feedpoint in directions other than into the dish. On transmission a spill-over from the feedpoint represents a loss of possible gain and, in reception, may result in space communication in the interception of

TABLE 13.12

Parabolic reflector

Band (MHz)	Dish diameter (ft)				
	10dB	15dB	20dB	25dB	30dB
432	3	5	10	15	30
1,296	1	2	3·5	6	12
2,400		1	2	3	4
10,000					1·5

Gain for various dish diameters. These are approximate figures. The actual gain achievable will depend upon the feed arrangements and upon the accuracy of the reflector relative to a true parabola.

unwanted hot body radiation from the earth, ie noise. A complete discussion of feed systems for parabolic reflectors is, however, outside the scope of this handbook.

Wide Band Aerial Structures

Many wide band aerial structures have been developed in recent years based upon the log periodic aerial. One of these has proved valuable as a very wide band radiating feed for a parabolic reflector covering the frequency band from 1,215MHz to 5,850MHz. It is called the trapezoidal tooth log periodic aerial and is shown in **Fig 13.39**. It can be constructed from the dimensions shown in **Fig 13.40** by accurately drawing out these shapes on paper, preferably using a drawing board, and pasting these on to 20swg brass sheet. The brass is then cut to the drawing and then the paper is removed by soaking. The two plates are then soldered to the feed line which projects through the reflector. Results obtained without a balun have proved satisfactory and the performance on the 1,296MHz and 2,450MHz bands found equal to the slot fed dipole with disc reflector shown in Fig 13.37.

The Helical Aerial

Another simple beam aerial possessing high gain and wide-band frequency characteristics simultaneously is the *helical aerial*: see **Fig 13.41** and **Table 13.13**. When the circumference of the helix is of the order of one wavelength axial radiation occurs; ie the maximum field strength is found to lie along the axis of the helix. This radiation is circularly polarized, the sense of the polarization depending

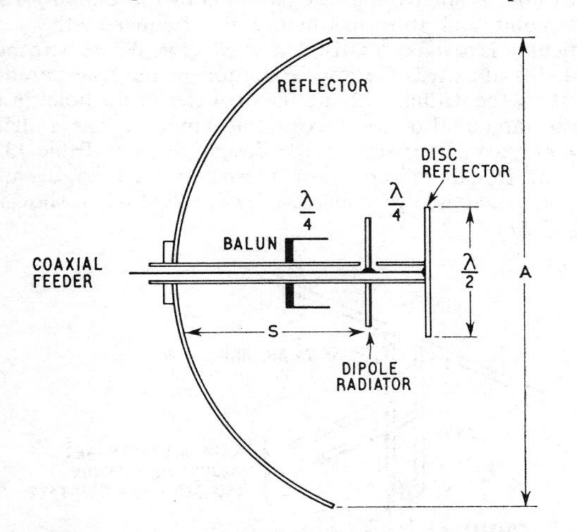

Fig 13.36. Typical design for a parabolic reflector feeder arrangement. The dipole should be at the focus of the parabola. To achieve maximum gain the distance S should be made adjustable. The coaxial-sleeve balun should be set back from the dipole.

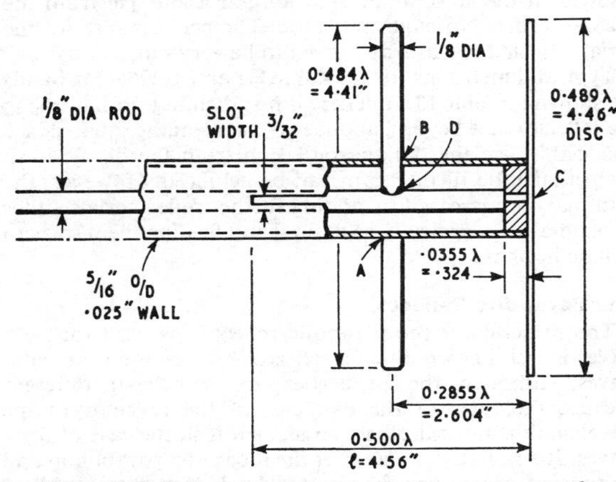

Fig 13.37. Dipole/reflector symmetric feed arrangement for a 1,296MHz parabolic reflector. One half of the dipole is soldered at A. The other half is threaded 6BA and is screwed into the centre conductor at D. It is finally soldered to the outer tube at B.

Example of an amateur-built parabolic aerial.

on whether the helix has a right- or left-hand thread viewed from the driven end.

If a pick-up dipole is used to explore the field in the direction of maximum radiation, the signal received by this dipole will show no change of amplitude as it is rotated through 360°, thus indicating true circular polarization. At any point to the side of the helix the wave will be elliptically polarized, ie the horizontal and vertical components will be of unequal strength.

A helix may be used to receive the circularly polarized waves radiated from a transmitting helix, but care must be taken to ensure that the receiving helix has a thread of the same sense as the radiator; if a thread of the wrong sense is used, the received signal will be very considerably weaker.

The properties of the helical aerial are determined by the diameter of the spiral D and the pitch P, and depend upon the resultant effect of the radiation taking place all along the helical conductor. The gain of the aerial depends on the number of turns in the helix. The diameter of the reflector R should be at least one half-wavelength. The diameter of the helix D should be about $\lambda/3$ and the pitch P about $\lambda/4$.

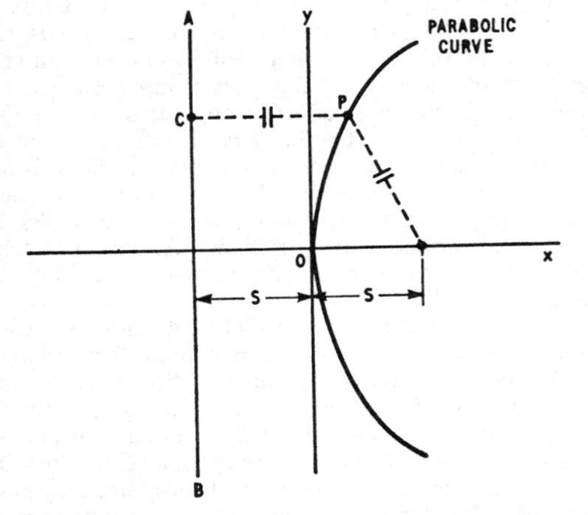

Fig 13.38. Parabolic curve: equation $y^2 = 4Sx$. The curve is the locus of points equi-distant from a fixed point, the focus F, and a fixed line AB, called the directrix, ie FP = PC. The focus F has the co-ordinates (S,O).

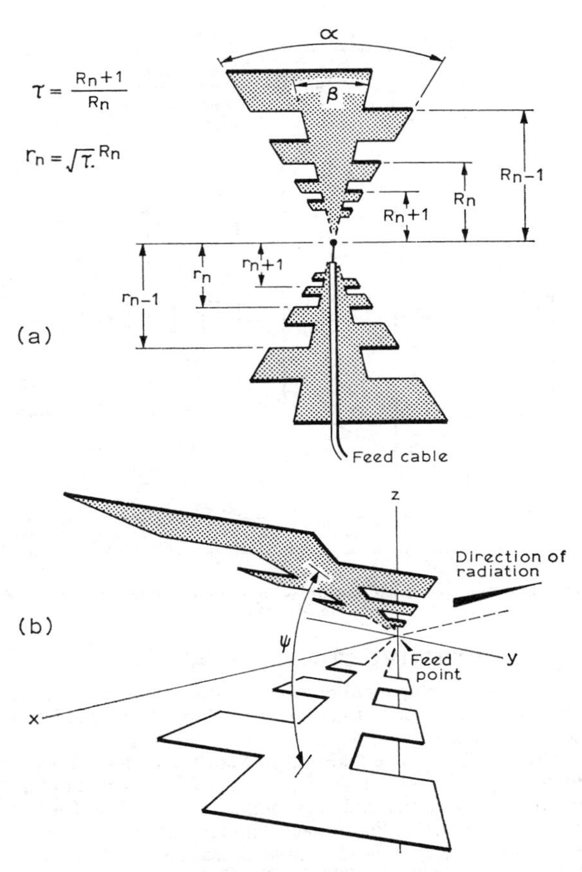

Fig 13.39. A trapezoidal-tooth log periodic aerial is shown in (a), the teeth of which may take on other shapes besides that pictured here. If the two halves are bent towards the reader the pyramidal structure of (b) is formed. The H plane is the xz plane. As an alternative to coaxial cable feed, twin transmission line may be placed along the axis.

It will be seen from Table 13.13 that the feed impedance is then about 140Ω unbalanced; this may be transformed to the feeder impedance by means of a quarter-wave transformer or a two-stub tuner.

It is important to note that the gain figures quoted will only be achieved if a circularly polarized aerial of the same sense, such as a helix, is used for reception. If a plane polarized aerial, such as a dipole, is used there will be a loss of 3dB.

The Long Helix

As with the Yagi, the gain of a helix aerial can be increased by extending the total effective length of the aerial. In the case of the helix, this is achieved by increasing the number of turns as shown in the second part of Table 13.13. The long helix has been used for frequencies as high as 11GHz with a gain of over 25dB and a bandwidth of approximately 5° between half power points. The helix had more than 500 turns. For amateur bands above 1,296MHz long helices provide an interesting alternative to the parabolic reflector. For example a helix having 69 turns has a gain of about 20dB at 10GHz.

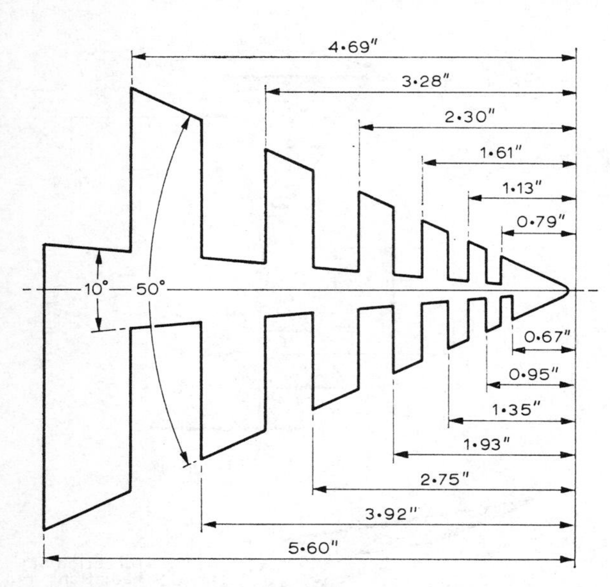

Fig 13.40. The structural outline, with dimensions, for the aerial of Fig 31.39(b).

TABLE 13.13

The helical aerial—dimensions and performance

Band	Dimensions				
	D	R	P	a	d
General (λ)	0·32	0·8	0·22	0·12	
432MHz (in)	8¾	22	6	3	¼
1,296MHz (in)	3	7	2	1⅛	¼ to ⅛

Turns	6	8	10	12	20
Gain (dB)	12	14	15	16	17
Beamwidth	47°	41°	36°	31°	24°

The gain and beamwidth of the helical aerial is dependent upon the tota number of turns as shown above.
Bandwidth = 0·75 to 1·3

$$\text{Feed Impedance} = 140 \times \frac{\textit{circumference}}{\lambda} \text{ ohms}$$

$$\text{Beamwidth} = \sqrt{\frac{12{,}300}{\text{No turns}}} \text{ degrees}$$

CONSTRUCTION OF AERIALS

Various constructional points have been mentioned in the text of this chapter but certain points are summarized and some are added here.

Great care must be taken to prevent the ingress of water to the coaxial cable at the feed point. Bostik sealing compound or Telecompound is recommended for this. The use of dissimilar metals in contact without weather protection must be avoided otherwise rapid corrosion due to electrolytic action will result. This is especially true of brass and aluminium. Even where protected it is still advisable to avoid dissimilar metals in contact if possible. Elements may be secured to the mounting booms by the use of proprietary cast alloy saddles supplied for television aerial use which are commonly available at a reasonable price. Similar clamps are available for attaching booms to masts. If the aerial is to be used out of doors, as most are, it is very desirable that it should be painted using a suitable primer; zinc chromate is ideal for aluminium and brass (adequate pre-cleaning with detergent or sugar soap is essential) with one or two coats of aluminium paint for finishing. Aerials protected in this way will last for ten years in a corrosive atmosphere whereas,

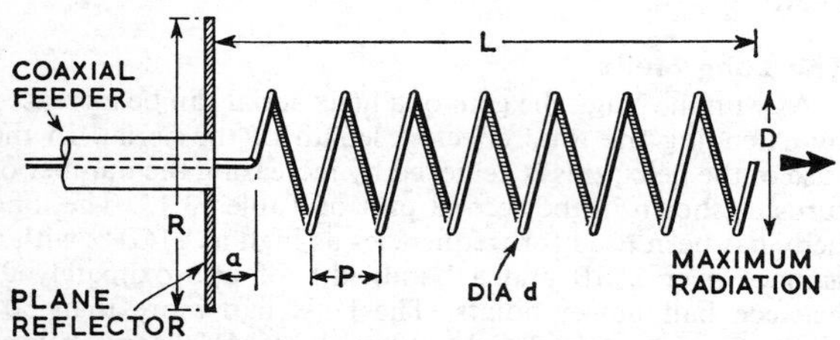

Fig 13.41. The helical aerial. The plane reflector may take the form of a dartboard type of wire grid. The dimensions given in Table 13.13 are based on a pitch angle of 12 degrees. The helix, which may be wound of copper tube or wire, the actual diameter of which is not critical, must be supported by low loss insulators.

unprotected, especially if made of some aluminium alloys, eg magnesium/aluminium, will only last a year or two or even less.

Certain alloys have, however, an inherently high resistance to corrosion. It is essential to have the specification of the material in detail and to have access to the manufacturer's data on corrosion before deciding on an alloy which will give a satisfactory life in an unpainted condition.

Particular care should be taken to ensure that all mechanical and electrical joints—especially sliding ones—are really good otherwise the aerial array will be unlikely to survive the first high wind.

TUNING ADJUSTMENTS

To tune up any aerial system it is essential to keep it away from large objects, such as buildings, sheds and trees. The array itself should be at least two wavelengths above the ground. It is useless to attempt any tuning indoors since the change in the surroundings will result in completely different performance when the array is taken outside.

Undoubtedly the most effective apparatus for tuning up any aerial system is a standing-wave indicator or reflectometer (see Chapter 18—*Measurements*). If there is zero reflection from the load, the standing-wave ratio on the aerial feeder is unity. Under this condition, known as a *flat line*, the maximum power is being radiated. All aerial matching adjustments should therefore be carried out to aim at a standing-wave ratio of unity.

If suitable apparatus is not available, the next best course of action is to tune the aerial for maximum forward radiation. A convenient device for this is a field-strength meter comprising a diode voltmeter connected to a half-wave dipole placed at least ten wavelengths from and at the same level as the aerial. When adjustments have resulted in a maximum reading on this voltmeter, the feed line may nevertheless not be "flat" and therefore some power will be wasted. However, if the best has been done with the resources available it is highly likely that good results will be achieved.

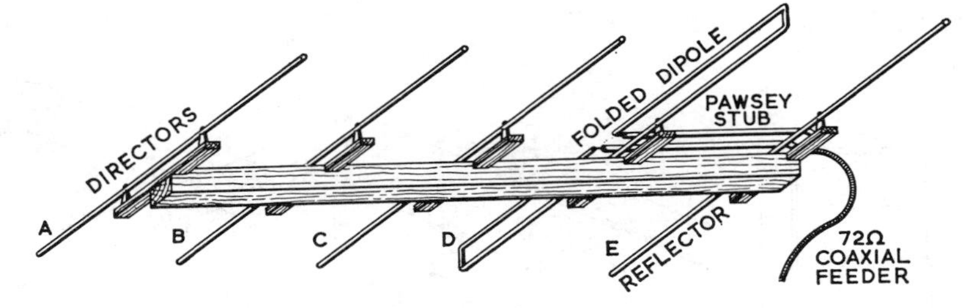

Fig 13.42. 5-element Yagi arrays for 70·3 and 145MHz with a folded dipole radiator. The gain obtainable is about 9dB.

TYPICAL AERIALS

To illustrate the principles of construction some aerial arrays which are typical of those in use at the present time are described below.

5-element Yagi Arrays for the 70 and 144MHz Bands

The 5-element Yagi array shown in **Fig 13.42** is a typical "flat-top" for vhf use. With the dimensions indicated in **Table 13.14** the centre frequencies are 70·3 and 145MHz.

The boom is made from wood and the elements are supported on polythene insulators mounted on wooden cross pieces. This form of construction is suitable where the workshop resources are limited and is quite as satisfactory as an all-metal construction although rather heavier.

The driven element is a folded dipole fed with 72Ω coaxial cable through a balun of the Pawsey-stub type. This balun is constructed from brass tube, the inside diameter of which is a sliding fit over the outer covering of the coaxial cable. The cable braiding is soldered at the aerial end of the balun to the outside of the balun tubing. See **Fig 13.43.**

The gain of the aerial is about 9dB, and the beamwidth is about 50° between half-power points.

A 6-over-6 Skeleton-slot Yagi Array for the 144MHz Band

This array can be constructed entirely from aluminium tubing, the elements being attached to the booms with suitable clamping devices.

With the dimensions given in **Fig 13.44**, the centre frequency is 145MHz and the bandwidth is approximately 3MHz for a standing-wave ratio of 1·2 : 1. The beamwidth for the half-power points is 45° and the forward gain approximately 11dB. The feed impedance is 72Ω and good-quality 72Ω coaxial cable should be used as the feeder. A balun is not essential, but it is recommended that one should be used and it may be either the Pawsey-stub or coaxial-sleeve type. For convenience in mounting this balun on the aerial mast, a short length of 72Ω balanced-twin feeder may be used to connect the output of the balun to the feed point. To prevent the ingress of water at the top end of the coaxial cable, the whole balun assembly should be enclosed in polythene film.

Provided the dimensions given are strictly adhered to, no matching adjustments will be necessary.

TABLE 13.14

Element lengths (in)			Spacings (in)		
	70·3MHz	145MHz		70·3MHz	145MHz
A	72	35	AB	25	12½
B	73	35½	BC	25	12¼
C	74	36	CD	25	12¼
D	79	38½	DE	42	20½
E	82½	40			
			balun length L	41	19¼

Materials
Elements: Aluminium rod or tube ⅜in dia (70·3), ¼in dia (145)
Folded Dipole (D): folded part ⅜in dia
fed part ¼in dia
centre-to-centre spacing 1in
Balun: brass tube to be a sliding fit over the coaxial feeder cable

An 11-element Long Yagi for the 144MHz Band

The 11-element long Yagi shown in **Fig 13.45** and **Table 13.15** is light in weight yet extremely effective.

The boom is drilled for the ⅛in diameter aluminium rod elements and the ¼in diameter folded dipole using a home workshop drill and drill stand located on the floor. The boom must be prevented from turning; this is probably best done by clamping it to a wooden stand or board. It is most important to ensure that the holes are drilled at right angles to the boom so that the elements are parallel to each other when fitted.

Elements are secured to the boom, having been inserted through the holes, either by centre punching the aluminium boom around the holes thus forming a compression joint for each element, or preferably by a clip of the type shown in **Fig 13.46**. The folded dipole must have one end bent after insertion in the boom. This can be done by forming the shape around a ⅞in diameter broom handle. The boom is secured to the mast by a mast clamp while the stay rods are secured by aluminium straps. The centre of the folded dipole is attached to a ⅝in thick polystyrene block which is drilled and tapped for the securing screws which are also the feed point. This block is screwed to the boom.

The folded dipole/feeder connections are waterproofed with Bostik sealing compound. Matching to the feeder is optimized, preferably using a reflectometer or, if not available, for maximum forward radiation by sliding the reflector assembly in and out of the boom. The resulting match is better than 1·2 to 1 at 145MHz. If greater gain is required additional directors may be added and the resulting slight change in matching can be effected by movement of the reflector as before. Finally, the reflector assembly should be clamped rigidly inside the boom by means of a hose clip.

An 18-element Parabeam-type Yagi for the 432MHz Band

The aerial shown in **Fig 13.47** is suitable for serious 432MHz band work, having a gain of about 15dB and horizontal beamwidth between half power points of 28°. The boom and elements are constructed of aluminium tubing, attachment being effected by the use of proprietary clamps. The centre feed impedance is 280Ω and a 4 to 1 balun can be used to match to a 72Ω coaxial cable.

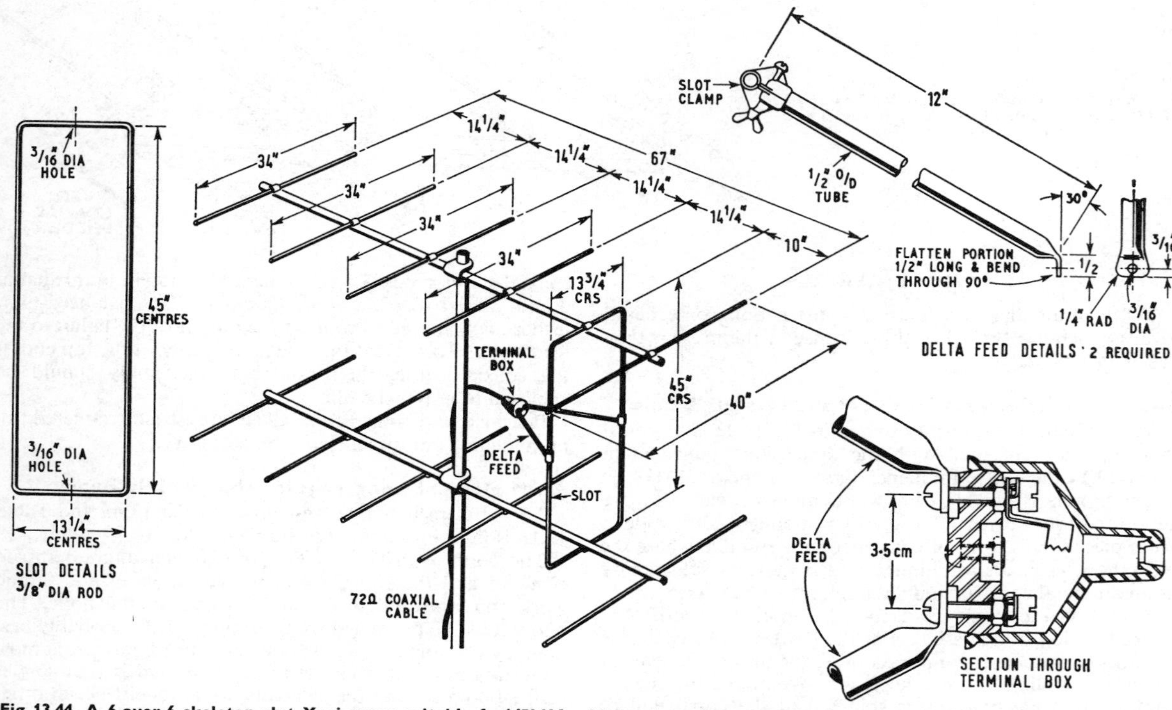

Fig 13.44. A 6-over-6 skeleton slot Yagi array suitable for 145MHz. Note slot-reflector spacing should be 20in, not 10in as shown. *With acknowledgements to Jaybeam Ltd.*

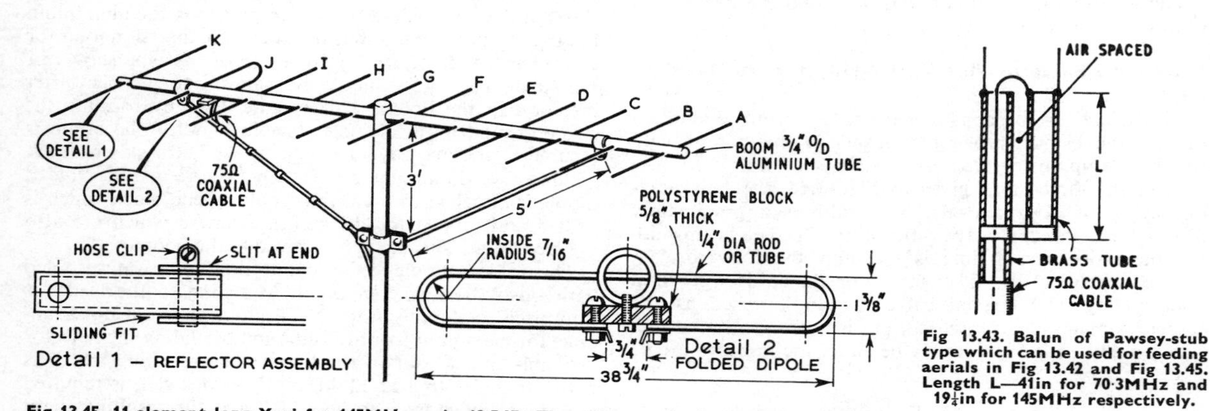

Fig 13.45. 11-element long Yagi for 145MHz, gain 13·5dB. The gain may be increased by adding additional directors, each $\frac{1}{8}$in less in length than its predecessors. The folded dipole is fed using a balun as shown in Fig 13.43.

Fig 13.43. Balun of Pawsey-stub type which can be used for feeding aerials in Fig 13.42 and Fig 13.45. Length L—41in for 70·3MHz and 19½in for 145MHz respectively.

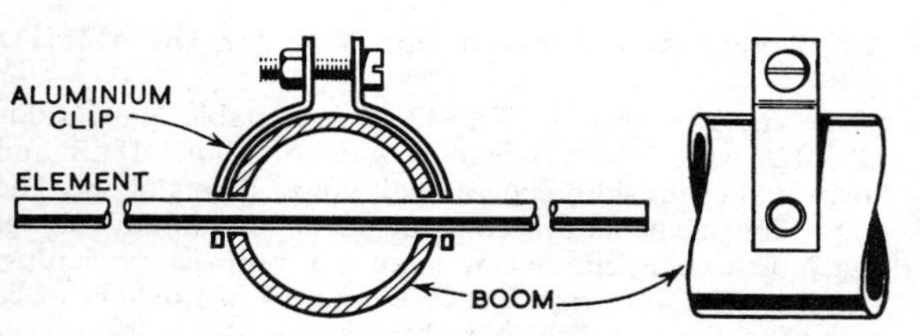

Fig 13.46. Element securing clip.

TABLE 13.15

Element	A	B	C	D	E	F	G	H	I	J	K
Element length (in)	$35\frac{3}{4}$	$35\frac{7}{8}$	36	$36\frac{1}{8}$	$36\frac{1}{4}$	$36\frac{3}{8}$	$36\frac{1}{2}$	$36\frac{5}{8}$	$36\frac{3}{4}$	detail 2	41
Parasitic element $\frac{1}{8}$in dia aluminium rod	AB	BC	CD	DE	EF	FG	GH	HI	IJ	JK	
Element spacing from (in)	16	16	16	16	16	16	16	20	8	19	

Feed details: as for the 5-element Yagi arrays for 70·3 and 145MHz.

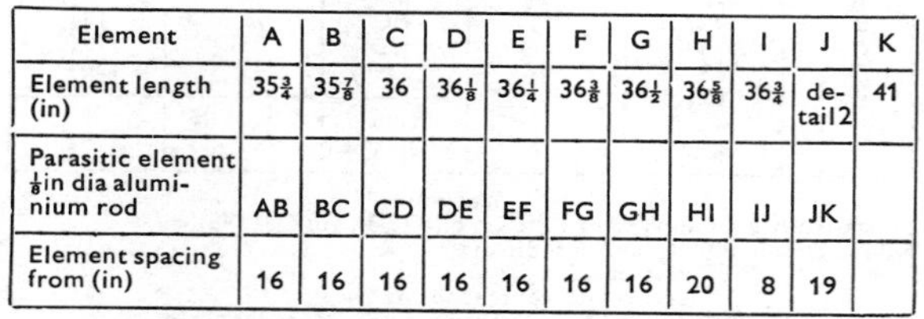

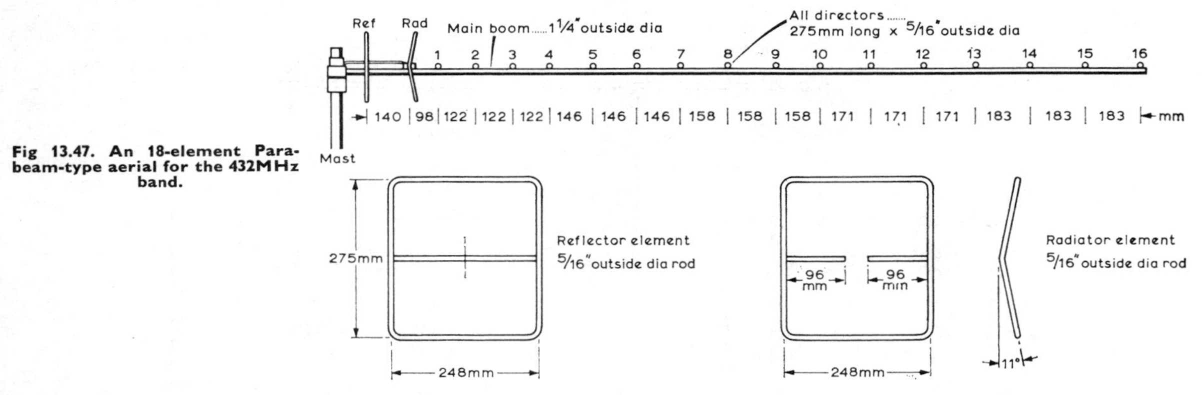

Fig 13.47. An 18-element Parabeam-type aerial for the 432MHz band.

A 10-over-10 Element Slot-fed Beam for the 1,296MHz Band

The aerial shown in **Fig 13.48** is ideal either for the newcomer to this increasingly popular band or as a permanent general purpose aerial. The gain is about 12dB and the vswr not worse than 1·2 : 1. It is constructed from ⅜in od brass or copper tube for the main booms and ⅛in od brass rod for the elements which are either soft or silver soldered to the booms. If desired an 8-over-8 beam can be constructed by removing directors D_7 and D_8 and shortening the booms accordingly. The gain will be reduced by about 2dB but the matching will not be affected.

A 36-element Parabeam-type Yagi for the 1,296MHz Band

The aerial shown in **Fig 13.49** is a 1,296MHz long Yagi developed by G8AZM and based upon the Parabeam. Its gain is roughly equivalent to that of a 3ft dish.

The method of construction is straightforward and evident from Fig 13.49. The critical dimensions marked should be observed, but other dimensions may be varied somewhat to suit the materials to hand. The ends of the boom should first be crimped and soldered to exclude moisture, and the mast extension and support arms then soldered into place. Starting with the smallest, the directors are soldered to the boom using a large (250W) soldering iron. The elements are located in small grooves filed in the boom, and are initially cut slightly longer than required and filed to length after soldering in place. In this way the required accuracy of construction, within a few tenths of a millimetre, can be achieved. The director lengths are given in the table. The spacing between the reflector and radiator should be 73mm, and between the radiator and first director 42·2mm. The directors are all spaced 57·7mm apart.

The radiator and reflector may best be made by soldering together two L-shaped pieces of rod, one limb of each of which is accurately filed to length after bending. It may be found necessary to use small clamps to hold joints in place while others are soldered. The radiator clamp is fabricated from polythene or nylon. The hole to take the boom may be cut with a heated piece of the boom material. The 4BA nut used to secure this clamp is melted into place.

The balun assembly is constructed from a 7·6mm-bore tube 120mm long. Approximately 130mm of the outer protective covering is stripped from the full length of uhf low-loss coaxial cable to be employed, which is pushed into the tube. The outer braid is then soldered at both ends taking extreme care not to melt the dielectric. The $\lambda/4$ coaxial sleeve is made from 12·5mm bore tube; the inside length is the critical dimension and should be 57·5mm.

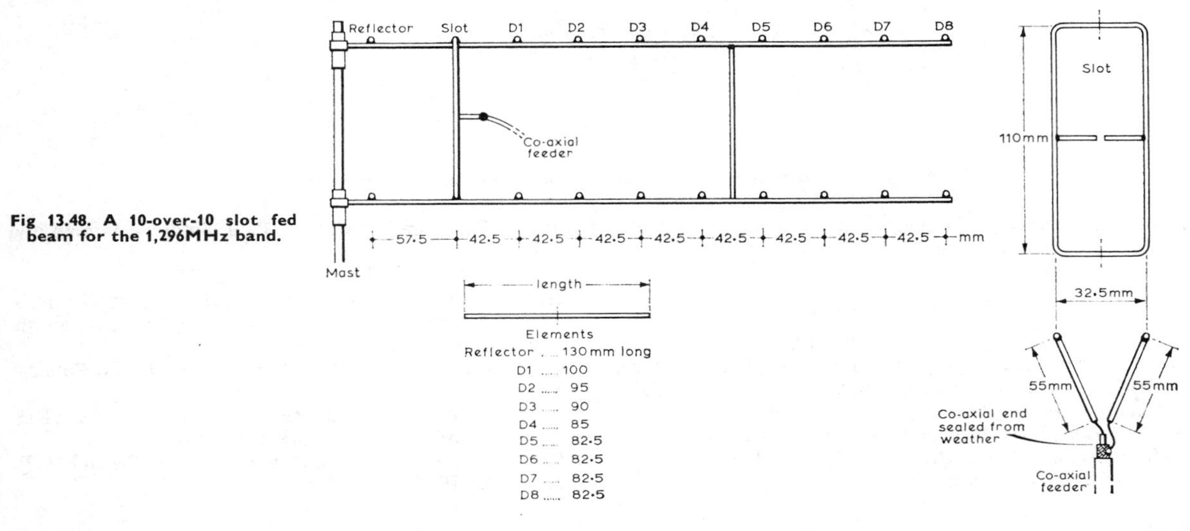

Fig 13.48. A 10-over-10 slot fed beam for the 1,296MHz band.

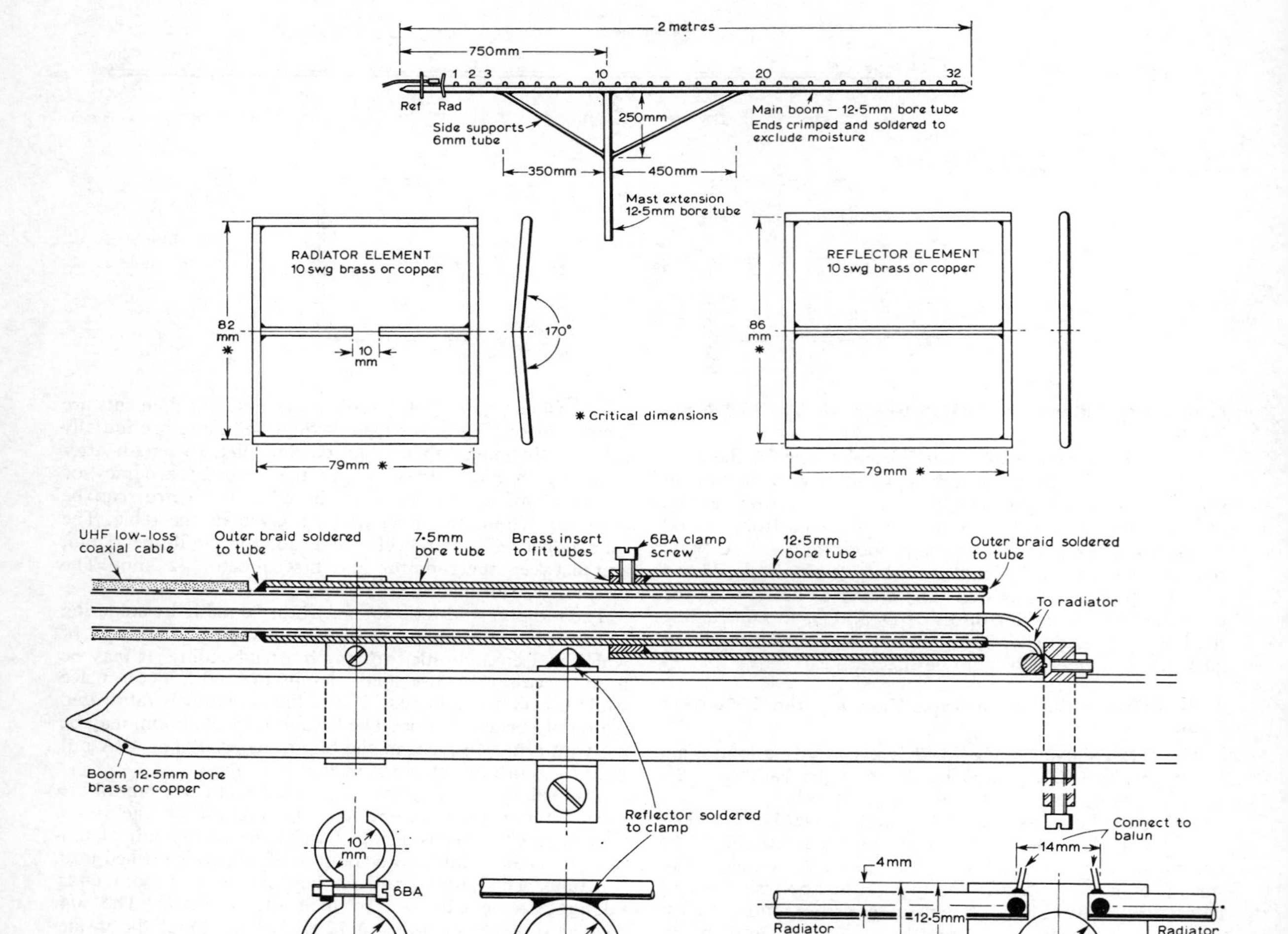

Fig 13.49. A 36-element Parabeam-type aerial for the 1,296MHz band.

After final assembly, the aerial may be painted with aluminium paint to reduce corrosion.

BIBLIOGRAPHY

Antennas, J. D. Kraus, McGraw-Hill, 1950.

"Measurements in high gain helical aerials and on helical aerials of triangular cross section", *Proc IEE*, Vol 3, July-August 1964.

Radio Aerials, E. B. Moulin, Oxford University Press.

VHF Handbook, W. I. Orr, W6SAI, and H. G. Johnson, W6QKI, Radio Publications.

Antenna Theory and Design, H. Paul, Sir Isaac Pitman and Sons.

The Radio Amateur's VHF Manual, ARRL.

VHF Antenna Handbook, J. Kyle, K5JKX, 73 Publications.

Aerials for Metre and Decimetre Wavelengths, R. A. Smith, Cambridge University Press.

Cubical Quad Antennas, W. I. Orr, W6SAI, Radio Publications.

"Wideband aerial structures", C. R. Fry, VE2ARO/G3MTI, *RSGB Bulletin* November 1966.

VHF/UHF Manual, 3rd edn, D. S. Evans, G3RPE and G. R. Jessop, G6JP, RSGB, 1976.

CHAPTER 14

MOBILE AND PORTABLE EQUIPMENT

THE advent of small factory-made transceivers designed specifically for the amateur bands, together with the establishment of an fm repeater network on vhf and uhf, has effected little short of a revolution in mobile and portable operation in the past few years.

While the older valve transceivers required a fairly substantial mounting underneath the vehicle dashboard or a boot mounting, the latest equipment will fit easily into most vehicles. Furthermore, current drain for the same rf power has been cut by up to half, and this, together with the recent change to more efficient alternators instead of generators in vehicles, means that the mobile operator need no longer be quite so careful about inadvertently discharging the vehicle battery.

The radical improvements in vhf and uhf equipment has been paralleled by a large swing to the fm mode used in conjunction with repeaters, which are now in operation on 144MHz and 432MHz in Europe.

As explained in Chapters 5 and 9, fm offers considerable advantages for mobile operation and, when used with a repeater, it permits improved mobile-to-mobile contacts over a substantial distance, even in built-up areas.

One requirement of mobile equipment has always been simplicity of operation, and consequently most vhf and uhf fm equipment is designed for fixed-channel crystal-controlled service. The simplex channels used in the UK and IARU Region 1 are given in **Table 14.1.**

Similar changes have occurred in portable operation, and hand-portable vhf or uhf transceivers working on the mobile simplex and repeater channels are now very popular. Facets of amateur radio which depend upon portable operation, such as emergency communication and df contests, have been greatly assisted by these trends.

Although most mobile and portable operation now takes place on vhf and uhf, the hf bands are still widely used for longer-distance contacts, and these days most commercially made hf transceivers are designed to operate directly from a 12V supply if required.

Almost irrespective of the band or mode, there are three key requirements for satisfactory mobile operation: a suitable transceiver, minimum electrical noise in the vehicle, and an efficient aerial. These aspects will now be considered in turn.

TABLE 14.1

IARU Region 1 vhf and uhf simplex channels

Channel	Frequency (MHz)	Notes
S20	145·500	Calling channel
S21	145·525	
S22	145·550	
S23	145·575	
SU15	433·375	
SU16	433·400	
SU17	433·425	
SU18	433·450	
SU19	433·475	
SU20	433·500	Calling channel

433·2MHz, which corresponds to channel RB8, is also in use as a simplex channel (SU8) in the UK.

MOBILE TRANSCEIVERS

The poor aerial systems, high noise levels, fluctuating supply voltages and constant vibration in mobile operation may mean that a transceiver which performs adequately in a base station may be very disappointing when installed in a vehicle. Many hf and vhf transceivers not specifically designed for mobile use will however be found satisfactory, the main requirements being the following:

(a) High-sensitivity receiver with efficient agc and a noise blanker (ssb) or with excellent limiting characteristics and a correctly adjusted squelch (fm).
(b) Adequate receiver audio output (at least 2W).
(c) Protection of the power amplifier against high vswr.
(d) A minimum of easy-to-operate controls with clear and unambiguous dials and meters, preferably illuminated at night.
(e) 12V power requirements not to exceed the spare capacity of the vehicle generation system (see later). Internal voltage regulation, with over-voltage protection.
(f) Small size, capable of being mounted within easy reach of the driver, preferably in or under the dashboard.
(g) Rugged mechanical construction, together with high stability of electronic circuits including the toneburst circuit if one is fitted.

Simple noise pulse limiter circuits were formerly used to suppress ignition noise in mobile a.m. and ssb receivers. These usually consisted of a two-diode clipper, often positioned immediately after the detector, which ensured that loud noise pulses did not overload the audio stages. However, the noise still remained audible, even if it was of a lower intensity, and the greater complexity of circuits made possible by using semiconductor devices has now allowed a more sophisticated system to be used—noise blanking. This mutes the receiver for the duration of each noise pulse, and has been found to be extremely effective against ignition noise, although not so much against other forms of interference. The principles of noise blanking are explained in Chapter 4.

The provision of a squelch circuit is desirable if not essential for mobile operation on fm. Such a circuit mutes the receiver audio unless a carrier is present and thus avoids subjecting the listener to a loud hiss (or roar) in the absence of a signal. The signal level at which the audio is switched on must, however, be carefully adjusted, or else the receiver will remain muted when weak signals are present. The control voltage necessary to operate the squelch switching is usually derived from the agc line (a.m. receivers) or from i.f. noise present at the discriminator (fm receivers).

When planning a mobile installation, consideration should be given to the possibility of using the transceiver in other modes of operation, particularly as a portable station, for in this case requirements are in many ways similar. For example, if a hand-portable ssb or fm transceiver is already available, all that needs to be done is to obtain an add-on rf amplifier and aerial, and leave these permanently in the vehicle (see p14.18).

Expensive radio equipment left permanently attached to the vehicle should always be insured against theft, and the driver's ordinary insurance may not cover this eventuality. If the vehicle is to be left unattended for long periods on a public highway it is advisable if possible to remove the transceiver from the car completely when it is not in use.

Power Supply

Although the nominal off-load voltage of a vehicle battery is 12V, in practice the voltage supplied by a vehicle electrical system will vary over a range from about 11·5 to 14·5V, depending on several factors such as whether or not the charging system is in operation and the other loads on the battery. Mobile transceivers are therefore designed and adjusted to work at an average voltage slightly higher than 12V, and 13·8V is an accepted standard in this respect. It follows that any alignment of such equipment should be undertaken with it connected to a well-regulated 13·8V power supply.

Due to power restrictions imposed by the vehicle's electrical system, mobile transmitters are usually operated with a dc input to the power amplifier not exceeding 50W. Higher powers can be used, but unless special provisions are made, transmitter operating time needs to be restricted to the capabilities of the storage and charging system, particularly during the hours of darkness when the vehicle lighting circuits are in operation.

While no hard and fast rules can be laid down concerning how much current can be consumed by the transceiver installed in a specific vehicle, certain assumptions can be made from which a reasonable maximum loading for that vehicle may be determined. The first and most important assumption is that the storage and charging system is in good order and maintains a state of balance between supply and demand when the load of the vehicle is at maximum. For all practical purposes, this load may be considered as the sum of the consumptions of the following items: (a) tail-lights; (b) side-lights; and (c) headlights dipped. In a typical case these amounted to (a) 12W; (b) 12W; and (c) 90W which added together produce a total of 114W, or a current of approximately 10A in a 12V system. With an equivalent radio equipment load, under daylight conditions, the electrical system of the vehicle should be capable of maintaining a state of balance. In fact, allowing for full headlight conditions, and adding in the loading of the ignition system, rear number plate light and a heater blower motor, the peak will

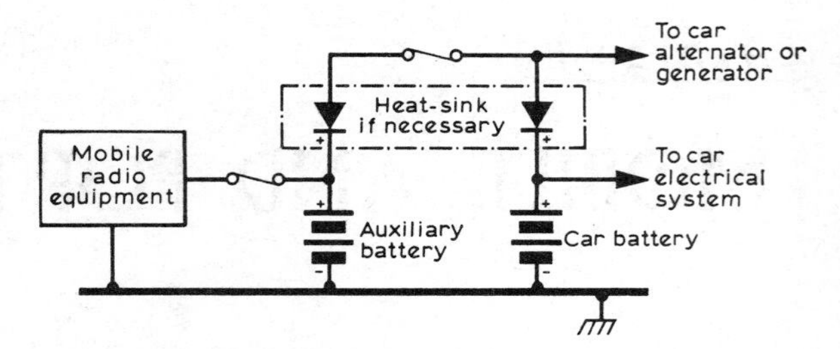

Fig 14.1. Showing how an auxiliary battery can be added to the charging system of a car using negative earth. For positive earth systems the polarity of the diodes should be reversed. Note that the diode current rating must be adequate in each case and a heat sink may be required. The diode supplying the auxiliary battery should have a current rating at least equal to the current drain of the mobile radio equipment.

be in the region of 160W for the vehicle cited. Thus, to assume a total available capacity of 120W maximum, ie a current of 10A on a 12V system, will err on the right side.

In relation to this loading it will be appreciated that the electrical system could be overtaxed if the radio equipment is used during darkness and wet weather, and in these cases either the storage battery will have to be charged from an external source while the car is at rest, or alternatively sufficient mileage covered during daylight without the equipment operating to allow the deficiency to be made up. To some extent, the extra loading can be compensated for by increasing the output of the generator or alternator, but where the loading is of the order of 120W, full compensation is not normally possible without the risk of overcharging the battery.

Where a great deal of high-power mobile working is envisaged, particularly with add-on PAS, and so long as there is adequate space under the bonnet or in the boot, consideration should be given to fitting an additional battery to run the installation. While this involves purchasing an additional battery, its cost may be preferable to being faced with a flat main battery. Where such a course is adopted, the auxiliary battery may be linked into the charging system of the car in the manner shown in **Fig 14.1.** If this circuit is installed, the output of the generator or alternator may be increased without running into difficulty since the charging current is shared by both batteries when their state of charge is equal, but when imbalance arises, the battery which is more discharged will take the greater current. Commercial kits are also available for these modifications (originally intended for additional batteries for caravans, etc).

In many mobile installations no additional battery or other modifications can be made to the system in order to reduce the extra load imposed on the vehicle battery, and in such cases (especially in older vehicles fitted with generators) it is important that extra care be devoted to that item to avoid premature failure.

The modern lead-acid battery is inherently reliable but because of its very reliability, it frequently fails to give good service simply because it is taken for granted. Just as a motor vehicle needs to be serviced regularly if it is to be maintained in good condition, so the battery needs a little attention now and then but, compared to the vehicle as a whole, its requirements are minor. The most important of its needs is distilled water.

Distilled water must always be used for topping-up the electrolyte, never tap water or boiled water. Both of these contain traces of mineral elements, some of which can react with the coating on the plates to produce either a semi-insulating compound, or a compound which is shed during the charging process.

Prolonged charging at an excessive rate is as damaging to a storage battery as is allowing the level of the electrolyte to fall below the level of the plates. For this reason, adjustments should never be made with a view to increasing the charging current without first determining that the higher rate is within the capabilities of the battery. It is usually permissible to increase the charging rate to some degree to compensate for the extra loading imposed by mobile radio equipment, but such adjustments should always be made by a competent automobile electrical engineer. Alternatively, if the extra loading is heavy, consideration might be given to the provision of an auxiliary battery in the manner described above.

Where high-power radio equipment is added to the circuit loading and no compensating output from the generator arranged, it is essential that provision is made to trickle charge the battery overnight after the equipment has been used.

The battery terminals should be kept clean and lightly coated with petroleum jelly, and the connecting lugs well tightened.

Should it be necessary to remove and store the battery, it should first be fully charged before being placed into store. It should be given a weekly boost charge to keep it in the fully charged condition.

The off-load voltage of a battery may not give a correct indication of its condition. As the state of charge of a battery varies so does the specific gravity of the electrolyte. For this reason, the most reliable indication of the state of a battery is given by a hydrometer, and this is an instrument which every mobile operator should possess. Normal sg levels lie between 1·150 and 1·250 and may vary from battery to battery of the same type. It is also wise to fit either a voltmeter or ammeter in the car so that charge rates and loads can be monitored.

ELECTRICAL NOISE SUPPRESSION

In the normal operation of many items of motor vehicle electrical equipment rf energy is generated which can be distributed over a wide range of frequencies, this energy being recognized in a receiver as interference or radio noise. While modern amateur mobile radio equipment is usually of advanced design with the necessary high orders of sensitivity, it can still be affected by radio interference generated by motor vehicle equipment. It is the purpose of this section to describe some of the methods that can be adopted to reduce this interference to acceptable levels.

The various forms of suppression and simple arrangements described below have been employed extensively to suppress mf, hf and vhf interference on vehicles fitted with commercial or amateur radio equipment and have given satisfactory performance. Interference to uhf reception is not often encountered, but where this occurs, adequate suppression can usually be obtained by fitting suppressors designed for operation in the range 30–400MHz.

Typical Sources of Interference

Abrupt changes or interruptions in the flow of direct or alternating currents occur in the normal operation of many items of vehicle electrical equipment, and it is the transient currents and associated electromagnetic fields arising from these changes or interruptions that cause radio interference.

The continuously generated transients constitute the most annoying sources of interference. Such transients are usually associated with the ignition system; machines having commutators or slip-ring brush gear such as dc generators, alternators, windscreen wiper motors and fan motors; voltage and current regulators, instrument voltage stabilizers and electric petrol pumps. In addition, semiconductor devices such as silicon diodes and transistors must not be overlooked since they perform functions similar to switches or vibrating contacts and are thus able to create just as much radio interference. These devices are used in automobile alternators, electrical output control units and electronic ignition systems.

On a petrol-engined vehicle the major source of interference is the ignition system. Energy is stored in the ignition secondary circuit capacitance (50–100pF) which is charged to a voltage in the region of 10-25kV. Each time a plug and distributor spark gap breaks down, this energy discharges in an oscillatory manner as a result of resonance between the secondary circuit capacitance of the ignition system (including spark plug, distribution cables and distributor) and the attached lead inductances. The magnitude of these oscillatory currents, which may attain peak values of 200A or more and flow for a few milliseconds, results in electromagnetic fields being set up which can cause serious interference to radio communication systems over a wide frequency range. The rf energy is radiated from the ht circuit and from the lt wiring due to mutual and direct coupling with the ht system. In a typical saloon vehicle the interference amplitude increases with frequency to a maximum in the region of 40-100MHz and then reduces with increasing frequency, although it may remain seriously appreciable up to at least 600MHz. If copper-cored ignition distribution cables are employed, then the interference field in the hf and vhf bands at a distance of 10m from the vehicle would be of the order 800-1,000μV/m: this value can readily exceed the field value set up by many radio transmitters.

Electronic motors and generator control units are further items of vehicle equipment which left unsuppressed can produce appreciable fields at a distance from the vehicle. Consequently, they can often be troublesome to radio equipment on the vehicle in close proximity.

Electrical equipment is not the only source of interference on a vehicle. Friction in body parts or wheels can build up static charges if there is poor earthing continuity in those areas.

Other sources of radio interference received via the vehicle aerial which are worthy of mention, but beyond the scope of this work to discuss at length, are overhead electrical transmission lines, some underground cables, neon signs, electrical equipment in factory areas, and some older vehicles with inadequate suppression schemes. While radio interference from these sources may be annoying to radio equipment users, it is usually only of a temporary nature in a moving vehicle.

Propagation of Interference

In addition to the source of radio interference, the magnitude of the field radiated from a motor vehicle depends on several other factors, for example the length and positioning of the vehicle wiring. Also, the screening effect of the vehicle body can have a pronounced influence. Its effectiveness is extremely variable and unpredictable, however, since it depends on the manner in which the various metal panels are joined and the nature of the joint. This in turn depends on how well the joints are welded; or if they are bolted, on the number and position of the bolts, whether serrated washers are used, and how much insulation in the form of paint is included in the joint. The effectiveness of the body as an electrical shield also varies over the rf spectrum due to resonances which can be set up within the panels. For example, the hinged bonnet of a typical saloon vehicle can resonate in the frequency range 80-120MHz, and if not effectively earthed (or bonded) will radiate interference which may cause serious problems in the reception of vhf signals.

Radio frequency interference associated with the electrical equipment on a vehicle can be introduced into a receiver in a number of ways. The most important of these can be summarized as follows:

(a) By direct radiation from the electrical equipment to the radio aerial.
(b) By conduction along the wiring from the equipment to the supply lead of the radio receiver or its power supply.
(c) By conduction along the vehicle wiring and then by radiation to the receiver aerial, or coupling to the aerial feeder or other receiver wiring if this is in close proximity.
(d) By radiation to mechanical devices such as control cables, rods, wires, tubes (exhaust pipe) which are not associated with the electrical system but located in close proximity to it, and then by re-radiation to the receiver aerial.
(e) By direct radiation from the vehicle wiring to the radio receiver circuits when receiver screening is inadequate.

Thus in any vehicle installation, apart from there being more than one source of interference which can create annoyance, an interference signal can arrive at the receiver by a number of different routes.

Installation of Equipment

When mobile radio equipment is installed in a vehicle, consideration must be given to the siting of the equipment and to the suppression requirements of each part of the vehicle electrical system, so that little or no interference is generated within the frequency range of the radio equipment.

The siting of the aerial and the routing of the feeder cable should be studied so that as far as possible they are clear of known sources of interference such as ignition and generator circuits and long cable runs in the vehicle wiring harness.

While a vhf aerial on a vehicle is usually sited on either a rear wing or the roof, an hf aerial, because of its larger physical size, is invariably sited at the rear of the vehicle (**Fig 14.2**). Normally this will mean the aerial is well away from the engine compartment and the ignition system, but in the vicinity of such secondary sources of interference as the wiring system to the rear lamp assemblies and the rear section of the exhaust pipe.

A good earth bond of the feeder cable screen direct to the vehicle chassis at the aerial base is important. Usually, poor feeder cable bonding at this point results in interference to the receiver from most items of ancillary equipment (starter direction indicators, windscreen wipers, horns, etc). Salt

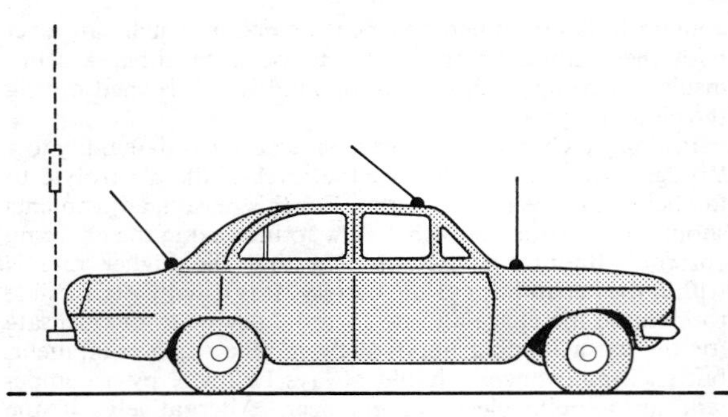

Fig 14.2. Alternative fixing positions for vhf aerials, and hf aerial position (shown dotted).

corrosion of earthing bonds can also be responsible for similar interference problems if the vehicle has been in use for some time during adverse winter weather conditions. A good protection for them is "Silcoset" by ICI, which is an air-drying silicone rubber. If possible, avoid any joints in the coaxial feeder between the aerial and the receiver.

It is vital that amateur mobile equipment is installed in accordance with any instructions supplied by the manufacturer, particularly with regard to receiver position and earthing to the vehicle chassis. An unsatisfactory installation in respect of high-frequency circulating currents set up by poor earthing connections, high-impedance supply lines to the equipment, and the use of inferior aerial feeder cable with poor screening can create many problems when undertaking interference suppression. With correctly installed equipment on the other hand, it is very much easier to identify each source of radio interference and determine how the interference reaches the equipment.

Location of Interference Sources

It is important to first ensure that no external source (see p14.3) or receiver malfunction is responsible for interference at the locations where an attempt is being made to diagnose vehicle equipment as the source of radio interference. For example, an apparently poor signal-to-noise ratio in mobile fm receivers is often caused simply by one or more receive channels not being accurately aligned.

Once this is done, the next step is to check that the vehicle ignition system is operating satisfactorily. With the engine switched off, check:

(i) that the spark plugs are clean and adjusted to the vehicle manufacturer's recommended gap. Wide gaps, especially when electrodes are worn and uneven, require an abnormally high ignition voltage which will generally radiate higher levels of interference and may affect engine performance;
(ii) that the distributor cap, rotor arm and points are clean and properly adjusted, and that the spark quench capacitor is connected. Never file or scrape a rotor tip when it is worn—replace it;
(iii) that there is continuity in the ignition secondary circuit (usually a few hundred ohms). An ignition coil with poor connections or a break in the secondary circuit will often work satisfactorily as far as the engine is concerned but will spark and *this cannot be suppressed;*

TABLE 14.2

Determination of major sources of vehicle interference. Vehicle stationary, receiver on with aerial disconnected.

Test conditions	Interference sound on receiver	Source
1. Ignition off	Regular ticking	Electric clock
2. Ignition on	Intermittent ticking or whine Slow crackling	Fuel pump Voltage stabilizer
3. Engine ticking over	Loud crackling proportional to engine speed	Ignition system
4. Engine speed increased (dc charging systems)	Whine proportional to engine speed Low-frequency crackle constant throughout charging speeds	Dynamo Control box
5. Engine speed increased (ac charging systems)	Whine varying with speed and load High-pitched whine constant throughout charging speed range	Alternator Alternator control
6. Wiper motor on	Whine in unison with wiper motor movement	Wiper motor
7. Heater motor on	Whine when heater is in operation	Heater motor

(iv) that all end terminals in the resistive ignition cable are tight. Never pull or strain these connections. Never cut a resistive cable to fit a screw-in suppressor. Never attempt to repair a loose end terminal—replace the lead.

These checks should ensure that the ignition system is in good order. Malfunctions in other electrical equipment or wiring may also cause high levels of interference and before and after fitting any suppression components to an item of equipment, it should be checked that it is operating satisfactorily.

The procedure shown in **Table 14.2** may now be followed to establish the major sources of interference. Any interference heard should be dealt with before moving on to the next step. Details of suppression procedures for the various items of vehicle equipment listed will be found in the following sections.

When the interference has been suppressed to an adequate level reconnect the aerial and start the engine. Any interference now heard is due to re-radiation into the aerial, probably from vehicle parts with poor earth connections, and should be dealt with by bonding (see p14.9).

Interference heard only while the car is on the move may be due to wheel static (see p14.10).

Methods of Interference Suppression

Since 1952 all new motor vehicles manufactured in the UK have been fitted with suppression components where necessary to comply with the statutory regulations, and more recently the relevant EEC regulations.

The suppression level to meet the above requirements ensures that the radiated interference from the vehicle will not cause undue interference to licensed television receivers operating in private dwellings. At this level very little interference is caused to domestic receivers operated in the long and medium wavebands. However, there will be few

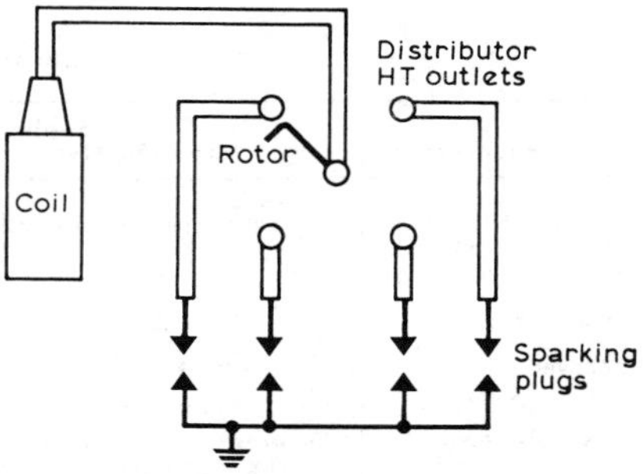

Fig 14.3. Typical arrangement for high-voltage ignition suppression using resistive or inductively wound cables.

amateur radio mobile installations where the level of interference from certain items of electrical equipment in the same vehicle will not warrant the fitting of suppression components. Moreover, these installations cannot be predicted with any certainty as conditions may vary between vehicles of a particular model.

The more involved suppression problems arise when the radio receiver is subjected to interference from several sources; for example, interference from the ignition system may be heard simultaneously with that from the screen-wiper and the instrument voltage stabilizer. It is therefore preferable to deal with the sources in a logical order, as explained above, and adopt a separate suppression scheme for each item of equipment. It is also preferable to arrange for most electrical equipment to be disconnected while assessing the suppression required by the ignition system; afterwards each piece of equipment can be separately reconnected to the supply and checked.

In general the aim is to reducc interference from each offending item of electrical equipment through the fitting of suitable suppression components as shown in **Table 14.3**, so that the ratio of signal to interference at the receiver is better than 30dB (approximately 30:1). Suppression components

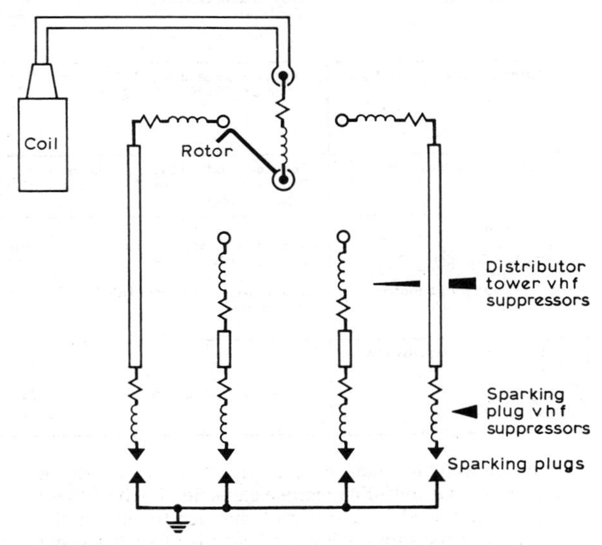

Fig 14.4. Typical arrangement for improved suppression at vhf using wirewound resistors and resistive cable.

TABLE 14.3

Methods of interference suppression

Equipment to be suppressed	Suppression components: Up to 30MHz	Suppression components: 30–400MHz	Comments
Ignition system			
HT cables (coil-distributor and distributor-sparking plugs)	Resistive core ht cable (Fig 14.3)	Resistive core ht cable (Fig 14.4)	Standard fitting on most production vehicles.
Sparking plugs		Screened plug-suppressor (Fig 14.4)	Screened wirewound resistor (Fig 14.15).
Distributor ht towers		Unscreened distributor suppressor (Fig 14.4)	Unscreened wirewound resistor (Fig 14.17).
Distributor ht cover		Metallic shield (if required)	A cylindrical metallic shield mounted to surround the distributor cap. It is attached to the metallic body of the distributor.
Ignition coil supply terminal ("SW" or "+")	Suppression capacitor 1μF 250V dc wkg (Fig 14.5)	None usually required: otherwise feedthrough suppression capacitor 1μF 250V dc wkg (Fig 14.5)	Suppression capacitor secured by earth terminal and ignition coil fixing bolt.
Charging system			
DC generator (dynamo)	Suppression capacitor 1-3μF 150-250V dc wkg (Fig 14.12)	Feedthrough suppression capacitor 1μF 250V dc wkg (Fig 14.12)	Fit close to "D" or output terminal of generator.
Alternator	Suppression capacitor 1-3μF 150-250V dc wkg (Fig 14.11)	None usually required	Fit in recess provided on slip-ring end casting. Connect between "+" output terminal and earth.
Control boxes for dc generators			
RB106-2	Filter unit	None usually required	
RB310	Filter unit	Special order only from manufacturer	Filter units available only from manufacturer of control box.
RB340	Filter unit	Special order only from manufacturer	Filter units available only from manufacturer of control box.
Electronic regulators for alternators			All electronic regulators incorporate some degree of rf suppression.
Types built-in to alternator	Suppression capacitor 1-3μF 150-250V dc wkg (Fig 14.11)	None required	Fit in recess provided on slip-ring end casting. Connect between "+" output terminal and earth.
Externally mounted types	Suppression capacitor 1μF	None required	Connect across "+" and "−" regulator terminals. Fit short earth lead to "−" terminal.
Screen wipers			
Permanent magnetic field	Suppression capacitor(s) 1μF (Fig 14.8)	Ferrite-cored inductors 30μH 7A (Fig 14.8)	Inductors fitted in supply leads. Capacitors connected between brushes and earth.
Wound field	Suppression capacitor(s) 1μF	Ferrite-cored inductors 10μH 3A	For frequencies up to 30MHz earthing straps may be required on motor body assembly.
Electric clocks (continuously driven types)	Suppression capacitor 1μF (Fig 14.10)	Ferrite-cored inductor 10μH 3A (Fig 14.10)	Fit components close to clock supply terminal: capacitor in parallel and inductor in series with clock.
Screen washer (pm field)	Suppression capacitor 1μF (Fig 14.13)—if required	Ferrite-cored inductors 10μH 3A (Fig 14.13)	Fit components close to motor terminals, capacitor in parallel and inductors (usually two) in series with motor.
Heater fan motor	Suppression capacitors 1μF (Fig 14.6)	Ferrite-cored inductors 30μH 7A (Fig 14.6)	Fit components close to motor terminals; inductors (usually two) in series with motor and capacitors between brushes and earth.
Instrument voltage stabilizer	Suppression capacitor 1μF and/or inductor 200μH 0·5A (Fig 14.7)	Ferrite-cored inductors 10μH 3A (Fig 14.7) and 5,000pF (maximum) capacitor	1μF suppression capacitor is connected between "B" terminal and earth, and inductors in "B" and "I" leads. The 5,000pF capacitor is connected across stabilizer contacts.
Oil pressure transducer	Ferrite-cored inductor 200μH 0·5A and (if required) a 0·2μF capacitor (Fig 14.9)	Ferrite-cored inductor 10μH 3A (Fig 14.9)	Fit components close to transducer terminals, inductor in series and capacitor in parallel with transducer.
Petrol pump			
Solenoid types	Suppression capacitor 1μF	None usually required	Fit capacitor close to pump supply terminal.
Immersed, motor driven	None usually required	Ferrite-cored inductors 30μH 7A (Fig 14.6)	Inductors connected in each supply lead and close to motor.
Petrol injection pump motor	Suppression capacitors 1μF (Fig 14.6)	Ferrite-cored inductors 30μH 7A (Fig 14.6)	Fit components close to motor terminals, inductors in each supply lead and capacitors between brushes and earth.
Other electrical equipment			
Starter	None usually required	None usually required	Suppression of items listed not usually attempted due to their low duty requirements.
Horn	None usually required	None usually required	Suppression of items listed not usually attempted due to their low duty requirements.
Flasher unit	None usually required	None usually required	Suppression of items listed not usually attempted due to their low duty requirements.
Switches	None usually required	None usually required	Suppression of items listed not usually attempted due to their low duty requirements.

vary considerably in their characteristics, and consist of resistive, inductive and/or capacitive elements. Each of these elements is employed in a different manner to obtain the desired interference suppression, and each is of such a value that optimum performance of the suppression component occurs in one of the two frequency ranges 0·15–30MHz and 30–400MHz.

For example, resistive or inductive suppressors having high ohmic values are inserted into ignition plug leads (**Figs 14.3** and **14.4**) to attenuate the high-frequency transient

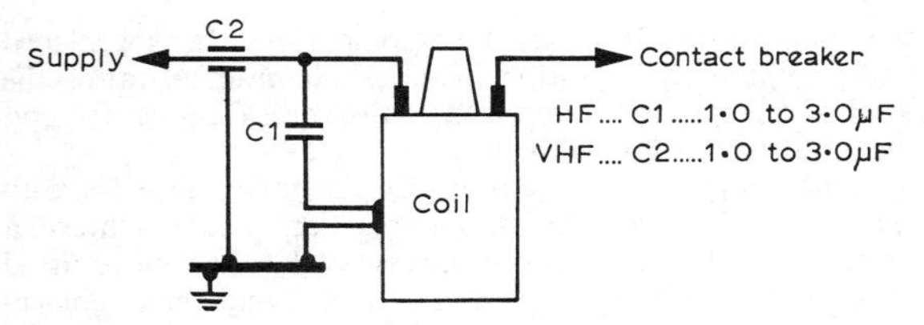

Fig 14.5. Arrangement of shunt and feedthrough vhf capacitors for ignition coil primary circuit.

currents responsible for interference. Such a suppressor will have little or no effect on the ignition function. On the other hand, where electrical circuits conduct heavier currents (eg battery charging systems) the introduction of such high resistance would be unacceptable. Instead, low-resistance inductors constructed on iron cores are used, designed to present a high impedance to either low-frequency currents (for example, due to alternator ripple) or rf currents which otherwise would flow into the vehicle's wiring system or radio equipment.

The capacitor presents a low impedance path to high-frequency currents and is of great value in restricting their flow to small non-radiating circuits (**Fig 14.5**).

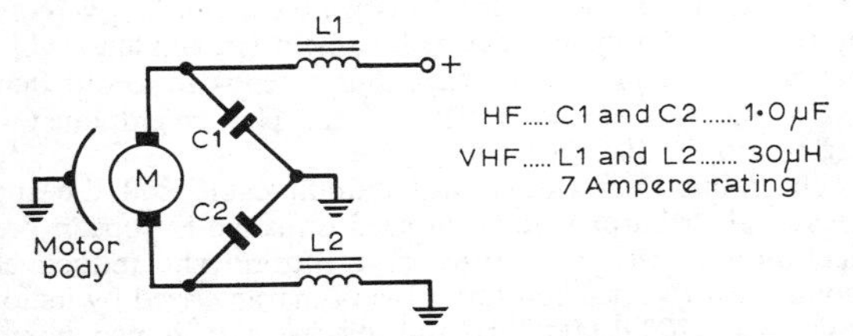

Fig 14.6. PM fan motor, electric petrol injection pump motor.

Circuits giving high attenuation of radio interference over certain frequency ranges can be constructed utilizing the characteristics of resistive, inductive and capacitive suppressors. These circuits are known as rf filters, examples of which are shown in **Figs 14.6–14.9** and also **Fig 14.15**.

When selecting such suppression components as inductors wound on ferrite cores, the current rating of the equipment to be suppressed is a guide to the size of suppressor required. Many of the commercially available inductors designed to be resonant at vhf communication frequencies, however, may have a greater current rating than that of the equipment with which it is intended to operate. Such suppressors produce an attenuation of interference of some 30dB when mounted close to the terminals of the offending equipment (Figs 14.6–14.9 and also **Fig 14.10**).

The complexity of suppression components and their arrangements will depend upon the interference frequencies,

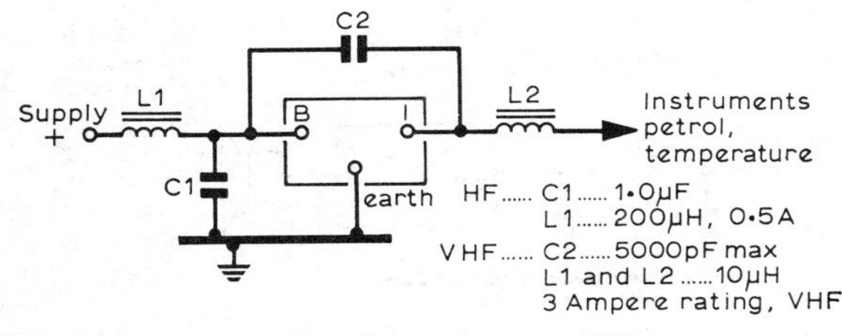

Fig 14.7. Instrument voltage stabilizer.

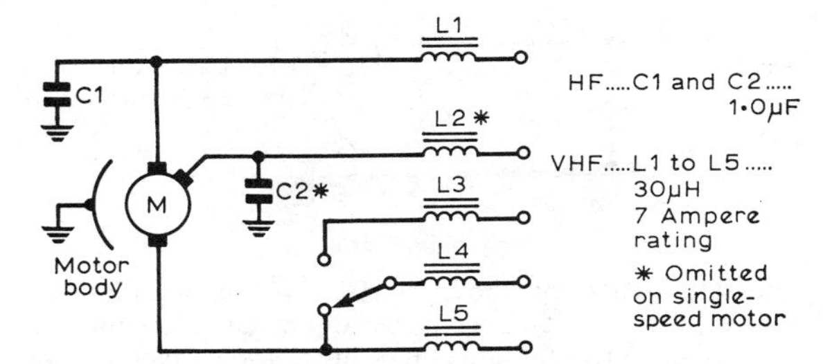

Fig 14.8. PM screen wiper, two-speed with regenerative braking.

the magnitude of the interference and the lowest value of radio signal which the receiving equipment can be expected to resolve. *It must always be ascertained that the suppression component does not impair the function of the equipment or circuit to which it is fitted.*

A reliable indication of the effectiveness of each suppression scheme can be made by detuning the main receiver to the upper or lower sidebands of the received signal to an extent where some considerable distortion occurs in the loudspeaker. When detuning is not possible, some indication of the suppression level achieved can usually be made by manually switching out the audio muting (squelch) circuit and advancing the volume control towards maximum.

A reliable assessment of the effectiveness of supply line filtering can be made by substituting the aerial with a dummy one inserted into the receiver aerial socket, and turning the audio gain control to near maximum. Some receiver supply line filters may need to be improved by fitting either an inductor and possibly a capacitor, or a pi-section filter (**Fig 14.14**) outboard of the receiver but close to the supply inlet terminal. Supply line filter components should not be mounted within the receiver housing unless they are adequately screened from other components and circuits in the receiver.

Ignition system

In some instances the amount of built-in suppression may be adequate for reception of hf and vhf signals in the vehicle but this depends on how well the vehicle bodywork is assembled to form an electrical screen around the ignition circuits. However, experience shows that more often than not the ex-works level of suppression needs to be improved, particularly where weaker hf and vhf signals are encountered.

Most vehicle manufacturers fit as initial equipment a resistive-type cable with distributed capacitance which operates as an RC filter network to produce a high attenuation of rf oscillatory currents. **Fig 14.16** shows resistive suppression cable to have a rising impedance characteristic with increasing frequency. This can be compared with the impedance characteristic of the lumped composite carbon type of suppressor, also shown in Fig 14.16, which falls with

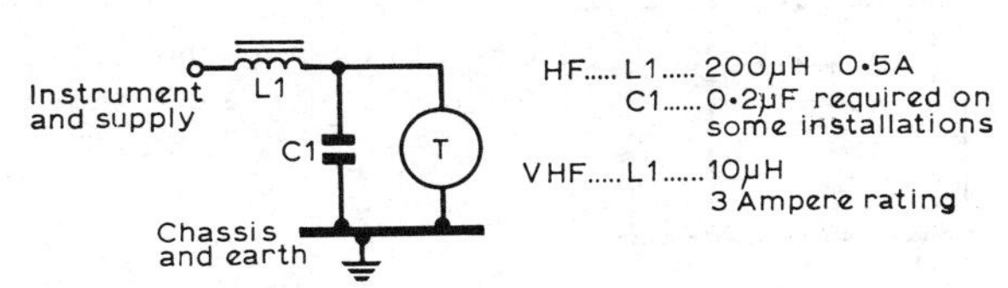

Fig 14.9. Oil pressure transducer.

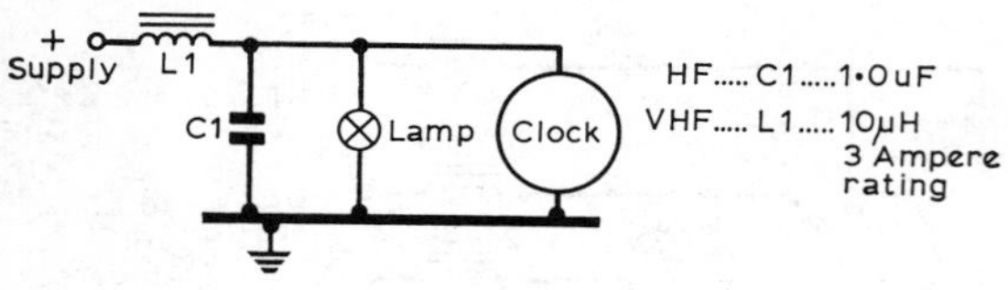

Fig 14.10. Clock.

increasing frequency above 40MHz. It follows that this latter suppression component, usually made in the form of spark plug "elbow" units and distributor "plug-in" or "screw-in" units, is generally unsatisfactory for suppressing vhf and any uhf interference, but can be useful on the hf bands.

Inductively wound resistors assembled into screened units for sparking plugs (Fig 14.15) and "push-in" units for distributor ht outlets (**Fig 14.17**) are further types of ignition suppressors. The impedance characteristics of these types is also shown in Fig 14.16. It will be seen that they give maximum attenuation of interference in the vhf broadcast frequency range, and adequate protection for frequencies up to approximately 300MHz. If suppression units of this design are used in conjunction with resistive cables, the arrangement exhibits wide-band attenuation characteristics of interference frequencies in the long-wave, medium-wave, hf and vhf bands. If copper-cored ignition cable is employed instead of the resistive type, a less effective result is obtained, although in most instances the vehicle will meet regulation standards at frequencies of 40–250MHz.

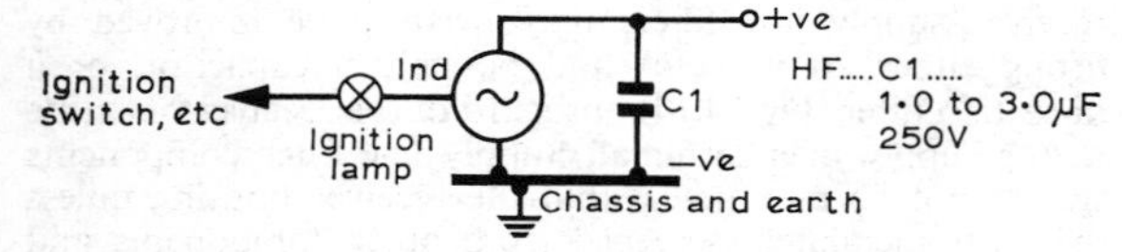

Fig 14.11. Alternator.

On no account should the lumped composite carbon-type suppressors be used with copper-cored ignition cable because their effective impedance reduces with increasing frequency. In some cases the radiated ignition interference from the vehicle will meet mandatory requirements (ie ECE Regulation No 10), but it is more usual for the specified limits at 90MHz and above to be exceeded. Also, the increase in radiation generally creates added problems for the satisfactory installation of hf, vhf and uhf radio communication equipment on the vehicle.

A common fallacy which exists in the motor vehicle service trade is that the increased ht resistance resulting from the use of resistive cables, and/or the insertion of suppressors in resistive cables, produces a deterioration in the performance of the ignition system and hence in that of the engine. Consequently, it is not uncommon to find that the manufacturer's original equipment has been replaced with copper-cored ignition cable. Such practice raises the level of radiated interference from the vehicle concerned and may infringe the law in the UK.

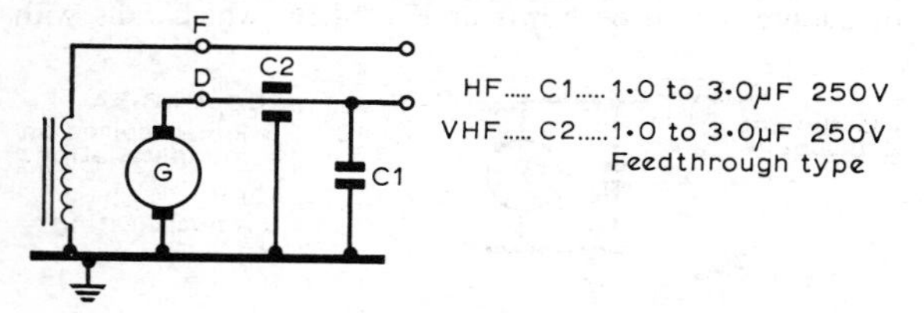

Fig 14.12. Dynamo.

Some vehicles are now fitted by the manufacturers with spark plugs having an internal resistor. These achieve a better level of interference suppression than the more usual plug/suppressor plug cover combination, and their replacement by non-resistor types may again increase the level of radiated interference beyond the statutory level.

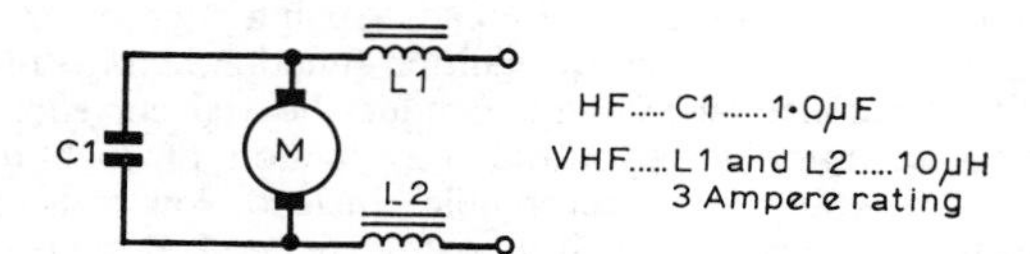

Fig 14.13. PM electric screen washer.

It is often found that ignition interference is evident even when suitable suppression methods are applied to ignition sparking plugs and their connecting leads. In many vehicles this can be shown to be due to radiation from the distributor on account of the jump-gaps between the rotor arm and outlet tower electrodes. This problem can be overcome by the use of a cylindrical metal shield (eg tinplate) which surrounds the distributor cap, and extends to about 0·5in above the top of the outlet towers. This shield must be earthed to the distributor base plate.

The inductively wound distribution cable and the suppressed distributor rotor arm used on some European production vehicles are worthy of mention. The inductively wound resistive distribution cable is characterized by its low resistance (200–1,000Ω/ft) and by its high impedance at frequencies of 30–300MHz (Fig 14.16). The cable has therefore found use where vhf and uhf communication equipment is carried on the vehicle as well as being extensively employed to meet legislative limits. Its low resistance is usually a disadvantage for suppression of interference at frequencies in the long- and medium-wave and hf bands, when it is necessary to introduce some additional impedance in the form of composite carbon suppressor units at the sparking plugs and distributor ht outlets. The suppressed distributor rotor arm has an inductively wound resistor mounted in the rotor moulding so that it is electrically in series with each distributor spark gap in turn during rotation. The suppressed rotor arm is used in conjunction with resistive cables to obtain improved suppression of vhf and hf interference emanating from the ignition distributor. However, neither is generally available in the UK.

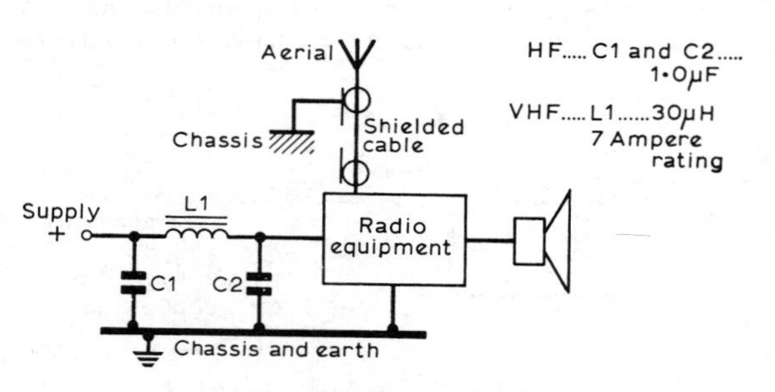

Fig 14.14. Radio equipment.

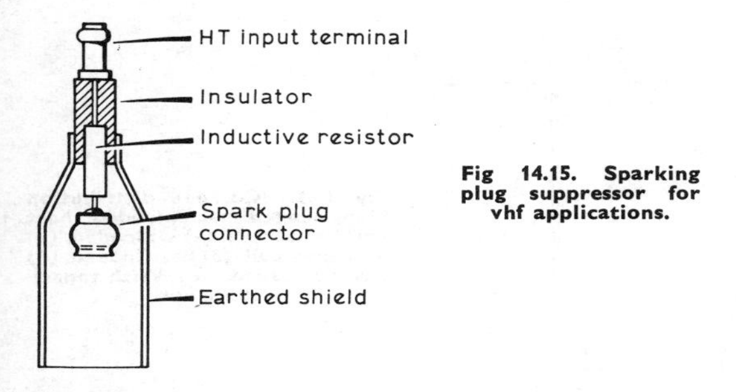

Fig 14.15. Sparking plug suppressor for vhf applications.

Ancillary equipment

Figs 14.5–14.14 show the suppression components it may be necessary to fit to various items of ancillary electrical equipment if both hf and vhf communication equipment is to be operated from the same vehicle. If either hf or vhf equipment only is to be operated from the same vehicle then not all the suppression components indicated in each figure will be required. Instead, it will be necessary to fit only those components which provide the appropriate hf or vhf suppression as specified in Table 14.3.

Suppression techniques used for electric motor-driven equipment are illustrated in Figs 14.6, 14.8, 14.10 and also **Fig 14.13**, and those for alternators and dynamos in **Figs 14.11** and **14.12** respectively. Alternators having built-in electronic circuits rarely cause interference problems at vhf although there may be diode commutation and regulator noise over the frequency range 0·15–10MHz. Diode commutation noise may also be present in the audio frequency range. The suppression arrangement shown in Fig 14.11 takes account of these eventualities, attenuating any interference at the above frequencies that may enter intermediate and audio frequency stages of a receiver by one or other of the routes suggested earlier.

Figs 14.7 and 14.9 show methods of suppressing such contact-operated devices as instrument voltage stabilizers and oil pressure transducers.

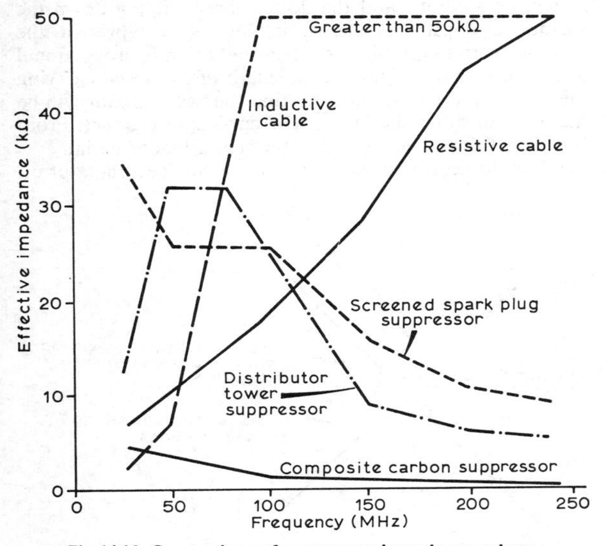

Fig 14.16. Comparison of suppressor impedance values.

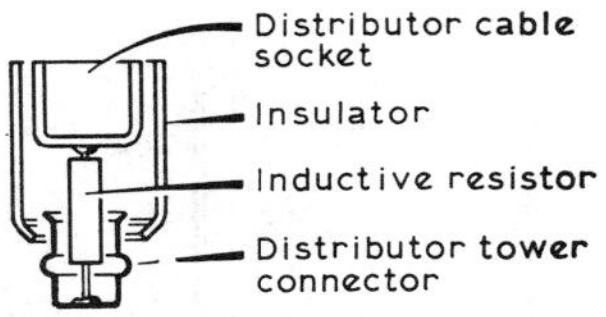

Fig 14.17. Distributor spark suppressor for vhf applications.

All inductors and capacitors used for suppression purposes should be connected as near to the source of interference as possible. In particular the leads of capacitors must be kept very short in order to keep the resonant frequencies of capacitor circuits as high as possible **(Fig 14.18)** and thereby obtain the best possible attenuation. The value of a feed-through capacitor for vhf and uhf suppression as opposed to a shunt type can be seen in Fig 14.18 where their frequency characteristics are compared. Each capacitor has a value of 1μF.

Some tachometers operate from ignition pulses picked up from the ignition coil and the wiring to the instrument radiates into wiring in the rest of the harness on the way to the instrument panel. This interference can only be suppressed by shielding the wire or fitting it with a low-capacitance screen. Some tachometers will not, however, maintain accuracy if the pick-up wire is screened. Other cases of interference introduced in the low-voltage wiring can often only be cured by rearranging the wiring to reduce the pick-up.

Vehicle body parts

Re-radiation of locally generated interference by various parts of the vehicle making poor electrical contact can often be suppressed by bonding at the following points:

(a) Corners of engine to chassis.
(b) Bonnet at both sides (ensure hinge and fastener are not insulated).
(c) Boot lid—as bonnet.
(d) Exhaust pipe to chassis and engine.
(e) Coil and distributor to engine frame.
(f) Battery earth to chassis.
(g) Air cleaner to engine frame.
(h) Steering column.
(i) Alternator to engine block.

Copper braiding of at least the size obtained from the outer of ½in coaxial cable should be used, together with tooth-type locking washers at each connection to ensure a good permanent contact.

Sometimes surfaces rubbing together produce static, and the cure for this is bonding to earth as described above.

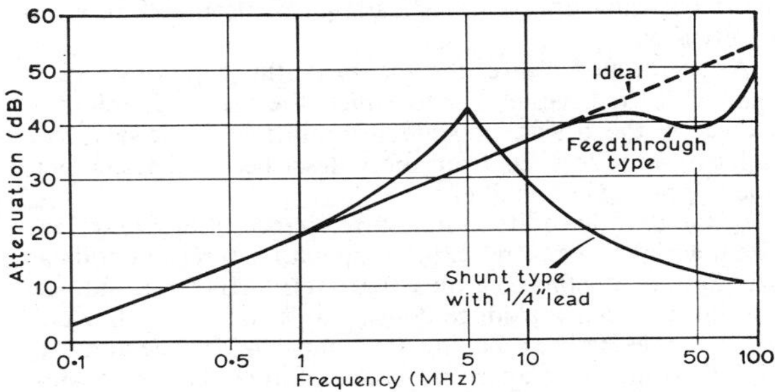

Fig 14.18. Suppression capacitor frequency characteristics.

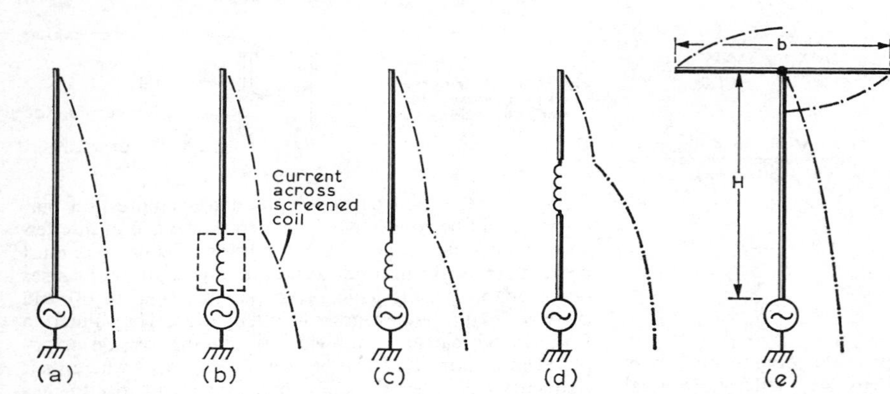

Fig 14.19. Current distribution along loaded and unloaded short vertical aerials. (a) Unloaded. (b) Screened coil. (c) Base loaded. (d) Centre loaded. (e) With capacitance hat.

Wheel static

Wheel static discharge can be suppressed on the front by installing a spring wiping device inside the hub cap which makes good electrical contact between the wheels and the axle. Somewhat similar devices which make electrical contact between the rear wheels and the chassis can be used.

A different type of "wheel static" which cannot be suppressed by earthing the wheels is sometimes encountered. This can be a static discharge within the tyre itself and can be eliminated by introducing a carbon powder inside the tyre or inner tube.

Coasting a car downhill with the engine off can often be the best way of identifying static type interference. However, a private road should of course be used for this exercise.

MOBILE AERIAL INSTALLATIONS

The success of any transmitter not only depends on the power which it has available, but the ability of the aerial system to radiate this power efficiently. A station with a 50W capability may in fact radiate a weaker signal than that achieved by a 10W station, simply due to an inefficient aerial.

The following discussion will mainly concern only those theoretical and practical aspects of aerials which are peculiar to mobile operation; readers are referred to Chapters 12 and 13 for the basic theory of aerials and transmission lines.

LF and HF Aerials

When operating on frequencies of 28MHz and above it is practical to use a $\lambda/4$ vertical whip, but on the lower frequencies such aerials are impractical due to their size. A short vertical aerial with a loading coil is therefore commonly used.

As mentioned above, the theory of this type of aerial is dealt with in Chapter 12, to which the reader is referred. However, the design of suitable mobile whip aerials does differ in some respects from those described in that chapter, due to practical difficulties.

In the first place the radiation resistance of any practical-size lf aerial is very low—a 6ft whip with a screened coil has a radiation resistance of only 0·2Ω at 3·5MHz. Second, the loading coil can introduce considerable losses—such a coil for the 3·5MHz band having a reactance of 1,400Ω and a *Q* of 140 would have an rf resistance of 10Ω. Third, because the feed-point impedance is low, a substantial proportion of the transmitter power may be lost in the matching system unless care is taken.

The position and construction of the loading coil affect all three of these factors, and this vital component will now be considered in more detail.

Loading Coil Position

The position of the loading coil of an lf vertical aerial has a pronounced effect on the radiation resistance (**Fig 14.19**). Where the coil is placed at the base of the aerial, and is fully screened as in Fig 14.19(b), the radiation resistance can be of the order of 0·2Ω. If the coil is now made a component part of the vertical system and allowed to radiate, the radiation resistance of the system rises to about 2Ω. This is illustrated in Fig 14.19(c). If the coil is now positioned in the centre of the whip as in Fig 14.19(d) both the lower section of the whip and the coil will have current flowing in them, and this again raises the radiation resistance, this time to about 4Ω. These figures relate to an aerial operating on 3·8MHz.

As is to be expected, the position of the loading coil has a pronounced effect on the current distribution in the aerial system as a whole, and the dotted lines running down the various arrangements shown in Fig 14.19 represent the current distribution. Since the field radiation is proportional to the current flowing, and the length of the aerial carrying this current, it will be apparent that the best radiator will be the one which has the highest current-length product. From Fig 14.19 this will be seen to be the centre-loaded aerial.

While the centre-loaded aerial may be the best radiator on

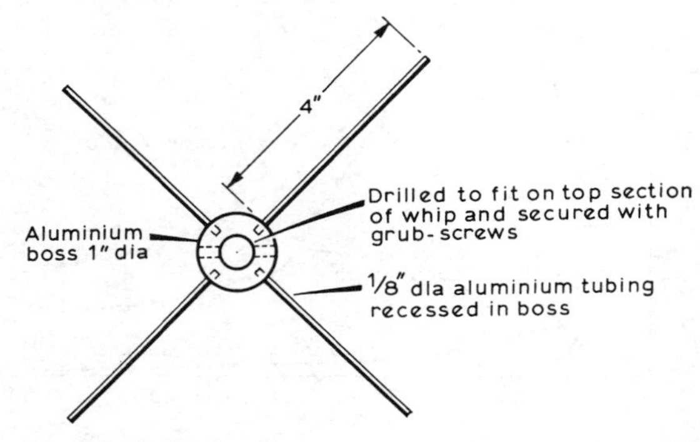

Fig 14.20. A capacitance hat suitable for vertical whip aerials.

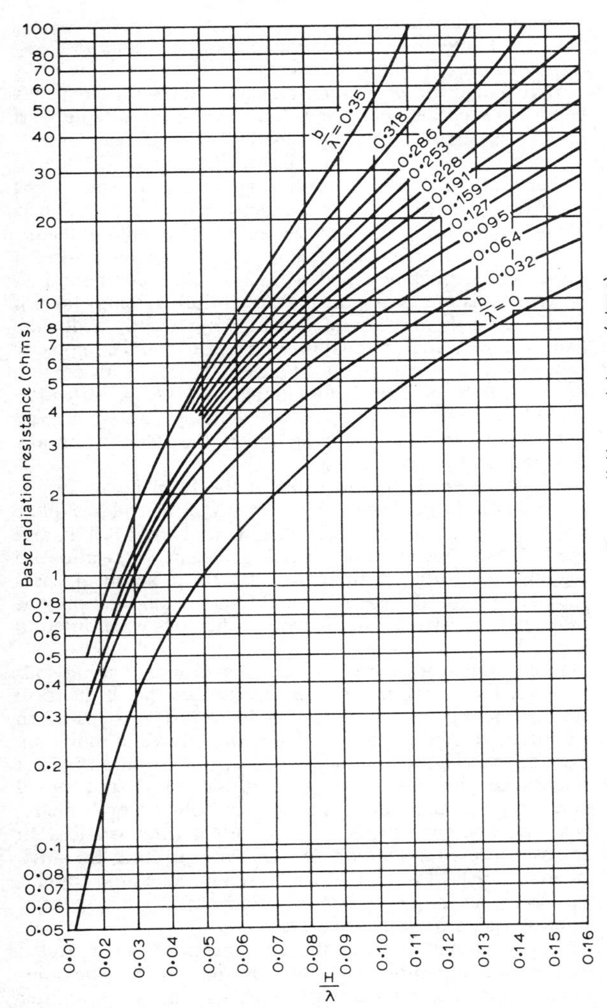

Fig 14.21. Base radiation resistance (Ω) of vertical aerials up to 0·16λ long with non-radiating top termination. Height of aerial = H, length of top = b. The special case of b = 0 represents an unloaded short vertical aerial.

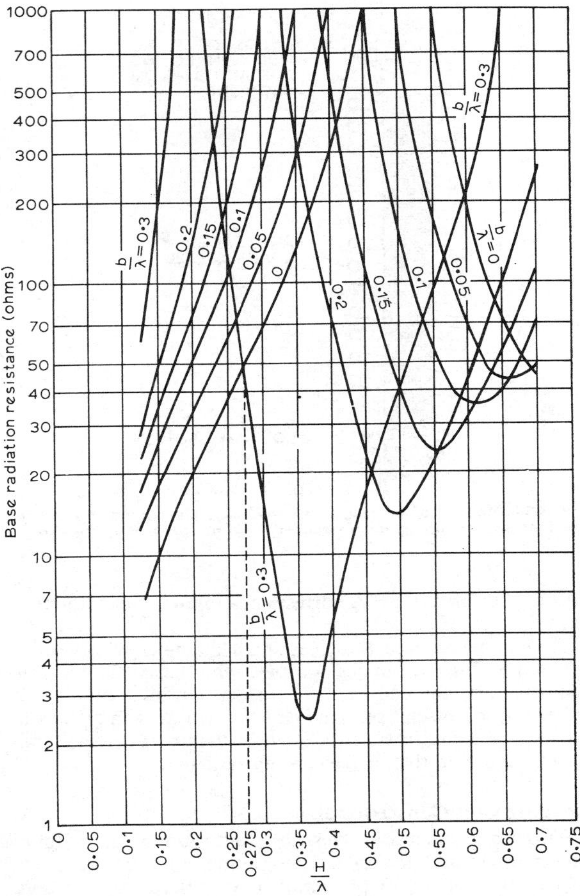

Fig 14.22. Base radiation resistance (Ω) of vertical aerials up to 0·75λ long, with non-radiating top termination.

the lower frequency bands, particularly 1·8MHz and 3·5MHz, mechanical considerations may influence the loading coil position. On these two bands it is more usual to employ a base-loaded whip. However, as will be seen later, it is possible to introduce some compensation by attention to the form of the loading coil.

Actually the statement that centre-loaded whips are the best radiators is not strictly correct since, for a given whip length and depending on frequency, there is an optimum position for the loading coil. The radiation resistance of an aerial, but more important its capacitance, are not things confined to one point on the whip, but rather are spread along its entire length.

When the coil is placed at the base of the whip it is resonated by the total capacitance of the whip but as it is raised the capacitance of the section of whip decreases. As the capacitance decreases, so the loading coil size has to be increased to restore resonance. Increasing the size of the loading coil also increases the losses in the coil, and eventually as the coil is further raised the law of diminishing returns takes over and the current-length product falls.

This effect may be offset to some extent by deliberately introducing capacitance into the top section of the whip above the coil. Such a method uses what is commonly described as a *capacitance hat*, and this device is shown in **Fig 14.20**. The current distribution in short vertical aerials with capacitance hats is shown in Fig 14.19(e). By introducing capacitance in this manner, the size of the loading coil may be reduced, and its losses also. However this system is not without its disadvantages. As capacitance is added above the coil, so the aerial becomes increasingly frequency selective. As an example, with a centre-loaded coil on 3·5MHz it may be possible to vary the transmitter frequency over about 25kHz before it becomes essential to re-adjust the loading. If this coil is replaced by another which is tuned by a capacitance hat, a frequency shift in excess of 10kHz may well entail retuning the aerial.

Base radiation resistances for varying aerial heights (*H*) and capacitance hat lengths *b* are given in **Figs 14.21** and

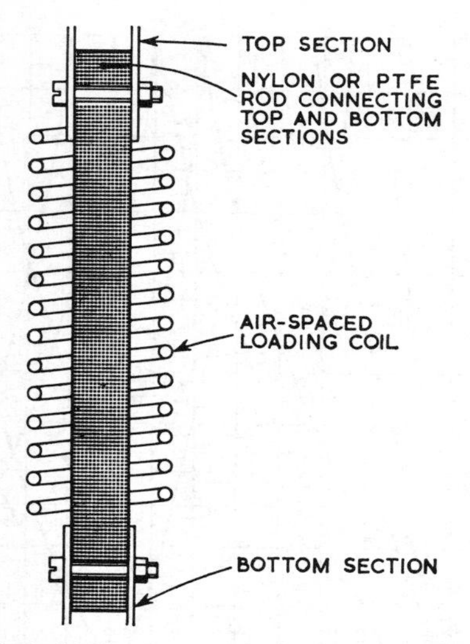

Fig 14.23. Method of bracing and connecting the upper and lower sections of a whip when air-spaced loading coils are employed.

14.22. The curve $b/\lambda = 0$ corresponds to an aerial without a capacitance hat.

It should be noted that if a small diameter wire with a pointed end is used at the top of a whip aerial the voltage gradient near the end of the wire may be high enough to cause corona discharge. This can be seen as a faint bluish glow due to ionization of the air near the tip of the wire and it will modulate the rf signal being radiated.

Loading Coil Construction

This is a key factor in the success or failure of any hf (and especially lf) mobile aerial. The loading coil must have the highest possible Q, or put another way, the resistance and losses of the coil must be as low as possible. As the Q of the coil is increased for a given frequency, so the losses in it will decrease and the strength of the radiated signal will increase.

Ideally, and to reduce dielectric losses, the loading coil should be self-supporting, but from the point of view of achieving a high degree of mechanical strength this is rarely possible except perhaps on the 14MHz and 21MHz bands. Where such a course is adopted, the strength of the actual whip must be maintained through the length of the coil winding. The most satisfactory method of achieving this is shown in **Fig 14.23** and consists of inserting an insulated rod into the bore of the whip top and bottom of the distance occupied by the loading coil, and running this rod through the centre of the coil. As the rod runs through the centre of the coil it must have a low dielectric loss, and ptfe or nylon are the most suitable materials.

It is essential to protect any loading coil from the effects of moisture, and where the coil is a self-supporting unit this is best accomplished by enclosing it in a shield constructed from polythene and which is fully sealed. The shield should clear the turns of the coil by about ½in. Freely available polythene bottles make a good basis for such shields.

Where formers are employed for the loading coil, ideal materials are again nylon, because of its high mechanical strength, or ptfe. Paxolin can be used on 1·8MHz always provided that its relatively high dielectric losses are acceptable. Using the first two mentioned materials, coils with a Q in the region of 400 are practical.

Wire gauge will be a compromise on the lower frequency bands between the final size of the coil and the acceptable rf resistance. On 1·8MHz the smallest wire gauge contemplated should be 18swg; on 3·5MHz 16swg; and 7MHz and higher 12swg. Whenever possible the wire should be Formvar insulated. This is a tough plastic coating impervious to moisture and, unlike enamel, will not chip or flake.

After the coil has been finally adjusted to resonance, it must be thoroughly doped with a suitable compound to prevent moisture getting at the turns. Polyurethane varnish seems to be the most suitable material currently available, and by using a coat-on-coat process, a tough and completely waterproof layer can be built up over the coil. This coating must include the coil terminations, and special attention should be paid to the top of the loading coil which will receive all the water running down the whip when the vehicle is stationary.

Finally, there is the question of the form factor of the loading coil. For optimum Q the length of the coil should be about twice its diameter. This means that the 1·8MHz and the 3·5MHz bands pose special problems in relation to loading coils for use with an 8ft whip, for in nearly all cases the size of a coil designed for optimum Q would not only be difficult to construct, but also look ridiculous when fitted to a vehicle.

In a number of experiments conducted with loading coils for these bands, the conclusion was reached that if the form factor requirement was disregarded and instead steps taken to produce a coil which exhibited a high degree of radiation, then the overall radiation efficiency was improved in quite a remarkable manner. These conclusions were not based solely on the readings of an adjacent field strength meter, but were also confirmed by stations at varying distances. In many respects the loading coil appeared to have the attributes of a helical whip, an aerial system at one time popular on the higher frequency bands but now mainly used for vhf hand-portable equipment (see later).

As the radiation from the coil increases, as indicated earlier, the radiation resistance of the whole system increases, and this in turn makes matching to the transmitter easier to accomplish, so reducing the losses which are inevitable where the feed impedance is very small. Quoting from the results of experiments with a whip tuned to 3·750MHz, the measured impedance was 16Ω and the calculated overall efficiency of the aerial 60 per cent. This used a coil 1½in in diameter and approximately 16½in long wound with 155 turns of 16swg, having a measured Q of 200. The coil was positioned 12in up from the feed point. Compared to a coil wound to the optimum form factor with a measured Q of 400, the long loading coil produced almost twice the field strength.

Although the effects of centre loading with such coils was not evaluated, it is evident from the base-loading experiments that this line of approach should be fruitful. It seems likely that the performance of a particular commercial centre-loaded whip, in which the loading coil has a long length compared to its diameter, arises from the radiation of the coil.

Table 14.4 gives approximate details of loading coils for the lower frequency bands. The details must be approximate since every aerial system will need to be tuned in relation to

TABLE 14.4

Starting details of loading coils for various frequency bands and positions on the whip. Overall length of whip is 8ft.

Band (MHz)	Position	Dia (in)	Turns	SWG	Length (in)	Notes
1·8	Bottom	3	140	18	7	Q 300
1·8	Bottom*	1½	260	16	17	Max radiation
1·8	Centre	1¼	475	22	27	Max radiation
3·5	Bottom	2½	75	14	6	Q 300
3·5	Bottom	2½	55	18	2½	Q 350
3·5	Bottom*	1½	155	16	16½	Max radiation
3·5	Centre	2½	105	16	10½	Q 300
14	Bottom	2	10	14	1½	Q 380
14	Centre	2	16	14	2	Q 350
21	Bottom	1¾	6	12	2	Q 300
21	Centre	2	8	12	2	Q 280

* Coil mounted 12in above base of aerial.

the vehicle with which it is to be employed, for it is impossible to estimate such factors as earth-loss resistance and car capacitance.

Adjustment of Loading Coils

Just as spending time on the construction of the loading coil will be worthwhile, so the care taken in tuning the system will pay dividends.

Tuning to frequency will be made a great deal easier if the section of the whip above the loading coil is constructed so that the length can be varied. If this cannot be arranged, then a variable-size capacitance hat will almost certainly be required to achieve precise resonance on any particular frequency. The aerial should only be trimmed in its final position on the vehicle.

The lower end of the whip should be earthed to the car body through a two-turn link, the length of wire between the whip and the link being made as short as possible. A grid dip oscillator is then offered up to the link and tuned to find the resonant frequency. The coupling between the gdo and the link should be as small as possible otherwise some inexplicable dips may be found. During this operation, the top section of the whip should be extended to within about 6in of its full length.

If the frequency is found to be very low, turns should be removed from the loading coil. If on the other hand the frequency is found to be high, this indicates inadequate turns. Where Table 14.4 is used, the frequency will be low in each case, for the specifications are on the basis that it is easier to remove turns rather than have to rewind the coil.

Accuracy in the gdo is essential and, if it is an unfamiliar instrument borrowed for the purpose of setting up the whip prior to use, its calibration should be checked against a receiver of known accuracy, and over the frequency range of interest. A mental note can then be made of any calibration error.

As the resonant frequency of the whip approaches the low side of the desired frequency, the frequency of the gdo should be continuously monitored if possible on a receiver. Normally the receiving equipment employed in the vehicle may be used. When within 10kHz of the low-frequency side of the desired resonant frequency, adjustment of the length of the top section of the whip should permit the whip to be precisely resonated to the desired frequency. To increase the frequency of resonance, the top section of the whip is reduced in length.

Since the feed impedance of a loaded whip will be substantially lower than the normal transmitter output impedance of 50Ω, the base impedance of the aerial must be matched to the feed line impedance. Experience indicates that this is best accomplished by the use of a variable series loading coil as shown in **Fig 14.24(a)**. This should preferably consist of a roller wheel making contact with the turns which, by rotating the coil, travels along the winding. Units employing this principle are sometimes available on the surplus market. To construct such a "roller-coaster" is usually hardly a practical proposition for the average radio amateur, and this being the case, a tapped coil will probably have to be employed.

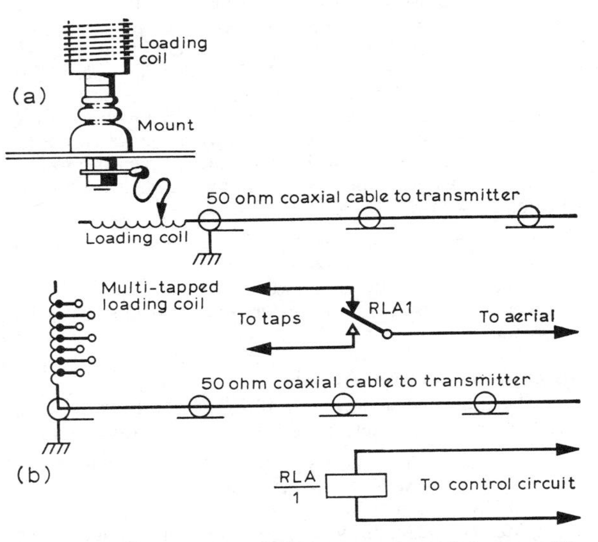

Fig 14.24. (a) The use of a series inductance to match the base of the aerial to the transmitter feeder. (b) For limited coverage a tapped inductance may be used in conjunction with a relay to adjust the inductance as the frequency of operation is changed.

If the coil is constructed from 16swg tinned copper wire, the turns of which are spaced by one wire diameter, connecting lugs may be soldered into place every one-third of a turn. Closely adjacent taps are essential to achieve correct loading and, while the coil will not be very beautiful when it is completed. it will be efficient. Typically such a coil consists of 30 turns on a 1½in dia former.

As the frequency of operation is changed from the resonant frequency of the whip, the whip should really be re-resonated by adjustment of the length of the top section. On 1·8MHz a change in excess of 10kHz may well reflect on the transmitter loading, and a change of 20kHz produce a serious drop in performance. To some extent such changes may be taken up in the loading system. With the tapped coil, it is possible to incorporate a relay actuated from the driving position which changes the tap on the coil and so compensates for the change in frequency. This is shown in **Fig 14.24(b)**. If this course is adopted, the control switch

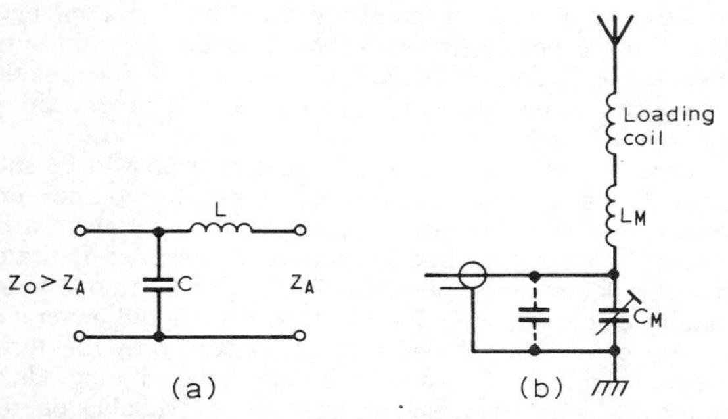

Fig 14.25. Use of a simple L network (a) to match a short vertical aerial to a feeder (b). Value of Cm is as follows: 160m, 2,000pF; 80m, 1,000pF; 40m, 220pF; 20m, 50pF. Cm is not required for the 10m and 15m bands. Cm can be made up of a mica compression trimmer in parallel with silver mica capacitors to the appropriate value.

is best marked with the actual resonant frequencies as a reminder of the range to which the aerial is tuned.

Matching a loaded whip to the feeder can also be achieved by means of a simple L network (**Fig 14.25**). The values of Lm and Cm can be calculated using the method described on p12.40. Lm can be combined with the loading coil if required. Cm can be made by using a mica compression trimmer in parallel with silver mica capacitors to the appropriate value. This type of matching network is adjusted as follows:

(i) Resonate the aerial with a loosely coupled gdo as described in the last section.
(ii) Adjust Cm for minimum swr.
(iii) Repeat (i) and (ii) until no further improvement can be made.

Installing LF and HF Aerials

In deciding the optimum position for mounting such aerials the following factors need to be taken into account:

(a) For maximum efficiency, the feed point should be as high as possible. However, adequate clearance must obviously be left for bridges etc or the garage.
(b) Mechanical stability requires a low fixing point.
(c) To avoid obstructing vision, the aerial should be at the rear of the vehicle.
(d) To avoid interference, the aerial should be as far as possible from the ignition system.

The best positions are clearly the rear wings or the rear bumper bar, and examples of both these types of mounting are shown in the photographs. Commercially made bumper mounts are available and can considerably lessen the risk of unsatisfactory mounting.

The aerial fixing must be mechanically strong enough not to be damaged when the vehicle is travelling at high speeds, or succumb to a blow caused by a low-hanging branch, or entering the garage before dismantling the aerial. Further, if the vehicle is not to lose some of its resale value, careful thought must be given to any idea of making fixing holes in the bodywork. Whatever fixing system is employed, it should be so arranged that the main support bracket can be removed, and the aerial system easily demounted.

Care should be taken that the aerial does not project horizontally from the vehicle (even at speed) so that it becomes a danger to other vehicles or pedestrians.

A centre-loaded whip for 3·5MHz: the G-whip Tribander with 3·5MHz loading coil.

VHF and UHF Aerials

On frequencies of 70MHz and above aerials become simpler since it is possible to fit a full $\lambda/4$ or even $5\lambda/8$ aerial. Under ideal conditions, and with the aerial mounted in the centre of the roof, it will act as a ground plane aerial, the feed impedance of which will be of the order of 35Ω. It will also be out of range of prying fingers. However, and understandably, there is usually considerable objection to cutting holes in the centre of the roof of a saloon car.

If a roof-mounted aerial is impractical, a $\lambda/4$ aerial may be mounted either on the apron surrounding the lid of the boot, or forward of the windscreen, without degrading the performance to any appreciable degree. If it is close to rising metalwork of the saloon, its polar diagram might be very distorted, and for this reason it should be positioned as far away from the rise of the saloon as possible. The polar diagram will still deviate substantially from a circle, but it will not have so many peaks and nulls in unexpected directions. A typical polar diagram of a 70MHz aerial installation is shown in **Fig 14.26**.

Sometimes it is necessary to fit an aerial without any drilling of the vehicle metalwork. There are four main alternatives possible—a gutter mount, a window clip mount, a magnetic mount or a boot-lip mount.

Gutter mounts take advantage of the fact that in most vehicles there is a gutter running the length of the roof above the window, and the mount is clamped to this ridge.

A window clip mount is useful for temporary mobile

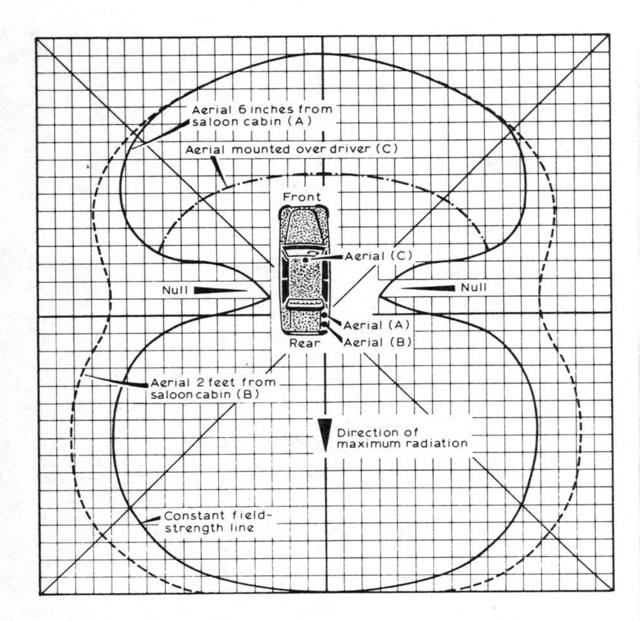

Fig 14.26. Polar diagram of a 70MHz λ/4 vertical aerial mounted on the rear of a small saloon car. Maximum radiation takes place towards the rear, but that towards the front is not greatly lower. Note the deep nulls off the sides of the vehicle when the aerial is within 6in of saloon cabin (aerial A). Moving it further back so that it is 2ft from cabin (aerial B) substantially reduces nulls. When the aerial is mounted above the driver (aerial C) the forward field becomes more restricted.

operation and is clamped to the top of the window pane.

Magnetic mounts utilize the holding power of a large magnet in their base. Such a device can provide a surprisingly strong fixing, often able to withstand motorway speeds, but there are three disadvantages to this type of mount. In the first place the efficiency is lower than that of a directly mounted aerial placed in the same position. This is due to the fact that the magnetic base must necessarily be insulated from the car body, and the optimum ground plane effect is not obtained. Second, although secure enough in normal circumstances, an emergency stop could conceivably cause the mount to break loose. Third, there is often an annoying tendency for the assembly to "walk" across the roof if the vehicle passes over rough ground.

A boot-lip mount is particularly useful when the transceiver is boot-mounted. It consists of a small clamp fitting on to the lip of the boot lid, with the cable feeder passing into the boot interior.

There is an additional problem common to all four types of non-drilling mount, and that is the question of the aerial feeder, which may have to go through an open window, thus rendering the car less secure when not in use, and also possibly letting rain into the interior. This problem can be minimized by using miniature coaxial feeder.

Where possible, 35Ω coaxial cable should be used to feed the whip as this will give an almost perfect match. As an alternative the length of the aerial may be adjusted to present an impedance of 50Ω at its base. As an aerial is made longer than λ/4, the base impedance rises, and at some point (about 0·275λ) can be made to look like 50Ω (Fig 14.22).

Two-aerial installation. Left: 70MHz λ/4 whip; right: 7MHz helical whip with a base-loading coil for 1·8MHz.

A halo aerial installation for 144MHz featuring the Jaybeam 2HM aerial.

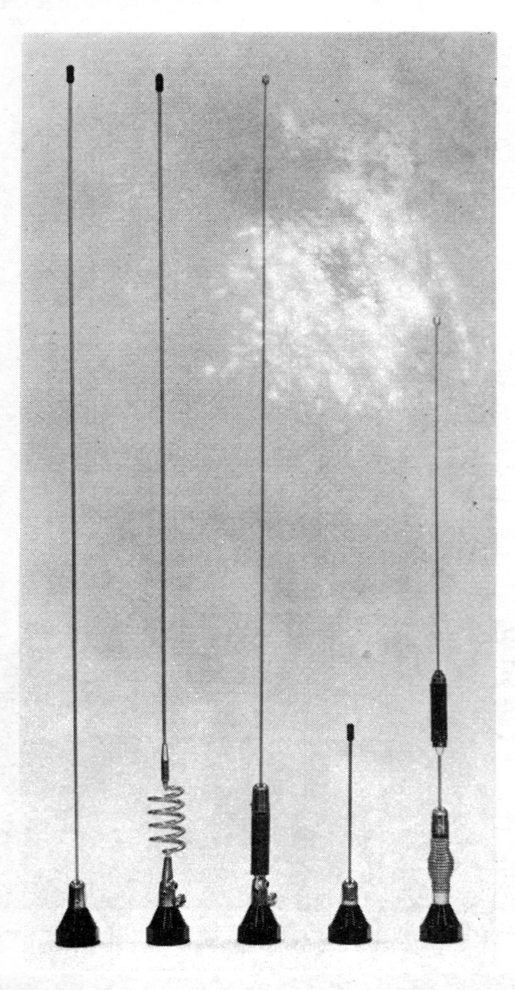

Whip aerials for mobile use at vhf and uhf. Left to right: 70MHz λ/4, 144MHz 5λ/8 with air-spaced coil, 144MHz 5λ/8 with coil on former, uhf λ/4, uhf collinear with spring-loaded base. *(Photo by courtesy of Pye Telecomms Ltd.)*

To arrive at this situation with any accuracy two instruments are required: an swr bridge for a line impedance of 50Ω and a field strength meter covering the band of interest. The swr bridge is connected in the feed line as near to the base of the aerial as possible, and the field strength meter is placed at a convenient distance. Do not stand the field strength meter on the vehicle itself, as rf currents in the car will result in misleading indications. Cut the aerial to a calculated 0·3 wavelength: 23in for 144MHz or 49in for 70MHz. Switch the swr bridge to FORWARD. Apply power to the transmitter, and adjust the transmitter output circuit for maximum forward reading. Slowly decrease the length of the whip to either (a) increase the forward reading on the swr bridge or (b) decrease the reflected indication. Once an apparent optimum point has been found, again adjust the transmitter output circuit for maximum forward indication of the bridge. Now carefully adjust the aerial length for minimum reflected power as indicated by the swr bridge.

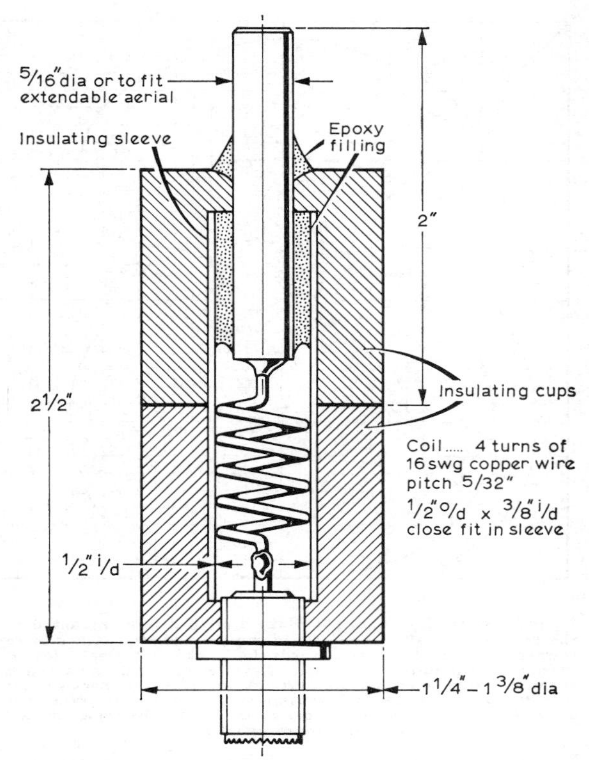

Fig 14.27. General arrangement of the 5λ/8 whip base and loading coil.

When correctly adjusted, minimum reflected power will coincide with maximum forward power and maximum indication on the field strength meter. Using this method, the swr on the feed line can be expected to be better than 1·25:1.

During the course of these adjustments it will be found that opening and closing doors and the lid of the boot will have a pronounced effect on the readings, as will walking near to the aerial or the vehicle. This is a practical demonstration of how surrounding objects affect the operation of a mobile aerial. In all cases the readings should be made from inside the car and with all the openings closed, and if possible, all adjustments conducted with the car in an open space.

5λ/8 WHIP AERIAL

The "5λ/8 whip" aerial is very popular among mobile operators, due to its low angle of radiation, and this is in reality a 3λ/4 aerial with the bottom λ/8 made in the form of a coil. This also allows a standard 50/75Ω feed to be used without matching problems.

Fig 14.27 shows a typical design for the base of a 2m version. The loading coil consists of four turns of 16swg wire wound ⅜in inside diameter and length approximately ⅝in, fitted as a close fit inside a short piece of thin plastic tube (such as pvc pipe). The whole assembly is enclosed in two cups made to fit, with the top terminal rod cemented into the top cup. Both cups should be cemented to the inner sleeve after the assembly has been checked for performance.

In Fig 14.27 a 2in top terminal is shown to mount a

telescopic aerial that will provide a 48in length. If preferred a single 49in length of whip may be used, in which case $\frac{1}{8}$ or $\frac{3}{16}$in diameter stainless steel would be suitable.

The most suitable connector is a uhf type with locking spigots to prevent it shaking loose in service.

Such an aerial is also suitable for base station use in conjunction with four or more $\lambda/4$ radials.

MINI-HALO FOR 144MHz

As an alternative to a vertical whip, a halo aerial may be employed on 144MHz. Being virtually a horizontal dipole, this aerial should show a performance superior to that of the $\lambda/4$ whip when working to ssb base stations using horizontal polarization. Standard halo aerials are often uncomfortably large even on 144MHz and this led to the design of a miniaturized capacitance-loaded halo, the construction of which is illustrated in **Figs 14.28–14.30**.

The most important part of the design is the tuning capacitor, and this must be constructed with care. To allow sufficient threads to be cut in the wall of the outer adjusting sleeve, this is tapped at 6BA. It should be noted that the inner of the two tube sections making up the capacitor is secured to one end of the circular element by a screw which force fits into the bore of the tube. The polythene or ptfe lining of this inner tube is fitted to the other end of the circular element by cutting a thread on its outside diameter and force screwing the lining over this thread. To assemble, the outer sleeve of the capacitor is slipped over the ptfe end of the element sprung away from the sleeve end, and then inserted into the bore of the capacitor inner sleeve.

The gamma match is of the same dimensions as would be used on a full-sized version, that is approximately 4½in long. In setting the system up the outer sleeve of the capacitor is used to bring the aerial to resonance, and then in conjunction with an swr bridge the gamma match is set for minimum swr.

Contrary to what might be expected, the mini-halo is not particularly frequency conscious, and tuning the 144MHz band does not show any decrease in signal strength from one end of the band to the other.

MOBILE SAFETY

While it is to the credit of mobile operators that no accidents have been attributed to the operation of mobile radio equipment, in the event of such an accident there could be no valid defence.

One item which could cause such an accident is the fist microphone and its trailing lead. Anyone who has foolishly tried to negotiate a sharp turn or a roundabout while clinging to a microphone with a press-to-talk switch will readily verify that the hand-held microphones can in these circumstances be particularly dangerous.

One microphone arrangement which goes a long way to solving this problem is that used in some radio-taxis and in which the microphone is fitted to a length of flexible tubing. Even this is not perfect for while it leaves both hands free for control of the vehicle, the head has to be maintained in almost a fixed position for close speaking, resulting in a restricted field of vision for the driver.

A solution to this is to employ a halter as shown in **Fig 14.31**. This fits over each shoulder and runs round the back of the neck of the wearer, and incorporates a boom upon

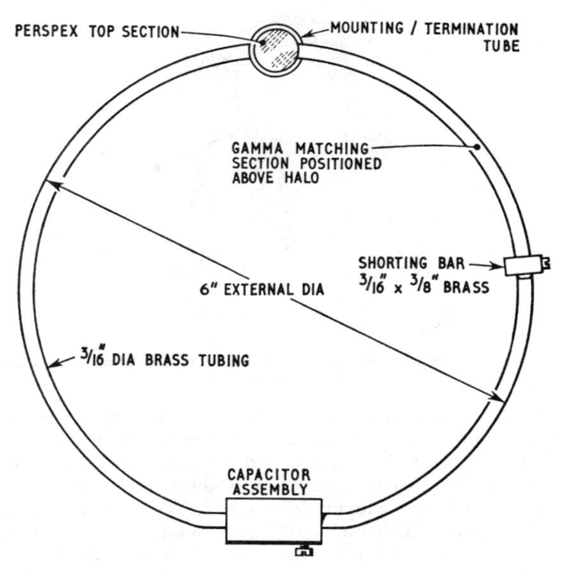

Fig 14.28. Looking down on top of the completed mini-halo aerial

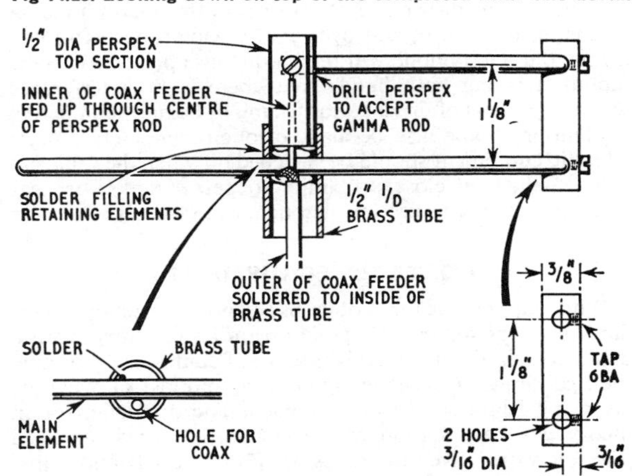

Fig 14.29. Side elevation of the completed halo, with drilling details of the gamma match support and Perspex mounting rod.

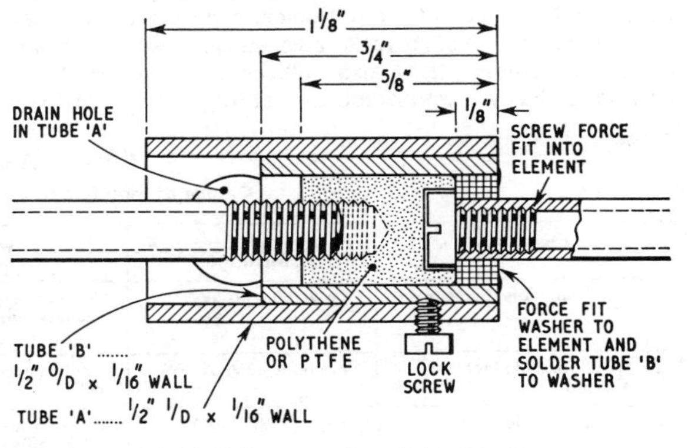

Fig 14.30. Cutaway view of the capacitor.

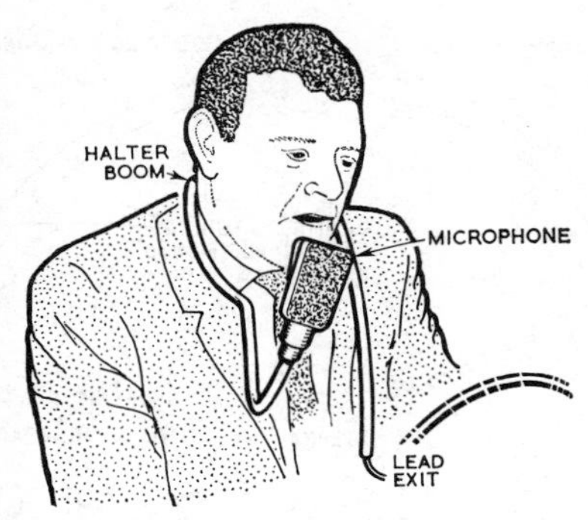

Fig 14.31. General view of the combined halter and microphone mounting boom.

which is mounted the microphone, and a section to carry the microphone lead clear of the user. Any of the usual microphone inserts may be fitted, and the lead run through the bore of the copper tube. If it is desired to incorporate transmit/receive switching with this halter, this can be accomplished by fitting a small box containing such a switch to the end of the halter where the microphone lead leaves the tube.

In use, as the driver/operator turns, so the microphone on the boom turns also, and over a very wide range of movements, the microphone will remain in close proximity to the mouth. Leaving both hands free, there is no interference with the process of driving, nor feeling of being encumbered.

Mention should also be made of one other safety aspect: all radio equipment should be switched off when the vehicle is being fuelled, in close proximity to petrol tanks, or near quarries where charges are detonated electrically.

PORTABLE EQUIPMENT

While equipment for mobile use is now reasonably standardized, that for true portable operation can vary from a matchbox-sized vhf transceiver to an hf contest rig "running the legal limit." Consequently no hard and fast rules can be given regarding choice of equipment and/or mode, for in theory *any* base station equipment can be used for fixed portable work, given the necessary power supply and transportation.

However, there are other practical considerations which limit the choice considerably. Portable equipment shares with mobile equipment the prolonged exposure to vibration in cars which soon reveals mechanical resonances and loosens fastenings. It is also subjected to the additional hazards of hasty erection and dismantling, of snatch loads when radiators become trip wires, and of fatigue failures in connections that have to be made and broken repeatedly.

Transceivers

Table 14.5 shows the three main classes of portable transmitting equipment, as used in practice. The first class consists generally of completely self-contained equipment in one enclosure and specially made for that particular type of operation. The second class covers a wide range of equipment, but the most suitable will often be mobile transceivers with add-on pre-amplifiers or power amplifiers if required, in which case mobile operation can also be readily utilized. The third class comprises main station equipment generally delivering close to the maximum legal power.

Careful thought should be given to which type of portable operation is really intended, and the station planned accordingly. For instance, if high-power portable operation is the eventual goal, it may be worth considering purchasing a generator and using base station equipment from the outset, rather than obtaining a low-power transmitter and adding on successively higher-power amplifiers. On the other hand, if the main interest is in hand-portable operation with occasional mobile work, an attractive possibility is an add-on power amplifier, which may be left permanently connected in the vehicle ready for service. Such amplifiers, which contain circuits that detect rf from the transmitter and then switch the amplifier into circuit, are now readily available. An example of a medium-power design for portable operation is given in Chapter 7.

DC-DC Voltage Conversion

Occasionally the need arises to use valve equipment which requires upwards of 250V ht, and to run this from a 12V supply. The conversion of 12V to 250V was formerly carried out using dynamotors, which are electric motor-driven generators, or vibrators, which use a vibrating reed contact to "chop" the 12V dc into pulses which are then stepped up to the higher voltage in a transformer and rectified. Both devices have now been superseded by the transistorized dc converter (or *inverter*). This utilizes one or more power transistors oscillating at a frequency of 500Hz or more, the resulting ac voltage being stepped up to the required voltage and rectified. Because the oscillation is at a relatively high frequency, the transformer can be efficient and of smaller size, and with suitable circuits conversion efficiency is at least 80 per cent.

The circuit of a simple transistorized dc converter is shown in **Fig 14.32**. The supply voltage is applied to the centre of the primary winding, the outside ends of which are connected to the collectors of TR1 and TR2 respectively. Switching within the transistors takes place between the

TABLE 14.5

Choice of mobile and portable equipment in the UK

Operation	Power output range (continuous)	Power supply	Transportation	Alternative operation
1. Hand-portable (pedestrian)	Low (50mW–3W)	Dry batteries or nicad rechargeable batteries (6–12V dc)	Pocket or shoulder	Mobile (add-on pa) Base station (add-on pa)
2. Fixed portable (vehicle mobile)	Medium (3W–50W)	Vehicle battery (12V or 13·8V dc)	Vehicle	Base station (12V or 13·8V dc psu)
3. Fixed portable	High (50W–100W)	Petrol or gas generator (110–250V ac)	Vehicle	Base station

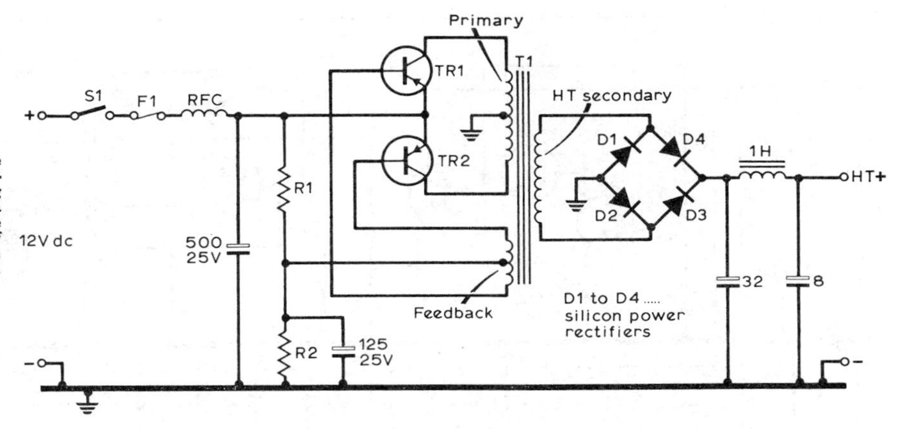

Fig 14.32. An example of the circuit of a high-efficiency transistorized dc converter operating around 1kHz. T1 is generally a toroidal transformer to minimize external fields. With circuits of this type efficiencies better than 80 per cent are to be expected.

collector and emitter connections and depends on the forward bias applied to the base; when the transistor is driven into saturation, the resistance between collector and emitter is effectively zero. Separate feedback windings feed the switching currents to the bases of the transistors.

Since transistors TR1 and TR2 are connected in push-pull they conduct on alternate half cycles of the primary oscillation. When the base of TR1 is driven negative, the base of TR2 will be driven positive, and hence TR2 becomes non-conducting. During the next half cycle TR2 is driven negative, and hence conducts, while TR1 is cut-off.

Such a circuit needs to be designed with a number of points in mind. First, the transistors selected must be able to switch the primary current without difficulty. Second, since the amount of feedback will decrease as the load on the transformer is increased, the feedback circuit must be designed so that when the transformer is supplying its rated load, the feedback level is sufficient to maintain oscillation. Third, under lightly loaded conditions, the feedback voltage to the bases of the transistors will be at its highest, and the transistors must be able to withstand the reverse base-emitter voltage which will appear across this junction during the cut-off condition. Fourth, the transistors must be rated to withstand any peak transient voltage occurring during the switching process, as any such spikes will cause failure of the transistors if they are not adequately rated. A good rule of thumb is to employ transistors in which the collector voltage is specified as at least three times the maximum supply voltage. In the case of mobile operation the line voltage may rise to 14·5V in a 12V nominal system when the generator is charging fully, and this implies using transistors with a V_{ce} of at least 45V.

From the foregoing it will have been noted that the feedback conditions are equated for full load, and from this it follows that under reduced loading conditions, the efficiency will be lower since TR1 and TR2 are being switched for a greater current than that which the primary will actually need to handle to meet the demand.

It is sometimes thought that transistorized dc converters do not cause interference, this conclusion being based on the premise that since there are no moving parts, and no contacts opening and closing and arcing in the process, there is nothing to be suppressed. This is a fallacy, as explained earlier in this chapter. While it is true that there is no interference due to contact arcing, the waveform within the generator is basically a square wave. Thus harmonics are present and these can produce a buzz similar to that of a vibrator unit. For this reason rf suppression is essential. Furthermore, there is often a strong magnetic field around the transformer, and the audio circuits should be positioned well clear of the converter if pick-up of the converter frequency is to be avoided.

Transistors used in dc converters must be mounted on an adequate heat sink, and if germanium types are used they must be operated in such a manner that under full load conditions their cases feel only warm to the touch. If there is any sign of overheating, either the load is too heavy for the type employed, or they are being overdriven. In the first case a larger transistor should be fitted, and in the second, the degree of base feedback should be investigated with the object of reducing it to a lower level.

Sometimes a transistor dc converter may operate satisfactorily for months and then, for no accountable reason, fail. Such failures are invariably due to transients on the 12V vehicle system created by some other electrical device. To absorb such spikes, it is essential to include a high-value capacitor across the supply line to the dc converter, as shown in Fig 14.32. Another line of approach is to clamp the collector-emitter potential with a zener diode having a breakdown voltage somewhat less than that of the V_{ce} rating of the transistors used. The arrangement is shown in **Fig 14.33.** It should be noted that a zener diode must be placed across the collector-emitter connections of each transistor.

One effect which is sometimes noted with transistorized dc converters is that they show a reluctance to start under load but, once running, perform in a satisfactory manner. This is invariably due to the feedback being set at a critical level. If

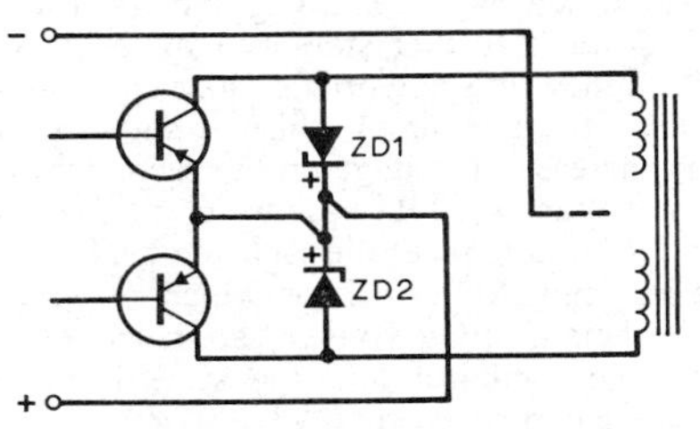

Fig 14.33. Use of zener diodes across the collector-emitter junction to limit transient voltage spikes. The diodes should be rated for a breakdown voltage slightly less than the V_{ce} rating of the transistors.

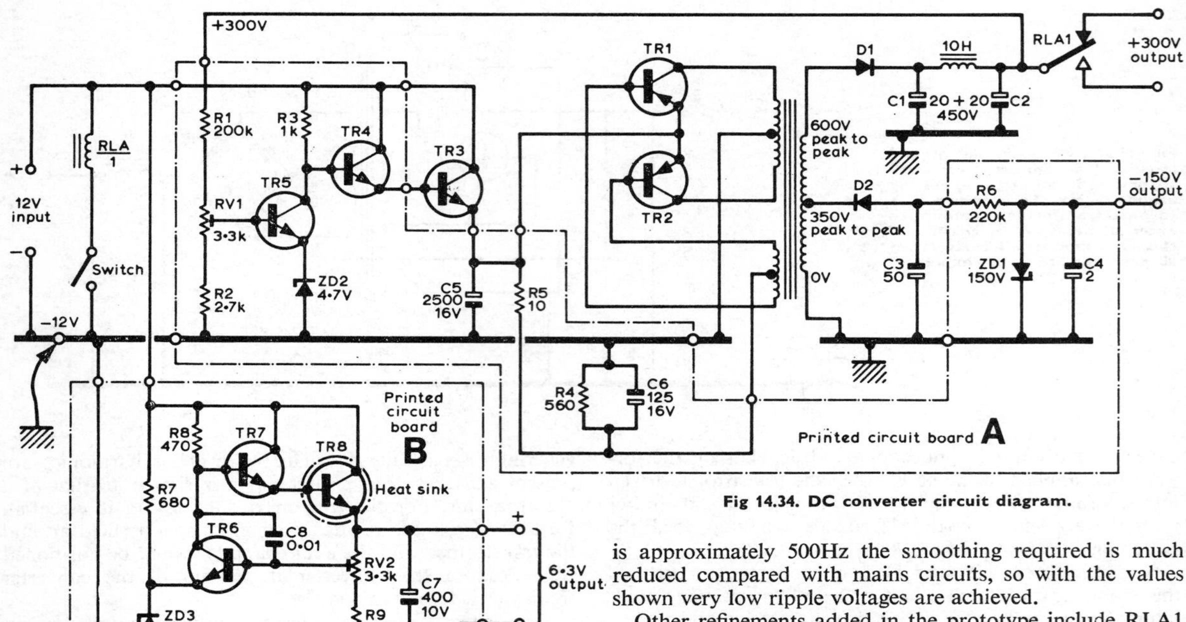

Fig 14.34. DC converter circuit diagram.

there is no way of increasing the level of the feedback in a convenient manner, sequential switching of the supply and the load will usually provide a solution to erratic starting. This entails employing a double-pole relay, one set of contacts of which are in series with the supply, and the other set in series with the ht output to the load. The contacts in series with the supply are carefully adjusted so that they close before the ht contacts. With this arrangement even the most reluctant circuit should start without difficulty.

GENERAL-PURPOSE REGULATED DC CONVERTER

The unit to be described was originally designed to run small klystrons from a car battery on portable excursions, but is easily adapted to supply small valve linear amplifiers requiring an anode supply of 250-350V.

Circuit Description

The circuit (**Fig 14.34**) is quite straightforward and consists of a basic oscillator (TR1 and TR2) driving a transformer. The secondary outputs are rectified by D1 and D2 and smoothed by C1, C2, C3, and the 10H choke; the negative output is further stabilized by ZD1. The positive voltage is applied to a potential divider chain and fed to the base of TR5, the emitter of which is stabilized to 4·7V by ZD2. Any increase of voltage on the secondary output thus reduces the voltage on TR4 base which acts as a Darlington pair with TR3 and reduces the voltage on TR3 emitter. TR3 is the series regulator transistor which provides TR1 and TR2 with their emitter voltage, so negative feedback is applied to the oscillator and the system stabilizes at an output voltage controlled by RV1.

TR6, TR7 and TR8 form a conventional series regulator to supply 6·3V dc for valve heaters. As the inverter frequency is approximately 500Hz the smoothing required is much reduced compared with mains circuits, so with the values shown very low ripple voltages are achieved.

Other refinements added in the prototype include RLA1 which gives a choice of two outputs for transmit/receive.

The unit can be used with either a positive or negative earth, the connection being made at the input terminal only. The output voltage will be slightly reduced on the positive earth version as the supply volts are effectively added to the feedback loop.

Construction

The unit can be constructed in a small box. The power transistors except for TR8 are mounted on the rear panel, and the transformers and the inverter board are mounted on a chassis within the unit. The valve heater regulator is built as a separate unit and fitted inside the box towards the front.

Printed circuit layouts of the regulator board and the inverter board are given in **Figs 14.35** and **14.36**.

The exact layout of the transformer and choke on the chassis will depend on the size and shape of the transformer purchased. The transformer used in the prototype was an ex-equipment potted toroidal type, probably from a Pye Ranger, with a tapped secondary. Such components are readily available on the surplus market. The requirement is

TABLE 14.6

Components list for regulated inverter

C1, C2	20 + 20μF 450V electrolytic	R1	200kΩ
		R2	2·7kΩ
C3	50μF 275V	R3	1kΩ
C4	2μF 150V	R4	560Ω 1W
C5	2,500μF 16V	R5	10Ω 1W
C6	125μF 16V	R6	220kΩ*
C7	100μF 10V	R7	680Ω
C8	0·01μF	R8	470Ω
D1	1,000V piv	R9	3·3kΩ
D2	BY100	RV1	3·3kΩ
ZD1	150V 400mW zener	RV2	3·3kΩ
ZD2	4·7V 400mW zener		
ZD3	3·3–3·9V 400mW zener		
TR1, TR2	NKT404 or 2G220		
TR3, TR8	2N3055		
TR4, TR7	2N3053		
TR5, TR6	2N3705 or BC109		

* R6 is chosen to drop the output voltage to the zener; in the prototype R6 dropped 25V.

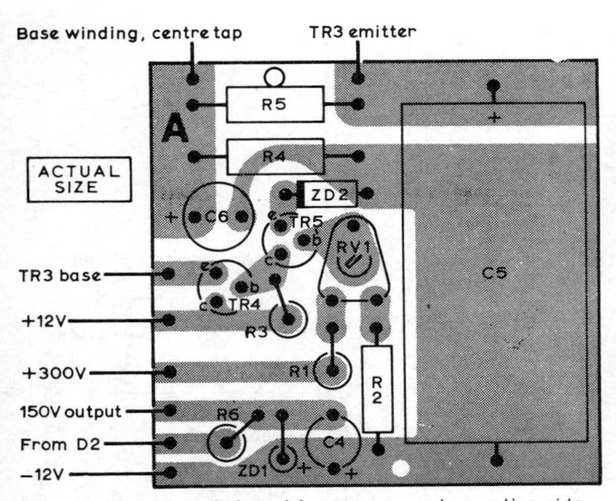

Printed circuit board viewed from component mounting side—copper track on reverse side

Fig 14.35. 300V regulator pcb.

for a transformer with 12V input and 400V and 250V outputs.

A 2,500μF 16V electrolytic capacitor across the input from the battery, and a 10A diode in one input lead, are useful additions to the basic inverter unit.

Testing and Performance

On completion, the 12V input should be applied with both outputs open circuit, whereupon the inverter should start immediately, and RV1 can be adjusted to provide the required output voltage.

The input voltage is then removed and a load applied to the outputs. The inverter should start and give the full output immediately the input is applied. Should this not be the case some adjustment of the ratio R4:R5 will probably cure this, although in the two versions built by the original author no such problems were encountered. RV2 is adjusted to give 6·3V from the output of the low voltage regulator.

The performance of the final prototype was as follows:

Input voltage (V)	13·5		
Output voltage (V)	250–380 open circuit, depending on setting of RV1		
Output noise (mV)	2		
Output current (mA)	0	30	60
Output voltage (V)	300	295	285
Input current (A)	0·5	1·5	2·25
Noise (mV)	2	5	10

Although the output noise is very low, great care in construction and filtering must be taken if sensitive vhf converters are to be used from the same 12V source as the inverter while it is running. On the original author's 6cm klystron receiver the low noise pre-amp picked up severe interference from the inverter until all inputs to and outputs from the unit were filtered and the unit was fully screened.

The above figures are typical of the performance when RV1 is set to its approximate mid-position. With an input of 12V, the maximum likely output is 380V when RV1 is adjusted to the limit. If the input falls to 10·4V, it is still

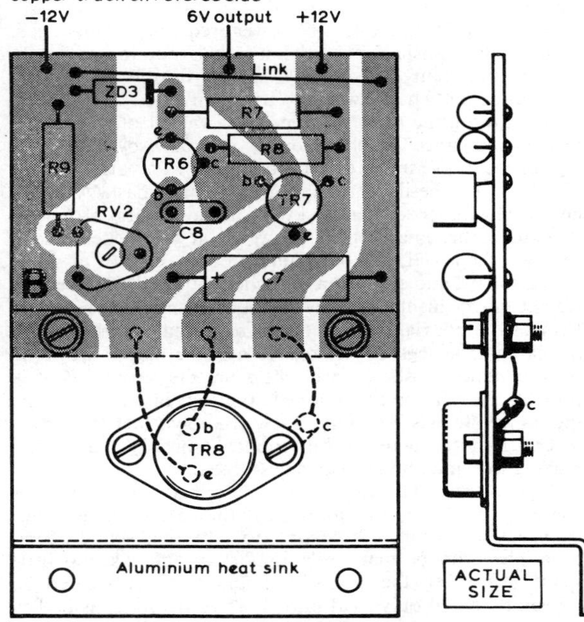

TR8 is insulated from heat sink which is isolated from printed circuit board

Fig 14.36. 6·3V regulator pcb.

possible to obtain 300V output by adjusting RV1, but in this case the regulation with varying loads will be poor.

With RV1 set to give an off-load output of 320V from a 12·8V input, the following figures show how the output voltage varies with fluctuations of input voltage.

Input (V)	Output (V)
11	280
12	305
12·8	320
14	340

By substituting a different transformer, and changing ZD1, other stabilized voltages may be obtained.

Aerials

It is an old adage that the aerial is the most important part of the station, and in portable operation with low-power equipment "the odd length of wire slung from a tree" is just not good enough if reliable communication is to be achieved. On the other hand, portable aerials have to be simpler and of lighter construction than those used in permanent installations: constant coiling and uncoiling of heavy-gauge cadmium-copper wire is not good for the wire—or the operator. Other factors include ease of tuning up and, perhaps surprisingly, space occupied, for sometimes the roadside is the only accessible common land.

As before with mobile aerials, the reader is referred to Chapters 12 and 13 for the theory and typical construction of aerials and their feeders, and only points peculiar to portable operation will be touched upon here.

Three popular aerials for portable work on the hf bands are the inverted-V dipole, the short loaded vertical and the ground-plane, and these will now be briefly discussed in turn.

A horizontal dipole for the lower frequency bands needs to be under considerable tension if it is not to sag in the centre, and many masts may be too flexible to support such an aerial. The answer is to use an inverted-V configuration, allowing the mast to support the weight of the coaxial feeder, and cancelling the sideways pull on it by having the legs of the dipole under equal gentle strain. This arrangement also has the merit of requiring only one mast. A dipole complete with feeder is an awkward aerial to carry around as it tends to become tangled. However, if the feeder is kept separate the aerial may be wound on to a piece of plywood or hardboard and easily run out when required. To keep the weight down, insulators may be made from small pieces of Perspex, a coaxial socket being mounted on the centre insulator. The legs of the dipole may be sloped down to bushes, fence posts or even pegs in the ground, but if possible they are best positioned about 7ft high; apart from the danger to passers-by it is a well-known fact that given several hundred acres of hill country, sheep will insist on grazing the small areas near one's dipole.

Short loaded vertical aerials, described earlier in this chapter, have the attraction that if they are made in interchangeable sections it is possible to achieve all-band operation. They can be made self-supporting and take up little space. However the pi-tank circuit commonly used in transmitters will only load into short vertical aerials if they are accurately resonated at the operating frequency and present a low impedance at the feedpoint. Unfortunately the process of resonating the aerial under portable conditions requires considerable patience if an atu is not available. Almost any variation in configuration will alter the resonant frequency; even pouring water over an earth spike in dry ground will alter the loading considerably, and it may be necessary to use the vehicle as an electrical counterpoise.

The ground-plane aerial is an attractive possibility for single-band hf operation. As discussed elsewhere in this book, it is a low-angle radiator which works perfectly well at ground level provided it has a satisfactory ground plane and, although this latter condition is comparatively rarely found in practice, useful amounts of low-angle radiation can be still achieved. The Z-match described in Chapter 12 may be used to match the low-impedance feedpoint to the transmitter or alternatively the electrical length of the radiator can be increased slightly (see Fig 14.22) and the reactance tuned out capacitively. This latter method also improves the low-angle radiation. On 14MHz any length from 19ft to 25ft can be tuned capacitively (or with a Z-match)—the most useful length being about 25ft. Beyond this the impedance rises rapidly and a low swr is unobtainable by simple means.

On the other hand the aerial can be shortened below λ/4 and tuned inductively—15ft is the limit that some Z-matches will tune on 14MHz, but this still produces low-angle radiation.

A portable ground-plane aerial of about 18ft can be constructed simply from lengths of scrap steel conduit and screwed connectors. If the mast is made up of five pieces of roughly equal length it will fit into the average car. It is inadvisable to try anything much more ambitious with this material, as the conduit snaps off readily at the start of the threaded sections. A glass jar will serve as a base insulator, and string and iron spikes form the four guys (somehow it is much easier to put a mast up single-handed with four guys

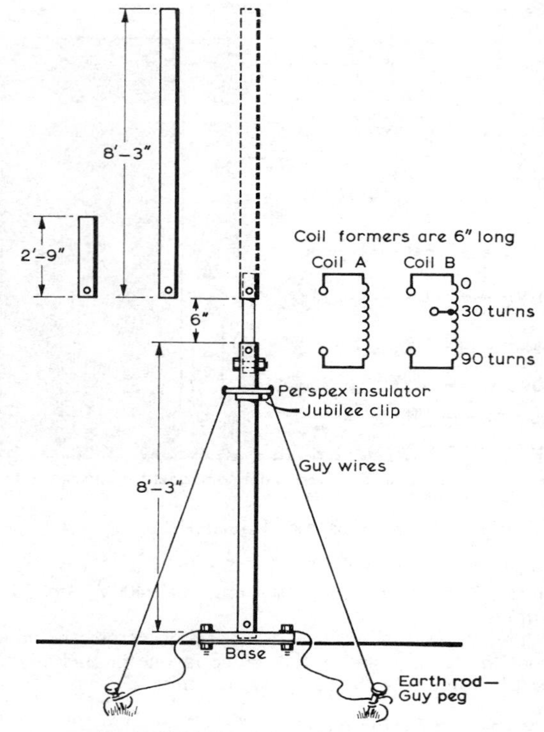

Fig 14.37. The complete all-band aerial.

than with three). In the UK, government surplus "29/41-foot aerials" have also been modified and utilized by some amateurs for ground planes and also short loaded verticals. If such an aerial is acquired it will be found that the rope guys and chain-link insulators supplied are extremely clumsy. These should be removed from the two stay plates and replaced with line made of polypropylene or similar insulating materials. Quite thin line, about ⅛in diameter, can be used and this will look much neater.

Four radials are the minimum for omnidirectional coverage and eight, at least in portable work, the maximum. Note that if the aerial is tuned capacitively the feeder will act as a radial, whereas if a Z-match or other transformer coupling is used, the feeder will make little contribution. The radials are probably best made from single-strand wire of sufficient stiffness to roll up easily, and with brightly coloured insulation (they are not earthed) so they show up on the ground. They can be pegged down with short lengths of fence wire.

Most vhf and uhf aerials for fixed portable operation consist of small Yagi or cubical quad arrays mounted on 15–20ft vertical masts, but the HB9CV aerial is also worth considering for this service. In the same way, a normal-mode helical aerial is an attractive alternative to the more usual λ/4 whip for hand-portable work.

ALL-BAND PORTABLE AERIAL

Primarily the aerial consists of three lengths of 20 gauge aluminium tubing of 1in diameter, two coils and a base plate that acts as insulator and earth connector. The earth

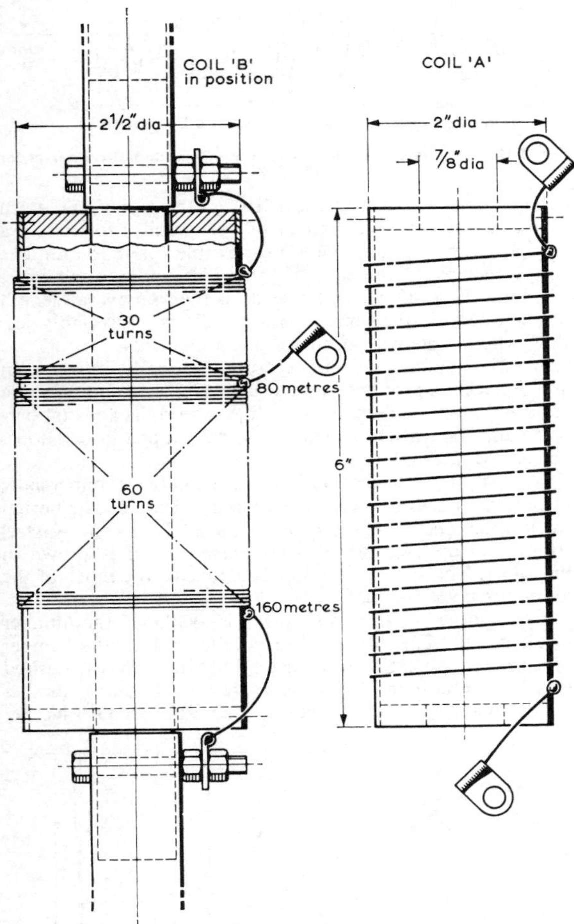

Fig 14.38. Coil construction details.

rods double as guy pegs for aerial support. The aerial can be erected by one person in approximately 10min. The overall construction is shown in **Fig 14.37**. It will be seen that the lengths of aluminium tubing are linked by a piece of ⅞in diameter Perspex, which is a close inside fit. The selected coils are constructed to fit over this link, and are electrically connected to the aluminium.

Tests with an impedance bridge indicate that the base impedance varies between 25 and 35Ω and vswr checks indicate that the worst figure is 1·3 to 1 using 50Ω coaxial cable as a feeder.

The aerial is adjusted to the desired band as follows:

28MHz .. 8ft 3in section only.
21MHz .. 8ft 3in + 2ft 9in sections (linked).
14MHz .. 8ft 3in + 8ft 3in sections (linked).
7MHz .. 8ft 3in + coil "A" + 8ft 3in sections.
3·5MHz .. 8ft 3in + part of coil "B" + 8ft 3in sections.
1·8MHz .. 8ft 3in + all of coil "B" + 8ft 3in sections.

Construction

The coils are constructed as follows:

Coil "A" consists of 24 turns of 10swg copper wire wound

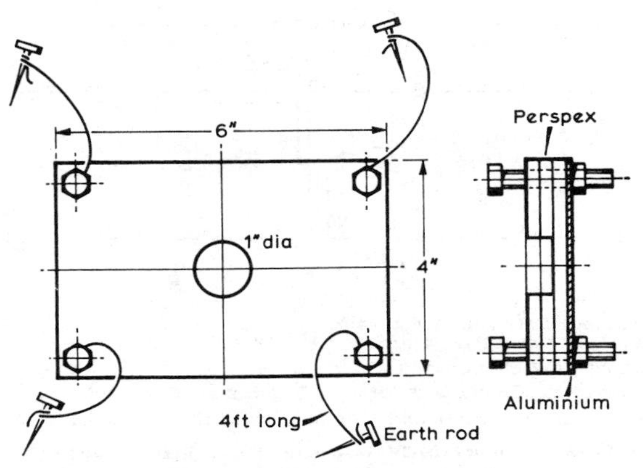

Fig 14.39. Detail of base plate construction.

on a 2in diameter former and the turns spaced over 5½in of the length.

Coil "B" consists of 90 turns of 20swg enamelled copper wire close wound on a 2½in diameter former and tapped at 30 turns for the 3·5MHz band. Both coil formers are exactly 6in long, made of Paxolin, and have discs in the end of the coils. The centres of the discs are drilled out to ⅞in diameter to fit over the ⅞in diameter centre Perspex rod of the aerial system (see Fig 14.37). The discs themselves are glued and pinned to the coil former.

Fig 14.38 shows the method of construction of the coils.

The base plate is made from ¼in or ½in Perspex. From **Fig 14.39** it can be seen that the top two pieces of Perspex have 1in diameter holes drilled into them but the third piece is left whole to act as the base insulator. The bottom sheet (of aluminium) acts as a connection between the four nuts and bolts and strengthens the base as a whole. The four bolts are left long in order that they press into the ground to prevent the base from moving. The 4ft lengths of flexible lead are made from thick copper braid covered with the outer sheath of some old coaxial cable.

A 2½in by 2½in piece of Perspex acts as a spreader and aerial insulator and has a 1in hole drilled into the centre and four small holes drilled in the corners to which are tied four 10ft lengths of nylon cord to act as guys. The aerial insulator is passed over the lower 8ft 3in section of the aerial and secured by a Jubilee clip at a height of about 8ft.

Assembly and Erection Procedure

To erect the aerial proceed as follows:

(a) Make up the aerial to the required band.
(b) Set the base plate in the desired position.
(c) Stretch out the four lengths of earth line and knock in the earth rod/guy pegs.
(d) Tie two of the nylon guys to two of the earth rod/guy pegs.
(e) Insert base of aerial into the base plate and hold it in the vertical position by pulling on the other two guys.
(f) Tie one nylon guy to the third earth rod/guy peg.
(g) Tie the fourth guy to the last peg and square all up as required to keep the aerial vertical.
(h) Connect the feeder inner to the aerial and the outer braid to one of the nuts and bolts of the base plate.

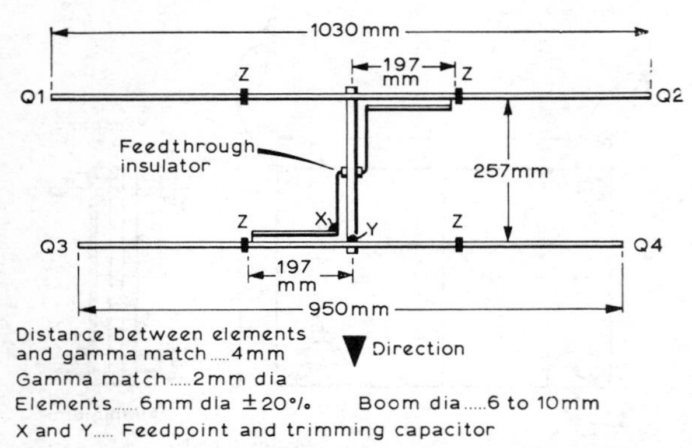

Fig 14.40. Details of HB9CV two-element beam designed by PA0TBE.

It can be observed that if additional guys are secured near the top of the 16ft 6in section, the aerial can double as a good mast for a 2m Yagi array.

The ⅞in diameter rod which joins the two aluminium parts of the aerial could equally well be made from ebonite, Paxolin, or any good insulating material.

Results obtained during portable operation with this aerial fed with approximately 20W of rf have been extremely satisfying and rewarding for the effort made in construction.

HB9CV AERIAL FOR 144MHz

Relatively few 144MHz enthusiasts know about the attractions of the HB9CV array, which is a development of the ZL-Special and has been known and used (mostly on hf) for many years. This two-element version, designed by PA0TBE, provides some 5dB gain and is very much easier to carry around than a three- or five-element Yagi or quad. Details are shown in **Fig 14.40.**

6mm diameter tubing is used for the centre parts of the elements (ie between points Z-Z which are approximately 400mm apart). Bolts that will take 4mm tubing are soldered at points Z; the four sections shown as Q1, Q2, Q3 and Q4 are then made of 4mm tubing threaded at one end to fit the bolts. A 30pF trimmer is connected between points X and Y and adjusted for minimum swr; for convenience this trimmer can then be replaced by a fixed capacitor of approximately the same value. The coaxial cable can be of 50, 60 or 70Ω impedance.

NORMAL-MODE HELICAL AERIAL

Helical aerials may be constructed for operation in one of two modes: the *axial mode* and the so-called *normal-mode*. When the circumference of each turn and the pitch (ie the distance between turns) are comparable to a wavelength, the helix functions as a beam aerial with maximum radiation along the axis of the helix. For this reason the helix is said to be operating in the axial mode, and the radiation is circularly polarized (see Chapter 13).

If the diameter and pitch of the helix are reduced below $\lambda/2$ the radiation pattern is drastically changed, and maximum radiation then occurs in a direction normal (ie at right angles) to the axis of the helix; hence its name. Axial radiation under these conditions is negligible. The polarization is in general elliptical but by appropriate choice of dimensions

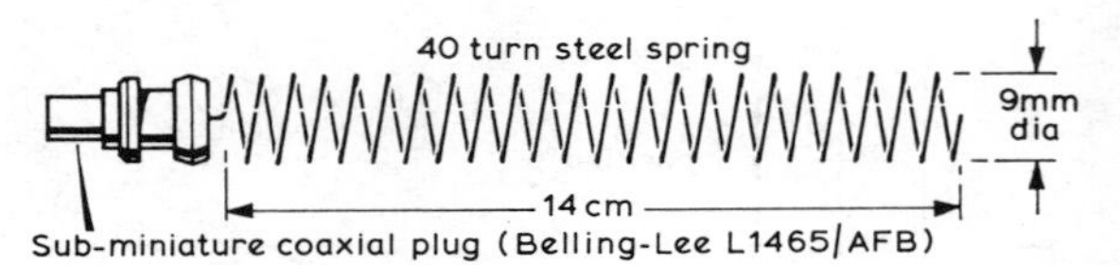

Fig 14.41. Example of a home-made normal-mode helical aerial for 145MHz.

can be made very closely linear. In fact the radiation pattern of the normal-mode helix can be very similar to that of a straight wire and it is therefore possible for it to simulate aerials such as $\lambda/2$ dipoles, $\lambda/4$ monopoles and ground planes. The important difference, however, is that the overall length of such aerials when made with a helix is appreciably less than when made with ordinary linear elements.

A recent application of the normal-mode helix has been as a replacement for telescopic whip aerials on hand-portable vhf transceivers. Telescopic $\lambda/4$ whips are fragile and cumbersome and have a nasty tendency to be trapped in car doors or poke people in the eye.

Normal-mode helical aerials are available commercially but these tend to be rather bulky and inflexible. The design to be described makes use of any available ordinary steel spring and an example of a completed aerial is shown in **Fig 14.41**. The spring is simply soldered to the inner of the miniature coaxial plug.

Assuming a suitable steel spring is available, the number of turns required is determined as follows. First the number of turns for a helix with diameter 1cm and with the desired length is determined from **Fig 14.42** and then this number is multiplied by a factor obtained from **Fig 14.43**. Five per

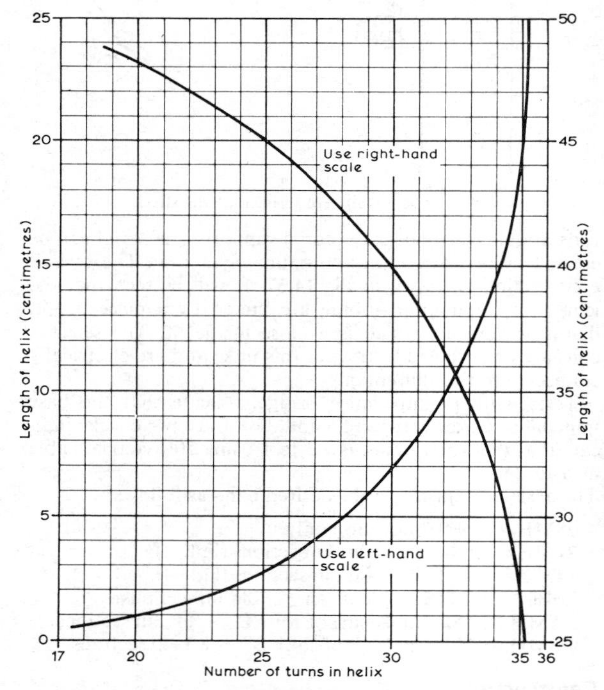

Fig 14.42. Graph showing the way in which the number of turns (N) is related to the overall length (h) for a normal-mdoe helix resonant at a constant frequency of 145MHz.

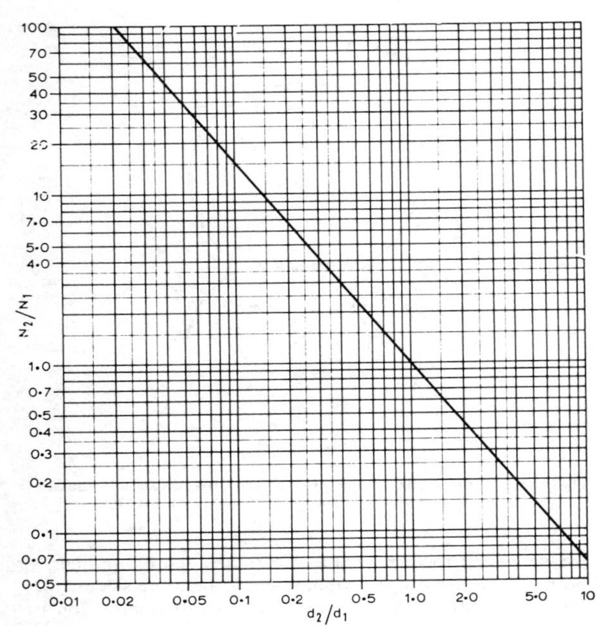

Fig 14.43. Graph relating the ratios N_2/N_1 and d_2/d_1.

cent extra turns are added to the calculated number and these turns are removed one at a time (and the tank circuit then adjusted each time) until maximum radiation occurs from the aerial as shown on a field strength meter.

As an example, suppose one requires a helix 20cm long with 1·2cm diameter to be resonant at 145MHz. From Fig 14.42 the number of turns required if the diameter were 1cm would be 35. The ratio of the actual diameter d_2 to the standard diameter d_1 is 1·2/1·0 and the ratio of the actual number of turns N_2 to the number for a 1cm diameter helix (N_1) is read off from Fig 14.43 as 0·81. Thus

$$N_2 = 35 \times 0{\cdot}81 = 28.$$

A suitable starting point for the aerial, allowing for trimming to resonance, would be a coil with 32 turns.

When an aerial is required for a frequency other than 145MHz, Figs 14.42 and 14.43 can still be used. Since shortened aerials of this type are especially convenient for the 4m band, a useful example would be a helix resonant at 70MHz with length 30cm and diameter 1·3cm. Steps in the calculation are:

1. Scale the length and diameter of the desired aerial (for frequency f_1) to find the dimensions of the required aerial at 145MHz. In this case the latter would have a length equal to

$$30 \times \frac{70}{145} = 14{\cdot}5\text{cm}$$

and diameter equal to

$$1{\cdot}3 \times \frac{70}{145} = 0{\cdot}63\text{cm}$$

2. Referring to Fig 14.42 again one finds a 145MHz helix 14·5cm long and 1cm diameter has 34 turns.

3. Finally, a helix with the same length but with diameter 0·63cm has 34 × 1·75 = 59 turns. (The factor 1·75 is the value of N_2/N_1 which corresponds to

$$\frac{d_2}{d_1} = \frac{0{\cdot}63}{1{\cdot}0}$$

and was obtained from Fig 14.43).

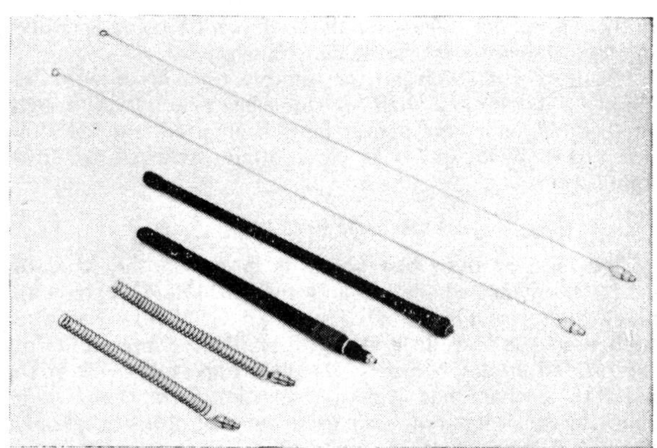

Various helical aerials contrasted. All except the longest spring aerial are designed to work at 145MHz, the latter operates at 85MHz. The two aerials at the rear are home-made $\frac{1}{4}\lambda$ whips, while the two in the foreground are home-made "spring" aerials.

A suitable starting point for the 4m aerial would thus be a helix with 63 turns, 1·3cm diameter and 30cm long.

Naturally one cannot expect the same performance as with a straight whip, but the difference is acceptably small. For example, compared with the straight $\lambda/4$ whip made from steel wire shown in the photograph, the steel spring helix also shown gives a loss of very roughly half an S-point.

DIRECTION-FINDING EQUIPMENT

The recent popularity of df contests has in turn led to the development of highly directional portable receivers for use on the amateur hf bands.

Two bands commonly used for this activity are 1·8MHz and 3·5MHz, and df receivers for these frequencies utilize the marked directional properties of a horizontal ferrite-rod aerial.

Fig 14.44 shows that two fairly sharp nulls exist along the axis of the rod, and because they are much easier to determine by ear than the corresponding maxima, it is these nulls that are used as the basis for direction finding.

Obviously the ferrite rod used alone gives ambiguous results in that the hidden transmitter could be in either of two opposing directions. To resolve this difficulty, a small non-directional *sense aerial* is placed on one side of the ferrite rod. The output of this aerial is connected, either directly or via an rf amplifier, to the ferrite rod in such a way that it modifies its overall directional pattern (Fig 14.44). A minimum of 6dB front-to-back ratio is required for reliable determination of the true direction of the hidden transmitter.

The sense aerial and/or amplifier is usually left switched off until required as it can affect the sharpness of the nulls.

Another necessary feature of a df receiver is some provision for rf attenuation, in order to prevent overloading of the rf amplifier stage when the equipment is being operated in close proximity to the hidden transmitter.

Ordinary vhf receivers are suitable for vhf df work in conjunction with small hand-portable beam aerials. Although small quad and Yagi arrays have been used, the HB9CV beam (see page 14.24) is now often preferred for this application.

1·8MHz DF RECEIVER

The receiver described below is based on the Mullard TAD100 integrated circuit and the LP1175 470kHz a.m. block filter (see **Fig 14.45**). The Toko CFU filter has also been used successfully and an alternative arrangement for the printed circuit board is shown in **Fig 14.46**. The Toko MFH41T, which has superior selectivity and bandwidth characteristics, has not been tried but will almost certainly produce a superior result, albeit at an increased cost.

General Details

The receiver must be housed in a metal or screened cabinet; the well-known double-U construction is easy to reproduce. The beauty of this method is that only one important dimension is associated with the bending process and small errors are usually easily corrected.

The TAD100 provides the frequency changer and oscillator, i.f. amplifier, detector and af driver. This is followed by a Class B output pair of transistors. A bfo is included in the main pcb. The sense amplifier is separate. Miniature components must be used unless the sizes of the pc boards are increased, but the latter practice is not recommended. Providing suitable components are used the boards will not be found to be overcrowded. The original receiver used one

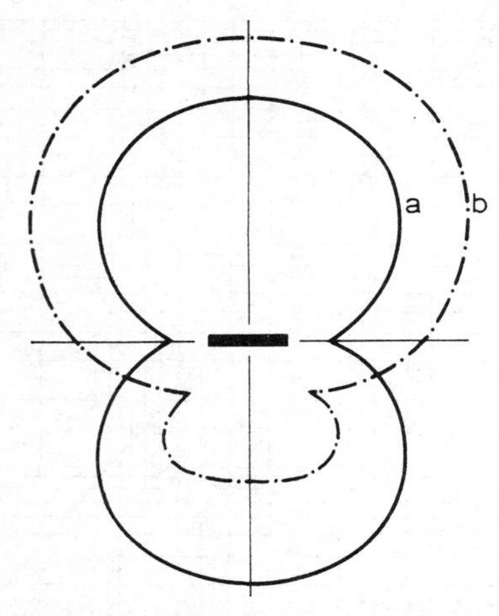

Fig 14.44. The directional properties of a ferrite rod aerial used alone (a) and in combination with a sense aerial (b).

rotary switch to turn on the bfo or the sense amplifier, or both. Miniature toggle switches would be more satisfactory, but are slightly more expensive. The rf attenuator consists simply of a loop of wire which is moved on to the rod aerial when required.

Construction

Starting with the largest pcb, use only a good, single-sided, heat-resisting material. Unless the etching process is to be

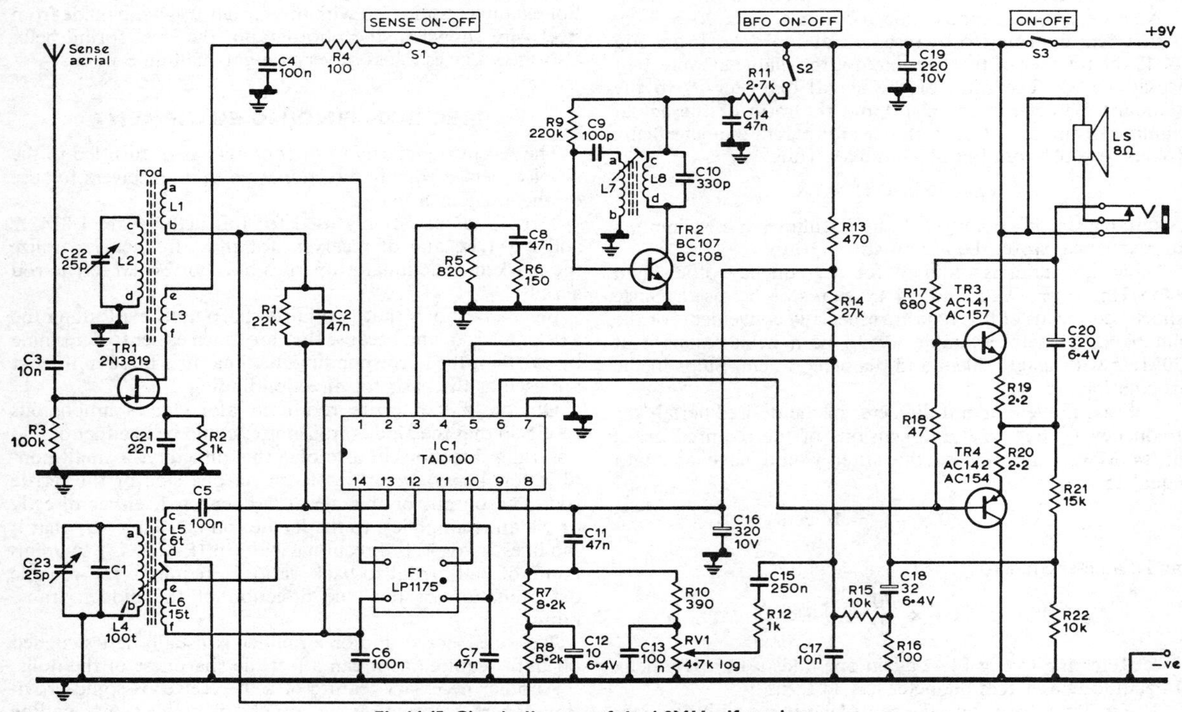

Fig 14.45. Circuit diagram of the 1·8MHz df receiver.

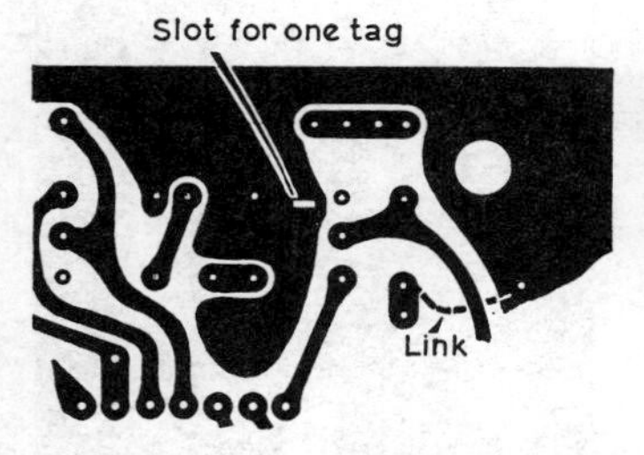

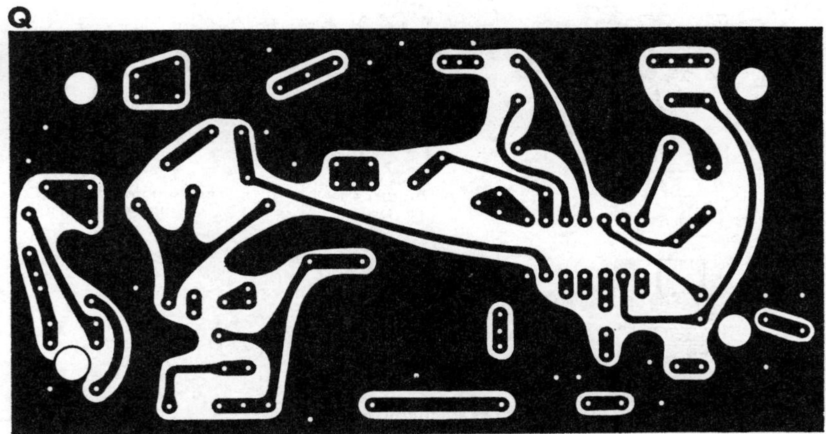

Fig 14.46. Main pc board, actual size. The section above shows the modifications required to part of the main pcb for the Toko CFU filter.

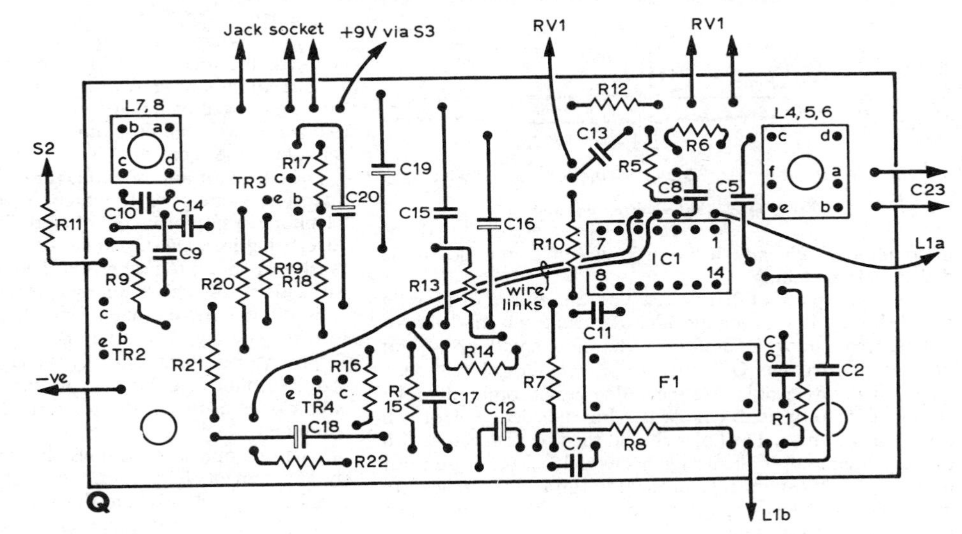

Fig 14.47. Main pcb connections.

TABLE 14.7

Components list

R1 22kΩ	R9 220kΩ	R17 680Ω
R2 1kΩ	R10 390Ω	R18 47Ω
R3 100kΩ	R11 2·7kΩ	R19 2·2Ω
R4 100Ω	R12 1kΩ	R20 2·2Ω
R5 820Ω	R13 470Ω	R21 15kΩ
R6 150Ω	R14 27kΩ	R22 10kΩ
R7 8·2kΩ	R15 10kΩ	All min ¼W
R8 8·2kΩ	R16 100Ω	RV1 6·7kΩ log
C1 Select	C9 100pF	C17 10nF
C2 47nF	C10 330pF	C18 32μF 6·4V
C3 10nF	C11 47nF	C19 220μF 10V
C4 100nF	C12 10μF 6·4V	C20 320μF 6·4V
C5 100nF	C13 100nF	C21 22nF
C6 100nF	C14 47nF	VC1,
C7 47nF	C15 250nF	VC2 25pF airspaced
C8 47nF	C16 320μF 10V	
IC1 TAD100		TR3 AC1[illegible]7, AC141
TR1 2N3819		TR4 AC154, AC142
TR2 BC107, BC108		Block filter LP1175
Ferrite rod ½in dia, 8 in long	Miniature toggle switch off/on, 3 off	
Speaker 8Ω, 2½in dia	$\frac{3}{16}$in dia coil formers and slugs	
Epicyclic slow motion drive	Battery clips	
Insulated jack socket, 3·5mm		

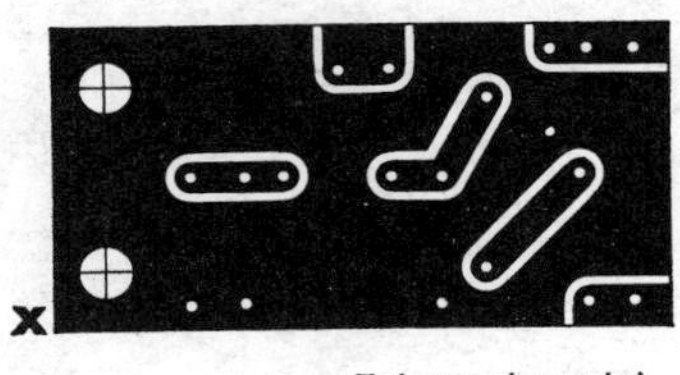

Fig 14.48. Sense amplifier pcb and connections.

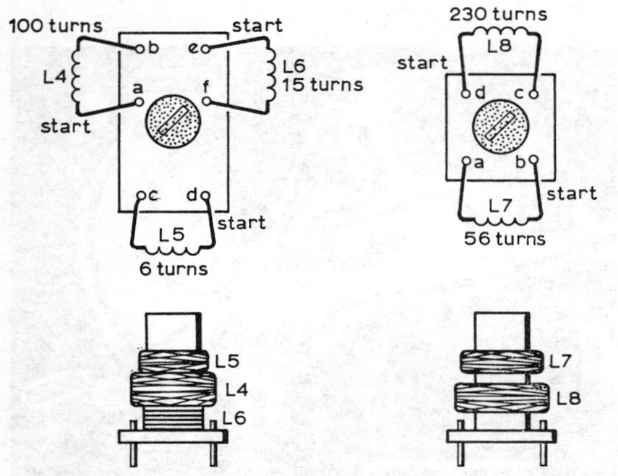

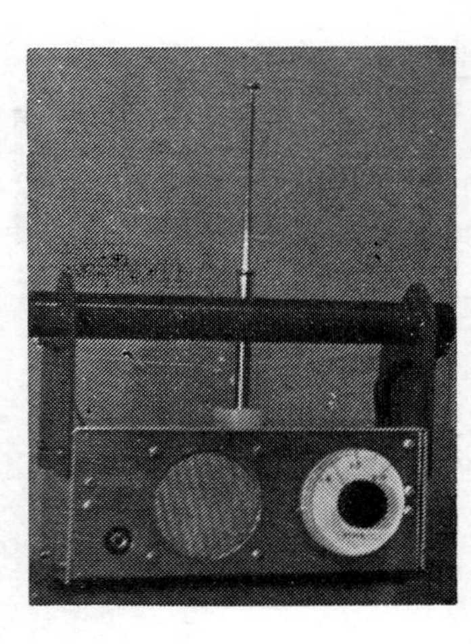

The completed 1·8MHz df receiver.

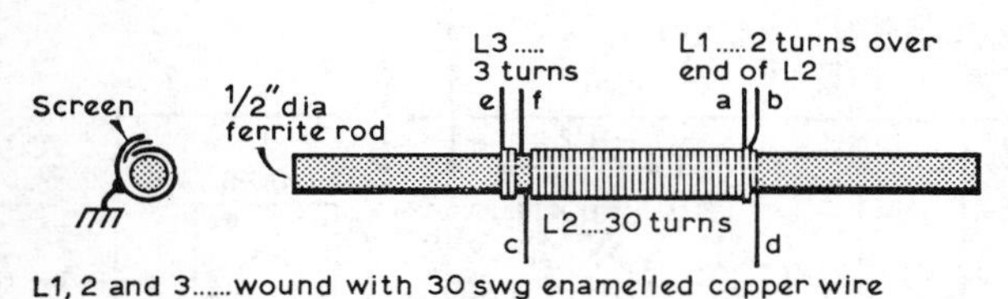

Fig 14.49. Coil winding details.

done photographically it is suggested that the board be drilled before etching.

First use a piece of 0·1in pitch Veroboard as a drilling jig and drill the 14 holes for the TAD100 using a 1mm drill. An Eclipse pin chuck is useful for holding small drills and preventing breakage. Only about a $\frac{3}{16}$in length of drill need be exposed. Other holes are then added to accommodate the components to be used. One or two holes drilled in error will not usually matter. The holes should then be connected with the resist material. In the designer's experience cellulose paint provides the most reliable protection, and if the holes are drilled first as suggested the use of one of the special resist pens is not recommended.

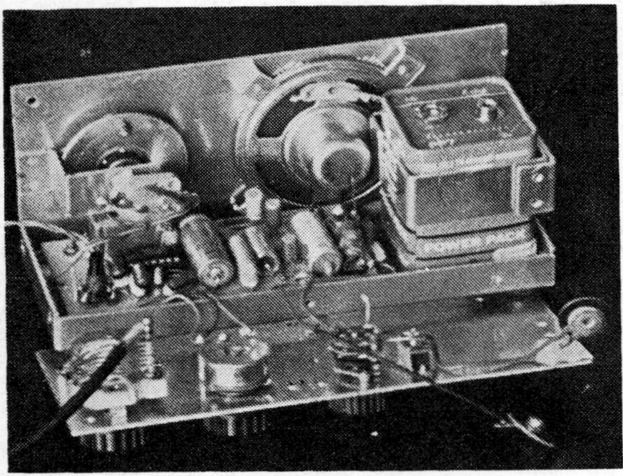

Internal view of the instrument.

The paint is applied with a fine artist's brush or a loop of fine wire. When the paint is dry and the circuit has been checked, the board should be floated upside down on the surface of a strong solution in water of ferric chloride. Etching is often completed in 10 to 15min. The resist is then cleaned off with cellulose thinners. The sense amplifier board is made in the same way.

The oscillator and bfo coils (**Fig 14.49**) are wound on $\frac{3}{16}$in diameter formers of the type often found in old radio-telephone equipment or offered for rewinding by firms advertising in *Radio Communication*. They should be wound as shown and preferably checked for resonance together with their respective tuning capacitors with a gdo. The oscillator coil has its tuned winding pile wound on top of the 15-turn winding. The other 6-turn winding is put on last. The number of turns made by the last winding may be modified to obtain the best result in individual cases. If the circuit does not oscillate, try reversing the direction of this winding.

When complete, both coils should be painted with polystyrene cement. The coils, other components and flyleads connecting pin 4 of IC1 to the junction of R14 and R15, and pin 5 of IC1 to the junction of R21 and R22, are then soldered on to the boards.

If V bending equipment is available it may be more convenient to fit the back of the cabinet separately—this was done in the original. Bending can be done very successfully over blocks of wood, providing that bending-quality aluminium is used. If the lid does not fit well, the inside dimension is easily adjusted as shown in **Fig 14.50**, providing there is a small radius at the bend. Precise dimensions of the front and back panel layouts have not been given because these will depend on the components available.

The directional aerial is wound on an 8in long, $\frac{1}{2}$in diameter ferrite rod. The number of turns for this diameter is shown but may be adjusted for other rods. Again, it is a good idea to check the result with a gdo. The screen is made from aluminium cooking foil and, by insulating the ends with varnished paper, care can be taken to avoid a short-circuited turn. The screen is then connected to the case. The rod is best protected in a handle made from an srbp tube (**Fig 14.51**). The sense aerial is made from a telescopic portable aerial and need only be 6-9in long.

The main pcb is mounted above the floor of the cabinet on

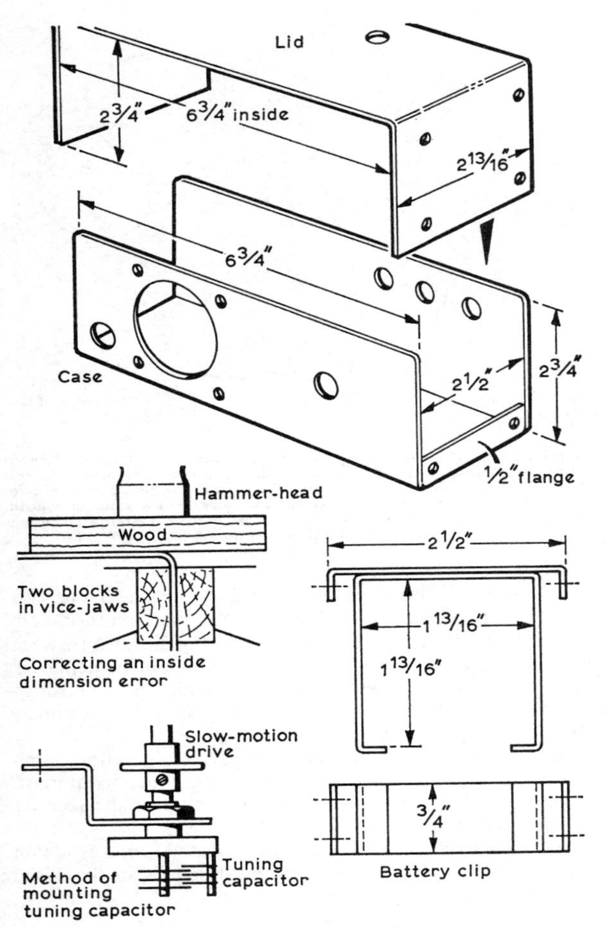

Fig 14.50. Constructional details.

two suitable spacers, and the sense amplifier is carried on a small bracket screwed to the back. A suitable epicyclic slow-motion drive is required for the tuning capacitor, and is probably best held in position with an aluminium ring and self-tapping screws. The tuning capacitor is mounted on a separate bracket as shown.

When the receiver is wired up and working, check the effectiveness of the sense aerial. There should be a minimum signal when the transmitting station is at right angles to the axis of the rod, and a maximum when the receiver is turned through 180°. With the sense aerial switched off a sharper null is found in the direction of the axis of the rod.

Sense Mechanism

A signal derived from the vertical aerial is amplified by a single fet before being added to the signal obtained from the ferrite rod. The effective strength of this received signal is almost independent of the horizontal position of the receiver, provided that the number of turns in the rod which forms the coupling winding are few in number (as indicated in the rod winding diagram).

The amplitude of the sense signal is adjusted by altering

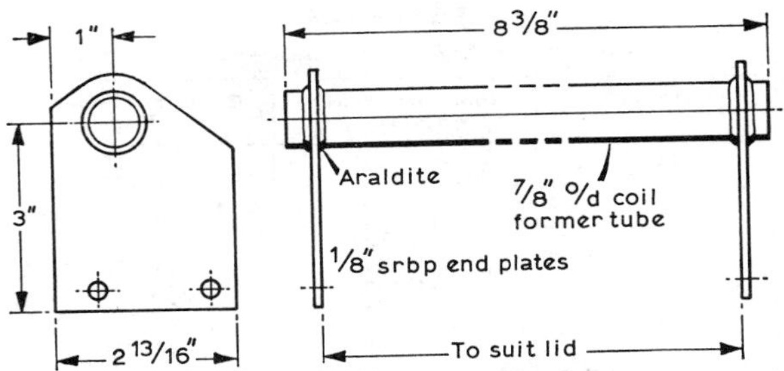

Fig 14.51. Details of directional aerial and mounting plate.

the length of the vertical whip, and when this is of the correct magnitude it will almost cancel the signal from the rod when the receiver is at right angles to the direction of propagation of the wanted signal in one position. The two signals are additive when the rod is turned through 180°. Thus it is possible to roughly determine the direction of the hidden station, but the null is not normally sharp enough to be used for taking an accurate bearing. Once a rough indication is obtained, the sense amplifier is switched off and the receiver is turned through 90° to obtain a more accurate result.

Using the Receiver

Results are sometimes more reliable when the receiver is near to the ground. Avoid wire fences and other metal objects which might distort the signal field pattern. In a walk-round hunt it is hardly worthwhile taking compass bearings, and a follow-your-nose technique may best be used; otherwise bearings may be taken with an oil-damped compass and plotted on an Ordnance Survey map. Note that the compass needle points to magnetic north, which does not correspond to true north and the grid lines on the map, but there is a simple method of making allowances for this on every map. The designer uses a Swiss Recta compass which is set to allow automatically for the error; no doubt others have the same facility.

REPEATERS

Over the past few years repeaters have had a considerable impact on vhf and uhf mobile (and low-power portable) operation. The purpose of this section is to describe the basic concepts of the system as used for mobile operation in the UK, together with the equipment required by the user. A note on operating procedures is also given.

Advantages of Repeater Operation

A repeater is a device which will receive a signal on one frequency and simultaneously transmit it on another frequency. Thus a low-power transmitter in a vehicle can transmit on the repeater's *input channel* and the signal will be faithfully re-transmitted on the repeater's *output channel*. Repeater channels used in Europe and the UK are given in **Tables 14.8** and **14.9**.

In effect, the receiving and transmitting coverage of the mobile station then becomes that of the repeater and, since the latter is favourably sited on top of a hill or high mast, the range is usually greatly improved over that of unassisted or *simplex* operation (see **Fig 14.52**). Typically, the effective range is increased from the order of 5–15 miles from the mobile station to something like 30 miles in any direction from the repeater depending upon terrain and band used.

TABLE 14.8

IARU Region 1 vhf and uhf repeater channels

Channel	Input frequency (MHz)	Output frequency (MHz)
R0	145·000	145·600
R1	145·025	145·625
R2	145·050	145·650
R3	145·075	145·675
R4	145·100	145·700
R5	145·125	145·725
R6	145·150	145·750
R7	145·175	145·775
R8	145·200	145·800
R9	145·225	145·825
RU0	433·000	434·600
RU1	433·025	434·625
RU2	433·050	434·650
RU3	433·075	434·675
RU4	433·100	434·700
RU5	433·125	434·725
RU6	433·150	434·750
RU7	433·175	434·775
RU8	433·200	434·800
RU9	433·225	434·825

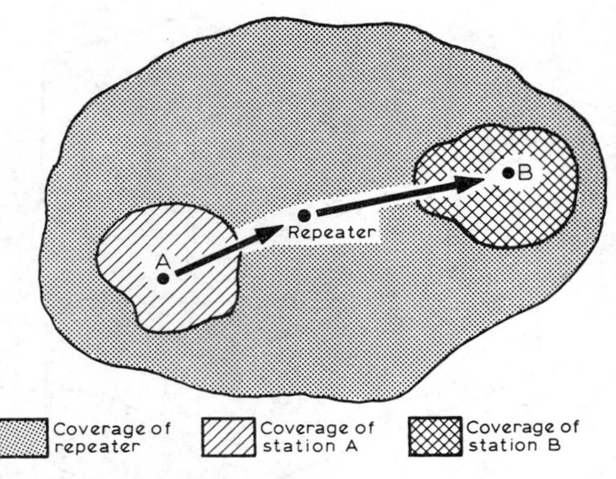

Fig 14.52. The improved range of communication available between mobile stations using a repeater. The simplex coverage areas of stations A and B shown are constantly changing shape as the two vehicles pass through different terrain, and thus station B would have to be quite close to station A before reliable simplex communication was possible.

Another advantage is that contact with other stations becomes more predictable. The coverages of stations A and B in Fig 14.52 continually change shape as the two stations pass through different terrain. Thus it is never easy for the mobile operator to estimate his or her simplex range. In contrast, the repeater service area is a known and much less variable factor.

One further advantage is that mobile "flutter" is usually diminished due to the superior aerial location at the repeater site.

It should be noted that these advantages are not gained at the expense of operating convenience, access to the repeater and the offset in transmit and receive frequencies being accomplished automatically by the user's transceiver.

How a Repeater Works

The major components of a repeater are an aerial system, a receiver, a transmitter and a control system. The receiver and transmitter do not differ greatly from those in use at a conventional amateur station, although a high professional standard is commonplace. The main differences lie instead in the aerial and control systems used, and these will now be discussed.

Aerial system

Because the same band is used for simultaneous transmission and reception, special arrangements have to be made so that the repeater transmitter does not seriously degrade the performance of the associated receiver.

TABLE 14.9

UK uhf repeater channels

Channel	Repeater input (MHz)	Repeater output (MHz)
RB0	434·60	433·00
RB2	434·65	433·05
RB4	434·70	433·10
RB6	434·75	433·15
RB8	434·80	433·20
RB10	434·85	433·25
RB14	434·95	433·35

These frequencies are the reverse of the IARU-recommended frequencies. Other channels have been allocated to repeaters but these are the only ones in use at the time of going to press. RB8 is reserved for Radio Amateur Emergency Network (Raynet) mobile repeaters. 433·2MHz is used in some areas as a simplex channel (SU8).

Fig 14.53(a) illustrates the two-aerial approach. The two aerials are physically separated from one another and, depending on the frequency in use and the distance between the aerials, a certain amount of isolation can be achieved. This is further improved by the use of high-pass and low-pass filters as shown, these usually being of extremely high-*Q* cavity construction.

Fig 14.53(b) shows the single aerial configuration, commonly used where space is at a premium on the aerial mast. A device called a *circulator* is used to prevent the receiver from "seeing" the transmitter and vice versa.

Whichever repeater configuration is used, good isolation between the transmitted and received signals in the aerial system is vital for satisfactory service.

Control system

A repeater is an unmanned relay station and therefore requires an automatic system to control its operation. This system must ensure as far as possible that the repeater only relays signals which are intended for that repeater, and that those signals it does relay come up to an acceptable standard in respect of frequency, strength and deviation.

To ensure that signals which are not intended for relaying cannot activate the repeater transmitter, UK repeaters require a short audio tone (or *toneburst*) of precise frequency and duration at the beginning of each transmitting over. This signals to the repeater that this signal is intended for relaying, and also sets in operation timing circuits within the control system. It should be noted that this requirement is by no means universal throughout the world. For example, Australian repeaters do not require a toneburst at all, while Continental European repeaters only require one initial toneburst to switch the repeater on.

A typical sequence of events during a transmitting over is as follows. The repeater receiver is switched on continually and monitors the input channel, using a squelch system. When a signal appears on the input the control system determines whether or not:

(i) the signal has tripped the squelch,

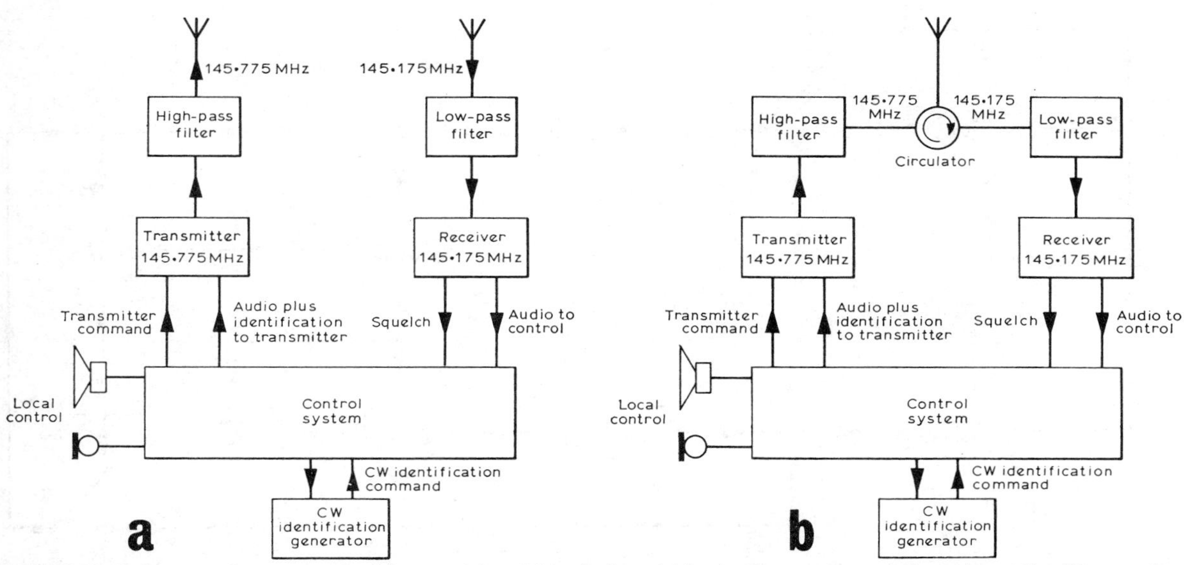

Fig 14.53. Block diagrams of repeaters using (a) two aerials and (b) a single aerial for simultaneous transmission and reception. The repeaters shown here are operating on channel R7 (see Table 14.8).

(ii) there was an audio tone of correct frequency and duration present,
(iii) the received signal is correctly deviated.

If these initial conditions are satisfied (known as a *valid access*) the control system will switch the repeater transmitter on and allow the receiver audio output to modulate it by opening what is known as the *talk-gate*. During the over, the control logic continuously monitors the incoming signal level and its deviation and, if either of these parameters fall at any time below the standard required for valid access, may switch off the transmitter or disconnect the receiver audio from the modulator by closing the talk-gate.

When the over is finished and the incoming signal disappears from the input channel, the repeater squelch will close and indicate to the control logic that it must ready itself for the next over. After a short delay the repeater will signal it is ready by transmitting either a "K" in morse code or a tone, and the next over may then commence.

The short delay between the end of an over and the "K" is quite important. During this interval the repeater will relay any new signals on the input channel (irrespective of whether they are preceded by a toneburst or not). Consequently this interval may be used by a third station to quickly announce its presence. This practice is termed *tail-ending* and is a good way of inserting urgent or emergency messages between overs (see later).

If the repeater has relayed a signal for more than a certain period of time (typically set between 60 and 90s) the control system will go into the *time-out* mode, close the talk-gate, and may transmit some form of *busy* signal until the incoming signal disappears off the input channel. This is done solely to prevent overs from being too long. When the input signal finally clears, the repeater will stop its busy signal, ready itself and respond with a "K" when it has done so.

Sometimes a perfectly valid access is made without being followed by a proper transmission. This is termed an *erroneous valid access* and the repeater may send its callsign and then switch off its transmitter.

The flow chart of **Fig 14.54** is a diagrammatic summary of the sequences described above and illustrates their relationships in a basic control system. Many repeaters have much more sophisticated control logic than is implied here. For example, some send a telemetric indication if the incoming signal is off-frequency or over-deviated. Again, some repeaters have slightly more stringent conditions for valid access and require the user to give his or her callsign immediately after the toneburst. Repeaters which have provision for automatic switchover to emergency power may indicate they have done so by some form of telemetry.

Further discussion of repeater logic and facilities is beyond the scope of this book, but most repeater groups have literature available which fully describes their repeater and prospective users are advised to study this carefully, and then spend some time listening to the repeater in use.

Toneburst Circuits

As stated above, UK repeaters require a short audio tone of precise frequency at the beginning of each transmitting over. Although the type of circuit required to produce this tone is quite simple, few things cause more discussion among repeater operators than toneburst circuits and their adjustment.

Fundamentally, there are only three requirements for a toneburst unit able to access all UK repeaters—it has to produce a tone of 1,750Hz ± 5Hz lasting for about 500ms and deviating the carrier by about 2·5kHz. The tone frequency of 1,750Hz is standard in Europe, but the duration of the tone required varies from repeater to repeater, some requiring as little as 200ms, others (including some in Continental Europe) requiring much longer. However 500ms will be sufficient in the UK.

These three requirements may be quite easy to achieve indoors at a fairly steady temperature of around 20°C, but

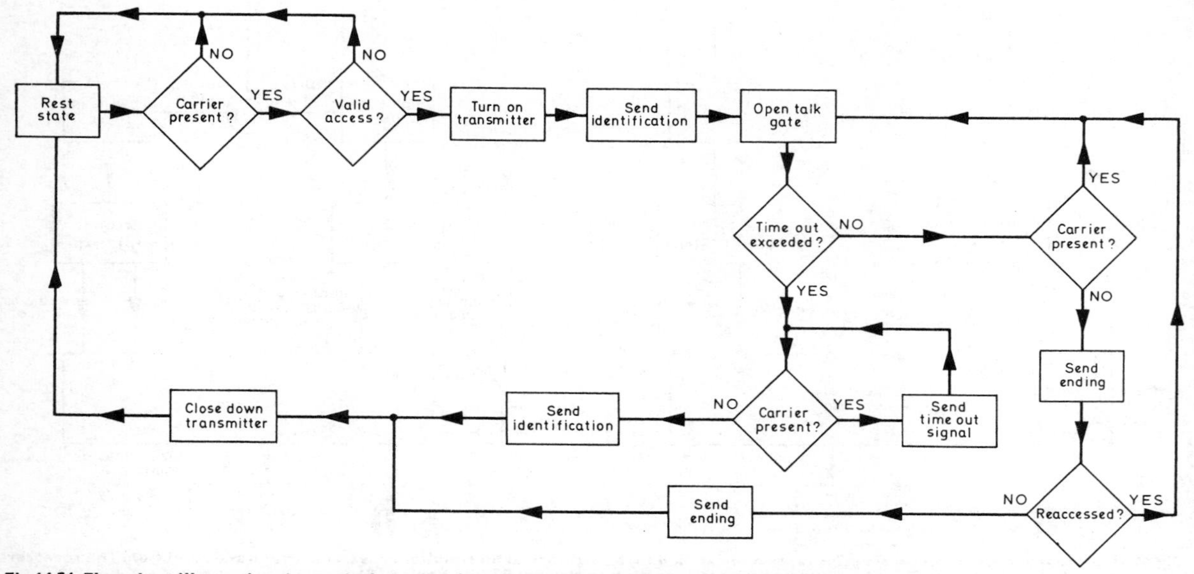

Fig 14.54. Flow chart illustrating the method of operation of a simple repeater control system. The starting point is the "Rest state" box (top left). Diamonds denote interrogation of the receiving system by the control logic and the answers it obtains, while rectangles denote the resulting action taken. In practice, most repeater control systems are more complex and may include facilities, for example, to indicate if the incoming user signal is off-frequency.

the interior temperature of a car can range from −10°C on a cold winter morning to 40°C when it has been standing out in the sun on a summer's day. Even some commercial units are incapable of meeting the requirements over this temperature range. Add to this the fluctuations in battery voltage and the constant vibration encountered in mobile service and it can be seen that problems can readily occur.

There are two approaches widely used to solve this difficulty. One is to use an RC oscillator but with high-quality components, the other to use a crystal-controlled oscillator and divide its frequency down to the toneburst frequency. An example of each type of circuit is described later.

If the toneburst unit is of the RC type it will require setting on to the correct frequency of 1,750Hz. This can easily be done with, for example, a digital frequency meter, but this type of test equipment may not be to hand. In this case one possible method is to ask a station with a correctly set toneburst to transmit the tone on a simplex channel and beat this against the tone from the circuit being adjusted, which for this purpose should have its output connected to an audio amplifier and loudspeaker. If possible both toneburst units should be set to run continuously as it is more difficult to adjust a half-second bleep accurately.

If another station's co-operation cannot be enlisted then the next best method is to listen on the input frequency of the local repeater and compare the toneburst frequency with those of the stations using the repeater.

Only when it is considered that the toneburst is on the right frequency should a test call be made into the repeater, eg "G4XYZ testing access through GB3ZZ". If access is achieved the repeater will respond with a "K" or some other indication. If it does not the toneburst should be re-adjusted *off-the-air*. Under no circumstances should a continuous transmission be made to the repeater while adjustment of the tone frequency is being carried out.

G3VEH "SUPER-BLEEP"

This circuit, shown in **Fig 14.55**, uses two μL741 operational amplifiers. IC1 is connected as an oscillator using a

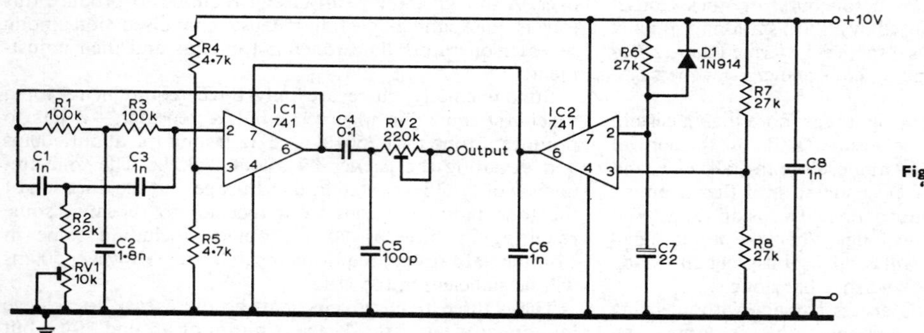

Fig 14.55. Circuit of G3VEH "Super-Bleep" toneburst unit.

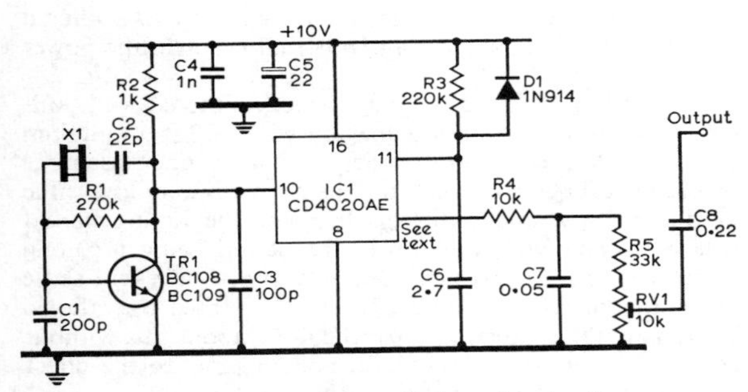

Fig 14.56. Crystal-controlled toneburst unit.

T network to derive the frequency, while IC2 is arranged as a timing device.

The frequency of the oscillator is set by RV1 and the output level by RV2. Keying is achieved by either switching the 12V positive line or by grounding the negative line on transmit. When this occurs the oscillator will start sending a tone, C7 will commence to charge up and the voltage at pin 2 will rise. When it passes the voltage at pin 3, set to half the supply voltage by potential divider R7 and R8, IC2 will switch off the oscillator. Thus the values of R6, C7 determine the duration of the tone—increasing the value of C7 will give a longer burst and vice versa. When at the end of the transmission the 10V supply is switched off, C7 will discharge through D1, thus resetting the timer ready for the next transmission.

There is no need to stabilize the power supply as the oscillator does not vary in frequency by more than a few hertz over a supply range of 5–15V. However, it is essential that C1, C2 and C3 are polystyrene capacitors and R1, R2, R3, R4 and R5 are metal film resistors.

When initial adjustments are being made to RV1 and RV2, it will be found convenient to temporarily ground pin 2 of IC2, thus ensuring a continuous tone.

CRYSTAL-CONTROLLED TONEBURST

Crystal-controlled tonebursts have two great advantages over those using RC oscillators: they are not so temperature-sensitive and they usually require no initial frequency adjustment before use.

This toneburst design **(Fig 14.56)** uses a simple crystal-controlled oscillator, the output of which is divided down to 1,750Hz in a cmos divider, type CD4020AE. This particular device is capable of dividing by 2^n, where $n = 7$ to 12, according to which output pin is selected. Consequently there is a choice of crystal frequencies possible; **Table 14.10** gives details. It will be seen that it is possible to use 450kHz FT241 crystals from the junk box and **Table 14.11** shows those channel numbers which will give the required 1,750Hz

TABLE 14.10

CD4020AE divider connections

Crystal frequency (MHz)	Division	Output pin for 1,750Hz
7·168	4,096 (2^{12})	1
3·584	2,048 (2^{11})	15
1·792	1,024 (2^{10})	14
0·896	512 (2^9)	12
0·448	256 (2^8)	13
0·224	128 (2^7)	6

TABLE 14.11

FT241 crystals suitable for toneburst use

	Channel number	Marked	Fundamental (kHz)	Toneburst (Hz)
54th harmonic series:	41	24·1	446·296	1,743·3
	42	24·2	448·148	1,750·6
	43	24·3	450·000	1,757·8
72nd harmonic series:	321	32·1	445·833	1,741·5
	322	32·2	447·222	1,746·9
	323	32·3	448·611	1,752·4
	324	32·4	450·000	1,757·8

when divided by 2^8. If these older crystals or other low-frequency crystals are used the oscillator may need to be modified by increasing its capacitances. A change of C1 to 470pF, C2 to 100pF and C3 to 1,000pF has been found to work well with FT241 crystals.

The reset facility connected to pin 11 on the ic permits an extremely simple timing circuit to be used. When C5 has charged up to above half the supply voltage (this takes about 400ms) the ic resets and mutes the tone automatically, although the oscillator continues to function. D1 provides a fast discharge path for C5.

The output from the divider is a square wave which is changed to triangular waveform by R4, C6. The output-voltage level is adjusted by RV1 to give about 2·5kHz deviation.

Power for the unit may be taken from any 10V+ regulated source in the transmitter, such as that which often feeds the rf oscillator. Power consumption is minimal. If no such regulated source is available a simple zener diode and series resistor regulator will be sufficient—good regulation is not required but the unit should be protected against voltage spikes or over-voltage.

TIME-OUT INDICATOR

Not everyone can mentally time overs accurately, especially while driving a car, and this simple indicator will be found a useful addition to some mobile installations.

The circuit is a standard monostable or pulse generator **(Fig 14.57)**. The only components which affect the time delay are C2 and R3, and since the time involved is fairly long, the capacitor should have low leakage and thus preferably be a tantalum type. Various combinations of R and C are

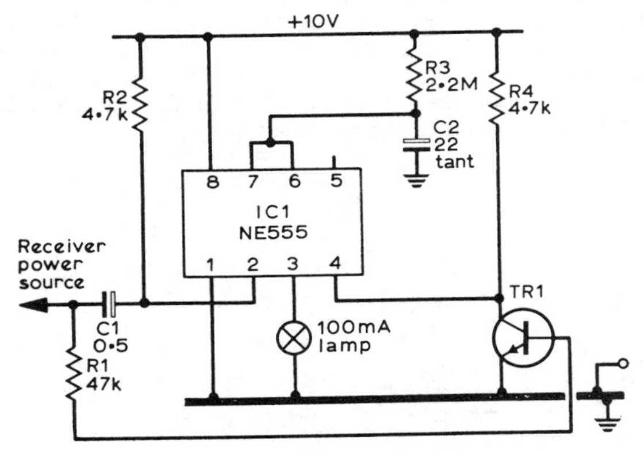

Fig 14.57. Simple time-out indicator. TR1 can be almost any npn transistor.

TABLE 14.12

Code of practice for repeater operation

1. Before attempting to transmit, ensure that:
   (a) your transmitter and receiver are on the correct frequency,
   (b) your toneburst is operating correctly,
   (c) your peak deviation is correctly set.
   Any adjustment to the transmitter or toneburst required should be carried out off-the-air, *not* through a repeater.

2. Check you will only access the repeater you wish to use (especially important in lift conditions).

3. Listen to the repeater before you transmit to make sure it is not in use or in beacon mode. If you hear a local station you wish to call, listen on the input frequency to check whether the station is within simplex range before calling.

4. Unless you are specifically calling another station, simply announce that you are "listening through", eg "G4XYZ listening through GB3ZZ". One announcement is sufficient. If you are calling another station, give its callsign followed by your own callsign, eg "G2XYZ from G4XYZ".

5. Once contact is established:
   (a) at the beginning and end of each over you need give only your own callsign, eg "From G4XYZ";
   (b) change frequency to a simplex channel at the first opportunity (especially if you are operating a fixed station);
   (c) keep your overs short and to the point or they will time-out and do not forget to wait for the "K";
   (d) do not monopolize the repeater as others may be waiting to use it;
   (e) if your signal is very noisy into the repeater, or if you are only opening the repeater squelch intermittently, finish the contact and try later when you are putting a better signal into the repeater.

6. If the repeater is busy emergency calls may be made by tail-ending before the "K", and announcing (a) that you have emergency traffic, and (b) which facilities you wish a station to provide. This will normally in most "risk-to-life" situations be a telephone so that the other station can alert the emergency services. Do not reply to an emergency call if you cannot provide the services requested.

usable, provided a limit of 3·3MΩ for R3 is not exceeded. The formula for on-time is T = 1·1 RC, R in megohms, C in microfarads, provided the leakage of C is negligible. The values given provide a time of 54s, but if a slightly shorter time is required reduce the value of R3 slightly.

Output voltage at pin 3 is high while timing is in progress, falling to zero at the end of the timing process, and is able to source up to 200mA. A suitable lamp would in many transceivers be the TRANSMIT lamp, and if this were used, no drilling of the front panel would be necessary.

The timer is switched on by a short negative pulse at pin 2, which is normally held at V_{cc}. This can usually be conveniently obtained from the receiver supply voltage, which will drop from 10–12V to zero on transmit. The pulse is formed by C1 and R2. The timer is switched off by grounding pin 4, the reset line, and this is accomplished by TR1, which may be almost any npn transistor, inverting the positive transition of the receiver supply.

The timer should be permanently connected to a filtered source of +10V to avoid transients that occur in the power supply.

The only problem that may be encountered with this circuit is that of capacitor leakage. Even if C2 is a tantalum type, it will need several cycles of charge and discharge before its leakage settles down to its normally very low value of under 1mA, and until this happens the timing period will vary. This will show up in the period being too long when the timer is first activated each day, but it will settle down to the correct value after two or three operations. There is nothing much that can be done about this without greatly complicating the circuit, and in practice the effect does not seem to matter very much.

Using a Repeater

The proper use of a repeater requires a high standard of operating ability and courtesy. Knowledge of the way in which repeaters work and confidence that one's own equipment is "spot-on" does help, but also required is an ability to think reasonably quickly and express oneself concisely, this being especially important on a repeater with a high level of activity. Such skills can be developed, but it does take time. If the reader is a newcomer to repeater operation, it will be found that there is no substitute for listening to a repeater and noting the best ways in which to operate.

A spell of listening will soon show that the two most common errors are "timing-out" and forgetting to wait for the "K". In each case the repeater will eventually interrupt communication and the user, quite unaware of this, may spend up to a minute or so blocking the repeater to no avail. If it is found difficult to time overs mentally it may be worthwhile buying or building a simple time-out warning device as described in the previous section.

Table 14.12 gives a suggested code of practice for repeater operation in general. Certain repeaters, especially those outside the UK, may have different requirements, but if the rule "listen before transmitting" is followed this should present no particular problem. Repeater groups usually have available literature giving full details of their repeater and its facilities, which is well worth studying.

It should be noted that although a repeater may provide a telemetric indication that a user's transmitting equipment is operating incorrectly this does not mean that equipment should be aligned through the repeater, nor is a lack of indication necessarily a cause for complacency. For example over-deviated signals having deviations in the range ±2·5kHz to ±5kHz will often pass through a repeater without actuating its audio blanking circuits. All indications should therefore be regarded as a warning, not as a substitute for off-air alignment.

CHAPTER 15

NOISE

IN every receiver there is a continuous background of noise which is present even when no signal is being received. This noise is caused partly by the valves and circuits of the receiver and partly by electrical impulses which reach the aerial from various sources: ie it is partly *internal* and partly *external*. In general, the more sensitive the receiver the higher the noise level, and increasing the sensitivity beyond a certain point brings no further increase in range. It is then possible to receive at adequate strength any signal not submerged in the noise, and further amplification will be of no assistance since both noise and signal will increase together. The sensitivity is then described as being *noise limited*, and this is the normal condition in modern receivers intended for weak-signal reception.

The purpose of this chapter is to discuss the characteristics of the various kinds of noise and to outline the broad principles of receiver design which must be followed if a high performance is to be achieved. Detailed treatment of special noise suppressors or noise limiters for reducing interference from ignition systems and other impulse generators will be found in the various chapters dealing with receivers.

An account is given here of the general characteristics of noise, the way it affects reception, the various sources of noise, and the principles involved in reducing its effects to a minimum. Finally the application of this knowledge to the needs of the average amateur is discussed, and it is hoped that this may prove of value to those who have not the time or are not sufficiently interested in the "whys and wherefores" to read the earlier parts of the chapter.

THE NATURE OF NOISE

The noise contributed by the receiver can be minimized by careful design, but it is important to appreciate that internal noise tends to increase with frequency whereas noise from external sources usually decreases. By careful design receiver noise can usually be made negligible compared with external noise below about 100MHz. Internal noise is caused by the unavoidable random movement of electrons in valves and circuits, and occasionally by avoidable defects such as dry joints in the wiring, faulty valves, and partial breakdowns of insulation. External noise can be caused by electrical machines and appliances (*man-made noise*), by atmospheric electrical discharges (*static* or *atmospherics*) and by radiations from outer space.

Most forms of electrical noise can be roughly pictured as a mixture of alternating currents of varying amplitudes and of all possible frequencies. Receivers, however, have a limited bandwidth and the observed noise level is a summation of all the components occurring within this band. As

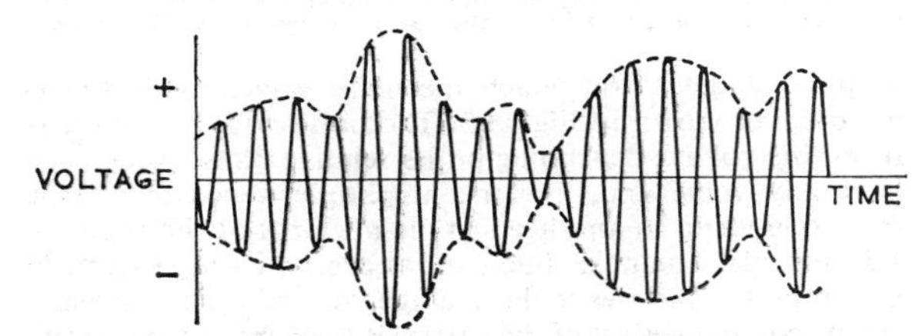

Fig 15.1. Noise waveform at the output of an i.f. amplifier. The oscillations, which are at the intermediate frequency, are approximately sinusoidal but have randomly varying amplitude.

might be expected from this picture, the mean noise power is proportional to bandwidth, although at any given instant the noise amplitude can have any value depending on the extent to which the various components happen to be assisting or cancelling each other.

Fig 15.1 shows a typical noise waveform such as might be produced at the output of an intermediate-frequency amplifier by random movements of electrons in early stages; the i.f. amplifier can only transmit frequencies within its passband, and the noise applied to the detector has therefore the character of an i.f. signal of fluctuating amplitude. **Fig 15.2** shows the same noise after detection by a linear rectifier; this consists of a positive (or negative) voltage which fluctuates in accordance with the amplitude of the input noise voltage. The important features of this waveform are its mean level, its statistical properties (summed up by the right-hand scale on the diagram) and its rate of fluctuation which is such that if the bandwidth of the circuits preceding the detector is B Hz it requires a time of about $1/B$ seconds for the amplitude to change significantly in value.

A steady carrier or the output from the beat-frequency oscillator when applied to the detector produces a steady dc

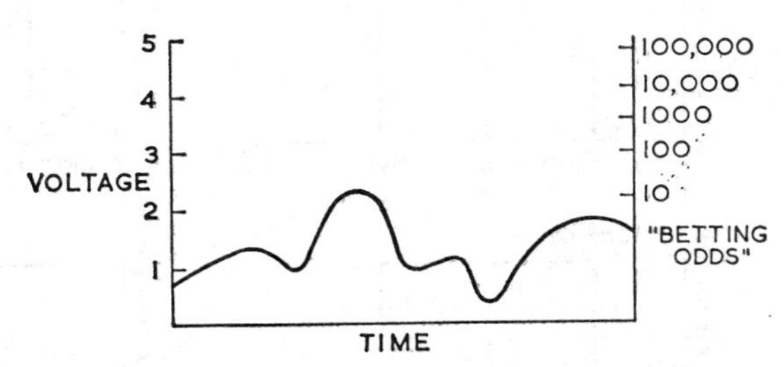

Fig 15.2. The same noise waveform after linear rectification. Note that it consists of the envelope of the original noise oscillations. The waveform as drawn has an average voltage amplitude of 1V and the "betting odds" against any particular instantaneous amplitude being exceeded are indicated by the right-hand scale: this applies to "white-noise" only.

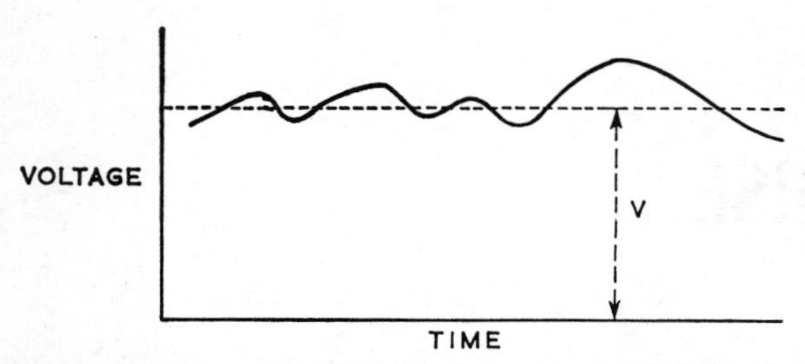

Fig 15.3. Output of linear detector when the input consists of a carrier (of voltage V) plus noise. Note that the average voltage remains equal to V as long as the noise peaks are less than V, ie as long as the noise-modulation of the carrier is less than 100 per cent.

output voltage V, on which the noise waveform is superimposed as shown in **Fig 15.3.** The noise adds or subtracts from this voltage depending on its relative phase, which can have any value and, like the amplitude, tends to change from one value to another in time-intervals of the order of $1/B$ seconds. The noise fluctuations are now symmetrical in the sense that they leave the average voltage V unchanged.

One common effect of the carrier is to change the apparent pitch of the noise. Thus the waveform of Fig 15.2 can be thought of as being produced by noise frequencies up to a bandwidth of B Hz beating together, while that of Fig 15.3 is due (in the case of normal double sideband reception of A3 signals) to a carrier in the centre of the passband beating with noise frequencies out to a distance of only $B/2$ away from it on either side. This causes a lowering of pitch, which rises again when the carrier or bfo is detuned as for ssb or single-signal cw reception. This behaviour is illustrated in **Fig 15.4.** With a little practice the operator can use this rise in pitch as a method of setting up the bfo.

Frequently the presence of the carrier or the output from the bfo increases the noise level, and this is commonly mistaken for noise on the carrier. The correct explanation is that when the signal level is low all detectors operate in a square-law manner, which is relatively inefficient; frequently the noise level is too low to cause the detector to operate on the linear part of its characteristic, so that the increase in noise level when the carrier is applied indicates more efficient detection and not a noisy carrier.

Noise originating externally to the receiver is often much more impulsive in character than the internal noise depicted in Fig 15.1. It can have a wide variety of characteristics, which makes it difficult to discuss in general terms, but the representation of Fig 15.1 can be applied to it, bearing in mind that (i) the dynamic range may be much greater and the "betting odds" of Fig 15.2 are not applicable, (ii) the envelope may show much slower variations, and (iii) there may be gaps in the noise waveform which can be exploited for reception particularly when, as is frequently the case in amateur working, low-grade communication is better than none.

THE EFFECT OF NOISE ON RECEPTION

The various random voltages originating in the receiver, together with certain kinds of externally-generated voltages, give rise to what is popularly known as *white noise.* It is described in this way because, like white light, its energy is uniformly distributed over the relevant frequency band. The manner in which such noise interferes with the reception of a signal is governed by certain elementary principles. Some discussion of these principles may prove interesting and helpful, although as will be seen their application to practical problems is not always straightforward.

It is useful to start by considering the nature of the signal. Any signal waveform can be considered as being built up from a succession of impulses, and it is obvious that the

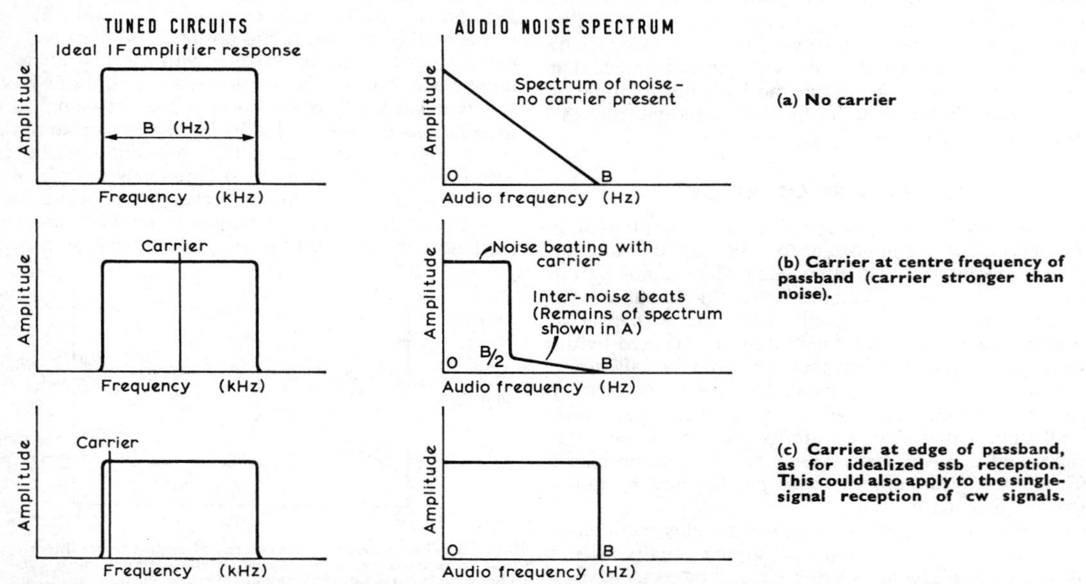

Fig 15.4. Frequency characteristics of noise in various receiving conditions. When the frequency of the carrier (or the bfo) is at the centre of the passband the noise-pitch is at its lowest, and when it is tuned to the edge of the passband the noise-pitch is at its highest. The bandwidth of the amplifier is assumed to be at least as great as the i.f. bandwidth.

more rapidly the voltage constituting the signal changes, ie the higher the frequencies in its modulation envelope, the shorter the impulses which must be used to obtain a reasonably accurate representation of it. A cw signal, for example, could be built up from impulses equal in length to a morse dot, say 0·1s, whereas a telephony signal having frequencies up to 3kHz would need to be regarded as a succession of impulses not more than 1/3,000s in length. If t is the duration of an impulse, an i.f. bandwidth of $1/t$ is needed for reproducing it, ie 3,000Hz in the case of a telephony signal. The use of the term *i.f. bandwidth* avoids any possible confusion due to factors arising in the detection process, such as the ratio of two between the overall and i.f. noise bandwidths which is illustrated in Fig 15.4(b), but it will be shown later that such distinctions are not important during the actual reception of normal amateur signals when these are interfered with by noise only. The rf bandwidth is usually much greater than the i.f. bandwidth and can therefore be ignored.

The optimum bandwidth must be in the region of $1/t$, since this value allows the wanted impulse to build up almost to full amplitude and a greater bandwidth can only increase the effective amount of noise and not the signal. On the other hand, a smaller bandwidth reduces the amplitude of the signal more rapidly than that of the noise.

The noise in a bandwidth of $1/t$ also consists of impulses of duration t, but these vary in amplitude and phase in a random manner. Each impulse, before detection, normally contains a considerable number of cycles of rf or i.f. oscillation, like those shown in Fig 15.1. When a waveform is represented by impulses in this way, the phase and amplitude are regarded as being constant for the duration of any one impulse but there need not be any similarity between consecutive impulses. It will be appreciated that this is merely a convenient approximation, since in practice there is usually a smooth transition from one value to the next.

The size of any one noise impulse is a matter of chance, some values being more likely to occur than others, and Fig 15.2 shows the "betting odds" against any given impulse exceeding the indicated values. To see what this means in practice, consider the reception of a single morse dot, the bandwidth being just wide enough and no more; Fig 15.2 shows that there is a reasonable chance of the noise amplitude reaching, say, three or four times the rms value, and unless the signal is even larger in amplitude it is likely to be mistaken for noise or vice versa. It is impossible to detect with certainty the presence of the morse dot in these circumstances, either aurally or mechanically or by any means whatsoever, unless it is stronger than any likely value of the noise. In aural reception, however, with the stipulated bandwidth (about 10Hz), the noise has a rough musical pitch similar to the wanted beat-note, resulting in an unpleasant overall effect, and the discriminatory powers of the ear are not used to best advantage. Under normal conditions it is found that the bandwidth can be extended indefinitely, and although the noise level rises the ability to copy a given weak signal through it is unaffected. This is because the ear itself, from the point of view of discrimination between signals and noise, acts as a filter having a passband of about 50Hz and thus tends to determine the overall receiving bandwidth. The noise which is effective in masking a wanted 1,000Hz beat note is mainly that part of the noise spectrum between 975 and 1,025Hz, and adding noise outside these limits has no adverse effect at normal audio levels. The usefulness of aural selectivity is not restricted to the separation of signals from noise, and most readers will be familiar with the fact that a weak morse signal can be read in the presence of a much stronger unwanted signal when the beat notes are well separated in pitch, whereas two signals producing nearly the same heterodyne frequency are much more difficult to copy.

It has been explained above that the problem of separating weak signals from a noise background is in principle one of deciding at each instant whether a true signal or only noise is present at the output of the receiver. This is a useful way of visualizing the basic problem, but in aural or visual reception the process is not carried out consciously and deliberately in the manner stated, and a variety of subjective or intellectual influences are usually at work. These are dependent partly on "integration" and partly on the fact that large portions of most messages are redundant so that missing words and phrases can often be filled in simply because nothing else would make sense. A missing dot in a morse transmission may be serious if it occurs in a callsign, but in a group of figures, known to contain no letters (eg a contest number or RST report), the figures can often be correctly deduced even if several of the dots are missing.

Integration is the process of adding together several signal impulses to give one larger one. This is largely the preserve of radio astronomers and radar engineers, but it also forms part of the subjective processes which are called into play while an attempt is being made to decipher a weak signal against a noise background.

Because of these and other complexities (eg fading) it is difficult to give any useful figures of signal-to-noise ratios required for amateur communication: more-or-less agreed figures exist for various types of commercial service, but amateur requirements tend to be both more modest and more elusive. The ability to read signals through white noise is not possessed in the same degree by all operators, but in general the human ear is a very efficient device for this purpose. When, as in amateur communication, messages are in plain language and tend to have standardized content such as RST reports, names and greetings, the aural selectivity tends to be supplemented by discrimination of an intellectual kind and no improvement is obtainable by the use of additional filters or other devices. It must be emphasized, however, that this applies only to white noise. It further assumes the absence of overloading in the receiver, adequate bfo injection voltage at the detector and a reasonable amount of audio gain. At very high volume levels the ear has a non-linear characteristic, reception is degraded and, if the bandwidth is increased at constant gain, the noise level will be further raised and the situation will be aggravated. On the other hand, if the volume is too low, signals comparable with the noise level may not be loud enough for good intelligibility. In the case of impulsive noise (eg atmospherics and ignition noise) the ear can often be given considerable assistance by devices such as limiters which prevent overloading and enable a signal to be copied through gaps in the noise (see Chapters 4 and 14).

An understanding of the mechanism of detection (including the effects shown in Fig 15.4 and the relatively complex phenomenon of modulation suppression) is important to the understanding of many noise problems and leads, as we have seen, to correct interpretation of the ways in which the amplitude and the pitch of the noise vary during the tuning-in of a signal. However, in regard to the actual copying of signals of the kinds met with in amateur radio, the problem can be very much simplified; this is because during such reception the detector is always operating *as a mixer*, ie there is always present a continuous signal or carrier which is strong compared with the noise, and the detector output is the beat frequency between the signal which conveys the intelligence and this strong oscillation. So long as this holds true, the detector, like any other frequency changer, is merely a device for shifting the frequency spectrum; it has no other effect on the signal or the noise, and no distinction need be made between post-detector and pre-detector selectivity, except where a factor of 2 must be introduced to allow for the superimposing of two similar sidebands in the process of detection, as indicated in Fig 15.4(b). This argument breaks down if, in a.m. reception, the carrier is not strong compared with the mean noise level, but in the case of the narrow band receivers usually employed for amateur reception the signal is then too weak for normal speech signals to be readable. If, however, the predetector bandwidth is increased to an amount greatly in excess of requirements, the noise voltage may be increased so that it exceeds the carrier level of a weak signal; it then exerts a "capture effect" and the wanted modulation is suppressed.

INTERNAL NOISE

Noise originating in the receiver can usually be distinguished from external noise by the fact that it persists after disconnecting the aerial, but this test can be misleading since the change in damping or tuning of the first circuit may alter the internal noise level, and it is better to make the experiment by changing over from the aerial to an equivalent dummy aerial. In the hf bands, where external noise levels are high, this gives a simple and accurate check of the noise performance of the receiver; thus a change of 6dB (ie a ratio of 4 : 1) in the noise level would mean that the receiver was responsible for a quarter of the total noise power, and the overall performance would be about 1dB worse than with a perfect receiver. These figures, of course, assume linear detection and no overloading in the receiver.

Provided that no part of the internal noise is due to faulty wiring or components, hf noise consists mainly of

(*a*) thermal noise,
(*b*) shot noise in valves and transistors,
(*c*) induced grid noise in valves.

Whichever variety of noise is being considered, it is of practical importance only when generated in the early stages of the receiver, since noise generated later in the receiving chain is not amplified to the same extent.

Thermal Noise

The electrons in a conductor are in continuous random motion at a rate which increases with the absolute temperature (obtained in degrees Kelvin by adding 273 to the temperature in degrees Centigrade). The voltage produced by the electrons add together like other ac voltages, ie as the square root of the sum of the squares, and therefore if two similar conductors having equal resistances are joined in series the resultant noise voltage produced is $\sqrt{2}$ times that developed across either of the conductors considered singly. For any given conductor the noise voltage corresponding to a bandwidth of B Hz is equal to $\sqrt{(4KTBR)}$ where K is Boltzmann's constant ($1{\cdot}37 \times 10^{-23}$ joules/deg K), T is the

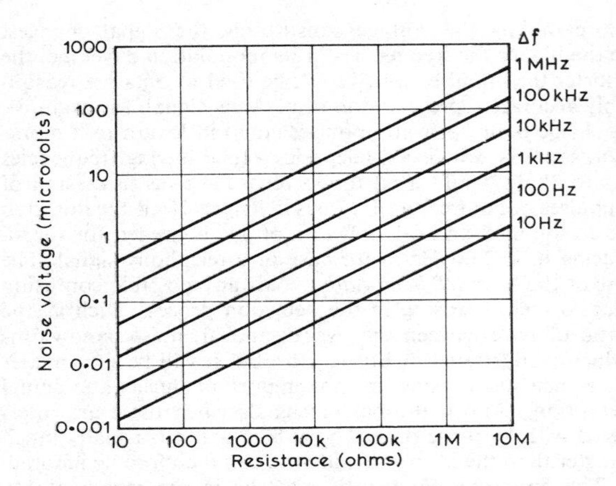

Fig 15.5. Chart for determining the noise voltage developed in a resistance for a specified bandwidth Δf at room temperature (20°C). Note that the scales are logarithmic.

absolute temperature (in deg K) and R is the resistance of the conductor (in ohms). This relationship is plotted graphically in **Fig 15.5**, but a handy figure to remember is $0{\cdot}13\mu V$ (approximately) for 1,000Ω at 1kHz bandwidth and at room temperature. The thermal noise generated in a tuned circuit can be calculated from the same expression if R is taken to represent the dynamic impedance: thus for a parallel resonant circuit R should be substituted by $Q\omega L$: see Chapter 1—*Principles*.

Thermal noise tends to be overshadowed in hf reception by noise picked up by the aerial, and in vhf/uhf reception by noise generated in valves or transistors.

The concept of noise temperature is helpful in assessing and comparing noise from all sources; thus it is now customary to express the external noise level by assigning an *equivalent temperature* to the radiation resistance of the aerial, ie the temperature it would be required to have if it were to produce the observed amount of noise merely as a consequence of thermal agitation. Similarly, valve shot noise is expressed in terms of an *equivalent* (room temperature) *noise resistance*, and induced grid noise in terms of an equivalent temperature attributed to the resistive component of the valve input impedance. The advantage of these concepts becomes apparent in dealing with *noise factor* as discussed later in this chapter.

Valve Shot Noise

This can be pictured by thinking of the anode current as being due to electrons hitting the anode like a stream of shot, or to the impact of hail against a window-pane, rather than as a continuous smooth flow. If the anode current I flows through a resistance R, the noise voltage developed is $\sqrt{(2eIBRF^2)}$ where e is the charge on an electron and F^2 is the space-charge smoothing factor. Comparison of this formula with the thermal-noise formula is the basis of the noise-diode method of measuring receiver noise performance, F^2 being unity when, as in the noise diode, all available electrons are drawn to the anode. This condition is particularly useful for the construction of standard noise sources for use in noise measurements, the noise output being accurately calculable. Almost any valve with a tungsten (including thoriated tungsten) filament can be used in this way but coated cathodes are unsuitable. Temperature limitation is ensured by increasing the anode voltage until there is no further increase of current, and assuming adequate voltage (typically about 50V) the current, and therefore the noise output, may be varied by alteration of filament current. For amplifying valves operating normally, F^2 is usually about 0·15 and it is convenient for comparison with aerial and thermal noise to represent the valve shot noise by a fictitious *equivalent noise resistance* (R_{eq}) such that a noise voltage $\sqrt{(4KTBR_{eq})}$ applied between grid and cathode would produce the equivalent fluctuation of anode current. It should be noted that the equivalent noise resistance is a mathematical fiction, convenient for calculations but not a true resistance to which Ohm's Law could be applied. Additional noise ("partition noise") is generated in tetrode and pentode valves due to random sharing of the electron stream between anode and screen. This is aggravated at high frequencies by inverse feedback due to cathode-lead inductance and is the largest source of noise in such valves, which are now obsolescent for low-noise applications.

Formulae for the calculation of the values of R_{eq} are given in Chapter 2—*Valves*.

Induced Grid Noise

At low frequencies the voltage on the grid of a valve due to an applied signal can be considered as being constant for the complete interval of time taken for one electron to leave the cathode space charge and arrive at the anode. As the frequency is raised this state of affairs ceases to exist and as the number of electrons approaching the grid is no longer equal to the number leaving it, a current is induced in the grid circuit giving the effect of a resistor R_t, which decreases as the square of the frequency in shunt with the grid circuit. This resistor also appears as a generator of thermal noise at an equivalent temperature comparable with that of the valve cathode, ie about five times the absolute room temperature. There is partial correlation in phase between the shot noise and the induced grid noise with the result that one can be partially balanced against the other by detuning the grid circuit and thereby altering the relative phases. In general the reduction in noise is small and the two sources can be treated as independent to a first approximation, although it is often found that the signal-to-noise ratio tends to vary in an unsymmetrical way as the tuning of the input circuit is varied.

Flicker (1/*f*) Noise

Amplifying devices normally exhibit additional noise at low frequencies, characterized by the property that noise power increases inversely as the frequency. It is negligible above some frequency, often known as the "corner frequency", and tends to be proportional to current, being described by such terms as flicker noise, $(1/f)$ noise and "current noise". The mechanisms are not all fully understood but "surface states" are often involved and there is some analogy between flicker noise and noise in carbon microphones where it is associated with variable contact between the granules. It can also be caused by temperature fluctuations, eg 2–3μV per 0·001°C for a semiconductor diode [1]. The "corner frequency" may vary between a few tens of hertz as for valves with tungsten filaments (eg a noise diode) and a megahertz or more for some semiconductor devices. It is not usually an important factor in amateur reception but will probably be encountered if an attempt is made to obtain as much as possible of the receiver gain at audio frequencies, as for example in direct-conversion receivers. In one instance it proved a major obstacle in the development of a transistorized version of an ssb transceiver having a filter constructed from "surplus" crystals; the high level of spurious responses required low level operation of the product detector which was followed by a high-gain audio amplifier, and satisfactory performance was not realizable until the advent of low-noise field-effect transistors.

Semiconductor Noise

Noise effects in semiconductors are largely analogous to those in valves though the situation is complicated by (*a*) base (or source) "spreading resistance" which is important mainly as a source of thermal noise in bipolar transistors but can also contribute to signal attenuation at very high frequencies, (*b*) the need in the case of bipolar transistors to take into account the shot noise associated with the base current and (*c*) a variety of effects difficult to take into account theoretically, such as leakage currents and the interdependence as well as temperature and frequency dependence of various parameters; fortunately these can usually be ignored as a first approximation, at least when dealing with the silicon planar and field-effect transistors most likely to be selected for low-noise applications. With transistors, unlike valves, it is impossible to make a direct connection to the "control electrode", and the base current has to find its way to the scene of action by spreading out from the base connection into the bulk of the semiconductor material, hence the term "spreading resistance". The capacitance is also a rather complex phenomenon, being attributable in part to phase shifts associated with the inability of electrons (or "holes") to move fast enough. An important practical aspect of this is the location of the capacitive reactance, which is more or less at the far end of the spreading resistance so that it is not accessible (even in part) for removal by tuning.

Additional sources of noise in transistors include leakage currents, but these tend to be unimportant in modern transistors intended for low-noise applications.

Semiconductor devices benefit from the absence of a hot cathode so that the temperature of the electrons is lower, and R_{eq} for an fet is of the order of $0{\cdot}7/g_m$ as compared with $2{\cdot}5/g_m$ for a triode valve, g_m being the mutual conductance. The noise of bipolar transistors may be represented [2] by a current generator $R_{eqc} = 2\beta/g$ plus a voltage generator $R_{eqv} = (r_b + 1/2g)$ where r_b is the base spreading resistance and g the emitter-to-base diffusion conductance, $1/g$ being given in ohms by $27/I_e$ where I_e is the emitter current in milliamps. The value of r_b may be obtained (sometimes) from manufacturers' data sheets. As in the case of R_{eq} these are not actual resistances but descriptions of the noise voltages and currents in terms of the equivalent thermal noise generators. In these expressions β is the current gain and β/g the input resistance of the active part of the transistor, ie the resistance which would be measured if it were possible to get at the junction directly, without having to go through the spreading resistance, assuming grounded-emitter operation. The use of these concepts in the design of receiver input stages is explained in a later section and it will be noted that for best noise performance g should be as small as possible. This in turn means that collector current must be kept as low as possible consistent with adequate gain, and values of 1mA or less are typical of medium frequencies, though higher values increase the upper frequency limit and tend to improve the noise performance as this limit is approached. On the other hand much lower values of current are usually desirable at low frequencies in cases where flicker noise is appreciable; as already noted, field-effect transistors are particularly good in this respect, but this applies only to junction types, not insulated-gate transistors, which suffer from surface-leakage problems at lf.

Due to the high input impedance of an fet the source spreading resistance r_s is of negligible consequence unless the frequency exceeds $1/(2\ Cr_s)$ where C is the input capacitance.

AERIAL NOISE

If a receiver is connected to a signal generator or to a dummy aerial a certain thermal noise voltage is fed into the receiver. This so-called *aerial noise* would determine the minimum usable signal level if the receiver itself generated no noise and if the real aerial produced the same noise as the dummy aerial. It has been found convenient to use the thermal noise level of the dummy aerial as a basis for the definition of *noise factor* as explained later in this chapter, because in this way the definition is directly related to the normal conditions of measurement. Unfortunately the noise levels of real and dummy aerials are usually quite different, so that both the definition and the measurements are not directly related to actuality and can be very misleading unless the appropriate allowances are made.

When the impedance of a real aerial is measured, for example with an impedance bridge, it is found to consist of resistance and probably some reactance as well. The reactance can be ignored for the purpose of the present discussion since it is usually tuned out in one way or another and does not itself contribute to the noise level. The resistance consists partly of radiation resistance and partly of loss resistance. The loss resistance is made up of the ohmic resistance of the aerial wire, leakage across insulators, losses in the immediate surroundings of the aerial, and also includes the resistance of the earth connection when a Marconi aerial is used; it is generally important in the case of lf reception and when using indoor and frame aerials, and is one of the two factors (the other being the bandwidth) which determine the extent to which it is practicable to reduce the size of an aerial system. The loss resistance is a source of thermal noise corresponding to a temperature which is usually taken as 300°K, although it can of course vary by 10 per cent or more from this figure depending on the local climatic conditions. In the higher frequency bands the loss resistance can normally be made negligible compared with the radiation resistance, which in practical cases can be considered to have any noise temperature from less than 10°K up to thousands or millions of degrees depending on the frequency and other factors.

Radiation Resistance

There is no connection between the noise associated with the radiation resistance of an aerial and the actual physical temperature of the aerial or its surroundings, and to appreciate why this must be so it is necessary to understand the meaning of radiation resistance.

When power is fed into a load, it is absorbed in the resistance of the load, and this is just as true for a transmitter feeding power into the radiation resistance of an aerial as it is for any other generator and load; but as is well known, power radiated from an aerial may travel immense distances, even to outer space, before being absorbed. The radiation resistance of an aerial therefore tends to be a property of the whole of space of which the local features occupy merely a trivial part. The radiation resistance as such depends only on the fact that power is dissipated, regardless of whether it is "consumed" by terrestrial objects or travels on into space to be eventually absorbed or "lost without trace". Any object capable of absorbing power from an aerial tends to radiate thermal noise power back to it, and "aerial noise" arises also due to radiation from a variety of terrestrial and extra-terrestrial processes.

EXTERNAL NOISE

The main sources of noise external to the receiver which affect communication may be grouped under three headings:

(*a*) man-made,
(*b*) atmospheric, and
(*c*) cosmic.

During periods of intense solar disturbance, radio noise from the sun also tends to interfere with normal reception.

Man-made Noise

In remote rural areas, the level of man-made noise may be pleasantly low or even negligible, whereas in congested towns it may be intolerably high. The various kinds are often identifiable by reason of their time characteristics. Most of them originate in domestic electrical appliances or in industrial equipment. Common sources of this kind of noise are high voltage power lines, light switches, vacuum cleaners, refrigerator motors, advertising signs, electric drills, petrol engines, trolley buses and electric trains.

Intermittent clicks, such as those produced by switches and well-designed thermostats—unless very frequent—do not cause serious trouble: isolated noise impulses destroy only occasional short elements of the signal and comparatively little of the information content is lost.

Continuous interference, such as that caused by commutators or faulty discharge-tube lighting, is much more damaging to good communications. Such interference spreads over a very wide frequency range, although some fluorescent tubes generate a broadly tunable noise band of unstable frequency.

The type of noise associated with commutators, often known as *hash*, is only partly impulsive in character and cannot be dealt with successfully by any noise-limiting device in the receiver. The true impulse type of noise, such as ignition interference, is much easier to suppress. One of the advantages of the frequency-modulation system is that if the receiver is correctly designed and adjusted, interference from high-amplitude impulse noise is virtually eliminated, though a similar result can be achieved in the reception of a.m. signals if sufficient i.f. bandwidth and appropriate limiters are used. Individual sources of man-made noise usually have a relatively small range, and considerable improvement may be obtained by raising the aerial system as high as possible and by using a screened feeder. In the

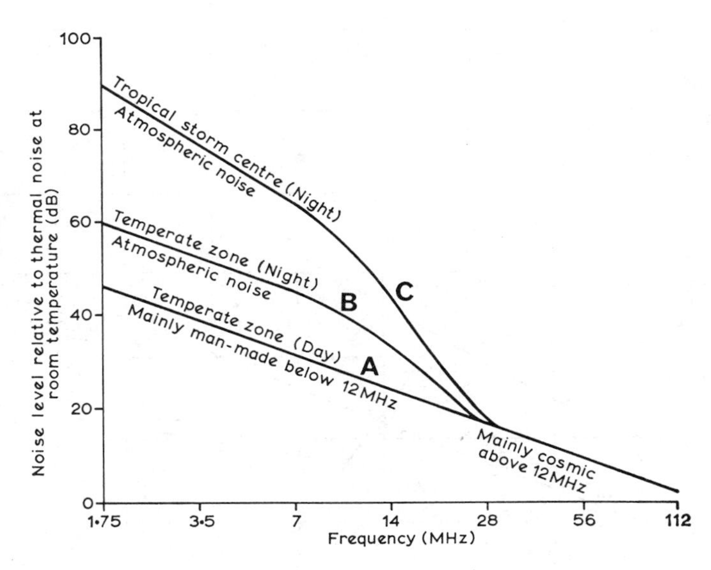

Fig 15.6. Variation of external noise level with frequency. Curve A shows that during the day in temperate zones the noise is mainly man-made at frequencies below about 12MHz. In these zones, atmospheric noise adds considerably to the total noise level at night (curve B). In tropical zones the atmospheric noise is relatively severe: curve C represents the worst conditions in these zones. The vertical scale indicates the number of decibels by which the noise level in a perfect receiver would increase if it were disconnected from a dummy aerial and fed from an efficient aerial of similar impedance.

lower hf range, however, man-made noise from the large numbers of distant sources tends to be a limiting factor even at some distance from main centres of population.

Atmospheric Noise

The primary source of atmospheric noise (or *static*) is the ordinary lightning discharge. When propagation is world-wide, noise may be received from large numbers of different thunderstorms all over the world, and if the number of discharges is large enough the general appearance and statistical properties will have some resemblance to those of white noise. On the other hand, a local thunderstorm will produce large "crashes" with intervals between them so that the ratio of peak-to-mean noise level can be very large. The noise from local storms is naturally much more intense and is considerably greater on the lower radio frequencies; of course this is only intermittent and does not require to be taken into account in the design of the "front end" of receivers, though some form of crash limitation is necessary to avoid "aural overload" and usually exists by virtue of agc action or audio limiting.

A not infrequent, but usually short-lived, form of interference is due to rain static, ie currents induced in the aerial by electrically charged raindrops.

Cosmic (Galactic) Noise

Considerable noise radiation is received from the Milky Way and other galaxies, some of which are so remote that they are not visible even in the most powerful optical telescopes. At the wavelengths mainly used for long-distance communication the ionosphere is transparent except at the fairly low angles of incidence responsible for the propagation of wanted signals. Over the range 14–28MHz, galactic noise is usually about 20–30dB greater than the internally generated noise of a triode or transistor rf stage designed for minimum noise factor (Fig 15.7). Apart from the variation with frequency, cosmic noise is similar in character to thermal noise.

Total Noise Level

A general idea of the variation of atmospheric, man-made and cosmic noise levels with frequency over the hf and vhf bands is given in **Fig 15.6**. These curves have been somewhat freely adapted from previously published data, bearing in mind the need for a simplified presentation. Below 14MHz the diagram should be regarded only as a rough guide, since atmospheric and man-made noise levels are extremely variable, and the noise is generally more impulsive in character. As explained earlier, impulsive noise tends to be relatively less serious the lower the grade of communication, and for amateur purposes the *effective* lower-frequency noise levels may be lower than indicated although some slight "weighting" in this direction has taken place in the course of adapting the data.

The marked difference in atmospheric noise levels between day and night is due to the more efficient propagation at night from distant storm centres located mainly in the tropics. Above 100MHz the external noise level continues to decrease, being much less than room-temperature thermal noise in the 432MHz band, and frequently equivalent to only a few degrees absolute in the microwave bands.

NOISE PERFORMANCE OF RECEIVERS

When it is desired to measure the sensitivity of a receiver, some form of test signal is fed in through an impedance which simulates that of the aerial from which the receiver is designed to work. Such an impedance is known as a *dummy aerial* and is a source of thermal noise at room temperature, T. If the receiver were perfect the noise at its output terminals would consist only of amplified noise from the dummy aerial, but in practice the receiver adds a certain amount of noise as previously explained.

Actual aerials may produce more or less noise than equivalent dummy aerials, depending on the frequency and other circumstances. This may be taken into account by assigning the appropriate "noise temperature" T_a to the aerial, eg from

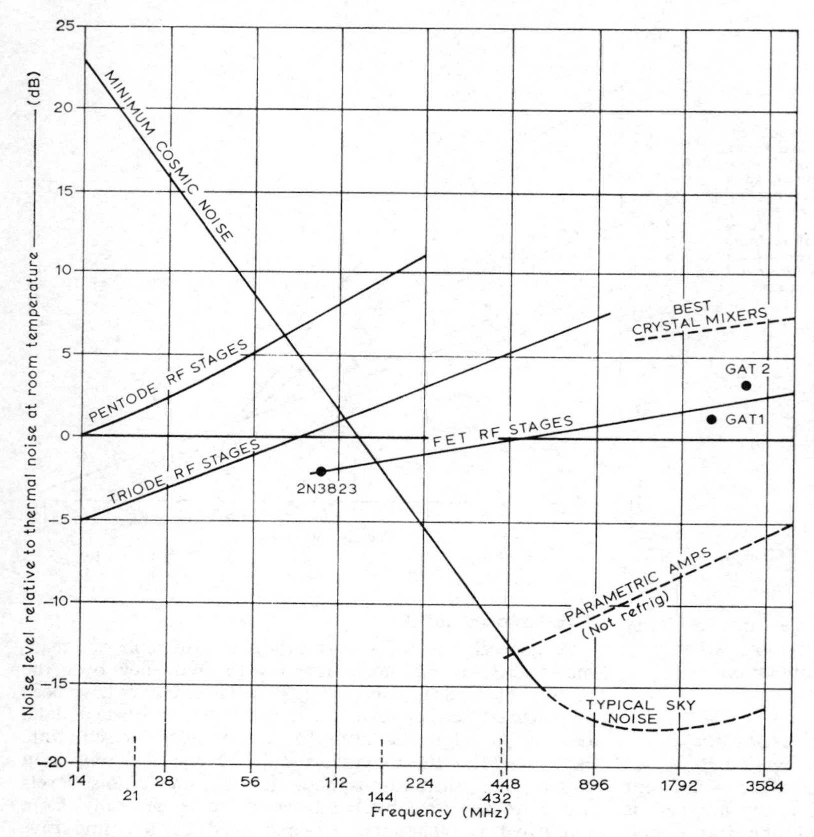

Fig 15.7. Approximate relationship between aerial noise and receiver noise over the amateur frequency bands. The valve and transistor noise curves represent typical good designs. Considerably higher noise levels may be observed with narrow-beam aerials directed at the sun or towards the centre of the galaxy.

Fig 15.6 or 15.7. Similarly any increase of noise due to imperfections in the receiver can be represented by an equivalent increase T_r in the aerial noise temperature. The ratio T_r/T_a is then the true measure of noise performance of the receiver, which is 3dB worse than perfect if $T_r = T_a$, or 1dB (ie 24 per cent) worse if T_r is 24 per cent of T_a. By expressing T_r in decibels relative to room temperature T, it can be directly compared with the noise levels indicated in Figs 15.6 or 15.7 to establish whether it is significant in relation to the external noise level.

The valve and transistor curves in Fig 15.7 illustrate the lowest values of T_r which can be achieved without undue difficulty compared with the lowest likely external noise levels, as a function of frequency. As explained later, however, it is not always desirable to aim at a low value of T_r, and hf receivers almost always have much higher values than vhf receivers.

Noise Factor

The equivalent-temperature concept of receiver noise is not yet universally adopted: the noise performance of receivers is usually specified in terms of *noise factor*, this being the number of times by which the total noise power effectively exceeds that part of it attributable to the aerial (which, for this purpose, is assumed to be at room temperature corresponding to the usual conditions of measurement). Unfortunately extreme care is required in specifying the conditions of measurement and in interpreting the figures obtained. The advice offered in Chapter 18 on the subject of noise measurement should be closely followed; the method of measurement described there is based on the definition

$$\textit{Noise Factor} = (T + T_r)/T$$

where T is the room temperature. This can of course be arranged in the form

$$T_r = (\textit{Noise Factor} - 1)T$$

to enable the noise factor as normally measured to be expressed directly as an equivalent aerial temperature.

The object of good receiver design is to reduce T_r until it is negligible compared with the aerial noise temperature but not necessarily to make T_r as small as possible; any further possibility of improvement can generally be used instead to improve the receiver in other respects as discussed later, or perhaps to reduce its cost. The value of T_r is obtained from the measurement in the form $T_r/T = (F - 1)$ where F is the noise factor expressed as a numerical ratio, in other words T_r is obtained in units of "room temperature" and there is no need to multiply by 290, thus converting it to degrees Kelvin, unless using data which is also expressed in degrees K. Thus if the receiver noise level read from Fig 15.6 is 3dB, ie a numerical ratio of 2, we have $T_r = 2$ "room temperatures" or $2 \times 290 = 580°K$ according to choice.

For the rest of this chapter "room temperature" will be used as the basic unit so that $T_r = (F - 1)$, ie $F = (1 + T_r)$.

It is important to note that in this case T_r and F are both numerical ratios, F being used directly as measured, ie without the complication of converting it to the more usual logarithmic form:

$$F(\text{dB}) = 10 \log (1 + T_r).$$

The advantage of using "temperature" rather than F is demonstrated by the following example. If T_a is nearly zero and T_r for two receivers is 0·3 and 0·6 respectively, it is immediately obvious that the second receiver produces twice as much noise power as the first, and the real difference in performance is 2 : 1 or 3dB. The ratio of the noise factors, however, is only 1·6/1·3 = 1·23 or a difference of only 0·9dB and, as a comparative figure, this is clearly misleading.

Use of "temperature" is now universal in professional low-noise applications such as satellite communications and radio astronomy but continued use of F(dB) in the literature and manufacturers' data sheets makes conversion necessary, eg with the aid of **Fig 15.8**. In plotting curves of T_r covering a wide range of values it becomes necessary to use a logarithmic scale as in Figs 15.7 and 15.8.

Optimization of Noise Performance

In designing a low-noise input stage for a receiver the basic problem is to amplify the aerial noise without adding noise. This may seem a curious way of stating the problem since noise, whatever the origin, is by definition unwanted; obviously, however, if aerial noise is amplified without any additions, so is the signal. Some readers will be familiar with definitions of noise factor in terms of signal-to-noise ratios and the complexity to which this can lead both in the definition itself and in its subsequent application to design problems. The approach used in this chapter is simplified by the fact that all the voltages to be compared are described "in the same language" so that bandwidth (which also involves problems of definition) becomes a common factor which can be eliminated. It is a basic assumption that all noise sources are subject to the same narrow-band amplification and caution is needed if, for example, an ssb filter is preceded only by a mixer and followed by a broad-band amplifier. It should also be noted that double-sideband reception of ssb or cw signals involves a 3dB increase of noise which is not reflected in the noise factor. A further 3dB loss would be incurred in the absence of image-rejection.

The relationship of noise sources and impedances at the input of a receiver can be represented by an equivalent circuit such as **Fig 15.9(a)**, which shows an aerial connected to a triode valve through a "transformer" (consisting, for example, of a tuned circuit) arranged so that the aerial impedance R_a as "seen" by the valve can be adjusted to any desired value. R_c is the parallel resonant impedance of the tuned circuit, including any damping due to the valve input impedance, and R_{eq} is the valve noise resistance as defined above. The noise produced by the resistances is represented by separate thermal noise generators, so that a noisy resistance such as R_c becomes a noiseless resistance in association with a voltage generator $v_n = \sqrt{(4kT_cBR_c)}$, the temperature T_c being assigned to allow for the possibility that R_c may be "hotter" than room temperature, as explained in the section on induced grid noise. To avoid confusing the issue by uncertainties in regard to the aerial temperature, R_a is assumed to be at "room temperature" which, apart from conforming to the usual conditions of measurement, provides as we have seen a convenient unit of temperature. Ignoring any large differences in "temperature" between the three resistances, it will be obvious from inspection that T_r is negligible provided $R_c \gg R_a \gg R_{eq}$; in this case R_c does not appreciably attenuate the aerial noise but is itself virtually short-circuited by the aerial. This only leaves R_{eq}, which is negligible compared with R_a, so that virtually all the noise comes from the aerial.

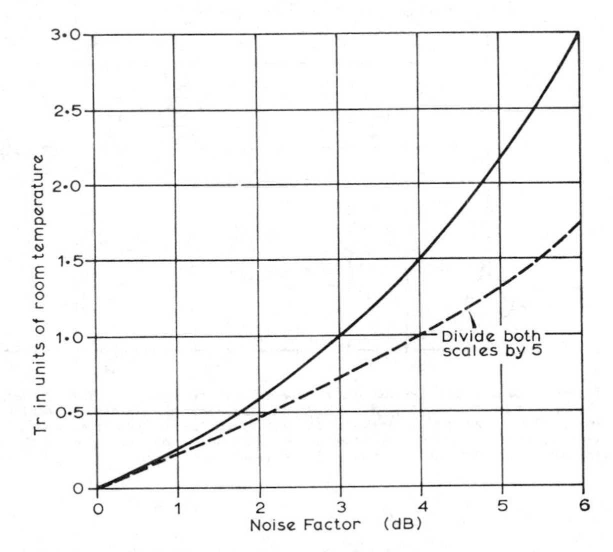

Fig 15.8. Comparison between noise temperature and noise factor. Use dotted curve for very low values. At higher values N and Tr become nearly equal when both are expressed in the same terms.

This is typical of what can be achieved with practical circuits at low frequencies, but note that the tuned circuit will be heavily damped because its natural impedance R_c is shunted by the relatively low resistance R_a. This is equivalent to "overcoupling" the aerial and produces a loss of selectivity which may not always be acceptable. The aerial itself is badly mismatched, but the consequent reflections do not affect reception of narrow-band signals. With a matched aerial, given that $R_a/2 \gg R_{eq}$ and $T_c = T$, we have $T_r = T$, ie F = 3dB. In this case the selectivity is only 3dB worse than that of the unloaded input circuit. At vhf and uhf, low noise is of primary importance and the design of the input circuit needs careful optimization. R_cT_c is then accounted for mainly by the transit time damping and associated induced-grid noise, typical values for R_c (approximately equal to R_t) and T_c being 1,000Ω and $5T$ respectively. There is obviously now an optimum value of R_a since if R_a is very small there is a large value of T_r given by R_{eq}/R_a, and if R_a is large the aerial noise voltage is attenuated by a factor R_a/R_c relative to the large noise voltage produced by $(R_cT_c + R_{eq})$, again giving a large value of T_r. Between these extremes there is an optimum value of R_a given by

$$R_c\sqrt{\frac{R_{eq}}{T_cR_c + \mathrm{R}_{eq}}}$$

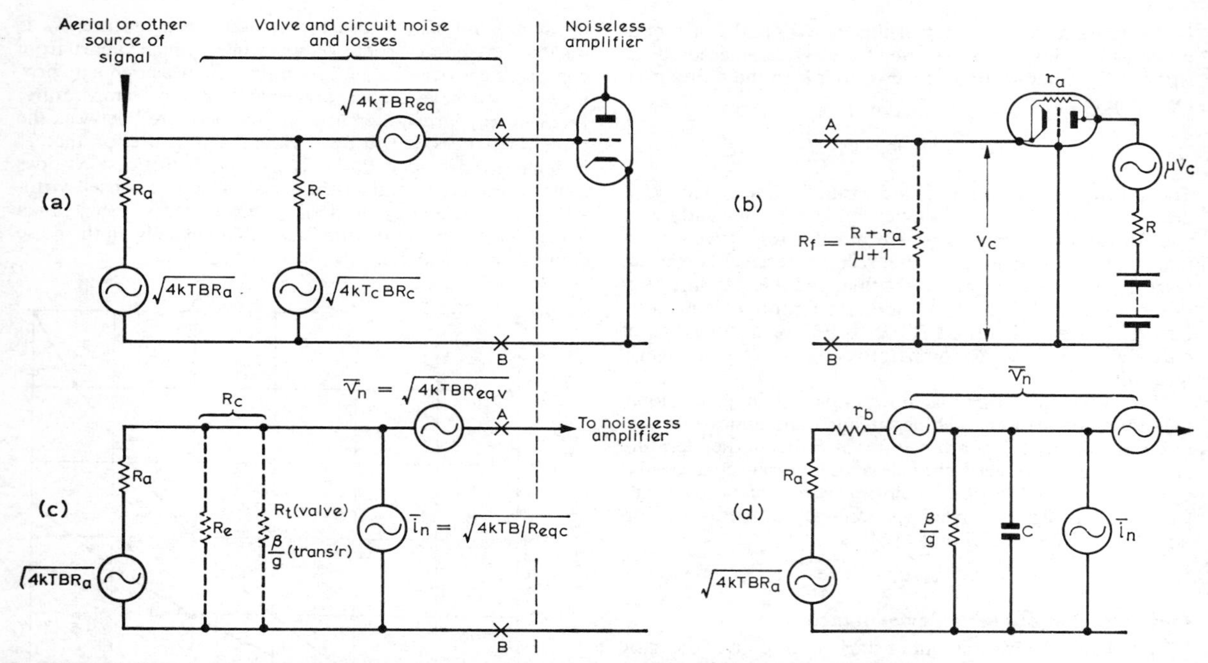

Fig 15.9. Equivalent circuits for receiver input stages. (a) Triode valve. R_c includes circuit losses R_e and transit-time damping R_t. The ratio T_c/T given by $(R_t + 5R_e)/(R_t + R_e)$ is approximately equal to one at hf and five at vhf/uhf. (b) Grounded-grid triode. Feedback through the internal resistance R_a causes the noiseless resistance Rf to appear across the input terminals AB. (c) General-purpose equivalent circuit using voltage and current noise generators. Applicable to bipolar transistors by including the thermal noise generator r_b in $Reqv$. (d) Representation of bipolar transistor showing spreading resistance and input capacitance.

and with this value of R_a

$$F_{(min)} = (T_r + 1)_{(min)} = 1 + 2\frac{R_{eq}}{R_c} + 2\sqrt{\frac{R_{eq}}{R_c}\left(T_c + \frac{R_{eq}}{R_c}\right)}$$

At large values of R_c this gives $T_r = 2\sqrt{T_c R_{eq}/R_c}$ and at small values (ie at very high frequencies) $T_r \approx 4R_{eq}/R_c$. The disappearance of T_c means that the valve input resistance, though now heavily attenuating the aerial noise (and of course any signals) relative to valve noise, is no longer itself a significant noise source. Since R_c at vhf is proportional to $1/F^2$, the value of T_r increases with frequency at 6dB per octave. The value of R_a is not critical and a 2 to 1 error increases F by a maximum of 0·5dB. Low values of F are reduced by much less than this.

In the case of a grounded-grid triode the inverse feedback from anode to cathode operates equally against signals and all the sources of noise, leaving their ratio unaffected. This means that $R_a(opt)$ and $F(min)$ are exactly the same as for grounded cathode operation, although the reduction of level is equivalent to the small resistance Rf in **Fig 15.9(b)** and the input resistance is very low. The input circuit is now of course heavily damped and provides little or no selectivity, but this makes no difference to the need for a high-Q circuit, the dependence of T_r on R_c being unchanged. So far, however, no account has been taken of stages after the first, and the reduced gain with grounded-grid operation means that noise from later stages is more likely to be important; it should also be noted that if resistance networks are used to provide feedback they may contribute some extra noise.

Use of Pentode RF Stages

In the case of pentode valves R_{eq} is greatly increased due to random sharing or "partition" of the electron stream between anode and screen, as explained earlier. The equivalent circuit of Fig 15.9(a) is applicable in this case at low and medium frequencies, but must be elaborated to take into account the inverse feedback due to cathode lead inductance which discriminates against the signal and in favour of the partition noise, thus even further degrading the noise factor at high frequencies. In the case of hf receivers it is still possible to obtain an acceptable noise factor but, in order to override the valve noise, signals have to be presented to the valve at a higher voltage level and, other things being equal, cross-modulation effects and intermodulation products are greatly increased.

Mixer Noise

When a valve or transistor is used as a mixer there is less gain for a given output current than when the same device is used as an amplifier, so that signal is reduced relative to noise. This is another way of saying that R_{eq} is increased, so that the input signal voltage must be stepped up to a higher value and the risk of generating intermodulation products is thereby increased. In the case of hf receivers this tends to make low-noise design of mixers more important than that of rf stages unless the interfering signals are far enough off-tune to be rejected by the extra rf selectivity made possible by use of the rf stage. This is unfortunately not the case at the hf end of the 7 and 21MHz bands where strong broadcasting signals are likely to be found in adjacent channels:

use of an rf stage is then a disadvantage depending on the rf gain setting, since it either adds to the noise level or alternatively amplifies unwanted signals, thus increasing the risk of intermodulation occurring at the mixer. Low-noise design of mixers is also a requirement at uhf and vhf, due to the relatively low gain of rf stages. The design of mixer stages for low noise is similar to that of rf stages, allowing of course for the higher values of R_{eq} and any relevant circuit differences. It is usually necessary for dc operating conditions and oscillator injection voltage to be optimized by trial and error.

Second-stage Noise

When two stages contribute to the internal noise level, the overall noise factor is given by

$$N_{ov} = N_1 + \frac{N_2 - 1}{G_1}$$

where N_1 and N_2 are the noise factors of the first and second stages and G_1 is the gain of the first stage. Similarly the contribution of a third stage would be $(N_3 - 1)/(G_1G_2)$ but this is usually negligible.

The lower N_2 is, the less is the need for a high value of G_1.

Bipolar Transistors

Fig 15.9(a) is not in a convenient form for evaluating the performance of bipolar transistors but this can easily be remedied by transforming it into **Fig 15.9(c)**, using the fact that the noise of a resistance may be represented either by a noise voltage generator of zero resistance, in series with the resistance, as in Fig 15.9(a) or a noise current generator of zero conductance (ie infinite resistance) in parallel with it as in Fig 15.9(c). Like the "noise resistance" R_{eq}, the current noise resistance R_{eqc} can exist as a generator without necessarily having an "ohmic" origin. To make Fig 15.9(c) into the exact equivalent of Fig 15.9(a) however, it is necessary to add the resistance R_c in parallel with R_{eqc}. The magnitude in this case of the current noise generator is given by $\bar{i}_n = \sqrt{4kT_cB/R_c}$, this being the current which would flow due to the thermal noise in the resistance R_c when R_c is short circuited, and R_{eqc} is identical with R_c/T_c. Analysis of Fig 15.9(c) leads to the simple results

$$R_a\,(opt) = \sqrt{R_{eqc}\,.\,R_{eqv}}$$

$$\text{and } T_r\,(min) = F(min) - 1 = \sqrt{R_{eqv}/R_{eqc}}$$

To apply these results to transistor input circuits at medium frequencies it is only necessary to substitute the values given in the section on semiconductor noise, in which case

$$R_a = \sqrt{\frac{2\beta}{g}}\,.\,\sqrt{r_b + \frac{1}{2g}} \approx \sqrt{\frac{2\beta\ r_b}{g}}$$

$$\text{and } T_r = \sqrt{(1 + 2gr_b)/\beta} \approx \sqrt{2gr_b/\beta}$$

With typical hf transistors operating at $I_B \approx$ 1mA, $2gr_b$ is about 4 and, using the simpler formula, T_r will be underestimated by about 10 per cent only. On the other hand low-noise transistors at lf may operate with currents of 100μA or less in which case it is the (gr_b) term which can be neglected.

These formulae allow for thermal noise in r_b but not for attenuation of signals which, as evident from inspection of **Fig 15.9(d)**, is negligible provided $r_b \ll \beta/g$. This condition

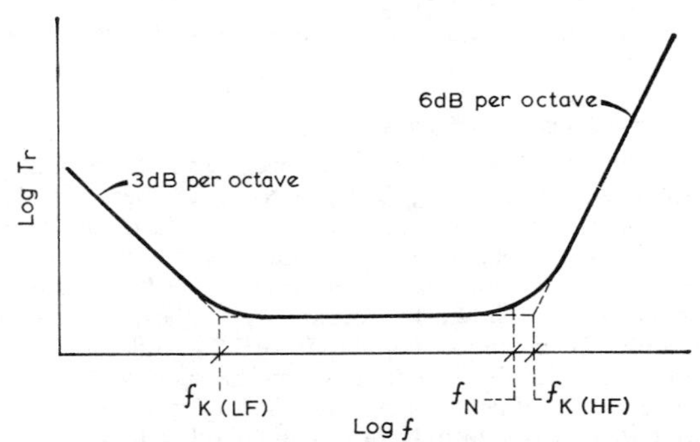

Fig 15.10. Variation of noise with frequency for bipolar transistors. General tendencies are indicated. Large variations are possible not only between types of transistor but for a given transistor with different operating conditions.

is normally met except at very high frequencies, ie near cut-off, where the effective grounded-emitter current-gain drops to a low value.

The grounded-base connection is closely analogous to grounded-grid operation of triodes, optimum values for R_a and T_r being identical with those for the grounded-emitter connection and in this case also there is considerable input mismatch due to the low impedance. The consequent reduction of gain will increase the importance of any second-stage noise. The hf noise performance of bipolar transistors cannot be calculated accurately owing to the large number of interdependent variables, r_b for example being to some extent frequency dependent. The general behaviour of T_r with frequency is shown by **Fig 15.10**, the corner frequency f_K being the point at which F is 3dB greater than its low frequency value. It is usually between 3 and 6 times f_N for modern transistors and the region from f_N to perhaps a little above f_K is likely to be the one of main interest since it includes the 144 and 432MHz bands, assuming appropriate choice of transistors.

The upper frequency limit for useful operation may be described in various ways as explained in Chapter 3, the most convenient for noise calculations being the gain-bandwidth product F_T given in terms of Fig 15.9(d) by $g/(2\pi C)$. The frequency f_N is the one at which the effective grounded-emitter current gain $\beta' = \beta/2$ and is given by $F_T/\sqrt{\beta}$, β' for any other frequency f being obtainable by dividing $1 + \left(\frac{f}{f_N}\right)^2$ into β. At f_N, the value of T_r is exactly double the midband value. The reduction in effective β corresponds more or less to the attenuation of signal relative to transistor shot noise and a rough estimate of optimum R_a and T_r, below f_K, may be obtained by using β' in place of β in the low-frequency formulae. Well above f_K, T_r tends towards the value $\frac{2gr_b}{\beta}\left(\frac{f}{f_N}\right)^2$ which is consistent with the 6dB per octave slope of Fig 15.10.

For greater accuracy, particularly near f_K, it is advisable to use the Nielson formula [3]. In its usual form this looks rather complicated, but can be simplified to

$$T_{r(min)} \approx D[1 + \sqrt{1 + A/D}]$$

by putting $D = (1 + gr_b)/\beta'$ and $A = (1 + 2gr_b)/(1 + gr_b)$.

The value of A must of course lie between 1 and 2, but with modern hf transistors there is a bias towards the upper limit.

Field-effect Transistors

In the case of the fet, data sheets give the actual noise voltages corresponding to R_{eq} or R_{eqv}, but these can also be obtained from the formula given earlier. In the hf region it would be a simple matter to make R_c large compared with R_{eq} so that T_r is negligible, but in practice R_c is likely to be reduced in the interest of selectivity to the point where T_r is only just negligible compared with T_a, and Fig 15.9(a) is appropriate to this situation. When the value of R_a is chosen on grounds other than optimization of noise performance and R_c consists entirely of circuit or other losses at room-temperature

$$T_r = \frac{R_a}{R_c}\left(1 + \frac{2R_{eq}}{R_a} + \frac{R_{eq}}{R_c}\right)$$

$\approx R_a/R_c$ if R_{eq} is small compared with R_a and R_c.

These conditions are representative for a field-effect transistor rf or mixer stage at hf.

The induced gate noise at high frequencies is accounted for [4] by putting $R_{eqc} = 4g_m/\omega^2C^2$ where C is the input capacitance, and since $R_{eqv} \approx 0{\cdot}7/g_m$ we have

$$R_{a(opt)} = 1{\cdot}65/\omega C$$

$$\text{and } (F - 1) = T_r = 0{\cdot}8\omega C/g_m.$$

Values of g_m and C are given in data sheets and g_m/C is the high-frequency figure-of-merit for an fet. In a typical case (2N3823), g_m = 5mmho, C = 6pF and at 100MHz $R_{a(opt)}$ = 430Ω and F_{min} = 2·1dB in good agreement with the specified upper limit of 2·5dB. There is some variation in figures quoted in the literature for the constant in the noise-factor formula, the lowest figure [6] being 0·52 which gives F = 4dB for the Plessey GAT1 gallium-arsenide fet at 2·2GHz in close agreement with the claimed performance. As with valves the input resistance of field-effect transistors decreases as $1/f^2$ at high frequencies, but, as the above examples demonstrate, it remains high enough to be ignored in typical applications.

Other Types of Low-noise Amplifier

(*a*) *Tunnel diodes:* These are in effect "negative resistances" which amplify by neutralizing some of the positive resistance in a tuned circuit, thereby increasing the circulating current, and hence the voltage across the circuit, at the expense of bandwidth. This is exactly the same process as regeneration (eg as used in Q-multipliers) and, since noise and signals are in general affected equally, the noise factor does not depend directly on the degree of amplification. Typically, the tunnel diode is used as a negative resistance in series with the aerial and load resistances and must be almost equal to their sum. The diode contributes shot noise and the best noise-factors so far obtained with tunnel diodes at vhf and uhf are of the order of 3dB (ie T_r = 1). In general, better results are likely to be achieved with field-effect transistors. Practical disadvantages of tunnel diodes include poor strong-signal performance, critical adjustments which are highly sensitive to aerial and load impedances (with possible instability) and the difficulty of using more than one stage.

(*b*) *The parametric amplifier:* Parametric amplifiers also work on the negative resistance principle, but this effect is achieved by the use of a capacitance which is "pumped" by means of an oscillator at a relatively high frequency. In one comparatively simple form [5] a cavity resonates at the signal frequency f_s and simultaneously at the much higher frequency f_i known as the idler frequency. A reverse-biased diode is connected across the resonator and acts as a voltage-dependent capacitance, the value of which is varied by means of a large signal from a pump oscillator at the frequency $f_p = f_i + f_s$. Some of the pump power is converted to energy at the signal frequency, thus providing amplification, and the mechanism is inherently low noise since a reactance is not a noise generator, though thermal noise is contributed by the signal and idler circuits. T_r values of about 1·0 were obtained in this way at 400MHz with pumping at 900–2,000MHz, but the limit of possible performance for more elaborate types of parametric amplifier with very high pump frequencies is in the region of T_r = 0·1 at 1,000MHz and 1·0 at 10,000MHz. Because of the critical adjustments required for stable amplification these very low-noise amplifiers are difficult to adapt to amateur requirements. In commercial use the value of T_r can be yet further reduced by a factor of 3 or 4 using an idler load resistance cooled down to liquid nitrogen temperature. Even lower values of T_r, down to about 0·04, can be obtained with masers, but these require refrigeration at liquid helium temperature and are restricted to such applications as satellite communication and radio astronomy.

Parametric amplifiers can take various other forms, the most important of these being the up-converter. This does not rely on negative resistance, is inherently stable, and provides even better noise performance, but practical difficulties arise from the high value of output frequency, the gain being determined by (and nearly equal to) the frequency ratio.

PRACTICAL ASPECTS OF NOISE

Every amateur will wish to ensure that the performance of his station is not handicapped by avoidable defects in the receiver. This situation can arise from the following causes:

(*a*) Gain is not sufficient to make certain that the useful sensitivity is noise-limited;

(*b*) Receiver contributes significantly and avoidably to the noise background level.

In the case of insufficient gain the trouble and the remedy are both obvious. Those designing their own receivers should have no difficulty in making a rough estimate, with the aid of Figs 15.5 and 15.6 and the foregoing explanations, of the amount of rf and i.f. gain required to give a reasonable working level (say 1V) at the detector, and hence of the amount of audio gain needed to bring the rectified noise (about 0·4V rms, assuming a diode detector with 1V input) up to a reasonable listening level.

Above about 100MHz, low noise factor is all-important and Fig 15.7 gives a guide to the performance which should be achievable by using various types of low-noise amplifier. The curves for valves and transistors are based on the most appropriate choice of type. For example, an expensive microwave device would not be used at 144MHz where the improvement, if any, is not likely to be cost-effective. On the other hand, to get the high cost of low-noise amplifiers for 1,296MHz and higher frequencies into perspective, the cost per dB should be compared to that of a similar increase of transmitter power. The dotted curve for parametric amplifiers [5] is unlikely to be realizable with amateur resources and it may be difficult to obtain much improvement compared with the best fet. The very low external noise levels at microwaves, indicated in Fig 15.7, will be swamped in practice by aerial thermal noise due to losses and "spillover." A feeder loss of 3dB brings the minimum value of T_a up to $T/2$ and at 144MHz increases the total noise level with a good fet input stage by 1dB, which is additional to the feeder loss.

On the hf bands it is the external noise which limits the ability of the receiver to deal with weak signals, provided of course that the aerial is a good one and that the receiver is functioning correctly. In this case, provided a low-noise mixer is used, there is usually no need for an rf stage and, as previously discussed, it may aggravate intermodulation interference. One danger, unless the rf gain is offset by additional preselection, is that of an apparent increase in the noise level caused by intermodulation at the mixer involving large numbers of moderately strong signals within the passband of the rf circuits. On the other hand, if it is necessary to use an inefficient aerial, such as a wire fastened to the picture rail, or a miniature beam having appreciable loss-resistance in the elements or the feeders, the external noise level will in most cases be lower; the receiver noise must therefore be reduced in proportion and an rf stage may be necessary. Indoor aerials are sometimes responsible for increased pick-up of man-made noise from local sources, and in such cases this reasoning will obviously not be applicable.

In defence of the standard practice of employing rf stages in hf receivers, it must be added that they provide a margin of safety against the consequences of inefficient coupling between the receiver and the aerial. Moreover, the disadvantages associated with rf stages can often be mitigated by making sure that rf gain control is not advanced beyond the point where external noise just swamps the internal noise as judged by a simple check, eg by disconnecting the aerial.

Aural Discrimination

When the above conditions have been satisfied, discrimination between signals and noise becomes mainly a job for the ear and brain working together, and except in a few special cases very little can be done in the receiver to assist this process. If the overall bandwidth is greater than necessary, decreasing it reduces the noise level but does not make copying any easier, because it is only the noise which occupies the same acoustic band as the signal that is effective is preventing it from being copied.

This argument, however, must not be carried too far; it can, for example, be used to justify the normal practice of using bandwidths of several hundred kilohertz for the sound channel of television receivers, but it is not valid when applied to amateur vhf receivers. The reason for this is that the amateur is able to make use of low signal levels, and it is possible by increasing the pre-detector bandwidth to reach a point where the noise voltage at the detector becomes comparable with the carrier voltage of, say, a just-readable telephony signal. The situation at the detector is then as depicted in Fig 15.4(c), but with the signal overmodulated by the noise, the wanted modulation of the carrier tends to be destroyed. The effect is the same as modulation suppression, a well-known property of linear detectors whereby the presence of a strong carrier destroys the modulation of a weaker one. There is no advantage in using a square-law detector, since this also by its nature discriminates in favour of strong signals.

Effects of Overloading

Another possible cause of deterioration in the ability to copy signals through noise is overloading, particularly of the i.f. stages, by noise peaks which may reach several times the mean noise level and will increase in amplitude if the bandwidth is increased. If a reduction of bandwidth appears to give improved reception of weak signals, it is possible that some such effect may be taking place and the gain should be reduced.

In the case of impulsive noise like that due to ignition systems or isolated atmospherics, the dynamic range is relatively large and noise limiters are useful in the prevention of overloading, particularly if the bandwidth is large so that the impulses are not lengthened and caused to overlap—as they may be by the ear itself—before being limited. The time intervals between atmospherics are frequently long enough to enable words or syllables, or morse characters, to be read through the gaps, and it may thus be possible to sustain low-grade communication through a relatively high noise level; in suitable circumstances this may allow some discounting of the lower-frequency noise levels plotted in Fig 15.6.

SSB Reception

In the reception of morse and sideband signals, the voltage injected into a conventional diode detector from the beat frequency oscillator must be large compared with the mean noise voltage, but a ratio of 8–10 times in voltage is sufficient in regard to the signal-to-noise ratio, whereas for maximum discrimination against strong signals an even higher injection level is necessary.

It is possible to employ ssb reception of ordinary amplitude modulated signals in order to remove interference which may be present on one sideband. This unfortunately entails a reduction in the signal-to-noise ratio since the two sidebands of a double-sideband signal add up in phase and therefore the removal of one of them halves the signal voltage; at the same time, the noise power which accompanies the rejected sideband is eliminated. This halves the total noise power so that there is a decrease of 3dB in the noise level to offset the decrease of 6dB in the signal level, leaving a penalty of 3dB. As a corollary to this, it follows that ssb, dsb and a.m. signals should give equal signal-to-noise ratios for equal *total* sideband power.

Effects of Aerial Characteristics

The gain of an aerial, in terms of the signal-to-noise ratio, is the same as its transmitting power gain if there are no losses in the aerial and if the noise is non-directional. In relatively rare circumstances the external noise will be stronger from one direction than another, and the effective gain of the aerial will then be reduced for signals in the same direction as the noise and increased for signals in other directions. The removal of high-angle lobes should tend to reduce the external (cosmic) noise level in the 14–28MHz bands. Losses in the aerial and feeder system are subtracted from the transmitting power gain but do not affect the gain in terms of the signal-to-noise ratio as long as the receiver noise can be kept well below the external noise level: thus, referring to Fig 15.6, the external noise level at 14MHz is 25dB so that an aerial loss of 18dB would still leave the external noise 7dB above thermal noise, ie 6dB above the receiver-plus-thermal noise for a noise factor of 1dB, and the signal-to-noise ratio would be degraded only by about 1dB compared with that for a loss-free aerial. This permits the use of much more compact aerial systems for reception than for transmission.

A very interesting point arises in the case of rhombic and certain other long-wire aerials, since the terminating resistance absorbs half the transmitter power and, of course, half the received noise power. On transmission the power which is absorbed is that which would otherwise be radiated in the backward direction, and the forward gain is unaffected. The elimination of noise from the backward direction, however, doubles the signal-to-noise ratio provided that the receiver noise level is low enough for advantage to be taken of the lower aerial noise level. Thus, in general, a terminated rhombic operating at frequencies below about 30MHz has an effective gain 3dB greater for receiving than for transmitting.

REFERENCES

[1] "Characteristics and Limitations of Transistors", R. D. Thornton *et al*, Vol 4 of *Semiconductor Electronics Education Committee* series, John Wiley 1966. Chapter 4.

[2] "Noise in Transistor Circuits", P. J. Baxandall, *Wireless World*, December 1968.

[3] "Behaviour of Noise Figure in Junction Transistors", E. G. Nielsen, *Proc IRE*, Vol 45, pp 957–963, July 1957.

[4] "Noise in Transistors", F. N. H. Robinson, *Wireless World*, July 1970.

[5] "Some Types of Low-Noise Amplifier", R. Hearn *et al*, *J Brit IRE*, Vol 22, pp 393–403, November 1961.

[6] *Field-Effect Transistors*, Ed Wallmark and Johnson, Prentice Hall, 1966.

[7] *Handbook of Semiconductor Electronics*, Lloyd P. Hunter. McGraw Hill, 1962.

CHAPTER 16

POWER SUPPLIES

AMATEUR radio equipment usually derives its power supply from one of three sources. These are:

(a) Public ac mains
(b) Primary or secondary batteries
(c) Engine-driven generators

For normal fixed station operation the ac mains is readily available and cheap, and is by far the most commonly-used system. Transformers, rectifiers and smoothing circuits can provide the wide range of dc voltages and currents required for amateur equipment. Various circuit configurations are described later in this chapter together with appropriate design data and component characteristics. Electronic regulators are also described which cope with cases requiring a high degree of voltage regulation.

Batteries have always provided a convenient source of power for low-power portable equipment and some items of test gear. It is not always convenient to use batteries of appropriate terminal voltage to supply power directly, and dc to dc converters are normally employed to change the voltage (see Chapter 14).

Engine-driven generators may be ac or dc. The most popular units, however, are those which produce mains voltage and frequency, and will therefore power fixed station equipment directly when used for portable operation. Present-day dc generators usually generate 12V or 24V and are normally used for battery charging on portable stations.

The safety of operation of power supplies and protection against misuse are important topics which are also discussed. Attention is also drawn to the Amateur Radio Safety Recommendations in Chapter 19.

RECTIFIER CIRCUITS

A rectifier is necessary when ac is to be converted to dc. The essential property of a rectifier is to pass current in one direction only. When an ac voltage is applied across such a device, current flows for alternate half-cycles, ie for one polarity of voltage. During the complementary half-cycles the rectifier exhibits a high resistance and current flow is negligible. The conventional symbol for a rectifier, as shown in **Fig 16.1**, includes an arrow head pointing in the direction of conventional current flow, ie anode to cathode. It is obvious that an ac voltage applied to the anode results in positive half-cycles only appearing at the cathode. For this reason semiconductor rectifiers often have a red mark or + sign at the cathode terminal.

Fig 16.1. Rectifier symbol.

Fig 16.2 shows three types of rectifier circuit which cover most of the applications in amateur equipment, together with curves indicating the shape of the current wave delivered by the rectifier to a resistive load. The output current is seen to be unidirectional but pulsating in character, and may be shown to consist of a dc component plus an ac component. The ac component is composed of the fundamental and harmonics of the supply frequency, the fundamental being predominant in the half-wave circuit and the second harmonic in the full-wave circuit. This ac component is termed *ripple* and may be removed or attenuated to any desired degree by the filter which follows the rectifier.

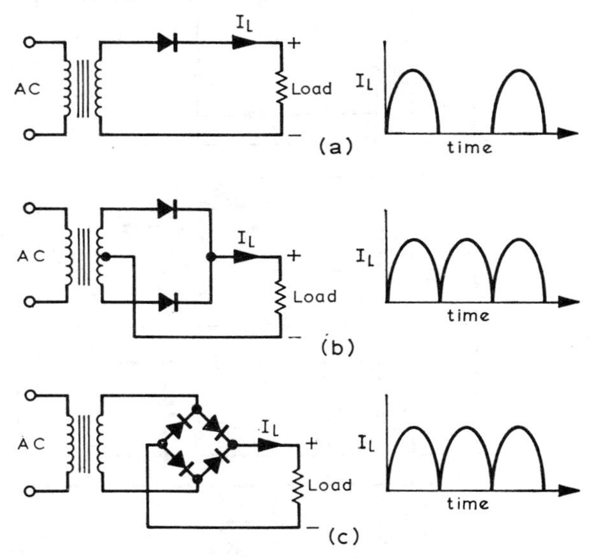

Fig 16.2. Rectifier circuits showing the output current waveforms with resistive loads: (a) half-wave (b) full-wave or bi-phase half-wave (c) bridge.

The half-wave circuit (Fig 16.2(a)) is best suited to low current applications such as grid bias supplies and the eht supply for cathode ray tubes. When used to supply current of the order of tens of milliamperes and higher, the half-wave circuit has disadvantages in that the ripple content is high, necessitating a large filter. In addition, the dc load current flows through the secondary winding of the transformer feeding the rectifier and can saturate the core, giving rise to low transformer efficiency. The *regulation* or variation of output voltage with load current is also poor.

The full-wave circuit (Fig 16.2(b)) is known more correctly as the bi-phase half-wave circuit. A centre-tapped winding is used to supply the rectifiers and must provide twice the ac voltage applied to each rectifier. The load is connected between the centre tap of the transformer and the strapped output terminals of the rectifiers. Each rectifier conducts on alternate half-cycles and contributes one half of the total load current. The dc component of the load current flows through each half of the transformer secondary in such a direction as to cancel the dc magnetization of the core, enabling a smaller and more efficient transformer to be used than is possible in a half-wave circuit providing the same output.

The bridge circuit (Fig 16.2 (c)) is preferred for higher voltage supplies since the peak inverse voltage across each rectifier is only half of the bi-phase half-wave arrangement for the same dc voltage output. The peak inverse voltage, referred to as piv, is the voltage appearing across the rectifier diode in the non-conducting condition. During each half-cycle of the input voltage, the rectifiers in opposite arms of the bridge are conducting and supply half the total load current. The dc magnetization of the transformer core is

TABLE 16.1

Operating conditions of single phase rectifier circuits

Circuit	DC output voltage	PIV across diode	Diode DC current	Diode peak current	Secondary RMS current
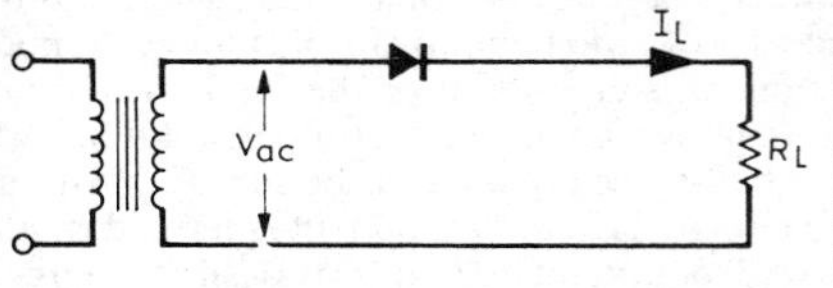	$0.45V_{ac}$	$1.4V_{ac}$	I_L	$3.14I_L$	$1.57I_L$
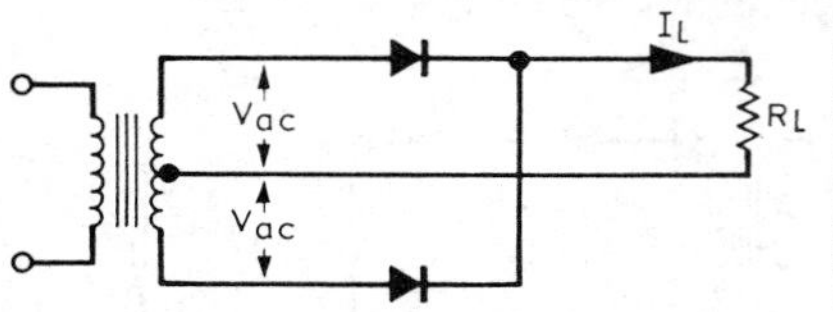	$0.9V_{ac}$	$2.8V_{ac}$	$0.5I_L$	$1.57I_L$	$0.785I_L$
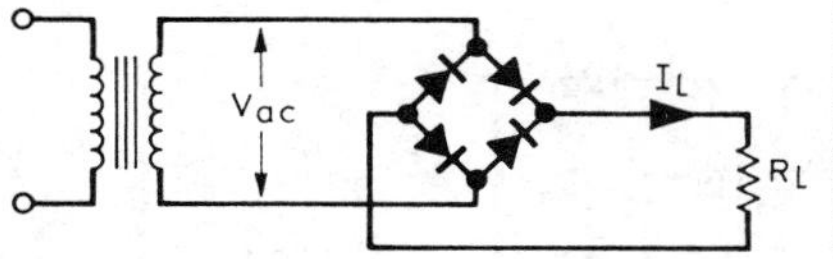	$0.9V_{ac}$	$1.4V_{ac}$	$0.5I_L$	$1.57I_L$	$1.11I_L$
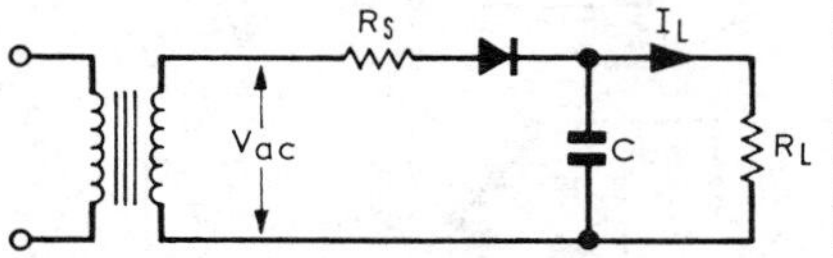	$1.4V_{ac}$ (No load)	$2.8V_{ac}$ Maximum	I_L	See Fig 16.8	= Diode RMS current See Fig 16.7
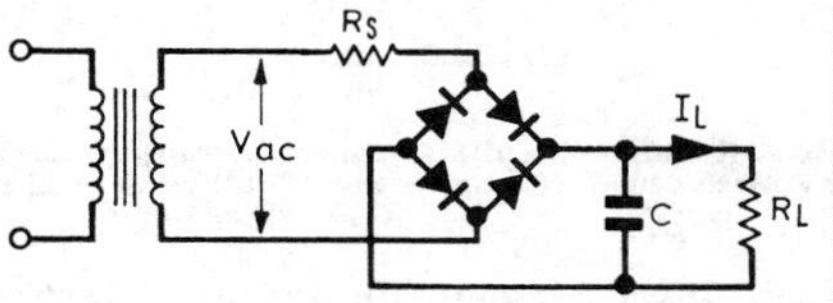	$1.4V_{ac}$ (No load) See Fig 16.6	$1.4V_{ac}$ Maximum	$0.5I_L$	See Fig 16.8	= Diode RMS current x 1.4 See Fig 16.7
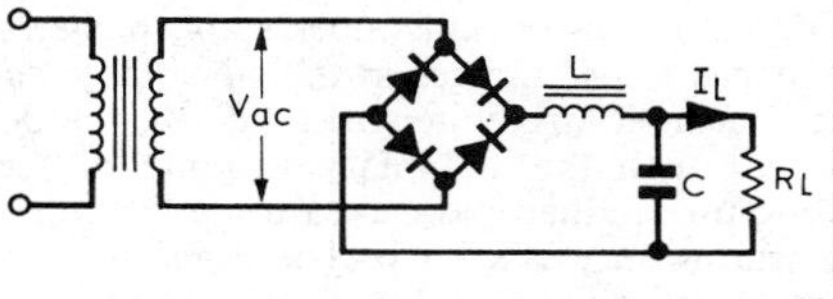	$0.9V_{ac}$	$1.4V_{ac}$	$0.5I_L$	$2I_L$ when $L=L_C$	$1.22I_L$ when $L=L_C$

effectively cancelled, as the dc component of the load current flows through the ht secondary winding in opposite directions during each half-cycle. The advantage of the bridge is seen in the simplification and reduction in cost of the transformer. Relative to the bi-phase half-wave circuit, the rms current rating is increased by about one-third, and the turns on the winding feeding the rectifiers are reduced by half. In practice this results in a smaller transformer used more efficiently.

Table 16.1 gives the operating conditions of rectifier circuits.

VOLTAGE MULTIPLIER CIRCUITS

The voltage multiplier is a form of rectifier circuit where the output voltage may be two or more times that obtained from the conventional half-wave or full-wave rectifier with capacitor input when supplied with the same ac input voltage. A combination of rectifiers and capacitors is used such that the capacitors are charged in parallel via the rectifiers and then discharged in series, the rectifiers acting as switches in addition to their normal function. The regulation of these circuits is poor as a result of this mode of operation, but the use of large value capacitors will assist in maintaining a reasonable regulation if the load varies appreciably.

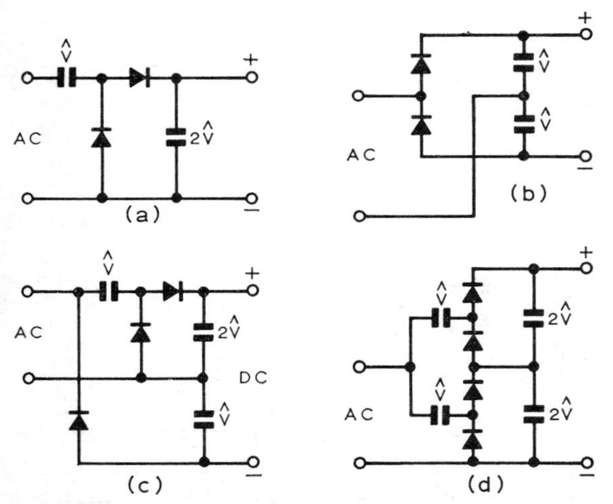

Fig 16.3. Voltage multiplier circuits: (a) half-wave voltage doubler; (b) full-wave voltage doubler; (c) voltage tripler; (d) voltage quadrupler. $\hat{V}$ = peak value of the ac input voltage. The working voltages of the capacitors should not be less than the values shown.

Half-wave and full-wave voltage doubler circuits are illustrated in **Fig 16.3(a)** and **(b)**. The full-wave circuit has better regulation but has the disadvantage of not having a common input and output terminal. As in all full-wave circuits, the ripple component is at twice the supply frequency and is therefore easier to remove. At zero load the output voltage is very nearly twice the peak value of the input ac voltage but falls rapidly to a value approximately twice the rms input voltage when appreciable current is drawn. The regulation may be improved by increasing the size of the capacitors, but care should be taken to ensure that the peak current ratings of the rectifiers are not exceeded.

Semiconductor rectifiers are more convenient for use in these circuits although valve rectifiers may be used if desired.

Voltage tripler and quadrupler circuits are illustrated in **Fig 16.3 (c)** and **(d)**. The tripler is a form of half-wave circuit where the output of a half-wave doubler is added to a normal half-wave rectifier. The quadrupler circuit is also known as the Cockcroft-Walton multiplier and, if required, additional stages may be added to provide higher output voltages.

When used to provide appreciable currents (up to 50–60mA) the capacitors shown in the circuits of Fig 16.3(a), (b) and (c) should be at least 8μF to provide reasonable regulation of the output voltage. Lower values (0·1 to 0·5μF) will be satisfactory when the load is not more than 1–2mA as in eht supplies. The polarity of the output voltage may be reversed in all the circuits shown by reversing the connections to each rectifier.

RECTIFIER CHARACTERISTICS

The following are the important parameters of rectifiers and the symbols shown comply with IEC recommendations and BS9000/1969:

I_{FSM} — maximum peak forward surge current (non-repetitive)
I_O — average rectified output current
I_{FRM} — maximum peak forward surge current (repetitive)
V_{RSM} — maximum reverse voltage (non-repetitive, maximum time duration 5ms)
V_{RRM} — maximum reverse voltage (repetitive), ie piv
T_{CASE} — case temperature for maximum rating.

It is important not to exceed any of these ratings. The characteristics of some typical modern silicon diodes are given in **Table 16.2.**

TABLE 16.2

Characteristics of some silicon diodes

Type	VRSM	VRRM	IFRM	IO	IFSM
1N4004		400	6A	1A	30A
1N4007		1000	6A	1A	30A
1S105		600	10A	750mA	40A
1S023		400	20A	1·5A	125A
1N2071	800	600	6A	700mA	25A
3F80	950	800	10A	4A	40A
4AF05	75	50	100A	30A	420A

SMOOTHING CIRCUITS

The behaviour of the rectifier circuit depends on the input element of the filter which follows the diodes. The two common types are the capacitor input filter and the choke input filter. The capacitor input filter has an output voltage which is equal to the peak value of the rectified voltage (or 1·4 times the rms voltage applied to the rectifier) on no load, and somewhat less when loaded. The choke input filter produces an output voltage which is the average value of the rectified voltage (or 0·9 times the rms voltage applied to a full-wave or bridge rectifier), but the voltage regulation is better. The choke input filter has the added weight and circuit complication of the additional inductor but it imposes lower peak currents on the diodes than the capacitor input filter and can therefore supply higher output currents without exceeding the peak current rating of the diodes.

Capacitor Input Filter

The operation of a rectifier circuit feeding a capacitor is very complex because of the complex nature of the waveforms involved. A bridge rectifier circuit with capacitor input filter is shown in **Fig 16.4,** and **Fig 16.5** shows the voltage and current waveforms. It will be seen that the current through the diodes is in the form of narrow pulses occurring near the peaks of the input ac voltage. It is also obvious from these waveforms that the dc output voltage is equal to the peak value of the rectifier output voltage when there is no output load, because the capacitor is not discharged at all between the voltage peaks. When loaded, the output voltage is governed by the total series resistance R_S, capacitance value

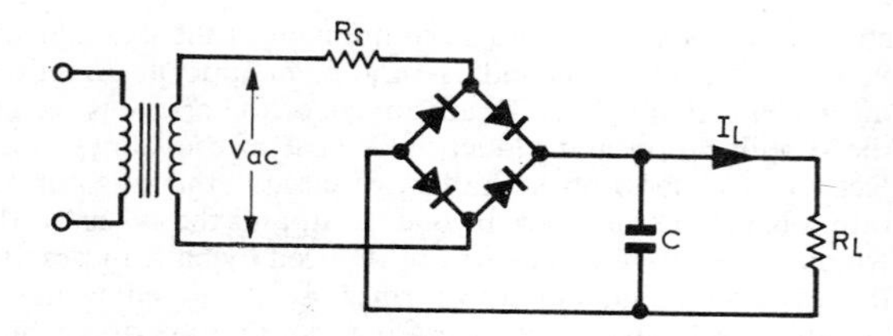

Fig 16.4. Bridge rectifier with capacitor input filter.

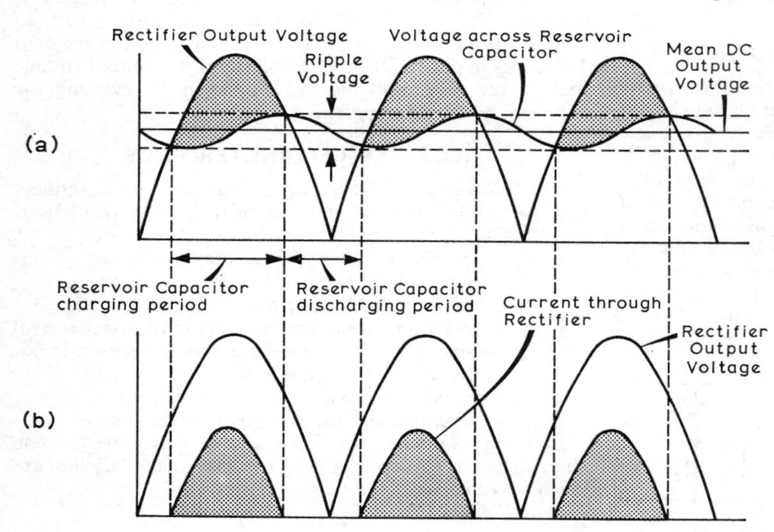

Fig 16.5. Curves illustrating the output voltage and current waveforms from a full-wave rectifier with capacitor input filter. The shaded portions in (a) represent periods during which the rectifier input voltage exceeds the voltage across the reservoir capacitor causing charging current to flow into it from the rectifier.

C and output load resistance R_L as well as the supply frequency f. The graphs in **Fig 16.6** show the dc output voltage as a percentage of the peak ac input voltage for a number of series resistance and load resistance combinations plotted against ωCR_L where $\omega = 2\pi f$. The series resistance R_S is made up of the effective transformer resistance, two diode resistances in series and any added resistance. The effective transformer resistance is:

$$\text{Secondary resistance} + n^2 \text{ (primary resistance)}$$

where n is the ratio of secondary turns to primary turns (or secondary voltage to primary voltage on no load).

The diode resistances may be determined from the forward voltage/forward current characteristics in the manufacturer's data sheets but for high voltage supplies where the load resistance R_L is greater than 500Ω the resistance of silicon diodes may be neglected. In the case of low voltage supplies, however,

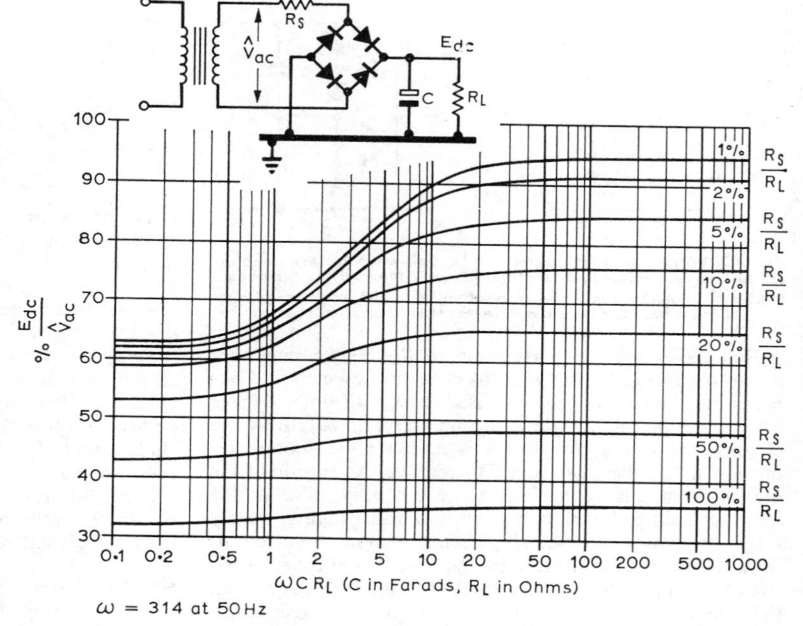

Fig 16.6. Output dc voltage as a percentage of peak ac input voltage for a bridge rectifier with capacitor input filter.

the resistance of a silicon diode can represent a significant percentage of the load resistance and should be included in the design. For example, in a power supply with a dc output of 20V at 1A (ie $R_L = 20\Omega$) a diode resistance of 1Ω represents five per cent of the R_L value and has a marked effect on the output voltage.

It is usual to select a value of capacitance such that ωCR_L lies well to the right of the upper knee on the curves of Fig 16.6. In this way operation is on a flat portion of the curves and the output voltage is not influenced by changes in capacitance but is determined entirely by the value of R_S.

The total series resistance R_S is also an important factor in determining the peak current which the rectifiers must handle and the rms current rating of the transformer for any given dc current output. These quantities increase with increasing capacitance up to ωCR_L values of approximately 10 but thereafter become independent of capacitance and relate only to the ratio of R_S to R_L. The ratio of diode rms current to diode dc current is shown in **Fig 16.7,** and the diode peak current as a ratio of diode dc current is shown in **Fig 16.8.** In the bridge rectifier circuit the diode dc current is half the load current since only one pair of diodes conducts during alternate half-cycles. For the same reason the total rms current supplied by the transformer secondary is the diode rms current multiplied by 1·4.

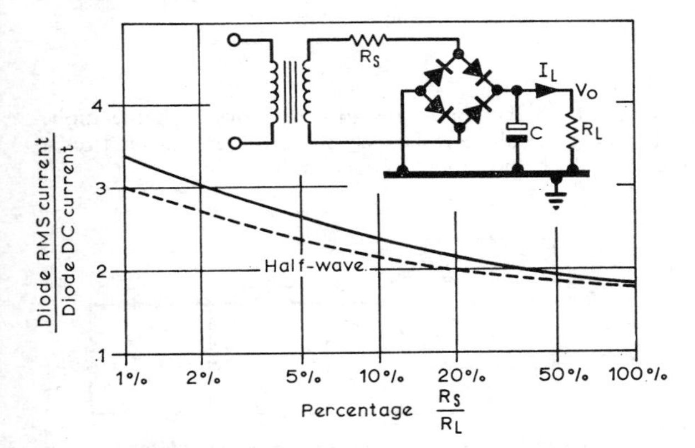

Fig 16.7. Relationship between diode rms current and percentage R_S/R_L for values of ωCR_L greater than 10. ($\omega = 314$ for 50Hz mains). The dotted line applies to half-wave rectifiers.

A typical design makes use of a value of R_S which is five per cent of R_L and a value of ωCR_L of 100. Reference to Fig 16.6 indicates that a power supply of such a design using a bridge rectifier produces a dc output voltage which is 84 per cent of the transformer peak ac voltage. Thus for any required output voltage, the transformer secondary voltage is specified. Fig 16.7 shows that the diode rms current is approximately 2·6 times the diode dc current, enabling the transformer secondary current rating to be specified, and Fig 16.8 shows that the diodes must handle a peak current which is eight times the diode dc current.

As an example, consider a power supply using a bridge rectifier and reservoir capacitor to supply 300V at 100mA.

$$R_L = \frac{300 \times 1{,}000}{100} = 3{,}000\Omega$$

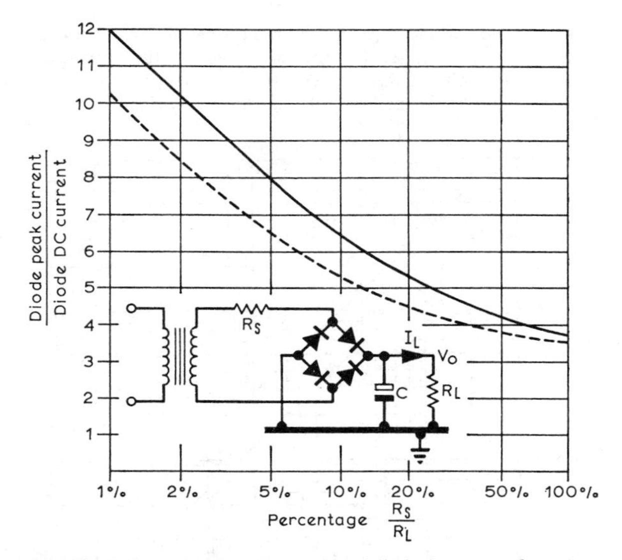

Fig 16.8. Diode peak current as a ratio of diode dc current for values of ωCR_L greater than 10. *Note:* **in a bridge rectifier circuit, diode dc current is half the load current. The dotted line applies to half-wave rectifiers; in this case the diode dc current is equal to the load current.**

Choosing $R_S/R_L = 5$ per cent then $R_S = 150\Omega$
Diode dc current $= 0{\cdot}5I_L = 50$mA
Diode peak current (see Fig 16.8) $= 8 \times 50 = 400$mA
Diode rms current (see Fig 16.7) $= 2{\cdot}6 \times 50 = 130$mA
Transformer rms current $= 1{\cdot}4 \times 130 = 182$mA
Transformer peak secondary voltage (see Fig 16.6)

$$= \frac{300 \times 100}{84} = 357\text{V}$$

$$\text{Transformer secondary rms voltage} = \frac{357}{1{\cdot}4} = 255\text{V}$$

Choose $\omega CR_L = 100$

$$2\pi \times 50 \times C \times 3{,}000 = 100 \quad (C \text{ in farads})$$

$$C\,(\mu F) = \frac{100 \times 10^6}{2\pi \times 50 \times 3{,}000} \doteqdot 100\mu F$$

A transformer should be selected which will supply 255V at a current of at least 182mA. The effective transformer resistance should be computed as previously described and subtracted from the R_S value of 150Ω above. A resistance of the value found should be connected in series with the transformer secondary. The circuit of the finished power supply is shown in **Fig 16.9.**

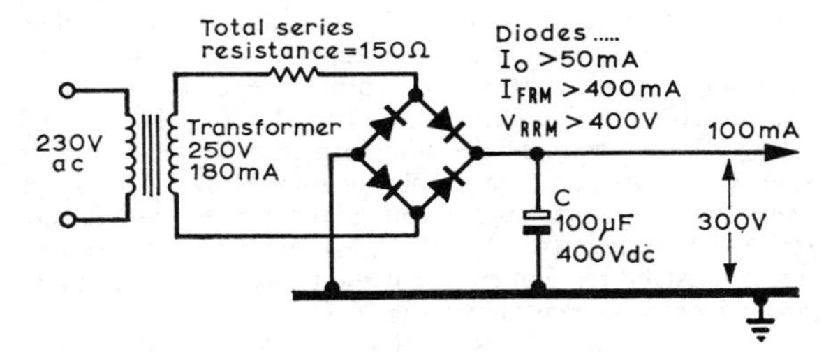

Fig 16.9. Circuit diagram of a power supply for 300V at 100mA.

The ripple voltage appearing at the reservoir capacitor may be found from **Fig 16.10.** In the bridge rectifier example above in which $\omega CR_L = 100$, the rms ripple voltage is approximately 0·7 per cent of the output voltage or 2·1V.

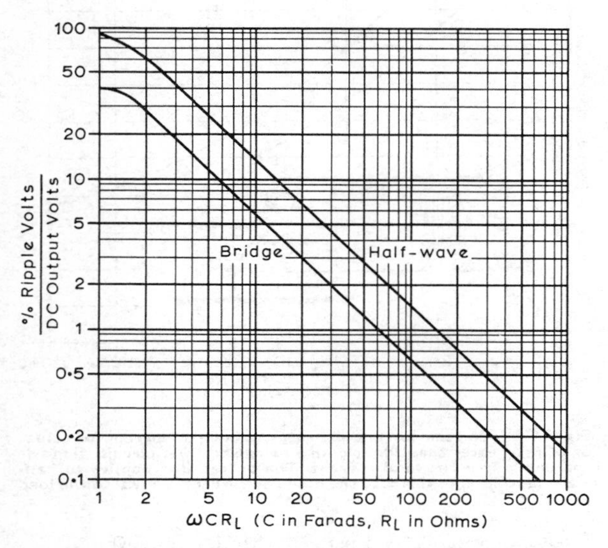

Fig 16.10. Percentage ripple voltage (rms) against values of ωCR_L. ($\omega = 2\pi f$ where f is the mains supply frequency).

The ripple current in the capacitor is obtained by dividing this voltage by the reactance of the capacitor ($1/\omega C$) at the ripple frequency of 100Hz.

$$\text{Ripple current} = \frac{2{\cdot}1 \times 2\pi \times 100 \times 100}{10^6} = 132\text{mA}$$

If greater smoothing is required a higher value of capacitance may be used. The ripple current however does not increase, since the decrease in ripple voltage is accompanied by the lower reactance of the larger capacitance. Doubling the capacitance, for example, halves the ripple voltage but of course the reactance is also halved, resulting in no change in ripple current. A good rule of thumb in full-wave or bridge-rectifier circuits operating with a 50Hz input is to regard the rms ripple current as 1·3 times the dc output current.

The requirements for the diode in this example, of 50mA dc current and 400mA peak repetitive current, put it in the low current class. It is likely that the amateur workshop component stock will contain some of the inexpensive general-purpose diodes such as the 1N4007 (see Table 16.2) with high peak repetitive current ratings. It is then possible, when using these devices in power supplies with relatively low load currents, to decrease the required series resistance R_S considerably without exceeding the diode peak current ratings. In many cases the transformer itself may provide sufficient series resistance. The advantages of a low series resistance are higher output voltage for a given transformer secondary voltage (see Fig 16.6), and improved regulation, but it must be remembered that the transformer secondary rms current is increased (see Fig 16.7).

It is important to ensure that the non-repetitive peak current rating I_{FSM} of the diode is not exceeded. A current surge occurs when the mains is switched on to a power supply with its reservoir capacitor discharged. The peak current in these conditions is found by dividing the transformer secondary peak voltage by the total series resistance. This current surge is exponential in character and has a time constant which is the product of the series resistance and filter input capacitance. In practice this pulse is likely to be of shorter duration than the one half-cycle for which the diode I_{FSM} is specified. The calculated value is therefore a pessimistic value. A diode with I_{FSM} high enough to cope with this current therefore has a safety margin.

Choke Input Filter

In the choke input filter rectifier circuit an inductor is connected in series between the output of the rectifier circuit and the shunt capacitor as shown in **Fig 16.11.** With very small values of inductance approaching zero it is obvious that the arrangement is similar to the capacitor input filter and the operation is the same. Current flow in this case is in the form of short duration pulses with no current flow between pulses (Fig 16.5). To make the current flow continuously the inductance requires to be greater than a certain value, usually called the *critical inductance*, and at least this inductance is required if the circuit is to have the properties of a choke input filter rectifier including good regulation. The critical inductance L_C may be calculated from the expression which relates to bridge or full-wave rectifier circuits:

$$L_C = \frac{R_S + R_L}{6\pi f}$$

where R_S is the total resistance in series with the diodes including the diode resistance, R_L is the output load resistance and f is the supply frequency.

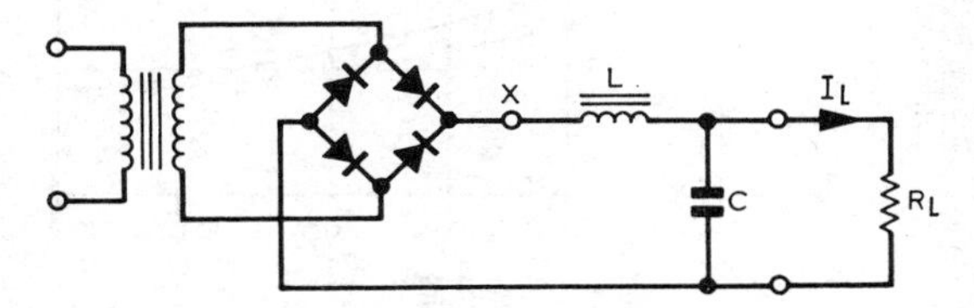

Fig 16.11. Bridge rectifier with choke input filter.

For 50Hz supply and when R_S is small compared with R_L:

$$L_C = \frac{R_L}{940}$$

The voltage and current waveforms in a choke input filter circuit with the critical value of inductance are shown in **Fig 16.12.**

With light loads (ie large values of R_L) the required critical inductance is large and approaches infinity as the output dc current approaches zero. In order to limit the required inductance to some maximum value it is usual to connect a bleed resistor across the output. The resistance may be calculated by re-arranging the critical inductance formula to read:

Bleed resistance = 940 × maximum inductance value.

In applications where there are large variations in output current, as in the case of a supply for a Class B amplifier, a special choke called a *swinging choke* with an unusually small airgap is used. The inductance of this type of choke decreases when the dc current through it increases so that the inductance varies with output load in keeping with the requirements of the critical inductance formula.

The peak diode current in a bridge circuit feeding a choke input filter with inductance $= L_C$ is $2I_L$, as can be seen in Fig 16.12, where I_L is the output load current including bleed current. The transformer rms current under these conditions is $1{\cdot}22\ I_L$ and the dc current per diode is $0{\cdot}5\ I_L$ as always in a bridge configuration.

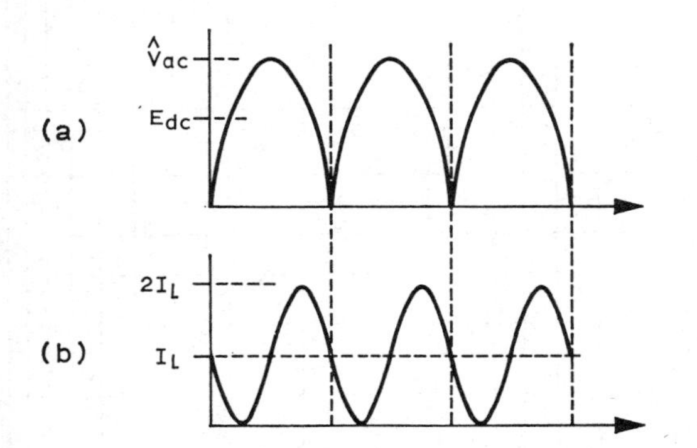

Fig 16.12. Waveforms at rectifier output (point X in Fig 16.11) in a choke input circuit: (a) voltage waveform; (b) current waveform ($L = L_C$).

The output dc voltage is the average value of the full-wave rectified voltage waveform, which is $0{\cdot}9\ V_{ac}$ if voltage drops are neglected. Including voltage drops, the output voltage E_{dc} may be calculated from:

$$E_{dc} = 0{\cdot}9\ V_{ac} - I_L (R_c + R_t) - E_r$$

where V_{ac} is the transformer rms secondary voltage, I_L is the output load current, R_c is the resistance of the choke, R_t is the resistance of the transformer secondary and E_r is the voltage drop of two diodes in series. In high-voltage supplies the voltage drop in the diodes may be neglected but in low-voltage high-current supplies the manufacturer's diode characteristics should be consulted to estimate this voltage.

The value of the filter capacitance is chosen to achieve some specified low level of rms ripple voltage E_R according to the formula:

$$E_R = \frac{E_{dc}}{0{\cdot}8LC}$$

where L is the inductance in henrys and C the capacitance in microfarads.

Care must be taken to avoid a condition of resonance of the series circuit composed of L and C. In a bridge system where the ripple frequency is 100Hz, the LC product for resonance is 2·53. However, it is usual to aim for a ripple voltage of less than 10 per cent of the dc output voltage so that LC is usually greater than 12 and well removed from the resonance value.

The rms ripple current in the capacitor is determined by dividing the ripple voltage by the reactance of the capacitor at the ripple frequency. However, for any given inductance value the ripple current is approximately constant and independent of the capacitance value, because an increase in capacitance results in a proportionate decrease in ripple voltage but the capacitive reactance decreases in the same ratio. This allows ripple current I_R to be found from the formula:

$$I_R = 0{\cdot}7I_L$$

This value of ripple current applies to cases where the inductance is the critical value. When larger values of inductance $= kL_C$ are used, the ripple current is further decreased and becomes:

$$I_R = \frac{0{\cdot}7I_L}{k}$$

Additional Smoothing

The simple capacitor input and choke input filters described will not reduce the ripple voltage to a sufficiently low level for some applications such as receivers, variable frequency oscillators, audio pre-amplifiers and the early stages of transmitters without the use of unreasonably large capacitors. For most cases one additional section comprising a series inductor and shunt capacitor as shown in **Fig 16.13(a)** will be adequate, and in low current supplies or in applications where large voltage drops can be tolerated the resistance-capacitance filter shown in **Fig 16.13(b)** may be used.

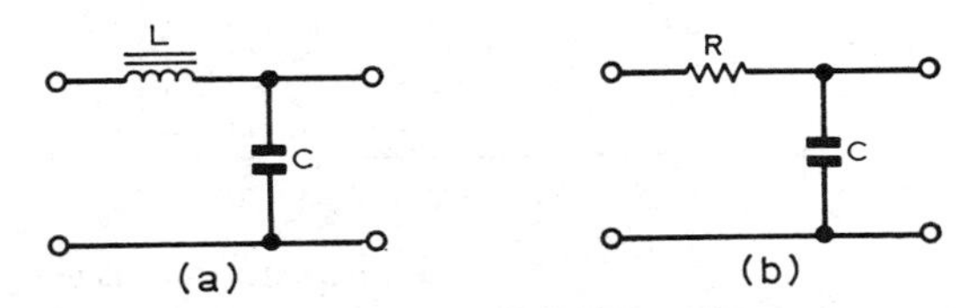

Fig 16.13. Additional smoothing sections to follow circuit of Fig 16.4 or Fig 16.11.

The ripple voltages at the outputs of these additional sections expressed as a percentage of their input ripple voltages are shown in **Figs 16.14** and **16.15** for various ripple frequencies. At a ripple frequency of 100Hz the ratio of output to input ripple voltage can also be calculated from the formula:

$$\frac{e_2}{e_1} = \frac{1}{0{\cdot}4LC} \quad \text{(L in henrys, C in microfarads)}$$

in the case of the LC filter and:

$$\frac{e_2}{e_1} = \frac{1}{0{\cdot}63RC} \quad \text{(R in k}\Omega\text{, C in microfarads)}$$

in the case of the RC filter, where e_1 and e_2 are the input and output ripple voltages respectively.

The capacitors in both types of filter sections should have a voltage rating at least equal to the peak ac voltage. The ripple currents are not high so the capacitors need not be as large physically as those immediately following the rectifiers. In the LC filter the choke is a constant inductance type rated to provide the required inductance at the maximum load current. The resistor in the RC filter requires a wattage rating sufficient to avoid overheating, and the required RC product should be obtained by using the largest practicable capacitance in order to minimize the resistance value and consequently the voltage drop.

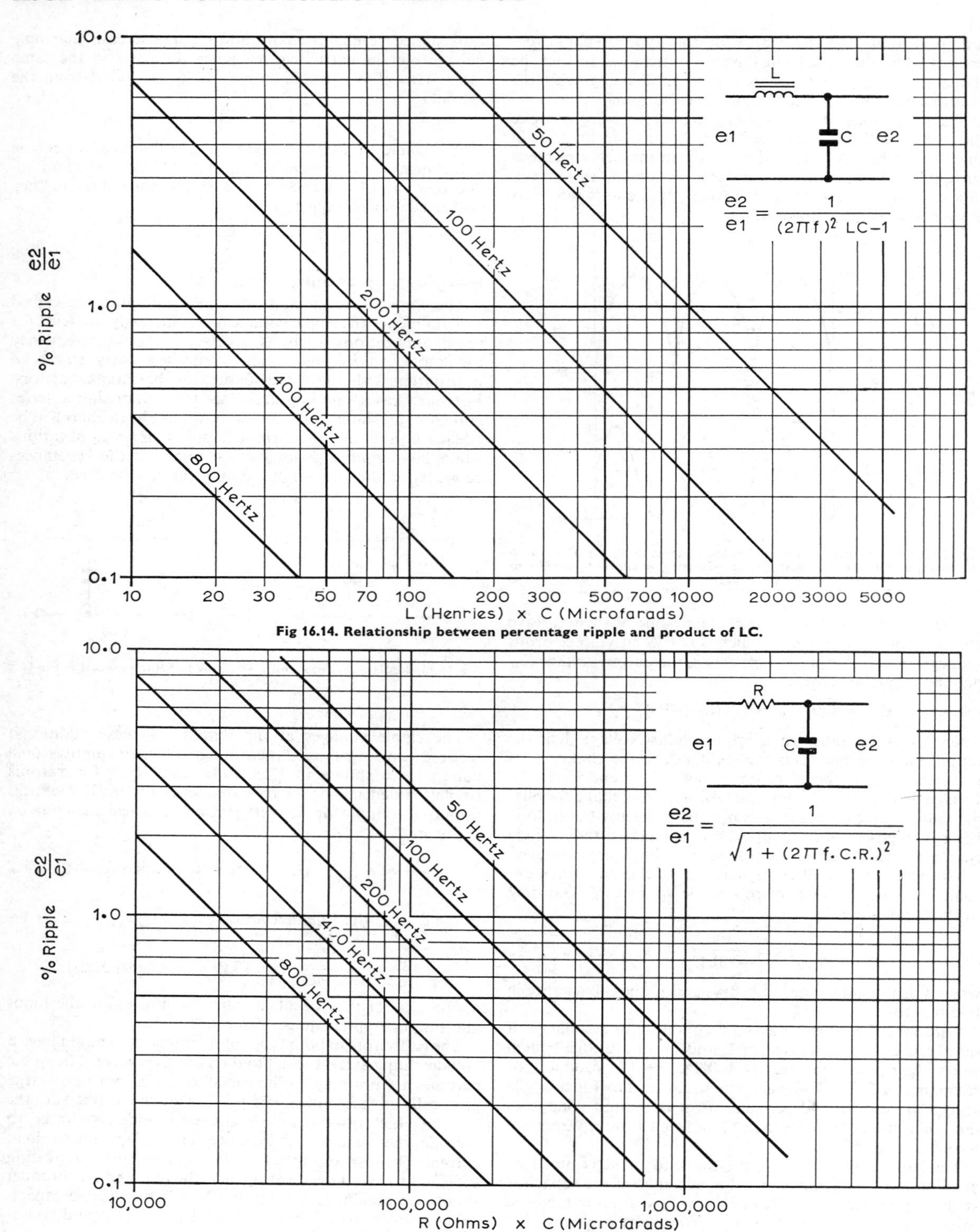

Fig 16.14. Relationship between percentage ripple and product of LC.

Fig 16.15. Relationship of percentage ripple and product of RC.

DUAL POWER SUPPLY

Transmitters and transceivers of the thermionic valve type generally require a high voltage supply for the pa anode and a lower voltage supply for the earlier stages. An attractive alternative to two separate supplies or to wasteful voltage dropping is the dual voltage supply shown in **Fig 16.16.** This type of circuit combines the principles of the bridge rectifier and the bi-phase half-wave rectifier. The diodes D1 and D2 provide the bi-phase half-wave rectification with the interconnection of the two diodes grounded and the output taken from the transformer centre tap. This arrangement is the reverse of normal but with the diode polarity shown a positive output is obtained. Since the diodes D1 and D2 function in both rectifier circuits, their currents are higher than the D3 and D4 currents, although this is only likely to be of practical interest if the diodes are operated near to their maximum ratings. With capacitor input filters at both outputs as shown in Fig 16.16, voltage V_2 is approximately half that of voltage V_1. Sometimes it is necessary to provide a low voltage which is less than half the high voltage, as for example in a transmitter requiring 900V for the pa anode and 300V for the driver stages. A choke input filter at the V_2 output is then more appropriate. This simply entails the removal of C2 shown in Fig 16.16 and ensuring that the choke L1 has an inductance which is at least the critical value for the load in question. (See page 16.6).

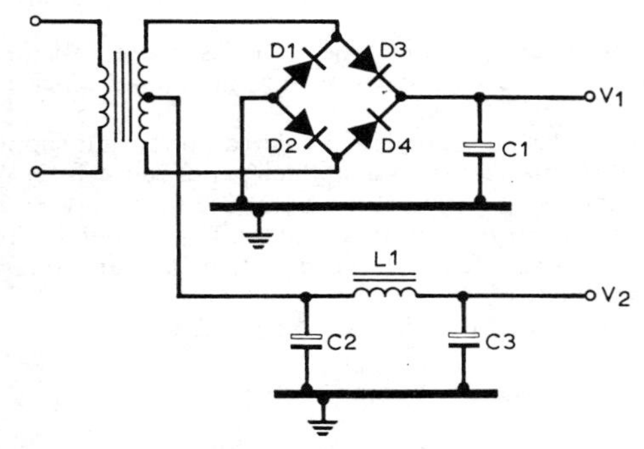

Fig 16.16. Circuit diagram of a dual voltage power supply.

VOLTAGE REGULATORS

There are many circuits which require a power supply of constant voltage for their proper operation. These include dc amplifiers, variable frequency oscillators, receiver local oscillators and some logic circuits. A regulated power supply is also very desirable for the bias supply for Class AB2 and Class B amplifiers used in modulators and ssb transmitters.

Zener Diode Regulator

The simplest form of voltage regulator, which is sufficient for many purposes, uses a zener diode as shown in **Fig 16.17.** Zener diodes have the characteristic property that they exhibit very little conduction when a reverse voltage is applied, as with normal diodes, but when the reverse voltage is increased the conduction increases markedly when a certain voltage is reached. This voltage, known as the zener voltage, is very well defined and very stable. Zener diodes are usually available with zener voltages in the EIA ± 5% series of preferred numbers, like resistors, starting at 3·3V and extending to 220V. When in the zener conducting state the impedance of the diode is very low and this zener impedance (Z_Z) is an important parameter since it determines the voltage change caused by a current change through the diode.

Fig 16.17. Circuit of a simple zener diode regulator.

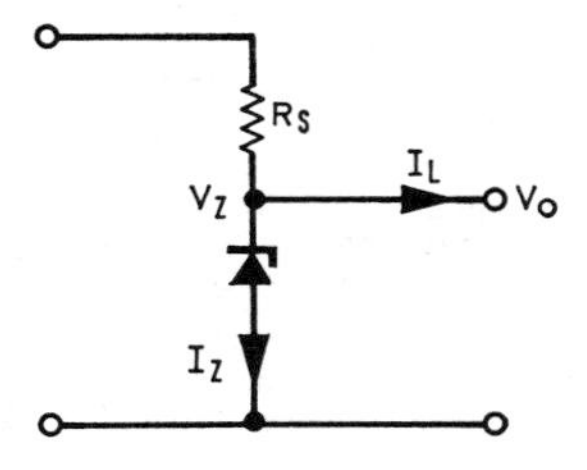

TABLE 16.3

Characteristics of some zener diodes from the Texas 1S3000 series

Type No.	Zener Voltage Vz	Test Current Iz mA	Zz Ohms at Iz
1S3006A	6·8	50	1·5
1S3010A	10	20	5
1S3020A	20	10	30
1S3039A	39	5	80

Table 16.3 shows the zener impedance of some regulator diodes from the Texas 1S3000 series and the current at which the impedance measurement is made. The effect of small changes in input voltage on the output voltage may be calculated by regarding the series resistor R_S and the zener impedance as the components of a potential divider. The fraction of input voltage change appearing at the output is:

$$\frac{Z_Z}{R_S + Z_Z}$$

Any ripple voltage on the input voltage is also decreased by the action of the zener diode.

In the circuit in Fig 16.17 the zener diode should provide a regulating action even when the load current (I_L) is maximum and the input voltage (V_{IN}) is minimum. In this extreme condition the zener diode current is minimum and the series resistance value must be such that this current is sufficient to ensure satisfactory regulation. This requirement enables the series resistance (R_S) to be calculated using the formula:

$$R_S = \frac{V_{IN\ (min)} - V_Z}{I_{L\ (max)} + I_{Z\ (min)}}$$

the units being volts, amps and ohms. A suitable minimum zener diode current is 10mA. The opposite extreme condition occurs when the input voltage is maximum and the load current minimum. This condition is used to calculate the power ratings of the series resistor and zener diode since the power dissipated in each is maximum. The total current (I_T) through the series resistor in these circumstances is:

$$I_T = \frac{V_{IN\ (max)} - V_Z}{R_S}$$

and the power dissipated by the resistor is:

$$\frac{(V_{IN\ (max)} - V_Z)^2}{R_S}$$

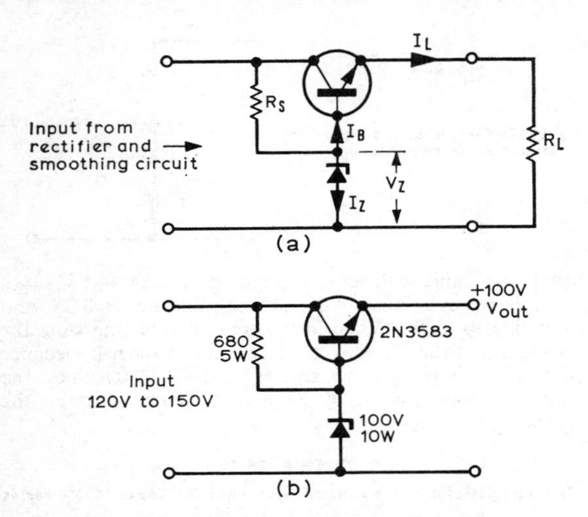

Fig 16.18. A series transistor regulator.

The total current divides between the load and the zener diode so the maximum diode current is:

$$I_{Z\,(max)} = I_T - I_{L\,(min)}$$

The power dissipation of the zener diode is then:

$$I_{Z\,(max)}\, V_Z \text{ watts.}$$

As an example, consider the design of a regulated supply of 10V derived from a power source with voltage varying from 22V to 28V. The full load current is 50mA but the load can be disconnected (ie $I_{L\,(min)}$ is zero).

First calculate R_S

$$R_S = \frac{22 - 10}{50 + 10} \times 1{,}000\Omega = 200\Omega$$

Then calculate power ratings

$$I_r = \frac{28 - 10}{200} = 0{\cdot}09\text{A} = 90\text{mA}$$

$$\text{Resistor power dissipation} = \frac{(28 - 10)^2}{200} = 1{\cdot}62\text{W}$$

$$\text{Zener diode dissipation} = \frac{90 \times 10}{1000} = 0{\cdot}9\text{W}$$

A 2W resistor and a 1W zener diode such as the 1S3010A would be suitable components.

Series Transistor Regulator

A regulator with performance superior to the simple shunt zener diode type just described uses a series transistor as shown in **Fig 16.18(a)** with a zener diode circuit to stabilize the base voltage. The zener diode circuit is only required to deal with the low base current since the load current flows in the emitter circuit of the transistor. The transistor behaves as an emitter follower and the output voltage is less than the stabilized base voltage by the base-to-emitter voltage (V_{BE}). Typical values of V_{BE} are in the range 0·5 to 1·0V and the main causes of change in the output voltage are changes in V_{BE} with current and temperature.

The design of the zener diode base regulator circuit is carried out in the manner already described but of course its load current is the relatively low base current. The base current

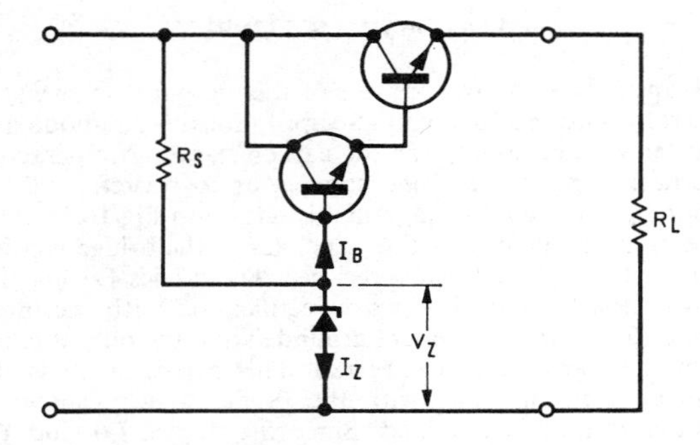

Fig 16.19. A series regulator using two compounded transistors as the series element.

(I_B) is related to the load current (I_L) by the forward current transfer ratio h_{FE} of the transistor at the load current in question, ie:

$$I_B = \frac{I_L}{h_{FE}}$$

The expression for the series resistance is then:

$$R_S = \frac{V_{IN\,(min)} - V_Z}{I_{B\,(max)} + I_{Z\,(min)}}$$

where $I_{B\,(max)} = \dfrac{I_{L\,(max)}}{h_{FE}}$ and the other terms have the same meaning as before. The wattages of the resistor and zener diode are calculated as for the simple zener diode regulator.

Consider as an example a regulated power unit supplying 100V at 400mA from a source which varies between 120V and 150V, and using a transistor type 2N3583 with $h_{FE} = 20$.

A 100V zener diode is used and the V_{BE} drop in the transistor may be neglected. (The actual output voltage is approximately 99V).

First calculate the series resistance value

$$I_{B\,(max)} = \frac{400}{20} = 20\text{mA or } 0{\cdot}02\text{A}$$

$$R_S = \frac{120 - 100}{0{\cdot}02 + 0{\cdot}01} = \frac{20}{0{\cdot}03} = 666\Omega$$

The preferred value of 680Ω is used.

The maximum zener diode current when the input voltage is maximum and the load current zero is

$$I_Z = \frac{150 - 100}{680} = 0{\cdot}073\text{A}$$

$$\text{Resistor dissipation} = \frac{50^2}{680} = 3{\cdot}6\text{W}$$

$$\text{Zener diode dissipation} = 0{\cdot}073 \times 100 = 7{\cdot}3\text{W}$$

A 5W resistor and 10W zener diode are used in practice. The transistor dissipation in the worst case of 150V input and 400mA load current is 50 × 0·4 = 20W. The power supply is shown in **Fig 16.18(b).** Although in normal operation the voltage across the transistor is 50V, it is a wise precaution to use a device such as the 2N3583 which has a maximum V_{CE} rating equal to the maximum input voltage of 150V, to cope with the case of switching on the power supply with an uncharged capacitive load.

It should be noted that the resistor and zener diode in this example require comparatively high power ratings. Components of lower ratings may be used if a second transistor is used as shown in **Fig 16.19.** The two transistors behave like a single device with h_{FE} which is the product of the forward current transfer ratios of the individual transistors, and repeating the calculation above with two transistors with h_{FE} of, say, 20 and 50 will show that a resistor of 2,000Ω dissipating 1·25W can be used. A disadvantage of the circuit is that the output voltage differs from the zener voltage by two base-to-emitter voltages and suffers the variations of these two.

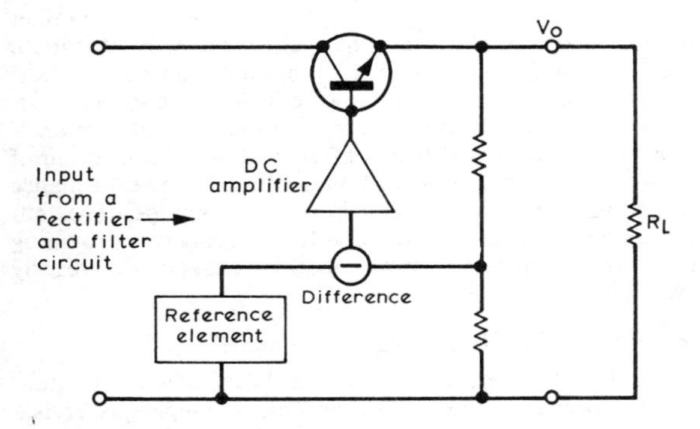

Fig 16.20. Diagram of a feedback-controlled series regulator.

A more precise regulator is achieved by the use of circuits of the form shown schematically in **Fig 16.20.** In this arrangement, a fraction of the output voltage is compared with a stable voltage established by a reference element (usually a zener diode) and the difference voltage is amplified by a dc amplifier. The amplifier provides the base current for the series transistor. Because of the high amplifier gain, only a very small decrease in regulator output voltage increases the transistor base current in response to a demand for increased load current. The power supply can therefore tolerate large changes in load current with very small resulting changes in output voltage. This will be recognized as a feedback control loop, and it is obvious that the greater the gain of the dc amplifier the smaller the changes in output voltage. It is usual to describe the performance of these regulated power supplies by the expressions *stabilization factor* and *output resistance*. The stabilization factor is a measure of how the output voltage is influenced by changes in the input voltage, and it is the ratio of input voltage change to the resulting change in output voltage, ie:

$$\text{Stabilization Factor} = \frac{\text{Change in input voltage}}{\text{Change in output voltage}}$$

when the output current is held constant. The output resistance may be regarded as the internal resistance of the regulator and, when an output current flows, the terminal voltage drops by the voltage drop in this resistance. The resistance value may be calculated by dividing the change in output voltage by the change in output load current which causes it, ie:

$$\text{Output Resistance} = \frac{\text{Change in output voltage}}{\text{Change in output load current}}$$

when the input voltage is constant. Stabilization factors of 5,000 and output resistances of 10mΩ are common.

In the simplest form the amplifier is just a single transistor stage and the reference zener diode is connected as shown in **Fig 16.21(a), Fig 16.21(b)** or **Fig 16.21(c).**

The collector resistor for TR2 (shown as R1 in the figures) has a value which is calculated by considering the situation when V_{IN} is minimum and the output load current I_L is maximum. In this extreme condition the output voltage is minimum and the collector current of TR2 is negligibly small. The base current of TR1 is therefore equal to the current through R1 and if the V_{BE} voltage of TR1 is ignored the voltage across R1 is $V_{IN\,(min)} - V_{OUT}$. The base current in TR1 necessary to produce an emitter current of $I_{L\,(max)}$ is $I_{L\,(max)}$ divided by the h_{FE} of TR1. This leads to the equation:

$$\frac{V_{IN\,(min)} - V_{OUT}}{R1} = \frac{I_{L\,(max)}}{h_{FE}}$$

and it follows that:

$$R1 = \frac{(V_{IN\,(min)} - V_{OUT})\,h_{FE}}{I_{L\,(max)}}$$

The other extreme condition, when V_{IN} is maximum and the load current is zero, is used to calculate the maximum current in the resistor R1 and in the collector of TR2 and hence to determine the wattage of these two components. This current $I_{C\,(max)}$ is found from the equation:

$$I_{C\,(max)} = \frac{V_{IN\,(max)} - V_{OUT}}{R1}$$

and the power dissipated in R1 is:

$$\frac{(V_{IN\,(max)} - V_{OUT})^2}{R1}\ \text{watts}$$

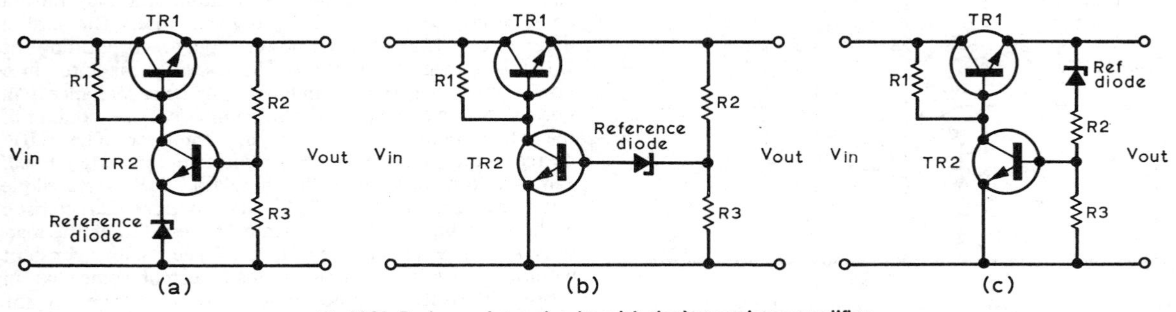

Fig 16.21. Basic regulator circuits with single transistor amplifier.

The collector-to-emitter voltage V_{CE} of TR2 is obviously V_{OUT} in the case of the circuits in Fig 16.21(b) and (c), and less than V_{OUT} by the reference diode voltage in the circuit of Fig 16.21(a). The required power rating of TR2 is found by multiplying V_{CE} by the calculated $I_{C\,(max)}$ value.

A practical circuit for a regulator operating from an input voltage of 55V and producing an output voltage adjustable from 27 to 35V is shown in **Fig 16.22.** The output current is 30mA, the output resistance is 1Ω and the stabilization factor is greater than 100.

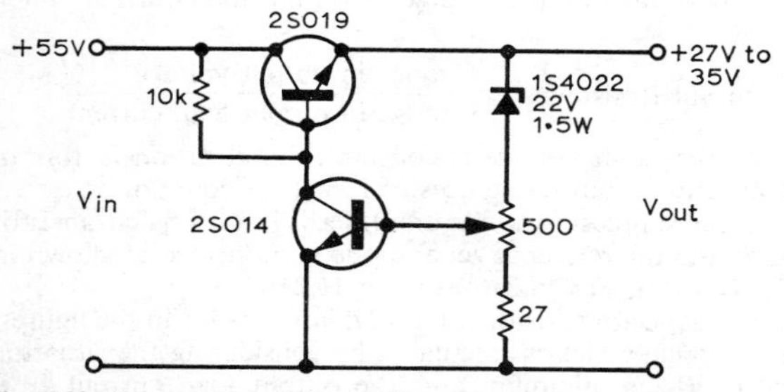

Fig 16.22. A regulator with variable output voltage (27 to 35V) at 30mA.

More elaborate regulators may contain an amplifier with two or more stages, sometimes combined with two or more transistors in a compound configuration for the series control element. An emitter-coupled pair is the preferred type of amplifier stage, at least at the input of the amplifier, because the drift with temperature is less than in a single transistor. A Texas Instruments design for a regulator to supply 50V at currents up to 500mA from an input which varies between 65 and 85V is shown in **Fig 16.23.** The zener diode D1 is a 5·1V device and is chosen because reference diodes at this voltage have a temperature coefficient of voltage which is near zero. Three diodes in series (each with a zener voltage of 5·6V) are chosen for the D2 position, rather than a single diode of 16V approximately, in order to provide a small temperature coefficient of opposite sign to the temperature coefficient of V_{BE} in the transistor TR3. Another interesting feature of this regulator is the stabilized supply for the collector of TR3 (and base of TR2) which is obtained by adding a stabilized 10V supply to the output rail. The zener diode D3 and 1·5kΩ resistor perform this function. The performance of the regulator is enhanced by this means because variations in the raw input voltage are not fed to the base of the series transistor as is the case in the simpler circuits of Fig 16.21. The stabilization factor of this regulator is greater than 4,000 and the output resistance less than 0·68Ω. The purpose of the 100μF capacitor is to prevent the high frequency oscillation which is a danger in any multi-stage feedback system.

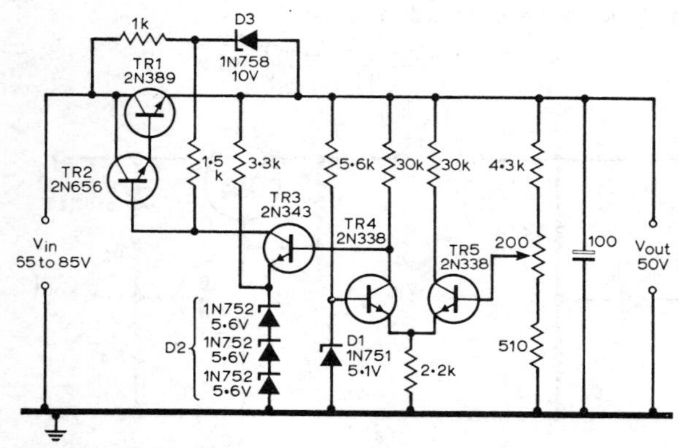

Fig 16.23. Circuit diagram of a regulator for 50V at 500mA (*Texas Instruments*).

Regulators are vulnerable to damage due to excessive load current, particularly an output short circuit. The series transistor is most likely to suffer because the amplifier supplies maximum base current to the series transistor in an effort to maintain the output voltage in spite of the short circuit! Most of the circuit refinements in modern regulators are aimed at protection against such damage. Protection is usually provided in the form of fast switch-off in the event of an overload, or by output current limiting. High performance regulators containing these features are complex units, but fortunately integrated circuit voltage regulators providing such performance in very small packages are readily available.

Integrated Circuit Voltage Regulators

The ic voltage regulators are rugged devices which contain in monolithic form the series transistors, amplifiers, reference diodes and the protection arrangements already mentioned.

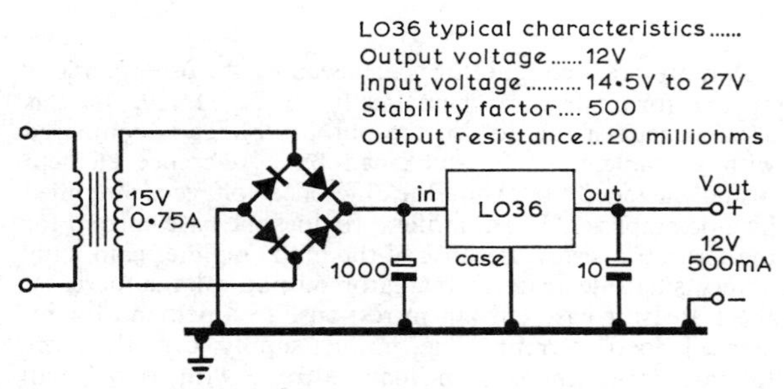

Fig 16.24. Circuit diagram of a 12V 500mA power unit using a regulator type L 036.

For example, the SGS-ATES L 036 voltage regulator has an output of 12V at 500mA and is designed to operate from an input which can vary from 14·5V to 27V. It is contained in a standard TO-3 case and has simply an input and an output connection apart from the grounded case. The circuit diagram of a 12V regulated power supply incorporating the L 036 is shown in **Fig 16.24.** The reservoir capacitor in a power supply of this type can have a smaller capacitance than normal, with a subsequent saving in physical size, because of the ripple reducing action of the regulator. The L 036 decreases the input ripple by a factor of approximately 1,000. For overload protection this regulator uses a technique known as *current fold-over*. The voltage-current characteristic is shown in **Fig 16.25** which demonstrates that, as the load resistance is decreased, the output voltage remains constant within narrow limits until a load current somewhat in excess of the full load value is reached. Further decreasing the load resistance causes the output voltage and current to

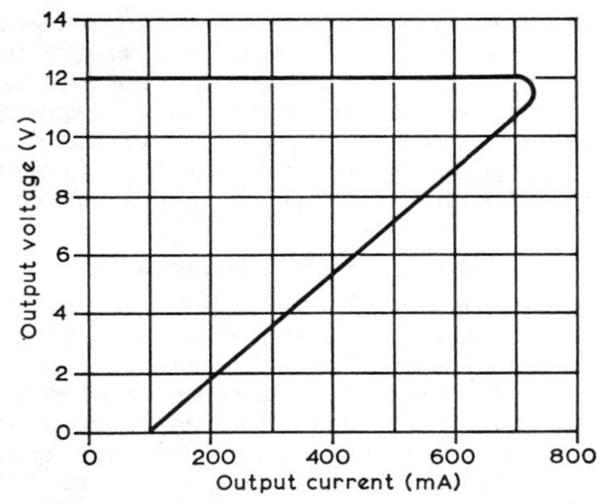

Fig 16.25. Voltage/current characteristic of the L 036.

decrease rapidly until the current flowing in a short-circuited output is only 100mA. The output resistance of the L 036 is approximately 20mΩ and the stabilization factor is about 500.

Another series of popular integrated circuit regulators is the Motorola MC 1460, MC 1461, MC 1560 and MC 1561. The first two will operate in any ambient temperature over the range 0°C to +75°C and the other two over the range −55°C to +125°C. All four are available with suffix G or suffix R indicating the type of package. Those in a type G package can deliver 250mA and in the type R package, which is a bolt-down flange type, up to 600mA. The maximum output voltages range from 17V for the MC 1460 and MC 1560 to 32V for the MC 1461 and 37V for the MC 1561. These integrated circuits incorporate facilities for short-circuit protection and electronic shut-down. The latter feature allows a number of control and protection techniques to be applied to circuits employing these devices. For example, it is possible to arrange that automatic shut-down occurs if the temperature of the chip becomes excessive or when the output is momentarily short-circuited. All of the devices in this series have an output impedance of approximately 20mΩ in a typical case and a stabilization factor of about 3,000.

A circuit for a power supply with output voltage adjustable from zero to 25V and capable of supplying up to 300mA is shown in **Fig 16.26.** The short-circuit current limit is determined by the resistor between pins 1 and 4. The value of 1Ω shown limits the short-circuited output current to 400mA.

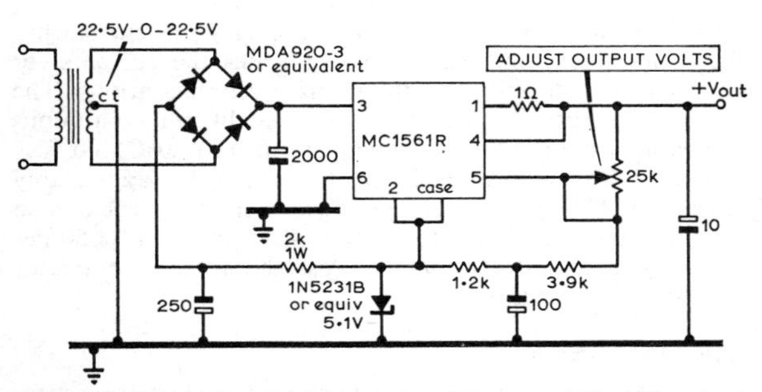

Fig 16.26. Circuit diagram of a regulated power supply with output voltage adjustable 0-25V (*Motorola*).

The output current capabilities of these devices may be enhanced by the use of a series-pass transistor as shown in **Fig 16.27.** In this circuit a 2N3055 series transistor boosts the output current to more than 4A. The 0·12Ω resistor at the emitter of the series transistor sets the short-circuit current at 5A. The output voltage is determined by the values of resistors R1 and R2. A value of 6·8kΩ is normally selected for R2. Resistor R1 is then calculated from the formula $R1 = (2V_O - 7)k\Omega$. The value of 20kΩ shown sets the output voltage at 13·5V and the unit is then useful for the bench testing of mobile equipment. A suitable input supply could comprise a transformer with a 15V 6A secondary, a bridge rectifier to supply 5A dc and a 20,000μF 25V wkg electrolytic capacitor with a ripple current rating of 6A. The 2N3055 must be mounted on a heat sink with thermal resistance less than 2·5°C/W. If the leads to the output load are long, the problem of voltage drop can be overcome by connecting the *sense* leads to the actual load, rather than to the output terminals of the power supply. The extended leads are shown by the dotted lines in the figure. In this way the regulator maintains close control over the load voltage instead of the terminal voltage. In commercial practice the sense leads are typically labelled V+ and V−, the current leads I+ and I−.

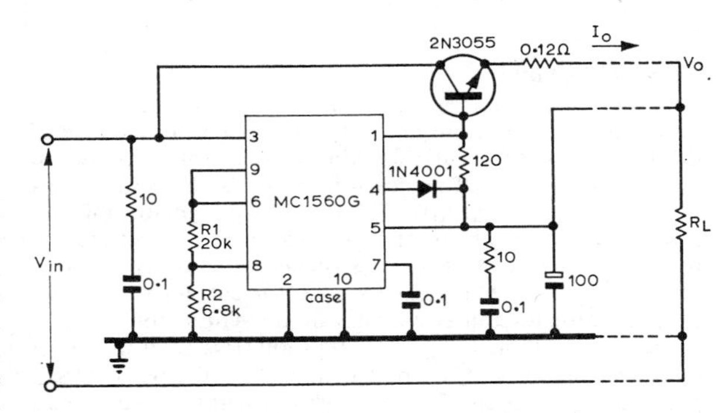

Fig 16.27. Circuit diagram of a regulator with maximum output current 4A (*Motorola*).

Regulated power supplies with output voltages higher than the specified maximum voltage of these integrated circuits can be constructed by the use of a series transistor with a zener diode to shift its required base voltage by the necessary amount from the ic output voltage. A circuit to produce 100V at 100mA is shown in **Fig 16.28.** The MC 1561 G operates from an input supply of nominally 30V to provide an output at pin 1 of approximately 25V. The 75V zener diode in series with a 1N4001 adds on just more than 75V to operate the base of TR1 at a little over 100V. The purpose of the 1N4001 is to prevent the MC 1561 G from supplying excessive current under output short-circuit conditions. The transistor TR2 in association with the 5·6Ω resistor limits the short-circuit current to 100mA.

It is important to remember that integrated circuits of this sort contain high-gain circuits of extremely small dimensions and consequently extremely small internal capacitances. This can lead to oscillations in the vhf region if care is not taken to use short leads and direct earthing, similar to the layout used in vhf amplifiers. Damping networks can be

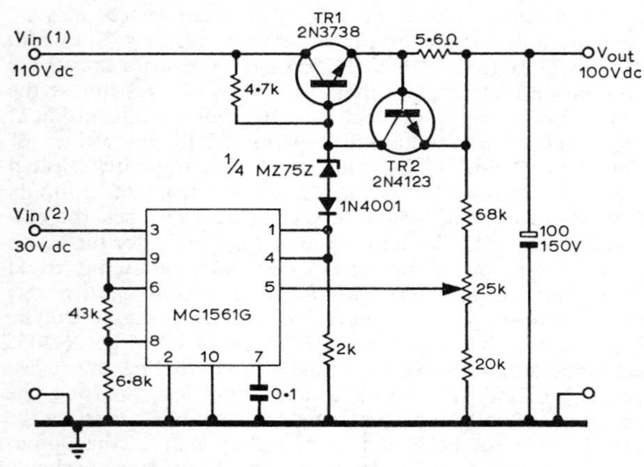

Fig 16.28. Circuit diagram of a 100V regulator *(Motorola)*.

used to inhibit such instability. These consist of a low resistance (2·7 to 10Ω) in series with a high quality 0·1μF capacitor. Two such networks are shown in the circuit in Fig 16.27, connected from pin 3 to ground and from pin 5 to ground.

Switching Regulators

The voltage regulators described earlier provide a high degree of regulation but are not very efficient. In the case of the series transistor regulator a considerable proportion of the input power is wasted as heat dissipated by the series transistor, ie $(V_{IN} - V_O) I_O$W. The switching regulator offers a solution to this problem of power waste. The current to the base of the series transistor is switched repetitively between zero and a value which causes collector current saturation. The collector-to-emitter voltage in the latter condition is extremely small (ie = $V_{CE\ (sat)}$, typically less than 0·5V), so although the collector current is maximum the power dissipated is very small. The transistor operates like a fast repetitive switch and the control of output current is achieved by altering the ratio of ON to OFF durations. A basic schematic diagram of a switching regulator is shown in **Fig 16.29.** The choke L and capacitor C are necessary to smooth the current pulses produced by the switch. The values of both L and C are quite small in practice because of the high frequency of the switching (eg 10kHz) as normally used. The control circuit comprises a multivibrator operating at the required switching frequency and a means of controlling the mark-space ratio of the generated pulses. The diode is necessary to protect the base-emitter junction of the transistor against the back emf developed by the choke when the current is switched off. The regulation obtained from a switching regulator is not as good as from a series type regulator, and the speed of response to input voltage and output current variations is also inferior. For these reasons a switching regulator is often followed by a conventional series regulator when highest performance is required.

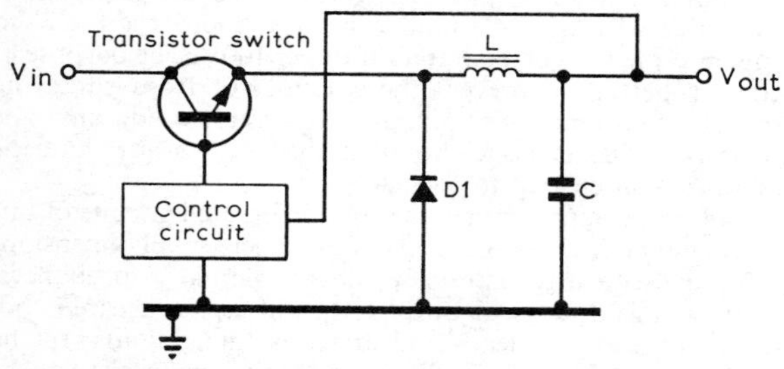

Fig 16.29. Schematic diagram of a switching regulator.

COMPONENT RATINGS

The components used in power supplies must be properly rated in the interests of reliability. In all cases increased reliability can be obtained by under-running the component in question. For example, the average life of electrolytic capacitors is more than doubled if they are operated at a voltage which is 50 to 75 per cent of the rated voltage, as compared with operation near the rated voltage. Ambient temperature also influences reliability, and operation of electrolytic capacitors in an ambient temperature of 45°C results in halving the average life obtaining when the ambient temperature is 25°C.

The voltage and current ratings of rectifier diodes has already been discussed and Table 16.1 may be used as a guide to design. Diodes of the same type may be connected in series to obtain a multiple of the peak inverse voltage rating but it is desirable to connect resistors of equal value across the diodes to ensure that the reverse voltage is shared equally. A commonly-recommended resistance value is 500Ω/V of reverse voltage. Diodes with a peak inverse voltage rating of 1,000V, for example, should be shunted with 500kΩ resistors or a near preferred value (ie 470kΩ or 560kΩ). It is also desirable to shunt each diode with a 0·01μF capacitor of adequate voltage rating to ensure that transient voltages are properly shared between diodes.

The smoothing capacitors must have voltage ratings at least as high as the voltages to which they are subjected under no-load conditions. They must also have ripple current ratings in excess of the maximum ripple currents which they are called upon to handle. The need sometimes arises for a high value electrolytic capacitor with a voltage rating greater than the usual available maximum of around 600V. A suitable component can be constructed from two or more capacitors of equal value connected in series. Resistors of a value about 100Ω/V of rated voltage must be connected across each capacitor to make the applied voltage divide equally. By this means a 16μF capacitor of 1,000V rating, for example, can be obtained by connecting two 32μF 500V capacitors in series and shunting each with a 50kΩ resistor. The capacitor connected to the 1,000V must have its can insulated from the chassis to withstand 500V.

The calculation of transformer secondary rms current has been described, and Table 16.1 indicates the currents for various rectifier and smoothing circuit configurations. The current ratings of transformers are usually for continuous operation at normal ambient temperatures (+20°C) but very often in amateur applications these can be appreciably exceeded. For transmitter applications the duty cycle is so low that the mains transformers may be over-run by 50 per cent (eg a 200mA transformer can be used in a circuit requiring 300mA rms).

Fuse Protection

A power supply can be adequately protected against harmful overload by a fuse of suitable rating connected in series with the transformer primary. The rms currents of all the secondary windings of the transformer should be ascertained under full load conditions and the current in each winding multiplied by the voltage of that winding. The VA figures of all the secondaries calculated in this manner should be added to give the total VA supplied by the transformer. (The magnetizing current can normally be neglected in this total.) The primary current can then be found by dividing this total VA figure by the primary voltage. A fuse with a rating 20 per cent greater than the primary current, or the nearest larger available rating, should be used. Anti-surge fuses are necessary to survive the high current transient at switch-on.

BATTERIES

Batteries are increasingly being used as a power source for amateur equipment because of their portability and the simplification possible in the supply circuitry of the equipment using them. The choice of a suitable battery depends on a number of factors.

Primary batteries have a single life-span, thus demanding periodic replacement. Many different types are now available and often the choice of a battery is governed more by what is readily available than by what is most suitable, and this can lead to dissatisfaction with the subsequent performance. Alternatively, secondary batteries, which can be recharged, may be the best choice for certain applications.

Battery Characteristics

In order to select a battery correctly, consideration must be given to the following factors:

Voltage. It is necessary to know both the initial and end point voltages. The lowest operating voltage of the equipment is an important factor in battery life.

Type of service. This could be intermittent communications use or constant current, and the operating cycle must be considered.

Load. The average current that the battery will be expected to deliver at the operating voltage.

Size. This will depend upon the size of the battery compartment in the equipment. To a lesser extent the weight may have some influence on the choice.

Storage life. This may or may not be a worthwhile consideration according to the usage of the equipment.

Temperature. Maximum performance is usually obtainable at about 20°C. Temperature around freezing point will reduce the efficiency of certain battery types.

Cost and availability. Obviously economy will be a major factor but it is necessary to consider both the initial cost and the battery life. It may be more economical to use a larger and initially more expensive battery but one which will give longer service and a lower running cost. When choosing a primary battery this should be of a type that is obtainable without difficulty.

Types Available

Primary cells

The *zinc carbon* or Leclanché battery is the most popular type available and is usually the least expensive. Among its characteristics are:

(*a*) It should not be discharged beyond the end of its useful life or left in equipment in a discharged state because of the possibility of leakage;

(*b*) a "rest" period is required. Continuous drain will effectively reduce battery life;

(*c*) the "high power" (HP) type is suitable for heavy current drain with reasonable voltage stability. It is used in most motor powered devices;

(*d*) the "power pack" (PP) range of batteries has been designed for use with transistor operated equipment and they are fitted with non-reversible contacts.

No attempt should be made to recharge zinc carbon batteries owing to the very real danger of explosion.

The *mercury* cell was first produced in quantity during the second world war when military requirements hastened its development.

Its characteristics include:

(*a*) high cathode efficiency providing a stable voltage during current discharge;

(*b*) a "rest" period is not required for maximum performance;

(*c*) a greater capacity/volume ratio than zinc carbon batteries;

(*d*) storage for long periods at temperatures around 20°C is possible without any appreciable loss of capacity. Typically 90 per cent of the capacity will be retained during a period of 30 months;

(*e*) the cell containers are usually of nickel plated steel, not forming part of an active electrode and resistant to corrosion.

The *alkaline manganese* cell has undergone improvement comparatively recently and in its present form is suitable for continuous discharge applications. However, its voltage is not as stable as that of the mercury cell. Characteristics include:

(*a*) a good storage life. Typically 95 per cent of the capacity will be retained during a period of 20 months at 20°C;

(*b*) the case is not an active element and is usually of electroplated steel which is resistant to corrosion;

(*c*) satisfactory operation is possible over the temperature range −20 to +70°C.

Rechargeable alkaline manganese cells are obtainable but at the present time do not appear to be readily available from UK retail sources.

TABLE 16.4
Primary cells

Type	Nominal voltage	Weight (g)	Suggested current range (mA)	Typical use
RM675H (mercury)	1·5	2·3	0–10	Hearing aid
RM401 (mercury)	1·5	11·3	0–80	Hearing aid
ZM9 (mercury)	1·5	29·8	0–200	Hearing aid
HP7	1·5	16·5	0–75	Radio
HP11	1·5	45·0	0–1,000	Motor
HP2	1·5	90·0	0–2,000	Motor
SP2	1·5	90·0	25–100	Radio
SP11	1·5	45·0	20–60	Radio
AD28	4·5	443·6	30–300	Radio
PP1	6·0	283·0	5–50	Radio
PP3	9·0	38·0	0–10	Radio
PP6	9·0	142·0	2·5–15	Radio
PP9	9·0	425·0	5–50	Radio
PP10	9·0	1,250·0	15–150	Radio
HP1	12·0	1,550·0	0–4,000	General purpose
B106	45·0	255.0	1–10	General purpose
B1702	60·0	2,720·0	5–50	General purpose
B126	90·0	454·0	1–10	Radio

The above types are representative of the range manufactured by the Ever Ready Company (GB) Ltd.

Rechargeable cells

The *lead acid* battery is mentioned for completeness and its ready availability. Although it may have limited use for mobile applications its use in portable equipment is not recommended owing to the weight and corrosive liquid electrolyte.

The type of rechargeable cell generally associated with communication equipment is the *nickel cadmium* battery. This is manufactured in several forms, eg button cell, cylindrical cell and rectangular battery. Although the cell construction may differ between these forms the basic principles of operation are identical. Nickel cadmium (abbreviated to "nicad") cells are referred to by different manufacturers as cells, accumulators or batteries. It is here assumed that the descriptions are interchangeable, although strictly the term "cell" applies to the individual unit, ie a nicad battery can consist of several separate cells.

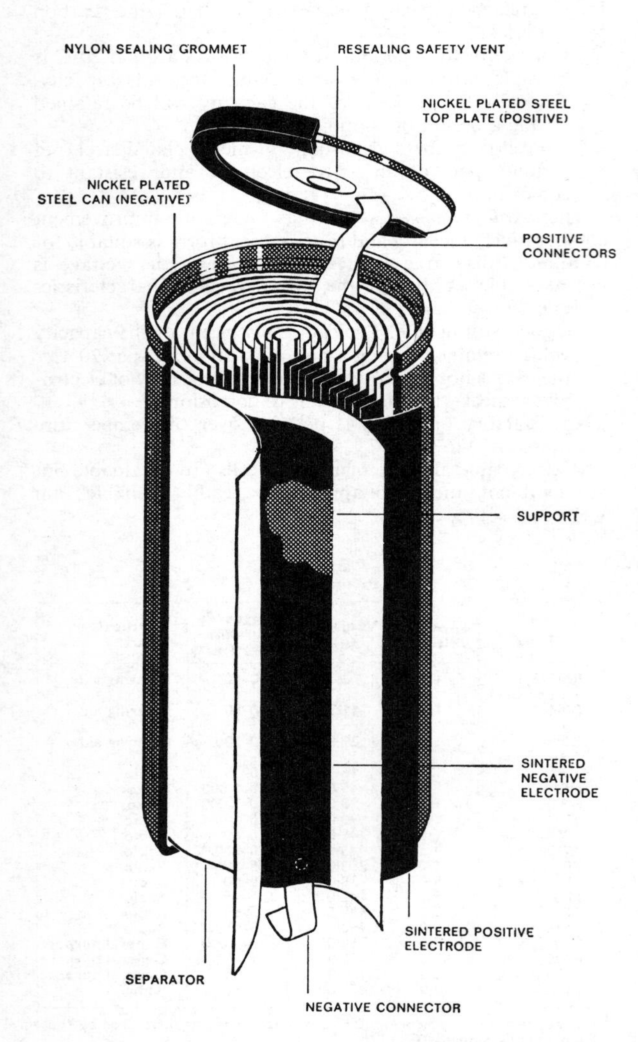

Fig 16.30. Typical nicad cylindrical cell construction.

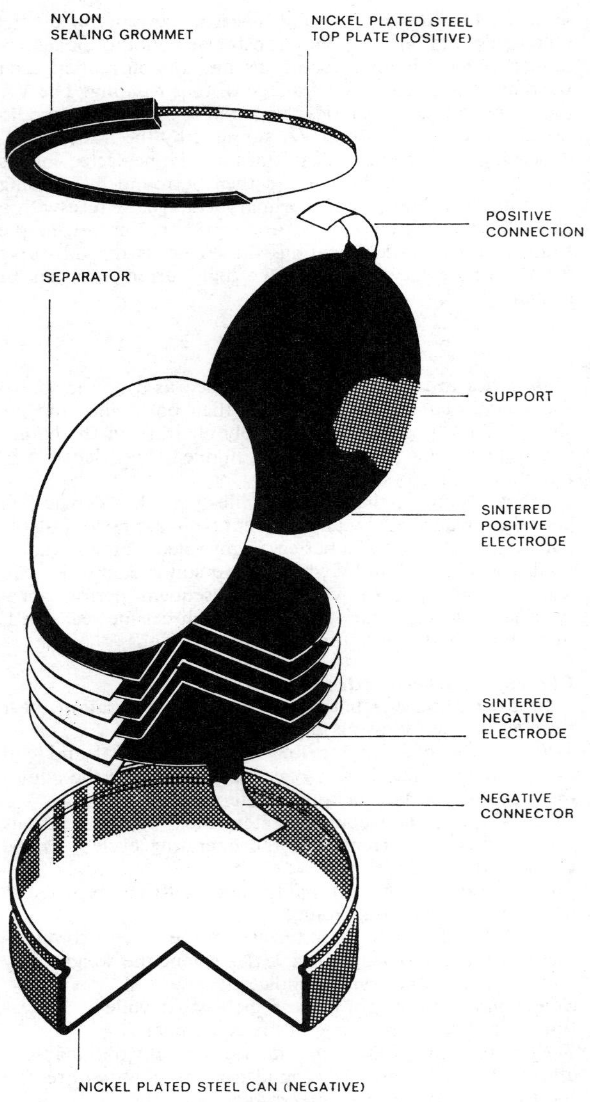

Fig 16.31. Typical button cell construction.

Characteristics of the nicad are:

(*a*) stable voltage during discharge cycle;
(*b*) suited to a continuous high discharge rate, a "rest' period is not required;
(*c*) unlike most secondary cells, no maintenance is required;
(*d*) cells may be charged and discharged hundreds or thousands of times according to the conditions of use;
(*e*) cells will operate over a wide temperature range, for discharge typically −20 to +45°C. For storage a greater range is possible, although the optimum figure for both charge and discharge is 20°C;
(*f*) nicads have a low self-discharge rate at normal temperatures. The curve for a typical cylindrical DEAC cell shows

that 70 per cent of the nominal capacity will be retained after a storage period of three months;

(*g*) the cells are fully sealed and may be used or stored in any position and are shock and vibration resistant. There is no corrosion problem.

The correct choice of a nicad battery is dependent on service requirements. The following points must be taken into consideration:

(i) nominal voltage;
(ii) upper and lower voltage limits;
(iii) maximum current and duration;
(iv) type of operation;
(v) space and weight limitations.

The above points, with the possible exception of (iv), are self-explanatory. For communications use the type of operation will usually be a charge-discharge sequence, rather than trickle charging or float operation.

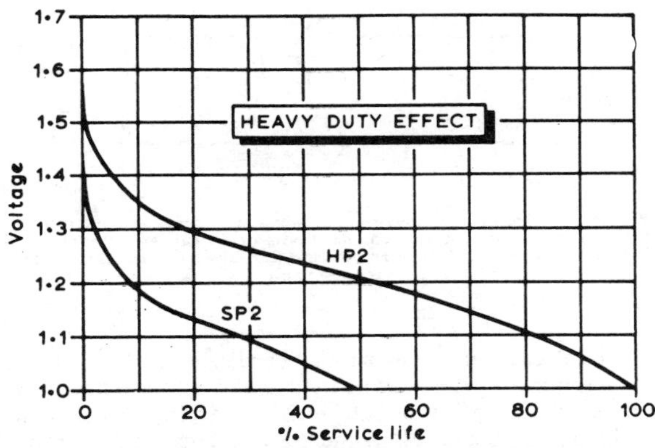

Fig 16.32. The graph illustrates the difference in service life between the specially designed high power battery and the standard battery on a tape recorder test. The discharge is for two hours a day through 5Ω for Ever Ready HP2 and SP2 batteries.

TABLE 16.5

Nickel cadmium rechargeable cells

Ever Ready Cat No.	Nominal capacity (Ah)	Weight (g)	Type
NCB28	0·28	16·5	Button cell 1·2V nominal
NCB55	0·55	28·5	Button cell 1·2V nominal
NCB175	1·75	100·0	Button cell 1·2V nominal
NCC60	0·60	30·0	Cylindrical cell, equivalent size HP7
NCC200	2·00	78·0	Cylindrical cell, equivalent size U11
NCC400	4·00	170·0	Cylindrical cell, equivalent size U2
NCB28/8	0·28	126·0	Button cell battery. 10V nom.
NCB55/8	0·55	232·0	Button cell battery. 10V nom.
DEAC type			
10/225DK	0·225	135·0	Button cell battery. 12V nom.
10/500DKZ	01·500	280·0	Button cell battery. 12V nom.
10/1000DK	0·000	610·0	Button cell battery. 12V nom.
TR7/8	0·070	45·0	Battery, equivalent to PP3. 9V nominal
501RS	0·500	30·0	Cylindrical cell, equivalent size H P7
RS1·8	1·800	65·0	Cylindrical cell, equivalent size U11
RS4	4·00	150·0	Cylindrical cell, equivalent size U2

The above types are representative of the extensive range of rechargeable cells produced by these two manufacturers.

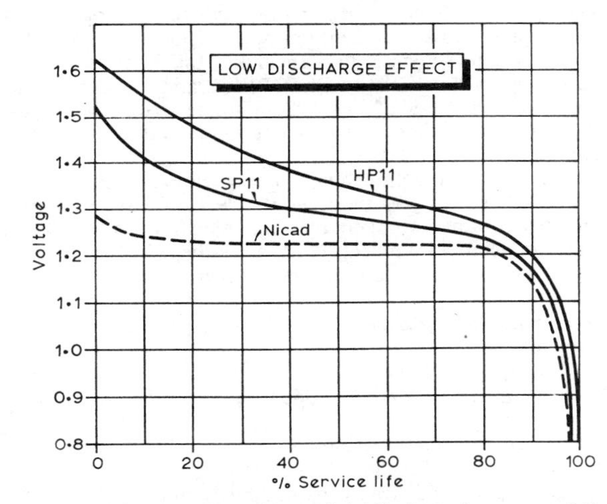

Fig 16.33. This shows that where a light discharge is concerned the heavy discharge type HP11 does not show any advantage over the SP11 when discharged through 300Ω for two hours a day. Note also the nicad discharge curve.

Charging Nickel Cadmium Cells

It is important that attention should be given to the correct methods of charging nicads. There is no particular difficulty involved but the methods are different from those employed with the usual automotive battery. Nicads are expensive and this is a very good reason for devoting some care to this aspect of their use. It is essential that the charging current should be kept to a constant known value. The use of constant voltage charging systems is not recommended due to the very low internal resistance of nicads, which could lead to the possibility of drawing high currents and consequent overheating.

The manufacturers of DEAC cells recommend that the charging current should be one tenth of the nominal capacity of the cell, ie a type 10/500DKZ with a capacity of 0·5Ah requires a charging current of 50mA. The charging factor of nicads is 1·4, ie in the case of a fully or partially discharged cell, 1·4 times the capacity taken out must be replaced. For normally discharged cells the charging time with the rated current is therefore 14h. This time can be reduced by fast charging, for details of which the manufacturer's literature should be consulted.

TABLE 16.6

Comparison of types

Type	Nominal voltage	Energy/density (Wh/weight)	Low temperature performance	Shelf life	Initial cost	Order of merit in cost per 100h
Zinc carbon	1·5	25	Poor	Fair	Lowest	2
Alkaline manganese	1·5	35	Fair	Good	Moderate	3
Mercury	1·4	45	Special type available	Good	High	4
Nickel cadmium (rechargeable)	1·25	*	Good	Very good	Expensive	1

* *Small size but moderate weight*

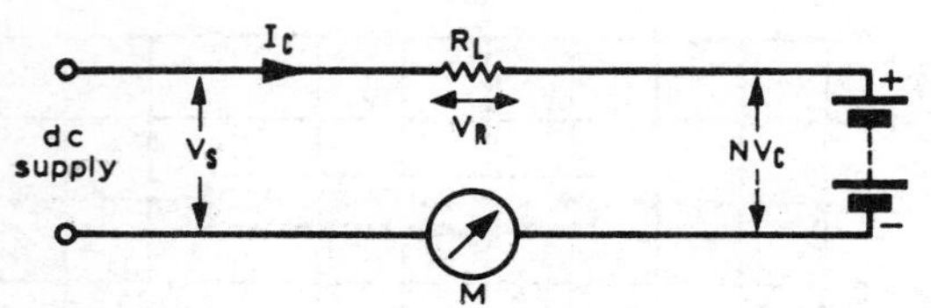

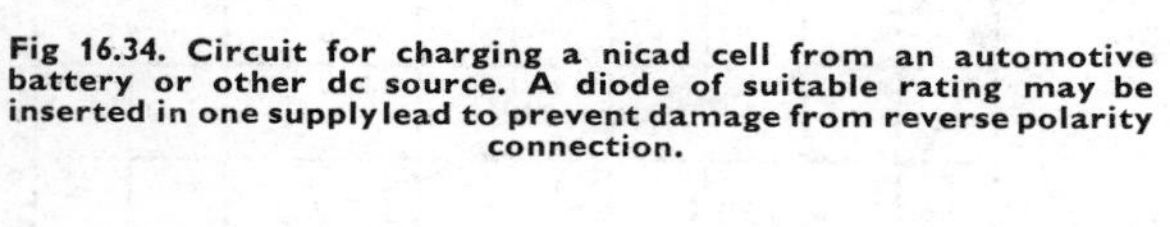

Fig 16.34. Circuit for charging a nicad cell from an automotive battery or other dc source. A diode of suitable rating may be inserted in one supply lead to prevent damage from reverse polarity connection.

Nickel cadmium cells should be discharged fully before recharging and 1·1V may be used as the practical knee point on a slow discharge cycle, at which discharging may cease and recharging commence. In the case of batteries, a drop of 10 per cent in terminal voltage may be considered discharged.

Charging of nicads can be carried out using automotive batteries, but usually it will be more convenient to use the ac mains supply. For this only a simple unit comprising a transformer, rectifier and current limiting resistance is required.

The circuits shown in **Figs 16.34** and **16.35** have been suggested by the Ever Ready Co (GB) Ltd and are suitable for charging their range of nickel cadmium batteries. The following sections outline how to determine values for these circuits.

A number of commercial chargers are available, both DEAC (GB) Ltd and the Ever Ready Co (GB) Ltd manufacture constant current chargers intended for nicad charging. The more elaborate units provide a relatively high maximum current with metered output and a timer.

Referring to Fig 16.34

Symbols

V_S = source voltage
V_C = on charge voltage of one nicad cell (approx 1·45V at 20°C)
V_R = voltage across resistor
N = number of nicad cells in battery
C = capacity of nicad cells in ampere hours
R_L = resistor required to limit current
I_C = charging current required (C/8 for Ever Ready cylindrical cells)
W_R = watts to be dissipated by resistor

Calculation of resistor R_L

$V_S - NV_C = R_L \times I_C$ (if this result is negative, V_S is too low). Assume charge rate of C/8, then $V_S - NV_C = R_L \times C/8$.

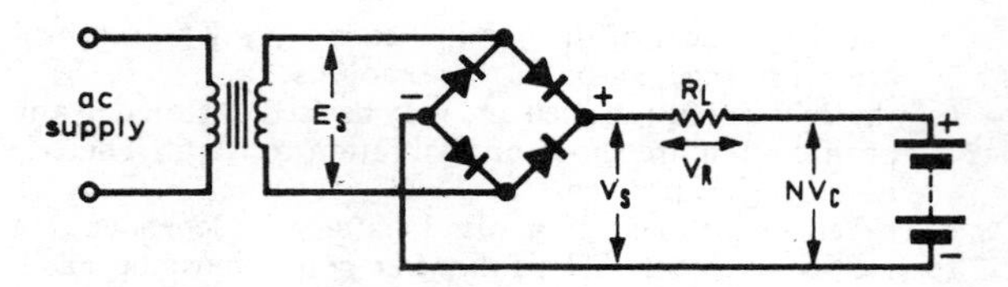

Fig 16.35. Circuit for charging a nicad cell from the domestic ac mains supply.

Therefore $R_L = \frac{8}{C}(V_S - NV_C)\ \Omega$.

Referring to Fig 16.35

Symbols

E_S = transformer secondary voltage
V_S = voltage of rectified dc
(remainder as in Fig 16.34)

Calculation of resistor R_L—as above

Rating of transformer

The transformer secondary voltage E_S should be at least twice NV_C and its current rating should be at least equal to I_C (in rms value).

Calculation of wattage of resistor R_L

$V_S = V_R + NV_C$
$V_R = V_S - NV_C = R_L \times I_C$
$W_R = V_R \times I_C$
$W_R = R_L \times I_C^2$ watts

Practical example

To charge one Ever Ready NCC400 from (*a*) a 12V car battery, and (*b*) 240V ac mains supply. Charging rate to be C/8.

(a) $V_S = NV_C = 1 \times 1·45V$
$C = 4Ah;\ I_C = 6/8 = 500mA$
Value of $R_L = \frac{8}{4}(12V - 1·45V)$
$R_L = 21·10\Omega$
$W_R = 21·10 \times 0·5 \times 0·5 = 5·2W$
A resistor of about 22Ω 6W is required.

(b) For rating of transformer see above.
Bridge rectifier: the piv should be greater than E_S (3V) and have a forward current of at least I_C (500mA).
$V_S = 3V;\ NV_C = 1 \times 1·45$
$C = 4Ah;\ I_C = C/8 = 500mA$
Value of $R_L = \frac{8}{4}(3V - 1·45V)$

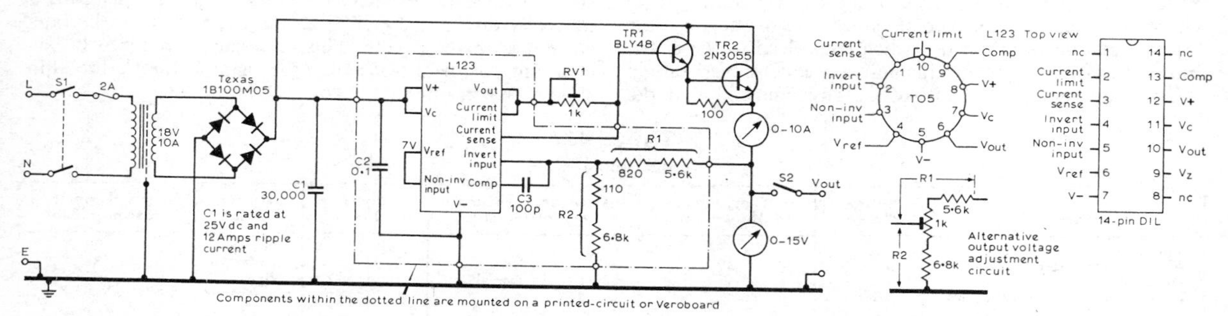

Fig 16.36. Circuit of the ac psu for mobile equipment. Suitable alternative transistors for TR1 are TIP29, TIP31, 2N3053 or 2N3055.

Front view of the unit.

$R_L = 3{\cdot}10\Omega$
$W_R = 3{\cdot}10 \times 0{\cdot}5 \times 0{\cdot}5 = 0{\cdot}77W$
A resistor of 3Ω 1W is required.

AC PSU FOR MOBILE EQUIPMENT

A regulated power supply with a terminal voltage of approximately 13·5V is useful for operating or testing mobile equipment. The circuit of a suitable unit designed by GI3HXV is shown in **Fig 16.36.** The circuit is based on an ic voltage regulator type L123, to which is added a current boost transistor TR2 and driver transistor TR1. Variable current limiting is provided by the variable resistor RV1, which senses the L123 output current to the base of TR1 and limits this current when the voltage drop across RV1 reaches approximately 0·65V. The output current of the psu is related to the L123 current by the current gain product of TR1 and TR2, so adjustment of RV1 gives output current limiting from about 2A upwards.

The output voltage V_o is determined by the resistances R1 and R2 according to the formula:

$$V_o = \frac{7\,(R1 + R2)}{R2}$$

The values shown give a nominal output of 13·5V but some trial and error adjustments may be required in practice due to the tolerance of the resistances and the L123 reference voltage. Two resistors are used for both R1 and R2 to permit selection for voltage variation. Alternatively single resistors may be used for both R1 and R2 with a 1kΩ wirewound potentiometer connected between them as shown in Fig 16.36. Adjusting the potentiometer from one extreme to the other gives a total output voltage variation of 1·0V either side of 13·0V. Note that the voltage feedback network R1, R2 is connected to the output load side of the ammeter. This arrangement includes the ammeter in the control loop and avoids the associated voltage drop which would otherwise degrade the output resistance of the psu. A load of 10A causes an output voltage drop of less than 50mV, and the ripple level is less than 5mV peak-to-peak at all load currents up to 10A.

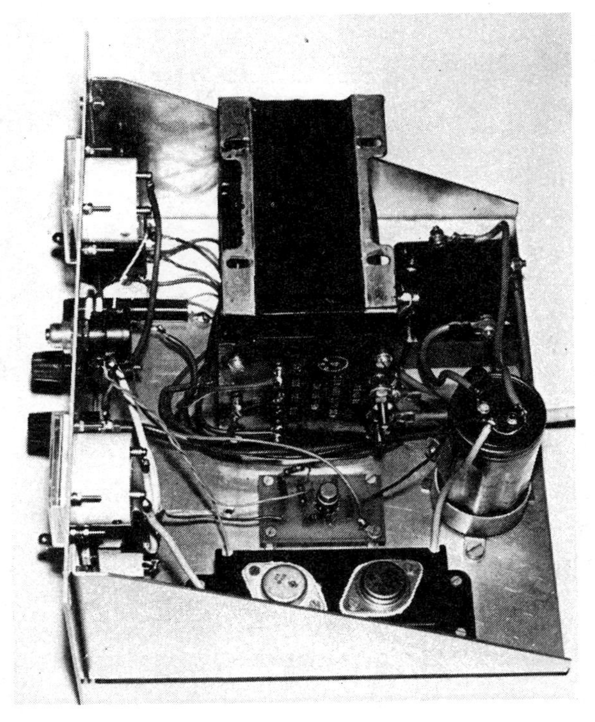

Internal layout of the unit.

The layout of the psu is not critical. The high-current parts of the circuit should be wired with 32/0·2 pvc covered wire or equivalent. The ic and its associated components may be wired on Veroboard or a small printed circuit board as shown in the photograph.

The setting of the current limit is carried out by turning RV1 to its maximum resistance position, connecting the output terminals together with a connection which will carry 10A and then adjusting RV1 to obtain the required current reading on the ammeter.

TR1 and TR2 should be bolted to a heat sink with a thermal resistance of less than 2°C/W. It is likely that an aluminium chassis large enough to accommodate the psu components will provide sufficient cooling. The unit should not be allowed to operate for more than a few minutes in the current-limiting mode if the current limit is greater than 5A, due to excessive heat dissipation in the output transistor.

CHAPTER 17

INTERFERENCE

INTERFERENCE is probably the most universal and challenging problem the radio amateur has to face today. He may be suffering from interference, in which case he will be irritated and frustrated that he is hearing an audio output other than the signal he is trying to listen to. More important, the transmitting amateur may be *causing* interference, not only to other (professional as well as amateur) users of the spectrum, but also to users of domestic receiving equipment (radio and tv). This type of interference can even be felt by users of audio equipment not even intended for reception of radio transmissions, and can lead to complaints to the authorities.

Interference can occur in almost any stage of a transmitting or receiving system, or other item of electronic equipment, such as a hi-fi amplifier. The primary cause is inadequate selectivity in some part of the equipment. In a transmitting context, the loaded Q of a transmitter output stage is typically 12, which means that it has poor selectivity and may not adequately attenuate harmonics of the wanted signal. On the other hand, audio amplifiers may suffer interference because some stage has excessive bandwidth and responds to rf signals.

In practice it is not possible to produce an ideal selectivity curve, so any system will respond to a certain extent to signals outside the required bandwidth, as shown in **Fig 17.1**, although in many instances this alone may not matter. A superheterodyne receiver, for example, has a relatively broad-band input stage followed by better selectivity in the i.f. stages, but because of non-linearities in the rf and mixer stages, signals on the skirts of the selectivity curve may generate new signals *inside* the required bandwidth which *cannot* be removed in later stages. Consequently, non-linearities are as important as lack of selectivity when considering possible causes of interference. The cure may lie in attention to either or both of these causes.

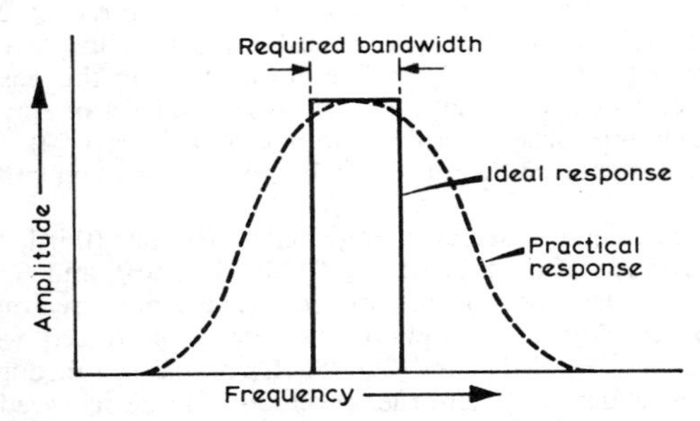

Fig 17.1. Typical response of a communication system.

SELECTIVITY

The selectivity requirements of tuned circuits in various pieces of equipment are dealt with in the relevant chapters of this book. In order to cure interference it is often necessary to improve the selectivity of certain stages, or the whole of the equipment, by additional filtering and screening.

NON-LINEARITIES (DISTORTION)

The effect produced by any electrical circuit may be expressed by the following equation:

$$V_2 = aV_1 + bV_1^2 + cV_1^3 + \ldots \text{(etc)}$$

which shows how the output voltage (V_2) depends on the input voltage (V_1). The circuit in **Fig 17.2(a)** is a simple potential divider where V_2 is directly proportional to V_1. If, for example, R1 = R2, then $V_2 = \frac{1}{2}V_1$; that is, the coefficient a is equal to $\frac{1}{2}$ and b, c, d etc are 0. Furthermore, the waveform of V_2 is exactly the same as that of V_1. If V_1 is a pure sinewave of frequency f_1, then V_2 is also a pure sinewave of the same frequency. In other words the circuit is perfectly linear.

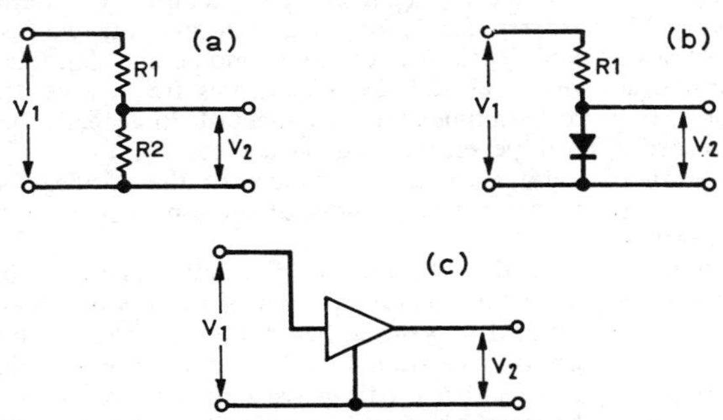

Fig 17.2. (a) V_2 is directly proportional to V_1, in other words the circuit is linear. (b) V_2 will be a distorted version of V_1. The degree of distortion will depend on the amplitude of V_1. (c) V_2 may be defined by an equation in terms of V_1 for both a linear or non-linear amplifier.

In **Fig 17.2(b)** the resistor R2 is replaced by a diode. If V_1 is very small and insufficient to cause the diode to conduct any appreciable current even when it is forward-biased (ie much less than the diode "knee" voltage) the circuit appears to be linear. If V_1 is large enough the diode will conduct when V_1 is positive and this prevents V_2 from following V_1. However, when V_1 is negative the diode is reverse-biased, does not conduct, and thus V_2 follows V_1. Hence V_2 is a greatly distorted form of V_1. As the circuit was apparently linear

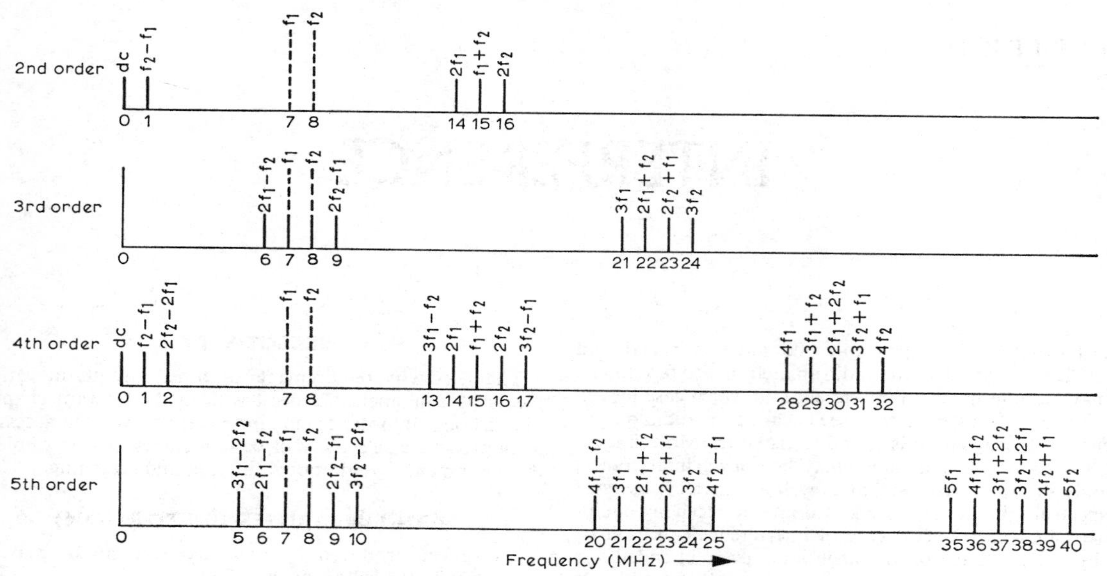

Fig 17.3. Build-up of distortion products in a non-linear system.

when V_1 was small, it is obvious that distortion depends upon signal levels.

The same equation may be applied to the amplifier in **Fig 17.2(c)**. If V_1 is a sinewave $V \sin 2\pi f_1 t$ of frequency f_1, the equation may be written:

$$V_2 = aV \sin 2\pi f_1 t + bV^2 \sin^2 2\pi f_1 t + cV^3 \sin^3 2\pi f_1 t + \ldots \text{(etc)}$$

The "first order" term $aV \sin 2\pi f_1 t$ is responsible for the linear amplification of the input signal. The second order term $bV^2 \sin^2 2\pi f_1 t$ is responsible for the generation of the two frequencies $2f_1$ and $0f_1$ (ie the second harmonic and dc). The third order term $cV^3 \sin^3 2\pi f_1 t$ generates frequencies $3f_1$ and f_1 (ie third harmonic and fundamental). In general, the *n*th order term generates frequencies of nf_1, $(n - 2)f_1$, $(n - 4)f_1$ etc, down to dc. The values of the coefficients *a*, *b*, *c*, etc, govern the amplitudes of these new signals so generated.

It will be seen that the level of the first order term is proportional to V (and hence V_1), that of the second order term is proportional to V^2 (and hence V_1^2) and that of the third order term is proportional to V^3 (and hence V_1^3). This means, for example, that a 1dB increase of the input signal level causes the second harmonic output to increase by 2dB and the third harmonic output by 3dB. The general rule is that for every dB increase in drive, the *n*th order product increases by *n*dB.

The third order term generates a signal on the fundamental frequency f_1. This may add to or subtract from that generated by the first order term, depending upon the signs of the coefficients *a* and *c*. When subtraction occurs, this causes a compression or limiting of the fundamental signal—a common occurrence in amplifiers operating at high signal levels. When addition occurs, this causes an expansion of the fundamental signal—an effect seen in Class B amplifiers suffering from crossover distortion when operated at low signal levels.

Intermodulation

If V_1 is not a single-frequency source at f_1, but consists of two frequencies f_1 and f_2, the second order term "bV_1^2" will not only produce $2f_1$ and $2f_2$, but will also produce $f_1 + f_2$ and $f_1 - f_2$, ie a total of four new signals are produced in addition to the original f_1 and f_2 input signals. The third-order term "cV_1^3" will produce $3f_1$, $3f_2$, $2f_1 + f_2$, $2f_1 - f_2$, $2f_2 + f_1$, and $2f_2 - f_1$ in addition to the original f_1 and f_2 signals, ie six extra signals. Fourth-order distortion produces 12 extra signals, fifth-order 16 extra, and so on. The generation of these signals is generally called *intermodulation*.

When three signals f_1, f_2 and f_3 are present at the input the number of products is even greater. Third-order distortion produces signals of the form $f_1 \pm f_2 \pm f_3$ in any combination, in other words a further eight signals in addition to the others outlined above.

For second-order intermodulation ($f_1 \pm f_2$) each signal contributes equally to the output signal. Changing either f_1 or f_2 alone by 1dB will produce a similar 1dB change in the output product levels, so that if both inputs are raised by 1dB then the outputs rise by 2dB.

For third-order intermodulation (eg $2f_1 + f_2$), if the level of f_1 is raised by 1dB the level of $(2f_1 + f_2)$ will rise by 2dB. However, if the level of f_2 is raised by 1dB then the level of $(2f_1 + f_2)$ will rise by 1dB. The output level in this case is twice as sensitive to the level of f_1 as to the level of f_2.

When third-order intermodulation is of the form $(f_1 \pm f_2 \pm f_3)$ each signal contributes equally to the resulting output level.

When f_1 and f_2 are close together, odd-order (third, fifth etc) products of the type $(2f_1 \pm f_2)$ fall close to f_1 and f_2, but even-order (second, fourth etc) products are more remote in frequency. **Fig 17.3** shows the spectrum produced when signals of f_1 = 7MHz and f_2 = 8MHz are passed through a non-linear device. Even-order products may be removed by moderate selectivity, but it is much more difficult to prevent

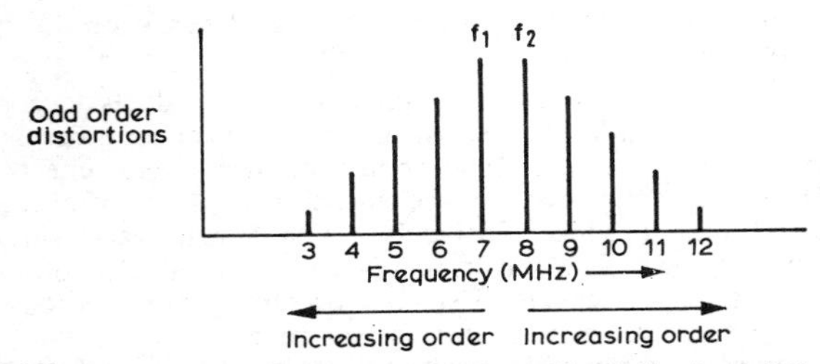

Fig 17.4. Spectrum of a typical wideband amplifier indicating relative amplitudes of distortion products.

interference due to the odd-order products. It is for this reason that odd-order distortion causes more problems, both on transmission and reception. The spectrum produced by a typical wideband amplifier is shown in **Fig 17.4.**

Harmonic Generation

Harmonic generation is a special case of intermodulation in which the frequencies f_1, f_2, f_3 etc are all the same (the input frequency). Second-order distortion will give $f_1 \pm f_1$, ie $2f_1$ and 0, third-order distortion will give $3f_1$ and f_1, fifth-order distortion will give $5f_1$, $3f_1$ and f_1, and so on.

Crossmodulation

Crossmodulation is a specific case of third-order intermodulation (in severe cases fifth-order and above), which results in the modulation of one signal being transferred to another. **Fig 17.5** shows the spectrum of two signals, f_1 modulated and f_2 unmodulated. The modulation frequency is f_a, so the lower sideband of the modulated signal is $f_1 - f_a$, and the upper sideband is $f_1 + f_a$. By manipulating three of the four signals present (according to the rules of third-order intermodulation), two new signals $f_2 - f_a$ and $f_2 + f_a$ are produced. These two signals are, in effect, modulation sidebands for signal f_2. Note that each sideband is produced twice, and the two signals add together.

Raising the modulated signal by 1dB raises both f_1 and $(f_1 \pm f_a)$ by 1dB, so the output $(f_2 \pm f_a)$ therefore rises by 2dB. This means that for a 1dB change of interfering modulated signal level, the depth of modulation impressed on the wanted signal increases by 2dB.

Raising f_2 by 1dB also raises $(f_2 - f_a)$ by 1dB, which means that increasing the wanted signal will have *no effect* on the *depth* of modulation impressed on it. So the general rule is that *crossmodulation depth depends only upon the strength of the interfering signal*, and not at all on the wanted signal. If the wanted signal is removed then the interfering signal will no longer be heard since any percentage modulation of zero carrier is zero!

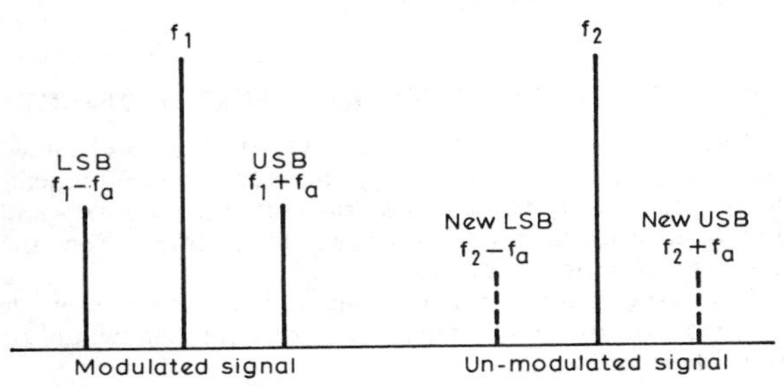

Fig 17.5. Production of unwanted sidebands by crossmodulation.

Non-linearities—Necessary and Unnecessary

While non-linearities are normally detrimental to the performance of electronic equipment, some are permissible and some are absolutely necessary. Examples are listed below:

Second-order—*Absolutely necessary* for frequency doublers, mixers, and for amplifiers where gain is controlled by a current or voltage (such as agc controlled stages). *Permissible* in selective amplifiers where second harmonics, or sum and difference frequencies can be filtered out. *Undesirable* in wideband amplifiers and oscillators.

Third-order—*Absolutely necessary* for frequency triplers, but otherwise *undesirable*. It produces third-order intermodulation and crossmodulation as described, and in many cases selectivity will be inadequate to prevent interference from occurring.

INTERFERENCE DUE TO RECEIVER DEFECTS

Many interfering signals are not present at the receiver aerial terminals on the frequency to which the receiver is tuned. Interference can be caused by:

Harmonics of the i.f.
Harmonics of the bfo and other oscillators.
Poor overall selectivity.
Image response.
Other responses (harmonic beats).
Signals which act as the receiver local oscillator.
I.F. breakthrough.
Intermodulation, crossmodulation and harmonic generation.
Detection in audio circuits, video circuits etc.

Harmonics of the I.F.

These are self-generated in the receiver, and may usually be identified by removing the aerial. With the bfo off, i.f. harmonics may not be apparent, but if they are sufficiently strong they will appear as a fast-tuning carrier. With the bfo on they may be detected as a beat note. The i.f. harmonics reach the receiver input or mixer by radiation or via the receiver internal wiring, and therefore suitable remedial action may be taken. However, if the mixer does not have a perfect square law (ie only second-order non-linearity), i.f. harmonics will be produced in the mixing device itself, and such signals then act as input signals, but cannot be removed. The only solution is to use a mixer where this does not occur.

Harmonics of the BFO and Other Oscillators

These are self-generated in the receiver, and like i.f. harmonics may be identified by removing the aerial. The bfo harmonics will be detected as a whistle as the receiver is tuned through the harmonic frequency. Again, these harmonics reach the receiver input by radiation or via the wiring, and remedial action may be taken.

Fortunately, i.f. and bfo harmonics usually only affect one spot frequency on the lf amateur bands, for example 4 × 465kHz interferes with 1,860kHz, and 8 × 465kHz interferes with 3,720kHz.

Poor Overall Selectivity

This results in signals on adjacent frequencies being heard at the same time as the wanted signal. The receiver bandwidth required depends upon the mode of transmission, and varies from several megahertz for television to tens of hertz for slow morse under noisy conditions. There is a greater chance of interference occurring to receivers of wide bandwidth, so the use of the minimum bandwidth for acceptable results is recommended.

Image Response

This topic is discussed in detail elsewhere in this book (see Chapter 4). It is due to inadequate pre-mixer selectivity and therefore tends to be worse at higher frequencies. Multiple conversion receivers are designed to overcome this problem, although domestic receivers operating on the medium waveband sometimes suffer interference from amateurs operating on the 160m band. For example, a receiver with a 465kHz i.f. tuned to 908kHz will have its local oscillator on 908 + 465 = 1,373kHz. A strong amateur signal on 1,838kHz may cause interference as this will also give an i.f. of 465kHz (see Fig 17.26).

Similarly, on an amateur receiver it may be possible to tune in a strong signal in two places on the 10m band. The higher signal (on the receiver dial) is usually the genuine one, as the local oscillator is normally above the wanted frequency. Image problems are increased if the rf stage does not track correctly with the local oscillator, and under certain circumstances the image signal may even be stronger than the correct one. This is more likely to occur in general coverage receivers.

Other Responses (Harmonic Beats)

These are produced when harmonics of the local oscillator beat with a signal to produce an i.f. signal. It is not uncommon with cheap transistor radios that an amateur operating on the hf bands can be tuned in on the medium waveband of the receiver. Examples are given in Fig 17.26.

Signals Which Act as the Receiver Local Oscillator

When the interfering signal is very strong and on approximately the same frequency as the local oscillator, the wanted signal may be heterodyned to the i.f. by either of these signals. This will produce a beat frequency, and the modulation of the interfering signal will also be heard. In fact, any two signals spaced by the i.f. and reaching the receiver mixer will produce an interfering i.f. signal. This is a case of second-order intermodulation. Again, this problem is more likely to occur in domestic or general coverage receivers (see Fig 17.26).

I.F. Breakthrough

This occurs when signals on the same frequency as the i.f. reach the mixer input or i.f. stages. Some coastal radio stations operate around 465kHz, and other stations operate on other common i.f. frequencies. When the interfering signal is entering the receiver via the aerial, a trap tuned to the i.f. and inserted in the aerial lead will effect a cure. I.F. breakthrough produces tunable whistles on all stations to which the receiver is tuned, and is again predominantly a domestic receiver problem.

Intermodulation, Crossmodulation and Harmonic Generation

Intermodulation and crossmodulation products occur because of non-linearities, usually in the mixer, but also in the rf and i.f. stages. Intermodulation has several effects. The speech components in the sidebands of the transmitted signal may produce sidebands which spread each side, similar to the effect shown in Fig 17.4. This effect is often responsible for reports of splatter on strong transmissions, whereas the true cause is the receiver mixer.

Similarly, strong signals near the wanted frequency may produce interfering signals either side of them. An example of this is on the 40m band where two strong commercial stations on 7,100kHz and 7,150kHz will produce third order beats on (2 × 7,100) − 7,150 = 7,050kHz (and other frequencies). Similar effects are produced by three signals on, for example, 7,100, 7,110 and 7,120kHz, producing third-order beats of the type 7,100 + 7,110 − 7,120 = 7,090kHz.

Crossmodulation is (as described) simply a special case of the above three-signal intermodulation which results in the modulation of a signal on (for example) 7,100kHz appearing on a signal on any nearby frequency.

These effects may be alleviated by increasing the pre-mixer selectivity so that only the wanted signal is selected and the rest are not, but this is difficult if the wanted and interfering signals are close together. The use of an atu and/or bandpass filter will improve the situation. Reducing the mixer input (with an aerial attenuator or rf gain control) will reduce the interference by 2dB for each 1dB of signal attenuation (see *Crossmodulation*), which will probably render the interference inaudible before the wanted signal is lost in the noise.

A more elegant solution is to ensure that the stages susceptible to third-order effects are either linear or have only second-order non-linearity. FETs are used increasingly in rf, mixer and other stages, using the square law action to give agc or second-order mixing action as desired.

Harmonics of a strong signal may be produced in the rf or mixer stages. If the receiver is tuned close to one of these harmonic frequencies, interference will occur due to the beat between the harmonic and the wanted signal. This is prevalent in television sets, especially on Band I.

Detection in Audio Circuits, Video Circuits etc

Detection of a strong rf signal may occur in af or video stages which are driven into non-linearity. No "receiver" may be involved, for example in hi-fi systems. This form of interference is prevalent in domestic equipment, and is discussed later in the chapter.

INTERFERENCE DUE TO TRANSMITTER DEFECTS

Interference due to transmitter defects may be classified into two groups: interference to users of frequencies immediately adjacent to the wanted transmitter frequency; and interference to users of frequencies quite remote from the wanted transmitter frequency.

The first group affects mainly other amateurs and users of "shared" bands, and is caused by excessive bandwidth or transmitter noise.

The second group, affecting mainly users of tv and radio

sets in the immediate vicinity, but occasionally more widespread, is caused by harmonics of the transmitter output frequency, and radiation of signals produced internally but not wanted at the output.

Common to both groups is interference caused by the radiation of unintentional frequencies, including parasitic oscillations.

Excessive Bandwidth

Any form of modulation produces sidebands. Morse conveyed by on-off keying of a carrier wave not only generates sidebands of the keying frequency (which is relatively low compared with voice modulation) but, as the keying waveform is effectively a rectangular shape (rich in harmonics), the sidebands may extend to a considerable width either side of the centre frequency and produce "key-clicks". Slowing down the transition time from on to off and vice versa will reduce these harmonics, thus reducing the sidebands and the corresponding interference to adjacent frequencies.

Frequency shift keying (fsk), either morse or rtty, may also produce key-clicks. FSK may be regarded as two separate transmissions, one on the "mark" frequency and one on the "space" frequency, which are keyed on and off alternately. The extent of the sidebands produced depends upon the rate of transition from one frequency to the other, the keying speed, and the frequency shift employed.

Tone modulation (mcw, afsk etc) may also produce excessive bandwidth if the tones are switched on and off suddenly. Such forms of modulation are sometimes used to send morse and fsk rtty on ssb transceivers.

"Chirp" is the term applied when the transmission frequency changes each time it is keyed. Slow chirp is usually produced by a change in the supply voltage to a vfo (when the transmitter is keyed on, the supply voltage falls). Fast chirp is usually caused by rf feedback from the pa stage or aerial into an oscillator. Because the transmitter is operating over a band of frequencies (by sliding from one to another), excessive bandwidth will be occupied.

Excessive bandwidth on a.m., ssb or fm transmissions may be caused by excessive bandwidth of the modulating audio frequencies. On congested amateur bands, a modulating frequency of 3kHz must be considered the maximum, so definite steps should be taken to tailor the audio frequency range. In this context, reliance should not be placed upon the microphone to cut off the higher frequencies.

The bandwidth of a.m. transmissions will increase if overmodulation occurs. Because the carrier is broken abruptly during part of the modulation cycle, the modulation takes on the form of a rectangular wave, and the resultant sidebands spread much wider than the original modulating frequency.

The bandwidth of ssb transmissions will be wide if the filter (or phase shift network in a phasing rig) response is incorrect, causing poor sideband suppression. Probably the most common reason for excessive bandwidth is odd-order distortion in the pa stage or linear amplifier. The speech components of the sideband signal intermodulate to produce third- and fifth-order products which are inside and immediately outside the required bandwidth (similar to Fig 17.4). The products falling inside cause poor quality, which is usually acceptable in practice if they are better than about 15dB down, but the splatter produced *outside* the required bandwidth may cause severe interference. Note that the

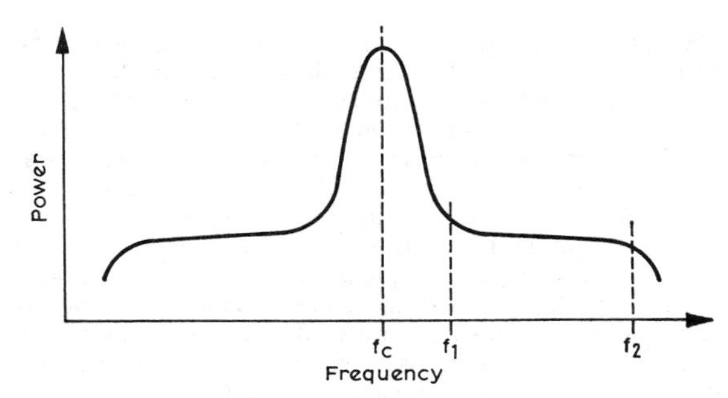

Fig 17.6. Typical transmitter noise spectrum.

unwanted products rise rapidly in level as the drive is increased (see *Non-linearities*).

Excessive bandwidth may be caused on fm or pm by overdeviation.

The remedy for all the above is similar. The audio gain control should be set so that overmodulation, overdriving or overdeviation cannot occur. To ensure that the best use is made of the transmitter peak modulation capability, use of a speech compressor and/or audio limiter is recommended. With volume compression techniques care must be taken with the attack time to avoid excessive splatter during the first speech syllables, before the compressor operates. This is particularly important if the modulator is so designed that it is capable (without the compressor operating) of modulating the transmitter greater than 100 per cent. In these cases it may therefore be necessary to follow the compression stages by some simple speech peak limiter. This in itself will result in distortion on peak speech, so a low-pass filter may be required after the limiter. It is fundamental that this low-pass filter is fitted into the circuit such that it is unaffected by any feedback loop forming part of the compression circuit. Examples of suitable circuits and filters are given in Chapter 9 (*Modulation Systems*).

Transmitter Noise

This will mainly affect operators trying to work duplex (ie transmitting and receiving simultaneously), or where two stations are being operated simultaneously on the same band and in close physical proximity. Assuming that there are no limitations due to the receiver performance, the bandwidth of the transmitted signal may still appear to be too wide. This may be due to the noise spectrum (which can also carry the modulation) emitted by the transmitter, typically as shown in **Fig 17.6**. The region f_c to f_1 is determined by the bandwidth or Q factor of oscillators or early stages in the transmitter. This is added to a wider band of noise f_c to f_2 generated mainly in the pa stage—f_2 may be well removed from f_c. With valve transmitters the noise sideband power radiated between f_1 and f_2 is in the order of 120dB below the main signal power, but transistors produce more noise, and a figure of 90dB below is typical for a transistor transmitter.

Spurious Emissions

An output on any frequency other than that which is intended is defined as a spurious emission. Spurious emissions include:

Harmonics of the transmitter output frequency. The problems of transmitter harmonics which interfere with neighbouring domestic equipment is perhaps one of the reasons why amateur operation is often restricted to "traditional" hours. This problem is dealt with separately later in the chapter.

Since most amateur bands are harmonically related, harmonics produced by a transmitter will inevitably fall in or just outside higher frequency bands. All transmitters produce some level of harmonic output, and when stations are operated in close proximity mutual interference may occur. The use of good transmitter circuit design and suitable filtering in the lead to the aerial (low-pass filter, atu etc) is recommended. Details are given later in this chapter.

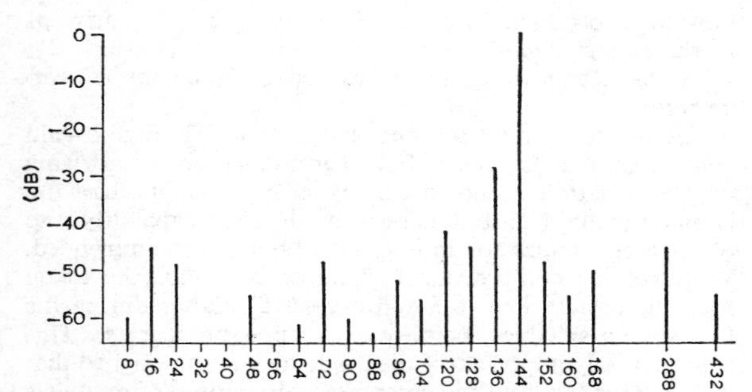

Fig 17.7. Spectrum analysis of a commercial amateur 2m transmitter.

Radiation of signals produced internally but not wanted at output. Frequencies produced intentionally in early stages of the transmitter may be radiated, such as signals from the multiplier chain, oscillators and mixers. A transmitter pi-tank is basically a low-pass filter tuned to match one impedance to another and so gives little attenuation of signals below the wanted transmitter output frequency. For example, a 3·5MHz transmitter with a 1·8MHz vfo may radiate quite strongly on 1·8MHz. To reduce this effect the pa stage should be driven by a buffer amplifier tuned to the output frequency, not by a frequency doubler. An atu may give considerable attenuation of such unwanted signals, but this depends upon its design.

A 144MHz transmitter may radiate a whole spectrum of harmonics of the original oscillator frequency. If this is 8MHz tripled to 24MHz, then tripled to 72MHz and doubled to 144MHz, strong signals may appear on 8, 24 and 72MHz, with other 8, 12 and 24MHz signals appearing as sidebands of these signals. The result is a continuous spectrum of signals appearing at 8MHz intervals extending well above 144MHz as shown in **Fig 17.7**. The aerial will not radiate effectively at the low frequencies, but severe interference may be caused to other services operating on frequencies near 144MHz. This effect may be reduced by bandpass coupling and screening between transmitter stages. The use of a bandpass filter in the aerial feeder, rather than a low-pass filter, is recommended for vhf and uhf transmitters. Similar considerations must be given to transmitters which use mixing to obtain the required output frequency.

Radiation of unintentional frequencies. Unwanted frequencies may be generated in mixer stages, for example image and other signals produced by harmonic beats (similar to the mechanism occurring in receivers) may be produced. If the ssb generation frequency is, say, 9MHz, a 5MHz vfo designed to produce an output on 14MHz may contain sufficient harmonic (or this may be generated in the mixer) to produce a strong output on $3 \times 5 = 15$MHz.

Parasitic Oscillations

Parasitic oscillations are oscillations which occur where no oscillation is intended. They are unpredictable and may be responsible for distortion, low efficiency and instability. These oscillations can be divided into three groups:

Oscillations much lower than the output frequency.
Oscillations on or near the output frequency.
Oscillations at vhf.

Oscillations much lower than the output frequency. These are usually recognized by the presence of a number of frequencies each side of the wanted frequency. In a.m. transmissions they may take the form of multiple sidebands spaced at intervals equal to the oscillation frequency. Such oscillations may occur in rf amplifiers which burst into oscillation at high frequency, at a low repetition rate ("squegging"), which can be due to poor neutralization.

Continuous low-frequency oscillation may occur due to the resonance of rf chokes with other circuit capacitances, when used both on the input and output circuits of an rf amplifier. It may be avoided if the input choke is made smaller than the output choke, so that the input choke resonates higher than the output, which prevents this type of "tuned anode—tuned grid" (tatg) oscillation. In the case of beam-power tetrodes it is better to avoid an input choke and use a fairly high value grid resistor, which causes only small rf power loss. However, care should be taken not to exceed the maximum value quoted for the valve type—typically 30kΩ. RF power transistors often have high gain at low frequencies and so are prone to low frequency oscillations. This may be avoided by incorporating additional low-frequency decoupling on such stages to ensure that the gain is low.

Oscillations in audio circuits may occur at supersonic frequencies and therefore be inaudible, although they may cause the production of parasitic sidebands due to the excessive modulation bandwidth, except perhaps with a filter-type ssb transmitter. Reduction of the audio quality will also result.

Oscillations on or near the output frequency. These are usually caused by rf feedback in one of the rf amplifier stages (usually the pa) due to lack of correct neutralization, poor screening or bad layout, so that the stage acts as a tatg oscillator. The oscillation may not be present all the time, and may occur only on peaks of modulation or when an amplifier has no drive (key-up condition on cw or between words on ssb). It will probably be unstable and change frequency if the amplifier input or output tuning is changed, but may lock onto the wanted signal when it is strong enough, such as on speech peaks. The result is often poor speech quality and "splash" on telephony or clicks on telegraphy. The situation is often aggravated by over-running amplifiers with excessive supply voltages, particularly with high-slope valves.

Oscillations at vhf. These are often due to an unfortunate choice of layout and may occur in almost any stage of a transmitter (or other equipment) including the audio stages. The oscillations arise from an unintentional form of tatg circuit, and the cure may lie in improved screening, layout, or decoupling of leads to vhf. As with all tatg circuits, beware of un-neutralized stages when the output could be tuned slightly above the input tuning frequency. A common practice is to use low-value resistors or vhf chokes in series with grid and anode connections of an hf transmitter pa stage. This reduces the stage gain at vhf and prevents oscillation. The chokes may consist of a few turns of wire wound on a suitable size of low-value resistor. Sometimes it is possible to eliminate vhf oscillations by connecting a small capacitor (about 30pF) directly between grid (or anode) and earth. This must be done using the shortest possible leads, and can only be done where this capacitance will not otherwise affect the performance of the circuit (usually below 28MHz).

Checking for parasitic oscillations. An essential procedure in the testing of a new transmitter is the running of the pa stage without drive (unless self-bias is used) and at the maximum permissible safe standing current. Under these circumstances the input and output tuning should be varied on all bands to ensure that no parasitic oscillations occur. They may be indicated by sudden changes in the standing anode (or collector) current, by the flow of grid current or by rf voltages appearing at the output. A transmitter should never be connected to an aerial until it has satisfactorily passed this test. During the test a suitable dummy load should be used. Neutralizing should be carried out near the maximum frequency used, although some compromise may be necessary, such as neutralizing at 21MHz on a 3·5–28MHz multiband transmitter. Maximum power output should be obtained near maximum dip in anode current in a conventional Class B or Class C pa stage when switched to the "tune" position.

TRANSMITTER HARMONICS

Perhaps the most common form of interference caused by transmitter shortcomings is the radiation of harmonics which fall on or near frequencies allocated to television and vhf radio. For trouble-free operation it is essential that such harmonics are suppressed to a very high degree. The suppression of harmonics which fall in higher frequency amateur bands is less demanding except where another amateur or other radio service is located nearby. In any case of harmonic (and other unwanted) radiation the responsibility for curing the interference rests with the amateur concerned. Hence the suppression of harmonics is vitally important in the design of a transmitter.

CW, a.m., fm and similar transmitters usually operate with a non-linear Class C pa stage (and other amplifier or multiplier stages), which must be expected to produce a high level of harmonic output. SSB transmitters usually use a more linear pa operating in Class A, AB1 or AB2, which, although it should produce less harmonic output than Class C, must still be considered as a harmonic generator. It is also quite common to drive an ssb pa into Class C for the purpose of tuning up.

There are several sophisticated methods of reducing the generation of transmitter harmonics, but satisfactory operation is usually obtained by paying attention to filtering, sometimes before, and invariably after the final amplifier stage. Although it is relatively easy to suppress harmonics by filtering the feeder to the aerial, for freedom from interference it is often necessary to ensure that the transmitter power and other interconnecting leads are adequately filtered and screened, and that the transmitter case allows no leakage of rf signals.

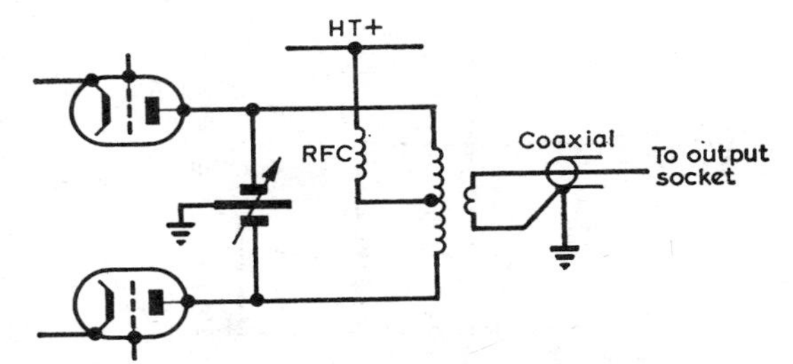

Fig 17.8. A push-pull anode tank circuit. The output link should be placed at the "earthy" section of the coil to minimize capacitive coupling of harmonic energy.

Tank Circuit Design

Transmitter output stages produce harmonics because the anode (or collector) current is distorted, a phenomenon which is most marked in a Class C pa stage where only short pulses of current flow. In order that only the component of current at the desired frequency generates appreciable voltage, the output tank circuit must be tuned and the Q of this determines the suppression of unwanted frequencies. A loaded Q of about 12 is a reasonable compromise between circuit losses and suppression. In the case of other stages in the transmitter, such as multipliers, mixers and drivers it is advantageous to operate at a higher Q since efficiency is not of prime importance. For detailed design procedure see Chapter 6—*HF Transmitters*.

Even in buffer amplifiers it is better to use a tuned circuit with a fairly low L/C ratio than the commonly used 2·5mH choke. The choke gives no selectivity, and has multiple resonances, whereas the tuned circuit reduces unwanted responses outside the desired frequency range. Better still, a wideband coupler may be used.

It is most important that the tank circuit is wired to provide an effective bypass from anode to cathode at the harmonic frequency. Short, stout leads should be used to minimize stray inductance throughout the whole of the rf path. Inductance between cathode and earth, being common to both input and output circuits can also cause unwanted feedback and this is doubly undesirable.

Some tank circuits may prove rather ineffective in reducing the harmonic output unless suitable precautions are taken. A typical example is that of a push-pull output circuit in which the split-stator tuning capacitor has its rotor connected to earth and the coil centre-tap connected to the ht line through an rf choke as shown in **Fig 17.8.** This arrangement is normally used to provide a better circuit balance and a low-impedance path to earth for harmonic frequencies. The even harmonic frequencies present at the anodes are in phase with each other and may be transferred by unintentional capacitive coupling to the output coupling coil. This effect can be minimized by making an earth connection to the coil, which should be positioned at the part of the tank circuit at the lowest rf potential, ie rf earth.

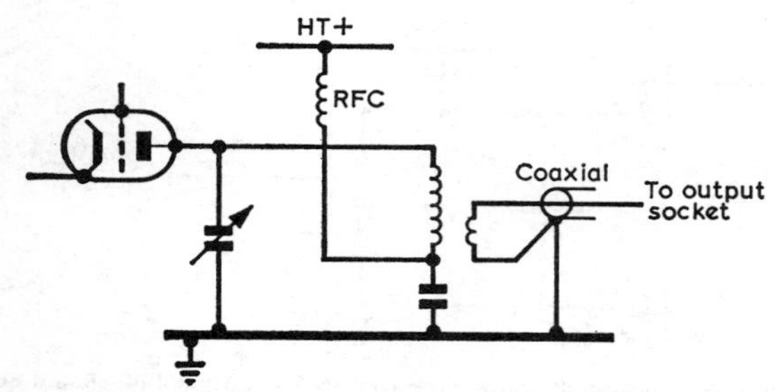

Fig 17.9. A single-ended tank circuit. As in Fig 17.8 the output link should be placed at the "earthy" end of the coil.

The same applies to the single-ended stage shown in **Fig 17.9**. The coil should feed into a coaxial feeder to simplify the screening problem. In addition the use of a Faraday screen between the tank circuit and the coupling coil is recommended since it will serve to eliminate the capacitive coupling without affecting the inductive coupling. This is desirable because it is the higher order harmonics which will be passed more readily by stray capacitive coupling. This point should be borne in mind when determining the layout. Where only one or two link turns are required a simple form of Faraday screen may be made from a piece of coaxial cable, as shown in **Fig 17.10**.

Pi-networks

In hf band transmitters the use of a pi-network for single-ended pa tank circuits is almost universal, and has the advantage of allowing the transmitter to operate into a range of loads (typically with a maximum swr of 2 : 1) for tolerable loaded Q factor. Transistor circuits also use pi-networks, or variations, especially in vhf transmitters. When used correctly the circuit gives good suppression of harmonics due to the frequency dependent action of the potential divider formed by L1 and C2 (see **Fig 17.11**). Compared with a simple tuned circuit, the pi-network gives four times more suppression of the second harmonic, nine times more suppression of the third harmonic, and so on, the general formula being n^2 times for the nth harmonic. However, on signals below the wanted frequency the attenuation is poor, as previously described.

Most transmitters using pi-tank output stages claim harmonic suppression of about 40dB, ie a 400W transmitter will emit harmonics of about 40mW, but this figure is quoted when the transmitter is operated into a constant impedance dummy load. When fed into an aerial having an impedance probably quite different on the unwanted frequencies, the harmonic suppression may be better or worse than this figure. It is unlikely that this level of harmonic will cause much trouble to other stations, but the use of an atu suitable for attenuating signals both below and above the wanted frequency is recommended.

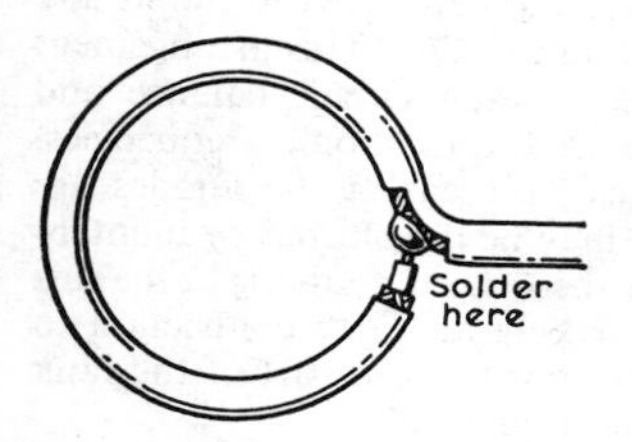

Fig 17.10. A shielded link coil made from coaxial cable. The outer braiding is left in position but does not form a closed loop. The inner conductor is soldered to the outer braiding at the point shown. This construction is not suitable for coils of more than a few turns.

In spite of the use of a correctly designed tank circuit and aerial coupling system it is often found that additional filtering-out of unwanted signals is required. Even if no interference to tv is apparent, it is good design practice to use an additional filter to reduce the possibility of it arising at a later date. It also reduces the risk of interference to other radio services, such as Police, Fire Brigade and Air Traffic Control.

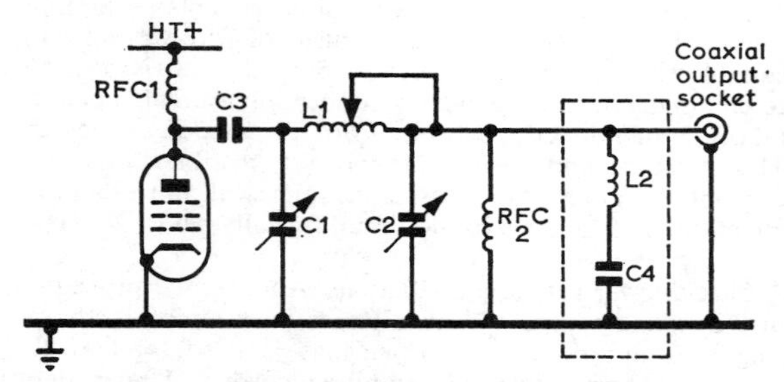

Fig 17.11. A pi-network tank circuit. C1—250pF tuning capacitor; C2—1,500pF loading capacitor; C3—1,000pF blocking capacitor; L1—inductance tapped as necessary; RFC1—anode choke; RFC2—safety choke to prevent ht voltage from appearing at the output in event of C3 shorting; L2-C4—harmonic trap (see text).

The effectiveness of a filter may be negated if unwanted signals are allowed to leak from the transmitter case by direct radiation or by being conveyed to the aerial via the earthy side of the system—it is very difficult to earth a transmitter case effectively at vhf or uhf because any earth lead is too long at such frequencies. Signals may also be re-radiated via the transmitter power and interconnecting leads if they are not correctly decoupled where they leave the case, so the necessity for proper transmitter screening and lead decoupling cannot be over-emphasized.

Screening and Decoupling

Poor screening and decoupling is often responsible for cases of "incurable" interference to television sets or other equipment. Unless the principles involved are fully understood the results may be disappointing. The basic principle can be explained with the aid of the simplified arrangement shown in **Fig 17.12(a)**. A transmitter T is enclosed with its load resistance R in a perfect screening box so that there will be no detectable radiation outside. For the sake of simplicity, it is assumed that the transmitter derives its power from batteries also contained within the box, so there can be no question of radiation from any of the power supply lines. The problem is to allow the transmitter to radiate its energy outside the box but only at the proper frequency; any other frequencies must be prevented from leaving the box. **Fig 17.12(b)** shows how this can be achieved. Here the transmitter is connected to an external load by screened cable and a screened low-pass filter. The load receives power from the transmitter at the wanted frequency while the filter stops the transmission of harmonic energy. It is important to note that the screening must be continuous to ensure that the only way out of the box is through the filter.

Although the arrangement of Fig 17.12(b) enables a carrier to be radiated with minimum harmonic content there still remains the practical problem of making connections to a modulator or keying circuit and also of connecting an external power source instead of the internal battery. While doing this it must be made impossible for any rf potential to exist between the leads and the outside of the box.

Fig 17.13 shows the relative effectiveness of various methods of decoupling. The capacitor values are not critical and usually lie between 100pF and 10,000pF, which are large enough values for adequate decoupling at rf but have negligible effect on the supply carried by the lead. The use of screening on all internal leads which are not normally carrying rf currents is strongly recommended. Chokes may be easily made with about five turns of wire through a suitable lossy ferrite bead, and will usually be free from spurious resonances up to uhf.

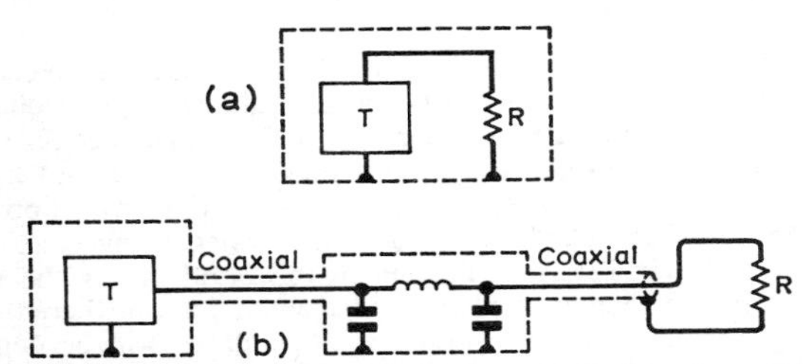

Fig 17.12. A simple way of looking at screening of transmitters. In (a) the perfect screening prevents any radiation from the transmitter or its load; in (b) a low-pass filter inserted in the line between transmitter and load eliminates harmonics while permitting radiation of the (wanted) fundamental.

Beware of paralleling decoupling capacitors to increase their effectiveness over a wide band. Because capacitors series-resonate with their own lead inductance (even if the leads are very short), it will be found that for a given type of capacitor certain values are optimum for decoupling certain frequencies. For example, 1,000pF may be best for 50MHz and 100pF for 200MHz. While good decoupling will be obtained at both these frequencies, at some other frequency (or frequencies) one capacitor may resonate with the effective inductance of the other. This will form a parallel tuned circuit (high impedance) across the line and result in *no* decoupling at all at that frequency. If such techniques are to be employed, it is safer to use an rf choke or low-value resistor between the capacitors.

The case which houses the transmitter should be as rf-tight as possible, and it is regrettable that many commercial units do not conform to this criterion. For best results the case should be made from a low-resistance material, such as brass or aluminium. Ventilation holes should not exceed ¼in. in any direction, and those which do may have to be screened with a fine mesh wire gauze. It is not always essential to screen meter holes, but where necessary the meters may be mounted in screened enclosures with their leads decoupled where they pass through the enclosure. Where there are joints in the metalwork there should be at least a ½in overlap of low-resistance contact (clean and paint-free), and surfaces should be bolted together with bolts not more than about 2in apart, or even less for vhf and uhf transmitters.

Transmitters screened and decoupled in accordance with the principles described should be adequately free of interference problems caused by rf leakage from the transmitter case and leads, thus enabling external filters to work effectively.

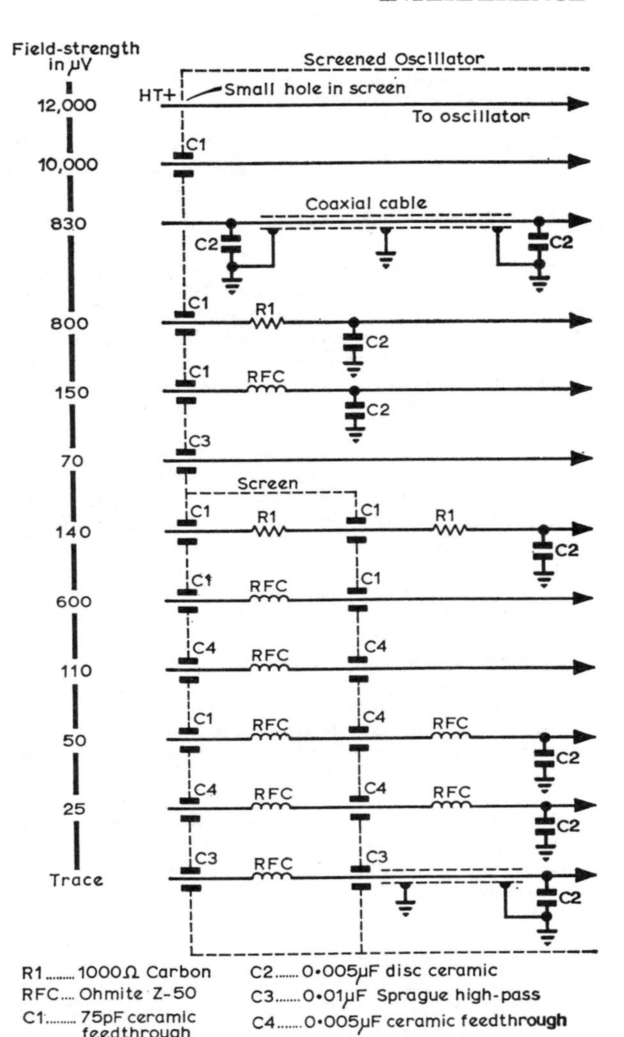

Fig 17.13. Comparison of the effectiveness of various methods of bypassing the leads emerging from a screened cabinet (measured at 80MHz).

Correct Station Layout

To ensure that the minimum of unwanted signal is radiated by the transmitter the layout shown in **Fig 17.14** is suggested. The function of each component is as follows:

Aerial tuning unit or aerial coupler. This ensures that the impedance of the aerial, including feeders, is transformed to the characteristic impedance (Z_0) of the filter. It enables the filter to work into the correct load on the wanted frequency and prevents damage to it due to high voltages produced by a high swr. Depending upon its design the atu may give extra rejection to unwanted signals both above and below the wanted frequency. When the impedance of the aerial system is close to the Z_0 of the filter the atu may be omitted, but other benefits will be lost. The atu should be fully screened.

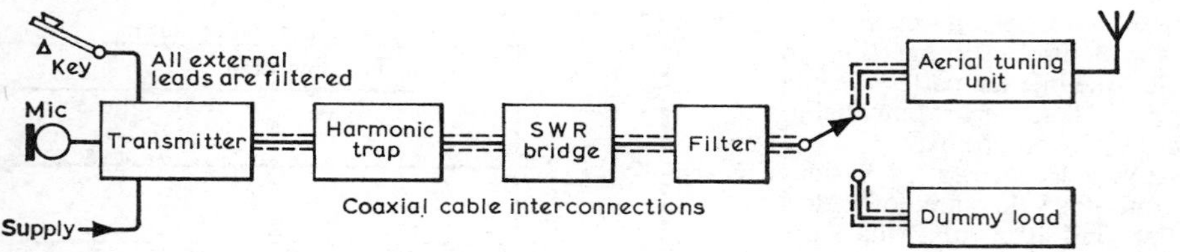

Fig 17.14. Suggested layout of equipment for minimum spurious radiation.

Filter—low-pass or band-pass. A low-pass filter which cuts off at about 30MHz is recommended for hf stations, while for vhf and uhf a band-pass filter for each band is preferable. The low-pass filter is designed to give high attenuation above 30MHz, often with even higher attenuation on a specified unwanted frequency which may cause interference to tv sets in Band I. This frequency is the one which could be radiated by the transmitter and *not* that of the tv channel. Because the input impedance of the atu plus aerial system will almost certainly not be correct for the filter on frequencies other than the wanted transmitter frequency, the attenuation of the filter may be quite different from the published figures which assume *correct* termination on *all* frequencies. This problem may be avoided if the filter is of the type designed to absorb signals above the cut-off frequency in a dummy load. The filter should be fully screened.

SWR bridge. This enables the atu to be adjusted so that the filter works into its correct impedance, as described. The bridge should be inserted as shown before the filter because the diodes in the bridge may generate harmonics. If the filter is correctly terminated the bridge should indicate 1 : 1. The bridge should be fully screened.

Harmonic trap. This is added to give even greater attenuation on a specific frequency, especially if the filter is not adequate on that frequency. It should be contained in a screened compartment, either inside the transmitter, or as an external unit.

Transmitter. The procedures described above should be noted. If a separate linear amplifier is used, then the same procedures should be adopted. A low-pass or band-pass filter between exciter and linear amplifier is desirable.

Choice and Location of Aerial and Transmitting Station

While strictly not a transmitting station fault, it may be possible to reduce interference to domestic equipment by taking measures which reduce the interfering field strength in the vicinity of the equipment. The interference may well be due to the strength of the amateur fundamental signal.

It is obvious that the transmitting aerial should be located as far as possible from domestic equipment and aerial installations. Vertical and end-fed aerials radiate strong local signals and may induce considerable voltages and currents into tv and fm radio downleads, which are predominantly vertical. It is advantageous to mount beam aerials well above roof level (sometimes a good argument to persuade neighbours not to raise objections to modest masts and towers).

It helps if the station "shack" is remote from the house, such as in a garden shed, garage, or outbuilding, to reduce the chances of trouble from the effects of radiation of signals from the transmitter case, leads and feeders. However, there is no reason why a transmitter should not be located right next to domestic equipment, provided that the transmitter is well screened and decoupled, and there is no radiation from the aerial feeder.

Feeders and Baluns

The connection of an unbalanced feeder (such as coaxial cable) to a balanced aerial (such as a dipole, beam, delta-loop etc) has little effect on the apparent performance of the aerial, but causes currents to flow on the outer braid of the cable. **Fig 17.15** shows the current distribution in a dipole both when using balanced feeder and when using coaxial cable. The feed current flowing in the braid brings the rf potential above that of earth and, depending upon where the earth reference for the station may be, there will be considerable radiation from both the feeder and even the casing of the transmitting equipment. Unless there is a very good rf earth on the transmitter the braid current will flow through the case and into the mains power wiring earth, which may be some distance from the transmitter as shown in **Fig 17.16.** As the line and neutral phases of the supply run parallel and are in close proximity to the earth wire there will be considerable radiation and direct pick-up of both signal frequency and harmonics on the power distribution wiring in the locality.

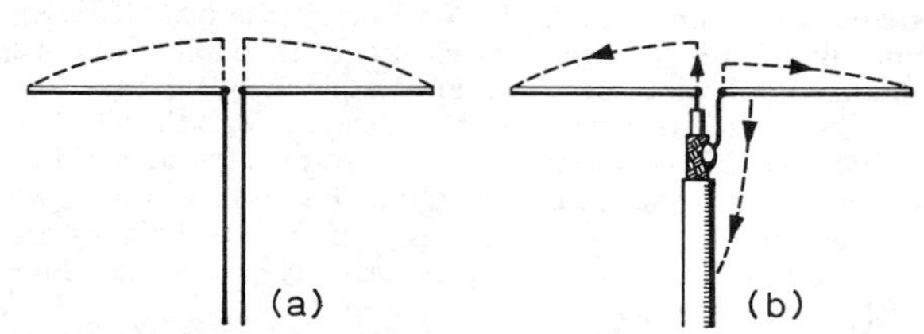

Fig 17.15. Comparison of current distributions in a dipole aerial fed by (a) a balanced feeder (b) a coaxial feeder.

As coaxial cable has the advantage of being able to be used without any ill effects due to the proximity of surrounding objects, it is a popular type of feeder. To connect it to a balanced aerial it is necessary to use a balun (ie a balance-to-unbalance transformer) at the point where the coaxial cable is joined to the aerial. Where proximity effects are no problem, balanced feeder may be used and can be coupled to the transmitter via a balun, or via an atu which has a balanced output (preferably "floating", ie it has no connection to the atu earth). The balanced feeder may be chosen to match the aerial impedance, although a suitable atu may be needed to transform this impedance to that of the filter. The absence of a direct connection between the aerial system and the atu earth may also reduce the flow of unwanted signals which may leak out of an imperfect transmitter. To prevent a build-up of static voltage on the feeder (due to electrically

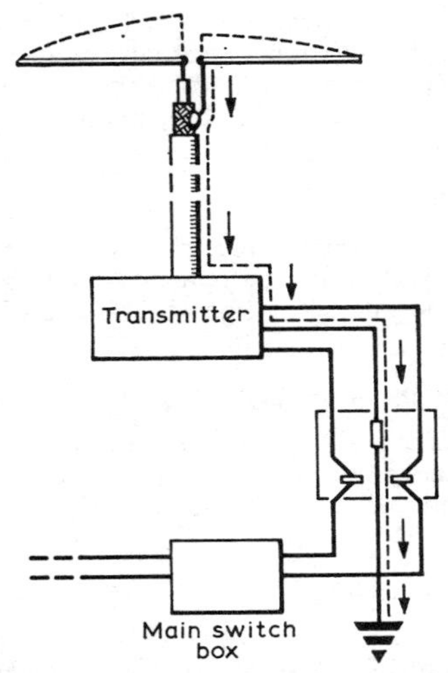

Fig 17.16. Due to the imbalance of aerial currents when a coaxial feeder is used, a proportion of the current on the leg connected to the braiding flows down the feeder and through thee quipment casing to earth. If there is no separate earth, these currents will flow via the power wiring to the house main earthing system. It is usual practice for the mains neutral to be connected to earth at the sub-station so the result can be considerable radiation on the mains distribution.

charged rain, for example), it is desirable to provide a leakage path, so a high-value resistor (about 100kΩ) should be connected from the aerial system to earth.

INTERFERENCE TO DOMESTIC EQUIPMENT

Because of the wide range of frequencies used by both amateurs and the broadcast television and radio services, the causes of interference are various, and may be categorized as follows:

Transmitting station faults.

TV and radio set faults.

External non-linear devices.

All three may occur simultaneously, but by a methodical process of elimination it is possible to determine the cause(s) of interference. Since many of the reasons for interference to tv, radio and audio equipment are the same, and to avoid repetition, tv interference is dealt with in most detail.

It is usually impracticable to improve the set performance by internal means, but where this is unavoidable, it should be undertaken with great caution and as a last resort. Attempting internal modifications may have unforeseen results, even if completely successful in curing the interference. The amateur may be continually held responsible for subsequent poor performance of the set, often with no justification. It should be remembered that the amateur is in no way obliged to provide any material assistance to the owners of the deficient equipment, although co-operation with all parties concerned is usually necessary to secure a satisfactory solution.

INTERFERENCE TO TV SETS (TVI)

To understand the problem, it is useful to know the frequency of the tv channel which is suffering the interference. The frequencies of various tv systems are given in **Table 17.1.**

TABLE 17.1

Television channel allocations. Note that cable tv frequencies may be considerably different from those stated. Continental European uhf vision frequencies are the same as the UK, but sound is 0·5MHz lower.

UNITED KINGDOM

Channel No	Vision carrier	Sound carrier
B1	45·00	41·50
B2	51·75	48·25
B3	56·75	53·25
B4	61·75	58·25
B5	66·75	63·25
B6	179·75	176·25
B7	184·75	181·25
B8	189·75	186·25
B9	194·75	191·25
B10	199·75	196·25
B11	204·75	201·25
B12	209·75	206·25
B13	214·75	211·25
B14	219·75	216·25
	System A 405 lines	

CONT. EUROPE

Channel No	Vision carrier	Sound carrier
E2	48·25	53·75
E2A	49·75	55·25
E3	55·25	60·75
E4	62·25	67·75
E5	175·25	180·75
E6	182·25	187·75
E7	189·25	194·75
E8	196·25	201·75
E9	203·25	202·75
E10	210·25	215·75
E11	217·25	222·75
E12	224·25	229·75
	System B 625 lines	

USA

Channel No	Vision carrier	Sound carrier
A1	—	—
A2	55·25	59·75
A3	61·25	65·75
A4	67·25	71·75
A5	77·25	81·75
A6	83·25	87·75
A7	175·25	179·75
A8	181·25	185·75
A9	187·25	191·75
A10	193·25	197·75
A11	199·25	203·75
A12	205·25	209·75
A13	211·25	215·75
	System M 525 lines	

REPUBLIC OF IRELAND

Channel No	Vision carrier	Sound carrier
1B	53·75	59·75
1D	175·25	181·25
1F	191·25	197·25
1H	207·25	213·25
1J	215·25	221·25
	System I 625 lines	

UK CABLE TV

Channel No	Vision carrier	Sound carrier
A	45·75	51·75
B	53·75	59·75
C	61·75	67·75
D	175·25	181·25
E	183·25	189·25
F	191·25	197·25
G	199·25	205·25
H	207·25	213·25
I	215·25	221·25
	System I 625 lines	

UK ALLOCATION UHF TELEVISION (BAND IV/V)

Channel No	Vision carrier	Sound carrier	Channel No	Vision carrier	Sound carrier	Channel No	Vision carrier	Sound carrier
21	471·25	477·25	40	623·25	629·25	55	743·25	749·25
22	479·25	485·25	41	631·25	637·25	56	751·25	757·25
23	487·25	493·25	42	639·25	645·25	57	759·25	765·25
24	495·25	501·25	43	647·25	653·25	58	767·25	773·25
25	503·25	509·25	44	655·25	661·25	59	775·25	781·25
26	511·25	517·25	45	663·26	669·25	60	783·25	789·25
27	519·25	525·25	46	671·25	677·25	61	791·25	797·25
28	527·25	533·25	47	679·25	685·25	62	799·25	805·25
29	535·25	541·25	48	687·25	693·25	63	807·25	813·25
30	543·25	549·25	49	695·25	701·25	64	815·25	821·25
31	551·25	557·25	50	703·25	709·25	65	823·25	829·25
32	559·25	565·25	51	711·25	717·25	66	831·25	837·25
33	567·25	573·25	52	719·25	725·25	67	839·25	845·25
34	575·25	581·25	53	727·25	733·25	68	847·25	853·25
39	615·25	621·25	54	735·25	741·25			

Until recently in the UK perhaps the most common cause of tvi has been the production of harmonics from hf band transmitters which affect Band I television. **Fig 17.17** indicates the harmonic relationship between these frequencies, and a similar chart can be drawn up for other tv and amateur frequency allocations. These harmonics may be generated either in the transmitting system, the receiving system, or by an external device. It will be seen that some channels are clear of possible harmonic troubles, yet interference on these channels is not uncommon, being often due to crossmodulation in the tv set rf, mixer or other stages.

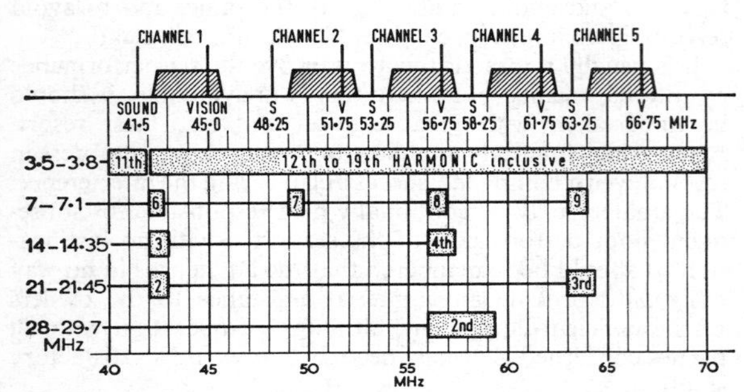

Fig 17.17. Harmonic relationship between the hf amateur bands and tv channels in the UK Band I.

With the growth of uhf television in the UK, and the corresponding decrease in the use of Bands I and III, harmonic interference from hf transmitters is much less prevalent, but crossmodulation and other forms of interference are much more common. In addition the increasing activity on the 2m and 70cm bands has resulted in more tvi to uhf tv.

Transmitting Station Faults

These are usually the radiation of an unwanted signal on or near the frequency to which the tv (or radio) is tuned, or, less frequently, on some other unauthorized frequency such as the tv i.f. The reasons for these faults have been listed in the previous section. Attention should be paid to all these possibilities to ensure that a clean signal is being radiated and that the local field strength is as low as possible.

TV Set at Fault

Television sets may suffer from all the shortcomings of receivers described earlier. As there is fierce commercial competition between manufacturers, some sets are designed with scant regard to the problems which occur when the set is operated in the presence of high rf fields, and the extra components required for filtering, screening and decoupling are often omitted in the interests of economy. Unfortunately, there are no regulations in the UK designed to protect tv sets from such interfering signals.

Interference is usually due to the following causes:
Self-generated harmonics.
Crossmodulation.
Detection of the interfering signal.
Direct reproduction.

Self-generated Harmonics

This is a common cause of tvi on Band I from hf transmissions (usually 14, 21 and 28MHz). VHF transmissions may also cause interference to uhf bands (eg 4 × 145 = 580MHz). Such interference is usually limited to low-order harmonics, ie second, third, fourth and fifth, which are generated in the rf, mixer or i.f. stages of the tv set because of non-linearities.

Crossmodulation

Crossmodulation may occur in the rf, mixer or i.f. stages of a tv set regardless of the frequency relationship between the wanted and interfering signals. This is a common cause of tvi to uhf tv.

Detection of the Interfering Signal

This occurs in audio or video stages and may take place in intentional diode detectors or because valves and transistors are driven into non-linearity by the strong interfering signal.

Direct Reproduction

This may occur because 1·8 and 3·5MHz are within the video spectrum of 0–5·5MHz, and even 7MHz can cause trouble. Colour decoder and other signal processing circuits are particularly susceptible to signals in the 80m band because they are designed to operate on the colour subcarrier frequency of 4·43MHz (625 line PAL system) or 3·57MHz (525 line NTSC system).

How Interference Enters the TV Set

There may well be other reasons for the interference, but the cure for all the mechanisms of interference is the same, that is, to prevent the interfering signal from entering the set.

The interfering signal can enter in four ways:
Via the coaxial feeder inner conductor;
Via the coaxial feeder braid;
Via the mains power leads;
By direct pick-up in the circuit.

Coaxial feeder inner conductor. When the tv aerial is large enough to act as a reasonably efficient aerial at the interfering frequency, it will feed a significant amount of interfering signal to the tv set via the inner conductor of the coaxial cable.

In the UK it is common practice to isolate the aerial input —both inner and outer conductors—from possible mains voltages by coupling the input socket to the set via capacitors of 270 or 470pF (see **Figs 17.18, 17.19**). These have negligible impedance at the tv frequencies, but not so at amateur frequencies. The coaxial cable braiding acts as an aerial for the interfering signal, causing a current to flow through the isolation capacitor to the tv set and on to earth. The current flowing through the capacitor impedance produces a voltage, so the braid becomes "live" at the interfering frequency.

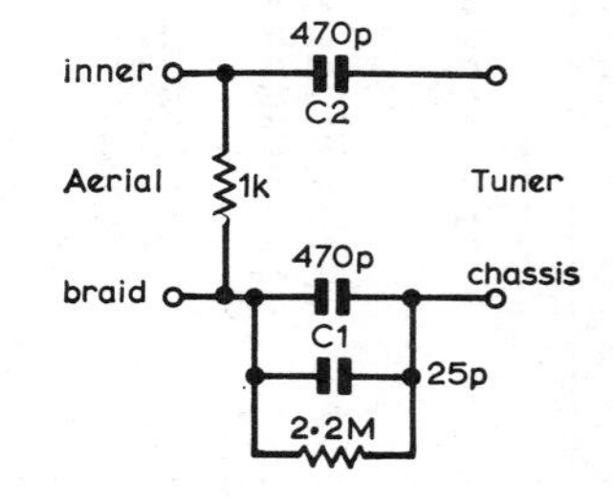

Fig 17.18. Typical aerial isolation network found in vhf tv sets in the UK.

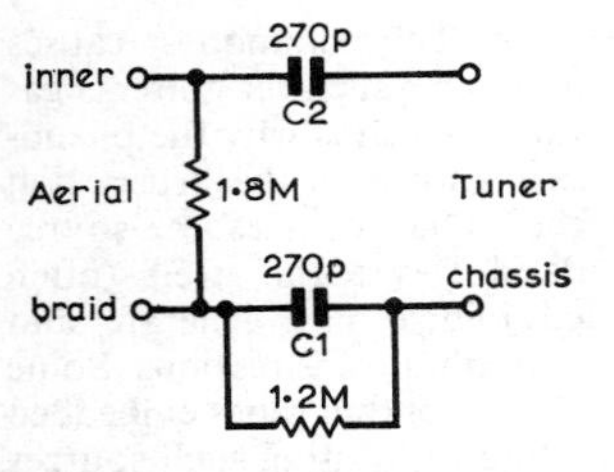

Fig 17.19. Typical aerial isolation network found in uhf tv sets in the UK.

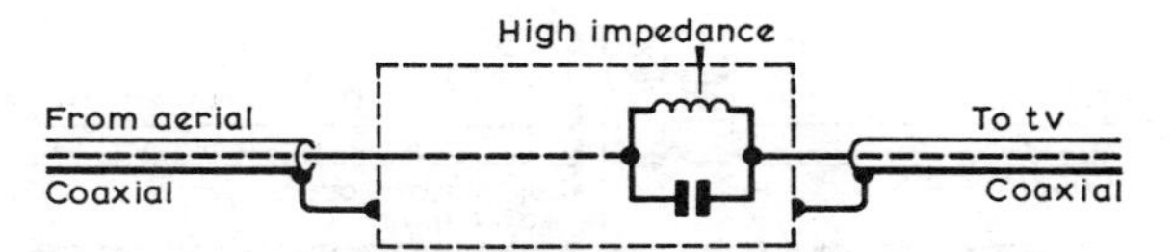

Fig 17.21. A common form of filter with a high impedance in the final element, which will prevent interfering signals from entering via the inner conductor.

The braid voltage is then transferred to the inner conductor via the effective impedance Z_f seen between outer and inner (see **Fig 17.20**).

If the braid of the coaxial cable is insubstantial it may not effectively screen the inner conductor, especially at low frequencies. This allows the inner conductor to pick up some interference, although the occurrence is probably rare.

With any of the three causes described above, a filter which prevents the interfering signal from entering via the inner conductor will be effective. However, the filter design must be chosen with care. When the first effect alone is present, almost any filter designed to attenuate the interfering frequency will be effective. Similarly with the third effect. The filter may contain inductors and capacitors in any suitable combination. However, when the second effect ("live" braid) is present a conventional filter (one with a common connection for the coaxial braid at both input and output) will only be satisfactory if the filter is as shown in **Fig 17.21**. The filter must contain immediately before its output a circuit which presents a high impedance to the interference. If the design is of the type shown in **Fig 17.22** where the intention is to bypass the interference to the earthy side of the filter, the reverse will happen, and the braid voltage will be transferred to the inner conductor, possibly increasing the interference. It is usually found that this type of filter will work satisfactorily only when inserted *inside* the tv set, between the aerial isolating network and tuner, where the true reference earth for the set will be found. A simpler cure is to use the "braid-breaking" type of filter described later.

Coaxial feeder braid. Again, the braid of the coaxial feeder may act as an end-fed aerial for the interfering signal. The capacitor isolating the braid, while presenting some impedance to the interference, still allows considerable current to flow into the tv set. Under certain circumstances the capacitor may actually help to resonate the system (see **Fig 17.23**). The current flowing through a printed circuit board conductor which has a relatively high impedance at rf may produce voltages across ostensibly equi-potential points in the circuit. A very small change in base-emitter voltage in a transistor circuit will cause a considerable change in base current and hence in collector current.

The cures for interference due to any braid current or voltage (as previously described) may be similar. One solution is to provide a good rf earth for the coaxial braid, as shown in **Fig 17.24**, so that the braid will then be at zero potential. No interference will be transferred to the inner conductor, and no current will flow into the set via the braid connection. If there is no suitable earth connection available, the mains earth on the tv set power plug may be tried, but this will probably be much less effective. Do not attempt to overcome problems caused by the set decoupling capacitors by connecting the braid of the feeder direct to the tv set; the set earth (ie the chassis) should *never* be connected to any other earth connection.

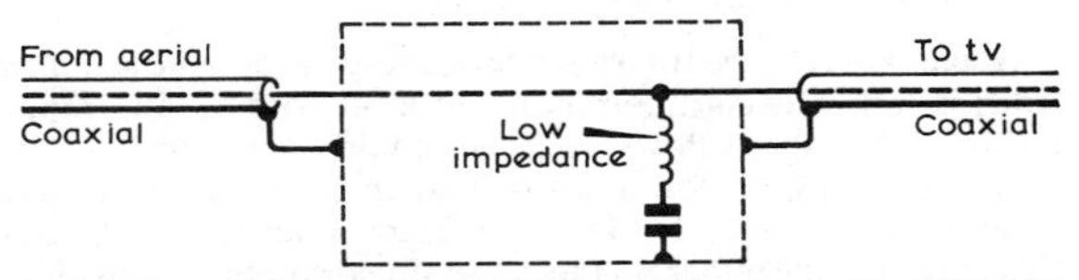

Fig 17.22. An alternative form of filter to the one in Fig 17.21. This type can suffer from the defect that any signals present on the braid will be passed on to the inner conductor. Only really suitable when mounted inside the set.

Another cure for these effects is to oblige the set to respond only to antiphase signals developed at the tv aerial and fed down via the coaxial cable, and to make the set immune to in-phase signals picked up on the braid. One method is to break the connection to the set by the use of a 1 : 1 rf transformer of low inter-winding capacitance as in Fig 17.46. Practical information is given in Figs 17.47–17.50. Where pick-up of the interference on the aerial is still significant, any conventional filter may be used as a back-up for the transformer.

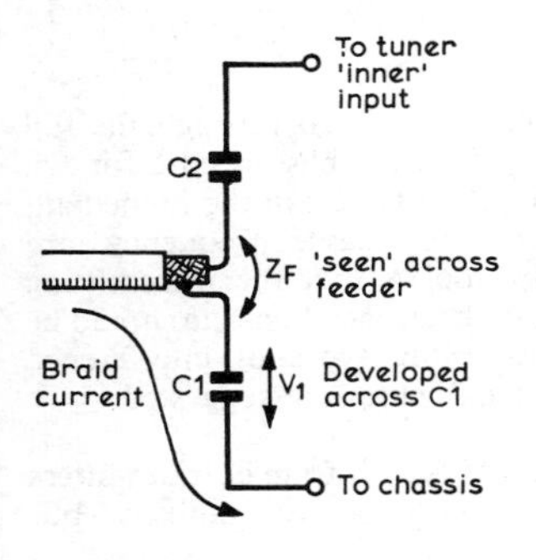

Fig 17.20. Current flowing in the outer braid of tv coaxial feeder can cause a signal to appear on the inner conductor as shown here. V1 is produced by the braid current flowing through C1, transferred to the coaxial inner via Z_f and thence passed via C2 to the tuner input.

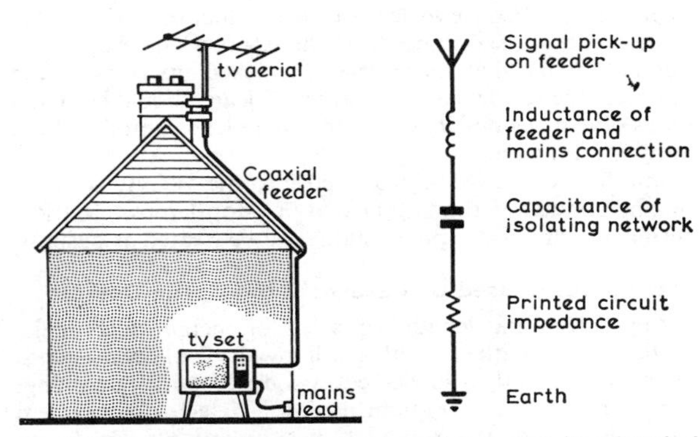

Fig 17.23. A typical domestic tv installation and its electrical equivalent.

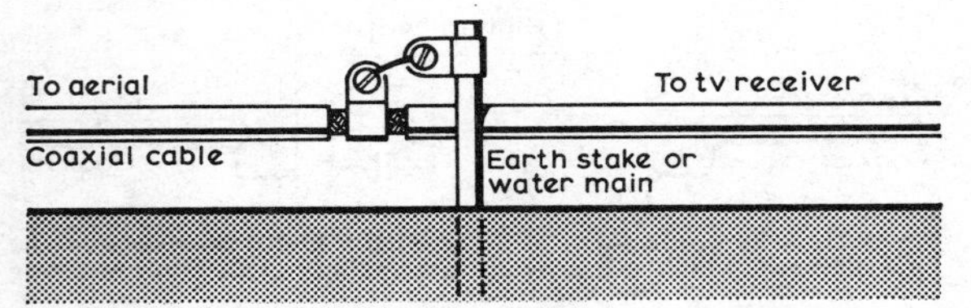

Fig 17.24. A simple solution to interference on the coaxial braid is to earth the braid at some point before it enters the tv set.

A second method is to wind the coaxial cable to form a choke. This presents a high impedance to interfering signals which are in phase with each other on the braid *and* the inner conductor, but has no impedance to antiphase currents. The loss of the wanted tv signal will therefore be low—only that of the short extra length of coaxial cable required (see Fig 17.45).

A third solution is to increase the impedance of the isolation capacitors, so that the relatively low-frequency interference cannot enter the set by either inner or outer conductor. This approach is particularly effective for uhf tv where interference is from amateur hf bands, and can be filtered out with ease. A suitable circuit is shown in Fig 17.51. There is some risk of pick-up of tv signals on the braiding but this appears to give no trouble.

Mains power leads. Interference due to pick-up on the mains leads is not in itself a common cause of trouble. The interfering signal usually flows from the coaxial braid, through the tv set circuits and power leads, and into the earthy power supply, as in Fig 17.23. The cure is to break this path, and this may be done either on the aerial side or on the mains side. Where mains filtering is necessary, the mains lead may be wound to form a bifilar rf choke, similar to the coaxial choke braid-breaking filter (Fig 17.45). Additional capacitors may be used as in Fig 17.53.

Direct pick-up in the circuit. Direct pick-up is an increasing problem due, to some extent, to the use of unscreened printed circuits. Apart from internal modifications to the tv set, the amateur is limited in what he can do to effect a cure. It is sometimes found that the use of all kinds of external filters proves ineffective, and when this occurs the set is almost certainly suffering from direct pick-up. It is possible that the interference is introduced into the vicinity of the set by the coaxial feeder, in which case earthing the coaxial cable as described, perhaps in conjunction with a braid-breaking filter, may be effective. Alternatively the filter may be inserted in the feeder some distance from the set—perhaps 8 or 10ft—but not so far that interference is picked up between the filter and the set. The feeder and mains leads should leave the set as directly as possible, and any spare lead should be kept away from the set.

Interference may be introduced into the vicinity of a set by any conductor running near it. TV-top lamps, electrical wiring, gas and water pipes, should all be treated as suspect.

Harmonics Caused by External Non-linear Devices

Any substantial lengths or areas of metal which make partial contact with each other will, by virtue of the existence of oxides and other substances associated with tarnishing, behave like an aerial system having a detector somewhere along its length. Intermodulation between any signals, or harmonic generation from one signal, will take place and may be radiated from the system. The commonest causes are rusty joints in domestic plumbing such as gutters, gas pipes and electrical wiring conduit, which is why the phenomenon has for many years been known as the "rusty bolt effect" or the "drain pipe effect". One troublesome source of such effects is the UK Band I tv aerial itself (often disused) which, with corroded joints, may generate and efficiently radiate harmonics of hf band transmissions. Some of the most obscure causes are rusty conduit pipes embedded in the plaster of walls. One method of locating such sources is with the use of a portable receiver, preferably with at least two tuned rf stages and having a loop aerial. The transmitter should be modulated at full power while the receiver is taken around the neighbourhood looking for points of origin of maximum harmonic. It is extremely important that the receiver should be fitted with a trap tuned to the transmitter fundamental frequency, to reduce further the possibility of self-generated harmonics within the receiver.

TVI Diagnosis

This section is intended to be a practical guide to an amateur confronted with a complaint of interference. The emphasis is therefore on logical procedures, rather than on repeating the particular application of topics dealt with elsewhere in this chapter. For example, the statement "Apply traps" implies "For details, see the section on traps".

It is not to be expected that each amateur will have full facilities for dealing with all tvi problems, and here the value of co-operative effort becomes apparent. All well-organized radio groups and clubs should have a small nucleus of members who can help with these problems. They should be reasonably well skilled in the diagnosis of tvi and should have available, as club property, test equipment such as low-pass filters, high-pass filters, tunable traps, coaxial couplers and adaptors. The services of the group should preferably be available to all local amateurs, whether members of the club or not, because a badly handled case of tvi directs public opinion against amateur radio as a whole.

It will be helpful to have a complete circuit diagram of the transmitter, including control and mains wiring, so that no potential radiators are neglected, and to keep a record of all tests, even those which do not work, since this will avoid repeating the same tests again later.

Before tvi can be cured, the nature of the trouble must be correctly diagnosed, and the chart shown in **Fig 17.25** sets out a series of logical tests which will enable this to be done expeditiously. Once the cause is known, reference to the appropriate sections of this chapter should be made for the necessary cures.

TVI Quick Cures

In some cases it is not necessary to go through the full procedure outlined in Fig 17.25 to establish a cure for tvi. This is so when, for example, other tv sets in the immediate neighbourhood operating on the same frequency are unaffected. This immediately implies that the transmitting station is not at fault (although beam headings etc should be taken into account). The following types of interference occur relatively often and "quick" cures are suggested.

Interference to uhf tv (Bands IV and V) from hf transmitters—transmitter harmonic interference is very unlikely, but

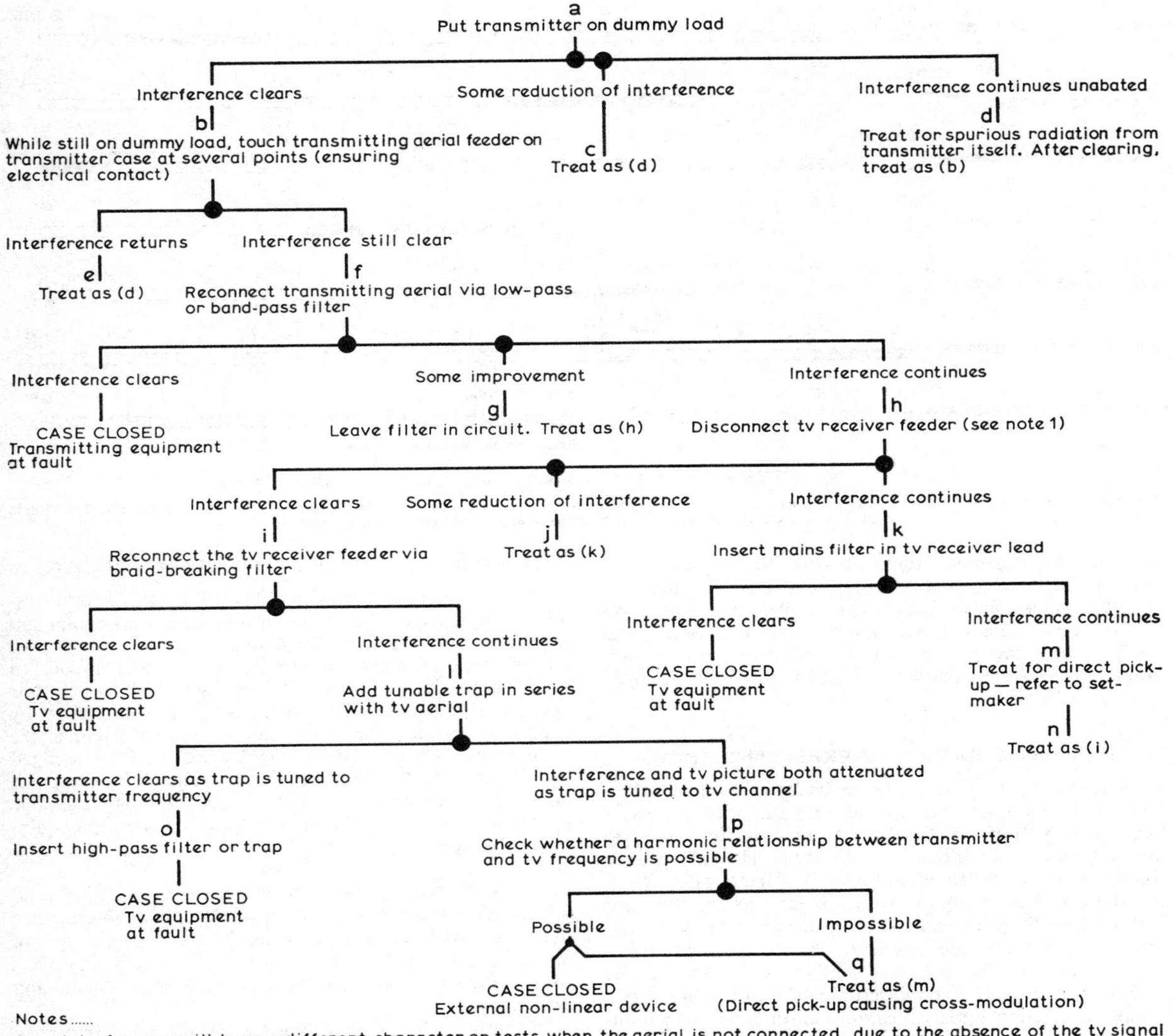

Fig 17.25. A systematic fault-finding chart for tracing tvi problems.

pick-up of the interfering signal on the braid of the tv coaxial feeder is common. Cure—filter tv as in Fig 17.45 or 17.51.

Loss of colour on tv when transmitting on the 80m band—the interfering signal is affecting the colour decoder circuits. Again, pick-up on the tv coaxial feeder is probable. Cure—filter tv as in Fig 17.45 or 17.51.

Interference to vhf tv (Bands I and III) from vhf or uhf transmitters—harmonic interference is unlikely (except perhaps from 70MHz). Significant pick-up of the interfering signal on the aerial is likely. Cure—trap or stub filter on tv set (Fig 17.32, 17.34, 17.36 or 17.39).

Improving Receiver Aerials

Where interference is caused by harmonic generation, the situation may be improved by increasing the strength of the wanted signal. This may be done by raising the receiving aerial or by using a higher gain and more directive array, which can improve the signal/interference ratio sufficiently to effect a cure.

However, it is important to note that where interference

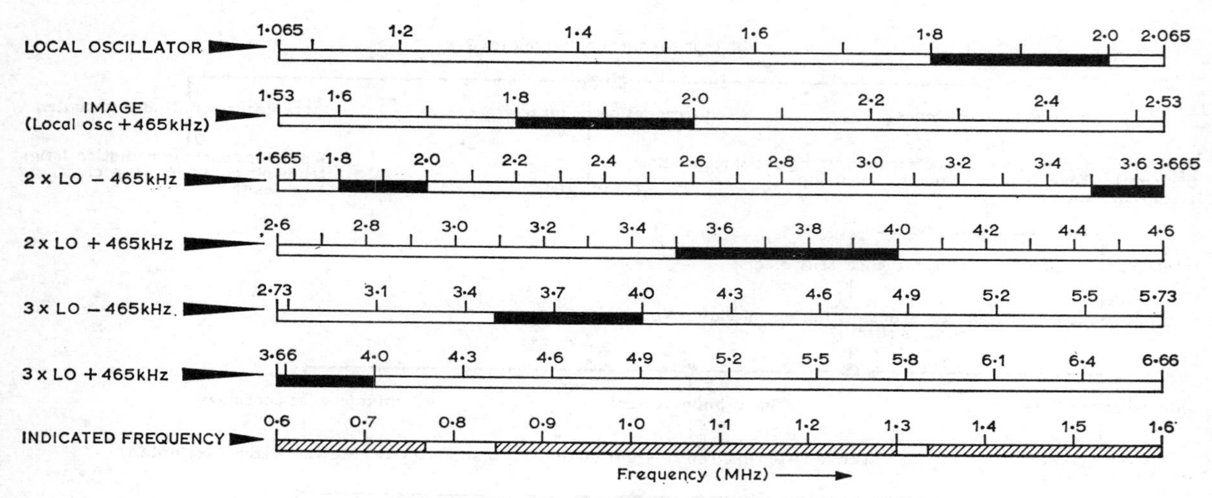

Fig 17.26. Some possible spurious responses of a superheterodyne receiver with 465kHz i.f. tuning the medium wave band. Similar charts can be drawn up for other combinations of i.f. and signal frequency.

is caused by crossmodulation (eg from hf transmitters to uhf tv), "improving" the aerial system may not alter the degree of interference—indeed it may worsen the situation if more aerial feeder is installed, due to increased braid pick-up. The percentage modulation impressed by crossmodulation is *independent* of the strength of the wanted signal (see *Crossmodulation*).

DOMESTIC RADIO INTERFERENCE (BCI)

Interference to domestic radio sets is rarely due to spurious outputs from transmitters, as the long and medium wavebands are both below the lowest amateur frequency, so interference from harmonics cannot occur. The vhf fm band (Band II) may suffer from harmonics of hf band transmitters, but these will normally be weak, except perhaps the third harmonic of the 10m band. The third harmonic of signals on the 80m band (which may be radiated by the transmitter or generated in the receiver) may cause trouble to the 10·7MHz i.f. However, the discrimination of the fm system against interfering signals ("capture effect") tends to minimize the effect of weak interfering signals.

The domestic receiver may suffer from any or all of the defects described in the section on receiver faults, and, as for tv sets, commercial pressures often dictate that sets are not adequate for operation in the presence of strong local signals. Domestic sets are usually superheterodyne designs, so a variety of spurious responses are possible. **Fig 17.26** indicates some of these.

BCI Diagnosis

Fig 17.27 indicates a logical approach to the diagnosis of bci. In many cases the receivers affected are portable types (ie they have no aerial or mains leads), and in these cases little can be done to effect a cure. Special attention should be paid to the section above on the location of transmitters and aerials. The effect of re-positioning the affected set should be investigated—well away from electrical wiring and plumbing.

AUDIO FREQUENCY INTERFERENCE (AFI)

There are an increasing number of cases where transmitting equipment is a source of interference to equipment which operates purely at audio frequencies, such as record players, tape recorders, electronic organs, and so on. In these cases the equipment affected is not designed (nor licensed) to operate as a radio receiver, and is therefore, by definition, at fault when interference occurs. Neither the amateur nor the Post Office Interference Branch is obliged to take any steps to cure such interference. However, this does not entirely absolve the amateur from taking reasonable precautions to prevent his signals from reaching the equipment. by reducing the local field strength as described earlier.

Almost all audio equipment now being produced is solid state, which appears to be more susceptible to interference than the earlier valve equipment. The widespread use of magnetic cartridges with low outputs requires amplifiers with higher gain than those needed for the earlier crystal and ceramic types.

In the majority of cases, the interfering signal is carried into the amplifier by the interconnecting leads. In practice it is found that the loudspeaker leads are the main offenders, often forming a dipole of considerable proportions in stereo systems. RF energy is fed back (by the internal negative feedback line) to earlier stages of the main amplifier, rectified by non-linearities, and then amplified. When interference enters the main amplifier, the volume control will have little effect on the interference level. Alternatively, rf currents may flow through the circuit board and be detected by earlier, more sensitive stages—in this case the volume control will affect the interference level.

The second major path for rf signals to enter the amplifier is via the mains leads. Alternatively, the signals may enter via other connecting leads, which, despite being normally shorter and often screened, may still cause trouble, particularly with vhf transmissions.

In the above cases, the trouble can normally be cured by the use of rf chokes in the leads (or rf decoupling). The

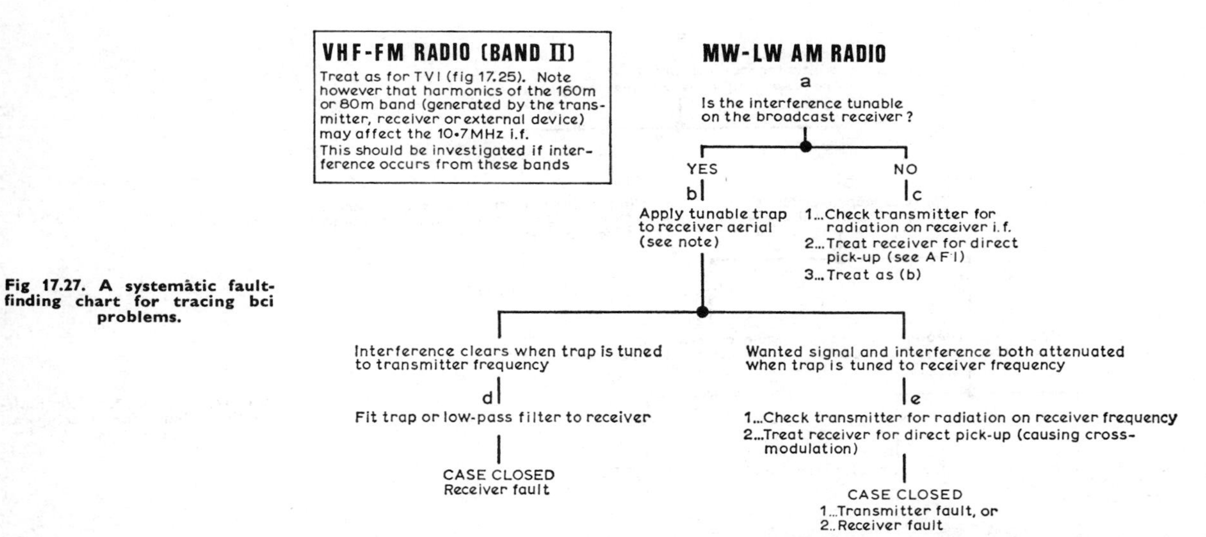

Fig 17.27. A systematic fault-finding chart for tracing bci problems.

simplest form of choke is to wrap the leads into several turns round a ferrite ring core (preferably) or a ferrite rod, which chokes the rf currents in a similar manner to the braid-breaking filter in Fig 17.45. In the case of loudspeaker leads it is possible that using screened or coaxial cable will effect a cure. Re-routing the leads or altering their length may help if the interference is confined to one frequency band.

If interference still persists when this has been done, then it is almost certain that the interference is being picked up directly internally.

Fig 17.28(a) shows a typical transistor amplifier stage. RF voltages appearing across the base/emitter junction will be amplified, or with large signals rectified and amplified, so it is necessary to decouple the rf from the stage. The simplest form of decoupling is shown in **Fig 17.28(b).** Note that if a resistor is used in series with the base its value must be low, such that it does not affect the transistor bias conditions. The capacitor may be large, but not so large that it affects the audio response, and in practice the use of a capacitor alone (typically 1,000pF) is usually adequate. A small rf choke is preferable to the resistor, as this will give greater rf attenuation and will not affect the bias conditions. A convenient form of choke is to use a small ferrite bead which may be slipped over a transistor lead or wire as in **Fig 17.29**. To increase its effectiveness, more than one turn of wire may be passed through the bead. Again, using a bead alone may well be adequate to cure interference.

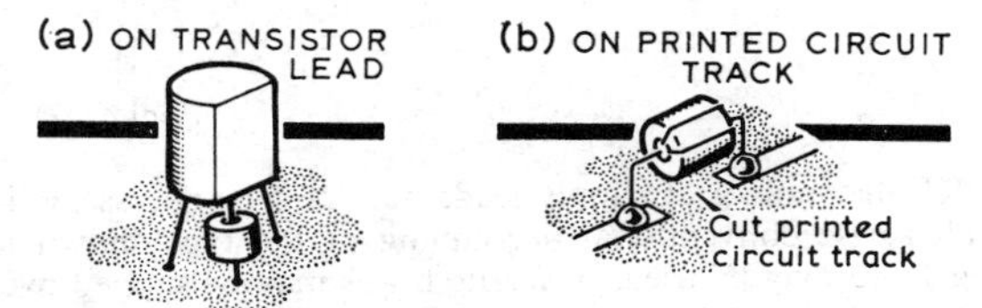

Fig 17.29. Use of a ferrite bead as an rf choke (a) on transistor leads, and (b) on a printed circuit.

One disadvantage exists in decoupling of the form shown in Fig 17.28(b). RF voltages introduced through the feedback line will not be effectively decoupled, and so it is preferable to decouple the base direct to the emitter as in **Fig 17.28(c)** and it is recommended that wherever possible this form of decoupling should be used. The capacitor should be connected as close as possible to the transistor, and if necessary a ferrite bead slipped over the base lead.

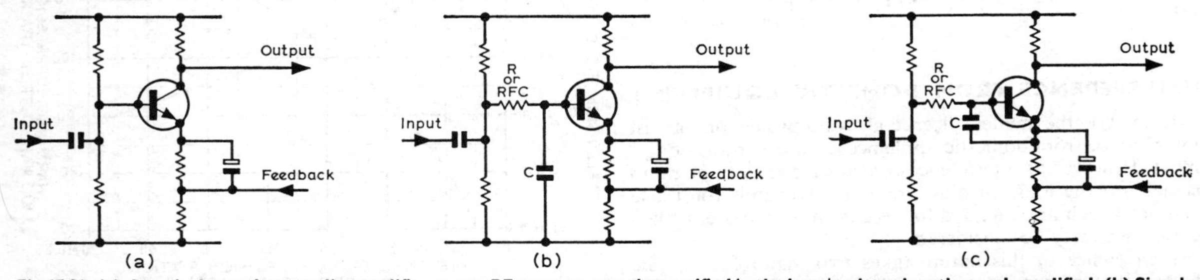

Fig 17.28. (a) A typical transistor audio amplifier stage. RF currents may be rectified in the base/emitter junction and amplified. (b) Simple rf filtering added to the basic stage. R should be low value to avoid upsetting the transistor bias conditions, and would be better replaced by an rfc. C is typically 1,000pF. (c) Decoupling the base to the emitter improves rejection of rf injected via the feedback line. C must be physically as close as possible to the transistor. This type of decoupling is preferable to that described in (b).

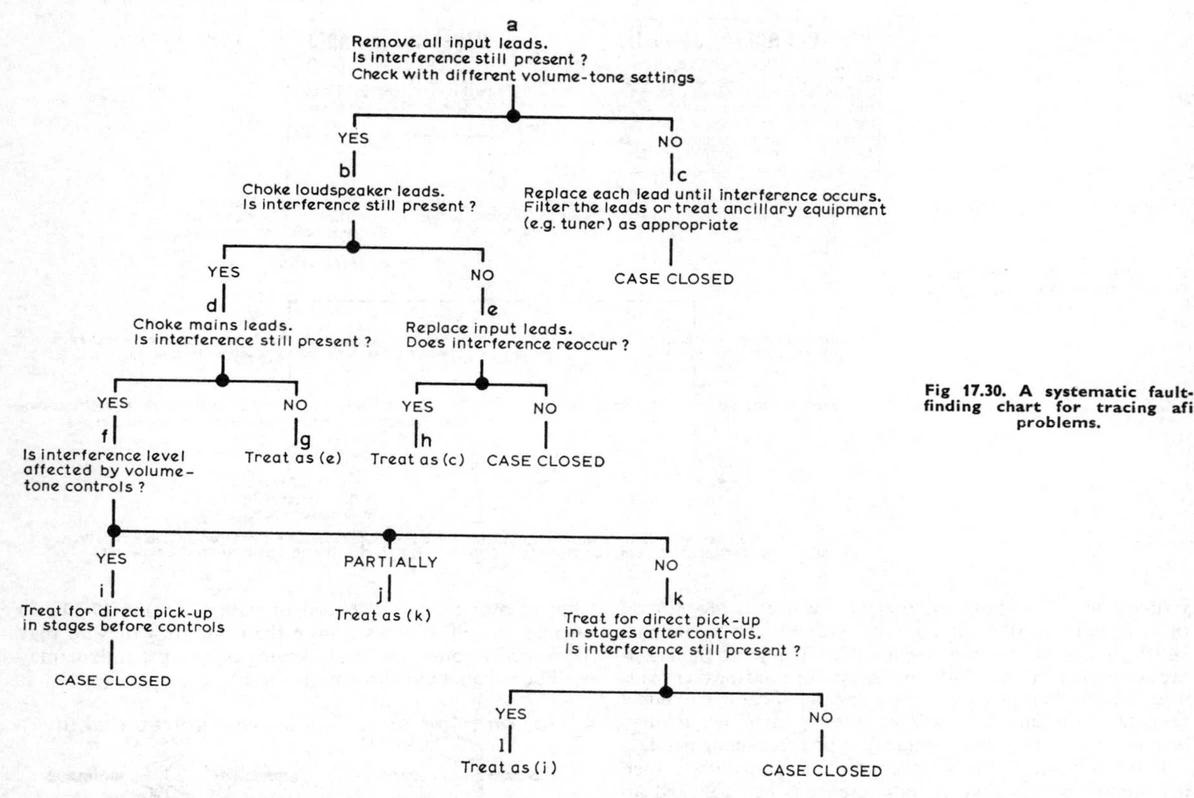

Fig 17.30. A systematic fault-finding chart for tracing afi problems.

RF decoupling of supply leads may also be necessary, in which case conventional decoupling of the type shown in Fig 17.13 may be used, or ferrite beads may be slipped over the supply lines between stages. Ferrite beads may also be found useful as chokes in such places as the leads attached to the cartridge in record players, to tape recorder heads and microphone leads.

AFI Diagnosis

Fig 17.30 shows a logical approach to the diagnosis of afi. If the audio equipment includes a radio tuner, it should be treated separately along the lines given for bci. The comments made on tvi with regard to the positioning of filters and spare leads, and placement of equipment away from wiring and plumbing to avoid direct pick-up, should also be noted for afi.

INTERFERENCE FROM DOMESTIC EQUIPMENT

In the UK the amateur licence gives no protection against interference from domestic appliances and electronic equipment. However, if interference is also caused to tv or radio signals on the long, medium, or vhf wavebands (on those stations which are intended for reception in that area), then complaints may be considered.

Interference of this nature takes two major forms. The first is that of occasional spark-based interference, usually from contactor devices such as thermostats in electric irons, refrigerators, central heating control circuits and so on, which give short bursts of interference while switching; or from electric motors in vacuum cleaners, hair dryers, food mixers, electric drills etc, which give interference for several minutes at a time. This type of interference can be reduced by the use of conventional mains filters and spark suppressors in the offending devices. A selection of suitable circuits is shown in Fig 17.52.

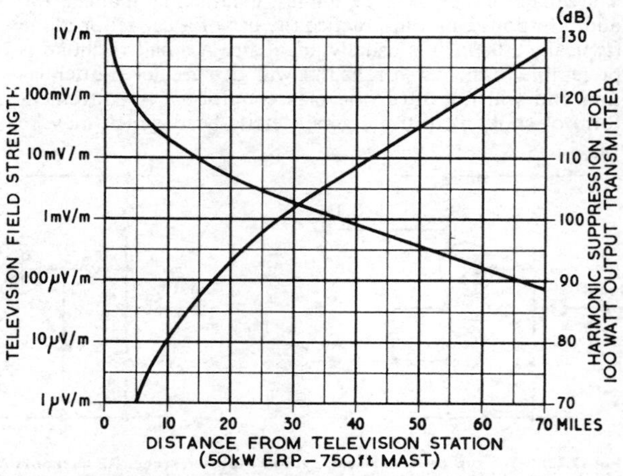

Fig 17.31. Showing the field strength of the tv signal and the consequent hf transmitter harmonic suppression required, as a function of the distance from the television station.

The second major source of interference is tv sets, and this can be particularly annoying as they are often on for long periods. Interference may be radiated from any stage in the receiver. The line output stage generates harmonics of about 16kHz (625- and 525-line systems) or about 10kHz (405-line system). In addition, the PAL switch in a colour tv, which operates at half the line frequency (about 7·8kHz), often radiates. The real cure for this form of interference is better design and screening in the tv sets themselves. Some of the radiation comes from the tv aerial coaxial feeder or mains leads, so that treating the set in the same manner as for tvi due to braid or mains pick-up may reduce the radiation from the set. Since the problems of interference from amateurs to domestic equipment are often reciprocal, the considerations for the location and types of aerials and feeders are appropriate.

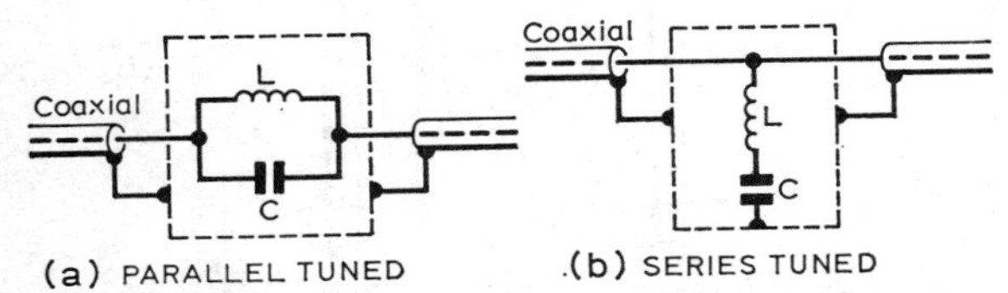

Fig 17.32. Single-frequency traps: (a) parallel-tuned. A tunable trap of this type is a valuable aid to diagnosis. Suggested values are: C—75pF variable, L—10 turns, 16swg, ½in dia, 1in long for 25/70MHz, or 22 turns 24swg ½in dia, 1in long for 12/35MHz. (b) Series-tuned. This type of trap must be used with care. In cases where interference is picked up on the coaxial cable braiding, this type of trap may worsen the situation.

PRACTICAL DESIGNS

Harmonic Suppression Calculations

Where transmitter harmonics are a problem, consideration must be given to the permissible level of radiated harmonic. **Fig 17.31** gives a guide to the suppression required for hf transmitters operating in the presence of Band I tv signals. The figure for suppression obtained from the chart cannot take account of local variations, so at least another 10 or 20dB should be allowed for an adequate safety margin.

At 60 miles from the transmitter the requirements may seem difficult to meet, but harmonic attenuation may be added up as follows:

30dB—basic transmitter
15dB—atu
50dB—low-pass filter as Fig 17.40
30dB—extra trap on output of transmitter

125dB—Total

Where the end sections of the low-pass filter are series-tuned as in Fig 17.40, the trap should be parallel-tuned (in series with the inner conductor), and vice-versa. Note that the figure given above for the low-pass filter is low, so the total rejection should be greater than the 125dB stated. *This will only be true if full attention is paid to the transmitter screening and decoupling considerations.*

Single Frequency Filters (Traps)

A trap may be used to prevent interference which occurs due to the presence of a single specific frequency, and also in the actual diagnosis of interference to determine which frequency is responsible. It may be in the form of either a parallel- or series-tuned circuit as shown in **Fig 17.32.** When the trap is intended for use in the tv coaxial feeder to reject the amateur fundamental frequency, it may be tuned by inserting it in the feeder to the amateur receiver and adjusting it for minimum received signal. Note that under certain circumstances a series-tuned circuit may cause increased interference unless used in conjunction with a braid-breaking filter. When the trap is to be used in the transmitting station to prevent radiation of a particular frequency which may not be on, or even close to, the frequency being interfered with, it may be necessary to carry out tuning either by trial and error, or by using a suitable signal generator and rf voltmeter. Final adjustment may be necessary when installed in the transmitter feeder to the aerial. A grid-dip oscillator (gdo) may be used as a signal generator by coupling it into the filter via a coupling coil as shown in **Fig 17.33**. This arrangement may be used for other types of filters, but harmonics from the gdo (or other signal generator) place a limit on the *measured* rejection of frequencies in the stop band.

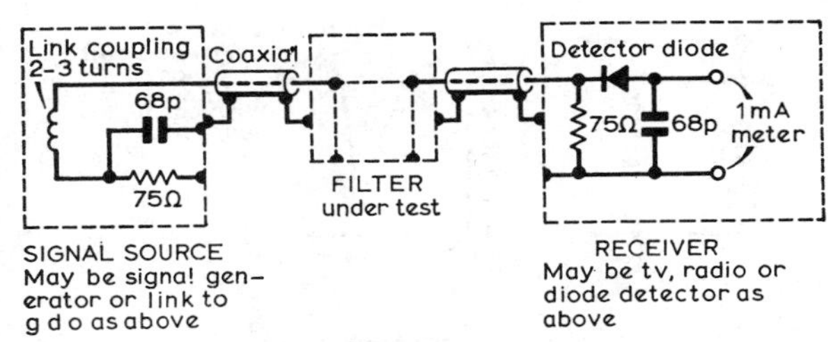

Fig 17.33. Arrangement of test equipment for filter alignment.

The L/C ratio of traps may be important. Parallel-tuned traps give the narrowest bandwidth and least attenuation with a low L/C ratio; series-tuned traps give the narrowest bandwidth and least attenuation with a high L/C ratio. Therefore, when it is required to trap, for example, 70MHz without altering 60MHz, a high L/C ratio series-tuned trap may be used, but the rejection at 70MHz will only be moderate. In this case it is possible to obtain greater attenuation by cascading traps, as shown in **Fig 17.34**. When the L/C ratios of each tuned circuit are chosen properly it will be found that a correctly designed stop-filter has been constructed, although it is necessary to ensure complete isolation between each tuned circuit to achieve maximum attenuation. When the trap is so wide that it also attenuates a wanted frequency on one side of the skirt (such as a trap for 35MHz in the transmitter feeder attenuating 28MHz),

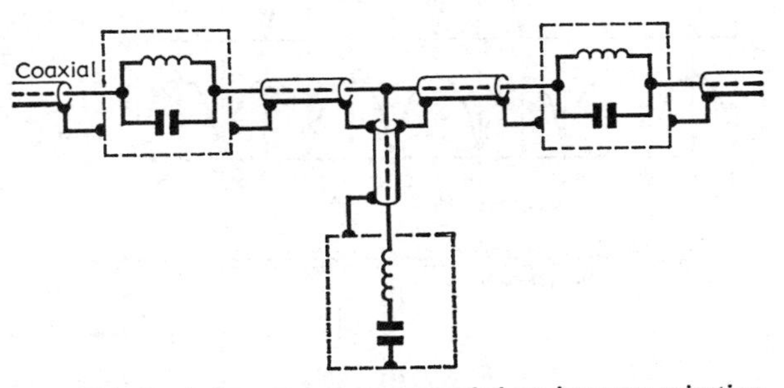

Fig 17.34. Single-frequency traps cascaded to increase rejection. They may be enclosed in one case to form a conventional stop-filter, provided each section is screened.

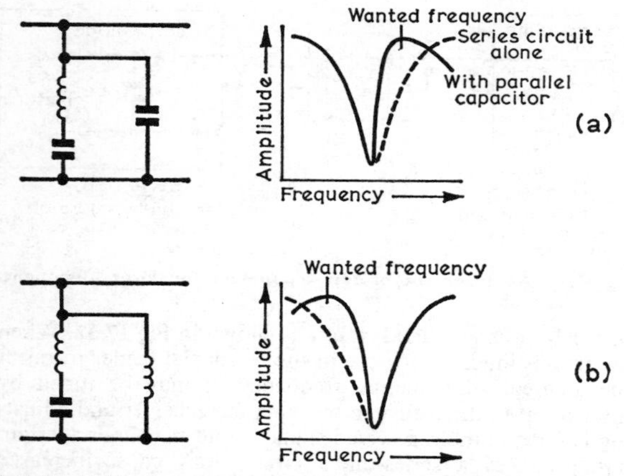

Fig 17.35. (a) The use of a capacitor across a trap or λ/4 stub sharpens the response on the high-frequency side. (b) The use of inductance across a trap or λ/4 stub sharpens the response on the low-frequency side.

this may be corrected by adding a suitable reactance across the trap as shown in **Fig 17.35**. When the wanted frequency is lower than the trap frequency an inductor must be added, and when the wanted frequency is higher a capacitor must be added.

Coaxial Stubs

A coaxial stub may be used as a trap by connecting it across a feeder. Its operation is based on the fact that a quarter-wave long stub with one end open-circuit presents a very low impedance at the other end, which is also true for odd multiples of the quarter-wave length. Conversely, when one end is shorted the other end presents a high impedance. When the stub is a half-wavelength long, any impedance at one end is reflected at the other end, which is also true for all multiples of the half-wave length. **Fig 17.36** shows these properties for 145MHz coaxial stubs.

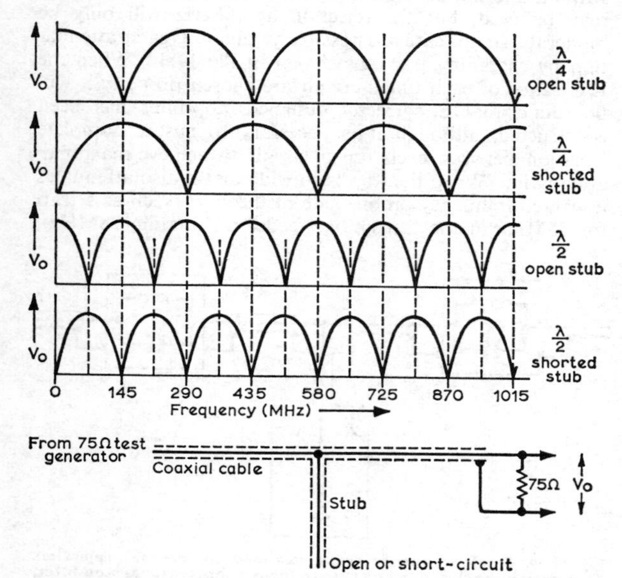

Fig 17.36. Attenuation properties of various 145MHz stubs.

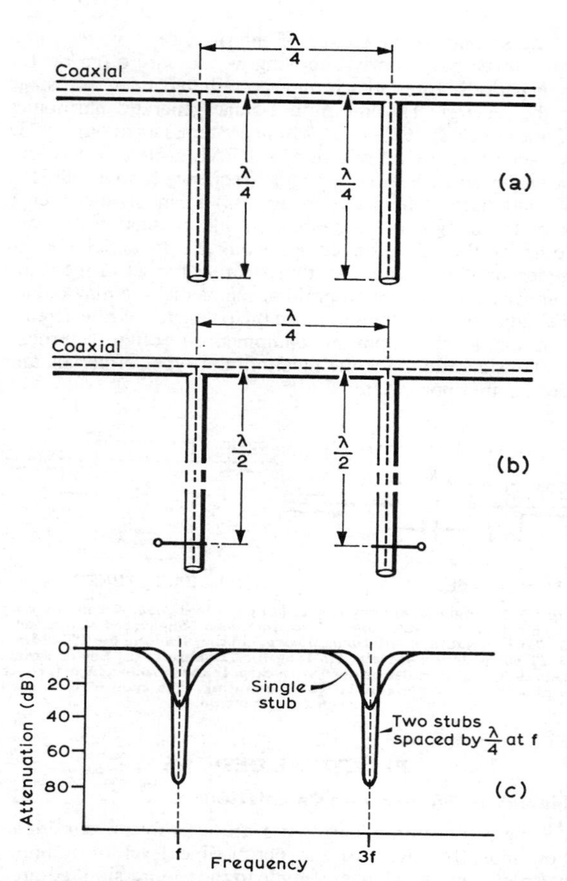

Fig 17.37. Cascading two stubs to achieve a notch of about 80dB at the centre frequency. In (a) open-circuit quarter-wave stubs are used, in (b) short-circuit half-wave stubs with a movable "pin" shorting device are used. A graph of the likely attenuation of a single and double stub notch filter is given in (c).

A quarter-wavelength open-circuit stub connected across a matched 75Ω feeder gives typically 30dB of attenuation, depending on the stub cable quality. This is comparable to a series-tuned trap of similar bandwidth.

A coaxial stub has the advantage of simple construction. No box is needed, and the length for a specific frequency is predictable—between 65 per cent of a "free-space" quarter-wave length for solid dielectric cable, and 85 per cent for semi-airspaced cable. The stub is also easily tuned—it is first made oversize, and gradually trimmed back until maximum attenuation is achieved. It may then be taped to the main feeder.

A single stub has fairly broad bandwidth, so to increase the attenuation and narrow the bandwidth, stubs may be cascaded like L-C traps, as shown in **Fig 17.37.** This form of filter is easy to construct, since there is no need for elaborate screening between each section, unlike the filter shown in Fig 17.34.

It will be seen from Fig 17.36 that the stub also attenuates odd multiples of the fundamental frequency, and in practice

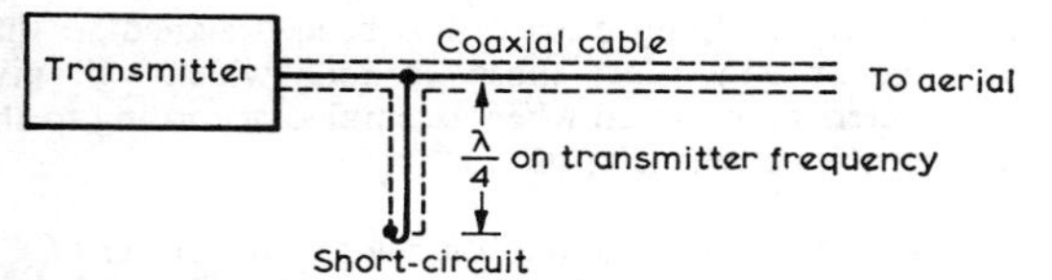

Fig 17.38. A short-circuit stub used between transmitter and aerial to trap even harmonics.

these overtone resonances may place some limitation on the usefulness of stubs. For example, a stub inserted in a tv feeder and cut to attenuate 145MHz will also attenuate 725MHz, so such a stub could not be used if the tv signals were around the UK channel 52. However, it is possible to shift the overtone resonances. If the stub is cut shorter than an electrical quarter-wavelength it may be tuned to resonance by a trimmer capacitor across the open end, in which case the overtone resonances move higher in frequency. Alternatively, the stub may be cut longer, and the trimmer inserted in series with the stub at the connection to the main feeder. In this case the overtone resonances will move lower in frequency. Similar procedures may be carried out with an inductor.

When the problems of multiple resonances are considered, it is doubtful whether single stubs have any real advantage over L-C circuits for single frequency applications. Note however that in some circumstances these multiple resonances may be used to advantage. A short-circuit quarter-wave stub for 145MHz will have maximum attenuation on even multiples of 145MHz (ie 290MHz, 580MHz, 870MHz, etc) as shown in Fig 17.36. This will provide useful harmonic attenuation when connected across the output of a 2m transmitter (see **Fig 17.38**).

About 14in of semi-airspaced coaxial cable is required for a stub on 2m. Connect it across the receiver and cut the stub for maximum attenuation (on a steady signal source) with its end open-circuit. Then short-circuit the end without altering the length of the stub, so the insertion loss at 145MHz should now be virtually zero. Then connect the stub across the transmitter output, again taking care not to alter its length.

A combined coaxial and L-C trap filter to filter out 70cm transmissions from tv receivers without affecting the uhf band is shown in **Fig 17.39**. The quarter-wavelength of coaxial cable is used to isolate the two tuned circuits from each other. The two coils should be screened from each other to avoid inductive coupling, and the tuning capacitors adjusted to give maximum rejection to the 70cm signal.

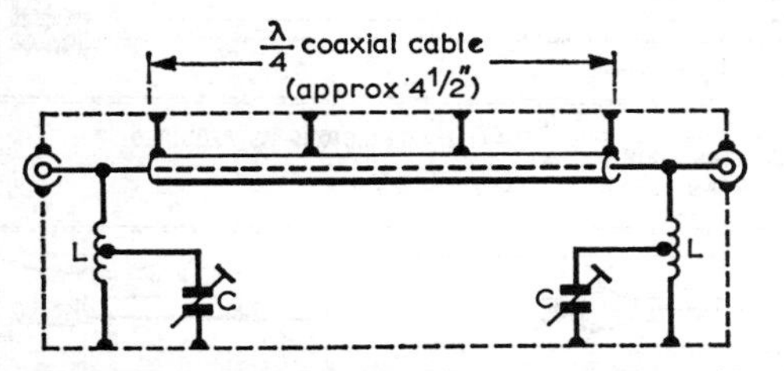

Fig 17.39. A 70cm trap for use with uhf tv sets. L is 8 turns 20swg ¼in inside dia, stretched to 0·6in long, centre-tapped; C is a 2·5–6pF low-inductance trimmer. The two tuned circuits must be screened from each other or kept well apart, and the braid of the cable must be soldered to the box at several points as shown. All leads must be very short.

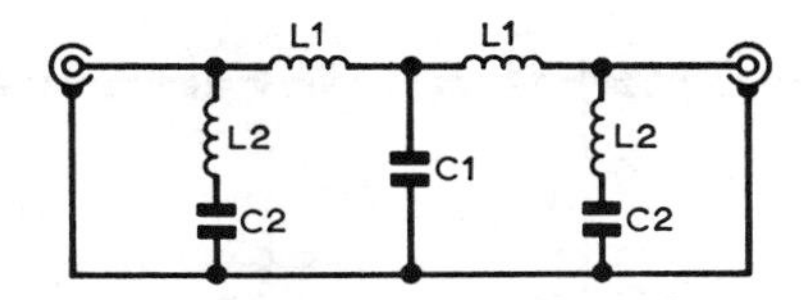

Chan No	C1 pF	C2 pF	L1 µH	L1 Turns	L1 Wdg length	L2 µH	L2 Turns	L2 Wdg length
1	110	30·2	0·55	9	7/8"	0·44	8	7/8"
2	110	38·2	0·56	9	7/8"	0·26	6	3/4"
3 & 4	103	38·3	0·58	9	7/8"	0·2	5	3/4"
5	93·5	34·8	0·52	9	7/8"	0·18	5	1"
Self-supporting coils					Internal diameter 1/2"			
Lead length 1"					16swg copper wire			

Fig 17.40. An easy-to-construct and align low-pass filter giving high attenuation on a specific frequency.

For use with 70cm tv transmission the two traps may be slightly staggered to give even rejection over the required bandwidth.

Low-pass Filters

The designs of low-pass filters are numerous. For hf amateur bands a typical filter has a cut-off frequency of 30MHz, with a very high attenuation of the specific frequency which may cause tvi to Band I tv. This is not the tv channel frequency, but that of the harmonic of the amateur signal. Only four variations of the basic design are required to give best results on Band I, depending on the actual frequency allocation. It is necessary to decide which harmonic is likely to cause the most trouble; Fig 17.17 gives the harmonic relationships involved, while the filter design is shown in **Fig 17.40**. The channel numbers refer to the UK Band I allocations. Capacitor values may be made up from paralleled capacitors to give approximately the correct value. All inductors should be screened from each other, kept away from surrounding metalwork by at least one coil diameter,

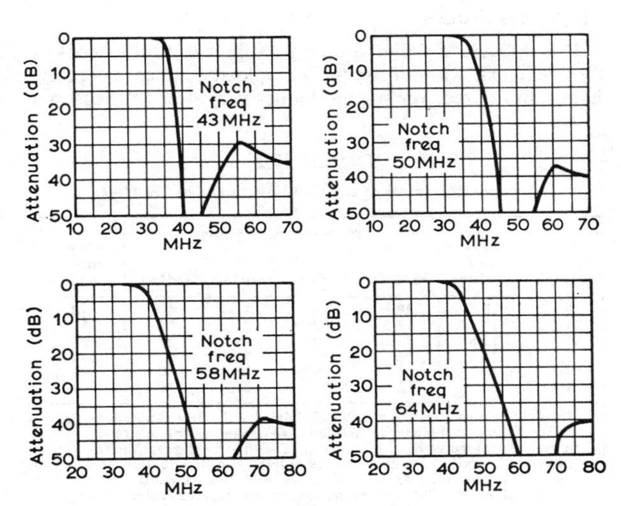

Fig 17.41. Performance curves of the filters shown in Fig 17.40.

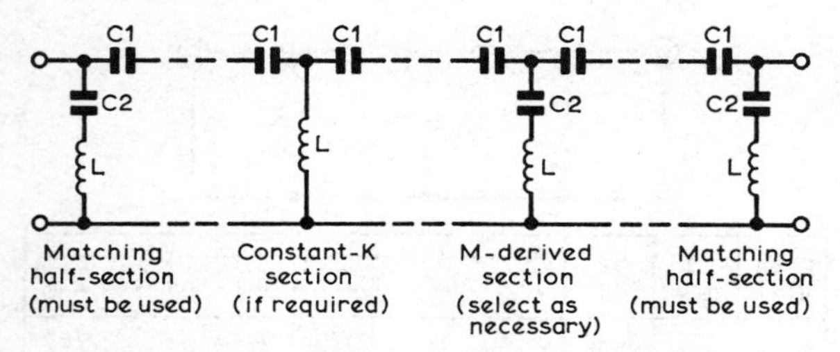

	Half-section	Constant K	M-derived 28MHz	21MHz	14MHz	7MHz
C1	98·5p	54p	98·5p	74p	64p	60p
C2	53p	—	107p	300p	700p	3000p
L	0·55μH	0·16μH	0·28μH	0·21μH	0·18μH	0·17μH
Turns	10	3½	5½	4	4	4
Internal dia ½"....wound on whitworth bolt with 16swg wire						

Fig 17.42. Building blocks for a 75Ω high-pass filter.

and all leads should be kept as short as possible. All capacitors should be of the low-inductance type, have short leads, and be rated to at least 750V.

Alignment may be carried out using a gdo as follows:

First short the series tuned circuits L2-C2 (conveniently with a short-circuit across the input and output terminals). Couple the gdo loosely to L2, and stretch or compress L2 so that the tuned circuit L2-C2 is resonant at the required frequency of maximum rejection (ie 42·5, 49·3, 58 or 63·6MHz). For more uniform rejection of the second harmonic of 28MHz the tuning of the 58MHz circuits may be staggered with one L2-C2 on about 56·5MHz, and the other on about 58·5MHz.

Now remove the short-circuits, and short C1 with a stout, short lead. Couple the gdo loosely into L1 or L2, and stretch or compress L1 so that the tuned circuits L1–L2–C2 are tuned to 28·5MHz.

Finally, remove the short from C1: the filter is now ready for use.

High-pass Filters

It is difficult to design a high-pass filter to fit the needs of all amateurs without undue complexity, since operating habits and tv receiver susceptibilities vary. For this reason a series of 75Ω "building blocks" has been designed so that each amateur can select those sections which will give the characteristic required when assembled according to the following instructions (**Fig 17.42**).

TABLE 17.2

Attenuation in dB of the building blocks of Fig 17.42.

Band (MHz)	Pair of matching half sections	M-derived sections 28MHz	21MHz	14MHz	7MHz	Con-stant-K section
28	28 *	28 *	15	12	10	10
21	17·5	17·5	40	23	19	19
14	13·5	13·5	23	40	27	27
7	12	12	19	30	40	40
3·5	12	12	19	28	40	40
1·6	12	12	19	28	40	40

*Due to the width of the 28–29·7MHz band, the attenuation falls to this figure at the edges if sections are peaked at the band centre. It may therefore be desirable to stagger-tune the end sections (or an additional section) to 28·5 and 29·5MHz. The attenuation at the series-resonant frequency in all cases will exceed 60dB, but the figures of 28 or 40dB represent the minimum attenuation at any point in the band if the circuit is resonated at mid-band.

From the bands used and the television receiver i.f. or image frequencies, decide which frequencies need maximum attenuation, some attenuation, and those which can be neglected.

For each band requiring maximum attenuation pick the corresponding *m*-derived section. Note that for 28MHz the matching half sections, which must be used, will contribute high attenuation (see **Table 17.2**).

From this table add up the attenuation given by the chosen *m*-derived sections plus the attenuation given by the matching half sections. Subtract 6dB for losses, and consider whether the attenuation given on each band is adequate. If not, the addition of one or two constant-*k* sections will probably help.

Draw out the circuit of the resulting filter. The chosen sections *must* be placed *between* the matching half sections. Pairs of capacitors in series may be replaced with their equivalent single value.

Build the filter. The series-tuned circuits may be tuned by temporarily shorting them, and resonating with a gdo. The filter may be constructed in a tin box (eg tobacco tin), and care must be taken to mount the coils at 90° to each other, or use screening to avoid stray coupling.

Bandpass Filters

Bandpass filters may be constructed for use on hf by using conventional double-tuned circuits (as described in Chapter 1—*Principles*). In practice, it is usually better to use a trap in a tv feeder, tuned to reject the interfering frequency, rather than a bandpass filter tuned to the wanted tv channel.

Bandpass filters may be constructed by cascading high-pass and low-pass filters with slightly staggered cut-off frequencies. Both filters must be *m*-derived to improve the matching but even so the resulting bandpass filter may introduce some mismatch into the feeder.

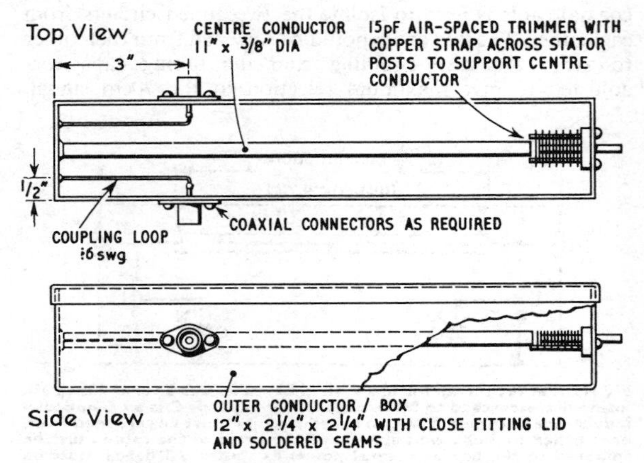

Fig 17.43. Construction of a 145MHz coaxial line filter.

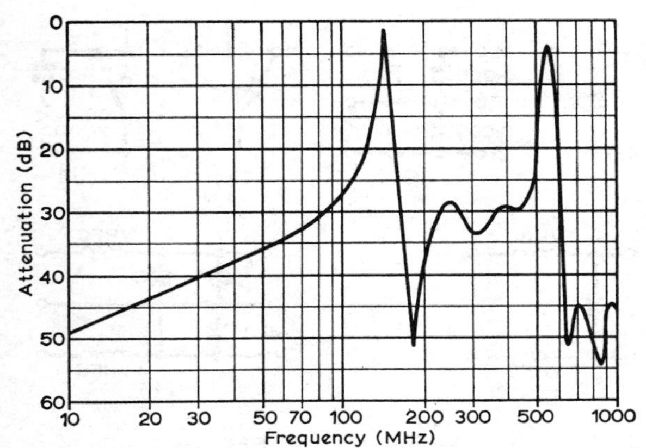

Fig 17.44. Response curve of the 145MHz coaxial line filter of Fig 17.43.

Coaxial Line Filters

As previously described, a bandpass filter is preferred to a low-pass filter for vhf transmitter use (see Fig 17.7). At vhf it becomes impractical to design a filter using conventional coils and capacitors, but fortunately the coaxial line technique allows filters to be made relatively simple for these frequencies. **Fig 17.43** shows a design for 2m use. This type of filter suffers from overtone resonances in the same way as coaxial stub filters, so end-capacitance is used to shift these resonances lower in frequency, and thus give adequate rejection to possible transmitter harmonics (see **Fig 17.44**). Two filters made to different dimensions may be coupled in series, spaced by a quarter-wavelength of coaxial cable to give good isolation. Because the overtone resonances will not coincide, the filters will then give much improved attenuation to all frequencies outside the required passband.

Braid-breaking Filters

Important. Braid-breaking filters are designed to present a large impedance to rf currents or voltages present on the braid of coaxial feeders. The performance of such filters will be adversely affected if the input and output leads are placed in close proximity to each other, thus capacitively bypassing the filter. For the same reason the use of metal boxes should be avoided.

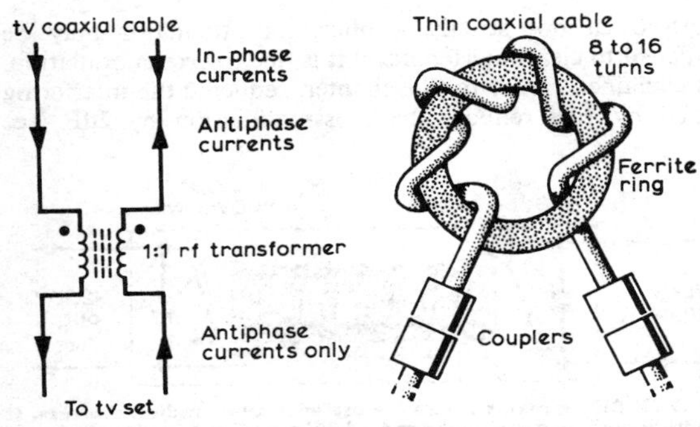

Fig 17.45. A coaxial rf choke for vhf and uhf tv/radio. In-phase currents are blocked while antiphase currents pass unhindered.

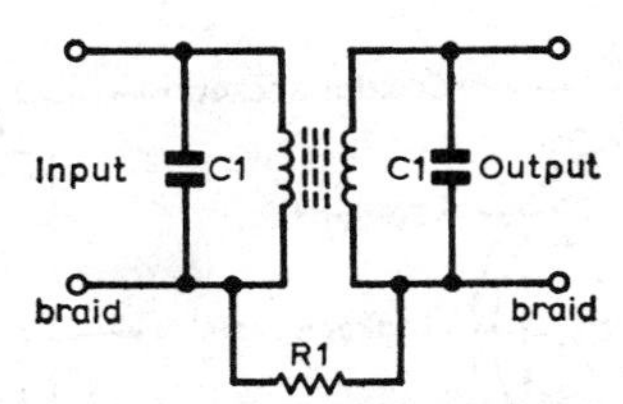

Fig 17.46. A 1 : 1 transformer braid-breaker for vhf tv. The capacitors across the input and output may be added to reduce loss at 216MHz to an absolute minimum. Typical values are 0–4·7pF. R1 is about 1MΩ.

Coaxial choke for vhf and uhf tv/radio (see **Fig 17.45**). The tv feeder is made to act as an rf choke to all but the aerial signals by winding it around a material of high magnetic permeability (conveniently on large ferrite ring cores, such as Mullard FX1588 or Neosid 4324R/1). A ferrite rod (transistor radio type) may also be used, wound with many turns of small-diameter wire.

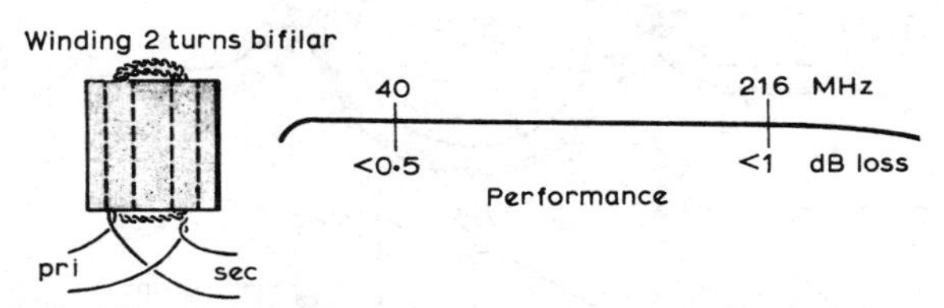

Fig 17.47. A braid-breaker using purpose-made baluns such as Neosid type 1050/1 or /2, material F14 (μ = 200); or Mullard type FX2249, material B2 (μ = 200).

1 : 1 rf transformers for vhf tv/radio. Fig 17.46 gives the circuit of a 1 : 1 rf transformer. The resistor provides a leakage path for static charges which would otherwise build up on the aerial system. The primary and secondary are closely coupled by using bifilar windings, ie the transformer is wound with a twisted pair of wires of suitable gauge. Use four to six twists per inch, depending on wire size. The core may be a purpose-made vhf balun core, such as Mullard FX2249 or Neosid 1050/1 or /2, a pair of ferrite beads, or i.f. transformer cores (with hexagonal holes) laid side by side and wound together as **Figs 17.47** and **17.48**. A suggested construction is given in **Fig 17.49**.

Two Faraday loops of 2 to 3in diameter may be used, taped together, ensuring that the screens do not touch. However, the loss is fairly high (**Fig 17.50**).

UHF braid-breaker/high-pass filter. Fig 17.51 gives the circuit of a uhf high-pass filter and braid-breaker combined.

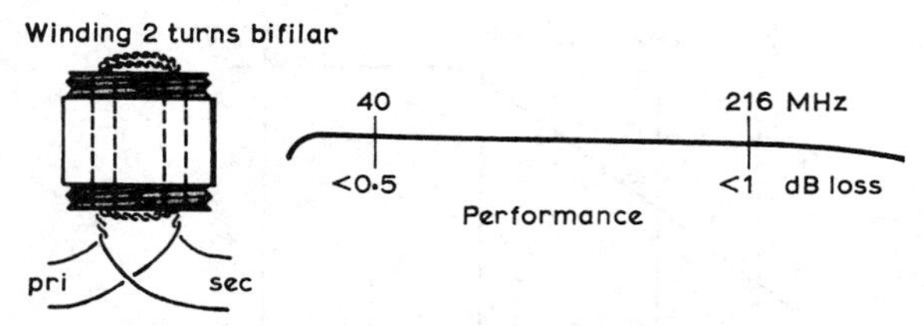

Fig 17.48. Alternative method using a pair of ferrite i.f. cores with hexagonal trimming holes, or ferrite tubes of similar size etc.

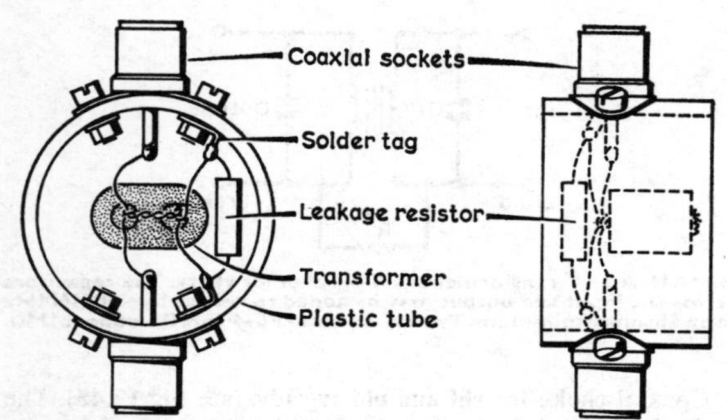

Fig 17.49. Suggested construction of a braid-breaker for vhf tv (40-216MHz).

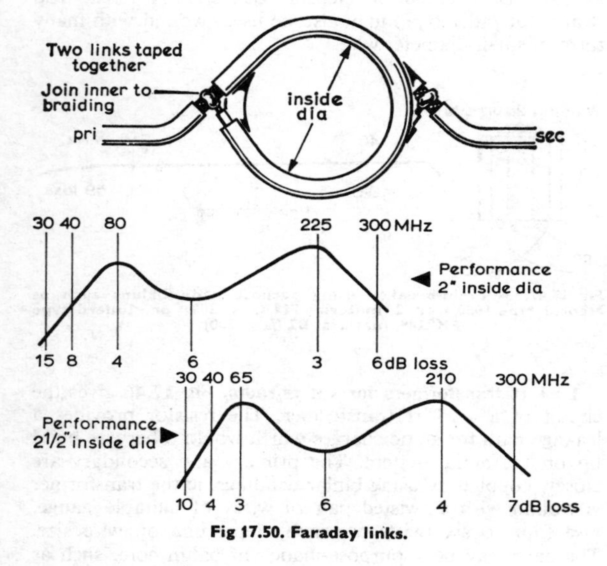

Fig 17.50. Faraday links.

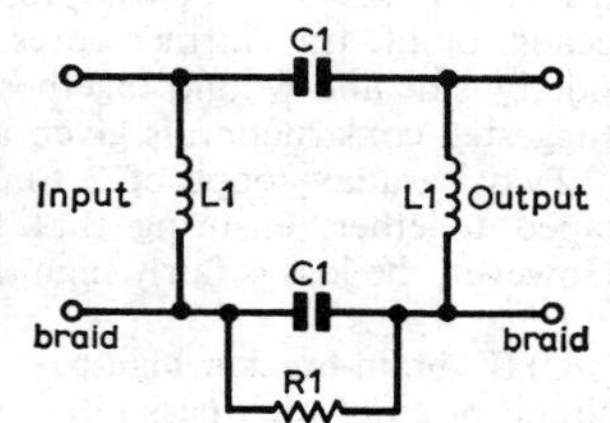

Fig 17.51. Circuit of a uhf tv braid-breaker. L1 is 4 turns 24swg ¼in inside dia, C1 is 5·6pF small disc or tubular. R1 is about 1MΩ. Construction must conform to good uhf practice to minimize losses, but values are not critical. Its performance is shown below.

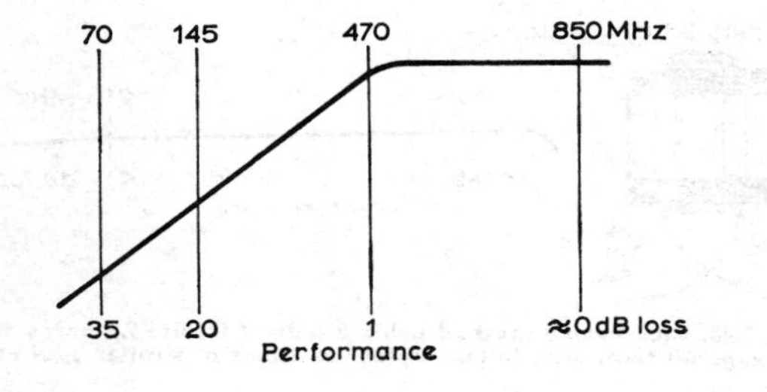

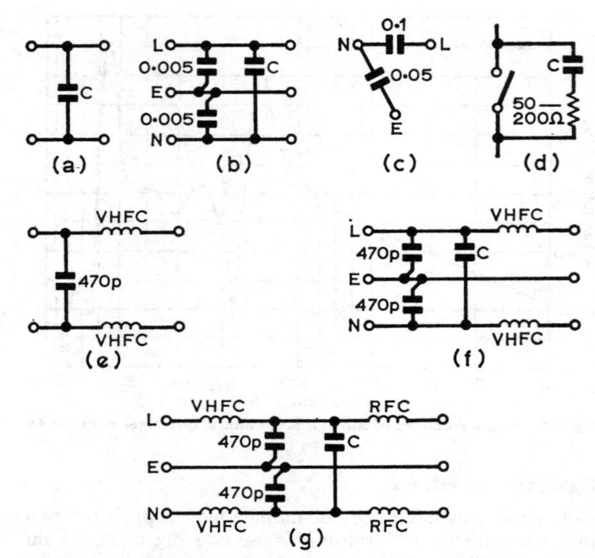

Fig 17.52. Some common forms of interference suppression filters (from *Radio Servicing Pocket Book*). (a) For two-core cable appliances, (b) for three-core appliances, (c) for three-pin sockets, (d) for thermostats and thyristor light dimmers etc. Types (e) and (f) are vhf types intended for use at tv frequencies. Type (g) is a combined mf/hf/vhf filter. The value of C may vary between 0·01 and 0·5μF. On type (c) the values given are the largest permissible. All capacitors must be rated for ac. L—line, N—neutral, E—earth.

Its performance is also given in Fig 17.51, and it may be constructed along the lines of the vhf braid-breaker given in Fig 17.49.

Mains Filters

Mains filters based on the low-pass filter design are shown in **Fig 17.52.** (Many are available as commercial units.) **Fig 17.53** shows a method of constructing a simple mains filter for use with tv, radio etc. Where three-core mains leads are used the capacitors may be made from two separate capacitors in series with the centre tap to the earth lead. Filters should, if possible, be contained in a screened box and placed close to the devices being filtered. Capacitors should always be adequately rated, and precautions taken to eliminate the risk of shock.

Attenuators

Although not strictly a filter, an attenuator may be sufficient to clear interference if it is due to crossmodulation. As explained earlier in this chapter, reducing the interfering signal by 1dB reduces the crossmodulation by 2dB (see

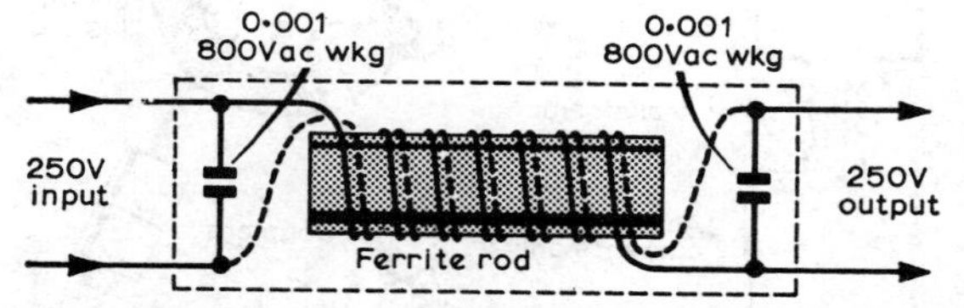

Fig 17.53. Bifilar mains choke for use with tv or radio receivers. It can be wound on a transistor radio ferrite rod or a ferrite ring core. For example, about 30 double turns of 22swg pvc-covered close-wound on a ⅜in dia rod about 3in long (not critical).

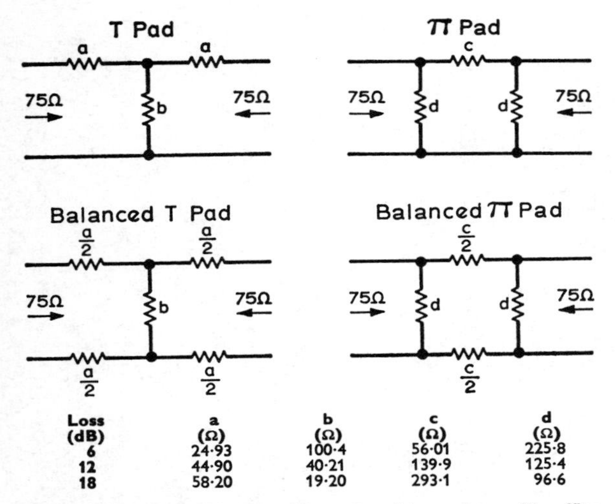

Loss (dB)	a (Ω)	b (Ω)	c (Ω)	d (Ω)
6	24·93	100·4	56·01	225·8
12	44·90	40·21	139·9	125·4
18	58·20	19·20	293·1	96·6

Fig 17.54. Simple attenuators. Nearest resistor values will suffice.

Crossmodulation). Commercial in-line attenuators are available. **Fig 17.54** gives details of useful attenuators, but it should be borne in mind that these can only be used when adequate wanted signal is available from the aerial.

Characteristic Impedance of Filters

The circuits of filters given are for 75Ω systems. For any other impedance Z_0, capacitance values should be multiplied by $\frac{75}{Z_0}$, and inductances and resistances by $\frac{Z_0}{75}$.

The Effect of Mis-termination on Filters

The input impedance of a tv or radio set is usually not a particularly good match to the feeder on the channel to which it is tuned, and is invariably a very bad reactive mismatch on the interfering frequency. The aerial itself is also a bad reactive mismatch on the interfering frequency, which means the filter is therefore not correctly terminated on either side. In unfortunate circumstances, rather than acting as a filter it may act as a matching section and make the interference worse, though the figures published for the filter indicate large attenuation—such figures are given for a correctly terminated filter. Such effects cannot occur if the filter design specifically open-circuits or short-circuits the interfering frequency, so that it is rejected whatever the source or load impedance may be. Such a filter in conjunction with a braid-breaker should always be effective *provided that* the interference is entering *via the aerial connection.*

If the tv/radio signal level is sufficient, an attenuator may be used between the filter and the tv/radio set. A 6dB pad will provide a reasonable match for the filter even if the tv or radio set is a complete mismatch, thus enabling a good approximation to the correct filter performance to be obtained.

Baluns

These take two basic forms. One type acts as a choke to rf currents on the coaxial braiding, and the other acts as a transformer which isolates rf from the braiding. A simple balun may be made by coiling a length of feeder, or by wrapping it round a ferrite ring as in Fig 17.45. This type of balun will operate over a wide range of frequencies.

Further information on baluns may be found in Chapters 12 and 13.

CHAPTER 18

MEASUREMENTS

CORRECT operation of amateur radio equipment involves measurements to ensure optimum performance, to comply with the terms of the amateur transmitting licence and to avoid interference to other users. The purpose of these measurements is to give the operator information regarding the conditions under which his equipment is functioning. Basically, they are concerned with voltage, current and frequency. For example, in even the simplest transmitter it is necessary to know the grid drive to the various stages (current measurements), the input power to the pa (current and voltage measurements) and the frequency of the radiated signal.

If apparatus is to be designed and constructed to give reliable and optimum performance, a considerable range of measuring equipment may be desirable. While it is possible to make do with only one or two simple items of test gear, this chapter describes equipment for dc, ac and rf voltage, current and power measurements, wavemeters and frequency standards and a number of more sophisticated items for both transmitter and receiver measurements.

DC MEASUREMENTS

The basis of most instruments for the measurement of voltage, current and resistance is the moving coil meter, in which a coil of wire, generally wound on a rectangular former, is mounted on pivots in the field of a permanent magnet (**Fig 18.1**). The coil experiences a torque proportional to (i) the current flowing through it and (ii) the strength of the field of the permanent magnet. Current is fed to the coil through two hairsprings mounted near to each end of the spindle. These springs also serve to return the pointer to the zero position (on the left hand side of its travel in standard meters) when the current ceases to flow. Provision for adjusting the position of the pointer is made by a zero adjuster accessible from the front of the instrument.

Since the movement of the coil and its associated pointer is proportional to the field of the magnet and that of the current being measured, the scale is linear. A minor disadvantage is that the instrument can only be used on dc, but it can be adapted to measure ac with a suitable rectifier.

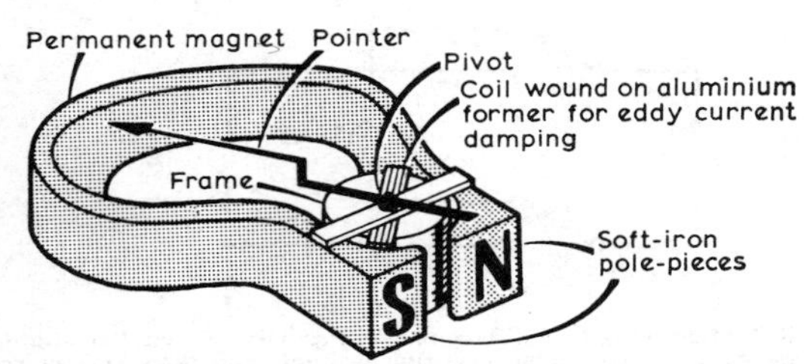

Fig 18.1. Construction of the moving coil meter.

It is usual to damp the coil system (ie prevent it swinging freely after a change of current), a common method being to wind the coil on an aluminium former, which then acts as a short-circuited single-turn coil in which the eddy currents serve to oppose the movement. The degree of eddy current damping is also dependent on the external resistance across the terminals of the moving coil and is greatest when the resistance is low. It is a wise precaution to protect sensitive instruments not in use by short circuiting the terminals.

When a moving coil meter is used for measuring current it is called an *ammeter*, *milliammeter* or *microammeter* depending on its full scale deflection (fsd). A dc voltmeter is a milliammeter or microammeter equipped with a voltage dropping (series) resistor.

Milliammeters

Milliammeters and microammeters are commonly manufactured with basic full scale deflections of 50μA, 100μA, 500μA, 1mA, 5mA and 10mA. For higher current ranges, a shunt resistor is connected across the meter (**Fig 18.2(a)**). The value of the shunt may be obtained from the formula

$$R_s = \frac{R_m}{n - 1}$$

where R_s is the resistance of the shunt, R_m is the resistance of the meter and n is the scale multiplying factor.

For example, if a milliammeter of 10Ω resistance and a fsd of 1mA is to be used to measure 100mA, a shunt must be provided to carry the excess current, that is 100 − 1 = 99mA. Thus the required resistance of the shunt is

$$R_s = \frac{10}{100 - 1} = \frac{10}{99} = 0{\cdot}101\Omega$$

When moving coil meters are used in circuits where high voltages are present (ie in pa anode circuits) care must be taken to avoid accidental electric shock from the zero adjuster, which may be live. If the instrument is of the flush mounting type the front of the meter may be covered with a piece of clear plastic about $\frac{1}{16}$in thick.

Voltmeters

A milliammeter may be used to read dc voltages by connecting a resistor, termed a *multiplier*, in series with it (**Fig 18.2(b)**). The value of the multiplier depends on the fsd of the meter and may be calculated from Ohm's Law.

For low voltage ranges, the value of the multiplier can be obtained from

$$R_s = R_m \left(\frac{V}{V_m} - 1\right)$$

where R_s is the resistance of the multiplier, R_m the resistance of the meter, V the required voltage, and V_m the voltage across the meter (this can be determined by applying Ohm's Law to the resistance of the meter and current flowing through it). In practice, however, the resistance of the meter can be ignored and the formula simplified to

$$R_s = \frac{1{,}000\,V}{I}$$

where V is the desired voltage range and I is the fsd of the meter in milliamps.

For example, assume a 0–5mA meter is to be used as the basis of a voltmeter to read 100V. Then

$$R_s = \frac{1{,}000 \times 100}{5} = 20{,}000\Omega$$

It is usual to describe the sensitivity of a voltmeter in ohms per volt; in the example considered above, the meter reading 100V for an fsd of 5mA would be said to have a sensitivity of 200Ω/V. In practice, such a sensitivity is too low for accurate measurements in radio and electronic equipment, due to the current drawn, and a more sensitive meter would therefore be chosen as the basis of the instrument. The lowest sensitivity which can be considered satisfactory for amateur purposes is 1kΩ/V (requiring a 0–1mA meter) but sensitivities of 5, 10, 20 and 100kΩ/V are common for accurate work.

Whenever a voltage measurement is made on a circuit of appreciable resistance, the current taken by the voltmeter should be considered, to ensure that the operating conditions are not significantly altered by connecting it.

The accuracy of a voltmeter depends largely on the accuracy of the multipliers. Precision resistors are the most suitable but they are rather costly and may be replaced in home-built equipment by one per cent close tolerance carbon resistors of adequate power rating. Alternatively, a supplier may be persuaded to select suitable values from normal stock.

It should be noted that even high-stability resistors may change in value by several per cent with time; if an instrument is required to maintain high accuracy for an extended period only wire-wound resistors of suitable low temperature coefficient material should be used.

AC MEASUREMENTS

The moving coil meter can be adapted to measure ac by the addition of a suitable rectifier such as the copper oxide type. Such a meter will read ac at audio frequencies, indicating the average value, 0·636 of the peak value of a sine wave

Commercial ac instruments of the rectifier type are calibrated in rms values assuming a sine wave and hence read incorrectly if used on any other wave form.

In conjunction with a thermocouple, a moving coil instrument can be used to read alternating currents at both audio and radio frequencies. The thermocouple is a junction of two dissimilar metals which, when heated, produces a dc voltage. The junction is heated by the current to be measured passing through a heater to which it is attached. A disadvantage of the arrangement is that low current readings are rather severely compressed and it is necessary therefore to have several instruments if widely different currents are to be measured. Thermocouple instruments read true rms values irrespective of wave form. Unless specially designed, such instruments become less accurate as the frequency increases due to the effect of the shunt capacitance.

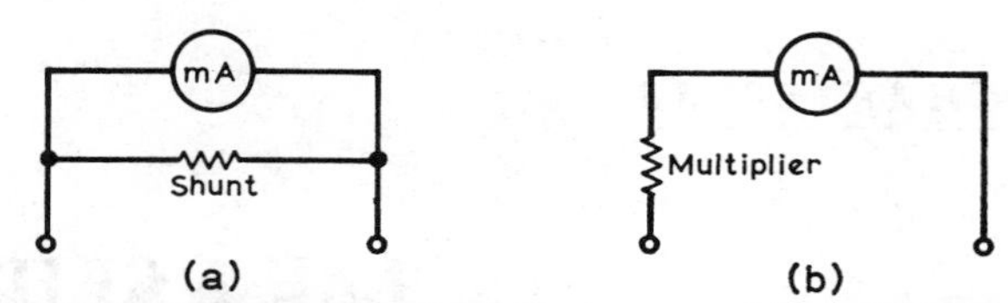

Fig 18.2. Extending the range of the mc meter: (a) to read higher current with a parallel shunt, (b) to measure voltages with a series resistor or multiplier.

The hot-wire ammeter is also designed to make use of the heating effect of a current passing through a wire. In this case, the wire is supported at both ends and kept under tension by a fibre attached to its centre and loaded by a spring. The current through the wire causes heating and expansion to take place so that a spindle, round which the fibre is wound, moves a pointer across a scale. The heating effect is independent of the type of current, and both ac and dc may therefore be measured. Despite its simplicity, the instrument is now rarely used as it is rather inaccurate.

The moving iron meter is useful for measurements on ac circuits (40–60Hz) where the current taken is unimportant. In these circumstances fixed iron and moving iron elements are mounted in the magnetic field at the centre of a coil through which the current to be measured is passing. Under these conditions, a mutual repulsion exists between the two parts, the moving iron being deflected, thus indicating on a scale the current measured.

The deflection obtained is approximately proportional to the square of the current or voltage measured (ie it is a square-law instrument). The scale divisions are thus close together at the low end of the scale and wide apart at the higher. This means that accurate readings cannot be taken at the low end, but the open scale at the top is sometimes useful. For example, a moving iron meter with an fsd of 250V would be excellent for reading small changes of voltage on a nominally 240V mains supply but would be useless for measuring voltages below about 50V.

AC voltage measurements of signal frequencies are generally concerned with peak values; electronic voltmeters or oscilloscopes are more suitable for such measurements.

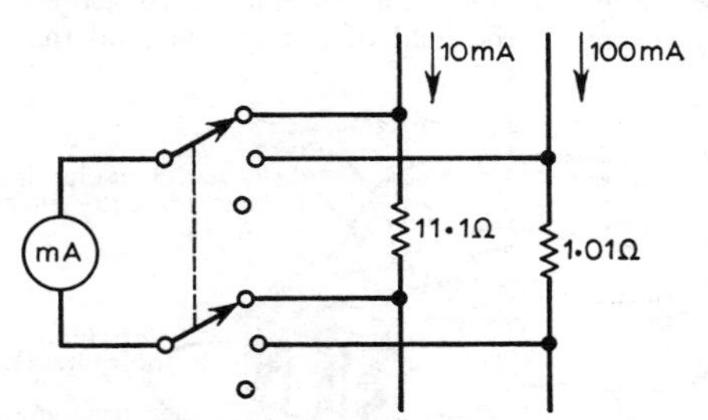

Fig 18.3. Measuring two different currents with a switched meter. In the case of the lower resistance shunt, the lead and switch resistance will cause errors.

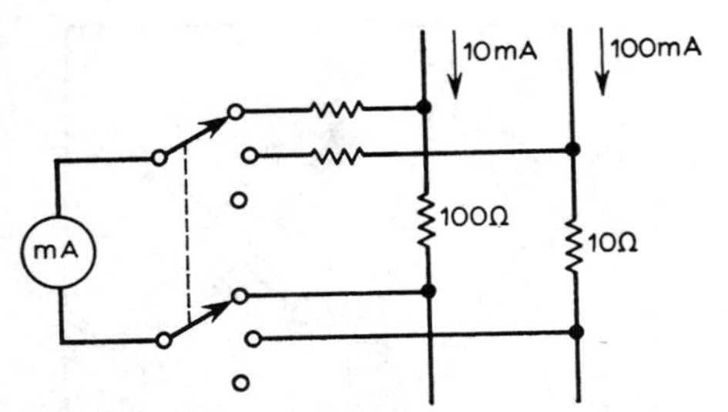

Fig 18.4. Using the milliammeter as a voltmeter allows shunt values to be large compared with possible lead and switch resistances.

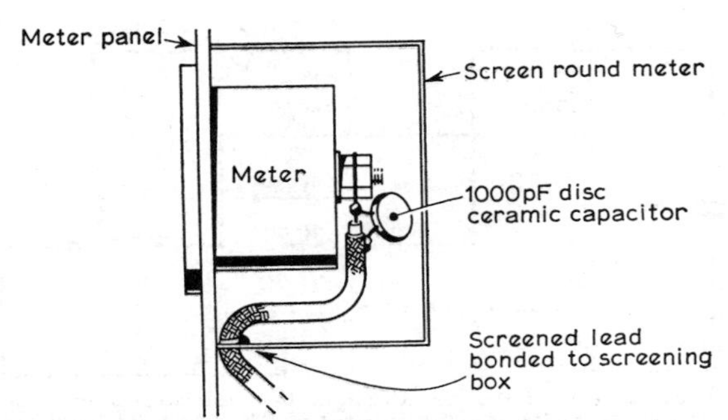

Fig 18.6. Screening and bypassing a meter in a transmitter.

METER SWITCHING

In order to save cost and panel space it is usual in complete equipments to switch a meter so that a number of different circuit currents can be measured. For example, in a transmitter it is desirable to be able to check the anode, screen and grid currents of the final valve amplifier as well as some of the supplies to earlier stages. Three alternative methods are possible.

First, suitable low-value resistors can be wired into the supply leads to act as meter shunts. For example, if the basic meter has an fsd of 1mA and a resistance of 100Ω, and if a range of 10mA fsd would be suitable to measure grid and screen currents and 100mA fsd for anode current, then shunts of 11·1Ω in the grid and screen dc paths and of 1·01Ω in the anode dc lead would produce the correct scales (see **Fig 18.3**). However, this method leads to considerable errors unless the resistance of the switch contacts and the connecting leads are very small compared with the resistance of the shunt.

A better method is to use shunts of 10 to 20 times the resistance of the meter. The meter will then read almost at its full sensitivity, subject to a reduction of 5–10 per cent which is often an acceptable error in this situation. If, for example, a 1mA 100Ω meter is taken and 11·1Ω is connected across its terminals the combination will have an fsd of 10mA and a resistance of almost 10Ω. Resistors of 200Ω in the circuits to be measured will only reduce this sensitivity by five per cent and the reading will be independent of practical values of lead or switch resistance. This method is only suitable when the currents to be measured are all of the same order.

A third method which permits a wide range of currents to be measured is to use a sensitive current meter as a voltmeter. If a 50μA fsd meter is made into a 1V fsd voltmeter by the addition of a multiplier of 20kΩ less the meter resistance, it may be switched across a 100Ω resistor carrying 10mA or a 10Ω resistor carrying 100mA and read full scale deflection for both currents (see **Fig 18.4**). By varying the series resistor and the shunt values, very wide current ranges can be measured.

Meter Protection

Meters are relatively expensive and are easily damaged if subjected to excessive current. Damage from this cause can be prevented by connecting two silicon diodes back to back across the meter terminals. These will have no effect until the voltage across the meter approaches some 400mV when they will conduct and effectively shunt it. A similar circuit is shown in **Fig 18.5** which includes an rf bypass capacitor.

When meters are built into transmitters, careful consideration should be given to shielding the meter movement from rf fields. Such fields can lead to heating of the hair springs and hence to permanent damage. Further, the hole in a metal panel can be a source of rf leakage. It is good practice to contain the meter in a screening can, to use shielded leads and to by-pass the meter to rf by a capacitor connected directly across its terminals (see also **Fig 18.6**).

MEASUREMENT OF RESISTANCE

The measurement of resistance is based on Ohm's Law, two common arrangements being shown in **Figs 18.7**(a) and (b). It will be seen that each circuit comprises a battery in series with a milliammeter, a resistor RV1 and an unknown resistor X. In practice, the terminals across which X is connected are first short circuited and RV1 adjusted until the meter reads full scale. When the resistor X is connected across the terminals, the meter reading will fall; calibration can therefore be carried out by connecting a number of resistors of known value across the terminals in turn and

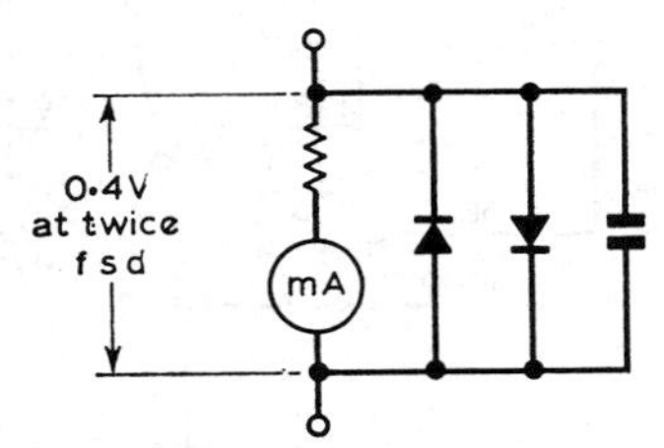

Fig 18.5. Voltage drop across meter increased by external resistance so that silicon protection diodes conduct at about twice full scale deflection. An rf bypass capacitor (say 1,000pF) should usually be included.

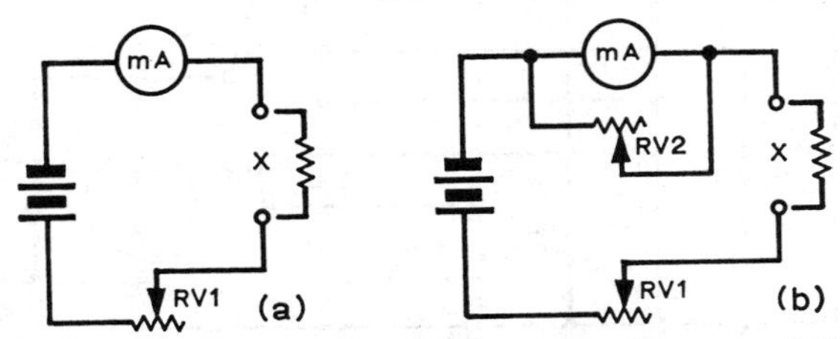

Fig 18.7. Measurement of resistance with a simple ohmmeter. In (b) RV2 is to adjust the sensitivity of the meter to compensate for a drop in battery voltage. In both circuits RV1 is for setting the meter to a full scale deflection with the terminals X short circuited.

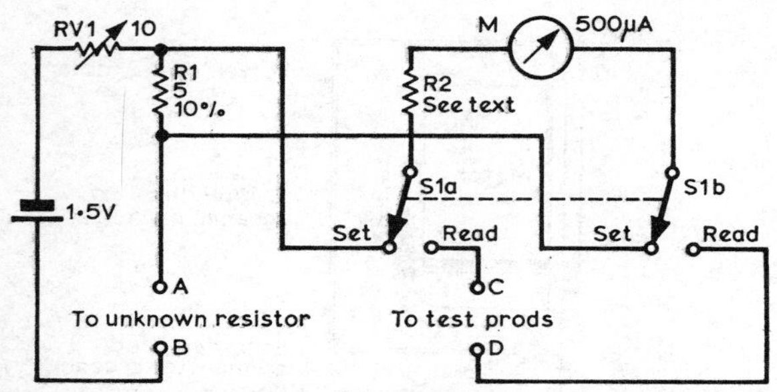

Fig 18.8. Circuit of the direct-reading low-resistance ohmmeter. R1 should be a high-stability resistor of ½W rating. R2 should be 1,000Ω less the internal resistance of the meter. The terminals A and B should be insulated and have large contact areas.

marking the scale accordingly. Alternatively, a graph relating meter current and resistance can be prepared.

An instrument designed for resistance measurements is termed an *ohmmeter* and may have several ranges. The resistance ranges in multimeters (see next section) are based on these simple circuits but are not very accurate below about 5Ω. Such inaccuracy is partly due to the resistance of the connecting leads.

A circuit arrangement suitable for measuring low resistances and differentiating between them and short circuits is shown in **Fig 18.8**. The instrument measures resistance up to 5Ω by comparing the voltage drops across a standard resistor and the unknown resistor when the same current flows through both. By selecting a 5Ω resistor as the standard (R1) and using a 0–500μA meter as a 0–0·5V voltmeter, it is possible to use the calibrated scale, each 100μA (0·1V) division representing 1Ω. The value of R2 should be 1,000Ω less the internal resistance of the meter.

The leads from A and B terminate in strong crocodile clips and those from C and D in sharp test prods.

The method of operation is as follows. With the meter switch at SET and the variable resistor RV1 at maximum, the unknown resistor is connected across terminals A and B.

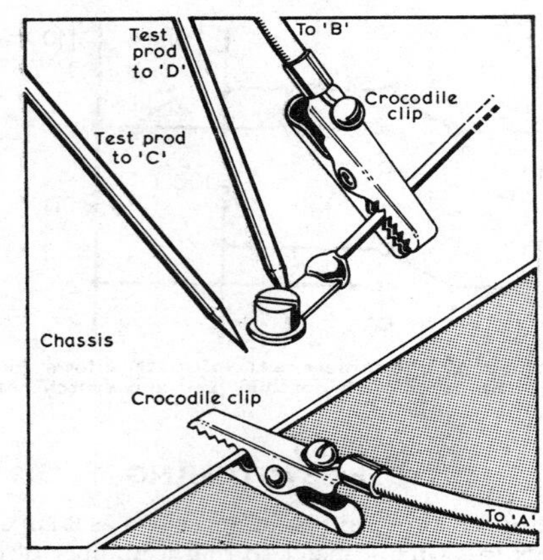

Fig 18.9. Use of the low-resistance ohmmeter of Fig 18.8.

RV1 is then adjusted so that the meter reads full scale. The switch is next moved to the READ position, the meter then indicating the resistance of the unknown resistor directly.

An example of the use of the instrument is illustrated in **Fig 18.9** where the resistance between a solder tag and chassis is to be found. Terminal A is connected to the chassis and B to the wire to the tag at any point; terminals C and D are connected across the joint to be measured and as close to it as possible. In this way, the same current flows through the standard resistor R1 as through the joint, and the resistance of the connecting leads to A and B does not affect the reading.

MULTI-RANGE METERS

A number of shunts and multipliers selected by a switch can be used in association with a single basic meter to form a multi-range meter (or multimeter), measuring current and voltage and, if containing an internal battery, resistance as well. A wide range of such instruments may be purchased at

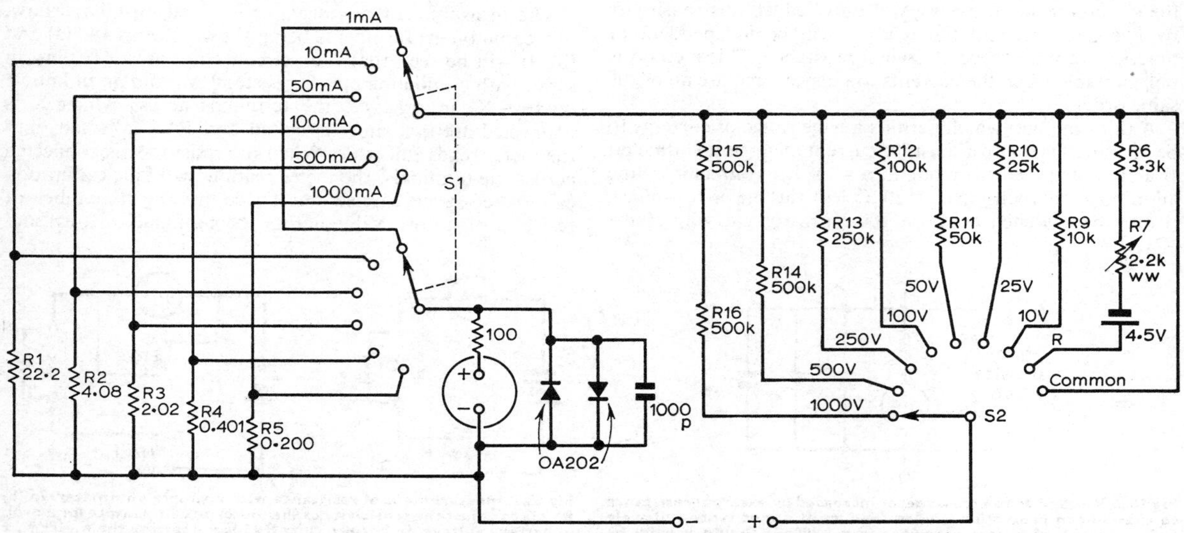

Fig 18.10. Basic multimeter unit reading current to 1A dc, voltages to 1,000V dc and resistance to 20kΩ.

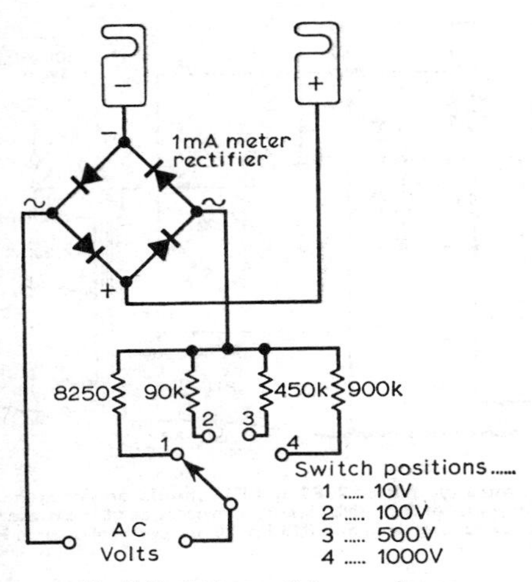

Fig 18.11. Add-on unit for ac voltages.

prices depending on the size of the scale and the sensitivity of the movement.

A circuit for a multi-range meter based on a 0–1mA meter of 100Ω resistance is shown in **Fig 18.10**. This gives six current ranges from 1mA to 1A fsd and dc voltage ranges from 10V to 1,000V fsd, as well as an ohms range readable up to 200kΩ. The meter is protected by silicon diodes. A separate add-on unit is also shown (**Fig 18.11**), containing a meter rectifier with associated switched series resistors to give four ac ranges up to 1,000V. When using the add-on unit, the main instrument is set to measure 1mA fsd and the add-on unit connected to its terminals by lugs.

The component values shown in Figs 18.10 and 18.11 are suitable for a meter having its resistance made up to 200Ω. The series resistors should be one per cent tin-oxide types, while the shunts may be made of Eureka wire.

ELECTRONIC INSTRUMENTS

The best-known instrument in this class is the valve voltmeter, although the field effect transistor is now frequently used instead of a thermionic valve. In conjunction with diode probes, useful ac measurements may be made at frequencies well above 100MHz.

DC VALVE VOLTMETER

In the simplest valve voltmeter, a dc voltage applied to the grid of a thermionic triode valve is measured by the resulting change in anode current. A practical circuit is shown in **Fig 18.12** in which a double triode is used as a differential amplifier. With the test prods shorted, the anode voltages of the two triodes may be made equal by the adjustment of RV1. The meter M1 will then read zero since there is no voltage across it. If a known dc voltage is then applied across the test prods, the anode currents of the two triodes will unbalance and M1 will deflect. By adjusting the series resistor RV2 the meter deflection may be calibrated in terms of the input voltage. The voltage applied at the grid must never be large enough to cause grid current to flow—higher voltages are measured by tapping the grid down a calibrated input potentiometer. An input probe consisting of a short length of coaxial cable and the input resistor R14 is desirable; this allows dc measurements in the presence of rf without significantly detuning the circuit under test. In operating, valve voltmeters should always be allowed to warm up for at least 10min before adjusting the zero control, and this adjustment should be checked from time to time.

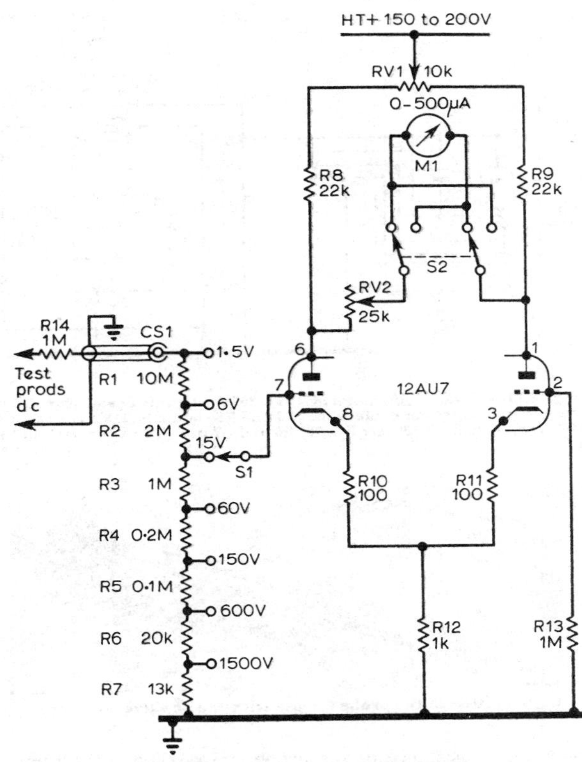

Fig 18.12. DC valve voltmeter. The meter M1 (0–500μA) should be calibrated 0–1·5 and 0–6V. Resistors R1-R7 should be one per cent tolerance; S1 should have ceramic insulators.

AC VALVE VOLTMETER

A more sophisticated valve voltmeter suitable for measuring af and rf voltage up to at least 50MHz and providing useful indications at 150MHz is illustrated in **Fig 18.13**. It consists of a diode detector producing a dc output voltage proportional to the peak value of the applied alternating voltage, followed by a differential dc amplifier using a double triode. Considerable dc negative feedback is applied to each triode which serves to linearize the input/output characteristic of the amplifier, and also improves the zero stability. Further improvement of zero stability is obtained by applying the output of a second diode to the other input of the differential amplifier. This type of amplifier responds to the difference between the potentials applied to its two inputs, and provided similar diodes are used for V1 and V3, changes in the contact potential of the diodes will not vary the zero setting appreciably.

The diode V1 and its associated components R1, C1 and C2 are built in the form of a shielded probe which is applied to the circuit under test when measurements are to be made.

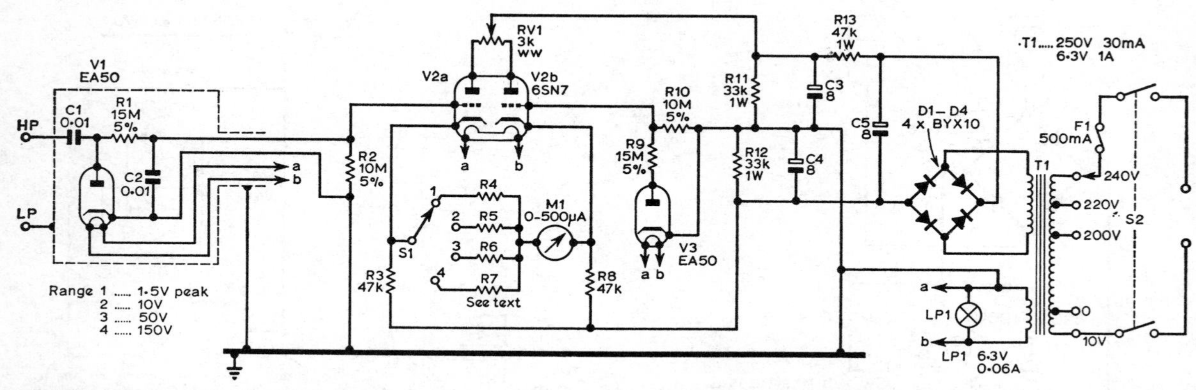

Fig 18.13. AC valve voltmeter. C1 is a 500V wkg mica capacitor. C2 is a 350V wkg mica type. R1, R2, R9 and R10 should be ½W high-stability resistors, five per cent tolerance; R3 and R8 should be ½W 10 per cent tolerance. The value of R4 is 300Ω less the resistance of the meter M1; R5 10kΩ; R6 47kΩ; R7 150kΩ; R4, R5, R6 and R7 are ¼W rating, R11 and R12 1W 10 per cent tolerance and R13 1W 20 per cent tolerance. Rectifier diodes are BYX10.

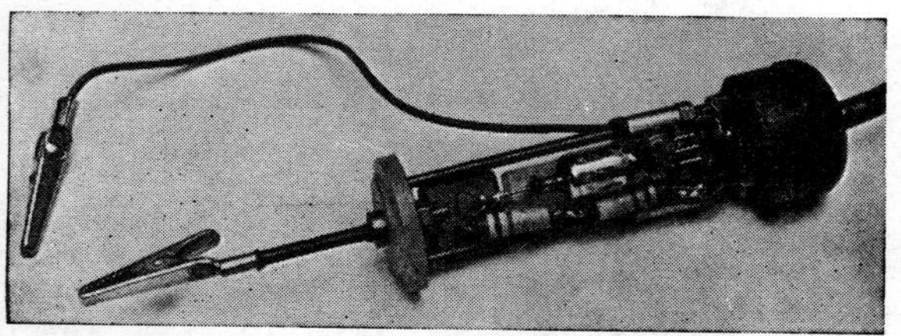

Construction of the probe for use with the ac valve voltmeter.

The input resistance of the probe is 3–4MΩ and the input capacitance approximately 8pF. The latter would be reduced by dispensing with the earthed shield but this would render the probe very sensitive to stray pickup, particularly when the instrument is used on its lowest range. A photograph of the probe is shown on this page. Connection of the probe to the remainder of the circuit is made by a four-way screened lead, terminated in a valve base which in turn fits a standard valveholder on the panel.

In use the high and low potential terminals of the probe are short circuited and the meter set to zero by means of the potentiometer RV1, with the range switch S1 at Range 1. The terminals are then separated and the high potential terminal applied to the point at which measurements are to be made, the low potential terminal being connected to chassis or earth. The instrument should be allowed to warm up for 5min or so before setting the zero.

Calibration may be carried out at mains frequency. Provided that the supply is sinusoidal the reading of a rectifier type voltmeter multiplied by 1·414 will give the peak value of the calibrating voltage. Exact adjustment of the fsd for each range is achieved by variation of the appropriate meter series resistances R4, R5, R6 and R7 by adding an additional resistor in series or parallel with the specified resistor. At least two separate scales on the meter will be required, as the 1·5V range is non-linear due to curvature of the diode characteristics at low input voltages.

A FET VOLTMETER

A solid state equivalent of the valve voltmeter is shown in **Fig 18.14** in which the triodes are replaced by field effect transistors. They should be selected for gate voltages within

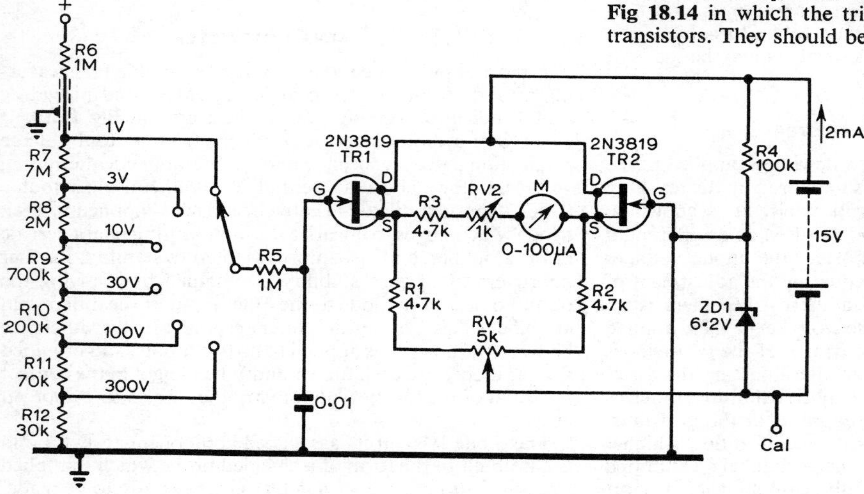

Fig 18.14. Circuit of the fet voltmeter. TR1, TR2 2N3819; TR3 6·2V zener; 0–100μA; R6 ½W carbon; R7–R12 one per cent high stability.

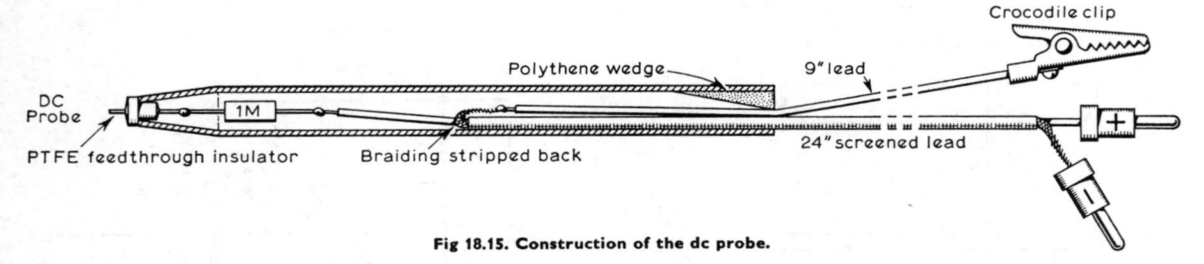

Fig 18.15. Construction of the dc probe.

20 per cent of each other at a drain current of 1mA but, if this is not possible, dissimilar values of the source resistors R1 and R2 will overcome the lack of balance. The gate voltage of TR2 is defined by the zener diode ZD1; this will be somewhat lower than its rated voltage because it is operated at a very low current in order to reduce battery consumption. Calibration will be slightly dependent on the battery voltage but a 15V battery should remain serviceable down to about 9V provided that the zero adjuster RV1 is corrected.

Calibration is by the variable resistor RV2 in series with the meter and is carried out by applying known voltages to the input terminals. A shielded input probe is desirable and a suitable arrangement using a ballpoint pen case is shown in **Fig 18.15**.

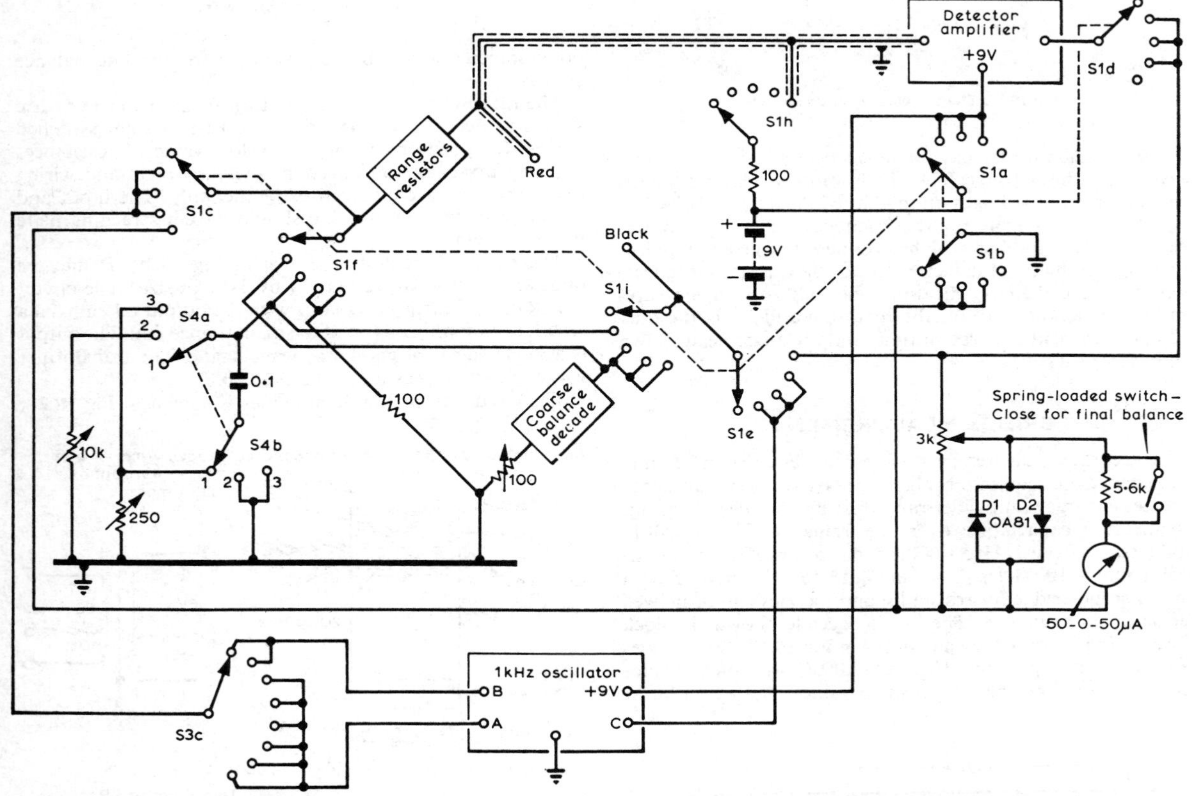

Fig 18.16. Switching layout of R, C and L component bridge. The 0·1μF capacitor shown is a standard and should have a tolerance of ±1 per cent.

S1 function switch
1 off
2 capacitance
3 inductance

S4 phase balance
1 capacitance and high-Q inductance
2 electrolytic capacitance and transformers
3 rf coils

S2 10-position decade switch—see Fig 18.17.

S3 range switch

	R	L	C
1	0–0·1Ω	0–1μH	0–100μF
2	0–1Ω	0–10μH	0–10μF
3	0–10Ω	0–100μH	0–1μF
4	0–100Ω	0–1mH	0–100nF
5	0–1kΩ	0–10mH	0–10nF
6	0–10kΩ	0–100mH	0–1nF
7	0–100kΩ	0–1H	0–100pF
8	0–1MΩ	0–10H	0–10pF
9	0–10MΩ	0–100H	0–1pF

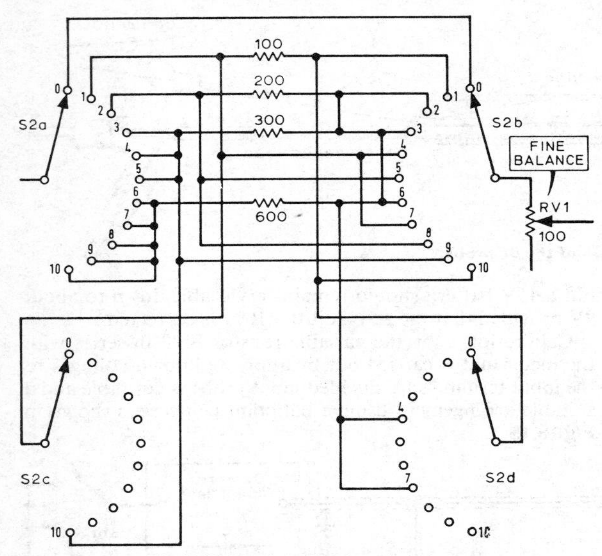

Fig 18.17. The 0–1,000Ω decade switch.

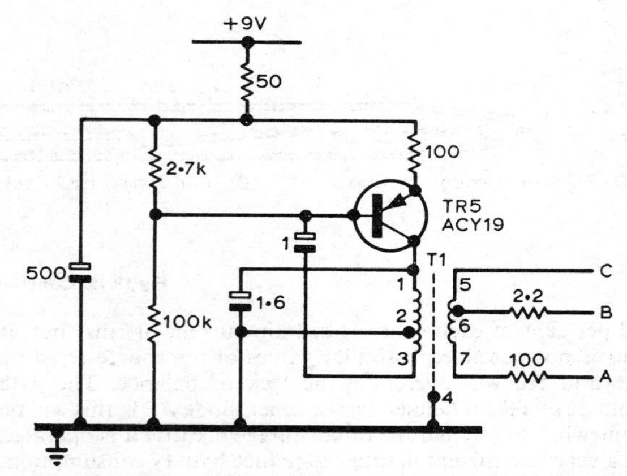

Fig 18.19. Circuit of oscillator; 0·5V rms across terminals 7–5 when terminated in a 100Ω resistor; 0·05V rms across terminals 6-5 when terminated in a 1·5Ω resistor.

AC measurements may be made by replacing the input probe by a diode probe. The 1N914 diode is suitable for the measurement of voltages up to about 75V at up to 100MHz. Junction-type silicon diodes intended for use as power rectifiers are suitable for higher voltages but give falsely low readings at high frequencies due to charge storage effects which reduce their rectification efficiency. For rf measurements above about 100V, the thermionic diode is the only reliable device at the present time, and the diode head shown in Fig 18.13 is suitable.

BRIDGE MEASUREMENTS

If accurate measurements of resistance, capacitance and inductance are required, bridge methods are preferable since greater accuracy and wider range can be obtained. A bridge to measure capacitance over the range 1pF to 1,000μF, inductance from 1μH to 1,000H and resistance from fractions of an ohm to 100MΩ is shown in **Fig 18.16**. It is derived from a Wheatstone bridge for resistance and the Heys and Maxwell bridges for reactance. The coarse decade shown in block form in Fig 18.16 is shown in detail in **Fig 18.17** and provides 10 switched steps from 100Ω to 1,000Ω, the fine balance control providing the intermediate values. A 10-turn linear potentiometer would be very suitable for the fine balance control.

The range resistors shown in block form in Fig 18.16 are shown in detail in **Fig 18.18** and consist of eight switched values. On positions 1 and 2, the low value of resistances required must take into account switch contact and wiring resistance, and are adjusted experimentally. The input and output switches S3a and S3b should have a screening plate between them.

The oscillator circuit is shown in **Fig 18.19**. It may be built on a Veroboard about 3½ by 2½in in size. The circuit is a standard Hartley oscillator with the output taken from a secondary winding on top of the main tapped coil. The output is a 1kHz signal of good sine wave shape. The transformer winding details are shown in **Fig 18.20**.

The circuit of a suitable amplifier is shown in **Fig 18.21**;

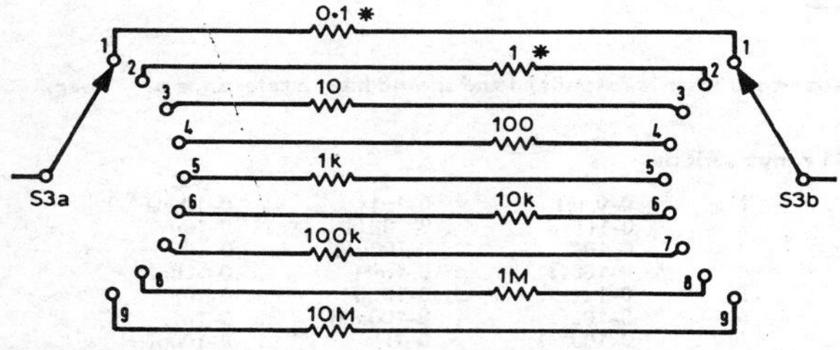

Fig 18.18. The range resistors. * indicates precise values adjusted experimentally.

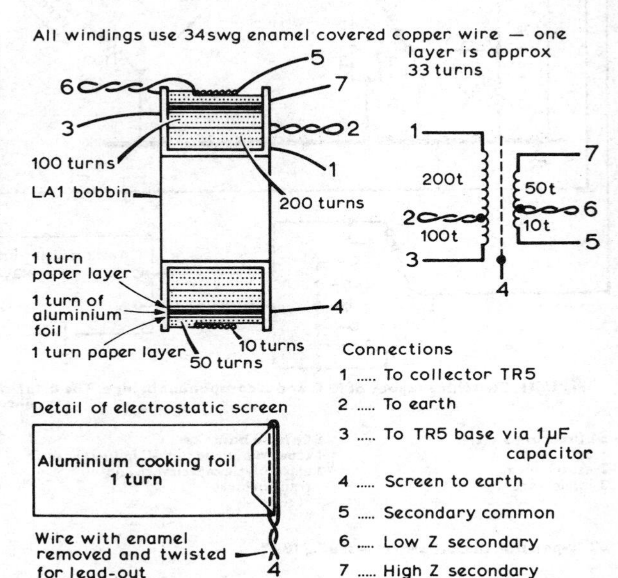

Fig 18.20. Constructional details of oscillator transformer T1.

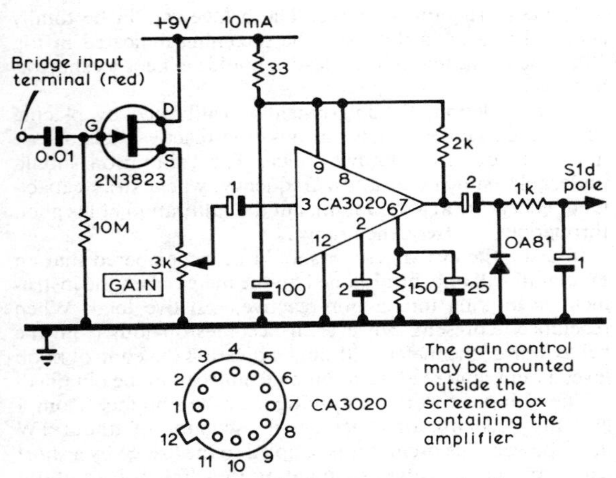

Fig 18.21. The amplifier and detector circuit.

the fet input stage provides an input impedance of several megohms. Silicon diodes across the 25–0–25μA meter protect it from damage when the bridge is out of balance by acting as shunts when the voltage across the meter exceeds some 500mV. The oscillator and the amplifier should be well separated but otherwise the layout is not critical. The variable resistors associated with S4 for phase balance dependent on circuit Q may be ganged if desired, as only one is adjusted at a time. Similarly the ac and dc sensitivity controls may be ganged.

Using the Bridge

To measure resistance, the coarse balance control is rotated with the gain control at minimum while searching for a null in the meter reading. As this is approached, the gain should be increased and the fine balance control used to obtain a complete null. The resistance value is then read off the decade switch and the calibrated scale of the fine balance control.

To measure capacitance, the same procedure is adopted except that the phase balance controls must also be adjusted. Position 1 of switch S4 should be used for low-loss capacitors and position 2 for low-Q capacitors such as electrolytics.

To measure inductance, the same method is followed using position 3 of S4 for air cored coils, position 1 for coils on ferrite cores and position 2 for mains frequency components.

The 0·1Ω and 1Ω range resistors may be constructed of resistance wire wound non-inductively and adjusted in value to correctly measure known resistors. A 7in length of 30swg Eureka wire for 1Ω and 2·5in of 24swg Eureka wire for 0·1Ω are suitable starting points.

RF CAPACITANCE BRIDGE

It is frequently necessary to measure small-value capacitors and a simple bridge operating at a relatively high frequency is convenient.

The following bridge is suitable for measuring capacitances down to 1pF and up to 1,500pF. The circuit (**Fig 18.22**) is a version of the Wheatstone bridge and is energized from a Colpitts oscillator at about 1·5MHz. The output of the bridge is rectified and applied to a "magic eye" indicator V2 which is arranged to be fully open when the bridge is balanced.

Calibration

The calibration of the bridge is quite easy, all that is required being a few accurate capacitors. The following capacitors are needed and they should be good quality silver

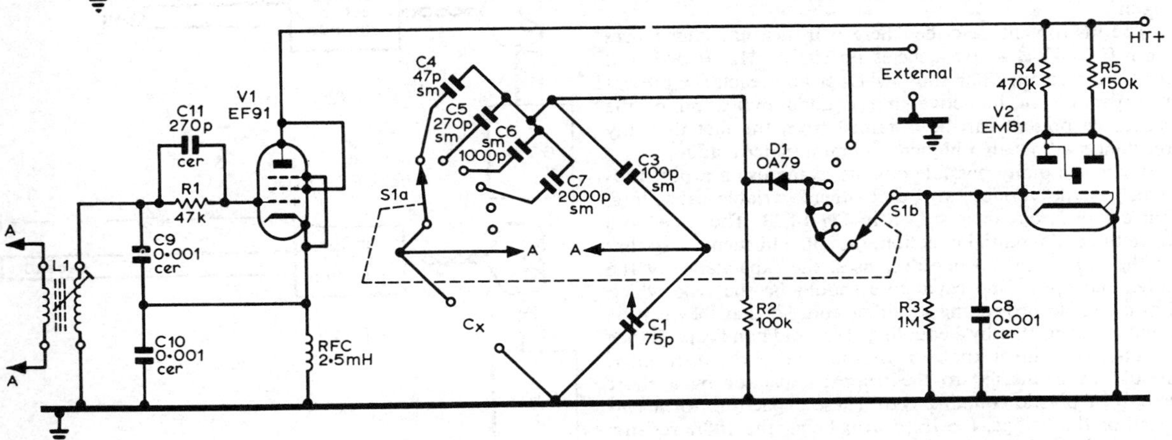

Fig 18.22. Circuit diagram of the 1·5MHz bridge. C3, C4, C5, C6, C7, silvered mica; C8, C9, C10, C11, ceramic; C12, 350V electrolytic; L1, Maxi-Q Red 2 miniature dual-purpose coil; MR1, 250V 15mA contact-cooled rectifier; S1, 2-pole 6-way; T1, 250V 15mA, 6·3V 1A. All resistors are ½W rating. L1, medium-wave oscillator coil.

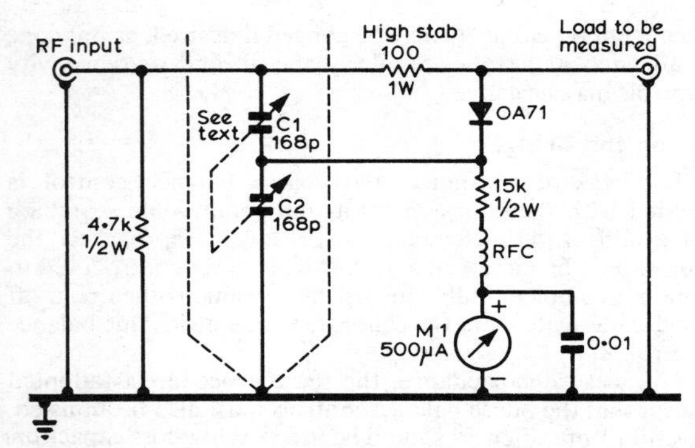

Fig 18.23. Simple rf bridge. C1, C2 form a differential capacitor with a maximum value of 168pF in each section.

mica with a tolerance of ± one per cent: 5, 10, 25, 50, 100, 250, 500, 1,000pF.

With these standards it is possible to calibrate the bridge throughout its range. Proceed as follows. Start on the lowest range and with no capacitor connected across the terminals, adjust C1 for balance. This point should be marked zero. (This is to balance out the stray capacitance. If it is not possible to obtain a balance—the "eye" must be fully open—solder a small capacitor, about 5pF, across the terminals until a balance can be obtained.) Connect the 5pF standard to the test terminals using the shortest possible leads and adjust C1 for balance, then mark the scale 5. Repeat the process using the 10pF standard. Next solder the two together and repeat, marking the scale 15.

RF IMPEDANCE BRIDGE

The need for an instrument which will measure impedance is felt at some time or other by every experimenting amateur. The instrument normally used is the full rf bridge, but commercial rf bridges are elaborate and expensive. On the other hand, it is possible to build a simple rf bridge which, provided its limitations are appreciated, can be an inexpensive and most useful adjunct in the amateur workshop. In fact, it is essential if experiments with aerials are undertaken.

The instrument described here will measure impedances from 0 to 400Ω at frequencies up to 30MHz. It does not measure reactance nor show whether any reactance present is capacitive or inductive, but a good indication of the reactance present can be obtained from the fact that any reactance will mean a higher minimum meter reading.

There are many possible circuits, some using a potentiometer as the variable arm and others variable capacitors, but a typical circuit is shown in **Fig 18.23**. The capacitors have to be differential in action, mounted in such a way that as the capacitance of one decreases, the capacitance of the other increases. The capacitors should be the type which has a spindle protruding at either end, so that they can be connected together by a coupling. To avoid hand capacitance effects, the control knob on the outside of the instrument should be connected to the nearest capacitor by a short length of plastic coupling rod. These capacitors form two arms of the bridge, the third arm being the 100Ω resistor and the fourth the load. Balance of the bridge is indicated by a zero reading on the meter M1.

The construction is simple. The bridge should be totally enclosed in a metal box and the screening indicated in Fig 18.23 incorporated. All the leads should be kept as short as possible.

On completion, the instrument is calibrated by placing across the load terminals various non-reactive resistors (ie not wire-wound) of known value. The calibration should preferably be made at a low frequency, where stray capacitance effects are at a minimum, but the calibration holds good throughout the frequency range.

In using the instrument, it should be remembered that an exact null will only be obtained on the meter when the instrument is looking into a non-reactive, resistive load. When reactance is present, however, it becomes obvious from the behaviour of the meter; although adjusting the control knob gives a minimum reading, a complete null cannot be obtained.

The rf input to drive the bridge can be obtained from a grid dip oscillator or other small oscillator of about 1W input power. The oscillator is coupled to the bridge by a short length of coaxial cable, terminating in a link coil of about four turns, which is placed on or near the gdo coil. Care should be taken not to overcouple or the gdo may change frequency or even stop oscillating. As the coupling is increased it will be seen that the meter reading of the bridge increases up to a certain point, after which further increase in coupling causes the meter reading to fall. A little less coupling than that which gives the maximum bridge meter reading is the best to use. The bridge can be used to find aerial impedance and can be used for many other purposes: for example, to find the input impedance of a receiver on a particular frequency.

One useful application of this type of simple bridge is to find the frequency at which a length of transmission line is $\lambda/4$ or $\lambda/2$ long electrically. If it is desired to find the frequency at which the transmission line is $\lambda/4$ long, the line is connected to the bridge and the far end of it left open-circuit. The bridge control is set at 0Ω. The dip oscillator is then adjusted until the lowest frequency is found at which the bridge shows a sharp null. This is the frequency at which

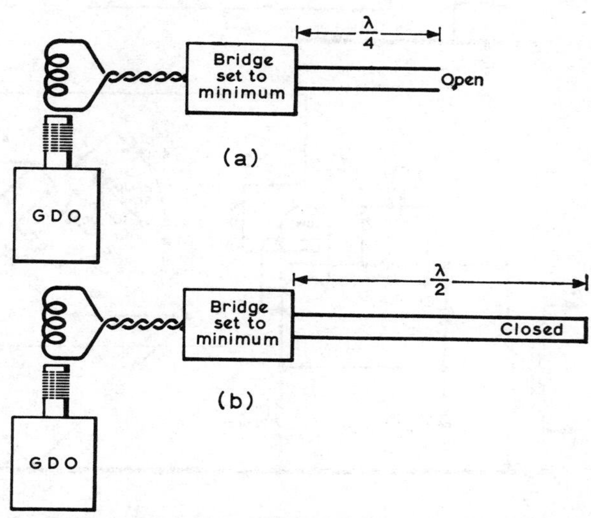

Fig 18.24. Method of using the rf bridge to determine the frequency at which a transmission line exhibits a $\lambda/4$ or $\lambda/2$ characteristic.

the piece of transmission line is λ/4 long. Odd multiples of this frequency can be checked in the same manner. In the same way, the frequency at which a piece of transmission line is λ/2 long can also be found. The procedure is the same, except that the far end of the transmission line is shorted instead of being left open. The method in both cases is illustrated in **Fig 18.24**.

The bridge can also be used to check the characteristic impedance of a transmission line. This is often a worthwhile exercise, since appearances can be misleading. The procedure is as follows:

(a) Find the frequency at which the length of transmission line under test is λ/4 long. Once this has been done, leave the oscillator on this frequency.
(b) Select a carbon resistor of approximately the same value as the probable characteristic impedance of the transmission line. Substitute this resistor for the transmission line as the bridge load and check its value at the frequency obtained in (a). (This will not necessarily agree with its dc value.)
(c) Disconnect the resistor from the bridge and reconnect the transmission line. Connect the resistor across the far end of the transmission line.
(d) Measure the impedance now presented by the transmission line at the frequency of (a). Then the characteristic impedance Z_o of the line is given by

$$Z_o = \sqrt{Z_s \times Z_r}$$

where Z_s = impedance presented by the line and Z_r = resistor value.

FREQUENCY MEASUREMENT

It is important to know the frequency to which a receiver is tuned and the frequency of a signal radiated by a transmitter; in particular, it is necessary to be quite sure that the signal is within the amateur band in which operation is taking place.

For this purpose, a crystal controlled oscillator is almost essential as a reference by which a vfo can be calibrated. Even if the transmitter is directly controlled by a crystal of certified accuracy, it is necessary to ensure that the transmitter is working on the correct harmonic of the crystal and for this purpose an absorption wavemeter is convenient. When constructing new equipment, it is most helpful to be able to check the resonant frequency of the tuned circuits before power is applied; this can be done with a dip oscillator.

Standard Frequency Services

Even when a crystal oscillator is used as a calibration source, it is necessary to set it against a frequency standard, as the actual frequency obtained from a crystal depends slightly on the circuit conditions. Standard frequency transmissions are provided in the United Kingdom by transmissions from MSF at Rugby on 2·5, 5 and 10MHz, while the BBC transmitter at Droitwich on 200kHz is also maintained at a very accurate frequency. Similar services in the USA are provided by WWV on 2·5, 5, 10, 15, 20 and 25MHz, and some of these signals are normally receivable in the UK. WWVH in Hawaii operates on 2·5, 5, 10 and 15MHz.

The transmissions from MSF on 5MHz are on a time sharing basis with HBN (Neuchatel) in accordance with

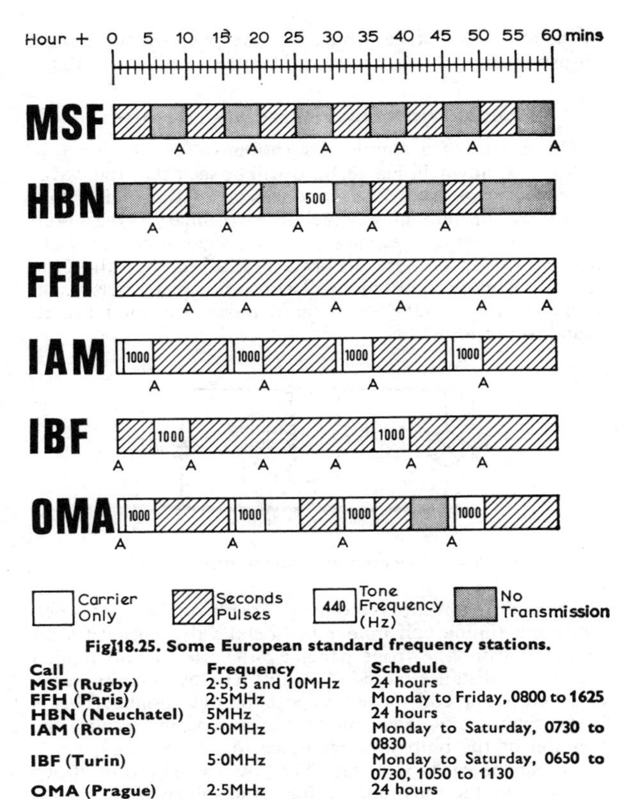

Fig 18.25. Some European standard frequency stations.

Call	Frequency	Schedule
MSF (Rugby)	2·5, 5 and 10MHz	24 hours
FFH (Paris)	2·5MHz	Monday to Friday, 0800 to 1625
HBN (Neuchatel)	5MHz	24 hours
IAM (Rome)	5·0MHz	Monday to Saturday, 0730 to 0830
IBF (Turin)	5·0MHz	Monday to Saturday, 0650 to 0730, 1050 to 1130
OMA (Prague)	2·5MHz	24 hours

the schedule shown in **Fig 18.25** which also shows the hourly schedules of other standard frequency transmissions useful in Europe. From 0–5min past each hour, MSF transmits carrier and seconds pulses, from 5–9½ min past each hour there is no transmission, and from 9½–10min past each hour, MSF transmits its callsign and the amount of the frequency offset (in parts in 10^{10}), each given three times in slow morse. The cycle is repeated six times in each hour. MSF also transmits on 60kHz from 1429 to 1530ut daily. Note that the schedules of standard frequency transmissions are subject to change. Details of these changes are published in *Radio Communication* and other magazines from time to time.

Absorption Wavemeters

An absorption wavemeter consists simply of a calibrated tuned circuit which absorbs power from the circuit being measured when the circuits are tuned to the same frequency.

The power collected by the wavemeter can be made to light a small lamp or operate a sensitive meter. With the wavemeter resonated to the circuit to which it is coupled, some energy is absorbed from that circuit. If the wavemeter is held close enough the rf current induced into it will be sufficient to light the bulb. Thus, provided that the wavemeter has been previously calibrated, it is only a matter of tuning for a resonance indication on the bulb and reading off the frequency.

A low-power stage may not be capable of providing enough rf power to light the bulb. Under such circumstances resonance indication can be obtained from the anode current of the low-power stage, which will rise when the wavemeter absorbs energy, or a dip in the grid current of the next stage.

ABSORPTION WAVEMETER FOR 1·5–30MHz

The circuit of a simple absorption wavemeter for 1·5–30MHz is shown in **Fig 18.26**. It will be seen that the wavemeter frame is connected to one side of the tuning coil, coupling coil, and tuning capacitor. An ordinary 6V 0·3A bulb is used to indicate resonance. A lower-consumption bulb would provide a more accurate indication, but the type specified is more robust and will withstand greater overloads, while showing a sufficiently sharp resonance point for all practical purposes.

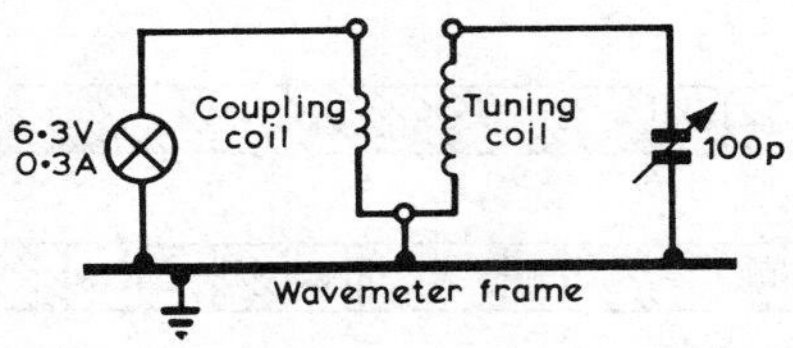

Fig 18.26. Absorption wavemeter for 1·5–30MHz.

The tuning capacitor is mounted directly on the front panel, while the tuning coil holder (an octal valve socket) is off mounted from the front panel by the use of two tapped aluminium distance pieces. Two stiff lengths of copper wire attached to the coil socket support the bulb holder. This is better than a more rigid method of mounting, because the location of the bulb with reference to the metal capacitor cover can be easily adjusted. The cover is cut from 20swg tin plate, bent to the required shape and soldered.

The coil formers are made from octal valve bases, force fitted into a bakelized paper tube 2½in long by 1⅛in in diameter (internal diameter ⅞in). After fitting, the tube and base should be cemented together.

TABLE 18.1

Range	Tuning coil	Coupling coil	Wire
1·5–4MHz	80 turns	6 turns	32swg enam
4–12MHz	29 turns	3 turns	22swg enam
12–30MHz	6¼ turns	2¼ turns	22swg enam

The coils should be wound in accordance with the data given in **Table 18.1**. The illustrations show the type of construction.

Calibration

Calibration may be carried out with the aid of an oscillator or a calibrated receiver. As most amateur stations have an accurately calibrated communications receiver, the latter method will be described.

With the receiver switched on and the aerial connected, a signal is tuned in at the low-frequency end of the band to be calibrated. A coupling coil consisting of a few turns of sufficient diameter to slide over the wavemeter tuning coil is connected in series with the aerial. The S-meter reading should be observed while the wavemeter is slowly tuned: at one point the reading will drop, indicating that signal frequency energy is being absorbed. This point can now be marked on the prepared dial of the wavemeter. The receiver is then tuned to the next signal higher in frequency, and the process is repeated until finally the whole dial is calibrated.

View of the wavemeter of Fig 18.26 showing the scale, cursor resonance indicator lamp and spare coils.

Some receivers are not fitted with an S-meter, in which case the bfo should be switched on, an ac rectifier type voltmeter being connected to the phones terminals of the set. A signal is then tuned in until the beat note provides a convenient reading on the voltmeter. As before, the wavemeter (coupled to the receiver aerial) should be tuned slowly until a dip occurs in the reading, indicating a calibration point.

When calibrating has been completed, the dial may be removed and permanently marked with Indian ink, after which it should be replaced and covered with a protective sheet of $\frac{1}{16}$in Perspex, held at the corners by 6BA screws tapped into the front panel. Perspex of the same thickness is also used for the dial cursor, a central hairline being engraved on it with a stylus.

The size of the front panel is 2½ by 3½in, and that of the capacitor cover box is 2⅛ by 3¼in. The resonance indicator bulb is pushed through a hole drilled in the top of the cover, and is protected from shock by a rubber grommet.

For checking any apparatus which has a power output of 500mW or more, the bulb will indicate resonance quite satisfactorily, but where output is low, as in the local oscillator circuit of a receiver, a meter inserted in the gridleak earth return will dip as the wavemeter is tuned through resonance. Alternatively, a milliammeter may be inserted in the anode circuit of the local oscillator valve, and this will give an increased reading as the wavemeter is tuned through resonance.

Internal construction of the absorption wavemeter.

SIMPLE ABSORPTION WAVEMETER FOR 65–230MHz

The absorption wavemeter circuit shown in **Fig 18.27** is an easily built unit covering 230MHz and can therefore be used to check frequencies in Band II (fm broadcasting) and Band III (television) in addition to 70 and 144MHz stages in amateur vhf transmitters. The appearance of the wavemeter with its associated indicating meter can be seen in the photograph.

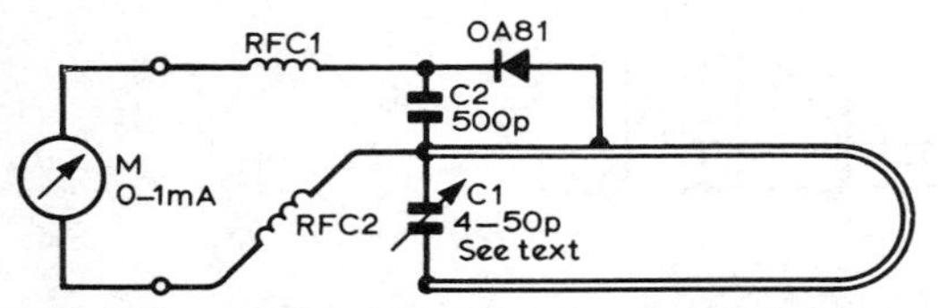

Fig 18.27. Circuit diagram of the vhf wavemeter. C1 (4–50pF) is a type C804 manufactured by Jackson Bros. M should have a fsd of 1-2mA. RFC1, RFC2 80 turns 40 swg enamelled wire wound on ½W resistor of 1kΩ or more and wax dipped.

Construction is straightforward and all the components apart from the meter are mounted on a Perspex plate measuring 7½ by 3 by ⅛in. Details of the tuned circuit are shown in **Fig 18.28(a)**, and should be closely followed. The layout of the other components is not critical provided they are kept away from the inductance loop. A heat shunt should be used when the diode is soldered in order to prevent any damage to it.

For accurate calibration a signal generator would be required but, provided the inductance loop is carefully constructed and the knob and scale are non-metallic, dial markings can be determined from **Fig 18.28(b)**. These should be accurate enough for most purposes.

In operation, the unit should be loosely coupled to the tuned circuit under test and the capacitor then tuned until the meter indicates resonance. For low-power oscillators a more sensitive meter should be used if available.

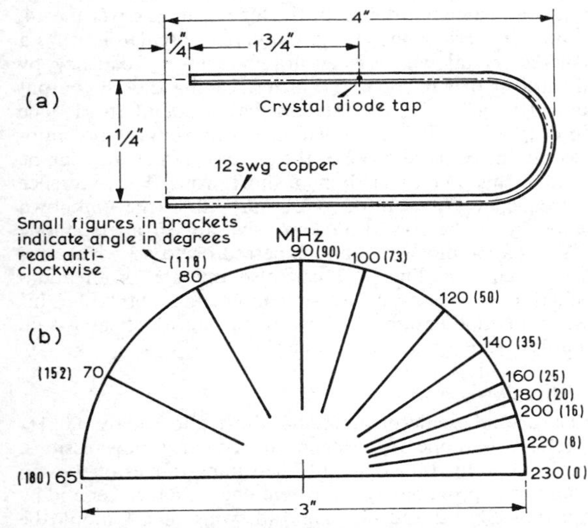

Fig 18.28. (a) Details of the inductance loop made of 12swg copper wire. The dimensions should be closely followed if the calibration of (b) is to be used. (b) Dial calibration. The calibration points relative to the base line (anti-clockwise) are: 230MHz 0°; 220MHz 8°; 200MHz 16°; 180MHz 20°; 160MHz 25°; 140MHz 35°; 120MHz 50°; 100MHz 73°; 90MHz 90°; 80MHz 118°; 70MHz 152°; 65MHz 180°.

VHF absorption wavemeter with its indicating meter.

The wavemeter can also be used as a field strength indicator when making adjustments to vhf aerial arrays. A single turn coil should be loosely coupled to the wavemeter loop and connected via a low impedance feeder to a dipole directed towards the aerial under test.

LECHER LINES

Above about 100MHz it is practicable to measure directly the wavelength at which a transmitter or oscillator is operating by using Lecher lines, which comprise a pair of taut parallel wires, spaced an inch or so apart to form an open-wire transmission line, and a bridge to short circuit the wires which can be moved along the line as required.

For transmitter frequency measurement, one end of the line is loosely coupled by a loop to the pa circuit. Starting near the coupling loop end of the line, the bridge is slowly moved towards the open end of the line until a point of maximum current in the bridge is found; this will be indicated by a deflection on the anode current meter, or by observing when a flash lamp and loop loosely coupled to the pa coil passes through a minimum in brightness. This position should be carefully noted and the bridge then moved further along the line until the next similar position is found. This should be repeated several times until the distance between successive positions is the same. The distance between the two points will be one half of the wavelength at which the transmitter is operating.

Application of the formula

$$\text{Frequency (MHz)} = \frac{15{,}000}{\text{Distance between bridge position (cm)}}$$

will enable the frequency of the oscillation to be determined. For example, if the distance between the two bridge positions is found to be 100cm, then by substituting in the formula

$$\text{Frequency} = \frac{15{,}000}{100}\text{MHz} = 150\text{MHz}$$

For the most sensitive condition of adjustment, the Lecher lines and the bulb and loop should be very loosely coupled to the tank circuit. This is especially important when measuring the frequency of a self-excited oscillator, the

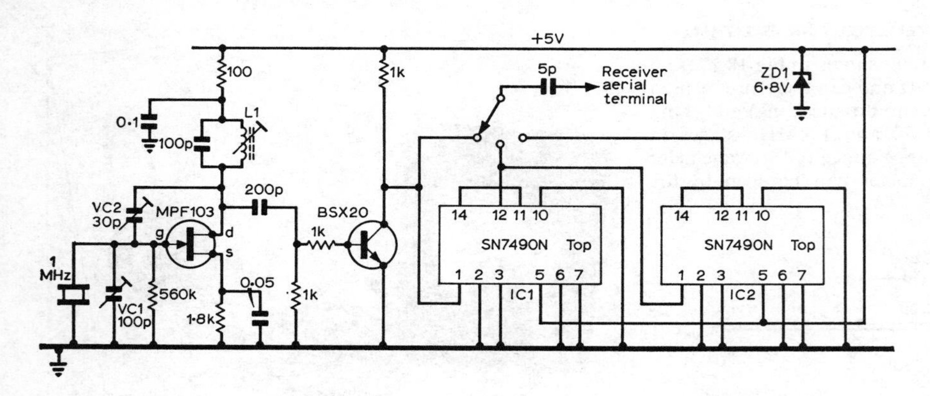

Fig 18.29. Integrated-circuit crystal calibrator. A medium-wave broadcast tuning coil is suitable for L1 (about 150μH). The SN7490N may be replaced by an FJJ141.

frequency of which may be altered if the coupling is too tight. Although the accuracy of frequency measurement by this method is not very high the method is of great use to the amateur as the only instrument required is a metre rule. The longer the Lecher lines and the further the two minima are from the coupling loop end, the more accurate will be the reading. An accuracy of 0·1 per cent can be attained with care.

CRYSTAL CALIBRATOR

A precision crystal oscillator with means of dividing the crystal frequency so as to produce more closely spaced reference points is generally known as a *crystal calibrator*. A convenient form uses a 1MHz crystal oscillator with dividers to produce 100kHz and 10kHz markers. For really accurate work, the oscillator could be phase locked on to a standard frequency transmission, the Droitwich 200kHz transmission being the most convenient.

A crystal calibrator using a fet 1MHz crystal oscillator and integrated circuits as dividers is shown in **Fig 18.29.** The rapid switching time of the ICs ensures useful output into the vhf region. Unlike multivibrator circuits normally used, the integrated circuits have the advantage that no setting up is required as each ic achieves its division of 10 by dividing first by 5 with output on pin 11 and then by 2 with output on pin 14. It is possible to take outputs of 200kHz and 20kHz from the junction of pins 11 and 14 on IC1 and IC2 respectively.

Integrated circuits are easily damaged and every care should be taken to ensure that their voltage ratings are never exceeded. Leakage potentials from soldering irons, voltages derived from test instruments, and pick-up from strong rf fields are among the sources which can destroy ICs. They are also intolerant of excessive soldering temperatures and the use of holders is a useful safeguard. The power supply for the calibrator should be stabilized at between 4·5V and 5·5V. The 6·8V zener diode shown in Fig 18.29 is intended to clamp the supply at just below the absolute maximum rating of the ICs in the event of excessive voltage being applied.

The output from the calibrator may be satisfactorily connected to a receiver aerial socket but must not be connected to a live circuit which might easily induce sufficient voltage to destroy the ICs. If an isolating amplifier is required to buffer the calibrator from a live circuit, it should use a transistor capable of fast switching if the vhf output is to be preserved.

The only setting-up required is the adjustment of VC1 to bring the crystal oscillator to zero beat with a standard frequency signal and possible adjustment of the feedback capacitor VC2. With active crystals, this may be omitted altogether.

HETERODYNE FREQUENCY METER

In order to interpolate between frequency markers, either a calibrated receiver or a calibrated vfo is required. An instrument having a calibrated vfo, a crystal reference and means of mixing the unknown and vfo frequencies to produce an audible note is known as a *heterodyne wavemeter*. Many amateurs use an excellent instrument of this type known as the BC221.

A circuit for a heterodyne wavemeter is given in **Fig 18.30.** Good mechanical stability and a dial capable of accurate and repeatable readings are essential.

A variable frequency oscillator V1 tunes over two ranges only: from 1·75MHz to 2MHz and from 7 to 7·5MHz. The second harmonic of the lower frequency range covers the 80m band while harmonics of the higher range cover the 14, 21 and 28MHz bands. V4 is the crystal oscillator using a 100kHz crystal which is accurately set on frequency by adjustment of C6. V2 is the mixer, having the vfo output applied to g3 and the crystal oscillator output to g1. The vfo output is available at low impedance from the potentiometer in the cathode of V2 or the unknown frequency can be fed in at this point, usually by a short probe. The difference in frequency between the vfo and either the unknown frequency or the crystal oscillator will appear at the anode of V2 as an audio signal, and is passed on to V3 which acts as an audio amplifier. V3 can also operate as an audio oscillator by means of S2 in conjunction with T1; this results in modulation of the vfo output which assists in identifying the signal in the receiver.

Calibration

The crystal oscillator should be adjusted to exactly 100kHz by reference to one of the standard frequency transmissions. This is done by adjusting C6. The tuning ranges of the vfo should be approximately set by reference to a receiver and by adjusting C1, C2 and C3. The full swing of C4 should be made to fully cover the bands 1·75 to 2MHz and 7 to 7·5MHz so as to have as open a calibration scale as possible. The vfo tuning scale should now be accurately calibrated at the 100kHz points provided by the crystal oscillator and it will

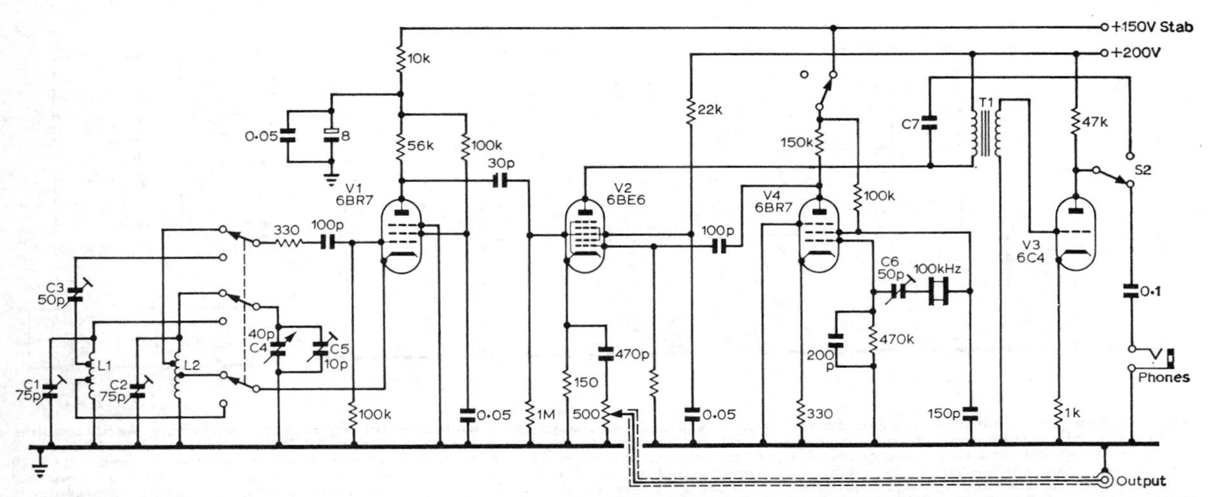

Fig 18.30. Circuit of the heterodyne wavemeter. L1: 16 turns 24swg 1in dia, ¾in long, cathode tap 2 turns, grid tap 11 turns from earthy end. L2: 67 turns 26swg enam, 1in dia, 1¼in long, cathode tap 5 turns, grid tap 41 turns from earthy end. C7: adjust for satisfactory audio note, typically 0·005 to 0·01μF. T1: any audio transformer of ratio 3:1 to 5:1. Reverse one winding if circuit does not oscillate.

be found that 50kHz points can also be identified. A smooth calibration curve can then be drawn for the vfo.

Each time the heterodyne wavemeter is used it should be allowed to warm up until the vfo is stable and one of the 100kHz calibration points checked against the crystal; any error may be corrected by adjusting C5.

An ht supply of about 200V at 50mA is required and the supply to the vfo and crystal oscillators should be stabilized. A glow discharge stabilizer such as the QS150/40 or a combination of zener diodes to give a stabilized supply of 150V would be suitable.

VHF Use of Frequency Markers

Although it is not difficult to produce frequency markers into the vhf region, the presence of strong lower-frequency signals is sometimes a disadvantage, particularly when double conversion receivers are used. A selective amplifier will increase the effective level of signals in the region of interest. Such an amplifier is shown in **Fig 18.31**. The signals from any standard source are amplified by the triode stage which drives a diode harmonic multiplier, followed by an amplifier having an anode circuit tuned to the desired band. As the harmonic generator is rather liable to pick up other frequencies, the two diodes and their associated resistor and capacitor should be carefully screened.

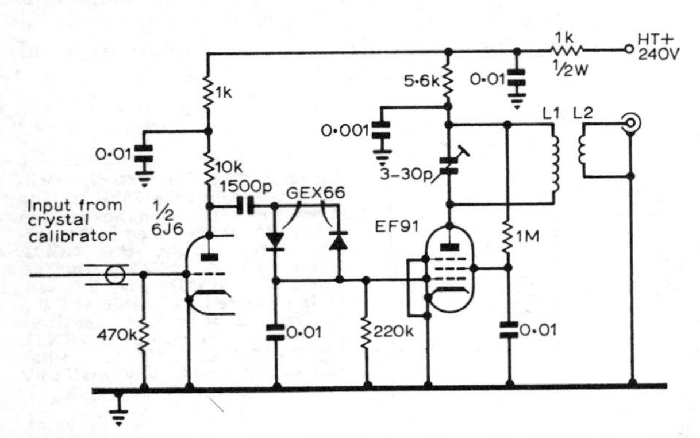

Fig 18.31. Selective amplifier. L1: 3 turns 18swg, ⅜in dia, ½in long for 145MHz. L2: single-turn coupling loop.

Harmonic Indicators

Although absorption wavemeters may be used to ensure that strong harmonics of the wanted frequency are not produced, they are not really sensitive enough when interference problems arise. For such cases, a tunable trf receiver covering 35 to 70MHz may be used for Band I television frequencies and also to cover the usual i.f. of about 38MHz. For television on higher frequencies, one of the television turret tuners available on the surplus market could be used, followed by the trf receiver as i.f. amplifier and detector. It is better to use valves rather than semiconductors throughout the rf and i.f. stages, since signals large enough to damage semiconductors may easily exist. The equipment required depends on the frequencies involved and the type of tuner available. An indicator of adequate sensitivity for Band I frequencies is shown in **Fig 18.32**; the output is shown on the meter M1 and can be heard on phones connected at J1. The meter should have an fsd of 50μA and should be provided with switched shunts to give ranges of, say, 50μA, 250μA and 1·25mA. The transmitter may either be fed into the input socket through a high-pass filter to remove the fundamental, or may be received by means of a rod aerial provided as an alternative input.

GRID DIP OSCILLATORS

Another type of frequency measuring device which is extremely useful is the grid dip oscillator. This is really nothing more than a calibrated oscillator which can be tuned over a wide range of frequencies and which has a moving coil meter indicating the grid current. If it is coupled to an external tuned circuit the grid current meter will dip when the oscillator is tuned to the resonant frequency of that circuit because the circuit will absorb energy from the oscillator, thus reducing the amount of rf feedback to the grid and, in consequence, reducing the grid current.

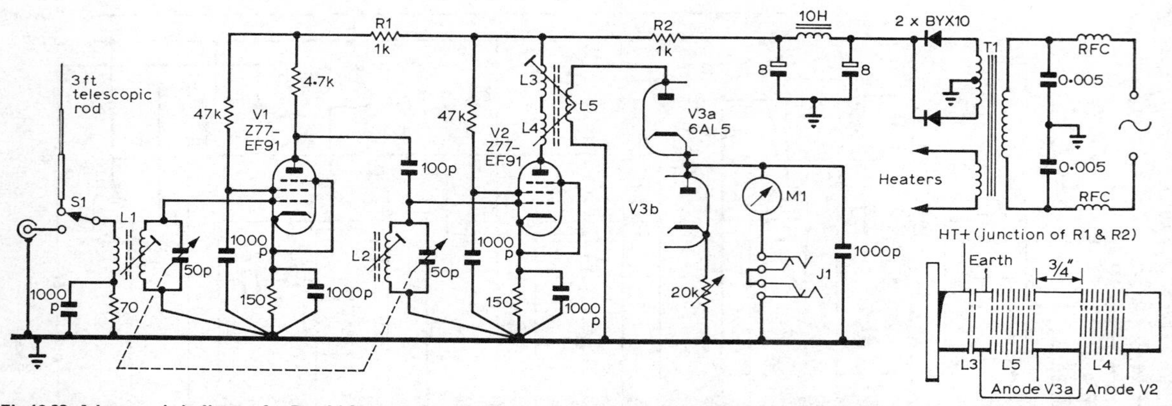

Fig 18.32. A harmonic indicator for Band 1 frequencies. L1: 5 turns 26swg spaced 1/16in, 1/4in dia, Aladdin PP5938 dust cored former. Aerial coupling coil consists of 2 turns wound at earthy end of grid coil. L2: 5 turns 26swg spaced 1/16in, 1/4in dia Aladdin PP5938 dust cored former. L3: 2 turns 26swg enam, wound close to the end of L5. L4, L5: 10 turns 26swg, close-wound on Aladdin 5937 former with dust cores at each end. L6, L8: 10 turns 18swg enam, close-wound, self-supporting, 3/8in dia, opened to 11/16in winding length. L7: 6 turns 18swg enam, self-supporting, close-wound 3/8in dia, opened to 1/2in winding length. Construction of the bandpass transformer L3, L4, L5 is shown on the right. The mains transformer is 250-0-250V at 25mA and 6·3V at 2A.

Since the grid dip oscillator provides its own rf energy it does not require the circuit being checked to be energized. It is therefore useful for checking the resonant frequency of tuned circuits, rf chokes and aerial systems. The gdo is also useful as an absorption wavemeter, signal monitor or simple signal generator.

Earlier versions of the gdo used valves in the oscillator but modern designs use transistor oscillators with a diode rectifier and dc amplifier to operate the meter. There is, of course, no grid current so that the title is somewhat misleading; the term *dip oscillator* is more appropriate.

A TRANSISTORIZED DIP OSCILLATOR FOR 0·85–150MHz

The circuit of a typical dip oscillator which is simple to build is shown in **Fig 18.33**.

Basically, the circuit comprises a multi-frequency range transistor oscillator covering 0·85 to 150MHz in seven ranges, using plug-in coils, a diode detector and a transistor dc amplifier operating a meter. The unit contains its own 9V battery, the dip oscillator actually running from a 6·8V zener-stabilized line, an arrangement which helps to reduce the effects of battery voltage variation. The total current consumption is 5mA. A generous overlap is provided between the frequency ranges, L2, L3, L4 and L7 covering two amateur bands each, and L5 three amateur bands (**Table 18.2**).

The oscillator circuit is a grounded-collector Colpitts, with only part of the tuning capacitance tapped for connection to the emitter. If the 15pF variable capacitor C1, shown ganged to the 150pF variable C2 in the circuit diagram, were in fact fixed, the coupling of the transistor input and output circuits to the tuned circuit would vary with rotation of the main 150pF tuning capacitor, the effective positive feedback falling as the value of the tuning capacitance was increased. The combination of the stray capacitances in the circuit with the ganged 15pF variable capacitor results in almost constant feedback being obtained over all ranges except the highest frequency one, where the feedback mechanism is somewhat different. The required value of the capacitance C_F varies with the frequency range in use and so the appropriate value is built into each coil range. Increasing the value of C_F reduces the feedback ratio.

The reactance of the 3·3pF coupling capacitor C6, and the essentially resistive impedance of the diode and its load, form a potentiometer which delivers a nearly constant dc voltage to the base of TR2, except on the highest frequency range, where the meter reading is reduced to about one-third of full scale deflection. The coil is used as part of the dc return circuit for the oscillator emitter current to avoid

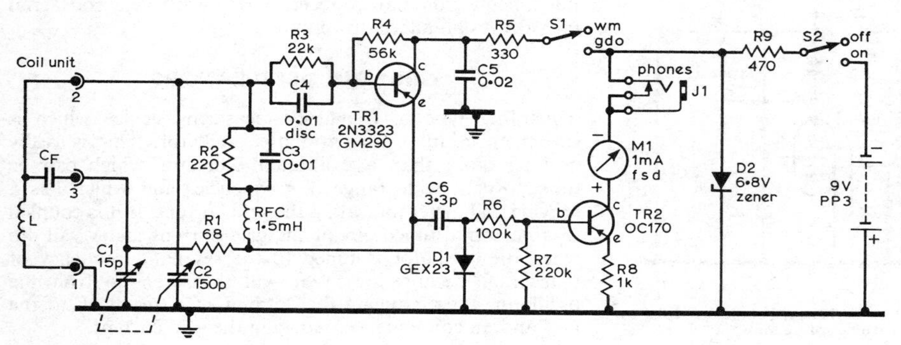

Fig 18.33. The G3HBW dip oscillator covering 0·85-150MHz. The rf transistor TR1 is an inexpensive vhf type manufactured by Motorola. Alternatively, the AF102, AF118, AF139, AFZ12, 2N1742, GM0290 or GM0378 may be used. S1 is provided to enable the dip oscillator to be used as a sensitive wavemeter. D1 may be a GEX23, GEX54, OA70, OA71 or similar diode. ZDI may be any small 6·8V zener diode.

TABLE 18.2

Coil winding details

Coil No	Range (MHz)	No or turns	SWG	Former od (in)	Former length (in)	Winding length (in)	CF total (pF)
1	0·85 to 2·0	180	28	$\frac{7}{8}$	$4\frac{1}{8}$	$3\frac{1}{16}$ (close wound)	3 × 1,000
2	1·8 to 4·0	54	28	$\frac{7}{8}$	$2\frac{1}{2}$	$\frac{7}{8}$ (close wound)	2,200 + 820
3	3·4 to 8·0	27	22	$\frac{7}{8}$	$2\frac{1}{2}$	$\frac{7}{8}$	2,200 + 680
4	6·7 to 16	13	22	$\frac{7}{8}$	$2\frac{1}{2}$	$\frac{7}{8}$	3,300
5	13·5 to 34	6	22	$\frac{7}{8}$	$2\frac{1}{2}$	$\frac{7}{8}$	2,200
6	33 to 85	3	18	Wind on $\frac{3}{8}$in drill (see Fig 18.36(d))	shroud $\frac{7}{8}$in long	coil $\frac{1}{2}$in long	330
7	50 to 150	1	18	(see Fig 18.36(c))			68

shunting the greater part of the tuned circuit with the 1·5mH rf choke.

Provision is made to remove the dc supply from the oscillator transistor when required so that the instrument may be used as a sensitive wavemeter and also as a monitor by employing the phones jack in the dc amplifier collector.

Construction

The dip oscillator is built into a small 18swg aluminium box, provided with a close fitting, flanged lid, the box and lid being available ready made from H. L. Smith and Co, 281 Edgware Road, London W2.

The coils for each range plug into a socket on one end of the box. Ordinary three-pin battery plugs and sockets are used, the plug being glued and then forced into a length of Paxolin tubing of suitable diameter on which the coil is wound.

The battery is mounted in a small clamp inside the lid and connected to the rest of the instrument by means of flying leads. A normal toggle switch is used for the main on-off function and a small slide switch to select either D.O. or WAVEMETER operation. This was done so that there would be no confusion as to whether the instrument was switched off or not, which might have occurred if two similar switches had been used. Any type of insulated jack socket may be utilized, provided that it is of the shorting variety.

First, the Perspex dial cursor and aluminium battery clamp should be made according to **Figs 18.34(a)** and **(b)** respectively. It is necessary to be very careful when cutting and drilling the $\frac{1}{16}$in thick Perspex sheet. A hand drill is to be preferred to a power-operated one to avoid the risk of splitting. The large centre hole should be opened out from a suitable smaller size with a reamer or a round file. A useful ancillary to be employed when marking the dial may be made from 18swg aluminium sheet of the same dimensions as the Perspex cursor. Mark and drill the holes as in the cursor proper, using the latter as a marking-out template. Draw the centre line of the aluminium strip along its greater dimension and then mark out and drill four $\frac{1}{16}$in diameter holes along this centre line, at distances of $\frac{7}{16}$, $\frac{5}{8}$, $\frac{13}{16}$ and 1in from one end. Finally, cut carefully along the centre line, through the four holes and as far as the centre hole with a fine hacksaw.

When making the battery clamp, preform the inner bends first and then the outer, finally marking and drilling the holes. The holes may now be drilled in the box and lid (see **Fig 18.35**). The four holes for fixing the lid should be marked out through the corresponding holes in the box and then drilled and tapped 6BA.

The shaft of the tuning capacitor is cut off so that only about $\frac{5}{16}$in of its length is left protruding. Both capacitor trimmers, if fitted, are then opened out to their fullest extent and the capacitor is mounted in the box by means of three 4BA countersunk head screws, *not more than* $\frac{1}{2}$*in long*, using two 2BA full nuts on each as spacing washers and screwing tightly into the tapped holes in the capacitor endplate. *If the screws are too long, they are liable to damage the vanes of the capacitor beyond repair.* Before the capacitor is bolted home, insert a $\frac{3}{8}$in long 6BA cheesehead screw in to the lug fixing hole for the slow-motion drive, the head of the screw being inside the box. Next, the operating shaft of the slow-motion drive is cut off so that only about $\frac{1}{2}$in of it remains, the drive is fitted over the capacitor shaft and the slotted lug is bolted to the box, using a single 6BA spacing washer and a 4BA nut as packing, between the lug and the box. The clamping screws may now be tightened on the capacitor shaft by manipulating a small screwdriver through the hole in the side of the box which has been provided for the purpose.

A scale disc $2\frac{5}{8}$in in diameter is made from thin card or, better, from $\frac{1}{16}$in thick white plastic sheet, such as Ivorine, and a $\frac{5}{16}$in diameter hole is made in the centre together with two 8BA clearance holes $\frac{5}{8}$in apart, so that the disc may be

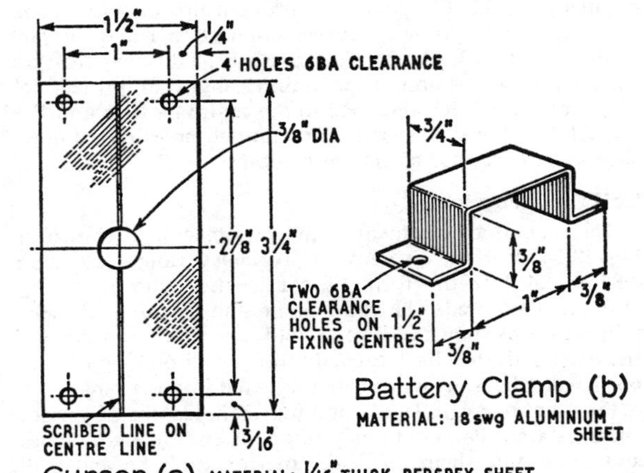

Fig 18.34. (a) The Perspex cursor dimensions and (b) the method of forming the battery clamp.

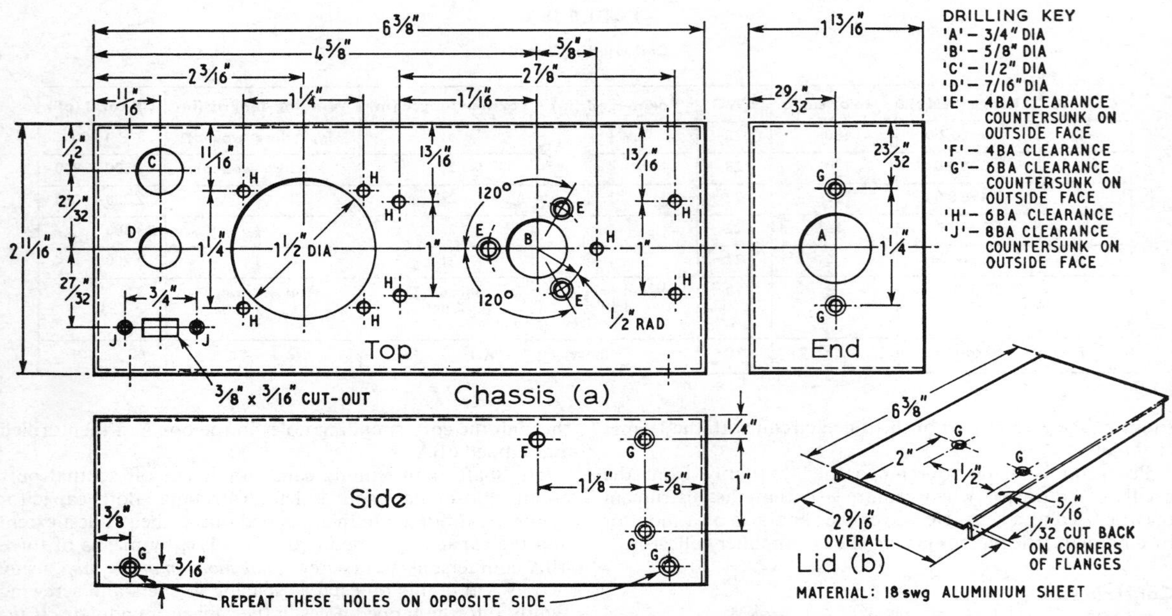

Fig 18.35. Drilling details suitable for a standard 6⅜ by 2 11/16 by 1 13/16 in aluminium box. The cursor is spaced above the top of the box, over the slow motion drive which is also mounted on the exterior of the box.

screwed to the drive flange, using two 8BA 3/16 in long countersunk head screws. The two tagstrip-mounting screws are ¾in long and serve also to support the lower end of the Perspex dial cursor, with two 19/32 in long pieces of 3/16 in od, 21swg wall, brass or aluminium tubing as packing pieces between the top of the box and the cursor. The other end of the cursor is supported in a similar way using two more 6BA screws, nuts and pieces of tubing.

Wiring up the instrument is a simple process. First, the double tagstrip arrangement should be partially completed as a unit, inserting R4, R5, R6, C5, C6 and D1 but not TR1. Then bolt the tagstrips in position in the box. Connect R1, R3, C4, R2, C3, RFC and also the interconnecting wires between the tuning capacitor, coil socket and tagstrips. The parallel pair, R2, C3, should be brought up vertically from coil socket connection 2, between connection 1 and the pin sockets of 2 and 3. The 2N3323 transistor, TR1, may then be soldered into position using a heat shunt to protect it. Next, attach R7, R8 and TR2 to the centre tagstrip and R9 and ZD1 to the slide switch. Two unused poles of the dpdt slide switch are used as anchoring points for R9.

Coil Construction

The general arrangement of the various coils is shown in **Fig 18.36**. For the five lowest frequency ranges covering 0·85 to 34MHz the form of construction shown in Fig 18.36(a) is adopted. The vhf coils L6 and L7 are made self-supporting, as shown in Figs 18.36(d) and 36(c) respectively. In all cases, the feedback capacitor (or capacitors) is mounted on the plug base, its wires being soldered into the pins.

First, cut one 4⅛in length and four 2½in lengths of ⅞in od, ⅛in thick wall, Paxolin tubing, filing the ends at right angles and smoothing them off. Take the longest tube and drill 1/16 in diameter holes 3/16 in and 3⅜in from one end on a line along the length of the tube. Do the same with the four

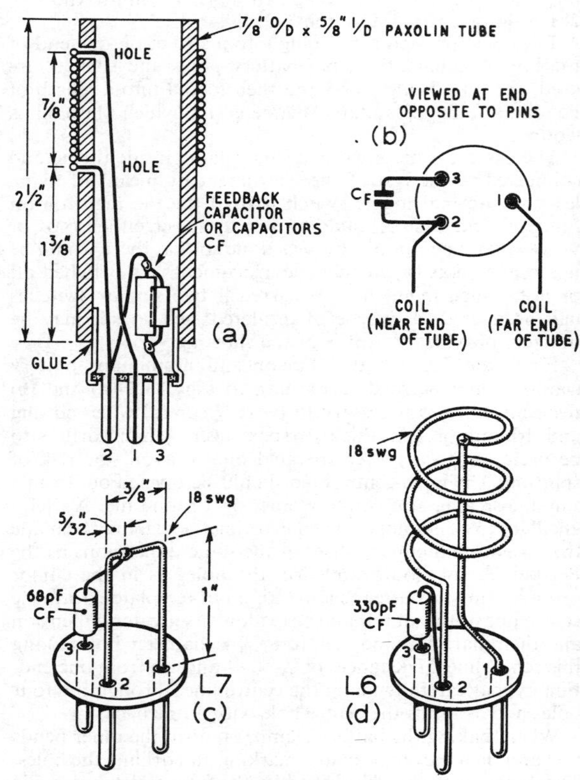

Fig 18.36. The coil assemblies. The five lower ranges are wound on Paxolin tubes, while ranges 6 and 7 are airwound.

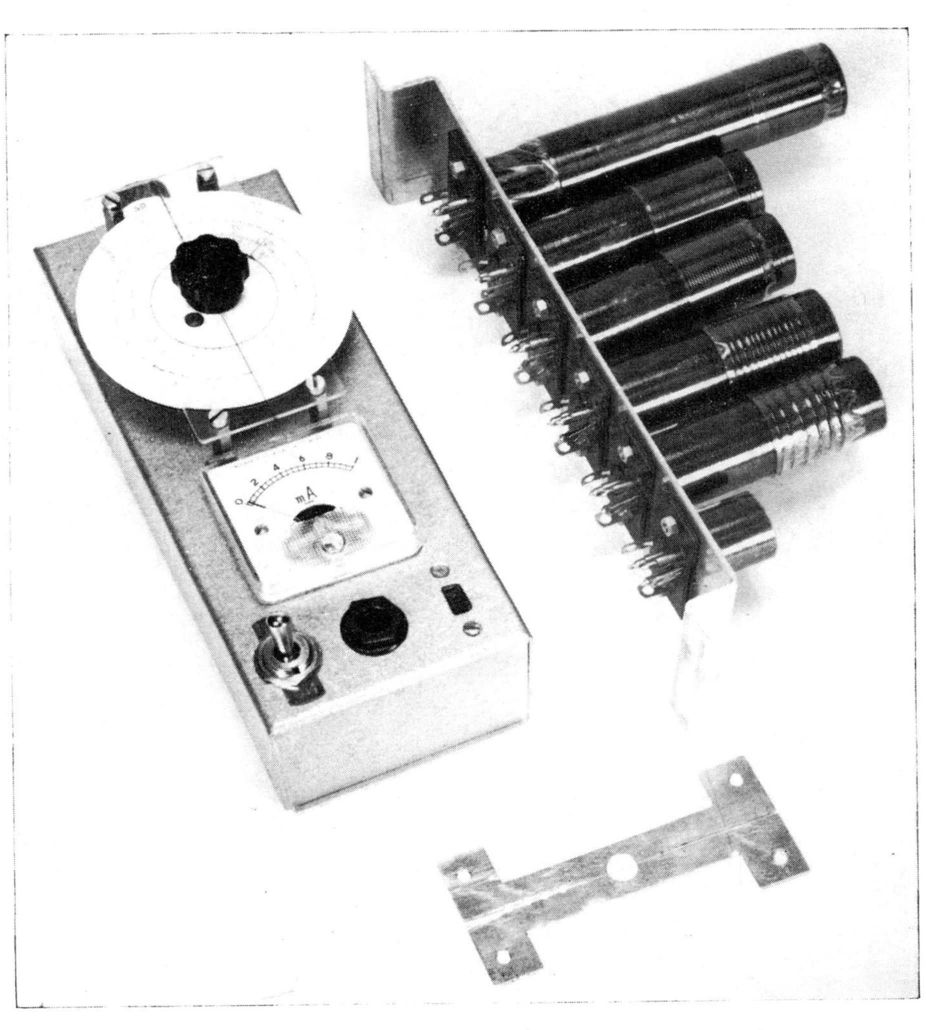

Complete transistorized dip oscillator. Coils should be handled with care, preferably at the base, and mounted on a stand as shown to avoid damage which could result in calibration errors.

shorter lengths, drilling the holes $\frac{5}{16}$in and $1\frac{1}{4}$in from one end. Remove the burrs and sharp edges from these holes both inside and out. Now cut off and file square two lengths of $\frac{3}{4}$in od, $\frac{1}{16}$in thick wall, Paxolin tube, both $\frac{7}{8}$in long (the other size of tubing will do if this is not available). Take the seven aluminium plug-shrouds and, supporting them carefully with a pair of narrow-nosed pliers, cut off the constricted portion with a fine hacksaw, leaving a $\frac{3}{16}$in length of the $\frac{5}{8}$in od aluminium tube portion with, of course, the four tabs still attached. Square off the end remote from the tabs and file a small chamfer on the outside at this end to provide a lead when inserting into the Paxolin tubes. Put a smear of Durofix on the outside of the aluminium tubes and then force each of them into one of the five longer Paxolin tubes, at the end remote from the holes, so that a $\frac{1}{16}$in length of the aluminium tubes, together with the four tabs, is left protruding. When the glue is dry, the coils may be completed.

The three lowest-frequency coils are close-wound (Table 18.2). Winding the coils for Ranges 1 and 2 is made easier if a simple procedure is adopted. Start at the plug end, passing the 28swg wire through the hole and then temporarily anchoring the free end with Sellotape. Hold the plug end of the former in the left hand and wind on about ten turns at a time, with a small spacing. Then, keeping the tension on the wire with the right hand, use the index finger of the left hand to push the turns together. When the coil is completed, pass the free end of the wire through the hole and anchor it temporarily with Sellotape, as at the start. The beginning and end turns of the coil may be secured with adhesive Melinex tape if desired, but ordinary Sellotape should not be used for a permanent job as it is hygroscopic.

The coils for Ranges 4 and 5 should be spaced out after winding to fill the available winding space. Take both ends of the coil down through the plug end of the tube, cut off so that only about 1in is left protruding and then bare the whole of this length, sliding an inch or so of loose-fitting sleeving on to each lead. Take the feedback capacitor appropriate to the range (Table 18.2). If two or three are required, parallel them as shown in Fig 18.36. Then pass one capacitor lead into pin 3 of a plug base, cut off and solder in

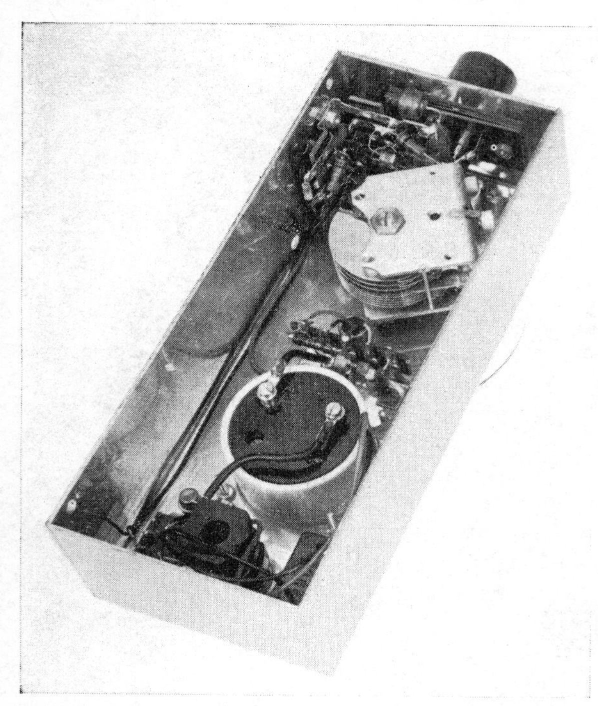

The underside of the transistor dip oscillator.

position, leaving the capacitor pointing away from the plug base as shown. Bend the other capacitor lead over and push into pin 2, but do not solder. Then offer up the plug base to the coil, pushing the coil leads into the appropriate pins, cut off the leads and solder. Test the coil in the dip oscillator before folding over the tabs.

The coil for Range 6 is wound on a $\frac{3}{8}$in diameter drill or rod as a mandrel. The forms of construction of L6 and L7 are self-evident from Figs 18.36(d) and 18.36(c). The two short pieces of $\frac{3}{4}$in od Paxolin tubing already prepared are used as protecting shrouds for these two vhf coils.

The finished appearance of the coils is shown on p18.19.

Testing and Calibration

The circuit connections should first be checked carefully. If all is well, plug in the Range 1 coil (0·85 to 2·0MHz) and, with the function switch set to D.O. and the tuning capacitor at mid-scale, switch on. A meter reading of from 0·5 to 0·7mA should result. If no meter reading is produced try the next Range 2 coil. Should success still not be obtained, check the polarity of the detector diode D1: its red (positive) end should be connected to the chassis. Correct operation of the dc amplifier may be ascertained by momentarily bridging the collector and base connections on the TR2 tagstrip with a 470kΩ resistor, which should give a reading of about 0·4mA on the meter. If this is successful, listen for the signal from the oscillator on a receiver. If nothing is heard, disconnect one end of D1 and listen again, as a diode with very poor reverse characteristics may prevent oscillation. Using methods such as these, the fault should be localized.

When all the ranges have been made to oscillate, check the frequency coverage, which should not differ by more than a few per cent from the frequencies quoted in Table 18.2. If the frequencies all seem too low, particularly at the hf ends, the trimmers on the tuning capacitor have probably not been unscrewed. It will usually be found easier to break the strip leads to the trimmers with a small pair of pliers, or unsolder them carefully.

The greatest meter deflection is usually obtained near the hf end of Range 3 or 4. This should be almost full scale. If the deflection is too large or too small, it may be adjusted by reducing or increasing respectively the value of R7.

It may perhaps be found that some coil ranges will not oscillate over the whole of the tuning capacitor travel; usually the hf ends are affected. If only one or two ranges are defective, the value of the built-in feedback capacitor should be reduced by about five or 10 per cent. However, if several ranges are unsatisfactory, a 2N3323 transistor of exceptionally low gain may have been used for TR1 and in this case it should be either changed, or the value of the negative feedback resistor R1 should be reduced to 56Ω or even lower until the trouble is cured. Check all the coil Ranges 1 to 7 in the oscillator, tuning right round on each range. The indicated currents should vary smoothly across each range without spurious dips or peaks. It will be found possible on any particular range to use the self-resonance of the coil two ranges lower to check the dip oscillator action. Oscillation should be completely stopped at resonance, with only moderately close coupling between the coils. The meter deflection on Range 7 becomes rather small below about 60MHz, but only the upper portion is required as the Range 6 coil covers the frequencies below 85MHz.

The instrument is now ready for calibration. Four concentric circles are drawn on the dial, using differently coloured ballpoint pens. The radii of the four circles are $\frac{5}{8}$, $\frac{13}{16}$, 1 and 1$\frac{3}{16}$in respectively. The capacitor is then completely unmeshed and a line is drawn along the diameter of the disc.

The easiest method of calibration is to listen for the oscillator signal on a general coverage receiver. Above 30MHz a 2 or 4m receiver may be used and, by listening to harmonics of the oscillator, the sub-multiple frequencies as well may be checked. A tv or fm receiver will provide further calibration points. Simple absorption wavemeters may be employed to check that spurious signals are not being received.

Both the upper and lower halves of the dial are calibrated, reversing the dial-marking template to suit. It is best to make the Range 1 and 2 calibrations on the outside of the $\frac{5}{8}$in radius circle, Ranges 3 and 4 on the outside of the $\frac{13}{16}$in radius circle and so on up to Range 7, leaving one space on the outside circle as a "spare". An "H" pencil is probably the best writing instrument for the job.

Using the Dip Oscillator

Determination of the resonant frequency of a tuned circuit. The resonant frequency of a tuned circuit is found by placing the d.o. coil close to that of the circuit and tuning for resonance. No power should be applied to the circuit under test, and the coupling should be as loose as possible consistent with a reasonable dip being obtained on the indicating meter. The size of the dip is proportional to the Q of the circuit under test, a circuit having a high Q producing a more pronounced dip than one having only a low or moderate Q.

Absorption wavemeter. By switching off the ht and coupling the d.o. in the usual manner, the instrument may be used as an absorption wavemeter. In this case power has to be applied to the circuit under test. Resonance is detected by a deflection on the meter due to rectified rf. It should be noted that an absorption wavemeter will respond to a harmonic if the wavemeter is tuned to its frequency.

Capacitance and inductance. Obviously, if an instrument has the ability to measure the frequency of a tuned circuit, it can also be used for the determination of capacitance and inductance. To measure capacitance, a close-tolerance capacitor (C_S) is first connected in parallel with a coil (any coil will do, providing it will resonate at a frequency within the range of the d.o.). This circuit is coupled to the oscillator and its frequency (F_1 MHz) noted. The unknown capacitor (C_X) is then connected in place of C_S and the resonant frequency again determined. If this is now F_2 MHz, the unknown capacitance is given by:

$$C_X = \frac{F_1^2 C_S}{F_2^2} \text{pF}$$

Similarly, inductances may be measured by connecting the unknown coil in parallel with a known capacitance and applying the formula

$$L = \frac{25{,}300}{CF^2}$$

where L is in μH, C is in pF and F_2 is the resonant frequency expressed in MHz.

Signal generator. For receiver testing the d.o. may be used to provide unmodulated cw signals by tuning the oscillator to the required frequency and placing it close to the aerial terminal of the receiver. The amplitude of the signal may be controlled by adjusting the distance between the d.o. coil and the aerial terminal.

AF signals can be injected at the jack socket on the d.o. to provide a modulated test signal, the modulation depth being dependent on the level of the modulation voltage, 5V being adequate.

CW and phone monitor. By connecting a pair of headphones to the instrument, it may be used as a cw monitor. It should be tuned to the transmitter frequency and the distance between the oscillator and transmitter adjusted for optimum signal strength. To monitor telephony transmissions the ht is switched off.

Checking crystals. It is possible to check the activity and frequency of oscillation of a crystal by inserting it in the coil socket of the d.o. and setting the frequency dial to maximum (ie minimum tuning capacitance). The meter reading will vary in accordance with the activity of the crystal and the frequency may be checked by locating the oscillation on a calibrated receiver. Generally speaking, this procedure may only be applied to fundamental mode crystals.

TRANSMITTER POWER OUTPUT MEASUREMENT

Although amateur licence conditions do not require the measurement of output except in the case of ssb transmitters, it is desirable to measure transmitter output if only to establish that the power amplifier is operating under correct conditions. Inadequate power output could easily be a sign of excessive dissipation in the pa valve or transistor. Power is normally measured when the transmitter is working into a dummy resistive load which presents the correct load impedance through the coupling system to the pa.

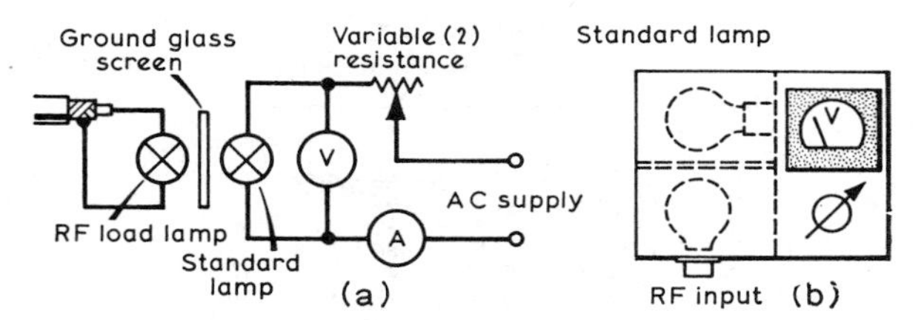

Fig 18.37. (a) Circuit arrangement of the two lamps. (b) Suggested layout. An effective "ground glass" screen may be made by using a sheet of Perspex, one side of which has been roughened with a household abrasive such as Vim. Suitable rotary variable resistors are available from P. X. Fox, Curtis and Berco.

Such a load is in any case required to permit non-radiating adjustments to be made to the transmitter. At frequencies below 30MHz an rf ammeter measuring the rms current into the load gives a satisfactory indication of power but its accuracy declines as frequency increases.

Effect of Modulation on Power Readings

When a carrier is 100 per cent sine-wave modulated the mean power increases by 50 per cent. This increase will be shown on output power measurements and may be used as a good indication of adequate upward modulation. It is useful to note that aerial current will increase by only $\sqrt{1{\cdot}5} = 1{\cdot}22$ for 100 per cent modulation.

Lamps as Dummy Loads

The rf output from a transmitter may be dissipated in a lamp and the brightness of the filament used as an indication of the power output. It is unlikely that the transmitter can be operated into exactly the load impedance offered by the aerial since the resistance of a lamp filament varies with its temperature. It is extremely difficult to guess the power in a lamp by observing the brightness and some form of comparison should be arranged. A good method is to use two identical lamps, one connected as the rf load and the other supplied from a metered variable power source. This is best arranged by fitting the two lamps side by side in a box having a ground glass screen over the lamps, and a dividing panel between them. The two lamps will then each illuminate the two separate halves of the ground glass screen. A suitable arrangement is shown in **Fig 18.37.**

It is obviously desirable to choose a lamp having about the required impedance and **Tables 18.3** and **18.4** and the curves of **Fig 18.38** will assist in choosing suitable lamps.

For powers in excess of about 50W lamps may still be used but a load consisting of a resistor is a more convenient way of dissipating the power. Power measurement would then have to be made by voltage measurement across the load, by current measurement using a thermal meter or thermocouple or by using a directional wattmeter.

TABLE 18.3

Lamp class	V	W	A	Ω
Aircraft (General)	24	6	0·25	96
Aircraft (General)	28	18	0·645	43·5(*a*)
Aircraft (General)	24	10	0·415	58
Small Projector	50	25	0·5	100
Indicator	16	3	0·188	85
Indicator	12	2·2	0·183	65·5

(*a*) Two in series for 36 watts at 87 ohms.

TABLE 18.4

Lamp, small projector type, 50V 25W

V	A	W	Ω	Lumens (%)
50	0·5	25	100	100
45	0·47	21·5	95	70
40	0·44	17·6	90	46
35	0·425	14·9	84	31
30	0·375	11·25	78	20
25	0·34	8·5	71	10

The type of resistor in which the resistive element is coated on the outside of a ceramic tube of 0·5in to 1in diameter is suitable. The resistor should be regarded as the outer section of a coaxial line and a centre conductor of copper mounted inside it. The diameter of the inner line should be chosen to give the correct characteristic impedance. For a 1in diameter resistor and an impedance of 75Ω, an inner conductor of about 0·3in diameter is needed. The inner conductor should be terminated in the centre connection of a suitable coaxial socket at one end and connected to the resistance by a low inductance path at the other end. The whole unit should be screened to prevent radiation.

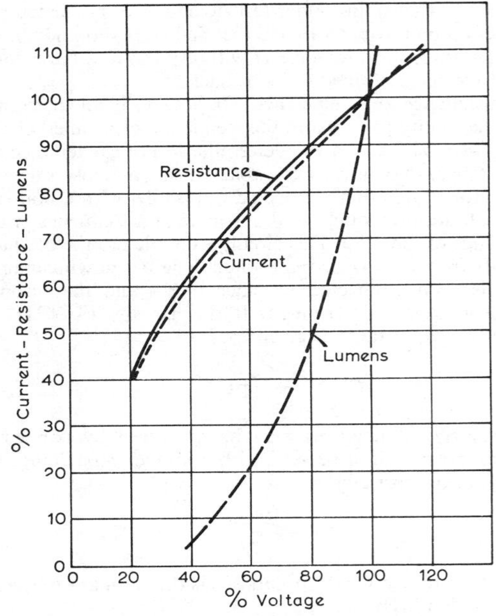

Fig 18.38. Characteristics of gas-filled lamps.

AERIAL AND FEEDER MEASUREMENTS

RF ammeters and voltmeters do not reveal the true power unless the line in which they are connected is correctly matched, but they may always be used as a means of tuning up the transmitter and the aerial coupler. The reflectometer is, however, an instrument which can distinguish between the forward and reflected waves in a standing-wave system, and can therefore be used to indicate a true match of impedance, because only the forward wave is radiated. Such an instrument can be calibrated to read true power flow into the aerial.

HF REFLECTOMETER

Reflectometers designed as vswr indicators have normally used sampling loops capacitively coupled to a length of transmission line. This results in a meter deflection roughly proportional to the frequency, and they are therefore unsuitable for power measurement unless calibrated for use over a narrow band.

By the use of lumped components this shortcoming can be largely eliminated and the following design may be regarded as independent of frequency up to 70MHz.

The circuit is shown in Fig 18.39 and uses a current transformer in which the low resistance at the secondary is split into two equal parts. The centre connection is taken to the voltage sampling network so that the sum and difference voltages are available at the ends of the transformer secondary winding.

Layout of the sampling circuit is fairly critical. The input and output sockets should be a few inches apart and connected together with a short length of coaxial cable. The coaxial outer must be earthed at one end only so that it acts as an electrostatic screen between the primary and secondary windings of the toroidal transformer. The layout of the sensing circuits in a similar instrument is shown in the photograph.

The primary of the toroidal transformer is formed by

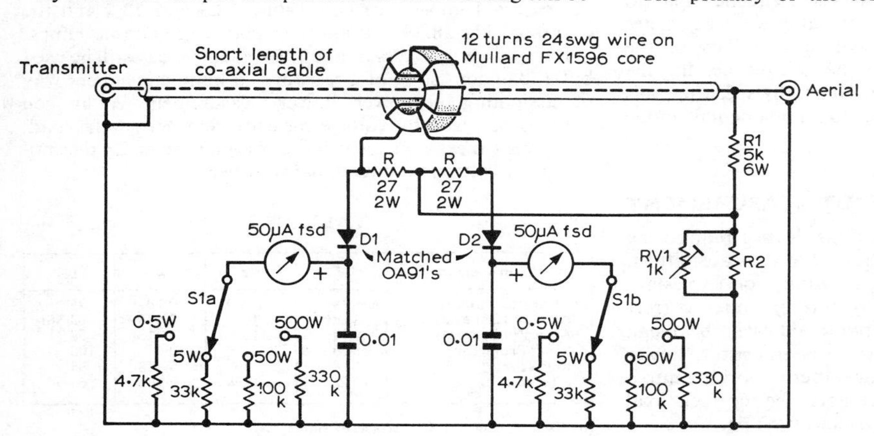

Fig 18.39. Circuit of the frequency-independent directional wattmeter, with four ranges corresponding to FSDs of 0·5, 5, 50 and 500W in 50Ω lines, when the value of R2 (including RV1, if fitted) should be 220Ω. For 75Ω systems R2 = 150Ω, and the calibration is different. The coaxial cable acts as an electrostatic screen between its centre conductor and the secondary winding of the toroidal transformer: the cable length is unimportant.

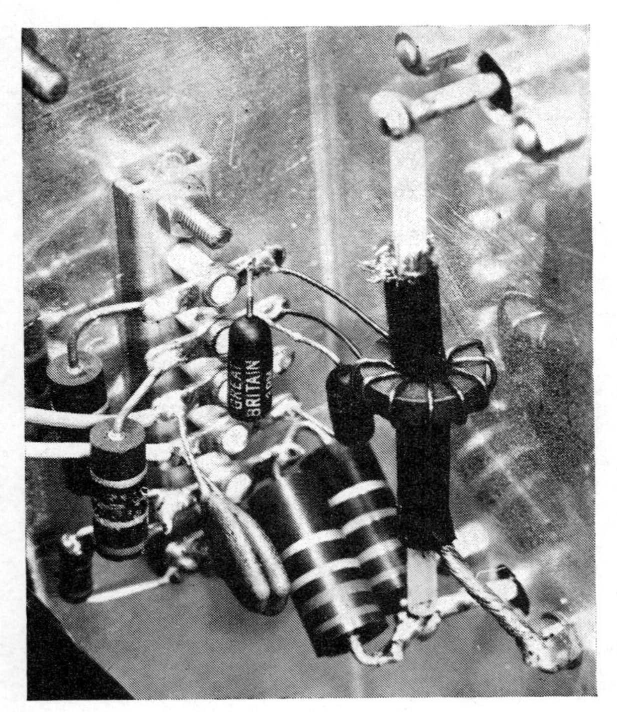

View of the sensing circuits of the frequency-independent directional wattmeter.

simply threading a ferrite ring on to the coaxial cable. Twelve turns of 24swg enamelled wire equally spaced around the entire circumference of the ring form the secondary winding. The ferrite material should maintain a high permeability over the frequency range to be used; a suitable ferrite ring is the Mullard FX1596.

Other components in the sampling circuits should have the shortest possible leads. R1 and R2 should be non-inductive carbon types. For powers above about 100W, R1 can consist of several 2W carbon resistors in parallel. RV1 should be a miniature skeleton potentiometer to keep stray reactance to a minimum. The detector diodes D1 and D2 should be matched point-contact types with a piv rating of about 50V. OA79 and OA91 diodes are suitable. The current transformer resistors should be matched to five per cent.

The ratio of the sampling resistors R1 and R2 is determined by the sensitivity of the current sensing circuit. As the two sampling voltages must be equal in magnitude under matched conditions, RV1 provides a fine adjustment of the ratio.

Calibration

Accurate calibration requires a transmitter and an rf voltmeter. The wattmeter is calibrated by feeding power through the meter into a dummy load of correct impedance. RV1 is adjusted for minimum reflected power indication and the power scale calibrated according to the rf voltage appearing across the load. The reflected power meter is calibrated by reversing the connections to the coaxial line.

This instrument has full scale deflections of 0·5, 5, 50 and 500W selected by the range switches. These should not be ganged since the reflected power will normally be much less than the forward power.

FIELD STRENGTH METER

Field strength meters are normally used as indicators to maximize the radiated power. An absorption wavemeter used with a simple aerial is adequate but the lack of linearity of a simple rectifier milliammeter indicator tends to distort the effects of aerial modifications. A field strength-meter with an untuned input system and a two-stage amplifier following the diode rectifier is shown in **Fig 18.40**.

NOISE BRIDGE

The noise bridge is an rf bridge energized by a wideband noise which derives from a zener diode operating at a low current level. This noise extends up to at least 200MHz. The noise source is followed by an amplifier to raise the level so that a receiver may be used as a null indicator. The noise output is applied to a quadrafilar-wound toroid providing two arms of a bridge circuit. A variable resistor in the third arm is used to obtain a balance against the aerial connected in the fourth arm. When the noise across the resistance arm and the aerial are equal, the bridge is balanced. Coincident adjustment of the receiver tuning and of the variable resistance is necessary to discover the point of minimum noise which occurs at the resonant frequency and radiation resistance of the aerial.

The basic circuits are shown in **Figs 18.41** and **18.42**. A ferrite core of about 0·25in internal diameter is suitable and the winding may consist of four lengths of 28swg enamelled copper wire twisted together to form one multistrand length. Five or six turns of this twisted group are spaced around the circumference of the core. The four wires are divided into two pairs and each pair connected in series so that the end of one wire is connected to the beginning of the other. The variable resistor should be a carbon type with a linear track and calibrated in ohms.

Operation of the Noise Bridge

With a receiver connected to the secondary winding centre point and a resistor of 50 or 75Ω across the aerial terminals, the noise minimum should be obtained when the variable resistor is set to the value of the test resistor.

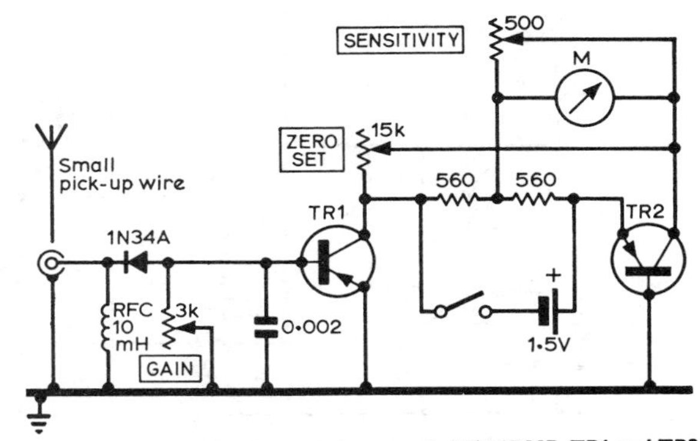

Fig 18.40. High-gain field-strength meter by WA4DXP. TR1 and TR2 may be any type of general purpose transistor.

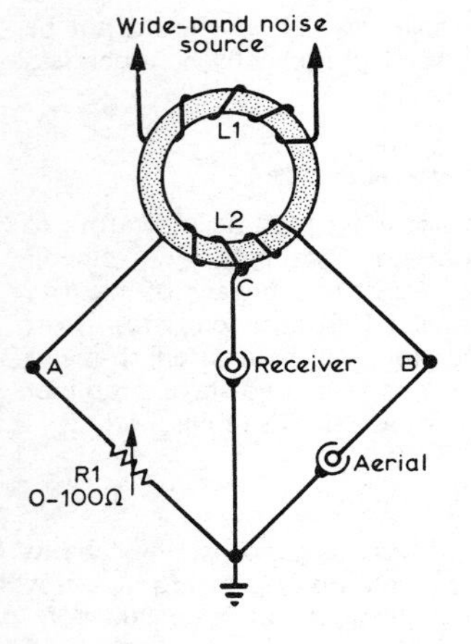

Fig 18.41. The bridge configuration of the aerial noise bridge. The two bifilar windings L1 and L2 are interwound on a toroid core.

With the aerial replacing the test resistor, a noise minimum should occur with the receiver tuned to the design frequency of the aerial and approach null as the resistor is set to the characteristic impedance of the line.

SIGNAL GENERATORS

In setting up and testing radio and electronic equipment it is frequently necessary to have an af or rf signal of suitable frequency. AF oscillators are used in the testing of speech amplifiers and modulators and in setting up ssb transmitters while rf signal generators are of great value in the alignment of communication receivers and converters.

SIMPLE AF OSCILLATOR

The generator shown in **Fig 18.43** will give an output of up to 1V rms over a frequency range of 15Hz to 200kHz in four switched ranges. A single range might be adequate for strictly amateur use but the extended range is useful if hi-fi equipment is also used. The power requirements are 9V at 20mA but the oscillator will continue to operate down to 6V. A simple switched output attenuator gives maximum output of 1V or 0·1V rms.

The frequency is adjusted by the two-gang potentiometer RV1/RV2 and switched by S1, S2 changing capacitors. The ranges are:

1. 15Hz to 200Hz
2. 150Hz to 2kHz
3. 1·5kHz to 20kHz
4. 15kHz to 200kHz

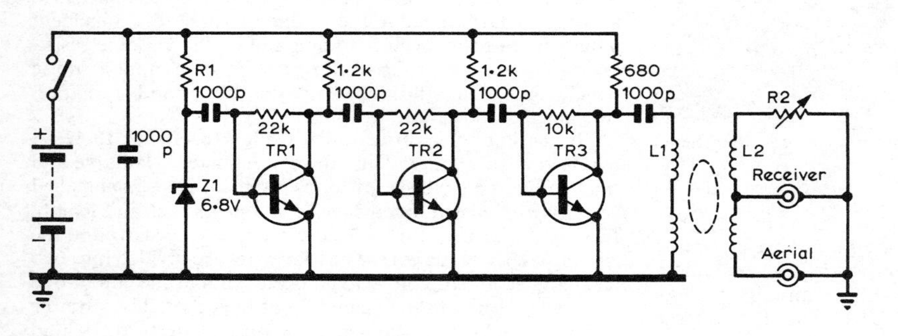

Fig 18.42. Basic circuit diagram of the noise bridge. Suggested equivalents for the diode and transistors are 2F6·8 and 2N708 respectively.

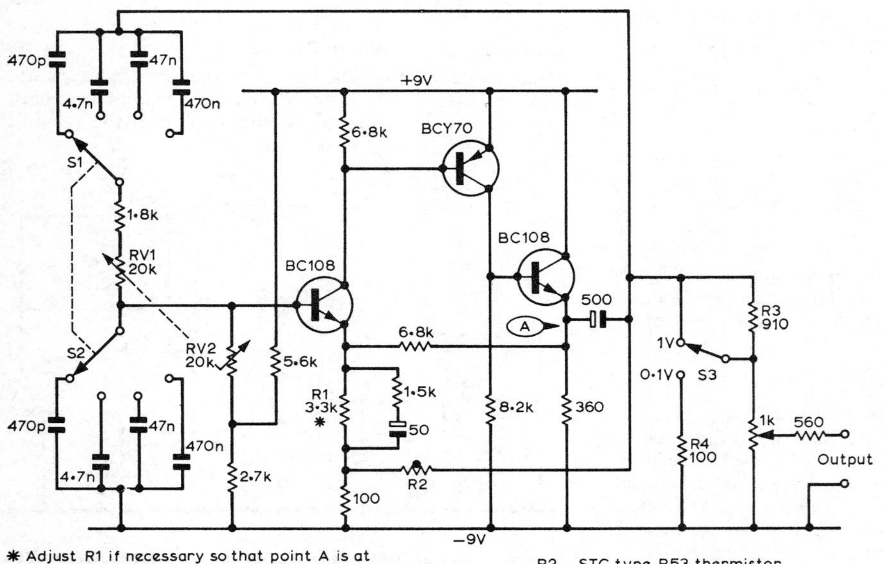

Fig 18.43. A wide-range audio frequency generator using a Wien bridge oscillator with a thermistor (R2) amplitude control device.

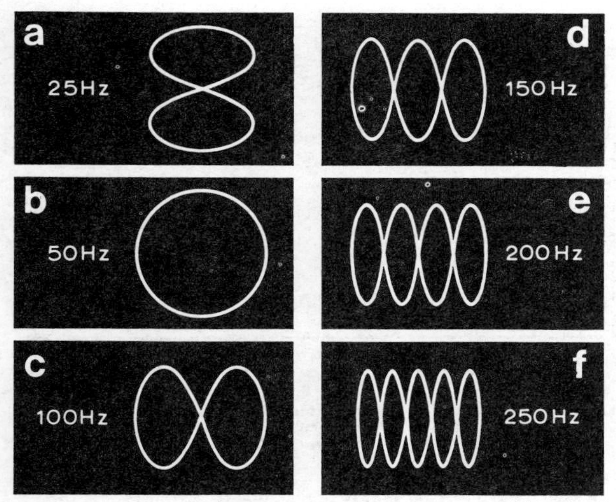

Fig 18.44. Calibration of an audio oscillator by means of Lissajous' figures using 50Hz ac mains.

Lissajous' figures are often suggested as the best means of calibrating an instrument of this type. The method of using them is as follows. With a 50Hz (mains frequency) ac voltage fed into the horizontal deflecting plates of an oscilloscope, the output of the audio oscillator should be applied to the vertical plates and the tuning control adjusted until the pattern of **Fig 18.44(a)** appears on the screen. The oscillator frequency is then 25Hz and the scale may be appropriately marked. The tuning control may then be rotated until the pattern of **Fig 18.44(b)** appears, which indicates a frequency of 50Hz. Similarly the calibration points relating to the range 100–250Hz may be fixed by reference to the patterns shown; finally the whole scale may be calibrated accurately by interpolation. There are of course many intermediate Lissajous' figures which may be obtained in addition to those shown. Readers who wish to work out the frequencies for which these apply may do so from the method given in **Fig 18.45**.

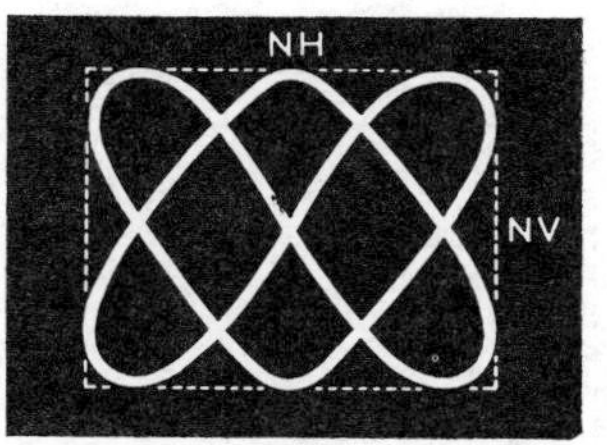

Fig 18.45. If a sine wave of unknown frequency is fed to the horizontal deflecting plates of an oscilloscope and a signal from a calibrated source is fed to the vertical plates, a stationary pattern such as that shown above may be obtained by adjusting the frequency of the calibrated source. The number of loops in both horizontal and vertical planes should be counted, and the unknown frequency may then be calculated from the following equation: Unknown frequency = NV/NH × frequency of calibrated source. In the diagram, the unknown frequency is $\frac{2}{3}$ of the calibrating frequency.

Above 400–600Hz calibration against the mains ceases to be practicable. One way to go from there is to use a second oscillator which may be set to an intermediate point (say 200Hz) and used to calibrate the first oscillator by means of Lissajous' figures up to approximately 1kHz. The auxiliary oscillator may thereupon be set to 1kHz and the process repeated in sequence until the full range has been covered. The principal requirement of the auxiliary oscillator is frequency stability; purity of waveform is unimportant.

RF SIGNAL GENERATOR

The rf signal generator shown in **Fig 18.46** covers 80kHz to 56MHz and employs a transitron oscillator (V1) modulated by a conventional af oscillator (V2).

Approximately 3–4mA anode current and 6–8mA screen current is drawn by V1, the sum of the two being practically constant for any setting of RV2. The value of R5 depends

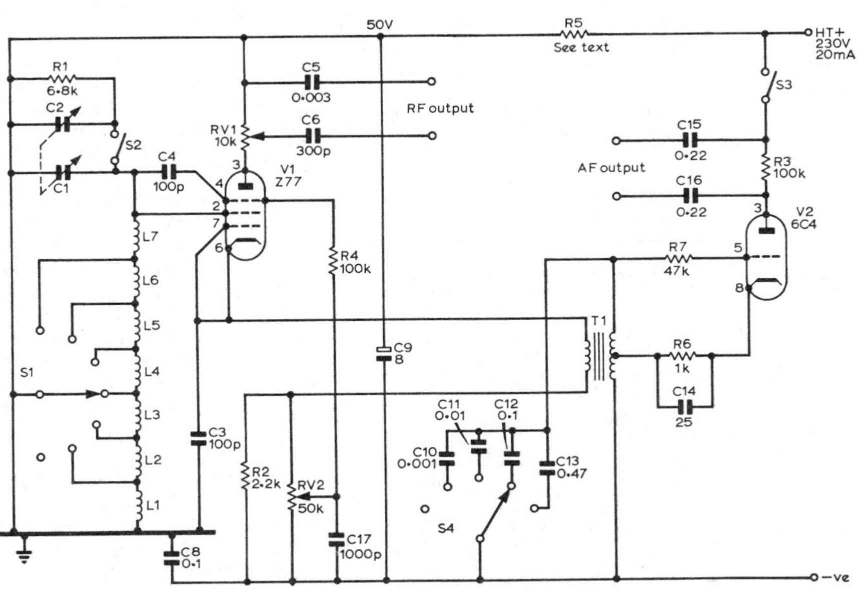

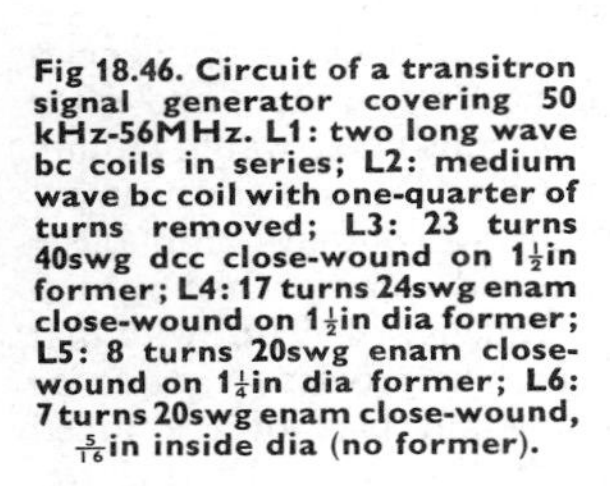
Fig 18.46. Circuit of a transitron signal generator covering 50 kHz-56MHz. L1: two long wave bc coils in series; L2: medium wave bc coil with one-quarter of turns removed; L3: 23 turns 40swg dcc close-wound on 1½in former; L4: 17 turns 24swg enam close-wound on 1¼in dia former; L5: 8 turns 20swg enam close-wound on 1¼in dia former; L6: 7 turns 20swg enam close-wound, $\frac{5}{16}$in inside dia (no former).

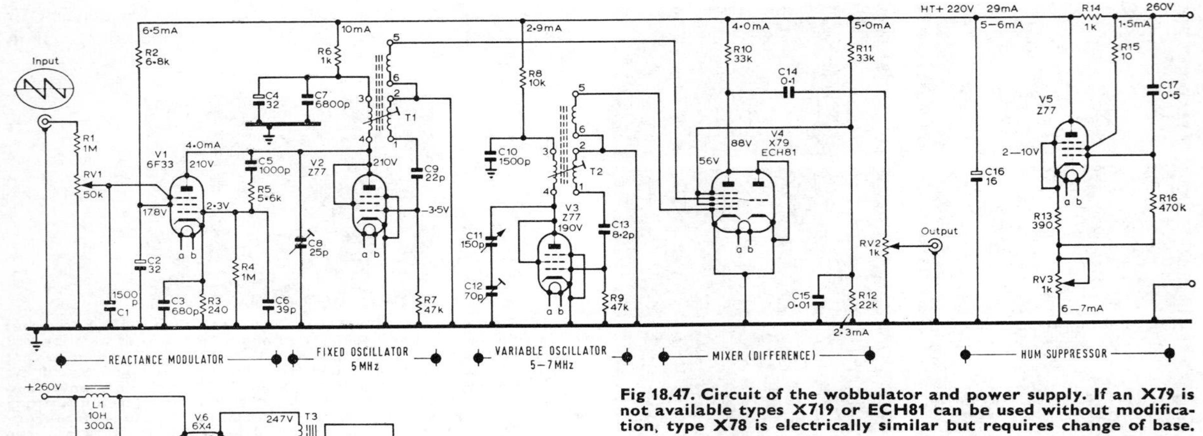

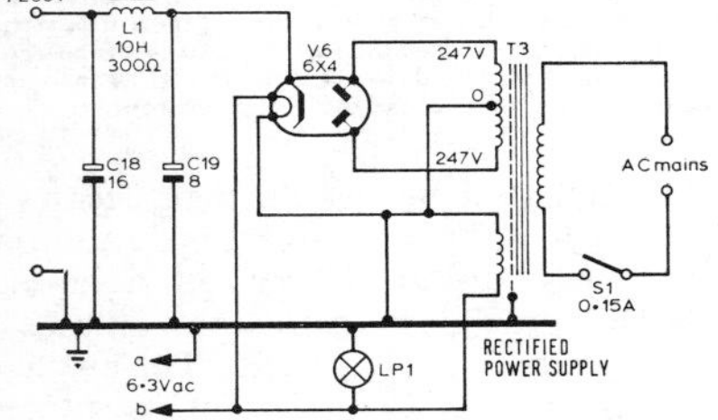

Fig 18.47. Circuit of the wobbulator and power supply. If an X79 is not available types X719 or ECH81 can be used without modification, type X78 is electrically similar but requires change of base.

on the ht supply and should be arranged to produce 50V at the point indicated. Construction is simplified if this 50V rail is earthed instead of the negative as is usual practice.

The seven-range coil assembly consists of inductances in series which are tapped by S1. One end of the highest frequency coil (L7) should be mounted directly on the screen grid pin of the valveholder, the switch being arranged close to it. Winding of the coils should commence from the L7 end to make sure that there are no gaps. Some experiment may be necessary with the number of turns.

A 500pF twin-gang capacitor with ceramic insulation is used for C1, C2. The C1 section is reduced to approximately 100pF by removing vanes; it is used on Ranges 4 to 7. For Ranges 1 to 3, C2 is switched in parallel with C1 by means of S2.

The frequency ranges covered by the capacitors suggested and the coils specified are as follows: Range 1, 80 to 220kHz; Range 2, 210 to 760kHz; Range 3, 660 to 2,700kHz; Range 4, 2·6 to 6·5MHz; Range 5, 5·6 to 15·8MHz; Range 6, 14 to 29MHz; Range 7, 24 to 56MHz.

Attenuation of the rf output is provided by RV1, a non-inductive carbon type potentiometer. The arrangement is unorthodox but it does provide an easy means of attenuating the output. Modulation is by cathode injection, the audio oscillator V2 providing four fixed tones selected by S4.

Screening of the entire unit, including the power pack, is desirable in order to avoid signal leakage. The tuning dial should be fitted with a vernier so that it may be read to one-tenth of a degree.

Calibration can be carried out with the aid of a frequency meter or by listening on a receiver for beats with known broadcast stations and short-wave stations.

TABLE 18.5

Details of oscillator coils

Fixed Frequency Oscillator

T1—$\frac{7}{16}$in dia Aladdin F804/PP5892 former with dust core.

Main tank winding	34 turns of 32swg
Feedback winding	16 turns of 32 swg
Coupling winding	4½ turns of 26swg

All windings single layer, close-wound, enamel. Feedback winding on top of main winding near earthy end.

Coupling winding spaced $\frac{1}{8}$in from main tank winding. Main tank winding inductance approximately 17μH, Q approximately 70.

Variable Frequency Oscillator

T2—$\frac{7}{16}$in dia Aladdin former with dust core.

Main tank winding	30 turns of 32swg
Feedback winding	12 turns of 32swg
Coupling winding	4¾ turns of 26swg

All windings single layer, close-wound, enamel. Feedback winding on top of main winding near earthy end.

Coupling winding spaced $\frac{1}{8}$in from main tank winding. Main tank winding inductance approximately 15·6μH, Q approximately 70.

WOBBULATOR FOR RECEIVER ALIGNMENT

The main function of a wobbulator is to simplify the alignment of i.f. strips and bandpass filters to produce a clean, flat-topped, steep skirted response with minimum side lobes. Using a signal generator and a vtvm, the time consumed can be considerable because a new graph has to be constructed following every major adjustment; with a wobbulator and oscilloscope, the pass-band can be observed at a glance, and even the effect of an adjustment actually being made can be continuously monitored.

As there are a number of different intermediate frequencies in general use, extending from 50kHz to 2MHz and often beyond, a simple oscillator to cover this range would require several stages of coil switching. An alternative arrangement was, however, chosen which produces an output continuously variable over the desired range. This is accomplished by mixing the output of a 5MHz fixed-frequency oscillator with that of a 5 to 7MHz variable oscillator. The process of frequency modulating the output occurs within part of the tuned circuit of the fixed oscillator, which ensures that the frequency deviation remains constant regardless of the output frequency, this being another reason for the choice of the particular method of generation. The maximum permissible sawtooth wave injection to the 6F33 reactance modulator V1 (**Fig 18.47**) to produce a linear frequency deviation is able to shift the nominal 5MHz by approximately 60kHz which should be adequate for all normal purposes. In order to present a single display on the oscilloscope, the signal injected into V1 should be of the same repetition frequency as the X scan on the oscilloscope, and also be a clean sawtooth. The wobbulator has therefore been designed without the inclusion of a sweep-frequency oscillator, for the output of the timebase of the oscilloscope in use should be perfectly satisfactory. Indeed, any irregularities in the wave shape become unimportant when the same oscillator is used in both cases. This method also eliminates the necessity for synchronization of oscillators.

The 6F33 (V1) or S6F33 was chosen as the reactance modulator because of its superior linear mutual conductance/suppressor grid voltage characteristic. It is important to see that both the anode and screen are adequately decoupled with respect to the sweep frequency.

The oscillator (V2, V3) frequencies were originally chosen as a compromise between stability of output centre frequency and ease of obtaining the required tuning range. As the output frequency is dependent on the difference frequency between the two oscillators 5MHz was not considered unduly high for stability. The oscillators being electrically similar, conditions causing a shift in one oscillator will similarly affect the other. The difference frequency should thus remain constant.

Should the necessity arise, the frequency range can be extended by the modification of one or both oscillators, with due consideration for unwanted mixing products.

The oscillator outputs are fed into the mixer (V4) and as more than sufficient output was available, the anode load of V4 was deliberately made low.

The instrument will work satisfactorily from an ht supply of 220 to 250V, and to avoid low-frequency cyclic distortion of the display, the smoothing must be of a high order. A conventional capacitance-input filter followed by a hum suppressor was found to give better results than two identical filter sections. The ac ripple is minimized by adjustment of RV3 in the cathode circuit of the hum suppressor, and in the case of the prototype was reduced to less than 2mV rms on full load (30mA) with no sweep voltage applied.

Establishing the correct nominal frequencies may be accomplished in one of three ways. After reducing the variable oscillator tuning capacitor to minimum, the fixed oscillator may be aligned by alteration of the slug of T1 and also C8, together with any one of the following three methods of frequency determination.

(a) Absorption wavemeter.
(b) Zero beat with a receiver tuned to 5MHz having its bfo on.
(c) Zero beat with a signal generator tuned to 5MHz, the outputs being mixed and displayed on an oscilloscope. The circuit of a suitable mixer is shown in **Fig 18.48.**

The variable oscillator must be set to cover the range 5MHz to 7MHz. For 5MHz, C11 should be positioned at maximum capacitance, and the slug of T2 adjusted for zero beat with the fixed oscillator. If with this setting, C1 is not capable of tuning the full 2MHz, alteration of the series padder C12 will improve the coverage, but the initial 5MHz adjustment will have to be repeated.

Calibration

The variable oscillator capacitor is calibrated directly in terms of difference frequency, and hence "centre frequency" when a sweep waveform is applied. It is recommended that the process of calibration be attempted when the outputs of the oscillators are not being mixed, as direct 5–7MHz frequency measurements leave less likelihood of unintentional measurement of spurious mixer products.

Calibration of the deviation control is the next step. With RV1 at minimum (zero sweep) the output can be mixed with a reference frequency from a signal generator, and tuned to zero beat. The combined output should for convenience be displayed on an oscilloscope. On variation of the generator frequency, by say, 10kHz, the zero beat pattern will immediately vanish until RV1 is altered to resume the condition. The process should be repeated until the deviation control is at maximum, calibration marks being applied to the dial of RV1 at each step. The deviation will always remain constant, regardless of the centre frequency, since the maximum output frequency of the frequency modulated oscillator cannot vary.

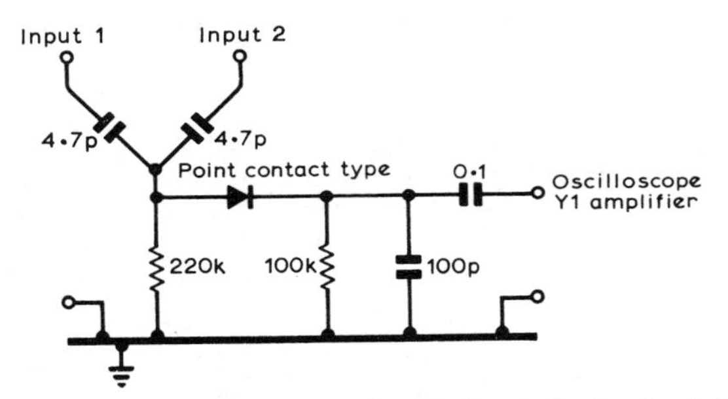

Fig 18.48. A simple mixer for use when aligning the fixed and variable frequency oscillators of the wobbulator.

Amplitude calibration of the output is unfortunately impracticable. The reason is that the output voltage varies in sympathy with variation of the centre frequency. Reference divisions of from, say, 1 to 10, however, are very useful.

A TWO-TONE TEST OSCILLATOR

A simple transistorized two-tone test oscillator for the alignment of ssb transmitters is shown in **Fig 18.49.**

The oscillators provide frequencies of 700kHz and 2kHz, the output from each being independently adjustable, giving a maximum of 1mW into 600Ω combined. The power supply required is 12–15V at 10mA.

The two test oscillators TR1 and TR4 are both single-stage phase shift oscillators using three-section ladder RC networks. This type of oscillator has been chosen for its simplicity and economy of components, though it does suffer from considerable dependence on transistor parameters. In many circuits embodying this type of oscillator, potentiometers are used to vary the conditions on the base or emitter in order to establish satisfactory oscillation. This should be unnecessary in this case, but should transistors other than those specified be used it is useful to remember that small changes to R3 or R15 will affect the waveform and amplitude of the oscillator outputs.

The signals at the collectors of the oscillators are coupled by the isolating resistors R6 and R13 to the gain controls RV1 and RV2. This dc coupling ensures that the controls vary both the signal amplitude and dc bias on the bases of

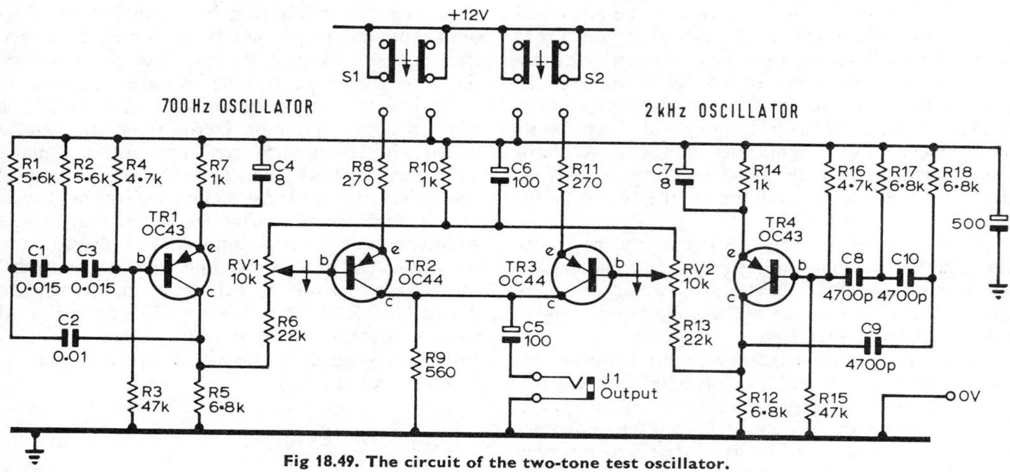

Fig 18.49. The circuit of the two-tone test oscillator.

the output amplifiers. Thus, when only a low output amplitude is required, the standing current of the Class A amplifiers is also low, ensuring maximum efficiency of these stages. The cold ends of the amplitude controls are taken to a fixed dc potential determined by R10 and C6. This prevents TR2 and TR3 from being cut off at very low amplitude settings.

The switches S1 and S2 each apply power to both oscillators and one of the output amplifiers. The output of each oscillator may therefore be obtained independently for setting up levels or response checks.

The common collector load for TR2 and TR3 ensures linear mixing of the two test tones and provides the required output impedance. Coupling to the load is through the 100μF capacitor C5. This is suitable for driving into the low impedances encountered in transistorized modulators, but if the application is restricted to valve amplifiers the use of a 0·01μF capacitor here may be necessary to reduce the effects due to the changing dc conditions in the output amplifier.

Pulsed Operation

If the prolonged application of sufficient drive to produce repetitive peak envelope power results in over-running the output devices of the linear amplifier, it becomes necessary to pulse the two-tone signal. An on/off ratio of about 1:1 should be sufficient and a pulse rate of several pulses a second makes observation on the oscilloscope quite easy. A suitable pulser for the two-tone source is shown in **Fig 18.50** in which TR1 and TR2 constitute a multivibrator

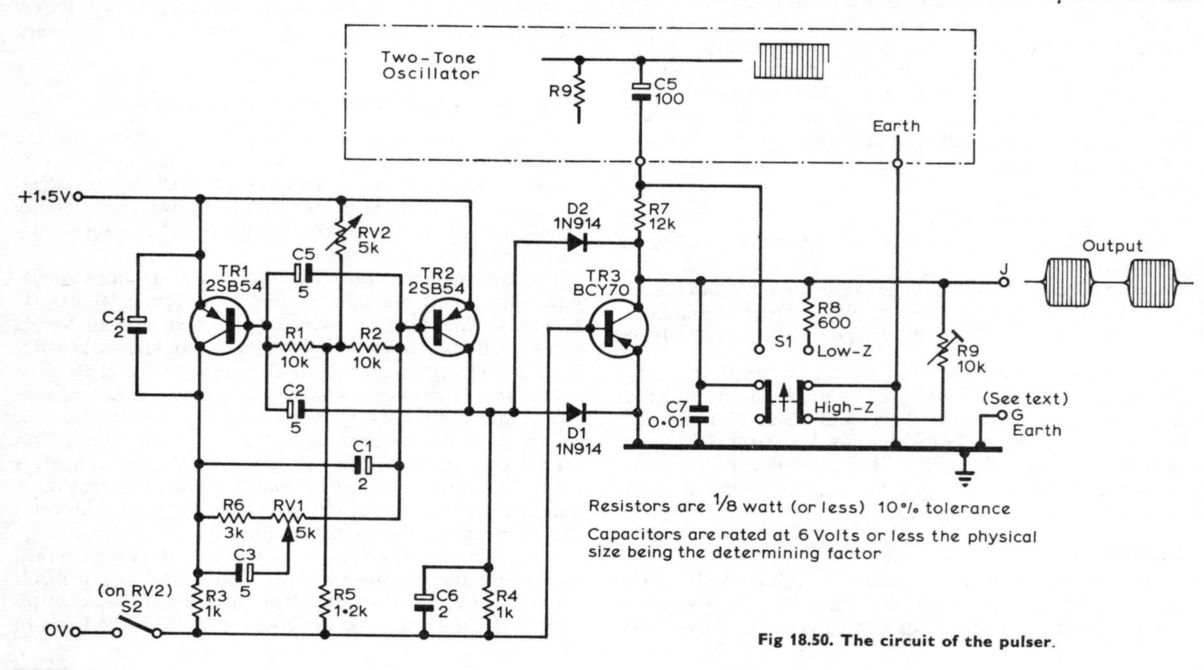

Fig 18.50. The circuit of the pulser.

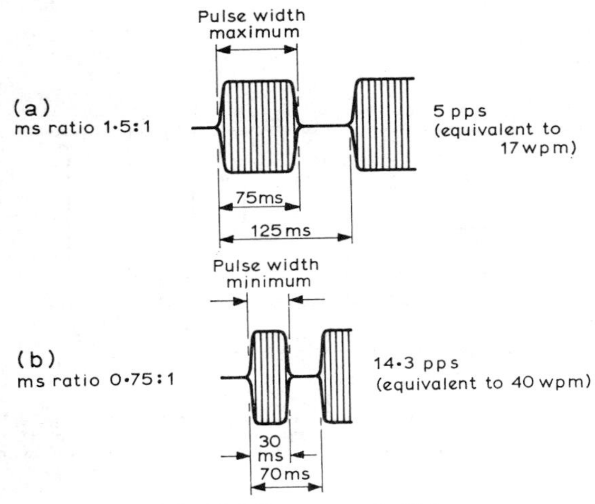

Fig 18.51. Pulse rates and widths.

driving the shunt transistor switch TR3. The pulse width is controlled by RV1 and the pulse rate by RV2. Switch S1 provides alternative high- or low-impedance outputs depending on the audio input requirements of the transmitter.

The performance of the pulser is shown in **Fig 18.51** when supplied with a single tone of 2kHz, with the pulse rate control set to maximum and the pulse width control set to (a) maximum and (b) minimum. The attenuation of the signal in the "off" condition is greater than 50dB in the high-impedance output condition and about 30dB for low impedance output. The battery consumption is 2·5mA at 1·5V and it should be noted that neither pole of the battery is earthed.

Use of the Two-tone Generator

Reference should be made to Chapter 6—*HF Transmitters* for details of the interpretation of two-tone tests. It should be noted that the clean cross-over at zero amplitude which is sought in transmitter adjustment is only possible if the two tones are of equal amplitude. A misleading display is likely if the beat note between the tones coincides with the frequency of one of the tones, and hence harmonically related tones should be avoided.

MEASUREMENT OF NOISE FACTOR

The performance of a receiver front-end or converter is largely dependent on its ability to amplify without seriously degrading the signal/noise ratio. This is usually expressed as the noise factor in decibels where

$$\text{Noise factor} = 10 \log_{10} \frac{S/N_{in}}{S/N_{out}} \text{ (dB)}$$

To make a measurement of noise factor, a source of noise of known amplitude is required; for hf and vhf frequencies the saturated thermionic diode is usually employed.

The anode current of a diode contains noise components due to shot effect, and the frequency spectrum covered by the noise output is exceedingly wide, from vlf up to the gigahertz region. When the diode is operated under temperature-limited conditions, ie the emission is controlled only by cathode temperature and not by anode-to-cathode potential, the noise power output may be shown to be directly proportional to anode current. Thus by varying the current taken by the filament, and hence its temperature, a close control of noise power output may be achieved.

A germanium diode is sometimes used as a noise source, since it generates considerable noise power when a current of a few milliamps is passed through it in the reverse conduction (high-resistance) direction. It has been found that low turnover voltage diodes produce the greatest noise output. Unfortunately the law relating diode current to noise output varies for each diode so that a germanium diode source requires calibration by a thermionic generator if it is to be used for absolute measurements.

VALVE-TYPE NOISE GENERATORS

The noise output of the generator is developed across a load resistor equal in value to the source resistance from which the receiver under test is designed to operate. The anode resistance and output capacitance of the noise diode appear in shunt with the load resistor, and care must be taken in the practical layout to keep the rf connections as short as possible in order to reduce the stray capacitance to a minimum. At the higher frequencies it is desirable to tune out this stray capacitance by means of shunt inductance if the greatest accuracy is required.

This problem arises because the value of the load resistor appears in the formula used to calculate the noise factor, and any shunt capacitance will lower the effective value of the load to an extent directly proportional to the frequency of measurement. The anode resistance of the diode will be much higher than the load resistor, as the diode is operated with sufficiently high anode voltage to give anode current saturation, and the resistive component of the shunt impedance may be neglected for all practical purposes. The dc anode supply to the diode must be fed in via rf chokes to ensure that all the noise generated is passed to the load resistor. It is not necessary for very elaborate smoothing to be used as the anode current will be independent of anode voltage variations under saturation conditions.

A suitable noise diode is the A2087 (CV2171) made by the M-O Valve Co Ltd. This is a B7G-based valve with a short mount, which minimizes the lead inductance associated with the valve itself. An ht supply of 100–150V at 20mA is required, together with a filament supply capable of varying the applied voltage up to a maximum of 4·4V. The tungsten filament takes 0·64A at the maximum filament voltage. The anode current of the CV2171 should not be allowed to exceed 20mA.

For use at 450MHz the M-OV CV2398 is preferable although the CV2171 is satisfactory at this frequency. The CV2398 is a noval based flying-lead valve, has a tungsten filament rated at 6·0V, 1·15A and requires an ht supply of 100–150V at 45mA maximum.

For ease of operation it is desirable to use coarse and fine variable resistors for the filament supply so that the noise diode anode current may be set accurately to any required value.

If equipment with both balanced and unbalanced input is used at the station, two separate diode circuits will be needed and these may be conveniently fed from a common power unit. Suitable circuit arrangements are shown in **Figs 18.52** and **18.53**. The meter M1 has three ranges 1,

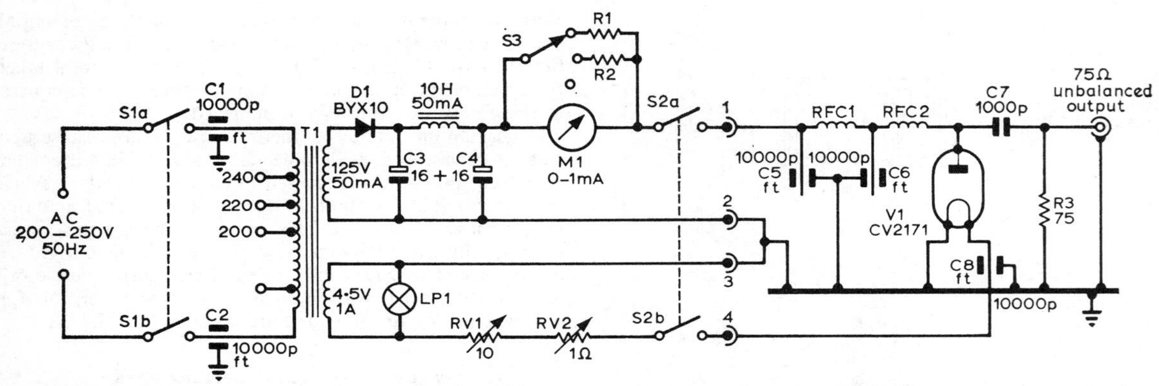

Fig 18.52. Noise generator using CV2171, suitable for unbalanced input. C1, C2, C5, C6, C8 are 10,000pF lead-through capacitors; C1,1,000pF tubular ceramic; RV1, 10Ω five per cent wire-wound linear (Spectral 132); RV2 1Ω 10 per cent wire-wound linear (Spectral 132); R3, 75Ω five per cent resistor. T1, 125V 50mA, 4·5V 1A; S1a,b, S2a, dpst toggle; S3, sp three-way wafer; RFC1, 2, 50 turns 30swg close-wound on ⅜in former.

5 and 25mA fsd obtained by switching in appropriate shunts across the basic 0–1mA meter. The maximum values of noise factor which may be measured on each range are shown in **Table 18.6.** When measurements at 75Ω only are required then a 0–5mA meter shunted to read 25mA will be adequate.

The figures given in Table 18.6 are calculated by substitution in the formula F (dB) $= 10 \log_{10} 20IR$, where R is the source resistance in ohms, and I is in amperes.

At frequencies up to 150MHz it is permissible to connect the noise generator to the receiver by a short coaxial or balanced line, but for the highest possible accuracy of measurement at 450MHz the noise diode and its associated rf circuit should be constructed in the form of a probe which can be connected directly to the input of the receiver under test.

The load resistor used should be of a type having minimum associated series inductance and parallel capacitance, and a tolerance of ± five per cent.

The output capacitance of the noise diode appears in shunt with the load resistor and may be tuned out by the arrangement shown in **Fig 18.54.** The shunt inductance should be adjusted for a rise in noise output from the generator at the centre of the band of interest; it is scarcely

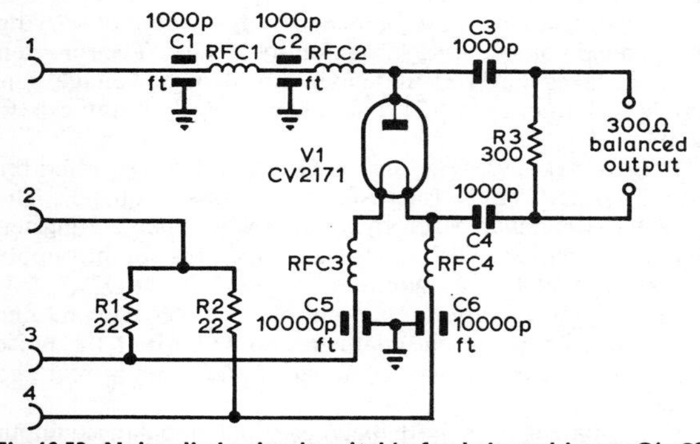

Fig 18.53. Noise diode circuit suitable for balanced input. C1, C2, 1,000pF ceramic lead-through capacitors; C3, C4, 1,000pF tubular ceramic, C5, C6, 10,000pF ceramic lead-through; R1, R2 22Ω ½W; R3, 300Ω five per cent (Erie type 5B); RFC1, RFC2, 50 turns 30swg close-wound on ⅜in dia former; RFC3, RFC4, 36 turns 18swg self-supporting ½in internal diameter.

worth doing except at frequencies of 400MHz and upwards. A practical point to observe when using a noise diode is to permit anode current to flow for the minimum time possible when making measurements, as the expected life of a valve of this type does not exceed a few hundred hours, and is much less if the filament voltage is kept at the maximum value for long periods.

TABLE 18.6

Meter FSD (mA)	Noise Factor (dB) 75Ω	Noise Factor (dB) 300Ω
1	1·76	7·78
5	8·75	14·77
25	15·74	21·76

Method of Using the Noise Generator to Measure the Noise Factor of a Receiver

When a noise factor measurement is to be made it is essential to ensure that the measuring system is linear. The receiver under test may be assumed to be linear for this purpose if doubling the input power produces a corresponding doubling of the output power.

Two methods of using the noise generator to measure the noise factor of a receiver will be described. The first is a simple method, involving one measurement only of noise diode anode current, while the other involves three such measurements and a little more calculation, but has the advantage of incorporating a linearity check.

Both methods require some form of power measuring device which is connected to the receiver at a suitable point to measure the af noise output. In the event that an af output meter is not available, a low-range ac voltmeter may be connected across the primary of the receiver output transformer. Alternatively an 0–100μA dc meter may be placed in series with the earthy end of the detector diode load. A fairly heavily damped meter will assist in reducing the fluctuations observed when reading noise output. It should be remembered that power is proportional to voltage or current squared so that to double the power output when reading in microamps or volts the meter reading should increase by a factor of $\sqrt{2}$.

In the first method of measuring the noise factor the

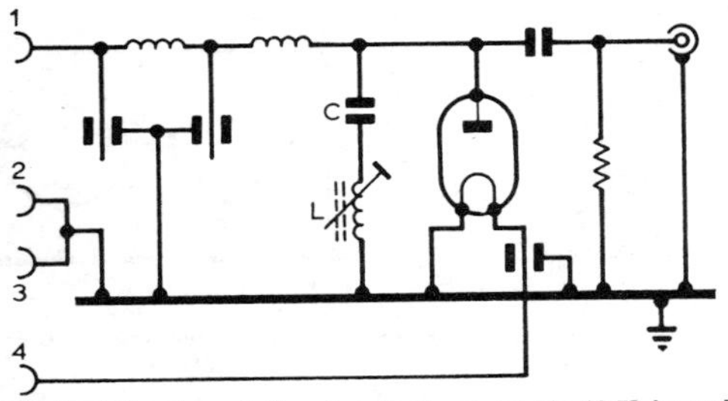

Fig 18.54. Modification of the circuit shown in Fig 18.53 in order to tune out capacitance of diode. C: 100–1,000pF blocking capacitor (disc); L: inductor resonated with strays and diode capacitance.

noise generator is connected to the input terminals of the receiver but not switched on. The i.f. and af gain controls are adjusted to give a convenient reading on the output meter. If a dc microammeter in the detector diode load is being used, the af gain control is ignored and the i.f. gain control adjusted to give a reading of 20–40μA. The noise diode is then switched on and the anode current increased until the receiver power output is doubled or the detector diode current increases by $\sqrt{2}$ times. The noise factor of the receiver in decibels is given by $F = 10 \log_{10} \frac{20IR}{1{,}000}$ where I is the noise diode current in milliamperes and R is the value of the noise generator source impedance.

The procedure described above is only valid if the receiver is known to be linear over the range of input levels involved in making the measurement. If there is any doubt then the three-measurement method should be used as follows.

With the noise generator connected but not switched on the receiver gain is adjusted to give a convenient meter reading, P_1, due to its own noise. The noise generator is then switched on and the noise diode anode current adjusted to give a considerable increase in receiver noise power output, but not enough to cause saturation. Let this output power be P_2 and the noise diode anode current required to produce it I_1. The gain is now reduced appreciably, and the noise diode currents I_2 (usually made equal to I_1 by suitable adjustment of the gain control) and I_3 required to give the original output readings P_1 and P_2 are determined, and the three values of I substituted in the following formula:

$$\mathrm{F\,(dB)} = 10 \log_{10} \frac{20\, I_1\, I_2\, R}{1{,}000\,(I_3 - I_2 - I_1)}$$

where R is the noise generator load impedance and I_1, I_2 and I_3 are in milliamps. It should be noted that if $I_1 + I_2$ is nearly equal to I_3 a small error in readings gives a large error in F, and this condition should be avoided by making P_2/P_1 as large as possible. If the receiver has a response at the image frequency a correction to the noise factor will be required depending on the relative response at the wanted and image frequencies. If the responses were equal, the correction would be 3dB.

If the equivalent noise temperature rather than the noise factor is required, it is given by $Tr = (F - 1)T_0$ where F is the numerical value of noise factor and T_0 is the equivalent noise temperature of the load resistor, which may be assumed to be 290°K (°K = °C + 273). The numerical value of the noise factor F is a ratio and not the figure given in decibels. In order to substitute for F in the equation for equivalent noise temperature, it is necessary to take the antilogarithm of the quoted noise factor in decibels divided by 10, ie

$$\text{numerical value} = \text{antilog}\ \frac{F(\text{dB})}{10}$$

CATHODE-RAY OSCILLOSCOPES

The oscilloscope is one of the most versatile instruments an amateur can possess and permits the visual display of af and rf signals. It is particularly useful for monitoring phone transmissions.

The heart of the oscilloscope is the cathode-ray tube. If it is desired to examine a waveform, the input is connected to the vertical plates while a sweep or time base circuit is connected to the horizontal deflection plates. For many applications it is sufficient to derive the sweep voltage from the ac mains.

By using readily available and inexpensive tubes, an instrument designed expressly for phone monitoring may be constructed quite cheaply. The 3BP1 is suitable as its screen is large enough to give an unambiguous picture, and although it is a short and handy tube, it has a reasonably high deflection sensitivity so that a high input is not required. This obviates the need for deflection amplifiers in monitor service. A number of tubes of this type have been tested and all were found to have the electron guns so well centred that the spot appeared in the centre of the screen without the assistance of a shift-control network. This simplifies construction and reduces cost. The anode voltage requirements are modest; many tubes will give a quite well focussed spot at voltages as low as 400–500, and nearly all will work excellently at 700V and above.

MONITORING DISPLAY UNIT

The circuit of a cathode-ray oscilloscope embodying the foregoing principles is given in **Fig 18.55.** This, though deliberately modest, is adequate for setting up ssb transmitters and monitoring a.m. transmitters.

An eht of 1,000/1,500V is obtained from a voltage doubler circuit and, if the ht secondary of the transformer is centre-tapped, a positive supply at 200/250V can readily be obtained by a single BYX10 rectifier taken from the centre tap for any time base which may be added.

Brilliance of the pattern is controlled by RV3, and focus by RV2. It should be noted that a high-intensity spot must not be permitted to remain stationary on the screen for any length of time, otherwise it will leave a permanent blemish on the fluorescent phosphor. Horizontal sweep is taken from the 50Hz ac mains. It is therefore necessary to make sure that C2 is connected to the live side of the mains supply. If no sweep is obtained on switching on, it is clear that C2 is connected to the neutral lead, and the fault should be rectified by transferring it to the other side of T1 primary.

A 50Hz sweep causes some distortion of the display, but this may be kept within reasonable limits by using a high sweep amplitude so that only the centre portion of the sweep appears on the screen. The resulting trace is linear enough for most monitoring work. The sweep amplitude may be reduced by RV1, but this control should be set at maximum when monitoring is being carried out.

If other work is contemplated, it is worth including some form of linear horizontal sweep. The Miller-transitron oscillator, shown in its most elementary form in **Fig 18.56,**

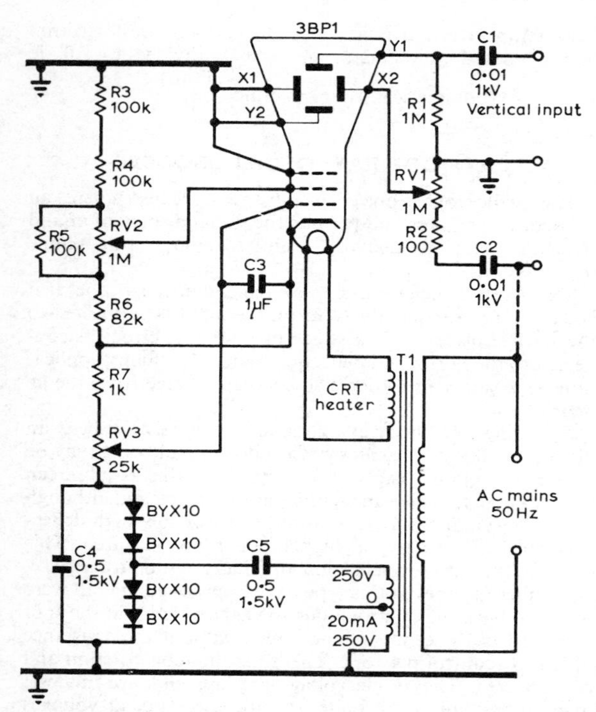

Fig 18.55. Oscilloscope display unit for monitoring telephony transmissions. The resistor chain R3-7 may need adjustment to suit particular CRTs or power supply voltages.

is suitable. Any sharp cut-off pentode may be used in this circuit, so long as it has its suppressor grid brought out to a separate pin. The required ht supply can be obtained from the centre tap of the ht winding as previously noted.

If the linear sweep of Fig 18.56 is preferred to the 50Hz mains sweep, C2 in Fig 18.55 should be disconnected from the primary of the heater transformer and connected to the anode of the oscillator valve.

This simple display unit can be employed for monitoring a.m. transmissions using the method described, with illustrations of typical patterns, in Chapter 9—*Modulation*. Connections for envelope and trapezoidal patterns are shown in **Fig 18.57**.

Calculation of modulation depth using the trapezoidal method is possible by measuring the sides of the trapezium

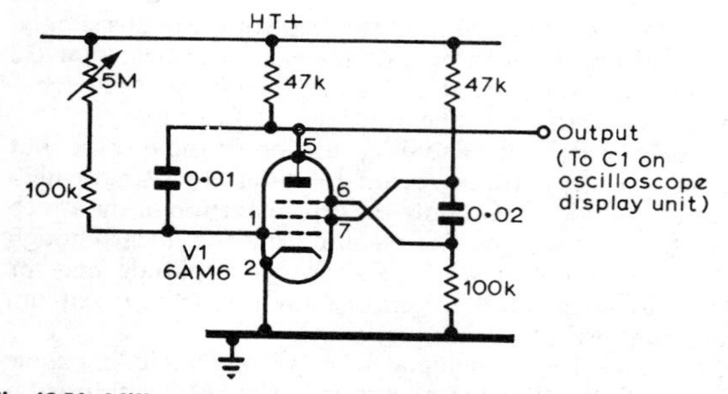

Fig 18.56. Miller-transitron sawtooth timebase generator for use with the oscilloscope of Fig 18.55.

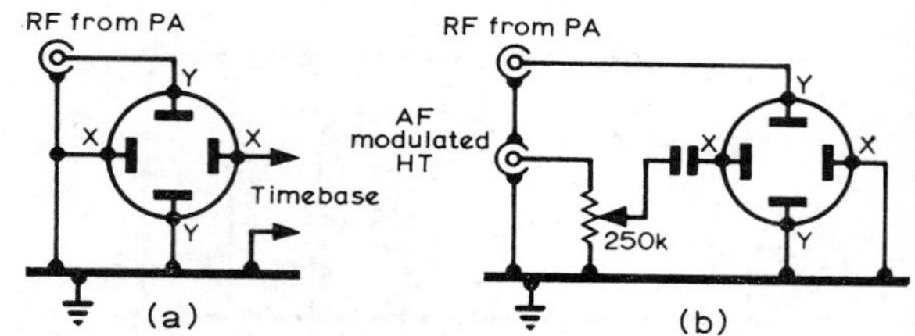

Fig 18.57. (a) Oscilloscope connections for envelope pattern. (b) Connections for trapezoidal pattern.

and using the abac of **Fig 18.58**. A disadvantage of the trapezoidal method is the necessity to tap off from the secondary of the modulation transformer an audio signal for the X (horizontal) plates. However, this can be overcome by building a small tuned circuit and demodulator unit (**Fig 18.59**) which can be connected to the output of the transmitter to provide the necessary audio signal. The unit should be made up on a small sub-chassis and fitted as near to the holder of the cathode ray tube as possible. If the signal is fed to the X amplifier input the trapezoidal display will be obtained.

SSB transmissions may be conveniently checked by the 45° method which calls for the application of similar voltage from the input and output of the linear amplifier. This necessitates the use of two detectors, as shown in **Fig 18.60**. The resistors R1 and R2 should be adjusted so that the voltage output from each is similar. The interpretation of the various displays obtained when a sinusoidal tone is fed into the ssb transmitter is shown in **Fig 18.61**. It should be noted that the use of the demodulator of Fig 18.59 does not permit checks on linearity to be made.

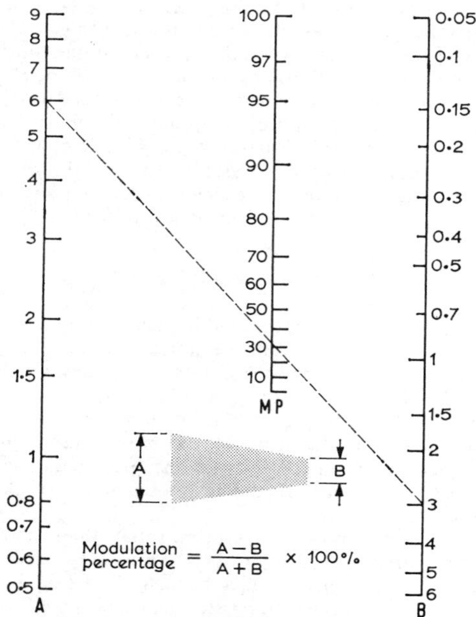

Fig 18.58. Abac for the calculation of modulation depth from the trapezoidal pattern. The dotted line illustrates an example in which the large side (a) is 6 units long and the shorter one (b) 0·3 units indicating a depth of modulation of just over 90 per cent.

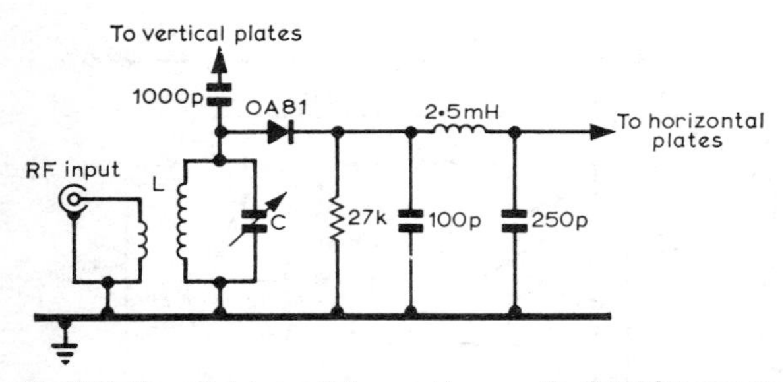

Fig 18.59. Demodulator unit to provide an audio signal for connection to the X plates of the oscilloscope in the trapezoidal method of measuring modulation depth.

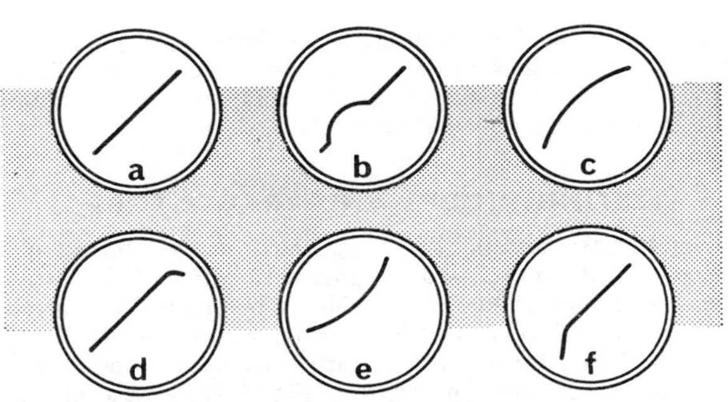

Fig 18.61. Interpretation of displays obtained with the arrangement of Fig 18.60. (a) Linear condition. (b) Incorrect grid circuit loading. (c) Incorrect bias. (d) Amplifier overloaded. (e) Insufficient standing current in amplifier. (f) Pattern obtained from speech input to a correctly adjusted linear amplifier.

APPLICATIONS OF THE OSCILLOSCOPE

It is important to realize that an oscilloscope is virtually a two-dimensional meter, ie it is intended for making measurements, not simply for displaying an interesting picture. Consequently the temptation to continually adjust the X and Y sensitivities to make the picture fill the screen should be firmly resisted. If the Y gain is continuously variable it is worthwhile calibrating it against a voltmeter on 50Hz ac (remembering that the meter reads rms, not peak-to-peak: 1V rms = $2\sqrt{2}$V peak-to-peak) so that the settings for, say, 1, 3, 10, 30 and 100V/cm can be found rapidly. Time calibration can be similarly checked against a crystal calibrator, with a divider to 10 and 1kHz and the 50Hz mains. If the oscilloscope is provided with a triggered time base it is convenient to calibrate it in terms of time/cm, but the older slightly less convenient synchronized time base has to be variable to synchronize, so a frequency (or period) calibration is more suitable.

The uses to which an oscilloscope can be put can be conveniently classified by the inputs to the Y and X plates.

(a) Signal to Y, nothing to X. This mode uses the electron beam as a meter pointer, and does not fully exploit an oscilloscope, but it is convenient when a valve voltmeter would otherwise be required, as for example measuring the dc anode potential of a high-gain audio amplifier. In practice there is no need to switch off the timebase for dc measurements, although this may be convenient to measure the peak-to-peak amplitude of an ac signal. Care should be taken to avoid burning the screen in either case.

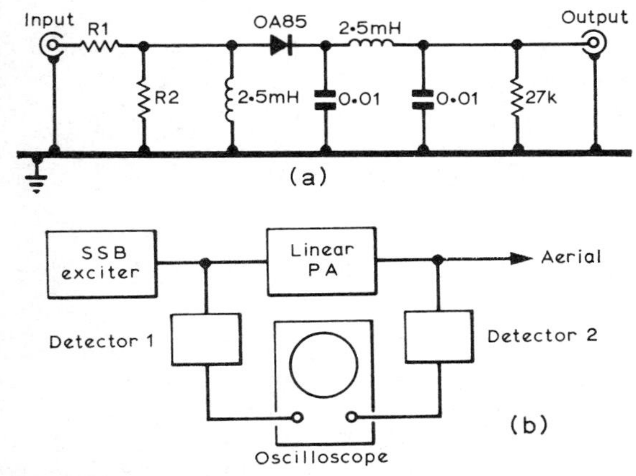

Fig 18.60. (a) Demodulator unit. (b) Circuit arrangement for checking ssb transmissions by the 45° method.

(b) Signal to Y, timebase to X. The additional dimension of time reveals much more information about a signal. The frequency can be measured against a calibrated timebase, and any distortion or unwanted signal traced. The input capacitance of the average Y amplifier is some 30pF and this, if connected to a high-impedance point, can severely restrict the bandwidth. The use of coaxial cable leads with an additional capacitance of 22pF per foot will make this much worse; sometimes this effect can be used to remove unwanted rf from a low-frequency signal. In general, however, it is preferable to use either a high-impedance probe unit, or failing this, short leads of unscreened wire.

The bandwidth of an oscilloscope amplifier may be quite low so that measurements above, say, 10MHz may not be possible. Often it is possible to work slightly outside the nominal frequency range if a reduced sensitivity is tolerable, but this should be done with caution since excessive signals outside the bandwidth can cause distortion and even modulate the timebase speed. Many oscilloscopes have, however, provision for inputs direct to the Y plates, and this permits higher-frequency signals to be applied, although the sensitivity is quite low. A modern cathode-ray tube with a side arm connection to the Y plates will give a useful deflection with an input frequency of 150MHz. Unless a very fast timebase is provided, it would not be possible to resolve the rf waveform at this frequency, but it would be possible to examine the modulation envelope as in the two-tone test for ssb transmitters.

(c) One signal to Y, another to X. The most familiar form of this mode is the trapezoidal display of a modulated amplifier; but the input and output of any device may be so displayed. The input/output linearity of a diode detector can therefore be displayed and compared with a product detector, taking care to avoid leakage of the bfo signal into the oscilloscope. The maximum undistorted output from an amplifier can be measured by increasing the drive till the diagonal line on the screen starts to curve.

The wobbulator, or swept alignment oscillator, is another application in which the X signal is related to the frequency of an oscillator, the output of which is fed to the rf amplifier under test, and the amplifier output (possibly after detection) is fed to the Y amplifier. Provided the rate of frequency sweeping is sufficiently slow, a graph of the amplifier response is presented on the cathode-ray tube.

Finally, a word of caution. It has been known for the timebase harmonics of an oscilloscope to cause tvi. This possibility should be borne in mind, and suitable screening and lead filtering applied if necessary.

TESTING DIODES AND TRANSISTORS

Since transistors and diodes are nearly always soldered in place and often purchased in "untested" batches it is worth making simple tests on all such devices before use. This also permits them to be graded for selective gain and leakage so that the better quality devices are reserved for the more onerous applications. The circuit of **Fig 18.62** shows a simple tester which will identify the polarity and measure the leakage and small-signal gain of transistors, and also the forward resistance of diodes.

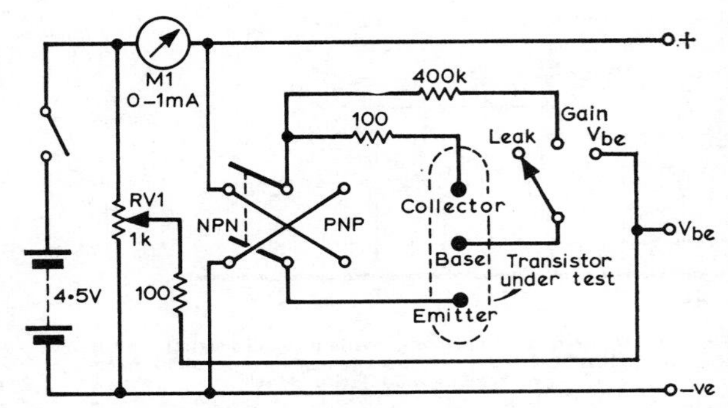

Fig 18.62. Simple transistor and diode tester.

Testing Transistors

To check the dc current gain (which approximates to the small-signal current gain β), the transistor is connected with collector to C, base to B and emitter to E. Moving S1 to the GAIN position applies 10μA of base current, and M1 will read the emitter current. With S1 at the LEAK position any common emitter leakage current is shown, which for silicon transistors should be barely perceptible. The difference between the two values of current divided by 10μA gives the value of $\beta + 1$ which is near enough to β for most purposes. A high value of leakage current probably indicates a short-circuited transistor, while absence of current in the gain position indicates either an open-circuited transistor or one of reversed polarity. No damage is done by reverse connection and pnp and npn transistors may be identified by finding the polarity which gives normal gain.

With S1 in the V_{be} position, the base emitter voltage is controlled by RV1 which should be near the negative end for npn and near the positive end for pnp devices. V_{be} may be measured by a voltmeter connected between the terminal marked "V_{be}" and either the positive or negative voltage rail, depending on the polarity of the device. This test position may be used for FETS but only positive or zero bias is possible.

Testing Diodes

The forward voltage drop across a diode may be measured by connecting it across the terminals "+" and "V_{be}" with a voltmeter in parallel. The forward current is set by RV1. Diodes may be matched for forward resistance, and by reversing the diode connections any reverse leakage can be seen. The value of forward voltage drop can be used to differentiate between germanium and silicon diodes.

CHAPTER 19

OPERATING TECHNIQUE AND STATION LAYOUT

HOWEVER much time and money have been spent on the construction or purchase of equipment, results are certain to depend to a large degree on the way in which the station is operated. A listening session on any of the amateur bands at almost any time will reveal many examples of bad behaviour and faulty operating techniques which are more usually the result of ignorance than a deliberate attempt to be a nuisance to other band users.

The two basic essentials of successful operating on any mode are judgment and courtesy. A third and most important factor is adequate knowledge of the various bands and their propagation habits. This can only come as a result of observation and practice and is the one great advantage which the former keen short-wave listener has over the pure technician when first coming on the air.

The newcomer may wonder exactly how he should begin his first contact and the wisest course for him to take is to first of all listen around carefully and establish exactly what signals are to be heard and the general level of activity. He can then decide whether to put out a "CQ" call or call a specific station at the appropriate time—this is when that station has finished his previous contact or has himself put out a "CQ" call. It is not good manners to try to "break in" on his frequency while he is still in contact.

Having established contact it is usual to exchange signal strength reports, names, locations, and details of equipment being used. Further general discussion should concern "matters of a personal nature in which the licensee, or the person with whom he is in communication, has been directly concerned . . ." (Wireless Telegraphy Act). It should be made a golden rule never to discuss politics, religion, or any matter which may offend the person to whom one is talking or anyone who may be listening.

Telephony Operation

Always speak clearly and not too quickly, especially when talking to someone whose native tongue is not one's own. Remember that many overseas amateurs deserve great credit for the trouble that they have taken to learn a foreign language such as English in order to be able to talk to those who do not speak their own language. Their ability to converse may well be limited to the more standard subjects involved in the usual simple contact and the use of basic words or even of some of the "Q" code (see Table 19.2) is helpful under such circumstances. The latter should never be used on phone under any other circumstances as nothing sounds more stupid to the casual listener than two local amateurs talking to each other by spelling out morse code abbreviations.

CW Operation

In general the good cw man is the one whose copy is easy to read and who does not try to impress by sending faster than he is really capable of doing properly. The speed of sending should partly depend on circumstances and when conditions are poor it is sensible to send more slowly than when signals are loud and clear. A good principle to follow is to send at the same speed as the operator at the other end of the contact—and to slow down if he indicates that he is having difficulty.

Morse from an electronic keyer can be a joy to the ear if the device is being operated correctly, and it should be the aim of the semi-automatic "bug" or straight key owner to send out signals just as readable. It is highly desirable to be able to listen to one's own transmission to know how one's style is developing.

There are five different ways of ending a cw transmission and each indicates to a listener exactly what the intentions of the sender are. These are as follows:

(1) ". . . AR." These letters are sent at the end of a call to a specific station before contact has been established eg "G3AAA de G3ZZZ AR".

(2) ". . . K." The letter K is sent at the end of a "CQ" call, or at the end of a transmission during an established contact where there is no objection to others joining in.

(3) ". . . KN." This is used when a call is made to a specific station and replies are not desired for any other station, or at the end of a transmission when already in contact where other calls are not wanted.

(4) ". . . SK." This combination is used at the end of a final transmission in a contact, immediately before the callsigns are given. It means that the sender is ready to receive other calls.

(5) ". . . CL." Denotes that the sender is closing down and will not answer any more calls.

Note that it is not correct to call "QRZ?" after finishing a transmission unless someone has already been heard calling—the right thing to do is to call "CQ". It should be borne in mind however that it is courteous to move off a frequency at the end of a contact if the station contacted was originally operating there. It is polite to move away from a frequency on which one has been fortunate enough to have been called by a rare dx station and leave it for others to use—there may be many others who would like a contact with the latter and it is not difficult to find a fresh frequency. Remember that no one has an absolute right to any frequency under any circumstances whatsoever and that the policy of "live and let live" is by far the best to adopt.

The "CQ" Call

There are many occasions when it is a good idea to put out a "CQ" call. One of these is when a band appears to be "dead". This can happen particularly on 21 and 28MHz when there can be suitable long distance propagation but the level of activity is low, most having left in the belief (after a quick listen around) that it is a waste of time to transmit.

There are important rules to be observed when putting out a "CQ" call. Some of the more vital ones are as follows:

(1) Listen on the chosen frequency to make sure that it is quite clear of other signals before starting to transmit.

(2) When using telephony confirm (1) by asking "Is anyone using this frequency please?" several times, and then listen very carefully for replies before commencing the call.

(3) Keep calls short and listen frequently for replies. Many contacts have never taken place because potential callers have become bored while waiting for an opportunity to identify and have moved away. The ideal arrangement is to use full break-in on cw and vox when on ssb so that it is possible to "listen through" one's calls. Not more than three "CQs" followed by one's own callsign given twice—this being repeated until a contact is established—is perhaps the best procedure to be followed. If no answers are obtained it may be because the frequency chosen is subject to interference elsewhere in the world and a change to another one is advisable.

If contact with a specific area is desired it is a good idea to put out a directional call. For example—"CQ VK, CQ VK, CQ VK de G——" etc (on cw) or "CQ Australia" on phone should produce replies from that part of the world only if conditions are suitable, and all other callers should be ignored. Likewise, if another station's directional CQ is heard and this does not apply to one's own area it should not be answered.

When answering a "CQ" call it is good practice to make transmissions short and to emphasize one's own call. The other man knows his own callsign well but not the caller's! Short calls mean minimum interference; if the contacts being made by the station being called are of the "rubber stamp" variety (RS/T, name, location, promise to QSL) it is quite possible that a persistent caller may find that a contact has been established and is half over by the time he finishes his transmission. It is a very good idea to listen to any station for a short while before calling; it will then be possible to estimate the speed at which contacts are being made and also where callers should transmit.

DX Working

This is a branch of amateur radio which holds a fascination of its own and commands a considerable following. The interpretation of the expression "dx" varies according to the frequency being used, the power and equipment of the stations involved, and even the scarcity of licensed amateurs in the country in which the dx station is located.

Without doubt the most important attribute which those looking for dx contacts need is patience. The more elusive stations are often running low power, may have simple equipment, and may be situated a very long way away. They may also not be particularly anxious to talk to yet another station in the British Isles! It is therefore all the more important to obey any calling instructions which they may give, and not to cause any interference problems.

On the whole it is more rewarding to seek out a station and call it at the correct time than to call "CQ" and expect a reply from a rare station. Many of the operators in dx locations have set operating habits and tend to appear at regular times and on regular frequencies, and the good listener-amateur will try to find out about these and then lie in wait for his quarry. The writer considers that dx working is very similar to fishing and requires similar study.

There are a number of news sheets published throughout the world at regular intervals which contain a great deal of valuable information on the latest expeditions and also on frequencies and times of operation of more unusual stations.

Many expedition stations make a practice of never listening on their own frequencies but indicate where they will seek replies, and the more practised mean exactly what they say. On telephony, for example, they may say "tuning 14,190 to 14,200kHz" and under these circumstances it is often better not to choose a calling frequency at the limits of the range indicated but to find a clear spot in between and call there. Some of the less experienced operators keep tuning to stations using the same frequency as that of the previous station contacted and if this is the case it is advisable to zero beat with the latter. A call just off the frequency of the last station contacted has often produced results for the writer. It is

TABLE 19.1

The Phonetic Alphabet

A	Alfa	J	Juliet	S	Sierra
B	Bravo	K	Kilo	T	Tango
C	Charlie	L	Lima	U	Uniform
D	Delta	M	Mike	V	Victor
E	Echo	N	November	W	Whisky
F	Foxtrot	O	Oscar	X	X-ray
G	Golf	P	Papa	Y	Yankee
H	Hotel	Q	Quebec	Z	Zulu
I	India	R	Romeo		

TABLE 19.2

The Q Code

QAV	I am calling . . .
QCM	There seems to be a defect in your transmission.
QIF	. . . (callsign of station) is using . . . (frequency).
QRA	The name of my station is . . .
QRB	The distance between our stations is . . .
QRG	Your exact frequency in kHz is . . .
QRH	Your frequency varies.
QRI	Your note varies.
QRJ	Your signals are very weak.
QRK	The intelligibility of your signals is (1 to 5)
QRL	I am busy.
QRM	There is interference.
QRN	I am being troubled by atmospheric noise.
QRO	Increase power.
QRP	Reduce power.
QRQ	Send faster (. . . words per minute).
QRS	Send more slowly (. . . words per minute).
QRT	Stop sending.
QRU	I have nothing for you.
QRV	I am ready.
QRW	Please tell . . . that I am calling him.
QRX	I will call you again.
QRZ	You are being called by . . .
QSK	I can hear between my signals (ie I am using break-in).
QSL	I give you acknowledgement of receipt.
QSM	Repeat the last message.
QSP	I will relay to . . .
QSV	Please send a series of Vs.
QSW	I will transmit on . . . kHz.
QSY	Move to . . . kHz.
QSZ	Send each word or group twice.
QTH	My location is . . .
QTR	The exact time is . . .

TABLE 19.3

The Z Code

ZAN	I am receiving absolutely nothing.
ZAP	Acknowledge please.
ZCK	Check your keying.
ZCL	Transmit your call letters intelligibly.
ZDF	Your frequency is drifting.
ZDM	Your dots are missing.
ZFO	Your signals have faded out.
ZGS	Your signals are getting stronger.
ZGW	Your signals are getting weaker.
ZOK	I am receiving OK.
ZRN	You have a rough note.
ZSU	Your signals are unreadable.
ZWO	Send words once.
ZWT	Send words twice.

common practice on cw to indicate tuning frequencies by sending "U 5" or "L 5" (for example) to indicate that the listening area is 5kHz higher or lower than the station's own transmitting frequency.

Under expedition conditions the dx station is usually trying to make as many contacts as possible in limited time and therefore tends to restrict exchanges to reports and callsigns only. Once again, adequate preparatory listening should alert the caller to what is required in order for him to avoid offending others who may be waiting for their turn.

Tail-ending

The use of this procedure, which consists of dropping one's own callsign in almost zero beat with a station just completing his last transmission, is accepted by some good operators but is open to grave abuse if used unskillfully and is usually best avoided. Likewise the temptation to break into a contact already established on ssb should be resisted unless one of the participants is well known to the caller. It should be remembered that joining into a conversation between two complete strangers is similar to joining a conversation anywhere else and that the rules of common courtesy should be exercised.

Phonetics

The use of internationally recognized words as phonetics is most useful under certain circumstances and the alphabet suggested in Appendix 16 of the Radio Regulations, Geneva, 1959 (as described on amateur licences in the UK) is shown in **Table 19.1**. Remember that "words used in this manner shall not be of a facetious or objectionable character" (British Amateur Licence).

The "Q" Code

The most common abbreviations employed by amateurs using morse code is the "Q" code **(Table 19.2)**. A much less frequently used code is the "Z" code—this has a number of combinations which could be of considerable use in amateur work (see **Table 19.3**).

Where a "Q" or "Z" group is followed by a question mark an answer is required and where appropriate this should be qualified by the addition of a number according to the following classification: 1—very slight; 2—slight; 3—moderate; 4—severe; 5—extreme. For example "QRM?" means "is there any interference" and "QRM 5" means "there is extremely severe interference".

A great deal of information can be transmitted in a short time on cw if full use is made of these standard abbreviations and codes. Their use may also be of great assistance at times during contacts with amateurs who are in difficulty on telephony because of language differences (but not otherwise).

TABLE 19.4

The RST System

Readability
1—Unreadable.
2—Barely readable, some words distinguishable.
3—Readable with considerable difficulty.
4—Readable with practically no difficulty.
5—Fully readable.

Signal strength
1—Faint signals, barely perceptible.
2—Very weak signals.
3—Weak signals.
4—Fair signals.
5—Fairly good signals.
6—Good signals.
7—Moderately strong signals.
8—Strong signals.
9—Extremely strong signals.

Tone
1—Extremely rough hissing note.
2—Very rough ac note.
3—Rough low pitched ac note, slightly musical.
4—Rough ac note, moderately musical.
5—Musically modulated note.
6—Modulated note, trace of whistle.
7—Near dc note, smooth ripple.
8—Good dc note, trace of ripple.
9—Pure dc note.

Originally the only code available for reporting signals was the one used to denote loudness and that was graded in the arbitrary scale R1–R9. Later, when interference became a problem it was necessary to indicate the readability of the signal and the scale QSA1–QSA5 was used to denote this; at the same time the signal strength was given as QRK1–QRK9. Later, when attention began to be paid to the quality of cw signals another code was evolved—this is the now universally used RST code **(Table 19.4)**. This shows Readability (R1 to R5), Signal strength (S1 to S9), and Tone (T1 to T9). On telephony it is usual to give the report as Readability (1 to 5) and Strength (1 to 9). A cw note which is chirpy is indicated by the addition of a "c" after the tone number. There is another code—the SINPO code—**(Table 19.5)** which is more frequently used for reporting broadcast transmissions.

TABLE 19.5

The SINPO Code

S	Signal strength	1	Barely audible.
		2	Poor.
		3	Fair.
		4	Good.
		5	Excellent.
I	Interference	1	Extreme.
		2	Severe.
		3	Moderate
		4	Slight.
		5	Nil.
N	Noise	1	Extreme.
		2	Severe.
		3	Moderate.
		4	Slight.
		5	Nil.
P	Fading	1	Extreme.
		2	Severe.
		3	Moderate.
		4	Slight
		5	Nil.
O	Overall rating	1	Unusable.
		2	Poor.
		3	Fair.
		4	Good.
		5	Excellent.

TABLE 19.6

Morse code and sound equivalents

A	di-dah	**S**	di-di-dit
B	dah-di-di-dit	**T**	dah
C	dah-di-dah-dit	**U**	di-di-dah
D	dah-di-dit	**V**	di-di-di-dah
E	dit	**W**	di-dah-dah
F	di-di-dah-dit	**X**	dah-di-di-dah
G	dah-dah-dit	**Y**	dah-di-dah-dah
H	di-di-di-dit	**Z**	dah-dah-di-dit
I	di-dit	**1**	di-dah-dah-dah-dah
J	di-dah-dah-dah	**2**	di-di-dah-dah-dah
K	dah-di-dah	**3**	di-di-di-dah-dah
L	di-dah-di-dit	**4**	di-di-di-di-dah
M	dah-dah	**5**	di-di-di-di-dit
N	dah-dit	**6**	dah-di-di-di-dit
O	dah-dah-dah	**7**	dah-dah-di-di-dit
P	di-dah-dah-dit	**8**	dah-dah-dah-di-dit
Q	dah-dah-di-dah	**9**	dah-dah-dah-dah-dit
R	di-dah-dit	**0**	dah-dah-dah-dah-dah

An abbreviated form of 0 (zero) is sometimes used and consists of one long dash. Similarly an abbreviated 9 in the form of an N (dah-dit) is in quite frequent use and can save considerable time during contest work where many numbers are being transmitted.

Punctuation

Question mark	di-di-dah-dah-di-dit
Full stop	di-dah-di-dah-di-dah
Comma	dah-dah-di-di-dah-dah

(The comma is frequently used to indicate an exclamation mark.)

Procedure signals

Stroke (/)	dah-di-di-dah-dit
Break sign (=)	dah-di-di-di-dah
End of message (+ or AR)	di-dah-di-dah-dit
End of work (SK)	di-di-di-dah-di-dah
Wait (AS)	di-dah-di-di-dit
Preliminary call (CT)	dah-di-dah-di-dah
Error	di-di-di-di-di-di-di-dit
Invitation to transmit (K)	dah-di-dah

One dah is equal to three di's (dits).
The space between part of the same letter should equal one dit.
The space between two letters should equal three dits.
The space between two words should equal between five and seven dits.

Learning Morse

It will be noticed that in **Table 19.6** the various morse code characters are described in dits and dahs. It is most important that anyone wishing to learn the code should not attempt to do so by memorizing first of all an image of dots and dashes which then have to be translated into sounds. It is far better to learn each character as a sound entity from the very beginning.

Many amateurs look upon the task of preparing themselves for a morse test as a great ordeal and never attempt it, thereby denying themselves the pleasure of owning a higher class licence with all its associated privileges. With determination and assistance from the various morse instruction tapes and records which are now on the market (and available from RSGB) a very useful degree of skill can be acquired. CW signals can often be copied over long distances even if very low power is all that is available, and under today's crowded band conditions they take up far less space and are less subject to being rendered unreadable by interference.

QSL Cards

Most amateurs and keen listeners experience pleasure when they receive written confirmation of enjoyable contacts or verification of reception of distant stations, and in former times the QSL card was considered to be the final courtesy of the contact. Present day costs have meant that many can no longer follow a policy of confirming every contact but it should be the aim of all those who make promises to send out QSL cards to do so, or to state the fact that they do not quite clearly at the time. In many countries the national radio society runs a central bureau where members may send their cards for world-wide distribution in bulk at reduced cost. They are also able to deposit envelopes for collection of their incoming cards from other QSL bureaux.

One of the uses of QSL cards is as proof of contact for the purpose of claiming the various certificates of operating proficiency which are offered by many of the world's national societies. The more important of these are fully described in the RSGB publication *Amateur Radio Awards*. It is therefore important to realize that the information given on cards should be accurate, and should contain the following data as an essential minimum:

(1) The fact that a contact took place must be mentioned.
(2) The date and time of the contact.
(3) The frequency band and mode used.
(4) The signal strength of the station contacted.

It is absolutely vital that no alterations whatever are made on cards which are likely to be used for the purposes of certain award applications (eg ARRL's DX Century Club). This applies whether such alterations are made by either the sender or the recipient.

It is not necessary to use elaborate cards and oversized ones or those weighing more than four grammes should not be sent via bureaux. A confirmation written in the form of a note or letter is equally acceptable and some enthusiasts have had blank cards printed with a suitable layout for the distant station to complete and return to the sender. A rubber stamping—suitably worded and applied to the applicant's own card—is also quite satisfactory.

Top Band Operation

Special care must be exercised when using the 160m band as the amateur service is only permitted to use it as a secondary user. This means that amateur privileges depend entirely on good behaviour and the utmost care being used to avoid interference to the primary users of the band. It is absolutely vital that such interference should not occur and any request from, for example, a coastal station, to close down or move frequency must be instantly obeyed. A list of some European coastal stations and their frequencies is given below. This was correct at the time of writing but changes may well have occurred and it is as well to check every frequency before transmitting. Under night time conditions even some of the more distant stations listed may be subject to interference from low power amateur signals from the British Isles.

European Special Service Stations using frequencies on the 160m band are as follows.

Patra (Greece)	1800kHz	Kotka (OFU)	1862kHz
Lyngby (OXZ)	1806kHz	Vaasa (OFW)	1862kHz
Brest (FFU)	1806kHz	Scheveningen (PCH)	1862kHz
Blavand (OXB)	1813kHz	Humber Radio (GKZ)	1869kHz
Gdynia (SPC)	1818kHz	Brest (FFU)	1876kHz
Bordeaux (FFC)	1820kHz	Reykjavik (TFA)	1876kHz
Wick (GKR)	1827kHz	Portpatrick (GPK)	1883kHz
Niton (GNI)	1834kHz	Civitavecchia (Italy)	1888kHz
Lands End (GLD)	1841kHz	Scheveningen (PCH)	1890kHz
North Foreland (GNF)	1848kHz	Goteborg (SAG)	1904kHz
Stonehaven (GND)	1856kHz	Marseilles (FFM)	1906kHz
Bordeaux (FFC)	1862kHz		

TABLE 19.7

Band (MHz)		Type of emission
3.50—3.60		cw (2)
3.60	± 20kHz	rtty (1)
3.60—3.80		cw and phone (2, 3)
7.00—7.04		cw
7.04	± 5kHz	rtty (1)
7.04—7.10		cw and phone
10.100—10.150		cw
10.145	± 5kHz	rtty (1)
14.00—14.10		cw
14.09	± 10kHz	rtty (1)
14.10—14.35		cw and phone
18.068—18.110		cw
18.105	± 5kHz	rtty (1)
18.110—18.168		cw and phone
21.00—21.15		cw
21.10	± 20kHz	rtty (1)
21.15—21.45		cw and phone
24.890—24.930		cw
24.925	± 5kHz	rtty (1)
24.930—24.990		cw and phone
28.00—28.20		cw
28.10	± 50kHz	rtty (1)
28.20—29.70		cw and phone

Notes

(1) For rtty, recommended section of operation shared with cw.
(2) 3,500-3,510 and 3,775-3,800kHz reserved for intercontinental working.
(3) 3,635-3,650kHz is used by USSR stations for intercontinental working.
(4) For sstv recommended operation frequencies are: 3,735, 7,040, 14,230, 21,340, 28,680kHz, all ± 5kHz.
(5) For beacons, 28.2-28.3MHz is recommended.
(6) For the downlink of amateur satellites, 29.3-29.55MHz is recommended
(7) The transmitter power on the 10MHz band should not exceed 250W mean output power, (NB: UK max carrier power is 20dBW).
(8) No contests should be organized on the 10MHz band.
(9) Credit for awards or diplomas should be accepted for contacts made on the 10MHz band.
(10) SSB may be used on the 10MHz band during emergencies involving the immediate safety of life and property, and only by stations actually involved in the handling of emergency traffic.
(11) Contest preferred segments for major contests: 3.5-3.56, 3.6-3.65, 3.7-3.8, 14-14.06, 14.125-14.3MHz.

This list is compiled from the ITU *List of Radiodetermination and Special Service Stations.*

The frequencies available to amateurs in various countries on 160m vary considerably and in some are very restricted. Different parts of the United States of America are allowed to use different parts of the band—eg stations in the Hawaiian Islands are allowed the 1900–2000kHz section whereas the rest of the USA uses the lower half of the band. Users of 160m who are looking for long distance contacts follow a special transmitting and listening pattern. For example it is the custom for European stations wishing to make transatlantic contacts to transmit in the 1825–1830kHz region and to listen for replies at the very low end of the band—this area is relatively free of other signals in Europe and the 1825–1830kHz band (known as the "dx window") is relatively clear in America. During the relatively few hours when propagation permits long distance contacts on 160m other stations are considered to be courteous if they avoid this part of the band.

Eighty Metres

The Region 1 allocation on the band covers 3500–3800kHz, but in Region 2 (which includes the American continent) the upper limit is 4000kHz. Australian amateurs are restricted to the portion below 3700kHz, and the portions which may be used by some other countries (eg India and Japan) are very small indeed. The whole band is available on a shared basis with other services. In order to facilitate intercontinental working it has been recommended by Region 1 IARU that 3500–3510kHz and 3790–3800kHz should be reserved for such purposes when propagation is favourable, and that the segment 3635–3650kHz be reserved for use by stations in the USSR for intercontinental contacts.

Forty Metres

Amateurs in Region 1 (which includes Britain) have the poorest slice of this band and are limited to the portion between 7000 and 7100kHz. Elsewhere the upper limit is 7300kHz. It should be noted that contacts with stations in other Regions (eg those in the USA) who are transmitting above 7100kHz should not be made as they come within the definition of "cross band" contacts according to ITU regulations.

Band Plans

On all bands there are recommended sections set aside for use by each mode. In some parts of the world (eg the USA) observance of these band sub-divisions is mandatory and their use also depends on the class of licence held by the operator. The Region 1 HF Band Plan (which is supported by all IARU member societies in Europe and Africa) is set out in **Table 19.7** and should be observed at all times even though its recommendations are only advisory as far as British amateurs are concerned.

Single-sideband Transmissions

This mode of operation has almost replaced amplitude modulation telephony and enables its users to carry on telephone type conversations if both stations are making full use of voice control (vox). This is most convenient as questions can be answered as they arise, interference detected as soon as it occurs, and long monologues avoided. Regulations in the UK provide that stations operating in this way should identify by giving their callsigns at intervals of not more than 15 minutes. Stations in the USA must identify every 10 minutes.

When transmitting on ssb (A3j) it is very important to avoid driving the final amplifier too hard and causing serious deterioration of the signal quality and severe interference to other band users. A good general rule to follow is to see that on speech peaks the needle of the meter which reads the anode current of the final stage in the transmitter should not exceed a point one third of the reading it gives when the stage is in the fully inserted carrier position. By far the best plan of all is to arrange for a sample of the transmitted signal to be viewed on a monitor oscilloscope at all times—in this instance "flat-topping" becomes immediately obvious.

Log Keeping

The Home Office requires all amateurs to keep a log book recording full details of all transmissions. A well-kept log is essential, not only because it may be required for inspection by the authorities, but also because it provides a permanent record of events, results and observations which may be interesting or useful later on.

Suitable log books which comply with regulations are readily available and may be purchased from RSGB Headquarters.

It should be made a practice to carry out regular checks for television interference, and such episodes should be duly entered into the log book. Few amateurs seem to realize that such tests are in fact stipulated in the Amateur Licence [section 4 (2)].

Power

The transmitting licence details the various power inputs permitted on the various bands, and the maximum power output permissible in the case of single sideband transmitters. The correct power to use for any contact is the lowest that will ensure satisfactory reception by the distant station, and

TABLE 19.8

UK 144MHz band plan

Band edge (MHz)	Section	Frequency	Use
144.000	CW only	144.000 to 144.025	Moonbounce
		144.050	CW calling frequency
		144.100	MS cw reference frequency
144.150	SSB and cw only	144.250	Used for GB2RS (ssb) and slow morse transmissions
		144.260 ±	Used by Raynet
		144.300	SSB calling frequency
		144.400	MS ssb reference frequency
144.500	All modes non-channelized	144.500	SSTV calling frequency
		144.600	RTTY calling frequency
		144.600 ±	RTTY working (fsk)
		144.675	Data and packet radio calling frequency
		144.700	FAX calling frequency
		144.750	ATV calling and talkback
		144.775	Raynet
		144.800	Raynet
		144.825	Raynet
144.845	Beacons	(144.850	Raynet)
144.990	FM repeater inputs	145.000 R0	
		145.025 R1	
		145.050 R2	
		145.075 R3	
		145.100 R4	
		145.125 R5	
		145.150 R6	
		145.175 R7	
145.200	FM simplex channels	145.200 S8	Raynet
		145.225 S9	Used by Raynet
		145.250 S10	Used for slow morse tone modulated transmissions
		145.275 S11	
		145.300 S12	RTTY-afsk
		145.325 S13	
		145.350 S14	
		145.375 S15	
		145.400 S16	
		145.425 S17	
		145.450 S18	
		145.475 S19	
		145.500 S20	FM calling channel
		145.525 S21	Used for GB2RS (fm) broadcast
		145.550 S22	Used for rally/exhibition talk-in
		145.575 S23	
145.600	FM repeater outputs	145.600 R0	
		145.625 R1	
		145.650 R2	
		145.675 R3	
		145.700 R4	
		145.725 R5	
		145.750 R6	
		145.775 R7	
145.800	Satellite service		
146.000			

UK 430-440MHz band plan

Band edge (MHz)	Section	Frequency	Use
430.000		NB: 431-432MHz not available within 100km of Charing Cross, London.	
432.000	CW only	432.000 to 432.025	Moonbounce
		432.050	CW centre of activity
432.150	SSB and cw only	432.200	SSB centre of activity
		432.350	Microwave talk-back
432.500	All modes non-channelized	432.600	RTTY calling frequency
		432.600 ±	RTTY working (fsk)
		432.675	Data transmission calling frequency
		432.700	FAX calling frequency
432.800	Beacons		
433.000	FM repeater outputs in UK only	433.000 RB0	
		433.025 RB1	
		433.050 RB2	
		433.075 RB3	
		433.100 RB4	
		433.125 RB5	
		433.150 RB6	
		433.175 RB7	
		433.200 RB8	
		433.225 RB9	
		433.250 RB10	
		433.275 RB11	
		433.300 RB12/SU12	RTTY repeater and rtty afsk working
		433.325 RB13	
		433.350 RB14	
		433.375 RB15	
433.400	FM simplex channels	433.400 SU16	
		433.425 SU17	
		433.450 SU18	
		433.475 SU19	
		433.500 SU20	FM calling channel
		433.600 SU24	RTTY-afsk
		433.700	Raynet
		433.725	Raynet
		433.750	Raynet
		433.775	Raynet
434.600	FM repeater inputs in UK only	434.600 RB0	
		434.625 RB1	
		434.650 RB2	
		434.675 RB3	
		434.700 RB4	
		434.725 RB5	
		434.750 RB6	
		434.775 RB7	
		434.800 RB8	
		434.825 RB9	
		434.850 RB10	
		434.875 RB11	
		434.900 RB12	RTTY repeater-afsk
		434.925 RB13	
		434.950 RB14	
		434.975 RB15	
435.000		434–440	ATV – frequencies chosen so as to avoid interference to other band users and, in particular, the amateur satellite service
		435–438	Amateur satellite service
440.000			

Notes on UK 144MHz and 430MHz band plans

MS operation can take place up to 26kHz higher than the reference frequency (see RSGB *Amateur Radio Operating Manual* p80).

The beacon and satellite service must be kept free of normal communication transmissions to prevent interference with these services. († – 144.850MHz in use by Raynet until further notice, subject to 25W erp max and vertical polarization).

The use of the fm mode within the ssb/cw section and cw and ssb in the fm-only sector is not recommended.

Repeater stations are primarily intended as an aid for mobile working and they are not intended to be used for dx communication. FM stations wishing to work dx should use the all-mode section, taking care to avoid frequencies allocated for specific purposes.

From January 1987, 433.200MHz will cease to be a simplex channel and will become a permanently designated repeater channel, RB8.

See page 19.10 for 50MHz and 70MHz band plans.

TABLE 19.9

Allocation of international callsign series

A2A–A2Z	Botswana
A3A–A3Z	Tonga
A4A–A4Z	Oman
A5A–A5Z	Bhutan
A6A–A6Z	United Arab Emirate
A7A–A7Z	Qatar
A8A–A8Z	Liberia
A9A–A9Z	Bahrain
C2A–C2Z	Nauru
C3A–C3Z	Andorra
C4A–C4Z	Cyprus
C5A–C5Z	Gambia
C6A–C6Z	Bahamas
C7A–C7Z	World Meteorological Service
C9A–C9Z	Mozambique
D2A–D3Z	Angola
D4A–D4Z	Cape Verde Republic
D5A–D4Z	Liberia
D6A–D6Z	State of the Comoros
D7A–D7Z	Republic of Korea
H3A–H3Z	Panama
L2A–L9Z	Argentina
P2A–P2Z	Papua New Guinea
S2A–S3Z	Bangladesh
S6A–S6Z	Singapore
S7A–S7Z	Republic of the Seychelles
S8A–S8Z	Transkei
AAA–ALZ	USA
AMA–AOZ	Spain
APA–ASZ	Pakistan
ATA–AWZ	India
AXA–AXZ	Australia
AYA–AZZ	Argentina
BAA–BZZ	China
CAA–CEZ	Chile
CFA–CKZ	Canada
CLA–CMZ	Cuba
CPA–CPZ	Bolivia
CQA–CRZ	Portuguese Overseas Provinces
CSA–CUZ	Portugal
CVA–CXZ	Uruguay
CYA–CZZ	Canada
DAA–DTZ	Germany
DUA–DZZ	Philippines
EAA–EHZ	Spain
EIA–EJZ	Eire
EKA–EKZ	USSR
ELA–ELZ	Liberia
EMA–EOZ	USSR
EPA–EQZ	Iran
ERA–ERZ	USSR
ESA–ESZ	Estonia
ETA–ETZ	Ethiopia
EUA–EWZ	Bielorussian SSR
EXA–EZZ	USSR
FAA–FZZ	France & Overseas Territories
GAA–GZZ	UK of Gt Britain & N Ireland
HAA–HAZ	Hungary
HBA–HBZ	Switzerland
HCA–HDZ	Ecuador
HEA–HEZ	Switzerland
HFA–HFZ	Poland
HGA–HGZ	Hungary
HHA–HHZ	Haiti
HIA–HIZ	Dominican Republic
HJA–HKZ	Colombia
HLA–HMZ	Korea
HNA–HNZ	Iraq
HOA–HPZ	Panama
HQA–HRZ	Honduras
HSA–HSZ	Thailand
HTA–HTZ	Nicaragua
HUA–HUZ	El Salvador
HVA–HVZ	Vatican City
HWA–HYZ	France & Overseas Territories
HZA–HZZ	Saudi Arabia
IAA–IZZ	Italy
JAA–JSZ	Japan
JTA–JVZ	Mongolia
JWA–JXZ	Norway
JYA–JYZ	Jordan
KAA–KZZ	USA
LAA–LNZ	Norway
LOA–LWZ	Argentina
LXA–LXZ	Luxembourg
LYA–LYZ	Lithuania
LZA–LZZ	Bulgaria
MAA–MZZ	UK of Gt Britain & N Ireland
NAA–NZZ	USA
OAA–OCZ	Peru
ODA–ODZ	Lebanon
OEA–OEZ	Austria
OFA–OJZ	Finland
OKA–OMZ	Czechoslovakia
ONA–OTZ	Belgium
OUA–OZZ	Denmark
PAA–PIZ	Netherlands
PJA–PJZ	Netherlands Antilles
PKA–POZ	Indonesia
PPA–PYZ	Brazil
PZA–PZZ	Surinam
RAA–RZZ	USSR
SAA–SMZ	Sweden
SNA–SRZ	Poland
SSA–SSM	Egypt
SSN–STZ	Sudan
SUA–SUZ	Egypt
SVA–SZZ	Greece
TAA–TCZ	Turkey
TDA–TDZ	Guatemala
TEA–TEZ	Costa Rica
TFA–TFZ	Iceland
TGA–TGZ	Guatemala
THA–THZ	France & Overseas Territories
TIA–TIZ	Costa Rica
TJA–TJZ	Cameroon
TLA–TLZ	Central African Republic
TNA–TNZ	Congo
TRA–TRZ	Gabon
TSN–TSZ	Tunisia
TTA–TTZ	Chad
TUA–TUZ	Ivory Coast
TYA–TYZ	Dahomey
TZA–TZZ	Mali
UAA–UQZ	USSR
URA–UTZ	Ukraine SSR
UUA–UZZ	USSR
VAA–VGZ	Canada
VHA–VNZ	Australia
VOA–VOZ	Canada
VPA–VSZ	British Overseas Territories
VTA–VWZ	India
VXA–VYZ	Canada
VZA–VZZ	Australia
WAA–WZZ	USA
XAA–XIZ	Mexico
XJA–XOZ	Canada
XPA–XPZ	Denmark
XQA–XRZ	Chile
XSA–XSZ	China
XTA–XTZ	Upper Volta
XUA–XUZ	Khmer Republic
XVA–XVZ	Viet-Nam
XWA–XWZ	Laos
XXA–XXZ	Portuguese Overseas Provinces
XYA–XZZ	Burma
YAA–YAZ	Afghanistan
YBA–YHZ	Indonesia
YIA–YIZ	Iraq
YJA–YJZ	New Hebrides
YKA–YKZ	Syria
YLA–YLZ	Latvian SSR
YMA–YMZ	Turkey
YNA–YNZ	Nicaragua
YOA–YRZ	Roumania
YSA–YSZ	El Salvador
YTA–YUZ	Yugoslavia
YVA–YVZ	Venezuela
YZA–YZZ	Yugoslavia
ZAA–ZAZ	Albania
ZBA–ZJZ	UK Overseas Territories
ZKA–ZMZ	New Zealand
ZNA–ZOZ	UK Overseas Territories
ZPA–ZPZ	Paraguay
ZQA–ZQZ	UK Overseas Territories
ZRA–ZUZ	Rep of South Africa
ZVA–ZZZ	Brazil
2AA–2ZZ	UK of Gt Britain & N Ireland
3AA–3AZ	Monaco
3BA–3BZ	Mauritius & Dependencies
3CA–3CZ	Equatorial Guinea
3DA–3DM	Swaziland
3DN–3DZ	Fiji
3EA–3FZ	Panama
3GA–3GZ	Chile
3HA–3UZ	China
3VA–3VZ	Tunisia
3WA–3WZ	Viet-Nam
3XA–3XZ	Guinea
3YA–3YZ	Norway
3ZA–3ZZ	Poland
4AA–4CZ	Mexico
4DA–4IZ	Philippines
4JA–4LZ	USSR
4MA–4MZ	Venezuela
4NA–4OZ	Yugoslavia
4PA–4SZ	Sri Lanka Republic
4TA–4TZ	Peru
4UA–4UZ	UN
5AA–5AZ	Libya
5BA–5BZ	Cyprus
5CA–5GZ	Morocco
5HA–5IZ	Tanzania
5JA–5KZ	Colombia
5LA–5MZ	Liberia
5NA–5OZ	Nigeria
5PA–5QZ	Denmark
5RA–5SZ	Malagasy Republic
5TA–5TZ	Mauritania
5UA–5UZ	Niger
5VA–5VZ	Togo
5WA–5WZ	Western Samoa
5XA–5XZ	Uganda
5YA–5ZZ	Kenya
6AA–6BZ	Egypt
6CA–6CZ	Syria
6DA–6JZ	Mexico
6KA–6NZ	Korea
6PA–6SZ	Pakistan
6TA–6UZ	Sudan
6VA–6WZ	Senegal
6XA–6XZ	Malagasy Republic
6YA–6YZ	Jamaica
6ZA–6ZZ	Liberia
7AA–7IZ	Indonesia
7JA–7NZ	Japan
7OA–7OZ	Yemen
7PA–7PZ	Lesotho
7QA–7QZ	Malawi
7RA–7RZ	Algeria
7SA–7SZ	Sweden
7TA–7YZ	Algeria
7ZA–7ZZ	Saudi Arabia
8AA–8IZ	Indonesia
8JA–8NZ	Japan
8OA–8OZ	Botswana
8PA–8PZ	Barbados
8QA–8QZ	Maldive Republic
8RA–8RZ	Guyana
8SA–8SZ	Sweden
8TA–8YZ	India
8ZA–8ZZ	Saudi Arabia
9AA–9AZ	San Marino
9BA–9DZ	Iran
9EA–9FZ	Ethiopia
9GA–9GZ	Ghana
9HA–9HZ	Malta
9IA–9JZ	Zambia
9LA–9LZ	Sierra Leone
9MA–9MZ	Malaysia
9NA–9NZ	Nepal
9OA–9TZ	Zaire
9UA–9UZ	Burundi
9VA–9VZ	Singapore
9WA–9WZ	Malaysia
9XA–9XZ	Rwanda
9YA–9ZZ	Trinidad & Tobago

except where the transmitter is designed only for low power it is desirable to have some easy means of reducing power input whenever conditions permit. Cross town contacts may be carried on with very low power indeed and a great deal of interference to other stations avoided. Local contacts should be arranged to take place on frequencies which are not open and in use for long distance working, and 28MHz is an ideal band to use for the purpose during sunspot minima.

STATION LAYOUT

Before starting to construct equipment it is desirable to have some idea of the ultimate layout of the completed station, for the physical size of the various units will, to some extent, depend on whether a table-top, bureau-bookcase, cupboard, rack and panel, or console assembly will be used. This in turn will depend on the amount of space that can be devoted to the station. Nowadays it is relatively easy to construct gear, capable of being operated at the full permissible input power, to occupy very little space, and the old rack and panel layout is virtually obsolete. This is the result of the availability of semiconductor devices and miniaturized components, and the need for effective screening of the modern transmitter as a precaution against tvi.

Choosing a Site

Amateur stations have been set up in many different places in and around the home, the location obviously

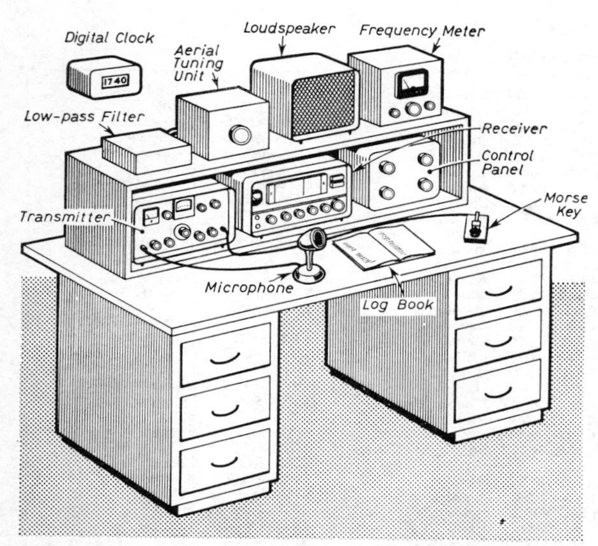

Fig 19.1. Typical table-top arrangement of an amateur station. A suitable operating table may be made from a standard flush door polished and supported on two cupboards or two two-drawer filing cabinets.

depending on domestic circumstances, the available accommodation, and the operator's ambitions. Ideally the best arrangement is for a small room to be set aside for the station. Not only does this provide the maximum comfort and quietness for operating, but also affords complete safety from danger for other members of the family. However, there are other quite suitable places in most homes.

Very efficient installations have been set up in cupboards under staircases, built into bureaux or cupboards in downstairs rooms or in sheds in gardens. All these places suffer from some drawback that does not exist for the lucky person who can devote a whole room to his station. The site under the staircase will inevitably be small, dark, and difficult to ventilate (a most important factor when extended operating periods are anticipated), and the station in the downstairs room shared with other members of the family will often be noisy, while the garden shed may be too cold and damp in the winter and too hot in the summer, besides being less accessible. The choice must be made by each individual, preferably in consultation with the rest of the family, after weighing the pros and cons of each alternative. The decision should take into account the siting of aerials—which should be as near to the transmitter as possible—and conversely as far from television aerials as can be arranged. The accessibility of power points should also be borne in mind, and also the availability of suitable earthing for the equipment. The aim should be to arrange that operating, even over long periods, is a pleasure so that maximum efficiency is obtained.

Station Wiring

The care that has gone into the construction of equipment should be continued when linking it together. Connecting cables should be short and concealed wherever possible. It is advisable to use connecting leads of differing colours since not only does this facilitate rapid servicing but it is also a safety precaution. A most important point to be watched where several plugs and sockets are to be used to carry different voltages is never to use identical components because it is only too easy to make an expensive mistake and put a wrong plug in a socket. In these days of the existence of a large amount of imported equipment, wired for 110V ac input, it is a good plan to feed the mains from a Variac or auto-transformer to a distribution board with three-pin sockets of a type different from all the 230V power sources. The 110V gear is then fitted with plugs which will only enter the 110V sockets and what would be an expensive disaster cannot occur!

Circuit diagrams of all equipment should be kept readily available, as this is an obvious aid to rapid and efficient servicing. Diagrams of all the interconnecting wiring will also be found to be invaluable.

Switching

The ideal number of switches to be used to change the equipment from the receive to the send position is none! The most convenient arrangement is to have vox operation when operating on ssb, and full break-in when using cw.

If the portion of the circuit where the transmitter is keyed carries a high voltage it is vital to key it through a relay. In any case a keying relay should be used whenever the leads from the keyed circuit to the key are of any appreciable length (see Chapter 8 for further details).

Arranging the Equipment

A typical arrangement of the equipment that goes to make up the average station is shown in **Fig 19.1**, but there are, of course, many other ways of assembling the various items according to personal preference. One of the most important features of the station is the operator's chair, which should be as comfortable as possible. From the operating position, all the main controls should be within easy reach, and all meters clearly in view.

The transmitter should be sited so that aerial feeders can be connected with the minimum of bends and of length within the room. If spaced feeder is being used this should be kept as far as possible from other objects. Ideally feeders should pass through a small window. Replacement of the glass with Perspex for easier drilling is not recommended (see Chapter 12).

The morse key is normally placed on the right side of the operating surface but should be located for maximum comfort during long periods of use. Small pieces of insulating tape folded beneath the feet of the key will help to stop any tendency for it to wander around on the bench. The microphone can be located in any convenient position on the desk. A good flat area, with sufficient space to accommodate the key, microphone, log, and a scribbling pad is essential for comfortable operating.

SAFETY PRECAUTIONS

Safety is of paramount importance and every precaution should be taken to ensure that the equipment is perfectly safe, not only for the operator himself but also for the other members of the household, or visitors. Double-pole switches should be used for all ac supply circuits, and interconnected switches should be fitted to home constructed equipment so that no part of it can have high voltage applied to it until the valve heaters and low power stages have been switched on. This precaution may not only save the life of the operator, but also protects the transmitter against damage.

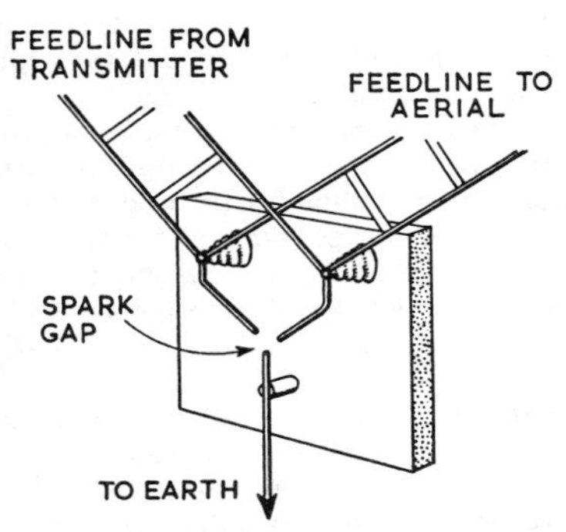

Fig 19.2. Lightning arrester. This arrangement is suitable for all types of twin-feeder systems. The mounting board should be of fireproof insulating material. Metal can be used provided that the spacing between the feeder and earth gaps is considerably smaller than the distance between the feeder gaps and the metal plate.

It should be possible to turn off power to the entire station by operating one master switch, located in a very prominent position, and all members of the household should know that in the event of an emergency this must be switched off before anything is touched.

All aerials should be protected against lightning, either by the provision of a lightning conductor in the immediate vicinity, or by the use of lightning arresters. The construction of a suitable spark-gap arrester for use with open wire feeders is shown in **Fig 19.2**. Arresters for coaxial cable are also available. Great care should be exercised before touching feeders which have been disconnected during a thunderstorm as they may hold a dangerous charge for a considerable time afterwards. Further information on lightning protection is given in Chapter 12.

An often overlooked precaution concerns the current carrying capability of the mains supply to the station. A 150 watt amateur station fully equipped with ancillary apparatus can draw quite a heavy current from the supply, and when assembling the equipment it is important to calculate the current that will be drawn when everything is in use and to check that the house wiring will carry this amount. If there is any doubt new wiring should be installed.

It is most important that every amateur should develop a strict code of safety discipline for use when handling his equipment. It should be the rule never to work on equipment which is plugged into the ac supply if this can possibly be avoided. However, there are occasions when this is unavoidable and under these circumstances the following precautions should be followed:

(1) Keep one hand in a pocket.
(2) Never wear headphones.
(3) Be certain that no part of the body is touching an object which is earthed and use a rubber or similar non-conductive covering over concrete floors.
(4) Use insulated tools.

Before working on equipment of any kind, plugged into the mains or not, it is vital to make sure that all filter capacitors are fully discharged—these are capable of retaining what could be a lethal charge for a considerable time.

The vast majority of shocks sustained from electrical apparatus are derived from the 240 volt mains line lead. Every year there are 100 or more deaths in the UK due to electrocution, mostly as a result of accidental contact with mains voltage. There is evidence to suggest that because of the different physiological effects, those who receive shocks from voltages of more than 1000 volts have a better chance of survival than those subjected to severe medium voltage shocks. Voltages as low as 32 volts have been known to cause death—as the jingle says: "It's volts that jolts, but mils that kills."

The danger of electrocution is increased where the victim's skin resistance is lowered by dampness or perspiration, or where he grips an extensive area of "live" metal while in good contact with earth. It is against this second possibility that particular care is needed in amateur stations.

A particular hazard is equipment which has a mains connected chassis which is being used under conditions for which it was not intended. All modern British television sets and a fair proportion of domestic broadcast receivers fall into this category; not only the "ac/dc" sets but also—and this is not always appreciated—a large proportion of "ac only" models. These sets are built to comply with the British Standards safety specification (BS 415) which, among other precautions, lays down that it should be impossible for the little finger to touch any part of the chassis, and also specifies that double pole on-off switches are used. Old radio sets, especially pre-war models, often do not comply with these standards, or may have damaged backs, or be used with exposed control spindles or grub screws. If there is any such equipment in your station, the exposed parts should be checked with a neon to make certain that they are not "live" with the on/off switch in either position. Remember that a single pole switch in the neutral lead will leave the chassis "live" even with the set apparently turned off. After such checks have been carried out non-reversible plugs should be fitted to ac supply leads.

It is wise to check all three-pin supply sockets in the house to see whether they have been correctly wired; all too often this is not the case. A three-pin socket with the "earth" socket at the top should have the "neutral" socket on the bottom left, and the live "line" socket on the bottom right—these directions apply when looking into the socket and must of course be reversed when looking at the back of the socket for wiring purposes (see **Fig 19.3**). Correct colour coding of leads is now: "Live", brown; "Neutral", blue; "Earth", yellow and green. It is very important to note that this coding may not apply to the wiring on some imported equipment, and the manufacturer's instructions should be very carefully studied before plugging into the supply. The use of modern fused plugs is recommended.

An even greater hazard, because it is seldom anticipated, can arise under fault conditions on equipment fitted with a double wound (ie "isolating") transformer of the type so often used in amateur equipment. It is by no means unusual or unknown for the primary winding to short circuit to the screening plate between the primary and secondary, the core, or to one of the secondary windings, so that the chassis of the equipment becomes "live". Such equipment will often continue to operate quite normally and can thus represent a very real danger over a considerable period. The best safeguard against this danger is to ensure that the screening plate between the primary and the other windings, the core, and the chassis, are all effectively earthed. The earth connection must be of very low resistance otherwise the supply fuses may not blow. These fuses should be of the minimum practicable rating—it is no use having a 50Ω resistance to earth and a 10A fuse—if this should be the case the size

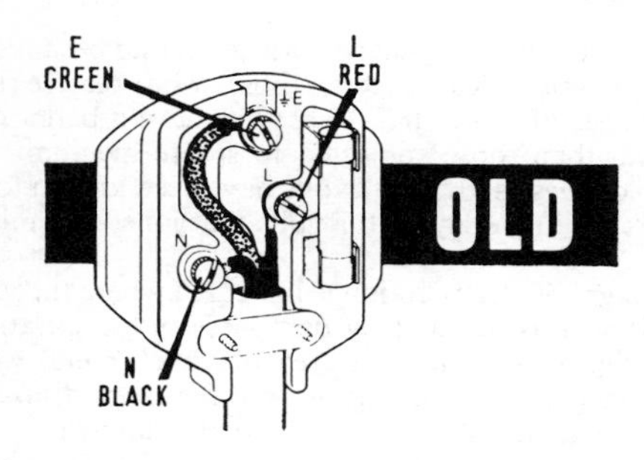

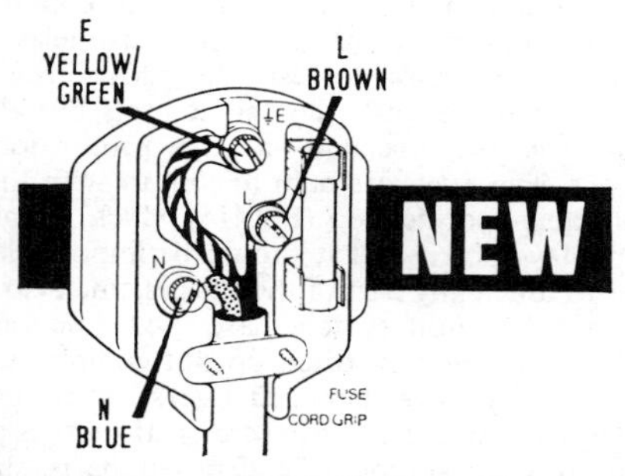

Fig 19.3. The correct wiring for three-pin plugs and sockets. To test that a socket is correctly wired, a lamp should light if connected between "L" and "N" or "L" and "E" but not when connected between "N" and "E". A neon bulb will glow when touched against "L".

of the electricity bills may be surprising, but the hazard is likely to remain undetected!

Another source of danger is the electric tool which has developed a fault and which has a "live" casing. This can happen, for example, with soldering irons and electric drills. The modern electric drill is designed with this danger in mind, but even so it should be remembered that in industry it is recommended that such tools should only be used in "earth free" areas—which is far from the state of affairs in the average amateur installation. A very careful check should be kept on the leads to all such tools, and any "tingles" felt when they are in use must be investigated immediately.

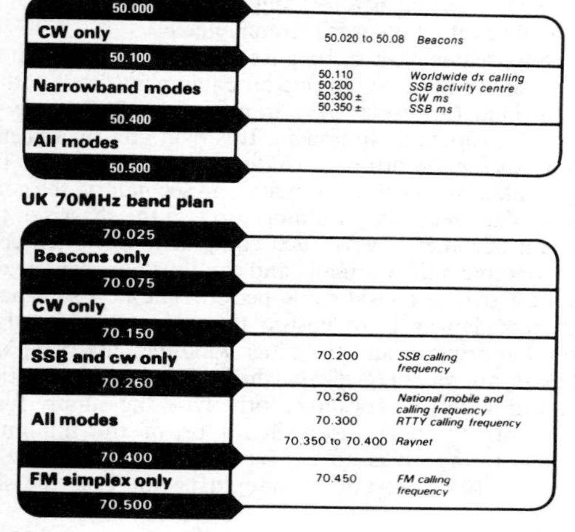

UK 50MHz band plan (Effective 1 February 1986)

Frequency	Segment	Details
50.000		
	CW only	50.020 to 50.08 *Beacons*
50.100		
	Narrowband modes	50.110 *Worldwide dx calling* 50.200 *SSB activity centre* 50.300 ± *CW ms* 50.350 ± *SSB ms*
50.400		
	All modes	
50.500		

UK 70MHz band plan

Frequency	Segment	Details
70.025		
	Beacons only	
70.075		
	CW only	
70.150		
	SSB and cw only	70.200 *SSB calling frequency*
70.260		
	All modes	70.260 *National mobile and calling frequency* 70.300 *RTTY calling frequency* 70.350 to 70.400 *Raynet*
70.400		
	FM simplex only	70.450 *FM calling frequency*
70.500		

Many amateurs fit extra power sockets in their stations and the control arrangements may call for quite a lot of semi-permanent ac wiring and switching.

In the UK these should always conform with the high standards laid down in the IEE Wiring Regulations. These are rather formidable reading for the non-professional but a number of books giving sound advice on modern wiring practice, based on the IEE recommendations, have been published and can often be obtained from local libraries. Advice can also be obtained from the offices of local electricity boards.

In most countries overseas, similar regulations exist and operators in these areas are recommended to obtain copies or seek the advice of the supply authorities. Finally, taking the worst possible event into consideration the operator and members of his household are advised to familiarize themselves with the procedures for the treatment of electric shock.

While every amateur will try to construct and maintain his station so that it is completely safe for himself and any others who may visit it, there is always the possibility that an accident may occur. Owing to a component failure a visitor may receive an electric shock, or an aerial or mast may fall and injure someone or damage property. Such an occurrence can result in a legal action, and in these days can result in the award of very substantial damages against the person held to be responsible for the accident. This risk can, and should, be insured against, either by an extension to the existing Householder's Comprehensive policy or by taking out a separate public liability policy if only fire insurance is held. The annual premium for this will only be quite a small amount and readers cannot be too strongly urged to consult their insurance advisers over this matter.

It is a wise precaution to have a fire extinguisher of the type suitable for use on electrical equipment in the radio room. The best type is that which directs a stream of carbon dioxide gas on to the burning area; the powder and carbon tetrachloride types may be used but are liable to cause further damage to such electrical equipment with which they come into contact.

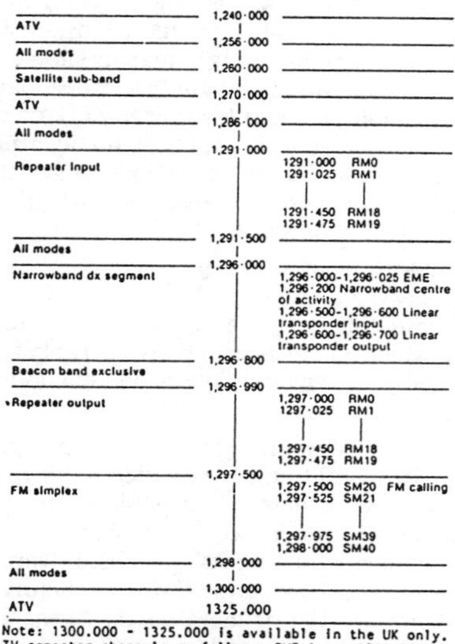

Provisional Region 1 1·3GHz band plan

Segment	Frequency	Details
	1,240·000	
ATV		
	1,256·000	
All modes		
	1,260·000	
Satellite sub-band		
	1,270·000	
ATV		
	1,286·000	
All modes		
	1,291·000	
Repeater input		1291·000 RM0 1291·025 RM1 | 1291·450 RM18 1291·475 RM19
	1,291·500	
All modes		
	1,296·000	
Narrowband dx segment		1,296·000-1,296·025 EME 1,296·200 Narrowband centre of activity 1,296·500-1,296·600 Linear transponder input 1,296·600-1,296·700 Linear transponder output
	1,296·800	
Beacon band exclusive		
	1,296·990	
•Repeater output		1,297·000 RM0 1297·025 RM1 | 1,297·450 RM18 1,297·475 RM19
	1,297·500	
FM simplex		1,297·500 SM20 FM calling 1,297·525 SM21 | 1,297·975 SM39 1,298·000 SM40
	1,298·000	
All modes		
	1,300·000	
ATV	1325.000	

Note: 1300.000 - 1325.000 is available in the UK only. TV repeater channels as follows: RMT 1 out 1311.5MHz - in 1276.5MHz; RMT 2 out 1318.5MHz - in 1249.0MHz; RMT 3 out 1308.0MHz - in 1248.0MHz.

SAFETY RECOMMENDATIONS FOR THE AMATEUR RADIO STATION

1. All equipment should be controlled by one master switch, the position of which should be well known to others in the house or club.
2. All equipment should be properly connected to a good and permanent earth. (*Note A.*)
3. Wiring should be adequately insulated, especially where voltages greater than 500V are used. Terminals should be suitably protected.
4. Transformers operating at more than 100V rms should be fitted with an earthed screen between the primary and secondary windings.
5. Capacitors of more than 0·01μF capacitance operating in power packs, modulators, etc (other than for rf bypass or coupling) should have a bleeder resistor connected directly across their terminals. The value of the bleeder resistor should be low enough to ensure rapid discharge. A value of 1/C megohms (where C is in microfarads) is recommended. The use of earthed probe leads for discharging capacitors in case the bleeder resistor is defective is also recommended. (*Note B*). Low leakage capacitors, such as paper and oil filled types, should be stored with their terminals short-circuited to prevent static charging.
6. Indicator lamps should be installed showing that the equipment is live. These should be clearly visible at the operating and test position. Faulty indicator lamps should be replaced immediately. Gas filled (neon) lamps are more reliable than filament types.
7. Double-pole switches should be used for breaking mains circuits on equipment. Fuses of correct rating should be connected to the equipment side of each switch. (*Note C.*) Always switch off before changing a fuse. The use of ac/dc equipment should be avoided.
8. In metal enclosed equipment install primary circuit breakers, such as micro-switches, which operate when the door or lid is opened. Check their operation frequently.
9. Test prods and test lamps should be of the insulated pattern.
10. A rubber mat should be used when the equipment is installed on a floor that is likely to become damp.
11. Switch off before making any adjustments. If adjustments must be made while the equipment is live, use one hand only and keep the other in your pocket. Never attempt two-handed work without switching off first. Use good quality insulated tools for adjustments.
12. Do not wear headphones while making internal adjustments on live equipment.
13. Ensure that the metal cases of microphones, morse keys, etc, are properly connected to the chassis.
14. Do not use meters with metal zero adjusting screws in high voltage circuits. Beware of live shafts projecting through panels particularly when metal grub screws are used in control knobs.
15. Aerials should not, under any circumstances, be connected to the mains or other ht source. Where feeders are connected through a capacitor, which may have ht on the other side, a low resistance dc path to earth should be provided (rf choke).

Note A.—Owing to the common use of plastic water main and sections of plastic pipe in effecting repairs, it is no longer safe to assume that a mains water pipe is effectively connected to earth. Steps must be taken, therefore, to ensure that the earth connection is of sufficiently low resistance to provide safety in the event of a fault. Checks should be made whenever repairs are made to the mains water system in the building.

Note B.—A " wandering earth lead " or an " insulated earthed probe lead " is an insulated lead permanently connected at one end to the chassis of the equipment; at the other end a suitable length of bare wire is provided for touch contacting the high potential terminals to be discharged.

Note C.—Where necessary, surge-proof fuses can be used.

CHAPTER 20

AMATEUR SATELLITE COMMUNICATION

RECENT years have seen a considerable growth in the use of amateur artificial satellites for long-distance vhf communication. Improvements in amateur equipment and the satellites themselves are bringing what was once a very specialized field into the mainstream of amateur activity. It is however possible that many amateurs are still suffering disappointment in their attempts to use this method of communication, due to an inadequate knowledge of the fundamentals of satellite orbital geometry. This topic is therefore dealt with in some detail below before the aspects of station equipment are considered.

Unlike the moon, amateur satellites in orbit relatively close to the earth's surface appear to move rather rapidly across the sky from horizon to horizon. How long the satellite will be within the range of a station is dependent on two factors: (i) the altitude of the satellite, and (ii) the distance it will be at the point of closest approach to the station. The longest duration at any altitude will occur on orbits that pass directly over the station location. The duration will decrease for orbits that pass further away from the station. For instance, a satellite in a 1,000-mile orbit would be within "line of sight" range for about 25min on an overhead pass, about 20min when it comes only within 1,000 miles of a ground station and only about 10min with a 2,000-mile distance of closest approach.

When using a satellite for two-way communication it is necessary to take into consideration the length of time the satellite will be within the simultaneous range of all the stations involved. The higher a satellite is the greater the effective range a ground station using it will have. Since higher satellites will be further away from the ground station, signal strengths will be smaller due to path losses unless either more powerful transmitters or higher gain receiving systems are used. The greater the point-to-point distance between the user and the satellite, the more power or gain will be needed to maintain adequate signal levels. Therefore, although high altitude satellites will allow contacts with more distant stations, a more elaborate station will be needed, or a satellite with larger transmitters and aerials.

GEOMETRY OF ORBITING SATELLITES

The laws governing the characteristics of one object rotating around another were established long before communication satellites were thought of. The motion of the planets around the sun has been a subject of study for a great many years, and the laws that have been derived from these observations are equally applicable to man-made satellites placed in orbit around the earth. It is not intended to derive these laws here, but merely to show how most of the information required for calculating orbital parameters for a satellite can be obtained from a few simple equations.

The force of attraction between any two objects is defined in the basic laws of physics as being proportional to the masses of the two objects divided by the square of the distance between them. This may be written as

$$F = \frac{m_1 \times m_2}{d^2}$$

where m_1 and m_2 are the two masses and d is the distance separating them.

This concept is analogous to the property of magnetism where the force of attraction between two magnets (with opposite polarity) increases as the strength of either, or both, magnets is increased, and decreases as they are moved apart.

The other main factor to be considered is that when an object rotates around a point, there is a force generated that tends to make the object move away from the centre of rotation. This can be likened to spinning a weight around on a piece of string; the heavier the weight and the faster it is spun the greater is the force pulling the weight away. This force is known as the centrifugal or centripetal force and can be written mathematically as

$$P = \frac{mv^2}{d}$$

where m is the mass of the object, v its velocity and d is the distance from the centre of the orbit to the object (or the length of string in the analogy).

Now consider the situation where there are two objects separated by a distance d, and one of the objects is rotating around the other, see **Fig 20.1**. If all extraneous effects are ignored, a condition of equilibrium will exist when the two forces previously mentioned (ie P and F), completely balance out. In this situation the outward, or centrifugal force, P, is equal to the force of attraction F, but in the opposite direction: thus there is no residual force causing the separation of the two objects to alter. The mathematical expression for this state of equilibrium is

$$\frac{m_1 v^2}{d} = \frac{m_1 m_2}{d^2} \qquad \ldots \text{(i)}$$

As before, m_1 and m_2 are the masses of the two objects separated by a distance d, and v is the velocity of the rotating object.

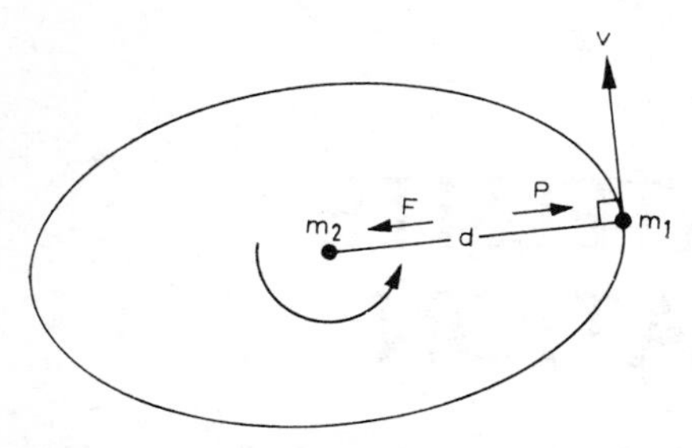

Fig 20.1. A state of equilibrium exists when the centrifugal force P is equal to the force of attraction.

Taking the situation of a satellite in orbit around the earth, d is the sum of the radius of the earth R (6,375km) and the height of the satellite orbit h.

Equation (i) can then be re-arranged to give

$$v^2 = \frac{m_2}{R + h} \quad \ldots \text{(ii)}$$

The circumference of the orbit is given by $l = 2\pi(R + h)$. If the time required to complete one revolution (known as the orbital period) is τ then from the general relationship of *velocity* = *distance* ÷ *time* it can be seen that

$$v = \frac{2\pi(R + h)}{\tau}$$

and that by substituting this in equation (ii) we can arrive at the following expression for the period:

$$\tau = 2\pi(R + h)\sqrt{\frac{R + h}{m_2}} \quad \ldots \text{(iii}$$

This shows that the orbital period of a satellite is completely independent of the weight of the satellite and depends only on the height of the orbit. The parameter m_2 is the gravitational mass of the earth and has a value of 398,600km^3/s^2.

In the following calculation we shall take as an example the orbit of Oscar 6 which had a mean altitude of 1,460km. In fact the orbit was not completely circular but the variations cause very minor errors to the calculations. By substituting this value of h in equation (iii) the period is

$$\tau = 2\pi(6{,}375 + 1{,}460)\sqrt{\frac{6{,}375 + 1{,}460}{398{,}600}} \text{ seconds}$$
$$= 6{,}900\text{s or } 115\text{min.}$$

Thus the satellite would complete about 12½ orbits each day. **Fig 20.2** shows how the satellite period varies for typical values of altitude. Another example that will be considered later is where the satellite altitude is 35,800km, which gives rise to an orbital period of 24h.

Many of the terms used in orbital geometry are self-explanatory but others may not be so obvious and the more important ones are listed below.

Sub-satellite point: this is the point on the earth that lies directly below the satellite. At this point the distance to the satellite is a minimum.

Ascending and descending nodes: a node is the point at

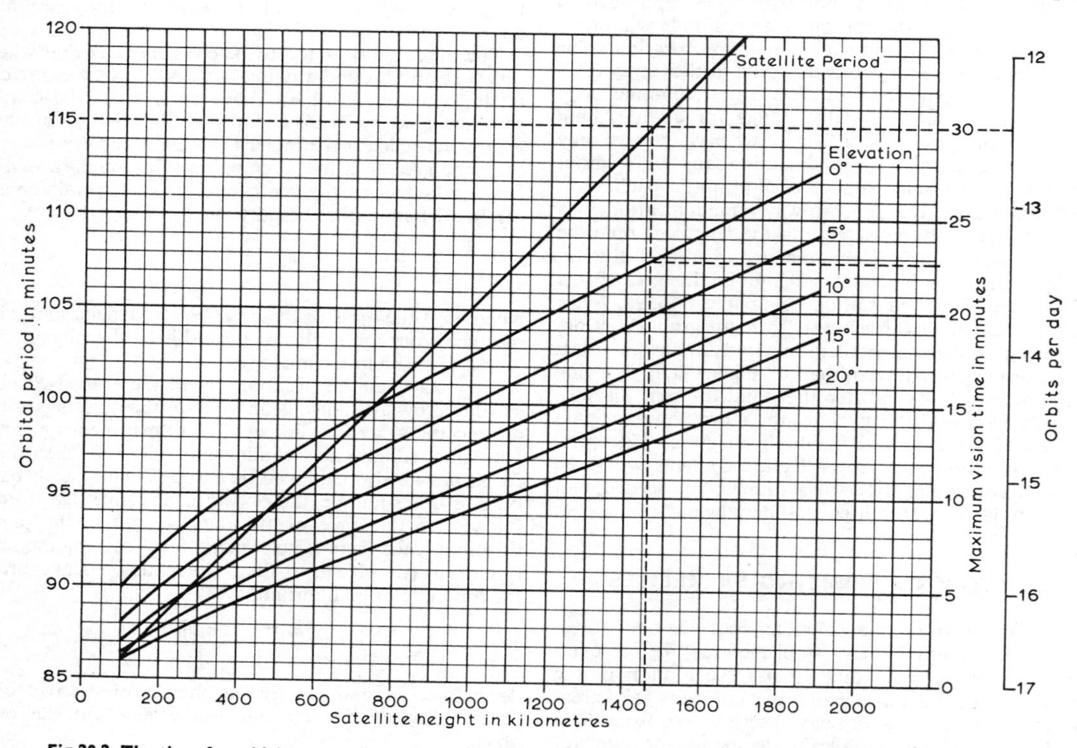

Fig 20.2. The time for which a satellite can be seen varies with the altitude of the satellite according to this chart.

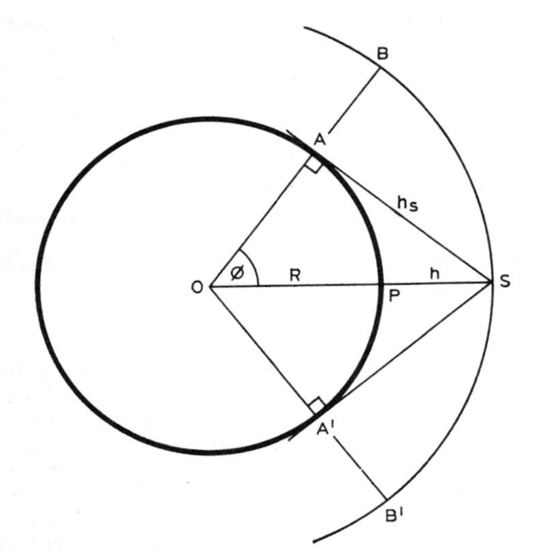

Fig 20.3. The geometry of a satellite orbit, showing the half angle of visibility (ϕ) and the slant height (h_s).

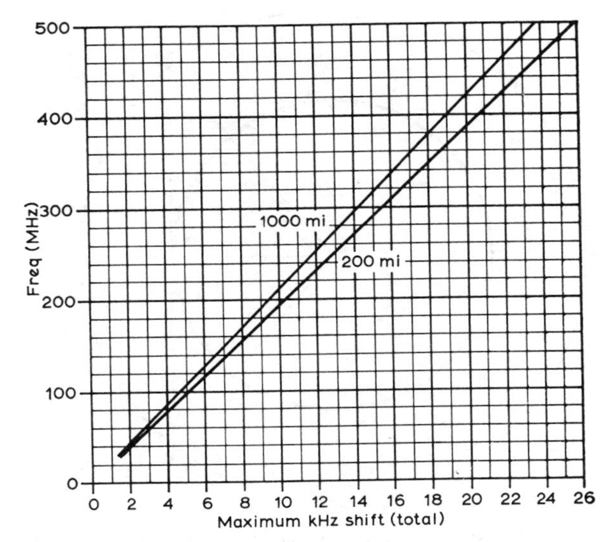

Fig 20.4. Doppler shift in relation to frequency and orbital height.

which a satellite's sub-satellite point crosses the earth's equator. If it crosses from south to north it is called an *ascending* node, from north to south a *descending* node.

Half angle of visibility: referring to **Fig 20.3**, this is the angle ϕ. It represents half the coverage angle of the satellite which is subtended at the centre of the earth. It can be calculated from the following expression.

$$\cos \phi = \frac{R}{R + h}$$

Slant height: the distance from a point on the earth to the satellite. The maximum slant height is when the location is at point A on Fig 20.3 and is given by

$$h = R \tan \phi$$

Slant range: the sum of the slant heights between two points on the earth, via the satellite. The maximum slant range will be twice the maximum slant height, ie $2 R \tan \phi$. It is necessary to know these distances when calculating the path losses for the up and down link paths.

Map range: the maximum great circle distance between two points on the earth that can simultaneously see the satellite. In Fig 20.3 this is the length of the arc APA′ and can be shown to be equal to $(220 \times \phi)$km, where ϕ is in degrees.

For Oscar 6 the above parameters are as follows:

Half angle of visibility	35·5°
Slant height	4,550km
Slant range	9,100km
Map range	7,900km

As stated before these values are all determined solely by the altitude of the satellite.

Doppler Shift

The movement of a satellite relative to the ground station results in a change in frequency of signals received in either direction. This change, known as Doppler shift, can be determined from the formula:

$$f_{dcom} = 3{\cdot}36\,(f_{up} - f_{down})\,v$$

where f_{dcom} is shift on each side of the centre frequency
f_{up} is frequency of ground station transmitter (uplink) in MHz
f_{down} is frequency of the satellite repeated signal (downlink) in MHz
v is speed in kilometres per second from equation (ii).

Maximum Doppler shift will occur on overhead passes. It can be seen in the formula that Doppler shift is a function of frequency as well as speed and is greater at higher frequencies. **Fig 20.4** indicates the total shift that may be expected at various altitudes and frequencies.

This shift in frequency of course must be taken into account when tuning receivers and transmitters. The frequency of a satellite moving toward a ground station will appear higher than the actual satellite transmission frequency and will drop as the satellite approaches until at the exact point of closest approach when it will be on the true frequency. Past this point the received signal will continue to drop lower in frequency as the satellite moves away from the ground station. Problems of tuning transmitter frequency are

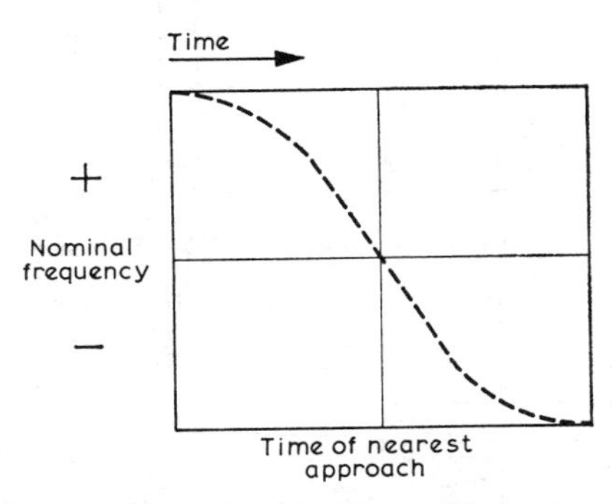

Fig 20.5. Variation of Doppler shift as the satellite passes a receiving station. The received frequency is at its nominal value when the satellite is at its nearest point, ie when its relative velocity is zero.

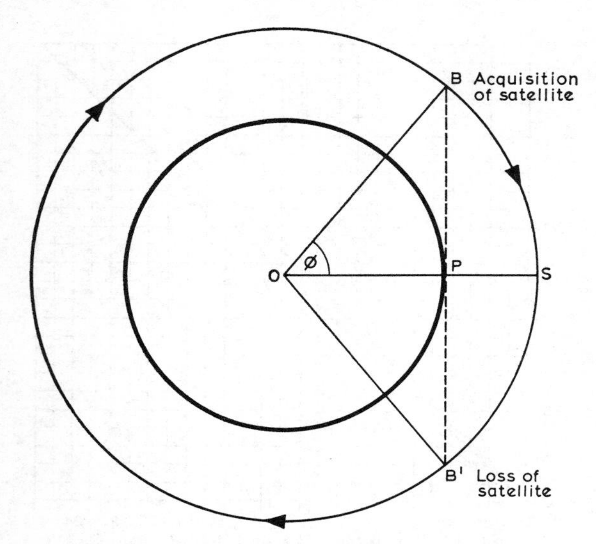

Fig 20.6. The satellite can be seen for its longest period when it passes directly overhead. Points B and B′ are the furthest at which the satellite can be seen.

reduced when a satellite receiver and transmitter frequencies are sufficiently far apart to permit a station to monitor its downlink while it is transmitting, for example, when the uplink is on 144MHz and the downlink is on 28MHz. This allows maximum efficiency of spectrum use since mutual interference by more than one station on the same frequency can be immediately detected.

The maximum Doppler shift of a signal from a satellite beacon is:

$$f_{dsat} = 3{\cdot}36(f_b\, v)$$

where f_{dsat} is the Doppler shift occurring each side of the beacon centre frequency (in Hz)
f_b is the frequency of the satellite beacon transmitter (in MHz)
v is the speed of the satellite (in kilometres per second). Values for v can be found from equation (ii).

Longitude Increment

After each orbit of the satellite the earth will have rotated a certain amount so the satellite will appear above a different point on the earth. It is convenient to choose a location on the equator as a reference point and determine how far the earth will have rotated during one complete satellite orbit. The earth completes one orbit in 24h, which is 1° every 4min; so if the period of the satellite is 115min, the earth will have rotated by 28·75° after each orbit. In other words the longitude of the point at which the satellite crosses the equator is increased by 28·75° per orbit.

Maximum Satellite Visibility Time

Fig 20.6 shows a satellite whose orbit passes directly over the earth station at point P; at this situation the satellite is visible for a maximum length of time. The actual duration of the pass depends on the velocity of the satellite which,

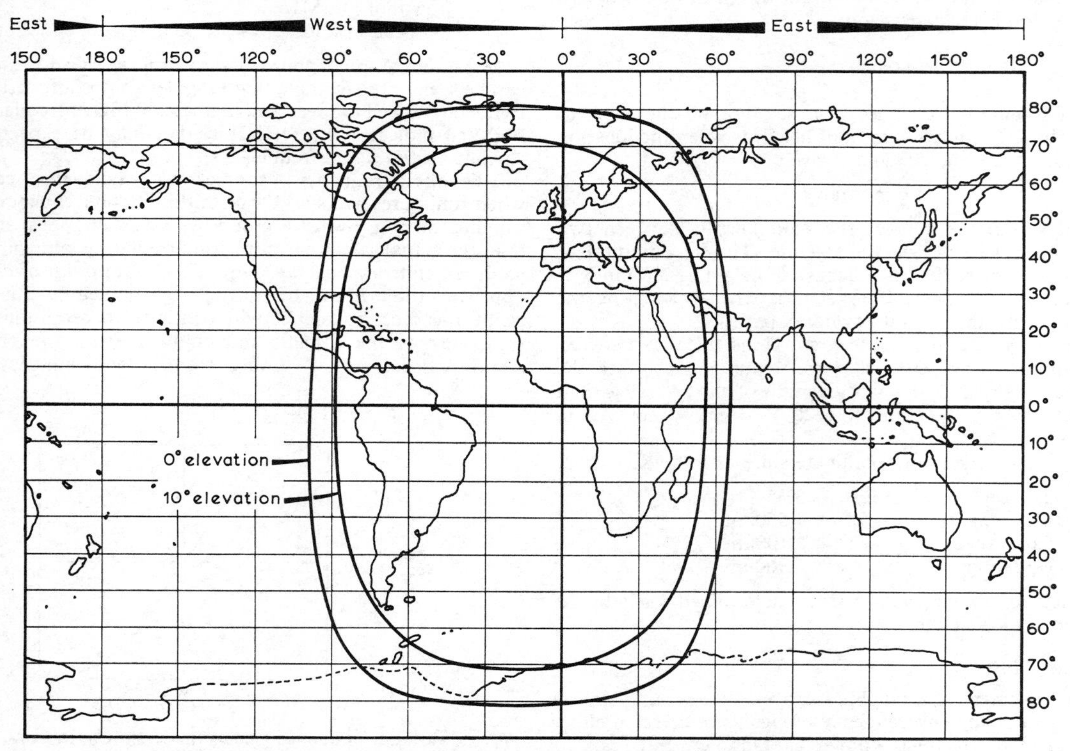

Fig 20.7. Showing the increased coverage area provided by a satellite in a geostationary orbit.

as has already been shown, is determined by its altitude. The satellite completes one complete orbit, or 360°, in its period τ: therefore to travel from the point B to B′ on the orbit will take $2\phi/360 \times \tau$ where ϕ is the half angle of visibility referred to earlier. For Oscar 6 ϕ was 35·5°, hence the maximum time for which the satellite was in direct line-of-sight was 22·7min.

In practice most satellite passes will not be overhead and the duration of the pass will consequently be reduced. Fig 20.2 shows how the time for which a satellite is visible varies with the altitude of the satellite. In calculating the time for which a given satellite may be seen, two satellite positions have been identified. These are B and B′ in Fig 20.6, which are known as the point of acquisition of the satellite (AOS) and the point of loss of the satellite (LOS) respectively. It should be remembered that these positions are based on line-of-sight criteria and observations have shown that satellite signals may not be heard until after, or in some cases before the satellite rises above, or falls below, the horizon.

Geosynchronous and Geostationary Orbits

It has been shown in the derivation of equation (iii), p20.2, that the period of a satellite is determined by its altitude, and it has been noted that for a satellite altitude of 35,800km the orbital period is 24h. Thus the satellite would appear above a certain point on the earth at the same time every day; this is known as a geosynchronous orbit.

If the orbit of the satellite is in the plane of the equator then both the earth and the satellite will rotate together and the satellite will always be above the same point on the earth; this is known as a geostationary orbit. One of the major advantages of this type of orbit is that the satellite always appears at the same point in the sky and hence there are no problems with orbit prediction and aerial tracking. With the much higher altitude of the satellite the coverage area is very greatly increased; a typical coverage area is given in **Fig 20.7**.

The main drawback with this type of orbit is that the path losses are much higher than those experienced with the subsynchronous orbits used by present amateur satellites. However, if it does become possible to put an amateur satellite into a geostationary orbit there would almost certainly be a sufficient number of stations able to use it to justify such a project. If it is possible to bounce signals off the moon at a distance of 380,000km, it should not be too difficult to relay them via a satellite at 36,000km.

Free Space Path Loss

The reduction in signal level between two points varies as the inverse of the square of the distance between the points. Considering an isotropic aerial, ie one that radiates equally in all directions, the field strength at a distance d from a source of P watts will be

$$E = \frac{P}{4\pi d^2}\ \text{W/m}^2$$

Now the gain of an aerial is given by

$$G = \frac{4\pi A}{\lambda^2}$$

where A is the effective aperture and λ is the wavelength. Hence, if an isotropic receiving aerial were placed at a distance d from the source the received power level would be

$$P_r = E \times A = \frac{P}{4\pi d^2} \times \frac{G\lambda^2}{4\pi}$$

But $G = 1$ for an isotropic aerial, hence

$$P_r = P \times \frac{\lambda^2}{(4\pi d)^2}.$$

The ratio of P to P_r is known as the loss between two isotropic aerials, or the *free space path loss*. **Fig 20.8** shows a nomogram which enables a free space loss to be calculated for various frequencies (see also Chapter 11).

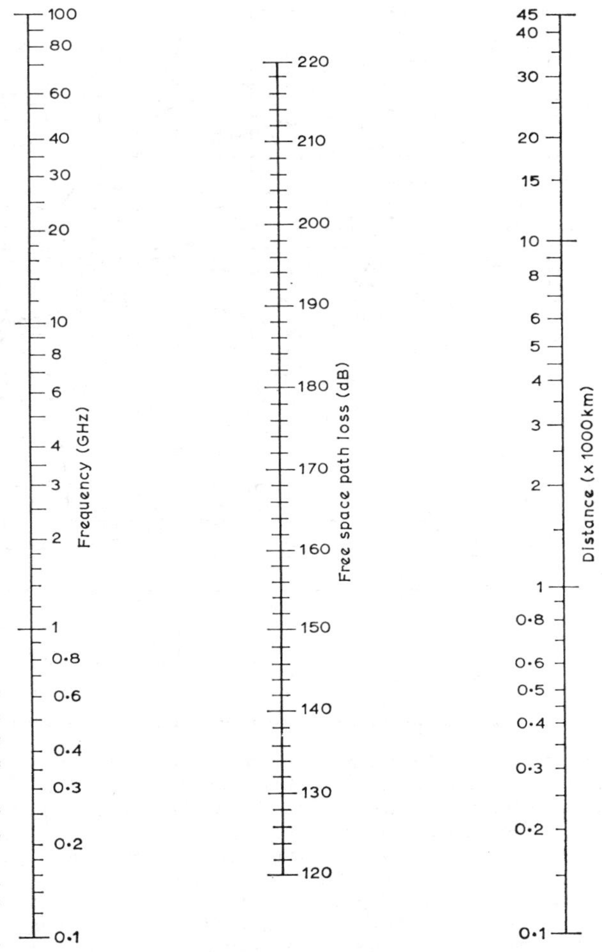

Fig 20.8. Nomogram for deriving the free space loss for various distances and frequencies.

TRACKING OF SATELLITES

At hf, and to some extent vhf, it is possible to use relatively low-gain aerials that have very wide beamwidths and consequently do not have to be directed towards the satellite. However, as the frequency is increased the losses become

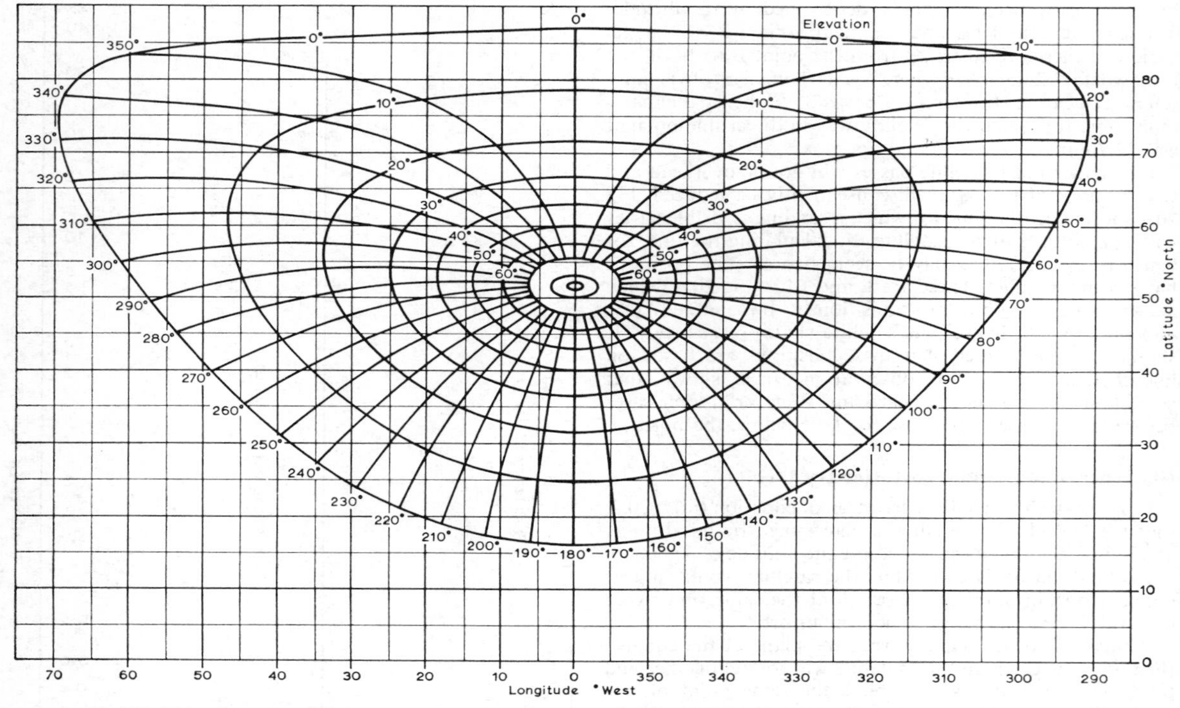

Fig 20.9. This chart gives the bearing necessary to direct an aerial towards a point of any given latitude and longitude.

greater and it becomes necessary to employ aerials with higher gains and hence smaller beamwidths. Thus, in order to obtain the maximum signal from the satellite it is necessary to point the aerial in the right direction.

To do this one needs to know the azimuth (heading) and elevation of the satellite as it progresses through the pass. A method for calculating this information for any satellite was published by W5PAG in the *AMSAT Newsletter* (December 1973) and this method is described below.

Azimuth—elevation Contour Chart

It is first necessary to draw a chart similar to **Fig 20.9** which gives the bearing required to direct the aerial towards a point of given latitude and longitude. The necessary steps are

1. The great circle angle (ie the angle subtended at the centre of the earth) between the receiving station and the point on the earth which is directly below the satellite should be calculated. **Fig 20.10** shows the great circle angle D and the other parameters required in the calculations. This angle can be calculated as follows:

$$D = \cos^{-1}\left(\frac{R}{R+h} \cdot \cos\gamma\right) - \gamma \text{ degrees}$$

where γ is the elevation angle of the satellite at the station, R is the radius of the earth (6,375km) and h is the altitude of the satellite (1,460km).

2. Next it is necessary to calculate the latitude of the point on the first bearing (say 0°) which corresponds to the elevation angle γ.

$$\sin B = \sin \alpha \cos D + \cos \alpha \sin D \cos C$$

where B is the latitude of the point below the satellite, ie the sub-satellite point, α is the latitude of the receiving station and C is the bearing to north (in this case 0°).

3. Finally, the corresponding longitude of the sub-satellite point should be calculated:

$$\sin L = \frac{\sin C \cdot \sin D}{\cos B}$$

where L is the difference in longitude between the sub-satellite point and the receiving station.

Thus the latitude and longitude of a point corresponding to a particular elevation angle have been calculated, on a

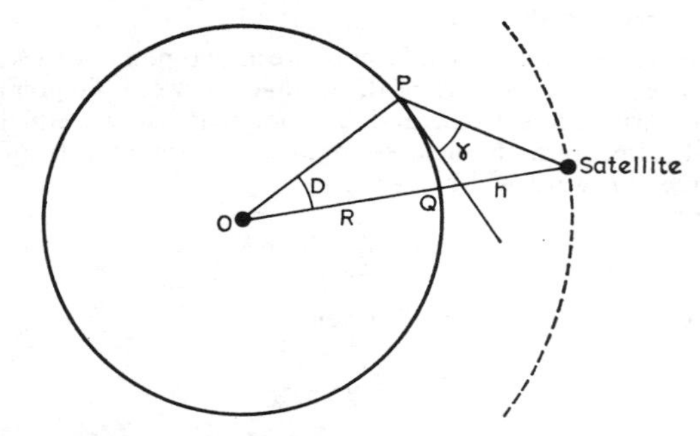

Fig 20.10. Q is the point on the surface of the earth which is directly beneath the satellite. The latitude is B° and the longitude relative to the receiving station at P is L°.

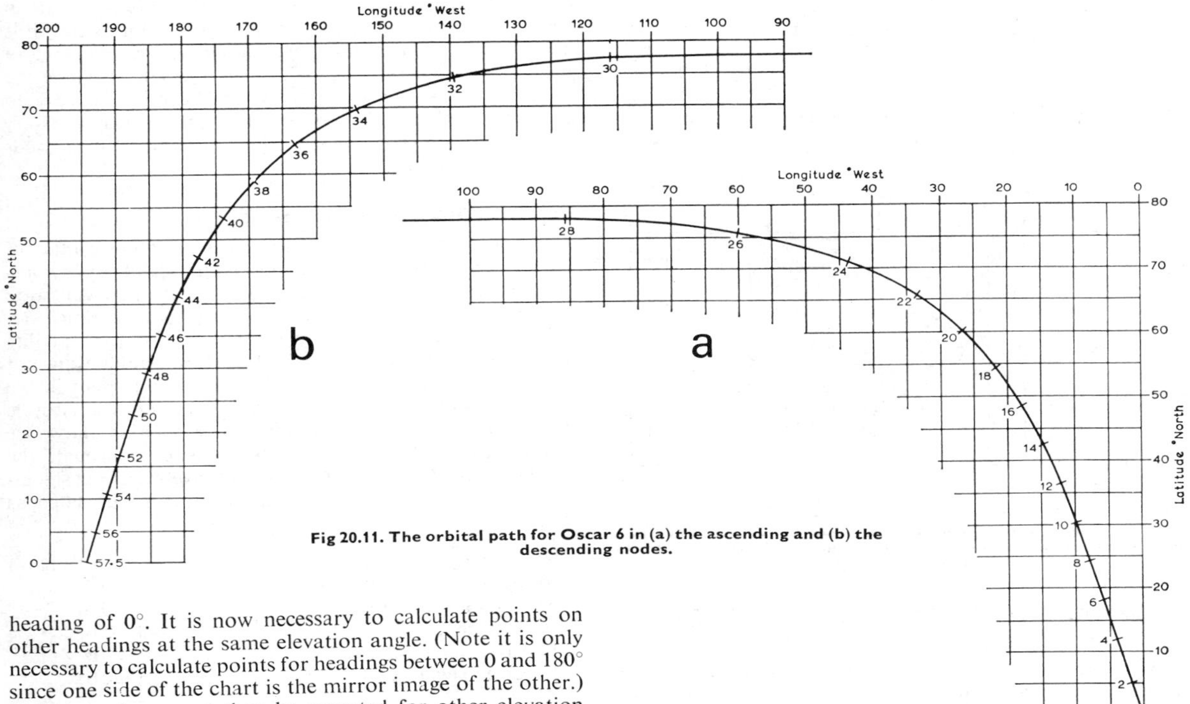

Fig 20.11. The orbital path for Oscar 6 in (a) the ascending and (b) the descending nodes.

heading of 0°. It is now necessary to calculate points on other headings at the same elevation angle. (Note it is only necessary to calculate points for headings between 0 and 180° since one side of the chart is the mirror image of the other.) This procedure must then be repeated for other elevation angles up to 90°.

Satellite Orbital Path

Having drawn the chart it only remains to determine where the satellite will be at any given time. The only information that is required about the satellite is the orbit period and the angle of inclination of the orbit to the equatorial plane. The latitude of the sub-satellite point is given by:

$$\sin b = \sin (360t/\tau) \times \sin \psi$$

where t is the length of time after the satellite has crossed the equator, τ is the satellite orbit period and ψ is the angle of inclination of the orbit. The values of ψ for Oscars 6 and 7 are 101·77° and 101·71° respectively. However, such a small difference is unlikely to cause any significant error. This angle is greater than 90° and the satellite orbit is referred to as retrograde, ie the orbit moves in a westerly direction as the earth moves from west to east.

The corresponding longitude of the sub-satellite point at a time t after the equator crossing is given by:

$$l = \cos^{-1} [\cos(360t/\tau) \div \cos b] \pm t/4$$

The factor $t/4$ is due to the rotation of the earth; the earth rotates by $\frac{1}{4}$° every minute. When the orbit is retrograde, ie when ψ is greater than 90°, $t/4$ is added, otherwise it is subtracted.

As an example, consider Oscar 7 which has a period of almost 115min and an orbital inclination of 101·7°. To find the location of the sub-satellite point these two values must be substituted in the equation above. The location at a time $t = 18$ min after the satellite crosses the equator is obtained from:

$$b = \sin^{-1} \{\sin [(360 \times 18)/115] \times \sin (101{\cdot}7)\} = 54{\cdot}6°$$

$$\text{and } l = \cos^{-1} \{\cos [(360 \times 18)/115] \div \cos (54{\cdot}6\} + 18/4 = 21{\cdot}5°$$

If the longitude of the equator crossing is, say, 15°W and the satellite is in the ascending node (ie moving in a direction towards the North Pole), the sub-satellite point at 18min after the equator crossing will be 54·6°N and 15 + 21·5 = 36·5°W.

To plot the complete orbital path as shown in **Fig 20.11** the above calculations must be repeated for convenient intervals of time, say one or two minutes. The orbital path should be plotted on a transparent sheet so that it can be positioned over the elevation contour diagram.

Using the Azimuth Elevation Chart

Having completed both parts of the chart, it may now be used to determine the appropriate azimuth and elevation of the satellite at any time during its pass in the following way.

The orbital path is first positioned on the elevation contour diagram and the equator lines on each are aligned. The equator crossing point on the orbital path overlay is positioned at the longitude of the equator crossing of the particular orbit. Most orbital data is given for the ascending node, although to calculate the descending node equator crossing all that is necessary is to subtract (180° − $x/2$) from the ascending node crossing point, where x is the longitude increment per orbit (28·7° for Oscar 7).

With the overlay in position it is now possible to read off the corresponding azimuth and elevation of the satellite to a sufficient degree of accuracy for most purposes.

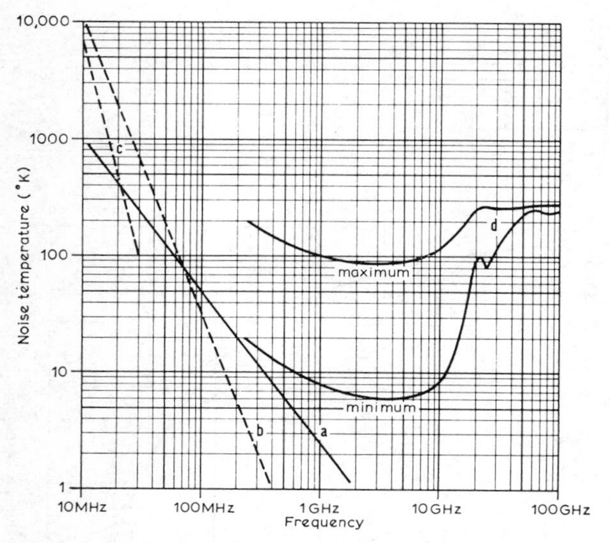

Fig 20.12. The overall noise figure comprises the noise temperatures of all components referred to a single point, usually the input to the first amplifier in the receiver.

TYPICAL LINK BUDGET FOR AN AMATEUR SATELLITE

All amateur satellites launched to date have been in sub-synchronous orbits and, at best, have had only single axis stabilization, ie one axis of the satellite has been kept constant in relation to the earth. Thus, it has only been possible to make use of simple non-directional aerials on board the satellite.

Transmitting Station Requirements

To calculate the amount of power required from an amateur station to access a satellite it is first necessary to know the signal level required at the input to the satellite receiver. The level at the input of Oscar 6 for full output was −100dBm, which should remain typical for future satellites. The required ground station power is calculated as follows:

Required signal level	−100dBm
Effective satellite aerial gain	• 0dB
Maximum slant range	4,550km
Free space path loss at maximum range	149dB
Required ground station erp	49dBm
	or 19dBW

This corresponds to an effective radiated power (erp) of about 80W. The erp is the product of the aerial gain and the amount of power actually fed to the aerial (not to the feeder). This figure can be achieved either by feeding 80W to an omnidirectional aerial, or a lower level of power to an aerial with some gain. From an operating point of view it is better to use an omnidirectional aerial since no tracking of the satellite is necessary. At higher frequencies where the signal losses are greater it may not be quite so easy to generate high levels of rf power and it may become necessary to use high-gain aerials which require quite accurate methods of satellite tracking.

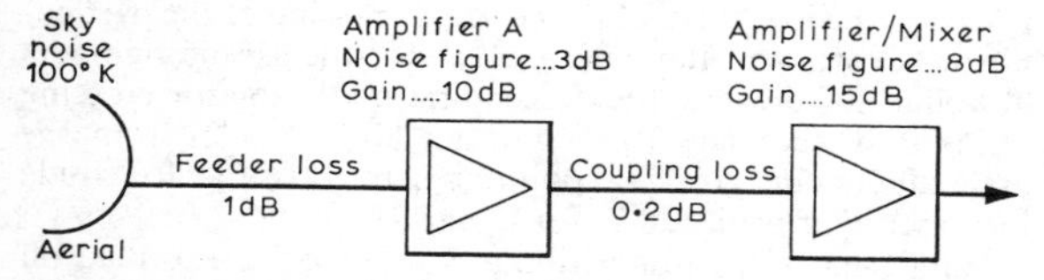

Fig 20.13. In the microwave bands the low signal levels used call for low-noise receivers. An outline of a suitable receiver is shown here.

Receiving Station Requirements

For single sideband operation a signal-to-noise ratio of 10dB is an acceptable working level, although for cw working much lower levels may be tolerated. A typical calculation is given below which derives the signal-to-noise ratio in a 3kHz bandwidth for a signal received from a satellite delivering a peak power of 1W at 29MHz. An allowance of 10dB is made for other signals that may be present in the input passband of the satellite receiver. With low duty cycle types of modulation such as ssb and cw such an allowance would provide for about 50 transmissions through the satellite provided there are no excessively strong signals present. The received signal-to-noise ratio is calculated as follows:

Satellite erp	0dBW (1W)
Multi channel loading	−10dB
Maximum slant range	4,550km
Free space loss	−135dB
Receiver aerial gain	0dB
Received signal level	−145dBW
Receiver noise bandwidth	3kHz
Receiver noise figure	15·0dB
Noise power in receiver bandwidth	−154·3dBW

hence signal-to-noise ratio = −145−(−154·3)
= 9·3dB

The receiver noise figure is mainly due to atmospheric noise. If the receiving station were situated directly below the satellite the path loss would be reduced by about 10dB and the signal-to-noise ratio correspondingly increased. A further improvement could be obtained by employing an aerial with some gain.

Optimizing the Receiving System

At hf the noise in a receiving system is generally swamped by the very high levels of atmospheric noise, sometimes known as sky noise (or temperature). The variation of this parameter with frequency is shown in **Fig 20.12**, which clearly shows that in the microwave bands the sky noise reduces to a level where the noise generated within the receiver has a significant effect on the overall system performance. Thus it is very important, for systems working at the low levels involved in satellite and moonbounce working, to reduce the internal receiver noise to the lowest possible value. In general this means using the minimum of feeder, and selecting amplifiers with the lowest noise figure consistent with useful gain.

Fig 20.13 shows an outline diagram for a typical low-level receiver. An input filter is not shown since the loss introduced is highly undesirable, and the use of a receive filter should be avoided unless there is the likelihood of crossmodulation from adjacent signals. (A filter will of course generally be required in the transmitter output.)

To find the overall noise figure of a receiver it is usual to find the noise contribution of each element and refer this to one point in the receiver, normally the input to the first amplifier. This is done as follows. Sky noise—a typical value might be 100°K, which is the equivalent noise temperature at the output of the aerial. This noise signal will be attenuated by the feeder, which has a loss of 1dB, or 1·26 as a ratio; therefore the equivalent sky noise temperature referred to the input of the amplifier is:

$$\frac{100}{1{\cdot}26} = 79{\cdot}4°\text{K}$$

The feeder not only has a loss but also increases the noise temperature.

The noise temperature at the input end of the feeder is given by $(1{\cdot}26 - 1) \times 290 = 75{\cdot}3°\text{K}$. The general expression is (loss as a ratio − 1) multiplied by the ambient temperature in degrees Kelvin. Hence the noise temperature referred to the amplifier input is:

$$\frac{75{\cdot}3}{1{\cdot}26} = 59{\cdot}8°\text{K}$$

The noise temperature of the amplifier is calculated from a similar expression to that for the feeder loss, ie (noise figure as a ratio − 1) multiplied by the ambient temperature in degrees Kelvin, which in this case is $(2 - 1) \times 290 = 290°\text{K}$. The coupling loss between the two amplifiers has a noise temperature of

$$\left(\text{antilog}\,\frac{0{\cdot}2}{10} - 1\right) \times 290 = 14{\cdot}5°\text{K}$$

This value is reduced by the amplifier gain when referred to the amplifier input, hence the equivalent noise temperature is 14·5/10 or 1·45°K. The mixer stage of the second amplifier usually has a much worse noise figure than the first amplifier, but the higher the gain of the first stage the lower the effect of following stages. An 8dB noise figure is equivalent to a noise temperature of (antilog 8/10 − 1) × 290 or 1,535°K, which is reduced to

$$\frac{1{,}535}{10} \times 1{\cdot}05 = 161°\text{K}$$

when referred to the input of the first amplifier. Thus the equivalent noise temperature at the input to the first has been calculated for each element. The total noise temperature is simply the sum of each of the individual values:

Sky noise	79·4°K
Feeder loss	59·8°K
1st amplifier	290°K
Coupling loss	1·45°K
2nd amplifier/mixer	161°K
	591·7°K

The noise figure is calculated from

$$N = 10 \log_{10}\left(\frac{T}{290} + 1\right)\text{dB}$$

where T is the total equivalent noise temperature. Hence for the receiver considered the noise figure is 4·8dB. By doing this type of calculation it is possible to locate those areas in a receiving system that have a significant effect on the overall performance. In particular it can be seen what the effect is if the second amplifier cannot be ignored.

AMATEUR SATELLITES LAUNCHED

Oscar 1

The first amateur satellite was designed and built by the Project Oscar group (Oscar is an acronym for Orbiting Satellite Carrying Amateur Radio). It was launched on 12 December 1961 and consisted of a 100mW telemetry beacon operating on 144·98MHz. More than 5,000 reports of the satellite were received from 600 amateur stations during its three-week life.

Oscar 2

This satellite was launched on 2 June 1962 and operated for 18 days. Like Oscar 1 it carried a telemetry beacon in the 2m band.

Oscar 3

This was the first "free access" satellite in the world, being launched on 9 March 1965, only one month before Early Bird. The transponder accepted signals within a 50kHz bandwidth centred on 144·1MHz and re-radiated them around 145·9MHz at a power level of 1W p.e.p. A 50mW telemetry beacon also radiated on 145·85MHz. Operation lasted for only two weeks, during which time more than 100 stations communicated through the satellite, including two-way transatlantic contacts.

Oscar 4

This was launched on 21 December 1965 and carried a 2m to 70cm transponder. Unfortunately the satellite failed to achieve its correct orbit and only operated long enough for about a dozen two-way contacts to be made. However, Oscar 4 does have the distinction of providing the first direct satellite communication link between the United States and the Soviet Union.

Australis Oscar 5 (AO-5)

The fifth amateur satellite was designed and constructed in Australia. A newly formed group, the Radio Amateur Satellite Corporation (AMSAT), checked the satellite for space operations before its launch on 23 January 1970. AO-5 carried telemetry beacons on 144·05MHz and 29·45MHz and the latter could be switched by ground command. Although no communication facility was provided, AO-5 did provide valuable experience in telecommand systems, which are now standard. Also, the two beacons allowed investigation of anomalous propagation.

AMSAT—Oscar 6

Oscar 6 was launched on 12 October 1972 as part of the payload of a Thor-Delta rocket carrying the NOAA-2 weather satellite. The orbital parameters were very similar to Oscar 5, allowing access to the satellite several times each day. The communication equipment comprised a 2m to 10m transponder with a 100kHz bandwidth. The input frequency range was 145·90 to 146·000MHz and the output range 29·45 to 29·55MHz.

Beacon signals were transmitted on 29·45MHz and 435·1 MHz, although the latter failed after a short period of

Oscar 7 during vibration tests. The 2,304MKz quadrifilar aerial furnished by RCA can be seen at the top.

operation. A new feature on Oscar 6 was the Codestore unit which was an 800-bit message storage unit capable of storing or playing back up to 18 words of morse code. The information from the Codestore unit was transmitted on the hf beacon signal by suitably keying the oscillator. The type of message transmitted via the Codestore ranged from satellite orbital data to telemetry parameters such as battery voltage and current drain. Oscar 6 was still functioning well after more than four years' operation, although times of operation were being restricted to conserve battery power.

AMSAT—Oscar 7

The next package from AMSAT was a satellite carrying two communications transponders and four beacons, three of which may be keyed by a Codestore unit similar to the one flown on Oscar 6. A brief summary of the characteristics of the satellite is given below. The satellite was launched on 15 November 1975 from the Western Test Range in the USA.

145 to 29MHz repeater. This is similar to the one flown on Oscar 6 except that the power output is increased to 2W p.e.p. The repeater will receive signals between 145·85 and 145·95 MHz and re-radiate them between 29·4 and 29·5MHz. There is also a beacon on 29·502MHz. The transmitter erp required to access the satellite is the same as for Oscar 6, ie about 80 to 100W.

432 to 145MHz repeater. This repeater, like that above, is also a linear device and hence is more efficient with low duty cycles types of modulation such as cw or ssb. The input frequency range is 432·125 to 432·175MHz and the output, which is inverted, is transmitted between 145·975 and 145·925MHz. There is also a telemetry beacon at 145·980 MHz. For reasons of loading of the 2m satellite receiver it is not possible to operate both repeaters simultaneously.

In order to access the repeater an effective radiated power of 80 or 100W is required, preferably by using a low-gain aerial and a high-level power amplifier, since this reduces the need for continual tracking of the satellite. The output of the repeater is about 8W p.e.p. fed to a circularly polarized aerial, which helps to reduce fading of the signals.

435·1MHz telemetry beacon. This beacon is similar to that flown on Oscar 6, but does not operate when the 432 to 145MHz transponder is in use. (The Doppler shift at this frequency will be up to $\pm 8{\cdot}4$kHz).

2,304MHz beacon. It was originally proposed to carry a beacon transmitter at the frequency, but due to authorization difficulties this was not possible.

Geostationary Satellites

Future amateur satellites planned for the immediate future are likely to use orbits similar to those of Oscar 5 or 6, however it is possible that within the next decade an amateur satellite could be placed in a geostationary orbit (see p20.5). As mentioned previously a satellite in such an orbit remains in an almost constant position relative to the earth. There are slight variations in the actual position due to the fact that the orbit may not be truly equatorial, but those are not likely to be signicant with amateur aerials.

The most appropriate frequency band for the downlink, ie satellite to earth, is probably 70cm (actually 435 to 438MHz), since at this frequency it is relatively easy to achieve low receiver noise figures (2 to 3dB) and high aerial gains (20 to 25dB). For the uplink the choice is somewhat under debate but due to the ease at which high power levels may be generated 145MHz seems to be the preferred solution.

A power link for the above suggested choice of frequencies is given below.

2m uplink

Input level at satellite for full output	−130dBW
Free space path loss	−168dB
Propagation loss	−1dB
Pointing and polarization loss	−2dB
Required ground station erp	41dBW
Ground station aerial gain	20dB
Transmitter power	21dBW
	or 126W

70cm downlink

Satellite erp	10dBW
Free space path loss	178dB
Multichannel loading factor	−10dB
Propagation loss	−1dB
Pointing and polarization loss	−2dB
Ground station aerial gain	20dB
Received signal level	−161dBW
Receiver noise figure	3dB
Signal bandwidth	3kHz
Signal-to-noise ratio	−168dB
Noise power	−169dBW

Thus it can be seen that communication via a geostationery satellite is quite within the existing capabilities of many amateur stations.

TABLE 20.3

Amateur satellite allocations

Frequency band	Status
7,000–7,100kHz	Equal primary region 1, 2 & 3
14,000–14,250kHz	Equal primary region 1, 2 & 3
21,000–21,450kHz	Equal primary region 1, 2 & 3
28·0–29·7MHz	Equal primary region 1, 2 & 3
144–146MHz	Equal primary region 1, 2 & 3
435–438MHz	Permitted subject to non-interference
24·0–24·05GHz	Equal primary. 24·125 ± 0·125GHz is also designated for industrial, scientific and medical purposes

FREQUENCY ALLOCATIONS TO THE AMATEUR SATELLITE SERVICE

In 1971 a world conference was convened by the International Telecommunication Union (ITU) in Geneva to allocate specific frequency bands to the various satellite services that either already existed or were proposed for the future. As far as amateurs were concerned this effectively meant designating certain existing allocations, or parts of them, as allocations to the Amateur Satellite Service. A complete list of the allocations is given in **Table 20.1**.

COMMUNICATING THROUGH OSCAR 7

The theoretical background to satellite orbital geometry and user station requirements has been given in some detail earlier in this chapter. The following section is intended to give the reader a brief summary of the way in which these considerations are applied in practice to communicating through a typical amateur satellite: Oscar 7.

Equipment Required

A receiver having good performance on 29·4–29·5MHz is essential for reception on Mode A. Most newer equipment is satisfactory but a preamplifier is desirable for older receivers. A 3N140, 40673 or similar mosfet should be used to give a gain of 10–20dB and a noise factor of about 3dB. A suitable design appears on p4.36.

Normally a triband beam will give good results up to an elevation angle of about 30°, but above this angle it will be found that crossed dipoles or a slant vertical will be superior and have the added advantage that tracking will not be required.

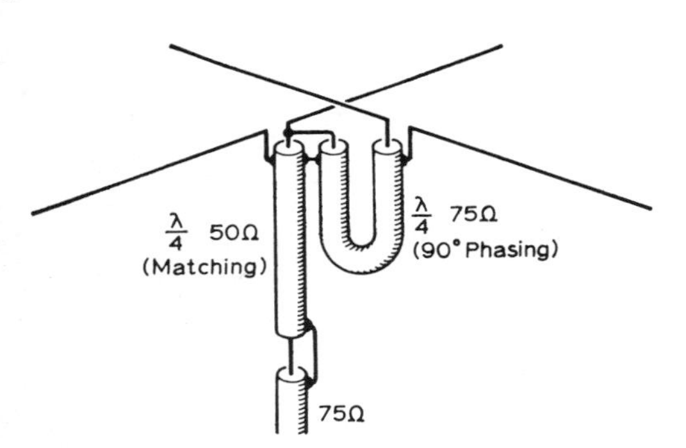

Fig 20.14. Crossed dipoles aerial, showing phasing and 75Ω matching arrangements. If the feeder run to the transmitter is short, or if the aerial is only being used for reception, 50Ω feeder may be used (without the λ/4 matching section shown). When calculating the length of cable required for the λ/4 sections, due allowance must be made for the velocity factor in each case (see Chapter 13).

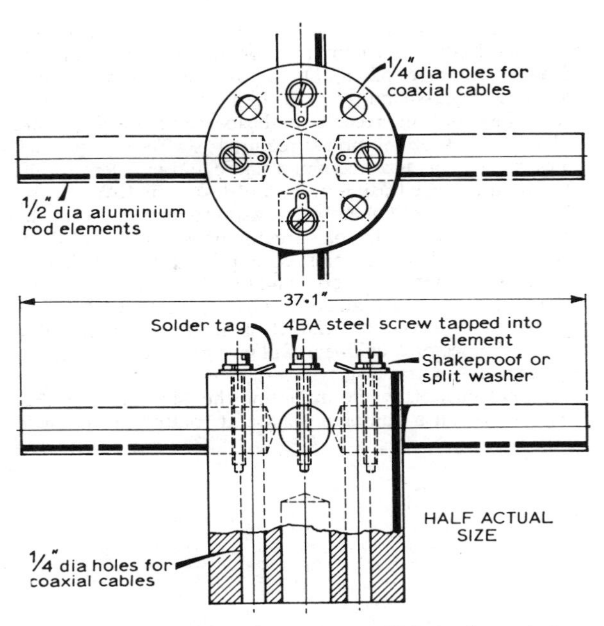

Fig 20.15. Details of the centre insulator, which may be made from 2in diameter nylon, Tufnol, ebonite etc.

The 144MHz uplink power requirements are 100W erp at 2,000 miles range, decreasing to 10W when near overhead. Most fm equipment can be modified for cw keying, and a vfo or vxo is well worthwhile. Aerials range from crossed Yagis for dx working at extreme ranges to simple crossed dipoles (also known as the *turnstile*) for higher angle communication.

Details of a crossed dipoles aerial for 144MHz is given in **Figs 20.14–20.16.** This aerial is also suitable for the downlink on Mode B.

The aerials suggested for the Mode A uplink are also suitable for reception on Mode B (145·925–145·975MHz).

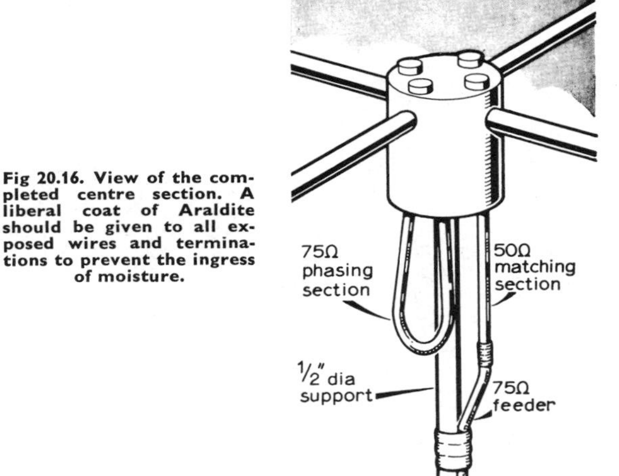

Fig 20.16. View of the completed centre section. A liberal coat of Araldite should be given to all exposed wires and terminations to prevent the ingress of moisture.

They must be followed by a low-noise system comprising a converter feeding a communication receiver providing the i.f./af functions (see p20.8). Do not overlook the inversion of the Mode B passband—cw will be transmitted at the high end of the uplink and received at the low end of the downlink.

The Mode B uplink power required is 80W erp at 2,000 miles range. This is a maximum figure and excellent results have been achieved using powers of less than 10W. RF on 432MHz may be obtained by a low-power transmitter or by the use of a transverter or varactor tripler from 144MHz. The latter is probably the most economical approach. A Yagi is suitable for the uplink aerial and this also provides enough radiated power at high angles to give adequate results without the use of a second aerial. If higher power is available crossed dipoles or a collinear will give good results without the necessity for tracking.

It might be worthwhile mounting the 144MHz and 432-MHz aerials on a common mast and rotator to minimize tracking problems. However, if the 432MHz output is derived by tripling from 144MHz, leakage may desensitize the receiver.

Orbital Predictions

Orbital prediction data usually give the longitude and time of the ascending node of a reference orbit for each day, together with the orbit number. Using this data it is possible to calculate the longitude and time of the next ascending node. This is done simply by adding the longitude increment (see p20.4) to the longitude given, and the satellite orbit period to the time given. Successive ascending nodes are determined in the same way.

Once the nearest ascending nodes to the user have been determined the method given on p20.6 can be used to plot orbital paths. Not only will these indicate the direction and elevation of the satellite at any given time, but also the usable time periods.

CHAPTER 21

IMAGE COMMUNICATION

THIS chapter is concerned with the transmission and reception of visual and graphic images using normal voice communication bandwidths and equipment. Two systems are in use in the UK at the present time: slow-scan television and facsimile, and these are described below. With either system it is possible to transmit and receive images on a world-wide basis, and these can provide a valuable adjunct to conventional voice or cw communication.

SLOW-SCAN TELEVISION

Introduction

One of the more recent aspects of amateur radio which has gained wide popularity in a relatively short space of time is slow-scan television. This popularity is in a sense the more remarkable as sstv is almost entirely an amateur development. While considerable experimentation went on in the field of low-definition tv systems in the early 'sixties, the rapid development of satellite transmission systems made most of the work in this field superfluous, and it was left to radio amateurs to recognize the potential of narrow-band picture transmission systems.

Slow-scan television, as its name implies, is a system by which a television picture is slowed down so that it may be contained radio the 2·8kHz audio bandwidth normally used in amateur radio equipment. The slowing-down process results in a 120-line picture which takes 7·2s to scan out, making one complete frame.

In theory the definition in terms of visible elements is around one-fourteenth of that obtainable with a 625-line tv picture, but this resolution is found to be adequate for the transmission of still pictures.

Since the total bandwidth lies well within the audio spectrum it becomes possible to transmit pictures using a normal amateur ssb transceiver and, similarly, an ordinary domestic tape recorder may be used to record the signals for playback at a later date.

Picture Composition

A composite video signal (ie video plus sync pulses) is used to frequency modulate a 1,200Hz audio subcarrier oscillator, and the resultant constant-amplitude fm signal is used to modulate an ssb transmitter.

The aspect ratio of the picture is usually 1:1. This square format was chosen mainly because the surplus cathode-ray tubes generally used for sstv are round, and therefore in order to utilize as much of the available screen area as possible a square picture was desirable.

Fig 21.1 shows the frequency composition of a single slow-scan line. An audio frequency of 1,500Hz is equivalent to "black" level and 2,300Hz is equivalent to "white". It follows that any intermediate shade of grey can be represented by a transmitted frequency between 1,500Hz and 2,300Hz.

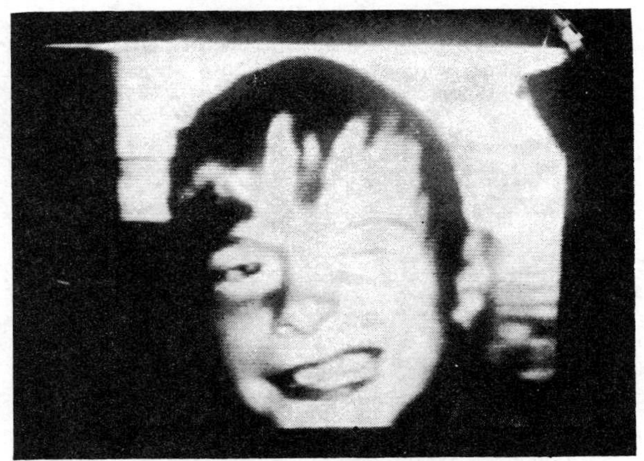

Typical sstv picture received on long-persistence cathode-ray tube.

Synchronization pulses are transmitted at the subcarrier frequency of 1,200Hz and, since these are "ultra" black, do not appear in the displayed image. The line scan shown in Fig 21.1 thus consists of a 5ms sync pulse at 1,200Hz followed by the varying frequencies which represent the light intensity of the scanned visual image. At the end of the line the scanning spot on the monitor screen "flies back" during the period of the next sync pulse and is then ready to scan the next line. The flyback is visually suppressed since the sync pulse is below black level.

At the beginning of each complete frame the first 5ms line

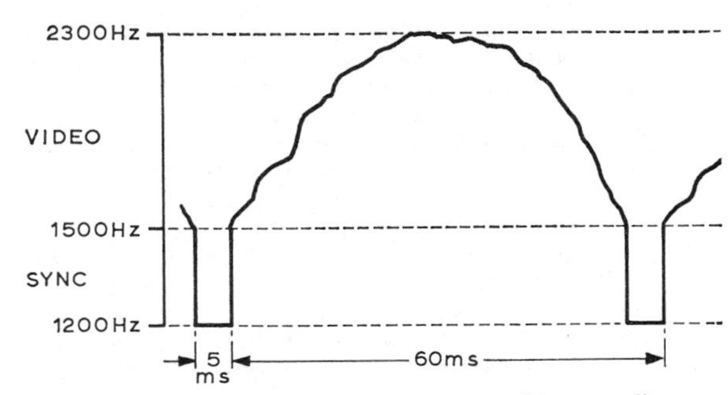

Fig 21.1. Frequency composition of a single slow-scan line.

sync pulse is replaced by a 30ms frame sync pulse during which the scanning spot resets from the bottom right to the top left of the displayed picture. Once again this flyback is visually suppressed.

Standards

The universally adopted standards for sstv are shown in **Table 21.1.** It can be seen that the line frequency of 16·6Hz is obtained by dividing the mains supply frequency by three (50Hz countries). Similarly if the mains frequency is divided by 360 one obtains the sstv frame frequency of 0·14Hz (7·2s). It will be apparent from Table 21.1 that in countries that use a 60Hz mains frequency a different divider network must be employed to arrive at approximately the same standards. Most sstv monitors will accept either standard, the only noticeable effect being a slightly larger picture on the monitor screen.

TABLE 21.1

SSTV standards

Parameter	50Hz Mains	60Hz Mains
Line speed	50Hz ÷ 3=16·6Hz (60ms)	60Hz ÷ 4=15Hz (66ms)
Line performance	120	120
Frame speed	7·2s	8s
Picture aspect ratio	1 to 1	1 to 1
Scanning direction		
Horizontal	left to right	left to right
Vertical	top to bottom	top to bottom
Sync pulse duration		
Horizontal	5ms	5ms
Vertical	30ms	30ms
Subcarrier frequency		
Sync	1,200Hz	1,200Hz
Black	1,500Hz	1,500Hz
White	2,300Hz	2,300Hz
Required transmission bandwidth	1·0 to 2·5kHz	1·0 to 2·5kHz

Equipment

A station equipped for sstv needs, in addition to the ordinary ssb equipment, a monitor and picture generator. No modification whatsoever is necessary to the existing transmitting and receiving equipment because the sstv signal is always at audio frequencies. However since the tv signal uses a constant-amplitude subcarrier, and the duty cycle is thus 100 per cent, care should be taken not to exceed the limitations of the transmitter power amplifier. In most modern transceivers this means a maximum input of around half the rated cw power. If long transmissions are to be made a cooling fan for the pa would be a worthwhile investment.

The monitor is normally plugged into the phones socket, extension speaker socket or phone-patch output of the receiver. The choice of monitor depends largely upon the individual. For those wishing to build their own, examples of high-quality monitors are those designed by SM0BUO [1] and W6MXV [2]. Less sophisticated (and cheaper) designs are those by W4TB [2] and G3RHI [3]. Full details of a design based on [1] are given later in this chapter.

The simpler designs often have a tendency to "false trigger" the monitor under conditions of weak signal strengths or strong adjacent-channel interference. These monitors should not be under-rated however, and under most operating conditions will give very good results.

There is no need at first to obtain picture generating equipment, and an ordinary domestic tape recorder or one of the popular cassette machines is all that is necessary. Most sstv amateurs will be glad to make short "programmes" from the reader's own caption cards and photographs, and record them on to tape. The output of the recorder is plugged directly into the microphone socket of the transceiver. Several methods exist for generating one's own pictures, but the simplest is probably the electronic pattern generator which is usually only used for test purposes.

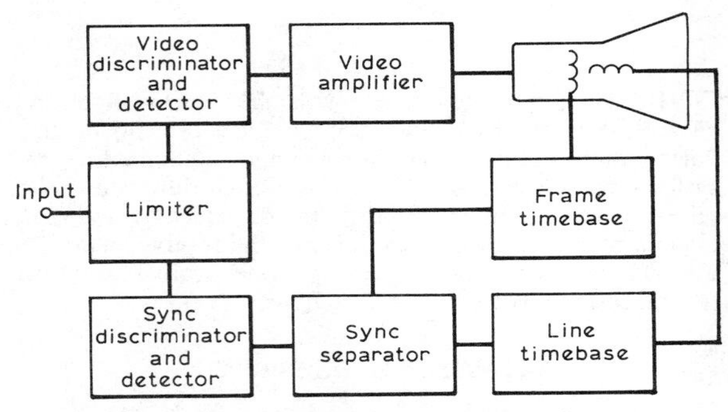

Fig 21.2. Block diagram of an sstv monitor.

The "flying-spot scanner", which is discussed later in this chapter, may be built for a few pounds, and needs no expensive camera tubes or lighting arrangements. It can only reproduce caption cards or photographs.

SSTV cameras fall into three main groups:

(a) A camera using a tube such as a "Plumbicon", which is scanned directly at slow-scan rate.

(b) An ordinary fast-scan (405 or 625-line) camera which is sampled at slow-scan rate and needs slight modification to the timebases.

(c) The fast-scan to slow-scan converter, which takes the output from an unmodified fast-scan camera and converts it digitally to slow scan.

Monitors

The heart of an sstv monitor is the cathode-ray tube which should be carefully chosen to suit the operator's requirements. The tube should have a long-persistence (P7) phosphor, the most popular types being the surplus 3FP7 and 5FP7 ex-radar tubes, these having 3in and 5in round screens respectively. This type of tube is however becoming rather scarce although suitable alternatives primarily intended for oscilloscopes can be found at reasonable prices.

The 3FP7, while permitting only a small picture display, has the advantage of being electrostatically deflected and it therefore requires no deflection coils. Tubes such as the larger 5FP7 require separate focus magnet and deflection assemblies. The deflection coils can be taken from an old domestic tv set having a tube with a narrow deflection angle. (Most modern tv tubes have a deflection angle of around 110° and the scan coils are therefore unsuitable.)

The focus assembly can use either permanent or electromagnets, and focus magnet assemblies can be taken from old domestic tv receivers having tetrode-type tubes. The electromagnetic types require however a fairly large dc source.

The basic block diagram for an sstv monitor is shown in

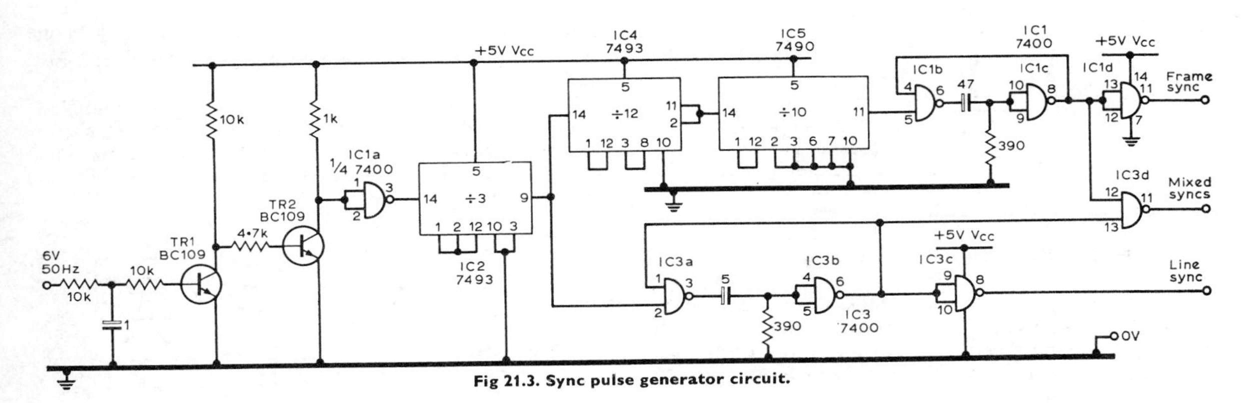

Fig 21.3. Sync pulse generator circuit.

Fig 21.2. The signal is applied to a limiter which eliminates any distortion due to amplitude modulation, and the signal is then passed to the video discriminator and detector. The discriminator output is arranged to give maximum response at 1,500Hz and minimum response at 2,300Hz, and the signal is then passed to the tube. Similarly the limited signal is passed to the sync discriminator and detector, the output of which should give maximum response at 1,200Hz. The sync separator routes the line and frame sync pulses to the two timebases which drive the crt scan coils.

Of considerable importance in any monitor are the various filters. These ensure that only the wanted information is fed to the video and sync stages and that all interference signals outside the passband are attenuated. Obviously the better the filter the less risk there is of adjacent-channel interference. Simple filters often employ the 88mH toroids commonly used in rtty equipment and suitable capacitors are chosen to achieve the correct centre frequency. More widely used today, however, is the "active" filter employing the popular 741 operational amplifiers. By cascading several sections together a very effective filter is obtained with steep sides and narrow skirt selectivity.

When constructing a monitor it is important to keep all ac fields such as those associated with the mains transformer well away from the tube, and failure to do so may result in vertical hum bars appearing in this display. The 5kV or so required on the tube's post deflection anode can be lethal, so proper safety precautions must be taken during construction and testing.

Sync Pulse Generators

The sync pulse generator provides line and frame pulses for the picture generating equipment. In the sstv camera or flying spot scanner the sync pulses are used to trigger the line and frame timebases, and these pulses are also routed to the modulator unit so that, when mixed with the picture information, they produce the required composite video output.

A typical sync pulse generator is shown in **Fig 21.3.** The pulses are derived from the 50Hz mains frequency. The sinusoidal input is squared and brought to ttl level by the stages TR1, TR2 and IC1a. This 50Hz square wave is divided by three in IC2 producing 16·6Hz pulses at pin 9. These are shaped and timed in IC3 to produce the line syncs. Further division by 120 takes place in IC4 and 5 to produce the frame rate of 0·14Hz (7·2s).

The shaper and timer IC1b,c produces the correctly timed frame syncs, and the pulse duration is set by adjusting the capacitors on IC1 and IC3. The frame and line syncs are combined in IC3d to produce a composite sync pulse output at pin 11.

Picture Generators

Perhaps the simplest picture generator, as previously mentioned, is the electronic pattern generator. Its main use is for setting up monitoring equipment, and of particular value is the grey scale which is used to check the monitor's tonal rendering and contrast.

A typical grey scale generator is shown in **Fig 21.4.** The oscillator IC1a drives a 7490 counter to produce square wave outputs at a, b and c. When the counter has counted to seven (logic 111), outputs a, b and c will all be at logic 1 and the output of IC1b (pin 6) will go to zero, thus disabling the oscillator. The line sync pulse will reset the counter to

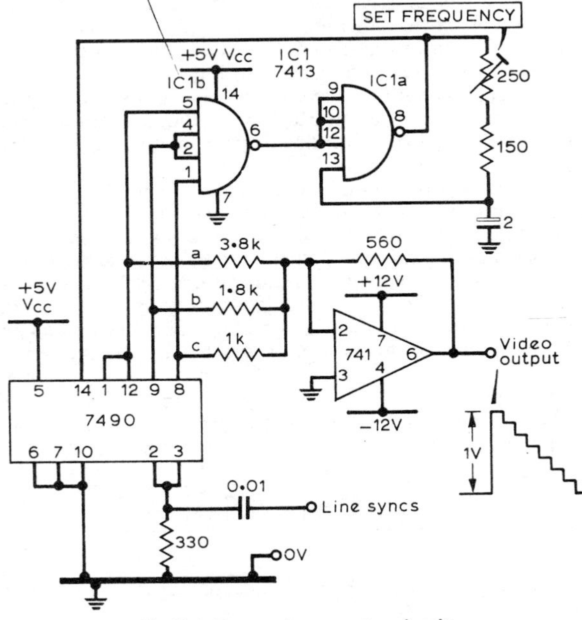

Fig 21.4. Grey scale generator circuit.

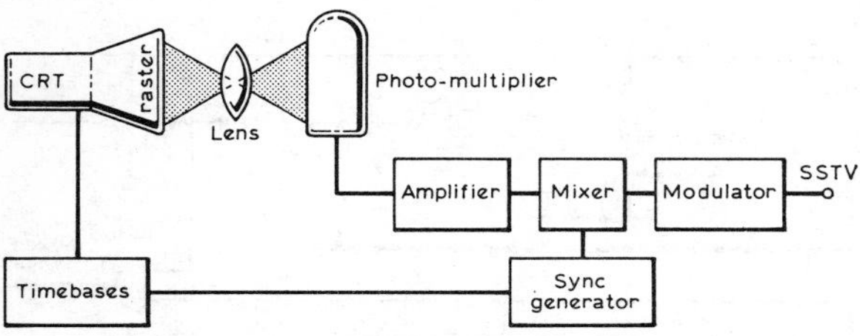

Fig 21.5. Block diagram of flying-spot scanner.

zero. IC1b (pin 6) will now go to logic 1 and the oscillator will re-start. Outputs a, b and c are fed through resistors to a current-summing junction employing a 741 operational amplifier. The 560Ω feedback resistor is adjusted to give an amplitude of 1V p-p.

The flying-spot scanner represents a most economical way of generating one's own pictures. In the block diagram of **Fig 21.5** a raster is scanned on to the front of a cathode-ray tube. The light from the raster is focused on to the photograph or drawing to be transmitted, and the reflected light from the picture is detected by the photomultiplier tube and converted into a low-level signal. The signal is amplified by the photomultiplier and the output fed to the video/sync mixer, the output of which shifts the voltage-controlled subcarrier oscillator from 1,500Hz to 2,300Hz depending on the level of light falling on the photocell. Sync signals from the pulse generator are combined with the video signal in the mixer stage which then shifts the subcarrier down to the sync frequency of 1,200Hz.

For the flying-spot source the station monitor may be used with the contrast level turned right down so that a constant brilliance raster is traced out. The output of the scanner is then fed to a tape recorder. The only problem with this method is that if only one monitor is used the results cannot be seen until the tape is replayed, and this makes adjustment rather tedious. A simplification is to draw the caption on to transparent sheet and place it against the tube face, because this does away with the need for a lens.

A block diagram of a slow-scan camera is shown in **Fig 21.6.** This system uses the sampling method of scan conversion, and any 405 or 625-line camera may be adapted for sstv.

The camera is laid on its right-hand side and the original frame timebase is used as the slow-scan line timebase. Dividing the fast-scan frame rate (50Hz) by three gives the required slow-scan rate, and the usual way to achieve this division is to replace the original frame sync pulse by a slow-scan line sync pulse, adding a capacitor in the camera to slow down the frame timebase. The original fast-scan line is still traced out at its usual speed but instead along the vertical axis. This line is "sampled" over a period of 7·2s, resulting in slow-scan video output which is fed to the sstv modulator.

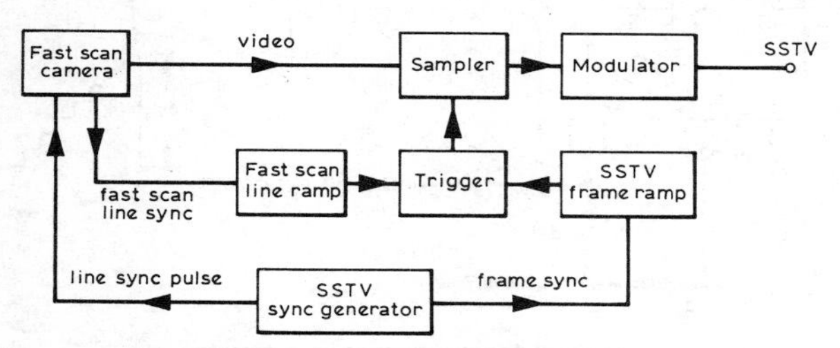

Fig 21.6. Block diagram of sstv sampling camera.

The second form of camera conversion uses a digital technique, and this method has the advantage that it requires only the standard composite video output from an ordinary tv camera operated the right way up and with no modifications whatsoever.

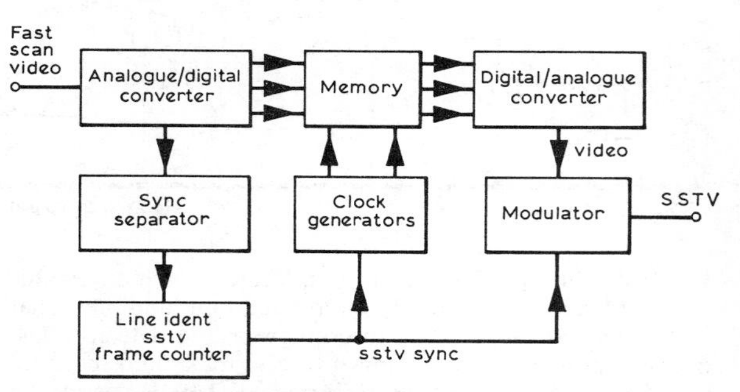

Fig 21.7. Block diagram of fast-scan to slow-scan converter.

A basic block diagram is shown in **Fig 21.7** but the whole system is somewhat complex and detailed information can be obtained from private [4] or published [5] designs.

The technique used is to convert the fast-scan tv signal into digital coded information (analogue-to-digital conversion) and load it a line at a time into a memory at fast-scan rate. The information from the memory is then recalled at a slow-scan rate, and the resulting digital information is fed to a digital-to-analogue converter and then to the sstv modulator in the normal way.

Advanced Techniques

One such technique is the sstv keyboard which was pioneered by W0LMD [6]. The keyboard is, as its name suggests, a method of writing information on to a slow-scan monitor by simply typing in the characters on a keyboard rather like that of a typewriter. The system normally writes five rows with up to six characters per row. This is useful for transmitting written information and for contest work, and the keyboard may also be used with a camera to superimpose titles over live shots.

Perhaps the most interesting development is a design for a slow-scan to fast-scan receive converter. This converter accepts a slow-scan picture, converts it to digital form, stores the complete frame in a memory bank, and then scans the memory at 625-line rate. The resultant picture is then viewed directly on a vision monitor or domestic tv set. The results on the screen are quite remarkable and far superior to those obtained with a conventional sstv monitor.

This system would seem to come at an opportune time since the supply of surplus long-persistence tubes now seems to be running out. The limiting factor at the moment, apart from the complexity, is the high cost of the ic memories, but hopefully these will become cheaper in due course.

There are a number of experiments being carried out on the production of colour pictures, and this is usually done photographically. The subject is recorded three times, each time with a different colour filter (red, blue or green) placed in front of the camera lens. The receiving station photographs each frame on to a single exposure of colour film

SSTV picture received on a domestic tv set after slow-scan to fast-scan conversion.

using the same sequence of filters in front of the camera lens. A Polaroid camera is best for this purpose as the results can be seen immediately.

A high-definition system is in use which utilizes a 256-line picture, taking 34s to scan one complete frame. Since the normal P7 phosphor crt does not have the required persistence (under average lighting conditions), perhaps the best method is to photograph the picture from the monitor with a Polaroid camera. The results are a much better picture which can be very useful for transmitting intricate circuit diagrams.

SSTV Operation

Licensing authorities in many countries throughout the world permit sstv transmissions, and present activity is concentrated around 3·740 and 14·230MHz, contacts being often established initially on ssb.

It should be borne in mind that sstv is a *picture* transmitting technique and while printed captions are necessary for "rubber stamp" information (name, QTH, rig, etc) and station identification, the real purpose is to exchange pictures with other amateurs to enhance normal communication. The printed word should therefore be kept to a minimum to maintain the interest of the receiving station, and in this way all language barriers are crossed and the potential of sstv is realized more fully.

SSTV MONITOR

This sstv monitor is representative of the more sophisticated designs, the performance of which compares favourably with commercially-built equipment. The slow-scan video bandwidth is 900Hz, ie the highest video frequency is 900Hz, consequently there are only 300Hz between this frequency and the sync frequency of 1,200Hz. To improve on this interference situation a linear integrated circuit working as a limiter is used in the input stage, **Fig 21.8(a)**. This ic has differential outputs, and the output pulses will therefore have opposite polarities. They are used to drive two monostable multivibrators, the digital ICS SN74121, which work as pulse-counting detectors and give an output signal that has a frequency range that is twice the range of the input signal. The frequency range of the input signal is 1,200Hz to 2,300 Hz, and thus the output signal will have a range of 2,400Hz to 4,600Hz. The difference between 900Hz and sync (2 × 1,200Hz = 2,400Hz) is now considerably greater and a much cleaner video signal is obtained.

The input ic limits, according to data, at an input signal of 100μV, while in this circuit an input signal of about 50mV is needed for full limiting. The output of the multivibrators goes to a low-pass filter with a flat characteristic from zero to about 900Hz, and the quality of the desired video signal is closely associated with the design of this filter. All signals up to about 900Hz should pass through the filter unaffected, while all higher frequencies should be greatly attenuated.

The output signal from the filter, which should be about 400mV p-p, is fed to video stages TR1, TR2 and TR3. The signal from TR3 intensity-modulates the cathode of the monitor tube. From the emitter of TR2, the signal will be connected to the active filter stages TR6 and TR7. Here all signals between approximately 2kHz and 3kHz will be greatly attenuated, ie noise and other interference will be considerably decreased. The attenuation will be about 60dB as shown in **Fig 21.9.**

From the emitter of TR7 the signal goes back to agc amplifiers TR5 and TR4 and further to the video input stage TR1. The level with respect to ground of the signal at TR7 emitter should be 1·9V. With the preset control RV2 the voltage on TR5 base is set to 6·2V, and RV1 is adjusted so that the signal on the emitter of TR4 will be at 1·7V for peak sync and 2·4V for black. Maximum white will then be at 3·5 to 3·8V depending on the quality of the incoming signal. By proper adjustment of the agc amplifier the frequency setting of the ssb receiver is no longer so critical.

The signal is fed from TR8 to TR9 where it is integrated by the 10kΩ resistor and the 27nF capacitor. The pulse measured at the collector of TR9 is set by RV3 to around 5V p-p. On the base of TR11 there will now be a positive-going square pulse with a width of 3ms. This pulse is integrated in the next stage and will appear as a sawtooth on the collector of TR12 from where it is fed to the horizontal sweep stages.

Other features of this design involve the increasing of the signal sensitivity and the decreasing of the sensitivity to interference. Earlier, the method used was that the incoming sync pulse triggered the sweep oscillators, but as the system could not differentiate between a sync pulse and an interference pulse the oscillator was triggered by any pulse that appeared. At strong interference levels this could result in beam absence during long periods—it was consequently impossible to get the monitor to synchronize. At weak signal levels the sync was not large enough and the sweeps did not operate for that reason.

In this design, local horizontal and vertical oscillators through which there will be a raster on the screen continuously, irrespective of whether a signal is being received or not, are used. Together with the action of the active filter stages TR6 and TR7, the result is such that it is possible to receive signals under rather strong interference conditions, and weak signals under fading conditions.

The sawtooth from TR12 will consequently control the frequency of the sawtooth oscillator TR13, the basic frequency of which is adjusted to a somewhat lower frequency by potentiometer RV4 on the front panel. This setting is not critical but is necessary as the sweep frequencies from various stations can differ.

From the emitter of TR15 the sawtooth is picked off through potentiometer RV5 which controls the sweep amplitude. To the same emitter is also connected transistor TR16 which centres the sweep on the screen, this setting being made by potentiometer RV6. The sawtooth is amplified in TR17 and TR18 and will drive the complementary stage TR19-TR20 which brings about a sawtooth sweep in the

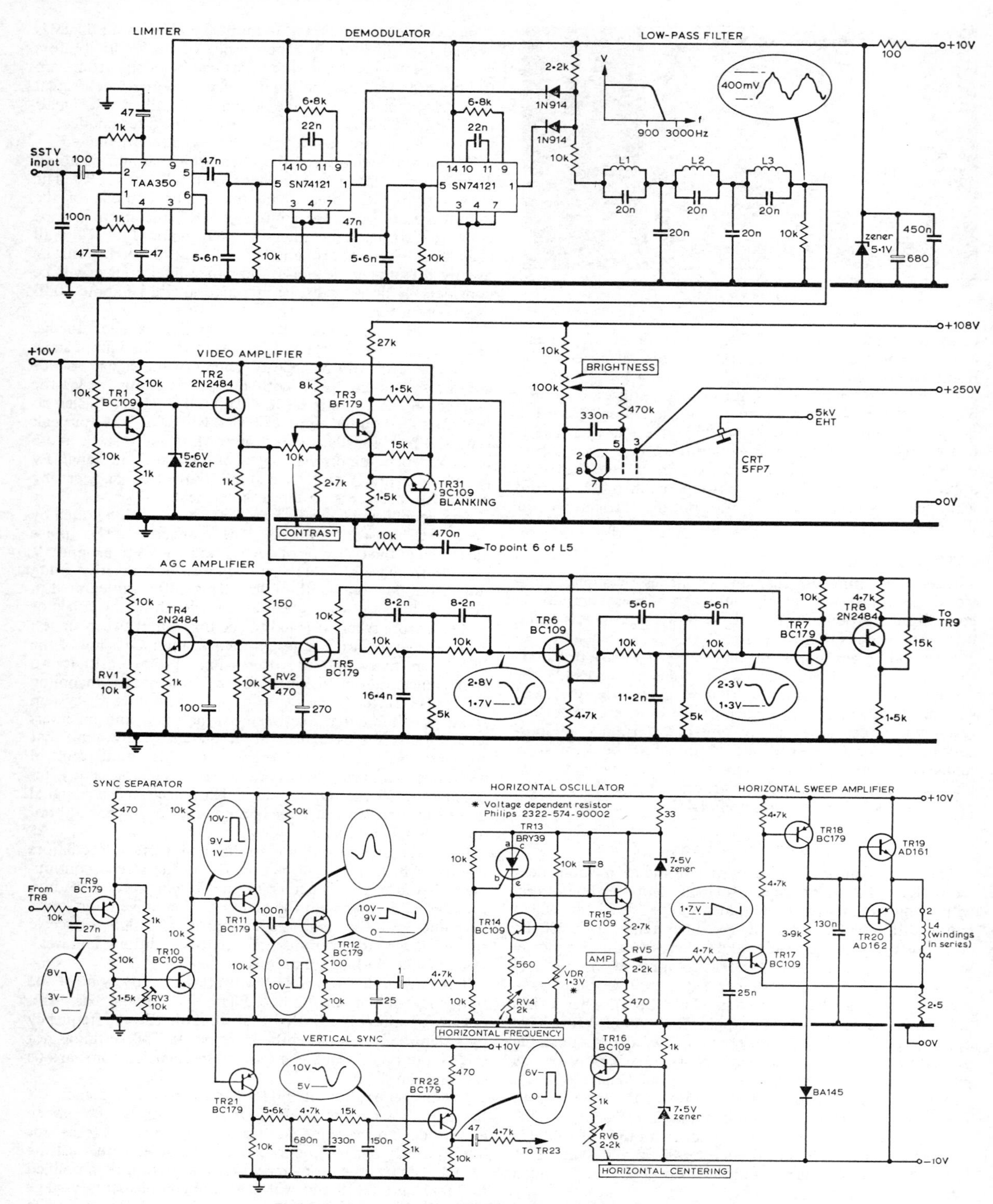

Fig 21.8(a). First part of two-part circuit diagram of sstv monitor.

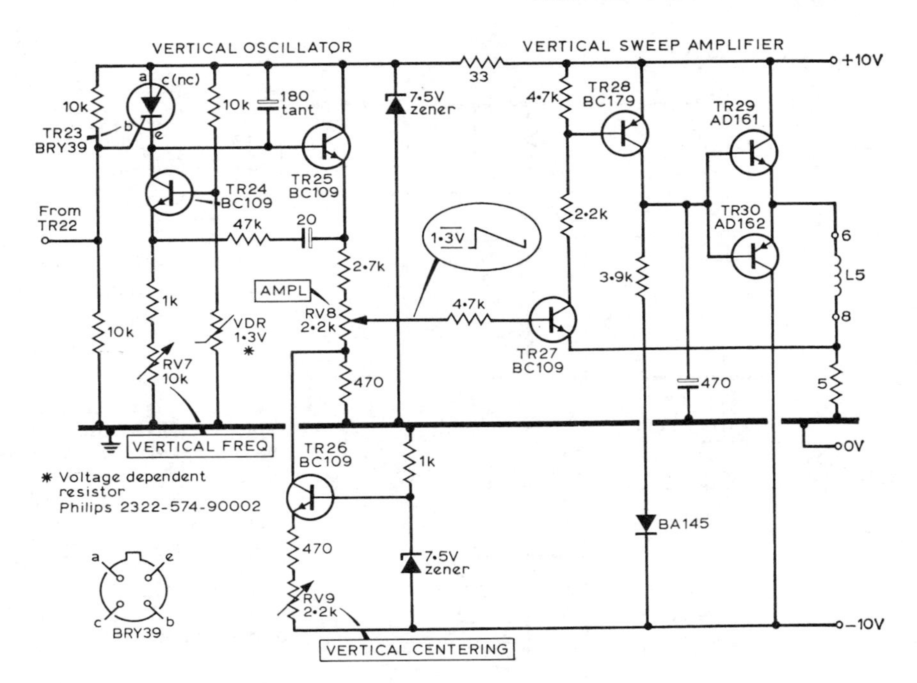

Fig 21.8(b). Second part of two-part circuit diagram of sstv monitor.

deflection coils. To improve the linearity a small signal is fed back to TR17.

From the base of TR11 the square pulse is fed to the vertical sync stage TR21 and is integrated in that collector circuit. On the collector of TR22 there will be a square pulse that controls the local oscillator TR23. The function of the vertical sweep stages is similar to that of the horizontal stages.

Between the emitters of the final amplifier stage and ground there should be a sweep of 15V p-p for the horizontal stage and 13V p-p for the vertical stage. Between the emitters of TR24 and TR25 there is a 47kΩ resistor and a 20μF capacitor in order to improve the linearity of the vertical sweep.

The capacitors on TR17 base and TR18 collector serve to decouple the vertical sweep which could be superimposed on the horizontal sweep. The capacitor on TR28 collector decouples the horizontal sweep from the vertical sweep. Transistor TR31 is a stage for blanking the beam during vertical flyback.

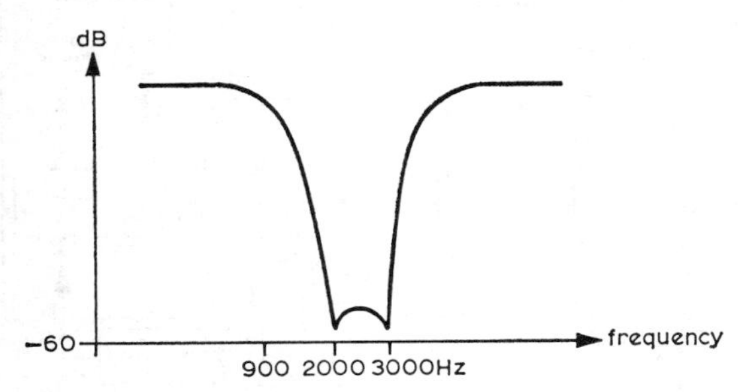

Fig 21.9. Frequency characteristics of active filter stages TR6-TR7.

Power Supplies

The low-voltage power supplies are shown in **Fig 21.10**. It is important that the + 10V and − 10V supplies be stabilized in order to maintain good linearity. The transformer used should have a current rating of 5A. The horizontal deflection coil requires a current of about 500mA p-p and the vertical coil about 300mA p-p.

The 250V supply for the eht generator uses conventional circuitry and is not shown. The + 108V should be dropped from the 250V supply using a suitable resistor and should be stabilized if possible.

A line-output transformer suitable for the eht supply may be obtained from a domestic tv set. The large overwind coil is used and is wired up according to the circuit in **Fig 21.11**. The eht diode should be mounted between two 1in ceramic stand-off insulators and covered to eliminate the possibility of being touched accidentally. This eht is lethal and adequate safety precautions should be taken.

Component Considerations

TAA350 may be replaced with a TAA350A, and in this case the pin numbers are all rotated clockwise by three pins.

The low-pass filter in Fig 21.8(a) uses Philips cores H20; an equivalent is the Mullard LA1302. A slightly lossy substitute can be wound using LA1 cores.

The 1·3V voltage-dependent resistors in the bases of TR14 and TR24 may be replaced by two silicon diodes in series (BA100) or possibly with a 1·3V zener diode.

A suitable scan coil assembly and permanent magnets for focussing may be obtained from an old tv set having a tube neck of similar size to the 5FP7. The scan coils should have their windings series or parallel connected as necessary to give the nearest resistance readings to those quoted in **Table 21.2**.

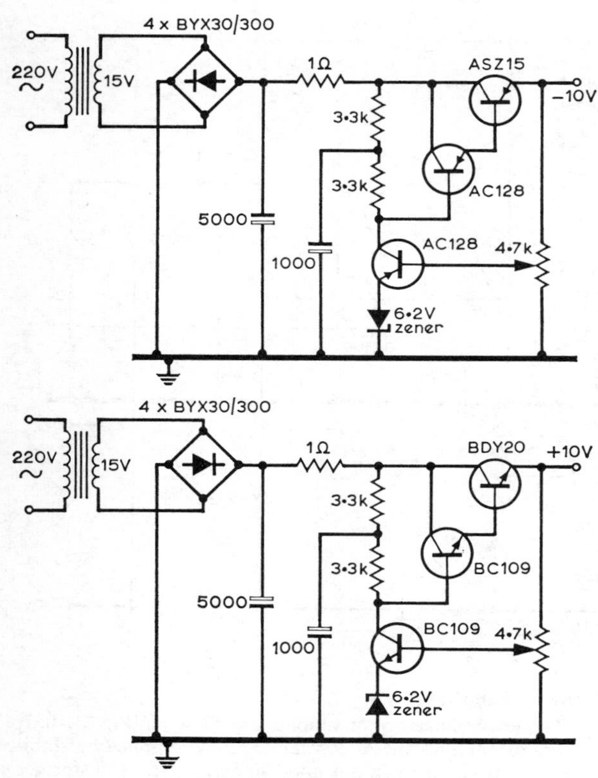

Fig 21.10. Low-voltage supplies.

Mechanical Design

The electronic and high-voltage circuits can be housed in a cabinet measuring 160mm wide by 240mm high by 300mm deep. However, it may be preferable to build the power supplies in a separate cabinet as trouble may be experienced with the ac fields surrounding the transformers.

The monitor tube is clamped against the front panel by fabricating a round clamp from a strip of 18swg aluminium. A pinch-bolt should be used for tightening on to the tube, and four lugs bent out at 90° to fix the clamp to the front panel.

The transistor stages may be mounted on two Vector-boards measuring 130mm by 220mm, or by using printed circuit boards. The input stages TR1 to TR12, TR21 and TR22 are mounted on one board with the sweep stages on the other. The boards are placed vertically on both sides of the monitor tube and fixed to small distance-spacers.

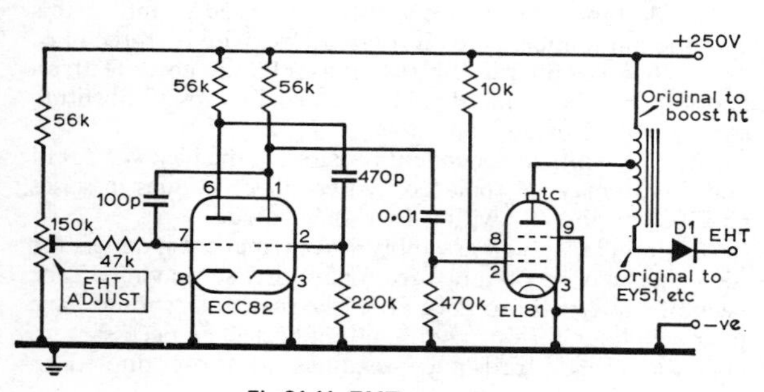

Fig 21.11. EHT generator.

TABLE 21.2

Component details

L1, L3	1H. Philips H core type H-20 or Mullard LA1302, 400 turns of 34 swg enamelled copper wire.
L2	2H. Philips H core type H-20 or Mullard LA1302, 550 turns 34 swg.
L4	Horizontal deflection coil, approximately 4·6Ω.
L5	Vertical deflection coil, approximately 38Ω.
D1	BY140, BY176 or BY182.

FACSIMILE

Facsimile (fax) is a process by which graphic information is converted into electrical signals which are transmitted by cable or radio to a receiver and then reproduced exactly as the original. The receiving systems usually available to the radio amateur operate on an electromechanical basis. A block diagram of a typical receiver system is shown in **Fig 21.12.** A.M. or fm signals are detected and the output in both cases is a varying dc signal, the value of which corresponds to the variation in the density of the material being processed. Transferring the varying dc potential into a printed record is usually achieved by the use of electrolytic or photo-sensitive paper.

Electrolytic paper presents a change of colour following the passage of a current through a metal stylus. The variations in the current caused by the received signal voltage appear as variations in the density, ie black, grey or white, of the paper. There are various types of electro-sensitive paper available and one often encountered is known as "Teledeltos".

When photo-sensitive paper is used a lamp replaces the stylus. The signal variations received by cable or radio are amplified and applied to the lamp, and the resulting beam is focused sharply on to the sensitive paper. The intensity of the beam determines the density of the final printed copy.

Standards

These can vary between the service involved and the type of machine. Generally, however, the characteristics of amateur service facsimile transmissions conform to CCIR recommendations.

There are two modes of emission in current use. The A4

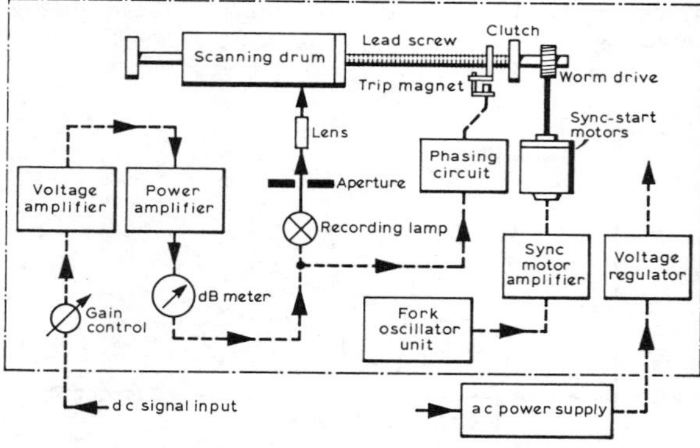

Fig 21.12. Block diagram of facsimile receiving unit.

mode denotes amplitude modulation of the carrier. The tone frequency is between 1,400 and 1,900Hz, and the upper limit is white and the lower limit black. The bandwidth using ssb is nominally 2·4kHz.

F4 denotes frequency modulation of the carrier. The centre frequency is 1,900Hz, white is +400Hz and black —400Hz.

For the amateur service it is recommended that the mode of transmissions shall be chosen so that the bandwidth on frequencies below 146MHz shall not exceed 4kHz.

Recommended mechanical standards are:

(a) *Drum speed.* This is generally a multiple of 60 revolutions per minute with a preferred speed of 120rpm.
(b) The *drum diameter* is 152mm.
(c) *Drum length.* This should be at least 550mm.
(d) *Index of co-operation.* This is the measure of the degree of the compatibility of different machines. It is the product of the total length of one scan line and the number of scan lines per unit length. The recommended figures are 576 for minimum black or white picture elements of 0·4mm and 288 for minimum picture elements of 0·7mm.
(e) *Direction of scanning* at the transmitter. The picture area is scanned from left to right and from top to bottom.
(f) *Synchronization.* The scanning speed should be maintained within 5 parts in 10^6 of the normal value.

Equipment

The cost of new facsimile equipment is very high and amateur operation invariably makes use of converted commercial or home-built equipment. In Europe equipment manufactured by Creed, Muirhead and Siemens may be found. Generally the conversion of commercial equipment for amateur operation is not difficult. There has been some interest shown in the conversion of equipment to receive weather photographs from the automatic picture transmission satellites and further information on this aspect of facsimile operation may be found in [7].

The setting up and testing of equipment can be facilitated by signals from three stations located at Bracknell which operate 24h a day. These are: GFA 21 on 3,289·5kHz, GFA 23 on 8,040kHz and GFA 24 on 11,086·5kHz. These stations use F4 emission and a drum speed of 120rpm.

Typical facsimile equipment as used at I1LCF. The scanning drum and lead screw are clearly visible at the bottom of the picture.

REFERENCES

[1] "Slow-scan television", by A. Backmann, SM0BUO, *Radio Communication* February 1971.
[2] "The W6MXV hi-performance magnetic deflection sstv monitor" in *Slow-Scan Television Handbook*, a *73 Magazine* publication.
[3] *Slow-Scan Television*, British Amateur Television Club.
[4] MXV–200 Scanverter, M. Tallent, W6MXV, 6941 Lenwood Way, San Jose, California 95133.
[5] "W0LMD scan converter", *73 Magazine* August 1974.
[6] "An sstv keyboard", by W0LMD, *CQ Magazine* September 1974.
[7] *A guide to the construction of inexpensive automatic picture transmission ground stations*, NASA, SP-50800, Greenbelt, Maryland, USA.

CHAPTER 22

THE RSGB AND THE RADIO AMATEUR

THE Radio Society of Great Britain (RSGB) is the national society of radio amateurs in the United Kingdom. The majority of its 37,000 members hold amateur transmitting licences; the others either hope to do so later or are interested primarily in the receiving side of amateur radio.

HRH The Prince Philip, Duke of Edinburgh, KG, is the Society's patron.

The Society was founded as the London Wireless Club more than 70 years ago in 1913 but soon attracted members throughout the country; the name of Radio Society of Great Britain was formally adopted in 1922. For many years its activities have been devoted almost entirely to the many aspects of amateur radio—that is, the transmission and reception of radio signals as a hobby pursued for the pleasure to be derived from an interest in radio techniques and construction, and for the ensuing friendships with those people throughout the world who possess similar interests.

The Society is proud of its role in the development of amateur radio and of the many eminent scientsists who have been connected with it, including Marconi, Lodge and Fleming. Members of the Society were prominent among those radio amateurs who were the first to demonstrate that wavelengths below 100m could provide world-wide communication on low power. But the Society looks to the future, rather than the past, and concentrates its efforts on the effective organization of amateur activity in order to provide the greatest opportunities for useful experimental work, and to encourage general interest in and enjoyment of the scientific hobby of amateur radio.

WHY YOU SHOULD JOIN

The following are just a few of the many reasons why, if you are really interested in amateur radio, you should join the RSGB immediately.

You will receive every month a copy of the Society's journal, *Radio Communication*. First published in 1925, this magazine is noted internationally for the high standard of its technical and constructional articles, written by Britain's leading radio amateurs, and for the wide scope of its news coverage. The needs of newcomers are not overlooked in the selection of technical articles.

You will be eligible to take part in many interesting contests and field days—some of which are open to listeners as well as licensed members. Trophies or certificates are awarded to leading entrants. You may also be eligible for a number of certificates representing graded degrees of achievement in amateur radio operating (receiving as well as transmitting) which are issued by the Society. These include the DX Listeners' Century Award, the Worked The British Commonwealth Award, Four Metres and Down Certificates and the Commonwealth DX Certificate. The rules governing all awards are available from RSGB Headquarters.

You will be able to use the world's largest and most comprehensive QSL Bureau, operated by the Society. Use of this efficient bureau will save the active amateur and listener a great deal of trouble. QSL cards are sent to and received from the bureau in batches, thus eliminating the need for stamping, addressing and posting individual cards. The bureau distributes the cards via other national societies' bureaux to amateurs throughout the world. Full details of how the QSL Bureau operates are sent to every member on election.

You will receive a certificate of membership and a lapel badge which identifies you as part of the amateur radio movement. Members who do not hold amateur transmitting licences are given special identification numbers for use in connection with the QSL Bureau, beginning RS followed by a number.

You will be encouraged to contribute, according to your interests, to the advancement of amateur radio. Many members serve as local representatives, or on local or national committees, or pass on to other members, through *Radio Communication* or by lectures, the results of their experiments and observations. The members in fact are the Society.

WHAT THE SOCIETY DOES

Some of the important activities of the Society have already been described, but there are many other ways in which the Society helps radio amateurs and all those interested in amateur radio. A few of these are outlined below.

Organization

The Society is administered by a Council elected by the Corporate membership, and a full-time staffed headquarters is maintained in London. There are also 20 elected Regional Representatives who, with the help of Area Representatives, arrange local meetings and other activities throughout the British Isles.

The Society maintains special committees dealing with such subjects as technical matters and publications, liaison with the licensing administration and other external organizations, contests, interference problems, propagation studies, vhf operation and exhibitions.

Meetings

In addition to the local meetings mentioned above, the Society organizes lectures, conferences and exhibitions on a national basis.

The Society also helps organize many special scientific studies and tests, some of which also involve regular meetings. Details of these and other forthcoming events are given in *Radio Communication.*

Publications

The main publishing activity of the Society is the monthly journal, *Radio Communication.* However, the Society also produces many books and other publications to help the radio amateur. Because the Society is anxious to disseminate sound technical information as widely as possible, many are issued at prices well below what a commercial publishing organization would have to charge. A notable example is this handbook.

Some publications of the Society are specially written for newcomers to help them obtain transmitting licences; these include *A Guide to Amateur Radio, The Radio Amateur's Examination Manual, How to Pass the RAE*, and *Morse Code for Radio Amateurs*.

The Society also provides facilities for obtaining a selection of the many amateur radio publications issued in the USA where there are over 250,000 radio amateurs. The most popular USA publications are generally available from Headquarters.

Amateur Radio Licences

The Society is recognized as the representative of the amateur service in all negotiations with the DTI on matters affecting the issue of amateur transmitting licences, including the frequencies assigned to the amateur service. It sends official representatives to the important World Administrative Radio Conferences of the International Telecommunication Union and other conferences where decisions vital to the future of amateur radio are taken. The RSGB is a founder member of the International Amateur Radio Union, the worldwide organization of national amateur radio societies.

News Bulletin Service

Every Sunday morning special news bulletins for radio amateurs are transmitted under the callsign GB2RS from stations throughout the British Isles, in the 3·6, 7, and 144MHz bands.

Slow Morse Transmissions

The Society sponsors the transmission by amateurs throughout the country of morse practice lessons intended for beginners. Details appear periodically in *Radio Communication.*

Beacon and Repeater Stations

The Society is the licensee and co-ordinator of stations operating in the 28, 70, 144 and 432MHz bands, and also the 1·3 and 10GHz bands. The beacon station transmissions are used extensively in propagation studies, while the repeater stations provide reliable communication for mobile and low-power users.

Radio Amateur Emergency Network

The Society has set up the Radio Amateur Emergency Network (Raynet), in collaboration with the British Red Cross Society, the St John Ambulance Brigade, the Police and **County Emergency Planning Officers to assist in providing communications during disaster relief operations.**

HOW TO JOIN THE RSGB

Joining the RSGB is simple, but there are of course certain formalities to be observed. As explained earlier, anyone with an active and genuine interest in amateur radio is warmly welcome to apply for membership.

If you are over 18 years of age, or hold an amateur transmitting licence, you are eligible to become a Corporate Member of the Society. You do not have to be engaged professionally in radio but equally this would not debar you from joining. Many members do in fact work in the electronic field, but for very many others radio is purely a spare-time hobby. Those under 18 who do not hold an amateur transmitting licence may become Associates, who have many of the privileges of full membership but do not vote in the annual Council election or on matters affecting the management of the Society. Associates must apply for transfer to Corporate membership on reaching 18 years of age or immediately they obtain a transmitting licence if under this age.

All applicants, for both Corporate and Associate membership, should be proposed by a Corporate Member of the Society to whom they are personally known. The member simply completes the proposal on the application form, and you will find that he or she will be glad to do this. Many newcomers to amateur radio (who are most welcome as members) may not know or be in touch with other members. In such cases a brief reference in writing should be submitted from a suitable person who can vouch for your interest in **amateur radio. All applications are placed before the Council at its regular meetings. All correspondence should be sent to the Radio Society of Great Britain, Lambda House, Cranborne Rd, Potters Bar, Herts EN6 3JW.**

* * *

The Society supports and encourages all activities "For the Advancement of Amateur Radio". It welcomes within its ranks all those who share this view.

CHAPTER 23

GENERAL DATA

Bias Resistor

The value of the resistance to be connected in the cathode lead for developing the required bias is—

$$R_c = \frac{E_c}{I_c} \times 1000 \text{ ohms}$$

where E_c = bias voltage required (volts)
I_c = total cathode current (mA).

Capacitance

The capacitance of a parallel-plate capacitor is—

$$C = \frac{0{\cdot}224\ KA}{d} \text{ picofarads}$$

where K = dielectric constant (air = 1·0)
A = area of dielectric (sq in)
d = thickness of dielectric (in)

If A is expressed in sq cm and d in cm,

$$C = \frac{0{\cdot}0885\ KA}{d} \text{ picofarads}$$

For multi-plate capacitors, multiply by the number of dielectric thicknesses.

Capacitance of a coaxial cylinder—

$$C = \frac{0{\cdot}242}{\log_{10}\ (D/d)} \text{ picofarads per cm length}$$

where D = inside dia of outer
d = outside dia of inner.

Capacitors in Series or Parallel

The effective capacitance of a number of capacitors in *series* is—

$$C = \frac{1}{\frac{1}{C_1} + \frac{1}{C_2} + \frac{1}{C_3} + \text{etc}}$$

The effective capacitance of a number of capacitors in *parallel* is—

$$C = C_1 + C_2 + C_3 + \text{etc}$$

Characteristic Impedance

The characteristic impedance Z_0 of a feeder or transmission line depends on its cross-sectional dimensions.

(i) Open-wire line:

$$Z_0 = 276 \log_{10} \frac{2\,D}{d} \text{ ohms}$$

where D = centre-to-centre spacing of wires (in)
d = wire diameter (in)

(ii) Coaxial line:

$$Z_0 = \frac{138}{\sqrt{K}} \log_{10} \frac{d_0}{d_i} \text{ ohms}$$

where K = dielectric constant of insulation between the conductors (eg 2·3 for polythene, 1·0 for air)
d_0 = inside diameter of outer conductor (in)
d_i = diameter of inner conductor (in)

Decibel

The decibel is the unit commonly used for expressing the relationship between two power levels (or between two voltages or two currents). A *decibel* (dB) is one-tenth of a *bel* (B). The number of decibels N representing the ratio of two power levels P_1 and P_2 is 10 times the common logarithm of the power ratio, thus—

$$\text{the } ratio\ N = 10 \log_{10} \frac{P_2}{P_1} \text{ decibels}$$

If it is required to express *voltage* (or *current*) ratios in this way, they must relate to similar impedance values; ie the two different voltages must appear across equal impedances (or the two different currents must flow through equal impedances). Under such conditions the *power* ratio is proportional to the square of the *voltage* (or the *current*) ratio, and hence—

$$N = 20 \log_{10} \frac{V_2}{V_1} \text{ decibels} \qquad N = 20 \log_{10} \frac{I_2}{I_1} \text{ decibels}$$

Dynamic Resistance

In a parallel-tuned circuit at resonance the dynamic resistance is—

$$R_D = \frac{L}{Cr} = Q\omega L = \frac{Q}{\omega C} \text{ ohms}$$

where L = inductance (henrys)
C = capacitance (farads)
r = effective series resistance (ohms)
Q = Q-value of coil
ω = 2π × frequency (hertz)

Frequency—Wavelength—Velocity

The velocity of propagation of a wave is—

$$v = f\lambda \text{ centimetres per second}$$

where f = frequency (hertz)
λ = wavelength (centimetres)

For electromagnetic waves in free space the velocity of

11—RCH2 • •

propagation v is approximately 3×10^{10} cm/s, and if f is expressed in kilohertz and λ in metres—

$$f = \frac{300{,}000}{\lambda} \text{ kilohertz}$$

$$\lambda = \frac{300{,}000}{f} \text{ metres}$$

$$\text{Free space } \frac{\lambda}{2} = \frac{492}{\text{MHz}} \text{ feet}$$

$$\text{Free space } \frac{\lambda}{4} = \frac{246}{\text{MHz}} \text{ feet}$$

Impedance

The impedance of a circuit comprising inductance, capacitance and resistance in series is—

$$Z = \sqrt{R^2 + \left(\omega L - \frac{1}{\omega C}\right)^2}$$

where R = resistance (ohms)
$\omega = 2\pi \times$ frequency (hertz)
L = inductance (henrys)
C = capacitance (farads)

Inductance of a Single-Layer Coil

For coils of ordinary proportions, the inductance is given approximately by—

$$L = \frac{a^2 n^2}{5(3a + 9b)} \text{ microhenrys} \qquad = \frac{r^2 n^2}{9r + 10l}$$

where a = diameter of coil (in)
n = number of turns
b = length of coil (in)
r = radius of coil
l = length of coil

By rearranging the formula—

$$n = \frac{1}{a}\sqrt{5L(3a + 9b)}$$

Slug Tuning. The variation in inductance obtainable with adjustable slugs depends on the winding length and the size and composition of the core and no universal correction factor can be given. For coils wound on Aladdin type F804 formers and having a winding length of 0·3–0·8in a dust-iron core will *increase* the inductance to about twice the air-core value: a brass core will *reduce* the inductance to a minimum of about 0·8 times the air-core value.

Inductances in Series or Parallel

The total effective value of a number of inductances connected in *series* (assuming that there is no mutual coupling) is given by—

$$L = L_1 + L_2 + L_3 + \text{etc}$$

If they are connected in *parallel*, the total effective value is—

$$L = \frac{1}{\frac{1}{L_1} + \frac{1}{L_2} + \frac{1}{L_3} + \text{etc}}$$

When there is mutual coupling M, the total effective value of two inductances connected in series is—

$$L = L_1 + L_2 + 2M \text{ (windings aiding)}$$

$$\text{or } L = L_1 + L_2 - 2M \text{ (windings opposing)}$$

Neon Stabilizer Dropper Resistance

The resistance to be connected in series with a neon stabilizer tube is—

$$R = \frac{E_s - E_r}{I} \times 1000 \text{ ohms}$$

where E_s = unregulated ht supply voltage (volts)
E_r = regulated ht supply voltage (volts)
I = maximum permissible current in regulator tube (milliamperes)

Ohm's Law

For a unidirectional current of constant magnitude flowing in a metallic conductor—

$$I = \frac{E}{R} \qquad E = IR \qquad R = \frac{E}{I}$$

where I = current (amperes)
E = voltage (volts)
R = resistance (ohms).

Power

In a dc circuit the power developed is given by—

$$W = EI = \frac{E^2}{R} = I^2 R \text{ watts}$$

where E = voltage (volts)
I = current (amperes)
R = resistance (ohms)

Q

The Q-value of an inductance is given by—

$$Q = \frac{\omega L}{R}$$

where $\omega = 2\pi \times$ frequency (hertz)
L = inductance (henrys)
R = effective resistance (ohms)

Reactance

The reactance of an inductance and a capacitance respectively is given by—

$$X_L = \omega L \text{ ohms} \qquad X_C = \frac{1}{\omega C} \text{ ohms}$$

where $\omega = 2\pi \times$ frequency (hertz)
L = inductance (henrys)
C = capacitance (farads)

The total reactance of an inductance and a capacitance in series is $X_L - X_C$.

Resistances in Series or Parallel

The effective value of several resistances connected in series is—

$$R = R_1 + R_2 + R_3 + \text{etc}$$

When several resistances are connected in parallel the effective total resistance is—

$$R = \frac{1}{\frac{1}{R_1} + \frac{1}{R_2} + \frac{1}{R_3} + \text{etc}}$$

Resonance

The resonant frequency of a tuned circuit is given by—

$$f = \frac{1}{2\pi\sqrt{LC}} \text{ hertz}$$

where L = inductance (henrys)
C = capacitance (farads)
If L is in microhenrys (μH) and C is in picofarads (pF) this formula becomes—

$$f = \frac{10^2}{2\pi\sqrt{LC}} \text{ kilohertz}$$

The basic formula can be rearranged thus:

$$L = \frac{1}{4\pi^2 f^2 C} \text{ henrys} \qquad C = \frac{1}{4\pi^2 f^2 L} \text{ farads}$$

Since $2\pi f$ is commonly represented by ω, these expressions can be written as—

$$L = \frac{1}{\omega^2 C} \text{ henrys} \qquad C = \frac{1}{\omega^2 L} \text{ farads}$$

Time Constant

For a combination of inductance and resistance in series the time constant (ie the time required for the current to reach $1/\epsilon$ or 63 per cent of its final value) is given by—

$$t = \frac{L}{R} \text{ seconds}$$

where L = inductance (henrys)
R = resistance (ohms)

For a combination of capacitance and resistance in series the time constant (ie the time required for the voltage across the capacitance to reach $1/\epsilon$ or 63 per cent of its final value) is given by—

$$t = C\,R \text{ seconds}$$

where C = capacitance (farads)
R = resistance (ohms)

Transformer Ratios

The ratio of a transformer refers to the ratio of the number of turns in one winding to the number of turns in the other winding. To avoid confusion it is always desirable to state in which sense the ratio is being expressed: eg the "primary-to-secondary" ratio n_p/n_s. The turns ratio is related to the impedance ratio thus—

$$\frac{n_p}{n_s} = \sqrt{\frac{Z_p}{Z_s}}$$

where n_p = number of primary turns
n_s = number of secondary turns
Z_p = impedance of primary circuit (ohms)
Z_s = impedance of secondary circuit (ohms)

BASIC SI UNITS

Quantity	Name of unit	Unit symbol
Electric current	ampere	A
Length	metre	m
Luminous intensity	candela	cd
Mass	kilogramme	kg
Thermodynamic temperature	degree Kelvin	°K
Time	second	s

DERIVED SI UNITS

Physical quantity	SI unit	Unit symbol
Electric capacitance	farad	F = A s/V
Electric charge	coulomb	C = A s
Electrical potential	volt	V = W/A
Electric resistance	ohm	Ω = V/A
Force	newton	N = kg m/s²
Frequency	hertz*	Hz = s^{-1}
Illumination	lux	lx = lm/m²
Inductance	henry	H = V s/A
Luminous flux	lumen	lm = cd sr
Magnetic flux	weber	Wb = V s
Magnetic flux density	tesla†	T = Wb/m²
Power	watt	W = J/s
Work, energy, quantity of heat	joule	J = N m

* Hertz is equivalent to cycle per second.
† Tesla is equivalent to weber per square metre.

CONVERSION FACTORS

To convert	into	Multiply by	Conversely
Amp hours	Coulombs	3600	$2{\cdot}778 \times 10^{-4}$
Atmospheres	Lb/sq in	14·70	0·068
Centigrade	Kelvin	°C+273=°K	°K − 273=°C
Cubic inches	Cubic feet	$5{\cdot}787 \times 10^{-4}$	1728
Cubic inches	Cubic metres	$1{\cdot}639 \times 10^{-5}$	$6{\cdot}102 \times 10^{4}$
Degrees (angular)	Radians	$1{\cdot}745 \times 10^{-2}$	57·3
Dynes	Pounds	$2{\cdot}248 \times 10^{-6}$	$4{\cdot}448 \times 10^{5}$
Ergs	Foot pounds	$7{\cdot}376 \times 10^{-8}$	$1{\cdot}356 \times 10^{7}$
Feet	Centimetres	30·48	$3{\cdot}281 \times 10^{-2}$
Foot pounds	Kilowatt hours	$3{\cdot}766 \times 10^{-7}$	$2{\cdot}655 \times 10^{6}$
Gausses	Lines per sq in	6·452	0·155
Grams	Dynes	980·7	$1{\cdot}02 \times 10^{-3}$
Grams per cm	Pounds per in	$5{\cdot}6 \times 10^{-3}$	178·6
Horse power	Kilowatts	0·746	1·341
Inches	Centimetres	2·54	0·3937
Kilograms	Pounds (lb)	2·205	0·454
Kilometres	Feet	3281	$3{\cdot}048 \times 10^{-4}$
Kilometres	Nautical miles	0·540	1·853
Kilometres	Statute miles	0·621	1·609
Kilowatt hours	Joules	$3{\cdot}6 \times 10^{6}$	$2{\cdot}778 \times 10^{-7}$
Kilowatt hours	HP hours	1·341	0·7457
Knots	Miles per hour	1·1508	0·869
Lamberts	Candles per sq cm	0·3183	3·142
Lamberts	Candles per sq in	2·054	0·4869
Lumens per sq ft	Foot candles	1	1
Lux	Foot candles	0·0929	10·764
Metres	Feet	3·28	0·3048
Metres	Yards	1·094	0·9144
Miles per hour	Feet per second	1·467	0·68182
Nepers	Decibels	8·686	0·1151
Tons	Pounds	2240	$4{\cdot}464 \times 10^{-4}$
Watts	Ergs per second	10^{7}	10^{-7}

WINDING COILS ON STANDARD FORMERS

Coil formers of the Aladdin type are widely used in modern radio equipment. There is, however, always the problem of calculating the number of turns required. For this reason two charts have been prepared by J. Greenwell (G3AEZ) to enable the amateur to calculate quickly and easily the necessary winding data.

Use of Chart 23.2

The use of **Chart 23.2** is best illustrated by describing a typical calculation.

Example: It is required to wind a coil on an Aladdin type F804 former which will resonate at 7MHz with a 50pF capacitor.

The method is as follows:

1. Draw a straight line through 50pF (axis *A*) and 7MHz (axis *B*).
2. Project the line to cut axis *C* and read off the required inductance, which in this case is 10·3μH.
3. Draw a horizontal line through 10·3μH on axis *D* and a vertical line through a reasonable winding length (say 0·5in) and determine the most suitable wire gauge to use, ie 32 swg.
4. From the 32swg curve determine the exact winding length to give an inductance of 10·3μH, ie 0·48in.

The coil required will therefore be close wound with 32 swg enamelled copper wire and 0·48in long.

If desired, the number of turns may be calculated using wire tables (such as **Table 23.1**) from which it will be found that the turns per inch for 32swg enamelled copper wire is 83. Hence, a winding 0·48in long will consist of (83 × 0·48) = 40 turns.

TABLE 23.1

The following table, prepared from information provided by the London Electric Wire Company, shows the minimum turns per inch for enamelled copper wire of the gauges most commonly used by amateurs.

SWG	Turns/in	SWG	Turns/in
20	26	32	82·6
22	33	34	96·2
24	41·5	36	116·3
26	50·3	38	144·9
28	61	40	178·6
30	72·5	42	212

For coils of low inductance, ie less than 1μH, it is advisable to space wind rather than close wind with a heavy gauge wire. Curves are, therefore, given in Chart 23.2 for pitches of 10, 15 and 20 turns per inch using 26swg enamelled copper wire. Other gauges may be used, however, without introducing significant errors.

The values shown in Chart 23.2 have been calculated for Aladdin F804 formers without cores. The variation in inductance obtainable with dust-iron or brass cores depends on the winding length and composition of the core material and therefore no simple correction factor may be quoted. However, experiments show that for coils between 0·3 and 0·8in long a dust-iron core will give a maximum possible inductance of about twice the "core-less" inductance and a brass core a minimum possible inductance of about 0·8 times the " core-less " inductance. These factors should be borne in mind when designing variable inductances from the charts.

Chart for 0·3in Diameter Formers

The inductance required is found from Chart 23.2 in the same way as for Aladdin F804 formers and the winding details determined from Chart 23.1.

Measurements show that the effect of a screening can on the average coil wound on 0·3in diameter formers is to reduce the inductance by about 5 per cent.

When designing very low inductance coils, an allowance of approximately 0·15μH should be made for the leads.

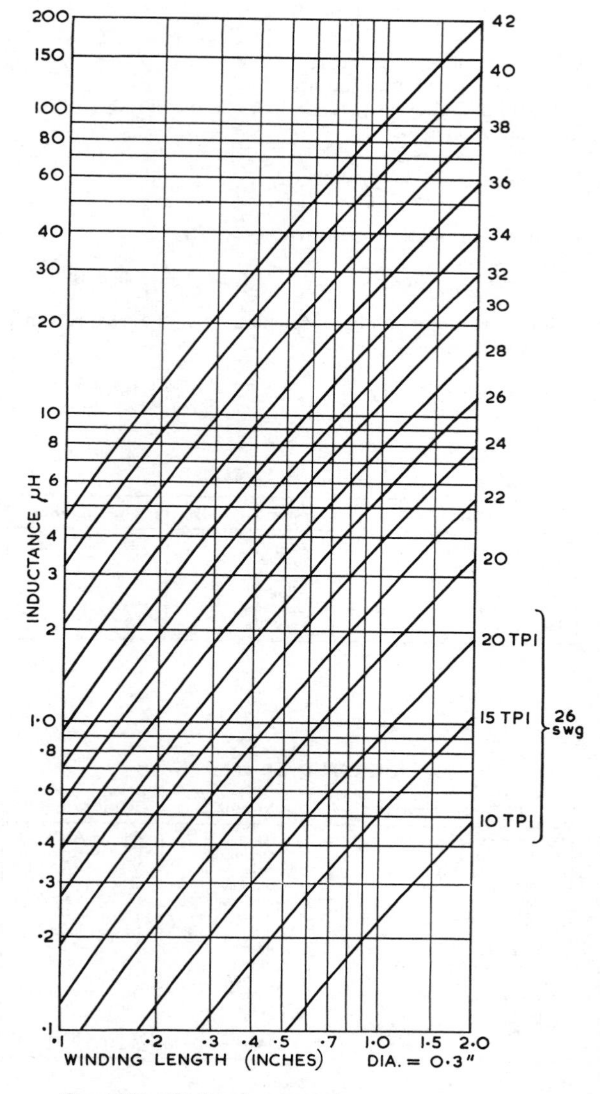

Chart 23.1. Winding data for 0·3in diameter coil forms.

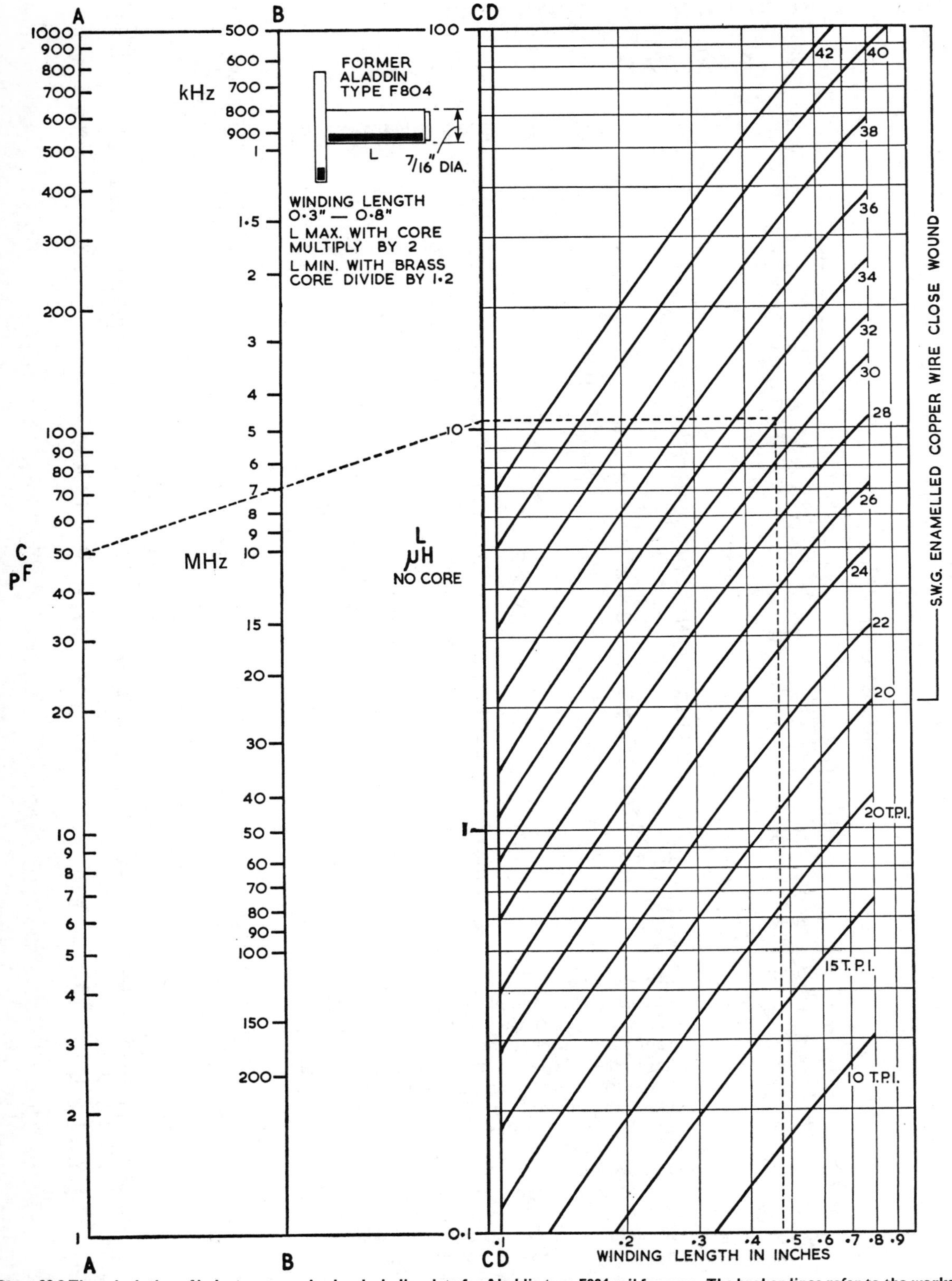

Chart 23.2 The calculation of inductance required and winding data for Aladdin type F804 coil formers. The broken lines refer to the worked example on page 23.4.

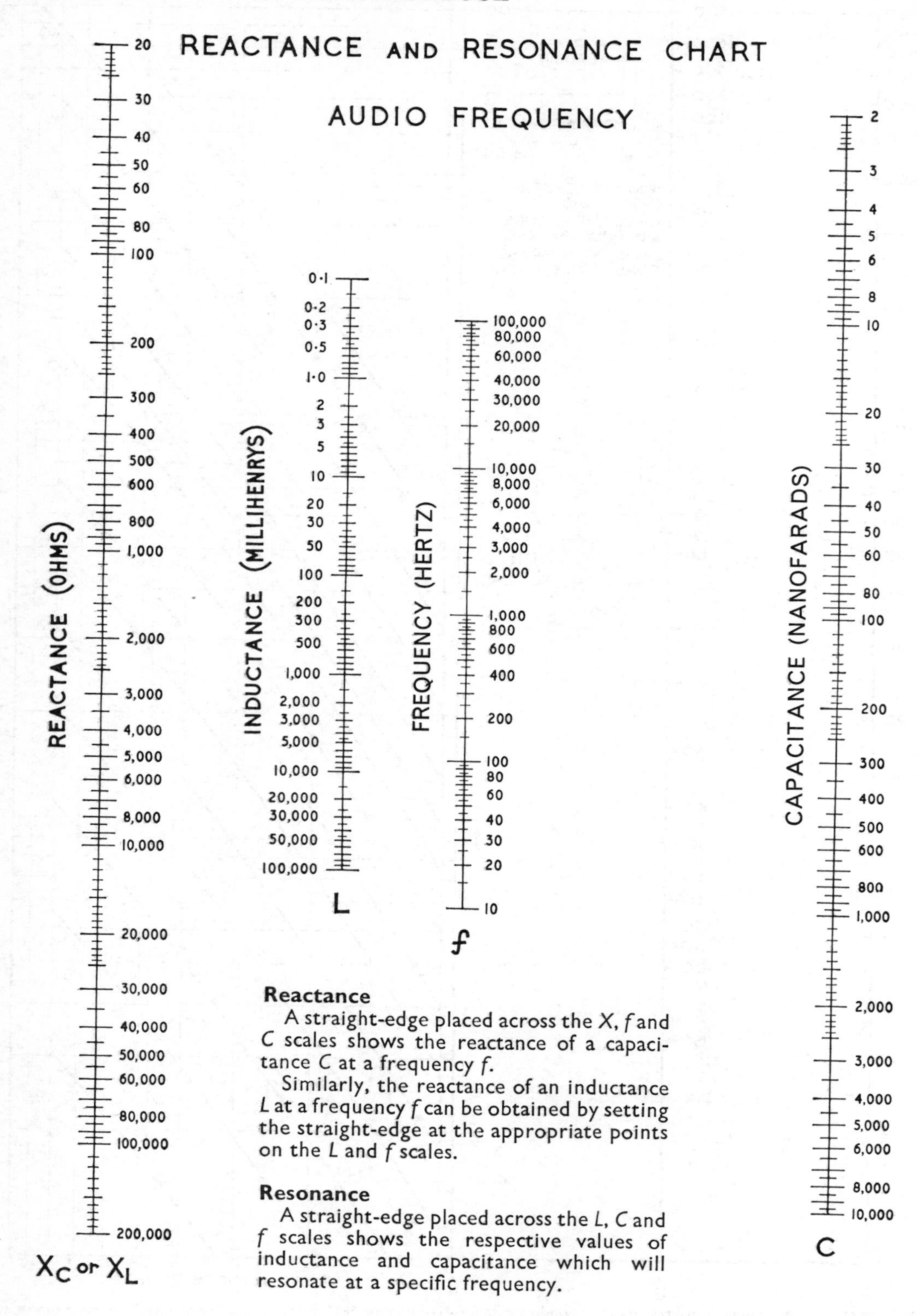

Reactance

A straight-edge placed across the *X*, *f* and *C* scales shows the reactance of a capacitance *C* at a frequency *f*.

Similarly, the reactance of an inductance *L* at a frequency *f* can be obtained by setting the straight-edge at the appropriate points on the *L* and *f* scales.

Resonance

A straight-edge placed across the *L*, *C* and *f* scales shows the respective values of inductance and capacitance which will resonate at a specific frequency.

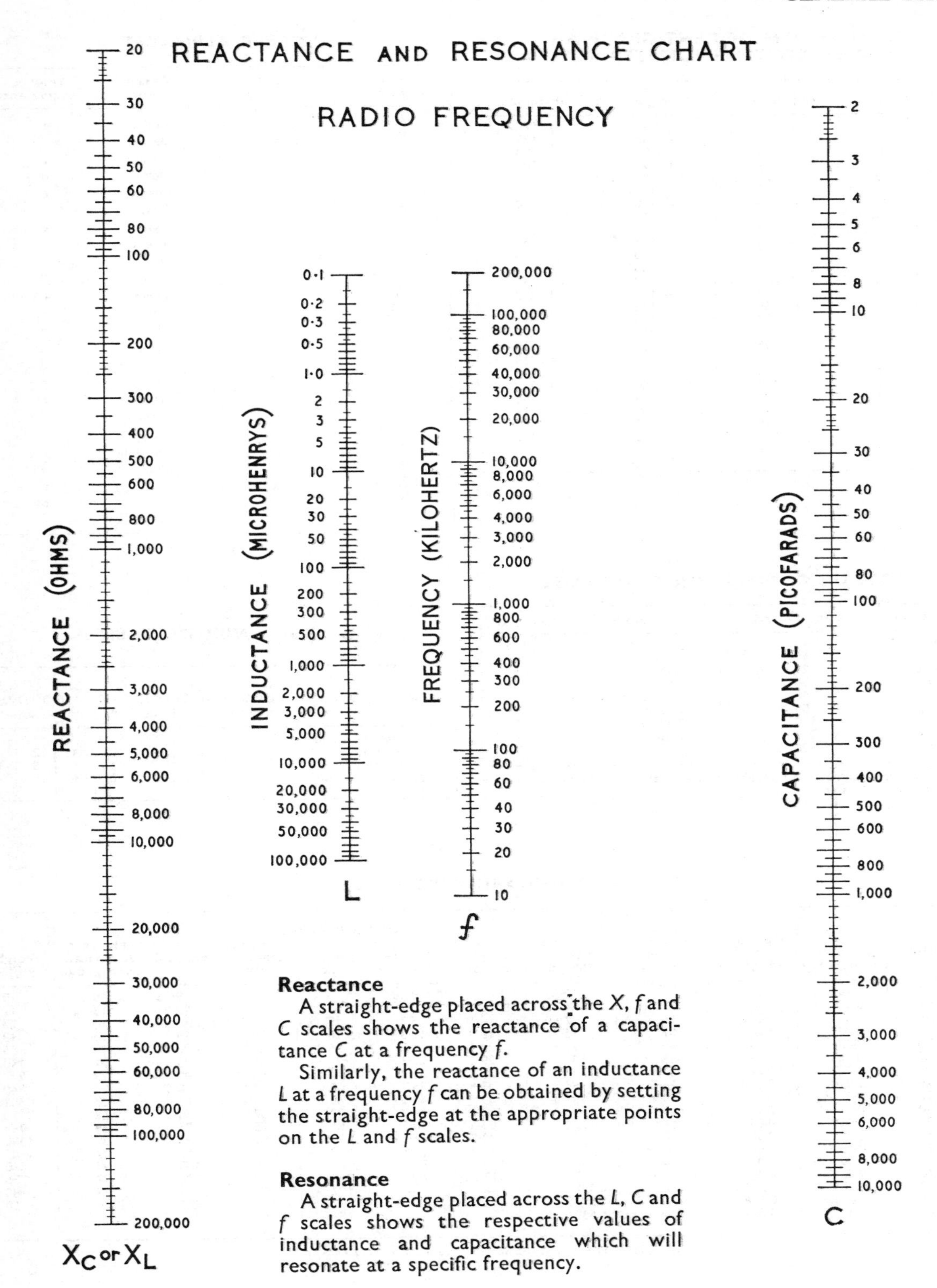

Reactance

A straight-edge placed across the *X*, *f* and *C* scales shows the reactance of a capacitance *C* at a frequency *f*.

Similarly, the reactance of an inductance *L* at a frequency *f* can be obtained by setting the straight-edge at the appropriate points on the *L* and *f* scales.

Resonance

A straight-edge placed across the *L*, *C* and *f* scales shows the respective values of inductance and capacitance which will resonate at a specific frequency.

COMPARISON OF CENTIGRADE AND FAHRENHEIT THERMOMETER SCALES

Centigrade	Fahrenheit	Centigrade	Fahrenheit
− 50	− 58	+ 80	+ 176
− 45	− 49	+ 85	+ 185
− 40	− 40	+ 90	+ 194
− 35	− 31	+ 95	+ 203
− 30	− 22	+ 100	+ 212
− 25	− 13	+ 105	+ 221
− 20	− 4	+ 110	+ 230
− 15	+ 5	+ 115	+ 239
− 10	+ 14	+ 120	+ 248
− 5	+ 23	+ 125	+ 257
0	+ 32	+ 130	+ 266
+ 5	+ 41	+ 135	+ 275
+ 10	+ 50	+ 140	+ 284
+ 15	+ 59	+ 145	+ 293
+ 20	+ 68	+ 150	+ 302
+ 25	+ 77	+ 155	+ 311
+ 30	+ 86	+ 160	+ 320
+ 35	+ 95	+ 165	+ 329
+ 40	+ 104	+ 170	+ 338
+ 45	+ 113	+ 175	+ 347
+ 50	+ 122	+ 180	+ 356
+ 55	+ 131	+ 185	+ 365
+ 60	+ 140	+ 190	+ 374
+ 65	+ 149	+ 195	+ 383
+ 70	+ 158	+ 200	+ 392
+ 75	+ 167		

GREEK ALPHABET

Capital letters	Small letters	Greek name	English equivalent
Α	α	Alpha	a
Β	β	Beta	b
Γ	γ	Gamma	g
Δ	δ	Delta	d
Ε	ε	Epsilon	e
Ζ	ζ	Zeta	z
Η	η	Eta	é
Θ	θ	Theta	th
Ι	ι	Iota	i
Κ	κ	Kappa	k
Λ	λ	Lambda	l
Μ	μ	Mu	m
Ν	ν	Nu	n
Ξ	ξ	Xi	x
Ο	ο	Omicron	ŏ
Π	π	Pi	p
Ρ	ρ	Rho	r
Σ	σ	Sigma	s
Τ	τ	Tau	t
Υ	υ	Upsilon	u
Φ	φ	Phi	ph
Χ	χ	Chi	ch
Ψ	ψ	Psi	ps
Ω	ω	Omega	ō

COLOUR CODING FOR GLASS FUSES

Colour	Rating (mA)	Colour	Rating (A)
Green/yellow	10	Green	0·75
Red/turquoise	15	Blue	1·0
Eau-de-Nil	25	Light blue	1·5
Salmon pink	50	Purple	2·0
Black	60	Yellow and purple	2·5
Grey	100	White	3·0
Red	150	Black and white	5·0
Brown	250	Orange	10·0
Yellow	500		

Note that this coding does not apply to the ceramic-bodied fuse commonly ou nd in 13A plugs etc.

USEFUL TWIST DRILL SIZES

Twist Drill No.	¼in 0·250 in	1 0·228 in	9 0·196 in	17 0·173 in	24 0·152 in	32 0·116 in	43 0·089 in	50 0·070 in
Clearance for wood-screw No.	14	12	10	8	6	4	2	0
Clearance for BA	0	1	2	3	4	6	8	10
Tapping size for BA	—	—	0	1	2	4	6	8

STANDARD WIRE TABLE

SWG	Diameter (inches)	Current* at 1000A/in²	Fusing current	Resistance ohms† 1000yd	TURNS PER INCH					Nearest American wire gauge	Weight yd/lb
					Single cotton	Double cotton	Single silk	Double silk	Enamel		
12	0·104	8·5	—	2·83	—	8·48	—	—	9·26	10	10
14	0·08	5·03	—	4·78	—	10·7	—	—	12·1	12	17
16	0·064	3·217	166	7·46	14·1	13·3	15	14·7	14·8	14	27
18	0·048	1·81	107	13·27	18·3	17·3	20	19·6	19·7	16	48
20	0·036	1·02	69·9	23·6	24·1	21·7	26·3	25·3	26·1	19	85
22	0·028	0·61	48	39	29·8	26·3	33·3	31·8	33·3	21	140
24	0·022	0·38	33·4	63·2	37	31·3	42·1	40	41·1	23	227
26	0·018	0·254	24·7	94·3	43·5	35·7	50·6	47·6	50·6	25	340
28	0·0148	0·17	18·4	139·5	50·5	40·2	60·4	56·2	61·4	27	503
30	0·0124	0·12	14·1	199·0	57·5	44·7	72	67·1	73·3	28	716
32	0·0108	0·09	11·5	262	63·5	50·5	81·3	75·2	83	29	944
34	0·0092	0·07	—	361	70·5	54·9	93·4	85·5	98	31	1300
36	0·0076	0·05	6·79	529	86·2	64·1	110	102	116	32	1905
38	0 006	0·03	—	849	100	71·4	133	121	143	34	3060
40	0·0048	0·018	3·41	1326	112·5	78·1	159	142	180	36	4775
42	0·004	0·0126	—	1910	—	—	192	161	217	38	6880
44	0·0032	0·008	—	2985	—	—	227	185	270	40	10,750
46	0·0024	0·0045	—	5307	—	—	278	217	357		

Note (*) Current loading may be used up to 2000A/in.²
(†) Some small variations depending on makers.

USA RG SERIES COAXIAL CABLES

Cable no.	Nominal impedance Z_o (ohms)	Cable outside diameter (in)	Velocity factor	Approximate attenuation (dB per 100ft)					Capacitance (pF/ft)	Maximum operating voltage rms
				1MHz	10MHz	100MHz	1,000MHz	3,000MHz		
RG-5/U	52·5	0·332	0·659	0·21	0·77	2·9	11·5	22·0	28·5	3,000
RG-5B/U	50·0	0·332	0·659	0·16	0·66	2·4	8·8	16·7	29·5	3,000
RG-6A/U	75·0	0·332	0·659	0·21	0·78	2·9	11·2	21·0	20·0	2,700
RG-8A/U	50·0	0·405	0·659	0·16	0·55	2·0	8·0	16·5	30·5	4,000
RG-9/U	51·0	0·420	0·659	0·16	0·57	2·0	7·3	15·5	30·0	4,000
RG-9B/U	50·0	0·425	0·659	0·175	0·61	2·1	9·0	18·0	30·5	4,000
RG-10A/U	50·0	0·475	0·659	0·16	0·55	2·0	8·0	16·5	30·5	4,000
RG-11A/U	75·0	0·405	0·66	0·18	0·7	2·3	7·8	16·5	20·5	5,000
RG-12A/U	75·0	0·475	0·659	0·18	0·66	2·3	8·0	16·5	20·5	4,000
RG-13A/U	75·0	0·425	0·659	0·18	0·66	2·3	8·0	16·5	20·5	4,000
RG-14A/U	50·0	0·545	0·659	0·12	0·41	1·4	5·5	12·0	30·0	5,500
RG-16/U	52·0	0·630	0·670	0·1	0·4	1·2	6·7	16·0	29·5	6,000
RG-17A/U	50·0	0·870	0·659	0·066	0·225	0·80	3·4	8·5	30·0	11,000
RG-18A/U	50·0	0·945	0·659	0·066	0·225	0·80	3·4	8·5	30·5	11,000
RG-19A/U	50·0	1·120	0·659	0·04	0·17	0·68	3·5	7·7	30·5	14,000
RG-20A/U	50·0	1·195	0·659	0·04	0·17	0·68	3·5	7·7	30·5	14,000
RG-21A/U	50·0	0·332	0·659	1·4	4·4	13·0	43·0	85·0	30·0	2,700
RG-29/U	53·5	0·184	0·659	0·33	1·2	4·4	16·0	30·0	28·5	1,900
RG-34A/U	75·0	0·630	0·659	0·065	0·29	1·3	6·0	12·5	20·5	5,200
RG-34B/U	75	0·630	0·66		0·3	1·4	5·8		21·5	6,500
RG-35A/U	75·0	0·945	0·659	0·07	0·235	0·85	3·5	8·60	20·5	10,000
RG-54A/U	58·0	0·250	0·659	0·18	0·74	3·1	11·5	21·5	26·5	3,000
RG-55B/U	53·5	0·206	0·659	0·36	1·3	4·8	17·0	32·0	28·5	1,900
RG-55A/U	50·0	0·216	0·659	0·36	1·3	4·8	17·0	32·0	29·5	1,900
RG-58/U	53·5	0·195	0·659	0·33	1·25	4·65	17·5	37·5	28·5	1,900
RG-58C/U	50·0	0·195	0·659	0·42	1·4	4·9	24·0	45·0	30·0	1,900
RG-59A/U	75·0	0·242	0·659	0·34	1·10	3·40	12·0	26·0	20·5	2,300
RG-59B/U	75	0·242	0·66		1·1	3·4	12		21	2,300
RG-62A/U	93·0	0·242	0·84	0·25	0·85	2·70	8·6	18·5	13·5	700
RG-74A/U	50·0	0·615	0·659	0·10	0·38	1·5	6·0	11·5	30·0	5,500
RG-83/U	35·0	0·405	0·66	0·23	0·80	2·8	9·6	24·0	44·0	2,000
*RG-213/U	50	0·405	0·66	0·16	0·6	1·9	8·0		29·5	5,000
†RG-218/U	50	0·870	0·66	0·066	0·2	1·0	4·4		29·5	11,000
‡RG-220/U	50	1·120	0·66	0·04	0·2	0·7	3·6		29·5	14,000

* Formerly RG8A/U † Formerly RG17A/U ‡ Formerly RG19A/U

CONSTANT-K FILTERS

LOW PASS FILTERS

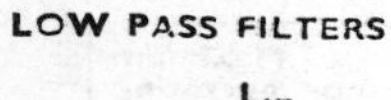

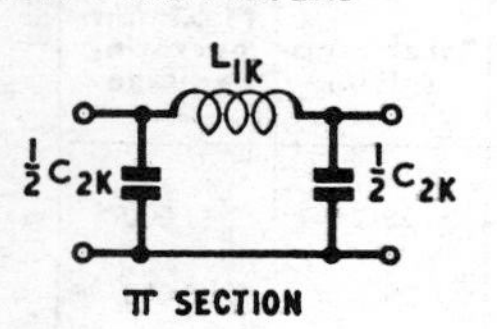

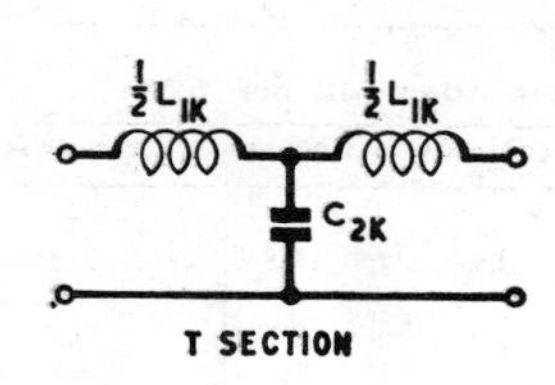

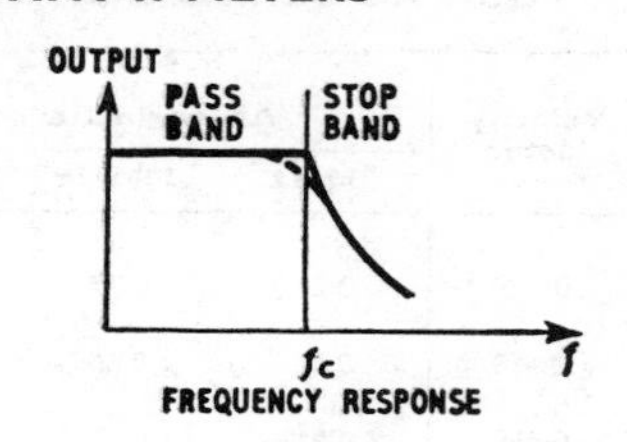

$$L_{1K} = \frac{R_o}{\pi f_c}\,;\quad C_{2K} = \frac{1}{\pi f_c R_o}$$

HIGH PASS FILTERS

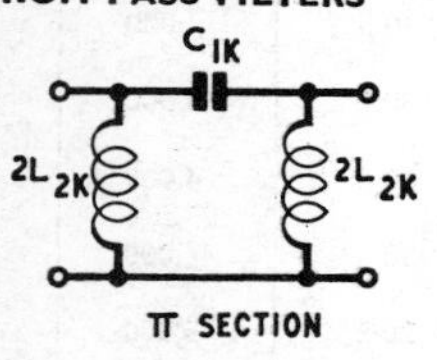

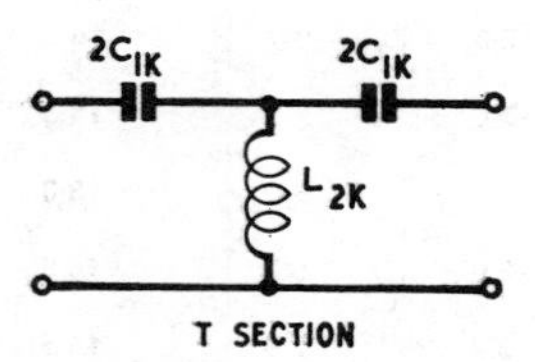

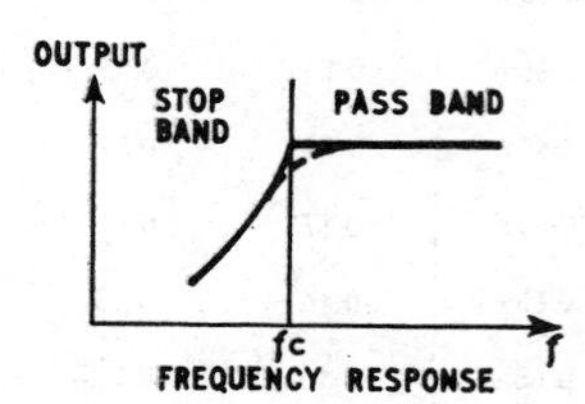

$$L_{2K} = \frac{R_o}{4\pi f_c}\,;\quad C_{1K} = \frac{1}{4\pi f_c R_o}$$

BAND PASS FILTERS

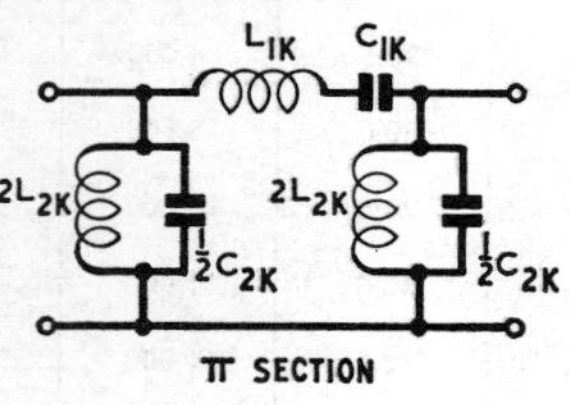

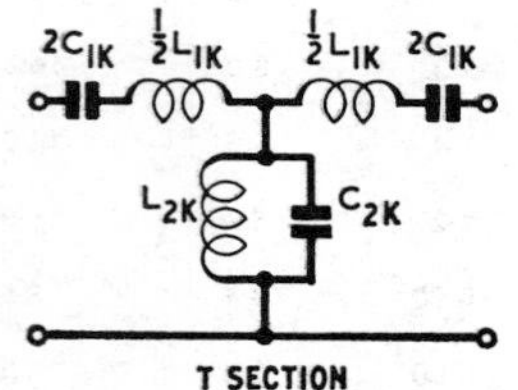

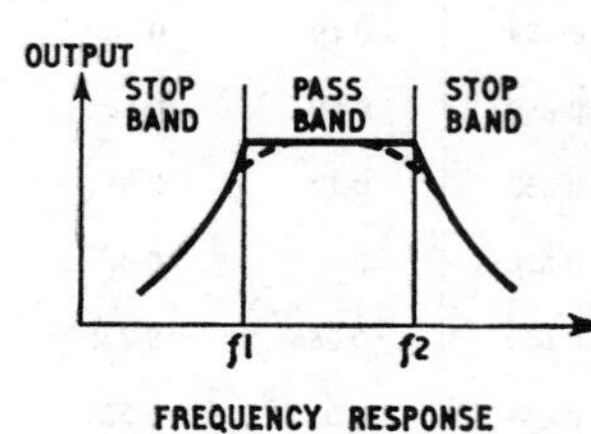

$$L_{1K} = \frac{R}{\pi(f_2 - f_1)}\,;\quad C_{1K} = \frac{(f_2 - f_1)}{4\pi R f_1 f_2}$$

$$L_{2K} = \frac{R(f_2 - f_1)}{4\pi f_1 f_2}\,;\quad C_{2K} = \frac{1}{\pi R(f_2 - f_1)}$$

M-DERIVED FILTERS

LOW PASS FILTERS

$$m = \sqrt{1 - \left(\frac{f_c}{f_\infty}\right)^2}$$

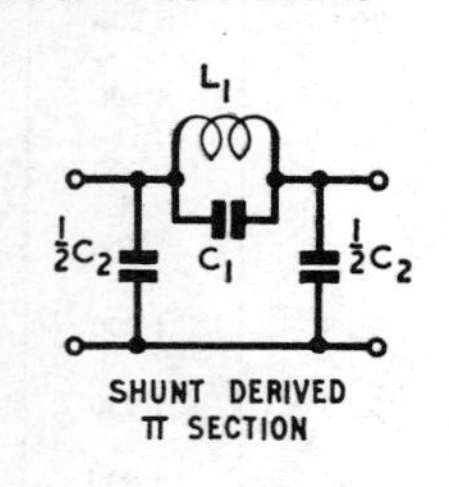

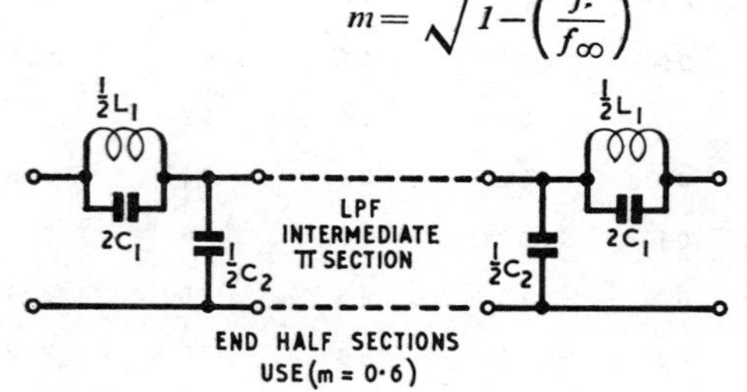

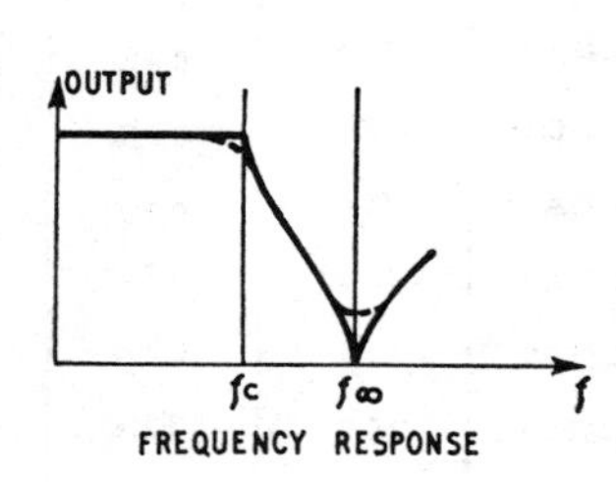

$$L_1 = m\,L_{1K}\,;\quad C_1 = \frac{1 - m^2}{4m}C_{2K}\,;\quad C_2 = mC_{2K},$$

HIGH PASS FILTERS

$$m = \sqrt{1 - \left(\frac{f_\infty}{f_c}\right)^2}$$

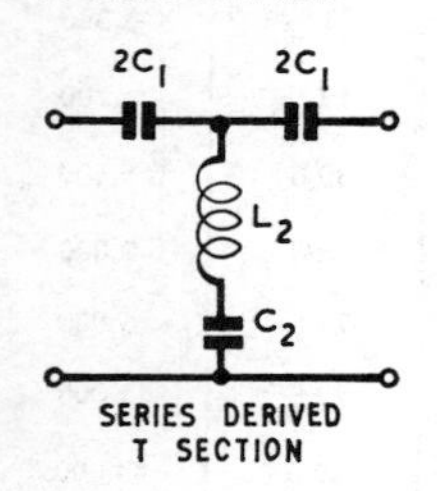

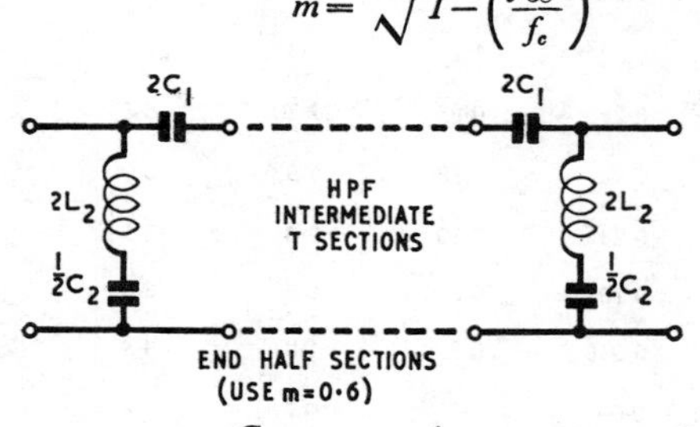

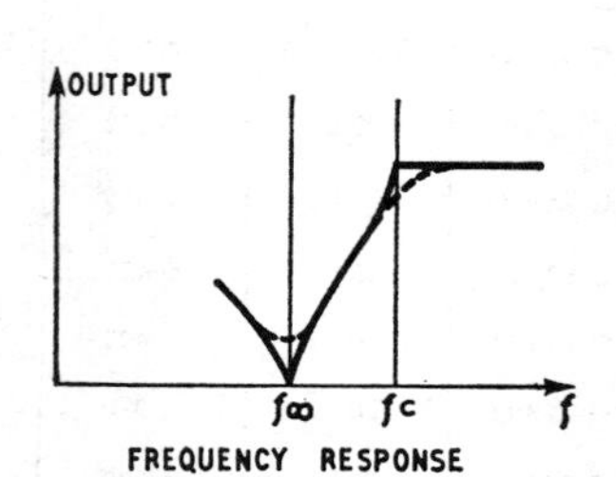

$$C_1 = \frac{C_{1K}}{m}\,;\quad C_2 = \frac{4m}{1 - m^2}\,C_{1K}\,;\quad L_2 = \frac{L_{2K}}{m}$$

COMPONENT COLOUR CODES

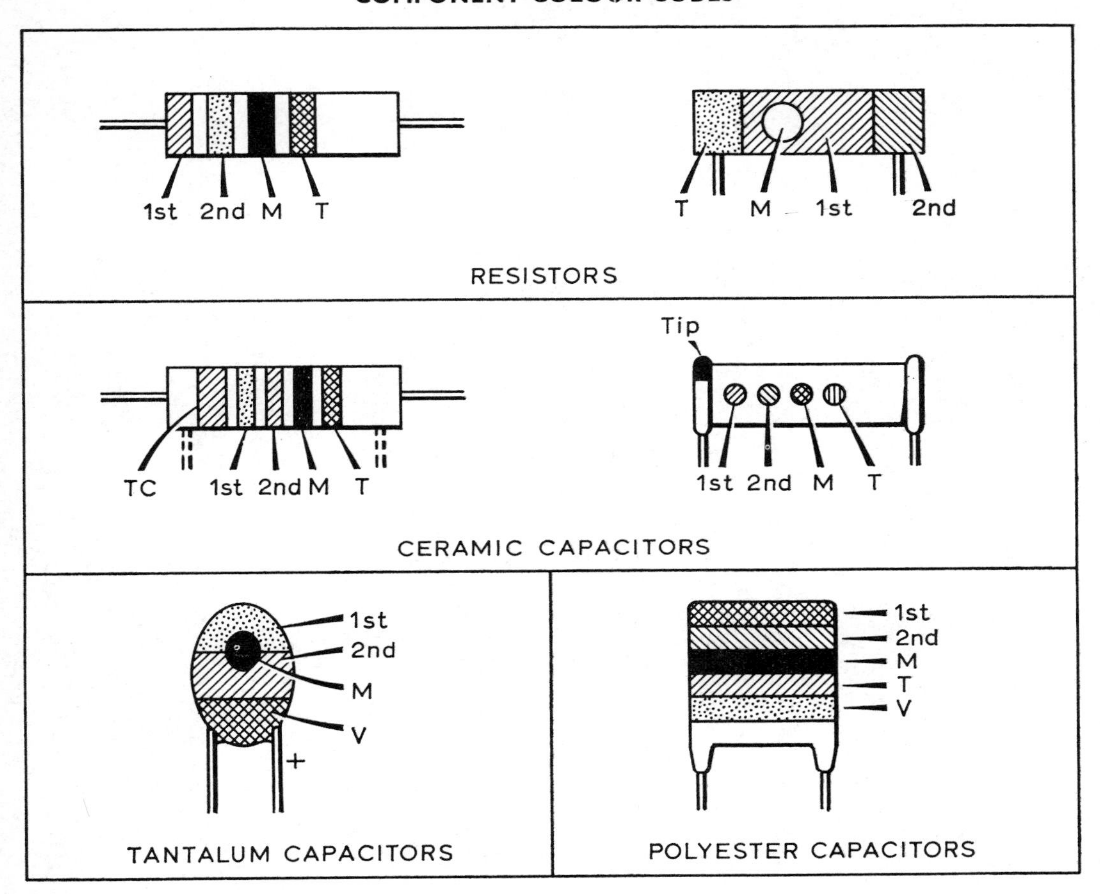

Colour	Significant figure (1st, 2nd)	Decimal multiplier (M)	Tolerance (T) (per cent)	Temp coeff (TC) (parts/10^6/°C)	Voltage (V) (tantalum cap)	Voltage (V) (polyester cap)
Black	0	1	±20	0	10	—
Brown	1	10	±1	−30	—	100
Red	2	100	±2	−80	—	250
Orange	3	1,000	±3	−150	—	—
Yellow	4	10,000	+100, −0	−220	6·3	400
Green	5	100,000	±5	−330	16	—
Blue	6	1,000,000	±6	−470	20	—
Violet	7	10,000,000	—	−750	—	—
Grey	8	100,000,000	—	+30	25	—
White	9	1,000,000,000	±10	+100 to −750	3	—
Gold	—	—	±5	—	—	—
Silver	—	—	±10	—	—	—
Pink	—	—	—	—	35	—
No colour	—	—	±20	—	—	—

Units used are ohms for resistors, picofarads for ceramic and polyester capacitors, and microfarads for tantalum capacitors.

INDEX

S

T

U

V

W

Z